EDITION 7

Beginning and Intermediate Algebra

Margaret L. Lial
American River College

John Hornsby
University of New Orleans

Terry McGinnis

Vice President, Courseware Portfolio Management: Chris Hoag
Director, Courseware Portfolio Management: Michael Hirsch
Courseware Portfolio Manager: Karen Montgomery
Courseware Portfolio Assistant: Kayla Shearns
Managing Producer: Scott Disanno
Content Producer: Lauren Morse
Producers: Stacey Miller and Noelle Saligumba
Managing Producer: Vicki Dreyfus
Associate Content Producer, TestGen: Rajinder Singh
Content Managers, MathXL: Eric Gregg and Dominick Franck
Manager, Courseware QA: Mary Durnwald
Senior Product Marketing Manager: Alicia Frankel
Product Marketing Assistant: Brooke Imbornone
Senior Author Support/Technology Specialist: Joe Vetere
Full Service Vendor, Cover Design, Composition: Pearson CSC
Full Service Project Management: Pearson CSC (Carol Merrigan)
Cover Image: Borchee/E+/Getty Images

Library of Congress Cataloging-in-Publication Data

Names: Lial, Margaret L., author. | Hornsby, John, 1949- author. | McGinnis, Terry, author.
Title: Beginning and intermediate algebra / Margaret L. Lial (American River College), John Hornsby (University of New Orleans), Terry McGinnis.
Description: 7th edition. | Boston : Pearson, [2020] | Includes index.
Identifiers: LCCN 2019000104 | ISBN 9780134895994 (student edition) | ISBN 0134895991 (student edition)
Subjects: LCSH: Algebra--Textbooks.
Classification: LCC QA152.3 .L52 2020 | DDC 512.9--dc23
LC record available at https://lccn.loc.gov/2019000104

1   19

Pearson

ISBN 13: 978-0-13-489599-4
ISBN 10: 0-13-489599-1

CONTENTS

Preface   viii

Study Skills   S-1

R   Prealgebra Review   1

1   The Real Number System   27

2   Linear Equations and Inequalities in One Variable   103

6   Rational Expressions and Applications   411

7   Linear Equations, Graphs, and Systems   489

8   Inequalities and Absolute Value   587

WELCOME TO THE 7TH EDITION

The first edition of Marge Lial's *Beginning and Intermediate Algebra* was published in 1996, and now we are pleased to present the 7th edition—with the same successful, well-rounded framework that was established 24 years ago and updated to meet the needs of today's students and professors. The names Lial and Miller, two faculty members from American River College in Sacramento, California, have become synonymous with excellence in Developmental Mathematics, Precalculus, Finite Mathematics, and Applications-Based Calculus.

With Chuck Miller's passing, Marge Lial was joined by a team of carefully selected coauthors who partnered with her. John Hornsby (University of New Orleans) joined Marge in this capacity in 1992, and in 1999, Terry McGinnis became part of this developmental author team. Since Marge's passing in 2012, John and Terry have dedicated themselves to carrying on the Lial/Miller legacy.

In the preface to the first edition of *Intermediate Algebra,* Marge Lial wrote

> *" . . . the strongest theme . . . is a combination of readability and suitability for the book's intended audience: students who are not completely self-confident in mathematics as they come to the course, but who must be self-confident and proficient . . . by the end of the course."*

Today's Lial author team upholds these same standards. With the publication of the 7th edition of *Beginning and Intermediate Algebra,* we proudly present a complete course program for students who need developmental algebra. Revisions to the core text, working in concert with such innovations in the MyLab Math course as Skill Builder and Learning Catalytics, combine to provide superior learning opportunities appropriate for all types of courses (traditional, hybrid, online).

We hope you enjoy using it as much as we have enjoyed writing it. We welcome any feedback that you have as you review and use this text.

WHAT'S NEW IN THIS EDITION?

We are pleased to offer the following new features and resources in the text and MyLab.

IMPROVED STUDY SKILLS These special activities are now grouped together at the front of the text, prior to Chapter R. **Study Skills Reminders** that refer students to specific Study Skills are found liberally throughout the text. Many Study Skills now include a *Now Try This* section to help students implement the specific skill.

REVISED EXPOSITION With each edition of the text, we continue to polish and improve discussions and presentations of topics to increase readability and student understanding. This edition is no exception.

NEW FIGURES AND DIAGRAMS For visual learners, we have included more than 50 new mathematical figures, graphs, and diagrams, including several new "hand drawn" style graphs. These are meant to suggest what a student who is graphing with paper and pencil should obtain. We use this style when introducing a particular type of graph for the first time.

ENHANCED USE OF PEDAGOGICAL COLOR We have thoroughly reviewed the use of pedagogical color in discussions and examples and have increased its use whenever doing so would enhance concept development, emphasize important steps, or highlight key procedures.

INCREASED Concept Check AND WHAT WENT WRONG? EXERCISES The number of Concept Check exercises, which facilitate students' mathematical thinking and conceptual understanding, and which begin each exercise set, has been increased. We have also more than doubled the number of WHAT WENT WRONG? exercises that highlight common student errors.

INCREASED RELATING CONCEPTS EXERCISES We have doubled the number of these flexible groups of exercises, which are located at the end of many exercise sets. These sets of problems were specifically written to help students tie concepts together, compare and contrast ideas, identify and describe patterns, and extend concepts to new situations. They may be used by individual students or by pairs or small groups working collaboratively. All answers to these exercises appear in the student answer section.

ENHANCED MYLAB MATH RESOURCES MyLab exercise coverage in the revision has been expanded, and video coverage has also been expanded and updated to a modern format for today's students. WHAT WENT WRONG? problems and all RELATING CONCEPTS exercise sets (both even- and odd-numbered problems) are now assignable in MyLab Math.

SKILL BUILDER These exercises offer just-in-time additional adaptive practice in MyLab Math. The adaptive engine tracks student performance and delivers, to each individual, questions that adapt to his or her level of understanding. This new feature enables instructors to assign fewer questions for homework, allowing students to complete as many or as few questions as they need.

LEARNING CATALYTICS This new student response tool uses students' own devices to engage them in the learning process. Problems that draw on prerequisite skills are included at the beginning of each section to gauge student readiness for the section. Accessible through MyLab Math and customizable to instructors' specific needs, these problems can be used to generate class discussion, promote peer-to-peer learning, and provide real-time feedback to instructors. More information can be found via the Learning Catalytics link in MyLab Math. Specific exercises notated in the text can be found by searching LialBegIntAlg# where the # is the chapter number.

CONTENT CHANGES

Specific content changes include the following:

- **Exercise sets** have been scrutinized and updated with a renewed focus on conceptual understanding and skill development. Even and odd pairing of the exercises, an important feature of the text, has been carefully reviewed.

- **Real-world data** in all examples and exercises and in their accompanying graphs has been updated.

- **An increased emphasis on fractions, decimals, and percents** appears throughout the text. We have **expanded Chapter R** to include new figures and revised explanations and examples on converting among fractions, decimals, and percents. And we have included an **all-new set of Cumulative Review Exercises,** many of which focus on fractions, decimals, and percents, at the end of Chapter 1. Sets of Cumulative Review Exercises in subsequent chapters now begin with new exercises that review skills related to these topics.

- **A new Section 2.4 provides expanded coverage of linear equations in one variable with fractional and decimal coefficients.** Two new examples have been included, and the number of exercises has been doubled.

- **Solution sets of linear inequalities in Section 2.9** are now graphed first, before they are written using interval notation.

- **Expanded Mid-Chapter Summary Exercises** in Chapter 2 continue our emphasis on the difference between simplifying an expression and solving an equation. New examples in the Summary Exercises in Chapters 5 and 7 illustrate and distinguish between solution methods.

- **Chapters 13 and 14 on Nonlinear Functions, Conic Sections, Nonlinear Systems, and Further Topics in Algebra,** previously available online in MyLab Math, are now included in the text. The material has been fully revised and updated.

- **Presentations of the following topics have been enhanced and expanded,** often including new examples and exercises.

  Order of operations involving absolute value expressions (Section 1.5)
  Solving linear equations in one variable (Sections 2.1, 2.2)
  Solving problems involving proportions and percent (Section 2.7)
  Writing an equation of a line from a graph (Section 3.4)
  Adding, subtracting, and dividing polynomials (Sections 4.4 and 4.7)
  Finding reciprocals of rational expressions (Section 6.2)
  Geometric interpretation of slope as rise/run (Section 7.1)
  Solving systems of equations using the elimination method (Section 7.5)
  Solving systems of linear equations in three variables (Section 7.6)
  Identifying functions and domains from equations (Section 9.1)
  Graphing polynomial functions (Section 9.3)
  Concepts and relationships among real numbers, non-real complex numbers, and imaginary numbers; simplifying powers of i (Section 10.7)
  Solving quadratic equations using the quadratic formula (Section 11.3)
  Solving exponential and logarithmic equations (Sections 12.2, 12.3)

LIAL DEVELOPMENTAL HALLMARK FEATURES

We have enhanced the following popular features, each of which is designed to increase ease of use by students and/or instructors.

- *Emphasis on Problem-Solving* We introduce our six-step problem-solving method in Chapter 2 and integrate it throughout the text. The six steps, *Read, Assign a Variable, Write an Equation, Solve, State the Answer,* and *Check,* are emphasized in boldface type and repeated in examples and exercises to reinforce the problem-solving process for students. We also provide students with PROBLEM-SOLVING HINT boxes that feature helpful problem-solving tips and strategies.

- *Helpful Learning Objectives* We begin each section with clearly stated, numbered objectives, and the included material is directly keyed to these objectives so that students and instructors know exactly what is covered in each section.

- *Cautions and Notes* One of the most popular features of previous editions is our inclusion of information marked ! CAUTION and NOTE to warn students about common errors and to emphasize important ideas throughout the exposition. The updated text design makes them easy to spot.

- *Comprehensive Examples* The new edition features a multitude of step-by-step, worked-out examples that include pedagogical color, helpful side comments, and special pointers. We give special attention to checking example solutions—more checks, designated using a special CHECK tag and ✓, are included than in past editions.

- *More Pointers* There are more pointers in examples and discussions throughout this edition of the text. They provide students with important on-the-spot reminders, as well as warnings about common pitfalls.

- *Numerous Now Try Problems* These margin exercises, with answers immediately available at the bottom of the page, have been carefully written to correspond to every example in the text. This key feature allows students to immediately practice the material in preparation for the exercise sets.

- *Updated Figures, Photos, and Hand-Drawn Graphs* Today's students are more visually oriented than ever. As a result, we provide detailed mathematical figures, diagrams, tables, and graphs, including a "hand-drawn" style of graphs, whenever possible. We have incorporated depictions of well-known mathematicians, as well as appealing photos to accompany applications in examples and exercises.

- *Relevant Real-Life Applications* We include many new or updated applications from fields such as business, pop culture, sports, technology, and the health sciences that show the relevance of algebra to daily life.

- *Extensive and Varied Exercise Sets* The text contains a wealth of exercises to provide students with opportunities to practice, apply, connect, review, and extend the skills they are learning. Numerous illustrations, tables, graphs, and photos help students visualize the problems they are solving. Problem types include skill building and writing exercises, as well as applications, matching, true/false, multiple-choice, and fill-in-the-blank problems. Special types of exercises include Concept Check, WHAT WENT WRONG? , Extending Skills, and RELATING CONCEPTS .

- *Special Summary Exercises* We include a set of these popular in-chapter exercises in every chapter. They provide students with the all-important *mixed review problems* they need to master topics and often include summaries of solution methods and/or additional examples.

- *Extensive Review Opportunities* We conclude each chapter with the following review components:

  A **Chapter Summary** that features a helpful list of **Key Terms** organized by section, **New Symbols,** a **Test Your Word Power** vocabulary quiz (with answers immediately following), and a **Quick Review** of each section's main concepts, complete with additional examples.

  A comprehensive set of **Chapter Review Exercises,** keyed to individual sections for easy student reference.

  A set of **Mixed Review Exercises** that helps students further synthesize concepts and skills.

  A **Chapter Test** that students can take under test conditions to see how well they have mastered the chapter material.

  A set of **Cumulative Review Exercises** for ongoing review that covers material going back to Chapter R.

- *Comprehensive Glossary* The online Glossary includes key terms and definitions (with section references) from throughout the text.

ACKNOWLEDGMENTS

The comments, criticisms, and suggestions of users, non-users, instructors, and students have positively shaped this text over the years, and we are most grateful for the many responses we have received. The feedback gathered for this edition was particularly helpful.

We especially wish to thank the following individuals who provided invaluable suggestions.

Barbara Aaker, *Community College of Denver*
Kim Bennekin, *Georgia Perimeter College*
Dixie Blackinton, *Weber State University*
Eun Cha, *College of Southern Nevada, Charleston*
Callie Daniels, *St. Charles Community College*
Cheryl Davids, *Central Carolina Technical College*
Robert Diaz, *Fullerton College*
Chris Diorietes, *Fayetteville Technical Community College*
Sylvia Dreyfus, *Meridian Community College*
Sabine Eggleston, *Edison State College*
LaTonya Ellis, *Bishop State Community College*
Beverly Hall, *Fayetteville Technical Community College*
Aaron Harris, *College of Southern Nevada, Charleston*
Loretta Hart, *NHTI, Concord's Community College*
Sandee House, *Georgia Perimeter College*
Joe Howe, *St. Charles Community College*
Lynette King, *Gadsden State Community College*
Linda Kodama, *Windward Community College*
Carlea McAvoy, *South Puget Sound Community College*
James Metz, *Kapi'olani Community College*
Jean Millen, *Georgia Perimeter College*
Molly Misko, *Gadsden State Community College*
Charles Patterson, *Louisiana Tech*
Jane Roads, *Moberly Area Community College*
Melanie Smith, *Bishop State Community College*
Erik Stubsten, *Chattanooga State Technical Community College*
Tong Wagner, *Greenville Technical College*
Rick Woodmansee, *Sacramento City College*
Sessia Wyche, *University of Texas at Brownsville*

Over the years, we have come to rely on an extensive team of experienced professionals. Our sincere thanks go to these dedicated individuals at Pearson who worked long and hard to make this revision a success.

We would like to thank Michael Hirsch, Matthew Summers, Karen Montgomery, Alicia Frankel, Lauren Morse, Vicki Dreyfus, Stacey Miller, Noelle Saligumba, Eric Gregg, and all of the Pearson math team for helping with the revision of the text.

We are especially pleased to welcome Callie Daniels, who has taught from our texts for many years, to our team. Her assistance has been invaluable. She thoroughly reviewed all chapters and helped extensively with manuscript preparation.

We are grateful to Carol Merrigan for her excellent production work. We appreciate her positive attitude, responsiveness, and expert skills. We would also like to thank Pearson CSC for their production work; Emily Keaton for her detailed help in updating real data applications; Connie Day for supplying her copyediting expertise; Pearson CSC for their photo research; and Lucie Haskins for producing another accurate, useful index. Paul Lorczak and Hal Whipple did a thorough, timely job accuracy-checking the page proofs and answers, and Sarah Sponholz checked the index.

We particularly thank the many students and instructors who have used this text over the years. You are the reason we do what we do. It is our hope that we have positively impacted your mathematics journey. We would welcome any comments or suggestions you might have via email to math@pearson.com.

John Hornsby
Terry McGinnis

DEDICATION

To BK and Vangie

E.J.H.

To Andrew and Tyler

Mom

Resources for Success

Pearson MyLab

Get the Most Out of MyLab Math for *Beginning and Intermediate Algebra,* Seventh Edition by Lial, Hornsby, McGinnis

The Lial team has helped thousands of students learn algebra with an approachable, teacherly writing style and balance of skill and concept development. With this revision, the series retains the hallmarks that have helped students succeed in math, and includes new and updated digital tools in the MyLab Math course.

Take advantage of the following resources to get the most out of your MyLab Math course.

Get Students Prepared with Integrated Review

Every student enters class with different levels of preparedness and prerequisite knowledge. To ensure students are caught up on prior skills, every Lial MyLab course now includes Integrated Review.

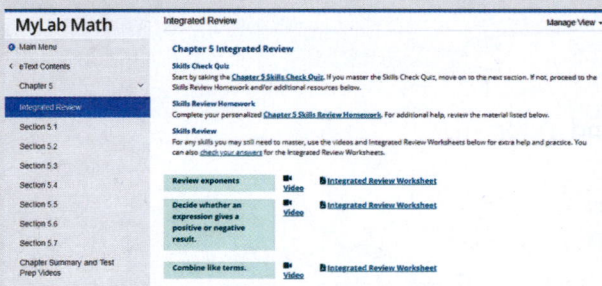

New! Integrated Review provides embedded and personalized review of prerequisite topics within relevant chapters. Students can check their prerequisite skills, and receive personalized practice on the topics they need to focus on, with study aids like worksheets and videos also available to help.

Integrated Review assignments are premade and available to assign in the Assignment Manager.

Personalize Learning

New! Skill Builder exercises offer just-in-time additional adaptive practice. The adaptive engine tracks student performance and delivers questions to each individual that adapt to his or her level of understanding. This new feature allows instructors to assign fewer questions for homework, allowing students to complete as many or as few questions as they need.

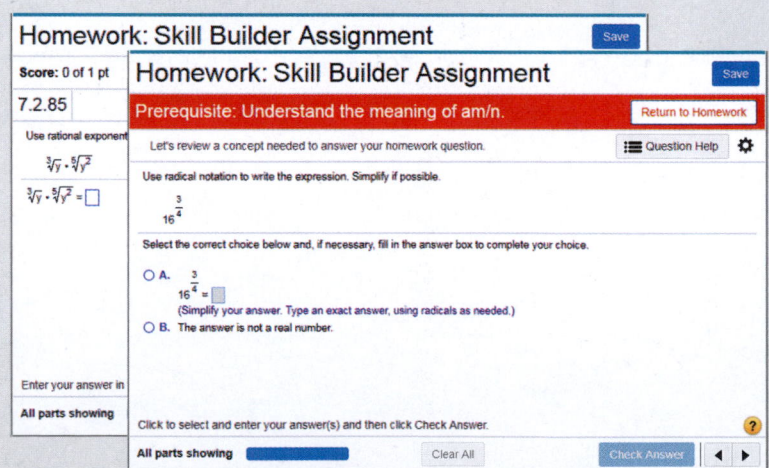

Resources for Success

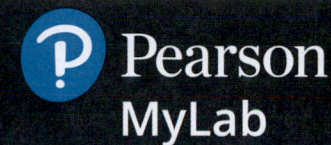

Support Students Whenever, Wherever

Updated! The **complete video program** for the Lial series includes:

- Full Section Lecture Videos
- Solution clips for select exercises
- Chapter Test Prep videos
- Short Quick Review videos that recap each section

Full Section Lecture Videos are also available as shorter, objective-level videos. No matter your students' needs—if they missed class, need help solving a problem, or want a short summary of a section's concepts—they can get support whenever they need it, wherever they need it. Much of the video series has been updated in a modern presentation format.

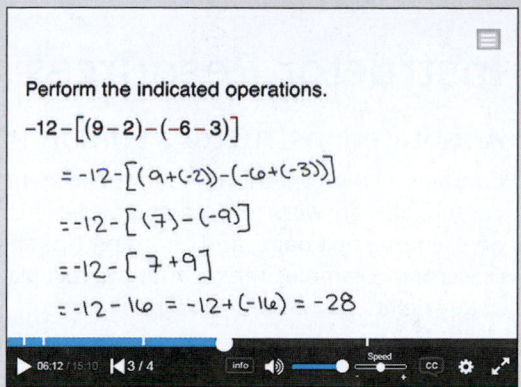

Foster a Growth Mindset

New! A **Mindset module** is available in the course, with mindset-focused videos and exercises that encourage students to maintain a positive attitude about learning, value their own ability to grow, and view mistakes as a learning opportunity.

Get Students Engaged

New! Learning Catalytics Learning Catalytics is an interactive student response tool that uses students' smartphones, tablets, or laptops to engage them in more sophisticated tasks and thinking.

In addition to a library of developmental math questions, Learning Catalytics questions created specifically for this text are pre-built to make it easy for instructors to begin using this tool! These questions, which cover prerequisite skills before each section, are noted in the margin of the Annotated Instructor's Edition, and can be found in Learning Catalytics by searching for "LialBegIntAlg#", where # is the chapter number.

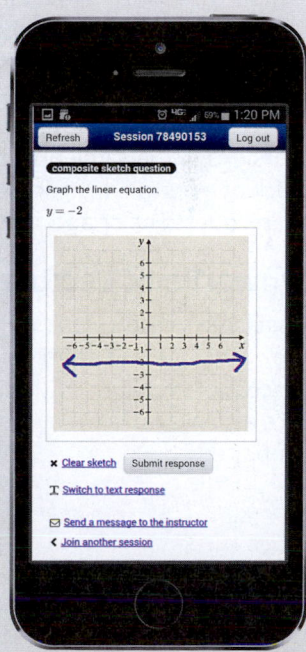

Resources for Success

Instructor Resources

Annotated Instructor's Edition

Contains all the content found in the student edition, plus answers to even and odd exercises on the same text page, and Teaching Tips and Classroom Examples throughout the text placed at key points.

The resources below are available through Pearson's Instructor Resource Center, or from MyLab Math.

Instructor's Resource Manual with Tests

Includes mini-lectures for each text section, several forms of tests per chapter—two diagnostic pretests, four free-response and two multiple-choice test forms per chapter, and two final exams.

Instructor's Solutions Manual

Contains detailed, worked-out solutions to all exercises in the text.

TestGen®

Enables instructors to build, edit, print, and administer tests using a computerized bank of questions developed to cover all the objectives of the text. TestGen is algorithmically based, allowing instructors to create multiple but equivalent versions of the same question or test with the click of a button. Instructors can also modify test bank questions or add new questions.

PowerPoint Lecture Slides

Available for download only, these slides present key concepts and definitions from the text. Accessible versions of the PowerPoint slides are also available for students who are vision-impaired.

Student Resources

Guided Notebook

This Guided Notebook helps students keep their work organized as they work through their course. The notebook includes:

- Guided Examples that are worked out for students, plus corresponding Now Try This exercises for each text objective.
- Extra practice exercises for every section of the text, with ample space for students to show their work.
- Learning objectives and key vocabulary terms for every text section, along with vocabulary practice problems.

Student Solutions Manual

Provides completely worked-out solutions to the odd-numbered section exercises and to all exercises in the Now Trys, Relating Concepts, Chapter Reviews, Mixed Reviews, Chapter Tests, and Cumulative Reviews. Available at no additional charge in the MyLab Math course.

Using Your Math Text

Your text is a valuable resource. You will learn more if you make full use of the features it offers.

Now **TRY THIS**

General Features of This Text

Locate each feature, and complete any blanks.

- **Table of Contents** This is located at the front of the text.

  Find it and mark the chapters and sections you will cover, as noted on your course syllabus.

- **Answer Section** This is located at the back of the text.

  Tab this section so you can easily refer to it when doing homework or reviewing for tests.

- **List of Formulas** This helpful list of geometric formulas, along with review information on triangles and angles, is found at the back of the text.

  The formula for the volume of a cube is _____.

Specific Features of This Text

Look through Chapter 1 or 2 and give the number of a page that includes an example of each of the following specific features.

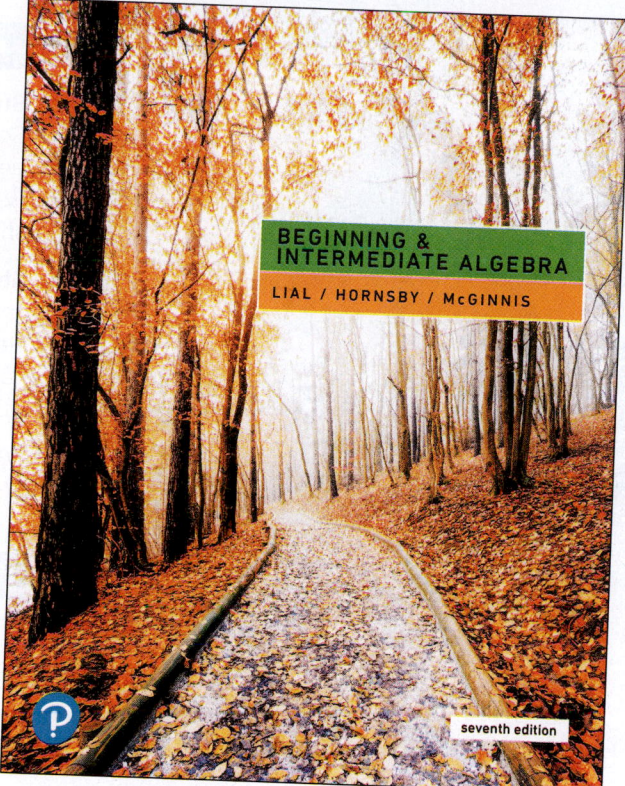

- **Objectives** The objectives are listed at the beginning of each section and again within the section as the corresponding material is presented. Once you finish a section, ask yourself if you have accomplished them. *See page _____.*

- **Vocabulary List** Important vocabulary is listed at the beginning of each section. You should be able to define these terms when you finish a section. *See page _____.*

- **Now Try Exercises** These margin exercises allow you to immediately practice the material covered in the examples and prepare you for the exercises. Check your results using the answers at the bottom of the page. *See page _____.*

- **Pointers** These small, shaded balloons provide on-the-spot warnings and reminders, point out key steps, and give other helpful tips. *See page _____.*

- **Cautions** These provide warnings about common errors that students often make or trouble spots to avoid. *See page _____.*

- **Notes** These provide additional explanations or emphasize other important ideas. *See page _____.*

- **Problem-Solving Hints** These boxes give helpful tips or strategies to use when you work applications. Look for them beginning in Chapter 2. *See page _____.*

Reading Your Math Text

Take time to read each section and its examples before doing your homework. You will learn more and be better prepared to work the exercises your instructor assigns.

Approaches to Reading Your Math Text

Student A learns best by listening to her teacher explain things. She "gets it" when she sees the instructor work problems. She previews the section before the lecture, so she knows generally what to expect. **Student A carefully reads the section in her text *AFTER* she hears the classroom lecture on the topic.**

Student B learns best by reading on his own. He reads the section and works through the examples before coming to class. That way, he knows what the teacher is going to talk about and what questions he wants to ask. **Student B carefully reads the section in his text *BEFORE* he hears the classroom lecture on the topic.**

Which of these reading approaches works best for you—that of Student A or Student B?

Tips for Reading Your Math Text

- **Turn off your cell phone and the TV.** You will be able to concentrate more fully on what you are reading.

- **Survey the material.** Glance over the assigned material to get an idea of the "big picture." Look at the list of objectives to see what you will be learning.

- **Read slowly.** Read only one section—or even part of a section—at a sitting, with paper and pencil in hand.

- **Pay special attention to important information given in colored boxes or set in boldface type.** Highlight any additional information you find helpful.

- **Study the examples carefully.** Pay particular attention to the blue side comments and any pointer balloons.

- **Do the Now Try exercises in the margin on separate paper as you go.** These problems mirror the examples and prepare you for the exercise set. Check your answers with those given at the bottom of the page.

- **Make study cards as you read.** Make cards for new vocabulary, rules, procedures, formulas, and sample problems.

- **Mark anything you don't understand. *ASK QUESTIONS* in class**—everyone will benefit. Follow up with your instructor, as needed.

Now TRY THIS

Think through and answer each question.

1. Which two or three reading tips given above will you try this week?

2. Did the tips you selected improve your ability to read and understand the material? Explain.

Taking Lecture Notes

Come to class prepared.

- Bring paper, pencils, notebook, text, completed homework, and any other materials you need.

- Arrive 10–15 minutes early if possible. Use the time before class to review your notes or study cards from the last class period.

- Select a seat carefully so that you can hear and see what is going on.

Study the set of sample math notes given at the right.

- **Include the date and the title** of the day's lecture topic.

- **Include definitions,** written here in parentheses—don't trust your memory.

- **Skip lines and write neatly** to make reading easier.

- **Emphasize direction words** (like *evaluate*, *simplify*, or *solve*) with their explanations.

- **Mark important concepts with stars, underlining, etc.**

- **Use two columns,** which allows an example and its explanation to be close together.

- **Use brackets and arrows** to clearly show steps, related material, etc.

- **Highlight any material and/or information that your instructor emphasizes.** Instructors often give "clues" about material that will definitely be on an exam.

January 12

Exponents

Exponents used to show repeated multiplication.

$3 \cdot 3 \cdot 3 \cdot 3$ can be written 3^4 — exponent (how many times it's multiplied)

base (the number being multiplied)

Read 3^2 as 3 to the 2nd power or 3 squared

3^3 as 3 to the 3rd power or 3 cubed

3^4 as 3 to the 4th power

etc.

Simplifying an expression with exponents → actually do the repeated multiplication

2^3 means $2 \cdot 2 \cdot 2$ and $2 \cdot 2 \cdot 2 = 8$

★ Careful! 5^2 means $5 \cdot 5$ NOT $5 \cdot 2$

so $5^2 = 5 \cdot 5 = 25$ BUT $5^2 \neq 10$

Example	Explanation
Simplify $2^4 \cdot 3^2$	Exponents mean multiplication.
$2 \cdot 2 \cdot 2 \cdot 2 \quad \cdot \quad 3 \cdot 3$	Use 2 as a factor 4 times. Use 3 as a factor 2 times.
$16 \quad \cdot \quad 9$	$2 \cdot 2 \cdot 2 \cdot 2$ is 16 $3 \cdot 3$ is 9 $\rangle$ $16 \cdot 9$ is 144
144	Simplified result is 144 (no exponents left)

Consider using a three-ring binder to organize your notes, class handouts, and completed homework.

Now TRY THIS

With a study partner or in a small group, compare lecture notes. Then answer each question.

1. What are you doing to show main points in your notes (such as boxing, using stars, etc.)?

2. In what ways do you set off explanations from worked problems and subpoints (such as indenting, using arrows, circling, etc.)?

3. What new ideas did you learn by examining your classmates' notes?

4. What new techniques will you try when taking notes in future lectures?

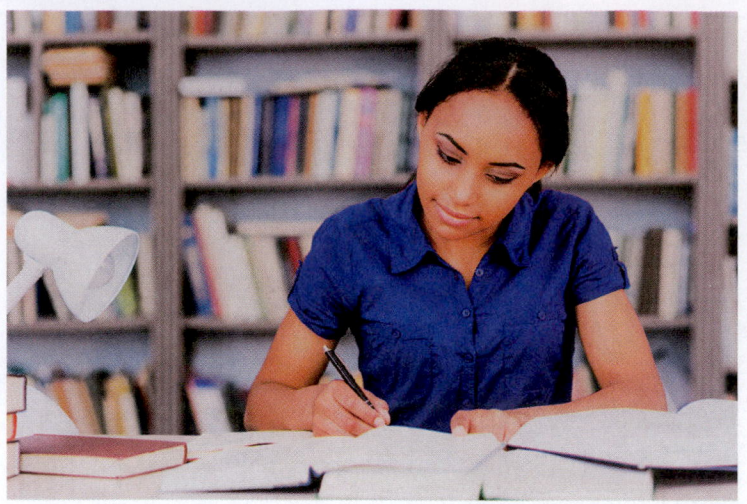

Completing Your Homework

You are ready to do your homework **AFTER** you have read the corresponding text section and worked through the examples and Now Try exercises.

Homework Tips

- **Keep distractions and potential interruptions to a minimum.** Turn off your cell phone and the TV. Find a quiet, comfortable place to work, away from a lot of other people, so you can concentrate on what you are doing.

- **Review your class notes.** Pay particular attention to anything your instructor emphasized during the lecture on this material.

- **Survey the exercise set.** Glance over the problems that your instructor has assigned to get a general idea of the types of exercises you will be working. Skim directions, and note any references to section examples.

- **Work problems neatly.** NEVER do your math homework in pen. Use pencil and write legibly, so others can read your work. Skip lines between steps. Clearly separate problems from each other.

- **Show all your work.** It is tempting to take shortcuts. Include ALL steps.

- **Check your work frequently to make sure you are on the right track.** It is hard to unlearn a mistake. For all odd-numbered problems, answers are given in the back of the text.

- **If you have trouble with a problem, refer to the corresponding worked example in the section.** The exercise directions will often reference specific examples to review. Pay attention to every line of the worked example to see how to get from step to step.

- **If you have trouble with an even-numbered problem, work the corresponding odd-numbered problem.** Check your answer in the back of the text, and apply the same steps to work the even-numbered problem.

- **If you have genuinely tried to work a problem but have not been able to complete it in a *reasonable* amount of time, it's ok to STOP.** Mark these problems. Ask for help at your school's tutor center or from fellow classmates, study partners, or your instructor.

- **Do some homework problems every day.** This is a good habit, even if your math class does not meet each day.

Now TRY THIS

Think through and answer each question.

1. What is your instructor's policy regarding homework?

2. Think about your current approach to doing homework. Be honest in your assessment.

    (a) What are you doing that is working well?

    (b) What improvements could you make?

3. Which one or two homework tips will you try this week?

4. In the event that you need help with homework, what resources are available? When does your instructor hold office hours?

Using Study Cards

You may have used "flash cards" in other classes. In math, "study cards" can help you remember terms and definitions, procedures, and concepts. Use study cards to

- Help you understand and learn the material;
- Quickly review when you have a few minutes;
- Review before a quiz or test.

One of the advantages of study cards is that you learn the material while you are making them.

Vocabulary Cards

Put the word and a page reference on the front of the card. On the back, write the definition, an example, any related words, and a sample problem (if appropriate).

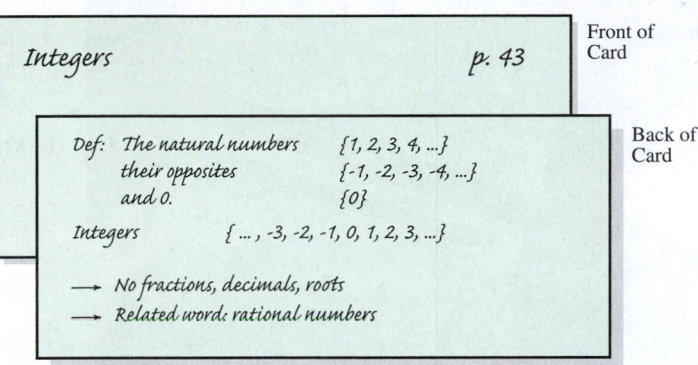

Front of Card

Integers   *p. 43*

Back of Card

Def: The natural numbers   {1, 2, 3, 4, ...}
their opposites   {-1, -2, -3, -4, ...}
and 0.   {0}

Integers   { ... , -3, -2, -1, 0, 1, 2, 3, ...}

⟶ *No fractions, decimals, roots*
⟶ *Related word: rational numbers*

Procedure ("Steps") Cards

Write the name of the procedure on the front of the card. Then write each step in words. On the back of the card, put an example showing each step.

Front of Card

Evaluating Absolute Value (Simplifying)   p. 48

1. Work inside absolute value bars first (like working inside parentheses).
2. Find the absolute value (never negative).
3. A negative sign in front of the absolute value bar is NOT affected, so keep it!

Back of Card

Examples:

Simplify |10 − 6|
                 |4| = 4

Work inside: 10 − 6 = 4
Absolute value of 4 is 4.

Simplify   − |−12|
                 −12

Absolute value of −12 is 12.
Keep negative sign that was in front.

Practice Problem Cards

Write a problem with direction words (like *solve, simplify*) on the front of the card, and work the problem on the back. Make one for each type of problem you learn.

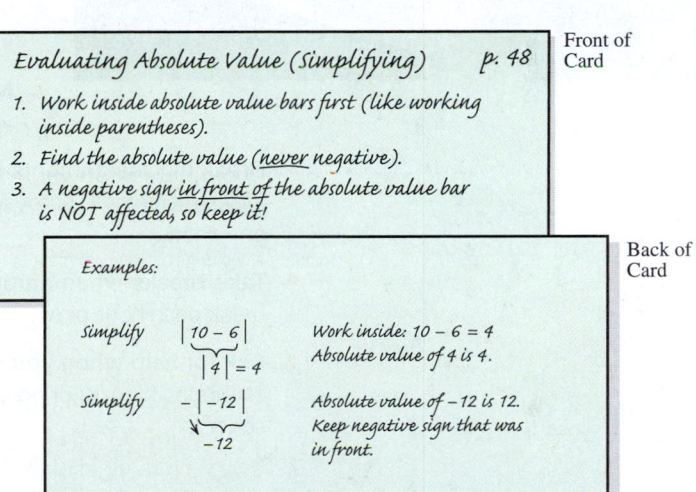

Front of Card

p. 121

Solve 4 (3x − 4) = 2 (6x − 9) + 2.

Back of Card

4 (3x −4) = 2 (6x − 9) + 2	
12x − 16 = 12x − 18 + 2	Distributive property
12x − 16 = 12x − 16	Combine like terms.
12x − 16 + 16 = 12x − 16 + 16	Add 16.
12x = 12x	Combine like terms.
12x − 12x = 12x − 12x	Subtract 12x.
0 = 0	True

When both sides of an equation are the same, it is called an identity.

Any real number will work, so the solution set is {all real numbers} (not just {0}).

Now **TRY THIS**

Make a vocabulary card, a procedure card, and a practice problem card for material that you are learning or reviewing.

Managing Your Time

Many college students juggle a busy schedule and multiple responsibilities, including school, work, and family demands.

Time Management Tips

- **Read the syllabus for each class.** Understand class policies, such as attendance, late homework, and make-up tests. Find out how you are graded.

- **Make a semester or quarter calendar.** Put test dates and major due dates for *all* your classes on the *same* calendar. Try using a different color for each class.

- **Make a weekly schedule.** After you fill in your classes and other regular responsibilities, block off some study periods. Aim for 2 hours of study for each 1 hour in class.

- **Choose a regular study time and place** (such as the campus library). Routine helps.

- **Keep distractions to a minimum.** Get the most out of the time you have set aside for studying by limiting interruptions. Turn off your cell phone. Take a break from social media. Avoid studying in front of the TV.

- **Make "to-do" lists.** Number tasks in order of importance. To see your progress, cross off tasks as you complete them.

- **Break big assignments into smaller chunks.** Don't wait until the last minute to begin big assignments or to study for tests. Make deadlines for each smaller chunk so that you stay on schedule.

- **Take breaks when studying.** Do not try to study for hours at a time. Take a 10-minute break each hour or so.

- **Ask for help when you need it.** Talk with your instructor during office hours. Make use of the learning/tutoring center, counseling office, or other resources available at your school.

Now TRY THIS

Work through the following, answering any questions.

1. Evaluate when and where you are currently studying. Are these places quiet and comfortable? Are you studying when you are most alert?

2. Which of the above tips will you try this week to improve your time management?

3. Create a weekly calendar that includes your class times, study times, and other family and/or work obligations.

4. Once the week is over, evaluate how these tips worked. Did you use your calendar and stick to it? What will you do differently next week?

5. Ask classmates, friends, and/or family members for tips on how they manage their time. Try any that you think might work for you.

Reviewing a Chapter

Your text provides extensive material to help you prepare for quizzes or tests in this course. Refer to the **Chapter 1 Summary** as you read through the following techniques.

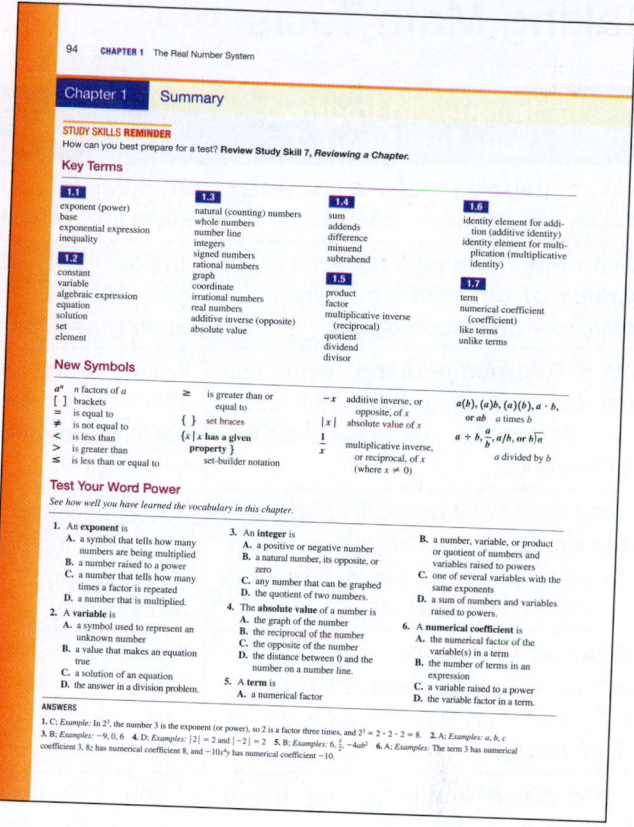

Techniques for Reviewing a Chapter

- **Review the Key Terms and any New Symbols.** Make a study card for each. Include a definition, an example, a sketch (if appropriate), and a section or page reference.

- **Take the Test Your Word Power quiz** to check your understanding of new vocabulary. The answers immediately follow.

- **Read the Quick Review.** Pay special attention to the headings. Study the explanations and examples given for each concept. Try to think about the whole chapter.

- **Reread your lecture notes.** Focus on what your instructor has emphasized in class, and review that material in your text.

- **Look over your homework.** Pay special attention to any trouble spots.

- **Work the Review Exercises.** They are grouped by section. Answers are included at the back of the text.

  ▸ Pay attention to direction words, such as *simplify, solve,* and *evaluate.*

  ▸ Are your answers exact and complete? Did you include the correct labels, such as $, cm², ft, etc.?

  ▸ Make study cards for difficult problems.

- **Work the Mixed Review Exercises.** They are in random order. Check your answers in the answer section at the back of the text.

- **Take the Chapter Test under test conditions.**

  ▸ Time yourself.

  ▸ Use a calculator or notes only if your instructor permits them on tests.

  ▸ Take the test in one sitting.

  ▸ Show all your work.

  ▸ Check your answers in the answer section. Section references are provided.

Reviewing a chapter takes time. Avoid rushing through your review in one night. Use the suggestions over a few days or evenings to better understand and remember the material.

Now TRY THIS

Follow these reviewing techniques to prepare for your next test. Then answer each question.

1. How much time did you spend reviewing for your test? Was it enough?

2. Which reviewing techniques worked best for you?

3. Are you investing enough time and effort to really *know* the material and set yourself up for success? Explain.

4. What will you do differently when reviewing for your next test?

Taking Math Tests

Techniques to Improve Your Test Score	Comments
Come prepared with a pencil, eraser, paper, and calculator, if allowed.	Working in pencil lets you erase, keeping your work neat.
Scan the entire test, note the point values of different problems, and plan your time accordingly.	To do 20 problems in 50 minutes, allow $50 \div 20 = 2.5$ minutes per problem. Spend less time on easier problems.
Do a "knowledge dump" when you get the test. Write important notes, such as formulas, in a corner of the test for reference.	Writing down tips and other special information that you've learned at the beginning allows you to relax as you take the test.
Read directions carefully, and circle any significant words. When you finish a problem, reread the directions. Did you do what was asked?	Pay attention to any announcements written on the board or made by your instructor. Ask if you don't understand something.
Show all your work. Many teachers give partial credit if some steps are correct, even if the final answer is wrong. **Write neatly.**	If your teacher can't read your writing, you won't get credit for it. If you need more space to work, ask to use extra paper.
Write down anything that might help solve a problem: a formula, a diagram, etc. If necessary, circle the problem and come back to it later. Do *not* erase anything you wrote down.	If you know even a little bit about a problem, write it down. The answer may come to you as you work on it, or you may get partial credit. Don't spend too long on any one problem.
If you can't solve a problem, make a guess. Do not change it unless you find an obvious mistake.	Have a good reason for changing an answer. Your first guess is usually your best bet.
Check that the answer to an application problem is reasonable and makes sense. Reread the problem to make sure you've answered the question.	Use common sense. Can the father really be seven years old? Would a month's rent be $32,140? Remember to label your answer if needed: $, years, inches, etc.
Check for careless errors. Rework each problem without looking at your previous work. Then compare the two answers.	Reworking a problem from the beginning forces you to rethink it. If possible, use a different method to solve the problem.

Now TRY THIS

Think through and answer each question.

1. What two or three tips will you try when you take your next math test?

2. How did the tips you selected work for you when you took your math test?

3. What will you do differently when taking your next math test?

4. Ask several classmates how they prepare for math tests. Did you learn any new preparation ideas?

Analyzing Your Test Results

An exam is a learning opportunity—learn from your mistakes. After a test is returned, do the following:

- **Note what you got wrong and why you had points deducted.**

- **Figure out how to solve the problems you missed.** Check your text or notes, or ask your instructor. Rework the problems correctly.

- **Keep all quizzes and tests that are returned to you.** Use them to study for future tests and the final exam.

Typical Reasons for Errors on Math Tests

1. You read the directions wrong.

2. You read the question wrong or skipped over something.

3. You made a computation error.

4. You made a careless error. (For example, you incorrectly copied a correct answer onto a separate answer sheet.)

5. Your answer was not complete.

6. You labeled your answer wrong. (For example, you labeled an answer "ft" instead of "ft^2.")

7. You didn't show your work.

These are test-taking errors. They are easy to correct if you read carefully, show all your work, proofread, and double-check units and labels.

8. You didn't understand a concept.

9. You were unable to set up the problem (in an application).

10. You were unable to apply a procedure.

These are test preparation errors. Be sure to practice all the kinds of problems that you will see on tests.

Now TRY THIS

Work through the following, answering any questions.

1. Use the sample charts at the right to track your test-taking progress. Refer to the tests you have taken so far in your course. For each test, check the appropriate box in the charts to indicate that you made an error in a particular category.

2. What test-taking errors did you make? Do you notice any patterns?

3. What test preparation errors did you make? Do you notice any patterns?

4. What will you do to avoid these kinds of errors on your next test?

▼ **Test-Taking Errors**

Test	Read directions wrong	Read question wrong	Made computation error	Made careless error	Answer not complete	Answer labeled wrong	Didn't show work
1							
2							
3							

▼ **Test Preparation Errors**

Test	Didn't understand concept	Didn't set up problem correctly	Couldn't apply a procedure
1			
2			
3			

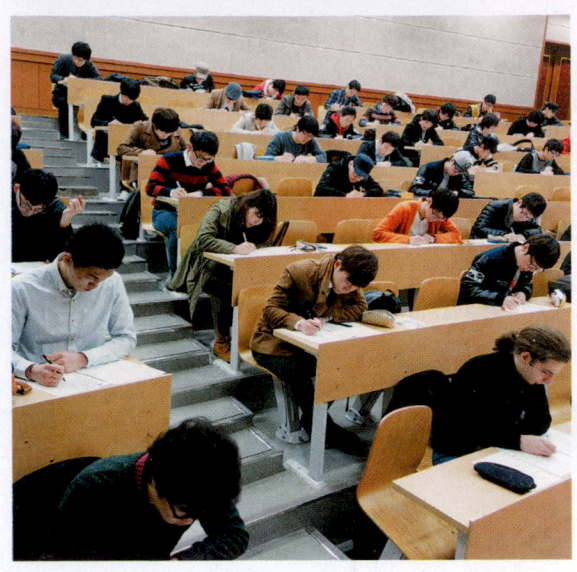

Preparing for Your Math Final Exam

Your math final exam is likely to be a comprehensive exam, which means it will cover material from the entire term. **One way to prepare for it now is by working a set of Cumulative Review Exercises** each time your class finishes a chapter. This continual review will help you remember concepts and procedures as you progress through the course.

Final Exam Preparation Suggestions

1. **Figure out the grade you need to earn on the final exam to get the course grade you want.** Check your course syllabus for grading policies, or ask your instructor if you are not sure.

2. **Create a final exam week plan.** Set priorities that allow you to spend extra time studying. This may mean making adjustments, in advance, in your work schedule or enlisting extra help with family responsibilities.

3. **Use the following suggestions to guide your studying.**

   - **Begin reviewing several days before the final exam.** DON'T wait until the last minute.

   - **Know exactly which chapters and sections will be covered on the exam.**

   - **Divide up the chapters.** Decide how much you will review each day.

   - **Keep returned quizzes and tests. Use them to review.**

   - **Practice all types of problems. Use the Cumulative Review Exercises** at the end of each chapter in your text beginning in Chapter 1. All answers, with section references, are given in the answer section at the back of the text.

   - **Review or rewrite your notes** to create summaries of important information.

   - **Make study cards for all types of problems.** Carry the cards with you, and review them whenever you have a few minutes.

   - **Take plenty of short breaks as you study to reduce physical and mental stress.** Exercising, listening to music, and enjoying a favorite activity are effective stress busters.

   Finally, *DON'T* stay up all night the night before an exam—*get a good night's sleep*.

Now TRY THIS

Think through and answer each question.

1. How many points do you need to earn on your math final exam to get the grade you want in your course?

2. What adjustments to your usual routine or schedule do you need to make for final exam week? List two or three.

3. Which of the suggestions for studying will you use as you prepare for your math final exam? List two or three.

4. Analyze your final exam results. How will you prepare differently next time?

PREALGEBRA REVIEW

R.1 Fractions
R.2 Decimals and Percents

OBJECTIVES

1. Write numbers in factored form.
2. Write fractions in lowest terms.
3. Convert between improper fractions and mixed numbers.
4. Multiply and divide fractions.
5. Add and subtract fractions.
6. Solve applied problems that involve fractions.
7. Interpret data in a circle graph.

The numbers used most often in everyday life are the **natural (counting) numbers,**

$$1, 2, 3, 4, \ldots,$$

the **whole numbers,**

$$0, 1, 2, 3, 4, \ldots,$$

and **fractions,** such as

$$\frac{1}{2}, \quad \frac{2}{3}, \quad \text{and} \quad \frac{11}{12}.$$

The three dots, or *ellipsis points,* indicate that each list of numbers continues in the same way indefinitely.

The parts of a fraction are named as shown.

Fraction bar → $\dfrac{3}{8}$ ← Numerator
     ← Denominator

The fraction bar represents division $\left(\frac{a}{b} = a \div b \right).$

> **NOTE**   Fractions are a way to represent parts of a whole. In a fraction, the **numerator** gives the number of parts being represented. The **denominator** gives the total number of equal parts in the whole. See **FIGURE 1.**
>
>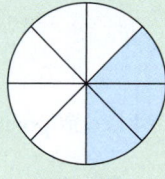
>
> The shaded region represents $\frac{3}{8}$ of the circle.
>
> **FIGURE 1**

VOCABULARY

☐ natural (counting) numbers
☐ whole numbers
☐ fractions
☐ numerator
☐ denominator
☐ proper fraction
☐ improper fraction
☐ factors
☐ product
☐ prime number
(continued)

A fraction is classified as being either a **proper fraction** or an **improper fraction.**

Proper fractions $\dfrac{1}{5}, \dfrac{2}{7}, \dfrac{9}{10}, \dfrac{23}{25}$ Numerator **is less than** denominator. Value is less than 1.

Improper fractions $\dfrac{3}{2}, \dfrac{5}{5}, \dfrac{11}{7}, \dfrac{28}{4}$ Numerator **is greater than or equal to** denominator. Value is greater than or equal to 1.

OBJECTIVE 1   Write numbers in factored form.

In the statement $3 \times 6 = 18$, the numbers 3 and 6 are **factors** of 18. Other factors of 18 include 1, 2, 9, and 18. The result of the multiplication, 18, is the **product.** We can represent the product of two numbers, such as 3 and 6, in several ways.

$$3 \times 6, \quad 3 \cdot 6, \quad (3)(6), \quad (3)6, \quad 3(6) \qquad \text{Products}$$

We *factor* a number by writing it as the product of two or more numbers.

Multiplication
$$\underset{\substack{\uparrow \quad \uparrow \quad \uparrow \\ \text{Factors Product}}}{3 \; \cdot \; 6 = 18}$$

Factoring
$$\underset{\substack{\uparrow \quad \uparrow \quad \uparrow \\ \text{Product Factors}}}{18 = 3 \; \cdot \; 6}$$

Factoring is the reverse of multiplying two numbers to get the product.

> **NOTE**   In algebra, a raised dot $\cdot$ is often used instead of the $\times$ symbol to indicate multiplication because $\times$ may be confused with the letter *x*.

A natural number greater than 1 is **prime** if it has only itself and 1 as factors. "Factors" are understood here to mean natural number factors.

$$2, 3, 5, 7, 11, 13, 17, 19, 23, 29, 31, 37 \qquad \text{First dozen prime numbers}$$

A natural number greater than 1 that is not prime is a **composite number.**

$$4, 6, 8, 9, 10, 12, 14, 15, 16, 18, 20, 21 \qquad \text{First dozen composite numbers}$$

The number 1 is considered to be neither prime nor composite.

⟳ NOW TRY EXERCISE 1

Identify the number 60 as *prime, composite,* or *neither.* If the number is composite, write it as a product of prime factors.

EXAMPLE 1   Writing Numbers in Prime Factored Form

Identify each number as *prime, composite,* or *neither.* If the number is composite, write it as a product of prime factors.

(a) 43

There are no natural numbers other than 1 and 43 itself that divide *evenly* into 43, so the number 43 is prime.

(b) 35

The number 35 is composite and can be written as the product of the prime factors 5 and 7.

$$35 = 5 \cdot 7$$

(c) 24

The number 24 is composite. We show a factor tree on the right, with prime factors circled.

Divide by the least prime factor of 24, which is 2. $24 = 2 \cdot 12$

Divide 12 by 2 to find two factors of 12. $24 = 2 \cdot 2 \cdot 6$

Now factor 6 as $2 \cdot 3$. $24 = \underbrace{2 \cdot 2 \cdot 2 \cdot 3}$
All factors are prime.

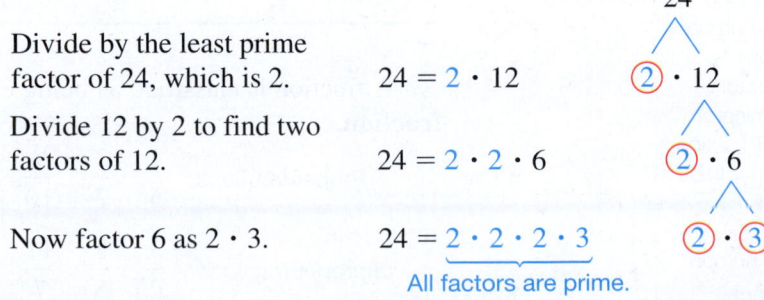

> **NOTE**   No matter which prime factor we start with when factoring, we will *always* obtain the same prime factorization. We verify this in **Example 1(c)** by starting with 3 instead of 2.
>
> Divide 24 by 3. $24 = 3 \cdot 8$
>
> Divide 8 by 2. $24 = 3 \cdot 2 \cdot 4$
>
> Divide 4 by 2. $24 = 3 \cdot 2 \cdot 2 \cdot 2$
>
> The same prime factors result.

OBJECTIVE 2 Write fractions in lowest terms.

The following properties are useful when writing a fraction in *lowest terms*.

Properties of 1

Any nonzero number divided by itself is equal to 1.   *Example:* $\frac{3}{3} = 1$

Any number multiplied by 1 remains the same.   *Example:* $\frac{2}{5} \cdot 1 = \frac{2}{5}$

A fraction is in **lowest terms** when the numerator and denominator have no factors in common (other than 1).

Writing a Fraction in Lowest Terms

Step 1   Write the numerator and denominator in factored form.

Step 2   Replace each pair of factors common to the numerator and denominator with 1.

Step 3   Multiply the remaining factors in the numerator and in the denominator.

(This procedure is sometimes called **"simplifying the fraction."**)

EXAMPLE 2   Writing Fractions in Lowest Terms

Write each fraction in lowest terms.

(a) $\dfrac{10}{15} = \dfrac{2 \cdot 5}{3 \cdot 5} = \dfrac{2}{3} \cdot \dfrac{5}{5} = \dfrac{2}{3} \cdot 1 = \dfrac{2}{3}$ Use the first property of 1 to replace $\frac{5}{5}$ with 1.

(b) $\dfrac{15}{45}$

By inspection, the greatest common factor of 15 and 45 is 15.

$$\frac{15}{45} = \frac{15}{3 \cdot 15} = \frac{1}{3 \cdot 1} = \frac{1}{3}$$

> Remember to write 1 in the numerator.

If the greatest common factor is not obvious, factor the numerator and denominator into prime factors.

$$\frac{15}{45} = \frac{3 \cdot 5}{3 \cdot 3 \cdot 5} = \frac{1 \cdot 1}{3 \cdot 1 \cdot 1} = \frac{1}{3}$$ The same answer results.

NOW TRY EXERCISE 2

Write each fraction in lowest terms.

(a) $\dfrac{30}{42}$ (b) $\dfrac{10}{70}$ (c) $\dfrac{72}{120}$

(c) $\dfrac{150}{200} = \dfrac{3 \cdot 50}{4 \cdot 50} = \dfrac{3}{4} \cdot 1 = \dfrac{3}{4}$ 50 is the greatest common factor of 150 and 200.

Another strategy is to choose *any* common factor and work in stages.

$$\dfrac{150}{200} = \dfrac{15 \cdot 10}{20 \cdot 10} = \dfrac{3 \cdot 5 \cdot 10}{4 \cdot 5 \cdot 10} = \dfrac{3}{4} \cdot 1 \cdot 1 = \dfrac{3}{4}$$ The same answer results.

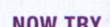 NOW TRY

OBJECTIVE 3   Convert between improper fractions and mixed numbers.

A **mixed number** is a single number that represents the sum of a natural number and a proper fraction. The mixed number $2\frac{3}{4}$ is illustrated in **FIGURE 2**.

Mixed number $\rightarrow$ $2\dfrac{3}{4} = 2 + \dfrac{3}{4}$

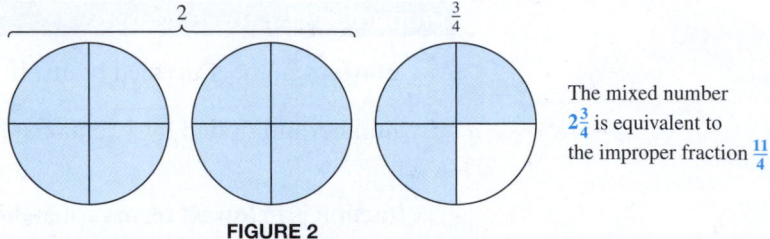

The mixed number $2\frac{3}{4}$ is equivalent to the improper fraction $\frac{11}{4}$.

FIGURE 2

NOW TRY EXERCISE 3

Write $\dfrac{92}{5}$ as a mixed number.

EXAMPLE 3   Converting an Improper Fraction to a Mixed Number

Write $\dfrac{59}{8}$ as a mixed number.

Because the fraction bar represents division $\left(\dfrac{a}{b} = a \div b, \text{ or } b\overline{)a}\,\right)$, divide the numerator of the improper fraction by the denominator.

Denominator of fraction $\rightarrow$ $8\overline{)59}$ $\begin{array}{r} 7 \leftarrow \text{Quotient} \\ \underline{56} \\ 3 \end{array}$ $\leftarrow$ Numerator of fraction $\dfrac{59}{8} = 7\dfrac{3}{8}$

$\quad$ $\leftarrow$ Remainder

NOW TRY

NOW TRY EXERCISE 4

Write $11\frac{2}{3}$ as an improper fraction.

EXAMPLE 4   Converting a Mixed Number to an Improper Fraction

Write $6\frac{4}{7}$ as an improper fraction.

Multiply the denominator of the fraction by the natural number and then add the numerator to obtain the numerator of the improper fraction.

$$7 \cdot 6 = 42 \quad \text{and} \quad 42 + 4 = 46$$

The denominator of the improper fraction is the same as the denominator in the mixed number, which is 7 here.

$$6\dfrac{4}{7} = \dfrac{7 \cdot 6 + 4}{7} = \dfrac{46}{7}$$

NOW TRY

NOW TRY ANSWERS

2. (a) $\frac{5}{7}$ (b) $\frac{1}{7}$ (c) $\frac{3}{5}$
3. $18\frac{2}{5}$
4. $\frac{35}{3}$

Multiplying Fractions

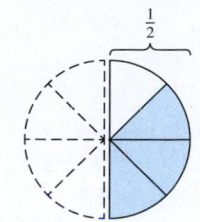

$\frac{3}{4}$ of $\frac{1}{2}$ is equivalent to $\frac{3}{4} \cdot \frac{1}{2}$, which equals $\frac{3}{8}$ of the circle.

FIGURE 3

NOW TRY EXERCISE 5

Find each product, and write it in lowest terms as needed.

(a) $\frac{4}{7} \cdot \frac{5}{8}$ (b) $3\frac{2}{5} \cdot 6\frac{2}{3}$

NOW TRY ANSWERS

5. (a) $\frac{5}{14}$ (b) $\frac{68}{3}$, or $22\frac{2}{3}$

OBJECTIVE 4 Multiply and divide fractions.

FIGURE 3 illustrates multiplying fractions.

Multiplying Fractions

If $\frac{a}{b}$ and $\frac{c}{d}$ are fractions, then $\frac{a}{b} \cdot \frac{c}{d} = \frac{a \cdot c}{b \cdot d}$.

That is, to multiply two fractions, multiply their numerators and then multiply their denominators.

EXAMPLE 5 Multiplying Fractions

Find each product, and write it in lowest terms as needed.

(a) $\frac{3}{8} \cdot \frac{4}{9}$

$= \frac{3 \cdot 4}{8 \cdot 9}$ Multiply numerators. Multiply denominators.

$= \frac{12}{72}$ Multiply.

$= \frac{1 \cdot 12}{6 \cdot 12}$ The greatest common factor of 12 and 72 is 12.

Make sure the product is in lowest terms.

$= \frac{1}{6}$ $\frac{1 \cdot 12}{6 \cdot 12} = \frac{1}{6} \cdot 1 = \frac{1}{6}$

Another strategy is to factor and divide out any common factors *before* multiplying.

$\frac{3}{8} \cdot \frac{4}{9}$

$= \frac{3}{2 \cdot 4} \cdot \frac{4}{3 \cdot 3}$ Factor.

$= \frac{1}{2 \cdot 3}$ Divide out common factors. Multiply.

$= \frac{1}{6}$ The same answer results.

(b) $2\frac{1}{3} \cdot 5\frac{1}{4}$

$= \frac{7}{3} \cdot \frac{21}{4}$ Write each mixed number as an improper fraction.

$= \frac{7 \cdot 21}{3 \cdot 4}$ Multiply numerators. Multiply denominators.

$= \frac{7 \cdot 3 \cdot 7}{3 \cdot 4}$ Factor the numerator.

Think: $\frac{49}{4}$ means $49 \div 4$.

$= \frac{49}{4}$, or $12\frac{1}{4}$ Write in lowest terms and as a mixed number.

NOW TRY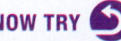

Two numbers are **reciprocals** of each other if their product is 1.

▼ Reciprocals

Number	Reciprocal
$\frac{3}{4}$	$\frac{4}{3}$
$\frac{11}{7}$	$\frac{7}{11}$
$\frac{1}{5}$	5, or $\frac{5}{1}$
10, or $\frac{10}{1}$	$\frac{1}{10}$

A number and its reciprocal have a product of 1. For example,

$$\frac{3}{4} \cdot \frac{4}{3} = \frac{12}{12}, \text{ or } 1.$$

Dividing Fractions

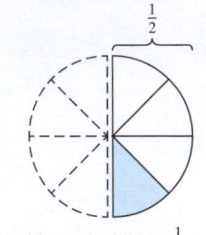

$\frac{1}{2} \div 4$ is equivalent to $\frac{1}{2} \cdot \frac{1}{4}$, which equals $\frac{1}{8}$ of the circle.

FIGURE 4

Division is the inverse or opposite of multiplication, and as a result we use reciprocals to divide fractions. **FIGURE 4** illustrates dividing fractions.

Dividing Fractions

If $\frac{a}{b}$ and $\frac{c}{d}$ are fractions, then $\qquad \frac{a}{b} \div \frac{c}{d} = \frac{a}{b} \cdot \frac{d}{c}.$

Multiply by the reciprocal.

That is, to divide by a fraction, multiply by its reciprocal.

As an example of why this procedure works, we know that

$$20 \div 10 = 2 \quad \text{and also that} \quad 20 \cdot \frac{1}{10} = 2.$$

The answer to a division problem is a **quotient**. In $\frac{a}{b} \div \frac{c}{d}$, the first fraction $\frac{a}{b}$ is the **dividend,** and the second fraction $\frac{c}{d}$ is the **divisor.**

EXAMPLE 6   Dividing Fractions

Find each quotient, and write it in lowest terms as needed.

(a) $\frac{3}{4} \div \frac{8}{5}$

$= \frac{3}{4} \cdot \frac{5}{8}$ Multiply by the reciprocal of the divisor.

$= \frac{3 \cdot 5}{4 \cdot 8}$ Multiply numerators. Multiply denominators.

$= \frac{15}{32}$ Make sure the quotient is in lowest terms.

(b) $\frac{3}{4} \div \frac{5}{8}$

$= \frac{3}{4} \cdot \frac{8}{5}$ Multiply by the reciprocal.

$= \frac{3 \cdot 4 \cdot 2}{4 \cdot 5}$ Multiply and factor.

$= \frac{6}{5}, \text{ or } 1\frac{1}{5}$

(c) $\frac{5}{8} \div 10$ Think of 10 as $\frac{10}{1}$ here.

$= \frac{5}{8} \cdot \frac{1}{10}$ Multiply by the reciprocal.

$= \frac{5 \cdot 1}{8 \cdot 2 \cdot 5}$ Multiply and factor.

$= \frac{1}{16}$ Remember to write 1 in the numerator.

NOW TRY EXERCISE 6

Find each quotient, and write it in lowest terms as needed.

(a) $\dfrac{2}{7} \div \dfrac{8}{9}$ **(b)** $3\dfrac{3}{4} \div 4\dfrac{2}{7}$

(d) $1\dfrac{2}{3} \div 4\dfrac{1}{2}$

$= \dfrac{5}{3} \div \dfrac{9}{2}$ Write each mixed number as an improper fraction.

$= \dfrac{5}{3} \cdot \dfrac{2}{9}$ Multiply by the reciprocal of the divisor.

$= \dfrac{10}{27}$ Multiply. The quotient is in lowest terms. **NOW TRY**

Adding Fractions

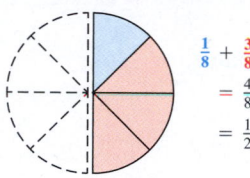

$\dfrac{1}{8} + \dfrac{3}{8}$

$= \dfrac{4}{8}$

$= \dfrac{1}{2}$

FIGURE 5

OBJECTIVE 5 Add and subtract fractions.

The result of adding two numbers is the **sum** of the numbers. For example, $2 + 3 = 5$, so 5 is the sum of 2 and 3.

FIGURE 5 illustrates adding fractions.

Adding Fractions

If $\dfrac{a}{b}$ and $\dfrac{c}{b}$ are fractions, then $\dfrac{a}{b} + \dfrac{c}{b} = \dfrac{a + c}{b}.$

That is, to find the sum of two fractions having the *same* denominator, add the numerators and *keep the same denominator.*

NOW TRY EXERCISE 7

Find the sum, and write it in lowest terms as needed.

$\dfrac{1}{8} + \dfrac{3}{8}$

EXAMPLE 7 Adding Fractions (Same Denominator)

Find each sum, and write it in lowest terms as needed.

(a) $\dfrac{3}{7} + \dfrac{2}{7}$

$= \dfrac{3 + 2}{7}$ Add numerators. Keep the same denominator.

$= \dfrac{5}{7}$ The sum is in lowest terms.

(b) $\dfrac{2}{10} + \dfrac{3}{10}$

$= \dfrac{2 + 3}{10}$ Add numerators. Keep the same denominator.

$= \dfrac{5}{10}$

$= \dfrac{1}{2}$ Write in lowest terms.

NOW TRY

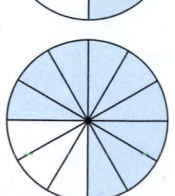

$\dfrac{3}{4}$ and $\dfrac{9}{12}$ are equivalent fractions.

FIGURE 6

If the fractions to be added do *not* have the same denominator, we must first rewrite them with a common denominator. For example, to rewrite $\dfrac{3}{4}$ as an equivalent fraction with denominator 12, think as follows.

$$\dfrac{3}{4} = \dfrac{?}{12}$$

NOW TRY ANSWERS

6. **(a)** $\dfrac{9}{28}$ **(b)** $\dfrac{7}{8}$

7. $\dfrac{1}{2}$

We must find the number that can be multiplied by 4 to give 12. Because $4 \cdot 3 = 12$, by the second property of 1 we multiply the numerator and the denominator by 3.

$$\dfrac{3}{4} = \dfrac{3}{4} \cdot 1 = \dfrac{3}{4} \cdot \dfrac{3}{3} = \dfrac{3 \cdot 3}{4 \cdot 3} = \dfrac{9}{12}$$

$\dfrac{3}{4}$ is equivalent to $\dfrac{9}{12}$. See **FIGURE 6.**

NOTE The process of writing an equivalent fraction is the reverse of writing a fraction in lowest terms.

Finding the Least Common Denominator (LCD)

To add or subtract fractions with different denominators, find the **least common denominator (LCD)** as follows.

Step 1 Factor each denominator using prime factors.

Step 2 The LCD is the product of every (different) factor that appears in any of the factored denominators. If a factor is repeated, use the greatest number of repeats as factors of the LCD.

Step 3 Write each fraction with the LCD as the denominator.

EXAMPLE 8 Adding Fractions (Different Denominators)

Find each sum, and write it in lowest terms as needed.

(a) $\dfrac{4}{15} + \dfrac{5}{9}$

Step 1 To find the LCD, factor each denominator using prime factors.

$$15 = 5 \cdot 3 \quad \text{and} \quad 9 = 3 \cdot 3 \qquad \text{The } \textit{different} \text{ factors are 3 and 5.}$$

Step 2
$$\overset{15 \quad 9}{\text{LCD} = 5 \cdot 3 \cdot 3 = 45}$$

In this example, the LCD needs one factor of 5 and two factors of 3 because the second denominator has two factors of 3.

Step 3 Write each fraction with 45 as denominator.

$$\dfrac{4}{15} = \dfrac{4}{15} \cdot \dfrac{3}{3} = \dfrac{12}{45} \quad \text{and} \quad \dfrac{5}{9} = \dfrac{5}{9} \cdot \dfrac{5}{5} = \dfrac{25}{45}$$

At this stage, the fractions are *not* in lowest terms.

$$\dfrac{4}{15} + \dfrac{5}{9}$$

$$= \dfrac{12}{45} + \dfrac{25}{45} \qquad \text{Use the equivalent fractions with the common denominator.}$$

Make sure the sum is in lowest terms.

$$= \dfrac{37}{45} \qquad \begin{array}{l}\text{Add numerators.}\\ \text{Keep the same denominator.}\end{array}$$

(b) $3\dfrac{1}{2} + 2\dfrac{3}{4}$

Method 1 $\qquad 3\dfrac{1}{2} + 2\dfrac{3}{4}$

$$= \dfrac{7}{2} + \dfrac{11}{4} \qquad \text{Write each mixed number as an improper fraction.}$$

Think: $\dfrac{7}{2} \cdot \dfrac{2}{2} = \dfrac{14}{4}$

$$= \dfrac{14}{4} + \dfrac{11}{4} \qquad \text{Find a common denominator. The LCD is 4.}$$

$$= \dfrac{25}{4}, \quad \text{or} \quad 6\dfrac{1}{4} \qquad \text{Add. Write as a mixed number.}$$

**NOW TRY
EXERCISE 8**
Find each sum, and write it in lowest terms as needed.

(a) $\dfrac{5}{12} + \dfrac{3}{8}$ (b) $3\dfrac{1}{4} + 5\dfrac{5}{8}$

Method 2

$$3\dfrac{1}{2} = 3\dfrac{2}{4}$$
$$+ 2\dfrac{3}{4} = 2\dfrac{3}{4}$$

Write $3\dfrac{1}{2}$ as $3\dfrac{2}{4}$. Then add vertically.

Add the whole numbers and the fractions separately.

$$5\dfrac{5}{4} = 5 + 1\dfrac{1}{4} = 6\dfrac{1}{4}, \quad \text{or} \quad \dfrac{25}{4} \qquad \text{The same answer results.}$$

 NOW TRY

Subtracting Fractions

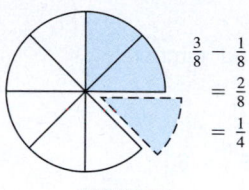

$$\dfrac{3}{8} - \dfrac{1}{8}$$
$$= \dfrac{2}{8}$$
$$= \dfrac{1}{4}$$

FIGURE 7

The result of subtracting one number from another number is the **difference** of the numbers. For example, $9 - 5 = 4$, so 4 is the difference of 9 and 5.

FIGURE 7 illustrates subtracting fractions.

Subtracting Fractions

If $\dfrac{a}{b}$ and $\dfrac{c}{b}$ are fractions, then $\qquad \dfrac{a}{b} - \dfrac{c}{b} = \dfrac{a - c}{b}.$

That is, to find the difference of two fractions having the *same* denominator, subtract the numerators and *keep the same denominator.*

EXAMPLE 9 Subtracting Fractions

Find each difference, and write it in lowest terms as needed.

(a) $\dfrac{15}{8} - \dfrac{3}{8}$

$\quad = \dfrac{15 - 3}{8}$ 　　Subtract numerators.
　　　　　　　　　Keep the same denominator.

$\quad = \dfrac{12}{8}$

Think: $\dfrac{12}{8} = \dfrac{3 \cdot 4}{2 \cdot 4} = \dfrac{3}{2}$ 　$= \dfrac{3}{2}, \quad \text{or} \quad 1\dfrac{1}{2}$ 　　Write in lowest terms and as a mixed number.

(b) $\dfrac{15}{16} - \dfrac{4}{9}$

$\quad = \dfrac{15}{16} \cdot \dfrac{9}{9} - \dfrac{4}{9} \cdot \dfrac{16}{16}$ 　　Because 16 and 9 have no common factors except 1, the LCD is $16 \cdot 9 = 144$.

$\quad = \dfrac{135}{144} - \dfrac{64}{144}$ 　　Write equivalent fractions.

$\quad = \dfrac{71}{144}$ 　　Subtract numerators.
　　　　　　Keep the common denominator.

(c) $\dfrac{7}{18} - \dfrac{4}{15}$

$\quad = \dfrac{7}{2 \cdot 3 \cdot 3} \cdot \dfrac{5}{5} - \dfrac{4}{3 \cdot 5} \cdot \dfrac{2 \cdot 3}{2 \cdot 3}$ 　　$18 = 2 \cdot 3 \cdot 3$ and $15 = 3 \cdot 5$, so the LCD is $2 \cdot 3 \cdot 3 \cdot 5 = 90$.

$\quad = \dfrac{35}{90} - \dfrac{24}{90}$ 　　Write equivalent fractions.

$\quad = \dfrac{11}{90}$ 　　Subtract. The answer is in lowest terms.

NOW TRY ANSWERS

8. (a) $\dfrac{19}{24}$ (b) $\dfrac{71}{8}$, or $8\dfrac{7}{8}$

NOW TRY
EXERCISE 9

Find each difference, and write it in lowest terms as needed.

(a) $\dfrac{5}{11} - \dfrac{2}{9}$ **(b)** $4\dfrac{1}{3} - 2\dfrac{5}{6}$

(d) $4\dfrac{1}{2} - 1\dfrac{3}{4}$

Method 1 $4\dfrac{1}{2} - 1\dfrac{3}{4}$

$$= \dfrac{9}{2} - \dfrac{7}{4}$$ Write each mixed number as an improper fraction.

$$= \dfrac{18}{4} - \dfrac{7}{4}$$ Find a common denominator. The LCD is 4.

Think: $\dfrac{9}{2} \cdot \dfrac{2}{2} = \dfrac{18}{4}$

$$= \dfrac{11}{4}, \quad \text{or} \quad 2\dfrac{3}{4}$$ Subtract. Write as a mixed number.

Method 2
$$4\dfrac{1}{2} = 4\dfrac{2}{4} = 3\dfrac{6}{4}$$ The LCD is 4.
$4\dfrac{2}{4} = 3 + 1 + \dfrac{2}{4} = 3 + \dfrac{4}{4} + \dfrac{2}{4} = 3\dfrac{6}{4}$

$$-1\dfrac{3}{4} = 1\dfrac{3}{4} = 1\dfrac{3}{4}$$

$$\overline{}$$

$$2\dfrac{3}{4}, \quad \text{or} \quad \dfrac{11}{4}$$ The same answer results.

NOW TRY

OBJECTIVE 6   Solve applied problems that involve fractions.

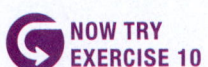

NOW TRY
EXERCISE 10

A board is $10\dfrac{1}{2}$ ft long. If it must be divided into four pieces of equal length for shelves, how long must each piece be?

EXAMPLE 10   Adding Fractions to Solve an Applied Problem

The diagram in **FIGURE 8** appears with directions for a woodworking project. Find the height of the desk to the top of the writing surface.

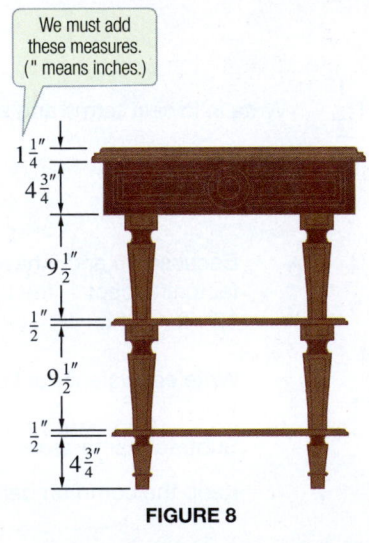

We must add these measures. (" means inches.)

FIGURE 8

Add the section and writing surface heights to obtain the total height. The common denominator is 4.

$$1\dfrac{1}{4} \rightarrow 1\dfrac{1}{4}$$

$$4\dfrac{3}{4} \rightarrow 4\dfrac{3}{4}$$

$$9\dfrac{1}{2} = 9\dfrac{2}{4}$$

$$\dfrac{1}{2} = \dfrac{2}{4}$$

$$9\dfrac{1}{2} = 9\dfrac{2}{4}$$

$$\dfrac{1}{2} = \dfrac{2}{4}$$

$$+\,4\dfrac{3}{4} \rightarrow 4\dfrac{3}{4}$$

$$\overline{}$$

$$27\dfrac{15}{4}$$

Because $\dfrac{15}{4}$ is an improper fraction, we simplify this answer.

Think: $\dfrac{15}{4}$ means $15 \div 4$.

Because $\dfrac{15}{4} = 3\dfrac{3}{4}$, we have $27\dfrac{15}{4} = 27 + 3\dfrac{3}{4} = 30\dfrac{3}{4}$. The height is $30\dfrac{3}{4}$ in.

NOW TRY

OBJECTIVE 7 Interpret data in a circle graph.

In a **circle graph,** or **pie chart,** a circle is used to indicate the total of all the data categories represented. The circle is divided into *sectors,* or wedges, whose sizes show the relative magnitudes of the categories. The sum of all the fractional parts must be 1 (for 1 whole circle).

NOW TRY
EXERCISE 11

Refer to the circle graph in **FIGURE 9.**

(a) Which region had the least number of Internet users?

(b) *Estimate* the number of Internet users in Asia.

(c) How many *actual* Internet users were there in Asia?

EXAMPLE 11 Using a Circle Graph to Interpret Information

In a recent year, there were about 3900 million (that is, 3.9 billion) Internet users worldwide. The circle graph in **FIGURE 9** shows the fractions of these users living in various regions of the world.

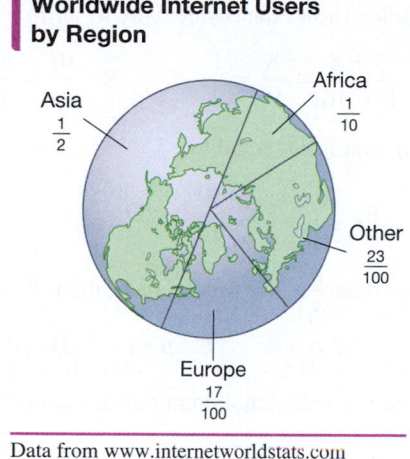

Worldwide Internet Users by Region

Asia $\frac{1}{2}$

Africa $\frac{1}{10}$

Other $\frac{23}{100}$

Europe $\frac{17}{100}$

Data from www.internetworldstats.com

FIGURE 9

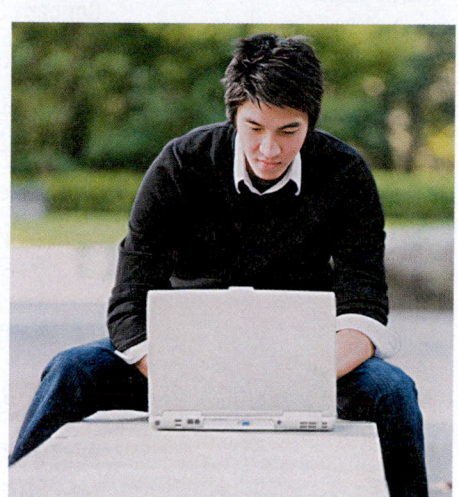

(a) Which region had the largest share of Internet users? What was that share?

The sector for Asia is the largest. Asia had the largest share of Internet users, $\frac{1}{2}$.

(b) *Estimate* the number of Internet users in Europe.

A share of $\frac{17}{100}$ can be rounded to $\frac{20}{100}$, or $\frac{1}{5}$, and the total number of Internet users, 3900 million, can be rounded to 4000 million. We multiply $\frac{1}{5}$ by 4000.

$$\frac{1}{5} \cdot 4000 = 800 \text{ million} \qquad \text{Approximate number of Internet users in Europe}$$

(c) How many *actual* Internet users were there in Europe?

$$\frac{17}{100} \cdot 3900 \qquad \text{Multiply the actual fraction from the graph for Europe by the number of users in millions.}$$

$$= \frac{17}{100} \cdot \frac{3900}{1} \qquad a = \frac{a}{1}, \text{ for all } a.$$

$$= \frac{66{,}300}{100} \qquad \begin{array}{l}\text{Multiply numerators.}\\ \text{Multiply denominators.}\end{array}$$

$$= 663 \qquad \text{Divide.}$$

NOW TRY ANSWERS

11. (a) Africa
  (b) 2000 million, or 2 billion
  (c) 1950 million, or 1.95 billion

Thus, 663 million people in Europe used the Internet.   **NOW TRY**

R.1 Exercises

FOR EXTRA HELP   ▶ **MyLab Math**

▶ *Video solutions for select problems available in MyLab Math*

STUDY SKILLS REMINDER
You will increase your chance of success in this course if you fully utilize your text. **Review Study Skill 1,** *Using Your Math Text.*

Concept Check   *Decide whether each statement is* true *or* false. *If it is false, explain why.*

1. In the fraction $\frac{5}{8}$, 5 is the numerator and 8 is the denominator.

2. The mixed number equivalent of the improper fraction $\frac{31}{5}$ is $6\frac{1}{5}$.

3. The fraction $\frac{7}{7}$ is proper.

4. The number 1 is prime.

5. The fraction $\frac{13}{39}$ is in lowest terms.

6. The reciprocal of $\frac{6}{2}$ is $\frac{3}{1}$.

7. The product of 10 and 2 is 12.

8. The difference of 10 and 2 is 5.

Concept Check   *Choose the letter of the correct response.*

9. Which choice shows the correct way to write $\frac{16}{24}$ in lowest terms?

A. $\frac{16}{24} = \frac{8+8}{8+16} = \frac{8}{16} = \frac{1}{2}$ **B.** $\frac{16}{24} = \frac{4 \cdot 4}{4 \cdot 6} = \frac{4}{6}$ **C.** $\frac{16}{24} = \frac{8 \cdot 2}{8 \cdot 3} = \frac{2}{3}$

10. Which fraction is *not* equal to $\frac{5}{9}$?

A. $\frac{15}{27}$ **B.** $\frac{30}{54}$ **C.** $\frac{40}{74}$ **D.** $\frac{55}{99}$

11. For the fractions $\frac{p}{q}$ and $\frac{r}{s}$, which one of the following can serve as a common denominator?

A. $q \cdot s$ **B.** $q + s$ **C.** $p \cdot r$ **D.** $p + r$

12. Which fraction with denominator 24 is equivalent to $\frac{5}{8}$?

A. $\frac{21}{24}$ **B.** $\frac{15}{24}$ **C.** $\frac{5}{24}$ **D.** $\frac{10}{24}$

Identify each number as prime, composite, *or* neither. *If the number is composite, write it as a product of prime factors.* **See Example 1.**

13. 19   **14.** 31   **15.** 30   **16.** 50

17. 64   **18.** 81   **19.** 1   **20.** 0

21. 57   **22.** 51   **23.** 79   **24.** 83   **25.** 124

26. 138   **27.** 500   **28.** 700   **29.** 3458   **30.** 1025

Write each fraction in lowest terms. **See Example 2.**

31. $\frac{8}{16}$ **32.** $\frac{4}{12}$ **33.** $\frac{15}{18}$ **34.** $\frac{16}{20}$ **35.** $\frac{90}{150}$

36. $\frac{100}{140}$ **37.** $\frac{18}{90}$ **38.** $\frac{16}{64}$ **39.** $\frac{144}{120}$ **40.** $\frac{132}{77}$

Write each improper fraction as a mixed number. **See Example 3.**

41. $\frac{12}{7}$ **42.** $\frac{16}{9}$ **43.** $\frac{77}{12}$ **44.** $\frac{101}{15}$ **45.** $\frac{83}{11}$ **46.** $\frac{67}{13}$

Write each mixed number as an improper fraction. **See Example 4.**

47. $2\frac{3}{5}$ **48.** $5\frac{6}{7}$ **49.** $10\frac{3}{8}$ **50.** $12\frac{2}{3}$ **51.** $10\frac{1}{5}$ **52.** $18\frac{1}{6}$

Find each product or quotient, and write it in lowest terms as needed. **See Examples 5 and 6.**

53. $\dfrac{4}{5} \cdot \dfrac{6}{7}$ **54.** $\dfrac{5}{9} \cdot \dfrac{2}{7}$ **55.** $\dfrac{2}{15} \cdot \dfrac{3}{8}$ **56.** $\dfrac{3}{20} \cdot \dfrac{5}{21}$

57. $\dfrac{1}{10} \cdot \dfrac{12}{5}$ **58.** $\dfrac{1}{8} \cdot \dfrac{10}{7}$ **59.** $\dfrac{15}{4} \cdot \dfrac{8}{25}$ **60.** $\dfrac{21}{8} \cdot \dfrac{4}{7}$

61. $21 \cdot \dfrac{3}{7}$ **62.** $36 \cdot \dfrac{4}{9}$ **63.** $3\dfrac{1}{4} \cdot 1\dfrac{2}{3}$ **64.** $2\dfrac{2}{3} \cdot 1\dfrac{3}{5}$

65. $2\dfrac{3}{8} \cdot 3\dfrac{1}{5}$ **66.** $3\dfrac{3}{5} \cdot 7\dfrac{1}{6}$ **67.** $5 \cdot 2\dfrac{1}{10}$ **68.** $3 \cdot 4\dfrac{2}{9}$

69. $\dfrac{7}{9} \div \dfrac{3}{2}$ **70.** $\dfrac{6}{11} \div \dfrac{5}{4}$ **71.** $\dfrac{5}{4} \div \dfrac{3}{8}$ **72.** $\dfrac{7}{5} \div \dfrac{3}{10}$

73. $\dfrac{32}{5} \div \dfrac{8}{15}$ **74.** $\dfrac{24}{7} \div \dfrac{6}{21}$ **75.** $\dfrac{3}{4} \div 12$ **76.** $\dfrac{2}{5} \div 30$

77. $6 \div \dfrac{3}{5}$ **78.** $8 \div \dfrac{4}{9}$ **79.** $6\dfrac{3}{4} \div \dfrac{3}{8}$ **80.** $5\dfrac{3}{5} \div \dfrac{7}{10}$

81. $2\dfrac{1}{2} \div 1\dfrac{5}{7}$ **82.** $2\dfrac{2}{9} \div 1\dfrac{2}{5}$ **83.** $2\dfrac{5}{8} \div 1\dfrac{15}{32}$ **84.** $2\dfrac{3}{10} \div 1\dfrac{4}{5}$

Find each sum or difference, and write it in lowest terms as needed. **See Examples 7–9.**

85. $\dfrac{7}{15} + \dfrac{4}{15}$ **86.** $\dfrac{2}{9} + \dfrac{5}{9}$ **87.** $\dfrac{7}{12} + \dfrac{1}{12}$ **88.** $\dfrac{3}{16} + \dfrac{5}{16}$

89. $\dfrac{5}{9} + \dfrac{1}{3}$ **90.** $\dfrac{4}{15} + \dfrac{1}{5}$ **91.** $\dfrac{3}{8} + \dfrac{5}{6}$ **92.** $\dfrac{5}{6} + \dfrac{2}{9}$

93. $\dfrac{5}{9} + \dfrac{3}{16}$ **94.** $\dfrac{3}{4} + \dfrac{6}{25}$ **95.** $3\dfrac{1}{8} + 2\dfrac{1}{4}$ **96.** $4\dfrac{2}{3} + 2\dfrac{1}{6}$

97. $3\dfrac{1}{4} + 1\dfrac{4}{5}$ **98.** $5\dfrac{3}{4} + 1\dfrac{1}{3}$ **99.** $\dfrac{7}{9} - \dfrac{2}{9}$ **100.** $\dfrac{8}{11} - \dfrac{3}{11}$

101. $\dfrac{13}{15} - \dfrac{3}{15}$ **102.** $\dfrac{11}{12} - \dfrac{3}{12}$ **103.** $\dfrac{7}{12} - \dfrac{1}{3}$ **104.** $\dfrac{5}{6} - \dfrac{1}{2}$

105. $\dfrac{7}{12} - \dfrac{1}{9}$ **106.** $\dfrac{11}{16} - \dfrac{1}{12}$ **107.** $4\dfrac{3}{4} - 1\dfrac{2}{5}$ **108.** $3\dfrac{4}{5} - 1\dfrac{4}{9}$

109. $6\dfrac{1}{4} - 5\dfrac{1}{3}$ **110.** $5\dfrac{1}{3} - 4\dfrac{1}{2}$ **111.** $8\dfrac{2}{9} - 4\dfrac{2}{3}$ **112.** $7\dfrac{5}{12} - 4\dfrac{5}{6}$

Work each problem involving fractions.

113. For each description, write a fraction in lowest terms that represents the region described.

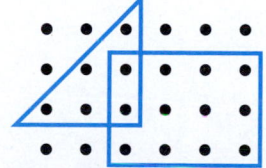

    (a) The dots in the rectangle as a part of the dots in the entire figure

    (b) The dots in the triangle as a part of the dots in the entire figure

    (c) The dots in the overlapping region of the triangle and the rectangle as a part of the dots in the triangle alone

    (d) The dots in the overlapping region of the triangle and the rectangle as a part of the dots in the rectangle alone

114. At the conclusion of the Pearson softball league season, batting statistics for five players were as shown in the table.

Player	At-Bats	Hits	Home Runs
Maureen	36	12	3
Christine	40	9	2
Chase	11	5	1
Joe	16	8	0
Greg	20	10	2

Use this information to answer each question. Estimate as necessary.

(a) Which player got a hit in exactly $\frac{1}{3}$ of his or her at-bats?

(b) Which player got a hit in just less than $\frac{1}{2}$ of his or her at-bats?

(c) Which player got a home run in just less than $\frac{1}{10}$ of his or her at-bats?

(d) Which player got a hit in just less than $\frac{1}{4}$ of his or her at-bats?

(e) Which two players got hits in exactly the same fractional part of their at-bats? What was the fractional part, expressed in lowest terms?

Use the table to work each problem.

115. How many cups of water would be needed for eight microwave servings?

116. How many teaspoons of salt would be needed for five stove-top servings? (*Hint:* 5 servings is halfway between 4 and 6 servings.)

	Microwave	Stove Top		
Servings	1	1	4	6
Water	$\frac{3}{4}$ cup	1 cup	3 cups	4 cups
Grits	3 Tbsp	3 Tbsp	$\frac{3}{4}$ cup	1 cup
Salt (optional)	Dash	Dash	$\frac{1}{4}$ tsp	$\frac{1}{2}$ tsp

Data from www.quakeroats.com

The Pride Golf Tee Company uses the Professional Tee System shown in the figure. Use the information given to work each problem. (Data from www.pridegolftee.com)

117. Find the difference in length between the ProLength Plus and the once-standard Shortee.

118. The ProLength Max tee is the longest tee allowed by the U.S. Golf Association's *Rules of Golf.* How much longer is the ProLength Max than the Shortee?

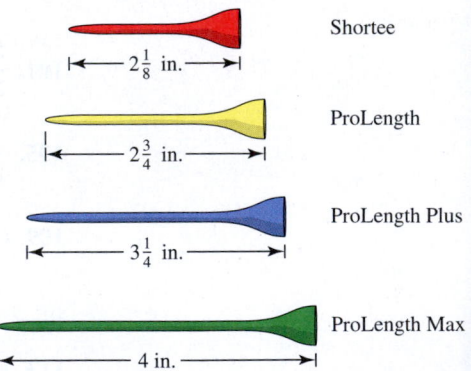

Solve each problem. **See Example 10.**

119. A hardware store sells a 40-piece socket wrench set. The measure of the largest socket is $\frac{3}{4}$ in., while the measure of the smallest is $\frac{3}{16}$ in. What is the difference between these measures?

120. Two sockets in a socket wrench set have measures of $\frac{9}{16}$ in. and $\frac{3}{8}$ in. What is the difference between these two measures?

121. A piece of property has an irregular shape, with five sides, as shown in the figure. Find the total distance around the piece of property. (This distance is the **perimeter** of the figure.)

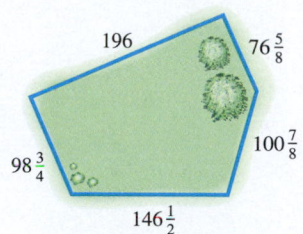

196    $76\frac{5}{8}$

$100\frac{7}{8}$

$98\frac{3}{4}$

$146\frac{1}{2}$

Measurements are in feet.

122. Find the perimeter of the triangle in the figure.

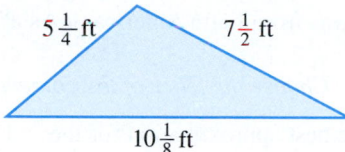

$5\frac{1}{4}$ ft $7\frac{1}{2}$ ft

$10\frac{1}{8}$ ft

123. A board is $15\frac{5}{8}$ in. long. If it must be divided into three pieces of equal length, how long must each piece be?

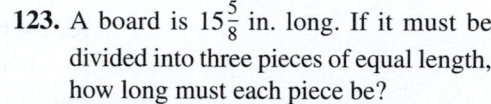

$15\frac{5}{8}$ in.

124. Paul's favorite recipe for barbecue sauce calls for $2\frac{1}{3}$ cups of tomato sauce. The recipe makes enough barbecue sauce to serve seven people. How much tomato sauce is needed for one serving?

125. A cake recipe calls for $1\frac{3}{4}$ cups of sugar. A caterer has $15\frac{1}{2}$ cups of sugar on hand. How many cakes can he make?

126. Kyla needs $2\frac{1}{4}$ yd of fabric to cover a chair. How many chairs can she cover with $23\frac{2}{3}$ yd of fabric?

127. It takes $2\frac{3}{8}$ yd of fabric to make a costume for a school play. How much fabric would be needed for seven costumes?

128. A cookie recipe calls for $2\frac{2}{3}$ cups of sugar. How much sugar would be needed to make four batches of cookies?

129. First published in 1953, the digest-sized *TV Guide* has changed to a full-sized magazine. The full-sized magazine is 3 in. wider than the old guide. What is the difference in their heights? (Data from *TV Guide*.)

$7\frac{1}{8}$ in.

$10\frac{1}{2}$ in.

⟵5 in.⟶   ⟵8 in.⟶

Old   New

130. Under existing standards, most of the holes in Swiss cheese must have diameters between $\frac{11}{16}$ and $\frac{13}{16}$ in. To accommodate new high-speed slicing machines, the U.S. Department of Agriculture wants to reduce the minimum size to $\frac{3}{8}$ in. How much smaller is $\frac{3}{8}$ in. than $\frac{11}{16}$ in.? (Data from U.S. Department of Agriculture.)

Approximately 40 million people living in the United States were born in other countries. The circle graph gives the fractional number from each region of birth for these people. Use the graph to work each problem. **See Example 11.**

131. *Estimate* the number of people living in the United States who were born in Europe. Then find the *actual* number who were born in Europe.

132. How many people were born in Latin America?

133. What fractional part of the foreign-born population was from Other regions?

134. What fractional part of the foreign-born population was from Latin America or Asia?

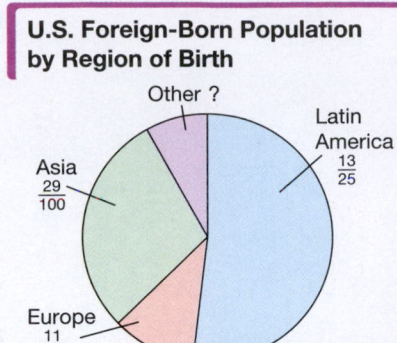

U.S. Foreign-Born Population by Region of Birth

Data from U.S. Census Bureau.

Extending Skills *Choose the letter of the correct response.*

135. Estimate the best approximation for the sum.

$$\frac{14}{26} + \frac{98}{99} + \frac{100}{51} + \frac{90}{31} + \frac{13}{27}$$

A. 5 **B.** 6 **C.** 7 **D.** 8

136. Estimate the best approximation for the product.

$$\frac{202}{50} \cdot \frac{99}{100} \cdot \frac{21}{40} \cdot \frac{75}{36}$$

A. 3 **B.** 4 **C.** 8 **D.** 16

R.2 Decimals and Percents

OBJECTIVES
1 Write decimals as fractions.
2 Add and subtract decimals.
3 Multiply and divide decimals.
4 Write fractions as decimals.
5 Write percents as decimals and decimals as percents.
6 Write percents as fractions and fractions as percents.
7 Solve applied problems that involve percents.

Fractions are one way to represent parts of a whole. Another way is with a decimal fraction or **decimal,** a number written with a decimal point.

9.25, 14.001, 0.3   Decimal numbers

In **FIGURE 10** at the right, 3 parts of the whole 10 are shaded. As a fraction, we say that $\frac{3}{10}$ of the figure is shaded. As a decimal, 0.3 is shaded. Both of these numbers are read *"three-tenths."*

Each digit in a decimal number has a place value, as shown below.

FIGURE 10

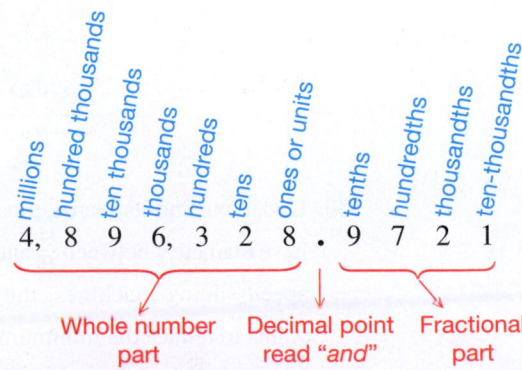

Each successive place value is ten times greater than the place value to its right and one-tenth as great as the place value to its left.

OBJECTIVE 1 Write decimals as fractions.

Place value is used to write a decimal number as a fraction.

VOCABULARY
- decimal
- decimal places
- divisor
- dividend
- quotient
- terminating decimal
- repeating decimal
- percent

> ### Converting a Decimal to a Fraction
>
> Read the decimal using the correct place value. Write it in fractional form just as it is read.
>
> - The numerator will be the digits to the right of the decimal point.
> - The denominator will be a power of 10—that is, 10 for tenths, 100 for hundredths, and so on.

NOW TRY EXERCISE 1

Write each decimal as a fraction. (Do not write in lowest terms.)

(a) 0.8   (b) 0.431   (c) 2.58

EXAMPLE 1 Writing Decimals as Fractions

Write each decimal as a fraction. (Do not write in lowest terms.)

(a) 0.95   We read 0.95 as "*ninety-five hundredths*."

$$0.95 = \frac{95}{100} \leftarrow \text{For hundredths}$$

(b) 0.056   We read 0.056 as "*fifty-six thousandths*."

> Do not confuse **0.056** with **0.56**, read "fifty-six *hundredths*," which is the fraction $\frac{56}{100}$.

$$0.056 = \frac{56}{1000} \leftarrow \text{For thousandths}$$

(c) 4.2095   We read this decimal number, which is greater than 1, as "*Four **and** two thousand ninety-five ten-thousandths*."

$$4.2095 = 4\frac{2095}{10,000} \quad \text{Write the decimal number as a mixed number.}$$

Think: $10,000 \cdot 4 + 2095$

$$= \frac{42,095}{10,000} \quad \text{Write the mixed number as an improper fraction.}$$

NOW TRY

OBJECTIVE 2 Add and subtract decimals.

EXAMPLE 2 Adding and Subtracting Decimals

Add or subtract as indicated.

(a) 6.92 + 14.8 + 3.217

Place the digits of the decimal numbers in columns by place value, so that tenths are in one column, hundredths in another column, and so on.

Be sure to line up decimal points.

```
  6.92      Decimal points are aligned.
 14.8
+ 3.217
```

To avoid errors, attach zeros as placeholders so that there are the same number of places to the right of each decimal point.

```
  6.92                    6.920     Attach 0s.
 14.8       becomes      14.800
+ 3.217                 + 3.217
                        ─────────
                         24.937
```

6.92 is equivalent to 6.920.
14.8 is equivalent to 14.800.

NOW TRY ANSWERS

1. (a) $\frac{8}{10}$ (b) $\frac{431}{1000}$ (c) $\frac{258}{100}$

NOW TRY EXERCISE 2

Add or subtract as indicated.

(a) 68.9 + 42.72 + 8.973

(b) 351.8 − 2.706

(b) 47.6 − 32.509

$$\begin{array}{r} 47.6 \\ -\ 32.509 \end{array}$$ becomes

$$\begin{array}{r} 47.600 \\ -\ 32.509 \\ \hline 15.091 \end{array}$$ Write the decimal numbers in columns, attaching 0s to 47.6.

(c) 3 − 0.253

$$\begin{array}{r} 3.000 \\ -\ 0.253 \\ \hline 2.747 \end{array}$$ A whole number is assumed to have the decimal point at the right of the number. Write 3 as 3.000.   **NOW TRY**

OBJECTIVE 3 Multiply and divide decimals.

> **Multiplying Decimals**
>
> **Step 1** Ignore the decimal points, and multiply as if the numbers were whole numbers.
>
> **Step 2** Add the number of **decimal places** (digits to the *right* of the decimal point) in each factor. Place the decimal point that many digits from the right in the product.

NOW TRY EXERCISE 3

Multiply.

(a) 9.32 × 1.4

(b) 0.6 × 0.004

EXAMPLE 3   Multiplying Decimals

Multiply.

(a) 29.3 × 4.52

$$\begin{array}{r} 29.3 \\ \times\ \ 4.52 \\ \hline 586 \\ 1465 \\ 1172 \\ \hline 132.436 \end{array}$$

1 decimal place
2 decimal places
1 + 2 = 3

3 decimal places

(b) 31.42 × 65

$$\begin{array}{r} 31.42 \\ \times\ \ \ \ 65 \\ \hline 15710 \\ 18852 \\ \hline 2042.30 \end{array}$$

2 decimal places
0 decimal places
2 + 0 = 2

2 decimal places

The final 0 can be dropped, and the product can be written 2042.3.

(c) 0.05 × 0.3   Here 5 × 3 = 15. Be careful placing the decimal point.

2 decimal places   1 decimal place

0.05   ×   0.3

= 0.015   ⟵ Do *not* write 0.150.

2 + 1 = 3 decimal places
Attach 0 as a placeholder
in the tenths place.   **NOW TRY**

> **NOTE**   Decimal numbers can be multiplied by converting to fractions first.
>
> 0.05 × 0.3   See Example 3(c).
>
> $= \dfrac{5}{100} \times \dfrac{3}{10}$ Write decimals as fractions.
>
> $= \dfrac{15}{1000}$ Multiply fractions.
>
> = 0.015   Write "*fifteen thousandths*" as a decimal. The same answer results.

SECTION R.2

$$5 \leftarrow \text{Quotient}$$

Divisor $\rightarrow 25\overline{)125}$

$\uparrow$

Dividend

Remember this terminology for the parts of a division problem.

Dividing Decimals

Step 1   Change the **divisor** (the number we are dividing *by*) into a whole number by moving the decimal point as many places as necessary to the right.

Step 2   Move the decimal point in the **dividend** (the number we are dividing *into*) to the right by the same number of places.

Step 3   Move the decimal point straight up, and then divide as with whole numbers to find the **quotient.**

NOW TRY EXERCISE 4

Divide.

(a) $451.47 \div 14.9$

(b) $7.334 \div 1.3$
  (Round the quotient to two decimal places.)

EXAMPLE 4   Dividing Decimals

Divide.

(a) $233.45 \div 11.5$

Write the problem as follows. $11.5\overline{)233.45}$

$11.5\overline{)233.45}$ To change the divisor 11.5 into a whole number, move *each* decimal point one place to the right.

To see why this works, write the division in fractional form and multiply by $\frac{10}{10}$. The result is the same as when we moved the decimal point one place to the right in the divisor and the dividend.

$$\frac{233.45}{11.5} \cdot \frac{10}{10} = \frac{2334.5}{115}$$ Multiplying by $\frac{10}{10}$ is equivalent to multiplying by 1.

Move the decimal point straight up and divide as with whole numbers.

```
        20.3
115)2334.5
    230
     345    115 does not divide into 34, so we used
     345    zero as a placeholder in the quotient.
       0
```
Move the decimal point straight up.

(b) $8.949 \div 1.25$ (Round the quotient to two decimal places.)

$1.25\overline{)8.949}$ Move each decimal point two places to the right.

```
         7.159
125)894.900
    875
     199
     125
     740
     625
    1150
    1125
      25
```
Move the decimal point straight up, and divide as with whole numbers. Attach 0s as placeholders.

We carried out the division to three decimal places so that we could round to two decimal places, obtaining the quotient 7.16.

NOW TRY ANSWERS
4. (a) 30.3   (b) 5.64

NOW TRY

> **NOTE** To round 7.159 in **Example 4(b)** to two decimal places (that is, to the nearest hundredth), we look at the digit to the *right* of the hundredths place. **If this digit is 5 or greater, we round up. If it is less than 5, we drop the digit(s) beyond the desired place.**
>
> Hundredths place
> ↓
> 7.159   9, the digit to the right of the hundredths place, is 5 or greater.
> ≈ 7.16   Round 5 up to 6. ≈ means "is approximately equal to."

Multiplying and Dividing by Powers of 10 (Shortcuts)

• To **multiply** by a power of 10, **move the decimal point to the right** as many places as the number of zeros.

• To **divide** by a power of 10, **move the decimal point to the left** as many places as the number of zeros.

In both cases, insert 0s as placeholders if necessary.

NOW TRY EXERCISE 5

Multiply or divide as indicated.

(a) 294.72×10

(b) $4.793 \div 100$

EXAMPLE 5   Multiplying and Dividing by Powers of 10

Multiply or divide as indicated.

(a) 48.731×10
$= 48.731$ Move the decimal point one place
$= 487.31$ to the *right*.

(b) 48.731×1000
$= 48.731$ Move the decimal point three places
$= 48,731$ to the *right*.

(c) $48.731 \div 10$
$= 48.731$ Move the decimal point one place
$= 4.8731$ to the *left*.

(d) $48.731 \div 1000$
$= 048.731$ Move the decimal point three places
$= 0.048731$ to the *left*.

NOW TRY

OBJECTIVE 4   Write fractions as decimals.

Converting a Fraction to a Decimal

Because a fraction bar indicates division, write a fraction as a decimal by dividing the numerator by the denominator.

EXAMPLE 6   Writing Fractions as Decimals

Write each fraction as a decimal.

(a) $\dfrac{19}{8}$

$$\begin{array}{r} 2.375 \\ 8\overline{)19.000} \\ 16 \\ \overline{30} \\ 24 \\ \overline{60} \\ 56 \\ \overline{40} \\ 40 \\ \overline{0} \end{array}$$

Divide 19 by 8. Add a decimal point and as many 0s as necessary to 19.

(b) $\dfrac{2}{3}$

$$\begin{array}{r} 0.6666\ldots \\ 3\overline{)2.0000\ldots} \\ 18 \\ \overline{20} \\ 18 \\ \overline{20} \\ 18 \\ \overline{20} \\ 18 \\ \overline{20} \\ 18 \\ \overline{2} \end{array}$$

$\dfrac{19}{8} = 2.375$ ← Terminating decimal

$\dfrac{2}{3} = 0.6666\ldots$ ← Repeating decimal

NOW TRY EXERCISE 6

Write each fraction as a decimal. For repeating decimals, write the answer by first using bar notation and then rounding to the nearest thousandth.

(a) $\dfrac{17}{20}$ (b) $\dfrac{2}{9}$

- The remainder in the division in part (a) is 0, so this quotient is a **terminating decimal.**

- The remainder in the division in part (b) is never 0. Because a number, in this case 2, is always left after the subtraction, this quotient is a **repeating decimal.** A convenient notation for a repeating decimal is a bar over the digit (or digits) that repeats.

$$\frac{2}{3} = 0.6666\ldots, \quad \text{or} \quad 0.\overline{6}$$

We often round repeating decimals to as many places as needed.

$$\frac{2}{3} \approx 0.667 \qquad \text{An approximation to the nearest thousandth} \qquad \textbf{NOW TRY} \ \text{}$$

OBJECTIVE 5 Write percents as decimals and decimals as percents.

The word **percent** means "*per 100*." Percent is written with the symbol **%**.

> *"One percent" means "one per one hundred," or "one one-hundredth."*

In **FIGURE 11**, 35 of the 100 squares are shaded—that is, $\dfrac{35}{100}$, or 35%, are shaded.

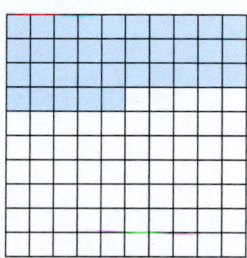

FIGURE 11

Percent, Fraction, and Decimal Equivalents

$$1\% = \frac{1}{100} = 0.01, \quad 10\% = \frac{10}{100} = 0.10, \quad 100\% = \frac{100}{100} = 1$$

NOW TRY EXERCISE 7

Write each percent as a decimal.

(a) 23%   (b) 350%

EXAMPLE 7 Writing Percents as Decimals

Write each percent as a decimal.

(a) 73%   73% means "73 *per one hundred*."

$$73\% = \frac{73}{100} = 0.73$$

Essentially, we are dropping the % symbol from 73% and dividing 73 by 100. Doing this moves the decimal point, which is understood to be after the 3, two places to the *left*.

(b) $125\% = \dfrac{125}{100} = 1.25$

> A percent greater than 100 represents a number greater than 1.

Drop the % symbol and move the decimal point, which is understood to be after the 5, two places to the *left*.

(c) $3\dfrac{1}{2}\%$

First write the fractional part as a decimal.

$$3\frac{1}{2}\% = (3 + 0.5)\% = 3.5\%$$

Now write the percent in decimal form.

$$3.5\% = \frac{3.5}{100} = 0.035$$

NOW TRY ANSWERS
6. (a) 0.85   (b) $0.\overline{2}$, 0.222
7. (a) 0.23   (b) 3.50, or 3.5

$\textbf{NOW TRY}$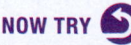

EXAMPLE 8   Writing Decimals as Percents

Write each decimal as a percent.

(a) 0.73

This conversion is the opposite of what we did in **Example 7(a)** when we wrote 73% as a decimal. There, we dropped the % symbol and divided 73 by 100. Here, we attach a % symbol and multiply 73 by 100. This is the same as multiplying by 1.

$$0.73 = 0.73 \cdot 100\% = 73\% \qquad 100\% = 1$$

Moving the decimal point two places to the *right* and attaching a % symbol gives the same result.

(b) $0.05 = 0.05 \cdot 100\% = 5\%$

(c) $2.63 = 2.63 \cdot 100\% = 263\%$

Move the decimal point two places to the *right* and attach a % symbol.

A number greater than 1 is more than 100%.

NOW TRY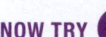

Converting Percents and Decimals (Shortcuts)

- To convert a percent to a decimal, move the decimal point two places to the *left* and drop the % symbol.

- To convert a decimal to a percent, move the decimal point two places to the *right* and attach a % symbol.

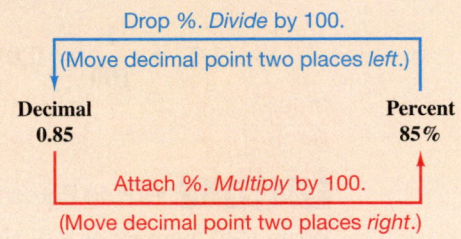

Drop %. *Divide* by 100.
(Move decimal point two places *left*.)

Decimal
0.85

Percent
85%

Attach %. *Multiply* by 100.
(Move decimal point two places *right*.)

EXAMPLE 9   Converting Percents and Decimals by Moving the Decimal Point

Convert each percent to a decimal and each decimal to a percent.

(a) $45\% = 0.45$ **(b)** $250\% = 2.50$, or 2.5 **(c)** $9\% = 09\% = 0.09$

(d) $0.57 = 57\%$ **(e)** $1.5 = 1.50 = 150\%$ **(f)** $0.007 = 0.7\%$

NOW TRY

OBJECTIVE 6   Write percents as fractions and fractions as percents.

EXAMPLE 10   Writing Percents as Fractions

Write each percent as a fraction. Give answers in lowest terms.

(a) 8%

We use the fact that percent means "*per one hundred*," and convert as follows.

$$8\% = \frac{8}{100}$$ As with converting percents to decimals, drop the % symbol and divide by 100.

NOW TRY EXERCISE 10

Write each percent as a fraction. Give answers in lowest terms.

(a) 20% **(b)** 160%

In lowest terms,

$$\frac{8}{100} = \frac{2 \cdot 4}{25 \cdot 4} = \frac{2}{25}.$$

Thus, $8\% = \frac{2}{25}$.

(b) $175\% = \frac{175}{100}$ Drop the % symbol, and divide by 100.

In lowest terms,

$$\frac{175}{100} = \frac{7 \cdot 25}{4 \cdot 25} = \frac{7}{4}, \quad \text{or} \quad 1\frac{3}{4}.$$

> A number greater than 1 is more than 100%.

(c) 13.5%

$$= \frac{13.5}{100}$$ Drop the % symbol. Divide by 100.

$$= \frac{13.5}{100} \cdot \frac{10}{10}$$ Use a property of 1 to write the numerator as a natural number.

$$= \frac{135}{1000}$$ Multiply.

$$= \frac{27}{200}$$ Write in lowest terms. $\frac{135}{1000} = \frac{27 \cdot 5}{200 \cdot 5} = \frac{27}{200}$

NOW TRY

We know that 100% of something is the whole thing. One way to convert a fraction to a percent is to multiply by 100%, which is equivalent to 1. This involves the same steps that are used for converting a decimal to a percent—*attach a % symbol and multiply by 100.*

NOW TRY EXERCISE 11

Write each fraction as a percent.

(a) $\frac{6}{25}$ **(b)** $\frac{7}{9}$

EXAMPLE 11 Writing Fractions as Percents

Write each fraction as a percent.

(a) $\frac{2}{5}$

$$= \frac{2}{5} \cdot 100\%$$ Multiply by 1 in the form 100%.

$$= \frac{2}{5} \cdot \frac{100}{1}\%$$

$$= \frac{2 \cdot 5 \cdot 20}{5 \cdot 1}\%$$ Multiply and factor.

$$= \frac{2 \cdot 20}{1}\%$$ Divide out the common factor.

$$= 40\%$$ Simplify.

(b) $\frac{1}{6}$

$$= \frac{1}{6} \cdot 100\%$$

$$= \frac{1}{6} \cdot \frac{100}{1}\%$$

$$= \frac{1 \cdot 2 \cdot 50}{2 \cdot 3 \cdot 1}\%$$

$$= \frac{50}{3}\%$$

$$= 16\frac{2}{3}\%, \quad \text{or} \quad 16.\overline{6}\%$$

NOW TRY

NOW TRY ANSWERS

10. (a) $\frac{1}{5}$ (b) $\frac{8}{5}$, or $1\frac{3}{5}$

11. (a) 24% (b) $77.\overline{7}\%$

OBJECTIVE 7 Solve applied problems that involve percents.

The decimal form of a percent is generally used in calculations.

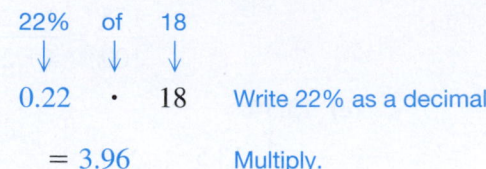

EXAMPLE 12   Using Percent to Solve an Applied Problem

A DVD with a regular price of $18 is on sale this week at 22% off. Find the amount of the discount and the sale price of the DVD.

The discount is 22% *of* 18. The word *of* here means multiply.

$$22\% \quad of \quad 18$$
$$\downarrow \qquad \downarrow \quad \downarrow$$
$$0.22 \quad \cdot \quad 18 \qquad \text{Write 22\% as a decimal.}$$

$$= 3.96 \qquad \text{Multiply.}$$

The discount is $3.96. The sale price is found by subtracting.

$$\$18.00 - \$3.96 = \$14.04 \qquad \text{Original price} - \text{discount} = \text{sale price}$$

NOW TRY

NOW TRY
EXERCISE 12

A winter coat is on sale for 60% off. The regular price is $120. Find the amount of the discount and the sale price of the coat.

NOW TRY ANSWER
12. $72; $48

R.2 Exercises

FOR EXTRA HELP

▶ **MyLab Math**

▶ *Video solutions for select problems available in MyLab Math*

STUDY SKILLS REMINDER
Be sure to read and work through the section material before working the exercises. **Review Study Skill 2, *Reading Your Math Text.***

Concept Check   *Provide the correct response.*

1. In the decimal number 367.9412, name the digit that has each place value.

   (a) tens   **(b)** tenths   **(c)** thousandths   **(d)** ones or units   **(e)** hundredths

2. Write a decimal number that has 5 in the thousands place, 0 in the tenths place, and 4 in the ten-thousandths place.

3. For the decimal number 46.249, round to the place value indicated.

   (a) hundredths   **(b)** tenths   **(c)** ones or units   **(d)** tens

4. Round each decimal to the nearest thousandth.

   (a) $0.\overline{8}$ **(b)** $0.\overline{4}$ **(c)** 0.9762   **(d)** 0.8645

*Write each decimal as a fraction. (Do not write in lowest terms.) **See Example 1.***

5. 0.4	**6.** 0.6	**7.** 0.64	**8.** 0.82	**9.** 0.138
10. 0.104	**11.** 0.043	**12.** 0.087	**13.** 3.805	**14.** 5.166

*Add or subtract as indicated. **See Example 2.***

15. 25.32 + 109.2 + 8.574	**16.** 90.527 + 32.43 + 589.8	**17.** 28.73 − 3.12
18. 46.88 − 13.45	**19.** 43.5 − 28.17	**20.** 345.1 − 56.31
21. 3.87 + 15 + 2.9	**22.** 8.2 + 1.09 + 12	**23.** 32.56 + 47.356 + 1.8
24. 75.2 + 123.96 + 3.897	**25.** 18 − 2.789	**26.** 29 − 8.582

*Multiply or divide as indicated. **See Examples 3 and 4.***

27. 12.8 × 9.1	**28.** 34.04 × 0.56	**29.** 22.41 × 33
30. 55.76 × 72	**31.** 0.2 × 0.03	**32.** 0.07 × 0.004

33. $78.65 \div 11$ **34.** $73.36 \div 14$ **35.** $32.48 \div 11.6$

36. $85.26 \div 17.4$ **37.** $19.967 \div 9.74$ **38.** $44.4788 \div 5.27$

*Multiply or divide as indicated. **See Example 5.***

39. 123.26×10 **40.** 785.91×10 **41.** 57.116×100

42. 82.053×100 **43.** 0.094×1000 **44.** 0.025×1000

45. $1.62 \div 10$ **46.** $8.04 \div 10$ **47.** $124.03 \div 100$

48. $490.35 \div 100$ **49.** $23.29 \div 1000$ **50.** $59.8 \div 1000$

Concept Check *Complete the table of fraction, decimal, and percent equivalents.*

	Fraction in Lowest Terms (or Whole Number)	Decimal	Percent
51.	$\frac{1}{100}$	0.01	
52.	$\frac{1}{50}$		2%
53.		0.05	5%
54.	$\frac{1}{10}$		
55.	$\frac{1}{8}$	0.125	
56.			20%
57.	$\frac{1}{4}$		
58.	$\frac{1}{3}$		
59.			50%
60.	$\frac{2}{3}$		$66\frac{2}{3}\%$, or $66.\overline{6}\%$
61.		0.75	
62.	1	1.0	

*Write each fraction as a decimal. For repeating decimals, write the answer by first using bar notation and then rounding to the nearest thousandth. **See Example 6.***

63. $\dfrac{21}{5}$ **64.** $\dfrac{9}{5}$ **65.** $\dfrac{9}{4}$ **66.** $\dfrac{15}{4}$ **67.** $\dfrac{3}{8}$

68. $\dfrac{7}{8}$ **69.** $\dfrac{5}{9}$ **70.** $\dfrac{8}{9}$ **71.** $\dfrac{1}{6}$ **72.** $\dfrac{5}{6}$

*Write each percent as a decimal. **See Examples 7 and 9(a)–9(c).***

73. 54%   **74.** 39%   **75.** 7%   **76.** 4%

77. 117%   **78.** 189%   **79.** 2.4%   **80.** 3.1%

81. $6\frac{1}{4}\%$ **82.** $5\frac{1}{2}\%$ **83.** 0.8%   **84.** 0.9%

*Write each decimal as a percent. **See Examples 8 and 9(d)–9(f).***

85. 0.79   **86.** 0.83   **87.** 0.02   **88.** 0.08

89. 0.004   **90.** 0.005   **91.** 1.28   **92.** 2.35

93. 0.4   **94.** 0.6   **95.** 6   **96.** 10

Write each percent as a fraction. Give answers in lowest terms. See Example 10.

97. 51%   **98.** 47%   **99.** 15%   **100.** 35%   **101.** 2%

102. 8%   **103.** 140%   **104.** 180%   **105.** 7.5%   **106.** 2.5%

Write each fraction as a percent. See Example 11.

107. $\dfrac{4}{5}$ **108.** $\dfrac{3}{25}$ **109.** $\dfrac{7}{50}$ **110.** $\dfrac{9}{20}$ **111.** $\dfrac{2}{11}$

112. $\dfrac{4}{9}$ **113.** $\dfrac{9}{4}$ **114.** $\dfrac{8}{5}$ **115.** $\dfrac{13}{6}$ **116.** $\dfrac{31}{9}$

Solve each problem. See Example 12.

117. What is 50% of 320?   **118.** What is 25% of 120?

119. What is 6% of 80?   **120.** What is 5% of 70?

121. What is 14% of 780?   **122.** What is 26% of 480?

Solve each problem. See Example 12.

123. Elwyn's bill for dinner at a restaurant was $89. He wants to leave a 20% tip. How much should he leave for the tip? What is his total bill for dinner and tip?

124. Gary earns $15 per hour at his job. He recently received a 7% raise. How much per hour was his raise? What is his new hourly rate?

125. Find the discount on a leather recliner with a regular price of $795 if the recliner is on sale at 15% off. What is the sale price of the recliner?

126. A laptop computer with a regular price of $597 is on sale at 20% off. Find the amount of the discount and the sale price of the computer.

In a recent year, approximately 76 million people from other countries visited the United States. The circle graph shows the distribution of these international visitors by country or region. Use the graph to work each problem.

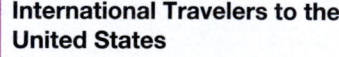

International Travelers to the United States

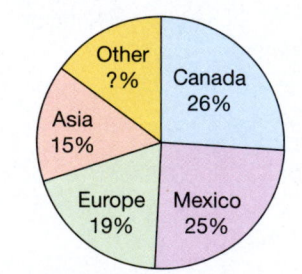

127. How many travelers visited the United States from Canada?

128. How many travelers visited the United States from Mexico?

129. What percent of travelers visited the United States from places other than Canada, Mexico, Europe, and Asia? (*Hint:* The sum of the parts of the graph must equal 1 whole, that is, 100%.)

Data from U.S. Department of Commerce.

130. How many travelers visited the United States from places other than Canada, Mexico, Europe, and Asia?

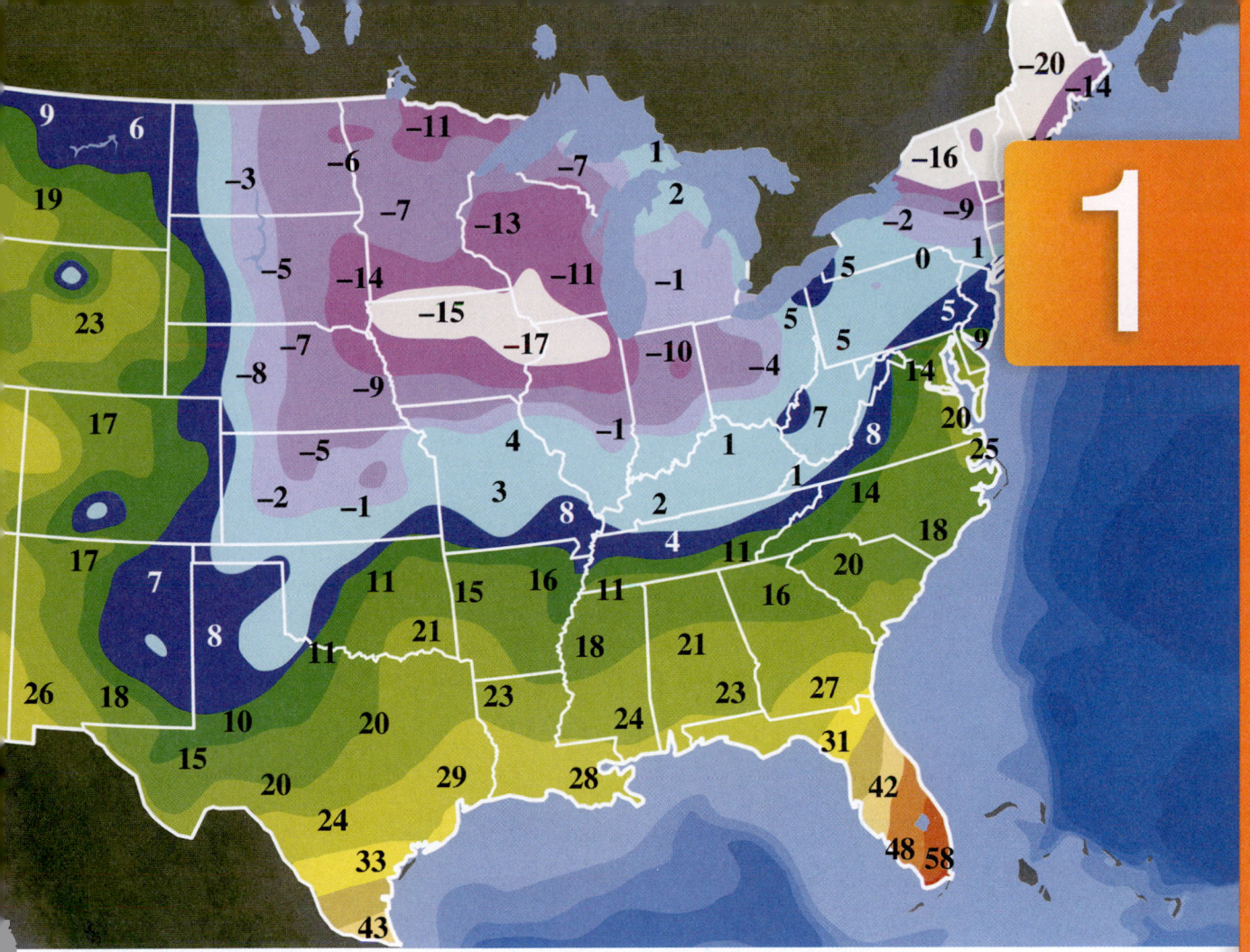

THE REAL NUMBER SYSTEM

Positive and *negative numbers,* used to indicate temperatures above and below zero, elevations above and below sea level, and gains and losses in the stock market or on a football field, are examples of *real numbers,* the subject of this chapter.

1.1 Exponents, Order of Operations, and Inequality

OBJECTIVES

1 Use exponents.
2 Use the rules for order of operations.
3 Use more than one grouping symbol.
4 Use inequality symbols in statements.
5 Translate word statements to symbols.
6 Write statements that change the direction of inequality symbols.

VOCABULARY

☐ exponent (power)
☐ base
☐ exponential expression
☐ inequality

**NOW TRY
EXERCISE 1**
Find the value of each exponential expression.

(a) 6^2 **(b)** $\left(\dfrac{4}{5}\right)^3$ **(c)** $(0.7)^2$

OBJECTIVE 1 Use exponents.

We can write a number as the product of its prime factors. For example,

$$81 \text{ can be written as } 3 \cdot 3 \cdot 3 \cdot 3. \quad \cdot \text{ indicates multiplication.}$$

Here the factor 3 appears four times. In algebra, repeated factors are often written using an *exponent*.

$$\underbrace{3 \cdot 3 \cdot 3 \cdot 3}_{\text{4 factors of } 3} = 3^4 \leftarrow \text{Exponent}$$
$$\uparrow \text{ Base}$$

In the **exponential expression** 3^4, the number 4 is the **exponent, or power,** and 3 is the **base.** An exponent tells how many times its base is used as a factor. We read 3^4 as **"3 to the fourth power,"** or **"3 to the fourth."**

    A number raised to the first power is simply that number.

$$5^1 = 5 \quad \text{and} \quad \left(\frac{1}{2}\right)^1 = \frac{1}{2} \qquad \text{In general, } a^1 = a.$$

EXAMPLE 1 Evaluating Exponential Expressions

Find the value of each exponential expression.

(a) 5^2 means $\underbrace{5 \cdot 5}_{\text{5 is used as a factor 2 times.}}$, which equals 25.

Read 5^2 as "5 *to the second power*" or, more commonly, "5 *squared.*"

(b) 6^3 means $\underbrace{6 \cdot 6 \cdot 6}_{\text{6 is used as a factor 3 times.}}$, which equals 216.

Read 6^3 as "6 *to the third power,*" or more commonly, "6 *cubed.*"

(c) 2^5 means $2 \cdot 2 \cdot 2 \cdot 2 \cdot 2$, which equals 32.   2 is used as a factor 5 times.

Read 2^5 as "2 *to the fifth power,*" or "2 *to the fifth.*"

$$2^5 \quad \textbf{does not equal} \quad \textbf{2} \cdot \textbf{5.} \longleftarrow \begin{array}{l} \text{Do } \textit{not} \text{ multiply the} \\ \text{base and exponent.} \end{array}$$

(d) $\left(\dfrac{2}{3}\right)^3$ means $\dfrac{2}{3} \cdot \dfrac{2}{3} \cdot \dfrac{2}{3}$, which equals $\dfrac{8}{27}$. $\frac{2}{3}$ is used as a factor 3 times.

(e) $(0.3)^2$ means $0.3 \cdot 0.3$, which equals 0.09.   0.3 is used as a factor 2 times.

NOW TRY

⚠ **CAUTION** *Squaring, or raising a number to the second power, is not the same as doubling the number.* In **Example 1(a),**

$$5^2 \quad \text{means} \quad 5 \cdot 5, \quad \textit{not} \quad 5 \cdot 2.$$

Thus $5^2 = 25$, *not* 10. Similarly, cubing, or raising a number to the third power, does *not* mean tripling the number. In **Example 1(b),** $6^3 = 216$, *not* 18.

NOW TRY ANSWERS
1. **(a)** 36 **(b)** $\frac{64}{125}$ **(c)** 0.49

OBJECTIVE 2 Use the rules for order of operations.

When an expression involves more than one operation, we often use **grouping symbols,** such as parentheses $(\quad)$, to indicate the order in which the operations should be performed.

Consider the following expression.

$$5 + 2 \cdot 3$$

To show that the multiplication should be performed before the addition, we could use parentheses to group $2 \cdot 3$.

$$5 + (2 \cdot 3) \quad \text{equals} \quad 5 + 6, \quad \text{which equals} \quad 11.$$

If the addition is to be performed first, the parentheses should group $5 + 2$.

$$(5 + 2) \cdot 3 \quad \text{equals} \quad 7 \cdot 3, \quad \text{which equals} \quad 21.$$

Other grouping symbols are brackets $[\quad]$, braces $\{\quad\}$, and fraction bars. (For example, in $\frac{8 - 2}{3}$, the expression $8 - 2$ is "grouped" in the numerator.)

To simplify an expression that involves more than one operation, we use the following rules for **order of operations.** This order is used by most calculators and computers.

Order of Operations

If grouping symbols are present, work within them, innermost first (and above and below fraction bars separately), in the following order.

Step 1 Apply all **exponents.**

Step 2 Do any **multiplications** and **divisions** in order from left to right.

Step 3 Do any **additions** and **subtractions** in order from left to right.

If no grouping symbols are present, start with Step 1.

NOTE Multiplication is understood in expressions with parentheses.

Examples: $3(7)$, $(6)2$, $(-5)(-4)$, $3(4 + 1)$

EXAMPLE 2 Using the Rules for Order of Operations

Find the value of each expression.

(a) $24 - 12 \div 3$ A helpful strategy is to label the order in which the
       ② ①   operations should be performed.

$= 24 - 4$ Divide.

$= 20$ Subtract.

(b) $48 \div 2 \cdot 3$
     ① ②   Divide, then multiply
            in order from left to right.

$= 24 \cdot 3$

$= 72$ Multiply.

(c) $6 \cdot 8 + 5 \cdot 2$
     ① ③ ②   Multiply, working
              from left to right.

$= 48 + 10$

$= 58$ Add.

**NOW TRY
EXERCISE 2**

Find the value of each expression.

(a) $15 - 2 \cdot 6$

(b) $8 + 2(5 - 1)$

(c) $6(2 + 4) - 7 \cdot 5$

(d) $8 \cdot 10 \div 4 - 2^3 + 3 \cdot 4^2$

(d) $16 - 3(2 + 3)$ ◁─ Do *not* subtract $16 - 3$ first.

  ③ ② ①

  $= 16 - 3(5)$ Add inside the parentheses.

  $= 16 - 15$ Multiply.

  $= 1$ Subtract.

(e) $2(5 + 6) + 7 \cdot 3$

  ② ① ④ ③

  $= 2(11) + 7 \cdot 3$ Add inside the parentheses.

  $= 22 + 21$ Multiply, working from left to right.

  $= 43$ Add.

     ⟨ $2^3 = 2 \cdot 2 \cdot 2$, not $2 \cdot 3$. ⟩

(f) $9 - 2^3 + 5$

  $= 9 - 2 \cdot 2 \cdot 2 + 5$ Apply the exponent.

  $= 9 - 8 + 5$ Multiply.

  $= 1 + 5$ Subtract.

  $= 6$ Add.

(g) $72 \div 2 \cdot 3 + 4 \cdot 2^3 - 3^3$ ◁─ Think: $3^3 = 3 \cdot 3 \cdot 3$

  $= 72 \div 2 \cdot 3 + 4 \cdot 8 - 27$ Apply the exponents.

  $= 36 \cdot 3 + 4 \cdot 8 - 27$ Divide.

  $= 108 + 32 - 27$ Multiply, working from left to right.

  $= 140 - 27$ Add.

  $= 113$ Subtract.

  Multiplications and divisions are done from left to right as they appear. Then additions and subtractions are done from left to right as they appear.

               NOW TRY ↻

OBJECTIVE 3 Use more than one grouping symbol.

In an expression such as $2(8 + 3(6 + 5))$, we often use brackets, $[\quad]$, in place of the outer pair of parentheses.

EXAMPLE 3 Using Brackets and Fraction Bars as Grouping Symbols

Simplify each expression.

(a) $2[8 + 3(6 + 5)]$ Work from the inside out.

  $= 2[8 + 3(11)]$ Add inside the parentheses.

  $= 2[8 + 33]$ Multiply inside the brackets.

  $= 2[41]$ Add inside the brackets.

  $= 82$ Multiply.

NOW TRY ANSWERS

2. (a) 3 (b) 16
 (c) 1 (d) 60

NOW TRY
EXERCISE 3

Simplify each expression.

(a) $7[3(3-1)+4]$

(b) $\dfrac{9(14-4)-2}{4+3\cdot6}$

(b) $\dfrac{4(5+3)+3}{2(3)-1}$ Simplify the numerator and denominator separately.

$=\dfrac{4(8)+3}{2(3)-1}$ Work inside the parentheses in the numerator.

$=\dfrac{32+3}{6-1}$ Multiply.

$=\dfrac{35}{5}$ Add and subtract.

$=7$ Divide.

NOW TRY

NOTE   The expression $\dfrac{4(5+3)+3}{2(3)-1}$ in **Example 3(b)** can be written as a quotient.

$$[4(5+3)+3]\div[2(3)-1]$$

The fraction bar "groups" the numerator and denominator separately.

OBJECTIVE 4 Use inequality symbols in statements.

So far, we have used the equality symbol $=$. The symbols

$$\neq,\ \ <,\ \ >,\ \ \leq,\ \ \text{and}\ \ \geq\ \ \ \ \text{Inequality symbols}$$

are used to express an **inequality,** a statement that two expressions may not be equal. The equality symbol with a slash through it, $\neq$, means "is *not* equal to."

$$7\neq8\ \ \ \ \text{7 is not equal to 8.}$$

If two numbers are not equal, then one of the numbers must be less than the other. Reading from left to right, the symbol $<$ represents "is less than."

$$7<8\ \ \ \ \text{7 is less than 8.}$$

Reading from left to right, the symbol $>$ means "is greater than."

$$8>2\ \ \ \ \text{8 is greater than 2.}$$

To keep the meanings of the symbols $<$ and $>$ clear, remember that the symbol always points to the lesser number.

Lesser number $\rightarrow$ $8<15$

$15>8$ $\leftarrow$ Lesser number

Reading from left to right, the symbol $\leq$ means "is less than or equal to."

$$5\leq9\ \ \ \ \text{5 is less than or equal to 9.}$$

If either the $<$ part or the $=$ part is true, then the inequality $\leq$ is true. The statement $5\leq9$ is true because $5<9$ is true. Also, the statement $8\leq8$ is true because $8=8$ is true.

Reading from left to right, the symbol $\geq$ means "is greater than or equal to."

NOW TRY ANSWERS

3. **(a)** 70 **(b)** 4

$$9\geq5\ \ \ \ \text{9 is greater than or equal to 5.}$$

NOW TRY
EXERCISE 4

Determine whether each statement is *true* or *false*.

(a) $12 \neq 10 - 2$

(b) $5 > 4 \cdot 2$

(c) $\dfrac{1}{4} \leq 0.25$

(d) $\dfrac{5}{9} > \dfrac{7}{11}$

EXAMPLE 4   Using Inequality Symbols

Determine whether each statement is *true* or *false*.

(a) $6 \neq 5 + 1$ This statement is false because $6 = 5 + 1$.

(b) $5 + 3 < 19$ The statement $5 + 3 < 19$ is true because $8 < 19$.

(c) $15 \leq 20 \cdot 2$ The statement $15 \leq 20 \cdot 2$ is true because $15 < 40$.

(d) $25 \geq 30$ Both $25 > 30$ and $25 = 30$ are false, so $25 \geq 30$ is false.

(e) $12 \geq 12$ Because $12 = 12$, this statement is true.

(f) $0.5 < \dfrac{1}{2}$ Because $0.5 = \frac{1}{2}$, this statement is false.

(g) $\dfrac{6}{15} \geq \dfrac{2}{3}$ Find a common denominator.

$\dfrac{6}{15} \geq \dfrac{10}{15}$ Both $\frac{6}{15} > \frac{10}{15}$ and $\frac{6}{15} = \frac{10}{15}$ are false, so $\frac{6}{15} \geq \frac{2}{3}$ is false.   **NOW TRY**

OBJECTIVE 5 Translate word statements to symbols.

NOW TRY
EXERCISE 5

Write each word statement in symbols.

(a) Ten is not equal to eight minus two.

(b) Fifty is greater than fifteen.

(c) Eleven is less than or equal to twenty.

EXAMPLE 5   Translating from Words to Symbols

Write each word statement in symbols.

(a) Twelve equals ten plus two.
$$12 = 10 + 2$$

(b) Nine is less than ten.
$$9 < 10$$

(c) Fifteen is not equal to eighteen.
$$15 \neq 18$$

(d) Seven is greater than four.
$$7 > 4$$

(e) Thirteen is less than or equal to forty.
$$13 \leq 40$$

(f) Eleven is greater than or equal to eleven.
$$11 \geq 11$$

NOW TRY

OBJECTIVE 6 Write statements that change the direction of inequality symbols.

Any statement involving $<$ can be converted to one with $>$, and any statement involving $>$ can be converted to one with $<$. *We do this by reversing the order of the numbers and the direction of the symbol.*

Interchange numbers.

$6 < 10$ becomes $10 > 6$

Reverse symbol.

EXAMPLE 6   Converting between Inequality Symbols

Write each statement as another true statement with the inequality symbol reversed.

(a) $5 > 2$ is equivalent to $2 < 5$.

(b) $\dfrac{1}{2} \leq \dfrac{3}{4}$ is equivalent to $\dfrac{3}{4} \geq \dfrac{1}{2}$.

NOW TRY

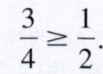
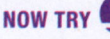

NOW TRY EXERCISE 6

Write the statement as another true statement with the inequality symbol reversed.

$$8 < 9$$

▼ **Summary of Equality and Inequality Symbols**

Symbol	Meaning	Example
$=$	Is equal to	$0.5 = \frac{1}{2}$ means 0.5 is equal to $\frac{1}{2}$.
$\neq$	Is not equal to	$3 \neq 7$ means 3 is not equal to 7.
$<$	Is less than	$6 < 10$ means 6 is less than 10.
$>$	Is greater than	$15 > 14$ means 15 is greater than 14.
$\leq$	Is less than or equal to	$4 \leq 8$ means 4 is less than or equal to 8.
$\geq$	Is greater than or equal to	$1 \geq 0$ means 1 is greater than or equal to 0.

NOW TRY ANSWER
6. $9 > 8$

⚠ **CAUTION** Equality and inequality symbols are used to write mathematical **sentences**. Operation symbols ($+$, $-$, $\cdot$, and $\div$) are used to write mathematical **expressions**.

Sentence: $4 < 10$ ← Gives the relationship between 4 and 10

Expression: $4 + 10$ ← Tells how to operate on 4 and 10 to get 14

1.1 Exercises

FOR EXTRA HELP

▶ **MyLab Math**

▶ *Video solutions for select problems available in MyLab Math*

STUDY SKILLS REMINDER
You will increase your chance of success in this course if you fully utilize your text. **Review Study Skill 1, *Using Your Math Text*.**

Concept Check *Decide whether each statement is* true *or* false. *If it is* false, *explain why.*

1. $3^2 = 6$ **2.** $1^3 = 3$ **3.** $3^1 = 1$

4. The expression 6^2 means that 2 is used as a factor 6 times.

5. When evaluated, $4 + 3(8 - 2)$ is equal to 42.

6. When evaluated, $12 \div 2 \cdot 3$ is equal to 2.

Concept Check *For each expression, label the order in which the operations should be performed. Do not actually perform them.*

7. $18 - 2 + 3$
  ○ ○

8. $28 - 6 \div 2$
  ○ ○

9. $2 \cdot 8 - 6 \div 3$
  ○ ○ ○

10. $40 + 6(3 - 1)$
  ○ ○ ○

11. $3 \cdot 5 - 2(4 + 2)$
  ○ ○ ○ ○

12. $9 - 2^3 + 3 \cdot 4$
  ○○○ ○

Find the value of each exponential expression. **See Example 1.**

13. 7^2 **14.** 8^2 **15.** 12^2 **16.** 14^2 **17.** 4^3 **18.** 5^3

19. 10^3 **20.** 11^3 **21.** 3^4 **22.** 6^4 **23.** 4^5 **24.** 3^5

25. $\left(\frac{1}{6}\right)^2$ **26.** $\left(\frac{1}{3}\right)^2$ **27.** $\left(\frac{2}{3}\right)^4$ **28.** $\left(\frac{3}{4}\right)^3$

29. $(0.6)^2$ **30.** $(0.9)^2$ **31.** $(0.4)^3$ **32.** $(0.5)^4$

33. Concept Check The value of an expression was found incorrectly as follows.

$$8 + 2 \cdot 3$$
$$= 10 \cdot 3$$
$$= 30$$

WHAT WENT WRONG? Find the correct value of the expression.

34. Concept Check The value of an expression was found incorrectly as follows.

$$16 - 2^3 + 5$$
$$= 16 - 6 + 5$$
$$= 10 + 5$$
$$= 15$$

WHAT WENT WRONG? Find the correct value of the expression.

*Find the value of each expression. **See Example 2 and 3.***

35. $64 \div 4 \cdot 2$

36. $250 \div 5 \cdot 2$

37. $13 + 9 \cdot 5$

38. $11 + 7 \cdot 6$

39. $25.2 - 12.6 \div 4.2$

40. $12.4 - 9.3 \div 3.1$

41. $9 \cdot 4 - 8 \cdot 3$

42. $11 \cdot 4 + 10 \cdot 3$

43. $\dfrac{1}{4} \cdot \dfrac{2}{3} + \dfrac{2}{5} \cdot \dfrac{11}{3}$

44. $\dfrac{9}{4} \cdot \dfrac{2}{3} + \dfrac{4}{5} \cdot \dfrac{5}{3}$

45. $20 - 4 \cdot 3 + 5$

46. $18 - 7 \cdot 2 + 6$

47. $10 + 40 \div 5 \cdot 2$

48. $12 + 64 \div 8 - 4$

49. $18 - 2(3 + 4)$

50. $30 - 3(4 + 2)$

51. $3(4 + 2) + 8 \cdot 3$

52. $9(1 + 7) + 2 \cdot 5$

53. $18 - 4^2 + 3$

54. $22 - 2^3 + 9$

55. $2 + 3[5 + 4(2)]$

56. $5 + 4[1 + 7(3)]$

57. $5[3 + 4(2^2)]$

58. $6[2 + 8(3^3)]$

59. $3^2[(11 + 3) - 4]$

60. $4^2[(13 + 4) - 8]$

61. $\dfrac{6(3^2 - 1) + 8}{8 - 2^2}$

62. $\dfrac{2(8^2 - 4) + 8}{29 - 3^3}$

63. $\dfrac{4(6 + 2) + 8(8 - 3)}{6(4 - 2) - 2^2}$

64. $\dfrac{6(5 + 1) - 9(1 + 1)}{5(8 - 6) - 2^3}$

Concept Check   *Insert one pair of parentheses in each expression so that the given value results when the operations are performed.*

65. $3 \cdot 6 + 4 \cdot 2$
$= 60$

66. $2 \cdot 8 - 1 \cdot 3$
$= 42$

67. $10 - 7 - 3$
$= 6$

68. $8 + 2^2$
$= 100$

*First simplify both sides of each inequality. Then determine whether the given statement is true or false. **See Examples 2–4.***

69. $9 \cdot 3 - 11 \leq 16$

70. $6 \cdot 5 - 12 \leq 18$

71. $5 \cdot 11 + 2 \cdot 3 \leq 60$

72. $9 \cdot 3 + 4 \cdot 5 \geq 48$

73. $0 \geq 12 \cdot 3 - 6 \cdot 6$

74. $10 \leq 13 \cdot 2 - 15 \cdot 1$

75. $45 \geq 2[2 + 3(2 + 5)]$

76. $55 \geq 3[4 + 3(4 + 1)]$

77. $[3 \cdot 4 + 5(2)] \cdot 3 > 72$

78. $2 \cdot [7 \cdot 5 - 3(2)] \leq 58$

79. $\dfrac{3 + 5(4 - 1)}{2 \cdot 4 + 1} \geq 3$

80. $\dfrac{7(3 + 1) - 2}{3 + 5 \cdot 2} \leq 2$

81. $3 \geq \dfrac{2(5 + 1) - 3(1 + 1)}{5(8 - 6) - 4 \cdot 2}$

82. $7 \leq \dfrac{3(8 - 3) + 2(4 - 1)}{9(6 - 2) - 11(5 - 2)}$

Write each statement in words, and determine whether it is true *or* false. ***See Examples 4 and 5.***

83. $5 < 17$

84. $8 < 12$

85. $5 \neq 8$

86. $6 \neq 9$

87. $7 \geq 14$

88. $6 \geq 12$

89. $15 \leq 15$

90. $21 \leq 21$

91. $\dfrac{1}{3} = \dfrac{3}{10}$

92. $\dfrac{10}{6} = \dfrac{3}{2}$

93. $2.5 > 2.50$

94. $1.80 > 1.8$

*Write each word statement in symbols. **See Example 5.***

95. Fifteen is equal to five plus ten.

96. Twelve is equal to twenty minus eight.

97. Nine is greater than five minus four.

98. Ten is greater than six plus one.

99. Sixteen is not equal to nineteen.

100. Three is not equal to four.

101. One-half is less than or equal to two-fourths.

102. One-third is less than or equal to three-ninths.

Write each statement as another true statement with the inequality symbol reversed. See Example 6.

103. $5 < 20$ **104.** $30 > 9$ **105.** $\dfrac{4}{5} > \dfrac{3}{4}$

106. $\dfrac{5}{4} < \dfrac{3}{2}$ **107.** $2.5 \geq 1.3$ **108.** $4.1 \leq 5.3$

One way to measure a person's cardiofitness is to calculate how many METs, or metabolic units, he or she can reach at peak exertion. One MET is the amount of energy used when sitting quietly. To calculate ideal METs, we can use the following expressions.

$$14.7 - age \cdot 0.13 \quad \text{For women}$$

$$14.7 - age \cdot 0.11 \quad \text{For men}$$

(Data from *New England Journal of Medicine.*)

109. A 40-yr-old woman wishes to calculate her ideal MET.

  (a) Write the expression, using her age.

  (b) Calculate her ideal MET.

  (c) Researchers recommend that a person reach approximately 85% of his or her MET when exercising. Calculate 85% of the ideal MET from part (b). Then refer to the following table. What activity listed in the table can the woman do that is approximately this value?

Activity	METs	Activity	METs
Golf (with cart)	2.5	Skiing (water or downhill)	6.8
Walking (3 mph)	3.3	Swimming	7.0
Mowing lawn (power mower)	4.5	Walking (5 mph)	8.0
Ballroom or square dancing	5.5	Jogging	10.2
Cycling	5.7	Skipping rope	12.0

Data from Harvard School of Public Health.

  (d) Repeat parts (a)–(c) for a 55–yr-old man.

110. Repeat parts (a)–(c) of **Exercise 109** using your age and gender.

The table shows the number of pupils per teacher in U.S. public schools in selected states.

111. Which states had a number greater than 12.6?

112. Which states had a number that was at most 15.2?

113. Which states had a number not less than 12.6?

114. Which states had a number less than 13.0?

State	Pupils per Teacher
Alaska	16.4
Texas	15.2
California	22.5
Virginia	12.6
Maine	12.4
Idaho	19.7
Missouri	12.1

Data from National Center for Education Statistics.

1.2   Variables, Expressions, and Equations

OBJECTIVES

1 Evaluate algebraic expressions, given values for the variables.
2 Translate word phrases to algebraic expressions.
3 Identify solutions of equations.
4 Identify solutions of equations from a set of numbers.
5 Distinguish between *expressions* and *equations*.

A **constant** is a fixed, unchanging number. A **variable** is a symbol, usually a letter, used to represent an unknown number.

Examples: $5,\ \dfrac{3}{4},\ 8\dfrac{1}{2},\ 10.8$ Constants $\Big|$ $a,\ x,\ y,\ z$ Variables

An **algebraic expression** is a sequence of constants, variables, operation symbols, and/or grouping symbols formed according to the rules of algebra.

Examples: $x + 5,\quad 2m - 9,\quad 8p^2 + 6(p - 2)$ Algebraic expressions

$2m$ means $2 \cdot m$, the product of 2 and m.

$6(p - 2)$ means the product of 6 and $p - 2$.

OBJECTIVE 1 Evaluate algebraic expressions, given values for the variables.

To *evaluate* an expression means to find its *value*. An algebraic expression can have different numerical values for different values of the variables.

EXAMPLE 1   Evaluating Algebraic Expressions

Evaluate each expression for $x = 5$.

(a) $8 + x$

$\qquad = 8 + 5$ Let $x = 5$.

$\qquad = 13$ Add.

(b) $2x - 9$

$\qquad = 2 \cdot 5 - 9$ Let $x = 5$.

$\qquad = 10 - 9$ Multiply.

$\qquad = 1$ Subtract.

(c) $3x^2$

$\qquad = 3 \cdot x^2$ $\boxed{5^2 = 5 \cdot 5}$

$\qquad = 3 \cdot 5^2$ Let $x = 5$.

$\qquad = 3 \cdot 25$ Square 5.

$\qquad = 75$ Multiply.

(d) $\dfrac{4x - 2}{7}$

$\qquad = \dfrac{4 \cdot 5 - 2}{7}$ Let $x = 5$.

$\qquad = \dfrac{20 - 2}{7}$ Multiply.

$\qquad = \dfrac{18}{7},\quad$ or $\quad 2\dfrac{4}{7}$

NOW TRY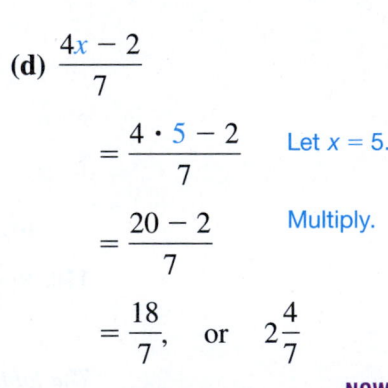

VOCABULARY

☐ constant
☐ variable
☐ algebraic expression
☐ equation
☐ solution
☐ set
☐ element

**NOW TRY
EXERCISE 1**

Evaluate each expression for
$x = 6$.

(a) $9x - 5$ **(b)** $4x^2$

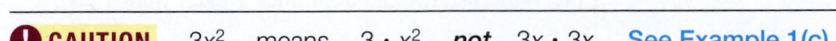

⚠ **CAUTION** $3x^2$ means $3 \cdot x^2$, **not** $3x \cdot 3x$.   See Example 1(c).

Unless parentheses are used, the exponent refers only to the variable or number just before it. We would need to use parentheses to write $3x \cdot 3x$ with exponents.

$$(3x)^2\quad \text{means}\quad 3x \cdot 3x.$$

NOW TRY ANSWERS
1. (a) 49   **(b)** 144

 **NOW TRY EXERCISE 2**

Evaluate each expression for $x = 4$ and $y = 7$.

(a) $3x + 4y$ **(b)** $\dfrac{6x - 2y}{2y - 9}$

(c) $4x^2 - y^2$

EXAMPLE 2 Evaluating Algebraic Expressions

Evaluate each expression for $x = 5$ and $y = 3$.

(a) $2x + 7y$ | We could use parentheses and write $2(5) + 7(3)$. |

[Follow the rules for order of operations.] $= 2 \cdot 5 + 7 \cdot 3$ Let $x = 5$ and $y = 3$.

$= 10 + 21$ Multiply.

$= 31$ Add.

(b) $\dfrac{9x - 8y}{2x - y}$

$= \dfrac{9 \cdot 5 - 8 \cdot 3}{2 \cdot 5 - 3}$ Let $x = 5$ and $y = 3$.

$= \dfrac{45 - 24}{10 - 3}$ Multiply.

$= \dfrac{21}{7}$ Subtract.

$= 3$ Divide.

(c) $x^2 - 2y^2$ | $3^2 = 3 \cdot 3$ |

$= 5^2 - 2 \cdot 3^2$ Let $x = 5$ and $y = 3$.

| $5^2 = 5 \cdot 5$ |

$= 25 - 2 \cdot 9$ Apply the exponents.

$= 25 - 18$ Multiply.

$= 7$ Subtract.

NOW TRY

OBJECTIVE 2 Translate word phrases to algebraic expressions.

EXAMPLE 3 Using Variables to Write Word Phrases as Algebraic Expressions

Write each word phrase as an algebraic expression, using x as the variable.

(a) The sum of a number and 9

$x + 9$,   or $9 + x$ "Sum" is the answer to an addition problem.

(b) 7 minus a number

$7 - x$ "Minus" indicates subtraction.

The expression $x - 7$ is incorrect. We cannot subtract in either order and obtain the same result.

(c) A number subtracted from 12

$12 - x$ | Be careful with order. |

Compare this result with "12 subtracted from a number," which is $x - 12$.

(d) The product of 11 and a number

$11 \cdot x$,   or $11x$

NOW TRY ANSWERS
2. (a) 40 **(b)** 2 **(c)** 15

**NOW TRY
EXERCISE 3**

Write each word phrase as an algebraic expression, using x as the variable.

(a) The sum of a number and 10

(b) A number divided by 7

(c) The product of 3 and the difference of 9 and a number

(e) 5 divided by a number

$$5 \div x, \quad \text{or} \quad \frac{5}{x} \quad \boxed{\frac{x}{5} \text{ is } not \text{ correct here.}}$$

(f) The **product** of 2 and the **difference** of a number and 8

We are multiplying 2 times "something." This "something" is the difference of a number and 8, written $x - 8$. We use parentheses around this difference.

$$2 \cdot (x - 8), \quad \text{or} \quad 2(x - 8) \quad \boxed{8 - x, \text{ which means the difference of 8 and a number, is } not \text{ correct.}}$$

NOW TRY

OBJECTIVE 3   Identify solutions of equations.

An **equation** is a statement that two algebraic expressions are equal. *An equation always includes the equality symbol,* $=$.

$$\left.\begin{array}{lll} x + 4 = 11, & 2y = 16, & 4p + 1 = 25 - p, \\[2mm] \dfrac{3}{4}x + \dfrac{1}{2} = 0, & z^2 = 4, & 4(m - 0.5) = 2m \end{array}\right\} \text{Equations}$$

To **solve an equation** means to find the value of the variable that makes the equation true. Such a value of the variable is a **solution** of the equation.

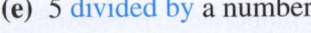

**NOW TRY
EXERCISE 4**

Determine whether the equation has the given number as a solution.

$$8k + 5 = 61; \quad 7$$

EXAMPLE 4   Deciding Whether a Number Is a Solution of an Equation

Determine whether each equation has the given number as a solution.

(a) $5p + 1 = 36; \quad 7$

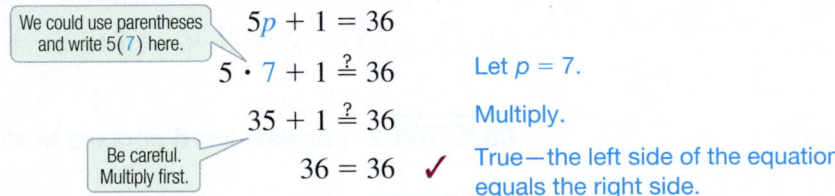

$$5p + 1 = 36$$
$$5 \cdot 7 + 1 \stackrel{?}{=} 36 \qquad \text{Let } p = 7.$$
$$35 + 1 \stackrel{?}{=} 36 \qquad \text{Multiply.}$$
$$36 = 36 \quad \checkmark \qquad \text{True—the left side of the equation equals the right side.}$$

The number 7 is a solution of the equation.

(b) $9m - 6 = 32; \quad \dfrac{14}{3}$

$$9m - 6 = 32$$
$$9 \cdot \frac{14}{3} - 6 \stackrel{?}{=} 32 \qquad \text{Let } m = \tfrac{14}{3}.$$
$$42 - 6 \stackrel{?}{=} 32 \qquad \text{Multiply: } 9 \cdot \frac{14}{3} = \frac{3 \cdot 3}{1} \cdot \frac{14}{3} = 42.$$
$$36 = 32 \qquad \text{False—the left side does } not \text{ equal the right side.}$$

The number $\dfrac{14}{3}$ is not a solution of the equation.   **NOW TRY**

OBJECTIVE 4   Identify solutions of equations from a set of numbers.

A **set** is a collection of objects. In mathematics, these objects are usually numbers. The objects that belong to a set are its **elements.** They are written between **braces** $\{ \quad \}$.

$$\{1, 2, 3, 4, 5\} \longleftarrow \text{The set containing the elements 1, 2, 3, 4, and 5}$$

NOW TRY ANSWERS
3. **(a)** $x + 10$, or $10 + x$ **(b)** $\frac{x}{7}$
   (c) $3(9 - x)$
4. yes

NOW TRY
EXERCISE 5

Write the word statement as an equation. Use x as the variable. Then find the solution of the equation from the set $\{0, 2, 4, 6, 8, 10\}$.

The sum of a number and nine is equal to the difference of 25 and the number.

EXAMPLE 5 Finding a Solution from a Given Set

Write each word statement as an equation. Use x as the variable. Then find the solution of the equation from the following set.

$$\{0, 2, 4, 6, 8, 10\}$$

(a) The sum of a number and four is six.

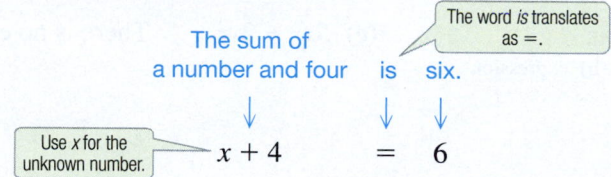

The sum of a number and four   is   six.

The word *is* translates as =.

Use x for the unknown number.

$$x + 4 = 6$$

One by one, mentally substitute each number from the given set $\{0, 2, 4, 6, 8, 10\}$ in $x + 4 = 6$. The only solution is 2 because

$$2 + 4 = 6 \text{ is true.}$$

(b) Nine more than five times a number is 49.

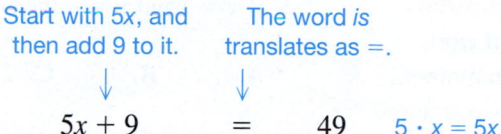

Start with $5x$, and then add 9 to it.   The word *is* translates as =.

$$5x + 9 = 49 \qquad 5 \cdot x = 5x$$

Substitute each of the given numbers. The solution is 8 because

$$5 \cdot 8 + 9 = 49 \text{ is true.}$$

(c) The sum of a number and 12 is equal to four times the number.

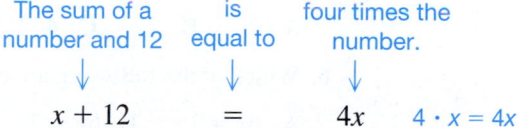

The sum of a number and 12   is equal to   four times the number.

$$x + 12 = 4x \qquad 4 \cdot x = 4x$$

Substituting each of the given numbers leads to a true statement only for $x = 4$ because

$$4 + 12 = 4(4) \text{ is true.}$$ **NOW TRY**

OBJECTIVE 5 Distinguish between *expressions* and *equations*.

Distinguishing between an Expression and an Equation

An **expression** is a phrase that represents a number.

An **equation** is a sentence—it has something on the left side, an = symbol, and something on the right side.

$$4x + 5 \qquad\qquad 4x + 5 = 9$$

Left side   Right side

Expression   Equation
(to simplify or evaluate)   (to solve)

NOW TRY ANSWER
5. $x + 9 = 25 - x$; 8

NOW TRY EXERCISE 6
Determine whether each of the following is an *expression* or an *equation*.
(a) $2x + 5 = 6$
(b) $2x + 5 - 6$

NOW TRY ANSWERS
6. (a) equation (b) expression

EXAMPLE 6 Distinguishing between Expressions and Equations

Determine whether each of the following is an *expression* or an *equation*.

(a) $2x - 3$ Ask, "*Is there an equality symbol?*" The answer is no, so this is an expression.

(b) $2x - 3 = 8$ Because there is an equality symbol with something on either side of it, this is an equation.

(c) $5x^2 + 2y^2$ There is no equality symbol. This is an expression.

NOW TRY

1.2 Exercises

 **MyLab Math**

▶ *Video solutions for select problems available in MyLab Math*

STUDY SKILLS REMINDER
Be sure to read and work through the section material before working the exercises.
Review Study Skill 2, *Reading Your Math Text*.

Concept Check   *Choose the letter(s) of the correct response.*

1. The expression $8x^2$ means _____.

   A. $8 \cdot x \cdot 2$ B. $8 \cdot x \cdot x$ C. $8 + x^2$ D. $8x \cdot 8x$

2. If $x = 2$ and $y = 1$, then the value of xy is _____.

   A. $\dfrac{1}{2}$ B. 1 C. 2 D. 3

3. The sum of 15 and a number x is represented by _____.

   A. $15 + x$ B. $15 - x$ C. $x - 15$ D. $15x$

4. 7 less than a number x is represented by _____.

   A. $7 - x$ B. $7 < x$ C. $7 + x$ D. $x - 7$

5. Which of the following is a solution of the equation $3x - 1 = 5$?

   A. 0 B. 2 C. 4 D. 6

6. Which of the following are expressions?

   A. $6x = 7$ B. $6x + 7$ C. $6x - 7$ D. $6x - 7 = 0$

7. **Concept Check**   The value of $5x^2$ for $x = 4$ was found incorrectly as follows.

$$5x^2$$
$$= 5 \cdot 4^2$$
$$= 20^2$$
$$= 400$$

WHAT WENT WRONG?   Find the correct value of the expression.

8. **Concept Check**   The value of $\dfrac{x+3}{5}$ for $x = 10$ was found incorrectly as follows.

$$\frac{x+3}{5}$$
$$= \frac{10+3}{5}$$
$$= 2 + 3$$
$$= 5$$

WHAT WENT WRONG?   Find the correct value of the expression.

Evaluate each expression for (a) x = 4 and (b) x = 6. See Example 1.

9. $x + 7$

10. $x - 3$

11. $4x$

12. $6x$

13. $5x - 4$

14. $7x - 9$

15. $4x^2$

16. $5x^2$

17. $\dfrac{x + 1}{3}$

18. $\dfrac{x + 2}{5}$

19. $\dfrac{3x - 5}{2x}$

20. $\dfrac{4x - 1}{3x}$

21. $3x^2 + x$

22. $2x + x^2$

23. $6.459x$

24. $3.275x$

Evaluate each expression for (a) x = 2 and y = 1 and (b) x = 1 and y = 5. See Example 2.

25. $8x + 3y + 5$

26. $4x + 2y + 7$

27. $3(x + 2y)$

28. $2(2x + y)$

29. $x + \dfrac{4}{y}$

30. $y + \dfrac{8}{x}$

31. $\dfrac{x}{2} + \dfrac{y}{3}$

32. $\dfrac{x}{5} + \dfrac{y}{4}$

33. $\dfrac{2x + 4y}{5x + 2y}$

34. $\dfrac{7x + 5y}{8x + y}$

35. $3x^2 + y^2$

36. $4x^2 + 2y^2$

37. $\dfrac{3x + y^2}{2x + 3y}$

38. $\dfrac{x^2 + 1}{4x + 5y}$

39. $0.841x^2 + 0.32y^2$ **40.** $0.941x^2 + 0.25y^2$

Write each word phrase as an algebraic expression, using x as the variable. See Example 3.

41. Twelve times a number

42. Fifteen times a number

43. Nine added to a number

44. Six added to a number

45. Two subtracted from a number

46. Seven subtracted from a number

47. A number subtracted from seven

48. A number subtracted from four

49. The difference of a number and 8

50. The difference of 8 and a number

51. 18 divided by a number

52. A number divided by 18

53. The product of 6 and four less than a number

54. The product of 9 and five more than a number

Determine whether each equation has the given number as a solution. See Example 4.

55. $4m + 2 = 6$; 1

56. $2r + 6 = 8$; 1

57. $2y + 3(y - 2) = 14$; 3

58. $6x + 2(x + 3) = 14$; 2

59. $6p + 4p + 9 = 11$; $\dfrac{1}{5}$

60. $2x + 3x + 8 = 20$; $\dfrac{12}{5}$

61. $3r^2 - 2 = 46$; 4

62. $2x^2 + 1 = 19$; 3

63. $\dfrac{3}{8}x + \dfrac{1}{4} = 1$; 2

64. $\dfrac{7}{10}x + \dfrac{1}{2} = 4$; 5

65. $0.5(x - 4) = 80$; 20

66. $0.2(x - 5) = 70$; 40

Write each word statement as an equation. Use x as the variable. Then find the solution of the equation from the set $\{2, 4, 6, 8, 10\}$. See Example 5.

67. The sum of a number and 8 is 18.

68. A number minus three equals 1.

69. One more than twice a number is 5.

70. The product of a number and 3 is 6.

71. Sixteen minus three-fourths of a number is 13.

72. The sum of six-fifths of a number and 2 is 14.

73. Three times a number is equal to 8 more than twice the number.

74. Twelve divided by a number equals $\frac{1}{3}$ times that number.

*Determine whether each of the following is an expression or an equation. **See Example 6.***

75. $3x + 2(x - 4)$ **76.** $8y - (3y + 5)$ **77.** $7t + 2(t + 1) = 4$

78. $9r + 3(r - 4) = 2$ **79.** $x + y = 9$ **80.** $x + y - 9$

One example of a **mathematical model** is an equation that describes the relationship between two quantities. For example, the life expectancy at birth of Americans can be approximated by the equation

$$y = 0.157x - 237,$$

where x is a year between 1990 and 2015 and y is age in years. (Data from Centers for Disease Control and Prevention.)

Use this model to approximate life expectancy (to the nearest year) in each of the following years.

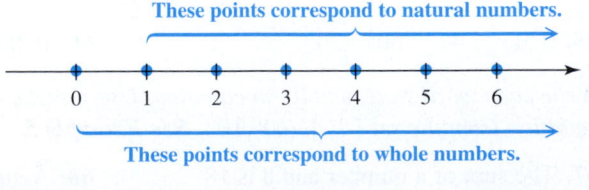

81. 1990   **82.** 1995   **83.** 2005   **84.** 2015

1.3   Real Numbers and the Number Line

OBJECTIVES

1. Classify numbers and graph them on number lines.
2. Use inequality symbols with real numbers.
3. Find the additive inverse of a real number.
4. Find the absolute value of a real number.
5. Interpret meanings of real numbers from a table of data.

OBJECTIVE 1 Classify numbers and graph them on number lines.

The set of numbers used for counting is the *natural numbers*. The set of *whole numbers* includes 0 with the natural numbers.

> **Natural Numbers and Whole Numbers**
>
> $\{1, 2, 3, 4, 5, \ldots\}$ is the set of **natural numbers** (or **counting numbers**).
>
> $\{0, 1, 2, 3, 4, 5 \ldots\}$ is the set of **whole numbers.**

We can represent numbers on a **number line** like the one in **FIGURE 1**.

These points correspond to natural numbers.

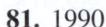

These points correspond to whole numbers.

FIGURE 1

To draw a number line, choose any point on the line and label it 0. Then choose any point to the right of 0 and label it 1. Use the distance between 0 and 1 as the scale to locate, and then label, other points.

The natural numbers are located to the right of 0 on the number line. For each natural number, we can place a corresponding number to the left of 0, labeling the points -1, -2, -3, and so on, as shown in **FIGURE 2** on the next page. Each is the **opposite,** or **negative,** of a natural number. The natural numbers, their opposites, and 0 form the set of *integers*.

Integers

$\{\ldots, -3, -2, -1, 0, 1, 2, 3, \ldots\}$ is the set of **integers.**

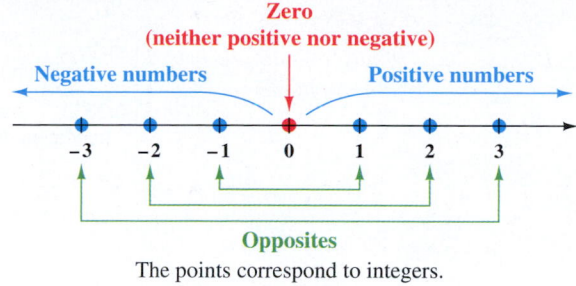

Opposites
The points correspond to integers.

FIGURE 2

Positive numbers and *negative numbers* are **signed numbers.**

NOW TRY EXERCISE 1

Use an integer to express the number in boldface italics in the following statement.

At its deepest point, the floor of West Okoboji Lake sits *136* ft below the water's surface. (Data from www.watersafetycouncil.org)

EXAMPLE 1 Using Signed Numbers

Use an integer to express the number in boldface italics in each statement.

(a) Through Internet advertising, a company increased its sales *$12,500.*

Use $12,500 because "increase" indicates a positive number.

(b) The shore surrounding the Dead Sea is *1348* ft below sea level. (Data from *The World Almanac and Book of Facts.*)

Here "below sea level" indicates a negative number, -1348. **NOW TRY**

NOTE **Set-builder notation** uses the set symbolism $\{x \,|\, x \text{ has a given property}\}$ to represent a set of numbers. For example, the set $\{-2, -1, 0, 1, 2\}$ can be described using set-builder notation as follows.

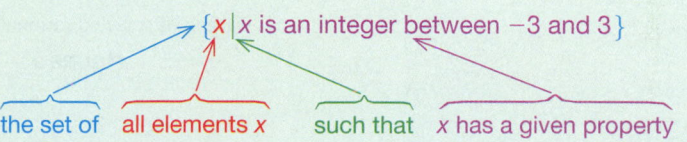

Fractions are *rational numbers.*

Rational Numbers

$\{x \,|\, x \text{ is a quotient of two integers, with denominator not } 0\}$ is the set of **rational numbers.**

Because any number that can be written as the quotient of two integers (that is, as a fraction) is a rational number, *all integers, mixed numbers, terminating (or ending) decimals, and repeating decimals are rational numbers.*

NOW TRY ANSWER
1. -136

▼ Rational Numbers

Rational Number	Equivalent Quotient of Two Integers
−5	$\frac{-5}{1}$ (means −5 ÷ 1)
$1\frac{3}{4}$	$\frac{7}{4}$ (means 7 ÷ 4)
0.23 (terminating decimal)	$\frac{23}{100}$ (means 23 ÷ 100)
0.3333 . . . , or $0.\overline{3}$ (repeating decimal)	$\frac{1}{3}$ (means 1 ÷ 3)
4.7	$\frac{47}{10}$ (means 47 ÷ 10)

To **graph** a number, we place a dot on a number line at the point that corresponds to the number. The number is the **coordinate** of the point.

NOW TRY EXERCISE 2

Graph each rational number on a number line.

$$-3, \quad \frac{17}{8}, \quad -2.75, \quad 1\frac{1}{2}, \quad -\frac{3}{4}$$

EXAMPLE 2   Graphing Rational Numbers

Graph each rational number on a number line.

$$-\frac{3}{2}, \quad -\frac{2}{3}, \quad 0.5, \quad 1\frac{1}{3}, \quad \frac{23}{8}, \quad 3.25, \quad 4$$

To locate the improper fractions on a number line, write them as mixed numbers or decimals. The graph is shown in **FIGURE 3**.

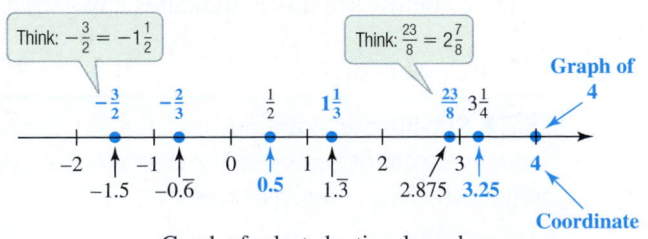

Graph of selected rational numbers

FIGURE 3

Think of the graph of a set of numbers as a picture of the set.

NOW TRY

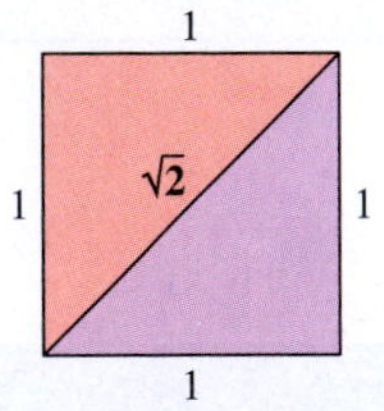

Not all numbers are rational. For example, the square root of 2, written $\sqrt{2}$, cannot be written as a quotient of two integers. Because of this, $\sqrt{2}$ is an *irrational number*. (See **FIGURE 4**.)

This square has diagonal of length $\sqrt{2}$. The number $\sqrt{2}$ is an irrational number.

FIGURE 4

Irrational Numbers

$\{x \mid x$ is a nonrational number represented by a point on a number line$\}$ is the set of **irrational numbers**.

NOW TRY ANSWER

2.

The decimal form of an irrational number neither terminates nor repeats.

Both rational and irrational numbers can be represented by points on a number line and together form the set of *real numbers*. See **FIGURE 5** on the next page.

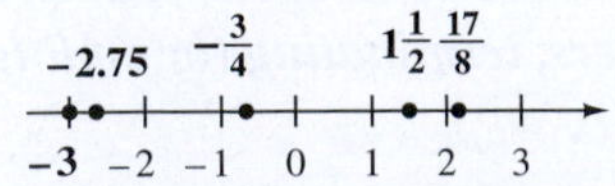

Real Numbers

$\{x \mid x \text{ is a rational or an irrational number}\}$ is the set of **real numbers.**

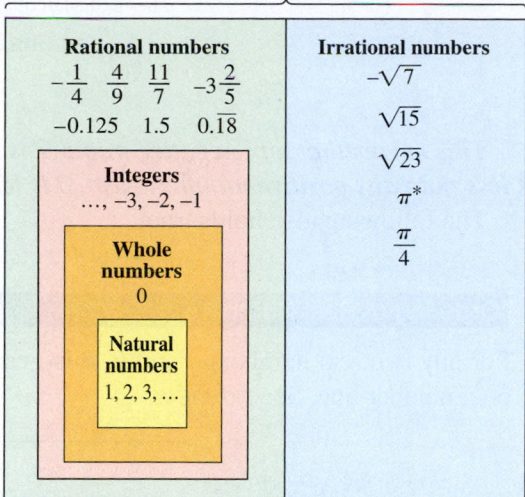

FIGURE 5

**NOW TRY
EXERCISE 3**

List the numbers in the following set that belong to each set of numbers.

$$\left\{-7, -\frac{4}{5}, 0, \sqrt{3}, 2.7, \pi, 13\right\}$$

(a) Whole numbers

(b) Integers

(c) Rational numbers

(d) Irrational numbers

EXAMPLE 3   Determining Whether a Number Belongs to a Set

List the numbers in the following set that belong to each set of numbers.

$$\left\{-5, -\frac{2}{3}, 0, 0.\overline{6}, \sqrt{2}, 3\frac{1}{4}, 5, 5.8\right\}$$

(a) Natural numbers: 5

(b) Whole numbers: 0 and 5

   The whole numbers consist of the natural (counting) numbers and 0.

(c) Integers: $-5, 0,$ and 5

(d) Rational numbers: $-5, -\frac{2}{3}, 0, 0.\overline{6}\left(\text{or }\frac{2}{3}\right), 3\frac{1}{4}\left(\text{or }\frac{13}{4}\right), 5,$ and $5.8\left(\text{or }\frac{58}{10}\right)$

   Each of these numbers can be written as the quotient of two integers.

(e) Irrational numbers: $\sqrt{2}$

(f) Real numbers: All the numbers in the set are real numbers.

NOW TRY

OBJECTIVE 2 Use inequality symbols with real numbers.

Given any two different positive integers, we can determine which number is less than the other. Positive numbers decrease as the corresponding points on a number line go to the left. For example,

$$8 < 12 \text{ because 8 is to the left of 12 on a number line.}$$

This ordering is extended to all real numbers by definition.

NOW TRY ANSWERS

3. (a) $0, 13$

   (b) $-7, 0, 13$

   (c) $-7, -\frac{4}{5}, 0, 2.7, 13$

   (d) $\sqrt{3}, \pi$

*The value of the irrational number π (pi) is approximately 3.14159265. The decimal digits continue forever with no repeated pattern.

Ordering of Real Numbers

For any two real numbers a and b, **a is less than b** if a lies to the left of b on a number line. See **FIGURE 6**.

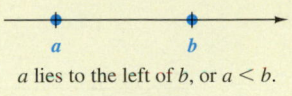

a lies to the left of b, or $a < b$.

FIGURE 6

This means that any negative number is less than 0, and any negative number is less than any positive number. Also, 0 is less than any positive number.

The following also holds true.

Ordering of Real Numbers

For any two real numbers a and b, **a is greater than b** if a lies to the right of b on a number line. See **FIGURE 7**.

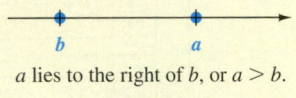

a lies to the right of b, or $a > b$.

FIGURE 7

NOW TRY
EXERCISE 4

Determine whether the statement is *true* or *false*.

$$-8 \leq -9$$

EXAMPLE 4   Determining the Order of Real Numbers

Determine whether the statement $-3 < -1$ is *true* or *false*.

Locate -3 and -1 on a number line. See **FIGURE 8**. Because -3 lies to the left of -1 on the number line, -3 is less than -1. The statement $-3 < -1$ is true.

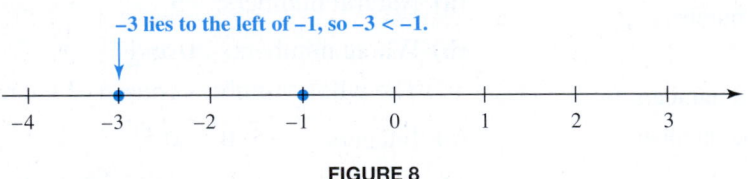

FIGURE 8

Also, the statement $-1 > -3$ is true because -1 lies to the right of -3 on the number line.

NOW TRY

OBJECTIVE 3   Find the additive inverse of a real number.

For any real number x (except 0), there is exactly one number on a number line the same distance from 0 as x, but on the *opposite* side of 0. See **FIGURE 9**. Such pairs of numbers are *additive inverses,* or *opposites,* of each other.

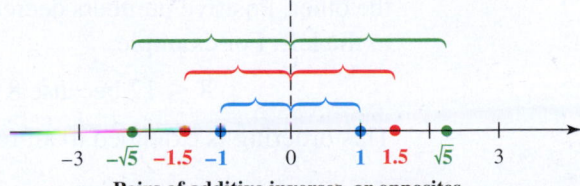

Pairs of additive inverses, or opposites

FIGURE 9

NOW TRY ANSWER
4. false

Additive Inverse

The **additive inverse** of a number x is the number that is the same distance from 0 on a number line as x, but on the *opposite* side of 0.

We indicate the additive inverse of a number by writing the symbol $-$ in front of the number. For example, the additive inverse of 7 is written -7 (read "*negative* 7"). We could write the additive inverse of -3 as $-(-3)$, but we know that 3 is the opposite of -3. Because a number can have only one additive inverse, 3 and $-(-3)$ must represent the same number.

$$-(-3) = 3$$

This idea can be generalized.

Double Negative Rule

For any real number x, the following holds true.

$$-(-x) = x$$

▼ **Additive Inverses**

Number	Additive Inverse
7	-7
-3	$-(-3)$, or 3
0	0
19	-19
$-\frac{2}{3}$	$\frac{2}{3}$
0.52	-0.52

Note that the number 0 is its own additive inverse.

The table in the margin gives examples of additive inverses.

Finding an Additive Inverse

The additive inverse of a nonzero number is found by changing the sign of the number. (A nonzero number and its additive inverse have opposite signs.)

OBJECTIVE 4 Find the absolute value of a real number.

Because additive inverses are the same distance from 0 on a number line, a number and its additive inverse have the same *absolute value*. The **absolute value** of a real number x, written $|x|$ and read "*the absolute value of x*," can be defined as the distance between 0 and the number on a number line.

$|2| = 2$ The distance between 2 and 0 on a number line is 2 units.

$|-2| = 2$ The distance between -2 and 0 on a number line is also 2 units.

Distance is a physical measurement, which is never negative. *Therefore, the absolute value of a number is never negative.*

Absolute Value

For any real number x, the absolute value of x is defined as follows.

$$|x| = \begin{cases} x & \text{if } x \geq 0 \\ -x & \text{if } x < 0 \end{cases}$$

By this definition, if x is a positive number or 0, then its absolute value is x itself.

$$|8| = 8 \quad \text{and} \quad |0| = 0$$

If x is a negative number, then its absolute value is the additive inverse of x.

$$|-8| = -(-8) = 8 \quad \text{The additive inverse of } -8 \text{ is } 8.$$

> **! CAUTION**
>
> $$|x| = \begin{cases} x & \text{if } x \geq 0 \\ -x & \text{if } x < 0 \end{cases} \qquad \text{Definition of absolute value}$$
>
> The "$-x$" in the second part of the definition of absolute value does *not* represent a negative number. Because x is negative in the second part, it follows that $-x$ represents the *opposite* of a negative number—that is, a positive number. ***The absolute value of a number is never negative.***

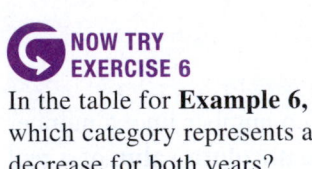
NOW TRY
EXERCISE 5

Find each absolute value.

(a) $|4|$ (b) $|-4|$

(c) $-|-4|$ (d) $|4-4|$

EXAMPLE 5   Finding Absolute Values

Find each absolute value.

(a) $|0| = 0$

(b) $|5| = 5$

(c) $|-5| = -(-5) = 5$

(d) $-|5| = -(5) = -5$

(e) $-|-5| = -(5) = -5$

(f) $|8 - 2| = |6| = 6$

(g) $-|8 - 2| = -|6| = -6$

Absolute value bars are grouping symbols. In parts (f) and (g), we perform any operations inside the absolute value bars *before* finding the absolute value.

NOW TRY

OBJECTIVE 5 Interpret meanings of real numbers from a table of data.

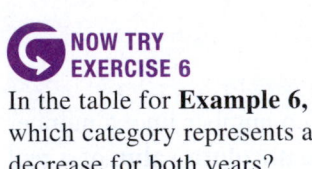
NOW TRY
EXERCISE 6

In the table for **Example 6,** which category represents a decrease for both years?

EXAMPLE 6   Interpreting Data

The Consumer Price Index (CPI) measures the average change in prices of goods and services purchased by urban consumers in the United States. The table shows the percent change in the CPI for selected categories of goods and services from 2013 to 2014 and from 2014 to 2015. Use the table to answer each question.

Category	Change from 2013 to 2014	Change from 2014 to 2015
Apparel	0.1	−1.3
Food and Beverage	2.3	1.8
Housing	2.6	2.1
Medical Care	2.4	2.6
Transportation	−0.7	−7.8

Data from U.S. Bureau of Labor Statistics.

(a) Which category in which year represents the greatest percent decrease?

We must find the negative number with the greatest absolute value. The number that satisfies this condition is -7.8, so the greatest percent decrease was shown by Transportation from 2014 to 2015.

(b) Which category in which year represents the least change?

We must find the number (either positive, negative, or zero) with the least absolute value. From 2013 to 2014, Apparel showed the least change, an increase of 0.1%.

NOW TRY ANSWERS

5. (a) 4 (b) 4 (c) −4 (d) 0
6. Transportation

NOW TRY

1.3 Exercises

FOR EXTRA HELP

 MyLab Math

▶ *Video solutions for select problems available in MyLab Math*

STUDY SKILLS REMINDER
Study cards are a great way to learn vocabulary, procedures, etc. **Review Study Skill 5, *Using Study Cards*.**

Concept Check *Complete each statement.*

1. The number _____ is a whole number, but not a natural number.

2. The natural numbers, their additive inverses, and 0 form the set of _____.

3. The additive inverse of every negative number is a (*negative / positive*) number.

4. If x and y are real numbers with $x > y$, then x lies to the (*left / right*) of y on a number line.

5. A rational number is the _____ of two integers, with the _____ not equal to 0.

6. Decimal numbers that neither terminate nor repeat are _____ numbers.

7. The additive inverse of -5 is _____, while the additive inverse of the absolute value of -5 is _____.

8. If a is negative, then $|a| =$ _____.

9. **Concept Check** Match each expression in Column I with its value in Column II. Choices in Column II may be used once, more than once, or not at all.

I	II
(a) $\lvert -9 \rvert$	**A.** 9
(b) $-(-9)$	**B.** -9
(c) $-\lvert -9 \rvert$	**C.** Neither A nor B
(d) $-\lvert -(-9) \rvert$	**D.** Both A and B

10. Students often say "*Absolute value is always positive.*" Is this true? Explain.

Concept Check *Give a number that satisfies the given condition.*

11. An integer between 3.6 and 4.6

12. A rational number between 2.8 and 2.9

13. A whole number that is not positive and is less than 1

14. A whole number greater than 3.5

15. An irrational number that is between $\sqrt{12}$ and $\sqrt{14}$

16. A real number that is neither negative nor positive

Concept Check *Determine whether each statement is* true *or* false.

17. Every natural number is positive.

18. Every whole number is positive.

19. Every integer is a rational number.

20. Every rational number is a real number.

21. Some numbers are both rational and irrational.

22. Every terminating decimal is a rational number.

Concept Check *Give three numbers between* -6 *and* 6 *that satisfy each given condition.*

23. Positive real numbers but not integers

24. Real numbers but not positive numbers

25. Real numbers but not whole numbers

26. Rational numbers but not integers

27. Real numbers but not rational numbers

28. Rational numbers but not negative numbers

Use a signed number to express each number in boldface italics. ***See Example 1.***

29. Between July 1, 2016, and July 1, 2017, the population of the United States increased by approximately **2,216,602.** (Data from U.S. Census Bureau.)

30. Between 2015 and 2016, the number of movie screens in the United States increased by **218.** (Data from Motion Picture Association of America.)

31. During the 2017 World Series, attendance went from 54,253 at the first game of the series to 43,282 at the third game, a decrease of **10,971**. (Data from Major League Baseball.)

32. In 1935, there were 15,295 banks in the United States. By 2016, the number was 6068, a decrease of **9227** banks. (Data from Federal Deposit Insurance Corporation.)

33. On Friday, November 10, 2017, the Dow Jones Industrial Average (DJIA) closed at 23,422.21. On the previous day it had closed at 23,461.94. Thus, on Friday, it closed down **39.73** points. (Data from *The Wall Street Journal*.)

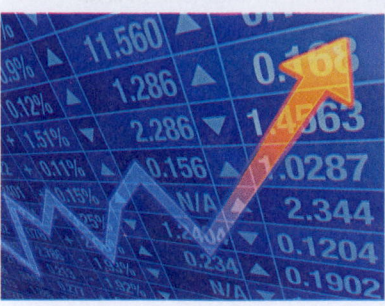

34. On Wednesday, November 8, 2017, the NASDAQ closed at 6789.12. On the previous day, it had closed at 6767.78. Thus, on Wednesday, it closed up **21.34** points. (Data from *Wall Street Journal*.)

Graph each number on a number line. See Example 2.

35. 0, 3, −5, −6

36. 2, 6, −2, −1

37. −2, −6, −4, 3, 4

38. −5, −3, −2, 0, 4

39. $\frac{1}{4}, 2\frac{1}{2}, -3.8, -4, -1\frac{5}{8}$

40. $5.25, 4\frac{5}{9}, -2\frac{1}{3}, 0, -3\frac{2}{5}$

List all numbers from each set that are the following. See Example 3.

(a) natural numbers   **(b)** whole numbers   **(c)** integers

(d) rational numbers   **(e)** irrational numbers   **(f)** real numbers

41. $\left\{-9, -\sqrt{7}, -1\frac{1}{4}, -\frac{3}{5}, 0, 0.\overline{1}, \sqrt{5}, 3, 5.9, 7\right\}$

42. $\left\{-5.3, -5, -\sqrt{3}, -1, -\frac{1}{9}, 0, 0.\overline{27}, 1.2, 3, \sqrt{11}\right\}$

43. $\left\{\frac{7}{9}, -2.\overline{3}, \sqrt{3}, 0, -8\frac{3}{4}, 11, -6, \pi\right\}$

44. $\left\{1\frac{5}{8}, -0.\overline{4}, \sqrt{6}, 9, -12, 0, \sqrt{10}, 0.026\right\}$

*For each number, find **(a)** the additive inverse and **(b)** the absolute value. See Objective 3 and Example 5.*

45. −2

46. −4

47. 8

48. 10

49. $-\frac{3}{4}$

50. $-\frac{2}{5}$

51. 5.6

52. 8.1

Find each absolute value. See Example 5.

53. $|-6|$

54. $|-14|$

55. $-|12|$

56. $-|19|$

57. $-\left|-\frac{2}{3}\right|$

58. $-\left|-\frac{4}{5}\right|$

59. $|6-3|$

60. $|9-4|$

61. $-|6-3|$

62. $-|9-4|$

Select the lesser of the two given numbers. See Examples 4 and 5.

63. −11, −4

64. −8, −13

65. $-\frac{2}{3}, -\frac{1}{4}$

66. $-\frac{3}{8}, -\frac{9}{16}$

67. 4, $|-5|$

68. 4, $|-3|$

69. $|-3.5|, |-4.5|$

70. $|-8.9|, |-9.8|$

71. $-|-6|, -|-4|$

72. $-|-2|, -|-3|$

73. $|5-3|, |6-2|$

74. $|7-2|, |8-1|$

Determine whether each statement is true *or* false. *See Examples 4 and 5.*

75. $-5 < -2$

76. $-8 > -2$

77. $-4 \le -(-5)$

78. $-6 \le -(-3)$

79. $|-6| < |-9|$

80. $|-12| < |-20|$

81. $-|8| > |-9|$

82. $-|12| > |-15|$

83. $-|-5| \ge -|-9|$

84. $-|-12| \le -|-15|$

85. $|6 - 5| \ge |6 - 2|$

86. $|13 - 8| \le |7 - 4|$

The table shows the percent change in the Consumer Price Index (CPI) for selected categories of goods and services from March 2017 to April 2017 and from April 2017 to May 2017. Use the table to answer each question. **See Example 6.**

87. Which category in which time period represents the greatest percent increase?

88. Which category in which time period represents the greatest percent decrease?

89. Which category in which time period represents the least change?

90. Which categories represent a decrease for both time periods?

Category	Change from March to April	Change from April to May
Apparel	−0.3	−0.8
Fuel oil	−0.3	−2.8
Gasoline	1.2	−6.4
Natural gas service	2.2	1.9
Shelter	0.3	0.2

Data from U.S. Bureau of Labor Statistics.

1.4 Adding and Subtracting Real Numbers

OBJECTIVES

1 Add two numbers with the same sign.

2 Add two numbers with different signs.

3 Use the definition of subtraction.

4 Use the rules for order of operations when adding and subtracting signed numbers.

5 Translate words and phrases involving addition and subtraction.

6 Use signed numbers to interpret data.

OBJECTIVE 1 Add two numbers with the same sign.

Recall that the answer to an addition problem is a **sum.** The numbers being added are the **addends.**

$$x + y = z \leftarrow \text{Sum}$$

$$\text{Addends}$$

EXAMPLE 1 Adding Numbers (Same Sign) on a Number Line

Use a number line to find each sum.

(a) $2 + 3$

Step 1 Start at 0 and draw an arrow 2 units to the *right*. See **FIGURE 10**.

Step 2 From the right end of that arrow, draw another arrow 3 units to the *right*.

The number below the end of the second arrow is 5, so $2 + 3 = 5$.

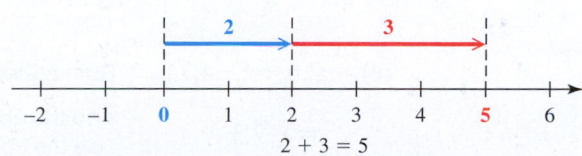

$2 + 3 = 5$

FIGURE 10

VOCABULARY
□ sum
□ addends
□ difference
□ minuend
□ subtrahend

NOW TRY
EXERCISE 1
Use a number line to find each sum.

(a) $3 + 5$ (b) $-1 + (-3)$

(b) $-2 + (-4)$

(We put parentheses around -4 due to the $+$ and $-$ symbols next to each other.)

Step 1 Start at 0 and draw an arrow 2 units to the *left*. See **FIGURE 11**.

Step 2 From the left end of the first arrow, draw a second arrow 4 units to the *left* to represent the addition of a *negative* number.

The number below the end of the second arrow is -6, so $-2 + (-4) = -6$.

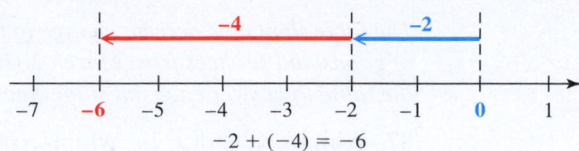

$$-2 + (-4) = -6$$

FIGURE 11

NOW TRY

In **Example 1(b)**, the sum of the two negative numbers -2 and -4 is a negative number whose distance from 0 is the sum of the distance of -2 from 0 and the distance of -4 from 0. *That is, the sum of two negative numbers is the negative of the sum of their absolute values.*

Adding Signed Numbers (Same Sign)

To add two numbers with the *same* sign, add their absolute values. The sum has the same sign as the addends.

Examples: $2 + 4 = 6$ and $-2 + (-4) = -6$

NOW TRY
EXERCISE 2
Find each sum.

(a) $-6 + (-11)$

(b) $-\dfrac{2}{5} + \left(-\dfrac{1}{2}\right)$

EXAMPLE 2 Adding Two Negative Numbers

Find each sum.

(a) $-9 + (-2)$ Both addends are negative.

$= \underset{\underset{\text{Sign of each addend}}{\uparrow}}{-}(|-9| + |-2|)$ Add the absolute values of the addends.

$= -(9 + 2)$ Take the absolute values.

$= -11$ The sum of two negative numbers is negative.

(b) $-\dfrac{1}{4} + \left(-\dfrac{2}{3}\right)$ Both addends are negative.

Think: $\left|-\dfrac{3}{12}\right| = \dfrac{3}{12}$
and $\left|-\dfrac{8}{12}\right| = \dfrac{8}{12}$ $= -\dfrac{3}{12} + \left(-\dfrac{8}{12}\right)$ Write equivalent fractions using the LCD, 12.

$= -\dfrac{11}{12}$ Add the absolute values of the addends. Use the common negative sign.

(c) $-2.6 + (-4.7)$ Both addends are negative.

$= -7.3$ Add the absolute values of the addends. Use the common negative sign.

NOW TRY

NOW TRY ANSWERS
1. (a) 8 (b) -4
2. (a) -17 (b) $-\dfrac{9}{10}$

OBJECTIVE 2 Add two numbers with different signs.

NOW TRY EXERCISE 3

Use a number line to find the sum.

$$4 + (-8)$$

EXAMPLE 3 Adding Numbers (Different Signs) on a Number Line

Use a number line to find the sum $-2 + 5$.

Step 1 Start at 0 and draw an arrow 2 units to the *left*. See **FIGURE 12**.

Step 2 From the left end of this arrow, draw a second arrow 5 units to the *right* to represent the addition of a *positive* number.

The number below the end of the second arrow is 3, so $-2 + 5 = 3$.

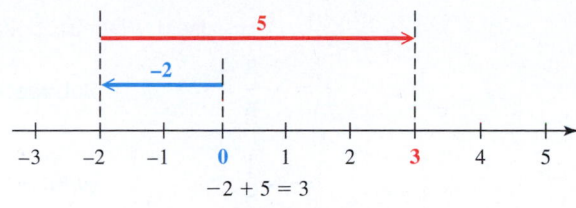

$$-2 + 5 = 3$$

FIGURE 12

 **NOW TRY**

Adding Signed Numbers (Different Signs)

To add two numbers with *different* signs, find their absolute values and subtract the lesser absolute value from the greater. The sum has the same sign as the addend with greater absolute value.

Examples: $-2 + 6 = 4$ and $2 + (-6) = -4$

EXAMPLE 4 Adding Signed Numbers (Different Signs)

Find each sum.

(a) $-12 + 5$

$|-12|$ $|5|$ Find the absolute value of each addend,
    ↓   ↓   and subtract the lesser from the greater.

$= -(12 - 5)$

  Use the sign of the addend with greater absolute value.

$= -7$

(b) $-8 + 12$ Find the absolute value of each addend, and subtract the lesser from the greater.

$= +(12 - 8)$

  Use the sign of the addend with greater absolute value.

$= 4$ The + symbol is understood.

(c) $6 + (-15)$ $|6| = 6$ and $|-15| = 15$;

$= -(15 - 6)$ Subtract the lesser from the greater.

  Use a − symbol because $|-15| > |6|$.

$= -9$

NOW TRY ANSWER
3. -4

**NOW TRY
EXERCISE 4**

Find each sum.

(a) $7 + (-4)$

(b) $\dfrac{2}{3} + \left(-2\dfrac{1}{9}\right)$

(c) $-5.7 + 3.7$

(d) $-10 + 10$

(d) $\dfrac{5}{6} + \left(-1\dfrac{1}{3}\right)$

$= \dfrac{5}{6} + \left(-\dfrac{4}{3}\right)$ Write the mixed number as an improper fraction.

$= \dfrac{5}{6} + \left(-\dfrac{8}{6}\right)$ Find a common denominator.

$= -\left(\dfrac{8}{6} - \dfrac{5}{6}\right)$ $\left|\dfrac{5}{6}\right| = \dfrac{5}{6}$ and $\left|-\dfrac{8}{6}\right| = \dfrac{8}{6}$;
Subtract the lesser absolute value from the greater.

Use a − symbol because $\left|-\dfrac{8}{6}\right| > \left|\dfrac{5}{6}\right|$.

$= -\dfrac{3}{6}$ Subtract the fractions.

$= -\dfrac{1}{2}$ Write in lowest terms.

(e) $8.1 + (-4.6)$ $|8.1| = 8.1$ and $|-4.6| = 4.6$;
Subtract the lesser absolute value from the greater.

$= +(8.1 - 4.6)$

$|8.1| > |-4.6|$

$= 3.5$

(f) $-16 + 16$ $|-16| = 16$ and $|16| = 16$;
The difference of the absolute values is 0,
which is neither positive nor negative.

$= 0$

(g) $42 + (-42)$

$= 0$ ***When additive inverses are added, the sum is 0.***   NOW TRY

OBJECTIVE 3 Use the definition of subtraction.

Recall that the answer to a subtraction problem is a **difference**. In the subtraction $x - y$, x is the **minuend** and y is the **subtrahend**.

$$x \quad - \quad y \quad = \quad z$$
Minuend   Subtrahend   Difference

**NOW TRY
EXERCISE 5**

Use a number line to find the difference.

$$6 - 2$$

EXAMPLE 5 Subtracting Numbers on a Number Line

Use a number line to find the difference $7 - 4$.

Step 1   Start at 0 and draw an arrow 7 units to the *right*. See **FIGURE 13**.

Step 2   From the right end of the first arrow, draw a second arrow 4 units to the *left* to represent the *subtraction*.

The number below the end of the second arrow is 3, so $7 - 4 = 3$.

NOW TRY ANSWERS

4. (a) 3   **(b)** $-\dfrac{13}{9}$, or $-1\dfrac{4}{9}$
   (c) -2 **(d)** 0
5. 4

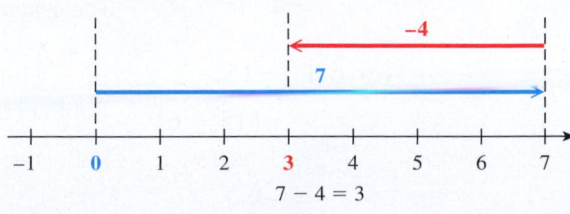

$$7 - 4 = 3$$

FIGURE 13

NOW TRY

The procedure used in **Example 5** to find the difference $7 - 4$ is exactly the same procedure that would be used to find the sum $7 + (-4)$.

$$7 - 4 \quad \text{is equal to} \quad 7 + (-4).$$

This suggests that *subtracting* a positive number from a greater positive number is the same as *adding* the additive inverse of the lesser number to the greater.

Definition of Subtraction

For any real numbers x and y, the following holds true.

$$x - y = x + (-y)$$

To subtract y from x, add the additive inverse (or opposite) of y to x. That is, change the subtrahend to its opposite and add.

Example: $4 - 9$
$$= 4 + (-9)$$
$$= -5$$

Subtracting Signed Numbers

Step 1 Change the subtraction symbol to an addition symbol, and change the sign of the subtrahend.

Step 2 Add the signed numbers.

EXAMPLE 6 Subtracting Signed Numbers

Find each difference.

(a) $12 - 3$

Change – to +.

$$= 12 + (-3)$$

No change ⟶   Additive inverse of 3

$$= 9$$

> 12 has the greater absolute value, so the sum is positive.

(b) $5 - 7$

Change – to +.

$$= 5 + (-7)$$

No change ⟶   Additive inverse of 7

$$= -2$$

> −7 has the greater absolute value, so the sum is negative.

(c) $-8 - 15$

Change – to +.

$$= -8 + (-15)$$

No change ⟶   Additive inverse of 15

$$= -23$$

> The sum of two negative numbers is negative.

NOW TRY
EXERCISE 6
Find each difference.
(a) $-5 - (-11)$
(b) $4 - 15$
(c) $-\dfrac{5}{7} - \dfrac{1}{3}$
(d) $5.25 - (-3.24)$

(d) $-3 - (-5)$

Change – to +.

$= -3 + 5$

No change ⎯↑   ↑⎯ Additive inverse of -5

$= 2$ ⎯⎯ 5 has the greater absolute value, so the sum is positive.

(e) $\dfrac{3}{8} - \left(-\dfrac{4}{5}\right)$

$= \dfrac{15}{40} - \left(-\dfrac{32}{40}\right)$ Write equivalent fractions using the LCD, 40.

$= \dfrac{15}{40} + \dfrac{32}{40}$ Definition of subtraction

$= \dfrac{47}{40},$ or $1\dfrac{7}{40}$ Add the fractions. Write as a mixed number.

(f) $-8.75 - (-2.41)$

$= -8.75 + 2.41$ Definition of subtraction

$= -6.34$ Add the decimals.   **NOW TRY**

Uses of the Symbol −

We use the symbol − for three purposes.

1. *It can represent subtraction,* as in $9 - 5 = 4$.

2. *It can represent negative numbers,* such as -10, -2, and -3.

3. *It can represent the additive inverse (or opposite) of a number,* as in "the additive inverse (or opposite) of 8 is -8."

We may see more than one use of − in the same expression, such as $-6 - (-9)$, where -9 is subtracted from -6. The meaning of the symbol − depends on its position in the algebraic expression.

OBJECTIVE 4 Use the rules for order of operations when adding and subtracting signed numbers.

EXAMPLE 7   Using the Rules for Order of Operations

Perform each indicated operation.

(a) $-6 - [2 - (8 + 3)]$ ⎯ Work from the inside out.

$= -6 - [2 - 11]$ Add inside the parentheses.

$= -6 - [2 + (-11)]$ Definition of subtraction

$= -6 - [-9]$ Add inside the brackets.

$= -6 + 9$ Definition of subtraction

$= 3$ Add.

NOW TRY ANSWERS
6. **(a)** 6 **(b)** -11
   (c) $-\dfrac{22}{21},$ or $-1\dfrac{1}{21}$
   (d) 8.49

**NOW TRY
EXERCISE 7**

Perform each indicated
operation.

(a) $8 - [(-3 + 7) - (3 - 9)]$

(b) $3|6 - 9| - |4 - 12|$

(b) $5 + [(-3 - 2) - (4 - 1)]$

$= 5 + [(-3 + (-2)) - 3]$ Work within each set of parentheses inside the brackets.

$= 5 + [(-5) - 3]$

$= 5 + [(-5) + (-3)]$ Show all steps to avoid sign errors.

$= 5 + [-8]$

$= -3$

(c) $\dfrac{2}{3} - \left[\dfrac{1}{12} - \left(-\dfrac{1}{4}\right) \right]$

$= \dfrac{8}{12} - \left[\dfrac{1}{12} - \left(-\dfrac{3}{12}\right) \right]$ Write equivalent fractions using the LCD, 12.

$= \dfrac{8}{12} - \left[\dfrac{1}{12} + \dfrac{3}{12} \right]$ Work inside the brackets.

$= \dfrac{8}{12} - \dfrac{4}{12}$ Add inside the brackets.

$= \dfrac{4}{12}$ Subtract.

$= \dfrac{1}{3}$ Write in lowest terms.

(d) $|4 - 7| + 2|6 - 3|$ The absolute value bars serve as grouping symbols.

$= |-3| + 2|3|$ Work within the absolute value bars.

$= 3 + 2 \cdot 3$ Find each absolute value.

$= 3 + 6$ Multiply.

Be careful.
Multiply first.

$= 9$ Add.

NOW TRY

OBJECTIVE 5 Translate words and phrases involving addition and subtraction.

▼ **Words and Phrases That Indicate Addition**

Word or Phrase	Example	Numerical Expression and Simplification
Sum of	The *sum of* −3 and 4	−3 + 4, which equals 1
Added to	5 *added to* −8	−8 + 5, which equals −3
More than	12 *more than* −5	−5 + 12, which equals 7
Increased by	−6 *increased by* 13	−6 + 13, which equals 7
Plus	3 *plus* 14	3 + 14, which equals 17

EXAMPLE 8 Translating Words and Phrases (Addition)

Write a numerical expression for each phrase, and simplify the expression.

(a) The sum of −8 and 4 and 6

Add in order from left to right. $-8 + 4 + 6$ simplifies to $-4 + 6$,   which equals   2.

NOW TRY EXERCISE 8

Write a numerical expression for the phrase, and simplify the expression.

The sum of -3 and 7, increased by 10

(b) 3 more than -5, increased by 12

$$(-5 + 3) + 12 \quad \text{simplifies to} \quad -2 + 12, \quad \text{which equals} \quad 10.$$

Here we *simplified* each expression by performing the operations.   **NOW TRY**

▼ **Words and Phrases That Indicate Subtraction**

Word, Phrase, or Sentence	Example	Numerical Expression and Simplification
Difference of	The *difference of* -3 and -8	$-3 - (-8)$ simplifies to $-3 + 8$, which equals 5
Subtracted from*	12 *subtracted from* 18	$18 - 12$, which equals 6
From . . . , subtract	From 12, subtract 8.	$12 - 8$ simplifies to $12 + (-8)$, which equals 4
Less	6 *less* 5	$6 - 5$, which equals 1
Less than*	6 *less than* 5	$5 - 6$ simplifies to $5 + (-6)$, which equals -1
Decreased by	9 *decreased by* -4	$9 - (-4)$ simplifies to $9 + 4$, which equals 13
Minus	8 *minus* 5	$8 - 5$, which equals 3

* Be careful with order when translating.

⚠️ **CAUTION**   When subtracting two numbers, be careful to write them in the correct order. In general,

$$x - y \neq y - x. \qquad \text{For example, } 5 - 3 \neq 3 - 5.$$

Think carefully before interpreting an expression involving subtraction.

NOW TRY EXERCISE 9

Write a numerical expression for each phrase, and simplify the expression.

(a) The difference of 5 and -8, decreased by 4

(b) 7 less than -2

EXAMPLE 9   Translating Words and Phrases (Subtraction)

Write a numerical expression for each phrase, and simplify the expression.

(a) The difference of -8 and 5

When "difference of" is used, write the numbers in the order given.

$$-8 - 5 \quad \text{simplifies to} \quad -8 + (-5), \quad \text{which equals} \quad -13.$$

(b) 4 subtracted from the sum of 8 and -3

Here the operation of addition is also used, as indicated by the words *sum of*. First, add 8 and -3. Next, subtract 4 from this sum.

$$[8 + (-3)] - 4 \quad \text{simplifies to} \quad 5 - 4, \quad \text{which equals} \quad 1.$$

(c) 4 less than -6

Here, 4 must be subtracted *from* -6, so write -6 first.

Be careful with order. $-6 - 4 \quad \text{simplifies to} \quad -6 + (-4), \quad \text{which equals} \quad -10.$

Notice that "4 less than -6" differs from "4 *is less than* -6." The second of these is symbolized $4 < -6$ (which is a false statement).

(d) 8, decreased by 5 less than 12

First, write "5 less than 12" as $12 - 5$. Next, subtract $12 - 5$ from 8.

$$8 - (12 - 5) \quad \text{simplifies to} \quad 8 - 7, \quad \text{which equals} \quad 1. \quad \text{NOW TRY}$$

NOW TRY ANSWERS
8. $(-3 + 7) + 10$; 14
9. (a) $[5 - (-8)] - 4$; 9
   (b) $-2 - 7$; -9

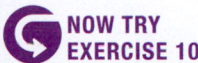

**NOW TRY
EXERCISE 10**

Find the difference between a gain of 226 yd on the football field by the Chesterfield Bears and a loss of 7 yd by the New London Wildcats.

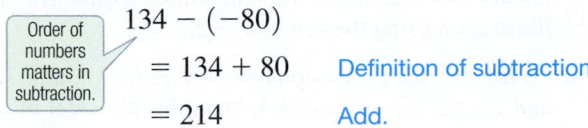

EXAMPLE 10 Solving an Application Involving Subtraction

The record-high temperature in the United States is 134°F, recorded at Death Valley, California, in 1913. The record low is −80°F, at Prospect Creek, Alaska, in 1971. See **FIGURE 14.** What is the difference between these highest and lowest temperatures? (Data from National Climatic Data Center.)

We must subtract the lowest temperature from the highest temperature.

> Order of numbers matters in subtraction.

$$134 - (-80)$$
$$= 134 + 80 \qquad \text{Definition of subtraction}$$
$$= 214 \qquad \text{Add.}$$

The difference between the two temperatures is 214°F.

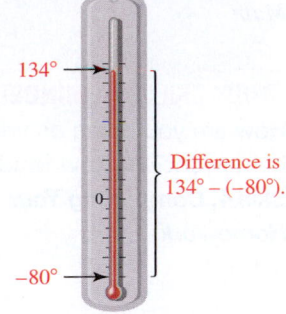

134° → Difference is
134° − (−80°).

0 −

−80° →

FIGURE 14

NOW TRY

OBJECTIVE 6 Use signed numbers to interpret data.

**NOW TRY
EXERCISE 11**

Refer to **FIGURE 15** and use a signed number to represent the change in enrollment from 1985 to 1990.

EXAMPLE 11 Using a Signed Number to Interpret Data

The bar graph in **FIGURE 15** shows public high school (grades 9–12) enrollment in millions of students for selected years from 1985 to 2015.

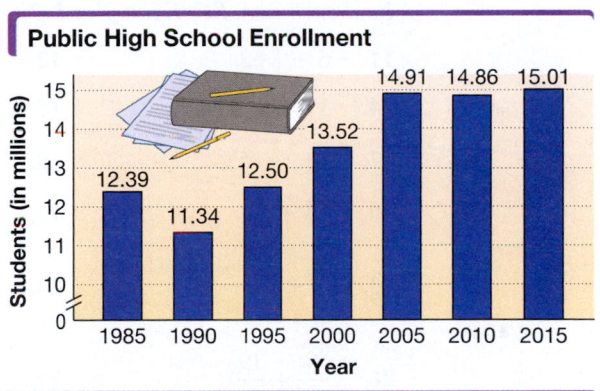

Public High School Enrollment

Data from U.S. National Center for Education Statistics.

FIGURE 15

(a) Use a signed number to represent the change in enrollment in millions from 2000 to 2005.

Start with the number for 2005. Subtract from it the number for 2000.

    2005   2000
     ↓   ↓
14.91 − 13.52 = +1.39 million students ← *A positive number indicates an increase. The bar for 2005 is "higher" than the bar for 2000.*

(b) Use a signed number to represent the change in enrollment in millions from 2005 to 2010.

Start with the number for 2010. Subtract from it the number for 2005.

    2010   2005
     ↓   ↓
14.86 − 14.91 = −0.05 million students ← *A negative number indicates a decrease. The bar for 2010 is "lower" than the bar for 2005.*

NOW TRY ANSWERS
10. 233 yd
11. −1.05 million students

NOW TRY

1.4 Exercises

FOR EXTRA HELP

 MyLab Math

Concept Check   *Complete each of the following.*

1. The sum of two negative numbers will always be a (*positive*/*negative*) number. Give a number-line illustration using the sum $-2 + (-3)$.

2. The sum of a number and its opposite will always be _____ .

3. When adding a positive number and a negative number, where the negative number has the greater absolute value, the sum will be a (*positive*/*negative*) number. Give a number-line illustration using the sum $-4 + 2$.

4. To simplify the expression $8 + [-2 + (-3 + 5)]$, one should begin by adding _____ and _____ , according to the rules for order of operations.

5. By the definition of subtraction, in order to perform the subtraction $-6 - (-8)$, we must add the opposite of _____ to _____ to obtain _____ .

6. "The difference of 7 and 12" translates as _____ , while "the difference of 12 and 7" translates as _____ .

Concept Check   *Suppose that x represents a positive number and y represents a negative number. Determine whether the given expression represents a* positive *or a* negative *number.*

7. $x - y$ 8. $y - x$ 9. $y - |x|$ 10. $x + |y|$

Find each sum. See Examples 1–7.

11. $-6 + (-2)$ 12. $-9 + (-2)$ 13. $-5 + (-7)$

14. $-11 + (-5)$ 15. $6 + (-4)$ 16. $11 + (-8)$

17. $12 + (-15)$ 18. $3 + (-7)$ 19. $-16 + 7$

20. $-13 + 6$ 21. $6 + (-6)$ 22. $-11 + 11$

23. $-\dfrac{1}{3} + \left(-\dfrac{4}{15}\right)$ 24. $-\dfrac{1}{4} + \left(-\dfrac{5}{12}\right)$ 25. $-\dfrac{1}{6} + \dfrac{2}{3}$

26. $-\dfrac{6}{25} + \dfrac{19}{20}$ 27. $\dfrac{5}{8} + \left(-\dfrac{17}{12}\right)$ 28. $\dfrac{9}{10} + \left(-\dfrac{11}{8}\right)$

29. $2\dfrac{1}{2} + \left(-3\dfrac{1}{4}\right)$ 30. $1\dfrac{3}{8} + \left(-2\dfrac{1}{4}\right)$ 31. $-3.5 + 12.4$

32. $-12.5 + 21.3$ 33. $-2.34 + (-3.67)$ 34. $-1.25 + (-6.88)$

35. $4 + [13 + (-5)]$ 36. $6 + [12 + (-3)]$ 37. $8 + [-2 + (-1)]$

38. $12 + [-3 + (-4)]$ 39. $-2 + [5 + (-1)]$ 40. $-8 + [9 + (-2)]$

41. $-6 + [6 + (-9)]$ 42. $-3 + [3 + (-8)]$ 43. $[(-9) + (-3)] + 12$

44. $[(-8) + (-6)] + 14$ 45. $-6.1 + [3.2 + (-4.8)]$ 46. $-9.4 + [5.8 + (-7.9)]$

47. $[-3 + (-4)] + [5 + (-6)]$ 48. $[-8 + (-3)] + [4 + (-6)]$

49. $[-4 + (-3)] + [8 + (-1)]$ 50. $[-5 + (-9)] + [16 + (-2)]$

51. $[-4 + (-6)] + [-3 + (-8)] + [12 + (-11)]$

52. $[-2 + (-11)] + [-12 + (-2)] + [18 + (-6)]$

Find each difference. See Examples 1–7.

53. $4 - 7$ **54.** $8 - 13$ **55.** $5 - 9$

56. $6 - 11$ **57.** $-7 - 1$ **58.** $-9 - 4$

59. $-8 - 6$ **60.** $-9 - 5$ **61.** $7 - (-3)$

62. $9 - (-2)$ **63.** $-6 - (-2)$ **64.** $-7 - (-5)$

65. $2 - (3 - 5)$ **66.** $-3 - (4 - 11)$ **67.** $\dfrac{1}{2} - \left(-\dfrac{1}{4}\right)$

68. $\dfrac{1}{3} - \left(-\dfrac{1}{12}\right)$ **69.** $-\dfrac{3}{4} - \dfrac{5}{8}$ **70.** $-\dfrac{5}{6} - \dfrac{1}{2}$

71. $\dfrac{5}{8} - \left(-\dfrac{1}{2} - \dfrac{3}{4}\right)$ **72.** $\dfrac{9}{10} - \left(\dfrac{1}{8} - \dfrac{3}{10}\right)$ **73.** $3.4 - (-8.2)$

74. $5.7 - (-11.6)$ **75.** $-6.4 - 3.5$ **76.** $-4.4 - 8.6$

Perform each indicated operation. See Examples 1–7.

77. $(4 - 6) + 12$ **78.** $(3 - 7) + 4$ **79.** $(8 - 1) - 12$

80. $(9 - 3) - 15$ **81.** $6 - (-8 + 3)$ **82.** $8 - (-9 + 5)$

83. $2 + (-4 - 8)$ **84.** $6 + (-9 - 2)$ **85.** $|-5 - 6| + |9 + 2|$

86. $|-4 + 8| + |6 - 1|$ **87.** $|-8 - 2| - |-9 - 3|$ **88.** $|-4 - 2| - |-8 - 1|$

89. $\left(-\dfrac{3}{8} - \dfrac{2}{3}\right) - \left(-\dfrac{9}{8} - 3\right)$ **90.** $\left(-\dfrac{3}{4} - \dfrac{5}{2}\right) - \left(-\dfrac{1}{8} - 1\right)$

91. $\left(-\dfrac{1}{2} + 0.25\right) - \left(-\dfrac{3}{4} + 0.75\right)$ **92.** $\left(-\dfrac{3}{2} - 0.75\right) - \left(0.5 - \dfrac{1}{2}\right)$

93. $-9 + [(3 - 2) - (-4 + 2)]$ **94.** $-8 - [(-4 - 1) + (9 - 2)]$

95. $-3 + [(-5 - 8) - (-6 + 2)]$ **96.** $-4 + [(-12 + 1) - (-1 - 9)]$

97. $-9.12 + [(-4.8 - 3.25) + 11.279]$ **98.** $-7.62 - [(-3.99 + 1.427) - (-2.8)]$

Write a numerical expression for each phrase, and simplify the expression. See Examples 8 and 9.

99. The sum of -5 and 12 and 6   **100.** The sum of -3 and 5 and -12

101. 14 added to the sum of -19 and -4 **102.** -2 added to the sum of -18 and 11

103. The sum of -4 and -10, increased by 12   **104.** The sum of -7 and -13, increased by 14

105. $\dfrac{2}{7}$ more than the sum of $\dfrac{5}{7}$ and $-\dfrac{9}{7}$ **106.** 1.85 more than the sum of -1.25 and -4.75

107. The difference of 4 and -8 **108.** The difference of 7 and -14

109. 8 less than -2 **110.** 9 less than -13

111. The sum of 9 and -4, decreased by 7   **112.** The sum of 12 and -7, decreased by 14

113. 12 less than the difference of 8 and -5 **114.** 19 less than the difference of 9 and -2

The table gives scores (above or below par—that is, above or below the score "standard") for selected golfers during the 2017 Blue Bay LPGA Tournament. Write a signed number that represents the total score (above or below par) for the four rounds for each golfer.

	Golfer	Round 1	Round 2	Round 3	Round 4
115.	Shanshan Feng	−3	−5	+1*	−2
116.	Jessica Korda	−1	0	−2	−1
117.	Pernilla Lindberg	−5	+7	−1	+2
118.	Brittany Lang	+3	+5	+2	−4

*Golf scoring commonly includes a + symbol with a score over par.
Data from LPGA.

Solve each problem. **See Example 10.**

119. Based on 2020 population projections, Illinois will probably lose 2 seats in the U.S. House of Representatives, Minnesota will lose 1 seat, and New York will lose 1. Write a signed number that represents the total number of seats these three states are projected to lose. (Data from Election Data Services.)

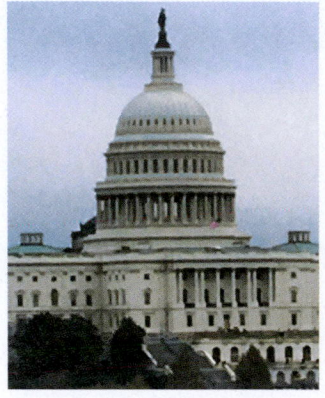

120. Both Alabama and Ohio are projected to lose 1 seat in the U.S. House of Representatives in 2020. The states projected to gain the most seats are Texas with 4 and Florida with 2. Write a signed number that represents the algebraic sum of these changes. (Data from Election Data Services.)

121. The surface, or rim, of a canyon is at altitude 0. On a hike down into the canyon, a party of hikers stops for a rest at 130 m below the surface. The hikers then descend another 54 m. Write the new altitude as a signed number.

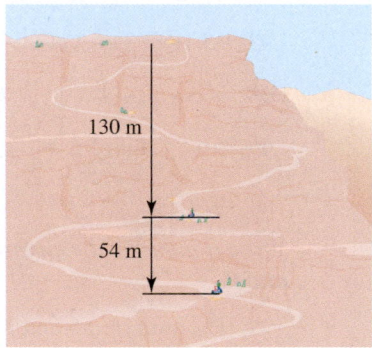

130 m

54 m

122. A pilot announces to the passengers that the current altitude of their plane is 34,000 ft. Because of turbulence, the pilot is forced to descend 2100 ft. Write the new altitude as a signed number.

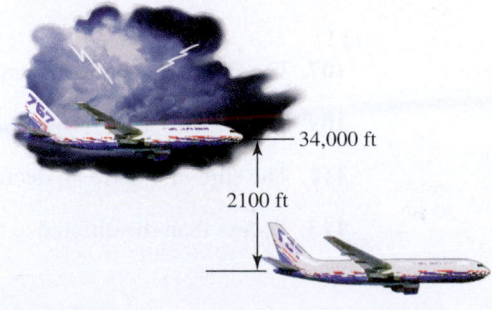

34,000 ft

2100 ft

123. The lowest temperature ever recorded in Arkansas was −29°F. The highest temperature ever recorded there was 149°F more than the lowest. What was this highest temperature? (Data from National Climatic Data Center.)

124. On January 23, 1943, the temperature rose 49°F in two minutes in Spearfish, South Dakota. If the starting temperature was −4°F, what was the temperature two minutes later? (Data from *Guinness World Records.*)

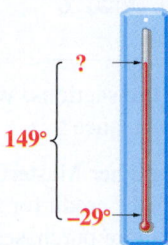

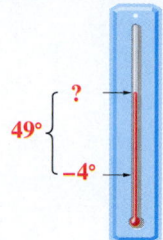

125. The lowest temperature ever recorded in Illinois was −36°F on January 5, 1999. The lowest temperature ever recorded in Utah was on February 1, 1985, and was 33°F lower than Illinois's record low. What is the record low temperature for Utah? (Data from National Climatic Data Center.)

126. The lowest temperature ever recorded in South Carolina was −19°F. The lowest temperature ever recorded in Wisconsin was 36° lower than South Carolina's record low. What is the record low temperature for Wisconsin? (Data from National Climatic Data Center.)

127. Nadine enjoys playing Triominoes every Wednesday night. Last Wednesday, on four successive turns, her scores were

$$-19, \quad 28, \quad -5, \quad \text{and} \quad 13.$$

What was her final score for the four turns?

128. Bruce enjoys playing Triominoes. On five successive turns, his scores were

$$-13, \quad 15, \quad -12, \quad 24, \quad \text{and} \quad 14.$$

What was his total score for the five turns?

129. In 2005, Americans saved −0.5% of their after-tax incomes. In September 2017, they saved 3.1%. (Data from U.S. Bureau of Economic Analysis.)

(a) Express the difference between these amounts as a positive number.

(b) How could Americans have a negative personal savings rate in 2005?

130. In 2000, the U.S. federal budget had a surplus of $236 billion. In 2016, the federal budget had a deficit of $616 billion. Express the difference between these amounts as a positive number. (Data from U.S. Office of Management and Budget.)

131. In 2006, bachelor's degree recipients from private four-year institutions had an average of $18,800 in total student debt. This average increased $1400 by 2011 and then dropped $300 by 2016. What was the average amount of total student debt in 2016? (Data from The College Board.)

132. The average annual spending per U.S. household on apparel and apparel services was $1786 in 2014. This amount increased $60 by 2015 and then decreased $43 by 2016. What was the average household expenditure for apparel and apparel services in 2016? (Data from U.S. Bureau of Labor Statistics.)

133. In August, Susan began with a checking account balance of $904.89. She made the following withdrawals and deposits.

Withdrawals	Deposits
$35.84	$85.00
$26.14	$120.76
$3.12	

Assuming no other transactions, what was her new account balance?

134. In September, Jeffery began with a checking account balance of $537.12. He made the following withdrawals and deposits.

Withdrawals	Deposits
$41.29	$80.59
$13.66	$276.13
$84.40	

Assuming no other transactions, what was his new account balance?

135. Linda owes $870.00 on her MasterCard account. She returns two items costing $35.90 and $150.00 and receives credit for these on the account. Next, she makes a purchase of $82.50 and then two more purchases of $10.00 each. She makes a payment of $500.00. She then incurs a finance charge of $37.23. How much does she still owe?

136. Marcial owes $679.00 on his Visa account. He returns three items costing $36.89, $29.40, and $113.55 and receives credit for these on the account. Next, he makes purchases of $135.78 and $412.88 and two purchases of $20.00 each. He makes a payment of $400. He then incurs a finance charge of $24.57. How much does he still owe?

The graph shows annual returns in percent for Class A shares of the Invesco S&P 500 Index Fund from 2012 to 2016. Use a signed number to represent the change in percent return for each period. **See Example 11.**

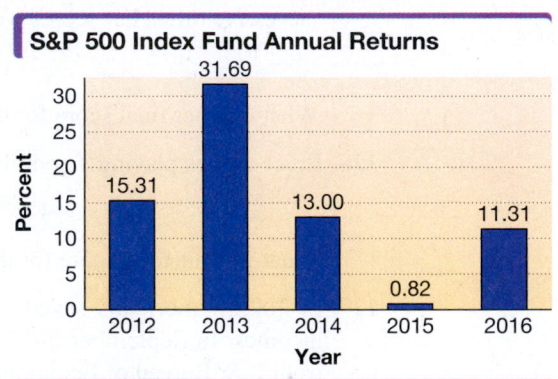

Data from Invesco.

137. 2012 to 2013   **138.** 2013 to 2014

139. 2014 to 2015   **140.** 2012 to 2016

The two tables show the heights of some selected mountains and the depths of some selected trenches. Use the information given to answer each question.

Mountain	Height (in feet, as a positive number)
Foraker	17,400
Wilson	14,246
Pikes Peak	14,110

Trench	Depth (in feet, as a negative number)
Philippine	−32,995
Cayman	−24,721
Java	−23,376

Data from *The World Almanac and Book of Facts.*

141. What is the difference between the height of Mt. Foraker and the depth of the Philippine Trench?

142. What is the difference between the height of Pikes Peak and the depth of the Java Trench?

143. How much deeper is the Cayman Trench than the Java Trench?

144. How much deeper is the Philippine Trench than the Cayman Trench?

145. How much higher is Mt. Wilson than Pikes Peak?

146. If Mt. Wilson and Pikes Peak were stacked one on top of the other, how much higher would they be than Mt. Foraker?

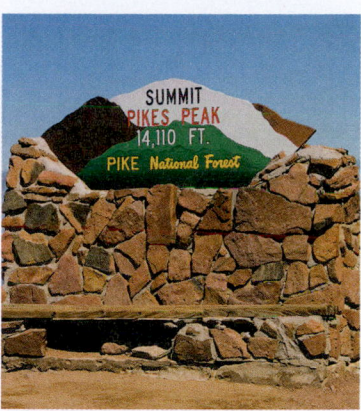

1.5 Multiplying and Dividing Real Numbers

OBJECTIVES

1 Find the product of a positive number and a negative number.

2 Find the product of two negative numbers.

3 Identify factors of integers.

4 Use the reciprocal of a number to apply the definition of division.

5 Use the rules for order of operations when multiplying and dividing signed numbers.

6 Evaluate algebraic expressions given values for the variables.

7 Translate words and phrases involving multiplication and division.

8 Translate simple sentences into equations.

The result of multiplication is a **product.** We know that the product of two positive numbers is positive. We also know that the product of 0 and any positive number is 0, so we extend that property to all real numbers.

> **Multiplication Property of 0**
>
> For any real number x, the following hold true.
> $$x \cdot 0 = 0 \quad \text{and} \quad 0 \cdot x = 0$$

OBJECTIVE 1 Find the product of a positive number and a negative number.

Observe the following pattern.

$$3 \cdot 5 = 15$$
$$3 \cdot 4 = 12$$
$$3 \cdot 3 = 9$$
$$3 \cdot 2 = 6$$
$$3 \cdot 1 = 3$$
$$3 \cdot 0 = 0$$
$$3 \cdot (-1) = ?$$

The products decrease by 3.

What should $3 \cdot (-1)$ equal? The product $3 \cdot (-1)$ represents the sum

$$-1 + (-1) + (-1), \quad \text{which equals} \quad -3,$$

so the product should be -3. Also, $3 \cdot (-2)$ and $3 \cdot (-3)$ represent the sums

$$-2 + (-2) + (-2), \quad \text{which equals} \quad -6$$

and

$$-3 + (-3) + (-3), \quad \text{which equals} \quad -9.$$

These results maintain the pattern in the list and suggest the following rules.

 **NOW TRY
EXERCISE 1**

Find each product.

(a) $-11(9)$ **(b)** $3.1(-2.5)$

Multiplying Signed Numbers (Different Signs)

For any positive real numbers x and y, the following hold true.

$$x(-y) = -(xy) \quad \text{and} \quad (-x)y = -(xy)$$

That is, the product of two numbers with different signs is negative.

Examples: $6(-3) = -18$ and $(-6)3 = -18$

EXAMPLE 1   **Multiplying Signed Numbers (Different Signs)**

Find each product.

(a) $8(-5)$

$= -(8 \cdot 5)$

$= -40$

The product of two numbers
with *different* signs is *negative.*

(b) $-9\left(\dfrac{1}{3}\right)$

$= -\left(9 \cdot \dfrac{1}{3}\right)$

$= -3$

(c) $-6.2(4.1)$

$= -(6.2 \cdot 4.1)$

$= -25.42$

NOW TRY

OBJECTIVE 2 Find the product of two negative numbers.

Look at another pattern.

$$-5(4) = -20$$
$$-5(3) = -15$$
$$-5(2) = -10$$
$$-5(1) = -5$$
$$-5(0) = 0$$
$$-5(-1) = ?$$

The products
increase by 5.

The numbers in color on the left side of the equality symbols decrease by 1 for each step down the list. The products on the right increase by 5 for each step down the list. To maintain this pattern, $-5(-1)$ should be 5 more than $-5(0)$, or 5 more than 0, so

$$-5(-1) = 5.$$

The pattern continues with

$$-5(-2) = 10$$
$$-5(-3) = 15$$
$$-5(-4) = 20$$
$$-5(-5) = 25, \quad \text{and so on.}$$

These results suggest the next rule.

Multiplying Two Negative Numbers

For any positive real numbers x and y, the following holds true.

$$-x(-y) = xy$$

That is, the product of two negative numbers is positive.

Example: $-5(-4) = 20$

NOW TRY ANSWERS
1. **(a)** -99 **(b)** -7.75

NOW TRY
EXERCISE 2
Find each product.
(a) $-8(-11)$
(b) $-\dfrac{1}{7}\left(-\dfrac{5}{2}\right)$

EXAMPLE 2 Multiplying Two Negative Numbers

Find each product.

(a) $-9(-2)$
$= 18$

(b) $-\dfrac{2}{3}\left(-\dfrac{3}{2}\right)$
$= 1$

(c) $-0.5(-1.25)$
$= 0.625$

The product of two numbers with the same sign is positive.   **NOW TRY**

The following box summarizes multiplying signed numbers.

Multiplying Signed Numbers

The product of two numbers with the *same* sign is *positive*.

The product of two numbers with *different* signs is *negative*.

Examples: $7(3) = 21,$ $-7(-3) = 21,$ $-7(3) = -21,$ and $7(-3) = -21$

OBJECTIVE 3 Identify factors of integers.

The definition of **factor** can be extended to integers. If the product of two integers is a third integer, then each of the two integers is a *factor* of the third.

▼ Integer Factors

Integer	18	20	15	7	1
	1, 18	1, 20	1, 15	1, 7	1, 1
	2, 9	2, 10	3, 5	−1, −7	−1, −1
Pairs of Factors	3, 6	4, 5	−1, −15		
	−1, −18	−1, −20	−3, −5		
	−2, −9	−2, −10			
	−3, −6	−4, −5			

▼ Reciprocals

Number	Reciprocal (Multiplicative Inverse)
4	$\dfrac{1}{4}$
0.3, or $\dfrac{3}{10}$	$\dfrac{10}{3}$
−5	$\dfrac{1}{-5}$, or $-\dfrac{1}{5}$
$-\dfrac{5}{8}$	$-\dfrac{8}{5}$

A number and its reciprocal have a product of **1.** For example,

$$4 \cdot \frac{1}{4} = \frac{4}{4}, \text{ or } 1.$$

0 has no reciprocal because the product of 0 and any number is 0, not 1.

OBJECTIVE 4 Use the reciprocal of a number to apply the definition of division.

The definition of division depends on the idea of a *reciprocal,* or *multiplicative inverse,* of a number.

Reciprocals, or Multiplicative Inverses

Pairs of numbers whose product is 1 are **reciprocals,** or **multiplicative inverses,** of each other.

The table in the margin gives examples of reciprocals.

Recall that the answer to a division problem is a **quotient.** For example, we can write the quotient of 15 and 3 as $15 \div 3$, which equals 5. We obtain the same answer if we multiply $15 \cdot \dfrac{1}{3}$, the reciprocal of 3.

This discussion suggests the next definition.

Definition of Division

For any real numbers x and y, where $y \neq 0$, the following holds true.

$$x \div y = x \cdot \frac{1}{y}$$

That is, to divide two numbers, multiply the first number (the **dividend**) by the reciprocal, or multiplicative inverse, of the second number (the **divisor**).

Example: $15 \div 3 = 15 \cdot \dfrac{1}{3} = 5$

Recall that an equivalent form of $x \div y$ is $\dfrac{x}{y}$, where the fraction bar represents division. ***In algebra, quotients are usually represented with a fraction bar.***

$$15 \div 3 \quad \text{is equivalent to} \quad \frac{15}{3}.$$

NOTE   The following all represent division of x by y, where $y \neq 0$.

$$x \div y, \quad \frac{x}{y}, \quad x/y, \quad \text{and} \quad y)\overline{x}$$

Example: $15 \div 3$, $\dfrac{15}{3}$, $15/3$, and $3)\overline{15}$ are equivalent forms of 15 divided by 3.

Because division is defined in terms of multiplication, the rules for multiplying signed numbers also apply to dividing them.

Dividing Signed Numbers

The quotient of two numbers with the *same* sign is *positive.*

The quotient of two numbers with *different* signs is *negative.*

Examples: $\dfrac{15}{3} = 5,$ $\dfrac{-15}{-3} = 5$ $\dfrac{15}{-3} = -5,$ and $\dfrac{-15}{3} = -5$

**NOW TRY
EXERCISE 3**

Find each quotient.

(a) $\dfrac{-10}{5}$ **(b)** $\dfrac{-1.44}{-0.12}$

(c) $-\dfrac{3}{8} \div \dfrac{7}{10}$

EXAMPLE 3   Dividing Signed Numbers

Find each quotient.

(a) $\dfrac{8}{-2} = -4$

(b) $\dfrac{-100}{5} = -20$

(c) $\dfrac{-4.5}{-0.09} = 50$

(d) $-\dfrac{1}{8} \div \left(-\dfrac{3}{4}\right)$

$= -\dfrac{1}{8} \cdot \left(-\dfrac{4}{3}\right)$ Multiply by the reciprocal of the divisor.

$= \dfrac{1}{6}$ Multiply the fractions. Write in lowest terms.

NOW TRY

NOW TRY ANSWERS
3. **(a)** -2 **(b)** 12 **(c)** $-\frac{15}{28}$

Consider the quotient $\frac{12}{3}$.

$$\frac{12}{3} = 4 \qquad \text{because} \qquad 4 \cdot 3 = 12. \quad \boxed{\text{Multiply to check a division problem.}}$$

Using this relationship between multiplication and division, we investigate division by 0. Consider the quotient $\frac{0}{3}$.

$$\frac{0}{3} = 0 \qquad \text{because} \qquad 0 \cdot 3 = 0.$$

Now consider $\frac{3}{0}$.

$$\frac{3}{0} = ?$$

We need to find a number that when multiplied by 0 will equal 3, that is, $? \cdot 0 = 3$. *No* real number satisfies this equation because the product of any real number and 0 must be 0. Thus,

$\frac{x}{0}$ **is not a number, and** *division by 0 is undefined.* **If a division problem involves division by 0, write "undefined."**

Division Involving 0

For any real number x, where $x \neq 0$, the following hold true.

$$\frac{0}{x} = 0 \quad \text{and} \quad \frac{x}{0} \text{ is undefined.}$$

Examples: $\quad \frac{0}{-10} = 0 \quad \text{and} \quad \frac{-10}{0} \text{ is undefined.}$

From the definitions of multiplication and division of real numbers,

$$\frac{-40}{8} = -5 \quad \text{and} \quad \frac{40}{-8} = -5, \quad \text{so} \quad \frac{-40}{8} = \frac{40}{-8}.$$

Based on this example, the quotient of a positive number and a negative number can be expressed in different, yet equivalent, forms.

Also, $\qquad \dfrac{-40}{-8} = 5 \quad \text{and} \quad \dfrac{40}{8} = 5, \quad \text{so} \quad \dfrac{-40}{-8} = \dfrac{40}{8}.$

Equivalent Forms

For any positive real numbers x and y, the following are equivalent.

$$\frac{-x}{y}, \quad \frac{x}{-y}, \quad \text{and} \quad -\frac{x}{y} \quad \leftarrow \text{We generally use this form for negative final answers.}$$

$$\frac{-x}{-y} \quad \text{and} \quad \frac{x}{y} \quad \leftarrow \text{We generally use this form for positive final answers.}$$

OBJECTIVE 5 Use the rules for order of operations when multiplying and dividing signed numbers.

NOW TRY
EXERCISE 4
Perform each indicated operation.
(a) $-4(6) - (-5)(5)$
(b) $\dfrac{12(-4) - 6(-3)}{-4(7 - 16)}$

EXAMPLE 4 Using the Rules for Order of Operations

Perform each indicated operation.

(a) $-9(2) - (-3)(2)$

$= -18 - (-6)$ Multiply.

$= -18 + 6$ Definition of subtraction

$= -12$ Add.

(b) $-6 + 2(3 - 5)$ Begin inside the parentheses.

Do *not* add first.

$= -6 + 2(-2)$ Subtract inside the parentheses.

$= -6 + (-4)$ Multiply.

$= -10$ Add.

(c) $|3 - 2(4)| - 2(-6)$

$= |3 - 8| - 2(-6)$ Multiply inside the absolute value bars.

$= |-5| - 2(-6)$ Subtract inside the absolute value bars.

$= 5 - 2(-6)$ Find the absolute value.

$= 5 - (-12)$ Multiply.

$= 17$ Subtract.

(d) $\dfrac{5(-2) - 3(4)}{2(1 - 6)}$ Simplify the numerator and denominator separately.

$= \dfrac{-10 - 12}{2(-5)}$ Multiply in the numerator.
Subtract inside the parentheses in the denominator.

$= \dfrac{-22}{-10}$ Subtract in the numerator.
Multiply in the denominator.

$= \dfrac{11}{5}$ Write in lowest terms.

NOW TRY

OBJECTIVE 6 Evaluate algebraic expressions given values for the variables.

EXAMPLE 5 Evaluating Algebraic Expressions

Evaluate each expression for $x = -1$, $y = -2$, and $m = -3$.

(a) $(3x + 4y)(-2m)$ Use parentheses around substituted negative values to avoid errors.

$= [3(-1) + 4(-2)][-2(-3)]$ Substitute the given values for the variables.

$= [-3 + (-8)][6]$ Multiply.

$= [-11]6$ Add inside the brackets.

$= -66$ Multiply.

NOW TRY ANSWERS
4. **(a)** 1 **(b)** $-\dfrac{5}{6}$

NOW TRY
EXERCISE 5

Evaluate $\dfrac{3x^2 - 12}{y}$ for $x = -4$ and $y = -3$.

(b)

$$2x^2 - 3y^2$$

> Think: $(-2)^2 = -2(-2) = 4$

$$= 2(-1)^2 - 3(-2)^2 \qquad \text{Substitute } -1 \text{ for } x \text{ and } -2 \text{ for } y.$$

> Think: $(-1)^2 = -1(-1) = 1$

$$= 2(1) - 3(4) \qquad \text{Apply the exponents.}$$

$$= 2 - 12 \qquad \text{Multiply.}$$

$$= -10 \qquad \text{Subtract.}$$

(c) $\dfrac{4y^2 + x}{m}$

$$= \frac{4(-2)^2 + (-1)}{-3} \qquad \text{Substitute } -2 \text{ for } y, -1 \text{ for } x, \text{ and } -3 \text{ for } m.$$

$$= \frac{4(4) + (-1)}{-3} \qquad \text{Apply the exponent.}$$

$$= \frac{16 + (-1)}{-3} \qquad \text{Multiply.}$$

$$= \frac{15}{-3} \qquad \text{Add.}$$

$$= -5 \qquad \text{Divide.}$$

NOW TRY

OBJECTIVE 7 Translate words and phrases involving multiplication and division.

▼ **Words and Phrases That Indicate Multiplication**

Word or Phrase	Example	Numerical Expression and Simplification
Product of	The *product of* −5 and −2	$-5(-2)$, which equals 10
Times	13 *times* −4	$13(-4)$, which equals −52
Twice (meaning "2 times")	*Twice* 6	$2(6)$, which equals 12
Triple (meaning "3 times")	*Triple* 4	$3(4)$, which equals 12
Of (used with fractions)	$\frac{1}{2}$ *of* 10	$\frac{1}{2}(10)$, which equals 5
Percent of	12*% of* −16	$0.12(-16)$, which equals −1.92
As much as	$\frac{2}{3}$ *as much as* 30	$\frac{2}{3}(30)$, which equals 20

EXAMPLE 6 Translating Words and Phrases (Multiplication)

Write a numerical expression for each phrase, and simplify the expression.

(a) The product of 12 and the sum of 3 and −6

$$12[3 + (-6)] \quad \text{simplifies to} \quad 12[-3], \quad \text{which equals} \quad -36.$$

(b) Twice the difference of 8 and −4

> The "difference of a and b" means $a - b$.

$$2[8 - (-4)] \quad \text{simplifies to} \quad 2[12], \quad \text{which equals} \quad 24.$$

(c) Two-thirds of the sum of −5 and −3

$$\frac{2}{3}[-5 + (-3)] \quad \text{simplifies to} \quad \frac{2}{3}[-8], \quad \text{which equals} \quad -\frac{16}{3}.$$

NOW TRY
EXERCISE 6

Write a numerical expression for each phrase, and simplify the expression.

(a) Twice the sum of -10 and 7

(b) 40% of the difference of 45 and 15

(d) 15% of the difference of 14 and -2

Remember that $15\% = 0.15$. $\;0.15[14 - (-2)]\;$ simplifies to $\;0.15[16]$, which equals $\;2.4$.

(e) Double the product of 3 and 4

Double means "2 times." $\;2 \cdot (3 \cdot 4)\;$ simplifies to $\;2(12)$, which equals $\;24$.

NOW TRY

When translating a phrase involving division into a fraction, we write the first number named as the numerator and the second as the denominator.

▼ Phrases That Indicate Division

Phrase	Example	Numerical Expression and Simplification
Quotient of	The *quotient of* -24 and 3	$\frac{-24}{3}$, which equals -8
Divided by	-16 *divided by* -4	$\frac{-16}{-4}$, which equals 4
Ratio of	The *ratio of* 2 to 3	$\frac{2}{3}$

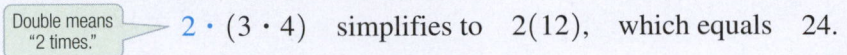

NOW TRY
EXERCISE 7

Write a numerical expression for the phrase, and simplify the expression.

The quotient of 21 and the sum of 10 and -7

EXAMPLE 7 Translating Words and Phrases (Division)

Write a numerical expression for each phrase, and simplify the expression.

(a) The quotient of 14 and the sum of -9 and 2

"Quotient" indicates division. $\dfrac{14}{-9 + 2}\;$ simplifies to $\;\dfrac{14}{-7}$, which equals $\;-2$.

(b) The product of 5 and -6, divided by the difference of -7 and 8

$\dfrac{5(-6)}{-7 - 8}\;$ simplifies to $\;\dfrac{-30}{-15}$, which equals $\;2$.   **NOW TRY**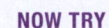

OBJECTIVE 8 Translate simple sentences into equations.

EXAMPLE 8 Translating Sentences into Equations

Write each sentence as an equation, using x as the variable. Then find the solution of the equation from the set of integers between -12 and 12, inclusive.

(a) Three times a number is -18.

The word *times* indicates multiplication.   The word *is* translates as $=$.

$3 \cdot x = -18$, or $3x = -18$ $3 \cdot x = 3x$

The integer between -12 and 12, inclusive, that makes this statement true is -6 because $3(-6) = -18$. The solution of the equation is -6.

(b) The sum of a number and 9 is 12.

$$x + 9 = 12$$

Because $3 + 9 = 12$, the solution of this equation is 3.

NOW TRY ANSWERS

6. **(a)** $2(-10 + 7)$; -6
(b) $0.40(45 - 15)$; 12

7. $\frac{21}{10 + (-7)}$; 7

NOW TRY EXERCISE 8

Write each sentence as an equation, using x as the variable. Then find the solution of the equation from the set of integers between -12 and 12, inclusive.

(a) The sum of a number and -4 is 7.

(b) The difference of -8 and a number is -11.

(c) The *difference of* a number and 5 *is* 0.

$$x - 5 = 0$$

Because $5 - 5 = 0$, the solution of this equation is 5.

(d) The *quotient of* 24 and a number *is* -2.

$$\frac{24}{x} = -2$$

Here, x must be a negative number because the numerator is positive and the quotient is negative. Because $\frac{24}{-12} = -2$, the solution is -12.   **NOW TRY**

⚠ **CAUTION**   In **Examples 6 and 7,** the *phrases* translate as *expressions,* while in **Example 8,** the *sentences* translate as *equations.*

- *An expression is a phrase.*

- *An equation is a sentence with something on the left side, an = symbol, and something on the right side.*

$$\underset{\underset{\text{Expression}}{\uparrow}}{\frac{5(-6)}{-7-8}} \qquad \underset{\underset{\text{Equation}}{\uparrow}}{3x = -18}$$

NOW TRY ANSWERS
8. (a) $x + (-4) = 7$; 11
   (b) $-8 - x = -11$; 3

1.5 Exercises

FOR EXTRA HELP   **MyLab Math**

▶ *Video solutions for select problems available in MyLab Math*

STUDY SKILLS REMINDER
Time management can be a challenge for students.
**Review Study Skill 6,
Managing Your Time.**

Concept Check   *Fill in each blank with one of the following.*

positive   negative   0

1. The product or the quotient of two numbers with the same sign is _____.

2. The product or the quotient of two numbers with different signs is _____.

3. If three negative numbers are multiplied, the product is _____.

4. If two negative numbers are multiplied and then their product is divided by a negative number, the result is _____.

5. If a negative number is squared and the result is added to a positive number, the result is _____.

6. The reciprocal of a negative number is _____.

7. If three positive numbers, five negative numbers, and zero are multiplied, the product is _____.

8. The cube of a negative number is _____.

9. Concept Check   Complete this statement: The quotient formed by any nonzero number divided by 0 is _____, and the quotient formed by 0 divided by any nonzero number is _____. Give an example of each quotient.

10. Concept Check   Which expression is undefined?

  A. $\dfrac{4+4}{4+4}$ **B.** $\dfrac{4-4}{4+4}$ **C.** $\dfrac{4-4}{4-4}$ **D.** $\dfrac{4-4}{4}$

*Find each product. **See Examples 1 and 2.***

11. $-7(4)$ **12.** $-8(5)$ **13.** $-5(-6)$ **14.** $-3(-4)$ **15.** $-10(-12)$

16. $-9(-5)$ **17.** $3(-11)$ **18.** $3(-15)$ **19.** $-0.5(0)$ **20.** $-0.3(0)$

21. $-6.8(0.35)$ **22.** $-4.6(0.24)$ **23.** $-\dfrac{3}{8}\left(-\dfrac{20}{9}\right)$ **24.** $-\dfrac{5}{6}\left(-\dfrac{16}{15}\right)$

25. $\dfrac{2}{15}\left(-1\dfrac{1}{4}\right)$ **26.** $\dfrac{3}{7}\left(-1\dfrac{5}{9}\right)$ **27.** $-8\left(-\dfrac{3}{4}\right)$ **28.** $-6\left(-\dfrac{2}{3}\right)$

*Find all integer factors of each number. **See Objective 3.***

29. 32   **30.** 36   **31.** 40   **32.** 50   **33.** 31   **34.** 17

*Find each quotient. **See Example 3** and the discussion of division involving 0.*

35. $\dfrac{15}{5}$ **36.** $\dfrac{35}{5}$ **37.** $\dfrac{-42}{6}$ **38.** $\dfrac{-28}{7}$ **39.** $\dfrac{-32}{-4}$

40. $\dfrac{-35}{-5}$ **41.** $\dfrac{96}{-16}$ **42.** $\dfrac{38}{-19}$ **43.** $\dfrac{-8.8}{2.2}$ **44.** $\dfrac{-4.6}{2.3}$

45. $-\dfrac{4}{3} \div \left(-\dfrac{1}{8}\right)$ **46.** $-\dfrac{6}{5} \div \left(-\dfrac{1}{3}\right)$ **47.** $-\dfrac{5}{6} \div \dfrac{8}{9}$ **48.** $-\dfrac{7}{10} \div \dfrac{3}{4}$

49. $\dfrac{0}{-5}$ **50.** $\dfrac{0}{-9}$ **51.** $\dfrac{11.5}{0}$ **52.** $\dfrac{15.2}{0}$

*Perform each indicated operation. **See Example 4.***

53. $7 - 3 \cdot 6$ **54.** $8 - 2 \cdot 5$ **55.** $-10 - (-4)(2)$

56. $-11 - (-3)(6)$ **57.** $-2(5) - (-4)(2)$ **58.** $-4(3) - (-3)(6)$

59. $-7(3 - 8)$ **60.** $-5(4 - 7)$ **61.** $7 + 2(4 - 1)$

62. $5 + 3(6 - 4)$ **63.** $-4 + 3(2 - 8)$ **64.** $-8 + 4(5 - 7)$

65. $(12 - 14)(1 - 4)$ **66.** $(8 - 9)(4 - 12)$ **67.** $(7 - 10)(10 - 4)$

68. $(5 - 12)(19 - 4)$ **69.** $(-2 - 8)(-6) + 7$ **70.** $(-9 - 4)(-2) + 10$

71. $3(-5) + |3 - 10|$ **72.** $4(-8) + |4 - 15|$ **73.** $|8 - 7(2)| - 6(-2)$

74. $|5 - 3(9)| - 7(-4)$ **75.** $\dfrac{-5(-6)}{9 - (-1)}$ **76.** $\dfrac{-12(-5)}{7 - (-5)}$

77. $\dfrac{-21(3)}{-3 - 6}$ **78.** $\dfrac{-40(3)}{-2 - 3}$ **79.** $\dfrac{-10(2) + 6(2)}{-3 - (-1)}$

80. $\dfrac{-12(4) + 5(3)}{-14 - (-3)}$ **81.** $\dfrac{-6 - |-9 + 5|}{2 - (-3)}$ **82.** $\dfrac{-8 - |-3 + 2|}{-3 - (-6)}$

83. $\dfrac{3^2 - 4^2}{7(-8 + 9)}$ **84.** $\dfrac{5^2 - 7^2}{2(3 + 3)}$ **85.** $\dfrac{8(-1) - |(-4)(-3)|}{-6 - (-1)}$

86. $\dfrac{-27(-2) - |6 \cdot 4|}{-2(3) - 2(2)}$ **87.** $\dfrac{-13(-4) - (-8)(-2)}{(-10)(2) - 4(-2)}$ **88.** $\dfrac{-5(2) + [3(-2) - 4]}{-3 - (-1)}$

Concept Check *A few years ago, the following question and expression appeared on boxes of Swiss Miss Hot Cocoa Mix:*

On average, how many mini-marshmallows are in one serving?

$$3 + 2 \times 4 \div 2 - 3 \times 7 - 4 + 47$$

89. The box gave 92 as the answer. What is the *correct* answer?

90. **WHAT WENT WRONG?** Explain the algebraic error that somebody at the company made in calculating the answer.

Evaluate each expression for $x = 6$, $y = -4$, and $a = 3$. **See Example 5.**

91. $5x - 2y + 3a$

92. $6x - 5y + 4a$

93. $(2x + y)(3a)$

94. $(5x - 2y)(-2a)$

95. $\left(\dfrac{1}{3}x - \dfrac{4}{5}y\right)\left(-\dfrac{1}{5}a\right)$

96. $\left(\dfrac{5}{6}x + \dfrac{3}{2}y\right)\left(-\dfrac{1}{3}a\right)$

97. $(6 - x)(5 + y)(3 + a)$

98. $(-5 + x)(-3 + y)(3 - a)$

99. $-2y^2 + 3a^2$

100. $5x^2 - 4y^2$

101. $\dfrac{5y^2 + 4}{x}$

102. $\dfrac{11 - 3a^2}{y}$

103. $\dfrac{2y - x}{a - 3}$

104. $\dfrac{xy + 8a}{x - 6}$

Write a numerical expression for each phrase, and simplify the expression. **See Examples 6 and 7.**

105. The product of -9 and 2, added to 9

106. The product of 4 and -7, added to -12

107. Twice the product of -1 and 6, subtracted from -4

108. Twice the product of -8 and 2, subtracted from -1

109. Nine subtracted from the product of 1.5 and -3.2

110. Three subtracted from the product of 4.2 and -8.5

111. The product of 12 and the difference of 9 and -8

112. The product of -3 and the difference of 3 and -7

113. The quotient of -12 and the sum of -5 and -1

114. The quotient of -20 and the sum of -8 and -2

115. The sum of 15 and -3, divided by the product of 4 and -3

116. The sum of -18 and -6, divided by the product of 2 and -4

117. Two-thirds of the difference of 8 and -1

118. Three-fourths of the sum of -8 and 12

119. 20% of the product of -5 and 6

120. 30% of the product of -8 and 5

121. The sum of $\dfrac{1}{2}$ and $\dfrac{5}{8}$, times the difference of $\dfrac{3}{5}$ and $\dfrac{1}{3}$

122. The sum of $\dfrac{3}{4}$ and $\dfrac{1}{2}$, times the difference of $\dfrac{2}{3}$ and $\dfrac{1}{6}$

123. The product of $-\dfrac{1}{2}$ and $\dfrac{3}{4}$, divided by $-\dfrac{2}{3}$

124. The product of $-\dfrac{2}{3}$ and $-\dfrac{1}{5}$, divided by $\dfrac{1}{7}$

Write each sentence as an equation, using x as the variable. Then find the solution of the equation from the set of integers between −12 and 12, inclusive. **See Example 8.**

125. The quotient of a number and 3 is −3.

126. The quotient of a number and 4 is −1.

127. 6 less than a number is 4.

128. 7 less than a number is 2.

129. When 5 is added to a number, the result is −5.

130. When 6 is added to a number, the result is −3.

The operation of division is used in **divisibility tests.** *A divisibility test allows us to determine whether a given number is divisible (without remainder) by another number.*

131. An integer is divisible by 2 if its last digit is divisible by 2, and not otherwise.

    (a) Is 3,473,986 divisible by 2?   **(b)** Is 4,336,879 divisible by 2?

132. An integer is divisible by 3 if the sum of its digits is divisible by 3, and not otherwise.

    (a) Is 4,799,232 divisible by 3?   **(b)** Is 2,443,871 divisible by 3?

133. An integer is divisible by 4 if its last two digits form a number divisible by 4, and not otherwise.

    (a) Is 2,876,335 divisible by 4?   **(b)** Is 6,221,464 divisible by 4?

134. An integer is divisible by 5 if its last digit is divisible by 5, and not otherwise.

    (a) Is 9,332,123 divisible by 5?   **(b)** Is 3,774,595 divisible by 5?

135. An integer is divisible by 6 if it is divisible by both 2 and 3, and not otherwise.

    (a) Is 1,524,822 divisible by 6?   **(b)** Is 2,873,590 divisible by 6?

136. An integer is divisible by 8 if its last three digits form a number divisible by 8, and not otherwise.

    (a) Is 2,923,296 divisible by 8?   **(b)** Is 7,291,623 divisible by 8?

137. An integer is divisible by 9 if the sum of its digits is divisible by 9, and not otherwise.

    (a) Is 2,287,321 divisible by 9?   **(b)** Is 4,114,107 divisible by 9?

138. An integer is divisible by 12 if it is divisible by both 3 and 4, and not otherwise.

    (a) Is 4,249,474 divisible by 12?   **(b)** Is 4,253,520 divisible by 12?

RELATING CONCEPTS For Individual or Group Work (Exercises 139–146)

To find the **average (mean)** *of a group of numbers, we add the numbers and then divide the sum by the number of terms added.* **Work Exercises 139–142 in order,** *to find the average of* 23, 18, 13, −4, *and* −8.

139. Find the sum of the given group of numbers.

140. How many numbers are in the group?

141. Divide the answer for **Exercise 139** by the answer for **Exercise 140.** Give the quotient as a mixed number.

142. What is the average of the given group of numbers?

Find the average of each group of numbers.

143. −15, 29, 8, −6

144. −17, 34, 9, −2

145. All integers between −10 and 14, including both −10 and 14

146. All integers between −15 and −10, including both −15 and −10

SUMMARY EXERCISES   Performing Operations with Real Numbers

Operations with Signed Numbers

Addition

Same sign   Add the absolute values of the numbers. The sum has the same sign as the addends.

Different signs   Find the absolute values of the numbers, and subtract the lesser absolute value from the greater. The sum has the same sign as the addend with greater absolute value.

Subtraction

Add the additive inverse (or opposite) of the subtrahend to the minuend.

Multiplication and Division

Same sign   The product or quotient of two numbers with the same sign is positive.

Different signs   The product or quotient of two numbers with different signs is negative.

Division by 0 is undefined.

Perform operations with signed numbers using the rules for order of operations.

Order of Operations

If grouping symbols are present, work within them, innermost first (and above and below fraction bars separately), in the following order.

Step 1   Apply all **exponents.**

Step 2   Do any **multiplications** and **divisions** in order from left to right.

Step 3   Do any **additions** and **subtractions** in order from left to right.

If no grouping symbols are present, start with Step 1.

Perform each indicated operation.

1. $14 - 3 \cdot 10$

2. $-3(8) - 4(-7)$

3. $(3 - 8)(-2) - 10$

4. $-6(7 - 3)$

5. $7 + 3(2 - 10)$

6. $-4[(-2)(6) - 7]$

7. $(-4)(7) - (-5)(2)$

8. $-5[-4 - (-2)(-7)]$

9. $40 - (-2)[8 - 9]$

10. $-5.1(-0.2)$

11. $-0.9(-3.7)$

12. $|-4(9)| - |-11|$

13. $|-2(3) + 4| - |-2|$

14. $|11 - 3(-4)| - 5(3)$

15. $\dfrac{1}{2} \div \left(-\dfrac{1}{2}\right)$

16. $-\dfrac{3}{4} \div \left(-\dfrac{5}{8}\right)$

17. $\left[\dfrac{5}{8} - \left(-\dfrac{1}{16}\right)\right] + \dfrac{3}{8}$

18. $\left(\dfrac{1}{2} - \dfrac{1}{3}\right) - \dfrac{5}{6}$

19. $\dfrac{5(-4)}{-7 - (-2)}$

20. $\dfrac{5(-8 + 3)}{13(-2) + (-7)(-3)}$

21. $\dfrac{-3 - (-9 + 1)}{-7 - (-6)}$

22. $\dfrac{2^2 + 4^2}{5^2 - 3^2}$ **23.** $\dfrac{(2+4)^2}{(5-3)^2}$ **24.** $\dfrac{4^2 - 3^2}{-5(-4+2)}$

25. $\dfrac{6^2 - 8}{-2(2) + 4(-1)}$ **26.** $\dfrac{6(-10+3)}{15(-2) - 3(-9)}$ **27.** $\dfrac{9(-6) - 3(8)}{4(-7) + (-2)(-11)}$

28. $\dfrac{3^2 - 5^2}{(-9)^2 - 9^2}$ **29.** $\dfrac{(-10)^2 + 10^2}{-10(5)}$ **30.** $\dfrac{8^2 - 12}{(-5)^2 + 2(6)}$

31. $\dfrac{-9(-6) + (-2)(27)}{3(8-9)}$ **32.** $\dfrac{16(-8+5)}{15(-3) + (-7-4)(-3)}$

Evaluate each expression for $x = -2$, $y = 3$, *and* $a = 4$.

33. $-x + y - 3a$ **34.** $(x - y) - (a - 2y)$ **35.** $(-8x - 3y)(-2a)$

36. $\dfrac{2x + 3y}{a - xy}$ **37.** $\dfrac{x^2 - y^2}{x^2 + y^2}$ **38.** $-x^2 + 3y$

39. $\left(\dfrac{1}{2}x + \dfrac{2}{3}y\right)\left(-\dfrac{1}{4}a\right)$ **40.** $\dfrac{3a + 6x}{-2y}$

1.6   Properties of Real Numbers

OBJECTIVES
1 Use the commutative properties.
2 Use the associative properties.
3 Use the identity properties.
4 Use the inverse properties.
5 Use the distributive property.

In the basic properties covered in this section, a, b, and c represent real numbers.

OBJECTIVE 1   Use the commutative properties.

The word *commute* means to go back and forth. We might commute to work or to school. If we travel from home to work and follow the same route from work to home, we travel the same distance each time. The **commutative properties** say that if two numbers are added or multiplied in either order, the result is the same.

> **Commutative Properties**
>
> $$a + b = b + a \qquad \text{Addition}$$
> $$ab = ba \qquad \text{Multiplication}$$

EXAMPLE 1   Using the Commutative Properties

Use a commutative property to complete each statement.

(a) $-8 + 5 = 5 + \underline{\ ?\ }$ *Notice that the "order" changed.*

$-8 + 5 = 5 + (-8)$ Commutative property of addition

(b) $(-2)7 = \underline{\ ?\ }\ (-2)$

$-2(7) = 7(-2)$ Commutative property of multiplication

VOCABULARY
☐ identity element for addition (additive identity)
☐ identity element for multiplication (multiplicative identity)

NOW TRY

NOW TRY
EXERCISE 1
Use a commutative property to complete each statement.
(a) $7 + (-3) = -3 +$ _____
(b) $(-5)4 = 4 \cdot$ _____

OBJECTIVE 2 Use the associative properties.

When we *associate* one object with another, we think of those objects as being grouped together. The **associative properties** say that when we add or multiply three numbers, we can group the first two together or the last two together and obtain the same answer.

Associative Properties

$$(a + b) + c = a + (b + c) \qquad \text{Addition}$$

$$(ab)c = a(bc) \qquad \text{Multiplication}$$

NOW TRY
EXERCISE 2
Use an associative property to complete each statement.
(a) $-9 + (3 + 7) =$ _____
(b) $5[(-4) \cdot 9] =$ _____

EXAMPLE 2 Using the Associative Properties

Use an associative property to complete each statement.

(a) $-8 + (1 + 4) = (-8 + \underline{\ ?\ }) + 4$ The "order" is the same. The "grouping" changed.

$-8 + (1 + 4) = (-8 + 1) + 4$ Associative property of addition

(b) $[2 \cdot (-7)] \cdot 6 = 2 \cdot \underline{\ ?\ }$

$[2 \cdot (-7)] \cdot 6 = 2 \cdot [(-7) \cdot 6]$ Associative property of multiplication

NOW TRY ↺

By the associative property, the sum (or product) of three numbers will be the same no matter how the numbers are "associated" in groups. Parentheses can be left out if a problem contains only addition (or multiplication). For example,

$$(-1 + 2) + 3 \quad \text{and} \quad -1 + (2 + 3) \quad \text{can be written as} \quad -1 + 2 + 3.$$

NOW TRY
EXERCISE 3
Is $5 + (7 + 6) = 5 + (6 + 7)$ an example of a *commutative property* or an *associative property*?

EXAMPLE 3 Distinguishing between Properties

Decide whether each statement is an example of a *commutative property*, an *associative property*, or *both*.

(a) $(2 + 4) + 5 = 2 + (4 + 5)$

The order of the three numbers is the same on both sides of the equality symbol. The only change is in the *grouping*, or association, of the numbers. This is an example of the *associative property*.

(b) $6 \cdot (3 \cdot 10) = 6 \cdot (10 \cdot 3)$

The same numbers, 3 and 10, are grouped on each side. On the left, the 3 appears first, but on the right, the 10 appears first. The only change involves the *order* of the numbers, so this is an example of the *commutative property*.

(c) $(8 + 1) + 7 = 8 + (7 + 1)$

Both the order and the grouping are changed. On the left, the order of the three numbers is 8, 1, and 7. On the right, it is 8, 7, and 1. On the left, the 8 and 1 are grouped. On the right, the 7 and 1 are grouped. Therefore, *both* properties are used.

NOW TRY ↺

NOW TRY ANSWERS
1. (a) 7   (b) −5
2. (a) $(-9 + 3) + 7$
   (b) $[5 \cdot (-4)] \cdot 9$
3. commutative property

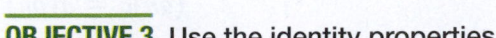

**NOW TRY
EXERCISE 4**

Find each sum or product.

(a) $8 + 54 + 7 + 6 + 32$

(b) $5(37)(20)$

EXAMPLE 4   Using the Commutative and Associative Properties

Find each sum or product.

(a) $23 + 41 + 2 + 9 + 25$

$= (41 + 9) + (23 + 2) + 25$

$= 50 + 25 + 25$

$= 100$

(b) $25(69)(4)$

$= 25(4)(69)$

$= 100(69)$

$= 6900$

Use the commutative and associative properties to rearrange and regroup numbers for easier computation.

NOW TRY

OBJECTIVE 3 Use the identity properties.

If a child wears a costume on Halloween, the child's appearance is changed, but his or her *identity* is unchanged. The identity of a real number is unchanged when identity properties are applied.

The **identity properties** say that

- the sum of 0 and any number equals that number, and

- the product of 1 and any number equals that number.

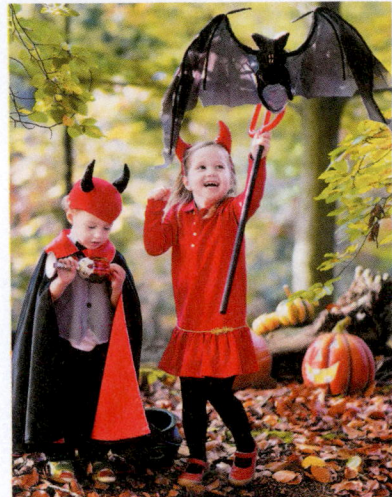

Identity Properties

$$a + 0 = a \quad \text{and} \quad 0 + a = a \quad \text{Addition}$$

$$a \cdot 1 = a \quad \text{and} \quad 1 \cdot a = a \quad \text{Multiplication}$$

The number 0 leaves the identity, or value, of any real number unchanged by addition, so **0** is the **identity element for addition,** or the **additive identity.** Because multiplication by 1 leaves any real number unchanged, the number **1** is the **identity element for multiplication,** or the **multiplicative identity.**

**NOW TRY
EXERCISE 5**

Use an identity property to complete each statement.

(a) $\dfrac{2}{5} \cdot \underline{\hspace{1cm}} = \dfrac{2}{5}$

(b) $8 + \underline{\hspace{1cm}} = 8$

EXAMPLE 5   Using the Identity Properties

Use an identity property to complete each statement.

(a) $-3 + \underline{\ ?\ } = -3$

$-3 + 0 = -3$

Identity property of addition

(b) $\underline{\ ?\ } \cdot \dfrac{1}{2} = \dfrac{1}{2}$

$1 \cdot \dfrac{1}{2} = \dfrac{1}{2}$

Identity property of multiplication

NOW TRY

NOW TRY ANSWERS

4. (a) 107 (b) 3700
5. (a) 1 (b) 0

We use the identity property of multiplication to write fractions in lowest terms and to find common denominators in **Example 6** on the next page.

**NOW TRY
EXERCISE 6**

Simplify.

(a) $\dfrac{16}{20}$ (b) $\dfrac{2}{5} + \dfrac{3}{20}$

EXAMPLE 6   Using the Identity Property to Simplify Expressions

Simplify. In part (a), write in lowest terms. In part (b), perform the operation.

(a) $\dfrac{49}{35}$

$= \dfrac{7 \cdot 7}{5 \cdot 7}$ Factor.

$= \dfrac{7}{5} \cdot \dfrac{7}{7}$ Write as a product.

$= \dfrac{7}{5} \cdot 1$ Property of 1

$= \dfrac{7}{5}$ Identity property

(b) $\dfrac{3}{4} + \dfrac{5}{24}$

$= \dfrac{3}{4} \cdot 1 + \dfrac{5}{24}$ Identity property

$= \dfrac{3}{4} \cdot \dfrac{6}{6} + \dfrac{5}{24}$ Use $1 = \dfrac{6}{6}$ to obtain a common denominator.

$= \dfrac{18}{24} + \dfrac{5}{24}$ Multiply.

$= \dfrac{23}{24}$ Add.   **NOW TRY**

OBJECTIVE 4   Use the inverse properties.

Each day before we go to work or school, we probably put on our shoes. Before we go to sleep at night, we probably take them off. These operations from everyday life are examples of *inverse* operations.

The **inverse properties** of addition and multiplication lead to the additive and multiplicative identities, respectively. Recall that $-a$ is the **additive inverse,** or **opposite,** of a and $\dfrac{1}{a}$ is the **multiplicative inverse,** or **reciprocal,** of the nonzero number a.

Inverse Properties

$$a + (-a) = 0 \quad \text{and} \quad -a + a = 0 \qquad \text{Addition}$$

$$a \cdot \dfrac{1}{a} = 1 \quad \text{and} \quad \dfrac{1}{a} \cdot a = 1 \quad (a \neq 0) \qquad \text{Multiplication}$$

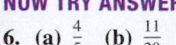

**NOW TRY
EXERCISE 7**

Use an inverse property to complete each statement.

(a) $10 + \underline{\hspace{1cm}} = 0$

(b) $-9 \cdot \underline{\hspace{1cm}} = 1$

EXAMPLE 7   Using the Inverse Properties

Use an inverse property to complete each statement.

(a) $\underline{\ ?\ } + \dfrac{1}{2} = 0$

$-\dfrac{1}{2} + \dfrac{1}{2} = 0$

(b) $4 + \underline{\ ?\ } = 0$

$4 + (-4) = 0$

(c) $-0.75 + \dfrac{3}{4} = \underline{\ ?\ }$

$-0.75 + \dfrac{3}{4} = 0$

The inverse property of addition is used in parts (a)–(c).

(d) $\underline{\ ?\ } \cdot \dfrac{5}{2} = 1$

$\dfrac{2}{5} \cdot \dfrac{5}{2} = 1$

(e) $-5(\underline{\ ?\ }) = 1$

$-5\left(-\dfrac{1}{5}\right) = 1$

(f) $4(0.25) = \underline{\ ?\ }$

$4(0.25) = 1$

The inverse property of multiplication is used in parts (d)–(f).   **NOW TRY**

NOW TRY ANSWERS

6. (a) $\dfrac{4}{5}$ (b) $\dfrac{11}{20}$

7. (a) -10 (b) $-\dfrac{1}{9}$

**NOW TRY
EXERCISE 8**
Simplify.

$$-\frac{1}{3}x + 7 + \frac{1}{3}x$$

EXAMPLE 8   Using Properties to Simplify an Expression

Simplify.

$$-2x + 10 + 2x$$

$$= (-2x + 10) + 2x \qquad \text{Order of operations}$$

$$= [10 + (-2x)] + 2x \qquad \text{Commutative property}$$

$$= 10 + [(-2x) + 2x] \qquad \text{Associative property}$$

$$= 10 + 0 \qquad \text{Inverse property}$$

*For any value of x,
$-2x$ and $2x$ are
additive inverses.*

$$= 10 \qquad \text{Identity property}$$ **NOW TRY**

NOTE   The steps of **Example 8** may be skipped when we actually do the simplification.

OBJECTIVE 5 Use the distributive property.

The word *distribute* means "to give out from one to several." Consider the following.

$$2(5 + 8) \quad \text{equals} \quad 2(13), \quad \text{which equals} \quad 26.$$
$$2(5) + 2(8) \quad \text{equals} \quad 10 + 16, \quad \text{which equals} \quad 26.$$

Both expressions equal 26.

Thus, $2(5 + 8) = 2(5) + 2(8).$

This is an example of the *distributive property of multiplication with respect to addition,* the only property involving *both* addition and multiplication. With this property, a product can be changed to a sum or difference. This idea is illustrated in **FIGURE 16**.

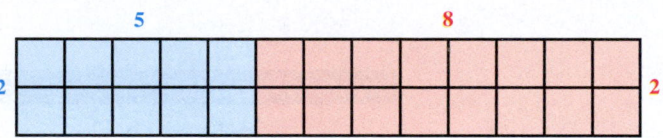

The area of the left part is 2(5) = 10.
The area of the right part is 2(8) = 16.
The total area is 2(5 + 8) = 2(13) = 26,
or the total area is 2(5) + 2(8) = 10 + 16 = 26.
Thus, 2(5 + 8) = 2(5) + 2(8).

FIGURE 16

The **distributive property** says that multiplying a number a by a sum of numbers $b + c$ gives the same result as multiplying a by b and a by c and adding the products.

Distributive Property

$$a(b + c) = ab + ac \qquad \text{and} \qquad (b + c)a = ba + ca$$

As the arrows show, the a outside the parentheses is "distributed" over the b and c inside. The distributive property is also valid for multiplication over subtraction.

$$a(b - c) = ab - ac \qquad \text{and} \qquad (b - c)a = ba - ca$$

The distributive property can be extended to more than two numbers.

$$a(b + c + d) = ab + ac + ad$$

NOW TRY ANSWER
8. 7

**NOW TRY
EXERCISE 9**

Use the distributive property
to rewrite each expression.

(a) $2(p + 5)$

(b) $-5(4x + 1)$

(c) $6(2r + t - 5z)$

EXAMPLE 9 Using the Distributive Property

Use the distributive property to rewrite each expression.

(a) $5(9 + 6)$ | We could write $5(9) + 5(6)$ here.

$= 5 \cdot 9 + 5 \cdot 6$ The factor 5 is "distributed" to the numbers 9 and 6.

Multiply first. $= 45 + 30$ Multiply.

$= 75$ Add.

(b) $4(x + 5 + y)$

$= 4x + 4 \cdot 5 + 4y$

$= 4x + 20 + 4y$

(c) $-2(x + 3)$

$= -2x + (-2)(3)$

$= -2x + (-6)$

$= -2x - 6$

(d) $-\dfrac{1}{2}(4x + 3)$ | Think: $-\frac{1}{2}(4x) = \left(-\frac{1}{2} \cdot 4\right)x = \left(-\frac{1}{2} \cdot \frac{4}{1}\right)x$

$= -\dfrac{1}{2}(4x) + \left(-\dfrac{1}{2}\right)(3)$ Distributive property

This step is often omitted. $= -2x + \left(-\dfrac{3}{2}\right)$ Multiply.

$= -2x - \dfrac{3}{2}$ Definition of subtraction

(e) $3(k - 9)$ | Be careful here.

$= 3[k + (-9)]$ Definition of subtraction

$= 3k + 3(-9)$ Distributive property

$= 3k - 27$ Multiply; definition of subtraction

(f) $-2(3x - 4)$

$= -2[3x + (-4)]$ Definition of subtraction

This step is often omitted. $= -2(3x) + (-2)(-4)$ Distributive property

$= (-2 \cdot 3)x + (-2)(-4)$ Associative property

$= -6x + 8$ Multiply.

(g) $8(3r + 11t + 5z)$

$= 8(3r) + 8(11t) + 8(5z)$ Distributive property

This step is often omitted. $= (8 \cdot 3)r + (8 \cdot 11)t + (8 \cdot 5)z$ Associative property

$= 24r + 88t + 40z$ Multiply.   **NOW TRY**

⚠️ **CAUTION**   We often omit writing subtraction as addition of the additive inverse.

$3(k - 9)$ **See Example 9(e).**

$= 3k - 3(9)$ Be careful not to make a sign error.

$= 3k - 27$ Multiply.

The expression $-a$ may be interpreted as $-1 \cdot a$. Using this result and the distributive property, we can *clear* (or *remove*) *parentheses*.

NOW TRY
EXERCISE 10

Write each expression without parentheses.

(a) $-(2 - r)$

(b) $-(2x - 5y - 7)$

EXAMPLE 10   Using the Distributive Property to Clear Parentheses

Write each expression without parentheses.

(a) $-(2y + 3)$

The $-$ symbol indicates a factor of -1.

$$= -1 \cdot (2y + 3) \qquad -a = -1 \cdot a$$

$$= -1 \cdot 2y + (-1) \cdot 3 \qquad \text{Distributive property}$$

$$= -2y - 3 \qquad \text{Multiply; definition of subtraction}$$

(b) $-(-9w - 2)$

$$= -1(-9w - 2)$$

$$= -1(-9w) - 1(-2)$$

$$= 9w + 2$$

We can also interpret the negative sign in front of the parentheses to mean the *opposite* of each of the terms within the parentheses.

$$-1(-9w - 2)$$

$$= +9w + 2$$

(c) $-(-x - 3y + 6z)$

$$= -1(-1x - 3y + 6z)$$

Be careful with signs.

$-1(-1x)$
$= 1x$
$= x$

$$= -1(-1x) - 1(-3y) - 1(6z) \qquad \text{Distributive property}$$

$$= x + 3y - 6z \qquad \text{Multiply.}$$

NOW TRY

Summary of the Properties of Addition and Multiplication

For any real numbers a, b, and c, the following properties hold true.

Commutative Properties $a + b = b + a \qquad ab = ba$

Associative Properties $(a + b) + c = a + (b + c)$

$$(ab)c = a(bc)$$

Identity Properties   There is a real number 0 such that

$$a + 0 = a \qquad \text{and} \qquad 0 + a = a.$$

There is a real number 1 such that

$$a \cdot 1 = a \qquad \text{and} \qquad 1 \cdot a = a.$$

Inverse Properties   For each real number a, there is a single real number $-a$ such that

$$a + (-a) = 0 \qquad \text{and} \qquad (-a) + a = 0.$$

For each nonzero real number a, there is a single real number $\frac{1}{a}$ such that

$$a \cdot \frac{1}{a} = 1 \qquad \text{and} \qquad \frac{1}{a} \cdot a = 1.$$

Distributive Properties $a(b + c) = ab + ac \qquad (b + c)a = ba + ca$

NOW TRY ANSWERS
10. (a) $-2 + r$
    (b) $-2x + 5y + 7$

1.6 Exercises

FOR EXTRA HELP

 MyLab Math

 Video solutions for select problems available in MyLab Math

STUDY SKILLS REMINDER
Are you fully utilizing the features of your text? **Review Study Skill 1,** *Using Your Math Text.*

1. Concept Check Match each item in Column I with the correct choice(s) from Column II. Choices may be used once, more than once, or not at all.

I	II
(a) Identity element for addition	**A.** $(5 \cdot 4) \cdot 3 = 5 \cdot (4 \cdot 3)$
(b) Identity element for multiplication	**B.** 0
(c) Additive inverse of a	**C.** $-a$
(d) Multiplicative inverse, or reciprocal, of the nonzero number a	**D.** -1
(e) The number that is its own additive inverse	**E.** $5 \cdot 4 \cdot 3 = 60$
(f) The two numbers that are their own multiplicative inverses	**F.** 1
(g) The only number that has no multiplicative inverse	**G.** $(5 \cdot 4) \cdot 3 = 3 \cdot (5 \cdot 4)$
(h) An example of the associative property	**H.** $5(4 + 3) = 5 \cdot 4 + 5 \cdot 3$
(i) An example of the commutative property	**I.** $\dfrac{1}{a}$
(j) An example of the distributive property	

2. Concept Check Complete each statement.

The commutative property allows us to change the (*order/grouping*) of the addends in a sum or the factors in a product.

The associative property allows us to change the (*order/grouping*) of the addends in a sum or the factors in a product.

Concept Check *Tell whether or not the following everyday activities are commutative.*

3. Washing your face and brushing your teeth

4. Putting on your left sock and putting on your right sock

5. Preparing a meal and eating a meal

6. Starting a car and driving away in a car

7. Putting on your socks and putting on your shoes

8. Getting undressed and taking a shower

Concept Check *Work each problem involving the properties of real numbers.*

9. Use parentheses to show how the associative property can be used to give two different meanings to the phrase "large deposit slip."

10. Use parentheses to show how the associative property can be used to give two different meanings to the phrase "defective merchandise counter."

11. Evaluate the following expressions.

$$25 - (6 - 2) \quad \text{and} \quad (25 - 6) - 2.$$

Does it appear that subtraction is associative?

12. Evaluate the following expressions.

$$180 \div (15 \div 3) \quad \text{and} \quad (180 \div 15) \div 3.$$

Does it appear that division is associative?

13. Complete the table and each statement beside it.

Number	Additive Inverse	Multiplicative Inverse
5		
-10		
$-\dfrac{1}{2}$		
$\dfrac{3}{8}$		
x		($x \neq 0$)
$-y$		($y \neq 0$)

A number and its additive inverse have (*the same* / *opposite*) signs.

A number and its multiplicative inverse have (*the same* / *opposite*) signs.

14. The following conversation took place between one of the authors of this text and his son Jack, when Jack was 4 years old.

> DADDY: "Jack, what is 3 + 0?"
>
> JACK: "3."
>
> DADDY: "Jack, what is 4 + 0?"
>
> JACK: "4. And Daddy, *string* plus zero equals *string*!"

What property of addition did Jack recognize?

Use a commutative or an associative property to complete each statement. State which property is used. **See Examples 1 and 2.**

15. $-15 + 9 = 9 +$ _____

16. $6 + (-2) = -2 +$ _____

17. $-8 \cdot 3 =$ _____ $\cdot (-8)$

18. $-12 \cdot 4 = 4 \cdot$ _____

19. $(3 + 6) + 7 = 3 + ($ _____ $+ 7)$

20. $(-2 + 3) + 6 = -2 + ($ _____ $+ 6)$

21. $7 \cdot (2 \cdot 5) = ($ _____ $\cdot 2) \cdot 5$

22. $8 \cdot (6 \cdot 4) = (8 \cdot$ _____ $) \cdot 4$

Decide whether each statement is an example of a commutative, *an* associative, *an* identity, *an* inverse, *or the* distributive property. **See Examples 1, 2, 3, 5, 6, 7, and 9.**

23. $4 + 15 = 15 + 4$

24. $3 + 12 = 12 + 3$

25. $5 \cdot (13 \cdot 7) = (5 \cdot 13) \cdot 7$

26. $-4 \cdot (2 \cdot 6) = (-4 \cdot 2) \cdot 6$

27. $-6 + (12 + 7) = (-6 + 12) + 7$

28. $(-8 + 13) + 2 = -8 + (13 + 2)$

29. $-9 + 9 = 0$

30. $1 + (-1) = 0$

31. $\dfrac{2}{3}\left(\dfrac{3}{2}\right) = 1$

32. $\dfrac{5}{8}\left(\dfrac{8}{5}\right) = 1$

33. $1.75 + 0 = 1.75$

34. $-8.45 + 0 = -8.45$

35. $(4 + 17) + 3 = 3 + (4 + 17)$

36. $(-8 + 4) + 12 = 12 + (-8 + 4)$

37. $2(x + y) = 2x + 2y$

38. $9(t + s) = 9t + 9s$

39. $-\dfrac{5}{9} = -\dfrac{5}{9} \cdot \dfrac{3}{3} = -\dfrac{15}{27}$

40. $-\dfrac{7}{12} = -\dfrac{7}{12} \cdot \dfrac{7}{7} = -\dfrac{49}{84}$

41. $4(2x) + 4(3y) = 4(2x + 3y)$

42. $6(5t) - 6(7r) = 6(5t - 7r)$

Find each sum or product. **See Example 4.**

43. $97 + 13 + 3 + 37$

44. $49 + 199 + 1 + 1$

45. $1999 + 2 + 1 + 8$

46. $2998 + 3 + 2 + 17$

47. $159 + 12 + 141 + 88$

48. $106 + 8 + 14 + 72$

49. $843 + 627 + (-43) + (-27)$ **50.** $1846 + 1293 + (-46) + (-93)$

51. $5(47)(2)$ **52.** $2(79)5$

53. $-4 \cdot 5 \cdot 93 \cdot 5$ **54.** $2 \cdot 25 \cdot 67 \cdot (-2)$

Simplify each expression. See Examples 7 and 8.

55. $6t + 8 - 6t + 3$ **56.** $9r + 12 - 9r + 1$ **57.** $\dfrac{2}{3}x - 11 + 11 - \dfrac{2}{3}x$

58. $\dfrac{1}{5}y + 4 - 4 - \dfrac{1}{5}y$ **59.** $\left(\dfrac{9}{7}\right)(-0.38)\left(\dfrac{7}{9}\right)$ **60.** $\left(\dfrac{4}{5}\right)(-0.73)\left(\dfrac{5}{4}\right)$

61. $t + (-t) + \dfrac{1}{2}(2)$ **62.** $w + (-w) + \dfrac{1}{4}(4)$

63. Concept Check A student used the distributive property to rewrite the expression $-3(4 - 6)$ as shown.

$$-3(4 - 6)$$
$$= -3(4) - 3(6)$$
$$= -12 - 18$$
$$= -30$$

This answer is incorrect. **WHAT WENT WRONG?** Rewrite the given expression correctly.

64. Concept Check A student cleared parentheses in the expression $-(3x + 4)$ as shown.

$$-(3x + 4)$$
$$= -1(3x + 4)$$
$$= -1(3x) + 4$$
$$= -3x + 4$$

This answer is incorrect. **WHAT WENT WRONG?** Rewrite the given expression correctly.

65. Explain how the procedure for changing $\dfrac{3}{4}$ to $\dfrac{9}{12}$ requires the use of the multiplicative identity element, 1.

66. Explain how the procedure for changing $\dfrac{9}{12}$ to $\dfrac{3}{4}$ requires the use of the multiplicative identity element, 1.

Use the distributive property to rewrite each expression. See Example 9.

67. $5(9 + 8)$ **68.** $6(11 + 8)$ **69.** $4(t + 3)$

70. $5(w + 4)$ **71.** $7(z - 8)$ **72.** $8(x - 6)$

73. $-8(r + 3)$ **74.** $-11(x + 4)$ **75.** $-\dfrac{1}{4}(8x + 3)$

76. $-\dfrac{1}{3}(9x + 5)$ **77.** $-\dfrac{1}{3}(9x - 4)$ **78.** $-\dfrac{1}{5}(5x - 7)$

79. $2(6x + 5)$ **80.** $3(3x + 4)$ **81.** $-3(2x - 5)$

82. $-4(3x - 2)$ **83.** $-0.6(8x + 1.2)$ **84.** $-5.2(4x + 2.3)$

85. $-\dfrac{4}{3}(12y + 15z)$ **86.** $-\dfrac{2}{5}(10b + 20a)$ **87.** $8(3r + 4s - 5y)$

88. $2(5u - 3v + 7w)$ **89.** $-3(8x + 3y + 4z)$ **90.** $-5(2x - 5y + 6z)$

Write each expression without parentheses. See Example 10.

91. $-(4t + 3m)$ **92.** $-(9x + 12y)$ **93.** $-(-5c - 4d)$

94. $-(-13x - 15y)$ **95.** $-(-3q + 5r - 8s)$ **96.** $-(-z + 5w - 9y)$

1.7   Simplifying Expressions

OBJECTIVES

1 Simplify expressions.
2 Identify terms and numerical coefficients.
3 Identify like terms.
4 Combine like terms.
5 Simplify expressions from word phrases.

OBJECTIVE 1 Simplify expressions.

We now simplify expressions using the properties of real numbers.

EXAMPLE 1   Simplifying Expressions

Simplify each expression.

(a) $4x + 8 + 9$ simplifies to $4x + 17$.

(b) $4(3m - 2n)$ ⟵ To simplify, we clear the parentheses.

$$= 4(3m) - 4(2n) \qquad \text{Distributive property}$$

$$= (4 \cdot 3)m - (4 \cdot 2)n \qquad \text{Associative property}$$

$$= 12m - 8n \qquad \text{Multiply.}$$

(c) $\qquad 6 + 3(4k + 5)$

Do *not* start by adding.

$$= 6 + 3(4k) + 3(5) \qquad \text{Distributive property}$$

$$= 6 + (3 \cdot 4)k + 3(5) \qquad \text{Associative property}$$

$$= 6 + 12k + 15 \qquad \text{Multiply.}$$

$$= 6 + 15 + 12k \qquad \text{Commutative property}$$

$$= 21 + 12k \qquad \text{Add.}$$

(d) $\qquad 5 - (2y - 8)$

$$= 5 - 1(2y - 8) \qquad -a = -1 \cdot a$$

$$= 5 - 1(2y) - 1(-8) \qquad \text{Distributive property}$$

Be careful with signs.

$$= 5 - 2y + 8 \qquad \text{Multiply.}$$

$$= 5 + 8 - 2y \qquad \text{Commutative property}$$

$$= 13 - 2y \qquad \text{Add.} \qquad \textbf{NOW TRY}$$

VOCABULARY

☐ term
☐ numerical coefficient (coefficient)
☐ like terms
☐ unlike terms

NOW TRY
EXERCISE 1
Simplify each expression.
(a) $3(2x - 4y)$
(b) $-4 - (-3y + 5)$

> **NOTE**   The steps using the commutative and associative properties will not be shown in the rest of the examples. However, be aware that they are usually involved.

OBJECTIVE 2 Identify terms and numerical coefficients.

A **term** is a number (constant), a variable, or a product or quotient of numbers and variables raised to powers.

$$9x, \quad 15y^2, \quad -3, \quad -8m^2n, \quad \frac{2}{p}, \quad \text{and} \quad k \qquad \text{Terms}$$

In the term $9x$, the **numerical coefficient,** or simply the **coefficient,** of the variable x is 9. Additional examples are shown in the table on the next page.

NOW TRY ANSWERS
1. **(a)** $6x - 12y$ **(b)** $3y - 9$

▼ Terms and Their Coefficients

Term	Numerical Coefficient
8	8
$-7y$	-7
$34r^3$	34
$-26x^5yz^4$	-26
$-k$, or $-1k$	-1
r, or $1r$	1
$\frac{3x}{8}$, or $\frac{3}{8}x$	$\frac{3}{8}$
$\frac{x}{3} = \frac{1x}{3}$, or $\frac{1}{3}x$	$\frac{1}{3}$

! **CAUTION** It is important to be able to distinguish between *terms* and *factors*. Consider the following expressions.

$8x^3 + 12x^2$ This expression has **two terms**, $8x^3$ and $12x^2$. Terms are separated by a $+$ or $-$ symbol.

$(8x^3)(12x^2)$ This is a **one-term** expression. The **factors** $8x^3$ and $12x^2$ are multiplied.

OBJECTIVE 3 Identify like terms.

Terms with exactly the same variables that have the same exponents on the variables are **like terms**.

Like Terms	**Unlike Terms**	
$9t$ and $4t$	$4y$ and $7t$	Different variables
$6x^2$ and $-5x^2$	$17x$ and $-8x^2$	Different exponents
$-2pq$ and $11pq$	$4xy^2$ and $4xy$	Different exponents
$3x^2y$ and $5x^2y$	$-7wz^3$ and $2xz^3$	Different variables

OBJECTIVE 4 Combine like terms.

The distributive property

$$a(b + c) = ab + ac \quad \text{can be written "in reverse" as} \quad ab + ac = a(b + c).$$

This last form, which may be used to find the sum or difference of like terms, provides justification for **combining like terms**.

NOW TRY EXERCISE 2

Simplify each expression.
(a) $4x + 6x - 7x$
(b) $z^2 + z^2$
(c) $4p^2 - 3p$

EXAMPLE 2 Combining Like Terms

Simplify each expression.

(a) $-9m + 5m$ Distributive property in reverse
$= (-9 + 5)m$
$= -4m$

(b) $6r + 3r + 2r$
$= (6 + 3 + 2)r$
$= 11r$

(c) $4x + x$
$= 4x + 1x$ $x = 1x$
$= (4 + 1)x$
$= 5x$

(d) $16y^2 - 9y^2$
$= (16 - 9)y^2$
$= 7y^2$

(e) $32y + 10y^2$ These unlike terms cannot be combined.   **NOW TRY**

! **CAUTION** Remember that only like terms may be combined.

NOW TRY ANSWERS
2. (a) $3x$ **(b)** $2z^2$
(c) cannot be combined

Simplifying an Expression

An expression has been simplified when the following conditions have been met.

• All grouping symbols have been removed.

• All like terms have been combined.

• Operations have been performed, when possible.

EXAMPLE 3   Simplifying Expressions

Simplify each expression.

(a) $14y + 2(6 + 3y)$ ⟵ *Start by distributing the 2.*

$$= 14y + 2(6) + 2(3y) \qquad \text{Distributive property}$$

*$14y + 6y$
$= (14 + 6)y$
$= 20y$*

$$= 14y + 12 + 6y \qquad \text{Multiply.}$$

$$= 20y + 12 \qquad \text{Combine like terms.}$$

(b) $9k - 6 - 3(2 - 5k)$ *Be careful with signs.*

$$= 9k - 6 - 3(2) - 3(-5k) \qquad \text{Distributive property}$$

$$= 9k - 6 - 6 + 15k \qquad \text{Multiply.}$$

$$= 24k - 12 \qquad \text{Combine like terms.}$$

(c) $-(2 - r) + 10r$

$$= -1(2 - r) + 10r \qquad -(2 - r) = -1(2 - r)$$

$$= -1(2) - 1(-r) + 10r \qquad \text{Distributive property}$$

Be careful with signs.

$$= -2 + 1r + 10r \qquad \text{Multiply.}$$

$$= -2 + 11r \qquad \text{Combine like terms.}$$

Alternatively, $-(2 - r)$ can be thought of as the *opposite* of $(2 - r)$—that is, $-2 + r$—which can then be added to $10r$ to obtain $-2 + 11r$.

By the commutative property, the final expression $-2 + 11r$ is equivalent to $11r - 2$.

(d) $100[0.03(x + 4)]$

$$= [100(0.03)](x + 4) \qquad \text{Associative property}$$

*Think:
$100(0.03)$
$= 3$*

$$= 3(x + 4) \qquad \text{Multiply.}$$

$$= 3x + 3(4) \qquad \text{Distributive property}$$

$$= 3x + 12 \qquad \text{Multiply.}$$

(e) $5(2a - 6) - 3(4a - 9)$

$$- 5(2a) + 5(-6) - 3(4a) - 3(-9) \qquad \text{Distributive property twice}$$

$$= 10a - 30 - 12a + 27 \qquad \text{Multiply.}$$

$$= -2a - 3 \qquad \text{Combine like terms.}$$

NOW TRY EXERCISE 3

Simplify each expression.

(a) $5k - 6 - (3 - 4k)$

(b) $\frac{1}{4}x - \frac{2}{3}(x - 9)$

(f) $-\frac{2}{3}(x - 6) - \frac{1}{6}x$

$$= -\frac{2}{3}x - \frac{2}{3}(-6) - \frac{1}{6}x \qquad \text{Distributive property}$$

$$= -\frac{2}{3}x + 4 - \frac{1}{6}x \qquad \text{Multiply.}$$

$$= -\frac{4}{6}x + 4 - \frac{1}{6}x \qquad \text{Find a common denominator.}$$

$$= -\frac{5}{6}x + 4 \qquad \text{Combine like terms.} \qquad \text{NOW TRY} \ \ \text{}$$

> **NOTE** **Examples 2 and 3** suggest that like terms may be combined by adding or subtracting the coefficients of the terms and keeping the same variable factors.

OBJECTIVE 5 Simplify expressions from word phrases.

NOW TRY EXERCISE 4

Write the following phrase as a mathematical expression using x as the variable, and simplify.

   Twice a number, subtracted from the sum of the number and 5

NOW TRY ANSWERS

3. (a) $9k - 9$ (b) $-\frac{5}{12}x + 6$

4. $(x + 5) - 2x; \ -x + 5$

EXAMPLE 4 Translating Words into a Mathematical Expression

Write the following phrase as a mathematical expression using x as the variable, and simplify.

The sum of 9, five times a number,
four times the number, and
six times the number

The word "sum" indicates that the terms should be added. Use x for the number.

$$9 + 5x + 4x + 6x \quad \text{simplifies to} \quad 9 + 15x. \qquad \text{Combine like terms.}$$

 This is an expression to be simplified, *not* an equation to be solved.

NOW TRY

1.7 Exercises

FOR EXTRA HELP   ▶ **MyLab Math**

▶ *Video solutions for select problems available in MyLab Math*

STUDY SKILLS REMINDER

Reread your class notes before working the assigned exercises. **Review Study Skill 3,** *Taking Lecture Notes.*

Concept Check *Choose the letter of the correct response.*

1. Which expression is a simplified form of $-(6x - 3)$?

   A. $-6x - 3$ **B.** $-6x + 3$ **C.** $6x - 3$ **D.** $6x + 3$

2. Which is an example of a term with numerical coefficient 5?

   A. $5x^3y^7$ **B.** x^5 **C.** $\frac{x}{5}$ **D.** $-5xy^3$

3. Which is an example of a pair of like terms?

   A. $6t, 6w$ **B.** $-8x^2y, 9xy^2$ **C.** $5ry, 6yr$ **D.** $-5x^2, 2x^3$

4. Which is a correct translation for "six times a number, subtracted from the product of eleven and the number" (if x represents the number)?

   A. $6x - 11x$ **B.** $11x - 6x$ **C.** $(11 + x) - 6x$ **D.** $6x - (11 + x)$

5. Concept Check   A student simplified the expression $7x - 2(3 - 2x)$ incorrectly as shown.

$$7x - 2(3 - 2x)$$
$$= 7x - 2(3) - 2(2x)$$
$$= 7x - 6 - 4x$$
$$= 3x - 6$$

WHAT WENT WRONG?   Find the correct simplified answer.

6. Concept Check   A student simplified the expression $3 + 2(4x - 5)$ incorrectly as shown.

$$3 + 2(4x - 5)$$
$$= 5(4x - 5)$$
$$= 5(4x) + 5(-5)$$
$$= 20x - 25$$

WHAT WENT WRONG?   Find the correct simplified answer.

Simplify each expression. See Example 1.

7. $4r + 19 - 8$ **8.** $7t + 18 - 4$ **9.** $7(3x - 4y)$

10. $8(2p - 9q)$ **11.** $5 + 2(x - 3y)$ **12.** $8 + 3(s - 6t)$

13. $8 + 4(3x + 6)$ **14.** $10 + 5(2y + 7)$ **15.** $-2 - (5 - 3p)$

16. $-10 - (7 - 14r)$ **17.** $6 + (4 - 3x) - 8$ **18.** $-12 + (7 - 8x) + 6$

In each term, give the numerical coefficient. See Objective 2.

19. $-12k$ **20.** $-11y$ **21.** $3m^2$ **22.** $9n^6$

23. xw **24.** pq **25.** $-x$ **26.** $-t$

27. $\dfrac{x}{2}$ **28.** $\dfrac{x}{6}$ **29.** $\dfrac{2x}{5}$ **30.** $\dfrac{8x}{9}$

31. $-0.5x^3$ **32.** $-1.75x^2$ **33.** 10 **34.** 15

Identify each group of terms as like *or* unlike. *See Objective 3.*

35. $8r, -13r$ **36.** $-7x, 12x$ **37.** $5z^4, 9z^3$ **38.** $8x^5, -10x^3$

39. $4, 9, -24$ **40.** $7, 17, -83$ **41.** x, y **42.** t, s

Simplify each expression. See Examples 1–3.

43. $7y + 6y$ **44.** $5m + 2m$ **45.** $-6x - 3x$

46. $-4z - 8z$ **47.** $12b + b$ **48.** $19x + x$

49. $3k + 8 + 4k + 7$ **50.** $15z + 1 + 4z + 2$

51. $-5y + 3 - 1 + 5 + y - 7$ **52.** $2k - 7 - 5k + 6 - 1 + 2$

53. $-2x + 3 + 4x - 17 + 20$ **54.** $r - 6 - 12r - 4 + 16$

55. $16 - 5m - 4m - 2 + 2m$ **56.** $6 - 3z - 2z - 5 - 2z$

57. $2.3x - 1.1 + 4.2x - 0.7$ **58.** $-3.4p - 0.8 + 2.5 + 7.2p$

59. $7.2x - 5.1 + 2.3x + 5.1$ **60.** $-9.6r + 2.7 - 8.5r - 2.7$

61. $-\dfrac{4}{3} + 2t + \dfrac{1}{3}t - 8 - \dfrac{8}{3}t$ **62.** $-\dfrac{5}{6} + 8x + \dfrac{1}{6}x - 7 - \dfrac{7}{6}$

63. $6y^2 + 11y^2 - 8y^2$ **64.** $-9m^3 + 3m^3 - 7m^3$

65. $2p^2 + 3p^2 - 8p^3 - 6p^3$ **66.** $5y^3 + 6y^3 - 3y^2 - 4y^2$

67. $2(4x + 6) + 3$ **68.** $4(6y + 9) + 7$

69. $-6 - 4(y - 7)$ **70.** $-4 - 5(t - 13)$

71. $13p + 4(4 - 8p)$ **72.** $5x + 3(7 - 2x)$

73. $3t - 5 - 2(2t - 4)$ **74.** $8p + 6 - 3(3p - 1)$

75. $100[0.05(x + 3)]$ **76.** $100[0.06(x + 5)]$

77. $10[0.3(5 - 3x)]$ **78.** $10[0.5(8 - 2z)]$

79. $-5(5y - 9) + 3(3y + 6)$ **80.** $-3(2t + 4) + 8(2t - 4)$

81. $2(5r + 3) - 3(2r - 3)$ **82.** $3(2y - 5) - 4(5y - 7)$

83. $8(2k - 1) - (4k - 3)$ **84.** $6(3p - 2) - (5p + 1)$

85. $-\dfrac{4}{3}(y - 12) - \dfrac{1}{6}y$ **86.** $-\dfrac{7}{5}(t - 15) - \dfrac{1}{2}t$

87. $\dfrac{1}{2}(2x + 4) - \dfrac{1}{3}(9x - 6)$ **88.** $\dfrac{1}{4}(8x + 16) - \dfrac{1}{5}(20x - 15)$

89. $-\dfrac{2}{3}(5x + 7) - \dfrac{1}{3}(4x + 8)$ **90.** $-\dfrac{3}{4}(7x + 9) - \dfrac{1}{4}(5x + 7)$

91. $-7.5(2y + 4) - 2.9(3y - 6)$ **92.** $8.4(6t - 6) + 2.4(9 - 3t)$

93. $-2(-3k + 2) - (5k - 6) - 3k - 5$ **94.** $-2(3r - 4) - (6 - r) + 2r - 5$

95. $-4(-3x + 3) - (6x - 4) - 2x + 1$ **96.** $-5(8x + 2) - (5x - 3) - 3x + 17$

Extending Skills   *Write each of the following as a mathematical expression, and simplify.*

97. Add $3x - 2$ to $4x + 8$.   **98.** Add $8t + 5$ to $10t - 8$.

99. Subtract $x - 7$ from $5x + 1$.   **100.** Subtract $3x - 5$ from $2x - 3$.

Write each phrase as a mathematical expression using x as the variable, and simplify.
See Example 4.

101. Five times a number, added to the sum of the number and three

102. Six times a number, added to the sum of the number and six

103. A number multiplied by -7, subtracted from the sum of 13 and six times the number

104. A number multiplied by 5, subtracted from the sum of 14 and eight times the number

105. Six times a number added to -4, subtracted from twice the sum of three times the number and 4 (*Hint: Twice* means two times.)

106. Nine times a number added to 6, subtracted from triple the sum of 12 and 8 times the number (*Hint: Triple* means three times.)

RELATING CONCEPTS   For Individual or Group Work   (Exercises 107–110)

A manufacturer has fixed costs of 1000 to produce gizmos. Each gizmo costs 5 to make. The fixed cost to produce gadgets is 750, and each gadget costs 3 to make.
Work Exercises 107–110 in order.

107. Write an expression for the cost to make x gizmos. (*Hint:* The cost will be the sum of the fixed cost and the cost per item times the number of items.)

108. Write an expression for the cost to make y gadgets.

109. Write an expression for the total cost to make x gizmos and y gadgets.

110. Simplify the expression from **Exercise 109.**

Chapter 1   Summary

STUDY SKILLS REMINDER
How can you best prepare for a test? **Review Study Skill 7, *Reviewing a Chapter.***

Key Terms

1.1

exponent (power)
base
exponential expression
inequality

1.2

constant
variable
algebraic expression
equation
solution
set
element

1.3

natural (counting) numbers
whole numbers
number line
integers
signed numbers
rational numbers
graph
coordinate
irrational numbers
real numbers
additive inverse (opposite)
absolute value

1.4

sum
addends
difference
minuend
subtrahend

1.5

product
factor
multiplicative inverse
  (reciprocal)
quotient
dividend
divisor

1.6

identity element for addition (additive identity)
identity element for multiplication (multiplicative
  identity)

1.7

term
numerical coefficient
  (coefficient)
like terms
unlike terms

New Symbols

a^n	n factors of a	$\geq$	is greater than or equal to	$-x$	additive inverse, or opposite, of x
$[\ ]$	brackets			$\lvert x \rvert$	absolute value of x
$=$	is equal to	$\{\ \}$	set braces		
$\neq$	is not equal to	$\{x \mid x$ **has a given**		$\dfrac{1}{x}$	multiplicative inverse,
$<$	is less than	**property** $\}$			or reciprocal, of x
$>$	is greater than		set-builder notation		(where $x \neq 0$)
$\leq$	is less than or equal to				

$a(b), (a)b, (a)(b), a \cdot b,$
or ab a times b

$a \div b, \dfrac{a}{b}, a/b,$ or $b\overline{)a}$

    a divided by b

Test Your Word Power

See how well you have learned the vocabulary in this chapter.

1. An **exponent** is
   A. a symbol that tells how many numbers are being multiplied
   B. a number raised to a power
   C. a number that tells how many times a factor is repeated
   D. a number that is multiplied.

2. A **variable** is
   A. a symbol used to represent an unknown number
   B. a value that makes an equation true
   C. a solution of an equation
   D. the answer in a division problem.

3. An **integer** is
   A. a positive or negative number
   B. a natural number, its opposite, or zero
   C. any number that can be graphed
   D. the quotient of two numbers.

4. The **absolute value** of a number is
   A. the graph of the number
   B. the reciprocal of the number
   C. the opposite of the number
   D. the distance between 0 and the number on a number line.

5. A **term** is
   A. a numerical factor

   B. a number, variable, or product or quotient of numbers and variables raised to powers
   C. one of several variables with the same exponents
   D. a sum of numbers and variables raised to powers.

6. A **numerical coefficient** is
   A. the numerical factor of the variable(s) in a term
   B. the number of terms in an expression
   C. a variable raised to a power
   D. the variable factor in a term.

ANSWERS

1. C; *Example:* In 2^3, the number 3 is the exponent (or power), so 2 is a factor three times, and $2^3 = 2 \cdot 2 \cdot 2 = 8$.   **2.** A; *Examples: a, b, c*
3. B; *Examples:* $-9, 0, 6$ **4.** D; *Examples:* $\lvert 2 \rvert = 2$ and $\lvert -2 \rvert = 2$ **5.** B; *Examples:* $6, \frac{x}{2}, -4ab^2$ **6.** A; *Examples:* The term 3 has numerical coefficient 3, $8z$ has numerical coefficient 8, and $-10x^4y$ has numerical coefficient -10.

Quick Review

CONCEPTS	EXAMPLES

1.1 Exponents, Order of Operations, and Inequality

Order of Operations
If grouping symbols are present, work within them, innermost first (and above and below fraction bars separately), in the following order.

Step 1 Apply all exponents.

Step 2 Do any multiplications and divisions in order from left to right.

Step 3 Do any additions and subtractions in order from left to right.

Simplify $36 - 4(2^2 + 3)$.

$36 - 4(2^2 + 3)$

$= 36 - 4(4 + 3)$ Apply the exponent.

$= 36 - 4(7)$ Add inside the parentheses.

$= 36 - 28$ Multiply.

$= 8$ Subtract.

1.2 Variables, Expressions, and Equations

To *evaluate* an expression means to find its *value*. Evaluate an expression with a variable by substituting a given number for the variable.

Evaluate $2x + y^2$ for $x = 3$ and $y = -4$.

$2x + y^2$

$= 2(3) + (-4)^2$ Substitute.

$= 6 + 16$ Multiply. Apply the exponent.

$= 22$ Add.

A value of a variable that makes an equation true is a solution of the equation.

Is 2 a solution of $5x + 3 = 18$?

$5(2) + 3 \stackrel{?}{=} 18$ Let $x = 2$.

$13 = 18$ False

2 is not a solution.

1.3 Real Numbers and the Number Line

Ordering Real Numbers
a is less than b if a lies to the left of b on a number line.

a is greater than b if a lies to the right of b on a number line.

The additive inverse, or opposite, of x is $-x$.

The absolute value of x, written $|x|$, is the distance between x and 0 on a number line.

Graph $-2, 0,$ and 3.

$-2 < 3$ $3 > 0$ $0 < 3$

$-(5) = -5$ $-(-7) = 7$ $-0 = 0$

$|13| = 13$ $|0| = 0$ $|-5| = 5$

1.4 Adding and Subtracting Real Numbers

Adding Two Signed Numbers
Same sign Add their absolute values. The sum has the same sign as the addends.

Different signs Subtract their absolute values. The sum has the sign of the addend with greater absolute value.

Definition of Subtraction

$$x - y = x + (-y)$$

Add. $9 + 4 = 13$

$-8 + (-5) = -13$

$7 + (-12) = -5$

$-5 + 13 = 8$

Subtract. $5 - (-2)$ $-3 - 4$ $-2 - (-6)$

$= 5 + 2$ $= -3 + (-4)$ $= -2 + 6$

$= 7$ $= -7$ $= 4$

CONCEPTS	EXAMPLES

1.5 Multiplying and Dividing Real Numbers

Multiplying and Dividing Two Signed Numbers
Same sign   The product (or quotient) is *positive*.

Different signs   The product (or quotient) is *negative*.

Definition of Division

$$x \div y = x \cdot \frac{1}{y} \quad \text{(where } y \neq 0\text{)}$$

0 divided by a nonzero number equals 0.
Division by 0 is undefined.

Multiply or divide.

$$6 \cdot 5 = 30 \qquad -7(-8) = 56 \qquad \frac{-24}{-6} = 4$$

$$-6(5) = -30 \qquad \frac{-18}{9} = -2 \qquad \frac{49}{-7} = -7$$

$$10 \div 2 = \frac{10}{2} = 10 \cdot \frac{1}{2} = 5$$

$$\frac{0}{5} = 0 \qquad \frac{5}{0} \text{ is undefined.}$$

1.6 Properties of Real Numbers

Commutative Properties

$$a + b = b + a$$
$$ab = ba$$

Associative Properties

$$(a + b) + c = a + (b + c)$$
$$(ab)c = a(bc)$$

Identity Properties

$$a + 0 = a \qquad 0 + a = a$$
$$a \cdot 1 = a \qquad 1 \cdot a = a$$

Inverse Properties

$$a + (-a) = 0 \qquad -a + a = 0$$
$$a \cdot \frac{1}{a} = 1 \qquad \frac{1}{a} \cdot a = 1 \quad \text{(where } a \neq 0\text{)}$$

Distributive Properties

$$a(b + c) = ab + ac \quad \text{and} \quad (b + c)a = ba + ca$$
$$a(b - c) = ab - ac \quad \text{and} \quad (b - c)a = ba - ca$$

$$7 + (-1) = -1 + 7$$
$$5(-3) = (-3)5$$

$$(3 + 4) + 8 = 3 + (4 + 8)$$
$$[-2(6)]4 = -2[(6)4]$$

$$-7 + 0 = -7 \qquad 0 + (-7) = -7$$
$$9 \cdot 1 = 9 \qquad 1 \cdot 9 = 9$$

$$7 + (-7) = 0 \qquad -7 + 7 = 0$$
$$-2\left(-\frac{1}{2}\right) = 1 \qquad -\frac{1}{2}(-2) = 1$$

$$5(x + 2) = 5x + 5(2) \qquad (x + 2)5 = x \cdot 5 + 2(5)$$
$$9(5y - 4) = 9(5y) - 9(4) \qquad (5y - 4)9 = 5y(9) - 4(9)$$

1.7 Simplifying Expressions

Only like terms may be combined. We use a form of the distributive property.

Simplify each expression.

$$-3y^2 + 6y^2 + 14y^2$$
$$= (-3 + 6 + 14)y^2$$
$$= 17y^2$$

$$4(3 + 2x) - 6(5 - x)$$
$$= 4(3) + 4(2x) - 6(5) - 6(-x)$$
$$= 12 + 8x - 30 + 6x$$
$$= 14x - 18$$

Chapter 1	Review Exercises

1.1 *Find the value of each exponential expression.*

1. 5^4 **2.** $\left(\dfrac{3}{5}\right)^3$ **3.** $\left(\dfrac{1}{8}\right)^2$ **4.** $(0.1)^3$

Evaluate each expression.

5. $8 \cdot 5 - 13$ **6.** $16 + 12 \div 4 - 2$ **7.** $20 - 2(5 + 3)$

8. $7[3 + 6(3^2)]$ **9.** $\dfrac{5(6^2 - 2^4)}{3 \cdot 5 + 10}$ **10.** $\dfrac{6(5 - 4) + 2(4 - 2)}{3^2 - (4 + 3)}$

Determine whether each statement is true *or* false.

11. $12 \cdot 3 - 6 \cdot 6 \le 0$ **12.** $3[5(2) - 3] > 20$ **13.** $9 \le 4^2 - 8$

Write each word statement in symbols.

14. Thirteen is less than seventeen.   **15.** Five plus two is not equal to ten.

16. Two-thirds is greater than or equal to four-sixths.

1.2 *Evaluate each expression for $x = 6$ and $y = 3$.*

17. $2x + 6y$ **18.** $4(3x - y)$ **19.** $\dfrac{x}{3} + 4y$ **20.** $\dfrac{x^2 + 3}{3y - x}$

Write each word phrase as an algebraic expression, using x as the variable.

21. Six added to a number   **22.** A number subtracted from eight

23. The difference of six times a number and nine   **24.** Three-fifths of a number added to 12

Determine whether each equation has the given number as a solution.

25. $5x + 3(x + 2) = 52;$ 2   **26.** $\dfrac{2}{9}t + \dfrac{1}{3} = 1;$ 3

Write each word statement as an equation. Use x as the variable. Then find the solution of the equation from the set $\{0, 2, 4, 6, 8, 10\}$.

27. Six less than twice a number is 10.   **28.** The product of a number and 4 is 8.

1.3 *Graph each number on a number line.*

29. $-4, -\dfrac{1}{2}, 0, 2.5, 5$ **30.** $-3, -1\dfrac{1}{2}, \dfrac{2}{3}, 2.25, 3$

Classify each number, using the sets natural numbers, whole numbers, integers, rational numbers, irrational numbers, *and* real numbers.

31. $\dfrac{4}{3}$ **32.** $0.\overline{63}$ **33.** 19   **34.** $\sqrt{6}$

Select the lesser of the two given numbers.

35. $-10, 5$ **36.** $-8, -9$ **37.** $-\dfrac{2}{3}, -\dfrac{3}{4}$ **38.** $0, -|23|$

Determine whether each statement is true *or* false.

39. $12 > -13$ **40.** $0 > -5$ **41.** $-9 < -7$ **42.** $-13 \ge -13$

*For each number, find (**a**) the additive inverse and (**b**) the absolute value.*

43. -9 **44.** 0 **45.** 6 **46.** $-\dfrac{5}{7}$

Find each absolute value.

47. $|-12|$ **48.** $-|3|$ **49.** $-|-19|$ **50.** $-|9-2|$

1.4 *Perform each indicated operation.*

51. $-10 + 4$ **52.** $14 + (-18)$ **53.** $-8 + (-9)$

54. $\dfrac{4}{9} + \left(-\dfrac{5}{4}\right)$ **55.** $-13.5 + (-8.3)$ **56.** $|-10 + 7| + |-11|$

57. $[-6 + (-8) + 8] + [9 + (-13)]$ **58.** $(-4 + 7) + (-11 + 3) + (-15 + 1)$

59. $-7 - 4$ **60.** $-12 - (-11)$

61. $5 - (-2)$ **62.** $-\dfrac{3}{7} - \dfrac{4}{5}$

63. $2.56 - (-7.75)$ **64.** $(-10 - 4) - (-2)$

65. $-5 - [(-7 - 4) + (8 - 12)]$ **66.** $-5.6 + [(-7.4 + 3.6) - 4.82]$

Write a numerical expression for each phrase, and simplify the expression.

67. 19 added to the sum of -31 and 12 **68.** 13 more than the sum of -4 and -8

69. The difference of -4 and -6 **70.** Five less than the difference of 7 and -5

Solve each problem.

71. George found that his checkbook balance was $-\$23.75$, so he deposited $\$50.00$. What is his new balance?

72. The low temperature in Yellowknife, in the Canadian Northwest Territories, one January day was $-26°F$. It rose $16°$ that day. What was the high temperature?

73. Reginald owed a friend $\$28$. He repaid $\$13$, but then borrowed another $\$14$. What positive or negative amount represents his present financial status?

74. If the temperature drops $7°$ below its previous level of $-3°$, what is the new temperature?

75. A quarterback passed for a gain of 8 yd, was sacked for a loss of 12 yd, and then threw a 42 yd touchdown pass. What positive or negative number represents the total net yardage for the plays?

76. On Wednesday, October 25, 2017, the Dow Jones Industrial Average closed at 23,329.46, down 112.30 from the previous day. What was the closing value the previous day? (Data from *The Wall Street Journal*.)

1.5 *Perform each indicated operation.*

77. $-12(-3)$ **78.** $15(-7)$ **79.** $-\dfrac{4}{3}\left(-\dfrac{3}{8}\right)$

80. $-4.8(-2.1)$ **81.** $5(8 - 12)$ **82.** $(5 - 7)(8 - 3)$

83. $2(-6) - (-4)(-3)$ **84.** $3(-10) - 5$ **85.** $\dfrac{-36}{-9}$

86. $\dfrac{220}{-11}$

87. $-\dfrac{1}{2} \div \dfrac{2}{3}$

88. $\dfrac{-33.9}{-3}$

89. $\dfrac{-5(3) - 1}{8 + 4(-2)}$

90. $\dfrac{5(-2) - 3(4)}{-2[3 - (-2)] - 1}$

91. $\dfrac{10^2 - 5^2}{8^2 + 3^2 - (-2)}$

92. $\dfrac{4(0.4)^2 - (0.8)^2}{(-1.2)^2 - (-0.56)}$

Evaluate each expression for $x = -5$, $y = 4$, and $z = -3$.

93. $6x - 4z$ **94.** $5x^2$ **95.** $z^2(3x - 8y)$ **96.** $\dfrac{3y - z}{x + 5}$

Write a numerical expression for each phrase, and simplify the expression.

97. Nine less than the product of -4 and 5

98. Five-sixths of the sum of 12 and -6

99. The quotient of 12 and the sum of 8 and -4

100. The product of -20 and 12, divided by the difference of 15 and -15

Write each sentence as an equation, using x as the variable. Then find the solution of the equation from the set of integers between -12 and 12, inclusive.

101. 8 times a number is -24.

102. The quotient of a number and 3 is -2.

1.6   *Decide whether each statement is an example of a* commutative, *an* associative, *an* identity, *an* inverse, *or the* distributive property.

103. $6 + 0 = 6$

104. $5 \cdot 1 = 5$

105. $-\dfrac{2}{3}\left(-\dfrac{3}{2}\right) = 1$

106. $17 + (-17) = 0$

107. $5 + (-9 + 2) = [5 + (-9)] + 2$

108. $w(xy) = (wx)y$

109. $3(x + y) = 3x + 3y$

110. $(1 + 2) + 3 = 3 + (1 + 2)$

Use the distributive property to rewrite each expression.

111. $7(y + 2)$ **112.** $-12(4 - t)$ **113.** $3(2s + 5y)$ **114.** $-(-4r + 5s)$

1.7   *Simplify each expression.*

115. $2m + 9m$

116. $15p^2 - 7p^2 + 8p^2$

117. $5p^2 - 4p + 6p + 11p^2$

118. $-2(3k - 5) + 2(k + 1)$

119. $7(2m + 3) - 2(8m - 4)$

120. $\dfrac{2}{5}(15x - 4) - \dfrac{1}{5}(10x + 7)$

Write each phrase as a mathematical expression using x as the variable, and simplify.

121. Seven times a number, subtracted from the product of -2 and three times the number

122. A number multiplied by 8, added to the sum of 5 and four times the number

Chapter 1	Mixed Review Exercises

Complete the table.

	Number	Absolute Value	Additive Inverse	Multiplicative Inverse
1.	-3			
2.	12			
3.				$-\dfrac{3}{2}$
4.			-0.2	

5. To which of the following sets does $0.\overline{6}$ belong: natural numbers, whole numbers, integers, rational numbers, irrational numbers, real numbers?

6. Evaluate $(x + 6)^3 - y^3$ for $x = -2$ and $y = 3$.

Perform each indicated operation.

7. $\dfrac{6(-4) + 2(-12)}{5(-3) + (-3)}$

8. $\dfrac{3}{8} - \dfrac{5}{12}$

9. $\dfrac{8^2 + 6^2}{7^2 + 1^2}$

10. $-\dfrac{12}{5} \div \dfrac{9}{7}$

11. $2\dfrac{5}{6} - 4\dfrac{1}{3}$

12. $\left(\dfrac{5}{6}\right)^2$

13. $[(-2) + 7 - (-5)] + [-4 - (-10)]$

14. $-16(-3.5) - 7.2(-3)$

15. $-8 + [(-4 + 17) - (-3 - 3)]$

16. $-4(2t + 1) - 8(-3t + 4)$

17. $5x^2 - 12y^2 + 3x^2 - 9y^2$

18. $(-8 - 3) - 5(2 - 9)$

Solve each problem.

19. The highest temperature ever recorded in Iowa was 118°F. The lowest temperature ever recorded in the state was 165° lower than the highest temperature. What is the record low temperature for Iowa? (*Data from* National Climatic Data Center.)

20. The top of Mt. Whitney, visible from Death Valley, has an altitude of 14,494 ft above sea level. The bottom of Death Valley is 282 ft below sea level. Using 0 as sea level, find the difference between these two elevations. (*Data from The World Almanac and Book of Facts.*)

Chapter 1	Test	FOR EXTRA HELP

Step-by-step test solutions are found on the Chapter Test Prep Videos available in **MyLab Math**.

▶ *View the complete solutions to all Chapter Test exercises in MyLab Math.*

STUDY SKILLS REMINDER

Using test-taking strategies can help you improve your test scores. **Review Study Skill 8, *Taking Math Tests.***

Determine whether each statement is true *or* false.

1. $4[-20 + 7(-2)] \le 135$

2. $\left(\dfrac{1}{2}\right)^2 + \left(\dfrac{2}{3}\right)^2 = \left(\dfrac{1}{2} + \dfrac{2}{3}\right)^2$

3. Graph the numbers -1, -3, $|-4|$, and $|-1|$ on a number line.

4. To which of the following sets does $-\dfrac{2}{3}$ belong: natural numbers, whole numbers, integers, rational numbers, irrational numbers, real numbers?

Select the lesser of the two given numbers.

5. 6, $-|-8|$

6. -0.742, -1.277

7. Write a numerical expression for the phrase, and simplify the expression.

The quotient of -6 and the sum of 2 and -8

Perform each indicated operation.

8. $-2 - (5 - 17) + (-6)$

9. $-5\frac{1}{2} + 2\frac{2}{3}$

10. $-6.2 - [-7.1 + (2.0 - 3.1)]$

11. $4^2 + (-8) - (2^3 - 6)$

12. $(-5)(-12) + 4(-4) - 8^2$

13. $25 - 15 \div 3 - 12$

14. $\dfrac{30(-1 - 2)}{-9[3 - (-2)] - 12(-2)}$

15. $\dfrac{-7 - |-6 + 2|}{-5 - (-4)}$

Evaluate each expression for $x = -2$ and $y = 4$.

16. $3x - 4y^2$

17. $\dfrac{5x + 7y}{3(x + y)}$

Solve each problem.

18. The highest elevation in Argentina is Mt. Aconcagua, which is 6960 m above sea level. The lowest point in Argentina is the Valdés Peninsula, 40 m below sea level. Find the difference between the highest and lowest elevations.

19. For a certain system of rating relief pitchers, 3 points are awarded for a save, 3 points are awarded for a win, 2 points are subtracted for a loss, and 2 points are subtracted for a blown save. If a pitcher has 4 saves, 3 wins, 2 losses, and 1 blown save, how many points does he have?

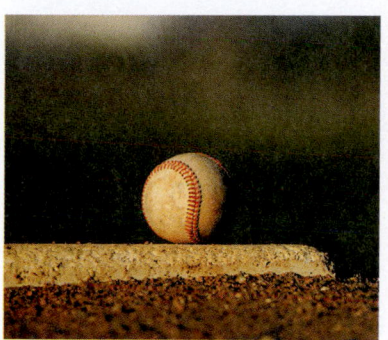

20. For 2016, the U.S. federal government collected $3.34 trillion in revenues, but spent $3.95 trillion. Write the federal budget deficit as a signed number. (Data from U.S. Office of Management and Budget.)

Match each statement in Column I with the property it illustrates in Column II.

I

II

21. $-\dfrac{2}{3} + \dfrac{2}{3} = 0$

A. Commutative property

22. $3x + 0 = 3x$

B. Associative property

23. $(5 + 2) + 8 = 8 + (5 + 2)$

C. Inverse property

24. $-3(x + y) = -3x + (-3y)$

D. Identity property

25. $-5 + (3 + 2) = (-5 + 3) + 2$

E. Distributive property

26. $-\dfrac{5}{3}\left(-\dfrac{3}{5}\right) = 1$

Simplify each expression.

27. $8x + 4x - 6x + x + 14x$

28. $-8.5t - 0.9 + 7.6 + 5.7t$

29. $-\dfrac{1}{6}(12x - 3) + 5x$

30. $5(2x - 1) - (x - 12) + 2(3x - 5)$

Chapters R and 1   Cumulative Review Exercises

Perform each indicated operation. Write answers in lowest terms as needed.

1. $\dfrac{5}{8} \cdot \dfrac{2}{7}$

2. $9 \div \dfrac{3}{2}$

3. $\dfrac{11}{16} - \dfrac{5}{12}$

4. $42.5 - 15.72$

5. 0.3×0.05

6. $9.26 \div 100$

Solve each problem.

7. A board is $12\frac{3}{4}$ in. long. If it must be divided into three pieces of equal length, how long must each piece be?

8. A smart watch with a regular price of $379 is on sale at 15% off. Find the amount of the discount and the sale price of the watch.

Simplify. Then determine whether each statement is true *or* false.

9. $105 \le 5[4 + 3(7 - 4)]$

10. $\dfrac{6(4^2 - 10) + 12}{15 - 3^2} > 2$

Determine whether each equation has the given number as a solution.

11. $5x + 4(2x - 7) = -19; \quad 3$

12. $\dfrac{5}{8}x - \dfrac{3}{2} = 1; \quad 4$

13. Graph each number on a number line.

$$\frac{1}{3}, \ 1\frac{1}{2}, \ -2.5, \ 0, \ -4$$

14. Evaluate the expression for $x = -4$ and $y = 5$.

$$\frac{5x + 4y}{8x^2 - y}$$

Perform each indicated operation.

15. $-5.37 + 2.76$

16. $-\dfrac{3}{4} - \dfrac{5}{8}$

17. $|-4 - 2| - |-9 + 1|$

18. $\dfrac{-12(-14)}{8 - (-6)}$

19. $\dfrac{3^2 - 6^2}{-3(12 - 9)}$

20. $\dfrac{-7(2) + [3(-5) - 6]}{-9 - (-3)(3)}$

Name the property illustrated by each equation.

21. $-4\left(-\dfrac{1}{4}\right) = 1$

22. $\left(-\dfrac{1}{2} + \dfrac{1}{3}\right) + \dfrac{2}{3} = -\dfrac{1}{2} + \left(\dfrac{1}{3} + \dfrac{2}{3}\right)$

Simplify each expression.

23. $8t + 4t - 12 + 7 - 9t$

24. $5(3x - 7) - (2x - 8)$

25. The highest elevation in Arkansas is Mount Magazine's Signal Hill, which is 2753 ft above sea level. The lowest point in Arkansas along the Ouachita River is 55 ft above sea level. Find the difference between the highest and lowest elevations. (Data from *The World Almanac and Book of Facts*.)

STUDY SKILLS REMINDER

It is not too soon to begin preparing for your final exam. **Review Study Skill 10,** *Preparing for Your Math Final Exam.*

LINEAR EQUATIONS AND INEQUALITIES IN ONE VARIABLE

Solving *linear equations,* the subject of this chapter, can be thought of in terms of the concept of balance.

2.1 The Addition Property of Equality

OBJECTIVES
1 Identify linear equations.
2 Use the addition property of equality.
3 Simplify, and then use the addition property of equality.

An **equation** is a statement asserting that two algebraic expressions are equal.

> ⚠ **CAUTION** *Remember that an equation always includes an equality symbol.*

Equation	Expression
↓	↓
Left side → $x - 5 = 2$ ← Right side	$x - 5$
An equation can be solved.	An expression **cannot** be solved. (It can be *evaluated* for a given value, or *simplified*.)

OBJECTIVE 1 Identify linear equations.

Linear Equation in One Variable

A **linear equation in one variable** (here x) is an equation that can be written in the form

$$ax + b = 0,$$

where a and b are real numbers and $a \neq 0$.

Examples: $4x + 9 = 0$, $\dfrac{1}{2}x + 3 = 1$, $2x - 1.5 = -5$, $x = 7$ Linear equations in one variable

VOCABULARY
☐ equation
☐ linear equation in one variable
☐ first-degree equation
☐ solution
☐ solution set
☐ equivalent equations

A linear equation in one variable has just one variable. Such an equation is also called a *first-degree equation* because its variable has an exponent of *one*. (Recall that $x = x^1$.)

$2x - 3y = 6$ (There are two distinct variables.)   Not linear equations in one variable
$x^2 + 4 = 0$ (The exponent on the variable is not 1.)

STUDY SKILLS REMINDER
Are you getting the most out of your class time?
Review Study Skill 3,
Taking Lecture Notes.

A **solution** of an equation is a number that makes the equation true when it replaces the variable. An equation is solved by finding its **solution set,** the set of all solutions. Equations with exactly the same solution sets are **equivalent equations.**

A linear equation in x is *solved* by using a series of steps to "isolate" the variable and obtain a simpler equivalent equation of the form

$$x = \textbf{a number} \quad \text{or} \quad \textbf{a number} = x.$$

OBJECTIVE 2 Use the addition property of equality.

In the linear equation $x - 5 = 2$, both $x - 5$ and 2 represent the same number because that is the meaning of the equality symbol. To solve the equation, we isolate x on the left side as follows.

$x - 5 = 2$ Given equation

$x - 5 + 5 = 2 + 5$ Add 5 to *each* side to keep them equal.

$x + 0 = 7$ Additive inverse property

$x = 7$ Additive identity property

> Add 5. It is the opposite **(additive inverse)** of -5, and $-5 + 5 = 0$.

To check that 7 is the solution, we replace x with 7 in the original equation.

CHECK $x - 5 = 2$ Original equation

$7 - 5 \stackrel{?}{=} 2$ Let $x = 7$.

> The left side equals
> the right side.

$2 = 2$ ✓   True

> We write a solution set
> using set braces { }.

The final equation is true, so 7 is the solution and $\{7\}$ is the solution set.

To solve the equation $x - 5 = 2$, we used the **addition property of equality.**

Addition Property of Equality

If a, b, and c represent real numbers, then the equations

$$a = b \quad \text{and} \quad a + c = b + c \quad \text{are equivalent.}$$

That is, the same number may be added to each side of an equation without changing the solution set.

In this property, any quantity that represents a real number c can be added to each side of an equation to obtain an equivalent equation.

NOTE   Equations can be thought of in terms of a balance. Thus, adding the *same* quantity to each side does not affect the balance. See **FIGURE 1**.

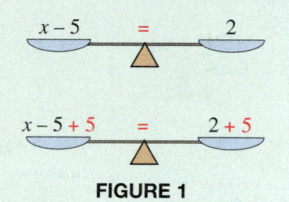

FIGURE 1

 **NOW TRY
EXERCISE 1**

Solve $x - 13 = 4$.

EXAMPLE 1   Applying the Addition Property of Equality

Solve $x - 16 = 7$.

Our goal is to obtain an equivalent equation of the form $x = $ a number.

$$x - 16 = 7$$ To isolate x on the left

$$x - 16 + 16 = 7 + 16$$ side, add 16 to each side.

$$x = 23$$ Combine like terms.

CHECK   Substitute 23 for x in the *original* equation.

$$x - 16 = 7$$ Original equation

$$23 - 16 \stackrel{?}{=} 7$$ Let $x = 23$.

> 7 is *not* the
> solution.

$$7 = 7 ✓$$ True

A true statement results, so 23 is the solution and $\{23\}$ is the solution set.

NOW TRY

NOW TRY ANSWER

1. $\{17\}$

⚠ **CAUTION**   *The final line of the* **CHECK** *does not give the solution of the equation.* It gives a confirmation that the value found is actually a solution.

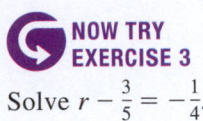
**NOW TRY
EXERCISE 2**
Solve $x - 5.7 = -7.2$.

EXAMPLE 2   Applying the Addition Property of Equality

Solve $x - 2.9 = -6.4$.

> Our goal is to isolate x.

$$x - 2.9 = -6.4$$

$$x - 2.9 + 2.9 = -6.4 + 2.9 \qquad \text{Add 2.9 to each side.}$$

$$x = -3.5$$

CHECK $\qquad x - 2.9 = -6.4 \qquad$ Original equation

$$-3.5 - 2.9 \overset{?}{=} -6.4 \qquad \text{Let } x = -3.5.$$

$$-6.4 = -6.4 \ \checkmark \qquad \text{True}$$

A true statement results, so the solution set is $\{-3.5\}$.   **NOW TRY**

**NOW TRY
EXERCISE 3**
Solve $r - \dfrac{3}{5} = -\dfrac{1}{4}$.

EXAMPLE 3   Applying the Addition Property of Equality

Solve $z - \dfrac{2}{3} = -\dfrac{1}{2}$.

> Other letters besides x can be used as variables.

$$z - \frac{2}{3} = -\frac{1}{2}$$

$$z - \frac{2}{3} + \frac{2}{3} = -\frac{1}{2} + \frac{2}{3} \qquad \text{Add } \tfrac{2}{3} \text{ to each side.}$$

$$z = -\frac{3}{6} + \frac{4}{6} \qquad \begin{array}{l}\text{Add fractions on the left. Write fractions with}\\\text{a common denominator on the right.}\end{array}$$

$$z = \frac{1}{6} \qquad \text{Add fractions.}$$

CHECK $\qquad z - \dfrac{2}{3} = -\dfrac{1}{2} \qquad$ Original equation

$$\frac{1}{6} - \frac{2}{3} \overset{?}{=} -\frac{1}{2} \qquad \text{Let } z = \tfrac{1}{6}.$$

$$\frac{1}{6} - \frac{4}{6} \overset{?}{=} -\frac{3}{6} \qquad \text{Write fractions with a common denominator.}$$

$$-\frac{3}{6} = -\frac{3}{6} \ \checkmark \qquad \text{True}$$

A true statement results, so the solution set is $\left\{\dfrac{1}{6}\right\}$.   **NOW TRY**

Subtraction was previously defined as addition of the opposite. Thus, we can also use the following rule when solving an equation.

Addition Property of Equality Extended to Subtraction

The same number may be *subtracted* from each side of an equation without changing the solution set.

NOW TRY ANSWERS
2. $\{-1.5\}$
3. $\left\{\dfrac{7}{20}\right\}$

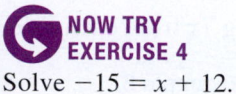 **NOW TRY EXERCISE 4**
Solve $-15 = x + 12$.

EXAMPLE 4 Applying the Addition Property of Equality

Solve $-7 = x + 22$.

Here, the variable x is on the right side of the equation.

$$-7 = x + 22$$ The variable can be isolated on *either* side.

$$-7 - 22 = x + 22 - 22$$ Subtract 22 from each side.

$$-29 = x$$ Rewrite; a number $= x$, or $x =$ a number.

$$x = -29$$

CHECK

$$-7 = x + 22$$ Original equation

$$-7 \stackrel{?}{=} -29 + 22$$ Let $x = -29$.

$$-7 = -7 \checkmark$$ True

A true statement results, so the solution set is $\{-29\}$. NOW TRY

NOTE See Example 4. What happens if we subtract $-7 - 22$ incorrectly, obtaining $x = -15$ (instead of $x = -29$) as the last line of the solution? A check should indicate an error.

CHECK

$$-7 = x + 22$$ Original equation from **Example 4**

$$-7 \stackrel{?}{=} -15 + 22$$ Let $x = -15$.

The left side does *not* equal the right side. → $-7 = 7$ False

The false statement indicates that -15 is *not* a solution of the equation. If this happens, rework the problem.

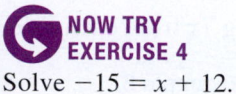 **NOW TRY EXERCISE 5**
Solve $x - 5 = 2x$.

EXAMPLE 5 Subtracting a Variable Term

Solve $6x - 8 = 7x$.

This equation has two variable terms.

$$6x - 8 = 7x$$ We must get the terms with x on the same side of the $=$ symbol.

$$6x - 8 - 6x = 7x - 6x$$ Subtract $6x$ from each side.

$$-8 = x$$ Combine like terms.

CHECK

$$6x - 8 = 7x$$ Original equation

$$6(-8) - 8 \stackrel{?}{=} 7(-8)$$ Let $x = -8$.

Use parentheses when substituting to avoid errors.

$$-48 - 8 \stackrel{?}{=} -56$$ Multiply.

$$-56 = -56 \checkmark$$ True

NOW TRY ANSWERS
4. $\{-27\}$
5. $\{-5\}$

A true statement results, so the solution set is $\{-8\}$. NOW TRY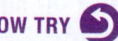

Extra work is required in **Example 5** if we start by subtracting $7x$ from each side.

$$6x - 8 = 7x \qquad \text{Original equation from Example 5}$$

$$6x - 8 - 7x = 7x - 7x \qquad \text{Subtract } 7x \text{ from each side.}$$

$$-8 - x = 0 \qquad \text{Combine like terms.}$$

$$-8 - x + 8 = 0 + 8 \qquad \text{Add 8 to each side.}$$

$$-x = 8 \qquad \text{Combine like terms.}$$

This result gives the value of $-x$, but not of x itself. However, it does say that the additive inverse of x is 8, which means that x must be -8.

$$x = -8 \qquad \text{Same result as in Example 5}$$

We can make the following generalization.

If a is a number and $-x = a$, then $x = -a$.

**NOW TRY
EXERCISE 6**

Solve $\frac{2}{3}x + 4 = \frac{5}{3}x$.

EXAMPLE 6 Subtracting a Variable Term (Fractional Coefficients)

Solve $\frac{3}{5}x + 15 = \frac{8}{5}x$.

$$\frac{3}{5}x + 15 = \frac{8}{5}x$$

$$\frac{3}{5}x + 15 - \frac{3}{5}x = \frac{8}{5}x - \frac{3}{5}x \qquad \text{Subtract } \frac{3}{5}x \text{ from each side.}$$

$$15 = 1x \qquad \frac{3}{5}x - \frac{3}{5}x = 0; \ \frac{8}{5}x - \frac{3}{5}x = \frac{5}{5}x = 1x$$

From now on we will skip this step.

$$15 = x \qquad \text{Multiplicative identity property}$$

Check by substituting 15 for x in the original equation. The solution set is $\{15\}$.

NOW TRY

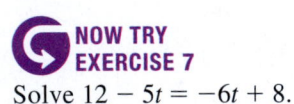
**NOW TRY
EXERCISE 7**

Solve $12 - 5t = -6t + 8$.

EXAMPLE 7 Applying the Addition Property of Equality Twice

Solve $8 - 6p = -7p + 5$.

$$8 - 6p = -7p + 5$$

$$8 - 6p + 7p = -7p + 5 + 7p \qquad \text{Add } 7p \text{ to each side.}$$

$$8 + p = 5 \qquad \text{Combine like terms.}$$

$$8 + p - 8 = 5 - 8 \qquad \text{Subtract 8 from each side.}$$

$$p = -3 \qquad \text{Combine like terms.}$$

CHECK

$$8 - 6p = -7p + 5 \qquad \text{Original equation}$$

$$8 - 6(-3) \stackrel{?}{=} -7(-3) + 5 \qquad \text{Let } p = -3.$$

$$8 + 18 \stackrel{?}{=} 21 + 5 \qquad \text{Multiply.}$$

$$26 = 26 \ \checkmark \qquad \text{True}$$

NOW TRY ANSWERS
6. $\{4\}$
7. $\{-4\}$

A true statement results, so the solution set is $\{-3\}$.

NOW TRY

> **NOTE** *There are often several correct ways to solve an equation.* In the equation
>
> $$8 - 6p = -7p + 5, \qquad \text{See Example 7.}$$
>
> we could begin by adding $6p$ (instead of $7p$) to each side. Combining like terms and sub-tracting 5 from each side gives $3 = -p$. (Try this.) If $3 = -p$, then $-3 = p$, and the variable has been isolated on the right side of the equation. The same solution results.

OBJECTIVE 3 Simplify, and then use the addition property of equality.

NOW TRY EXERCISE 8

Solve.

$$4w - 6 + 2w + 3$$
$$= 7 + 5w - 10$$

EXAMPLE 8 Combining Like Terms When Solving

Solve $3t - 12 + t + 2 = 5 + 3t - 15$.

$$3t - 12 + t + 2 = 5 + 3t - 15$$

$$4t - 10 = -10 + 3t \qquad \text{Combine like terms on each side.}$$

$$4t - 10 - 3t = -10 + 3t - 3t \qquad \text{Subtract } 3t \text{ from each side.}$$

$$t - 10 = -10 \qquad \text{Combine like terms.}$$

$$t - 10 + 10 = -10 + 10 \qquad \text{Add 10 to each side.}$$

The real number 0 can be a solution of an equation. ⟶ $t = 0$ Combine like terms.

CHECK
$$3t - 12 + t + 2 = 5 + 3t - 15 \qquad \text{Original equation}$$

$$3(0) - 12 + 0 + 2 \overset{?}{=} 5 + 3(0) - 15 \qquad \text{Let } t = 0.$$

$$0 - 12 + 0 + 2 \overset{?}{=} 5 + 0 - 15 \qquad \text{Multiply.}$$

$$-10 = -10 \;\checkmark \qquad \text{True}$$

The check confirms that the solution set is $\{0\}$.   **NOW TRY**

NOW TRY EXERCISE 9

Solve.

$$4(3x - 2) - (11x - 4) = 3$$

EXAMPLE 9 Using the Distributive Property When Solving

Solve $3(2 + 5x) - (1 + 14x) = 6$.

Begin by *clearing* (or *removing*) the parentheses.

$$3(2 + 5x) - (1 + 14x) = 6$$

Be sure to distribute to *all* terms within the parentheses.

$$3(2 + 5x) - 1(1 + 14x) = 6 \qquad -(1 + 14x) = -1(1 + 14x)$$

$$3(2) + 3(5x) - 1(1) - 1(14x) = 6 \qquad \text{Distributive property}$$

Be careful here, or a sign error may result.

$$6 + 15x - 1 - 14x = 6 \qquad \text{Multiply.}$$

$$x + 5 = 6 \qquad \text{Combine like terms.}$$

$$x + 5 - 5 = 6 - 5 \qquad \text{Subtract 5 from each side.}$$

$$x = 1 \qquad \text{Combine like terms.}$$

Check by substituting 1 for x in the original equation. The solution set is $\{1\}$.

NOW TRY

NOW TRY ANSWERS
8. $\{0\}$
9. $\{7\}$

2.1 Exercises

     MyLab Math

▶ *Video solutions for select problems available in MyLab Math*

STUDY SKILLS REMINDER
Be sure to read and work through the section material before working the exercises. **Review Study Skill 2, *Reading Your Math Text*.**

Concept Check   *Complete each statement with the correct response. The following terms may be used once, more than once, or not at all.*

linear	expression	solution set	multiplication
equation	addition	equivalent equations	variable

1. A(n) _____ includes an equality symbol, while a(n) _____ does not.

2. A(n) _____ equation in one _____ (here x) is an equation that can be written in the form $ax + b \, (= / \neq) \, 0$.

3. Equations that have exactly the same solution set are _____ .

4. The _____ property of equality states that the same expression may be added to or subtracted from each side of an equation without changing the _____ .

5. Concept Check   Decide whether each of the following is an *expression* or an *equation*. If it is an expression, simplify it. If it is an equation, solve it.

(a) $5x + 8 - 4x + 7$ (b) $-6z + 12 + 7z - 5$

(c) $5x + 8 - 4x = 7$ (d) $-6z + 12 + 7z = -5$

6. Concept Check   Which pairs of equations are equivalent equations?

A. $x + 2 = 6$ and $x = 4$ **B.** $10 - x = 5$ and $x = -5$

C. $x + 3 = 9$ and $x = 6$ **D.** $4 + x = 8$ and $x = -4$

7. Concept Check   Which of the following are *not* linear equations in one variable?

A. $x^2 - 5x + 6 = 0$ **B.** $4x + 3y = 12$

C. $3x - 4 = 0$ **D.** $7x - 6x = 3 + 9x$

8. Explain how to check a solution of an equation.

Solve each equation, and check the solution. **See Examples 1–7.**

9. $x - 3 = 9$ **10.** $x - 9 = 8$ **11.** $x - 12 = 19$

12. $x - 18 = 22$ **13.** $x - 6 = -9$ **14.** $x - 5 = -7$

15. $r + 8 = 12$ **16.** $x + 7 = 11$ **17.** $x + 26 = 17$

18. $x + 47 = 26$ **19.** $x - 8.4 = -2.1$ **20.** $x - 15.5 = -5.1$

21. $t + 12.3 = -4.6$ **22.** $x + 21.5 = -13.4$ **23.** $x + \dfrac{1}{4} = -\dfrac{1}{2}$

24. $x + \dfrac{2}{3} = -\dfrac{1}{6}$ **25.** $k - \dfrac{3}{4} = -\dfrac{2}{3}$ **26.** $p - \dfrac{5}{9} = -\dfrac{1}{2}$

27. $7 + r = -3$ **28.** $8 + k = -4$ **29.** $2 = p + 15$

30. $5 = z + 19$ **31.** $-4 = x - 14$ **32.** $-7 = x - 22$

33. $5.2 = z - 4.9$ **34.** $11.8 = z - 3.6$ **35.** $-\dfrac{1}{3} = x - \dfrac{3}{5}$

36. $-\dfrac{1}{4} = x - \dfrac{2}{3}$ **37.** $8x - 3 = 9x$ **38.** $6x - 4 = 7x$

39. $6t - 2 = 5t$

40. $4z - 6 = 3z$

41. $3x = 2x + 7$

42. $5x = 4x + 9$

43. $10x + 4 = 9x$

44. $8t + 5 = 7t$

45. $\dfrac{2}{5}w - 6 = \dfrac{7}{5}w$

46. $\dfrac{2}{7}z - 2 = \dfrac{9}{7}z$

47. $\dfrac{1}{2}x + 5 = -\dfrac{1}{2}x$

48. $\dfrac{1}{5}x + 7 = -\dfrac{4}{5}x$

49. $5.6x + 2 = 4.6x$

50. $9.1x + 5 = 8.1x$

51. $1.4x - 3 = 0.4x$

52. $1.9t - 6 = 0.9t$

53. $3x + 7 = 2x + 4$

54. $9x + 5 = 8x + 4$

55. $8t + 6 = 7t + 6$

56. $13t + 9 = 12t + 9$

57. $7 - 4x = -5x + 9$

58. $3 - 6x = -7x + 10$

59. $5 - x = -2x - 11$

60. $3 - 8x = -9x - 1$

61. $1.2y - 4 = 0.2y - 4$

62. $7.7r - 6 = 6.7r - 6$

Solve each equation, and check the solution. **See Examples 8 and 9.**

63. $4x - 6 - 3x = 3$

64. $7x - 8 - 6x = 4$

65. $3x + 7 - 2x = 0$

66. $5x + 4 - 4x = 0$

67. $3x + 6 - 10 = 2x - 2$

68. $8x + 4 - 8 = 7x - 1$

69. $5t + 3 + 2t - 6t = 4 + 12$

70. $4x - 6 + 3x - 6x = 3 + 10$

71. $6x + 5 + 7x + 3 = 12x + 4$

72. $4x + 3 + 8x + 1 = 11x + 2$

73. $10x + 5x + 7 - 4 = 12x + 3 + 2x$

74. $7p + 4p + 13 - 7 = 7p + 6 + 3p$

75. $5.2q - 4.6 - 7.1q = -0.9q - 4.6$

76. $4.0x + 2.7 - 9.6x = -4.6x + 2.7$

77. $\dfrac{5}{7}x + \dfrac{1}{3} = \dfrac{2}{5} - \dfrac{2}{7}x + \dfrac{2}{5}$

78. $\dfrac{6}{7}s - \dfrac{3}{4} = \dfrac{4}{5} - \dfrac{1}{7}s + \dfrac{1}{6}$

79. $5(3x - 2) - 14x = 3$

80. $2(3x - 4) - 5x = 7$

81. $13p + 3(2 - 4p) = 4$

82. $21w + 4(6 - 5w) = 8$

83. $(5y + 6) - (3 + 4y) = 10$

84. $(8r + 3) - (1 + 7r) = 6$

85. $2(p + 5) - (9 + p) = -3$

86. $4(k + 6) - (8 + 3k) = -5$

87. $-6(2b + 1) + (13b - 7) = 0$

88. $-5(3w - 3) + (16w + 1) = 0$

89. $10(-2x + 1) = -19(x + 1)$

90. $2(-3r + 2) = -5(r - 3)$

Extending Skills *Solve each equation, and check the solution.*

91. $-2(8p + 2) - 3(2 - 7p) - 2(4 + 2p) = 0$

92. $-5(1 - 2z) + 4(3 - z) - 7(3 + z) = 0$

93. $4(7x - 1) + 3(2 - 5x) - 4(3x + 5) = -6$

94. $9(2m - 3) - 4(5 + 3m) - 5(4 + m) = -3$

Concept Check *Work each problem.*

95. Write an equation that requires the use of the addition property of equality, where 6 must be added to each side and the solution is a negative number.

96. Write an equation that requires the use of the addition property of equality, where $\dfrac{1}{2}$ must be subtracted from each side and the solution is a positive number.

2.2   The Multiplication Property of Equality

OBJECTIVES

1 Use the multiplication property of equality.

2 Simplify, and then use the multiplication property of equality.

OBJECTIVE 1   Use the multiplication property of equality.

The addition property of equality is not sufficient to solve some equations. Consider the following.

$$3x + 2 = 17$$

$$3x + 2 - 2 = 17 - 2 \qquad \text{Subtract 2 from each side.}$$

$$3x = 15 \qquad \text{Combine like terms.}$$

The coefficient of x is 3, not 1 as desired. The **multiplication property of equality** is needed to change $3x = 15$ to an equation of the form

$$x = \textbf{a number.}$$

Because $3x = 15$, both $3x$ and 15 must represent the same number. Multiplying both $3x$ and 15 by the same number will result in an equivalent equation.

Multiplication Property of Equality

If a, b, and c represent real numbers, where $c \neq 0$, then the equations

$$a = b \quad \text{and} \quad ac = bc \quad \text{are equivalent.}$$

That is, each side of an equation may be multiplied by the same nonzero number without changing the solution set.

To isolate the variable in $3x = 15$, we must change $3x$ to $1x$, or x. To do this, we multiply each side of the equation by $\frac{1}{3}$, the *reciprocal* of 3, because $\frac{1}{3} \cdot 3 = \frac{3}{3} = 1$.

$$3x = 15$$

$$\frac{1}{3} \cdot 3x = \frac{1}{3} \cdot 15 \qquad \text{Multiply each side by } \frac{1}{3}, \text{ the reciprocal of 3.}$$

$$\left(\frac{1}{3} \cdot 3 \right) x = \frac{1}{3} \cdot 15 \qquad \text{Associative property}$$

The product of a number and its reciprocal is **1**.

$$1x = 5 \qquad \text{Multiplicative inverse property}$$

$$x = 5 \qquad \text{Multiplicative identity property}$$

The solution is 5. We can check this result in the original equation.

Just as the addition property of equality permits *subtracting* the same number from each side of an equation, the multiplication property of equality permits *dividing* each side of an equation by the same nonzero number.

$$3x = 15$$

$$\frac{3x}{3} = \frac{15}{3} \qquad \text{Divide each side by 3.}$$

$$x = 5 \qquad \text{Same result as above}$$

Multiplication Property of Equality Extended to Division

We can divide each side of an equation by the same nonzero number without changing the solution. *Do not, however, divide each side by a variable, because the variable might be equal to 0.*

> **NOTE**   It is usually easier to multiply on each side of an equation if the coefficient of the variable is a fraction, and to divide on each side if the coefficient is an integer.
>
> To solve $\frac{3}{4}x = 12$,   it is easier to multiply by $\frac{4}{3}$ than to divide by $\frac{3}{4}$.
>
> To solve $5x = 20$,   it is easier to divide by 5 than to multiply by $\frac{1}{5}$.

**NOW TRY
EXERCISE 1**

Solve $8x = 80$.

EXAMPLE 1   Applying the Multiplication Property of Equality

Solve $5x = 60$.

$$5x = 60 \qquad \text{Our goal is to isolate } x.$$

$$\frac{5x}{5} = \frac{60}{5} \qquad \text{Divide each side by 5, the coefficient of } x.$$

Dividing by 5 is the same as multiplying by $\frac{1}{5}$.

$$x = 12 \qquad \frac{5x}{5} = \frac{5}{5}x = 1x = x$$

CHECK   Substitute 12 for x in the original equation.

$$5x = 60 \qquad \text{Original equation}$$

$$5(12) \stackrel{?}{=} 60 \qquad \text{Let } x = 12.$$

60 is *not* the solution.

$$60 = 60 \quad \checkmark \quad \text{True}$$

A true statement results, so the solution set is $\{12\}$.

NOW TRY

**NOW TRY
EXERCISE 2**

Solve $-10x = 24$.

EXAMPLE 2   Applying the Multiplication Property of Equality

Solve $-25x = 30$.

$$-25x = 30$$

To avoid errors later, show the division as a separate step.

$$\frac{-25x}{-25} = \frac{30}{-25} \qquad \text{Divide each side by } -25, \text{ the coefficient of } x.$$

$$x = -\frac{30}{25} \qquad \frac{a}{-b} = -\frac{a}{b}$$

$$x = -\frac{6}{5} \qquad \text{Write in lowest terms.}$$

CHECK

$$-25x = 30 \qquad \text{Original equation}$$

$$\frac{-25}{1}\left(-\frac{6}{5}\right) \stackrel{?}{=} 30 \qquad \text{Let } x = -\frac{6}{5}.$$

$$30 = 30 \quad \checkmark \quad \text{True}$$

NOW TRY ANSWERS

1. $\{10\}$

2. $\left\{-\frac{12}{5}\right\}$

The check confirms that the solution set is $\left\{-\frac{6}{5}\right\}$.

NOW TRY

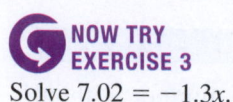

**NOW TRY
EXERCISE 3**

Solve $7.02 = -1.3x$.

EXAMPLE 3 Solving a Linear Equation (Decimal Coefficient)

Solve $6.09 = -2.1x$.

$$6.09 = -2.1x \quad \boxed{\text{Isolate } x \text{ on the right.}}$$

$$\frac{6.09}{-2.1} = \frac{-2.1x}{-2.1} \qquad \text{Divide each side by } -2.1.$$

$$-2.9 = x, \quad \text{or} \quad x = -2.9$$

Check by replacing x with -2.9 in the original equation. The solution set is $\{-2.9\}$.

NOW TRY

**NOW TRY
EXERCISE 4**

Solve $\frac{x}{5} = -7$.

EXAMPLE 4 Solving a Linear Equation (Fractional Coefficient)

Solve $\frac{x}{4} = 3$.

$$\frac{x}{4} = 3$$

$$\frac{1}{4}x = 3 \qquad\qquad \frac{x}{4} = \frac{1x}{4} = \frac{1}{4}x$$

$$4 \cdot \frac{1}{4}x = 4 \cdot 3 \qquad \begin{array}{l}\text{Multiply each side by } \frac{4}{1}, \text{ or } 4,\\ \text{the reciprocal of } \frac{1}{4}.\end{array}$$

$$\boxed{4 \cdot \frac{1}{4}x = 1x = x} \quad x = 12 \qquad \begin{array}{l}\text{Multiplicative inverse property;}\\ \text{multiplicative identity property}\end{array}$$

CHECK

$$\frac{x}{4} = 3 \qquad\qquad \text{Original equation}$$

$$\frac{12}{4} \overset{?}{=} 3 \qquad\qquad \text{Let } x = 12.$$

$$3 = 3 \quad \checkmark \quad \text{True}$$

A true statement results, so the solution set is $\{12\}$.

NOW TRY

**NOW TRY
EXERCISE 5**

Solve $-\frac{4}{7}z = -16$.

EXAMPLE 5 Solving a Linear Equation (Fractional Coefficient)

Solve $-\frac{3}{4}w = -6$.

$$-\frac{3}{4}w = -6$$

$$-\frac{4}{3} \cdot \left(-\frac{3}{4}w\right) = -\frac{4}{3} \cdot (-6) \qquad \begin{array}{l}\text{Multiply each side by } -\frac{4}{3},\\ \text{the reciprocal of } -\frac{3}{4}.\end{array}$$

$$\boxed{\text{Reciprocals have the }\textit{same}\text{ sign.}}$$

$$1 \cdot w = -\frac{4}{3} \cdot \left(-\frac{6}{1}\right) \qquad \text{Multiplicative inverse property}$$

$$w = 8 \qquad \begin{array}{l}\text{Multiplicative inverse property;}\\ \text{Multiply fractions.}\end{array}$$

Check to confirm that the solution set is $\{8\}$.

NOW TRY

NOW TRY ANSWERS

3. $\{-5.4\}$
4. $\{-35\}$
5. $\{28\}$

NOTE We can use reasoning to solve an equation such as $-x = 8$. Because this equation says that the additive inverse (or opposite) of x is 8, x must equal -8. We can also use the multiplication property of equality to obtain the same result. This is done in **Example 6.**

 **NOW TRY**
EXERCISE 6
Solve $-x = -9$.

EXAMPLE 6   Applying the Multiplication Property of Equality

Solve $-x = 8$.

$$-x = 8$$
$$-1x = 8 \qquad \qquad -x = -1x$$
$$-1(-1x) = -1(8) \qquad \text{Multiply each side by } -1.$$
$$[-1(-1)]x = -8 \qquad \text{Associative property; Multiply.}$$

> These steps are usually omitted.

$$1x = -8 \qquad \text{Multiplicative inverse property}$$
$$x = -8 \qquad \text{Multiplicative identity property}$$

CHECK
$$-x = 8 \qquad \text{Original equation}$$
$$-(-8) \stackrel{?}{=} 8 \qquad \text{Let } x = -8.$$
$$8 = 8 \quad \checkmark \qquad \text{True}$$

A true statement results, so $\{-8\}$ is the solution set.   **NOW TRY**

OBJECTIVE 2   Simplify, and then use the multiplication property of equality.

 **NOW TRY**
EXERCISE 7
Solve $9n - 6n = 21$.

EXAMPLE 7   Combining Like Terms When Solving

Solve $5m + 6m = 33$.

$$5m + 6m = 33$$
$$11m = 33 \qquad \text{Combine like terms.}$$
$$\frac{11m}{11} = \frac{33}{11} \qquad \text{Divide by 11.}$$
$$m = 3 \qquad \text{Multiplicative identity property; Divide.}$$

CHECK
$$5m + 6m = 33 \qquad \text{Original equation}$$
$$5(3) + 6(3) \stackrel{?}{=} 33 \qquad \text{Let } m = 3.$$
$$15 + 18 \stackrel{?}{=} 33 \qquad \text{Multiply.}$$
$$33 = 33 \quad \checkmark \qquad \text{True}$$

The check confirms that the solution set is $\{3\}$.   **NOW TRY**

EXAMPLE 8   Using the Distributive Property When Solving

Solve $2(6x - 5) + 10 = -24$.

Begin by clearing the parentheses.

$$2(6x - 5) + 10 = -24$$
$$2(6x) + 2(-5) + 10 = -24 \qquad \text{Distributive property}$$
$$12x - 10 + 10 = -24 \qquad \text{Multiply.}$$
$$12x = -24 \qquad \text{Combine like terms.}$$
$$\frac{12x}{12} = \frac{-24}{12} \qquad \text{Divide each side by 12.}$$
$$x = -2$$

NOW TRY ANSWERS
6. $\{9\}$
7. $\{7\}$

NOW TRY
EXERCISE 8
Solve $3(5x - 4) + 12 = -15$.

CHECK

$$2(6x - 5) + 10 = -24 \qquad \text{Original equation}$$

$$2[6(-2) - 5] + 10 \overset{?}{=} -24 \qquad \text{Let } x = -2.$$

$$2[-17] + 10 \overset{?}{=} -24 \qquad \begin{array}{l}\text{Multiply; then add} \\ \text{within the brackets.}\end{array}$$

$$-34 + 10 \overset{?}{=} -24 \qquad \text{Multiply.}$$

$$-24 = -24 \quad \checkmark \quad \text{True}$$

NOW TRY ANSWER
8. $\{-1\}$

The check confirms that the solution set is $\{-2\}$.

NOW TRY

2.2 Exercises

FOR EXTRA HELP   **MyLab Math**

 Video solutions for select problems available in MyLab Math

1. Concept Check Tell whether to use the *addition property of equality* or the *multiplication property of equality* to solve each equation. *Do not actually solve.*

(a) $3x = 12$ **(b)** $3 + x = 12$ **(c)** $-x = 4$ **(d)** $-12 = 6 + x$

Concept Check *Choose the letter of the correct response.*

2. Which equation does *not* require the use of the multiplication property of equality?

A. $3x - 5x = 6$ **B.** $-\dfrac{1}{4}x = 12$ **C.** $5x - 4x = 7$ **D.** $\dfrac{x}{3} = -2$

3. In the solution of a linear equation, the next-to-the-last step reads "$-x = -\dfrac{3}{4}$." Which of the following is the solution of this equation?

A. $-\dfrac{3}{4}$ **B.** $\dfrac{3}{4}$ **C.** -1 **D.** $\dfrac{4}{3}$

4. Which of the following is the solution of the equation $-x = -24$?

A. 24 **B.** -24 **C.** 1 **D.** -1

Concept Check *By what number is it necessary to multiply both sides of each equation to isolate x on the left side? Do not actually solve.*

5. $\dfrac{4}{5}x = 8$ **6.** $\dfrac{2}{3}x = 6$ **7.** $\dfrac{x}{10} = 5$ **8.** $\dfrac{x}{100} = 10$

9. $-\dfrac{9}{2}x = -4$ **10.** $-\dfrac{8}{3}x = -11$ **11.** $-x = 0.75$ **12.** $-x = 0.48$

Concept Check *By what number is it necessary to divide both sides of each equation to isolate x on the left side? Do not actually solve.*

13. $6x = 12$ **14.** $7x = 35$ **15.** $-4x = 16$ **16.** $-13x = 26$

17. $0.12x = 48$ **18.** $0.21x = 63$ **19.** $-x = 25$ **20.** $-x = 50$

Solve each equation, and check the solution. See Examples 1–6.

21. $6x = 36$ **22.** $8x = 64$ **23.** $2m = 15$ **24.** $3m = 10$

25. $3a = -15$ **26.** $5x = -60$ **27.** $-7x = 28$ **28.** $-9x = 36$

29. $10t = -36$ **30.** $10s = -54$ **31.** $-6x = -72$ **32.** $-4x = -64$

33. $4r = 0$ **34.** $7x = 0$ **35.** $-x = 12$ **36.** $-t = 14$

37. $-x = -\dfrac{3}{4}$ **38.** $-x = -\dfrac{1}{2}$ **39.** $0.2t = 8$ **40.** $0.9x = 18$

41. $-0.3x = 9$ **42.** $-0.5x = 20$ **43.** $0.6x = -1.44$ **44.** $0.8x = -2.96$

45. $-9.1 = -2.6x$ **46.** $-7.2 = -4.5x$ **47.** $-2.1m = 25.62$ **48.** $-3.9x = 32.76$

49. $\dfrac{1}{4}x = -12$ **50.** $\dfrac{1}{5}p = -3$ **51.** $\dfrac{z}{6} = 12$ **52.** $\dfrac{x}{5} = 15$

53. $\dfrac{x}{7} = -5$ **54.** $\dfrac{r}{8} = -3$ **55.** $\dfrac{2}{7}p = 4$ **56.** $\dfrac{3}{8}x = 9$

57. $-\dfrac{5}{6}t = -15$ **58.** $-\dfrac{3}{4}z = -21$ **59.** $-\dfrac{7}{9}x = \dfrac{3}{5}$ **60.** $-\dfrac{5}{6}x = \dfrac{4}{9}$

Solve each equation, and check the solution. **See Examples 7 and 8.**

61. $4x + 3x = 21$ **62.** $8x + 3x = 121$ **63.** $6r - 8r = 10$

64. $3p - 7p = 24$ **65.** $\dfrac{2}{5}x - \dfrac{3}{10}x = 2$ **66.** $\dfrac{2}{3}x - \dfrac{5}{9}x = 4$

67. $5m + 6m - 2m = 63$ **68.** $9r + 2r - 7r = 68$ **69.** $-6x + 4x - 7x = 0$

70. $-5x + 4x - 8x = 0$ **71.** $8w - 4w + w = -3$ **72.** $9x - 3x + x = -4$

73. $\dfrac{1}{3}x - \dfrac{1}{4}x + \dfrac{1}{12}x = 3$ **74.** $\dfrac{2}{5}x + \dfrac{1}{10}x - \dfrac{1}{20}x = 18$

75. $0.9w - 0.5w + 0.1w = -3$ **76.** $0.5x - 0.6x + 0.3x = -1$

77. $4(3x - 1) + 4 = -36$ **78.** $5(3x + 2) - 10 = 30$

79. $-4(x + 2) + 8 = 16$ **80.** $-6(4x + 3) + 18 = -48$

Concept Check   *Work each problem.*

81. Write an equation that requires the use of the multiplication property of equality, where each side must be multiplied by $\frac{2}{3}$ and the solution is a negative number.

82. Write an equation that requires the use of the multiplication property of equality, where each side must be divided by 100 and the solution is not an integer.

2.3 Solving Linear Equations Using Both Properties of Equality

OBJECTIVES

1 Use the four steps for solving a linear equation.

2 Solve equations with no solution or infinitely many solutions.

3 Write expressions for two related unknown quantities.

OBJECTIVE 1 Use the four steps for solving a linear equation.

We now apply *both* properties of equality to solve linear equations.

Solving a Linear Equation in One Variable

Step 1 **Simplify each side separately.** Use the distributive property as needed.
- Clear any parentheses.
- Combine like terms.

Step 2 **Isolate the variable terms on one side.** Use the addition property of equality so that all terms with variables are on one side of the equation and all constants (numbers) are on the other side.

Step 3 **Isolate the variable.** Use the multiplication property of equality to obtain an equation that has just the variable with coefficient 1 on one side.

Step 4 **Check.** Substitute the value found into the *original* equation. If a true statement results, write the solution set. If not, rework the problem.

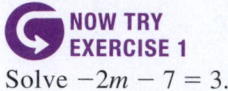

**NOW TRY
EXERCISE 1**

Solve $-2m - 7 = 3$.

Remember that when we solve an equation, our primary goal is to isolate the variable on one side of the equation.

EXAMPLE 1   Solving a Linear Equation

Solve $-6x + 5 = 17$.

Step 1   There are no parentheses, nor are there like terms to combine on one side in this equation. This step is not necessary.

> Our goal is to isolate x. $-6x + 5 = 17$

Step 2 $-6x + 5 - 5 = 17 - 5$ Subtract 5 from each side.

$-6x = 12$ Combine like terms.

Step 3 $\dfrac{-6x}{-6} = \dfrac{12}{-6}$ Divide each side by -6.

$x = -2$

Step 4   Check by substituting -2 for x in the original equation.

CHECK $-6x + 5 = 17$ Original equation

$-6(-2) + 5 \overset{?}{=} 17$ Let $x = -2$.

$12 + 5 \overset{?}{=} 17$ Multiply.

> 17 is *not* the solution. $17 = 17$ ✓   True

The check confirms that -2 is the solution. The solution set is $\{-2\}$.   **NOW TRY**

EXAMPLE 2   Solving a Linear Equation

Solve $3x + 2 = 5x - 8$.

Step 1   There are no parentheses, nor are there like terms to combine on one side in this equation. We begin with Step 2.

Step 2   To isolate the variable terms on one side of the equation, we can subtract either $3x$ or $5x$ from each side. We suggest subtracting $3x$ so that the coefficient of the variable remains positive.

> Our goal is to isolate x. $3x + 2 = 5x - 8$

$3x + 2 - 3x = 5x - 8 - 3x$ Subtract $3x$ from each side.

$2 = 2x - 8$ Combine like terms.

$2 + 8 = 2x - 8 + 8$ Add 8 to each side.

$10 = 2x$ Combine like terms.

Step 3 $\dfrac{10}{2} = \dfrac{2x}{2}$ Divide each side by 2.

$5 = x$

NOW TRY ANSWER
1. $\{-5\}$

Step 4   Check by substituting 5 for x in the original equation.

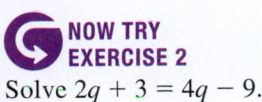
NOW TRY
EXERCISE 2
Solve $2q + 3 = 4q - 9$.

CHECK

$$3x + 2 = 5x - 8 \quad \text{Original equation}$$

$$3(5) + 2 \overset{?}{=} 5(5) - 8 \quad \text{Let } x = 5.$$

$$15 + 2 \overset{?}{=} 25 - 8 \quad \text{Multiply.}$$

$$17 = 17 \ \checkmark \quad \text{True}$$

The check confirms that 5 is the solution. The solution set is $\{5\}$.   **NOW TRY**

NOTE *Remember that the variable can be isolated on either side of an equation. In* **Example 2**, *x will be isolated on the left if we begin by subtracting* $5x$.

$$3x + 2 = 5x - 8 \quad \text{Equation from Example 2}$$

$$3x + 2 - 5x = 5x - 8 - 5x \quad \text{Subtract } 5x \text{ from each side.}$$

$$-2x + 2 = -8 \quad \text{Combine like terms.}$$

$$-2x + 2 - 2 = -8 - 2 \quad \text{Subtract 2 from each side.}$$

$$-2x = -10 \quad \text{Combine like terms.}$$

$$\frac{-2x}{-2} = \frac{-10}{-2} \quad \text{Divide each side by } -2.$$

$$x = 5 \quad \text{The same solution results.}$$

There are often several equally correct ways to solve an equation.

NOW TRY
EXERCISE 3
Solve.

$3(z - 6) - 5z = -7z + 7$

EXAMPLE 3 Solving a Linear Equation

Solve $4(k - 3) - k = k - 6$.

Step 1 Clear parentheses using the distributive property.

$$4(k - 3) - k = k - 6$$

$$4(k) + 4(-3) - k = k - 6 \quad \text{Distributive property}$$

$$4k - 12 - k = k - 6 \quad \text{Multiply.}$$

$$3k - 12 = k - 6 \quad \text{Combine like terms.}$$

Step 2
$$3k - 12 - k = k - 6 - k \quad \text{Subtract } k.$$

$$2k - 12 = -6 \quad \text{Combine like terms.}$$

$$2k - 12 + 12 = -6 + 12 \quad \text{Add 12.}$$

$$2k = 6 \quad \text{Combine like terms.}$$

Step 3
$$\frac{2k}{2} = \frac{6}{2} \quad \text{Divide by 2.}$$

$$k = 3$$

Step 4 CHECK
$$4(k - 3) - k = k - 6 \quad \text{Original equation}$$

$$4(3 - 3) - 3 \overset{?}{=} 3 - 6 \quad \text{Let } k = 3.$$

$$4(0) - 3 \overset{?}{=} 3 - 6 \quad \text{Work inside the parentheses.}$$

$$-3 = -3 \ \checkmark \quad \text{True}$$

NOW TRY ANSWERS
2. $\{6\}$
3. $\{5\}$

A true statement results, so the solution set is $\{3\}$.   **NOW TRY**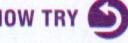

**NOW TRY
EXERCISE 4**

Solve.

$$5x - (x + 9) = x - 4$$

EXAMPLE 4   Solving a Linear Equation

Solve $8z - (3 + 2z) = 3z + 1$.

Step 1

$$8z - (3 + 2z) = 3z + 1$$

$$8z - 1(3 + 2z) = 3z + 1 \qquad \text{Multiplicative identity property}$$

$$8z - 1(3) - 1(2z) = 3z + 1 \qquad \text{Distributive property}$$

$$8z - 3 - 2z = 3z + 1 \qquad \text{Multiply.}$$

> Be careful with signs.

$$6z - 3 = 3z + 1 \qquad \text{Combine like terms.}$$

Step 2

$$6z - 3 - 3z = 3z + 1 - 3z \qquad \text{Subtract } 3z.$$

$$3z - 3 = 1 \qquad \text{Combine like terms.}$$

$$3z - 3 + 3 = 1 + 3 \qquad \text{Add 3.}$$

$$3z = 4 \qquad \text{Combine like terms.}$$

Step 3

$$\frac{3z}{3} = \frac{4}{3} \qquad \text{Divide by 3.}$$

$$z = \frac{4}{3}$$

Step 4   Check by substituting $\frac{4}{3}$ for z in the original equation.

CHECK

$$8z - (3 + 2z) = 3z + 1 \qquad \text{Original equation}$$

$$8\left(\frac{4}{3}\right) - \left[3 + 2\left(\frac{4}{3}\right)\right] \overset{?}{=} 3\left(\frac{4}{3}\right) + 1 \qquad \text{Let } z = \frac{4}{3}.$$

$$\frac{32}{3} - \left[3 + \frac{8}{3}\right] \overset{?}{=} 4 + 1 \qquad \text{Multiply.}$$

$$\frac{32}{3} - \frac{17}{3} \overset{?}{=} 5 \qquad 3 = \frac{9}{3}; \frac{9}{3} + \frac{8}{3} = \frac{17}{3}$$

$$5 = 5 \quad \checkmark \qquad \text{True}$$

A true statement results, so the solution set is $\left\{\frac{4}{3}\right\}$.   **NOW TRY**

⚠ **CAUTION**   In an expression such as $8z - (3 + 2z)$, the $-$ sign acts like a factor of -1 and affects the sign of *every* term within the parentheses.

$$8z - (3 + 2z) \leftarrow \text{Left side of the equation in } \textbf{Example 4}$$

$$= 8z - 1(3 + 2z)$$

$$= 8z + (-1)(3 + 2z)$$

$$= 8z - 3 - 2z$$

Change to $-$ in *both* terms.

NOW TRY
EXERCISE 5

Solve.

$4(t - 1) = 24 - 4(7 - 2t)$

EXAMPLE 5 Solving a Linear Equation

Solve $4(4 - 3x) = 32 - 8(x + 2)$.

> *Do not subtract 8 from 32 here.*

Step 1

$4(4 - 3x) = 32 - 8(x + 2)$ *Be careful with signs.*

Multiply mentally. $16 - 12x = 32 - 8x - 16$ Distributive property

$16 - 12x = 16 - 8x$ Combine like terms.

Step 2

$16 - 12x + 12x = 16 - 8x + 12x$ Add $12x$.

Adding $12x$ keeps a positive coefficient on x.

$16 = 16 + 4x$ Combine like terms.

$16 - 16 = 16 + 4x - 16$ Subtract 16.

$0 = 4x$ Combine like terms.

Step 3

$\dfrac{0}{4} = \dfrac{4x}{4}$ Divide by 4.

$0 = x$

Step 4   **CHECK**

$4(4 - 3x) = 32 - 8(x + 2)$ Original equation

$4[4 - 3(0)] \overset{?}{=} 32 - 8(0 + 2)$ Let $x = 0$.

$4(4 - 0) \overset{?}{=} 32 - 8(2)$ Multiply and add.

$4(4) \overset{?}{=} 32 - 16$ Subtract and multiply.

$16 = 16$ ✓   True

A true statement results, so the solution set is $\{0\}$.   **NOW TRY**

> **NOTE** It is perfectly acceptable for an equation to have solution set $\{0\}$.

OBJECTIVE 2 Solve equations with no solution or infinitely many solutions.

Each equation so far has had exactly one solution. An equation with exactly one solution is a **conditional equation** because it is true only under certain conditions. Some equations may have no solution or infinitely many solutions.

EXAMPLE 6 Solving an Equation That Has Infinitely Many Solutions

Solve $5x - 15 = 5(x - 3)$.

$5x - 15 = 5(x - 3)$

$5x - 15 = 5x - 15$ Distributive property

$5x - 15 - 5x = 5x - 15 - 5x$ Subtract $5x$.

Notice that the variable "disappeared." $-15 = -15$ Combine like terms.

$-15 + 15 = -15 + 15$ Add 15.

$0 = 0$ True

Solution set: {all real numbers}

NOW TRY ANSWER
5. $\{0\}$

NOW TRY
EXERCISE 6
Solve.

$$-3(x - 7) = -3x + 21$$

Because the last statement $(0 = 0)$ in **Example 6** is true, *any* real number is a solution. We could have predicted this from the second line in the solution.

$$5x - 15 = 5x - 15 \;\longleftarrow\; \text{This is true for } any \text{ value of } x.$$

Try several values for x in the original equation to see that they all satisfy it.

An equation with both sides exactly the same, like $0 = 0$, is an **identity.** An identity is true for *all* replacements of the variables. As shown above, we write the solution set as **{all real numbers}.**   **NOW TRY**

⚠ CAUTION   In **Example 6,** do not write $\{0\}$ as the solution set. While 0 is a solution, there are infinitely many other solutions. *For the solution set to be {0}, the last line must include a variable, such as x, and read x = 0 (as in Example 5), not 0 = 0.*

NOW TRY
EXERCISE 7
Solve.

$$x + 4(x - 3) = 3 + 5x$$

| **EXAMPLE 7**   Solving an Equation That Has No Solution |

Solve $2x + 3(x + 1) = 5x + 4$.

$$2x + 3(x + 1) = 5x + 4$$

$$2x + 3x + 3 = 5x + 4 \qquad \text{Distributive property}$$

$$5x + 3 = 5x + 4 \qquad \text{Combine like terms.}$$

$$5x + 3 - 5x = 5x + 4 - 5x \qquad \text{Subtract } 5x.$$

Again, the variable "disappeared." → $3 = 4 \qquad$ **False**

There is no solution.   Solution set: ∅

A false statement $(3 = 4)$ results, so there is no solution. We could have predicted this from the third line in the solution.

$$5x + 3 = 5x + 4 \;\longleftarrow\; \text{This can } never \text{ be true.}$$

A **contradiction** is an equation that has no solution. Its solution set is the **empty set,** or **null set,** symbolized ∅.   **NOW TRY**

⚠ CAUTION   Do not write $\{\varnothing\}$ to represent the empty set. $\{\varnothing\}$ is not empty. It is the set containing the symbol ∅.

▼ **Summary of Solution Sets of Equations**

Type of Equation	Final Equation in Solution	Number of Solutions	Solution Set
Conditional (See Examples 1–5.)	$x =$ a number	One	{a number}
Identity (See Example 6.)	A true statement with no variable, such as $0 = 0$	Infinitely many	{all real numbers}
Contradiction (See Example 7.)	A false statement with no variable, such as $3 = 4$	None	∅

NOW TRY ANSWERS
6. {all real numbers}
7. ∅

OBJECTIVE 3 Write expressions for two related unknown quantities.

NOW TRY
EXERCISE 8

Two numbers have a sum of 18. If one of the numbers is represented by m, find an expression for the other number.

EXAMPLE 8 Translating a Phrase into an Algebraic Expression

Perform each translation.

(a) Two numbers have a sum of 23. If one of the numbers is represented by x, find an expression for the other number.

First, suppose that the sum of two numbers is 23, and one of the numbers is 10. To find the other number, we would subtract 10 from 23.

$$23 - 10 \leftarrow \text{This gives 13 as the other number.}$$

Instead of using 10 as one of the numbers, we use x. The other number would be obtained in the same way—by subtracting x from 23.

$$23 - x. \quad \boxed{x - 23 \text{ is not correct.} \atop \text{Subtraction is not commutative.}}$$

CHECK We find the sum of the two numbers.

$$x + (23 - x)$$
$$= 23, \quad \text{as required.} \quad \checkmark$$

(b) Two numbers have a product of 24. If one of the numbers is represented by x, find an expression for the other number.

Suppose that one of the numbers is 4. To find the other number, we would divide 24 by 4.

$$\frac{24}{4} \quad \leftarrow \text{This gives 6 as the other number.} \atop \text{The product } 6 \cdot 4 \text{ is 24.}$$

In the same way, if x is one of the numbers, then we divide 24 by x to find the other number.

NOW TRY ANSWER
8. $18 - m$

$$\frac{24}{x} \quad \leftarrow \text{The other number}$$ **NOW TRY**

2.3 Exercises

FOR EXTRA HELP   ▶ **MyLab Math**

▶ *Video solutions for select problems available in MyLab Math*

STUDY SKILLS REMINDER
Study cards are a great way to learn vocabulary, procedures, etc. **Review Study Skill 5,**
Using Study Cards.

Concept Check *Using the methods of this section, what should we do first when solving each equation? Do not actually solve.*

1. $7x + 8 = 1$ **2.** $7x - 5x + 15 = 8 + x$

3. $3(2t - 4) = 20 - 2t$ **4.** $-5z = -15$

5. Concept Check Suppose that when solving three linear equations, we obtain the final results shown in parts (a)–(c). Fill in the blanks in parts (a)–(c), and then match each result with the solution set in choices A–C for the *original* equation.

(a) $6 = 6$ (The original equation is a(n) _____.)   **A.** $\{0\}$

(b) $x = 0$ (The original equation is a(n) _____ equation.)   **B.** {all real numbers}

(c) $-5 = 0$ (The original equation is a(n) _____.)   **C.** $\varnothing$

6. Concept Check Which equation does *not* have {all real numbers} as its solution set?

A. $5x = 4x + x$ **B.** $2(x + 6) = 2x + 12$ **C.** $\frac{1}{2}x = 0.5x$ **D.** $3x = 2x$

Solve each equation, and check the solution. ***See Examples 1–7.***

7. $3x + 2 = 14$

8. $4x + 3 = 27$

9. $-5z - 4 = 21$

10. $-7w - 4 = 10$

11. $4p - 5 = 2p$

12. $6q - 2 = 3q$

13. $2x + 5 = 4x - 3$

14. $7p + 8 = 9p - 2$

15. $5m + 8 = 7 + 3m$

16. $4r + 9 = 5 + r$

17. $-12x - 5 = 10 - 7x$

18. $-16w - 3 = 13 - 8w$

19. $12h - 5 = 11h + 5 - h$

20. $-4x - 1 = -5x + 1 + 3x$

21. $7r - 5r + 2 = 5r + 2 - r$

22. $9p - 4p + 6 = 7p + 6 - 3p$

23. $5(p - 1) - p = 2p - 7$

24. $6(z - 3) - 2z = z - 27$

25. $3(4x + 2) + 5x = 30 - x$

26. $5(2m + 3) - 4m = 2m + 25$

27. $6x - (2 + 3x) = x + 1$

28. $7x - (3 + 2x) = 2x + 5$

29. $-2p + 7 = 3 - (5p + 1)$

30. $4x + 9 = 3 - (x - 2)$

31. $-(8x - 2) + 5x - 6 = -4$

32. $-(7x - 5) + 4x - 7 = -2$

33. $6(3 - x) = -6x + 18$

34. $9(2 - p) = -9p + 18$

35. $11x - 5(x + 2) = 6x + 5$

36. $6x - 4(x + 1) = 2x + 4$

37. $24 - 4(7 - 2t) = 4(t - 1)$

38. $8 - 2(2 - x) = 4(x + 1)$

39. $6(3w + 5) = 2(10w + 10)$

40. $4(2x - 1) = -6(x + 3)$

41. $10(2x - 1) = 8(2x + 1) + 14$

42. $9(3k - 5) = 12(3k - 1) - 51$

43. $-(4x + 2) - (-3x - 5) = 3$

44. $-(6k - 5) - (-5k + 8) = -3$

45. $3(2x - 4) = 6(x - 2)$

46. $3(6 - 4x) = 2(-6x + 9)$

47. $6(4x - 1) = 12(2x + 3)$

48. $6(2x + 8) = 4(3x - 6)$

49. $4(x + 8) = 2(2x + 6) + 20$

50. $4(x + 3) = 2(2x + 8) - 4$

51. $9(v + 1) - 3v = 2(3v + 1) - 8$

52. $8(t - 3) + 4t = 6(2t + 1) - 10$

Perform each translation. ***See Example 8.***

53. Two numbers have a sum of 11. One of the numbers is q. What expression represents the other number?

54. Two numbers have a sum of 34. One of the numbers is r. What expression represents the other number?

55. The product of two numbers is 9. One of the numbers is x. What expression represents the other number?

56. The product of two numbers is -6. One of the numbers is m. What expression represents the other number?

57. A baseball player got 65 hits one season. He got h of the hits in one game. What expression represents the number of hits he got in the rest of the games?

58. A hockey player scored 42 goals in one season. He scored n goals in one game. What expression represents the number of goals he scored in the rest of the games?

59. Monica is x years old. What expression represents her age 15 yr from now?   5 yr ago?

60. Chandler is y years old. What expression represents his age 4 yr ago?   11 yr from now?

61. Cliff has r quarters. Express the value of the quarters in cents.

62. Claire has y dimes. Express the value of the dimes in cents.

63. A clerk has t dollars, all in \$5 bills. What expression represents the number of \$5 bills?

64. A clerk has v dollars, all in \$10 bills. What expression represents the number of \$10 bills?

2.4 Clearing Fractions and Decimals When Solving Linear Equations

OBJECTIVES

1 Solve equations with fractions as coefficients.

2 Solve equations with decimals as coefficients.

OBJECTIVE 1 Solve equations with fractions as coefficients.

We solved some simple equations that involved fractions, two of which follow, in earlier sections.

$$z - \frac{2}{3} = -\frac{1}{2}$$

$$z - \frac{2}{3} + \frac{2}{3} = -\frac{1}{2} + \frac{2}{3} \qquad \text{Add } \frac{2}{3}.$$

$$z = -\frac{3}{6} + \frac{4}{6} \qquad \begin{array}{l} \text{Write} \\ \text{fractions with} \\ \text{a common} \\ \text{denominator.} \end{array}$$

$$z = \frac{1}{6}$$

The solution set is $\left\{ \frac{1}{6} \right\}$.

$$-\frac{3}{4}w = -6$$

$$-\frac{4}{3} \cdot \left(-\frac{3}{4}w \right) = -\frac{4}{3} \cdot (-6) \qquad \begin{array}{l} \text{Multiply} \\ \text{by } -\frac{4}{3}\text{, the} \\ \text{reciprocal} \\ \text{of } -\frac{3}{4}. \end{array}$$

$$1 \cdot w = -\frac{4}{3} \cdot \left(-\frac{6}{1} \right)$$

$$w = 8$$

The solution set is $\{8\}$.

In more complicated equations that involve fractions, such as

$$\frac{x}{4} - 7 = -\frac{x}{3},$$

we can avoid messy computations and potential errors by *clearing* (or *eliminating*) the fractions. We do this as part of Step 1, multiplying each side by the least common denominator (LCD) of all the fractions in the equation. The result is an equation with only *integer* coefficients.

Solving a Linear Equation in One Variable

Step 1 **Simplify each side separately.** Use the distributive property as needed.
- Clear any parentheses.
- **Clear any fractions or decimals.**
- Combine like terms.

Step 2 **Isolate the variable terms on one side.** Use the addition property of equality so that all terms with variables are on one side of the equation and all constants (numbers) are on the other side.

Step 3 **Isolate the variable.** Use the multiplication property of equality to obtain an equation that has just the variable with coefficient 1 on one side.

Step 4 **Check.** Substitute the value found into the *original* equation. If a true statement results, write the solution set. If not, rework the problem.

**NOW TRY
EXERCISE 1**

Solve.

$$\frac{x}{2} - 7 = -\frac{x}{5}$$

EXAMPLE 1   Solving a Linear Equation (Fractional Coefficients)

Solve $\frac{x}{4} - 7 = -\frac{x}{3}$.

Step 1

$$\frac{x}{4} - 7 = -\frac{x}{3}$$

The LCD of all the fractions in the equation is 12.

$$12\left(\frac{x}{4} - 7\right) = 12\left(-\frac{x}{3}\right)$$

Multiply each side by 12, the LCD.

Think: $12\left(\frac{x}{4}\right)$
$= \frac{12}{1} \cdot \frac{x}{4}$
$= \frac{12x}{4}$
$= 3x$

$$12\left(\frac{x}{4}\right) + 12(-7) = 12\left(-\frac{x}{3}\right)$$

Distributive property

$$3x - 84 = -4x$$

Multiply.

Step 2

$$3x - 84 - 3x = -4x - 3x$$

Subtract 3x.

$$-84 = -7x$$

Combine like terms.

Step 3

$$\frac{-84}{-7} = \frac{-7x}{-7}$$

Divide by −7.

$$12 = x$$

Step 4   CHECK

$$\frac{x}{4} - 7 = -\frac{x}{3}$$

Original equation

$$\frac{12}{4} - 7 \overset{?}{=} -\frac{12}{3}$$

Let x = 12.

$$3 - 7 \overset{?}{=} -4$$

Simplify fractions.

$$-4 = -4 \quad ✓$$

True

The check confirms that the solution set is $\{12\}$.

NOW TRY

EXAMPLE 2   Solving a Linear Equation (Fractional Coefficients)

Solve $\frac{2}{3}x - \frac{1}{2}x = -\frac{1}{6}x - 2$.

Step 1

$$\frac{2}{3}x - \frac{1}{2}x = -\frac{1}{6}x - 2$$

The LCD of all the fractions in the equation is 6.

Use parentheses to "group" the original terms.

$$6\left(\frac{2}{3}x - \frac{1}{2}x\right) = 6\left(-\frac{1}{6}x - 2\right)$$

Multiply each side by 6, the LCD.

$$6\left(\frac{2}{3}x\right) + 6\left(-\frac{1}{2}x\right) = 6\left(-\frac{1}{6}x\right) + 6(-2)$$

Distributive property; Multiply *each* term inside the parentheses by 6.

The fractions have been cleared.

$$4x - 3x = -x - 12$$

Multiply.

$$x = -x - 12$$

Combine like terms.

Step 2

$$x + x = -x - 12 + x$$

Add x.

$$2x = -12$$

Combine like terms.

Step 3

$$\frac{2x}{2} = \frac{-12}{2}$$

Divide by 2.

$$x = -6$$

 **NOW TRY EXERCISE 2**

Solve.

$$\frac{1}{2}x + \frac{5}{8}x = \frac{3}{4}x - 6$$

Step 4 **CHECK**

$\dfrac{2}{3}x - \dfrac{1}{2}x = -\dfrac{1}{6}x - 2$	Original equation
$\dfrac{2}{3}(-6) - \dfrac{1}{2}(-6) \overset{?}{=} -\dfrac{1}{6}(-6) - 2$	Let $x = -6$.
$-4 + 3 \overset{?}{=} 1 - 2$	Multiply.
$-1 = -1$ ✓	True

The check confirms that the solution set is $\{-6\}$.   **NOW TRY**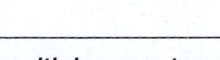

⚠ **CAUTION** *When clearing an equation of fractions, be sure to multiply every term on each side of the equation by the LCD.*

 **NOW TRY EXERCISE 3**

Solve.

$$\frac{2}{3}(x + 2) - \frac{1}{2}(3x + 4) = -4$$

EXAMPLE 3 **Solving a Linear Equation (Fractional Coefficients)**

Solve $\dfrac{1}{3}(x + 5) - \dfrac{3}{5}(x + 2) = 1$.

Step 1 We clear the parentheses first. Then we clear the fractions.

$$\frac{1}{3}(x + 5) - \frac{3}{5}(x + 2) = 1 \quad \boxed{\text{Study Step 1 carefully.}}$$

$\dfrac{1}{3}(x) + \dfrac{1}{3}(5) - \dfrac{3}{5}(x) - \dfrac{3}{5}(2) = 1$	Distributive property
$\dfrac{1}{3}x + \dfrac{5}{3} - \dfrac{3}{5}x - \dfrac{6}{5} = 1$	Multiply.
$15\left(\dfrac{1}{3}x + \dfrac{5}{3} - \dfrac{3}{5}x - \dfrac{6}{5}\right) = 15(1)$	Multiply each side by 15, the LCD.
$15\left(\dfrac{1}{3}x\right) + 15\left(\dfrac{5}{3}\right) + 15\left(-\dfrac{3}{5}x\right) + 15\left(-\dfrac{6}{5}\right) = 15(1)$	Distributive property

$\boxed{\text{Think: } 15\left(\frac{1}{3}x\right) = \left(\frac{15}{1} \cdot \frac{1}{3}\right)x = 5x}$

	$5x + 25 - 9x - 18 = 15$	Multiply.
	$-4x + 7 = 15$	Combine like terms.
Step 2	$-4x + 7 - 7 = 15 - 7$	Subtract 7.
	$-4x = 8$	Combine like terms.
Step 3	$\dfrac{-4x}{-4} = \dfrac{8}{-4}$	Divide by -4.
	$x = -2$	

Step 4 **CHECK**

$\dfrac{1}{3}(x + 5) - \dfrac{3}{5}(x + 2) = 1$	Original equation
$\dfrac{1}{3}(-2 + 5) - \dfrac{3}{5}(-2 + 2) \overset{?}{=} 1$	Let $x = -2$.
$\dfrac{1}{3}(3) - \dfrac{3}{5}(0) \overset{?}{=} 1$	Add.
$1 = 1$ ✓	True

NOW TRY ANSWERS
2. $\{-16\}$
3. $\{4\}$

The check confirms that the solution set is $\{-2\}$.   **NOW TRY**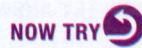

OBJECTIVE 2 Solve equations with decimals as coefficients.

Because decimals can be written as fractions, the same procedure used to clear an equation of fractions can be used to clear decimals. Consider the following equation.

$$1.5x - 3 = 2.2x + 11$$

Writing the decimal coefficients as fractions gives an equivalent equation.

$$\frac{15}{10}x - 3 = \frac{22}{10}x + 11$$

We can clear the fractional coefficients by multiplying each side of the equation by the LCD, 10. It follows then that we can clear the decimal coefficients in the original equation by multiplying each side by 10.

NOW TRY EXERCISE 4
Solve.

$$2.3x - 1 = 3.1x + 7$$

EXAMPLE 4 Solving a Linear Equation (Decimal Coefficients)

Solve $1.5x - 3 = 2.2x + 11$.

Step 1

$$1.5x - 3 = 2.2x + 11$$

$$10(1.5x - 3) = 10(2.2x + 11) \qquad \text{Multiply by 10.}$$

$$10(1.5x) + 10(-3) = 10(2.2x) + 10(11) \qquad \text{Distributive property}$$

To multiply a decimal by 10, move the decimal point one place to the right.

Think: $10(1.5x)$ = $15x$

$$15x - 30 = 22x + 110 \qquad \text{Multiply.}$$

Step 2

$$15x - 30 - 15x = 22x + 110 - 15x \qquad \text{Subtract } 15x.$$

$$-30 = 7x + 110 \qquad \text{Combine like terms.}$$

$$-30 - 110 = 7x + 110 - 110 \qquad \text{Subtract 110.}$$

$$-140 = 7x \qquad \text{Combine like terms.}$$

Step 3

$$\frac{-140}{7} = \frac{7x}{7} \qquad \text{Divide by 7.}$$

$$-20 = x$$

Step 4   CHECK

$$1.5x - 3 = 2.2x + 11 \qquad \text{Original equation}$$

$$1.5(-20) - 3 \overset{?}{=} 2.2(-20) + 11 \qquad \text{Let } x = -20.$$

$$-30 - 3 \overset{?}{=} -44 + 11 \qquad \text{Multiply.}$$

$$-33 = -33 \quad \checkmark \qquad \text{True}$$

The check confirms that the solution set is $\{-20\}$.   **NOW TRY**

In general, to clear decimals in an equation, we multiply each side of the *original* equation by a power of 10.

EXAMPLE 5 Solving a Linear Equation (Decimal Coefficients)

Solve $0.1t + 0.05(20 - t) = 0.09(20)$.

Step 1

$$0.1t + 0.05(20 - t) = 0.09(20) \qquad \text{Clear the parentheses first.}$$

$$0.1t + 0.05(20) + 0.05(-t) = 0.09(20) \qquad \text{Distributive property}$$

$$0.1t + 1 - 0.05t = 1.8 \qquad (*) \quad \text{Multiply.}$$

NOW TRY ANSWER
4. $\{-10\}$

NOW TRY EXERCISE 5

Solve.

$-0.2t + 0.05(13 - t) = 0.08(30)$

The decimals here are expressed as tenths (0.1 and 1.8) and hundredths (0.05). We choose the *least* exponent on 10 to eliminate the decimal points, which will make all coefficients integers. Here, we multiply by 10^2 — that is, 100.

> Now clear the decimals.

$$100(0.1t + 1 - 0.05t) = 100(1.8) \qquad \text{Multiply by 100.}$$

$$100(0.1t) + 100(1) + 100(-0.05t) = 100(1.8) \qquad \text{Distributive property}$$

$$10t + 100 - 5t = 180 \qquad (**) \quad \text{Multiply.}$$

$$5t + 100 = 180 \qquad \text{Combine like terms.}$$

Step 2
$$5t + 100 - 100 = 180 - 100 \qquad \text{Subtract 100.}$$

$$5t = 80 \qquad \text{Combine like terms.}$$

Step 3
$$\frac{5t}{5} = \frac{80}{5} \qquad \text{Divide by 5.}$$

$$t = 16$$

Step 4 Check to confirm that $\{16\}$ is the solution set.   **NOW TRY**

NOW TRY ANSWER

5. $\{-7\}$

NOTE In **Example 5**, multiplying by 100 is the same as moving the decimal points *two* places to the right.

$$0.10\,t + 1.00 - 0.05\,t = 1.80 \qquad \text{Equation (*) from Example 5 with 0s included as placeholders}$$

$$10t + 100 - 5t = 180 \qquad \text{Multiply by 100. Equation (**) results.}$$

2.4 Exercises

FOR EXTRA HELP   **MyLab Math**

▶ *Video solutions for select problems available in MyLab Math*

STUDY SKILLS REMINDER
How are you doing on your homework? **Review Study Skill 4, Completing Your Homework.**

Concept Check *Find the least common denominator (LCD) for the fractions in each list.*

1. $\dfrac{3}{8}, \dfrac{1}{4}$

2. $\dfrac{7}{10}, \dfrac{1}{5}$

3. $\dfrac{1}{2}, \dfrac{2}{3}$

4. $\dfrac{3}{5}, \dfrac{1}{2}$

5. $\dfrac{5}{6}, \dfrac{1}{2}, \dfrac{3}{4}$

6. $\dfrac{1}{9}, \dfrac{2}{3}, \dfrac{5}{6}$

7. $\dfrac{4}{5}, \dfrac{1}{3}, \dfrac{3}{10}$

8. $\dfrac{3}{4}, \dfrac{1}{6}, \dfrac{5}{8}$

Concept Check *Multiply as indicated.*

9. 3.5×10 **10.** 2.0×10 **11.** 0.1×10 **12.** 0.5×10

13. 0.72×100 **14.** 0.45×100 **15.** 0.07×100 **16.** 0.01×100

17. 0.1×100 **18.** 0.3×100 **19.** 2.5×100 **20.** 1.2×100

21. 0.009×1000 **22.** 0.001×1000 **23.** 0.48×1000 **24.** 0.84×1000

Concept Check *Give the letter of the correct response.*

25. The expression $12\left(\dfrac{1}{6}x + \dfrac{1}{3} - \dfrac{2}{3}x - \dfrac{2}{3}\right)$ is equivalent to which of the following?

A. $-6x - 4$ **B.** $2x + 4$ **C.** $10x - 4$ **D.** $-6x + 4$

26. The expression $100(0.03x - 0.3)$ is equivalent to which of the following?

A. $0.03x - 0.3$ **B.** $3x - 3$ **C.** $3x - 10$ **D.** $3x - 30$

Solve each equation, and check the solution. See Examples 1–3.

27. $\dfrac{x}{2} - 5 = \dfrac{x}{7}$

28. $\dfrac{x}{3} - 2 = \dfrac{x}{5}$

29. $\dfrac{3z}{4} + \dfrac{z}{3} = \dfrac{13}{2}$

30. $\dfrac{4p}{9} + \dfrac{p}{6} = \dfrac{11}{3}$

31. $\dfrac{1}{3}x - \dfrac{5}{6} = \dfrac{3}{4}x$

32. $\dfrac{5}{8}x - \dfrac{7}{12} = \dfrac{1}{3}x$

33. $\dfrac{3}{4}x - \dfrac{1}{2}x = -\dfrac{1}{8}x - 12$

34. $\dfrac{3}{5}x - \dfrac{1}{2}x = \dfrac{7}{10}x + 3$

35. $\dfrac{3}{5}t - \dfrac{1}{10}t = t - \dfrac{5}{2}$

36. $-\dfrac{2}{7}r + 2r = \dfrac{1}{2}r + \dfrac{17}{2}$

37. $\dfrac{3}{4}x - \dfrac{1}{3}x + 5 = \dfrac{5}{6}x$

38. $\dfrac{1}{5}x - \dfrac{2}{3}x - 2 = -\dfrac{2}{5}x$

39. $\dfrac{1}{6}x - \dfrac{1}{2}x = 2 - \dfrac{1}{3}x$

40. $\dfrac{1}{2}x - \dfrac{1}{5}x = 1 + \dfrac{3}{10}x$

41. $\dfrac{2}{3}w + 4 = \dfrac{2}{3}(w + 3) + 2$

42. $\dfrac{3}{4}z - 1 = \dfrac{3}{4}(z - 8) + 5$

43. $\dfrac{1}{4}(p + 3) - \dfrac{2}{5}(p + 1) = -1$

44. $\dfrac{1}{7}(n + 5) - \dfrac{1}{2}(n + 2) = -1$

45. $\dfrac{1}{7}(3x + 2) - \dfrac{1}{5}(x + 4) = 2$

46. $\dfrac{1}{4}(3x - 1) + \dfrac{1}{6}(x + 3) = 3$

47. $\dfrac{1}{2}(x + 2) + \dfrac{3}{4}(x + 4) = x + 5$

48. $\dfrac{1}{3}(x + 3) + \dfrac{1}{6}(x - 6) = x + 3$

49. $-\dfrac{1}{4}(x - 12) + \dfrac{1}{2}(x + 2) = x + 4$

50. $\dfrac{1}{9}(p + 18) + \dfrac{1}{3}(2p + 3) = p + 3$

51. $\dfrac{2}{3}k - \left(k - \dfrac{1}{2}\right) = \dfrac{1}{6}(k - 51)$

52. $-\dfrac{5}{6}q - (q - 1) = \dfrac{1}{4}(-q + 80)$

Solve each equation, and check the solution. See Examples 4 and 5.

53. $0.6x + 2.2 = 7$

54. $0.8x + 3.6 = 6$

55. $2.4t - 9.4 = -7$

56. $1.6k - 1.8 = -5$

57. $2.7x - 3 = 5.2x + 2$

58. $3.6x - 1 = 5.1x + 5$

59. $0.25x - 1 = 0.32x - 0.3$

60. $0.45x - 0.2 = 0.29x - 1$

61. $0.75x - 3.2 = 0.55 - 0.5x$

62. $1.35x - 0.6 = 1.65 + 2.1x$

63. $0.8t + 0.15 = 2t - 1.35$

64. $-0.12p + 3.4 = 0.84 + 5p$

65. $0.1(x + 80) + 0.2x = 14$

66. $0.3(x + 12) + 0.9x = 6$

67. $0.05(x - 6) = 0.05x - 3$

68. $0.04(x + 5) = 0.04x + 2$

69. $0.5q + 0.04(5 - 8q) = 0.14(4)$

70. $0.2z + 0.04(10 - z) = -0.08(3)$

71. $2.4x - 1.4 = 0.8(3x - 2) + 0.2$

72. $1.8x + 1.2 = 0.9(2x + 3) - 1.5$

73. $0.3(z - 25) + 0.6(z + 15) = -12$

74. $0.3(x + 15) + 0.4(x + 25) = 25$

75. $0.2(60) + 0.05x = 0.1(60 + x)$

76. $0.3(30) + 0.15x = 0.2(30 + x)$

77. $1.00x + 0.05(12 - x) = 0.10(63)$

78. $0.92x + 0.98(12 - x) = 0.96(12)$

79. $0.6(10,000) + 0.8x = 0.72(10,000 + x)$ **80.** $0.2(5000) + 0.3x = 0.25(5000 + x)$

SUMMARY EXERCISES   Applying Methods for Solving Linear Equations

Students often confuse *simplifying an expression* with *solving an equation*. Compare these two processes in the example that follows.

NOW TRY
EXERCISE
Simplify.

$8 + 2(4x - 1) + 2x$

Solve.

$8 + 2(4x - 1) = 2x$

EXAMPLE   Simplifying an Expression vs. Solving an Equation

Simplify.

$5(x - 1) - 3x + 1$	This is an expression.
$= 5x - 5 - 3x + 1$	Distributive property
$= 2x - 4$	Combine like terms.

We *simplify an expression* by performing the operations and combining like terms. The result is usually a simpler expression that still contains the variable (unless the variable terms add to 0).

Solve.

$5(x - 1) - 3x = 1$	This is an equation.
$5x - 5 - 3x = 1$	Distributive property
$2x - 5 = 1$	Combine like terms.
$2x = 6$	Add 5 to each side.
$x = 3$	Divide each side by 2.

Check to confirm that the solution set is $\{3\}$.

We *solve an equation* by isolating the variable on one side to find a value of the variable. The result is a potential solution that we check by substituting in the original equation.

NOW TRY

NOW TRY ANSWER
$10x + 6; \{-1\}$

Concept Check   *Decide whether each of the following is an* expression *or an* equation. *If it is an expression, simplify it. If it is an equation, solve it.*

1. $x + 2 = -3$

2. $4p - 6 + 3p - 8$

3. $-(m - 1) - (3 + 2m)$

4. $6q - 9 = 12 + 3q$

5. $5x - 9 = 3(x - 3)$

6. $\dfrac{2}{3}x + 8 = \dfrac{1}{4}x$

7. $2 - 6(z + 1) - 4(z - 2) - 10$

8. $7(p - 2) + p = 2(p + 2)$

9. $\dfrac{1}{2}(x + 10) - \dfrac{2}{3}x$

10. $-4(k + 2) + 3(2k - 1)$

Solve each equation, and check the solution.

11. $-6z = -14$

12. $2m + 8 = 16$

13. $12.5x = -63.75$

14. $-x = -12$

15. $\dfrac{4}{5}x = -20$

16. $7m - 5m = -12$

17. $-x = 6$

18. $\dfrac{x}{-2} = 8$

19. $4x + 2(3 - 2x) = 6$

20. $x - 16.2 = 7.5$

21. $7m - (2m - 9) = 39$

22. $2 - (m + 4) = 3m - 2$

23. $-3(m - 4) + 2(5 + 2m) = 29$

24. $-0.3x + 2.1(x - 4) = -6.6$

25. $0.08x + 0.06(x + 9) = 1.24$

26. $3(m + 5) - 1 + 2m = 5(m + 2)$

27. $-2t + 5t - 9 = 3(t - 4) - 5$

28. $2.3x + 13.7 = 1.3x + 2.9$

29. $0.2(50) + 0.8r = 0.4(50 + r)$

30. $r + 9 + 7r = 4(3 + 2r) - 3$

31. $2(3 + 7x) - (1 + 15x) = 2$

32. $0.06(10 - x) + 0.4x = 0.02(47)$

33. $\dfrac{1}{4}x - 4 = \dfrac{3}{2}x + \dfrac{3}{4}x$

34. $\dfrac{3}{4}(z - 2) - \dfrac{1}{3}(5 - 2z) = -2$

2.5 Applications of Linear Equations

OBJECTIVES

1 Learn the six steps for solving applied problems.

2 Solve problems involving unknown numbers.

3 Solve problems involving sums of quantities.

4 Solve problems involving consecutive integers.

5 Solve problems involving complementary and supplementary angles.

VOCABULARY

☐ consecutive integers
☐ consecutive even (or odd) integers
☐ degree
☐ complementary angles
☐ right angle
☐ supplementary angles
☐ straight angle

OBJECTIVE 1 Learn the six steps for solving applied problems.

While there is not one specific method, we suggest the following.

Solving an Applied Problem

Step 1 **Read** the problem carefully, several times if necessary. *What information is given? What is to be found?*

Step 2 **Assign a variable** to represent the unknown value. Make a sketch, diagram, or table, as needed. If necessary, express any other unknown values in terms of the variable.

Step 3 **Write an equation** using the variable expression(s).

Step 4 **Solve** the equation.

Step 5 **State the answer.** Label it appropriately. *Does it seem reasonable?*

Step 6 **Check** the answer in the words of the *original* problem.

OBJECTIVE 2 Solve problems involving unknown numbers.

EXAMPLE 1 Finding the Value of an Unknown Number

The product of 4, and a number decreased by 7, is 100. What is the number?

Step 1 **Read** the problem carefully. We are asked to find a number.

Step 2 **Assign a variable** to represent the unknown quantity.

Let x = the number.

Step 3 **Write an equation.**

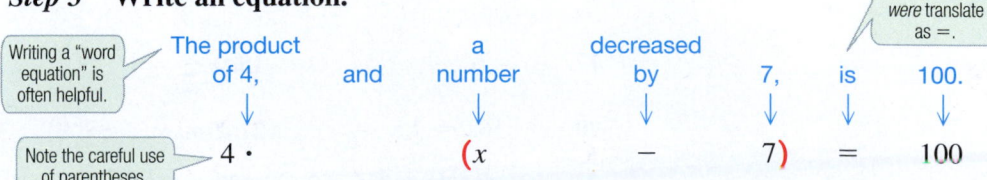

Because of the commas in the given problem, writing the equation as $4x - 7 = 100$ is *incorrect*. This equation corresponds to the statement *"The product of 4 and a number, decreased by 7, is 100."*

**NOW TRY
EXERCISE 1**

The product of 7, and a number increased by 3, is −63. What is the number?

Step 4 **Solve.**

$$4(x - 7) = 100 \qquad \text{Equation from Step 3}$$
$$4x - 28 = 100 \qquad \text{Distributive property}$$
$$4x - 28 + 28 = 100 + 28 \qquad \text{Add 28.}$$
$$4x = 128 \qquad \text{Combine like terms.}$$
$$\frac{4x}{4} = \frac{128}{4} \qquad \text{Divide by 4.}$$
$$x = 32$$

Step 5 **State the answer.** The number is 32.

Step 6 **Check.** The number 32 decreased by 7 is 25. The product of 4 and 25 is 100, as required. The answer, 32, is correct.

NOW TRY

**NOW TRY
EXERCISE 2**

If 5 is added to a number, the result is 7 less than three times the number. Find the number.

EXAMPLE 2 Finding the Value of an Unknown Number

If 6 is subtracted from five times a number, the result is 9 more than twice the number. Find the number.

Step 1 **Read** the problem. We are asked to find a number.

Step 2 **Assign a variable** to represent the unknown quantity.

Let x = the number.

Step 3 **Write an equation.**

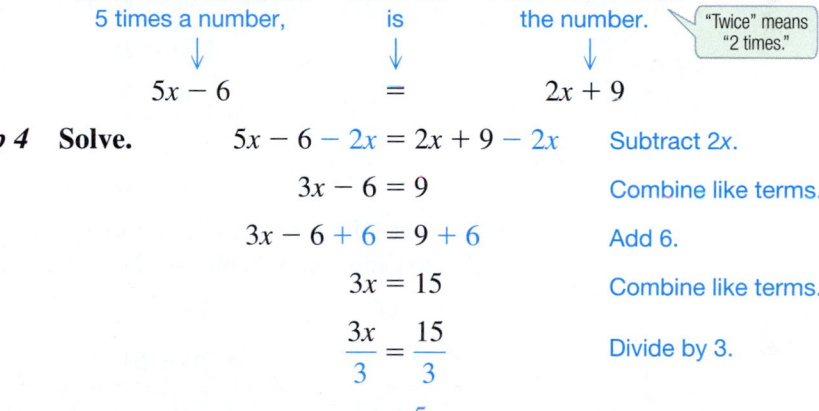

Step 4 **Solve.**

$$5x - 6 - 2x = 2x + 9 - 2x \qquad \text{Subtract } 2x.$$
$$3x - 6 = 9 \qquad \text{Combine like terms.}$$
$$3x - 6 + 6 = 9 + 6 \qquad \text{Add 6.}$$
$$3x = 15 \qquad \text{Combine like terms.}$$
$$\frac{3x}{3} = \frac{15}{3} \qquad \text{Divide by 3.}$$
$$x = 5$$

Step 5 **State the answer.** The number is 5.

Step 6 **Check.** If 6 is subtracted from 5 times the number, we obtain

$$5 \cdot 5 - 6, \quad \text{which equals 19.}$$

Nine more than twice the number would be

$$2 \cdot 5 + 9, \quad \text{which also equals 19.}$$

The answer, 5, checks.

NOW TRY

OBJECTIVE 3 Solve problems involving sums of quantities.

NOW TRY ANSWERS
1. −12
2. 6

PROBLEM-SOLVING HINT In general, to solve problems involving sums of quantities, choose a variable to represent one of the unknowns. *Then represent the other quantity in terms of the same variable.*

NOW TRY EXERCISE 3

In the 2016 Summer Olympics in Rio de Janeiro, Brazil, Germany won 25 fewer medals than Great Britain. The two countries won a total of 109 medals. How many medals did each country win? (Data from *The World Almanac and Book of Facts*.)

EXAMPLE 3   Finding Numbers of Olympic Medals

In the 2016 Summer Olympics in Rio de Janeiro, Brazil, the United States won 51 more medals than China. The two countries won a total of 191 medals. How many medals did each country win? (Data from *The World Almanac and Book of Facts*.)

Step 1   **Read** the problem. We are given the total number of medals and asked to find the number each country won.

Step 2   **Assign a variable.**

Let x = the number of medals China won.

Then $x + 51$ = the number of medals the United States won.

Step 3   **Write an equation.**

The total	is	the number of medals China won	plus	the number of medals the United States won.
↓	↓	↓	↓	↓
191	=	x	+	$(x + 51)$

Step 4   **Solve.**

$$191 = 2x + 51 \qquad \text{Combine like terms.}$$
$$191 - 51 = 2x + 51 - 51 \qquad \text{Subtract 51.}$$
$$140 = 2x \qquad \text{Combine like terms.}$$
$$\frac{140}{2} = \frac{2x}{2} \qquad \text{Divide by 2.}$$
$$70 = x, \quad \text{or} \quad x = 70 \quad \leftarrow \text{Medals China won}$$

Step 5   **State the answer.** The variable x represents the number of medals China won, so China won 70 medals. Now determine the number of U.S. medals.

$$x + 51$$
$$= 70 + 51$$
$$= 121 \leftarrow \text{Medals the United States won}$$

Step 6   **Check.** The United States won 121 medals and China won 70, so the difference was

$$121 - 70 = 51, \quad \text{as required.}$$

The total number of medals was

$$121 + 70 = 191, \quad \text{as required.}$$

All conditions of the problem are satisfied. The answer checks.

NOW TRY

> ⚠ **CAUTION**   The nature of an applied problem may restrict the set of possible solutions. An answer such as −33 medals or $25\frac{1}{2}$ medals would be inappropriate in **Example 3**. *Be sure that an answer is reasonable given the context of the problem.*

NOW TRY ANSWER

3. Great Britain: 67 medals; Germany: 42 medals

NOTE   The problem in **Example 3** could also be solved by letting x represent the number of medals the United States won. Then $x - 51$ would represent the number of medals China won. The equation would be different.

$$191 = x + (x - 51)$$ Alternative equation for **Example 3**

The solution of this equation is 121, which is the number of U.S. medals. The number of Chinese medals would be $121 - 51 = 70$. **The answers are the same,** whichever approach is used, even though the equation and its solution are different.

EXAMPLE 4   Finding the Number of Orders for Tea

The owner of Terry's Coffeehouse found that on one day the number of orders for tea was $\frac{1}{3}$ the number of orders for coffee. If the total number of orders for the two drinks was 76, how many orders were placed for tea?

Step 1   **Read** the problem. It asks for the number of orders for tea.

Step 2   **Assign a variable.** Because of the way the problem is stated, let the variable represent the number of orders for *coffee*.

Let x = the number of orders for coffee.

Then $\frac{1}{3}x$ = the number of orders for tea.

A sketch is helpful. See **FIGURE 2**.

Coffee orders x

Tea orders $\frac{1}{3}x$

FIGURE 2

Step 3   **Write an equation.** Use the fact that the total number of orders was 76.

The total	is	orders for coffee	plus	orders for tea.
↓	↓	↓	↓	↓
76	=	x	+	$\frac{1}{3}x$

Remember the x in $\frac{1}{3}x$.

Step 4   **Solve.** $76 = \frac{4}{3}x$ $x = 1x = \frac{3}{3}x; \frac{3}{3}x + \frac{1}{3}x = \frac{4}{3}x$

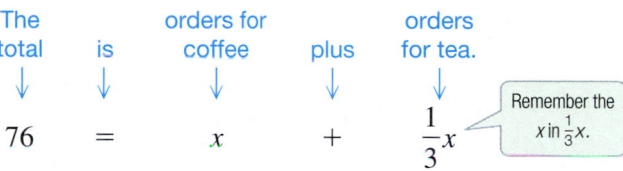

$$\frac{3}{4}(76) = \frac{3}{4}\left(\frac{4}{3}x\right)$$ Multiply by $\frac{3}{4}$, the reciprocal of $\frac{4}{3}$.

Be careful. This is *not* the answer.

$57 = x$ ← Number of orders for coffee

Step 5   **State the answer.** In this problem, *x does not represent the quantity that we must find.* The number of orders for tea was $\frac{1}{3}x$.

$$\frac{1}{3}(57) = 19$$ ← Number of orders for tea

Step 6   **Check.** The number of orders for tea, 19, is one-third the number of orders for coffee, 57, and $19 + 57 = 76$, as required. This agrees with the information given in the problem. The answer checks.   **NOW TRY**

PROBLEM-SOLVING HINT   In **Example 4**, it was easier to let the variable represent the quantity that was *not* specified. This required extra work in Step 5 to find the number of orders for tea. In some cases, this approach is easier.

NOW TRY EXERCISE 4

In one week, the owner of Carly's Coffeehouse found that the number of orders for bagels was $\frac{2}{3}$ the number of orders for chocolate scones. If the total number of orders for the two items was 525, how many orders were placed for bagels?

NOW TRY ANSWER
4. 210 bagel orders

**NOW TRY
EXERCISE 5**

A fly spray mixture requires 7 oz of water for each 1 oz of essential oil. To fill a quart bottle (32 oz), how many ounces of water and how many ounces of essential oil are required?

EXAMPLE 5   Analyzing a Gasoline-Oil Mixture

A lawn trimmer uses a mixture of gasoline and oil. The mixture contains 16 oz of gasoline for each 1 ounce of oil. If the tank holds 68 oz of the mixture, how many ounces of oil and how many ounces of gasoline does it require when it is full?

Step 1   **Read** the problem. We must find how many ounces of oil and gasoline are needed to fill the tank.

Step 2   **Assign a variable.**

Let x = the number of ounces of oil required.

Then $16x$ = the number of ounces of gasoline required.

A sketch like the one in **FIGURE 3** is helpful.

| Oil x |
| Gasoline $16x$ |

FIGURE 3

Step 3   **Write an equation.**

Amount of gasoline	plus	amount of oil	is	total amount in tank.
↓	↓	↓	↓	↓
$16x$	$+$	x	$=$	68

Step 4   **Solve.**

$$17x = 68 \qquad \text{Combine like terms.}$$

$$\frac{17x}{17} = \frac{68}{17} \qquad \text{Divide by 17.}$$

$$x = 4$$

Step 5   **State the answer.** When full, the lawn trimmer requires 4 oz of oil, and

$$16x = 16(4)$$

$$= 64 \text{ oz of gasoline.}$$

Step 6   **Check.** Because $4 + 64 = 68$, and 64 is 16 times 4, the answer checks.

NOW TRY

PROBLEM-SOLVING HINT   Sometimes we must find three unknown quantities. *When the three unknowns are compared in pairs, let the variable represent the unknown found in both pairs.*

EXAMPLE 6   Dividing a Board into Pieces

A project calls for three pieces of wood. The longest piece must be twice the length of the middle-sized piece. The shortest piece must be 10 in. shorter than the middle-sized piece. If a board 70 in. long is to be used, how long must each piece be?

Step 1   **Read** the problem. Three lengths must be found.

Step 2   **Assign a variable.** The middle-sized piece appears in both pairs of comparisons, so let x represent the length, in inches, of the middle-sized piece.

Let x = the length of the middle-sized piece.

Then $2x$ = the length of the longest piece,

and $x - 10$ = the length of the shortest piece.

NOW TRY ANSWER
5. water: 28 oz; essential oil: 4 oz

NOW TRY EXERCISE 6

Over a 6-hr period, a basketball player spent twice as much time lifting weights as practicing free throws and 2 hr longer watching game films than practicing free throws. How many hours did he spend on each task?

A sketch is helpful. See **FIGURE 4**.

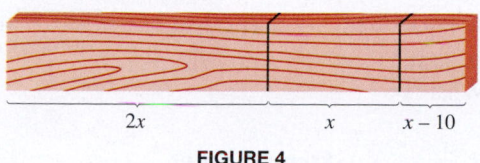

FIGURE 4

Step 3 **Write an equation.**

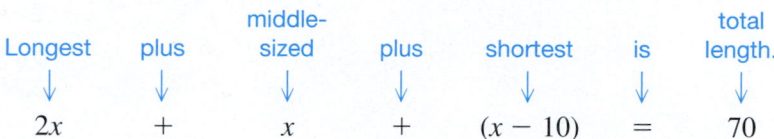

$$2x + x + (x - 10) = 70$$

Step 4 **Solve.**

$$4x - 10 = 70 \quad \text{Combine like terms.}$$
$$4x - 10 + 10 = 70 + 10 \quad \text{Add 10.}$$
$$4x = 80 \quad \text{Combine like terms.}$$
$$\frac{4x}{4} = \frac{80}{4} \quad \text{Divide by 4.}$$
$$x = 20$$

Step 5 **State the answer.** The middle-sized piece is 20 in. long, the longest piece is

$$2(20) = 40 \text{ in. long,}$$

and the shortest piece is

$$20 - 10 = 10 \text{ in. long.}$$

Step 6 **Check.** The sum of the lengths is $20 + 40 + 10 = 70$ in. All conditions of the problem are satisfied.   **NOW TRY**

Consecutive integers

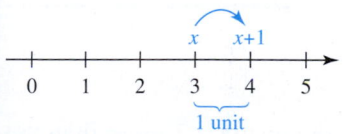

FIGURE 5

OBJECTIVE 4 Solve problems involving consecutive integers.

Two integers that differ by 1 are **consecutive integers.** For example, 3 and 4, 16 and 17, and −2 and −1 are pairs of consecutive integers. See **FIGURE 5**.

In general, if x represents an integer, then x + 1 represents the next greater consecutive integer.

EXAMPLE 7 Finding Consecutive Integers

Two pages that face each other in the print edition of this book have 277 as the sum of their page numbers. What are the page numbers?

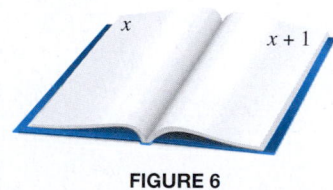

FIGURE 6

Step 1 **Read** the problem. Because the two pages face each other, they must have page numbers that are consecutive integers.

Step 2 **Assign a variable.**

Let $x =$ the lesser page number.

Then $x + 1 =$ the greater page number.

FIGURE 6 illustrates the situation.

NOW TRY
EXERCISE 7

Two pages that face each other in a book have 593 as the sum of their page numbers. What are the page numbers?

Step 3   **Write an equation.** The sum of the page numbers is 277.

Lesser page number	plus	greater page number	is	the sum.
↓	↓	↓	↓	↓
x	$+$	$(x + 1)$	$=$	277

Step 4   **Solve.**

$$2x + 1 = 277 \qquad \text{Combine like terms.}$$
$$2x + 1 - 1 = 277 - 1 \qquad \text{Subtract 1.}$$
$$2x = 276 \qquad \text{Combine like terms.}$$
$$\frac{2x}{2} = \frac{276}{2} \qquad \text{Divide by 2.}$$
$$x = 138$$

Step 5   **State the answer.** The lesser page number is 138, and the greater page number is $138 + 1 = 139$. (If you are using a print text, it is opened to these two pages.)

Step 6   **Check.** The sum of 138 and 139 is 277. The answer checks.

NOW TRY ↺

Consecutive *even* integers, such as 2 and 4, and 8 and 10, differ by 2. Similarly, consecutive *odd* integers, such as 1 and 3, and 9 and 11, also differ by 2. See **FIGURE 7.**

In general, if x represents an even (or odd) integer, then x + 2 represents the next greater consecutive even (or odd) integer, respectively.

Consecutive even integers

2 units
x ⟶ $x+2$

| 0 | 1 | 2 | 3 | 4 | 5 |

x ⟵ $x+2$
2 units

Consecutive odd integers

FIGURE 7

> **PROBLEM-SOLVING HINT**   If x = the lesser (least) integer in a consecutive integer problem, then the following apply.
> - For two consecutive integers, use $\quad x, \quad x + 1.$
> - For two consecutive *even* integers, use $\quad x, \quad x + 2.$
> - For two consecutive *odd* integers, use $\quad x, \quad x + 2.$

EXAMPLE 8   **Finding Consecutive Odd Integers**

If the lesser of two consecutive odd integers is doubled, the result is 7 more than the greater of the two integers. Find the two integers.

Step 1   **Read** the problem. We must find two consecutive odd integers.

Step 2   **Assign a variable.**

Let $\qquad x =$ the lesser consecutive odd integer.

Then $\quad x + 2 =$ the greater consecutive odd integer.

Step 3   **Write an equation.**

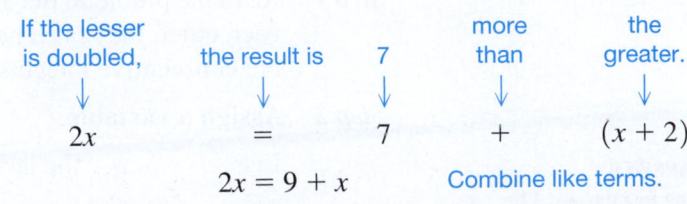

If the lesser is doubled,	the result is	7	more than	the greater.
↓	↓	↓	↓	↓
$2x$	$=$	7	$+$	$(x + 2)$

Step 4   **Solve.**

$$2x = 9 + x \qquad \text{Combine like terms.}$$
$$2x - x = 9 + x - x \qquad \text{Subtract } x.$$
$$x = 9 \qquad \text{Combine like terms.}$$

NOW TRY ANSWER
7. 296, 297

**NOW TRY
EXERCISE 8**

Find two consecutive odd
integers such that the sum
of twice the lesser and three
times the greater is 191.

Step 5 **State the answer.** The lesser integer is 9. The greater is $9 + 2 = 11$.

Step 6 **Check.** When 9 is doubled, we obtain 18, which is 7 more than the greater
odd integer, 11. The answer checks.

NOW TRY

OBJECTIVE 5 Solve problems involving complementary and supplementary angles.

An angle can be measured using a unit called the **degree** (°), which is $\frac{1}{360}$ of a complete
rotation. Some special angles are shown in **FIGURE 8**.

- Two angles whose sum is 90° are **complementary,** or *complements* of each other.
- An angle that measures 90° is a **right angle.**
- Two angles whose sum is 180° are **supplementary,** or *supplements* of each other.
- An angle that measures 180° is a **straight angle.**

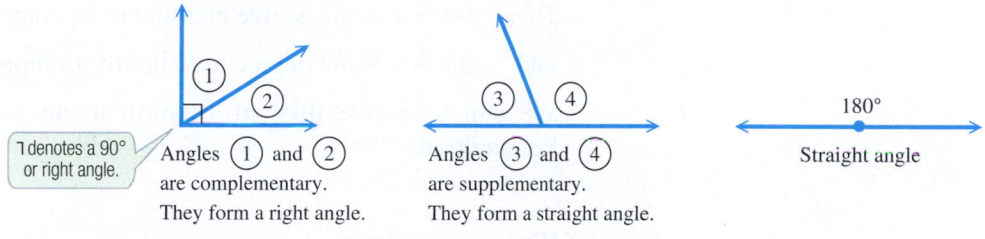

1 denotes a 90°
or right angle.

Angles ① and ② | Angles ③ and ④ | Straight angle
are complementary. | are supplementary.
They form a right angle. | They form a straight angle.

FIGURE 8

**NOW TRY
EXERCISE 9**

Find the measure of an angle
whose complement is twice
its measure.

EXAMPLE 9 Finding the Measure of an Angle

Find the measure of an angle whose complement is five times its measure.

Step 1 **Read** the problem. We must find the measure of an angle, given information
about the measure of its complement.

Step 2 **Assign a variable.**

Let $\qquad x =$ the degree measure of the angle.

Then $\quad 90 - x =$ the degree measure of its complement.

Step 3 **Write an equation.**

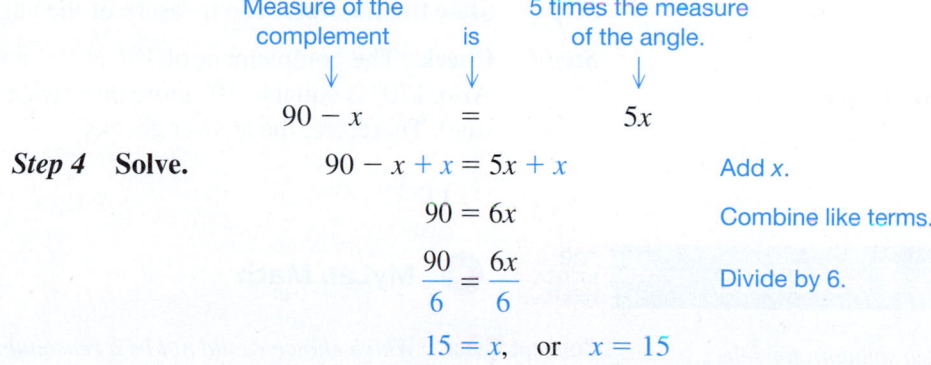

Measure of the   5 times the measure
complement   is   of the angle.
↓   ↓   ↓
$90 - x$ $=$ $5x$

Step 4 **Solve.** $\qquad 90 - x + x = 5x + x \qquad$ Add x.

$$90 = 6x \qquad \text{Combine like terms.}$$

$$\frac{90}{6} = \frac{6x}{6} \qquad \text{Divide by 6.}$$

$$15 = x, \quad \text{or} \quad x = 15$$

Step 5 **State the answer.** The measure of the angle is 15°.

Step 6 **Check.** If the angle measures 15°, then the measure of its complement is

$$90° - 15° = 75°,$$

which is equal to five times 15°, as required.

NOW TRY

NOW TRY ANSWERS
8. 37, 39
9. 30°

> **PROBLEM-SOLVING HINT**   Let x represent the degree measure of an angle.
>
> **90 − x** represents the degree measure of its complement.
>
> **180 − x** represents the degree measure of its supplement.

**NOW TRY
EXERCISE 10**
Find the measure of an
angle whose supplement is
46° less than three times its
complement.

EXAMPLE 10 Finding the Measure of an Angle

Find the measure of an angle whose supplement is 10° more than twice its complement.

Step 1 **Read** the problem. We are to find the measure of an angle, given information about its complement and its supplement.

Step 2 **Assign a variable.**

Let x = the degree measure of the angle.

Then $90 - x$ = the degree measure of its complement,

and $180 - x$ = the degree measure of its supplement.

We can visualize this information using a sketch. See **FIGURE 9**.

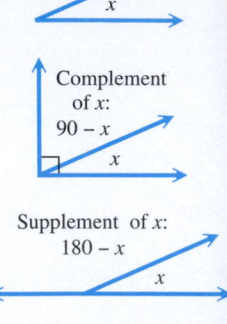

All measures are in degrees.

FIGURE 9

Step 3 **Write an equation.**

Supplement   is   10   more than   twice   its complement.

$$180 - x \;=\; 10 \;+\; 2 \cdot \;(90 - x)$$

> Be sure to use parentheses here.

Step 4 **Solve.**

$$180 - x = 10 + 180 - 2x \qquad \text{Distributive property}$$
$$180 - x = 190 - 2x \qquad \text{Combine like terms.}$$
$$180 - x + 2x = 190 - 2x + 2x \qquad \text{Add } 2x.$$
$$180 + x = 190 \qquad \text{Combine like terms.}$$
$$180 + x - 180 = 190 - 180 \qquad \text{Subtract } 180.$$
$$x = 10$$

Step 5 **State the answer.** The measure of the angle is 10°.

Step 6 **Check.** The complement of 10° is 80° and the supplement of 10° is 170°. Also, 170° is equal to 10° more than twice 80° (that is, $170 = 10 + 2(80)$ is true). Therefore, the answer checks.

NOW TRY

NOW TRY ANSWER
10. 22°

2.5 Exercises

FOR EXTRA HELP

 MyLab Math

▶ *Video solutions for select
problems available in MyLab
Math*

Concept Check   *Which choice would **not** be a reasonable answer? Justify your response.*

1. A problem requires finding the number of cars on a dealer's lot.

   A. 0   **B.** 1   **C.** $6\frac{1}{2}$ **D.** 45

2. A problem requires finding the number of hours a light bulb is on during a day.

   A. 0   **B.** 4.5   **C.** 13   **D.** 25

STUDY SKILLS REMINDER
Are you fully utilizing the features of you text? **Review Study Skill 1,** *Using Your Math Text*.

3. A problem requires finding the distance traveled in miles.

    A. -10 **B.** 1.8   **C.** $10\frac{1}{2}$ **D.** 50

4. A problem requires finding elapsed time in minutes.

    A. -5 **B.** 0   **C.** 10.5   **D.** 90

Concept Check   *Fill in each blank with the correct response.*

5. Consecutive integers differ by _____, such as 15 and _____, and -8 and _____. If x represents an integer, then _____ represents the next greater integer.

6. Consecutive odd integers are odd integers that differ by _____, such as _____ and 13. Consecutive even integers are even integers that differ by _____, such as 12 and _____.

7. Two angles whose measures sum to 90° are _____ angles. Two angles whose measures sum to 180° are _____ angles.

8. A right angle has measure _____. A straight angle has measure _____.

Solve each problem. In each case, give an equation using x as the variable and give the answer. **See Examples 1 and 2.**

9. The sum of a number and 9 is -26. What is the number?

10. The difference of a number and 11 is -31. What is the number?

11. The product of 8, and a number increased by 6, is 104. What is the number?

12. The product of 5, and 3 more than twice a number, is 85. What is the number?

13. If 2 is added to five times a number, the result is equal to 5 more than four times the number. Find the number.

14. If four times a number is added to 8, the result is equal to three times the number, added to 5. Find the number.

15. If 2 is subtracted from a number and this difference is tripled, the result is 6 more than the number. Find the number.

16. If 3 is added to a number and this sum is doubled, the result is 2 more than the number. Find the number.

17. When 6 is added to $\frac{3}{4}$ of a number, the result is 4 less than the number. Find the number.

18. When $\frac{2}{3}$ of a number is added to 10, the result is 5 more than the number. Find the number.

19. The sum of three times a number and 7 more than the number is the same as the difference of -11 and twice the number. What is the number?

20. If 4 is added to twice a number and this sum is multiplied by 2, the result is the same as if the number is multiplied by 3 and 4 is added to the product. What is the number?

Complete the six suggested problem-solving steps to solve each problem.

21. The 50-member California State Senate includes 14 more Democrats than Republicans. (No other parties are represented.) How many Democrats and how many Republicans are there in the legislature? (Data from www.senate.ca.gov) **(See Example 3.)**

    Step 1 **Read** the problem carefully.

            We must find the number of Democrats and the number of _____.

    Step 2 **Assign a variable.**

            Let x = the number of Republicans.

            Then _____ = the number of _____.

Step 3   **Write an equation.**

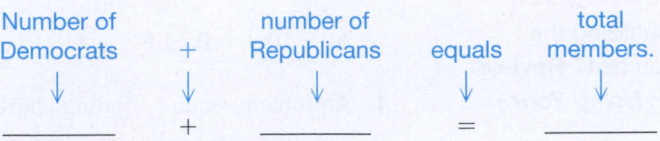

_____ + _____ = _____

Complete Steps 4–6 to solve the problem.

22. The sum of two consecutive even integers is 254. Find the integers. **(See Example 8.)**

Step 1   **Read** the problem carefully.
We must find two consecutive _____ .

Step 2   **Assign a variable.**
Let x = the lesser of the two _____ even integers.
Then _____ = the greater of the two consecutive even integers.

Step 3   **Write an equation.**

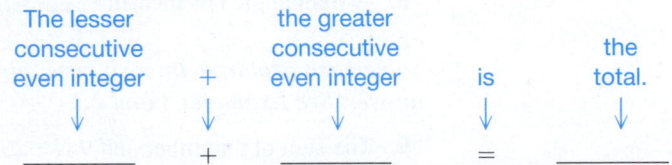

_____ + _____ = _____

Complete Steps 4–6 to solve the problem.

Solve each problem. ***See Example 3.***

23. New York and Ohio are among the states with the most remaining drive-in movie screens. New York has 4 more screens than Ohio, and there are 52 screens total in the two states. How many drive-in movie screens remain in each state? (Data from www. drive-ins.com)

24. Two of the most watched episodes in television were the final episodes of *M*A*S*H* and *Cheers*. The total number of viewers for these two episodes was about 92 million, with 8 million more people watching the *M*A*S*H* episode than the *Cheers* episode. How many people watched each episode? (Data from Nielsen Media Research.)

25. During the 115th session, the U.S. Senate had a total of 98 Democrats and Republians. There were 6 fewer Democrats than Republicans. How many members of each party were there? (Data from www. senate.gov)

26. During the 115th session, the total number of Democrats and Republicans in the U.S. House of Representatives was 434. There were 46 more Republicans than Democrats. How many members of each party were there? (Data from www.house.gov)

27. Beyoncé and Guns N' Roses had the two top-grossing concert tours for 2016, together generating $300 million in ticket sales. Guns N' Roses took in $38 million less than Beyoncé. How much did each tour generate? (Data from Pollstar.)

28. The Toyota Camry and the Honda Civic were the top-selling passenger cars in the United States in 2016. Civic sales were 22 thousand less than Camry sales, and 756 thousand of the two cars were sold. How many of each car were sold? (Data from www.businessinsider.com)

29. In the 2016–2017 NBA regular season, the Golden State Warriors won 7 more than four times as many games as they lost. The Warriors played 82 games. How many wins and losses did the team have? (Data from www.NBA.com)

30. In the 2017 regular baseball season, the Cleveland Indians won 18 fewer than twice as many games as they lost. They played 162 regular-season games. How many wins and losses did the team have? (Data from www.MLB.com)

31. A one-cup serving of orange juice contains 3 mg less than four times the amount of vitamin C as a one-cup serving of pineapple juice. Servings of the two juices contain a total of 122 mg of vitamin C. How many milligrams of vitamin C are in a serving of each type of juice? (Data from U.S. Agriculture Department.)

32. A one-cup serving of pineapple juice has 9 more than three times as many calories as a one-cup serving of tomato juice. Servings of the two juices contain a total of 173 calories. How many calories are in a serving of each type of juice? (Data from U.S. Agriculture Department.)

*Solve each problem. **See Examples 4 and 5.***

33. In one day, a store sold $\frac{2}{3}$ as many DVDs as Blu-ray discs. The total number of DVDs and Blu-ray discs sold that day was 280. How many DVDs were sold?

34. A workout that combines weight training and aerobics burns a total of 371 calories. If weight training burns $\frac{2}{5}$ as many calories as aerobics, how many calories does weight training burn?

35. The world's largest taco contained approximately 1 kg of onion for every 6.6 kg of grilled steak. The total weight of these two ingredients was 617.6 kg. To the nearest tenth of a kilogram, how many kilograms of each ingredient were used to make the taco? (Data from *Guinness World Records*.)

36. As of 2017, the combined population of China and India was estimated at 2.7 billion. If there were about 0.9 as many people living in India as China, what was the population of each country, to the nearest tenth of a billion? (Data from U.S. Census Bureau.)

37. The value of a "Mint State-63" (uncirculated) 1950 Jefferson nickel minted at Denver is twice the value of a 1945 nickel in similar condition minted at Philadelphia. Together, the total value of the two coins is $24.00. What is the value of each coin? (Data from Yeoman, R., *A Guide Book of United States Coins*.)

38. U.S. five-cent coins are made from a combination of nickel and copper. For every 1 lb of nickel, 3 lb of copper are used. How many pounds of each metal would be needed to make 560 lb of five-cent coins? (Data from The United States Mint.)

39. A recipe for whole-grain bread calls for 1 oz of rye flour for every 4 oz of whole-wheat flour. How many ounces of each kind of flour should be used to make a loaf of bread weighing 32 oz?

40. A medication contains 9 mg of active ingredients for every 1 mg of inert ingredients. How much of each kind of ingredient would be contained in a single 250-mg caplet?

*Solve each problem. **See Example 6.***

41. An office manager booked 55 airline tickets. He booked 7 more tickets on American Airlines than United Airlines. On Southwest Airlines, he booked 4 more than twice as many tickets as on United. How many tickets did he book on each airline?

42. A mathematics textbook editor spent 7.5 hr making telephone calls, writing e-mails, and attending meetings. She spent twice as much time attending meetings as making telephone calls and 0.5 hr longer writing e-mails than making telephone calls. How many hours did she spend on each task?

43. A party-length submarine sandwich that is 59 in. long is cut into three pieces. The middle piece is 5 in. longer than the shortest piece, and the shortest piece is 9 in. shorter than the longest piece. How long is each piece?

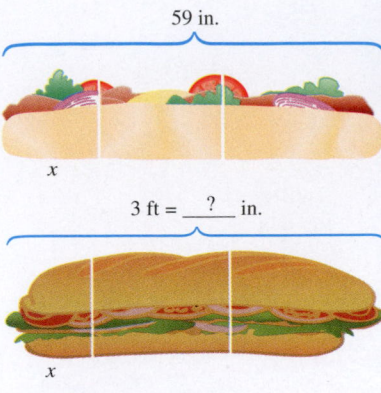

59 in.

x

44. A three-foot-long deli sandwich must be split into three pieces so that the middle piece is twice as long as the shortest piece and the shortest piece is 8 in. shorter than the longest piece. How long should the three pieces be?

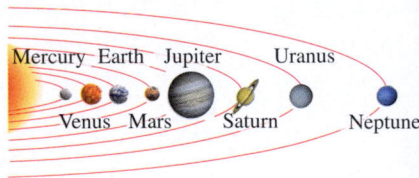

3 ft = ___?___ in.

x

45. The United States earned 121 medals at the 2016 Summer Olympics. The number of silver medals earned was 1 fewer than the number of bronze medals. The number of gold medals earned was 8 more than the number of bronze medals. How many of each kind of medal did the United States earn? (Data from *The World Almanac and Book of Facts*.)

46. Norway earned 39 medals at the 2018 Winter Olympics. The number of gold medals earned was the same as the number of silver medals. The number of bronze medals earned was 3 fewer than the number of silver medals. How many of each kind of medal did Norway earn? (Data from *The Gazette*.)

47. Venus is 31.2 million mi farther from the sun than Mercury. Earth is 57 million mi farther from the sun than Mercury. The total of the distances from these three planets to the sun is 196.2 million mi. How far from the sun is Mercury? (Distances are *mean (average)* distances.) (Data from *The New York Times Almanac*.)

Mercury Earth Jupiter Uranus

Venus Mars Saturn Neptune

48. Saturn, Jupiter, and Uranus have a total of 167 known satellites (moons). Jupiter has 18 more satellites than Saturn, and Uranus has 34 fewer satellites than Saturn. How many known satellites does Uranus have? (Data from http://solarsystem.nasa.gov)

49. The sum of the measures of the angles of any triangle is 180°. In triangle *ABC*, angles *A* and *B* have the same measure, while the measure of angle *C* is 60° greater than each of angles *A* and *B*. What are the measures of the three angles?

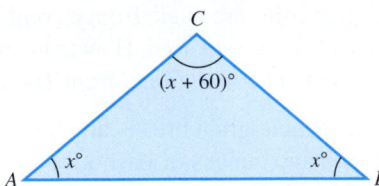

C

$(x + 60)°$

A $x°$ $x°$ *B*

50. The sum of the measures of the angles of any triangle is 180°. In triangle *ABC*, the measure of angle *A* is 141° greater than the measure of angle *B*. The measure of angle *B* is the same as the measure of angle *C*. Find the measure of each angle.

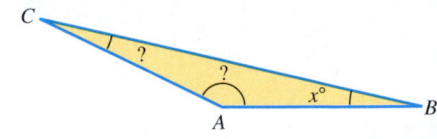

C

?

?

A $x°$ *B*

Solve each problem. ***See Examples 7 and 8.***

51. The numbers on two consecutively numbered gym lockers have a sum of 137. What are the locker numbers?

x $x + 1$

52. The numbers on two consecutive checkbook checks have a sum of 357. What are the numbers?

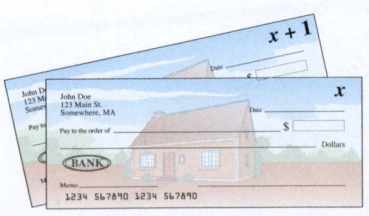

$x + 1$

x

John Doe
123 Main St.
Somewhere, MA

BANK

1234 567890 1234 567890

53. Two pages that are back-to-back in this text have 203 as the sum of their page numbers. What are the page numbers?

54. Two apartments have numbers that are consecutive integers. The sum of the numbers is 59. What are the two apartment numbers?

55. Find two consecutive even integers such that the lesser added to three times the greater gives a sum of 46.

56. Find two consecutive even integers such that six times the lesser added to the greater gives a sum of 86.

57. Find two consecutive odd integers such that 59 more than the lesser is four times the greater.

58. Find two consecutive odd integers such that twice the greater is 17 more than the lesser.

59. When the lesser of two consecutive integers is added to three times the greater, the result is 43. Find the integers.

60. If five times the lesser of two consecutive integers is added to three times the greater, the result is 59. Find the integers.

Extending Skills   *Solve each problem.*

61. If the sum of three consecutive even integers is 60, what is the first of the three even integers? (*Hint:* If x and $x + 2$ represent the first two consecutive even integers, how would we represent the third consecutive even integer?)

62. If the sum of three consecutive odd integers is 69, what is the third of the three odd integers?

63. If 6 is subtracted from the third of three consecutive odd integers and the result is multiplied by 2, the answer is 23 less than the sum of the first and twice the second of the integers. Find the integers.

64. If the first and third of three consecutive even integers are added, the result is 22 less than three times the second integer. Find the integers.

Solve each problem. See Examples 9 and 10.

65. Find the measure of an angle whose complement is four times its measure. (*Hint:* If x represents the measure of the unknown angle, how would we represent its complement?)

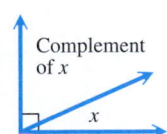

66. Find the measure of an angle whose complement is five times its measure.

67. Find the measure of an angle whose supplement is eight times its measure. (*Hint:* If x represents the measure of the unknown angle, how would we represent its supplement?)

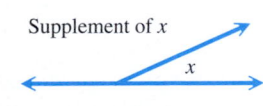

68. Find the measure of an angle whose supplement is three times its measure.

69. Find the measure of an angle whose supplement measures 39° more than twice its complement.

70. Find the measure of an angle whose supplement measures 38° less than three times its complement.

71. Find the measure of an angle such that the difference between the measures of its supplement and three times its complement is 10°.

72. Find the measure of an angle such that the sum of the measures of its complement and its supplement is 160°.

2.6   Formulas and Additional Applications from Geometry

OBJECTIVES

1 Solve a formula for one variable, given values of the other variables.

2 Use a formula to solve an applied problem.

3 Solve problems involving vertical angles and straight angles.

4 Solve a formula for a specified variable.

A **formula** is an equation in which variables are used to describe a relationship. For example, formulas exist for finding perimeters and areas of geometric figures, calculating money earned on bank savings, and converting among measurements.

$$P = 4s, \quad \mathcal{A}^* = \pi r^2, \quad I = prt, \quad F = \frac{9}{5}C + 32 \qquad \text{Formulas}$$

Many of the formulas used in this text are given at the back of the text.

OBJECTIVE 1   Solve a formula for one variable, given values of the other variables.

The **area** of a plane (two-dimensional) geometric figure is a measure of the surface covered or enclosed by the figure. Area is measured in square units.

EXAMPLE 1   Using Formulas to Evaluate Variables

Find the value of the remaining variable in each formula.

(a) $\mathcal{A} = LW; \quad \mathcal{A} = 64, L = 10$

This formula gives the area of a rectangle. See **FIGURE 10**.

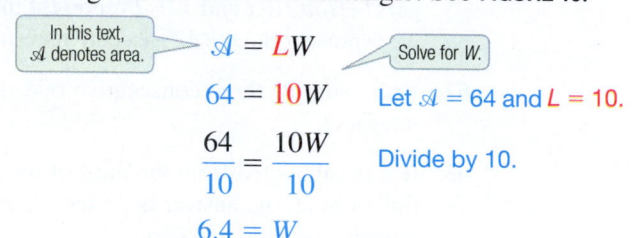

In this text, $\mathcal{A}$ denotes area.

$$\mathcal{A} = LW \qquad \text{Solve for } W.$$
$$64 = 10W \qquad \text{Let } \mathcal{A} = 64 \text{ and } L = 10.$$
$$\frac{64}{10} = \frac{10W}{10} \qquad \text{Divide by 10.}$$
$$6.4 = W$$

VOCABULARY

☐ formula
☐ area
☐ perimeter
☐ vertical angles

L

W

Rectangle
$\mathcal{A} = LW$

FIGURE 10

NOW TRY EXERCISE 1

Find the value of the remaining variable.

$$P = 2a + 2b;$$
$$P = 78, a = 12$$

The width is 6.4. Because $10(6.4) = 64$, the given area, the answer checks.

(b) $\mathcal{A} = \frac{1}{2}h(b + B); \quad \mathcal{A} = 210, B = 27, h = 10$

This formula gives the area of a trapezoid. See **FIGURE 11**.

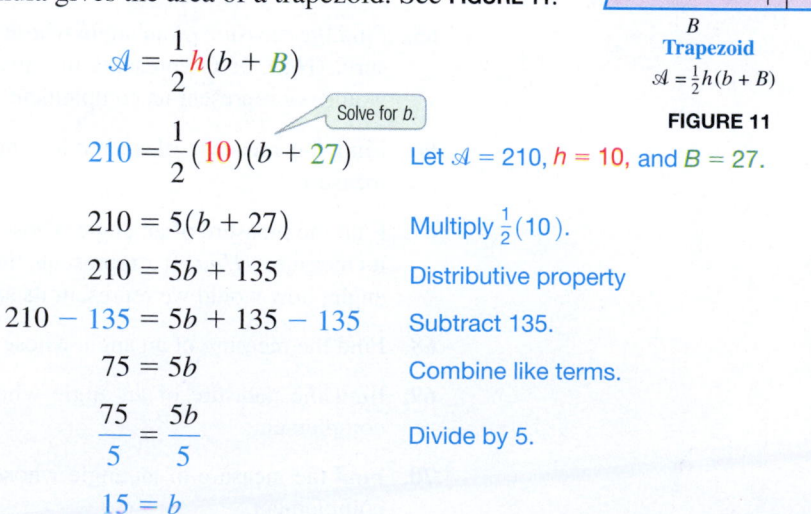

b

h

B

Trapezoid
$\mathcal{A} = \frac{1}{2}h(b + B)$

FIGURE 11

$$\mathcal{A} = \frac{1}{2}h(b + B)$$

$$210 = \frac{1}{2}(10)(b + 27) \qquad \text{Solve for } b. \qquad \text{Let } \mathcal{A} = 210, h = 10, \text{ and } B = 27.$$

$$210 = 5(b + 27) \qquad \text{Multiply } \frac{1}{2}(10).$$

$$210 = 5b + 135 \qquad \text{Distributive property}$$

$$210 - 135 = 5b + 135 - 135 \qquad \text{Subtract 135.}$$

$$75 = 5b \qquad \text{Combine like terms.}$$

$$\frac{75}{5} = \frac{5b}{5} \qquad \text{Divide by 5.}$$

$$15 = b$$

The length of the shorter parallel side, b, is 15. Because $\frac{1}{2}(10)(15 + 27) = 210$, as required, the answer checks.

NOW TRY ANSWER
1. $b = 27$

NOW TRY

---

* In this text, we use $\mathcal{A}$ to denote area.

OBJECTIVE 2 Use a formula to solve an applied problem.

The **perimeter** of a plane (two-dimensional) geometric figure is the measure of the outer boundary of the figure. For a polygon (e.g., a rectangle, square, or triangle), it is the sum of the lengths of the sides.

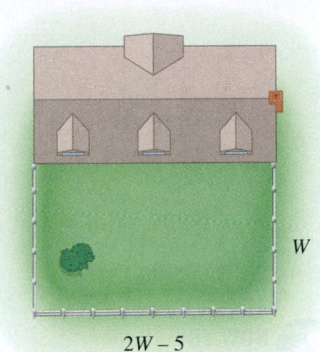

NOW TRY EXERCISE 2

A garden is in the shape of a rectangle. The length is 10 ft less than twice the width, and the perimeter is 160 ft. Find the dimensions of the garden.

EXAMPLE 2 Finding the Dimensions of a Rectangular Yard

A backyard is in the shape of a rectangle. The length is 5 m less than twice the width, and the perimeter is 80 m. Find the dimensions of the yard.

Step 1   **Read** the problem. We must find the dimensions of the yard.

Step 2   **Assign a variable.** Let W = the width of the lot, in meters. The length is 5 meters less than twice the width, so the length is $L = 2W - 5$. See **FIGURE 12**.

FIGURE 12

Step 3   **Write an equation.** Use the formula for the perimeter of a rectangle.

$$P = 2L + 2W \qquad \text{Perimeter of a rectangle}$$

Perimeter   = 2 · Length   + 2 · Width

$$80 \qquad = 2(2W - 5) \quad + \quad 2W \qquad \begin{array}{l}\text{Substitute 80 for perimeter } P \\ \text{and } 2W - 5 \text{ for length } L.\end{array}$$

Step 4   **Solve.**

$$80 = 4W - 10 + 2W \qquad \text{Distributive property}$$

$$80 = 6W - 10 \qquad \text{Combine like terms.}$$

$$80 + 10 = 6W - 10 + 10 \qquad \text{Add 10.}$$

$$90 = 6W \qquad \text{Combine like terms.}$$

$$\frac{90}{6} = \frac{6W}{6} \qquad \text{Divide by 6.}$$

We must also find the length.

$$15 = W$$

Step 5   **State the answer.** The width is 15 m and the length is $2(15) - 5 = 25$ m.

Step 6   **Check.** If the width is 15 m and the length is 25 m, the perimeter is

$$2(25) + 2(15) = 80 \text{ m}, \quad \text{as required.} \qquad \text{NOW TRY} \; \text{}$$

EXAMPLE 3 Finding the Dimensions of a Triangle

The longest side of a triangle is 3 ft longer than the shortest side. The medium side is 1 ft longer than the shortest side. If the perimeter of the triangle is 16 ft, what are the lengths of the three sides?

Step 1   **Read** the problem. We must find the lengths of the sides of a triangle.

Step 2   **Assign a variable.**

Let s = the length of the shortest side, in feet,

$s + 1$ = the length of the medium side, in feet, and

$s + 3$ = the length of the longest side, in feet.

It is a good idea to draw a sketch. See **FIGURE 13**.

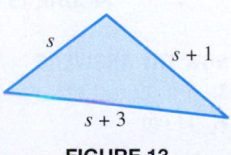

FIGURE 13

NOW TRY ANSWER
2. width: 30 ft; length: 50 ft

 **NOW TRY EXERCISE 3**

The perimeter of a triangle is 30 ft. The longest side is 1 ft longer than the medium side, and the shortest side is 7 ft shorter than the medium side. What are the lengths of the three sides of the triangle?

Step 3 **Write an equation.** Use the formula for the perimeter of a triangle.

$$P = a + b + c$$ Perimeter of a triangle

$$16 = s + (s + 1) + (s + 3)$$ Substitute for P, a, b, and c.

Step 4 **Solve.** $16 = 3s + 4$ Combine like terms.

We mentally subtracted 4 from each side, then divided each side by 3.

$$12 = 3s$$ Subtract 4.

$$4 = s$$ Divide by 3.

Step 5 **State the answer.** The shortest side, s, has length 4 ft.

$$s + 1 = 4 + 1 = 5 \text{ ft}$$ Length of medium side

$$s + 3 = 4 + 3 = 7 \text{ ft}$$ Length of longest side

Step 6 **Check.** The medium side, 5 ft, is 1 ft longer than the shortest side, and the longest side, 7 ft, is 3 ft longer than the shortest side. The perimeter is

$$4 + 5 + 7 = 16 \text{ ft}, \quad \text{as required.}$$ **NOW TRY**

 **NOW TRY EXERCISE 4**

The area of a triangle is 77 cm². The base is 14 cm. Find the height of the triangle.

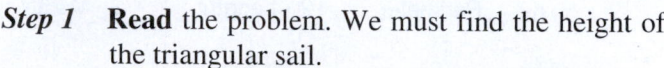 **EXAMPLE 4**   Finding the Height of a Triangular Sail

The area of a triangular sail of a sailboat is 126 ft². (Recall that "ft²" means "square feet.") The base of the sail is 12 ft. Find the height of the sail.

Step 1 **Read** the problem. We must find the height of the triangular sail.

Step 2 **Assign a variable.** Let h = the height of the sail, in feet. See **FIGURE 14**.

Step 3 **Write an equation.** Use the formula for the area of a triangle.

$$\mathcal{A} = \frac{1}{2}bh$$ $\mathcal{A}$ is area, b is base, and h is height.

$$126 = \frac{1}{2}(12)h$$ Let $\mathcal{A} = 126$ and $b = 12$.

Step 4 **Solve.** $126 = 6h$ Multiply.

$$21 = h$$ Divide by 6.

Step 5 **State the answer.** The height of the sail is 21 ft.

Step 6 **Check** to see that the values $\mathcal{A} = 126$, $b = 12$, and $h = 21$ satisfy the formula for the area of a triangle.

$$126 = \frac{1}{2}(12)(21) \text{ is true.}$$ **NOW TRY**

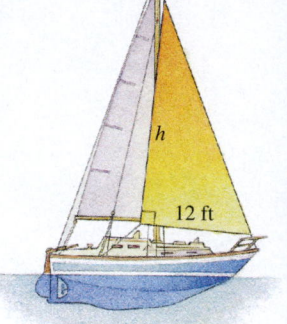

FIGURE 14

OBJECTIVE 3 Solve problems involving vertical angles and straight angles.

FIGURE 15 shows two intersecting lines forming angles that are numbered, ①, ②, ③, and ④. Angles ① and ③ lie "opposite" each other. They are **vertical angles.** Another pair of vertical angles is ② and ④. *Vertical angles have equal measures.*

Consider angles ① and ②. When their measures are added, we obtain 180°, the measure of a **straight angle.** There are three other angle pairs that form straight angles: ② and ③, ③ and ④, and ① and ④.

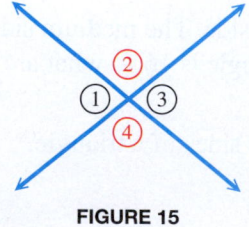

FIGURE 15

NOW TRY ANSWERS
3. 5 ft, 12 ft, 13 ft
4. 11 cm

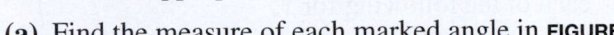

EXAMPLE 5 Finding Angle Measures

Refer to the appropriate figure in each part.

(a) Find the measure of each marked angle in **FIGURE 16**.

The marked angles are vertical angles, so they have equal measures.

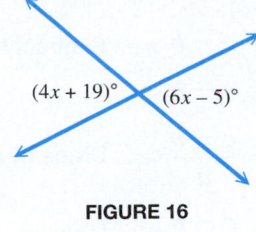

FIGURE 16

$$4x + 19 = 6x - 5 \qquad \text{Set } 4x + 19 \text{ equal to } 6x - 5.$$
$$19 = 2x - 5 \qquad \text{Subtract } 4x.$$
$$24 = 2x \qquad \text{Add 5.}$$
$$12 = x \qquad \text{Divide by 2.}$$

[This is *not* the answer.]

Replace x with 12 in the expression for the measure of each angle.

$4x + 19$		$6x - 5$	
$= 4(12) + 19$	Let $x = 12$.	$= 6(12) - 5$	Let $x = 12$.
$= 48 + 19$	Multiply.	$= 72 - 5$	Multiply.
$= 67$	Add.	$= 67$	Subtract.

The angles have equal measures, as required. Each measures $67°$.

(b) Find the measure of each marked angle in **FIGURE 17**.

The measures of the marked angles must add to 180° because together they form a straight angle. (They are also *supplements* of each other.)

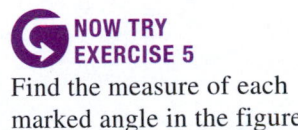

FIGURE 17

$$(3x - 30) + 4x = 180 \qquad \text{Supplementary angles sum to } 180°.$$
$$7x - 30 = 180 \qquad \text{Combine like terms.}$$
$$7x = 210 \qquad \text{Add 30.}$$
$$x = 30 \qquad \text{Divide by 7.}$$

[Don't stop here.]

NOW TRY EXERCISE 5

Find the measure of each marked angle in the figure.

Replace x with 30 in the expression for the measure of each angle.

$3x - 30$		$4x$	
$= 3(30) - 30$	Let $x = 30$.	$= 4(30)$	Let $x = 30$.
$= 90 - 30$	Multiply.	$= 120$	Multiply.
$= 60$	Subtract.		

The two angles measure $60°$ and $120°$, which add to $180°$, as required. **NOW TRY**

! CAUTION In **Example 5**, the answer is *not* the value of *x*. *Remember to substitute the value of the variable into the expression given for each angle.*

OBJECTIVE 4 Solve a formula for a specified variable.

Sometimes we want to rewrite a formula in terms of a *different* variable in the formula. For example, consider $\mathcal{A} = LW$, the formula for the area of a rectangle.

How can we rewrite $\mathcal{A} = LW$ in terms of W?

The process whereby we do this involves **solving for a specified variable,** or **solving a literal equation.**

NOW TRY ANSWER
5. $32°, 32°$

To solve a formula for a specified variable, we use the *same* steps that we used to solve an equation with just one variable. Consider the parallel reasoning to solve each of the following for x.

$$3x + 4 = 13$$
$$3x + 4 - 4 = 13 - 4 \qquad \text{Subtract 4.}$$
$$3x = 9$$
$$\frac{3x}{3} = \frac{9}{3} \qquad \text{Divide by 3.}$$
$$x = 3 \qquad \begin{array}{l}\text{Equation}\\ \text{solved for } x\end{array}$$

$$ax + b = c$$
$$ax + b - b = c - b \qquad \text{Subtract } b.$$
$$ax = c - b$$
$$\frac{ax}{a} = \frac{c - b}{a} \qquad \text{Divide by } a.$$
$$x = \frac{c - b}{a} \qquad \begin{array}{l}\text{Formula}\\ \text{solved for } x\end{array}$$

When we solve a formula for a specified variable, we treat the specified variable as if it were the ONLY variable in the equation, and we treat the other variables as if they were constants (numbers).

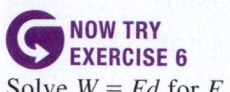
NOW TRY
EXERCISE 6
Solve $W = Fd$ for F.

EXAMPLE 6   Solving for a Specified Variable

Solve $\mathcal{A} = LW$ for W.

W is multiplied by L, so we undo the multiplication by dividing each side by L.

$$\mathcal{A} = LW \qquad \text{\small Our goal is to isolate } W.$$

$$\frac{\mathcal{A}}{L} = \frac{LW}{L} \qquad \text{Divide by } L.$$

$$\frac{\mathcal{A}}{L} = W, \quad \text{or} \quad W = \frac{\mathcal{A}}{L} \qquad \frac{LW}{L} = \frac{L}{L} \cdot W = 1 \cdot W = W$$

CHECK   Substitute $\frac{\mathcal{A}}{L}$ for W in the original equation.

$$\mathcal{A} = LW \qquad \text{Original equation}$$
$$\mathcal{A} = L\left(\frac{\mathcal{A}}{L}\right) \qquad \text{Let } W = \frac{\mathcal{A}}{L}.$$
$$\mathcal{A} = \mathcal{A} \quad \checkmark \qquad \text{True}$$

A true statement results, so $W = \frac{\mathcal{A}}{L}$.

NOW TRY

NOW TRY
EXERCISE 7
Solve $Ax + By = C$ for A.

EXAMPLE 7   Solving for a Specified Variable

Solve $P = 2L + 2W$ for L.

$$\text{\small Our goal is to isolate } L.$$

$$P = 2L + 2W$$
$$P - 2W = 2L + 2W - 2W \qquad \text{Subtract } 2W.$$
$$P - 2W = 2L \qquad \text{Combine like terms.}$$
$$\frac{P - 2W}{2} = \frac{2L}{2} \qquad \text{Divide by 2.}$$
$$\frac{P - 2W}{2} = L, \quad \text{or} \quad L = \frac{P - 2W}{2} \qquad \frac{2L}{2} = \frac{2}{2} \cdot L = 1 \cdot L = L$$

$$\frac{P - 2W}{2} \neq P - W$$

NOW TRY ANSWERS
6. $F = \frac{W}{d}$
7. $A = \frac{C - By}{x}$

NOW TRY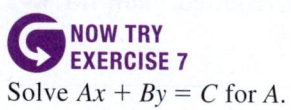

NOW TRY
EXERCISE 8

Solve $S = \frac{1}{2}(a + b + c)$ for a.

EXAMPLE 8 Solving for a Specified Variable

Solve $\mathcal{A} = \frac{1}{2}h(b + B)$ for B.

> Our goal is to isolate B.

$$\mathcal{A} = \frac{1}{2}h(b + B)$$

$$\mathcal{A} = \frac{1}{2}hb + \frac{1}{2}hB \qquad \text{Clear the parentheses using the distributive property.}$$

$$2 \cdot \mathcal{A} = 2\left(\frac{1}{2}hb + \frac{1}{2}hB\right) \qquad \text{Multiply each side by 2 to clear the fractions.}$$

$$2 \cdot \mathcal{A} = 2 \cdot \frac{1}{2}hb + 2 \cdot \frac{1}{2}hB \qquad \text{Distributive property}$$

$$2\mathcal{A} = hb + hB \qquad \text{Multiply; } 2 \cdot \frac{1}{2} = \frac{2}{2} = 1$$

$$2\mathcal{A} - hb = hb + hB - hb \qquad \text{Subtract } hb.$$

$$2\mathcal{A} - hb = hB \qquad \text{Combine like terms.}$$

$$\frac{2\mathcal{A} - hb}{h} = \frac{hB}{h} \qquad \text{Divide by } h.$$

$$\frac{2\mathcal{A} - hb}{h} = B, \quad \text{or} \quad B = \frac{2\mathcal{A} - hb}{h}$$

NOW TRY

NOW TRY
EXERCISE 9

Solve each equation for y.

(a) $5x + y = 3$

(b) $x - 2y = 8$

EXAMPLE 9 Solving for a Specified Variable

Solve each equation for y.

(a) $\qquad 2x - y = 7$

> Our goal is to isolate y.

$$2x - y - 2x = 7 - 2x \qquad \text{Subtract } 2x.$$

> Be careful to retain the $-$ symbol.

$$-y = 7 - 2x \qquad \text{Combine like terms.}$$

$$-1(-y) = -1(7 - 2x) \qquad \text{Multiply by } -1.$$

$$y = -7 + 2x \qquad \text{Multiply; distributive property}$$

$$y = 2x - 7 \qquad -a + b = b - a$$

We could have added y and subtracted 7 from each side of the equation to isolate y on the right, giving $2x - 7 = y$, a different form of the same result.

(b) $\qquad -3x + 2y = 6$

$$-3x + 2y + 3x = 6 + 3x \qquad \text{Add } 3x.$$

$$2y = 3x + 6 \qquad \text{Combine like terms; commutative property}$$

$$\frac{2y}{2} = \frac{3x + 6}{2} \qquad \text{Divide by 2.}$$

> Be careful here.

$$y = \frac{3x}{2} + \frac{6}{2} \qquad \frac{a + b}{c} = \frac{a}{c} + \frac{b}{c}$$

> $\frac{3x}{2} = \frac{3}{2} \cdot \frac{x}{1} = \frac{3}{2}x$

$$y = \frac{3}{2}x + 3 \qquad \text{Simplify.}$$

NOW TRY ANSWERS

8. $a = 2S - b - c$
9. (a) $y = -5x + 3$
    (b) $y = \frac{1}{2}x - 4$

Although we could have given our answer as $y = \frac{3x + 6}{2}$, we simplified further in preparation for later work.   NOW TRY

2.6 Exercises

FOR EXTRA HELP **MyLab Math**

 Video solutions for select problems available in MyLab Math

STUDY SKILLS REMINDER
Time management can be a challenge for students. **Review Study Skill 6, *Managing Your Time*.**

Concept Check *Give a one-sentence definition of each term.*

1. Perimeter of a plane geometric figure

2. Area of a plane geometric figure

Concept Check *Decide whether* perimeter *or* area *would be used to solve a problem concerning the measure of the quantity.*

3. Carpeting for a bedroom

4. Sod for a lawn

5. Fencing for a yard

6. Baseboards for a living room

7. Tile for a bathroom

8. Fertilizer for a garden

9. Determining the cost of replacing a linoleum floor with a wood floor

10. Determining the cost of planting rye grass in a lawn for the winter

Find the value of the remaining variable in each formula. Use 3.14 as an approximation for π (pi). **See Example 1.**

11. $P = 2L + 2W$ (perimeter of a rectangle); $L = 8, W = 5$

12. $P = 2L + 2W$; $L = 6, W = 4$

13. $A = \frac{1}{2}bh$ (area of a triangle); $b = 8, h = 16$

14. $A = \frac{1}{2}bh$; $b = 10, h = 14$

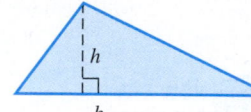

15. $P = a + b + c$ (perimeter of a triangle); $P = 12, a = 3, c = 5$

16. $P = a + b + c$; $P = 15, a = 3, b = 7$

17. $d = rt$ (distance formula); $d = 252, r = 45$

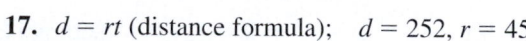

18. $d = rt$; $d = 100, t = 2.5$

19. $A = \frac{1}{2}h(b + B)$ (area of a trapezoid); $A = 91, h = 7, b = 12$

20. $A = \frac{1}{2}h(b + B)$; $A = 75, b = 19, B = 31$

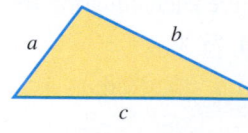

21. $C = 2\pi r$ (circumference of a circle); $C = 16.328$

22. $C = 2\pi r$; $C = 8.164$

23. $C = 2\pi r$; $C = 20\pi$

24. $C = 2\pi r$; $C = 100\pi$

25. $A = \pi r^2$ (area of a circle); $r = 4$

26. $A = \pi r^2$; $r = 12$

27. $S = 2\pi rh$; $S = 120\pi, h = 10$

28. $S = 2\pi rh$; $S = 720\pi, h = 30$

The **volume** of a three-dimensional geometric figure is a measure of the space occupied by the figure. For example, we would need to know the volume of a gasoline tank in order to determine how many gallons of gasoline would completely fill the tank. Volume is measured in cubic units.

In each exercise, a formula for the volume (V) of a three-dimensional figure is given, along with values for the other variables. Evaluate V. (Use 3.14 as an approximation for π.)
See Example 1.

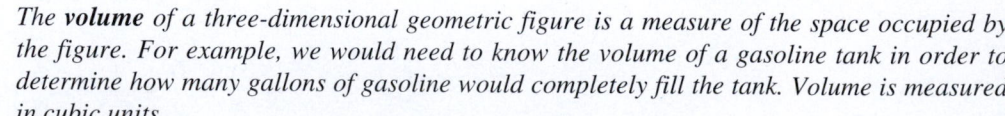
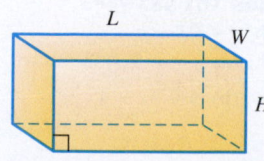

29. $V = LWH$ (volume of a rectangular box); $L = 10, W = 5, H = 3$

30. $V = LWH$; $L = 12, W = 8, H = 4$

31. $V = \dfrac{1}{3}Bh$ (volume of a pyramid); $B = 12, h = 13$

32. $V = \dfrac{1}{3}Bh$; $B = 36, h = 4$

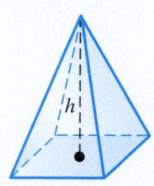

33. $V = \dfrac{4}{3}\pi r^3$ (volume of a sphere); $r = 12$

34. $V = \dfrac{4}{3}\pi r^3$; $r = 6$

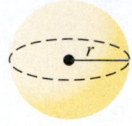

Simple interest I in dollars is calculated using the formula

$$I = prt, \qquad \text{Simple interest formula}$$

where p represents the principal, or amount, in dollars that is invested or borrowed, r represents the annual interest rate, expressed as a decimal, and t represents time, in years.
    In each exercise, find the value of the remaining variable in the simple interest formula.
See Example 1. *(Hint: Write percents as decimals.* **Recall that** $1\% = \dfrac{1}{100} = \mathbf{0.01}.)$

35. $p = \$7500, r = 4\%, t = 2$ yr   **36.** $p = \$3600, r = 3\%, t = 4$ yr

37. $I = \$33, r = 2\%, t = 3$ yr   **38.** $I = \$270, r = 5\%, t = 6$ yr

39. $I = \$180, p = \$4800, r = 2.5\%$ **40.** $I = \$162, p = \$2400, r = 1.5\%$

Solve each problem. **See Examples 2 and 3.**

41. The length of a rectangle is 9 in. more than the width. The perimeter is 54 in. Find the length and the width of the rectangle.

42. The width of a rectangle is 3 ft less than the length. The perimeter is 62 ft. Find the length and the width of the rectangle.

43. The perimeter of a rectangle is 36 m. The length is 2 m more than three times the width. Find the length and the width of the rectangle.

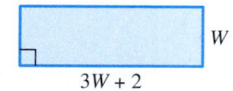

44. The perimeter of a rectangle is 36 yd. The width is 18 yd less than twice the length. Find the length and the width of the rectangle.

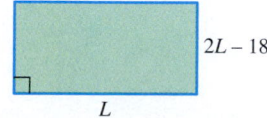

45. The longest side of a triangle is 3 in. longer than the shortest side. The medium side is 2 in. longer than the shortest side. If the perimeter of the triangle is 20 in., what are the lengths of the three sides?

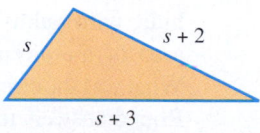

46. The medium side of a triangle is 4 ft longer than the shortest side. The longest side is twice as long as the shortest side. If the perimeter of the triangle is 28 ft, what are the lengths of the three sides?

47. Two sides of a triangle have the same length. The third side measures 4 m less than twice that length. The perimeter of the triangle is 24 m. Find the lengths of the three sides.

48. Two sides of a triangle have the same length. The third side measures 6 yd less than three times that length. The perimeter of the triangle is 19 yd. Find the lengths of the three sides.

Use a formula to write an equation for each application, and then solve. (Use 3.14 as an approximation for π.) **Formulas are found at the back of this text. See Examples 2–4.**

49. One of the largest fashion catalogues in the world was published in Hamburg, Germany. Each of the 212 pages in the catalogue measured 1.2 m by 1.5 m. What was the perimeter of a page? What was the area? (Data from *Guinness World Records.*)

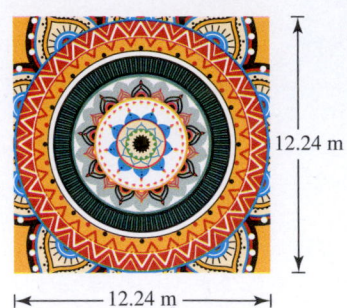

50. One of the world's largest mandalas (sand paintings) measures 12.24 m by 12.24 m. What is the perimeter of the sand painting? To the nearest hundredth of a square meter, what is the area? (Data from *Guinness World Records.*)

51. The area of a triangular road sign is 70 ft². If the base of the sign measures 14 ft, what is the height of the sign?

52. The area of a triangular advertising banner is 96 ft². If the height of the banner measures 12 ft, what is the measure of the base?

53. A prehistoric ceremonial site dating to 3000 B.C. was discovered in southwestern England. The site is a nearly perfect circle, consisting of nine concentric rings that probably held upright wooden posts. Around this timber temple is a wide, encircling ditch enclosing an area with a diameter of 443 ft. Find this enclosed area to the nearest thousand square feet. (*Hint:* Find the radius. Then use $\mathcal{A} = \pi r^2$.) (Data from *Archaeology.*)

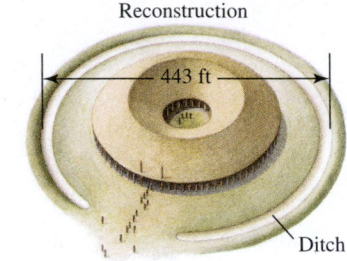

Reconstruction

Ditch

54. The Rogers Centre in Toronto, Canada, is the first stadium with a hard-shell, retractable roof. The steel dome is 630 ft in diameter. To the nearest foot, what is the circumference of this dome? (Data from www.ballparks.com)

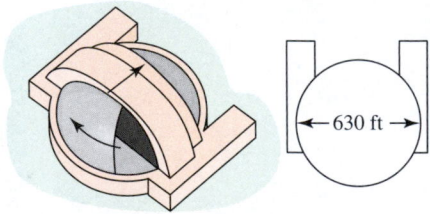

55. One of the largest drums ever constructed was made from Japanese cedar and cowhide, with radius 7.87 ft. What was the area of the circular face of the drum? What was the circumference of the drum? Round answers to the nearest hundredth. (Data from *Guinness World Records.*)

56. A drum played at the Royal Festival Hall in London had radius 6.5 ft. What was the area of the circular face of the drum? What was the circumference of the drum? (Data from *Guinness World Records.*)

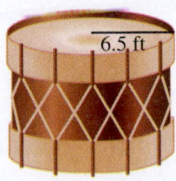

57. The survey plat depicted here shows two lots that form a trapezoid. The measures of the parallel sides are 115.80 ft and 171.00 ft. The height of the trapezoid is 165.97 ft. Find the combined area of the two lots. Round the answer to the nearest hundredth of a square foot.

58. Lot A in the survey plat is in the shape of a trapezoid. The parallel sides measure 26.84 ft and 82.05 ft. The height of the trapezoid is 165.97 ft. Find the area of Lot A. Round the answer to the nearest hundredth of a square foot.

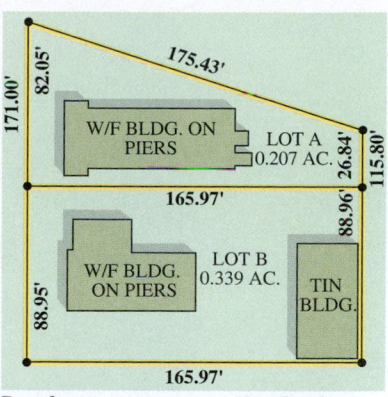

Data from property survey, New Roads, Louisiana.

59. The U.S. Postal Service requires that any box sent by Priority Mail® have length plus girth (distance around) totaling no more than 108 in. The maximum volume that meets this condition is contained by a box with a square end 18 in. on each side. What is the length of the box? What is the maximum volume? (Data from United States Postal Service.)

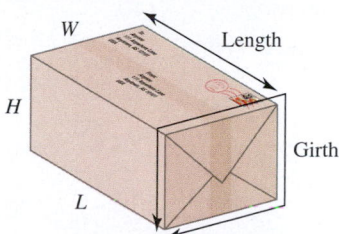

60. One of the world's largest sandwiches, made by Wild Woody's Chill and Grill in Roseville, Michigan, was 12 ft long, 12 ft wide, and $17\frac{1}{2}$ in. ($1\frac{11}{24}$ ft) thick. What was the volume of the sandwich? (Data from *Guinness World Records*.)

Not to scale

Find the measure of each marked angle. See Example 5.

61.
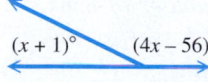
$(x + 1)°$ $(4x - 56)°$

62.
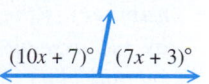
$(10x + 7)°$ $(7x + 3)°$

63.
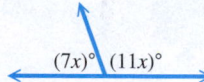
$(7x)°$ $(11x)°$

64.
$(20x + 10)°$ $(3x + 9)°$

65.
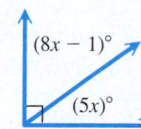
$(8x - 1)°$
$(5x)°$

66.
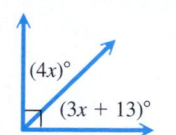
$(4x)°$
$(3x + 13)°$

67.
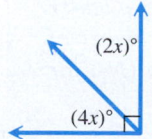
$(2x)°$
$(4x)°$

68.
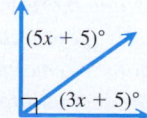
$(5x + 5)°$
$(3x + 5)°$

69.
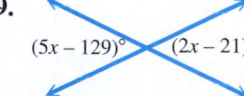
$(5x - 129)°$ $(2x - 21)°$

70.
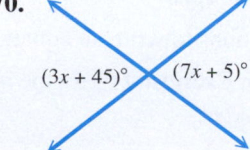
$(3x + 45)°$ $(7x + 5)°$

71.
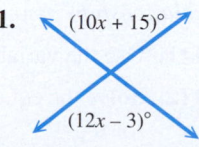
$(10x + 15)°$
$(12x - 3)°$

72.
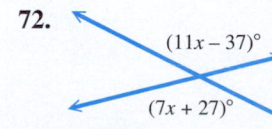
$(11x - 37)°$
$(7x + 27)°$

Solve each formula for the specified variable. See Examples 6–8.

73. $d = rt$ for t **74.** $d = rt$ for r **75.** $A = bh$ for b

76. $A = LW$ for L **77.** $C = \pi d$ for d **78.** $P = 4s$ for s

79. $V = LWH$ for H **80.** $V = LWH$ for W **81.** $I = prt$ for r

82. $I = prt$ for p **83.** $A = \frac{1}{2}bh$ for h **84.** $A = \frac{1}{2}bh$ for b

85. $V = \frac{1}{3}\pi r^2 h$ for h **86.** $V = \pi r^2 h$ for h **87.** $P = a + b + c$ for b

88. $P = a + b + c$ for a **89.** $P = 2L + 2W$ for W **90.** $A = p + prt$ for r

91. $y = mx + b$ for m **92.** $y = mx + b$ for x **93.** $Ax + By = C$ for y

94. $Ax + By = C$ for x **95.** $M = C(1 + r)$ for r **96.** $A = p(1 + rt)$ for t

97. $P = 2(a + b)$ for a **98.** $P = 2(a + b)$ for b **99.** $S = \frac{1}{2}(a + b + c)$ for b

100. $S = \frac{1}{2}(a + b + c)$ for c **101.** $C = \frac{5}{9}(F - 32)$ for F **102.** $A = \frac{1}{2}h(b + B)$ for b

Solve each equation for y. See Example 9.

103. $6x + y = 4$ **104.** $3x + y = 6$ **105.** $5x - y = 2$ **106.** $4x - y = 1$

107. $-3x + 5y = -15$ **108.** $-2x + 3y = -9$ **109.** $x - 3y = 12$ **110.** $x - 5y = 10$

RELATING CONCEPTS   For Individual or Group Work   (Exercises 111–114)

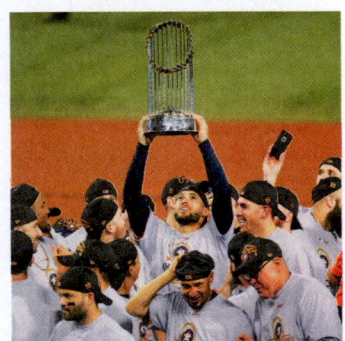

The climax of any sports season is the playoffs. Baseball fans eagerly debate predictions of which team will win the pennant for their division. The magic number for each first-place team is often reported in media outlets. The **magic number** (sometimes called the **elimination number**) is the combined number of wins by the first-place team and losses by the second-place team that would clinch the title for the first-place team.

To calculate the magic number, consider the following conditions.

The number of wins for the first-place team (W_1) plus the magic number (M) is one more than the sum of the number of wins to date (W_2) and the number of games remaining in the season (N_2) for the second-place team.

▼ **American League**

East	W	L	PCT	GB
Boston Red Sox	81	62	.566	—
New York Yankees	78	65	.545	3.0
Baltimore Orioles	71	73	.493	10.5
Tampa Bay Rays	71	74	.490	11.0
Toronto Blue Jays	67	77	.465	14.5
Central	**W**	**L**	**PCT**	**GB**
Cleveland Indians	88	56	.611	—
Minnesota Twins	74	69	.517	13.5
Kansas City Royals	71	72	.497	16.5
Detroit Tigers	60	83	.420	27.5
Chicago White Sox	57	86	.399	30.5
West	**W**	**L**	**PCT**	**GB**
Houston Astros	88	57	.601	—
Los Angeles Angels	73	70	.510	13.0
Texas Rangers	72	71	.503	14.0
Seattle Mariners	71	73	.493	15.5
Oakland Athletics	63	80	.441	23.0

Data from MLB.com

Work Exercises 111–114 in order, to see the relationships among these concepts.

111. Use the variable definitions to write an equation involving the magic number.

112. Solve the equation from **Exercise 111** for the magic number M and write a formula for it.

113. The American League standings on September 11, 2017, are shown in the table on the preceding page. There were 162 regulation games in the 2017 season. Find the magic number for each first-place team. The number of games remaining in the season for the second-place team is calculated as

$$N_2 = 162 - (W_2 + L_2),$$

where L_2 represents the number of losses for the second-place team.

(a) AL East: Boston vs New York

(b) AL Central: Cleveland vs Minnesota

(c) AL West: Houston vs Los Angeles

114. Calculate the magic number for Oakland vs Houston. (Treat Oakland as though it were the second-place team.) How can we interpret the result?

2.7 Ratio, Proportion, and Percent

OBJECTIVES

1 Write ratios.
2 Solve proportions.
3 Solve applied problems using proportions.
4 Find percents and percentages.

VOCABULARY

☐ ratio
☐ proportion
☐ terms of a proportion
☐ extremes
☐ means
☐ cross products of a proportion
☐ percent
☐ percentage
☐ base

NOW TRY EXERCISE 1

Write a ratio for each word phrase. Express fractions in lowest terms.

(a) 7 in. to 4 in.

(b) 45 sec to 2 min

NOW TRY ANSWERS

1. (a) $\frac{7}{4}$ (b) $\frac{3}{8}$

OBJECTIVE 1 Write ratios.

A **ratio** is a comparison of two quantities using a quotient.

Ratio

The ratio of a number a to a number b (where $b \neq 0$) is written as follows.

$$a \text{ to } b, \quad a:b, \quad \text{or} \quad \frac{a}{b}$$

Writing a ratio as a quotient $\frac{a}{b}$ is most common in algebra.

Examples: 2 to 3,   2:3, $\frac{2}{3}$

EXAMPLE 1 Writing Word Phrases as Ratios

Write a ratio for each word phrase. Express fractions in lowest terms.

(a) 5 hr to 3 hr $\dfrac{5 \text{ hr}}{3 \text{ hr}} = \dfrac{5}{3}$

(b) 6 in. to 14 in. $\dfrac{6 \text{ in.}}{14 \text{ in.}} = \dfrac{6}{14} = \dfrac{3}{7}$ Write in lowest terms.

(c) 6 hr to 3 days

First convert 3 days to hours.

$$3 \text{ days} = 3 \cdot 24 = 72 \text{ hr} \qquad \text{1 day = 24 hr}$$

Now write the ratio using the common unit of measure, hours.

$$\dfrac{6 \text{ hr}}{3 \text{ days}} = \dfrac{6 \text{ hr}}{72 \text{ hr}} = \dfrac{6}{72} = \dfrac{1}{12} \qquad \text{Write in lowest terms.}$$

NOW TRY

One application of ratios is in *unit pricing,* to see which size of an item offered in different sizes produces the best price per unit.

NOW TRY
EXERCISE 2

A supermarket charges the following prices for a certain brand of liquid detergent.

Size	Price
75 oz	$8.94
100 oz	$13.97
150 oz	$19.97

Which size is the best buy? What is the unit price (to the nearest thousandth) for that size?

EXAMPLE 2   Finding Price per Unit

A supermarket charges the following prices for a jar of crunchy peanut butter.

Peanut Butter

Size	Price
18 oz	$3.49
28 oz	$4.99
40 oz	$6.79

Which size is the best buy? That is, which size has the lowest unit price?

To find the best buy, write ratios comparing the price for each size jar to the number of units (ounces) per jar.

Peanut Butter

Size	Price	Unit Price (dollars per ounce)
18 oz	$3.49	$\dfrac{\$3.49}{18} = \$0.194$
28 oz	$4.99	$\dfrac{\$4.99}{28} = \$0.178$
40 oz	$6.79	$\dfrac{\$6.79}{40} = \$0.170$ ← Best buy

> To find the price per ounce, the number of ounces goes in the denominator.

(Results are rounded to the nearest thousandth.)

Because the 40-oz size produces the lowest unit price, it is the best buy. Buying the largest size does not always provide the best buy, although it often does, as in this case.

NOW TRY

OBJECTIVE 2 Solve proportions.

A ratio is used to compare two numbers or amounts. A **proportion** says that two ratios are equal. For example, the proportion

$$\frac{3}{4} = \frac{15}{20}$$

> A proportion is a special type of equation.

says that the ratios $\frac{3}{4}$ and $\frac{15}{20}$ are equal. In the proportion

$$\frac{a}{b} = \frac{c}{d} \quad (\text{where } b, d \neq 0),$$

$a, b, c,$ and d are the **terms** of the proportion. The terms a and d are the **extremes,** and the terms b and c are the **means.** We read the proportion $\frac{a}{b} = \frac{c}{d}$ as

"a is to b as c is to d."

NOW TRY ANSWER
2. 75 oz; $0.119 per oz

Multiplying each side of this proportion by the common denominator, bd, gives the following.

$$\frac{a}{b} = \frac{c}{d}$$

$$bd \cdot \frac{a}{b} = bd \cdot \frac{c}{d} \qquad \text{Multiply each side by } bd.$$

$$\frac{b}{b}(d \cdot a) = \frac{d}{d}(b \cdot c) \qquad \text{Associative and commutative properties}$$

$$ad = bc \qquad \text{Commutative and identity properties}$$

We can also find the products ad and bc by multiplying diagonally.

$$ad = bc$$

$$\frac{a}{b} \times \frac{c}{d}$$

For this reason, ad and bc are the **cross products of the proportion.**

Cross Products of a Proportion

If $\frac{a}{b} = \frac{c}{d}$, then the cross products ad and bc are equal—that is, ***the product of the extremes equals the product of the means.***

Also, if $ad = bc$, then $\frac{a}{b} = \frac{c}{d}$ (where $b, d \neq 0$).

NOTE If $\frac{a}{c} = \frac{b}{d}$, then $ad = cb$, or $ad = bc$. This means that the two proportions are equivalent, and the proportion

$$\frac{a}{b} = \frac{c}{d} \qquad \text{can also be written as} \qquad \frac{a}{c} = \frac{b}{d} \qquad (\text{where } c, d \neq 0).$$

Sometimes one form is more convenient to work with than the other.

NOW TRY EXERCISE 3

Determine whether each proportion is *true* or *false*.

(a) $\frac{1}{3} = \frac{33}{100}$ (b) $\frac{4}{13} = \frac{16}{52}$

EXAMPLE 3 Determining Whether Proportions Are True

Determine whether each proportion is *true* or *false*.

(a) $\frac{3}{4} = \frac{15}{20}$

Check to see whether the cross products are equal.

$$4 \cdot 15 = 60$$

$$\frac{3}{4} \times \frac{15}{20}$$

$$3 \cdot 20 = 60$$

The cross products are equal, so the proportion is true.

(b) $\frac{6}{7} = \frac{30}{32}$ The cross products, $6 \cdot 32 = 192$ and $7 \cdot 30 = 210$, are not equal, so the proportion is false.

NOW TRY

Four numbers are used in a proportion. If any three of these numbers are known, the fourth can be found.

NOW TRY EXERCISE 4

Solve.

$$\frac{9}{7} = \frac{x}{56}$$

EXAMPLE 4   Finding an Unknown in a Proportion

Solve $\frac{5}{9} = \frac{x}{63}$.

$$\frac{5}{9} = \frac{x}{63} \quad \text{Solve for } x.$$

$$5 \cdot 63 = 9 \cdot x \quad \text{Cross products must be equal.}$$

$$315 = 9x \quad \text{Multiply.}$$

$$35 = x \quad \text{Divide by 9.}$$

CHECK   Because $\frac{5}{9} = \frac{35}{63}$ is a true statement, the solution set is $\{35\}$.   **NOW TRY**

NOW TRY EXERCISE 5

Solve.

$$\frac{k-3}{6} = \frac{3k+2}{4}$$

EXAMPLE 5   Solving an Equation Using Cross Products

Solve $\frac{m-2}{5} = \frac{m+1}{3}$.

$$\frac{m-2}{5} = \frac{m+1}{3} \quad \boxed{\text{Be sure to use parentheses.}}$$

$$3(m-2) = 5(m+1) \quad (\ast) \quad \text{Cross products}$$

$$3m - 6 = 5m + 5 \quad \text{Distributive property}$$

$$-2m - 6 = 5 \quad \text{Subtract } 5m.$$

$$-2m = 11 \quad \text{Add 6.}$$

$$m = -\frac{11}{2} \quad \text{Divide by } -2.$$

Check by substituting $-\frac{11}{2}$ for m in the proportion. The solution set is $\left\{-\frac{11}{2}\right\}$.

NOW TRY

> **NOTE**   When we set cross products equal to each other, we are actually multiplying each ratio in the proportion by a common denominator.
>
> $$\frac{m-2}{5} = \frac{m+1}{3} \quad \text{See Example 5.}$$
>
> $$15\left(\frac{m-2}{5}\right) = 15\left(\frac{m+1}{3}\right) \quad \text{Multiply each ratio by 15, the LCD.}$$
>
> $$15\left(\frac{m-2}{5}\right)$$
> $$= 15 \cdot \frac{1}{5}(m-2)$$
> $$= 3(m-2)$$
>
> $$3(m-2) = 5(m+1) \quad \text{This is equation }(\ast)\text{ from Example 5.}$$

> ⊘ **CAUTION**   *The cross-product method cannot be used directly if there is more than one term on either side of the equality symbol.*
>
> $$\underbrace{\frac{m-1}{5} = \frac{m+1}{3} - 4}_{2 \text{ terms}}, \qquad \underbrace{\frac{x}{3} + \frac{5}{4} = \frac{1}{2}}_{2 \text{ terms}}$$
>
> Do **not** use the cross-product method to solve equations in this form.

NOW TRY ANSWERS

4. $\{72\}$

5. $\left\{-\frac{12}{7}\right\}$

OBJECTIVE 3 Solve applied problems using proportions.

NOW TRY EXERCISE 6

Twenty gallons of gasoline costs $69.80. How much would 27 gal of the same gasoline cost?

EXAMPLE 6 Applying Proportions

After Lee Ann pumped 5.0 gal of gasoline, the display showing the price read $18.10. When she finished pumping the gasoline, the price display read $52.49. How many gallons did she pump?

To solve this problem, set up a proportion, with prices in the numerators and gallons in the denominators. Let x = the number of gallons she pumped.

Price $\longrightarrow$ $\dfrac{\$18.10}{5.0} = \dfrac{\$52.49}{x}$ $\longleftarrow$ Price
Gallons $\longrightarrow$ $\qquad\qquad$ $\longleftarrow$ Gallons

> Be sure that numerators represent the *same* quantities and denominators represent the *same* quantities.

$18.10x = 5.0(52.49)$ $\qquad$ Cross products

$18.10x = 262.45$ $\qquad$ Multiply.

$x = 14.5$ $\qquad$ Divide by 18.10.

She pumped 14.5 gal. Check this answer. Notice that the way the proportion was set up uses the fact that the unit price is the same, no matter how many gallons are purchased.

NOW TRY

NOW TRY EXERCISE 7

Suppose you are in the 50th car in line to park your car for a concert.

(a) In 75 sec, 3 cars are able to park. Assuming this rate stays the same, how long (to the nearest minute) will you be waiting to park your car?

(b) It takes 10 min to get from the parking lot to your seat. If the concert begins in 45 min, will you get your car parked in time to see the beginning of the concert?

EXAMPLE 7 Applying Proportions

Suppose you are the 100th person in line to buy tickets for the blockbuster movie *Star Wars: The Last Jedi*.

(a) In 90 sec, 4 people were able to buy tickets. Assuming this rate stays the same, how long (in minutes) will you be waiting in line to purchase your ticket?

Let x = the number of seconds you will be waiting in line.

Number of seconds $\longrightarrow$ $\dfrac{90}{4} = \dfrac{x}{100}$ $\longleftarrow$ Number of seconds
People buying tickets $\longrightarrow$ $\qquad\qquad$ $\longleftarrow$ People buying tickets

$90(100) = 4x$ $\qquad$ Cross products

$9000 = 4x$ $\qquad$ Multiply.

$2250 = x$ $\qquad$ Divide by 4.

Because 60 sec = 1 min, the number of minutes waiting in line will be

$$\dfrac{2250}{60}, \quad \text{or} \quad 37.5 \text{ min.}$$

(b) In 30 min the previews end and the movie begins. Will you be able to purchase your ticket in time to see the beginning of the movie?

No. Based on the answer to part (a), you are going to miss the first

$$37.5 - 30 = 7.5 \text{ min of the movie.}$$

NOW TRY

NOW TRY ANSWERS
6. $94.23
7. (a) 21 min **(b)** yes

OBJECTIVE 4 Find percents and percentages.

A ***percent*** *is a ratio where the second number is always 100.* For example,

50% represents the ratio of 50 to 100, that is, $\frac{50}{100}$, or, as a decimal, 0.50.

The word **percent** means *"per 100."* One percent means *"one per 100."*

$$1\% = \frac{1}{100} = 0.01, \quad 10\% = \frac{10}{100} = 0.10, \quad 100\% = \frac{100}{100} = 1 \qquad \text{Percent, fraction, and decimal equivalents}$$

We can solve a percent problem involving $x\%$ by writing it as a proportion. The amount, or **percentage,** is compared to the **base** (the whole amount).

$$\frac{x}{100} = \frac{amount}{base}$$

We can also write this proportion as follows.

$$\text{percent (as a decimal)} = \frac{amount}{base} \qquad \frac{x}{100} \text{ or } 0.01x \text{ is equivalent to } x \text{ percent.}$$

percent (as a decimal) · base = amount   Basic percent equation

**NOW TRY
EXERCISE 8**

Solve each problem.
(a) 20% of 70 is what number?
(b) 40% of what number is 130?
(c) What percent of 484 is 121?

EXAMPLE 8 Solving Percent Equations

Solve each problem.
(a) 15% of 600 is what number?

Let n = the number. The word *of* indicates multiplication, and the word *is* indicates equality.

  15%   of   600   is   what number?   *Translate each word or phrase to write the equation.*

  0.15   ·   600   = n Percent · base = amount

Write 15% as a decimal.

             $90 = n$ Multiply.

Thus, 15% of 600 is 90.

(b) 32% of what number is 64?

  32%   of   what number   is   64?

  0.32   · n =   64   Percent · base = amount

Write 32% as a demical.

                 $n = \dfrac{64}{0.32}$ Divide by 0.32.

                 $n = 200$ Simplify.

Thus, 32% of 200 is 64.

(c) What percent of 360 is 90?

  What percent   of   360   is   90?

         p ·   360   =   90   Percent · base = amount

                  $p = \dfrac{90}{360}$ Divide by 360.

Also, $\frac{90}{360} = \frac{1}{4} = \frac{25}{100}$, or 25%.

              $p = 0.25$,   or $p = 25\%$ Simplify. Write 0.25 as a percent.

Thus, 25% of 360 is 90.

NOW TRY ANSWERS
8. (a) 14   **(b)** 325   **(c)** 25%

NOW TRY

> **NOTE**   We can also solve percent problems using proportions. To answer the question "15% of 600 is what number?" in **Example 8(a),** we could work as follows.
>
> $$\frac{x}{100} = \frac{\text{amount}}{\text{base}}$$ Percent proportion
>
> $$\frac{15}{100} = \frac{n}{600}$$ Substitute the percent for x and **600** for the **base.** We must find the amount n.
>
> $$15 \cdot 600 = 100 \cdot n$$ Cross products must be equal.
>
> $$9000 = 100n$$ Multiply.
>
> $$90 = n$$ Divide by 100.
>
> The same answer results—that is, 15% of 600 is 90.

 **NOW TRY EXERCISE 9**

A winter coat is on a clearance sale for $48. The regular price is $120. What percent of the regular price is the savings?

EXAMPLE 9   Solving an Applied Percent Problem

A newspaper ad offered a set of tires at a sales price of $258. The regular price was $300. What percent of the regular price was the savings?

The savings on the tires amounted to $300 − $258 = $42. We can now restate the problem: *What percent of 300 is 42?*

What percent   of   300   is   42?
↓   ↓   ↓   ↓   ↓
p ·   300   =   42   Write the percent equation.

$$p = \frac{42}{300}$$ Divide by 300.

$$p = 0.14, \quad \text{or} \quad 14\%$$ Simplify. Write 0.14 as a percent.

The sale price represents a 14% savings.   **NOW TRY**

 **NOW TRY EXERCISE 10**

A wash, cut, and style at a hair salon costs $38.00. Maria wants to leave a 20% tip. Estimate how much she should leave to the nearest dollar.

EXAMPLE 10   Solving an Applied Percent Problem

A restaurant bill (before tax) is $18.15. If Debbie wants to leave a 15% tip, estimate how much she should leave to the nearest dollar.

Round $18.15 to $20.00. We can restate the problem:

"15% of $20.00 *is what number?*"

15%   of   20   is   what number?
↓   ↓   ↓   ↓   ↓
0.15 ·   20   = n

$$3 = n$$ Multiply.

She should leave a $3.00 tip.

To check mentally, 15% of $10 is $1.50. So 15% of $20 is twice as much—that is, $3.00.   **NOW TRY**

2.7 Exercises

FOR EXTRA HELP   ▶ **MyLab Math**

Video solutions for select problems available in MyLab Math

1. **Concept Check**   Ratios are used to _____ two numbers or quantities. Which of the following indicate the ratio of a to b?

   A. $\dfrac{a}{b}$ **B.** $\dfrac{b}{a}$ **C.** $a \cdot b$ **D.** $a:b$

2. **Concept Check**   A proportion says that two _____ are equal. The equation

$$\frac{a}{b} = \frac{c}{d} \quad (\text{where } b, d \neq 0)$$

   is a _____, where ad and bc are the _____.

3. **Concept Check**   Match each ratio in Column I with the ratio equivalent to it in Column II.

I	II
(a) 75 to 100	**A.** 80 to 100
(b) 5 to 4	**B.** 50 to 100
(c) $\dfrac{1}{2}$	**C.** 3 to 4
(d) 4 to 5	**D.** 15 to 12

4. **Concept Check**   Which of the following represent a ratio of 4 days to 2 weeks?

   A. $\dfrac{4}{2}$ **B.** $\dfrac{4}{7}$ **C.** $\dfrac{4}{14}$

   D. $\dfrac{2}{1}$ **E.** $\dfrac{2}{7}$ **F.** $\dfrac{1}{2}$

   G. $\dfrac{2}{4}$ **H.** $\dfrac{7}{2}$ **I.** 2

Concept Check   *Write each percent as a ratio, a fraction (with denominator 100), and a decimal. For example,*

   24% represents the *ratio* 24 to 100, the *fraction* $\dfrac{24}{100}$, and the *decimal* 0.24.

   5. 1%   **6.** 5%   **7.** 75%   **8.** 30%   **9.** 100%   **10.** 125%

*Write a ratio for each word phrase. Express fractions in lowest terms. **See Example 1.***

11. 40 mi to 30 mi   **12.** 60 ft to 70 ft

13. 72 dollars to 220 dollars   **14.** 120 people to 90 people

15. 20 yd to 8 ft   **16.** 30 in. to 8 ft

17. 16 min to 1 hr   **18.** 24 min to 2 hr

19. 60 in. to 2 yd   **20.** 720 sec to 1 hr

Find the best buy for each item. Give the unit price to the nearest thousandth for that size. ***See Example 2.*** *(Data from various grocery stores.)*

21. Granulated Sugar

Size	Price
4 lb	$3.29
10 lb	$7.49

22. Applesauce

Size	Price
23 oz	$1.99
48 oz	$3.49

23. Orange Juice

Size	Price
64 oz	$2.99
89 oz	$4.79
128 oz	$6.49

24. Salad Dressing

Size	Price
8 oz	$1.69
16 oz	$1.97
36 oz	$5.99

25. Maple Syrup

Size	Price
8.5 oz	$5.79
12.5 oz	$7.99
32 oz	$16.99

26. Mouthwash

Size	Price
16.9 oz	$3.39
33.8 oz	$3.49
50.7 oz	$5.29

27. Tomato Ketchup

Size	Price
32 oz	$1.79
36 oz	$2.69
40 oz	$2.49
64 oz	$4.38

28. Grape Jelly

Size	Price
12 oz	$1.05
18 oz	$1.73
32 oz	$1.84
48 oz	$2.88

29. Laundry Detergent

Size	Price
87 oz	$7.88
131 oz	$10.98
263 oz	$19.96

30. Spaghetti Sauce

Size	Price
14 oz	$1.79
24 oz	$1.77
48 oz	$3.65

Determine whether each proportion is true *or false.* **See Example 3.**

31. $\dfrac{5}{35} = \dfrac{8}{56}$

32. $\dfrac{4}{12} = \dfrac{7}{21}$

33. $\dfrac{120}{82} = \dfrac{7}{10}$

34. $\dfrac{27}{160} = \dfrac{18}{110}$

35. $\dfrac{\frac{1}{2}}{5} = \dfrac{1}{10}$

36. $\dfrac{\frac{1}{3}}{6} = \dfrac{1}{18}$

Solve each equation. **See Examples 4 and 5.**

37. $\dfrac{k}{4} = \dfrac{175}{20}$

38. $\dfrac{x}{6} = \dfrac{18}{4}$

39. $\dfrac{49}{56} = \dfrac{z}{8}$

40. $\dfrac{20}{100} = \dfrac{z}{80}$

41. $\dfrac{x}{24} = \dfrac{15}{16}$

42. $\dfrac{x}{4} = \dfrac{12}{30}$

43. $\dfrac{8}{12} = \dfrac{12k}{18}$

44. $\dfrac{14}{10} = \dfrac{21t}{15}$

45. $\dfrac{z}{2} = \dfrac{z+1}{3}$

46. $\dfrac{m}{5} = \dfrac{m-2}{2}$

47. $\dfrac{3y-2}{5} = \dfrac{6y-5}{11}$

48. $\dfrac{2r+8}{4} = \dfrac{3r-9}{3}$

49. $\dfrac{5k+1}{6} = \dfrac{3k-2}{3}$

50. $\dfrac{x+4}{6} = \dfrac{x+10}{8}$

51. $\dfrac{2p+7}{3} = \dfrac{p-1}{4}$

52. $\dfrac{3m-2}{5} = \dfrac{4-m}{3}$

53. $\dfrac{2(x-4)}{3} = \dfrac{4(x-3)}{5}$

54. $\dfrac{9(x-3)}{6} = \dfrac{6(x-2)}{2}$

Solve each problem. (Assume that all items are equally priced.) **See Example 6.**

55. If 16 candy bars cost $20.00, how much do 24 candy bars cost?

56. If 12 ring tones cost $30.00, how much do 8 ring tones cost?

57. Eight quarts of oil cost $14.00. How much do 5 qt of oil cost?

58. Four tires cost $398.00. How much do 7 tires cost?

59. If 9 pairs of jeans cost $121.50, find the cost of 5 pairs.

60. If 7 shirts cost $87.50, find the cost of 11 shirts.

61. If 6 gal of premium unleaded gasoline costs $22.56, how much would it cost to completely fill a 15-gal tank?

62. If sales tax on a $16.00 DVD is $1.32, find the sales tax on a $120.00 DVD player.

Solve each problem. (Round answers to the nearest tenth as necessary.) ***See Examples 6 and 7.***

63. Biologists tagged 500 fish in North Bay. At a later date, they found 7 tagged fish in a sample of 700. Estimate the total number of fish in North Bay to the nearest hundred.

64. Researchers at West Okoboji Lake tagged 840 fish. A later sample of 1000 fish contained 18 that were tagged. Approximate the fish population in West Okoboji Lake to the nearest hundred.

65. Suppose you are the 120th person in line at airport security.

    (a) In 75 sec, 5 people are able to pass through security. Assuming this rate stays the same, how long (in minutes) will you be waiting in the security line?

    (b) It takes 15 min to walk from security to the gate for your flight. If the flight stops boarding in 60 min, will you arrive at the gate in time?

66. Suppose you are the 30th person in line to renew your driver's license at the Department of Transportation.

    (a) In 150 sec, 2 people are helped. Assuming this rate stays the same, how long (in minutes) will it take to reach the service counter?

    (b) It takes 15 min to drive back to work. If 45 min of your lunch hour remain, will you arrive back at work in time?

67. The distance between Kansas City, Missouri, and Denver is 600 mi. On a certain wall map, this is represented by a length of 2.4 ft. On the map, how many feet would there be between Memphis and Philadelphia, two cities that are actually 1000 mi apart?

68. The distance between Singapore and Tokyo is 3300 mi. On a certain wall map, this distance is represented by 11 in. The actual distance between Mexico City and Cairo is 7700 mi. How far apart are they on the same map?

69. A wall map of the United States has a distance of 8.5 in. between Memphis and Denver, two cities that are actually 1040 mi apart. The actual distance between St. Louis and Des Moines is 333 mi. How far apart are St. Louis and Des Moines on the map?

70. A wall map of the United States has a distance of 8.0 in. between New Orleans and Chicago, two cities that are actually 912 mi apart. The actual distance between Milwaukee and Seattle is 1940 mi. How far apart are Milwaukee and Seattle on the map?

71. On a world globe, the distance between Capetown and Bangkok, two cities that are actually 10,080 km apart, is 12.4 in. The actual distance between Moscow and Berlin is 1610 km. How far apart are Moscow and Berlin on this globe?

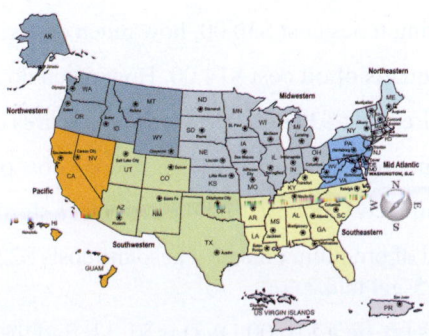

72. On a world globe, the distance between Rio de Janeiro and Hong Kong, two cities that are actually 17,615 km apart, is 21.5 in. The actual distance between Paris and Stockholm is 1605 km. How far apart are Paris and Stockholm on this globe?

73. According to the directions on a bottle of Armstrong® Concentrated Floor Cleaner, for routine cleaning, $\frac{1}{4}$ cup of cleaner should be mixed with 1 gal of warm water. How much cleaner should be mixed with $10\frac{1}{2}$ gal of water?

74. The directions on a bottle of Armstrong® Concentrated Floor Cleaner also specify that, for extra-strength cleaning, $\frac{1}{2}$ cup of cleaner should be used for each gallon of water. How much cleaner should be mixed with $15\frac{1}{2}$ gal of water for extra-strength cleaning?

75. On November 21, 2017, the exchange rate between euros and U.S. dollars was 1 euro to $1.1739. Ashley went to Rome and exchanged her U.S. currency for euros, receiving 300 euros. How much in U.S. dollars did she exchange? (Data from www.xe.com)

76. If 8 U.S. dollars can be exchanged for 150.775 Mexican pesos, how many pesos can be obtained for $65? (Round to the nearest tenth.)

*Two triangles are **similar** if they have the same shape (but not necessarily the same size). Similar triangles have sides that are proportional. The figure shows two similar triangles. Notice that the ratios of the corresponding sides all equal $\frac{3}{2}$.*

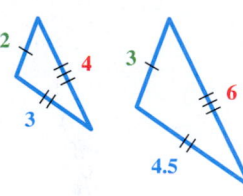

$$\frac{3}{2} = \frac{3}{2}, \quad \frac{4.5}{3} = \frac{3}{2}, \quad \frac{6}{4} = \frac{3}{2}$$

If we know that two triangles are similar, we can set up a proportion to solve for the length of an unknown side.

   Find the lengths x and y as needed in each pair of similar triangles.

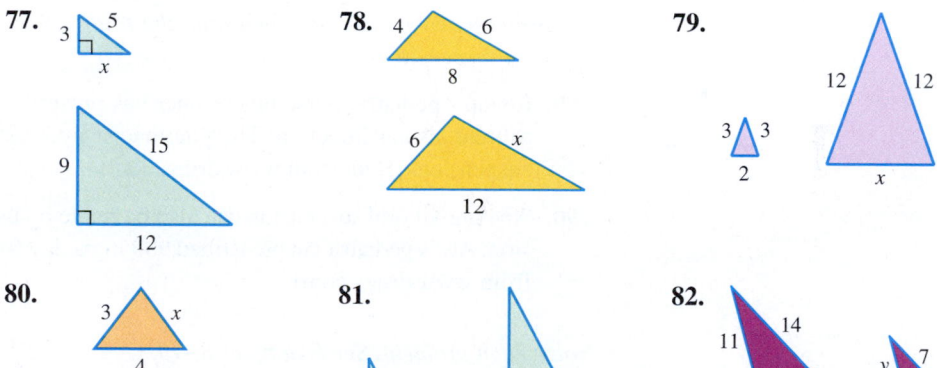

77.

78.

79.

80.

81.

82.

*For each exercise, **(a)** draw a sketch consisting of two right triangles depicting the situation described, and **(b)** solve the problem. (Data from Guinness World Records.)*

83. An enlarged version of the chair used by George Washington at the Constitutional Convention casts a shadow 18 ft long at the same time a vertical pole 12 ft high casts a shadow 4 ft long. How tall is the chair?

84. One of the tallest candles ever constructed was exhibited at the 1897 Stockholm Exhibition. If it cast a shadow 5 ft long at the same time a vertical pole 32 ft high cast a shadow 2 ft long, how tall was the candle?

The Consumer Price Index (CPI) provides a means of determining the purchasing power of the U.S. dollar from one year to the next. Using the period from 1982 to 1984 as a measure of 100.0, the CPI for selected years from 2003 through 2015 is shown in the table. To use the CPI to predict a price in a particular year, we set up a proportion and compare it with a known price in another year.

Year	Consumer Price Index
2003	184.0
2005	195.3
2007	207.3
2009	214.5
2011	224.9
2013	233.0
2015	237.0

Data from U.S. Bureau of Labor Statistics.

$$\frac{\text{price in year } A}{\text{index in year } A} = \frac{\text{price in year } B}{\text{index in year } B}$$

Use the CPI figures in the table to find the amount that would be charged for using the same amount of electricity that cost $225 in 2003. Give answers to the nearest dollar.

85. in 2005　　　　**86.** in 2007　　　　**87.** in 2013　　　　**88.** in 2015

Children are often given antibiotics in liquid form, called an oral suspension. Pharmacists make up these suspensions by mixing medication in powder form with water. They use proportions to calculate the volume of the suspension for the amount of medication that has been prescribed. For each exercise, do the following.

(a) Find the total amount of medication in milligrams to be given over the full course of treatment.

(b) Write a proportion that can be solved to find the total volume of the liquid suspension that the pharmacist will prepare. Use x as the variable.

(c) Solve the proportion to determine the total volume of the oral suspension.

89. Logan's pediatric nurse practitioner has prescribed 375 mg of Amoxil a day for 7 days to treat his ear infection. The pharmacist uses 125 mg of Amoxil in each 5 mL of the suspension. (Data from www.drugs.com)

90. An Amoxil oral suspension can also be made by using 250 mg for each 5 mL of suspension. Ava's pediatrician prescribed 900 mg a day for 10 days to treat her bronchitis. (Data from www.drugs.com)

Solve each problem. See Examples 8–10.

91. 18% of 780 is what number?　　　　**92.** 23% of 480 is what number?

93. 42% of what number is 294?　　　　**94.** 18% of what number is 108?

95. 120% of what number is 510?　　　　**96.** 140% of what number is 315?

97. What percent of 50 is 4?　　　　**98.** What percent of 64 is 8?

99. What percent of 30 is 36?　　　　**100.** What percent of 48 is 96?

101. 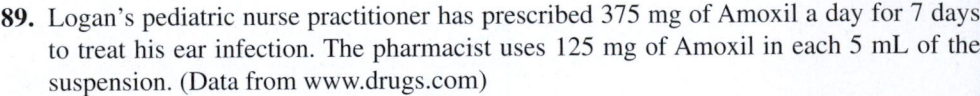 Clayton earned 48 points on a 60-point geometry project. What percent of the total points did he earn?

102. On a 75-point algebra test, Grady scored 63 points. What percent of the total points did he score?

103. A laptop computer that has a regular price of $700 is on sale for $504. What percent of the regular price is the savings?

104. An all-in-one desktop computer that has a regular price of $980 is on sale for $833. What percent of the regular price is the savings?

105. Tyler has a monthly income of $1500. His rent is $480 per month. What percent of his monthly income is his rent?

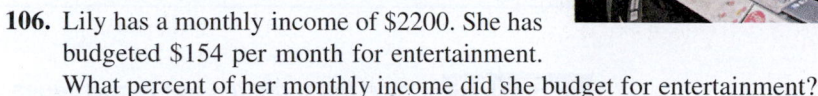

106. Lily has a monthly income of $2200. She has budgeted $154 per month for entertainment. What percent of her monthly income did she budget for entertainment?

107. Anna saved $1950, which was 65% of the amount she needed for a used car. What was the total amount she needed for the car?

108. Bryn had $525, which was 70% of the total amount she needed for a deposit on an apartment. What was the total deposit she needed?

109. A restaurant bill (before tax) was $58.26. Megan wants to leave a 25% tip for exceptional service. Estimate how much she should leave to the nearest dollar.

110. An Uber ride cost $21.89. Matt wants to leave a 10% tip. Estimate how much he should leave to the nearest dollar.

RELATING CONCEPTS For Individual or Group Work (Exercises 111–114)

Work Exercises 111–114 in order. The steps justify the method of solving a proportion using cross products.

111. What is the LCD of the fractions in the following equation? $\dfrac{x}{6} = \dfrac{2}{5}$

112. Solve the equation in **Exercise 111** as follows.

   (a) Multiply each side by the LCD. What equation results?

   (b) Solve the equation from part (a) by dividing each side by the coefficient of x.

113. Solve the equation in **Exercise 111** using cross products.

114. Compare the answers from **Exercises 112(b) and 113.** What do you notice?

2.8 Further Applications of Linear Equations

OBJECTIVES

1 Solve percent problems involving rates.

2 Solve problems involving mixtures.

3 Solve problems involving simple interest.

4 Solve problems involving denominations of money.

5 Solve problems involving distance, rate, and time.

OBJECTIVE 1 Solve percent problems involving rates.

Recall that the word *percent* means "*per* 100." One percent means "*one per* 100."

$$1\% = \frac{1}{100} = 0.01, \quad 10\% = \frac{10}{100} = 0.10, \quad 100\% = \frac{100}{100} = 1 \qquad \text{Percent, fraction, and decimal equivalents}$$

Example:

27% represents the ratio of 27 to 100, that is, $\frac{27}{100}$, or, as a decimal, 0.27.

> **PROBLEM-SOLVING HINT**   Mixing different concentrations of a substance or different interest rates involves percents. To obtain the amount of pure substance or the interest, we multiply as follows.
>
> | Mixture Problems | Interest Problems (annual) |
> |---|---|
> | base · rate (%) = percentage | principal · rate (%) = interest |
> | b · r = p | p · r = I |
>
> *In an equation, percent is always written as a decimal (or a fraction).*

NOW TRY
EXERCISE 1

Answer each question.

(a) How much pure alcohol is in 70 L of a 20% alcohol solution?

(b) If $3200 is invested for 1 yr at 2% simple interest, how much interest is earned?

EXAMPLE 1   Using Percents to Find Percentages

Answer each question.

(a) How much pure acid is in 40 L of a 35% acid solution?

The amount of pure acid is found by multiplying.

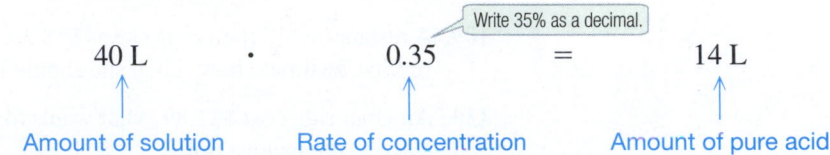

Write 35% as a decimal.

$$40 \text{ L} \quad \cdot \quad 0.35 \quad = \quad 14 \text{ L}$$

Amount of solution   Rate of concentration   Amount of pure acid

(b) If $1300 is invested for 1 yr at 3% simple interest, how much interest is earned?

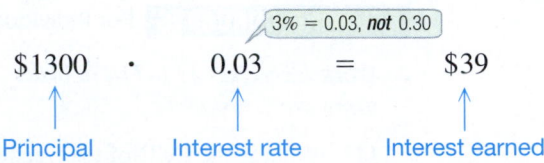

3% = 0.03, *not* 0.30

$$\$1300 \quad \cdot \quad 0.03 \quad = \quad \$39$$

Principal   Interest rate   Interest earned   **NOW TRY**

> **PROBLEM-SOLVING HINT**   In the applications that follow, using a table helps organize the information in a problem. This allows us to more easily set up an equation, which is usually the most difficult step.

OBJECTIVE 2 Solve problems involving mixtures.

EXAMPLE 2   Solving a Mixture Problem

A chemist mixes 20 L of a 40% acid solution with some 70% acid solution to obtain a mixture that is 50% acid. How many liters of the 70% acid solution should she use?

Step 1   **Read** the problem. Note the percent of each solution and of the mixture.

Step 2   **Assign a variable** to represent the unknown quantity.

Let x = the number of liters of 70% acid solution needed.

As in **Example 1(a),** the amount of pure acid in this solution is the product of the percent of strength and the number of liters of solution.

$0.70x$ Liters of pure acid in x liters of 70% solution

The amount of pure acid in the 20 L of 40% solution is found similarly.

$0.40(20)$ Liters of pure acid in the 40% solution

NOW TRY ANSWERS
1. (a) 14 L   **(b)** $64

**NOW TRY
EXERCISE 2**

A certain seasoning is 70% salt. How many ounces of this seasoning must be mixed with 30 oz of dried herbs containing 10% salt to obtain a seasoning that is 50% salt?

The new solution will contain $(x + 20)$ liters of 50% solution. The amount of pure acid in this solution is again found by multiplying.

$0.50(x + 20)$ Liters of pure acid in the 50% solution

FIGURE 18 illustrates this information, which is organized in the table.

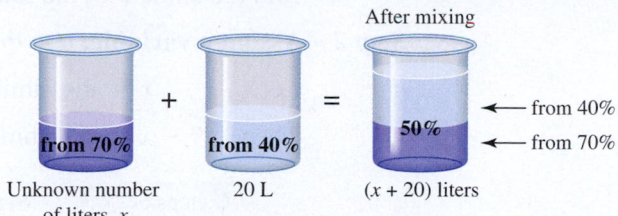

FIGURE 18

Liters of Solution	Rate (as a decimal)	Liters of Pure Acid
x	0.70	$0.70x$
20	0.40	$0.40(20)$
$x + 20$	0.50	$0.50(x + 20)$

Sum must equal

Step 3   **Write an equation.** The number of liters of pure acid in the 70% solution added to the number of liters of pure acid in the 40% solution will equal the number of liters of pure acid in the final mixture.

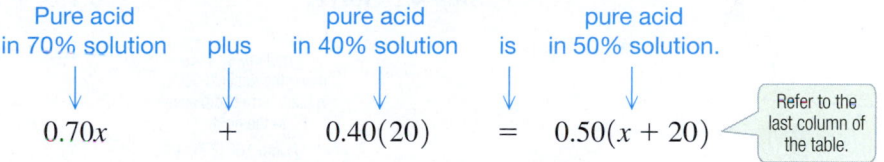

Pure acid in 70% solution plus pure acid in 40% solution is pure acid in 50% solution.

$$0.70x + 0.40(20) = 0.50(x + 20)$$

Refer to the last column of the table.

Step 4   **Solve.** We clear the parentheses first. Then we clear the decimals.

$0.70x = 0.7x$ and $0.50x = 0.5x$.

Think: $10(0.7x) = 0.7x = 7x$

$$0.7x + 8 = 0.5x + 10$$ Multiply; distributive property
$$10(0.7x + 8) = 10(0.5x + 10)$$ Multiply by 10.
$$7x + 80 = 5x + 100$$ Distributive property
$$2x + 80 = 100$$ Subtract 5x.
$$2x = 20$$ Subtract 80.
$$x = 10$$ Divide by 2.

Step 5   **State the answer.** The chemist needs to use 10 L of 70% solution.

Step 6   **Check.** If 10 L of 70% solution are used, the amounts of pure acid are the same.

$0.70(10) + 0.40(20)$ Sum of the two solutions
$= 7 + 8$
$= 15$

$0.50(10 + 20)$ Mixture
$= 0.50(30)$
$= 15$ NOW TRY

⚠️ **CAUTION** In a mixture problem, the concentration of the final mixture must be *between* the concentrations of the two solutions making up the mixture.

NOW TRY ANSWER
2. 60 oz

NOW TRY EXERCISE 3

How many liters of a 25% saline solution must be mixed with a 10% saline solution to obtain 15 L of a 15% solution?

EXAMPLE 3   Solving a Mixture Problem

How many ounces of a seasoning that is 15% pepper must be mixed with a version that is 30% pepper to obtain 9 oz of a seasoning that is 20% pepper?

Step 1   **Read** the problem. We are given the *total* amount of the mixture. We must find the amount of the seasoning that is 15% pepper.

Step 2   **Assign a variable.** Use the fact that the total mixture is 9 oz.

Let $x =$ the number of ounces of seasoning that is 15% pepper.

Then $9 - x =$ the number of ounces of seasoning that is 30% pepper.

Ounces of Seasoning	Rate (as a decimal)	Ounces of Pepper
x	0.15	$0.15x$
$9 - x$	0.30	$0.30(9 - x)$
9	0.20	$0.20(9)$

Use a table to organize the given information.

Step 3   **Write an equation.** Refer to the last column of the table.

Pepper in 15% seasoning   plus   pepper in 30% seasoning   is   pepper in 20% seasoning.

$$0.15x + 0.30(9 - x) = 0.20(9)$$

Step 4   **Solve.**

$$0.15x + 2.7 - 0.3x = 1.8 \qquad \text{Distributive property; multiply.}$$

> To multiply by 100, move the decimal point in each term two places to the right.
> Think: $100(2.7) = 2.70$

$$15x + 270 - 30x = 180 \qquad \text{Multiply by 100.}$$

$$-15x + 270 = 180 \qquad \text{Combine like terms.}$$

$$-15x = -90 \qquad \text{Subtract 270.}$$

$$x = 6 \qquad \text{Divide by } -15.$$

Step 5   **State the answer.** 6 oz of seasoning that is 15% pepper is needed. (This means that $9 - 6 = 3$ oz of the 30% pepper seasoning is needed, although the problem does not specifically ask for this amount.)

Step 6   **Check.** The ounces of pepper before and after mixing are the same.

$0.15(6) + 0.30(9 - 6)$ Sum of the two seasonings $\qquad 0.20(9)$ Mixture

$= 0.9 + 0.9$ $\qquad = 1.8$

$= 1.8$

NOW TRY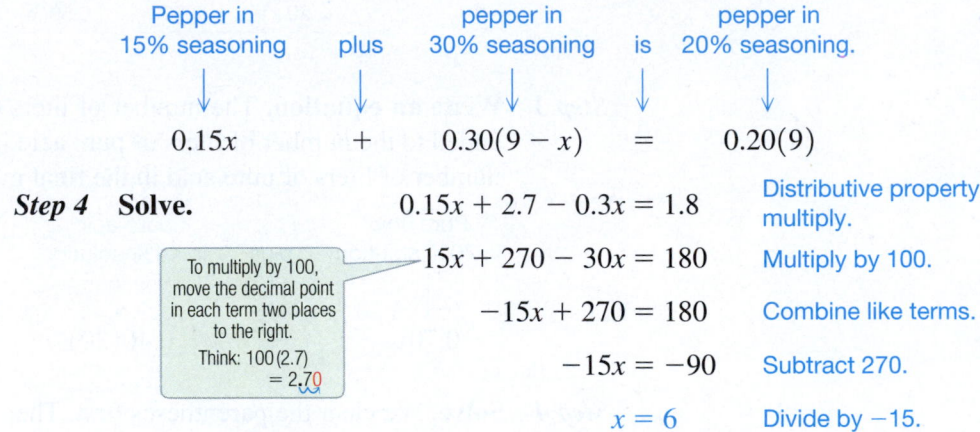

OBJECTIVE 3 Solve problems involving simple interest.

The formula for simple interest

$$I = prt \quad \text{becomes} \quad I = pr \quad \text{when time } t = 1 \text{ (for annual interest)},$$

as shown in the Problem-Solving Hint at the beginning of this section. Multiplying the total amount (principal) by the rate (rate of interest) gives the percentage (amount of interest).

NOW TRY ANSWER
3. 5 L

NOW TRY EXERCISE 4

A financial advisor invests some money in a municipal bond paying 3% annual interest and $5000 more than that amount in a certificate of deposit paying 2% annual interest. To earn $400 per year in interest, how much should he invest at each rate?

EXAMPLE 4   Solving a Simple Interest Problem

Susan plans to invest some money at 2% and $2000 more than this amount at 4%. To earn $380 per year in interest, how much should she invest at each rate?

Step 1   **Read** the problem. There will be two answers.

Step 2   **Assign a variable.**

Let $x =$ the amount invested at 2% (in dollars).

Then $x + 2000 =$ the amount invested at 4% (in dollars).

Amount Invested (in dollars)	Rate (as a decimal)	Interest for One Year (in dollars)
x	0.02	0.02x
$x + 2000$	0.04	0.04($x + 2000$)

Use a table to organize the given information.

Step 3   **Write an equation.** Multiply amount by rate to obtain interest earned. The two amounts of interest must total $380.

Interest at 2%	plus	interest at 4%	is	total interest.
↓	↓	↓	↓	↓
0.02x	+	0.04($x + 2000$)	=	380

Step 4   **Solve.**

$$0.02x + 0.04x + 80 = 380 \qquad \text{Distributive property}$$
$$2x + 4x + 8000 = 38{,}000 \qquad \text{Multiply by 100.}$$
$$6x + 8000 = 38{,}000 \qquad \text{Combine like terms.}$$
$$6x = 30{,}000 \qquad \text{Subtract 8000.}$$
$$x = 5000 \qquad \text{Divide by 6.}$$

Step 5   **State the answer.** At 2%, she should invest $5000. At 4%, she should invest

$$\$5000 + \$2000 = \$7000.$$

Step 6   **Check.** The sum of the two interest amounts is

$$0.02(\$5000) + 0.04(\$7000)$$
$$= \$100 + \$280$$
$$= \$380, \quad \text{as required.} \qquad \text{NOW TRY}$$

OBJECTIVE 4 Solve problems involving denominations of money.

PROBLEM-SOLVING HINT   To obtain the total value in problems that involve different denominations of money or items with different monetary values, we multiply as follows.

Money Denominations Problems

number · value of one item = total value

Examples: 30 dimes have a value of 30($0.10) = $3.

15 five-dollar bills have a value of 15($5) = $75.

NOW TRY ANSWER
4. $11,000 at 2%; $6000 at 3%

**NOW TRY
EXERCISE 5**
A boy has saved $5.65 in dimes and quarters. He has 10 more quarters than dimes. How many of each denomination of coin does he have?

EXAMPLE 5   Solving a Money Denominations Problem

A bank teller has 25 more $5 bills than $10 bills. The total value of the money is $200. How many of each denomination of bill does she have?

Step 1   **Read** the problem. We must find the number of each denomination of bill.

Step 2   **Assign a variable.**

Let $x =$ the number of $10 bills.

Then $x + 25 =$ the number of $5 bills.

Number of Bills	Denomination (in dollars)	Total Value (in dollars)
x	10	$10x$
$x + 25$	5	$5(x + 25)$

Step 3   **Write an equation.** Multiplying the number of bills by the denomination gives the monetary value. The value of the tens added to the value of the fives must be $200.

$$\begin{array}{ccccc} \text{Value of} & \text{plus} & \text{value of} & \text{is} & \$200. \\ \text{tens} & & \text{fives} & & \\ \downarrow & \downarrow & \downarrow & \downarrow & \downarrow \\ 10x & + & 5(x+25) & = & 200 \end{array}$$

Step 4   **Solve.**

$$10x + 5x + 125 = 200 \qquad \text{Distributive property}$$
$$15x + 125 = 200 \qquad \text{Combine like terms.}$$
$$15x = 75 \qquad \text{Subtract 125.}$$
$$x = 5 \qquad \text{Divide by 15.}$$

Step 5   **State the answer.** The teller has 5 tens and $5 + 25 = 30$ fives.

Step 6   **Check.** The teller has $30 - 5 = 25$ more fives than tens. The value is

$$5(\$10) + 30(\$5) = \$200, \quad \text{as required.} \qquad \textbf{NOW TRY} \,\, \text{⟲}$$

OBJECTIVE 5 Solve problems involving distance, rate, and time.

If a car travels at an average rate of 50 mph for 2 hr, then it travels

$$50 \cdot 2 = 100 \text{ mi.}$$

This is an example of the basic relationship among distance, rate, and time.

$$\textbf{distance} = \textbf{rate} \cdot \textbf{time,} \quad \text{or} \quad \textbf{\textit{d} = \textit{rt}}$$

By solving, in turn, for r and t in the formula $d = rt$, we obtain two other equivalent forms of the formula.

Forms of the Distance Formula

$$d = rt \qquad r = \frac{d}{t} \qquad t = \frac{d}{r}$$

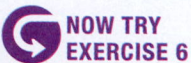

**NOW TRY
EXERCISE 6**

It took a driver 6 hr to travel from St. Louis to Fort Smith, a distance of 400 mi. What was the driver's rate, to the nearest hundredth?

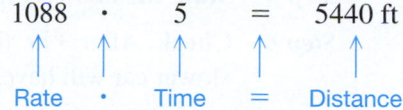

EXAMPLE 6 Finding Distance, Rate, or Time

Solve each problem using a form of the distance formula.

(a) The speed (rate) of sound is 1088 ft per sec at sea level at 32°F. Find the distance sound travels in 5 sec under these conditions.

We must find distance, given rate and time, using $d = rt$ (or $rt = d$).

$$1088 \cdot 5 = 5440 \text{ ft}$$

Rate · Time = Distance

(b) The winner of the first Indianapolis 500 race (in 1911) was Ray Harroun, driving a Marmon Wasp at an average rate of 74.59 mph. (Data from *The Universal Almanac.*) How long did it take him to complete the 500 mi?

We must find time, given rate and distance, using $t = \frac{d}{r}$ $\left(\text{or } \frac{d}{r} = t\right)$.

Distance ⟶ $\dfrac{500}{74.59} = 6.70$ hr (rounded) ⟵ Time
Rate ⟶

To convert 0.70 hr to minutes, we multiply by 60 to obtain $0.70(60) = 42$. It took Harroun about 6 hr, 42 min to complete the race.

(c) At the 2016 Olympic Games, Hungarian swimmer Katinka Hosszú won a gold medal with a time of 58.45 sec in the women's 100-m backstroke swimming event. (Data from *The World Almanac and Book of Facts.*) Find her rate.

We must find rate, given distance and time, using $r = \frac{d}{t}$ (or $\frac{d}{t} = r$).

Distance ⟶ $\dfrac{100}{58.45} = 1.71$ m per sec (rounded) ⟵ Rate
Time ⟶

NOW TRY

EXAMPLE 7 Solving a Distance-Rate-Time Problem

Two cars leave Iowa City, Iowa, at the same time and travel east on Interstate 80. One travels at a constant rate of 55 mph. The other travels at a constant rate of 63 mph. In how many hours will the distance between them be 24 mi?

Step 1 **Read** the problem carefully.

Step 2 **Assign a variable.** We must find a time.

Let $t =$ the number of hours until the distance between them is 24 mi.

The sketch in **FIGURE 19** shows what is happening in the problem.

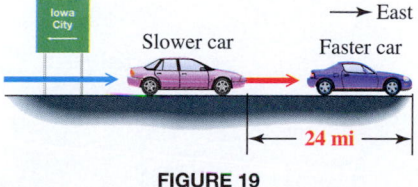

FIGURE 19

To construct a table, we fill in the rates given in the problem, using t for the time traveled by each car. Because $d = rt$, or $rt = d$, we multiply rate by time to find expressions for the distances traveled.

	Rate	Time	Distance
Faster Car	63	t	$63t$
Slower Car	55	t	$55t$

⟵ The quantities $63t$ and $55t$ represent the two distances.

NOW TRY ANSWER

6. 66.67 mph

NOW TRY EXERCISE 7

From a point on a straight road, two bicyclists ride in the same direction. One travels at a rate of 18 mph. The other travels at a rate of 20 mph. In how many hours will they be 5 mi apart?

Step 3 **Write an equation.**

$$63t - 55t = 24$$

The *difference* between the larger distance and the smaller distance is 24 mi.

Step 4 **Solve.** $8t = 24$ Combine like terms.

$t = 3$ Divide by 8.

Step 5 **State the answer.** It will take the cars 3 hr to be 24 mi apart.

Step 6 **Check.** After 3 hr, the faster car will have traveled $63 \cdot 3 = 189$ mi and the slower car will have traveled $55 \cdot 3 = 165$ mi. The difference is

$$189 - 165 = 24, \quad \text{as required.}$$

 NOW TRY

PROBLEM-SOLVING HINT   In distance-rate-time problems, once we have filled in two pieces of information in each row of a table, we can automatically fill in the third piece of information, using the appropriate form of the distance formula. Then we set up the equation based on a sketch and the information in the table.

NOW TRY EXERCISE 8

Two cars leave a parking lot at the same time, one traveling east and the other traveling west. The westbound car travels 6 mph faster than the eastbound car. In $\frac{1}{4}$ hr, they are 35 mi apart. What are their rates?

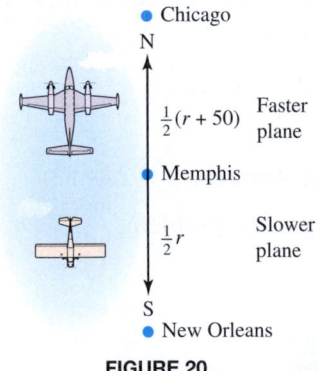

• Chicago

N

$\frac{1}{2}(r + 50)$ Faster plane

• Memphis

$\frac{1}{2}r$ Slower plane

S

• New Orleans

FIGURE 20

EXAMPLE 8   Solving a Distance-Rate-Time Problem

Two planes leave Memphis at the same time. One heads south to New Orleans. The other heads north to Chicago. The Chicago plane flies 50 mph faster than the New Orleans plane. In $\frac{1}{2}$ hr, the planes are 275 mi apart. What are their rates?

Step 1 **Read** the problem carefully.

Step 2 **Assign a variable.**

Let r = the rate of the slower plane.

Then $r + 50$ = the rate of the faster plane.

	Rate	Time	Distance
Slower Plane	r	$\frac{1}{2}$	$\frac{1}{2}r$
Faster Plane	$r + 50$	$\frac{1}{2}$	$\frac{1}{2}(r + 50)$

Sum is 275 mi.

Step 3 **Write an equation.** As **FIGURE 20** shows, the planes are headed in *opposite* directions. The *sum* of their distances equals 275 mi.

$$\frac{1}{2}r + \frac{1}{2}(r + 50) = 275$$

Step 4 **Solve.**

$$\frac{1}{2}r + \frac{1}{2}r + 25 = 275 \quad \text{Distributive property}$$

$$r + 25 = 275 \quad \text{Combine like terms.}$$

Rate of slower plane ⟶ $r = 250$ Subtract 25.

Step 5 **State the answer.** The slower plane (headed south) has a rate of 250 mph.

$$250 + 50 = 300 \text{ mph} \longleftarrow \text{Rate of faster plane}$$

Step 6 **Check.** Verify that $\frac{1}{2}(250) + \frac{1}{2}(300) = 275$ mi, as required.

NOW TRY ANSWERS
7. 2.5 hr
8. slower car: 67 mph; faster car: 73 mph

 NOW TRY

2.8 Exercises

FOR EXTRA HELP

▶ **MyLab Math**

▶ *Video solutions for select problems available in MyLab Math*

Concept Check *Choose the letter of the correct response.*

1. Which expression represents the amount of pure alcohol in x liters of a 75% alcohol solution?

 A. $0.75x$ liters   **B.** $(75 + x)$ liters   **C.** $(75 - x)$ liters   **D.** $75x$ liters

2. Which expression represents the value of x quarters?

 A. $25x$ dollars   **B.** $\dfrac{25}{x}$ dollars   **C.** $0.25x$ dollars   **D.** $(x + 0.25)$ dollars

3. If a minivan travels at 55 mph for t hours, which expression represents the distance traveled?

 A. $(t + 55)$ miles   **B.** $(t - 55)$ miles   **C.** $55t$ miles   **D.** $\dfrac{55}{t}$ miles

4. If a car travels at r mph for 6 hr, which expression represents the distance traveled?

 A. $\dfrac{r}{6}$ miles   **B.** $(r - 6)$ miles   **C.** $(r + 6)$ miles   **D.** $6r$ miles

5. Suppose that a chemist is mixing two acid solutions, one of 20% concentration and the other of 30% concentration. Which concentration could *not* be obtained?

 A. 22%   **B.** 24%   **C.** 28%   **D.** 32%

6. Suppose that water is added to a 24% alcohol mixture. Which concentration could be obtained? (*Hint:* The solution is being diluted.)

 A. 22%   **B.** 26%   **C.** 28%   **D.** 30%

7. Which choice is the best estimate for the average rate of a bus trip of 405 mi that lasted 8.2 hr?

 A. 30 mph   **B.** 40 mph   **C.** 50 mph   **D.** 60 mph

8. Which choice is the best estimate for the time an automobile trip of 185 mi driven at an average rate of 58 mph would take?

 A. 3 hr   **B.** 4 hr   **C.** 4.5 hr   **D.** 5 hr

9. Concept Check An automobile averages 45 mph and travels for 30 min. A student claimed that the distance traveled was

$$45 \cdot 30 = 1350 \text{ mi.}$$

 This is incorrect. **WHAT WENT WRONG?** Give the correct distance.

10. Concept Check A principal of $1000 is invested for 1 yr at 2% simple interest. A student claims that the interest earned will be

$$\$1000 \cdot 0.2 = \$200.$$

 This is incorrect. **WHAT WENT WRONG?** Give the correct amount of interest earned.

Answer each question. **See Example 1 and the Problem-Solving Hint preceding Example 5.**

11. How much pure alcohol is in 150 L of a 30% alcohol solution?

12. How much pure acid is in 250 mL of a 14% acid solution?

13. If $25,000 is invested for 1 yr at 3% simple interest, how much interest is earned?

14. If $10,000 is invested for 1 yr at 3.5% simple interest, how much interest is earned?

15. What is the monetary value of 35 half-dollars?

16. What is the monetary value of 283 nickels?

Solve each problem. ***See Examples 2 and 3.***

17. How many liters of 25% acid solution must a chemist add to 80 L of 40% acid solution to obtain a mixture that is 30% acid?

Liters of Solution	Rate (as a decimal)	Liters of Pure Acid
x	0.25	0.25x
80	0.40	0.40(80)
x + 80	0.30	0.30(x + 80)

18. How many gallons of 50% antifreeze must be mixed with 80 gal of 20% anti-freeze to obtain a mixture that is 40% antifreeze?

Gallons of Mixture	Rate (as a decimal)	Gallons of Pure Antifreeze
x	0.50	0.50x
80	0.20	0.20(80)
x + 80	0.40	0.40(x + 80)

19. A pharmacist has 20 L of a 10% drug solution. How many liters of a 5% drug solution must be added to obtain a mixture that is 8%?

Liters of Solution	Rate (as a decimal)	Liters of Pure Drug
20		20(0.10)
	0.05	
	0.08	

20. A certain metal is 20% tin. How many kilograms of this metal must be mixed with 80 kg of a metal that is 70% tin to obtain a metal that is 50% tin?

Kilograms of Metal	Rate (as a decimal)	Kilograms of Pure Tin
x	0.20	
	0.70	
	0.50	

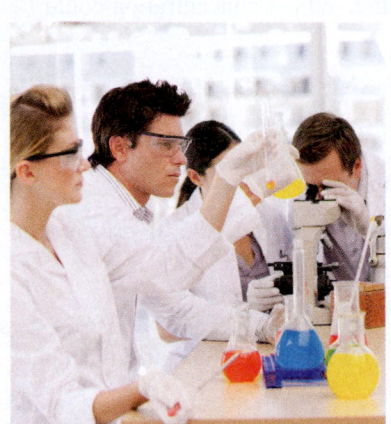

21. In a chemistry class, 12 L of a 12% alcohol solution must be mixed with a 20% solution to obtain a 14% solution. How many liters of the 20% solution are needed?

22. How many liters of a 10% alcohol solution must be mixed with 40 L of a 50% solution to obtain a 40% solution?

23. Minoxidil is a drug that has proven to be effective in treating male pattern baldness. Water must be added to 20 mL of a 4% minoxidil solution to dilute it to a 2% solution. How many milliliters of water should be used? (*Hint:* Water is 0% minoxidil.)

24. Water must be added to 150 mL of a 10% essential oil solution to dilute it to a 5% solution. How many milliliters of water should be used? (*Hint:* Water is 0% essential oil.)

25. How many liters of a 60% acid solution must be mixed with a 75% acid solution to obtain 20 L of a 72% solution?

26. How many gallons of a fruit drink that is 50% real juice must be mixed with a fruit drink that is 20% real juice to obtain 12 gal of a fruit drink that is 40% real juice?

Solve each problem. ***See Example 4.***

27. Arlene is saving money for her college education. She deposited some money in a savings account paying 5% and $1200 less than that amount in a second account paying 4%. The two accounts produced a total of $141 interest in 1 yr. How much did she invest at each rate?

28. Margaret won a prize for her work. She invested part of the money in a certificate of deposit at 2% and $3000 more than that amount in a bond paying 3%. Her annual interest income was $390. How much did Margaret invest at each rate?

29. An artist invests in a mutual fund account paying 6%, and $6000 more than three times as much in a mutual fund account paying 5%. Her total annual interest income from the investments is $825. How much does she invest at each rate?

30. With income earned by selling the rights to his life story, an actor invests some of the money at 3% and $30,000 more than twice as much at 4%. The total annual interest earned from the investments is $5600. How much is invested at each rate?

31. Jamal had $2500, some of which he deposited in a mutual fund account paying 8%. The rest he deposited in a money market account paying 2%. How much did he deposit in each account if the total annual interest was $152?

Amount Invested (in dollars)	Rate (as a decimal)	Interest for One Year (in dollars)
x	0.08	
	0.02	

32. Carter invested a total of $9000 in two accounts, one paying 1% and the other paying 4%. If he earned total annual interest of $285, how much did he deposit in each account?

Amount Invested (in dollars)	Rate (as a decimal)	Interest for One Year (in dollars)
x	0.01	
	0.04	

*Solve each problem. **See Example 5.***

33. A coin collector has $1.70 in dimes and nickels. She has two more dimes than nickels. How many nickels does she have?

Number of Coins	Denomination (in dollars)	Total Value (in dollars)
x	0.05	0.05x
	0.10	

34. A bank teller has $725 in $5 bills and $20 bills. The teller has five more twenties than fives. How many $5 bills does the teller have?

Number of Bills	Denomination (in dollars)	Total Value (in dollars)
x	5	
x + 5	20	

35. In January 2017, U.S. first-class mail rates increased to 49 cents for the first ounce, and 21 cents for each additional ounce. If Sabrina spent $16.45 for a total of 45 stamps of these two denominations, how many stamps of each denomination did she buy? (Data from U.S. Postal Service.)

36. A movie theater has two ticket prices: $8 for adults and $5 for children. If the box office took in $4116 from the sale of 600 tickets, how many tickets of each kind were sold?

37. Harriet operates a coffee shop. One of her customers wants to buy two kinds of beans: Arabian Mocha and Colombian Decaf. If she wants twice as much Arabian Mocha as Colombian Decaf, how much of each can she buy for a total of $87.50? (Prices are listed on the sign.)

38. A customer at Harriet's coffee shop wants to buy Italian Espresso beans and Kona Deluxe beans. If he wants four times as much Kona Deluxe as Italian Espresso, how much of each can he buy for a total of $247.50? (Prices are listed on the sign.)

Arabian Mocha	$8.50/lb
Chocolate Mint	$10.50/lb
Colombian Decaf	$8.00/lb
French Roast	$7.50/lb
Guatemalan Spice	$9.50/lb
Hazelnut Decaf	$10.00/lb
Italian Espresso	$9.00/lb
Kona Deluxe	$11.50/lb

*Solve each problem. **See Example 6.***

39. A driver averaged 53 mph and took 10 hr to travel from Memphis to Chicago. What is the distance between Memphis and Chicago?

40. A small plane traveled from Warsaw to Rome, averaging 164 mph. The trip took 2 hr. What is the distance from Warsaw to Rome?

41. The winner of the 2016 Indianapolis 500 (mile) race was Alexander Rossi, who drove his Dellara-Honda at a rate of 166.634 mph. What was his time (to the nearest thousandth of an hour)? (Data from *The World Almanac and Book of Facts*.)

42. In 2016, Kyle Busch drove his Toyota to victory in the Brickyard 400 (mile) race at a rate of 128.940 mph. What was his time (to the nearest thousandth of an hour)? (Data from *The World Almanac and Book of Facts.*)

In each exercise, find the rate on the basis of the information provided. Round answers to the nearest hundredth. All events were at the 2016 Olympics. (Data from The World Almanac and Book of Facts.) **See Example 6.**

	Event	Participant	Time
43.	200-m run, women	Elaine Thompson, Jamaica	21.78 sec
44.	400-m hurdles, women	Dalilah Muhammed, USA	53.13 sec
45.	400-m hurdles, men	Kerron Clement, USA	47.73 sec
46.	200-m run, men	Usain Bolt, Jamaica	19.78 sec

Solve each problem. **See Examples 7 and 8.**

47. From a point on a straight road, Marco and Celeste ride bicycles in the same direction. Marco rides at 10 mph and Celeste rides at 12 mph. In how many hours will they be 15 mi apart?

48. At a given hour, two steamboats leave a city in the same direction on a straight canal. One travels at 18 mph and the other travels at 24 mph. In how many hours will the boats be 9 mi apart?

49. Atlanta and Cincinnati are 440 mi apart. John leaves Cincinnati, driving toward Atlanta at an average rate of 60 mph. Pat leaves Atlanta at the same time, driving toward Cincinnati in her antique auto, averaging 28 mph. How long will it take them to meet?

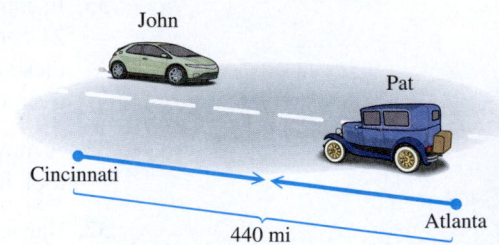

	r	t	d
John	60	t	60t
Pat	28	t	28t

50. St. Louis and Portland are 2060 mi apart. A small plane leaves Portland, traveling toward St. Louis at an average rate of 90 mph. Another plane leaves St. Louis at the same time, traveling toward Portland and averaging 116 mph. How long will it take them to meet?

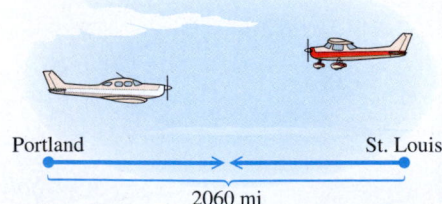

	r	t	d
Plane Leaving Portland	90	t	90t
Plane Leaving St. Louis	116	t	116t

51. A train leaves Kansas City, Kansas, and travels north at 85 km per hr. Another train leaves at the same time and travels south at 95 km per hour. How long will it take before they are 315 km apart?

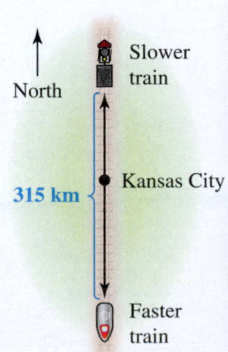

52. Two planes leave Boston at the same time and fly in opposite directions. If one flies at 410 mph and the other flies at 530 mph, how long will it take them to be 3290 mi apart?

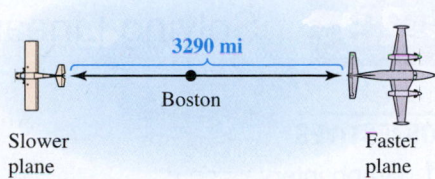

53. Two planes leave an airport at the same time, one flying east, the other flying west. The eastbound plane travels 150 mph slower. They are 2250 mi apart after 3 hr. Find the rate of each plane.

	r	t	d
Eastbound	$x - 150$	3	
Westbound	x	3	

54. Two trains leave a city at the same time. One travels north, and the other travels south 20 mph faster. In 2 hr, the trains are 280 mi apart. Find their rates.

	r	t	d
Northbound	x	2	
Southbound	$x + 20$	2	

55. Two cars start from towns 400 mi apart and travel toward each other. They meet after 4 hr. Find the rate of each car if one travels 20 mph faster than the other.

56. Two cars leave towns 230 km apart at the same time, traveling directly toward one another. One car travels 15 km per hr slower than the other. They pass one another 2 hr later. What are their rates?

Extending Skills *Solve each problem.*

57. Elena works for $8 an hour. A total of 25% of her salary is deducted for taxes and insurance. How many hours must she work to take home $450?

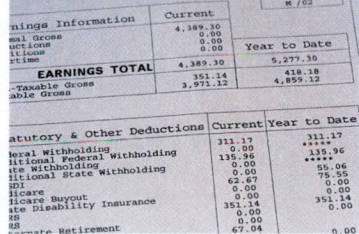

58. Paula received a paycheck for $585 for her weekly wages less 10% deductions. How much was she paid before the deductions were made?

59. At the end of a day, the owner of a gift shop had $2394 in the cash register. This amount included sales tax of 5% on all sales. Find the amount of the sales.

60. A restaurant bill totaled $180, which included a 20% tip. Find the actual price of the food and drinks served.

61. Carter received a credit card bill for $104. This included interest of 1.5% per month for one month, plus a $5 late charge. To the nearest cent, how much were his actual purchases?

62. Ricardo paid $7000 for a used car. This included 10% sales tax, less a trade-in allowance of $1250 for his old car. Find the actual price of the car he purchased.

63. Kevin is three times as old as Bob. Three years ago the sum of their ages was 22 yr. How old is each now? (*Hint:* Write an expression first for the age of each now and then for the age of each three years ago.)

64. A store has 39 qt of milk, some in pint cartons and some in quart cartons. There are six times as many quart cartons as pint cartons. How many quart cartons are there? (*Hint:* 1 qt = 2 pt)

65. A table is three times as long as it is wide. If it were 3 ft shorter and 3 ft wider, it would be square (with all sides equal). How long and how wide is the table?

2.9   Solving Linear Inequalities

An **inequality** relates algebraic expressions using the following symbols.

$<$ Is less than $\leq$ Is less than or equal to

$>$ Is greater than $\geq$ Is greater than or equal to

Linear Inequality in One Variable

A **linear inequality in one variable** (here x) is an inequality that can be written in the form

$$ax + b < 0, \quad ax + b \leq 0, \quad ax + b > 0, \quad \text{or} \quad ax + b \geq 0,$$

where a and b are real numbers and $a \neq 0$.

Examples: $x + 5 < 0$, $z - \dfrac{3}{4} \geq 5$, $2k + 5 \leq 10$, $x > -1$ Linear inequalities in one variable

We solve a linear inequality by finding all of its real number solutions. For example, the solution set $\{x \mid x \leq 2\}$ includes *all real numbers* that are less than or equal to 2, not just the *integers* less than or equal to 2.

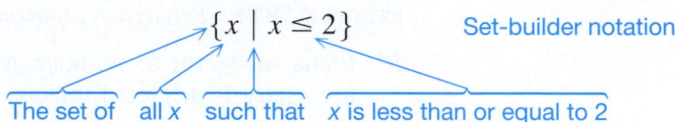

OBJECTIVE 1   Graph intervals on a number line.

Graphing is a good way to show the solution set of an inequality. To graph all real numbers belonging to the set

$$\{x \mid x \leq 2\},$$

we place a square bracket at 2 on a number line to show that it is included and then draw an arrow extending from the bracket to the left (because all numbers *less than* 2 are also part of the graph). See **FIGURE 21**.

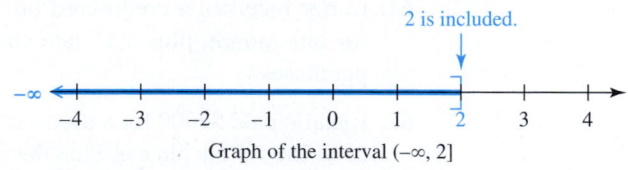

Graph of the interval $(-\infty, 2]$

FIGURE 21

The set of numbers less than or equal to 2 is an example of an **interval** on a number line. We can write this interval using **interval notation** as follows.

$$(-\infty, 2]$$ Interval notation

The **negative infinity symbol** $-\infty$ does not indicate a number, but shows that the interval includes *all* real numbers less than 2. Again, the square bracket indicates that 2 is part of the solution. Intervals that continue indefinitely in the positive direction are written with the **positive infinity symbol** ∞.

**NOW TRY
EXERCISE 1**
Graph each inequality,
and write it using interval
notation.

(a) $x < -1$ (b) $-2 \leq x$

EXAMPLE 1   Graphing Intervals on a Number Line

Graph each inequality, and write it using interval notation.

(a) $x > -5$

The statement $x > -5$ says that x can represent any number greater than -5 but cannot equal -5. We graph this interval by placing a parenthesis at -5 and drawing an arrow to the right, as in **FIGURE 22**. The parenthesis at -5 indicates that -5 is *not* part of the graph. The interval is written $(-5, \infty)$.

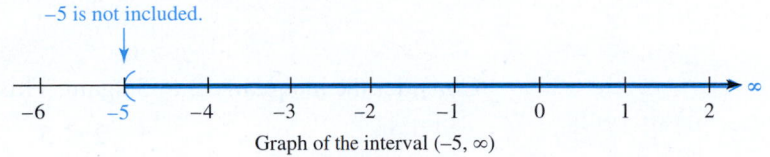

Graph of the interval $(-5, \infty)$

FIGURE 22

(b) $3 \geq x$

The statement $3 \geq x$ means the same as $x \leq 3$. **The inequality symbol continues to point toward the lesser number.** We place a square bracket at 3 (because 3 is part of the graph) and draw an arrow to the left, as in **FIGURE 23**. The interval is $(-\infty, 3]$.

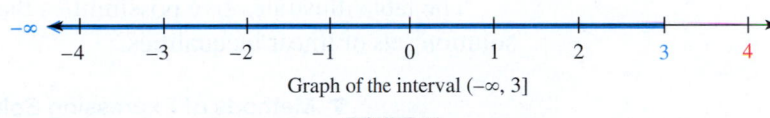

Graph of the interval $(-\infty, 3]$

FIGURE 23

CHECK   To confirm that the interval in **FIGURE 23** is graphed in the proper direction, select a value that is part of the graph and substitute it into the given inequality $3 \geq x$. For example, we select 0 and substitute to obtain $3 \geq 0$, a true statement.

If we were to select a value that is *not* part of the graph, such as 4, we would obtain $3 \geq 4$, a false statement, further confirming our result. ✓   **NOW TRY** ↻

Important Concepts Regarding Interval Notation

1. A parenthesis indicates that an endpoint is *not included* in a solution set.

2. A bracket indicates that an endpoint is *included* in a solution set.

3. A parenthesis is *always* used next to an infinity symbol, $-\infty$ or ∞.

4. The set of all real numbers is written in interval notation as $(-\infty, \infty)$.

NOTE   Some texts use a solid circle ●, rather than a square bracket, to indicate that an endpoint is included in a number line graph. An open circle ○, rather than a parenthesis, is used to indicate noninclusion.

OBJECTIVE 2   Use the addition property of inequality.

Consider the true inequality $2 < 5$. Add 4 to each side.

$$2 < 5$$
$$2 + 4 < 5 + 4 \qquad \text{Add 4.}$$
$$6 < 9 \qquad \text{True}$$

The result is a true statement. This suggests the **addition property of inequality.**

NOW TRY ANSWERS
1. (a)

-3 -2 -1 0 1 2 3

$(-\infty, -1)$

(b)

-3 -2 -1 0 1 2 3

$[-2, \infty)$

Addition Property of Inequality

If a, b, and c represent real numbers, then the inequalities

$$a < b \quad \text{and} \quad a + c < b + c \quad \text{are equivalent.*}$$

That is, the same number may be added to each side of an inequality without changing the solution set.

*This also applies to $a \le b$, $a > b$, and $a \ge b$.

Consider the inequality $2 < 5$ again. This time subtract 4 from each side.

$$2 < 5$$
$$2 - 4 < 5 - 4 \qquad \text{Subtract 4.}$$
$$-2 < 1 \qquad \text{True}$$

Again, a true statement results. *As with the addition property of equality, the same number may be* **subtracted** *from each side of an inequality.*

The table illustrates five possibilities that may occur when writing and graphing solution sets of linear inequalities.

▼ Methods of Expressing Solution Sets of Linear Inequalities

Set-Builder Notation	Graph	Interval Notation
$\{x \mid x < a\}$		$(-\infty, a)$
$\{x \mid x \le a\}$		$(-\infty, a]$
$\{x \mid x > a\}$		(a, ∞)
$\{x \mid x \ge a\}$		$[a, \infty)$
$\{x \mid x \text{ is a real number}\}$		$(-\infty, \infty)$

NOW TRY EXERCISE 2

Solve the inequality, and graph the solution set.

$$5 + 5x \ge 4x + 3$$

EXAMPLE 2   Using the Addition Property of Inequality

Solve $7 + 3x \ge 2x - 5$, and graph the solution set.

$$7 + 3x \ge 2x - 5 \qquad \text{As with equations, our goal is to isolate } x.$$
$$7 + 3x - 2x \ge 2x - 5 - 2x \qquad \text{Subtract } 2x.$$
$$7 + x \ge -5 \qquad \text{Combine like terms.}$$
$$7 + x - 7 \ge -5 - 7 \qquad \text{Subtract 7.}$$
$$x \ge -12 \qquad \text{Combine like terms.}$$

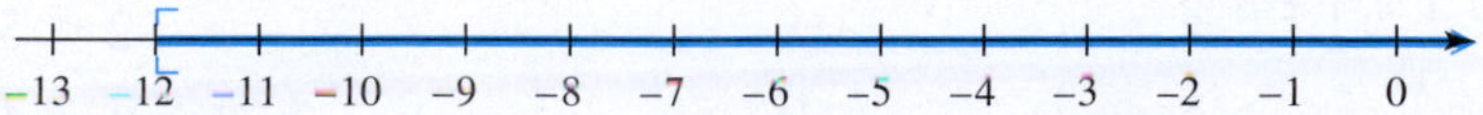

FIGURE 24

The solution set is graphed in **FIGURE 24** and written $[-12, \infty)$.

NOW TRY ANSWER

2.

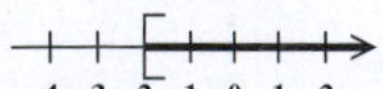

$[-2, \infty)$

NOW TRY

NOTE Because an inequality has many solutions, we cannot check all of them by substitution as we did with the single solution of an equation. To check the solutions in the interval $[-12, \infty)$ in **Example 2,** we first substitute -12 for x in the related *equation.*

CHECK

$$7 + 3x = 2x - 5 \qquad \text{Related equation}$$
$$7 + 3(-12) \overset{?}{=} 2(-12) - 5 \qquad \text{Let } x = -12.$$
$$7 - 36 \overset{?}{=} -24 - 5 \qquad \text{Multiply.}$$
$$-29 = -29 \quad \checkmark \qquad \text{True}$$

A true statement results, so -12 is indeed the "boundary" point. Next we test a number other than -12 from the interval $[-12, \infty)$. We choose 0.

CHECK

$$7 + 3x \geq 2x - 5 \qquad \text{Original inequality}$$
$$7 + 3(0) \overset{?}{\geq} 2(0) - 5 \qquad \text{Let } x = 0.$$
$$7 + 0 \overset{?}{\geq} 0 - 5 \qquad \text{Multiply.}$$
$$7 \geq -5 \quad \checkmark \qquad \text{True}$$

0 is easy to substitute.

Again, a true statement results. The checks confirm that solutions to the inequality are in the interval $[-12, \infty)$. Any number "outside" the interval $[-12, \infty)$—that is, any number in $(-\infty, -12)$—will give a false statement when tested. (Try this with -13. A false statement, $-32 \geq -31$, results.)

OBJECTIVE 3 Use the multiplication property of inequality.

Consider the true inequality $3 < 7$. Multiply each side by the positive number 2.

$$3 < 7$$
$$2(3) < 2(7) \qquad \text{Multiply each side by 2.}$$
$$6 < 14 \qquad \text{True}$$

The result is a true statement. Now multiply each side of $3 < 7$ by the negative number -5.

$$3 < 7$$
$$-5(3) < -5(7) \qquad \text{Multiply each side by } -5.$$
$$-15 < -35 \qquad \text{False}$$

To obtain a true statement when multiplying each side by -5, *we must reverse the direction of the inequality symbol.*

$$3 < 7$$
$$-5(3) > -5(7) \qquad \text{Multiply by } -5. \text{ Reverse the direction of the symbol.}$$
$$-15 > -35 \qquad \text{True}$$

NOTE The above illustrations began with the inequality $3 < 7$, a true statement involving two positive numbers. Similar results occur when one or both of the numbers are negative. Verify this by multiplying each of the following inequalities first by 2 and then by -5.

$$-3 < 7, \quad 3 > -7, \quad \text{and} \quad -7 < -3$$

These observations suggest the **multiplication property of inequality.**

Multiplication Property of Inequality

Let a, b, and c represent real numbers, where $c \neq 0$.

1. If c is *positive,* then the inequalities

$$a < b \quad \text{and} \quad ac < bc \quad \text{are equivalent.}^*$$

2. If c is *negative,* then the inequalities

$$a < b \quad \text{and} \quad ac > bc \quad \text{are equivalent.}^*$$

That is, each side of an inequality may be multiplied by the same positive number without changing the direction of the inequality symbol. *If the multiplier is negative, we must reverse the direction of the inequality symbol.*

*This also applies to $a \leq b$, $a > b$, and $a \geq b$.

As with the multiplication property of equality, the same nonzero number may be divided into each side of an inequality.

Note the following differences for positive and negative numbers.

1. When each side of an inequality is multiplied or divided by a *positive number,* the direction of the inequality symbol *does not change.*

2. When each side of an inequality is multiplied or divided by a *negative number, reverse the direction of the inequality symbol.*

EXAMPLE 3   Using the Multiplication Property of Inequality

Solve each inequality, and graph the solution set.

(a) $3x < -18$

We divide each side by 3, a positive number, so the direction of the inequality symbol *does not* change. **(It does not matter that the number on the right side of the inequality is negative.)**

$$3x < -18$$

3 is *positivie.* Do NOT reverse the direction of the symbol.

$$\frac{3x}{3} < \frac{-18}{3} \qquad \text{Divide by 3.}$$

$$x < -6$$

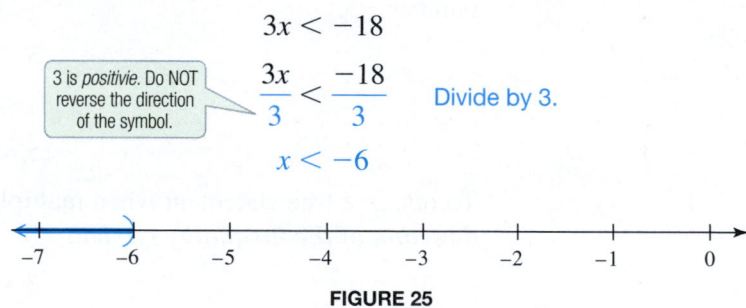

FIGURE 25

The solution set is graphed in **FIGURE 25** and written $(-\infty, -6)$.

(b) $-4x \geq 8$

Here, each side of the inequality must be divided by -4, a negative number, which *does* require changing the direction of the inequality symbol.

$$-4x \geq 8 \qquad \text{To avoid errors, show the division as a separate step.}$$

-4 is *negative.* Change $\geq$ to $\leq$.

$$\frac{-4x}{-4} \leq \frac{8}{-4} \qquad \text{Divide by } -4. \text{ Reverse the symbol.}$$

$$x \leq -2$$

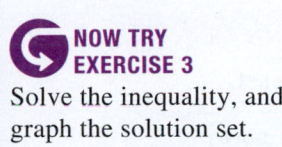
NOW TRY EXERCISE 3

Solve the inequality, and graph the solution set.

$$-5k \geq 15$$

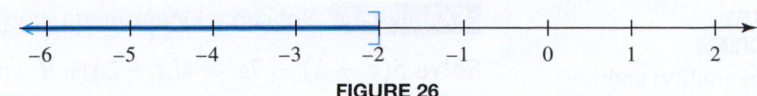

FIGURE 26

The solution set is graphed in **FIGURE 26** and written $(-\infty, -2]$.   **NOW TRY**

OBJECTIVE 4 Solve linear inequalities using both properties of inequality.

Solving a Linear Inequality in One Variable

Step 1 **Simplify each side separately.** Use the distributive property as needed.

- Clear any parentheses.
- Clear any fractions or decimals.
- Combine like terms.

Step 2 **Isolate the variable terms on one side.** Use the addition property of inequality so that all terms with variables are on one side of the inequality and all constants (numbers) are on the other side.

Step 3 **Isolate the variable.** Use the multiplication property of inequality to obtain an inequality in one of the following forms, where k is a constant (number).

$$\text{variable} < k, \quad \text{variable} \leq k, \quad \text{variable} > k, \quad \text{or} \quad \text{variable} \geq k$$

Remember: *Reverse the direction of the inequality symbol only when multiplying or dividing each side of an inequality by a negative number.*

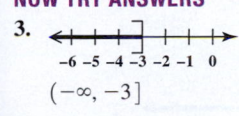
NOW TRY EXERCISE 4

Solve the inequality, and graph the solution set.

$$6 - 2x + 5x < 8x - 4$$

EXAMPLE 4 Solving a Linear Inequality

Solve $3x + 2 - 5 > -x + 7 + 2x$, and graph the solution set.

Step 1 Simplify by combining like terms.

$$3x + 2 - 5 > -x + 7 + 2x$$

$$3x - 3 > x + 7$$

Step 2 Isolate the variable term using the addition property of inequality.

$3x - 3 - x > x + 7 - x$	Subtract x.
$2x - 3 > 7$	Combine like terms.
$2x - 3 + 3 > 7 + 3$	Add 3.
$2x > 10$	Combine like terms.

Step 3 Isolate the variable using the multiplication property of inequality.

$$\frac{2x}{2} > \frac{10}{2} \qquad \text{Divide by 2.}$$

Because 2 is positive, keep the symbol $>$.

$$x > 5$$

NOW TRY ANSWERS

3.
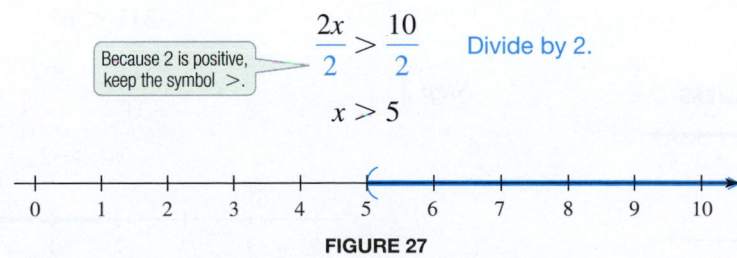
$(-\infty, -3]$

4.

$(2, \infty)$

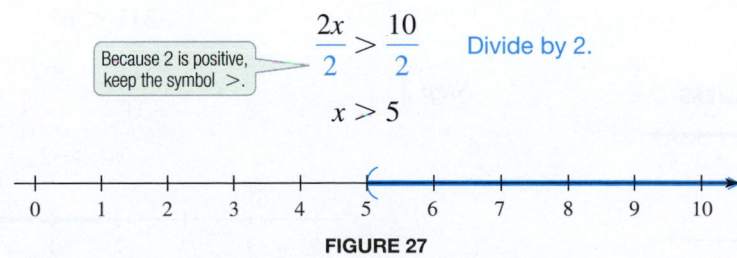

FIGURE 27

The solution set is graphed in **FIGURE 27** and written $(5, \infty)$.   **NOW TRY**

 **NOW TRY
EXERCISE 5**

Solve the inequality, and graph the solution set.

$$2t - 3(t - 6) \leq 4(t + 7)$$

EXAMPLE 5   Solving a Linear Inequality

Solve $5(z - 3) - 7z \geq 4(z - 3) + 9$, and graph the solution set.

Step 1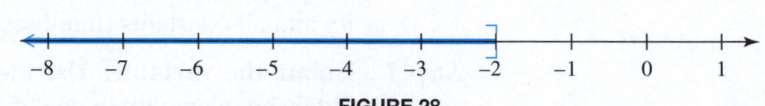

$$5(z - 3) - 7z \geq 4(z - 3) + 9$$ [Start by clearing parentheses.]

$$5z - 15 - 7z \geq 4z - 12 + 9$$ Distributive property

$$-2z - 15 \geq 4z - 3$$ Combine like terms.

Step 2

$$-2z - 15 - 4z \geq 4z - 3 - 4z$$ Subtract 4z.

$$-6z - 15 \geq -3$$ Combine like terms.

$$-6z - 15 + 15 \geq -3 + 15$$ Add 15.

$$-6z \geq 12$$ Combine like terms.

Step 3   [Because −6 is negative, change ≥ to ≤.]

$$\frac{-6z}{-6} \leq \frac{12}{-6}$$ Divide by −6. Reverse the symbol.

$$z \leq -2$$

```
←——+——+——+——+——+——+——[——+——+——+——→
  -8  -7  -6  -5  -4  -3  -2  -1   0   1
```

FIGURE 28

The solution set is graphed in **FIGURE 28** and written $(-\infty, -2]$.   NOW TRY

 **NOW TRY
EXERCISE 6**

Solve the inequality, and graph the solution set.

$$\frac{1}{8}(x + 4) > \frac{1}{6}(2x + 8)$$

EXAMPLE 6   Solving a Linear Inequality (Fractional Coefficients)

Solve $\frac{3}{4}(x - 6) < \frac{2}{3}(5x + 1)$, and graph the solution set.

Step 1

$$\frac{3}{4}(x - 6) < \frac{2}{3}(5x + 1)$$ [Clear the parentheses first. Then clear the fractions.]

$$\frac{3}{4}x - \frac{9}{2} < \frac{10}{3}x + \frac{2}{3}$$ Distributive property

$$12\left(\frac{3}{4}x - \frac{9}{2}\right) < 12\left(\frac{10}{3}x + \frac{2}{3}\right)$$ Multiply each side by the LCD, 12.

$$12\left(\frac{3}{4}x\right) + 12\left(-\frac{9}{2}\right) < 12\left(\frac{10}{3}x\right) + 12\left(\frac{2}{3}\right)$$ Distributive property

$$9x - 54 < 40x + 8$$ Multiply.

Step 2

$$9x - 54 - 40x < 40x + 8 - 40x$$ Subtract 40x.

$$-31x - 54 < 8$$ Combine like terms.

$$-31x - 54 + 54 < 8 + 54$$ Add 54.

$$-31x < 62$$ Combine like terms.

Step 3

$$\frac{-31x}{-31} > \frac{62}{-31}$$ Divide by −31. Reverse the symbol.

$$x > -2$$

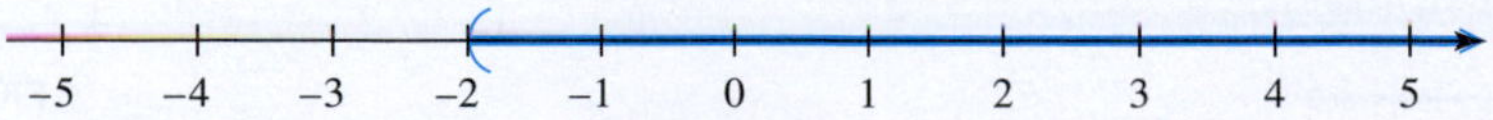

NOW TRY ANSWERS

5.
```
++——[—+——+——+——+——→
-4 -3 -2 -1  0  1  2
```
$[-2, \infty)$

6.
```
←—+——)—+——+——+——+——→
-6 -5 -4 -3 -2 -1  0
```
$(-\infty, -4)$

```
←——+——+——+——(——+——+——+——+——+——+——→
  -5  -4  -3  -2  -1   0   1   2   3   4   5
```

FIGURE 29

The solution set is graphed in **FIGURE 29** and written $(-2, \infty)$.   NOW TRY

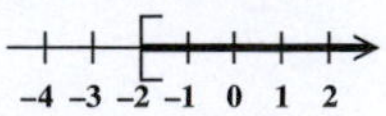

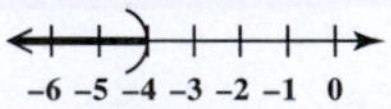

OBJECTIVE 5 Solve applied problems using inequalities.

▼ **Words and Phrases That Indicate Inequality**

Word or Phrase	Example	Inequality
Is more than	A number *is more than* 4	$x > 4$
Is less than	A number *is less than* -12	$x < -12$
Exceeds	A number *exceeds* 3.5	$x > 3.5$
Is at least	A number *is at least* 6	$x \geq 6$
Is at most	A number *is at most* 8	$x \leq 8$

! CAUTION Do not confuse a statement such as "5 is more than a number," expressed as $5 > x$, with the phrase "5 more than a number," expressed as $x + 5$ or $5 + x$.

The next example uses the idea of finding the *average* of a number of scores. ***To find the average of n numbers, add the numbers and divide by n.*** We continue to use the six problem-solving steps, changing *Step 3* to **"Write an inequality."**

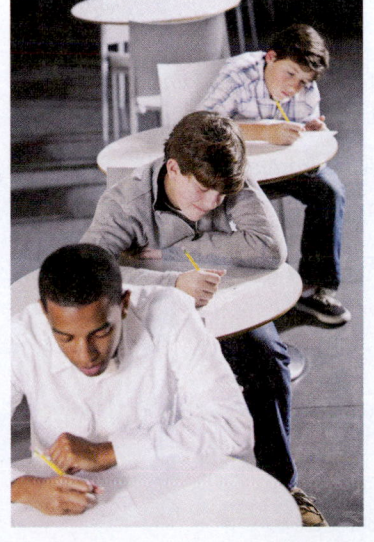

NOW TRY EXERCISE 7

Will has grades of 98 and 85 on his first two tests in algebra. If he wants an average of at least 90 after his third test, what possible scores may he make on that test?

EXAMPLE 7   Finding an Average Test Score

John has grades of 86, 88, and 78 on his first three tests in geometry. If he wants an average of at least 80 after his fourth test, what possible scores may he make on that test?

Step 1   **Read** the problem again.

Step 2   **Assign a variable.** Let x = John's score on his fourth test.

Step 3   **Write an inequality.**

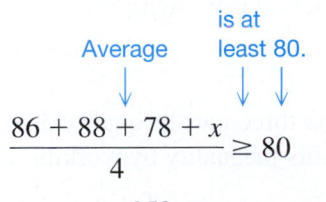

$$\frac{86 + 88 + 78 + x}{4} \geq 80$$

To find his average after four tests, add the test scores and divide by 4.

Step 4   **Solve.**

$$\frac{252 + x}{4} \geq 80 \qquad \text{Add in the numerator.}$$

$$4\left(\frac{252 + x}{4}\right) \geq 4(80) \qquad \text{Multiply by 4.}$$

$$252 + x \geq 320$$

$$252 + x - 252 \geq 320 - 252 \qquad \text{Subtract 252.}$$

$$x \geq 68 \qquad \text{Combine like terms.}$$

Step 5   **State the answer.** He must score 68 or more on the fourth test to have an average of *at least* 80.

Step 6   **Check.**

$$\frac{86 + 88 + 78 + 68}{4} = \frac{320}{4} = 80$$

To complete the check, also show that any number greater than 68 (but less than or equal to 100) makes the average greater than 80.

NOW TRY ANSWER

7. 87 or more

NOW TRY

> ⚠ **CAUTION** In applied problems, remember that
>
> is at least　translates as　is greater than or equal to
>
> and　is at most　translates as　is less than or equal to.

OBJECTIVE 6　Solve linear inequalities with three parts.

An inequality that says that one number is *between* two other numbers is a **three-part inequality.** For example,

$$-3 < 5 < 7 \quad \text{says that} \quad 5 \text{ is } between -3 \text{ and } 7.$$

↻ NOW TRY EXERCISE 8

Graph the inequality, and write it using interval notation.

$$0 \le x < 2$$

EXAMPLE 8　Graphing a Three-Part Inequality

Graph the inequality $-1 \le x < 3$, and write it using interval notation.

The statement is read "-1 *is less than or equal to x **and** x is less than* 3." We want the set of numbers *between* -1 and 3, with -1 included and 3 excluded. We use a square bracket at -1 because -1 is part of the graph and a parenthesis at 3 because 3 is not part of the graph. See **FIGURE 30.**

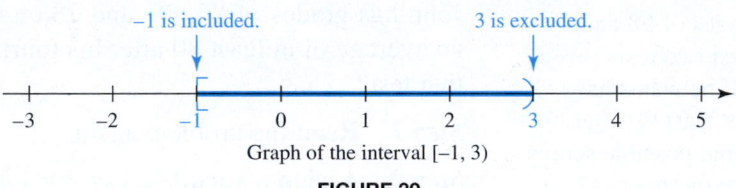

Graph of the interval $[-1, 3)$

FIGURE 30

The interval is written $[-1, 3)$.　　**NOW TRY ↻**

The three-part inequality $3 < x + 2 < 8$ says that $x + 2$ is *between* 3 and 8. We solve this inequality by working with all three parts at the same time.

$$3 - 2 < x + 2 - 2 < 8 - 2 \qquad \text{Subtract 2 from } each \text{ part.}$$
$$1 < \quad x \quad < 6$$

The idea is to obtain an inequality in the following form.

$$\text{a number} < x < \text{ another number}$$

EXAMPLE 9　Solving Three-Part Inequalities

Solve each inequality, and graph the solution set.

(a)
$$4 < \quad 3x - 5 \quad \le 10 \qquad \text{⟨ Work with all three parts at the same time.}$$
$$4 + 5 < 3x - 5 + 5 \le 10 + 5 \qquad \text{Add 5 to each part.}$$
$$9 < \quad 3x \quad \le 15 \qquad \text{Combine like terms.}$$
$$\frac{9}{3} < \quad \frac{3x}{3} \quad \le \frac{15}{3} \qquad \text{Divide each part by 3.}$$
$$3 < \quad x \quad \le 5$$

Remember to divide all *three* parts by 3.

NOW TRY ANSWER

8.
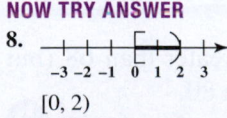

$[0, 2)$

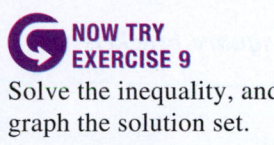

NOW TRY
EXERCISE 9

Solve the inequality, and graph the solution set.

$$-4 \leq \frac{3}{2}x - 1 \leq 0$$

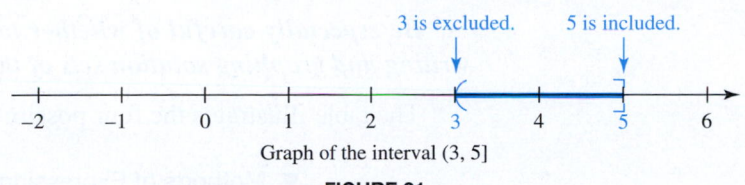

3 is excluded.   5 is included.

Graph of the interval (3, 5]

FIGURE 31

The solution set is graphed in **FIGURE 31** and written (3, 5].

(b)

$$-4 \leq \frac{2}{3}m - 1 \leq 8$$ ← Work with all three parts at the same time.

$$3(-4) \leq 3\left(\frac{2}{3}m - 1\right) \leq 3(8)$$ Multiply each part by 3 to clear the fraction.

$$-12 \leq 2m - 3 \leq 24$$ Multiply; distributive property

$$-12 + 3 \leq 2m - 3 + 3 \leq 24 + 3$$ Add 3 to each part.

$$-9 \leq 2m \leq 27$$ Combine like terms.

$$\frac{-9}{2} \leq \frac{2m}{2} \leq \frac{27}{2}$$ Divide each part by 2.

$$-\frac{9}{2} \leq m \leq \frac{27}{2}$$

Think: $-\frac{9}{2} = -4\frac{1}{2}$

Think: $\frac{27}{2} = 13\frac{1}{2}$

FIGURE 32

The solution set is graphed in **FIGURE 32** and written $\left[-\frac{9}{2}, \frac{27}{2}\right]$.   **NOW TRY**

NOTE The inequality in **Example 9(b)** could also be solved as follows.

$$-4 \leq \frac{2}{3}m - 1 \leq 8$$ See Example 9(b).

$$-4 + 1 \leq \frac{2}{3}m - 1 + 1 \leq 8 + 1$$ Add 1 to each part.

$$-3 \leq \frac{2}{3}m \leq 9$$

$$\frac{3}{2}(-3) \leq \frac{3}{2}\left(\frac{2}{3}m\right) \leq \frac{3}{2}(9)$$ Multiply each part by $\frac{3}{2}$.

$$-\frac{9}{2} \leq m \leq \frac{27}{2}$$ The same solution set $\left[-\frac{9}{2}, \frac{27}{2}\right]$ results.

NOW TRY ANSWER

9.

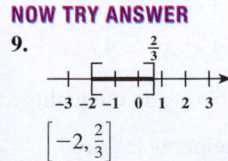

$$\left[-2, \frac{2}{3}\right]$$

⚠ **CAUTION** *Three-part inequalities are written so that the symbols point in the same direction and both point toward the lesser number.*

$3 < x + 2 < 8$ is equivalent to $8 > x + 2 > 3$.

It would be *wrong* to write $8 < x + 2 < 3$, which would imply that $8 < 3$, a *false* statement.

Be especially careful of whether to use parentheses or square brackets when writing and graphing solution sets of three-part inequalities.

The table illustrates the four possibilities that may occur.

▼ **Methods of Expressing Solution Sets of Three-Part Inequalities**

Set-Builder Notation	Graph	Interval Notation
$\{x \mid a < x < b\}$		(a, b)
$\{x \mid a < x \le b\}$		$(a, b]$
$\{x \mid a \le x < b\}$		$[a, b)$
$\{x \mid a \le x \le b\}$		$[a, b]$

2.9 Exercises

FOR EXTRA HELP   ▶ **MyLab Math**

▶ *Video solutions for select problems available in MyLab Math*

Concept Check   *Work each problem.*

1. When graphing an inequality, use a parenthesis if the inequality symbol is _____ or _____ . Use a square bracket if the inequality symbol is _____ or _____ .

2. *True* or *false?*   In interval notation, a square bracket is sometimes used next to an infinity symbol.

3. In interval notation, the set $\{x \mid x > 0\}$ is written _____ .

4. In interval notation, the set of all real numbers is written _____ .

Concept Check   *Write an inequality using the variable x that corresponds to each graph of solutions on a number line.*

5.

6.

7.

8.

Graph each inequality, and write it using interval notation. See Example 1.

9. $k \le 4$	**10.** $x \le 3$	**11.** $x < -3$	**12.** $r < -11$
13. $t > 4$	**14.** $m > 5$	**15.** $-\dfrac{1}{2} \le x$	**16.** $-\dfrac{3}{4} \le x$
17. $0 \ge x$	**18.** $1 \ge x$	**19.** $-2 < x$	**20.** $-1 < x$

Solve each inequality. Graph the solution set, and write it using interval notation. See Example 2.

21. $z - 8 \ge -7$	**22.** $p - 3 \ge -11$	**23.** $2k + 3 \ge k + 8$
24. $3x + 7 \ge 2x + 11$	**25.** $3n + 5 < 2n - 6$	**26.** $5x - 2 < 4x - 5$

27. Under what conditions must the inequality symbol be reversed when solving an inequality?

28. **Concept Check**   If $p < q$ and $r < 0$, which one of the following statements is false?

   A. $pr < qr$ **B.** $pr > qr$ **C.** $p + r < q + r$ **D.** $p - r < q - r$

29. Concept Check A student solved the following inequality incorrectly as shown.

$$-5x < 15$$

$$\frac{-5x}{-5} < \frac{15}{-5}$$

$$x < -3$$

WHAT WENT WRONG? Give the correct solution set.

30. Concept Check A student solved the following inequality incorrectly as shown.

$$\frac{2}{3}x \geq -6$$

$$\frac{3}{2}\left(\frac{2}{3}x\right) \leq \frac{3}{2}(-6)$$

$$x \leq -9$$

WHAT WENT WRONG? Give the correct solution set.

Solve each inequality. Graph the solution set, and write it using interval notation. **See Example 3.**

31. $3x < 18$ **32.** $5x < 35$ **33.** $2y \geq -20$

34. $6m \geq -24$ **35.** $-8t > 24$ **36.** $-7x > 49$

37. $-x \geq 0$ **38.** $-k < 0$ **39.** $-\frac{3}{4}r < -15$

40. $-\frac{7}{8}t < -14$ **41.** $-0.02x \leq 0.06$ **42.** $-0.03v \geq -0.12$

Solve each inequality. Graph the solution set, and write it using interval notation. **See Examples 4–6.**

43. $8x + 9 \leq -15$ **44.** $6x + 7 \leq -17$

45. $-4x - 3 < 1$ **46.** $-5x - 4 < 6$

47. $5r + 1 \geq 3r - 9$ **48.** $6t + 3 < 3t + 12$

49. $5x - 2 \leq -x + 10$ **50.** $3x - 9 \geq -2x + 6$

51. $-7x + 4 > -3x - 2$ **52.** $-8x + 1 < -4x + 11$

53. $6x + 3 + x < 2 + 4x + 4$ **54.** $-4w + 12 + 9w \geq w + 9 + w$

55. $-x + 4 + 7x \leq -2 + 3x + 6$ **56.** $14x - 6 + 7x > 4 + 10x - 10$

57. $5(t - 1) > 3(t - 2)$ **58.** $7(m - 2) < 4(m - 4)$

59. $5(x + 3) - 6x \leq 3(2x + 1) - 4x$ **60.** $2(x - 5) + 3x > 4(x - 6) + 1$

61. $\frac{1}{3}(5x - 4) \geq \frac{2}{5}(x + 3)$ **62.** $\frac{5}{12}(5x - 7) < \frac{5}{6}(x - 5)$

63. $\frac{2}{3}(p + 3) > \frac{5}{6}(p - 4)$ **64.** $\frac{7}{9}(x - 4) \leq \frac{4}{3}(x + 5)$

65. $\frac{4}{5}x - \frac{1}{2}(x + 3) \leq \frac{3}{10}$ **66.** $\frac{1}{6}x + \frac{1}{3}(x - 1) > \frac{1}{2}$

67. $4x - (6x + 1) \leq 8x + 2(x - 3)$ **68.** $2z - (4z + 3) > 6z + 3(z + 4)$

69. $5(2k + 3) - 2(k - 8) > 3(2k + 4) + k - 2$

70. $2(3z - 5) + 4(z + 6) \geq 2(3z + 2) + 3z - 15$

Concept Check *Translate each statement into an inequality. Use x as the variable.*

71. You must be at least 18 yr old to vote.

72. Less than 1 in. of rain fell.

73. Chicago received more than 5 in. of snow.

74. A full-time student must take at least 12 credits.

75. Tracy could spend at most $20 on a gift.

76. The car's speed exceeded 60 mph.

*Solve each problem. **See Example 7.***

77. Christy has scores of 76 and 81 on her first two algebra tests. If she wants an average of at least 80 after her third test, what possible scores may she make on that test?

78. Joseph has scores of 96 and 86 on his first two geometry tests. What possible scores may he make on his third test so that his average is at least 90?

79. A student has scores of 87, 84, 95, and 79 on four quizzes. What scores may she make on the fifth quiz to have an average of at least 85?

80. Another student has scores of 82, 93, 94, and 86 on four quizzes. What scores may he make on the fifth quiz to have an average of at least 90?

81. The average monthly precipitation in Houston, Texas, for October, November, and December is 4.6 in. If 5.7 in. falls in October and 4.3 in. falls in November, how many inches must fall in December so that the average monthly precipitation for these months exceeds 4.6 in.? (Data from National Climatic Data Center.)

82. The average monthly precipitation in New Orleans, Louisiana, for June, July, and August is 6.7 in. If 8.1 in. falls in June and 5.7 in. falls in July, how many inches must fall in August so that the average monthly precipitation for these months exceeds 6.7 in.? (Data from National Climatic Data Center.)

83. When 2 is added to the difference of six times a number and 5, the result is greater than 13 added to five times the number. Find all such numbers.

84. When 8 is subtracted from the sum of three times a number and 6, the result is less than 4 more than the number. Find all such numbers.

85. The formula for converting Fahrenheit temperature to Celsius is

$$C = \frac{5}{9}(F - 32).$$

If the Celsius temperature on a certain winter day in Minneapolis is never less than $-25°$, how would we describe the corresponding Fahrenheit temperatures? (Data from National Climatic Data Center.)

86. The formula for converting Celius temperature to Fahrenheit is

$$F = \frac{9}{5}C + 32.$$

The Fahrenheit temperature of Phoenix has never exceeded 122°. How would we describe this using Celsius temperature? (Data from National Climatic Data Center.)

87. For what values of x would the rectangle have a perimeter of at least 400?

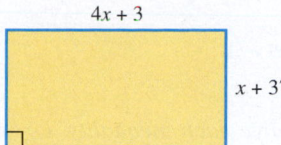

4x + 3

x + 37

88. For what values of x would the triangle have a perimeter of at least 72?

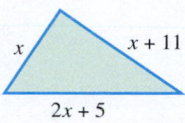

x

x + 11

2x + 5

89. A phone call costs \$2.00, plus \$0.30 per minute or fractional part of a minute. If x represents the number of minutes of the length of the call, then $2 + 0.30x$ represents the cost of the call. If Alan has \$5.60 to spend on a call, what is the maximum total time he can use the phone?

90. At the Speedy Gas'n Go, a car wash costs \$3.00 and gasoline is selling for \$3.60 per gallon. Carla has \$48.00 to spend, and her car is so dirty that she must have it washed. What is the maximum number of gallons of gasoline that she can purchase?

A company that produces DVDs has found that revenue from sales of DVDs is \$5 per DVD, less sales costs of \$100. Production costs are \$125, plus \$4 per DVD. Profit (P) is given by revenue (R) less cost (C), so the company must find the production level x that makes

$$P > 0, \quad \text{that is,} \quad R - C > 0. \qquad P = R - C$$

91. Write an expression for revenue R, letting x represent the production level (number of DVDs to be produced).

92. Write an expression for production costs C in terms of x.

93. Write an expression for profit P, and then solve the inequality $P > 0$.

94. Describe the solution in terms of the problem.

Concept Check *Write a three-part inequality using the variable x that corresponds to each graph of solutions on a number line.*

95.

-1 0 1 2

96.

-1 0 1 2

97.

-1 0 1 2

98.

-1 0 1 2

Graph each inequality, and write it using interval notation. See Example 8.

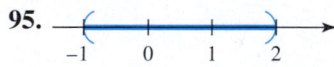

99. $8 \le x \le 10$

100. $3 \le x \le 5$

101. $0 < y \le 10$

102. $-3 \le x < 0$

103. $4 > x > -3$

104. $6 \ge x \ge -4$

Solve each inequality. Graph the solution set, and write it using interval notation. See Example 9.

105. $-8 < 4x \le 4$

106. $-3 \le 3x < 12$

107. $-5 \le 2x - 3 \le 9$

108. $-7 \le 3x - 4 \le 8$

109. $10 < 7p + 3 < 24$

110. $-8 \le 3r - 1 \le -1$

111. $6 \le 3(x - 1) < 18$

112. $-4 < 2(x + 1) \le 6$

113. $-12 \le \frac{1}{2}z + 1 \le 4$

114. $-6 \le \frac{1}{3}x + 3 \le 5$

115. $1 \le 3 + \dfrac{2}{3}p \le 7$

116. $2 < 6 + \dfrac{3}{4}x < 12$

117. $-7 \le \dfrac{5}{4}r - 1 \le -1$

118. $-12 \le \dfrac{3}{7}x + 2 \le -4$

Extending Skills *Solve each inequality. Graph the solution set, and write it using interval notation.*

119. $-4 < -2x < 12$

120. $9 < -3x < 15$

121. $5 < 1 - 6m < 12$

122. $-1 \le 1 - 5q \le 16$

RELATING CONCEPTS   For Individual or Group Work   (Exercises 123–126)

Work Exercises 123–126 in order, to see the connection between the solution of an equation and the solutions of the corresponding inequalities.

123. Solve the following equation, and graph the solution set on a number line.

$$3x + 2 = 14$$

124. Solve the following inequality, and graph the solution set on a number line.

$$3x + 2 > 14$$

125. Solve the following inequality, and graph the solution set on a number line.

$$3x + 2 < 14$$

126. If we were to graph all the solution sets from **Exercises 123–125** on the same number line, describe the graph that we would obtain. (This is the **union** of all the solution sets.)

Chapter 2	Summary

STUDY SKILLS REMINDER

How do you best prepare for a test? **Review Study Skill 7,** *Reviewing a Chapter.*

Key Terms

2.1

equation
linear equation in one
 variable
first-degree equation
solution
solution set
equivalent equations

2.3

conditional equation
identity

contradiction
empty (null) set

2.5

consecutive integers
consecutive even (or odd)
 integers
degree
complementary angles
right angle
supplementary angles
straight angle

2.6

formula
area
perimeter
vertical angles

2.7

ratio
proportion
terms of a proportion
extremes
means

cross products of
 a proportion
percent
percentage
base

2.9

inequality
linear inequality in one
 variable
interval
three-part inequality

New Symbols

∅ empty set a to b, $a:b$,   or $\frac{a}{b}$ ratio of a to b ∞   infinity

1° one degree (a, b) interval notation for $a < x < b$ −∞   negative infinity

⌐ right angle $[a, b]$ interval notation for $a \le x \le b$ $(−∞, ∞)$ set of all real numbers

Test Your Word Power

See how well you have learned the vocabulary in this chapter.

1. A **solution set** is the set of numbers that
   A. make an expression undefined
   B. make an equation false
   C. make an equation true
   D. make an expression equal to 0.

2. **Complementary angles** are angles
   A. formed by two parallel lines
   B. whose sum is 90°
   C. whose sum is 180°
   D. formed by perpendicular lines.

3. **Supplementary angles** are angles
   A. formed by two parallel lines
   B. whose sum is 90°
   C. whose sum is 180°
   D. formed by perpendicular lines.

4. A **ratio**
   A. compares two quantities using a quotient
   B. says that two quotients are equal
   C. is a product of two quantities
   D. is a difference of two quantities.

5. A **proportion**
   A. compares two quantities using a quotient
   B. says that two quotients are equal
   C. is a product of two quantities
   D. is a difference of two quantities.

6. An **inequality** is
   A. a statement that two algebraic expressions are equal
   B. a point on a number line
   C. an equation with no solutions
   D. a statement that relates algebraic expressions using $<, \le, >$, or $\ge$.

ANSWERS

1. C; *Example:* {8} is the solution set of $2x + 5 = 21$.   **2.** B; *Example:* Angles with measures 35° and 55° are complementary angles.
3. C; *Example:* Angles with measures 112° and 68° are supplementary angles.   **4.** A; *Example:* $\frac{7 \text{ in.}}{12 \text{ in.}}$, or $\frac{7}{12}$ **5.** B; *Example:* $\frac{2}{3} = \frac{8}{12}$
6. D; *Examples:* $x < 5, 7 + 2y \ge 11, -5 < 2z - 1 \le 3$

Quick Review

CONCEPTS	EXAMPLES
2.1 **The Addition Property of Equality** The same number may be added to (or subtracted from) each side of an equation without changing the solution set.	Solve. $x - 6 = 12$ $x - 6 + 6 = 12 + 6$ Add 6. $x = 18$ Combine like terms. Solution set: $\{18\}$
2.2 **The Multiplication Property of Equality** Each side of an equation may be multiplied (or divided) by the same nonzero number without changing the solution set.	Solve. $\frac{3}{4}x = -9$ $\frac{4}{3} \cdot \frac{3}{4}x = \frac{4}{3} \cdot (-9)$ Multiply by $\frac{4}{3}$, the reciprocal of $\frac{3}{4}$. $x = -12$ Solution set: $\{-12\}$

CONCEPTS	EXAMPLES
2.3 Solving Linear Equations Using Both Properties of Equality **2.4** Clearing Fractions and Decimals When Solving Linear Equations **Solving a Linear Equation in One Variable** *Step 1* Simplify each side separately. • Clear any parentheses. • Clear any fractions or decimals. • Combine like terms. *Step 2* Isolate the variable terms on one side. *Step 3* Isolate the variable. *Step 4* Check.	Solve. $2x + 2(x + 1) = 14 + x$ $2x + 2x + 2 = 14 + x$ Distributive property $4x + 2 = 14 + x$ Combine like terms. $4x + 2 - x - 2 = 14 + x - x - 2$ Subtract x. Subtract 2. $3x = 12$ Combine like terms. $\dfrac{3x}{3} = \dfrac{12}{3}$ Divide by 3. $x = 4$ **CHECK** $2(4) + 2(4 + 1) \stackrel{?}{=} 14 + 4$ Let $x = 4$. $18 = 18$ ✓ True Solution set: $\{4\}$
2.5 Applications of Linear Equations **Solving an Applied Problem** *Step 1* Read. *Step 2* Assign a variable. *Step 3* Write an equation. *Step 4* Solve the equation. *Step 5* State the answer. *Step 6* Check.	One number is five more than another. Their sum is 21. What are the numbers? Let x = the lesser number. Then $x + 5$ = the greater number. $x + (x + 5) = 21$ $2x + 5 = 21$ Combine like terms. $2x = 16$ Subtract 5. $x = 8$ Divide by 2. The numbers are 8 and $8 + 5 = 13$. 13 is five more than 8, and $8 + 13 = 21$. The answer checks.
2.6 Formulas and Additional Applications from Geometry To find the value of one of the variables in a formula, given values for the others, substitute the known values into the formula.	Find L if $\mathscr{A} = LW$, given that $\mathscr{A} = 24$ and $W = 3$. $\mathscr{A} = LW$ $24 = L \cdot 3$ $\mathscr{A} = 24, W = 3$ $\dfrac{24}{3} = \dfrac{L \cdot 3}{3}$ Divide by 3. $8 = L$
Solving a Formula for a Specified Variable (Solving a Literal Equation) To solve a formula for one of the variables, isolate that variable by treating the other variables as constants (numbers) and using the steps for solving equations.	Solve $P = 2a + 2b$ for b. $P = 2a + 2b$ $P - 2a = 2a + 2b - 2a$ Subtract $2a$. $P - 2a = 2b$ Combine like terms. $\dfrac{P - 2a}{2} = \dfrac{2b}{2}$ Divide by 2. $\dfrac{P - 2a}{2} = b,$ or $b = \dfrac{P - 2a}{2}$

CONCEPTS	EXAMPLES

2.7 Ratio, Proportion, and Percent

To write a ratio, express quantities using the same units.

4 ft to 8 in. can be written 48 in. to 8 in., which is the ratio

$$\frac{48}{8}, \quad \text{or} \quad \frac{6}{1}.$$

To solve a proportion, use the method of cross products.

Solve.

$$\frac{x}{12} = \frac{35}{60}$$

$$60x = 12 \cdot 35 \qquad \text{Cross products}$$

$$60x = 420 \qquad \text{Multiply.}$$

$$x = 7 \qquad \text{Divide by 60.}$$

Solution set: $\{7\}$

Percent, Fraction, and Decimal Equivalents

$$1\% = \frac{1}{100} = 0.01, \quad 10\% = \frac{10}{100} = 0.10, \quad 100\% = \frac{100}{100} = 1$$

$$5\% = \frac{5}{100} = 0.05, \quad 35\% = \frac{35}{100} = 0.35, \quad 200\% = \frac{200}{100} = 2$$

To solve a percent problem, use the percent equation.

percent (as a decimal) · base = amount

What percent of 325 is 65?

$$p \cdot 325 = 65$$

$$p = \frac{65}{325}$$

$$p = 0.2, \quad \text{or} \quad p = 20\%$$

Thus, 20% of 325 is 65.

2.8 Further Applications of Linear Equations

Solving an Applied Problem

Step 1 Read.

Step 2 Assign a variable. Make a sketch, diagram, or table, as needed.

There are three forms of the formula relating distance, rate, and time.

$$d = rt, \quad r = \frac{d}{t}, \quad t = \frac{d}{r}$$

Step 3 Write an equation.

Step 4 Solve the equation.

Step 5 State the answer.

Step 6 Check.

Two cars leave from the same point, traveling in opposite directions. One travels at 45 mph and the other at 60 mph. How long will it take them to be 210 mi apart?

Let t = time it takes for the two cars to be 210 mi apart.

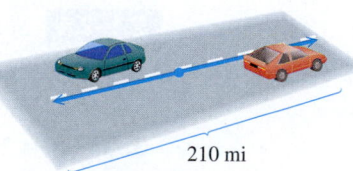

210 mi

	Rate	Time	Distance
Slower Car	45	t	$45t$
Faster Car	60	t	$60t$

The sum of the distance is 210 mi.

$$45t + 60t = 210$$

$$105t = 210 \qquad \text{Combine like terms.}$$

$$t = 2 \qquad \text{Divide by 105.}$$

It will take the cars 2 hr to be 210 mi apart.

CONCEPTS	EXAMPLES

2.9 **Solving Linear Inequalities**

Solving a Linear Inequality in One Variable

Step 1   Simplify each side separately.
- Clear any parentheses.
- Clear any fractions or decimals.
- Combine like terms.

Step 2   Isolate the variable terms on one side.

Step 3   Isolate the variable.

Be sure to reverse the direction of the inequality symbol when multiplying or dividing by a negative number.

To solve a three-part inequality such as

$$4 < 2x + 6 \le 8,$$

work with all three parts at the same time.

Solve the inequality, and graph the solution set.

$$3(1 - x) + 5 - 2x > 9 - 6$$

$3 - 3x + 5 - 2x > 9 - 6$ Distributive property

$8 - 5x > 3$ Combine like terms.

$8 - 5x - 8 > 3 - 8$ Subtract 8.

$-5x > -5$ Combine like terms.

$\dfrac{-5x}{-5} < \dfrac{-5}{-5}$ Divide by -5.
  Change $>$ to $<$.

$x < 1$

Solution set: 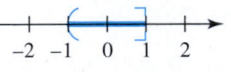 $(-\infty, 1)$

Solve the inequality, and graph the solution set.

$$4 < \quad 2x + 6 \quad \le 8$$

$4 - 6 < 2x + 6 - 6 \le 8 - 6$ Subtract 6.

$-2 < \quad 2x \quad \le 2$ Combine like terms.

$\dfrac{-2}{2} < \quad \dfrac{2x}{2} \quad \le \dfrac{2}{2}$ Divide by 2.

$-1 < \quad x \quad \le 1$

Solution set: $(-1, 1]$

Chapter 2   Review Exercises

2.1–2.4 *Solve each equation.*

1. $x - 5 = 1$

2. $x + 8 = -4$

3. $3t + 1 = 2t + 8$

4. $5z = 4z + \dfrac{2}{3}$

5. $(4r - 2) - (3r + 1) = 8$

6. $3(2x - 5) = 2 + 5x$

7. $7x = 35$

8. $12r = -48$

9. $2p - 7p + 8p = 15$

10. $\dfrac{x}{12} = -1$

11. $-\dfrac{6}{5}q = -18$

12. $12m + 11 = 59$

13. $3(2x + 6) - 5(x + 8) = x - 22$

14. $5x + 9 - (2x - 3) = 2x - 7$

15. $2(3t - 1) = 10 - 4(t + 3)$

16. $0.1(x + 80) + 0.2x = 14$

17. $3x - (-2x + 6) = 4(x - 4) + x$

18. $\dfrac{1}{2}(x + 3) - \dfrac{2}{3}(x - 2) = 3$

2.5 *Solve each problem.*

19. If 7 is added to five times a number, the result is equal to three times the number. Find the number.

20. In 2017, Illinois had 118 members in its House of Representatives, consisting of only Democrats and Republicans. There were 16 more Democrats than Republicans. How many representatives from each party were there? (Data from www.ilga.gov)

21. The land area of Hawaii is 5213 mi^2 greater than the area of Rhode Island. Together, the areas total 7637 mi^2. What is the area of each state?

22. The height of Seven Falls in Colorado is $\frac{5}{2}$ the height of Twin Falls in Idaho. The sum of the heights is 420 ft. Find the height of each. (Data from *The World Almanac and Book of Facts.*)

23. A lawn mower uses a mixture of gasoline and oil. The mixture contains 1 oz of oil for every 32 oz of gasoline. How many ounces of oil and how many ounces of gasoline are required to fill a 132-oz tank completely?

24. Find two consecutive odd integers such that when the lesser is added to twice the greater, the result is 24 more than the greater integer.

25. A 72-in. board is to be cut into three pieces. The longest piece must be three times as long as the shortest piece. The middle-sized piece must be 7 in. longer than the shortest piece. How long must each piece be?

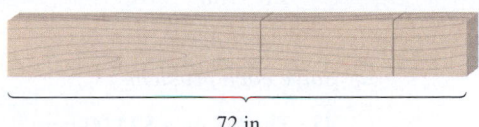

72 in.

26. The supplement of an angle measures 10 times the measure of its complement. What is the measure of the angle?

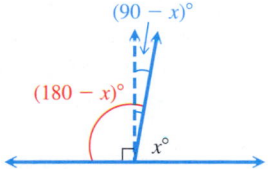

2.6 *Find the value of the variable in each formula. Use 3.14 as an approximation for π.*

27. $A = \dfrac{1}{2}bh$; $A = 44, b = 8$

28. $A = \dfrac{1}{2}h(b + B)$; $h = 8, b = 3, B = 4$

29. $C = 2\pi r$; $C = 29.83$

30. $V = \dfrac{4}{3}\pi r^3$; $r = 6$

Solve each problem.

31. The perimeter of a certain rectangle is 16 times the width. The length is 12 cm more than the width. Find the width of the rectangle.

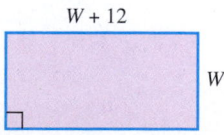

32. A baseball diamond is a square with a side of 90 ft. The pitcher's position is 60.5 ft from home plate, as shown in the figure. Find the measures of the angles marked in the figure. (*Hint:* Recall that the sum of the measures of the angles of any triangle is 180°.)

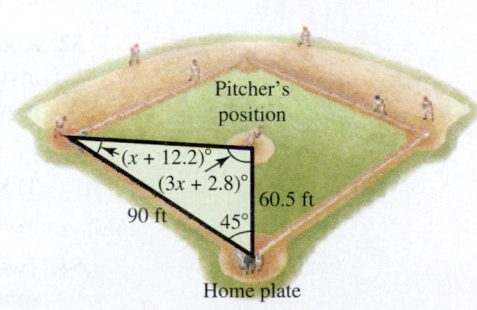

Find the measure of each marked angle.

33.

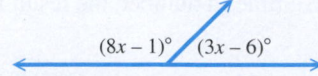

$(8x - 1)°$ $(3x - 6)°$

34.

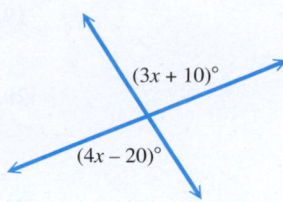

$(3x + 10)°$

$(4x - 20)°$

Solve each formula for the specified variable.

35. $A = bh$ for h

36. $A = \dfrac{1}{2}h(b + B)$ for h

Solve each equation for y.

37. $x + y = 11$

38. $3x - 2y = 12$

2.7 *Write a ratio for each word phrase. Express fractions in lowest terms.*

39. 60 cm to 40 cm

40. 5 days to 2 weeks

41. 90 in. to 10 ft

Solve each equation.

42. $\dfrac{p}{21} = \dfrac{5}{30}$

43. $\dfrac{5 + x}{3} = \dfrac{2 - x}{6}$

44. $\dfrac{z}{5} = \dfrac{6z - 5}{11}$

Solve each problem.

45. The tax on a $24.00 item is $2.04. How much tax would be paid on a $36.00 item?

46. The distance between two cities on a road map is 32 cm. The two cities are actually 150 km apart. The distance on the map between two other cities is 80 cm. How far apart are these cities?

47. In the 2016 Summer Olympics in Rio de Janeiro, Brazil, Italian athletes earned 28 medals. Two of every 7 medals were gold. How many gold medals did Italy earn? (Data from *The World Almanac and Book of Facts.*)

48. Find the best buy. Give the unit price to the nearest thousandth for that size. (Data from Jewel-Osco.)

Cereal

Size	Price
9 oz	$3.49
14 oz	$3.99
18 oz	$4.49

49. What percent of 12 is 21?

50. 36% of what number is 900?

2.8 *Solve each problem.*

51. A nurse must mix 15 L of a 10% solution of a drug with some 60% solution to obtain a 20% mixture. How many liters of the 60% solution will be needed?

52. Robert invested $10,000, from which he earns an annual income of $400 per year. He invested part of the $10,000 at 5% annual interest and the remainder paying 3% interest. How much did he invest at each rate?

53. In 1846, the vessel *Yorkshire* traveled from Liverpool to New York, a distance of 3150 mi, in 384 hr. What was the *Yorkshire's* average rate? Round the answer to the nearest tenth.

54. Two planes leave St. Louis at the same time. One flies north at 350 mph and the other flies south at 420 mph. In how many hours will they be 1925 mi apart?

2.9 *Graph each inequality, and write it using interval notation.*

55. $x \geq -4$ **56.** $x < 7$ **57.** $-5 \leq x < 6$

58. Which inequality requires reversing the direction of the inequality symbol when it is solved?

    A. $4x \geq -36$ **B.** $-4x \leq 36$ **C.** $4x < 36$ **D.** $4x > 36$

Solve each inequality. Graph the solution set, and write it using interval notation.

59. $x + 6 \geq 3$ **60.** $5x < 4x + 2$

61. $-6x \leq -18$ **62.** $8(x - 5) - (2 + 7x) \geq 4$

63. $4x - 3x > 10 - 4x + 7x$ **64.** $3(2x + 5) + 4(8 + 3x) < 5(3x + 7)$

65. $-3 \leq 2x + 1 \leq 4$ **66.** $9 < 3x + 5 \leq 20$

Solve each problem.

67. Awilda has grades of 94 and 88 on her first two calculus tests. What possible scores on a third test will give her an average of at least 90?

68. If nine times a number is added to 6, the result is at most 3. Find all such numbers.

Chapter 2   Mixed Review Exercises

Solve.

1. $\dfrac{x}{7} = \dfrac{x - 5}{2}$ **2.** $I = prt$ for r

3. $-2x > -4$ **4.** $2k - 5 = 4k + 13$

5. $0.05x + 0.02x = 4.9$ **6.** $2 - 3(x - 5) = 4 + x$

7. $9x - (7x + 2) = 3x + (2 - x)$ **8.** $\dfrac{1}{3}s + \dfrac{1}{2}s + 7 = \dfrac{5}{6}s + 5 + 2$

9. Athletes in vigorous training programs can eat 50 calories per day for every 2.2 lb of body weight. To the nearest hundred, how many calories can a 175-lb athlete consume per day? (Data from *The Gazette.*)

10. In 2017, the top-selling frozen pizza brands, DiGiorno and Red Baron, together had sales of $1586.9 million. Red Baron's sales were $442.3 million less than DiGiorno's. What were sales in millions for each brand? (Data from IRI.)

11. Find the best buy. Give the unit price to the nearest thousandth for that size. (Data from Jewel-Osco.)

Laundry Detergent

Size	Price
50 oz	$ 3.99
100 oz	$ 7.29
160 oz	$ 9.99

12. Find the measure of each marked angle.

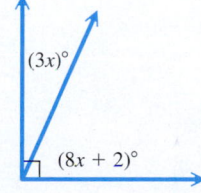

$(3x)^\circ$

$(8x + 2)^\circ$

13. Janet drove from Louisville to Dallas, a distance of 819 mi, averaging 63 mph. What was her driving time?

14. Two trains are 390 mi apart. They start at the same time and travel toward one another, meeting 3 hr later. If the rate of one train is 30 mph more than the rate of the other train, find the rate of each train.

15. The perimeter of a triangle is 96 m. One side is twice as long as another, and the third side is 30 m long. What is the length of the longest side?

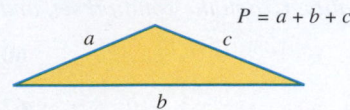

$$P = a + b + c$$

16. The perimeter of a certain square cannot be greater than 200 m. Find the possible values for the length of a side.

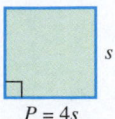

$$P = 4s$$

| Chapter 2 | Test | FOR EXTRA HELP | *Step-by-step test solutions are found on the Chapter Test Prep Videos available in* **MyLab Math**. |

▶ *View the complete solutions to all Chapter Test exercises in* MyLab Math.

Solve each equation.

1. $5x + 9 = 7x + 21$

2. $-\dfrac{4}{7}x = -12$

3. $7 - (x - 4) = -3x + 2(x + 1)$

4. $0.06(x + 20) + 0.08(x - 10) = 4.6$

5. $-8(2x + 4) = -4(4x + 8)$

6. $2 - 3(x - 5) = 3 + (x + 1)$

Solve each problem.

7. In the 2017 baseball season, the Los Angeles Dodgers won 12 less than twice as many games as they lost. They played 162 regular-season games. How many wins and losses did the Dodgers have? (Data from www.MLB.com)

8. Three islands in the Hawaiian island chain are Hawaii (the Big Island), Maui, and Kauai. Together, their areas total 5300 mi². The island of Hawaii is 3293 mi² larger than the island of Maui, and Maui is 177 mi² larger than Kauai. What is the area of each island?

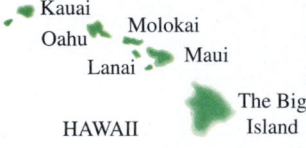

9. Find the measure of an angle if its supplement measures 10° more than three times its complement.

10. If the lesser of two consecutive even integers is tripled, the result is 20 more than twice the greater integer. Find the two integers.

11. The formula for the perimeter of a rectangle is $P = 2L + 2W$.

   (a) Solve for W.   **(b)** If $P = 116$ and $L = 40$, find the value of W.

12. Solve the equation $5x - 4y = 8$ for y.

Find the measure of each marked angle.

13.

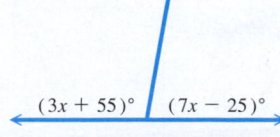

$(3x + 55)°$ $(7x - 25)°$

14.

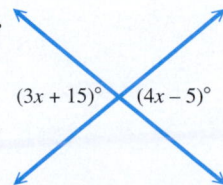

$(3x + 15)°$ $(4x - 5)°$

Solve each equation.

15. $\dfrac{z}{8} = \dfrac{12}{16}$

16. $\dfrac{x + 5}{3} = \dfrac{x - 3}{4}$

Solve each problem.

17. What percent of 65 is 26?

18. Dawn has a monthly income of $2200 and plans to spend 12% of this amount on groceries. How much will she spend on groceries?

19. Find the best buy. Give the unit price to the nearest thousandth for that size. (Data from Jewel-Osco.)

Cheese Slices

Size	Price
8 oz	$ 2.99
16 oz	$ 3.99
48 oz	$14.69

20. The distance between Milwaukee and Boston is 1050 mi. On a certain map, this distance is represented by 42 in. On the same map, Seattle and Cincinnati are 92 in. apart. What is the actual distance between Seattle and Cincinnati?

21. Carlos invested some money at 3% simple interest and $6000 more than that amount at 4.5% simple interest. After 1 yr, his total interest from the two accounts was $870. How much did he invest at each rate?

22. Two cars leave from the same point, traveling in opposite directions. One travels at a constant rate of 50 mph, while the other travels at a constant rate of 65 mph. How long will it take for them to be 460 mi apart?

23. Write an inequality using the variable x that corresponds to each graph of solutions on a number line.

(a) 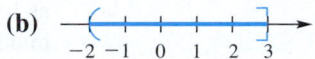   **(b)**

Solve each inequality. Graph the solution set, and write it using interval notation.

24. $-3x > -33$

25. $-0.04x \leq 0.12$

26. $-4x + 2(x - 3) \geq 4x - (3 + 5x) - 7$

27. $-10 < 3x - 4 \leq 14$

28. Susan has grades of 76 and 81 on her first two algebra tests. If she wants an average of at least 80 after her third test, what scores may she make on that test?

Chapters R–2 Cumulative Review Exercises

STUDY SKILLS REMINDER
It is not too soon to begin preparing for your final exam.
Review Study Skill 10,
Preparing for Your Math Final Exam.

Write each fraction in lowest terms.

1. $\dfrac{15}{40}$

2. $\dfrac{108}{144}$

Perform the indicated operations.

3. $\dfrac{5}{6} + \dfrac{1}{4} + \dfrac{7}{15}$

4. $16\dfrac{7}{8} - 3\dfrac{1}{10}$

5. $\dfrac{9}{8} \cdot \dfrac{16}{3}$

6. $\dfrac{3}{4} \div \dfrac{5}{8}$

7. $4.8 + 16.73$

8. $56.3 - 28.99$

9. $67.8 \, (0.45)$

10. $236.46 \div 4.2$

11. $13 + (-19) + 7$

12. $(-7 - 1)(-4) + (-4)$

13. $\dfrac{-3 - (-5)}{1 - (-1)}$

14. $\dfrac{6(-4) - (-2)(12)}{3^2 + 7^2}$

15. Find the value of $\dfrac{3x^2 - y^3}{-4z}$ for $x = -2$, $y = -4$, and $z = 3$.

Name the property illustrated by each equation.

16. $7(p + q) = 7p + 7q$ **17.** $7 + (-7) = 0$ **18.** $3.5(1) = 3.5$

Simplify each expression.

19. $-(m - 1) - 3 + 2m$ **20.** $2(p - 1) - 3p + 2$

Solve each equation, and check the solution.

21. $2r - 6 = 8r$ **22.** $4 - 5(s + 2) = 3(s + 1) - 1$

23. $\dfrac{2}{3}x + \dfrac{3}{4}x = -17$ **24.** $\dfrac{2x + 3}{5} = \dfrac{x - 4}{2}$

25. Solve $3x + 4y = 24$ for y.   **26.** Solve $P = a + b + c + B$ for c.

Solve each inequality. Graph the solution set, and write it using interval notation.

27. $6(r - 1) + 2(3r - 5) \leq -4$ **28.** $-18 \leq -9z < 9$

Solve each problem.

29. A 40-cm piece of yarn must be cut into three pieces. The longest piece is to be three times as long as the middle-sized piece, and the shortest piece is to be 5 cm shorter than the middle-sized piece. Find the length of each piece.

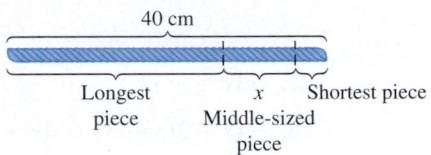

30. A fully inflated professional basketball has a circumference of 78 cm. What is the radius of a circular cross section through the center of the ball? (Use 3.14 as an approximation for π.) Round the answer to the nearest hundredth.

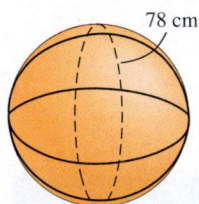

31. Two cars are 400 mi apart. Both start at the same time and travel toward one another. They meet 4 hr later. If the rate of one car is 20 mph faster than the other, what is the rate of each car?

32. The graph shows the breakdown of the colors chosen for new 2016 model-year midsize cars sold in the United States. If approximately 2.1 million of these cars were sold, about how many were each color? (Data from www.GoodCarBadCar.net)

(a) White

(b) Silver

(c) Red

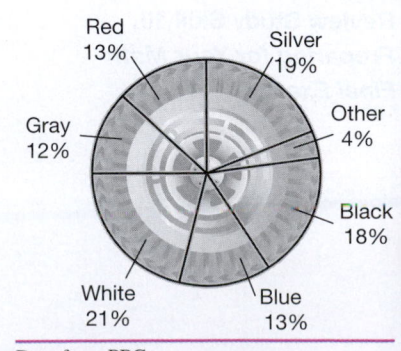

Most Popular Automobile Colors

Data from PPG.

STUDY SKILLS REMINDER
We can learn from the mistakes we make. **Review Study Skill 9,** *Analyzing Your Test Results.*

LINEAR EQUATIONS IN TWO VARIABLES

We determine location on a map using *coordinates,* a concept that is based on a *rectangular coordinate system,* one of the topics of this chapter.

3.1   Linear Equations and Rectangular Coordinates

VOCABULARY

☐ line graph
☐ linear equation in two variables
☐ ordered pair
☐ table of values
☐ *x*-axis
☐ *y*-axis
☐ origin
☐ rectangular (Cartesian) coordinate system
☐ quadrant
☐ plane
☐ coordinates
☐ plot
☐ scatter diagram

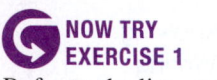 **NOW TRY EXERCISE 1**

Refer to the line graph in **FIGURE 1**.

(a) Estimate the average price of a gallon of gasoline in 2010.

(b) About how much did the average price of a gallon of gasoline increase from 2010 to 2012?

NOW TRY ANSWERS
1. **(a)** about $2.80
   (b) about $0.90

OBJECTIVE 1 Interpret line graphs.

A line graph is used to show changes or trends in data over time. To form a **line graph,** we connect a series of points representing data with line segments.

EXAMPLE 1   Interpreting a Line Graph

The line graph in **FIGURE 1** shows average prices of a gallon of regular unleaded gasoline in the United States for the years 2009 through 2016.

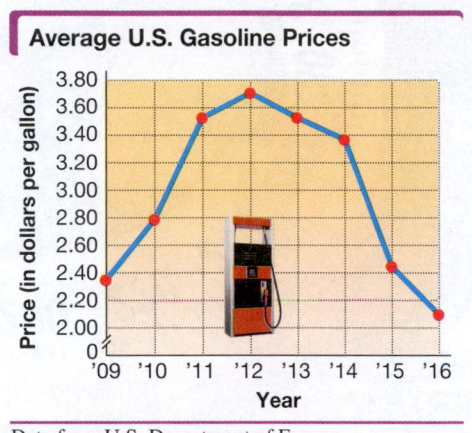

Average U.S. Gasoline Prices

Data from U.S. Department of Energy.

FIGURE 1

(a) During what year from 2009 to 2016 was the price of a gallon of regular unleaded gasoline the greatest?

The highest point on the graph indicates that the price was greatest in 2012.

(b) What was the general trend in the average price of a gallon of gasoline from 2012 through 2016?

The line graph *falls* from left to right from 2012 to 2016, so the average price of a gallon of gasoline decreased over those years.

(c) Estimate the average price of a gallon of gasoline in 2009 and 2012. About how much did the price increase between 2009 and 2012?

Move up from 2009 on the horizontal scale to the point plotted for 2009. This point is about three-fourths of the way between the lines on the vertical scale for $2.20 and $2.40—that is, about $2.35 per gallon.

Locate the point plotted for 2012. Moving across to the vertical scale, this point is about halfway between the lines for $3.60 and $3.80—that is, about $3.70 per gallon.

Between 2009 and 2012, the average price increased about

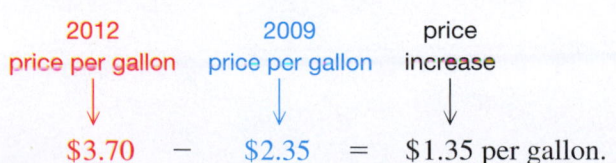

$$\underset{\substack{\text{2012}\\\text{price per gallon}}}{\$3.70} - \underset{\substack{\text{2009}\\\text{price per gallon}}}{\$2.35} = \underset{\substack{\text{price}\\\text{increase}}}{\$1.35 \text{ per gallon.}}$$

NOW TRY

Year	Average Price (in dollars per gallon)
2009	2.35
2010	2.79
2011	3.53
2012	3.71
2013	3.53
2014	3.37
2015	2.45
2016	2.14

Data from U.S. Department of Energy.

STUDY SKILLS REMINDER
Make study cards to help you learn and remember the material in this chapter. **Review Study Skill 5, Using Study Cards.**

The line graph in **FIGURE 1** relates years to average prices for a gallon of gasoline. We can also represent these two related quantities using a table of data, as shown in the margin. In table form, we can see more precise data rather than estimating it. Trends in the data are easier to see from the graph, which gives a "picture" of the data.

We can extend these ideas to the subject of this chapter, *linear equations in two variables*. A linear equation in two variables, one for each of the quantities being related, can be used to represent the data in a table or graph. ***The graph of a linear equation in two variables is a line.***

Linear Equation in Two Variables

A **linear equation in two variables** (here x and y) is an equation that can be written in the form

$$Ax + By = C,$$

where A, B, and C are real numbers and A and B are not both 0. This form is called *standard form*.

Examples: $3x + 4y = 9$, $x - y = 0$, $x + 2y = -8$ Linear equations in two variables in standard form

NOTE Linear equations in two variables that are not written in standard form, such as

$$y = 4x + 5 \quad \text{and} \quad 3x = 7 - 2y,$$

can be algebraically rewritten in this form, as we will discuss later.

OBJECTIVE 2 Write a solution as an ordered pair.

Recall that a *solution* of an equation is a number that makes the equation true when it replaces the variable. For example, the linear equation in *one* variable

$$x - 2 = 5 \quad \text{has solution} \quad 7$$

because replacing x with 7 gives a true statement.

*A solution of a linear equation in **two** variables requires **two** numbers, one for each variable.* For example, a true statement results when we replace x with 2 and y with 13 in the equation $y = 4x + 5$.

$$13 = 4(2) + 5 \qquad \text{Let } x = 2 \text{ and } y = 13.$$

The pair of numbers $x = 2$ and $y = 13$ gives one solution of the equation $y = 4x + 5$. The phrase "$x = 2$ and $y = 13$" is abbreviated as a pair of numbers written inside parentheses.

x-value —↓ ↓— y-value
$$(2, 13)$$
Ordered pair

The x-value is always given first. Such a pair of numbers is an **ordered pair**.

⚠ CAUTION The ordered pairs $(2, 13)$ and $(13, 2)$ are *not* the same. In the first pair, $x = 2$ and $y = 13$. In the second pair, $x = 13$ and $y = 2$. ***The order in which the numbers are written in an ordered pair is important.***

OBJECTIVE 3 Determine whether a given ordered pair is a solution of a given equation.

We substitute the x- and y-values of an ordered pair into a linear equation in two variables to see whether the ordered pair is a solution. An ordered pair that is a solution of an equation is said to *satisfy* the equation.

NOW TRY EXERCISE 2

Determine whether each ordered pair is a solution of the equation.

$$3x - 7y = 19$$

(a) $(3, 4)$ **(b)** $(-3, -4)$

EXAMPLE 2 Determining Whether Ordered Pairs Are Solutions of an Equation

Determine whether each ordered pair is a solution of the equation $2x + 3y = 12$.

(a) $(3, 2)$

Substitute 3 for x and 2 for y in the given equation.

$$2x + 3y = 12$$
$$2(3) + 3(2) \overset{?}{=} 12 \qquad \text{Let } x = 3 \text{ and } y = 2.$$
$$6 + 6 \overset{?}{=} 12 \qquad \text{Multiply.}$$
$$12 = 12 \checkmark \qquad \text{True}$$

This result is true, so $(3, 2)$ is a solution of $2x + 3y = 12$.

(b) $(-2, -7)$

Substitute -2 for x and -7 for y in the given equation.

$$2x + 3y = 12$$
$$2(-2) + 3(-7) \overset{?}{=} 12 \qquad \text{Let } x = -2 \text{ and } y = -7.$$

Use parentheses to avoid errors.

$$-4 + (-21) \overset{?}{=} 12 \qquad \text{Multiply.}$$
$$-25 = 12 \qquad \text{False}$$

This result is false, so $(-2, -7)$ is *not* a solution of $2x + 3y = 12$.   **NOW TRY**

OBJECTIVE 4 Complete ordered pairs for a given equation.

Substituting a number for one variable in a linear equation makes it possible to find the value of the other variable.

EXAMPLE 3 Completing Ordered Pairs

Complete each ordered pair for the equation $y = 4x + 5$.

(a) $(7, \underline{\quad})$

In this ordered pair, $x = 7$. To find the corresponding value of y, replace x with 7 in the given equation.

$$y = 4x + 5$$

Solve for the value of y.

$$y = 4(7) + 5 \qquad \text{Let } x = 7.$$
$$y = 28 + 5 \qquad \text{Multiply.}$$
$$y = 33 \qquad \text{Add.}$$

The x-value always comes first.

NOW TRY ANSWERS
2. (a) no **(b)** yes

The ordered pair is $(7, 33)$.

NOW TRY
EXERCISE 3
Complete each ordered pair for the equation.

$$y = 3x - 12$$

(a) $(3, \underline{\quad})$ **(b)** $(\underline{\quad}, 3)$

(b) $(\underline{\quad}, -3)$

In this ordered pair, $y = -3$. Find the corresponding value of x by replacing y with -3 in the given equation.

$$y = 4x + 5$$

> Solve for the value of *x*.

$$-3 = 4x + 5 \qquad \text{Let } y = -3.$$

$$-8 = 4x \qquad \text{Subtract 5 from each side.}$$

$$-2 = x \qquad \text{Divide each side by 4.}$$

The ordered pair is $(-2, -3)$.   **NOW TRY** ↻

OBJECTIVE 5 Complete a table of values.

Ordered pairs are often displayed in a **table of values.** Although we usually write tables of values vertically, they may be written horizontally.

EXAMPLE 4 Completing Tables of Values

Complete the table of values for each equation. Write the results as ordered pairs.

(a) $x - 2y = 8$

x	y	Ordered Pairs
0		⟶ $(0, \underline{\quad})$
10		⟶ $(10, \underline{\quad})$
	0	⟶ $(\underline{\quad}, 0)$
	-2	⟶ $(\underline{\quad}, -2)$

From the first row of the table, let $x = 0$ in the equation. From the second row of the table, let $x = 10$.

If $x = 0,$	If $x = 10,$
then $x - 2y = 8$	then $x - 2y = 8$
becomes $0 - 2y = 8$	becomes $10 - 2y = 8$
$-2y = 8$	$-2y = -2$
$y = -4.$	$y = 1.$

The first two ordered pairs are $(0, -4)$ and $(10, 1)$. From the third and fourth rows of the table, let $y = 0$ and $y = -2$, respectively.

If $y = 0,$	If $y = -2,$
then $x - 2y = 8$	then $x - 2y = 8$
becomes $x - 2(0) = 8$	becomes $x - 2(-2) = 8$
$x - 0 = 8$	$x + 4 = 8$
$x = 8.$	$x = 4.$

The last two ordered pairs are $(8, 0)$ and $(4, -2)$. The completed table of values and corresponding ordered pairs follow.

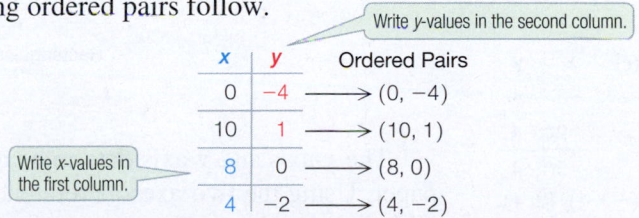

> Write *y*-values in the second column.

x	y	Ordered Pairs
0	-4	⟶ $(0, -4)$
10	1	⟶ $(10, 1)$
8	0	⟶ $(8, 0)$
4	-2	⟶ $(4, -2)$

> Write *x*-values in the first column.

NOW TRY ANSWERS
3. (a) $(3, -3)$ **(b)** $(5, 3)$

Each ordered pair is a solution of the given equation $x - 2y = 8$.

NOW TRY
EXERCISE 4
Complete the table of values for each equation. Write the results as ordered pairs.

(a) $5x - 4y = 20$

x	y
0	
	0
2	

(b) $x = -1$

x	y
	-4
	0
	2

(c) $y = 4$

x	y
-3	
2	
5	

(b) $x = 5$ (Using two variables, $x = 5$ could be written $x + 0y = 5$.)

x	y
	0
	6
	-2

The given equation is $x = 5$. No matter which value of y is chosen, the value of x is *always* 5.

x	y	Ordered Pairs
5	0	$\longrightarrow (5, 0)$
5	6	$\longrightarrow (5, 6)$
5	-2	$\longrightarrow (5, -2)$

(c) $y = -3$ (Using two variables, $y = -3$ could be written $0x + y = -3$.)

x	y
-5	
0	
2	

The given equation is $y = -3$. No matter which value of x is chosen, the value of y is *always* -3.

x	y	Ordered Pairs
-5	-3	$\longrightarrow (-5, -3)$
0	-3	$\longrightarrow (0, -3)$
2	-3	$\longrightarrow (2, -3)$

NOW TRY

OBJECTIVE 6 Plot ordered pairs.

Recall that a linear equation in *one* variable can have zero, one, or an infinite number of real number solutions. These solutions can be graphed on *one* number line. For example, the linear equation in one variable

$$x - 2 = 5 \text{ has solution 7.}$$

The solution 7 is graphed on the number line in **FIGURE 2**.

FIGURE 2

Every linear equation in *two* variables has an infinite number of ordered pairs (x, y) as solutions. To graph these solutions, we need *two* number lines, one for each variable, drawn at right angles as in **FIGURE 3**. The horizontal number line is the **x-axis,** and the vertical line is the **y-axis.** The point at which the x-axis and y-axis intersect is the **origin.** Together, the x-axis and y-axis form a **rectangular coordinate system.**

The rectangular coordinate system is divided into four regions, or **quadrants.** These quadrants are numbered counterclockwise, as shown in **FIGURE 3**.

René Descartes (1596–1650)
The rectangular coordinate system is also called the **Cartesian coordinate system,** in honor of René Descartes, the French mathematician credited with its invention.

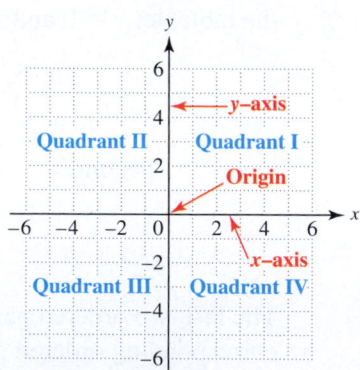

Rectangular Coordinate System
FIGURE 3

NOW TRY ANSWERS

4. (a)

x	y
0	-5
4	0
2	$-\frac{5}{2}$

$(0, -5), (4, 0), \left(2, -\frac{5}{2}\right)$

(b)

x	y
-1	-4
-1	0
-1	2

$(-1, -4), (-1, 0),$
$(-1, 2)$

(c)

x	y
-3	4
2	4
5	4

$(-3, 4), (2, 4),$
$(5, 4)$

The x-axis and y-axis determine a **plane**—a flat surface illustrated by a sheet of paper. Using the two axes as references, every point in the plane can be associated with an ordered pair. The numbers in the ordered pair are the **coordinates** of the point.

> **NOTE** In a plane, *both* numbers in the ordered pair are needed to locate a point. The ordered pair is a name for the point.

NOW TRY
EXERCISE 5

Plot each point in a rectangular coordinate system.

$(-3, 1)$, $(2, -4)$, $(0, -1)$, $\left(\frac{5}{2}, 3\right)$, $(-4, -3)$, $(-4, 0)$

EXAMPLE 5 Plotting Ordered Pairs

Plot each point in a rectangular coordinate system.

(a) $(2, 3)$ **(b)** $(-1, -4)$ **(c)** $(-2, 3)$ **(d)** $(3, -2)$ **(e)** $\left(\frac{3}{2}, 2\right)$

(f) $(4, -3.75)$ **(g)** $(5, 0)$ **(h)** $(0, -3)$ **(i)** $(0, 0)$

The point $(2, 3)$ from part (a) is **plotted** (graphed) in **FIGURE 4**. The other points are plotted in **FIGURE 5**. In each case, we begin at the origin.

Step 1 Move right or left the number of units that corresponds to the *x*-coordinate in the ordered pair—*right if the x-coordinate is positive or left if the x-coordinate is negative.*

Step 2 Then turn and move up or down the number of units that corresponds to the *y*-coordinate in the ordered pair—*up if the y-coordinate is positive or down if the y-coordinate is negative.*

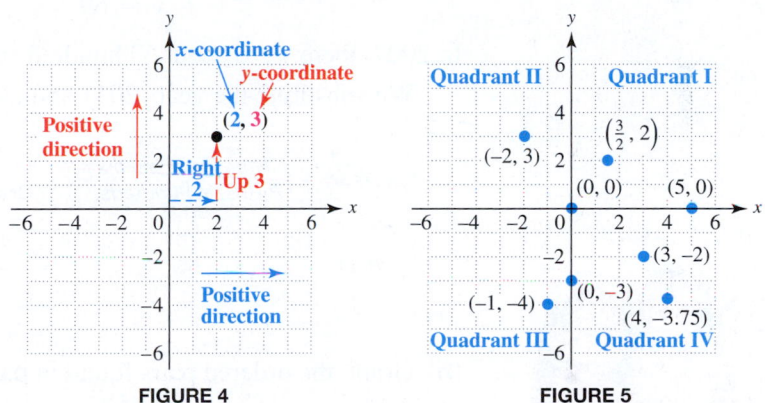

FIGURE 4   FIGURE 5

Notice in **FIGURE 5** that the point $(-2, 3)$ is in quadrant II, whereas the point $(3, -2)$ is in quadrant IV.

The order of the coordinates is important. The x-coordinate is always given first in an ordered pair.

To plot the point $\left(\frac{3}{2}, 2\right)$, think of the improper fraction $\frac{3}{2}$ as the mixed number $1\frac{1}{2}$ and move to the right $\frac{3}{2}$ $\left(\text{or } 1\frac{1}{2}\right)$ units along the *x*-axis. Then turn and go up 2 units, parallel to the *y*-axis. The point $(4, -3.75)$ is plotted similarly, by approximating the location of the decimal *y*-coordinate.

The point $(5, 0)$ lies on the *x*-axis because the *y*-coordinate is 0. The point $(0, -3)$ lies on the *y*-axis because the *x*-coordinate is 0. The point $(0, 0)$ is at the origin.

Points on the axes themselves are not in any quadrant. **NOW TRY**

NOW TRY ANSWER
5.

We can use a linear equation in two variables to mathematically describe, or *model*, certain real-life situations, as shown in the next example.

NOW TRY
EXERCISE 6

Use the linear equation in **Example 6** to approximate the number of multiple births, to the nearest hundred, in 2016. Interpret the results.

EXAMPLE 6   Using a Linear Equation to Model Multiple Births

The annual number of multiple births—that is, triplet and higher-order births—in the United States from 2007 through 2015 can be approximated by the linear equation

Number of multiple births ⌐   ⌐ Year

$$y = -3.05x + 6190,$$

which relates x, the year, and y, the number of multiple births in hundreds. (Data from National Center for Health Statistics.)

(a) Complete the table of values for the given linear equation.

x (Year)	y (Number of Multiple Births, in hundreds)
2007	
2011	
2015	

To find y when $x = 2007$, we substitute into the equation.

$$y = -3.05x + 6190$$

$\approx$ means "is approximately equal to."

$$y = -3.05(2007) + 6190 \quad \text{Let } x = 2007.$$

$$y \approx 69 \quad \text{Use a calculator.}$$

In 2007, there were about 69 hundred (or 6900) multiple births.

We substitute the years 2011 and 2015 in the same way to complete the table.

x (Year)	y (Number of Multiple Births, in hundreds)	Ordered Pairs (x, y)
2007	69	→ (2007, 69)
2011	56	→ (2011, 56)
2015	44	→ (2015, 44)

Here each year x is paired with a number of multiple births y (in hundreds).

(b) Graph the ordered pairs found in part (a).

See **FIGURE 6.** A graph of ordered pairs of data is a **scatter diagram.** A scatter diagram enables us to describe how the two quantities are related to each other. In **FIGURE 6**, the plotted points could be connected to approximate a straight *line,* so the variables x (year) and y (number of multiple births) have a *line*ar relationship. The decrease in the number of multiple births is also reflected.

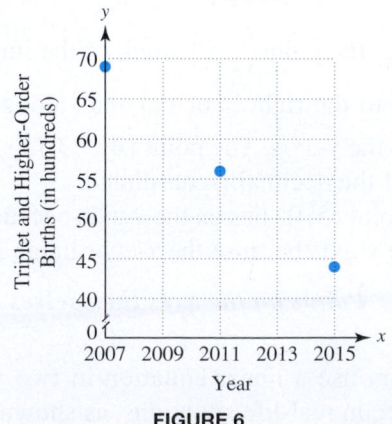

Number of Multiple Births

Notice the axis labels and scales. Each grid square represents 2 units in the horizontal direction and 5 units in the vertical direction. We show a break in the y-axis, to indicate the jump from 0 to 40.

NOW TRY ANSWER
6. $y \approx 41$; There were approximately 41 hundred (or 4100) multiple births in the U.S. in 2016.

FIGURE 6

NOW TRY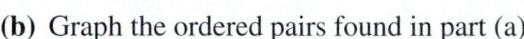

3.1 Exercises

FOR EXTRA HELP

 MyLab Math

▶ *Video solutions for select problems available in MyLab Math*

STUDY SKILLS REMINDER
We can learn from the mistakes we make. **Review Study Skill 9, *Analyzing Your Test Results.***

Concept Check *Complete each statement.*

1. The symbol (x, y) (*does/does not*) represent an ordered pair, while the symbols $[x, y]$ and $\{x, y\}$ (*do/do not*) represent ordered pairs.

2. The origin is represented by the ordered pair ____.

3. The point whose graph has coordinates $(-4, 2)$ is in quadrant ____.

4. The point whose graph has coordinates $(0, 5)$ lies on the ____-axis.

5. The ordered pair $(4, \underline{})$ is a solution of the equation $y = 3$.

6. The ordered pair $(\underline{}, -2)$ is a solution of the equation $x = 6$.

Concept Check *Fill in each blank with the word* positive *or the word* negative.

The point with coordinates (x, y) is in

7. quadrant III if x is _____ and y is _____.

8. quadrant II if x is _____ and y is _____.

9. quadrant IV if x is _____ and y is _____.

10. quadrant I if x is _____ and y is _____.

11. A point (x, y) has the property that $xy < 0$. In which quadrant(s) must the point lie? Explain.

12. A point (x, y) has the property that $xy > 0$. In which quadrant(s) must the point lie? Explain.

The line graph shows the overall unemployment rate in the U.S. civilian labor force for the years 2009 through 2016. Use the graph to work each problem. **See Example 1.**

13. Between which pairs of consecutive years did the unemployment rate increase?

14. What was the general trend in the unemployment rate between 2010 and 2016?

15. Estimate the overall unemployment rate in 2011 and 2012. About how much did the unemployment rate decline between 2011 and 2012?

16. During which year(s)

    (a) was the unemployment rate greater than 9%, but less than 10%?

    (b) was the unemployment rate less than 7%?

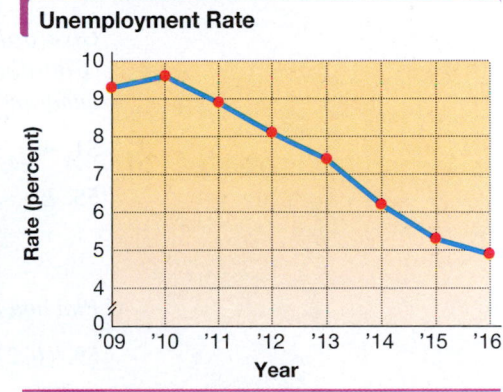

Data from Bureau of Labor Statistics.

Determine whether each ordered pair is a solution of the given equation. **See Example 2.**

17. $x + y = 8$; $(0, 8)$ **18.** $x + y = 9$; $(0, 9)$ **19.** $2x + y = 5$; $(3, -1)$

20. $2x - y = 6$; $(4, 2)$ **21.** $5x - 3y = 15$; $(5, 2)$ **22.** $4x - 3y = 6$; $(2, 1)$

23. $x = -4y$; $(-8, 2)$ **24.** $y = 3x$; $(2, 6)$ **25.** $x = -6$; $(-6, 5)$

26. $x - 9 = 0$; $(9, 2)$ **27.** $y = 2$; $(2, 4)$ **28.** $y + 4 = 0$; $(-6, 2)$

Complete each ordered pair for the equation $y = 2x + 7$. **See Example 3.**

29. $(5, \underline{})$ **30.** $(0, \underline{})$ **31.** $(\underline{}, -3)$ **32.** $(\underline{}, 0)$

Complete each ordered pair for the equation $y = -4x - 4$. See Example 3.

33. $(0, \underline{\hspace{0.8cm}})$ **34.** $(\underline{\hspace{0.8cm}}, 0)$ **35.** $(\underline{\hspace{0.8cm}}, 24)$ **36.** $(\underline{\hspace{0.8cm}}, 16)$

Complete the table of values for each equation. Write the results as ordered pairs. See Example 4.

37. $4x + 3y = 24$

x	y
0	
	0
4	

38. $2x + 3y = 12$

x	y
0	
	0
	8

39. $4x - 9y = -36$

x	y
	0
0	
	8

40. $3x - 5y = -15$

x	y
	0
0	
	-6

41. $x = 12$

x	y
	3
	8
	0

42. $x = -9$

x	y
	6
	2
	-3

43. $y = -10$

x	y
4	
0	
-4	

44. $y = 6$

x	y
8	
4	
-2	

45. $y - 2 = 0$

x	y
9	
2	
0	

46. $y + 6 = 0$

x	y
6	
3	
0	

47. $x + 4 = 0$

x	y
	4
	0
	-4

48. $x - 8 = 0$

x	y
	8
	3
	0

49. Concept Check   A student incorrectly claimed that $(3, 4)$ and $(4, 3)$ correspond to the same point. **WHAT WENT WRONG?** Explain.

50. Concept Check   A student incorrectly claimed that the point $(6, 0)$ lies on the *y*-axis because the *y*-coordinate is 0. **WHAT WENT WRONG?** Explain.

Give ordered pairs for the points labeled A–H in the figure. (Coordinates of the points shown are integers.) Identify the quadrant in which each point is located. See Example 5.

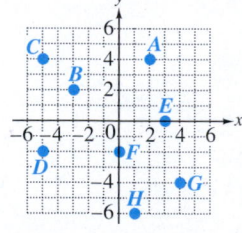

51. *A*   **52.** *B*   **53.** *C*   **54.** *D*

55. *E*   **56.** *F*   **57.** *G*   **58.** *H*

Plot and label each point in a rectangular coordinate system. See Example 5.

59. $(6, 2)$ **60.** $(5, 3)$ **61.** $(-4, 2)$ **62.** $(-3, 5)$

63. $\left(-\dfrac{4}{5}, -1\right)$ **64.** $\left(-\dfrac{3}{2}, -4\right)$ **65.** $(3, -1.75)$ **66.** $(5, -4.25)$

67. $(0, 4)$ **68.** $(0, -3)$ **69.** $(4, 0)$ **70.** $(-3, 0)$

Complete the table of values for each equation. Then plot and label the ordered pairs. See Examples 4 and 5.

71. $x - 2y = 6$

x	y
0	
	0
2	
	-1

72. $2x - y = 4$

x	y
0	
	0
1	
	-6

73. $3x - 4y = 12$

x	y
0	
	0
-4	
	-4

74. $2x - 5y = 10$

x	y
0	
	0
-5	
	-3

75. $y + 4 = 0$

x	y
0	
5	
-2	
-3	

76. $x - 5 = 0$

x	y
	1
	0
	6
	-4

77. Describe the pattern indicated by the plotted points in **Exercises 71–76.**

78. Answer each question.

   (a) A line through the plotted points in **Exercise 75** would be horizontal. What do you notice about the y-coordinates of the ordered pairs?

   (b) A line through the plotted points in **Exercise 76** would be vertical. What do you notice about the x-coordinates of the ordered pairs?

*Solve each problem. **See Example 6.***

79. Suppose that it costs a flat fee of $20 plus $15 per day to rent a pressure washer. Therefore, the cost y in dollars to rent the pressure washer for x days is given by the linear equation

$$y = 15x + 20.$$

Express each of the following as an ordered pair.

   (a) When the washer is rented for 5 days, the cost is $95.

   (b) When the cost is $110, the washer is rented for 6 days.

80. Suppose that it costs $5000 to start up a business selling snow cones. Furthermore, it costs $0.50 per cone in labor, ice, syrup, and overhead. Then the cost y in dollars to make x snow cones is given by the linear equation

$$y = 0.50x + 5000.$$

Express each of the following as an ordered pair.

   (a) When 100 snow cones are made, the cost is $5050.

   (b) When the cost is $6000, the number of snow cones made is 2000.

81. The table shows the worldwide number of monthly active Facebook users at year end in millions.

Monthly Active Facebook Users at Year End

Year	Facebook Users (in millions)
2010	608
2011	845
2012	1056
2013	1228
2014	1393
2015	1591
2016	1860

Data from Facebook.

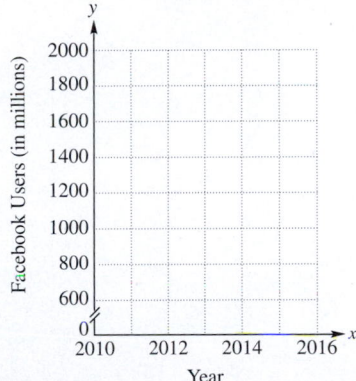

   (a) Write the data from the table as ordered pairs (x, y), where x represents the year and y represents the number of Facebook users in millions.

   (b) What does the ordered pair (2016, 1860) mean in the context of this problem?

   (c) Make a scatter diagram of the ordered pairs of data. Use the given grid.

   (d) Describe the pattern indicated by the points on the scatter diagram. What happened to the worldwide number of monthly active Facebook users?

82. The table shows the number of U.S. students who studied abroad (in thousands).

Academic Year	Number of Students (in thousands)
2009	270
2010	274
2011	283
2012	289
2013	304
2014	313
2015	325

Data from Institute of International Education.

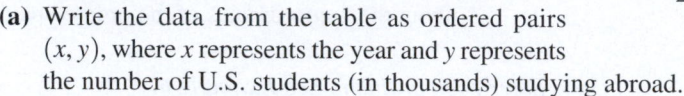

U.S. Students Studying Abroad

(a) Write the data from the table as ordered pairs (x, y), where x represents the year and y represents the number of U.S. students (in thousands) studying abroad.

(b) What does the ordered pair $(2015, 325)$ mean in the context of this problem?

(c) Make a scatter diagram of the ordered pairs of data. Use the given grid.

(d) Describe the pattern indicated by the points on the scatter diagram. What was the trend in the number of U.S. students studying abroad during these years?

83. The maximum benefit for the heart from exercising occurs when the heart rate is in the target heart rate zone. The lower limit of this zone can be approximated by the linear equation

$$y = -0.65x + 143,$$

where x represents age and y represents heartbeats per minute. (Data from *The Gazette.*)

Age	Heartbeats (per minute)
20	
40	
60	
80	

(a) Complete the table of values for this linear equation.

(b) Write the data from the table of values as ordered pairs.

(c) Make a scatter diagram of the data. Do the points lie in a linear pattern?

84. The maximum benefit for the heart from exercising occurs when the heart rate is in the target heart rate zone. The upper limit of this zone can be approximated by the linear equation

$$y = -0.85x + 187,$$

where x represents age and y represents heartbeats per minute. (Data from *The Gazette.*)

Age	Heartbeats (per minute)
20	
40	
60	
80	

(a) Complete the table of values for this linear equation.

(b) Write the data from the table of values as ordered pairs.

(c) Make a scatter diagram of the data. Describe the pattern indicated by the data.

85. See **Exercises 83 and 84.** What is the target heart rate zone for age 20? Age 40?

86. See **Exercises 83 and 84.** What is the target heart rate zone for age 60? Age 80?

3.2 Graphing Linear Equations in Two Variables

OBJECTIVES

1 Graph linear equations by plotting ordered pairs.

2 Find intercepts.

3 Graph linear equations of the form $Ax + By = 0$.

4 Graph linear equations of the form $y = b$ or $x = a$.

5 Use a linear equation to model data.

VOCABULARY

☐ graph, graphing
☐ x-intercept
☐ y-intercept
☐ horizontal line
☐ vertical line

STUDY SKILLS REMINDER

Time management can be a challenge for students. **Review Study Skill 6, *Managing your Time.***

OBJECTIVE 1 Graph linear equations by plotting ordered pairs.

There are infinitely many ordered pairs that satisfy a linear equation in two variables. We find these ordered-pair solutions by choosing as many values of x (or y) as we wish and then completing each ordered pair.

For example, consider the equation $x + 2y = 7$. If we choose $x = 1$, then we can substitute to find the corresponding value of y.

$$x + 2y = 7 \quad \text{Given equation.}$$
$$1 + 2y = 7 \quad \text{Let } x = 1.$$
$$2y = 6 \quad \text{Subtract 1.}$$
$$y = 3 \quad \text{Divide by 2.}$$

If $x = 1$, then $y = 3$, so the ordered pair $(1, 3)$ is a solution of the equation.

$$1 + 2(3) = 7 \quad \begin{array}{l} \text{A true statement results, so} \\ (1, 3) \text{ is a solution.} \end{array}$$

This ordered pair and other solutions of $x + 2y = 7$ are graphed in **FIGURE 7**.

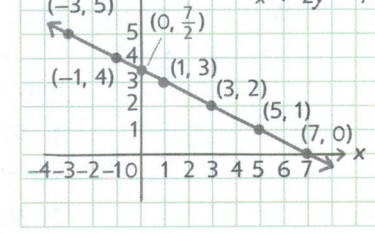

FIGURE 7   **FIGURE 8**

Notice that the points plotted in **FIGURE 7** all appear to lie on a straight line, as shown in **FIGURE 8**. In fact, the following is true.

> *Every point on the line represents a solution of the equation $x + 2y = 7$, and every solution of the equation corresponds to a point on the line.*

The line gives a "picture" of all the solutions of the equation

$$x + 2y = 7.$$

The line extends indefinitely in both directions, as suggested by the arrowhead on each end, and is the **graph** of the equation $x + 2y = 7$. **Graphing** is the process of plotting ordered pairs and drawing a line through the corresponding points.

Graph of a Linear Equation

The graph of any linear equation in two variables is a straight line.

Notice that the word *line* appears in the name "*line*ar equation."

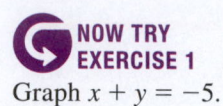

NOW TRY
EXERCISE 1
Graph $x + y = -5$.

EXAMPLE 1   Graphing a Linear Equation

Graph $x + y = 5$.

Two ordered pairs are needed to draw the graph of a line. We generally find at least one additional ordered-pair solution to check our work.

The equation $x + y = 5$ can be read as *"The sum of two numbers is 5."* Ordered pairs that make this statement true include the following.

$(0, 5)$ because $0 + 5 = 5.$

$(2, 3)$ because $2 + 3 = 5.$

$(5, 0)$ because $5 + 0 = 5.$

$(-2, 7)$ because $-2 + 7 = 5,$ and so on.

> Ordered pairs with x- and y-values that sum to 5 are solutions of the equation $x + y = 5$.

We can also find ordered-pair solutions by choosing a value for x (or y) and substituting it into the equation.

Choose $x = 0$ and solve for y.	Choose $y = 0$ and solve for x.	Choose $x = -2$ and solve for y.
$x + y = 5$	$x + y = 5$	$x + y = 5$
$0 + y = 5$	$x + 0 = 5$	$-2 + y = 5$
$y = 5$	$x = 5$	$y = 7$
This gives the ordered pair $(0, 5)$.	This gives the ordered pair $(5, 0)$.	This gives the ordered pair $(-2, 7)$.

We plot the ordered-pair solutions found above and draw a line through them. See **FIGURE 9.**

The linear equation $x + y = 5$ may be represented in three ways:

1. *Verbally* as *"The sum of two numbers is 5,"*

2. *Numerically* as a set of ordered pairs (or a table of values),

3. *Graphically* as a picture of the solutions.

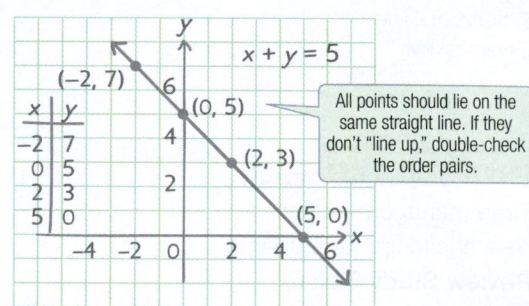

> All points should lie on the same straight line. If they don't "line up," double-check the order pairs.

FIGURE 9

NOW TRY

EXAMPLE 2   Graphing a Linear Equation

Graph $4x - 5y = 20$.

To find three ordered pairs that are solutions of $4x - 5y = 20$, we choose three arbitrary values for x or y that we think will be easy to substitute.

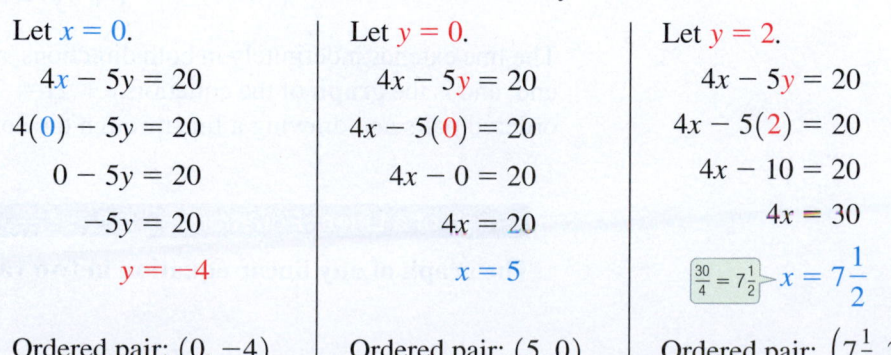

Let $x = 0$.	Let $y = 0$.	Let $y = 2$.
$4x - 5y = 20$	$4x - 5y = 20$	$4x - 5y = 20$
$4(0) - 5y = 20$	$4x - 5(0) = 20$	$4x - 5(2) = 20$
$0 - 5y = 20$	$4x - 0 = 20$	$4x - 10 = 20$
$-5y = 20$	$4x = 20$	$4x = 30$
$y = -4$	$x = 5$	$\frac{30}{4} = 7\frac{1}{2} \rightarrow x = 7\frac{1}{2}$
Ordered pair: $(0, -4)$	Ordered pair: $(5, 0)$	Ordered pair: $\left(7\frac{1}{2}, 2\right)$

NOW TRY ANSWER

1.

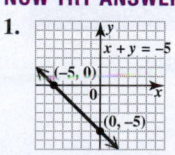

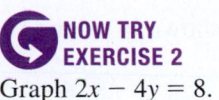

NOW TRY EXERCISE 2

Graph $2x - 4y = 8$.

We plot the three ordered-pair solutions and draw a line through them. See **FIGURE 10**. Two points determine the line, and the third point is used to check that no errors have been made.

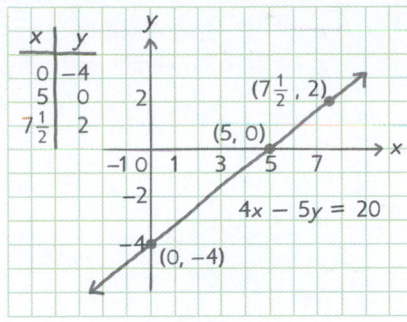

FIGURE 10

NOW TRY

NOTE   The ordered pairs that we find and use to graph an equation are *solutions* of the equation. For each value of x, there will be a corresponding value of y, and for each value of y, there will be a corresponding value of x. Substituting each ordered pair into the equation should produce a true statement.

OBJECTIVE 2   Find intercepts.

In **FIGURE 10**, the graph intersects (crosses) the x-axis at $(5, 0)$ and the y-axis at $(0, -4)$. For this reason, $(5, 0)$ is the **x-intercept** and $(0, -4)$ is the **y-intercept** of the graph. The intercepts are often convenient points to use when graphing.

Finding Intercepts

To find the x-intercept, let $y = 0$ in the given equation and solve for x. Then $(x, 0)$ is the x-intercept.

To find the y-intercept, let $x = 0$ in the given equation and solve for y. Then $(0, y)$ is the y-intercept.

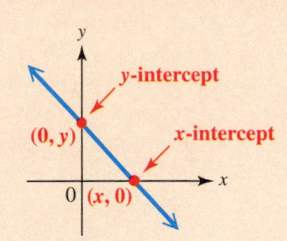

EXAMPLE 3   Graphing a Linear Equation Using Intercepts

Find the intercepts for the graph of $2x + y = 4$. Then draw the graph.

To find the intercepts, we first let $x = 0$ and then let $y = 0$. To find a third point, we arbitrarily let $x = 4$.

Let $x = 0$.	Let $y = 0$.	Let $x = 4$.
$2x + y = 4$	$2x + y = 4$	$2x + y = 4$
$2(0) + y = 4$	$2x + 0 = 4$	$2(4) + y = 4$
$0 + y = 4$	$2x = 4$	$8 + y = 4$
$y = 4$	$x = 2$	$y = -4$
y-intercept: $(0, 4)$	x-intercept: $(2, 0)$	Third point: $(4, -4)$

NOW TRY ANSWER

2.
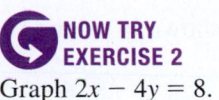

**NOW TRY
EXERCISE 3**

Find the intercepts for the graph of $x + 2y = 2$. Then draw the graph.

The graph, with the two intercepts in red, and a table of values are shown in **FIGURE 11**.

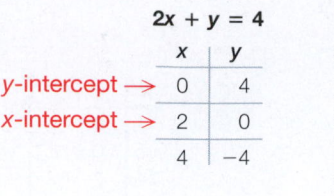

$2x + y = 4$

x	y
0	4
2	0
4	−4

y-intercept →
x-intercept →

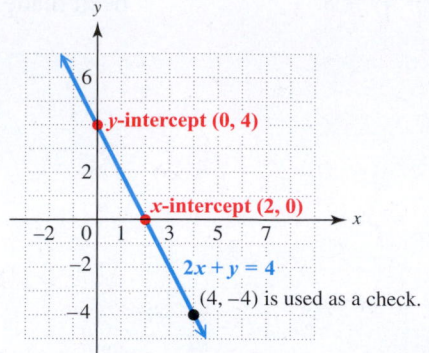

y-intercept (0, 4)

x-intercept (2, 0)

$2x + y = 4$

(4, −4) is used as a check.

FIGURE 11

NOW TRY

⚠ **CAUTION**   *When choosing x- or y-values to find ordered pairs to plot, be careful to choose so that the resulting points are not too close together.* For example, using $(-1, -1)$, $(0, 0)$, and $(1, 1)$ to graph $x - y = 0$ may result in an inaccurate line. It is better to choose points whose *x*-values differ by at least 2.

EXAMPLE 4   Graphing a Linear Equation Using Intercepts

Graph $y = -\dfrac{3}{2}x + 3$.

Although this linear equation is not in standard form $(Ax + By = C)$, it *could* be written in that form. To find the intercepts, we first let $x = 0$ and then let $y = 0$.

$$y = -\frac{3}{2}x + 3$$

$$y = -\frac{3}{2}(0) + 3 \qquad \text{Let } x = 0.$$

$$y = 0 + 3 \qquad \text{Multiply.}$$

$$y = 3 \qquad \text{Add.}$$

y-intercept: $(0, 3)$

$$y = -\frac{3}{2}x + 3$$

$$0 = -\frac{3}{2}x + 3 \qquad \text{Let } y = 0.$$

$$\frac{3}{2}x = 3 \qquad \text{Add } \tfrac{3}{2}x.$$

$$x = 2 \qquad \text{Multiply by } \tfrac{2}{3}.$$

x-intercept: $(2, 0)$

To find a third point, we arbitrarily let $x = -2$.

$$y = -\frac{3}{2}x + 3 \qquad \boxed{\text{Choosing a multiple of 2 makes multiplying by } -\tfrac{3}{2} \text{ easier.}}$$

$$y = -\frac{3}{2}(-2) + 3 \qquad \text{Let } x = -2.$$

$$y = 3 + 3 \qquad \text{Multiply.}$$

$$y = 6 \qquad \text{Add.}$$

Third point: $(-2, 6)$

We plot the three ordered-pair solutions and draw a line through them, as shown in **FIGURE 12** on the next page.

NOW TRY ANSWER

3. *x*-intercept: $(2, 0)$; *y*-intercept: $(0, 1)$

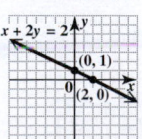

NOW TRY EXERCISE 4
Graph $y = \frac{1}{3}x + 1$.

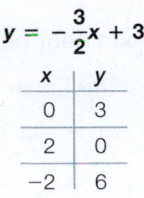

$y = -\frac{3}{2}x + 3$

x	y
0	3
2	0
-2	6

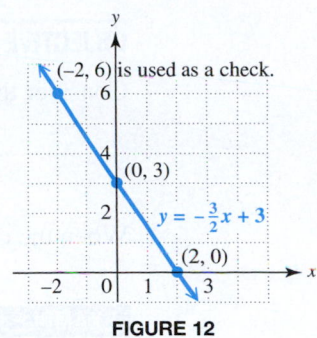

FIGURE 12

NOW TRY

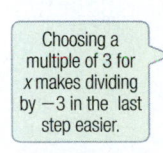
NOW TRY EXERCISE 5
Graph $2x + y = 0$.

OBJECTIVE 3 Graph linear equations of the form $Ax + By = 0$.

EXAMPLE 5 Graphing an Equation with x- and y-Intercepts (0, 0)

Graph $x - 3y = 0$.

To find the *y*-intercept, let $x = 0$.

$x - 3y = 0$
$0 - 3y = 0$ Let $x = 0$.
$-3y = 0$ Subtract.
$y = 0$ Divide by -3.
y-intercept: $(0, 0)$

To find the *x*-intercept, let $y = 0$.

$x - 3y = 0$
$x - 3(0) = 0$ Let $y = 0$.
$x - 0 = 0$ Multiply.
$x = 0$ Subtract.
x-intercept: $(0, 0)$

The *x*- and *y*-intercepts are the *same* point, $(0, 0)$. We must select *two other values* for *x* or *y* to find two other points on the graph. We choose $x = 6$ and $x = -3$.

Choosing a multiple of 3 for x makes dividing by −3 in the last step easier.

$x - 3y = 0$
$6 - 3y = 0$ Let $x = 6$.
$-3y = -6$ Subtract 6.
$y = 2$ Divide by -3.
Ordered pair: $(6, 2)$

$x - 3y = 0$
$-3 - 3y = 0$ Let $x = -3$.
$-3y = 3$ Add 3.
$y = -1$ Divide by -3.
Ordered pair: $(-3, -1)$

We use the three ordered-pair solutions to draw the graph in **FIGURE 13**.

$x - 3y = 0$

x	y
0	0
6	2
-3	-1

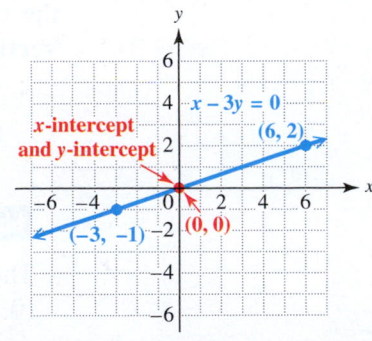

FIGURE 13

NOW TRY

NOW TRY ANSWERS
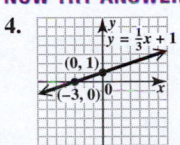
4.
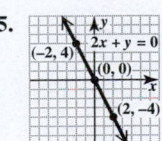
5.

Line through the Origin

The graph of a linear equation of the form

$$Ax + By = 0,$$

where *A* and *B* are nonzero real numbers, passes through the origin $(0, 0)$.

Examples: $2x + 3y = 0$, $x = y$, $-4y = 3x$ The last two can be written in $Ax + By = 0$ form.

OBJECTIVE 4 Graph linear equations of the form $y = b$ or $x = a$.

Consider the following linear equations.

$$y = -4 \quad \text{can be written as} \quad 0x + y = -4.$$
$$x = 3 \quad \text{can be written as} \quad x + 0y = 3.$$

When the coefficient of x or y is 0, the graph is a horizontal or vertical line.

NOW TRY
EXERCISE 6
Graph $y = 2$.

EXAMPLE 6 Graphing a Horizontal Line ($y = b$)

Graph $y = -4$.

For any value of x, the value of y is always -4. Three ordered pairs that satisfy the equation are given in the table of values. Drawing a line through these points gives the **horizontal line** in FIGURE 14.

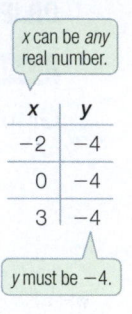

x	y
-2	-4
0	-4
3	-4

x can be any real number.

y must be -4.

The y-intercept is $(0, -4)$.

There is no x-intercept.

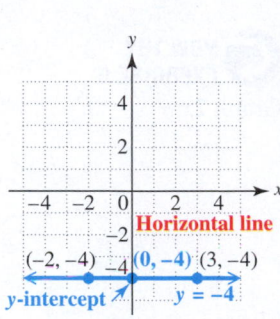

FIGURE 14

NOW TRY

NOW TRY
EXERCISE 7
Graph $x + 4 = 0$.

EXAMPLE 7 Graphing a Vertical Line ($x = a$)

Graph $x - 3 = 0$.

First we add 3 to each side of the equation $x - 3 = 0$ to obtain $x = 3$. All ordered-pair solutions of this equation have x-coordinate 3. Any number can be used for y. Three ordered pairs that satisfy the equation are given in the table of values. The graph is the **vertical line** in FIGURE 15.

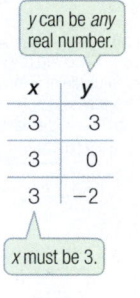

x	y
3	3
3	0
3	-2

y can be any real number.

x must be 3.

The x-intercept is $(3, 0)$.

There is no y-intercept.

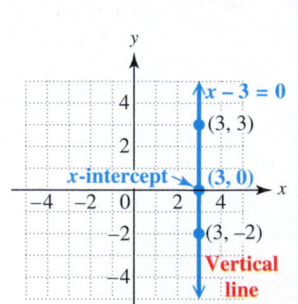

FIGURE 15

NOW TRY

Horizontal and Vertical Lines

The graph of **$y = b$,** where b is a real number, is a **horizontal line** with y-intercept $(0, b)$ and no x-intercept (unless the horizontal line is the x-axis itself).

Examples: $y = 5$ and $y + 2 = 0$ (which can be written $y = -2$)

The graph of **$x = a$,** where a is a real number, is a **vertical line** with x-intercept $(a, 0)$ and no y-intercept (unless the vertical line is the y-axis itself).

Examples: $x = -1$ and $x - 6 = 0$ (which can be written $x = 6$)

NOW TRY ANSWERS
6.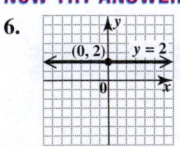
7.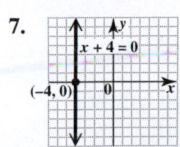

Keep the following in mind regarding the x- and y-axes.

- *The x-axis is the horizontal line given by the equation $y = 0$.*
- *The y-axis is the vertical line given by the equation $x = 0$.*

> **!CAUTION** The equations of horizontal and vertical lines are often confused with each other.
>
> - The graph of $y = b$ is parallel to the x-axis (for $b \neq 0$).
> - The graph of $x = a$ is parallel to the y-axis (for $a \neq 0$).

▼ **Forms of Linear Equations**

Equation	To Graph	Example
$Ax + By = C$ (where A, B, and C are real numbers not equal to 0)	Find any two points on the line. A good choice is to find the intercepts. Let $x = 0$, and find the corresponding value of y. Then let $y = 0$, and find x. As a check, find a third point by choosing a value for x or y that has not yet been used.	$3x - 2y = 6$; points $(4, 3)$, $(2, 0)$, $(0, -3)$
$Ax + By = 0$	The graph passes through the point $(0, 0)$. To find additional points that lie on the graph, choose any values for x or y, except 0.	$x - 2y = 0$; points $(0, 0)$, $(4, 2)$, $(-4, -2)$
$y = b$	Draw a horizontal line through the point $(0, b)$.	$y = -2$; point $(0, -2)$
$x = a$	Draw a vertical line through the point $(a, 0)$.	$x = 4$; point $(4, 0)$

OBJECTIVE 5 Use a linear equation to model data.

EXAMPLE 8 Using a Linear Equation to Model Internet Use

In the United States, the weekly time spent online y in hours can be modeled by the linear equation

$$y = 0.91x + 9.3,$$

where $x = 0$ represents 2000, $x = 1$ represents 2001, and so on. (Data from *The 2017 Digital Future Report,* USC.)

(a) Use the equation to approximate weekly time spent online in the years 2000, 2008, and 2016.

Substitute the appropriate value for each year x to find weekly time spent online that year.

$$y = 0.91x + 9.3 \qquad \text{Given linear equation}$$

For 2000: $\quad y = 0.91(0) + 9.3 \qquad$ Replace x with 0.

$$y = 9.3 \text{ hr} \qquad \text{Multiply, and then add.}$$

NOW TRY
EXERCISE 8

Use **(a)** the graph and **(b)** the equation in **Example 8** to approximate weekly time spent online in 2010. (Round the answer in part (b) to the nearest tenth.)

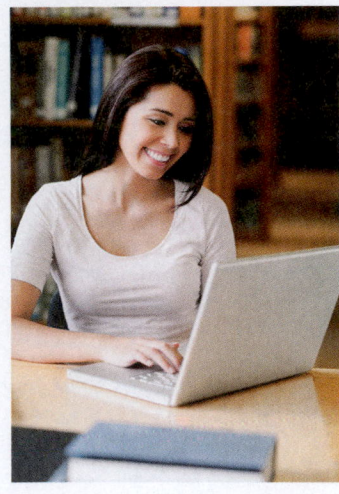

$$y = 0.91x + 9.3 \qquad \text{Given linear equation}$$

For 2008: $\quad y = 0.91(8) + 9.3 \qquad$ 2008 − 2000 = 8

$\qquad\qquad\quad y \approx 16.6 \text{ hr} \qquad$ Replace x with 8.

For 2016: $\quad y = 0.91(16) + 9.3 \qquad$ 2016 − 2000 = 16

$\qquad\qquad\quad y \approx 23.9 \text{ hr} \qquad$ Replace x with 16.

(b) Write the information from part (a) as three ordered pairs, and use them to graph the given linear equation.

Because x represents the year and y represents the time, the ordered pairs are

$$(0, 9.3), \quad (8, 16.6), \quad \text{and} \quad (16, 23.9).$$

See **FIGURE 16**. (Arrowheads are not included with the graphed line because the data are for the years 2000 to 2016 only—that is, from $x = 0$ to $x = 16$.)

x	y
0	9.3
8	16.6
16	23.9

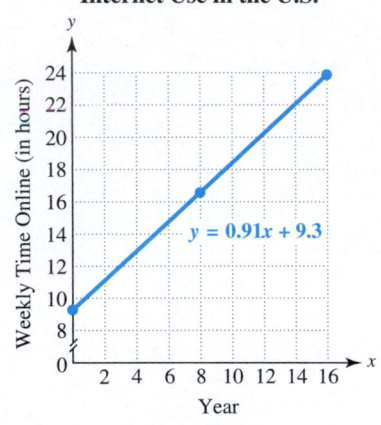

Internet Use in the U.S.

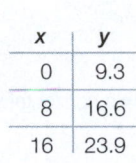

FIGURE 16

(c) Use the graph and then the equation to determine the year when the weekly time spent online was about 20 hours.

Locate 20 on the vertical axis in **FIGURE 16**. Move horizontally from 20 to the graphed line, then down from the line to the value of x on the horizontal axis to find the year, here 12. Because $x = 0$ represents 2000, $x = 12$ represents the year

$$2000 + 12 = 2012. \quad\leftarrow \text{Year when the weekly time spent online was about 20 hours}$$

To use the equation to find the year, substitute 20 for y and solve for x.

$$y = 0.91x + 9.3 \qquad \text{Given linear equation}$$

$$20 = 0.91x + 9.3 \qquad \text{Let } y = 20.$$

$$10.7 = 0.91x \qquad \text{Subtract 9.3.}$$

$$11.8 \approx x \qquad \text{Divide by 0.91.}$$

Rounding up to the nearest year gives $x = 12$—that is, 2012—which agrees with the graph.

NOW TRY ↺

NOW TRY ANSWERS
8. **(a)** about 18 hr
   (b) 18.4 hr

3.2 Exercises

FOR EXTRA HELP

 MyLab Math

Video solutions to select problems available in MyLab Math

Concept Check *Fill in each blank with the correct response.*

1. A linear equation in two variables x and y is an equation that can be written in the form $Ax +$ ____ = ____ , where A, B, and C are real numbers and A and B are not both _____.

2. The graph of any linear equation in two variables is a straight _____. Every point on the line represents a _____ of the equation.

3. **Concept Check** Match the information about each graph in Column I with the correct linear equation in Column II.

	I		**II**
(a)	The graph of the equation has y-intercept $(0, -4)$.	**A.**	$3x + y = -4$
(b)	The graph of the equation has $(0, 0)$ as x-intercept and y-intercept.	**B.**	$x - 4 = 0$
		C.	$y = 4x$
(c)	The graph of the equation does not have an x-intercept.	**D.**	$y = 4$
(d)	The graph of the equation has x-intercept $(4, 0)$.		

4. **Concept Check** Which of these equations have a graph with only one intercept?

   A. $x + 8 = 0$ **B.** $x - y = 3$ **C.** $x + y = 0$ **D.** $y = 4$

Concept Check *Identify the intercepts of each graph. (Coordinates of the points shown are integers.)*

5.
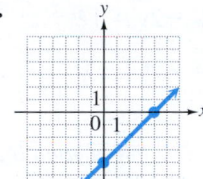

6.
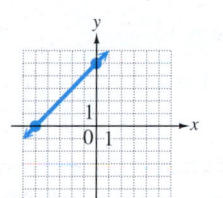

7.
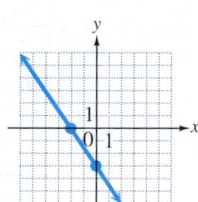

8.
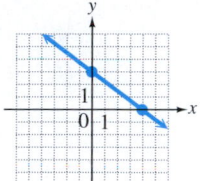

9. **Concept Check** Match each equation in (a)–(d) with its graph in A–D.

   (a) $x = -2$ **(b)** $y = -2$ **(c)** $x = 2$ **(d)** $y = 2$

A.
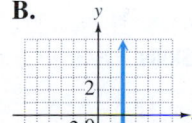

B.
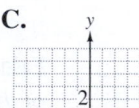

C.

D.
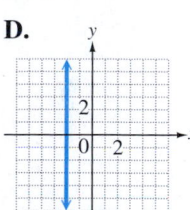

10. **Concept Check** Using x and y, write a linear equation in two variables to represent each statement.

   (a) The sum of two numbers is 8.   **(b)** The sum of two numbers is -2.

*Complete the given ordered-pair solutions for each equation. Then graph each equation by plotting the points and drawing a line through them. **See Examples 1–4.***

11. $x + y = 4$

   $(0,$ ____ $)$, $($ ____ $, 0)$, $(2,$ ____ $)$

12. $x - y = 2$

   $(0,$ ____ $)$, $($ ____ $, 0)$, $(5,$ ____ $)$

13. $y = \dfrac{2}{3}x + 1$

   $(0,$ ____ $)$, $(3,$ ____ $)$, $(-3,$ ____ $)$

14. $y = -\dfrac{3}{4}x + 2$

   $(0,$ ____ $)$, $(4,$ ____ $)$, $(-4,$ ____ $)$

15. $3x = -y - 6$

   $(0,$ ____ $)$, $($ ____ $, 0)$, $\left(-\dfrac{1}{3},$ ____ $\right)$

16. $x = 2y + 3$

   $($ ____ $, 0)$, $(0,$ ____ $)$, $\left($ ____ $, \dfrac{1}{2}\right)$

Find the x- and y-intercepts for the graph of each equation. **See Examples 1–7.**

17. $x - y = 8$ **18.** $x - y = 7$ **19.** $5x - 2y = 20$ **20.** $-3x + 2y = 12$

21. $x + 6y = 0$ **22.** $3x + y = 0$ **23.** $y = -2x + 4$ **24.** $y = 3x + 6$

25. $y = \dfrac{1}{3}x - 2$ **26.** $y = \dfrac{1}{4}x - 1$ **27.** $2x - 3y = 0$ **28.** $4x - 5y = 0$

29. $x - 4 = 0$ **30.** $x - 5 = 0$ **31.** $y = 2.5$ **32.** $y = -1.5$

Graph each linear equation. **See Examples 1–7.**

33. $x - y = 4$ **34.** $x - y = 5$ **35.** $2x + y = 6$

36. $-3x + y = -6$ **37.** $y = 2x - 5$ **38.** $y = 4x + 3$

39. $x = y + 2$ **40.** $x = -y + 6$ **41.** $2x - 5y = 10$

42. $3x + 2y = 6$ **43.** $3x + 7y = 14$ **44.** $6x - 5y = 18$

45. $y = -\dfrac{3}{4}x + 3$ **46.** $y = -\dfrac{2}{3}x - 2$ **47.** $y - 2x = 0$

48. $y + 3x = 0$ **49.** $y = -6x$ **50.** $y = 4x$

51. $y = -1$ **52.** $y = 3$ **53.** $x = 5$

54. $x = -1$ **55.** $x + 2 = 0$ **56.** $x - 4 = 0$

57. $-3y = 15$ **58.** $-2y = 12$ **59.** $x + 2 = 8$

60. $x - 1 = -4$ **61.** $y - 2 = 0$ **62.** $y + 3 = 0$

Concept Check *Describe what the graph of each linear equation will look like in the coordinate plane. (Hint: Rewrite the equation if necessary so that it is in a more recognizable form.)*

63. $3x = y - 9$ **64.** $2x = y - 4$ **65.** $x - 10 = 1$ **66.** $x + 4 = 3$

67. $3y = -6$ **68.** $5y = -15$ **69.** $2x = 4y$ **70.** $3x = 9y$

Extending Skills *Plot each set of points, and draw a line through them. Then give an equation of the line.*

71. $(3, 5)$, $(3, 0)$, and $(3, -3)$ **72.** $(1, 3)$, $(1, 0)$, and $(1, -1)$

73. $(-3, -3)$, $(0, -3)$, and $(4, -3)$ **74.** $(-5, 5)$, $(0, 5)$ and $(3, 5)$

Solve each problem. **See Example 8.**

75. The weight y (in pounds) of a man taller than 60 in. can be approximated by the linear equation

$$y = 5.5x - 220,$$

where x is the height of the man in inches.

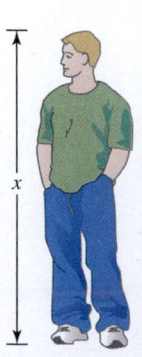

(a) Use the equation to approximate the weights of men whose heights are 62 in., 66 in., and 72 in.

(b) Write the information from part (a) as three ordered pairs.

(c) Graph the equation for $x \geq 62$, using the data from part (b).

(d) Use the graph to estimate the height of a man who weighs 155 lb. Then use the equation to find the height of this man to the nearest inch.

76. The height y (in centimeters) of a woman can be approximated by the linear equation

$$y = 3.9x + 73.5,$$

where x is the length of her radius bone in centimeters.

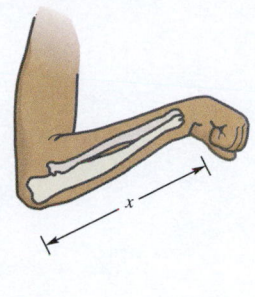

(a) Use the equation to approximate the heights of women with radius bones of lengths 20 cm, 22 cm, and 26 cm.

(b) Write the information from part (a) as three ordered pairs.

(c) Graph the equation for $x \geq 20$, using the data from part (b).

(d) Use the graph to estimate the length of the radius bone in a woman who is 167 cm tall. Then use the equation to find the length of the radius bone to the nearest centimeter.

77. As a fundraiser, a club is selling posters. The printer charges a $25 set-up fee, plus $0.75 for each poster. The cost y in dollars to print x posters is given by the linear equation

$$y = 0.75x + 25.$$

(a) What is the cost y in dollars to print 50 posters? To print 100 posters?

(b) Find the number of posters x if the printer billed the club for costs of $175.

(c) Write the information from parts (a) and (b) as three ordered pairs.

(d) Use the data from part (c) and the given grid to graph the equation for $x \geq 0$.

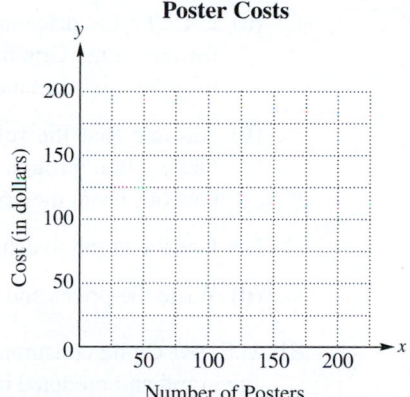

78. A gas station is selling gasoline for $3.50 per gallon and charges $7 for a car wash. The cost y in dollars for x gallons of gasoline and a car wash is given by the linear equation

$$y = 3.50x + 7.$$

(a) What is the cost y in dollars for 9 gal of gasoline and a car wash?   For 4 gal of gasoline and a car wash?

(b) Find the number of gallons of gasoline x if the cost for gasoline and a car wash is $35.

(c) Write the information from parts (a) and (b) as three ordered pairs.

(d) Use the data from part (c) and the given grid to graph the equation for $x \geq 0$.

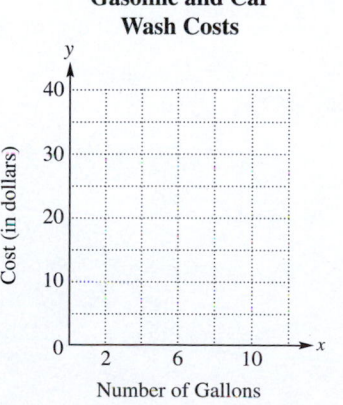

79. The graph shows the value of a sport-utility vehicle (SUV) over the first 5 yr of ownership.

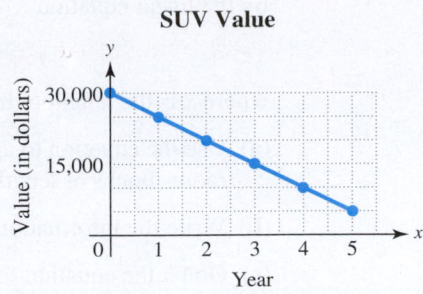

Use the graph to do the following.

(a) Determine the initial value of the SUV.

(b) Find the **depreciation** (loss in value) from the original value after the first 3 yr.

(c) What is the annual or yearly depreciation in each of the first 5 yr?

(d) What does the ordered pair (5, 5000) mean in the context of this problem?

80. Demand for an item is often closely related to its price. As price increases, demand decreases, and as price decreases, demand increases. Suppose demand for a video game is 2000 units when the price is $40, and demand is 2500 units when the price is $30.

(a) Let x be the price and y be the demand for the game. Graph the two given pairs of prices and demands on the given grid.

(b) Assume that the relationship is linear. Draw a line through the two points from part (a). From the graph, estimate the demand if the price drops to $20.

(c) Use the graph to estimate the price if the demand is 3500 units.

(d) Write the prices and demands from parts (b) and (c) as ordered pairs.

81. U.S. per capita consumption y of cheese in pounds from 2000 through 2016 is shown in the graph and modeled by the linear equation

$$y = 0.356x + 29.7,$$

where $x = 0$ represents 2000, $x = 2$ represents 2002, and so on.

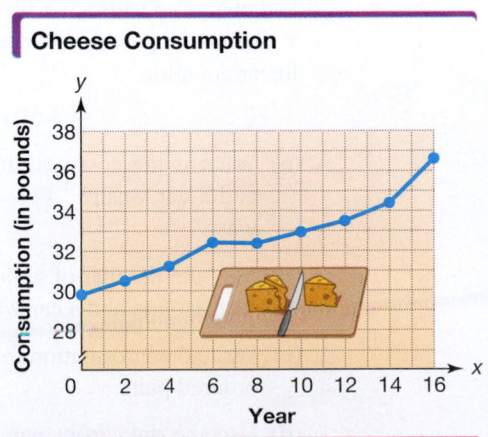

Data from U.S. Department of Agriculture.

(a) Use the equation to approximate cheese consumption (to the nearest tenth) in 2000, 2008, and 2016.

(b) Use the graph to estimate cheese consumption for the same years.

(c) How do the approximations using the equation compare with the estimates from the graph?

(d) The USDA projects that per capita consumption of cheese in 2022 will be 36.8 lb. Use the equation to approximate per capita cheese consumption (to the nearest tenth) in 2022. How does the approximation using the equation compare to the USDA projection?

82. The numbers of U.S. marathon finishers y, in thousands, from 2000 through 2016 are shown in the graph and modeled by the linear equation

$$y = 12.68x + 340.9,$$

where $x = 0$ represents 2000, $x = 4$ represents 2004, and so on.

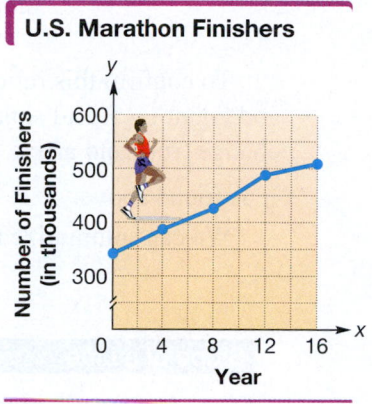

U.S. Marathon Finishers

Data from Running U.S.A.

(a) Use the equation to approximate the number of U.S. marathon finishers (to the nearest thousand) in 2004, 2012, and 2016.

(b) Use the graph to estimate the number of U.S. marathon finishers for the same years.

(c) How do the approximations using the equation compare with the estimates from the graph?

3.3   The Slope of a Line

OBJECTIVES

1 Find the slope of a line given two points.

2 Find the slope from the equation of a line.

3 Use slopes to determine whether two lines are parallel, perpendicular, or neither.

An important characteristic of the lines we graphed in the previous section is their slant, or "steepness" as viewed from *left to right*. See **FIGURE 17**.

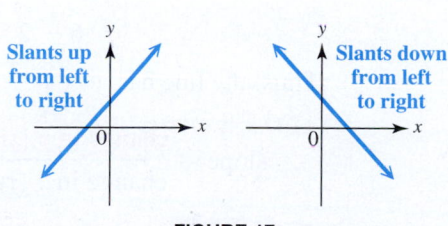

FIGURE 17

One way to measure the steepness of a line is to compare the vertical change in the line with the horizontal change while moving along the line from one fixed point to another. This measure of steepness is the *slope* of the line.

VOCABULARY
☐ rise
☐ run
☐ slope
☐ parallel lines
☐ perpendicular lines

STUDY SKILLS REMINDER
Are you fully utilizing the features of your text? **Review Study Skill 1, *Using Your Math Text*.**

OBJECTIVE 1 Find the slope of a line given two points.

To find the steepness, or slope, of the line in **FIGURE 18**, we begin at point Q and move to point P. The vertical change, or **rise,** is the change in the y-values, which is the difference

$$6 - 1 = 5 \text{ units.}$$

The horizontal change, or **run,** from Q to P is the change in the x-values, which is the difference

$$5 - 2 = 3 \text{ units.}$$

One way to compare two numbers is by using a ratio. **Slope** is a ratio of the vertical change in y to the horizontal change in x. The line in **FIGURE 18** has

$$\text{slope} = \frac{\text{vertical change in } y \text{ (rise)}}{\text{horizontal change in } x \text{ (run)}} = \frac{5}{3}.$$

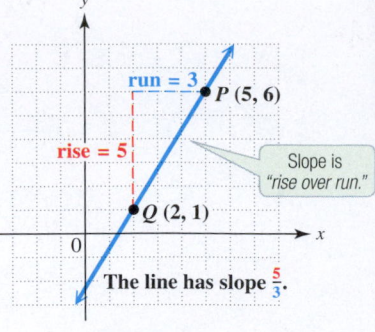
FIGURE 18

To confirm this ratio, we can count grid squares. We start at point Q in **FIGURE 18** and count *up* 5 grid squares to find the vertical change (rise). To find the horizontal change (run) and arrive at point P, we count to the *right* 3 grid squares. The slope is $\frac{5}{3}$, as found above.

We can summarize this discussion as follows.

Slope of a Line

Slope is a single number that allows us to determine the direction in which a line is slanting from left to right, as well as how much slant there is to the line.

NOW TRY EXERCISE 1
Find the slope of the line.

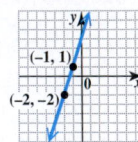

EXAMPLE 1 Finding the Slope of a Line

Find the slope of the line in **FIGURE 19**.

We use the coordinates of the two points shown on the line. The vertical change is the difference of the y-values.

$$-1 - 3 = -4$$

The horizontal change is the difference of the x-values.

$$6 - 2 = 4$$

Thus, the line has

$$\text{slope} = \frac{\text{change in } y \text{ (rise)}}{\text{change in } x \text{ (run)}} = \frac{-4}{4}, \text{ or } -1.$$

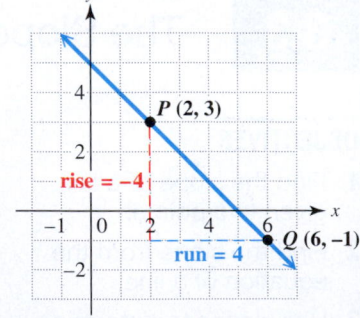
FIGURE 19

Counting grid squares, we begin at point P and count *down* 4 grid squares. Because we counted down, we write the vertical change as a negative number, -4 here. Then we count to the *right* 4 grid squares to reach point Q. The slope is $\frac{-4}{4}$, or -1.

NOW TRY ANSWER
1. 3

NOW TRY ↺

NOTE *The slope of a line is the same for any two points on the line.* In **FIGURE 19**, locate the points (3, 2) and (5, 0), which also lie on the line. Start at (3, 2) and count *down* 2 units and then to the *right* 2 units to arrive at the point (5, 0).

$$\text{The slope is } \frac{-2}{2}, \quad \text{or} \quad -1.$$

This is the same slope found in **Example 1** using the points (2, 3) and (6, −1).

The concept of slope is used in many everyday situations. See **FIGURE 20**.

- A highway with a 10%, or $\frac{1}{10}$, grade (or slope) rises 1 m for every 10 m horizontally.

- A roof with pitch (or slope) $\frac{5}{12}$ rises 5 ft for every 12 ft that it runs horizontally.

- A stairwell with slope $\frac{8}{12}$ $\left(\text{or } \frac{2}{3}\right)$ indicates a vertical rise of 8 ft for a horizontal run of 12 ft.

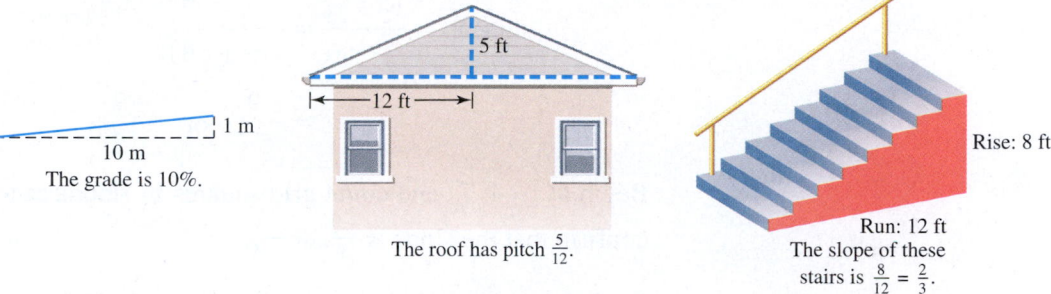

5 ft

12 ft

1 m

10 m

The grade is 10%.

The roof has pitch $\frac{5}{12}$.

Rise: 8 ft

Run: 12 ft

The slope of these stairs is $\frac{8}{12} = \frac{2}{3}$.

FIGURE 20

We can generalize the preceding discussion and find the slope of a line through two nonspecific points (x_1, y_1) and (x_2, y_2). This notation is **subscript notation.** Read x_1 as **"x-sub-one"** and x_2 as **"x-sub-two."**

Moving along the line from the point (x_1, y_1) in **FIGURE 21** to the point (x_2, y_2), we see that y changes by $y_2 - y_1$ units. This is the vertical change (rise). Similarly, x changes by $x_2 - x_1$ units, which is the horizontal change (run). The slope of the line is the ratio of $y_2 - y_1$ to $x_2 - x_1$.

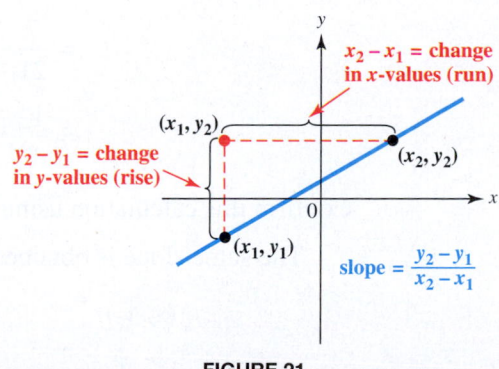

$x_2 - x_1$ = change in x-values (run)

(x_1, y_2)

(x_2, y_2)

$y_2 - y_1$ = change in y-values (rise)

(x_1, y_1)

slope = $\dfrac{y_2 - y_1}{x_2 - x_1}$

FIGURE 21

Slope Formula

The **slope** m of the line passing through the points (x_1, y_1) and (x_2, y_2) is defined as follows. (Traditionally, the letter m represents slope.)

$$m = \frac{\text{change in } y}{\text{change in } x} = \frac{y_2 - y_1}{x_2 - x_1} \quad (\text{where } x_1 \neq x_2)$$

The slope gives the change in y for each unit of change in x.

NOTE   Subscript notation is used to identify a point. It does *not* indicate an operation. *Note the difference between x_2, which represents a nonspecific value, and x^2, which means $x \cdot x$. Read x_2 as "x-sub-two," not "x squared."*

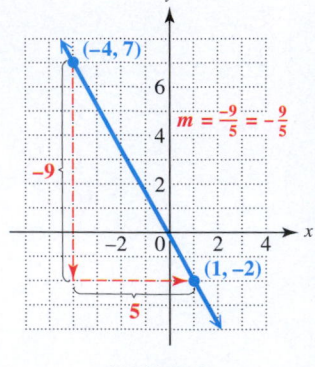

NOW TRY EXERCISE 2

Find the slope of the line passing through $(4, -5)$ and $(-2, -4)$.

EXAMPLE 2   Finding Slopes of Lines

Find the slope of each line.

(a) The line passing through $(-4, 7)$ and $(1, -2)$

Label the points. Then apply the slope formula.

$$\underset{\underset{(-4, 7)}{\downarrow\downarrow}}{(x_1, y_1)} \quad \text{and} \quad \underset{\underset{(1, -2)}{\downarrow\downarrow}}{(x_2, y_2)}$$

$$\text{slope } m = \frac{y_2 - y_1}{x_2 - x_1} = \frac{-2 - 7}{1 - (-4)} \quad \boxed{\text{Substitute carefully.}}$$

$$= \frac{-9}{5}, \quad \text{or} \quad -\frac{9}{5}$$

FIGURE 22

Begin at $(-4, 7)$ and count grid squares in **FIGURE 22** to confirm that the slope is $\frac{-9}{5}$, or $-\frac{9}{5}$.

(b) The line passing through $(-9, -2)$ and $(12, 5)$

Label the points. Then apply the slope formula.

$$\underset{\underset{(-9, -2)}{\downarrow\downarrow}}{(x_1, y_1)} \quad \text{and} \quad \underset{\underset{(12, 5)}{\downarrow\downarrow}}{(x_2, y_2)}$$

$$\text{slope } m = \frac{y_2 - y_1}{x_2 - x_1} = \frac{5 - (-2)}{12 - (-9)}$$

$$= \frac{7}{21} \qquad \text{Subtract.}$$

$$= \frac{1}{3} \qquad \text{Write in lowest terms.}$$

FIGURE 23

Confirm this calculation using **FIGURE 23**. (Note the scale on the *x*- and *y*-axes.)

The same slope is obtained if we label the points in reverse order.

$$\underset{\underset{(-9, -2)}{\downarrow\downarrow}}{(x_2, y_2)} \quad \text{and} \quad \underset{\underset{(12, 5)}{\downarrow\downarrow}}{(x_1, y_1)}$$

It makes no difference which point is identified as (x_1, y_1) or (x_2, y_2).

$$\text{slope } m = \frac{y_2 - y_1}{x_2 - x_1} = \frac{-2 - 5}{-9 - 12} \qquad \text{Substitute.}$$

$\boxed{\text{Start with the values of the \textit{same} point. Subtract the values of the other point.}}$

$$= \frac{-7}{-21} \qquad \text{Subtract.}$$

$$= \frac{1}{3} \qquad \text{The same slope results.}$$

NOW TRY ANSWER

2. $-\frac{1}{6}$

NOW TRY

The slopes of the lines in **FIGURES 22** and **23** suggest the following.

> **Orientation of Lines with Positive and Negative Slopes**
>
> A line with positive slope rises (slants up) from left to right.
>
> A line with negative slope falls (slants down) from left to right.

NOW TRY
EXERCISE 3

Find the slope of the line passing through $(1, -3)$ and $(4, -3)$.

EXAMPLE 3   Finding the Slope of a Horizontal Line

Find the slope of the line passing through $(-5, 4)$ and $(2, 4)$.

$$m = \frac{y_2 - y_1}{x_2 - x_1} = \frac{4 - 4}{2 - (-5)} \qquad \text{Subtract } y\text{-values.}$$
$$\text{Subtract } x\text{-values in the } same \text{ order.}$$

$$= \frac{0}{7} \qquad \text{Subtract.}$$

$$= \mathbf{0} \qquad \text{Slope 0}$$

As shown in **FIGURE 24**, the line passing through these two points is horizontal, with equation $y = 4$.

All horizontal lines have slope 0.

This is because the difference of the y-values for any two points on a horizontal line is always 0, which results in a 0 *numerator* when we are calculating slope.

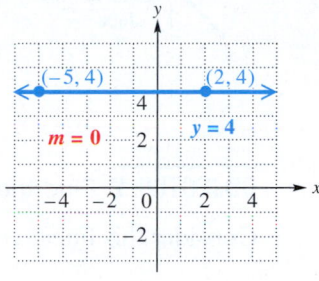

FIGURE 24

NOW TRY

NOW TRY
EXERCISE 4

Find the slope of the line passing through $(-2, 1)$ and $(-2, -4)$.

EXAMPLE 4   Applying the Slope Concept to a Vertical Line

Find the slope of the line passing through $(6, 2)$ and $(6, -4)$.

$$m = \frac{y_2 - y_1}{x_2 - x_1} = \frac{-4 - 2}{6 - 6} \qquad \text{Subtract } y\text{-values.}$$
$$\text{Subtract } x\text{-values in the } same \text{ order.}$$

$$= \frac{-6}{0} \qquad \text{Undefined slope}$$

Because division by 0 is undefined, this line has undefined slope. (This is why the slope formula has the restriction $x_1 \neq x_2$.)

The graph in **FIGURE 25** shows that this line is vertical, with equation $x = 6$.

The slope of any vertical line is undefined.

This is because the difference of the x-values for any two points on a vertical line is always 0, which results in a 0 *denominator* when we are calculating slope.

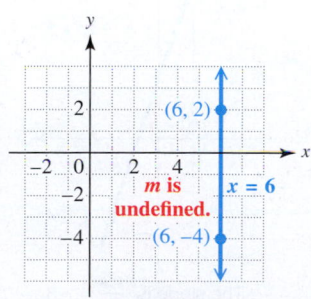

FIGURE 25

NOW TRY

NOW TRY ANSWERS

3. 0
4. undefined slope

Slopes of Horizontal and Vertical Lines

A **horizontal line,** which has an equation of the form

$$y = b \text{ (where } b \text{ is a constant (number)),}$$

has **slope 0.**

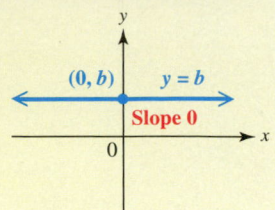

A **vertical line,** which has an equation of the form

$$x = a \text{ (where } a \text{ is a constant (number)),}$$

has **undefined slope.**

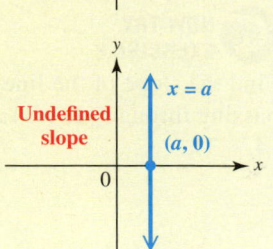

FIGURE 26 summarizes the four cases for slopes of lines.

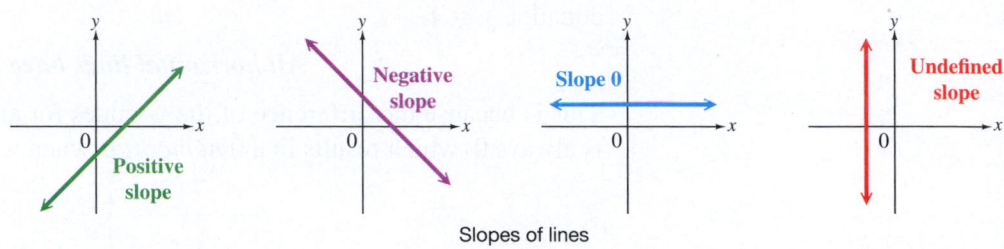

Slopes of lines

FIGURE 26

OBJECTIVE 2 Find the slope from the equation of a line.

Consider this linear equation.

$$y = -3x + 5$$

We can find the slope of this line using any two points on the line. Because the equation is solved for y, it involves less work to choose two different values of x and then find the corresponding values of y. We arbitrarily choose $x = -2$ and $x = 4$.

$y = -3x + 5$		$y = -3x + 5$	
$y = -3(-2) + 5$	Let $x = -2$.	$y = -3(4) + 5$	Let $x = 4$.
$y = 6 + 5$	Multiply.	$y = -12 + 5$	Multiply.
$y = 11$	Add.	$y = -7$	Add.
Ordered pair: $(-2, 11)$		Ordered pair: $(4, -7)$	

Now we apply the slope formula using the ordered pairs $(-2, 11)$ and $(4, -7)$.

$$m = \frac{-7 - 11}{4 - (-2)} = \frac{-18}{6} = -3$$

See the graph of this line in **FIGURE 27.** Counting grid squares from $(-2, 11)$ to $(4, -7)$ confirms that the slope is $\frac{-18}{6}$, which is equivalent to $\frac{-3}{1}$ or -3. **_This is the same number as the coefficient of x in the given equation $y = -3x + 5$._** It can be shown that this always happens, *as long as the equation is solved for y.*

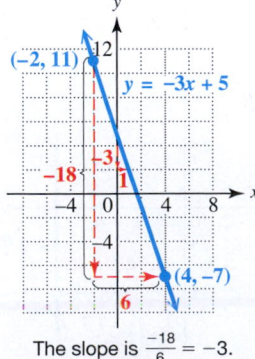

The slope is $\frac{-18}{6} = -3$.

FIGURE 27

> **Finding the Slope of a Line from Its Equation**
>
> **Step 1** Solve the equation for y.
>
> **Step 2** The slope is given by the coefficient of x.

NOW TRY
EXERCISE 5
Find the slope of the line.
$$3x + 5y = -1$$

EXAMPLE 5 Finding Slopes from Equations

Find the slope of each line.

(a) $2x - 5y = 4$

Step 1 Solve the equation for y.

$$2x - 5y = 4 \quad \text{[Isolate } y \text{ on one side.]}$$

$$-5y = 4 - 2x \qquad \text{Subtract } 2x.$$

$$-5y = -2x + 4 \qquad \text{Commutative property}$$

$$\frac{-5y}{-5} = \frac{-2x + 4}{-5} \qquad \text{Divide by } -5.$$

$$y = \frac{-2x}{-5} + \frac{4}{-5} \qquad \frac{a+b}{c} = \frac{a}{c} + \frac{b}{c}$$

$$\frac{-2x}{-5} = \frac{-2}{-5}x = \frac{2}{5}x \qquad y = \frac{2}{5}x - \frac{4}{5}$$

$$\uparrow \text{ Slope}$$

Step 2 The slope is given by the coefficient of x, so the slope is $\frac{2}{5}$.

(b) $$8x + 4y = 1$$

[Solve for y.] $$4y = 1 - 8x \qquad \text{Subtract } 8x.$$

$$4y = -8x + 1 \qquad \text{Commutative property}$$

$$y = -2x + \frac{1}{4} \qquad \text{Divide } each \text{ term by 4.}$$

The slope is given by the coefficient of x, which is -2.

(c) $$3y + x = -3 \quad \text{[We omit the step showing the commutative property.]}$$

$$3y = -x - 3 \qquad \text{Subtract } x.$$

$$y = \frac{-x}{3} - 1 \qquad \text{Divide } each \text{ term by 3.}$$

[The slope is $-\frac{1}{3}$, **not** $\frac{-x}{3}$ or $-\frac{x}{3}$.] $$y = -\frac{1}{3}x - 1 \qquad \frac{-x}{3} = \frac{-1x}{3} = -\frac{1}{3}x$$

The coefficient of x is $-\frac{1}{3}$, so the slope of this line is $-\frac{1}{3}$. **NOW TRY**

> **NOTE** We can solve the linear equation $Ax + By = C$ (where $B \neq 0$) for y to show that, in general, the slope of a line is $m = -\frac{A}{B}$.

NOW TRY ANSWER
5. $-\frac{3}{5}$

OBJECTIVE 3 Use slopes to determine whether two lines are parallel, perpendicular, or neither.

Two lines in a plane that never intersect are **parallel.** We use slopes to determine whether two lines are parallel.

FIGURE 28 shows the graphs of $x + 2y = 4$ and $x + 2y = -6$. These lines appear to be parallel. We solve each equation for y to find the slope.

$$x + 2y = 4$$

$$2y = -x + 4 \qquad \text{Subtract } x.$$

$$y = \frac{-x}{2} + 2 \qquad \text{Divide by 2.}$$

$$y = -\frac{1}{2}x + 2 \qquad \frac{-x}{2} = \frac{-1x}{2} = -\frac{1}{2}x$$

> The slope is $-\frac{1}{2}$, not $-\frac{x}{2}$.

↑ Slope

$$x + 2y = -6$$

$$2y = -x - 6 \qquad \text{Subtract } x.$$

$$y = \frac{-x}{2} - 3 \qquad \text{Divide by 2.}$$

$$y = -\frac{1}{2}x - 3 \qquad \frac{-x}{2} = \frac{-1x}{2} = -\frac{1}{2}x$$

↑ Slope

Both lines have slope $-\frac{1}{2}$. *Nonvertical parallel lines always have equal slopes.*

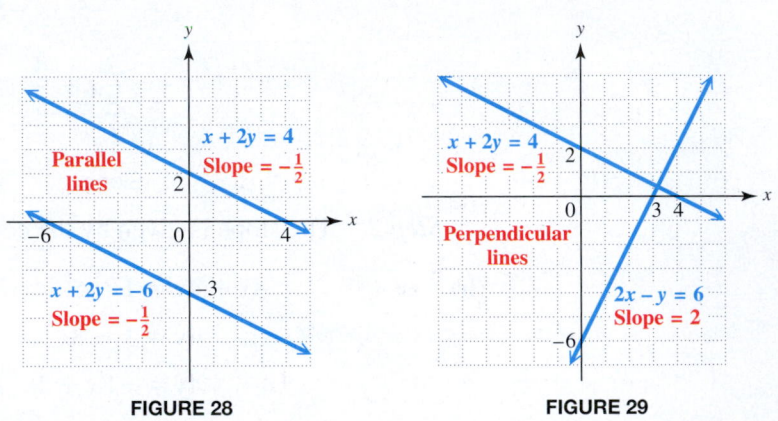

FIGURE 28   **FIGURE 29**

FIGURE 29 shows the graphs of $x + 2y = 4$ and $2x - y = 6$. These lines appear to be **perpendicular** (that is, they intersect at a 90° angle). As we have seen, solving $x + 2y = 4$ for y gives $y = -\frac{1}{2}x + 2$, with slope $-\frac{1}{2}$. We solve $2x - y = 6$ for y to find the slope.

$$2x - y = 6$$

$$-y = -2x + 6 \qquad \text{Subtract } 2x.$$

$$y = 2x - 6 \qquad \text{Multiply by } -1.$$

↑ Slope

The product of the slopes of the two lines is

$$-\frac{1}{2}(2) = -1.$$

This condition is true in general.

▼ Negative Reciprocals

Number	Negative Reciprocal
$\frac{3}{4}$	$-\frac{4}{3}$
$\frac{1}{2}$	$-\frac{2}{1}$, or -2
-6, or $-\frac{6}{1}$	$\frac{1}{6}$
-0.4, or $-\frac{4}{10}$	$\frac{10}{4}$, or 2.5

The product of each number and its negative reciprocal is -1.

The product of the slopes of two perpendicular lines, neither of which is vertical, is always -1.

This means that the slopes of perpendicular lines are negative (or opposite) reciprocals—if one slope is the nonzero number a, then the other slope is $-\frac{1}{a}$. The table in the margin shows several examples.

Slopes of Parallel and Perpendicular Lines

Two lines with the same slope are parallel.

Two lines whose slopes have a product of -1 are perpendicular.

EXAMPLE 6 Determining Whether Two Lines Are Parallel or Perpendicular

Determine whether each pair of lines is *parallel, perpendicular,* or *neither.*

(a) $x + 3y = 5$ and $-3x + y = 3$

Find the slope of each line by first solving each equation for y.

$$x + 3y = 5$$
$$3y = -x + 5 \qquad \text{Subtract } x.$$
$$y = -\frac{1}{3}x + \frac{5}{3} \qquad \text{Divide by 3.}$$

The slope is $-\frac{1}{3}$.

$$-3x + y = 3$$
$$y = 3x + 3 \qquad \text{Add } 3x.$$

The slope is 3.

Because the slopes are not equal, the lines are not parallel. Check the product of the slopes: $-\frac{1}{3}(3) = -1$. The two lines are *perpendicular* because the product of their slopes is -1. See **FIGURE 30**.

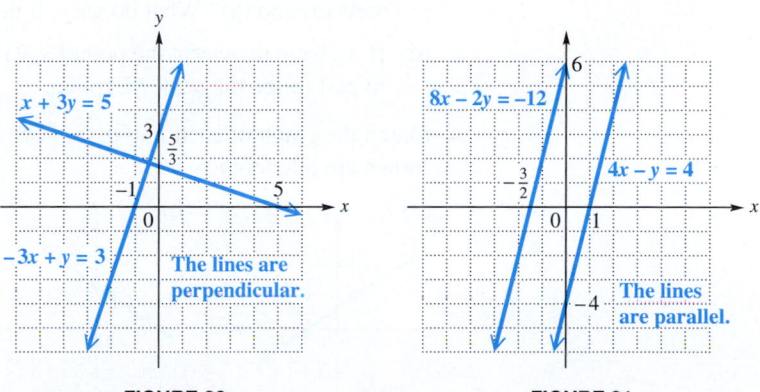

FIGURE 30   FIGURE 31

(b) $4x - y = 4$ $y = 4x - 4$
         $8x - 2y = -12$ $y = 4x + 6$

Solve each equation for y.

We see when each equation is solved for y that both lines have slope 4 and different y-intercepts. Therefore, the two lines are *parallel*. See **FIGURE 31**.

**NOW TRY
EXERCISE 6**

Determine whether the pair of lines is *parallel*, *perpendicular*, or *neither*.

$$2x - 3y = 1$$
$$4x + 6y = 5$$

NOW TRY ANSWER

6. neither

(c) $4x + 3y = 6$ Solve each $y = -\dfrac{4}{3}x + 2$

   $2x - y = 5$ equation for *y*. $y = 2x - 5$

The slopes, $-\dfrac{4}{3}$ and 2, are not the same $\left(-\dfrac{4}{3} \neq 2\right)$, nor are they negative reciprocals $\left(-\dfrac{4}{3}(2) \neq -1\right)$. The two lines are *neither* parallel nor perpendicular.

(d) $5x - y = 1$ Solve each $y = 5x - 1$

   $x - 5y = -10$ equation for *y*. $y = \dfrac{1}{5}x + 2$ $5\left(\dfrac{1}{5}\right) = 1, \textit{not} -1.$

The slopes, 5 and $\dfrac{1}{5}$, are neither the same, nor are they negative reciprocals. The two lines are *neither* parallel nor perpendicular.   **NOW TRY**

3.3 Exercises

FOR EXTRA HELP   ▶ **MyLab Math**

▶ *Video solutions for select problems available in MyLab Math*

STUDY SKILLS REMINDER

How are you doing on your homework? **Review Study Skill 4, *Completing Your Homework.***

Concept Check   *Work each problem involving slope.*

1. Slope is a measure of the _____ of a line. Slope is the (*horizontal / vertical*) change compared to the (*horizontal / vertical*) change while moving along the line from one point to another.

2. Slope is the _____ of the vertical change in _____, called the (*rise / run*), to the horizontal change in _____, called the (*rise / run*).

3. Use the graph at the right to answer the following.

   (a) Start at the point $(-1, -4)$ and count vertically up to the horizontal line that goes through the other plotted point. What is this vertical change? (Remember: "up" means positive, "down" means negative.)

   (b) From this new position, count horizontally to the other plotted point. What is this horizontal change? (Remember: "right" means positive, "left" means negative.)

   (c) What is the ratio (quotient) of the numbers found in parts (a) and (b)? What do we call this number?

   (d) If we were to *start* at the point $(3, 2)$ and *end* at the point $(-1, -4)$ would the answer to part (c) be the same? Explain.

4. Match the graph of each line in (a)–(d) with its slope in A–D. (Coordinates of the points shown are integers.)

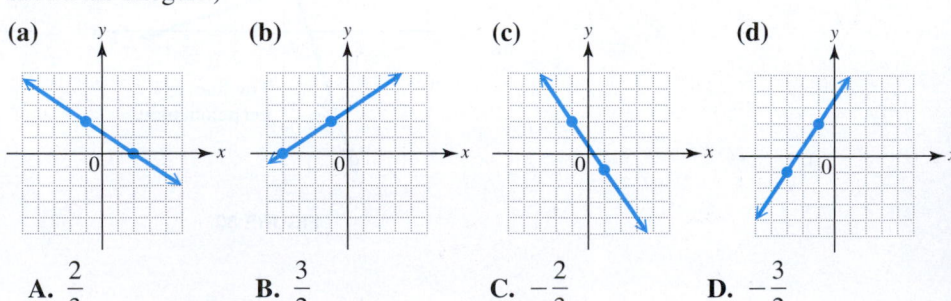

(a)   **(b)**   **(c)**   **(d)**

A. $\dfrac{2}{3}$ B. $\dfrac{3}{2}$ C. $-\dfrac{2}{3}$ D. $-\dfrac{3}{2}$

5. Determine whether the line with the given slope *m* rises from left to right, falls from left to right, is *horizontal*, or is *vertical*.

   (a) $m = -4$ **(b)** $m = 0$ **(c)** *m* is undefined.   **(d)** $m = \dfrac{3}{7}$

6. On a pair of axes similar to the one shown, sketch the graph of a straight line having the indicated slope.

(a) Negative

(b) Positive

(c) Undefined

(d) 0

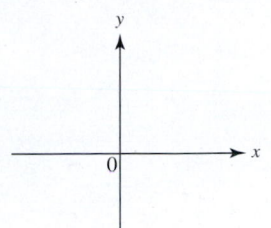

Concept Check *The figure at the right shows a line that has a positive slope (because it rises from left to right) and a positive y-value for the y-intercept (because it intersects the y-axis above the origin).*

For each line graphed, answer the following.

(a) Is the slope positive, negative, or 0?

(b) Is the y-value of the y-intercept positive, negative, or 0?

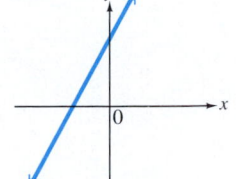

7.

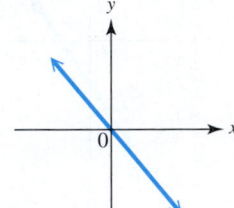

8.

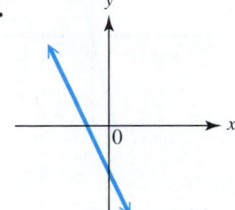

9.

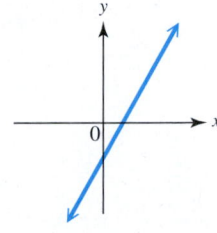

10.

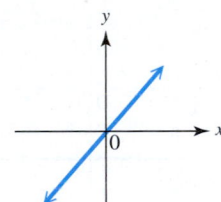

11.

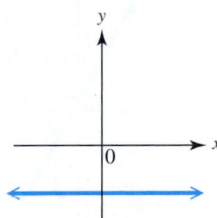

12.

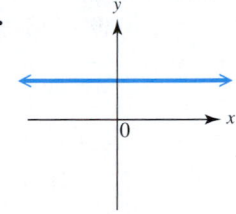

13. Concept Check   A student was asked to find the slope of the line through the points $(2, 5)$ and $(-1, 3)$. His answer, $-\frac{2}{3}$, was incorrect. He showed his work as

$$\frac{3 - 5}{2 - (-1)} = \frac{-2}{3} = -\frac{2}{3}.$$

WHAT WENT WRONG? Give the correct slope.

14. Concept Check   A student was asked to find the slope of the line through the points $(-2, 4)$ and $(6, -1)$. Her answer, $-\frac{8}{5}$, was incorrect. She showed her work as

$$\frac{6 - (-2)}{-1 - 4} = \frac{8}{-5} = -\frac{8}{5}.$$

WHAT WENT WRONG? Give the correct slope.

Concept Check *Answer each question.*

15. What is the slope (or grade) of this hill?

16. What is the slope (or pitch) of this roof?

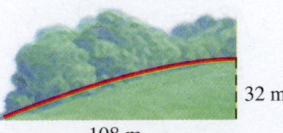

108 m

17. What is the slope of the slide? (*Hint:* The slide *drops* 8 ft vertically as it extends 12 ft horizontally.)

18. What is the slope (or grade) of this ski slope? (*Hint:* The ski slope *drops* 25 ft vertically for every 100 horizontal feet.)

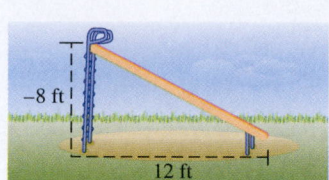

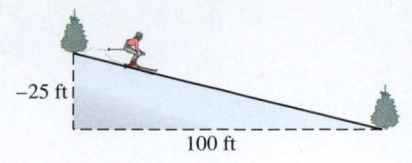

Use the coordinates of the indicated points to find the slope of each line. (Coordinates of the points shown are integers.) ***See Examples 1–4.***

19.

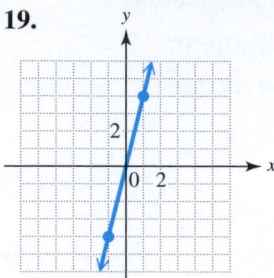

20.

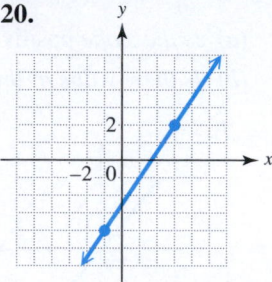

21.

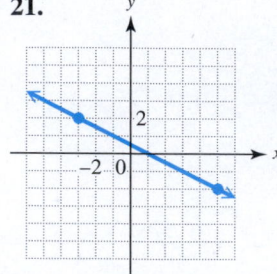

22.

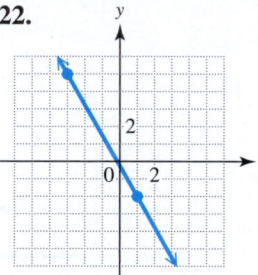

23.

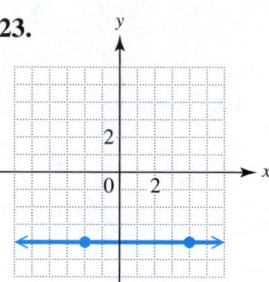

24.

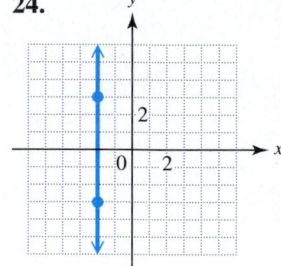

Find the slope of the line passing through each pair of points. ***See Examples 2–4.***

25. $(1, -2)$ and $(-3, -7)$ **26.** $(4, -1)$ and $(-2, -8)$ **27.** $(0, 3)$ and $(-2, 0)$

28. $(8, 0)$ and $(0, -5)$ **29.** $(4, 3)$ and $(-6, 3)$ **30.** $(6, 5)$ and $(-12, 5)$

31. $(-2, 4)$ and $(-3, 7)$ **32.** $(-4, 5)$ and $(-5, 8)$ **33.** $(-12, 3)$ and $(-12, -7)$

34. $(-8, 6)$ and $(-8, -1)$ **35.** $(4.8, 2.5)$ and $(3.6, 2.2)$ **36.** $(3.1, 2.6)$ and $(1.6, 2.1)$

37. $\left(-\dfrac{7}{5}, \dfrac{3}{10}\right)$ and $\left(\dfrac{1}{5}, -\dfrac{1}{2}\right)$ **38.** $\left(-\dfrac{4}{3}, \dfrac{1}{2}\right)$ and $\left(\dfrac{1}{3}, -\dfrac{5}{6}\right)$

Find the slope of each line. ***See Example 5.***

39. $y = 5x + 12$ **40.** $y = 2x + 3$ **41.** $4y = x + 1$ **42.** $2y = x + 4$

43. $3x - 2y = 3$ **44.** $6x - 4y = 4$ **45.** $-3x + 2y = 5$ **46.** $-2x + 4y = 5$

47. $x + y = -4$ **48.** $x - y = -2$ **49.** $x = 2y$ **50.** $x = -5y$

51. $y = -5$ **52.** $y = 4$ **53.** $x = 6$ **54.** $x = -2$

Find the slope of each line in two ways by doing the following.

(a) Give any two points that lie on the line, and use them to determine the slope.

(b) Solve the equation for y, and identify the slope from the equation.

See Objective 2 and Example 5.

55. $2x + y = 10$ **56.** $-4x + y = -8$ **57.** $5x - 3y = 15$ **58.** $3x + 2y = 12$

Each table of values gives several points that lie on a line.

(a) Use any two of the ordered pairs to find the slope of the line.

(b) What is the x-intercept of the line?   The y-intercept?

(c) Graph the line.

59.

x	y
−4	0
−2	2
0	4
1	5

60.

x	y
−4	3
−1	0
0	−1
2	−3

61.

x	y
3	−3
0	−2
−3	−1
−6	0

62.

x	y
−1	−6
0	−4
2	0
5	6

Concept Check *Answer each question.*

63. What is the slope of a line whose graph is

    (a) parallel to the graph of $3x + y = 7$?

    (b) perpendicular to the graph of $3x + y = 7$?

64. What is the slope of a line whose graph is

    (a) parallel to the graph of $-5x + y = -3$?

    (b) perpendicular to the graph of $-5x + y = -3$?

65. If two lines are both vertical or both horizontal, which of the following are they?

    A. Parallel

    B. Perpendicular

    C. Neither parallel nor perpendicular

66. If a line is vertical, what is true of any line that is perpendicular to it?

For each pair of equations, give the slopes of the lines and then determine whether the two lines are parallel, perpendicular, *or* neither. ***See Example 6.***

67. $2x + 5y = 4$

$4x + 10y = 1$

68. $3x + 2y = 6$

$9x + 6y = 11$

69. $8x - 9y = 6$

$8x + 6y = -5$

70. $5x - 3y = -2$

$3x - 5y = -8$

71. $3x - 5y = -1$

$5x + 3y = 2$

72. $3x - 2y = 6$

$2x + 3y = 3$

73. $-4x + 3y = 4$

$-8x + 6y = 0$

74. $9x - 2y = 0$

$-18x + 4y = 3$

75. $5x - y = 1$

$x - 5y = -10$

76. $x + 7y = -4$

$7x + y = -3$

77. $2x - 5y = 10$

$5x + 2y = 12$

78. $3x - 4y = 12$

$4x + 3y = 12$

The graph shows global shipments of desktop PCs and tablets, in millions of units, from 2010 through 2016. Use the graph to work each problem.

79. Locate the line on the graph that represents tablets.

   (a) Write two ordered pairs (x, y), where x is the year and y is shipments in millions of units, to represent the data for the years 2010 and 2016.

   (b) Use the ordered pairs from part (a) to find the slope of the line.

   (c) Interpret the meaning of the slope in the context of this problem.

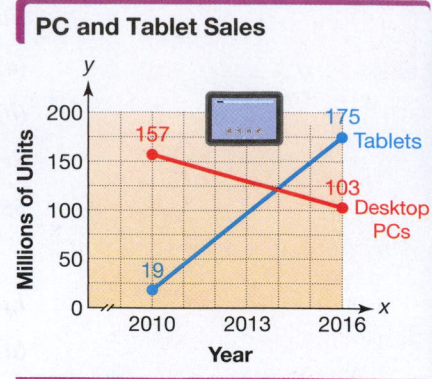

PC and Tablet Sales

Data from IDC.

80. Locate the line on the graph that represents PCs. Repeat parts (a)–(c) of **Exercise 79.**

RELATING CONCEPTS   For Individual or Group Work   (Exercises 81–86)

FIGURE A *gives the percent of freshmen at 4-year colleges and universities who planned to major in the Biological Sciences.* **FIGURE B** *shows the percent of the same group of students who planned to major in Business.* **Work Exercises 81–86 in order.**

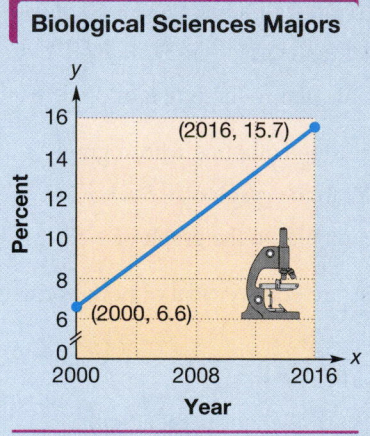

Biological Sciences Majors

Data from Higher Education Research Institute.

FIGURE A

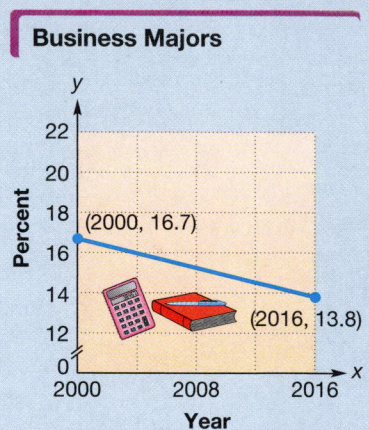

Business Majors

Data from Higher Education Research Institute.

FIGURE B

81. Find the slope of the line in **FIGURE A** to the nearest hundredth.

82. The slope of the line in **FIGURE A** is (*positive/negative*). This means that during the period represented, the percent of freshmen planning to major in the Biological Sciences (*increased/decreased*).

83. The slope of a line represents the *rate of change*. Based on **FIGURE A**, what was the increase in the percent of freshmen *per year* who planned to major in the Biological Sciences during the period shown?

84. Find the slope of the line in **FIGURE B** to the nearest hundredth.

85. The slope of the line in **FIGURE B** is (*positive/negative*). This means that during the period represented, the percent of freshmen planning to major in Business (*increased/decreased*).

86. Based on **FIGURE B**, what was the decrease in the percent of freshmen *per year* who planned to major in Business?

3.4 Slope-Intercept Form of a Linear Equation

OBJECTIVES

1 Use slope-intercept form of the equation of a line.
2 Graph a line using its slope and a point on the line.
3 Write an equation of a line using its slope and any point on the line.
4 Graph and write equations of horizontal and vertical lines.

OBJECTIVE 1 Use slope-intercept form of the equation of a line.

Recall that we can find the slope (steepness) of a line by solving the equation of the line for y. In that form, the slope is the coefficient of x. For example, the line with equation

$$y = 2x + 3 \quad \text{has slope} \quad 2.$$

What does the number 3 *represent?* To find out, suppose a line has slope m and y-intercept $(0, b)$. We can find an equation of this line by choosing another point (x, y) on the line, as shown in **FIGURE 32**. Then we apply the slope formula.

$$m = \frac{y - b}{x - 0} \quad \leftarrow \text{Change in } y\text{-values}$$
$$\phantom{m = \frac{y - b}{x - 0}} \leftarrow \text{Change in } x\text{-values}$$

$$m = \frac{y - b}{x} \quad \text{Subtract in the denominator.}$$

$$mx = y - b \quad \text{Multiply by } x.$$

$$mx + b = y \quad \text{Add } b.$$

$$y = mx + b \quad \text{Rewrite.}$$

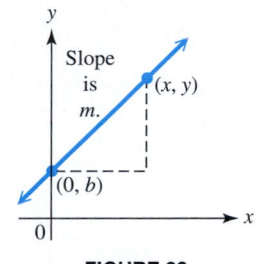

FIGURE 32

This result is the *slope-intercept form* of the equation of a line. Both the *slope* and the *y-intercept* of the line can be read directly from this form. For the line with equation $y = 2x + 3$, the number 3 gives the y-intercept $(0, 3)$.

Slope-Intercept Form

The **slope-intercept form** of the equation of a line with slope m and y-intercept $(0, b)$ is

$$y = mx + b.$$

Slope ⌐   ⌐ $(0, b)$ is the y-intercept.

The intercept given is the y-intercept.

**NOW TRY
EXERCISE 1**

Identify the slope and y-intercept of the line with each equation.

(a) $y = -\dfrac{3}{5}x - 9$

(b) $y = -\dfrac{x}{3} + \dfrac{7}{3}$

NOW TRY ANSWERS

1. (a) slope: $-\dfrac{3}{5}$; y-intercept: $(0, -9)$

(b) slope: $-\dfrac{1}{3}$; y-intercept: $\left(0, \dfrac{7}{3}\right)$

EXAMPLE 1 Identifying Slopes and y-Intercepts

Identify the slope and y-intercept of the line with each equation.

(a) $y = -4x + 1$

Slope ⌐   ⌐ y-intercept $(0, 1)$

(b) $y = x - 8$ can be written as $y = 1x + (-8)$.

Slope ⌐   ⌐ y-intercept $(0, -8)$

(c) $y = 6x$ can be written as $y = 6x + 0$.

Slope ⌐   ⌐ y-intercept $(0, 0)$

(d) $y = \dfrac{x}{4} - \dfrac{3}{4}$ can be written as $y = \dfrac{1}{4}x + \left(-\dfrac{3}{4}\right)$.

Slope ⌐   ⌐ y-intercept $\left(0, -\dfrac{3}{4}\right)$ **NOW TRY**

> **NOTE** Slope-intercept form is an especially useful form for a linear equation because of the information we can determine from it. It is also the form used by graphing calculators and the one that describes a *linear function*.

OBJECTIVE 2 Graph a line using its slope and a point on the line.

We can use the slope and the point represented by the y-intercept to graph a line.

Graphing a Line Using the Slope and y-Intercept

Step 1 Write the equation in slope-intercept form $y = mx + b$, if necessary, by solving for y.

Step 2 Identify the y-intercept. Plot the point $(0, b)$.

Step 3 Identify the slope m of the line. Use the geometric interpretation of slope ("*rise over run*") to find another point on the graph by counting from the y-intercept.

Step 4 Join the two points with a line to obtain the graph. (If desired, obtain a third point, such as the x-intercept, as a check.)

EXAMPLE 2 Graphing Lines Using Slopes and y-Intercepts

Graph the equation of each line using the slope and y-intercept.

(a) $y = \dfrac{2}{3}x - 1$

Step 1 The equation is in slope-intercept form.

$$y = \frac{2}{3}x - 1$$

$\qquad\qquad\uparrow\qquad\quad\uparrow$

$\qquad$ Slope $\quad$ Value of b in y-intercept $(0, b)$

Step 2 The y-intercept is $(0, -1)$. Plot this point. See **FIGURE 33**.

Step 3 The slope is $\dfrac{2}{3}$. By definition,

$$\text{slope } m = \frac{\text{change in } y \text{ (rise)}}{\text{change in } x \text{ (run)}} = \frac{2}{3}.$$

From the y-intercept, count *up* 2 units and to the *right* 3 units to the point $(3, 1)$.

Step 4 Draw the line through the points $(0, -1)$ and $(3, 1)$ to obtain the graph in **FIGURE 33**.

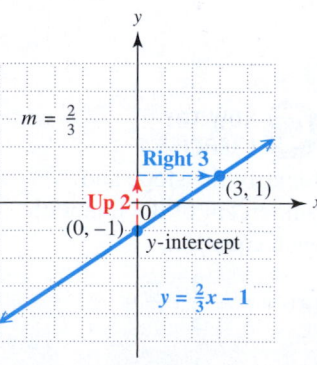

FIGURE 33

(b) $3x + 4y = 8$

Step 1 Solve for y to write the equation in slope-intercept form.

$$3x + 4y = 8$$

Isolate y on one side.

$$4y = -3x + 8 \qquad \text{Subtract } 3x.$$

Slope-intercept form $\longrightarrow$ $y = -\dfrac{3}{4}x + 2 \qquad$ Divide by 4.

**NOW TRY
EXERCISE 2**

Graph $3x + 2y = 8$ using the slope and y-intercept.

Step 2 The y-intercept in $y = -\frac{3}{4}x + 2$ is $(0, 2)$. Plot this point. See **FIGURE 34**.

Step 3 The slope is $-\frac{3}{4}$, which can be written as either $\frac{-3}{4}$ or $\frac{3}{-4}$. We use $\frac{-3}{4}$ here.

$$m = \frac{\text{change in } y \text{ (rise)}}{\text{change in } x \text{ (run)}} = \frac{-3}{4}$$

From the y-intercept, count *down* 3 units (because of the negative sign) and to the *right* 4 units to the point $(4, -1)$.

Step 4 Draw the line through the two points $(0, 2)$ and $(4, -1)$ to obtain the graph in **FIGURE 34**.

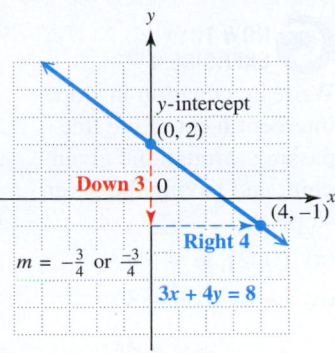

FIGURE 34

NOW TRY

> **NOTE** In Step 3 of **Example 2(b)**, we could use $\frac{3}{-4}$ for the slope. From the y-intercept $(0, 2)$ in **FIGURE 34**, count *up* 3 units and to the *left* 4 units (because of the negative sign) to the point $(-4, 5)$. Verify that this produces the same line.

**NOW TRY
EXERCISE 3**

Graph the line passing through the point $(-3, -4)$, with slope $\frac{5}{2}$.

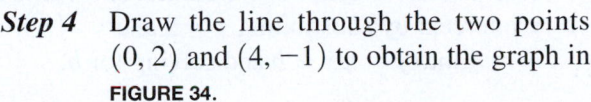
EXAMPLE 3   Graphing a Line Using Its Slope and a Point

Graph the line passing through the point $(-2, 3)$, with slope -4.

First, plot the point $(-2, 3)$. See **FIGURE 35**. Then write the slope -4 as

$$m = \frac{\text{change in } y \text{ (rise)}}{\text{change in } x \text{ (run)}} = \frac{-4}{1}.$$

Locate another point on the line by counting *down* 4 units from $(-2, 3)$ and then to the *right* 1 unit. Finally, draw the line through this new point $(-1, -1)$ and the given point $(-2, 3)$. See **FIGURE 35**.

We could have written the slope as $\frac{4}{-1}$ instead. In this case, we would move *up* 4 units from $(-2, 3)$ and then to the *left* 1 unit. Verify that this produces the same line.

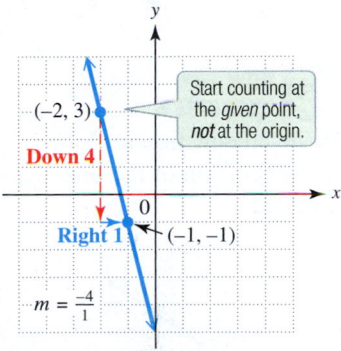

FIGURE 35

NOW TRY

OBJECTIVE 3 Write an equation of a line using its slope and any point on the line.

We can use the slope-intercept form to write the equation of a line if we know the slope and any point on the line.

NOW TRY ANSWERS

2.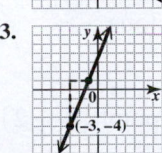

3.

EXAMPLE 4   Using Slope-Intercept Form to Write Equations

Write an equation in slope-intercept form of the line passing through the given point and having the given slope.

(a) $(0, -1)$, $m = \frac{2}{3}$

Because the point $(0, -1)$ is the y-intercept, $b = -1$. We can substitute this value for b and the given slope $m = \frac{2}{3}$ directly into slope-intercept form $y = mx + b$ to write an equation.

NOW TRY
EXERCISE 4

Write an equation in slope-intercept form of the line passing through the given point and having the given slope.

(a) $(0, 2)$, $m = -4$

(b) $(-2, 1)$, $m = 3$

Slope —↓   ↓— y-intercept is $(0, b)$.

$$y = mx + b \qquad \text{Slope-intercept form}$$

$$y = \frac{2}{3}x + (-1) \qquad \text{Substitute for } m \text{ and } b.$$

$$y = \frac{2}{3}x - 1 \qquad \text{Definition of subtraction}$$

(b) $(2, 5)$, $m = 4$

This line passes through the point $(2, 5)$, which is **not** the y-intercept because the x-coordinate is 2, **not** 0. **We cannot substitute directly as in part (a).** We can find the y-intercept by substituting $x = 2$ and $y = 5$ from the given point and the given slope $m = 4$ into $y = mx + b$ and solving for b.

$$y = mx + b \qquad \text{Slope-intercept form}$$

$$5 = 4(2) + b \qquad \text{Let } y = 5,\ m = 4,\ \text{and } x = 2.$$

$(0, b)$ is the y-intercept. Don't stop here.

$$5 = 8 + b \qquad \text{Multiply.}$$

$$-3 = b \qquad \text{Subtract 8.}$$

Now substitute the values of m and b into slope-intercept form.

$$y = mx + b \qquad \text{Slope-intercept form}$$

$$y = 4x - 3 \qquad \text{Let } m = 4 \text{ and } b = -3.$$

NOW TRY

EXAMPLE 5   Writing an Equation of a Line from a Graph

Write an equation in slope-intercept form of the line graphed in **FIGURE 36**.

NOW TRY
EXERCISE 5

Write an equation in slope-intercept form of the line graphed.

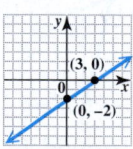

From the graph in **FIGURE 36**, identify the y-intercept $(0, 4)$. Thus, $b = 4$. To find the slope, count grid squares *down* 4 units from the y-intercept and to the *right* 2 units to the x-intercept $(2, 0)$.

$$\text{slope } m = \frac{-4}{2}, \quad \text{or} \quad -2.$$

We now have the information needed to write the equation of the line.

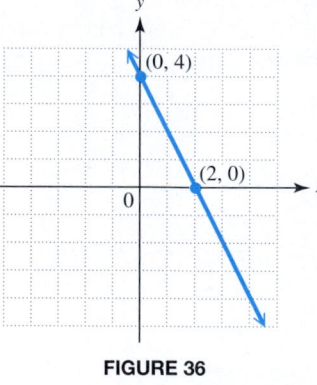

FIGURE 36

Slope —↓   ↓— y-intercept is $(0, b)$.

$$y = mx + b \qquad \text{Slope-intercept form}$$

$$y = -2x + 4 \qquad \text{Let } m = -2 \text{ and } b = 4.$$

NOW TRY

OBJECTIVE 4 Graph and write equations of horizontal and vertical lines.

EXAMPLE 6   Graphing Horizontal and Vertical Lines Using Slope and a Point

Graph each line passing through the given point and having the given slope.

(a) $(4, -2)$, $m = 0$

Recall that a horizontal line has slope 0. To graph this line, plot the point $(4, -2)$ and draw the horizontal line through it. See **FIGURE 37** on the next page.

NOW TRY ANSWERS

4. (a) $y = -4x + 2$

   (b) $y = 3x + 7$

5. $y = \frac{2}{3}x - 2$

**NOW TRY
EXERCISE 6**

Graph each line passing through the given point and having the given slope.

(a) $(-3, 3)$, undefined slope

(b) $(3, -3)$, slope 0

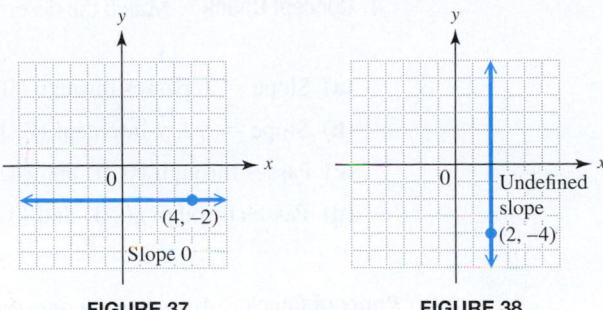

FIGURE 37   **FIGURE 38**

(b) $(2, -4)$, undefined slope

Recall that a vertical line has undefined slope. To graph this line, plot the point $(2, -4)$ and draw the vertical line through it. See **FIGURE 38**.   **NOW TRY**

**NOW TRY
EXERCISE 7**

Write an equation of the line passing through the point $(-1, 1)$ and having the given slope.

(a) Undefined slope

(b) $m = 0$

NOW TRY ANSWERS

6. (a)   (b)

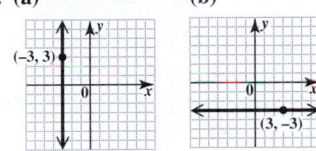
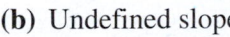

7. (a) $x = -1$ (b) $y = 1$

EXAMPLE 7   Writing Equations of Horizontal and Vertical Lines

Write an equation of the line passing through the point $(2, -2)$ and having the given slope.

(a) Slope 0

This line is horizontal because it has slope 0. Recall that a horizontal line through the point (a, b) has equation $y = b$. The y-coordinate of the point $(2, -2)$ is -2, so the equation is $y = -2$. See **FIGURE 39**.

(b) Undefined slope

This line is vertical because it has undefined slope. A vertical line through the point (a, b) has equation $x = a$. The x-coordinate of $(2, -2)$ is 2, so the equation is $x = 2$. See **FIGURE 39**.   **NOW TRY**

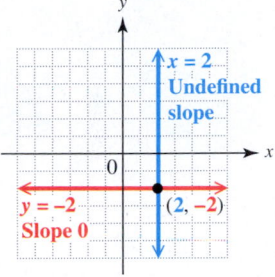

FIGURE 39

3.4 Exercises

FOR EXTRA HELP   ▶ **MyLab Math**

▶ *Video solutions to select problems available in MyLab Math*

STUDY SKILLS REMINDER

Reread your class notes before working the assigned exercises. **Review Study Skill 3,** *Taking Lecture Notes.*

Concept Check   *Fill in each blank with the correct response.*

1. In slope-intercept form $y = mx + b$ of the equation of a line, the slope is _____ and the y-intercept is _____.

2. The line with equation $y = -\dfrac{x}{2} - 3$ has slope _____ and y-intercept _____.

3. Concept Check   Match each equation in parts (a)–(d) with the graph in A–D that would most closely resemble its graph.

(a) $y = x + 3$ (b) $y = -x + 3$ (c) $y = x - 3$ (d) $y = -x - 3$

A. 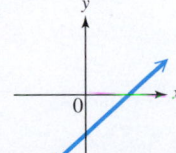   **B.** 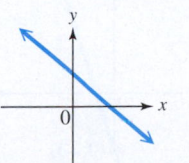   **C.** 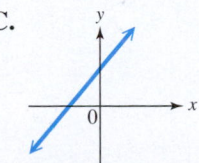   **D.**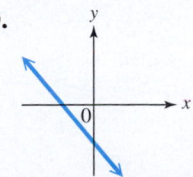

4. Concept Check   Match the description in Column I with the correct equation in Column II.

I	II
(a) Slope = 2, passes through $(0, 4)$	**A.** $y = 4x$ **B.** $y = -4x$
(b) Slope = -2, y-intercept $(0, 1)$	**C.** $y = \frac{1}{4}x$ **D.** $y = -\frac{1}{4}x$
(c) Passes through $(0, 0)$ and $(4, 1)$	**E.** $y = -2x + 1$ **F.** $y = 2x + 4$
(d) Passes through $(0, 0)$ and $(1, 4)$	

Concept Check   *Answer each question.*

5. What is the common name given to a vertical line whose x-intercept is the origin?

6. What is the common name given to a line with slope 0 whose y-intercept is the origin?

Identify the slope and y-intercept of the line with each equation. ***See Example 1.***

7. $y = \frac{5}{2}x - 4$

8. $y = \frac{7}{3}x - 6$

9. $y = -x + 9$

10. $y = x + 1$

11. $y = \frac{x}{5} - \frac{3}{10}$

12. $y = \frac{x}{7} - \frac{5}{14}$

Graph the equation of each line using the slope and y-intercept. ***See Example 2.***

13. $y = 3x + 2$

14. $y = 4x - 4$

15. $y = -\frac{1}{3}x + 4$

16. $y = -\frac{1}{2}x + 2$

17. $y = 2x$

18. $y = -3x$

19. $2x + y = -5$

20. $3x + y = -2$

21. $4x - 5y = 20$

22. $6x - 5y = 30$

23. $x + 2y = 0$

24. $x - 3y = 0$

Graph each line passing through the given point and having the given slope. ***See Examples 3 and 6.***

25. $(0, 1)$, $m = 4$

26. $(0, -3)$, $m = \frac{1}{2}$

27. $(0, 4)$, $m = -4$

28. $(0, -5)$, $m = -2$

29. $(1, -5)$, $m = -\frac{2}{5}$

30. $(2, -1)$, $m = -\frac{1}{3}$

31. $(-1, 4)$, $m = \frac{2}{5}$

32. $(-2, 2)$, $m = \frac{3}{2}$

33. $(0, 0)$, $m = -2$

34. $(0, 0)$, $m = -3$

35. $(-2, 3)$, $m = 0$

36. $(3, 2)$, $m = 0$

37. $(2, 4)$, undefined slope

38. $(3, -2)$, undefined slope

39. $(5, -5)$, slope 0

40. $(-4, 4)$, slope 0

41. $(-2, 2)$, undefined slope

42. $(-1, -3)$, undefined slope

Write an equation in slope-intercept form of each line graphed. ***See Example 5.***

43.

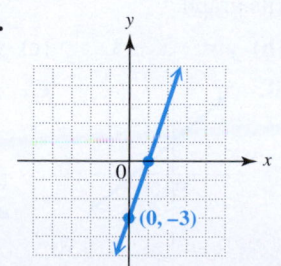

44.

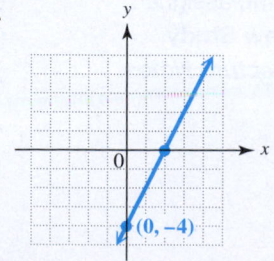

45.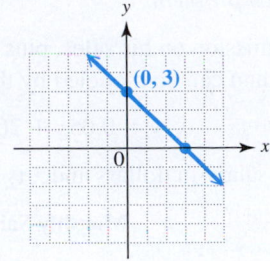

46.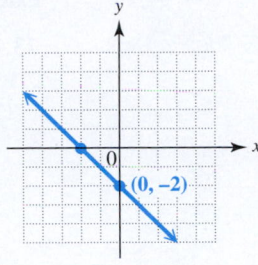

47.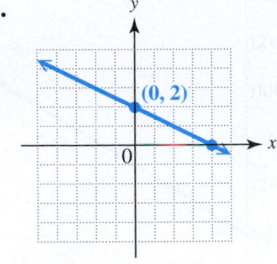

48.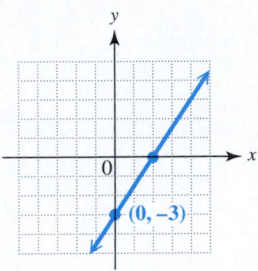

*Write an equation in slope-intercept form (if possible) of the line passing through the given point and having the given slope. **See Examples 4 and 7.***

49. slope 4, y-intercept $(0, -3)$

50. slope -5, y-intercept $(0, 6)$

51. $(0, -7), m = -1$

52. $(0, -9), m = 1$

53. $(4, 1), m = 2$

54. $(2, 7), m = 3$

55. $(-1, 3), m = -4$

56. $(-3, 1), m = -2$

57. $(9, 3), m = 1$

58. $(8, 4), m = 1$

59. $(-4, 1), m = \dfrac{3}{4}$

60. $(2, 1), m = \dfrac{5}{2}$

61. $(0, 3), m = 0$

62. $(0, -4), m = 0$

63. $(2, -6)$, undefined slope

64. $(-1, 7)$, undefined slope

65. $(0, -2)$, undefined slope

66. $(0, 5)$, undefined slope

67. $(6, -6)$, slope 0

68. $(-3, 3)$, slope 0

Each table of values gives several points that lie on a line.

(a) *Use any two of the ordered pairs to find the slope of the line.*

(b) *Identify the y-intercept of the line.*

(c) *Use the slope and y-intercept from parts (a) and (b) to write an equation of the line in slope-intercept form.*

(d) *Graph the equation.*

69.

x	y
0	-1
3	5
5	9

70.

x	y
0	4
2	2
4	0

71.

x	y
-9	1
-6	0
0	-2

72.

x	y
-10	-1
0	3
5	5

Extending Skills   *Solve each problem.*

73. Andrew earns 5% commission on his sales, plus a base salary of $2000 per month. This is illustrated in the graph and can be modeled by the linear equation

$$y = 0.05x + 2000,$$

where *y* is his monthly salary in dollars and *x* is his sales, also in dollars.

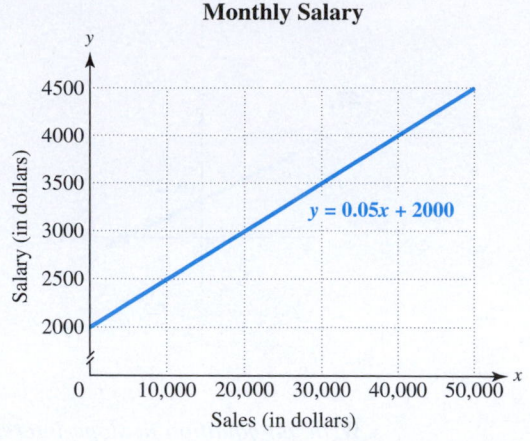

Monthly Salary

(a) What is the slope? With what does the slope correspond in the problem?

(b) What is the *y*-intercept? With what does the *y*-value of the *y*-intercept correspond in the problem?

(c) Use the equation to determine Andrew's monthly salary if his sales are $10,000. Confirm this using the graph.

(d) Use the graph to determine his sales if he wants to earn a monthly salary of $3500. Confirm this using the equation.

74. The cost to rent a moving van is $0.50 per mile, plus a flat fee of $100. This is illustrated in the graph and can be modeled by the linear equation

$$y = 0.50x + 100,$$

where *y* is the total rental cost in dollars and *x* is the number of miles driven.

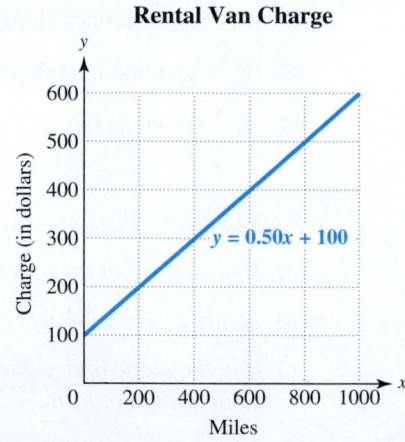

Rental Van Charge

(a) What is the slope? With what does the slope correspond in the problem?

(b) What is the *y*-intercept? With what does the *y*-value of the *y*-intercept correspond in the problem?

(c) Use the equation to determine the total charge if 400 mi are driven. Confirm this using the graph.

(d) Use the graph to determine the number of miles driven if the charge is $500. Confirm this using the equation.

Extending Skills *The cost y of producing x items is, in some cases, expressed in the form* **y = mx + b.** *The value of b gives the* **fixed cost** *(the cost that is the same no matter how many items are produced), and the value of m is the* **variable cost** *(the cost of producing an additional item). Use this information to work each problem.*

75. It costs $400 to start a business selling campaign buttons. Each button costs $0.25 to make.

   (a) What is the fixed cost?   **(b)** What is the variable cost?

   (c) Write the cost equation.

   (d) Find the cost of producing 100 buttons.

   (e) How many campaign buttons will be produced if the total cost is $775?

76. It costs $2000 to purchase a copier, and each copy costs $0.02 to make.

   (a) What is the fixed cost?   **(b)** What is the variable cost?

   (c) Write the cost equation.

   (d) Find the cost of producing 10,000 copies.

   (e) How many copies will be produced if the total cost is $2600?

RELATING CONCEPTS For Individual or Group Work (Exercises 77–80)

A line with equation written in slope-intercept form $y = mx + b$ *has slope m and y-intercept* $(0, b)$. *The standard form of a linear equation in two variables is*

$$Ax + By = C, \qquad \text{Standard form}$$

where A, B, and C are real numbers and A and B are not both 0. **Work Exercises 77–80 in order.**

77. Write the standard form of a linear equation in slope-intercept form—that is, solved for y—to show that, in general, the slope is given by $-\frac{A}{B}$ (where $B \neq 0$).

78. Use the fact that $m = -\frac{A}{B}$ to find the slope of the line with each equation.

   (a) $2x + 3y = 18$ **(b)** $4x - 2y = -1$ **(c)** $3x - 7y = 21$

79. Refer to the slope-intercept form found in **Exercise 77.** What is the y-intercept?

80. Use the result of **Exercise 79** to find the y-intercept of each line in **Exercise 78.**

3.5   Point-Slope Form of a Linear Equation and Modeling

OBJECTIVES

1 Use point-slope form to write an equation of a line.

2 Write an equation of a line using two points on the line.

3 Write an equation of a line that fits a data set.

OBJECTIVE 1 Use point-slope form to write an equation of a line.

There is another form that can be used to write an equation of a line. To develop this form, let m represent the slope of a line and let (x_1, y_1) represent a given point on the line. We let (x, y) represent any other point on the line. See **FIGURE 40.**

$$m = \frac{y - y_1}{x - x_1} \qquad \text{Definition of slope}$$

$$m(x - x_1) = y - y_1 \qquad \text{Multiply each side by } x - x_1.$$

$$y - y_1 = m(x - x_1) \qquad \text{Interchange sides.}$$

This result is the *point-slope form* of the equation of a line.

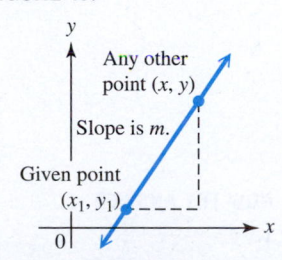

FIGURE 40

Point-Slope Form

The **point-slope form** of the equation of a line with slope m passing through the point (x_1, y_1) is

$$\underset{\uparrow}{y - y_1} = m\underset{\uparrow}{(x - x_1)}.$$

Slope

Given point

NOW TRY
EXERCISE 1

Write an equation of the line passing through $(3, -1)$, with slope $-\frac{2}{5}$. Give the final answer in slope-intercept form.

EXAMPLE 1   Using Point-Slope Form to Write Equations

Write an equation of each line. Give the final answer in slope-intercept form.

(a) The line passing through $(-2, 4)$, with slope -3

Here $x_1 = -2$, $y_1 = 4$, and $m = -3$. Substitute into the point-slope form.

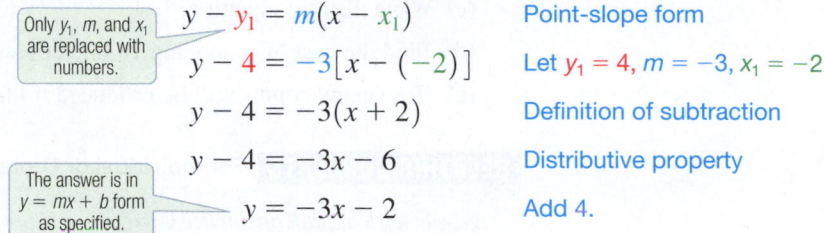

Only y_1, m, and x_1 are replaced with numbers. → $y - y_1 = m(x - x_1)$	Point-slope form
$y - 4 = -3[x - (-2)]$	Let $y_1 = 4$, $m = -3$, $x_1 = -2$.
$y - 4 = -3(x + 2)$	Definition of subtraction
$y - 4 = -3x - 6$	Distributive property
The answer is in $y = mx + b$ form as specified. → $y = -3x - 2$	Add 4.

(b) The line passing through $(4, 2)$, with slope $\frac{3}{5}$

$y - y_1 = m(x - x_1)$	Point-slope form
$y - 2 = \dfrac{3}{5}(x - 4)$	Let $y_1 = 2$, $m = \frac{3}{5}$, $x_1 = 4$.
$y - 2 = \dfrac{3}{5}x - \dfrac{12}{5}$	Distributive property
Do not clear fractions here because we want the answer in slope-intercept form—that is, solved for y. → $y = \dfrac{3}{5}x - \dfrac{12}{5} + \dfrac{10}{5}$	Add $2 = \frac{10}{5}$ to each side.
$y = \dfrac{3}{5}x - \dfrac{2}{5}$	Combine like terms. **NOW TRY**

OBJECTIVE 2 Write an equation of a line using two points on the line.

Many of the linear equations we have worked with were given in **standard form**

$$Ax + By = C, \qquad \text{Standard form}$$

where A, B, and C are real numbers and A and B are not both 0. In most cases, A, B, and C are rational numbers. ***For consistency in this text, we give answers so that A, B, and C are integers with greatest common factor 1 and $A \geq 0$. (If $A = 0$, then we give $B > 0$.)***

> **NOTE**   The definition of standard form is not the same in all texts. A linear equation can be written in many different, yet equally correct, ways. For example,
>
> $$3x + 4y = 12, \quad 6x + 8y = 24, \quad \text{and} \quad -9x - 12y = -36$$
>
> all represent the same set of ordered pairs. When answers are given in standard form, the form $3x + 4y = 12$ is preferable to the other forms because the greatest common factor of 3, 4, and 12 is 1 and $A \geq 0$.

NOW TRY ANSWER
1. $y = -\frac{2}{5}x + \frac{1}{5}$

NOW TRY
EXERCISE 2

Write an equation of the line passing through the points $(4, 1)$ and $(6, -2)$. Give the final answer in

(a) slope-intercept form and

(b) standard form.

EXAMPLE 2 Writing an Equation of a Line Using Two Points

Write an equation of the line passing through the points $(-2, 5)$ and $(3, 4)$. Give the final answer in slope-intercept form and then in standard form.

First, find the slope of the line using the given points.

$$\text{slope } m = \frac{4 - 5}{3 - (-2)} \qquad \text{Subtract } y\text{-values.}$$
$$\text{Subtract } x\text{-values in the same order.}$$

$$= \frac{-1}{5} \qquad \text{Simplify the fraction.}$$

$$= -\frac{1}{5} \qquad \frac{-a}{b} = -\frac{a}{b}$$

Use $m = -\frac{1}{5}$ and $(-2, 5)$ or $(3, 4)$ as (x_1, y_1) in point-slope form. We choose $(3, 4)$.

$$y - y_1 = m(x - x_1) \qquad \text{Point-slope form}$$

$$y - 4 = -\frac{1}{5}(x - 3) \qquad \text{Let } y_1 = 4, \ m = -\frac{1}{5}, \ x_1 = 3.$$

$$y - 4 = -\frac{1}{5}x + \frac{3}{5} \qquad \text{Distributive property}$$

$$y = -\frac{1}{5}x + \frac{3}{5} + \frac{20}{5} \qquad \text{Add } 4 = \frac{20}{5} \text{ to each side.}$$

Slope-intercept form $\longrightarrow y = -\frac{1}{5}x + \frac{23}{5} \qquad$ Combine like terms.

$$5y = -x + 23 \qquad \text{Multiply by 5 to clear fractions.}$$

Standard form $\longrightarrow x + 5y = 23 \qquad$ Add x. **NOW TRY**

NOTE There is often more than one way to write an equation of a line. In **Example 2**, the same equation will result using the point $(-2, 5)$ for (x_1, y_1) in point-slope form. We could also use slope-intercept form $y = mx + b$ and substitute the slope and either given point, solving for b.

▼ Summary of the Forms of Linear Equations

Equation	Description	Example
$y = mx + b$	**Slope-intercept form** Slope is m. y-intercept is $(0, b)$.	$y = \frac{3}{2}x - 6$
$y - y_1 = m(x - x_1)$	**Point-slope form** Slope is m. Line passes through (x_1, y_1).	$y + 3 = \frac{3}{2}(x - 2)$
$Ax + By = C$ (where A, B, and C are real numbers and A and B are not both 0)	**Standard form** Slope is $-\frac{A}{B}$ $(B \neq 0)$. x-intercept is $\left(\frac{C}{A}, 0\right)$ $(A \neq 0)$. y-intercept is $\left(0, \frac{C}{B}\right)$ $(B \neq 0)$.	$3x - 2y = 12$
$x = a$	**Vertical line** Slope is undefined. x-intercept is $(a, 0)$.	$x = 3$
$y = b$	**Horizontal line** Slope is 0. y-intercept is $(0, b)$.	$y = 3$

NOW TRY ANSWERS

2. (a) $y = -\frac{3}{2}x + 7$
(b) $3x + 2y = 14$

OBJECTIVE 3 Write an equation of a line that fits a data set.

If a given set of data fits a linear pattern—that is, if its graph consists of points lying close to a straight line—we can write a linear equation that models the data.

EXAMPLE 3 Writing an Equation of a Line That Models Data

The table lists average annual cost y (in dollars) of tuition and fees for instate students at public 4-year colleges and universities for selected years, where $x = 1$ represents 2011, $x = 2$ represents 2012, and so on.

Year	x	Cost y (in dollars)
2011	1	8820
2012	2	9080
2013	3	9150
2014	4	9240
2015	5	9500
2016	6	9650

Data from The College Board.

(a) Plot the data and write an equation that approximates it.

We plot the data as shown in **FIGURE 41**.

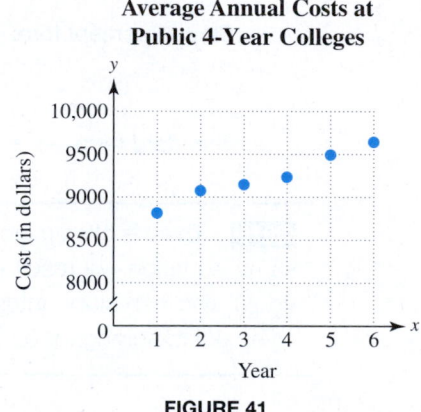

Average Annual Costs at Public 4-Year Colleges

FIGURE 41

The points appear to lie approximately in a straight line. To find an equation of the line, we choose two ordered pairs $(1, 8820)$ and $(6, 9650)$ from the table and determine the slope of the line through these points.

$$m = \frac{y_2 - y_1}{x_2 - x_1} = \frac{9650 - 8820}{6 - 1} = 166 \qquad \text{Let } (6, 9650) = (x_2, y_2) \text{ and } (1, 8820) = (x_1, y_1).$$

The slope, 166, is positive, indicating that tuition and fees *increased* \$166 each year. Now substitute this slope and the point $(1, 8820)$ in the point-slope form to find an equation of the line.

$$y - y_1 = m\,(x - x_1) \qquad \text{Point-slope form}$$

$$y - 8820 = 166(x - 1) \qquad \text{Let } (x_1, y_1) = (1, 8820), m = 166.$$

$$y - 8820 = 166x - 166 \qquad \text{Distributive property}$$

$$y = 166x + 8654 \qquad \text{Add 8820.}$$

Thus, the equation $y = 166x + 8654$ can be used to model the data.

NOW TRY EXERCISE 3

Use the points $(3, 9150)$ and $(5, 9500)$ to write an equation in slope-intercept form that approximates the data of **Example 3.** How well does this equation approximate the cost in 2016?

(b) Use the equation found in part (a) to determine the cost of tuition and fees in 2015.

We let $x = 5$ (for 2015) in the equation from part (a), and solve for y.

$$y = 166x + 8654 \qquad \text{Equation of the line}$$
$$y = 166(5) + 8654 \qquad \text{Substitute 5 for } x.$$
$$y = 9484 \qquad \text{Multiply, and then add.}$$

Using the equation, tuition and fees in 2015 were \$9484. The corresponding value in the table for $x = 15$ is 9500, so the equation approximates the data reasonably well.

NOW TRY

NOW TRY ANSWER

3. $y = 175x + 8625$; The equation gives $y = 9675$ when $x = 6$, which approximates the data reasonably well.

> **NOTE** In **Example 3,** if we had chosen two different data points, we would have obtained a slightly different equation. See **Now Try Exercise 3.**
> Also, we could have used slope-intercept form $y = mx + b$ (instead of point-slope form) to write an equation that models the data.

3.5 Exercises

FOR EXTRA HELP   ▶ **MyLab Math**

▶ *Video solutions for select problems available in MyLab Math*

Concept Check   *Fill in each blank with the correct response.*

1. In point-slope form $y - y_1 = m(x - x_1)$ of the equation of a line, the slope is _____ and the line passes through the point _____.

2. The line with equation $y - 4 = \frac{1}{2}(x + 2)$ has slope _____ and passes through the point _____.

Concept Check   *Work each problem.*

3. Match each form or description in Column I with the corresponding equation in Column II.

I	II
(a) Point-slope form	**A.** $x = a$
(b) Horizontal line	**B.** $y = mx + b$
(c) Slope-intercept form	**C.** $y = b$
(d) Standard form	**D.** $y - y_1 = m(x - x_1)$
(e) Vertical line	**E.** $Ax + By = C$

4. Write the equation $y + 1 = -2(x - 5)$ in slope-intercept form and then in standard form.

5. Which equations are equivalent to $2x - 3y = 6$?

   A. $y = \frac{2}{3}x - 2$ **B.** $-2x + 3y = -6$

   C. $y = -\frac{3}{2}x + 3$ **D.** $y - 2 = \frac{2}{3}(x - 6)$

6. In the summary box following **Example 2,** we give the equations

$$y = \frac{3}{2}x - 6 \quad \text{and} \quad y + 3 = \frac{3}{2}(x - 2)$$

   as examples of equations in slope-intercept form and point-slope form, respectively. Write each of these equations in standard form. What do you notice?

*Write an equation of the line passing through the given point and having the given slope. Give the final answer in slope-intercept form. See **Example 1**.*

7. $(1, 7)$, $m = 5$

8. $(2, 9)$, $m = 6$

9. $(6, -3)$, $m = 1$

10. $(-4, 4)$, $m = 1$

11. $(1, -7)$, $m = -3$

12. $(1, -5)$, $m = -7$

13. $(3, -2)$, $m = -1$

14. $(-5, 4)$, $m = -1$

15. $(-2, 5)$, $m = \dfrac{2}{3}$

16. $(4, 2)$, $m = -\dfrac{1}{3}$

17. $(6, -3)$, $m = -\dfrac{4}{5}$

18. $(7, -2)$, $m = -\dfrac{7}{2}$

19. Concept Check   When asked to write the equation $y = 2x + 4$ in standard form, a student wrote the following equation, which did not agree with the answer in the answer section.

$$-2x + y = 4$$

WHAT WENT WRONG?   Write the equation in standard form as defined in this text.

20. Concept Check   When asked to write the equation $6y = 12 - 3x$ in standard form, a student wrote the following equation, which did not agree with the answer in the answer section.

$$3x + 6y = 12$$

WHAT WENT WRONG?   Write the equation in standard form as defined in this text.

*Write an equation of the line passing through the given pair of points. Give the final answer in **(a)** slope-intercept form and **(b)** standard form. See **Example 2**.*

21. $(4, 10)$ and $(6, 12)$

22. $(8, 5)$ and $(9, 6)$

23. $(-4, 0)$ and $(0, 2)$

24. $(0, -2)$ and $(-3, 0)$

25. $(-2, -1)$ and $(3, -4)$

26. $(-1, -7)$ and $(-8, -2)$

27. $(5, -7)$ and $(-3, 2)$

28. $(-4, 6)$ and $(9, -1)$

29. $\left(-\dfrac{2}{3}, \dfrac{8}{3}\right)$ and $\left(\dfrac{1}{3}, \dfrac{7}{3}\right)$

30. $\left(\dfrac{1}{4}, \dfrac{9}{4}\right)$ and $\left(\dfrac{5}{4}, -\dfrac{1}{4}\right)$

31. $\left(\dfrac{1}{2}, \dfrac{3}{2}\right)$ and $\left(-\dfrac{1}{4}, \dfrac{5}{4}\right)$

32. $\left(\dfrac{2}{3}, -\dfrac{1}{3}\right)$ and $\left(-\dfrac{1}{3}, \dfrac{1}{3}\right)$

*Write an equation of the given line through the given points. Give the final answer in **(a)** slope-intercept form and **(b)** standard form.*

33.
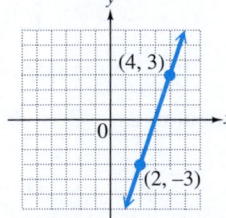

34.
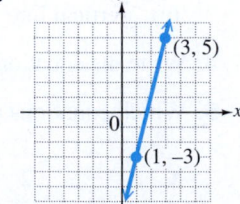

35.
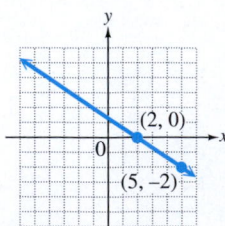

36.
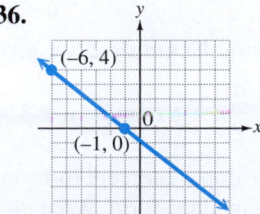

37.
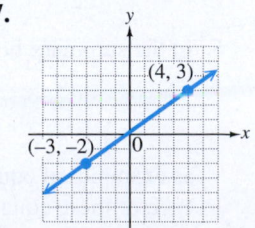

38.
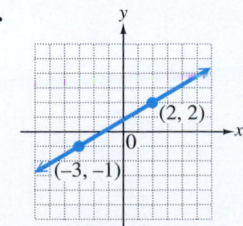

Extending Skills *Write an equation of the line satisfying the given conditions. Give the final answer in slope-intercept form. (Hint: Recall the relationships among slopes of parallel and perpendicular lines.)*

39. Parallel to $5x - y = 10$;
y-intercept $(0, -2)$

40. Parallel to $3x + y = 7$;
y-intercept $(0, 4)$

41. Perpendicular to $x - 2y = 7$;
y-intercept $(0, -3)$

42. Perpendicular to $x + 4y = -5$;
y-intercept $(0, -1)$

43. Passes through $(2, 3)$;
parallel to $4x - y = -2$

44. Passes through $(2, -3)$;
parallel to $3x - 4y = 5$

45. Passes through $(4, 2)$;
perpendicular to $x - 3y = 7$

46. Passes through $(-1, 4)$;
perpendicular to $2x + 3y = 8$

Solve each problem. See Example 3.

47. The table lists the average annual cost y (in dollars) of tuition and fees at 2-year colleges for selected years x, where year 1 represents 2012, year 2 represents 2013, and so on.

Year	x	Cost (in dollars)
2012	1	3310
2013	2	3340
2014	3	3370
2015	4	3460
2016	5	3520

Data from The College Board.

(a) Write five ordered pairs (x, y) for the data.

(b) Plot the ordered pairs (x, y). Do the points lie approximately in a straight line?

(c) Use the ordered pairs $(2, 3340)$ and $(4, 3460)$ to write an equation of a line that approximates the data. Give the final equation in slope-intercept form.

(d) Use the equation from part (c) to estimate the average annual cost at 2-year colleges in 2017 to the nearest dollar. (*Hint:* What is the value of x for 2017?)

48. The table gives the worldwide number y of monthly active Twitter users in millions for selected years, where $x = 1$ represents 2014, $x = 2$ represents 2015, and so on.

Year	x	Twitter Users (in millions)
2014	1	271
2015	2	304
2016	3	313
2017	4	326

Data from Twitter.

(a) Write four ordered pairs (x, y) for the data.

(b) Plot the ordered pairs (x, y). Do the points lie approximately in a straight line?

(c) Use the ordered pairs $(1, 271)$ and $(3, 313)$ to write an equation of a line that approximates the data. Give the final equation in slope-intercept form.

(d) Use the equation from part (c) to estimate the worldwide number of monthly active Twitter users in 2018. (*Hint:* What is the value of x for 2018?)

The points on the graph show the annual average number y of paid music streaming subscriptions (in millions) in the United States for recent years x. Here x = 1 represents 2011, x = 2 represents 2012, and so on. The graph of a linear equation that models the data is also shown.

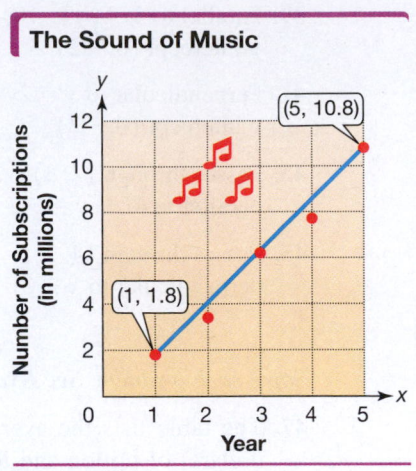

The Sound of Music

Data from RIAA.

49. Use the ordered pairs shown on the graph to write an equation of the line that models the data. Give the final equation in slope-intercept form.

50. Use the equation from **Exercise 49** to do the following.

   (a) Estimate the annual average number of paid music streaming subscriptions in the United States in 2016. (*Hint:* What is the value of x for 2016?)

   (b) How does the answer in part (a) compare to the actual value of 22.6 million subscriptions in 2016?

The points on the graph indicate years of life expected at birth y in the United States for selected years x. Here x = 0 represents 1930, x = 10 represents 1940, and so on. The graph of a linear equation that models the data is also shown.

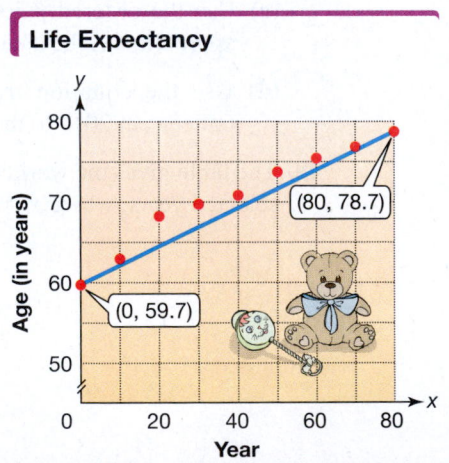

Life Expectancy

Data from National Center for Health Statistics.

51. Use the ordered pairs shown on the graph to write an equation of the line that models the data. Give the final equation in slope-intercept form.

52. Use the equation from **Exercise 51** to do the following.

   (a) Find years of life expected at birth in 2000. (*Hint:* What is the value of x for 2000?) Round the answer to the nearest tenth.

   (b) How does the answer in part (a) compare to the actual value of 76.8 yr?

RELATING CONCEPTS For Individual or Group Work (Exercises 53–60)

If we think of ordered pairs of the form (C, F), then the two most common methods of measuring temperature, Celsius and Fahrenheit, can be related as follows:

When $C = 0, F = 32,$ and when $C = 100, F = 212.$

Work Exercises 53–60 in order.

53. Write two ordered pairs relating these two temperature scales.

54. Find the slope of the line through the two points.

55. Use the point-slope form to find an equation of the line. (Use C and F as variables rather than x and y.)

56. Write an equation for F in terms of C.

57. Use the equation from **Exercise 56** to write an equation for C in terms of F.

58. Use the equation from **Exercise 56** to find the Fahrenheit temperature when $C = 30$.

59. Use the equation from **Exercise 57** to find the Celsius temperature when $F = 50$.

60. For what temperature is $F = C$? (Use the thermometer above to check the answer.)

SUMMARY EXERCISES   Applying Graphing and Equation-Writing Techniques for Lines

1. Concept Check Match the description in Column I with the correct equation in Column II.

I	II
(a) Slope $= -0.5, b = -2$	**A.** $y = -\frac{1}{2}x$
(b) x-intercept $(4, 0)$, y-intercept $(0, 2)$	**B.** $y = -\frac{1}{2}x - 2$
(c) Passes through $(4, -2)$ and $(0, 0)$	**C.** $x - 2y = 2$
(d) $m = \frac{1}{2}$, passes through $(-2, -2)$	**D.** $x + 2y = 4$
	E. $x = 2y$

2. Concept Check Which equations are equivalent to $2x + 5y = 20$?

A. $y = -\frac{2}{5}x + 4$ **B.** $y - 2 = -\frac{2}{5}(x - 5)$

C. $y = \frac{5}{2}x - 4$ **D.** $2x = 5y - 20$

Graph each line, using the given information or equation.

3. $x - 2y = -4$ **4.** $2x + 3y = 12$

5. $m = 1$, y-intercept $(0, -2)$ **6.** $y - 4 = -9$

7. $m = -\frac{2}{3}$, passes through $(3, -4)$ **8.** $4x = 3y + 12$

9. $x - 4y = 0$ **10.** $m = -\frac{3}{4}$, passes through $(4, -4)$

11. $5x + 2y = 10$ **12.** $x + 5y = 0$

13. $x + 4 = 0$ **14.** $y = -x + 6$

Write an equation in slope-intercept form of each line represented by the table of ordered pairs or the graph.

15.

x	y
3	0
1	4
−1	8

16.

x	y
−6	0
0	8
3	12

17.

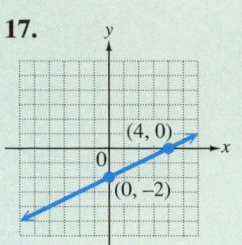

18.

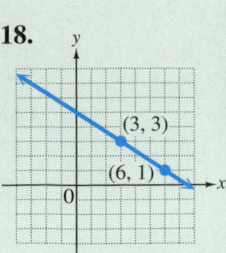

Write an equation of each line. Give the final answer in slope-intercept form if possible.

19. $m = -3, b = -6$

20. $m = \frac{3}{2}$, passes through $(-4, 6)$

21. Passes through $(1, -7)$ and $(-2, 5)$

22. Passes through $(0, 0)$, undefined slope

23. Passes through $(0, 0)$ and $(3, 2)$

24. $m = -1, b = -4$

25. Passes through $(5, 0)$ and $(0, -5)$

26. Passes through $(0, 0)$, $m = 0$

27. $m = \frac{5}{3}$, passes through $(-3, 0)$

28. Passes through $(1, -13)$ and $(-2, 2)$

Chapter 3   Summary

STUDY SKILLS REMINDER

Be prepared for your math test on this chapter. **Review Study Skills 7 and 8,** *Reviewing a Chapter* **and** *Taking Math Tests.*

Key Terms

3.1

line graph
linear equation in two
    variables
ordered pair
table of values
x-axis
y-axis

origin
rectangular (Cartesian)
    coordinate system
quadrant
plane
coordinates
plot
scatter diagram

3.2

graph, graphing
x-intercept
y-intercept
horizontal line
vertical line

3.3

rise
run
slope
parallel lines
perpendicular lines

New Symbols

(x, y) ordered pair

m slope

(x_1, y_1) subscript notation
        (read "x-sub-one,
        y-sub-one")

Test Your Word Power

See how well you have learned the vocabulary in this chapter.

1. An **ordered pair** is a pair of numbers written
   A. in numerical order between brackets
   B. between parentheses or brackets
   C. between parentheses in which order is important
   D. between parentheses in which order does not matter.

2. An **intercept** is
   A. the point where the *x*-axis and *y*-axis intersect
   B. a pair of numbers written in parentheses in which order matters

   C. one of the four regions determined by a rectangular coordinate system
   D. the point where a graph intersects the *x*-axis or the *y*-axis.

3. The **slope** of a line is
   A. the measure of the run over the rise of the line
   B. the distance between two points on the line
   C. the ratio of the change in *y* to the change in *x* along the line
   D. the horizontal change compared to the vertical change of two points on the line.

4. Two lines in a plane are **parallel** if
   A. they represent the same line
   B. they never intersect
   C. they intersect at a 90° angle
   D. one has a positive slope and one has a negative slope.

5. Two lines in a plane are **perpendicular** if
   A. they represent the same line
   B. they never intersect
   C. they intersect at a 90° angle
   D. one has a positive slope and one has a negative slope.

ANSWERS

1. C; *Examples:* $(0, 3)(-3, 8)$, $(4, 0)$ 2. D; *Example:* The graph of the equation $4x - 3y = 12$ has *x*-intercept $(3, 0)$ and *y*-intercept $(0, -4)$.   3. C; *Example:* The line through $(3, 6)$ and $(5, 4)$ has slope $\frac{4-6}{5-3} = \frac{-2}{2} = -1$.   4. B; *Example:* See **FIGURE A**.   5. C; *Example:* See **FIGURE B**.

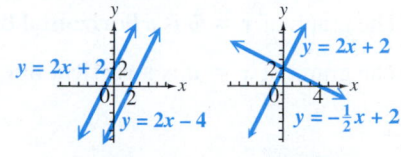

FIGURE A   FIGURE B

Quick Review

CONCEPTS	EXAMPLES

3.1 Linear Equations and Rectangular Coordinates

An ordered pair is a solution of an equation if it makes the equation a true statement.

Are $(2, -5)$ and $(0, -6)$ solutions of $4x - 3y = 18$?

$4(2) - 3(-5) \overset{?}{=} 18$

$8 + 15 \overset{?}{=} 18$

$23 = 18$ False

$(2, -5)$ is not a solution.

$4(0) - 3(-6) \overset{?}{=} 18$

$0 + 18 \overset{?}{=} 18$

$18 = 18$ ✓ True

$(0, -6)$ is a solution.

If a value of either variable in an equation is given, then the value of the other variable can be found by substitution.

Complete the ordered pair $(0, \underline{\quad})$ for the given equation.

$$3x = y + 4$$

$$3(0) = y + 4 \quad \text{Let } x = 0.$$

$$0 = y + 4 \quad \text{Multiply.}$$

$$-4 = y \quad \text{Subtract 4.}$$

The ordered pair is $(0, -4)$.

To plot an ordered pair, begin at the origin.

Step 1 Move right or left the number of units corresponding to the *x*-coordinate—right if it is positive or left if it is negative.

Step 2 Then turn and move up or down the number of units corresponding to the *y*-coordinate—up if it is positive or down if it is negative.

Plot the ordered pair $(-3, 4)$.

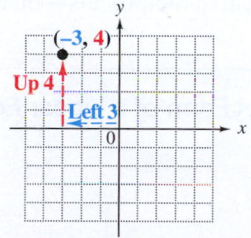

CONCEPTS	**EXAMPLES**

3.2 Graphing Linear Equations in Two Variables

Graphing a Linear Equation in Two Variables

Step 1 Find at least two ordered pairs that are solutions of the equation. (The intercepts are good choices.) It is good practice to find a third ordered pair as a check.

Step 2 Plot the corresponding points.

Step 3 Draw a straight line through the points.

Graph $x - 2y = 4$.

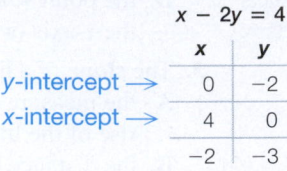

$x - 2y = 4$	
x	**y**
0	−2
4	0
−2	−3

y-intercept → (0, −2)
x-intercept → (4, 0)

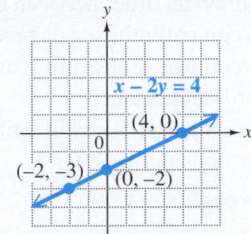

The graph of $Ax + By = 0$ passes through the origin. In this case, find and plot at least one other point that satisfies the equation. Then draw the line through these points.

The graph of $y = b$ is a **horizontal line** through $(0, b)$.

The graph of $x = a$ is a **vertical line** through $(a, 0)$.

Graph $2x + 3y = 0$.

$2x + 3y = 0$	
x	**y**
−3	2
0	0
3	−2

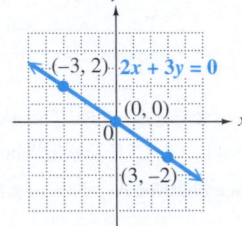

Graph $y = -3$ and $x = -3$.

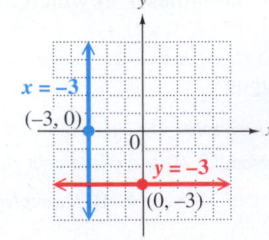

3.3 The Slope of a Line

The slope m of the line passing through the points (x_1, y_1) and (x_2, y_2) is defined as follows.

$$m = \frac{\text{change in } y}{\text{change in } x} = \frac{y_2 - y_1}{x_2 - x_1} \quad (\text{where } x_1 \neq x_2)$$

Find the slope m of the line passing through the points $(-2, 3)$ and $(4, -5)$.

$$m = \frac{-5 - 3}{4 - (-2)} = \frac{-8}{6} = -\frac{4}{3}$$

A horizontal line has slope **0**.

A vertical line has **undefined slope**.

The line $y = -2$ has slope 0.

The line $x = 4$ has undefined slope.

Finding the Slope of a Line from Its Equation

Step 1 Solve the equation for y.

Step 2 The slope is given by the coefficient of x.

Find the slope of the line with the following equation.

$$3x - 4y = 12$$
$$-4y = -3x + 12 \qquad \text{Subtract } 3x.$$
$$y = \frac{3}{4}x - 3 \qquad \text{Divide by } -4.$$

↑ Slope

Parallel lines have the same slope.

The lines $y = 3x - 1$ and $y = 3x + 4$ are parallel because both have slope 3.

Perpendicular lines (neither of which is vertical) have slopes that are negative reciprocals—that is, their product is −1.

The lines $y = -3x - 1$ and $y = \frac{1}{3}x + 4$ are perpendicular because their slopes are −3 and $\frac{1}{3}$, and $-3\left(\frac{1}{3}\right) = -1$.

3.4 Slope-Intercept Form of a Linear Equation

Slope-Intercept Form

$y = mx + b$

m is the slope.

$(0, b)$ is the y-intercept.

Write an equation in slope-intercept form of the line with slope 2 and y-intercept $(0, -5)$.

$$y = mx + b \qquad \text{Slope-intercept form}$$
$$y = 2x - 5 \qquad \text{Let } m = 2 \text{ and } b = -5.$$

CONCEPTS	EXAMPLES

3.5 Point-Slope Form of a Linear Equation and Modeling

Point-Slope Form

$y - y_1 = m(x - x_1)$

m is the slope.

(x_1, y_1) is a point on the line.

Write an equation of the line passing through $(-4, 5)$ with slope $-\frac{1}{2}$.

$$y - y_1 = m(x - x_1) \qquad \text{Point-slope form}$$

$$y - 5 = -\frac{1}{2}[x - (-4)] \qquad \text{Substitute } y_1, m, \text{ and } x_1.$$

$$y - 5 = -\frac{1}{2}(x + 4) \qquad \text{Definition of subtraction}$$

$$y - 5 = -\frac{1}{2}x - 2 \qquad \text{Distributive property}$$

$$y = -\frac{1}{2}x + 3 \qquad \text{Add 5.}$$

Standard Form

$Ax + By = C$

A, B, and C are real numbers and A and B are not both 0. (In answers, we give A, B, and C as integers with greatest common factor 1 and $A \geq 0$.)

Write the equation $y = -\frac{1}{2}x + 3$ in standard form.

$$y = -\frac{1}{2}x + 3$$

$$2y = -x + 6 \qquad \text{Multiply each term by 2.}$$

$$x + 2y = 6 \qquad \text{Add } x.$$

Chapter 3   Review Exercises

3.1 *The line graph shows the number, in millions, of real Christmas trees purchased for the years 2011 through 2016.*

1. Between which years did the number of real trees purchased increase?

2. Between which years did the number of real trees purchased remain the same?

3. Estimate the number of real trees purchased in 2013 and 2014.

4. By about how much did the number of real trees purchased between 2013 and 2014 decrease?

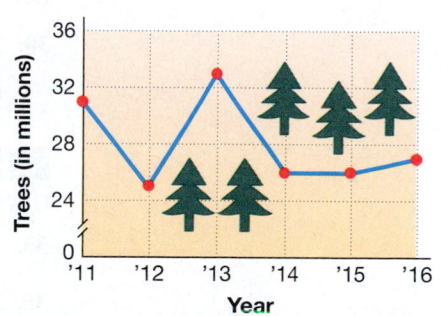

Purchases of Real Christmas Trees

Data from National Christmas Tree Association.

Complete the given ordered pairs for each equation.

5. $y = 3x + 2$; $(-1, \underline{\quad}), (0, \underline{\quad}), (\underline{\quad}, 5)$

6. $4x + 3y = 6$; $(0, \underline{\quad}), (\underline{\quad}, 0), (-2, \underline{\quad})$

7. $x = 3y$; $(0, \underline{\quad}), (8, \underline{\quad}), (\underline{\quad}, -3),$

8. $x - 7 = 0$; $(\underline{\quad}, -3), (\underline{\quad}, 0), (\underline{\quad}, 5)$

Determine whether each ordered pair is a solution of the given equation.

9. $x + y = 7$; $(2, 5)$

10. $2x + y = 5$; $(-1, 3)$

11. $3x - y = 4$; $\left(\frac{1}{3}, -3\right)$

12. $x = -1$; $(0, -1)$

Identify the quadrant in which each point is located. Then plot and label each point in a rectangular coordinate system.

13. $(2, 3)$

14. $(-4, 2)$

15. $(3, 0)$

16. $(0, -6)$

3.2 *Find the x- and y-intercepts for the graph of each equation. Then draw the graph.*

17. $y = 2x + 5$

18. $3x + 2y = 8$

19. $x + 2y = -4$

20. $x = -6$

3.3 *Find the slope of each line. (In Exercises 26 and 27, coordinates of the points shown are integers.)*

21. The line passing through $(2, 3)$ and $(-4, 6)$

22. The line passing through $(2, 5)$ and $(2, 8)$

23. $y = 3x - 4$

24. $y = 5$

25. $x = -7$

26.

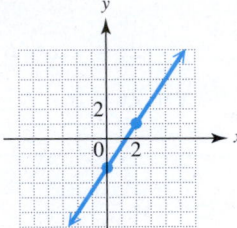

27.

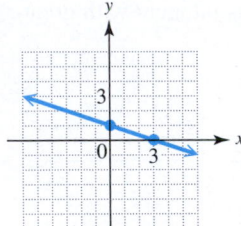

28. The line passing through these points

x	y
0	1
2	4
6	10

29. Find each slope.

    (a) A line whose graph is parallel to the graph of $y = 2x + 3$

    (b) A line whose graph is perpendicular to the graph of $y = -3x + 3$

Determine whether each pair of lines is parallel, perpendicular, *or* neither.

30. $3x + 2y = 6$

    $6x + 4y = 8$

31. $x - 3y = 1$

    $3x + y = 4$

32. $x - 2y = 8$

    $x + 2y = 8$

3.4, 3.5 *Write an equation of each line. Give the final answer in slope-intercept form (if possible).*

33. The line with $m = -1$, $b = \frac{2}{3}$

34. The line passing through $(2, 3)$ and $(-4, 6)$

35. The line passing through $(4, -3)$, $m = 1$

36. The line passing through $(-1, 4)$, $m = \frac{2}{3}$

37. The line passing through $(1, -1)$, $m = -\frac{3}{4}$

38. The line with $m = -\frac{1}{4}$, $b = \frac{3}{2}$

39. The line passing through $(-4, 1)$, slope 0

40. The line passing through $(-4, 1)$, undefined slope

41. The line graphed in **Exercise 26**

42. The line graphed in **Exercise 27**

Chapter 3   Mixed Review Exercises

Match each statement to the appropriate graph or graphs in choices A–D. Graphs may be used more than once.

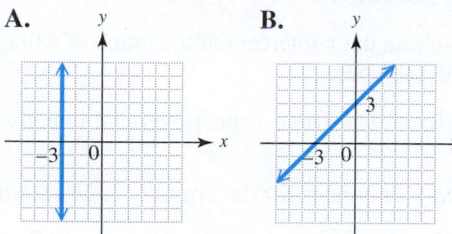

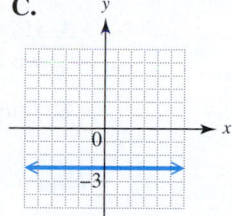

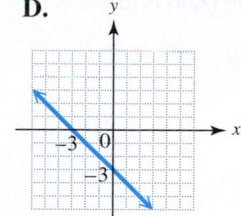

1. The line shown in the graph has undefined slope.

2. The graph of the equation has *y*-intercept $(0, -3)$.

3. The graph of the equation has *x*-intercept $(-3, 0)$.

4. The line shown in the graph has negative slope.

5. The graph is that of the equation $y = -3$.

6. The line shown in the graph has slope 1.

Find the x- and y-intercepts and the slope of each line. Then graph the line.

7. $y = -2x - 5$

8. $x + 3y = 0$

9. $y - 5 = 0$

10. $x = -5$

*Write an equation of each line. Give the final answer in **(a)** slope-intercept form and **(b)** standard form.*

11. The line with $m = -\frac{1}{4}$, $b = -\frac{5}{4}$

12. The line passing through $(8, 6)$, $m = -3$

13. The line passing through $(3, -5)$ and $(-4, -1)$

14. The line passing through $(5, -5)$, slope 0

The points on the graph show the number y of members of the professional social networking service LinkedIn (in millions) for selected years x. Here x = 0 represents 2010, x = 1 represents 2011, and so on. The graph of a linear equation that models the data is also shown.

15. Because the points of the graph lie approximately in a linear pattern, a straight line can be used to model the data. Will this line have positive or negative slope? Explain.

16. Write two ordered pairs (x, y) for the data for 2010 and 2016.

17. Use the two ordered pairs from **Exercise 16** to write an equation of a line that models the data. Give the final equation in slope-intercept form. Round the slope to the nearest tenth.

18. Use the equation from **Exercise 17** to approximate the number of LinkedIn members in 2014 to the nearest million. (What is the value of *x* for 2014?) How does the answer compare to the actual value of 332 million?

LinkedIn Members

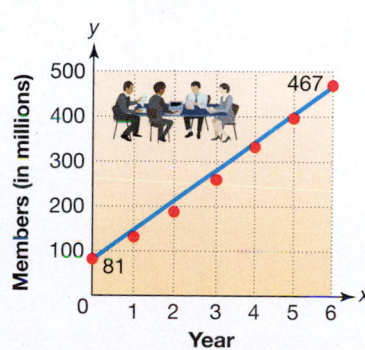

Data from LinkedIn.

| Chapter 3 | Test | FOR EXTRA HELP | *Step-by-step test solutions are found on the Chapter Test Prep Videos available in* **MyLab Math**. |

▶ *View the complete solutions to all Chapter Test exercises in MyLab Math.*

1. Complete the ordered pairs $(0, \underline{\quad}), (\underline{\quad}, 0), (\underline{\quad}, -3)$ for the equation $3x + 5y = -30$.

2. Is $(4, -1)$ a solution of $4x - 7y = 9$?

3. How do we find the x-intercept of the graph of a linear equation in two variables? How do we find the y-intercept?

4. *True* or *false:* The x-axis is the horizontal line given by the equation $y = 0$.

Find the x- and y-intercepts for the graph of each equation. Then draw the graph.

5. $3x + y = 6$ 6. $y - 2x = 0$ 7. $x + 3 = 0$ 8. $y = 1$

9. Give the slope and y-intercept of the graph of $y = x - 4$. Use them to graph the equation.

10. Graph the line passing through the point $(-2, 3)$, with slope $-\frac{1}{2}$.

Find the slope of each line. (In Exercise 13, coordinates of the points shown are integers.)

11. The line passing through $(-4, 6)$ and $(-1, -2)$

12. $2x + y = 10$

13.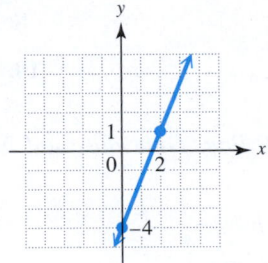

14. A line whose graph is parallel to the graph of $y - 4 = 6$

15. $x + 12 = 0$

Write an equation of each line. Give the final answer in slope-intercept form.

16. The line passing through $(-1, 4)$, $m = 2$ 17. The line graphed in **Exercise 13**

18. The line passing through $(2, -6)$ and $(1, 3)$

19. The line with x-intercept $(3, 0)$, y-intercept $\left(0, \frac{9}{2}\right)$

The graph shows worldwide snowmobile sales y for selected years x, where x = 0 represents 2000, x = 1 represents 2001, and so on. Use the graph to work each problem.

20. Is the slope of the line in the graph positive or negative? Explain.

21. Two ordered pairs (x, y) are shown on the graph.

    (a) Use the ordered pairs to find the slope of the line to the nearest tenth.

    (b) Write an equation of a line that models the data. Give the final equation in slope-intercept form.

    (c) Use the equation from part (b) to approximate worldwide snowmobile sales for 2016. How does the answer compare to the actual sales of 127.0 thousand?

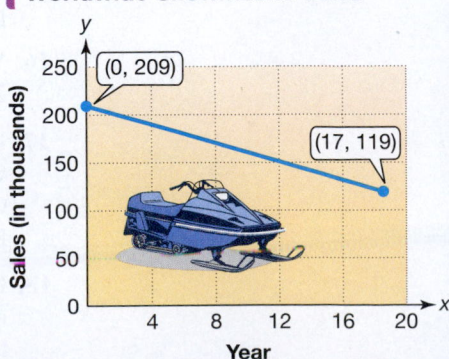

Worldwide Snowmobile Sales

Data from www.snowmobile.org

22. What does the ordered pair $(15, 129.5)$ mean in the context of this problem?

1. For the decimal number 135.264, round to the place value indicated.

    (a) hundredths   **(b)** tenths   **(c)** ones or units   **(d)** tens

Perform each indicated operation.

2. $\dfrac{2}{9} + \dfrac{5}{6}$ **3.** $10\dfrac{5}{8} - 3\dfrac{1}{10}$ **4.** $\dfrac{3}{4} \div \dfrac{1}{8}$

5. $8.3 + 2.09$ **6.** 0.2×0.08 **7.** $530.26 \div 100$

8. $5 - (-4) + (-2)$ **9.** $\dfrac{-13(-5)}{41 - 6^2}$ **10.** $\dfrac{(-3)^2 - (-4)(2^4)}{5(2) - (-2)^3}$

11. *True* or *false?* $\dfrac{4(3 - 9)}{2 - 6} \geq 6$

12. Find the value of $xz^3 - 5y^2$ for $x = -2$, $y = -3$, and $z = -1$.

13. What property does $3(-2 + x) = -6 + 3x$ illustrate?

14. Simplify $-4p - 6 + 3p + 8$ by combining like terms.

Solve.

15. $V = \dfrac{1}{3}\pi r^2 h$ for h **16.** $6 - 3(1 + x) = 2(x + 5) - 2$

17. $-(m - 3) = 5 - 2m$ **18.** $\dfrac{x - 2}{3} = \dfrac{2x + 1}{5}$

Solve each inequality. Graph the solution set, and write it using interval notation.

19. $-2.5x < 6.5$ **20.** $4(x + 3) - 5x < 12$ **21.** $\dfrac{2}{3}x - \dfrac{1}{6}x \leq -2$

Solve each problem.

22. Find the measure of each marked angle.

    (a)
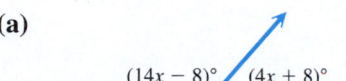
$(14x - 8)°$ $(4x + 8)°$

    (b)
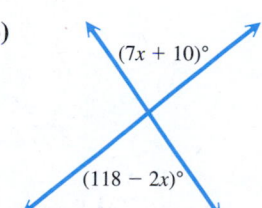
$(7x + 10)°$

$(118 - 2x)°$

In 2016, a total of 2,014,220 electric cars were on the road worldwide. The circle graph shows the distribution of these electric cars by country.

23. How many electric cars could be found in China in 2016? Round to the nearest whole number.

24. How many electric cars could be found in the Netherlands in 2016? Round to the nearest whole number.

Global Electric Car Distribution

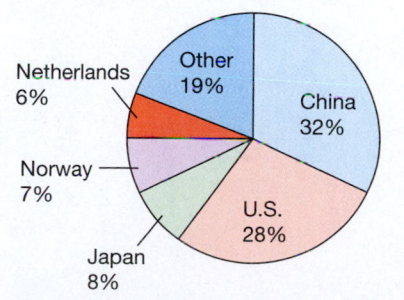

Netherlands 6%

Other 19%

China 32%

Norway 7%

U.S. 28%

Japan 8%

Data from International Energy Agency.

Consider the linear equation $-3x + 4y = 12$*. Find the following.*

25. The x- and y-intercepts   **26.** The slope   **27.** The graph

28. Are the lines with equations $x + 5y = -6$ and $y = 5x - 8$ *parallel, perpendicular,* or *neither*?

Write an equation of each line. Give the final answer in slope-intercept form.

29. The line passing through $(2, -5)$, slope 3

30. The line passing through $(0, 4)$ and $(2, 4)$

STUDY SKILLS REMINDER

It is not too soon to begin preparing for your final exam. **Review Study Skill 10, *Preparing for Your Math Final Exam.***

EXPONENTS AND POLYNOMIALS

Exponents and *scientific notation,* two of the topics of this chapter, are often used to express very large or very small numbers. Using this notation, one light-year, which is about 6 trillion miles, is written

$$6 \times 10^{12}.$$

4.1 The Product Rule and Power Rules for Exponents

OBJECTIVES

1 Use exponents.
2 Use the product rule for exponents.
3 Use the rule $(a^m)^n = a^{mn}$.
4 Use the rule $(ab)^m = a^m b^m$.
5 Use the rule $\left(\dfrac{a}{b}\right)^m = \dfrac{a^m}{b^m}$.
6 Use combinations of the rules for exponents.
7 Apply the rules for exponents in a geometry application.

VOCABULARY

☐ base
☐ exponent (power)
☐ exponential expression

NOW TRY
EXERCISE 1
Write $4 \cdot 4 \cdot 4$ in exponential form and evaluate.

NOW TRY
EXERCISE 2
Identify the base and the exponent of each expression. Then evaluate.
(a) 3^4 **(b)** -3^4 **(c)** $(-3)^4$

NOW TRY ANSWERS
1. 4^3; 64
2. **(a)** 3; 4; 81 **(b)** 3; 4; −81
   (c) −3; 4; 81

OBJECTIVE 1 Use exponents.

Recall from earlier work that in the expression 5^2, the number 5 is the **base** and 2 is the **exponent,** or **power.** The expression 5^2 is an **exponential expression.** Although we do not usually write the exponent when it is 1, in general

$$a^1 = a, \quad \text{for any quantity } a.$$

EXAMPLE 1 Using Exponents

Write $3 \cdot 3 \cdot 3 \cdot 3$ in exponential form and evaluate.

Because 3 occurs as a factor four times, the base is 3 and the exponent is 4. The exponential expression is 3^4, read "3 *to the fourth power*" or simply "3 *to the fourth.*"

$$\underbrace{3 \cdot 3 \cdot 3 \cdot 3}_{\text{4 factors of 3}} \quad \text{means} \quad 3^4, \quad \text{which equals} \quad 81.$$

NOW TRY

EXAMPLE 2 Evaluating Exponential Expressions

Identify the base and the exponent of each expression. Then evaluate.

Expression	Base	Exponent	Value
(a) 5^4	5	4	$5 \cdot 5 \cdot 5 \cdot 5$, which equals 625
(b) -5^4	5	4	$-1 \cdot (5 \cdot 5 \cdot 5 \cdot 5)$, which equals −625
(c) $(-5)^4$	−5	4	$(-5)(-5)(-5)(-5)$, which equals 625

NOW TRY

⚠ CAUTION Compare Examples 2(b) and 2(c). In -5^4, the absence of parentheses means that the exponent 4 applies only to the base 5, not −5. In $(-5)^4$, the parentheses mean that the exponent 4 applies to the base −5.

In summary, $-a^n$ and $(-a)^n$ are not necessarily the same.

Expression	Base	Exponent	Example
$-a^n$	a	n	$-3^2 = -(3 \cdot 3) = -9$
$(-a)^n$	$-a$	n	$(-3)^2 = (-3)(-3) = 9$

OBJECTIVE 2 Use the product rule for exponents.

To develop the product rule, we use the definition of exponents.

$$2^4 \cdot 2^3$$

$$= \underbrace{(2 \cdot 2 \cdot 2 \cdot 2)}_{\text{4 factors}} \underbrace{(2 \cdot 2 \cdot 2)}_{\text{3 factors}}$$

$$= \underbrace{2 \cdot 2 \cdot 2 \cdot 2 \cdot 2 \cdot 2 \cdot 2}_{4 + 3 = 7 \text{ factors}}$$

$$= 2^7$$

Also,
$$6^2 \cdot 6^3$$
$$= (6 \cdot 6)(6 \cdot 6 \cdot 6)$$
$$= 6 \cdot 6 \cdot 6 \cdot 6 \cdot 6$$
$$= 6^5.$$

Generalizing from these examples, we have the following.

$2^4 \cdot 2^3$ is equal to 2^{4+3}, which equals 2^7.

$6^2 \cdot 6^3$ is equal to 6^{2+3}, which equals 6^5.

In each case, adding the exponents gives the exponent of the product, suggesting the **product rule for exponents.**

Product Rule for Exponents

For any positive integers m and n, $a^m \cdot a^n = a^{m+n}$.
(Keep the same base and add the exponents.)

Example: $6^2 \cdot 6^5 = 6^{2+5} = 6^7$

! CAUTION When using the product rule, *keep the same base and add the exponents.* Do **not** multiply the bases.

$6^2 \cdot 6^5$ is equal to 6^7, *not* 36^7.

NOW TRY EXERCISE 3

Use the product rule for exponents to simplify each expression, if possible.

(a) $y^2 \cdot y \cdot y^5$

(b) $(2x^3)(4x^6)$

(c) $2^4 \cdot 5^3$

(d) $3^2 + 3^3$

EXAMPLE 3 Using the Product Rule

Use the product rule for exponents to simplify each expression, if possible.

(a) $6^3 \cdot 6^5$
$= 6^{3+5}$ Product rule
$= 6^8$ Add the exponents.

(b) $x^2 \cdot x$
$= x^2 \cdot x^1$ $a = a^1$, for all a.
$= x^{2+1}$ Product rule
$= x^3$ Add the exponents.

(c) $m^4 \cdot m^3 \cdot m^5$ — This can be written as $m^4 m^3 m^5$.
$= m^{4+3+5}$ Product rule
$= m^{12}$ Add the exponents.

(d) $2^3 \cdot 3^2$ — Think: 2^3 means $2 \cdot 2 \cdot 2$. Think: 3^2 means $3 \cdot 3$.   The product rule does not apply. *The bases are different.*
$= 8 \cdot 9$ Evaluate 2^3 and 3^2.
$= 72$ Multiply.

(e) $2^3 + 2^4$ The product rule does not apply. *This is a sum, not a product.*
$= 8 + 16$ Evaluate 2^3 and 2^4.
$= 24$ Add.

(f) $(2x^3)(3x^7)$ — $2x^3$ means $2 \cdot x^3$ and $3x^7$ means $3 \cdot x^7$.
$= (2 \cdot 3) \cdot (x^3 \cdot x^7)$ Commutative and associative properties
$= 6x^{3+7}$ Multiply; product rule
$= 6x^{10}$ Add the exponents.   NOW TRY

NOW TRY ANSWERS
3. (a) y^8 (b) $8x^9$
(c) The product rule does not apply; 2000
(d) The product rule does not apply; 36

! **CAUTION**   Note the important difference between *adding* and *multiplying* exponential expressions.

$$8x^3 + 5x^3 \quad \text{means} \quad (8+5)x^3, \quad \text{which equals} \quad 13x^3.$$

$$(8x^3)(5x^3) \quad \text{means} \quad (8 \cdot 5)x^{3+3}, \quad \text{which equals} \quad 40x^6.$$

OBJECTIVE 3 Use the rule $(a^m)^n = a^{mn}$.

We can simplify an expression such as $(5^2)^4$ with the product rule for exponents, as follows.

$$(5^2)^4$$

$= 5^2 \cdot 5^2 \cdot 5^2 \cdot 5^2$	Definition of exponent
$= 5^{2+2+2+2}$	Product rule
$= 5^8$	Add.

Observe that $2 \cdot 4 = 8$. This example suggests **power rule (a) for exponents.**

> ### Power Rule (a) for Exponents
>
> For any positive integers m and n, $(a^m)^n = a^{mn}$.
> (Raise a power to a power by multiplying exponents.)
>
> *Example:* $(3^4)^2 = 3^{4 \cdot 2} = 3^8$

**NOW TRY
EXERCISE 4**

Use power rule (a) for exponents to simplify.

(a) $(4^7)^5$ **(b)** $(y^4)^7$

EXAMPLE 4   Using Power Rule (a)

Use power rule (a) for exponents to simplify.

(a) $(2^5)^3$

$= 2^{5 \cdot 3}$

$= 2^{15}$

(b) $(5^7)^2$

$= 5^{7 \cdot 2}$

$= 5^{14}$

(c) $(x^2)^5$

$= x^{2 \cdot 5}$ Power rule (a)

$= x^{10}$ Multiply.

NOW TRY

OBJECTIVE 4 Use the rule $(ab)^m = a^m b^m$.

Consider the following.

$$(4x)^3$$

$= (4x)(4x)(4x)$	Definition of exponent
$= (4 \cdot 4 \cdot 4)(x \cdot x \cdot x)$	Commutative and associative properties
$= 4^3 x^3$	Definition of exponent

This example suggests **power rule (b) for exponents.**

> ### Power Rule (b) for Exponents
>
> For any positive integer m, $(ab)^m = a^m b^m$.
> (Raise a product to a power by raising each factor to the power.)
>
> *Example:* $(2p)^5 = 2^5 p^5$

NOW TRY ANSWERS
4. **(a)** 4^{35} **(b)** y^{28}

NOW TRY EXERCISE 5

Use power rule (b) for exponents to simplify.

(a) $(-5ab)^3$ (b) $(4t^3p^5)^2$
(c) $(-3)^6$

EXAMPLE 5 Using Power Rule (b)

Use power rule (b) for exponents to simplify.

(a) $(3xy)^2$
$= 3^2x^2y^2$ Power rule (b)
$= 9x^2y^2$ $3^2 = 3 \cdot 3 = 9$

(b) $3(xy)^2$
$= 3(x^2y^2)$ Power rule (b)
$= 3x^2y^2$ Multiply.

(c) $(2m^2p^3)^4$
$= 2^4(m^2)^4(p^3)^4$ Power rule (b)
$= 2^4m^8p^{12}$ Power rule (a)
$= 16m^8p^{12}$ $2^4 = 2 \cdot 2 \cdot 2 \cdot 2 = 16$

Compare parts (a) and (b). Pay attention to the use of parentheses.

(d) $(-5^6)^3$
$= (-1 \cdot 5^6)^3$ $-a = -1 \cdot a$
$= (-1)^3 \cdot (5^6)^3$ Power rule (b)
$= -1 \cdot 5^{18}$ Power rule (a)
$= -5^{18}$ Multiply.

Raise -1 to the designated power.

(e) $(-4)^8$
$= (-1 \cdot 4)^8$ $-a = -1 \cdot a$
$= (-1)^8(4^8)$ Power rule (b)
$= 1(4^8)$ $(-1)^8 = 1$
$= 4^8$ $1 \cdot a = a$

NOW TRY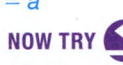

> **! CAUTION** *Power rule (b) does not apply to a sum.*
> $$(4x)^2 = 4^2x^2, \quad \text{but} \quad (4+x)^2 \neq 4^2 + x^2.$$

OBJECTIVE 5 Use the rule $\left(\frac{a}{b}\right)^m = \frac{a^m}{b^m}$.

Consider the following.

$$\left(\frac{2}{3}\right)^4 = \left(\frac{2}{3}\right)\left(\frac{2}{3}\right)\left(\frac{2}{3}\right)\left(\frac{2}{3}\right) \quad \text{Definition of exponent}$$
$$= \frac{2 \cdot 2 \cdot 2 \cdot 2}{3 \cdot 3 \cdot 3 \cdot 3} \quad \text{Multiply fractions.}$$
$$= \frac{2^4}{3^4} \quad \text{Definition of exponent}$$

This example suggests **power rule (c) for exponents.**

Power Rule (c) for Exponents

For any positive integer m, $\left(\frac{a}{b}\right)^m = \frac{a^m}{b^m}$ (where $b \neq 0$).

(Raise a quotient to a power by raising both numerator and denominator to that power.)

Example: $\left(\frac{5}{3}\right)^2 = \frac{5^2}{3^2}$

NOW TRY ANSWERS
5. (a) $-125a^3b^3$ (b) $16t^6p^{10}$
   (c) 3^6

NOW TRY EXERCISE 6

Use power rule (c) for exponents to simplify.

(a) $\left(\dfrac{3}{4}\right)^3$ **(b)** $\left(\dfrac{p}{q}\right)^5$

$(q \neq 0)$

EXAMPLE 6 Using Power Rule (c)

Use power rule (c) for exponents to simplify.

(a) $\left(\dfrac{2}{3}\right)^5$

$= \dfrac{2^5}{3^5}$ Power rule (c)

$= \dfrac{32}{243}$ Simplify.

(b) $\left(\dfrac{1}{5}\right)^4$

$= \dfrac{1^4}{5^4}$

$= \dfrac{1}{625}$

(c) $\left(\dfrac{m}{n}\right)^4$

$= \dfrac{m^4}{n^4}$

(where $n \neq 0$)

NOW TRY

NOTE In **Example 6(b),** we used the fact that $1^4 = 1$ because $1 \cdot 1 \cdot 1 \cdot 1 = 1$.

In general, $1^n = 1$, *for any integer n.*

Rules for Exponents

For positive integers m and n, the following hold true.

			Examples
Product rule	$a^m \cdot a^n = a^{m+n}$		$6^2 \cdot 6^5 = 6^{2+5} = 6^7$
Power rules (a)	$(a^m)^n = a^{mn}$		$(3^4)^2 = 3^{4 \cdot 2} = 3^8$
(b)	$(ab)^m = a^m b^m$		$(2p)^5 = 2^5 p^5$
(c)	$\left(\dfrac{a}{b}\right)^m = \dfrac{a^m}{b^m}$ (where $b \neq 0$)		$\left(\dfrac{5}{3}\right)^2 = \dfrac{5^2}{3^2}$

OBJECTIVE 6 Use combinations of the rules for exponents.

EXAMPLE 7 Using Combinations of the Rules

Simplify.

(a) $\left(\dfrac{2}{3}\right)^2 \cdot 2^3$

$= \dfrac{2^2}{3^2} \cdot \dfrac{2^3}{1}$ Power rule (c)

$= \dfrac{2^2 \cdot 2^3}{3^2 \cdot 1}$ Multiply fractions.

$= \dfrac{2^{2+3}}{3^2}$ Product rule

$= \dfrac{2^5}{3^2}$ Add.

$= \dfrac{32}{9}$ Apply the exponents.

(b) $(5x)^3(5x)^4$

$= (5x)^7$ Product rule

$= 5^7 x^7$ Power rule (b)

Another correct way to simplify this expression follows.

$(5x)^3(5x)^4$ **Alternative solution**

$= 5^3 x^3 5^4 x^4$ Power rule (b)

$= 5^3 \cdot 5^4 \cdot x^3 x^4$ Commutative property

$= 5^{3+4} x^{3+4}$ Product rule

$= 5^7 x^7$ Add the exponents.

NOW TRY ANSWERS

6. (a) $\dfrac{27}{64}$ **(b)** $\dfrac{p^5}{q^5}$

**NOW TRY
EXERCISE 7**

Simplify.

(a) $\left(\dfrac{3}{5}\right)^3 \cdot 3^2$ **(b)** $(8k)^5(8k)^4$

(c) $(x^4y)^5(-2x^2y^5)^3$

(c) $(2x^2y^3)^4(3xy^2)^3$

$= 2^4(x^2)^4(y^3)^4 \cdot 3^3x^3(y^2)^3$ Power rule (b)

$= 2^4x^8y^{12} \cdot 3^3x^3y^6$ Power rule (a)

$= 2^4 \cdot 3^3 \cdot x^8x^3y^{12}y^6$ Commutative and associative properties

$= 16 \cdot 27 \cdot x^{11}y^{18}$ Apply the exponents; product rule

$= 432x^{11}y^{18}$ Multiply.

Notice that $(2x^2y^3)^4$ means $2^4x^{2\cdot4}y^{3\cdot4}$,   ***not*** $(2 \cdot 4)x^{2\cdot4}y^{3\cdot4}$.

(d) $(-x^3y)^2(-x^5y^4)^3$ | Think of the negative sign as a factor of -1.

$= (-1 \cdot x^3y)^2(-1 \cdot x^5y^4)^3$ $-a = -1 \cdot a$

$= (-1)^2(x^3)^2y^2 \cdot (-1)^3(x^5)^3(y^4)^3$ Power rule (b)

$= (-1)^2x^6y^2 \cdot (-1)^3x^{15}y^{12}$ Power rule (a)

$= (-1)^5x^{21}y^{14}$ Product rule

$= -x^{21}y^{14}$ Simplify; $(-1)^5 = -1$ **NOW TRY**

! CAUTION   Be aware of the distinction between $(2y)^3$ and $2y^3$.

$(2y)^3 = 2y \cdot 2y \cdot 2y = 8y^3$ The base is $2y$.

$2y^3 = 2 \cdot y \cdot y \cdot y$ The base is y.

OBJECTIVE 7 Apply the rules for exponents in a geometry application.

EXAMPLE 8   Using Area Formulas

Find an expression that represents the area, in square units, of **(a)** **FIGURE 1** and **(b)** **FIGURE 2.**

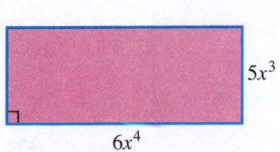

$5x^3$

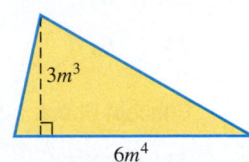

$3m^3$ Assume $x > 0$, $m > 0$.

$6x^4$ $6m^4$

FIGURE 1   **FIGURE 2**

(a) For **FIGURE 1,** use the formula for the area of a rectangle.

$\mathcal{A} = LW$ Area formula

$\mathcal{A} = (6x^4)(5x^3)$ Substitute.

$\mathcal{A} = 6 \cdot 5 \cdot x^{4+3}$ Commutative property; product rule

$\mathcal{A} = 30x^7$ Multiply. Add the exponents.

NOW TRY ANSWERS

7. (a) $\dfrac{243}{125}$ **(b)** 8^9k^9

   (c) $-8x^{26}y^{20}$

**NOW TRY
EXERCISE 8**
Find an expression that represents the area, in square units, of the figure. Assume $x > 0$.

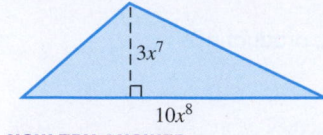

NOW TRY ANSWER

8. $15x^{15}$

(b) **FIGURE 2** is a triangle with base $6m^4$ and height $3m^3$.

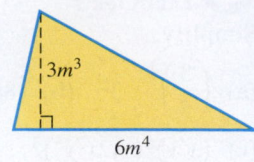

$$\mathcal{A} = \frac{1}{2}bh \qquad \text{Area formula}$$

$$\mathcal{A} = \frac{1}{2}(6m^4)(3m^3) \qquad \text{Substitute.}$$

$$\mathcal{A} = \frac{1}{2}(6 \cdot 3 \cdot m^{4+3}) \qquad \text{Properties of real numbers; product rule}$$

$$\mathcal{A} = 9m^7 \qquad \text{Multiply. Add the exponents.} \qquad \textbf{NOW TRY} \;\circlearrowleft$$

FIGURE 2 (repeated)

4.1 Exercises

FOR EXTRA HELP   ▶ **MyLab Math**

▶ *Video solutions for select problems available in MyLab Math*

Concept Check   *Determine whether each statement is* true *or* false. *If false, correct the right-hand side of the statement.*

1. $3^3 = 9$

2. $(-2)^4 = 2^4$

3. $(a^2)^3 = a^5$

4. $\left(\frac{1}{4}\right)^2 = \frac{1}{4^2}$

5. $-2^2 = 4$

6. $2^3 \cdot 2^4 = 4^7$

7. $(3x)^2 = 6x$

8. $(-x)^2 = x^2$

Write each expression in exponential form. ***See Example 1.***

9. $w \cdot w \cdot w \cdot w \cdot w \cdot w$

10. $t \cdot t \cdot t \cdot t \cdot t \cdot t \cdot t$

11. $\left(\frac{1}{2}\right)\left(\frac{1}{2}\right)\left(\frac{1}{2}\right)\left(\frac{1}{2}\right)\left(\frac{1}{2}\right)\left(\frac{1}{2}\right)$

12. $\left(\frac{1}{4}\right)\left(\frac{1}{4}\right)\left(\frac{1}{4}\right)\left(\frac{1}{4}\right)\left(\frac{1}{4}\right)$

13. $(-4)(-4)(-4)(-4)$

14. $(-3)(-3)(-3)(-3)(-3)(-3)$

15. $(-7y)(-7y)(-7y)(-7y)$

16. $(-8p)(-8p)(-8p)(-8p)(-8p)$

17. Explain how the expressions $(-3)^4$ and -3^4 are different.

18. Explain how the expressions $(5x)^3$ and $5x^3$ are different.

Identify the base and the exponent of each expression. In Exercises 19–26, also evaluate. ***See Example 2.***

19. 3^5

20. 2^7

21. $(-3)^5$

22. $(-2)^7$

23. $(-6)^2$

24. $(-9)^2$

25. -6^2

26. -9^2

27. $(-6x)^4$

28. $(-8x)^4$

29. $-6x^4$

30. $-8x^4$

Concept Check   *Simplify each expression.*

31. $8^2 \cdot 8^5$

$= 8^{—+—}$

$= \underline{\hspace{1cm}}$

32. $5m^2 \cdot 2m^6$

$= (5 \cdot \underline{\hspace{1cm}}) \cdot (m^— \cdot m^—)$

$= \underline{\hspace{1cm}} m^{—+—}$

$= \underline{\hspace{1cm}}$

Use the product rule for exponents to simplify each expression, if possible. Write each answer in exponential form. ***See Example 3.***

33. $5^2 \cdot 5^6$

34. $3^6 \cdot 3^7$

35. $4^2 \cdot 4^7 \cdot 4^3$

36. $5^3 \cdot 5^8 \cdot 5^2$

37. $t^3 \cdot t^8 \cdot t^{13}$

38. $n^5 \cdot n^6 \cdot n^9$

39. $(-8r^4)(7r^3)$ **40.** $(10a^7)(-4a^3)$ **41.** $(-6p^5)(-7p^5)$

42. $(-5w^8)(-9w^8)$ **43.** $(5x^2)(-2x^3)(3x^4)$ **44.** $(12y^3)(4y)(-3y^5)$

45. $(-2x^4)(-5x^6)(-3x)$ **46.** $(-10y^2)(-2y^5)(-8y)$ **47.** $3^8 + 3^9$

48. $4^{12} + 4^5$ **49.** $5^8 \cdot 3^9$ **50.** $6^3 \cdot 8^9$

Use the power rules for exponents to simplify each expression. Leave answers in exponential form in Exercises 51–66. Assume that variables in denominators are not zero. See Examples 4–6.

51. $(4^3)^2$ **52.** $(8^3)^6$ **53.** $(t^4)^5$ **54.** $(y^6)^5$

55. $(7r)^3$ **56.** $(11x)^4$ **57.** $(5xy)^5$ **58.** $(9pq)^6$

59. $(-5)^4$ **60.** $(-6)^6$ **61.** $(-5)^5$ **62.** $(-6)^3$

63. $8(qr)^3$ **64.** $4(vw)^5$ **65.** $\left(\dfrac{9}{5}\right)^8$ **66.** $\left(\dfrac{12}{7}\right)^3$

67. $\left(\dfrac{1}{2}\right)^3$ **68.** $\left(\dfrac{1}{3}\right)^5$ **69.** $\left(\dfrac{a}{b}\right)^3$ **70.** $\left(\dfrac{r}{t}\right)^4$

71. $\left(\dfrac{x}{4}\right)^3$ **72.** $\left(\dfrac{y}{3}\right)^4$ **73.** $\left(-\dfrac{2x}{y}\right)^5$ **74.** $\left(-\dfrac{4p}{q}\right)^3$

75. $(-2x^2y)^3$ **76.** $(-5m^4p^2)^3$ **77.** $(3a^3b^2)^2$ **78.** $(4x^3y^5)^4$

Concept Check *If a negative number is raised to an even power, the result is positive. For example,*

$$(-3)^4 = (-3)(-3)(-3)(-3) = +81\text{—that is, }81.$$

If a negative number is raised to an odd power, the result is negative. For example,

$$(-3)^5 = (-3)(-3)(-3)(-3)(-3) = -243.$$

Without actually performing the computation, state whether the power is positive *or* negative.

79. (a) $(-4)^3$ **(b)** $(-4)^4$ **(c)** $(-4)^{12}$ **(d)** $(-4)^{13}$

80. (a) $(-4)^4(-4)^2$ **(b)** $(-4)^3(-4)^8$ **(c)** $(-4)^{12}(-4)^{16}$ **(d)** $(-4)(-4)^{96}$

Simplify each expression. Leave answers in exponential form in Exercises 81–90. **See Example 7.**

81. $\left(\dfrac{5}{2}\right)^3 \cdot \left(\dfrac{5}{2}\right)^2$ **82.** $\left(\dfrac{3}{4}\right)^5 \cdot \left(\dfrac{3}{4}\right)^6$ **83.** $\left(\dfrac{9}{8}\right)^3 \cdot 9^2$

84. $\left(\dfrac{8}{5}\right)^4 \cdot 8^3$ **85.** $(2x)^9(2x)^3$ **86.** $(6y)^5(6y)^8$

87. $(6x^2y^3)^5$ **88.** $(5r^5t^6)^7$ **89.** $(x^2)^3(x^3)^5$

90. $(y^4)^5(y^3)^5$ **91.** $(2w^2x^3y)^2(x^4y)^5$ **92.** $(3x^4y^2z)^3(yz^4)^5$

93. $(-r^4s)^2(-r^2s^3)^5$ **94.** $(-ts^6)^4(-t^3s^5)^3$

95. $\left(\dfrac{5a^2b^5}{c^6}\right)^3$ $(c \neq 0)$ **96.** $\left(\dfrac{6x^3y^9}{z^5}\right)^4$ $(z \neq 0)$

Concept Check *Determine whether each statement is* true *or* false.

97. (a) $(-3)^3 = 3^3$ **98. (a)** $(-2)^2(-2)^4 = 2^6$

 (b) $(-3)^4 = 3^4$ **(b)** $(-2)^5(-2)^5 = 2^{10}$

 (c) $(-3)^5 = 3^5$ **(c)** $(-2)^8(-2)^9 = 2^{17}$

 (d) $(-3)^6 = 3^6$ **(d)** $(-2)^{10}(-2)^{20} = 2^{30}$

99. Concept Check A student simplified the expression $(10^2)^3$ as 1000^6. This is incorrect. **WHAT WENT WRONG?** Simplify correctly.

100. Concept Check A student simplified the expression $(3x^2y^3)^4$ as $3 \cdot 4x^8y^{12}$, or $12x^8y^{12}$. This is incorrect. **WHAT WENT WRONG?** Simplify correctly.

*Find an expression that represents the area, in square units, of each figure. **See Example 8.** (If necessary, refer to the formulas at the back of this text. The symbol ⌐ in the figures indicates 90° angles.)*

101. 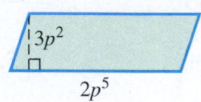 $3x^2$ $4x^3$

102. 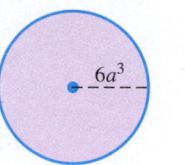 m^2 $4m^4$

103. 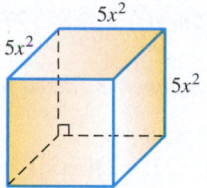 $3p^2$ $2p^5$

104. 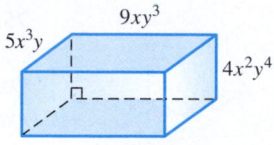 $6a^3$

Find an expression that represents the volume, in cubic units, of each figure. (If necessary, refer to the formulas at the back of this text.)

105. $5x^2$ $5x^2$ $5x^2$

106. $9xy^3$ $5x^3y$ $4x^2y^4$

Compound interest is interest paid on both the principal and the interest earned earlier. The formula for compound interest is

$$A = P(1 + r)^n,$$

where A is the amount accumulated from a principal of P dollars left untouched for n years with an annual interest rate r (expressed as a decimal).

Use the preceding formula and a calculator to find A to the nearest cent.

107. $P = \$250$, $r = 0.04$, $n = 5$

108. $P = \$400$, $r = 0.04$, $n = 3$

109. $P = \$1500$, $r = 0.015$, $n = 6$

110. $P = \$2000$, $r = 0.015$, $n = 4$

4.2 Integer Exponents and the Quotient Rule

OBJECTIVES

1 Use 0 as an exponent.
2 Use negative numbers as exponents.
3 Use the quotient rule for exponents.
4 Use combinations of the rules for exponents.

Consider the following list.

$2^4 = 16$
$2^3 = 8$
$2^2 = 4$

As exponents decrease by 1, the results are divided by 2 each time.

Each time we decrease the exponent by 1, the value is divided by 2 (the base). Using this pattern, we can continue the list to lesser and lesser integer exponents.

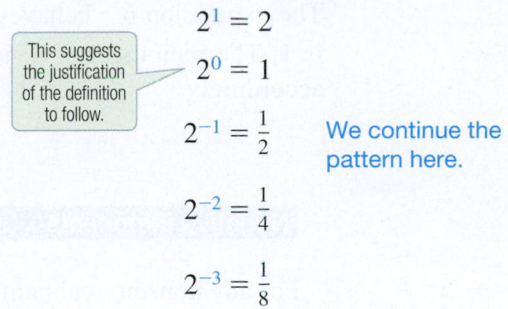

This suggests the justification of the definition to follow.

$$2^1 = 2$$
$$2^0 = 1$$
$$2^{-1} = \tfrac{1}{2}$$ We continue the pattern here.
$$2^{-2} = \tfrac{1}{4}$$
$$2^{-3} = \tfrac{1}{8}$$

From this list, it appears that we should define 2^0 as 1 and bases raised to negative exponents as reciprocals of those bases.

___

OBJECTIVE 1   Use 0 as an exponent.

The definitions of 0 and negative exponents must be consistent with the rules for exponents from the preceding section. For example, if we define 6^0 to be 1, then

$$6^0 \cdot 6^2 = 1 \cdot 6^2 = 6^2 \quad \text{and} \quad 6^0 \cdot 6^2 = 6^{0+2} = 6^2,$$

and we see that the product rule is satisfied. The power rules are also valid for a 0 exponent. Thus, we define a 0 exponent as follows.

> **Zero Exponent**
>
> For any nonzero real number a, $a^0 = 1$.
>
> *Example:* $17^0 = 1$

NOW TRY
EXERCISE 1
Evaluate.
(a) 6^0
(b) -12^0
(c) $(-12x)^0$ $(x \neq 0)$
(d) $14^0 - 12^0$

EXAMPLE 1   Using Zero Exponents

Evaluate.

(a) $60^0 = 1$

(b) $(-60)^0 = 1$

(c) $-60^0 = -(1) = -1$

(d) $y^0 = 1$ $(y \neq 0)$

(e) $6y^0 = 6(1) = 6$ $(y \neq 0)$

(f) $(6y)^0 = 1$ $(y \neq 0)$

(g) $8^0 + 11^0 = 1 + 1 = 2$

(h) $-8^0 - 11^0 = -1 - 1 = -2$

NOW TRY

___

⚠ **CAUTION**   Look again at Examples 1(b) and 1(c). In $(-60)^0$, the base is -60, and because any nonzero base raised to the 0 exponent is 1, $(-60)^0 = 1$. In -60^0, which can be written $-(60)^0$, the base is 60, so $-60^0 = -1$.

___

OBJECTIVE 2   Use negative numbers as exponents.

From the lists at the beginning of this section, $2^{-1} = \tfrac{1}{2}$, $2^{-2} = \tfrac{1}{4}$, and $2^{-3} = \tfrac{1}{8}$. We can make a conjecture that 2^{-n} should equal $\tfrac{1}{2^n}$. *Is the product rule valid in such cases?* For example,

$$6^{-2} \cdot 6^2 = 6^{-2+2} = 6^0 = 1.$$

NOW TRY ANSWERS
1. (a) 1   **(b)** -1 **(c)** 1   **(d)** 0

The expression 6^{-2} behaves as if it were the reciprocal of 6^2, because their product is 1. The reciprocal of 6^2 is also $\frac{1}{6^2}$, leading us to define 6^{-2} as $\frac{1}{6^2}$, and generalize accordingly.

Negative Exponents

For any nonzero real number a and any integer n, $\quad a^{-n} = \dfrac{1}{a^n}.$

Example: $\quad 3^{-2} = \dfrac{1}{3^2}$

By definition, a^{-n} and a^n are reciprocals.

$$a^n \cdot a^{-n} = a^n \cdot \frac{1}{a^n} = 1$$

Because $1^n = 1$, the definition of a^{-n} can also be written as follows.

$$a^{-n} = \frac{1}{a^n} = \frac{1^n}{a^n} = \left(\frac{1}{a}\right)^n$$

For example, $\qquad 6^{-3} = \left(\dfrac{1}{6}\right)^3 \quad$ and $\quad \left(\dfrac{1}{3}\right)^{-2} = 3^2.$

EXAMPLE 2   Using Negative Exponents

Write with positive exponents and simplify. Assume that all variables represent non-zero real numbers.

(a) $3^{-2} = \dfrac{1}{3^2} = \dfrac{1}{9} \qquad a^{-n} = \frac{1}{a^n}$

(b) $5^{-3} = \dfrac{1}{5^3} = \dfrac{1}{125} \qquad a^{-n} = \frac{1}{a^n}$

(c) $\left(\dfrac{1}{2}\right)^{-3} = 2^3 = 8 \qquad$ $\frac{1}{2}$ and 2 are reciprocals.
(Reciprocals have a product of 1.)

Notice that we can change the base to its reciprocal if we also change the sign of the exponent.

(d) $\left(\dfrac{2}{5}\right)^{-4}$

$= \left(\dfrac{5}{2}\right)^4 \qquad$ $\frac{2}{5}$ and $\frac{5}{2}$ are reciprocals.

$= \dfrac{5^4}{2^4} \qquad$ Power rule (c)

$= \dfrac{625}{16} \qquad$ Apply the exponents.

(e) $\left(\dfrac{4}{3}\right)^{-5}$

$= \left(\dfrac{3}{4}\right)^5 \qquad$ $\frac{4}{3}$ and $\frac{3}{4}$ are reciprocals.

$= \dfrac{3^5}{4^5} \qquad$ Power rule (c)

$= \dfrac{243}{1024} \qquad$ Apply the exponents.

**NOW TRY
EXERCISE 2**

Write with positive exponents and simplify. Assume that all variables represent nonzero real numbers.

(a) 2^{-3} **(b)** $\left(\dfrac{1}{7}\right)^{-2}$

(c) $\left(\dfrac{3}{2}\right)^{-4}$ **(d)** $3^{-2} + 4^{-2}$

(e) p^{-4} **(f)** $x^{-6}y^{3}$

(f) $4^{-1} - 2^{-1}$

$= \dfrac{1}{4} - \dfrac{1}{2}$ Apply the exponents.

$= \dfrac{1}{4} - \dfrac{2}{4}$ Find a common denominator.

$= -\dfrac{1}{4}$ Subtract.

(h) $\dfrac{1}{x^{-4}}$

$= \dfrac{1^{-4}}{x^{-4}}$ $1^n = 1$, for any integer n.

$= \left(\dfrac{1}{x}\right)^{-4}$ Power rule (c)

$= x^4$ $\dfrac{1}{x}$ and x are reciprocals.

(g) $3p^{-2}$

$= \dfrac{3}{1} \cdot \dfrac{1}{p^2}$ $a^{-n} = \dfrac{1}{a^n}$

$= \dfrac{3}{p^2}$ Multiply.

(i) $x^3 y^{-4}$

$= \dfrac{x^3}{1} \cdot \dfrac{1}{y^4}$ $a^{-n} = \dfrac{1}{a^n}$

$= \dfrac{x^3}{y^4}$ Multiply.

In general, $\dfrac{1}{a^{-n}} = a^n.$

NOW TRY

> ⚠ **CAUTION** *A negative exponent does not indicate a negative number. Negative exponents lead to reciprocals.*

Expression	Example	
a^{-n}	$3^{-2} = \dfrac{1}{3^2} = \dfrac{1}{9}$	Not negative
$-a^{-n}$	$-3^{-2} = -\dfrac{1}{3^2} = -\dfrac{1}{9}$	Negative

Consider the following.

$$\dfrac{2^{-3}}{3^{-4}}$$

$$= \dfrac{\dfrac{1}{2^3}}{\dfrac{1}{3^4}}$$ Definition of negative exponent

$$= \dfrac{1}{2^3} \div \dfrac{1}{3^4}$$ $\dfrac{a}{b}$ means $a \div b$.

$$= \dfrac{1}{2^3} \cdot \dfrac{3^4}{1}$$ To divide, multiply by the reciprocal of the divisor.

$$= \dfrac{3^4}{2^3}$$ Multiply.

NOW TRY ANSWERS

2. **(a)** $\dfrac{1}{8}$ **(b)** 49 **(c)** $\dfrac{16}{81}$

(d) $\dfrac{25}{144}$ **(e)** $\dfrac{1}{p^4}$ **(f)** $\dfrac{y^3}{x^6}$

Therefore, $\dfrac{2^{-3}}{3^{-4}} = \dfrac{3^4}{2^3}.$

> ## Negative-to-Positive Rules for Exponents
>
> For any nonzero numbers a and b and any integers m and n, the following hold true.
>
> $$\frac{a^{-m}}{b^{-n}} = \frac{b^n}{a^m} \quad \text{and} \quad \left(\frac{a}{b}\right)^{-m} = \left(\frac{b}{a}\right)^m$$
>
> *Examples:* $\dfrac{3^{-5}}{2^{-4}} = \dfrac{2^4}{3^5}$ and $\left(\dfrac{4}{5}\right)^{-3} = \left(\dfrac{5}{4}\right)^3$

**NOW TRY
EXERCISE 3**

Write with positive exponents and simplify. Assume that all variables represent nonzero real numbers.

(a) $\dfrac{5^{-3}}{6^{-2}}$ **(b)** $\dfrac{n^{-4}}{m^{-2}}$

(c) $\dfrac{x^2 y^{-3}}{5z^{-4}}$ **(d)** $\left(\dfrac{r}{3s}\right)^{-3}$

EXAMPLE 3   Changing from Negative to Positive Exponents

Write with positive exponents and simplify. Assume that all variables represent non-zero real numbers.

(a) $\dfrac{4^{-2}}{5^{-3}} = \dfrac{5^3}{4^2} = \dfrac{125}{16}$

(b) $\dfrac{m^{-5}}{p^{-1}} = \dfrac{p^1}{m^5} = \dfrac{p}{m^5}$

(c) $\dfrac{a^{-2}b}{3d^{-3}} = \dfrac{bd^3}{3a^2}$ Notice that b in the numerator and the coefficient 3 in the denominator are not affected.

(d) $\left(\dfrac{x}{2y}\right)^{-4}$

$= \left(\dfrac{2y}{x}\right)^4$ Negative-to-positive rule

$= \dfrac{2^4 y^4}{x^4}$ Power rules (b) and (c)

$= \dfrac{16y^4}{x^4}$ Apply the exponent.

NOW TRY

⚠ **CAUTION**   Be careful. We cannot use the rule $\frac{a^{-m}}{b^{-n}} = \frac{b^n}{a^m}$ to change negative exponents to positive exponents if the exponents occur in a *sum* or *difference* of terms.

Example: $\dfrac{5^{-2} + 3^{-1}}{7 - 2^{-3}}$ written with positive exponents is $\dfrac{\dfrac{1}{5^2} + \dfrac{1}{3}}{7 - \dfrac{1}{2^3}}$.

OBJECTIVE 3 Use the quotient rule for exponents.

Consider the following.

$$\frac{6^5}{6^3} = \frac{6 \cdot 6 \cdot 6 \cdot 6 \cdot 6}{6 \cdot 6 \cdot 6} = 6^2 \quad \begin{array}{l}\text{Divide out the}\\\text{common factors.}\end{array}$$

The difference of the exponents, $5 - 3 = 2$, is the exponent in the quotient.

Also, $\dfrac{6^2}{6^4} = \dfrac{6 \cdot 6}{6 \cdot 6 \cdot 6 \cdot 6} = \dfrac{1}{6^2} = 6^{-2}$. $\begin{array}{l}\text{Divide out the}\\\text{common factors.}\end{array}$

Here, $2 - 4 = -2$. These examples suggest the **quotient rule for exponents.**

NOW TRY ANSWERS

3. **(a)** $\dfrac{36}{125}$ **(b)** $\dfrac{m^2}{n^4}$ **(c)** $\dfrac{x^2 z^4}{5y^3}$

(d) $\dfrac{27s^3}{r^3}$

Quotient Rule for Exponents

For any nonzero real number a and any integers m and n,

$$\frac{a^m}{a^n} = a^{m-n}.$$

(Keep the same base and subtract the exponents.)

Example: $\dfrac{5^8}{5^4} = 5^{8-4} = 5^4$

⚠ **CAUTION**   A common **error** is to write $\frac{5^8}{5^4} = 1^{8-4} = 1^4$. **This is incorrect.** By the quotient rule, the quotient must have the *same base,* 5, just like the product in the product rule.

$$\frac{5^8}{5^4} = 5^{8-4} = 5^4$$

We can confirm this by writing out the factors.

$$\frac{5^8}{5^4} = \frac{5 \cdot 5 \cdot 5 \cdot 5 \cdot 5 \cdot 5 \cdot 5 \cdot 5}{5 \cdot 5 \cdot 5 \cdot 5} = 5^4$$

🔄 **NOW TRY**
EXERCISE 4

Simplify. Assume that all variables represent nonzero real numbers.

(a) $\dfrac{6^3}{6^4}$ **(b)** $\dfrac{t^4}{t^{-5}}$

(c) $\dfrac{(p+q)^{-3}}{(p+q)^{-7}}$ $(p \neq -q)$

(d) $\dfrac{5^2 x y^{-3}}{3^{-1} x^{-2} y^2}$

NOW TRY ANSWERS

4. (a) $\dfrac{1}{6}$ **(b)** t^9 **(c)** $(p+q)^4$

(d) $\dfrac{75x^3}{y^5}$

EXAMPLE 4   Using the Quotient Rule

Simplify. Assume that all variables represent nonzero real numbers.

(a) $\dfrac{5^8}{5^6} = 5^{8-6} = 5^2 = 25$

> Keep the same base.

(b) $\dfrac{4^2}{4^5} = 4^{2-5} = 4^{-3} = \dfrac{1}{4^3} = \dfrac{1}{64}$

(c) $\dfrac{5^{-3}}{5^{-7}} = 5^{-3-(-7)} = 5^4 = 625$

> Be careful with signs.

(d) $\dfrac{q^5}{q^{-3}} = q^{5-(-3)} = q^8$

(e) $\dfrac{3^2 x^5}{3^4 x^3}$

$= \dfrac{3^2}{3^4} \cdot \dfrac{x^5}{x^3}$

$= 3^{2-4} \cdot x^{5-3}$ Quotient rule

$= 3^{-2} x^2$ Subtract.

$= \dfrac{x^2}{3^2}$ Definition of negative exponent

$= \dfrac{x^2}{9}$ Apply the exponent.

(f) $\dfrac{(m+n)^{-2}}{(m+n)^{-4}}$

$= (m+n)^{-2-(-4)}$

$= (m+n)^{-2+4}$

$= (m+n)^2$ $(m \neq -n)$

The restriction $m \neq -n$ is necessary to prevent a denominator of 0 in the original expression. Division by 0 is undefined.

(g) $3x^{-5}$

> Avoid the error of applying -5 to 3.

$= 3 \cdot \dfrac{1}{x^5}$ -5 applies *only* to x.

$= \dfrac{3}{x^5}$ Multiply.

(h) $\dfrac{7x^{-3} y^2}{2^{-1} x^2 y^{-5}}$

$= \dfrac{7 \cdot 2^1 y^2 y^5}{x^2 x^3}$ Negative-to-positive rule

$= \dfrac{14y^7}{x^5}$ Multiply; product rule

🔄 **NOW TRY**

The definitions and rules for exponents are summarized here.

Summary of Definitions and Rules for Exponents

For all integers m and n, and all real numbers a and b for which the following are defined, these definitions and rules hold true.

Examples

Product rule	$a^m \cdot a^n = a^{m+n}$	$7^4 \cdot 7^5 = 7^{4+5} = 7^9$
Zero exponent	$a^0 = 1$	$(-3)^0 = 1$
Negative exponent	$a^{-n} = \dfrac{1}{a^n}$	$5^{-3} = \dfrac{1}{5^3}$
Quotient rule	$\dfrac{a^m}{a^n} = a^{m-n}$	$\dfrac{2^2}{2^5} = 2^{2-5} = 2^{-3} = \dfrac{1}{2^3}$
Power rules (a)	$(a^m)^n = a^{mn}$	$(4^2)^3 = 4^{2 \cdot 3} = 4^6$
(b)	$(ab)^m = a^m b^m$	$(3k)^4 = 3^4 k^4$
(c)	$\left(\dfrac{a}{b}\right)^m = \dfrac{a^m}{b^m}$	$\left(\dfrac{2}{3}\right)^2 = \dfrac{2^2}{3^2}$
Negative-to-positive rules	$\dfrac{a^{-m}}{b^{-n}} = \dfrac{b^n}{a^m}$	$\dfrac{2^{-4}}{5^{-3}} = \dfrac{5^3}{2^4}$
	$\left(\dfrac{a}{b}\right)^{-m} = \left(\dfrac{b}{a}\right)^m$	$\left(\dfrac{4}{7}\right)^{-2} = \left(\dfrac{7}{4}\right)^2$

OBJECTIVE 4 Use combinations of the rules for exponents.

EXAMPLE 5 Using Combinations of Rules

Simplify. Assume that all variables represent nonzero real numbers.

(a) $\dfrac{(4^2)^3}{4^5}$

$= \dfrac{4^6}{4^5}$ Power rule (a)

$= 4^{6-5}$ Quotient rule

$= 4^1$ Subtract.

$= 4$ $a^1 = a$, for all a.

(b) $x^2 \cdot x^{-6} \cdot x^{-1}$

$= x^{2+(-6)+(-1)}$ Product rule

$= x^{-5}$ Add.

$= \dfrac{1}{x^5}$ Definition of negative exponent

(c) $(2x)^3 (2x)^2$

$= (2x)^5$ Product rule

$= 2^5 x^5$ Power rule (b)

$= 32x^5$ $2^5 = 32$

$(2x)^3 (2x)^2$ **Alternative solution**

$= 2^3 x^3 \cdot 2^2 x^2$ Power rule (b)

$= 2^{3+2} x^{3+2}$ Product rule

$= 2^5 x^5$ Add exponents.

$= 32x^5$ $2^5 = 32$

NOW TRY EXERCISE 5

Simplify. Assume that all variables represent nonzero real numbers.

(a) $\dfrac{3^{15}}{(3^3)^4}$ **(b)** $(4t)^5(4t)^{-3}$

(c) $\left(\dfrac{7y^4}{10}\right)^{-3}$ **(d)** $\dfrac{(a^2b^{-2}c)^{-3}}{(2ab^3c^{-4})^5}$

(e) $\dfrac{(5k)^{-6}(5k)^8}{(5k)^7(5k)^{-4}}$

(d) $\left(\dfrac{2x^3}{5}\right)^{-4}$

$= \left(\dfrac{5}{2x^3}\right)^4$ Negative-to-positive rule

$= \dfrac{5^4}{2^4x^{12}}$ Power rules (a)–(c)

$= \dfrac{625}{16x^{12}}$ Apply the exponents.

(f) $\dfrac{(4m)^{-3}}{(3m)^{-4}}$

$= \dfrac{4^{-3}m^{-3}}{3^{-4}m^{-4}}$ Power rule (b)

$= \dfrac{3^4m^4}{4^3m^3}$ Negative-to-positive rule

$= \dfrac{3^4m^{4-3}}{4^3}$ Quotient rule

$= \dfrac{3^4m^1}{4^3}$ Subtract.

$= \dfrac{81m}{64}$ Apply the exponents.

(e) $\left(\dfrac{3x^{-2}}{4^{-1}y^3}\right)^{-3}$

$= \dfrac{3^{-3}x^6}{4^3y^{-9}}$ Power rules (a)–(c)

$= \dfrac{x^6y^9}{4^3 \cdot 3^3}$ Negative-to-positive rule

$= \dfrac{x^6y^9}{1728}$ $4^3 \cdot 3^3 = 64 \cdot 27 = 1728$

(g) $\dfrac{(7y)^{-3}(7y)^4}{(7y)^{12}(7y)^{-10}}$

$= \dfrac{(7y)^{-3+4}}{(7y)^{12+(-10)}}$ Product rule

$= \dfrac{(7y)^1}{(7y)^2}$ Add.

$= (7y)^{1-2}$ Quotient rule

$= (7y)^{-1}$ Subtract.

$= \dfrac{1}{7y}$ $a^{-1} = \dfrac{1}{a}$, for $a \neq 0$.

NOW TRY ↩

NOW TRY ANSWERS

5. (a) 27 **(b)** $16t^2$ **(c)** $\dfrac{1000}{343y^{12}}$

(d) $\dfrac{c^{17}}{32a^{11}b^9}$ **(e)** $\dfrac{1}{5k}$

NOTE Because the steps can be done in several different orders, there are many correct ways to simplify expressions like those in **Example 5**. For instance, see the alternative solution for **Example 5(c)**.

4.2 Exercises

FOR EXTRA HELP

▶ **MyLab Math**

▶ *Video solutions for select problems available in MyLab Math*

Concept Check   *Determine whether each expression is* positive, negative, *or* 0.

1. $(-2)^{-3}$ **2.** $(-3)^{-2}$ **3.** -2^4 **4.** -3^6

5. $\left(\dfrac{1}{4}\right)^{-2}$ **6.** $\left(\dfrac{1}{5}\right)^{-2}$ **7.** $1 - 5^0$ **8.** $1 - 7^0$

Concept Check   *Simplify each expression.*

9. $\dfrac{5^{11}}{5^8}$

$= 5\underline{\ \ }{}^{-}\underline{\ \ }$

$= 5\underline{\ \ }$

$= \underline{\ \ \ \ }$

10. $\dfrac{6^{-5}}{6^{-2}}$

$= 6\underline{\ \ }{}^{-(\underline{\ \ })}$

$= 6\underline{\ \ }$

$= \dfrac{1}{6\underline{\ \ }}$

$= \underline{\ \ \ \ }$

Decide whether each expression is equal to 0, 1, or −1. *See Example 1.*

11. 9^0

12. 3^0

13. $(-2)^0$

14. $(-12)^0$

15. -8^0

16. -6^0

17. $-(-6)^0$

18. $-(-13)^0$

19. $(-4)^0 - 4^0$

20. $(-11)^0 - 11^0$

21. $(3x)^0$ $(x \neq 0)$

22. $(-5t)^0$ $(t \neq 0)$

23. $8^0 - 12^0$

24. $6^0 - 13^0$

25. $\dfrac{0^{10}}{12^0}$

26. $\dfrac{0^5}{2^0}$

Concept Check *Match each expression in Column I with the equivalent expression in Column II. Choices in Column II may be used once, more than once, or not at all. (In Exercise 27, $x \neq 0$.)*

	I		II
27.	**(a)** x^0	**A.**	0
	(b) $-x^0$	**B.**	1
	(c) $7x^0$	**C.**	-1
	(d) $(7x)^0$	**D.**	7
	(e) $-7x^0$	**E.**	-7
	(f) $(-7x)^0$	**F.**	$\dfrac{1}{7}$

	I		II
28.	**(a)** -2^{-4}	**A.**	8
	(b) $(-2)^{-4}$	**B.**	16
	(c) 2^{-4}	**C.**	$-\dfrac{1}{16}$
	(d) $\dfrac{1}{2^{-4}}$	**D.**	-8
	(e) $\dfrac{1}{-2^{-4}}$	**E.**	-16
	(f) $\dfrac{1}{(-2)^{-4}}$	**F.**	$\dfrac{1}{16}$

Evaluate each expression. See Examples 1 and 2.

29. $6^0 + 8^0$

30. $4^0 + 2^0$

31. 4^{-3}

32. 5^{-4}

33. $\left(\dfrac{1}{2}\right)^{-4}$

34. $\left(\dfrac{1}{3}\right)^{-3}$

35. $\left(\dfrac{6}{7}\right)^{-2}$

36. $\left(\dfrac{2}{3}\right)^{-3}$

37. $(-3)^{-4}$

38. $(-4)^{-3}$

39. $3x^0$ $(x \neq 0)$

40. $-5t^0$ $(t \neq 0)$

41. $5^{-1} + 3^{-1}$

42. $6^{-1} + 2^{-1}$

43. $3^{-2} - 2^{-1}$

44. $6^{-2} - 3^{-1}$

45. $\left(\dfrac{1}{2}\right)^{-1} + \left(\dfrac{2}{3}\right)^{-1}$

46. $\left(\dfrac{1}{3}\right)^{-1} + \left(\dfrac{4}{3}\right)^{-1}$

Simplify each expression. Assume that all variables represent nonzero real numbers and no denominators are zero. See Examples 2–4.

47. $\dfrac{5^8}{5^5}$

48. $\dfrac{10^6}{10^3}$

49. $\dfrac{9^4}{9^5}$

50. $\dfrac{7^3}{7^4}$

51. $\dfrac{3^{-2}}{5^{-3}}$

52. $\dfrac{4^{-3}}{3^{-2}}$

53. $\dfrac{5}{5^{-1}}$

54. $\dfrac{6}{6^{-2}}$

55. $\dfrac{6^{-3}}{6^2}$

56. $\dfrac{4^{-2}}{4^3}$

57. $\dfrac{x^{12}}{x^{-3}}$

58. $\dfrac{y^4}{y^{-6}}$

59. $\dfrac{1}{6^{-3}}$

60. $\dfrac{1}{5^{-2}}$

61. $\dfrac{2}{r^{-4}}$

62. $\dfrac{3}{s^{-8}}$

63. $\dfrac{4^{-3}}{5^{-2}}$ **64.** $\dfrac{6^{-2}}{5^{-4}}$ **65.** $p^5 q^{-8}$ **66.** $x^{-8} y^4$

67. $\dfrac{r^5}{r^{-4}}$ **68.** $\dfrac{a^6}{a^{-4}}$ **69.** $\dfrac{6^4 x^8}{6^5 x^3}$ **70.** $\dfrac{3^8 y^5}{3^{10} y^2}$

71. $\dfrac{x^{-3} y}{4 z^{-2}}$ **72.** $\dfrac{p^{-5} q^{-4}}{9 r^{-3}}$ **73.** $\dfrac{(a+b)^{-3}}{(a+b)^{-4}}$

74. $\dfrac{(x+y)^{-8}}{(x+y)^{-9}}$ **75.** $\dfrac{(x+2y)^{-3}}{(x+2y)^{-5}}$ **76.** $\dfrac{(p-3q)^{-2}}{(p-3q)^{-4}}$

Simplify each expression. Assume that all variables represent nonzero real numbers. See **Example 5.**

77. $\dfrac{(7^4)^3}{7^9}$ **78.** $\dfrac{(5^3)^2}{5^2}$ **79.** $x^{-3} \cdot x^5 \cdot x^{-4}$ **80.** $y^{-8} \cdot y^5 \cdot y^{-2}$

81. $\dfrac{(3x)^{-2}}{(4x)^{-3}}$ **82.** $\dfrac{(2y)^{-3}}{(5y)^{-4}}$ **83.** $\left(\dfrac{x^{-1} y}{z^2}\right)^{-2}$ **84.** $\left(\dfrac{p^{-4} q}{r^{-3}}\right)^{-3}$

85. $(6x)^4 (6x)^{-3}$ **86.** $(10y)^9 (10y)^{-8}$ **87.** $\dfrac{(m^7 n)^{-2}}{m^{-4} n^3}$

88. $\dfrac{(m^{-8} n^{-4})^2}{m^2 n^5}$ **89.** $\dfrac{(x^{-1} y^2 z)^{-2}}{(x^{-3} y^3 z)^{-1}}$ **90.** $\dfrac{(a^2 b^3 c^4)^{-4}}{(a^{-2} b^{-3} c^{-4})^{-5}}$

91. $\left(\dfrac{xy^{-2}}{x^2 y}\right)^{-3}$ **92.** $\left(\dfrac{wz^{-5}}{w^3 z}\right)^{-2}$ **93.** $\dfrac{(2r)^{-4} (2r)^5}{(2r)^9 (2r)^{-7}}$

94. $\dfrac{(8x)^{-8} (8x)^9}{(8x)^{13} (8x)^{-11}}$ **95.** $\dfrac{(-4y)^8 (-4y)^{-8}}{(-4y)^{-26} (-4y)^{27}}$ **96.** $\dfrac{(-9p)^{16} (-9p)^{-16}}{(-9p)^{-41} (-9p)^{42}}$

97. Concept Check   A student simplified $\dfrac{16^3}{2^2}$ incorrectly as shown.

$$\dfrac{16^3}{2^2} = \left(\dfrac{16}{2}\right)^{3-2} = 8^1 = 8$$

WHAT WENT WRONG? Give the correct answer.

98. Concept Check   A student simplified -5^4 incorrectly as shown.

$$-5^4 = (-5)^4 = 625$$

WHAT WENT WRONG? Give the correct answer.

Extending Skills   *Simplify each expression. Assume that all variables represent nonzero real numbers.*

99. $\dfrac{(4a^2 b^3)^{-2} (2ab^{-1})^3}{(a^3 b)^{-4}}$ **100.** $\dfrac{(m^6 n)^{-2} (m^2 n^{-2})^3}{m^{-1} n^{-2}}$

101. $\dfrac{(2y^{-1} z^2)^2 (3y^{-2} z^{-3})^3}{(y^3 z^2)^{-1}}$ **102.** $\dfrac{(3p^{-2} q^3)^2 (5p^{-1} q^{-4})^{-1}}{(p^2 q^{-2})^{-3}}$

103. $\dfrac{(9^{-1} z^{-2} x)^{-1} (4z^2 x^4)^{-2}}{(5z^{-2} x^{-3})^2}$ **104.** $\dfrac{(4^{-1} a^{-1} b^{-2})^{-2} (5a^{-3} b^4)^{-2}}{(3a^{-3} b^{-5})^2}$

SUMMARY EXERCISES Applying the Rules for Exponents

1. **Concept Check** *Match each expression (a)–(j) in Column I with the equivalent expression A–J in Column II. Choices in Column II may be used once, more than once, or not at all.*

<table>
<tr><td colspan="2" align="center">**I**</td><td colspan="2" align="center">**II**</td></tr>
<tr><td>**(a)** $2^0 + 2^0$</td><td>**(b)** $2^1 \cdot 2^0$</td><td>**A.** 0</td><td>**B.** 1</td></tr>
<tr><td>**(c)** $2^0 - 2^{-1}$</td><td>**(d)** $2^1 - 2^0$</td><td>**C.** -1</td><td>**D.** 2</td></tr>
<tr><td>**(e)** $2^0 \cdot 2^{-2}$</td><td>**(f)** $2^1 \cdot 2^1$</td><td>**E.** $\dfrac{1}{2}$</td><td>**F.** 4</td></tr>
<tr><td>**(g)** $2^{-2} - 2^{-1}$</td><td>**(h)** $2^0 \cdot 2^0$</td><td>**G.** -2</td><td>**H.** -4</td></tr>
<tr><td>**(i)** $2^{-2} \div 2^{-1}$</td><td>**(j)** $2^0 \div 2^{-2}$</td><td>**I.** $-\dfrac{1}{4}$</td><td>**J.** $\dfrac{1}{4}$</td></tr>
</table>

Simplify each expression. Assume that all variables represent nonzero real numbers.

2. $5^{-1} + 6^{-1}$

3. $-4^0 + (-4)^0$

4. $5^{-2} + 6^{-2}$

5. $-(-19^0)$

6. $-(-13)^0$

7. $\dfrac{0^{13}}{13^0}$

8. $8^{-1} + 6^{-1}$

9. $52^0 - (-8)^0$

10. $-(-8^0)^0$

11. $(10x^2y^4)^2(10xy^2)^3$

12. $(-2ab^3c)^4(-2a^2b)^3$

13. $\left(\dfrac{9wx^3}{y^4}\right)^3$

14. $(4x^{-2}y^{-3})^{-2}$

15. $\dfrac{c^{11}(c^2)^4}{(c^3)^3(c^2)^{-6}}$

16. $\left(\dfrac{k^4t^2}{k^2t^{-4}}\right)^{-2}$

17. $\dfrac{(3y^{-1}z^3)^{-1}(3y^2)}{(y^3z^2)^{-3}}$

18. $\dfrac{(2xy^{-1})^3}{2^3x^{-3}y^2}$

19. $(z^4)^{-3}(z^{-2})^{-5}$

20. $\left(\dfrac{r^2st^5}{3r}\right)^{-2}$

21. $\dfrac{(3^{-1}x^{-3}y)^{-1}(2x^2y^{-3})^2}{(5x^{-2}y^2)^{-2}}$

22. $\left(\dfrac{5x^2}{3x^{-4}}\right)^{-1}$

23. $\left(\dfrac{-9x^{-2}}{9x^2}\right)^{-2}$

24. $\dfrac{(x^{-4}y^2)^3(x^2y)^{-1}}{(xy^2)^{-3}}$

25. $\dfrac{(a^{-2}b^3)^{-4}}{(a^{-3}b^2)^{-2}(ab)^{-4}}$

26. $(2a^{-30}b^{-29})(3a^{31}b^{30})$

27. $\left(\dfrac{(x^{43}y^{23})^2}{x^{-26}y^{-42}}\right)^0$

28. $\left(\dfrac{7a^2b^3}{2}\right)^3$

29. $\dfrac{(2xy^{-3})^{-2}}{(3x^{-2}y^4)^{-3}}$

30. $\left(\dfrac{a^2b^3c^4}{a^{-2}b^{-3}c^{-4}}\right)^{-2}$

31. $(6x^{-5}z^3)^{-3}$

32. $(2p^{-2}qr^{-3})(2p)^{-4}$

33. $\dfrac{(xy)^{-3}(xy)^5}{(xy)^{-4}}$

34. $\dfrac{(7^{-1}x^{-3})^{-2}(x^4)^{-6}}{7^{-1}x^{-3}}$

35. $\left(\dfrac{3^{-4}x^{-3}}{3^{-3}x^{-6}}\right)^{-2}$

36. $(5p^{-2}q)^{-3}(5pq^3)^4$

37. $\left(\dfrac{4r^{-6}s^{-2}t}{2r^8s^{-4}t^2}\right)^{-1}$

38. $(13x^{-6}y)(13x^{-6}y)^{-1}$

39. $\dfrac{(8pq^{-2})^4}{(8p^{-2}q^{-3})^3}$

40. $\left(\dfrac{mn^{-2}p}{m^2np^4}\right)^{-2}\left(\dfrac{mn^{-2}p}{m^2np^4}\right)^3$

4.3 Scientific Notation

OBJECTIVES

1 Express numbers in scientific notation.
2 Convert numbers in scientific notation to standard notation.
3 Use scientific notation in calculations.

OBJECTIVE 1 Express numbers in scientific notation.

Numbers occurring in science are often extremely large (such as the distance from Earth to the sun, 93,000,000 mi) or extremely small (the wavelength of blue light, approximately 0.000000475 m). Because of the difficulty of working with many zeros, scientists often express such numbers with exponents using *scientific notation.*

> **Scientific Notation**
>
> A number is written in **scientific notation** when it is expressed in the form
>
> $$a \times 10^n,$$
>
> where $1 \le |a| < 10$ and n is an integer.

VOCABULARY

☐ scientific notation
☐ standard notation

In scientific notation, there is *always* one nonzero digit before the decimal point.

Scientific notation

$3.19 \times 10^1 = 3.19 \times 10 = 31.9$	Decimal point moves 1 place to the right.
$3.19 \times 10^2 = 3.19 \times 100 = 319.$	Decimal point moves 2 places to the right.
$3.19 \times 10^3 = 3.19 \times 1000 = 3190.$	Decimal point moves 3 places to the right.
$3.19 \times 10^{-1} = 3.19 \times 0.1 = 0.319$	Decimal point moves 1 place to the left.
$3.19 \times 10^{-2} = 3.19 \times 0.01 = 0.0319$	Decimal point moves 2 places to the left.
$3.19 \times 10^{-3} = 3.19 \times 0.001 = 0.00319$	Decimal point moves 3 places to the left.

NOTE In scientific notation, the multiplication cross $\times$ is commonly used.

A number in scientific notation is always written with the decimal point after the first nonzero digit and then multiplied by the appropriate power of 10.

Example: 56,200 is written 5.62×10^4 because

$$56,200 = 5.62 \times 10,000 = 5.62 \times 10^4.$$

Additional examples:

42,000,000	is written	$4.2 \times 10^7.$
0.000586	is written	$5.86 \times 10^{-4}.$
and 2,000,000,000	is written	$2 \times 10^9.$

It is not necessary to write 2.0.

To write a number in scientific notation, follow these steps. (For a negative number, follow these steps using the *absolute value* of the number. Then make the result negative.)

Converting a Positive Number to Scientific Notation

Step 1 **Position the decimal point.** Place a caret $^\wedge$ to the right of the first non-zero digit, where the decimal point will be placed.

Step 2 **Determine the numeral for the exponent.** Count the number of digits from the decimal point to the caret. This number gives the absolute value of the exponent on 10.

Step 3 **Determine the sign for the exponent.** Decide whether multiplying by 10^n should make the result of Step 1 greater or less.

- The exponent should be positive to make the result greater.

- The exponent should be negative to make the result less.

**NOW TRY
EXERCISE 1**
Write each number in
scientific notation.

(a) 12,600,000

(b) 0.00027

(c) −0.0000341

EXAMPLE 1   Using Scientific Notation

Write each number in scientific notation.

(a) 93,000,000

Step 1 Place a caret to the right of the 9 (the first nonzero digit) to mark the new location of the decimal point.

$$9_\wedge 3,000,000$$

Step 2 Count from the decimal point, which is understood to be after the last 0, to the caret.

$$9.3,000,000. \leftarrow \text{Decimal point}$$

Count 7 places.

Step 3 Because 9.3 is to be made greater, the exponent on 10 is positive.

$$93,000,000 = 9.3 \times 10^7$$

(b) $63,200,000,000 = 6.3200000000 = 6.32 \times 10^{10}$

10 places

(c) 0.00462

Move the decimal point to the right of the first nonzero digit, and count the number of places the decimal point was moved.

$$0.00462 \quad \text{3 places}$$

Because 0.00462 is *less* than 4.62, the exponent must be *negative*.

$$0.00462 = 4.62 \times 10^{-3}$$

(d) $-0.0000762 = -7.62 \times 10^{-5}$ Remember the negative sign.   **NOW TRY**

5 places

NOTE   When writing a positive number in scientific notation, think as follows.

- If the original number is "large," like 93,000,000, use a *positive* exponent on 10 because positive is greater than negative.

- If the original number is "small," like 0.00462, use a *negative* exponent on 10 because negative is less than positive.

NOW TRY ANSWERS
1. **(a)** 1.26×10^7
   (b) 2.7×10^{-4}
   (c) -3.41×10^{-5}

OBJECTIVE 2 Convert numbers in scientific notation to standard notation.

Multiplying a number by a positive power of 10 will make the number greater. Multiplying by a negative power of 10 will make the number less.

We refer to a number such as 475 as the **standard notation** of 4.75×10^2.

**NOW TRY
EXERCISE 2**

Write each number in
standard notation.

(a) 5.71×10^4

(b) 2.72×10^{-5}

(c) -8.81×10^{-4}

EXAMPLE 2 Writing Numbers in Standard Notation

Write each number in standard notation.

(a) 6.2×10^3

Because the exponent is positive, we make 6.2 greater by moving the decimal point 3 places to the right. We attach two zeros.

$$6.2 \times 10^3 = 6.200 = 6200$$

(b) $4.283 \times 10^6 = 4.283000 = 4,283,000$ Move 6 places to the right.
Attach zeros as necessary.

(c) $-7.04 \times 10^{-3} = -0.00704$ Move 3 places to the left.

The exponent tells the number of places and the direction in which the decimal point is moved.

NOW TRY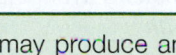

OBJECTIVE 3 Use scientific notation in calculations.

**NOW TRY
EXERCISE 3**

Perform each calculation.
Write answers in both
scientific and standard
notation.

(a) $(6 \times 10^7)(7 \times 10^{-4})$

(b) $\dfrac{18 \times 10^{-3}}{6 \times 10^4}$

EXAMPLE 3 Multiplying and Dividing with Scientific Notation

Perform each calculation. Write answers in both scientific and standard notation.

(a) $(7 \times 10^3)(5 \times 10^4)$

$= (7 \times 5)(10^3 \times 10^4)$ Commutative and associative properties

$= 35 \times 10^7$ Multiply. Use the product rule.

> Don't stop. This number is *not* in scientific notation because 35 is not between 1 and 10.

$= (3.5 \times 10^1) \times 10^7$ Write 35 in scientific notation.

$= 3.5 \times (10^1 \times 10^7)$ Associative property

$= 3.5 \times 10^8$ Product rule; Answer in scientific notation

$= 350,000,000$ Answer in standard notation

(b) $\dfrac{4 \times 10^{-5}}{2 \times 10^3}$

$= \dfrac{4}{2} \times \dfrac{10^{-5}}{10^3}$

$= 2 \times 10^{-8}$ Divide. Use the quotient rule.
Answer in scientific notation

$= 0.00000002$ Answer in standard notation

NOW TRY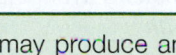

NOW TRY ANSWERS

2. (a) 57,100 (b) 0.0000272

   (c) -0.000881

3. (a) 4.2×10^4, or 42,000

   (b) 3×10^{-7}, or 0.0000003

NOTE Multiplying or dividing numbers written in scientific notation may produce an answer in the form $a \times 10^0$. Because $10^0 = 1$, $a \times 10^0 = a$.

Example: $(8 \times 10^{-4})(5 \times 10^4) = 40 \times 10^0 = 40$ $10^0 = 1$

Also, if $a = 1$, then $a \times 10^n = 10^n$.

Example: 1,000,000 could be written as 10^6 instead of 1×10^6.

NOW TRY EXERCISE 4

See **Example 4.** About how much would 8,000,000 nanometers measure in inches?

NOW TRY EXERCISE 5

The land area of California is approximately 1.6×10^5 mi², and the 2016 population of California was approximately 3.9×10^7 people. Use this information to estimate the number of square miles per California resident in 2016. (Data from U.S. Census Bureau.)

EXAMPLE 4   Using Scientific Notation to Solve an Application

A *nanometer* is a very small unit of measure that is equivalent to about 0.00000003937 in. About how much would 700,000 nanometers measure in inches? (Data from *The World Almanac and Book of Facts.*)

Write each number in scientific notation, and then multiply.

$$700{,}000(0.00000003937)$$

$$= (7 \times 10^5)(3.937 \times 10^{-8}) \qquad \text{Write in scientific notation.}$$

$$= (7 \times 3.937)(10^5 \times 10^{-8}) \qquad \text{Properties of real numbers}$$

$$= 27.559 \times 10^{-3} \qquad \text{Multiply. Use the product rule.}$$

Don't stop here.

$$= (2.7559 \times 10^1) \times 10^{-3} \qquad \text{Write 27.559 in scientific notation.}$$

$$= 2.7559 \times 10^{-2} \qquad \text{Product rule}$$

$$= 0.027559 \qquad \text{Write in standard notation.}$$

Thus, 700,000 nanometers would measure

$$2.7559 \times 10^{-2} \text{ in., } \quad \text{or} \quad 0.027559 \text{ in.} \qquad \textbf{NOW TRY} \; \text{}$$

EXAMPLE 5   Using Scientific Notation to Solve an Application

As of November 30, 2017, the gross federal debt was about $\$2.0590 \times 10^{13}$ (which is more than \$20 trillion). The population of the United States was approximately 326 million that year. About how much would each person have had to contribute in order to pay off the federal debt? (Data from www.usgovernmentdebt.us; www.census.gov)

Divide to obtain the per person contribution.

$$\frac{2.0590 \times 10^{13}}{326{,}000{,}000}$$

$$= \frac{2.0590 \times 10^{13}}{3.26 \times 10^8} \qquad \text{Write 326 million in scientific notation.}$$

$$= \frac{2.0590}{3.26} \times 10^5 \qquad \text{Quotient rule}$$

$$= 0.63160 \times 10^5 \qquad \text{Divide. Round to 5 decimal places.}$$

$$= 63{,}160 \qquad \text{Write in standard notation.}$$

NOW TRY ANSWERS
4. 3.1496×10^{-1} in., or 0.31496 in.
5. 4.1×10^{-3} mi², or 0.0041 mi²

Each person would have to pay about \$63,160.   **NOW TRY**

4.3 Exercises

FOR EXTRA HELP   **MyLab Math**

 Video solutions for select problems available in MyLab Math

Concept Check   *Match each number written in scientific notation in Column I with the correct choice from Column II. Not all choices in Column II will be used.*

	I	II		I	II
1.	**(a)** 4.6×10^{-4}	**A.** 46,000	**2.**	**(a)** 1×10^9	**A.** 1 billion
	(b) 4.6×10^4	**B.** 460,000		**(b)** 1×10^6	**B.** 100 million
	(c) 4.6×10^5	**C.** 0.00046		**(c)** 1×10^8	**C.** 1 million
	(d) 4.6×10^{-5}	**D.** 0.000046		**(d)** 1×10^{10}	**D.** 10 billion
		E. 4600			**E.** 100 billion

Concept Check   *Determine whether or not each number is written in scientific notation as defined in **Objective 1**. If it is not, write it as such.*

3. 4.56×10^4 **4.** 7.34×10^6 **5.** 5,600,000   **6.** 34,000

7. 0.8×10^2 **8.** 0.9×10^3 **9.** 0.004   **10.** 0.0007

11. Concept Check   Write each number in scientific notation.

   (a) 63,000

      The first nonzero digit is _____ . The decimal point should be moved _____ places.

$$63,000 = \underline{\hspace{1cm}} \times 10^{\overline{\hspace{0.5cm}}}$$

   (b) 0.0571

      The first nonzero digit is _____ . The decimal point should be moved _____ places.

$$0.0571 = \underline{\hspace{1cm}} \times 10^{\overline{\hspace{0.5cm}}}$$

12. Concept Check   Write each number in standard notation.

   (a) 4.2×10^3

      Move the decimal point _____ places to the _____ .

$$4.2 \times 10^3 = \underline{\hspace{1cm}}$$

   (b) 6.42×10^{-3}

      Move the decimal point _____ places to the _____ .

$$6.42 \times 10^{-3} = \underline{\hspace{1cm}}$$

*Write each number in scientific notation. **See Example 1**.*

13. 5,876,000,000   **14.** 9,994,000,000   **15.** 82,350   **16.** 78,330

17. 0.000007   **18.** 0.0000004   **19.** 0.00203   **20.** 0.0000578

21. −13,000,000   **22.** −25,000,000,000   **23.** −0.006   **24.** −0.01234

*Write each number in standard notation. **See Example 2**.*

25. 7.5×10^5 **26.** 8.8×10^6 **27.** 5.677×10^{12} **28.** 8.766×10^9

29. 1×10^{12} **30.** 1×10^7 **31.** 6.21×10^0 **32.** 8.56×10^0

33. 7.8×10^{-4} **34.** 8.9×10^{-5} **35.** 5.134×10^{-9} **36.** 7.123×10^{-10}

37. -4×10^{-3} **38.** -6×10^{-4} **39.** -8.1×10^5 **40.** -9.6×10^6

*Each statement contains a number in **boldface italic** type. If the number is in scientific notation, write it in standard notation. If the number is not in scientific notation, write it as such. **See Examples 1 and 2**.*

41. A *muon* is an atomic particle closely related to an electron. The half-life of a muon is about 2 millionths $(\mathbf{2 \times 10^{-6}})$ of a second. (Data from www2.fisica.unlp.edu.ar)

42. There are 13 red balls and 39 black balls in a box. Mix them up and draw 13 out one at a time without returning any ball . . . the probability that the 13 drawings each will produce a red ball is . . . $\mathbf{1.6 \times 10^{-12}}$. (Data from Weaver, W., *Lady Luck*.)

43. An electron and a positron attract each other in two ways: the electromagnetic attraction of their opposite electric charges, and the gravitational attraction of their two masses. The electromagnetic attraction is

$$\mathbf{4,200,000,000,000,000,000,000,000,000,000,000,000,000,000}$$

times as strong as the gravitational. (Data from Asimov, I., *Isaac Asimov's Book of Facts*.)

44. The name "googol" applies to the number

$$10,000,000,000,000,000,000,000,000,000,000,000,000,000,000,000,000,000,$$
$$000,000,000,000,000,000,000,000,000,000,000,000,000,000.$$

The Web search engine Google honors this number. Sergey Brin, president and cofounder of Google, Inc., was a mathematics major. He chose the name Google to describe the vast reach of this search engine. (Data from *The Gazette*.)

*Perform the indicated operations. Write each answer (a) in scientific notation and (b) in standard notation. **See Example 3.***

45. $(2 \times 10^8)(3 \times 10^3)$ **46.** $(4 \times 10^7)(3 \times 10^3)$

47. $(5 \times 10^4)(3 \times 10^2)$ **48.** $(8 \times 10^5)(2 \times 10^3)$

49. $(3 \times 10^{-4})(-2 \times 10^8)$ **50.** $(4 \times 10^{-3})(-2 \times 10^7)$

51. $(6 \times 10^3)(4 \times 10^{-2})$ **52.** $(7 \times 10^5)(3 \times 10^{-4})$

53. $(9 \times 10^4)(7 \times 10^{-7})$ **54.** $(6 \times 10^4)(8 \times 10^{-8})$

55. $\dfrac{9 \times 10^{-5}}{3 \times 10^{-1}}$ **56.** $\dfrac{12 \times 10^{-4}}{4 \times 10^{-3}}$ **57.** $\dfrac{8 \times 10^3}{-2 \times 10^2}$

58. $\dfrac{15 \times 10^4}{-3 \times 10^3}$ **59.** $\dfrac{2.6 \times 10^{-3}}{2 \times 10^2}$ **60.** $\dfrac{9.5 \times 10^{-1}}{5 \times 10^3}$

61. $\dfrac{4 \times 10^5}{8 \times 10^2}$ **62.** $\dfrac{3 \times 10^9}{6 \times 10^5}$ **63.** $\dfrac{-4.5 \times 10^4}{1.5 \times 10^{-2}}$

64. $\dfrac{-7.2 \times 10^3}{6.0 \times 10^{-1}}$ **65.** $\dfrac{-8 \times 10^{-4}}{-4 \times 10^3}$ **66.** $\dfrac{-5 \times 10^{-6}}{-2 \times 10^2}$

Calculators can express numbers in scientific notation. The displays often use notation such as

$$5.4\text{E}3 \quad \text{to represent} \quad 5.4 \times 10^3.$$

Similarly, $5.4\text{E}-3$ represents 5.4×10^{-3}. Predict the display the calculator would give for the expression shown in each screen.

67.
```
NORMAL FLOAT AUTO REAL RADIAN MP
0.00000047
```

68.
```
NORMAL FLOAT AUTO REAL RADIAN MP
0.000021
```

69.
```
NORMAL FLOAT AUTO REAL RADIAN MP
(8E5)/(4E-2)
```

70.
```
NORMAL FLOAT AUTO REAL RADIAN MP
(9E-4)/(3E3)
```

71.
```
NORMAL FLOAT AUTO REAL RADIAN MP
(2E6)*(2E-3)/(4E2)
```

72.
```
NORMAL FLOAT AUTO REAL RADIAN MP
(5E-3)*(1E9)/(5E3)
```

Extending Skills   *Use scientific notation to calculate the result in each expression. Write answers in scientific notation.*

73. $\dfrac{650,000,000(0.0000032)}{0.00002}$ **74.** $\dfrac{3,400,000,000(0.000075)}{0.00025}$

75. $\dfrac{0.00000072(0.00023)}{0.000000018}$ **76.** $\dfrac{0.000000081(0.000036)}{0.00000048}$

77. $\dfrac{0.0000016(240,000,000)}{0.00002(0.0032)}$ **78.** $\dfrac{0.000015(42,000,000)}{0.000009(0.000005)}$

Use scientific notation to calculate the answer to each problem. ***See Examples 3–5.***

79. The Double Helix Nebula, a conglomeration of dust and gas stretching across the center of the Milky Way galaxy, is 25,000 light-years from Earth. If one light-year is about 6,000,000,000,000 mi, about how many miles is the Double Helix Nebula from Earth? (Data from www.spitzer.caltech.edu)

80. Pollux, one of the brightest stars in the night sky, is 33.7 light-years from Earth. If one light-year is about 6,000,000,000,000 mi (that is, 6 trillion mi), about how many miles is Pollux from Earth? (Data from *The World Almanac and Book of Facts.*)

81. In 2017, the population of the United States was about 326.4 million. To the nearest dollar, calculate how much each person in the United States would have had to contribute in order to make one person a trillionaire (that is, to give that person $1,000,000,000,000). (Data from U.S. Census Bureau.)

82. In 2016, the U.S. government collected about $5712 per person in individual income taxes. If the population of the United States at that time was 323,000,000, how much did the government collect in taxes for 2016? (Data from www.usgovernmentrevenue.com)

83. Before Congress raised the debt limit in 2017, it was 1.98×10^{13}. To the nearest dollar, how much was this for every man, woman, and child in the country? Use 326 million as the population of the United States. (Data from www.washingtonpost.com; www.census.gov)

84. In 2016, the state of Minnesota had about 7.33×10^4 farms with an average of 3.53×10^2 acres per farm. What was the total number of acres devoted to farmland in Minnesota that year? (Data from U.S. Department of Agriculture.)

85. Light travels at a speed of 1.86×10^5 mi per sec. When Venus is 6.68×10^7 mi from the sun, how long (in seconds) does it take light to travel from the sun to Venus? (Data from *The World Almanac and Book of Facts.*)

86. The distance to Earth from Pluto is 4.58×10^9 km. *Pioneer 10* transmitted radio signals from Pluto to Earth at the speed of light, 3.00×10^5 km per sec. About how long (in seconds) did it take for the signals to reach Earth?

87. During the 2016–2017 season, Broadway shows grossed a total of 1.45×10^9. Total attendance for the season was 1.33×10^7. What was the average ticket price (to the nearest cent) for a Broadway show? (Data from The Broadway League.)

88. In 2016, 1.14×10^{10} was spent to attend motion pictures in the United States and Canada. The total number of tickets sold was 1.32 billion. What was the average ticket price (to the nearest cent) for a movie? (Data from Motion Picture Association of America.)

89. In 2017, the world's fastest computer could perform 93,014,600,000,000,000 calculations per second. How many calculations could it perform per minute?   Per hour? (Data from www.top500.org)

90. In 2017, the fastest computer in the United States could handle 17.59 quadrillion calculations per second. (*Hint:* 1 quadrillion $= 1 \times 10^{15}$.) How many calculations could it perform per minute?   Per hour? (Data from www.top500.org)

RELATING CONCEPTS For Individual or Group Work (Exercises 91–95)

In 1935, Charles F. Richter devised a scale to compare the intensities of earthquakes. The **intensity** of an earthquake is measured relative to the intensity of a standard **zero-level earthquake** of intensity I_0. The relationship is equivalent to

$$I = I_0 \times 10^R, \quad \text{where } R \text{ is the \textbf{Richter scale} measure.}$$

For example, if an earthquake has magnitude 5.0 on the Richter scale, then its intensity is calculated as

$$I = I_0 \times 10^{5.0} = I_0 \times 100{,}000,$$

which is 100,000 times as intense as a zero-level earthquake.

Intensity	$I_0 \times 10^0$	$I_0 \times 10^1$	$I_0 \times 10^2$	$I_0 \times 10^3$	$I_0 \times 10^4$	$I_0 \times 10^5$	$I_0 \times 10^6$	$I_0 \times 10^7$	$I_0 \times 10^8$
Richter Scale	0	1	2	3	4	5	6	7	8

To compare two earthquakes, such as one that measures 8.0 to one that measures 5.0, calculate the ratio of their intensities.

$$\frac{\text{intensity 8.0}}{\text{intensity 5.0}} = \frac{I_0 \times 10^{8.0}}{I_0 \times 10^{5.0}} = \frac{10^8}{10^5} = 10^{8-5} = 10^3 = 1000$$

An earthquake that measures 8.0 is 1000 times as intense as one that measures 5.0.

Use the information in the table to **work Exercises 91–95 in order.**

Year	Earthquake Location	Richter Scale Measurement
1960	Valdivia, Chile	9.5
2010	Maule region, Chile	8.8
2007	Southern Sumatra, Indonesia	8.5
2015	Gorkha district, Nepal	7.8
2015	Farkhar, Afghanistan	7.5
2018	Mooreland, Oklahoma	3.8

Data from earthquake.usgs.gov

91. Compare the intensity of the 1960 Valdivia earthquake with that of the 2007 Southern Sumatra earthquake.

92. Compare the intensity of the 2010 Maule earthquake with that of 2015 Gorkha earthquake.

93. Compare the intensity of the 1960 Valdivia earthquake with that of the 2015 Farkhar earthquake.

94. Compare the intensity of the 2010 Maule earthquake with that of the 2018 Mooreland earthquake.

95. Compare the intensity of the 2015 Gorkha earthquake with that of the 2018 Mooreland earthquake.

4.4   Adding, Subtracting, and Graphing Polynomials

OBJECTIVES

1. Identify terms and coefficients.
2. Combine like terms.
3. Describe polynomials using appropriate vocabulary.
4. Evaluate polynomials.
5. Add and subtract polynomials.
6. Graph equations defined by polynomials of degree 2.

VOCABULARY

☐ term
☐ leading term
☐ numerical coefficient (coefficient)
☐ like terms
☐ unlike terms
☐ polynomial
☐ descending powers
☐ degree of a term
☐ degree of a polynomial
☐ monomial
☐ binomial
☐ trinomial
☐ parabola
☐ vertex
☐ axis of symmetry (axis)

NOW TRY EXERCISE 1

Identify the coefficient of each term in the expression. Give the number of terms.

$$t - 10t^2$$

NOW TRY ANSWER
1. $1; -10$; two terms

OBJECTIVE 1   Identify terms and coefficients.

Recall that in an expression such as

$$4x^3 + 6x^2 + 5x + 8,$$

the quantities $4x^3$, $6x^2$, $5x$, and 8 are **terms.** In the **leading** (or first) **term** $4x^3$, the number 4 is the **numerical coefficient,** or simply the **coefficient,** of x^3. In the same way, 6 is the coefficient of x^2 in the term $6x^2$, and 5 is the coefficient of x in the term $5x$. The constant term 8 can be thought of as

$$8 \cdot 1 = 8x^0 \quad \text{because} \quad x^0 = 1,$$

so 8 is the coefficient in the term 8. Other examples are given in the table.

▼ Terms and Their Coefficients

Term	Numerical Coefficient
6	6
$-7y$	-7
$34r^3$	34
$-26x^5yz^4$	-26
$-k = -1k$	-1
$r = 1r$	1
$\frac{3x}{8} = \frac{3}{8}x$	$\frac{3}{8}$
$\frac{x}{3} = \frac{1x}{3} = \frac{1}{3}x$	$\frac{1}{3}$

EXAMPLE 1   Identifying Coefficients

Identify the coefficient of each term in the expression. Give the number of terms.

(a) $x - 6x^4 + 3$ can be written as $1x + (-6x^4) + 3x^0$

There are three terms:

$$x, \quad -6x^4, \quad \text{and} \quad 3.$$

The coefficients are 1, -6, and 3.

(b) $5 - v^3$ can be written as $5v^0 + (-1v^3)$.

There are two terms.   The coefficients are 5 and -1.

NOW TRY

OBJECTIVE 2   Combine like terms.

Recall that **like terms** have exactly the same variables, with the same exponents on the variables. *Only the coefficients may differ.*

Like terms		Unlike terms	
$19m^5$ and $14m^5$		$7x$ and $7y$	
$6y^9$, $-37y^9$, and y^9	Examples of **like terms**	z^4 and z	Examples of **unlike terms**
$3pq$ and $-2pq$		$2pq$ and $2p$	
$2xy^2$ and $-xy^2$		$-4xy^2$ and $5x^2y$	

Using the distributive property, we combine like terms by adding or subtracting their coefficients.

NOW TRY
EXERCISE 2

Simplify by combining like terms.

(a) $-7x^6 + 12x^6$

(b) $x - \dfrac{2}{5}x$

(c) $3x^2 - x^2 + 2x$

EXAMPLE 2 Combining Like Terms

Simplify by combining like terms.

(a) $-4x^3 + 6x^3$

$$= (-4 + 6)x^3 \qquad \begin{array}{l} ac + bc \\ = (a+b)c \end{array}$$

$$= 2x^3$$

(b) $9x^6 - 14x^6 + x^6 \qquad \boxed{x^6 = 1x^6}$

$$= (9 - 14 + 1)x^6$$

$$= -4x^6$$

(c) $y + \dfrac{2}{3}y$

$$= 1y + \dfrac{2}{3}y \qquad y = 1y$$

$$= \left(\dfrac{3}{3} + \dfrac{2}{3}\right)y \qquad \begin{array}{l} 1 = \frac{3}{3}; \text{ Distributive} \\ \text{property} \end{array}$$

$$= \dfrac{5}{3}y \qquad \text{Add the fractions.}$$

(d) $8rs - 13rs + 9rs$

$$= (8 - 13 + 9)rs$$

$$= 4rs$$

> Note how the distributive property justifies the procedure for combining like terms.

(e) $12m^2 + 5m + 4m^2$

$$= (12 + 4)m^2 + 5m$$

$$= 16m^2 + 5m \qquad \boxed{\text{Stop here. These are unlike terms.}}$$

(f) $5u + 11v$

These are unlike terms. They cannot be combined.

NOW TRY

⚠️ **CAUTION**   In **Example 2(e),** we cannot combine $16m^2$ and $5m$ because the exponents on the variables are different. *Unlike terms have different variables or different exponents on the same variables.*

OBJECTIVE 3 Describe polynomials using appropriate vocabulary.

Polynomial in x

A **polynomial in x** is a term or the sum of a finite number of terms of the form

$$ax^n, \quad \text{for any real number } a \text{ and any whole number } n.$$

For example, the expression

$$16x^8 - 7x^6 + 5x^4 - 3x^2 + 4 \qquad \begin{array}{l} \text{Polynomial in } x \\ \text{(The 4 can be written as } 4x^0.) \end{array}$$

is a polynomial in x. This polynomial is written in **descending powers** of the variable because the exponents on x decrease from left to right. By contrast,

$$2x^3 - x^2 + \dfrac{4}{x}, \quad \text{or} \quad 2x^3 - x^2 + 4x^{-1}, \qquad \text{Not a polynomial}$$

NOW TRY ANSWERS

2. (a) $5x^6$ **(b)** $\frac{3}{5}x$ **(c)** $2x^2 + 2x$

is not a polynomial in x. A variable appears in a denominator or as a factor to a negative power in a numerator.

> **NOTE** We can define a polynomial using any variable, not just x, as in **Example 2(e).** A polynomial may have terms with more than one variable, as in **Example 2(d).**

The **degree of a term** is the sum of the exponents on the variables. The **degree of a polynomial** is the greatest degree of any nonzero term of the polynomial.

▼ Degrees of Terms and Polynomials

Term	Degree	Polynomial	Degree
$3x^4$	4	$3x^4 - 5x^2 + 6$	4
$5x$, or $5x^1$	1	$5x + 7$	1
-7, or $-7x^0$	0	$x^5 + 3x^6 - 7$	6
$2x^2y$, or $2x^2y^1$	$2 + 1 = 3$	$5y^2 + xy - 2x^2y$	3

Some polynomials with a specific number of terms have special names.

- A polynomial with exactly one term is a **monomial.** (*Mono-* means "one," as in *mono*rail.)

$$9m, \quad -6y^5, \quad x^2, \quad \text{and} \quad 6 \qquad \text{Monomials}$$

- A polynomial with exactly two terms is a **binomial.** (*Bi-* means "two," as in *bi*cycle.)

$$-9x^4 + 9x^3, \quad 8m^2 + 6m, \quad \text{and} \quad 3t - 10 \qquad \text{Binomials}$$

- A polynomial with exactly three terms is a **trinomial.** (*Tri-* means "three," as in *tri*angle.)

$$9m^3 - 4m^2 + 6, \quad \frac{19}{3}y^2 + \frac{8}{3}y + 5, \quad \text{and} \quad -3z^5 - z^2 + z \qquad \text{Trinomials}$$

EXAMPLE 3 Classifying Polynomials

For each polynomial, first simplify, if possible. Then give the degree and tell whether the simplified polynomial is a *monomial*, a *binomial*, a *trinomial*, or *none of these*.

(a) $2x^3 + 5$ The polynomial cannot be simplified. It is a *binomial* of degree 3.

(b) $6x - 8x + 13x$

$= 11x$ Combine like terms to simplify.

The degree is 1 (because $x = x^1$). The simplified polynomial is a *monomial*.

(c) $4xy - 5xy + 2xy$

$= xy$ Combine like terms to simplify.

The degree is 2 (because $xy = x^1y^1$, and $1 + 1 = 2$). The simplified polynomial is a *monomial*.

(d) $2x^2 - 3x + 8x - 12$

$= 2x^2 + 5x - 12$ Combine like terms.

The degree is 2. The simplified polynomial is a *trinomial*.   **NOW TRY** ↻

NOW TRY EXERCISE 3

Simplify, give the degree, and tell whether the simplified polynomial is a *monomial*, a *binomial*, a *trinomial*, or *none of these.*

(a) $3x^2 + 2x - 4$

(b) $x^3 + 4x^3$

(c) $x^8 - x^7 + 2x^8$

(d) $-3x^2 + 8x - 6x + 4$

NOW TRY ANSWERS

3. (a) The polynomial cannot be simplified; degree 2; trinomial

(b) $5x^3$; degree 3; monomial

(c) $3x^8 - x^7$; degree 8; binomial

(d) $-3x^2 + 2x + 4$; degree 2; trinomial

OBJECTIVE 4   Evaluate polynomials.

When we *evaluate* a polynomial, we find its *value*. A polynomial usually represents different numbers for different values of the variable.

NOW TRY
EXERCISE 4
Find the value for $t = -3$.
$$4t^3 - t^2 - t$$

EXAMPLE 4   Evaluating a Polynomial

Find the value of $3x^4 + 5x^3 - 4x - 4$ for **(a)** $x = -2$ and **(b)** $x = 3$.

(a)

$$3x^4 + 5x^3 - 4x - 4$$

> Use parentheses to avoid errors.

$$= 3(-2)^4 + 5(-2)^3 - 4(-2) - 4 \qquad \text{Substitute } -2 \text{ for } x.$$

$$= 3(16) + 5(-8) - 4(-2) - 4 \qquad \text{Apply the exponents.}$$

$$= 48 - 40 + 8 - 4 \qquad \text{Multiply.}$$

$$= 12 \qquad \text{Add and subtract.}$$

(b)

$$3x^4 + 5x^3 - 4x - 4$$

> Replace x with 3.

$$= 3(3)^4 + 5(3)^3 - 4(3) - 4 \qquad \text{Let } x = 3.$$

$$= 3(81) + 5(27) - 4(3) - 4 \qquad \text{Apply the exponents.}$$

$$= 243 + 135 - 12 - 4 \qquad \text{Multiply.}$$

$$= 362 \qquad \text{Add and subtract.} \qquad \text{NOW TRY}$$

> **! CAUTION**   Use parentheses around the numbers that are substituted for the variable, as in Example 4. *Be particularly careful when substituting a negative number for a variable that is raised to a power, or a sign error may result.*

OBJECTIVE 5   Add and subtract polynomials.

Adding Polynomials

To add two polynomials, combine (add) like terms.

EXAMPLE 5   Adding Polynomials Vertically

Find each sum.

(a) Add $6x^3 - 4x^2 + 3$ and $-2x^3 + 7x^2 - 5$.

$$6x^3 - 4x^2 + 3 \qquad \text{Write like terms in columns.}$$
$$+ \ (-2x^3 + 7x^2 - 5)$$

Now add, column by column.

> Add the coefficients only. Do *not* add the exponents.

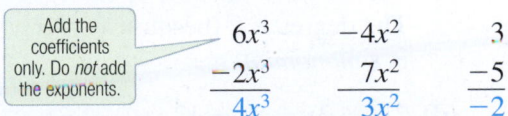

Add the three sums together to obtain the answer.

$$4x^3 + 3x^2 + (-2) = 4x^3 + 3x^2 - 2 \leftarrow \text{Final sum}$$

NOW TRY ANSWER
4. -114

**NOW TRY
EXERCISE 5**

Find each sum.

(a) Add $7y^3 - 4y^2 + 2$ and $-6y^3 + 5y^2 - 3$.

(b) Add $-5x^4 - 2x + 3$ and $x^3 - 5x$.

(b) Add $2x^2 - 4x + 3$ and $x^3 + 5x$.

Write like terms in columns and add column by column.

$$
\begin{array}{r}
2x^2 - 4x + 3 \\
+\ (x^3 \qquad + 5x \qquad) \\
\hline
x^3 + 2x^2 + \ x + 3
\end{array}
$$

Leave spaces for missing terms.

NOW TRY

The polynomials in **Example 5** also can be added horizontally.

**NOW TRY
EXERCISE 6**

Add $10x^4 - 3x^2 - x$ and $x^4 - 3x^2 + 5x$ horizontally.

EXAMPLE 6 Adding Polynomials Horizontally

Find each sum.

(a) Add $6x^3 - 4x^2 + 3$ and $-2x^3 + 7x^2 - 5$.

$$(6x^3 - 4x^2 + 3) + (-2x^3 + 7x^2 - 5) = 4x^3 + 3x^2 - 2$$

Same answer as found in **Example 5(a)**

(b) Add $2x^2 - 4x + 3$ and $x^3 + 5x$.

$$
\begin{aligned}
(2x^2 - 4x + 3) &+ (x^3 + 5x) \\
&= x^3 + 2x^2 - 4x + 5x + 3 \qquad \text{Commutative property} \\
&= x^3 + 2x^2 + x + 3 \qquad\qquad \text{See Example 5(b).}
\end{aligned}
$$

NOW TRY

The difference $x - y$ is defined as $x + (-y)$. (We find the difference $x - y$ by adding x and the opposite of y.)

Examples: $7 - 2$ is equivalent to $7 + (-2)$,   which equals 5.

$-8 - (-2)$ is equivalent to $-8 + 2$,   which equals -6.

A similar method is used to subtract polynomials.

Subtracting Polynomials

To subtract two polynomials, change the sign of each term in the subtrahend (second polynomial) and add the result to the minuend (first polynomial)—that is, add the *opposite* of each term of the second polynomial to the first polynomial.

EXAMPLE 7 Subtracting Polynomials Horizontally

Perform each subtraction.

(a) $(5x - 2) - (3x - 8)$

$$
\begin{aligned}
&= (5x - 2) + [-(3x - 8)] \qquad \text{Definition of subtraction} \\
&= (5x - 2) + [-1(3x - 8)] \qquad -a = -1a \\
&= (5x - 2) + (-3x + 8) \qquad\quad \text{Distributive property} \\
&= 2x + 6 \qquad\qquad\qquad\qquad\quad \text{Combine like terms.}
\end{aligned}
$$

NOW TRY ANSWERS
5. (a) $y^3 + y^2 - 1$
   (b) $-5x^4 + x^3 - 7x + 3$
6. $11x^4 - 6x^2 + 4x$

 **NOW TRY EXERCISE 7**

Perform each subtraction.

(a) $(3x - 8) - (5x - 9)$

(b) $(4t^4 - t^2 + 7)$
   $- (5t^4 - 3t^2 + 1)$

CHECK To check a subtraction problem, use the following fact.

$$\text{If} \quad a - b = c, \quad \text{then} \quad a = b + c.$$

We found that $(5x - 2) - (3x - 8) = 2x + 6$. Check as follows.

$$(3x - 8) + (2x + 6)$$
$$= 5x - 2 \checkmark$$

(b) Subtract $6x^3 - 4x^2 + 2$ from $11x^3 + 2x^2 - 8$.

> Be careful to write the problem in the correct order.

$$(11x^3 + 2x^2 - 8) - (6x^3 - 4x^2 + 2)$$
$$= (11x^3 + 2x^2 - 8) + (-6x^3 + 4x^2 - 2)$$
$$= 5x^3 + 6x^2 - 10 \quad \text{Combine like terms.}$$

NOW TRY

Subtraction can also be done in columns.

 **NOW TRY EXERCISE 8**

Subtract by columns.

$(12x^2 - 9x + 4)$
$- (-10x^2 - 3x + 7)$

EXAMPLE 8 Subtracting Polynomials Vertically

Subtract by columns: $(14y^3 - 6y^2 + 2y - 5) - (2y^3 - 7y^2 - 4y + 6)$.

$$14y^3 - 6y^2 + 2y - 5$$
$$- \underline{(2y^3 - 7y^2 - 4y + 6)} \quad \text{Arrange like terms in columns.}$$

Change all signs in the second row (the subtrahend), and then add.

$$14y^3 - 6y^2 + 2y - 5$$
$$+ \underline{(-2y^3 + 7y^2 + 4y - 6)} \quad \text{Change all signs.}$$
$$12y^3 + y^2 + 6y - 11 \quad \text{Add.}$$

NOW TRY

 NOW TRY EXERCISE 9

Perform the indicated operations.

$(6p^4 - 8p^3 + 2p - 1)$
$- (-7p^4 + 6p^2 - 12)$
$+ (p^4 - 3p + 8)$

EXAMPLE 9 Adding and Subtracting More Than Two Polynomials

Perform the indicated operations.

$$(4 - x + 3x^2) - (2 - 3x + 5x^2) + (8 + 2x - 4x^2)$$

Rewrite, using the definition of subtraction.

$$(4 - x + 3x^2) - (2 - 3x + 5x^2) + (8 + 2x - 4x^2)$$
$$= (4 - x + 3x^2) + (-2 + 3x - 5x^2) + (8 + 2x - 4x^2)$$
$$= (2 + 2x - 2x^2) + (8 + 2x - 4x^2) \quad \text{Combine like terms.}$$
$$= 10 + 4x - 6x^2 \quad \text{Combine like terms.}$$

NOW TRY

NOW TRY EXERCISE 10

Subtract.

$(4x^2 - 2xy + y^2)$
$- (6x^2 - 7xy + 2y^2)$

EXAMPLE 10 Adding and Subtracting Multivariable Polynomials

Add or subtract as indicated.

(a) $(4a + 2ab - b) + (3a - ab + b)$

$$= 4a + 2ab - b + 3a - ab + b$$
$$= 7a + ab \quad \text{Combine like terms.}$$

NOW TRY ANSWERS

7. (a) $-2x + 1$
   (b) $-t^4 + 2t^2 + 6$
8. $22x^2 - 6x - 3$
9. $14p^4 - 8p^3 - 6p^2 - p + 19$
10. $-2x^2 + 5xy - y^2$

(b) $(2x^2y + 3xy + y^2) - (3x^2y - xy - 2y^2)$

$$= 2x^2y + 3xy + y^2 - 3x^2y + xy + 2y^2$$
$$= -x^2y + 4xy + 3y^2$$

> Be careful with signs. The coefficient of xy is 1.

NOW TRY

OBJECTIVE 6 Graph equations defined by polynomials of degree 2.

Earlier we graphed linear equations (which are actually polynomial equations of degree 1). By plotting points, we can graph polynomial equations of degree 2.

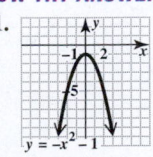

**NOW TRY
EXERCISE 11**

Graph $y = -x^2 - 1$.

EXAMPLE 11 Graphing Equations Defined by Polynomials of Degree 2

Graph each equation.

(a) $y = x^2$

It is easier to select values for x and find corresponding y-values. Selecting $x = 2$ and substituting in $y = x^2$ gives

$$y = 2^2 = 4.$$

The point $(2, 4)$ is on the graph of $y = x^2$. (Recall that in an ordered pair such as $(2, 4)$, *the x-value comes first and the y-value second.*) We show some ordered pairs that satisfy $y = x^2$ in the table with **FIGURE 3.** If we plot the ordered pairs from the table on a coordinate system and draw a smooth curve through them, we obtain the graph shown in **FIGURE 3.**

The graph of $y = x^2$ is the graph of a function, because each input x is related to just one output y. The curve in **FIGURE 3** is a **parabola.**

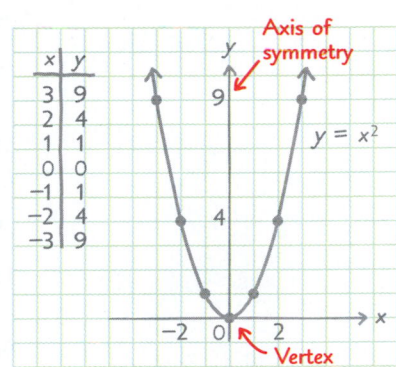

FIGURE 3

- The point $(0, 0)$, the *lowest* point on this graph, is the **vertex** of the parabola.
- The vertical line through the vertex (the y-axis here) is the **axis of symmetry,** or simply the **axis,** of the parabola. This axis is a line of symmetry for the graph. If the graph is folded on this line, the two halves will coincide.

(b) $y = -x^2 + 3$

Plot points to obtain the graph. For example, let $x = -2$ and $x = 0$.

$$y = -x^2 + 3 \qquad\qquad y = -x^2 + 3$$
$$y = -(-2)^2 + 3 \qquad\qquad y = -0^2 + 3$$
$$y = -4 + 3 \qquad\qquad y = 0 + 3$$
$$y = -1 \qquad\qquad y = 3$$

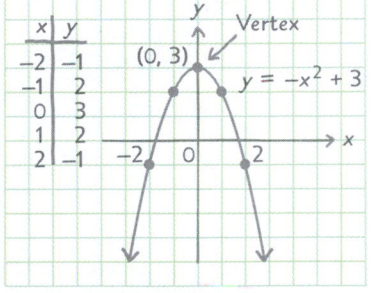

FIGURE 4

The points $(-2, -1)$, $(0, 3)$, and several others are shown in the table that accompanies the graph in **FIGURE 4.** The vertex $(0, 3)$ is the *highest* point on this graph. The graph opens downward because x^2 has a negative coefficient.

NOW TRY

NOW TRY ANSWER

11.

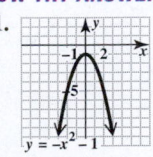

$y = -x^2 - 1$

NOTE   *All polynomials of degree 2 have parabolas as their graphs.* When graphing, find points until the vertex and points on either side of it are located. (In this section, all parabolas have their vertices on the x-axis or the y-axis.)

4.4 Exercises

FOR EXTRA HELP

 MyLab Math

▶ *Video solutions for select problems available in MyLab Math*

Concept Check *Complete each statement.*

1. In the term $4x^6$, the coefficient of x^6 is _____ and the exponent is _____.

2. The expression $4x^3 - 5x^2$ has exactly (*one / two / three*) term(s).

3. The degree of the term $-3x^9$ is _____.

4. The polynomial $4x^2 + y^2$ (*is / is not*) an example of a trinomial.

5. When $x^2 + 10$ is evaluated for $x = 3$, the result is _____. For $x = -3$, the result is _____.

6. $5x^{\underline{}} + 3x^3 - 7x$ is a trinomial of degree 6.

7. Combining like terms in $-3xy - 2xy + 5xy$ gives _____.

8. _____ is a monomial with coefficient 8, in the variable x, having degree 5.

Identify the coefficient of each term in the expression, and give the number of terms. See Example 1.

9. $6x^4$ **10.** $-9y^5$ **11.** t^4 **12.** s^7

13. $-19r^2 - r$ **14.** $2y^3 - y$ **15.** $x + 8x^2 + 5x^3$ **16.** $v - 2v^3 - v^7$

In each polynomial, simplify by combining like terms whenever possible. Write results that have more than one term in descending powers of the variable. See Example 2 and Objective 3.

17. $-3m^5 + 5m^5$ **18.** $-4y^3 + 3y^3$ **19.** $2r^5 + (-3r^5)$

20. $9y^2 + (-19y^2)$ **21.** $0.2m^5 - 0.5m^2$ **22.** $-0.9y + 0.9y^2$

23. $-3x^5 + 3x^5 - 5x^5$ **24.** $6x^3 - 9x^3 + 10x^3$ **25.** $-4p^7 + 8p^7 + 5p^9$

26. $-3a^8 + 4a^8 - 3a^2$ **27.** $-1.5x^2 + 5.3x^2 - 3.8x^2$ **28.** $8.6y^4 - 10.3y^4 + 1.7y^4$

29. $-4y^2 + 3y^2 - 2y^2 + y^2$ **30.** $3r^5 - 8r^5 + r^5 + 2r^5$

31. $-\dfrac{1}{3}tu^7 + \dfrac{2}{5}tu^7 + \dfrac{1}{15}tu^7 - \dfrac{8}{5}tu^7$ **32.** $-\dfrac{3}{4}p^2q - \dfrac{1}{3}p^2q + \dfrac{7}{12}p^2q - \dfrac{1}{6}p^2q$

For each polynomial, first simplify, if possible, and write the result in descending powers of the variable. Then give the degree and tell whether the simplified polynomial is a monomial, *a* binomial, *a* trinomial, *or none of these. See Example 3.*

33. $6x^4 - 9x$ **34.** $7t^3 - 3t$

35. $5m^4 - 3m^2 + 6m^4 - 7m^3$ **36.** $6p^5 + 4p^3 - 8p^5 + 10p^2$

37. $\dfrac{5}{3}x^4 - \dfrac{2}{3}x^4$ **38.** $\dfrac{4}{5}r^6 + \dfrac{1}{5}r^6$

39. $0.8x^4 - 0.3x^4 - 0.5x^4 + 7$ **40.** $1.2t^3 - 0.9t^3 - 0.3t^3 + 9$

41. $-11ab + 2ab - 4ab$ **42.** $5xy + 13xy - 12xy$

43. Concept Check A student incorrectly evaluated the polynomial below for $x = -2$. **WHAT WENT WRONG?** Give the correct answer.

$$-x^2 + 3x + 4$$
$$= [-(-2)]^2 + 3(-2) + 4$$
$$= 4 - 6 + 4$$
$$= 2$$

44. Concept Check A student incorrectly evaluated the polynomial below for $x = -1$. **WHAT WENT WRONG?** Give the correct answer.

$$2x^3 + 8x - 5$$
$$= [2(-1)]^3 + 8(-1) - 5$$
$$= (-2)^3 - 8 - 5$$
$$= -8 - 8 - 5$$
$$= -21$$

Find the value of each polynomial for **(a)** $x = 2$ *and* **(b)** $x = -1$. *See Example 4.*

45. $2x^2 + 5x + 1$

46. $x^2 + 5x - 10$

47. $-3x^2 + 14x - 2$

48. $-2x^2 + 5x - 1$

49. $2x^5 - 4x^4 + 5x^3 - x^2$

50. $x^4 - 6x^3 + x^2 - x$

Add. See Examples 5 and 6.

51. $2x^2 - 4x$
$+(3x^2 + 2x)$

52. $-5y^3 + 3y$
$+(8y^3 - 4y)$

53. $3m^2 + 5m + 6$
$+(2m^2 - 2m - 4)$

54. $4a^3 - 4a^2 - 4$
$+(6a^3 + 5a^2 - 8)$

55. $\frac{2}{3}x^2 + \frac{1}{5}x + \frac{1}{6}$
$+\left(\frac{1}{2}x^2 - \frac{1}{3}x + \frac{2}{3}\right)$

56. $\frac{4}{7}y^2 - \frac{1}{5}y + \frac{7}{9}$
$+\left(\frac{1}{3}y^2 - \frac{1}{3}y + \frac{2}{5}\right)$

57. $9m^3 - 5m^2 + 4m - 8$ and $-3m^3 + 6m^2 - 6$

58. $12r^5 + 11r^4 - 7r^3 - 2r^2$ and $-8r^5 + 3r^3 + 2r^2$

Subtract. See Example 8.

59. $5y^3 - 3y^2$
$-(2y^3 + 8y^2)$

60. $-6t^3 + 4t^2$
$-(8t^3 - 6t^2)$

61. $12x^4 - x^2 + x$
$-(8x^4 + 3x^2 - 3x)$

62. $13y^5 - y^3 - 8y^2$
$-(7y^5 + 5y^3 + y^2)$

63. $12m^3 - 8m^2 + 6m + 7$
$-(-3m^3 + 5m^2 - 2m - 4)$

64. $5a^4 - 3a^3 + 2a^2 - a + 6$
$-(-6a^4 + a^3 - a^2 + a - 1)$

Perform each indicated operation. See Examples 6, 7, and 9.

65. $(8m^2 - 7m) - (3m^2 + 7m - 6)$

66. $(x^2 + x) - (3x^2 + 2x - 1)$

67. $(16x^3 - x^2 + 3x) + (-12x^3 + 3x^2 + 2x)$

68. $(-2b^6 + 3b^4 - b^2) + (b^6 + 2b^4 + 2b^2)$

69. Subtract $9x^2 - 3x + 7$ from $-2x^2 - 6x + 4$.

70. Subtract $-5w^3 + 5w^2 - 7$ from $6w^3 + 8w + 5$.

71. $(9a^4 - 3a^2 + 2) + (4a^4 - 4a^2 + 2) + (-12a^4 + 6a^2 - 3)$

72. $(4m^2 - 3m + 2) + (5m^2 + 13m - 4) + (-16m^2 - 4m + 3)$

73. $[(8m^2 + 4m - 7) - (2m^2 - 5m + 2)] - (m^2 + m + 1)$

74. $[(9b^3 - 4b^2 + 3b + 2) - (-2b^3 - 3b^2 + b)] - (8b^3 + 6b + 4)$

75. $[(3x^2 - 2x + 7) - (4x^2 + 2x - 3)] - [(9x^2 + 4x - 6) + (-4x^2 + 4x + 4)]$

76. $[(6t^2 - 3t + 1) - (12t^2 + 2t - 6)] - [(4t^2 - 3t - 8) + (-6t^2 + 10t - 12)]$

Concept Check   *Answer each of the following.*

77. Without actually performing the operations, mentally determine the coefficient of the x^2-term in the simplified form of

$$(-4x^2 + 2x - 3) - (-2x^2 + x - 1) + (-8x^2 + 3x - 4).$$

78. Without actually performing the operations, mentally determine the coefficient of the x-term in the simplified form of

$$(-8x^2 - 3x + 2) - (4x^2 - 3x + 8) - (-2x^2 - x + 7).$$

Add or subtract as indicated. See Example 10.

79. $(6b + 3c) + (-2b - 8c)$ **80.** $(-5t + 13s) + (8t - 3s)$

81. $(4x + 2xy - 3) - (-2x + 3xy + 4)$ **82.** $(8ab + 2a - 3b) - (6ab - 2a + 3b)$

83. $(2c^4d + 3c^2d^2 - 4d^2) - (c^4d + 8c^2d^2 - 5d^2)$

84. $(3k^2h^3 + 5kh + 6k^3h^2) - (2k^2h^3 - 9kh + k^3h^2)$

Find a polynomial that represents the perimeter of each rectangle, square, or triangle.

85.
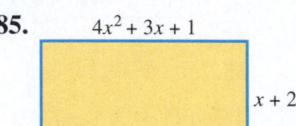
$4x^2 + 3x + 1$
$x + 2$

86.
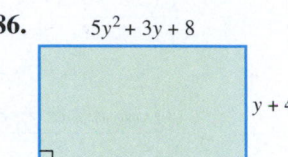
$5y^2 + 3y + 8$
$y + 4$

87.
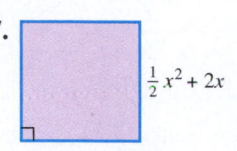
$\frac{1}{2}x^2 + 2x$

88.
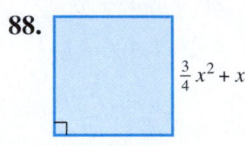
$\frac{3}{4}x^2 + x$

89.
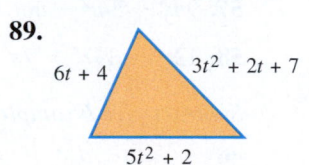
$6t + 4$ $3t^2 + 2t + 7$
$5t^2 + 2$

90.
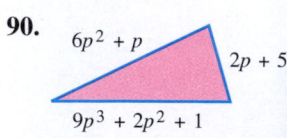
$6p^2 + p$ $2p + 5$
$9p^3 + 2p^2 + 1$

*Find **(a)** a polynomial that represents the perimeter of each triangle and **(b)** the degree measures of the angles of the triangle. (Hint: The sum of the measures of the angles of any triangle is 180°.)*

91.
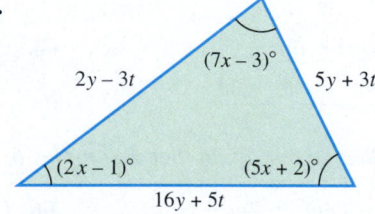
$(7x - 3)°$
$2y - 3t$ $5y + 3t$
$(2x - 1)°$ $(5x + 2)°$
$16y + 5t$

92.
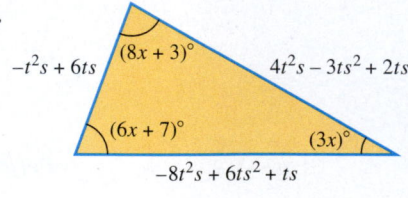
$-t^2s + 6ts$ $(8x + 3)°$ $4t^2s - 3ts^2 + 2ts$
$(6x + 7)°$ $(3x)°$
$-8t^2s + 6ts^2 + ts$

Extending Skills   *Perform each indicated operation.*

93. Find the difference of the sum of $5x^2 + 2x - 3$ and $x^2 - 8x + 2$ and the sum of $7x^2 - 3x + 6$ and $-x^2 + 4x - 6$.

94. Subtract the sum of $9t^3 - 3t + 8$ and $t^2 - 8t + 4$ from the sum of $12t + 8$ and $t^2 - 10t + 3$.

Graph each equation by completing the table of values. See Example 11.

95. $y = x^2 - 4$ **96.** $y = x^2 - 6$ **97.** $y = 2x^2 - 1$

x	y
-2	
-1	
0	
1	
2	

x	y
-2	
-1	
0	
1	
2	

x	y
-2	
-1	
0	
1	
2	

98. $y = 2x^2 + 2$

x	y
-2	
-1	
0	
1	
2	

99. $y = -x^2 + 4$

x	y
-2	
-1	
0	
1	
2	

100. $y = -x^2 + 2$

x	y
-2	
-1	
0	
1	
2	

101. $y = (x + 3)^2$

x	-5	-4	-3	-2	-1
y					

102. $y = (x - 4)^2$

x	2	3	4	5	6
y					

RELATING CONCEPTS For Individual or Group Work (Exercises 103–106)

The following binomial models the distance in feet that a car going approximately 68 mph *will skid in t seconds.*

$$100t - 13t^2$$

When we evaluate this binomial for a value of t, we obtain a value for distance. This illustrates the concept of a **function**—*for each input of a time, we obtain one and only one output for distance.*

   Exercises 103–106 further illustrate the function concept with polynomials. **Work them in order.**

103. Evaluate the given binomial $100t - 13t^2$ for $t = 5$. Use the result to fill in the blanks:

In _____ seconds, the car will skid _____ feet.

104. If one "dog" year is estimated to be about seven "human" years, the monomial $7x$ gives the dog's age in human years for x dog years. Evaluate this monomial for $x = 9$. Use the result to fill in the blanks:

If a dog is _____ in dog years,

then it is _____ in human years.

105. If it costs $15 plus $2 per day to rent a chain saw, the binomial

$$2x + 15$$

gives the cost in dollars to rent the chain saw for x days. Evaluate this binomial for $x = 6$. Use the result to fill in the blanks:

If the saw is rented for _____ days,

then the cost is _____ dollars.

106. If an object is projected upward under certain conditions, its height in feet is given by the trinomial

$$-16t^2 + 60t + 80,$$

where t is in seconds. Evaluate this trinomial for $t = 2.5$. Use the result to fill in the blanks:

If _____ seconds have elapsed, then the height of the object is _____ feet.

4.5   Multiplying Polynomials

OBJECTIVES

1 Multiply monomials.
2 Multiply a monomial and a polynomial.
3 Multiply two polynomials.
4 Multiply binomials using the FOIL method.

VOCABULARY

☐ FOIL method
☐ outer product
☐ inner product

NOW TRY EXERCISE 1

Find each product.
(a) $-6x(5x^3)$
(b) $6mn^3(12mn)$

NOW TRY EXERCISE 2

Find the product.

$-3x^5(2x^3 - 5x^2 + 10)$

OBJECTIVE 1 Multiply monomials.

Recall that we multiply monomials using the product rule for exponents.

| EXAMPLE 1 | Multiplying Monomials |

Find each product.

(a) $8m^2(-9m)$ (b) $4x^3y^2(2x^2y)$

$= 8(-9) \cdot m^2m^1$ $= 4(2) \cdot x^3x^2 \cdot y^2y^1$ Commutative and associative properties

$= -72m^{2+1}$ $= 8x^{3+2}y^{2+1}$ Multiply; product rule

$= -72m^3$ $= 8x^5y^3$ Add.   NOW TRY

⚠ **CAUTION**   *Do not confuse addition of terms with multiplication of terms.*

$7q^5 + 2q^5$		$(7q^5)(2q^5)$	
$= (7 + 2)q^5$	Distributive property	$= 7 \cdot 2q^{5+5}$	Commutative property; product rule
$= 9q^5$	Add.	$= 14q^{10}$	Multiply. Add.

OBJECTIVE 2 Multiply a monomial and a polynomial.

We use the distributive property and multiplication of monomials.

| EXAMPLE 2 | Multiplying Monomials and Polynomials |

Find each product.

(a) $4x^2(3x + 5)$ $a(b + c) = ab + ac$

$= 4x^2(3x) + 4x^2(5)$ Distributive property

$= 12x^3 + 20x^2$ Multiply monomials.

(b) $-8m^3(4m^3 + 3m^2 + 2m - 1)$

$= -8m^3(4m^3) + (-8m^3)(3m^2)$

$+ (-8m^3)(2m) + (-8m^3)(-1)$ Distributive property

$= -32m^6 - 24m^5 - 16m^4 + 8m^3$ Multiply monomials.   NOW TRY

OBJECTIVE 3 Multiply two polynomials.

To find the product of the polynomials $x^2 + 3x + 5$ and $x - 4$, we can think of $x - 4$ as a single quantity and use the distributive property as follows.

$(x^2 + 3x + 5)(x - 4)$

$= x^2(x - 4) + 3x(x - 4) + 5(x - 4)$ Distributive property

$= x^2(x) + x^2(-4) + 3x(x) + 3x(-4) + 5(x) + 5(-4)$

Distributive property again

$= x^3 - 4x^2 + 3x^2 - 12x + 5x - 20$ Multiply monomials.

$= x^3 - x^2 - 7x - 20$ Combine like terms.

NOW TRY ANSWERS

1. (a) $-30x^4$ (b) $72m^2n^4$
2. $-6x^8 + 15x^7 - 30x^5$

Multiplying Polynomials

To multiply two polynomials, multiply each term of the first polynomial by each term of the second polynomial. Then combine like terms.

NOW TRY
EXERCISE 3
Multiply.

$(x^2 - 4)(2x^2 - 5x + 3)$

EXAMPLE 3 Multiplying Two Polynomials

Multiply $(m^2 + 5)(4m^3 - 2m^2 + 4m)$.

$(m^2 + 5)(4m^3 - 2m^2 + 4m)$ Multiply each term of the first polynomial by each term of the second.

$= m^2(4m^3) + m^2(-2m^2) + m^2(4m) + 5(4m^3) + 5(-2m^2) + 5(4m)$ Distributive properly

$= 4m^5 - 2m^4 + 4m^3 + 20m^3 - 10m^2 + 20m$ Multiply monomials.

$= 4m^5 - 2m^4 + 24m^3 - 10m^2 + 20m$ Combine like terms.

NOW TRY

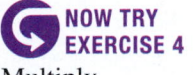
NOW TRY
EXERCISE 4
Multiply.

$5t^2 - 7t + 4$
$\underline{\qquad 2t - 6}$

EXAMPLE 4 Multiplying Polynomials Vertically

Multiply $(x^3 + 2x^2 + 4x + 1)(3x + 5)$ vertically.

$x^3 + 2x^2 + 4x + 1$ Write the polynomials vertically.
$\underline{\qquad\qquad 3x + 5}$

Begin by multiplying each of the terms in the top row by 5.

$x^3 + 2x^2 + 4x + 1$
$\underline{\qquad\qquad 3x + 5}$
$5x^3 + 10x^2 + 20x + 5$ $5(x^3 + 2x^2 + 4x + 1)$

Now multiply each term in the top row by $3x$. Then add like terms.

$x^3 + 2x^2 + 4x + 1$
$\underline{\qquad\qquad 3x + 5}$
$5x^3 + 10x^2 + 20x + 5$ This process is similar to multiplication of whole numbers.

Place *like* terms in columns so they can be combined.

$\underline{3x^4 + 6x^3 + 12x^2 + 3x}$ $3x(x^3 + 2x^2 + 4x + 1)$
$3x^4 + 11x^3 + 22x^2 + 23x + 5$ Add in columns.   **NOW TRY**

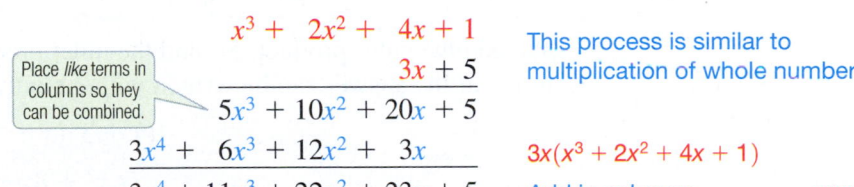

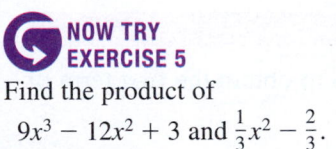
NOW TRY
EXERCISE 5
Find the product of
$9x^3 - 12x^2 + 3$ and $\frac{1}{3}x^2 - \frac{2}{3}$.

EXAMPLE 5 Multiplying Polynomials Vertically (Fractional Coefficients)

Find the product of $4m^3 - 2m^2 + 4m$ and $\frac{1}{2}m^2 + \frac{5}{2}$.

$4m^3 - 2m^2 + 4m$
$\underline{\qquad \frac{1}{2}m^2 + \frac{5}{2}}$
$10m^3 - 5m^2 + 10m$ Terms of top row are multiplied by $\frac{5}{2}$.
$\underline{2m^5 - m^4 + 2m^3}$ Terms of top row are multiplied by $\frac{1}{2}m^2$.
$2m^5 - m^4 + 12m^3 - 5m^2 + 10m$ Add in columns.   **NOW TRY**

We can use a rectangle to model polynomial multiplication. For example, to find $(2x + 1)(3x + 2)$, we label a rectangle, as shown below on the left. Then we put the product of each pair of monomials in the appropriate box, as shown on the right.

NOW TRY ANSWERS

3. $2x^4 - 5x^3 - 5x^2 + 20x - 12$
4. $10t^3 - 44t^2 + 50t - 24$
5. $3x^5 - 4x^4 - 6x^3 + 9x^2 - 2$

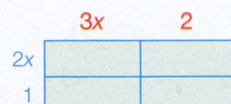

	3x	2
2x	6x²	4x
1	3x	2

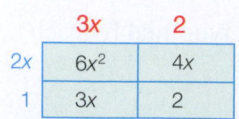

This approach can be extended to polynomials with any number of terms.

The product of the binomials is the sum of the four monomial products.

$$(2x + 1)(3x + 2)$$
$$= 6x^2 + 4x + 3x + 2$$
$$= 6x^2 + 7x + 2 \qquad \text{Combine like terms.}$$

OBJECTIVE 4 Multiply binomials using the FOIL method.

When multiplying binomials, the **FOIL method** reduces the rectangle method to a systematic approach without the rectangle. Consider this example.

$$(x + 3)(x + 5)$$
$$= x(x + 5) + 3(x + 5) \qquad \text{Distributive property}$$
$$= x(x) + x(5) + 3(x) + 3(5) \qquad \text{Distributive property again}$$
$$= x^2 + 5x + 3x + 15 \qquad \text{Multiply.}$$
$$= x^2 + 8x + 15 \qquad \text{Combine like terms.}$$

The letters of the word FOIL refer to the positions of the terms.

$(x + 3)(x + 5)$ Multiply the **First terms:** $x(x)$.   **F**

$(x + 3)(x + 5)$ Multiply the **Outer terms:** $x(5)$.   **O**
This is the **outer product.**

$(x + 3)(x + 5)$ Multiply the **Inner terms:** $3(x)$.   **I**
This is the **inner product.**

$(x + 3)(x + 5)$ Multiply the **Last terms:** $3(5)$.   **L**

We add the outer product, $5x$, and the inner product, $3x$, to obtain $8x$ so that the three terms of the answer can be written without extra steps.

$$(x + 3)(x + 5)$$
$$= x^2 + 8x + 15$$

FOIL Method for Multiplying Binomials

Step 1   Multiply the two **First** terms of the binomials to obtain the first term of the product.

Step 2   Find the **Outer** product and the **Inner** product and combine them (when possible) to obtain the middle term of the product.

Step 3   Multiply the two **Last** terms of the binomials to obtain the last term of the product.

Step 4   Add the terms found in Steps 1–3.

$$\mathbf{F} = x^2 \qquad \mathbf{L} = 15$$

Example: $(x + 3)(x + 5)$ $(x + 3)(x + 5)$
$$= x^2 + 8x + 15$$

$$\mathbf{I} = 3x$$
$$\mathbf{O} = 5x$$
$$8x \qquad \text{Combine like terms.}$$

NOW TRY
EXERCISE 6
Use the FOIL method to find the product.

$$(t - 6)(t + 5)$$

EXAMPLE 6   Using the FOIL Method

Use the FOIL method to find the product $(x + 8)(x - 6)$.

Step 1 **F** Multiply the First terms: $x(x) = x^2$.

Step 2 **O** Find the Outer product: $x(-6) = -6x$.   ⎫ Combine:
            I Find the Inner product: $8(x) = 8x$.   ⎭ $-6x + 8x = 2x$

Step 3 **L** Multiply the Last terms: $8(-6) = -48$.

Step 4 The product $(x + 8)(x - 6)$ is $x^2 + 2x - 48$.   Add the terms found in Steps 1–3.

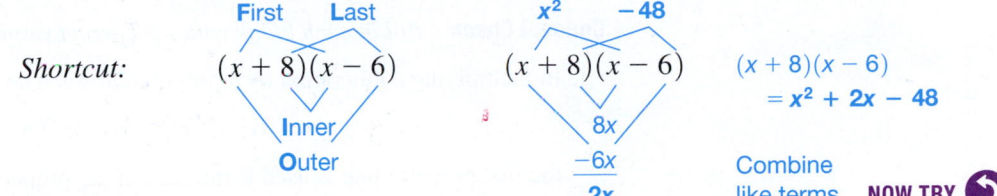

Shortcut:

$$(x + 8)(x - 6)$$

First   Last   Inner   Outer

$$(x + 8)(x - 6)$$

x^2 -48 $8x$ $-6x$ $2x$

$$(x + 8)(x - 6) = x^2 + 2x - 48$$

Combine like terms.   **NOW TRY**

NOW TRY
EXERCISE 7
Multiply.

$$(7y - 3)(2x + 5)$$

EXAMPLE 7   Using the FOIL Method

Multiply $(9x - 2)(3y + 1)$.

First $(9x - 2)(3y + 1)$ **27xy**

Outer $(9x - 2)(3y + 1)$ **9x** ⎱ These unlike terms *cannot* be combined.

Inner $(9x - 2)(3y + 1)$ **−6y**

Last $(9x - 2)(3y + 1)$ **−2**

F   O   I   L

The product $(9x - 2)(3y + 1)$ is $27xy + 9x - 6y - 2$.   **NOW TRY**

NOW TRY
EXERCISE 8
Find each product.
(a) $(3p - 5q)(4p - q)$
(b) $5x^2(3x + 1)(x - 5)$

EXAMPLE 8   Using the FOIL Method

Find each product.

(a) $(2k + 5y)(k + 3y)$

F   O   I   L
$= 2k(k) + 2k(3y) + 5y(k) + 5y(3y)$

$= 2k^2 + 6ky + 5ky + 15y^2$ Multiply.

$= 2k^2 + 11ky + 15y^2$ Combine like terms.

(b) $(7p + 2q)(3p - q)$

$= 21p^2 - pq - 2q^2$ FOIL method

(c) $2x^2(x - 3)(3x + 4)$

$= 2x^2(3x^2 - 5x - 12)$ FOIL method

$= 6x^4 - 10x^3 - 24x^2$ Distributive property

NOW TRY

NOW TRY ANSWERS
6. $t^2 - t - 30$
7. $14yx + 35y - 6x - 15$
8. **(a)** $12p^2 - 23pq + 5q^2$
   (b) $15x^4 - 70x^3 - 25x^2$

NOTE   Alternatively, the factors in **Example 8(c)** can be multiplied as follows.

$2x^2(x - 3)(3x + 4)$ Multiply $2x^2$ and $x - 3$ first.

$= (2x^3 - 6x^2)(3x + 4)$ Multiply that product and $3x + 4$.

$= 6x^4 - 10x^3 - 24x^2$ FOIL method; The same answer results.

4.5 Exercises

FOR EXTRA HELP   **MyLab Math**

▶ *Video solutions for select problems available in MyLab Math*

Concept Check   *Match each product in Column I with the correct polynomial in Column II.*

I	II		I	II
1. (a) $5x^3(6x^7)$	**A.** $125x^{21}$		**2. (a)** $(x-5)(x+4)$	**A.** $x^2+9x+20$
(b) $-5x^7(6x^3)$	**B.** $30x^{10}$		**(b)** $(x+5)(x+4)$	**B.** $x^2-9x+20$
(c) $(5x^7)^3$	**C.** $-216x^9$		**(c)** $(x-5)(x-4)$	**C.** x^2-x-20
(d) $(-6x^3)^3$	**D.** $-30x^{10}$		**(d)** $(x+5)(x-4)$	**D.** x^2+x-20

Concept Check   *Fill in each blank with the correct response.*

3. In multiplying a monomial by a polynomial, such as in

$$4x(3x^2+7x^3)=4x(3x^2)+4x(7x^3),$$

the first property that is used is the _____ property.

4. The FOIL method can only be used to multiply two polynomials when both polynomials are _____ .

5. The product $2x^2(-3x^5)$ has exactly _____ term(s) after the multiplication is performed.

6. The product $(a+b)(c+d)$ has exactly _____ term(s) after the multiplication is performed.

Find each product. ***See Example 1.***

7. $5y^4(3y^7)$ **8.** $10p^2(5p^3)$ **9.** $-15a^4(-2a^5)$

10. $-3m^6(-5m^4)$ **11.** $5p(3q^2)$ **12.** $4a^3(3b^2)$

13. $-6m^3(3n^2)$ **14.** $9r^3(-2s^2)$ **15.** $y^5 \cdot 9y \cdot y^4$

16. $x^2 \cdot 3x^3 \cdot 2x$ **17.** $(4x^3)(2x^2)(-x^5)$ **18.** $(7t^5)(3t^4)(-t^8)$

Find each product. ***See Example 2.***

19. $2m(3m+2)$ **20.** $4x(5x+3)$ **21.** $3p(-2p^3+4p^2)$

22. $4x(3+2x+5x^3)$ **23.** $-8z(2z+3z^2+3z^3)$ **24.** $-7y(3+5y^2-2y^3)$

25. $2y^3(3+2y+5y^4)$ **26.** $2m^4(6+5m+3m^2)$

27. $-4r^3(-7r^2+8r-9)$ **28.** $-9a^5(-3a^6-2a^4+8a^2)$

29. $3a^2(2a^2-4ab+5b^2)$ **30.** $4z^3(8z^2+5zy-3y^2)$

Concept Check   *Multiply.*

31. $7m^3n^2(3m^2+2mn-n^3)$

$= 7m^3n^2(\underline{\quad}) + 7m^3n^2(\underline{\quad})$

$+ 7m^3n^2(\underline{\quad})$

$= \underline{\qquad\qquad}$

32. $2p^2q(3p^2q^2-5p+2q^2)$

$= \underline{\quad}(3p^2q^2) + \underline{\quad}(-5p)$

$+ \underline{\quad}(2q^2)$

$= \underline{\qquad\qquad}$

Find each product. ***See Examples 3–5.***

33. $(6x+1)(2x^2+4x+1)$ **34.** $(9a+2)(9a^2+a+1)$

35. $(9y-2)(8y^2-6y+1)$ **36.** $(2r-1)(3r^2+4r-4)$

37. $(4m + 3)(5m^3 - 4m^2 + m - 5)$

38. $(2y + 8)(3y^4 - 2y^2 + 1)$

39. $(2x - 1)(3x^5 - 2x^3 + x^2 - 2x + 3)$

40. $(2a + 3)(a^4 - a^3 + a^2 - a + 1)$

41. $(5x^2 + 2x + 1)(x^2 - 3x + 5)$

42. $(2m^2 + m - 3)(m^2 - 4m + 5)$

43. $(6x^4 - 4x^2 + 8x)\left(\frac{1}{2}x + 3\right)$

44. $(8y^6 + 4y^4 - 12y^2)\left(\frac{3}{4}y^2 + 2\right)$

Find each product using the rectangle method shown in the text. Determine the individual terms that should appear on the blanks or in the rectangles, and then give the final product.

45. $(x + 3)(x + 4)$

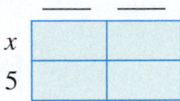

Product: _____

46. $(x + 5)(x + 2)$

Product: _____

47. $(2x + 1)(x^2 + 3x + 2)$

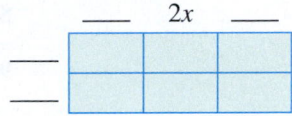

Product: _____

48. $(x + 4)(3x^2 + 2x + 1)$

Product: _____

Concept Check   *For each product, find and simplify the following.*

(a) *Product of first terms*

____ (____)

= ____

(b) *Outer product*

____ (____)

= ____

(c) *Inner product*

____ (____)

= ____

(d) *Product of last terms*

____ (____)

= ____

(e) *Complete product in simplified form*

_____

49. $(2p - 5)(3p + 7)$

50. $(2p - 5)(2p + 5)$

*Find each product. **See Examples 6–8.***

51. $(m + 7)(m + 5)$

52. $(n + 9)(n + 3)$

53. $(n - 1)(n + 4)$

54. $(t - 3)(t + 8)$

55. $(2x + 3)(6x - 4)$

56. $(3y + 5)(8y - 6)$

57. $(9 + t)(9 - t)$

58. $(10 + r)(10 - r)$

59. $(3x - 2)(3x - 2)$

60. $(4m + 3)(4m + 3)$

61. $(5a + 1)(2a + 7)$

62. $(b + 8)(6b - 2)$

63. $(6 - 5m)(2 + 3m)$

64. $(8 - 3a)(2 + a)$

65. $(5 - 3x)(4 + x)$

66. $(6 - 5x)(2 + x)$

67. $(3t - 4s)(t + 3s)$

68. $(2m - 3n)(m + 5n)$

69. $(4x + 3)(2y - 1)$

70. $(5x + 7)(3y - 8)$

71. $(3x + 2y)(5x - 3y)$

72. $(5a + 3b)(5a - 4b)$

73. $3y^3(2y + 3)(y - 5)$

74. $2x^2(2x - 5)(x + 3)$

75. $-8r^3(5r^2 + 2)(5r^2 - 2)$

76. $-5t^4(2t^4 + 1)(2t^4 - 1)$

77. $-3r(r - 1)(r^2 + r + 1)$

78. Concept Check Multiply each of the following.

(a) $(x - 1)(x^2 + x + 1)$ (b) $(y - 2)(y^2 + 2y + 4)$

Describe the final product as "the _____ of two _____."

Find polynomials that represent, in appropriate units, (a) the area and (b) the perimeter of each square or rectangle. (If necessary, refer to the formulas at the back of this text.)

79.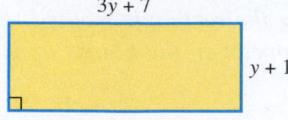

80.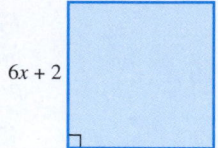

Extending Skills *Find each product. Recall that $a^2 = a \cdot a$ and $a^3 = a^2 \cdot a$.*

81. $(x + 7)^2$

82. $(m + 6)^2$

83. $(a - 4)(a + 4)$

84. $(b - 10)(b + 10)$

85. $(2p - 5)^2$

86. $(3m - 1)^2$

87. $(5k + 3q)^2$

88. $(8m + 3n)^2$

89. $(m - 5)^3$

90. $(p - 3)^3$

91. $(2a + 1)^3$

92. $(3m + 1)^3$

93. $-3a(3a + 1)(a - 4)$

94. $-4r(3r + 2)(2r - 5)$

95. $7(4m - 3)(2m + 1)$

96. $5(3k - 7)(5k + 2)$

97. $(3r - 2s)^4$

98. $(2z - 5y)^4$

99. $3p^3(2p^2 + 5p)(p^3 + 2p + 1)$

100. $5k^2(k^3 - 3)(k^2 - k + 4)$

101. $-2x^5(3x^2 + 2x - 5)(4x + 2)$

102. $-4x^3(3x^4 + 2x^2 - x)(-2x + 1)$

103. $\left(3p^2 + \dfrac{5}{4}q\right)\left(2p^2 - \dfrac{5}{3}q\right)$

104. $\left(2x^2 + \dfrac{2}{3}y\right)\left(3x^2 - \dfrac{3}{4}y\right)$

The figures in Exercises 105–108 are composed of triangles, squares, rectangles, and circles. Find a polynomial that represents the area, in square units, of each shaded region. In Exercises 107 and 108, leave π in the answers. (If necessary, refer to the formulas at the back of this text.)

105.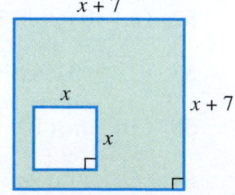

106.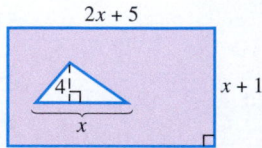

107.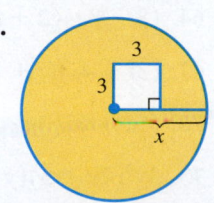

108.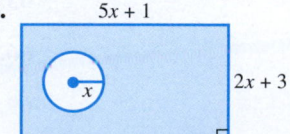

RELATING CONCEPTS For Individual or Group Work (Exercises 109–114)

Work Exercises 109–114 in order. (All units are in feet.)

109. Find a polynomial that represents the area, in square feet, of the rectangle.

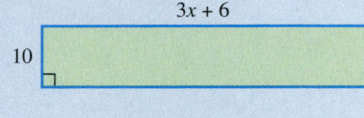

110. Suppose we know that the area of the rectangle is 600 ft^2. Use this information and the polynomial from **Exercise 109** to write an equation in x, and solve it.

111. Refer to **Exercise 110**. What are the dimensions of the rectangle?

112. Use the result of **Exercise 111** to find the perimeter of the rectangle.

113. Suppose the rectangle represents a strip of lawn and it costs $0.75 per square foot to lay sod on the lawn. How much will it cost to sod the entire lawn?

114. Suppose it costs $20.50 per linear foot for fencing. How much will it cost to fence the entire lawn?

4.6 Special Products

OBJECTIVES

1 Square binomials.

2 Find the product of the sum and difference of two terms.

3 Find greater powers of binomials.

VOCABULARY

☐ difference of two squares
☐ conjugates

NOW TRY EXERCISE 1

Find $(x + 5)^2$.

OBJECTIVE 1 Square binomials.

EXAMPLE 1 Squaring a Binomial

Find $(m + 3)^2$.

$$(m + 3)(m + 3) \quad \text{[}(m + 3)^2 \text{ means } (m + 3)(m + 3).\text{]}$$

$$= m^2 + 3m + 3m + 9 \qquad \text{FOIL method}$$

$$\text{This is the answer.} \quad = m^2 + 6m + 9 \qquad \text{Combine like terms.}$$

This result has the squares of the first and the last terms of the binomial.

$$m^2 = m^2 \quad \text{and} \quad 3^2 = 9$$

The middle term of the trinomial, 6m, is twice the product of the two terms of the binomial, m and 3. This is true because when we used the FOIL method above, the outer and inner products were $m(3)$ and $3(m)$, and

$$m(3) + 3(m) \quad \text{equals} \quad 2(m)(3).$$

Thus, $\quad (m + 3)^2 \quad = \quad m^2 \quad + \quad 6m \quad + \quad 9.$

Square m. $2(m)(3)$ Square 3.   **NOW TRY**

Square of a Binomial

The square of a binomial is a *trinomial* consisting of

the square of the first term $+$ twice the product of the two terms $+$ the square of the last term.

For x and y, the following hold true.

$$(x + y)^2 = x^2 + 2xy + y^2$$

$$(x - y)^2 = x^2 - 2xy + y^2$$

NOW TRY ANSWER

1. $x^2 + 10x + 25$

Square each binomial.

(a) $(3x - 1)^2$

(b) $(4p - 5q)^2$

(c) $\left(6t - \frac{1}{3}\right)^2$

(d) $-(3y + 2)^2$

(e) $m(2m + 3)^2$

EXAMPLE 2 Squaring Binomials

Square each binomial.

$$(x - y)^2 \;=\; x^2 \;-\; 2 \cdot x \cdot y \;+\; y^2$$

(a) $(t - 8)^2 = t^2 - 2(t)(8) + 8^2$

$$= t^2 - 16t + 64$$

(b) $(5z - 1)^2$

$$= (5z)^2 - 2(5z)(1) + 1^2 \qquad (x - y)^2 = x^2 - 2xy + y^2$$

> Be careful to square 5z correctly.

$$= 5^2 z^2 - 10z + 1 \qquad (5z)^2 = 5^2 z^2 = 25z^2 \text{ by power rule (b)}$$

$$= 25z^2 - 10z + 1$$

(c) $(3b + 5r)^2$

$$= (3b)^2 + 2(3b)(5r) + (5r)^2$$

> Be careful to square 3b and 5r correctly.

$$= 9b^2 + 30br + 25r^2$$

(d) $(2a - 9x)^2$

$$= (2a)^2 - 2(2a)(9x) + (9x)^2$$

$$= 4a^2 - 36ax + 81x^2$$

(e) $\left(4m + \frac{1}{2}\right)^2$

$$= (4m)^2 + 2(4m)\left(\frac{1}{2}\right) + \left(\frac{1}{2}\right)^2$$

$$= 16m^2 + 4m + \frac{1}{4}$$

(f) $-(2x - 3)^2$

$$= -\left[(2x)^2 - 2(2x)(3) + 3^2\right]$$

$$= -(4x^2 - 12x + 9)$$

$$= -4x^2 + 12x - 9$$

(g) $x(4x - 3)^2$

> Remember the middle term, $2(4x)(3)$.

$$= x(16x^2 - 24x + 9) \qquad \text{Square the binomial.}$$

$$= 16x^3 - 24x^2 + 9x \qquad \text{Distributive property}$$

NOW TRY

In the square of a sum, all of the terms are positive. (See Example 2(c).) In the square of a difference, the middle term is negative. (See Example 2(a).)

⚠ CAUTION A common error in squaring a binomial is to forget the middle term of the product. In general, remember the following.

$$(x + y)^2 = x^2 + 2xy + y^2, \quad \textbf{\textit{not}} \quad x^2 + y^2.$$

$$(x - y)^2 = x^2 - 2xy + y^2, \quad \textbf{\textit{not}} \quad x^2 - y^2.$$

OBJECTIVE 2 Find the product of the sum and difference of two terms.

In binomial products of the form $(x + y)(x - y)$, one binomial is a sum of two terms. The other is a difference of the *same* two terms. Consider the following.

$$(x + 2)(x - 2)$$

$$= x^2 - 2x + 2x - 4 \qquad \text{FOIL method}$$

$$= x^2 - 4 \qquad \text{Combine like terms.}$$

Thus, the product of $x + y$ and $x - y$ is a **difference of two squares.**

Product of a Sum and Difference of Two Terms

The product of a sum and difference of two terms is a *binomial* consisting of

the square of the first term $-$ the square of the second term.

For x and y, the following holds true.

$$(x + y)(x - y) = x^2 - y^2$$

NOW TRY EXERCISE 3

Find the product.

$(t + 10)(t - 10)$

EXAMPLE 3 Finding the Product of a Sum and Difference of Two Terms

Find each product.

(a) $(x + 4)(x - 4)$ ← This is the product of a sum and difference of two terms.

$= x^2 - 4^2$ $(x + y)(x - y) = x^2 - y^2$

$= x^2 - 16$ Square 4.

(b) $\left(\dfrac{2}{3} - w\right)\left(\dfrac{2}{3} + w\right)$

$= \left(\dfrac{2}{3} + w\right)\left(\dfrac{2}{3} - w\right)$ Commutative property

$= \left(\dfrac{2}{3}\right)^2 - w^2$ $(x + y)(x - y) = x^2 - y^2$

$= \dfrac{4}{9} - w^2$ Square $\frac{2}{3}$.

(c) $x(x + 2)(x - 2)$ Find the product of the sum and difference of two terms.

$= x(x^2 - 4)$

$= x^3 - 4x$ Distributive property

NOW TRY EXERCISE 4

Find each product.

(a) $(4x - 6)(4x + 6)$

(b) $\left(5r - \frac{4}{5}\right)\left(5r + \frac{4}{5}\right)$

(c) $y(3y + 1)(3y - 1)$

(d) $-5(p + q^2)(p - q^2)$

EXAMPLE 4 Finding the Product of a Sum and Difference of Two Terms

Find each product.

$(x + y)(x - y)$

(a) $(5m + 3)(5m - 3)$ ← This is the product of a sum and difference of two terms.

$= (5m)^2 - 3^2$ $(x + y)(x - y) = x^2 - y^2$

$= 25m^2 - 9$ Apply the exponents.

Be careful to square 5m correctly.

(b) $(4x + y)(4x - y)$

$= (4x)^2 - y^2$

$= 16x^2 - y^2$ $(4x)^2 = 4^2x^2 = 16x^2$

(c) $\left(z - \dfrac{1}{4}\right)\left(z + \dfrac{1}{4}\right)$

$= z^2 - \dfrac{1}{16}$

(d) $p(2p + 1)(2p - 1)$

$= p(4p^2 - 1)$

$= 4p^3 - p$ Distributive property

(e) $-3(x + y^2)(x - y^2)$

$= -3(x^2 - y^4)$

$= -3x^2 + 3y^4$

NOW TRY ANSWERS

3. $t^2 - 100$

4. (a) $16x^2 - 36$
   (b) $25r^2 - \frac{16}{25}$
   (c) $9y^3 - y$
   (d) $-5p^2 + 5q^4$

> **NOTE** The expressions $x + y$ and $x - y$, the sum and difference of the *same* two terms, are **conjugates**.
>
> *Example:* $x + 2$ and $x - 2$ are conjugates.

OBJECTIVE 3 Find greater powers of binomials.

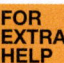 **NOW TRY EXERCISE 5**

Find the product.

$$(2m - 1)^3$$

EXAMPLE 5 Finding Greater Powers of Binomials

Find each product.

(a) $(x + 5)^3$

$$= (x + 5)(x + 5)^2 \qquad \qquad a^3 = a \cdot a^2$$
$$= (x + 5)(x^2 + 10x + 25) \qquad \text{Square the binomial.}$$
$$= x^3 + 10x^2 + 25x + 5x^2 + 50x + 125 \qquad \text{Multiply polynomials.}$$
$$= x^3 + 15x^2 + 75x + 125 \qquad \text{Combine like terms.}$$

(b) $(2y - 3)^4$

$$= (2y - 3)^2(2y - 3)^2 \qquad \qquad a^4 = a^2 \cdot a^2$$
$$= (4y^2 - 12y + 9)(4y^2 - 12y + 9) \qquad \text{Square each binomial.}$$
$$= 16y^4 - 48y^3 + 36y^2 - 48y^3 + 144y^2 \qquad \text{Multiply polynomials.}$$
$$\quad - 108y + 36y^2 - 108y + 81$$
$$= 16y^4 - 96y^3 + 216y^2 - 216y + 81 \qquad \text{Combine like terms.}$$

(c) $-2r(r + 2)^3$

$$= -2r(r + 2)(r + 2)^2 \qquad \qquad a^3 = a \cdot a^2$$
$$= -2r(r + 2)(r^2 + 4r + 4) \qquad \text{Square the binomial.}$$
$$= -2r(r^3 + 4r^2 + 4r + 2r^2 + 8r + 8) \qquad \text{Multiply polynomials.}$$
$$= -2r(r^3 + 6r^2 + 12r + 8) \qquad \text{Combine like terms.}$$
$$= -2r^4 - 12r^3 - 24r^2 - 16r \qquad \text{Distributive property} \qquad \text{NOW TRY} \, \circlearrowleft$$

NOW TRY ANSWER

5. $8m^3 - 12m^2 + 6m - 1$

4.6 Exercises

FOR EXTRA HELP    **MyLab Math**

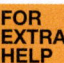 *Video solutions for select problems available in MyLab Math*

1. Concept Check A student incorrectly squared $(a + b)$ as follows.
$$(a + b)^2 = a^2 + b^2 \qquad \text{WRONG}$$
WHAT WENT WRONG? Give the correct answer.

2. Concept Check A student incorrectly squared $(x - y)$ as follows.
$$(x - y)^2 = x^2 + y^2 \qquad \text{WRONG}$$
WHAT WENT WRONG? Give the correct answer.

3. Concept Check Consider the square of the binomial $4x + 3$: $(4x + 3)^2$.

(a) What is the first term of the binomial? Square it.

(b) Find twice the product of the two terms of the binomial: $2(\_\_\_)(\_\_\_) = \_\_\_$.

(c) What is the last term of the binomial? Square it.

(d) Use the results of parts (a)–(c) to find $(4x + 3)^2$.

4. Concept Check Consider the product of $(7x + 3y)$ and $(7x - 3y)$:

$$(7x + 3y)(7x - 3y).$$

(a) What is the first term of each binomial factor? Square it.

(b) What is the product of the outer terms? The inner terms? Add them.

(c) What are the last terms of the binomial factors? Multiply them.

(d) Use the results of parts (a)–(c) to find $(7x + 3y)(7x - 3y)$.

Find each product. See Examples 1 and 2.

5. $(m + 2)^2$ **6.** $(x + 8)^2$ **7.** $(r - 3)^2$ **8.** $(z - 5)^2$

9. $(x + 2y)^2$ **10.** $(p - 3m)^2$ **11.** $(5p + 2q)^2$ **12.** $(8a + 3b)^2$

13. $(4x - 3)^2$ **14.** $(9x - 4)^2$ **15.** $(4a + 5b)^2$ **16.** $(9y + 4z)^2$

17. $\left(6m - \dfrac{4}{5}n\right)^2$ **18.** $\left(5x + \dfrac{2}{5}y\right)^2$ **19.** $\left(\dfrac{1}{2}x + \dfrac{1}{3}\right)^2$ **20.** $\left(\dfrac{1}{4}x + \dfrac{1}{5}\right)^2$

21. $2(x + 6)^2$ **22.** $4(x + 3)^2$ **23.** $t(3t - 1)^2$ **24.** $x(2x + 5)^2$

25. $3t(4t + 1)^2$ **26.** $2x(7x - 2)^2$ **27.** $-(4r - 2)^2$ **28.** $-(3y - 8)^2$

Find each product. See Examples 3 and 4.

29. $(k + 5)(k - 5)$ **30.** $(a + 8)(a - 8)$ **31.** $(4 - 3t)(4 + 3t)$

32. $(7 - 2x)(7 + 2x)$ **33.** $(5x + 2)(5x - 2)$ **34.** $(2m + 5)(2m - 5)$

35. $(5y + 3x)(5y - 3x)$ **36.** $(3x + 4y)(3x - 4y)$ **37.** $(10x + 3y)(10x - 3y)$

38. $(13r + 2z)(13r - 2z)$ **39.** $(2x^2 - 5)(2x^2 + 5)$ **40.** $(9y^2 - 2)(9y^2 + 2)$

41. $\left(\dfrac{3}{4} - x\right)\left(\dfrac{3}{4} + x\right)$ **42.** $\left(\dfrac{2}{3} + r\right)\left(\dfrac{2}{3} - r\right)$

43. $\left(9y + \dfrac{2}{3}\right)\left(9y - \dfrac{2}{3}\right)$ **44.** $\left(7x + \dfrac{3}{7}\right)\left(7x - \dfrac{3}{7}\right)$

45. $q(5q - 1)(5q + 1)$ **46.** $p(3p + 7)(3p - 7)$

47. $-5(a - b^3)(a + b^3)$ **48.** $-6(r - s^4)(r + s^4)$

49. $\dfrac{1}{2}(2k - 1)(2k + 1)$ **50.** $\dfrac{1}{3}(3m - 5)(3m + 5)$

51. $-\dfrac{1}{100}(10x + 10)(10x - 10)$ **52.** $-\dfrac{1}{200}(20y + 20)(20y - 20)$

Find each product. See Example 5.

53. $(x + 1)^3$ **54.** $(y + 2)^3$ **55.** $(t - 3)^3$ **56.** $(m - 5)^3$

57. $(r + 5)^3$ **58.** $(p + 3)^3$ **59.** $(2a + 1)^3$ **60.** $(3m + 1)^3$

61. $(4x - 1)^4$ **62.** $(2x - 1)^4$ **63.** $(3r - 2t)^4$ **64.** $(2z + 5y)^4$

65. $2x(x + 1)^3$ **66.** $3y(y + 2)^3$ **67.** $-4t(t + 3)^3$

68. $-5r(r + 1)^3$ **69.** $(x + y)^2(x - y)^2$ **70.** $(s + 2)^2(s - 2)^2$

The special product $(x + y)(x - y) = x^2 - y^2$ *can be used to perform some multiplications.*

Examples:

51×49	102×98
$= (50 + 1)(50 - 1)$	$= (100 + 2)(100 - 2)$
$= 50^2 - 1^2$	$= 100^2 - 2^2$
$= 2500 - 1$	$= 10{,}000 - 4$
$= 2499$	$= 9996$

Use this method to calculate each product mentally.

71. 101×99 **72.** 103×97 **73.** 201×199

74. 301×299 **75.** $20\frac{1}{2} \times 19\frac{1}{2}$ **76.** $30\frac{1}{3} \times 29\frac{2}{3}$

Find a polynomial that represents the area, in square units, of each figure. (If necessary, refer to the formulas at the back of this text.)

77.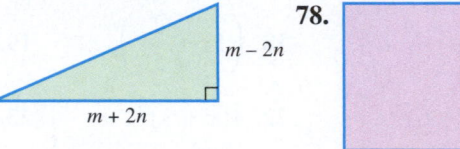

78.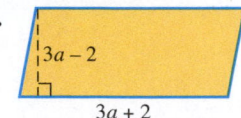

79.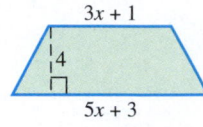

80.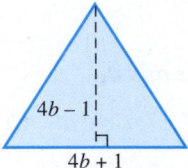

81.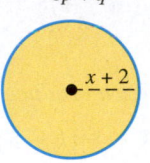

82.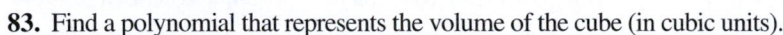

Refer to the cube shown here.

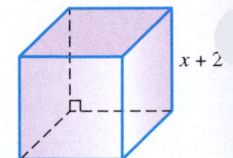

83. Find a polynomial that represents the volume of the cube (in cubic units).

84. If the value of x is 6, what is the volume of the cube (in cubic units)?

RELATING CONCEPTS **For Individual or Group Work** (Exercises 85–94)

Refer to the figure, and justify the special product

$$(a + b)^2 = a^2 + 2ab + b^2.$$

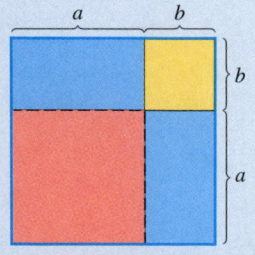

85. Express the area of the large square as the square of a binomial.

86. Give the monomial that represents the area of the red square.

87. Give the monomial that represents the sum of the areas of the blue rectangles.

88. Give the monomial that represents the area of the yellow square.

89. What is the sum of the monomials obtained in **Exercises 86–88?**

90. Why must the binomial square (**Exercise 85**) equal the polynomial (**Exercise 89**)?

Apply the special product $(a + b)^2 = a^2 + 2ab + b^2$ *to a purely numerical problem.*

91. Evaluate 35^2, using either traditional paper-and-pencil methods or a calculator.

92. The number 35 can be written as $30 + 5$. Therefore, $35^2 = (30 + 5)^2$. Use the special product for squaring a binomial with $a = 30$ and $b = 5$ to write an expression for $(30 + 5)^2$. Do not simplify at this time.

93. Use the order of operations to simplify the expression found in **Exercise 92.**

94. How do the answers in **Exercises 91 and 93** compare?

4.7 Dividing Polynomials

OBJECTIVES

1 Divide a polynomial by a monomial.
2 Divide a polynomial by a polynomial.
3 Apply polynomial division in a geometry problem.

OBJECTIVE 1 Divide a polynomial by a monomial.

We add two fractions with a common denominator as follows.

$$\frac{a}{c} + \frac{b}{c} = \frac{a+b}{c}$$

In reverse, this statement gives a rule for dividing a polynomial by a monomial.

Dividing a Polynomial by a Monomial

To divide a polynomial by a monomial, divide each term of the polynomial by the monomial.

$$\frac{a+b}{c} = \frac{a}{c} + \frac{b}{c} \quad \textbf{(where } c \neq 0\textbf{)}$$

Examples: $\frac{2+5}{3} = \frac{2}{3} + \frac{5}{3}$ and $\frac{x+3z}{2y} = \frac{x}{2y} + \frac{3z}{2y}$ $(y \neq 0)$

The parts of a division problem are named as follows.

Dividend $\longrightarrow$ $\dfrac{12x^2 + 6x}{6x} = 2x + 1$ $\longleftarrow$ Quotient
Divisor $\longrightarrow$

NOW TRY
EXERCISE 1
Divide $16a^6 - 12a^4$ by $4a^2$.

EXAMPLE 1 Dividing a Polynomial by a Monomial

Divide $5m^5 - 10m^3$ by $5m^2$.

$$\frac{5m^5 - 10m^3}{5m^2}$$ *A fraction bar means division.*

$$= \frac{5m^5}{5m^2} - \frac{10m^3}{5m^2}$$ Use the preceding rule, with + replaced by −.

$$= m^3 - 2m$$ Quotient rule

CHECK Multiply $5m^2 \cdot (m^3 - 2m) = 5m^5 - 10m^3$ ✔
          Divisor   Quotient   Original polynomial (Dividend)

Because division by 0 is undefined, the quotient $\frac{5m^5 - 10m^3}{5m^2}$ is undefined if $5m^2 = 0$, or $m = 0$. From now on, we assume that no denominators are 0.   **NOW TRY**

EXAMPLE 2 Dividing a Polynomial by a Monomial

Divide.

$$\frac{16a^5 - 12a^4 + 8a^2}{4a^3}$$ This becomes $\frac{2}{a}$, not $2a$.

$$= \frac{16a^5}{4a^3} - \frac{12a^4}{4a^3} + \frac{8a^2}{4a^3}$$ Divide each term by $4a^3$.

$$= 4a^2 - 3a + \frac{2}{a}$$ $\frac{8a^2}{4a^3} = \frac{8}{4}a^{2-3} = 2a^{-1} = \frac{2}{a}$

NOW TRY ANSWER
1. $4a^4 - 3a^2$

NOW TRY
EXERCISE 2
Divide.

$$\frac{36x^5 + 24x^4 - 12x^3}{6x^4}$$

The quotient $4a^2 - 3a + \frac{2}{a}$ is *not* a polynomial because $\frac{2}{a}$ has a variable in the denominator. While the sum, difference, and product of two polynomials are always polynomials, the quotient of two polynomials may not be a polynomial.

CHECK $4a^3\left(4a^2 - 3a + \frac{2}{a}\right)$ Divisor × Quotient should equal Dividend.

$$= 4a^3(4a^2) + 4a^3(-3a) + 4a^3\left(\frac{2}{a}\right)$$ Distributive property

$$= 16a^5 - 12a^4 + 8a^2 \checkmark$$ Dividend   **NOW TRY**

NOW TRY
EXERCISE 3
Divide $7y^4 - 40y^5 + 100y^2$
by $-5y^2$.

EXAMPLE 3   Dividing a Polynomial by a Monomial (Negative Coefficient)

Divide $-7x^3 + 12x^4 - 4x$ by $-4x$.

Write the dividend polynomial in descending powers.

$$\frac{12x^4 - 7x^3 - 4x}{-4x}$$ Write in descending powers before dividing.

$$= \frac{12x^4}{-4x} - \frac{7x^3}{-4x} - \frac{4x}{-4x}$$ Divide each term by $-4x$.

$$= -3x^3 - \frac{7x^2}{-4} - (-1)$$ Quotient rule

$$= -3x^3 + \frac{7x^2}{4} + 1$$ Be careful with signs, and be sure to include 1 in the answer.

CHECK $-4x\left(-3x^3 + \frac{7x^2}{4} + 1\right)$ Divisor × Quotient should equal Dividend.

$$= -4x(-3x^3) - 4x\left(\frac{7x^2}{4}\right) - 4x(1)$$ Distributive property

$$= 12x^4 - 7x^3 - 4x \checkmark$$ Dividend   **NOW TRY**

NOW TRY
EXERCISE 4
Divide

$$35m^5n^4 - 49m^2n^3 + 12mn$$

by $7m^2n$.

EXAMPLE 4   Dividing a Polynomial by a Monomial

Divide $-180x^4y^{10} + 150x^3y^8 - 120x^2y^6 + 90xy^4 - 100y$ by $-30xy^2$.

$$\frac{-180x^4y^{10} + 150x^3y^8 - 120x^2y^6 + 90xy^4 - 100y}{-30xy^2}$$

$$= \frac{-180x^4y^{10}}{-30xy^2} + \frac{150x^3y^8}{-30xy^2} - \frac{120x^2y^6}{-30xy^2} + \frac{90xy^4}{-30xy^2} - \frac{100y}{-30xy^2}$$ Divide each term by $-30xy^2$.

$$= 6x^3y^8 - 5x^2y^6 + 4xy^4 - 3y^2 + \frac{10}{3xy}$$

NOW TRY ANSWERS

2. $6x + 4 - \frac{2}{x}$

3. $8y^3 - \frac{7y^2}{5} - 20$

4. $5m^3n^3 - 7n^2 + \frac{12}{7m}$

Check by multiplying the divisor by the quotient.   **NOW TRY**

OBJECTIVE 2   Divide a polynomial by a polynomial.

We use a method of "long division" to divide a polynomial by a polynomial (other than a monomial). ***Both polynomials must first be written in descending powers.***

Dividing Whole Numbers	Dividing Polynomials

Step 1

Divide 6696 by 27.

$$27\overline{)6696}$$

Divide $8x^3 - 4x^2 - 14x + 15$ by $2x + 3$.

$$2x + 3\overline{)8x^3 - 4x^2 - 14x + 15}$$

Step 2

66 divided by $27 = 2$.

$2 \cdot 27 = 54$

$$\begin{array}{r} 2 \\ 27\overline{)6696} \\ 54 \end{array}$$

$8x^3$ divided by $2x = 4x^2$.

$4x^2(2x + 3) = 8x^3 + 12x^2$

$$\begin{array}{r} 4x^2 \\ 2x + 3\overline{)8x^3 - 4x^2 - 14x + 15} \\ 8x^3 + 12x^2 \end{array}$$

Step 3

Subtract.

$$\begin{array}{r} 2 \\ 27\overline{)6696} \\ -54 \\ \hline 12 \end{array}$$

Subtract.

$$\begin{array}{r} 4x^2 \\ 2x + 3\overline{)8x^3 - 4x^2 - 14x + 15} \\ -(8x^3 + 12x^2) \\ \hline -16x^2 \end{array}$$

Bring down the next digit.

$$\begin{array}{r} 2 \\ 27\overline{)6696} \\ -54\downarrow \\ \hline 129 \end{array}$$

Bring down the next term.

$$\begin{array}{r} 4x^2 \\ 2x + 3\overline{)8x^3 - 4x^2 - 14x + 15} \\ -(8x^3 + 12x^2)\downarrow \\ \hline -16x^2 - 14x \end{array}$$

Step 4

129 divided by $27 = 4$.

$4 \cdot 27 = 108$

$$\begin{array}{r} 24 \\ 27\overline{)6696} \\ -54 \\ \hline 129 \\ 108 \end{array}$$

$-16x^2$ divided by $2x = -8x$.

$-8x(2x + 3) = -16x^2 - 24x$

$$\begin{array}{r} 4x^2 - 8x \\ 2x + 3\overline{)8x^3 - 4x^2 - 14x + 15} \\ -(8x^3 + 12x^2) \\ \hline -16x^2 - 14x \\ -16x^2 - 24x \end{array}$$

Step 5

Subtract. Bring down.

$$\begin{array}{r} 24 \\ 27\overline{)6696} \\ -54 \\ \hline 129 \\ -108\downarrow \\ \hline 216 \end{array}$$

Subtract. Bring down.

$$\begin{array}{r} 4x^2 - 8x \\ 2x + 3\overline{)8x^3 - 4x^2 - 14x + 15} \\ -(8x^3 + 12x^2) \\ \hline -16x^2 - 14x \\ -(-16x^2 - 24x)\downarrow \\ \hline 10x + 15 \end{array}$$

(continued)

Dividing Whole Numbers	Dividing Polynomials
Step 6	
216 divided by 27 = 8.	$10x$ divided by $2x = 5.$
$8 \cdot 27 = 216$	$5(2x + 3) = 10x + 15$

$$
\begin{array}{r}
248 \\
27\overline{)6696} \\
-54 \\
\hline
129 \\
-108 \\
\hline
216 \\
-216 \\
\hline
\end{array}
$$

Remainder $\longrightarrow 0$

$$
\begin{array}{r}
4x^2 - 8x + 5 \\
2x + 3\overline{)8x^3 - 4x^2 - 14x + 15} \\
-(8x^3 + 12x^2) \\
\hline
-16x^2 - 14x \\
-(-16x^2 - 24x) \\
\hline
10x + 15 \\
-(10x + 15) \\
\hline
\end{array}
$$

Remainder $\longrightarrow 0$

6696 divided by 27 is 248.	$8x^3 - 4x^2 - 14x + 15$ divided by $2x + 3$ is $4x^2 - 8x + 5.$
Step 7 Multiply to check.	Multiply to check.
CHECK $27 \cdot 248 = 6696$ ✓	**CHECK** $(2x + 3)(4x^2 - 8x + 5)$ $= 8x^3 - 4x^2 - 14x + 15$ ✓

NOW TRY
EXERCISE 5
Divide.

$$\frac{4x^2 + x - 18}{x - 2}$$

EXAMPLE 5   Dividing a Polynomial by a Polynomial

Divide. $\dfrac{3x^2 - 5x - 28}{x - 4}$

Divisor

$$
\begin{array}{r}
3x + 7 \quad \longleftarrow \text{Quotient} \\
x - 4\overline{)3x^2 - 5x - 28} \quad \longleftarrow \text{Dividend} \\
-(3x^2 - 12x) \quad \longleftarrow (3x^2 - 5x) - (3x^2 - 12x) = 7x \\
\hline
7x - 28 \\
-(7x - 28) \quad \longleftarrow (7x - 28) - (7x - 28) = 0 \\
\hline
0
\end{array}
$$

Step 1 $3x^2$ divided by x is $3x$. $3x(x - 4) = 3x^2 - 12x$

Step 2   Subtract $3x^2 - 12x$ from $3x^2 - 5x$. Bring down -28.

Step 3 $7x$ divided by x is 7. $7(x - 4) = 7x - 28$

Step 4   Subtract $7x - 28$ from $7x - 28$. The remainder is 0.

CHECK   Multiply the divisor, $x - 4$, by the quotient, $3x + 7$. The product must be the original dividend, $3x^2 - 5x - 28$.

$$(x - 4)(3x + 7) = 3x^2 + 7x - 12x - 28 \qquad \text{FOIL method}$$
$$= 3x^2 - 5x - 28 \ ✓ \qquad \text{Combine like terms.}$$

Divisor   Quotient   Dividend

**NOW TRY
EXERCISE 6**

Divide.

$$\frac{6k^3 - 20k - k^2 + 1}{2k - 3}$$

EXAMPLE 6   Dividing a Polynomial by a Polynomial

Divide. $\dfrac{5x + 4x^3 - 8 - 4x^2}{2x - 1}$

The first polynomial must be written in descending powers as

$$4x^3 - 4x^2 + 5x - 8.$$

Then divide by $2x - 1$.

$$
\begin{array}{r}
2x^2 - x + 2 \\
2x - 1\overline{\smash{)}4x^3 - 4x^2 + 5x - 8} \\
\underline{-(4x^3 - 2x^2)} \\
-2x^2 + 5x \\
\underline{-(-2x^2 + x)} \\
4x - 8 \\
\underline{-(4x - 2)} \\
-6 \leftarrow \text{Remainder}
\end{array}
$$

Write in descending powers.

In each subtraction, add the opposite.

Step 1 $4x^3$ divided by $2x$ is $2x^2$. $2x^2(2x - 1) = 4x^3 - 2x^2$

Step 2   Subtract. Bring down the next term.

Step 3 $-2x^2$ divided by $2x$ is $-x$. $-x(2x - 1) = -2x^2 + x$

Step 4   Subtract. Bring down the next term.

Step 5 $4x$ divided by $2x$ is 2. $2(2x - 1) = 4x - 2$

Step 6   Subtract. The remainder is -6. Write the remainder as the numerator of a fraction that has $2x - 1$ as its denominator. Because there is a nonzero remainder, the answer is not a polynomial.

Remember to add $\frac{\text{remainder}}{\text{divisor}}$. Don't forget the $+$ sign.

Dividend $\longrightarrow$
Divisor $\longrightarrow$
$$\frac{4x^3 - 4x^2 + 5x - 8}{2x - 1} = \underbrace{2x^2 - x + 2}_{\substack{\text{Quotient} \\ \text{polynomial}}} + \underbrace{\frac{-6}{2x - 1}}_{\substack{\text{Fractional part} \\ \text{of quotient}}}$$
$\leftarrow$ Remainder
$\leftarrow$ Divisor

Step 7   CHECK

$$(2x - 1)\left(2x^2 - x + 2 + \frac{-6}{2x - 1}\right) \quad \text{Multiply Divisor} \times \text{(Quotient including the Remainder).}$$

$$= (2x - 1)(2x^2) + (2x - 1)(-x) + (2x - 1)(2) + (2x - 1)\left(\frac{-6}{2x - 1}\right)$$

$$= 4x^3 - 2x^2 - 2x^2 + x + 4x - 2 - 6$$

$$= 4x^3 - 4x^2 + 5x - 8 \quad \checkmark \qquad \text{Divisor} \times \text{Quotient} = \text{Dividend} \qquad \textbf{NOW TRY}$$

NOW TRY ANSWER

6. $3k^2 + 4k - 4 + \dfrac{-11}{2k - 3}$

! CAUTION   Remember to include "$+ \dfrac{\text{remainder}}{\text{divisor}}$" as part of the answer.

 **NOW TRY EXERCISE 7**

Divide $m^3 - 1000$ by $m - 10$.

EXAMPLE 7 Dividing into a Polynomial with Missing Terms

Divide $x^3 - 1$ by $x - 1$.

Here, the dividend, $x^3 - 1$, is missing the x^2-term and the x-term.

$$
\begin{array}{r}
x^2 + x + 1 \\
x - 1 \overline{)\ x^3 + 0x^2 + 0x - 1} \\
\underline{-(x^3 - x^2)} \\
x^2 + 0x \\
\underline{-(x^2 - x)} \\
x - 1 \\
\underline{-(x - 1)} \\
0
\end{array}
$$

> Insert placeholders for the missing terms.

The remainder is 0. The quotient is $x^2 + x + 1$.

CHECK $(x - 1)(x^2 + x + 1)$

$ = x^3 + x^2 + x - x^2 - x - 1$

$ = x^3 - 1$ ✓ Divisor × Quotient = Dividend

NOW TRY

 **NOW TRY EXERCISE 8**

Divide

$y^4 - 5y^3 + 6y^2 + y - 4$

by $y^2 + 2$.

EXAMPLE 8 Dividing by a Polynomial with Missing Terms

Divide $x^4 + 2x^3 + 2x^2 - x - 1$ by $x^2 + 1$.

Because the divisor, $x^2 + 1$, has a missing x-term, write it as $x^2 + 0x + 1$.

$$
\begin{array}{r}
x^2 + 2x + 1 \\
x^2 + 0x + 1 \overline{)\ x^4 + 2x^3 + 2x^2 - x - 1} \\
\underline{-(x^4 + 0x^3 + x^2)} \\
2x^3 + x^2 - x \\
\underline{-(2x^3 + 0x^2 + 2x)} \\
x^2 - 3x - 1 \\
\underline{-(x^2 + 0x + 1)} \\
-3x - 2 \ \leftarrow \text{Remainder}
\end{array}
$$

> Insert a placeholder for the missing term.

When the result of subtracting ($-3x - 2$ here) is a constant or a polynomial of degree *less* than the divisor ($x^2 + 0x + 1$ in this case), that constant or polynomial is the remainder. We write the answer as follows.

$$x^2 + 2x + 1 + \frac{-3x - 2}{x^2 + 1}$$

> Remember to write " + $\frac{\text{remainder}}{\text{divisor}}$."

NOW TRY

 **NOW TRY EXERCISE 9**

Divide $10x^3 + 21x^2 + 5x - 8$ by $2x + 4$.

EXAMPLE 9 Dividing a Polynomial When the Quotient Has Fractional Coefficients

Divide $4x^3 + 2x^2 + 3x + 2$ by $4x - 4$.

$$\frac{6x^2}{4x} = \frac{3}{2}x$$

$$\frac{9x}{4x} = \frac{9}{4}$$

$$
\begin{array}{r}
x^2 + \frac{3}{2}x + \frac{9}{4} \\
4x - 4 \overline{)\ 4x^3 + 2x^2 + 3x + 2} \\
\underline{-(4x^3 - 4x^2)} \\
6x^2 + 3x \\
\underline{-(6x^2 - 6x)} \\
9x + 2 \\
\underline{-(9x - 9)} \\
11
\end{array}
$$

NOW TRY ANSWERS

7. $m^2 + 10m + 100$

8. $y^2 - 5y + 4 + \dfrac{11y - 12}{y^2 + 2}$

9. $5x^2 + \dfrac{1}{2}x + \dfrac{3}{2} + \dfrac{-14}{2x + 4}$

The answer is $x^2 + \frac{3}{2}x + \frac{9}{4} + \dfrac{11}{4x - 4}$.

NOW TRY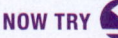

NOW TRY EXERCISE 10

The area of a rectangle is given by $(x^3 + 7x^2 + 17x + 20)$ sq. units. The width is given by $(x + 4)$ units. What is its length?

Length = ?

Width = $x + 4$

Area = $x^3 + 7x^2 + 17x + 20$

OBJECTIVE 3 Apply polynomial division in a geometry problem.

EXAMPLE 10 Using an Area Formula

The area of the rectangle in **FIGURE 5** is given by $(x^3 + 4x^2 + 8x + 8)$ sq. units. The width is given by $(x + 2)$ units. What is its length?

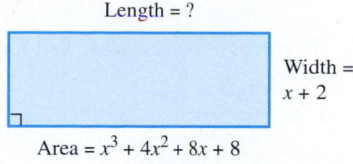

Length = ?

Width = $x + 2$

Area = $x^3 + 4x^2 + 8x + 8$

FIGURE 5

For a rectangle, $\mathcal{A} = LW$. Solving for L gives $L = \frac{\mathcal{A}}{W}$. Divide the area, $x^3 + 4x^2 + 8x + 8$, by the width, $x + 2$, to find the length.

$$
\begin{array}{r}
x^2 + 2x + 4 \\
x + 2 \overline{\smash{)}\ x^3 + 4x^2 + 8x + 8} \\
-(x^3 + 2x^2) \\
\hline
2x^2 + 8x \\
-(2x^2 + 4x) \\
\hline
4x + 8 \\
-(4x + 8) \\
\hline
0
\end{array}
$$

NOW TRY ANSWER
10. $(x^2 + 3x + 5)$ units

The quotient $(x^2 + 2x + 4)$ units represents the length.

NOW TRY

4.7 Exercises

FOR EXTRA HELP

 MyLab Math

 Video solutions for select problems available in MyLab Math

Concept Check *Complete each statement.*

1. In the following statement, _____ is the dividend, _____ is the divisor, and _____ is the quotient.

$$\frac{10x^2 + 8}{2} = 5x^2 + 4$$

2. To check the division shown in **Exercise 1**, multiply _____ by _____ and show that the product is _____.

3. **Concept Check** A student incorrectly divided

$$\frac{6x^2 - 12x}{6} \quad \text{to obtain} \quad x^2 - 12x.$$

WHAT WENT WRONG? Give the correct quotient.

4. **Concept Check** A student incorrectly divided

$$\frac{8x^6 + 6x^3}{2x^4} \quad \text{to obtain} \quad 4x^2 + 3x.$$

WHAT WENT WRONG? Give the correct quotient.

Concept Check   *Divide.*

5. $\dfrac{6p^4 + 18p^7}{3p^2}$

$= \dfrac{}{3p^2} + \dfrac{}{3p^2}$

$= \underline{}$

6. $\dfrac{20x^4 - 25x^3 + 5x}{5x^2}$

$= \dfrac{}{} - \dfrac{}{} + \dfrac{}{}$

$= \underline{}$

Perform each division. **See Examples 1–3.**

7. $\dfrac{12m^4 - 6m^3}{6m^2}$

8. $\dfrac{35n^5 - 5n^2}{5n}$

9. $\dfrac{60x^4 - 20x^2 + 10x}{2x}$

10. $\dfrac{120x^6 - 60x^3 + 80x^2}{2x}$

11. $\dfrac{20m^5 - 10m^4 + 5m^2}{5m^2}$

12. $\dfrac{12t^5 - 6t^3 + 6t^2}{6t^2}$

13. $\dfrac{8t^5 - 4t^3 + 4t^2}{2t}$

14. $\dfrac{8r^4 - 4r^3 + 6r^2}{2r}$

15. $\dfrac{4a^5 - 4a^2 + 8}{4a}$

16. $\dfrac{5t^8 + 5t^7 + 15}{5t}$

17. $\dfrac{18p^5 + 12p^3 - 6p^2}{-6p^3}$

18. $\dfrac{32x^8 + 24x^5 - 8x}{-8x^2}$

19. $\dfrac{-7r^7 + 6r^5 - r^4}{-r^5}$

20. $\dfrac{-13t^9 + 8t^6 - t^5}{-t^6}$

Divide each polynomial by $3x^2$. **See Examples 1–3.**

21. $12x^5 - 9x^4 + 6x^3$

22. $24x^6 - 12x^5 + 30x^4$

23. $3x^2 + 15x^3 - 27x^4$

24. $3x^2 - 18x^4 + 30x^5$

25. $36x + 24x^2 + 6x^3$

26. $9x - 12x^2 + 9x^3$

27. $4x^4 + 3x^3 + 2x$

28. $5x^4 - 6x^3 + 8x$

Perform each division. **See Examples 1–4.**

29. $\dfrac{-27r^4 + 36r^3 - 6r^2 - 26r + 2}{-3r}$

30. $\dfrac{-8k^4 + 12k^3 + 2k^2 - 7k + 3}{-2k}$

31. $\dfrac{2m^5 - 6m^4 + 8m^2}{-2m^3}$

32. $\dfrac{6r^5 - 8r^4 + 10r^2}{-2r^3}$

33. $(120x^{11} - 60x^{10} + 140x^9 - 100x^8) \div (10x^{12})$

34. $(45y^7 + 9y^6 - 6y^5 + 12y^4) \div (3y^8)$

35. $(20a^4b^3 - 15a^5b^2 + 25a^3b) \div (-5a^4b)$

36. $(16y^5z - 8y^2z^2 + 12yz^3) \div (-4y^2z^2)$

37. $(120x^5y^4 - 80x^2y^3 + 40x^2y^4 - 20x^5y^3) \div (20xy^2)$

38. $(200a^5b^6 - 160a^4b^7 - 120a^3b^9 + 40a^2b^2) \div (40a^2b)$

Perform each division using the "long division" process. **See Examples 5 and 6.**

39. $\dfrac{x^2 - x - 6}{x - 3}$

40. $\dfrac{m^2 - 2m - 24}{m - 6}$

41. $\dfrac{2y^2 + 9y - 35}{y + 7}$

42. $\dfrac{2y^2 + 9y + 7}{y + 1}$

43. $\dfrac{p^2 + 2p + 20}{p + 6}$

44. $\dfrac{x^2 + 11x + 16}{x + 8}$

45. $\dfrac{12m^2 - 20m + 3}{2m - 3}$

46. $\dfrac{12y^2 + 20y + 7}{2y + 1}$

47. $\dfrac{4a^2 - 22a + 32}{2a + 3}$

48. $\dfrac{9w^2 + 6w + 10}{3w - 2}$

49. $\dfrac{8x^3 - 10x^2 - x + 3}{2x + 1}$

50. $\dfrac{12t^3 - 11t^2 + 9t + 18}{4t + 3}$

51. $\dfrac{8k^4 - 12k^3 - 2k^2 + 7k - 6}{2k - 3}$

52. $\dfrac{27r^4 - 36r^3 - 6r^2 + 26r - 24}{3r - 4}$

53. $\dfrac{5y^4 + 5y^3 + 2y^2 - y - 8}{y + 1}$

54. $\dfrac{2r^3 - 5r^2 - 6r + 15}{r - 3}$

55. $\dfrac{3k^3 - 4k^2 - 6k + 10}{k - 2}$

56. $\dfrac{5z^3 - z^2 + 10z + 2}{z + 2}$

57. $\dfrac{6p^4 - 15p^3 + 14p^2 - 5p + 10}{3p^2 + 1}$

58. $\dfrac{6r^4 - 10r^3 - r^2 + 15r - 8}{2r^2 - 3}$

Perform each division. **See Examples 6–9.**

59. $(x^3 + 2x^2 - 3) \div (x - 1)$

60. $(x^3 - 2x^2 - 9) \div (x - 3)$

61. $\dfrac{3y^3 + y^2 + 2}{y + 1}$

62. $\dfrac{2r^3 - 6r - 36}{r - 3}$

63. $(2x^3 + x + 2) \div (x + 3)$

64. $(3x^3 + x + 5) \div (x + 1)$

65. $\dfrac{5 - 2r^2 + r^4}{r^2 - 4}$

66. $\dfrac{4t^2 + t^4 + 7}{t^2 - 4}$

67. $\dfrac{-4x + 3x^3 + 2}{x - 1}$

68. $\dfrac{-5x + 6x^3 + 5}{x - 1}$

69. $\dfrac{y^3 + 27}{y + 3}$

70. $\dfrac{y^3 - 64}{y - 4}$

71. $\dfrac{a^4 - 25}{a^2 - 5}$

72. $\dfrac{a^4 - 36}{a^2 + 6}$

73. $\dfrac{x^4 - 4x^3 + 5x^2 - 3x + 2}{x^2 + 3}$

74. $\dfrac{3t^4 + 5t^3 - 8t^2 - 13t + 2}{t^2 - 5}$

75. $\dfrac{2x^5 + 9x^4 + 8x^3 + 10x^2 + 14x + 5}{2x^2 + 3x + 1}$

76. $\dfrac{4t^5 - 11t^4 - 6t^3 + 5t^2 - t + 3}{4t^2 + t - 3}$

77. $(3a^2 - 11a + 17) \div (2a + 6)$

78. $(4x^2 + 11x - 8) \div (3x + 6)$

79. $\dfrac{3x^3 + 5x^2 - 9x + 5}{3x - 3}$

80. $\dfrac{5x^3 + 4x^2 + 10x + 20}{5x + 5}$

Work each problem. Give answers in units (or as specified). If necessary, refer to the formulas found at the back of this text. **See Example 10.**

81. What expression represents the length of the rectangle?

Area = $(5x^3 + 7x^2 - 13x - 6)$ sq. units

82. What expression represents the length of the rectangle?

Area = $(15x^3 + 12x^2 - 9x + 3)$ sq. units

83. What expression represents the length of the base of the triangle?

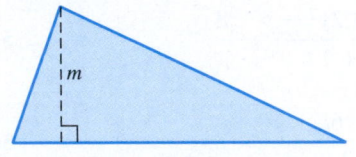

Area = $(24m^3 + 48m^2 + 12m)$ sq. units

84. What expression represents the length of the base of the parallelogram?

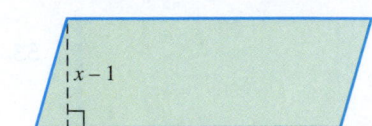

Area = $(2x^3 + 2x^2 - 3x - 1)$ sq. units

85. If the distance traveled is $(5x^3 - 6x^2 + 3x + 14)$ miles and the rate is $(x + 1)$ mph, write an expression, in hours, for the time traveled.

86. If it costs $(4x^5 + 3x^4 + 2x^3 + 9x^2 - 29x + 2)$ dollars to fertilize a garden, and fertilizer costs $(x + 2)$ dollars per square yard, write an expression, in square yards, for the area of the garden.

Chapter 4   Summary

Key Terms

4.1

base
exponent (power)
exponential expression

4.3

scientific notation
standard notation

4.4

term
leading term
numerical coefficient
  (coefficient)
like terms
unlike terms
polynomial

descending powers
degree of a term
degree of a polynomial
monomial
binomial
trinomial
parabola
vertex
axis of symmetry (axis)

4.5

FOIL method
outer product
inner product

4.6

difference of two squares
conjugates

New Symbols

x^{-n} x to the negative n
      power

Test Your Word Power

See how well you have learned the vocabulary in this chapter.

1. A **polynomial** is an algebraic expression made up of
   A. a term or a finite product of terms with positive coefficients and exponents
   B. a term or a finite sum of terms with real coefficients and whole number exponents
   C. the product of two or more terms with positive exponents
   D. the sum of two or more terms with whole number coefficients and exponents.

2. The **degree of a term** is the
   A. number of variables in the term
   B. product of the exponents on the variables
   C. least exponent on the variables
   D. sum of the exponents on the variables.

3. The **FOIL** method is used when
   A. adding two binomials
   B. adding two trinomials
   C. multiplying two binomials
   D. multiplying two trinomials.

4. A **binomial** is a polynomial with
   A. only one term
   B. exactly two terms
   C. exactly three terms
   D. more than three terms.

5. A **monomial** is a polynomial with
   A. only one term
   B. exactly two terms
   C. exactly three terms
   D. more than three terms.

6. A **trinomial** is a polynomial with
   A. only one term
   B. exactly two terms
   C. exactly three terms
   D. more than three terms.

ANSWERS

1. B; *Example:* $5x^3 + 2x^2 - 7$ 2. D; *Examples:* The term 6 has degree 0, $3x$ has degree 1, $-2x^8$ has degree 8, and $5x^2y^4$ has degree 6.

$$\underset{F}{} \quad \underset{O}{} \quad \underset{I}{} \quad \underset{L}{}$$

3. C; *Example:* $(m + 4)(m - 3) = m(m) - 3m + 4m + 4(-3) = m^2 + m - 12$ 4. B; *Example:* $3t^3 + 5t$ 5. A; *Examples:* -5 and $4xy^5$

6. C; *Example:* $2a^2 - 3ab + b^2$

Quick Review

CONCEPTS	EXAMPLES
4.1 The Product Rule and Power Rules for Exponents For any integers m and n, the following hold true. **Product Rule** $a^m \cdot a^n = a^{m+n}$ **Power Rules** (a) $(a^m)^n = a^{mn}$ (b) $(ab)^m = a^m b^m$ (c) $\left(\dfrac{a}{b}\right)^m = \dfrac{a^m}{b^m}$ (where $b \neq 0$)	Simplify using the rules for exponents. $2^4 \cdot 2^5 = 2^{4+5} = 2^9$ $(3^4)^2 = 3^{4\cdot 2} = 3^8$ $(6a)^5 = 6^5 a^5$ $\left(\dfrac{2}{3}\right)^4 = \dfrac{2^4}{3^4}$
4.2 Integer Exponents and the Quotient Rule For any nonzero real numbers a and b and any integers m and n, the following hold true. **Zero Exponent** $a^0 = 1$ **Negative Exponent** $a^{-n} = \dfrac{1}{a^n}$ **Quotient Rule** $\dfrac{a^m}{a^n} = a^{m-n}$ **Negative-to-Positive Rules** $\dfrac{a^{-m}}{b^{-n}} = \dfrac{b^n}{a^m}$ $\left(\dfrac{a}{b}\right)^{-m} = \left(\dfrac{b}{a}\right)^m$	Simplify using the rules for exponents. $15^0 = 1$ $5^{-2} = \dfrac{1}{5^2} = \dfrac{1}{25}$ $\dfrac{4^8}{4^3} = 4^{8-3} = 4^5$ $\dfrac{4^{-2}}{3^{-5}} = \dfrac{3^5}{4^2}$ $\left(\dfrac{6}{5}\right)^{-3} = \left(\dfrac{5}{6}\right)^3$

CONCEPTS	EXAMPLES

4.3 Scientific Notation

To write a positive number in scientific notation

$$a \times 10^n, \quad \text{where} \quad 1 \leq |a| < 10,$$

move the decimal point to follow the first nonzero digit.

1. If moving the decimal point makes the number less, then n is positive.

2. If moving the decimal point makes the number greater, then n is negative.

3. If the decimal point is not moved, then n is 0.

For a negative number, follow these steps using the *absolute value* of the number. Then make the result negative.

Write in scientific notation.

$$247 = 2.47 \times 10^2$$
$$0.0051 = 5.1 \times 10^{-3}$$
$$-4.8 = -4.8 \times 10^0$$

Write in standard notation.

$$3.25 \times 10^5 = 325{,}000$$
$$8.44 \times 10^{-6} = 0.00000844$$

4.4 Adding, Subtracting, and Graphing Polynomials

Adding Polynomials
Combine (add) like terms.

Add.
$$\begin{array}{r} 2x^2 + 5x - 3 \\ + (5x^2 - 2x + 7) \\ \hline 7x^2 + 3x + 4 \end{array}$$

Subtracting Polynomials
Change the sign of each term in the subtrahend (second polynomial) and add the result to the minuend (first polynomial).

Subtract.
$$(2x^2 + 5x - 3) - (5x^2 - 2x + 7)$$
$$= (2x^2 + 5x - 3) + (-5x^2 + 2x - 7)$$
$$= -3x^2 + 7x - 10$$

Graphing Polynomials
To graph a simple polynomial equation of degree 2, plot points near the vertex. (In this chapter, all parabolas have a vertex on the x-axis or the y-axis.)

Graph $y = x^2 - 2$.

x	y
-2	2
-1	-1
0	-2
1	-1
2	2

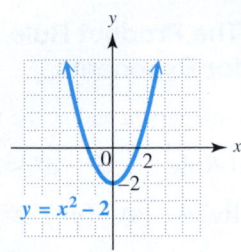

$y = x^2 - 2$

4.5 Multiplying Polynomials

General Method for Multiplying Polynomials
Multiply each term of the first polynomial by each term of the second polynomial. Then combine like terms.

Multiply.
$$\begin{array}{r} 3x^3 - 4x^2 + 2x - 7 \\ 4x + 3 \\ \hline 9x^3 - 12x^2 + 6x - 21 \\ 12x^4 - 16x^3 + 8x^2 - 28x \\ \hline 12x^4 - 7x^3 - 4x^2 - 22x - 21 \end{array}$$

FOIL Method for Multiplying Binomials

Step 1 Multiply the two **F**irst terms to obtain the first term of the product.

Step 2 Find the **O**uter product and the **I**nner product and combine them (when possible) to obtain the middle term of the product.

Step 3 Multiply the two **L**ast terms to obtain the last term of the product.

Step 4 Add the terms found in Steps 1–3.

Multiply. $(2x + 3)(5x - 4)$

$$2x(5x) = 10x^2 \quad \textbf{F}$$
$$2x(-4) + 3(5x) = 7x \quad \textbf{O, I}$$
$$3(-4) = -12 \quad \textbf{L}$$

The product is $10x^2 + 7x - 12$.

CONCEPTS	EXAMPLES

4.6   Special Products

Square of a Binomial

$$(x + y)^2 = x^2 + 2xy + y^2$$
$$(x - y)^2 = x^2 - 2xy + y^2$$

Multiply.

$(3x + 1)^2$

$= (3x)^2 + 2(3x)(1) + 1^2$

$= 9x^2 + 6x + 1$

$(2m - 5n)^2$

$= (2m)^2 - 2(2m)(5n) + (5n)^2$

$= 4m^2 - 20mn + 25n^2$

Product of a Sum and Difference of Two Terms

$$(x + y)(x - y) = x^2 - y^2$$

$(4a + 3)(4a - 3)$

$= (4a)^2 - 3^2$

$= 16a^2 - 9$ $(4a)^2 = 4^2a^2 = 16a^2$

4.7   Dividing Polynomials

Dividing a Polynomial by a Monomial
Divide each term of the polynomial by the monomial.

$$\frac{a + b}{c} = \frac{a}{c} + \frac{b}{c} \quad \text{(where } c \neq 0\text{)}$$

Divide.

$$\frac{4x^3 - 2x^2 + 6x - 8}{2x} = \frac{4x^3}{2x} - \frac{2x^2}{2x} + \frac{6x}{2x} - \frac{8}{2x}$$

$$= 2x^2 - x + 3 - \frac{4}{x} \qquad \text{Divide each term by } 2x.$$

Dividing a Polynomial by a Polynomial
Use "long division."

$$\begin{array}{r} 2x - 5 \\ 3x + 4\overline{)6x^2 - 7x - 21} \\ -(6x^2 + 8x) \\ \hline -15x - 21 \\ -(-15x - 20) \\ \hline -1 \leftarrow \text{Remainder} \end{array}$$

The answer is $2x - 5 + \frac{-1}{3x + 4}$.

Chapter 4   Review Exercises

4.1 *Use the product rule, power rules, or both to simplify each expression. Leave answers in exponential form in Exercises 1–8.*

1. $4^3 \cdot 4^8$

2. $(-5)^6(-5)^5$

3. $(-8x^4)(9x^3)$

4. $(2x^2)(5x^3)(x^9)$

5. $(19x)^5$

6. $(-4y)^7$

7. $5(pt)^4$

8. $\left(\frac{7}{5}\right)^6$

9. $(3x^2y^3)^3$

10. $(t^4)^8(t^2)^5$

11. $(6x^2z^4)^2(x^3yz^2)^4$

12. $\left(\frac{2m^3n}{p^2}\right)^3$

4.2 *Evaluate each expression.*

13. -10^0

14. $-(-23)^0$

15. $6^0 + (-6)^0$

16. $-3^0 - 2^0$

Simplify each expression. Assume that all variables represent nonzero real numbers.

17. -7^{-2}

18. $\left(\frac{5}{8}\right)^{-2}$

19. $(2^{-2})^{-3}$

20. $9^3 \cdot 9^{-5}$

21. $2^{-1} + 4^{-1}$

22. $\dfrac{6^{-5}}{6^{-3}}$

23. $\dfrac{x^{-7}}{x^{-9}}$

24. $\dfrac{y^4 \cdot y^{-2}}{y^{-5}}$

25. $(3r^{-2})^{-4}$

26. $(3p)^4(3p^{-7})$

27. $\dfrac{ab^{-3}}{a^4b^2}$

28. $\dfrac{(6r^{-1})^2(2r^{-4})}{r^{-5}(r^2)^{-3}}$

4.3 *Write each number in scientific notation.*

29. 48,000,000

30. 28,988,000,000

31. 0.0000000824

32. −4,820,000

Write each number in standard notation.

33. 2.4×10^4

34. 7.83×10^7

35. 8.97×10^{-7}

36. -7.6×10^{-4}

Perform the indicated operations. Write each answer in (a) scientific notation and (b) standard notation.

37. $(2 \times 10^{-3})(4 \times 10^5)$

38. $(2.5 \times 10^{-51})(2.0 \times 10^{51})$

39. $\dfrac{8 \times 10^4}{2 \times 10^{-2}}$

40. $\dfrac{60 \times 10^{-1}}{24 \times 10}$

*Each statement contains a number in **boldface italic** type. If the number is in scientific notation, write it in standard notation. If the number is not in scientific notation, write it as such.*

41. In 2016, China was the world's largest energy producer. China accounted for ***1.0072 × 10^{17}*** Btu. (Data from *Enerdata.*)

42. The 2016 population of Tokyo, Japan, was ***3.8140 × 10^7*** (Data from United Nations, Population Division.)

43. The energy consumption of India in 2016 was ***35,079,955,200,000,000*** Btu. (Data from *Enerdata.*)

44. In 2016, the budget of the U.S. Department of Defense was ***585,300,000,000*** dollars. (Data from www.defense.gov)

4.4 *For each polynomial, first simplify, if possible, and write the result in descending powers of the variable. Then give the degree and tell whether the simplified polynomial is a monomial, a binomial, a trinomial, or none of these.*

45. $9m^2 + 11m^2$

46. $-4p + p^3 - p^2$

47. $-7y^5 - 8y^4 - y^5 + y^4$

Add or subtract as indicated.

48. $(12r^4 - 7r^3 + 2r^2) - (5r^4 - 3r^3 + 2r^2 - 1) - (7r^4 - 3r^3 + 2r - 6)$

49. $(5x^3y^2 - 3xy^5 + 12x^2) - (-9x^2 - 8x^3y^2 + 2xy^5)$

50. $\begin{array}{r} -2a^3 + 5a^2 \\ + \underline{(3a^3 - a^2)} \end{array}$

51. $\begin{array}{r} 6y^2 - 8y + 2 \\ - \underline{(5y^2 + 2y - 7)} \end{array}$

52. $\begin{array}{r} -12k^4 - 8k^2 + 7k \\ - \underline{(k^4 + 7k^2 - 11k)} \end{array}$

Graph each equation by completing the table of values.

53. $y = -x^2 + 5$

x	−2	−1	0	1	2
y					

54. $y = 3x^2 - 2$

x	−2	−1	0	1	2
y					

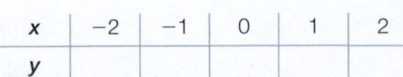

4.5 *Find each product.*

55. $(a + 2)(a^2 - 4a + 1)$

56. $(3r - 2)(2r^2 + 4r - 3)$

57. $(5p^2 + 3p)(p^3 - p^2 + 5)$

58. $(m - 9)(m + 2)$

59. $(3k - 6)(2k + 1)$

60. $(a + 3b)(2a - b)$

61. $(6k + 5q)(2k - 7q)$

62. $(s - 1)^3$

4.6 *Find each product.*

63. $(a + 4)^2$

64. $(2r + 5t)^2$

65. $(6m - 5)(6m + 5)$

66. $(5a + 6b)(5a - 6b)$

67. $(r + 2)^3$

68. $t(5t - 3)^2$

69. Choose values for x and y to show that, in general, the following hold true.

(a) $(x + y)^2 \neq x^2 + y^2$

(b) $(x + y)^3 \neq x^3 + y^3$

70. Explain how to raise a binomial to the third power. Give an example.

Solve each problem. (If necessary, refer to the formulas at the back of this text.)

71. Find a polynomial that represents, in cubic centimeters, the volume of a cube with one side having length $(x^2 + 2)$ centimeters.

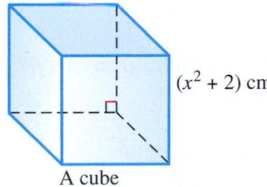

A cube

72. Find a polynomial that represents, in cubic inches, the volume of a sphere with radius $(x + 1)$ inches.

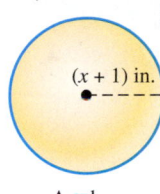

$(x + 1)$ in.

A sphere

4.7 *Perform each division.*

73. $\dfrac{-15y^4}{9y^2}$

74. $(-10m^4n^2 + 5m^3n^2 + 6m^2n^4) \div (5m^2n)$

75. $\dfrac{6y^4 - 12y^2 + 18y}{6y}$

76. $\dfrac{24r^8s^6 + 12r^7s^5 - 8r}{-4r^3s^5}$

77. $\dfrac{2r^2 + 3r - 14}{r - 2}$

78. $\dfrac{10a^3 + 9a^2 - 14a + 9}{5a - 3}$

79. $\dfrac{x^4 - 5x^2 + 3x^3 - 3x + 4}{x^2 - 1}$

80. $\dfrac{m^4 + 4m^3 - 12m - 5m^2 + 6}{m^2 - 3}$

81. $\dfrac{16x^2 - 25}{4x + 5}$

82. $\dfrac{25y^2 - 100}{5y + 10}$

83. $\dfrac{y^3 - 8}{y - 2}$

84. $\dfrac{1000x^6 + 1}{10x^2 + 1}$

85. $\dfrac{6y^4 - 15y^3 + 14y^2 - 5y - 1}{3y^2 + 1}$

86. $\dfrac{4x^5 - 8x^4 - 3x^3 + 22x^2 - 15}{4x^2 - 3}$

Chapter 4   Mixed Review Exercises

Perform each indicated operation, or simplify each expression. Assume that all variables represent nonzero real numbers.

1. $5^0 + 7^0$

2. $\left(\dfrac{6r^2p}{5}\right)^3$

3. $(12a + 1)(12a - 1)$

4. 2^{-4}

5. $(4^{-2})^2$

6. $\dfrac{2p^3 - 6p^2 + 5p}{2p^2}$

7. $\dfrac{(2m^{-5})(3m^2)^{-1}}{m^{-2}(m^{-1})^2}$

8. $(3k - 6)(2k^2 + 4k + 1)$

9. $\dfrac{r^9 \cdot r^{-5}}{r^{-2} \cdot r^{-7}}$

10. $(2r + 5s)^2$

11. $\dfrac{2y^3 + 17y^2 + 37y + 7}{2y + 7}$

12. $(2r + 5)(5r - 2)$

13. $(-5y^2 + 3y - 11) + (4y^2 - 7y + 15)$

14. $(25x^2y^3 - 8xy^2 + 15x^3y) \div (10x^2y^3)$

15. $(6p^2 - p - 8) - (-4p^2 + 2p - 3)$

16. $\dfrac{3x^3 - 2x + 5}{x - 3}$

17. $(-7 + 2k)^2$

18. $\left(\dfrac{x}{y^{-3}}\right)^{-4}$

Find polynomials that represent, in appropriate units, (a) the perimeter and (b) the area of each square or rectangle.

19.

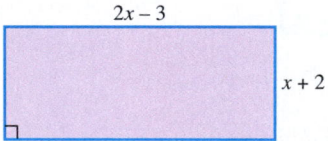

$2x - 3$

$x + 2$

20.

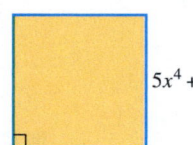

$5x^4 + 2x^2$

Chapter 4   Test   FOR EXTRA HELP

Step-by-step test solutions are found on the Chapter Test Prep Videos available in **MyLab Math.**

 View the complete solutions to all Chapter Test exercises in MyLab Math.

Evaluate each expression.

1. 5^{-4}

2. $(-3)^0 + 4^0$

3. $4^{-1} + 3^{-1}$

4. $(-2)^3(-2)^2$

Simplify each expression. Assume that all variables represent nonzero real numbers.

5. $\left(\dfrac{6}{m^2}\right)^3$

6. $\dfrac{(3x^2y)^2(xy^3)^2}{(xy)^3}$

7. $\dfrac{8^{-1} \cdot 8^4}{8^{-2}}$

8. $\dfrac{(x^{-3})^{-2}(x^{-1}y)^2}{(xy^{-2})^2}$

9. Determine whether each expression represents a number that is *positive*, *negative*, or *zero*.

    (a) 3^{-4} **(b)** $(-3)^4$ **(c)** -3^4 **(d)** 3^0 **(e)** $(-3)^0 - 3^0$ **(f)** $(-3)^{-3}$

10. (a) Write 45,000,000,000 using scientific notation.

   (b) Write 3.6×10^{-6} using standard notation.

   (c) Write the quotient $\dfrac{9.5 \times 10^{-1}}{5 \times 10^{3}}$ using standard notation.

11. A satellite galaxy of the Milky Way, known as the Large Magellanic Cloud, is *1000* light-years across. A *light-year* is equal to *5,890,000,000,000* mi. (Data from *USA Today*.)

   (a) Write the two boldface italic numbers in scientific notation.

   (b) How many miles across is the Large Magellanic Cloud?

For each polynomial, first simplify, if possible, and write the result in descending powers of the variable. Then give the degree and tell whether the simplified polynomial is a monomial, a binomial, a trinomial, or none of these.

12. $5x^2 + 8x - 12x^2$

13. $13n^3 - n^2 + n^4 + 3n^4 - 9n^2$

14. Graph the equation $y = 2x^2 - 4$ by completing the table of values.

x	-2	-1	0	1	2
y					

Perform each indicated operation.

15. $(2y^2 - 8y + 8) + (-3y^2 + 2y + 3) - (y^2 + 3y - 6)$

16. $(-9a^3b^2 + 13ab^5 + 5a^2b^2) - (6ab^5 + 12a^3b^2 + 10a^2b^2)$

17. $-6r^5 + 4r^2 - 3$
     $+ (6r^5 + 12r^2 - 16)$

18. $9t^3 - 4t^2 + 2t + 2$
     $- (9t^3 + 8t^2 - 3t - 6)$

19. $3x^2(-9x^3 + 6x^2 - 2x + 1)$

20. $(t - 8)(t + 3)$

21. $(4x + 3y)(2x - y)$

22. $(5x - 2y)^2$

23. $(10v + 3w)(10v - 3w)$

24. $(2r - 3)(r^2 + 2r - 5)$

Refer to the square below. Find polynomials, in appropriate units, that represent the following.

25. The perimeter

26. The area

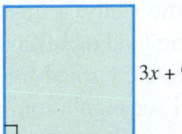

$3x + 9$

Perform each division.

27. $\dfrac{8y^3 - 6y^2 + 4y + 10}{2y}$

28. $(-9x^2y^3 + 6x^4y^3 + 12xy^3) \div (3xy)$

29. $\dfrac{5x^2 - x - 18}{5x + 9}$

30. $(3x^3 - x + 4) \div (x - 2)$

| Chapters R–4 | Cumulative Review Exercises |

Perform each operation.

1. $\dfrac{2}{3} + \dfrac{1}{8}$

2. $\dfrac{7}{4} - \dfrac{9}{5}$

3. 0.07×0.0006

Solve each problem.

4. A contractor installs sheds. Each requires $1\frac{1}{4}$ yd³ of concrete. How much concrete would be needed for 25 sheds?

5. A retailer has $34,000 invested in her business. She finds that last year she earned 5.4% on this investment. How much did she earn?

6. Find the value of $\dfrac{4x - 2y}{x + y}$ for $x = -2$ and $y = 4$.

Perform each indicated operation.

7. $\dfrac{(-13 + 15) - (3 + 2)}{6 - 12}$

8. $-7 - 3[2 + (5 - 8)]$

Name the property illustrated.

9. $(9 + 2) + 3 = 9 + (2 + 3)$

10. $6(4 + 2) = 6(4) + 6(2)$

11. Simplify the expression $-3(2x^2 - 8x + 9) - (4x^2 + 3x + 2)$.

Solve each equation.

12. $2 - 3(t - 5) = 4 + t$

13. $2(5x + 1) = 10x + 4$

14. $d = rt$ for r

15. $\dfrac{x}{5} = \dfrac{x - 2}{7}$

16. $3x - (4 + 2x) = -4$

17. $0.05x + 0.15(50 - x) = 5.50$

18. $\dfrac{1}{3}p - \dfrac{1}{6}p = -2$

19. $4 - (3x + 12) = -7 - (3x + 1)$

Solve each problem.

20. A husky running the Iditarod in Alaska burns $5\frac{3}{8}$ calories in exertion for every 1 calorie burned in thermoregulation in extreme cold. According to one scientific study, a husky in top condition burns an amazing total of 11,200 calories per day. How many calories are burned for exertion, and how many are burned for regulation of body temperature? Round answers to the nearest whole number.

21. One side of a triangle is twice as long as a second side. The third side of the triangle is 17 ft long. The perimeter of the triangle cannot be more than 50 ft. Find the longest possible values for the other two sides of the triangle.

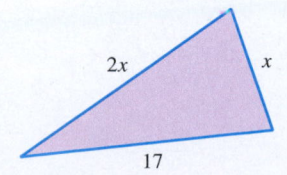

Solve each inequality.

22. $-2(x + 4) > 3x + 6$

23. $-3 \leq 2x + 5 < 9$

24. Graph $y = -3x + 6$.

25. Consider the two points $(-1, 5)$ and $(2, 8)$.

    (a) Find the slope of the line passing through them.

    (b) Find the equation of the line passing through them.

Evaluate each expression.

26. $(-5)^2$

27. -4^2

28. $4^{-1} + 3^0$

29. $\dfrac{8^{-5} \cdot 8^7}{8^2}$

30. Write $\dfrac{(a^{-3}b^2)^2}{(2a^{-4}b^{-3})^{-1}}$ with positive exponents only.

31. It takes about 3.6×10^1 sec at a speed of 3.0×10^5 km per sec for light from the sun to reach Venus. How far is Venus from the sun? (Data from *The World Almanac and Book of Facts.*)

32. Graph the equation $y = (x + 4)^2$ by completing the table of values.

x	-6	-5	-4	-3	-2
y					

Perform each indicated operation.

33. $(7x^3 - 12x^2 - 3x + 8) - (-4x^3 + 8x^2 - 2x - 2)$

34. $(7x + 4)(9x + 3)$

35. $\dfrac{y^3 - 3y^2 + 8y - 6}{y - 1}$

STUDY SKILLS REMINDER

Have you begun to prepare for your final exam? **Review Study Skill 10, *Preparing for Your Math Final Exam.***

FACTORING AND APPLICATIONS

The motion of a freely falling object or of an object that is projected upward can be described using a *quadratic equation*. A key topic of this chapter, *factoring,* is used in solving some such equations.

5.1   Greatest Common Factors; Factoring by Grouping

OBJECTIVES

1 Find the greatest common factor of a list of numbers.
2 Find the greatest common factor of a list of variable terms.
3 Factor out the greatest common factor.
4 Factor by grouping.

To **factor** a number means to write it as a product of two or more numbers. The product is a **factored form** of the number.

Factors   Factors

Example: $6 \cdot 2 = 12$ $12 = 6 \cdot 2$ $6 \cdot 2$ is a **factored form** of 12.

Multiplying   Factoring

Factoring is a process that "undoes" multiplying. We multiply $6 \cdot 2$ to obtain 12, but we factor 12 by writing it as $6 \cdot 2$. Other factored forms of 12 are

$$-6(-2), \quad 3 \cdot 4, \quad -3(-4), \quad 12 \cdot 1, \quad -12(-1), \quad \text{and} \quad 2 \cdot 2 \cdot 3.$$

OBJECTIVE 1 Find the greatest common factor of a list of numbers.

An integer that is a factor of two or more integers is a **common factor** of those integers. Consider the positive integer factors of 18 and 24.

Factors of 18:   1, 2, 3, 6, 9, 18

Factors of 24:   1, 2, 3, 4, 6, 8, 12, 24

The common factors of 18 and 24 are 1, 2, 3, and 6.

The **greatest common factor (GCF)** of a list of integers is the largest common factor of those integers. Thus, 6 is the greatest common factor of 18 and 24.

VOCABULARY

☐ factor
☐ factored form
☐ common factor
☐ greatest common factor (GCF)

STUDY SKILLS REMINDER

Make study cards to help you learn and remember the material in this chapter. **Review Study Skill 5, Using Study Cards.**

NOTE *Factors* of a number are also *divisors* of the number. The *greatest common factor* is the same as the *greatest common divisor.* Divisibility tests are useful for deciding what numbers divide into a given number.

▼ **Divisibility Tests**

A Whole Number Divisible by	Must Have the Following Property:
2	Ends in 0, 2, 4, 6, or 8
3	Sum of digits divisible by 3
4	Last two digits form a number divisible by 4
5	Ends in 0 or 5
6	Divisible by both 2 and 3
8	Last three digits form a number divisible by 8
9	Sum of digits divisible by 9
10	Ends in 0

EXAMPLE 1   Finding the Greatest Common Factor (Numbers)

Find the greatest common factor for each list of numbers.

(a) 30, 45

Although we could find the GCF by listing the common factors (as we did above with 18 and 24), we would run the risk of leaving out factors. Another strategy is to write the prime factored form of each number. We can do this using factor trees.

NOW TRY
EXERCISE 1

Find the greatest common factor for each list of numbers.

(a) 24, 36

(b) 54, 90, 108

(c) 15, 19, 25

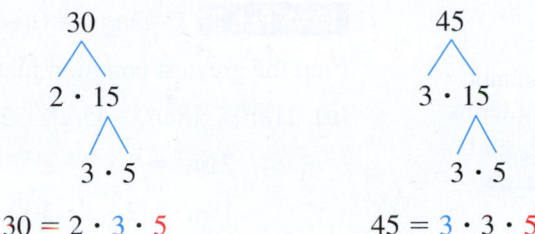

$$30 = 2 \cdot 3 \cdot 5 \qquad 45 = 3 \cdot 3 \cdot 5$$

To find the GCF, use each prime the **least** ***number of times it appears in*** **all** ***the factored forms.*** There is no 2 in the prime factored form of 45, so there will be no 2 in the greatest common factor. The least number of times 3 appears in all the factored forms is 1. The least number of times 5 appears is also 1.

$$\text{GCF} = 3^1 \cdot 5^1 = 15 \qquad 3^1 = 3 \text{ and } 5^1 = 5.$$

(b) 72, 120, 432

$$72 = 2 \cdot 2 \cdot 2 \cdot 3 \cdot 3$$
$$120 = 2 \cdot 2 \cdot 2 \cdot 3 \cdot 5$$
$$432 = 2 \cdot 2 \cdot 2 \cdot 2 \cdot 3 \cdot 3 \cdot 3$$

Write the prime factored form of each number.

The least number of times 2 appears in all the factored forms is 3, and the least number of times 3 appears is 1. There is no 5 in the prime factored form of either 72 or 432.

$$\text{GCF} = 2^3 \cdot 3^1 = 24 \qquad 2^3 = 8 \text{ and } 3^1 = 3.$$

(c) 10, 11, 14

$$10 = 2 \cdot 5$$
$$11 = 11$$
$$14 = 2 \cdot 7$$

Write the prime factored form of each number.

There are no primes common to all three numbers, so the GCF is 1.   **NOW TRY**

OBJECTIVE 2 Find the greatest common factor of a list of variable terms.

The terms x^4, x^5, x^6, and x^7 have x^4 as the greatest common factor because the ***least exponent*** on the variable x in the factored forms is 4.

$$x^4 = 1 \cdot x^4, \quad x^5 = x \cdot x^4, \quad x^6 = x^2 \cdot x^4, \quad x^7 = x^3 \cdot x^4$$

$$\text{GCF} = x^4$$

Finding the Greatest Common Factor (GCF)

Step 1 **Factor.** Write each number in prime factored form.

Step 2 **List common factors.** List each prime number or each variable that is a factor of every term in the list. (If a prime does not appear in one of the prime factored forms, it *cannot* appear in the greatest common factor.)

Step 3 **Choose least exponents.** Use as exponents on the common prime factors the *least* exponents from the prime factored forms.

Step 4 **Multiply** the primes from Step 3. If there are no primes left after Step 3, the greatest common factor is 1.

NOW TRY ANSWERS
1. **(a)** 12 **(b)** 18 **(c)** 1

**NOW TRY
EXERCISE 2**

Find the greatest common factor for each list of terms.

(a) $25k^3, 15k^2, 35k^5, 50k^4$

(b) $m^3n^5, m^4n^4, m^5n^2, m^6$

EXAMPLE 2   Finding the Greatest Common Factor (Variable Terms)

Find the greatest common factor for each list of terms.

(a) $21m^7, \quad 18m^6, \quad 45m^8, \quad 24m^5$

$$21m^7 = 3 \cdot 7 \cdot m^7$$
$$18m^6 = 2 \cdot 3 \cdot 3 \cdot m^6$$
$$45m^8 = 3 \cdot 3 \cdot 5 \cdot m^8$$
$$24m^5 = 2 \cdot 2 \cdot 2 \cdot 3 \cdot m^5$$

Here, 3 is the greatest common factor of the coefficients 21, 18, 45, and 24. The least exponent on m is 5.

$$\text{GCF} = 3m^5$$

(b) $x^4y^2, \quad x^7y^5, \quad x^3y^7, \quad y^{15}$

$$x^4y^2 = x^4 \cdot y^2$$
$$x^7y^5 = x^7 \cdot y^5$$
$$x^3y^7 = x^3 \cdot y^7$$
$$y^{15} = y^{15}$$

There is no x in the last term, y^{15}, so x will not appear in the greatest common factor. There is a y in each term, however, and 2 is the least exponent on y.

$$\text{GCF} = y^2$$

NOW TRY

OBJECTIVE 3 Factor out the greatest common factor.

Factoring a polynomial is the process of writing a polynomial sum in factored form as a product. For example, the polynomial

$$3m + 12$$

has two terms, $3m$ and 12. The greatest common factor of these two terms is 3. We can write $3m + 12$ so that each term is a product with 3 as one factor.

$$3m + 12$$
$$= 3 \cdot m + 3 \cdot 4 \qquad \text{GCF} = 3$$
$$= 3(m + 4) \qquad \text{Distributive property,}$$
$$a \cdot b + a \cdot c = a(b + c)$$

The factored form of $3m + 12$ is $3(m + 4)$. This process is called **factoring out the greatest common factor.**

⚠ **CAUTION**   The polynomial $3m + 12$ is *not* in factored form when written as

$$3 \cdot m + 3 \cdot 4. \qquad \text{Not in factored form}$$

The terms are factored, but the polynomial is not. The factored form of $3m + 12$ is the *product*

> The factors here are 3 and $m + 4$.

$3(m + 4).$ In factored form

EXAMPLE 3   Factoring Out the Greatest Common Factor

Write in factored form by factoring out the greatest common factor.

(a) $5y^2 + 10y$

$$= 5y(y) + 5y(2) \qquad \text{GCF} = 5y$$
$$= 5y(y + 2) \qquad \text{Distributive property}$$

CHECK $5y(y + 2)$ Multiply the factored form.

$$= 5y(y) + 5y(2) \qquad \text{Distributive property}$$
$$= 5y^2 + 10y \checkmark \qquad \text{Original polynomial}$$

NOW TRY ANSWERS
2. **(a)** $5k^2$ **(b)** m^3

NOW TRY
EXERCISE 3

Write in factored form by factoring out the greatest common factor.

(a) $7t^4 - 14t^3$

(b) $8x^6 - 20x^5 + 28x^4$

(c) $m^7 + m^9$

(d) $30m^4n^3 - 42m^2n^2$

(b) $20m^5 + 10m^4 - 15m^3$

$= 5m^3(4m^2) + 5m^3(2m) - 5m^3(3)$ GCF $= 5m^3$

$= 5m^3(4m^2 + 2m - 3)$ Factor out $5m^3$.

CHECK $5m^3(4m^2 + 2m - 3)$

$= 5m^3(4m^2) + 5m^3(2m) + 5m^3(-3)$ Distributive property

$= 20m^5 + 10m^4 - 15m^3$ ✓   Original polynomial

(c) $x^5 + x^3$

$= x^3(x^2) + x^3(1)$ GCF $= x^3$

$= x^3(x^2 + 1)$ ◁— Don't forget the 1.

CHECK   Mentally distribute x^3 over each term inside the parentheses. ✓

(d) $20m^7p^2 - 36m^3p^4$

$= 4m^3p^2(5m^4) - 4m^3p^2(9p^2)$ GCF $= 4m^3p^2$

$= 4m^3p^2(5m^4 - 9p^2)$ Factor out $4m^3p^2$.

CHECK   Mentally distribute $4m^3p^2$ over each term inside the parentheses. ✓

NOW TRY

⚠ CAUTION   Be sure to include the 1 in a problem like **Example 3(c)**. *Check that the factored form can be multiplied out to give the original polynomial.*

NOW TRY
EXERCISE 4

Write

$-14b^2 - 21b^3 + 7b$

in factored form by factoring out a negative common factor.

EXAMPLE 4   Factoring Out a Negative Common Factor

Write $-8x^4 + 16x^3 - 4x^2$ in factored form.

We can factor out either $4x^2$ or $-4x^2$ here. So that the coefficient of the leading (first) term in the trinomial factor will be positive, we factor out $-4x^2$.

$-8x^4 + 16x^3 - 4x^2$ Be careful with signs.

$= -4x^2(2x^2) - 4x^2(-4x) - 4x^2(1)$ $-4x^2$ is a common factor.

$= -4x^2(2x^2 - 4x + 1)$ Factor out $-4x^2$.
↑—— Positive coefficient

CHECK $-4x^2(2x^2 - 4x + 1)$

$= -4x^2(2x^2) - 4x^2(-4x) - 4x^2(1)$ Distributive property

$= -8x^4 + 16x^3 - 4x^2$ ✓   Original polynomial

NOW TRY

NOW TRY ANSWERS

3. (a) $7t^3(t - 2)$

(b) $4x^4(2x^2 - 5x + 7)$

(c) $m^7(1 + m^2)$

(d) $6m^2n^2(5m^2n - 7)$

4. $-7b(2b + 3b^2 - 1)$

NOTE   Whenever we factor a polynomial in which the coefficient of the leading term is negative, we will factor out the negative common factor. However, it would also be correct to factor out $4x^2$ in **Example 4.**

$4x^2(-2x^2 + 4x - 1)$ Equivalent form of the answer
↑—— Negative coefficient

NOW TRY
EXERCISE 5

Write in factored form by factoring out the greatest common factor.

(a) $x(x + 2) + 5(x + 2)$

(b) $a(t + 10) - b(t + 10)$

EXAMPLE 5   Factoring Out the Greatest Common Factor

Write in factored form by factoring out the greatest common factor.

Same

(a) $a(a + 3) + 4(a + 3)$ The binomial $a + 3$ is the greatest common factor.

$= (a + 3)(a + 4)$ Factor out $a + 3$.

(b) $x^2(x + 1) - 5(x + 1)$

$= (x + 1)(x^2 - 5)$ Factor out $x + 1$.   **NOW TRY**

> **NOTE**   In factored forms, the order of the factors does not matter because of the commutative property of multiplication, **ab = ba.**
>
> $(x + 1)(x^2 - 5)$ can also be written $(x^2 - 5)(x + 1)$.   **See Example 5(b).**

OBJECTIVE 4   Factor by grouping.

When a polynomial has four terms, common factors can sometimes be used to factor by grouping.

EXAMPLE 6   Factoring by Grouping

Factor by grouping.

(a) $2x + 6 + ax + 3a$

Group the first two terms and the last two terms because the first two terms have a common factor of 2 and the last two terms have a common factor of a.

$$2x + 6 + ax + 3a$$

$$= (2x + 6) + (ax + 3a)   \text{Group the terms.}$$

$$= 2(x + 3) + a(x + 3)   \text{Factor each group.}$$

The expression is still not in factored form because it is the *sum* of two terms. Now, however, $x + 3$ is a common factor and can be factored out.

$$= 2(x + 3) + a(x + 3)   x + 3 \text{ is a common factor.}$$

$(2 + a)(x + 3)$ is also correct.

$$= (x + 3)(2 + a)   \text{Factor out } x + 3.$$

The final result $(x + 3)(2 + a)$ is in factored form because it is a ***product.***

CHECK $(x + 3)(2 + a)$

$$= x(2) + x(a) + 3(2) + 3(a)   \text{Multiply using the FOIL method.}$$
$$\;\; \text{F} \qquad \text{O} \qquad \text{I} \qquad \text{L}$$

$$= 2x + ax + 6 + 3a   \text{Simplify.}$$

$$= 2x + 6 + ax + 3a \;\checkmark   \text{Rearrange terms to obtain the original polynomial.}$$

(b) $6ax + 24x + a + 4$

$$= (6ax + 24x) + (a + 4)   \text{Group the terms.}$$

$$= 6x(a + 4) + 1(a + 4)   \text{Factor each group. Remember the 1.}$$

$$= (a + 4)(6x + 1)   \text{Factor out } a + 4.$$

NOW TRY ANSWERS

5. (a) $(x + 2)(x + 5)$
   (b) $(t + 10)(a - b)$

**NOW TRY
EXERCISE 6**

Factor by grouping.

(a) $ab + 3a + 5b + 15$

(b) $12xy + 3x + 4y + 1$

(c) $x^3 + 5x^2 - 8x - 40$

CHECK $(a + 4)(6x + 1)$

$= 6ax + a + 24x + 4$ FOIL method

$= 6ax + 24x + a + 4$ ✓   Rearrange terms to obtain the original polynomial.

(c) $2x^2 - 10x + 3xy - 15y$

$= (2x^2 - 10x) + (3xy - 15y)$ Group the terms.

$= 2x(x - 5) + 3y(x - 5)$ Factor each group.

$= (x - 5)(2x + 3y)$ Factor out $x - 5$.

CHECK $(x - 5)(2x + 3y)$

$= 2x^2 + 3xy - 10x - 15y$ FOIL method

$= 2x^2 - 10x + 3xy - 15y$ ✓   Original polynomial

(d) $t^3 + 2t^2 - 3t - 6$ ⟵ Write a + sign between the groups.

$= (t^3 + 2t^2) + (-3t - 6)$ Group the terms.

$= t^2(t + 2) - 3(t + 2)$ Factor out -3 so there is a common factor, $t + 2$. *Check:* $-3(t + 2) = -3t - 6$
⟵ Be careful with signs.

$= (t + 2)(t^2 - 3)$ Factor out $t + 2$.

CHECK   Multiply using the FOIL method. ✓   **NOW TRY** ↺

⚠ **CAUTION**   *Be careful with signs when grouping.* In a problem like **Example 6(d),** it is wise to check the factoring in the second step, as shown in the example side comment.

Factoring a Polynomial with Four Terms by Grouping

Step 1 **Group the terms.** Collect the terms into two groups so that each group has a common factor.

Step 2 **Factor within the groups.** Factor out the greatest common factor from each group.

Step 3 **If possible, factor the entire polynomial.** Factor out a common binomial factor from the results of Step 2.

Step 4 **If necessary, rearrange terms.** If Step 2 does not result in a common binomial factor, try a different grouping.

Always check the factored form by multiplying.

EXAMPLE 7   Rearranging Terms before Factoring by Grouping

Factor by grouping.

(a) $10x^2 - 12y + 15x - 8xy$

Factoring out a common factor of 2 from the first two terms and a common factor of x from the last two terms gives the following.

$(10x^2 - 12y) + (15x - 8xy)$ Group the terms.

$= 2(5x^2 - 6y) + x(15 - 8y)$ No common factor results.

NOW TRY ANSWERS
6. **(a)** $(b + 3)(a + 5)$
  (b) $(4y + 1)(3x + 1)$
  (c) $(x + 5)(x^2 - 8)$

NOW TRY
EXERCISE 7

Factor by grouping.

(a) $12p^2 - 28q - 16pq + 21p$

(b) $5xy - 6 - 15x + 2y$

This grouping does not lead to a common factor, so we try rearranging the terms. There is usually more than one way to do this. We try the following.

$$10x^2 - 12y + 15x - 8xy \qquad \text{Original polynomial}$$
$$= 10x^2 - 8xy - 12y + 15x \qquad \text{Commutative property}$$
$$= (10x^2 - 8xy) + (-12y + 15x) \qquad \text{Group the terms}$$
$$= 2x(5x - 4y) + 3(-4y + 5x) \qquad \text{Factor each group}$$
$$= 2x(5x - 4y) + 3(5x - 4y) \qquad \text{Rewrite } -4y + 5x.$$
$$= (5x - 4y)(2x + 3) \qquad \text{Factor out } 5x - 4y.$$

CHECK $(5x - 4y)(2x + 3)$
$$= 10x^2 + 15x - 8xy - 12y \qquad \text{FOIL method}$$
$$= 10x^2 - 12y + 15x - 8xy \; \checkmark \quad \text{Original polynomial}$$

(b) $2xy + 12 - 3y - 8x$

We must rearrange these terms to obtain two groups that each have a common factor. Trial and error suggests the following grouping.

$$2xy + 12 - 3y - 8x \qquad \boxed{\text{Write a } + \text{ sign between the groups.}}$$
$$= (2xy - 3y) + (-8x + 12) \qquad \text{Rearrange and group the terms.}$$
$$= y(2x - 3) - 4(2x - 3) \qquad \begin{array}{l}\text{Factor each group.}\\ \textit{Check: } -4(2x - 3) = -8x + 12\end{array}$$

$\boxed{\text{Be careful with signs.}}$

$$= (2x - 3)(y - 4) \qquad \text{Factor out } 2x - 3.$$

Because the quantities in parentheses in the second step must be the same, we factored out -4 rather than 4.

CHECK $(2x - 3)(y - 4)$
$$= 2xy - 8x - 3y + 12 \qquad \text{FOIL method}$$
$$= 2xy + 12 - 3y - 8x \; \checkmark \quad \text{Original polynomial} \qquad \textbf{NOW TRY} \; \text{}$$

NOW TRY ANSWERS
7. **(a)** $(3p - 4q)(4p + 7)$
   (b) $(5x + 2)(y - 3)$

> ⚠ **CAUTION** Use negative signs carefully when grouping, as in **Example 7(b)**, or a sign error will occur. *Always check by multiplying.*

5.1 Exercises

FOR EXTRA HELP   **MyLab Math**

▶ *Video solutions for select problems available in MyLab Math*

Concept Check *Complete each statement.*

1. To factor a number or quantity means to write it as a(n) _____. Factoring is the opposite, or inverse, process of _____.

2. An integer or variable expression that is a factor of two or more terms is a(n) _____. For example, 12 (*is / is not*) a common factor of both 36 and 72 because it _____ evenly into both integers.

Find the greatest common factor for each list of numbers. See Example 1.

3. 12, 16   **4.** 18, 24   **5.** 40, 20, 4   **6.** 50, 30, 5

7. 18, 24, 36, 48   **8.** 15, 30, 45, 75   **9.** 4, 9, 12   **10.** 9, 16, 24

Find the greatest common factor for each list of terms. ***See Examples 1 and 2.***

11. $16y$, 24

12. $18w$, 27

13. $30x^3$, $40x^6$, $50x^7$

14. $60z^4$, $70z^8$, $90z^9$

15. x^4y^3, xy^2

16. a^4b^5, a^3b

17. $42ab^3$, $36a$, $90b$, $48ab$

18. $45c^3d$, $75c$, $90d$, $105cd$

19. $12m^3n^2$, $18m^5n^4$, $36m^8n^3$

20. $25p^5r^7$, $30p^7r^8$, $50p^5r^3$

Concept Check   *An expression is factored when it is written as a product, not a sum. Decide whether each expression is* factored *or* not factored.

21. $2k^2(5k)$

22. $2k^2(5k + 1)$

23. $2k^2 + (5k + 1)$

24. $(2k^2 + 5k) + 1$

25. Concept Check   A student incorrectly factored as follows.

$$18x^3y^2 + 9xy$$
$$= 9xy(2x^2y)$$

WHAT WENT WRONG?   Factor correctly.

26. Concept Check   When asked to factor a polynomial completely, a student answered incorrectly as follows.

$$12x^2y - 24xy$$
$$= 3xy(4x - 8)$$

WHAT WENT WRONG?   Factor completely.

Complete each factorization by writing the polynomial as the product of two factors. ***See Example 3.***

27. $9m^4$
   $= 3m^2(\underline{\hspace{1cm}})$

28. $12p^5$
   $= 6p^3(\underline{\hspace{1cm}})$

29. $-8z^9$
   $= -4z^5(\underline{\hspace{1cm}})$

30. $-15k^{11}$
   $= -5k^8(\underline{\hspace{1cm}})$

31. $6m^4n^5$
   $= 3m^3n(\underline{\hspace{1cm}})$

32. $27a^3b^2$
   $= 9a^2b(\underline{\hspace{1cm}})$

33. $12y + 24$
   $= 12(\underline{\hspace{1.5cm}})$

34. $18p + 36$
   $= 18(\underline{\hspace{1.5cm}})$

35. $10a^2 - 20a$
   $= 10a(\underline{\hspace{1.5cm}})$

36. $15x^2 - 30x$
   $= 15x(\underline{\hspace{1.5cm}})$

37. $8x^2y + 12x^3y^2$
   $= 4x^2y(\underline{\hspace{1.5cm}})$

38. $10st + 18s^2t^2$
   $= 2st(\underline{\hspace{1.5cm}})$

Write in factored form by factoring out the greatest common factor. ***See Examples 3–5.***

39. $x^2 - 4x$

40. $m^2 - 7m$

41. $6t^2 + 15t$

42. $8x^2 + 6x$

43. $27m^3 - 9m$

44. $12p^3 - 4p$

45. $m^3 - m^2$

46. $p^3 - p^2$

47. $16z^4 + 24z^2$

48. $25k^4 + 15k^2$

49. $-12x^3 - 6x^2$

50. $-21b^3 - 7b^2$

51. $65y^{10} + 35y^6$

52. $100a^5 + 16a^3$

53. $11w^3 - 100$

54. $13z^5 - 80$

55. $8mn^3 + 24m^2n^3$

56. $19p^2y + 38p^2y^3$

57. $13y^8 + 26y^4 - 39y^2$

58. $5x^5 + 25x^4 - 20x^3$

59. $-4x^3 + 10x^2 - 6x$

60. $-9z^3 + 6z^2 - 12z$

61. $36p^6q + 45p^5q^4 + 81p^3q^2$

62. $125a^3z^5 + 60a^4z^4 + 85a^5z^2$

63. $a^5 + 2a^3b^2 - 3a^5b^2 + 4a^4b^3$

64. $x^6 + 5x^4y^3 - 6xy^4 + 10xy$

65. $c(x + 2) - d(x + 2)$

66. $r(x + 5) - t(x + 5)$

67. $m(m + 2n) + n(m + 2n)$

68. $q(q + 4p) + p(q + 4p)$

69. $q^2(p - 4) - 3(p - 4)$

70. $y^2(x - 9) - 2(x - 9)$

Students often have difficulty when factoring by grouping because they are not able to tell when a polynomial is completely factored. For example,

$$5y(2x - 3) + 8t(2x - 3) \qquad \text{Not in factored form}$$

is not in factored form, because it is the **sum** of two terms: $5y(2x - 3)$ and $8t(2x - 3)$. However, because $2x - 3$ is a common factor of these two terms, the expression can now be factored.

$$(2x - 3)(5y + 8t) \qquad \text{In factored form}$$

The factored form is a **product** of the two factors $2x - 3$ and $5y + 8t$.

Concept Check *Determine whether each expression is* in factored form *or is* not in factored form. *If it is not in factored form, factor it if possible.*

71. $8(7t + 4) + x(7t + 4)$

72. $3r(5x - 1) + 7(5x - 1)$

73. $(8 + x)(7t + 4)$

74. $(3r + 7)(5x - 1)$

75. $18x^2(y + 4) + 7(y - 4)$

76. $12k^3(s - 3) + 7(s + 3)$

77. Concept Check A student factored as follows.

$$x^3 + 4x^2 - 2x - 8$$
$$= (x^3 + 4x^2) + (-2x - 8)$$
$$= x^2(x + 4) + 2(-x - 4)$$

The student could not find a common factor of the two terms.

WHAT WENT WRONG? Complete the factoring.

78. Concept Check A student factored as follows.

$$10xy + 18 + 12x + 15y$$
$$= (10xy + 18) + (12x + 15y)$$
$$= 2(5xy + 9) + 3(4x + 5y)$$

The student could not find a common factor of the two terms.

WHAT WENT WRONG? Complete the factoring.

Factor by grouping. ***See Examples 6 and 7.***

79. $5m + mn + 20 + 4n$

80. $ts + 5t + 2s + 10$

81. $6xy - 21x + 8y - 28$

82. $2mn - 8n + 3m - 12$

83. $p^2 + 4p + pq + 4q$

84. $m^2 + 2m + mn + 2n$

85. $a^2 - 2a + ab - 2b$

86. $y^2 - 6y + yw - 6w$

87. $7z^2 + 14z - az - 2a$

88. $2b^2 + 3b - 8ab - 12a$

89. $18r^2 + 12ry - 3xr - 2xy$

90. $8s^2 + 6sy - 4st - 3yt$

91. $w^3 + w^2 + 9w + 9$

92. $y^3 + y^2 + 6y + 6$

93. $3a^3 + 6a^2 - 2a - 4$

94. $10x^3 + 15x^2 - 8x - 12$

95. $10xy + 50y + x + 5$

96. $9ab + 27a + b + 3$

97. $12 - 4a - 3b + ab$ **98.** $6 - 3x - 2y + xy$

99. $16m^3 - 4m^2p^2 - 4mp + p^3$ **100.** $10t^3 - 2t^2s^2 - 5ts + s^3$

101. $y^2 + 3x + 3y + xy$ **102.** $m^2 + 14p + 7m + 2mp$

103. $5m - 6p - 2mp + 15$ **104.** $7y - 9x - 3xy + 21$

105. $2z^2 + 6w - 4z - 3wz$ **106.** $2a^2 + 20b - 8a - 5ab$

107. $18r^2 - 2ty + 12ry - 3rt$ **108.** $12a^2 - 4bc + 16ac - 3ab$

109. $a^5 - 3 + 2a^5b - 6b$ **110.** $b^3 - 2 + 5ab^3 - 10a$

RELATING CONCEPTS **For Individual or Group Work** (Exercises 111–114)

*In many cases, the choice of which pairs of terms to group when factoring by grouping can be made in different ways. To see this, **work Exercises 111–114 in order.***

111. Start with the polynomial from **Example 7(b)**,

$$2xy + 12 - 3y - 8x, \quad \text{and rearrange the terms to obtain} \quad 2xy - 8x - 3y + 12.$$

What property allows this?

112. Group the first two terms and the last two terms of the rearranged polynomial in **Exercise 111.** Then factor each group.

113. Is the result from **Exercise 112** in factored form? Explain.

114. If the answer to **Exercise 113** is "No," factor the polynomial. Is the result the same as the one shown for **Example 7(b)?**

| 5.2 | Factoring Trinomials |

OBJECTIVES

1 Factor trinomials with coefficient 1 for the second-degree term.

2 Factor out the greatest common factor first.

Using the FOIL method, we can find the product of the binomials $k - 3$ and $k + 1$.

$$(k - 3)(k + 1) = k^2 - 2k - 3 \qquad \text{\color{blue}Multiplying}$$

Suppose instead that we are given the polynomial $k^2 - 2k - 3$ and want to write it as the product $(k - 3)(k + 1)$.

$$k^2 - 2k - 3 = (k - 3)(k + 1) \qquad \text{\color{blue}Factoring}$$

Recall that *factoring* is a process that reverses, or "undoes," multiplying.

OBJECTIVE 1 **Factor trinomials with coefficient 1 for the second-degree term.**

When factoring polynomials with integer coefficients, we use only integers in the factors. For example, we can factor $x^2 + 5x + 6$ by finding integers m and n such that

$$x^2 + 5x + 6 \quad \text{is written as} \quad (x + m)(x + n).$$

To find these integers m and n, we multiply the two binomials on the right.

VOCABULARY
☐ prime polynomial

$$(x + m)(x + n)$$

$$= x^2 + nx + mx + mn \qquad \text{\color{blue}FOIL method}$$

$$= x^2 + (n + m)x + mn \qquad \text{\color{blue}Distributive property}$$

Comparing this result with $x^2 + 5x + 6$ shows that we must find integers m and n that have a sum of 5 and a product of 6.

Product of m and n is 6.

$$x^2 + 5x + 6 = x^2 + (n + m)x + mn$$

Sum of m and n is 5.

Because many pairs of integers have a sum of 5, it is best to begin by listing those pairs of integers whose product is 6. Both 5 and 6 are positive, so we consider only pairs in which both integers are positive.

Factors of 6	Sums of Factors
6, 1	$6 + 1 = 7$
3, 2	$3 + 2 = 5$

Sum is 5.

Both pairs have a product of 6, but only the pair 3 and 2 has a sum of 5. So 3 and 2 are the required integers.

$$x^2 + 5x + 6 \quad \text{factors as} \quad (x + 3)(x + 2).$$

Check by using the FOIL method to multiply the binomials. *Make sure that the sum of the outer and inner products produces the correct middle term.*

CHECK $(x + 3)(x + 2) = x^2 + 5x + 6$ ✓ Correct

$3x$
$2x$
$5x$ Add.

This method can be used only to factor trinomials that have 1 as the coefficient of the second-degree (squared variable) term.

> **NOTE** In the examples that follow, we list factors in descending order (disregarding signs) when looking for the required pair of integers. This helps to avoid skipping the correct combination.

NOW TRY
EXERCISE 1

Factor $p^2 + 7p + 10$.

EXAMPLE 1 Factoring a Trinomial (All Positive Terms)

Factor $m^2 + 9m + 14$.

Look for two integers whose product is 14 and whose sum is 9. List pairs of integers whose product is 14, and examine the sums. Only positive integers are needed because all signs in $m^2 + 9m + 14$ are positive.

Factors of 14	Sums of Factors
14, 1	$14 + 1 = 15$
7, 2	$7 + 2 = 9$

Sum is 9.

The required integers are 7 and 2 because $7 \cdot 2 = 14$ and $7 + 2 = 9$.

$$m^2 + 9m + 14 \quad \text{factors as} \quad (m + 7)(m + 2).$$

$(m + 2)(m + 7)$ is also correct.

CHECK $(m + 7)(m + 2)$

$\qquad = m^2 + 2m + 7m + 14 \qquad$ FOIL method

$\qquad = m^2 + 9m + 14$ ✓ $\qquad$ Original polynomial $\qquad$ **NOW TRY**

NOW TRY ANSWER
1. $(p + 5)(p + 2)$

NOW TRY
EXERCISE 2

Factor $t^2 - 9t + 18$.

EXAMPLE 2 Factoring a Trinomial (Negative Middle Term)

Factor $x^2 - 9x + 20$.

We must find two integers whose product is 20 and whose sum is -9. Because the numbers we are looking for have a *positive product* and a *negative sum,* we consider only pairs of negative integers.

Factors of 20	Sums of Factors
$-20, -1$	$-20 + (-1) = -21$
$-10, -2$	$-10 + (-2) = -12$
$-5, -4$	$-5 + (-4) = -9$

Sum is -9.

The required integers are -5 and -4.

$$x^2 - 9x + 20 \quad \text{factors as} \quad (x - 5)(x - 4).$$

> The order of the factors does not matter.

CHECK

$$(x - 5)(x - 4)$$

$$= x^2 - 4x - 5x + 20 \qquad \text{FOIL method}$$

$$= x^2 - 9x + 20 \ \checkmark \qquad \text{Original polynomial}$$

NOW TRY

NOW TRY
EXERCISE 3

Factor $x^2 + x - 42$.

EXAMPLE 3 Factoring a Trinomial (Negative Last (Constant) Term)

Factor $x^2 + x - 6$.

We must find two integers whose product is -6 and whose sum is 1 (because the coefficient of x, or $1x$, is 1). To obtain a *negative product,* the pairs of integers must have different signs.

> Once we find the required pair, we can stop listing factors.

Factors of -6	Sums of Factors
$6, -1$	$6 + (-1) = 5$
$-6, 1$	$-6 + 1 = -5$
$3, -2$	$3 + (-2) = 1$

Sum is 1.

The required integers are 3 and -2.

$$x^2 + x - 6 \quad \text{factors as} \quad (x + 3)(x - 2).$$

CHECK $\qquad (x + 3)(x - 2)$

$$= x^2 - 2x + 3x - 6 \qquad \text{FOIL method}$$

$$= x^2 + x - 6 \ \checkmark \qquad \text{Original polynomial}$$

NOW TRY

EXAMPLE 4 Factoring a Trinomial (Two Negative Terms)

Factor $p^2 - 2p - 15$.

Find two integers whose product is -15 and whose sum is -2. Because the constant term, -15, is negative, list pairs of integers with different signs.

NOW TRY ANSWERS
2. $(t - 6)(t - 3)$
3. $(x + 7)(x - 6)$

Factors of -15	Sums of Factors
$15, -1$	$15 + (-1) = 14$
$-15, 1$	$-15 + 1 = -14$
$5, -3$	$5 + (-3) = 2$
$-5, 3$	$-5 + 3 = -2$

Sum is -2.

**NOW TRY
EXERCISE 4**
Factor $x^2 - 4x - 21$.

The required integers are -5 and 3.

$$p^2 - 2p - 15 \quad \text{factors as} \quad (p - 5)(p + 3).$$

CHECK   Multiply $(p - 5)(p + 3)$ to obtain $p^2 - 2p - 15$. ✓   **NOW TRY** ↻

Guidelines for Factoring $x^2 + bx + c$

Find two integers whose product is c and whose sum is b.

1. Both integers must be positive if b and c are positive. (See **Example 1.**)

2. Both integers must be negative if c is positive and b is negative. (See **Example 2.**)

3. One integer must be positive and one must be negative if c is negative. (See **Examples 3 and 4.**)

NOTE   Consider all possible sign combinations for multiplying two bionomials to understand the resulting signs of the terms of a trinomial.

$(x + 3)(x + 2) = x^2 + 5x + 6$ All signs in the trinomial product are
Both signs positive   positive.

$(x - 3)(x - 2) = x^2 - 5x + 6$ The last term of the trinomial product is
Both signs negative   positive and the middle term is negative.

$(x + 3)(x - 2) = x^2 + x - 6$ The last term of the trinomial product is
                                                                   negative, and the middle term has the same
$(x - 3)(x + 2) = x^2 - x - 6$ sign as the number in the binomials with the
Different signs   greater absolute value.

Some trinomials cannot be factored using only integers. Such trinomials are **prime polynomials.**

**NOW TRY
EXERCISE 5**
Factor each trinomial if possible.

(a) $m^2 + 5m + 8$

(b) $t^2 + 11t - 24$

EXAMPLE 5   Determining Whether Polynomials Are Prime

Factor each trinomial if possible.

(a) $x^2 - 5x + 12$

As in **Example 2,** both factors must be negative to give a positive product and a negative sum. List pairs of negative integers whose product is 12, and examine the sums.

Factors of 12	Sums of Factors
$-12, -1$	$-12 + (-1) = -13$
$-6, -2$	$-6 + (-2) = -8$
$-4, -3$	$-4 + (-3) = -7$

No sum is -5.

None of the pairs of integers has a sum of -5. Therefore, the trinomial $x^2 - 5x + 12$ *cannot be factored using only integers.* It is a prime polynomial.

NOW TRY ANSWERS
4. $(x - 7)(x + 3)$
5. (a) prime   **(b)** prime

(b) $k^2 - 8k + 11$

There is no pair of integers whose product is 11 and whose sum is -8, so $k^2 - 8k + 11$ is a prime polynomial.

NOW TRY ↻

 **NOW TRY
EXERCISE 6**

Factor $a^2 + 2ab - 15b^2$.

EXAMPLE 6 Factoring a Multivariable Trinomial

Factor $z^2 - 2bz - 3b^2$.

Here, the coefficient of z in the middle term is $-2b$, so we need to find two expressions whose product is $-3b^2$ and whose sum is $-2b$.

Factors of $-3b^2$	Sums of Factors
$3b, -b$	$3b + (-b) = 2b$
$-3b, b$	$-3b + b = -2b$

Sum is $-2b$.

$$z^2 - 2bz - 3b^2 \quad \text{factors as} \quad (z - 3b)(z + b).$$

CHECK

$$(z - 3b)(z + b)$$

$$= z^2 + zb - 3bz - 3b^2 \qquad \text{FOIL method}$$

$$= z^2 + 1bz - 3bz - 3b^2 \qquad \text{Identity and commutative properties}$$

$$= z^2 - 2bz - 3b^2 \ \checkmark \qquad \text{Combine like terms.} \qquad \text{NOW TRY}$$

OBJECTIVE 2 Factor out the greatest common factor first.

If the terms of a trinomial have a common factor greater than 1, first factor it out. Then factor the remaining trinomial.

**NOW TRY
EXERCISE 7**

Factor each trinomial completely.

(a) $3y^4 - 27y^3 + 60y^2$

(b) $-5x^5 - 15x^4 + 90x^3$

EXAMPLE 7 Factoring Trinomials with Common Factors

Factor each trinomial completely.

The terms have a common factor.

(a) $\qquad\qquad 4x^5 - 28x^4 + 40x^3$

$$= 4x^3(x^2 - 7x + 10) \qquad \text{Factor out the greatest common factor, } 4x^3.$$

Factor $x^2 - 7x + 10$. The integers -5 and -2 have a product of 10 and a sum of -7.

Include $4x^3$.

$$= 4x^3(x - 5)(x - 2) \qquad \text{Completely factored form}$$

CHECK

$$4x^3(x - 5)(x - 2)$$

$$= 4x^3(x^2 - 2x - 5x + 10) \qquad \text{FOIL method}$$

$$= 4x^3(x^2 - 7x + 10) \qquad \text{Combine like terms.}$$

$$= 4x^5 - 28x^4 + 40x^3 \ \checkmark \qquad \text{Distributive property}$$

(b) $-3y^8 - 18y^7 + 21y^6$

The coefficient of the first term is negative, so we factor out $-3y^6$.

$$-3y^8 - 18y^7 + 21y^6$$

$$= -3y^6(y^2 + 6y - 7) \qquad \text{Factor out the greatest common factor, } -3y^6.$$

Factor $y^2 + 6y - 7$. The integers 7 and -1 have a product of -7 and a sum of 6.

$$= -3y^6(y + 7)(y - 1) \qquad \text{Completely factored form}$$

CHECK

$$-3y^6(y + 7)(y - 1)$$

$$= -3y^6(y^2 + 6y - 7) \qquad \text{FOIL method}$$

$$= -3y^8 - 18y^7 + 21y^6 \ \checkmark \qquad \text{Distributive property}$$

NOW TRY

NOW TRY ANSWERS

6. $(a + 5b)(a - 3b)$

7. (a) $3y^2(y - 5)(y - 4)$

 (b) $-5x^3(x + 6)(x - 3)$

> ⚠ **CAUTION** *When factoring, always look for a common factor first.* If the coefficient of the leading term is negative, factor out the negative common factor.
> Remember to include the common factor as part of the answer, and check by multiplying out the completely factored form.

5.2 Exercises

FOR EXTRA HELP ▶ **MyLab Math**

▶ *Video solutions for select problems available in MyLab Math*

Concept Check *Answer each question.*

1. When factoring a trinomial in x as

$$(x + a)(x + b),$$

what must be true of a and b if the coefficient of the constant term of the trinomial is negative?

2. When factoring a trinomial in x as

$$(x + a)(x + b),$$

what must be true of a and b if the coefficient of the constant term of the trinomial is positive?

3. Which is the correct factored form of $x^2 - 12x + 32$?

A. $(x - 8)(x + 4)$ **B.** $(x + 8)(x - 4)$

C. $(x - 8)(x - 4)$ **D.** $(x + 8)(x + 4)$

4. Which is the correct factored form of $x^2 - 5x - 14$?

A. $(x - 2)(x - 7)$ **B.** $(x + 2)(x + 7)$

C. $(x - 2)(x + 7)$ **D.** $(x + 2)(x - 7)$

5. What polynomial can be factored as $(a + 9)(a + 4)$?

6. What polynomial can be factored as $(y - 7)(y + 3)$?

7. Concept Check A student factored as follows.

$$x^3 + 3x^2 - 28x$$
$$= x(x^2 + 3x - 28x)$$
$$= (x + 7)(x - 4) \qquad \text{Incorrect}$$

WHAT WENT WRONG? Factor correctly.

8. Concept Check A student factored as follows.

$$x^2 + x + 6$$
$$= (x + 6)(x - 1) \qquad \text{Incorrect}$$

WHAT WENT WRONG? Factor correctly if possible.

List all pairs of integers with the given product. Then find the pair whose sum is given. **See the tables in Examples 1–4.**

9. Product: 48; Sum: -19

10. Product: 18; Sum: 9

11. Product: -24; Sum: -5

12. Product: -36; Sum: -16

Concept Check   *Complete each factoring.*

13. To factor $y^2 + 12y + 20$, find two integers whose product is _____ and whose sum is _____. Complete the table.

Factors of 20	Sums of Factors
20, 1	20 + 1 = 21
10, ___	10 + ___ = ___
5, ___	5 + ___ = ___

Which pair of factors has the required sum? _____

Now factor the trinomial.

14. To factor $t^2 - 12t + 32$, find two integers whose product is _____ and whose sum is _____. Complete the table.

Factors of 32	Sums of Factors
−32, −1	−32 + (−1) = −33
−16, ___	−16 + (___) = ___
−8, ___	−8 + (___) = ___

Which pair of factors has the required sum? _____

Now factor the trinomial.

Complete each factorization. **See Examples 1–4.**

15. $p^2 + 11p + 30$

$= (p + 5)(\underline{\hspace{1cm}})$

16. $x^2 + 10x + 21$

$= (x + 7)(\underline{\hspace{1cm}})$

17. $x^2 + 15x + 44$

$= (x + 4)(\underline{\hspace{1cm}})$

18. $r^2 + 15r + 56$

$= (r + 7)(\underline{\hspace{1cm}})$

19. $x^2 - 9x + 8$

$= (x - 1)(\underline{\hspace{1cm}})$

20. $t^2 - 14t + 24$

$= (t - 2)(\underline{\hspace{1cm}})$

21. $y^2 - 2y - 15$

$= (y + 3)(\underline{\hspace{1cm}})$

22. $t^2 - t - 42$

$= (t + 6)(\underline{\hspace{1cm}})$

23. $x^2 + 9x - 22$

$= (x - 2)(\underline{\hspace{1cm}})$

24. $x^2 + 6x - 27$

$= (x - 3)(\underline{\hspace{1cm}})$

25. $y^2 - 7y - 18$

$= (y + 2)(\underline{\hspace{1cm}})$

26. $y^2 - 2y - 24$

$= (y + 4)(\underline{\hspace{1cm}})$

Factor completely. If a polynomial cannot be factored using integers, write prime. **See Examples 1–5.**

27. $y^2 + 9y + 8$

28. $a^2 + 9a + 20$

29. $b^2 + 8b + 15$

30. $x^2 + 6x + 8$

31. $m^2 + m - 20$

32. $p^2 + 4p - 5$

33. $y^2 - 8y + 15$

34. $y^2 - 6y + 8$

35. $x^2 + 4x + 5$

36. $t^2 + 11t + 12$

37. $z^2 - 15z + 56$

38. $x^2 - 13x + 36$

39. $r^2 - r - 30$

40. $q^2 - q - 42$

41. $a^2 - 8a - 48$

42. $d^2 - 4d - 45$

43. $x^2 + 3x - 39$

44. $m^2 + 10m - 30$

45. $-32 + 14x + x^2$

(*Hint:* First write in descending powers.)

46. $-39 + 10x + x^2$

(*Hint:* First write in descending powers.)

Factor completely. **See Example 6.**

47. $r^2 + 3ra + 2a^2$ **48.** $x^2 + 5xa + 4a^2$

49. $x^2 + 4xy + 3y^2$ **50.** $p^2 + 9pq + 8q^2$

51. $t^2 - tz - 6z^2$ **52.** $a^2 - ab - 12b^2$

53. $v^2 - 11vw + 30w^2$ **54.** $v^2 - 11vx + 24x^2$

55. $m^2 + 4mn - 12n^2$ **56.** $x^2 + 6xy - 16y^2$

57. $a^2 - 9ab + 18b^2$ **58.** $h^2 - 11hk + 28k^2$

Factor completely. **See Example 7.**

59. $4x^2 + 12x - 40$ **60.** $5y^2 + 5y - 30$

61. $2t^3 + 8t^2 + 6t$ **62.** $3t^3 + 27t^2 + 24t$

63. $6z^4 - 24z^3 + 18z^2$ **64.** $5x^4 - 35x^3 + 30x^2$

65. $-2x^6 - 8x^5 + 42x^4$ **66.** $-4y^5 - 12y^4 + 40y^3$

67. $5m^5 - 25m^4 + 40m^2$ **68.** $12k^5 - 6k^3 + 10k^2$

69. $x^3 - 7x^2y + 12xy^2$ **70.** $p^3 - 8p^2q + 15pq^2$

71. $z^{10} - 4z^9y - 21z^8y^2$ **72.** $k^7 - 2k^6m - 15k^5m^2$

73. $-a^5 - 3a^4b + 4a^3b^2$ **74.** $-x^9 - 5x^8w + 24x^7w^2$

75. $m^3n - 10m^2n^2 + 24mn^3$ **76.** $p^3q - 11p^2q^2 + 18pq^3$

77. $y^3z + y^2z^2 - 6yz^3$ **78.** $a^3b + 3a^2b^2 - 54ab^3$

Extending Skills   *Factor each polynomial.*

79. $(a + b)x^2 + (a + b)x - 12(a + b)$

80. $(x + y)n^2 + (x + y)n - 20(x + y)$

81. $(2p + q)r^2 - 12(2p + q)r + 27(2p + q)$

82. $(3m - n)k^2 - 13(3m - n)k + 40(3m - n)$

5.3   More on Factoring Trinomials

OBJECTIVES

1 Factor trinomials by grouping when the coefficient of the second-degree term is not 1.

2 Factor trinomials using the FOIL method.

OBJECTIVE 1 Factor trinomials by grouping when the coefficient of the second-degree term is not 1.

We now extend our work to factor a trinomial in which the coefficient of the second-degree term is *not* 1, such as

$$2x^2 + 7x + 6.$$

There are two methods for factoring this trinomail, as shown in **Examples 1 and 4.** The first method uses factoring by grouping.

EXAMPLE 1   Factoring by Grouping (Coefficient of the Second-Degree Term Not 1)

Factor $2x^2 + 7x + 6$.

To factor this trinomial, we look for two positive integers whose product is $2 \cdot 6 = 12$ and whose sum is 7.

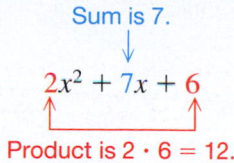

Sum is 7.

$$2x^2 + 7x + 6$$

Product is $2 \cdot 6 = 12$.

The required integers are 3 and 4, since $3 \cdot 4 = 12$ and $3 + 4 = 7$. We use these integers to write the middle term $7x$ as $3x + 4x$.

$$2x^2 + 7x + 6$$
$$= 2x^2 + \underbrace{3x + 4x}_{7x} + 6$$
$$= (2x^2 + 3x) + (4x + 6) \qquad \text{Group the terms.}$$
$$= x(2x + 3) + 2(2x + 3) \qquad \text{Factor each group.}$$

Must be the same factor

$$= (2x + 3)(x + 2) \qquad \text{Factor out } 2x + 3.$$

CHECK   Multiply $(2x + 3)(x + 2)$ to obtain $2x^2 + 7x + 6$. ✓   **NOW TRY**

> **NOTE**   In **Example 1**, we could have written $7x$ as $4x + 3x$, rather than as $3x + 4x$. Factoring by grouping would give the same answer. Try this.

EXAMPLE 2   Factoring Trinomials by Grouping

Factor each trinomial.

(a) $6r^2 + r - 1$

We must find two integers with a product of $6(-1) = -6$ and a sum of 1.

Sum is 1.

$$6r^2 + 1r - 1 \qquad \text{The coefficient of } r, \text{ or } 1r, \text{ is } 1.$$

Product is $6(-1) = -6$.

The integers -2 and 3 have a product of -6 and a sum of 1. We write the middle term r as $-2r + 3r$.

$$6r^2 + r - 1$$
$$= 6r^2 - 2r + 3r - 1 \qquad r = -2r + 3r$$
$$= (6r^2 - 2r) + (3r - 1) \qquad \text{Group the terms.}$$
$$= 2r(3r - 1) + 1(3r - 1) \qquad \text{The binomials must be the same.}$$

Remember the 1.

$$= (3r - 1)(2r + 1) \qquad \text{Factor out } 3r - 1.$$

CHECK   Multiply $(3r - 1)(2r + 1)$ to obtain $6r^2 + r - 1$. ✓

NOW TRY EXERCISE 1

Factor $2m^2 + 7m + 3$.

NOW TRY ANSWER

1. $(2m + 1)(m + 3)$

NOW TRY
EXERCISE 2

Factor each trinomial.

(a) $2z^2 + 5z + 3$

(b) $15m^2 + m - 2$

(c) $8x^2 - 2xy - 3y^2$

(b) $12z^2 - 5z - 2$

Look for two integers whose product is $12(-2) = -24$ and whose sum is -5. The required integers are 3 and -8.

$$12z^2 - 5z - 2$$
$$= 12z^2 + 3z - 8z - 2 \qquad -5z = 3z - 8z$$
$$= (12z^2 + 3z) + (-8z - 2) \qquad \text{Group the terms.}$$
$$= 3z(4z + 1) - 2(4z + 1) \qquad \text{Factor each group.}$$
$$= (4z + 1)(3z - 2) \qquad \text{Factor out } 4z + 1.$$

> Be careful with signs.

CHECK Multiply $(4z + 1)(3z - 2)$ to obtain $12z^2 - 5z - 2$. ✓

(c) $10m^2 + mn - 3n^2$

Two integers whose product is $10(-3) = -30$ and whose sum is 1 are -5 and 6.

$$10m^2 + mn - 3n^2 \qquad \text{The coefficient of } mn \text{ is 1.}$$
$$= 10m^2 - 5mn + 6mn - 3n^2 \qquad mn = -5mn + 6mn$$
$$= (10m^2 - 5mn) + (6mn - 3n^2) \qquad \text{Group the terms.}$$
$$= 5m(2m - n) + 3n(2m - n) \qquad \text{Factor each group.}$$
$$= (2m - n)(5m + 3n) \qquad \text{Factor out } 2m - n.$$

CHECK Multiply $(2m - n)(5m + 3n)$ to obtain $10m^2 + mn - 3n^2$. ✓

 **NOW TRY**

NOW TRY
EXERCISE 3

Factor $15z^6 + 18z^5 - 24z^4$.

EXAMPLE 3 Factoring a Trinomial with a Common Factor by Grouping

Factor $28x^5 - 58x^4 - 30x^3$.

$$28x^5 - 58x^4 - 30x^3$$
$$= 2x^3(14x^2 - 29x - 15) \qquad \text{Factor out the greatest common factor, } 2x^3.$$

To factor $14x^2 - 29x - 15$, find two integers whose product is $14(-15) = -210$ and whose sum is -29. Factoring 210 into prime factors helps find these integers.

$$210 = 2 \cdot 3 \cdot 5 \cdot 7$$

Combine the prime factors into pairs in different ways, using one positive factor and one negative factor to obtain -210. The factors 6 and -35 have the correct sum, -29.

$$28x^5 - 58x^4 - 30x^3$$
$$= 2x^3(14x^2 - 29x - 15) \qquad \text{Factor out the GCF.}$$
$$= 2x^3(14x^2 + 6x - 35x - 15) \qquad -29x = 6x - 35x$$
$$= 2x^3[(14x^2 + 6x) + (-35x - 15)] \qquad \text{Group the terms.}$$
$$= 2x^3[2x(7x + 3) - 5(7x + 3)] \qquad \text{Factor each group.}$$
$$= 2x^3[(7x + 3)(2x - 5)] \qquad \text{Factor out } 7x + 3.$$
$$= 2x^3(7x + 3)(2x - 5)$$

> Remember the common factor.

NOW TRY ANSWERS
2. (a) $(2z + 3)(z + 1)$
 (b) $(3m - 1)(5m + 2)$
 (c) $(4x - 3y)(2x + y)$
3. $3z^4(5z - 4)(z + 2)$

CHECK One way to check is to first multiply $2x^3(7x + 3)$ to obtain $(14x^4 + 6x^3)$. Then multiply

$$(14x^4 + 6x^3)(2x - 5) \quad \text{to obtain} \quad 28x^5 - 58x^4 - 30x^3. \text{ ✓}$$ **NOW TRY**

OBJECTIVE 2 Factor trinomials using the FOIL method.

There is an alternative method of factoring trinomials that uses trial and error.

**NOW TRY
EXERCISE 4**

Factor $2p^2 + 9p + 9$.

> **EXAMPLE 4** Factoring Using FOIL (Coefficient of the Second-Degree Term Not 1)

Factor $2x^2 + 7x + 6$. (We factored this trinomial by grouping in **Example 1.**)

We want to write $2x^2 + 7x + 6$ as the product of two binomials.

$$2x^2 + 7x + 6$$
$$= (\underline{\hspace{1cm}})(\underline{\hspace{1cm}})$$

We use the FOIL method in reverse.

The product of the two first terms of the binomials must be $2x^2$. The possible factors of $2x^2$ are $2x$ and x, or $-2x$ and $-x$. Because all terms of the trinomial are positive, we consider only positive factors. Thus, we have the following.

$$2x^2 + 7x + 6$$
$$= (2x\underline{\hspace{1cm}})(x\underline{\hspace{1cm}})$$

The product of the two last terms of the binomials must be 6. It can be factored as

$$1 \cdot 6, \quad 6 \cdot 1, \quad 2 \cdot 3, \quad \text{or} \quad 3 \cdot 2.$$

Beginning with 1 and 6, we try each pair of factors in $(2x\underline{\hspace{1cm}})(x\underline{\hspace{1cm}})$ to find the pair that gives the correct middle term, $7x$.

Try 1 and 6.   Try 6 and 1.

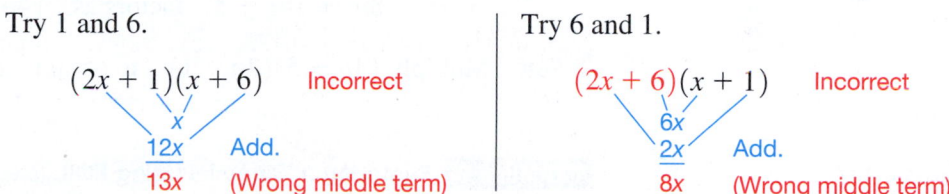

On the right above, $2x + 6 = 2(x + 3)$. The terms of the binomial $2x + 6$ have a common factor of 2, while the terms of the given trinomial $2x^2 + 7x + 6$ have no common factor other than 1. Thus, the product $(2x + 6)(x + 1)$ cannot be correct.

> *If the terms of the original polynomial have greatest common factor 1, then each of its factors will also have terms with GCF 1.*

Try 2 and 3.   Try 3 and 2.

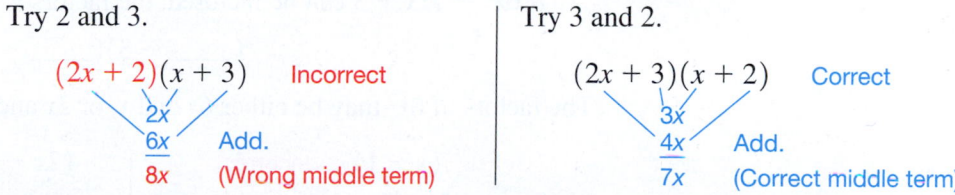

We might have recognized that the product on the left above would be incorrect because the terms of $2x + 2$ have a common factor of 2, while the terms of the given trinomial do not. The last combination produces $7x$, the correct middle term.

$$2x^2 + 7x + 6 \quad \text{factors as} \quad (2x + 3)(x + 2).$$

CHECK Multiply $(2x + 3)(x + 2)$ to obtain $2x^2 + 7x + 6$. ✓   **NOW TRY** ↻

**NOW TRY
EXERCISE 5**

Factor $8y^2 + 22y + 5$.

EXAMPLE 5   Factoring a Trinomial Using FOIL (All Positive Terms)

Factor $8p^2 + 14p + 5$.

The number 8 has several possible pairs of factors, but 5 has only 1 and 5, or -1 and -5, so we begin by considering the factors of 5. We ignore the negative factors because all coefficients in the trinomial are positive. If $8p^2 + 14p + 5$ can be factored, the factors will have this form.

$$(\underline{} + 5)(\underline{} + 1)$$

The possible pairs of factors of $8p^2$ are $8p$ and p, or $4p$ and $2p$. We try various combinations, checking to see if the middle term is $14p$.

Try $8p$ and p.

$(8p + 5)(p + 1)$ Incorrect

$5p$
$8p$
$13p$ Add.

Try p and $8p$.

$(p + 5)(8p + 1)$ Incorrect

$40p$
p
$41p$ Add.

Try $2p$ and $4p$.

$(2p + 5)(4p + 1)$ Incorrect

$20p$
$2p$
$22p$ Add.

Try $4p$ and $2p$.

$(4p + 5)(2p + 1)$ Correct

$10p$
$4p$
$14p$ Add.

The last combination produces $14p$, the correct middle term.

$$8p^2 + 14p + 5 \quad \text{factors as} \quad (4p + 5)(2p + 1).$$

CHECK   Multiply $(4p + 5)(2p + 1)$ to obtain $8p^2 + 14p + 5$. ✓   NOW TRY

**NOW TRY
EXERCISE 6**

Factor $10x^2 - 9x + 2$.

EXAMPLE 6   Factoring a Trinomial Using FOIL (Negative Middle Term)

Factor $6x^2 - 11x + 3$.

Because 3 has as factors only 1 and 3, or -1 and -3, it is better here to begin by considering the factors of 3. The last (constant) term of the trinomial $6x^2 - 11x + 3$ is positive and the middle term has a negative coefficient, so we consider only negative factors. We need two negative factors, because the *product* of two negative factors is positive and their *sum* is negative, as required.

If $6x^2 - 11x + 3$ can be factored, the factors will have this form.

$$(\underline{} - 3)(\underline{} - 1)$$

The factors of $6x^2$ may be either $6x$ and x, or $2x$ and $3x$.

$(6x - 3)(x - 1)$ Incorrect

$-3x$
$-6x$
$-9x$ Add.

$(2x - 3)(3x - 1)$ Correct

$-9x$
$-2x$
$-11x$ Add.

We can save some work by recognizing that the product on the left cannot be correct—the terms of $6x - 3$ have a common factor of 3, while those of the given trinomial do not. This is also why we used $2x$ and $3x$ on the right, rather than $3x$ and $2x$ (which would have given $3x - 3$ as a factor).

The factors $2x$ and $3x$ produce $-11x$, the correct middle term.

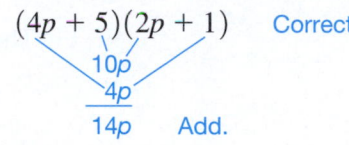

Check by multiplying.

$$6x^2 - 11x + 3 \quad \text{factors as} \quad (2x - 3)(3x - 1).$$ NOW TRY

NOW TRY ANSWERS
5. $(4y + 1)(2y + 5)$
6. $(5x - 2)(2x - 1)$

**NOW TRY
EXERCISE 7**

Factor $10a^2 + 31a - 14$.

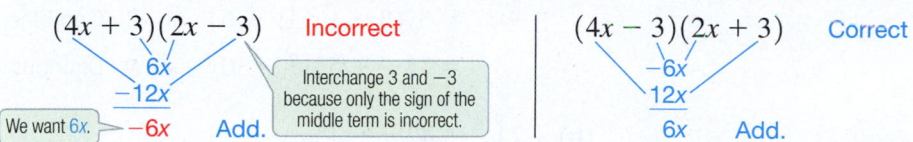

EXAMPLE 7 Factoring a Trinomial Using FOIL (Negative Constant Term)

Factor $8x^2 + 6x - 9$.

The integer 8 has several possible pairs of factors, as does -9. Because the constant term is negative, one positive factor and one negative factor of -9 are needed. The coefficient of the middle term is relatively small, so we avoid large factors such as 8 or 9. We try $4x$ and $2x$ as factors of $8x^2$, and 3 and -3 as factors of -9.

$(4x + 3)(2x - 3)$ Incorrect

$6x$

$-12x$

We want $6x$. $-6x$ Add.

Interchange 3 and -3 because only the sign of the middle term is incorrect.

$(4x - 3)(2x + 3)$ Correct

$-6x$

$12x$

$6x$ Add.

The combination on the right produces $6x$, the correct middle term.

$$8x^2 + 6x - 9 \quad \text{factors as} \quad (4x - 3)(2x + 3).$$

Check by multiplying.

 NOW TRY

**NOW TRY
EXERCISE 8**

Factor $8z^2 + 2wz - 15w^2$.

EXAMPLE 8 Factoring a Multivariable Trinomial

Factor $12a^2 - ab - 20b^2$.

There are several pairs of factors of $12a^2$, including

$$12a \text{ and } a, \quad 6a \text{ and } 2a, \quad \text{and} \quad 3a \text{ and } 4a.$$

There are also many pairs of factors of $-20b^2$, including

$$20b \text{ and } -b, \quad -20b \text{ and } b, \quad 10b \text{ and } -2b, \quad -10b \text{ and } 2b,$$

$$4b \text{ and } -5b, \quad \text{and} \quad -4b \text{ and } 5b.$$

Once again, because the coefficient of the middle term is relatively small, avoid the larger factors. Try the factors $3a$ and $4a$, and $4b$ and $-5b$.

$(3a - 5b)(4a + 4b)$ Incorrect

This cannot be correct because the terms of $4a + 4b$ have 4 as a common factor, while the terms of the given trinomial do not.

$(3a + 4b)(4a - 5b)$ Incorrect

$16ab$

$-15ab$

ab Add.

In the factorization attempt on the right, the middle term is ab, rather than $-ab$. We interchange the signs of the last two terms in the factors to obtain the correct result.

$$(3a - 4b)(4a + 5b)$$

$$= 12a^2 - ab - 20b^2 \quad \text{Correct}$$

Thus, $12a^2 - ab - 20b^2$ factors as $(3a - 4b)(4a + 5b)$.

 NOW TRY

EXAMPLE 9 Factoring Trinomials with Common Factors

Factor each trinomial.

(a) $\qquad 15y^3 + 55y^2 + 30y$

$$= 5y(3y^2 + 11y + 6) \quad \text{Factor out the greatest common factor, } 5y.$$

Now factor $3y^2 + 11y + 6$. Try $3y$ and y as factors of $3y^2$, and 2 and 3 as factors of 6. We know that $(3y + 3)(y + 2)$ is incorrect because the terms of $3y + 3$ have a common factor of 3. So we switch the 3 and 2 and try the following factors to see if we obtain the correct middle term.

NOW TRY ANSWERS

7. $(5a - 2)(2a + 7)$

8. $(4z - 5w)(2z + 3w)$

$$(3y + 2)(y + 3)$$

$$= 3y^2 + 11y + 6 \quad \text{Correct}$$

 **NOW TRY
EXERCISE 9**

Factor each trinomial.

(a) $36z^3 + 10z^2 + 7z$

(b) $-10x^3 - 45x^2 + 90x$

This leads to the completely factored form.

$$15y^3 + 55y^2 + 30y$$

Remember the common factor. $= 5y(3y + 2)(y + 3)$

CHECK $5y(3y + 2)(y + 3)$

$\quad = 5y(3y^2 + 9y + 2y + 6)$ FOIL method

$\quad = 5y(3y^2 + 11y + 6)$ Combine like terms.

$\quad = 15y^3 + 55y^2 + 30y$ ✓   Distributive property

(b) $-24a^3 - 42a^2 + 45a$

The common factor could be $3a$ or $-3a$. If we factor out $-3a$, the leading term of the trinomial will be positive, which makes it easier to factor the remaining trinomial.

$$-24a^3 - 42a^2 + 45a$$

It is easier to factor if the first term is $8a^2$ rather than $-8a^2$.

$\quad = -3a(8a^2 + 14a - 15)$ Factor out $-3a$.

$\quad = -3a(4a - 3)(2a + 5)$ Factor the trinomial.

CHECK $-3a(4a - 3)(2a + 5)$

We can multiply $-3a(4a - 3)$ first.

$\quad = (-12a^2 + 9a)(2a + 5)$ Distributive property

$\quad = -24a^3 - 42a^2 + 45a$ ✓   FOIL method   **NOW TRY**

NOW TRY ANSWERS

9. (a) $6z(3z + 4)(2z + 3)$

   (b) $-5x(2x - 3)(x + 6)$

⚠ **CAUTION**   Remember to include the common factor in the final factored form.

5.3 Exercises

FOR EXTRA HELP   ▶ **MyLab Math**

▶ *Video solutions for select problems available in MyLab Math*

Concept Check   *The middle term of each trinomial has been rewritten. Now factor by grouping.*

1. $10t^2 + 9t + 2$

$\quad = 10t^2 + 5t + 4t + 2$

2. $6x^2 + 13x + 6$

$\quad = 6x^2 + 9x + 4x + 6$

3. $15z^2 - 19z + 6$

$\quad = 15z^2 - 10z - 9z + 6$

4. $12p^2 - 17p + 6$

$\quad = 12p^2 - 9p - 8p + 6$

5. $8s^2 + 2st - 3t^2$

$\quad = 8s^2 - 4st + 6st - 3t^2$

6. $3x^2 - xy - 14y^2$

$\quad = 3x^2 - 7xy + 6xy - 14y^2$

Concept Check   *Complete the steps to factor each trinomial by grouping.*

7. $2m^2 + 11m + 12$

   (a) Find two integers whose product is _____ · _____ = _____ and whose sum is _____.

   (b) The required integers are _____ and _____.

   (c) Now write the middle term, $11m$, as _____ + _____.

   (d) Rewrite the given trinomial using four terms as _____.

   (e) Factor the polynomial in part (d) by grouping.

   (f) Check by multiplying.

8. $6y^2 - 19y + 10$

(a) Find two integers whose product is _____ · _____ = _____ and whose sum is _____.

(b) The required integers are _____ and _____.

(c) Now write the middle term, $-19y$, as _____ + _____.

(d) Rewrite the given trinomial using four terms as _____.

(e) Factor the polynomial in part (d) by grouping.

(f) Check by multiplying.

Concept Check *Answer each question.*

9. Which pair of integers would be used to rewrite the middle term of $12y^2 + 5y - 2$ when factoring by grouping?

A. $-8, 3$ **B.** $8, -3$ **C.** $-6, 4$ **D.** $6, -4$

10. Which pair of integers would be used to rewrite the middle term of $20b^2 - 13b + 2$ when factoring by grouping?

A. $10, 3$ **B.** $-10, -3$ **C.** $8, 5$ **D.** $-8, -5$

11. Concept Check On a quiz, a student factored $16x^2 - 24x + 5$ by grouping as follows, but he did not receive credit for his answer.

$$16x^2 - 24x + 5$$

$$= 16x^2 - 4x - 20x + 5$$

$$= 4x(4x - 1) - 5(4x - 1) \qquad \text{His answer}$$

WHAT WENT WRONG? Give the correct factored form.

12. Concept Check A student factored $3k^3 - 12k^2 - 15k$ by first factoring out the common factor $3k$ to obtain $3k(k^2 - 4k - 5)$. Then she wrote the following.

$$k^2 - 4k - 5$$

$$= k^2 - 5k + k - 5$$

$$= k(k - 5) + 1(k - 5)$$

$$= (k - 5)(k + 1) \qquad \text{Her answer}$$

WHAT WENT WRONG? Give the correct factored form.

Concept Check *Which is the correct factored form of the given polynomial?*

13. $2x^2 - x - 1$

 A. $(2x - 1)(x + 1)$

 B. $(2x + 1)(x - 1)$

14. $3a^2 - 5a - 2$

 A. $(3a + 1)(a - 2)$

 B. $(3a - 1)(a + 2)$

15. $4y^2 + 17y - 15$

 A. $(y + 5)(4y - 3)$

 B. $(2y - 5)(2y + 3)$

16. $12c^2 - 7c - 12$

 A. $(6c - 2)(2c + 6)$

 B. $(4c + 3)(3c - 4)$

17. Concept Check A student factoring the trinomial

$$12x^2 + 7x - 12$$

wrote $(4x + 4)$ as one binomial factor. **WHAT WENT WRONG?** Factor correctly.

18. Concept Check A student factoring the trinomial

$$4x^2 + 10x - 6$$

wrote $(4x - 2)(x + 3)$. **WHAT WENT WRONG?** Factor correctly.

Complete each factorization. **See Examples 1–9.**

19. $6a^2 + 7ab - 20b^2$

$= (3a - 4b)(\underline{\hspace{2cm}})$

20. $9m^2 + 6mn - 8n^2$

$= (3m - 2n)(\underline{\hspace{2cm}})$

21. $2x^2 + 6x - 8$

$= 2(\underline{\hspace{3cm}})$

$= 2(\underline{\hspace{1.5cm}})(\underline{\hspace{1.5cm}})$

22. $3x^2 + 9x - 30$

$= 3(\underline{\hspace{3cm}})$

$= 3(\underline{\hspace{1.5cm}})(\underline{\hspace{1.5cm}})$

23. $-4z^3 + 10z^2 + 6z$

$= -2z(\underline{\hspace{3cm}})$

$= -2z(\underline{\hspace{1.5cm}})(\underline{\hspace{1.5cm}})$

24. $-15r^3 + 39r^2 + 18r$

$= -3r(\underline{\hspace{3cm}})$

$= -3r(\underline{\hspace{1.5cm}})(\underline{\hspace{1.5cm}})$

Factor each trinomial completely. **See Examples 1–9.**

25. $3a^2 + 10a + 7$

26. $7r^2 + 8r + 1$

27. $2y^2 + 7y + 6$

28. $5z^2 + 12z + 4$

29. $15m^2 + m - 2$

30. $6x^2 + x - 1$

31. $12s^2 + 11s - 5$

32. $20x^2 + 11x - 3$

33. $10m^2 - 23m + 12$

34. $6x^2 - 17x + 12$

35. $8w^2 - 14w + 3$

36. $9p^2 - 18p + 8$

37. $20y^2 - 39y - 11$

38. $10x^2 - 11x - 6$

39. $3x^2 - 15x + 16$

40. $2t^2 - 14t + 15$

41. $20x^2 + 22x + 6$

42. $36y^2 + 81y + 45$

43. $24x^2 - 42x + 9$

44. $48b^2 - 74b - 10$

45. $4z^2 - 5z + 7$

46. $6x^2 - 8x + 5$

47. $16 + 16x + 3x^2$

48. $18 + 65x + 7x^2$

49. $12x^2 - 7x - 4$

50. $12x^2 - 9x - 10$

51. $24x^2 - 46x + 15$

52. $24x^2 - 94x + 35$

53. $32x^3 + 44x^2 + 12x$

54. $30t^3 + 55t^2 + 25t$

55. $-10x^3 + 5x^2 + 140x$

56. $-18k^3 - 48k^2 + 66k$

57. $15n^4 - 39n^3 + 18n^2$

58. $24a^4 + 10a^3 - 4a^2$

59. $-40m^2q - mq + 6q$

60. $-15a^2b - 22ab - 8b$

61. $15x^2y^2 - 7xy^2 - 4y^2$

62. $14a^2b^3 + 15ab^3 - 9b^3$

63. $5a^2 - 7ab - 6b^2$

64. $6x^2 - 5xy - y^2$

65. $12s^2 + 11st - 5t^2$

66. $25a^2 + 25ab + 6b^2$

67. $12p^2 + 7pq - 12q^2$

68. $6m^2 + 5mn - 6n^2$

69. $24y^2 - 41xy - 14x^2$

70. $24x^2 + 19xy - 5y^2$

71. $48a^2 - 94ab - 4b^2$

72. $48t^2 - 147ts + 9s^2$

73. $24x^2 + 38xy + 15y^2$

74. $24x^2 + 62xy + 33y^2$

75. $10x^4y^5 + 39x^3y^5 - 4x^2y^5$

76. $14x^7y^4 - 31x^6y^4 + 6x^5y^4$

77. $36a^3b^2 - 104a^2b^2 - 12ab^2$

78. $36p^4q + 129p^3q - 60p^2q$

79. $6m^6n + 7m^5n^2 + 2m^4n^3$

80. $12k^3q^4 - 4k^2q^5 - kq^6$

If a trinomial has a negative coefficient for the second-degree term, such as $-2x^2 + 11x - 12$, it is usually easier to factor by first factoring out the common factor -1.

$$-2x^2 + 11x - 12$$

$$= -1(2x^2 - 11x + 12) \quad \text{Factor out } -1.$$

$$= -1(2x - 3)(x - 4) \quad \text{Factor the trinomial.}$$

Use this method to factor each trinomial.

81. $-x^2 - 4x + 21$

82. $-x^2 + x + 72$

83. $-3x^2 - x + 4$

84. $-5x^2 + 2x + 16$

85. $-2a^2 - 5ab - 2b^2$

86. $-3p^2 + 13pq - 4q^2$

Extending Skills *Factor each trinomial.*

87. $24x^4 + 55x^2 - 24$ **88.** $24x^4 + 17x^2 - 20$

89. $36x^4 - 64x^2y + 15y^2$ **90.** $36x^4 + 59x^2y + 24y^2$

Extending Skills *Factor each polynomial. (Hint: As the first step, factor out the greatest common factor.)*

91. $25q^2(m + 1)^3 - 5q(m + 1)^3 - 2(m + 1)^3$

92. $18x^2(y - 3)^2 - 21x(y - 3)^2 - 4(y - 3)^2$

93. $9x^2(r + 3)^3 + 12xy(r + 3)^3 + 4y^2(r + 3)^3$

94. $4t^2(k + 9)^7 + 20ts(k + 9)^7 + 25s^2(k + 9)^7$

Extending Skills *Find all integers k so that the trinomial can be factored using the methods of this section.*

95. $5x^2 + kx - 1$ **96.** $2x^2 + kx - 3$ **97.** $2m^2 + km + 5$ **98.** $3y^2 + ky + 4$

RELATING CONCEPTS For Individual or Group Work (Exercises 99–104)

Often there are several different equivalent forms of an answer that are all correct. Work Exercises 99–104 in order, to see this for factoring problems.

99. Factor the integer 35 as the product of two prime numbers.

100. Factor the integer 35 as the product of the negatives of two prime numbers.

101. Verify that $6x^2 - 11x + 4$ factors as $(3x - 4)(2x - 1)$.

102. Verify that $6x^2 - 11x + 4$ factors as $(4 - 3x)(1 - 2x)$.

103. Compare the two valid factored forms in **Exercises 101 and 102.** How do the factors in each case compare?

104. Suppose we know that the correct factored form of a particular trinomial is $(7t - 3)(2t - 5)$. From **Exercises 99–103,** what is another valid factored form?

5.4 Special Factoring Techniques

OBJECTIVES

1 Factor a difference of squares.

2 Factor a perfect square trinomial.

3 Factor a difference of cubes.

4 Factor a sum of cubes.

OBJECTIVE 1 Factor a difference of squares.

The rule for finding the product of a sum and difference of the same two terms is

$$(x + y)(x - y) = x^2 - y^2.$$

Reversing this rule leads to the following special factoring rule.

Factoring a Difference of Squares

$$x^2 - y^2 = (x + y)(x - y)$$

Example:

$$m^2 - 4$$
$$= m^2 - 2^2$$
$$= (m + 2)(m - 2) \qquad \text{Difference of squares}$$

VOCABULARY
☐ perfect square
☐ perfect square trinomial
☐ perfect cube
☐ difference of cubes
☐ sum of cubes

Two conditions must be true for a binomial to be a difference of squares.

1. Both terms of the binomial must be **perfect squares,** such as

$$x^2, \quad 9y^2 = (3y)^2, \quad m^4 = (m^2)^2, \quad 1 = 1^2, \quad 25 = 5^2, \quad 144 = 12^2.$$

2. The terms of the binomial must have different signs, one positive and one negative.

> **NOTE**   Although some polynomials can be factored using rational or irrational numbers, recall that we are factoring using only integers.

**NOW TRY
EXERCISE 1**

Factor each binomial if possible.

(a) $x^2 - 100$

(b) $x^2 + 100$

(c) $x^2 - 32$

EXAMPLE 1   Factoring Binomials

Factor each binomial if possible.

$$x^2 - y^2 = (x + y)(x - y)$$
$$\downarrow \quad \downarrow \quad \quad \downarrow \quad \downarrow \quad \downarrow \quad \downarrow$$

(a) $p^2 - 16 = p^2 - 4^2 = (p + 4)(p - 4)$

(b) $x^2 - 8$

Because 8 is not a perfect square, this binomial does not satisfy Condition 1 above. It cannot be factored using integers, so it is a prime polynomial.

(c) $p^2 + 16$

The binomial $p^2 + 16$ does not satisfy Condition 2 above. It is a *sum* of squares—it is *not* equal to $(p + 4)(p - 4)$. (See part (a).) We can use the FOIL method and try the following.

$$(p - 4)(p - 4) \qquad\qquad\qquad (p + 4)(p + 4)$$
$$= p^2 - 8p + 16, \quad \text{not} \quad p^2 + 16. \qquad = p^2 + 8p + 16, \quad \text{not} \quad p^2 + 16.$$

Thus, $p^2 + 16$ cannot be factored using integers and is a prime polynomial.

NOW TRY ↺

> ## Sum of Squares
>
> **If x and y have no common factors (except 1),** the following holds true.
>
> **A sum of squares $x^2 + y^2$ cannot be factored using real numbers.**
>
> Thus, $x^2 + y^2$ is prime. **(See Example 1(c).)**

EXAMPLE 2   Factoring Differences of Squares

Factor each binomial.

$$x^2 \quad - y^2 = (x \quad + y)(x \quad - y)$$
$$\downarrow \quad\quad \downarrow \quad\quad \downarrow \quad \downarrow \quad \downarrow \quad\quad \downarrow$$

(a) $25m^2 - 4 = (5m)^2 - 2^2 = (5m + 2)(5m - 2)$

(b) $49z^2 - 64t^2$

$$= (7z)^2 - (8t)^2 \qquad \text{Write each term as a square.}$$

$$= (7z + 8t)(7z - 8t) \qquad \text{Factor the difference of squares.}$$

NOW TRY ANSWERS

1. (a) $(x + 10)(x - 10)$
  (b) prime **(c)** prime

**NOW TRY
EXERCISE 2**

Factor each binomial.

(a) $9t^2 - 100$

(b) $36a^2 - 49b^2$

CHECK $(7z + 8t)(7z - 8t)$

$= 49z^2 - 56zt + 56tz - 64t^2$ FOIL method

$= 49z^2 - 64t^2$ ✓   Commutative property; Combine like terms.

NOW TRY 

NOTE *Always check a factored form by multiplying.*

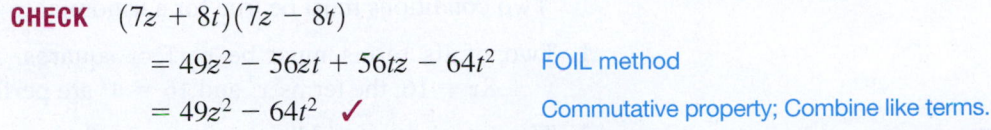

**NOW TRY
EXERCISE 3**

Factor each binomial
completely.

(a) $16k^2 - 64$

(b) $m^4 - 144$

(c) $v^4 - 625$

EXAMPLE 3 Factoring More Complex Differences of Squares

Factor each binomial completely.

(a) $81y^2 - 36$ Always check for a common factor first.

$= 9(9y^2 - 4)$ Factor out the GCF, 9.

$= 9[(3y)^2 - 2^2]$ Write each term as a square.

$= 9(3y + 2)(3y - 2)$ Factor the difference of squares.

(b) $p^4 - 36$

$= (p^2)^2 - 6^2$ Write each term as a square.

Neither binomial can be factored further. $= (p^2 + 6)(p^2 - 6)$ Factor the difference of squares.

(c) $m^4 - 16$

$= (m^2)^2 - 4^2$ Write each term as a square.

$= (m^2 + 4)(m^2 - 4)$ Factor the difference of squares.

Don't stop here! $= (m^2 + 4)(m + 2)(m - 2)$ Factor the difference of squares again.

NOW TRY 

! CAUTION *Factor again when any of the factors is a difference of squares, as in* **Example 3(c).** Check by multiplying.

OBJECTIVE 2 Factor a perfect square trinomial.

Recall the rules for squaring binomials.

$(x + y)^2$ Squared binomial $(x - y)^2$ Squared binomial

$= (x + y)(x + y)$ $= (x - y)(x - y)$

$= x^2 + 2xy + y^2$ Perfect square trinomial $= x^2 - 2xy + y^2$ Perfect square trinomial

A **perfect square trinomial** is a trinomial that is the square of a binomial. For example, $x^2 + 8x + 16$ is a perfect square trinomial because it is the square of the binomial $x + 4$.

$(x + 4)^2$ Squared binomial

$= (x + 4)(x + 4)$

$= x^2 + 8x + 16$ Perfect square trinomial

NOW TRY ANSWERS

2. (a) $(3t + 10)(3t - 10)$

   (b) $(6a + 7b)(6a - 7b)$

3. (a) $16(k + 2)(k - 2)$

   (b) $(m^2 + 12)(m^2 - 12)$

   (c) $(v^2 + 25)(v + 5)(v - 5)$

Two conditions must be true for a trinomial to be a perfect square trinomial.

1. Two of its terms must be perfect squares. In the perfect square trinomial $x^2 + 8x + 16$, the terms x^2 and $16 = 4^2$ are perfect squares.

2. *The remaining (middle) term of a perfect square trinomial is always twice the product of the two terms in the squared binomial.*

$$x^2 + 8x + 16$$
$$= x^2 + 2(x)(4) + 4^2 \qquad 8x = 2(x)(4)$$
$$= (x + 4)^2 \qquad \text{Factor.}$$

The following are *not* perfect square trinomials.

$16x^2 + 4x + 15$ Violates Condition 1 (Only $16x^2 = (4x)^2$ is a perfect square; 15 is not.)

$x^2 + 6x + 36$ Violates Condition 2 (x^2 and $36 = 6^2$ are perfect squares, but $2(x)(6) = 12x$, *not* 6x.)

Reversing the rules for squaring binomials leads to the following special factoring rules.

Factoring Perfect Square Trinomials

$$x^2 + 2xy + y^2 = (x + y)^2$$
$$x^2 - 2xy + y^2 = (x - y)^2$$

**NOW TRY
EXERCISE 4**

Factor $y^2 + 14y + 49$.

EXAMPLE 4 Factoring a Perfect Square Trinomial

Factor $x^2 + 10x + 25$.

The x^2-term is a perfect square, and so is 25, which equals 5^2.

Try to factor $x^2 + 10x + 25$ as the squared binomial $(x + 5)^2$.

To check, take twice the product of the two terms in the squared binomial.

$$2 \cdot x \cdot 5 = 10x \leftarrow \text{Middle term of } x^2 + 10x + 25$$

Twice   First term   Last term
of binomial   of binomial

Because $10x$ is the middle term of the trinomial, the trinomial is a perfect square.

$$x^2 + 10x + 25 \quad \text{factors as} \quad (x + 5)^2. \qquad \text{NOW TRY}$$

EXAMPLE 5 Factoring Perfect Square Trinomials

Factor each trinomial.

(a) $x^2 - 22x + 121$

The first and last terms are perfect squares ($121 = 11^2$ or $(-11)^2$). Check to see whether the middle term of $x^2 - 22x + 121$ is twice the product of the first and last terms of the binomial $x - 11$.

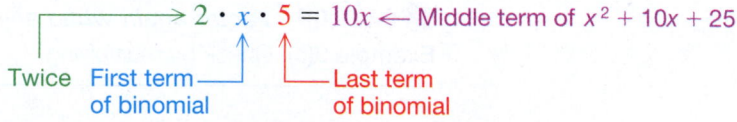

$$2 \cdot x \cdot (-11) = -22x \leftarrow \text{Middle term of } x^2 - 22x + 121$$

Twice   First
term   Last
term

NOW TRY ANSWER
4. $(y + 7)^2$

 **NOW TRY
EXERCISE 5**

Factor each trinomial.

(a) $t^2 - 18t + 81$

(b) $4p^2 - 28p + 49$

(c) $9x^2 + 6x + 4$

(d) $80x^3 + 120x^2 + 45x$

(e) $64x^2 + 80xy + 25y^2$

Thus, $x^2 - 22x + 121$ is a perfect square trinomial.

> Check by squaring the binomial.

$$x^2 - 22x + 121 \quad \text{factors as} \quad (x - 11)^2.$$

Same sign

Notice that the sign of the second term in the squared binomial is the same as the sign of the middle term in the trinomial.

(b) $9m^2 - 24m + 16 = (3m)^2 + 2(3m)(-4) + (-4)^2$

Perfect squares   Twice   First term   Last term

$$= (3m - 4)^2$$

(c) $25y^2 + 20y + 16$

The first and last terms are perfect squares.

$$25y^2 = (5y)^2 \quad \text{and} \quad 16 = 4^2$$

Twice the product of the first and last terms of the binomial $5y + 4$ is

$$2 \cdot 5y \cdot 4 = 40y,$$

which is *not* the middle term of

$$25y^2 + 20y + 16.$$

This trinomial is not a perfect square. In fact, the trinomial cannot be factored even with the methods of the previous sections. It is a prime polynomial.

(d) $12z^3 + 60z^2 + 75z$

$$= 3z(4z^2 + 20z + 25) \qquad \text{Factor out the common factor, 3z.}$$

$$= 3z[(2z)^2 + 2(2z)(5) + 5^2] \qquad 4z^2 + 20z + 25 \text{ is a perfect square trinomial.}$$

$$= 3z(2z + 5)^2 \qquad \text{Factor.}$$

(e) $4x^2 + 36xy + 81y^2 = (2x)^2 + 2(2x)(9y) + (9y)^2$

Perfect squares   Twice   First term   Last term

$$= (2x + 9y)^2$$

NOW TRY

NOTE   Keep the following in mind when factoring perfect square trinomials.

1. The sign of the second term in the squared binomial is always the same as the sign of the middle term in the trinomial.

2. The first and last terms of a perfect square trinomial must be *positive* because they are squares. For example, $x^2 - 2x - 1$ cannot be a perfect square trinomial because the last term is negative.

3. Perfect square trinomials can also be factored by grouping or the FOIL method. Using the method of this section is often easier.

4. It is wise to check factored forms. One way to do this is to write the squared binomial as the product of two terms and multiply using the FOIL method.

$$(x - 11)^2 = (x - 11)(x - 11) = x^2 - 22x + 121 \qquad \text{See Example 5(a).}$$

NOW TRY ANSWERS

5. (a) $(t - 9)^2$

   (b) $(2p - 7)^2$

   (c) prime

   (d) $5x(4x + 3)^2$

   (e) $(8x + 5y)^2$

OBJECTIVE 3 Factor a difference of cubes.

In a difference of cubes $x^3 - y^3$, both terms of the binomial must be **perfect cubes,** such as

$$x^3, \quad 8p^3 = (2p)^3, \quad s^6 = (s^2)^3, \quad 1 = 1^3, \quad 27 = 3^3, \quad 216 = 6^3.$$

We can factor a **difference of cubes** using the following rule.

> **Factoring a Difference of Cubes**
>
> $$x^3 - y^3 = (x - y)(x^2 + xy + y^2)$$

This rule for factoring a difference of cubes should be memorized. To see that the rule is correct, multiply $(x - y)(x^2 + xy + y^2)$.

$$
\begin{array}{rl}
x^2 + xy \ + y^2 & \text{Multiply vertically.} \\
\underline{x \ - y} & \\
-x^2y - xy^2 - y^3 & -y(x^2 + xy + y^2) \\
\underline{x^3 + x^2y + xy^2 } & x(x^2 + xy + y^2) \\
x^3 - y^3 & \text{Add.}
\end{array}
$$

Notice the pattern of the terms in the factored form of $x^3 - y^3$.

- $x^3 - y^3$ factors as (a binomial factor) · (a trinomial factor).

- The binomial factor has the difference of the cube roots of the given terms.
  (*Note:* A cube root of 1 is 1 because $1^3 = 1$, a cube root of 8 is 2 because $2^3 = 8$, and so on.)

- The terms in the trinomial factor are all positive.

- The terms in the binomial factor help to determine the trinomial factor.

$$
x^3 - y^3 = (x - y)(\underset{\substack{\text{First term} \\ \text{squared}}}{x^2} + \underset{\substack{\text{positive} \\ \text{product of} \\ \text{the terms}}}{xy} + \underset{\substack{\text{second term} \\ \text{squared}}}{y^2})
$$

⚠ **CAUTION**   The polynomial $x^3 - y^3$ is **not** equivalent to $(x - y)^3$.

$$
\begin{array}{l|l}
x^3 - y^3 & (x - y)^3 \\
= (x - y)(x^2 + xy + y^2) & = (x - y)(x - y)(x - y) \\
& = (x - y)(x^2 - 2xy + y^2)
\end{array}
$$

EXAMPLE 6 Factoring Differences of Cubes

Factor each binomial.

$$x^3 \ - \ y^3 \ = \ (x \ - \ y) \ (x^2 \ + \ xy \ + \ y^2)$$

(a) $m^3 - 125 = m^3 - 5^3 = (m - 5)(m^2 + 5m + 5^2)$ Let $x = m$ and $y = 5$.

$$= (m - 5)(m^2 + 5m + 25)$$ $5^2 = 25$

**NOW TRY
EXERCISE 6**

Factor each binomial.

(a) $a^3 - 27$

(b) $8t^3 - 125$

(c) $3k^3 - 192$

(d) $125x^3 - 343y^6$

(b) $8p^3 - 27$

$= (2p)^3 - 3^3$ $8p^3 = (2p)^3$ and $27 = 3^3$.

$= (2p - 3)[(2p)^2 + (2p)3 + 3^2]$ Let $x = 2p$ and $y = 3$.

$= (2p - 3)(4p^2 + 6p + 9)$ Apply the exponents. Multiply.

> $(2p)^2 = 2^2p^2 = 4p^2$,
> *not* $2p^2$.

(c) $4m^3 - 32$

$= 4(m^3 - 8)$ Factor out the common factor, 4.

$= 4(m^3 - 2^3)$ $8 = 2^3$

$= 4(m - 2)(m^2 + 2m + 4)$ Factor the difference of cubes.

(d) $125t^3 - 216s^6$

$= (5t)^3 - (6s^2)^3$ Write each term as a cube.

$= (5t - 6s^2)[(5t)^2 + 5t(6s^2) + (6s^2)^2]$ Factor the difference of cubes.

$= (5t - 6s^2)(25t^2 + 30ts^2 + 36s^4)$
> Square carefully.
> $(6s^2)^2 = 6^2(s^2)^2 = 36s^4$

NOW TRY

⚠ **CAUTION** A common error when factoring $x^3 - y^3 = (x - y)(x^2 + xy + y^2)$ is to try to factor $x^2 + xy + y^2$. This is usually not possible.

OBJECTIVE 4 Factor a sum of cubes.

A *sum of squares,* such as $m^2 + 25$, *cannot* be factored using real numbers, but a **sum of cubes** can.

> **Factoring a Sum of Cubes**
>
> $$x^3 + y^3 = (x + y)(x^2 - xy + y^2)$$

Compare the rule for the *sum* of cubes with that for the *difference* of cubes.

$x^3 - y^3 = (x - y)(x^2 + xy + y^2)$ Difference of cubes

 └─Same─┘   Opposite   Always
  sign   sign   positive

       The only difference between the rules is the positive and negative signs.

$x^3 + y^3 = (x + y)(x^2 - xy + y^2)$ Sum of cubes

 └─Same─┘   Opposite   Always
  sign   sign   positive

EXAMPLE 7 Factoring Sums of Cubes

Factor each binomial.

NOW TRY ANSWERS

6. (a) $(a - 3)(a^2 + 3a + 9)$
   (b) $(2t - 5)(4t^2 + 10t + 25)$
   (c) $3(k - 4)(k^2 + 4k + 16)$
   (d) $(5x - 7y^2) \cdot$
      $(25x^2 + 35xy^2 + 49y^4)$

(a) $k^3 + 27$

$= k^3 + 3^3$ $27 = 3^3$

$= (k + 3)(k^2 - 3k + 3^2)$ Factor the sum of cubes.

$= (k + 3)(k^2 - 3k + 9)$ Apply the exponent.

NOW TRY EXERCISE 7

Factor each binomial.

(a) $x^3 + 125$

(b) $27a^3 + 8b^3$

(b) $8m^3 + 125n^3$

$= (2m)^3 + (5n)^3$ $8m^3 = (2m)^3$ and $125n^3 = (5n)^3$.

$= (2m + 5n)[(2m)^2 - 2m(5n) + (5n)^2]$ Factor the sum of cubes.

$= (2m + 5n)(4m^2 - 10mn + 25n^2)$ Be careful: $(2m)^2 = 2^2 m^2$ and $(5n)^2 = 5^2 n^2$.

(c) $1000a^6 + 27b^3$

$= (10a^2)^3 + (3b)^3$

$= (10a^2 + 3b)[(10a^2)^2 - (10a^2)(3b) + (3b)^2]$ Factor the sum of cubes.

$= (10a^2 + 3b)(100a^4 - 30a^2b + 9b^2)$ $(10a^2)^2 = 10^2(a^2)^2 = 100a^4$

NOW TRY

The methods of factoring discussed in this section are summarized here.

NOW TRY ANSWERS

7. (a) $(x + 5)(x^2 - 5x + 25)$

   (b) $(3a + 2b)(9a^2 - 6ab + 4b^2)$

Special Factoring Rules

Difference of squares	$x^2 - y^2 = (x + y)(x - y)$
Perfect square trinomials	$x^2 + 2xy + y^2 = (x + y)^2$
	$x^2 - 2xy + y^2 = (x - y)^2$
Difference of cubes	$x^3 - y^3 = (x - y)(x^2 + xy + y^2)$
Sum of cubes	$x^3 + y^3 = (x + y)(x^2 - xy + y^2)$

A sum of squares can be factored only if the terms have a common factor.

5.4 Exercises

FOR EXTRA HELP

▶ **MyLab Math**

▶ *Video solutions for select problems available in MyLab Math*

Concept Check   *Work each problem.*

1. To help factor differences of squares, complete the following list of perfect squares.

$1^2 = $ _____ $2^2 = $ _____ $3^2 = $ _____ $4^2 = $ _____ $5^2 = $ _____

$6^2 = $ _____ $7^2 = $ _____ $8^2 = $ _____ $9^2 = $ _____ $10^2 = $ _____

$11^2 = $ _____ $12^2 = $ _____ $13^2 = $ _____ $14^2 = $ _____ $15^2 = $ _____

$16^2 = $ _____ $17^2 = $ _____ $18^2 = $ _____ $19^2 = $ _____ $20^2 = $ _____

2. The following powers of x are all perfect squares:

$$x^2, \quad x^4, \quad x^6, \quad x^8, \quad x^{10}.$$

On the basis of this observation, we may make a conjecture (an educated guess) that if the power of a variable is divisible by ____ (with 0 remainder), then we have a perfect square.

3. Which of the following are differences of squares?

   A. $x^2 - 4$ **B.** $y^2 + 9$ **C.** $2a^2 - 25$ **D.** $9m^2 - 1$

4. Which of the following binomial sums can be factored?

   A. $x^2 + 36$ **B.** $x^3 + x$ **C.** $3x^2 + 12$ **D.** $25x^2 + 49$

5. Concept Check   On a quiz, a student indicated *prime* when asked to factor $4x^2 + 16$ because she said that a sum of a squares cannot be factored.   **WHAT WENT WRONG?**

6. **Concept Check**   When directed to factor $k^4 - 81$ completely, a student did not earn full credit.

$$(k^2 + 9)(k^2 - 9) \qquad \text{His answer}$$

The student argued that because his answer does indeed give $k^4 - 81$ when multiplied out, he should be given full credit. **WHAT WENT WRONG?** Give the correct factored form.

Factor each binomial completely. If the binomial is prime, say so. See Examples 1–3.

7. $y^2 - 25$

8. $t^2 - 36$

9. $x^2 - 144$

10. $x^2 - 400$

11. $m^2 - 12$

12. $k^2 - 18$

13. $m^2 + 64$

14. $k^2 + 49$

15. $4m^2 + 16$

16. $9x^2 + 81$

17. $9r^2 - 4$

18. $4x^2 - 9$

19. $36x^2 - 16$

20. $32a^2 - 8$

21. $196p^2 - 225$

22. $361q^2 - 400$

23. $16r^2 - 25a^2$

24. $49m^2 - 100p^2$

25. $81x^2 - 49y^2$

26. $36y^2 - 121z^2$

27. $54x^2 - 6y^2$

28. $48m^2 - 75n^2$

29. $100x^2 + 49$

30. $81w^2 + 16$

31. $4 - x^2$

32. $25 - x^2$

33. $36 - 25t^2$

34. $16 - 49p^2$

35. $x^3 + 4x$

36. $z^3 + 25z$

37. $x^4 - x^2$

38. $y^4 - 9y^2$

39. $p^4 - 49$

40. $r^4 - 25$

41. $x^4 - 1$

42. $y^4 - 10,000$

43. $p^4 - 256$

44. $k^4 - 81$

Concept Check   *Work each problem.*

45. Which of the following are perfect square trinomials?

   A. $y^2 - 13y + 36$ **B.** $x^2 + 6x + 9$ **C.** $4z^2 - 4z + 1$ **D.** $16m^2 + 10m + 1$

46. In the polynomial $9y^2 + 14y + 25$, the first and last terms are perfect squares. Can the polynomial be factored? If it can, factor it. If it cannot, explain why it is not a perfect square trinomial.

Concept Check   *Find the value of the indicated variable.*

47. Find b so that $x^2 + bx + 25$ factors as $(x + 5)^2$.

48. Find c so that $4m^2 - 12m + c$ factors as $(2m - 3)^2$.

49. Find a so that $ay^2 - 12y + 4$ factors as $(3y - 2)^2$.

50. Find b so that $100a^2 + ba + 9$ factors as $(10a + 3)^2$.

Factor each trinomial completely. See Examples 4 and 5.

51. $w^2 + 2w + 1$

52. $p^2 + 4p + 4$

53. $x^2 - 8x + 16$

54. $x^2 - 10x + 25$

55. $x^2 - 10x + 100$

56. $x^2 - 18x + 36$

57. $2x^2 + 24x + 72$

58. $3y^2 + 48y + 192$

59. $4x^2 + 12x + 9$

60. $25x^2 + 10x + 1$

61. $16x^2 - 40x + 25$

62. $36y^2 - 60y + 25$

63. $36r^3 + 96r^2 + 64r$

64. $12k^3 + 12k^2 + 3k$

65. $49x^2 - 28xy + 4y^2$

66. $4z^2 - 12zw + 9w^2$

67. $64x^2 + 48xy + 9y^2$

68. $9t^2 + 24tr + 16r^2$

69. $50h^2 - 40hy + 8y^2$

70. $18x^2 - 48xy + 32y^2$

71. $25z^4 + 5z^3 + z^2$

72. $4x^4 + 2x^3 + x^2$

Concept Check   *Work each problem.*

73. To help factor sums or differences of cubes, complete the following list of perfect cubes.

$1^3 = $ _____ $2^3 = $ _____ $3^3 = $ _____ $4^3 = $ _____ $5^3 = $ _____

$6^3 = $ _____ $7^3 = $ _____ $8^3 = $ _____ $9^3 = $ _____ $10^3 = $ _____

74. The following powers of x are all perfect cubes:

$$x^3, \quad x^6, \quad x^9, \quad x^{12}, \quad x^{15}.$$

On the basis of this observation, we may make a conjecture that if the power of a variable is divisible by _____ (with 0 remainder), then we have a perfect cube.

75. Which of the following are differences of cubes?

   A. $9x^3 - 125$ **B.** $x^3 - 16$ **C.** $x^3 - 1$ **D.** $8x^3 - 27y^3$

76. Which of the following are sums of cubes?

   A. $x^3 + 1$ **B.** $x^3 + 36$ **C.** $12x^3 + 27$ **D.** $64x^3 + 216y^3$

77. Identify each monomial as a *perfect square*, a *perfect cube*, *both of these*, or *neither of these*.

   (a) $4x^3$ **(b)** $8y^6$ **(c)** $49x^{12}$ **(d)** $81r^{10}$ **(e)** $64x^6y^{12}$ **(f)** $125t^6$

78. What must be true for x^n to be both a perfect square and a perfect cube?

Factor each binomial completely. ***See Examples 6 and 7.***

79. $a^3 - 1$

80. $m^3 - 8$

81. $m^3 + 8$

82. $b^3 + 1$

83. $y^3 - 216$

84. $x^3 - 343$

85. $k^3 + 1000$

86. $p^3 + 512$

87. $27x^3 - 64$

88. $64y^3 - 27$

89. $125x^3 + 8$

90. $216x^3 + 125$

91. $6p^3 + 6$

92. $81x^3 + 3$

93. $5x^3 + 40$

94. $128y^3 + 54$

95. $y^3 - 8x^3$

96. $w^3 - 216z^3$

97. $2x^3 - 16y^3$

98. $27w^3 - 216z^3$

99. $8p^3 + 729q^3$

100. $64x^3 + 125y^3$

101. $27a^3 + 64b^3$

102. $125m^3 + 8p^3$

103. $125t^3 + 8s^3$

104. $27r^3 + 1000s^3$

105. $8x^3 - 125y^6$

106. $27t^3 - 64s^6$

107. $27m^6 + 8n^3$

108. $1000r^6 + 27s^3$

109. $125k^3 - 8m^9$

110. $125c^6 - 216d^3$

111. $x^9 + y^9$

112. $x^9 - y^9$

Extending Skills *Although we usually factor polynomials using integers, we can apply the same concepts to factoring using fractions and decimals.*

$$z^2 - \frac{9}{16}$$

$$= z^2 - \left(\frac{3}{4}\right)^2 \qquad \frac{9}{16} = \left(\frac{3}{4}\right)^2$$

$$= \left(z + \frac{3}{4}\right)\left(z - \frac{3}{4}\right) \qquad \text{Factor the difference of squares.}$$

$$x^2 + \frac{2}{5}x + \frac{1}{25}$$

$$= x^2 + 2(x)\left(\frac{1}{5}\right) + \left(\frac{1}{5}\right)^2 \qquad \frac{1}{25} = \left(\frac{1}{5}\right)^2$$

$$= \left(x + \frac{1}{5}\right)^2 \qquad \text{Factor the perfect square trinomial.}$$

Apply the special factoring rules of this section to factor each polynomial.

113. $p^2 - \dfrac{1}{9}$

114. $q^2 - \dfrac{1}{4}$

115. $36m^2 - \dfrac{16}{25}$

116. $100b^2 - \dfrac{4}{49}$

117. $x^2 - 0.64$

118. $y^2 - 0.36$

119. $t^2 + t + \dfrac{1}{4}$

120. $m^2 + \dfrac{2}{3}m + \dfrac{1}{9}$

121. $x^2 - 1.8x + 0.81$

122. $y^2 - 1.4y + 0.49$

123. $x^3 + \dfrac{1}{8}$

124. $x^3 + \dfrac{1}{64}$

Extending Skills *Factor each polynomial completely.*

125. $(m + n)^2 - (m - n)^2$

126. $(a - b)^3 - (a + b)^3$

127. $m^2 - p^2 + 2m + 2p$

128. $3r - 3k + 3r^2 - 3k^2$

129. $4x^2 + 4x + 1 - y^2$

130. $16y^2 + 24y + 9 - z^2$

131. $9a^2 - 48ab + 64b^2 - 4c^2$

132. $25x^2 - 10xy + y^2 - 36z^2$

SUMMARY EXERCISES Recognizing and Applying Factoring Strategies

When factoring a polynomial, ask these questions to determine a suitable factoring technique.

Factoring a Polynomial

Question 1 **Is there a common factor other than 1?** If so, factor it out.

Question 2 **How many terms are in the polynomial?**

Two terms: Is it a difference of squares or a sum or difference of cubes? If so, factor as in **Section 5.4.**

Three terms: Is it a perfect square trinomial? In this case, factor as in **Section 5.4.**

If the trinomial is not a perfect square trinomial, what is the coefficient of the second-degree term?

- If it is 1, use the factoring method of **Section 5.2.**
- If it is not 1, use the general factoring methods of **Section 5.3.**

Four terms: Try to factor by grouping, as in **Section 5.1.**

Question 3 **Can any factors be factored further?** If so, factor them.

⊕ **CAUTION**   It is wise to check the answer to a factoring problem.

1. Check that the product of all the factors does indeed yield the original polynomial.

2. Check that the original polynomial has been factored **completely.**

↻ **NOW TRY
EXERCISE 1**
Factor completely.

$$100a^2 - 49$$

EXAMPLE 1 Applying Factoring Strategies

Factor $144x^2 - 169$ completely.

Question 1   **Is there a common factor other than 1?**

There is no common factor other than 1. Proceed to Question 2.

Question 2   **How many terms are there in the polynomial?**

The polynomial $144x^2 - 169$ has two terms. It is a difference of squares because $144x^2 = (12x)^2$ and $169 = 13^2$.

$$144x^2 - 169$$

$$= (12x)^2 - 13^2 \qquad \text{Write each term as a square.}$$

$$= (12x + 13)(12x - 13) \qquad \text{Factor the difference of squares.}$$

Question 3   **Can any factors be factored further?**

No. The original polynomial has been factored completely.

NOW TRY ↻

EXAMPLE 2 Applying Factoring Strategies

Factor $12x^2 + 26xy + 12y^2$ completely.

↻ **NOW TRY
EXERCISE 2**
Factor completely.

$$24m^2 - 42my + 9y^2$$

Question 1   **Is there a common factor other than 1?**

Yes, 2 is a common factor, so factor it out.

$$12x^2 + 26xy + 12y^2$$

$$= 2(6x^2 + 13xy + 6y^2)$$

Question 2   **How many terms are in the polynomial?**

The polynomial $6x^2 + 13xy + 6y^2$ has three terms. It is not a perfect square trinomial. To factor by grouping, we find two integers with a product of $6 \cdot 6$, or 36, and a sum of 13. These integers are 4 and 9.

$$12x^2 + 26xy + 12y^2$$

$$= 2(6x^2 + 13xy + 6y^2) \qquad \text{Factor out the GCF, 2.}$$

$$= 2(6x^2 + 4xy + 9xy + 6y^2) \qquad 4 \cdot 9 = 36;\ 4 + 9 = 13$$

$$= 2[(6x^2 + 4xy) + (9xy + 6y^2)] \qquad \text{Group the terms.}$$

$$= 2[2x(3x + 2y) + 3y(3x + 2y)] \qquad \text{Factor each group.}$$

$$= 2(3x + 2y)(2x + 3y) \qquad \text{Factor out the common factor, } 3x + 2y.$$

We could also have factored the trinomial $6x^2 + 13xy + 6y^2$ by trial and error, using the FOIL method in reverse.

↻ **NOW TRY ANSWERS**
1. $(10a + 7)(10a - 7)$
2. $3(4m - y)(2m - 3y)$

Question 3   **Can any factors be factored further?**

No. The original polynomial has been factored completely.

NOW TRY ↻

Match each polynomial in Column I with the best choice for factoring it in Column II. The choices in Column II may be used once, more than once, or not at all.

I

1. $12x^2 + 20x + 8$

2. $x^2 - 17x + 72$

3. $16m^2n + 24mn - 40mn^2$

4. $64a^2 - 121b^2$

5. $36p^2 - 60pq + 25q^2$

6. $z^2 - 4z + 6$

7. $8r^3 - 125$

8. $x^6 + 4x^4 - 3x^2 - 12$

9. $4w^2 + 49$

10. $z^2 - 24z + 144$

II

A. Factor out the GCF. No further factoring is possible.

B. Factor a difference of squares.

C. Factor a difference of cubes.

D. Factor a sum of cubes.

E. Factor a perfect square trinomial.

F. Factor by grouping.

G. Factor out the GCF. Then factor a trinomial by grouping or trial and error.

H. Factor into two binomials by finding two integers whose product is the constant in the trinomial and whose sum is the coefficient of the middle term.

I. The polynomial is prime.

Factor each polynomial completely.

11. $a^2 - 4a - 12$

12. $a^2 + 17a + 72$

13. $6y^2 - 6y - 12$

14. $7y^6 + 14y^5 - 168y^4$

15. $6a + 12b + 18c$

16. $m^2 - 3mn - 4n^2$

17. $p^2 - 17p + 66$

18. $z^2 - 6z + 7z - 42$

19. $10z^2 - 7z - 6$

20. $2m^2 - 10m - 48$

21. $17x^3y^2 + 51xy$

22. $15y + 5$

23. $8a^5 - 8a^4 - 48a^3$

24. $8k^2 - 10k - 3$

25. $z^2 - 3za - 10a^2$

26. $50z^2 - 100$

27. $x^2 - 4x - 5x + 20$

28. $x^2 + 2x + 16$

29. $6n^2 - 19n + 10$

30. $9y^2 + 12y - 5$

31. $16x + 20$

32. $m^2 + 2m - 15$

33. $6y^2 - 5y - 4$

34. $m^2 - 81$

35. $6z^2 + 31z + 5$

36. $12x^2 + 47x - 4$

37. $4k^2 - 12k + 9$

38. $8p^2 + 23p - 3$

39. $54m^2 - 24z^2$

40. $8m^2 - 2m - 3$

41. $3k^2 + 4k - 4$

42. $8a^3 - 27$

43. $14k^3 + 7k^2 - 70k$

44. $5 + r - 5s - rs$

45. $y^4 - 16$

46. $9z^2 + 64$

47. $8m - 16m^2$

48. $k^2 - 16$

49. $z^3 - 8$

50. $y^2 - y - 56$

51. $k^2 + 9$

52. $27p^{10} - 45p^9 - 252p^8$

53. $32m^9 + 16m^5 + 24m^3$

54. $8m^3 + 125$

55. $16r^2 + 24rm + 9m^2$

56. $z^2 - 12z + 36$

57. $15h^2 + 11hg - 14g^2$

58. $5z^3 - 45z^2 + 70z$

59. $k^2 - 11k + 30$

60. $64p^2 - 100m^2$

61. $3k^3 - 12k^2 - 15k$

62. $y^2 - 4yk - 12k^2$

63. $1000p^3 + 27$

64. $64r^3 - 343$

65. $6 + 3m + 2p + mp$

66. $2m^2 + 7mn - 15n^2$

67. $16z^2 - 8z + 1$

68. $a^4 - 625$

69. $108m^2 - 36m + 3$

70. $100a^2 - 81y^2$

71. $x^2 - xy + y^2$

72. $4y^2 - 25$

73. $32z^3 + 56z^2 - 16z$

74. $10m^2 + 25m - 60$

75. $64m^2 - 80mn + 25n^2$

76. $4 - 2q - 6p + 3pq$

77. $6a^2 + 10a - 4$

78. $36y^6 - 42y^5 - 120y^4$

79. $36x^2 + 32x + 9$

80. $16k^2 - 48k + 36$

81. $20 + 5m + 12n + 3mn$

82. $72y^3z^2 + 12y^2 - 24y^4z^2$

83. $8k^2 - 2kh - 3h^2$

84. $2a^2 - 7a - 30$

85. $2x^3 + 128$

86. $45a^3b^5 - 60a^4b^2 + 75a^6b^4$

87. $10y^2 - 7yz - 6z^2$

88. $m^2 - 4m + 4$

89. $8a^2 + 23ab - 3b^2$

90. $125m^4 - 400m^3n + 195m^2n^2$

5.5 Solving Quadratic Equations Using the Zero-Factor Property

OBJECTIVES

1 Solve quadratic equations using the zero-factor property.

2 Solve other equations using the zero-factor property.

Galileo Galilei developed theories to explain physical phenomena. According to legend, Galileo dropped objects of different weights from the Leaning Tower of Pisa to disprove the belief that heavier objects fall faster than lighter objects. He developed a formula that describes the motion of freely falling objects,

$$d = 16t^2,$$

where d is the distance in feet that an object falls (disregarding air resistance) in t seconds, regardless of weight. The equation $d = 16t^2$ is a *quadratic equation*.

Quadratic Equation

A **quadratic equation** (in x here) is an equation that can be written in the form

$$ax^2 + bx + c = 0,$$

where a, b, and c are real numbers and $a \neq 0$. The given form is called **standard form**.

Examples: $x^2 + 5x + 6 = 0$, $2x^2 - 5x = 3$, $x^2 = 4$ Quadratic equations

---

A quadratic equation has a second-degree term and no terms of greater degree. Such an equation is called a **second-degree equation.** Of the above examples, only $x^2 + 5x + 6 = 0$ is in standard form.

We have factored many quadratic *expressions* of the form

$$ax^2 + bx + c.$$

Galileo Galilei (1564–1642)

In this section, we use factored quadratic expressions to solve quadratic *equations*.

VOCABULARY
☐ quadratic equation
☐ second-degree equation
☐ double solution

OBJECTIVE 1 Solve quadratic equations using the zero-factor property.

We use the following property to solve some quadratic equations.

> ### Zero-Factor Property
> **If a and b are real numbers and if $ab = 0$, then $a = 0$ or $b = 0$.**
>
> That is, if the product of two numbers is 0, then at least one of the numbers must be 0. One number *must* be 0, but both *may* be 0.

NOW TRY EXERCISE 1

Solve each equation.
(a) $(x - 4)(3x + 1) = 0$
(b) $y(4y - 5) = 0$

EXAMPLE 1 Using the Zero-Factor Property

Solve each equation.

(a) $(x + 3)(2x - 1) = 0$

The product $(x + 3)(2x - 1)$ is equal to 0. By the zero-factor property, the product of these two factors will equal 0 only if at least one of the factors equals 0. Therefore, either $x + 3 = 0$ or $2x - 1 = 0$.

$$x + 3 = 0 \quad \text{or} \quad 2x - 1 = 0 \qquad \text{Zero-factor property}$$
$$x = -3 \qquad\qquad 2x = 1 \qquad \text{Solve each equation.}$$
$$x = \frac{1}{2} \qquad \text{Divide each side by 2.}$$

Check these values by substituting -3 for x in the original equation. ***Then start over*** and substitute $\frac{1}{2}$ for x.

CHECK

$$(x + 3)(2x - 1) = 0$$
$$(-3 + 3)[2(-3) - 1] \stackrel{?}{=} 0 \qquad \text{Let } x = -3.$$
$$0(-7) \stackrel{?}{=} 0$$
$$0 = 0 \checkmark \text{ True}$$

$$(x + 3)(2x - 1) = 0$$
$$\left(\frac{1}{2} + 3\right)\left(2 \cdot \frac{1}{2} - 1\right) \stackrel{?}{=} 0 \qquad \text{Let } x = \frac{1}{2}.$$
$$\frac{7}{2}(0) \stackrel{?}{=} 0$$
$$0 = 0 \checkmark \text{ True}$$

Because true statements result, the solution set is $\left\{-3, \frac{1}{2}\right\}$. ◁ *Include* both *solutions in the solution set.*

(b)
$$y(3y - 4) = 0$$
$$y = 0 \quad \text{or} \quad 3y - 4 = 0 \qquad \text{Zero-factor property}$$

Don't forget that 0 can be a solution.

$$3y = 4 \qquad \text{Add 4.}$$
$$y = \frac{4}{3} \qquad \text{Divide by 3.}$$

CHECK

$$y(3y - 4) = 0$$
$$0(3 \cdot 0 - 4) \stackrel{?}{=} 0 \qquad \text{Let } y = 0.$$
$$0(-4) \stackrel{?}{=} 0$$
$$0 = 0 \checkmark \text{ True}$$

$$y(3y - 4) = 0$$
$$\frac{4}{3}\left(3 \cdot \frac{4}{3} - 4\right) \stackrel{?}{=} 0 \qquad \text{Let } y = \frac{4}{3}.$$
$$\frac{4}{3}(0) \stackrel{?}{=} 0$$
$$0 = 0 \checkmark \text{ True}$$

NOW TRY ANSWERS
1. (a) $\left\{-\frac{1}{3}, 4\right\}$ **(b)** $\left\{0, \frac{5}{4}\right\}$

True statements result. The solution set is $\left\{0, \frac{4}{3}\right\}$.

NOW TRY

NOTE The word *or* as used in the zero-factor property means *"one or the other or both"* factors may be 0.

NOW TRY EXERCISE 2

Solve $t^2 = -3t + 18$.

EXAMPLE 2 Solving Quadratic Equations

Solve each equation.

(a) $x^2 - 5x = -6$

First, write the equation in standard form $ax^2 + bx + c = 0$.

> *Don't factor x out at this step.*

$$x^2 - 5x = -6$$

$$x^2 - 5x + 6 = 0 \qquad \text{Add 6 to each side.}$$

Now factor $x^2 - 5x + 6$. Find two numbers whose product is 6 and whose sum is -5. These two numbers are -2 and -3, so we factor as follows.

$$(x - 2)(x - 3) = 0 \qquad \text{Factor.}$$

$$x - 2 = 0 \quad \text{or} \quad x - 3 = 0 \qquad \text{Zero-factor property}$$

$$x = 2 \quad \text{or} \qquad x = 3 \qquad \text{Solve each equation.}$$

CHECK

$$x^2 - 5x = -6 \qquad\qquad\qquad x^2 - 5x = -6$$

$$2^2 - 5(2) \overset{?}{=} -6 \quad \text{Let } x = 2. \qquad 3^2 - 5(3) \overset{?}{=} -6 \quad \text{Let } x = 3.$$

$$4 - 10 \overset{?}{=} -6 \qquad\qquad\qquad 9 - 15 \overset{?}{=} -6$$

$$-6 = -6 \ \checkmark \ \text{True} \qquad\qquad -6 = -6 \ \checkmark \ \text{True}$$

Both values check, so the solution set is $\{2, 3\}$.

(b)

$$y^2 = y + 20 \qquad \text{Write this equation in standard form.}$$

$$\text{Standard form} \rightarrow \quad y^2 - y - 20 = 0 \qquad \text{Subtract } y \text{ and 20.}$$

$$(y - 5)(y + 4) = 0 \qquad \text{Factor.}$$

$$y - 5 = 0 \quad \text{or} \quad y + 4 = 0 \qquad \text{Zero-factor property}$$

$$y = 5 \quad \text{or} \qquad y = -4 \qquad \text{Solve each equation.}$$

Check each value to verify that the solution set is $\{-4, 5\}$.

NOW TRY

Solving a Quadratic Equation Using the Zero-Factor Property

Step 1 **Write the equation in standard form**—that is, with all terms on one side of the equality symbol in descending powers of the variable and 0 on the other side.

Step 2 **Factor** completely.

Step 3 **Apply the zero-factor property.** Set each factor with a variable equal to 0.

Step 4 **Solve** the resulting equations.

Step 5 **Check** each value in the original equation. Write the solution set.

NOW TRY ANSWER
2. $\{-6, 3\}$

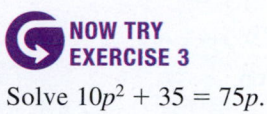

**NOW TRY
EXERCISE 3**

Solve $10p^2 + 35 = 75p$.

EXAMPLE 3 Solving a Quadratic Equation (Common Factor)

Solve $4x^2 + 40 = 26x$.

$$4x^2 + 40 = 26x$$

> Write this equation in the form $ax^2 + bx + c = 0$.

Step 1

$$4x^2 - 26x + 40 = 0$$ Standard form

> This 2 is *not* a solution of the equation.

$$2(2x^2 - 13x + 20) = 0$$ Factor out 2.

$$2x^2 - 13x + 20 = 0$$ Divide each side by 2.

Step 2 $$(2x - 5)(x - 4) = 0$$ Factor.

Step 3 $$2x - 5 = 0 \quad \text{or} \quad x - 4 = 0$$ Zero-factor property

Step 4 $$2x = 5 \qquad\qquad x = 4$$ Solve each equation.

$$x = \frac{5}{2}$$

Step 5   *Check* each value to verify that the solution set is $\left\{\frac{5}{2}, 4\right\}$.   **NOW TRY**

⚠ **CAUTION** A common error is to include the common factor 2 as a solution in **Example 3.** *Only factors containing variables lead to solutions,* such as the factor y in the equation $y(3y - 4) = 0$ in **Example 1(b).**

EXAMPLE 4 Solving Quadratic Equations

Solve each equation.

(a) $$16m^2 - 25 = 0$$

> This equation is in standard form $ax^2 + bx + c = 0$. There is no first-degree term because $b = 0$.

$$(4m + 5)(4m - 5) = 0$$ Factor the difference of squares.

$$4m + 5 = 0 \quad \text{or} \quad 4m - 5 = 0$$ Zero-factor property

$$4m = -5 \quad \text{or} \quad 4m = 5$$ Solve each equation.

$$m = -\frac{5}{4} \quad \text{or} \quad m = \frac{5}{4}$$

Check $-\frac{5}{4}$ and $\frac{5}{4}$ in the original equation. The solution set is $\left\{-\frac{5}{4}, \frac{5}{4}\right\}$.

(b) $$y^2 = 2y$$

> This equation is in the form $ax^2 + bx + c = 0$. Here, $c = 0$.

$$y^2 - 2y = 0$$

$$y(y - 2) = 0$$ Factor.

> Don't forget to set the variable factor y equal to 0.

$$y = 0 \quad \text{or} \quad y - 2 = 0$$ Zero-factor property

$$y = 2$$ Solve.

A check confirms that the solution set is $\{0, 2\}$.

(c) $$k(2k + 1) = 3$$

> To be in standard form, 0 must be on the right side.

$$2k^2 + k = 3$$ Distributive property

Standard form $\rightarrow$ $$2k^2 + k - 3 = 0$$ Subtract 3.

$$(2k + 3)(k - 1) = 0$$ Factor.

NOW TRY ANSWER

3. $\left\{\frac{1}{2}, 7\right\}$

**NOW TRY
EXERCISE 4**

Solve each equation.

(a) $9x^2 - 64 = 0$

(b) $m^2 = 5m$

(c) $p(6p - 1) = 2$

$$2k + 3 = 0 \quad \text{or} \quad k - 1 = 0 \qquad \text{Zero-factor property}$$

$$2k = -3 \qquad\qquad k = 1 \qquad \text{Solve each equation.}$$

$$k = -\frac{3}{2}$$

A check confirms that the solution set is $\left\{-\frac{3}{2}, 1\right\}$.

⚠ **CAUTION**   In **Example 4(b)**, it is tempting to begin by dividing both sides of

$$y^2 = 2y$$

by y to obtain $y = 2$. Note, however, that we do not find the other solution, 0, if we divide by a variable. (We *may* divide each side of an equation by a *nonzero* real number, however. In **Example 3** we divided each side by 2.)

In **Example 4(c)**, we cannot directly apply the zero-factor property to solve

$$k(2k + 1) = 3$$

in its given form because of the 3 on the right side of the equation. **We can apply the zero-factor property only to a product that equals 0.**

EXAMPLE 5   Solving Quadratic Equations (Double Solutions)

Solve each equation.

(a)
$$z^2 - 22z + 121 = 0 \qquad \text{This is a perfect square trinomial.}$$

$$(z - 11)^2 = 0 \qquad \text{Factor.}$$

$$(z - 11)(z - 11) = 0 \qquad a^2 = a \cdot a$$

$$z - 11 = 0 \quad \text{or} \quad z - 11 = 0 \qquad \text{Zero-factor property}$$

$$z = 11 \qquad \text{Add 11.}$$

Because the two factors are identical, they both lead to the same solution, which is called a **double solution**.

CHECK $z^2 - 22z + 121 = 0$ \qquad Original equation

$$11^2 - 22(11) + 121 \stackrel{?}{=} 0 \qquad \text{Let } z = 11.$$

$$121 - 242 + 121 \stackrel{?}{=} 0 \qquad \text{Apply the exponent. Multiply.}$$

$$0 = 0 \quad ✓ \qquad \text{True}$$

The solution set is $\{11\}$.

(b)
$$9t^2 + 30t = -25$$

$$9t^2 + 30t + 25 = 0 \qquad \text{Standard form}$$

$$(3t + 5)^2 = 0 \qquad \text{Factor the perfect square trinomial.}$$

$$3t + 5 = 0 \quad \text{or} \quad 3t + 5 = 0 \qquad \text{Zero-factor property}$$

$$3t = -5 \qquad \text{Solve the equation.}$$

$$t = -\frac{5}{3} \qquad -\frac{5}{3} \text{ is a double solution.}$$

NOW TRY ANSWERS

4. **(a)** $\left\{-\frac{8}{3}, \frac{8}{3}\right\}$ **(b)** $\{0, 5\}$

  (c) $\left\{-\frac{1}{2}, \frac{2}{3}\right\}$

**NOW TRY
EXERCISE 5**

Solve $4x^2 + 4x + 1 = 0$.

CHECK

$$9t^2 + 30t = -25 \qquad \text{Original equation}$$

$$9\left(-\frac{5}{3}\right)^2 + 30\left(-\frac{5}{3}\right) \overset{?}{=} -25 \qquad \text{Let } t = -\frac{5}{3}.$$

$$9\left(\frac{25}{9}\right) + 30\left(-\frac{5}{3}\right) \overset{?}{=} -25 \qquad \text{Apply the exponent.}$$

$$25 - 50 \overset{?}{=} -25 \qquad \text{Multiply.}$$

$$-25 = -25 \quad \checkmark \qquad \text{True}$$

The solution set is $\left\{-\frac{5}{3}\right\}$. **NOW TRY**

⚠ **CAUTION** Each of the equations in **Example 5** has only *one* distinct solution. *We write a double solution only once in a solution set.*

OBJECTIVE 2 Solve other equations using the zero-factor property.

We can also use the zero-factor property to solve equations that involve more than two factors with variables. (These equations will have at least one term greater than second degree. They are *not* quadratic equations.)

**NOW TRY
EXERCISE 6**

Solve $3x^3 - 27x = 0$.

EXAMPLE 6 Solving an Equation with More Than Two Variable Factors

Solve $6z^3 - 6z = 0$.

$$6z^3 - 6z = 0$$

$$6z(z^2 - 1) = 0 \qquad \text{Factor out } 6z.$$

$$6z(z + 1)(z - 1) = 0 \qquad \text{Factor } z^2 - 1.$$

By an extension of the zero-factor property, this product can equal 0 only if at least one of the factors equals 0. We solve three equations, one for each factor with a variable.

$$6z = 0 \quad \text{or} \quad z + 1 = 0 \quad \text{or} \quad z - 1 = 0$$

$$z = 0 \quad \text{or} \quad z = -1 \quad \text{or} \quad z = 1$$

Check by substituting, in turn, 0, -1, and 1 into the original equation. The solution set is $\{-1, 0, 1\}$. **NOW TRY**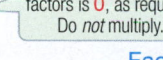

**NOW TRY
EXERCISE 7**

Solve.

$(3a - 1)(2a^2 - 5a - 12) = 0$

EXAMPLE 7 Solving an Equation with a Quadratic Factor

Solve $(3x - 1)(x^2 - 9x + 20) = 0$.

$$(3x - 1)(x^2 - 9x + 20) = 0 \qquad \text{The product of the factors is 0, as required. Do \textit{not} multiply.}$$

$$(3x - 1)(x - 5)(x - 4) = 0 \qquad \text{Factor the trinomial.}$$

$$3x - 1 = 0 \quad \text{or} \quad x - 5 = 0 \quad \text{or} \quad x - 4 = 0 \qquad \text{Zero-factor property}$$

$$x = \frac{1}{3} \quad \text{or} \quad x = 5 \quad \text{or} \quad x = 4 \qquad \text{Solve each equation.}$$

Check to verify that the solution set is $\left\{\frac{1}{3}, 4, 5\right\}$. **NOW TRY**

NOW TRY ANSWERS

5. $\left\{-\frac{1}{2}\right\}$

6. $\{-3, 0, 3\}$

7. $\left\{-\frac{3}{2}, \frac{1}{3}, 4\right\}$

⚠ **CAUTION**   In **Example 7**, it would be unproductive to begin by multiplying the two factors together. The zero-factor property requires the *product* of two or more factors to equal 0. *Always consider first whether an equation is given in an appropriate form to apply the zero-factor property.*

NOW TRY EXERCISE 8

Solve.

$x(4x - 9) = (x - 2)^2 + 24$

EXAMPLE 8   Solving an Equation Requiring Multiplication before Factoring

Solve $(3x + 1)x = (x + 1)^2 + 5$.

The zero-factor property requires the *product* of two or more factors to equal 0.

$$(3x + 1)x = (x + 1)^2 + 5 \quad \text{← This equation is *not* in the correct form.}$$

$$3x^2 + x = x^2 + 2x + 1 + 5 \qquad \text{Multiply on the left. Square } x + 1 \text{ on the right.}$$

$$3x^2 + x = x^2 + 2x + 6 \qquad \text{Combine like terms.}$$

$$2x^2 - x - 6 = 0 \qquad \text{Standard form}$$

$$(2x + 3)(x - 2) = 0 \qquad \text{Factor.} \quad \text{← The product of the factors is now 0.}$$

$$2x + 3 = 0 \quad \text{or} \quad x - 2 = 0 \qquad \text{Zero-factor property}$$

$$x = -\frac{3}{2} \quad \text{or} \quad x = 2 \qquad \text{Solve each equation.}$$

NOW TRY ANSWER

8. $\left\{ -\frac{7}{3}, 4 \right\}$

Check to verify that the solution set is $\left\{ -\frac{3}{2}, 2 \right\}$.

NOW TRY ↻

5.5 Exercises

FOR EXTRA HELP   ▶ **MyLab Math**

▶ *Video solutions for select problems available in MyLab Math*

Concept Check   *Fill in each blank with the correct response.*

1. A quadratic equation in x can be written in the form _____ = 0.

2. The form $ax^2 + bx + c = 0$ is called _____ form.

3. If the product of two numbers is 0, then at least one of the numbers is _____ . This is the _____ property.

4. If a quadratic equation is in standard form, to solve the equation we should begin by attempting to _____ the polynomial.

5. The equation $x^3 + x^2 + x = 0$ is not a quadratic equation because _____ .

6. If a quadratic equation $ax^2 + bx + c = 0$ has $c = 0$, then _____ *must* be a solution because _____ is a factor of the polynomial.

Concept Check   *Work each problem.*

7. Identify each equation as *linear* or *quadratic*.

    (a) $2x - 5 = 6$ **(b)** $x^2 - 5 = -4$

    (c) $x^2 + 2x - 3 = 2x^2 - 2$ **(d)** $5^2x + 2 = 0$

8. The number 9 is a *double solution* of the following equation. Why is this so?

$$(x - 9)^2 = 0$$

9. Concept Check Consider this incorrect "solution."

$$2x(3x - 4) = 0$$
$$x = 2 \quad \text{or} \quad x = 0 \quad \text{or} \quad 3x - 4 = 0$$
$$x = \frac{4}{3}$$

The solution set is $\left\{2, 0, \frac{4}{3}\right\}$.

WHAT WENT WRONG? Solve the equation correctly.

10. Concept Check Consider this incorrect "solution."

$$7x^2 - x = 0$$
$$7x - 1 = 0 \qquad \text{Divide by } x.$$
$$x = \frac{1}{7} \qquad \text{Solve the equation.}$$

The solution set is $\left\{\frac{1}{7}\right\}$.

WHAT WENT WRONG? Solve the equation correctly.

*Solve each equation, and check the solutions. **See Example 1.***

11. $(x + 5)(x - 2) = 0$

12. $(x - 1)(x + 8) = 0$

13. $(2m - 7)(m - 3) = 0$

14. $(6x + 5)(x + 4) = 0$

15. $(2x + 1)(6x - 1) = 0$

16. $(3x + 2)(10x - 1) = 0$

17. $t(6t + 5) = 0$

18. $w(4w + 1) = 0$

19. $2x(3x - 4) = 0$

20. $6y(4y + 9) = 0$

21. $(x - 6)(x - 6) = 0$

22. $(y + 1)(y + 1) = 0$

*Solve each equation, and check the solutions. **See Examples 2–8.***

23. $y^2 + 3y + 2 = 0$

24. $p^2 + 8p + 7 = 0$

25. $y^2 - 3y + 2 = 0$

26. $r^2 - 4r + 3 = 0$

27. $x^2 = 24 - 5x$

28. $t^2 = 2t + 15$

29. $x^2 = 3 + 2x$

30. $x^2 = 4 + 3x$

31. $z^2 + 3z = -2$

32. $p^2 - 2p = 3$

33. $m^2 + 8m + 16 = 0$

34. $x^2 + 6x + 9 = 0$

35. $x^2 - 20x + 100 = 0$

36. $t^2 - 26t + 169 = 0$

37. $3x^2 + 5x - 2 = 0$

38. $6r^2 - r - 2 = 0$

39. $12p^2 = 8 - 10p$

40. $18x^2 = 12 + 15x$

41. $9s^2 + 12s = -4$

42. $36x^2 + 60x = -25$

43. $25x^2 - 10x + 1 = 0$

44. $81x^2 - 18x + 1 = 0$

45. $y^2 - 9 = 0$

46. $m^2 - 100 = 0$

47. $16x^2 - 49 = 0$

48. $4w^2 - 9 = 0$

49. $n^2 = 121$

50. $x^2 = 400$

51. $x^2 + 6x = 0$

52. $x^2 + 4x = 0$

53. $x^2 = 7x$

54. $t^2 = 9t$

55. $6r^2 = 3r$

56. $10y^2 = -5y$

57. $x(x - 7) = -10$

58. $r(r - 5) = -6$

59. $3z(2z + 7) = 12$

60. $4x(2x + 3) = 36$

61. $2y(y + 13) = 136$

62. $t(3t - 20) = -12$

63. $(x - 8)(x + 6) = 6x$

64. $(x - 2)(x + 9) = 4x$

65. $(x + 4)(x + 7) = 10$

66. $(x + 2)(x + 5) = 4$

67. $9y^3 - 49y = 0$

68. $16r^3 - 9r = 0$

69. $r^3 - 2r^2 - 8r = 0$

70. $x^3 - x^2 - 6x = 0$

71. $x^3 + x^2 - 20x = 0$

72. $y^3 - 6y^2 + 8y = 0$

73. $4x^3 - 18x^2 + 8x = 0$

74. $9x^3 - 24x^2 + 12x = 0$

75. $r^4 = 2r^3 + 15r^2$

76. $x^4 = 3x^2 + 2x^3$

77. $(2r + 5)(3r^2 - 16r + 5) = 0$

78. $(3m + 4)(6m^2 + m - 2) = 0$

79. $(2x + 7)(x^2 + 2x - 3) = 0$

80. $(x + 1)(6x^2 + x - 12) = 0$

81. $3x(x + 1) = (2x + 3)(x + 1)$

82. $2x(x + 3) = (3x + 1)(x + 3)$

83. $x^2 + (x + 1)^2 = (x + 2)^2$

84. $x^2 + (x - 7)^2 = (x + 1)^2$

Extending Skills *Solve each equation, and check the solutions.*

85. $(2x)^2 = (2x + 4)^2 - (x + 5)^2$

86. $5 - (x - 1)^2 = (x - 2)^2$

87. $(x + 3)^2 - (2x - 1)^2 = 0$

88. $(4y - 3)^3 - 9(4y - 3) = 0$

89. $6p^2(p + 1) = 4(p + 1) - 5p(p + 1)$

90. $6x^2(2x + 3) = 4(2x + 3) + 5x(2x + 3)$

Galileo's formula describing the motion of freely falling objects is

$$d = 16t^2.$$

*The distance d in feet an object falls depends on the time t elapsed, in seconds. (This is an example of a **function**, a key mathematical concept.)*

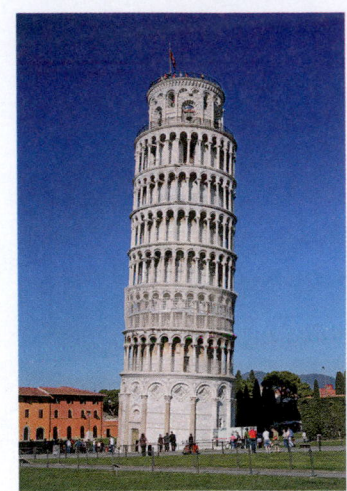

91. (a) Use Galileo's formula and complete the following table. (*Hint:* Substitute each given value into the formula and solve for the unknown value.)

t in seconds	0	1	2	3		
d in feet	0	16			256	576

(b) When $t = 0$, we find that $d = 0$. Explain this in the context of the problem.

92. Refer to **Exercise 91.** When 256 was substituted for d and the formula was solved for t, there should have been two solutions, 4 and -4. Why doesn't -4 make sense as an answer?

5.6 Applications of Quadratic Equations

OBJECTIVES

1 Solve problems involving geometric figures.

2 Solve problems involving consecutive integers.

3 Solve problems by applying the Pythagorean theorem.

4 Solve problems using given quadratic models.

We use the zero-factor property to solve quadratic equations that arise in application problems. We follow the same six problem-solving steps as earlier.

Solving an Applied Problem

Step 1 **Read** the problem carefully. *What information is given? What is to be found?*

Step 2 **Assign a variable** to represent the unknown value. Make a sketch, diagram, or table, as needed. If necessary, express any other unknown values in terms of the variable.

Step 3 **Write an equation** using the variable expression(s).

Step 4 **Solve** the equation.

Step 5 **State the answer.** Label it appropriately. *Does it seem reasonable?*

Step 6 **Check** the answer in the words of the *original* problem.

VOCABULARY

☐ consecutive integers
☐ consecutive even (odd) integers
☐ legs
☐ hypotenuse

PROBLEM-SOLVING HINT Refer to the formulas at the back of the text as needed when solving application problems.

OBJECTIVE 1 Solve problems involving geometric figures.

EXAMPLE 1   Solving an Area Problem

Abe wants to plant a triangular flower bed in a corner of his garden. One leg of the right-triangular flower bed will be 2 m shorter than the other leg. He wants the bed to have an area of 24 m². Find the lengths of the legs.

Step 1   **Read** the problem. We need to find the lengths of the legs of a right triangle with area 24 m².

Step 2   **Assign a variable.**

Let x = the length of one leg.

Then $x - 2$ = the length of the other leg.

See **FIGURE 1**.

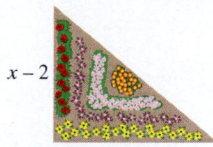

FIGURE 1

Step 3   **Write an equation.** In a right triangle, the legs are the base and height, so we substitute 24 for the area, x for the base, and $x - 2$ for the height in the formula for the area of a triangle.

$$\mathscr{A} = \frac{1}{2}bh \qquad \text{Formula for the area of a triangle}$$

$$24 = \frac{1}{2}x(x - 2) \qquad \text{Let } \mathscr{A} = 24, b = x, \text{ and } h = x - 2.$$

Step 4   **Solve.**

$$24 = \frac{1}{2}x^2 - x \qquad \text{Distributive property}$$

$$48 = x^2 - 2x \qquad \text{Multiply each term by 2.}$$

$$x^2 - 2x - 48 = 0 \qquad \text{Standard form}$$

$$(x + 6)(x - 8) = 0 \qquad \text{Factor.}$$

$$x + 6 = 0 \quad \text{or} \quad x - 8 = 0 \qquad \text{Zero-factor property}$$

$$x = -6 \quad \text{or} \quad x = 8 \qquad \text{Solve each equation.}$$

Step 5   **State the answer.** The solutions are -6 and 8. Because a triangle cannot have a side of negative length, we discard the solution -6. Then the lengths of the legs will be 8 m and $8 - 2 = 6$ m.

Step 6   **Check.** The length of one leg is 2 m less than the length of the other leg, and the area is

$$\frac{1}{2}(8)(6) = 24 \text{ m}^2, \quad \text{as required.}$$

NOW TRY

! CAUTION *When solving applied problems, always check solutions against physical facts and discard any answers that are not appropriate.*

OBJECTIVE 2 Solve problems involving consecutive integers.

Recall that **consecutive integers,** illustrated in **FIGURE 2**, are integers that are next to each other on a number line, such as

3 and 4,   or -11 and -10.

NOW TRY EXERCISE 1

A right triangle has one leg that is 4 ft shorter than the other leg. The area of the triangle is 6 ft². Determine the lengths of the legs.

x $x+1$ $x+2$

0   1   2   3   4   5   6

Consecutive integers

FIGURE 2

NOW TRY ANSWER
1. 2 ft, 6 ft

Consecutive even integers

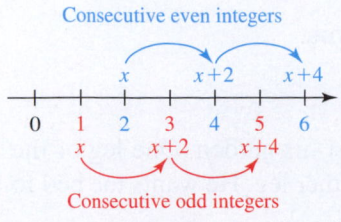

Consecutive odd integers

FIGURE 3

Consecutive even integers are *even* integers that are next to each other on a number line, such as

$$4 \text{ and } 6, \quad \text{or} \quad -10 \text{ and } -8.$$

Consecutive odd integers are defined similarly—for example, 3 and 5 are consecutive *odd* integers, as are -13 and -11. See **FIGURE 3**.

PROBLEM-SOLVING HINT   If $x =$ the lesser (least) integer in a consecutive integer problem, then the following apply.
• For two consecutive integers, use $\quad$ $x, \; x + 1.$
• For three consecutive integers, use $\quad$ $x, \; x + 1, \; x + 2.$
• For two consecutive even or odd integers, use $\quad$ $x, \; x + 2.$
• For three consecutive even or odd integers, use $\quad$ $x, \; x + 2, \; x + 4.$

In this text, we list consecutive integers in increasing order.

EXAMPLE 2   Solving a Consecutive Integer Problem

The product of the second and third of three consecutive integers is 2 more than 7 times the first integer. Find the integers.

Step 1   **Read** the problem. Note that the integers are consecutive.

Step 2   **Assign a variable.**

Let $\qquad x =$ the first integer.

Then $\quad x + 1 =$ the second integer,

and $\quad x + 2 =$ the third integer.

Step 3   **Write an equation.**

The product of the second and third $\qquad$ is $\qquad$ 2 more than 7 times the first.

$$(x + 1)(x + 2) \qquad\qquad = \qquad\qquad 7x + 2$$

Step 4   **Solve.**

$$x^2 + 3x + 2 = 7x + 2 \qquad \text{Multiply.}$$
$$x^2 - 4x = 0 \qquad \text{Standard form}$$
$$x(x - 4) = 0 \qquad \text{Factor.}$$
$$x = 0 \quad \text{or} \quad x - 4 = 0 \qquad \text{Zero-factor property}$$
$$x = 4 \qquad \text{Add 4.}$$

Step 5   **State the answer.** The values 0 and 4 each lead to a distinct answer.

If $x = 0$, then $x + 1 = 1$ and $x + 2 = 2$. The integers are $0, 1, 2$.

If $x = 4$, then $x + 1 = 5$ and $x + 2 = 6$. The integers are $4, 5, 6$.

Step 6   **Check.** The product of the second and third integers must equal 2 more than 7 times the first. Because

$$1 \cdot 2 = 7 \cdot 0 + 2 \quad \text{and} \quad 5 \cdot 6 = 7 \cdot 4 + 2 \quad \text{are both true,}$$

both sets of consecutive integers satisfy the statement of the problem.

NOW TRY
EXERCISE 2

The product of the first and second of three consecutive integers is 2 more than 8 times the third integer. Find the integers.

NOW TRY ANSWER
2. $9, 10, 11$ or $-2, -1, 0$

NOW TRY

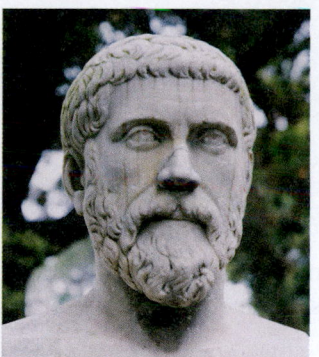

Pythagoras (c. 580–500 B.C.)

OBJECTIVE 3 Solve problems by applying the Pythagorean theorem.

Although there is evidence of the discovery of the Pythagorean relationship in earlier civilizations, the Greek mathematician and philosopher Pythagoras generally receives credit for being the first to prove it. This famous theorem from geometry relates the lengths of the sides of a right triangle.

Pythagorean Theorem

If a and b are the lengths of the shorter sides of a right triangle (a triangle with a 90° angle) and c is the length of the longest side, then

$$a^2 + b^2 = c^2.$$

The two shorter sides are the **legs** of the triangle, and the longest side, opposite the right angle, is the **hypotenuse.**

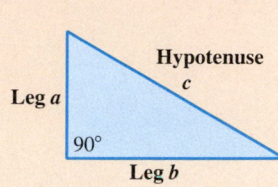

PROBLEM-SOLVING HINT In solving a problem involving the Pythagorean theorem, be sure that the expressions for the sides are properly placed.

$$(\text{one leg})^2 + (\text{other leg})^2 = \text{hypotenuse}^2$$

EXAMPLE 3 Applying the Pythagorean Theorem

Patricia and Ali leave their office, with Patricia traveling north and Ali traveling east. When Ali is 1 mi farther than Patricia from the office, the distance between them is 2 mi more than Patricia's distance from the office. Find their distances from the office and the distance between them.

Step 1 **Read** the problem again. We must find three distances.

Step 2 **Assign a variable.**

Let x = Patricia's distance from the office.

Then $x + 1$ = Ali's distance from the office,

and $x + 2$ = the distance between them.

Place these expressions on a right triangle, as in **FIGURE 4.**

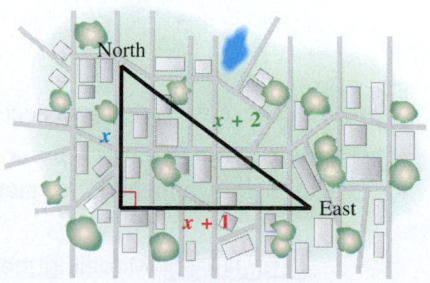

The symbol ⌐ indicates a right, or 90°, angle.

FIGURE 4

Step 3 **Write an equation.** Substitute into the Pythagorean theorem.

$$a^2 + b^2 = c^2$$

$$x^2 + (x + 1)^2 = (x + 2)^2 \quad \text{Be careful to substitute properly.}$$

NOW TRY EXERCISE 3

The longer leg of a right triangle is 7 ft longer than the shorter leg and the hypotenuse is 8 ft longer than the shorter leg. Find the lengths of the sides of the triangle.

Step 4 **Solve.**

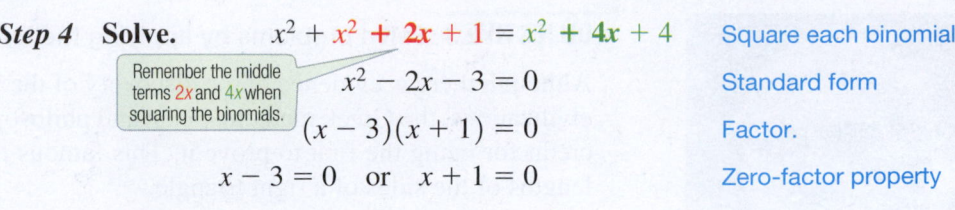

$$x^2 + x^2 + 2x + 1 = x^2 + 4x + 4 \qquad \text{Square each binomial.}$$

Remember the middle terms $2x$ and $4x$ when squaring the binomials.

$$x^2 - 2x - 3 = 0 \qquad \text{Standard form}$$
$$(x - 3)(x + 1) = 0 \qquad \text{Factor.}$$
$$x - 3 = 0 \quad \text{or} \quad x + 1 = 0 \qquad \text{Zero-factor property}$$
$$x = 3 \quad \text{or} \qquad x = -1 \qquad \text{Solve each equation.}$$

Step 5 **State the answer.** Because -1 cannot represent a distance, 3 is the only possible answer. Patricia's distance is 3 mi, Ali's distance is $3 + 1 = 4$ mi, and the distance between them is $3 + 2 = 5$ mi.

Step 6 **Check.** Because $3^2 + 4^2 = 5^2$ is true, the answer is correct. **NOW TRY**

OBJECTIVE 4 Solve problems using given quadratic models.

In **Examples 1–3,** we wrote quadratic equations to model, or mathematically describe, various situations and then solved the equations. In the remaining examples, we are given quadratic models and must use them to determine data.

NOW TRY EXERCISE 4

Refer to **Example 4.** How long will it take for the rocket to reach a height of 50 ft?

EXAMPLE 4 Finding the Height of a Rocket

A rocket is projected upward from ground level with an initial velocity of 180 ft per sec. The height h of the rocket in feet at time t in seconds is modeled by the quadratic equation

$$h = -16t^2 + 180t + 6.$$

How long will it take for the rocket to reach a height of 206 ft?

A height of 206 ft means that $h = 206$, so we substitute 206 for h in the equation and solve for t.

$$h = -16t^2 + 180t + 6$$
$$206 = -16t^2 + 180t + 6 \qquad \text{Let } h = 206.$$
$$-16t^2 + 180t + 6 = 206 \qquad \text{Interchange sides.}$$
$$-16t^2 + 180t - 200 = 0 \qquad \text{Standard form}$$
$$4t^2 - 45t + 50 = 0 \qquad \text{Divide by } -4.$$
$$(4t - 5)(t - 10) = 0 \qquad \text{Factor.}$$
$$4t - 5 = 0 \quad \text{or} \quad t - 10 = 0 \qquad \text{Zero-factor property}$$
$$4t = 5 \quad \text{or} \qquad t = 10 \qquad \text{Solve each equation.}$$
$$t = \frac{5}{4}$$

206 ft

FIGURE 5

Because we found two acceptable answers, the rocket will be 206 ft above the ground twice—once on its way up and once on its way down—at $\frac{5}{4}$ or $1\frac{1}{4}$ sec and at 10 sec after it is launched. See **FIGURE 5**. **NOW TRY**

EXAMPLE 5 Modeling the Foreign-Born Population of the United States

The foreign-born population of the United States over the years 1930–2016 can be modeled by the quadratic equation

$$y = 0.009578x^2 - 0.4886x + 15.07,$$

NOW TRY ANSWERS
3. 5 ft, 12 ft, 13 ft
4. $\frac{1}{4}$ sec and 11 sec

where $x = 0$ represents 1930, $x = 10$ represents 1940, and so on, and y is the number of people in millions. (Data from U.S. Census Bureau.)

NOW TRY EXERCISE 5

Use the model in **Example 5** to find the foreign-born population of the United States in the year 2000. Give the answer to the nearest tenth of a million. How does it compare to the actual value from the table?

(a) Use the model to find the foreign-born population in 1980 to the nearest tenth of a million.

Because $x = 0$ represents 1930, $x = 50$ represents 1980. Substitute 50 for x in the given equation.

$$y = 0.009578x^2 - 0.4886x + 15.07 \qquad \text{Given quadratic model}$$
$$y = 0.009578(50)^2 - 0.4886(50) + 15.07 \qquad \text{Let } x = 50.$$
$$y = 14.6 \qquad \text{Round to the nearest tenth.}$$

In 1980, the U.S. foreign-born population was about 14.6 million people.

(b) Repeat part (a) for 2010.

$$y = 0.009578(80)^2 - 0.4886(80) + 15.07 \qquad \text{For 2010, let } x = 80.$$
$$y = 37.3 \qquad \text{Round to the nearest tenth.}$$

In 2010, the U.S. foreign-born population was about 37.3 million people.

(c) The model used above was developed from the data in the table. How do the results in parts (a) and (b) compare to the actual data from the table?

Year	Foreign-Born Population (in millions)
1930	14.2
1940	11.6
1950	10.3
1960	9.7
1970	9.6
1980	14.1
1990	19.8
2000	28.4
2010	37.6
2016	43.7

NOW TRY ANSWER

5. 27.8 million people; The actual value is 28.4 million. The answer using the model is slightly low.

From the table, the actual value for 1980 is 14.1 million. Our answer in part (a), 14.6 million, is slightly high. For 2010, the actual value is 37.6 million, so our answer of 37.3 million in part (b) is slightly low, but a good estimate. **NOW TRY**

5.6 Exercises

FOR EXTRA HELP

 MyLab Math

▶ *Video solutions for select problems available in MyLab Math*

1. Concept Check Complete each statement to review the six problem-solving steps.

Step 1: _____ the problem carefully.

Step 2: Assign a _____ to represent the unknown value.

Step 3: Write a(n) _____ using the variable expression(s).

Step 4: _____ the equation.

Step 5: State the _____ .

Step 6: _____ the answer in the words of the _____ problem.

2. Concept Check A student solves an applied problem and gets 6 or -3 for the length of the side of a square. Which of these answers is reasonable? Why?

*A geometric figure is given in each exercise. Write the indicated formula. Then, using x as the variable, complete Steps 3–6 for each problem. (Refer to the steps in **Exercise 1** as needed.)*

3.

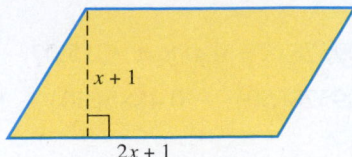

The area of this parallelogram is 45 sq. units. Find its base and height.

Formula for the area of a parallelogram:

_____

Step 3:   45 = _____

Step 4: $x =$ _____ or $x =$ _____

Step 5:   base: _____ units;

            height: _____ units

Step 6:   _____ = 45

4.

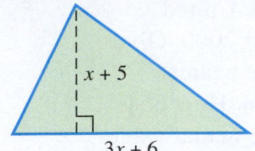

The area of this triangle is 60 sq. units. Find its base and height.

Formula for the area of a triangle:

_____

Step 3:   60 = _____

Step 4: $x =$ _____ or $x =$ _____

Step 5:   base: _____ units;

            height: _____ units

Step 6:   _____ = 60

5.

The area of this rug is 80 sq. units. Find its length and width.

Formula for the area of a rectangle:

_____

Step 3:   _____ = $(x + 8)$ _____

Step 4: $x =$ _____ or $x =$ _____

Step 5:   length: _____ units;

            width: _____ units

Step 6:   _____ = 80

6.

The volume of this box is 192 cu. units. Find its length and width.

Formula for the volume of a rectangular solid: _____

Step 3:   _____ = _____ $(x + 2)$

Step 4: $x =$ _____ or $x =$ _____

Step 5:   length: _____ units;

            width: _____ units

Step 6:   _____ $\cdot\, 4 =$ _____

Solve each problem. Check answers to be sure that they are reasonable. Refer to the formulas at the back of this text as needed. See Example 1.

7. The length of a standard jewel case is 2 cm more than its width. The area of the rectangular top of the case is 168 cm^2. Find the length and width of the jewel case.

8. A standard DVD case is 6 cm longer than it is wide. The area of the rectangular top of the case is 247 cm^2. Find the length and width of the case.

9. The area of a triangle is 30 in.^{2}. The base of the triangle measures 2 in. more than twice the height of the triangle. Find the measures of the base and the height.

10. A certain triangle has its base equal in measure to its height. The area of the triangle is 72 m^2. Find the equal base and height measure.

11. A 10-gal aquarium is 3 in. higher than it is wide. Its length is 21 in., and its volume is 2730 in.^{3}. What are the height and width of the aquarium?

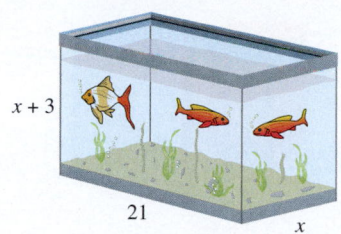

12. A toolbox is 2 ft high, and its width is 3 ft less than its length. If its volume is 80 ft^3, find the length and width of the box.

13. The dimensions of a rectangular monitor screen are such that its length is 3 in. more than its width. If the length were doubled and if the width were decreased by 1 in., the area would be increased by 150 in.^{2}. What are the length and width of the screen?

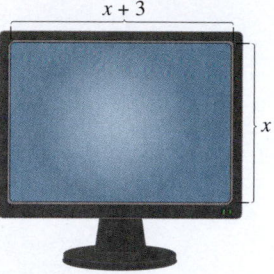

14. A computer keyboard is 11 in. longer than it is wide. If the length were doubled and if 2 in. were added to the width, the area would be increased by 198 in.^{2}. What are the length and width of the keyboard?

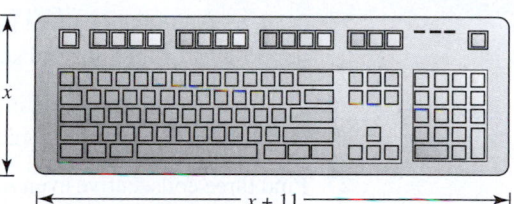

15. A square mirror has sides measuring 2 ft less than the sides of a square painting. If the difference between their areas is 32 ft^2, find the lengths of the sides of the mirror and the painting.

16. The sides of one square have length 3 m more than the sides of a second square. If the area of the larger square is subtracted from 4 times the area of the smaller square, the result is 36 m^2. What are the lengths of the sides of each square?

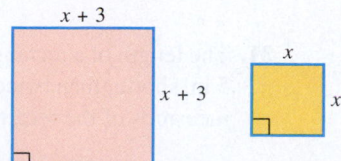

Solve each problem. **See Example 2.**

17. The product of the numbers on two consecutive volumes of research data is 420. Find the volume numbers.

18. The product of the page numbers on two facing pages of a book is 600. Find the page numbers.

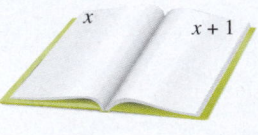

19. The product of two consecutive integers is 11 more than their sum. Find the integers.

20. The product of two consecutive integers is 4 less than four times their sum. Find the integers.

21. The product of the second and third of three consecutive integers is 2 more than 10 times the first integer. Find the integers.

22. The product of the first and third of three consecutive integers is 3 more than 3 times the second integer. Find the integers.

23. Find two consecutive odd integers such that their product is 15 more than three times their sum.

24. Find two consecutive odd integers such that five times their sum is 23 less than their product.

25. Find three consecutive odd integers such that 3 times the sum of all three is 18 more than the product of the first and second integers.

26. Find three consecutive odd integers such that the sum of all three is 42 less than the product of the second and third integers.

27. Find three consecutive even integers such that the sum of the squares of the first and second integers is equal to the square of the third integer.

28. Find three consecutive even integers such that the square of the sum of the first and second integers is equal to twice the third integer.

Solve each problem. **See Example 3.**

29. The hypotenuse of a right triangle is 1 cm longer than the longer leg. The shorter leg is 7 cm shorter than the longer leg. Find the length of the longer leg of the triangle.

30. The longer leg of a right triangle is 1 m longer than the shorter leg. The hypotenuse is 1 m shorter than twice the shorter leg. Find the length of the shorter leg of the triangle.

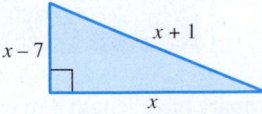

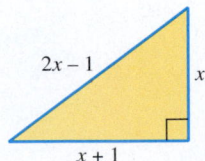

31. The length of a rectangle is 5 in. longer than its width. The diagonal is 5 in. shorter than twice the width. Find the length, width, and diagonal measures of the rectangle.

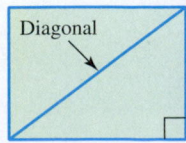

32. The length of a rectangle is 4 in. longer than its width. The diagonal is 8 in. longer than the width. Find the length, width, and diagonal measures of the rectangle.

33. Tram works due north of home. Her husband Alan works due east. They leave for work at the same time. By the time Tram is 5 mi from home, the distance between them is 1 mi more than Alan's distance from home. How far from home is Alan?

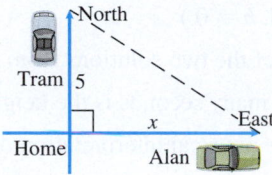

34. Two cars left an intersection at the same time. One traveled north. The other traveled 14 mi farther, but to the east. How far apart were they at that time if the distance between them was 4 mi more than the distance traveled east?

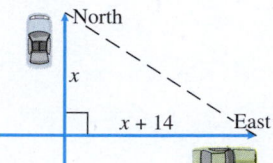

35. A ladder is leaning against a building. The distance from the bottom of the ladder to the building is 4 ft less than the length of the ladder. How high up the side of the building is the top of the ladder if that distance is 2 ft less than the length of the ladder?

36. A lot has the shape of a right triangle with one leg 2 m longer than the other. The hypotenuse is 2 m less than twice the length of the shorter leg. Find the length of the shorter leg.

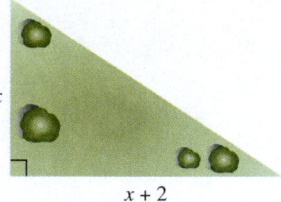

If an object is projected upward with an initial velocity of 128 ft per sec, its height h in feet after t seconds is given by the quadratic equation

$$h = -16t^2 + 128t.$$

*Find the height of the object after each time listed. See **Example 4**.*

37. 1 sec   **38.** 2 sec   **39.** 4 sec

40. How long does it take the object just described to return to the ground? (*Hint:* When the object hits the ground, $h = 0$.)

*Solve each problem. See **Example 4**.*

41. If an object is projected from a height of 48 ft with an initial velocity of 32 ft per sec, its height h in feet after t seconds is given by

$$h = -16t^2 + 32t + 48.$$

(a) After how many seconds is the height 64 ft? (*Hint:* Let $h = 64$ and solve.)

(b) After how many seconds is the height 60 ft?

(c) After how many seconds does the object hit the ground? (*Hint:* When the object hits the ground, $h = 0$.)

(d) Only one of the two solutions from part (c) is appropriate here. Why?

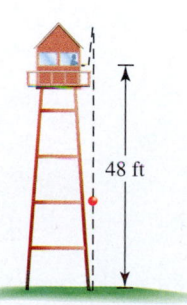

42. If an object is projected upward from ground level with an initial velocity of 64 ft per sec, its height h in feet t seconds later is given by

$$h = -16t^2 + 64t.$$

(a) After how many seconds is the height 48 ft?

(b) The object reaches its maximum height 2 sec after it is projected. What is this maximum height?

(c) After how many seconds does the object hit the ground? (*Hint:* When the object hits the ground, $h = 0$.)

(d) Only one of the two solutions from part (c) is appropriate here. Why?

(e) After how many seconds is the height 60 ft.

(f) What is the physical interpretation of why part (e) has two answers?

43. A cell phone is dropped off a nature observatory that is 144 ft above the ground. The height h in feet of the cell phone after t seconds is given by

$$h = -16t^2 + 144.$$

After how many seconds does the cell phone hit the ground?

44. A rock is dropped off a cliff that is 256 ft above a river. The height h in feet of the rock after t seconds is given by

$$h = -16t^2 + 256.$$

After how many seconds does the rock hit the water?

144 ft

*Solve each problem. **See Example 5.***

45. The table shows the number of cellular phone subscribers (in millions) in the United States.

Year	Subscribers (in millions)
2000	109
2002	141
2004	182
2006	233
2008	270
2010	296
2012	326
2014	355
2016	396

Data from CTIA.

We used the data to develop the quadratic equation

$$y = -0.229x^2 + 21.5x + 105,$$

which models the number of cellular phone subscribers y (in millions) in the year x, where $x = 0$ represents 2000, $x = 2$ represents 2002, and so on.

(a) What value of x corresponds to 2010?

(b) Use the model to find the number of subscribers in 2010, to the nearest million. How does the result compare with the actual data in the table?

(c) What value of x corresponds to 2016?

(d) Use the model to find the number of cellular phone subscribers in 2016, to the nearest million. How does the result compare with the actual data in the table?

(e) Assuming that the trend in the data continues, what value of x would correspond to 2020?

(f) Use the model to find the number of cellular phone subscribers in 2020, to the nearest million.

46. World population (in billions) is shown in the table.

Year	Population (in billions)
1950	2.5
1960	3.0
1970	3.7
1980	4.5
1990	5.3
2000	6.1
2010	7.0
2016	7.5

Data from www.worldpopulationstatistics.com

Using the data, we developed the quadratic equation

$$y = 0.00025x^2 + 0.0600x + 2.4420,$$

which models population y (in billions) in the year x, where $x = 0$ represents 1950, $x = 10$ represents 1960, and so on.

(a) What value of x corresponds to the year 2000? To the year 2016?

(b) Use the model to find world population in 2000 and 2016, to the nearest tenth. How do the results compare with the actual data in the table?

(c) World population is projected to reach 8.2 billion in 2025. What value of x corresponds to the year 2025?

(d) Use the model to project world population in 2025, to the nearest tenth. How does the result compare to the projection given in part (c)?

RELATING CONCEPTS For Individual or Group Work (Exercises 47–50)

A proof of the Pythagorean theorem is based on the figures shown. **Work Exercises 47–50 in order.**

47. What is an expression for the area of the dark square labeled ③ in **FIGURE A**?

48. The five regions in **FIGURE A** are equal in area to the six regions in **FIGURE B**. What is an expression for the area of the square labeled ① in **FIGURE B**?

49. What is an expression for the area of the square labeled ② in **FIGURE B**?

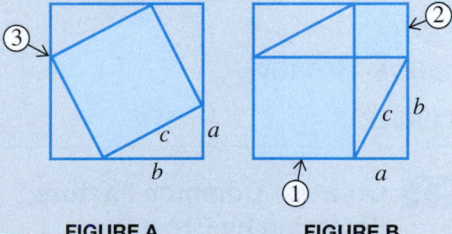

FIGURE A   FIGURE B

50. Represent this statement using algebraic expressions:

The sum of the areas of the shaded regions in **FIGURE B** *is equal to the area of the shaded region in* **FIGURE A**.

What does this equation represent?

Chapter 5   Summary

Key Terms

5.1

factor
factored form
common factor
greatest common factor
   (GCF)

5.2

prime polynomial

5.4

perfect square
perfect square trinomial
perfect cube
difference of cubes
sum of cubes

5.5

quadratic equation
second-degree equation
double solution

5.6

consecutive integers
consecutive even (odd)
   integers
legs
hypotenuse

Test Your Word Power

See how well you have learned the vocabulary in this chapter.

1. **Factoring** is
   A. a method of multiplying
      polynomials
   B. the process of writing a
      polynomial as a product
   C. the answer in a multiplication
      problem
   D. a way to add the terms of a
      polynomial.

2. A polynomial is in **factored form**
   when
   A. it is prime
   B. it is written as a sum

   C. the second-degree term has a
      coefficient of 1
   D. it is written as a product.

3. A **perfect square trinomial** is a
   trinomial
   A. that can be factored as the square
      of a binomial
   B. that cannot be factored
   C. that is multiplied by a binomial
   D. where all terms are perfect
      squares.

4. A **quadratic equation** is an equation
   that can be written in the form
   A. $y = mx + b$
   B. $ax^2 + bx + c = 0$ $(a \neq 0)$
   C. $Ax + By = C$
   D. $ax + b = 0.$

5. A **hypotenuse** is
   A. either of the two shorter sides of
      a triangle
   B. the shortest side of a triangle
   C. the side opposite the right angle
      in a triangle
   D. the longest side in any triangle.

ANSWERS

1. B; *Example:* $x^2 - 5x - 14$ factors as $(x - 7)(x + 2)$.   2. D; *Example:* The factored form of $x^2 - 5x - 14$ is $(x - 7)(x + 2)$.   3. A; *Example:*
$a^2 + 2a + 1$ is a perfect square trinomial. Its factored form is $(a + 1)^2$.   4. B; *Examples:* $y^2 - 3y + 2 = 0, x^2 - 9 = 0, 2m^2 = 6m + 8$ 5. C; *Example:*
See the triangle at the end of the Quick Review that follows.

Quick Review

CONCEPTS	EXAMPLES

**5.1   Greatest Common Factors;
Factoring by Grouping**

Finding the Greatest Common Factor (GCF)

Step 1   Write each number in prime factored form.

Step 2   List each prime number or each variable that is a factor
of every term in the list.

Step 3   Use as exponents on the common prime factors the
least exponents from the prime factored forms.

Step 4   Multiply the primes from Step 3.

Find the greatest common factor of $4x^2y$, $6x^2y^3$, and $2xy^2$.

$$4x^2y = 2 \cdot 2 \cdot x^2 \cdot y$$

$$6x^2y^3 = 2 \cdot 3 \cdot x^2 \cdot y^3$$

$$2xy^2 = 2 \cdot x \cdot y^2$$

$$\text{GCF} = 2xy$$

CONCEPTS	EXAMPLES
Factoring by Grouping	Factor by grouping.

Factoring by Grouping

Step 1 Collect the terms into two groups so that each group has a common factor.

Step 2 Factor out the greatest common factor from each group.

Step 3 Factor out a common binomial factor from the results of Step 2.

Step 4 If necessary, rearrange terms and try a different grouping.

Factor by grouping.

$$3x^2 + 5x - 24xy - 40y$$
$$= (3x^2 + 5x) + (-24xy - 40y) \quad \text{Group the terms.}$$
$$= x(3x + 5) - 8y(3x + 5) \quad \text{Factor each group.}$$
$$= (3x + 5)(x - 8y) \quad \text{Factor out } 3x + 5.$$

5.2 Factoring Trinomials

Factoring $x^2 + bx + c$

Find two integers m and n such that $mn = c$ and $m + n = b$.

$$mn = c$$
$$\downarrow$$
$$x^2 + bx + c$$
$$\uparrow$$
$$m + n = b$$

Then $x^2 + bx + c$ factors as $(x + m)(x + n)$.

Check by multiplying.

Factor $x^2 + 6x + 8$.

$$mn = 8$$
$$\downarrow$$
$$x^2 + 6x + 8 \qquad \text{Find two integers } m \text{ and } n \text{ whose}$$
$$\uparrow \qquad\qquad \text{product is } 8 \text{ and whose sum is } 6.$$
$$m + n = 6 \qquad \text{Here, } m = 2 \text{ and } n = 4.$$

$x^2 + 6x + 8$ factors as $(x + 2)(x + 4)$.

CHECK $(x + 2)(x + 4)$

$$= x^2 + 4x + 2x + 8 \quad \text{FOIL method}$$
$$= x^2 + 6x + 8 \checkmark \quad \text{Combine like terms.}$$

5.3 More on Factoring Trinomials

Factoring $ax^2 + bx + c$

Use one of the following methods.

Factoring by Grouping

Find m and n such that $mn = ac$ and $m + n = b$.

$$mn = ac$$
$$\downarrow \qquad \downarrow$$
$$ax^2 + bx + c$$
$$\uparrow$$
$$m + n = b$$

Then factor $ax^2 + mx + nx + c$ by grouping.

Factor $3x^2 + 14x - 5$ by grouping.

$$\overset{\displaystyle -15}{3x^2 + 14x - 5} \qquad \text{Here, } mn = -15 \text{ and } m + n = 14.$$

Find two integers with a product of $3(-5) = -15$ and a sum of 14. The integers are -1 and 15.

$$3x^2 + 14x - 5$$
$$= 3x^2 - x + 15x - 5 \qquad 14x = -x + 15x$$
$$= (3x^2 - x) + (15x - 5) \quad \text{Group the terms.}$$
$$= x(3x - 1) + 5(3x - 1) \quad \text{Factor each group.}$$
$$= (3x - 1)(x + 5) \qquad \text{Factor out } 3x - 1.$$

Factoring by Trial and Error

Use the FOIL method in reverse.

Factor $3x^2 + 14x - 5$ by trial and error.

Because the only positive factors of 3 are 3 and 1, and -5 has possible factors of 1 and -5, or -1 and 5, the possible factored forms for this trinomial follow.

$(3x - 5)(x + 1)$ Incorrect $(3x + 5)(x - 1)$ Incorrect

$(3x + 1)(x - 5)$ Incorrect $(3x - 1)(x + 5)$ Correct

Using grouping or trial and error,

$$3x^2 + 14x - 5 \quad \text{factors as} \quad (3x - 1)(x + 5).$$

CONCEPTS	EXAMPLES

5.4   Special Factoring Techniques

Difference of Squares

$$x^2 - y^2 = (x + y)(x - y)$$

Perfect Square Trinomials

$$x^2 + 2xy + y^2 = (x + y)^2$$
$$x^2 - 2xy + y^2 = (x - y)^2$$

Difference of Cubes

$$x^3 - y^3 = (x - y)(x^2 + xy + y^2)$$

Sum of Cubes

$$x^3 + y^3 = (x + y)(x^2 - xy + y^2)$$

Factor.

$$4x^2 - 9 \qquad\qquad 100y^4 - 49$$
$$= (2x + 3)(2x - 3) \qquad = (10y^2 + 7)(10y^2 - 7)$$

$$9x^2 + 6x + 1 \qquad\qquad 4x^2 - 20x + 25$$
$$= (3x + 1)^2 \qquad\qquad = (2x - 5)^2$$

$$m^3 - 8 \qquad\qquad\qquad z^3 + 27$$
$$= m^3 - 2^3 \qquad\qquad\qquad = z^3 + 3^3$$
$$= (m - 2)(m^2 + 2m + 4) \qquad = (z + 3)(z^2 - 3z + 9)$$

5.5   Solving Quadratic Equations Using the Zero-Factor Property

Zero-Factor Property

If a and b are real numbers and if $ab = 0$, then

$$a = 0 \quad \text{or} \quad b = 0.$$

Solving a Quadratic Equation Using the Zero-Factor Property

Step 1   Write the equation in standard form.

Step 2   Factor.

Step 3   Apply the zero-factor property.

Step 4   Solve the resulting equations.

Step 5   Check. Write the solution set.

If $(x - 2)(x + 3) = 0$, then

$$x - 2 = 0 \quad \text{or} \quad x + 3 = 0,$$

leading to $\qquad x = 2 \quad \text{or} \qquad x = -3.$

Solve. $\qquad\qquad\qquad 2x^2 = 7x + 15$

$$2x^2 - 7x - 15 = 0 \qquad \text{Standard form}$$
$$(2x + 3)(x - 5) = 0 \qquad \text{Factor.}$$
$$2x + 3 = 0 \quad \text{or} \quad x - 5 = 0 \qquad \text{Zero-factor property}$$
$$2x = -3 \qquad\qquad x = 5 \qquad \text{Solve each equation.}$$
$$x = -\frac{3}{2}$$

CHECK $\qquad\qquad 2x^2 = 7x + 15$

$$2(5)^2 \overset{?}{=} 7(5) + 15 \qquad \text{Let } x = 5.$$
$$50 \overset{?}{=} 35 + 15$$
$$50 = 50 \quad \checkmark \qquad \text{True}$$

The other value also checks. The solution set is $\left\{ -\frac{3}{2}, 5 \right\}$.

5.6   Applications of Quadratic Equations

Solving an Applied Problem

Step 1   Read the problem.

Step 2   Assign a variable.

Step 3   Write an equation.

Step 4   Solve the equation.

Step 5   State the answer.

Step 6   Check.

The longer leg of a right triangle is 2 ft longer than the shorter leg. The hypotenuse is 4 ft longer than the shorter leg. Find the lengths of the sides of the triangle.

Let $\qquad x =$ the length of the shorter leg.

Then $x + 2 =$ the length of the longer leg,

and $\quad x + 4 =$ the length of the hypotenuse.

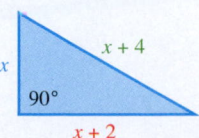

CONCEPTS	EXAMPLES

Pythagorean Theorem

In a right triangle, the sum of the squares of the legs equals the square of the hypotenuse.

$$a^2 + b^2 = c^2$$

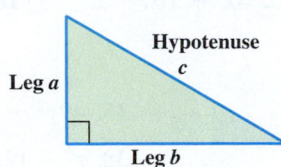

Leg a

Hypotenuse c

Leg b

$$\underset{\downarrow}{a^2} + \underset{\downarrow}{b^2} = \underset{\downarrow}{c^2}$$

$x^2 + (x + 2)^2 = (x + 4)^2$	Pythagorean theorem
$x^2 + x^2 + 4x + 4 = x^2 + 8x + 16$	Square each binomial.
$x^2 - 4x - 12 = 0$	Standard form
$(x - 6)(x + 2) = 0$	Factor.
$x - 6 = 0$ or $x + 2 = 0$	Zero-factor property
$x = 6$ or $x = -2$	Solve each equation.

Because -2 cannot represent a length of a triangle, 6 is the only possible answer. The sides have lengths

$$6 \text{ ft}, \quad 6 + 2 = 8 \text{ ft}, \quad \text{and} \quad 6 + 4 = 10 \text{ ft}.$$

Because $6^2 + 8^2 = 10^2$ is true, the answer is correct.

Chapter 5   Review Exercises

5.1 *Factor out the greatest common factor, or factor by grouping.*

1. $7t + 14$

2. $60z^3 + 30z$

3. $-3x^3 + 6x^2 + 3x$

4. $100m^2n^3 - 50m^3n^4 + 150m^2n^2$

5. $2xy - 8y + 3x - 12$

6. $6y^2 + 9y + 4xy + 6x$

5.2 *Factor completely.*

7. $x^2 + 5x + 6$

8. $y^2 - 13y + 40$

9. $q^2 + 6q - 27$

10. $r^2 - r - 56$

11. $x^2 + x + 1$

12. $3x^4 + 30x^3 + 48x^2$

13. $-8p^5 + 24p^4 + 80p^3$

14. $m^2 - 3mn - 18n^2$

15. $y^2 - 8yz + 15z^2$

16. $p^2 + 2pq - 120q^2$

17. $p^7 - p^6q - 2p^5q^2$

18. $-3r^5 + 6r^4s + 45r^3s^2$

5.3 *Answer each question.*

19. To begin factoring $6r^2 - 5r - 6$, what are the possible first terms of the two binomial factors if we consider only positive integer coefficients?

20. What is the first step to factor $2z^3 + 9z^2 - 5z$?

Factor completely.

21. $2k^2 - 5k + 2$

22. $3r^2 + 11r - 4$

23. $6r^2 - 5r - 6$

24. $10z^2 - 3z - 1$

25. $5t^2 - 11t + 12$

26. $24x^5 - 20x^4 + 4x^3$

27. $-30y^3 - 5y^2 + 10y$

28. $14a^2 - 27ab - 20b^2$

29. $3m^2 + 19mn - 40n^2$

30. $10r^3s + 17r^2s^2 + 6rs^3$

5.4 *Answer each question.*

31. Which one of the following is a difference of squares?

    A. $32x^2 - 1$ **B.** $4x^2y^2 - 25z^2$ **C.** $x^2 + 36$ **D.** $25y^3 - 1$

32. Which one of the following is a perfect square trinomial?

    A. $x^2 + x + 1$ **B.** $y^2 - 4y + 9$ **C.** $4x^2 + 10x + 25$ **D.** $x^2 - 20x + 100$

Factor completely.

33. $n^2 - 49$ **34.** $25b^2 - 121$ **35.** $49y^2 - 25w^2$

36. $144p^2 - 36q^2$ **37.** $x^2 + 100$ **38.** $r^2 - 12r + 36$

39. $9t^2 - 42t + 49$ **40.** $m^3 + 1000$ **41.** $125k^3 + 64x^3$

42. $343x^3 - 64$ **43.** $1000 - 27x^6$ **44.** $x^6 - y^6$

5.5 *Solve each equation, and check the solutions.*

45. $(4t + 3)(t - 1) = 0$ **46.** $x(2x - 5) = 0$

47. $z^2 + 4z + 3 = 0$ **48.** $m^2 - 5m + 4 = 0$

49. $x^2 = -15 + 8x$ **50.** $3z^2 - 11z - 20 = 0$

51. $81t^2 - 64 = 0$ **52.** $y^2 = 8y$

53. $n(n - 5) = 6$ **54.** $t^2 - 14t + 49 = 0$

55. $t^2 = 12(t - 3)$ **56.** $x^2 = 9$

57. $(5z + 2)(z^2 + 3z + 2) = 0$ **58.** $64x^3 - 9x = 0$

5.6 *Solve each problem.*

59. The length of a rectangular rug is 6 ft more than the width. The area is 40 ft². Find the length and width of the rug.

60. A treasure chest from a sunken galleon has dimensions (in feet) as shown in the figure. Its surface area is 650 ft². Find its width.

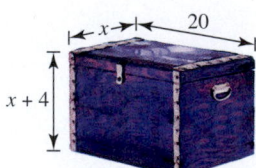

61. Two cars left an intersection at the same time. One traveled west, and the other traveled 14 mi less, but to the south. How far apart were they at that time, if the distance between them was 16 mi more than the distance traveled south?

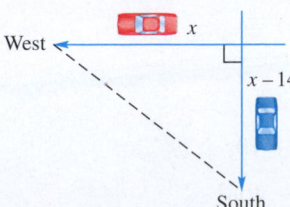

62. A pyramid has a rectangular base with a length that is 2 m more than its width. The height of the pyramid is 6 m, and its volume is 48 m³. Find the length and width of the base.

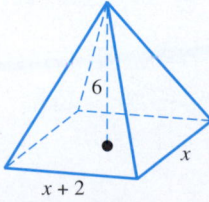

63. The product of the lesser two of three consecutive integers is equal to 23 plus the greatest. Find the integers.

64. If an object is dropped, the distance d in feet it falls in t seconds (disregarding air resistance) is given by the quadratic equation

$$d = 16t^2.$$

Find the distance an object would fall in **(a)** 4 sec and **(b)** 8 sec.

65. The numbers of electric cars in use worldwide are given in the table.

Year	Electric Cars (in thousands)
2010	17
2011	65
2012	183
2013	388
2014	715
2015	1263
2016	2014

Data from International Energy Agency.

Using the data, we developed the quadratic equation

$$y = 70.35x^2 - 103.54x + 59.69,$$

which models the number of electric cars y (in thousands) in the year x, where $x = 0$ represents 2010, $x = 1$ represents 2011, and so on.

(a) Use the model to find the number of electric cars in 2013 and 2016, to the nearest thousand.

(b) How do the results in part (a) compare with the actual data in the table?

Chapter 5   Mixed Review Exercises

1. Which one of the following is *not* factored completely?

  A. $3(7t)$ **B.** $3x(7t + 4)$ **C.** $(3 + x)(7t + 4)$ **D.** $3(7t + 4) + x(7t + 4)$

2. Which one of the following is the correct, completely factored form of $6x^2 + 16x - 32$?

  A. $(2x + 8)(3x - 4)$ **B.** $(6x - 8)(x + 4)$
  C. $2(3x - 4)(x + 4)$ **D.** $2(x - 4)(3x + 4)$

Factor completely.

3. $3k^2 + 11k + 10$ **4.** $z^2 - 11zx + 10x^2$

5. $y^4 - 625$ **6.** $6m^3 - 21m^2 - 45m$

7. $25a^2 + 15ab + 9b^2$ **8.** $2a^5 - 8a^4 - 24a^3$

9. $15m^2 + 20m - 12mp - 16p$ **10.** $24ab^3c^2 - 56a^2bc^3 + 72a^2b^2c$

11. $12x^2yz^3 + 12xy^2z - 30x^3y^2z^4$ **12.** $12r^2 + 18rq - 10r - 15q$

13. $49t^2 + 56t + 16$ **14.** $1000a^3 + 27$

Solve each equation.

15. $t(t - 7) = 0$ **16.** $x^2 + 3x = 10$ **17.** $25x^2 + 20x + 4 = 0$

Solve each problem.

18. A lot is in the shape of a right triangle. The hypotenuse is 3 m longer than the longer leg. The longer leg is 6 m longer than twice the length of the shorter leg. Find the lengths of the sides of the lot.

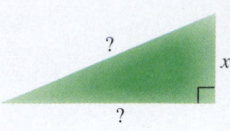

19. The triangular sail of a schooner has an area of 30 m². The height of the sail is 4 m more than the base. Find the base of the sail.

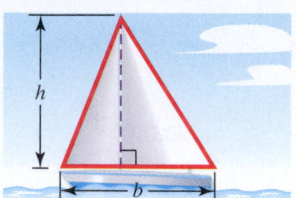

20. The floor plan for a house is a rectangle with length 7 m more than its width. The area is 170 m². Find the width and length of the house.

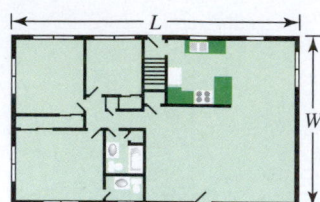

| Chapter 5 | Test | FOR EXTRA HELP | *Step-by-step test solutions are found on the Chapter Test Prep Videos available in* **MyLab Math**. |

▶ *View the complete solutions to all Chapter Test exercises in MyLab Math.*

1. Which one of the following is the correct, completely factored form of $2x^2 - 2x - 24$?

   A. $(2x + 6)(x - 4)$ **B.** $(x + 3)(2x - 8)$

   C. $2(x + 4)(x - 3)$ **D.** $2(x + 3)(x - 4)$

Factor completely. If the polynomial is prime, say so.

2. $12x^2 - 30x$ **3.** $2m^3n^2 + 3m^3n - 5m^2n^2$ **4.** $2ax - 2bx + ay - by$

5. $x^2 - 5x - 24$ **6.** $2x^2 + x - 3$ **7.** $10z^2 - 17z + 3$

8. $3x^2 - 12x - 15$ **9.** $t^2 + 2t + 3$ **10.** $x^2 + 36$

11. $12 - 6a + 2b - ab$ **12.** $9y^2 - 64$ **13.** $81a^2 - 121b^2$

14. $x^2 + 16x + 64$ **15.** $4x^2 - 28xy + 49y^2$ **16.** $6t^4 + 3t^3 - 108t^2$

17. $r^3 - 125$ **18.** $8k^3 + 64$ **19.** $x^4 - 81$ **20.** $81x^4 - 16y^4$

Solve each equation.

21. $(x + 3)(x - 9) = 0$ **22.** $2r^2 - 13r + 6 = 0$

23. $25x^2 - 4 = 0$ **24.** $t^2 = 9t$

25. $x(x - 20) = -100$ **26.** $(s + 8)(6s^2 + 13s - 5) = 0$

Solve each problem.

27. Find two consecutive integers such that the square of the sum of the two integers is 11 more than the lesser integer.

28. The length of a rectangular flower bed is 3 ft less than twice its width. The area of the bed is 54 ft². Find the dimensions of the flower bed.

29. A carpenter needs to cut a brace to support a wall stud, as shown in the figure. The brace should be 7 ft less than three times the length of the stud. If the brace will be anchored on the floor 15 ft away from the stud, how long should the brace be?

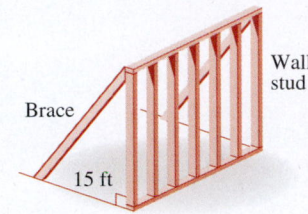

30. The public debt y (in billions of dollars) of the United States from 2010 through 2016 can be approximated by the quadratic equation

$$y = -49.77x^2 + 1117x + 9077,$$

where $x = 0$ represents 2010, $x = 1$ represents 2011, and so on. Use the model to estimate the public debt, to the nearest billion dollars, in the year 2016. (Data from Bureau of Public Debt.)

| Chapters R–5 | # Cumulative Review Exercises |

Perform the indicated operations. Write answers in lowest terms as needed.

1. $\dfrac{1}{10} + \dfrac{2}{5}$ **2.** $\dfrac{7}{9} - \dfrac{1}{6}$ **3.** $\dfrac{2}{3} \cdot \dfrac{5}{6}$ **4.** $\dfrac{5}{12} \div \dfrac{5}{6}$

Solve.

5. $3x + 2(x - 4) = 4(x - 2)$ **6.** $0.3x + 0.9x = 0.06$

7. Determine whether each of the following is an *expression* or an *equation*. If it is an expression, simplify it. If it is an equation, solve it.

    (a) $\dfrac{2}{3}m - \dfrac{1}{2}(m - 4) = 3$ **(b)** $\dfrac{2}{3}m - \dfrac{1}{2}(m - 4) - 3$

8. Solve for P: $A = P + Prt$.

9. Find the measures of the marked angles.

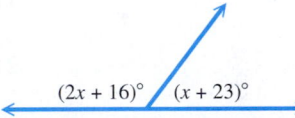

10. At the 2018 Winter Olympics in Pyeongchang, South Korea, the United States won a total of 23 medals. The United States won 1 more gold medal than silver and 2 fewer bronze medals than silver. Find the number of each type of medal won. (Data from *The Gazette*.)

11. In a recent survey, 1000 American adults were asked which tech companies they associate with online payment services. Complete the results shown in the table.

Tech Company	Percent That Associate the Company with Online Payments	Actual Number That Associate the Company with Online Payments
Apple	41%	
Facebook	34%	
Google		230
Amazon		160

Data from Statista.

12. Fill in each blank with *positive* or *negative*. The point with coordinates (a, b) is in

    (a) quadrant II if a is _____ and b is _____.

    (b) quadrant III if a is _____ and b is _____.

13. Consider the equation $y = -2x - 4$. Find the following.

    (a) The x- and y-intercepts   **(b)** The slope   **(c)** The graph

14. The points on the graph show the number of digital photos estimated to have been taken worldwide in the years 2013–2017, along with a graph of a linear equation that models the data.

    (a) Use the ordered pairs shown on the graph to find the slope of the line to the nearest whole number. Interpret the slope.

    (b) Use the graph to estimate the number of digital photos taken worldwide in the year 2014. Write the answer as an ordered pair of the form (year, number of digital photos taken in billions).

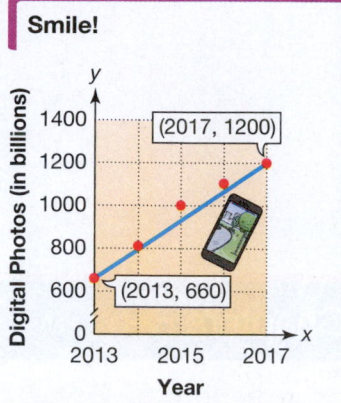

Smile!

Data from InfoTrends via Bitkom.

Evaluate each expression.

15. $\left(\dfrac{3}{4}\right)^{-2}$

16. $\left(\dfrac{4^{-3} \cdot 4^4}{4^5}\right)^{-1}$

Simplify each expression, and write the answer using only positive exponents. Assume that no denominators are 0.

17. $\dfrac{(p^2)^3 p^{-4}}{(p^{-3})^{-1} p}$

18. $\dfrac{(m^{-2})^3 m}{m^5 m^{-4}}$

Perform each indicated operation.

19. $(2k^2 + 4k) - (5k^2 - 2) - (k^2 + 8k - 6)$

20. $(9x + 6)(5x - 3)$

21. $(3p + 2)^2$

22. $\dfrac{8x^4 + 12x^3 - 6x^2 + 20x}{2x}$

Factor completely.

23. $2a^2 + 7a - 4$

24. $8t^2 + 10tv + 3v^2$

25. $4p^2 - 12p + 9$

26. $25r^2 - 81t^2$

Solve.

27. $6m^2 + m - 2 = 0$

28. The length of the hypotenuse of a right triangle is twice the length of the shorter leg, plus 3 m. The longer leg is 7 m longer than the shorter leg. Find the lengths of the sides.

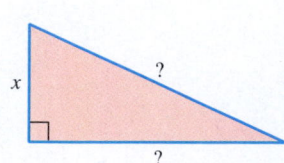

STUDY SKILLS REMINDER

Have you begun to prepare for your final exam? **Review Study Skill 10, *Preparing for Your Math Final Exam.***

6

RATIONAL EXPRESSIONS AND APPLICATIONS

The formula $r = \frac{d}{t}$ gives the rate r of a speeding car in terms of its distance d and its time traveled t. That formula involves a *rational expression* (or *algebraic fraction*), the subject of this chapter.

6.1   The Fundamental Property of Rational Expressions

OBJECTIVES

1 Find the numerical value of a rational expression.

2 Find the values of the variable for which a rational expression is undefined.

3 Write rational expressions in lowest terms.

4 Recognize equivalent forms of rational expressions.

The quotient of two integers (with denominator not 0), such as $\frac{2}{3}$ or $-\frac{3}{4}$, is a *rational number*. In the same way, the quotient of two polynomials with denominator not equal to 0 is a *rational expression*.

Rational Expression

A **rational expression** is an expression of the form $\frac{P}{Q}$, where P and Q are polynomials and $Q \neq 0$.

Examples: $\dfrac{-6x}{x^3 + 8}$, $\dfrac{9x}{y + 3}$, $\dfrac{2m^3}{8}$ Rational expressions

---

Our work with rational expressions requires much of what we have learned earlier about polynomials and factoring, as well as the rules for arithmetic fractions.

VOCABULARY

☐ rational expression
☐ lowest terms

OBJECTIVE 1 Find the numerical value of a rational expression.

Remember that to *evaluate* an expression means to find its *value*. We use substitution to evaluate a rational expression for a given value of the variable.

EXAMPLE 1 Evaluating Rational Expressions

Find the numerical value of $\frac{3x + 6}{2x - 4}$ for each value of x.

(a) $\dfrac{3x + 6}{2x - 4}$ for $x = 1$

$= \dfrac{3(1) + 6}{2(1) - 4}$ Let $x = 1$.

$= \dfrac{9}{-2}$

$= -\dfrac{9}{2}$ $\dfrac{a}{-b} = -\dfrac{a}{b}$

(b) $\dfrac{3x + 6}{2x - 4}$ for $x = 0$

$= \dfrac{3(0) + 6}{2(0) - 4}$ Let $x = 0$.

$= \dfrac{6}{-4}$

$= -\dfrac{3}{2}$ Lowest terms

(c) $\dfrac{3x + 6}{2x - 4}$ for $x = 2$

$= \dfrac{3(2) + 6}{2(2) - 4}$ Let $x = 2$.

$= \dfrac{12}{0}$ The expression is **undefined** for $x = 2$.

(d) $\dfrac{3x + 6}{2x - 4}$ for $x = -2$

$= \dfrac{3(-2) + 6}{2(-2) - 4}$ Let $x = -2$.

$= \dfrac{0}{-8}$

$= 0$ $\dfrac{0}{b} = 0$

NOW TRY

**NOW TRY
EXERCISE 1**

Find the numerical value of each expression for $x = -3$.

(a) $\dfrac{2x - 1}{x + 4}$ **(b)** $\dfrac{x + 3}{4}$

(c) $\dfrac{4}{x + 3}$

NOW TRY ANSWERS

1. **(a)** -7 **(b)** 0
   (c) The expression is undefined for $x = -3$.

NOTE   *The numerator of a rational expression may be any real number.* If the numerator equals 0 and the denominator does not equal 0, then the rational expression equals 0. See Example 1(d).

OBJECTIVE 2 Find the values of the variable for which a rational expression is undefined.

In the definition of a rational expression $\frac{P}{Q}$, Q cannot equal 0. *The denominator of a rational expression cannot equal 0 because division by 0 is undefined.*

For example, in the rational expression

$$\frac{8x^2}{x-3}, \leftarrow \text{Denominator cannot equal 0.}$$

the variable x can take on any real number value except 3. If x is 3, then the denominator becomes $3 - 3 = 0$, making the expression undefined. Thus, x cannot equal 3. We indicate this restriction by writing $x \neq 3$.

Determining When a Rational Expression Is Undefined

Step 1 Set the denominator of the rational expression equal to 0.

Step 2 Solve this equation.

Step 3 The solutions of the equation are the values that make the rational expression undefined. The variable *cannot* equal these values.

NOW TRY EXERCISE 2

Find any values of the variable for which each rational expression is undefined.

(a) $\dfrac{k-4}{2k-1}$

(b) $\dfrac{2x}{x^2 + 5x - 14}$

(c) $\dfrac{y+10}{y^2 + 10}$

EXAMPLE 2 Finding Values That Make Rational Expressions Undefined

Find any values of the variable for which each rational expression is undefined.

(a) $\dfrac{x+5}{3x+2}$ We must find any value of x that makes the *denominator* equal to 0 because division by 0 is undefined.

Step 1 Set the denominator equal to 0.

$$3x + 2 = 0$$

Step 2 Solve. $3x = -2$ Subtract 2.

$$x = -\frac{2}{3}$$ Divide by 3.

Step 3 The given expression is undefined for $-\frac{2}{3}$, so $x \neq -\frac{2}{3}$.

(b) $\dfrac{8x^2 + 1}{x - 3}$ The denominator $x - 3 = 0$ when x is 3. The given expression is undefined for 3, so $x \neq 3$.

(c) $\dfrac{9m^2}{m^2 - 5m + 6}$

$$m^2 - 5m + 6 = 0 \qquad \text{Set the denominator equal to 0.}$$
$$(m-2)(m-3) = 0 \qquad \text{Factor.}$$
$$m - 2 = 0 \quad \text{or} \quad m - 3 = 0 \qquad \text{Zero-factor property}$$
$$m = 2 \quad \text{or} \qquad m = 3 \qquad \text{Solve for } m.$$

The given expression is undefined for 2 and 3, so $m \neq 2$, $m \neq 3$.

(d) $\dfrac{2r}{r^2 + 1}$ This denominator will not equal 0 for any value of r, because r^2 is always greater than or equal to 0, and adding 1 makes the sum greater than or equal to 1. There are no values for which this expression is undefined.

NOW TRY

NOW TRY ANSWERS

2. **(a)** $k \neq \frac{1}{2}$ **(b)** $x \neq -7, x \neq 2$
 (c) It is never undefined.

OBJECTIVE 3 Write rational expressions in lowest terms.

A fraction such as $\frac{2}{3}$ is said to be in *lowest terms.*

Lowest Terms

A rational expression $\frac{P}{Q}$ (where $Q \neq 0$) is in **lowest terms** if the greatest common factor of its numerator and denominator is 1.

We use the **fundamental property of rational expressions** to write a rational expression in lowest terms.

Fundamental Property of Rational Expressions

If $\frac{P}{Q}$ (where $Q \neq 0$) is a rational expression and if K represents any polynomial (where $K \neq 0$), then the following holds true.

$$\frac{PK}{QK} = \frac{P}{Q}$$

This property is based on the identity property of multiplication.

$$\frac{PK}{QK} = \frac{P}{Q} \cdot \frac{K}{K} = \frac{P}{Q} \cdot 1 = \frac{P}{Q}$$

**NOW TRY
EXERCISE 3**
Write each rational expression
in lowest terms.

(a) $\dfrac{20}{48}$ **(b)** $\dfrac{21y^5}{7y^2}$

EXAMPLE 3 Writing in Lowest Terms

Write each rational expression in lowest terms.

(a) $\dfrac{30}{72}$ Rational number

Begin by factoring.

$$\frac{30}{72} = \frac{2 \cdot 3 \cdot 5}{2 \cdot 2 \cdot 2 \cdot 3 \cdot 3}$$

(b) $\dfrac{14k^2}{2k^3}$ Rational expression

Write k^2 as $k \cdot k$ and k^3 as $k \cdot k \cdot k$.

$$\frac{14k^2}{2k^3} = \frac{2 \cdot 7 \cdot k \cdot k}{2 \cdot k \cdot k \cdot k}$$

Group any factors common to the numerator and denominator.

$$= \frac{5 \cdot (2 \cdot 3)}{2 \cdot 2 \cdot 3 \cdot (2 \cdot 3)}$$

$$= \frac{7(2 \cdot k \cdot k)}{k(2 \cdot k \cdot k)}$$

Use the fundamental property.

$$= \frac{5}{2 \cdot 2 \cdot 3}$$

$$= \frac{5}{12}$$

$$= \frac{7}{k}$$

NOW TRY

Writing a Rational Expression in Lowest Terms

Step 1 **Factor** the numerator and denominator completely.

Step 2 **Use the fundamental property** to divide out any common factors.

NOW TRY ANSWERS
3. (a) $\frac{5}{12}$ **(b)** $3y^3$

 **NOW TRY**
EXERCISE 4

Write each rational expression in lowest terms.

(a) $\dfrac{3x + 15}{5x + 25}$

(b) $\dfrac{k^2 - 36}{k^2 + 8k + 12}$

(c) $\dfrac{xy + 5y - 2x - 10}{xy + 3x + 5y + 15}$

EXAMPLE 4   Writing in Lowest Terms

Write each rational expression in lowest terms.

(a) $\dfrac{3x - 12}{5x - 20}$

> $x \neq 4$ because the denominator is 0 for this value.

Step 1 $\quad = \dfrac{3(x - 4)}{5(x - 4)}$ — Factor.

Step 2 $\quad = \dfrac{3}{5}$ — $\dfrac{x-4}{x-4} = 1$; Fundamental property

The given expression is equal to $\frac{3}{5}$ for all values of x, where $x \neq 4$ (because the denominator of the original rational expression is 0 when x is 4).

(b) $\dfrac{2y^2 - 8}{2y + 4}$

> $y \neq -2$ because the denominator is 0 for this value.

Step 1 $\quad = \dfrac{2(y^2 - 4)}{2(y + 2)}$ — Factor.

$\quad = \dfrac{2(y + 2)(y - 2)}{2(y + 2)}$ — Factor the numerator completely. $y^2 - 4$ is a difference of squares.

Step 2 $\quad = y - 2$ — $\dfrac{2(y+2)}{2(y+2)} = 1$; Fundamental property

(c) $\dfrac{m^2 + 2m - 8}{2m^2 - m - 6}$

Step 1 $\quad = \dfrac{(m + 4)(m - 2)}{(2m + 3)(m - 2)}$

> $m \neq -\frac{3}{2}, m \neq 2$ Factor.

Step 2 $\quad = \dfrac{m + 4}{2m + 3}$ — $\dfrac{m-2}{m-2} = 1$; Fundamental property

From now on, we write statements of equality of rational expressions with the understanding that they apply only to real numbers that make neither denominator equal to 0.

(d) $\dfrac{4m + 12 - mp - 3p}{5m + mp + 15 + 3p}$ — The numerator and denominator each have four terms, so try to factor by grouping.

Step 1 $\quad = \dfrac{(4m + 12) + (-mp - 3p)}{(5m + mp) + (15 + 3p)}$ — Group terms in the numerator. Group terms in the denominator.

$\quad = \dfrac{4(m + 3) - p(m + 3)}{m(5 + p) + 3(5 + p)}$ — Factor each group.

$\quad = \dfrac{(m + 3)(4 - p)}{(5 + p)(m + 3)}$ — Factor out $m + 3$. Factor out $5 + p$.

Step 2 $\quad = \dfrac{4 - p}{5 + p}$ — $\dfrac{m+3}{m+3} = 1$; Fundamental property

> **! CAUTION**   *Rational expressions cannot be written in lowest terms until after the numerator and denominator have been factored. Only common factors can be divided out, not common terms.*

$$\frac{6x + 9}{4x + 6} = \frac{3(2x + 3)}{2(2x + 3)} = \frac{3}{2} \qquad \frac{6 + x}{4x} \leftarrow \text{Numerator cannot be factored.}$$

↑ Divide out the common factor.   Already in lowest terms

NOW TRY
EXERCISE 5
Write in lowest terms.

$$\frac{10 - a^2}{a^2 - 10}$$

EXAMPLE 5   Writing in Lowest Terms (Factors Are Opposites)

Write $\dfrac{x - y}{y - x}$ in lowest terms.

To find a common factor, the denominator $y - x$ can be factored as follows.

$$y - x \quad \boxed{\text{We are factoring out } -1, \text{ NOT multiplying by it.}}$$

$$= -1(-y + x) \qquad \text{Factor out } -1.$$

$$= -1(x - y) \qquad \text{Commutative property, } a + b = b + a$$

With this result in mind, we simplify as follows.

$$\frac{x - y}{y - x}$$

$$= \frac{1(x - y)}{-1(x - y)} \qquad y - x = -1(x - y) \text{ from above.}$$

$$= \frac{1}{-1} \qquad \text{Fundamental property}$$

$$= -1 \qquad \text{Lowest terms}$$

Alternatively, we could factor -1 from the numerator $x - y$.

$$\frac{x - y}{y - x} \qquad \text{Alternative solution}$$

$$= \frac{-1(-x + y)}{y - x} \qquad \text{Factor out } -1 \text{ in the numerator.}$$

$$= \frac{-1(y - x)}{y - x} \qquad \text{Commutative property}$$

$$= -1 \qquad \text{The result is the same.} \qquad \text{NOW TRY} \circlearrowright$$

> **! CAUTION**   Although *x* and *y* appear in both the numerator and denominator in **Example 5**, we cannot use the fundamental property right away because they are *terms*, not *factors*. *Terms are added, while factors are multiplied.*

NOW TRY ANSWER
5. -1

In **Example 5,** notice that $y - x$ is the **opposite** (or **additive inverse**) of $x - y$. A general rule for this situation follows.

Quotient of Opposites

If the numerator and the denominator of a rational expression are opposites, such as in $\dfrac{x-y}{y-x}$, then the rational expression is equal to -1.

Based on this result, the following are true.

Numerator and denominator are opposites. $\longrightarrow$ $\dfrac{q-7}{7-q} = -1$ and $\dfrac{-5a+2b}{5a-2b} = -1$

However, the following expression cannot be simplified further.

$$\dfrac{x-2}{x+2} \longleftarrow \text{ Numerator and denominator are } not \text{ opposites.}$$

NOW TRY EXERCISE 6

Write each rational expression in lowest terms.

(a) $\dfrac{p-4}{4-p}$ (b) $\dfrac{4m^2-n^2}{2n-4m}$

(c) $\dfrac{x+y}{x-y}$

EXAMPLE 6   Writing in Lowest Terms (Factors Are Opposites)

Write each rational expression in lowest terms.

(a) $\dfrac{2-m}{m-2}$

Because $2-m$ and $m-2$ are opposites, this expression equals -1.

(b) $\dfrac{4x^2-9}{6-4x}$

$= \dfrac{(2x+3)(2x-3)}{2(3-2x)}$ Factor the numerator and denominator.

$= \dfrac{(2x+3)(2x-3)}{2(-1)(2x-3)}$ Write $3-2x$ in the denominator as $-1(2x-3)$.

$= \dfrac{2x+3}{2(-1)}$ Fundamental property

$= \dfrac{2x+3}{-2}$ Multiply in the denominator.

$= -\dfrac{2x+3}{2}$ $\dfrac{a}{-b} = -\dfrac{a}{b}$

(c) $\dfrac{3+r}{3-r}$ $3-r$ is *not* the opposite of $3+r$.

This rational expression is already in lowest terms.   **NOW TRY**

OBJECTIVE 4 Recognize equivalent forms of rational expressions.

It is important in algebra to recognize equivalent forms of expressions. For example,

$$0.5, \quad \frac{1}{2}, \quad 50\%, \quad \text{and} \quad \frac{50}{100} \qquad \text{Equivalent expressions}$$

all represent the *same* real number. On a number line, the exact same point would apply to all four of them.

A similar situation exists with negative common fractions. The common fraction $-\dfrac{5}{6}$ can also be written $\dfrac{-5}{6}$ and $\dfrac{5}{-6}$, with the negative sign appearing in any of three different positions. All three forms represent the *same* rational number.

NOW TRY ANSWERS

6. (a) -1
   (b) $\dfrac{2m+n}{-2}$, or $-\dfrac{2m+n}{2}$
   (c) It is already in lowest terms.

Consider the following rational expression.

$$-\frac{2x+3}{2} \quad \text{Final result from \textbf{Example 6(b)}} \atop \text{in the form } -\frac{a}{b}$$

The $-$ sign representing the factor -1 is in front of the expression, aligned with the fraction bar. Although we usually give answers in this form, it is important to be able to recognize other equivalent forms of a rational expression. The factor -1 may instead be placed in the numerator or in the denominator.

Use parentheses.

$$\frac{-(2x+3)}{2} \quad \text{and} \quad \frac{2x+3}{-2} \qquad \text{Equivalent forms of } -\frac{2x+3}{2}$$

In the first of these two expressions, the distributive property can be applied.

Multiply *each* term in the binomial by -1.

$$\frac{-(2x+3)}{2} \quad \text{can also be written} \quad \frac{-2x-3}{2}.$$

> ⚠️ **CAUTION** $\frac{-2x+3}{2}$ is *not* an equivalent form of $\frac{-(2x+3)}{2}$. *Be careful to apply the distributive property correctly.*

NOW TRY
EXERCISE 7
Write four equivalent forms of the rational expression.

$$-\frac{4k-9}{k+3}$$

EXAMPLE 7   Writing Equivalent Forms of a Rational Expression

Write four equivalent forms of the rational expression.

$$-\frac{3x+2}{x-6}$$

If we apply the negative sign to the numerator, we obtain these equivalent forms.

$$①\rightarrow \frac{-(3x+2)}{x-6} \quad \text{and, by the distributive property,} \quad \frac{-3x-2}{x-6} \leftarrow②$$

If we apply the negative sign to the denominator, we obtain two more forms.

$$③\rightarrow \frac{3x+2}{-(x-6)} \quad \text{and, by distributing once again,} \quad \frac{3x+2}{-x+6} \leftarrow④$$

NOW TRY

NOW TRY ANSWER

7. $\dfrac{-(4k-9)}{k+3}$, $\dfrac{-4k+9}{k+3}$, $\dfrac{4k-9}{-(k+3)}$, $\dfrac{4k-9}{-k-3}$

> ⚠️ **CAUTION**   Recall that $-\frac{5}{6} \neq \frac{-5}{-6}$. Thus, in **Example 7**, it would be incorrect to distribute the negative sign in $-\frac{3x+2}{x-6}$ to *both* the numerator *and* the denominator. (Doing this would actually lead to the *opposite* of the original expression.)

6.1 Exercises

FOR EXTRA HELP   ▶ **MyLab Math**

▶ *Video solutions for select problems available in MyLab Math*

Concept Check   *Work each problem.*

1. Fill in each blank with the correct response: The rational expression $\frac{x+5}{x-3}$ is undefined when x is _____, so $x \neq$ _____. This rational expression is equal to 0 when $x =$ _____.

2. Which one of the following rational expressions can be simplified?

   A. $\dfrac{x^2+2}{x^2}$ B. $\dfrac{x^2+2}{2}$ C. $\dfrac{x^2+y^2}{y^2}$ D. $\dfrac{x^2-5x}{x}$

3. Which two of the following rational expressions equal -1?

    A. $\dfrac{2x + 3}{2x - 3}$ **B.** $\dfrac{2x - 3}{3 - 2x}$ **C.** $\dfrac{2x + 3}{3 + 2x}$ **D.** $\dfrac{2x + 3}{-2x - 3}$

4. Make the correct choice: $\dfrac{4 - r^2}{4 + r^2}$ (*is / is not*) equal to -1.

5. Which one of the following rational expressions is *not* equivalent to $\dfrac{x - 3}{4 - x}$?

    A. $\dfrac{3 - x}{x - 4}$ **B.** $\dfrac{x + 3}{4 + x}$ **C.** $-\dfrac{3 - x}{4 - x}$ **D.** $-\dfrac{x - 3}{x - 4}$

6. Make the correct choice: $\dfrac{5 + 2x}{3 - x}$ and $\dfrac{-5 - 2x}{x - 3}$ (*are / are not*) equivalent rational expressions.

7. Find the numerical value of the rational expression for $x = -3$.

$$\frac{x}{2x + 1}$$

$$= \frac{\overline{}}{2(\underline{}) + 1}\qquad \text{Let } x = -3.$$

$$= \frac{\overline{}}{\underline{} + 1}$$

$$= \underline{}$$

8. Find any values of the variable for which the rational expression is undefined.

$$\frac{x + 2}{x - 5}$$

Step 1 $\underline{} = 0$

Step 2 $x = \underline{}$

Step 3 The given expression is undefined for $\underline{}$. Thus, $x \,(= / \neq)\, 5.$

*Find the numerical value of each rational expression for (**a**) $x = 2$ and (**b**) $x = -3$. See Example 1.*

9. $\dfrac{3x + 1}{5x}$ 10. $\dfrac{5x - 2}{4x}$ 11. $\dfrac{x^2 - 4}{2x + 1}$ 12. $\dfrac{2x^2 - 4x}{3x - 1}$

13. $\dfrac{(-2x)^3}{3x + 9}$ 14. $\dfrac{(-3x)^2}{4x + 12}$ 15. $\dfrac{7 - 3x}{3x^2 - 7x + 2}$ 16. $\dfrac{5x + 2}{4x^2 - 5x - 6}$

17. $\dfrac{(x + 3)(x - 2)}{500x}$ 18. $\dfrac{(x - 2)(x + 3)}{1000x}$ 19. $\dfrac{x^2 - 4}{x^2 - 9}$ 20. $\dfrac{x^2 - 9}{x^2 - 4}$

Find any values of the variable for which each rational expression is undefined. Write answers with the symbol $\neq$. See Example 2.

21. $-\dfrac{5}{x}$ 22. $-\dfrac{2}{y}$ 23. $\dfrac{12}{5y}$ 24. $\dfrac{-7}{3z}$ 25. $\dfrac{x + 1}{x - 6}$ 26. $\dfrac{m - 2}{m - 5}$

27. $\dfrac{4x^2}{3x + 5}$ 28. $\dfrac{2x^3}{3x + 4}$ 29. $\dfrac{5m + 2}{m^2 + m - 6}$ 30. $\dfrac{2r - 5}{r^2 - 5r + 4}$

31. $\dfrac{x^2 + 3x}{4}$ 32. $\dfrac{x^2 - 4x}{6}$ 33. $\dfrac{3x - 1}{x^2 + 2}$ 34. $\dfrac{4q + 2}{q^2 + 9}$

Concept Check *Work each problem.*

35. Identify the two *terms* in the numerator and the two *terms* in the denominator of the rational expression $\dfrac{x^2 + 4x}{x + 4}$.

36. Describe the steps you would use to write the rational expression in **Exercise 35** in lowest terms. (*Hint:* It simplifies to x.)

Write each rational number in lowest terms. See Example 3(a).

37. $\dfrac{36}{84}$ 38. $\dfrac{16}{60}$ 39. $\dfrac{54}{198}$ 40. $\dfrac{48}{108}$

Write each rational expression in lowest terms. **See Examples 3(b) and 4.**

41. $\dfrac{18r^3}{6r}$

42. $\dfrac{27p^4}{3p}$

43. $\dfrac{4(y-2)}{10(y-2)}$

44. $\dfrac{15(m-1)}{9(m-1)}$

45. $\dfrac{(x+1)(x-1)}{(x+1)^2}$

46. $\dfrac{(t+5)(t-3)}{(t+5)^2}$

47. $\dfrac{7m+14}{5m+10}$

48. $\dfrac{16x+8}{14x+7}$

49. $\dfrac{6m-18}{7m-21}$

50. $\dfrac{5r+20}{3r+12}$

51. $\dfrac{m^2-n^2}{m+n}$

52. $\dfrac{a^2-b^2}{a-b}$

53. $\dfrac{2t+6}{t^2-9}$

54. $\dfrac{5s-25}{s^2-25}$

55. $\dfrac{12m^2-3}{8m-4}$

56. $\dfrac{20p^2-45}{6p-9}$

57. $\dfrac{3m^2-3m}{5m-5}$

58. $\dfrac{6t^2-6t}{5t-5}$

59. $\dfrac{9r^2-4s^2}{9r+6s}$

60. $\dfrac{16x^2-9y^2}{12x-9y}$

61. $\dfrac{x-6}{x^2-36}$

62. $\dfrac{x-8}{x^2-64}$

63. $\dfrac{x^2-9}{x^2-6x+9}$

64. $\dfrac{x^2-16}{x^2-8x+16}$

65. $\dfrac{13x^2-39x^3}{7x-21x^2}$

66. $\dfrac{30x^3-15x^5}{22x^2-11x^4}$

67. $\dfrac{5k^2-13k-6}{5k+2}$

68. $\dfrac{7t^2-31t-20}{7t+4}$

69. $\dfrac{x^2+2x-15}{x^2+6x+5}$

70. $\dfrac{y^2-5y-14}{y^2+y-2}$

71. $\dfrac{2x^2-3x-5}{2x^2-7x+5}$

72. $\dfrac{3x^2+8x+4}{3x^2-4x-4}$

73. $\dfrac{3x^3+13x^2+14x}{3x^3-5x^2-28x}$

74. $\dfrac{2x^3+7x^2-30x}{2x^3-11x^2+15x}$

75. $\dfrac{-3t+6t^2-3t^3}{7t^2-14t^3+7t^4}$

76. $\dfrac{-20r-20r^2-5r^3}{24r^2+24r^3+6r^4}$

77. $\dfrac{zw+4z-3w-12}{zw+4z+5w+20}$

78. $\dfrac{km+4k-4m-16}{km+4k+5m+20}$

79. $\dfrac{pr+qr+ps+qs}{pr+qr-ps-qs}$

80. $\dfrac{wt+ws+xt+xs}{wt-xs-xt+ws}$

81. $\dfrac{ac-ad+bc-bd}{ac-ad-bc+bd}$

82. $\dfrac{ac-bc-ad+bd}{ac-ad-bd+bc}$

83. $\dfrac{m^2-n^2-4m-4n}{2m-2n-8}$

84. $\dfrac{x^2-y^2-7y-7x}{3x-3y-21}$

85. $\dfrac{x^2y+y+x^2z+z}{xy+xz}$

86. $\dfrac{y^2k+pk-y^2z-pz}{yk-yz}$

Extending Skills   *These exercises involve factoring sums and differences of cubes. Write each rational expression in lowest terms.*

87. $\dfrac{1+p^3}{1+p}$

88. $\dfrac{8+x^3}{2+x}$

89. $\dfrac{x^3-27}{x-3}$

90. $\dfrac{r^3-1000}{r-10}$

91. $\dfrac{b^3-a^3}{a^2-b^2}$

92. $\dfrac{8y^3-27z^3}{9z^2-4y^2}$

93. $\dfrac{k^3+8}{k^2-4}$

94. $\dfrac{r^3+27}{r^2-9}$

95. $\dfrac{z^3+27}{z^3-3z^2+9z}$

96. $\dfrac{t^3 + 64}{t^3 - 4t^2 + 16t}$ **97.** $\dfrac{1 - 8r^3}{8r^2 + 4r + 2}$ **98.** $\dfrac{8 - 27x^3}{27x^2 + 18x + 12}$

*Write each rational expression in lowest terms. **See Examples 5 and 6.***

99. $\dfrac{6 - t}{t - 6}$ **100.** $\dfrac{2 - k}{k - 2}$ **101.** $\dfrac{m^2 - 1}{1 - m}$

102. $\dfrac{x^2 - 100}{10 - x}$ **103.** $\dfrac{q^2 - 4q}{4q - q^2}$ **104.** $\dfrac{z^2 - 5z}{5z - z^2}$

105. $\dfrac{p + 6}{p - 6}$ **106.** $\dfrac{5 - x}{5 + x}$ **107.** $\dfrac{-2m + 2n}{m - n}$

108. $\dfrac{-5p + 5q}{p - q}$ **109.** $\dfrac{3x - x^2}{x - 3}$ **110.** $\dfrac{6y - 2y^2}{y - 3}$

*Write four equivalent forms for each rational expression. **See Example 7.***

111. $-\dfrac{x + 4}{x - 3}$ **112.** $-\dfrac{x + 6}{x - 1}$ **113.** $-\dfrac{2x - 3}{x + 3}$

114. $-\dfrac{5x - 6}{x + 4}$ **115.** $-\dfrac{3x - 1}{5x - 6}$ **116.** $-\dfrac{2x - 9}{7x - 1}$

Solve each problem.

117. The average number of vehicles waiting in line to enter a sports arena parking area is approximated by the rational expression

$$\frac{x^2}{2(1 - x)},$$

where x is a quantity between 0 and 1 known as the **traffic intensity.** (Data from Mannering, F., and W. Kilareski, *Principles of Highway Engineering and Traffic Control,* John Wiley and Sons.)

     To the nearest tenth, find the average number of vehicles waiting if the traffic intensity is the given number.

(a) 0.1   **(b)** 0.8   **(c)** 0.9

(d) What happens to the number of vehicles waiting as traffic intensity increases?

118. The percent of deaths caused by smoking is modeled by the rational expression

$$\frac{x - 1}{x},$$

where x is the number of times a smoker is more likely than a nonsmoker to die of lung cancer. This is called the **incidence rate.** (Data from Walker, A., *Observation and Inference: An Introduction to the Methods of Epidemiology,* Epidemiology Resources Inc.) For example, $x = 10$ means that a smoker is 10 times more likely than a nonsmoker to die of lung cancer.

     Find the percent of deaths if the incidence rate is the given number.

(a) 5   **(b)** 10   **(c)** 20   **(d)** Can the incidence rate equal 0? Explain.

119. The area of the rectangle is represented by

$$x^4 + 10x^2 + 21.$$

Find the polynomial that represents the width of the rectangle.

$\left(\textit{Hint: } \text{Use } W = \dfrac{A}{L}.\right)$

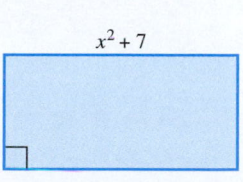

$x^2 + 7$

120. The volume of the box is represented by

$$(x^2 + 8x + 15)(x + 4).$$

Find the polynomial that represents the area of the bottom of the box.

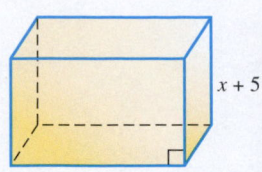

$x + 5$

RELATING CONCEPTS For Individual or Group Work (Exercises 121–124)

Earlier we used long division to find a quotient of two polynomials. We obtain the same quotient by expressing a division problem as a rational expression (fraction) and writing this rational expression in lowest terms, as shown below.

$$
\begin{array}{r}
x + 4 \\
2x - 3\overline{)2x^2 + 5x - 12} \\
\underline{2x^2 - 3x} \\
8x - 12 \\
\underline{8x - 12} \\
0
\end{array}
$$

$$\dfrac{2x^2 + 5x - 12}{2x - 3}$$

$$= \dfrac{(2x - 3)(x + 4)}{2x - 3}$$ Factor.

$$= x + 4$$ Fundamental property

Perform the long division. Then simplify the rational expression to show that the result is the same.

121. $4x + 7\overline{)8x^2 + 26x + 21}$

and $\dfrac{8x^2 + 26x + 21}{4x + 7}$

122. $6x + 5\overline{)12x^2 + 16x + 5}$

and $\dfrac{12x^2 + 16x + 5}{6x + 5}$

123. $x + 1\overline{)x^3 + x^2 + x + 1}$

and $\dfrac{x^3 + x^2 + x + 1}{x + 1}$

124. $x + 1\overline{)x^3 + x^2 + 2x + 2}$

and $\dfrac{x^3 + x^2 + 2x + 2}{x + 1}$

6.2 Multiplying and Dividing Rational Expressions

OBJECTIVES

1 Multiply rational expressions.
2 Find reciprocals of rational expressions.
3 Divide rational expressions.

OBJECTIVE 1 Multiply rational expressions.

The product of two fractions is found by multiplying the numerators and multiplying the denominators. Rational expressions are multiplied in the same way.

Multiplying Rational Expressions

The product of the rational expressions $\dfrac{P}{Q}$ and $\dfrac{R}{S}$ is defined as follows.

$$\frac{P}{Q} \cdot \frac{R}{S} = \frac{PR}{QS}$$

That is, to multiply rational expressions, multiply the numerators and multiply the denominators.

Example: $\dfrac{4}{7} \cdot \dfrac{2}{5} = \dfrac{4 \cdot 2}{7 \cdot 5} = \dfrac{8}{35}$

VOCABULARY

☐ reciprocals
(multiplicative inverses)

 **NOW TRY
EXERCISE 1**

Multiply. Write each answer
in lowest terms.

(a) $\dfrac{7}{18} \cdot \dfrac{9}{14}$ (b) $\dfrac{4k^2}{7} \cdot \dfrac{14}{11k}$

EXAMPLE 1   Multiplying Rational Expressions

Multiply. Write each answer in lowest terms.

(a) $\dfrac{3}{10} \cdot \dfrac{5}{9}$ Rational numbers   (b) $\dfrac{6}{x} \cdot \dfrac{x^2}{12}$ Rational expressions

Indicate the product of the numerators and the product of the denominators.

$$= \frac{3 \cdot 5}{10 \cdot 9} \qquad\qquad\qquad = \frac{6 \cdot x^2}{x \cdot 12}$$

Leave the products in factored form. Factor the numerator and denominator to further
identify any common factors. Then use the fundamental property to divide out any
common factors and write each product in lowest terms.

$$= \frac{3 \cdot 5}{2 \cdot 5 \cdot 3 \cdot 3} \qquad\qquad = \frac{6 \cdot x \cdot x}{x \cdot 2 \cdot 6}$$

$$= \frac{1}{6} \qquad \boxed{\text{Remember to write 1 in the numerator.}} \qquad = \frac{x}{2}$$

NOW TRY

> **NOTE**   It is also possible to divide out common factors in the numerator and denominator
> *before* multiplying the rational expressions. Consider the following.
>
> $$\frac{3}{10} \cdot \frac{5}{9} \qquad\qquad \text{Example 1(a)}$$
>
> $$= \frac{3}{5 \cdot 2} \cdot \frac{5}{3 \cdot 3} \qquad \text{Identify the common factors.}$$
>
> $$= \frac{1}{2 \cdot 3} \qquad\qquad \begin{array}{l}\text{Divide out the common factors.}\\ \text{Insert a factor of 1 in the numerator.}\end{array}$$
>
> $$= \frac{1}{6} \qquad\qquad \text{Multiply.}$$

 **NOW TRY
EXERCISE 2**

Multiply. Write the answer in
lowest terms.

$$\frac{m-3}{3m} \cdot \frac{9m^2}{8(m-3)^2}$$

EXAMPLE 2   Multiplying Rational Expressions

Multiply. Write the answer in lowest terms.

$$\frac{x+y}{2x} \cdot \frac{x^2}{(x+y)^2}$$

$$\boxed{\begin{array}{c}\text{Use parentheses}\\ \text{here around}\\ x + y.\end{array}} = \frac{(x+y)x^2}{2x(x+y)^2} \qquad \begin{array}{l}\text{Multiply numerators.}\\ \text{Multiply denominators.}\end{array}$$

$$= \frac{(x+y)x \cdot x}{2x(x+y)(x+y)} \qquad \text{Factor. Identify the common factors.}$$

NOW TRY ANSWERS

1. (a) $\frac{1}{4}$ (b) $\frac{8k}{11}$

2. $\frac{3m}{8(m-3)}$

$$= \frac{x}{2(x+y)} \qquad\qquad \frac{(x+y)x}{x(x+y)} = 1; \text{ Lowest terms}$$

NOW TRY

**NOW TRY
EXERCISE 3**
Multiply. Write the answer in lowest terms.

$$\frac{y^2 - 3y - 28}{y^2 - 9y + 14} \cdot \frac{y^2 - 7y + 10}{y^2 + 4y}$$

EXAMPLE 3   Multiplying Rational Expressions

Multiply. Write the answer in lowest terms.

$$\frac{x^2 + 3x}{x^2 - 3x - 4} \cdot \frac{x^2 - 5x + 4}{x^2 + 2x - 3}$$

$$= \frac{(x^2 + 3x)(x^2 - 5x + 4)}{(x^2 - 3x - 4)(x^2 + 2x - 3)} \qquad \text{Definition of multiplication}$$

$$= \frac{x(x + 3)(x - 4)(x - 1)}{(x - 4)(x + 1)(x + 3)(x - 1)} \qquad \text{Factor.}$$

$$= \frac{x}{x + 1} \qquad \begin{array}{l}\text{Divide out the common factors.}\\ \text{The result is in lowest terms.}\end{array}$$

The quotients $\frac{x + 3}{x + 3}, \frac{x - 4}{x - 4}$, and $\frac{x - 1}{x - 1}$ all equal 1, justifying the final product $\frac{x}{x + 1}$.

NOW TRY

OBJECTIVE 2   Find reciprocals of rational expressions.

If the product of two rational expressions is 1, the rational expressions are **reciprocals** (or **multiplicative inverses**) of each other. As with arithmetic fractions, the reciprocal of a rational expression is found by interchanging the numerator and denominator.

Examples: $\frac{2}{3}$ has reciprocal $\frac{3}{2}$. $\frac{2x - 1}{x - 5}$ has reciprocal $\frac{x - 5}{2x - 1}$.

**NOW TRY
EXERCISE 4**
Find the reciprocal of each rational expression.

(a) $\frac{5}{8}$ **(b)** $\frac{6b^5}{3r^2b}$

(c) $\frac{t^2 - 4t}{t^2 + 2t - 3}$

EXAMPLE 4   Finding Reciprocals of Rational Expressions

Find the reciprocal of each rational expression.

(a) $\frac{4p^3}{9q}$ has reciprocal $\frac{9q}{4p^3}$.   Interchange the numerator and denominator.

(b) $\frac{k^2 - 9}{k^2 - k - 20}$ has reciprocal $\frac{k^2 - k - 20}{k^2 - 9}$.   Reciprocals have product 1.

NOW TRY

OBJECTIVE 3   Divide rational expressions.

Suppose we have $\frac{7}{8}$ gal of milk and want to find how many quarts we have. Because 1 qt is $\frac{1}{4}$ gal, we ask, *"How many $\frac{1}{4}$s are there in $\frac{7}{8}$?"* This would be interpreted as follows.

$$\frac{7}{8} \div \frac{1}{4}, \quad \text{or} \quad \frac{\frac{7}{8}}{\frac{1}{4}} \leftarrow \text{The fraction bar means division.}$$

The fundamental property of rational expressions discussed earlier can be applied to rational number values of P, Q, and K.

$$\frac{P}{Q} = \frac{P \cdot K}{Q \cdot K} = \frac{\frac{7}{8} \cdot \frac{4}{1}}{\frac{1}{4} \cdot \frac{4}{1}} = \frac{\frac{7}{8} \cdot \frac{4}{1}}{1} = \frac{7}{8} \cdot \frac{4}{1} \qquad \begin{array}{l}\text{Let } P = \frac{7}{8}, Q = \frac{1}{4}, \text{ and } K = \frac{4}{1}.\\ (K \text{ is the reciprocal of } Q.)\end{array}$$

NOW TRY ANSWERS
3. $\frac{y - 5}{y}$

4. (a) $\frac{8}{5}$ (b) $\frac{3r^2b}{6b^5}$ (c) $\frac{t^2 + 2t - 3}{t^2 - 4t}$

Therefore, to divide $\frac{7}{8}$ by $\frac{1}{4}$, we multiply $\frac{7}{8}$ by the reciprocal of $\frac{1}{4}$, namely $\frac{4}{1}$ (or 4). Because $\frac{7}{8} \cdot \frac{4}{1} = \frac{7}{2}$, there are $\frac{7}{2}$ qt, or $3\frac{1}{2}$ qt, in $\frac{7}{8}$ gal.

Division of rational expressions is defined in the same way.

Dividing Rational Expressions

If $\frac{P}{Q}$ and $\frac{R}{S}$ are any two rational expressions where $\frac{R}{S} \neq 0$, then their quotient is defined as follows.

$$\frac{P}{Q} \div \frac{R}{S} = \frac{P}{Q} \cdot \frac{S}{R} = \frac{PS}{QR}$$

That is, to divide one rational expression by another rational expression, multiply the first expression (dividend) by the reciprocal of the second (divisor).

Example: $\quad \frac{2}{3} \div \frac{7}{10} = \frac{2}{3} \cdot \frac{10}{7} = \frac{2 \cdot 10}{3 \cdot 7} = \frac{20}{21}$

NOW TRY EXERCISE 5

Divide. Write each answer in lowest terms.

(a) $\dfrac{3}{10} \div \dfrac{11}{20}$

(b) $\dfrac{2x - 5}{3x^2} \div \dfrac{2x - 5}{12x}$

EXAMPLE 5 Dividing Rational Expressions

Divide. Write each answer in lowest terms.

(a) $\dfrac{5}{8} \div \dfrac{7}{16}$ Rational numbers

(b) $\dfrac{2y}{y + 3} \div \dfrac{4y}{y + 5}$ Rational expressions

Multiply the dividend by the reciprocal of the divisor.

$= \dfrac{5}{8} \cdot \dfrac{16}{7}$ ← Reciprocal of $\frac{7}{16}$

$= \dfrac{5 \cdot 16}{8 \cdot 7}$ Multiply.

$= \dfrac{5 \cdot 8 \cdot 2}{8 \cdot 7}$ Factor 16.

$= \dfrac{10}{7}$ Lowest terms

$= \dfrac{2y}{y + 3} \cdot \dfrac{y + 5}{4y}$ ← Reciprocal of $\frac{4y}{y + 5}$

$= \dfrac{2y(y + 5)}{(y + 3)(4y)}$ Multiply.

$= \dfrac{2y(y + 5)}{(y + 3)(2 \cdot 2y)}$ Factor.

$= \dfrac{y + 5}{2(y + 3)}$ Lowest terms

NOW TRY

NOW TRY EXERCISE 6

Divide. Write the answer in lowest terms.

$$\frac{(3k)^3}{2j^4} \div \frac{9k^2}{6j}$$

EXAMPLE 6 Dividing Rational Expressions

Divide. Write the answer in lowest terms.

$$\frac{(3m)^2}{(2p)^3} \div \frac{6m^3}{16p^2}$$

$= \dfrac{(3m)^2}{(2p)^3} \cdot \dfrac{16p^2}{6m^3}$ Multiply by the reciprocal of the divisor.

$(3m)^2 = 3^2m^2$; $(2p)^3 = 2^3p^3$
$= \dfrac{9m^2}{8p^3} \cdot \dfrac{16p^2}{6m^3}$ Power rule for exponents, $(ab)^2 = a^2b^2$

$= \dfrac{9 \cdot 16m^2p^2}{8 \cdot 6p^3m^3}$ Multiply numerators.
Multiply denominators.

$= \dfrac{3 \cdot 3 \cdot 8 \cdot 2 \cdot m^2 \cdot p^2}{8 \cdot 3 \cdot 2 \cdot p^2 \cdot p \cdot m^2 \cdot m}$ Factor.

$= \dfrac{3}{pm}, \quad \text{or} \quad \dfrac{3}{mp}$ Lowest terms; Either form is correct.

NOW TRY

NOW TRY ANSWERS

5. (a) $\dfrac{6}{11}$ (b) $\dfrac{4}{x}$

6. $\dfrac{9k}{j^3}$

NOW TRY
EXERCISE 7
Divide. Write the answer in lowest terms.

$$\frac{\dfrac{(t+2)(t-5)}{-4t}}{\dfrac{t^2-25}{(t+5)(t+2)}}$$

EXAMPLE 7 Dividing Rational Expressions

Divide. Write the answer in lowest terms.

$$\frac{x^2-4}{(x+3)(x-2)} \div \frac{(x+2)(x+3)}{-2x}$$

$$= \frac{x^2-4}{(x+3)(x-2)} \cdot \frac{-2x}{(x+2)(x+3)} \qquad \text{Multiply by the reciprocal of the divisor.}$$

$$= \frac{-2x(x^2-4)}{(x+3)(x-2)(x+2)(x+3)} \qquad \begin{array}{l}\text{Multiply numerators.}\\ \text{Multiply denominators.}\end{array}$$

$$= \frac{-2x(x+2)(x-2)}{(x+3)(x-2)(x+2)(x+3)} \qquad \text{Factor the numerator.}$$

$$= \frac{-2x}{(x+3)^2} \qquad \begin{array}{l}\text{Divide out the common factors;}\\ a \cdot a = a^2\end{array}$$

$$= -\frac{2x}{(x+3)^2} \qquad \frac{-a}{b} = -\frac{a}{b}; \text{ Lowest terms}$$

NOW TRY

NOW TRY
EXERCISE 8
Divide. Write the answer in lowest terms.

$$\frac{7-x}{2x+6} \div \frac{x^2-49}{x^2+6x+9}$$

EXAMPLE 8 Dividing Rational Expressions (Factors Are Opposites)

Divide. Write the answer in lowest terms.

$$\frac{m^2-4}{m^2-1} \div \frac{2m^2+4m}{1-m}$$

$$= \frac{m^2-4}{m^2-1} \cdot \frac{1-m}{2m^2+4m} \qquad \begin{array}{l}\text{Multiply by the reciprocal of the divisor.}\end{array}$$

$$= \frac{(m^2-4)(1-m)}{(m^2-1)(2m^2+4m)} \qquad \begin{array}{l}\text{Multiply numerators}\\ \text{Multiply denominators.}\end{array}$$

$$= \frac{(m+2)(m-2)(1-m)}{(m+1)(m-1)(2m)(m+2)} \qquad \begin{array}{l}\text{Factor. } 1-m \text{ and } m-1 \text{ are opposites.}\end{array}$$

$$= \frac{-1(m-2)}{2m(m+1)} \qquad \begin{array}{l}\text{Divide out the common factors.}\\ \text{Recall that } \frac{1-m}{m-1} = -1.\end{array}$$

$$= \frac{-m+2}{2m(m+1)} \qquad \text{Distribute } -1 \text{ in the numerator.}$$

$$= \frac{2-m}{2m(m+1)} \qquad \text{Rewrite } -m+2 \text{ as } 2-m.$$

NOW TRY

NOW TRY ANSWERS

7. $-\dfrac{(t+2)^2}{4t}$

8. $-\dfrac{x+3}{2(x+7)}$

Multiplying or Dividing Rational Expressions

Step 1 **Note the operation.** If the operation is division, use the definition of division to rewrite it as multiplication.

Step 2 **Multiply** numerators and multiply denominators.

Step 3 **Factor** all numerators and denominators completely.

Step 4 **Write in lowest terms** using the fundamental property.

Note: Steps 2 and 3 may be interchanged based on personal preference.

6.2 Exercises

FOR EXTRA HELP   **MyLab Math**

▶ *Video solutions for select problems available in MyLab Math*

1. Concept Check Match each multiplication problem in Column I with the correct product in Column II.

I		II
(a) $\dfrac{5x^3}{10x^4} \cdot \dfrac{10x^7}{4x}$	**A.** $\dfrac{4}{5x^5}$	
(b) $\dfrac{10x^4}{5x^3} \cdot \dfrac{10x^7}{4x}$	**B.** $\dfrac{5x^5}{4}$	
(c) $\dfrac{5x^3}{10x^4} \cdot \dfrac{4x}{10x^7}$	**C.** $\dfrac{1}{5x^7}$	
(d) $\dfrac{10x^4}{5x^3} \cdot \dfrac{4x}{10x^7}$	**D.** $5x^7$	

2. Concept Check Match each division problem in Column I with the correct quotient in Column II.

I		II
(a) $\dfrac{5x^3}{10x^4} \div \dfrac{10x^7}{4x}$	**A.** $\dfrac{5x^5}{4}$	
(b) $\dfrac{10x^4}{5x^3} \div \dfrac{10x^7}{4x}$	**B.** $5x^7$	
(c) $\dfrac{5x^3}{10x^4} \div \dfrac{4x}{10x^7}$	**C.** $\dfrac{4}{5x^5}$	
(d) $\dfrac{10x^4}{5x^3} \div \dfrac{4x}{10x^7}$	**D.** $\dfrac{1}{5x^7}$	

Multiply. Write each answer in lowest terms. See Examples 1 and 2.

3. $\dfrac{4}{9} \cdot \dfrac{15}{16}$

4. $\dfrac{10}{21} \cdot \dfrac{3}{5}$

5. $\dfrac{15a^2}{14} \cdot \dfrac{7}{5a}$

6. $\dfrac{21b^6}{18} \cdot \dfrac{9}{7b^4}$

7. $\dfrac{16y^4}{18y^5} \cdot \dfrac{15y^5}{y^2}$

8. $\dfrac{12m^5}{2m^2} \cdot \dfrac{6m^6}{28m^3}$

9. $\dfrac{2(c+d)}{3} \cdot \dfrac{18}{6(c+d)^2}$

10. $\dfrac{4(y-2)}{x} \cdot \dfrac{3x}{6(y-2)^2}$

11. $\dfrac{(x-y)^2}{2} \cdot \dfrac{24}{3(x-y)}$

12. $\dfrac{(a+b)^2}{5} \cdot \dfrac{30}{2(a+b)}$

13. $\dfrac{t-4}{8} \cdot \dfrac{4t^2}{t-4}$

14. $\dfrac{z+9}{12} \cdot \dfrac{3z^2}{z+9}$

15. $\dfrac{3x}{x+3} \cdot \dfrac{(x+3)^2}{6x^2}$

16. $\dfrac{(t-2)^2}{4t^2} \cdot \dfrac{2t}{t-2}$

Find the reciprocal of each rational expression. See Example 4.

17. $\dfrac{3p^3}{16q}$

18. $\dfrac{6x^4}{9y^2}$

19. $\dfrac{r^2+rp}{7}$

20. $\dfrac{16}{9a^2+36a}$

21. $\dfrac{z^2+7z+12}{z^2-9}$

22. $\dfrac{p^2-4p+3}{p^2-3p}$

Concept Check *Multiply or divide. Write each answer in lowest terms.*

23. $\dfrac{5x-10}{6} \cdot \dfrac{9}{10x-20}$

$= \dfrac{5(\underline{\quad})}{6} \cdot \dfrac{3 \cdot \underline{\quad}}{10(\underline{\quad})}$

$= \dfrac{5(x-2) \cdot 3 \cdot 3}{2 \cdot 3 \cdot 2 \cdot \underline{\quad} \cdot (x-2)}$

$= \underline{\quad}$

24. $\dfrac{6x-4}{3} \div \dfrac{15x-10}{9}$

$= \dfrac{6x-4}{3} \cdot \dfrac{\underline{\quad}}{\underline{\quad}}$

$= \dfrac{2(\underline{\quad})}{3} \cdot \dfrac{9}{5(\underline{\quad})}$

$= \dfrac{2(3x-2) \cdot 3 \cdot 3}{\underline{\quad} \cdot 5(3x-2)}$

$= \underline{\quad}$

Divide. Write each answer in lowest terms. See Examples 5 and 6.

25. $\dfrac{4}{5} \div \dfrac{13}{20}$

26. $\dfrac{7}{8} \div \dfrac{3}{4}$

27. $\dfrac{9z^4}{3z^5} \div \dfrac{3z^2}{5z^3}$

28. $\dfrac{35x^8}{7x^9} \div \dfrac{5x^5}{9x^6}$

29. $\dfrac{4t^4}{2t^5} \div \dfrac{(2t)^3}{-6}$

30. $\dfrac{-12a^6}{3a^2} \div \dfrac{(2a)^3}{27a}$

31. $\dfrac{3}{2y-6} \div \dfrac{6}{y-3}$

32. $\dfrac{4m+16}{10} \div \dfrac{3m+12}{18}$

33. $\dfrac{7t+7}{-6} \div \dfrac{4t+4}{15}$

34. $\dfrac{8z-16}{-20} \div \dfrac{3z-6}{40}$

35. $\dfrac{2x}{x-1} \div \dfrac{x^2}{x+2}$

36. $\dfrac{y^2}{y+1} \div \dfrac{3y}{y-3}$

37. $\dfrac{(x-3)^2}{6x} \div \dfrac{x-3}{x^2}$

38. $\dfrac{2a}{a+4} \div \dfrac{a^2}{(a+4)^2}$

39. $\dfrac{5x^3}{x^2-16} \div \dfrac{x^5}{(x-4)^2}$

40. $\dfrac{8x^4}{x^2-25} \div \dfrac{x^7}{(x-5)^2}$

41. $\dfrac{-4t^3}{t^2-1} \div \dfrac{t^2}{(t+1)^2}$

42. $\dfrac{-8x^5}{x^2-4} \div \dfrac{x^3}{(x+2)^2}$

Multiply or divide. Write each answer in lowest terms. **See Examples 3, 7, and 8.**

43. $\dfrac{5x-15}{3x+9} \cdot \dfrac{4x+12}{6x-18}$

44. $\dfrac{8r+16}{24r-24} \cdot \dfrac{6r-6}{3r+6}$

45. $\dfrac{2-t}{8} \div \dfrac{t-2}{6}$

46. $\dfrac{m-2}{4} \div \dfrac{2-m}{6}$

47. $\dfrac{27-3z}{4} \cdot \dfrac{12}{2z-18}$

48. $\dfrac{35-5x}{6} \cdot \dfrac{12}{3x-21}$

49. $\dfrac{p^2+4p-5}{p^2+7p+10} \div \dfrac{p-1}{p+4}$

50. $\dfrac{z^2-3z+2}{z^2+4z+3} \div \dfrac{z-1}{z+1}$

51. $\dfrac{m^2-4}{16-8m} \div \dfrac{m+2}{8}$

52. $\dfrac{r^2-36}{54-9r} \div \dfrac{r+6}{9}$

53. $\dfrac{m^2-4}{16-8m} \div \dfrac{m^2+3m+2}{8m+16}$

54. $\dfrac{t^2-49}{42-6t} \div \dfrac{t^2+10t+21}{6t+42}$

55. $\dfrac{2x^2-7x+3}{x-3} \cdot \dfrac{x+2}{x-1}$

56. $\dfrac{3x^2-5x-2}{x-2} \cdot \dfrac{x-3}{x+1}$

57. $\dfrac{2k^2-k-1}{2k^2+5k+3} \div \dfrac{4k^2-1}{2k^2+k-3}$

58. $\dfrac{3t^2-4t-4}{3t^2+10t+8} \div \dfrac{9t^2+21t+10}{3t^2-t-10}$

59. $\dfrac{2k^2+3k-2}{6k^2-7k+2} \cdot \dfrac{4k^2-5k+1}{k^2+k-2}$

60. $\dfrac{2m^2-5m-12}{m^2-10m+24} \cdot \dfrac{m^2-9m+18}{4m^2-9}$

61. $\dfrac{m^2+2mp-3p^2}{m^2-3mp+2p^2} \div \dfrac{m^2+4mp+3p^2}{m^2+2mp-8p^2}$

62. $\dfrac{x^2-2xy-3y^2}{x^2+xy-30y^2} \div \dfrac{x^2+xy-12y^2}{x^2-xy-20y^2}$

63. $\dfrac{m^2+3m+2}{m^2+5m+4} \cdot \dfrac{m^2+10m+24}{m^2+5m+6}$

64. $\dfrac{z^2-z-6}{z^2-2z-8} \cdot \dfrac{z^2+7z+12}{z^2-9}$

65. $\dfrac{y^2+y-2}{y^2+3y-4} \div \dfrac{y+2}{y+3}$

66. $\dfrac{r^2+r-6}{r^2+4r-12} \div \dfrac{r+3}{r-1}$

67. $\dfrac{2m^2+7m+3}{m^2-9} \cdot \dfrac{m^2-3m}{2m^2+11m+5}$

68. $\dfrac{6s^2+17s+10}{s^2-4} \cdot \dfrac{s^2-2s}{6s^2+29s+20}$

69. $\dfrac{r^2+rs-12s^2}{r^2-rs-20s^2} \div \dfrac{r^2-2rs-3s^2}{r^2+rs-30s^2}$

70. $\dfrac{m^2+8mn+7n^2}{m^2+mn-42n^2} \div \dfrac{m^23mn-4n^2}{m^2-mn-30n^2}$

71. $\dfrac{(q-3)^4(q+2)}{q^2+3q+2} \div \dfrac{q^2-6q+9}{q^2+4q+4}$

72. $\dfrac{(x+4)^3(x-3)}{x^2-9} \div \dfrac{x^2+8x+16}{x^2+6x+9}$

Extending Skills *These exercises involve grouping symbols, factoring by grouping, and factoring sums and differences of cubes. Multiply or divide as indicated. Write each answer in lowest terms.*

73. $\dfrac{3a - 3b - a^2 + b^2}{4a^2 - 4ab + b^2} \cdot \dfrac{4a^2 - b^2}{2a^2 - ab - b^2}$

74. $\dfrac{4r^2 - t^2 + 10r - 5t}{2r^2 + rt + 5r} \cdot \dfrac{4r^3 + 4r^2t + rt^2}{2r + t}$

75. $\dfrac{-x^3 - y^3}{x^2 - 2xy + y^2} \div \dfrac{3y^2 - 3xy}{x^2 - y^2}$

76. $\dfrac{b^3 - 8a^3}{4a^3 + 4a^2b + ab^2} \div \dfrac{4a^2 + 2ab + b^2}{-a^3 - ab^3}$

77. $\dfrac{x + 5}{x + 10} \div \left(\dfrac{x^2 + 10x + 25}{x^2 + 10x} \cdot \dfrac{10x}{x^2 + 15x + 50} \right)$

78. $\dfrac{m - 8}{m - 4} \div \left(\dfrac{m^2 - 12m + 32}{8m} \cdot \dfrac{m^2 - 8m}{m^2 - 8m + 16} \right)$

Find a rational expression that represents the unknown dimension of each rectangle. (Assume all measures are given in appropriate units.)

79. 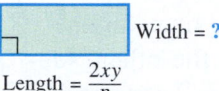 Width = ?

Length = $\dfrac{2xy}{p}$

The area is $\dfrac{5x^2y^3}{2pq}$.

80. Width = $\dfrac{9ab^2}{c}$

Length = ?

The area is $\dfrac{12a^3b^4}{5cd}$.

<table>
<tr><td>**6.3**</td><td></td></tr>
</table>

6.3 Least Common Denominators

OBJECTIVES

1 Find the least common denominator for a list of fractions.

2 Write equivalent rational expressions.

OBJECTIVE 1 Find the least common denominator for a list of fractions.

Adding or subtracting rational expressions often requires finding the **least common denominator (LCD)**. The LCD is the simplest expression that is divisible by all of the denominators in all of the expressions. For example, the fractions

$$\frac{2}{9} \quad \text{and} \quad \frac{5}{12} \quad \text{have LCD} \quad 36,$$

because 36 is the least positive number divisible by both 9 and 12.

We can often find least common denominators by inspection. In other cases, we find the LCD using the following procedure.

VOCABULARY

☐ least common denominator (LCD)

Finding the Least Common Denominator (LCD)

Step 1 **Factor** each denominator into prime factors.

Step 2 **List each different denominator factor** the *greatest* number of times it appears in any of the denominators.

Step 3 **Multiply** the denominator factors from Step 2 to find the LCD.

When each denominator is factored into prime factors, every prime factor must be a factor of the LCD.

> ! **CAUTION**   When finding the LCD, use each factor the **greatest** number of times it appears in any *single* denominator, not the **total** number of times it appears.

 **NOW TRY EXERCISE 1**

Find the LCD for each pair of fractions.

(a) $\dfrac{5}{48}, \dfrac{1}{30}$ **(b)** $\dfrac{3}{10y}, \dfrac{1}{6y}$

EXAMPLE 1   Finding Least Common Denominators

Find the LCD for each pair of fractions.

(a) $\dfrac{1}{24}, \dfrac{7}{15}$ Rational numbers   **(b)** $\dfrac{1}{8x}, \dfrac{3}{10x}$ Rational expressions

Step 1   Write each denominator in factored form with numerical coefficients in prime factored form.

$$24 = 2 \cdot 2 \cdot 2 \cdot 3 = 2^3 \cdot 3$$
$$15 = 3 \cdot 5$$

$$8x = 2 \cdot 2 \cdot 2 \cdot x = 2^3 \cdot x$$
$$10x = 2 \cdot 5 \cdot x$$

Step 2   Find the LCD by taking each different factor the *greatest* number of times it appears as a factor in any of the denominators.

The factor 2 appears three times in one product and not at all in the other, so the greatest number of times 2 appears is three. The greatest number of times both 3 and 5 appear is one.

Here, 2 appears three times in one product and once in the other, so the greatest number of times 2 appears is three. The greatest number of times 5 appears is one. The greatest number of times x appears in either product is one.

Step 3

$$\begin{aligned} LCD &= 2 \cdot 2 \cdot 2 \cdot 3 \cdot 5 \\ &= 2^3 \cdot 3 \cdot 5 \\ &= 120 \end{aligned}$$

$$\begin{aligned} LCD &= 2 \cdot 2 \cdot 2 \cdot 5 \cdot x \\ &= 2^3 \cdot 5 \cdot x \\ &= 40x \end{aligned}$$ **NOW TRY**

 **NOW TRY EXERCISE 2**

Find the LCD for the pair of fractions.

$$\dfrac{5}{6x^4} \quad \text{and} \quad \dfrac{7}{8x^3}$$

EXAMPLE 2   Finding the LCD

Find the LCD for $\dfrac{5}{6r^2}$ and $\dfrac{3}{4r^3}$.

Step 1

$$\left.\begin{aligned} 6r^2 &= 2 \cdot 3 \cdot r^2 \\ 4r^3 &= 2 \cdot 2 \cdot r^3 = 2^2 \cdot r^3 \end{aligned}\right\}$$ Factor each denominator.

Step 2   The greatest number of times 2 appears is two, the greatest number of times 3 appears is one, and the greatest number of times r appears is three.

Step 3 $LCD = 2^2 \cdot 3 \cdot r^3 = 12r^3$ **NOW TRY**

EXAMPLE 3   Finding LCDs

Find the LCD for the fractions in each list.

(a) $\dfrac{6}{5m}, \dfrac{4}{m^2 - 3m}$

$$\left.\begin{aligned} 5m &= 5 \cdot m \\ m^2 - 3m &= m(m - 3) \end{aligned}\right\}$$ Factor each denominator.

Use each different factor the greatest number of times it appears.

$$LCD = 5 \cdot m \cdot (m - 3) = 5m(m - 3)$$

> Be sure to include m as a factor in the LCD.

Because m is not a *factor* of $m - 3$, **both** m and $m - 3$ must appear in the LCD.

NOW TRY EXERCISE 3

Find the LCD for the fractions in each list.

(a) $\dfrac{3t}{2t^2 - 10t}, \dfrac{t+4}{t^2 - 25}$

(b) $\dfrac{1}{x^2 + 7x + 12}$, $\dfrac{2}{x^2 + 6x + 9}, \dfrac{5}{x^2 + 2x - 8}$

(c) $\dfrac{2}{a-4}, \dfrac{1}{4-a}$

(b) $\dfrac{1}{r^2 - 4r - 5}, \dfrac{3}{r^2 - r - 20}, \dfrac{1}{r^2 - 10r + 25}$

$$\left.\begin{array}{l} r^2 - 4r - 5 = (r-5)(r+1) \\ r^2 - r - 20 = (r-5)(r+4) \\ r^2 - 10r + 25 = (r-5)^2 \end{array}\right\}$$ Factor each denominator.

Use each different factor the greatest number of times it appears as a factor.

$$\text{LCD} = (r+1)(r+4)(r-5)^2$$ Be sure to include the exponent 2 on the factor $(r-5)$.

(c) $\dfrac{1}{q-5}, \dfrac{3}{5-q}$

The expressions $q - 5$ and $5 - q$ are opposites of each other. This means that if we multiply $q - 5$ by -1, we will obtain $5 - q$.

$$-(q-5) = -q + 5 = 5 - q$$

Therefore, either $q - 5$ or $5 - q$ can be used as the LCD.

NOW TRY

OBJECTIVE 2 Write equivalent rational expressions.

Once we have the LCD, the next step in preparing to add or subtract two rational expressions is to use the fundamental property to write equivalent expressions.

> **Writing a Rational Expression with a Specified Denominator**
>
> **Step 1** **Factor** both denominators.
>
> **Step 2** **Decide what factor(s) the denominator must be multiplied by** in order to equal the specified denominator.
>
> **Step 3** **Multiply** the rational expression by that factor divided by itself. (That is, multiply by 1.)

NOW TRY EXERCISE 4

Write each rational expression with the indicated denominator.

(a) $\dfrac{2}{9} = \dfrac{?}{27}$ (b) $\dfrac{4t}{11} = \dfrac{?}{33t}$

EXAMPLE 4 Writing Equivalent Rational Expressions

Write each rational expression with the indicated denominator.

(a) $\dfrac{3}{8} = \dfrac{?}{40}$ Rational numbers   (b) $\dfrac{9k}{25} = \dfrac{?}{50k}$ Rational expressions

Step 1 For each example, first factor the denominator on the right. Then compare the denominator on the left with the one on the right to decide what factors are missing. (It may sometimes be necessary to factor both denominators.)

$$\dfrac{3}{8} = \dfrac{?}{5 \cdot 8} \qquad\qquad \dfrac{9k}{25} = \dfrac{?}{25 \cdot 2k}$$

Step 2 A factor of 5 is missing.   Factors of 2 and k are missing.

Step 3 Multiply $\dfrac{3}{8}$ by $\dfrac{5}{5}$.   Multiply $\dfrac{9k}{25}$ by $\dfrac{2k}{2k}$.

$$\dfrac{3}{8} = \dfrac{3}{8} \cdot \dfrac{5}{5} = \dfrac{15}{40} \qquad\qquad \dfrac{9k}{25} = \dfrac{9k}{25} \cdot \dfrac{2k}{2k} = \dfrac{18k^2}{50k}$$

$$\dfrac{5}{5} = 1 \uparrow \qquad\qquad\qquad \dfrac{2k}{2k} = 1 \uparrow$$

NOW TRY

NOW TRY ANSWERS

3. (a) $2t(t-5)(t+5)$
   (b) $(x+3)^2(x+4)(x-2)$
   (c) either $a - 4$ or $4 - a$

4. (a) $\dfrac{6}{27}$ (b) $\dfrac{12t^2}{33t}$

 **NOW TRY EXERCISE 5**

Write each rational expression with the indicated denominator.

(a) $\dfrac{8k}{5k - 2} = \dfrac{?}{25k - 10}$

(b) $\dfrac{2t - 1}{t^2 + 4t} = \dfrac{?}{t^3 + 12t^2 + 32t}$

EXAMPLE 5 **Writing Equivalent Rational Expressions**

Write each rational expression with the indicated denominator.

(a) $\qquad \dfrac{8}{3x + 1} = \dfrac{?}{12x + 4}$

$\dfrac{8}{3x + 1} = \dfrac{?}{4(3x + 1)} \quad \longleftarrow$ Factor the denominator on the right.

The missing factor is 4, so multiply the fraction on the left by $\frac{4}{4}$.

$$\dfrac{8}{3x + 1} \cdot \dfrac{4}{4} = \dfrac{32}{12x + 4} \qquad \text{Fundamental property}$$

(b) $\qquad \dfrac{12p}{p^2 + 8p} = \dfrac{?}{p^3 + 4p^2 - 32p}$

Factor the denominator in each rational expression.

$$\dfrac{12p}{p(p + 8)} = \dfrac{?}{p(p + 8)(p - 4)}$$

$\begin{aligned} p^3 + 4p^2 - 32p \\ = p(p^2 + 4p - 32) \\ = p(p + 8)(p - 4) \end{aligned}$

The factor $p - 4$ is missing, so multiply $\frac{12p}{p(p + 8)}$ by $\frac{p - 4}{p - 4}$.

$$\dfrac{12p}{p^2 + 8p} = \dfrac{12p}{p(p + 8)} \cdot \dfrac{p - 4}{p - 4} \qquad \text{Fundamental property}$$

$$= \dfrac{12p(p - 4)}{p(p + 8)(p - 4)} \qquad \begin{aligned}&\text{Multiply numerators.}\\&\text{Multiply denominators.}\end{aligned}$$

$$= \dfrac{12p^2 - 48p}{p^3 + 4p^2 - 32p} \qquad \text{Multiply the factors.}$$

NOW TRY

NOTE In the last step in **Example 5(b)**, we multiplied the factors of the numerator and denominator in

$$\dfrac{12p(p - 4)}{p(p + 8)(p - 4)} \quad \text{to obtain} \quad \dfrac{12p^2 - 48p}{p^3 + 4p^2 - 32p}.$$

We did this to match the form of the fractions in the original statement of the problem. These are equivalent expressions, and either form is acceptable as the answer.

NOW TRY ANSWERS

5. (a) $\dfrac{40k}{25k - 10}$

(b) $\dfrac{2t^2 + 15t - 8}{t^3 + 12t^2 + 32t}$

6.3 Exercises

FOR EXTRA HELP

▶ **MyLab Math**

▶ *Video solutions for select problems available in MyLab Math*

Concept Check *Choose the correct response.*

1. The least common denominator for $\frac{11}{20}$ and $\frac{1}{2}$ is

    A. 2   **B.** 20   **C.** 40   **D.** none of these.

2. The least common denominator for $\frac{1}{a}$ and $\frac{1}{5a}$ is

    A. a **B.** $5a$ **C.** $5a^2$ **D.** 5.

3. **Concept Check** To find the LCD for $\frac{4}{25x^2}$ and $\frac{7}{10x^4}$, a student factored each denominator as follows.

$$25x^2 = 5^2 \cdot x^2$$

$$10x^4 = 2 \cdot 5 \cdot x^4$$

He multiplied the factors $2 \cdot 5^2 \cdot x^6$ to obtain $50x^6$ as the LCD. This is incorrect. **WHAT WENT WRONG?** Give the correct LCD.

4. **Concept Check** A student was asked to find the LCD for

$$\frac{2}{x-1} \quad \text{and} \quad \frac{5}{x+1}.$$

Her reasoning was that because the denominator expressions are opposites, either $x-1$ or $x+1$ could be used as the LCD. This is incorrect. **WHAT WENT WRONG?** Give the correct LCD.

5. **Concept Check** Find the LCD for the pair of fractions.

$$\frac{7}{10}, \frac{1}{25}$$

Step 1 $10 = 2 \cdot \underline{\hspace{1cm}}$

$25 = \underline{\hspace{1cm}} \cdot 5$

Step 2 The greatest number of times 2 appears is ____. The greatest number of times ____ appears is two.

Step 3 LCD $= \underline{\hspace{0.5cm}} \cdot \underline{\hspace{0.5cm}} \cdot 5$

LCD $= \underline{\hspace{1cm}}$

6. **Concept Check** Write the rational expression as an equivalent expression with the indicated denominator.

$$\frac{7k}{5} = \frac{?}{30p}$$

Step 1 Factor the denominator on the right.

$$\frac{7k}{5} = \frac{?}{5 \cdot \underline{\hspace{1cm}}}$$

Step 2 Factors of ____ and ____ are missing.

Step 3 $\dfrac{7k}{5} \cdot \dfrac{\underline{\hspace{1cm}}}{\underline{\hspace{1cm}}} = \dfrac{\underline{\hspace{1cm}}}{30p}$

Find the LCD for the fractions in each list. **See Examples 1 and 2.**

7. $\dfrac{7}{15}, \dfrac{21}{20}$

8. $\dfrac{9}{10}, \dfrac{13}{25}$

9. $\dfrac{17}{100}, \dfrac{23}{120}, \dfrac{43}{180}$

10. $\dfrac{17}{250}, \dfrac{21}{300}, \dfrac{1}{360}$

11. $\dfrac{9}{x^2}, \dfrac{8}{x^5}$

12. $\dfrac{12}{m^7}, \dfrac{14}{m^8}$

13. $\dfrac{-2}{5p}, \dfrac{13}{6p}$

14. $\dfrac{-14}{15k}, \dfrac{11}{4k}$

15. $\dfrac{6}{21r^3}, \dfrac{8}{12r^5}$

16. $\dfrac{2}{15x^7}, \dfrac{7}{18x^3}$

17. $\dfrac{4}{25m^3}, \dfrac{7}{10m^4}$

18. $\dfrac{6}{35t^2}, \dfrac{5}{49t^6}$

19. $\dfrac{13}{5a^2b^3}, \dfrac{29}{15a^5b}$

20. $\dfrac{7}{3r^4s^5}, \dfrac{23}{9r^6s^8}$

21. $\dfrac{1}{r^2t^3}, \dfrac{1}{r^5t}, \dfrac{1}{r^9t^2}$

22. $\dfrac{5}{x^8y^4}, \dfrac{5}{x^9y^3}, \dfrac{5}{xy^2}$

23. $\dfrac{7}{x+1}, \dfrac{9}{x-1}$

24. $\dfrac{3}{y+3}, \dfrac{2}{y-3}$

Find the LCD for the fractions in each list. **See Example 3.**

25. $\dfrac{7}{6p}, \dfrac{15}{4p - 8}$

26. $\dfrac{7}{8k}, \dfrac{28}{12k - 24}$

27. $\dfrac{9}{28m^2}, \dfrac{3}{12m - 20}$

28. $\dfrac{14}{27a^3}, \dfrac{7}{9a - 45}$

29. $\dfrac{7}{5b - 10}, \dfrac{11}{6b - 12}$

30. $\dfrac{3}{7x^2 + 21x}, \dfrac{2}{5x^2 + 15x}$

31. $\dfrac{37}{6r - 12}, \dfrac{25}{9r - 18}$

32. $\dfrac{14}{5p - 30}, \dfrac{11}{6p - 36}$

33. $\dfrac{5}{c - d}, \dfrac{8}{d - c}$

34. $\dfrac{4}{y - x}, \dfrac{8}{x - y}$

35. $\dfrac{12}{m - 3}, \dfrac{-4}{3 - m}$

36. $\dfrac{3}{a - 8}, \dfrac{-17}{8 - a}$

37. $\dfrac{29}{p - q}, \dfrac{18}{q - p}$

38. $\dfrac{16}{z - x}, \dfrac{9}{x - z}$

39. $\dfrac{13}{x^2 - 1}, \dfrac{-5}{2x + 2}$

40. $\dfrac{9}{y^2 - 9}, \dfrac{-2}{2y + 6}$

41. $\dfrac{4x^2}{(x - 4)^2}, \dfrac{17x}{3x - 12}$

42. $\dfrac{3y^2}{(y + 6)^2}, \dfrac{5y}{2y + 12}$

43. $\dfrac{5}{12p + 60}, \dfrac{-17}{p^2 + 5p}, \dfrac{-16}{p^2 + 10p + 25}$

44. $\dfrac{13}{r^2 + 7r}, \dfrac{-3}{5r + 35}, \dfrac{-4}{r^2 + 14r + 49}$

45. $\dfrac{-3}{8y + 16}, \dfrac{-22}{y^2 + 3y + 2}$

46. $\dfrac{-2}{9m - 18}, \dfrac{-6}{m^2 - 7m + 10}$

47. $\dfrac{3}{k^2 + 5k}, \dfrac{2}{k^2 + 3k - 10}$

48. $\dfrac{1}{z^2 - 4z}, \dfrac{9}{z^2 - 3z - 4}$

49. $\dfrac{6}{a^2 + 6a}, \dfrac{-5}{a^2 + 3a - 18}$

50. $\dfrac{8}{y^2 - 5y}, \dfrac{-5}{y^2 - 2y - 15}$

51. $\dfrac{5}{p^2 + 8p + 15}, \dfrac{3}{p^2 - 3p - 18}, \dfrac{12}{p^2 - p - 30}$

52. $\dfrac{10}{y^2 - 10y + 21}, \dfrac{2}{y^2 - 2y - 3}, \dfrac{15}{y^2 - 6y - 7}$

53. $\dfrac{-5}{k^2 + 2k - 35}, \dfrac{-8}{k^2 + 3k - 40}, \dfrac{19}{k^2 - 2k - 15}$

54. $\dfrac{-19}{z^2 + 4z - 12}, \dfrac{-16}{z^2 + z - 30}, \dfrac{16}{z^2 + 2z - 24}$

Write each rational expression as an equivalent expression with the indicated denominator. **See Examples 4 and 5.**

55. $\dfrac{4}{11} = \dfrac{?}{55}$

56. $\dfrac{8}{7} = \dfrac{?}{42}$

57. $\dfrac{-5}{k} = \dfrac{?}{9k}$

58. $\dfrac{-4}{q} = \dfrac{?}{6q}$

59. $\dfrac{15m^2}{8k} = \dfrac{?}{32k^4}$

60. $\dfrac{7t^2}{3y} = \dfrac{?}{9y^2}$

61. $\dfrac{19z}{2z-6} = \dfrac{?}{6z-18}$

62. $\dfrac{3r}{5r-5} = \dfrac{?}{15r-15}$

63. $\dfrac{-2a}{9a-18} = \dfrac{?}{18a-36}$

64. $\dfrac{-7y}{6y+18} = \dfrac{?}{24y+72}$

65. $\dfrac{14}{z^2-3z} = \dfrac{?}{z(z-3)(z-2)}$

66. $\dfrac{25}{m^2-9m} = \dfrac{?}{m(m-9)(m+8)}$

67. $\dfrac{4r-t}{r^2+rt+t^2} = \dfrac{?}{t^3-r^3}$

68. $\dfrac{3x-1}{x^2+2x+4} = \dfrac{?}{x^3-8}$

69. $\dfrac{2(z-y)}{y^2+yz+z^2} = \dfrac{?}{y^4-z^3y}$

70. $\dfrac{2p+3q}{p^2+2pq+q^2} = \dfrac{?}{(p+q)(p^3+q^3)}$

Extending Skills *Write each rational expression as an equivalent expression with the indicated denominator.*

71. $\dfrac{36r}{r^2-r-6} = \dfrac{?}{(r-3)(r+2)(r+1)}$

72. $\dfrac{4m}{m^2+m-2} = \dfrac{?}{(m-1)(m-3)(m+2)}$

73. $\dfrac{a+2b}{2a^2+ab-b^2} = \dfrac{?}{2a^3b+a^2b^2-ab^3}$

74. $\dfrac{m-4}{6m^2+7m-3} = \dfrac{?}{12m^3+14m^2-6m}$

RELATING CONCEPTS For Individual or Group Work (Exercises 75–80)

Work Exercises 75–80 in order.

75. Suppose that we want to write $\dfrac{3}{4}$ as an equivalent fraction with denominator 28. By what number must we multiply both the numerator and the denominator?

76. If we write $\dfrac{3}{4}$ as an equivalent fraction with denominator 28, by what number are we actually multiplying the fraction?

77. What property of multiplication is being used when we write a common fraction as an equivalent one with a larger denominator?

78. Suppose that we want to write $\dfrac{2x+5}{x-4}$ as an equivalent fraction with denominator $7x-28$. By what number must we multiply both the numerator and the denominator?

79. If we write $\dfrac{2x+5}{x-4}$ as an equivalent fraction with denominator $7x-28$, by what number are we actually multiplying the fraction?

80. Repeat **Exercise 77**, changing "a common" to "an algebraic."

6.4   Adding and Subtracting Rational Expressions

OBJECTIVE 1 Add rational expressions having the same denominator.

We find the sum of two such rational expressions using the procedure for adding two common fractions that have the same denominator.

> **Adding Rational Expressions (Same Denominator)**
>
> The rational expressions $\frac{P}{Q}$ and $\frac{R}{Q}$ (where $Q \neq 0$) are added as follows.
>
> $$\frac{P}{Q} + \frac{R}{Q} = \frac{P + R}{Q}$$
>
> That is, to add rational expressions with the same denominator, add the numerators and keep the same denominator.
>
> *Example:* $\frac{2}{7} + \frac{3}{7} = \frac{2 + 3}{7} = \frac{5}{7}$

NOW TRY EXERCISE 1

Add. Write each answer in lowest terms.

(a) $\dfrac{5}{12} + \dfrac{3}{12}$

(b) $\dfrac{4y}{y + 3} + \dfrac{12}{y + 3}$

EXAMPLE 1   Adding Rational Expressions (Same Denominator)

Add. Write each answer in lowest terms.

(a) $\dfrac{4}{9} + \dfrac{2}{9}$ Rational numbers

(b) $\dfrac{3x}{x + 1} + \dfrac{3}{x + 1}$ Rational expressions

The denominators are the same, so the sum is found by adding the two numerators and keeping the same (common) denominator.

$$= \frac{4 + 2}{9} \qquad \text{Add.}$$

$$= \frac{6}{9}$$

$$= \frac{2 \cdot 3}{3 \cdot 3} \qquad \text{Factor.}$$

$$= \frac{2}{3} \qquad \text{Lowest terms}$$

$$= \frac{3x + 3}{x + 1} \qquad \text{Add.}$$

$$= \frac{3(x + 1)}{x + 1} \qquad \text{Factor.}$$

$$= 3 \qquad \text{Lowest terms}$$

NOW TRY

OBJECTIVE 2 Add rational expressions having different denominators.

We use the following steps to add rational expressions having different denominators.

> **Adding Rational Expressions (Different Denominators)**
>
> *Step 1* **Find the least common denominator (LCD).**
>
> *Step 2* **Write each rational expression** as an equivalent rational expression with the LCD as the denominator.
>
> *Step 3* **Add** the numerators to obtain the numerator of the sum. The LCD is the denominator of the sum.
>
> *Step 4* **Write in lowest terms** using the fundamental property.

NOW TRY ANSWERS

1. (a) $\frac{2}{3}$ (b) 4

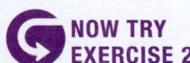 **NOW TRY
EXERCISE 2**

Add. Write each answer in
lowest terms.

(a) $\dfrac{5}{12} + \dfrac{3}{20}$ (b) $\dfrac{3}{5x} + \dfrac{2}{7x}$

EXAMPLE 2 Adding Rational Expressions (Different Denominators)

Add. Write each answer in lowest terms.

(a) $\dfrac{1}{12} + \dfrac{7}{15}$ (b) $\dfrac{2}{3y} + \dfrac{1}{4y}$

Step 1 Find the LCD, using the methods of the previous section.

$$12 = 2 \cdot 2 \cdot 3 = 2^2 \cdot 3 \qquad\qquad 3y = 3 \cdot y$$

$$15 = 3 \cdot 5 \qquad\qquad 4y = 2 \cdot 2 \cdot y = 2^2 \cdot y$$

$$\text{LCD} = 2^2 \cdot 3 \cdot 5 = 60 \qquad\qquad \text{LCD} = 2^2 \cdot 3 \cdot y = 12y$$

Step 2 Write each rational expression as a fraction with the LCD (60 and $12y$, respectively) as the denominator.

$$\dfrac{1}{12} + \dfrac{7}{15} \qquad \text{The LCD is 60.} \qquad\qquad \dfrac{2}{3y} + \dfrac{1}{4y} \qquad \text{The LCD is } 12y.$$

$$= \dfrac{1}{12} \cdot \dfrac{5}{5} + \dfrac{7}{15} \cdot \dfrac{4}{4} \qquad\qquad = \dfrac{2}{3y} \cdot \dfrac{4}{4} + \dfrac{1}{4y} \cdot \dfrac{3}{3}$$

$$= \dfrac{5}{60} + \dfrac{28}{60} \qquad\qquad = \dfrac{8}{12y} + \dfrac{3}{12y}$$

Step 3 Add the numerators. The LCD is the denominator.

Step 4 Write in lowest terms if necessary.

$$= \dfrac{5 + 28}{60} \qquad\qquad = \dfrac{8 + 3}{12y}$$

$$= \dfrac{33}{60} \qquad\qquad = \dfrac{11}{12y}$$

$$= \dfrac{11}{20}$$

NOW TRY

EXAMPLE 3 Adding Rational Expressions

Add. Write the answer in lowest terms.

$$\dfrac{2x}{x^2 - 1} + \dfrac{-1}{x + 1}$$

Step 1 The denominators are different, so find the LCD.

$$\left.\begin{array}{l} x^2 - 1 = (x + 1)(x - 1) \\ x + 1 \text{ is prime.} \end{array}\right\} \quad \text{The LCD is } (x + 1)(x - 1).$$

Step 2 Write each rational expression with the LCD as the denominator.

$$\dfrac{2x}{x^2 - 1} + \dfrac{-1}{x + 1} \qquad\qquad\qquad \text{The LCD is } (x + 1)(x - 1).$$

$$= \dfrac{2x}{(x + 1)(x - 1)} + \dfrac{-1(x - 1)}{(x + 1)(x - 1)} \qquad \begin{array}{l}\text{Multiply the second} \\ \text{fraction by } \frac{x-1}{x-1}.\end{array}$$

$$= \dfrac{2x}{(x + 1)(x - 1)} + \dfrac{-x + 1}{(x + 1)(x - 1)} \qquad \text{Distributive property}$$

NOW TRY ANSWERS

2. (a) $\frac{17}{30}$ (b) $\frac{31}{35x}$

NOW TRY EXERCISE 3

Add. Write the answer in lowest terms.

$$\frac{6t}{t^2 - 9} + \frac{-3}{t + 3}$$

Step 3 $= \dfrac{2x - x + 1}{(x + 1)(x - 1)}$ Add numerators.

             Keep the same denominator.

        $= \dfrac{x + 1}{(x + 1)(x - 1)}$ Combine like terms in the numerator.

Step 4 $= \dfrac{1(x + 1)}{(x + 1)(x - 1)}$ Identity property of multiplication

Remember to write 1 in the numerator.

        $= \dfrac{1}{x - 1}$ Divide out the common factor.
The result is in lowest terms.   **NOW TRY**

NOW TRY EXERCISE 4

Add. Write the answer in lowest terms.

$$\frac{x - 1}{x^2 + 6x + 8} + \frac{4x}{x^2 + x - 12}$$

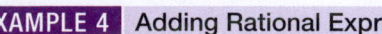

EXAMPLE 4   Adding Rational Expressions

Add. Write the answer in lowest terms.

$$\frac{2x}{x^2 + 5x + 6} + \frac{x + 1}{x^2 + 2x - 3}$$

$= \dfrac{2x}{(x + 2)(x + 3)} + \dfrac{x + 1}{(x + 3)(x - 1)}$ Factor the denominators.

$= \dfrac{2x(x - 1)}{(x + 2)(x + 3)(x - 1)} + \dfrac{(x + 1)(x + 2)}{(x + 2)(x + 3)(x - 1)}$ The LCD is $(x + 2)(x + 3)(x - 1)$.

$= \dfrac{2x(x - 1) + (x + 1)(x + 2)}{(x + 2)(x + 3)(x - 1)}$ Add numerators.
Keep the same denominator.

$= \dfrac{2x^2 - 2x + x^2 + 3x + 2}{(x + 2)(x + 3)(x - 1)}$ Multiply.

$= \dfrac{3x^2 + x + 2}{(x + 2)(x + 3)(x - 1)}$ Combine like terms.

The numerator cannot be factored. The expression is in lowest terms.   **NOW TRY**

NOTE   If the final expression in **Example 4** could be written in lower terms, the numerator would have a factor of $x + 2$, $x + 3$, or $x - 1$. Therefore, it is only necessary to check for possible factored forms of the numerator that would contain one of these binomials.

NOW TRY EXERCISE 5

Add. Write the answer in lowest terms.

$$\frac{2k}{k - 7} + \frac{5}{7 - k}$$

EXAMPLE 5   Adding Rational Expressions (Denominators Are Opposites)

Add. Write the answer in lowest terms.

$$\frac{y}{y - 2} + \frac{8}{2 - y}$$ The denominators are opposites.

$= \dfrac{y}{y - 2} + \dfrac{8(-1)}{(2 - y)(-1)}$ Multiply $\frac{8}{2 - y}$ by 1 in the form $\frac{-1}{-1}$ to find a common denominator.

$= \dfrac{y}{y - 2} + \dfrac{-8}{2 + y}$ Distributive property

$= \dfrac{y}{y - 2} + \dfrac{-8}{y - 2}$ Rewrite $-2 + y$ as $y - 2$.

$= \dfrac{y - 8}{y - 2}$ Add numerators.
Keep the same denominator.   **NOW TRY**

NOW TRY ANSWERS

3. $\dfrac{3}{t - 3}$

4. $\dfrac{5x^2 + 4x + 3}{(x + 4)(x + 2)(x - 3)}$

5. $\dfrac{2k - 5}{k - 7}$,   or $\dfrac{5 - 2k}{7 - k}$

NOTE If we had chosen to use $2 - y$ as the common denominator in **Example 5**, we would have obtained a different, yet equivalent, form of the answer $\frac{y - 8}{y - 2}$.

$$\frac{y}{y - 2} + \frac{8}{2 - y}$$ See Example 5.

$$= \frac{y(-1)}{(y - 2)(-1)} + \frac{8}{2 - y}$$ Multiply $\frac{y}{y - 2}$ by $\frac{-1}{-1}$.

$$= \frac{-y}{2 - y} + \frac{8}{2 - y}$$ In the first denominator, $(y - 2)(-1) = -y + 2 = 2 - y$.

$$= \frac{-y + 8}{2 - y}$$ Add numerators. Keep the same denominator.

$$= \frac{8 - y}{2 - y}$$

Multiply $\frac{8 - y}{2 - y}$ by $\frac{-1}{-1}$ to confirm that it is equivalent to $\frac{y - 8}{y - 2}$.

OBJECTIVE 3 Subtract rational expressions.

> **Subtracting Rational Expressions (Same Denominator)**
>
> The rational expressions $\frac{P}{Q}$ and $\frac{R}{Q}$ (where $Q \neq 0$) are subtracted as follows.
>
> $$\frac{P}{Q} - \frac{R}{Q} = \frac{P - R}{Q}$$
>
> That is, to subtract rational expressions with the same denominator, subtract the numerators and keep the same denominator.
>
> *Example:* $\frac{4}{5} - \frac{1}{5} = \frac{4 - 1}{5} = \frac{3}{5}$

**NOW TRY
EXERCISE 6**

Subtract. Write the answer in lowest terms.

$$\frac{2x}{x + 5} - \frac{x + 1}{x + 5}$$

EXAMPLE 6 Subtracting Rational Expressions (Same Denominator)

Subtract. Write the answer in lowest terms.

$$\frac{2m}{m - 1} - \frac{m + 3}{m - 1}$$ Use parentheses around the numerator of the subtrahend.

$$= \frac{2m - (m + 3)}{m - 1}$$ Subtract numerators. Keep the same denominator.

Be careful with signs. $$= \frac{2m - m - 3}{m - 1}$$ Distributive property

$$= \frac{m - 3}{m - 1}$$ Combine like terms.

NOW TRY

! CAUTION In subtraction problems like the one in **Example 6**, the numerator of the fraction being subtracted must be treated as a single quantity. *Be sure to use parentheses after the subtraction symbol. Subtract each term within the parentheses, or a sign error may occur.*

NOW TRY ANSWER

6. $\frac{x - 1}{x + 5}$

NOW TRY
EXERCISE 7
Subtract. Write the answer in lowest terms.

$$\frac{6}{y-6} - \frac{2}{y}$$

EXAMPLE 7 Subtracting Rational Expressions (Different Denominators)

Subtract. Write the answer in lowest terms.

$$\frac{9}{x-2} - \frac{3}{x}$$
The LCD is $x(x-2)$.

$$= \frac{9x}{x(x-2)} - \frac{3(x-2)}{x(x-2)}$$
Write each expression with the LCD.

$$= \frac{9x - 3(x-2)}{x(x-2)}$$
Subtract numerators.
Keep the same denominator.

Be careful with signs. $$= \frac{9x - 3x + 6}{x(x-2)}$$
Distributive property

$$= \frac{6x+6}{x(x-2)}, \quad \text{or} \quad \frac{6(x+1)}{x(x-2)}$$
Combine like terms.
Factor the numerator.

We factored in the last step to determine whether there were any common factors to divide out. There were not, so the expression is in lowest terms. The two final forms are equivalent. Either form can be given as the answer. **NOW TRY**

EXAMPLE 8 Subtracting Rational Expressions (Denominators Are Opposites)

NOW TRY
EXERCISE 8
Subtract. Write the answer in lowest terms.

$$\frac{2m}{m-4} - \frac{m-12}{4-m}$$

Subtract. Write the answer in lowest terms.

$$\frac{3x}{x-5} - \frac{2x-25}{5-x}$$
The denominators are opposites. We choose $x-5$ as the common denominator.

$$= \frac{3x}{x-5} - \frac{(2x-25)(-1)}{(5-x)(-1)}$$
Multiply $\frac{2x-25}{5-x}$ by $\frac{-1}{-1}$ to find a common denominator·

$$= \frac{3x}{x-5} - \frac{-2x+25}{x-5}$$
$(2x-25)(-1) = -2x+25$;
$(5-x)(-1) = -5+x = x-5$

Subtract the *entire* numerator. Use parentheses to show this. $$= \frac{3x - (-2x+25)}{x-5}$$
Subtract numerators.

Be careful with signs. $$= \frac{3x + 2x - 25}{x-5}$$
Distributive property

$$= \frac{5x-25}{x-5}$$
Combine like terms.

$$= \frac{5(x-5)}{x-5}$$
Factor.

$$= 5$$
Lowest terms **NOW TRY**

EXAMPLE 9 Subtracting Rational Expressions

Subtract. Write the answer in lowest terms.

NOW TRY ANSWERS
7. $\frac{4(y+3)}{y(y-6)}$
8. 3

$$\frac{6x}{x^2 - 2x + 1} - \frac{1}{x^2 - 1}$$

$$= \frac{6x}{(x-1)^2} - \frac{1}{(x-1)(x+1)}$$
Factor the denominators.
LCD $= (x-1)(x-1)(x+1)$

**NOW TRY
EXERCISE 9**

Subtract. Write the answer in lowest terms.

$$\frac{5}{t^2 - 6t + 9} - \frac{2t}{t^2 - 9}$$

From the factored denominators, we identify the LCD as $(x-1)^2(x+1)$. **We use the factor $x - 1$ twice** because it appears twice in the first denominator.

$$= \frac{6x(x+1)}{(x-1)^2(x+1)} - \frac{1(x-1)}{(x-1)(x-1)(x+1)} \qquad \text{Fundamental property}$$

$$= \frac{6x(x+1) - 1(x-1)}{(x-1)^2(x+1)} \qquad \text{Subtract numerators.}$$

$$= \frac{6x^2 + 6x - x + 1}{(x-1)^2(x+1)} \qquad \text{Distributive property}$$

$$= \frac{6x^2 + 5x + 1}{(x-1)^2(x+1)}, \quad \text{or} \quad \frac{(3x+1)(2x+1)}{(x-1)^2(x+1)} \qquad \begin{array}{l}\text{Combine like terms.}\\ \text{Factor the numerator.}\end{array}$$

The factored form indicates that the expression is in lowest terms—there are no common factors to divide out. Either form can be given as the answer.   **NOW TRY**

**NOW TRY
EXERCISE 10**

Subtract. Write the answer in lowest terms.

$$\frac{q}{2q^2 + 5q - 3} - \frac{3q+4}{3q^2 + 10q + 3}$$

EXAMPLE 10 Subtracting Rational Expressions

Subtract. Write the answer in lowest terms.

$$\frac{q}{q^2 - 4q - 5} - \frac{3}{2q^2 - 13q + 15}$$

$$= \frac{q}{(q+1)(q-5)} - \frac{3}{(q-5)(2q-3)} \qquad \begin{array}{l}\text{Factor the denominators. The LCD is}\\ (q+1)(q-5)(2q-3).\end{array}$$

$$= \frac{q(2q-3)}{(q+1)(q-5)(2q-3)} - \frac{3(q+1)}{(q+1)(q-5)(2q-3)} \qquad \text{Fundamental property}$$

$$= \frac{q(2q-3) - 3(q+1)}{(q+1)(q-5)(2q-3)} \qquad \text{Subtract numerators.}$$

$$= \frac{2q^2 - 3q - 3q - 3}{(q+1)(q-5)(2q-3)} \qquad \text{Distributive property}$$

$$= \frac{2q^2 - 6q - 3}{(q+1)(q-5)(2q-3)} \qquad \text{Combine like terms.}$$

The numerator cannot be factored. The final answer is in lowest terms.   **NOW TRY**

NOW TRY ANSWERS

9. $\dfrac{-2t^2 + 11t + 15}{(t-3)^2(t+3)}$

10. $\dfrac{-3q^2 - 4q + 4}{(2q-1)(q+3)(3q+1)}$

6.4 Exercises

FOR EXTRA HELP   **MyLab Math**

 Video solutions for select problems available in MyLab Math

Concept Check *Match each expression in Column I with the correct sum or difference in Column II.*

I

1. $\dfrac{x}{x+8} + \dfrac{8}{x+8}$ 2. $\dfrac{2x}{x-8} - \dfrac{16}{x-8}$

3. $\dfrac{8}{x-8} - \dfrac{x}{x-8}$ 4. $\dfrac{8}{x+8} - \dfrac{x}{x+8}$

5. $\dfrac{x}{x+8} - \dfrac{8}{x+8}$ 6. $\dfrac{1}{x} + \dfrac{1}{8}$

7. $\dfrac{1}{8} - \dfrac{1}{x}$ 8. $\dfrac{1}{8x} - \dfrac{1}{8x}$

II

A. 2   B. $\dfrac{x-8}{x+8}$

C. -1 D. $\dfrac{8+x}{8x}$

E. 1   F. 0

G. $\dfrac{x-8}{8x}$ H. $\dfrac{8-x}{x+8}$

9. Concept Check A student subtracted the following rational expressions incorrectly as shown.

$$\frac{2x}{x+5} - \frac{x+1}{x+5}$$

$$= \frac{2x - x + 1}{x+5}$$

$$= \frac{x+1}{x+5}$$

WHAT WENT WRONG? Give the correct answer.

10. Concept Check A student subtracted the following rational expressions incorrectly as shown.

$$\frac{7x}{2x-3} - \frac{3x-5}{2x-3}$$

$$= \frac{7x - (3x-5)}{2x-3}$$

$$= \frac{7x - 3x - 5}{2x-3}$$

$$= \frac{4x-5}{2x-3}$$

WHAT WENT WRONG? Give the correct answer.

Note: When adding and subtracting rational expressions, several different equivalent forms of the answer often exist. If your answer does not look exactly like the one given in the back of the book, check to see whether you have written an equivalent form.

Add or subtract. Write each answer in lowest terms. **See Examples 1 and 6.**

11. $\dfrac{5}{18} + \dfrac{7}{18}$

12. $\dfrac{11}{24} + \dfrac{7}{24}$

13. $\dfrac{4}{m} + \dfrac{7}{m}$

14. $\dfrac{5}{p} + \dfrac{12}{p}$

15. $\dfrac{5}{y+4} - \dfrac{1}{y+4}$

16. $\dfrac{6}{t+3} - \dfrac{3}{t+3}$

17. $\dfrac{x}{x+y} + \dfrac{y}{x+y}$

18. $\dfrac{a}{a+b} + \dfrac{b}{a+b}$

19. $\dfrac{4y}{y+3} + \dfrac{12}{y+3}$

20. $\dfrac{2x}{x+4} + \dfrac{8}{x+4}$

21. $\dfrac{5m}{m+1} - \dfrac{1+4m}{m+1}$

22. $\dfrac{4x}{x+2} - \dfrac{2+3x}{x+2}$

23. $\dfrac{a+b}{2} - \dfrac{a-b}{2}$

24. $\dfrac{x-y}{2} - \dfrac{x+y}{2}$

25. $\dfrac{x^2}{x+5} + \dfrac{5x}{x+5}$

26. $\dfrac{t^2}{t-3} + \dfrac{-3t}{t-3}$

27. $\dfrac{y^2 - 3y}{y+3} + \dfrac{-18}{y+3}$

28. $\dfrac{r^2 - 8r}{r-5} + \dfrac{15}{r-5}$

29. $\dfrac{x}{x^2-9} - \dfrac{-3}{x^2-9}$

30. $\dfrac{y}{y^2-16} - \dfrac{-4}{y^2-16}$

31. $\dfrac{y^2+x^2}{x^2-y^2} - \dfrac{2x^2}{x^2-y^2}$

32. $\dfrac{3a^2+b^2}{a^2-b^2} - \dfrac{4a^2}{a^2-b^2}$

Add or subtract. Write each answer in lowest terms. **See Examples 2, 3, 4, and 7.**

33. $\dfrac{5}{12} + \dfrac{3}{20}$

34. $\dfrac{7}{30} + \dfrac{2}{45}$

35. $\dfrac{z}{5} + \dfrac{1}{3}$

36. $\dfrac{p}{8} + \dfrac{4}{5}$

37. $\dfrac{5}{7} - \dfrac{r}{2}$

38. $\dfrac{20}{9} - \dfrac{z}{3}$

39. $-\dfrac{3}{4} - \dfrac{1}{2x}$

40. $-\dfrac{7}{8} - \dfrac{3}{2a}$

41. $\dfrac{7}{4t} + \dfrac{3}{7t}$

42. $\dfrac{8}{3r} + \dfrac{2}{5r}$

43. $\dfrac{x+1}{6} + \dfrac{3x+3}{9}$

44. $\dfrac{2x-6}{4} + \dfrac{x+5}{6}$

45. $\dfrac{x+3}{3x} + \dfrac{2x+2}{4x}$

46. $\dfrac{x+2}{5x} + \dfrac{6x+3}{3x}$

47. $\dfrac{7}{3p^2} - \dfrac{2}{p}$

48. $\dfrac{12}{5m^2} - \dfrac{5}{m}$

49. $\dfrac{1}{k+4} - \dfrac{2}{k}$

50. $\dfrac{3}{m+1} - \dfrac{4}{m}$

51. $\dfrac{x}{x-2} + \dfrac{-8}{x^2-4}$

52. $\dfrac{2x}{x-1} + \dfrac{-4}{x^2-1}$

53. $\dfrac{4m}{m^2+3m+2} + \dfrac{2m-1}{m^2+6m+5}$

54. $\dfrac{a}{a^2+3a-4} + \dfrac{4a}{a^2+7a+12}$

55. $\dfrac{4y}{y^2-1} - \dfrac{5}{y^2+2y+1}$

56. $\dfrac{2x}{x^2-16} - \dfrac{3}{x^2+8x+16}$

57. $\dfrac{t}{t+2} + \dfrac{5-t}{t} - \dfrac{4}{t^2+2t}$

58. $\dfrac{2p}{p-3} + \dfrac{2+p}{p} - \dfrac{-6}{p^2-3p}$

Concept Check *Answer each question.*

59. What are the *two* possible LCDs that could be used for the sum $\dfrac{10}{m-2} + \dfrac{5}{2-m}$?

60. If one form of the answer to a sum or difference of rational expressions is $\dfrac{4}{k-3}$, what would an alternative form of the answer be if the denominator is $3-k$?

Add or subtract. Write each answer in lowest terms. See Examples 5 and 8.

61. $\dfrac{4}{x-5} + \dfrac{6}{5-x}$

62. $\dfrac{10}{m-2} + \dfrac{5}{2-m}$

63. $\dfrac{-1}{1-y} - \dfrac{4y-3}{y-1}$

64. $\dfrac{-4}{p-3} - \dfrac{p+1}{3-p}$

65. $\dfrac{2}{x-y^2} + \dfrac{7}{y^2-x}$

66. $\dfrac{-8}{p-q^2} + \dfrac{3}{q^2-p}$

67. $\dfrac{x}{5x-3y} - \dfrac{y}{3y-5x}$

68. $\dfrac{t}{8t-9s} - \dfrac{s}{9s-8t}$

69. $\dfrac{3}{4p-5} + \dfrac{9}{5-4p}$

70. $\dfrac{8}{3-7y} - \dfrac{2}{7y-3}$

71. $\dfrac{15x}{5x-7} - \dfrac{-21}{7-5x}$

72. $\dfrac{24y}{6y-5} - \dfrac{-20}{5-6y}$

*In these subtraction problems, the rational expression that follows the subtraction sign has a numerator with more than one term. **Be careful with signs** and find each difference. See Examples 6–10.*

73. $\dfrac{2m}{m-n} - \dfrac{5m+n}{2m-2n}$

74. $\dfrac{5p}{p-q} - \dfrac{3p+1}{4p-4q}$

75. $\dfrac{5}{x^2-9} - \dfrac{x+2}{x^2+4x+3}$

76. $\dfrac{1}{a^2-1} - \dfrac{a-1}{a^2+3a-4}$

77. $\dfrac{2q+1}{3q^2+10q-8} - \dfrac{3q+5}{2q^2+5q-12}$

78. $\dfrac{4y-1}{2y^2+5y-3} - \dfrac{y+3}{6y^2+y-2}$

Perform each indicated operation. See Examples 1–10.

79. $\dfrac{y^2}{y-2} - \dfrac{9y-14}{y-2}$

80. $\dfrac{y^2}{y-4} - \dfrac{y+12}{y-4}$

81. $\dfrac{3}{x+4} + 7$

82. $\dfrac{9}{x+7} + 2$

83. $\dfrac{-x+2}{x} - \dfrac{x-5}{4x}$

84. $\dfrac{-y+3}{y} - \dfrac{y+4}{3y}$

85. $\dfrac{5x}{x-7} - \dfrac{3x}{x-3}$

86. $\dfrac{6t}{t+4} - \dfrac{2t}{t+1}$

87. $\dfrac{5a}{3a-6} - \dfrac{a-7}{a-2}$

88. $\dfrac{4a}{5a-15} - \dfrac{a-1}{a-3}$

89. $\dfrac{4}{3-x} + \dfrac{x}{2x-6}$

90. $\dfrac{5}{4-x} + \dfrac{x}{2x-8}$

91. $\dfrac{5x + 11}{x^2 - 11x + 18} - \dfrac{4x + 20}{x^2 - 11x + 18}$

92. $\dfrac{4x + 7}{x^2 + 2x - 3} - \dfrac{3x + 4}{x^2 + 2x - 3}$

93. $\dfrac{4}{r^2 - r} + \dfrac{6}{r^2 + 2r} - \dfrac{1}{r^2 + r - 2}$

94. $\dfrac{6}{k^2 + 3k} - \dfrac{1}{k^2 - k} + \dfrac{2}{k^2 + 2k - 3}$

95. $\dfrac{x + 3y}{x^2 + 2xy + y^2} + \dfrac{x - y}{x^2 + 4xy + 3y^2}$

96. $\dfrac{m}{m^2 - 1} + \dfrac{m - 1}{m^2 + 2m + 1}$

97. $\dfrac{r + y}{18r^2 + 9ry - 2y^2} + \dfrac{3r - y}{36r^2 - y^2}$

98. $\dfrac{2x - z}{2x^2 + xz - 10z^2} - \dfrac{x + z}{x^2 - 4z^2}$

*Find an expression that represents (**a**) the perimeter and (**b**) the area of each figure. Give answers in simplified form. (Assume all measures are given in appropriate units. If necessary, refer to the formulas at the back of this text.)*

99.

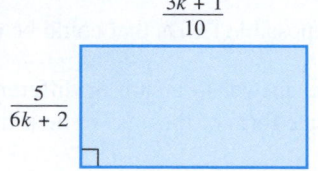

100.

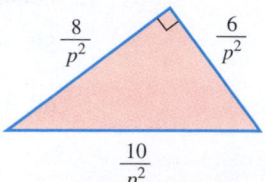

*A **Concours d'Elegance** is a competition in which a maximum of 100 points is awarded to a car based on its general attractiveness. The rational expression*

$$\dfrac{9010}{49(101 - x)} - \dfrac{10}{49}$$

approximates the cost, in thousands of dollars, of restoring a car so that it will win x points.

101. Simplify the given expression by performing the indicated subtraction.

102. Use the simplified expression from **Exercise 101** to determine, to two decimal places, how much it would cost to win 95 points.

6.5   Complex Fractions

OBJECTIVES

1 Define and recognize a complex fraction.

2 Simplify a complex fraction by writing it as a division problem (Method 1).

3 Simplify a complex fraction by multiplying numerator and denominator by the LCD (Method 2).

4 Simplify rational expressions with negative exponents.

VOCABULARY
☐ complex fraction

OBJECTIVE 1 Define and recognize a complex fraction.

The quotient of two mixed numbers can be written as a fraction.

Example: $2\frac{1}{2} \div 3\frac{1}{4}$ can be written as $\dfrac{2\frac{1}{2}}{3\frac{1}{4}}$,   or as $\dfrac{2 + \frac{1}{2}}{3 + \frac{1}{4}}$.

Complex Fraction

A quotient with one or more fractions in the numerator, or denominator, or both, is a **complex fraction.**

Examples: $\dfrac{2 + \frac{1}{2}}{3 + \frac{1}{4}}$, $\dfrac{\frac{3x^2 - 5x}{6x^2}}{2x - \frac{1}{x}}$, $\dfrac{3 + x}{5 - \frac{2}{x}}$ Complex fractions

The parts of a complex fraction are named as follows.

$$\left.\begin{matrix} \dfrac{2}{p} - \dfrac{1}{q} \\[2mm] \rule{3cm}{0.4pt} \\[1mm] \dfrac{3}{p} + \dfrac{5}{q} \end{matrix}\right\}$$

← Numerator of complex fraction

← Main fraction bar

← Denominator of complex fraction

OBJECTIVE 2 Simplify a complex fraction by writing it as a division problem (Method 1).

Because the main fraction bar represents division in a complex fraction, one method of simplifying a complex fraction involves division.

Simplifying a Complex Fraction (Method 1)

Step 1   Write both the numerator and the denominator as single fractions.

Step 2   Change the complex fraction to a division problem.

Step 3   Perform the indicated division.

EXAMPLE 1   Simplifying Complex Fractions (Method 1)

Simplify each complex fraction.

(a) $\dfrac{\dfrac{2}{3} + \dfrac{5}{9}}{\dfrac{1}{4} + \dfrac{1}{12}}$

(b) $\dfrac{6 + \dfrac{3}{x}}{\dfrac{x}{4} + \dfrac{1}{8}}$

Step 1   First, write each numerator as a single fraction.

$$\frac{2}{3} + \frac{5}{9}$$

$$= \frac{2}{3} \cdot \frac{3}{3} + \frac{5}{9}$$

$$= \frac{6}{9} + \frac{5}{9}$$

$$= \frac{11}{9}$$

$$6 + \frac{3}{x}$$

$$= \frac{6}{1} \cdot \frac{x}{x} + \frac{3}{x}$$

$$= \frac{6x}{x} + \frac{3}{x}$$

$$= \frac{6x + 3}{x}$$

Repeat the process for each denominator.

$$\frac{1}{4} + \frac{1}{12}$$

$$= \frac{1}{4} \cdot \frac{3}{3} + \frac{1}{12}$$

$$= \frac{3}{12} + \frac{1}{12}$$

$$= \frac{4}{12}$$

$$\frac{x}{4} + \frac{1}{8}$$

$$= \frac{x}{4} \cdot \frac{2}{2} + \frac{1}{8}$$

$$= \frac{2x}{8} + \frac{1}{8}$$

$$= \frac{2x + 1}{8}$$

Simplify each complex
fraction.

(a) $\dfrac{\frac{2}{5}+\frac{1}{4}}{\frac{1}{6}+\frac{3}{8}}$ (b) $\dfrac{2+\frac{4}{x}}{\frac{5}{6}+\frac{5x}{12}}$

Step 2   Write the equivalent complex fraction as a division problem.

$$\dfrac{\frac{11}{9}}{\frac{4}{12}} \qquad\qquad \dfrac{\frac{6x+3}{x}}{\frac{2x+1}{8}}$$

$$=\frac{11}{9}\div\frac{4}{12} \qquad\qquad =\frac{6x+3}{x}\div\frac{2x+1}{8}$$

Step 3   Now use the definition of division and multiply by the reciprocal. Then write in lowest terms using the fundamental property.

$$=\frac{11}{9}\cdot\frac{12}{4} \qquad\qquad =\frac{6x+3}{x}\cdot\frac{8}{2x+1}$$

$$=\frac{11\cdot3\cdot4}{3\cdot3\cdot4} \qquad\qquad =\frac{3(2x+1)}{x}\cdot\frac{8}{2x+1}$$

$$=\frac{11}{3} \qquad\qquad =\frac{24}{x}$$

Simplify the complex
fraction.

$$\dfrac{\frac{a^2b}{c}}{\frac{ab^2}{c^3}}$$

EXAMPLE 2   Simplifying a Complex Fraction (Method 1)

Simplify the complex fraction.

$$\dfrac{\frac{xp}{q^3}}{\frac{p^2}{qx^2}}\quad$$ The numerator and denominator are already single fractions.

$$=\frac{xp}{q^3}\div\frac{p^2}{qx^2}\quad$$ Write as a division problem.

$$=\frac{xp}{q^3}\cdot\frac{qx^2}{p^2}\quad$$ (*)   Multiply by the reciprocal.

$$=\frac{x\cdot p\cdot q\cdot x^2}{q\cdot q^2\cdot p\cdot p}\quad$$ Multiply. Factor.

$$=\frac{x^3}{q^2p}\quad$$ Divide out the common factors.

NOTE   Alternatively, we can simplify equation (*) in **Example 2** as follows.

$$\frac{xp}{q^3}\cdot\frac{qx^2}{p^2}$$

$$=x^{1+2}p^{1-2}q^{1-3}\quad$$ Product and quotient rules for exponents

$$=x^3p^{-1}q^{-2}\quad$$ Add and subtract exponents.

$$=\frac{x^3}{pq^2}\quad$$ Definition of negative exponent

NOW TRY ANSWERS

1. (a) $\frac{6}{5}$ (b) $\frac{24}{5x}$

2. $\frac{ac^2}{b}$

**NOW TRY
EXERCISE 3**
Simplify the complex
fraction.

$$\dfrac{5 + \dfrac{2}{a-3}}{\dfrac{1}{a-3} - 2}$$

EXAMPLE 3 **Simplifying a Complex Fraction (Method 1)**

Simplify the complex fraction.

$$\dfrac{\dfrac{3}{x+2} - 4}{\dfrac{2}{x+2} + 1}$$

> Find a common denominator before subtracting in the numerator or adding in the denominator.

$$= \dfrac{\dfrac{3}{x+2} - \dfrac{4(x+2)}{x+2}}{\dfrac{2}{x+2} + \dfrac{1(x+2)}{x+2}}$$ Write both second terms with a denominator of $x+2$.

$$= \dfrac{\dfrac{3 - 4(x+2)}{x+2}}{\dfrac{2 + 1(x+2)}{x+2}}$$ Subtract in the numerator.

Add in the denominator.

> Be careful with signs.

$$= \dfrac{\dfrac{3 - 4x - 8}{x+2}}{\dfrac{2 + x + 2}{x+2}}$$ Distributive property

$$= \dfrac{\dfrac{-5 - 4x}{x+2}}{\dfrac{4 + x}{x+2}}$$ Combine like terms.

$$= \dfrac{-5 - 4x}{x+2} \div \dfrac{4+x}{x+2}$$ Write as a division problem.

$$= \dfrac{-5 - 4x}{x+2} \cdot \dfrac{x+2}{4+x}$$ Multiply by the reciprocal of the denominator (divisor).

$$= \dfrac{-5 - 4x}{4 + x}$$ Divide out the common factor.   **NOW TRY** ↻

OBJECTIVE 3 Simplify a complex fraction by multiplying numerator and denominator by the LCD (Method 2).

If we multiply both the numerator and the denominator of a complex fraction by the LCD of all the fractions within the complex fraction, the result will no longer be complex. This is Method 2.

Simplifying a Complex Fraction (Method 2)

Step 1 Find the LCD of all fractions within the complex fraction.

Step 2 Multiply both the numerator and the denominator of the complex fraction by this LCD using the distributive property as necessary. Write in lowest terms.

NOW TRY ANSWER
3. $\dfrac{5a - 13}{7 - 2a}$

**NOW TRY
EXERCISE 4**

Simplify each complex fraction.

(a) $\dfrac{\dfrac{3}{5} - \dfrac{1}{4}}{\dfrac{1}{8} + \dfrac{3}{20}}$ **(b)** $\dfrac{\dfrac{2}{x} - 3}{7 + \dfrac{x}{5}}$

EXAMPLE 4   Simplifying Complex Fractions (Method 2)

Simplify each complex fraction.

(a) $\dfrac{\dfrac{2}{3} + \dfrac{5}{9}}{\dfrac{1}{4} + \dfrac{1}{12}}$ (In **Example 1**, we simplified these same fractions using Method 1.)   **(b)** $\dfrac{6 + \dfrac{3}{x}}{\dfrac{x}{4} + \dfrac{1}{8}}$

Step 1   Find the LCD for all denominators in the complex fraction.

The LCD for 3, 9, 4, and 12 is 36. | The LCD for x, 4, and 8 is $8x$.

Step 2   Multiply numerator and denominator of the complex fraction by the LCD.

$$= \dfrac{36\left(\dfrac{2}{3} + \dfrac{5}{9}\right)}{36\left(\dfrac{1}{4} + \dfrac{1}{12}\right)}$$

Multiply each term by 36.

$$= \dfrac{36\left(\dfrac{2}{3}\right) + 36\left(\dfrac{5}{9}\right)}{36\left(\dfrac{1}{4}\right) + 36\left(\dfrac{1}{12}\right)}$$

$$= \dfrac{24 + 20}{9 + 3} \quad \text{Multiply.}$$

$$= \dfrac{44}{12} \quad \text{Add.}$$

$$= \dfrac{4 \cdot 11}{4 \cdot 3} \quad \text{Factor.}$$

$$= \dfrac{11}{3} \quad \text{Same answer as in Example 1(a)}$$

$$= \dfrac{8x\left(6 + \dfrac{3}{x}\right)}{8x\left(\dfrac{x}{4} + \dfrac{1}{8}\right)}$$

Multiply each term by $8x$.

$$= \dfrac{8x(6) + 8x\left(\dfrac{3}{x}\right)}{8x\left(\dfrac{x}{4}\right) + 8x\left(\dfrac{1}{8}\right)} \quad \text{Distributive property}$$

$$= \dfrac{48x + 24}{2x^2 + x} \quad \text{Multiply.}$$

$$= \dfrac{24(2x + 1)}{x(2x + 1)} \quad \text{Factor.}$$

$$= \dfrac{24}{x} \quad \text{Same answer as in Example 1(b)}$$

NOW TRY

**NOW TRY
EXERCISE 5**

Simplify the complex fraction.

$$\dfrac{\dfrac{1}{y} + \dfrac{2}{3y^2}}{\dfrac{5}{4y^2} - \dfrac{3}{2y^3}}$$

EXAMPLE 5   Simplifying a Complex Fraction (Method 2)

Simplify the complex fraction.

$$\dfrac{\dfrac{3}{5m} - \dfrac{2}{m^2}}{\dfrac{9}{2m} + \dfrac{3}{4m^2}} \quad \text{The LCD for } 5m, m^2, 2m, \text{ and } 4m^2 \text{ is } 20m^2.$$

$$= \dfrac{20m^2\left(\dfrac{3}{5m} - \dfrac{2}{m^2}\right)}{20m^2\left(\dfrac{9}{2m} + \dfrac{3}{4m^2}\right)} \quad \text{Multiply numerator and denominator by } 20m^2.$$

$$= \dfrac{20m^2\left(\dfrac{3}{5m}\right) - 20m^2\left(\dfrac{2}{m^2}\right)}{20m^2\left(\dfrac{9}{2m}\right) + 20m^2\left(\dfrac{3}{4m^2}\right)} \quad \text{Distributive property}$$

$$= \dfrac{12m - 40}{90m + 15}, \quad \text{or} \quad \dfrac{4(3m - 10)}{5(18m + 3)} \quad \text{Multiply and factor.}$$

Either form can be given as the answer.

NOW TRY

NOW TRY ANSWERS

4. (a) $\dfrac{14}{11}$ **(b)** $\dfrac{10 - 15x}{x^2 + 35x}$

5. $\dfrac{12y^2 + 8y}{15y - 18}$, or $\dfrac{4y(3y + 2)}{3(5y - 6)}$

> **NOTE** *Either method can be used to simplify a complex fraction.* A little more or less work may be involved based on the method selected, but the same answer will result if the method is applied correctly **(Examples 1 and 4).**
>
> - We prefer Method 1 for problems that involve the quotient of two fractions, like **Example 2.**
>
> - We prefer Method 2 for complex fractions that have sums or differences in the numerators or denominators, like **Examples 3 and 5.**

EXAMPLE 6 Simplifying Complex Fractions

Simplify each complex fraction. Use either method.

(a) $\dfrac{\dfrac{x+2}{x-3}}{\dfrac{x^2-4}{x^2-9}}$ This is a quotient of two rational expressions. We use Method 1.

$$= \frac{x+2}{x-3} \div \frac{x^2-4}{x^2-9} \qquad \text{Write as a division problem.}$$

$$= \frac{x+2}{x-3} \cdot \frac{x^2-9}{x^2-4} \qquad \text{Multiply by the reciprocal of the divisor.}$$

$$= \frac{(x+2)(x+3)(x-3)}{(x-3)(x+2)(x-2)} \qquad \text{Multiply, and then factor.}$$

$$= \frac{x+3}{x-2} \qquad \text{Divide out the common factors.}$$

(b) $\dfrac{\dfrac{1}{y} + \dfrac{2}{y+2}}{\dfrac{4}{y} - \dfrac{3}{y+2}}$ There are sums and differences in the numerator and denominator. We use Method 2.

$$= \frac{\left(\dfrac{1}{y} + \dfrac{2}{y+2}\right) \cdot y(y+2)}{\left(\dfrac{4}{y} - \dfrac{3}{y+2}\right) \cdot y(y+2)} \qquad \begin{array}{l}\text{Multiply numerator and denominator} \\ \text{by the LCD, } y(y+2)\text{. Because } y \\ \text{appears in two denominators, it} \\ \text{must be a factor in the LCD.}\end{array}$$

$$= \frac{\left(\dfrac{1}{y}\right)y(y+2) + \left(\dfrac{2}{y+2}\right)y(y+2)}{\left(\dfrac{4}{y}\right)y(y+2) - \left(\dfrac{3}{y+2}\right)y(y+2)} \qquad \text{Distributive property}$$

$$= \frac{1(y+2) + 2y}{4(y+2) - 3y} \qquad \begin{array}{l}\text{Multiply. Divide out the} \\ \text{common factors.}\end{array}$$

$$= \frac{y+2+2y}{4y+8-3y} \qquad \text{Distributive property}$$

$$= \frac{3y+2}{y+8} \qquad \text{Combine like terms.}$$

**NOW TRY
EXERCISE 6**
Simplify each complex
fraction.

(a) $\dfrac{\dfrac{9y^2 - 16}{y^2 - 100}}{\dfrac{3y - 4}{y + 10}}$ **(b)** $\dfrac{1 - \dfrac{2}{x} - \dfrac{15}{x^2}}{1 + \dfrac{5}{x} + \dfrac{6}{x^2}}$

(c) $\dfrac{1 - \dfrac{2}{x} - \dfrac{3}{x^2}}{1 - \dfrac{5}{x} + \dfrac{6}{x^2}}$ As in **Example 6(b)**, there are sums and differences in the numerator and denominator. We use Method 2.

$$= \dfrac{\left(1 - \dfrac{2}{x} - \dfrac{3}{x^2}\right)x^2}{\left(1 - \dfrac{5}{x} + \dfrac{6}{x^2}\right)x^2}$$ Multiply numerator and denominator by the LCD, x^2.

$$= \dfrac{x^2 - 2x - 3}{x^2 - 5x + 6}$$ Distributive property

$$= \dfrac{(x - 3)(x + 1)}{(x - 3)(x - 2)}$$ Factor.

$$= \dfrac{x + 1}{x - 2}$$ Divide out the common factor.

NOW TRY

OBJECTIVE 4 Simplify rational expressions with negative exponents.

We begin by rewriting the expressions with only positive exponents. Recall that for any nonzero real number a and any integer n,

$$a^{-n} = \dfrac{1}{a^n}.$$ Definition of negative exponent

⊘ CAUTION $a^{-1} + b^{-1} = \dfrac{1}{a} + \dfrac{1}{b}$, **not** $\dfrac{1}{a + b}$.   Avoid this common error.

EXAMPLE 7   **Simplifying Rational Expressions with Negative Exponents**

Simplify each expression, using only positive exponents in the answer.

(a) $\dfrac{m^{-1} + p^{-2}}{2m^{-2} - p^{-1}}$ $a^{-n} = \dfrac{1}{a^n}$

$$= \dfrac{\dfrac{1}{m} + \dfrac{1}{p^2}}{\dfrac{2}{m^2} - \dfrac{1}{p}}$$ Write with positive exponents.
    $2m^{-2} = 2 \cdot m^{-2} = \dfrac{2}{1} \cdot \dfrac{1}{m^2} = \dfrac{2}{m^2}$

The base of $2m^{-2}$ is m, not $2m$:
$2m^{-2} = \dfrac{2}{m^2}.$

$$= \dfrac{m^2p^2\left(\dfrac{1}{m} + \dfrac{1}{p^2}\right)}{m^2p^2\left(\dfrac{2}{m^2} - \dfrac{1}{p}\right)}$$ Simplify by Method 2. Multiply the numerator and denominator by the LCD, m^2p^2.

$$= \dfrac{m^2p^2 \cdot \dfrac{1}{m} + m^2p^2 \cdot \dfrac{1}{p^2}}{m^2p^2 \cdot \dfrac{2}{m^2} - m^2p^2 \cdot \dfrac{1}{p}}$$ Distributive property

$$= \dfrac{mp^2 + m^2}{2p^2 - m^2p}$$ Write in lowest terms.

NOW TRY ANSWERS
6. **(a)** $\dfrac{3y + 4}{y - 10}$ **(b)** $\dfrac{x - 5}{x + 2}$

**NOW TRY
EXERCISE 7**

Simplify each expression, using only positive exponents in the answer.

(a) $\dfrac{r^{-2} - s^{-1}}{4r^{-1} + s^{-2}}$

(b) $\dfrac{2y^{-1} - 3y^{-2}}{y^{-2} + 3x^{-1}}$

NOW TRY ANSWERS

7. (a) $\dfrac{s^2 - r^2 s}{4rs^2 + r^2}$ (b) $\dfrac{2xy - 3x}{x + 3y^2}$

(b) $\dfrac{x^{-2} - 2y^{-1}}{y - 2x^2}$

> The 2 does *not* go in the denominator of this fraction.

$= \dfrac{\dfrac{1}{x^2} - \dfrac{2}{y}}{y - 2x^2}$ Write with positive exponents.

$= \dfrac{\left(\dfrac{1}{x^2} - \dfrac{2}{y}\right)x^2 y}{(y - 2x^2)x^2 y}$ Use Method 2. Multiply by the LCD, $x^2 y$.

$= \dfrac{y - 2x^2}{(y - 2x^2)x^2 y}$ Use the distributive property in the numerator.

$= \dfrac{1}{x^2 y}$ Write in lowest terms.

> Remember to write 1 in the numerator.

NOW TRY

6.5 Exercises

FOR EXTRA HELP

 MyLab Math

Video solutions for select problems available in MyLab Math

Concept Check *Answer each question.*

1. In a fraction, what operation does the fraction bar represent?

2. What property of real numbers justifies Method 2 of simplifying complex fractions?

Concept Check *In Exercises 3 and 4, consider the following complex fraction.*

$$\dfrac{\dfrac{1}{2} - \dfrac{1}{3}}{\dfrac{5}{6} - \dfrac{1}{12}}$$

3. Answer each part, outlining Method 1 for simplifying the complex fraction.

(a) To combine the terms in the numerator, we must find the LCD of $\frac{1}{2}$ and $\frac{1}{3}$.

What is this LCD?

Determine the simplified form of the numerator of the complex fraction.

(b) To combine the terms in the denominator, we must find the LCD of $\frac{5}{6}$ and $\frac{1}{12}$.

What is this LCD?

Determine the simplified form of the denominator of the complex fraction.

(c) Now use the results from parts (a) and (b) to write the complex fraction as a division problem using the symbol ÷.

(d) Perform the operation from part (c) to obtain the final simplification.

4. Answer each part, outlining Method 2 for simplifying the complex fraction.

(a) We must determine the LCD of all the fractions within the complex fraction.

What is this LCD?

(b) Multiply every term in the complex fraction by the LCD found in part (a), but at this time do not combine the terms in the numerator and the denominator.

(c) Now combine the terms from part (b) to obtain the simplified form of the complex fraction.

Concept Check   *Work each problem.*

5. Which complex fraction is equivalent to $\dfrac{2-\frac{1}{4}}{3-\frac{1}{2}}$? Answer this question without showing any work, and explain your reasoning.

   A. $\dfrac{2+\frac{1}{4}}{3+\frac{1}{2}}$ **B.** $\dfrac{2-\frac{1}{4}}{-3+\frac{1}{2}}$ **C.** $\dfrac{-2-\frac{1}{4}}{-3-\frac{1}{2}}$ **D.** $\dfrac{-2+\frac{1}{4}}{-3+\frac{1}{2}}$

6. Only one of these choices is equal to $\dfrac{\frac{1}{3}+\frac{1}{12}}{\frac{1}{2}+\frac{1}{4}}$. Which one is it? Answer this question without showing any work, and explain your reasoning.

   A. $\dfrac{5}{9}$ **B.** $-\dfrac{5}{9}$ **C.** $-\dfrac{9}{5}$ **D.** $-\dfrac{1}{12}$

*Simplify each complex fraction. Use either method. **See Examples 1–6.***

7. $\dfrac{-\frac{4}{3}}{\frac{2}{9}}$ 8. $\dfrac{-\frac{5}{6}}{\frac{5}{4}}$ 9. $\dfrac{\frac{5}{8}+\frac{2}{3}}{\frac{7}{3}-\frac{1}{4}}$ 10. $\dfrac{\frac{6}{5}-\frac{1}{9}}{\frac{2}{5}+\frac{5}{3}}$

11. $\dfrac{\frac{x}{y^2}}{\frac{x^2}{y}}$ 12. $\dfrac{\frac{p^4}{r}}{\frac{p^2}{r^2}}$ 13. $\dfrac{\frac{p}{6q^2}}{\frac{p^2}{q}}$ 14. $\dfrac{\frac{a}{x}}{\frac{a^2}{2x}}$

15. $\dfrac{\frac{4a^4b^3}{3a}}{\frac{2ab^4}{b^2}}$ 16. $\dfrac{\frac{2r^4t^2}{3t}}{\frac{5r^2t^5}{3r}}$ 17. $\dfrac{\frac{m+2}{3}}{\frac{m-4}{m}}$ 18. $\dfrac{\frac{q-5}{q}}{\frac{q+5}{3}}$

19. $\dfrac{\frac{2}{x}-3}{\frac{2-3x}{2}}$ 20. $\dfrac{6+\frac{2}{r}}{\frac{3r+1}{4}}$ 21. $\dfrac{\frac{1}{x}+x}{\frac{x^2+1}{8}}$ 22. $\dfrac{\frac{3}{m}-m}{\frac{3-m^2}{4}}$

23. $\dfrac{a-\frac{5}{a}}{a+\frac{1}{a}}$ 24. $\dfrac{q+\frac{1}{q}}{q+\frac{4}{q}}$ 25. $\dfrac{\frac{5}{8}+\frac{2}{3}}{\frac{7}{3}-\frac{1}{4}}$ 26. $\dfrac{\frac{6}{5}-\frac{1}{9}}{\frac{2}{5}+\frac{5}{3}}$

27. $\dfrac{\frac{1}{x^2}+\frac{1}{y^2}}{\frac{1}{x}-\frac{1}{y}}$ 28. $\dfrac{\frac{1}{a^2}-\frac{1}{b^2}}{\frac{1}{a}-\frac{1}{b}}$ 29. $\dfrac{\frac{2}{p^2}-\frac{3}{5p}}{\frac{4}{p}+\frac{1}{4p}}$ 30. $\dfrac{\frac{2}{m^2}-\frac{3}{m}}{\frac{2}{5m^2}+\frac{1}{3m}}$

31. $\dfrac{\frac{5}{x^2y}-\frac{2}{xy^2}}{\frac{3}{x^2y^2}+\frac{4}{xy}}$ 32. $\dfrac{\frac{1}{m^3p}+\frac{2}{mp^2}}{\frac{4}{mp}+\frac{1}{m^2p}}$ 33. $\dfrac{\frac{1}{4}-\frac{1}{a^2}}{\frac{1}{2}+\frac{1}{a}}$ 34. $\dfrac{\frac{1}{9}-\frac{1}{m^2}}{\frac{1}{3}+\frac{1}{m}}$

35. $\dfrac{\dfrac{1}{z+5}}{\dfrac{4}{z^2-25}}$

36. $\dfrac{\dfrac{1}{a+1}}{\dfrac{2}{a^2-1}}$

37. $\dfrac{\dfrac{1}{m+1}-1}{\dfrac{1}{m+1}+1}$

38. $\dfrac{\dfrac{2}{x-1}+2}{\dfrac{2}{x-1}-2}$

39. $\dfrac{\dfrac{12}{x+2}+2}{\dfrac{18}{x+2}-2}$

40. $\dfrac{\dfrac{6}{x+3}+3}{\dfrac{9}{x+3}-3}$

41. $\dfrac{\dfrac{x}{y}+\dfrac{y}{x}}{\dfrac{x}{y}-\dfrac{y}{x}}$

42. $\dfrac{\dfrac{x}{y}-\dfrac{y}{x}}{\dfrac{x}{y}+\dfrac{y}{x}}$

43. $\dfrac{1}{\dfrac{1}{a}+\dfrac{1}{b}}$

44. $\dfrac{-1}{\dfrac{1}{a}-\dfrac{1}{b}}$

45. $\dfrac{\dfrac{1}{m-1}+\dfrac{2}{m+2}}{\dfrac{2}{m+2}-\dfrac{1}{m-3}}$

46. $\dfrac{\dfrac{5}{r+3}-\dfrac{1}{r-1}}{\dfrac{2}{r+2}+\dfrac{3}{r+3}}$

47. $\dfrac{2+\dfrac{1}{x}-\dfrac{28}{x^2}}{3+\dfrac{13}{x}+\dfrac{4}{x^2}}$

48. $\dfrac{4-\dfrac{11}{x}-\dfrac{3}{x^2}}{2-\dfrac{1}{x}-\dfrac{15}{x^2}}$

49. $\dfrac{\dfrac{y+8}{y-4}}{\dfrac{y^2-64}{y^2-16}}$

50. $\dfrac{\dfrac{t+5}{t-8}}{\dfrac{t^2-25}{t^2-64}}$

51. $\dfrac{\dfrac{15a^2+15b^2}{5}}{\dfrac{a^4-b^4}{10}}$

52. $\dfrac{\dfrac{14x^2+14y^2}{21}}{\dfrac{x^4-y^4}{27}}$

53. $\dfrac{\dfrac{1}{x^3-y^3}}{\dfrac{1}{x^2-y^2}}$

54. $\dfrac{\dfrac{1}{x^3+y^3}}{\dfrac{1}{x^2+2xy+y^2}}$

55. Concept Check When simplifying the expression

$$\frac{1}{x^{-2}+y^{-2}},$$

a student noted the negative exponents in the denominator and wrote his answer as x^2+y^2. This is incorrect. **WHAT WENT WRONG?**

56. Concept Check A student simplifying the expression

$$\frac{a^{-2}-4b^{-2}}{3b-6a} \quad \text{wrote} \quad \frac{\dfrac{1}{a^2}-\dfrac{1}{4b^2}}{3b-6a}$$

as her first step. This is a common error. **WHAT WENT WRONG?**

Simplify each expression, using only positive exponents in the answer. ***See Example 7.***

57. $\dfrac{1}{x^{-2}+y^{-2}}$

58. $\dfrac{1}{p^{-2}-q^{-2}}$

59. $\dfrac{x^{-2}+y^{-2}}{x^{-1}+y^{-1}}$

60. $\dfrac{x^{-1}-y^{-1}}{x^{-2}-y^{-2}}$

61. $\dfrac{k^{-1}+p^{-2}}{k^{-1}-3p^{-2}}$

62. $\dfrac{x^{-2}-y^{-1}}{x^{-2}+4y^{-1}}$

63. $\dfrac{x^{-1}+2y^{-1}}{2y+4x}$

64. $\dfrac{a^{-2}-4b^{-2}}{3b-6a}$

Extending Skills *Simplify each fraction.*

65. $\dfrac{1+x^{-1}-12x^{-2}}{1-x^{-1}-20x^{-2}}$

66. $\dfrac{1+t^{-1}-56t^{-2}}{1-t^{-1}-72t^{-2}}$

Extending Skills *The fractions here are **continued fractions**. Simplify by starting at "the bottom" and working upward.*

67. $1+\dfrac{1}{1+\dfrac{1}{1+1}}$

68. $5+\dfrac{5}{5+\dfrac{5}{5+5}}$

69. $7-\dfrac{3}{5+\dfrac{2}{4-2}}$

70. $3-\dfrac{2}{4+\dfrac{2}{4-2}}$

71. $r+\dfrac{r}{4-\dfrac{2}{6+2}}$

72. $\dfrac{2q}{7}-\dfrac{q}{6+\dfrac{8}{4+4}}$

RELATING CONCEPTS   **For Individual or Group Work**   (Exercises 73–80)

*Complex fractions have various applications. To find the average of two numbers, we add them and divide by 2. Suppose that we wish to find the **average** of $\frac{3}{8}$ and $\frac{5}{6}$. **Work Exercises 73–76 in order,** to see how a complex fraction occurs in a problem like this.*

73. Write in symbols: "*The sum of $\frac{3}{8}$ and $\frac{5}{6}$, divided by 2.*" The result should be a complex fraction.

74. Use Method 1 to simplify the complex fraction from **Exercise 73.**

75. Use Method 2 to simplify the complex fraction from **Exercise 73.**

76. The answers in **Exercises 74 and 75** should be the same. Which method did you prefer? Why?

*Recall that **slope** measures the steepness of a line and is calculated using the formula*

$$\text{slope } m = \frac{\text{change in } y}{\text{change in } x} = \frac{y_2 - y_1}{x_2 - x_1} \quad (\text{where } x_1 \neq x_2).$$

*Work **Exercises 77–80** to find the slope of the line that passes through each pair of points. This will involve simplifying complex fractions.*

77. $\left(\frac{3}{4}, \frac{1}{3}\right)$ and $\left(\frac{5}{4}, \frac{10}{3}\right)$

78. $\left(\frac{1}{2}, \frac{5}{12}\right)$ and $\left(\frac{1}{4}, \frac{1}{3}\right)$

79. $\left(-\frac{2}{9}, \frac{5}{18}\right)$ and $\left(\frac{1}{18}, -\frac{5}{9}\right)$

80. $\left(-\frac{4}{5}, \frac{1}{2}\right)$ and $\left(-\frac{3}{10}, -\frac{1}{5}\right)$

6.6   Solving Equations with Rational Expressions

OBJECTIVES

1 Distinguish between operations with rational expressions and equations with terms that are rational expressions.

2 Solve equations with rational expressions.

3 Solve a formula for a specified variable.

VOCABULARY

☐ proposed solution
☐ extraneous solution
   (extraneous value)

OBJECTIVE 1   Distinguish between operations with rational expressions and equations with terms that are rational expressions.

Before solving equations with rational expressions, we emphasize the distinction between sums and differences of terms with rational coefficients—that is, rational *expressions*—and *equations* with terms that are rational expressions.

Sums and differences are expressions to simplify. Equations are solved.

EXAMPLE 1   Distinguishing between Expressions and Equations

Identify each of the following as an *expression* or an *equation*. Then simplify the expression or solve the equation.

(a) $\dfrac{3}{4}x - \dfrac{2}{3}x$
This is a difference of two terms. It represents an *expression* to simplify because there is no equality symbol.

$= \dfrac{3 \cdot 3}{3 \cdot 4}x - \dfrac{4 \cdot 2}{4 \cdot 3}x$
The LCD is 12. Write each coefficient with this LCD.

$= \dfrac{9}{12}x - \dfrac{8}{12}x$
Multiply.

$= \dfrac{1}{12}x$
Combine like terms, using the distributive property: $\frac{9}{12}x - \frac{8}{12}x = \left(\frac{9}{12} - \frac{8}{12}\right)x$.

 **NOW TRY**
EXERCISE 1

Identify each of the following as an *expression* or an *equation*. Then simplify the expression or solve the equation.

(a) $\dfrac{3}{2}t - \dfrac{5}{7}t$

(b) $\dfrac{3}{2}t - \dfrac{5}{7}t = \dfrac{11}{7}$

(b)

$$\frac{3}{4}x - \frac{2}{3}x = \frac{1}{2}$$ Because there is an equality symbol, this is an *equation* to be solved.

$$12\left(\frac{3}{4}x - \frac{2}{3}x\right) = 12\left(\frac{1}{2}\right)$$ Use the multiplication property of equality to clear fractions. Multiply by 12, the LCD.

Multiply *each* term by 12.

$$12\left(\frac{3}{4}x\right) - 12\left(\frac{2}{3}x\right) = 12\left(\frac{1}{2}\right)$$ Distributive property

$$9x - 8x = 6$$ Multiply.

$$x = 6$$ Combine like terms.

CHECK

$$\frac{3}{4}x - \frac{2}{3}x = \frac{1}{2}$$ Original equation

$$\frac{3}{4}(6) - \frac{2}{3}(6) \stackrel{?}{=} \frac{1}{2}$$ Let $x = 6$.

$$\frac{9}{2} - 4 \stackrel{?}{=} \frac{1}{2}$$ Multiply.

$$\frac{1}{2} = \frac{1}{2} \checkmark$$ True

A true statement results, so $\{6\}$ is the solution set of the equation.   **NOW TRY**

Uses of the LCD

When adding or subtracting rational expressions, keep the LCD throughout the simplification. (**See Example 1(a).**)

When solving an equation with terms that are rational expressions, multiply each side by the LCD so that denominators are eliminated. (**See Example 1(b).**)

OBJECTIVE 2 Solve equations with rational expressions.

When an equation involves fractions, as in **Example 1(b)**, we multiply by the LCD of all denominators so that the resulting equation contains no fractions.

EXAMPLE 2 Solving an Equation with Rational Expressions

Solve, and check the solution.

$$\frac{x}{3} + \frac{x}{4} = 10 + x$$

$$12\left(\frac{x}{3} + \frac{x}{4}\right) = 12(10 + x)$$ Multiply by the LCD, 12, to clear fractions.

$$12\left(\frac{x}{3}\right) + 12\left(\frac{x}{4}\right) = 12(10) + 12x$$ Distributive property

$$4x + 3x = 120 + 12x$$ Multiply.

$$7x = 120 + 12x$$ Combine like terms.

$$-5x = 120$$ Subtract 12x.

$$x = -24$$ Divide by −5.

NOW TRY ANSWERS

1. (a) expression; $\frac{11}{14}t$

   (b) equation; $\{2\}$

NOW TRY
EXERCISE 2

Solve, and check the solution.

$$\frac{x}{6} + \frac{x}{3} = 6 + x$$

CHECK

$$\frac{x}{3} + \frac{x}{4} = 10 + x \qquad \text{Original equation}$$

$$\frac{-24}{3} + \frac{-24}{4} \stackrel{?}{=} 10 + (-24) \qquad \text{Let } x = -24.$$

$$-8 + (-6) \stackrel{?}{=} -14 \qquad \text{Divide. Add.}$$

$$-14 = -14 \; \checkmark \qquad \text{True}$$

A true statement results, so the solution set is $\{-24\}$.

NOW TRY

❗ **CAUTION** *Be careful not to confuse the following procedures.*

- In **Examples 2 and 3,** we use the multiplication property of equality to multiply each side of an *equation* by the LCD.

- In our work with complex fractions, we used the fundamental property to multiply a *fraction* (an **expression**) by another fraction that had the LCD as both its numerator and its denominator.

NOW TRY
EXERCISE 3

Solve, and check the solution.

$$\frac{x}{7} - \frac{x+5}{5} = -\frac{3}{7}$$

EXAMPLE 3 Solving an Equation with Rational Expressions

Solve, and check the solution.

$$\frac{p}{2} - \frac{p-1}{3} = 1$$

$$6\left(\frac{p}{2} - \frac{p-1}{3}\right) = 6(1) \qquad \begin{array}{l}\text{Multiply each side by the LCD, 6,}\\ \text{to clear fractions.}\end{array}$$

$$6\left(\frac{p}{2}\right) - 6\left(\frac{p-1}{3}\right) = 6(1) \qquad \text{Distributive property}$$

$$3p - 2(p-1) = 6 \quad \longleftarrow \begin{array}{l}\text{Use parentheses around}\\ p-1 \text{ to avoid errors.}\end{array}$$

$$3p - 2(p) - 2(-1) = 6 \qquad \text{Distributive property}$$

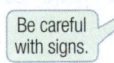
Be careful
with signs.

$$3p - 2p + 2 = 6 \qquad \text{Multiply.}$$

$$p + 2 = 6 \qquad \text{Combine like terms.}$$

$$p = 4 \qquad \text{Subtract 2.}$$

CHECK

$$\frac{p}{2} - \frac{p-1}{3} = 1 \qquad \text{Original equation}$$

$$\frac{4}{2} - \frac{4-1}{3} \stackrel{?}{=} 1 \qquad \text{Let } p = 4.$$

$$2 - 1 \stackrel{?}{=} 1 \qquad \text{Simplify.}$$

$$1 = 1 \; \checkmark \qquad \text{True}$$

NOW TRY ANSWERS
2. $\{-12\}$
3. $\{-10\}$

The solution set is $\{4\}$.

NOW TRY

Recall that division by 0 is undefined. *When we solve an equation with rational expressions that have variables in the denominator, the solution cannot be a number that makes the denominator equal 0.*

A value of the variable that appears to be a solution after both sides of a rational equation are multiplied by a variable expression is a **proposed solution.** *All proposed solutions must be checked in the original equation.*

NOW TRY EXERCISE 4

Solve, and check the proposed solution.

$$4 + \frac{6}{x-3} = \frac{2x}{x-3}$$

EXAMPLE 4   Solving an Equation with Rational Expressions

Solve, and check the proposed solution.

$$\frac{x}{x-2} = \frac{2}{x-2} + 2 \qquad \begin{array}{l}\textit{x cannot equal 2 because 2}\\ \textit{causes both denominators}\\ \textit{to equal 0.}\end{array}$$

$$(x-2)\left(\frac{x}{x-2}\right) = (x-2)\left(\frac{2}{x-2} + 2\right) \qquad \begin{array}{l}\text{Multiply each side by the LCD,}\\ x-2.\end{array}$$

$$(x-2)\left(\frac{x}{x-2}\right) = (x-2)\left(\frac{2}{x-2}\right) + (x-2)2 \qquad \text{Distributive property}$$

$$x = 2 + 2x - 4 \qquad \text{Simplify.}$$

$$x = -2 + 2x \qquad \text{Combine like terms.}$$

$$-x = -2 \qquad \text{Subtract } 2x.$$

$$\begin{array}{l}\text{Proposed}\\ \text{solution}\end{array} \rightarrow x = 2 \qquad \text{Multiply by } -1.$$

As noted, x cannot equal 2 because replacing x with 2 in the original equation causes the denominators to equal 0. We see this in the following check.

CHECK

$$\frac{x}{x-2} = \frac{2}{x-2} + 2 \qquad \text{Original equation}$$

$$\frac{2}{2-2} \overset{?}{=} \frac{2}{2-2} + 2 \qquad \text{Let } x = 2.$$

$$\frac{2}{0} \overset{?}{=} \frac{2}{0} + 2 \qquad \text{Subtract in the denominators.}$$

Division by 0 is undefined.

Thus, the proposed solution 2 must be rejected. The solution set is $\varnothing$.   **NOW TRY**

A proposed solution that is not an actual solution of the original equation, such as 2 in **Example 4,** is an **extraneous solution,** or **extraneous value.** Some students like to determine which numbers cannot be solutions *before* solving the equation, as we did at the beginning of **Example 4.**

Solving an Equation with Rational Expressions

Step 1   **Multiply each side of the equation by the LCD.** (This clears the equation of fractions.) Be sure to distribute to *every* term on *both* sides of the equation.

Step 2   **Solve** the resulting equation for proposed solutions.

Step 3   **Check** each proposed solution by substituting it into the original equation. Reject any values that cause a denominator to equal 0.

NOW TRY ANSWER

4. $\varnothing$

NOW TRY
EXERCISE 5
Solve, and check the proposed solution.

$$\frac{3}{2x^2 - 8x} = \frac{1}{x^2 - 16}$$

EXAMPLE 5   Solving an Equation with Rational Expressions

Solve, and check the proposed solution.

$$\frac{2}{x^2 - x} = \frac{1}{x^2 - 1}$$

Step 1 $\dfrac{2}{x(x - 1)} = \dfrac{1}{(x + 1)(x - 1)}$ Factor the denominators to find the LCD.

The LCD is $x(x + 1)(x - 1)$. *Notice that 0, 1, and −1 cannot be solutions.* Otherwise, a denominator will equal 0. Multiply both sides of the equation by the LCD to clear the fractions.

$$x(x + 1)(x - 1)\left(\frac{2}{x(x - 1)}\right) = x(x + 1)(x - 1)\left(\frac{1}{(x + 1)(x - 1)}\right)$$

Multiply by the LCD to clear the fractions.

Step 2 $(x + 1)2 = x$ Divide out the common factors.

$$2x + 2 = x$$ Distributive property

$$x + 2 = 0$$ Subtract x.

Proposed solution → $x = -2$ Subtract 2.

Step 3   The proposed solution −2 does not make any denominator equal 0.

CHECK $\dfrac{2}{x^2 - x} = \dfrac{1}{x^2 - 1}$ Original equation

$$\frac{2}{(-2)^2 - (-2)} \stackrel{?}{=} \frac{1}{(-2)^2 - 1}$$ Let $x = -2$.

$$\frac{2}{4 + 2} \stackrel{?}{=} \frac{1}{4 - 1}$$ Apply the exponents; Definition of subtraction

$$\frac{1}{3} = \frac{1}{3}\ \checkmark$$ True

A true statement results, so the solution set is $\{-2\}$.   **NOW TRY**

EXAMPLE 6   Solving an Equation with Rational Expressions

Solve, and check the proposed solution.

$$\frac{2m}{m^2 - 4} + \frac{1}{m - 2} = \frac{2}{m + 2}$$

$$\frac{2m}{(m + 2)(m - 2)} + \frac{1}{m - 2} = \frac{2}{m + 2}$$ Factor the first denominator on the left to find the LCD, $(m + 2)(m - 2)$.

Notice that −2 and 2 cannot be solutions of this equation.

$$(m + 2)(m - 2)\left(\frac{2m}{(m + 2)(m - 2)} + \frac{1}{m - 2}\right)$$

Multiply by the LCD.

$$= (m + 2)(m - 2)\left(\frac{2}{m + 2}\right)$$

NOW TRY ANSWER
5. $\{-12\}$

NOW TRY
EXERCISE 6

Solve, and check the proposed solution.

$$\frac{2y}{y^2 - 25} = \frac{8}{y + 5} - \frac{1}{y - 5}$$

$$(m + 2)(m - 2)\left(\frac{2m}{(m + 2)(m - 2)}\right) + (m + 2)(m - 2)\left(\frac{1}{m - 2}\right)$$

$$= (m + 2)(m - 2)\left(\frac{2}{m + 2}\right) \quad \text{Distributive property}$$

$$2m + m + 2 = (m - 2)2 \qquad \text{Divide out the common factors.}$$

$$3m + 2 = 2m - 4 \qquad \text{Combine like terms; distributive property}$$

$$m + 2 = -4 \qquad \text{Subtract } 2m.$$

$$m = -6 \qquad \text{Subtract 2.}$$

CHECK $\qquad \dfrac{2m}{m^2 - 4} + \dfrac{1}{m - 2} = \dfrac{2}{m + 2} \qquad$ Original equation

$$\frac{2(-6)}{(-6)^2 - 4} + \frac{1}{-6 - 2} \stackrel{?}{=} \frac{2}{-6 + 2} \qquad \text{Let } m = -6.$$

$$\frac{-12}{32} + \frac{1}{-8} \stackrel{?}{=} \frac{2}{-4} \qquad \text{Apply the exponent. Subtract and add.}$$

$$-\frac{1}{2} = -\frac{1}{2} \quad \checkmark \qquad \text{True}$$

A true statement results, so the solution set is $\{-6\}$.   **NOW TRY**

EXAMPLE 7 Solving an Equation with Rational Expressions

Solve, and check the proposed solution(s).

$$\frac{1}{x - 1} + \frac{1}{2} = \frac{2}{x^2 - 1}$$

$x \neq 1, -1.$ Otherwise, a denominator is 0.

$$\frac{1}{x - 1} + \frac{1}{2} = \frac{2}{(x + 1)(x - 1)} \qquad \begin{array}{l}\text{Factor the denominator} \\ \text{on the right. The LCD} \\ \text{is } 2(x + 1)(x - 1).\end{array}$$

$$2(x + 1)(x - 1)\left(\frac{1}{x - 1} + \frac{1}{2}\right) = 2(x + 1)(x - 1)\left(\frac{2}{(x + 1)(x - 1)}\right)$$

Multiply by the LCD.

$$2(x + 1)(x - 1)\left(\frac{1}{x - 1}\right) + 2(x + 1)(x - 1)\left(\frac{1}{2}\right) \qquad \text{Distributive property}$$

$$= 2(x + 1)(x - 1)\left(\frac{2}{(x + 1)(x - 1)}\right)$$

$$2(x + 1) + (x + 1)(x - 1) = 2(2) \qquad \text{Divide out the common factors.}$$

$$2x + 2 + x^2 - 1 = 4 \qquad \text{Distributive property; Multiply.}$$

Write in standard form.

$$x^2 + 2x - 3 = 0 \qquad \text{Subtract 4. Combine like terms.}$$

$$(x + 3)(x - 1) = 0 \qquad \text{Factor.}$$

$$x + 3 = 0 \quad \text{or} \quad x - 1 = 0 \qquad \text{Zero-factor property}$$

$$x = -3 \quad \text{or} \qquad x = 1 \quad \longleftarrow \quad \text{Proposed solutions}$$

NOW TRY ANSWER

6. $\{9\}$

Because 1 makes a denominator equal 0, the proposed solution 1 is an extraneous value. Check that -3 is a solution.

**NOW TRY
EXERCISE 7**

Solve, and check the proposed solution(s).

$$\frac{3}{m^2 - 9} = \frac{1}{2(m - 3)} - \frac{1}{4}$$

CHECK

$$\frac{1}{x - 1} + \frac{1}{2} = \frac{2}{x^2 - 1} \qquad \text{Original equation}$$

$$\frac{1}{-3 - 1} + \frac{1}{2} \stackrel{?}{=} \frac{2}{(-3)^2 - 1} \qquad \text{Let } x = -3.$$

$$\frac{1}{-4} + \frac{1}{2} \stackrel{?}{=} \frac{2}{9 - 1} \qquad \text{Subtract. Apply the exponent.}$$

$$\frac{1}{4} = \frac{1}{4} \checkmark \qquad \text{True}$$

A true statement results, so the solution set is $\{-3\}$.   **NOW TRY**

**NOW TRY
EXERCISE 8**

Solve, and check the proposed solution.

$$\frac{5}{k^2 + k - 2} = \frac{1}{3k - 3} - \frac{1}{k + 2}$$

EXAMPLE 8   Solving an Equation with Rational Expressions

Solve, and check the proposed solution.

$$\frac{1}{k^2 + 4k + 3} + \frac{1}{2k + 2} = \frac{3}{4k + 12}$$

$$\frac{1}{(k + 1)(k + 3)} + \frac{1}{2(k + 1)} = \frac{3}{4(k + 3)} \qquad \begin{array}{l}\text{Factor each denominator.} \\ \text{The LCD is } 4(k + 1)(k + 3).\end{array}$$

$k \neq -1, -3$

$$4(k + 1)(k + 3)\left(\frac{1}{(k + 1)(k + 3)} + \frac{1}{2(k + 1)}\right)$$

$$= 4(k + 1)(k + 3)\left(\frac{3}{4(k + 3)}\right) \qquad \begin{array}{l}\text{Multiply by} \\ \text{the LCD.}\end{array}$$

$$4(k + 1)(k + 3)\left(\frac{1}{(k + 1)(k + 3)}\right) + 2 \cdot 2(k + 1)(k + 3)\left(\frac{1}{2(k + 1)}\right)$$

Do *not* add $4 + 2$ here.
$$= 4(k + 1)(k + 3)\left(\frac{3}{4(k + 3)}\right) \qquad \begin{array}{l}\text{Distributive} \\ \text{property}\end{array}$$

$$4 + 2(k + 3) = (k + 1)3 \qquad \text{Divide out the common factors.}$$

$$4 + 2k + 6 = 3k + 3 \qquad \text{Distributive property}$$

$$2k + 10 = 3k + 3 \qquad \text{Combine like terms.}$$

$$10 = k + 3 \qquad \text{Subtract } 2k.$$

$$7 = k \qquad \text{Subtract } 3.$$

CHECK

$$\frac{1}{k^2 + 4k + 3} + \frac{1}{2k + 2} = \frac{3}{4k + 12} \qquad \text{Original equation}$$

$$\frac{1}{7^2 + 4(7) + 3} + \frac{1}{2(7) + 2} \stackrel{?}{=} \frac{3}{4(7) + 12} \qquad \text{Let } k = 7.$$

$$\frac{1}{80} + \frac{1}{16} \stackrel{?}{=} \frac{3}{40} \qquad \begin{array}{l}\text{Apply the exponent.} \\ \text{Multiply and add.}\end{array}$$

$$\frac{3}{40} = \frac{3}{40} \checkmark \qquad \text{True}$$

NOW TRY ANSWERS
7. $\{-1\}$
8. $\{-5\}$

A true statement results, so the solution set is $\{7\}$.   **NOW TRY**

OBJECTIVE 3 Solve a formula for a specified variable.

When solving a formula for a specified variable, remember to treat the variable for which you are solving as if it were the only variable, and all others as if they were constants.

NOW TRY
EXERCISE 9

Solve the following formula for x.

$$p = \frac{x - y}{z}$$

EXAMPLE 9 Solving for a Specified Variable

Solve the following formula for v.

$$a = \frac{v - w}{t} \quad \text{Our goal is to isolate } v.$$

$$at = \left(\frac{v - w}{t}\right) t \qquad \text{Multiply by } t \text{ to clear the fraction.}$$

$$at = v - w \qquad \text{Divide out the common factor.}$$

$$at + w = v \qquad \text{Add } w.$$

$$v = at + w \qquad \text{Interchange sides.}$$

CHECK

$$a = \frac{v - w}{t} \qquad \text{Original equation}$$

$$a \stackrel{?}{=} \frac{at + w - w}{t} \qquad \text{Let } v = at + w.$$

$$a \stackrel{?}{=} \frac{at}{t} \qquad \text{Combine like terms.}$$

$$a = a \checkmark \qquad \text{True}$$

A true statement results, so $v = at + w$. **NOW TRY**

NOW TRY
EXERCISE 10

Solve the following formula for d.

$$a = \frac{b}{c + d}$$

EXAMPLE 10 Solving for a Specified Variable

Solve the following formula for d.

$$F = \frac{k}{d - D} \quad \text{We must isolate } d.$$

$$F(d - D) = \frac{k}{d - D}(d - D) \qquad \text{Multiply by } d - D \text{ to clear the fraction.}$$

$$F(d - D) = k \qquad \text{Divide out the common factor.}$$

$$Fd - FD = k \qquad \text{Distributive property}$$

$$Fd = k + FD \qquad \text{Add } FD.$$

$$d = \frac{k + FD}{F} \qquad \text{Divide by } F.$$

We can write an equivalent form of this answer as follows.

$$d = \frac{k + FD}{F} \qquad \text{Answer from above}$$

$$d = \frac{k}{F} + \frac{FD}{F} \qquad \text{Definition of addition of fractions, } \frac{a + b}{c} = \frac{a}{c} + \frac{b}{c}$$

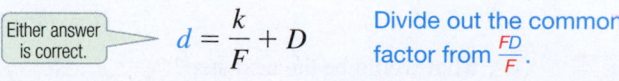

$$d = \frac{k}{F} + D \qquad \text{Divide out the common factor from } \frac{FD}{F}.$$

NOW TRY

NOW TRY ANSWERS

9. $x = pz + y$

10. $d = \frac{b - ac}{a}$, or $d = \frac{b}{a} - c$

NOW TRY EXERCISE 11

Solve the following formula for x.

$$\frac{2}{w} = \frac{1}{x} - \frac{3}{y}$$

EXAMPLE 11 Solving for a Specified Variable

Solve the following formula for c.

$$\frac{1}{a} = \frac{1}{b} + \frac{1}{c}$$

Goal: Isolate c, the specified variable.

$$abc\left(\frac{1}{a}\right) = abc\left(\frac{1}{b} + \frac{1}{c}\right)$$ Multiply by the LCD, abc, to clear the fractions.

$$abc\left(\frac{1}{a}\right) = abc\left(\frac{1}{b}\right) + abc\left(\frac{1}{c}\right)$$ Distributive property

$$bc = ac + ab$$ Divide out the common factors.

$$bc - ac = ab$$ (*) Subtract ac so that both terms with c are on the same side.

Pay careful attention here.

$$c(b - a) = ab$$ Factor out c.

$$c = \frac{ab}{b - a}$$ Divide by $b - a$.

NOW TRY

⚠ **CAUTION** In **Example 11**, we transformed to obtain equation (*), which has *both* terms with c on the same side of the equality symbol. This key step enabled us to factor out c on the left and ultimately isolate it.

When solving an equation for a specified variable, be sure that the specified variable appears alone on only one side of the equality symbol in the final equation.

NOW TRY ANSWER
11. $x = \frac{wy}{2y + 3w}$

6.6 Exercises

FOR EXTRA HELP

▶ **MyLab Math**

▶ *Video solutions for select problems available in MyLab Math*

Concept Check *Answer each question.*

1. What is the least positive whole number by which we can multiply both sides of the equation

$$\frac{2}{3}x + \frac{1}{4}x = 6$$

to obtain an equation with only integer coefficients?

2. Before even beginning to solve the equation

$$\frac{1}{x - 3} + \frac{2}{3 - x} = 4,$$

what number do we know cannot be a solution? Why?

3. What is the simplest monomial by which we can multiply both sides of the equation

$$\frac{1}{x} - \frac{1}{y} = \frac{1}{z}$$

so that there are no variables in the denominators?

4. If we were solving an equation for the variable k, and our steps led to the equation

$$kr - mr = km,$$

what would be the next step?

Identify each as an expression *or an* equation. *Then* simplify the expression *or* solve the equation. *See Example 1.*

5. $\dfrac{7}{8}x + \dfrac{1}{5}x$

6. $\dfrac{4}{7}x + \dfrac{3}{5}x$

7. $\dfrac{7}{8}x + \dfrac{1}{5}x = 1$

8. $\dfrac{4}{7}x + \dfrac{3}{5}x = 1$

9. $\dfrac{3}{5}y - \dfrac{7}{10}y$

10. $\dfrac{2}{3}y - \dfrac{9}{4}y$

11. $\dfrac{3}{5}x - \dfrac{7}{10}x = 1$

12. $\dfrac{2}{3}x - \dfrac{9}{4}x = -19$

Solve each equation, and check the solutions. See Examples 2 and 3.

13. $\dfrac{2}{3}x + \dfrac{1}{2}x = -7$

14. $\dfrac{1}{4}x - \dfrac{1}{3}x = 1$

15. $\dfrac{3x}{5} - 6 = x$

16. $\dfrac{5t}{4} + t = 9$

17. $\dfrac{4m}{7} + m = 11$

18. $x - \dfrac{3x}{2} = 1$

19. $\dfrac{2x + 3}{-6} = \dfrac{3}{2}$

20. $\dfrac{4x + 3}{6} = \dfrac{5}{2}$

21. $\dfrac{z - 1}{4} = \dfrac{z + 3}{3}$

22. $\dfrac{r - 5}{2} = \dfrac{r + 2}{3}$

23. $\dfrac{3p + 6}{8} = \dfrac{3p - 3}{16}$

24. $\dfrac{2z + 1}{5} = \dfrac{7z + 5}{15}$

25. $\dfrac{x}{2} + \dfrac{x}{7} = 5 + x$

26. $\dfrac{x}{5} + \dfrac{x}{2} = 6 + x$

27. $\dfrac{2x}{3} - \dfrac{x}{9} = x - 4$

28. $\dfrac{3x}{5} - \dfrac{x}{10} = x - 6$

29. $\dfrac{r}{6} - \dfrac{r - 2}{3} = -\dfrac{4}{3}$

30. $\dfrac{p}{2} - \dfrac{p - 1}{4} = \dfrac{5}{4}$

31. $\dfrac{q + 2}{3} + \dfrac{q - 5}{5} = \dfrac{7}{3}$

32. $\dfrac{x - 6}{6} + \dfrac{x + 2}{8} = \dfrac{11}{4}$

33. $\dfrac{a + 7}{8} - \dfrac{a - 2}{3} = \dfrac{4}{3}$

34. $\dfrac{x + 3}{7} - \dfrac{x + 2}{6} = \dfrac{1}{6}$

35. $\dfrac{3m}{5} - \dfrac{3m - 2}{4} = \dfrac{1}{5}$

36. $\dfrac{8p}{5} - \dfrac{3p - 4}{2} = \dfrac{5}{2}$

When solving an equation with variables in denominators, we must determine the values that cause these denominators to equal 0, so that we can reject these values if they appear as proposed solutions. Find all values for which at least one denominator is equal to 0. Write answers using the symbol ≠. Do not solve. See Examples 4–8.

37. $\dfrac{3}{x + 2} - \dfrac{5}{x} = 1$

38. $\dfrac{7}{x} + \dfrac{9}{x - 4} = 5$

39. $\dfrac{-1}{(x + 3)(x - 4)} = \dfrac{1}{2x + 1}$

40. $\dfrac{8}{(x - 7)(x + 3)} = \dfrac{7}{3x - 10}$

41. $\dfrac{4}{x^2 + 8x - 9} + \dfrac{1}{x^2 - 4} = 0$

42. $\dfrac{-3}{x^2 + 9x - 10} - \dfrac{12}{x^2 - 49} = 0$

Solve each equation, and check the solutions. See Examples 4–8.

43. $\dfrac{5}{m} - \dfrac{3}{m} = 8$

44. $\dfrac{4}{x} + \dfrac{1}{x} = 2$

45. $\dfrac{5}{x} + 4 = \dfrac{2}{x}$

46. $\dfrac{11}{q} - 3 = \dfrac{1}{q}$

47. $\dfrac{2x - 7}{x} = \dfrac{17}{5}$

48. $\dfrac{2x + 3}{x} = \dfrac{3}{2}$

49. $\dfrac{k}{k - 4} - 5 = \dfrac{4}{k - 4}$

50. $\dfrac{-5}{x + 5} = \dfrac{x}{x + 5} + 2$

51. $\dfrac{3}{x - 1} + \dfrac{2}{4x - 4} = \dfrac{7}{4}$

52. $\dfrac{2}{p + 3} + \dfrac{3}{8} = \dfrac{5}{4p + 12}$

53. $\dfrac{2}{m} = \dfrac{m}{5m + 12}$

54. $\dfrac{x}{4 - x} = \dfrac{2}{x}$

55. $\dfrac{5x}{14x + 3} = \dfrac{1}{x}$

56. $\dfrac{m}{8m + 3} = \dfrac{1}{3m}$

57. $\dfrac{2}{z - 1} - \dfrac{5}{4} = \dfrac{-1}{z + 1}$

58. $\dfrac{5}{p - 2} = 7 - \dfrac{10}{p + 2}$

59. $\dfrac{x}{3x + 3} = \dfrac{2x - 3}{x + 1} - \dfrac{2x}{3x + 3}$

60. $\dfrac{2k + 3}{k + 1} - \dfrac{3k}{2k + 2} = \dfrac{-2k}{2k + 2}$

61. $\dfrac{4}{x^2 - 3x} = \dfrac{1}{x^2 - 9}$

62. $\dfrac{2}{t^2 - 4} = \dfrac{3}{t^2 - 2t}$

63. $\dfrac{-2}{z + 5} + \dfrac{3}{z - 5} = \dfrac{20}{z^2 - 25}$

64. $\dfrac{3}{r + 3} - \dfrac{2}{r - 3} = \dfrac{-12}{r^2 - 9}$

65. $\dfrac{1}{x + 4} + \dfrac{x}{x - 4} = \dfrac{-8}{x^2 - 16}$

66. $\dfrac{x}{x - 3} + \dfrac{4}{x + 3} = \dfrac{18}{x^2 - 9}$

67. $\dfrac{2p}{p^2 - 1} = \dfrac{2}{p + 1} - \dfrac{1}{p - 1}$

68. $\dfrac{2x}{x^2 - 16} - \dfrac{2}{x - 4} = \dfrac{4}{x + 4}$

69. $\dfrac{9}{3x + 4} = \dfrac{36 - 27x}{16 - 9x^2}$

70. $\dfrac{25}{5x - 6} = \dfrac{-150 - 125x}{36 - 25x^2}$

71. $\dfrac{4}{3x + 6} - \dfrac{3}{x + 3} = \dfrac{8}{x^2 + 5x + 6}$

72. $\dfrac{-13}{t^2 + 6t + 8} + \dfrac{4}{t + 2} = \dfrac{3}{2t + 8}$

73. $\dfrac{1}{x^2 - 1} = \dfrac{2}{x - 1} - \dfrac{1}{x - 1}$

74. $\dfrac{3}{x^2 - 9} = \dfrac{3}{x - 3} - \dfrac{2}{x - 3}$

75. $\dfrac{2}{x + 2} + \dfrac{1}{x^2 - 2x + 4} = \dfrac{24}{x^3 + 8}$

76. $\dfrac{2}{x - 3} + \dfrac{1}{x^2 + 3x + 9} = \dfrac{54}{x^3 - 27}$

77. $\dfrac{3x}{x^2 + 5x + 6} = \dfrac{5x}{x^2 + 2x - 3} - \dfrac{2}{x^2 + x - 2}$

78. $\dfrac{m}{m^2 + m - 2} = \dfrac{m}{m^2 + 3m + 2} - \dfrac{m}{m^2 - 1}$

79. $\dfrac{x + 4}{x^2 - 3x + 2} - \dfrac{5}{x^2 - 4x + 3} = \dfrac{x - 4}{x^2 - 5x + 6}$

80. $\dfrac{3}{r^2 + r - 2} - \dfrac{1}{r^2 - 1} = \dfrac{7}{2r^2 + 6r + 4}$

81. Concept Check   A student simplified the following expression as shown.

$$\dfrac{3}{2}t + \dfrac{5}{7}t$$

$$= 14\left(\dfrac{3}{2}t + \dfrac{5}{7}t\right)$$

$$= 21t + 10t$$

$$= 31t \qquad \text{Incorrect}$$

WHAT WENT WRONG? Give the correct answer.

82. Concept Check   A student solved the following formula for r as shown.

$$\dfrac{1}{r} - \dfrac{1}{m} = \dfrac{1}{k}$$

$$rmk\left(\dfrac{1}{r} - \dfrac{1}{m}\right) = rmk\left(\dfrac{1}{k}\right)$$

$$mk - rk = rm$$

$$\dfrac{mk - rk}{m} = r \qquad \text{Incorrect}$$

WHAT WENT WRONG? Give the correct answer.

Solve each formula or equation for the specified variable. See Examples 9–11.

83. $m = \dfrac{kF}{a}$ for F

84. $I = \dfrac{kE}{R}$ for E

85. $m = \dfrac{kF}{a}$ for a

86. $I = \dfrac{kE}{R}$ for R

87. $m = \dfrac{y - b}{x}$ for y

88. $y = \dfrac{C - Ax}{B}$ for C

89. $I = \dfrac{E}{R + r}$ for R

90. $I = \dfrac{E}{R + r}$ for r

91. $h = \dfrac{2\mathcal{A}}{B + b}$ for $\mathcal{A}$

92. $d = \dfrac{2S}{n(a + L)}$ for S

93. $d = \dfrac{2S}{n(a + L)}$ for a

94. $h = \dfrac{2\mathcal{A}}{B + b}$ for B

95. $\dfrac{1}{x} = \dfrac{1}{y} - \dfrac{1}{z}$ for y

96. $\dfrac{3}{k} = \dfrac{1}{p} + \dfrac{1}{q}$ for q

97. $\dfrac{2}{r} + \dfrac{3}{s} + \dfrac{1}{t} = 1$ for t

98. $\dfrac{5}{p} + \dfrac{2}{q} + \dfrac{3}{r} = 1$ for r

99. $9x + \dfrac{3}{z} = \dfrac{5}{y}$ for z

100. $-3t - \dfrac{4}{p} = \dfrac{6}{s}$ for p

101. $\dfrac{t}{x - 1} - \dfrac{2}{x + 1} = \dfrac{1}{x^2 - 1}$ for t

102. $\dfrac{5}{y + 2} - \dfrac{r}{y - 2} = \dfrac{3}{y^2 - 4}$ for r

RELATING CONCEPTS For Individual or Group Work (Exercises 103–110)

In these exercises, we summarize various concepts involving rational expressions. **Work Exercises 103–110 in order.**

Let P, Q, and R be rational expressions defined as follows.

$$P = \frac{6}{x + 3}, \qquad Q = \frac{5}{x + 1}, \qquad R = \frac{4x}{x^2 + 4x + 3}$$

103. Find the values for which each expression is undefined. Write answers using the symbol $\neq$.

  (a) P **(b)** Q **(c)** R

104. Find and express $(P \cdot Q) \div R$ in lowest terms.

105. Why is $(P \cdot Q) \div R$ not defined if $x = 0$?

106. Find the LCD for P, Q, and R.

107. Perform the operations and express $P + Q - R$ in lowest terms.

108. Simplify the complex fraction $\dfrac{P + Q}{R}$.

109. Solve the equation $P + Q = R$.

110. How does the answer to **Exercise 103** help in working **Exercise 109**?

SUMMARY EXERCISES Simplifying Rational Expressions vs. Solving Rational Equations

Students often confuse *simplifying an expression* with *solving an equation*. We review the four operations to *simplify* the following rational *expressions*.

Add: $\dfrac{1}{x} + \dfrac{1}{x-2}$

$\qquad = \dfrac{1(x-2)}{x(x-2)} + \dfrac{x(1)}{x(x-2)}$ Write with a common denominator.

$\qquad = \dfrac{x-2+x}{x(x-2)}$ Add numerators.
Keep the same denominator.

$\qquad = \dfrac{2x-2}{x(x-2)}$ Combine like terms.

Subtract: $\dfrac{1}{x} - \dfrac{1}{x-2}$

$\qquad = \dfrac{1(x-2)}{x(x-2)} - \dfrac{x(1)}{x(x-2)}$ Write with a common denominator.

$\qquad = \dfrac{x-2-x}{x(x-2)}$ Subtract numerators.
Keep the same denominator.

$\qquad = \dfrac{-2}{x(x-2)}$ Combine like terms.

Multiply: $\dfrac{1}{x} \cdot \dfrac{1}{x-2}$

$\qquad = \dfrac{1}{x(x-2)}$ Multiply numerators.
Multiply denominators.

Divide: $\dfrac{1}{x} \div \dfrac{1}{x-2}$

$\qquad = \dfrac{1}{x} \cdot \dfrac{x-2}{1}$ Multiply by the reciprocal of the divisor.

$\qquad = \dfrac{x-2}{x}$ Multiply numerators.
Multiply denominators.

By contrast, we *solve* the following rational *equation*.

Solve. $\qquad \dfrac{1}{x} + \dfrac{1}{x-2} = \dfrac{3}{4}$ $x \neq 0, 2$ because a denominator is 0 for these values.

$$4x(x-2)\left(\dfrac{1}{x} + \dfrac{1}{x-2}\right) = 4x(x-2)\left(\dfrac{3}{4}\right)$$ Multiply each side by the LCD, $4x(x-2)$, to clear fractions.

$$4x(x-2)\left(\dfrac{1}{x}\right) + 4x(x-2)\left(\dfrac{1}{x-2}\right) = 4x(x-2)\left(\dfrac{3}{4}\right)$$ Distributive property

$$4(x-2) + 4x = x(x-2)(3)$$ Divide out the common factors.

$$4x - 8 + 4x = 3x^2 - 6x$$ Distributive property

$$3x^2 - 14x + 8 = 0 \qquad \text{Standard form}$$

$$(3x - 2)(x - 4) = 0 \qquad \text{Factor.}$$

$$3x - 2 = 0 \quad \text{or} \quad x - 4 = 0 \quad \text{Zero-factor property}$$

$$\text{Proposed solutions} \rightarrow x = \frac{2}{3} \quad \text{or} \qquad x = 4 \quad \text{Solve for } x.$$

Neither proposed solution makes a denominator equal 0. Check each proposed solution in the original equation to confirm that the solution set is $\left\{ \frac{2}{3}, 4 \right\}$.

Points to Remember

1. When simplifying rational expressions, the fundamental property is applied only after numerators and denominators have been *factored*.

2. When adding and subtracting rational expressions, the common denominator must be kept throughout the problem and in the final result.

3. When simplifying rational expressions, always check to see if the answer is in lowest terms. If it is not, use the fundamental property.

4. When solving equations with rational expressions, the LCD is used to clear fractions. Multiply each side by the LCD. (Notice how this use differs from that of the LCD in Point 2.)

5. When solving equations with rational expressions, reject any proposed solution that causes an original denominator to equal 0.

For each exercise, indicate "expression" if an expression is to be simplified or "equation" if an equation is to be solved. Then simplify the expression or solve the equation.

1. $\dfrac{4}{p} + \dfrac{6}{p}$

2. $\dfrac{x^3 y^2}{x^2 y^4} \cdot \dfrac{y^5}{x^4}$

3. $\dfrac{1}{x^2 + x - 2} \div \dfrac{4x^2}{2x - 2}$

4. $\dfrac{8}{t - 5} = 2$

5. $\dfrac{x - 4}{5} = \dfrac{x + 3}{6}$

6. $\dfrac{2}{k^2 - 4k} + \dfrac{3}{k^2 - 16}$

7. $\dfrac{2y^2 + y - 6}{2y^2 - 9y + 9} \cdot \dfrac{y^2 - 2y - 3}{y^2 - 1}$

8. $\dfrac{3t^2 - t}{6t^2 + 15t} \div \dfrac{6t^2 + t - 1}{2t^2 - 5t - 25}$

9. $\dfrac{4}{p + 2} + \dfrac{1}{3p + 6}$

10. $\dfrac{1}{x} + \dfrac{1}{x - 3} = -\dfrac{5}{4}$

11. $\dfrac{3}{t - 1} + \dfrac{1}{t} = \dfrac{7}{2}$

12. $\dfrac{6}{k} - \dfrac{2}{3k}$

13. $\dfrac{5}{4z} - \dfrac{2}{3z}$

14. $\dfrac{x + 2}{3} = \dfrac{2x - 1}{5}$

15. $\dfrac{1}{m^2 + 5m + 6} + \dfrac{2}{m^2 + 4m + 3}$

16. $\dfrac{2k^2 - 3k}{20k^2 - 5k} \div \dfrac{2k^2 - 5k + 3}{4k^2 + 11k - 3}$

17. $\dfrac{2}{x + 1} + \dfrac{5}{x - 1} = \dfrac{10}{x^2 - 1}$

18. $\dfrac{3}{x + 3} + \dfrac{4}{x + 6} = \dfrac{9}{x^2 + 9x + 18}$

19. $\dfrac{4t^2 - t}{6t^2 + 10t} \div \dfrac{8t^2 + 2t - 1}{3t^2 + 11t + 10}$

20. $\dfrac{x}{x - 2} + \dfrac{3}{x + 2} = \dfrac{8}{x^2 - 4}$

6.7   Applications of Rational Expressions

OBJECTIVES

1 Solve problems about numbers.

2 Solve problems about distance, rate, and time.

3 Solve problems about work.

We continue to use the six-step problem-solving method introduced earlier.

Solving an Applied Problem

Step 1   **Read** the problem, several times if necessary. *What information is given? What is to be found?*

Step 2   **Assign a variable** to represent the unknown value. Use a sketch, diagram, or table, as needed. Express any other unknown values in terms of the variable.

Step 3   **Write an equation** using the variable expression(s).

Step 4   **Solve** the equation.

Step 5   **State the answer.** Label it appropriately. *Does the answer seem reasonable?*

Step 6   **Check** the answer in the words of the *original* problem.

OBJECTIVE 1   Solve problems about numbers.

EXAMPLE 1   Solving a Problem about an Unknown Number

NOW TRY EXERCISE 1

In a certain fraction, the numerator is 4 less than the denominator. If 7 is added to both the numerator and denominator, the resulting fraction is equivalent to $\frac{7}{8}$. What is the original fraction?

If the same number is added to both the numerator and the denominator of the fraction $\frac{2}{5}$, the result is equivalent to $\frac{2}{3}$. Find the number.

Step 1   **Read** the problem carefully. We are trying to find a number.

Step 2   **Assign a variable.**

Let $x =$ the number added to the numerator and the denominator.

Step 3   **Write an equation.** The fraction

$$\frac{2 + x}{5 + x}$$

represents the result of adding the same number to both the numerator and the denominator. This result is equivalent to $\frac{2}{3}$.

$$\frac{2 + x}{5 + x} = \frac{2}{3}$$

Step 4   **Solve.** $3(5 + x)\left(\dfrac{2 + x}{5 + x}\right) = 3(5 + x)\left(\dfrac{2}{3}\right)$ Multiply by the LCD, $3(5 + x)$.

$$3(2 + x) = (5 + x)2$$ Divide out the common factors.

$$6 + 3x = 10 + 2x$$ Distributive property

$$x = 4$$ Subtract $2x$. Subtract 6.

Step 5   **State the answer.** The number is 4.

NOW TRY ANSWER

1. $\frac{21}{25}$

Step 6   **Check** the solution in the words of the original problem. If 4 is added to both the numerator and the denominator of $\frac{2}{5}$, the result is $\frac{2 + 4}{5 + 4} = \frac{6}{9} = \frac{2}{3}$, as required.

NOW TRY

OBJECTIVE 2 Solve problems about distance, rate, and time.

Recall the following formulas relating distance, rate, and time.

Forms of the Distance Formula

$$d = rt \qquad r = \frac{d}{t} \qquad t = \frac{d}{r}$$

PROBLEM-SOLVING HINT Many applied problems use forms of the distance formula. The following two strategies are especially helpful in setting up equations to solve such problems.

- *Make a sketch* to visualize what is happening in the problem.

- *Make a table* to organize the information given in the problem and the unknown quantities.

EXAMPLE 2 Solving a Problem about Distance, Rate, and Time

The Tickfaw River has a current of 3 mph. A motorboat takes as long to travel 12 mi downstream as to travel 8 mi upstream. What is the rate of the boat in still water?

Step 1 **Read** the problem again. We must find the rate (speed) of the boat in still water.

Step 2 **Assign a variable.**

Let $x =$ the rate of the boat in still water.

Because the current pushes the boat when the boat is going downstream, the rate of the boat downstream will be the *sum* of the rate of the boat and the rate of the current, $(x + 3)$ mph.

Because the current slows the boat down when the boat is going upstream, the boat's rate going upstream will be the *difference* between the rate of the boat in still water and the rate of the current, $(x - 3)$ mph. See **FIGURE 1**.

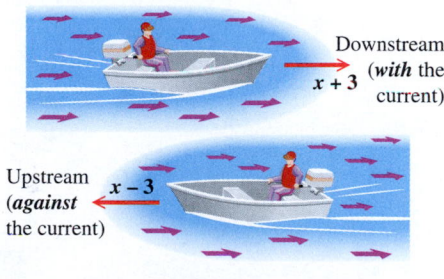

FIGURE 1

This information is summarized in the following table.

	d	r	t
Downstream	12	$x + 3$	
Upstream	8	$x - 3$	

Fill in the times using the formula $t = \frac{d}{r}$.

The time downstream is the distance divided by the rate.

$$t = \frac{d}{r} = \frac{12}{x + 3} \qquad \text{Time downstream}$$

The time upstream is that distance divided by that rate.

$$t = \frac{d}{r} = \frac{8}{x - 3} \qquad \text{Time upstream}$$

**NOW TRY
EXERCISE 2**
In her small boat, Jennifer can travel 12 mi downstream in the same amount of time that she can travel 4 mi upstream. The rate of the current is 2 mph. Find the rate of Jennifer's boat in still water.

	d	r	t
Downstream	12	$x + 3$	$\frac{12}{x+3}$
Upstream	8	$x - 3$	$\frac{8}{x-3}$

Enter each time, using $t = \frac{d}{r}$.

Step 3 **Write an equation.**

$$\frac{12}{x+3} = \frac{8}{x-3}$$

The time downstream equals the time upstream, so the two times from the table must be equal.

Step 4 **Solve.**

$$(x+3)(x-3)\left(\frac{12}{x+3}\right) = (x+3)(x-3)\left(\frac{8}{x-3}\right) \qquad \text{Multiply by the LCD, } (x+3)(x-3).$$

$$(x-3)12 = (x+3)8 \qquad \text{Divide out the common factors.}$$

$$12x - 36 = 8x + 24 \qquad \text{Distributive property}$$

$$4x = 60 \qquad \text{Subtract } 8x. \text{ Add 36.}$$

$$x = 15 \qquad \text{Divide by 4.}$$

Step 5 **State the answer.** The rate of the boat in still water is 15 mph.

Step 6 **Check.** The rate of the boat downstream is $(x + 3)$ mph, which is $15 + 3 = 18$ mph. Divide 12 mi by 18 mph to find the time.

$$t = \frac{d}{r} = \frac{12}{18} = \frac{2}{3} \text{ hr}$$

The rate of the boat upstream is $(x - 3)$ mph, which is $15 - 3 = 12$ mph. Divide 8 mi by 12 mph to find the time.

$$t = \frac{d}{r} = \frac{8}{12} = \frac{2}{3} \text{ hr}$$

The time upstream equals the time downstream, as required.

NOW TRY

> **NOTE** In distance-rate-time problems like the one in **Example 2**, once we have filled in two pieces of information in each row of a table, we can automatically fill in the third piece of information, using the appropriate form of the distance formula. Then we set up an equation based on our sketch and the information in the table.

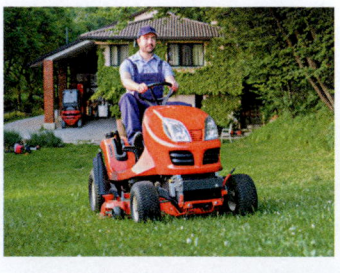

OBJECTIVE 3 **Solve problems about work.**

Suppose that we can mow a lawn in 4 hr. Then after 1 hr, we will have mowed $\frac{1}{4}$ of the lawn. After 2 hr, we will have mowed $\frac{2}{4}$, or $\frac{1}{2}$, of the lawn, and so on. This idea is generalized as follows.

Rate of Work

If a job can be completed in t units of time, then the rate of work is

$$\frac{1}{t} \text{ job per unit of time.}$$

NOW TRY ANSWER
2. 4 mph

PROBLEM-SOLVING HINT The amount of work W is equal to the rate of work r multiplied by the time worked t. Note the similarity to the distance formula $d = rt$.

In the lawn-mowing example, after 3 hr, the fractional part of the job done is as follows.

$$\underbrace{\frac{1}{4}}_{\substack{\text{Rate of} \\ \text{work}}} \cdot \underbrace{3}_{\substack{\text{Time} \\ \text{worked}}} = \underbrace{\frac{3}{4}}_{\substack{\text{Fractional part} \\ \text{of job done}}}$$

After 4 hr, $\frac{1}{4}(4) = 1$ whole job has been done.

EXAMPLE 3 Solving a Problem about Work Rates

Mateo can paint the trim on a small house in 10 hr. Chet needs 15 hr to complete the same job by hand. If both Mateo and Chet work together, how long will it take them to paint the trim?

Step 1 **Read** the problem again. We are looking for time working together.

Step 2 **Assign a variable.**

Let x = the number of hours it will take for Mateo and Chet to paint the trim, working together.

Making a table is helpful. Based on the previous discussion on work rates, Mateo's rate alone is $\frac{1}{10}$ job per hour. Chet's rate alone is $\frac{1}{15}$ job per hour.

	Rate	Time Working Together	Fractional Part of the Job Done When Working Together
Mateo	$\frac{1}{10}$	x	$\frac{1}{10}x$
Chet	$\frac{1}{15}$	x	$\frac{1}{15}x$

Because $rt = W$, the quantities $\frac{1}{10}x$ and $\frac{1}{15}x$ represent the two amounts of work.

Step 3 **Write an equation.**

$$\underbrace{\frac{1}{10}x}_{\substack{\text{Fractional part} \\ \text{done by Mateo}}} + \underbrace{\frac{1}{15}x}_{\substack{\text{Fractional part} \\ \text{done by Chet}}} = \underbrace{1}_{\text{1 whole job}}$$

Together, Mateo and Chet complete **1 whole job.** Add the fractional parts and set the sum equal to **1.**

Step 4 **Solve.**

$$30\left(\frac{1}{10}x + \frac{1}{15}x\right) = 30(1) \qquad \text{Multiply by the LCD, 30.}$$

$$30\left(\frac{1}{10}x\right) + 30\left(\frac{1}{15}x\right) = 30(1) \qquad \text{Distributive property}$$

$$3x + 2x = 30 \qquad \text{(*)} \quad \text{Multiply.}$$

$$5x = 30 \qquad \text{Combine like terms.}$$

$$x = 6 \qquad \text{Divide by 6.}$$

Step 5 **State the answer.** Working together, Mateo and Chet can paint the trim in 6 hr.

**NOW TRY
EXERCISE 3**

Sarah can proofread a
manuscript in 6 hr, and Joyce
can proofread the same
manuscript in 12 hr. How
long will it take them to
proofread the manuscript if
they work together?

Step 6 **Check.** The value of *x* must be *less than* 10 due to the fact that Mateo can
complete the job *alone* in 10 hr. So 6 hr seems reasonable.

In 6 hr, Mateo completes

$$\frac{1}{10}x = \frac{1}{10}(6) = \frac{6}{10} = \frac{3}{5} \text{ of the job.}$$

In 6 hr, Chet completes

$$\frac{1}{15}x = \frac{1}{15}(6) = \frac{6}{15} = \frac{2}{5} \text{ of the job.}$$

Working together, they complete $\frac{3}{5} + \frac{2}{5} = 1$ whole job, as required.
The answer, 6 hr, is correct.

NOW TRY ↻

⚠ **CAUTION**   A common error students make when solving a work problem like that in
Example 3 is to add the two times.

10 hr + 15 hr = 25 hr ← Incorrect answer

The answer 25 hr is unreasonable, because the slower worker (Chet) can do the job *alone*
in 15 hr.

Another common error students make is to add the two times and divide by 2—that is,
average the times.

$$\frac{10 + 15}{2} = \frac{25}{2} = 12\frac{1}{2} \text{ hr} \leftarrow \text{Incorrect answer}$$

The answer $12\frac{1}{2}$ hr is also unreasonable, because the faster worker (Mateo) can do the job
alone in 10 hr.

The correct time for the two workers together must be *less than* the time for the faster
worker alone (in this case Mateo, at 10 hr).

An alternative approach when solving work problems is to consider the part of the
job that can be done in 1 hr. For instance, in **Example 3** Mateo can do the entire job
in 10 hr, and Chet can do it in 15 hr. Thus, their work rates, as we saw in **Example 3,**
are $\frac{1}{10}$ and $\frac{1}{15}$, respectively. Since it takes them *x* hours to complete the job working
together, in 1 hr they can paint $\frac{1}{x}$ of the trim.

The amount painted by Mateo in 1 hr plus the amount painted by Chet in 1 hr
must equal the amount they can paint *together* in 1 hr. This leads to the following
alternative equation.

Amount by Chet
↓

Amount by Mateo → $\dfrac{1}{10} + \dfrac{1}{15} = \dfrac{1}{x}$ ← Amount together

Compare this alternative equation with the equation

$$\frac{1}{10}x + \frac{1}{15}x = 1$$

in Step 3 of **Example 3.** If we multiply each side of the alternative equation by the
LCD 30*x*, we obtain the following.

$$\frac{1}{10} + \frac{1}{15} = \frac{1}{x} \qquad \text{Alternative equation}$$

$$30x\left(\frac{1}{10} + \frac{1}{15}\right) = 30x\left(\frac{1}{x}\right) \qquad \text{Multiply by the LCD, } 30x.$$

$$30x\left(\frac{1}{10}\right) + 30x\left(\frac{1}{15}\right) = 30x\left(\frac{1}{x}\right) \qquad \text{Distributive property}$$

$$3x + 2x = 30 \qquad \text{Multiply.}$$

This is equation (*) in **Example 3.** The same solution, $x = 6$, results.

6.7 Exercises

FOR EXTRA HELP ▶ **MyLab Math**

▶ *Video solutions for select problems available in MyLab Math*

Concept Check *Answer each question.*

1. If a migrating hawk travels m mph in still air, what is its rate when it flies into a steady headwind of 5 mph? What is its rate with a tailwind of 5 mph?

2. Suppose Stephanie walks D miles at R mph in the same time that Wally walks d miles at r mph. What is an equation relating D, R, d, and r?

3. If it takes Katherine 10 hr to do a job, what is her rate?

4. If it takes Clayton 12 hr to do a job, how much of the job does he do in 8 hr?

Use Steps 2 and 3 of the six-step problem-solving method to set up an equation to use in solving each problem. (Remember that Step 1 is to read the problem carefully.) Do not actually solve the equation. **See Example 1.**

5. The numerator of the fraction $\frac{5}{6}$ is increased by an amount so that the value of the resulting fraction is equivalent to $\frac{13}{3}$. By what amount was the numerator increased?

   (a) Let $x =$ _____ . (*Step 2*)

   (b) Write an expression for "the numerator of the fraction $\frac{5}{6}$ is increased by an amount."

   (c) Set up an equation to solve the problem. (*Step 3*)

6. If the same number is added to the numerator and subtracted from the denominator of the fraction $\frac{23}{12}$, the resulting fraction is equivalent to $\frac{3}{2}$. What is the number?

   (a) Let $x =$ _____ . (*Step 2*)

   (b) Write an expression for "a number is added to the numerator of $\frac{23}{12}$." Then write an expression for "the same number is subtracted from the denominator of $\frac{23}{12}$."

   (c) Set up an equation to solve the problem. (*Step 3*)

In each problem, state what x represents, write an equation, and answer the question. **See Example 1.**

7. In a certain fraction, the denominator is 6 more than the numerator. If 3 is added to both the numerator and the denominator, the resulting fraction is equivalent to $\frac{5}{7}$. What was the original fraction (*not* written in lowest terms)?

8. In a certain fraction, the denominator is 4 less than the numerator. If 3 is added to both the numerator and the denominator, the resulting fraction is equivalent to $\frac{3}{2}$. What was the original fraction?

9. The denominator of a certain fraction is three times the numerator. If 2 is added to the numerator and subtracted from the denominator, the resulting fraction is equivalent to 1. What was the original fraction (*not* written in lowest terms)?

10. The numerator of a certain fraction is four times the denominator. If 6 is added to both the numerator and the denominator, the resulting fraction is equivalent to 2. What was the original fraction (*not* written in lowest terms)?

11. One-sixth of a number is 5 more than the same number. What is the number?

12. One-third of a number is 2 more than one-sixth of the same number. What is the number?

13. A quantity, its $\frac{3}{4}$, its $\frac{1}{2}$, and its $\frac{1}{3}$, added together, becomes 93. What is the quantity? (Data from *Rhind Mathematical Papyrus*.)

14. A quantity, its $\frac{2}{3}$, its $\frac{1}{2}$, and its $\frac{1}{7}$, added together, becomes 33. What is the quantity? (Data from *Rhind Mathematical Papyrus*.)

Solve each problem using a form of the distance formula.

15. British explorer and endurance swimmer Lewis Gordon Pugh was the first person to swim at the North Pole. He swam 0.6 mi at 0.0319 mi per min in waters created by melted sea ice. What was his time (to three decimal places)? (Data from *The Gazette*.)

16. In the 2016 Summer Olympics, Penny Oleksiak of Canada won the women's 100-m free-style swimming event. Her rate was 1.8975 m per sec. What was her time (to two decimal places)? (Data from www.olympic.org)

17. Hellen Onsando Obiri of Kenya won the women's 5000-m run in the 2016 Olympics with a time of 15.323 min. What was her rate (to three decimal places)? (Data from www.olympic.org)

18. Faith Chepngetich Kipyegon of Kenya won the women's 1500-m run in the 2016 Olympics with a time of 4.066 min. What was her rate (to three decimal places)? (Data from www.olympic.org)

19. In 2018, Austin Dillon drove his Chevrolet to victory in the Daytona 500 (mile) race with a rate of 150.545 mph. The actual distance traveled was 507.5 miles. What was his time (to the nearest thousandth of an hour)? (Data from *The World Almanac and Book of Facts*.)

20. In 2017, Kurt Busch drove his Ford to victory in the Daytona 500 (mile) race. His rate was 143.187 mph. What was his time (to the nearest thousandth of an hour)? (Data from *The World Almanac and Book of Facts*.)

Set up an equation to solve each problem. Do not actually solve the equation. See Example 2.

21. Luvenia can row her boat 4 mph in still water. She takes as long to row 8 mi upstream as 24 mi downstream. What is the rate of the current? (Let x = rate of the current.)

	d	r	t
Upstream	8	$4 - x$	
Downstream	24	$4 + x$	

22. Wayne flew his airplane 500 mi against the wind in the same time it took him to fly 600 mi with the wind. If the speed of the wind was 10 mph, what was the rate of his plane in still air? (Let x = rate of the plane in still air.)

	d	r	t
Against the Wind	500	$x - 10$	
With the Wind	600	$x + 10$	

Solve each problem. See Example 2.

23. A boat can travel 20 mi against a current in the same time that it can travel 60 mi with the current. The rate of the current is 4 mph. Find the rate of the boat in still water.

24. Vince can fly his plane 200 mi against the wind in the same time it takes him to fly 300 mi with the wind. The wind blows at 30 mph. Find the rate of his plane in still air.

25. The sanderling is a small shorebird about 6.5 in. long, with a thin, dark bill and a wide, white wing stripe. If a sanderling can fly 30 mi with the wind in the same time it can fly 18 mi against the wind when the wind speed is 8 mph, what is the rate of the bird in still air? (Data from U.S. Geological Survey.)

26. Airplanes usually fly faster from west to east than from east to west because the prevailing winds go from west to east. The air distance between Chicago and London is about 4000 mi, while the air distance between New York and London is about 3500 mi. If a jet can fly eastbound from Chicago to London in the same time it can fly westbound from London to New York in a 35-mph wind, what is the rate of the plane in still air? (Data from www.geobytes.com)

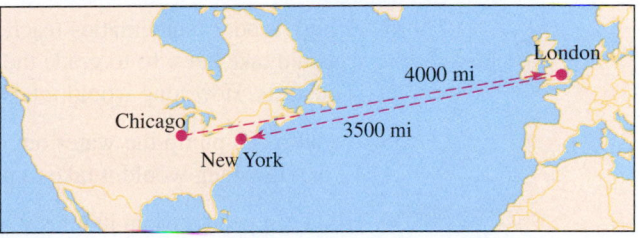

27. An airplane maintaining a constant airspeed takes as long to travel 450 mi with the wind as it does to travel 375 mi against the wind. If the wind is blowing at 15 mph, what is the rate of the plane in still air?

	d	r	t
Against the Wind			
With the Wind			

28. A river has a current of 4 km per hr. Find the rate of Jai's boat in still water if it travels 40 km downstream in the same time that it takes to travel 24 km upstream.

	d	r	t
Upstream			
Downstream			

29. Connie's boat travels at 12 mph. Find the rate of the current of the river if she can travel 6 mi upstream in the same amount of time it takes her to travel 10 mi downstream.

30. Mohammed can travel 8 mi upstream in the same time it takes him to travel 12 mi downstream. His boat travels 15 mph in still water. What is the rate of the current?

31. The distance from Seattle, Washington, to Victoria, British Columbia, is about 148 mi by ferry. It takes about 4 hr less to travel by the same ferry from Victoria to Vancouver, British Columbia, a distance of about 74 mi. What is the average rate of the ferry?

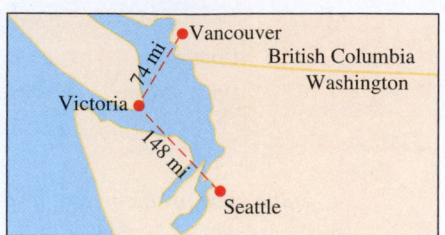

32. Driving from Tulsa to Detroit, Dean averaged 50 mph. He figured that if he had averaged 60 mph, his driving time would have decreased 3 hr. How far is it from Tulsa to Detroit?

Set up an equation to solve each problem. Do not actually solve the equation. **See Example 3.**

33. Working alone, Edward can paint a room in 8 hr. Abdalla can paint the same room working alone in 6 hr. How long will it take them if they work together? (Let t represent the time they work together.)

	r	t	w
Edward		t	
Abdalla		t	

34. Donald can tune up his Chevy in 2 hr working alone. Jeff can do the same job in 3 hr working alone. How long would it take them if they worked together? (Let t represent the time they work together.)

	r	t	w
Donald		t	
Jeff		t	

Solve each problem. **See Example 3.**

35. A copier can do a large printing job in 20 hr. An older model can do the same job in 30 hr. How long would it take to do the job using both copiers?

36. A company can prepare customer statements in 8 hr using a new computer. Using an older computer requires 24 hr to do the same job. How long would it take to prepare the statements using both computers?

37. A high school mathematics teacher gave a test to her geometry classes. Working alone, it would take her 4 hr to grade the tests. Her student teacher would take 6 hr to grade the same tests. How long would it take them to grade these tests if they work together?

38. A pump can pump the water out of a flooded basement in 10 hr. A smaller pump takes 12 hr. How long would it take to pump the water from the basement with both pumps?

39. Hilda can paint a room in 6 hr. Working together with Brenda, they can paint the room in $3\frac{3}{4}$ hr. How long would it take Brenda to paint the room by herself?

40. Grant can completely mess up his room in 15 min. If his cousin Wade helps him, they can completely mess up the room in $8\frac{4}{7}$ min. How long would it take Wade to mess up the room by himself?

41. An experienced employee can enter tax data into a computer twice as fast as a new employee. Working together, it takes the employees 2 hr. How long would it take the experienced employee working alone?

42. One roofer can put a roof on a house three times faster than another. Working together, they can roof a house in 4 days. How long would it take the faster roofer working alone?

Solve each problem. (Data from Mary Jo Boyer, Math for Nurses, *Wolter Kluwer.)*

43. Nurses use Young's Rule to calculate the pediatric (child's) dose P of a medication, given a child's age c in years and a normal adult dose a.

$$P = \frac{c}{c + 12} \cdot a$$

The normal adult dose for milk of magnesia is 30 mL. Use Young's Rule to calculate the correct dose to give a 6-yr-old boy.

44. Nurses use Clark's Rule to calculate the pediatric (child's) dose P of a medication, given a child's weight w in pounds and a normal adult dose a.

$$P = \frac{w}{150} \cdot a$$

The normal adult dose for ibuprofen is 200 mg. Use Clark's Rule to calculate the correct dose to give a 4-yr-old girl who weighs 30 lb.

RELATING CONCEPTS   For Individual or Group Work   (Exercises 45–48)

In the movie Little Big League, *young Billy Heywood inherits the Minnesota Twins baseball team and becomes its manager. Before the biggest game of the year, he can't keep his mind on his job because a homework problem is giving him trouble.*

If Joe can paint a house in 3 hr, and Sam can paint the same house in 5 hr, how long does it take for them to do it together?

With the help of one of his players, Billy solves the problem.

45. Use the method of **Example 3** of this section to solve this problem.

46. Billy got "help" from some of the other players. The incorrect answers they gave him follow. Explain the faulty reasoning behind each of these answers.

   (a) 15 hr   **(b)** 8 hr   **(c)** 4 hr

47. The player who gave Billy the correct answer solved the problem as follows:

Using the simple formula a times b over a plus b, we get our answer of one and seven-eighths.

Show that if it takes one person a hours to complete a job and another b hours to complete the same job, then the expression stated by the player,

$$\frac{a \cdot b}{a + b},$$

actually does give the number of hours it would take them to do the job together. (*Hint:* Refer to **Example 3** and use a and b rather than 10 and 15 to write a formula. Then solve the formula for x.)

48. Solve the following problem using the method of **Example 3.** Then solve it using the formula obtained in **Exercise 47.** How do the answers compare?

A screen printer can complete a t-shirt order for a Little League baseball organization in 15 hr using a large machine. The same order would take 30 hr using a smaller machine. How long would it take to complete the order using both machines together?

Chapter 6   Summary

Key Terms

6.1

rational expression
lowest terms

6.1

rational expression
lowest terms

6.3

least common denominator
  (LCD)

6.6

proposed solution
extraneous solution
  (extraneous value)

6.2

reciprocals
  (multiplicative inverses)

6.5

complex fraction

Test Your Word Power

See how well you have learned the vocabulary in this chapter.

1. A **rational expression** is
   A. an algebraic expression made up of a term or the sum of a finite number of terms with real coefficients and whole number exponents
   B. a polynomial equation of degree 2
   C. an expression with one or more fractions in the numerator, or denominator, or both
   D. the quotient of two polynomials with denominator not 0.

2. In a given set of fractions, the **least common denominator** is
   A. the smallest denominator of all the denominators
   B. the product of the factors from each denominator, with each factor raised to the greatest power that occurs
   C. the largest integer that evenly divides the numerator and denominator of all the fractions
   D. the sum of the factors raised to the greatest power that occurs in each denominator.

3. A **complex fraction** is
   A. an algebraic expression made up of a term or the sum of a finite number of terms with real coefficients and whole number exponents
   B. a polynomial equation of degree 2
   C. a quotient with one or more fractions in the numerator, or denominator, or both
   D. the quotient of two polynomials with denominator not 0.

ANSWERS

1. D; *Examples:* $-\dfrac{3}{4y}, \dfrac{5x^3}{x+2}, \dfrac{a+3}{a^2-4a-5}$ **2.** B; *Example:* The LCD of $\dfrac{1}{x}, \dfrac{2}{3}$, and $\dfrac{5}{x+1}$ is $3x(x+1)$. **3.** C; *Examples:* $\dfrac{\frac{2}{3}}{\frac{4}{7}}, \dfrac{x-\frac{1}{y}}{x+\frac{1}{y}}, \dfrac{\frac{2}{a+1}}{a^2-1}$

Quick Review

CONCEPTS	EXAMPLES
6.1 **The Fundamental Property of Rational Expressions**	Find the values for which the expression $\dfrac{x-4}{x^2-16}$ is undefined.
Determining When a Rational Expression Is Undefined	$x^2-16=0$
Step 1 Set the denominator equal to 0.	$(x-4)(x+4)=0$ Factor.
Step 2 Solve this equation.	$x-4=0$ or $x+4=0$ Zero-factor property
Step 3 The solutions of the equation are the values that make the rational expression undefined. The variable *cannot* equal these values.	$x=4$ or $x=-4$ Solve for x.
	The rational expression is undefined for 4 and -4, so $x\neq4$ and $x\neq-4$.

CONCEPTS	EXAMPLES

Writing a Rational Expression in Lowest Terms

Step 1 Factor the numerator and denominator completely

Step 2 Use the fundamental property to divide out any common factors.

Write in lowest terms.

$$\frac{x^2 - 1}{(x - 1)^2}$$

$$= \frac{(x - 1)(x + 1)}{(x - 1)(x - 1)} \qquad \text{Factor.}$$

$$= \frac{x + 1}{x - 1} \qquad \text{Lowest terms}$$

Writing Equivalent Forms of a Rational Expression
There are often several different equivalent forms of a rational expression.

Give four equivalent forms of $-\dfrac{x - 1}{x + 2}$.

① $\rightarrow \dfrac{-(x - 1)}{x + 2}$, or $\dfrac{-x + 1}{x + 2}$ ← ②   Distribute the negative sign in the numerator.

③ $\rightarrow \dfrac{x - 1}{-(x + 2)}$, or $\dfrac{x - 1}{-x - 2}$ ← ④   Distribute the negative sign in the denominator.

6.2 Multiplying and Dividing Rational Expressions

Multiplying or Dividing Rational Expressions

Step 1 Note the operation. If the operation is division, use the definition of division to rewrite it as multiplication.

Step 2 Multiply numerators and multiply denominators.

Step 3 Factor numerators and denominators completely.

Step 4 Write in lowest terms using the fundamental property.

Note: Steps 2 and 3 may be interchanged based on personal preference.

Multiply. $\dfrac{3x + 9}{x - 5} \cdot \dfrac{x^2 - 3x - 10}{x^2 - 9}$

$$= \frac{(3x + 9)(x^2 - 3x - 10)}{(x - 5)(x^2 - 9)} \qquad \begin{array}{l}\text{Multiply}\\\text{numerators and}\\\text{denominators.}\end{array}$$

$$= \frac{3(x + 3)(x - 5)(x + 2)}{(x - 5)(x + 3)(x - 3)} \qquad \text{Factor.}$$

$$= \frac{3(x + 2)}{x - 3} \qquad \text{Lowest terms}$$

Divide. $\dfrac{2x + 1}{x + 5} \div \dfrac{6x^2 - x - 2}{x^2 - 25}$

$$= \frac{2x + 1}{x + 5} \cdot \frac{x^2 - 25}{6x^2 - x - 2} \qquad \begin{array}{l}\text{Multiply by the}\\\text{reciprocal of}\\\text{the divisor.}\end{array}$$

$$= \frac{(2x + 1)(x^2 - 25)}{(x + 5)(6x^2 - x - 2)} \qquad \begin{array}{l}\text{Multiply}\\\text{numerators and}\\\text{denominators.}\end{array}$$

$$= \frac{(2x + 1)(x + 5)(x - 5)}{(x + 5)(2x + 1)(3x - 2)} \qquad \text{Factor.}$$

$$= \frac{x - 5}{3x - 2} \qquad \text{Lowest terms}$$

6.3 Least Common Denominators

Finding the LCD

Step 1 Factor each denominator into prime factors.

Step 2 List each different factor the *greatest* number of times it appears in any of the denominators.

Step 3 Multiply the factors from Step 2 to find the LCD.

Find the LCD for $\dfrac{3}{k^2 - 8k + 16}$ and $\dfrac{1}{4k^2 - 16k}$.

$$\left.\begin{array}{l}k^2 - 8k + 16 = (k - 4)^2\\[4pt]4k^2 - 16k = 4k(k - 4)\end{array}\right\} \begin{array}{l}\text{Factor each}\\\text{denominator.}\end{array}$$

$$\text{LCD} = (k - 4)^2 \cdot 4 \cdot k$$

$$= 4k(k - 4)^2$$

CONCEPTS	EXAMPLES

Writing a Rational Expression with a Specified Denominator

Step 1 Factor both denominators.

Step 2 Decide what factor(s) the denominator must be multiplied by in order to equal the specified denominator.

Step 3 Multiply the rational expression by that factor divided by itself. (That is, multiply by 1.)

Write the rational expression as an equivalent expression with the indicated denominator.

$$\frac{5}{2z^2 - 6z} = \frac{?}{4z^3 - 12z^2}$$

$$\frac{5}{2z(z - 3)} = \frac{?}{4z^2(z - 3)}$$

$2z(z - 3)$ must be multiplied by $2z$ to obtain $4z^2(z - 3)$.

$$\frac{5}{2z(z - 3)} \cdot \frac{2z}{2z} \qquad \frac{2z}{2z} = 1$$

$$= \frac{10z}{4z^2(z - 3)}, \quad \text{or} \quad \frac{10z}{4z^3 - 12z^2}$$

6.4 Adding and Subtracting Rational Expressions

Adding Rational Expressions

Step 1 Find the LCD.

Step 2 Write each rational expression as an equivalent rational expression with the LCD as denominator.

Step 3 Add the numerators to obtain the numerator of the sum. The LCD is the denominator of the sum.

Step 4 Write in lowest terms using the fundamental property.

Add. $\dfrac{2}{3m + 6} + \dfrac{m}{m^2 - 4}$

$\left.\begin{array}{l} 3m + 6 = 3(m + 2) \\ m^2 - 4 = (m + 2)(m - 2) \end{array}\right\}$ The LCD is $3(m + 2)(m - 2)$.

$$= \frac{2(m - 2)}{3(m + 2)(m - 2)} + \frac{3m}{3(m + 2)(m - 2)} \quad \text{Write with the LCD.}$$

$$= \frac{2m - 4 + 3m}{3(m + 2)(m - 2)} \quad \begin{array}{l}\text{Add numerators}\\\text{Keep the same demominator.}\end{array}$$

$$= \frac{5m - 4}{3(m + 2)(m - 2)} \quad \text{Combine like terms.}$$

Subtracting Rational Expressions

Follow the same steps as for addition, but subtract in Step 3.

Subtract. $\dfrac{6}{k + 4} - \dfrac{2}{k}$ The LCD is $k(k + 4)$.

$$= \frac{6k}{(k + 4)k} - \frac{2(k + 4)}{k(k + 4)} \quad \text{Write with the LCD.}$$

$$= \frac{6k - 2(k + 4)}{k(k + 4)} \quad \begin{array}{l}\text{Subtract numerators.}\\\text{Keep the same}\\\text{denominator.}\end{array}$$

$$= \frac{6k - 2k - 8}{k(k + 4)} \quad \text{Distributive property}$$

$$= \frac{4k - 8}{k(k + 4)} \quad \text{Combine like terms.}$$

$$= \frac{4(k - 2)}{k(k + 4)} \quad \text{Factor.}$$

CONCEPTS	EXAMPLES

6.5 Complex Fractions

Simplifying a Complex Fraction

Method 1

Step 1 Write both the numerator and the denominator as single fractions.

Step 2 Change the complex fraction to a division problem.

Step 3 Perform the indicated division.

Method 2

Step 1 Find the LCD of all fractions within the complex fraction.

Step 2 Multiply both the numerator and the denominator of the complex fraction by this LCD using the distributive property as necessary. Write in lowest terms.

Which method to use is a matter of individual preference.

Simplify.   **Method 1**   **Method 2**

Method 1

$$\frac{\frac{1}{a} - a}{1 - a}$$

$$= \frac{\frac{1}{a} - \frac{a^2}{a}}{1 - a}$$

$$= \frac{\frac{1 - a^2}{a}}{1 - a}$$

$$= \frac{1 - a^2}{a} \div (1 - a)$$

$$= \frac{1 - a^2}{a} \cdot \frac{1}{1 - a}$$

$$= \frac{(1 - a)(1 + a)}{a(1 - a)}$$

$$= \frac{1 + a}{a}$$

Method 2

$$\frac{\frac{1}{a} - a}{1 - a}$$

$$= \frac{a\left(\frac{1}{a} - a\right)}{a(1 - a)}$$

$$= \frac{\frac{a}{a} - a^2}{(1 - a)a}$$

$$= \frac{1 - a^2}{(1 - a)a}$$

$$= \frac{(1 + a)(1 - a)}{(1 - a)a}$$

$$= \frac{1 + a}{a}$$

The same answer results using either method.

6.6 Solving Equations with Rational Expressions

Solving an Equation with Rational Expressions

Step 1 Multiply each side of the equation by the LCD to clear the equation of fractions. Be sure to distribute to *every* term on *both* sides of the equation.

Step 2 Solve the resulting equation for proposed solutions.

Step 3 Check each proposed solution by substituting it in the original equation. Reject any value that causes a denominator to equal 0.

Solve. $\dfrac{x}{x - 3} + \dfrac{4}{x + 3} = \dfrac{18}{x^2 - 9}$

$$\frac{x}{x - 3} + \frac{4}{x + 3} = \frac{18}{(x - 3)(x + 3)} \quad \text{Factor.}$$

The LCD is $(x - 3)(x + 3)$. *Note that 3 and -3 cannot be solutions, as they cause a denominator to equal 0.*

$$(x - 3)(x + 3)\left(\frac{x}{x - 3} + \frac{4}{x + 3}\right) \quad \text{Multiply by the LCD.}$$

$$= (x - 3)(x + 3)\left(\frac{18}{(x - 3)(x + 3)}\right)$$

$$(x + 3)x + (x - 3)4 = 18 \qquad \text{Distributive property}$$

$$x^2 + 3x + 4x - 12 = 18 \qquad \text{Distributive property}$$

$$x^2 + 7x - 30 = 0 \qquad \text{Standard form}$$

$$(x - 3)(x + 10) = 0 \qquad \text{Factor.}$$

$$x - 3 = 0 \quad \text{or} \quad x + 10 = 0 \qquad \text{Zero-factor property}$$

$$\text{Reject} \rightarrow x = 3 \quad \text{or} \quad x = -10 \qquad \text{Solve.}$$

Because 3 causes denominators to equal 0, it is an extraneous value. Check that the only solution is -10. Thus, $\{-10\}$ is the solution set.

CONCEPTS

EXAMPLES

6.7 Applications of Rational Expressions

Solving a Distance-Rate-Time Problem

Use the formulas relating d, r, and t.

$$d = rt, \quad r = \frac{d}{t}, \quad t = \frac{d}{r}$$

A small plane flew from Chicago to Kansas City averaging 145 mph. The trip took 3.5 hr. What is the distance between Chicago and Kansas City?

$$145 \quad \cdot \quad 3.5 \quad = \quad 507.5 \text{ mi}$$
$$\uparrow \qquad \uparrow \qquad \qquad \uparrow$$
$$\text{Rate} \quad \text{Time} \quad \text{Distance}$$

Solving a Work Problem

Step 1 Read the problem, several times if necessary.

Step 2 Assign a variable. State what the variable represents. Organize the information from the problem in a table. If a job is done in t units of time, then the rate of work is

$$\frac{1}{t} \text{ job per unit of time.}$$

Step 3 Write an equation. The sum of the fractional parts should equal 1 (whole job).

Step 4 Solve the equation.

Step 5 State the answer.

Step 6 Check.

It takes the regular mail carrier 6 hr to cover her route. A substitute takes 8 hr to cover the same route. How long would it take them to cover the route together?

Let x = the number of hours to cover the route together.

The rate of the regular carrier is $\frac{1}{6}$ job per hour, and the rate of the substitute is $\frac{1}{8}$ job per hour.

	Rate	Time	Part of the Job Done
Regular	$\frac{1}{6}$	x	$\frac{1}{6}x$
Substitute	$\frac{1}{8}$	x	$\frac{1}{8}x$

Multiply rate by time to find the fractional part done.

$$\frac{1}{6}x + \frac{1}{8}x = 1 \qquad \text{The parts add to 1 whole job.}$$

$$24\left(\frac{1}{6}x + \frac{1}{8}x\right) = 24(1) \qquad \text{The LCD is 24.}$$

$$24\left(\frac{1}{6}x\right) + 24\left(\frac{1}{8}x\right) = 24 \qquad \begin{array}{l}\text{Distributive property;} \\ \text{Multiply.}\end{array}$$

$$4x + 3x = 24 \qquad \text{Multiply.}$$

$$7x = 24 \qquad \text{Combine like terms.}$$

$$x = \frac{24}{7} \qquad \text{Divide by 7.}$$

To cover the route together, it would take them

$$\frac{24}{7} \text{ hr,} \quad \text{or} \quad 3\frac{3}{7} \text{ hr.}$$

This makes sense. The time together is *less than* the time of the regular carrier working alone. Also,

$$\frac{1}{6}\left(\frac{24}{7}\right) + \frac{1}{8}\left(\frac{24}{7}\right) = 1 \quad \text{is true.}$$

Chapter 6   Review Exercises

6.1 *Find the numerical value of each rational expression for (a) $x = -2$ and (b) $x = 4$.*

1. $\dfrac{x^2}{x - 5}$ 2. $\dfrac{4x - 3}{5x + 2}$ 3. $\dfrac{3x}{x^2 - 4}$ 4. $\dfrac{x - 1}{x + 2}$

Find any values of the variable for which each rational expression is undefined. Write answers with the symbol $\neq$.

5. $\dfrac{4}{x - 3}$ 6. $\dfrac{y + 3}{2y}$ 7. $\dfrac{2k + 1}{3k^2 + 17k + 10}$ 8. $\dfrac{m - 2}{m^2 - 2m - 3}$

Write each rational expression in lowest terms.

9. $\dfrac{5a^3b^3}{15a^4b^2}$ 10. $\dfrac{m - 4}{4 - m}$ 11. $\dfrac{4x^2 - 9}{6 - 4x}$

12. $\dfrac{4p^2 + 8pq - 5q^2}{10p^2 - 3pq - q^2}$ 13. $\dfrac{x^2 - 9}{xy + 4x - 3y - 12}$ 14. $\dfrac{x^3 - 1}{x - 1}$

Write four equivalent forms for each rational expression.

15. $-\dfrac{4x - 9}{2x + 3}$ 16. $-\dfrac{8 - 3x}{3 - 6x}$

6.2 *Multiply or divide. Write each answer in lowest terms.*

17. $\dfrac{18p^3}{6} \cdot \dfrac{24}{p^4}$ 18. $\dfrac{8x^2}{12x^5} \cdot \dfrac{6x^4}{2x}$

19. $\dfrac{9m^2}{(3m)^4} \div \dfrac{6m^5}{36m}$ 20. $\dfrac{3q + 3}{5 - 6q} \div \dfrac{4q + 4}{2(5 - 6q)}$

21. $\dfrac{x - 3}{4} \cdot \dfrac{5}{2x - 6}$ 22. $\dfrac{2r + 3}{r - 4} \cdot \dfrac{r^2 - 16}{6r + 9}$

23. $\dfrac{6a^2 + 7a - 3}{2a^2 - a - 6} \div \dfrac{a + 5}{a - 2}$ 24. $\dfrac{y^2 - 6y + 8}{y^2 + 3y - 18} \div \dfrac{y - 4}{y + 6}$

25. $\dfrac{2p^2 + 13p + 20}{p^2 + p - 12} \cdot \dfrac{p^2 + 2p - 15}{2p^2 + 7p + 5}$ 26. $\dfrac{3z^2 + 5z - 2}{9z^2 - 1} \cdot \dfrac{9z^2 + 6z + 1}{z^2 + 5z + 6}$

27. $\dfrac{p^2 + 3p + 2pq + 6q}{p - 2q} \cdot \dfrac{2p + 6}{p^2 + 6p + 9}$ 28. $\dfrac{x^3 - 64}{x^2 - 16} \div \dfrac{x^2 + 4x + 16}{2x + 8}$

6.3 *Find the LCD for the fractions in each list.*

29. $\dfrac{1}{8}, \dfrac{5}{12}, \dfrac{7}{32}$ 30. $\dfrac{4}{9y}, \dfrac{7}{12y^2}, \dfrac{5}{27y^4}$

31. $\dfrac{1}{m^2 + 2m}, \dfrac{4}{m^2 + 7m + 10}$ 32. $\dfrac{3}{x^2 + 4x + 3}, \dfrac{5}{x^2 + 5x + 4}, \dfrac{2}{x^2 + 7x + 12}$

Write each rational expression as an equivalent expression with the indicated denominator.

33. $\dfrac{5}{8} = \dfrac{?}{56}$ 34. $\dfrac{10}{k} = \dfrac{?}{4k}$ 35. $\dfrac{3}{2a^3} = \dfrac{?}{10a^4}$ 36. $\dfrac{9}{x - 3} = \dfrac{?}{18 - 6x}$

37. $\dfrac{-3y}{2y - 10} = \dfrac{?}{50 - 10y}$ **38.** $\dfrac{4b}{b^2 + 2b - 3} = \dfrac{?}{(b + 3)(b - 1)(b + 2)}$

6.4 *Add or subtract. Write each answer in lowest terms.*

39. $\dfrac{10}{x} + \dfrac{5}{x}$ **40.** $\dfrac{6}{3p} - \dfrac{12}{3p}$

41. $\dfrac{9}{k} - \dfrac{5}{k - 5}$ **42.** $\dfrac{4}{y} + \dfrac{7}{7 + y}$

43. $\dfrac{m}{3} - \dfrac{2 + 5m}{6}$ **44.** $\dfrac{12}{x^2} - \dfrac{3}{4x}$

45. $\dfrac{5}{a - 2b} + \dfrac{2}{a + 2b}$ **46.** $\dfrac{4}{k^2 - 9} - \dfrac{k + 3}{3k - 9}$

47. $\dfrac{8}{z^2 + 6z} - \dfrac{3}{z^2 + 4z - 12}$ **48.** $\dfrac{11}{2p - p^2} - \dfrac{2}{p^2 - 5p + 6}$

6.5 *Simplify each complex fraction.*

49. $\dfrac{\dfrac{2}{3} - \dfrac{1}{6}}{\dfrac{1}{4} + \dfrac{2}{5}}$ **50.** $\dfrac{\dfrac{a^4}{b^2}}{\dfrac{a^3}{b}}$ **51.** $\dfrac{\dfrac{y - 3}{y}}{\dfrac{y + 3}{4y}}$ **52.** $\dfrac{x + \dfrac{1}{w}}{x - \dfrac{1}{w}}$

53. $\dfrac{\dfrac{1}{p} - \dfrac{1}{q}}{\dfrac{1}{q - p}}$ **54.** $\dfrac{\dfrac{x^2 - 16}{x - 3}}{\dfrac{x - 4}{x^2 - 9}}$ **55.** $\dfrac{\dfrac{1}{r + t} - 1}{\dfrac{1}{r + t} + 1}$ **56.** $\dfrac{x^{-2} - y^{-2}}{x^{-1} - y^{-1}}$

6.6 *Solve each equation, and check the solutions.*

57. $\dfrac{k}{5} - \dfrac{2}{3} = \dfrac{1}{2}$ **58.** $\dfrac{4 - z}{z} + \dfrac{3}{2} = \dfrac{-4}{z}$

59. $\dfrac{x}{2} - \dfrac{x - 3}{7} = -1$ **60.** $\dfrac{3x - 1}{x - 2} = \dfrac{5}{x - 2} + 1$

61. $\dfrac{3}{x + 4} - \dfrac{2x}{5} = \dfrac{3}{x + 4}$ **62.** $\dfrac{3}{m - 2} + \dfrac{1}{m - 1} = \dfrac{7}{m^2 - 3m + 2}$

Solve each formula or equation for the specified variable.

63. $m = \dfrac{Ry}{t}$ for t **64.** $b = \dfrac{s + t}{r}$ for s

65. $a = \dfrac{b}{c + d}$ for d **66.** $\dfrac{1}{r} - \dfrac{1}{s} = \dfrac{1}{t}$ for t

6.7 *Solve each problem.*

67. In a certain fraction, the denominator is 5 less than the numerator. If 5 is added to both the numerator and the denominator, the resulting fraction is equivalent to $\frac{5}{4}$. Find the original fraction (*not* written in lowest terms).

68. The denominator of a certain fraction is six times the numerator. If 3 is added to the numerator and subtracted from the denominator, the resulting fraction is equivalent to $\frac{2}{5}$. Find the original fraction (*not* written in lowest terms).

69. Ryan Hunter-Reay won the Iowa Corn Indy 300. He drove a Dallara-Honda the 262.5-mi distance with an average rate of 129.943 mph. What was his time (to the nearest thousandth of an hour)? (Data from www.indycar.com)

70. A plane flies 350 mi with the wind in the same time it takes to fly 310 mi against the wind. The plane has a speed of 165 mph in still air. Find the speed of the wind.

71. Sarita can plant her garden in 5 hr working alone. A friend can do the same job in 8 hr. How long would it take them if they worked together?

72. The head gardener can mow the lawns in the city park twice as fast as his assistant. Working together, they can complete the job in $1\frac{1}{3}$ hr. How long would it take the head gardener working alone?

Chapter 6   Mixed Review Exercises

Perform the indicated operation. Write each answer in lowest terms.

1. $\dfrac{4}{m-1} - \dfrac{3}{m+1}$

2. $\dfrac{8p^5}{5} \div \dfrac{2p^3}{10}$

3. $\dfrac{r-3}{8} \div \dfrac{3r-9}{4}$

4. $\dfrac{\dfrac{5}{x} - 1}{\dfrac{5-x}{3x}}$

5. $\dfrac{4}{z^2 - 2z + 1} - \dfrac{3}{z^2 - 1}$

6. $\dfrac{1}{t^2 - 4} + \dfrac{1}{2-t}$

7. $\dfrac{2x^2 + 5x - 12}{4x^2 - 9} \cdot \dfrac{x^2 - 3x - 28}{x^2 + 8x + 16}$

Solve each equation, and check the solutions.

8. $\dfrac{5t}{6} = \dfrac{2t-1}{3} + 1$

9. $\dfrac{2}{z} - \dfrac{z}{z+3} = \dfrac{1}{z+3}$

10. $\dfrac{1}{x^2 - 1} = \dfrac{1}{x+1}$

11. $\dfrac{2x}{x^2 - 16} - \dfrac{2}{x-4} = \dfrac{4}{x+4}$

12. $a = \dfrac{v-w}{t}$ for v

Solve each problem.

13. If the same number is added to both the numerator and the denominator of the fraction $\frac{4}{11}$, the result is equivalent to $\frac{1}{2}$. Find the number.

14. Seema can clean the house in 3 hr. Satish can clean the house in 6 hr. Working together, how long will it take them to clean the house?

15. Anne flew her plane 400 km with the wind in the same time it took her to go 200 km against the wind. The wind speed is 50 km per hr. Find the rate of the plane in still air.

	d	r	t
With the Wind	400		
Against the Wind	200		

| Chapter 6 | Test | **FOR EXTRA HELP** | Step-by-step test solutions are found on the Chapter Test Prep Videos available in **MyLab Math**. |

▶ *View the complete solutions to all Chapter Test exercises in MyLab Math.*

1. Find the numerical value of $\dfrac{6r + 1}{2r^2 - 3r - 20}$ for **(a)** $r = -2$ and **(b)** $r = 4$.

2. Find any values for which $\dfrac{3x - 1}{x^2 - 2x - 8}$ is undefined. Write the answer with the symbol $\neq$.

3. Write four equivalent forms of the rational expression $-\dfrac{6x - 5}{2x + 3}$.

Write each rational expression in lowest terms.

4. $\dfrac{-15x^6 y^4}{5x^4 y}$

5. $\dfrac{6a^2 + a - 2}{2a^2 - 3a + 1}$

Multiply or divide. Write each answer in lowest terms.

6. $\dfrac{5(d - 2)}{9} \div \dfrac{3(d - 2)}{5}$

7. $\dfrac{6k^2 - k - 2}{8k^2 + 10k + 3} \cdot \dfrac{4k^2 + 7k + 3}{3k^2 + 5k + 2}$

8. $\dfrac{4a^2 + 9a + 2}{3a^2 + 11a + 10} \div \dfrac{4a^2 + 17a + 4}{3a^2 + 2a - 5}$

9. $\dfrac{x^2 - 10x + 25}{9 - 6x + x^2} \cdot \dfrac{x - 3}{5 - x}$

Find the LCD for the fractions in each list.

10. $\dfrac{-3}{10p^2}, \dfrac{21}{25p^3}, \dfrac{-7}{30p^5}$

11. $\dfrac{r + 1}{2r^2 + 7r + 6}, \dfrac{-2r + 1}{2r^2 - 7r - 15}$

Write each rational expression as an equivalent expression with the indicated denominator.

12. $\dfrac{15}{4p} = \dfrac{?}{64p^3}$

13. $\dfrac{3}{6m - 12} = \dfrac{?}{42m - 84}$

Add or subtract. Write each answer in lowest terms.

14. $\dfrac{4x + 2}{x + 5} + \dfrac{-2x + 8}{x + 5}$

15. $\dfrac{-4}{y + 2} + \dfrac{6}{5y + 10}$

16. $\dfrac{x + 1}{3 - x} + \dfrac{x^2}{x - 3}$

17. $\dfrac{3}{2m^2 - 9m - 5} - \dfrac{m + 1}{2m^2 - m - 1}$

Simplify each complex fraction.

18. $\dfrac{\dfrac{2p}{k^2}}{\dfrac{3p^2}{k^3}}$

19. $\dfrac{\dfrac{x^2 - 25}{x + 3}}{\dfrac{x + 5}{x^2 - 9}}$

20. $\dfrac{\dfrac{1}{x + 3} - 1}{1 + \dfrac{1}{x + 3}}$

Solve each equation.

21. $\dfrac{3x}{x + 1} = \dfrac{3}{2x}$

22. $\dfrac{2}{x - 1} - \dfrac{2}{3} = \dfrac{-1}{x + 1}$

23. $4 + \dfrac{6}{x - 3} = \dfrac{2x}{x - 3}$

24. $\dfrac{2x}{x - 3} + \dfrac{1}{x + 3} = \dfrac{-6}{x^2 - 9}$

25. Solve $F = \dfrac{k}{d - D}$ for D.

Solve each problem.

26. A man can paint a room in his house, working alone, in 5 hr. His wife can do the job in 4 hr. How long will it take them to paint the room if they work together?

	Rate	Time Working Together	Fractional Part of the Job Done When Working Together
Man			
Wife			

27. A boat travels 7 mph in still water. It takes as long to go 20 mi upstream as 50 mi downstream. Find the rate of the current.

28. If the same number is added to the numerator and subtracted from the denominator of $\frac{5}{6}$, the resulting fraction is equivalent to $\frac{1}{10}$. What is the number?

Chapters R–6 Cumulative Review Exercises

1. Add. $\dfrac{2}{7} + \dfrac{9}{4}$

2. Divide. $\dfrac{7}{12} \div \dfrac{14}{5}$

3. Multiply. -5.62×4.3

4. Find 32% of 18.

5. Evaluate $3 + 4\left(\dfrac{1}{2} - \dfrac{3}{4}\right)$.

Solve.

6. $3(2t - 5) = 2 + 5t$

7. $\mathcal{A} = \dfrac{1}{2}bh$ for b

8. $\dfrac{2 + m}{3} = \dfrac{2 - m}{4}$

Solve each inequality. Graph the solution set and write it using interval notation.

9. $5x \le 6x + 8$

10. $5m - 9 > 2m + 3$

11. Consider the graph of $4x + 3y = -12$.

(a) What is the x-intercept? **(b)** What is the y-intercept?

12. What is the slope of the line with equation $4x + 3y = -12$?

Graph each equation.

13. $y = -3x + 2$

14. $y = -x^2 + 1$

Simplify each expression.

15. $\dfrac{(2x^3)^{-1} \cdot x}{2^3 x^5}$

16. $\dfrac{(m^{-2})^3 m}{m^5 m^{-4}}$

Perform each indicated operation.

17. $(2k^2 + 3k) - (k^2 + k - 1)$

18. $(2a - b)^2$

19. $(y^2 + 3y + 5)(3y - 1)$

20. $\dfrac{12p^3 + 2p^2 - 12p + 4}{2p - 2}$

Factor completely.

21. $8t^2 + 10tv + 3v^2$

22. $16x^4 - 1$

Solve each equation.

23. $r^2 = 2r + 15$

24. $(r - 5)(2r + 1)(3r - 2) = 0$

Solve each problem.

25. The length of a rectangle is 2 m less than twice the width. The area is 60 m². Find the width of the rectangle.

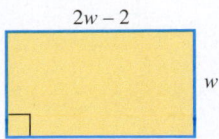

26. One number is 4 greater than another. The product of the numbers is 2 less than the lesser number. Find the lesser number.

Answer each question.

27. Which one of the following rational expressions is equal to 1 for *all* real numbers?

A. $\dfrac{k^2 + 2}{k^2 + 2}$ **B.** $\dfrac{4 - m}{4 - m}$ **C.** $\dfrac{2x + 9}{2x + 9}$ **D.** $\dfrac{x^2 - 1}{x^2 - 1}$

28. Which one of the following rational expressions is *not* equivalent to $\dfrac{4 - 3x}{7}$?

A. $-\dfrac{-4 + 3x}{7}$ **B.** $-\dfrac{4 - 3x}{-7}$ **C.** $\dfrac{-4 + 3x}{-7}$ **D.** $\dfrac{-(3x + 4)}{7}$

Perform each operation, and write the answer in lowest terms.

29. $\dfrac{3}{7} + \dfrac{4}{r}$

30. $\dfrac{4}{5q - 20} - \dfrac{1}{3q - 12}$

31. $\dfrac{7z^2 + 49z + 70}{16z^2 + 72z - 40} \div \dfrac{3z + 6}{4z^2 - 1}$

32. $\dfrac{\dfrac{4}{a} + \dfrac{5}{2a}}{\dfrac{7}{6a} - \dfrac{1}{5a}}$

Solve each equation. Check the solutions.

33. $\dfrac{r + 2}{5} = \dfrac{r - 3}{3}$

34. $\dfrac{1}{x} = \dfrac{1}{x + 1} + \dfrac{1}{2}$

35. Jody can weed the yard in 3 hr. Pat can weed the same yard in 2 hr. How long will it take them if they work together?

	Rate	Time Working Together	Fractional Part of the Job Done
Jody			
Pat			

7

LINEAR EQUATIONS, GRAPHS, AND SYSTEMS

The point of intersection of two lines can be found using a *system of linear equations,* one of the topics covered in this chapter.

7.1 Review of Graphs and Slopes of Lines

This section and the next review and extend some of the main topics of linear equations in two variables introduced earlier in the text.

OBJECTIVE 1 Plot ordered pairs.

Each of the pairs of numbers $(3, 2)$, $(-5, 6)$, and $(4, -1)$ is an example of an **ordered pair**—that is, a pair of numbers written within parentheses. The *order* in which the numbers are written is important. We graph an ordered pair using two perpendicular number lines that intersect at their 0 points. See **FIGURE 1**. The common 0 point is the **origin.**

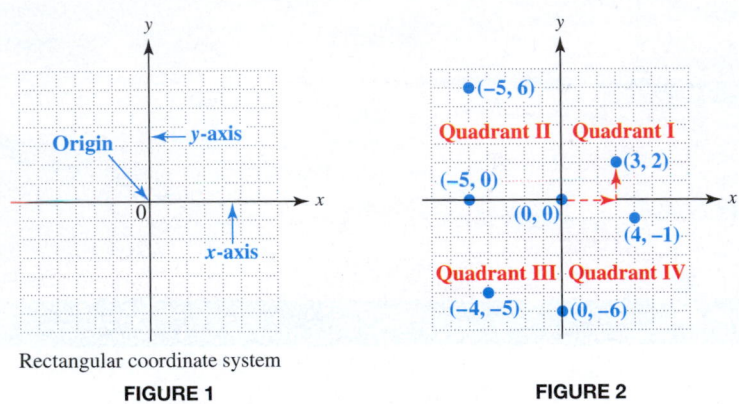

Rectangular coordinate system

FIGURE 1　　　　　　**FIGURE 2**

The position of any point in this coordinate plane is determined by referring to the horizontal number line, or **x-axis,** and the vertical number line, or **y-axis.** The *x*-axis and the *y*-axis make up a **rectangular** (or **Cartesian,** for Descartes) **coordinate system.**

The numbers in an ordered pair (x, y) are its **components.** The first component indicates position relative to the *x*-axis, and the second component indicates position relative to the *y*-axis. For example, to locate, or **plot,** the point on the graph that corresponds to the ordered pair $(3, 2)$, we move three units from 0 to the right along the *x*-axis and then two units up parallel to the *y*-axis. See **FIGURE 2**. The numbers in an ordered pair are the **coordinates** of the corresponding point.

The four regions of the graph shown in **FIGURE 2** are **quadrants I, II, III,** and **IV,** reading counterclockwise from the upper right quadrant. *The points on the x-axis and y-axis do not belong to any quadrant.*

OBJECTIVE 2 Graph lines and find intercepts.

Each solution of an equation with two variables, such as

$$2x + 3y = 6, \quad \text{Equation with two variables } x \text{ and } y$$

will include two numbers, one for each variable. To keep track of which number goes with which variable, we write the solutions as ordered pairs. *(If x and y are used as the variables, the x-value is given first.)*

We can show that $(6, -2)$ is a solution of $2x + 3y = 6$ by substitution.

$$2x + 3y = 6 \qquad \text{Equation with two variables}$$
$$2(6) + 3(-2) \overset{?}{=} 6 \qquad \text{Let } x = 6, y = -2.$$
$$12 - 6 \overset{?}{=} 6 \qquad \text{Multiply.}$$
$$6 = 6 \ \checkmark \qquad \text{True}$$

Use parentheses to avoid errors.

Because the ordered pair $(6, -2)$ makes the equation true, it is a solution. On the other hand, $(5, 1)$ is *not* a solution of $2x + 3y = 6$.

$$2x + 3y = 6 \qquad \text{Equation with two variables}$$
$$2(5) + 3(1) \overset{?}{=} 6 \qquad \text{Let } x = 5, y = 1.$$
$$10 + 3 \overset{?}{=} 6 \qquad \text{Multiply.}$$
$$13 = 6 \qquad \text{False}$$

To find ordered pairs that satisfy an equation, we select any number for one of the variables, substitute it into the equation for that variable, and then solve for the other variable.

> *Because any real number could be selected for one variable and would lead to a real number for the other variable, an equation with two variables such as $2x + 3y = 6$ has an infinite number of solutions.*

The **graph of an equation** is the set of points corresponding to *all* ordered pairs that satisfy the equation. It gives a "picture" of the equation. The graph of $2x + 3y = 6$ is shown in **FIGURE 3** along with a **table of ordered pairs** (or **table of values**).

The equation $2x + 3y = 6$ is a **first-degree equation** because it has no term with a variable to a power greater than 1.

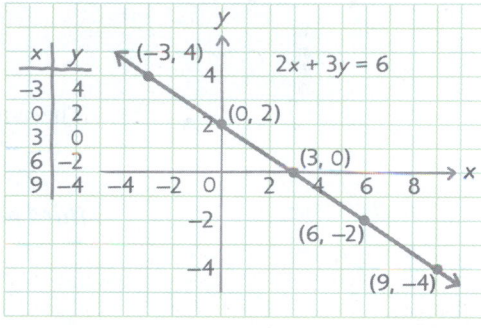

x	y
-3	4
0	2
3	0
6	-2
9	-4

FIGURE 3

> *The graph of a first-degree equation in two variables is a straight line.*

Because first-degree equations with two variables have straight-line graphs, they are called *linear equations in two variables.*

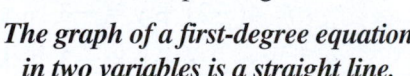

Linear Equation in Two Variables

A **linear equation in two variables** (here x and y) is an equation that can be written in the form

$$Ax + By = C,$$

where A, B, and C are real numbers and A and B are not both 0. This form is called **standard form.**

Examples: $3x + 4y = 9, \quad x - y = 0, \quad x + 2y = -8$

A straight line is determined if any two different points on the line are known. Therefore, finding two different points is sufficient to graph the line.

Two useful points for graphing are the x- and y-intercepts. The **x-intercept** is the point (if any) where the line intersects the x-axis. The **y-intercept** is the point (if any) where the line intersects the y-axis. See **FIGURE 4**.

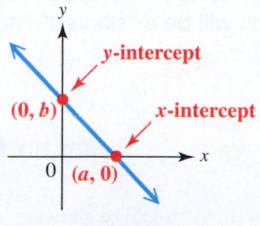

FIGURE 4

Remember the following regarding intercepts.

- The y-value of the point where the line intersects the x-axis is always 0.
- The x-value of the point where the line intersects the y-axis is always 0.

This suggests a method for finding the x- and y-intercepts.*

Finding Intercepts

When graphing the equation of a line, find the intercepts as follows.

$$\text{Let } y = 0 \text{ to find the } x\text{-intercept.}$$

$$\text{Let } x = 0 \text{ to find the } y\text{-intercept.}$$

NOW TRY
EXERCISE 1
Find the x- and y-intercepts, and graph the equation.

$$x - 2y = 4$$

EXAMPLE 1   Finding Intercepts

Find the x- and y-intercepts of $4x - y = -3$, and graph the equation.

To find the x-intercept, let $y = 0$.

$$4x - y = -3$$
$$4x - 0 = -3 \quad \text{Let } y = 0.$$
$$4x = -3 \quad \text{Subtract.}$$
$$x = -\frac{3}{4} \quad \text{Divide by 4.}$$

The x-intercept is $\left(-\frac{3}{4}, 0\right)$.

To find the y-intercept, let $x = 0$.

$$4x - y = -3$$
$$4(0) - y = -3 \quad \text{Let } x = 0.$$
$$-y = -3 \quad \text{Multiply. Subtract.}$$
$$y = 3 \quad \text{Multiply by } -1.$$

The y-intercept is $(0, 3)$.

To guard against errors when graphing the equation, it is a good idea to find a third point. We arbitrarily choose $x = -2$, and substitute this value in the equation.

$$4x - y = -3$$
$$4(-2) - y = -3 \quad \text{Let } x = -2.$$
$$-8 - y = -3 \quad \text{Multiply.}$$
$$-y = 5 \quad \text{Add 8.}$$
$$y = -5 \quad \text{Multiply by } -1.$$

The ordered pair $(-2, -5)$ lies on the graph. We plot the three ordered pairs and draw a line through them. See **FIGURE 5**.

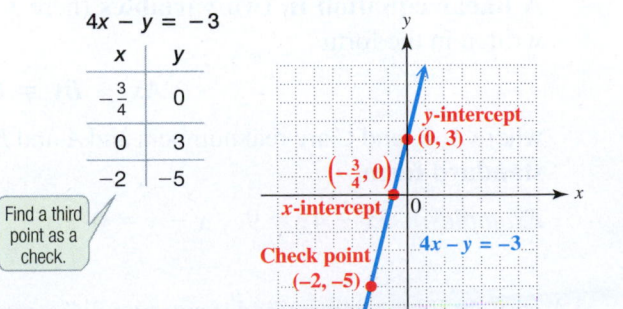

A linear equation with both x and y variables will have both x- and y-intercepts. Its graph will be a "slanted" line.

FIGURE 5

NOW TRY

NOW TRY ANSWER
1. x-intercept: $(4, 0)$;
   y-intercept: $(0, -2)$

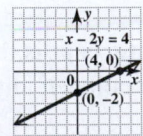

---

*Some texts define an intercept as a number, not a point. For example, "y-intercept $(0, 4)$" would be given as "y-intercept 4."

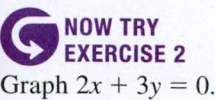

NOW TRY EXERCISE 2
Graph $2x + 3y = 0$.

EXAMPLE 2 Graphing a Line That Passes through the Origin

Graph $x + 2y = 0$.

Find the x-intercept.	Find the y-intercept.
$x + 2y = 0$	$x + 2y = 0$
$x + 2(0) = 0$ Let $y = 0$.	$0 + 2y = 0$ Let $x = 0$.
$x + 0 = 0$ Multiply.	$2y = 0$ Add.
$x = 0$ x-intercept is $(0, 0)$.	$y = 0$ y-intercept is $(0, 0)$.

Both intercepts are the *same* point, $(0, 0)$, which means that the graph passes through the origin. To find a second point so we can graph the line, we choose any nonzero number for x or y and solve for the other variable. We arbitrarily choose $x = 4$.

$x + 2y = 0$

$4 + 2y = 0$ Let $x = 4$.

$2y = -4$ Subtract 4.

$y = -2$ Divide by 2.

$x + 2y = 0$	
x	y
-2	1
0	0
4	-2

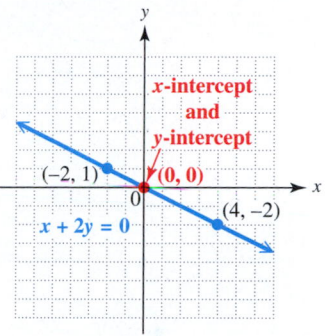

This gives the ordered pair $(4, -2)$. As a final check, substitute 1 for y in the equation and verify that $(-2, 1)$ also lies on the line. The graph is shown in **FIGURE 6**.

FIGURE 6

NOW TRY

OBJECTIVE 3 Graph equations of horizontal and vertical lines.

NOW TRY EXERCISE 3
Graph each equation.
(a) $y = -2$ **(b)** $x + 3 = 0$

EXAMPLE 3 Graphing Horizontal and Vertical Lines

Graph each equation.

(a) $y = 2$ (This equation can be written as $0x + y = 2$.)

Because y *always* equals 2, there is no value of x corresponding to $y = 0$, and the graph has no x-intercept. One value where $y = 2$ is on the y-axis, so the y-intercept is $(0, 2)$. Plot any two other points with y-coordinate 2, such as $(-1, 2)$ and $(3, 2)$.

The graph is shown in **FIGURE 7**. It is a horizontal line.

$y = 2$	
x	y
-1	2
0	2
3	2

$x + 1 = 0$	
x	y
-1	-4
-1	0
-1	5

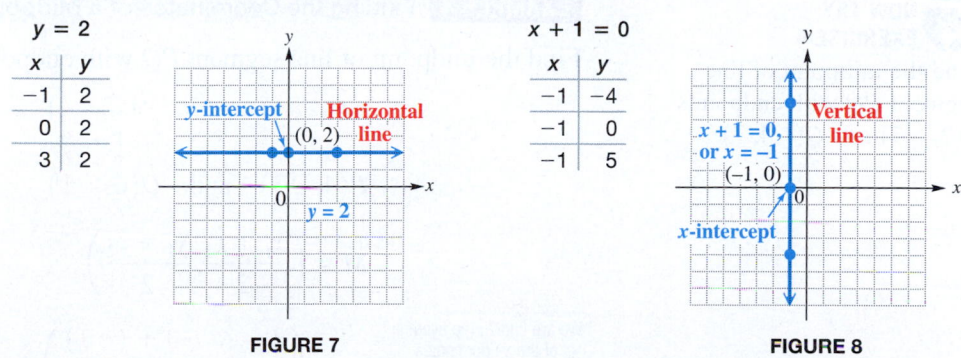

NOW TRY ANSWERS
2.
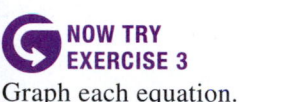

3.
(a)   **(b)**
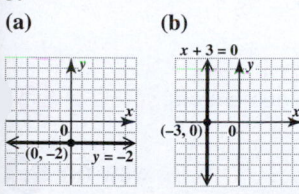

FIGURE 7   **FIGURE 8**

(b) $x + 1 = 0$ (This equation can be written as $x = -1$ or $x + 0y = -1$.)

Because x *always* equals -1, there is no value of y that makes $x = 0$, and the graph has no y-intercept. One value where $x = -1$ is on the x-axis, so the x-intercept is $(-1, 0)$. Plot any two other points with x-coordinate -1, such as $(-1, -4)$ and $(-1, 5)$.

The graph is shown in **FIGURE 8**. It is a vertical line.   **NOW TRY**

OBJECTIVE 4   Find the midpoint of a line segment.

If the coordinates of the endpoints of a line segment are known, then the coordinates of the *midpoint* of the segment can be found.

FIGURE 9 shows a line segment PQ with endpoints $P(-8, 4)$ and $Q(3, -2)$. R is the point with the same x-coordinate as P and the same y-coordinate as Q. So the coordinates of R are $(-8, -2)$.

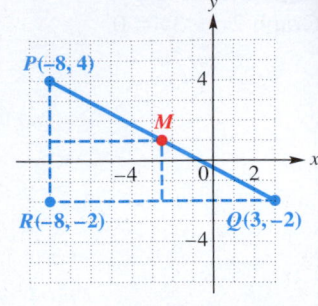

FIGURE 9

The x-coordinate of the midpoint M of PQ is the same as the x-coordinate of the midpoint of RQ. Because RQ is horizontal, the x-coordinate of its midpoint is the *average* (or *mean*) of the x-coordinates of its endpoints.

$$\frac{1}{2}(-8 + 3) = -\frac{5}{2}$$

The y-coordinate of M is the average (or *mean*) of the y-coordinates of the midpoint of PR.

$$\frac{1}{2}(4 + (-2)) = 1$$

The midpoint of PQ is $M\left(-\frac{5}{2}, 1\right)$.

This discussion leads to the *midpoint formula.*

Midpoint Formula

The midpoint M of a line segment PQ with endpoints $P(x_1, y_1)$ and $Q(x_2, y_2)$ is found as follows.

$$M = \left(\frac{x_1 + x_2}{2}, \frac{y_1 + y_2}{2}\right)$$

The two nonspecific points (x_1, y_1) and (x_2, y_2) use **subscript notation.** Read (x_1, y_1) as "*x-sub-one, y-sub-one.*"

NOW TRY
EXERCISE 4
Find the midpoint of line segment PQ with endpoints $P(2, -5)$ and $Q(-4, 7)$.

EXAMPLE 4   Finding the Coordinates of a Midpoint

Find the midpoint of line segment PQ with endpoints $P(4, -3)$ and $Q(6, -1)$.

$$\overset{(x_1, \ y_1)}{\underset{\downarrow \ \ \downarrow}{P(4, -3)}} \quad \text{and} \quad \overset{(x_2, \ y_2)}{\underset{\downarrow \ \ \downarrow}{Q(6, -1)}} \qquad \text{Label the points.}$$

$$M = \left(\frac{x_1 + x_2}{2}, \frac{y_1 + y_2}{2}\right) \qquad \text{Midpoint formula}$$

We are finding the average of the x-coordinates and the average of the y-coordinates.

$$= \left(\frac{4 + 6}{2}, \frac{-3 + (-1)}{2}\right) \qquad \text{Substitute.}$$

$$= \left(\frac{10}{2}, \frac{-4}{2}\right) \qquad \text{Add in the numerators.}$$

$$= (5, -2) \leftarrow \text{Midpoint of segment } PQ$$

NOW TRY ANSWER
4. $(-1, 1)$

NOW TRY

NOTE When graphing with a graphing calculator, we "set up" a rectangular coordinate system. The screen in **FIGURE 10** shows the **standard viewing window**. Minimum x- and y-values are -10, and maximum x- and y-values are 10. The scale on each axis, here 1, determines the distance between the tick marks.

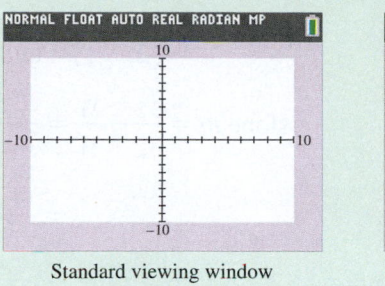

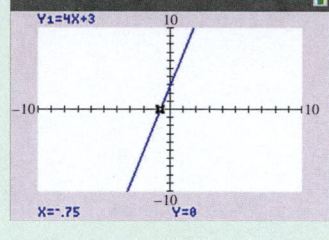

Standard viewing window

FIGURE 10   **FIGURE 11**

To graph an equation such as $4x - y = -3$, we use the intercepts $(-0.75, 0)$ and $(0, 3)$ to determine an appropriate window. Here, we choose the standard viewing window. We solve the equation for y to obtain $y = 4x + 3$ and enter it into the calculator. The graph in **FIGURE 11** also gives the coordinates of the x-intercept at the bottom of the screen.

OBJECTIVE 5 Find the slope of a line.

Slope (steepness) is used in many practical ways, as shown in **FIGURE 12**.

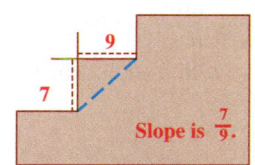

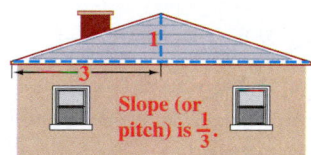

FIGURE 12

Slope is the ratio of vertical change, or **rise,** to horizontal change, or **run.** A simple way to remember this is to think, *"Slope is rise over run."*

To obtain a formal definition of the slope of a line, we designate two different points on the line as (x_1, y_1) and (x_2, y_2). See **FIGURE 13**. As we move along the line in **FIGURE 13** from (x_1, y_1) to (x_2, y_2), the y-value changes (vertically) from y_1 to y_2, an amount equal to $y_2 - y_1$. As y changes from y_1 to y_2, the value of x changes (horizontally) from x_1 to x_2 by the amount $x_2 - x_1$.

The ratio of the change in y to the change in x ("rise over run," or $\frac{\text{rise}}{\text{run}}$) is the *slope* of the line, with the letter m traditionally used for slope.

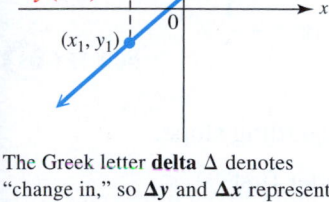

The Greek letter **delta** Δ denotes "change in," so Δy and Δx represent the change in y and the change in x, respectively.

FIGURE 13

Slope Formula

The **slope** m of the line passing through the distinct points (x_1, y_1) and (x_2, y_2) is defined as follows.

$$m = \frac{\text{rise}}{\text{run}} = \frac{\text{change in } y}{\text{change in } x} = \frac{\Delta y}{\Delta x} = \frac{y_2 - y_1}{x_2 - x_1} \quad (\text{where } x_1 \neq x_2)$$

NOW TRY
EXERCISE 5
Find the slope of the line passing through the points $(2, -6)$ and $(-3, 5)$.

EXAMPLE 5   Finding the Slope of a Line

Find the slope of the line passing through the points $(2, -1)$ and $(-5, 3)$.

Label the points, and then apply the slope formula.

$$(x_1, \ y_1) \qquad\qquad (x_2, \ y_2)$$
$$\downarrow \ \downarrow \qquad\qquad\quad \downarrow \ \downarrow$$
$$(2, -1) \quad \text{and} \quad (-5, 3)$$

$$\text{slope } m = \frac{y_2 - y_1}{x_2 - x_1} = \frac{3 - (-1)}{-5 - 2} \qquad \text{Substitute.}$$

$$= \frac{4}{-7}, \quad \text{or} \quad -\frac{4}{7} \qquad \text{Subtract; } \frac{a}{-b} = -\frac{a}{b}$$

The slope is $-\frac{4}{7}$.

The same slope is obtained if we label the points in reverse order. *It makes no difference which point is identified as (x_1, y_1) and which as (x_2, y_2).*

$$(x_2, \ y_2) \qquad\qquad (x_1, \ y_1)$$
$$\downarrow \ \downarrow \qquad\qquad\quad \downarrow \ \downarrow$$
$$(2, -1) \quad \text{and} \quad (-5, 3)$$

y-values are in the numerator, *x*-values in the denominator.

$$\text{slope } m = \frac{y_2 - y_1}{x_2 - x_1} = \frac{-1 - 3}{2 - (-5)} \qquad \text{Substitute.}$$

$$= \frac{-4}{7}, \quad \text{or} \quad -\frac{4}{7} \qquad \text{Subtract; } \frac{-a}{b} = -\frac{a}{b}$$

To confirm this slope, see **FIGURE 14**. Using the geometric interpretation of slope $\left(\text{as } \frac{\text{rise}}{\text{run}} \right)$ and beginning at the point $(2, -1)$, we move *up* 4 units and to the *left* 7 units to reach the point $(-5, 3)$.

$$\text{slope } m = \frac{\text{change in } y \,(\text{rise})}{\text{change in } x \,(\text{run})} = \frac{4}{-7}, \quad \text{or} \quad -\frac{4}{7}$$

The same slope results if we begin at the point $(-5, 3)$ and move *down* 4 units and to the *right* 7 units to reach the point $(2, -1)$. See **FIGURE 14**.

$$\text{slope } m = \frac{\text{change in } y \,(\text{rise})}{\text{change in } x \,(\text{run})} = \frac{-4}{7}, \quad \text{or} \quad -\frac{4}{7}$$

NOW TRY

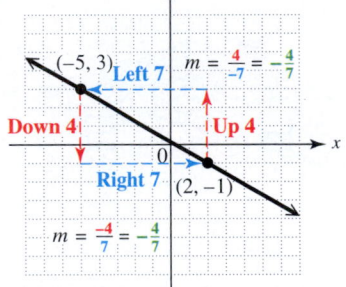

$m = \frac{4}{-7} = -\frac{4}{7}$

Left 7   Down 4   Up 4   Right 7 $(2, -1)$ $(-5, 3)$

$m = \frac{-4}{7} = -\frac{4}{7}$

Moves *down* or to the *left* are denoted as negative numbers.

FIGURE 14

Example 5 suggests the following important ideas regarding slope.

1. The slope is the same no matter which point we consider first.

2. Using similar triangles from geometry, we can show that the slope is the same no matter which two different points on the line we choose.

⚠ **CAUTION** *When calculating slope, remember that the change in y (rise) is the numerator and the change in x (run) is the denominator.*

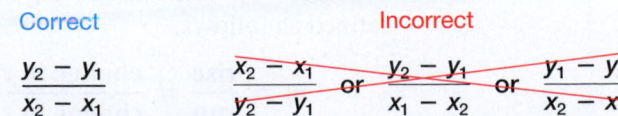

Correct

$$\frac{y_2 - y_1}{x_2 - x_1}$$

Incorrect

$$\frac{x_2 - x_1}{y_2 - y_1} \quad \text{or} \quad \frac{y_2 - y_1}{x_1 - x_2} \quad \text{or} \quad \frac{y_1 - y_2}{x_2 - x_1}$$

Be careful to subtract the y-values and the x-values in the same order.

NOW TRY ANSWER
5. $-\frac{11}{5}$

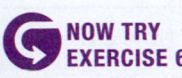

**NOW TRY
EXERCISE 6**

Find the slope of the line
$3x - 7y = 21$.

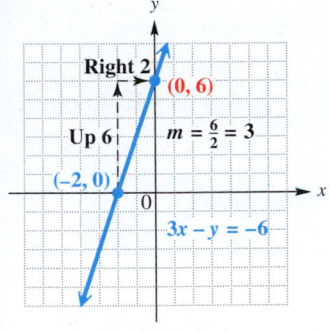

FIGURE 15

EXAMPLE 6 Finding the Slope of a Line from Its Equation

Find the slope of the line $3x - y = -6$.

One way to find the slope is to use the definition of slope with two different points on the line, such as the intercepts.

Let $y = 0$ to find that the x-intercept is $(-2, 0)$.

Let $x = 0$ to find that the y-intercept is $(0, 6)$.

Use these two points in the slope formula.

$$\text{slope } m = \frac{y_2 - y_1}{x_2 - x_1} = \frac{6 - 0}{0 - (-2)} \qquad \begin{array}{l}(x_1, y_1) = (-2, 0) \\ (x_2, y_2) = (0, 6)\end{array}$$

$$= \frac{6}{2} \qquad \text{Subtract.}$$

$$= 3 \qquad \text{Divide.}$$

The graph of $3x - y = -6$ in **FIGURE 15** confirms that the slope of the line is 3.

NOW TRY

The slope of a line can also be found directly from its equation. Consider the equation $3x - y = -6$ from **Example 6.** Solve this equation for y.

$$3x - y = -6 \qquad \text{Equation from \textbf{Example 6}}$$

$$-y = -6 - 3x \qquad \text{Subtract } 3x.$$

$$y = 3x + 6 \qquad \text{Multiply by } -1; \text{ Commutative property}$$

The slope, 3, found using the slope formula in **Example 6,** is the same number as the coefficient of x in the equation $y = 3x + 6$. We will see in the next section that this is true in general, *as long as the equation is solved for y.*

**NOW TRY
EXERCISE 7**

Find the slope of the line
$5x - 4y = 7$.

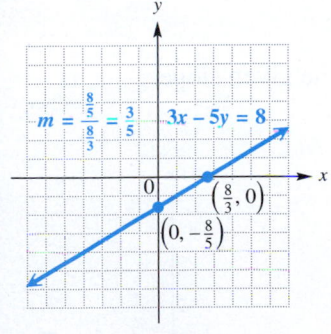

FIGURE 16

NOW TRY ANSWERS

6. $\frac{3}{7}$

7. $\frac{5}{4}$

EXAMPLE 7 Finding the Slope of a Line from Its Equation

Find the slope of the line $3x - 5y = 8$.

Many of the ordered-pair solutions of this equation have coordinates that are fractions, including the intercepts $\left(\frac{8}{3}, 0\right)$ and $\left(0, -\frac{8}{5}\right)$. See **FIGURE 16**. This makes calculations using the slope formula more difficult.

Instead, we solve the equation for y as discussed above.

$$3x - 5y = 8 \quad \boxed{\text{Solve for } y.}$$

$$-5y = 8 - 3x \qquad \text{Subtract } 3x.$$

$$\frac{-5y}{-5} = \frac{-3x + 8}{-5} \qquad \begin{array}{l}\text{Commutative property;} \\ \text{Divide each side by } -5.\end{array}$$

$$\boxed{\frac{-3x}{-5} = \frac{-3}{-5} \cdot \frac{x}{1} = \frac{3}{5}x} \quad y = \frac{3}{5}x - \frac{8}{5} \qquad \frac{a + b}{c} = \frac{a}{c} + \frac{b}{c}$$

The slope, given by the coefficient of x, is $\frac{3}{5}$.

We can confirm this slope using the intercepts and the slope formula, which involves simplifying a complex fraction.

$$m = \frac{0 - \left(-\frac{8}{5}\right)}{\frac{8}{3} - 0} = \frac{\frac{8}{5}}{\frac{8}{3}} = \frac{8}{5} \div \frac{8}{3} = \frac{8}{5} \cdot \frac{3}{8} = \frac{3}{5} \qquad \text{The same slope results.}$$

NOW TRY

NOTE We can solve the standard form of a linear equation $Ax + By = C$ (where $B \neq 0$) for y to show that, in general, **the slope of a line in this form is $-\frac{A}{B}$.**

$$Ax + By = C \qquad \text{Standard form}$$

$$By = -Ax + C \qquad \text{Subtract } Ax.$$

$$y = -\frac{A}{B}x + \frac{C}{B} \qquad \text{Divide each term by } B.$$

The slope is given by the coefficient of x, $-\frac{A}{B}$. In the equation $3x - 5y = 8$ from **Example 7,** $A = 3$ and $B = -5$, so the slope is

$$-\frac{A}{B} = -\frac{3}{-5} = \frac{3}{5}. \qquad \text{The same slope results.}$$

We review the following special cases of slope.

Horizontal and Vertical Lines

- An equation of the form $y = b$ always intersects the y-axis at the point $(0, b)$. A line with this equation is **horizontal** and has **slope 0.** See **FIGURE 17.**

- An equation of the form $x = a$ always intersects the x-axis at the point $(a, 0)$. A line with this equation is **vertical** and has **undefined slope.** See **FIGURE 18.**

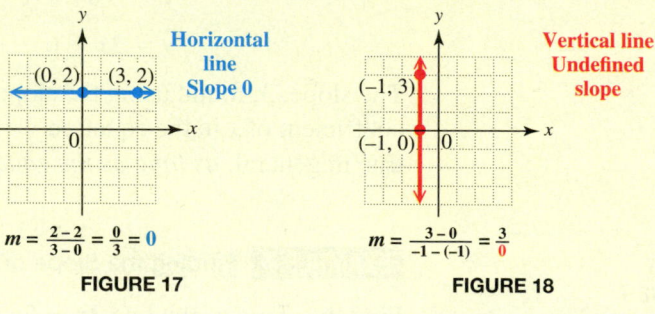

$$m = \frac{2-2}{3-0} = \frac{0}{3} = 0$$

FIGURE 17

$$m = \frac{3-0}{-1-(-1)} = \frac{3}{0}$$

FIGURE 18

OBJECTIVE 6 Graph a line given its slope and a point on the line.

EXAMPLE 8 Using the Slope and a Point to Graph Lines

Graph each line passing through the given point having the given slope.

(a) $(0, -4)$; slope $\frac{2}{3}$

Begin by plotting the point $P(0, -4)$, as shown in **FIGURE 19.** Then use the geometric interpretation of slope to find a second point.

$$m = \frac{\text{change in } y}{\text{change in } x} = \frac{2}{3} \begin{matrix} \leftarrow \text{rise} \\ \leftarrow \text{run} \end{matrix}$$

We move *up* 2 units from $(0, -4)$ and to the *right* 3 units to locate a second point on the graph, $R(3, -2)$. The line through $P(0, -4)$ and R is the required graph.

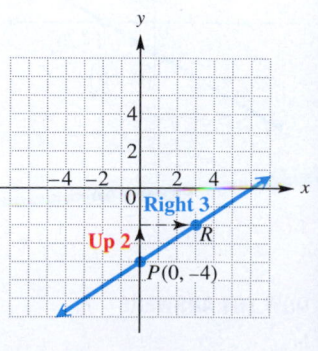

FIGURE 19

**NOW TRY
EXERCISE 8**

Graph the line passing through $(-4, 1)$ that has slope $-\frac{2}{3}$.

(b) $(3, 1)$; slope -4

Start by plotting the point $P(3, 1)$. See **FIGURE 20**. Find a second point R on the line by writing the slope -4 as $\frac{-4}{1}$ and using the geometric interpretation of slope.

$$m = \frac{\text{change in } y}{\text{change in } x} = \frac{-4}{1} \begin{matrix} \leftarrow \text{rise} \\ \leftarrow \text{run} \end{matrix}$$

We move *down* 4 units from $(3, 1)$ and to the *right* 1 unit to locate a second point, $R(4, -3)$. The line through $P(3, 1)$ and R is the required graph.

The slope -4 also could be written as

$$m = \frac{\text{change in } y}{\text{change in } x} = \frac{4}{-1}.$$

In this case, the second point R is located *up* 4 units and to the *left* 1 unit. Verify that this approach also produces the line in **FIGURE 20**.

NOW TRY

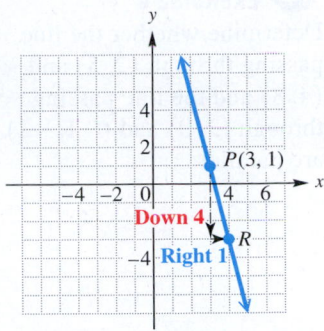

FIGURE 20

Orientation of a Line in the Plane

A *positive slope* indicates that the line *slants up (rises)* from left to right. See **FIGURE 19**.

A *negative slope* indicates that the line *slants down (falls)* from left to right. See **FIGURE 20**.

FIGURE 21 summarizes the four cases for slopes of lines.

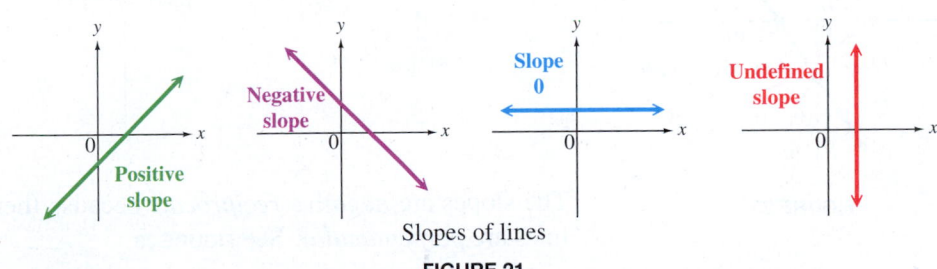

Slopes of lines
FIGURE 21

OBJECTIVE 7 Determine whether two lines are parallel, perpendicular, or neither using slope.

Recall that the slopes of a pair of parallel or perpendicular lines are related.

Slopes of Parallel Lines and Perpendicular Lines

- Two nonvertical lines with the same slope are parallel. Two nonvertical parallel lines have the same slope.

- Two perpendicular lines, neither of which is vertical, have slopes that are negative reciprocals—that is, their product is -1. Also, lines with slopes that are negative reciprocals are perpendicular.

- A line with slope 0 is perpendicular to a line with undefined slope.

NOW TRY ANSWER

8.

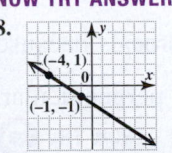

NOW TRY EXERCISE 9

Determine whether the line passing through $(2, 5)$ and $(4, 8)$ and the line passing through $(2, 0)$ and $(-1, -2)$ are parallel.

EXAMPLE 9   Determining Whether Two Lines Are Parallel

Determine whether the lines L_1, passing through $(-2, 1)$ and $(4, 5)$, and L_2, passing through $(3, 0)$ and $(0, -2)$, are parallel.

Slope of L_1:

$$m_1 = \frac{5 - 1}{4 - (-2)}$$

$$= \frac{4}{6}$$

$$= \frac{2}{3}$$

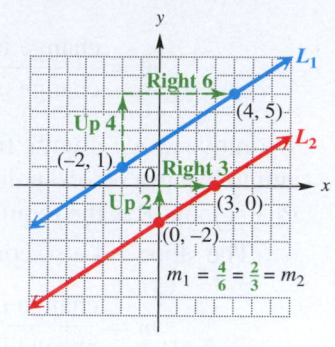

Slope of L_2:

$$m_2 = \frac{0 - (-2)}{3 - 0}$$

$$= \frac{2}{3}$$

FIGURE 22

The slopes are the *same*, both $\frac{2}{3}$, so the two lines are *parallel*. See **FIGURE 22**.

NOW TRY

NOW TRY EXERCISE 10

Determine whether each pair of lines is *parallel*, *perpendicular*, or *neither*.

(a) $x + 2y = 7$ and $2x = y - 4$

(b) $2x - y = 4$ and $2x + y = 6$

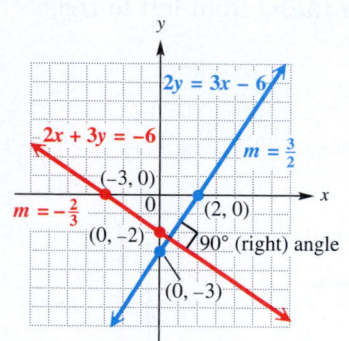

FIGURE 23

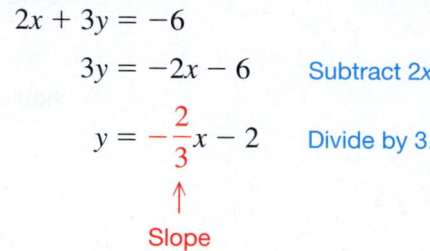

FIGURE 24

EXAMPLE 10   Determining Whether Two Lines Are Parallel, Perpendicular, or Neither

Determine whether each pair of lines is *parallel*, *perpendicular*, or *neither*.

(a) $2y = 3x - 6$ and $2x + 3y = -6$

Find the slope of each line by solving each equation for y.

$$2y = 3x - 6$$
$$y = \frac{3}{2}x - 3 \quad \text{Divide by 2.}$$
$$\uparrow$$
Slope

$$2x + 3y = -6$$
$$3y = -2x - 6 \quad \text{Subtract } 2x.$$
$$y = -\frac{2}{3}x - 2 \quad \text{Divide by 3.}$$
$$\uparrow$$
Slope

The slopes are *negative reciprocals* because their product is $\frac{3}{2}\left(-\frac{2}{3}\right) = -1$. The two lines are *perpendicular*. See **FIGURE 23**.

(b) $2x - 5y = 8$ and $2x + 5y = 8$

Find the slope of each line by solving each equation for y.

$$2x - 5y = 8$$
$$-5y = -2x + 8 \quad \text{Subtract } 2x.$$
$$y = \frac{2}{5}x - \frac{8}{5} \quad \text{Divide by } -5.$$
$$\uparrow$$
Slope

$$2x + 5y = 8$$
$$5y = -2x + 8 \quad \text{Subtract } 2x.$$
$$y = -\frac{2}{5}x + \frac{8}{5} \quad \text{Divide by 5.}$$
$$\uparrow$$
Slope

The slopes, $\frac{2}{5}$ and $-\frac{2}{5}$, are not the same. They are not negative reciprocals because their product is $-\frac{4}{25}$, *not* -1. The two lines are *neither* parallel nor perpendicular. See **FIGURE 24**.

NOW TRY

NOW TRY ANSWERS

9. not parallel
10. (a) perpendicular
    (b) neither

OBJECTIVE 8 Solve problems involving average rate of change.

The slope formula applied to any two points on a line gives the **average rate of change** in *y* per unit change in *x*.

For example, suppose the height of a boy increased from 60 to 68 in. between the ages of 12 and 16, as shown in **FIGURE 25**.

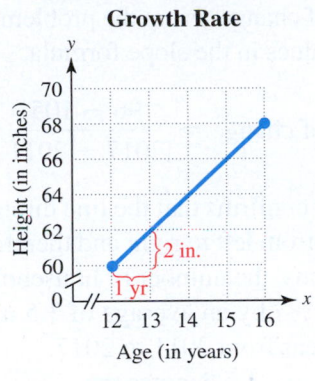

Growth Rate

Change in height $y \rightarrow \dfrac{68 - 60}{16 - 12} = \dfrac{8}{4} = 2$ in. per yr
Change in age $x \longrightarrow$

FIGURE 25

The boy may actually have grown more (or less) than 2 in. during some years. If we plotted ordered pairs (age, height) for those years and drew a line connecting any two of the points, the average rate of change would likely be slightly different. However, using the data for ages 12 and 16, the boy's *average* change in height was 2 in. per year over these years.

NOW TRY EXERCISE 11

There were approximately 84 million high-speed Internet subscribers in 2013. Using this number for 2013 and the number for 2017 from the graph in **FIGURE 26**, find the average rate of change per year from 2013 to 2017. How does it compare with the average rate of change found in **Example 11?**

EXAMPLE 11 Interpreting Slope as Average Rate of Change

The graph in **FIGURE 26** shows the number of high-speed Internet subscribers in the United States from 2012 to 2017. Find the average rate of change in number of subscribers per year.

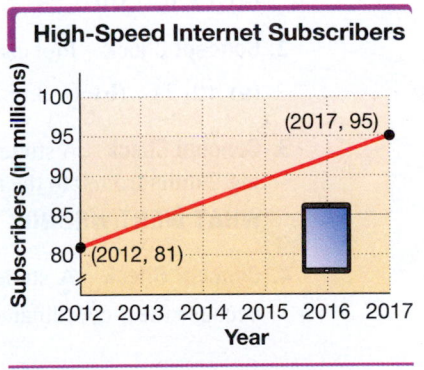

Data from Leichtman Research Group.

FIGURE 26

To find the average rate of change, we need two pairs of data. From the graph, we have the ordered pairs (2012, 81) and (2017, 95). We use the slope formula.

$$\text{average rate of change} = \frac{95 - 81}{2017 - 2012} = \frac{14}{5} = 2.8$$

A positive slope indicates an increase.

This means that the number of high-speed Internet subscribers *increased* by an average of 2.8 million subscribers per year from 2012 to 2017.

NOW TRY

NOW TRY ANSWER
11. 2.75 million subscribers per year; It is less than the average rate of change per year from 2012 to 2017.

NOW TRY EXERCISE 12

In 2011, global sales of digital cameras totaled 33 billion euros. In 2016, sales totaled 17 billion euros. Find the average rate of change in sales of digital cameras per year. (Data from Photoindustrie–Verband.)

NOW TRY ANSWER

12. -3.2 billion euros per year

EXAMPLE 12 Interpreting Slope as Average Rate of Change

In 2011, there were 105 million households with cable TV in the United States. There were 96 million such households in 2017. Find the average rate of change in the number of households per year. (Data from Nielson Media Research.)

We must write two ordered pairs of the form (year, number of households) to use to find the average rate of change. From the problem, we have (2011, 105) and (2017, 96). We substitute these values in the slope formula.

$$\text{average rate of change} = \frac{96 - 105}{2017 - 2011} = \frac{-9}{6} = -1.5$$

A negative slope indicates a decrease.

The graph in **FIGURE 27** confirms that the line through the ordered pairs falls from left to right and therefore has negative slope. Thus, the number of households with cable TV *decreased* by an average of 1.5 million households per year from 2011 to 2017.

The negative sign in -1.5 denotes the *decrease*. (We say "*The number of customers decreased by 1.5 million per year.*" It is **incorrect** to say "*The number of customers decreased by -1.5 million per year.*")

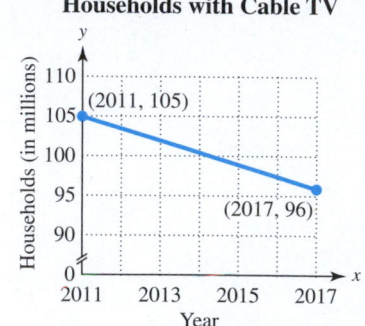

Households with Cable TV

FIGURE 27

NOW TRY

7.1 Exercises

FOR EXTRA HELP

▶ **MyLab Math**

▶ *Video solutions for select problems available in MyLab Math*

STUDY SKILLS REMINDER

Are you fully utilizing the features of your text? **Review Study Skill 1,** *Using Your Math Text.*

1. Concept Check Name the quadrant, if any, in which each point is located. **See Objective 1.**

(a) $(1, 6)$ **(b)** $(-4, -2)$ **(c)** $(-3, 6)$ **(d)** $(7, -5)$ **(e)** $(-3, 0)$ **(f)** $(0, -8)$

2. Concept Check Plot each point in a rectangular coordinate system. **See Objective 1.**

(a) $(2, 3)$ **(b)** $(-3, -2)$ **(c)** $(0, 5)$ **(d)** $(-2, 4)$ **(e)** $(-2, 0)$ **(f)** $(3, -3)$

3. Concept Check A student plotted the point with coordinates $(-4, 2)$ incorrectly by moving 2 units from 0 to the right along the x-axis and then 4 units down parallel to the y-axis. **WHAT WENT WRONG?**

4. Concept Check A student incorrectly claimed that the point $(0, -4)$ lies on the x-axis because the x-coordinate is 0. **WHAT WENT WRONG?**

Complete the given table for each equation and then graph the equation. **See Objective 2 and** **.**

5. $y = x - 4$

x	y
0	
1	
2	
3	
4	

6. $x - y = 3$

x	y
0	
	0
5	
2	

7. $x + 2y = 5$

x	y
0	
	0
2	
	2

8. $4x - 5y = 20$

x	y
0	
	0
2	
	-3

9. **Concept Check**   Match each equation in parts (a)–(d) with its graph in choices A–D. (Coordinates of the points shown are integers.)

(a) $x + 3y = 3$ (b) $x - 3y = -3$ (c) $x - 3y = 3$ (d) $x + 3y = -3$

A. 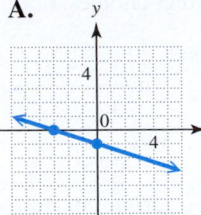   **B.** 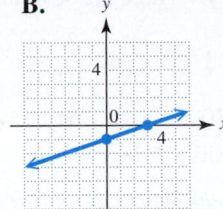   **C.** 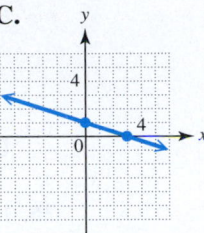   **D.**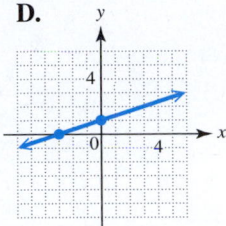

10. **Concept Check**   Which of the following equations have a graph that is a horizontal line? A vertical line?

A. $x - 6 = 0$ **B.** $x + y = 0$ **C.** $y + 3 = 0$ **D.** $y = -10$ **E.** $x + 1 = 5$

Find the x- and y-intercepts. Then graph each equation. See Examples 1–3.

11. $2x + 3y = 12$

12. $5x + 2y = 10$

13. $x - 3y = 6$

14. $x - 2y = -4$

15. $x + 5y = 0$

16. $x - 3y = 0$

17. $2x = 3y$

18. $4y = 3x$

19. $y = 5$

20. $y = -3$

21. $x = 2$

22. $x = -3$

Each table of values gives several points that lie on a line.

(a)   What is the x-intercept of the line?   The y-intercept?

(b)   Which equation in choices A–D corresponds to the given table of values?

(c)   Graph the equation.

23.

x	y
-4	-3
-2	0
0	3
2	6

A. $3x + 2y = 6$
B. $3x - 2y = -6$
C. $3x + 2y = -6$
D. $3x - 2y = 6$

24.

x	y
-1	6
0	4
1	2
2	0

A. $2x - y = 4$
B. $2x + y = -4$
C. $2x + y = 4$
D. $2x - y = -4$

25.

x	y
-2	-1
0	-1
2	-1
4	-1

A. $y = -1$
B. $y = 1$
C. $x = 1$
D. $x = -1$

26.

x	y
6	-1
6	0
6	1
6	2

A. $x = -6$
B. $y = 0$
C. $y = 6$
D. $x = 6$

Find the midpoint of each segment with the given endpoints. See Example 4.

27. $(-8, 4)$ and $(-2, -6)$

28. $(5, 2)$ and $(-1, 8)$

29. $(3, -6)$ and $(6, 3)$

30. $(-10, 4)$ and $(7, 1)$

31. $(-9, 3)$ and $(9, 8)$

32. $(4, -3)$ and $(-1, 3)$

33. $(2.5, 3.1)$ and $(1.7, -1.3)$

34. $(6.2, 5.8)$ and $(1.4, -0.6)$

Concept Check *Answer each question.*

35. A hill rises 30 ft for every horizontal 100 ft. Which of the following express its slope (or grade)? (There are several correct choices.)

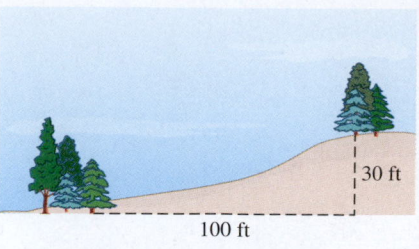

A. 0.3   **B.** $\dfrac{3}{10}$ **C.** $3\dfrac{1}{3}$

D. $\dfrac{30}{100}$ **E.** $\dfrac{10}{3}$ **F.** 30%

30 ft

100 ft

36. If a walkway rises 2 ft for every 24 ft on the horizontal, which of the following express its slope (or grade)? (There are several correct choices.)

A. 12%   **B.** $\dfrac{2}{24}$ **C.** $\dfrac{1}{12}$

D. 12   **E.** $8.\overline{3}\%$ **F.** $\dfrac{24}{2}$

2 ft

24 ft   (not to scale)

37. Concept Check Determine the slope of each line segment in the given figure.

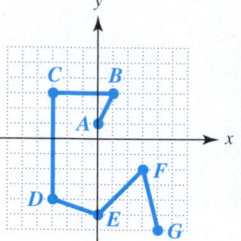

(a) *AB*   (b) *BC*   (c) *CD*

(d) *DE*   (e) *EF*   (f) *FG*

(g) If *A* and *F* were joined by a line segment in the figure, what would be its slope?

(h) If *B* and *D* were joined by a line segment in the figure, what would be its slope?

38. Concept Check Which forms of the slope formula are correct? Explain.

A. $m = \dfrac{y_1 - y_2}{x_2 - x_1}$ **B.** $m = \dfrac{y_1 - y_2}{x_1 - x_2}$ **C.** $m = \dfrac{x_2 - x_1}{y_2 - y_1}$ **D.** $m = \dfrac{y_2 - y_1}{x_2 - x_1}$

39. Concept Check A student found the slope of the line through the points $(-4, 5)$ and $(2, 7)$ as follows.

$$m = \frac{7 - 5}{-4 - 2} = \frac{2}{-6} = -\frac{1}{3}$$

This is incorrect. **WHAT WENT WRONG?** Give the correct slope.

40. Concept Check A student found the slope of the line through the points $(-3, 4)$ and $(-5, -1)$ as follows.

$$m = \frac{-3 - (-5)}{4 - (-1)} = \frac{2}{5}$$

This is incorrect. **WHAT WENT WRONG?** Give the correct slope.

*In the following exercises, **(a)** find the slope of the line passing through each pair of points, if possible, and **(b)** based on the slope, indicate whether the line* rises *from left to right,* falls *from left to right, is* horizontal, *or is* vertical. ***See Example 5 and* FIGURE 21**.

41. $(-2, -3)$ and $(-1, 5)$ **42.** $(-4, 1)$ and $(-3, 4)$ **43.** $(-4, 1)$ and $(2, 6)$

44. $(-3, -3)$ and $(5, 6)$ **45.** $(2, 4)$ and $(-4, 4)$ **46.** $(-6, 3)$ and $(2, 3)$

47. $(-2, 2)$ and $(4, -1)$ **48.** $(-3, 1)$ and $(6, -2)$ **49.** $(5, -3)$ and $(5, 2)$

50. $(4, -1)$ and $(4, 3)$ **51.** $(1.5, 2.6)$ and $(0.5, 3.6)$ **52.** $(3.4, 4.2)$ and $(1.4, 10.2)$

Each table of values gives several points that lie on a line. Find the slope of the line.

53.

x	y
−1	8
0	6
2	2
3	0

54.

x	y
−3	6
−1	0
0	−3
2	−9

55.

x	y
−6	−4
−3	0
0	4
3	8

56.

x	y
−5	−4
0	−2
5	0
10	2

Use the geometric interpretation of slope ("rise over run") to find the slope of each line. (Coordinates of the points shown are integers.)

57.
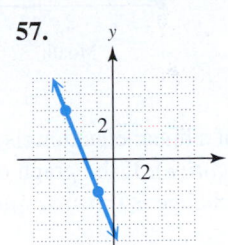

58.
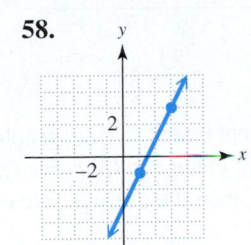

59.
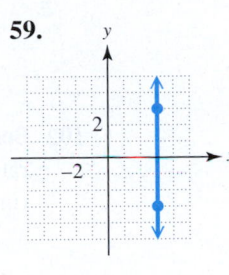

60.
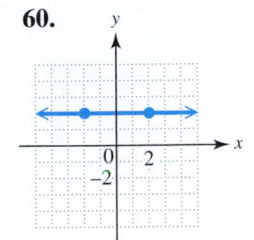

Find the slope of each line in three ways by doing the following.

(a) *Give any two points that lie on the line, and use them to determine the slope.* **See Example 6.**

(b) *Solve the equation for y, and identify the slope from the equation.* **See Example 7.**

(c) *For the form $Ax + By = C$, calculate $-\frac{A}{B}$.* **See the Note following Example 7.**

61. $2x - y = 8$ **62.** $3x + 4y = 12$ **63.** $x + y = -3$ **64.** $x - y = 4$

Find the slope of each line, and sketch its graph. **See Examples 5–7.**

65. $x + 2y = 4$ **66.** $4x - y = 4$ **67.** $y = 4x$ **68.** $y = -3x$

69. $x - 3 = 0$ **70.** $x + 2 = 0$ **71.** $y = -5$ **72.** $y = -4$

Graph each line passing through the given point and having the given slope. **See Example 8.**

73. $(-4, 2); m = \frac{1}{2}$ **74.** $(-2, -3); m = \frac{5}{4}$ **75.** $(0, -2); m = -\frac{2}{3}$

76. $(0, -4); m = -\frac{3}{2}$ **77.** $(-1, -2); m = 3$ **78.** $(-2, -4); m = 4$

79. $(0, 0); m = \frac{1}{5}$ **80.** $(0, 0); m = \frac{5}{3}$ **81.** $(2, -5); m = 0$

82. $(5, 3); m = 0$ **83.** $(-3, 1);$ undefined slope   **84.** $(-4, 1);$ undefined slope

Determine whether each pair of lines is parallel, perpendicular, or neither. **See Examples 9 and 10.**

85. The line passing through (15, 9) and (12, −7) and the line passing through (8, −4) and (5, −20)

86. The line passing through (4, 6) and (−8, 7) and the line passing through (−5, 5) and (7, 4)

87. $x + 4y = 7$ and $4x - y = 3$

88. $2x + 5y = -7$ and $5x - 2y = 1$

89. $4x - 3y = 6$ and $3x - 4y = 2$

90. $2x + y = 6$ and $x - y = 4$

91. $x = 6$ and $6 - x = 8$

92. $3x = y$ and $2y - 6x = 5$

93. $4x + y = 0$ and $5x - 8 = 2y$

94. $2x + 5y = -8$ and $6 + 2x = 5y$

95. $4x - 3y = 8$ and $4y + 3x = 12$

96. $2x = y + 3$ and $2y + x = 3$

97. $x = 6$ and $y = 4$

98. $x + 1 = 0$ and $y - 2 = 0$

Find and interpret the average rate of change illustrated in each graph. **See Objective 8 and FIGURE 25.**

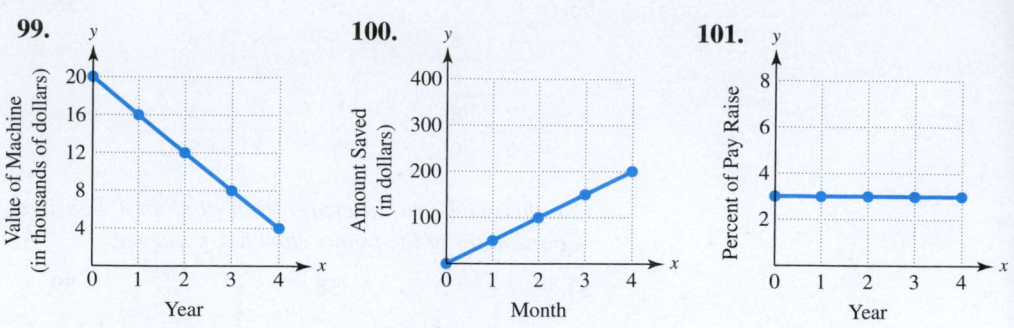

99.   **100.**   **101.**

102. Concept Check   If the graph of a linear equation rises from left to right, then the average rate of change is (*positive / negative*). If the graph of a linear equation falls from left to right, then the average rate of change is (*positive / negative*).

Solve each problem. **See Examples 11 and 12.**

103. The graph shows the number of wireless subscriber connections (that is, active devices, including smartphones, feature phones, tablets, etc.) in millions in the United States for the years 2011 to 2016.

(a) In the context of this graph, what does the ordered pair (2016, 396) mean?

(b) Use the given ordered pairs to find the slope of the line.

(c) Interpret the slope in the context of this problem.

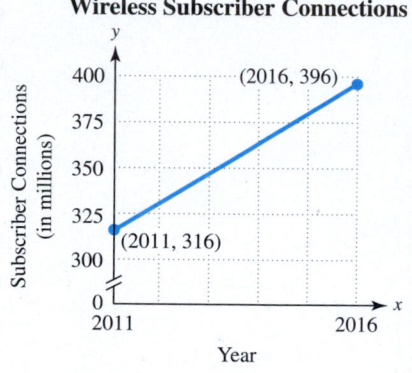

Wireless Subscriber Connections

Data from CTIA.

104. The graph shows the percent of households in the United States that were wireless-only households for the years 2011 to 2016.

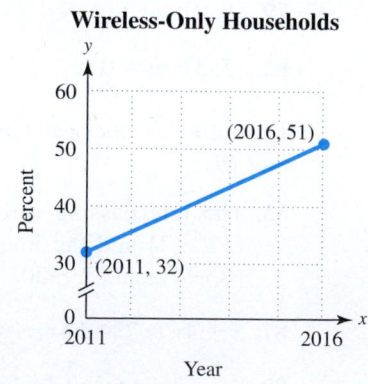

Wireless-Only Households

Data from CTIA.

(a) In the context of this graph, what does the ordered pair (2016, 51) mean?

(b) Use the given ordered pairs to find the slope of the line.

(c) Interpret the slope in the context of this problem.

105. The graph shows the number of drive-in movie theaters in the United States from 2010 through 2017.

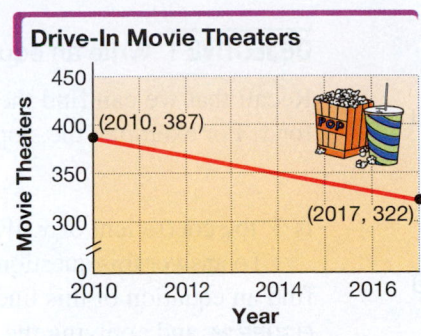

Drive-In Movie Theaters

Data from www.drive-ins.com

(a) Use the given ordered pairs to find the average rate of change in the number of drive-in theaters per year during this period. Round the answer to the nearest whole number.

(b) Explain how a negative slope is interpreted in this situation.

106. The graph shows the number of U.S. travelers to Canada (in thousands) from 2000 through 2016.

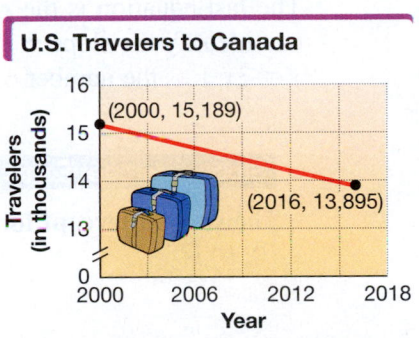

U.S. Travelers to Canada

Data from U.S. Department of Commerce.

(a) Use the given ordered pairs to find the average rate of change in the number of U.S. travelers to Canada per year during this period. Round the answer to the nearest thousand.

(b) Explain how a negative slope is interpreted in this situation.

107. The average price of a gallon of unleaded gasoline in 2000 was $1.51. In 2016, the average price was $2.14. Find and interpret the average rate of change in the price of a gallon of gasoline per year to the nearest cent. (Data from Energy Information Administration.)

108. The average price of a movie ticket in 2004 was $6.21. In 2016, the average price was $8.65. Find and interpret the average rate of change in the price of a movie ticket per year to the nearest cent. (Data from Motion Picture Association of America.)

109. In 2010, the number of digital cameras shipped worldwide totaled 122 million. There were 24 million shipped in 2016. Find and interpret the average rate of change in the number of digital cameras shipped worldwide per year to the nearest million. (Data from CIPA.)

110. In 2010, worldwide shipments of desktop computers totaled 157 million units. In 2016, 103 million units were shipped. Find and interpret the average rate of change in worldwide shipments of desktop computers per year. (Data from IDC.)

7.2   Review of Equations of Lines; Linear Models

OBJECTIVES

1 Write an equation of a line given its slope and y-intercept.

2 Graph a line using its slope and y-intercept.

3 Write an equation of a line given its slope and a point on the line.

4 Write an equation of a line given two points on the line.

5 Write equations of horizontal and vertical lines.

6 Write an equation of a line parallel or perpendicular to a given line.

7 Write an equation of a line that models real data.

OBJECTIVE 1 Write an equation of a line given its slope and y-intercept.

Recall that we can find the slope of a line from its equation by solving the equation for y. For example, the slope of the line with equation

$$y = 3x + 6$$

is 3, the coefficient of x. *What does the number 6 represent?*

To answer this question, suppose a line has slope m and y-intercept $(0, b)$. We can find an equation of this line by choosing another point (x, y) on the line, as shown in **FIGURE 28**, and applying the slope formula.

$$m = \frac{y - b}{x - 0} \leftarrow \text{Change in } y$$
$$\qquad\qquad\qquad \leftarrow \text{Change in } x$$

$$m = \frac{y - b}{x} \qquad \text{Subtract.}$$

$$mx = y - b \qquad \text{Multiply by } x.$$

$$mx + b = y \qquad \text{Add } b.$$

$$y = mx + b \qquad \text{Interchange sides.}$$

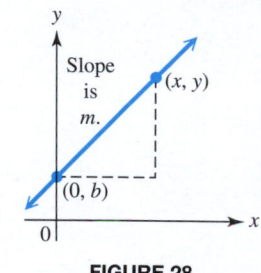

FIGURE 28

The last equation is the *slope-intercept form* of the equation of a line, because we can identify the slope m and y-intercept $(0, b)$ at a glance. In the line with equation $y = 3x + 6$, the number 6 indicates that the y-intercept is $(0, 6)$.

Slope-Intercept Form

The **slope-intercept form** of the equation of a line with slope m and y-intercept $(0, b)$ is

$$y = mx + b.$$

Slope ⤴   ⤴ $(0, b)$ is the y-intercept.

NOW TRY EXERCISE 1

Write an equation of the line with slope $\frac{2}{3}$ and y-intercept $(0, 1)$.

EXAMPLE 1   Writing an Equation of a Line

Write an equation of the line with slope $-\frac{4}{5}$ and y-intercept $(0, -2)$.

Here, $m = -\frac{4}{5}$ and $b = -2$. Substitute these values into the slope-intercept form.

$$y = mx + b \qquad \text{Slope-intercept form}$$

$$y = -\frac{4}{5}x + (-2) \qquad \text{Let } m = -\frac{4}{5} \text{ and } b = -2.$$

$$y = -\frac{4}{5}x - 2 \qquad \text{Definition of subtraction} \qquad \text{NOW TRY} \; \circlearrowright$$

NOTE Every linear equation (of a nonvertical line) has a *unique* (one and only one) slope-intercept form. *Linear functions* are defined using slope-intercept form. Also, this form is used in graphing a line with a graphing calculator.

NOW TRY ANSWER
1. $y = \frac{2}{3}x + 1$

OBJECTIVE 2 Graph a line using its slope and *y*-intercept.

We first saw this approach in the previous section.

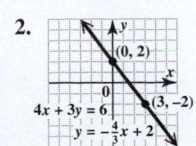

NOW TRY EXERCISE 2

Graph the line using its slope and *y*-intercept.

$$4x + 3y = 6$$

EXAMPLE 2 Graphing Lines Using Slope and *y*-Intercept

Graph each line using its slope and *y*-intercept.

(a) $y = 3x - 6$ (In slope-intercept form)

Here, $m = 3$ and $b = -6$. Plot the *y*-intercept $(0, -6)$. The slope 3 can be interpreted geometrically as follows.

$$m = \frac{\text{rise}}{\text{run}} = \frac{\text{change in } y}{\text{change in } x} = \frac{3}{1}$$

From $(0, -6)$, move *up* 3 units and to the *right* 1 unit, and plot a second point at $(1, -3)$. Join the two points with a straight line. See **FIGURE 29**.

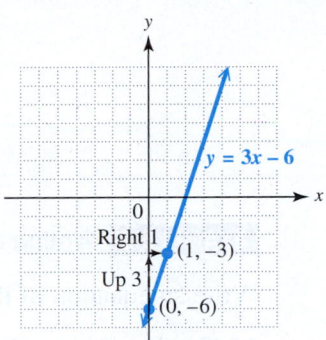

FIGURE 29

(b) $3y + 2x = 9$ (*Not* in slope-intercept form)

Write the equation in slope-intercept form by solving for *y*.

$$3y + 2x = 9$$

$$3y = -2x + 9 \qquad \text{Subtract } 2x.$$

Slope-intercept form $\longrightarrow y = -\frac{2}{3}x + 3 \qquad \text{Divide by 3.}$

Slope ⌐ ⌐ *y*-intercept is $(0, 3)$.

To graph this equation, plot the *y*-intercept $(0, 3)$. The slope can be interpreted as either $\frac{-2}{3}$ or $\frac{2}{-3}$. Using $\frac{-2}{3}$, begin at $(0, 3)$ and move *down* 2 units and to the *right* 3 units to locate the point $(3, 1)$. The line through these two points is the required graph. See **FIGURE 30**. (Verify that the point obtained using $\frac{2}{-3}$ as the slope is also on this line.)

NOW TRY ANSWER

2.

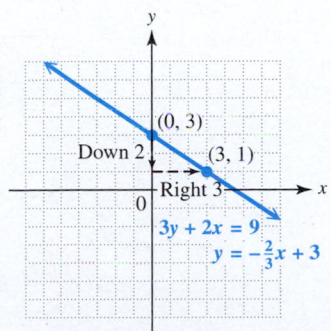

FIGURE 30

NOW TRY

OBJECTIVE 3   Write an equation of a line given its slope and a point on the line.

Let m represent the slope of a line and (x_1, y_1) represent a given point on the line. Let (x, y) represent any other point on the line. See **FIGURE 31**.

$$m = \frac{y - y_1}{x - x_1} \qquad \text{Slope formula}$$

$$m(x - x_1) = y - y_1 \qquad \text{Multiply each side by } x - x_1.$$

$$y - y_1 = m(x - x_1) \qquad \text{Interchange sides.}$$

This is the *point-slope form* of the equation of a line.

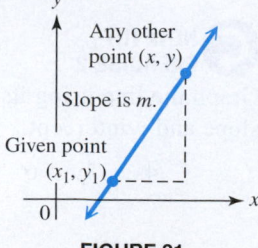

FIGURE 31

Point-Slope Form

The **point-slope form** of the equation of a line with slope m passing through the point (x_1, y_1) is

$$y - y_1 = m(x - x_1).$$

Slope ↓

↳ Given point ↲

NOW TRY EXERCISE 3

Write an equation of the line with slope $-\frac{1}{5}$ passing through the point $(5, -3)$.

EXAMPLE 3   Writing an Equation of a Line Given Its Slope and a Point

Write an equation of the line with slope $\frac{1}{3}$ passing through the point $(-2, 5)$.

Method 1   **Use point-slope form.** Let $(x_1, y_1) = (-2, 5)$ and $m = \frac{1}{3}$.

$$y - y_1 = m(x - x_1) \qquad \text{Point-slope form}$$

Leave x and y as variables.
$$y - 5 = \frac{1}{3}[x - (-2)] \qquad \text{Let } y_1 = 5, m = \frac{1}{3}, \text{ and } x_1 = -2.$$

$$y - 5 = \frac{1}{3}(x + 2) \qquad \text{Definition of subtraction}$$

$$y - 5 = \frac{1}{3}x + \frac{2}{3} \qquad \text{(*)} \quad \text{Distributive property}$$

$$y = \frac{1}{3}x + \frac{2}{3} + \frac{15}{3} \qquad \text{Add } 5 = \frac{15}{3}.$$

Slope-intercept form ⟶ $y = \frac{1}{3}x + \frac{17}{3}$ $\qquad \frac{2}{3} + \frac{15}{3} = \frac{17}{3}$

Method 2   **Use slope-intercept form.** Let $(x, y) = (-2, 5)$ and $m = \frac{1}{3}$.

$$y = mx + b \qquad \text{Slope-intercept form}$$

$$5 = \frac{1}{3}(-2) + b \qquad \text{Let } y = 5, m = \frac{1}{3}, \text{ and } x = -2.$$

Solve for b.
$$5 = -\frac{2}{3} + b \qquad \text{Multiply.}$$

$5 = \frac{15}{3}$

$$\frac{17}{3} = b, \quad \text{or} \quad b = \frac{17}{3} \qquad \text{Add } \frac{2}{3}; \frac{15}{3} + \frac{2}{3} = \frac{17}{3}$$

NOW TRY ANSWER

3. $y = -\frac{1}{5}x - 2$

Substitute $m = \frac{1}{3}$ and $b = \frac{17}{3}$ to obtain $y = \frac{1}{3}x + \frac{17}{3}$, as above.

NOW TRY

Previously we defined *standard form* for a linear equation.

$$Ax + By = C \qquad \text{Standard form}$$

Here A, B, and C are real numbers and A and B are not both 0. *For consistency, we give A, B, and C as integers with greatest common factor 1, and $A \geq 0$.* (If $A = 0$, then we give $B > 0$.) The equation in **Example 3** is written in standard form as follows.

$$y - 5 = \frac{1}{3}x + \frac{2}{3} \qquad \text{Equation (*) from Example 3}$$

$$3y - 15 = x + 2 \qquad \text{Multiply each term by 3.}$$

$$-x + 3y = 17 \qquad \text{Subtract } x. \text{ Add 15.}$$

Standard form $\longrightarrow x - 3y = -17 \qquad$ Multiply by -1.

NOTE "Standard form" is not standard among texts. A linear equation can be written correctly in many different ways.

Example:

$$\underbrace{2x + 3y = 8,}_{\substack{\text{Standard form} \\ \text{as described above}}} \qquad \underbrace{2x = 8 - 3y, \quad 3y = 8 - 2x, \quad x + \frac{3}{2}y = 4, \quad 4x + 6y = 16}_{\text{Equivalent forms of } 2x + 3y = 8}$$

OBJECTIVE 4 Write an equation of a line given two points on the line.

NOW TRY
EXERCISE 4
Write an equation of the line passing through the points $(3, -4)$ and $(-2, -1)$. Give the final answer in standard form.

EXAMPLE 4 Writing an Equation of a Line Given Two Points

Write an equation of the line passing through the points $(-4, 3)$ and $(5, -7)$. Give the final answer in standard form.

First find the slope using the slope formula.

$$m = \frac{-7 - 3}{5 - (-4)} = -\frac{10}{9}$$

Use either $(-4, 3)$ or $(5, -7)$ as (x_1, y_1) in the point-slope form of the equation of a line. We choose $(-4, 3)$, so $-4 = x_1$ and $3 = y_1$.

$$y - y_1 = m(x - x_1) \qquad \text{Point-slope form}$$

$$y - 3 = -\frac{10}{9}[x - (-4)] \qquad \text{Let } y_1 = 3, m = -\frac{10}{9}, \text{ and } x_1 = -4.$$

$$y - 3 = -\frac{10}{9}(x + 4) \qquad \text{Definition of subtraction}$$

$$y - 3 = -\frac{10}{9}x - \frac{40}{9} \qquad \text{Distributive property}$$

$$9y - 27 = -10x - 40 \qquad \text{Multiply each term by 9.}$$

Standard form $\longrightarrow 10x + 9y = -13 \qquad$ Add $10x$. Add 27.

NOW TRY ANSWER
4. $3x + 5y = -11$

Verify that if $(5, -7)$ were used, the same equation would result. **NOW TRY**

> **NOTE** Once the slope is found in **Example 4,** the equation of the line could also be determined using slope-intercept form.

OBJECTIVE 5 Write equations of horizontal and vertical lines.

A horizontal line has slope 0. Using point-slope form, we can find the equation of a horizontal line through the point (a, b).

$$y - y_1 = m(x - x_1) \qquad \text{Point-slope form}$$
$$y - b = 0(x - a) \qquad y_1 = b, m = 0, x_1 = a$$
$$y - b = 0 \qquad \text{Multiplication property of 0}$$
$$\text{Horizontal line} \longrightarrow y = b \qquad \text{Add } b.$$

Point-slope form does not apply to a vertical line because the slope of a vertical line is undefined. A vertical line through the point (a, b) has equation $x = a.$

> **Equations of Horizontal and Vertical Lines**
>
> A **horizontal line** through the point (a, b) has equation $y = b.$
>
> A **vertical line** through the point (a, b) has equation $x = a.$

**NOW TRY
EXERCISE 5**

Write an equation of the line passing through the point $(4, -4)$ and satisfying the given condition.

(a) The line has undefined slope.

(b) The line has slope 0.

EXAMPLE 5 Writing Equations of Horizontal and Vertical Lines

Write an equation of the line passing through the point $(-3, 3)$ and satisfying the given condition.

(a) The line has slope 0.

   Because the slope is 0, this is a horizontal line. A horizontal line through the point (a, b) has equation $y = b.$ In $(-3, 3)$, the y-coordinate is 3, so the equation is $y = 3.$

(b) The line has undefined slope.

   This is a vertical line because the slope is undefined. A vertical line through the point (a, b) has equation $x = a.$ In $(-3, 3)$, the x-coordinate is -3, so the equation is $x = -3.$

   Both lines are graphed in **FIGURE 32.**

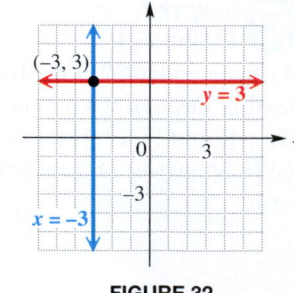

FIGURE 32

NOW TRY

OBJECTIVE 6 Write an equation of a line parallel or perpendicular to a given line.

EXAMPLE 6 Writing Equations of Parallel or Perpendicular Lines

Write an equation of the line passing through the point $(-3, 6)$ and satisfying the given condition. Give final answers in slope-intercept form.

(a) The line is parallel to the line $2x + 3y = 6.$

   We can find the slope of the given line by solving for y.

$$2x + 3y = 6$$
$$3y = -2x + 6 \qquad \text{Subtract } 2x.$$
$$y = -\frac{2}{3}x + 2 \qquad \text{Divide by 3.}$$
$$\uparrow\!\!\!\_\text{ Slope}$$

NOW TRY ANSWERS
5. **(a)** $x = 4$ **(b)** $y = -4$

NOW TRY
EXERCISE 6

Write an equation of the line passing through the point $(6, -1)$ and satisfying the given condition. Give final answers in slope-intercept form.

(a) The line is parallel to the line $3x - 5y = 7$.

(b) The line is perpendicular to the line $3x - 5y = 7$.

The slope of the line is given by the coefficient of x, so $m = -\frac{2}{3}$. See **FIGURE 33**.

The required equation of the line through $(-3, 6)$ and *parallel* to $2x + 3y = 6$ must have the *same* slope $-\frac{2}{3}$. To find this equation, we use the point-slope form with $(x_1, y_1) = (-3, 6)$ and $m = -\frac{2}{3}$.

$$y - y_1 = m(x - x_1) \qquad \text{Point-slope form}$$

$$y - 6 = -\frac{2}{3}[x - (-3)] \qquad y_1 = 6, m = -\frac{2}{3}, x_1 = -3$$

$$y - 6 = -\frac{2}{3}(x + 3) \qquad \begin{array}{l}\text{Definition of}\\ \text{subtraction}\end{array}$$

$$y - 6 = -\frac{2}{3}x - 2 \qquad \begin{array}{l}\text{Distributive}\\ \text{property}\end{array}$$

$$y = -\frac{2}{3}x + 4 \qquad \text{Add 6.}$$

We did not clear the fraction because we want the final equation in slope-intercept form—that is, solved for y. Both lines are shown in **FIGURE 34**.

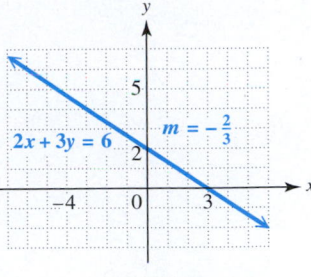

FIGURE 33

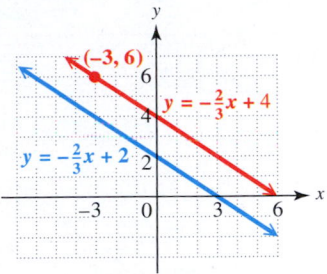

Both lines have slope $-\frac{2}{3}$. The lines are parallel.

FIGURE 34

(b) The line is perpendicular to the line $2x + 3y = 6$.

In part (a), we wrote the equation of the given line in slope-intercept form.

$$2x + 3y = 6$$

$$y = -\frac{2}{3}x + 2 \qquad \text{Slope-intercept form}$$

$$\uparrow\!\!\!\!\!\!\!\!\!\!\!\!\underline{\quad}\ \text{Slope}$$

To be *perpendicular* to the line $2x + 3y = 6$, a line must have slope $\frac{3}{2}$, the *negative reciprocal* of $-\frac{2}{3}$.

We use $(-3, 6)$ and slope $\frac{3}{2}$ in the point-slope form to find the equation of the perpendicular line shown in **FIGURE 35**.

$$y - y_1 = m(x - x_1) \qquad \text{Point-slope form}$$

$$y - 6 = \frac{3}{2}[x - (-3)] \qquad y_1 = 6, m = \frac{3}{2}, x_1 = -3$$

$$y - 6 = \frac{3}{2}(x + 3) \qquad \begin{array}{l}\text{Definition of}\\ \text{subtraction}\end{array}$$

$$y - 6 = \frac{3}{2}x + \frac{9}{2} \qquad \text{Distributive property}$$

$$y = \frac{3}{2}x + \frac{9}{2} + \frac{12}{2} \qquad \text{Add } 6 = \frac{12}{2}.$$

$$y = \frac{3}{2}x + \frac{21}{2} \qquad \frac{9}{2} + \frac{12}{2} = \frac{21}{2}$$

The slopes are negative reciprocals, $-\frac{2}{3}\left(\frac{3}{2}\right) = -1$. The lines are perpendicular.

FIGURE 35

NOW TRY ANSWERS

6. **(a)** $y = \frac{3}{5}x - \frac{23}{5}$

  (b) $y = -\frac{5}{3}x + 9$

NOW TRY

▼ Summary of Forms of Linear Equations

Equation	Description	When to Use
$y = mx + b$	**Slope-Intercept Form** Slope is m. y-intercept is $(0, b)$.	The slope and y-intercept can be easily identified and used to quickly graph the equation.
$y - y_1 = m(x - x_1)$	**Point-Slope Form** Slope is m. Line passes through (x_1, y_1).	This form is ideal for finding the equation of a line if the slope and a point on the line or two points on the line are known.
$Ax + By = C$	**Standard Form** (A, B, and C integers, $A \geq 0$) Slope is $-\frac{A}{B}$ $(B \neq 0)$. x-intercept is $\left(\frac{C}{A}, 0\right)$ $(A \neq 0)$. y-intercept is $\left(0, \frac{C}{B}\right)$ $(B \neq 0)$.	The x- and y-intercepts can be found quickly and used to graph the equation. The slope must be calculated.
$y = b$	**Horizontal Line** Slope is 0. y-intercept is $(0, b)$.	If the graph intersects only the y-axis, then y is the only variable in the equation.
$x = a$	**Vertical Line** Slope is undefined. x-intercept is $(a, 0)$.	If the graph intersects only the x-axis, then x is the only variable in the equation.

OBJECTIVE 7 Write an equation of a line that models real data.

If a given set of data changes at a fairly constant rate, the data may fit a linear pattern, where the rate of change is the slope of the line.

EXAMPLE 7 Writing a Linear Equation to Describe Real Data

A local gasoline station is selling 89-octane gas for $3.50 per gal.

(a) Write an equation that describes the cost y to buy x gallons of gas.

The total cost is determined by the number of gallons multiplied by the price per gallon (in this case, $3.50). As the gas is pumped, two sets of numbers spin by: the number of gallons pumped and the cost of that number of gallons.

The table illustrates this situation.

Number of Gallons Pumped	Cost of This Number of Gallons
0	0($3.50) = $ 0.00
1	1($3.50) = $ 3.50
2	2($3.50) = $ 7.00
3	3($3.50) = $10.50
4	4($3.50) = $14.00

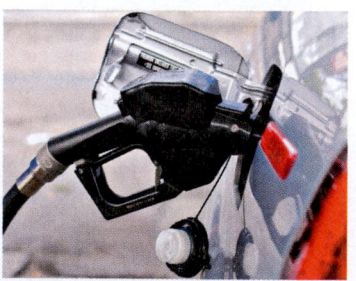

If we let x denote the number of gallons pumped, then the total cost y in dollars can be found using the following linear equation.

Total cost ↘   ↙ Number of gallons

$$y = 3.50x$$

Theoretically, there are infinitely many ordered pairs (x, y) that satisfy this equation, but here we are limited to nonnegative values for x because we cannot have a negative number of gallons. In this situation, there is also a practical maximum value for x, which varies from one car to another—the size of the gas tank.

**NOW TRY
EXERCISE 7**

A cell phone plan costs $100 for the telephone plus $85 per month for service. Write an equation that gives the cost y in dollars for x months of cell phone service using this plan.

(b) A car wash at this gas station costs an additional $3.00. Write an equation that defines the cost of gas and a car wash.

The cost will be $3.50x + 3.00$ dollars for x gallons of gas and a car wash.

$$y = 3.5x + 3 \qquad \text{Final 0's need not be included.}$$

(c) Interpret the ordered pairs $(5, 20.5)$ and $(10, 38)$ in the context of the equation from part (b).

$(5, 20.5)$ indicates that 5 gal of gas and a car wash cost $20.50.

$(10, 38)$ indicates that 10 gal of gas and a car wash cost $38.00.

 **NOW TRY**

NOTE In **Example 7(a)**, the ordered pair $(0, 0)$ satisfied the equation $y = 3.50x$, so the linear equation has the form

$$y = mx, \quad \text{where } b = 0.$$

If a realistic situation involves an initial charge b plus a charge per unit m, as in the equation $y = 3.5x + 3$ in **Example 7(b)**, then the equation has the form

$$y = mx + b, \quad \text{where } b \neq 0.$$

EXAMPLE 8 Writing an Equation of a Line That Models Data

Average annual tuition and fees for in-state students at public two-year colleges are shown in the table for selected years and graphed as ordered pairs of points in the scatter diagram in **FIGURE 36**, where $x = 0$ represents 2012, $x = 1$ represents 2013, and so on, and y represents the cost in dollars.

Year	Cost (in dollars)
2012 $(x = 0)$	3216
2013 $(x = 1)$	3264
2014 $(x = 2)$	3347
2015 $(x = 3)$	3435
2016 $(x = 4)$	3520
2017 $(x = 5)$	3570

Data from The College Board.

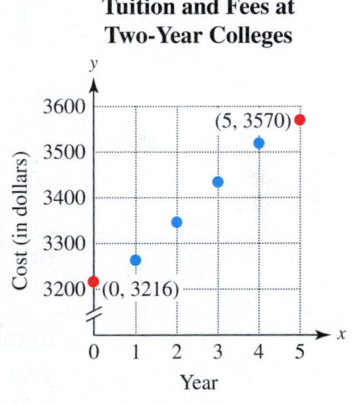

FIGURE 36

(a) Write an equation that models the data.

Because the points in **FIGURE 36** lie approximately on a straight line, we can write a linear equation that models the relationship between year x and cost y. We choose two data points, $(0, 3216)$ and $(5, 3570)$, to find the slope of the line.

$$m = \frac{3570 - 3216}{5 - 0} = \frac{354}{5} = 70.8$$

The slope 70.8 indicates that the cost of tuition and fees increased by about $71 per year from 2012 to 2017. We use this slope and the y-intercept $(0, 3216)$ to write an equation of the line in slope-intercept form.

$$y = 70.8x + 3216$$

NOW TRY ANSWER
7. $y = 85x + 100$

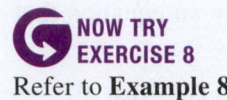

**NOW TRY
EXERCISE 8**

Refer to **Example 8.**

(a) Using the data values for the years 2012 and 2016, write an equation that models the data.

(b) Use the equation from part (a) to approximate the cost of tuition and fees in 2015.

(b) Use the equation from part (a) to approximate the cost of tuition and fees in 2018.

The value $x = 6$ corresponds to the year 2018.

$$y = 70.8x + 3216 \qquad \text{Equation from part (a)}$$

$$y = 70.8(6) + 3216 \qquad \text{Substitute 6 for } x.$$

$$y = 3640.8 \qquad \text{Multiply, and then add.}$$

According to the model, average tuition and fees for in-state students at public two-year colleges in 2018 were about \$3641.   **NOW TRY**

> **NOTE**   Choosing different data points in **Example 8** would result in a slightly different line (particularly in regard to its slope) and, hence, a slightly different equation. However, all such equations should yield similar results. See **Now Try Exercise 8.**

EXAMPLE 9   Writing an Equation of a Line That Models Data

Hospital expenditures (in billions of dollars) in the United States are shown in the graph in **FIGURE 37**.

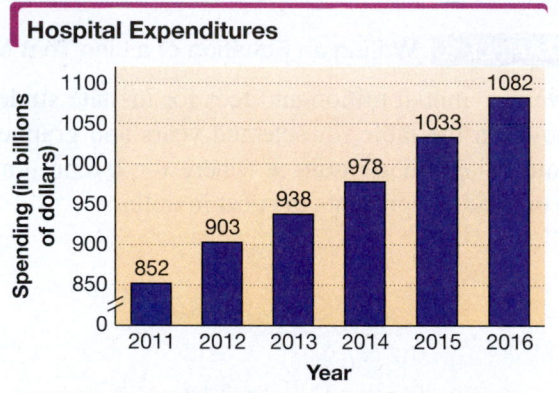
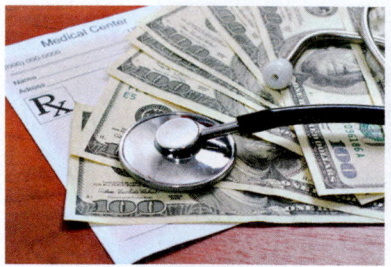

Data from Centers for Medicare & Medicaid Services.

FIGURE 37

(a) Write an equation that models the data.

The data increase linearly—that is, a straight line through the tops of any two bars in the graph would be close to the top of each bar. To model the relationship between year x and hospital expenditures y, we let $x = 1$ represent 2011, $x = 2$ represent 2012, and so on. The given data for 2011 and 2016 can be written as the ordered pairs $(1, 852)$ and $(6, 1082)$.

$$m = \frac{1082 - 852}{6 - 1} = \frac{230}{5} = 46 \qquad \begin{array}{l}\text{Find the slope of the line}\\ \text{through } (1, 852) \text{ and } (6, 1082).\end{array}$$

Thus, hospital expenditures increased by \$46 billion per year. To write an equation, we substitute this slope and the point $(1, 852)$ into the point-slope form.

$$y - y_1 = m(x - x_1) \qquad \text{Point-slope form}$$

$$y - 852 = 46(x - 1) \qquad (x_1, y_1) = (1, 852); m = 46$$

> Either point can be used here. (6, 1082) provides the same answer.

$$y - 852 = 46x - 46 \qquad \text{Distributive property}$$

$$y = 46x + 806 \qquad \text{Add 852.}$$

NOW TRY ANSWERS

8. (a) $y = 76x + 3216$
   (b) \$3444

NOW TRY EXERCISE 9

Refer to **Example 9.**

(a) Use the ordered pairs (4, 978) and (6, 1082) to write an equation that models the data.

(b) Use the equation from part (a) to estimate hospital expenditures in 2019.

NOW TRY ANSWERS

9. **(a)** $y = 52x + 770$
   (b) $1238 billion

(b) Use the equation from part (a) to estimate hospital expenditures in the United States in 2019. (Assume a constant rate of change.)

Because we let $x = 1$ represent 2011, $x = 9$ represents 2019.

$$y = 46x + 806 \qquad \text{Equation from part (a)}$$
$$y = 46(9) + 806 \qquad \text{Substitute 9 for } x.$$
$$y = 1220 \qquad \text{Multiply, and then add.}$$

Hospital expenditures in the United States in 2019 were about $1220 billion.

NOW TRY

7.2 Exercises

FOR EXTRA HELP

 MyLab Math

▶ *Video solutions for select problems available in MyLab Math*

STUDY SKILLS REMINDER

Be sure to read and work through the section material before working the exercises. **Review Study Skill 2, *Reading Your Math Text.***

Concept Check *Provide the appropriate response.*

1. The following equations all represent the same line. Which one is in standard form as specified in this section?

   A. $3x - 2y = 5$ **B.** $y = \dfrac{3}{2}x - \dfrac{5}{2}$ **C.** $\dfrac{3}{5}x - \dfrac{2}{5}y = 1$ **D.** $3x = 2y + 5$

2. Which equation is in point-slope form?

   A. $y = 6x + 2$ **B.** $4x + y = 9$ **C.** $y - 3 = 2(x - 1)$ **D.** $2y = 3x - 7$

3. Which equation in **Exercise 2** is in slope-intercept form?

4. Write the equation $y + 2 = -3(x - 4)$ in slope-intercept form.

5. Write the equation $y + 2 = -3(x - 4)$ in standard form.

6. Write the equation $10x - 7y = 70$ in slope-intercept form.

Concept Check *Match each equation with the graph that it most closely resembles. (Hint: Determining the signs of m and b will help in each case.)*

7. $y = 2x + 3$

8. $y = -2x + 3$

9. $y = -2x - 3$

10. $y = 2x - 3$

11. $y = 2x$

12. $y = -2x$

13. $y = 3$

14. $y = -3$

A.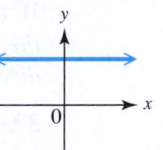

B.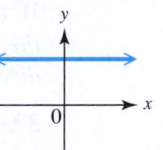

C.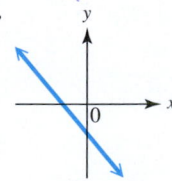

D.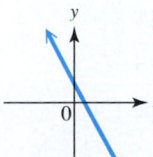

E.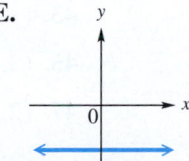

F.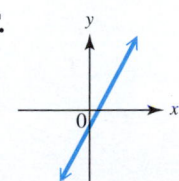

G.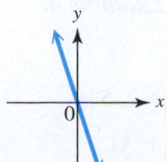

H.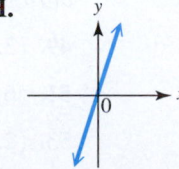

I.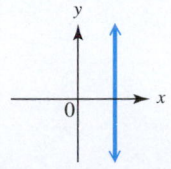

Write an equation in slope-intercept form of the line that satisfies the given conditions. **See Example 1.**

15. $m = 5; b = 15$

16. $m = 2; b = 12$

17. $m = -\frac{2}{3}; b = \frac{4}{5}$

18. $m = -\frac{5}{8}; b = -\frac{1}{3}$

19. Slope 1; y-intercept $(0, -1)$

20. Slope -1; y-intercept $(0, -3)$

21. Slope $\frac{2}{5}$; y-intercept $(0, 5)$

22. Slope $-\frac{3}{4}$; y-intercept $(0, 7)$

Write an equation in slope-intercept form of the line shown in each graph. (Coordinates of the points shown are integers.)

23.

24.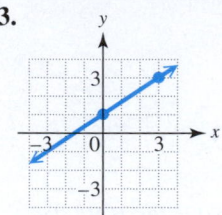

25.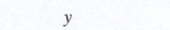

26.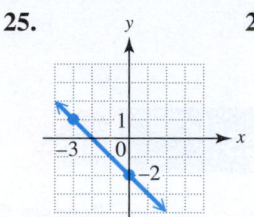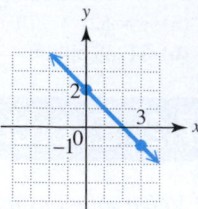

Each table of values gives several points that lie on a line. Write an equation in slope-intercept form of the line.

27.

x	y
-2	-8
0	-4
1	-2
3	2

28.

x	y
-2	-3
0	3
2	9
3	12

29.

x	y
-5	6
0	3
5	0
10	-3

30.

x	y
-4	5
-2	0
0	-5
2	-10

*For each equation, (**a**) write it in slope-intercept form, (**b**) give the slope of the line, (**c**) give the y-intercept, and (**d**) graph the line.* **See Example 2.**

31. $-x + y = 4$

32. $-x + y = 6$

33. $6x + 5y = 30$

34. $3x + 4y = 12$

35. $4x - 5y = 20$

36. $7x - 3y = 3$

37. $x + 2y = -4$

38. $x + 3y = -9$

*Write an equation of the line passing through the given point and having the given slope. Give the equation (**a**) in slope-intercept form and (**b**) in standard form.* **See Example 3 and the discussion on standard form.**

39. $(5, 8)$; slope -2

40. $(12, 10)$; slope 1

41. $(-2, 4)$; slope $-\frac{3}{4}$

42. $(-1, 6)$; slope $-\frac{5}{6}$

43. $(-5, 4)$; slope $\frac{1}{2}$

44. $(7, -2)$; slope $\frac{1}{4}$

45. $(3, 0)$; slope 4

46. $(-2, 0)$; slope -5

47. $(2, 6.8)$; slope 1.4

48. $(6, -1.2)$; slope 0.8

Write an equation of the line passing through the given points. Give the final answer in standard form. **See Example 4.**

49. $(3, 4)$ and $(5, 8)$

50. $(5, -2)$ and $(-3, 14)$

51. $(6, 1)$ and $(-2, 5)$

52. $(-2, 5)$ and $(-8, 1)$

53. $(2, 5)$ and $(1, 5)$

54. $(-2, 2)$ and $(4, 2)$

55. $(7, 6)$ and $(7, -8)$

56. $(13, 5)$ and $(13, -1)$

57. $\left(\frac{1}{2}, -3\right)$ and $\left(-\frac{2}{3}, -3\right)$

58. $\left(-\frac{4}{9}, -6\right)$ and $\left(\frac{12}{7}, -6\right)$

59. $\left(-\frac{2}{5}, \frac{2}{5}\right)$ and $\left(\frac{4}{3}, \frac{2}{3}\right)$

60. $\left(\frac{3}{4}, \frac{8}{3}\right)$ and $\left(\frac{2}{5}, \frac{2}{3}\right)$

Write an equation of the line passing through the given point and satisfying the given condition. See Example 5.

61. $(9, 5)$; slope 0

62. $(-4, -2)$; slope 0

63. $(9, 10)$; undefined slope

64. $(-2, 8)$; undefined slope

65. $\left(-\frac{3}{4}, -\frac{3}{2}\right)$; slope 0

66. $\left(-\frac{5}{8}, -\frac{9}{2}\right)$; slope 0

67. $(-7, 8)$; horizontal

68. $(2, -7)$; horizontal

69. $(0.5, 0.2)$; vertical

70. $(0.1, 0.4)$; vertical

71. $(0, 0)$; horizontal

72. $(0, 0)$; vertical

Write an equation of the line passing through the given point and satisfying the given condition. Give the equation (a) in slope-intercept form and (b) in standard form. See Example 6.

73. $(7, 2)$; parallel to $3x - y = 8$

74. $(4, 1)$; parallel to $2x + 5y = 10$

75. $(-2, -2)$; parallel to $-x + 2y = 10$

76. $(-1, 3)$; parallel to $-x + 3y = 12$

77. $(8, 5)$; perpendicular to $2x - y = 7$

78. $(2, -7)$; perpendicular to $5x + 2y = 18$

79. $(-2, 7)$; perpendicular to $x = 9$

80. $(8, 4)$; perpendicular to $x = -3$

Write an equation in the form $y = mx$ for each situation. Then give the three ordered pairs associated with the equation for x-values 0, 5, and 10. See Example 7(a).

81. x represents the number of hours traveling at 45 mph, and y represents the distance traveled (in miles).

82. x represents the number of caps sold at $26 each, and y represents the total cost of the caps (in dollars).

83. x represents the number of gallons of gas sold at $3.75 per gal, and y represents the total cost of the gasoline (in dollars).

84. x represents the number of hours a bicycle is rented at $7.50 per hour, and y represents the total charge for the rental (in dollars).

85. x represents the number of credit hours taken at a community college at $150 per credit hour, and y represents the total tuition paid for the credit hours (in dollars).

86. x represents the number of tickets to a performance of *Hamilton* purchased at $250 per ticket, and y represents the total paid for the tickets (in dollars).

For each situation, do the following.

(a) *Write an equation in the form y = mx + b.*

(b) *Find and interpret the ordered pair associated with the equation for x = 5.*

(c) *Answer the question posed in the problem.*

See Examples 7(b) and 7(c).

87. A ticket for Taylor Swift's Reputation Stadium Tour costs $140. There is a ticket order delivery fee of $18.50. Let x represent the number of tickets and y represent the cost in dollars. How much does it cost to purchase 2 tickets and have them delivered? (Data from Ticketmaster.)

88. Resident tuition at North Shore Community College is $206 per credit hour. There is also a $300 program fee for physical therapy. Let x represent the number of credit hours and y represent the cost in dollars. How much does it cost for a student in physical therapy to take 15 credit hours? (Data from www.northshore.edu)

89. A health club membership costs $99, plus $41 per month. Let x represent the number of months and y represent the cost in dollars. How much does the first year's membership cost? (Data from Midwest Athletic Club.)

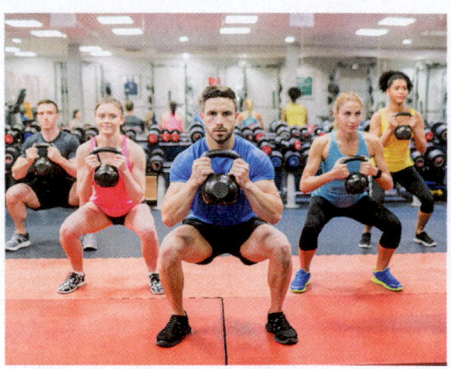

90. An Executive VIP/Gold membership to a health club costs $159, plus $57 per month. Let x represent the number of months and y represent the cost in dollars. How much does a one-year membership cost? (Data from Midwest Athletic Club.)

91. A wireless plan includes unlimited talk, text, and data for $90 per month. There is a $25 activation fee. Let x represent the number of months and y represent the cost in dollars. Over two years, how much will this plan cost?

92. A wireless plan includes unlimited talk, text, and data for $75 per month. There is a $350 charge for a phone. Let x represent the number of months and y represent the cost in dollars. Over two years, how much will this plan cost?

Solve each problem. Give equations in slope-intercept form. **See Examples 8 and 9.**

93. Total sales of e-readers in the United States (in millions of dollars) are shown in the graph, where the year 2013 corresponds to $x = 0$.

(a) Use the ordered pairs from the graph to write an equation that models the data. (Round the slope to the nearest tenth.) Interpret the slope in the context of this problem.

(b) Use the equation from part (a) to approximate sales of e-readers in the United States in 2014, the year data was unavailable.

(c) Use the equation from part (a) to approximate sales of e-readers in the United States in 2015. How does the result compare to the actual value, $527 million? Explain.

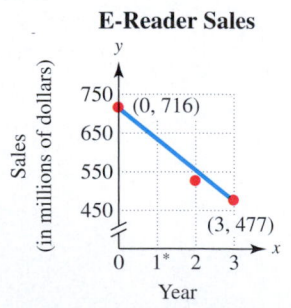

E-Reader Sales

Sales (in millions of dollars)

(0, 716)

(3, 477)

Year

*Data for this year are unavailable.

Data from Consumer Technology Association.

94. Total sales of smartphones in the United States (in billions of dollars) are shown in the graph, where the year 2013 corresponds to $x = 0$.

(a) Use the ordered pairs from the graph to write an equation that models the data. (Round the slope to the nearest tenth.) Interpret the slope in the context of this problem.

(b) Use the equation from part (a) to approximate smartphone sales in the United States in 2014, the year data was unavailable.

(c) Use the equation from part (a) to approximate sales of smartphones in the United States in 2015. How does the result compare to the actual value, $52.9 billion? Explain.

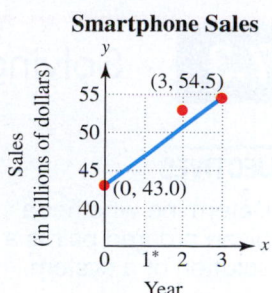

Smartphone Sales

*Data for this year are unavailable.
Data from Consumer Technology Association.

95. Expenditures for home health care in the United States are shown in the graph.

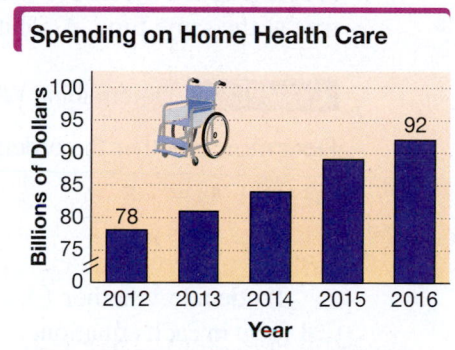

Spending on Home Health Care

Data from Centers for Medicare & Medicaid Services.

(a) If a straight line were drawn through the tops of the bars for 2012 and 2016, would this line have positive or negative slope? Explain.

(b) Use the information given for the years 2012 and 2016, letting $x = 12$ represent 2012 and $x = 16$ represent 2016, and letting y represent spending (in billions of dollars), to write an equation that models the data.

(c) Use the equation from part (b) to approximate the amount spent on home health care in 2015. How does the result compare with the actual value, $89 billion?

96. The number of pieces of first-class mail delivered in the United States is shown in the graph.

(a) If a straight line were drawn through the tops of the bars for 2011 and 2016, would this line have positive or negative slope? Explain.

(b) Use the information given for the years 2011 and 2016, letting $x = 11$ represent 2011 and $x = 16$ represent 2016, and letting y represent the number of pieces of mail (in billions), to write an equation that models the data.

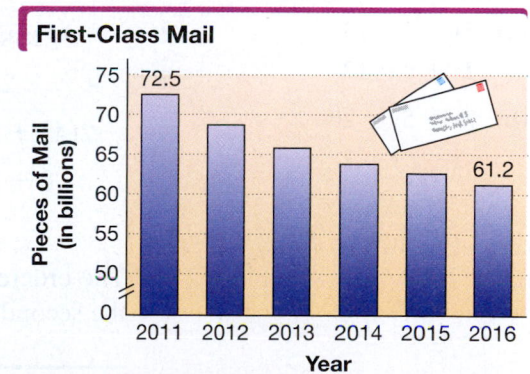

First-Class Mail

Data from U.S. Postal Service.

(c) Use the equation from part (b) to approximate the number of pieces of first-class mail delivered in 2015 to the nearest tenth. How does this result compare to the actual value, 62.6 billion?

7.3   Solving Systems of Linear Equations by Graphing

OBJECTIVES

1 Determine whether a given ordered pair is a solution of a system.

2 Solve linear systems by graphing.

3 Solve special systems by graphing.

4 Identify special systems without graphing.

A **system of linear equations,** or **linear system,** consists of two or more linear equations with the same variables.

Examples:

$$2x + 3y = 4 \qquad\qquad x + 3y = 1 \qquad\qquad x - y = 1$$
$$3x - y = -5 \qquad\quad -y = 4 - 2x \qquad\qquad y = 3$$

Systems of linear equations

In the system on the right, think of $y = 3$ as an equation in two variables by writing it as $0x + y = 3$.

OBJECTIVE 1 Determine whether a given ordered pair is a solution of a system.

A **solution of a system** of linear equations is an ordered pair that makes both equations true at the same time. A solution of an equation is said to *satisfy* the equation.

EXAMPLE 1 Determining Whether an Ordered Pair Is a Solution

Determine whether the ordered pair $(4, -3)$ is a solution of each system.

(a) $x + 4y = -8$

 $3x + 2y = 6$

To decide whether $(4, -3)$ is a solution of the system, substitute 4 for x and -3 for y in each equation.

$x + 4y = -8$	$3x + 2y = 6$
$4 + 4(-3) \overset{?}{=} -8$ Substitute.	$3(4) + 2(-3) \overset{?}{=} 6$ Substitute.
$4 + (-12) \overset{?}{=} -8$ Multiply.	$12 + (-6) \overset{?}{=} 6$ Multiply.
$-8 = -8$ ✓ True	$6 = 6$ ✓ True

Because $(4, -3)$ satisfies *both* equations, it is a solution of the system.

(b) $2x + 5y = -7$

 $3x + 4y = 2$

Again, substitute 4 for x and -3 for y in each equation.

$2x + 5y = -7$	$3x + 4y = 2$
$2(4) + 5(-3) \overset{?}{=} -7$ Substitute.	$3(4) + 4(-3) \overset{?}{=} 2$ Substitute.
$8 + (-15) \overset{?}{=} -7$ Multiply.	$12 + (-12) \overset{?}{=} 2$ Multiply.
$-7 = -7$ ✓ True	$0 = 2$ False

The ordered pair $(4, -3)$ is not a solution of this system because it does not satisfy the second equation.

NOW TRY

NOW TRY EXERCISE 1

Determine whether the ordered pair $(5, 2)$ is a solution of each system.

(a) $3x - y = 13$
 $2x + y = 12$

(b) $2x + 5y = 20$
 $x - y = 7$

OBJECTIVE 2 Solve linear systems by graphing.

The set of all ordered pairs that are solutions of a system is its **solution set.** One way to find the solution set of a system of two linear equations is to graph both equations on the same axes. Any intersection point would be on both lines and would therefore be a solution of *both* equations.

NOW TRY ANSWERS
1. **(a)** yes **(b)** no

Thus, the coordinates of any point at which the lines intersect give a solution of the system.

The graph in **FIGURE 38** shows that the solution of the system in **Example 1(a)** is the intersection point $(4, -3)$.

> *The solution (point of intersection) is always written as an ordered pair.*

Because *two different* straight lines can intersect at no more than one point, there can never be more than one solution for such a system.

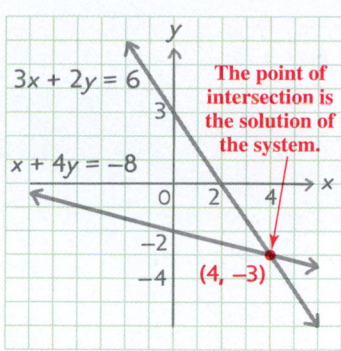

FIGURE 38

NOW TRY EXERCISE 2

Solve the system by graphing.

$$x - 2y = 4$$
$$2x + y = 3$$

EXAMPLE 2 Solving a System by Graphing (One Solution)

Solve the system by graphing.

$$2x + 3y = 4$$
$$3x - y = -5$$

We graph these two lines by plotting several points for each line. The intercepts are often convenient choices. We show finding the intercepts for $2x + 3y = 4$.

To find the *y*-intercept, let $x = 0$.

$$2x + 3y = 4$$
$$2(0) + 3y = 4 \qquad \text{Let } x = 0.$$
$$3y = 4$$
$$y = \frac{4}{3} \qquad \text{y-intercept } \left(0, \frac{4}{3}\right)$$

To find the *x*-intercept, let $y = 0$.

$$2x + 3y = 4$$
$$2x + 3(0) = 4 \qquad \text{Let } y = 0.$$
$$2x = 4$$
$$x = 2 \qquad \text{x-intercept } (2, 0)$$

The tables show these intercepts and a check point for each graph.

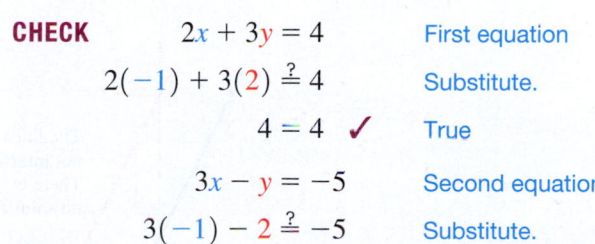

We recommend finding a third ordered pair as a check.

The lines in **FIGURE 39** suggest that the graphs intersect at the point $(-1, 2)$. We check by substituting -1 for x and 2 for y in *both* equations.

CHECK

$$2x + 3y = 4 \qquad \text{First equation}$$
$$2(-1) + 3(2) \overset{?}{=} 4 \qquad \text{Substitute.}$$
$$4 = 4 \checkmark \qquad \text{True}$$

$$3x - y = -5 \qquad \text{Second equation}$$
$$3(-1) - 2 \overset{?}{=} -5 \qquad \text{Substitute.}$$
$$-5 = -5 \checkmark \qquad \text{True}$$

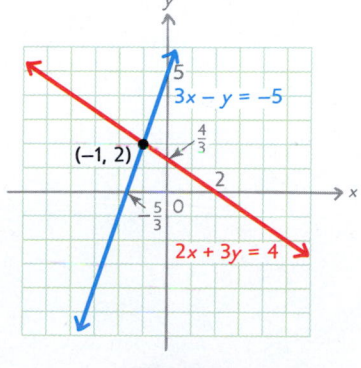

FIGURE 39

NOW TRY ANSWER
2. $\{(2, -1)\}$

Because $(-1, 2)$ satisfies both equations, the solution set of the system is $\{(-1, 2)\}$.

NOW TRY

> **NOTE**   We can also write each equation in a system in slope-intercept form and use the slope and y-intercept to graph each line. **See Example 2.**
>
> $2x + 3y = 4$ becomes $y = -\dfrac{2}{3}x + \dfrac{4}{3}.$ y-intercept $\left(0, \dfrac{4}{3}\right)$; slope $-\dfrac{2}{3}$
>
> $3x - y = -5$ becomes $y = 3x + 5.$ y-intercept $(0, 5)$; slope 3, or $\dfrac{3}{1}$
>
> Confirm that graphing these equations results in the same lines and the same solution shown in **FIGURE 39** on the preceding page.

Solving a Linear System by Graphing

Step 1   **Graph each equation** of the system on the same coordinate axes.

Step 2   **Find the coordinates of the point of intersection** of the graphs if possible, and write it as an ordered pair.

Step 3   **Check** that the ordered pair is the solution by substituting it in *both* of the *original* equations. If it satisfies *both* equations, write the solution set.

> ⚠ **CAUTION**   We recommend using graph paper and a straightedge when solving systems of equations graphically. It may not be possible to determine from the graph the exact coordinates of the point that represents the solution, particularly if those coordinates are not integers. The graphing method does, however, show geometrically how solutions are found and is useful when approximate answers will suffice.

OBJECTIVE 3   Solve special systems by graphing.

The graphs of the equations in a system may not intersect at all or may be the same line.

NOW TRY EXERCISE 3

Solve the system by graphing.

$4x + y = 7$
$12x + 3y = 10$

EXAMPLE 3   Solving a System by Graphing (No Solution)

Solve the system by graphing.

$$2x + y = 2$$
$$2x + y = 8$$

The graphs of these lines are shown in **FIGURE 40**. The two lines are parallel and have no points in common. For such a system, there is no solution. We write the solution set as $\varnothing$.

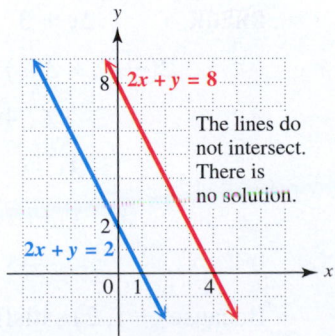

FIGURE 40

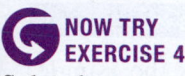

**NOW TRY
EXERCISE 4**

Solve the system by graphing.

$$5x - 3y = 2$$
$$10x - 6y = 4$$

EXAMPLE 4 Solving a System by Graphing (Infinite Number of Solutions)

Solve the system by graphing.

$$2x + 5y = 1$$
$$6x + 15y = 3$$

The graphs of these two equations are the same line. See **FIGURE 41**. We can obtain the second equation by multiplying each side of the first equation by 3. In this case, every point on the line is a solution of the system, and the solution set contains an infinite number of ordered pairs, each of which satisfies both equations of the system.

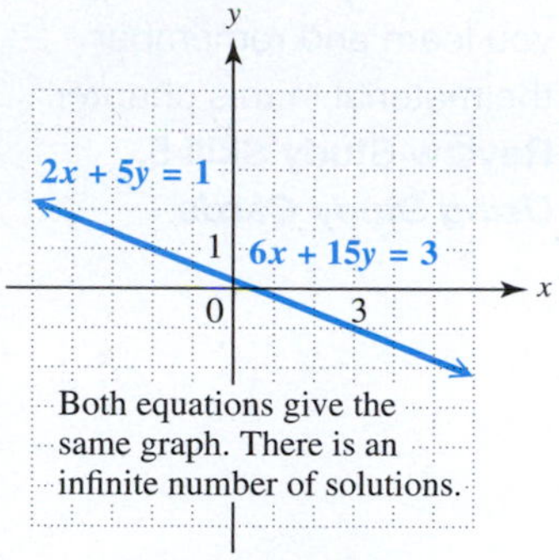

We write the solution set as

$$\{(x, y) \mid 2x + 5y = 1\},$$

> This is the first equation in the system. See the Note below.

read "*the set of ordered pairs* (x, y) *such that* $2x + 5y = 1$." Recall that this notation is called **set-builder notation**.

NOW TRY

Both equations give the same graph. There is an infinite number of solutions.

FIGURE 41

NOTE When a system has an infinite number of solutions, as in **Example 4,** either equation of the system can be used to write the solution set.

We prefer to use the equation in Ax + By = C form with integer coefficients having greatest common factor 1 and positive coefficient A of x.

If neither of the given equations is in this form, use an *equivalent* equation that is written this way.

Examples: For the system

$$-6x + 2y = -4$$
$$3x - y = 2,$$

This system has an infinite number of solutions.

we write the solution set using the second equation.

$$\{(x, y) \mid 3x - y = 2\}$$

For the system

$$2x - 4y = 8$$
$$-4x + 8y = -16,$$

This system has an infinite number of solutions.

we divide each term of the first equation by the common factor 2 (so that the coefficients have greatest common factor 1) and write the solution set as follows.

$$\{(x, y) \mid x - 2y = 4\}$$

The system in **Example 2** has exactly one solution. A system with at least one solution is a **consistent system.** A system with no solution, such as the one in **Example 3,** is an **inconsistent system.**

The equations in **Example 2** are **independent equations** with different graphs. The equations of the system in **Example 4** have the same graph and are equivalent. Because they are different forms of the same equation, these equations are **dependent equations.**

Examples 2–4 illustrate the three cases that may occur when solving a system of equations with two variables.

NOW TRY ANSWER

4. $\{(x, y) \mid 5x - 3y = 2\}$

Three Cases for Solutions of Linear Systems with Two Variables

Case 1   The graphs intersect at exactly one point, which gives the (single) ordered-pair solution of the system. The **system is consistent** and the **equations are independent.** See FIGURE 42(a).

Case 2   The graphs are parallel lines, so there is no solution and the solution set is $\varnothing$. The **system is inconsistent** and the **equations are independent.** See FIGURE 42(b).

Case 3   The graphs are the same line. There is an infinite number of solutions, and the solution set is written in set-builder notation as

$$\{(x, y) \mid \underline{\hspace{2cm}}\},$$

where one of the equations follows the $\mid$ symbol. The **system is consistent** and the **equations are dependent.** See FIGURE 42(c).

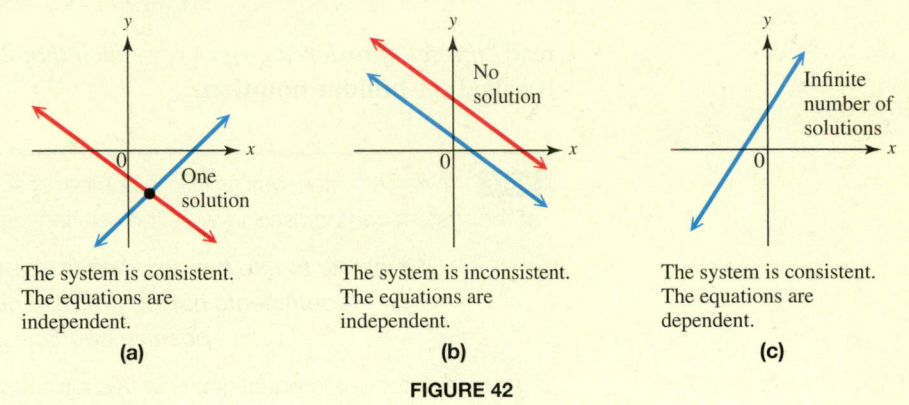

The system is consistent.
The equations are
independent.

(a)

The system is inconsistent.
The equations are
independent.

(b)

The system is consistent.
The equations are
dependent.

(c)

FIGURE 42

OBJECTIVE 4 Identify special systems without graphing.

We can recognize special systems without graphing by comparing their slopes and y-intercepts. We do this by writing each equation in slope-intercept form, solving for y.

EXAMPLE 5   Identifying the Three Cases Using Slopes

Describe each system without graphing. State the number of solutions.

(a) $3x + 2y = 6$

$-2y = 3x - 5$

Write each equation in slope-intercept form $y = mx + b$.

$3x + 2y = 6$ ◁ Solve for y.

$2y = -3x + 6$ Subtract $3x$.

$y = -\dfrac{3}{2}x + 3$ Divide *each* term by 2.

$-2y = 3x - 5$ ◁ Solve for y.

$y = -\dfrac{3}{2}x + \dfrac{5}{2}$ Divide *each* term by -2.

$\dfrac{3x}{-2}$ and $-\dfrac{3}{2}x$ are equivalent.

Both equations have slope $-\dfrac{3}{2}$, but they have different y-intercepts, $(0, 3)$ and $\left(0, \dfrac{5}{2}\right)$.

Lines with the same slope are parallel, so these equations have graphs that are parallel lines, which do not intersect. Thus, the system has no solution and is inconsistent.

NOW TRY
EXERCISE 5
Describe each system without graphing. State the number of solutions. If the system is *inconsistent* or the equations are *dependent,* say so.

(a) $2x + y = 7$
$3y = -6x - 12$

(b) $5x - 8y = 4$
$x - \frac{8}{5}y = \frac{4}{5}$

(c) $y - 3x = 7$
$3y - x = 0$

NOW TRY ANSWERS

5. (a) The equations have graphs that are parallel lines. The system has no solution and is inconsistent.
(b) The equations have graphs that are the same line. The system has an infinite number of solutions. The equations are dependent.
(c) The equations have graphs that intersect in one point. The system has exactly one solution.

(b) $2x - y = 4$

$x = \frac{y}{2} + 2$

Write each equation in slope-intercept form.

$2x - y = 4$

Be careful to retain the $-$ symbol. $\quad -y = -2x + 4$ Subtract 2x.

$y = 2x - 4$ Multiply by -1.

$\frac{y}{2} + 2 = x$ Interchange sides.

$\frac{y}{2} = x - 2$ Subtract 2.

$y = 2x - 4$ Multiply by 2.

The equations are exactly the same—their graphs are the same line. Any ordered-pair solution of one equation is also a solution of the other equation. Thus, the system has an infinite number of solutions. The equations are dependent.

(c) $x - 3y = 5$

$2x + y = 8$

In slope-intercept form, the equations are as follows.

$x - 3y = 5$

$-3y = -x + 5$ Subtract x.

  $y = \frac{1}{3}x - \frac{5}{3}$ Divide by -3.

↑
Slope

$2x + y = 8$

$y = -2x + 8$ Subtract 2x.

↑
Slope

The graphs of these equations are neither parallel nor the same line because the slopes are different. The graphs will intersect in one point—thus, the system has exactly one solution.

NOW TRY

7.3 Exercises

FOR EXTRA HELP

▶ **MyLab Math**

▶ *Video solutions for select problems available in MyLab Math*

STUDY SKILLS REMINDER
Are you getting the most out of your class time?
Review Study Skill 3, Taking Lecture Notes.

Concept Check *Complete each statement. The following terms may be used once, more than once, or not at all.*

consistent	system of linear equations	inconsistent	solution
ordered pair	independent	linear equation	dependent

1. A(n) _____ consists of two or more linear equations with the (*same/different*) variables.

2. A solution of a system of linear equations is a(n) _____ that makes all equations of the system (*true/false*) at the same time.

3. The equations of two parallel lines form a(n) _____ system that has (*one/no/infinitely many*) solution(s). The equations are _____ because their graphs are different.

4. If the graphs of a linear system intersect in one point, the point of intersection is the _____ of the system. The system is _____ and the equations are independent.

5. If two equations of a linear system have the same graph, the equations are _____. The system is _____ and has (*one/no/infinitely many*) solution(s).

6. If a linear system is inconsistent, the graphs of the two equations are (*intersecting/parallel/ the same*) line(s). The system has no _____.

7. Concept Check   A student determined incorrectly that the ordered pair $(1, -2)$ is a solution of the following system. His reasoning was that the ordered pair satisfies the equation $x + y = -1$ because $1 + (-2) = -1$. **WHAT WENT WRONG?**

$$x + y = -1$$
$$2x + y = 4$$

8. Concept Check   A student determined that the ordered pair $(0, 0)$ is the only solution of the following system. Her reasoning was that the ordered pair satisfies both equations of the system. This is not entirely correct. **WHAT WENT WRONG?**

$$x + y = 0$$
$$4x = -4y$$

Concept Check   *Work each problem.*

9. Which ordered pair could be a solution of the system graphed? Why is it the only valid choice?

   A. $(2, 2)$ **B.** $(-2, 2)$

   C. $(-2, -2)$ **D.** $(2, -2)$

10. Which ordered pair could be a solution of the system graphed? Why is it the only valid choice?

   A. $(2, 0)$ **B.** $(0, 2)$

   C. $(-2, 0)$ **D.** $(0, -2)$

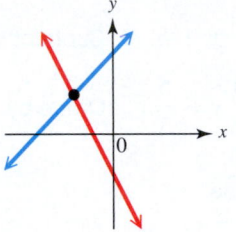
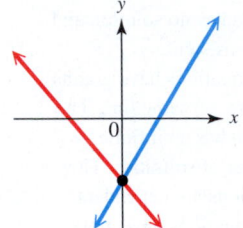

11. Each ordered pair in parts (a)–(d) is a solution of one of the systems graphed in choices A–D. Use the location of the point of intersection to determine the correct system for each solution. Match each system from A–D with its solution from (a)–(d).

(a) $(3, 4)$

A.

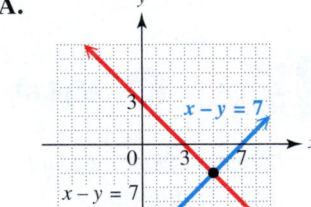

B.

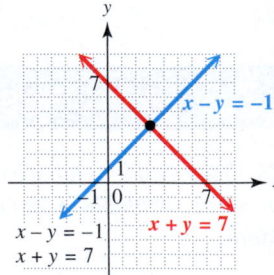

(b) $(-2, 3)$

(c) $(-3, 2)$

C.

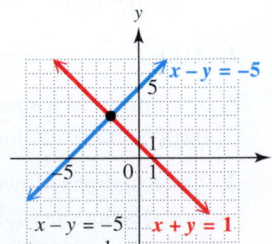

D.

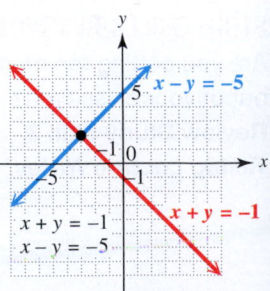

(d) $(5, -2)$

12. Each system has infinitely many solutions. Write its solution set using set-builder notation as described in the Note following **Example 4.**

   (a) $6x - 4y = 8$

       $3x - 2y = 4$

   (b) $3x - 9y = -6$

       $-6x + 18y = 12$

   (c) $x = 4 - y$

       $2x + 2y = 8$

13. We described systems without graphing in **Example 5.** Based on these descriptions, state the solution set of each system of equations.

    (a) $3x + 2y = 6$

         $-2y = 3x - 5$ See Example 5(a).

    (b) $2x - y = 4$

         $x = \dfrac{y}{2} + 2$ See Example 5(b).

14. Graph the lines in the following system of equations. Explain the difficulty in identifying a solution.

$$x - 3y = 5$$
$$2x + \ \ y = 8 \quad \text{See Example 5(c).}$$

Determine whether the given ordered pair is a solution of the given system. **See Example 1.**

15. $(2, -3)$

     $x + \ \ y = -1$

     $2x + 5y = 19$

16. $(4, 3)$

     $x + 2y = 10$

     $3x + 5y = 3$

17. $(-1, -3)$

     $3x + 5y = -18$

     $4x + 2y = -10$

18. $(-9, -2)$

     $2x - 5y = -8$

     $3x + 6y = -39$

19. $(7, -2)$

     $4x = 26 - \ \ y$

     $3x = 29 + 4y$

20. $(9, 1)$

     $2x = 23 - 5y$

     $3x = 24 + 3y$

21. $(6, -8)$

     $-2y = \ \ x + 10$

     $3y = 2x + 30$

22. $(-5, 2)$

     $5y = \ \ 3x + 20$

     $3y = -2x - 4$

23. $(0, 0)$

     $4x + 2y = 0$

     $x + \ \ y = 0$

24. $(-1, -1)$

     $-4x + 4y = 0$

     $x - \ \ y = 0$

25. $(1, 1)$

     $y = \dfrac{2}{3}x$

     $y = \dfrac{1}{2}x$

26. $(-2, 2)$

     $y = -\dfrac{3}{2}x$

     $y = -\dfrac{1}{3}x$

Solve each system by graphing. If the system is inconsistent or the equations are dependent, say so. **See Examples 2–4.**

27. $x - y = 2$

     $x + y = 6$

28. $x - y = 3$

     $x + y = -1$

29. $x + y = 4$

     $y - x = 4$

30. $x + y = -5$

     $y - x = -5$

31. $x - 2y = 6$

     $x + 2y = 2$

32. $2x - y = 4$

     $4x + y = 2$

33. $\ \ 3x - 2y = -3$

     $-3x - \ \ y = -6$

34. $2x - \ \ y = 4$

     $2x + 3y = 12$

35. $3x + \ \ y = 5$

     $6x + 2y = 10$

36. $2x - \ \ y = 4$

     $4x - 2y = 8$

37. $-3x + y = -3$

     $y = x - 3$

38. $2x - 3y = -6$

     $y = -3x + 2$

39. $y = 2x + 8$

     $x + y = -4$

40. $y = 4x - 4$

     $3x - 2y = 3$

41. $2x - \ \ y = 6$

     $4x - 2y = 8$

42. $\ \ x + 2y = 4$

     $2x + 4y = 12$

43. $2y - 6x = 12$

     $3x - \ \ y = -6$

44. $-8y - 2x = -8$

     $x + 4y = 4$

45. $3x - 4y = 24$

     $y = -\dfrac{3}{2}x + 3$

46. $4x + y = 5$

     $y = \dfrac{3}{2}x - 6$

47. $2x = y - 4$

     $4x + 4 = 2y$

48. $3x = y + 5$

     $6x - 5 = 2y$

49. $2x = -3y$

     $4x - y = 0$

50. $-x = y$

     $5x - 2y = 0$

Without graphing, do the following for each system of equations. **See Example 5.**

(a) *Describe each system.*

(b) *State the number of solutions.*

(c) *Is the system inconsistent, are the equations dependent, or neither?*

51. $y - x = -5$
$x + y = 1$

52. $y + 2x = 6$
$x - 3y = -4$

53. $x + 2y = 0$
$4y = -2x$

54. $2x - y = 4$
$y + 4 = 2x$

55. $x - 3y = 5$
$2x + y = 8$

56. $2x + 3y = 12$
$2x - y = 4$

57. $5x + 4y = 7$
$10x + 8y = 4$

58. $3x + 2y = 5$
$6x + 4y = 3$

59. $3x + 2y = 5$
$-6x - 4y = 10$

60. $2x - y = 6$
$-10x + 5y = 30$

61. $5x = 10y$
$\frac{1}{2}x - y = 0$

62. $y = -3x$
$x + \frac{1}{3}y = 0$

Work each problem using the graph provided.

63. The graph shows how revenues (in millions of dollars) from digital downloads of single songs and from paid music subscription services in the United States changed over the years 2012 through 2016.

(a) For which years was revenue from download singles greater than revenue from paid subscriptions?

(b) Estimate the year in which the revenues from download singles and paid subscriptions were closest to the same. About how much was the revenue from each format in that year?

U.S. Music Revenues

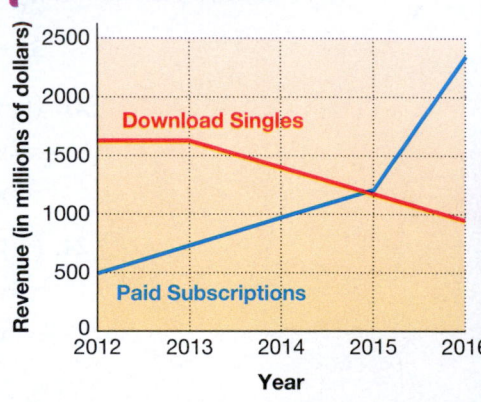

Data from RIAA.

(c) Express the point of intersection of the two graphs as an ordered pair written in the form (year, revenue in millions of dollars).

64. The graph shows global shipments of desktop PCs and tablet computers in millions of units from 2011 through 2016.

(a) Estimate the year in which the numbers of global shipments of desktop PCs and tablets were about the same. How many units was this?

(b) Express the point of intersection of the two graphs as an ordered pair written in the form (year, units shipped in millions).

(c) Describe the trend in shipments of desktop PCs over the years 2011 through 2016. Is the slope of the graph *positive, negative,* or *zero*? Explain.

Global Computer Shipments

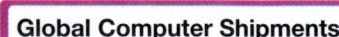
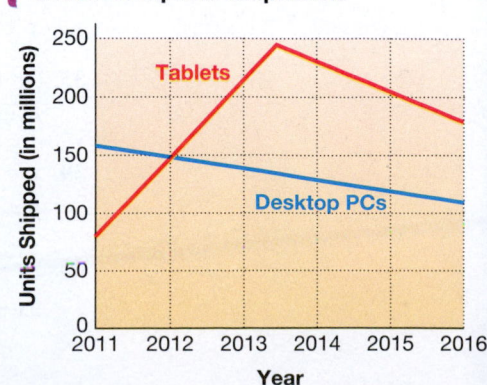

(d) Describe the trend in shipments of tablets over the years 2011 through 2016.

Data from IDC.

RELATING CONCEPTS For Individual or Group Work (Exercises 65–68)

*Economics deals with **supply** and **demand**. Typically, as the price of an item increases, the demand for the item decreases and the supply increases. If supply and demand can be described by straight-line equations, the point at which the lines intersect determines the **equilibrium supply** and **equilibrium demand**.*

The price per unit, p, and the demand, x, for aluminum siding are related by the linear equation

$$p = 60 - \frac{3}{4}x,$$

while the price and the supply are related by

$$p = \frac{3}{4}x.$$

Supply and Demand

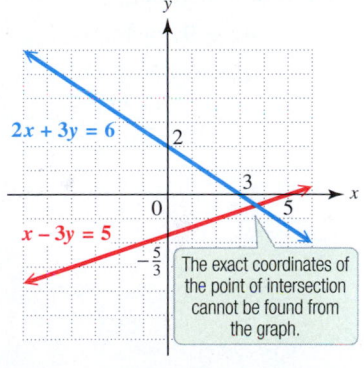

*Use the graph to work **Exercises 65–68 in order.***

65. At what value of x does supply equal demand? At what value of p does supply equal demand?

66. Express the equilibrium supply and equilibrium demand as an ordered pair of the form (quantity, price).

67. When $x > 40$, does demand exceed supply or does supply exceed demand?

68. When $x < 40$, does demand exceed supply or does supply exceed demand?

7.4 Solving Systems of Linear Equations by Substitution

OBJECTIVES

1 Solve linear systems by substitution.

2 Solve special systems by substitution.

3 Solve linear systems with fractions and decimals.

OBJECTIVE 1 Solve linear systems by substitution.

Graphing to solve a system of equations has a serious drawback. For example, consider the system graphed in **FIGURE 43**. It is difficult to determine an accurate solution of the system from the graph.

As a result, there are algebraic methods for solving systems of equations. The **substitution method,** which gets its name from the fact that an expression in one variable is *substituted* for the other variable, is one such method.

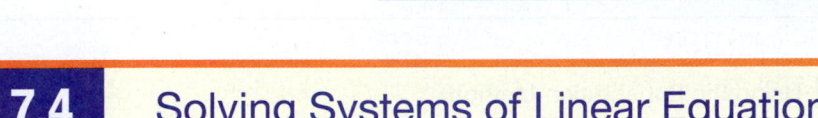

The exact coordinates of the point of intersection cannot be found from the graph.

FIGURE 43

EXAMPLE 1 Using the Substitution Method

Solve the system by the substitution method.

$$3x + 5y = 26 \qquad (1)$$
$$y = 2x \qquad (2)$$

We number the equations for reference in our discussion.

Equation (2) is already solved for y. This equation says that y is equal to $2x$, so we substitute $2x$ for y in equation (1).

$$3x + 5y = 26 \qquad (1)$$
$$3x + 5(2x) = 26 \qquad \text{Let } y = 2x.$$
$$3x + 10x = 26 \qquad \text{Multiply.}$$
$$13x = 26 \qquad \text{Combine like terms.}$$
$$x = 2 \qquad \text{Divide by 13.}$$

Don't stop here.

 **NOW TRY**
EXERCISE 1
Solve the system by the substitution method.

$$2x - 4y = 28$$
$$y = -3x$$

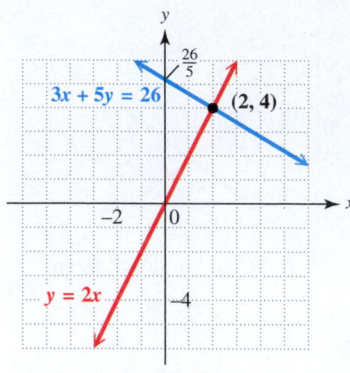

FIGURE 44

Now we can find the value of y by substituting 2 for x in either equation. We choose equation (2) because the substitution is easier.

$y = 2x$	(2)
$y = 2(2)$	Let $x = 2$.
$y = 4$	Multiply.

We check that the ordered pair $(2, 4)$ is the solution by substituting 2 for x and 4 for y in *both* equations.

CHECK

$3x + 5y = 26$	(1)		$y = 2x$	(2)
$3(2) + 5(4) \stackrel{?}{=} 26$	Substitute.		$4 \stackrel{?}{=} 2(2)$	Substitute.
$6 + 20 \stackrel{?}{=} 26$	Multiply.		$4 = 4$ ✓	True
$26 = 26$ ✓ True				

Because $(2, 4)$ satisfies both equations, the solution set of the system is $\{(2, 4)\}$.

The graph in **FIGURE 44**, which shows the point $(2, 4)$ as the intersection of the equations of this system, confirms our algebraic work. **NOW TRY** 🔄

> ⚠ **CAUTION** *A system is not completely solved until values for both x and y are found.* Write the solution set as a set containing an ordered pair.

 **NOW TRY**
EXERCISE 2
Solve the system by the substitution method.

$$4x + 9y = 1$$
$$x = y - 3$$

EXAMPLE 2 Using the Substitution Method

Solve the system by the substitution method.

$$2x + 5y = 7 \quad (1)$$
$$x = -1 - y \quad (2)$$

Equation (2) gives x in terms of y. Substitute $-1 - y$ for x in equation (1).

$2x + 5y = 7$	(1) ← Be sure to substitute in the *other* equation.
$2(-1 - y) + 5y = 7$	Let $x = -1 - y$.
(Distribute 2 to both -1 and $-y$.) $-2 - 2y + 5y = 7$	Distributive property
$-2 + 3y = 7$	Combine like terms.
$3y = 9$	Add 2.
$y = 3$	Divide by 3.

To find x, substitute 3 for y in equation (2).

$x = -1 - y$	(2)
$x = -1 - 3$	Let $y = 3$.
(Write the x-coordinate first.) $x = -4$	Subtract.

Check that $(-4, 3)$ is the solution.

CHECK

$2x + 5y = 7$	(1)		$x = -1 - y$	(2)
$2(-4) + 5(3) \stackrel{?}{=} 7$	Substitute.		$-4 \stackrel{?}{=} -1 - 3$	Substitute.
$-8 + 15 \stackrel{?}{=} 7$	Multiply.		$-4 = -4$ ✓	True
$7 = 7$ ✓ True				

Both results are true, so the solution set of the system is $\{(-4, 3)\}$. **NOW TRY** 🔄

NOW TRY ANSWERS
1. $\{(2, -6)\}$
2. $\{(-2, 1)\}$

> ⚠ **CAUTION** Even though we found y first in **Example 2**, *the x-coordinate is always written first in the ordered-pair solution of a system.* The ordered pair $(-4, 3)$ is **not** the same as $(3, -4)$.

Solving a Linear System by Substitution

Step 1 **Solve one equation for either variable.** If one of the equations has a variable term with coefficient 1 or -1, choose it because the substitution method is usually easier.

Step 2 **Substitute** for that variable in the other equation. The result should be an equation with just one variable.

Step 3 **Solve** the equation from Step 2.

Step 4 **Find the other value.** Substitute the result from Step 3 into the equation from Step 1 and solve for the other variable.

Step 5 **Check** the values in *both* of the *original* equations. Then write the solution set as a set containing an ordered pair.

EXAMPLE 3   Using the Substitution Method

Solve the system by the substitution method.

$$2x = 4 - y \qquad (1)$$
$$5x + 3y = 10 \qquad (2)$$

Step 1 We must solve one of the equations for either x or y. Because the coefficient of y in equation (1) is -1, we avoid fractions by solving this equation for y.

$$2x = 4 - y \qquad (1)$$
$$y + 2x = 4 \qquad \text{Add } y.$$
$$y = 4 - 2x \qquad \text{Subtract } 2x.$$

Step 2 Now substitute $4 - 2x$ for y in equation (2).

$$5x + 3y = 10 \qquad (2)$$
$$5x + 3(4 - 2x) = 10 \qquad \text{Let } y = 4 - 2x.$$

Step 3 Solve the equation from Step 2.

$$5x + 12 - 6x = 10 \qquad \text{Distributive property}$$
$$-x + 12 = 10 \qquad \text{Combine like terms.}$$
$$-x = -2 \qquad \text{Subtract } 12.$$
$$x = 2 \qquad \text{Multiply by } -1.$$

Step 4 We solved equation (1) for y in Step 1. Substitute 2 for x in this equation to find y.

$$y = 4 - 2x \qquad \text{Equation (1) solved for } y$$
$$y = 4 - 2(2) \qquad \text{Let } x = 2.$$
$$y = 4 - 4 \qquad \text{Multiply.}$$
$$y = 0 \qquad \text{Subtract.}$$

**NOW TRY
EXERCISE 3**
Solve the system by the substitution method.
$$2y = x + 6$$
$$4x - 5y = -24$$

Step 5   Check that $(2, 0)$ is the solution.

CHECK
$$2x = 4 - y \quad (1)$$
$$2(2) \stackrel{?}{=} 4 - 0 \quad \text{Substitute.}$$
$$4 = 4 \ \checkmark \quad \text{True}$$

$$5x + 3y = 10 \quad (2)$$
$$5(2) + 3(0) \stackrel{?}{=} 10 \quad \text{Substitute.}$$
$$10 = 10 \ \checkmark \quad \text{True}$$

Both results are true, so the solution set of the system is $\{(2, 0)\}$.   **NOW TRY**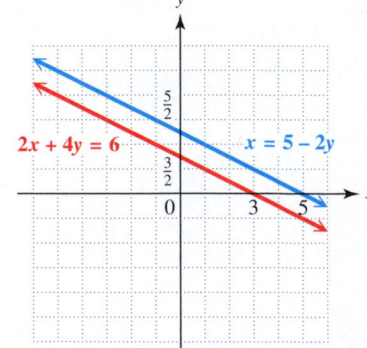

_____

OBJECTIVE 2 Solve special systems by substitution.

**NOW TRY
EXERCISE 4**
Solve the system by the substitution method.
$$8x - 2y = 1$$
$$y = 4x - 8$$

| **EXAMPLE 4** | Solving an Inconsistent System Using Substitution |

Solve the system by the substitution method.

$$x = 5 - 2y \quad (1)$$
$$2x + 4y = 6 \quad (2)$$

Because equation (1) is solved for x, we substitute $5 - 2y$ for x in equation (2).

$$2x + 4y = 6 \quad (2)$$
$$2(5 - 2y) + 4y = 6 \quad \text{Let } x = 5 - 2y.$$
$$10 - 4y + 4y = 6 \quad \text{Distributive property}$$
$$10 = 6 \quad \text{False (contradiction)}$$

The false result means that the equations in the system have graphs that are parallel lines. The system is inconsistent and has no solution, so the solution set is $\varnothing$.

CHECK   We can confirm the solution set by writing each equation in slope-intercept form—that is, solved for y.

$$x = 5 - 2y \quad (1)$$
$$2y = -x + 5$$
$$y = -\frac{1}{2}x + \frac{5}{2}$$

$$2x + 4y = 6 \quad (2)$$
$$4y = -2x + 6$$
$$y = -\frac{1}{2}x + \frac{3}{2}$$

The two lines have the same slope, $-\frac{1}{2}$, but different y-intercepts, $\left(0, \frac{5}{2}\right)$ and $\left(0, \frac{3}{2}\right)$. Therefore, they are parallel and do not intersect, confirming that the solution set is $\varnothing$. See **FIGURE 45.** $\checkmark$

FIGURE 45

NOW TRY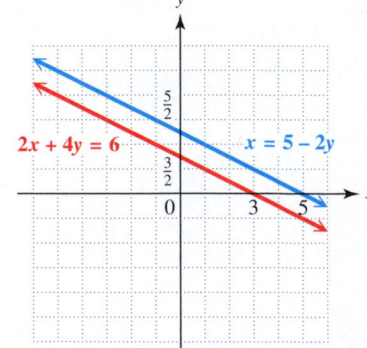

| **EXAMPLE 5** | Solving a System with Dependent Equations Using Substitution |

Solve the system by the substitution method.

$$3x - y = 4 \quad (1)$$
$$-9x + 3y = -12 \quad (2)$$

Begin by solving equation (1) for y.

$$y = 3x - 4 \quad \text{Equation (1) solved for } y$$

We substitute $3x - 4$ for y in equation (2) and solve the resulting equation.

NOW TRY ANSWERS
3. $\{(-6, 0)\}$
4. $\varnothing$

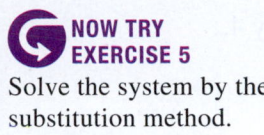

**NOW TRY
EXERCISE 5**

Solve the system by the substitution method.

$$5x - y = 6$$
$$-10x + 2y = -12$$

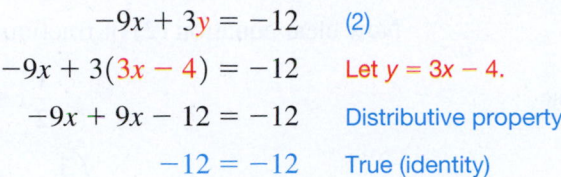

$$-9x + 3y = -12 \quad \text{(2)}$$
$$-9x + 3(3x - 4) = -12 \qquad \text{Let } y = 3x - 4.$$
$$-9x + 9x - 12 = -12 \qquad \text{Distributive property}$$
$$-12 = -12 \qquad \text{True (identity)}$$

This true result means that every solution of one equation is also a solution of the other, so the system has an infinite number of solutions. The solution set, written in set-builder notation using equation (1), is

$$\{(x, y) \mid 3x - y = 4\}.$$

CHECK If we multiply equation (1) by -3, we obtain equation (2). Therefore,

$$3x - y = 4 \quad \text{and} \quad -9x + 3y = -12$$

are equivalent equations. They represent the same line. All of the ordered pairs corresponding to points that lie on the common graph are solutions. See **FIGURE 46**. ✓

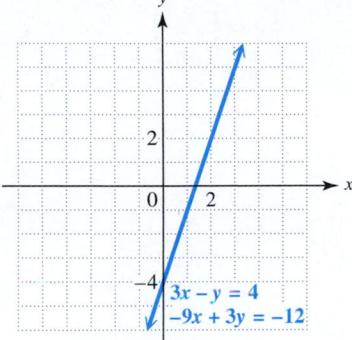

FIGURE 46

NOW TRY

⚠ **CAUTION** Avoid these common mistakes.

1. Do not give "false" as the solution of an inconsistent system. The correct response is ∅. **(See Example 4.)**

2. Do not give "true" as the solution of a system of dependent equations. In this text, we write the solution set in set-builder notation using the equation in the system (or an equivalent equation) that is in $Ax + By = C$ form with integer coefficients having greatest common factor 1 and positive coefficient A of x. **(See Example 5.)**

OBJECTIVE 3 Solve linear systems with fractions and decimals.

EXAMPLE 6 Using the Substitution Method (Fractional Coefficients)

Solve the system by the substitution method.

$$3x + \frac{1}{4}y = 2 \qquad \text{(1)}$$

$$\frac{1}{2}x + \frac{3}{4}y = -\frac{5}{2} \qquad \text{(2)}$$

Clear equation (1) of fractions by multiplying each side by 4.

$$3x + \frac{1}{4}y = 2 \qquad \text{(1)}$$

$$4\left(3x + \frac{1}{4}y\right) = 4(2) \qquad \text{Multiply by 4.}$$

$$4(3x) + 4\left(\frac{1}{4}y\right) = 4(2) \qquad \text{Distributive property}$$

$$12x + y = 8 \qquad \text{(3)}$$

NOW TRY ANSWER
5. $\{(x, y) \mid 5x - y = 6\}$

**NOW TRY
EXERCISE 6**
Solve the system by the substitution method.

$$x + \frac{1}{2}y = \frac{1}{2}$$

$$\frac{1}{6}x - \frac{1}{3}y = \frac{4}{3}$$

Now clear equation (2) of fractions by multiplying each side by 4.

$$\frac{1}{2}x + \frac{3}{4}y = -\frac{5}{2} \qquad \text{(2)}$$

$$4\left(\frac{1}{2}x + \frac{3}{4}y\right) = 4\left(-\frac{5}{2}\right) \qquad \text{Multiply by 4, the common denominator.}$$

$$4\left(\frac{1}{2}x\right) + 4\left(\frac{3}{4}y\right) = 4\left(-\frac{5}{2}\right) \qquad \text{Distributive property}$$

$$2x + 3y = -10 \qquad \text{(4)}$$

The given system of equations can be written as an equivalent system.

$$12x + y = 8 \qquad \text{(3) from the previous page}$$

$$2x + 3y = -10 \qquad \text{(4)}$$

To solve this system by substitution, solve equation (3) for y.

$$12x + y = 8 \qquad \text{(3)}$$

$$y = 8 - 12x \qquad \text{Subtract 12}x.$$

Now substitute this result for y in equation (4).

$$2x + 3y = -10 \qquad \text{(4)}$$

$$2x + 3(8 - 12x) = -10 \qquad \text{Let } y = 8 - 12x.$$

$$2x + 24 - 36x = -10 \qquad \text{Distributive property}$$

$$-34x = -34 \qquad \text{Combine like terms. Subtract 24.}$$

$$x = 1 \qquad \text{Divide by } -34.$$

Substitute 1 for x in $y = 8 - 12x$ (equation (3) solved for y).

$$y = 8 - 12(1) \qquad \text{Let } x = 1.$$

$$y = -4 \qquad \text{Multiply, and then subtract.}$$

Check $(1, -4)$ in both of the original equations. The solution set is $\{(1, -4)\}$.

NOW TRY

EXAMPLE 7   Using the Substitution Method (Decimal Coefficients)

Solve the system by the substitution method.

$$0.5x + 2.4y = 4.2 \qquad \text{(1)}$$

$$-0.1x + 1.5y = 5.1 \qquad \text{(2)}$$

Clear each equation of decimals by multiplying by 10. Recall that multiplying by 10 is the same as moving each decimal point one place to the right.

$$10(0.5x + 2.4y) = 10(4.2) \qquad \text{Multiply equation (1) by 10.}$$

$$10(0.5x) + 10(2.4y) = 10(4.2) \qquad \text{Distributive property}$$

$$5x + 24y = 42 \qquad \text{(3)}$$

$$10(-0.1x + 1.5y) = 10(5.1) \qquad \text{Multiply equation (2) by 10.}$$

$$10(-0.1x) + 10(1.5y) = 10(5.1) \qquad \text{Distributive property}$$

$$-x + 15y = 51 \qquad \text{(4)}$$

$$10(-0.1x)$$
$$= -1x$$
$$= -x$$

 **NOW TRY EXERCISE 7**

Solve the system by the substitution method.

$0.2x + 0.3y = 0.5$
$0.3x - 0.1y = 1.3$

Now solve the equivalent system of equations by substitution.

$$5x + 24y = 42 \qquad \text{(3)}$$
$$-x + 15y = 51 \qquad \text{(4)}$$

Equation (4) can be solved for x.

$$x = 15y - 51 \qquad \text{Equation (4) solved for } x$$

Substitute this result for x in equation (3).

$$5x + 24y = 42 \qquad \text{(3)}$$
$$5(15y - 51) + 24y = 42 \qquad \text{Let } x = 15y - 51.$$
$$75y - 255 + 24y = 42 \qquad \text{Distributive property}$$
$$99y = 297 \qquad \text{Combine like terms. Add 255.}$$
$$y = 3 \qquad \text{Divide by 99.}$$

Equation (4) solved for x is $x = 15y - 51$. Substitute 3 for y.

$$x = 15(3) - 51 \qquad \text{Let } y = 3.$$
$$x = -6 \qquad \text{Multiply, and then subtract.}$$

Check $(-6, 3)$ in both of the original equations. The solution set is $\{(-6, 3)\}$.

NOW TRY ANSWER
7. $\{(4, -1)\}$

NOW TRY

7.4 Exercises

FOR EXTRA HELP

 MyLab Math

 Video solutions for select problems available in MyLab Math

STUDY SKILLS REMINDER
How are you doing on your homework? **Review Study Skill 4, *Completing Your Homework.***

1. Concept Check A student solves the following system and finds that $x = 3$, which is correct. The student gives the solution set incorrectly as $\{3\}$. **WHAT WENT WRONG?** Give the correct solution set.

$$5x - y = 15$$
$$7x + y = 21$$

2. Concept Check A student solves the following system and obtains the statement $8 = 8$. The student gives the solution set incorrectly as $\{(8, 8)\}$. **WHAT WENT WRONG?** Give the correct solution set.

$$x + y = 4$$
$$2x + 2y = 8$$

Concept Check *Answer each question.*

3. When we use the substitution method, how can we tell that a system has no solution?

4. When we use the substitution method, how can we tell that a system has an infinite number of solutions?

Solve each system by the substitution method. Check each solution. **See Examples 1–5.**

5. $x + y = 12$
$\quad y = 3x$

6. $x + 3y = -28$
$\quad y = -5x$

7. $3x + 2y = 27$
$\quad x = y + 4$

8. $4x + 3y = -5$
$\quad x = y - 3$

9. $3x + 4 = -y$
$\quad 2x + y = 0$

10. $2x - 5 = -y$
$\quad x + 3y = 0$

11. $7x + 4y = 13$
$x + y = 1$

12. $3x - 2y = 19$
$x + y = 8$

13. $5x + 2y = -15$
$2x = y - 6$

14. $2x + 3y = 4$
$4x = y + 8$

15. $3x + 5y = 25$
$x - 2y = -10$

16. $4x + 3y = -15$
$x - 4y = 20$

17. $3x - y = 5$
$y = 3x - 5$

18. $4x - y = -3$
$y = 4x + 3$

19. $2x + 8y = 3$
$x = 8 - 4y$

20. $2x + 10y = 3$
$x = 1 - 5y$

21. $2y = 4x + 24$
$2x - y = -12$

22. $2y = 14 - 6x$
$3x + y = 7$

23. $x = y - 4$
$x - y = 1$

24. $x = 2 - y$
$x + y = -5$

25. $x + y = 0$
$3x - 3y = 0$

26. $5x + y = 0$
$x - y = 0$

27. $2x + y = 0$
$4x - 2y = 2$

28. $x + y = 0$
$4x + 2y = 3$

Solve each system by the substitution method. Check each solution. **See Examples 6 and 7.**

29. $\frac{1}{5}x + \frac{2}{3}y = -\frac{8}{5}$
$3x - y = 9$

30. $\frac{1}{3}x - \frac{1}{2}y = -\frac{2}{3}$
$4x + y = 6$

31. $\frac{1}{2}x + \frac{1}{3}y = -\frac{1}{3}$
$\frac{1}{2}x + 2y = -7$

32. $\frac{1}{6}x + \frac{1}{6}y = 1$
$-\frac{1}{2}x - \frac{1}{3}y = -5$

33. $\frac{x}{5} + 2y = \frac{8}{5}$
$\frac{3x}{5} + \frac{y}{2} = -\frac{7}{10}$

34. $\frac{x}{2} + \frac{y}{3} = \frac{7}{6}$
$\frac{x}{4} - \frac{3y}{2} = \frac{9}{4}$

35. $0.2x - 1.3y = -3.2$
$-0.1x + 2.7y = 9.8$

36. $0.1x + 0.9y = -2$
$0.5x - 0.2y = 4.1$

37. $0.3x - 0.1y = 2.1$
$0.6x + 0.3y = -0.3$

38. $0.8x - 0.1y = 1.3$
$2.2x + 1.5y = 8.9$

39. $-0.3x + 0.5y = -1.5$
$0.4x + 0.5y = 2$

40. $-0.1x + 0.1y = 0.2$
$0.3x + 0.1y = 0.2$

RELATING CONCEPTS   For Individual or Group Work (Exercises 41–44)

A system of linear equations can be used to model the cost and the revenue of a business. **Work Exercises 41–44 in order.**

41. Suppose that it costs $5000 to start a business manufacturing and selling bicycles. Each bicycle will cost $400 to manufacture. Explain why the linear equation

$$y_1 = 400x + 5000 \quad (y_1 \text{ in dollars})$$

gives the *total* cost to manufacture x bicycles.

42. Each bicycle will sell for $600. Write an equation using y_2 (in dollars) to express the revenue from the sale of x bicycles.

43. Form a system from the two equations in **Exercises 41 and 42.** Then solve the system, assuming $y_1 = y_2$—that is, cost = revenue.

44. The value of x from **Exercise 43** is the number of bicycles it takes to *break even*. Fill in the blanks: When _____ bicycles are sold, the break-even point is reached. At that point, we have spent _____ dollars and taken in _____ dollars.

7.5 Solving Systems of Linear Equations by Elimination

OBJECTIVES

1 Solve linear systems by elimination.
2 Multiply when using the elimination method.
3 Use an alternative method to find the second value in a solution.
4 Solve special systems by elimination.

OBJECTIVE 1 Solve linear systems by elimination.

Adding the same quantity to each side of an equation results in equal sums.

$$\text{If } A = B, \text{ then } A + C = B + C.$$

We can take this addition a step further. Adding *equal* quantities, rather than the *same* quantity, to each side of an equation also results in equal sums.

$$\text{If } A = B \text{ and } C = D, \text{ then } A + C = B + D.$$

The **elimination method** uses the addition property of equality to solve systems of equations.

EXAMPLE 1 Using the Elimination Method

Solve the system by the elimination method.

$$x + y = 5 \quad (1)$$
$$x - y = 3 \quad (2)$$

Each equation in this system is a statement of equality, so the sum of the left sides equals the sum of the right sides. Adding vertically in this way gives the following.

 The goal is to **eliminate** a variable.

$$x + y = 5 \quad (1)$$
$$\underline{x - y = 3} \quad (2)$$
$$2x \quad = 8 \quad \text{Add left sides and add right sides.}$$
$$x = 4 \quad \text{Divide by 2.}$$

Notice that y has been eliminated. The result, $x = 4$, gives the x-value of the ordered-pair solution of the given system. To find the y-value of the solution, substitute 4 for x in either of the two equations of the system. We choose equation (1).

$$x + y = 5 \quad (1)$$
$$4 + y = 5 \quad \text{Let } x = 4.$$
$$y = 1 \quad \text{Subtract 4.}$$

Check the ordered pair $(4, 1)$ in both equations of the given system.

CHECK

$$x + y = 5 \quad (1) \qquad\qquad x - y = 3 \quad (2)$$
$$4 + 1 \overset{?}{=} 5 \quad \text{Substitute.} \qquad 4 - 1 \overset{?}{=} 3 \quad \text{Substitute.}$$
$$5 = 5 \ \checkmark \quad \text{True} \qquad\qquad 3 = 3 \ \checkmark \quad \text{True}$$

Both results are true, so the solution set of the system is $\{(4, 1)\}$.

The graph in **FIGURE 47** confirms our algebraic work. We could also solve the system using the substitution method as an alternative confirmation. (Try this.)

NOW TRY

NOW TRY
EXERCISE 1
Solve the system by the elimination method.

$$x - y = 4$$
$$3x + y = 8$$

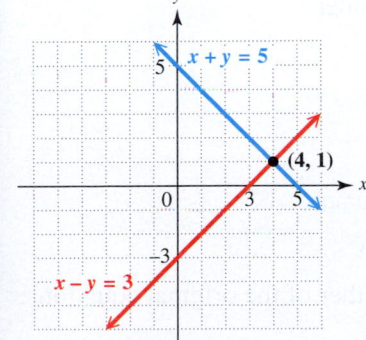

FIGURE 47

> ⚠️ **CAUTION** *A system is not completely solved until values for both x and y are found.*
> Do not stop after finding the value of only one variable. Remember to write the solution set as a set containing an ordered pair.

NOW TRY ANSWER
1. $\{(3, -1)\}$

With the elimination method, the idea is to *eliminate* one of the two variables in a system.

To do this, one pair of variable terms in the two equations must have coefficients that are opposites (additive inverses).

Solving a Linear System by Elimination

Step 1 **Write both equations in the form $Ax + By = C$.**

Step 2 **Transform the equations as needed so that the coefficients of one pair of variable terms are opposites.** Multiply one or both equations by appropriate numbers so that the sum of the coefficients of either the x- or y-terms is 0.

Step 3 **Add** the new equations to eliminate a variable. The sum should be an equation with just one variable.

Step 4 **Solve** the equation from Step 3 for the remaining variable.

Step 5 **Find the other value.** Substitute the result from Step 4 into either of the original equations, and solve for the other variable.

Step 6 **Check** the values in *both* of the *original* equations. Then write the solution set as a set containing an ordered pair.

It does not matter which variable is eliminated first. Usually, we choose the one that is more convenient to work with.

EXAMPLE 2   Using the Elimination Method

Solve the system by the elimination method.

$$y + 11 = 2x \quad \text{(1)}$$
$$5x = y + 26 \quad \text{(2)}$$

Step 1 Write both equations in the form $Ax + By = C$.

$$-2x + y = -11 \qquad \text{Subtract } 2x \text{ and } 11 \text{ in equation (1).}$$
$$5x - y = 26 \qquad \text{Subtract } y \text{ in equation (2).}$$

Step 2 Because the coefficients of y are 1 and -1, adding will **eliminate y.** It is not necessary to multiply either equation by a number.

Step 3 Add the two equations.

$$
\begin{array}{rcl}
-2x + y & = & -11 \\
5x - y & = & 26 \\
\hline
3x & = & 15 \qquad \text{Add in columns.}
\end{array}
$$

Step 4 Solve. $x = 5$ Divide by 3.

Step 5 Find the value of y by substituting 5 for x in either of the original equations.

$$y + 11 = 2x \qquad \text{(1)}$$
$$y + 11 = 2(5) \qquad \text{Let } x = 5.$$
$$y + 11 = 10 \qquad \text{Multiply.}$$
$$y = -1 \qquad \text{Subtract 11.}$$

NOW TRY
EXERCISE 2
Solve the system by the elimination method.
$$2x - 6 = -3y$$
$$5x - 3y = -27$$

Step 6 Check the ordered pair $(5, -1)$ in both of the original equations.

CHECK

$y + 11 = 2x$	(1)		$5x = y + 26$	(2)
$(-1) + 11 \overset{?}{=} 2(5)$	Substitute.		$5(5) = -1 + 26$	Substitute.
$10 = 10$ ✓	True		$25 = 25$ ✓	True

Both results are true, so the solution set is $\{(5, -1)\}$.

NOW TRY

OBJECTIVE 2 Multiply when using the elimination method.

Sometimes we need to multiply each side of one or both equations in a system by a number before adding will eliminate a variable.

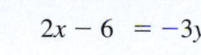

NOW TRY
EXERCISE 3
Solve the system by the elimination method.
$$4x - 3y = 6$$
$$x - 7y = -11$$

EXAMPLE 3	Using the Elimination Method

Solve the system by the elimination method.

$$3x - 2y = 10 \qquad (1)$$
$$x + 5y = -8 \qquad (2)$$

Step 1 The equations are already written in $Ax + By = C$ form.

Step 2 Adding the two equations gives $4x + 3y = 2$, which does not eliminate either variable. However, multiplying equation (2) by -3 and then adding will **eliminate x.**

Step 3 Add the two equations.

$$3x - 2y = 10 \qquad (1)$$
$$\underline{-3x - 15y = 24} \qquad \text{Multiply } both \text{ sides of equation (2) by } -3.$$
$$-17y = 34 \qquad \text{Add.}$$

Step 4 Solve. $y = -2$ Divide by -17.

Step 5 Find the value of x by substituting -2 for y in either equation (1) or equation (2).

$$x + 5y = -8 \qquad (2)$$
$$x + 5(-2) = -8 \qquad \text{Let } y = -2.$$
$$x - 10 = -8 \qquad \text{Multiply.}$$
$$x = 2 \qquad \text{Add 10.}$$

Step 6 Check the ordered pair $(2, -2)$ in both of the original equations.

CHECK

$3x - 2y = 10$	(1)		$x + 5y = -8$	(2)
$3(2) - 2(-2) \overset{?}{=} 10$	Substitute.		$2 + 5(-2) \overset{?}{=} -8$	Substitute.
$10 = 10$ ✓	True		$-8 = -8$ ✓	True

Both results are true, so the solution set is $\{(2, -2)\}$.

NOW TRY

NOW TRY ANSWERS
2. $\{(-3, 4)\}$
3. $\{(3, 2)\}$

 **CAUTION** When using the elimination method, remember to multiply *both* sides (not just one side) of an equation by the same nonzero number.

NOW TRY EXERCISE 4

Solve the system by the elimination method.

$$3x - 5y = 25$$
$$2x + 8y = -6$$

EXAMPLE 4 Using the Elimination Method

Solve the system by the elimination method.

$$2x + 3y = -15 \quad (1)$$
$$5x + 2y = 1 \quad (2)$$

Adding the two equations gives $7x + 5y = -14$, which does not eliminate either variable. However, we can multiply each equation by a suitable number so that the coefficients of one of the two variables are opposites. For example, to **eliminate x,** we multiply each side of $2x + 3y = -15$ (equation (1)) by 5 and each side of $5x + 2y = 1$ (equation (2)) by -2.

$$10x + 15y = -75 \qquad \text{Multiply } both\ sides \text{ of equation (1) by 5.}$$
$$\underline{-10x - 4y = -2} \qquad \text{Multiply } both\ sides \text{ of equation (2) by } -2.$$

The coefficients of x are opposites.

$$11y = -77 \qquad \text{Add.}$$
$$y = -7 \qquad \text{Divide by 11.}$$

Find the value of x by substituting -7 for y in either equation (1) or (2).

$$5x + 2y = 1 \qquad (2)$$
$$5x + 2(-7) = 1 \qquad \text{Let } y = -7.$$
$$5x - 14 = 1 \qquad \text{Multiply.}$$
$$5x = 15 \qquad \text{Add 14.}$$
$$x = 3 \qquad \text{Divide by 5.}$$

CHECK $2x + 3y = -15 \quad (1)$ $5x + 2y = 1 \quad (2)$

$$2(3) + 3(-7) \overset{?}{=} -15 \quad \text{Substitute.} \qquad 5(3) + 2(-7) \overset{?}{=} 1 \quad \text{Substitute.}$$
$$-15 = -15 \ \checkmark \ \text{True} \qquad\qquad\qquad 1 = 1 \ \checkmark \ \text{True}$$

The solution set of the system is $\{(3, -7)\}$. ← Write the x-value first.   **NOW TRY**

> **NOTE** In **Example 4**, we eliminated the variable x. Alternatively, we could multiply each equation of the system by a suitable number so that the variable y is eliminated.
>
> $$2x + 3y = -15 \quad (1) \xrightarrow{\text{Multiply by 2.}} 4x + 6y = -30$$
> $$5x + 2y = 1 \quad (2) \xrightarrow{\text{Multiply by } -3.} -15x - 6y = -3$$
>
> Complete this approach and confirm that the same solution results.

OBJECTIVE 3 Use an alternative method to find the second value in a solution.

Sometimes it is easier to find the value of the second variable in a solution using the elimination method twice.

> **NOTE** When the value of the first variable in a solution is a fraction, this alternative method helps avoid arithmetic errors. See **Example 5.** This method could be used to solve any system.

NOW TRY ANSWER
4. $\{(5, -2)\}$

NOW TRY
EXERCISE 5

Solve the system by the elimination method.

$$4x + 9y = 3$$
$$5y = 6 - 3x$$

EXAMPLE 5 Finding the Second Value Using an Alternative Method

Solve the system by the elimination method.

$$4x = 9 - 3y \qquad (1)$$
$$5x - 2y = 8 \qquad (2)$$

Write equation (1) in $Ax + By = C$ form by adding $3y$ to each side.

$$4x + 3y = 9 \qquad (3)$$
$$5x - 2y = 8 \qquad (2)$$

One way to proceed is to **eliminate y** by multiplying each side of equation (3) by 2 and each side of equation (2) by 3 and then adding.

$$8x + 6y = 18 \qquad \text{Multiply equation (3) by 2.}$$
$$\underline{15x - 6y = 24} \qquad \text{Multiply equation (2) by 3.}$$
$$23x \qquad = 42 \qquad \text{Add.}$$

The coefficients of y are opposites.

$$x = \frac{42}{23} \qquad \text{Divide by 23.}$$

Substituting $\frac{42}{23}$ for x in one of the given equations would give y, but the arithmetic would be complicated. Instead, solve for y by starting over again with the original equations written in $Ax + By = C$ form (equations (3) and (2)) and **eliminating x.**

$$20x + 15y = 45 \qquad \text{Multiply equation (3) by 5.}$$
$$\underline{-20x + 8y = -32} \qquad \text{Multiply equation (2) by } -4.$$
$$23y = 13 \qquad \text{Add.}$$

The coefficients of x are opposites.

$$y = \frac{13}{23} \qquad \text{Divide by 23.}$$

Check that the solution set is $\left\{\left(\frac{42}{23}, \frac{13}{23}\right)\right\}$.

NOW TRY

OBJECTIVE 4 Solve special systems by elimination.

EXAMPLE 6 Solving Special Systems Using the Elimination Method

Solve each system by the elimination method.

(a)
$$2x + 4y = 5 \qquad (1)$$
$$4x + 8y = -9 \qquad (2)$$

Multiply each side of equation (1) by -2. Then add the two equations.

$$-4x - 8y = -10 \qquad \text{Multiply equation (1) by } -2.$$
$$\underline{4x + 8y = -9} \qquad (2)$$
$$0 = -19 \qquad \text{False (contradiction)}$$

NOW TRY ANSWER

5. $\left\{\left(\frac{39}{7}, -\frac{15}{7}\right)\right\}$

The false statement $0 = -19$ indicates that the equations in the system have graphs that are parallel lines. The system has no solution, so the solution set is $\varnothing$.

NOW TRY
EXERCISE 6

Solve each system by the elimination method.

(a) $4x + 3y = 0$
     $-4x - 3y = -1$

(b) $x - y = 2$
     $5x - 5y = 10$

NOW TRY ANSWERS

6. (a) $\varnothing$ **(b)** $\{(x, y) \mid x - y = 2\}$

(b) $3x - y = 4$ (1)

                            $-9x + 3y = -12$ (2)

Multiply each side of equation (1) by 3. Then add the two equations.

               $9x - 3y = 12$ Multiply equation (1) by 3.

               $\underline{-9x + 3y = -12}$ (2)

                     $0 = 0$ True (identity)

A true statement occurs when the equations are equivalent. This indicates that every solution of one equation is also a solution of the other, so the system has an infinite number of solutions. The solution set is

$$\{(x, y) \mid 3x - y = 4\}.$$ **NOW TRY**

7.5 Exercises

FOR EXTRA HELP

▶ **MyLab Math**

▶ *Video solutions for select problems available in MyLab Math*

STUDY SKILLS REMINDER

Time management can be a challenge for students. **Review Study Skill 6, *Managing Your Time.***

Concept Check   *Answer* true *or* false. *If false, tell why.*

1. To eliminate the x-terms in this system, we multiply equation (1) by -2 and then add the result to equation (2).

               $3x + 5y = 7$ (1)
               $6x + 3y = -10$ (2)

2. To eliminate the y-terms in this system, we multiply equation (2) by 3 and then add the result to equation (1).

               $2x + 12y = 7$ (1)
               $3x + 4y = 1$ (2)

3. Concept Check   A student solved the following system by elimination.

               $x + y = 1$ (1)
               $-x - y = 2$ (2)

He obtained the false statement $0 = 3$ and incorrectly gave $\{(0, 3)\}$ as the solution set.   **WHAT WENT WRONG?**   Give the correct solution set.

4. Concept Check   To eliminate the y-terms in the system

               $2x - y = 5$ (1)
               $-6x + 3y = -15,$ (2)

a student began by multiplying equation (1) by 3 to obtain $6x - 3y = 5$.   **WHAT WENT WRONG?**   Solve correctly and give the solution set.

Solve each system by the elimination method. Check each solution. ***See Examples 1 and 2.***

5. $x - y = -2$
   $x + y = 10$

6. $x + y = 10$
   $x - y = -6$

7. $2x + y = -5$
   $x - y = 2$

8. $2x + y = -15$
   $-x - y = 10$

9. $x + 2y = 11$
   $-x + 3y = 4$

10. $-x - 4y = 10$
   $x - 4y = 14$

11. $2y = -3x$
   $-3x - y = 3$

12. $5x = y + 5$
   $-5x + 2y = 0$

13. $6x - y = -1$
   $5y = 17 + 6x$

14. $y = 9 - 6x$
   $-6x + 3y = 15$

15. $2x - 6 = -3y$
   $5x - 3y = -27$

16. $x - 2 = -y$
   $2x = y + 10$

Solve each system by the elimination method. Check each solution. ***See Examples 3–6.***

17. $2x - y = 12$
   $3x + 2y = -3$

18. $x + y = 3$
   $-3x + 2y = -19$

19. $3x - 4y = -17$
   $x + 3y = 3$

20. $5x - 4y = -1$
   $x + 8y = -9$

21. $x + 4y = 16$
   $3x + 5y = 20$

22. $2x + y = 8$
   $5x - 2y = -16$

23. $5x - 3y = -20$
   $-3x + 6y = 12$

24. $4x + 3y = -28$
   $5x - 6y = -35$

25. $2x - 8y = 0$
$4x + 5y = 0$

26. $3x - 15y = 0$
$6x + 10y = 0$

27. $3x + 3y = 33$
$5x - 2y = 27$

28. $4x - 3y = -19$
$3x + 2y = 24$

29. $3x - 7 = -5y$
$5x + 4y = -10$

30. $2x + 3y = 13$
$6 + 2y = -5x$

31. $5x + 4y = 12$
$3x + 5y = 15$

32. $2x + 3y = 21$
$5x - 2y = -14$

33. $5x - 4y = 15$
$-3x + 6y = -9$

34. $4x + 5y = -16$
$5x - 6y = -20$

35. $5x - 2y = 3$
$10x - 4y = 5$

36. $3x - 5y = 1$
$6x - 10y = 4$

37. $-x + 3y = 4$
$-2x + 6y = 8$

38. $6x - 2y = 24$
$-3x + y = -12$

39. $x + 3y = 6$
$-2x + 12 = 6y$

40. $7x + 2y = 0$
$4y = -14x$

41. $4x - 3y = 1$
$8x = 3 + 6y$

42. $5x + 8y = 10$
$24y = -15x - 10$

43. $4x = 3y - 2$
$5x + 3 = 2y$

44. $2x + 3y = 0$
$4x + 12 = 9y$

45. $24x + 12y = -7$
$16x - 18y = 17$

46. $9x + 4y = -3$
$6x + 6y = -7$

47. $3x = 3 + 2y$
$-\dfrac{4}{3}x + y = \dfrac{1}{3}$

48. $3x = 27 + 2y$
$x - \dfrac{7}{2}y = -25$

49. $\dfrac{1}{5}x + y = \dfrac{6}{5}$
$\dfrac{1}{10}x + \dfrac{1}{3}y = \dfrac{5}{6}$

50. $\dfrac{1}{3}x + \dfrac{1}{2}y = \dfrac{13}{6}$
$\dfrac{1}{2}x - \dfrac{1}{4}y = -\dfrac{3}{4}$

51. $2.4x + 1.7y = 7.6$
$1.2x - 0.5y = 9.2$

52. $0.5x + 3.4y = 13$
$1.5x - 2.6y = -25$

RELATING CONCEPTS For Individual or Group Work (Exercises 53–56)

The graph shows average U.S. movie theater ticket prices from 2006 through 2016. In 2006, the average price was \$6.55, as represented on the graph by the point P(2006, 6.55). In 2016, the average price was \$8.65, as represented on the graph by the point Q(2016, 8.65). **Work Exercises 53–56 in order.**

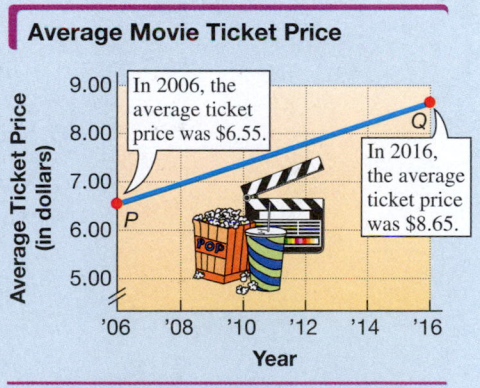

Average Movie Ticket Price

In 2006, the average ticket price was \$6.55.

In 2016, the average ticket price was \$8.65.

Data from National Association of Theatre Owners.

53. Line segment PQ has an equation that can be written in the form $y = ax + b$. Using the coordinates of point P with $x = 2006$ and $y = 6.55$, write an equation in the variables a and b.

54. Using the coordinates of point Q with $x = 2016$ and $y = 8.65$, write a second equation in the variables a and b.

55. Write the system of equations formed from the two equations in **Exercises 53 and 54.** Solve the system, giving the values of a and b. (*Hint:* Eliminate b using the elimination method.)

56. Answer each of the following.

  (a) What is the equation of the line on which segment PQ lies?

  (b) Let $x = 2015$ in the equation from part (a), and solve for y. How does the result compare with the actual figure of \$8.43?

SUMMARY EXERCISES   Applying Techniques for Solving Systems of Linear Equations

We have introduced two algebraic methods for solving systems of linear equations. We compare them in the next example.

NOW TRY EXERCISE

Consider the following system.

$$x - 2y = 2$$
$$4x - 5y = -4$$

(a) Solve by the substitution method.

(b) Solve by the elimination method.

EXAMPLE   Comparing the Substitution and Elimination Methods

Consider the following system.

$$3x + y = -2 \quad (1)$$
$$5x + 2y = 4 \quad (2)$$

(a) Solve by the substitution method.

Solve equation (1) for y.

$$y = -2 - 3x \qquad \text{Subtract } 3x.$$

Substitute $-2 - 3x$ for y in equation (2).

$$5x + 2y = 4 \qquad (2)$$
$$5x + 2(-2 - 3x) = 4 \qquad \text{Let } y = -2 - 3x.$$
$$5x - 4 - 6x = 4 \qquad \text{Multiply.}$$
$$-x = 8 \qquad \text{Combine like terms. Add 4.}$$
$$x = -8 \qquad \text{Multiply by } -1.$$

Substitute -8 for x in $y = -2 - 3x$ (that is, equation (1) solved for y).

$$y = -2 - 3(-8) \qquad \text{Let } x = -8.$$
$$y = -2 + 24 \qquad \text{Multiply.}$$
$$y = 22 \qquad \text{Subtract.}$$

Check that the solution set is $\{(-8, 22)\}$.

(b) Solve by the elimination method.

Multiply equation (1) by -2 and add the result to equation (2) to eliminate the y-terms.

$$-6x - 2y = 4 \qquad \text{Multiply equation (1) by } -2.$$
$$\underline{5x + 2y = 4} \qquad (2)$$
$$-x = 8 \qquad \text{Add.}$$
$$x = -8 \qquad \text{Multiply by } -1.$$

Substitute -8 for x in either of the original equations.

$$3x + y = -2 \qquad (1)$$
$$3(-8) + y = -2 \qquad \text{Let } x = -8.$$
$$-24 + y = -2 \qquad \text{Multiply.}$$
$$y = 22 \qquad \text{Add 24.}$$

Check that the solution set is $\{(-8, 22)\}$.   **NOW TRY**

Some systems are more easily solved using one method than the other.

Guidelines for Choosing a Method to Solve a System of Linear Equations

1. If one of the equations of the system is already solved for one of the variables, the substitution method is the better choice.

$$3x + 4y = 9 \qquad\qquad x = 3y - 7$$
$$\text{and}$$
$$y = 2x - 6 \qquad\qquad -5x + 3y = 9$$

2. If both equations are in $Ax + By = C$ form and none of the variables has coefficient -1 or 1, the elimination method is the better choice.

$$4x - 11y = 3$$
$$-2x + 3y = 4$$

3. If one or both of the equations are in $Ax + By = C$ form and the coefficient of one of the variables is -1 or 1, either method is appropriate.

$$3x + y = -2$$
$$5x + 2y = 4$$

This system is solved by both methods in the preceding example for comparison.

Concept Check   *Use the preceding guidelines to solve each problem.*

1. To minimize the amount of work required, determine whether to use the substitution or elimination method to solve each system, and why. *Do not actually solve.*

   (a) $3x + 5y = 69$
   $\quad\quad y = 4x$

   (b) $3x + y = -7$
   $\quad\quad x - y = -5$

   (c) $3x - 2y = 0$
   $\quad\quad 9x + 8y = 7$

2. Which system would be easier to solve with the substitution method? Why?

   System A: $5x - 3y = 7$
   $\quad\quad\quad\quad 2x + 8y = 3$

   System B: $7x + 2y = 4$
   $\quad\quad\quad\quad y = -3x + 1$

*Solve each system by any method. (For Exercises 3–5, see the answers to **Exercise 1**.)*

3. $3x + 5y = 69$
   $\quad y = 4x$

4. $3x + y = -7$
   $\quad x - y = -5$

5. $3x - 2y = 0$
   $\quad 9x + 8y = 7$

6. $x + y = 7$
   $\quad x = -3 - y$

7. $6x + 7y = 4$
   $\quad 5x + 8y = -1$

8. $6x - y = 5$
   $\quad y = 11x$

9. $\quad 4x - 6y = 10$
   $-10x + 15y = -25$

10. $3x - 5y = 7$
    $\quad 2x + 3y = 30$

11. $5x = 7 + 2y$
    $\quad 5y = 5 - 3x$

12. $4x + 3y = 1$
    $\quad 3x + 2y = 2$

13. $\quad 2x - 3y = 7$
    $-4x + 6y = 14$

14. $\quad x - 3y = 7$
    $\quad 4x + y = 5$

15. $7x - 4y = 0$
    $\quad 3x = 2y$

16. $\quad 5x - 4y = 15$
    $-3x + 6y = -9$

17. $3x = 7 - y$
    $\quad 2y = 14 - 6x$

18. $\dfrac{1}{5}x + \dfrac{2}{3}y = -\dfrac{8}{5}$
    $\quad 3x - y = 9$

19. $\dfrac{1}{6}x + \dfrac{1}{6}y = 2$
    $\quad -\dfrac{1}{2}x - \dfrac{1}{3}y = -8$

20. $\dfrac{x}{3} - \dfrac{3y}{4} = -\dfrac{1}{2}$
    $\quad \dfrac{x}{6} + \dfrac{y}{8} = \dfrac{3}{4}$

21. $0.1x + y = 1.6$
    $\quad 0.6x + 0.5y = -1.4$

22. $0.2x - 0.3y = 0.1$
    $\quad 0.3x - 0.2y = 0.9$

7.6   Systems of Linear Equations in Three Variables

A solution of an equation in three variables, such as

$$2x + 3y - z = 4, \quad \text{Linear equation in three variables}$$

is an **ordered triple** and is written (x, y, z). For example, the ordered triple $(0, 1, -1)$ is a solution of the preceding equation, because

$$2(0) + 3(1) - (-1) = 4 \quad \text{is a true statement.}$$

Verify that another solution of this equation is $(10, -3, 7)$.

We now extend the term *linear equation* to equations of the form

$$Ax + By + Cz + \cdots + Dw = K,$$

where not all the coefficients $A, B, C, \ldots, D$ equal 0. For example,

$$2x + 3y - 5z = 7 \quad \text{and} \quad x - 2y - z + 3w = 8$$

are linear equations, the first with three variables and the second with four.

OBJECTIVE 1 Understand the geometry of systems of three equations in three variables.

Consider the solution of a system such as the following.

$$\begin{aligned} 4x + 8y + z &= 2 \\ x + 7y - 3z &= -14 \\ 2x - 3y + 2z &= 3 \end{aligned} \quad \begin{array}{l}\text{System of linear equations} \\ \text{in three variables}\end{array}$$

Theoretically, a system of this type can be solved by graphing. However, the graph of a linear equation with three variables is a *plane*, not a line. Because visualizing a plane requires three-dimensional graphing, the method of graphing is not practical with these systems. However, it does illustrate the number of solutions possible for such systems, as shown in **FIGURE 48**.

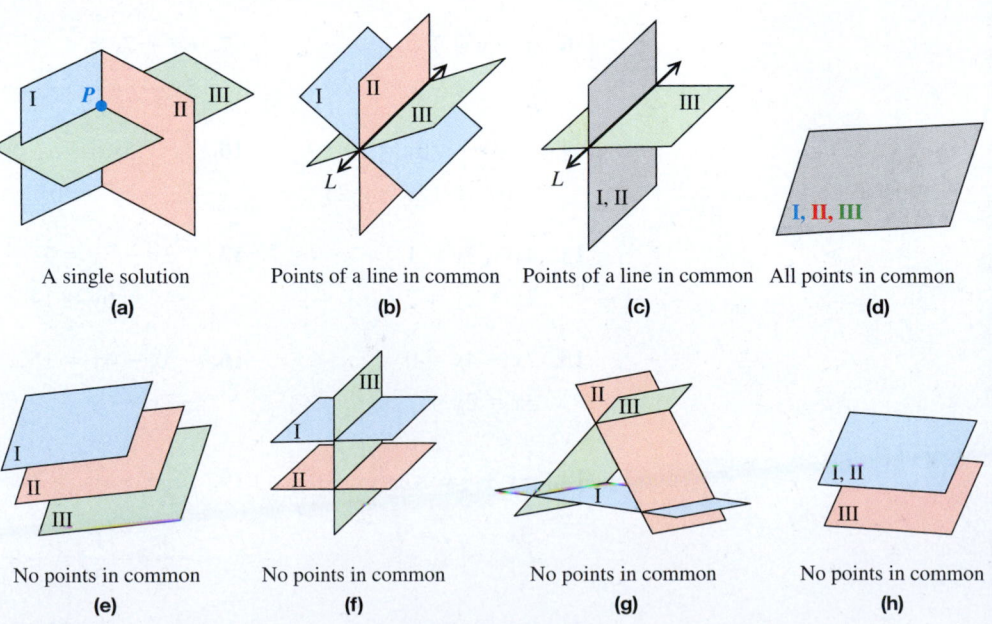

A single solution

(a)

Points of a line in common

(b)

Points of a line in common

(c)

All points in common

(d)

No points in common

(e)

No points in common

(f)

No points in common

(g)

No points in common

(h)

FIGURE 48

FIGURE 48 illustrates the following cases.

Graphs of Linear Systems in Three Variables

Case 1 **The three planes may meet at a single, common point.**

This point is the solution of the system. See **FIGURE 48(a)**.

Case 2 **The three planes may have the points of a line in common.**

The infinite set of points that satisfy the equation of the line is the solution of the system. See **FIGURES 48(b) and (c)**.

Case 3 **The three planes may coincide.**

The solution of the system is the set of all points on a plane. See **FIGURE 48(d)**.

Case 4 **The planes may have no points common to all three.**

There is no solution of the system. See **FIGURES 48(e)–(h)**.

OBJECTIVE 2 Solve linear systems (with three equations and three variables) by elimination.

Because graphing to find the solution set of a system of three equations in three variables is impractical, these systems are solved with an extension of the elimination method.

In the steps that follow, we use the term **focus variable** to identify the first variable to be eliminated. The focus variable will always be present in the **working equation**, which will be used twice to eliminate this variable.

Solving a Linear System in Three Variables

Step 1 **Select a variable and an equation.** A good choice for the variable, which we call the *focus variable,* is one that has coefficient 1 or -1. Then select an equation, one that contains the focus variable, as the *working equation.*

Step 2 **Eliminate the focus variable.** Use the working equation and one of the other two equations of the original system. The result is an equation in two variables.

Step 3 **Eliminate the focus variable again.** Use the working equation and the remaining equation of the original system. The result is another equation in two variables.

Step 4 **Write the equations in two variables that result from Steps 2 and 3 as a system, and solve it.** Doing this gives the values of two of the variables.

Step 5 **Find the value of the remaining variable.** Substitute the values of the two variables found in Step 4 into the working equation to obtain the value of the focus variable.

Step 6 **Check** the three values in *each* of the *original* equations of the system. Then write the solution set as a set containing an ordered triple.

EXAMPLE 1   Solving a System in Three Variables

Solve the system.

$$
\begin{aligned}
4x + 8y + z &= 2 \qquad &(1)\\
x + 7y - 3z &= -14 \qquad &(2)\\
2x - 3y + 2z &= 3 \qquad &(3)
\end{aligned}
$$

Step 1   Because z in equation (1) has coefficient 1, we choose z as the focus variable and (1) as the working equation. (Another option would be to choose x as the focus variable—it also has coefficient 1—and use (2) as the working equation.)

Focus variable

$$4x + 8y + z = 2 \qquad (1) \leftarrow \text{Working equation}$$

Step 2   Multiply working equation (1) by 3 and add the result to equation (2).

$$
\begin{aligned}
12x + 24y + 3z &= 6 \qquad &\text{Multiply each side of (1) by 3.}\\
x + 7y - 3z &= -14 \qquad &(2)\\
\hline
13x + 31y \phantom{{}+3z} &= -8 \qquad &\text{Add.} \quad (4)
\end{aligned}
$$

Focus variable z was eliminated.

Step 3   Multiply working equation (1) by -2 and add the result to remaining equation (3) to again eliminate focus variable z.

$$
\begin{aligned}
-8x - 16y - 2z &= -4 \qquad &\text{Multiply each side of (1) by } -2.\\
2x - 3y + 2z &= 3 \qquad &(3)\\
\hline
-6x - 19y \phantom{{}+2z} &= -1 \qquad &\text{Add.} \quad (5)
\end{aligned}
$$

Focus variable z was eliminated.

Step 4   Write the equations in two variables that result in Steps 2 and 3 as a system.

Make sure these equations have the same two variables.

$$
\begin{aligned}
13x + 31y &= -8 \qquad &(4) \quad \text{The result from Step 2}\\
-6x - 19y &= -1 \qquad &(5) \quad \text{The result from Step 3}
\end{aligned}
$$

Now solve this system. We choose to eliminate x.

$$
\begin{aligned}
78x + 186y &= -48 \qquad &\text{Multiply each side of (4) by 6.}\\
-78x - 247y &= -13 \qquad &\text{Multiply each side of (5) by 13.}\\
\hline
-61y &= -61 \qquad &\text{Add.}\\
y &= 1 \qquad &\text{Divide by } -61.
\end{aligned}
$$

Substitute 1 for y in either equation (4) or equation (5) to find x.

$$
\begin{aligned}
-6x - 19y &= -1 \qquad &(5)\\
-6x - 19(1) &= -1 \qquad &\text{Let } y = 1.\\
-6x - 19 &= -1 \qquad &\text{Multiply.}\\
-6x &= 18 \qquad &\text{Add 19.}\\
x &= -3 \qquad &\text{Divide by } -6.
\end{aligned}
$$

Step 5   Now substitute the two values we found in Step 4 in working equation (1) to find the value of the remaining variable, focus variable z.

$$
\begin{aligned}
4x + 8y + z &= 2 \qquad &(1)\\
4(-3) + 8(1) + z &= 2 \qquad &\text{Let } x = -3 \text{ and } y = 1.\\
-4 + z &= 2 \qquad &\text{Multiply, and then add.}\\
z &= 6 \qquad &\text{Add 4.}
\end{aligned}
$$

NOW TRY
EXERCISE 1
Solve the system.

$$x - y + 2z = 1$$
$$3x + 2y + 7z = 8$$
$$-3x - 4y + 9z = -10$$

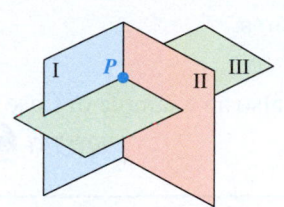

A single solution

FIGURE 48(a) (repeated)

> Write the values of x, y, and z in the correct order.

Step 6 It appears that $(-3, 1, 6)$ is the only solution of the system. We must check that this ordered triple satisfies all three original equations of the system. We begin with equation (1).

CHECK

$$4x + 8y + z = 2 \qquad (1)$$
$$4(-3) + 8(1) + 6 \overset{?}{=} 2 \qquad \text{Substitute.}$$
$$-12 + 8 + 6 \overset{?}{=} 2 \qquad \text{Multiply.}$$
$$2 = 2 \ \checkmark \ \text{True}$$

Because $(-3, 1, 6)$ also satisfies equations (2) and (3), the solution set contains a single ordered triple, $\{(-3, 1, 6)\}$. This is Case 1, shown earlier in **FIGURE 48(a)**.

NOW TRY

OBJECTIVE 3 Solve linear systems (with three equations and three variables) in which some of the equations have missing terms.

If a linear system includes an equation that is missing a term or terms, one elimination step can be omitted.

EXAMPLE 2 Solving a System of Equations with Missing Terms

Solve the system.

$$6x - 12y = -5 \qquad (1) \quad \text{Missing } z$$
$$8y + z = 0 \qquad (2) \quad \text{Missing } x$$
$$9x - z = 12 \qquad (3) \quad \text{Missing } y$$

Equation (1) is missing the variable z, so one way to begin is to use z as the focus variable and eliminate z again using equations (2) and (3).

$$8y + z = 0 \qquad (2)$$
$$\underline{9x \qquad - z = 12} \qquad (3)$$
$$9x + 8y \qquad = 12 \qquad \text{Add.} \quad (4)$$

Use resulting equation (4) in x and y, together with equation (1), $6x - 12y = -5$, to eliminate x.

$$18x - 36y = -15 \qquad \text{Multiply each side of (1) by 3.}$$
$$\underline{-18x - 16y = -24} \qquad \text{Multiply each side of (4) by } -2.$$
$$-52y = -39 \qquad \text{Add.}$$
$$y = \frac{-39}{-52} \qquad \text{Divide by } -52.$$
$$y = \frac{3}{4} \qquad \text{Write in lowest terms.}$$

We can find z by substituting this value for y in equation (2).

$$8y + z = 0 \qquad (2)$$
$$8\left(\frac{3}{4}\right) + z = 0 \qquad \text{Let } y = \tfrac{3}{4}.$$
$$6 + z = 0 \qquad \text{Multiply.}$$
$$z = -6 \qquad \text{Subtract 6.}$$

NOW TRY ANSWER
1. $\{(2, 1, 0)\}$

**NOW TRY
EXERCISE 2**

Solve the system.

$3x - z = -10$
$4y + 5z = 24$
$x - 6y = -8$

We can find x by substituting -6 for z in equation (3).

$$9x - z = 12 \quad \text{(3)}$$
$$9x - (-6) = 12 \quad \text{Let } z = -6.$$
$$9x + 6 = 12 \quad -(-a) = a$$
$$x = \frac{6}{9} \quad \text{Subtract 6. Divide by 9.}$$
$$x = \frac{2}{3} \quad \text{Write in lowest terms.}$$

Check to verify that the solution set is $\left\{\left(\frac{2}{3}, \frac{3}{4}, -6\right)\right\}$. This is also an example of Case 1.

NOW TRY

> **NOTE** Another way to solve the system in **Example 2** is to begin by eliminating the variable y from equations (1) and (2). The resulting equation together with equation (3) forms a system of two equations in the variables x and z. Try working **Example 2** this way to see that the same solution results. There are often multiple ways to solve a system of equations. Some ways may involve more work than others.

OBJECTIVE 4 Solve special systems.

**NOW TRY
EXERCISE 3**

Solve the system.

$x - 3y + 2z = 10$
$-2x + 6y - 4z = -20$
$\frac{1}{2}x - \frac{3}{2}y + z = 5$

EXAMPLE 3   Solving a System of Dependent Equations with Three Variables

Solve the system.

$$2x - 3y + 4z = 8 \quad \text{(1)}$$
$$-x + \frac{3}{2}y - 2z = -4 \quad \text{(2)}$$
$$6x - 9y + 12z = 24 \quad \text{(3)}$$

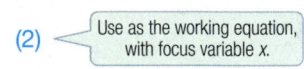
Use as the working equation, with focus variable x.

Eliminate focus variable x using equations (1) and (2).

$$2x - 3y + 4z = 8 \quad \text{(1)}$$
$$\underline{-2x + 3y - 4z = -8} \quad \text{Multiply each side of (2) by 2.}$$
$$0 = 0 \quad \text{True (identity)}$$

Eliminating x from equations (2) and (3) gives the same result.

$$-6x + 9y - 12z = -24 \quad \text{Multiply each side of (2) by 6.}$$
$$\underline{6x - 9y + 12z = 24} \quad \text{(3)}$$
$$0 = 0 \quad \text{True (identity)}$$

I, II, III

All points in common

FIGURE 48(d)   (repeated)

When solving a system such as this, attempting to eliminate one variable results in elimination of *all* variables. The equations are dependent—that is, they are equivalent forms of the *same* equation—and have the same graph. This is Case 3, as illustrated in **FIGURE 48(d)**. The solution set is written

$$\{(x, y, z) \mid 2x - 3y + 4z = 8\}. \quad \text{Set-builder notation}$$

Although any one of the three equations could be used to write the solution set, we use the equation with integer coefficients having greatest common factor 1 and positive coefficient of x, as in the previous section.

NOW TRY

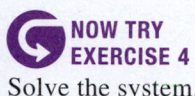

NOW TRY
EXERCISE 4

Solve the system.

$$x - 5y + 2z = 4$$
$$3x + y - z = 6$$
$$-2x + 10y - 4z = 7$$

EXAMPLE 4 Solving an Inconsistent System with Three Variables

Solve the system.

$$2x - 4y + 6z = 5 \quad \text{(1)}$$
$$-x + 3y - 2z = -1 \quad \text{(2)}$$
$$x - 2y + 3z = 1 \quad \text{(3)}$$

Use as the working equation, with focus variable x.

Eliminate the focus variable, x, using equations (1) and (3).

$$-2x + 4y - 6z = -2 \quad \text{Multiply each side of (3) by } -2.$$
$$\underline{2x - 4y + 6z = 5 \quad \text{(1)}}$$
$$0 = 3 \quad \text{False (contradiction)}$$

The resulting false statement indicates that equations (1) and (3) have no common solution. Thus, the system is inconsistent and the solution set is $\varnothing$. The graph of this system would show two planes parallel to one another, and a third plane that intersects both, as in **FIGURE 48(f)**. This is Case 4.

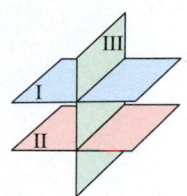

No points in common

FIGURE 48(f) (repeated)

NOW TRY

> **NOTE** If a false statement results when adding as in **Example 4,** it is not necessary to go any further with the solution. Because two of the three planes are parallel, it is not possible for the three planes to have any points in common.

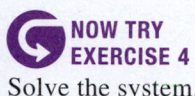

NOW TRY
EXERCISE 5

Solve the system.

$$x - 3y + 2z = 4$$
$$\frac{1}{3}x - y + \frac{2}{3}z = 7$$
$$\frac{1}{2}x - \frac{3}{2}y + z = 2$$

EXAMPLE 5 Solving Another Special System

Solve the system.

$$2x - y + 3z = 6 \quad \text{(1)}$$
$$x - \frac{1}{2}y + \frac{3}{2}z = 3 \quad \text{(2)}$$
$$4x - 2y + 6z = 1 \quad \text{(3)}$$

Equations (1) and (2) are equivalent. If we multiply each side of equation (2) by 2, we obtain equation (1). These two equations are dependent and have the same graph.

Equations (1) and (3) are *not* equivalent, however. If we multiply each side of equation (3) by $\frac{1}{2}$, we obtain

$$2x - y + 3z = \frac{1}{2}.$$

This equation has the same coefficients as equation (1), but a different constant term. Therefore, the graphs of equations (1) and (3) have *no* points in common—that is, the planes are parallel.

Thus, this system is inconsistent and the solution set is $\varnothing$, as illustrated in **FIGURE 48(h)**. This is another example of Case 4.

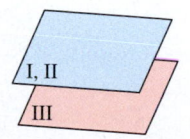

No points in common

FIGURE 48(h) (repeated)

NOW TRY

NOW TRY ANSWERS

4. $\varnothing$

5. $\varnothing$

7.6 Exercises

  MyLab Math

▶ *Video solutions for select problems available in MyLab Math*

Concept Check *Answer each of the following.*

1. Using your immediate surroundings, give an example of three planes that satisfy the condition.

   (a) They intersect in a single point.

   (b) They do not intersect.

   (c) They intersect in infinitely many points.

2. Suppose that a system has infinitely many ordered-triple solutions of the form (x, y, z) such that

$$x + y + 2z = 1.$$

Give three specific ordered triples that are solutions of the system.

3. Explain what the following statement means: "*The solution set of the following system is* $\{(-1, 2, 3)\}$."

$$2x + y + z = 3$$
$$3x - y + z = -2$$
$$4x - y + 2z = 0$$

4. The following two equations have a common solution of $(1, 2, 3)$.

$$x + y + z = 6$$
$$2x - y + z = 3$$

Which equation would complete a system of three linear equations in three variables having solution set $\{(1, 2, 3)\}$?

   A. $3x + 2y - z = 1$ **B.** $3x + 2y - z = 4$

   C. $3x + 2y - z = 5$ **D.** $3x + 2y - z = 6$

5. What constant should replace the question mark in this system so that the solution set is $\{(1, 1, 1)\}$?

$$2x - 3y + z = 0$$
$$-5x + 2y - z = -4$$
$$x + y + 2z = \;?$$

6. Complete the work of **Example 1** and show that the ordered triple $(-3, 1, 6)$ is also a solution of equations (2) and (3).

$$x + 7y - 3z = -14 \qquad \text{Equation (2)}$$
$$2x - 3y + 2z = 3 \qquad \text{Equation (3)}$$

*Solve each system. **See Example 1.***

7. $2x - 5y + 3z = -1$
     $x + 4y - 2z = 9$
     $x - 2y - 4z = -5$

8. $x + 3y - 6z = 1$
     $2x - y + z = 7$
     $x + 2y + 2z = 14$

9. $3x + 2y + z = 8$
     $2x - 3y + 2z = -16$
     $x + 4y - z = 20$

10. $-3x + y - z = -10$
     $-4x + 2y + 3z = -1$
     $2x + 3y - 2z = -5$

11. $2x + 5y + 2z = 0$
     $4x - 7y - 3z = 1$
     $3x - 8y - 2z = -6$

12. $5x - 2y + 3z = -9$
     $4x + 3y + 5z = 4$
     $2x + 4y - 2z = 14$

13. $x + 2y + z = 4$
     $2x + y - z = -1$
     $x - y - z = -2$

14. $x - 2y + 5z = -7$
     $-2x - 3y + 4z = -14$
     $-3x + 5y - z = -7$

15. $-x + 2y + 6z = 2$
     $3x + 2y + 6z = 6$
     $x + 4y - 3z = 1$

16. $2x + y + 2z = 1$
$x + 2y + z = 2$
$x - y - z = 0$

17. $x + y - z = -2$
$2x - y + z = -5$
$-x + 2y - 3z = -4$

18. $x + 2y + 3z = 1$
$-x - y + 3z = 2$
$-6x + y + z = -2$

19. $\dfrac{1}{3}x + \dfrac{1}{6}y - \dfrac{2}{3}z = -1$
$-\dfrac{3}{4}x - \dfrac{1}{3}y - \dfrac{1}{4}z = 3$
$\dfrac{1}{2}x + \dfrac{3}{2}y + \dfrac{3}{4}z = 21$

20. $\dfrac{2}{3}x - \dfrac{1}{4}y + \dfrac{5}{8}z = 0$
$\dfrac{1}{5}x + \dfrac{2}{3}y - \dfrac{1}{4}z = -7$
$-\dfrac{3}{5}x + \dfrac{4}{3}y - \dfrac{7}{8}z = -5$

21. $5.5x - 2.5y + 1.6z = 11.83$
$2.2x + 5.0y - 0.1z = -5.97$
$3.3x - 7.5y + 3.2z = 21.25$

22. $6.2x - 1.4y + 2.4z = -1.80$
$3.1x + 2.8y - 0.2z = 5.68$
$9.3x - 8.4y - 4.8z = -34.20$

Solve each system. ***See Example 2.***

23. $2x - 3y + 2z = -1$
$x + 2y + z = 17$
$2y - z = 7$

24. $2x - y + 3z = 6$
$x + 2y - z = 8$
$2y + z = 1$

25. $4x + 2y - 3z = 6$
$x - 4y + z = -4$
$-x + 2z = 2$

26. $2x + 3y - 4z = 4$
$x - 6y + z = -16$
$-x + 3z = 8$

27. $2x + y = 6$
$3y - 2z = -4$
$3x - 5z = -7$

28. $4x - 8y = -7$
$4y + z = 7$
$-8x + z = -4$

29. $-5x + 2y + z = 5$
$-3x - 2y - z = 3$
$-x + 6y = 1$

30. $-4x + 3y - z = 4$
$-5x - 3y + z = -4$
$-2x - 3z = 12$

31. $7x - 3z = -34$
$2y + 4z = 20$
$\dfrac{3}{4}x + \dfrac{1}{6}y = -2$

32. $5x - 2z = 8$
$4y + 3z = -9$
$\dfrac{1}{2}x + \dfrac{2}{3}y = -1$

33. $4x - z = -6$
$\dfrac{3}{5}y + \dfrac{1}{2}z = 0$
$\dfrac{1}{3}x + \dfrac{2}{3}z = -5$

34. $5x - z = 38$
$\dfrac{2}{3}y + \dfrac{1}{4}z = -17$
$\dfrac{1}{5}y + \dfrac{5}{6}z = 4$

Solve each system. If the system is inconsistent or has dependent equations, say so. ***See Examples 1, 3, 4, and 5.***

35. $2x + 2y - 6z = 5$
$-3x + y - z = -2$
$-x - y + 3z = 4$

36. $-2x + 5y + z = -3$
$5x + 14y - z = -11$
$7x + 9y - 2z = -5$

37. $-5x + 5y - 20z = -40$
$x - y + 4z = 8$
$3x - 3y + 12z = 24$

38. $x + 4y - z = 3$
$-2x - 8y + 2z = -6$
$3x + 12y - 3z = 9$

39. $x + 5y - 2z = -1$
$-2x + 8y + z = -4$
$3x - y + 5z = 19$

40. $x + 3y + z = 2$
$4x + y + 2z = -4$
$5x + 2y + 3z = -2$

41. $2x + y - z = 6$
$4x + 2y - 2z = 12$
$-x - \dfrac{1}{2}y + \dfrac{1}{2}z = -3$

42. $2x - 8y + 2z = -10$
$-x + 4y - z = 5$
$\dfrac{1}{8}x - \dfrac{1}{2}y + \dfrac{1}{8}z = -\dfrac{5}{8}$

43. $x + y - 2z = 0$
$3x - y + z = 0$
$4x + 2y - z = 0$

44. $2x + 3y - z = 0$

$x - 4y + 2z = 0$

$3x - 5y - z = 0$

45. $x - 2y + \dfrac{1}{3}z = 4$

$3x - 6y + z = 12$

$-6x + 12y - 2z = -3$

46. $4x + y - 2z = 3$

$x + \dfrac{1}{4}y - \dfrac{1}{2}z = \dfrac{3}{4}$

$2x + \dfrac{1}{2}y - z = 1$

Extending Skills *Extend the method of this section to solve each system. Express the solution in the form* (x, y, z, w).

47. $x + y + z - w = 5$

$2x + y - z + w = 3$

$x - 2y + 3z + w = 18$

$-x - y + z + 2w = 8$

48. $3x + y - z + 2w = 9$

$x + y + 2z - w = 10$

$x - y - z + 3w = -2$

$-x + y - z + w = -6$

49. $3x + y - z + w = -3$

$2x + 4y + z - w = -7$

$-2x + 3y - 5z + w = 3$

$5x + 4y - 5z + 2w = -7$

50. $x - 3y + 7z + w = 11$

$2x + 4y + 6z - 3w = -3$

$3x + 2y + z + 2w = 19$

$4x + y - 3z + w = 22$

RELATING CONCEPTS For Individual or Group Work (Exercises 51–55)

A circle *has an equation of the following form.*

$$x^2 + y^2 + ax + by + c = 0 \qquad \text{Equation of a circle}$$

*It is a fact from geometry that given three **non-collinear** points—that is, points that do not all lie on the same straight line—there will be a circle that contains them. For example, the points* $(4, 2), (-5, -2),$ *and* $(0, 3)$ *lie on the circle whose equation is shown in the figure.*

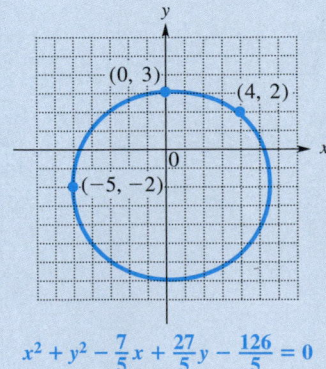

Work Exercises 51–55 in order, *to find an equation of the circle passing through the points*

$$(2, 1), \quad (-1, 0), \quad \text{and} \quad (3, 3).$$

51. Let $x = 2$ and $y = 1$ in the general equation $x^2 + y^2 + ax + by + c = 0$ to find an equation in a, b, and c.

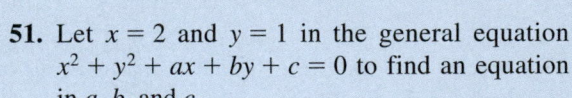

$$x^2 + y^2 - \tfrac{7}{5}x + \tfrac{27}{5}y - \tfrac{126}{5} = 0$$

52. Let $x = -1$ and $y = 0$ to find a second equation in a, b, and c.

53. Let $x = 3$ and $y = 3$ to find a third equation in a, b, and c.

54. Form a system of three equations using the answers from **Exercises 51–53.** Solve the system to find the values of a, b, and c.

55. Use the values of a, b, and c from **Exercise 54** and the form of the equation of a circle given above to write an equation of the circle passing through the given points.

7.7 Applications of Systems of Linear Equations

OBJECTIVES

1 Solve geometry problems using two variables.

2 Solve money problems using two variables.

3 Solve mixture problems using two variables.

4 Solve distance-rate-time problems using two variables.

5 Solve problems with three variables using a system of three equations.

Although some problems with two unknowns can be solved using just one variable, it is often easier to use two variables and a system of equations. The following problem, which can be solved with a system, appeared in a Hindu work that dates back to about a.d. 850. **(See Exercise 24.)**

The mixed price of 9 citrons (a lemonlike fruit) and 7 fragrant wood apples is 107; again, the mixed price of 7 citrons and 9 fragrant wood apples is 101. O you arithmetician, tell me quickly the price of a citron and the price of a wood apple here, having distinctly separated those prices well.

PROBLEM-SOLVING HINT When solving an applied problem using two variables, it is a good idea to pick letters that correspond to the descriptions of the unknown quantities. In the example above, we could choose

$$c \text{ to represent the number of citrons,}$$

and $$w \text{ to represent the number of wood apples.}$$

The following steps are based on the problem-solving method presented earlier in the text.

Solving an Applied Problem Using a System of Equations

Step 1 **Read** the problem carefully. *What information is given? What is to be found?*

Step 2 **Assign variables** to represent the unknown values. Write down what each variable represents. Make a sketch, diagram, or table, as needed.

Step 3 **Write a system of equations** using all the variables.

Step 4 **Solve** the system of equations.

Step 5 **State the answer**. Label it appropriately. *Does it seem reasonable?*

Step 6 **Check** the answer in the words of the *original* problem.

OBJECTIVE 1 Solve geometry problems using two variables.

EXAMPLE 1 Finding the Dimensions of a Soccer Field

A rectangular soccer field may have a width between 50 and 100 yd and a length between 100 and 130 yd. One particular soccer field has a perimeter of 320 yd. Its length measures 40 yd more than its width. What are the dimensions of this field? (Data from www.soccer-training-guide.com)

Step 1 **Read** the problem again. We must find the dimensions of the field.

Step 2 **Assign variables.** A sketch may be helpful. See **FIGURE 49** on the next page.

$$\text{Let } L = \text{the length} \quad \text{and} \quad W = \text{the width.}$$

A rectangular parking lot has a length that is 10 ft more than twice its width. The perimeter of the parking lot is 620 ft. What are the dimensions of the parking lot?

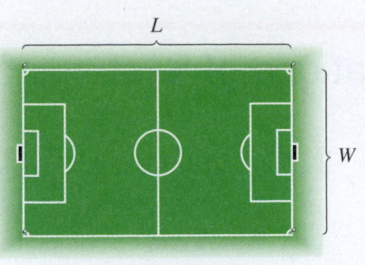

L

W

FIGURE 49

Step 3 **Write a system of equations.** Because the perimeter is 320 yd, we find one equation by using the perimeter formula.

$$2L + 2W = 320 \qquad 2L + 2W = P$$

For a second equation, use the information given about the length.

$$L = W + 40 \qquad \text{The length is 40 yd more than the width.}$$

These two equations form a system of equations.

$$2L + 2W = 320 \qquad (1)$$

$$L = W + 40 \qquad (2)$$

Step 4 **Solve** the system. Equation (2) is solved for L, so we use the substitution method, substituting $W + 40$ for L in equation (1), and solving for W.

$$2L + 2W = 320 \qquad (1)$$

$$2(W + 40) + 2W = 320 \qquad \text{Let } L = W + 40.$$

> Be sure to use parentheses around $W + 40$.

$$2W + 80 + 2W = 320 \qquad \text{Distributive property}$$

$$4W + 80 = 320 \qquad \text{Combine like terms.}$$

$$4W = 240 \qquad \text{Subtract 80.}$$

> Don't stop here.

$$W = 60 \qquad \text{Divide by 4.}$$

Let $W = 60$ in the equation $L = W + 40$ to find L.

$$L = 60 + 40 = 100$$

Step 5 **State the answer.** The length is 100 yd, and the width is 60 yd. Both dimensions are within the ranges given in the problem.

Step 6 **Check** using the words of the *original* problem.

$$2(100) + 2(60) = 320 \qquad \text{The perimeter is 320 yd, as required.}$$

$$100 = 60 + 40 \qquad \text{Length is 40 yd more than width, as required.}$$

NOW TRY

PROBLEM-SOLVING HINT   There is often more than one way to write the equations in a system used to solve an application. In **Example 1,** we might write the second equation as

$$W = L - 40. \qquad (2)$$

In this case, we would substitute $L - 40$ for W in equation (1) to obtain

$$2L + 2(L - 40) = 320 \qquad \text{Let } W = L - 40.$$

and solve for L first (instead of W, as in Step 4 above). The *same* answer results.

OBJECTIVE 2 Solve money problems using two variables.

EXAMPLE 2 Solving a Problem about Ticket Prices

For the 2015–2016 National Football League and National Basketball Association seasons, two football tickets and one basketball ticket purchased at their average prices cost $241.84. One football ticket and two basketball tickets cost $204.74. What were the average ticket prices for the two sports? (Data from Team Marketing Report.)

Step 1 **Read** the problem again. There are two unknowns.

Step 2 **Assign variables.**

Let f = the average price for a football ticket

and b = the average price for a basketball ticket.

Step 3 **Write a system of equations.** We write one equation using the fact that two football tickets and one basketball ticket cost $241.84.

$$2f + b = 241.84$$

By similar reasoning, we can write a second equation.

$$f + 2b = 204.74$$

These two equations form a system of equations.

$$2f + b = 241.84 \quad \text{(1)}$$
$$f + 2b = 204.74 \quad \text{(2)}$$

Step 4 **Solve** the system. Either the substitution or the elimination method can be used, based on personal preference. We choose the elimination method. To eliminate f, multiply equation (2) by -2 and add.

$$2f + b = 241.84 \quad \text{(1)}$$
$$\underline{-2f - 4b = -409.48} \quad \text{Multiply each side of (2) by } -2.$$
$$-3b = -167.64 \quad \text{Add.}$$
$$b = 55.88 \quad \text{Divide by } -3.$$

To find the value of f, let $b = 55.88$ in equation (2).

$$f + 2b = 204.74 \quad \text{(2)}$$
$$f + 2(55.88) = 204.74 \quad \text{Let } b = 55.88.$$
$$f + 111.76 = 204.74 \quad \text{Multiply.}$$
$$f = 92.98 \quad \text{Subtract 111.76.}$$

Step 5 **State the answer.** The average price for one basketball ticket was $55.88. For one football ticket, the average price was $92.98.

Step 6 **Check** that these values satisfy the problem conditions.

$$2(\$92.98) + \$55.88 = \$241.84, \quad \text{as required.}$$
$$\$92.98 + 2(\$55.88) = \$204.74, \quad \text{as required.}$$

NOW TRY

NOW TRY EXERCISE 2

Two general admission tickets and three tickets for children to a theme park cost $239.95. One general admission ticket and four tickets for children cost $224.95. Determine the ticket prices for general admission and for children.

PROBLEM-SOLVING HINT Rework Step 4 of **Example 2** using the substitution method to confirm that the same answer results. (There are two ways to do this—either solve equation (1) for b or solve equation (2) for f.)

Solving using a different method can be a good strategy for checking your answer.

NOW TRY ANSWER

2. general admission: $56.99; children: $41.99

OBJECTIVE 3 Solve mixture problems using two variables.

Some mixture problems can be solved using one variable. For many mixture problems, we can use more than one variable and a system of equations.

EXAMPLE 3   Solving a Mixture Problem

How many ounces each of 5% hydrochloric acid and 20% hydrochloric acid must be combined to obtain 10 oz of solution that is 12.5% hydrochloric acid?

Step 1   **Read** the problem. Two solutions of different strengths are being mixed to obtain a specific amount of a solution with an "in-between" strength.

Step 2   **Assign variables.**

Let x = the number of ounces of 5% solution

and y = the number of ounces of 20% solution.

Use a table to summarize the information from the problem.

Ounces of Solution	Percent (as a decimal)	Ounces of Pure Acid
x	5% = 0.05	$0.05x$
y	20% = 0.20	$0.20y$
10	12.5% = 0.125	$(0.125)10$

Multiply the amount of each solution (given in the first column) by its concentration of acid (given in the second column) to find the amount of acid in that solution (given in the third column).

Gives equation (1)   Gives equation (2)

FIGURE 50 illustrates what is happening in the problem.

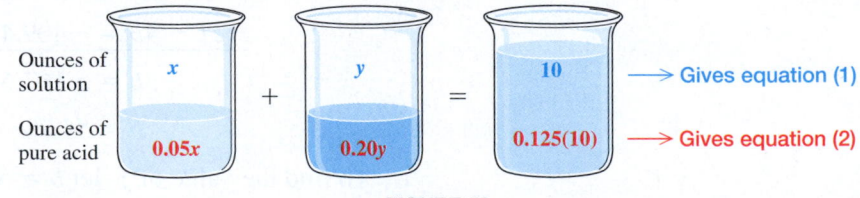

Ounces of solution x + y =   10   ⟶ Gives equation (1)

Ounces of pure acid $0.05x$ $0.20y$ $0.125(10)$ ⟶ Gives equation (2)

FIGURE 50

Step 3   **Write a system of equations.** When x ounces of 5% solution and y ounces of 20% solution are combined, the total number of ounces is 10.

$$x + y = 10$$

The ounces of acid in the 5% solution $(0.05x)$ added to the ounces of acid in the 20% solution $(0.20y)$ must equal the total ounces of acid in the mixture, which is $(0.125)10$, or 1.25.

$$0.05x + 0.20y = 1.25$$

Notice that these equations can be quickly determined by reading down the table or using the labels in **FIGURE 50.**

Multiply the second equation by 100 to clear the decimals and obtain an equivalent system of equations.

$$x + \quad y = 10 \qquad (1)$$
$$5x + 20y = 125 \qquad (2)$$

This is the system to solve.

**NOW TRY
EXERCISE 3**
How many liters each of a 15% acid solution and a 25% acid solution should be mixed to obtain 30 L of an 18% acid solution?

Step 4   **Solve** the system. Again, either the substitution or the elimination method can be used. We choose to eliminate x.

$$-5x - 5y = -50 \qquad \text{Multiply each side of (1) by } -5.$$
$$\underline{5x + 20y = 125} \qquad \text{(2)}$$
$$15y = 75 \qquad \text{Add.}$$

Ounces of 20% solution $\rightarrow y = 5 \qquad$ Divide by 15.

Substitute 5 for y in equation (1) to find the value of x.

$$x + y = 10 \qquad \text{(1)}$$
$$x + 5 = 10 \qquad \text{Let } y = 5.$$

Ounces of 5% solution $\rightarrow x = 5 \qquad$ Subtract 5.

Step 5   **State the answer.** The desired mixture will require 5 oz of the 5% solution and 5 oz of the 20% solution.

Step 6   **Check.**

Total amount of solution: $\qquad x + y = 5 \text{ oz} + 5 \text{ oz}$
$$= 10 \text{ oz}, \quad \text{as required.}$$

Total amount of acid: $\qquad 5\% \text{ of } 5 \text{ oz} + 20\% \text{ of } 5 \text{ oz}$
$$= 0.05(5) + 0.20(5)$$
$$= 1.25 \text{ oz}.$$

Percent of acid in solution:

Total acid $\longrightarrow$ $\dfrac{1.25}{10} = 0.125, \quad \text{or} \quad 12.5\%, \quad \text{as required.}$ **NOW TRY**
Total solution $\longrightarrow$

OBJECTIVE 4 Solve distance-rate-time problems using two variables.

Motion problems require a form of the distance formula $d = rt$, where d is distance, r is rate (or speed), and t is time.

EXAMPLE 4 Solving a Motion Problem

A car travels 250 km in the same time that a truck travels 225 km. If the rate of the car is 8 km per hr faster than the rate of the truck, find both rates.

Step 1   **Read** the problem again. Given the distances traveled, we need to find the rate of each vehicle.

Step 2   **Assign variables.**

Let x = the rate of the car,   and y = the rate of the truck.

As in **Example 3,** a table helps organize the information. Fill in the distance for each vehicle, and the variables for the unknown rates.

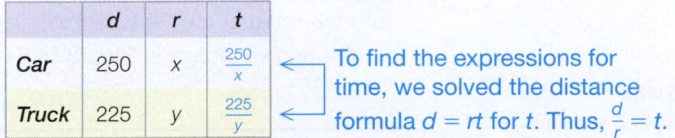

	d	r	t
Car	250	x	$\dfrac{250}{x}$
Truck	225	y	$\dfrac{225}{y}$

To find the expressions for time, we solved the distance formula $d = rt$ for t. Thus, $\dfrac{d}{r} = t$.

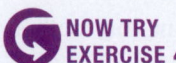
NOW TRY
EXERCISE 4
On a bicycle ride, Vann can travel 50 mi in the same amount of time that Ivy can travel 40 mi. Determine each bicyclist's rate, if Vann's rate is 2 mph faster than Ivy's.

Step 3   **Write a system of equations.** The rate x of the car is 8 km per hr faster than the rate y of the truck.

> Be careful writing this equation. *The car is faster.* $x = y + 8$ (1)

Both vehicles travel for the *same* time, so the times must be equal.

Time for car ⟶ $\dfrac{250}{x} = \dfrac{225}{y}$ ⟵ Time for truck

Multiply both sides by xy to obtain an equivalent equation with no variable denominators.

$$xy \cdot \frac{250}{x} = \frac{225}{y} \cdot xy \qquad \text{Multiply by the LCD, } xy.$$

$$\frac{250xy}{x} = \frac{225xy}{y} \qquad \text{Multiply; } xy = \frac{xy}{1}$$

$$250y = 225x \qquad \text{Divide out the common factors.}\quad (2)$$

We now have a system of linear equations.

$$x = y + 8 \qquad (1)$$
$$250y = 225x \qquad (2)$$

Step 4   **Solve** the system by substitution. Replace x with $y + 8$ in equation (2).

$$250y = 225x \qquad\qquad (2)$$

> Be sure to use parentheses around $y + 8$. $250y = 225(y + 8)$ Let $x = y + 8$.

$$250y = 225y + 1800 \qquad \text{Distributive property}$$
$$25y = 1800 \qquad \text{Subtract } 225y.$$

Truck's rate → $y = 72$ Divide by 25.

Let $y = 72$ in the equation $x = y + 8$ to find x.

Car's rate → $x = 72 + 8 = 80$

Step 5   **State the answer.** The rate of the car is 80 km per hr, and the rate of the truck is 72 km per hr.

Step 6   **Check.**

$$Car: \quad t = \frac{d}{r} = \frac{250}{80} = 3.125$$
$$Truck: \quad t = \frac{d}{r} = \frac{225}{72} = 3.125$$

Times are equal, as required.

The rate of the car, 80 km per hr, is 8 km per hour greater than that of the truck, 72 km per hr, as required.

NOW TRY

PROBLEM-SOLVING HINT   When one quantity in an application is compared to another (as are the rates of the car and truck in **Example 4**), be careful to translate correctly in terms of the two variables. In Step 3 of **Example 4,**

$$y = x - 8 \text{ is an alternative for equation (1).}$$

NOW TRY ANSWER
4. Vann: 10 mph; Ivy: 8 mph

**NOW TRY
EXERCISE 5**

In his motorboat, Ed traveled 42 mi upstream at top speed in 2.1 hr. Still at top speed, his return trip to the same spot took only 1.5 hr. Find the rate of Ed's boat in still water and the rate of the current.

Downstream
(with the current)

Upstream
(against the current)

FIGURE 51

EXAMPLE 5   Solving a Motion Problem

While kayaking on the Blackledge River, Rebecca traveled 9 mi upstream (against the current) in 2.25 hr. It took her only 1 hr paddling downstream (with the current) back to the spot where she started. Find Rebecca's kayaking rate in still water and the rate of the current.

Step 1   **Read** the problem. We must find two rates—Rebecca's kayaking rate in still water and the rate of the current.

Step 2   **Assign variables.**

Let x = Rebecca's kayaking rate in still water

and y = the rate of the current.

When the kayak is traveling *against* the current, the current slows it down. The rate of the kayak is the *difference* between its rate in still water and the rate of the current, which is $(x - y)$ mph.

When the kayak is traveling *with* the current, the current speeds it up. The rate of the kayak is the *sum* of its rate in still water and the rate of the current, which is $(x + y)$ mph.

Thus, $x - y$ = the rate of the kayak upstream (*against* the current),

and $x + y$ = the rate of the kayak downstream (*with* the current).

See **FIGURE 51**. Make a table. Use the formula $d = rt$, or $rt = d$.

	r	t	d
Upstream	$x - y$	2.25	$2.25(x - y)$
Downstream	$x + y$	1	$1(x + y)$

The distance is the same in each direction, 9 mi.

Step 3   **Write a system of equations.**

$$2.25(x - y) = 9 \quad \text{Upstream}$$
$$1(x + y) = 9 \quad \text{Downstream}$$

Clear parentheses in each equation, and then divide each term in the first equation by 2.25 to obtain an equivalent system.

$$x - y = 4 \quad (1)$$
$$\underline{x + y = 9} \quad (2)$$

Step 4   **Solve.**
$$2x = 13 \quad \text{Add.}$$

Rebecca's kayaking rate → $x = 6.5$ Divide by 2.

Substitute 6.5 for x in equation (2) and solve for y.

$$x + y = 9 \quad (2)$$
$$6.5 + y = 9 \quad \text{Let } x = 6.5.$$

Rate of current → $y = 2.5$ Subtract 6.5.

Step 5   **State the answer.** Rebecca's kayaking rate in still water was 6.5 mph, and the rate of the current was 2.5 mph.

Step 6   **Check.**

Distance upstream: $2.25(6.5 - 2.5) = 9$ True statements

Distance downstream: $1(6.5 + 2.5) = 9$ result.   **NOW TRY**

NOW TRY ANSWER
5. boat: 24 mph; current: 4 mph

OBJECTIVE 5 Solve problems with three variables using a system of three equations.

EXAMPLE 6 Solving a Geometry Problem

The sum of the measures of the angles of any triangle is 180°. The smallest angle of a certain triangle measures 36° less than the middle-sized angle. The largest angle measures 16° more than twice the smallest angle. Find the measure of each angle.

Step 1   **Read** the problem again. There are three unknowns in this problem.

Step 2   **Assign variables.** Make a sketch, as in **FIGURE 52**.

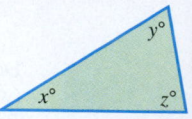

Let x = the measure of the smallest angle,

    y = the measure of the middle-sized angle,

and z = the measure of the largest angle.

FIGURE 52

Step 3   **Write a system of three equations.** The sum of the measures of the angles of any triangle is 180°.

$$x + y + z = 180$$

We can write two more equations from the information given.

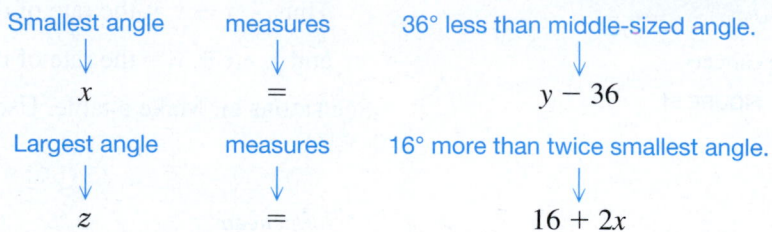

These three equations form a system.

$$x + y + z = 180 \qquad (1)$$

$$x = y - 36 \qquad (2)$$

$$z = 16 + 2x \qquad (3)$$

Step 4   **Solve** the system of equations. If we write equation (2) in terms of y, then equations (2) and (3) will give both y and z in terms of x.

$$x = y - 36 \qquad (2)$$

$$y = x + 36 \qquad \text{Add 36. Interchange sides.}$$

Now substitute $x + 36$ for y and $16 + 2x$ (from equation (3)) for z in equation (1).

$$x + y + z = 180 \qquad (1)$$

$$x + (x + 36) + (16 + 2x) = 180 \qquad \text{Let } y = x + 36 \text{ and } z = 16 + 2x.$$

$$4x + 52 = 180 \qquad \text{Combine like terms.}$$

$$x = 32 \qquad \text{Subtract 52. Divide by 4.}$$

Substitute 32 for x in $y = x + 36$ (equation (2) solved for y) to find y.

$$y = 32 + 36 \qquad \text{Let } x = 32.$$

$$y = 68 \qquad \text{Add.}$$

Substitute 32 for x in $z = 16 + 2x$ (equation (3)) to find z.

$$z = 16 + 2(32) \qquad \text{Let } x = 32.$$

$$z = 80 \qquad \text{Multiply, and then add.}$$

**NOW TRY
EXERCISE 6**

The sum of the measures of the angles of any triangle is 180°. The measure of the third angle of a certain triangle is four times that of the first angle. The sum of the measures of the first and second angles is 84°. Find the measure of each angle.

Step 5   **State the answer.** The three angles measure 32°, 68°, and 80°.

Step 6   **Check.** The sum of the measures of the three angles is

$$32° + 68° + 80° = 180°, \quad \text{as required.}$$

Also, each of the following is a true statement.

$32° = 68° - 36°$ Smallest measures 36° less than middle-sized.

$80° = 16° + 2(32°)$ Largest measures 16° more than twice smallest.

NOW TRY

PROBLEM-SOLVING HINT   In Step 4 of **Example 6,** we could also have substituted $16 + 2x$ (from equation (3)) for z in equation (1) as follows.

$$x + y + z = 180 \quad (1)$$

$$x + y + (16 + 2x) = 180 \quad \text{Let } z = 16 + 2x.$$

$$3x + y = 164 \quad \text{Combine like terms. Subtract 16.} \quad (4)$$

This gives a system of two equations in x and y.

$$x - y = -36 \quad \text{(2) in standard form}$$

$$3x + y = 164 \quad (4)$$

Eliminating y gives $x = 32$, which can be used to obtain $y = 68$ and $z = 80$. There is often more than one way to solve applications involving systems of three equations.

EXAMPLE 7   Solving a Problem Involving Prices

At Panera Bread, a loaf of honey wheat bread costs $3.99, a loaf of tomato basil bread costs $4.99, and a loaf of French bread costs $3.19. On a recent day, three times as many loaves of honey wheat were sold as loaves of tomato basil. The number of loaves of French bread sold was 5 less than the number of loaves of honey wheat sold. Total receipts for these breads were $90.17. How many loaves of each type of bread were sold? (Data from Panera Bread menu.)

Step 1   **Read** the problem again. There are three unknowns in this problem.

Step 2   **Assign variables** to represent the three unknowns.

Let $x =$ the number of loaves of honey wheat bread,

$y =$ the number of loaves of tomato basil bread,

and $z =$ the number of loaves of French bread.

Step 3   **Write a system of three equations.** Three times as many loaves of honey wheat bread were sold as loaves of tomato basil bread.

$$x = 3y, \quad \text{or} \quad x - 3y = 0 \quad \text{Subtract 3y.} \quad (1)$$

Also, we have the information needed for another equation.

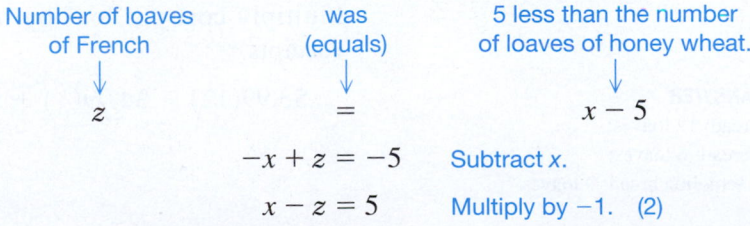

Number of loaves of French	was (equals)	5 less than the number of loaves of honey wheat.
↓	↓	↓
z	$=$	$x - 5$

$$-x + z = -5 \quad \text{Subtract } x.$$

$$x - z = 5 \quad \text{Multiply by } -1. \quad (2)$$

NOW TRY ANSWER
6. 24°, 60°, 96°

NOW TRY
EXERCISE 7

At Panera Bread, a loaf of white bread costs $3.99, a loaf of cheese bread costs $4.79, and a loaf of sesame semolina bread costs $8.19. On a recent day, twice as many loaves of white bread were sold as loaves of cheese bread. The number of loaves of sesame semolina sold was 3 less than the number of loaves of white sold. Total receipts for these breads were $150.33. How many loaves of each type of bread were sold? (Data from Panera Bread menu.)

Multiplying the cost of a loaf of each kind of bread by the number of loaves of that kind sold and adding gives an equation for the total receipts.

$$3.99x + 4.99y + 3.19z = 90.17$$

$$399x + 499y + 319z = 9017 \quad \text{Multiply each term by 100 to clear decimals.} \quad (3)$$

These three equations form a system.

$$x - 3y = 0 \quad (1)$$

$$x - z = 5 \quad (2)$$

$$399x + 499y + 319z = 9017 \quad (3)$$

Step 4   **Solve** the system of equations. Equation (1) is missing the variable z, so one way to begin is to eliminate z again, using equations (2) and (3).

$$319x \qquad - 319z = 1595 \quad \text{Multiply (2) by 319.}$$

$$\underline{399x + 499y + 319z = 9017 \quad (3)}$$

$$718x + 499y \qquad = 10{,}612 \quad \text{Add.} \quad (4)$$

Use resulting equation (4) in x and y, together with equation (1), $x - 3y = 0$, to eliminate x.

$$-718x + 2154y = 0 \qquad \text{Multiply (1) by } -718.$$

$$\underline{718x + 499y = 10{,}612 \quad (4)}$$

$$2653y = 10{,}612 \quad \text{Add.}$$

$$y = 4 \qquad \text{Divide by 2653.}$$

We can find x by substituting this value for y in equation (1).

$$x - 3y = 0 \qquad (1)$$

$$x - 3(4) = 0 \qquad \text{Let } y = 4.$$

$$x - 12 = 0 \qquad \text{Multiply.}$$

$$x = 12 \qquad \text{Add 12.}$$

We can find z by substituting this value for x in equation (2).

$$x - z = 5 \qquad (2)$$

$$12 - z = 5 \qquad \text{Let } x = 12.$$

$$z = 7 \qquad \text{Subtract 12. Multiply by } -1.$$

Step 5   **State the answer.** There were 12 loaves of honey wheat bread, 4 loaves of tomato basil bread, and 7 loaves of French bread sold.

Step 6   **Check.** Each of the following is a true statement.

$$12 = 3 \cdot 4 \qquad \text{Honey wheat is three times tomato basil.}$$

$$7 = 12 - 5 \qquad \text{French is 5 less than honey wheat.}$$

Multiply cost per loaf by number of loaves and add to confirm total receipts.

$$\$3.99(12) + \$4.99(4) + \$3.19(7) = \$90.17, \quad \text{as required.}$$

NOW TRY

NOW TRY ANSWER

7. white bread: 12 loaves;
cheese bread: 6 loaves;
sesame semolina bread: 9 loaves

7.7 Exercises

FOR EXTRA HELP

 MyLab Math

 Video solutions for select problems available in MyLab Math

Concept Check   *Answer each question.*

1. If a container of liquid contains 60 oz of solution, what is the number of ounces of pure acid if the given solution contains the following acid concentrations?

   (a) 10%   **(b)** 25%   **(c)** 40%   **(d)** 50%

2. If $5000 is invested in an account paying simple annual interest, how much interest will be earned during the first year at the following rates?

   (a) 2%   **(b)** 3%   **(c)** 4%   **(d)** 3.5%

3. If one pound of turkey costs $1.99, how much will x pounds cost?

4. If one movie ticket costs $13.50, how much will y tickets cost?

5. If the rate of a boat in still water is 10 mph and the rate of the current of a river is x mph, what is the rate of the boat in each case?

   (a) The boat is going upstream (that is, against the current, which slows the boat down).

   (b) The boat is going downstream (that is, with the current, which speeds the boat up).

6. If the rate of a plane in still air is x mph and the rate of a steady wind is 20 mph, what is the rate of the plane in each case?

   (a) The plane is flying into the wind (that is, into a headwind, which slows the plane down).

   (b) The plane is flying with the wind (that is, with a tailwind, which speeds the plane up).

7. A bear is running at a rate of 44 ft per sec.

   (a) If the bear runs for 5 sec, what is its distance?

   (b) If the bear travels d ft, what is its time?

8. The swimming rate of a whale is 25 mph.

   (a) If the whale swims for y hours, what is its distance?

   (b) If the whale travels 10 mi, what is its time?

Solve each problem. **See Example 1.**

9. Venus and Serena measured a tennis court and found that it was 42 ft longer than it was wide and had a perimeter of 228 ft. What were the length and the width of the tennis court?

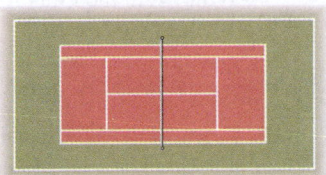

10. LeBron and Jose measured a basketball court and found that the width of the court was 44 ft less than the length. If the perimeter was 288 ft, what were the length and the width of the basketball court?

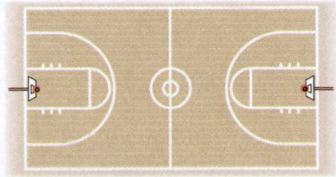

11. During the 2017 Major League Baseball season, the Cleveland Indians played 162 games. They won 42 more games than they lost. What was the team's win-loss record that year?

Team	W	L
Cleveland	—	—
Minnesota	85	77
Kansas City	80	82
Chicago White Sox	67	95
Detroit	—	—

Data from Major League Baseball.

12. During the 2017 Major League Baseball season, the Detroit Tigers played 162 games. They lost 34 more games than they won. What was the team's win-loss record?

13. In 2016, the two American telecommunication companies with the greatest revenues were AT&T and Verizon. The two companies had combined revenues of $288.9 billion. AT&T's revenue was $38.7 billion more than that of Verizon. What was the revenue for each company? (Data from Verizon and AT&T Annual Reports.)

14. In 2017, U.S. exports to Canada were $39.4 billion more than exports to Mexico. Together, exports to these two countries totaled $525.4 billion. How much were exports to each country? (Data from U.S. Census Bureau.)

Find the measures of angles x and y. Remember that (1) the sum of the measures of the angles of any triangle is 180°, (2) supplementary angles have a sum of 180°, and (3) vertical angles have equal measures.

15.

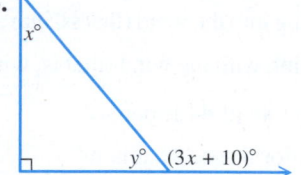

16.

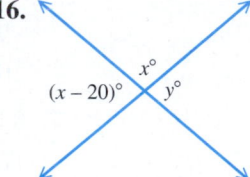

The Fan Cost Index (FCI) represents the cost of four average-price tickets to a sporting event (two adult, two child), four small soft drinks, two small beers, four hot dogs, parking for one car, two game programs, and two souvenir caps. (Data from Team Marketing Report.)
    *Use the concept of FCI to solve each problem. **See Example 2.***

17. For the 2016 Major League Baseball season, the FCI prices for the Cleveland Indians and the Boston Red Sox totaled $540.10. The Boston FCI was $181.22 more than that of Cleveland. What were the FCIs for these teams?

18. In 2016, the FCI prices for Major League Baseball and the National Football League totaled $722.37. The football FCI was $283.31 more than that of baseball. What were the FCIs for these sports?

*Solve each problem. **See Example 2.***

19. Leanna is a waitress at Bonefish Grill. During one particular day she sold 15 ribeye steak dinners and 20 grilled salmon dinners, totaling $886.50. Another day she sold 25 ribeye steak dinners and 10 grilled salmon dinners, totaling $973.50. How much did each type of dinner cost? (Data from Bonefish Grill menu.)

20. At a business meeting at Panera Bread, the bill for two cappuccinos and three caffe lattes was $19.75. At another table, the bill for one cappuccino and two caffe lattes was $11.97. How much did each type of beverage cost? (Data from Panera Bread menu.)

21. Two days at Busch Gardens (Tampa Bay) and 3 days at Universal Studios Florida (Orlando) cost $510, while 4 days at Busch Gardens and 2 days at Universal Studios cost $580. (Prices are based on single-day admissions.) What was the cost per day for each park? (Data from Busch Gardens and Universal Studios.)

22. On the basis of average total costs per day for business travel to New York City and Washington, DC (which include a hotel room, car rental, and three meals), 2 days in New York and 3 days in Washington cost $2484, while 4 days in New York and 2 days in Washington cost $3120. What was the average cost per day in each city? (Data from *Business Travel News*.)

23. Tickets to a production of *A Midsummer Night's Dream* at Broward College cost $5 for general admission or $4 with a student ID. If 184 people paid to see a performance and $812 was collected, how many of each type of ticket were sold?

24. The mixed price of 9 citrons and 7 fragrant wood apples is 107; again, the mixed price of 7 citrons and 9 fragrant wood apples is 101. O you arithmetician, tell me quickly the price of a citron and the price of a wood apple here, having distinctly separated those prices well. (*Source:* Hindu work, A.D. 850.)

Solve each problem. See Example 3.

25. How many gallons each of 25% alcohol and 35% alcohol should be mixed to obtain 20 gal of 32% alcohol?

Gallons of Solution	Percent (as a decimal)	Gallons of Pure Alcohol
x	25% = 0.25	
y	35% = 0.35	
20	32% =	

26. How many liters each of 15% acid and 33% acid should be mixed to obtain 120 L of 21% acid?

Liters of Solution	Percent (as a decimal)	Liters of Pure Acid
x	15% = 0.15	
y	33% =	
120	21% =	

27. A party mix is made by adding nuts that sell for $2.50 per kg to a cereal mixture that sells for $1 per kg. How much of each should be added to obtain 30 kg of a mix that will sell for $1.70 per kg?

	Number of Kilograms	Price per Kilogram (in dollars)	Value (in dollars)
Nuts	x	2.50	
Cereal	y	1.00	
Mixture		1.70	

28. A fruit drink is made by mixing juices. Such a drink with 50% juice is to be mixed with a drink that is 30% juice to obtain 200 L of a drink that is 45% juice. How much of each should be used?

	Liters of Drink	Percent (as a decimal)	Liters of Pure Juice
50% Juice	x	0.50	
30% Juice	y	0.30	
Mixture		0.45	

29. A total of $3000 is invested, part at 2% simple interest and part at 4%. If the total annual return from the two investments is $100, how much is invested at each rate?

Principal (in dollars)	Rate (as a decimal)	Interest (in dollars)
x	0.02	0.02x
y	0.04	0.04y
3000	XXXXX	100

30. An investor will invest a total of $15,000 in two accounts, one paying 4% annual simple interest and the other 3%. If he wants to earn $550 annual interest, how much should he invest at each rate?

Principal (in dollars)	Rate (as a decimal)	Interest (in dollars)
x	0.04	
y	0.03	
15,000	XXXXX	

31. Pure acid is to be added to a 10% acid solution to obtain 54 L of a 20% acid solution. What amounts of each should be used? (*Hint:* Pure acid is 100% acid.)

32. A truck radiator holds 36 L of fluid. How much pure antifreeze must be added to a mixture that is 4% antifreeze to fill the radiator with a mixture that is 20% antifreeze?

33. How many pounds of candy that sells for $1.75 per lb must be mixed with candy that sells for $1.25 per lb to obtain 10 lb of a mixture that should sell for $1.60 per lb?

34. How many pounds of candy that sells for $2.50 per lb must be mixed with candy that sells for $1.75 per lb to obtain 6 lb of a mixture that sells for $2.10 per lb?

Solve each problem. ***See Examples 4 and 5.***

35. A train travels 150 km in the same time that a plane travels 400 km. If the rate of the plane is 20 km per hr less than three times the rate of the train, find both rates.

	r	t	d
Train	x		150
Plane	y		400

36. A motor scooter travels 20 mi in the same time that a bicycle travels 8 mi. If the rate of the scooter is 5 mph more than twice the rate of the bicycle, find both rates.

37. A freight train and an express train leave towns 390 km apart, traveling toward one another. The freight train travels 30 km per hr slower than the express train. They pass one another 3 hr later. What are their rates?

	r	t	d
Freight Train	x	3	
Express Train	y	3	

38. A car and a truck leave towns 230 mi apart, traveling toward each other. The car travels 15 mph faster than the truck. They pass each other 2 hr later. What are their rates?

39. In his motorboat, Bill travels upstream at top speed to his favorite fishing spot, a distance of 36 mi, in 2 hr. Returning, he finds that the trip downstream, still at top speed, takes only 1.5 hr. Find the rate of Bill's boat and the rate of the current. Let $x =$ the rate of the boat and $y =$ the rate of the current.

	r	t	d
Upstream	$x - y$	2	
Downstream	$x + y$		

40. Traveling for 3 hr into a steady headwind, a plane flies 1650 mi. The pilot determines that flying *with* the same wind for 2 hr, he could make a trip of 1300 mi. Find the rate of the plane and the wind speed.

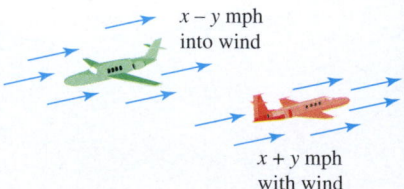

$x - y$ mph
into wind

$x + y$ mph
with wind

41. A plane flies 560 mi in 1.75 hr traveling with the wind. The return trip later against the same wind takes the plane 2 hr. Find the rate of the plane and the wind speed.

42. Braving blizzard conditions on the planet Hoth, Luke Skywalker sets out in his snow speeder for a rebel base 4800 mi away. He travels into a steady headwind and makes the trip in 3 hr. Returning, he finds that the trip back, now with a tailwind, takes only 2 hr. Find the rate of Luke's snow speeder and the wind speed.

Solve each problem. ***See Examples 6 and 7.***

43. In the triangle below, $z = x + 10$ and $x + y = 100$. Determine a third equation involving x, y, and z, and then find the measures of the three angles.

44. In the triangle below, x is 10 less than y and 20 less than z. Write a system of equations and find the measures of the three angles.

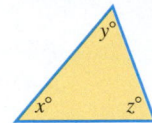

45. In a certain triangle, the measure of the second angle is 10° greater than three times the first. The third angle measure is equal to the sum of the measures of the other two. Find the measures of the three angles.

46. The measure of the largest angle of a triangle is 12° less than the sum of the measures of the other two. The smallest angle measures 58° less than the largest. Find the measures of the angles.

47. The perimeter of a triangle is 70 cm. The longest side is 4 cm less than the sum of the other two sides. Twice the shortest side is 9 cm less than the longest side. Find the length of each side of the triangle.

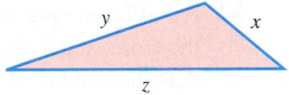

48. The perimeter of a triangle is 56 in. The longest side measures 4 in. less than the sum of the other two sides. Three times the shortest side is 4 in. more than the longest side. Find the lengths of the three sides.

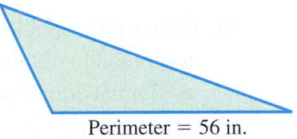

Perimeter = 56 in.

49. In the 2016 Summer Olympics, host Brazil earned 1 more gold medal than silver. The number of silver medals that Brazil earned was the same as the number of its bronze medals. Brazil earned a total of 19 medals. How many of each kind of medal did Brazil earn? (Data from *The World Almanac and Book of Facts.*)

50. In 2017, the average Facebook user in the United States spent 693 more minutes per month using Facebook than the average Instagram user spent using Instagram. The average Instagram user spent 36 fewer minutes per month using Instagram than the average Snapchat user spent using Snapchat. The total amount of time that average users of each social network spent per month using their respective networks was 1347 min. How much time did the average user spend on each social network? (Data from Verto Analytics.)

51. Tickets for a Harlem Globetrotters show cost $30 for upper level, $51 for center court, or $76 for floor seats. Nine times as many upper level tickets were sold as floor tickets, and the number of upper level tickets sold was 55 more than the sum of the number of center court tickets and floor tickets. Sales of all three kinds of tickets totaled $95,215. How many of each kind of ticket were sold? (Data from www.harlemglobetrotters.com)

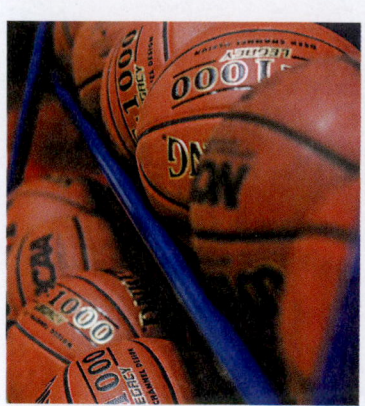

52. Three kinds of tickets are available for a rock concert: "up close," "in the middle," and "far out." "Up close" tickets cost $10 more than "in the middle" tickets. "In the middle" tickets cost $10 more than "far out" tickets. Twice the cost of an "up close" ticket is $20 more than three times the cost of a "far out" ticket. Find the price of each kind of ticket.

53. A wholesaler supplies college t-shirts to three college bookstores: A, B, and C. The wholesaler recently shipped a total of 800 t-shirts to the three bookstores. Twice as many t-shirts were shipped to bookstore B as to bookstore A, and the number shipped to bookstore C was 40 less than the sum of the numbers shipped to the other two bookstores. How many t-shirts were shipped to each bookstore?

54. An office supply store sells three models of computer desks: A, B, and C. In one month, the store sold a total of 85 computer desks. The number of model B desks was five more than the number of model C desks. The number of model A desks was four more than twice the number of model C desks. How many of each model did the store sell that month?

55. A plant food is to be made from three chemicals. The mix must include 60% of the first and second chemicals. The second and third chemicals must be in the ratio of 4 to 3 by weight. How much of each chemical is needed to make 750 kg of the plant food?

56. How many ounces of 5% hydrochloric acid, 20% hydrochloric acid, and water must be combined to obtain 10 oz of solution that is 8.5% hydrochloric acid if the amount of water used must equal the total amount of the other two solutions?

The National Hockey League uses a point system to determine team standings. A team is awarded 2 *points for a win* (W), 0 *points for a loss in regulation play* (L), *and* 1 *point for an overtime loss* (OTL). *Use this information to solve each problem.*

Team	GP	W	L	OTL	Points
Anaheim	82	——	——	——	105
Edmonton	82	47	26	9	103
San Jose	82	46	29	7	99
Calgary	82	45	33	4	94
Los Angeles	82	——	——	——	86

Data from www.nhl.com

57. During the 2016–2017 NHL regular season, the Anaheim Ducks played 82 games. Their wins and overtime losses resulted in a total of 105 points. They had 10 more losses in regulation play than overtimes losses. How many wins, losses, and overtime losses did they have that season?

58. During the same NHL regular season, the Los Angeles Kings also played 82 games. Their wins and overtimes losses resulted in a total of 86 points. They had 4 more total losses (in regulation play and overtime) than wins. How many wins, losses, and overtime losses did they have that season?

The following exercises are based on the "Bait Box Puzzle," featured in an "Ask Marilyn" column in Parade Magazine.

59. There are three kinds of bait boxes—those containing fish, those containing bugs, and those containing worms. Although the bait boxes look alike, each kind has a different weight. Use the information in **FIGURE A** to write a system of three equations, and solve it to determine the weight of each kind of bait box.

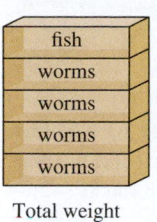

Total weight
28 oz

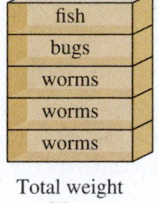

Total weight
25 oz

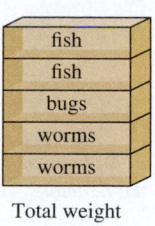

Total weight
28 oz

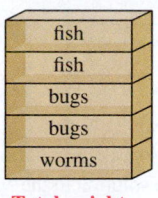

Total weight
?

FIGURE A   **FIGURE B**

60. Write an equation that describes the situation depicted in **FIGURE B**. Use the same three variables used in **Exercise 59**. Then determine the total weight of the bait boxes in **FIGURE B**.

Chapter 7   Summary

Key Terms

7.1
ordered pair
origin
x-axis
y-axis
rectangular (Cartesian) coordinate system
components
plot
coordinate
quadrant

graph of an equation
table of ordered pairs (table of values)
first-degree equation
linear equation in two variables
x-intercept
y-intercept
rise
run
slope

7.3
system of linear equations (linear system)
solution of a system
solution set of a system
consistent system
inconsistent system
independent equations
dependent equations

7.6
ordered triple
focus variable
working equation

New Symbols

(x, y) ordered pair

(x_1, y_1) subscript notation (read "x-sub-one, y-sub-one")

Δ Greek letter delta
m slope

(x, y, z) ordered triple

Test Your Word Power

See how well you have learned the vocabulary in this chapter.

1. A **linear equation in two variables** is an equation that can be written in the form
   A. $Ax + By < C$
   B. $ax = b$
   C. $y = x^2$
   D. $Ax + By = C.$

2. The **slope** of a line is
   A. the measure of the run over the rise of the line
   B. the distance between two points on the line
   C. the ratio of the change in y to the change in x along the line
   D. the horizontal change compared to the vertical change of two points on the line.

3. A **system of linear equations** consists of
   A. at least two linear equations with different variables
   B. two or more linear equations that have an infinite number of solutions
   C. two or more linear equations with the same variables
   D. two or more linear inequalities.

4. A **consistent system** is a system of equations
  A. with at least one solution
  B. with no solution
  C. with graphs that do not intersect
  D. with solution set $\varnothing$.

5. An **inconsistent system** is a system of equations
  A. with one solution
  B. with no solution
  C. with an infinite number of solutions
  D. that have the same graph.

6. Dependent equations
  A. have different graphs
  B. have no solution
  C. have one solution
  D. are different forms of the same equation.

ANSWERS

1. D; *Examples:* $3x + 2y = 6$, $x = y - 7$ **2.** C; *Example:* The line through $(3, 6)$ and $(5, 4)$ has slope $\frac{4-6}{5-3} = \frac{-2}{2} = -1$.
3. C; *Example:* $2x + y = 7$, $3x - y = 3$ **4.** A; *Example:* The system in **Answer 3** is consistent. The graphs of the equations intersect at exactly one point—in this case, the solution $(2, 3)$.   **5.** B; *Example:* The equations of two parallel lines make up an inconsistent system. Their graphs never intersect, so there is no solution to the system.   **6.** D; *Example:* The equations $4x - y = 8$ and $8x - 2y = 16$ are dependent because their graphs are the same line.

Quick Review

CONCEPTS	EXAMPLES

7.1   Review of Graphs and Slopes of Lines

Finding Intercepts

To find the *x*-intercept, let $y = 0$ and solve for x.

To find the *y*-intercept, let $x = 0$ and solve for y.

Find the intercepts of the graph of $2x + 3y = 12$.

$2x + 3(0) = 12$ Let $y = 0$. $\quad$ $2(0) + 3y = 12$ Let $x = 0$.
$\qquad 2x = 12$ $\qquad\qquad\qquad 3y = 12$
$\qquad\quad x = 6$ $\qquad\qquad\qquad\quad y = 4$

The *x*-intercept is $(6, 0)$. $\quad$ The *y*-intercept is $(0, 4)$.

Midpoint Formula

The midpoint M of a line segment PQ with endpoints $P(x_1, y_1)$ and $Q(x_2, y_2)$ is found as follows.

$$M = \left(\frac{x_1 + x_2}{2}, \frac{y_1 + y_2}{2}\right)$$

Find the midpoint of the segment with endpoints $(4, -7)$ and $(-10, -13)$.

$$M = \left(\frac{4 + (-10)}{2}, \frac{-7 + (-13)}{2}\right) = (-3, -10)$$

Slope Formula

The slope m of the line passing through the points (x_1, y_1) and (x_2, y_2) is defined as follows.

$$\text{slope } m = \frac{\text{rise}}{\text{run}} = \frac{\text{change in } y}{\text{change in } x} = \frac{\Delta y}{\Delta x} = \frac{y_2 - y_1}{x_2 - x_1}$$

$$(\text{where } x_1 \neq x_2)$$

Find the slope of the graph of $2x + 3y = 12$.
  Use the intercepts $(6, 0)$ and $(0, 4)$ and the slope formula.

$$m = \frac{4 - 0}{0 - 6} = \frac{4}{-6} = -\frac{2}{3} \qquad \begin{array}{l}(x_1, y_1) = (6, 0)\\ (x_2, y_2) = (0, 4)\end{array}$$

A horizontal line has slope 0.

A vertical line has undefined slope.

Parallel lines have the same slope.

The graph of the horizontal line $y = -5$ has slope $m = 0$.

The graph of the vertical line $x = 3$ has undefined slope.

The lines $y = 2x + 3$ and $4x - 2y = 6$ are parallel. Both have slope 2.

$\quad y = 2x + 3$ $\quad$ | $\quad$ $4x - 2y = 6$ $\leftarrow$ Solve for *y*.
$\qquad\qquad\qquad\qquad\qquad -2y = -4x + 6$
$\qquad\qquad\qquad\qquad\qquad\quad y = 2x - 3$

Perpendicular lines, neither of which is vertical, **have slopes that are negative reciprocals** (that is, have a product of -1).

The lines $y = 3x - 1$ and $x + 3y = 4$ are perpendicular. Their slopes are negative reciprocals because $3\left(-\frac{1}{3}\right) = -1$.

$\quad y = 3x - 1$ $\quad$ | $\quad$ $x + 3y = 4$ $\leftarrow$ Solve for *y*.
$\qquad\qquad\qquad\qquad\qquad 3y = -x + 4$
$\qquad\qquad\qquad\qquad\qquad\quad y = -\frac{1}{3}x + \frac{4}{3}$

CONCEPTS	EXAMPLES
Slope gives the **average rate of change** in y per unit change in x, where the value of y depends on the value of x.	The weight of a young child increased from 30 lb to 60 lb between the ages of 3 and 8. What was the child's average change in weight per year over these years? $$\frac{60 - 30}{8 - 3} = \frac{30}{5} = 6 \text{ lb per yr}$$

7.2 Review of Equations of Lines; Linear Models

Slope-Intercept Form $y = mx + b$	$y = 2x + 3$ $m = 2$; y-intercept is $(0, 3)$.
Point-Slope Form $y - y_1 = m(x - x_1)$	$y - 3 = 4(x - 5)$ $m = 4$; The point $(5, 3)$ is on the line.
Standard Form $Ax + By = C$, where A, B, and C are real numbers and A and B are not both 0. (We give A, B, and C as integers, with greatest common factor 1, and $A \geq 0$.)	$2x - 5y = 8$ Standard form
Horizontal Line $y = b$	$y = 4$ Horizontal line
Vertical Line $x = a$	$x = -1$ Vertical line

7.3 Solving Systems of Linear Equations by Graphing

An ordered pair is a solution of a system if it makes all equations of the system true at the same time.	Is $(4, -1)$ a solution of the following system? $$x + y = 3$$ $$2x - y = 9$$ Because $4 + (-1) = 3$ and $2(4) - (-1) = 9$ are both true, the ordered pair $(4, -1)$ is a solution.

Solving a Linear System by Graphing	Solve the system by graphing.
Step 1 Graph each equation of the system on the same axes.	$$x + y = 5$$ $$2x - y = 4$$
Step 2 Find the coordinates of the point of intersection.	The ordered pair $(3, 2)$ satisfies *both* equations, so $\{(3, 2)\}$ is the solution set.
Step 3 Check. Write the solution set.	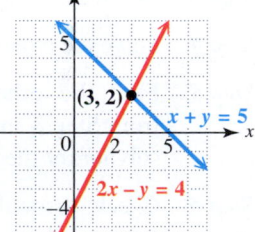

If the graphs of the equations do not intersect (that is, the lines are parallel), then the system has *no solution* and the solution set is $\varnothing$.

If the graphs of the equations are the same line, then the system has an *infinite number of solutions*. Use set-builder notation to write the solution set as

$$\{(x, y) \mid \underline{\hspace{1cm}}\},$$

where a form of the equation is written on the blank.

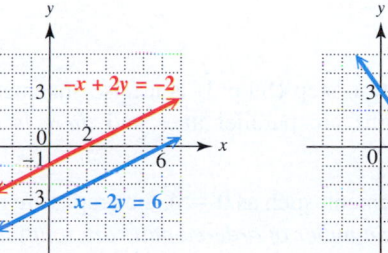

Solution set: $\varnothing$

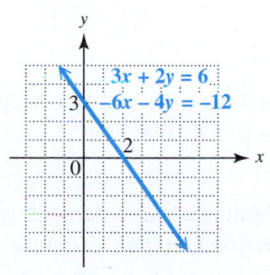

Solution set: $\{(x, y) \mid 3x + 2y = 6\}$

CONCEPTS	EXAMPLES

7.4 Solving Systems of Linear Equations by Substitution

Solving a Linear System by Substitution

Step 1 Solve one equation for either variable.

Step 2 Substitute for that variable in the other equation to obtain an equation in one variable.

Step 3 Solve the equation from Step 2.

Step 4 Find the other value by substituting the result from Step 3 into the equation from Step 1 and solving for the remaining variable.

Step 5 Check. Write the solution set.

Solve the system by substitution.

$$x + 2y = -5 \quad (1)$$
$$y = -2x - 1 \quad (2)$$

Equation (2) is already solved for y.

Substitute $-2x - 1$ for y in equation (1).

$$x + 2y = -5 \quad (1)$$
$$x + 2(-2x - 1) = -5 \quad \text{Let } y = -2x - 1.$$
$$x - 4x - 2 = -5 \quad \text{Distributive property}$$
$$-3x - 2 = -5 \quad \text{Combine like terms.}$$
$$x = 1 \quad \text{Add 2. Divide by } -3.$$

To find y, let $x = 1$ in equation (2).

$$y = -2x - 1 \quad (2)$$
$$y = -2(1) - 1 \quad \text{Let } x = 1.$$
$$y = -3 \quad \text{Multiply, and then subtract.}$$

A check confirms that $\{(1, -3)\}$ is the solution set.

7.5 Solving Systems of Linear Equations by Elimination

Solving a Linear System by Elimination

Step 1 Write both equations in the form $Ax + By = C$.

Step 2 Multiply to transform the equations so that the coefficients of one pair of variable terms are opposites.

Step 3 Add the equations to eliminate a variable.

Step 4 Solve the equation from Step 3.

Step 5 Find the other value by substituting the result from Step 4 into either of the original equations and solving for the remaining variable.

Step 6 Check. Write the solution set.

Solve the system by elimination.

$$x + 3y = 7 \quad (1)$$
$$3x - y = 1 \quad (2)$$

Multiply equation (1) by -3 to eliminate the x-terms.

$$-3x - 9y = -21 \quad \text{Multiply equation (1) by } -3.$$
$$\underline{3x - y = 1 \quad (2)}$$
$$-10y = -20 \quad \text{Add.}$$
$$y = 2 \quad \text{Divide by } -10.$$

Substitute to find the value of x.

$$x + 3y = 7 \quad (1)$$
$$x + 3(2) = 7 \quad \text{Let } y = 2.$$
$$x = 1 \quad \text{Multiply. Subtract 6.}$$

A check confirms that $\{(1, 2)\}$ is the solution set.

If the result of the addition step (Step 3) is a false statement, such as $0 = 4$, the graphs are parallel lines and *there is no solution*.

If the result is a true statement, such as $0 = 0$, the graphs are the same line, and an *infinite number of ordered pairs are solutions*.

$$x - 2y = 6$$
$$\underline{-x + 2y = -2}$$
$$0 = 4$$
Solution set: $\varnothing$

$$x - 2y = 6$$
$$\underline{-x + 2y = -6}$$
$$0 = 0$$
Solution set:
$\{(x, y) \mid x - 2y = 6\}$

CONCEPTS	EXAMPLES
7.6 Systems of Linear Equations in Three Variables	Solve the system.

Solving a Linear System in Three Variables

Step 1 Select a focus variable, preferably one with coefficient 1 or −1, and a working equation.

Step 2 Eliminate the focus variable, using the working equation and one of the equations of the system.

Step 3 Eliminate the focus variable again, using the working equation and the remaining equation of the system.

Step 4 Solve the system of two equations in two variables formed by the equations from Steps 2 and 3.

Step 5 Find the value of the remaining variable.

Step 6 Check the values in *each* of the *original* equations of the system. Then write the solution set as a set containing an ordered triple.

$$x + 2y - z = 6 \quad (1)$$
$$x + y + z = 6 \quad (2)$$
$$2x + y - z = 7 \quad (3)$$

We choose z as the focus variable and (2) as the working equation. Add equations (1) and (2).
$$2x + 3y = 12 \quad (4)$$
Add equations (2) and (3).
$$3x + 2y = 13 \quad (5)$$
Use equations (4) and (5) to eliminate x.

$-6x - 9y = -36$	Multiply (4) by −3.
$6x + 4y = 26$	Multiply (5) by 2.
$-5y = -10$	Add.
$y = 2$	Divide by −5.

To find x, substitute 2 for y in equation (4).

$2x + 3(2) = 12$	Let $y = 2$ in (4).
$2x + 6 = 12$	Multiply.
$2x = 6$	Subtract 6.
$x = 3$	Divide by 2.

Substitute 3 for x and 2 for y in working equation (2).

$x + y + z = 6$	(2)
$3 + 2 + z = 6$	Substitute.
$z = 1$	Subtract 5.

Check to verify that $\{(3, 2, 1)\}$ is the solution set.

7.7 Applications of Systems of Linear Equations

Use the six-step problem-solving method.

Step 1 Read the problem carefully.

Step 2 Assign variables.

Step 3 Write a system of equations.

Step 4 Solve the system.

Step 5 State the answer.

Step 6 Check.

The sum of two numbers is 30. Their difference is 6. Find the numbers.

Let $x =$ one number, and let $y =$ the other number.

$x + y = 30$	(1)
$x - y = 6$	(2)
$2x = 36$	Add.
$x = 18$	Divide by 2.

To find y, let $x = 18$ in equation (1).

$x + y = 30$	(1)
$18 + y = 30$	Let $x = 18$.
$y = 12$	Subtract 18.

The two numbers are 18 and 12. Their sum and difference are
$$18 + 12 = 30 \quad \text{and} \quad 18 - 12 = 6, \quad \text{as required.}$$

Chapter 7   Review Exercises

7.1 *Complete the given table of ordered pairs for each equation, and then graph the equation.*

1. $3x + 2y = 10$

x	y
0	
	0
2	
	-2

2. $x - y = 8$

x	y
2	
	-3
3	
	-2

Find the x- and y-intercepts. Then graph each equation.

3. $4x - 3y = 12$ **4.** $5x + 7y = 28$ **5.** $2x + 5y = 20$ **6.** $x - 4y = 8$

Find the midpoint of each segment with the given endpoints.

7. $(-8, -12)$ and $(8, 16)$ **8.** $(0, -5)$ and $(-9, 8)$

Find the slope of each line. (In Exercises 17 and 18, coordinates of the points shown are integers.)

9. Through $(-1, 2)$ and $(4, -5)$ **10.** Through $(0, 3)$ and $(-2, 4)$

11. Through $(-1, 5)$ and $(-1, -4)$ **12.** $y = 2x + 3$

13. $3x - 4y = 5$ **14.** $y = 4$

15. Perpendicular to $3x - y = 4$ **16.** Parallel to $3y = 2x + 5$

17. 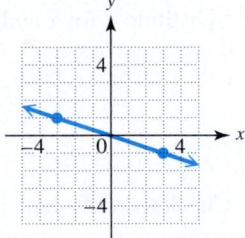   **18.** 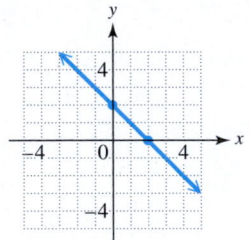   **19.**

x	y
-2	6
-1	3
0	0
1	-3

20. Determine whether the slope of the line is *positive, negative, 0,* or *undefined.*

(a) 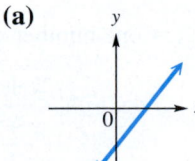   **(b)** 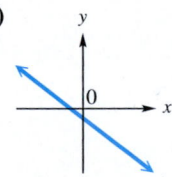   **(c)** 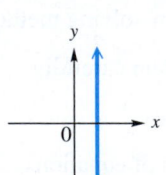   **(d)**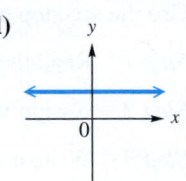

Determine whether each pair of lines is parallel, perpendicular, *or neither.*

21. $3x - y = 6$ and $x + 3y = 12$ **22.** $3x - y = 4$ and $6x + 12 = 2y$

Solve each problem.

23. If the pitch of a roof is $\frac{1}{4}$, how many feet in the horizontal direction correspond to a rise of 3 ft?

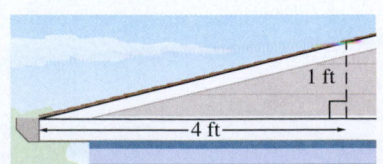

24. In 1980, the median family income in the United States was about $21,000 per year. In 2016, it was about $59,039 per year. Find the average rate of change in median family income per year to the nearest dollar during this period. (Data from U.S. Census Bureau.)

7.2 *Write an equation of the line passing through the given point(s) and satisfying any given condition. Give the equation* **(a)** *in slope-intercept form and* **(b)** *in standard form.*

25. $(0, -8)$; slope $\frac{3}{5}$

26. $(0, 5)$; slope $-\frac{1}{3}$

27. $(2, -5)$ and $(1, 4)$

28. $(-3, -1)$ and $(2, 6)$

29. $(6, -2)$; parallel to $4x - y = 3$

30. $(0, 1)$; perpendicular to $2x - 5y = 7$

Write an equation of the line passing through the given point and satisfying the given condition.

31. $(0, 12)$; slope 0

32. $(2, 7)$; undefined slope

33. $(0.3, 0.6)$; vertical

34. $(-1, 4)$; horizontal

Solve each problem.

35. A lease on a 2018 Chevrolet Sonic LT hatchback costs $2829 at signing, plus $229 per month. Let x represent the number of months since the signing and y represent the amount paid (in dollars). (Data from www.carsdirect.com)

 (a) Write an equation in slope-intercept form that represents this situation.

 (b) How much will be paid over the lease's full term of 39 months?

36. Average annual charges for tuition, fees, room, and board (in dollars) at private nonprofit 4-year universities in the United States are shown in the graph.

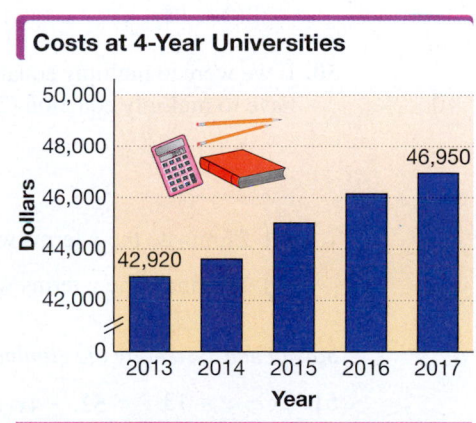

Costs at 4-Year Universities

Data from The College Board.

 (a) Use the information given for the years 2013 and 2017, letting $x = 13$ represent 2013 and $x = 17$ represent 2017, and letting y represent cost (in dollars), to write an equation that models the data. Write the equation in slope-intercept form. Interpret the slope.

 (b) Use the equation from part (a) to approximate costs at private 4-year universities in 2015.

7.3 *Determine whether the given ordered pair is a solution of the given system.*

37. $(3, 4)$

$$4x - 2y = 4$$
$$5x + \ y = 19$$

38. $(-5, 2)$

$$x - 4y = -13$$
$$2x + 3y = 4$$

Solve each system by graphing.

39. $x + y = 4$
$\ \ \ 2x - y = 5$

40. $x - 2y = 4$
$\ \ \ 2x + \ y = -2$

41. $2x + 4 = 2y$
$\ \ \ y - x = -3$

42. $x - 2\ = 2y$
$\ \ \ 2x - 4y = 4$

7.4 *Answer each question.*

43. To solve the following system by substitution, which variable in which equation would be easiest to solve for in the first step?

$$5x - 3y = 7$$
$$-x + 2y = 4$$

44. After solving a system of linear equations by the substitution method, a student obtained the false statement "$0 = 5$." What is the solution set of this system?

Solve each system by the substitution method.

45. $3x + y = 7$
$\ \ \ x = 2y$

46. $2x - 5y = -19$
$\ \ \ y = x + 2$

47. $5x + 15y = 30$
$\ \ \ x + \ 3y = 6$

48. $4x + 5y = 44$
$\ \ \ x + 2 = 2y$

7.5 *Answer each question.*

49. Which system does not require that we multiply one or both equations by a constant to solve the system by the elimination method?

A. $-4x + 3y = 7$
$\ \ \ \ \ \ 3x - 4y = 4$

B. $\ \ 5x + \ 8y = 13$
$\ \ \ \ \ \ 12x + 24y = 36$

C. $2x + 3y = 5$
$\ \ \ \ \ \ x - 3y = 12$

D. $\ \ x + 2y = 9$
$\ \ \ \ \ \ 3x - \ y = 6$

50. If we were to multiply equation (1) by -3 in the system below, by what number would we have to multiply equation (2) in order to do the following?

$$2x + 12y = 7 \quad (1)$$
$$3x + \ \ 4y = 1 \quad (2)$$

(a) Eliminate the x-terms when solving by the elimination method.

(b) Eliminate the y-terms when solving by the elimination method.

Solve each system by the elimination method.

51. $2x - y = 13$
$\ \ \ x + y = 8$

52. $-4x + 3y = 25$
$\ \ \ 6x - 5y = -39$

53. $5x - 3y = 11$
$\ \ \ x - 2y = 4$

54. $\ \ 2x + \ y = 3$
$\ \ \ \ \ \ -4x - 2y = 6$

7.3–7.5 *Solve each system by any method.*

55. $2x + 3y = -5$
$\ \ \ 3x + 4y = -8$

56. $6x - 9y = 0$
$\ \ \ 2x - 3y = 0$

57. $x - 2y = 5$
$\ \ \ y = x - 7$

58. $\dfrac{x}{2} + \dfrac{y}{3} = 7$

$\ \ \ \dfrac{x}{4} + \dfrac{2y}{3} = 8$

59. $\dfrac{3}{4}x - \dfrac{1}{3}y = \dfrac{7}{6}$

$\ \ \ \dfrac{1}{2}x + \dfrac{2}{3}y = \dfrac{5}{3}$

60. $0.3x + 0.2y = -0.5$
$\ \ \ 0.4x - 0.5y = -2.2$

7.6 *Solve each system. If a system is inconsistent or has dependent equations, say so.*

61. $2x + 3y - z = -16$
$x + 2y + 2z = -3$
$-3x + y + z = -5$

62. $3x - y - z = -8$
$4x + 2y + 3z = 15$
$-6x + 2y + 2z = 10$

63. $4x - y = 2$
$3y + z = 9$
$x + 2z = 7$

64. $5x - y = 26$
$4y + 3z = -4$
$x + z = 5$

65. $3x - 4y + z = 8$
$-6x + 8y - 2z = -16$
$\frac{3}{2}x - 2y + \frac{1}{2}z = 4$

66. $2x - y + 3z = 0$
$5x + y - z = 0$
$-2x + 3y + 4z = 0$

7.7 *Solve each problem.*

67. A regulation National Hockey League ice rink has perimeter 570 ft. The length of the rink is 30 ft longer than twice the width. What are the dimensions of an NHL ice rink? (Data from www.nhl.com)

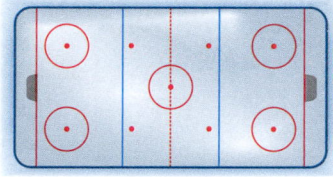

68. A plane flies 560 mi in 1.75 hr traveling with the wind. The return trip later against the same wind takes the plane 2 hr. Find the rate of the plane and the speed of the wind. Let x = the rate of the plane and y = the speed of the wind.

	r	t	d
With Wind	$x + y$	1.75	
Against Wind		2	

69. For Valentine's Day, Ms. Sweet will mix some $6-per-lb nuts with some $3-per-lb choco-late candy to obtain 100 lb of mix, which she will sell at $3.90 per lb. How many pounds of each should she use?

	Number of Pounds	Price per Pound (in dollars)	Value (in dollars)
Nuts	x		
Chocolate	y		
Mixture	100		

70. How many liters each of 8% and 20% hydrogen peroxide should be mixed together to obtain 8 L of 12.5% solution?

71. In the 2016 Summer Olympics in Rio de Janeiro, Japan earned 9 fewer gold medals than bronze. The number of silver medals earned was 34 less than twice the number of bronze medals. Japan earned a total of 41 medals. How many of each kind of medal did Japan earn? (Data from *The World Almanac and Book of Facts.*)

72. The sum of the measures of the angles of any triangle is 180°. The largest angle measures 10° less than the sum of the other two. The measure of the middle-sized angle is the average of the other two. Find the measures of the three angles.

73. Noemi sells real estate. On three recent sales, she made 10% commission, 6% commission, and 5% commission. Her total commissions on these sales were $17,000, and she sold property worth $280,000. If the 5% sale amounted to the sum of the other two, what were the three sales prices?

74. In the great baseball year of 1961, Yankee teammates Mickey Mantle, Roger Maris, and Yogi Berra combined for 137 home runs. Mantle hit 7 fewer than Maris, and Maris hit 39 more than Berra. What were the home run totals for each player? (Data from www.mlb.com)

| Chapter 7 | **Mixed Review Exercises** |

Determine whether each pair of lines is parallel, perpendicular, *or* neither.

1. $4x + 3y = 8$ and $6y = 7 - 8x$
2. $3x + y = 4$ and $3y = x - 6$

The graph shows per capita consumption of potatoes (in pounds) in the United States from 2008 to 2016.

3. Use the given ordered pairs to find and interpret the average rate of change in per capita potato consumption per year to the nearest tenth during this period.

4. Write an equation in slope-intercept form that models per capita consumption of potatoes y (in pounds) in year x, where $x = 0$ represents 2008.

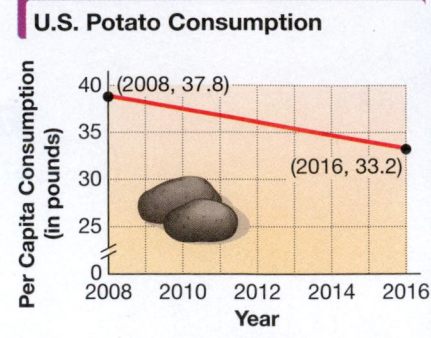

U.S. Potato Consumption

Data from U.S. Department of Agriculture.

Write an equation of a line (in the form specified, if given) passing through the given point(s) and satisfying any given conditions.

5. $(0, 0)$; $m = 3$ (slope-intercept form)
6. $(0, 3)$ and $(-2, 4)$ (standard form)

7. Suppose that two linear equations are graphed on the same set of coordinate axes. Sketch what the graph might look like if the system has the given description.

(a) The system has a single solution.

(b) The system has no solution.

(c) The system has infinitely many solutions.

8. Which system would be easier to solve using the substitution method? Why?

$$\text{System A:} \quad 5x - 3y = 7 \qquad \text{System B:} \quad 7x + 2y = 4$$
$$2x + 8y = 3 \qquad\qquad\qquad y = -3x + 1$$

Solve each system by any method.

9. $-7x + 3y = 12$
$5x + 2y = 8$

10. $x + 4y = 17$
$-3x + 2y = -9$

11. $x = 7y + 10$
$2x + 3y = 3$

12. $x + 2y = 48$
$\dfrac{1}{4}x + \dfrac{1}{2}y = 12$

13. $x + y - z = 0$
$2y - z = 1$
$2x + 3y - 4z = -4$

14. $x + 3y - 6z = 7$
$2x - y + z = 1$
$x + 2y + 2z = -1$

Solve each problem.

15. To make a 10% acid solution, Jeffrey wants to mix some 5% solution with 10 L of 20% solution. How many liters of 5% solution should he use?

16. In the 2016 Summer Olympics, China, the United States, and Great Britain won a combined total of 258 medals. China won 51 fewer medals than the United States. Great Britain won 54 fewer medals than the United States. How many medals did each country win? (Data from *The World Almanac and Book of Facts*.)

Chapter 7   Test

FOR EXTRA HELP *Step-by-step test solutions are found on the Chapter Test Prep Videos available in* **MyLab Math**.

▶ *View the complete solutions to all Chapter Test exercises in MyLab Math.*

Find the x- and y-intercepts. Then graph each equation.

1. $4x - 3y = -12$ **2.** $x = 2$ **3.** $y = -2x$

4. Find the slope of the line passing through the points $(6, 4)$ and $(-4, -1)$.

5. Find the slope and the x- and y-intercepts of the graph of $3x - 2y = 13$.

6. Which line has positive slope and negative y-coordinate for its y-intercept?

A.   **B.** 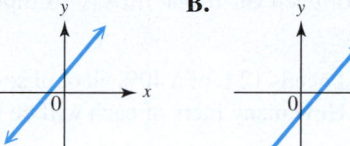   **C.** 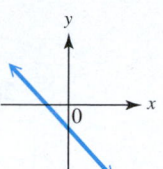   **D.**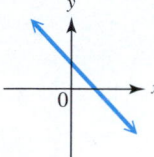

7. Determine whether the pair of lines $5x - y = 8$ and $5y = -x + 3$ is *parallel, perpendicular,* or *neither.*

8. In 2000, there were 94,000 farms in Iowa. As of 2016, there were 87,000. Find and interpret the average rate of change in the number of farms per year, to the nearest whole number. (Data from U.S. Department of Agriculture.)

*Write an equation of the line passing through the given point(s) and satisfying any given conditions. Give the equation (**a**) in slope-intercept form and (**b**) in standard form.*

9. $(0, 3)$; slope $-\dfrac{2}{5}$ **10.** $(-2, 3)$ and $(6, -1)$ **11.** $(4, -1)$; $m = -5$

Write an equation of the line passing through the given point and satisfying the given condition.

12. $(-3, 14)$; horizontal   **13.** $(5, -6)$; vertical

14. Write an equation in slope-intercept form of the line passing through the point $(-7, 2)$ and satisfying the given condition.

   (a) parallel to $3x + 5y = 6$ **(b)** perpendicular to $y = 2x$

15. A ticket to a rock concert costs \$142.75. An advance parking pass costs \$45. Let x represent the number of tickets purchased and y represent the total cost in dollars.

   (a) Write an equation in slope-intercept form that represents this situation.

   (b) How much does it cost for 6 tickets and a parking pass?

16. Solve the system by graphing. $\begin{aligned} x + y &= 7 \\ x - y &= 5 \end{aligned}$

Solve each system. If a system is inconsistent or has dependent equations, say so.

17. $2x - 3y = 24$
$\quad y = -\dfrac{2}{3}x$

18. $3x - y = -8$
$\quad 2x + 6y = 3$

19. $12x - 5y = 8$
$\quad 3x = \dfrac{5}{4}y + 2$

20. $3x + y = 12$
$\quad 2x - y = 3$

21. $-5x + 2y = -4$
$\quad 6x + 3y = -6$

22. $3x + 4y = 8$
$\quad 8y = 7 - 6x$

23. $3x + 5y + 3z = 2$
$6x + 5y + z = 0$
$3x + 10y - 2z = 6$

24. $4x + y + z = 11$
$x - y - z = 4$
$y + 2z = 0$

25. $-2x + 4y + 10z = 3$
$x - 2y - 5z = 1$
$x + 2y + z = -1$

Solve each problem.

26. Two top-grossing super hero films, *Captain America: Civil War* and *Deadpool,* earned $771.2 million together. If *Deadpool* grossed $45 million less than *Captain America,* how much did each film gross? (Data from comScore, Inc.)

27. Two cars start from points 420 mi apart and travel toward each other. They meet after 3.5 hr. Find the average rate of each car if one travels 30 mph slower than the other.

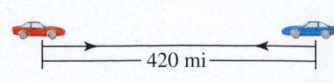

28. A chemist needs 12 L of a 40% alcohol solution. She must mix a 20% solution and a 50% solution. How many liters of each will be required to obtain what she needs?

Liters of Solution	Percent (as a decimal)	Liters of Pure Alcohol

29. The largest angle of a triangle measures 10° less than twice the middle-sized angle. The middle-sized angle measures 5° more than twice the smallest angle. The sum of the measures of the angles of any triangle is 180°. Find the measure of each angle.

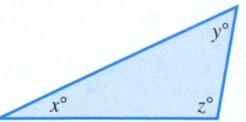

30. The owner of a tea shop wants to mix three kinds of tea to make 100 oz of a mixture that will sell for $0.83 per oz. He uses Orange Pekoe, which sells for $0.80 per oz, Irish Breakfast, for $0.85 per oz, and Earl Grey, for $0.95 per oz. If he wants to use twice as much Orange Pekoe as Irish Breakfast, how much of each kind of tea should he use?

Chapters R–7 | Cumulative Review Exercises

1. Add or subtract as indicated.

   (a) $\dfrac{1}{6} + \dfrac{3}{4}$ **(b)** $\dfrac{11}{12} - \dfrac{1}{4}$ **(c)** $7.5 - 2.75$

2. Multiply or divide as indicated.

   (a) $6 \div \dfrac{3}{4}$ **(b)** 1.5×100 **(c)** $1.25 \div 10$

3. Complete the table of fraction, decimal, and percent equivalents.

	Fraction in Lowest Terms (or Whole Number)	Decimal	Percent
(a)	$\frac{1}{100}$		
(b)		0.5	
(c)			75%
(d)		1.0	

Determine whether each statement is always true, sometimes true, *or* never true. *If the statement is* sometimes true, *give examples for which it is true and for which it is false.*

4. The absolute value of a negative number equals the additive inverse of the number.

5. The sum of two negative numbers is positive.

6. The sum of a positive number and a negative number is 0.

Evaluate.

7. 3^2

8. -3^2

9. $(-3)^2$

10. Evaluate $-3(2q - 3p)$ for $p = -4$, $q = \frac{1}{2}$, and $r = 16$.

Solve each equation or inequality.

11. $2z - 5 + 3z = 4 - (z + 2)$

12. $0.04x + 0.06(x - 1) = 1.04$

13. $ax + by = d$ for x

14. $2x + 3 > 5$

Solve each problem.

15. A jar contains only pennies, nickels, and dimes. The number of dimes is one more than the number of nickels, and the number of pennies is six more than the number of nickels. How many of each denomination are in the jar, if the total value is $4.80?

16. Two angles of a triangle have the same measure. The measure of the third angle is 4° less than twice the measure of each of the equal angles. Find the measures of the three angles. (*Hint:* The sum of the measures of the angles of any triangle is 180°.)

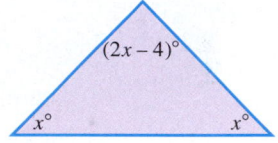

In Exercises 17-21, Point A has coordinates $(-2, 6)$ and point B has coordinates $(4, -2)$.

17. What is the equation of the horizontal line through *A*?

18. What is the equation of the vertical line through *B*?

19. What is the slope of line *AB*?

20. What is the slope of a line perpendicular to line *AB*?

21. What is the standard form of the equation of line *AB*?

22. Graph the line having slope $\frac{2}{3}$ and passing through the point $(-1, -3)$.

Perform the indicated operations.

23. $(7x + 3y)^2$

24. $\dfrac{2x^4 + x^3 + 7x^2 + 2x + 6}{x^2 + 2}$

25. $(3x^3 + 4x^2 - 7) - (2x^3 - 8x^2 + 3x)$

26. (a) Write 0.0004638 using scientific notation.

   (b) Write 5.66×10^5 in standard notation.

Factor.

27. $16w^2 + 50w - 21$

28. $100x^4 - 81$

29. $8p^3 + 27$

30. Solve $9x^2 = 6x - 1$.

Solve each problem.

31. A sign is to have the shape of a triangle with a height 3 ft greater than the length of the base. How long should the base be if the area is to be 14 ft²?

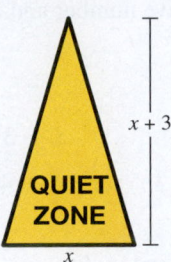

QUIET ZONE

$x + 3$

x

32. A game board has the shape of a rectangle. The longer sides are each 2 in. longer than the distance between them. The area of the board is 288 in.². Find the length of the longer sides and the distance between them.

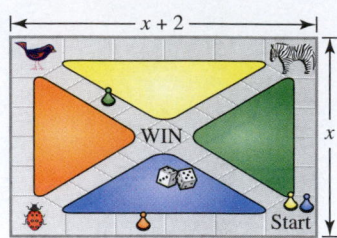

$x + 2$

WIN

Start

x

Perform each indicated operation. Write answers in lowest terms.

33. $\dfrac{8}{x + 1} - \dfrac{2}{x + 3}$

34. $\dfrac{x^2 + 5x + 6}{3x} \div \dfrac{x^2 - 4}{x^2 + x - 6}$

35. Simplify $\dfrac{\dfrac{12}{x + 6}}{\dfrac{4}{2x + 12}}$.

36. Solve $\dfrac{2}{x - 1} = \dfrac{5}{x - 1} - \dfrac{3}{4}$.

Solve each system.

37. $-2x + 3y = -15$
$\quad\ 4x -\ \ y = 15$

38. $\ \ x - 3y = 7$
$\quad 2x - 6y = 14$

39. $\ \ x + y + z = 10$
$\quad x - y - z = 0$
$\ -x + y - z = -4$

40. The graph shows a company's costs to produce computer parts and the revenue from the sale of computer parts.

(a) At what production level does the cost equal the revenue? What is the revenue at that point?

(b) Profit is revenue less cost. Estimate the profit on the sale of 1100 parts.

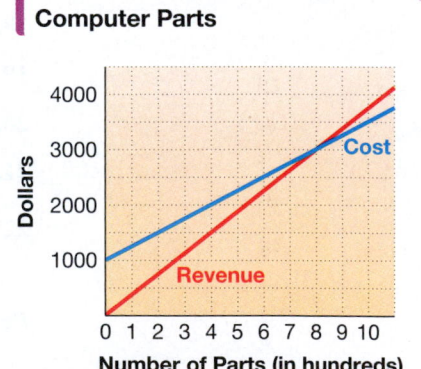

Computer Parts

Dollars (4000, 3000, 2000, 1000)

Cost

Revenue

0 1 2 3 4 5 6 7 8 9 10
Number of Parts (in hundreds)

STUDY SKILLS REMINDER

It is not too soon to begin preparing for your final exam. **Review Study Skill 10,** ***Preparing for Your Math Final Exam.***

INEQUALITIES AND ABSOLUTE VALUE

In manufacturing and quality control situations, relative error, or tolerance, in measurements uses *inequality* and *absolute value*, two topics covered in this chapter.

8

8.1   Review of Linear Inequalities in One Variable

OBJECTIVES

1 Review inequalities and interval notation.
2 Solve linear inequalities using the addition property.
3 Solve linear inequalities using the multiplication property.
4 Solve linear inequalities with three parts.

OBJECTIVE 1 Review inequalities and interval notation.

An **inequality** consists of algebraic expressions related by one of the symbols $<$ ("*is less than*"), $\leq$ ("*is less than or equal to*"), $>$ ("*is greater than*"), or $\geq$ ("*is greater than or equal to*"). We *solve an inequality* by finding all real number solutions for it.

For example, the solution set of $x \leq 2$ includes *all* real numbers that are less than or equal to 2. This set of numbers is an example of an **interval** on a number line. We write this interval using **interval notation** as $(-\infty, 2]$. The negative infinity symbol $-\infty$ does not indicate a number, but shows that the interval includes *all* real numbers less than 2. The square bracket indicates that 2 is included in the solution set, which is graphed in **FIGURE 1**.

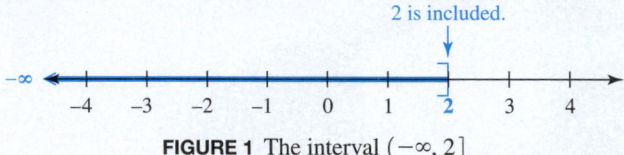

FIGURE 1 The interval $(-\infty, 2]$

Remember the following important concepts regarding interval notation.

- A parenthesis indicates that an endpoint is *not* included.

- A square bracket indicates that an endpoint is included.

- A parenthesis is always used next to an infinity symbol, $-\infty$ or ∞.

- The set of real numbers is written in interval notation as $(-\infty, \infty)$.

VOCABULARY

☐ inequality
☐ interval
☐ linear inequality in one variable
☐ three-part inequality

▼ Summary of Types of Intervals

Type of Interval	Set-Builder Notation	Graph	Interval Notation
Open interval	$\{x \mid a < x < b\}$		(a, b)
Closed interval	$\{x \mid a \leq x \leq b\}$		$[a, b]$
Half-open (or half-closed) interval	$\{x \mid a \leq x < b\}$		$[a, b)$
	$\{x \mid a < x \leq b\}$		$(a, b]$
Disjoint interval	$\{x \mid x < a \text{ or } x > b\}$		$(-\infty, a) \cup (b, \infty)$
	$\{x \mid x > a\}$		(a, ∞)
	$\{x \mid x \geq a\}$		$[a, \infty)$
Infinite interval	$\{x \mid x < a\}$		$(-\infty, a)$
	$\{x \mid x \leq a\}$		$(-\infty, a]$
	$\{x \mid x \text{ is a real number}\}$		$(-\infty, \infty)$

STUDY SKILLS REMINDER

Study cards are a great way to learn vocabulary, procedures, and so on.
Review Study Skill 5, Using Study Cards.

Linear Inequality in One Variable

A **linear inequality in one variable** (here x) is an inequality that can be written in the form

$$ax + b < 0, \quad ax + b \le 0, \quad ax + b > 0, \quad \text{or} \quad ax + b \ge 0,$$

where a and b are real numbers and $a \ne 0$.

Examples: $x + 5 < 0$, $2x + 5 \le 10$, $x > -1$ — Linear inequalities in one variable

OBJECTIVE 2 Solve linear inequalities using the addition property.

We solve an inequality by finding all numbers that make the inequality true. We use two important properties to solve inequalities.

Addition Property of Inequality

If a, b, and c represent real numbers, then the inequalities

$$a < b \quad \text{and} \quad a + c < b + c \quad \text{are equivalent.*}$$

That is, the same number may be added to (or subtracted from) each side of an inequality without changing the solution set.

*This also applies to $a \le b$, $a > b$, and $a \ge b$.

**NOW TRY
EXERCISE 1**

Solve $x - 10 > -7$, and graph the solution set.

EXAMPLE 1   Using the Addition Property of Inequality

Solve $x - 7 < -12$, and graph the solution set.

$$x - 7 < -12$$
$$x - 7 + 7 < -12 + 7 \qquad \text{Add 7.}$$
$$x < -5 \qquad \text{Combine like terms.}$$

CHECK   Substitute -5 for x in the *equation* $x - 7 = -12$.

$$x - 7 = -12$$
$$-5 - 7 \overset{?}{=} -12 \qquad \text{Let } x = -5.$$
$$-12 = -12 \quad \checkmark \qquad \text{True}$$

The result, a true statement, shows that -5 is the boundary point. Now test a number on each side of -5 to verify that numbers *less than* -5 make the inequality true. We choose -6 and -4.

$$x - 7 < -12$$

$-6 - 7 \overset{?}{<} -12$ Let $x = -6$.	$-4 - 7 \overset{?}{<} -12$ Let $x = -4$.
$-13 < -12$ $\checkmark$ True	$-11 < -12$ False
-6 *is* in the solution set.	-4 is *not* in the solution set.

The check confirms that the solution set is the graph shown in **FIGURE 2**.

NOW TRY ANSWER

1.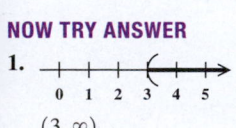

$(3, \infty)$

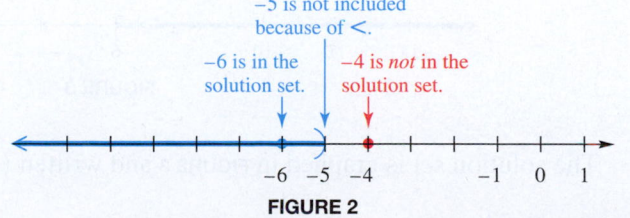

-5 is not included because of $<$.

-6 is in the solution set. -4 is *not* in the solution set.

FIGURE 2

Using interval notation, the solution set is the infinite interval $(-\infty, -5)$.

NOW TRY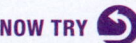

OBJECTIVE 3 Solve linear inequalities using the multiplication property.

Consider the following true statement.

$$-2 < 5$$

Multiply each side by some positive number—for example, 8.

$$-2(8) < 5(8) \qquad \text{Multiply by 8.}$$

$$-16 < 40 \qquad \text{True}$$

The result is true. Start again with $-2 < 5$, and multiply each side by some negative number—for example, -8.

$$-2(-8) < 5(-8) \qquad \text{Multiply by } -8.$$

$$16 < -40 \qquad \text{False}$$

The result is false. To make it true, we must change the direction of the inequality symbol.

$$16 > -40 \qquad \text{True}$$

As these examples suggest, multiplying each side of an inequality by a *negative* number requires reversing the direction of the inequality symbol.

Multiplication Property of Inequality

Let a, b, and c represent real numbers, where $c \neq 0$.

(a) If c is *positive*, then the inequalities

$$a < b \quad \text{and} \quad ac < bc \quad \text{are equivalent.*}$$

(b) If c is *negative*, then the inequalities

$$a < b \quad \text{and} \quad ac > bc \quad \text{are equivalent.*}$$

That is, each side of an inequality may be multiplied (or divided) by the same *positive* number without changing the direction of the inequality symbol. *If the multiplier is negative, we must reverse the direction of the inequality symbol.*

*This also applies to $a \leq b$, $a > b$, and $a \geq b$.

EXAMPLE 2   Using the Multiplication Property of Inequality

Solve each inequality, and graph the solution set.

(a) $5x \leq -30$

Divide each side by 5. *Because $5 > 0$, do not reverse the direction of the inequality symbol.*

$$5x \leq -30$$

$$\frac{5x}{5} \leq \frac{-30}{5} \qquad \text{Divide by 5.}$$

$$x \leq -6$$

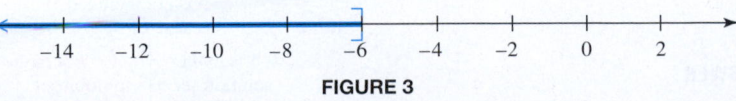

FIGURE 3

The solution set is graphed in **FIGURE 3** and written $(-\infty, -6]$.

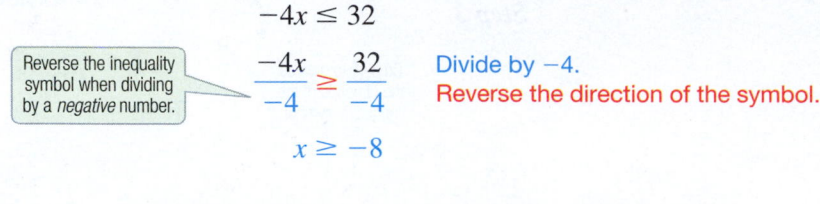

**NOW TRY
EXERCISE 2**

Solve each inequality, and graph the solution set.

(a) $8x \geq -40$

(b) $-20x > -60$

(b) $-4x \leq 32$

Divide each side by -4. *Because* $-4 < 0$, *reverse the direction of the inequality symbol.*

$$-4x \leq 32$$

Reverse the inequality symbol when dividing by a negative *number.*

$$\frac{-4x}{-4} \geq \frac{32}{-4}$$ Divide by -4.
Reverse the direction of the symbol.

$$x \geq -8$$

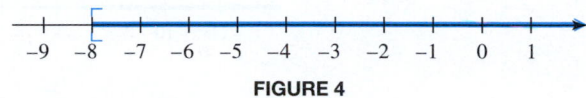

FIGURE 4

FIGURE 4 shows the graph of the solution set, $[-8, \infty)$.

NOW TRY

To solve a linear inequality in one variable, use the following steps.

Solving a Linear Inequality in One Variable

Step 1 **Simplify each side separately.** Use the distributive property as needed.

- Clear any parentheses.
- Clear any fractions or decimals.
- Combine like terms.

Step 2 **Isolate the variable terms on one side.** Use the addition property of inequality so that all terms with variables are on one side of the inequality and all constants (numbers) are on the other side.

Step 3 **Isolate the variable.** Use the multiplication property of inequality to obtain an inequality in one of the following forms, where k is a constant (number).

$$\text{variable} < k, \quad \text{variable} \leq k, \quad \text{variable} > k, \quad \text{or} \quad \text{variable} \geq k$$

Remember: Reverse the direction of the inequality symbol only when multiplying or dividing each side of an inequality by a negative number.

EXAMPLE 3 Solving a Linear Inequality

Solve $-3(x + 4) + 2 \geq 7 - x$, and graph the solution set.

Step 1 $\quad -3(x + 4) + 2 \geq 7 - x$

$\qquad -3x - 3(4) + 2 \geq 7 - x$ Distributive property

$\qquad -3x - 12 + 2 \geq 7 - x$ Multiply.

$\qquad -3x - 10 \geq 7 - x$ (*) Combine like terms.

Step 2 $\quad -3x - 10 + x \geq 7 - x + x$ Add x.

$\qquad -2x - 10 \geq 7$ Combine like terms.

NOW TRY ANSWERS

2. (a)

$[-5, \infty)$

(b)

$(-\infty, 3)$

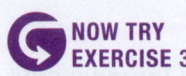
NOW TRY EXERCISE 3

Solve and graph the solution set.

$5 - 2(x - 4) \le 11 - 4x$

$$-2x - 10 + 10 \ge 7 + 10 \qquad \text{Add 10.}$$

$$-2x \ge 17 \qquad \text{Combine like terms.}$$

Step 3 $\qquad \dfrac{-2x}{-2} \le \dfrac{17}{-2} \qquad$ Divide by -2.
Change $\ge$ to $\le$.

> Be sure to reverse the direction of the inequality symbol.

$$x \le -\dfrac{17}{2}$$

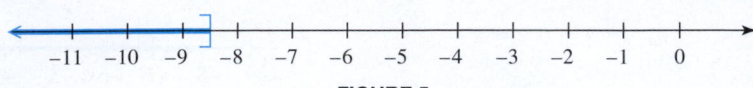

FIGURE 5

FIGURE 5 shows the graph of the solution set, $\left(-\infty, -\dfrac{17}{2}\right]$.

NOW TRY

NOW TRY EXERCISE 4

Solve and graph the solution set.

$\dfrac{3}{4}(x - 2) + \dfrac{1}{2} > \dfrac{1}{5}(x - 8)$

EXAMPLE 4   Solving a Linear Inequality with Fractions

Solve $-\dfrac{2}{3}(x - 3) - \dfrac{1}{2} < \dfrac{1}{2}(5 - x)$, and graph the solution set.

$$-\dfrac{2}{3}(x - 3) - \dfrac{1}{2} < \dfrac{1}{2}(5 - x)$$

Step 1 $\qquad -\dfrac{2}{3}x + 2 - \dfrac{1}{2} < \dfrac{5}{2} - \dfrac{1}{2}x \qquad$ Clear parentheses.

$$6\left(-\dfrac{2}{3}x + 2 - \dfrac{1}{2}\right) < 6\left(\dfrac{5}{2} - \dfrac{1}{2}x\right) \qquad \begin{array}{l}\text{To clear the fractions}\\\text{multiply by 6, the LCD.}\end{array}$$

$$6\left(-\dfrac{2}{3}x\right) + 6(2) + 6\left(-\dfrac{1}{2}\right) < 6\left(\dfrac{5}{2}\right) + 6\left(-\dfrac{1}{2}x\right) \qquad \text{Distributive property}$$

$$-4x + 12 - 3 < 15 - 3x \qquad \text{Multiply.}$$

$$-4x + 9 < 15 - 3x \qquad \text{Combine like terms.}$$

Step 2 $\qquad -4x + 9 + 3x < 15 - 3x + 3x \qquad$ Add $3x$.

$$-x + 9 < 15 \qquad \text{Combine like terms.}$$

$$-x + 9 - 9 < 15 - 9 \qquad \text{Subtract 9.}$$

$$-x < 6 \qquad \text{Combine like terms.}$$

Step 3 $\qquad -1(-x) > -1(6) \qquad$ Multiply by -1.
Change $<$ to $>$.

> Reverse the inequality symbol when multiplying by a *negative* number.

$$x > -6$$

NOW TRY ANSWERS

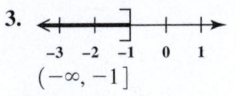
3.
$(-\infty, -1]$

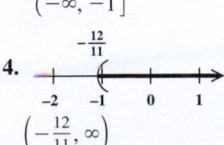
4.
$\left(-\dfrac{12}{11}, \infty\right)$

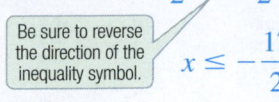

FIGURE 6

FIGURE 6 shows the graph of the solution set, $(-6, \infty)$.

NOW TRY

OBJECTIVE 4 Solve linear inequalities with three parts.

Some applications involve a **three-part inequality** such as

$$3 < x + 2 < 8, \quad \text{where } x + 2 \text{ is } \textit{between } 3 \text{ and } 8.$$

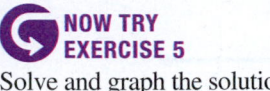

NOW TRY EXERCISE 5

Solve and graph the solution set.

$$-1 < x - 2 < 3$$

EXAMPLE 5 Solving a Three-Part Inequality

Solve $3 < x + 2 < 8$, and graph the solution set.

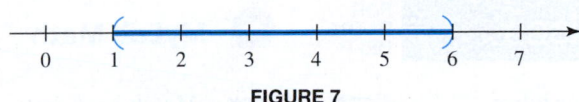
Do not write $8 < x + 2 < 3$, which implies $8 < 3$, a false statement.

$$3 < \quad x + 2 \quad < 8$$

$$3 - 2 < x + 2 - 2 < 8 - 2 \qquad \text{Subtract 2 from all three parts.}$$

$$1 < \quad x \quad < 6$$

Thus, x must be between 1 and 6 so that $x + 2$ will be between 3 and 8.

(number line: 0 1 2 3 4 5 6 7, open interval from 1 to 6)

FIGURE 7

The solution set is graphed in **FIGURE 7** and written as the open interval $(1, 6)$.

NOW TRY

⚠ **CAUTION**   *Write three-part inequalities so that the symbols point in the same direction, and both point toward the lesser number.*

NOW TRY EXERCISE 6

Solve and graph the solution set.

$$-2 < -4x - 5 \le 7$$

EXAMPLE 6 Solving a Three-Part Inequality

Solve $-2 \le -3x - 1 \le 5$, and graph the solution set.

$$-2 \le \quad -3x - 1 \quad \le 5$$

$$-2 + 1 \le -3x - 1 + 1 \le 5 + 1 \qquad \text{Add 1 to each part.}$$

$$-1 \le \quad -3x \quad \le 6$$

$$\frac{-1}{-3} \ge \frac{-3x}{-3} \ge \frac{6}{-3} \qquad \begin{array}{l}\text{Divide each part by } -3.\\ \text{Reverse the direction of the}\\ \text{inequality symbols.}\end{array}$$

$$\frac{1}{3} \ge \quad x \quad \ge -2$$

$$-2 \le \quad x \quad \le \frac{1}{3} \qquad \begin{array}{l}\text{Rewrite in the}\\ \text{order on the}\\ \text{number line.}\end{array}$$

NOW TRY ANSWERS

5. (number line: -1 0 1 2 3 4 5, open interval from 1 to 5)
$(1, 5)$

6. (number line: -4 -3 -2 -1 0 1, with $-\frac{3}{4}$ marked)
$\left[-3, -\frac{3}{4}\right)$

(number line: -3 -2 -1 0 $\frac{1}{3}$ 1 2, closed interval from -2 to $\frac{1}{3}$)

FIGURE 8

The solution set is graphed in **FIGURE 8** and written as the closed interval $\left[-2, \frac{1}{3}\right]$.

NOW TRY

▼ **Solution Sets of Equations and Inequalities**

Equation or Inequality	Typical Solution Set	Graph of Solution Set
Linear equation $5x + 4 = 14$	$\{2\}$	
Linear inequality $5x + 4 < 14$	$(-\infty, 2)$	
$5x + 4 > 14$	$(2, \infty)$	
Three-part inequality $-1 \leq 5x + 4 \leq 14$	$[-1, 2]$	

8.1 Exercises

FOR EXTRA HELP   ▶ **MyLab Math**

▶ *Video solutions for select problems available in MyLab Math*

Concept Check   *Match each inequality in Column I with the correct graph or interval in Column II.*

I

1. $x \leq 3$ **2.** $x > 3$

3. $x < 3$ **4.** $x \geq 3$

5. $-3 \leq x \leq 3$

6. $-3 < x < 3$

II

A.

B.

C. $(3, \infty)$ **D.** $(-\infty, 3]$

E. $(-3, 3)$ **F.** $[-3, 3]$

Concept Check   *Work each problem involving inequalities.*

7. A high level of LDL cholesterol ("bad cholesterol") in the blood increases a person's risk of heart disease. The table shows how LDL levels affect risk.

   If x represents the LDL cholesterol number, write a linear inequality or three-part inequality for each category. Use x as the variable.

   (a) Optimal

   (b) Near optimal/above optimal

   (c) Borderline high

   (d) High

   (e) Very high

LDL Cholesterol	Risk Category
Less than 100	Optimal
100–129	Near optimal/above optimal
130–159	Borderline high
160–189	High
190 and above	Very high

Data from WebMD.

8. A high level of triglycerides in the blood also increases a person's risk of heart disease. The table shows how triglyceride levels affect risk.

   If x represents the triglycerides number, write a linear inequality or three-part inequality for each category. Use x as the variable.

   (a) Normal   **(b)** Mildly high

   (c) High   **(d)** Very high

Triglycerides	Risk Category
Less than 100	Normal
100–199	Mildly high
200–499	High
500 or higher	Very high

Data from WebMD.

9. Concept Check A student solved the following inequality incorrectly as shown.

$$4x \geq -64$$

$$\frac{4x}{4} \leq \frac{-64}{4}$$

$$x \leq -16 \qquad (-\infty, -16]$$

Solution set: $(-\infty, -16]$

WHAT WENT WRONG? Give the correct solution set.

10. Concept Check A student solved the following inequality incorrectly as shown.

$$-2x < -18$$

$$\frac{-2x}{-2} < \frac{-18}{-2}$$

$$x < 9 \qquad (-\infty, 9)$$

Solution set: $(-\infty, 9)$

WHAT WENT WRONG? Give the correct solution set.

Solve each inequality. Graph the solution set, and write it using interval notation. See Examples 1–4.

11. $x - 4 \geq 12$

12. $x - 3 \geq 7$

13. $3k + 1 > 22$

14. $5x + 6 < 76$

15. $4x < -16$

16. $2x > -10$

17. $-4x < 16$

18. $-5x > 25$

19. $-\dfrac{3}{4}x \geq 30$

20. $-\dfrac{2}{3}x \leq 12$

21. $-1.3x \geq -5.2$

22. $-2.5x \leq -1.25$

23. $5x + 2 \leq -48$

24. $4x + 1 \leq -31$

25. $\dfrac{3k - 1}{4} > 5$

26. $\dfrac{5x - 6}{8} < 8$

27. $\dfrac{2x - 5}{-4} > 5$

28. $\dfrac{3x - 2}{-5} < 6$

29. $3k + 1 < -20$

30. $5z + 6 > -29$

31. $6x - 4 \geq -2x$

32. $2x - 8 \geq -2x$

33. $x - 2(x - 4) \leq 3x$

34. $x - 3(x + 1) \leq 4x$

35. $-(4 + r) + 2 - 3r < -14$

36. $-(9 + x) - 5 + 4x \geq 4$

37. $-3(x - 6) > 2x - 2$

38. $-2(x + 4) \leq 6x + 16$

39. $\dfrac{2}{3}(3x - 1) \geq \dfrac{3}{2}(2x - 3)$

40. $\dfrac{7}{5}(10x - 1) < \dfrac{2}{3}(6x + 5)$

41. $-\dfrac{1}{4}(p + 6) + \dfrac{3}{2}(2p - 5) < 10$

42. $\dfrac{3}{5}(t - 2) - \dfrac{1}{4}(2t - 7) \leq 3$

43. $3(2x - 4) - 4x < 2x + 3$

44. $7(4 - x) + 5x < 2(16 - x)$

45. $8\left(\dfrac{1}{2}x + 3\right) < 8\left(\dfrac{1}{2}x - 1\right)$

46. $10\left(\dfrac{1}{5}x + 2\right) < 10\left(\dfrac{1}{5}x + 1\right)$

Solve each inequality. Graph the solution set, and write it using interval notation. See Examples 5 and 6.

47. $-4 < x - 5 < 6$

48. $-1 < x + 1 < 8$

49. $-9 \leq x + 5 \leq 15$

50. $-4 \leq x + 3 \leq 10$

51. $-6 \leq 2x + 4 \leq 16$

52. $-15 < 3x + 6 < -12$

53. $-19 \leq 3x - 5 \leq 1$

54. $-16 < 3x + 2 < -10$

55. $4 \leq -9x + 5 < 8$

56. $4 \leq -2x + 3 < 8$

57. $-8 \leq -4x + 2 \leq 6$

58. $-12 \leq -6x + 3 \leq 15$

59. $-3 < \dfrac{3}{4}x < 6$

60. $-4 < \dfrac{2}{3}x < 12$

61. $-1 \leq \dfrac{2x - 5}{6} \leq 5$

62. $-3 \leq \dfrac{3x + 1}{4} \leq 3$

63. $-5 \leq \dfrac{6 - 5x}{2} \leq 0$

64. $-4 \leq \dfrac{2 - 4x}{3} \leq 0$

8.2   Set Operations and Compound Inequalities

OBJECTIVES

1 Recognize set intersection and union.
2 Find the intersection of two sets.
3 Solve compound inequalities with the word *and*.
4 Find the union of two sets.
5 Solve compound inequalities with the word *or*.

VOCABULARY
☐ intersection
☐ compound inequality
☐ union

OBJECTIVE 1 Recognize set intersection and union.

Consider the two sets A and B defined as follows.

$$A = \{1, 2, 3\}, \qquad B = \{2, 3, 4\}$$

The set of all elements that belong to both A **and** B, called their *intersection* and symbolized $A \cap B$, is given by

$$A \cap B = \{2, 3\}. \qquad \text{Intersection}$$

The set of all elements that belong to either A **or** B, or both, called their *union* and symbolized $A \cup B$, is given by

$$A \cup B = \{1, 2, 3, 4\}. \qquad \text{Union}$$

OBJECTIVE 2 Find the intersection of two sets.

The intersection of two sets is defined with the word *and*.

Intersection of Sets

For any two sets A and B, the **intersection** of A and B, symbolized $A \cap B$, is defined as follows.

$$A \cap B = \{x \mid x \text{ is an element of } A \text{ and } x \text{ is an element of } B\}$$

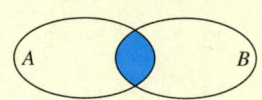

NOW TRY
EXERCISE 1

Let $A = \{2, 4, 6, 8\}$ and $B = \{0, 2, 6, 8\}$. Find $A \cap B$.

EXAMPLE 1 Finding the Intersection of Two Sets

Let $A = \{1, 2, 3, 4\}$ and $B = \{2, 4, 6\}$. Find $A \cap B$.

The set $A \cap B$ contains those elements that belong to both A *and* B.

$$A \cap B = \{1, 2, 3, 4\} \cap \{2, 4, 6\}$$
$$= \{2, 4\}$$

NOW TRY

OBJECTIVE 3 Solve compound inequalities with the word *and*.

A **compound inequality** consists of two inequalities linked by a connective word.

$$x + 1 \leq 9 \quad \text{and} \quad x - 2 \geq 3$$
$$2x > 4 \quad \text{or} \quad 3x - 6 < 5$$

Compound inequalities

Solving a Compound Inequality with *and*

Step 1   Solve each inequality individually.

Step 2   The solution set of the compound inequality includes all numbers that satisfy both inequalities in Step 1—that is, the *intersection* of the solution sets.

NOW TRY
EXERCISE 2

Solve the compound inequality, and graph the solution set.

$x - 2 \le 5$ and $x + 5 \ge 9$

EXAMPLE 2 Solving a Compound Inequality with *and*

Solve the compound inequality, and graph the solution set.

$$x + 1 \le 9 \quad \text{and} \quad x - 2 \ge 3$$

Step 1 Solve each inequality individually.

$$x + 1 \le 9 \quad \text{and} \quad x - 2 \ge 3$$
$$x + 1 - 1 \le 9 - 1 \quad \text{and} \quad x - 2 + 2 \ge 3 + 2$$
$$x \le 8 \quad \text{and} \quad x \ge 5$$

Step 2 The solution set will include all numbers that satisfy *both* inequalities in Step 1 at the same time. The compound inequality is true whenever $x \le 8$ and $x \ge 5$ are both true. See the graphs in **FIGURE 9**.

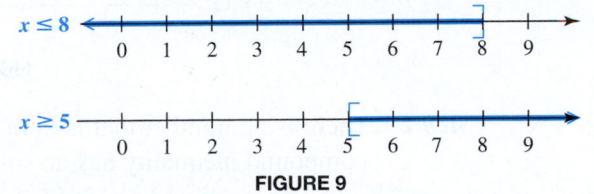

The set of points where the graphs "overlap" represents the intersection.

FIGURE 9

The intersection of the two graphs in **FIGURE 9** is the solution set. **FIGURE 10** shows the graph of the solution set, written using interval notation as the closed interval $[5, 8]$.

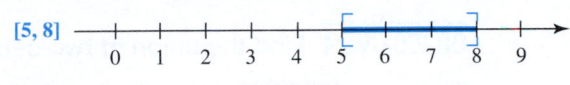

FIGURE 10

NOW TRY

NOW TRY
EXERCISE 3

Solve and graph.

$-4x - 1 < 7$ and
$\qquad 3x + 4 \ge -5$

EXAMPLE 3 Solving a Compound Inequality with *and*

Solve the compound inequality, and graph the solution set.

$$-3x - 2 > 5 \quad \text{and} \quad 5x - 1 \le -21$$

Step 1 Solve each inequality individually.

$$-3x - 2 > 5 \quad \text{and} \quad 5x - 1 \le -21$$
$$-3x > 7 \quad \text{and} \quad 5x \le -20$$
$$x < -\frac{7}{3} \quad \text{and} \quad x \le -4$$

Remember to reverse the direction of the inequality symbol.

The graphs of $x < -\frac{7}{3}$ and $x \le -4$ are shown in **FIGURE 11**.

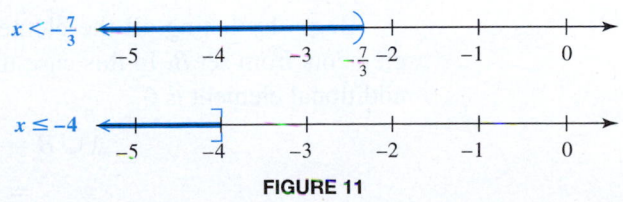

FIGURE 11

Step 2 Now find all values of x that are less than $-\frac{7}{3}$ and also less than or equal to -4. This is shown in **FIGURE 12**. The solution set is written as the infinite interval $(-\infty, -4]$.

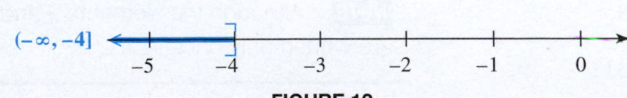

FIGURE 12

NOW TRY

NOW TRY ANSWERS

2.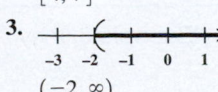
$[4, 7]$

3.

$(-2, \infty)$

NOW TRY
EXERCISE 4
Solve and graph.

$x - 7 < -12$ and

$2x + 1 > 5$

EXAMPLE 4 Solving a Compound Inequality with *and*

Solve the compound inequality, and graph the solution set.

$$x + 2 < 5 \quad \text{and} \quad x - 10 > 2$$

Step 1 Solve each inequality individually.

$$x + 2 < 5 \quad \text{and} \quad x - 10 > 2$$
$$x < 3 \quad \text{and} \quad x > 12$$

The graphs of $x < 3$ and $x > 12$ are shown in **FIGURE 13**.

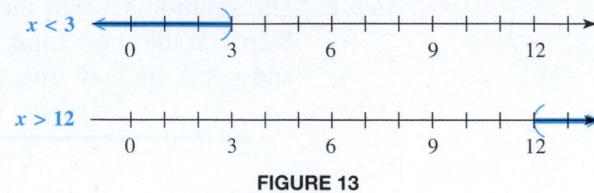

FIGURE 13

Step 2 There is no number that is both less than 3 *and* greater than 12, so the given compound inequality has no solution. See **FIGURE 14**. The solution set is $\varnothing$.

FIGURE 14   **NOW TRY**

OBJECTIVE 4 Find the union of two sets.

The union of two sets is defined with the word *or*.

> **Union of Sets**
>
> For any two sets A and B, the **union** of A and B, symbolized $A \cup B$, is defined as follows.
>
> $$A \cup B = \{x \mid x \text{ is an element of } A \textbf{ or } x \text{ is an element of } B\}$$
>
>

NOW TRY
EXERCISE 5
Let $A = \{5, 10, 15, 20\}$
and $B = \{5, 15, 25\}$.
Find $A \cup B$.

EXAMPLE 5 Finding the Union of Two Sets

Let $A = \{1, 2, 3, 4\}$ and $B = \{2, 4, 6\}$. Find $A \cup B$.

Begin by listing all the elements of set A: 1, 2, 3, 4. Then list any additional elements from set B. In this case the elements 2 and 4 are already listed, so the only additional element is 6.

$$A \cup B = \{1, 2, 3, 4\} \cup \{2, 4, 6\}$$
$$= \{1, 2, 3, 4, 6\}$$

The union consists of all elements in either A *or* B (or both).   **NOW TRY**

NOW TRY ANSWERS
4. $\varnothing$
5. $\{5, 10, 15, 20, 25\}$

NOTE Although the elements 2 and 4 appeared in both sets A and B in **Example 5,** they are written only once in $A \cup B$.

OBJECTIVE 5 Solve compound inequalities with the word *or*.

> **Solving a Compound Inequality with *or***
>
> **Step 1** Solve each inequality individually.
>
> **Step 2** The solution set of the compound inequality includes all numbers that satisfy either one or the other (or both) of the inequalities in Step 1—that is, the *union* of the solution sets.

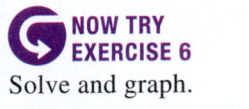
NOW TRY
EXERCISE 6
Solve and graph.

$-12x \le -24$ or $x + 9 < 8$

EXAMPLE 6 Solving a Compound Inequality with *or*

Solve the compound inequality, and graph the solution set.

$$6x - 4 < 2x \quad \text{or} \quad -3x \le -9$$

Step 1 Solve each inequality individually.

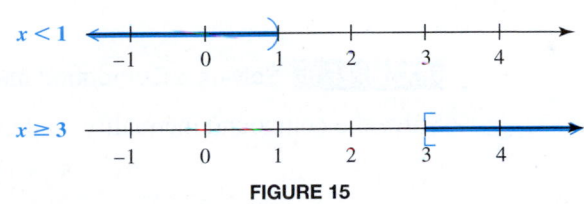

$$6x - 4 < 2x \quad \text{or} \quad -3x \le -9$$
$$4x < 4$$

> Remember to reverse the inequality symbol.

$$x < 1 \quad \text{or} \quad x \ge 3$$

The graphs of these two inequalities are shown in **FIGURE 15**.

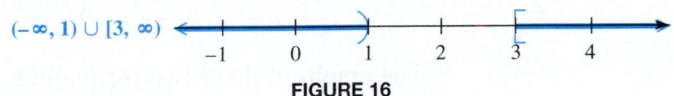

$x < 1$

$x \ge 3$

The set of points in *either* of the graphs represents the union.

FIGURE 15

Step 2 Because the inequalities are joined with *or*, find the union of the two solution sets. The union is the disjoint interval in **FIGURE 16**.

$(-\infty, 1) \cup [3, \infty)$

FIGURE 16

In interval notation, the solution set is $(-\infty, 1) \cup [3, \infty)$. Always pay particular attention to the end points of the solution sets and whether parentheses, brackets, or one of each should be used.   **NOW TRY**

⚠️ **CAUTION** When inequalities are used to write the solution set in **Example 6**, it *must* be written using two separate inequalities.

$$x < 1 \quad \text{or} \quad x \ge 3$$

Writing $3 \le x < 1$, which translates using *and*, would imply that

$$3 \le 1, \quad \text{which is } \textbf{FALSE.}$$

NOW TRY ANSWER

6.

$(-\infty, -1) \cup [2, \infty)$

**NOW TRY
EXERCISE 7**

Solve and graph.

$-x + 2 < 6$ or $6x - 8 \geq 10$

EXAMPLE 7   Solving a Compound Inequality with *or*

Solve the compound inequality, and graph the solution set.

$$-4x + 1 \geq 9 \quad \text{or} \quad 5x + 3 \leq -12$$

Step 1   Solve each inequality individually.

$$-4x + 1 \geq 9 \quad \text{or} \quad 5x + 3 \leq -12$$
$$-4x \geq 8 \quad \text{or} \quad 5x \leq -15$$
$$x \leq -2 \quad \text{or} \quad x \leq -3$$

The graphs of these two inequalities are shown in **FIGURE 17**.

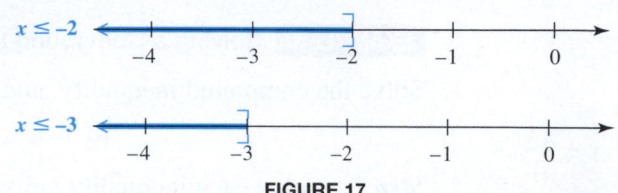

FIGURE 17

Step 2   **FIGURE 18** shows the union, the infinite interval $(-\infty, -2]$.

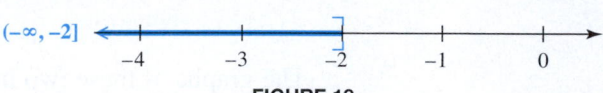

FIGURE 18

NOW TRY

**NOW TRY
EXERCISE 8**

Solve and graph.

$$8x - 4 \geq 20 \quad \text{or}$$
$$-2x + 1 > -9$$

EXAMPLE 8   Solving a Compound Inequality with *or*

Solve the compound inequality, and graph the solution set.

$$-2x + 5 \geq 11 \quad \text{or} \quad 4x - 7 \geq -27$$

Step 1   Solve each inequality individually.

$$-2x + 5 \geq 11 \quad \text{or} \quad 4x - 7 \geq -27$$
$$-2x \geq 6 \quad \text{or} \quad 4x \geq -20$$
$$x \leq -3 \quad \text{or} \quad x \geq -5$$

The graphs of these two inequalities are shown in **FIGURE 19**.

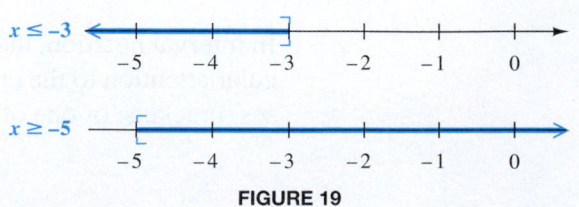

FIGURE 19

NOW TRY ANSWERS

7.
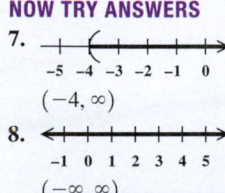
$(-4, \infty)$

8.
$(-\infty, \infty)$

Step 2   By taking the union, we obtain every real number as a solution because every real number satisfies at least one of the two inequalities. The set of all real numbers is graphed in **FIGURE 20**. It is written in interval notation as $(-\infty, \infty)$.

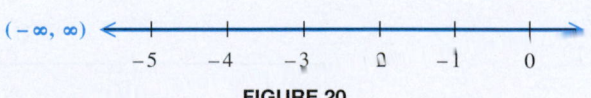

FIGURE 20

NOW TRY

NOW TRY EXERCISE 9

In **Example 9,** list the elements of each set.

(a) The set of countries to which exports were greater than $100,000 million and from which imports were less than $400,000 million

(b) The set of countries to which exports were less than $200,000 million or from which imports were greater than $250,000 million

EXAMPLE 9 Applying Intersection and Union

The five top U.S. trading partners for 2016 are listed in the table. Amounts are in millions of dollars.

Country	U.S. Exports to Country	U.S. Imports from Country
China	115,602	462,618
Canada	266,797	277,756
Mexico	229,702	294,056
Japan	63,236	132,046
Germany	49,363	114,099

Data from U.S. Census Bureau.

List the elements of the following sets.

(a) The set of countries to which exports were greater than $200,000 million *and* from which imports were less than $300,000 million

The countries that satisfy both conditions are Canada and Mexico, so the set is

$$\{\text{Canada, Mexico}\}.$$

(b) The set of countries to which exports were less than $100,000 million *or* from which imports were greater than $200,000 million

Here, any country that satisfies at least one of the conditions is in the set. This set includes all five countries:

$$\{\text{China, Canada, Mexico, Japan, Germany}\}. \qquad \textbf{NOW TRY} \ \text{}$$

NOW TRY ANSWERS
9. **(a)** {Canada, Mexico}
   (b) {China, Canada, Mexico, Japan, Germany}

8.2 Exercises

FOR EXTRA HELP

 MyLab Math

Video solutions for select problems available in MyLab Math

STUDY SKILLS REMINDER
Reread your class notes before working the assigned exercises.
Review Study Skill 3, Taking Lecture Notes.

Concept Check *Determine whether each statement is* true *or* false. *If it is false, explain why.*

1. The union of the solution sets of $x + 1 = 6$, $x + 1 < 6$, and $x + 1 > 6$ is $(-\infty, \infty)$.

2. The intersection of the sets $\{x \mid x \geq 9\}$ and $\{x \mid x \leq 9\}$ is $\varnothing$.

3. The union of the sets $(-\infty, 7)$ and $(7, \infty)$ is $\{7\}$.

4. The intersection of the sets $(-\infty, 7]$ and $[7, \infty)$ is $\{7\}$.

Let $A = \{1, 2, 3, 4, 5, 6\}$, $B = \{1, 3, 5\}$, $C = \{1, 6\}$, *and* $D = \{4\}$. *Find each set. See Examples 1 and 5.*

5. $A \cap D$ **6.** $B \cap C$ **7.** $B \cap A$ **8.** $C \cap A$ **9.** $B \cap \varnothing$

10. $A \cap \varnothing$ **11.** $A \cup B$ **12.** $B \cup D$ **13.** $B \cup C$ **14.** $C \cup D$

Concept Check *Two sets are specified by graphs. Graph the intersection of the two sets.*

15.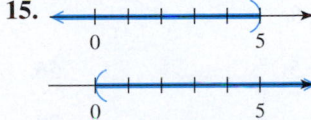

16.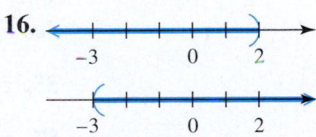

17.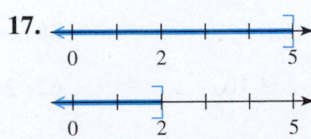

18.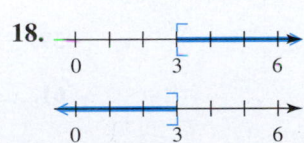

Solve each compound inequality. Graph the solution set, and write it using interval notation.
See Examples 2–4.

19. $x < 2$ and $x > -3$

20. $x < 5$ and $x > 0$

21. $x \leq 2$ and $x \leq 5$

22. $x \geq 3$ and $x \geq 6$

23. $x \leq 3$ and $x \geq 6$

24. $x \leq -1$ and $x \geq 3$

25. $x - 3 \leq 6$ and $x + 2 \geq 7$

26. $x + 5 \leq 11$ and $x - 3 \geq -1$

27. $-3x > 3$ and $x + 3 > 0$

28. $-3x < 3$ and $x + 2 < 6$

29. $3x - 4 \leq 8$ and $-4x + 1 \geq -15$

30. $7x + 6 \leq 48$ and $-4x + 3 \geq -21$

Concept Check   *Two sets are specified by graphs. Graph the union of the two sets.*

31.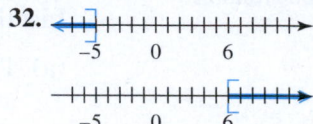

32.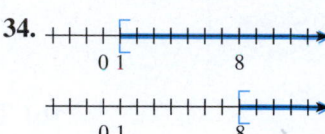

33.

34.

Solve each compound inequality. Graph the solution set, and write it using interval notation.
See Examples 6–8.

35. $x \leq 1$ or $x \leq 8$

36. $x \geq 1$ or $x \geq 8$

37. $x \geq -2$ or $x \geq 5$

38. $x \leq -2$ or $x \leq 6$

39. $x \geq -2$ or $x \leq 4$

40. $x \geq 5$ or $x \leq 7$

41. $x + 2 > 7$ or $1 - x > 6$

42. $x + 1 > 3$ or $x + 4 < 2$

43. $x + 1 > 3$ or $-4x + 1 > 5$

44. $-2x + 1 > -11$ or $x + 1 > 10$

45. $4x + 1 \geq -7$ or $-2x + 3 \geq 5$

46. $3x + 2 \leq -7$ or $-2x + 1 \leq 9$

Concept Check   *Express each set in simplest interval form. (Hint: Graph each set and look for the intersection or union.)*

47. $(-\infty, -1] \cap [-4, \infty)$

48. $[-1, \infty) \cap (-\infty, 9]$

49. $(-\infty, -6] \cap [-9, \infty)$

50. $(5, 11] \cap [6, \infty)$

51. $(-\infty, 3) \cup (-\infty, -2)$

52. $[-9, 1] \cup (-\infty, -3)$

53. $[3, 6] \cup (4, 9)$

54. $[-1, 2] \cup (0, 5)$

Solve each compound inequality. Graph the solution set, and write it using interval notation.
See Examples 2–4 and 6–8.

55. $x < -1$ and $x > -5$

56. $x > -1$ and $x < 7$

57. $x < 4$ or $x < -2$

58. $x < 5$ or $x < -3$

59. $-3x \leq -6$ or $-3x \geq 0$

60. $-8x \leq -24$ or $-5x \geq 15$

61. $x + 1 \geq 5$ and $x - 2 \leq 10$

62. $2x - 6 \leq -18$ and $2x \geq -18$

Average expenses for full-time resident college students at 4-year institutions during the 2015–2016 academic year are shown in the table.

▼ **College Expenses (in Dollars)**

Type of Expense	Public Schools (in-state)	Private Schools
Tuition and fees	8778	27,951
Board rates	4561	5116
Dormitory charges	5850	6463

Data from National Center for Education Statistics.

List the elements of each set. **See Example 9.**

63. The set of expenses that are less than $9000 for public schools *and* are greater than $15,000 for private schools

64. The set of expenses that are greater than $4000 for public schools *and* are less than $6000 for private schools

65. The set of expenses that are less than $9000 for public schools *or* are greater than $15,000 for private schools

66. The set of expenses that are greater than $15,000 *or* are between $7000 and $8000

The figures represent the backyards of neighbors Luigi, Mario, Than, and Joe. Suppose that each resident has 150 ft *of fencing and enough sod to cover* 1400 ft² *of lawn.*

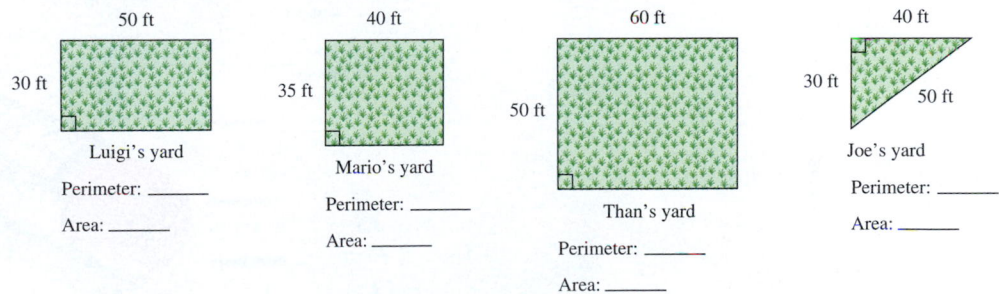

Luigi's yard — 50 ft, 30 ft
Perimeter: _____
Area: _____

Mario's yard — 40 ft, 35 ft
Perimeter: _____
Area: _____

Than's yard — 60 ft, 50 ft
Perimeter: _____
Area: _____

Joe's yard — 40 ft, 30 ft, 50 ft
Perimeter: _____
Area: _____

67. Determine the perimeters of the four yards.

68. Determine the areas of the four yards.

Give the name or names of the residents whose yards satisfy each description. (Hint: Use the perimeters and areas found in **Exercises 67 and 68.**)

69. The yard can be fenced *and* the yard can be sodded.

70. The yard can be fenced *and* the yard cannot be sodded.

71. The yard cannot be fenced *and* the yard can be sodded.

72. The yard cannot be fenced *and* the yard cannot be sodded.

73. The yard can be fenced *or* the yard can be sodded.

74. The yard cannot be fenced *or* the yard can be sodded.

RELATING CONCEPTS   For Individual or Group Work (Exercises 75–78)

An intermediate algebra teacher bases final grades on points earned for activities as given in the Graded Classwork table on the left. To determine final grades, the teacher strictly adheres to the point ranges given in the Grade Distribution table on the right.

▼ **Graded Classwork**

Activity	Points Available
Homework and vocabulary	45
Daily activities (scaled)	55
Lab participation and completion	100
Major exams (3 at 100 points)	300
Final exam	150
Total points	**650**

▼ **Grade Distribution**

Grade	Points Required
A	585–650
B	520–584
C	455–519
IP*	< 455 and active
F	< 455 and inactive

* In Progress

Use this information to **work Exercises 75–78 in order.**

75. Suppose Lauren earns all of the homework and vocabulary points, 50 points for daily activities, and 90 points for lab participation and completion.

Let x = points to be earned on exams.

(a) Write three inequalities to find the minimum number of points she needs on exams to earn grades no lower than A, B, and C.

(b) Solve each inequality from part (a) to find the minimum number of points she needs for each grade. What "test average" (as a percent) corresponds to this number of points, given that exams account for 450 possible points? (Round up to the nearest whole number.)

76. See Exercise 75. Write and solve a compound inequality to find the range of points Lauren needs in exam scores to earn a B average. What range of "test averages" (as percents) correspond to these scores, given that exams account for 450 possible points? (Round up to the nearest whole number.)

77. Suppose Mark earns only 15 points in homework and vocabulary, 40 points in daily activities, and 50 points in lab participation. Repeat **Exercise 75** using these values.

78. Repeat **Exercise 76** given that Mark wants to earn a C average. (Use his classwork points given in **Exercise 77.**)

8.3 Absolute Value Equations and Inequalities

OBJECTIVES

1 Use the distance definition of absolute value.

2 Solve equations of the form $|ax + b| = k$, for $k > 0$.

3 Solve inequalities of the form $|ax + b| < k$ and of the form $|ax + b| > k$, for $k > 0$.

4 Solve absolute value equations that involve rewriting.

5 Solve equations of the form $|ax + b| = |cx + d|$.

6 Solve special cases of absolute value equations and inequalities.

7 Solve an application involving relative error.

Suppose the government of a country decides that it will comply with a restriction on greenhouse gas emissions *within* 3 years of 2025. This means that the *difference* between the year it will comply and 2025 is less than 3, *without regard to sign*. We state this mathematically as follows, where x represents the year in which it complies.

$$|x - 2025| < 3 \qquad \text{Absolute value inequality}$$

We can intuitively reason that the year must be between 2022 and 2028, and thus values that satisfy $2022 < x < 2028$ make this inequality true.

OBJECTIVE 1 Use the distance definition of absolute value.

The **absolute value** of a number x, written $|x|$, represents the undirected distance from x to 0 on a number line. For example, the solution set of $|x| = 4$ is $\{-4, 4\}$, which means $x = -4$ or $x = 4$, as shown in **FIGURE 21**.

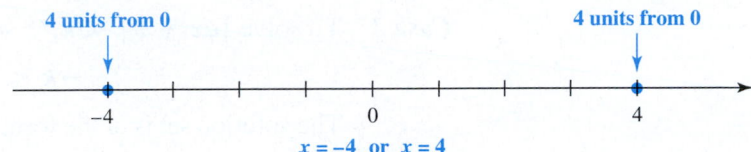

$$x = -4 \text{ or } x = 4$$

FIGURE 21

Because absolute value represents distance from 0, the solution set of $|x| > 4$ consists of all numbers that are *more* than four units from 0 on a number line. The set $(-\infty, -4) \cup (4, \infty)$ fits this description. The graph consists of disjoint intervals, which means $x < -4$ *or* $x > 4$, as shown in **FIGURE 22**.

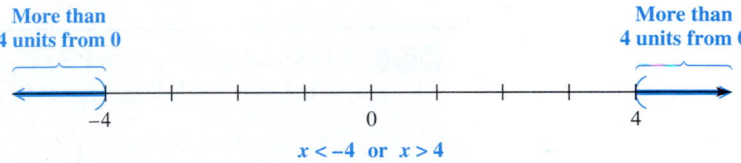

$$x < -4 \text{ or } x > 4$$

FIGURE 22

The solution set of $|x| < 4$ consists of all numbers that are *less* than 4 units from 0 on a number line. This is represented by all numbers *between* -4 and 4, which is given by the open interval $(-4, 4)$, as shown in **FIGURE 23**. Here, $-4 < x < 4$, which means $x > -4$ *and* $x < 4$.

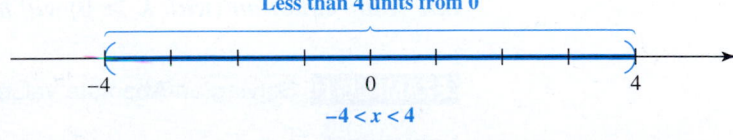

$$-4 < x < 4$$

FIGURE 23

Absolute value equations and inequalities generally take the form

$$|ax + b| = k, \qquad |ax + b| > k, \qquad \text{or} \qquad |ax + b| < k,$$

where k is a positive number. From **FIGURES 21–23**, we see that

| $|x| = 4$ | has the same solution set as | $x = -4$ or $x = 4,$ |

| $|x| > 4$ | has the same solution set as | $x < -4$ or $x > 4,$ |

| $|x| < 4$ | has the same solution set as | $x > -4$ and $x < 4.$ |

> This is equivalent to $-4 < x < 4.$

VOCABULARY
☐ absolute value equation
☐ absolute value inequality

Solving Absolute Value Equations and Inequalities

Let k be a positive real number, and p and q be real numbers.

Case 1   To solve $|ax + b| = k$, solve the compound equation

$$ax + b = k \quad \text{or} \quad ax + b = -k.$$

The solution set is usually of the form $\{p, q\}$, which includes two numbers.

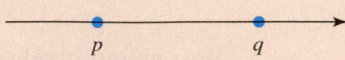

Case 2   To solve $|ax + b| > k$,* solve the compound inequality

$$ax + b > k \quad \text{or} \quad ax + b < -k.$$

The solution set is of the form $(-\infty, p) \cup (q, \infty)$, which consists of disjoint intervals.

Case 3   To solve $|ax + b| < k$,** solve the three-part inequality

$$-k < ax + b < k.$$

The solution set is of the form (p, q), which consists of a single interval.

*This also applies to $|ax + b| \geq k$. The solution set *includes* the endpoints, using brackets rather than parentheses.
**This also applies to $|ax + b| \leq k$. The solution set *includes* the endpoints, using brackets rather than parentheses.

NOTE   It is acceptable to write the compound statements in Cases 1 and 2 of the preceding box in the following equivalent forms.

$$ax + b = k \quad \text{or} \quad -(ax + b) = k \qquad \text{Alternative for Case 1}$$

$$ax + b > k \quad \text{or} \quad -(ax + b) > k \qquad \text{Alternative for Case 2}$$

OBJECTIVE 2   Solve equations of the form $|ax + b| = k$, for $k > 0$.

Remember that because absolute value refers to distance from the origin, an absolute value equation (with $k > 0$) will have two parts.

EXAMPLE 1   Solving an Absolute Value Equation (Case 1)

Solve $|2x + 1| = 7$.

For $|2x + 1|$ to equal 7, $2x + 1$ must be 7 units from 0 on a number line. This can happen only when $2x + 1 = 7$ or $2x + 1 = -7$. This is *Case 1* in the box above. Solve this compound equation as follows.

$$2x + 1 = 7 \quad \text{or} \quad 2x + 1 = -7$$
$$2x = 6 \quad \text{or} \quad 2x = -8 \qquad \text{Subtract 1.}$$
$$x = 3 \quad \text{or} \quad x = -4 \qquad \text{Divide by 2.}$$

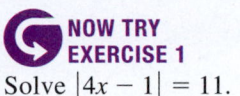

**NOW TRY
EXERCISE 1**
Solve $|4x - 1| = 11$.

CHECK $\qquad |2x + 1| = 7$

$|2(3) + 1| \overset{?}{=} 7$ Let $x = 3$. $\qquad |2(-4) + 1| \overset{?}{=} 7$ Let $x = -4$.

$|6 + 1| \overset{?}{=} 7 \qquad\qquad\qquad |-8 + 1| \overset{?}{=} 7$

$|7| \overset{?}{=} 7 \qquad\qquad\qquad\qquad |-7| \overset{?}{=} 7$

$7 = 7 \checkmark$ True $\qquad\qquad\qquad 7 = 7 \checkmark$ True

The solution set is $\{-4, 3\}$. See **FIGURE 24**.

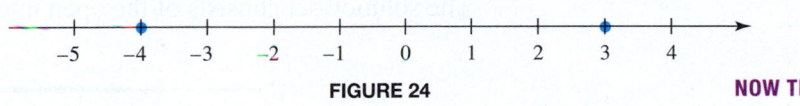

FIGURE 24 NOW TRY

OBJECTIVE 3 Solve inequalities of the form $|ax + b| < k$ and of the form $|ax + b| > k$, for $k > 0$.

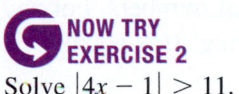

**NOW TRY
EXERCISE 2**
Solve $|4x - 1| > 11$.

EXAMPLE 2 Solving an Absolute Value Inequality (Case 2)

Solve $|2x + 1| > 7$.

Because $2x + 1$ must represent a number that is *more* than 7 units from 0 on either side of a number line, this absolute value inequality is rewritten as the following compound inequality. This is **Case 2**.

$$2x + 1 > 7 \quad \text{or} \quad 2x + 1 < -7$$

$$2x > 6 \quad \text{or} \qquad 2x < -8 \qquad \text{Subtract 1.}$$

$$x > 3 \quad \text{or} \qquad x < -4 \qquad \text{Divide by 2.}$$

The solution set consists of the disjoint intervals shown in **FIGURE 25** and is written $(-\infty, -4) \cup (3, \infty)$.

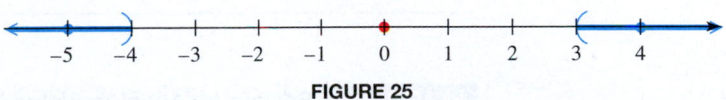

FIGURE 25

CHECK The excluded endpoints -4 and 3 are correct because from **Example 1** we know that -4 and 3 are the solutions of the related equation. Referring to **FIGURE 25**, we choose a test point in each of the three intervals $(-\infty, -4)$, $(-4, 3)$, and $(3, \infty)$.

For $(-\infty, -4)$, let $x = -5$. | For $(-4, 3)$, let $x = 0$. | For $(3, \infty)$, let $x = 4$.

$|2x + 1| > 7 \qquad\qquad |2x + 1| > 7 \qquad\qquad |2x + 1| > 7$

$|2(-5) + 1| \overset{?}{>} 7 \qquad |2(0) + 1| \overset{?}{>} 7 \qquad |2(4) + 1| \overset{?}{>} 7$

$|-9| \overset{?}{>} 7 \qquad\qquad |1| \overset{?}{>} 7 \qquad\qquad |9| \overset{?}{>} 7$

$9 > 7 \checkmark$ True $\qquad 1 > 7$ False $\qquad 9 > 7 \checkmark$ True

NOW TRY ANSWERS
1. $\left\{-\frac{5}{2}, 3\right\}$

2. $\left(-\infty, -\frac{5}{2}\right) \cup (3, \infty)$

The check confirms that the solution set is $(-\infty, -4) \cup (3, \infty)$. NOW TRY

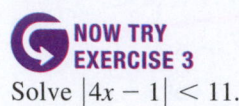
NOW TRY
EXERCISE 3
Solve $|4x - 1| < 11$.

EXAMPLE 3 Solving an Absolute Value Inequality (Case 3)

Solve $|2x + 1| < 7$. Graph the solution set.

The expression $2x + 1$ must represent a number that is less than 7 units from 0 on either side of a number line. That is, $2x + 1$ must be between -7 and 7, which is written as a three-part inequality. This is **Case 3**.

$$-7 < 2x + 1 < 7$$

$$-8 < \quad 2x \quad < 6 \qquad \text{Subtract 1 from each part.}$$

$$-4 < \quad x \quad < 3 \qquad \text{Divide each part by 2.}$$

The solution set consists of the open interval $(-4, 3)$ shown in **FIGURE 26**.

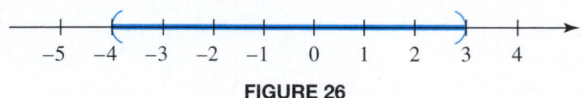

FIGURE 26

 NOW TRY

Look back at **FIGURES 24, 25, AND 26**, with the graphs of

$$|2x + 1| = 7, \quad |2x + 1| > 7, \quad \text{and} \quad |2x + 1| < 7.$$

If we find the union of the three sets, we obtain the set of all real numbers. For any value of x, $|2x + 1|$ will satisfy *one and only one* of the following: It is equal to 7, greater than 7, or less than 7.

⚠ **CAUTION** Remember the following important concepts.

1. The methods described apply when the constant is alone on one side of the absolute value equation or inequality and is *positive*.

2. Absolute value equations $|ax + b| = k$ and inequalities of the form $|ax + b| > k$ translate into **"or"** compound statements.

3. Absolute value inequalities of the form $|ax + b| < k$ translate into **"and"** compound statements. *Only "and" compound statements may be written as three-part inequalities.*

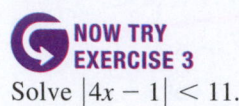
NOW TRY
EXERCISE 4
Solve $|7 - 4x| \geq 7$.

EXAMPLE 4 Solving an Absolute Value Inequality (Case 2, for ≥)

Solve $|5 - 2x| \geq 5$.

Case 2 is applied. Notice that the endpoints are included because equality is part of the symbol $\geq$.

$$5 - 2x \geq 5 \quad \text{or} \quad 5 - 2x \leq -5$$

$$-2x \geq 0 \quad \text{or} \quad -2x \leq -10 \qquad \text{Subtract 5.}$$

$$x \leq 0 \quad \text{or} \quad x \geq 5 \qquad \begin{array}{l}\text{Divide by } -2. \text{ Reverse the} \\ \text{direction of the inequality symbols.}\end{array}$$

See the graph of the disjoint intervals in **FIGURE 27**. The solution set is written $(-\infty, 0] \cup [5, \infty)$.

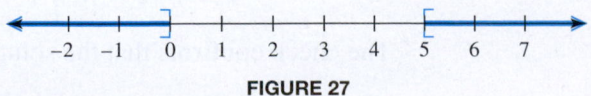

FIGURE 27

 NOW TRY

NOW TRY ANSWERS

3. $\left(-\frac{5}{2}, 3\right)$

4. $(-\infty, 0] \cup \left[\frac{7}{2}, \infty\right)$

OBJECTIVE 4 Solve absolute value equations that involve rewriting.

**NOW TRY
EXERCISE 5**

Solve $|10x - 2| - 2 = 12$.

EXAMPLE 5 Solving an Absolute Value Equation That Requires Rewriting

Solve $|x + 3| + 5 = 12$.

Isolate the absolute value expression on one side of the equality symbol.

$$|x + 3| + 5 = 12$$

$$|x + 3| + 5 - 5 = 12 - 5 \qquad \text{Subtract 5.}$$

$$|x + 3| = 7 \qquad \text{Combine like terms.}$$

$$x + 3 = 7 \quad \text{or} \quad x + 3 = -7 \qquad \text{Case 1}$$

$$x = 4 \quad \text{or} \quad x = -10 \qquad \text{Subtract 3.}$$

CHECK $\qquad\qquad |x + 3| + 5 = 12$

$$|4 + 3| + 5 \stackrel{?}{=} 12 \qquad \text{Let } x = 4. \qquad\qquad |-10 + 3| + 5 \stackrel{?}{=} 12 \qquad \text{Let } x = -10.$$

$$|7| + 5 \stackrel{?}{=} 12 \qquad\qquad\qquad\qquad\qquad |-7| + 5 \stackrel{?}{=} 12$$

$$12 = 12 \;\checkmark\; \text{True} \qquad\qquad\qquad\qquad\qquad 12 = 12 \;\checkmark\; \text{True}$$

The check confirms that the solution set is $\{-10, 4\}$. **NOW TRY**

**NOW TRY
EXERCISE 6**

Solve each inequality.

(a) $|x - 1| - 4 \leq 2$

(b) $|x - 1| - 4 \geq 2$

EXAMPLE 6 Solving Absolute Value Inequalities That Require Rewriting

Solve each inequality.

(a) $\quad |x + 3| + 5 \geq 12$

$$|x + 3| \geq 7 \qquad \text{Case 2}$$

$$x + 3 \geq 7 \quad \text{or} \quad x + 3 \leq -7$$

$$x \geq 4 \quad \text{or} \quad x \leq -10$$

The solution set is $(-\infty, -10] \cup [4, \infty)$.

(b) $\quad |x + 3| + 5 \leq 12$

$$|x + 3| \leq 7 \qquad \text{Case 3}$$

$$-7 \leq x + 3 \leq 7$$

$$-10 \leq \quad x \quad \leq 4$$

The solution set is $[-10, 4]$.

NOW TRY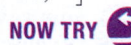

OBJECTIVE 5 Solve equations of the form $|ax + b| = |cx + d|$.

If two expressions have the same absolute value, they must either be equal or be negatives of each other.

Solving $|ax + b| = |cx + d|$

To solve an absolute value equation of the form

$$|ax + b| = |cx + d|,$$

solve the compound equation

$$ax + b = cx + d \quad \text{or} \quad ax + b = -(cx + d).$$

NOW TRY ANSWERS

5. $\left\{-\frac{6}{5}, \frac{8}{5}\right\}$

6. (a) $[-5, 7]$

$\quad$ **(b)** $(-\infty, -5] \cup [7, \infty)$

**NOW TRY
EXERCISE 7**

Solve

$|3x - 4| = |5x + 12|.$

EXAMPLE 7   Solving an Equation Involving Two Absolute Values

Solve $|x + 6| = |2x - 3|$.

This equation is satisfied either if $x + 6$ and $2x - 3$ are equal to each other, or if $x + 6$ and $2x - 3$ are negatives of each other.

$$
\begin{array}{rcl}
x + 6 = 2x - 3 & \text{or} & x + 6 = -(2x - 3) \\
x + 9 = 2x & \text{or} & x + 6 = -2x + 3 \\
9 = x & \text{or} & 3x = -3 \\
x = 9 & \text{or} & x = -1
\end{array}
$$

CHECK $|x + 6| = |2x - 3|$

$$
\begin{array}{ll}
|9 + 6| \stackrel{?}{=} |2(9) - 3| \quad \text{Let } x = 9. & \quad |-1 + 6| \stackrel{?}{=} |2(-1) - 3| \quad \text{Let } x = -1. \\
|15| \stackrel{?}{=} |18 - 3| & \quad |5| \stackrel{?}{=} |-2 - 3| \\
|15| \stackrel{?}{=} |15| & \quad |5| \stackrel{?}{=} |-5| \\
15 = 15 \ \checkmark \quad \text{True} & \quad 5 = 5 \ \checkmark \quad \text{True}
\end{array}
$$

The check confirms that the solution set is $\{-1, 9\}$.   **NOW TRY**

OBJECTIVE 6 Solve special cases of absolute value equations and inequalities.

When an absolute value equation or inequality involves a *negative constant or 0* alone on one side, we use the following properties to solve it.

> **Special Properties of Absolute Value**
>
> ***Property 1*** The absolute value of an expression can never be negative—that is, $|a| \geq 0$ for all real numbers a.
>
> ***Property 2*** The absolute value of an expression equals 0 only when the expression is equal to 0.

**NOW TRY
EXERCISE 8**

Solve each equation.

(a) $|3x - 8| = -2$

(b) $|7x + 12| = 0$

EXAMPLE 8   Solving Special Cases of Absolute Value Equations

Solve each equation.

(a) $|5x - 3| = -4$

See ***Property 1*** in the preceding box. ***The absolute value of an expression can never be negative,*** so there are no solutions for this equation. The solution set is $\varnothing$.

(b) $|7x - 3| = 0$

See ***Property 2*** in the preceding box. The expression $|7x - 3|$ will equal 0 *only* if $7x - 3 = 0$.

$$
\begin{array}{ll}
7x - 3 = 0 & \\
7x = 3 & \text{Add 3.} \\
x = \dfrac{3}{7} & \text{Divide by 7.}
\end{array}
$$

> Check by substituting in the original equation.

NOW TRY ANSWERS

7. $\{-8, -1\}$

8. **(a)** $\varnothing$ **(b)** $\left\{-\frac{12}{7}\right\}$

The solution set $\left\{\dfrac{3}{7}\right\}$ consists of just one element.   **NOW TRY**

**NOW TRY
EXERCISE 9**

Solve each inequality.

(a) $|x| > -10$

(b) $|4x + 1| + 5 < 4$

(c) $|x - 2| - 3 \leq -3$

EXAMPLE 9   Solving Special Cases of Absolute Value Inequalities

Solve each inequality.

(a) $|x| \geq -4$

  The absolute value of a number is always greater than or equal to 0. (Property 1)
Thus, $|x| \geq -4$ is true for *all* real numbers. The solution set is $(-\infty, \infty)$.

(b)
$$|x + 6| - 3 < -5$$
$$|x + 6| < -2 \qquad \text{Add 3 to each side.}$$

There is no number whose absolute value is less than -2, so this inequality has no solution. The solution set is $\varnothing$.

(c)
$$|x - 7| + 4 \leq 4$$
$$|x - 7| \leq 0 \qquad \text{Subtract 4 from each side.}$$

The value of $|x - 7|$ will never be less than 0. However, $|x - 7|$ will *equal* 0 when $x = 7$. Therefore, the solution set is $\{7\}$.

NOW TRY

OBJECTIVE 7 Solve an application involving relative error.

Absolute value is used to find the **relative error,** or **tolerance,** in a measurement. If x represents the actual measurement and x_t represents the expected measurement, then taking the absolute value of the difference of x and x_t, divided by x_t, gives the relative error in x.

$$\textbf{relative error in } \textit{x} = \left| \frac{x - x_t}{x_t} \right|$$

In quality control situations, the relative error must often be less than some predetermined amount.

**NOW TRY
EXERCISE 10**

Suppose a machine filling *quart* milk cartons is set for a relative error that *is no greater than* 0.03. How many ounces may a filled carton contain?

EXAMPLE 10   Solving an Application Involving Relative Error

Suppose a machine filling *quart* milk cartons is set for a relative error that *is no greater than* 0.05. How many ounces may a filled carton contain?

  Here $x_t = 32$ oz (because 1 qt = 32 oz) and the relative error = 0.05. We must find x, the actual measure of a filled carton.

$$\left| \frac{x - 32}{32} \right| \leq 0.05 \qquad \begin{array}{l} \text{Substitute the given values.} \\ \textit{Is no greater than} \text{ translates as } \leq. \end{array}$$

$$-0.05 \leq \frac{x - 32}{32} \leq 0.05 \qquad \text{Case 3}$$

$$-1.6 \leq x - 32 \leq 1.6 \qquad \text{Multiply by 32.}$$

$$30.4 \leq \quad x \quad \leq 33.6 \qquad \text{Add 32.}$$

The filled carton may contain between 30.4 and 33.6 oz, inclusive.   **NOW TRY**

NOW TRY ANSWERS
9. (a) $(-\infty, \infty)$ (b) $\varnothing$ (c) $\{2\}$
10. between 31.04 and 32.96 oz, inclusive

8.3 Exercises

    MyLab Math

▶ *Video solutions for select problems available in MyLab Math*

STUDY SKILLS REMINDER
Time management can be a challenge for students.
Review Study Skill 6,
Managing Your Time.

Concept Check *Match each absolute value equation or inequality in Column I with the graph of its solution set in Column II.*

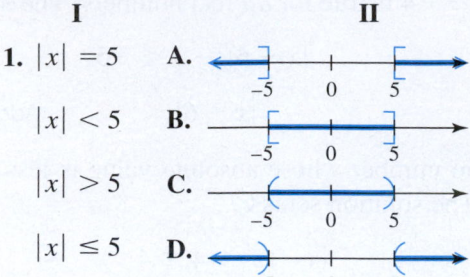

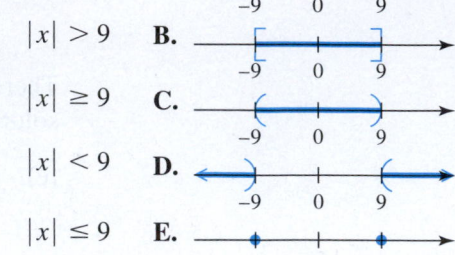

3. **Concept Check** How many solutions will $|ax + b| = k$ have for each situation?

    (a) $k = 0$ **(b)** $k > 0$ **(c)** $k < 0$

4. **Concept Check** If $k < 0$, what is the solution set of each of the following?

    (a) $|x - 1| < k$ **(b)** $|x - 1| > k$ **(c)** $|x - 1| = k$

Solve each equation. See Example 1.

5. $|x| = 12$ 6. $|x| = 14$ 7. $|4x| = 20$

8. $|5x| = 30$ 9. $|x - 3| = 9$ 10. $|x - 5| = 13$

11. $|2x - 1| = 11$ 12. $|2x + 3| = 19$ 13. $|4x - 5| = 17$

14. $|5x - 1| = 21$ 15. $|2x + 5| = 14$ 16. $|2x - 9| = 18$

17. $|-3x + 8| = 1$ 18. $|-6x + 5| = 4$ 19. $\left|12 - \dfrac{1}{2}x\right| = 6$

20. $\left|14 - \dfrac{1}{3}x\right| = 8$ 21. $|0.5x| = 6$ 22. $|0.3x| = 9$

23. $\left|\dfrac{1}{2}x + 3\right| = 2$ 24. $\left|\dfrac{2}{3}x - 1\right| = 5$ 25. $\left|1 + \dfrac{3}{4}x\right| = 7$

26. $\left|2 - \dfrac{5}{2}x\right| = 14$ 27. $|0.02x - 1| = 2.50$ 28. $|0.04x - 3| = 5.96$

Solve each inequality. Graph the solution set, and write it using interval notation. See Example 2.

29. $|x| > 3$ 30. $|x| > 5$ 31. $|x| \geq 4$

32. $|x| \geq 6$ 33. $|r + 5| \geq 20$ 34. $|x + 4| \geq 8$

35. $|5x + 2| > 10$ 36. $|4x + 1| \geq 21$ 37. $|3 - x| > 5$

38. $|5 - x| > 3$ 39. $|-5x + 3| \geq 12$ 40. $|-2x - 4| \geq 5$

41. Concept Check The graph of the solution set of $|2x + 1| = 9$ is given here.

Without doing any algebraic work, graph the solution set of each inequality, referring to the graph shown.

(a) $|2x + 1| < 9$ (b) $|2x + 1| > 9$

42. Concept Check The graph of the solution set of $|3x - 4| < 5$ is given here.

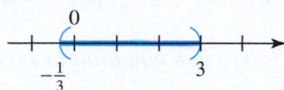

Without doing any algebraic work, graph the solution set of the following, referring to the graph shown.

(a) $|3x - 4| = 5$ (b) $|3x - 4| > 5$

Solve each inequality. Graph the solution set, and write it using interval notation. See ***Example 3.*** *(Hint: Compare the answers with those in* ***Exercises 29–40.****)*

43. $|x| \le 3$ **44.** $|x| \le 5$ **45.** $|x| < 4$

46. $|x| < 6$ **47.** $|r + 5| < 20$ **48.** $|x + 4| < 8$

49. $|5x + 2| \le 10$ **50.** $|4x + 1| < 21$ **51.** $|3 - x| \le 5$

52. $|5 - x| \le 3$ **53.** $|-5x + 3| < 12$ **54.** $|-2x - 4| < 5$

Solve each equation or inequality. In Exercises 55–66, graph the solution set. See ***Examples 1–4.***

55. $|-4 + x| > 9$ **56.** $|-3 + x| > 8$ **57.** $|3x + 2| < 11$

58. $|2x - 1| < 7$ **59.** $|7 + 2x| = 5$ **60.** $|9 - 3x| = 3$

61. $|3x - 1| \le 11$ **62.** $|2x - 6| \le 6$ **63.** $|-6x - 6| \le 1$

64. $|-2x - 6| \le 5$ **65.** $|-8 + x| \le 5$ **66.** $|-4 + x| \le 9$

67. $|10 - 12x| \ge 4$ **68.** $|8 - 10x| \ge 2$ **69.** $|3(x - 1)| = 8$

70. $|7(x - 2)| = 4$ **71.** $|0.1x - 1| > 3$ **72.** $|0.1x + 1| > 2$

73. $|x + 2| = 5 - 2$ **74.** $|x + 3| = 12 - 2$ **75.** $3|x - 6| = 9$

76. $5|x - 4| = 5$ **77.** $|2 - 0.2x| = 2$ **78.** $|5 - 0.5x| = 4$

Solve each equation or inequality. See ***Examples 5 and 6.***

79. $|x| - 1 = 4$ **80.** $|x| + 3 = 10$

81. $|x + 4| + 1 = 2$ **82.** $|x + 5| - 2 = 12$

83. $|2x + 1| + 3 > 8$ **84.** $|6x - 1| - 2 > 6$

85. $|x + 5| - 6 \le -1$ **86.** $|x - 2| - 3 \le 4$

87. $|0.1x - 2.5| + 0.3 \ge 0.8$ **88.** $|0.5x - 3.5| + 0.2 \ge 0.6$

89. $\left| \dfrac{1}{2}x + \dfrac{1}{3} \right| + \dfrac{1}{4} = \dfrac{3}{4}$ **90.** $\left| \dfrac{2}{3}x + \dfrac{1}{6} \right| + \dfrac{1}{2} = \dfrac{5}{2}$

Solve each equation. See Example 7.

91. $|3x + 1| = |2x + 4|$ **92.** $|7x + 12| = |x - 8|$

93. $\left|x - \dfrac{1}{2}\right| = \left|\dfrac{1}{2}x - 2\right|$ **94.** $\left|\dfrac{2}{3}x - 2\right| = \left|\dfrac{1}{3}x + 3\right|$

95. $|6x| = |9x + 1|$ **96.** $|13x| = |2x + 1|$

97. $|2x - 6| = |2x + 11|$ **98.** $|3x - 1| = |3x + 9|$

Solve each equation or inequality. See Examples 8 and 9.

99. $|x| \geq -10$ **100.** $|x| \geq -15$ **101.** $|12t - 3| = -8$

102. $|13x + 1| = -3$ **103.** $|4x + 1| = 0$ **104.** $|6x - 2| = 0$

105. $|2x - 1| = -6$ **106.** $|8x + 4| = -4$ **107.** $|x + 5| > -9$

108. $|x + 9| > -3$ **109.** $|7x + 3| \leq 0$ **110.** $|4x - 1| \leq 0$

111. $|5x - 2| = 0$ **112.** $|7x + 4| = 0$ **113.** $|x - 2| + 3 \geq 2$

114. $|x - 4| + 5 \geq 4$ **115.** $|10x + 7| + 3 < 1$ **116.** $|4x + 1| - 2 < -5$

Determine the number of ounces a filled carton of the given size may contain for the given relative error. See Example 10.

$$\left|\frac{x - x_t}{x_t}\right| = \textbf{relative error in } x$$

x represents actual measurement.
x_t represents expected measurement.

117. 64-oz carton; relative error no greater than 0.05

118. 24-oz carton; relative error no greater than 0.05

119. 32-oz carton; relative error no greater than 0.02

120. 36-oz carton; relative error no greater than 0.03

In later courses in mathematics, it is sometimes necessary to find an interval in which x must lie in order to keep y within a given difference of some number. For example, suppose

$$y = 2x + 1$$

and we want y to be within 0.01 *unit of* 4. *This criterion can be written as*

$$|y - 4| < 0.01.$$

Solving this inequality shows that x must lie in the interval (1.495, 1.505) *to satisfy the requirement.*

    Find the open interval in which x must lie in order for the given condition to hold.

121. $y = 2x + 1$, and the difference of y and 1 is less than 0.1.

122. $y = 4x - 6$, and the difference of y and 2 is less than 0.02.

123. $y = 4x - 8$, and the difference of y and 3 is less than 0.001.

124. $y = 5x + 12$, and the difference of y and 4 is less than 0.0001.

Work each problem.

125. Dr. Mosely has determined that 99% of the babies he has delivered have weighed x pounds, where

$$|x - 8.3| < 1.5.$$

What range of weights corresponds to this inequality?

126. The Celsius temperatures x on Mars approximately satisfy the inequality

$$|x + 85| \le 55.$$

What range of temperatures corresponds to this inequality?

127. The recommended daily intake (RDI) of calcium for females aged 19–50 is 1000 mg. Write this statement as an absolute value inequality, with x representing the RDI, to express the RDI plus or minus 100 mg. Solve the inequality. (Data from National Academy of Sciences—Institute of Medicine.)

128. The average clotting time of blood is 7.45 sec, with a variation of plus or minus 3.6 sec. Write this statement as an absolute value inequality, with x representing the time. Solve the inequality.

RELATING CONCEPTS **For Individual or Group Work** (Exercises 129–132)

The 10 tallest buildings in Houston, Texas, are listed along with their heights.

Building	Height (in feet)
JPMorgan Chase Tower	1002
Wells Fargo Plaza	992
Williams Tower	901
Bank of America Center	780
Texaco Heritage Plaza	762
609 Main at Texas	757
Enterprise Plaza	756
Centerpoint Energy Plaza	741
1600 Smith St.	732
Fulbright Tower	725

Data from *The World Almanac and Book of Facts.*

*Use this information to **work Exercises 129–132 in order.***

129. To find the average of a group of numbers, we add the numbers and then divide by the number of numbers. Use a calculator to find the average of the heights.

130. Let k represent the average height of these buildings. If a height x satisfies the inequality

$$|x - k| < t,$$

then the height is said to be within t feet of the average. Using the result from **Exercise 129,** list the buildings that are within 50 ft of the average.

131. Repeat **Exercise 130,** but list the buildings that are within 95 ft of the average.

132. Work each of the following.

   (a) Write an absolute value inequality that describes the height of a building that is *not* within 95 ft of the average.

   (b) Solve the inequality from part (a).

   (c) Use the result of part (b) to list the buildings that are not within 95 ft of the average.

   (d) Confirm that the answer to part (c) makes sense by comparing it with the answer to **Exercise 131.**

SUMMARY EXERCISES   Solving Linear and Absolute Value Equations and Inequalities

Solve each equation or inequality. Give the solution set in set notation for equations and in interval notation for inequalities.

1. $4x + 1 = 49$

2. $|x - 1| = 6$

3. $6x - 9 = 12 + 3x$

4. $3x + 7 = 9 + 8x$

5. $|x + 3| = -4$

6. $2x + 1 \leq x$

7. $8x + 2 \geq 5x$

8. $4(x - 11) + 3x = 20x - 31$

9. $2x - 1 = -7$

10. $|3x - 7| - 4 = 0$

11. $6x - 5 \leq 3x + 10$

12. $|5x - 8| + 9 \geq 7$

13. $9x - 3(x + 1) = 8x - 7$

14. $|x| \geq 8$

15. $9x - 5 \geq 9x + 3$

16. $13x - 5 > 13x - 8$

17. $|x| < 5.5$

18. $4x - 1 = 12 + x$

19. $\dfrac{2}{3}x + 8 = \dfrac{1}{4}x$

20. $-\dfrac{5}{8}x \geq -20$

21. $\dfrac{1}{4}x < -6$

22. $\dfrac{1}{2} \leq \dfrac{2}{3}x \leq \dfrac{5}{4}$

23. $\dfrac{3}{5}x - \dfrac{1}{10} = 2$

24. $\dfrac{x}{6} - \dfrac{3x}{5} = x - 86$

25. $x + 9 + 7x = 4(3 + 2x) - 3$

26. $6 - 3(2 - x) < 2(1 + x) + 3$

27. $-6 \leq \dfrac{3}{2} - x \leq 6$

28. $\dfrac{x}{4} - \dfrac{2x}{3} = -10$

29. $|5x + 1| \leq 0$

30. $5x - (3 + x) \geq 2(3x + 1)$

31. $-2 \leq 3x - 1 \leq 8$

32. $-1 \leq 6 - x \leq 5$

33. $|7x - 1| = |5x + 3|$

34. $|x + 2| = |x + 4|$

35. $|1 - 3x| \geq 4$

36. $7x - 3 + 2x = 9x - 8x$

37. $-(x + 4) + 2 = 3x + 8$

38. $|x - 1| < 7$

39. $|2x - 3| > 11$

40. $|5 - x| < 4$

41. $|x - 1| \geq -6$

42. $|2x - 5| = |x + 4|$

43. $8x - (1 - x) = 3(1 + 3x) - 4$

44. $8x - (x + 3) = -(2x + 1) - 12$

45. $|x - 5| = |x + 9|$

46. $|x + 2| < -3$

47. $2x + 1 > 5$ or $3x + 4 < 1$

48. $1 - 2x \geq 5$ and $7 + 3x \geq -2$

8.4 Linear Inequalities and Systems in Two Variables

OBJECTIVE 1 Graph linear inequalities in two variables.

Earlier we graphed linear inequalities in *one* variable on a number line. In this section, we graph linear inequalities in *two* variables on a rectangular coordinate system.

> ### Linear Inequality in Two Variables
>
> A **linear inequality in two variables** (here x and y) is an inequality that can be written in the form
>
> $$Ax + By < C, \quad Ax + By \leq C, \quad Ax + By > C, \quad \text{or} \quad Ax + By \geq C,$$
>
> where A, B, and C are real numbers and A and B are not both 0.
>
> *Examples:*
>
> $$2x + 5y < 10, \quad x - y > 0, \quad y \geq 1 \text{ (here } A = 0), \quad x \leq -6 \text{ (here } B = 0)$$

Consider the graph in **FIGURE 28**. The graph of the line $x + y = 5$ divides the points in the rectangular coordinate system into three sets of points.

1. Those points that lie *on* the line itself and satisfy the equation $x + y = 5$, such as $(0, 5)$, $(2, 3)$, and $(5, 0)$

2. Those points that lie in the region *above* the line and satisfy the inequality $x + y > 5$, such as $(5, 3)$ and $(2, 4)$

3. Those points that lie in the region *below* the line and satisfy the inequality $x + y < 5$, such as $(0, 0)$ and $(-3, -1)$

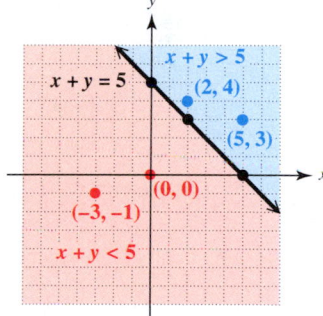

FIGURE 28

The graph of the line $x + y = 5$ is the **boundary line** for the two inequalities

$$x + y > 5 \quad \text{and} \quad x + y < 5.$$

A graph of a linear inequality in two variables is a region in the real number plane that may or may not include the boundary line.

To graph a linear inequality in two variables, follow these steps.

> ### Graphing a Linear Inequality in Two Variables
>
> **Step 1 Draw the graph of the straight line that is the boundary.**
>
> - Make the line solid if the inequality involves $\leq$ or $\geq$.
>
> - Make the line dashed if the inequality involves $<$ or $>$.
>
> **Step 2 Choose a test point.** Choose any point not on the line, and substitute the coordinates of that point in the inequality.
>
> **Step 3 Shade the appropriate region.** Shade the region that includes the test point if it satisfies the original inequality. Otherwise, shade the region on the other side of the boundary line.

🔄 **NOW TRY**
EXERCISE 1
Graph $-x + 2y \geq 4$.

EXAMPLE 1 Graphing a Linear Inequality

Graph $3x + 2y \geq 6$.

Step 1 First graph the boundary line $3x + 2y = 6$, which has intercepts $(2, 0)$ and $(0, 3)$, as shown in **FIGURE 29**.

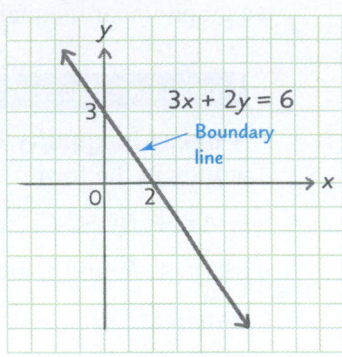

FIGURE 29

Step 2 The graph of the inequality $3x + 2y \geq 6$ includes the points of the boundary line $3x + 2y = 6$ (because the inequality symbol $\geq$ includes equality) and either the points *above* that line or the points *below* it. To decide which, select any point *not* on the boundary line to use as a test point. Substitute the values from the test point, here $(0, 0)$, for x and y in the inequality.

$$3x + 2y > 6 \quad \boxed{\text{We are testing the region.}}$$

$$\boxed{(0, 0) \text{ is a convenient test point.}} \quad 3(0) + 2(0) \overset{?}{>} 6 \quad \text{Let } x = 0 \text{ and } y = 0.$$

$$0 > 6 \quad \text{False}$$

Step 3 Because the result is false, $(0, 0)$ does *not* satisfy the inequality. The solution set includes all points in the region on the *other* side of the line. This region is shaded in **FIGURE 30**.

As a further check, select a test point in the shaded region, say $(3, 0)$.

$$3x + 2y > 6 \quad \text{Test the region.}$$

$$3(3) + 2(0) \overset{?}{>} 6 \quad \text{Let } x = 3 \text{ and } y = 0.$$

$$9 + 0 \overset{?}{>} 6 \quad \text{Multiply.}$$

$$9 > 6 \checkmark \quad \text{True}$$

The check confirms that the correct region is graphed in **FIGURE 30**.

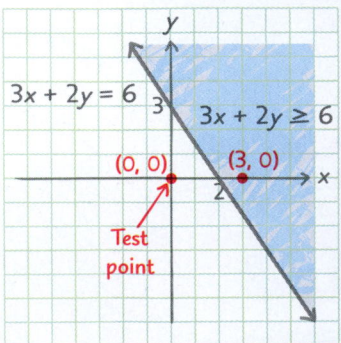

FIGURE 30

NOW TRY 🔄

NOW TRY ANSWER
1.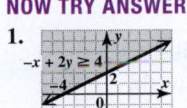

⚠️ **CAUTION** When drawing the boundary line in Step 1, be careful to draw a solid line if the inequality includes equality ($\leq$, $\geq$) or a dashed line if equality is not included ($<$, $>$).

If an inequality is written in the form $y > mx + b$ or $y < mx + b$, then the inequality symbol indicates which region to shade.

> If $y > mx + b$, then shade **above** the boundary line.

> If $y < mx + b$, then shade **below** the boundary line.

This method works only if the inequality is solved for y.

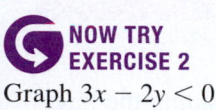

**NOW TRY
EXERCISE 2**

Graph $3x - 2y < 0$.

EXAMPLE 2   Graphing a Linear Inequality with Boundary Passing through the Origin

Graph $3x - 4y > 0$.

   Graph the boundary line. The x- and y-intercepts are the same point, $(0, 0)$, so this line passes through the origin. Two other points on the line are $(4, 3)$ and $(-4, -3)$. The points of the boundary line do *not* belong to the inequality $3x - 4y > 0$ (because the inequality symbol is $>$, *not* $\geq$). For this reason, the line is dashed.

   To use the method explained above, we solve the inequality for y.

$$3x - 4y > 0 \qquad \text{Original inequality}$$

$$-4y > -3x \qquad \text{Subtract } 3x.$$

> Use this equivalent inequality to decide which region to shade.

$$y < \frac{3}{4}x \qquad \begin{array}{l}\text{Divide by } -4.\\ \text{Change } > \text{ to } <.\end{array}$$

Because the *is less than* symbol occurs **when the original inequality is solved for y,** shade the region *below* the boundary line. See **FIGURE 31**.

CHECK   Choose a test point not on the line, which rules out the origin. We choose $(2, -1)$.

$$3x - 4y > 0 \qquad \text{Original inequality}$$

$$3(2) - 4(-1) \overset{?}{>} 0 \qquad \text{Let } x = 2 \text{ and } y = -1.$$

$$6 + 4 \overset{?}{>} 0 \qquad \text{Multiply.}$$

$$10 > 0 \ \checkmark \qquad \text{True}$$

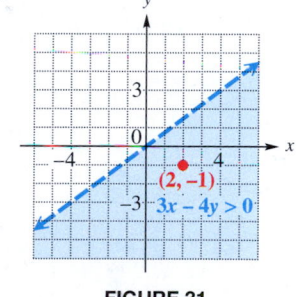

FIGURE 31

This result agrees with the decision to shade below the line. The solution set, graphed in **FIGURE 31**, includes only those points in the shaded region (and *not* those on the line).

NOW TRY

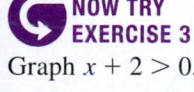

**NOW TRY
EXERCISE 3**

Graph $x + 2 > 0$.

EXAMPLE 3   Graphing a Linear Inequality

Graph $x - 3 < 1$.

   We graph $x - 3 = 1$, which is equivalent to $x = 4$, as a dashed vertical line passing through the point $(4, 0)$. To determine which region to shade, we choose $(0, 0)$ as a test point.

$$x - 3 < 1 \qquad \text{Original inequality}$$

$$0 - 3 \overset{?}{<} 1 \qquad \text{Let } x = 0.$$

$$-3 < 1 \qquad \text{True}$$

Because a true statement results, we shade the region containing $(0, 0)$. See **FIGURE 32**.

NOW TRY ANSWERS

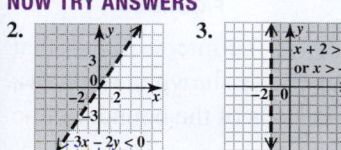

FIGURE 32

NOW TRY

OBJECTIVE 2 Solve systems of linear inequalities by graphing.

A **system of linear inequalities** consists of two or more linear inequalities. The **solution set of a system of linear inequalities** includes all ordered pairs that make all inequalities of the system true at the same time.

Solving a System of Linear Inequalities

Step 1   **Graph each linear inequality on the same axes.**

Step 2   **Choose the intersection.** Indicate the solution set by shading the intersection of the graphs—that is, the region where the graphs overlap.

EXAMPLE 4   Solving a System of Linear Inequalities

Graph the solution set of the system.

$$3x + 2y \le 6$$

$$2x - 5y > 10$$

Step 1   Graph $3x + 2y \le 6$ with the solid boundary line $3x + 2y = 6$ using the intercepts $(0, 3)$ and $(0, 2)$. Determine the region to shade.

$$3x + 2y < 6 \quad \text{We are testing the region.}$$

$$3(0) + 2(0) \overset{?}{<} 6 \quad \text{Use (0, 0) as a test point.}$$

$$0 < 6 \quad \text{True}$$

Shade the region that includes $(0, 0)$. See **FIGURE 33(a)**.

Now graph $2x - 5y > 10$ with dashed boundary line $2x - 5y = 10$ using the intercepts $(0, -2)$ and $(5, 0)$. Determine the region to shade.

$$2x - 5y > 10$$

$$2(0) - 5(0) \overset{?}{>} 10 \quad \text{Use (0, 0) as a test point.}$$

$$0 > 10 \quad \text{False}$$

Shade the region that does *not* include $(0, 0)$. See **FIGURE 33(b)**.

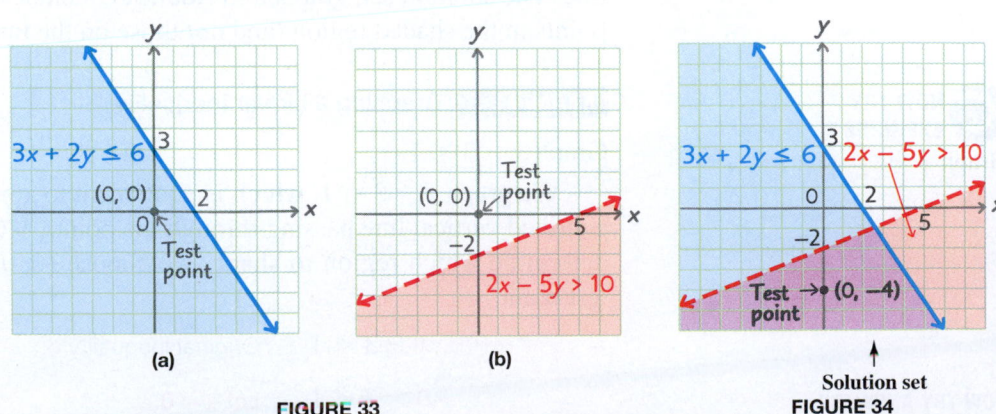

(a)   (b)

FIGURE 33

Solution set

FIGURE 34

Step 2   The solution set of this system includes all points in the intersection—that is, the overlap—of the graphs of the two inequalities. As shown in **FIGURE 34**, this intersection is the purple shaded region and portions of the boundary line $3x + 2y = 6$ that surrounds it.

NOW TRY
EXERCISE 4

Graph the solution set of
the system.

$$4x - 2y < 8$$
$$x + 3y \geq 3$$

CHECK To confirm the solution set in **FIGURE 34**, select a test point in the purple shaded region, such as $(0, -4)$, and substitute it into *both* inequalities to make sure that true statements result. (Using an ordered pair that has one coordinate 0 makes the substitution easier.)

$$3x + 2y < 6 \qquad\qquad 2x - 5y > 10$$
$$3(0) + 2(-4) \overset{?}{<} 6 \quad \text{Test } (0, -4). \qquad 2(0) - 5(-4) \overset{?}{>} 10 \quad \text{Test } (0, -4).$$
$$-8 < 6 \quad \text{True} \qquad\qquad 20 > 10 \quad \text{True}$$

We have shaded the correct region in **FIGURE 34**. Test points selected in the other three regions will satisfy only one of the inequalities or neither of them. (Verify this.) ✓

NOW TRY

> **NOTE** We usually do all the work on one set of axes. Be sure that the region of the final solution set is clearly indicated.

NOW TRY
EXERCISE 5

Graph the solution set of the system.

$$2x + 5y > 10$$
$$x - 2y < 0$$

EXAMPLE 5 Solving a System of Linear Inequalities

Graph the solution set of the system.

$$x - y > 0$$
$$2x + y > 2$$

FIGURE 35 shows the graphs of both $x - y > 0$ and $2x + y > 2$. Dashed lines indicate that the graphs of the inequalities do not include their boundary lines.

For $x - y > 0$, the dashed boundary line passes through the origin. To determine the region to shade, we must select a test point other than $(0, 0)$—that is, a point *off* the line. We use $(2, 0)$, which gives the true statement $2 > 0$, and so we shade the region that includes this point.

For $2x + y > 2$, we can use $(0, 0)$ as the test point. The false statement $0 > 2$ results, and so we shade the region that does *not* include this point.

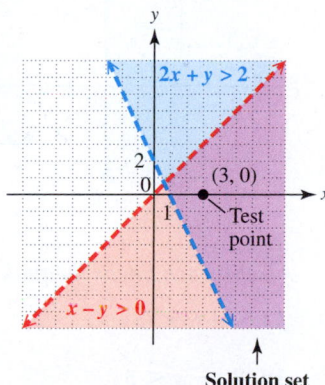

FIGURE 35

The solution set of the system is the region with purple shading in **FIGURE 35**. Neither boundary line is included.

CHECK To confirm the solution set in **FIGURE 35**, select a test point in the purple shaded region, such as $(3, 0)$, and substitute it into *both* inequalities. The resulting statements, $3 > 0$ and $6 > 2$, are true, so we have shaded the correct region as the solution set. ✓

NOW TRY

NOW TRY ANSWERS

4.
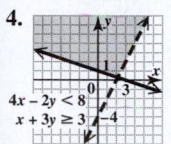

5.
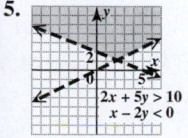

8.4 Exercises

 MyLab Math

▶ *Video solutions for select problems available in MyLab Math*

Concept Check   *Determine whether each ordered pair is a solution of the given inequality.*

1. $x - 2y \le 4$

  (a) $(0, 0)$ **(b)** $(2, -1)$

  (c) $(7, 1)$ **(d)** $(0, 2)$

2. $x + y > 0$

  (a) $(0, 0)$ **(b)** $(-2, 1)$

  (c) $(2, -1)$ **(d)** $(-4, 6)$

3. $x - 5 > 0$

  (a) $(0, 0)$ **(b)** $(5, 0)$

  (c) $(-1, 3)$ **(d)** $(6, 2)$

4. $y \le 1$

  (a) $(0, 0)$ **(b)** $(3, 1)$

  (c) $(2, -1)$ **(d)** $(-3, 3)$

Concept Check   *In each statement, fill in the first blank with either* solid *or* dashed. *Fill in the second blank with either* above *or* below.

5. The boundary of the graph of $y \le -x + 2$ will be a _____ line, and the shading will be _____ the line.

6. The boundary of the graph of $y < -x + 2$ will be a _____ line, and the shading will be _____ the line.

7. The boundary of the graph of $y > -x + 2$ will be a _____ line, and the shading will be _____ the line.

8. The boundary of the graph of $y \ge -x + 2$ will be a _____ line, and the shading will be _____ the line.

Concept Check   *Refer to the given graph, and complete each statement with the correct inequality symbol* $<$, $\le$, $>$, *or* $\ge$.

9. x _____ 4 **10.** y _____ -3 **11.** y _____ $3x - 2$ **12.** y _____ $-x + 3$

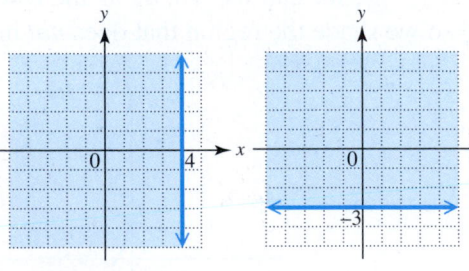
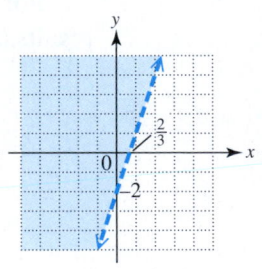
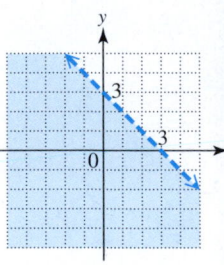

Graph each linear inequality. See Examples 1–3.

13. $x + y \le 2$ **14.** $x + y \le -3$ **15.** $4x - y < 4$

16. $3x - y < 3$ **17.** $x + 3y \ge -2$ **18.** $x + 4y \ge -3$

19. $y < \frac{1}{2}x + 3$ **20.** $y < \frac{1}{3}x - 2$ **21.** $y \ge -\frac{2}{5}x + 2$

22. $y \ge -\frac{3}{2}x + 3$ **23.** $2x + 3y \ge 6$ **24.** $3x + 4y \ge 12$

25. $5x - 3y > 15$ **26.** $4x - 5y > 20$ **27.** $x + y > 0$

28. $x + 2y > 0$ **29.** $x - 3y \le 0$ **30.** $x - 5y \le 0$

31. $y < x$ **32.** $y \le 4x$ **33.** $x + 3 \ge 0$

34. $x - 1 \le 0$ **35.** $y + 5 < 2$ **36.** $y - 1 > 3$

Extending Skills   *Complete each of the following to write an inequality for the graph shown.*

37. Determine the following for the boundary line.

Slope: _____

y-intercept: _____

Equation: $y =$ _____

The boundary line here is (*solid / dashed*), and the region (*above / below*) it is shaded.

The inequality symbol to indicate this is ($< / \le / > / \ge$).

Inequality for the graph: y _____

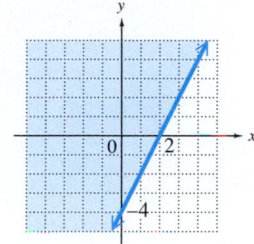

38. Determine the following for the boundary line.

Slope: _____

y-intercept: _____

Equation: $y =$ _____

The boundary line here is (*solid / dashed*), and the region (*above / below*) it is shaded.

The inequality symbol to indicate this is ($< / \le / > / \ge$).

Inequality for the graph: y _____

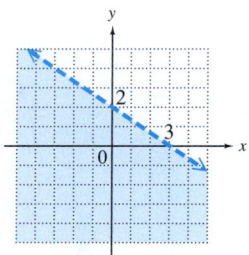

Concept Check   *Match each system of inequalities with the correct graph from choices A–D.*

39. $x \ge 5$
$\quad\ \ y \le -3$

A.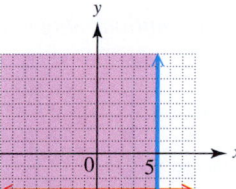

B.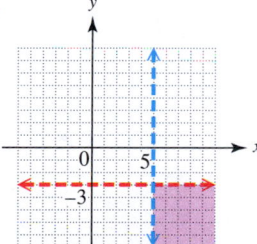

40. $x \le 5$
$\quad\ \ y \ge -3$

41. $x > 5$
$\quad\ \ y < -3$

C.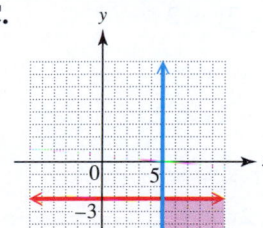

D.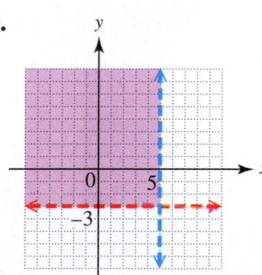

42. $x < 5$
$\quad\ \ y > -3$

Concept Check   *Describe the region that is the solution set of each system of inequalities.*

43. $x \ge 0$ **44.** $x \le 0$
$\quad\ \ y \ge 0$ $\qquad\ \ y \le 0$

45. $x > 0$ **46.** $x < 0$
$\quad\ \ y < 0$ $\qquad\ \ y > 0$

Concept Check   *Determine whether each ordered pair is a solution of the given system of inequalities. Then shade the solution set of each system. Boundary lines are already graphed.*

47. $x - 3y \leq 6$

    $x \geq -4$

  (a) $(-5, -4)$ **(b)** $(0, -4)$ **(c)** $(0, 0)$

48. $x - 2y \geq 4$

    $x \leq -2$

  (a) $(-3, 0)$ **(b)** $(0, 0)$ **(c)** $(-4, -5)$

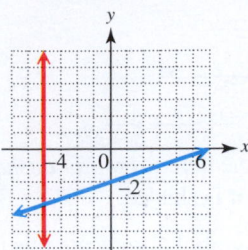

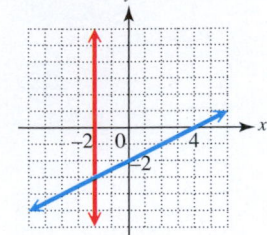

49. $x + y > 4$

    $5x - 3y < 15$

  (a) $(0, 0)$ **(b)** $(3, 3)$ **(c)** $(5, 0)$

50. $3x - 2y > 12$

    $4x + 3y < 12$

  (a) $(0, 0)$ **(b)** $(3, -3)$ **(c)** $(6, 0)$

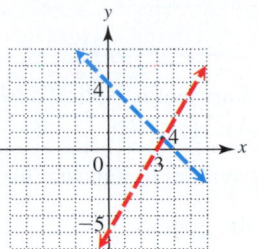

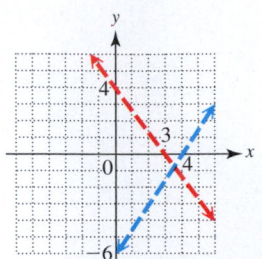

Graph the solution set of each system of linear inequalities. **See Examples 4 and 5.**

51. $x + y \leq 6$
    $x - y \geq 1$

52. $x + y \leq 2$
    $x - y \geq 3$

53. $4x + 5y \geq 20$
    $x - 2y \leq 5$

54. $x + 4y \leq 8$
    $2x - y \geq 4$

55. $2x + 3y < 6$
    $x - y < 5$

56. $x + 2y < 4$
    $x - y < -1$

57. $y \leq 2x - 5$
    $x < 3y + 2$

58. $x \geq 2y + 6$
    $y > -2x + 4$

59. $4x + 3y < 6$
    $x - 2y > 4$

60. $3x + y > 4$
    $x + 2y < 2$

61. $x \leq 2y + 3$
    $x + y < 0$

62. $x \leq 4y + 3$
    $x + y > 0$

63. $x + y < 3$
    $2x > y$

64. $x + y > 2$
    $3x < y$

65. $y \leq x + 4$
    $y \leq 4$

66. $y \geq x - 1$
    $y \geq 1$

67. $x - 3y \leq 6$
    $x \geq -5$

68. $x - 2y \geq 2$
    $x \leq -3$

69. $-3x + y \geq 1$
    $6x - 2y \geq -10$

70. $x + y < 4$
    $-2x - 2y < 4$

71. $2x + 3y < 6$
    $4x + 6y > 18$

72. $2x - y < -3$
    $6x - 3y > 9$

Extending Skills   *Graph the solution set of each system of linear inequalities.*

73. $4x + 5y < 8$
$y > -2$
$x > -4$

74. $x - 2y \geq -2$
$y \geq -2$
$x \leq 3$

75. $x + y \geq -3$
$x - y \leq 3$
$y \leq 3$

76. $x + y < 4$
$x - y > -4$
$y > -1$

77. $3x - 2y \geq 6$
$x + y \leq 4$
$x \geq 0$
$y \geq -4$

78. $2x - 3y < 6$
$x + y > 3$
$x < 4$
$y < 4$

RELATING CONCEPTS   **For Individual or Group Work** (Exercises 79–84)

Linear programming is a method for finding the optimal (best possible) solution that meets all the conditions for a problem such as the following.

> *A factory can have* no more than 200 *workers on a shift, but must have* at least 100 *and must manufacture* at least 3000 *units at minimum cost. How many workers should be on a shift in order to produce the required units at minimal cost?*

Let x represent the number of workers and y represent the number of units manufactured. **Work Exercises 79–84 in order.**

79. Write three inequalities expressing the problem conditions.

80. Graph the inequalities from **Exercise 79** using the axes below, and shade the intersection.

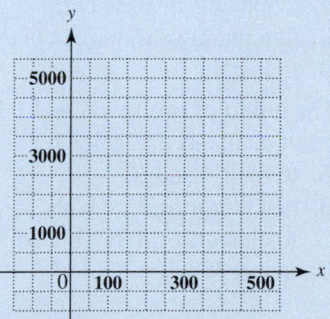

81. The cost per worker is $50 per day and the cost to manufacture 1 unit is $100. Write an equation in x, y, and C representing the total daily cost C.

82. Find values of x and y for several points in or on the boundary of the shaded region. Include any "corner points," where C is maximized or minimized.

83. Of the values of x and y found in **Exercise 82,** which ones give the least value when substituted in the cost equation from **Exercise 81?**

84. What does the answer in **Exercise 83** mean in terms of the given problem?

Chapter 8   Summary

STUDY SKILLS REMINDER
How do you best prepare for a test? **Review Study Skill 7,**
Reviewing a Chapter.

Key Terms

8.1	8.2	8.3	8.4
inequality	intersection	absolute value equation	linear inequality in two
interval	compound inequality	absolute value inequality	variables
linear inequality in one	union		boundary line
variable			system of linear inequalities
three-part inequality			solution set of a system of
			linear inequalities

New Symbols

∞ infinity

$-\infty$ negative infinity

$(-\infty, \infty)$ the set of real numbers

(a, b) interval notation for $a < x < b$

$[a, b]$ interval notation for $a \le x \le b$

$\cap$ set intersection

$\cup$ set union

Test Your Word Power

See how well you have learned the vocabulary in this chapter.

1. An **inequality** is
   A. a statement that two algebraic expressions are equal
   B. a point on a number line
   C. an equation with no solutions
   D. a statement consisting of algebraic expressions related by $<, \le, >,$ or $\ge$.

2. **Interval notation** is
   A. a point on a number line
   B. a special notation for describing a point on a number line
   C. a way to use symbols to describe an interval on a number line
   D. a notation to describe unequal quantities.

3. The **intersection** of two sets A and B is the set of elements that belong
   A. to both A and B
   B. to either A or B, or both
   C. to either A or B, but not both
   D. to just A.

4. The **union** of two sets A and B is the set of elements that belong
   A. to both A and B
   B. to either A or B, or both
   C. to either A or B, but not both
   D. to just B.

5. A **linear inequality in two variables** is an inequality that can be written in the form
   A. $Ax + By < C$ or $Ax + By > C$ ($\le$ or $\ge$ can be used)
   B. $ax < b$
   C. $y \ge x^2$
   D. $Ax + By = C$.

ANSWERS

1. D; *Examples:* $2 > -2, x < 5, 7 + 2k \ge 11$ 2. C; *Examples:* $(-\infty, 5], (1, \infty), [-3, 3)$

3. A; *Example:* If $A = \{2, 4, 6, 8\}$ and $B = \{1, 2, 3\}$, then $A \cap B = \{2\}$.

4. B; *Example:* If $A = \{2, 4, 6, 8\}$ and $B = \{1, 2, 3\}$, then $A \cup B = \{1, 2, 3, 4, 6, 8\}$.

5. A; *Examples:* $4x + 3y < 12, x > 6y, 2x \ge 4y + 5$

CONCEPTS	EXAMPLES

8.1   Review of Linear Inequalities in One Variable

Solving a Linear Inequality in One Variable

Step 1   Simplify each side separately.

- Clear any parentheses.
- Clear any fractions or decimals.
- Combine like terms.

Step 2   Isolate the variable terms on one side.

Step 3   Isolate the variable (here x) to write the inequality in one of these forms.

$$x < k, \quad x \le k, \quad x > k, \quad \text{or} \quad x \ge k$$

If an inequality is multiplied or divided by a negative number, the direction of the inequality symbol must be reversed.

Solve each inequality.

$$3(x + 2) - 5x \le 12$$

$3x + 6 - 5x \le 12$	Distributive property
$-2x + 6 \le 12$	Combine like terms.
$-2x + 6 - 6 \le 12 - 6$	Subtract 6.
$-2x \le 6$	Combine like terms.
$\dfrac{-2x}{-2} \ge \dfrac{6}{-2}$	Divide by -2. Change $\le$ to $\ge$.
$x \ge -3$	

Solution set: $[-3, \infty)$

To solve a three-part inequality, work with all three parts at the same time.

$$-4 < 2x + 3 \le 7$$

$-4 - 3 < 2x + 3 - 3 \le 7 - 3$	Subtract 3.
$-7 < 2x \le 4$	
$\dfrac{-7}{2} < \dfrac{2x}{2} \le \dfrac{4}{2}$	Divide by 2.
$-\dfrac{7}{2} < x \le 2$	

Solution set: $\left(-\dfrac{7}{2}, 2\right]$

8.2   Set Operations and Compound Inequalities

Solving a Compound Inequality

Step 1   Solve each inequality in the compound inequality individually.

Step 2   If the inequalities are joined with *and*, then the solution set is the intersection of the two individual solution sets.

If the inequalities are joined with *or*, then the solution set is the union of the two individual solution sets.

Solve each compound inequality.

$$x + 1 > 2 \quad \text{and} \quad 2x < 6$$
$$x > 1 \quad \text{and} \quad x < 3$$

Solution set: $(1, 3)$

$$2(x + 3) - 2 \le 4 \quad \text{or} \quad -4x \le -16$$
$$2x + 6 - 2 \le 4 \quad \text{or} \quad \dfrac{-4x}{-4} \ge \dfrac{-16}{-4}$$
$$2x + 4 \le 4 \quad \text{or} \quad x \ge 4$$
$$x \le 0$$

Solution set:

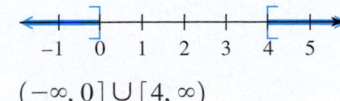

$$(-\infty, 0] \cup [4, \infty)$$

CONCEPTS

EXAMPLES

8.3 Absolute Value Equations and Inequalities

Solving Absolute Value Equations and Inequalities
Let k be a positive number.

Case 1 To solve $|ax + b| = k$, solve the compound equation

$$ax + b = k \quad \text{or} \quad ax + b = -k.$$

Case 2 To solve $|ax + b| > k$, solve the compound inequality

$$ax + b > k \quad \text{or} \quad ax + b < -k.$$

Case 3 To solve $|ax + b| < k$, solve the compound inequality

$$-k < ax + b < k.$$

Solve each equation or inequality.

$$|x - 7| = 3 \qquad \text{Case 1}$$

$$x - 7 = 3 \quad \text{or} \quad x - 7 = -3$$

$$x = 10 \quad \text{or} \qquad x = 4 \qquad \text{Add 7.}$$

Solution set: $\{4, 10\}$

$$|x - 7| > 3 \qquad \text{Case 2}$$

$$x - 7 > 3 \quad \text{or} \quad x - 7 < -3$$

$$x > 10 \quad \text{or} \qquad x < 4 \qquad \text{Add 7.}$$

Solution set:

$$(-\infty, 4) \cup (10, \infty)$$

$$|x - 7| < 3 \qquad \text{Case 3}$$

$$-3 < x - 7 < 3$$

$$4 < \quad x \quad < 10 \qquad \text{Add 7.}$$

Solution set: $(4, 10)$

To solve an absolute value equation of the form

$$|ax + b| = |cx + d|,$$

solve the compound equation

$$ax + b = cx + d \quad \text{or} \quad ax + b = -(cx + d).$$

$$|x + 2| = |2x - 6|$$

$$x + 2 = 2x - 6 \quad \text{or} \quad x + 2 = -(2x - 6)$$

$$-x = -8 \qquad \text{or} \quad x + 2 = -2x + 6$$

$$x = 8 \qquad \text{or} \qquad 3x = 4$$

$$x = \frac{4}{3}$$

Solution set: $\left\{\frac{4}{3}, 8\right\}$

Special Properties of Absolute Value

Property 1 The absolute value of an expression can never be negative—that is, $|a| \geq 0$ for all real numbers a.

Property 2 The absolute value of an expression equals 0 only when the expression is equal to 0.

$$|x - 2| = -4 \qquad \text{Solution set: } \varnothing$$

$$|x| \geq -1 \qquad \text{Solution set: } (-\infty, \infty) \quad \text{Property 1}$$

$$|2x - 3| < -6 \qquad \text{Solution set: } \varnothing$$

$$|3x - 9| = 0 \qquad \text{Property 2}$$

$$3x - 9 = 0 \qquad |a| = 0 \text{ implies } a = 0.$$

$$3x = 9 \qquad \text{Add 9.}$$

$$x = 3 \qquad \text{Divide by 3.}$$

Solution set: $\{3\}$

CONCEPTS	EXAMPLES

8.4 Linear Inequalities and Systems in Two Variables

Graphing a Linear Inequality

Step 1 Draw the graph of the straight line that is the boundary.
- Make the line solid if the inequality involves $\leq$ or $\geq$.
- Make the line dashed if the inequality involves $<$ or $>$.

Step 2 Choose any point not on the line as a test point. Substitute the coordinates of that point in the inequality.

Step 3 Shade the region that includes the test point if it satisfies the original inequality. Otherwise, shade the region on the other side of the boundary line.

Graph $2x - 3y \leq 6$.

Draw the graph of $2x - 3y = 6$ using the intercepts $(3, 0)$ and $(0, -2)$. Draw a solid line because of the inclusion of equality in the symbol $\leq$.

Choose $(0, 0)$ as a test point.

$$2(0) - 3(0) \overset{?}{<} 6$$
$$0 < 6 \qquad \text{True}$$

Shade the region that includes $(0, 0)$.

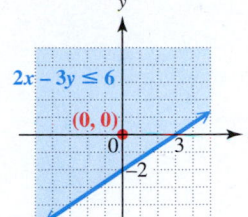

Solving a System of Linear Inequalities

Step 1 Graph each inequality on the same axes.

Step 2 Graph the intersection. Indicate the solution set by shading the intersection of the graphs—that is, the regions where the graphs overlap.

Graph the solution set of the system.

$$2x + 4y \geq 5$$
$$x \geq 1$$

First graph the solid boundary lines

$$2x + 4y = 5 \quad \text{and} \quad x = 1.$$

Then use a test point, such as $(0, 0)$, to determine the region to shade for each inequality. Using $(0, 0)$, two false statements result.

For $2x + 4y \geq 5$, shade the region "above" the line.

For $x \geq 1$, shade the region to the "right" of the line.

The intersection, the purple shaded region, is the solution set of the system.

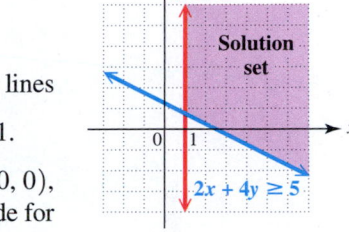

Chapter 8   Review Exercises

8.1 *Solve each inequality. Give the solution set in interval form.*

1. $-\dfrac{2}{3}x < 6$

2. $-5x - 4 \geq 11$

3. $\dfrac{6x + 3}{-4} < -3$

4. $5 - (6 - 4x) \geq 2x - 7$

5. $8 \leq 3x - 1 < 14$

6. $\dfrac{5}{3}(x - 2) + \dfrac{2}{5}(x + 1) > 1$

8.2 *Let $A = \{a, b, c, d\}$, $B = \{a, c, e, f\}$, and $C = \{a, e, f, g\}$. Find each set.*

7. $A \cap B$ **8.** $A \cap C$ **9.** $B \cup C$ **10.** $A \cup C$

Solve each compound inequality. Graph the solution set, and write it using interval notation.

11. $x > 6$ and $x < 9$

12. $x + 4 > 12$ and $x - 2 < 12$

13. $x > 5$ or $x \leq -3$

14. $x \geq -2$ or $x < 2$

15. $x - 4 > 6$ and $x + 3 \leq 10$

16. $-5x + 1 \geq 11$ or $3x + 5 \geq 26$

Express each set in simplest interval form.

17. $(-3, \infty) \cap (-\infty, 4]$

18. $(-\infty, 6] \cap (-\infty, 2]$

19. $(4, \infty) \cup (9, \infty)$

20. $(1, 2) \cup (1, \infty)$

8.3 *Solve each equation or inequality.*

21. $|x| = 7$

22. $|x + 2| = 9$

23. $|3x - 7| = 8$

24. $|x - 4| = -12$

25. $|2x - 7| + 4 = 11$

26. $|4x + 2| - 7 = -3$

27. $|3x + 1| = |x + 2|$

28. $|2x - 1| = |2x + 3|$

29. $|x| < 14$

30. $|-x + 6| \leq 7$

31. $|2x + 5| \leq 1$

32. $|x + 1| \geq -3$

33. $|3 - 4x| + 7 < -4$

34. $|-8 - 3x| - 7 > -8$

Determine the number of ounces a filled carton of the given size may contain for the given relative error.

$$\left| \frac{x - x_t}{x_t} \right| = \textbf{relative error in } x$$

x represents actual measurement.
x_t represents expected measurement.

35. 48-oz carton; relative error no greater than 0.03

36. 32-oz carton; relative error no greater than 0.01

8.4 *Graph each linear inequality.*

37. $3x - 2y \leq 12$

38. $5x - y > 6$

39. $x \geq 2y$

40. $x \geq 2$

Graph the solution set of each system of linear inequalities.

41. $x + y \geq 2$
$x - y \leq 4$

42. $y \geq 2x$
$2x + 3y \leq 6$

43. $3x - y \leq 3$
$y < 2$

44. Which system of linear inequalities is graphed in the figure?

A. $x \leq 3$
$\quad y \leq 1$

B. $x \leq 3$
$\quad y \geq 1$

C. $x \geq 3$
$\quad y \leq 1$

D. $x \geq 3$
$\quad y \geq 1$

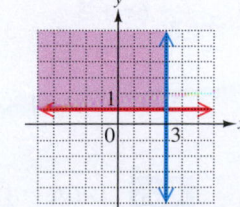

| Chapter 8 | Mixed Review Exercises |

Solve.

1. $5 - (6 - 4x) > 2x - 5$

2. $x + 4 < 7$ and $x + 5 \geq 3$

3. $|3x + 6| \geq 0$

4. $-5x \geq -10$

5. $|3x + 2| + 4 = 9$

6. $|x + 3| \leq 13$

7. $\dfrac{3}{4}(x - 2) - \dfrac{1}{3}(5 - 2x) < -2$

8. $-4 < 3 - 2x < 9$

9. $|5x - 1| > 14$

10. $x \geq -2$ or $x < 4$

11. $|x - 1| = |2x + 3|$

12. $|3x - 7| = 4$

13. $-5x < -30$ and $-7x > -56$

14. $-5x + 1 \geq 11$ or $3x + 5 \geq 26$

15. Which inequality has as its graph a dashed boundary line and shading below the line?

 A. $y \geq 4x + 3$ **B.** $y > 4x + 3$ **C.** $y \leq 4x + 3$ **D.** $y < 4x + 3$

16. Graph the solution set of the system of linear inequalities.

$$x + y < 5$$
$$x - y \geq 2$$

| Chapter 8 | Test | FOR EXTRA HELP | *Step-by-step test solutions are found on the Chapter Test Prep Videos available in* **MyLab Math**. |

 View the complete solutions to all Chapter Test exercises in MyLab Math.

Solve each inequality. Graph the solution set, and write it using interval notation.

1. $4 - 6(x + 3) \leq -2 - 3(x + 6) + 3x$

2. $-\dfrac{4}{7}x > -16$

3. $-1 < 3x - 4 < 2$

4. Which one of the following inequalities is equivalent to $x < -3$?

 A. $-3x < 9$ **B.** $-3x > -9$ **C.** $-3x > 9$ **D.** $-3x < -9$

5. Let $A = \{1, 2, 5, 7\}$ and $B = \{1, 5, 9, 12\}$. Find each set.

 (a) $A \cap B$ **(b)** $A \cup B$

Solve each compound or absolute value equation or inequality.

6. $3x \geq 6$ and $x < 9$

7. $-4x \leq -24$ or $4x < 12$

8. $|4x + 3| \leq 7$

9. $|5 - 6x| > 12$

10. $|3x - 9| = 6$

11. $|-3x + 4| - 4 < -1$

12. $|7 - x| \leq -1$

13. $|3x - 2| + 1 = 8$

14. $|3 - 5x| = |2x + 8|$

15. If $k < 0$, what is the solution set of each of the following?

 (a) $|8x - 5| < k$ **(b)** $|8x - 5| > k$ **(c)** $|8x - 5| = k$

Graph each linear inequality.

16. $3x - 2y > 6$

17. $y < 2x - 1$

18. Which system of linear inequalities is graphed in the figure?

A. $x > 4$
$y > 3$

B. $x < 4$
$y < 3$

C. $x > 4$
$y < 3$

D. $x < 3$
$y > 4$

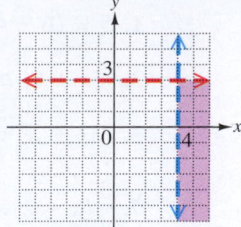

Graph the solution set of each system of linear inequalities.

19. $2x + 7y \leq 14$
$x - y \geq 1$

20. $2x - y > 6$
$4y + 12 \geq -3x$

Chapters R–8 **Cumulative Review Exercises**

1. Match each number in Column I with the equivalent number(s) in Column II. Choices may be used more than once.

I

(a) $\dfrac{1}{100}$ **(b)** $\dfrac{1}{10}$

(c) 1 **(d)** $\dfrac{1}{2}$

(e) $\dfrac{1}{4}$ **(f)** $\dfrac{5}{4}$

II

A. 0.5 **B.** 0.25 **C.** 0.1

D. 0.01 **E.** 125% **F.** 50%

G. 100% **H.** 10% **I.** 1%

2. Match each number in Column I with the choice or choices of sets of numbers in Column II to which the number belongs.

I

(a) 34 **(b)** 0

(c) 2.16 **(d)** $-\sqrt{36}$

(e) $\sqrt{13}$ **(f)** $-\dfrac{4}{5}$

II

A. Natural numbers **B.** Whole numbers

C. Integers **D.** Rational numbers

E. Irrational numbers **F.** Real numbers

Evaluate.

3. $9 \cdot 4 - 16 \div 4$

4. $-|8 - 13| - |-4| + |-9|$

5. Simplify $0.5(8x - 10) - 0.4(5 + 10x)$.

Solve.

6. $-5(8 - 2z) + 4(7 - z) = 7(8 + z) - 3$ **7.** $3(x + 2) - 5(x + 2) = -2x - 4$

8. $A = p + prt$ for t

9. $2(m + 5) - 3m + 1 > 5$

10. A survey polled Internet users age 12 and older about the Internet activities they engaged in daily. Complete the results shown in the table if 5000 such users were surveyed.

Daily Internet Activity	Percent	Actual Number
Browse the web	70%	
Download/listen to music	39%	
Download/watch videos		1800
Play games		1550

Data from the 2017 Center for the Digital Future report.

Find the slope of each line described.

11. Through $(-4, 5)$ and $(2, -3)$

12. Through $(4, 5)$; horizontal

*Write an equation of each line that satisfies the given conditions. Give the equation (**a**) in slope-intercept form and (**b**) in standard form.*

13. Through $(4, -1)$; $m = -4$

14. Through $(0, 0)$ and $(1, 4)$

Graph each equation or inequality.

15. $-3x + 4y = 12$

16. $3x + 2y < 0$

17. Per capita consumption of whole milk in the United States (in gallons) is shown in the graph, where $x = 0$ represents 1970.

(a) Use the given ordered pairs to find the average rate of change in per capita consumption of whole milk (in gallons) per year during this period. Interpret the answer.

(b) Use the answer from part (a) to write an equation of the line in slope-intercept form that models per capita consumption of whole milk y (in gallons).

(c) Use the equation from part (b) to approximate per capita consumption of whole milk in 2000.

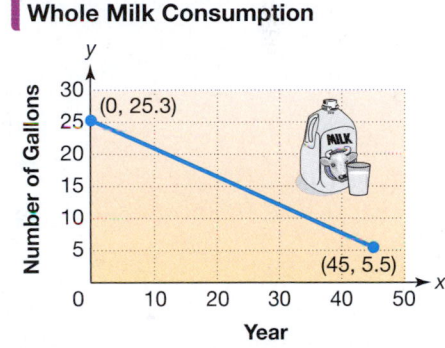

Whole Milk Consumption

Data from U.S. Department of Agriculture.

Simplify. Write answers with only positive exponents. Assume that all variables represent positive real numbers.

18. $\left(\dfrac{2m^3n}{p^2}\right)^3$

19. $\dfrac{x^{-6}y^3z^{-1}}{x^7y^{-4}z}$

Perform the indicated operations.

20. $(3x^2 - 8x + 1) - (x^2 - 3x - 9)$

21. $(3x + 2y)(5x - y)$

22. $\dfrac{16x^3y^5 - 8x^2y^2 + 4}{4x^2y}$

23. $\dfrac{m^3 - 3m^2 + 5m - 3}{m - 1}$

Factor each polyomial completely.

24. $m^2 + 12m + 32$

25. $25t^4 - 36$

26. $81z^2 + 72z + 16$

Perform each indicated operation. Express answers in lowest terms.

27. $\dfrac{x^2 - 3x - 4}{x^2 + 3x} \cdot \dfrac{x^2 + 2x - 3}{x^2 - 5x + 4}$

28. $\dfrac{t^2 + 4t - 5}{t + 5} \div \dfrac{t - 1}{t^2 + 8t + 15}$

29. $\dfrac{2}{x + 3} - \dfrac{4}{x - 1}$

30. $\dfrac{\dfrac{2}{3} + \dfrac{1}{2}}{\dfrac{1}{9} - \dfrac{1}{6}}$

Solve each equation.

31. $(x + 4)(x - 1) = -6$

32. $\dfrac{x}{x + 8} - \dfrac{3}{x - 8} = \dfrac{128}{x^2 - 64}$

Solve each system.

33. $3x - 4y = 1$
$2x + 3y = 12$

34. $3x - 2y = 4$
$-6x + 4y = 7$

35. $x + 3y - 6z = 7$
$2x - y + z = 1$
$x + 2y + 2z = -1$

36. The Star-Spangled Banner that flew over Fort McHenry during the War of 1812 had a perimeter of 144 ft. Its length measured 12 ft more than its width. Find the dimensions of this flag, which is displayed in the Smithsonian Institution's Museum of American History in Washington, DC. (Data from National Park Service brochure.)

Solve each equation or inequality.

37. $x > -4$ and $x < 4$

38. $2x + 1 > 5$ or $2 - x \geq 2$

39. $|3x - 1| = 2$

40. $|3z + 1| \geq 7$

STUDY SKILLS REMINDER

It is not too soon to begin preparing for your final exam. **Review Study Skill 10, *Preparing for Your Math Final Exam.***

RELATIONS AND FUNCTIONS

Linear equations whose graphs are straight (nonvertical) lines define *linear functions*, one of the topics of this chapter. We use the concept of slope, or steepness, to graph such functions.

| 9.1 | Introduction to Relations and Functions |

OBJECTIVES

1 Define and identify relations and functions.
2 Find the domain and range.
3 Identify functions defined by graphs and equations.

OBJECTIVE 1 Define and identify relations and functions.

Consider the relationship illustrated in the following table between number of hours worked and paycheck amount for an hourly worker.

Number of Hours Worked	Paycheck Amount (in dollars)	Ordered pairs
5	50	⟶ (5, 50)
10	100	⟶ (10, 100)
20	200	⟶ (20, 200)
40	400	⟶ (40, 400)

The data from the table can be represented by a set of ordered pairs.

$$\{(5, 50), (10, 100), (20, 200), (40, 400)\}$$

Number of hours worked ↑   ↑ Paycheck amount in dollars

Each first component of the ordered pairs represents a number of hours worked, and each second component represents the corresponding paycheck amount. Such a set of ordered pairs is a *relation*.

VOCABULARY

☐ relation
☐ function
☐ dependent variable
☐ independent variable
☐ domain
☐ range

Relation

A **relation** is any set of ordered pairs.

NOW TRY
EXERCISE 1

Write the relation as a set of ordered pairs.

Year	Average Gas Price per Gallon (in dollars)
2010	2.79
2012	3.64
2014	3.37
2016	2.14

Data from Energy Information Administration.

NOW TRY ANSWER
1. {(2010, 2.79), (2012, 3.64), (2014, 3.37), (2016, 2.14)}

EXAMPLE 1   Writing Ordered Pairs for a Relation

Write the relation as a set of ordered pairs.

Number of Gallons of Gas	Cost (in dollars)
0	0
1	3.50
2	7.00
3	10.50
4	14.00

The data in the table defines a relation between number of gallons of gas and cost and can be written as the following set of ordered pairs.

$$\{(0, 0), (1, 3.50), (2, 7.00), (3, 10.50), (4, 14.00)\}$$

Number of gallons of gas ↑   ↑ Cost in dollars

NOW TRY

A *function* is a special kind of relation.

Function

A **function** is a relation in which, for each distinct value of the first component of the ordered pairs, there is *exactly one* value of the second component.

NOW TRY
EXERCISE 2

Determine whether each relation defines a function.

(a) $\{(1, 5), (3, 5), (5, 5)\}$
(b) $\{(-1, -3), (0, 2), (-1, 6)\}$

EXAMPLE 2 Determining Whether Relations Are Functions

Determine whether each relation defines a function.

(a) $F = \{(1, 2), (-2, 4), (3, -1)\}$

Look at the ordered pairs that define this relation. For each distinct x-value, there is *exactly one* y-value. We can show this correspondence as follows.

$$\{1, \quad -2, \quad 3\} \quad x\text{-values of } F$$
$$\downarrow \quad \downarrow \quad \downarrow$$
$$\{2, \quad 4, \quad -1\} \quad y\text{-values of } F$$

Therefore, relation F is a function.

(b) $G = \{(-2, -1), (-1, 0), (0, 1), (1, 2), (2, 2)\}$

Relation G is also a function. Although the last two ordered pairs have the same y-value (1 is paired with 2, and 2 is paired with 2), this does not violate the definition of a function.

$$\{-2, \quad -1, \quad 0, \quad 1, \quad 2\} \quad x\text{-values of } G$$
$$\downarrow \quad \downarrow \quad \downarrow \quad \downarrow\swarrow$$
$$\{-1, \quad 0, \quad 1, \quad 2\} \quad y\text{-values of } G$$

The first components (x-values) are distinct, and each is paired with only one second component (y-value).

(c) $H = \{(-4, 1), (-2, 1), (-2, 0)\}$

In relation H, the last two ordered pairs have the ***same*** x-value paired with ***two different*** y-values (-2 is paired with both 1 and 0). H is a relation, but *not* a function.

$$\{-4, \quad -2\} \quad x\text{-values of } H$$
$$\downarrow\swarrow \quad \downarrow$$
$$\{1, \quad 0\} \quad y\text{-values of } H$$

In a function, no two ordered pairs have the same first component and different second components.

Different y-values

$$\text{Relation } H = \{(-4, 1), (-2, 1), (-2, 0)\} \quad \text{Not a function}$$

Same x-value

NOW TRY

Relations may be defined in several different ways.

• **A relation may be defined as a set of ordered pairs.**

$$\text{Relation } F = \{(1, 2), (-2, 4), (3, -1)\} \quad \text{Function (Example 2(a))}$$
$$\text{Relation } H = \{(-4, 1), (-2, 1), (-2, 0)\} \quad \text{Not a function (Example 2(c))}$$

- **A relation may be defined as a correspondence or *mapping*.**

See **FIGURE 1**. In the mapping for relation F, the arrow from 1 to 2 indicates that the ordered pair $(1, 2)$ belongs to F. Also, -2 is mapped to 4, and 3 is mapped to -1. Thus, F is a function—each first component is paired with exactly one second component.

In the mapping for relation H, which is *not* a function, the first component -2 is paired with two different second components, 1 and 0.

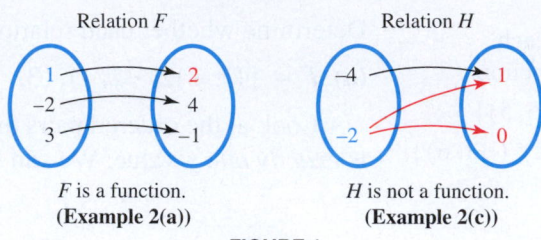

Relation F Relation H

F is a function. H is not a function.
(Example 2(a))   **(Example 2(c))**

FIGURE 1

- **A relation may be defined as a table.**

- **A relation may be defined as a graph.**

  FIGURE 2 includes a table and graph for relation F, which is a function.

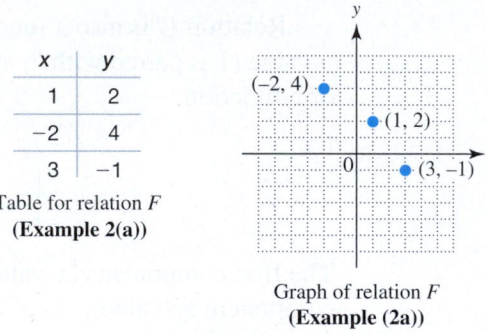

x	y
1	2
-2	4
3	-1

Table for relation F
(Example 2(a))

Graph of relation F
(Example (2a))

FIGURE 2

- **A relation may be defined as a rule.**

The rule may be given in words, such as "y is twice x." Usually the rule is an equation, such as

$$y = 2x.$$

The infinite number of ordered-pair solutions (x, y) can be represented by the graph in **FIGURE 3**.

In the equation $y = 2x$, the value of y *depends* on the value of x. Thus, the variable y is the **dependent variable.** The variable x is the **independent variable.**

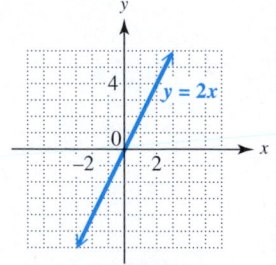

Graph of the relation $y = 2x$

FIGURE 3

Dependent variable $\longrightarrow y = 2x \longleftarrow$ Independent variable

An equation tells how to determine the value of the dependent variable for a specific value of the independent variable.

NOTE An equation that describes the relationship given at the beginning of this section between number of hours worked and paycheck amount is

$$y = 10x. \qquad \text{10 represents the hourly rate, \$10.}$$

Paycheck amount y *depends* on number of hours worked x. Thus, *paycheck amount* is the dependent variable, and *number of hours worked* is the independent variable.

In a function, there is exactly one value of the dependent variable, the second component, for each value of the independent variable, the first component.

> **NOTE**   Another way to think of a function relationship is to visualize the independent variable as an **input** and the dependent variable as an **output**. This **input-output (function) machine** illustrates the relationship between number of hours worked and paycheck amount.

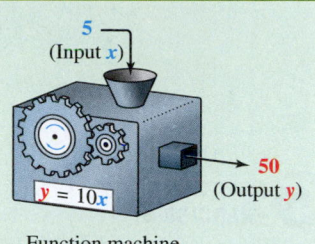

Function machine

OBJECTIVE 2 Find the domain and range.

Domain and Range

For every relation defined by a set of ordered pairs (x, y), there are two important sets of elements.

- The set of all values of the independent variable (x) is the **domain.**
- The set of all values of the dependent variable (y) is the **range.**

NOW TRY
EXERCISE 3
Give the domain and range of each relation. Determine whether the relation defines a function.

(a) $\{(2, 2), (2, 5), (4, 8)\}$

(b) The table from **Objective 1**

Number of Hours Worked	Paycheck Amount (in dollars)
5	50
10	100
20	200
40	400

NOW TRY ANSWERS
3. (a) domain: $\{2, 4\}$;
    range: $\{2, 5, 8\}$;
    not a function
(b) domain: $\{5, 10, 20, 40\}$;
    range: $\{50, 100, 200, 400\}$;
    function

EXAMPLE 3   Finding Domains and Ranges of Relations

Give the domain and range of each relation. Determine whether the relation defines a function.

(a) $\{(3, -1), (4, 2), (4, 5), (6, 8)\}$ List 4 only once.

Domain: $\{3, 4, 6\}$ Set of x-values

Range: $\{-1, 2, 5, 8\}$ Set of y-values

This relation is not a function because the same x-value 4 is paired with two different y-values, 2 and 5.

(b)

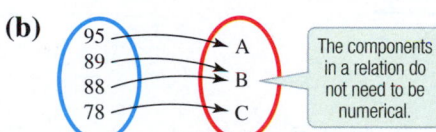

The components in a relation do not need to be numerical.

This mapping represents the following set of ordered pairs.

$$\{(95, A), (89, B), (88, B), (78, C)\}$$

Domain: $\{95, 89, 88, 78\}$ Set of first components

Range: $\{A, B, C\}$ Set of second components

The mapping defines a function—each domain value corresponds to exactly one range value.

(c)

x	y
-5	2
0	2
5	2

This table represents the following set of ordered pairs.

$$\{(-5, 2), (0, 2), (5, 2)\}$$

Domain: $\{-5, 0, 5\}$ Set of x-values

Range: $\{2\}$ Set of y-values

The table defines a function—each distinct x-value corresponds to exactly one y-value (even though it is the same y-value).

NOW TRY

A graph gives a "picture" of a relation and can be used to determine its domain and range.

> **NOTE** Pay particular attention to the use of color to interpret domain and range in **Example 4**—blue for domain and red for range.

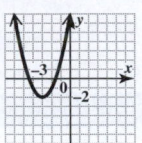

NOW TRY EXERCISE 4

Give the domain and range of the relation.

EXAMPLE 4 Finding Domains and Ranges from Graphs

Give the domain and range of each relation.

(a)

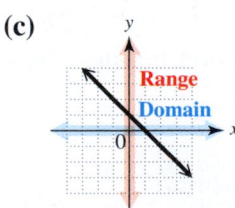

This relation includes the five ordered pairs that are graphed.

$$\{(-1, 1), (0, -1), (1, 2), (4, -3), (5, 2)\}$$

Domain: $\{-1, 0, 1, 4, 5\}$ Set of *x*-values

Range: $\{1, -1, 2, -3\}$ Set of *y*-values

> List 2 only once.

(b)

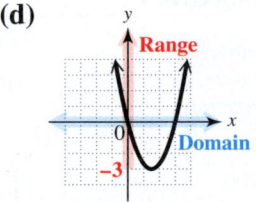

The *x*-values of the ordered pairs that form the graph include all numbers between -4 and 4, inclusive, as shown in blue. The *y*-values include all numbers between -6 and 6, inclusive, as shown in red.

Domain: $[-4, 4]$

Range: $[-6, 6]$ Use interval notation.

(c)

The arrowheads on the graphed line indicate that the line extends indefinitely left and right, as well as up and down. Therefore, both the domain (set of *x*-values), shown in blue, and the range (set of *y*-values), shown in red, include all real numbers.

Domain: $(-\infty, \infty)$ Range: $(-\infty, \infty)$

(d)

The graphed curve extends indefinitely left and right, as well as upward. The domain, shown in blue, includes all real numbers. Because there is a least *y*-value, -3, the range, shown in red, includes all numbers greater than or equal to -3.

Domain: $(-\infty, \infty)$ Range: $[-3, \infty)$

NOW TRY

OBJECTIVE 3 Identify functions defined by graphs and equations.

Because each value of *x* in a function corresponds to only one value of *y*, any vertical line drawn through the graph of a function must intersect the graph in at most one point. This is the *vertical line test* for a function.

FIGURE 4 on the next page illustrates this test with the graphs of two relations.

NOW TRY ANSWER

4. domain: $(-\infty, \infty)$; range: $[-2, \infty)$

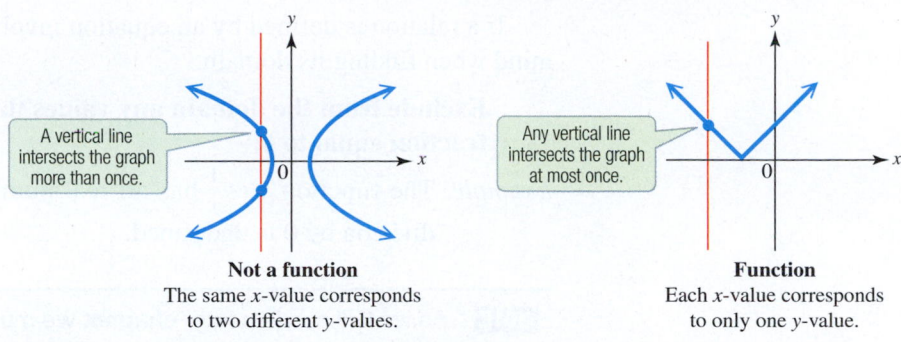

Not a function
The same *x*-value corresponds
to two different *y*-values.

Function
Each *x*-value corresponds
to only one *y*-value.

FIGURE 4

Vertical Line Test

If every vertical line intersects the graph of a relation in no more than one point, then the relation represents a function.

NOW TRY EXERCISE 5

Use the vertical line test to determine whether the relation is a function.

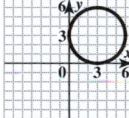

EXAMPLE 5 Using the Vertical Line Test

Use the vertical line test to determine whether each relation graphed in **Example 4** is a function. (We repeat the graphs here.)

(a)

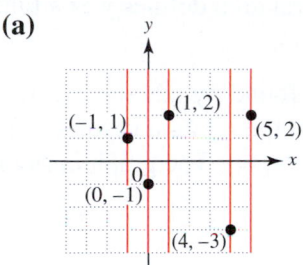

Function

(b)

Not a function

(c)

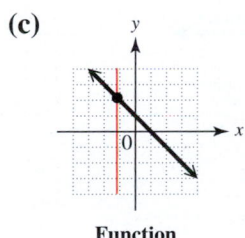

Function

(d)

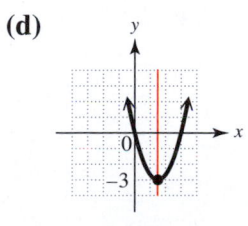

Function

The graphs in (a), (c), and (d) satisfy the vertical line test because every vertical line intersects each graph no more than once. These graphs represent functions.

The graph in (b) fails the vertical line test because a vertical line intersects the graph more than once—that is, the same *x*-value corresponds to two different *y*-values. This is not the graph of a function. **NOW TRY**

NOTE Graphs that do not represent functions are still relations. *All equations and graphs represent relations, and all relations have a domain and range.*

NOW TRY ANSWER
5. not a function

If a relation is defined by an equation involving a fraction, keep the following in mind when finding its domain.

> **Exclude from the domain any values that make the denominator of a fraction equal to 0.**

Example: The function $y = \frac{1}{x}$ has all real numbers *except* 0 as its domain because division by 0 is undefined.

> **NOTE** As we will see in a later chapter, we must also exclude from the domain any values that result in an even root of a negative number.
>
> *Example:* The function $y = \sqrt{x}$ has all *nonnegative* real numbers as its domain because the square root of a negative number is not real.

Agreement on Domain

Unless specified otherwise, the domain of a relation is assumed to be all real numbers that produce real numbers when substituted for the independent variable.

EXAMPLE 6 Identifying Functions and Domains from Equations

Determine whether each relation defines y as a function of x. Give the domain.

(a) $y = x + 4$

In this equation, y is found by adding 4 to x. Thus, each value of x corresponds to just one value of y, and the relation defines a function. Because x can be any real number, the domain is $(-\infty, \infty)$. The graph in **FIGURE 5** confirms this reasoning.

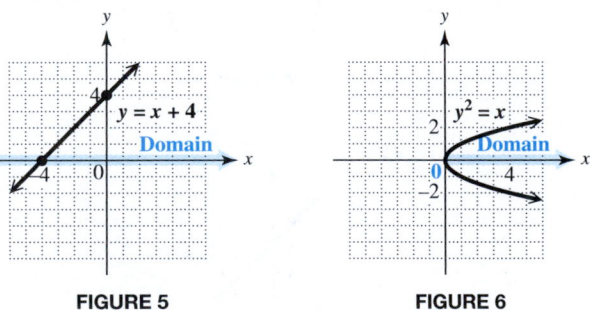

FIGURE 5   FIGURE 6

(b) $y^2 = x$

The ordered pairs $(4, 2)$ and $(4, -2)$ both satisfy this equation. One value of x, 4, corresponds to two values of y, 2 and -2, so this equation does not define a function. Because x is equal to the square of y, the values of x must always be nonnegative. The domain of the relation is $[0, \infty)$. See **FIGURE 6**.

(c) $y \leq x - 1$

By definition, y is a function of x if every value of x leads to exactly one value of y. Here, a particular value of x, such as 1, corresponds to many values of y. The ordered pairs

$$(1, 0), \quad (1, -1), \quad (1, -2), \quad (1, -3), \quad \text{and so on}$$

all satisfy the inequality. This relation does not define a function. Any number can be used for x, so the domain is the set of all real numbers, $(-\infty, \infty)$. See **FIGURE 7**.

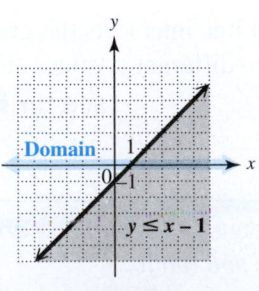

FIGURE 7

**NOW TRY
EXERCISE 6**

Determine whether each relation defines y as a function of x. Give the domain.

(a) $y = 4x - 3$

(b) $y = \dfrac{1}{x - 2}$

(c) $y < 3x + 1$

(d) $y = \dfrac{5}{x - 1}$

Given any value of x in the domain, we find y by subtracting 1 and then dividing the result into 5. This process produces exactly one value of y for each value in the domain, so the given equation defines a function.

The domain includes all real numbers except those which make the denominator 0.

$$x - 1 = 0 \qquad \text{Set the denominator equal to 0.}$$

$$x = 1 \qquad \text{Add 1.}$$

The domain includes all real numbers *except* 1, written $(-\infty, 1) \cup (1, \infty)$. In **FIGURE 8**, the open circle on the graph indicates that 1 is excluded from the domain.

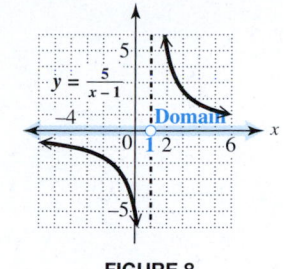

FIGURE 8

NOW TRY

In summary, we give three variations of the definition of a function.

Variations of the Definition of a Function

1. A **function** is a relation in which, for each distinct value of the first component of the ordered pairs, there is exactly one value of the second component.

2. A **function** is a set of distinct ordered pairs in which no first component is repeated.

3. A **function** is a correspondence (mapping) or an equation (rule) that assigns exactly one range value to each distinct domain value.

9.1 Exercises

FOR EXTRA HELP

 MyLab Math

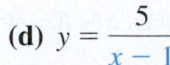
Video solutions for select problems available in MyLab Math

STUDY SKILLS REMINDER

How are you doing on your homework? **Review Study Skill 4, *Completing Your Homework*.**

Concept Check *Complete each statement. Choices may be used more than once.*

function	independent variable	vertical line test	relation
domain	ordered pairs	dependent variable	range

1. A(n) _____ is any set of _____ $\{(x, y)\}$.

2. A(n) _____ is a relation in which, for each distinct value of the first component of the _____ , there is exactly one value of the second component.

3. In a relation $\{(x, y)\}$, the _____ is the set of x-values, and the _____ is the set of y-values.

4. The relation $\{(0, -2), (2, -1), (2, -4), (5, 3)\}$ (*does / does not*) define a function. The set $\{0, 2, 5\}$ is its _____ , and the set $\{-2, -1, -4, 3\}$ is its _____ .

5. Consider the function $d = 50t$, where d represents distance and t represents time. The value of d depends on the value of t, so the variable t is the _____ , and the variable d is the _____ .

6. The _____ is used to determine whether a graph is that of a function. It says that any vertical line can intersect the graph of a(n) _____ in no more than (*zero / one / two*) point(s).

Write each relation as a set of ordered pairs. See Example 1.

7.

x	y
2	−2
2	0
2	1

8.

x	y
−1	−1
0	−1
1	−1

9.

Year	Average Movie Ticket Price (in dollars)
1960	0.76
1980	2.69
2000	5.39
2016	8.65

Data from Motion Picture Association of America.

10.

Year	Average ACT Composite Score
2010	21.0
2012	21.1
2014	21.0
2016	20.8

Data from ACT.

11.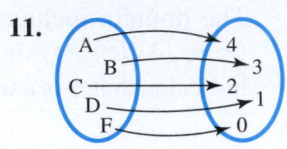

12.

Concept Check *Express each relation using a different form. (For example, if the given form is a set of ordered pairs, use a graph.) There is more than one correct way to do this. See Objective 1.*

13. $\{(0, 2), (2, 4), (4, 6)\}$

14. *y* is half of *x*.

15.

x	y
−1	−3
0	−1
1	1
3	3

16.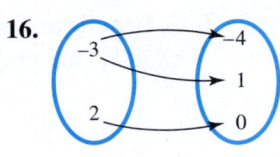

17.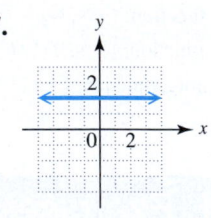

18. Concept Check Which of the relations represented in **Exercises 13–17** does *not* define a function? Explain.

Determine whether each relation defines a function, and give the domain and range. See Examples 2–5.

19. $\{(5, 1), (3, 2), (4, 9), (7, 6)\}$

20. $\{(8, 0), (5, 4), (9, 3), (3, 8)\}$

21. $\{(2, 4), (0, 2), (2, 5)\}$

22. $\{(9, -2), (-3, 5), (9, 2)\}$

23. $\{(-3, 1), (4, 1), (-2, 7)\}$

24. $\{(-12, 5), (-10, 3), (8, 3)\}$

25. $\{(1, 1), (1, -1), (0, 0), (2, 4), (2, -4)\}$

26. $\{(2, 5), (3, 7), (4, 9), (5, 11)\}$

27.

x	y
1	5
1	2
1	−1
1	−4

28.

x	y
−4	−4
−4	0
−4	4
−4	8

29.

x	y
4	−3
2	−3
0	−3
−2	−3

30.

x	y
−3	−6
−1	−6
1	−6
3	−6

31.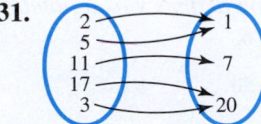

32.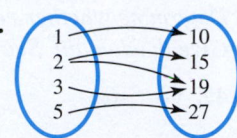

33.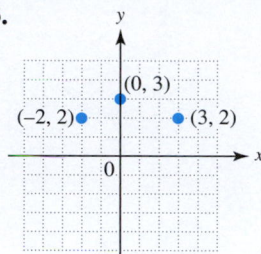

34.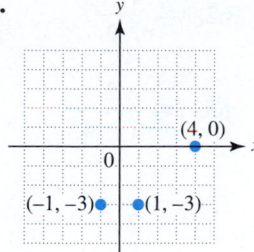

35.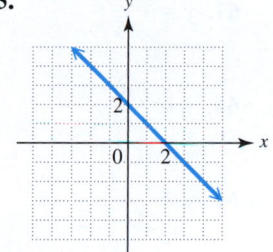

36.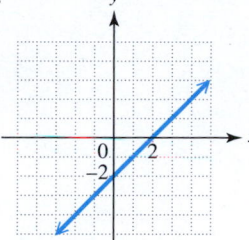

37.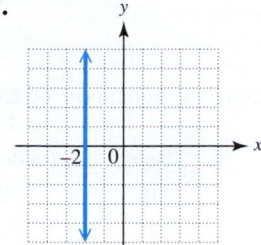

38.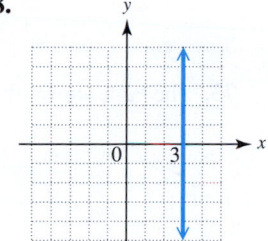

39.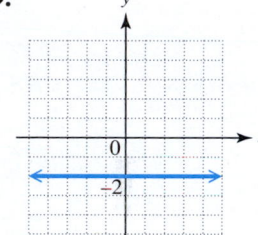

40.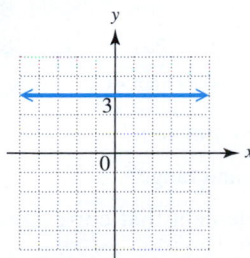

41.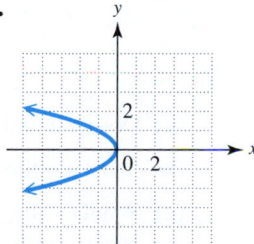

42.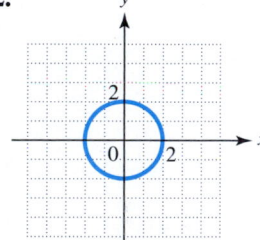

43.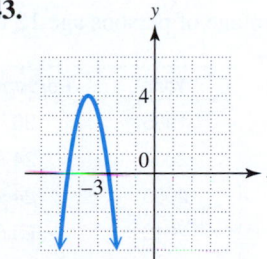

44.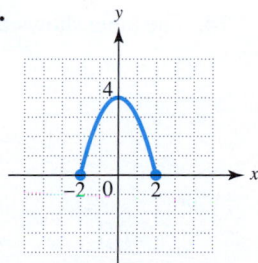

45.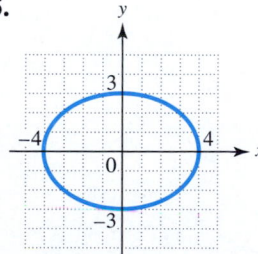

46.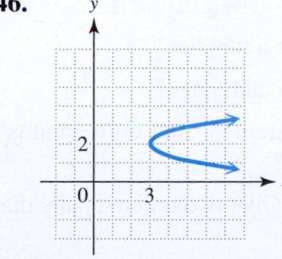

47.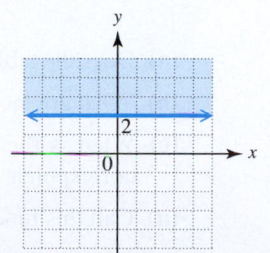

48.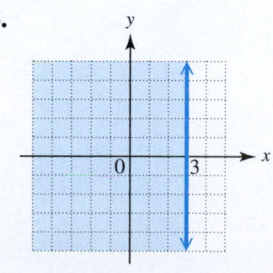

Determine whether each relation defines y as a function of x. (Solve for y first if necessary.) Give the domain. ***See Example 6.***

49. $y = -6x$

50. $y = -9x$

51. $y = 2x - 6$

52. $y = 6x + 8$

53. $y = x^2$

54. $y = x^3$

55. $x = y^6$

56. $x = y^4$

57. $x + y < 4$

58. $x - y < 3$

59. $y = x$

60. $y = -x$

61. $y = \dfrac{x + 4}{5}$

62. $y = \dfrac{x - 3}{2}$

63. $y = -\dfrac{2}{x}$

64. $y = -\dfrac{6}{x}$

65. $y = \dfrac{2}{x - 4}$

66. $y = \dfrac{7}{x - 2}$

67. $y = \dfrac{1}{4x + 2}$

68. $y = \dfrac{1}{2x + 9}$

69. $x = y^2 + 1$

70. $x = y^2 - 3$

71. $xy = 1$

72. $xy = 3$

Solve each problem.

73. The table shows the percentage of students at 4-year colleges who graduated within 5 years.

Year	Percentage
2013	52.8
2014	52.6
2015	52.6
2016	53.2
2017	53.7

Data from ACT.

(a) Does the table define a function?

(b) What are the domain and range?

(c) What is the range element that corresponds to 2015? The domain element that corresponds to 53.7?

(d) Call this function f. Give two ordered pairs that belong to f.

74. The table shows the percentage of persons age 12 or older who smoked cigarettes.

Year	Percentage
1985	38.7
2000	24.9
2005	24.9
2010	23.0
2015	19.4

Data from National Survey on Drug Use and Health.

(a) Does the table define a function?

(b) What are the domain and range?

(c) What is the range element that corresponds to 2015? The domain element that corresponds to 23.0?

(d) Call this function g. Give two ordered pairs that belong to g.

9.2   Function Notation and Linear Functions

OBJECTIVES

1 Use function notation.
2 Graph linear and constant functions.

VOCABULARY
☐ linear function
☐ constant function

OBJECTIVE 1 Use function notation.

When a function f is defined with a rule or an equation using x and y for the independent and dependent variables, we say, "y *is a function of* x" to emphasize that y *depends on* x. We use the notation

$$y = f(x),$$

The parentheses here do *not* indicate multiplication.

called **function notation,** to express this and read $f(x)$ as "*f of x*," or "*f at x*." The letter f is a name for this particular function. For example, if $y = 3x - 5$, we can name this function f and write the following.

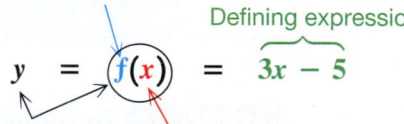

Name of the function

Defining expression

$$y \;=\; f(x) \;=\; 3x - 5$$

Value of the function   Value of the independent variable

$f(x)$ *is just another name for the dependent variable* y.

We evaluate a function at different values of x by substituting x-values from the domain into the function.

NOW TRY EXERCISE 1
Let $f(x) = 4x + 3$. Find the value of function f for each value of x.
(a) $x = -2$ (b) $x = 0$

EXAMPLE 1   Evaluating a Function

Let $f(x) = 3x - 5$. Find the value of function f for each value of x.

(a) $x = 2$

$$f(x) = 3x - 5$$

Read $f(2)$ as "f of 2" or "f at 2."

$$f(2) = 3 \cdot 2 - 5 \qquad \text{Replace } x \text{ with 2.}$$
$$f(2) = 6 - 5 \qquad \text{Multiply.}$$
$$f(2) = 1 \qquad \text{Subtract.}$$

For $x = 2$, the corresponding function value (or y-value) is 1. $f(2) = 1$ symbolizes the statement

"If $x = 2$ in the function f, then $y = 1$"

and is represented by the ordered pair $(2, 1)$.

(b) $x = -1$

$$f(x) = 3x - 5$$

Use parentheses to avoid errors.

$$f(-1) = 3(-1) - 5 \qquad \text{Replace } x \text{ with } -1.$$
$$f(-1) = -3 - 5 \qquad \text{Multiply.}$$
$$f(-1) = -8 \qquad \text{Subtract.}$$

NOW TRY ANSWERS
1. (a) -5 (b) 3

Thus, $f(-1) = -8$ and the ordered pair $(-1, -8)$ belongs to f.

NOW TRY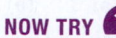

! CAUTION   The symbol $f(x)$ does *not* indicate "f times x," but represents the y-value associated with the indicated x-value. As shown in **Example 1(a)**, $f(2)$ is the y-value that corresponds to the x-value 2 in f.

NOTE   In the function $f(x) = 3x - 5$ in **Example 1,** $f(2) = 1$ and $f(-1) = -8$ correspond to the ordered pairs $(2, 1)$ and $(-1, -8)$. Because the domain of f is $(-\infty, \infty)$—that is, x can be any real number—this function defines an infinite set of ordered pairs whose graph is a line with slope 3 and y-intercept $(0, -5)$. See **FIGURE 9**. This makes sense because $f(x)$ is another name for y, and

$$f(x) = 3x - 5 \quad \text{is equivalent to} \quad y = 3x - 5.$$ f is a *linear function.*

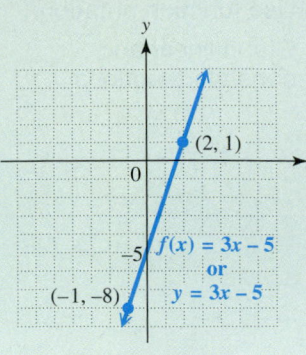

FIGURE 9

NOW TRY EXERCISE 2

Let $f(x) = 2x^2 - 4x + 1$. Find the following.

(a) $f(-2)$ **(b)** $f(a)$

EXAMPLE 2   Evaluating a Function

Let $f(x) = -x^2 + 5x - 3$. Find the following.

(a) $f(4)$

> Do *not* read this as "f times 4." Read it as "f of 4," or "f at 4."

$f(x) = -x^2 + 5x - 3$ The base in $-x^2$ is x, *not* $(-x)$.

$f(4) = -4^2 + 5 \cdot 4 - 3$ Replace x with 4.

$f(4) = -16 + 20 - 3$ Apply the exponent. Multiply.

$f(4) = 1$ Add and subtract.

Thus, $f(4) = 1$, and the ordered pair $(4, 1)$ belongs to f.

(b) $f(q)$

$f(x) = -x^2 + 5x - 3$

$f(q) = -q^2 + 5q - 3$ Replace x with q.

The replacement of one variable with another is important in later courses.

NOW TRY

Sometimes letters other than f, such as g, h, or capital letters F, G, and H are used to name functions.

NOW TRY EXERCISE 3

Let $g(x) = 8x - 5$. Find and simplify $g(a - 2)$.

EXAMPLE 3   Evaluating a Function

Let $g(x) = 2x + 3$. Find and simplify $g(a + 1)$.

$g(x) = 2x + 3$

$g(a + 1) = 2(a + 1) + 3$ Replace x with $a + 1$.

$g(a + 1) = 2a + 2 + 3$ Distributive property

$g(a + 1) = 2a + 5$ Add.   NOW TRY

NOW TRY ANSWERS

2. (a) 17   **(b)** $2a^2 - 4a + 1$
3. $8a - 21$

NOW TRY
EXERCISE 4
For each function, find $f(-1)$.
(a) $f = \{(-5, -1), (-3, 2),$
$(-1, 4)\}$
(b) $f(x) = x^2 - 12$

EXAMPLE 4 Evaluating Functions

For each function, find $f(3)$.

(a) $f(x) = 3x - 7$

$f(3) = 3(3) - 7$ Replace x with 3.

$f(3) = 9 - 7$ Multiply.

$f(3) = 2$ Subtract.

(b)

x	$y = f(x)$
6	-12
3	-6
0	0
-3	6

← Here, $f(3) = -6$.

(c) $f = \{(-3, 5), (0, 3), (3, 1), (6, -1)\}$

We want $f(3)$, the y-value of the ordered pair whose first component is $x = 3$. As indicated by the ordered pair $(3, 1)$, for $x = 3$, $y = 1$. Thus, $f(3) = 1$.

(d)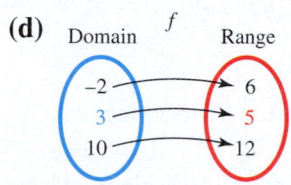

The domain element 3 is paired with 5 in the range, so

$$f(3) = 5.$$

NOW TRY

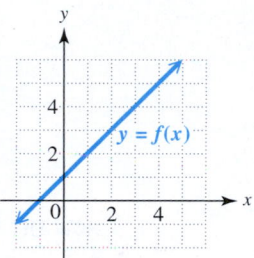

FIGURE 10

EXAMPLE 5 Finding Function Values from a Graph

Refer to the function graphed in **FIGURE 10**.

(a) Find $f(3)$.

Locate 3 on the x-axis. See **FIGURE 11**. Moving up to the graph of f and over to the y-axis gives 4 for the corresponding y-value. Thus, $f(3) = 4$, which corresponds to the ordered pair $(3, 4)$.

(b) Find $f(0)$.

Refer to **FIGURE 11** to see that $f(0) = 1$.

NOW TRY
EXERCISE 5
Refer to the function graphed in **FIGURE 10**.

(a) Find $f(-1)$.

(b) For what value of x is $f(x) = 2$?

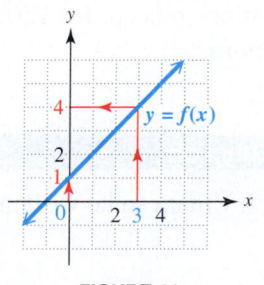

FIGURE 11

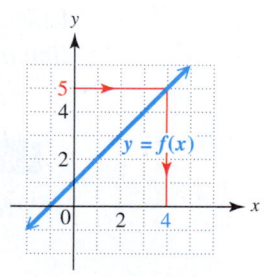

FIGURE 12

(c) For what value of x is $f(x) = 5$?

Because $f(x) = y$, we want the value of x that corresponds to $y = 5$. See **FIGURE 12**, and locate 5 on the y-axis. Moving across to the graph of f and down to the x-axis gives $x = 4$. Thus, $f(4) = 5$, which corresponds to the ordered pair $(4, 5)$. NOW TRY

If a function f is defined by an equation in x and y instead of function notation, use the following steps to find $f(x)$.

Writing an Equation Using Function Notation

Step 1 Solve the equation for y if it is not given in that form.

Step 2 Replace y with $f(x)$.

NOW TRY ANSWERS
4. **(a)** 4 **(b)** -11
5. **(a)** 0 **(b)** 1

NOW TRY
EXERCISE 6
Write the equation using function notation $f(x)$. Then find $f(-3)$.

$$-4x^2 + y = 5$$

EXAMPLE 6   Writing Equations Using Function Notation

Write each equation using function notation $f(x)$. Then find $f(-2)$.

(a) $\quad\quad y = x^2 + 1$ ← This equation is already solved for y.   (Step 1)

$\quad\quad\quad f(x) = x^2 + 1$ Replace y with $f(x)$.   (Step 2)

To find $f(-2)$, let $x = -2$.

$$f(x) = x^2 + 1$$

$$f(-2) = (-2)^2 + 1 \quad\quad \text{Let } x = -2.$$

$$f(-2) = 4 + 1 \quad\quad\quad (-2)^2 = (-2)(-2)$$

$$f(-2) = 5 \quad\quad\quad\quad \text{Add.}$$

(b) $\quad\quad\quad x - 4y = 5$

Step 1 $\quad -4y = -x + 5$ Subtract x.

$$y = \frac{1}{4}x - \frac{5}{4} \quad\quad \text{Divide by } -4.$$

Step 2 $\quad f(x) = \frac{1}{4}x - \frac{5}{4}$ Replace y with $f(x)$.

Now find $f(-2)$.

$$f(-2) = \frac{1}{4}(-2) - \frac{5}{4} = -\frac{7}{4} \quad\quad \text{Let } x = -2.$$

NOW TRY

OBJECTIVE 2 Graph linear and constant functions.

Linear equations (except for vertical lines with equations of the form $x = a$) define *linear functions*.

> **Linear Function**
>
> A function f that can be written in the form
>
> $$f(x) = ax + b,$$
>
> where a and b are real numbers, is a **linear function.** The value of a is the slope m of the graph of the function. The domain of a linear function is $(-\infty, \infty)$, unless specified otherwise.
>
> *Examples:*
>
> $$f(x) = 2x + 4, \quad f(x) = -5x, \quad f(x) = -\frac{1}{2}x - \frac{5}{4}, \quad f(x) = 0.75x + 10$$

A linear function whose graph is a horizontal line has the form

$$f(x) = b \quad\quad \text{Constant function}$$

NOW TRY ANSWER
6. $f(x) = 4x^2 + 5$;
   $f(-3) = 41$

and is a **constant function.** The range of any nonconstant linear function is $(-\infty, \infty)$, but the range of a constant function $f(x) = b$ is $\{b\}$.

 **NOW TRY EXERCISE 7**

Graph each function. Give the domain and range.

(a) $f(x) = \dfrac{1}{3}x - 2$

(b) $f(x) = -3$

EXAMPLE 7 Graphing Linear and Constant Functions

Graph each function. Give the domain and range.

(a) $f(x) = \dfrac{1}{4}x - \dfrac{5}{4}$ (from **Example 6(b)**)

Slope ↗ ↑ — y-intercept is $\left(0, -\dfrac{5}{4}\right)$.

To graph this function, plot the y-intercept $\left(0, -\dfrac{5}{4}\right)$. Use the geometric definition of slope as $\dfrac{\text{rise}}{\text{run}}$ to find a second point on the line. The slope is $\dfrac{1}{4}$, so we move up 1 unit from $\left(0, -\dfrac{5}{4}\right)$ and to the right 4 units to the point $\left(4, -\dfrac{1}{4}\right)$. Draw the straight line through these points. See **FIGURE 13**. The domain and range are both $(-\infty, \infty)$.

NOW TRY ANSWERS

7. (a)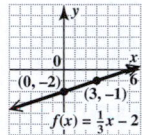

domain: $(-\infty, \infty)$;
range: $(-\infty, \infty)$

(b)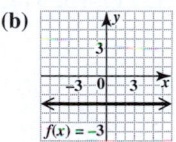

domain: $(-\infty, \infty)$;
range: $\{-3\}$

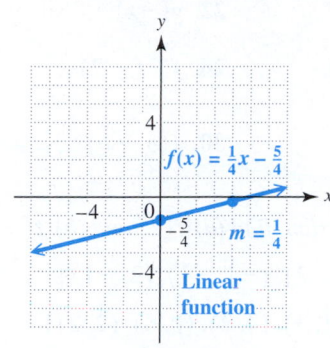

FIGURE 13

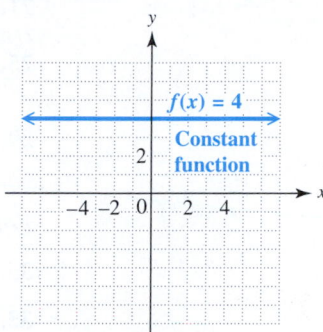

FIGURE 14

(b) $f(x) = 4$

The graph of this constant function is the horizontal line containing all points with y-coordinate 4. See **FIGURE 14**. The domain is $(-\infty, \infty)$. The value of $f(x)$—that is, y—is 4 for *every* value of x, so the range is $\{4\}$.

NOW TRY ↻

9.2 Exercises

FOR EXTRA HELP   ▶ **MyLab Math**

▶ *Video solutions for select problems available in MyLab Math*

Concept Check *Work each problem.*

1. To emphasize that "y is a function of x" for a given function f, we use function notation and write $y =$ _____. Here, f is the name of the _____, x is a value from the _____, and $f(x)$ is the function value (or y-value) that corresponds to _____. We read $f(x)$ as "_____."

2. Choose the correct response.

For a function f, the notation $f(3)$ means _____.

A. the variable f times 3, or $3f$.

B. the value of the dependent variable when the independent variable is 3.

C. the value of the independent variable when the dependent variable is 3.

D. f equals 3.

3. Fill in each blank with the correct response.

The equation $2x + y = 4$ has a straight _____ as its graph. One point that lies on the graph is $(3,$ _____$)$. If we solve the equation for y and use function notation, we have a(n) _____ function $f(x) =$ _____. For this function, $f(3) =$ _____, meaning that the point $($_____$,$ _____$)$ lies on the graph of the function.

4. Which of the following defines y as a linear function of x?

    A. $y = \dfrac{1}{4}x - \dfrac{5}{4}$ **B.** $y = \dfrac{1}{x}$ **C.** $y = x^2$ **D.** $y = x^3$

*Let $f(x) = -3x + 4$ and $g(x) = -x^2 + 4x + 1$. Find the following. **See Examples 1–3.***

5. $f(0)$ **6.** $g(0)$ **7.** $f(-3)$ **8.** $f(-5)$

9. $g(-2)$ **10.** $g(-1)$ **11.** $g(3)$ **12.** $g(10)$

13. $f(100)$ **14.** $f(-100)$ **15.** $f\left(\dfrac{1}{3}\right)$ **16.** $f\left(\dfrac{7}{3}\right)$

17. $g(0.5)$ **18.** $g(1.5)$ **19.** $f(p)$ **20.** $g(k)$

21. $f(-x)$ **22.** $g(-x)$ **23.** $f(x + 2)$ **24.** $f(x - 2)$

25. $f(2t + 1)$ **26.** $f(3t - 2)$ **27.** $g(\pi)$ **28.** $g(t)$

29. $f(x + h)$ **30.** $f(a + b)$ **31.** $g\left(\dfrac{p}{3}\right)$ **32.** $g\left(\dfrac{1}{x}\right)$

*For each function, find (a) $f(2)$ and (b) $f(-1)$. **See Examples 4, 5(a), and 5(b).***

33. $f = \{(-2, 2), (-1, -1), (2, -1)\}$ **34.** $f = \{(-1, -5), (0, 5), (2, -5)\}$

35. $f = \{(-1, 3), (4, 7), (0, 6), (2, 2)\}$ **36.** $f = \{(2, 5), (3, 9), (-1, 11), (5, 3)\}$

37.

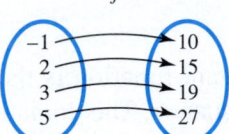

38.

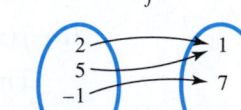

39.

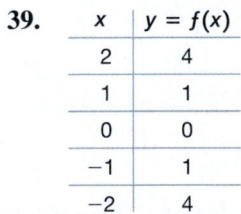

x	$y = f(x)$
2	4
1	1
0	0
−1	1
−2	4

40.

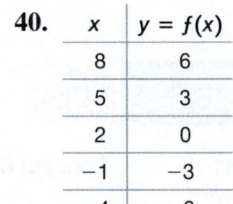

x	$y = f(x)$
8	6
5	3
2	0
−1	−3
−4	−6

41.

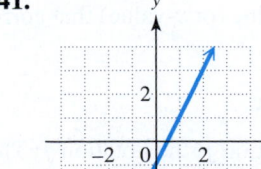

42.

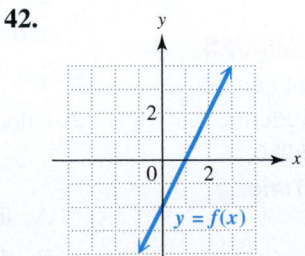

43.

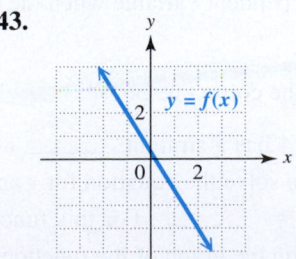

44.

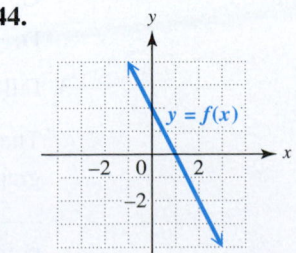

Refer to the given graph. Find the value of x for each value of f(x). **See Example 5(c).**

45. (a) $f(x) = 3$

(b) $f(x) = -1$

(c) $f(x) = -3$

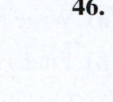

46. (a) $f(x) = 4$

(b) $f(x) = -2$

(c) $f(x) = 0$

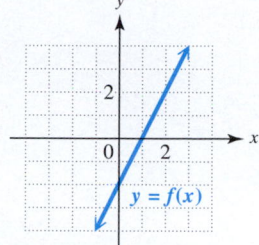

An equation that defines y as a function f of x is given. **(a)** *Solve for y in terms of x, and write each equation using function notation f(x).* **(b)** *Find f(3).* **See Example 6.**

47. $x + 3y = 12$

48. $x - 4y = 8$

49. $y + 2x^2 = 3$

50. $y - 3x^2 = 2$

51. $4x - 3y = 8$

52. $-2x + 5y = 9$

Graph each linear or constant function. Give the domain and range. **See Example 7.**

53. $f(x) = -2x + 5$

54. $g(x) = 4x - 1$

55. $h(x) = \dfrac{1}{2}x + 2$

56. $F(x) = -\dfrac{1}{4}x + 1$

57. $f(x) = x$

58. $f(x) = -x$

59. $H(x) = -3x$

60. $G(x) = 2x$

61. $g(x) = -4$

62. $f(x) = 5$

63. $f(x) = 0$

64. $f(x) = -2.5$

65. Concept Check What is the name that is usually given to the graph of $f(x) = 0$?

66. Concept Check Can the graph of a linear function have an undefined slope? Explain.

Solve each problem.

67. A package weighing x pounds costs $f(x)$ dollars to mail to a given location, where

$$f(x) = 3.75x.$$

(a) Evaluate $f(3)$.

(b) Describe what 3 and the value $f(3)$ mean in part (a), using the terms *independent variable* and *dependent variable*.

(c) How much would it cost to mail a 5-lb package? Express this situation using function notation.

68. A taxicab driver charges $2.50 per mile.

(a) Fill in the table with the correct response for the price $f(x)$ the driver charges for a trip of x miles.

(b) The linear function that gives a rule for the amount charged is $f(x) =$ _____.

(c) Graph this function for the domain $\{0, 1, 2, 3\}$ using the set of axes at the right.

x	f(x)
0	
1	
2	
3	

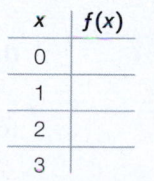

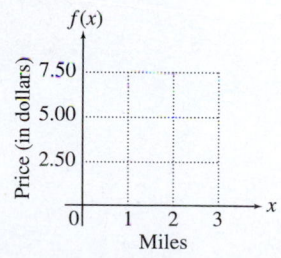

69. To print t-shirts, there is a $100 set-up fee, plus a $12 charge per t-shirt. Let x represent the number of t-shirts printed and $f(x)$ represent the total charge.

   (a) Write a linear function that models this situation.

   (b) Find $f(125)$. Interpret the answer in the context of this problem.

   (c) Find the value of x if $f(x) = 1000$. Express this situation using function notation, and interpret it in the context of this problem.

70. Rental on a car is $150, plus $0.50 per mile. Let x represent the number of miles driven and $f(x)$ represent the total cost to rent the car.

   (a) Write a linear function that models this situation.

   (b) Find $f(250)$. Interpret the answer in the context of this problem.

   (c) Find the value of x if $f(x) = 400$. Express this situation using function notation, and interpret it in the context of this problem.

71. The table represents a linear function.

   (a) What is $f(2)$?

   (b) If $f(x) = -1.3$, what is the value of x?

   (c) What is the slope of the line?

   (d) What is the y-intercept of the line?

   (e) Using the answers from parts (c) and (d), write an equation for $f(x)$.

x	$y = f(x)$
0	3.5
1	2.3
2	1.1
3	−0.1
4	−1.3

72. The table represents a linear function.

   (a) What is $f(2)$?

   (b) If $f(x) = 2.1$, what is the value of x?

   (c) What is the slope of the line?

   (d) What is the y-intercept of the line?

   (e) Using the answers from parts (c) and (d), write an equation for $f(x)$.

x	$y = f(x)$
−1	−3.9
0	−2.4
1	−0.9
2	0.6
3	2.1

73. The graph shows water in a swimming pool over time.

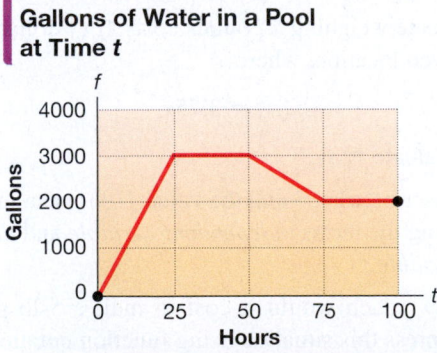

Gallons of Water in a Pool at Time _t_

   (a) What numbers are possible values of the independent variable? The dependent variable?

   (b) For how long is the water level increasing? Decreasing?

   (c) How many gallons of water are in the pool after 90 hr?

   (d) Call this function f. What is $f(0)$? What does it mean?

   (e) What is $f(25)$? What does it mean?

74. The graph shows electricity use on a summer day.

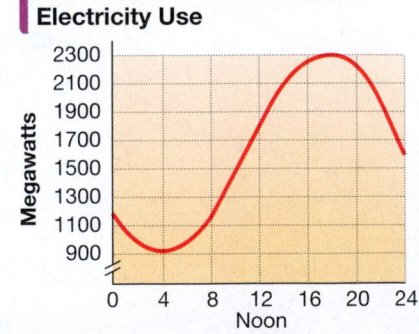

Electricity Use

   (a) Is this the graph of a function? Why or why not?

   (b) What is the domain?

   (c) Estimate the number of megawatts used at 8 A.M.

   (d) At what time was the most electricity used? The least electricity?

   (e) Call this function f. What is $f(12)$? What does it mean?

75. Forensic scientists use the lengths of certain bones to calculate the height of a person. Two such bones are the tibia (t), the bone from the ankle to the knee, and the femur (r), the bone from the knee to the hip socket. A person's height (h) in centimeters is determined from the lengths of these bones using the following functions.

For men: $h(r) = 69.09 + 2.24r$ or $h(t) = 81.69 + 2.39t$

For women: $h(r) = 61.41 + 2.32r$ or $h(t) = 72.57 + 2.53t$

Femur

Tibia

(a) Find the height of a man with a femur measuring 56 cm.

(b) Find the height of a man with a tibia measuring 40 cm.

(c) Find the height of a woman with a femur measuring 50 cm.

(d) Find the height of a woman with a tibia measuring 36 cm.

76. Based on federal regulations, a pool to house sea otters must have a volume that is "the square of the sea otter's average adult length (in meters) multiplied by 3.14 and by 0.91 meter." If x represents the sea otter's average adult length and $f(x)$ represents the volume (in cubic meters) of the corresponding pool size, this formula can be written as the function

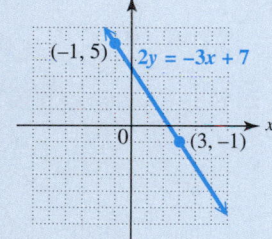

$$f(x) = 0.91(3.14)x^2.$$

Find the volume of the pool for each adult sea otter length (in meters). Round answers to the nearest hundredth.

(a) 0.8   (b) 1.0   (c) 1.2   (d) 1.5

RELATING CONCEPTS   For Individual or Group Work (Exercises 77–85)

Refer to the straight-line graph and **work Exercises 77–85 in order.**

77. By just looking at the graph, how can we tell whether the slope is *positive, negative, 0,* or *undefined*?

78. Apply the slope formula to find the slope of the line.

79. What is the slope of any line parallel to the line shown? Perpendicular to the line shown?

80. Find the x-intercept of the graph.

81. Find the y-intercept of the graph.

82. Use function notation to write the equation of the graphed line. Use f to designate the function.

83. Find $f(8)$.

84. If $f(x) = -8$, what is the value of x?

85. From the graph, find $f(1)$. Then confirm this value using the equation written in **Exercise 82.**

$(-1, 5)$ $2y = -3x + 7$

$(3, -1)$

9.3 Polynomial Functions, Graphs, Operations, and Composition

OBJECTIVES

1 Recognize and evaluate polynomial functions.

2 Graph basic polynomial functions.

3 Perform operations on polynomial functions.

4 Find the composition of functions.

VOCABULARY

☐ polynomial function
☐ identity function
☐ squaring function
☐ cubing function
☐ composite function (composition of functions)

 **NOW TRY EXERCISE 1**

Let $f(x) = x^3 - 2x^2 + 7$. Find $f(-3)$.

OBJECTIVE 1 Recognize and evaluate polynomial functions.

We have studied linear (first-degree polynomial) functions $f(x) = ax + b$. Now we consider more general polynomial functions.

> **Polynomial Function**
>
> A **polynomial function of degree** n is defined by
> $$f(x) = a_n x^n + a_{n-1} x^{n-1} + \cdots + a_1 x + a_0,$$
> for real numbers $a_n, a_{n-1}, \ldots, a_1$, and a_0, where $a_n \neq 0$ and n is a whole number.
>
> *Examples:* $f(x) = 2$, $f(x) = -6x + 1$, $f(x) = 5x^2 - x + 4$,
> $f(x) = x^3 + 2x^2 - 5x - 3$

Another way of describing a polynomial function is to say that it is a function defined by a polynomial in one variable, consisting of one or more terms. It is usually written in descending powers of the variable, and its degree is the degree of the polynomial that defines it.

We can evaluate a polynomial function $f(x)$ at different values of the variable x.

EXAMPLE 1 Evaluating Polynomial Functions

Let $f(x) = 4x^3 - x^2 + 5$. Find each value.

(a) $f(3)$

> Read this as "f of 3," not "f times 3."

$$f(x) = 4x^3 - x^2 + 5 \qquad \text{Given function}$$
$$f(3) = 4(3)^3 - 3^2 + 5 \qquad \text{Substitute 3 for } x.$$
$$f(3) = 4(27) - 9 + 5 \qquad \text{Apply the exponents.}$$
$$f(3) = 108 - 9 + 5 \qquad \text{Multiply.}$$
$$f(3) = 104 \qquad \text{Subtract, and then add.}$$

Thus, $f(3) = 104$ and the ordered pair $(3, 104)$ belongs to f.

(b) $f(-4)$

$$f(x) = 4x^3 - x^2 + 5 \qquad \text{Use parentheses.}$$
$$f(-4) = 4(-4)^3 - (-4)^2 + 5 \qquad \text{Let } x = -4.$$
$$f(-4) = 4(-64) - 16 + 5 \qquad \text{Be careful with signs.}$$
$$f(-4) = -256 - 16 + 5 \qquad \text{Multiply.}$$
$$f(-4) = -267 \qquad \text{Subtract, and then add.}$$

So, $f(-4) = -267$. The ordered pair $(-4, -267)$ belongs to f.   **NOW TRY**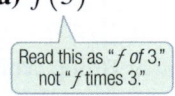

The capital letter P is sometimes used for polynomial functions. The function
$$P(x) = 4x^3 - x^2 + 5$$
yields the same ordered pairs as the function f in **Example 1.**

NOW TRY ANSWER
1. -38

OBJECTIVE 2 Graph basic polynomial functions.

Recall that each input (or x-value) of a function results in one output (or y-value). The set of input values (for x) defines the domain of the function, and the set of output values (for y) defines the range.

The simplest polynomial function is the **identity function** $f(x) = x$, graphed in **FIGURE 15**. This function pairs each real number with itself.

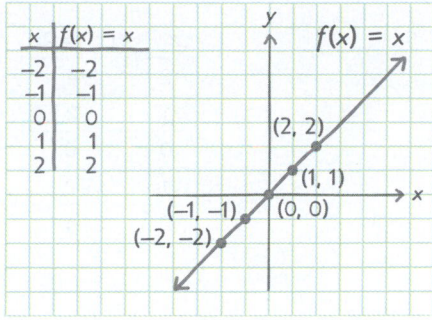

Identity function

$$f(x) = x$$

Domain: $(-\infty, \infty)$
Range: $(-\infty, \infty)$

FIGURE 15

> **NOTE** The identity function $f(x) = x$ shown in **FIGURE 15** is a *linear function* of the form $f(x) = ax + b$, where the slope a is 1 and the y-value of the y-intercept b is 0.

Another polynomial function, the **squaring function** $f(x) = x^2$, is graphed in **FIGURE 16**. For this function, every real number is paired with its square. The graph of the squaring function is a *parabola*.

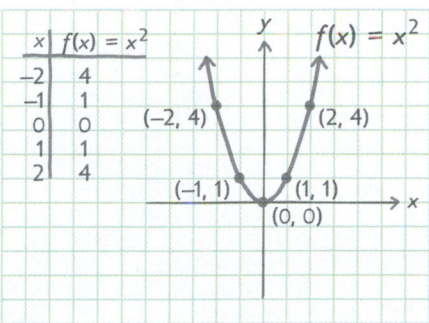

Squaring function

$$f(x) = x^2$$

Domain: $(-\infty, \infty)$
Range: $[0, \infty)$

FIGURE 16

The **cubing function** $f(x) = x^3$ is graphed in **FIGURE 17**. This function pairs every real number with its cube.

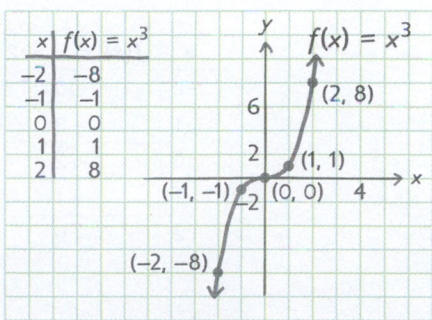

Cubing function

$$f(x) = x^3$$

Domain: $(-\infty, \infty)$
Range: $(-\infty, \infty)$

FIGURE 17

**NOW TRY
EXERCISE 2**

Graph $f(x) = x^2 - 4$. Give the domain and range.

EXAMPLE 2   Graphing Variations of Polynomial Functions

Graph each function by creating a table of ordered pairs. Give the domain and range of each function by observing its graph.

(a) $f(x) = 2x$

To find each range value, multiply the domain value by 2. Plot the points and join them with a straight line. See **FIGURE 18**. Both the domain and the range are $(-\infty, \infty)$.

x	$f(x) = 2x$
-2	-4
-1	-2
0	0
1	2
2	4

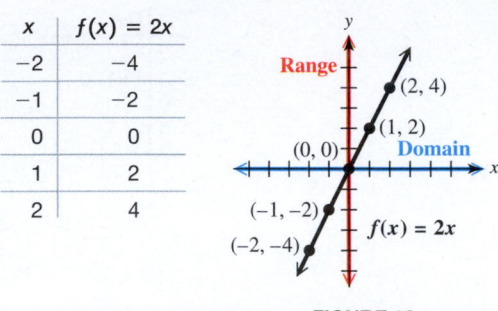

FIGURE 18

(b) $f(x) = -x^2$

For each input x, square it and then take its opposite. Plotting and joining the points gives a parabola that opens down. It is a *reflection* of the graph of the squaring function across the x-axis. See the table and **FIGURE 19**. The domain is $(-\infty, \infty)$ and the range is $(-\infty, 0]$.

x	$f(x) = -x^2$
-2	-4
-1	-1
0	0
1	-1
2	-4

FIGURE 19

(c) $f(x) = x^3 - 2$

For this function, cube the input and then subtract 2 from the result. The graph is that of the cubing function *shifted* down 2 units. See the table and **FIGURE 20**. The domain and range are both $(-\infty, \infty)$.

x	$f(x) = x^3 - 2$
-2	-10
-1	-3
0	-2
1	-1
2	6

NOW TRY ANSWER

2.

domain: $(-\infty, \infty)$;
range: $[-4, \infty)$

FIGURE 20

NOW TRY

OBJECTIVE 3 Perform operations on polynomial functions.

The operations of addition, subtraction, multiplication, and division are also defined for functions. For example, the graph in **FIGURE 21** shows dollars (in billions) spent for general science and for space/other technologies over a 20-year period.

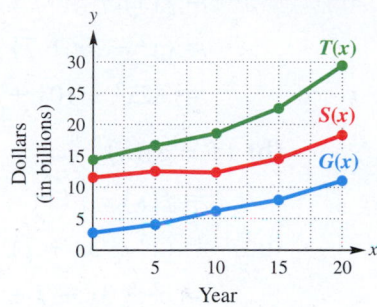

Science and Space Spending

Data from U.S. Office of Management and Budget.

FIGURE 21

$G(x)$ represents dollars spent for general science.

$S(x)$ represents dollars spent for space/other technologies.

$T(x)$ represents total expenditures for these two categories.

The total expenditures function can be found by *adding* the spending functions for the two individual categories.

$$T(x) = G(x) + S(x)$$

As another example, businesses use the equation "profit equals revenue minus cost," which can be written using function notation.

$$P(x) = R(x) - C(x)$$

↑   ↑   ↑   *x* is the number of items produced and sold.

Profit   Revenue   Cost
function   function   function

The profit function is found by *subtracting* the cost function from the revenue function.

We define the following **operations on functions.**

Operations on Functions

If $f(x)$ and $g(x)$ define functions, then

$$(f + g)(x) = f(x) + g(x), \qquad \text{Sum function}$$

$$(f - g)(x) = f(x) - g(x), \qquad \text{Difference function}$$

$$(fg)(x) = f(x) \cdot g(x), \qquad \text{Product function}$$

and

$$\left(\frac{f}{g}\right)x = \frac{f(x)}{g(x)}, \quad g(x) \neq 0. \qquad \text{Quotient function}$$

In each case, the domain of the new function is the intersection of the domains of $f(x)$ and $g(x)$. Additionally, the domain of the quotient function must exclude any values of x for which $g(x) = 0$.

NOW TRY
EXERCISE 3
For $f(x) = x^3 - 3x^2 + 4$
and $g(x) = -2x^3 + x^2 - 12$,
find each of the following.
(a) $(f + g)(x)$
(b) $(f - g)(x)$

EXAMPLE 3   Adding and Subtracting Functions

Find each of the following for the polynomial functions f and g as defined.
$$f(x) = x^2 - 3x + 7 \quad \text{and} \quad g(x) = -3x^2 - 7x + 7$$

(a) $(f + g)(x)$ *This notation does* ***not*** *indicate the distributive property.*

$\quad = f(x) + g(x)$ Sum function

$\quad = (x^2 - 3x + 7) + (-3x^2 - 7x + 7)$ Substitute.

$\quad = -2x^2 - 10x + 14$ Add the polynomials.

(b) $(f - g)(x)$

$\quad = f(x) - g(x)$ Difference function

$\quad = (x^2 - 3x + 7) - (-3x^2 - 7x + 7)$ Substitute.

$\quad = x^2 - 3x + 7 + 3x^2 + 7x - 7$ Change subtraction to addition.

$\quad = 4x^2 + 4x$ Add the polynomials.   **NOW TRY**

NOW TRY
EXERCISE 4
For $f(x) = x^2 - 4$
and $g(x) = -6x^2$,
find each of the following.
(a) $(f + g)(x)$
(b) $(f - g)(-4)$

EXAMPLE 4   Adding and Subtracting Functions

Find each of the following for the polynomial functions f and g as defined.
$$f(x) = 10x^2 - 2x \quad \text{and} \quad g(x) = 2x$$

(a) $(f + g)(2)$

$\quad = f(2) + g(2)$ Sum function

     $\overbrace{f(x) = 10x^2 - 2x}$ $\overbrace{g(x) = 2x}$

This is a key step. $\quad = [10(2)^2 - 2(2)] + 2(2)$ Substitute.

$\quad = [40 - 4] + 4$ Order of operations

$\quad = 40$ Subtract, and then add.

Alternative method: $(f + g)(x)$ Find $(f + g)(x)$.

$\quad = f(x) + g(x)$ Sum function

$\quad = (10x^2 - 2x) + 2x$ Substitute.

$\quad = 10x^2$ Combine like terms.

$\quad\quad (f + g)(2)$ Now find $(f + g)(2)$.

$\quad = 10(2)^2$ $(f + g)(x) = 10x^2$; Substitute.

The result is the same. $\longrightarrow = 40$ Apply the exponent. Multiply.

(b) $(f - g)(x)$ and $(f - g)(1)$

$\quad\quad (f - g)(x)$

$\quad = f(x) - g(x)$ Difference function

$\quad = (10x^2 - 2x) - 2x$ Substitute.

$\quad = 10x^2 - 4x$ Combine like terms.

NOW TRY ANSWERS
3. (a) $-x^3 - 2x^2 - 8$
   (b) $3x^3 - 4x^2 + 16$
4. (a) $-5x^2 - 4$ (b) 108

$\quad\quad (f - g)(1)$ Now find $(f - g)(1)$.

Confirm that $f(1) - g(1)$ gives the same result.

$\quad = 10(1)^2 - 4(1)$ $(f - g)(x) = 10x^2 - 4x$; Substitute.

$\quad = 6$ Perform the operations.   **NOW TRY**

NOW TRY
EXERCISE 5

For $f(x) = 3x^2 - 1$ and $g(x) = 8x + 7$, find $(fg)(x)$ and $(fg)(-2)$.

EXAMPLE 5 **Multiplying Polynomial Functions**

For $f(x) = 3x + 4$ and $g(x) = 2x^2 + x$, find $(fg)(x)$ and $(fg)(-1)$.

$(fg)(x)$

This notation indicates function multiplication.

$= f(x) \cdot g(x)$ Use the definition.

$= (3x + 4)(2x^2 + x)$ Substitute.

$= 6x^3 + 3x^2 + 8x^2 + 4x$ FOIL method

$= 6x^3 + 11x^2 + 4x$ Combine like terms.

$(fg)(-1)$ $(fg)(x) = 6x^3 + 11x^2 + 4x$

$= 6(-1)^3 + 11(-1)^2 + 4(-1)$ Let $x = -1$ in $(fg)(x)$.

Be careful with signs.

$= -6 + 11 - 4$ Apply the exponents. Multiply.

$= 1$ Add and subtract.

An alternative method for finding $(fg)(-1)$ is to find $f(-1)$ and $g(-1)$ and then multiply the results. Verify that $f(-1) \cdot g(-1) = 1$. This follows from the definition.

NOW TRY

NOW TRY
EXERCISE 6

For $f(x) = 8x^2 + 2x - 3$ and $g(x) = 2x - 1$, find $\left(\frac{f}{g}\right)(x)$ and $\left(\frac{f}{g}\right)(8)$.

EXAMPLE 6 **Dividing Polynomial Functions**

For $f(x) = 2x^2 + x - 10$ and $g(x) = x - 2$, find $\left(\frac{f}{g}\right)(x)$ and $\left(\frac{f}{g}\right)(-3)$.

$$\left(\frac{f}{g}\right)(x) = \frac{f(x)}{g(x)} = \frac{2x^2 + x - 10}{x - 2}$$

To find the quotient, divide as follows.

$$\begin{array}{r} 2x + 5 \\ x - 2 \overline{\smash{\big)}2x^2 + x - 10} \\ \underline{2x^2 - 4x} \quad \leftarrow 2x(x-2) \\ 5x - 10 \quad \text{Subtract.} \\ \underline{5x - 10} \quad \leftarrow 5(x-2) \\ 0 \end{array}$$

To subtract, add the opposite.

The quotient is $2x + 5$. Thus,

$$\left(\frac{f}{g}\right)(x) = 2x + 5, \quad x \neq 2. \quad \leftarrow \begin{array}{l}\text{2 is not in the domain.}\\ \text{It causes denominator}\\ g(x) = x - 2 \text{ to equal 0.}\end{array}$$

$$\left(\frac{f}{g}\right)(-3) = 2(-3) + 5 = -1 \quad \text{Let } x = -3 \text{ in } \left(\frac{f}{g}\right)(x) = 2x + 5.$$

Alternative method: The same result is found by evaluating $\frac{f(-3)}{g(-3)}$.

$$\frac{f(-3)}{g(-3)} \qquad \left(\frac{f}{g}\right)(-3) = \frac{f(-3)}{g(-3)}$$

$$= \frac{2(-3)^2 + (-3) - 10}{-3 - 2} \quad \begin{array}{l}\leftarrow f(x) = 2x^2 + x - 10\\ \leftarrow g(x) = x - 2\end{array}$$

$$= \frac{5}{-5}$$

$$= -1 \quad \text{The result is the same.}$$ **NOW TRY**

NOW TRY ANSWERS
5. $24x^3 + 21x^2 - 8x - 7$; -99
6. $4x + 3$, $x \neq \frac{1}{2}$; 35

OBJECTIVE 4 Find the composition of functions.

The diagram in **FIGURE 22** shows a function g that assigns, to each element x of set X, some element y of set Y. Suppose that a function f takes each element of set Y and assigns a value z of set Z. Then f and g together assign an element x in X to an element z in Z.

The result of this process is a new function h that takes an element x in X and assigns it an element z in Z.

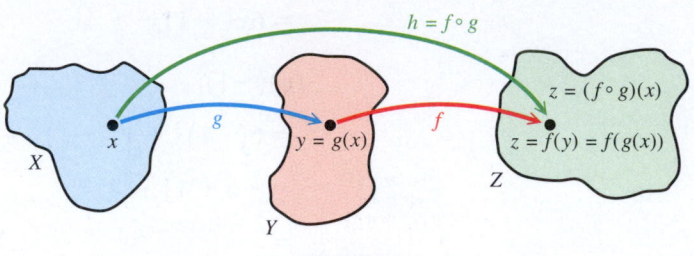

FIGURE 22

Function h is the *composition* of functions f and g, written $f \circ g$.

Composition of Functions

The **composite function**, or **composition**, of functions f and g is defined by

$$(f \circ g)(x) = f(g(x)),$$

for all x in the domain of g such that $g(x)$ is in the domain of f.

Read $f \circ g$ as "f of g" (or "f compose g").

As a real-life example of how composite functions occur, consider the following retail situation.

A $40 pair of blue jeans is on sale for 25% off. If we purchase the jeans before noon, the retailer offers an additional 10% off. What is the final sale price of the blue jeans?

We might be tempted to say that the jeans are 35% off and calculate as follows.

$40 − 0.35($40)   Original price − Discount = Sale price

= $40 − $14

= $26 ⟵——————— This is not correct.

To find the correct final sale price, we must first find the price after taking 25% off, and then take an additional 10% off *that* price.

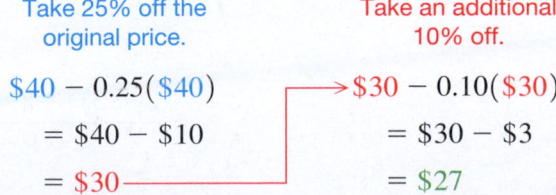

Take 25% off the original price.   Take an additional 10% off.

$40 − 0.25($40)   →$30 − 0.10($30)

= $40 − $10   = $30 − $3

= $30————   = $27

This is the idea behind composition of functions.

As another real-life example of composition, suppose an oil well off the coast is leaking, with the leak spreading oil in a circular layer over the surface. See **FIGURE 23**.

At any time t, in minutes, after the beginning of the leak, the radius of the circular oil slick is given by $r(t) = 5t$ feet. Because $\mathcal{A}(r) = \pi r^2$ gives the area of a circle of radius r, the area can be expressed as a function of time.

FIGURE 23

$$\mathcal{A}(r) = \pi r^2 \qquad \text{Area of a circle}$$
$$\mathcal{A}(r(t)) = \pi(5t)^2 \qquad \text{Substitute } 5t \text{ for } r.$$
$$\mathcal{A}(r(t)) = 25\pi t^2 \qquad \begin{array}{l}(5t)^2 = 5^2 t^2 = 25t^2;\\ \text{Commutative property}\end{array}$$

The function $\mathcal{A}(r(t))$ is a composite function of the functions $\mathcal{A}$ and r.

NOW TRY
EXERCISE 7

Let $f(x) = 3x + 7$ and $g(x) = x - 2$.
Find $(f \circ g)(7)$.

EXAMPLE 7   Finding a Composite Function

Let $f(x) = x^2$ and $g(x) = x + 3$. Find $(f \circ g)(4)$.

$$(f \circ g)(4) \qquad \boxed{\text{Evaluate the "inside" function value first.}}$$
$$= f(g(4)) \qquad \text{Definition of composition}$$
$$= f(4 + 3) \qquad \text{Use the rule for } g(x); g(4) = 4 + 3$$
$$= f(7) \qquad \text{Add.}$$
$$\boxed{\text{Now evaluate the "outside" function.}} \quad = 7^2 \qquad \text{Use the rule for } f(x); f(7) = 7^2$$
$$= 49 \qquad \text{Square 7.}$$

In this composition, g is the innermost "operation" and acts on x (here 4) first. Then the output value of g (here 7) becomes the input (domain) value of f.   **NOW TRY**

If we interchange the order of functions f and g, the composition $g \circ f$, read "g of f" (or "g compose f"), is defined by

$$(g \circ f)(x) = g(f(x)), \qquad \text{for all } x \text{ in the domain of } f \text{ such that } f(x)\text{ is in the domain of } g.$$

NOW TRY
EXERCISE 8

Let $f(x) = 3x + 7$ and $g(x) = x - 2$ as in **Now Try Exercise 7.**
Find $(g \circ f)(7)$.

EXAMPLE 8   Finding a Composite Function

Let $f(x) = x^2$ and $g(x) = x + 3$ as in **Example 7**. Find $(g \circ f)(4)$.

$$(g \circ f)(4) \qquad \boxed{\text{Evaluate the "inside" function value first.}}$$
$$= g(f(4)) \qquad \text{Definition of composition}$$
$$= g(4^2) \qquad \text{Use the rule for } f(x); f(4) = 4^2$$
$$= g(16) \qquad \text{Square 4.}$$
$$\boxed{\text{Now evaluate the "outside" function.}} \quad = 16 + 3 \qquad \text{Use the rule for } g(x); g(16) = 16 + 3$$
$$= 19 \qquad \text{Add.}$$

In this composition, f is the innermost "operation" and acts on x (again 4) first. Then the output value of f (here 16) becomes the input (domain) value of g.   **NOW TRY**

NOW TRY ANSWERS
7. 22
8. 26

We see in **Examples 7 and 8** that $(f \circ g)(4) \neq (g \circ f)(4)$ because $49 \neq 19$. In general, $(f \circ g)(x) \neq (g \circ f)(x)$.

NOW TRY
EXERCISE 9

Let $f(x) = x - 5$
and $g(x) = -x^2 + 2$.
Find each of the following.
(a) $(g \circ f)(-1)$
(b) $(f \circ g)(x)$

EXAMPLE 9 Finding Composite Functions

Let $f(x) = 4x - 1$ and $g(x) = x^2 + 5$. Find each of the following.

(a) $(f \circ g)(2)$

$\qquad = f(g(2))$ Definition of composition

$\qquad = f(2^2 + 5)$ $g(x) = x^2 + 5$

$\qquad = f(9)$ Work inside the parentheses.

$\qquad = 4(9) - 1$ $f(x) = 4x - 1$

$\qquad = 35$ Multiply, and then subtract.

(b) $(f \circ g)(x)$

$\qquad = f(g(x))$ Use $g(x)$ as the input for function f.

$\qquad = 4(g(x)) - 1$ Use the rule for $f(x)$; $f(x) = 4x - 1$

$\qquad = 4(x^2 + 5) - 1$ $g(x) = x^2 + 5$

$\qquad = 4x^2 + 20 - 1$ Distributive property

$\qquad = 4x^2 + 19$ Combine like terms.

(c) Find $(f \circ g)(2)$ again, this time using the rule obtained in part (b).

$$(f \circ g)(x) = 4x^2 + 19 \qquad \text{From part (b)}$$
$$(f \circ g)(2) = 4(2)^2 + 19 \qquad \text{Let } x = 2.$$
$$= 4(4) + 19 \qquad \text{Square 2.}$$
$$= 16 + 19 \qquad \text{Multiply.}$$

NOW TRY ANSWERS
9. **(a)** -34 **(b)** $-x^2 - 3$

Same result as in part (a) $\rightarrow$ $= 35$ Add.   **NOW TRY**

9.3 Exercises

FOR EXTRA HELP

 MyLab Math

▶ *Video solutions for select problems available in MyLab Math*

Concept Check *Work each problem involving polynomial functions and function notation.*

1. A polynomial function is a function defined by a _____ in (*one / two / three*) variable(s), consisting of one or more (*factors / terms*) and usually written in descending _____ of the variable.

2. Which of the following are *not* polynomial functions?

  A. $P(x) = x^{-2} - 2x$ **B.** $f(x) = \frac{1}{2}x^2 + x - 1$

  C. $g(x) = -4x + 1.5$ **D.** $p(x) = x^3 - x^2 - \frac{5}{x}$

3. For a function f, the notation $f(5)$ means _____.

  A. the variable f times 5, or $5f$ **B.** f equals 5

  C. the range value when the domain value is 5   **D.** the domain value when the range value is 5

4. If $f(x) = 2x$, then $f(5) = $ _____. Here $f(5) = 10$ is an abbreviation for the statement "If $x = $ _____ in the function f, then $y = $ _____." The ordered pair _____ belongs to f.

5. For $f(x) = x$, find $f(0)$, $f(1)$, and $f(2)$. Use the results to write three ordered pairs that belong to f.

6. For $f(x) = x^2$, find $f(-2)$, $f(0)$, and $f(2)$. Use the results to write three ordered pairs that belong to f.

7. Concept Check For the function
$$f(x) = -x^2 + 4,$$
a student claimed that $f(-2) = 8$. This is incorrect. **WHAT WENT WRONG?** Find the correct value of $f(-2)$.

8. Concept Check For the function
$$f(x) = x^2 - 1,$$
a student claimed that $f(-1) = -2$. This is incorrect. **WHAT WENT WRONG?** Find the correct value of $f(-1)$.

*For each polynomial function, find (**a**) $f(-1)$, (**b**) $f(2)$, and (**c**) $f(0)$. See Example 1.*

9. $f(x) = 6x - 4$ **10.** $f(x) = -2x + 5$ **11.** $f(x) = x^2 - 7x$

12. $f(x) = x^2 + 5x$ **13.** $f(x) = x^2 - 3x + 4$ **14.** $f(x) = x^2 - 5x - 4$

15. $f(x) = 2x^2 - 4x + 1$ **16.** $f(x) = 3x^2 + x - 5$ **17.** $f(x) = 5x^4 - 3x^2 + 6$

18. $f(x) = 4x^4 + 2x^2 - 1$ **19.** $f(x) = -x^2 + 2x^3 - 8$ **20.** $f(x) = -x^2 - x^3 + 11$

Graph each polynomial function. Give the domain and range. See Example 2.

21. $f(x) = 3x$ **22.** $f(x) = -4x$ **23.** $f(x) = -2x + 1$

24. $f(x) = 3x + 2$ **25.** $f(x) = -3x^2$ **26.** $f(x) = \dfrac{1}{2}x^2$

27. $f(x) = x^2 - 2$ **28.** $f(x) = -x^2 + 2$ **29.** $f(x) = x^3 + 1$

30. $f(x) = -x^3 + 2$ **31.** $f(x) = -2x^3 - 1$ **32.** $f(x) = \dfrac{1}{2}x^3 + 3$

*For each pair of functions, find (**a**) $(f + g)(x)$ and (**b**) $(f - g)(x)$. See Example 3.*

33. $f(x) = 5x - 10$, $g(x) = 3x + 7$

34. $f(x) = -4x + 1$, $g(x) = 6x + 2$

35. $f(x) = 4x^2 + 8x - 3$, $g(x) = -5x^2 + 4x - 9$

36. $f(x) = 3x^2 - 9x + 10$, $g(x) = -4x^2 + 2x + 12$

Concept Check *Find two polynomial functions defined by $f(x)$ and $g(x)$ such that each statement is true.*

37. $(f + g)(x) = 3x^3 - x + 3$ **38.** $(f - g)(x) = -x^2 + x - 5$

Let $f(x) = x^2 - 9$, $g(x) = 2x$, and $h(x) = x - 3$. Find each of the following. See Example 4.

39. $(f + g)(x)$ **40.** $(f - g)(x)$ **41.** $(f + g)(3)$ **42.** $(f - g)(-3)$

43. $(f - h)(x)$ **44.** $(f + h)(x)$ **45.** $(f - h)(-3)$ **46.** $(f + h)(-2)$

47. $(g + h)(-10)$ **48.** $(g - h)(10)$ **49.** $(g - h)(-3)$ **50.** $(g + h)(1)$

51. $(g + h)\left(\dfrac{1}{4}\right)$ **52.** $(g + h)\left(\dfrac{1}{3}\right)$ **53.** $(g + h)\left(-\dfrac{1}{2}\right)$ **54.** $(g + h)\left(-\dfrac{1}{4}\right)$

Solve each problem. See Objective 3.

55. The cost in dollars to produce x t-shirts is $C(x) = 2.5x + 50$. The revenue in dollars from sales of x t-shirts is $R(x) = 10.99x$.

   (a) Write and simplify a function P that gives profit in terms of x.

   (b) Find the profit if 100 t-shirts are produced and sold.

56. The cost in dollars to produce x youth baseball caps is $C(x) = 4.3x + 75$. The revenue in dollars from sales of x caps is $R(x) = 25x$.

   (a) Write and simplify a function P that gives profit in terms of x.

   (b) Find the profit if 50 caps are produced and sold.

For each pair of functions, find $(fg)(x)$. See Example 5.

57. $f(x) = 2x, \quad g(x) = 5x - 1$ **58.** $f(x) = 3x, \quad g(x) = 6x - 8$

59. $f(x) = x + 1, \quad g(x) = 2x - 3$ **60.** $f(x) = x - 7, \quad g(x) = 4x + 5$

61. $f(x) = 2x - 3, \quad g(x) = 4x^2 + 6x + 9$

62. $f(x) = 3x + 4, \quad g(x) = 9x^2 - 12x + 16$

Let $f(x) = x^2 - 9$, $g(x) = 2x$, and $h(x) = x - 3$. Find each of the following. See Example 5.

63. $(fg)(x)$ **64.** $(fh)(x)$ **65.** $(fg)(2)$

66. $(fh)(1)$ **67.** $(fh)(-1)$ **68.** $(gh)(-2)$

69. $(gh)(-3)$ **70.** $(fg)(-2)$ **71.** $(fg)(0)$

72. $(fh)(0)$ **73.** $(fg)\left(-\dfrac{1}{2}\right)$ **74.** $(fg)\left(-\dfrac{1}{3}\right)$

For each pair of functions, find $\left(\dfrac{f}{g}\right)(x)$ and give any x-values that are not in the domain of the quotient function. See Example 6.

75. $f(x) = 10x^2 - 2x, \quad g(x) = 2x$ **76.** $f(x) = 18x^2 - 24x, \quad g(x) = 3x$

77. $f(x) = 2x^2 - x - 3, \quad g(x) = x + 1$ **78.** $f(x) = 4x^2 - 23x - 35, \quad g(x) = x - 7$

79. $f(x) = 8x^3 - 27, \quad g(x) = 2x - 3$ **80.** $f(x) = 27x^3 + 64, \quad g(x) = 3x + 4$

Let $f(x) = x^2 - 9$, $g(x) = 2x$, and $h(x) = x - 3$. Find each of the following. See Example 6.

81. $\left(\dfrac{f}{g}\right)(x)$ **82.** $\left(\dfrac{f}{h}\right)(x)$ **83.** $\left(\dfrac{f}{g}\right)(2)$

84. $\left(\dfrac{f}{h}\right)(1)$ **85.** $\left(\dfrac{h}{g}\right)(x)$ **86.** $\left(\dfrac{g}{h}\right)(x)$

87. $\left(\dfrac{h}{g}\right)(3)$ **88.** $\left(\dfrac{g}{h}\right)(-1)$ **89.** $\left(\dfrac{f}{g}\right)\left(\dfrac{1}{2}\right)$

90. $\left(\dfrac{f}{g}\right)\left(\dfrac{3}{2}\right)$ **91.** $\left(\dfrac{h}{g}\right)\left(-\dfrac{1}{2}\right)$ **92.** $\left(\dfrac{h}{g}\right)\left(-\dfrac{3}{2}\right)$

Concept Check *Let $f(x) = x^2$ and $g(x) = 2x - 1$. Match each expression in Column I with the description of how to evaluate it in Column II.*

I	**II**

93. $(f \circ g)(5)$ **A.** Square 5. Take the result and square it.

94. $(g \circ f)(5)$ **B.** Double 5 and subtract 1. Take the result and square it.

95. $(f \circ f)(5)$ **C.** Double 5 and subtract 1. Take the result, double it, and subtract 1.

96. $(g \circ g)(5)$ **D.** Square 5. Take the result, double it, and subtract 1.

97. Concept Check For $f(x) = x - 8$ and $g(x) = 2$, a student found $(f \circ g)(x)$ incorrectly as follows.

$$(f \circ g)(x) = 2(x - 8)$$
$$= 2x - 16$$

WHAT WENT WRONG? Give the correct composition.

98. Concept Check For $f(x) = 2x + 3$ and $g(x) = x + 5$, a student found $(f \circ g)(x)$ correctly as

$$(f \circ g)(x) = 2x + 13.$$

When asked to find $(g \circ f)(x)$, he gave the same result. **WHAT WENT WRONG?** Give the correct composition.

Let $f(x) = x^2 + 4$, $g(x) = 2x + 3$, and $h(x) = x - 5$. Find each of the following. See Examples 7–9.

99. $(h \circ g)(4)$ **100.** $(f \circ g)(4)$ **101.** $(g \circ f)(6)$ **102.** $(h \circ f)(6)$

103. $(f \circ h)(-2)$ **104.** $(h \circ g)(-2)$ **105.** $(f \circ g)(0)$ **106.** $(f \circ h)(0)$

107. $(g \circ f)(x)$ **108.** $(g \circ h)(x)$ **109.** $(h \circ g)(x)$ **110.** $(h \circ f)(x)$

111. $(f \circ h)\left(\dfrac{1}{2}\right)$ **112.** $(h \circ f)\left(\dfrac{1}{2}\right)$ **113.** $(f \circ g)\left(-\dfrac{1}{2}\right)$ **114.** $(g \circ f)\left(-\dfrac{1}{2}\right)$

Extending Skills *The tables give some selected ordered pairs for functions f and g.*

x	3	4	6	8
$f(x)$	1	3	9	2

x	2	7	1	9
$g(x)$	3	6	9	12

Tables like these can be used to evaluate composite functions. For example, to evaluate $(g \circ f)(6)$, use the first table to find $f(6) = 9$. Then use the second table to find

$$(g \circ f)(6) = g(f(6)) = g(9) = 12.$$

Find each of the following.

115. $(f \circ g)(2)$ **116.** $(f \circ g)(7)$ **117.** $(g \circ f)(3)$

118. $(g \circ f)(8)$ **119.** $(f \circ f)(4)$ **120.** $(g \circ g)(1)$

Solve each problem. ***See Objective 4.***

121. The function $f(x) = 12x$ computes the number of inches in x feet, and the function $g(x) = 5280x$ computes the number of feet in x miles. Find and simplify $(f \circ g)(x)$. What does it compute?

122. The function $f(x) = 60x$ computes the number of minutes in x hours, and the function $g(x) = 24x$ computes the number of hours in x days. Find and simplify $(f \circ g)(x)$. What does it compute?

123. The perimeter x of a square with sides of length s is given by the formula $x = 4s$.

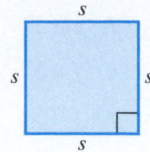

(a) Solve for s in terms of x.

(b) If y represents the area of this square, write y as a function of the perimeter x.

(c) Use the composite function of part (b) to find the area of a square with perimeter 6.

124. The perimeter x of an equilateral triangle with sides of length s is given by the formula $x = 3s$.

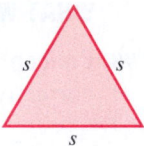

(a) Solve for s in terms of x.

(b) The area y of an equilateral triangle with sides of length s is given by the formula $y = \frac{s^2\sqrt{3}}{4}$. Write y as a function of the perimeter x.

(c) Use the composite function of part (b) to find the area of an equilateral triangle with perimeter 12.

125. In a sale room at a clothing store, every item is on sale for half its original price, plus $1.

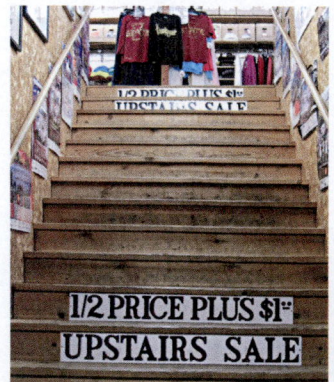

(a) Write a function g that finds half of x.

(b) Write a function f that adds 1 to x.

(c) Write and simplify the function $(f \circ g)(x)$.

(d) Use the function from part (c) to find the sale price of a shirt that has an original price of $60.

126. A pair of shoes is marked 50% off. A customer has a coupon for an additional $10 off.

(a) Write a function g that finds 50% of x.

(b) Write a function f that subtracts 10 from x.

(c) Write and simplify the function $(f \circ g)(x)$.

(d) Use the function from part (c) to find the sale price of a pair of shoes that has an original price of $100.

127. When a thermal inversion layer is over a city (as happens often in Los Angeles), pollutants cannot rise vertically, but are trapped below the layer and must disperse horizontally.

Assume that a factory smokestack begins emitting a pollutant at 8 A.M. and that the pollutant disperses horizontally over a circular area. Suppose that t represents the time, in hours, since the factory began emitting pollutants ($t = 0$ represents 8 A.M.), and assume that the radius of the circle of pollution is $r(t) = 2t$ miles. Let $\mathcal{A}(r) = \pi r^2$ represent the area of a circle of radius r. Find and interpret $(\mathcal{A} \circ r)(t)$.

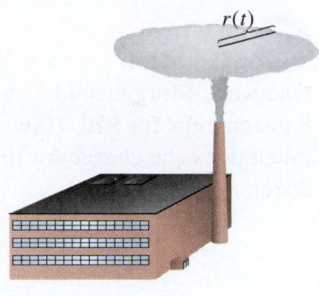

128. An oil well is leaking, with the leak spreading oil over the surface as a circle. At any time t, in minutes, after the beginning of the leak, the radius of the circular oil slick on the surface is $r(t) = 4t$ feet. Let $\mathcal{A}(r) = \pi r^2$ represent the area of a circle of radius r. Find and interpret $(\mathcal{A} \circ r)(t)$.

9.4 Variation

OBJECTIVES

1 Write an equation expressing direct variation.

2 Find the constant of variation, and solve direct variation problems.

3 Solve inverse variation problems.

4 Solve joint variation problems.

5 Solve combined variation problems.

VOCABULARY

☐ direct variation
☐ constant of variation
☐ inverse variation
☐ joint variation
☐ combined variation

OBJECTIVE 1 Write an equation expressing direct variation.

The circumference of a circle is given by the formula $C = 2\pi r$, where r is the radius of the circle. See **FIGURE 24**. The circumference is always a constant multiple of the radius—that is, C is always found by multiplying r by the constant 2π.

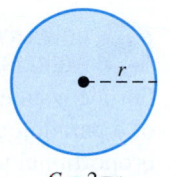

$C = 2\pi r$

FIGURE 24

As the ***radius increases,*** the ***circumference increases.***

As the ***radius decreases,*** the ***circumference decreases.***

Because of these relationships, the circumference is said to *vary directly* as the radius.

> **Direct Variation**
>
> ***y* varies directly as *x*** if there exists a real number *k* such that
>
> $$y = kx.$$
>
> Stated another way, ***y* is *directly proportional to x*.** The number *k* is the **constant of variation.**

In direct variation, for $k > 0$, as the value of x increases, the value of y increases. Similarly, as x decreases, y decreases.

OBJECTIVE 2 Find the constant of variation, and solve direct variation problems.

The direct variation equation $y = kx$ defines a linear function, where the constant of variation k is the slope of the line. For example, the following equation describes the cost y to buy x gallons of gasoline.

$$y = 3.50x$$

The cost varies directly as the number of gallons purchased.

As the *number of gallons increases,* the *cost increases.*

As the *number of gallons decreases,* the *cost decreases.*

The constant of variation k is 3.50, the cost of 1 gal of gasoline.

**NOW TRY
EXERCISE 1**

One week Morgan sold 8 dozen eggs for $20. How much does she charge for one dozen eggs?

EXAMPLE 1 Solving a Direct Variation Problem

Eva is paid an hourly wage. One week she worked 43 hr and was paid $795.50. How much does she earn per hour?

$$\text{Let } h = \text{the number of hours she works}$$

$$\text{and } P = \text{her corresponding pay.}$$

Write a variation equation.

k represents Eva's hourly wage. →	$P = kh$	P varies directly as h.
	$795.50 = k \cdot 43$	Let $P = 795.50$ and $h = 43$.
This is the constant of variation. →	$k = 18.50$	Use a calculator.

Her hourly wage is $18.50, and P and h are related by the equation

$$P = 18.50h.$$

We can use this equation to find her pay for any number of hours worked.

NOW TRY

**NOW TRY
EXERCISE 2**

For a constant height, the area of a parallelogram is directly proportional to its base. If the area is 20 cm² when the base is 4 cm, find the area when the base is 7 cm.

EXAMPLE 2 Solving a Direct Variation Problem

Hooke's law for an elastic spring states that the distance a spring stretches is directly proportional to the force applied. If a force of 150 newtons* stretches a certain spring 8 cm, how much will a force of 400 newtons stretch the spring? See **FIGURE 25**.

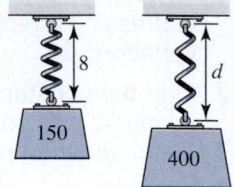

FIGURE 25

$$\text{Let } d = \text{the distance the spring stretches}$$

$$\text{and } f = \text{the force applied.}$$

Then $d = kf$ for some constant k. A force of 150 newtons stretches the spring 8 cm, so we use these values to find k.

	$d = kf$	Variation equation
Solve for *k*. →	$8 = k \cdot 150$	Let $d = 8$ and $f = 150$.
	$k = \dfrac{8}{150}$	Solve for *k*.
	$k = \dfrac{4}{75}$	Write in lowest terms.

Now we rewrite the variation equation $d = kf$ using $\frac{4}{75}$ for k.

$$d = \frac{4}{75}f \qquad \text{Let } k = \tfrac{4}{75}.$$

For a force of 400 newtons, substitute 400 for f.

$$d = \frac{4}{75}(400) = \frac{64}{3} \qquad \text{Let } f = 400.$$

The spring will stretch $\frac{64}{3}$ cm, or $21\frac{1}{3}$ cm, if a force of 400 newtons is applied.

NOW TRY

NOW TRY ANSWERS
1. $2.50
2. 35 cm²

*A newton is a unit of measure of force used in physics.

Solving a Variation Problem

Step 1 Write a variation equation.

Step 2 Substitute the initial values and solve for k.

Step 3 Rewrite the variation equation with the value of k from Step 2.

Step 4 Substitute the remaining values, solve for the unknown, and find the required answer.

One variable can be directly proportional to a power of another variable.

Direct Variation as a Power

y **varies directly as the *n*th power of** *x* if there exists a real number k such that

$$y = kx^n.$$

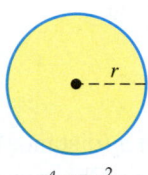

$\mathcal{A} = \pi r^2$

FIGURE 26

An example of direct variation as a power is the formula for the area of a circle, $\mathcal{A} = \pi r^2$. See **FIGURE 26**. Here, π is the constant of variation, and the area $\mathcal{A}$ varies directly as the *square* of the radius r.

**NOW TRY
EXERCISE 3**
Suppose y varies directly as the square of x, and $y = 200$ when $x = 5$. Find y when $x = 7$.

EXAMPLE 3 Solving a Direct Variation Problem

The distance a body falls from rest varies directly as the square of the time it falls (disregarding air resistance). If a skydiver falls 64 ft in 2 sec, how far will she fall in 8 sec?

Step 1 Let d = the distance the skydiver falls

and t = the time it takes to fall.

Then d is a function of t for some constant k.

$d = kt^2$ *d* varies directly as the square of *t*.

Step 2 To find the value of k, use the fact that the skydiver falls 64 ft in 2 sec.

$d = kt^2$ Variation equation

$64 = k(2)^2$ Let *d* = 64 and *t* = 2.

$k = 16$ Find *k*.

Step 3 Now we rewrite the variation equation $d = kt^2$ using 16 for k.

$d = 16t^2$ Let *k* = 16.

Step 4 Let $t = 8$ to find the number of feet the skydiver will fall in 8 sec.

$d = 16(8)^2 = 1024$ Let *t* = 8.

The skydiver will fall 1024 ft in 8 sec.   **NOW TRY**

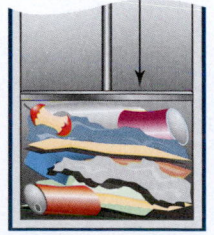
As pressure on trash increases, volume of trash decreases.

FIGURE 27

OBJECTIVE 3 Solve inverse variation problems.

Another type of variation is *inverse variation*. **With inverse variation, for $k > 0$, *as one variable increases, the other variable decreases.***

For example, in a closed space, volume decreases as pressure increases, which can be illustrated by a trash compactor. See **FIGURE 27**. As the compactor presses down, the pressure on the trash increases, and in turn, the trash occupies a smaller space.

NOW TRY ANSWER
3. 392

Inverse Variation

y varies inversely as x if there exists a real number k such that

$$y = \frac{k}{x}.$$

y varies inversely as the nth power of x if there exists a real number k such that

$$y = \frac{k}{x^n}.$$

The inverse variation equation $y = \frac{k}{x}$ defines a rational function.

Another example of inverse variation comes from the distance formula.

$$d = rt \qquad \text{Distance formula}$$

$$t = \frac{d}{r} \qquad \text{Divide each side by } r.$$

In the form $t = \frac{d}{r}$, t (time) varies inversely as r (rate or speed), with d (distance) serving as the constant of variation. For example, if the distance between Chicago and Des Moines is 300 mi, then

$$t = \frac{300}{r}.$$

The values of r and t might be any of the following.

$$\left.\begin{array}{l} r = 50, t = 6 \\ r = 60, t = 5 \\ r = 75, t = 4 \end{array}\right\} \begin{array}{l} \text{As } r \text{ increases,} \\ t \text{ decreases.} \end{array} \qquad \left.\begin{array}{l} r = 30, t = 10 \\ r = 25, t = 12 \\ r = 20, t = 15 \end{array}\right\} \begin{array}{l} \text{As } r \text{ decreases,} \\ t \text{ increases.} \end{array}$$

If we *increase* the rate (speed) at which we drive, time *decreases*. If we *decrease* the rate (speed) at which we drive, time *increases*.

NOW TRY
EXERCISE 4

For a constant area, the height of a triangle varies inversely as the base. If the height is 7 cm when the base is 8 cm, find the height when the base is 14 cm.

EXAMPLE 4   Solving an Inverse Variation Problem

In the manufacture of a phone-charging device, the cost of producing the device varies inversely as the number produced. If 10,000 units are produced, the cost is \$2 per unit. Find the cost per unit to produce 25,000 units.

$$\text{Let } x = \text{the number of units produced}$$

$$\text{and } c = \text{the cost per unit.}$$

Here, as production increases, cost decreases, and as production decreases, cost increases. We write a variation equation using the variables c and x and the constant k.

$$c = \frac{k}{x} \qquad c \text{ varies inversely as } x.$$

$$2 = \frac{k}{10,000} \qquad \text{Let } c = 2 \text{ and } x = 10,000.$$

$$20,000 = k \qquad \text{Multiply by } 10,000.$$

Thus, the variation equation $c = \frac{k}{x}$ becomes $c = \frac{20,000}{x}$.

$$c = \frac{20,000}{25,000} = 0.80 \qquad \text{Let } x = 25,000.$$

The cost per unit to make 25,000 units is \$0.80.

NOW TRY ANSWER
4. 4 cm

NOW TRY

**NOW TRY
EXERCISE 5**

The weight of an object above Earth varies inversely as the square of its distance from the center of Earth. If an object weighs 150 lb on the surface of Earth, and the radius of Earth is about 3960 mi, how much does it weigh when it is 1000 mi above Earth's surface? Round to the nearest pound.

EXAMPLE 5 Solving an Inverse Variation Problem

The weight of an object above Earth varies inversely as the square of its distance from the center of Earth. A space shuttle in an elliptical orbit has a maximum distance from the center of Earth (**apogee**) of 6700 mi. Its minimum distance from the center of Earth (**perigee**) is 4090 mi. See **FIGURE 28**. If an astronaut in the shuttle weighs 57 lb at its apogee, what does the astronaut weigh at its perigee?

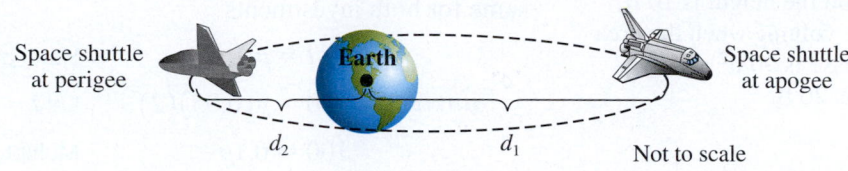

FIGURE 28

Let w = the astronaut's weight and d = the distance from the center of Earth, for some constant k. We write a variation equation using these variables.

$$w = \frac{k}{d^2} \qquad \text{\textit{w} varies inversely as the square of \textit{d}.}$$

At the apogee, the astronaut weighs 57 lb, and the distance from the center of Earth is 6700 mi. Use these values to find k.

$$57 = \frac{k}{(6700)^2} \qquad \text{Let } w = 57 \text{ and } d = 6700.$$

$$k = 57(6700)^2 \qquad \text{Solve for } k.$$

Substitute $k = 57(6700)^2$ and $d = 4090$ in the variation equation $w = \frac{k}{d^2}$ to find the astronaut's weight at the perigee.

$$w = \frac{57(6700)^2}{(4090)^2} = 153 \text{ lb} \qquad \begin{array}{l}\text{Use a calculator.}\\\text{Round to the nearest pound.}\end{array}$$

NOW TRY

OBJECTIVE 4 Solve joint variation problems.

If one variable varies directly as the *product* of several other variables (perhaps raised to powers), the first variable is said to *vary jointly* as the others.

Joint Variation

y **varies jointly as** *x* **and** *z* if there exists a real number k such that
$$y = kxz.$$

An example of joint variation is the formula for the volume of a right pyramid, $V = \frac{1}{3}Bh$. See **FIGURE 29**. Here, $\frac{1}{3}$ is the constant of variation, and the volume V varies jointly as the area of the base B and the height h.

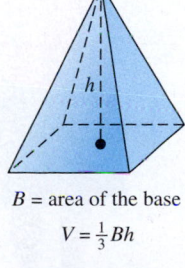

B = area of the base
$V = \frac{1}{3}Bh$

FIGURE 29

! CAUTION Note that *and* in the expression "*y* varies jointly as *x and z*" translates as a product in the variation equation $y = kxz$. The word *and* does **not** indicate addition here.

NOW TRY ANSWER
5. 96 lb

 **NOW TRY
EXERCISE 6**

The volume of a right pyramid varies jointly as the height and the area of the base. If the volume is 100 ft³ when the area of the base is 30 ft² and the height is 10 ft, find the volume when the area of the base is 90 ft² and the height is 20 ft.

EXAMPLE 6 Solving a Joint Variation Problem

The interest on a loan or an investment is given by the formula $I = prt$. Here, for a given principal p, the interest earned I varies jointly as the interest rate r and the time t the principal is left earning interest. If an investment earns $100 interest at 5% for 2 yr, how much interest will the same principal earn at 4.5% for 3 yr?

We use the formula $I = prt$, where p is the constant of variation because it is the same for both investments.

$$I = prt$$ Here, p is the constant of variation.

Solve for p. → $$100 = p(0.05)(2)$$ Let $I = 100$, $r = 5\% = 0.05$, and $t = 2$.

$$100 = 0.1p$$ Multiply.

$$p = 1000$$ Divide by 0.1. Rewrite.

Now we find I when $p = 1000$, $r = 4.5\% = 0.045$, and $t = 3$.

$$I = 1000(0.045)(3) = 135$$ Let $p = 1000$, $r = 4.5\% = 0.045$, and $t = 3$.

The interest will be $135.   **NOW TRY**

OBJECTIVE 5 Solve combined variation problems.

There are combinations of direct and inverse variation, called **combined variation.**

 **NOW TRY
EXERCISE 7**

In statistics, the sample size used to estimate a population mean varies directly as the variance and inversely as the square of the maximum error of the estimate. If the sample size is 200 when the variance is 25 m² and the maximum error of the estimate is 0.5 m, find the sample size when the variance is 25 m² and the maximum error of the estimate is 0.1 m.

EXAMPLE 7 Solving a Combined Variation Problem

Body mass index (BMI) is used to assess whether a person's weight is healthy. A BMI from 19 through 25 is considered desirable. BMI varies directly as an individual's weight in pounds and inversely as the square of his or her height in inches.

A person who weighs 118 lb and is 64 in. tall has a BMI of 20. (BMI is rounded to the nearest whole number.) Find the BMI of a man who weighs 165 lb and is 70 in. tall. (Data from *Washington Post*.)

Let B = BMI, w = weight, and h = height. Write a variation equation.

$$B = \frac{kw}{h^2}$$ ← BMI varies directly as the weight.
← BMI varies inversely as the square of the height.

To find k, let $B = 20$, $w = 118$, and $h = 64$.

$$20 = \frac{k(118)}{64^2}$$ $B = \frac{kw}{h^2}$

$$k = \frac{20(64^2)}{118}$$ Multiply by 64^2. Divide by 118.

$$k = 694$$ Use a calculator. Round to the nearest whole number.

Now find B when $k = 694$, $w = 165$, and $h = 70$.

$$B = \frac{694(165)}{70^2} = 23$$ Round to the nearest whole number.

NOW TRY ANSWERS
6. 600 ft³
7. 5000

The man's BMI is 23.   **NOW TRY**

9.4 Exercises

  MyLab Math

▶ *Video solutions for select problems available in MyLab Math*

Concept Check *Fill in each blank with the correct response.*

1. For $k > 0$, if y varies directly as x, then when x increases, y _____, and when x decreases, y _____.

2. For $k > 0$, if y varies inversely as x, then when x increases, y _____, and when x decreases, y _____.

Concept Check *Use personal experience or intuition to determine whether the situation suggests* direct *or* inverse *variation.*

3. The number of movie tickets purchased and the total price for the tickets

4. The rate and the distance traveled by a pickup truck in 3 hr

5. The amount of pressure put on the accelerator of a car and the speed of the car

6. The percentage off an item that is on sale and the price of the item

7. Your age and the probability that you believe in the tooth fairy

8. The surface area of a balloon and its diameter

9. The demand for an item and the price of the item

10. The number of hours worked by an hourly worker and the amount of money earned

Concept Check *Determine whether each equation represents* direct, inverse, joint, *or* combined *variation.*

11. $y = \dfrac{3}{x}$

12. $y = \dfrac{8}{x}$

13. $y = 10x^2$

14. $y = 2x^3$

15. $y = 3xz^4$

16. $y = 6x^3z^2$

17. $y = \dfrac{4x}{wz}$

18. $y = \dfrac{6x}{st}$

Concept Check *Write each formula using the "language" of variation. For example, the formula for the circumference of a circle, $C = 2\pi r$, can be written as*

"The circumference of a circle varies directly as the length of its radius."

19. $P = 4s$, where P is the perimeter of a square with side of length s

20. $d = 2r$, where d is the diameter of a circle with radius r

21. $S = 4\pi r^2$, where S is the surface area of a sphere with radius r

22. $V = \frac{4}{3}\pi r^3$, where V is the volume of a sphere with radius r

23. $\mathcal{A} = \frac{1}{2}bh$, where $\mathcal{A}$ is the area of a triangle with base b and height h

24. $V = \frac{1}{3}\pi r^2 h$, where V is the volume of a cone with radius r and height h

25. **Concept Check** What is the constant of variation in each of the variation equations in **Exercises 19–24?**

26. **Concept Check** What is meant by the constant of variation in a direct variation problem? If we were to graph the linear equation $y = kx$ for some nonnegative constant k, what role would k play in the graph?

Write a variation equation for each situation. Use k as the constant of variation. **See Examples 1–6.**

27. A varies directly as b.

28. W varies directly as f.

29. h varies inversely as t.

30. p varies inversely as s.

31. M varies directly as the square of d.

32. P varies inversely as the square of x.

33. I varies jointly as g and h.

34. C varies jointly as a and the square of b.

Solve each problem. See Examples 1–7.

35. If x varies directly as y, and $x = 9$ when $y = 3$, find x when $y = 12$.

36. If x varies directly as y, and $x = 10$ when $y = 7$, find y when $x = 50$.

37. If a varies directly as the square of b, and $a = 4$ when $b = 3$, find a when $b = 2$.

38. If h varies directly as the square of m, and $h = 15$ when $m = 5$, find h when $m = 7$.

39. If z varies inversely as w, and $z = 10$ when $w = 0.5$, find z when $w = 8$.

40. If t varies inversely as s, and $t = 3$ when $s = 5$, find s when $t = 5$.

41. If m varies inversely as the square of p, and $m = 20$ when $p = 2$, find m when $p = 5$.

42. If a varies inversely as the square of b, and $a = 48$ when $b = 4$, find a when $b = 7$.

43. p varies jointly as q and the square of r, and $p = 200$ when $q = 2$ and $r = 3$. Find p when $q = 5$ and $r = 2$.

44. f varies jointly as h and the square of g, and $f = 50$ when $h = 2$ and $g = 4$. Find f when $h = 6$ and $g = 3$.

Solve each problem. See Examples 1–7.

45. Ben bought 8.5 gal of gasoline and paid $33.32. What is the price of gasoline per gallon?

46. Sara gives horseback rides at Shadow Mountain Ranch. A 2.5-hr ride costs $50.00. What is the price per hour?

47. The weight of an object on Earth is directly proportional to the weight of that same object on the moon. A 200-lb astronaut would weigh 32 lb on the moon. How much would a 50-lb dog weigh on the moon?

48. The pressure exerted by a certain liquid at a given point is directly proportional to the depth of the point beneath the surface of the liquid. The pressure at 30 m is 80 newtons. What pressure is exerted at 50 m?

49. The volume of a can of tomatoes is directly proportional to the height of the can. If the volume of the can is 300 cm³ when its height is 10.62 cm, find the volume to the nearest whole number of a can with height 15.92 cm.

50. The force required to compress a spring is directly proportional to the change in length of the spring. If a force of 20 newtons is required to compress a certain spring 2 cm, how much force is required to compress the spring from 20 cm to 8 cm?

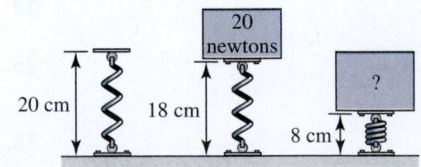

51. For a body falling freely from rest (disregarding air resistance), the distance the body falls varies directly as the square of the time. If an object is dropped from the top of a tower 576 ft high and hits the ground in 6 sec, how far did it fall in the first 4 sec?

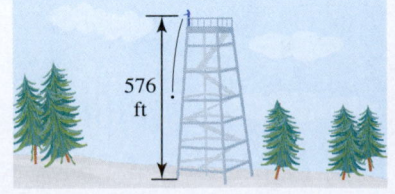

52. The amount of water emptied by a pipe varies directly as the square of the diameter of the pipe. For a certain constant water flow, a pipe emptying into a canal will allow 200 gal of water to escape in an hour. The diameter of the pipe is 6 in. How much water would a 12-in. pipe empty into the canal in an hour, assuming the same water flow?

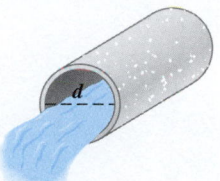

53. Over a specified distance, rate varies inversely as time. If a Dodge Viper on a test track goes a certain distance in one-half minute at 160 mph, what rate is needed to go the same distance in three-fourths minute?

54. For a constant area, the length of a rectangle varies inversely as the width. The length of a rectangle is 27 ft when the width is 10 ft. Find the width of a rectangle with the same area if the length is 18 ft.

55. The frequency of a vibrating string varies inversely as its length. That is, a longer string vibrates fewer times in a second than a shorter string. Suppose a piano string 2 ft long vibrates at 250 cycles per sec. What frequency would a string 5 ft long have?

56. The current in a simple electrical circuit varies inversely as the resistance. If the current is 20 amps when the resistance is 5 ohms, find the current when the resistance is 7.5 ohms.

57. The amount of light (measured in foot-candles) produced by a light source varies inversely as the square of the distance from the source. If the illumination produced 1 m from a light source is 768 foot-candles, find the illumination produced 6 m from the same source.

58. The force with which Earth attracts an object above Earth's surface varies inversely as the square of the distance of the object from the center of Earth. If an object 4000 mi from the center of Earth is attracted with a force of 160 lb, find the force of attraction if the object were 6000 mi from the center of Earth.

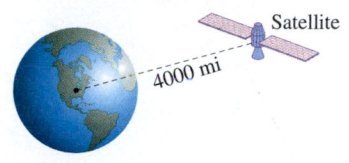

59. For a given interest rate, simple interest varies jointly as principal and time. If $2000 left in an account for 4 yr earned interest of $280, how much interest would be earned in 6 yr?

60. The collision impact of an automobile varies jointly as its mass and the square of its speed. Suppose a 2000-lb car traveling at 55 mph has a collision impact of 6.1. What is the collision impact (to the nearest tenth) of the same car at 65 mph?

61. The weight of a bass varies jointly as its girth and the square of its length. (**Girth** is the distance around the body of the fish.) A prize-winning bass weighed in at 22.7 lb and measured 36 in. long with a 21-in. girth. How much (to the nearest tenth of a pound) would a bass 28 in. long with an 18-in. girth weigh?

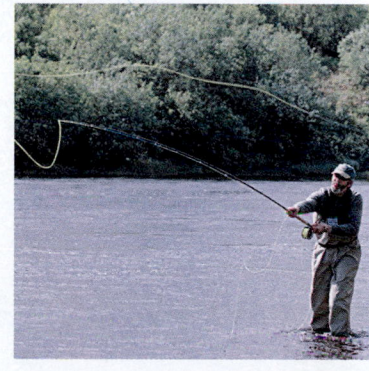

62. The weight of a trout varies jointly as its length and the square of its girth. One angler caught a trout that weighed 10.5 lb and measured 26 in. long with an 18-in. girth. Find the weight (to the nearest tenth of a pound) of a trout that is 22 in. long with a 15-in. girth.

63. The force needed to keep a car from skidding on a curve varies inversely as the radius of the curve and jointly as the weight of the car and the square of the speed. If 242 lb of force keeps a 2000-lb car from skidding on a curve of radius 500 ft at 30 mph, what force (to the nearest tenth of a pound) would keep the same car from skidding on a curve of radius 750 ft at 50 mph?

64. The maximum load that a cylindrical column with a circular cross section can hold varies directly as the fourth power of the diameter of the cross section and inversely as the square of the height. A 9-m column 1 m in diameter will support 8 metric tons. How many metric tons can be supported by a column 12 m high and $\frac{2}{3}$ m in diameter?

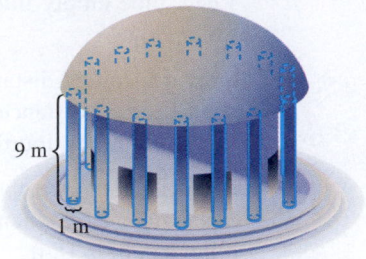

9 m

1 m

Load = 8 metric tons

65. The number of long-distance phone calls between two cities during a certain period varies jointly as the populations of the cities, p_1 and p_2, and inversely as the distance between them, in miles. If 80,000 calls are made between two cities 400 mi apart, with populations of 70,000 and 100,000, how many calls (to the nearest hundred) are made between cities with populations of 50,000 and 75,000 that are 250 mi apart?

66. The volume of gas varies inversely as the pressure and directly as the temperature. (Temperature must be measured in *kelvins* (K), a unit of measurement used in physics.) If a certain gas occupies a volume of 1.3 L at 300 K and a pressure of 18 newtons, find the volume at 340 K and a pressure of 24 newtons.

67. A body mass index from 27 through 29 carries a slight risk of weight-related health problems, while a BMI of 30 or more indicates a great increase in risk. Use your own height and weight and the information in **Example 7** to determine your BMI and whether you are at risk.

68. The maximum load of a horizontal beam that is supported at both ends varies directly as the width and the square of the height and inversely as the length between the supports. A beam 6 m long, 0.1 m wide, and 0.06 m high supports a load of 360 kg. What is the maximum load supported by a beam 16 m long, 0.2 m wide, and 0.08 m high?

RELATING CONCEPTS For Individual or Group Work (Exercises 69–74)

A routine activity such as pumping gasoline can be related to many of the concepts studied in this text. Suppose that premium unleaded costs $3.75 per gal. **Work Exercises 69–74 in order.**

69. 0 gal of gasoline cost $0.00, while 1 gal costs $3.75. Represent these two pieces of information as ordered pairs of the form (gallons, price).

70. Use the information from **Exercise 69** to find the slope of the line on which the two points lie.

71. Write the slope-intercept form of the equation of the line on which the two points lie.

72. Using function notation, if $f(x) = ax + b$ represents the line from **Exercise 71,** what are the values of a and b?

73. How is the value of a from **Exercise 72** related to gasoline in this situation? With relationship to the line, what do we call this number?

74. Why does the equation from **Exercise 72** satisfy the conditions for direct variation? In the context of variation, what do we call the value of a?

Chapter 9 Summary

Key Terms

9.1

relation
function
dependent variable
independent variable
domain
range

9.2

linear function
constant function

9.3

polynomial function
identity function
squaring function
cubing function
composite function
(composition of functions)

9.4

direct variation
constant of variation
inverse variation
joint variation
combined variation

New Symbols

$f(x)$ function notation; function of x (read "f of x" or "f at x")

$(f \circ g)(x) = f(g(x))$ composite function

Test Your Word Power

See how well you have learned the vocabulary in this chapter.

1. A **relation** is
  A. a set of ordered pairs
  B. the ratio of the change in y to the change in x along a line
  C. the set of all possible values of the independent variable
  D. all the second components of a set of ordered pairs.

2. A **function** is
  A. a pair of numbers in an ordered pair
  B. a set of ordered pairs in which each x-value corresponds to exactly one y-value
  C. a pair of numbers written between parentheses
  D. the set of all ordered pairs that satisfy an equation.

3. The **domain** of a relation is
  A. the set of all possible values of the dependent variable y
  B. a set of ordered pairs
  C. the difference between the x-values
  D. the set of all possible values of the independent variable x.

4. The **range** of a relation is
  A. the set of all possible values of the dependent variable y
  B. a set of ordered pairs
  C. the difference between the y-values
  D. the set of all possible values of the independent variable x.

5. If two positive quantities x and y are in **direct variation,** and the constant of variation is positive, then
  A. as x increases, y decreases
  B. as x increases, y increases
  C. as x increases, y remains constant
  D. as x decreases, y remains constant.

6. If two positive quantities x and y are in **inverse variation,** and the constant of variation is positive, then
  A. as x increases, y decreases
  B. as x increases, y increases
  C. as x increases, y remains constant
  D. as x decreases, y remains constant.

ANSWERS

1. A; *Example:* The set $\{(2, 0), (4, 3), (6, 6)\}$ defines a relation. **2.** B; *Example:* The relation given in Answer 1 is a function.
3. D; *Example:* In the relation in Answer 1, the domain is the set of x-values, $\{2, 4, 6\}$. **4.** A; *Example:* In the relation in
Answer 1, the range is the set of y-values, $\{0, 3, 6\}$. **5.** B; *Example:* The equation $y = 3x$ represents direct variation. When $x = 2$, $y = 6$.
If x increases to 3, then y increases to $3(3) = 9$. **6.** A; *Example:* The equation $y = \frac{3}{x}$ represents inverse variation. When $x = 1$, $y = 3$.
If x increases to 2, then y decreases to $\frac{3}{2}$, or $1\frac{1}{2}$.

Quick Review

CONCEPTS	**EXAMPLES**

9.1 **Introduction to Relations and Functions**

A **relation** is any set of ordered pairs. A **function** is a set of ordered pairs in which, for each distinct value of the first component, there is *exactly one* value of the second component.

The set of first components (x-values) is the **domain.**

The set of second components (y-values) is the **range.**

The set of ordered pairs $\{(-1, 4), (0, 6), (1, 4)\}$ defines a function.

Domain: $\{-1, 0, 1\}$ Set of x-values

Range: $\{4, 6\}$ Set of y-values

The equation $y = x^2$ defines a function.

Domain: $(-\infty, \infty)$ Range: $[0, \infty)$

9.2 **Function Notation and Linear Functions**

To evaluate a function f, where $f(x)$ defines the range value for a given value of x in the domain, substitute the domain value wherever x appears.

To write an equation that defines a function f in function notation, follow these steps.

Step 1 Solve the equation for y if it is not given in that form.

Step 2 Replace y with $f(x)$.

Let $f(x) = x^2 - 7x + 12$. Find $f(1)$.

$$f(x) = x^2 - 7x + 12$$
$$f(1) = 1^2 - 7(1) + 12 \qquad \text{Let } x = 1.$$
$$f(1) = 6$$

Write $2x + 3y = 12$ using function notation $f(x)$.

$$2x + 3y = 12$$
$$3y = -2x + 12 \qquad \text{Subtract } 2x.$$
$$y = -\frac{2}{3}x + 4 \qquad \text{Divide by 3.}$$
$$f(x) = -\frac{2}{3}x + 4 \qquad \text{Replace } y \text{ with } f(x).$$

9.3 **Polynomial Functions, Graphs, Operations, and Composition**

Graphs of Basic Polynomial Functions

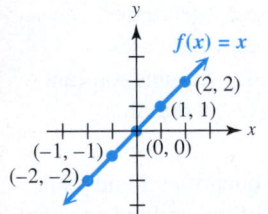

Identity function
$f(x) = x$
Domain: $(-\infty, \infty)$
Range: $(-\infty, \infty)$

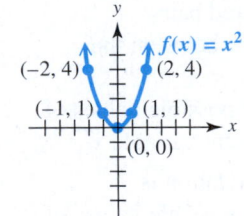

Squaring function
$f(x) = x^2$
Domain: $(-\infty, \infty)$
Range: $[0, \infty)$

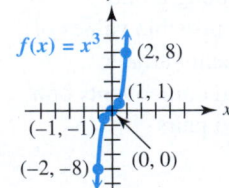

Cubing function
$f(x) = x^3$
Domain: $(-\infty, \infty)$
Range: $(-\infty, \infty)$

Operations on Functions

If $f(x)$ and $g(x)$ define functions, then

$$(f + g)(x) = f(x) + g(x),$$
$$(f - g)(x) = f(x) - g(x),$$
$$(fg)(x) = f(x) \cdot g(x),$$

and

$$\left(\frac{f}{g}\right)(x) = \frac{f(x)}{g(x)}, \quad g(x) \neq 0.$$

Let $f(x) = x^2$ and $g(x) = 2x + 1$. Find the following.

$(f + g)(x)$
$$= f(x) + g(x)$$
$$= x^2 + 2x + 1$$

$(f - g)(x)$
$$= f(x) - g(x)$$
$$= x^2 - (2x + 1)$$
$$= x^2 - 2x - 1$$

$(fg)(x)$
$$= f(x) \cdot g(x)$$
$$= x^2(2x + 1)$$
$$= 2x^3 + x^2$$

$\left(\dfrac{f}{g}\right)(x)$
$$= \frac{f(x)}{g(x)}$$
$$= \frac{x^2}{2x + 1}, \quad x \neq -\frac{1}{2}$$

CONCEPTS	EXAMPLES
Composition of f and g $$(f \circ g)(x) = f(g(x))$$ $$(g \circ f)(x) = g(f(x))$$	Let $f(x) = x^2$ and $g(x) = 2x + 1$. Find $(f \circ g)(x)$ and $(g \circ f)(x)$.

$(f \circ g)(x)$	$(g \circ f)(x)$
$= f(g(x))$	$= g(f(x))$
$= f(2x + 1)$	$= g(x^2)$
$= (2x + 1)^2$	$= 2x^2 + 1$

9.4 Variation

Let k be a real number.

If $\quad y = kx,\quad$ then y varies directly as x.

The diameter of a circle varies directly as the radius.

$$d = kr \quad \text{Here, } k = 2.$$

If $\quad y = kx^n,\quad$ then y varies directly as x^n.

The area of a circle varies directly as the square of the radius.

$$\mathcal{A} = kr^2 \quad \text{Here, } k = \pi.$$

If $\quad y = \dfrac{k}{x},\quad$ then y varies inversely as x.

Pressure varies inversely as volume.

$$p = \frac{k}{V}$$

If $\quad y = kxz,\quad$ then y varies jointly as x and z.

For a given principal, interest varies jointly as interest rate and time.

$$I = krt \quad k \text{ is the given principal.}$$

Chapter 9 Review Exercises

9.1 *Decide whether each relation defines a function, and give the domain and range.*

1. $\{(-4, 2), (-4, -2), (1, 5), (1, -5)\}$

2. 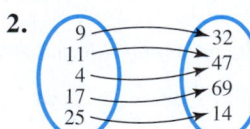   **3.** 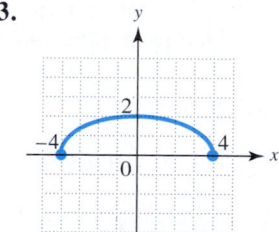   **4.**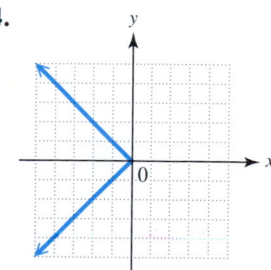

9.1, 9.2 *Decide whether each relation defines y as a function of x. Give the domain. Identify any linear functions.*

5. $y = 3x - 3$ **6.** $y < x + 2$ **7.** $x = y^2$ **8.** $y = \dfrac{7}{x - 6}$

9.2 *Let $f(x) = -2x^2 + 3x - 6$. Find the following.*

9. $f(0)$ **10.** $f(2.1)$ **11.** $f\left(-\dfrac{1}{2}\right)$ **12.** $f(k)$

Solve each problem.

13. The equation $2x^2 - y = 0$ defines y as a function f of x. Write it using function notation, and find $f(3)$.

14. Suppose that $2x - 5y = 7$ defines y as a function f of x. If $y = f(x)$, which one of the following defines the same function?

   A. $f(x) = -\dfrac{2}{5}x + \dfrac{7}{5}$ **B.** $f(x) = -\dfrac{2}{5}x - \dfrac{7}{5}$

   C. $f(x) = \dfrac{2}{5}x - \dfrac{7}{5}$ **D.** $f(x) = \dfrac{2}{5}x + \dfrac{7}{5}$

15. Describe the graph of a constant function.

16. The table shows life expectancy at birth in the United States for selected years.

   (a) Does the table define a function?

   (b) What are the domain and range?

   (c) Call this function f. Give two ordered pairs that belong to f.

   (d) Find $f(1980)$. What does this mean?

   (e) If $f(x) = 76.8$, what does x equal?

Year	Life Expectancy at Birth (years)
1960	69.7
1970	70.8
1980	73.7
1990	75.4
2000	76.8
2010	78.7
2015	78.8

Data from National Center for Health Statistics.

9.3   *For the polynomial function $f(x) = -2x^2 + 5x + 7$, find each value.*

17. $f(-2)$ 18. $f(3)$ 19. $f(0)$

Graph each polynomial function. Give the domain and range.

20. $f(x) = -2x + 5$ 21. $f(x) = x^2 - 6$ 22. $f(x) = -x^3 + 1$

Find each of the following for the polynomial functions

$$f(x) = 2x + 3 \quad and \quad g(x) = 5x^2 - 3x + 2.$$

23. **(a)** $(f + g)(x)$ **(b)** $(f - g)(-1)$ 24. **(a)** $(f - g)(x)$ **(b)** $(f + g)(-1)$

Find each of the following for the polynomial functions

$$f(x) = 12x^2 - 3x \quad and \quad g(x) = 3x.$$

25. **(a)** $(fg)(x)$ **(b)** $(fg)(-1)$ 26. **(a)** $\left(\dfrac{f}{g}\right)(x)$ **(b)** $\left(\dfrac{f}{g}\right)(2)$

Find each of the following for the polynomial functions

$$f(x) = 3x^2 + 2x - 1 \quad and \quad g(x) = 5x + 7.$$

27. **(a)** $(f \circ g)(x)$ **(b)** $(f \circ g)(3)$ **(c)** $(f \circ g)(-2)$

28. **(a)** $(g \circ f)(x)$ **(b)** $(g \circ f)(3)$ **(c)** $(g \circ f)(-2)$

9.4

29. In which one of the following does y vary inversely as x?

   A. $y = 2x$ **B.** $y = \dfrac{x}{3}$ **C.** $y = \dfrac{3}{x}$ **D.** $y = x^2$

Solve each problem.

30. For a particular camera, the viewing distance varies directly as the amount of enlargement. A picture that is taken with this camera and enlarged 5 times should be viewed from a distance of 250 mm. Suppose a print 8.6 times the size of the negative is made. From what distance should it be viewed?

31. The frequency (number of vibrations per second) of a vibrating guitar string varies inversely as its length. That is, a longer string vibrates fewer times in a second than a shorter string. Suppose a guitar string 0.65 m long vibrates 4.3 times per sec. What frequency would a string 0.5 m long have?

32. The volume of a rectangular box of a given height is jointly proportional to its width and length. A box with width 2 ft and length 4 ft has volume 12 ft³. Find the volume of a box with the same height that is 3 ft wide and 5 ft long.

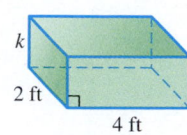

Volume: 12 ft³

Chapter 9   Mixed Review Exercises

Work each problem.

1. Give the domain and range of the relation. Does it define a function? Why or why not?

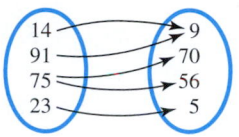

2. Refer to the graph of function f.

   (a) Find $f(-2)$.

   (b) Find $f(0)$.

   (c) For what value of x is $f(x) = -3$?

   (d) Give the domain and range.

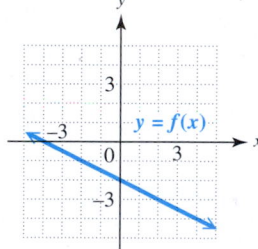

Let $f(x) = 2x^2 - 4$ and $g(x) = 3x + 1$. Find each of the following.

3. $f(-1)$ **4.** $(f + g)(3)$ **5.** $(gf)(-3)$

6. $\left(\dfrac{g}{f}\right)(-1)$ **7.** $(f \circ g)(2)$ **8.** $(g \circ f)(2)$

Solve each problem.

9. The area of a triangle varies jointly as the lengths of the base and height. A triangle with base 10 ft and height 4 ft has area 20 ft². Find the area of a triangle with base 3 ft and height 8 ft.

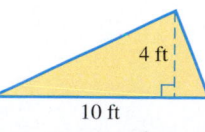

Area: 20 ft²

10. The circumference of a circle varies directly as its radius. A circle with circumference 9.42 in. has radius approximately 1.5 in. Find the circumference of a circle with radius 5.25 in. Give the answer to the nearest hundredth.

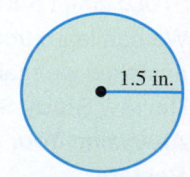

$C = 9.42$ in.

<table>
<tr><td>Chapter 9</td><td>Test</td></tr>
</table>

FOR EXTRA HELP

Step-by-step test solutions are found on the Chapter Test Prep Videos available in **MyLab Math**.

▶ *View the complete solutions to all Chapter Test exercises in* MyLab Math.

1. Which one of the following is the graph of a function?

A. 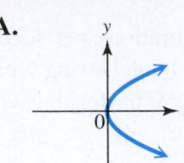 **B.** 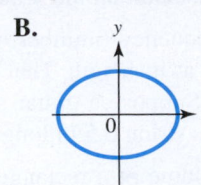 **C.** 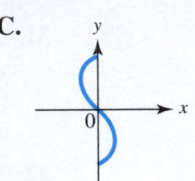 **D.**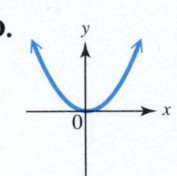

2. Which of the following does not define *y* as a function of *x*?

A. $\{(0, 1), (-2, 3), (4, 8)\}$

B. $y = 2x - 6$

C.

x	y
0	1
3	2
0	2
6	3

Give the domain and range of each relation.

3. Choice A of **Exercise 1**

4. Choice A of **Exercise 2**

5. For $f(x) = -x^2 + 2x - 1$, find $f(1)$ and $f(a)$.

Graph each polynomial function. Give the domain and range.

6. $f(x) = \dfrac{2}{3}x - 1$

7. $f(x) = -2x^2 + 3$

8. $f(x) = -x^3 + 3$

Solve each problem.

9. Find each of the following for the polynomial functions

$$f(x) = -2x^2 + 5x - 6 \quad \text{and} \quad g(x) = 7x - 3.$$

(a) $f(4)$ **(b)** $(f + g)(x)$ **(c)** $(f - g)(x)$ **(d)** $(f - g)(-2)$

10. Find each of the following for the polynomial functions

$$f(x) = x^2 + 3x + 2 \quad \text{and} \quad g(x) = x + 1.$$

(a) $(fg)(x)$ **(b)** $(fg)(-2)$

11. Use $f(x)$ and $g(x)$ from **Exercise 10** to find each of the following.

(a) $\left(\dfrac{f}{g}\right)(x)$ **(b)** $\left(\dfrac{f}{g}\right)(-2)$

12. Find each of the following for the polynomial functions

$$f(x) = 3x + 5 \quad \text{and} \quad g(x) = x^2 + 2.$$

(a) $(f \circ g)(-2)$ **(b)** $(f \circ g)(x)$ **(c)** $(g \circ f)(x)$

STUDY SKILLS REMINDER

We can learn from the mistakes we make. **Review Study Skill 9,** *Analyzing Your Test Results.*

13. The current in a simple electrical circuit is inversely proportional to the resistance. If the current is 80 amps when the resistance is 30 ohms, find the current when the resistance is 12 ohms.

14. The force of the wind blowing on a vertical surface varies jointly as the area of the surface and the square of the velocity. If a wind blowing at 40 mph exerts a force of 50 lb on a surface of 500 ft², how much force will a wind of 80 mph place on a surface of 2 ft²?

Chapters R–9   Cumulative Review Exercises

Perform the indicated operations.

1. $-\dfrac{2}{3} - \dfrac{1}{4}$

2. $-\dfrac{3}{4}\left(-\dfrac{5}{6}\right)$

3. $-\dfrac{3}{5} \div \dfrac{1}{2}$

4. $-8.75 + 3.08$

5. $5.25 \div 100$

6. 43.6×100

Evaluate.

7. (a) 5^2 (b) -5^2 (c) $(-5)^2$

8. $|2x| + 3y - z^3$ for $x = -4$, $y = 3$, and $z = 6$

Solve each equation or inequality.

9. $7(2x + 3) - 4(2x + 1) = 2(x + 1)$

10. $\dfrac{2}{3}y + \dfrac{5}{12}y \le 20$

Solve each problem.

11. A triangle has area 42 m². The base is 14 m long. Find the height of the triangle.

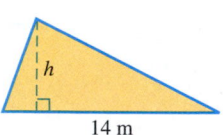

14 m

12. Abushieba invested some money at 4% interest and twice as much at 3% interest. His interest for the first year was $400. How much did he invest at each rate?

Find the slope of each line described.

13. Through $(-5, 8)$ and $(-1, 2)$

14. Perpendicular to $4x - 3y = 12$

15. Write an equation of the line in **Exercise 13.** Give the equation in the form $y = mx + b$.

Graph each equation or inequality.

16. $-4x + 2y = 8$ (Give the intercepts.)

17. $2x + 5y > 10$

18. Simplify $\left(\dfrac{m^{-4}n^2}{m^2n^{-3}}\right) \cdot \left(\dfrac{m^5n^{-1}}{m^{-2}n^5}\right)$. Write the answer with only positive exponents. Assume that all variables represent nonzero real numbers.

Perform the indicated operations.

19. $(3y^2 - 2y + 6) - (-y^2 + 5y + 12)$

20. $(4f + 3)(3f - 1)$

21. $\left(\dfrac{1}{4}x + 5\right)^2$

22. $(3x^3 + 13x^2 - 17x - 7) \div (3x + 1)$

Factor each polynomial completely.

23. $2x^2 - 13x - 45$

24. $8p^3 + 125$

25. Write $\dfrac{y^2 - 16}{y^2 - 8y + 16}$ in lowest terms.

Perform the indicated operations. Express answers in lowest terms.

26. $\dfrac{2a^2}{a + b} \cdot \dfrac{a - b}{4a}$

27. $\dfrac{x^2 - 9}{2x + 4} \div \dfrac{x^3 - 27}{4}$

28. $\dfrac{x + 4}{x - 2} + \dfrac{2x - 10}{x - 2}$

Solve each equation or inequality.

29. $3x^2 + 4x = 7$

30. $\dfrac{-3x}{x+1} + \dfrac{4x+1}{x} = \dfrac{-3}{x^2+x}$

31. $|6x - 8| - 4 = 0$

32. $|3x + 2| \geq 4$

Solve each system.

33. $4x - y = -7$
$5x + 2y = 1$

34. $x + y - 2z = -1$
$2x - y + z = -6$
$3x + 2y - 3z = -3$

Solve each problem.

35. Decide whether the relation $y = -x + 2$ is a function, and give its domain and range.

36. Suppose that $y = f(x)$ and $5x - 3y = 8$.

   (a) Find an equation that defines $f(x)$. That is, $f(x) = $ _____.

   (b) Find $f(1)$.

37. Find each of the following for the polynomial functions

$$f(x) = x^2 + 2x - 3, \quad g(x) = 2x^3 - 3x^2 + 4x - 1, \quad \text{and} \quad h(x) = x^2.$$

   (a) $(f + g)(x)$ **(b)** $(g - f)(x)$ **(c)** $(f + g)(-1)$ **(d)** $(f \circ h)(x)$

38. The cost of a pizza varies directly as the square of its radius. If a pizza with a 7-in. radius costs $6.00, how much should a pizza with a 9-in. radius cost?

STUDY SKILLS REMINDER

Have you begun to prepare for your final exam? **Review Study Skill 10,** *Preparing for Your Math Final Exam.*

10

ROOTS, RADICALS, AND ROOT FUNCTIONS

The formula for calculating the distance one can see to the horizon from the top of a tall building involves a *square root radical,* one of the topics covered in this chapter.

10.1 Radical Expressions and Graphs

OBJECTIVES

1 Find square roots.

2 Determine whether a given root is rational, irrational, or not a real number.

3 Find cube, fourth, and other roots.

4 Graph functions defined by radical expressions.

5 Find *n*th roots of *n*th powers.

6 Use a calculator to find roots.

OBJECTIVE 1 Find square roots.

Recall that *squaring* a number means multiplying the number by itself.

$$7^2 \text{ means } 7 \cdot 7, \text{ which equals } 49. \quad \text{The square of 7 is 49.}$$

The opposite (inverse) of squaring a number is taking its *square root*. This is equivalent to asking

"What number when multiplied by itself equals 49?"

From the example above, one answer is 7 because $7 \cdot 7 = 49$.

> **Square Root**
>
> A number b is a **square root** of a if $b^2 = a$ (that is, $b \cdot b = a$).

EXAMPLE 1 Finding All Square Roots of a Number

Find all square roots of 49.

We ask, *"What number when multiplied by itself equals 49?"* As mentioned above, one square root is 7. Another square root of 49 is −7 because

$$(-7)(-7) = 49.$$

Thus, the number 49 has *two* square roots, 7 and −7. One square root is positive, and one is negative.

NOW TRY

The **positive** or **principal square root** of a number is written with the symbol $\sqrt{}$. For example, the positive square root of 121 is 11.

$$\sqrt{121} = 11 \qquad 11^2 = 121$$

The symbol $-\sqrt{}$ is used for the **negative square root** of a number. For example, the negative square root of 121 is −11.

$$-\sqrt{121} = -11 \qquad (-11)^2 = 121$$

The **radical symbol** $\sqrt{}$ always represents the positive square root (except that $\sqrt{0} = 0$). The number inside the radical symbol is the **radicand,** and the entire expression—radical symbol and radicand—is a **radical.**

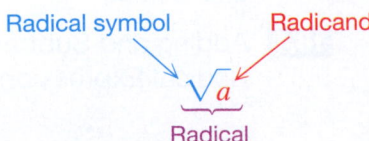

An algebraic expression containing a radical is a **radical expression.**

The radical symbol $\sqrt{}$ has been used since 16th-century Germany and was likely derived from the letter *R*. The radical symbol at the right comes from the Latin word *radix,* for "root." It was first used by Leonardo of Pisa (Fibonacci) in 1220.

Early radical symbol

VOCABULARY

☐ square root
☐ positive (principal) square root
☐ negative square root
☐ radicand
☐ radical
☐ radical expression
☐ perfect square
☐ irrational number
☐ cube root
☐ fourth root
☐ index (order)
☐ perfect cube
☐ perfect fourth power
☐ square root function
☐ cube root function

 **NOW TRY EXERCISE 1**

Find all square roots of 81.

NOW TRY ANSWER

1. 9, −9

We summarize our discussion of square roots as follows.

Square Roots of *a*

Let *a* be a positive real number.

$\sqrt{a}$ is the positive or principal square root of *a*.

$-\sqrt{a}$ is the negative square root of *a*.

For nonnegative *a*, the following hold true.

$$\sqrt{a} \cdot \sqrt{a} = \left(\sqrt{a}\right)^2 = a \quad \text{and} \quad -\sqrt{a} \cdot \left(-\sqrt{a}\right) = \left(-\sqrt{a}\right)^2 = a$$

Also, $\sqrt{0} = 0$.

NOW TRY EXERCISE 2

Find each square root.

(a) $\sqrt{400}$ (b) $-\sqrt{64}$

(c) $\sqrt{\dfrac{100}{121}}$ (d) $\sqrt{0.09}$

EXAMPLE 2 Finding Square Roots

Find each square root.

(a) $\sqrt{144}$

The radical $\sqrt{144}$ represents the positive or principal square root of 144. Think of a positive number whose square is 144.

$$12^2 = 144, \quad \text{so} \quad \sqrt{144} = 12.$$

(b) $-\sqrt{81}$

This symbol represents the negative square root of 81. Because $\sqrt{81} = 9$,

$$-\sqrt{81} = -9.$$

(c) $\sqrt{\dfrac{4}{9}} = \dfrac{2}{3}$ (d) $-\sqrt{\dfrac{16}{49}} = -\dfrac{4}{7}$ (e) $\sqrt{0.81} = 0.9$

$(0.9)^2 = 0.9 \cdot 0.9$
$= 0.81$

NOW TRY

! CAUTION By definition, $\sqrt{4} = 2$ because $2^2 = 4$. *In general, however, the square root of a number does not equal half the number.*

As noted above, when the square root of a positive real number is squared, the result is that positive real number. $\left(\text{Also, } \left(\sqrt{0}\right)^2 = 0.\right)$

NOW TRY EXERCISE 3

Find the square of each radical expression.

(a) $\sqrt{15}$ (b) $-\sqrt{23}$

(c) $\sqrt{2k^2 + 5}$

EXAMPLE 3 Squaring Radical Expressions

Find the square of each radical expression.

(a) $\sqrt{13}$

The square of $\sqrt{13}$ is $\left(\sqrt{13}\right)^2 = 13.$ Definition of square root

(b) $-\sqrt{29}$

$$\left(-\sqrt{29}\right)^2 = 29$$ The square of a *negative* number is positive.

(c) $\sqrt{p^2 + 1}$

$$\left(\sqrt{p^2 + 1}\right)^2 = p^2 + 1$$

NOW TRY

NOW TRY ANSWERS

2. (a) 20   (b) -8
   (c) $\dfrac{10}{11}$ (d) 0.3

3. (a) 15   (b) 23   (c) $2k^2 + 5$

OBJECTIVE 2 Determine whether a given root is rational, irrational, or not a real number.

Numbers with square roots that are rational are **perfect squares.**

Perfect squares   Rational square roots

$$25$$ $$\sqrt{25} = 5$$

$$144 \quad \text{are perfect squares because} \quad \sqrt{144} = 12$$

$$\frac{4}{9} \qquad\qquad\qquad\qquad\qquad \sqrt{\frac{4}{9}} = \frac{2}{3}$$

A number that is not a perfect square has a square root that is not a rational number. For example, $\sqrt{5}$ is not a rational number because it cannot be written as the ratio of two integers. Its decimal equivalent neither terminates nor repeats. However, $\sqrt{5}$ is a real number and corresponds to a point on the number line.

A real number that is not rational is an **irrational number.** The number $\sqrt{5}$ is irrational. *Many square roots of integers are irrational.*

> If a is a *positive* real number that is *not* a perfect square, then $\sqrt{a}$ is irrational.

Not every number has a real number square root. For example, there is no real number that can be squared to obtain -36. (The square of a real number can never be negative.) Because of this, $\mathbf{\sqrt{-36}}$ *is not a real number.*

> If a is a *negative* real number, then $\sqrt{a}$ is *not* a real number.

⊘ **CAUTION**   Do not confuse $\sqrt{-36}$ and $-\sqrt{36}$. $\sqrt{-36}$ is not a real number because there is no real number that can be squared to obtain -36. However, $-\sqrt{36}$ is the negative square root of 36, which is -6.

**NOW TRY
EXERCISE 4**

Determine whether each number is *rational, irrational,* or *not a real number.*

(a) $\sqrt{31}$ **(b)** $\sqrt{900}$
(c) $\sqrt{-16}$

EXAMPLE 4   Identifying Types of Square Roots

Determine whether each number is *rational, irrational,* or *not a real number.*

(a) $\sqrt{17}$ Because 17 is not a perfect square, $\sqrt{17}$ is irrational.

(b) $\sqrt{64}$ The number 64 is a perfect square, 8^2, so $\sqrt{64} = 8$, a rational number.

(c) $\sqrt{-25}$ There is no real number whose square is -25. Therefore, $\sqrt{-25}$ is not a real number.   **NOW TRY** ↻

NOTE   Not all irrational numbers are square roots of integers. For example, the number π (approximately 3.14159) is an irrational number that is not a square root of any integer.

OBJECTIVE 3 Find cube, fourth, and other roots.

Finding the square root of a number is the inverse (opposite) of squaring a number. In a similar way, there are inverses to finding the cube of a number and to finding the fourth or greater power of a number. These inverses are, respectively, the **cube root**, $\sqrt[3]{a}$, and the **fourth root**, $\sqrt[4]{a}$. Similar symbols are used for other roots.

> ### $\sqrt[n]{a}$
>
> The *n*th root of *a*, written $\sqrt[n]{a}$, is a number whose *n*th power equals *a*. That is,
>
> $$\sqrt[n]{a} = b \quad \text{means} \quad b^n = a.$$

In $\sqrt[n]{a}$, the number *n* is the **index,** or **order,** of the radical.

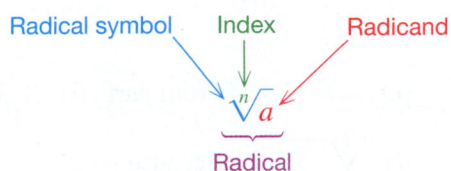

We could write $\sqrt[2]{a}$ instead of $\sqrt{a}$, but the simpler symbol $\sqrt{a}$ is customary because the square root is the most commonly used root.

> **NOTE** When working with cube roots or fourth roots, it is helpful to learn the first few **perfect cubes** ($1^3 = 1$, $2^3 = 8$, $3^3 = 27$, and so on) and the first few **perfect fourth powers** ($1^4 = 1$, $2^4 = 16$, $3^4 = 81$, and so on). See **Exercises 75 and 76.**

NOW TRY EXERCISE 5

Find each cube root.

(a) $\sqrt[3]{343}$ (b) $\sqrt[3]{-1000}$

(c) $\sqrt[3]{27}$ (d) $\sqrt[3]{\dfrac{125}{216}}$

EXAMPLE 5 Finding Cube Roots

Find each cube root.

$\boxed{2^3 = 2 \cdot 2 \cdot 2}$

(a) $\sqrt[3]{8}$ *"What number can be cubed to give* 8*?"* Because $2^3 = 8$, $\sqrt[3]{8} = 2$.

(b) $\sqrt[3]{-8}$ Because $(-2)^3 = -8$, $\sqrt[3]{-8} = -2$.

(c) $\sqrt[3]{216}$ Because $6^3 = 216$, $\sqrt[3]{216} = 6$.

(d) $\sqrt[3]{\dfrac{27}{125}}$ Because $\left(\dfrac{3}{5}\right)^3 = \dfrac{27}{125}$, $\sqrt[3]{\dfrac{27}{125}} = \dfrac{3}{5}$.

NOW TRY

In **Example 5(b),** $\sqrt[3]{-8} = -2$—that is, we can find the *cube root* of a negative number. (Contrast this with the *square root* of a negative number, which is not real.) The cube root of a positive number is positive. The cube root of a negative number is negative.

> *There is only one real number cube root for each real number.*

When a radical has an *even index* (square root, fourth root, and so on), *the radicand must be nonnegative* to yield a real number root. Also, for $a > 0$,

$$\sqrt{a}, \ \sqrt[4]{a}, \ \sqrt[6]{a}, \ \text{and so on are positive (principal) roots.}$$

$$-\sqrt{a}, \ -\sqrt[4]{a}, \ -\sqrt[6]{a}, \ \text{and so on are negative roots.}$$

NOW TRY ANSWERS

5. (a) 7 (b) −10

(c) 3 (d) $\frac{5}{6}$

**NOW TRY
EXERCISE 6**

Find each root.

(a) $\sqrt[4]{625}$ (b) $\sqrt[4]{-625}$

(c) $-\sqrt[4]{625}$ (d) $\sqrt[5]{3125}$

(e) $\sqrt[5]{-3125}$

EXAMPLE 6   Finding Other Roots

Find each root.

$$2^4 = 2 \cdot 2 \cdot 2 \cdot 2$$

(a) $\sqrt[4]{16}$ Because 2 is positive and $2^4 = 16$, $\sqrt[4]{16} = 2$.

(b) $-\sqrt[4]{16}$ From part (a), $\sqrt[4]{16} = 2$, so the negative root is $-\sqrt[4]{16} = -2$.

(c) $\sqrt[4]{-16}$

For a fourth root to be a real number, the radicand must be nonnegative. There is no real number that equals $\sqrt[4]{-16}$.

(d) $\sqrt[5]{32}$

Here 2 is the number whose fifth power is 32.

$$\sqrt[5]{32} = 2 \quad \text{because} \quad 2^5 = 32.$$

(e) $-\sqrt[5]{32}$ From part (d), $\sqrt[5]{32} = 2$, so it follows that $-\sqrt[5]{32} = -2$.

(f) $\sqrt[5]{-32}$ Because $(-2)^5 = -32$, $\sqrt[5]{-32} = -2$.   **NOW TRY**

OBJECTIVE 4   Graph functions defined by radical expressions.

A **radical expression** is an algebraic expression that contains radicals.

$$3 - \sqrt{x}, \quad \sqrt[3]{x}, \quad \sqrt{2x - 1} \qquad \text{Radical expressions}$$

In earlier chapters, we graphed functions defined by polynomial and rational expressions. Now we graph functions defined by basic radical expressions, such as $f(x) = \sqrt{x}$ and $f(x) = \sqrt[3]{x}$.

FIGURE 1 shows the graph of the **square root function,**

$$f(x) = \sqrt{x},$$

together with a table of selected points. Only nonnegative values can be used for x, so the domain is $[0, \infty)$. Because $\sqrt{x}$ is the principal square root of x, it always has a nonnegative value, so the range is also $[0, \infty)$.

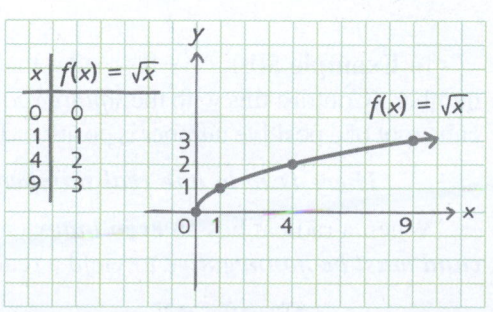

Square root function

$$f(x) = \sqrt{x}$$

Domain: $[0, \infty)$

Range: $[0, \infty)$

FIGURE 1

FIGURE 2 shows the graph of the **cube root function**

$$f(x) = \sqrt[3]{x}.$$

Any real number (positive, negative, or 0) can be used for x in the cube root function, so $\sqrt[3]{x}$ can be positive, negative, or 0. Thus, both the domain and the range of the cube root function are $(-\infty, \infty)$.

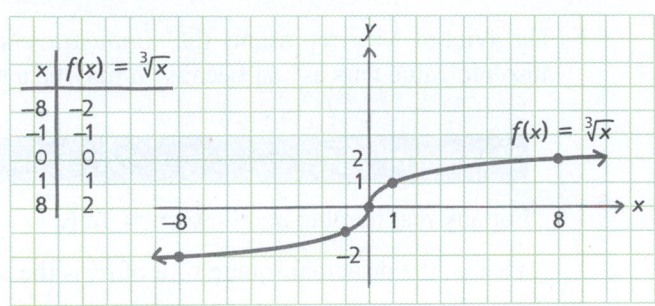

x	$f(x) = \sqrt[3]{x}$
-8	-2
-1	-1
0	0
1	1
8	2

Cube root function

$$f(x) = \sqrt[3]{x}$$

Domain: $(-\infty, \infty)$

Range: $(-\infty, \infty)$

FIGURE 2

NOW TRY
EXERCISE 7

Graph each function, and give its domain and range.

(a) $f(x) = \sqrt{x + 1}$

(b) $f(x) = \sqrt[3]{x} - 1$

EXAMPLE 7   Graphing Functions Defined with Radicals

Graph each function, and give its domain and range.

(a) $f(x) = \sqrt{x - 3}$

Create a table of values as given with the graph in **FIGURE 3**. The x-values were chosen in such a way that the function values are all integers. For the radicand to be nonnegative, we must have

$$x - 3 \geq 0, \quad \text{or} \quad x \geq 3.$$

Therefore, the domain of this function is $[3, \infty)$. Function values are positive or 0, so the range is $[0, \infty)$.

x	$f(x) = \sqrt{x - 3}$
3	$\sqrt{3 - 3} = 0$
4	$\sqrt{4 - 3} = 1$
7	$\sqrt{7 - 3} = 2$

This graph is shifted to the right 3 units compared to the graph of $y = \sqrt{x}$.

FIGURE 3

NOW TRY ANSWERS

7. **(a)**

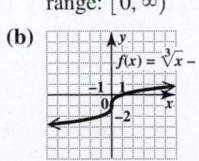

domain: $[-1, \infty)$;
range: $[0, \infty)$

(b)

domain: $(-\infty, \infty)$;
range: $(-\infty, \infty)$

(b) $f(x) = \sqrt[3]{x} + 2$

See **FIGURE 4**. Both the domain and the range are $(-\infty, \infty)$.

x	$f(x) = \sqrt[3]{x} + 2$
-8	$\sqrt[3]{-8} + 2 = 0$
-1	$\sqrt[3]{-1} + 2 = 1$
0	$\sqrt[3]{0} + 2 = 2$
1	$\sqrt[3]{1} + 2 = 3$
8	$\sqrt[3]{8} + 2 = 4$

This graph is shifted up 2 units compared to the graph of $y = \sqrt[3]{x}$.

FIGURE 4

NOW TRY

OBJECTIVE 5 Find *n*th roots of *n*th powers.

Consider the expression $\sqrt{a^2}$. At first glance, we might think that it is equivalent to *a*. However, this is not necessarily true. For example, consider the following.

$$\text{If } a = 6, \quad \text{then} \quad \sqrt{a^2} = \sqrt{6^2} = \sqrt{36} = 6.$$

$$\text{If } a = -6, \quad \text{then} \quad \sqrt{a^2} = \sqrt{(-6)^2} = \sqrt{36} = 6. \leftarrow \text{Instead of } -6, \text{ we get } 6, \text{ the } \textit{absolute value} \text{ of } -6.$$

The symbol $\sqrt{a^2}$ represents the *nonnegative* square root, so we express $\sqrt{a^2}$ with absolute value bars as $|a|$ because *a* may be a negative number.

Meaning of $\sqrt{a^2}$

For any real number *a*, $\qquad \sqrt{a^2} = |a|.$

That is, the principal square root of a^2 is the absolute value of *a*.

NOW TRY EXERCISE 8

Find each square root.

(a) $\sqrt{11^2}$ **(b)** $\sqrt{(-11)^2}$

(c) $\sqrt{z^2}$ **(d)** $\sqrt{(-z)^2}$

EXAMPLE 8 Simplifying Square Roots Using Absolute Value

Find each square root.

(a) $\sqrt{7^2} = |7| = 7$ **(b)** $\sqrt{(-7)^2} = |-7| = 7$

(c) $\sqrt{k^2} = |k|$ **(d)** $\sqrt{(-k)^2} = |-k| = |k|$ **NOW TRY**

We can generalize this idea to any *n*th root.

Meaning of $\sqrt[n]{a^n}$

If *n* is an *even* positive integer, then $\quad \sqrt[n]{a^n} = |a|.$

If *n* is an *odd* positive integer, then $\quad \sqrt[n]{a^n} = a.$

That is, use the absolute value symbol when *n* is even.

NOW TRY EXERCISE 9

Simplify each root.

(a) $\sqrt[8]{(-2)^8}$ **(b)** $\sqrt[3]{(-9)^3}$

(c) $-\sqrt[4]{(-10)^4}$ **(d)** $-\sqrt{m^8}$

(e) $\sqrt[3]{x^{18}}$ **(f)** $\sqrt[4]{t^{20}}$

EXAMPLE 9 Simplifying Higher Roots Using Absolute Value

Simplify each root.

(a) $\sqrt[6]{(-3)^6} = |-3| = 3$ *n* is even. Use absolute value.

(b) $\sqrt[5]{(-4)^5} = -4$ *n* is odd. Absolute value is not necessary.

(c) $-\sqrt[4]{(-9)^4} = -|-9| = -9$ *n* is even. Use absolute value.

(d) $-\sqrt{m^4} = -|m^2| = -m^2$ For all *m*, $|m^2| = m^2$.

No absolute value bars are needed here, because m^2 is nonnegative for any real number value of *m*.

(e) $\sqrt[3]{a^{12}} = a^4$, because $a^{12} = (a^4)^3$.

NOW TRY ANSWERS
8. **(a)** 11   **(b)** 11   **(c)** $|z|$ **(d)** $|z|$
9. **(a)** 2   **(b)** -9 **(c)** -10 **(d)** $-m^4$ **(e)** x^6 **(f)** $|t^5|$

(f) $\sqrt[4]{x^{12}} = |x^3|$

We use absolute value to guarantee that the result is not negative (because x^3 is negative when *x* is negative). If desired, $|x^3|$ can be written as $x^2 \cdot |x|$.   **NOW TRY**

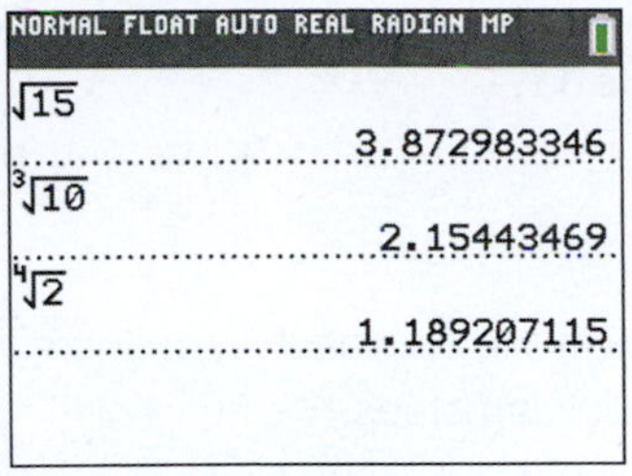

(a)

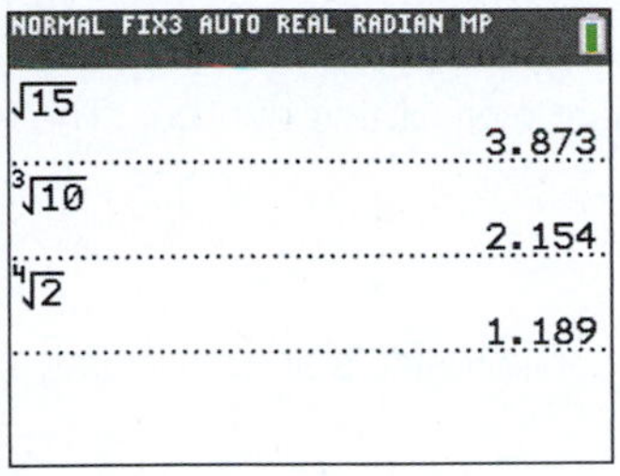

(b)

FIGURE 5

OBJECTIVE 6 Use a calculator to find roots.

While numbers such as $\sqrt{9}$ and $\sqrt[3]{-8}$ are rational, radicals are often irrational numbers. To find approximations of such radicals, we usually use a calculator. For example,

$$\sqrt{15} \approx 3.872983346, \quad \sqrt[3]{10} \approx 2.15443469, \quad \text{and} \quad \sqrt[4]{2} \approx 1.189207115,$$

where the symbol $\approx$ means "is approximately equal to." In this text, we often show approximations rounded to three decimal places. Thus,

$$\sqrt{15} \approx 3.873, \quad \sqrt[3]{10} \approx 2.154, \quad \text{and} \quad \sqrt[4]{2} \approx 1.189.$$

FIGURE 5 shows how the preceding approximations are displayed on a TI-83/84 Plus C graphing calculator. In **FIGURE 5(a)**, eight or nine decimal places are shown, while in **FIGURE 5(b)**, the number of decimal places is fixed at three.

There is a simple way to check that a calculator approximation is "in the ballpark." For example, because 16 is a little larger than 15, $\sqrt{16} = 4$ should be a little larger than $\sqrt{15}$. Thus, 3.873 is reasonable as an approximation for $\sqrt{15}$.

> **NOTE** Methods for finding approximations differ among makes and models of calculators. *Consult your owner's manual for keystroke instructions.* Be aware that graphing calculators often differ from scientific calculators in the order in which keystrokes are made.

NOW TRY EXERCISE 10
Use a calculator to approximate each radical to three decimal places.

(a) $\sqrt{73}$ (b) $-\sqrt{92}$

(c) $\sqrt[4]{92}$ (d) $\sqrt[5]{33}$

EXAMPLE 10 Finding Approximations for Roots

Use a calculator to approximate each radical to three decimal places.

(a) $\sqrt{39} \approx 6.245$

(b) $-\sqrt{72} \approx -8.485$

(c) $\sqrt[3]{93} \approx 4.531$

(d) $\sqrt[4]{39} \approx 2.499$ **NOW TRY**

NOW TRY EXERCISE 11
Use the formula in **Example 11** to approximate f to the nearest thousand if

$$L = 7 \times 10^{-5}$$
and $$C = 3 \times 10^{-9}.$$

EXAMPLE 11 Using Roots to Calculate Resonant Frequency

In electronics, the resonant frequency f of a circuit may be found using the formula

$$f = \frac{1}{2\pi\sqrt{LC}}, \qquad \text{Electronics formula}$$

where f is in cycles per second, L is in henrys, and C is in farads. (Henrys and farads are units of measure in electronics.) Find the resonant frequency f if $L = 5 \times 10^{-4}$ henry and $C = 3 \times 10^{-10}$ farad. Give the answer to the nearest thousand.

Find the value of f when $L = 5 \times 10^{-4}$ and $C = 3 \times 10^{-10}$.

$$f = \frac{1}{2\pi\sqrt{LC}} \qquad \text{Given formula}$$

$$f = \frac{1}{2\pi\sqrt{(5 \times 10^{-4})(3 \times 10^{-10})}} \qquad \text{Substitute for } L \text{ and } C.$$

$$f \approx 411{,}000 \qquad \text{Use a calculator.}$$

The resonant frequency f is approximately 411,000 cycles per sec.   **NOW TRY**

NOW TRY ANSWERS
10. (a) 8.544 (b) −9.592
    (c) 3.097 (d) 2.012
11. 347,000 cycles per sec

NOTE The expression in the second-to-last line of **Example 11,**

$$\frac{1}{2\pi \sqrt{(5 \times 10^{-4})(3 \times 10^{-10})}},$$

can be difficult to compute on a calculator. There are several ways to approach it, but here is one way that works nicely. Use a calculator to verify the following steps.

Step 1 Calculate the product under the radical in the denominator. The result is

$$1.5 \times 10^{-13}.$$

Step 2 Use the square root function ($\sqrt{x}$) to find the square root of this number. The result is approximately

$$3.872983346 \times 10^{-7}.$$

Step 3 Multiply by 2π, using the π function of the calculator. The result is approximately

$$2.433467206 \times 10^{-6}.$$

Step 4 Now use the reciprocal function (labeled $\frac{1}{x}$ or x^{-1}) of the calculator. The result should be 410936.296, which, rounded to the nearest thousand, is 411,000.

If the numerator had not been 1, and perhaps an expression to evaluate, one approach would be to save the result found in Step 3 in the calculator memory, evaluate the numerator, and then divide by the number saved in the memory.

10.1 Exercises

FOR EXTRA HELP

▶ **MyLab Math**

▶ *Video solutions for select problems available in MyLab Math*

STUDY SKILLS REMINDER
We can learn from the mistakes we make.
Review Study Skill 9,
Analyzing Your Test Results.

1. Concept Check   To help find square roots, complete this list of perfect squares.

$1^2 = $ _____ $2^2 = $ _____ $3^2 = $ _____ $4^2 = $ _____ $5^2 = $ _____

$6^2 = $ _____ $7^2 = $ _____ $8^2 = $ _____ $9^2 = $ _____ $10^2 = $ _____

$11^2 = $ _____ $12^2 = $ _____ $13^2 = $ _____ $14^2 = $ _____ $15^2 = $ _____

$16^2 = $ _____ $20^2 = $ _____ $25^2 = $ _____ $30^2 = $ _____ $50^2 = $ _____

2. Concept Check   Fill in each blank with the correct response.

The opposite (inverse) of squaring a number is taking its _____.

A number b is a square root of a if _____ $= a$. That is, $b \cdot$ _____ $= a$.

Concept Check   *Determine whether each statement is* true *or* false. *If false, tell why.*

3. Every positive number has two real square roots.

4. A negative number has negative real square roots.

5. Every nonnegative number has two real square roots.

6. The positive square root of a positive number is its principal square root.

7. The cube root of every nonzero real number has the same sign as the number itself.

8. Every positive number has three real cube roots.

Concept Check   *What must be true about the value of the variable a for each statement to be true?*

9. $\sqrt{a}$ represents a positive number.

10. $-\sqrt{a}$ represents a negative number.

11. $\sqrt{a}$ is not a real number.

12. $\sqrt{a}$ is not a real number.

Find all square roots of each number. See Example 1.

13. 9   **14.** 16   **15.** 64   **16.** 100   **17.** 169

18. 225   **19.** $\dfrac{25}{196}$ **20.** $\dfrac{81}{400}$ **21.** 900   **22.** 1600

23. Concept Check A student who was directed to find $\sqrt{-4}$ incorrectly obtained -2. **WHAT WENT WRONG?** Give the correct answer.

24. Concept Check A student who was directed to find $\sqrt{16}$ incorrectly obtained 8. **WHAT WENT WRONG?** Give the correct answer.

Find each square root. See Examples 2 and 4(c).

25. $\sqrt{1}$ **26.** $\sqrt{4}$ **27.** $\sqrt{49}$ **28.** $\sqrt{81}$ **29.** $\sqrt{100}$

30. $\sqrt{400}$ **31.** $-\sqrt{16}$ **32.** $-\sqrt{64}$ **33.** $-\sqrt{256}$ **34.** $-\sqrt{196}$

35. $\sqrt{\dfrac{4}{25}}$ **36.** $\sqrt{\dfrac{9}{100}}$ **37.** $-\sqrt{\dfrac{144}{121}}$ **38.** $-\sqrt{\dfrac{49}{36}}$

39. $\sqrt{0.64}$ **40.** $\sqrt{0.16}$ **41.** $-\sqrt{0.04}$ **42.** $-\sqrt{0.01}$

43. $\sqrt{-121}$ **44.** $\sqrt{-64}$ **45.** $-\sqrt{-49}$ **46.** $-\sqrt{-100}$

Find the square of each radical expression. See Example 3.

47. $\sqrt{19}$ **48.** $\sqrt{59}$ **49.** $-\sqrt{19}$ **50.** $-\sqrt{59}$

51. $\sqrt{\dfrac{2}{3}}$ **52.** $\sqrt{\dfrac{5}{7}}$ **53.** $\sqrt{3x^2 + 4}$ **54.** $\sqrt{9y^2 + 3}$

Determine whether each number is rational, irrational, *or* not a real number. *If a number is rational, give its exact value. If a number is irrational, give a decimal approximation to the nearest thousandth. See Examples 4 and 10.*

55. $\sqrt{25}$ **56.** $\sqrt{169}$ **57.** $\sqrt{29}$ **58.** $\sqrt{33}$

59. $-\sqrt{64}$ **60.** $-\sqrt{81}$ **61.** $-\sqrt{300}$ **62.** $-\sqrt{500}$

63. $\sqrt{-29}$ **64.** $\sqrt{-47}$ **65.** $\sqrt{1200}$ **66.** $\sqrt{1500}$

Concept Check *Without using a calculator, determine between which two consecutive integers each square root lies. For example,*

$\sqrt{75}$ *is between 8 and 9, because* $\sqrt{64} = 8$, $\sqrt{81} = 9$, *and* $64 < 75 < 81$.

67. $\sqrt{94}$ **68.** $\sqrt{43}$ **69.** $\sqrt{51}$ **70.** $\sqrt{30}$

71. $-\sqrt{40}$ **72.** $-\sqrt{63}$ **73.** $\sqrt{23.2}$ **74.** $\sqrt{10.3}$

75. Concept Check To help find cube roots, complete this list of perfect cubes.

$1^3 = $ _____ $2^3 = $ _____ $3^3 = $ _____ $4^3 = $ _____ $5^3 = $ _____

$6^3 = $ _____ $7^3 = $ _____ $8^3 = $ _____ $9^3 = $ _____ $10^3 = $ _____

76. Concept Check To help find fourth roots, complete this list of perfect fourth powers.

$1^4 = \underline{\hspace{1cm}}$ $2^4 = \underline{\hspace{1cm}}$ $3^4 = \underline{\hspace{1cm}}$ $4^4 = \underline{\hspace{1cm}}$ $5^4 = \underline{\hspace{1cm}}$

$6^4 = \underline{\hspace{1cm}}$ $7^4 = \underline{\hspace{1cm}}$ $8^4 = \underline{\hspace{1cm}}$ $9^4 = \underline{\hspace{1cm}}$ $10^4 = \underline{\hspace{1cm}}$

77. Concept Check A student who was directed to find $\sqrt[3]{27}$ incorrectly obtained 9. **WHAT WENT WRONG?** Give the correct answer.

78. Concept Check A student incorrectly claimed that $\sqrt[3]{-125}$ is not a real number. **WHAT WENT WRONG?** Give the correct answer.

Find each root. See Examples 5 and 6.

79. $\sqrt[3]{1}$ **80.** $\sqrt[3]{27}$ **81.** $\sqrt[3]{125}$ **82.** $\sqrt[3]{64}$

83. $\sqrt[3]{729}$ **84.** $\sqrt[3]{1000}$ **85.** $\sqrt[3]{-27}$ **86.** $\sqrt[3]{-64}$

87. $\sqrt[3]{-216}$ **88.** $\sqrt[3]{-343}$ **89.** $-\sqrt[3]{-8}$ **90.** $-\sqrt[3]{-343}$

91. $\sqrt[3]{\dfrac{8}{27}}$ **92.** $\sqrt[3]{\dfrac{64}{125}}$ **93.** $\sqrt[3]{-\dfrac{216}{125}}$ **94.** $\sqrt[3]{-\dfrac{1}{64}}$

95. $\sqrt[3]{0.125}$ **96.** $\sqrt[3]{0.008}$ **97.** $\sqrt[4]{81}$ **98.** $\sqrt[4]{16}$

99. $\sqrt[4]{625}$ **100.** $\sqrt[4]{256}$ **101.** $\sqrt[4]{1296}$ **102.** $\sqrt[4]{10,000}$

103. $\sqrt[4]{-1}$ **104.** $\sqrt[4]{-625}$ **105.** $-\sqrt[4]{81}$ **106.** $-\sqrt[4]{256}$

107. $\sqrt[5]{32}$ **108.** $\sqrt[5]{1}$ **109.** $\sqrt[5]{-1024}$ **110.** $\sqrt[5]{-100,000}$

Graph each function, and give its domain and range. See Example 7.

111. $f(x) = \sqrt{x+3}$ **112.** $f(x) = \sqrt{x-5}$ **113.** $f(x) = \sqrt{x-2}$

114. $f(x) = \sqrt{x+4}$ **115.** $f(x) = \sqrt[3]{x-3}$ **116.** $f(x) = \sqrt[3]{x+1}$

117. $f(x) = \sqrt[3]{x-3}$ **118.** $f(x) = \sqrt[3]{x+1}$

Simplify each root. See Examples 8 and 9.

119. $\sqrt{12^2}$ **120.** $\sqrt{19^2}$ **121.** $\sqrt{(-10)^2}$ **122.** $\sqrt{(-13)^2}$

123. $\sqrt[6]{(-2)^6}$ **124.** $\sqrt[6]{(-4)^6}$ **125.** $\sqrt[5]{(-9)^5}$ **126.** $\sqrt[5]{(-8)^5}$

127. $-\sqrt[6]{(-5)^6}$ **128.** $-\sqrt[6]{(-7)^6}$ **129.** $\sqrt{x^2}$ **130.** $-\sqrt{x^2}$

131. $\sqrt{(-z)^2}$ **132.** $\sqrt{(-q)^2}$ **133.** $\sqrt[3]{x^3}$ **134.** $-\sqrt[3]{x^3}$

135. $\sqrt[3]{x^{15}}$ **136.** $\sqrt[3]{m^9}$ **137.** $\sqrt[6]{x^{30}}$ **138.** $\sqrt[4]{k^{20}}$

139. Concept Check When approximating $\sqrt{19}$ to three decimal places, a student incorrectly rounded $\sqrt{19} \approx 4.358898944$ to 4.358. **WHAT WENT WRONG?** Give the correct approximation.

140. Concept Check When approximating $\sqrt{40}$ to three decimal places, a student incorrectly rounded $\sqrt{40} \approx 6.32455532$ to 6.324. **WHAT WENT WRONG?** Give the correct approximation.

Use a calculator to approximate each radical to three decimal places. **See Example 10.**

141. $\sqrt{9483}$　　　　**142.** $\sqrt{6825}$　　　　**143.** $\sqrt{284.361}$　　　　**144.** $\sqrt{846.104}$

145. $-\sqrt{82}$　　　　**146.** $-\sqrt{91}$　　　　**147.** $\sqrt[3]{423}$　　　　**148.** $\sqrt[3]{555}$

149. $\sqrt[4]{100}$　　　　**150.** $\sqrt[4]{250}$　　　　**151.** $\sqrt[5]{23.8}$　　　　**152.** $\sqrt[5]{98.4}$

Solve each problem. **See Example 11.**

153. Use the electronics formula

$$f = \frac{1}{2\pi\sqrt{LC}}$$

to calculate the resonant frequency f of a circuit, in cycles per second, to the nearest thousand for the following values of L and C.

(a) $L = 7.237 \times 10^{-5}$ and $C = 2.5 \times 10^{-10}$

(b) $L = 5.582 \times 10^{-4}$ and $C = 3.245 \times 10^{-9}$

154. The threshold weight T for a person is the weight above which the risk of death increases greatly. The threshold weight in pounds for men aged 40–49 is related to height h in inches by the formula

$$h = 12.3\sqrt[3]{T}.$$

What height corresponds to a threshold weight of 216 lb for a 43-year-old man? Round the answer to the nearest inch and then to the nearest tenth of a foot.

155. According to an article in *The World Scanner Report,* the distance D, in miles, to the horizon from an observer's point of view over water or "flat" earth is given by

$$D = \sqrt{2H},$$

where H is the height of the point of view, in feet. If a person whose eyes are 6 ft above ground level is standing at the top of a hill 44 ft above "flat" earth, approximately how far to the horizon will she be able to see?

156. The time for one complete swing of a simple pendulum is given by

$$t = 2\pi\sqrt{\frac{L}{g}},$$

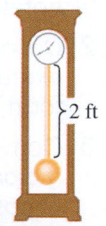

where t is time in seconds, L is the length of the pendulum in feet, and g, the force due to gravity, is about 32 ft per sec^2. Find the time of a complete swing of a 2-ft pendulum to the nearest tenth of a second.

The coefficient of self-induction L (in henrys), the energy P stored in an electronic circuit (in joules), and the current I (in amps) are related by the formula

$$I = \sqrt{\frac{2P}{L}}.$$

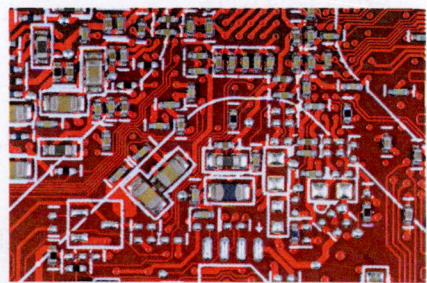

157. Find I if $P = 120$ and $L = 80$.

158. Find I if $P = 100$ and $L = 40$.

Heron's formula gives a method of finding the area of a triangle if the lengths of its sides are known. Suppose that a, b, and c are the lengths of the sides. Let s denote one-half of the perimeter of the triangle (called the **semiperimeter**)—that is,

$$s = \frac{1}{2}(a + b + c).$$

Then the area of the triangle is given by

$$\mathscr{A} = \sqrt{s(s - a)(s - b)(s - c)}.$$

Use Heron's formula to solve each problem.

159. Find the area of the Bermuda Triangle, to the nearest thousand square miles, if the "sides" of this triangle measure approximately 960 mi, 1030 mi, and 1030 mi.

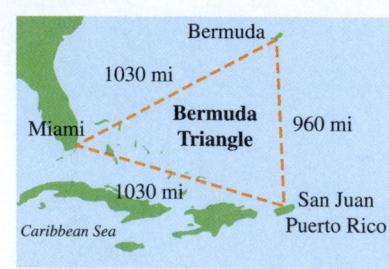

160. The Vietnam Veterans Memorial in Washington, DC, is in the shape of an unenclosed isosceles triangle with equal sides of length 246.75 ft. If the triangle were enclosed, the third side would have length 438.14 ft. Find the area of this enclosure to the nearest hundred square feet. (Data from information pamphlet obtained at the Vietnam Veterans Memorial.)

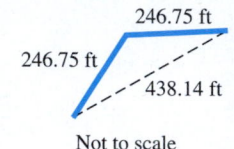

Not to scale

161. Find the area of a triangle with sides of lengths $a = 11$ m, $b = 60$ m, and $c = 61$ m.

162. Find the area of a triangle with sides of lengths $a = 20$ ft, $b = 34$ ft, and $c = 42$ ft.

10.2   Rational Exponents

OBJECTIVES

1 Use exponential notation for *n*th roots.

2 Define and use expressions of the form $a^{m/n}$.

3 Convert between radicals and rational exponents.

4 Use the rules for exponents with rational exponents.

OBJECTIVE 1   Use exponential notation for *n*th roots.

Consider the expression $(3^{1/2})^2$. We can simplify it as follows.

$$(3^{1/2})^2$$

$$= 3^{1/2} \cdot 3^{1/2} \qquad a^2 = a \cdot a$$

$$= 3^{1/2+1/2} \qquad \text{Product rule: } a^m \cdot a^n = a^{m+n}$$

$$= 3^1 \qquad \text{Add exponents.}$$

$$= 3 \qquad a^1 = a$$

Also, by definition,

$$(\sqrt{3})^2 = \sqrt{3} \cdot \sqrt{3} = 3.$$

Because both $(3^{1/2})^2$ and $(\sqrt{3})^2$ are equal to 3, it seems reasonable to define

$$3^{1/2} = \sqrt{3}.$$

This discussion suggests the following generalization.

Meaning of $a^{1/n}$

If $\sqrt[n]{a}$ is a real number, then $\quad a^{1/n} = \sqrt[n]{a}.$

Examples: $\quad 4^{1/2} = \sqrt{4}, \quad 8^{1/3} = \sqrt[3]{8}, \quad 16^{1/4} = \sqrt[4]{16}$

Notice that the denominator of the rational exponent is the index of the radical.

NOW TRY
EXERCISE 1

Evaluate each exponential.

(a) $81^{1/2}$ **(b)** $125^{1/3}$

(c) $-625^{1/4}$ **(d)** $(-625)^{1/4}$

(e) $(-125)^{1/3}$ **(f)** $\left(\dfrac{1}{16}\right)^{1/4}$

EXAMPLE 1 Evaluating Exponentials of the Form $a^{1/n}$

Evaluate each exponential.

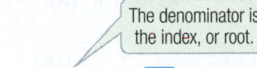 The denominator is the index, or root.

 The denominator is the index, or root. $\sqrt{}$ means $\sqrt[2]{}$.

(a) $64^{1/3} = \sqrt[3]{64} = 4$

(b) $100^{1/2} = \sqrt{100} = 10$

(c) $-256^{1/4} = -\sqrt[4]{256} = -4$

(d) $(-256)^{1/4} = \sqrt[4]{-256}$ is not a real number because the radicand, -256, is negative and the index is even.

(e) $(-32)^{1/5} = \sqrt[5]{-32} = -2$

(f) $\left(\dfrac{1}{8}\right)^{1/3} = \sqrt[3]{\dfrac{1}{8}} = \dfrac{1}{2}$ **NOW TRY**

⚠ **CAUTION** Notice the distinction between **Examples 1(c) and 1(d).** The radical in part (c) is the *negative fourth root of a positive number,* while the radical in part (d) is the *principal fourth root of a negative number, which is not a real number.*

OBJECTIVE 2 Define and use expressions of the form $a^{m/n}$.

We know that $8^{1/3} = \sqrt[3]{8}$. Now we can define a number like $8^{2/3}$, where the numerator of the exponent is not 1. For past rules of exponents to be valid,

$$8^{2/3} = 8^{(1/3)2} = \left(8^{1/3}\right)^2.$$

Because $8^{1/3} = \sqrt[3]{8}$,

$$8^{2/3} = \left(\sqrt[3]{8}\right)^2 = 2^2 = 4.$$

Generalizing from this example, we define $a^{m/n}$ as follows.

Meaning of $a^{m/n}$

If m and n are positive integers with m/n in lowest terms, then

$$a^{m/n} = \left(a^{1/n}\right)^m,$$

provided that $a^{1/n}$ is a real number. If $a^{1/n}$ is not a real number, then $a^{m/n}$ is not a real number.

EXAMPLE 2 Evaluating Exponentials of the Form $a^{m/n}$

Evaluate each exponential.

NOW TRY ANSWERS
1. (a) 9 **(b)** 5 **(c)** -5
  (d) It is not a real number.
  (e) -5 **(f)** $\frac{1}{2}$

Think: $36^{1/2} = \sqrt{36} = 6$

Think: $125^{1/3} = \sqrt[3]{125} = 5$

(a) $36^{3/2} = (36^{1/2})^3 = 6^3 = 216$

(b) $125^{2/3} = (125^{1/3})^2 = 5^2 = 25$

**NOW TRY
EXERCISE 2**
Evaluate each exponential.

(a) $32^{2/5}$ **(b)** $8^{5/3}$

(c) $-100^{3/2}$ **(d)** $(-125)^{4/3}$

(e) $(-121)^{3/2}$

> Be careful.
> The base is 4.

(c) $-4^{5/2} = -(4^{5/2}) = -(4^{1/2})^5 = -(2)^5 = -32$

Because the base is 4, the negative sign is *not* affected by the exponent.

(d) $(-27)^{2/3} = [(-27)^{1/3}]^2 = (-3)^2 = 9$

Notice that in part (c) we first evaluate the exponential and then find its negative. In part (d), the $-$ sign is part of the base, -27.

(e) $(-100)^{3/2} = [(-100)^{1/2}]^3$, which is not a real number, because

$$(-100)^{1/2}, \quad \text{or} \quad \sqrt{-100}, \quad \text{is not a real number.} \qquad \textbf{NOW TRY} \; \circlearrowleft$$

Recall that for any natural number n,

$$a^{-n} = \frac{1}{a^n} \quad (\text{where } a \ne 0). \qquad \text{Definition of negative exponent}$$

When a rational exponent is negative, this earlier interpretation is applied.

> **Meaning of $a^{-m/n}$**
>
> If $a^{m/n}$ is a real number, then
>
> $$a^{-m/n} = \frac{1}{a^{m/n}} \quad (\text{where } a \ne 0).$$

**NOW TRY
EXERCISE 3**
Evaluate each exponential.

(a) $243^{-3/5}$ **(b)** $4^{-5/2}$

(c) $\left(\dfrac{216}{125}\right)^{-2/3}$

EXAMPLE 3   **Evaluating Exponentials of the Form $a^{-m/n}$**

Evaluate each exponential.

(a) $16^{-3/4} = \dfrac{1}{16^{3/4}} = \dfrac{1}{\left(16^{1/4}\right)^3} = \dfrac{1}{\left(\sqrt[4]{16}\right)^3} = \dfrac{1}{2^3} = \dfrac{1}{8}$

> The denominator of 3/4 is the index and the numerator is the exponent.

(b) $25^{-3/2} = \dfrac{1}{25^{3/2}} = \dfrac{1}{\left(25^{1/2}\right)^3} = \dfrac{1}{\left(\sqrt{25}\right)^3} = \dfrac{1}{5^3} = \dfrac{1}{125}$

(c) $\left(\dfrac{8}{27}\right)^{-2/3} = \dfrac{1}{\left(\dfrac{8}{27}\right)^{2/3}} = \dfrac{1}{\left(\sqrt[3]{\dfrac{8}{27}}\right)^2} = \dfrac{1}{\left(\dfrac{2}{3}\right)^2} = \dfrac{1}{\dfrac{4}{9}} = \dfrac{9}{4}$

> $\dfrac{1}{\frac{4}{9}} = 1 \div \dfrac{4}{9} = 1 \cdot \dfrac{9}{4}$

We can also use the rule $\left(\dfrac{b}{a}\right)^{-m} = \left(\dfrac{a}{b}\right)^{m}$ here, as follows.

$$\left(\dfrac{8}{27}\right)^{-2/3} = \left(\dfrac{27}{8}\right)^{2/3} = \left(\sqrt[3]{\dfrac{27}{8}}\right)^2 = \left(\dfrac{3}{2}\right)^2 = \dfrac{9}{4} \qquad \text{The result is the same.}$$

> Take the reciprocal of only the base, *not* the exponent.

\qquad **NOW TRY** $\circlearrowleft$

NOW TRY ANSWERS
2. **(a)** 4 **(b)** 32 **(c)** -1000
(d) 625
(e) It is not a real number.
3. **(a)** $\frac{1}{27}$ **(b)** $\frac{1}{32}$ **(c)** $\frac{25}{36}$

> ⚠ **CAUTION** Be careful to distinguish between exponential expressions like the following.
>
> $16^{-1/4}$, which equals $\frac{1}{2}$, $-16^{1/4}$, which equals -2,   and $-16^{-1/4}$, which equals $-\frac{1}{2}$
>
> **A negative exponent does not necessarily lead to a negative result. Negative exponents lead to reciprocals, which may be positive.**

We obtain an alternative definition of $a^{m/n}$ by applying the power rule for exponents differently than in the earlier definition.

> **Alternative Meaning of $a^{m/n}$**
>
> If all indicated roots are real numbers, then
>
> $$a^{m/n} = (a^{1/n})^m = (a^m)^{1/n}.$$

As a result, we can evaluate an expression such as $27^{2/3}$ in two ways.

$$27^{2/3} = (27^{1/3})^2 = 3^2 = 9$$

The result is the same.

or $\qquad 27^{2/3} = (27^2)^{1/3} = 729^{1/3} = 9$

In most cases, it is easier to use $(a^{1/n})^m$.

> **Radical Form of $a^{m/n}$**
>
> If all indicated roots are real numbers, then
>
> $$a^{m/n} = \sqrt[n]{a^m} = \left(\sqrt[n]{a}\right)^m.$$
>
> That is, raise a to the mth power and then take the nth root, or take the nth root of a and then raise it to the mth power.

For example, $8^{2/3} = \sqrt[3]{8^2} = \sqrt[3]{64} = 4,$ and $8^{2/3} = \left(\sqrt[3]{8}\right)^2 = 2^2 = 4,$

so $\qquad\qquad\qquad\qquad 8^{2/3} = \sqrt[3]{8^2} = \left(\sqrt[3]{8}\right)^2.$

OBJECTIVE 3 Convert between radicals and rational exponents.

Using the definition of rational exponents, we can simplify many problems involving radicals by converting the radicals to numbers with rational exponents. After simplifying, we can convert the answer back to radical form if required.

EXAMPLE 4 **Converting Exponentials to Radicals**

Write each exponential as a radical. Assume that all variables represent positive real numbers. Use the definition that takes the root first.

(a) $13^{1/2} = \sqrt{13}$ **(b)** $6^{3/4} = \left(\sqrt[4]{6}\right)^3$ **(c)** $9m^{5/8} = 9\left(\sqrt[8]{m}\right)^5$

(d) $6x^{2/3} - (4x)^{3/5} = 6\left(\sqrt[3]{x}\right)^2 - \left(\sqrt[5]{4x}\right)^3$

(e) $r^{-2/3} = \dfrac{1}{r^{2/3}} = \dfrac{1}{\left(\sqrt[3]{r}\right)^2}$

(f) $(a^2 + b^2)^{1/2} = \sqrt{a^2 + b^2}$ $\boxed{\sqrt{a^2 + b^2} \neq a + b}$

**🔄 NOW TRY
EXERCISE 4**

Write each exponential as a radical. Assume that all variables represent positive real numbers. Use the definition that takes the root first.

(a) $21^{1/2}$ **(b)** $17^{5/4}$

(c) $4t^{3/5} + (4t)^{2/3}$

(d) $w^{-2/5}$ **(e)** $(a^2 - b^2)^{1/4}$

NOW TRY ANSWERS

4. **(a)** $\sqrt{21}$ **(b)** $\left(\sqrt[4]{17}\right)^5$
   (c) $4\left(\sqrt[5]{t}\right)^3 + \left(\sqrt[3]{4t}\right)^2$
   (d) $\dfrac{1}{\left(\sqrt[5]{w}\right)^2}$ **(e)** $\sqrt[4]{a^2 - b^2}$

NOW TRY

 **NOW TRY
EXERCISE 5**
Write each radical as an exponential and simplify. Assume that all variables represent positive real numbers.

(a) $\sqrt[3]{15}$ **(b)** $\sqrt[4]{4^2}$

(c) $\sqrt[4]{x^4}$

EXAMPLE 5 Simplifying Radicals Using Rational Exponents

Write each radical as an exponential and simplify. Assume that all variables represent positive real numbers.

(a) $\sqrt{10} = 10^{1/2}$

(b) $\sqrt{5^4} = 5^{4/2} = 5^2 = 25$

(c) $\sqrt[4]{3^8} = 3^{8/4} = 3^2 = 9$

(d) $\sqrt[6]{z^6} = z^{6/6} = z^1 = z$, because z is positive.

NOW TRY

> **NOTE** In **Example 5(d),** it is not necessary to use absolute value bars because the directions specifically state that the variable represents a positive real number. The absolute value of the positive real number z is z itself, so the answer is simply z.

OBJECTIVE 4 Use the rules for exponents with rational exponents.

The definition of rational exponents allows us to apply the rules for exponents.

Rules for Rational Exponents

Let r and s be rational numbers. For all real numbers a and b for which the following are defined, these rules hold true.

Rules	Examples	Rules	Examples
$a^r \cdot a^s = a^{r+s}$	$5^{2/3} \cdot 5^{5/3} = 5^{2/3+5/3} = 5^{7/3}$	$\left(\dfrac{a}{b}\right)^r = \dfrac{a^r}{b^r}$	$\left(\dfrac{5}{7}\right)^{2/3} = \dfrac{5^{2/3}}{7^{2/3}}$
$\dfrac{a^r}{a^s} = a^{r-s}$	$\dfrac{4^{5/3}}{4^{4/3}} = 4^{5/3-4/3} = 4^{1/3}$	$a^{-r} = \dfrac{1}{a^r}$	$4^{-3/7} = \dfrac{1}{4^{3/7}}$
$(a^r)^s = a^{rs}$	$(5^{1/2})^{2/5} = 5^{1/2 \cdot 2/5} = 5^{1/5}$	$\dfrac{a^{-r}}{b^{-s}} = \dfrac{b^s}{a^r}$	$\dfrac{5^{-1/2}}{3^{-3/4}} = \dfrac{3^{3/4}}{5^{1/2}}$
$(ab)^r = a^r b^r$	$(3x)^{1/2} = 3^{1/2}x^{1/2}$		

EXAMPLE 6 Applying Rules for Rational Exponents

Simplify each expression. Assume that all variables represent positive real numbers.

(a) $2^{1/2} \cdot 2^{1/4}$

$= 2^{1/2+1/4}$ Product rule

$= 2^{2/4+1/4}$ Write exponents with a common denominator.

$= 2^{3/4}$ Add exponents.

(b) $\dfrac{5^{2/3}}{5^{7/3}}$

$= 5^{2/3-7/3}$ Quotient rule

$= 5^{-5/3}$ Subtract exponents.

$= \dfrac{1}{5^{5/3}}$ $a^{-r} = \dfrac{1}{a^r}$

NOW TRY ANSWERS
5. **(a)** $15^{1/3}$ **(b)** 2 **(c)** x

NOW TRY
EXERCISE 6
Simplify each expression. Assume that all variables represent positive real numbers.

(a) $5^{1/4} \cdot 5^{2/3}$ **(b)** $\dfrac{9^{3/5}}{9^{7/5}}$

(c) $\dfrac{(r^{2/3}t^{1/4})^8}{t}$

(d) $\left(\dfrac{2x^{1/2}y^{-2/3}}{x^{-3/5}y^{-1/5}}\right)^{-3}$

(e) $y^{2/3}(y^{1/3} + y^{5/3})$

(c) $\dfrac{(x^{1/2}y^{2/3})^4}{y}$

$= \dfrac{(x^{1/2})^4(y^{2/3})^4}{y}$ Power rule

$= \dfrac{x^2 y^{8/3}}{y^1}$ Power rule; $y = y^1$

$= x^2 y^{8/3-1}$ Quotient rule

$= x^2 y^{5/3}$ $\dfrac{8}{3} - 1 = \dfrac{8}{3} - \dfrac{3}{3} = \dfrac{5}{3}$

(d) $\left(\dfrac{x^4 y^{-6}}{x^{-2}y^{1/3}}\right)^{-2/3}$

$= \dfrac{(x^4)^{-2/3}(y^{-6})^{-2/3}}{(x^{-2})^{-2/3}(y^{1/3})^{-2/3}}$ Power rule

$= \dfrac{x^{-8/3}y^4}{x^{4/3}y^{-2/9}}$ Power rule

$= x^{-8/3-4/3}y^{4-(-2/9)}$ Quotient rule

$= x^{-4}y^{38/9}$ *Use parentheses to avoid errors.* $4 - \left(-\dfrac{2}{9}\right) = \dfrac{36}{9} + \dfrac{2}{9} = \dfrac{38}{9}$

$= \dfrac{y^{38/9}}{x^4}$ Definition of negative exponent

Alternative method: We can simplify within the parentheses first, as follows.

$\left(\dfrac{x^4 y^{-6}}{x^{-2}y^{1/3}}\right)^{-2/3}$

$= \left(x^{4-(-2)}y^{-6-1/3}\right)^{-2/3}$ Quotient rule

$= \left(x^6 y^{-19/3}\right)^{-2/3}$ $-6 - \dfrac{1}{3} = -\dfrac{18}{3} - \dfrac{1}{3} = -\dfrac{19}{3}$

$= (x^6)^{-2/3}(y^{-19/3})^{-2/3}$ Power rule

$= x^{-4}y^{38/9}$ Power rule

$= \dfrac{y^{38/9}}{x^4}$ Definition of negative exponent; The result is the same.

(e) $m^{3/4}(m^{5/4} - m^{1/4})$

$= m^{3/4}(m^{5/4}) - m^{3/4}(m^{1/4})$ Distributive property

Do not make the common mistake of multiplying exponents in the first step. $= m^{3/4+5/4} - m^{3/4+1/4}$ Product rule

$= m^{8/4} - m^{4/4}$ Add exponents.

$= m^2 - m$ Write the exponents in lowest terms.

NOW TRY

NOW TRY ANSWERS
6. (a) $5^{11/12}$ **(b)** $\dfrac{1}{9^{4/5}}$

(c) $r^{16/3}t$ **(d)** $\dfrac{y^{7/5}}{8x^{33/10}}$

(e) $y + y^{7/3}$

 **CAUTION**   Use the rules of exponents in problems like those in **Example 6.** Do not convert the expressions to radical form.

NOW TRY
EXERCISE 7

Write each radical as an exponential and simplify. Leave answers in exponential form. Assume that all variables represent positive real numbers.

(a) $\sqrt[5]{y^3} \cdot \sqrt[3]{y}$ **(b)** $\dfrac{\sqrt[4]{y^3}}{\sqrt{y^5}}$

(c) $\sqrt{\sqrt[3]{y}}$

EXAMPLE 7   Applying Rules for Rational Exponents

Write each radical as an exponential and simplify. Leave answers in exponential form. Assume that all variables represent positive real numbers.

(a) $\sqrt[3]{x^2} \cdot \sqrt[4]{x}$

$= x^{2/3} \cdot x^{1/4}$	Convert to rational exponents.
$= x^{2/3 + 1/4}$	Product rule
$= x^{8/12 + 3/12}$	Write exponents with a common denominator.
$= x^{11/12}$	Add exponents.

(b) $\dfrac{\sqrt{x^3}}{\sqrt[3]{x^2}}$

$= \dfrac{x^{3/2}}{x^{2/3}}$	Convert to rational exponents.
$= x^{3/2 - 2/3}$	Quotient rule
$= x^{9/6 - 4/6}$	Write exponents with a common denominator.
$= x^{5/6}$	Subtract exponents.

(c) $\sqrt{\sqrt[4]{z}}$

$= \sqrt{z^{1/4}}$	Convert the inside radical to a rational exponent.
$= \left(z^{1/4}\right)^{1/2}$	Convert the square root to a rational exponent.
$= z^{1/8}$	Power rule

NOW TRY

NOW TRY ANSWERS

7. **(a)** $y^{14/15}$ **(b)** $\dfrac{1}{y^{7/4}}$ **(c)** $y^{1/6}$

10.2 Exercises

FOR EXTRA HELP

▶ **MyLab Math**

▶ *Video solutions for select problems available in MyLab Math*

Concept Check   *Match each expression in Column I with the equivalent choice in Column II.*

I

1. $3^{1/2}$ **2.** $(-27)^{1/3}$

3. $-16^{1/2}$ **4.** $(-25)^{1/2}$

5. $(-32)^{1/5}$ **6.** $(-32)^{2/5}$

7. $4^{3/2}$ **8.** $6^{2/4}$

9. $-6^{2/4}$ **10.** $36^{0.5}$

II

A. -4 **B.** 8

C. $\sqrt{3}$ **D.** $-\sqrt{6}$

E. -3 **F.** $\sqrt{6}$

G. 4 **H.** -2

I. 6 **J.** Not a real number

11. Concept Check   A student incorrectly evaluated $27^{1/3}$ as 9.   **WHAT WENT WRONG?**   Evaluate $27^{1/3}$ correctly.

12. Concept Check   A student evaluating $16^{-3/2}$ incorrectly suggested that the result is a negative number because the exponent is negative.   **WHAT WENT WRONG?**   Evaluate $16^{-3/2}$ correctly.

*Evaluate each exponential. **See Examples 1–3.***

13. $169^{1/2}$ **14.** $121^{1/2}$ **15.** $729^{1/3}$ **16.** $512^{1/3}$

17. $16^{1/4}$ **18.** $625^{1/4}$ **19.** $\left(\dfrac{64}{81}\right)^{1/2}$ **20.** $\left(\dfrac{8}{27}\right)^{1/3}$

21. $(-27)^{1/3}$ **22.** $(-32)^{1/5}$ **23.** $(-144)^{1/2}$ **24.** $(-36)^{1/2}$

25. $100^{3/2}$ **26.** $64^{3/2}$ **27.** $81^{3/4}$ **28.** $216^{2/3}$

29. $-16^{5/2}$ **30.** $-32^{3/5}$ **31.** $(-8)^{4/3}$ **32.** $(-243)^{2/5}$

33. $32^{-3/5}$ **34.** $27^{-4/3}$ **35.** $64^{-3/2}$ **36.** $81^{-3/2}$

37. $\left(\dfrac{125}{27}\right)^{-2/3}$ **38.** $\left(\dfrac{64}{125}\right)^{-2/3}$ **39.** $\left(\dfrac{16}{81}\right)^{-3/4}$ **40.** $\left(\dfrac{729}{64}\right)^{-5/6}$

Write each exponential as a radical. Assume that all variables represent positive real numbers. Use the definition that takes the root first. **See Example 4.**

41. $10^{1/2}$ **42.** $3^{1/2}$ **43.** $8^{3/4}$

44. $7^{2/3}$ **45.** $5x^{2/3}$ **46.** $8x^{3/4}$

47. $9q^{5/8} - (2x)^{2/3}$ **48.** $(3p)^{3/4} + 4x^{1/3}$ **49.** $x^{-3/5}$

50. $z^{-4/9}$ **51.** $(2y + x)^{2/3}$ **52.** $(r + 2z)^{3/2}$

Write each radical as an exponential and simplify. Assume that all variables represent positive real numbers. **See Example 5.**

53. $\sqrt{15}$ **54.** $\sqrt{26}$ **55.** $\sqrt{2^{12}}$ **56.** $\sqrt{5^{10}}$ **57.** $\sqrt[3]{4^9}$

58. $\sqrt[4]{6^8}$ **59.** $\sqrt[8]{x^8}$ **60.** $\sqrt{t^2}$ **61.** $\sqrt{x^{20}}$ **62.** $\sqrt{r^{50}}$

Simplify each expression. Assume that all variables represent positive real numbers. **See Example 6.**

63. $3^{1/2} \cdot 3^{3/2}$ **64.** $6^{4/3} \cdot 6^{2/3}$ **65.** $\dfrac{64^{5/3}}{64^{4/3}}$

66. $\dfrac{125^{7/3}}{125^{5/3}}$ **67.** $y^{7/3} \cdot y^{-4/3}$ **68.** $r^{-8/9} \cdot r^{17/9}$

69. $x^{2/3} \cdot x^{-1/4}$ **70.** $x^{2/5} \cdot x^{-1/3}$ **71.** $\dfrac{k^{1/3}}{k^{2/3} \cdot k^{-1}}$

72. $\dfrac{z^{3/4}}{z^{5/4} \cdot z^{-2}}$ **73.** $\dfrac{\left(x^{1/4}y^{2/5}\right)^{20}}{x^2}$ **74.** $\dfrac{\left(r^{1/5}s^{2/3}\right)^{15}}{r^2}$

75. $\dfrac{\left(x^{2/3}\right)^2}{\left(x^2\right)^{7/3}}$ **76.** $\dfrac{\left(p^3\right)^{1/4}}{\left(p^{5/4}\right)^2}$ **77.** $\dfrac{m^{3/4}n^{-1/4}}{\left(m^2n\right)^{1/2}}$

78. $\dfrac{a^{-1/2}b^{-5/4}}{\left(a^{-3}b^2\right)^{1/6}}$ **79.** $\left(\dfrac{x^8y^{-8}}{x^{-4}y^{1/2}}\right)^{-2/3}$ **80.** $\left(\dfrac{x^{2/3}y^{-6}}{x^{-1/12}y^{1/4}}\right)^{-4/5}$

81. $\left(\dfrac{b^{-3/2}}{c^{-5/3}}\right)^2 \left(b^{-1/4}c^{-1/3}\right)^{-1}$ **82.** $\left(\dfrac{m^{-2/3}}{a^{-3/4}}\right)^4 \left(m^{-3/8}a^{1/4}\right)^{-2}$ **83.** $\left(\dfrac{p^{-1/4}q^{-3/2}}{3^{-1}p^{-2}q^{-2/3}}\right)^{-2}$

84. $\left(\dfrac{2^{-2}w^{-3/4}x^{-5/8}}{w^{3/4}x^{-1/2}}\right)^{-3}$ **85.** $p^{2/3}\left(p^{1/3} + 2p^{4/3}\right)$ **86.** $z^{5/8}\left(3z^{5/8} + 5z^{11/8}\right)$

87. $k^{1/4}\left(k^{3/2} - k^{1/2}\right)$ **88.** $r^{3/5}\left(r^{1/2} + r^{3/4}\right)$ **89.** $6a^{7/4}\left(a^{-7/4} + 3a^{-3/4}\right)$

90. $4m^{5/3}\left(m^{-2/3} - 4m^{-5/3}\right)$ **91.** $-5x^{7/6}\left(x^{5/6} - x^{-1/6}\right)$ **92.** $-8y^{11/7}\left(y^{3/7} - y^{-4/7}\right)$

Write each radical as an exponential and simplify. Leave answers in exponential form. Assume that all variables represent positive numbers. **See Example 7.**

93. $\sqrt[5]{x^3} \cdot \sqrt[4]{x}$ **94.** $\sqrt[6]{y^5} \cdot \sqrt[3]{y^2}$ **95.** $\dfrac{\sqrt[3]{t^4}}{\sqrt[5]{t^4}}$ **96.** $\dfrac{\sqrt[4]{w^3}}{\sqrt[6]{w}}$

97. $\dfrac{\sqrt{x^5}}{\sqrt{x^8}}$ **98.** $\dfrac{\sqrt[3]{k^5}}{\sqrt[3]{k^7}}$ **99.** $\sqrt{y} \cdot \sqrt[3]{yz}$ **100.** $\sqrt[3]{xz} \cdot \sqrt{z}$

101. $\sqrt[4]{\sqrt[3]{m}}$ **102.** $\sqrt[3]{\sqrt{k}}$ **103.** $\sqrt{\sqrt[3]{\sqrt[4]{x}}}$ **104.** $\sqrt[3]{\sqrt[5]{\sqrt{y}}}$

Concept Check *Work each problem.*

105. Replace a with 3 and b with 4 to show that, in general,

$$\sqrt{a^2 + b^2} \neq a + b.$$

106. Suppose someone claims that $\sqrt[n]{a^n + b^n}$ must equal $a + b$, because when $a = 1$ and $b = 0$, a true statement results:

$$\sqrt[n]{a^n + b^n} = \sqrt[n]{1^n + 0^n} = \sqrt[n]{1^n} = 1 = 1 + 0 = a + b.$$

Explain why this is faulty reasoning.

Solve each problem.

107. Meteorologists can determine the duration of a storm using the function

$$T(d) = 0.07d^{3/2},$$

where d is the diameter of the storm in miles and T is the time in hours. Find the duration of a storm with a diameter of 16 mi. Round the answer to the nearest tenth of an hour.

108. The threshold weight t, in pounds, for a person is the weight above which the risk of death increases greatly. The threshold weight in pounds for men aged 40–49 is related to height H in inches by the function

$$H(t) = (1860.867t)^{1/3}.$$

What height corresponds to a threshold weight of 200 lb for a 46-yr-old man? Round the answer to the nearest inch and then to the nearest tenth of a foot.

*The **windchill factor** is a measure of the cooling effect that the wind has on a person's skin. It calculates the equivalent cooling temperature if there were no wind. The National Weather Service uses the formula*

Windchill temperature $= 35.74 + 0.6215T - 35.75V^{4/25} + 0.4275TV^{4/25}$,

where T is the temperature in °F and V is the wind speed in miles per hour, to calculate windchill. The table gives the windchill factor for various wind speeds and temperatures at which frostbite is a risk, and how quickly it may occur.

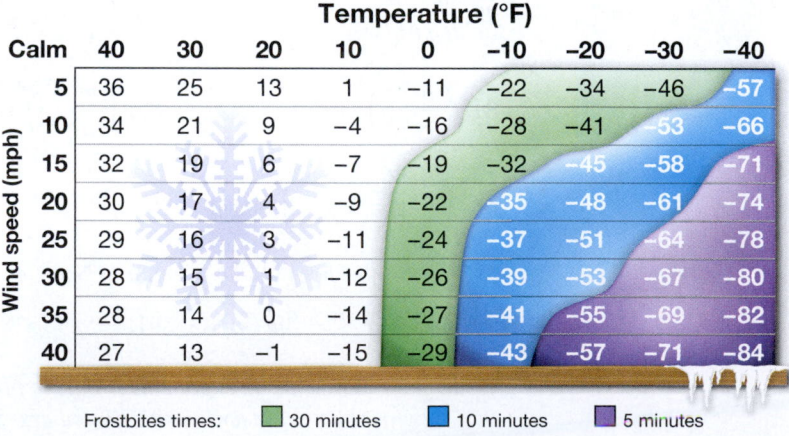

Temperature (°F)

Calm	40	30	20	10	0	−10	−20	−30	−40
5	36	25	13	1	−11	−22	−34	−46	−57
10	34	21	9	−4	−16	−28	−41	−53	−66
15	32	19	6	−7	−19	−32	−45	−58	−71
20	30	17	4	−9	−22	−35	−48	−61	−74
25	29	16	3	−11	−24	−37	−51	−64	−78
30	28	15	1	−12	−26	−39	−53	−67	−80
35	28	14	0	−14	−27	−41	−55	−69	−82
40	27	13	−1	−15	−29	−43	−57	−71	−84

(Wind speed (mph))

Frostbites times: □ 30 minutes □ 10 minutes □ 5 minutes

Data from National Oceanic and Atmospheric Administration, National Weather Service.

Use the formula to determine the windchill temperature to the nearest tenth of a degree, given the following conditions. Compare answers with the appropriate entries in the table.

109. 30°F, 15-mph wind **110.** 10°F, 30-mph wind

111. 20°F, 20-mph wind **112.** 40°F, 10-mph wind

10.3 Simplifying Radicals, the Distance Formula, and Circles

OBJECTIVES

1 Use the product rule for radicals.

2 Use the quotient rule for radicals.

3 Simplify radicals.

4 Simplify products and quotients of radicals with different indexes.

5 Use the Pythagorean theorem.

6 Use the distance formula.

7 Find an equation of a circle given its center and radius.

OBJECTIVE 1 Use the product rule for radicals.

Consider the expressions $\sqrt{36 \cdot 4}$ and $\sqrt{36} \cdot \sqrt{4}$.

$$\sqrt{36 \cdot 4} = \sqrt{144} = 12$$

The result is the same.

$$\sqrt{36} \cdot \sqrt{4} = 6 \cdot 2 = 12$$

This is an example of the **product rule for radicals.**

Product Rule for Radicals

If n is a natural number and $\sqrt[n]{a}$ and $\sqrt[n]{b}$ are real numbers, then the following holds true.

$$\sqrt[n]{a} \cdot \sqrt[n]{b} = \sqrt[n]{ab}$$

That is, the product of two nth roots is the nth root of the product.

We justify the product rule using the rules for rational exponents. Because $\sqrt[n]{a} = a^{1/n}$ and $\sqrt[n]{b} = b^{1/n}$,

$$\sqrt[n]{a} \cdot \sqrt[n]{b} = a^{1/n} \cdot b^{1/n} = (ab)^{1/n} = \sqrt[n]{ab}.$$

> **! CAUTION** *Use the product rule only when the radicals have the same index.*

 **NOW TRY EXERCISE 1**

Multiply. Assume that all variables represent positive real numbers.

(a) $\sqrt{7} \cdot \sqrt{11}$

(b) $\sqrt{2mn} \cdot \sqrt{15}$

EXAMPLE 1 Using the Product Rule

Multiply. Assume that all variables represent positive real numbers.

(a) $\sqrt{5} \cdot \sqrt{7}$

$= \sqrt{5 \cdot 7}$

$= \sqrt{35}$

(b) $\sqrt{11} \cdot \sqrt{p}$

$= \sqrt{11p}$

(c) $\sqrt{7} \cdot \sqrt{11xyz}$

$= \sqrt{77xyz}$

NOW TRY

NOW TRY EXERCISE 2

Multiply. Assume that all variables represent positive real numbers.

(a) $\sqrt[3]{4} \cdot \sqrt[3]{5}$ **(b)** $\sqrt[4]{5t} \cdot \sqrt[4]{6r^3}$

(c) $\sqrt[3]{20x} \cdot \sqrt[3]{3xy^3}$

(d) $\sqrt[3]{5} \cdot \sqrt[4]{9}$

EXAMPLE 2 Using the Product Rule

Multiply. Assume that all variables represent positive real numbers.

(a) $\sqrt[3]{3} \cdot \sqrt[3]{12}$

$= \sqrt[3]{3 \cdot 12}$

$= \sqrt[3]{36}$ ← Remember to write the index.

(b) $\sqrt[4]{8y} \cdot \sqrt[4]{3r^2}$

$= \sqrt[4]{24yr^2}$

(c) $\sqrt[6]{10m^4} \cdot \sqrt[6]{5m}$

$= \sqrt[6]{50m^5}$

NOW TRY ANSWERS

1. (a) $\sqrt{77}$ **(b)** $\sqrt{30mn}$

2. (a) $\sqrt[3]{20}$ **(b)** $\sqrt[4]{30tr^3}$

(c) $\sqrt[3]{60x^2y^3}$

(d) This expression cannot be simplified using the product rule.

(d) $\sqrt[4]{2} \cdot \sqrt[5]{2}$ This product cannot be simplified using the product rule for radicals because the indexes (4 and 5) are different.

NOW TRY

OBJECTIVE 2 Use the quotient rule for radicals.

The **quotient rule for radicals** is similar to the product rule.

> ### Quotient Rule for Radicals
>
> If n is a natural number and $\sqrt[n]{a}$ and $\sqrt[n]{b}$ are real numbers, then the following holds true.
>
> $$\sqrt[n]{\frac{a}{b}} = \frac{\sqrt[n]{a}}{\sqrt[n]{b}} \quad (\text{where } b \neq 0)$$
>
> That is, the nth root of a quotient is the quotient of the nth roots.

**NOW TRY
EXERCISE 3**

Simplify. Assume that all variables represent positive real numbers.

(a) $\sqrt{\dfrac{49}{36}}$ (b) $\sqrt{\dfrac{5}{144}}$

(c) $\sqrt[3]{-\dfrac{27}{1000}}$ (d) $\sqrt[4]{\dfrac{t}{16}}$

(e) $-\sqrt[5]{\dfrac{m^{15}}{243}}$

EXAMPLE 3 Using the Quotient Rule

Simplify. Assume that all variables represent positive real numbers.

(a) $\sqrt{\dfrac{16}{25}} = \dfrac{\sqrt{16}}{\sqrt{25}} = \dfrac{4}{5}$

(b) $\sqrt{\dfrac{7}{36}} = \dfrac{\sqrt{7}}{\sqrt{36}} = \dfrac{\sqrt{7}}{6}$

(c) $\sqrt[3]{-\dfrac{8}{125}} = \sqrt[3]{\dfrac{-8}{125}} = \dfrac{\sqrt[3]{-8}}{\sqrt[3]{125}} = \dfrac{-2}{5} = -\dfrac{2}{5}$ $\dfrac{-a}{b} = -\dfrac{a}{b}$

(d) $\sqrt[3]{\dfrac{7}{216}} = \dfrac{\sqrt[3]{7}}{\sqrt[3]{216}} = \dfrac{\sqrt[3]{7}}{6}$

(e) $\sqrt[5]{\dfrac{x}{32}} = \dfrac{\sqrt[5]{x}}{\sqrt[5]{32}} = \dfrac{\sqrt[5]{x}}{2}$

(f) $-\sqrt[3]{\dfrac{m^6}{125}} = -\dfrac{\sqrt[3]{m^6}}{\sqrt[3]{125}} = -\dfrac{m^2}{5}$ Think: $\sqrt[3]{m^6} = m^{6/3} = m^2$

NOW TRY

OBJECTIVE 3 Simplify radicals.

We use the product and quotient rules to simplify radicals. A radical is **simplified** if the following four conditions are met.

> ### Conditions for a Simplified Radical
>
> 1. The radicand has no factor raised to a power greater than or equal to the index.
> 2. The radicand has no fractions.
> 3. No denominator contains a radical.
> 4. Exponents in the radicand and the index of the radical have greatest common factor 1.
>
> *Examples:* $\sqrt{22}, \quad \sqrt{15xy}, \quad \sqrt[3]{18}, \quad \dfrac{\sqrt[4]{m^3}}{m}$ These radicals are simplified.
>
> $\sqrt{28}, \quad \sqrt[3]{\dfrac{3}{5}}, \quad \dfrac{7}{\sqrt{7}}, \quad \sqrt[3]{r^{12}}$ These radicals are not simplified. Each violates one of the above conditions.

NOW TRY ANSWERS

3. (a) $\dfrac{7}{6}$ (b) $\dfrac{\sqrt{5}}{12}$ (c) $-\dfrac{3}{10}$

    (d) $\dfrac{\sqrt[4]{t}}{2}$ (e) $-\dfrac{m^3}{3}$

 **NOW TRY EXERCISE 4**

Simplify.

(a) $\sqrt{50}$ (b) $\sqrt{192}$

(c) $\sqrt{42}$ (d) $\sqrt[3]{108}$

(e) $-\sqrt[4]{80}$

EXAMPLE 4 Simplifying Roots of Numbers

Simplify.

(a) $\sqrt{24}$

Check to see whether 24 is divisible by a perfect square (the square of a natural number) such as 4, 9, 16, ... The greatest perfect square that divides into 24 is 4.

$$\sqrt{24}$$
$$= \sqrt{4 \cdot 6} \qquad \text{Factor. 4 is a perfect square.}$$
$$= \sqrt{4} \cdot \sqrt{6} \qquad \text{Product rule}$$
$$= 2\sqrt{6} \qquad \sqrt{4} = 2$$

(b) $\sqrt{108}$

As shown on the left, the number 108 is divisible by the perfect square 36. If this perfect square is not immediately clear, try factoring 108 into its prime factors, as shown on the right.

$$\sqrt{108}$$
$$= \sqrt{36 \cdot 3} \qquad \text{Factor.}$$
$$= \sqrt{36} \cdot \sqrt{3} \qquad \text{Product rule}$$
$$= 6\sqrt{3} \qquad \sqrt{36} = 6$$

$$\sqrt{108}$$
$$= \sqrt{2^2 \cdot 3^3} \qquad \text{Factor into prime factors.}$$
$$= \sqrt{2^2 \cdot 3^2 \cdot 3} \qquad a^3 = a^2 \cdot a$$
$$= \sqrt{2^2} \cdot \sqrt{3^2} \cdot \sqrt{3} \qquad \text{Product rule}$$
$$= 2 \cdot 3 \cdot \sqrt{3} \qquad \sqrt{2^2} = 2, \sqrt{3^2} = 3$$
$$= 6\sqrt{3} \qquad \text{Multiply.}$$

(c) $\sqrt{10}$ No perfect square (other than 1) divides into 10, so $\sqrt{10}$ cannot be simplified further.

(d) $\sqrt[3]{16}$

$$\sqrt[3]{16} \qquad \text{Look for the greatest perfect } \textit{cube} \text{ that divides into 16.}$$

Remember to write the index.

$$= \sqrt[3]{8 \cdot 2} \qquad \text{Factor. 8 is a perfect cube.}$$
$$= \sqrt[3]{8} \cdot \sqrt[3]{2} \qquad \text{Product rule}$$
$$= 2\sqrt[3]{2} \qquad \sqrt[3]{8} = 2$$

(e)

$$-\sqrt[4]{162}$$

Remember the negative sign in each line.

$$= -\sqrt[4]{81 \cdot 2} \qquad \text{81 is a perfect fourth power.}$$
$$= -\sqrt[4]{81} \cdot \sqrt[4]{2} \qquad \text{Product rule}$$
$$= -3\sqrt[4]{2} \qquad \sqrt[4]{81} = 3$$

NOW TRY

NOW TRY ANSWERS

4. (a) $5\sqrt{2}$ (b) $8\sqrt{3}$

(c) $\sqrt{42}$ cannot be simplified further.

(d) $3\sqrt[3]{4}$ (e) $-2\sqrt[4]{5}$

⚠ **CAUTION** *Be careful with which factors belong outside the radical symbol and which belong inside.* In **Example 4(b)**, the $2 \cdot 3$ is written **outside** because $\sqrt{2^2} = 2$ and $\sqrt{3^2} = 3$. The remaining 3 is left inside the radical.

NOW TRY
EXERCISE 5

Simplify. Assume that all variables represent positive real numbers.

(a) $\sqrt{36x^5}$ **(b)** $\sqrt{32m^5n^4}$

(c) $\sqrt[3]{-125k^3p^7}$

(d) $-\sqrt[4]{162x^7y^8}$

EXAMPLE 5   Simplifying Radicals Involving Variables

Simplify. Assume that all variables represent positive real numbers.

(a) $\sqrt{16m^3}$

$\qquad = \sqrt{16m^2 \cdot m}$ \qquad\qquad *Factor.*

$\qquad = \sqrt{16m^2} \cdot \sqrt{m}$ \qquad *Product rule*

$\qquad = 4m\sqrt{m}$ \qquad\qquad *Take the square root.*

Absolute value bars are not needed around the m in color because all the variables represent *positive* real numbers.

(b) $\sqrt{200k^7q^8}$

$\qquad = \sqrt{10^2 \cdot 2 \cdot (k^3)^2 \cdot k \cdot (q^4)^2}$ \qquad *Factor into perfect squares.*

$\qquad = 10k^3q^4\sqrt{2k}$ \qquad\qquad\qquad *Take the square root.*

(c) $\sqrt[3]{-8x^4y^5}$

$\qquad = \sqrt[3]{(-8x^3y^3)(xy^2)}$ \qquad *Choose $-8x^3y^3$ as the perfect cube that divides into $-8x^4y^5$.*

$\qquad = \sqrt[3]{-8x^3y^3} \cdot \sqrt[3]{xy^2}$ \qquad *Product rule*

$\qquad = -2xy\sqrt[3]{xy^2}$ \qquad\qquad *Take the cube root.*

(d) $-\sqrt[4]{32y^9}$

$\qquad = -\sqrt[4]{(16y^8)(2y)}$ \qquad *$16y^8$ is the greatest fourth power that divides into $32y^9$.*

$\qquad = -\sqrt[4]{16y^8} \cdot \sqrt[4]{2y}$ \qquad *Product rule*

$\qquad = -2y^2\sqrt[4]{2y}$ \qquad\qquad *Take the fourth root.* \qquad **NOW TRY**

> **NOTE**   From **Example 5,** we see that if a variable is raised to a power with an exponent divisible by 2, it is a perfect square. If it is raised to a power with an exponent divisible by 3, it is a perfect cube. *In general, if it is raised to a power with an exponent divisible by n, it is a perfect nth power.*

NOW TRY
EXERCISE 6

Simplify. Assume that all variables represent positive real numbers.

(a) $\sqrt[6]{7^2}$ **(b)** $\sqrt[6]{y^4}$

(c) $\sqrt[4]{x^{10}}$

EXAMPLE 6   Simplifying Radicals Using Lesser Indexes

Simplify. Assume that all variables represent positive real numbers.

(a) $\sqrt[9]{5^6}$ *Exponents in the radicand and the index must have GCF 1.*

We write this radical using rational exponents and then write the exponent in lowest terms. We then express the answer as a radical.

$$\sqrt[9]{5^6} = (5^6)^{1/9} = 5^{6/9} = 5^{2/3} = \sqrt[3]{5^2} = \sqrt[3]{25}$$

(b) $\sqrt[4]{p^2} = (p^2)^{1/4} = p^{2/4} = p^{1/2} = \sqrt{p}$ (Recall the assumption that $p > 0$.)

(c) $\sqrt[4]{x^{18}} = (x^{18})^{1/4} = x^{18/4} = x^{9/2} = \sqrt{x^9} = \sqrt{x^8 \cdot x} = \sqrt{x^8} \cdot \sqrt{x} = x^4\sqrt{x}$

\qquad\qquad\qquad\qquad\qquad\qquad\qquad\qquad\qquad\qquad\qquad\qquad\qquad\qquad\qquad\qquad\qquad\qquad\qquad **NOW TRY**

NOW TRY ANSWERS
5. (a) $6x^2\sqrt{x}$ **(b)** $4m^2n^2\sqrt{2m}$
  (c) $-5kp^2\sqrt[3]{p}$ **(d)** $-3xy^2\sqrt[4]{2x^3}$
6. (a) $\sqrt[3]{7}$ **(b)** $\sqrt[3]{y^2}$ **(c)** $x^2\sqrt{x}$

These examples suggest the following rule.

> ## Meaning of $\sqrt[kn]{a^{km}}$
>
> If m is an integer, n and k are natural numbers, and all indicated roots exist, then the following holds true.
>
> $$\sqrt[kn]{a^{km}} = \sqrt[n]{a^m}$$

OBJECTIVE 4 Simplify products and quotients of radicals with different indexes.

We multiply and divide radicals with different indexes using rational exponents.

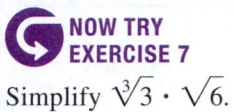

NOW TRY EXERCISE 7

Simplify $\sqrt[3]{3} \cdot \sqrt{6}$.

EXAMPLE 7 Multiplying Radicals with Different Indexes

Simplify $\sqrt{7} \cdot \sqrt[3]{2}$.

The indexes, 2 and 3, have a least common multiple of 6, so we use rational exponents to write each radical as a *sixth* root.

$$\sqrt{7} = 7^{1/2} = 7^{3/6} = \sqrt[6]{7^3} = \sqrt[6]{343}$$

$$\sqrt[3]{2} = 2^{1/3} = 2^{2/6} = \sqrt[6]{2^2} = \sqrt[6]{4}$$

Now we can multiply.

$$\sqrt{7} \cdot \sqrt[3]{2}$$

$$= \sqrt[6]{343} \cdot \sqrt[6]{4} \qquad \text{Substitute; } \sqrt{7} = \sqrt[6]{343},\ \sqrt[3]{2} = \sqrt[6]{4}$$

$$= \sqrt[6]{1372} \qquad \text{Product rule} \qquad \text{NOW TRY}$$

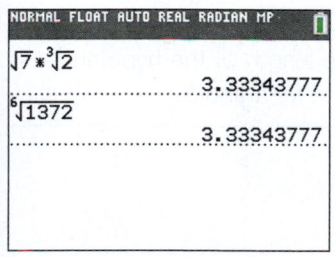

FIGURE 6

Results such as the one in **Example 7** can be supported with a calculator, as shown in **FIGURE 6**. Notice that the calculator gives the same *approximation* for the initial product and the final radical that we obtained.

⚠ **CAUTION**   The computation in **FIGURE 6** is not *proof* that the two expressions are equal. The algebra in **Example 7**, however, is valid proof of their equality.

OBJECTIVE 5 Use the Pythagorean theorem.

The **Pythagorean theorem** provides an equation that relates the lengths of the three sides of a right triangle.

> ## Pythagorean Theorem
>
> If a and b are the lengths of the shorter sides of a right triangle and c is the length of the longest side, then the following holds true.
>
> Hypotenuse c
> Leg a
> ⌐ denotes a 90° or right angle.
> Leg b
>
> $$a^2 + b^2 = c^2$$
>
> The two shorter sides are the **legs** of the triangle, and the longest side is the **hypotenuse.** The hypotenuse is the side opposite the right angle.
>
> $$\textbf{leg}^2 + \textbf{leg}^2 = \textbf{hypotenuse}^2$$

Later we will see that an equation such as $x^2 = 7$ has two solutions: $\sqrt{7}$ (the principal, or positive, square root of 7) and $-\sqrt{7}$. Similarly, $c^2 = 15$ has two solutions: $\pm\sqrt{15}$. In applications we often choose only the principal (positive) square root.

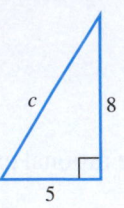

NOW TRY
EXERCISE 8
Find the length of the unknown side in each right triangle.

(a)

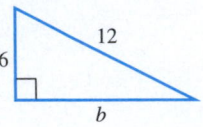

(b)

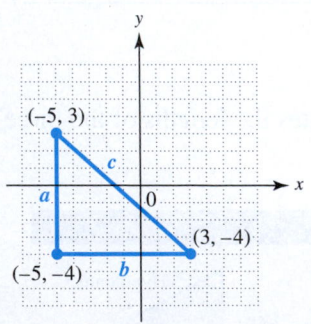

EXAMPLE 8   Using the Pythagorean Theorem

Find the length of the unknown side in the right triangle in **FIGURE 7**.

Substitute carefully. $\quad a^2 + b^2 = c^2$	Pythagorean theorem
$4^2 + 6^2 = c^2$	Let $a = 4$ and $b = 6$.
$16 + 36 = c^2$	Apply the exponents.
$c^2 = 52$	Add. Interchange sides.
$c = \sqrt{52}$	Choose the principal root.
$c = \sqrt{4 \cdot 13}$	Factor.
$c = \sqrt{4} \cdot \sqrt{13}$	Product rule
$c = 2\sqrt{13}$	Simplify.

FIGURE 7

The length of the hypotenuse is $2\sqrt{13}$.

NOW TRY

! CAUTION   In the equation $a^2 + b^2 = c^2$, be sure that the length of the hypotenuse is substituted for c and that the lengths of the legs are substituted for a and b.

OBJECTIVE 6   Use the distance formula.

The *distance formula* enables us to find the distance between two points in the coordinate plane, or the length of the line segment joining those two points.

FIGURE 8 shows the points $(3, -4)$ and $(-5, 3)$. The vertical line through $(-5, 3)$ and the horizontal line through $(3, -4)$ intersect at the point $(-5, -4)$. Thus, the point $(-5, -4)$ becomes the vertex of the right angle in a right triangle.

By the Pythagorean theorem, the sum of the squares of the lengths of the two legs a and b of the right triangle in **FIGURE 8** is equal to the square of the length of the hypotenuse, c.

$$a^2 + b^2 = c^2, \quad \text{or} \quad c^2 = a^2 + b^2.$$

The length a is the difference of the y-coordinates of the endpoints. The x-coordinate of both points in **FIGURE 8** is -5, so the side is vertical, and we can find a by finding the difference of the y-coordinates. We subtract -4 from 3 to obtain a positive value for a.

$$a = 3 - (-4) = 7$$

Similarly, we find b by subtracting -5 from 3.

$$b = 3 - (-5) = 8$$

FIGURE 8

NOW TRY ANSWERS
8. (a) $\sqrt{89}$ **(b)** $6\sqrt{3}$

Now substitute these values into the equation.

$$c^2 = a^2 + b^2 \qquad \text{Pythagorean theorem}$$

$$c^2 = 7^2 + 8^2 \qquad \text{Let } a = 7 \text{ and } b = 8.$$

$$c^2 = 49 + 64 \qquad \text{Apply the exponents.}$$

$$c^2 = 113 \qquad \text{Add.}$$

$$c = \sqrt{113} \qquad \text{Choose the principal root.}$$

We choose the principal root because the distance cannot be negative. Therefore, the distance between $(-5, 3)$ and $(3, -4)$ is $\sqrt{113}$.

NOTE It is customary to leave the distance in simplified radical form. Do not use a calculator to find an approximation unless specifically directed to do so.

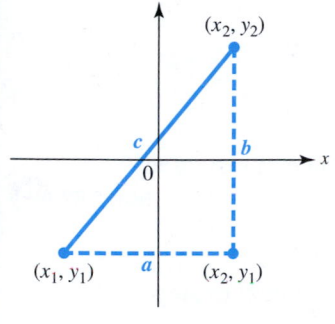

FIGURE 9

This work can be generalized. **FIGURE 9** shows the two points (x_1, y_1) and (x_2, y_2). The distance a between (x_1, y_1) and (x_2, y_1) is given by

$$a = |x_2 - x_1|,$$

and the distance b between (x_2, y_2) and (x_2, y_1) is given by

$$b = |y_2 - y_1|.$$

From the Pythagorean theorem, we obtain the following.

$$c^2 = a^2 + b^2$$

$$c^2 = (x_2 - x_1)^2 + (y_2 - y_1)^2 \qquad \begin{array}{l}\text{For all real numbers } a, \\ |a|^2 = a^2.\end{array}$$

Choosing the principal square root gives the **distance formula.** In this formula, we use d (to denote distance) rather than c.

Distance Formula

The distance d between the points (x_1, y_1) and (x_2, y_2) is given by the following.

$$d = \sqrt{(x_2 - x_1)^2 + (y_2 - y_1)^2}$$

NOW TRY
EXERCISE 9
Find the distance between the points $(-4, -3)$ and $(-8, 6)$.

EXAMPLE 9 Using the Distance Formula

Find the distance between the points $(-3, 5)$ and $(6, 4)$.

We arbitrarily choose to let $(x_1, y_1) = (-3, 5)$ and $(x_2, y_2) = (6, 4)$.

$$d = \sqrt{(x_2 - x_1)^2 + (y_2 - y_1)^2} \qquad \text{Distance formula}$$

$$d = \sqrt{[6 - (-3)]^2 + (4 - 5)^2} \qquad \text{Let } x_2 = 6, y_2 = 4, x_1 = -3, y_1 = 5.$$

Substitute carefully.

$$d = \sqrt{9^2 + (-1)^2}$$

$$d = \sqrt{82} \qquad \begin{array}{l}\text{Leave in radical form.} \\ \text{The distance is } \sqrt{82}.\end{array} \qquad \text{NOW TRY} \quad$$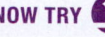

NOW TRY ANSWER
9. $\sqrt{97}$

OBJECTIVE 7 Find an equation of a circle given its center and radius.

A **circle** is the set of all points in a plane that lie a fixed distance from a fixed point. The fixed point is the **center,** and the fixed distance is the **radius.** We use the distance formula to find an equation of a circle.

**NOW TRY
EXERCISE 10**
Find an equation of the circle
with center $(0, 0)$ and
radius 6, and graph it.

EXAMPLE 10 Finding an Equation of a Circle and Graphing It

Find an equation of the circle with center $(0, 0)$ and radius 3, and graph it.

If the point (x, y) is on the circle, then the distance from (x, y) to the center $(0, 0)$ is the radius 3.

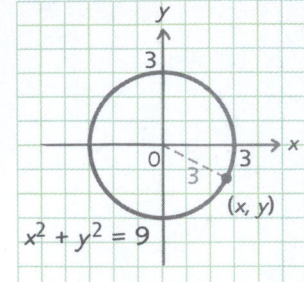

$$\sqrt{(x_2 - x_1)^2 + (y_2 - y_1)^2} = d \qquad \text{Distance formula}$$

$$\sqrt{(x - 0)^2 + (y - 0)^2} = 3 \qquad \text{Let } x_1 = 0, y_1 = 0,$$
$$\text{and } d = 3.$$

$$\sqrt{x^2 + y^2} = 3 \qquad \text{Simplify.}$$

$$\left(\sqrt{x^2 + y^2}\right)^2 = 3^2 \qquad \text{Square each side.}$$

$$x^2 + y^2 = 9 \qquad (\sqrt{a})^2 = a; 3^2 = 9$$

FIGURE 10

An equation of this circle is $x^2 + y^2 = 9$. The graph is shown in **FIGURE 10.**

NOW TRY

A circle may not be centered at the origin, as seen in the next example.

**NOW TRY
EXERCISE 11**
Find an equation of the circle
with center $(-2, 2)$ and
radius 3, and graph it.

EXAMPLE 11 Finding an Equation of a Circle and Graphing It

Find an equation of the circle with center $(4, -3)$ and radius 5, and graph it.

$$\sqrt{(x_2 - x_1)^2 + (y_2 - y_1)^2} = d \qquad \text{Distance formula}$$

$$\sqrt{(x - 4)^2 + [y - (-3)]^2} = 5 \qquad \text{Let } x_1 = 4, y_1 = -3, \text{ and } d = 5.$$

$$(x - 4)^2 + (y + 3)^2 = 25 \qquad \text{Square each side.}$$

To graph the circle

$$(x - 4)^2 + (y + 3)^2 = 25,$$

plot the center $(4, -3)$, and then, because the radius is 5, move right, left, up, and down 5 units from the center, plotting the points

$$(9, -3), \quad (-1, -3), \quad (4, 2), \quad \text{and} \quad (4, -8).$$

Draw a smooth curve through these four points. When graphing by hand, it is helpful to sketch one quarter of the circle at a time. See **FIGURE 11.** (Note that the center is not part of the actual graph, but provides help in drawing a more accurate graph.)

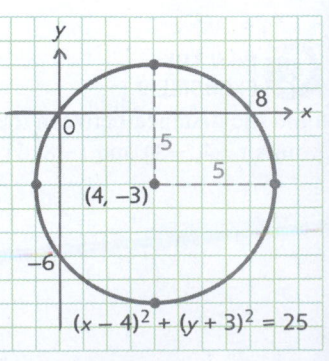

FIGURE 11

NOW TRY

NOW TRY ANSWERS
10. $x^2 + y^2 = 36$

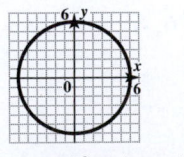

11. $(x + 2)^2 + (y - 2)^2 = 9$

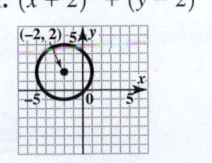

Examples 10 and 11 suggest the form of an equation of a circle with radius r and center (h, k). If (x, y) is a point on the circle, then the distance from the center (h, k) to the point (x, y) is r. See **FIGURE 12**. By the distance formula,

$$\sqrt{(x - h)^2 + (y - k)^2} = r.$$

Squaring each side gives the **center-radius form** of the equation of a circle.

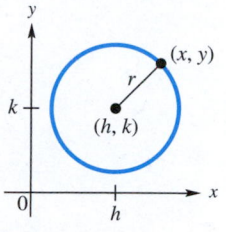

FIGURE 12

Equation of a Circle (Center-Radius Form)

A circle with center (h, k) and radius $r > 0$ has an equation of the form

$$(x - h)^2 + (y - k)^2 = r^2.$$

As a special case, a circle with center at the origin $(0, 0)$ and radius $r > 0$ has the following equation.

$$x^2 + y^2 = r^2$$

NOW TRY
EXERCISE 12

Find an equation of the circle with center $(-5, 4)$ and radius $\sqrt{6}$.

NOW TRY ANSWER
12. $(x + 5)^2 + (y - 4)^2 = 6$

EXAMPLE 12   Using the Center-Radius Form of the Equation of a Circle

Find an equation of the circle with center $(-1, 2)$ and radius $\sqrt{7}$.

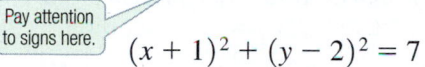   Center-radius form

$$[x - (-1)]^2 + (y - 2)^2 = \left(\sqrt{7}\right)^2$$ Let $h = -1$, $k = 2$, and $r = \sqrt{7}$.

Pay attention to signs here.

$$(x + 1)^2 + (y - 2)^2 = 7$$ Simplify; $\left(\sqrt{a}\right)^2 = a$ **NOW TRY**

10.3 Exercises

FOR EXTRA HELP   ▶ **MyLab Math**

▶ *Video solutions for select problems available in MyLab Math*

Concept Check   *Choose the correct response.*

1. Which is the greatest perfect square factor of 128?

  A. 12   **B.** 16   **C.** 32   **D.** 64

2. Which is the greatest perfect cube factor of $81a^7$?

  A. $8a^3$ **B.** $27a^3$ **C.** $81a^6$ **D.** $27a^6$

3. Which radical can be simplified?

  A. $\sqrt{21}$ **B.** $\sqrt{48}$ **C.** $\sqrt[3]{12}$ **D.** $\sqrt[4]{10}$

4. Which radical *cannot* be simplified?

  A. $\sqrt[3]{30}$ **B.** $\sqrt[3]{27a^2b}$ **C.** $\sqrt{\dfrac{25}{81}}$ **D.** $\dfrac{2}{\sqrt{7}}$

5. Which one of the following is *not* equal to $\sqrt{\dfrac{1}{2}}$? (Do not use calculator approximations.)

  A. $\sqrt{0.5}$ **B.** $\sqrt{\dfrac{2}{4}}$ **C.** $\sqrt{\dfrac{3}{6}}$ **D.** $\dfrac{\sqrt{4}}{\sqrt{16}}$

6. Which one of the following is *not* equal to $\sqrt[3]{\frac{2}{5}}$? (Do not use calculator approximations.)

A. $\sqrt[3]{\frac{6}{15}}$ **B.** $\frac{\sqrt[3]{50}}{5}$ **C.** $\frac{\sqrt[3]{10}}{\sqrt[3]{25}}$ **D.** $\frac{\sqrt[3]{10}}{5}$

7. Concept Check   A student multiplied incorrectly as follows.

$$\sqrt[3]{13} \cdot \sqrt[3]{5}$$
$$= \sqrt{13 \cdot 5} \qquad \text{Product rule}$$
$$= \sqrt{65} \qquad \text{Multiply.}$$

WHAT WENT WRONG?   Give the correct product.

8. Concept Check   A student multiplied incorrectly as follows.

$$\sqrt[3]{x} \cdot \sqrt[3]{x} = x$$

WHAT WENT WRONG?   Give the correct product.

Multiply. Assume that all variables represent positive real numbers. See Examples 1 and 2.

9. $\sqrt{3} \cdot \sqrt{3}$ **10.** $\sqrt{5} \cdot \sqrt{5}$ **11.** $\sqrt{18} \cdot \sqrt{2}$ **12.** $\sqrt{12} \cdot \sqrt{3}$

13. $\sqrt{5} \cdot \sqrt{6}$ **14.** $\sqrt{10} \cdot \sqrt{3}$ **15.** $\sqrt{14} \cdot \sqrt{x}$ **16.** $\sqrt{23} \cdot \sqrt{t}$

17. $\sqrt{14} \cdot \sqrt{3pqr}$ **18.** $\sqrt{7} \cdot \sqrt{5xt}$ **19.** $\sqrt[3]{2} \cdot \sqrt[3]{5}$ **20.** $\sqrt[3]{3} \cdot \sqrt[3]{6}$

21. $\sqrt[3]{7x} \cdot \sqrt[3]{2y}$ **22.** $\sqrt[3]{9x} \cdot \sqrt[3]{4y}$ **23.** $\sqrt[4]{11} \cdot \sqrt[4]{3}$ **24.** $\sqrt[4]{6} \cdot \sqrt[4]{9}$

25. $\sqrt[4]{2x} \cdot \sqrt[4]{3y^2}$ **26.** $\sqrt[4]{3y^2} \cdot \sqrt[4]{6yz}$ **27.** $\sqrt[3]{7} \cdot \sqrt[4]{3}$ **28.** $\sqrt[5]{8} \cdot \sqrt[6]{12}$

Simplify. Assume that all variables represent positive real numbers. See Example 3.

29. $\sqrt{\frac{64}{121}}$ **30.** $\sqrt{\frac{16}{49}}$ **31.** $\sqrt{\frac{3}{25}}$ **32.** $\sqrt{\frac{13}{49}}$

33. $\sqrt{\frac{x}{25}}$ **34.** $\sqrt{\frac{k}{100}}$ **35.** $\sqrt{\frac{p^6}{81}}$ **36.** $\sqrt{\frac{w^{10}}{36}}$

37. $\sqrt[3]{-\frac{27}{64}}$ **38.** $\sqrt[3]{-\frac{216}{125}}$ **39.** $\sqrt[3]{\frac{r^2}{8}}$ **40.** $\sqrt[3]{\frac{t}{125}}$

41. $-\sqrt[4]{\frac{81}{x^4}}$ **42.** $-\sqrt[4]{\frac{625}{y^4}}$ **43.** $\sqrt[5]{\frac{1}{x^{15}}}$ **44.** $\sqrt[5]{\frac{32}{y^{20}}}$

45. Concept Check   A student simplified $\sqrt{48}$ as follows and did not receive credit for his answer.

$$\sqrt{48}$$
$$= \sqrt{4 \cdot 12}$$
$$= \sqrt{4} \cdot \sqrt{12}$$
$$= 2\sqrt{12}$$

WHAT WENT WRONG?   Give the correct simplified form.

46. Concept Check   A student incorrectly claimed that the following radical is in simplified form.

$$\sqrt[3]{k^4}$$

WHAT WENT WRONG?   Give the correct simplified form.

Simplify. **See Example 4.**

47. $\sqrt{12}$ **48.** $\sqrt{18}$ **49.** $\sqrt{288}$ **50.** $\sqrt{72}$

51. $-\sqrt{32}$ **52.** $-\sqrt{48}$ **53.** $-\sqrt{28}$ **54.** $-\sqrt{24}$

55. $\sqrt{30}$ **56.** $\sqrt{46}$ **57.** $\sqrt[3]{128}$ **58.** $\sqrt[3]{24}$

59. $\sqrt[3]{40}$ **60.** $\sqrt[3]{375}$ **61.** $\sqrt[3]{-16}$ **62.** $\sqrt[3]{-250}$

63. $-\sqrt[4]{512}$ **64.** $-\sqrt[4]{1250}$ **65.** $\sqrt[5]{64}$ **66.** $\sqrt[5]{128}$

67. $-\sqrt[5]{486}$ **68.** $-\sqrt[5]{2048}$ **69.** $\sqrt[6]{128}$ **70.** $\sqrt[6]{1458}$

Simplify. Assume that all variables represent positive real numbers. **See Example 5.**

71. $\sqrt{72k^2}$ **72.** $\sqrt{18m^2}$ **73.** $\sqrt{144x^3y^9}$

74. $\sqrt{169s^5t^{10}}$ **75.** $\sqrt{121x^6}$ **76.** $\sqrt{256z^{12}}$

77. $-\sqrt[3]{27t^{12}}$ **78.** $-\sqrt[3]{64y^{18}}$ **79.** $-\sqrt{100m^8z^4}$

80. $-\sqrt{25t^6s^{20}}$ **81.** $-\sqrt[3]{-125a^6b^9c^{12}}$ **82.** $-\sqrt[3]{-216y^{15}x^6z^3}$

83. $\sqrt[4]{\dfrac{1}{16}r^8t^{20}}$ **84.** $\sqrt[4]{\dfrac{81}{256}t^{12}u^8}$ **85.** $\sqrt{50x^3}$ **86.** $\sqrt{300z^3}$

87. $-\sqrt{500r^{11}}$ **88.** $-\sqrt{200p^{13}}$ **89.** $\sqrt{13x^7y^8}$ **90.** $\sqrt{23k^9p^{14}}$

91. $\sqrt[3]{8z^6w^9}$ **92.** $\sqrt[3]{64a^{15}b^{12}}$ **93.** $\sqrt[3]{-16z^5t^7}$ **94.** $\sqrt[3]{-81m^4n^{10}}$

95. $\sqrt[4]{81x^{12}y^{16}}$ **96.** $\sqrt[4]{81t^8u^{28}}$ **97.** $-\sqrt[4]{162r^{15}s^9}$ **98.** $-\sqrt[4]{32k^5m^{11}}$

99. $\sqrt{\dfrac{y^{11}}{36}}$ **100.** $\sqrt{\dfrac{v^{13}}{49}}$ **101.** $\sqrt[3]{\dfrac{x^{16}}{27}}$ **102.** $\sqrt[3]{\dfrac{y^{17}}{125}}$

Simplify. Assume that $x \geq 0$. **See Example 6.**

103. $\sqrt[4]{48^2}$ **104.** $\sqrt[4]{50^2}$ **105.** $\sqrt[4]{25}$ **106.** $\sqrt[6]{8}$

107. $\sqrt[10]{x^{25}}$ **108.** $\sqrt[12]{x^{44}}$ **109.** $\sqrt[10]{x^{16}}$ **110.** $\sqrt[12]{x^{38}}$

Simplify. Assume that all variables represent positive real numbers. **See Example 7.**

111. $\sqrt[3]{4} \cdot \sqrt{3}$ **112.** $\sqrt[3]{5} \cdot \sqrt{6}$ **113.** $\sqrt[4]{3} \cdot \sqrt[3]{4}$

114. $\sqrt[3]{2} \cdot \sqrt[5]{3}$ **115.** $\sqrt{x} \cdot \sqrt[3]{x}$ **116.** $\sqrt[3]{y} \cdot \sqrt[4]{y}$

Find the length of the unknown side in each right triangle. Simplify answers if possible. **See Example 8.**

117.

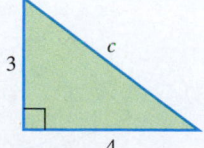

118.

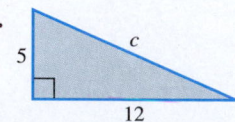

119.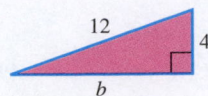

120.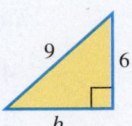

121.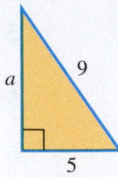

122.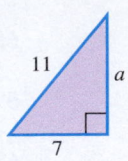

Find the distance between each pair of points. See Example 9.

123. $(6, 13)$ and $(1, 1)$

124. $(8, 13)$ and $(2, 5)$

125. $(-6, 5)$ and $(3, -4)$

126. $(-1, 5)$ and $(-7, 7)$

127. $(-8, 2)$ and $(-4, 1)$

128. $(-1, 2)$ and $(5, 3)$

129. $(4.7, 2.3)$ and $(1.7, -1.7)$

130. $(-2.9, 18.2)$ and $(2.1, 6.2)$

131. $\left(\sqrt{2}, \sqrt{6}\right)$ and $\left(-2\sqrt{2}, 4\sqrt{6}\right)$

132. $\left(\sqrt{7}, 9\sqrt{3}\right)$ and $\left(-\sqrt{7}, 4\sqrt{3}\right)$

133. $(x + y, y)$ and $(x - y, x)$

134. $(c, c - d)$ and $(d, c + d)$

Concept Check   *Work each problem.*

135. Match each equation with the correct graph.

(a) $(x - 3)^2 + (y - 2)^2 = 25$

A.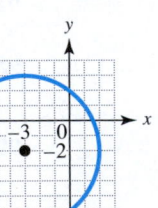

B.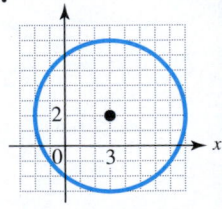

(b) $(x - 3)^2 + (y + 2)^2 = 25$

(c) $(x + 3)^2 + (y - 2)^2 = 25$

C.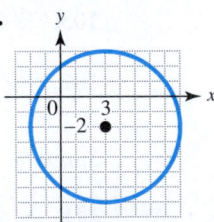

D.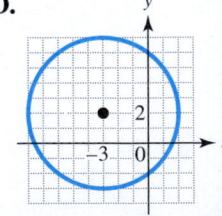

(d) $(x + 3)^2 + (y + 2)^2 = 25$

136. A circle can be drawn on a piece of posterboard by fastening one end of a string with a thumbtack, pulling the string taut with a pencil, and tracing a curve, as shown in the figure. Explain why this method works.

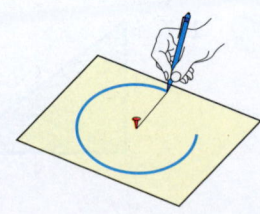

Find the equation of a circle satisfying the given conditions. See Examples 10–12.

137. Center: $(0, 0)$; radius: 12

138. Center: $(0, 0)$; radius: 9

139. Center: $(-4, 3)$; radius: 2

140. Center: $(5, -2)$; radius: 4

141. Center: $(-8, -5)$; radius: $\sqrt{5}$

142. Center: $(-12, 13)$; radius: $\sqrt{7}$

Graph each circle. Identify the center and the radius. See Examples 10–12.

143. $x^2 + y^2 = 9$

144. $x^2 + y^2 = 4$

145. $x^2 + y^2 = 16$

146. $x^2 + y^2 = 25$

147. $(x + 3)^2 + (y - 2)^2 = 9$

148. $(x - 1)^2 + (y + 3)^2 = 16$

149. $(x - 2)^2 + (y - 3)^2 = 4$

150. $(x + 4)^2 + (y + 1)^2 = 25$

Solve each problem.

151. A Panasonic Smart Viera E50 LCD HDTV has a rectangular screen with a 36.5-in. width. Its height is 20.8 in. What is the length of the diagonal of the screen to the nearest tenth of an inch? (Data from measurements of the author's television.)

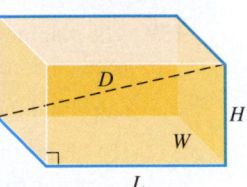

36.5 in.

20.8 in.

d

152. The length of the diagonal of a box is given by

$$D = \sqrt{L^2 + W^2 + H^2},$$

where L, W, and H are, respectively, the length, width, and height of the box. Find the length of the diagonal D of a box that is 4 ft long, 2 ft wide, and 3 ft high. Give the exact value, and then round to the nearest tenth of a foot.

D

H

W

L

153. In the study of sound, one version of the law of tensions is

$$f_1 = f_2 \sqrt{\frac{F_1}{F_2}}.$$

If $F_1 = 300$, $F_2 = 60$, and $f_2 = 260$, find f_1 to the nearest unit.

154. The illumination I, in foot-candles, produced by a light source is related to the distance d, in feet, from the light source by the equation

$$d = \sqrt{\frac{k}{I}},$$

where k is a constant. If $k = 640$, how far from the light source will the illumination be 2 foot-candles? Give the exact value, and then round to the nearest tenth of a foot.

Tom owns a condominium in a high rise building on the shore of Lake Michigan, and has a beautiful view of the lake from his window. He discovered that he can find the number of miles to the horizon by multiplying 1.224 by the square root of his eye level in feet from the ground.

Use Tom's discovery to do the following.

155. Write a formula that could be used to calculate the distance d in miles to the horizon from a height h in feet from the ground.

156. Tom lives on the 14th floor, which is 150 ft above the ground. His eyes are 6 ft above his floor. Use the formula from **Exercise 155** to calculate the distance, to the nearest tenth of a mile, that Tom can see to the horizon from his condominium window.

157. Tom's neighbor Sheri lives on a floor that is 100 ft above the ground. Assuming that her eyes are 5 ft above the ground, to the nearest tenth of a mile, how far can she see to the horizon?

RELATING CONCEPTS For Individual or Group Work (Exercises 158–160)

Pythagoras may have written the first proof of the Pythagorean relationship, but many other proofs have been given. Three of them follow. (If necessary, refer to the formulas at the back of the text.)

158. The Babylonians may have used a tile pattern like that shown here to illustrate the Pythagorean theorem.

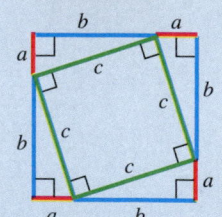

    (a) The side of the square along the hypotenuse measures 5 units. What are the measures of the sides along the legs?

    (b) Using the measures from part (a), show that the Pythagorean theorem is satisfied.

159. The diagram shown at the right can be used to verify the Pythagorean theorem.

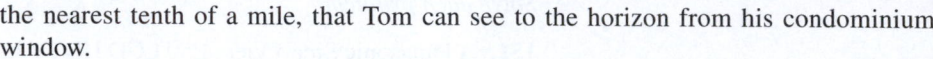

    (a) Find the area of the large square.

    (b) Find the sum of the areas of the smaller square and the four right triangles.

    (c) Set the areas equal and write a simpler, equivalent equation.

160. James A. Garfield, the twentieth president of the United States, provided a proof of the Pythagorean theorem using the given figure.

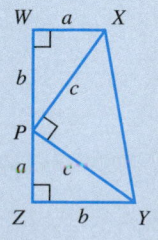

    (a) Find the area of the trapezoid $WXYZ$ using the formula for the area of a trapezoid.

    (b) Find the area of each of the right triangles PWX, PZY, and PXY.

    (c) The sum of the areas of the three right triangles must equal the area of the trapezoid. Set the expression from part (a) equal to the sum of the three expressions from part (b). Perform the operations, clear fractions, and write a simpler equivalent equation.

10.4   Adding and Subtracting Radical Expressions

OBJECTIVE

1 Simplify radical expressions involving addition and subtraction.

OBJECTIVE 1 Simplify radical expressions involving addition and subtraction.

We do so using the distributive property,

$$ac + bc = (a + b)c.$$

EXAMPLE 1   Adding and Subtracting Radicals

Add or subtract to simplify each radical expression.

(a) $4\sqrt{2} + 3\sqrt{2}$

$= (4 + 3)\sqrt{2}$ This is similar to simplifying $4x + 3x$ as $7x$.

$= 7\sqrt{2}$

(b) $2\sqrt{3} - 5\sqrt{3}$

$= (2 - 5)\sqrt{3}$ This is similar to simplifying $2x - 5x$ as $-3x$.

$= -3\sqrt{3}$

NOW TRY EXERCISE 1

Add or subtract to simplify each radical expression.

(a) $3\sqrt{5} + 7\sqrt{5}$

(b) $2\sqrt{7} - 3\sqrt{7}$

(c) $\sqrt{12} + \sqrt{75}$

(d) $-\sqrt{63t} + 3\sqrt{28t}, \quad t \geq 0$

(e) $6\sqrt{7} - 2\sqrt{3}$

(c) $3\sqrt{24} + \sqrt{54}$ Simplify each individual radical.

$= 3\sqrt{4 \cdot 6} + \sqrt{9 \cdot 6}$ Factor the radicands so that one factor is a perfect square.

$= 3\sqrt{4} \cdot \sqrt{6} + \sqrt{9} \cdot \sqrt{6}$ Product rule

$= 3 \cdot 2\sqrt{6} + 3\sqrt{6}$ Find the square roots.

$= 6\sqrt{6} + 3\sqrt{6}$ Multiply.

$= (6 + 3)\sqrt{6}$ Distributive property

$= 9\sqrt{6}$ Add.

(d) $2\sqrt{20x} - \sqrt{45x}, \quad x \geq 0$

$= 2\sqrt{4} \cdot \sqrt{5x} - \sqrt{9} \cdot \sqrt{5x}$ Product rule

$= 2 \cdot 2\sqrt{5x} - 3\sqrt{5x}$ Find the square roots.

$= 4\sqrt{5x} - 3\sqrt{5x}$ Multiply.

$= (4 - 3)\sqrt{5x}$ Distributive property

$= \sqrt{5x}$ $1\sqrt{5x} = \sqrt{5x}$ Subtract.

(e) $2\sqrt{3} - 4\sqrt{5}$

The radicands differ and are already simplified, so this expression cannot be simplified further.

NOW TRY

NOW TRY ANSWERS

1. **(a)** $10\sqrt{5}$ **(b)** $-\sqrt{7}$
 (c) $7\sqrt{3}$ **(d)** $3\sqrt{7t}$
 (e) The expression cannot be simplified further.

❗ **CAUTION**   *Only radical expressions with the same index and the same radicand may be combined.*

NOW TRY
EXERCISE 2

Add or subtract to simplify each radical expression. Assume that all variables represent positive real numbers.

(a) $3\sqrt[3]{2000} - 4\sqrt[3]{128}$

(b) $5\sqrt[4]{a^5b^3} + \sqrt[4]{81ab^7}$

(c) $\sqrt[3]{128t^4} - 2\sqrt{72t^3}$

EXAMPLE 2   Adding and Subtracting Radicals with Higher Indexes

Add or subtract to simplify each radical expression. Assume that all variables represent positive real numbers.

(a) $2\sqrt[3]{16} - 5\sqrt[3]{54}$ *Remember to write the index with each radical.*

$= 2\sqrt[3]{8 \cdot 2} - 5\sqrt[3]{27 \cdot 2}$ Factor.

$= 2\sqrt[3]{8} \cdot \sqrt[3]{2} - 5\sqrt[3]{27} \cdot \sqrt[3]{2}$ Product rule

$= 2 \cdot 2 \cdot \sqrt[3]{2} - 5 \cdot 3 \cdot \sqrt[3]{2}$ Find the cube roots.

$= 4\sqrt[3]{2} - 15\sqrt[3]{2}$ Multiply.

$= (4 - 15)\sqrt[3]{2}$ Distributive property

$= -11\sqrt[3]{2}$ Combine like terms.

In practice, the step indicating $(4 - 15)\sqrt[3]{2}$ can be done mentally, giving the final answer $-11\sqrt[3]{2}$ directly.

(b) $2\sqrt[3]{x^2y} + \sqrt[3]{8x^5y^4}$

$= 2\sqrt[3]{x^2y} + \sqrt[3]{8x^3y^3 \cdot x^2y}$ Factor.

$= 2\sqrt[3]{x^2y} + \sqrt[3]{8x^3y^3} \cdot \sqrt[3]{x^2y}$ Product rule

$= 2\sqrt[3]{x^2y} + 2xy\sqrt[3]{x^2y}$ Find the cube root.

This result cannot be simplified further.

$= (2 + 2xy)\sqrt[3]{x^2y}$ Distributive property

Although we were able to use the distributive property in the last step, 2 and $2xy$ are not like terms and cannot be combined into a single term.

(c) $5\sqrt{4x^3} + 3\sqrt[3]{64x^4}$ *Be careful. The indexes are different.*

$= 5\sqrt{4x^2 \cdot x} + 3\sqrt[3]{64x^3 \cdot x}$ Factor.

$= 5\sqrt{4x^2} \cdot \sqrt{x} + 3\sqrt[3]{64x^3} \cdot \sqrt[3]{x}$ Product rule

$= 5 \cdot 2x\sqrt{x} + 3 \cdot 4x\sqrt[3]{x}$ *Keep track of the indexes.*

$= 10x\sqrt{x} + 12x\sqrt[3]{x}$ Multiply.

The two terms in the final expression cannot be combined into a single term—the indexes are different.

NOW TRY

NOW TRY ANSWERS
2. **(a)** $14\sqrt[3]{2}$

   (b) $(5a + 3b)\sqrt[4]{ab^3}$

   (c) $4t\sqrt[3]{2t} - 12t\sqrt{2t}$

⚠ **CAUTION**   *The root of a sum does not equal the sum of the roots.*

Example: $\sqrt{9 + 16} \neq \sqrt{9} + \sqrt{16}$

because $\sqrt{9 + 16} = \sqrt{25} = 5$,   but $\sqrt{9} + \sqrt{16} = 3 + 4 = 7$.

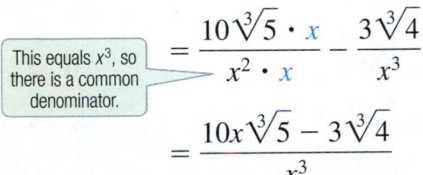

**NOW TRY
EXERCISE 3**

Perform the indicated
operations. Assume that all
variables represent positive
real numbers.

(a) $5\dfrac{\sqrt{5}}{\sqrt{45}} - 4\sqrt{\dfrac{28}{9}}$

(b) $6\sqrt[3]{\dfrac{16}{x^{12}}} + 7\sqrt[3]{\dfrac{9}{x^9}}$

EXAMPLE 3 Adding and Subtracting Radicals with Fractions

Perform the indicated operations. Assume that all variables represent positive real numbers.

(a) $2\sqrt{\dfrac{75}{16}} + 4\dfrac{\sqrt{8}}{\sqrt{32}}$

$= 2\dfrac{\sqrt{25 \cdot 3}}{\sqrt{16}} + 4\dfrac{\sqrt{4 \cdot 2}}{\sqrt{16 \cdot 2}}$ Quotient rule; Factor.

$= 2\left(\dfrac{5\sqrt{3}}{4}\right) + 4\left(\dfrac{2\sqrt{2}}{4\sqrt{2}}\right)$ Product rule; Find the square roots.

$= \dfrac{5\sqrt{3}}{2} + 2$ Multiply; $\dfrac{\sqrt{2}}{\sqrt{2}} = 1$

$= \dfrac{5\sqrt{3}}{2} + \dfrac{4}{2}$ Write with a common denominator; $2 = \frac{4}{2}$

$= \dfrac{5\sqrt{3} + 4}{2}$ $\dfrac{a}{c} + \dfrac{b}{c} = \dfrac{a+b}{c}$

(b) $10\sqrt[3]{\dfrac{5}{x^6}} - 3\sqrt[3]{\dfrac{4}{x^9}}$

$= 10\dfrac{\sqrt[3]{5}}{\sqrt[3]{x^6}} - 3\dfrac{\sqrt[3]{4}}{\sqrt[3]{x^9}}$ Quotient rule

$= \dfrac{10\sqrt[3]{5}}{x^2} - \dfrac{3\sqrt[3]{4}}{x^3}$ Simplify denominators.

$= \dfrac{10\sqrt[3]{5} \cdot x}{x^2 \cdot x} - \dfrac{3\sqrt[3]{4}}{x^3}$ Write with a common denominator.

This equals x^3, so there is a common denominator.

$= \dfrac{10x\sqrt[3]{5} - 3\sqrt[3]{4}}{x^3}$ $\dfrac{a}{c} - \dfrac{b}{c} = \dfrac{a-b}{c}$

NOW TRY

NOW TRY ANSWERS

3. (a) $\dfrac{5 - 8\sqrt{7}}{3}$

(b) $\dfrac{12\sqrt[3]{2} + 7x\sqrt[3]{9}}{x^4}$

10.4 Exercises

FOR EXTRA HELP

 MyLab Math

 Video solutions for select problems available in MyLab Math

Concept Check *Choose the correct response.*

1. Which sum can be simplified without first simplifying the individual radical expressions?

   A. $\sqrt{50} + \sqrt{32}$ **B.** $3\sqrt{6} + 9\sqrt{6}$

   C. $\sqrt[3]{32} - \sqrt[3]{108}$ **D.** $\sqrt[5]{6} - \sqrt[5]{192}$

2. Which difference can be simplified without first simplifying the individual radical expressions?

   A. $\sqrt{81} - \sqrt{18}$ **B.** $\sqrt[3]{8} - \sqrt[3]{16}$

   C. $4\sqrt[3]{7} - 9\sqrt[3]{7}$ **D.** $\sqrt{75} - \sqrt{12}$

3. Concept Check   A student incorrectly simplified

$$(3 + 3xy)\sqrt[3]{xy^2} \quad \text{as} \quad 6xy\sqrt[3]{xy^2}.$$

His teacher did not give him any credit for this answer. **WHAT WENT WRONG?**

4. Concept Check   A student incorrectly gave the difference

$$28 - 4\sqrt{2} \quad \text{as} \quad 24\sqrt{2}.$$

Her teacher did not give her any credit for this answer. **WHAT WENT WRONG?**

Add or subtract to simplify each radical expression. Assume that all variables represent positive real numbers. ***See Examples 1 and 2.***

5. $\sqrt{36} - \sqrt{100}$

6. $\sqrt{25} - \sqrt{81}$

7. $6\sqrt{10} + 2\sqrt{10}$

8. $5\sqrt{6} + 4\sqrt{6}$

9. $6\sqrt{5} - 7\sqrt{5}$

10. $3\sqrt{2} - 4\sqrt{2}$

11. $-2\sqrt{48} + 3\sqrt{75}$

12. $4\sqrt{32} - 2\sqrt{8}$

13. $5\sqrt{6} + 2\sqrt{10}$

14. $3\sqrt{11} - 5\sqrt{13}$

15. $\sqrt[3]{16} + 4\sqrt[3]{54}$

16. $3\sqrt[3]{24} - 2\sqrt[3]{192}$

17. $6\sqrt{18} - \sqrt{32} + 2\sqrt{50}$

18. $5\sqrt{8} + 3\sqrt{72} - 3\sqrt{50}$

19. $2\sqrt{5} + 3\sqrt{20} + 4\sqrt{45}$

20. $5\sqrt{54} - 2\sqrt{24} - 2\sqrt{96}$

21. $\sqrt{72x} - \sqrt{8x}$

22. $\sqrt{18k} - \sqrt{72k}$

23. $3\sqrt{72m^2} - 5\sqrt{32m^2} - 3\sqrt{18m^2}$

24. $9\sqrt{27p^2} - 14\sqrt{108p^2} + 2\sqrt{48p^2}$

25. $\sqrt[4]{32} + 3\sqrt[4]{2}$

26. $\sqrt[4]{405} - 2\sqrt[4]{5}$

27. $2\sqrt[3]{16} + \sqrt[3]{54}$

28. $15\sqrt[3]{81} + 4\sqrt[3]{24}$

29. $2\sqrt[3]{27x} - 2\sqrt[3]{8x}$

30. $6\sqrt[3]{128m} - 3\sqrt[3]{16m}$

31. $3\sqrt[3]{x^2y} - 5\sqrt[3]{8x^2y}$

32. $3\sqrt[3]{x^2y^2} - 2\sqrt[3]{64x^2y^2}$

33. $3x\sqrt[3]{xy^2} - 2\sqrt[3]{8x^4y^2}$

34. $6q^2\sqrt[3]{5q} - 2q\sqrt[3]{40q^4}$

35. $5\sqrt[4]{32} + 3\sqrt[4]{162}$

36. $2\sqrt[4]{512} + 4\sqrt[4]{32}$

37. $3\sqrt[4]{x^5y} - 2x\sqrt[4]{xy}$

38. $2\sqrt[4]{m^9p^6} - 3m^2p\sqrt[4]{mp^2}$

39. $2\sqrt[4]{32a^3} + 5\sqrt[4]{2a^3}$

40. $5\sqrt[4]{243x^3} + 2\sqrt[4]{3x^3}$

41. $\sqrt[3]{64xy^2} + \sqrt[3]{27x^4y^5}$

42. $\sqrt[4]{625s^3t} + \sqrt[4]{81s^7t^5}$

43. $\sqrt[3]{192st^4} - \sqrt{27s^3t}$

44. $\sqrt{125a^5b^5} + \sqrt[3]{125a^4b^4}$

45. $2\sqrt[3]{8x^4} + 3\sqrt[4]{16x^5}$

46. $3\sqrt[3]{64m^4} + 5\sqrt[4]{81m^5}$

Perform the indicated operations. Assume that all variables represent positive real numbers. ***See Example 3.***

47. $\sqrt{8} - \dfrac{\sqrt{64}}{\sqrt{16}}$

48. $\sqrt{48} - \dfrac{\sqrt{81}}{\sqrt{9}}$

49. $\dfrac{2\sqrt{5}}{3} + \dfrac{\sqrt{5}}{6}$

50. $\dfrac{4\sqrt{3}}{3} + \dfrac{2\sqrt{3}}{9}$

51. $\sqrt{\dfrac{8}{9}} + \sqrt{\dfrac{18}{36}}$

52. $\sqrt{\dfrac{12}{16}} + \sqrt{\dfrac{48}{64}}$

53. $\dfrac{\sqrt{32}}{3} + \dfrac{2\sqrt{2}}{3} - \dfrac{\sqrt{2}}{\sqrt{9}}$ **54.** $\dfrac{\sqrt{27}}{2} - \dfrac{3\sqrt{3}}{2} + \dfrac{\sqrt{3}}{\sqrt{4}}$ **55.** $3\sqrt{\dfrac{50}{9}} + 8\dfrac{\sqrt{2}}{\sqrt{8}}$

56. $5\sqrt{\dfrac{288}{25}} + 21\dfrac{\sqrt{2}}{\sqrt{18}}$ **57.** $\sqrt{\dfrac{25}{x^8}} + \sqrt{\dfrac{9}{x^6}}$ **58.** $\sqrt{\dfrac{100}{y^4}} + \sqrt{\dfrac{81}{y^{10}}}$

59. $3\sqrt{\dfrac{50}{49}} - \dfrac{\sqrt{27}}{\sqrt{12}}$ **60.** $9\sqrt{\dfrac{48}{25}} - 2\dfrac{\sqrt{2}}{\sqrt{98}}$ **61.** $3\sqrt[3]{\dfrac{m^5}{27}} - 2m\sqrt[3]{\dfrac{m^2}{64}}$

62. $2a\sqrt[4]{\dfrac{a}{16}} - 5a\sqrt[4]{\dfrac{a}{81}}$ **63.** $3\sqrt[3]{\dfrac{2}{x^6}} - 4\sqrt[3]{\dfrac{5}{x^9}}$ **64.** $-4\sqrt[3]{\dfrac{4}{t^9}} + 3\sqrt[3]{\dfrac{9}{t^{12}}}$

65. Concept Check   Consider the expression

$$\sqrt{63} + \sqrt{112} - \sqrt{252}.$$

   (a) Simplify this expression using the methods of this section.

   (b) Use a calculator to approximate the given expression.

   (c) Use a calculator to approximate the simplified expression in part (a).

   (d) Complete the following: Assuming the work in part (a) is correct, the approximations in parts (b) and (c) should be (*equal / unequal*).

66. Concept Check   Let $a = 1$ and let $b = 64$.

   (a) Evaluate $\sqrt{a} + \sqrt{b}$. Then find $\sqrt{a + b}$. Are they equal?

   (b) Evaluate $\sqrt[3]{a} + \sqrt[3]{b}$. Then find $\sqrt[3]{a + b}$. Are they equal?

   (c) Complete the following: In general,

$$\sqrt[n]{a} + \sqrt[n]{b} \neq \underline{\hspace{3cm}},$$

   based on the observations in parts (a) and (b) of this exercise.

Solve each problem.

67. A rectangular yard has a length of $\sqrt{192}$ m and a width of $\sqrt{48}$ m. Choose the best estimate of its dimensions. Then estimate the perimeter.

   A. 14 m by 7 m   **B.** 5 m by 7 m   **C.** 14 m by 8 m   **D.** 15 m by 8 m

68. If the sides of a triangle are $\sqrt{65}$ in., $\sqrt{35}$ in., and $\sqrt{26}$ in., which one of the following is the best estimate of its perimeter?

   A. 19 in.   **B.** 20 in.   **C.** 24 in.   **D.** 26 in.

Find the perimeter of each figure. Give answers as simplified radical expressions.

69.

$3\sqrt{20}$ in. $2\sqrt{45}$ in.

$\sqrt{75}$ in.

70.

$\sqrt{192}$ m

$\sqrt{48}$ m

71.

$4\sqrt{18}$ in.

$3\sqrt{12}$ in. $\sqrt{108}$ in.

$2\sqrt{72}$ in.

72. Find the area of the trapezoid. Give the answer as a simplified radical.

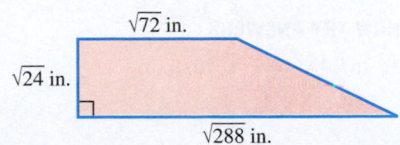

$\sqrt{72}$ in.

$\sqrt{24}$ in.

$\sqrt{288}$ in.

Extending Skills *Find the perimeter of each triangle.*

73.

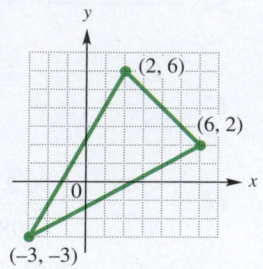

74.

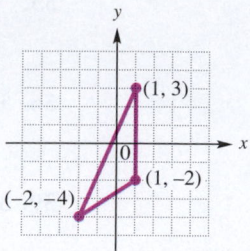

10.5 Multiplying and Dividing Radical Expressions

OBJECTIVES

1. Multiply radical expressions.

2. Rationalize denominators with one radical term.

3. Rationalize denominators with binomials involving radicals.

4. Write radical quotients in lowest terms.

OBJECTIVE 1 Multiply radical expressions.

The distributive property may be used when multiplying radical expressions.

EXAMPLE 1   Using the Distributive Property with Radicals

Multiply.

(a) $\sqrt{5}(2 + \sqrt{6})$

$\qquad = \sqrt{5} \cdot 2 + \sqrt{5} \cdot \sqrt{6}$ Distributive property: $a(b + c) = ab + ac$

$\qquad = 2\sqrt{5} + \sqrt{30}$ Commutative property; product rule

(b) $4(\sqrt{12} - \sqrt{27})$

$\qquad = 4\sqrt{12} - 4\sqrt{27}$ Distributive property

$\qquad = 4\sqrt{4 \cdot 3} - 4\sqrt{9 \cdot 3}$ Factor the radicands so that one factor is a perfect square.

$\qquad = 4 \cdot 2\sqrt{3} - 4 \cdot 3\sqrt{3}$ $\sqrt{4} = 2; \sqrt{9} = 3$

$\qquad = 8\sqrt{3} - 12\sqrt{3}$ Multiply.

$\qquad = -4\sqrt{3}$ $8\sqrt{3} - 12\sqrt{3} = (8 - 12)\sqrt{3}$ **NOW TRY**

VOCABULARY

☐ conjugates

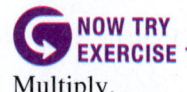
NOW TRY
EXERCISE 1
Multiply.

(a) $\sqrt{10}(4 + \sqrt{7})$

(b) $3(\sqrt{20} - \sqrt{45})$

We multiply binomial expressions involving radicals using the **FOIL** method. Recall that the acronym **FOIL** refers to the positions of the terms. We multiply the **F**irst terms, **O**uter terms, **I**nner terms, and **L**ast terms of the binomials.

EXAMPLE 2   Multiplying Binomials Involving Radical Expressions

Multiply, using the FOIL method.

(a) $(\sqrt{5} + 3)(\sqrt{6} + 1)$

$\qquad\qquad\quad$ First$\qquad$Outer$\qquad$Inner$\qquad$Last

$\qquad = \sqrt{5} \cdot \sqrt{6} + \sqrt{5} \cdot 1 + 3 \cdot \sqrt{6} + 3 \cdot 1$ FOIL method

$\qquad = \sqrt{30} + \sqrt{5} + 3\sqrt{6} + 3$ This result cannot be simplified further.

NOW TRY ANSWERS

1. **(a)** $4\sqrt{10} + \sqrt{70}$

   (b) $-3\sqrt{5}$

NOW TRY
EXERCISE 2

Multiply, using the FOIL method.

(a) $(8 - \sqrt{5})(9 - \sqrt{2})$

(b) $(\sqrt{7} + \sqrt{5})(\sqrt{7} - \sqrt{5})$

(c) $(\sqrt{15} - 4)^2$

(d) $(8 + \sqrt[3]{5})(8 - \sqrt[3]{5})$

(e) $(\sqrt{m} - \sqrt{n})(\sqrt{m} + \sqrt{n}),$
$\quad m \geq 0$ and $n \geq 0$

(b) $(7 - \sqrt{3})(\sqrt{5} + \sqrt{2})$

$\qquad$ F $\qquad$ O $\qquad$ I $\qquad$ L

$= 7\sqrt{5} + 7\sqrt{2} - \sqrt{3} \cdot \sqrt{5} - \sqrt{3} \cdot \sqrt{2}$ $\qquad$ FOIL method

$= 7\sqrt{5} + 7\sqrt{2} - \sqrt{15} - \sqrt{6}$ $\qquad$ Product rule

(c) $(\sqrt{10} + \sqrt{3})(\sqrt{10} - \sqrt{3})$

$= \sqrt{10} \cdot \sqrt{10} - \sqrt{10} \cdot \sqrt{3} + \sqrt{3} \cdot \sqrt{10} - \sqrt{3} \cdot \sqrt{3}$ $\quad$ FOIL method

$= 10 - 3$ $\qquad$ Product rule; $-\sqrt{30} + \sqrt{30} = 0$

$= 7$ $\qquad$ Subtract.

A product such as $(\sqrt{10} + \sqrt{3})(\sqrt{10} - \sqrt{3}) = (\sqrt{10})^2 - (\sqrt{3})^2$ is a difference of squares.

$$(x + y)(x - y) = x^2 - y^2 \qquad \text{Here, } x = \sqrt{10} \text{ and } y = \sqrt{3}.$$

(d) $(\sqrt{7} - 3)^2$

$= (\sqrt{7} - 3)(\sqrt{7} - 3)$ $\qquad$ $a^2 = a \cdot a$

$= \sqrt{7} \cdot \sqrt{7} - 3\sqrt{7} - 3\sqrt{7} + 3 \cdot 3$ $\qquad$ FOIL method

$= 7 - 6\sqrt{7} + 9$ $\qquad$ Multiply. Combine like terms.

$= 16 - 6\sqrt{7}$ $\qquad$ Add.

Be careful. These terms cannot be combined.

(e) $(5 - \sqrt[3]{3})(5 + \sqrt[3]{3})$

Remember to write the index 3 in each radical.

$= 5 \cdot 5 + 5\sqrt[3]{3} - 5\sqrt[3]{3} - \sqrt[3]{3} \cdot \sqrt[3]{3}$ $\qquad$ FOIL method

$= 25 - \sqrt[3]{3^2}$ $\qquad$ Multiply. Combine like terms.

$= 25 - \sqrt[3]{9}$ $\qquad$ Apply the exponent.

(f) $(\sqrt{k} + \sqrt{y})(\sqrt{k} - \sqrt{y}), \quad k \geq 0$ and $y \geq 0$

$= (\sqrt{k})^2 - (\sqrt{y})^2$ $\qquad$ Difference of squares

$= k - y$

NOW TRY

NOTE In **Example 2(d)**, we could have used the formula for the square of a binomial to obtain the same result.

$(\sqrt{7} - 3)^2$

$= (\sqrt{7})^2 - 2(\sqrt{7})(3) + 3^2$ $\qquad$ $(x - y)^2 = x^2 - 2xy + y^2$

$= 7 - 6\sqrt{7} + 9$ $\qquad$ Apply the exponents. Multiply.

$= 16 - 6\sqrt{7}$ $\qquad$ Add.

NOW TRY ANSWERS

2. (a) $72 - 8\sqrt{2} - 9\sqrt{5} + \sqrt{10}$
(b) 2 **(c)** $31 - 8\sqrt{15}$
(d) $64 - \sqrt[3]{25}$ **(e)** $m - n$

OBJECTIVE 2 Rationalize denominators with one radical term.

A simplified radical expression has no radical in the denominator. The origin of this agreement no doubt occurred before the days of high-speed calculation, when computation was a tedious process performed by hand.

Consider the expression $\frac{1}{\sqrt{2}}$. To find a decimal approximation by hand, it is necessary to divide 1 by a decimal approximation for $\sqrt{2}$, such as 1.414. It is much easier if the divisor is a whole number. This can be accomplished by multiplying $\frac{1}{\sqrt{2}}$ by 1 in the form $\frac{\sqrt{2}}{\sqrt{2}}$. ***Multiplying by 1 in any form does not change the value of the original expression.***

$$\frac{1}{\sqrt{2}} \cdot \frac{\sqrt{2}}{\sqrt{2}} = \frac{\sqrt{2}}{2} \qquad \text{Multiply by 1; } \tfrac{\sqrt{2}}{\sqrt{2}} = 1$$

Now the computation requires dividing 1.414 by 2 to obtain 0.707, which is easier.

With current technology, either form $\frac{1}{\sqrt{2}}$ or $\frac{\sqrt{2}}{2}$ can be approximated with the same number of keystrokes. See **FIGURE 13**, which shows that a calculator gives the same approximation for both forms of the expression.

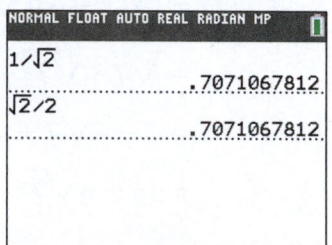

FIGURE 13

Rationalizing the Denominator

The process of removing radicals from a denominator so that the denominator contains only rational numbers is called **rationalizing the denominator**. This is done by multiplying by a form of 1.

EXAMPLE 3   Rationalizing Denominators with Square Roots

Rationalize each denominator.

(a) $\dfrac{3}{\sqrt{7}}$

Multiply by $\frac{\sqrt{7}}{\sqrt{7}}$. This is an application of the identity property of multiplication—in effect, we are multiplying by 1.

$$\frac{3}{\sqrt{7}} = \frac{3 \cdot \sqrt{7}}{\sqrt{7} \cdot \sqrt{7}} = \frac{3\sqrt{7}}{7} \qquad \begin{array}{l} \text{In the denominator,} \\ \sqrt{7} \cdot \sqrt{7} = \sqrt{7 \cdot 7} = \sqrt{49} = 7. \\ \text{The final denominator is now a rational number.} \end{array}$$

(b) $\dfrac{5\sqrt{2}}{\sqrt{5}} = \dfrac{5\sqrt{2} \cdot \sqrt{5}}{\sqrt{5} \cdot \sqrt{5}} = \dfrac{5\sqrt{10}}{5} = \sqrt{10} \qquad \text{Divide out the common factor, 5.}$

 **NOW TRY EXERCISE 3**

Rationalize each denominator.

(a) $\dfrac{8}{\sqrt{13}}$ **(b)** $\dfrac{9\sqrt{7}}{\sqrt{3}}$

(c) $\dfrac{-10}{\sqrt{20}}$

(c) $\dfrac{-6}{\sqrt{12}}$

Less work is involved if the radical in the denominator is simplified first.

$$\frac{-6}{\sqrt{12}} = \frac{-6}{\sqrt{4 \cdot 3}} = \frac{-6}{2\sqrt{3}} = \frac{-3}{\sqrt{3}}$$

Now we rationalize the denominator.

$$\frac{-3}{\sqrt{3}} = \frac{-3 \cdot \sqrt{3}}{\sqrt{3} \cdot \sqrt{3}} = \frac{-3\sqrt{3}}{3} = -\sqrt{3}$$

 **NOW TRY**

 **NOW TRY EXERCISE 4**

Simplify.

(a) $-\sqrt{\dfrac{27}{80}}$

(b) $\sqrt{\dfrac{48x^8}{y^3}}, \quad y > 0$

EXAMPLE 4 Rationalizing Denominators in Roots of Fractions

Simplify.

(a) $-\sqrt{\dfrac{18}{125}}$ **(b)** $\sqrt{\dfrac{50m^4}{p^5}}, \quad p > 0$

$= -\dfrac{\sqrt{18}}{\sqrt{125}}$ Quotient rule	$= \dfrac{\sqrt{50m^4}}{\sqrt{p^5}}$ Quotient rule
$= -\dfrac{\sqrt{9 \cdot 2}}{\sqrt{25 \cdot 5}}$ Factor.	$= \dfrac{\sqrt{25m^4 \cdot 2}}{\sqrt{p^4 \cdot p}}$ Factor.
$= -\dfrac{3\sqrt{2}}{5\sqrt{5}}$ Product rule	$= \dfrac{5m^2\sqrt{2}}{p^2\sqrt{p}}$ Product rule
$= -\dfrac{3\sqrt{2} \cdot \sqrt{5}}{5\sqrt{5} \cdot \sqrt{5}}$ Multiply by $\dfrac{\sqrt{5}}{\sqrt{5}}$.	$= \dfrac{5m^2\sqrt{2} \cdot \sqrt{p}}{p^2\sqrt{p} \cdot \sqrt{p}}$ Multiply by $\dfrac{\sqrt{p}}{\sqrt{p}}$.
$= -\dfrac{3\sqrt{10}}{5 \cdot 5}$ Product rule	$= \dfrac{5m^2\sqrt{2p}}{p^2 \cdot p}$ Product rule
$= -\dfrac{3\sqrt{10}}{25}$ Multiply.	$= \dfrac{5m^2\sqrt{2p}}{p^3}$ Multiply.

 **NOW TRY**

EXAMPLE 5 Rationalizing Denominators with Higher Roots

Simplify.

(a) $\sqrt[3]{\dfrac{27}{16}}$

Use the quotient rule, and simplify the numerator and denominator.

$$\sqrt[3]{\frac{27}{16}} = \frac{\sqrt[3]{27}}{\sqrt[3]{16}} = \frac{3}{\sqrt[3]{8} \cdot \sqrt[3]{2}} = \frac{3}{2\sqrt[3]{2}}$$

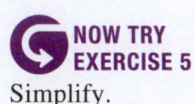

NOW TRY
EXERCISE 5

Simplify.

(a) $\sqrt[3]{\dfrac{8}{81}}$

(b) $\sqrt[4]{\dfrac{7x}{y}}, \quad x \ge 0, y > 0$

Because $2 \cdot 4 = 8$ is a perfect cube, multiply the numerator and denominator of the simplified expression by $\sqrt[3]{4}$.

$$\dfrac{3}{2\sqrt[3]{2}} \qquad \sqrt[3]{\dfrac{27}{16}} = \dfrac{3}{2\sqrt[3]{2}} \text{ from above}$$

$$= \dfrac{3 \cdot \sqrt[3]{4}}{2\sqrt[3]{2} \cdot \sqrt[3]{4}} \qquad \begin{array}{l}\text{Multiply by } \sqrt[3]{4} \text{ in the numerator}\\ \text{and denominator. This will give}\\ \sqrt[3]{8} = 2 \text{ in the denominator.}\end{array}$$

$$= \dfrac{3\sqrt[3]{4}}{2\sqrt[3]{8}} \qquad \text{Multiply.}$$

$$= \dfrac{3\sqrt[3]{4}}{2 \cdot 2} \qquad \sqrt[3]{8} = 2$$

$$= \dfrac{3\sqrt[3]{4}}{4} \qquad \text{Multiply.}$$

(b) $\qquad \sqrt[4]{\dfrac{5x}{z}}, \quad x \ge 0, z > 0$

$$= \dfrac{\sqrt[4]{5x}}{\sqrt[4]{z}} \qquad \text{Quotient rule}$$

$$= \dfrac{\sqrt[4]{5x} \cdot \sqrt[4]{z^3}}{\sqrt[4]{z} \cdot \sqrt[4]{z^3}} \qquad \begin{array}{l}\text{Multiply by } \sqrt[4]{z^3} \text{ in the numerator}\\ \text{and denominator.}\end{array}$$

$\sqrt[4]{z} \cdot \sqrt[4]{z^3}$ will give $\sqrt[4]{z^4}$.

$$= \dfrac{\sqrt[4]{5xz^3}}{\sqrt[4]{z^4}} \qquad \text{Product rule}$$

$$= \dfrac{\sqrt[4]{5xz^3}}{z}$$

NOW TRY

! CAUTION   In **Example 5(a),** a typical error is to multiply the numerator and denominator by $\sqrt[3]{2}$, forgetting that

$$\sqrt[3]{2} \cdot \sqrt[3]{2} = \sqrt[3]{2^2}, \quad \text{which does } \textbf{\textit{not}} \text{ equal} \quad 2.$$

We need **three** factors of 2 to obtain 2^3 under the radical.

$$\sqrt[3]{2} \cdot \sqrt[3]{2} \cdot \sqrt[3]{2} = \sqrt[3]{2^3}, \quad \text{which does equal} \quad 2.$$

OBJECTIVE 3   Rationalize denominators with binomials involving radicals.

To rationalize a denominator that contains a binomial expression (one that contains exactly two terms) involving radicals, such as

$$\dfrac{3}{1 + \sqrt{2}},$$

NOW TRY ANSWERS

5. (a) $\dfrac{2\sqrt[3]{9}}{9}$ (b) $\dfrac{\sqrt[4]{7xy^3}}{y}$

we use the special product $(x + y)(x - y) = x^2 - y^2$ and the concept of *conjugates*. The conjugate of $1 + \sqrt{2}$ is $1 - \sqrt{2}$.

In general, $x + y$ and $x - y$ are **conjugates.** Specifically, if x and y represent non-negative rational numbers, the product

$$\left(\sqrt{x} + \sqrt{y}\right)\left(\sqrt{x} - \sqrt{y}\right) \quad \text{produces the rational number} \quad x - y.$$

Rationalizing a Binomial Denominator

Whenever a radical expression has a sum or difference with square root radicals in the denominator, rationalize the denominator by multiplying both the numerator and denominator by the conjugate of the denominator.

EXAMPLE 6 Rationalizing Binomial Denominators

Rationalize each denominator.

(a) $\dfrac{3}{1 + \sqrt{2}}$

> Again, we are multiplying by a form of 1.

$$= \frac{3\left(1 - \sqrt{2}\right)}{\left(1 + \sqrt{2}\right)\left(1 - \sqrt{2}\right)}$$

Multiply the numerator and denominator by $1 - \sqrt{2}$, the conjugate of the denominator.

$$\begin{aligned}\left(1 + \sqrt{2}\right)\left(1 - \sqrt{2}\right) &= 1^2 - \left(\sqrt{2}\right)^2 \\ &= 1 - 2 \\ &= -1\end{aligned}$$

$$= \frac{3\left(1 - \sqrt{2}\right)}{-1}$$

> The denominator is now a rational number.

$$= \frac{3}{-1}\left(1 - \sqrt{2}\right) \qquad \frac{a \cdot b}{c} = \frac{a}{c} \cdot b$$

$$= -3\left(1 - \sqrt{2}\right) \qquad \frac{a}{-1} = -a$$

$$= -3 + 3\sqrt{2} \qquad \text{Distributive property}$$

Either of the forms in the last two lines can be given as the answer.

(b) $\dfrac{5}{4 - \sqrt{3}}$

$$= \frac{5\left(4 + \sqrt{3}\right)}{\left(4 - \sqrt{3}\right)\left(4 + \sqrt{3}\right)}$$

Multiply the numerator and denominator by $4 + \sqrt{3}$.

$$= \frac{5\left(4 + \sqrt{3}\right)}{16 - 3}$$

Multiply in the denominator.

$$= \frac{5\left(4 + \sqrt{3}\right)}{13}$$

Subtract in the denominator.

We leave the numerator in factored form. This makes it easier to determine whether the expression is written in lowest terms.

 **NOW TRY**
EXERCISE 6

Rationalize each denominator.

(a) $\dfrac{4}{1 + \sqrt{3}}$ **(b)** $\dfrac{4}{5 - \sqrt{7}}$

(c) $\dfrac{\sqrt{3} + \sqrt{7}}{\sqrt{5} - \sqrt{2}}$

(d) $\dfrac{8}{\sqrt{3x} + \sqrt{y}}$,

$3x \neq y, x > 0, y > 0$

(c) $\dfrac{\sqrt{2} - \sqrt{3}}{\sqrt{5} + \sqrt{3}}$

$= \dfrac{(\sqrt{2} - \sqrt{3})(\sqrt{5} - \sqrt{3})}{(\sqrt{5} + \sqrt{3})(\sqrt{5} - \sqrt{3})}$ Multiply the numerator and denominator by $\sqrt{5} - \sqrt{3}$.

$= \dfrac{\sqrt{10} - \sqrt{6} - \sqrt{15} + 3}{5 - 3}$ Multiply.

$= \dfrac{\sqrt{10} - \sqrt{6} - \sqrt{15} + 3}{2}$ Subtract in the denominator.

(d) $\dfrac{3}{\sqrt{5m} - \sqrt{p}}$, $5m \neq p, m > 0, p > 0$

$= \dfrac{3(\sqrt{5m} + \sqrt{p})}{(\sqrt{5m} - \sqrt{p})(\sqrt{5m} + \sqrt{p})}$ Multiply the numerator and denominator by $\sqrt{5m} + \sqrt{p}$.

$= \dfrac{3(\sqrt{5m} + \sqrt{p})}{5m - p}$ Multiply in the denominator.   **NOW TRY**

OBJECTIVE 4 Write radical quotients in lowest terms.

 NOW TRY
EXERCISE 7

Write each quotient in lowest terms.

(a) $\dfrac{15 - 6\sqrt{2}}{18}$

(b) $\dfrac{15k + \sqrt{50k^2}}{20k}$, $k > 0$

EXAMPLE 7   Writing Radical Quotients in Lowest Terms

Write each quotient in lowest terms.

(a) $\dfrac{6 + 2\sqrt{5}}{4}$ *This is a key step.*

$= \dfrac{2(3 + \sqrt{5})}{2 \cdot 2}$ Factor the numerator and denominator.

$= \dfrac{3 + \sqrt{5}}{2}$ Divide out the common factor.

Alternative method: $\dfrac{6 + 2\sqrt{5}}{4} = \dfrac{6}{4} + \dfrac{2\sqrt{5}}{4} = \dfrac{3}{2} + \dfrac{\sqrt{5}}{2} = \dfrac{3 + \sqrt{5}}{2}$

(b) $\dfrac{5y - \sqrt{8y^2}}{6y}$, $y > 0$

$= \dfrac{5y - 2y\sqrt{2}}{6y}$ $\sqrt{8y^2} = \sqrt{4y^2 \cdot 2} = 2y\sqrt{2}$

$= \dfrac{y(5 - 2\sqrt{2})}{6y}$ Factor the numerator.

$= \dfrac{5 - 2\sqrt{2}}{6}$ Divide out the common factor.   **NOW TRY**

NOW TRY ANSWERS

6. (a) $-2 + 2\sqrt{3}$

(b) $\dfrac{2(5 + \sqrt{7})}{9}$

(c) $\dfrac{\sqrt{15} + \sqrt{6} + \sqrt{35} + \sqrt{14}}{3}$

(d) $\dfrac{8(\sqrt{3x} - \sqrt{y})}{3x - y}$

7. (a) $\dfrac{5 - 2\sqrt{2}}{6}$ **(b)** $\dfrac{3 + \sqrt{2}}{4}$

⚠ CAUTION   *Be careful to factor before writing a quotient in lowest terms.*

10.5 Exercises

FOR EXTRA HELP

 MyLab Math

▶ *Video solutions for select problems available in MyLab Math*

Concept Check *Match each part of a rule for a special product in Column I with the part it equals in Column II. Assume that A and B represent positive real numbers.*

I	II
1. $(x + \sqrt{y})(x - \sqrt{y})$	**A.** $x - y$
2. $(\sqrt{x} + y)(\sqrt{x} - y)$	**B.** $x + 2y\sqrt{x} + y^2$
3. $(\sqrt{x} + \sqrt{y})(\sqrt{x} - \sqrt{y})$	**C.** $x - y^2$
4. $(\sqrt{x} + \sqrt{y})^2$	**D.** $x - 2\sqrt{xy} + y$
5. $(\sqrt{x} - \sqrt{y})^2$	**E.** $x^2 - y$
6. $(\sqrt{x} + y)^2$	**F.** $x + 2\sqrt{xy} + y$

Multiply, and then simplify each product. Assume that all variables represent positive real numbers. **See Examples 1 and 2.**

7. $\sqrt{6}(3 + \sqrt{2})$

8. $\sqrt{10}(5 + \sqrt{3})$

9. $\sqrt{3}(\sqrt{12} - 4)$

10. $\sqrt{5}(\sqrt{125} - 6)$

11. $5(\sqrt{72} - \sqrt{8})$

12. $7(\sqrt{50} - \sqrt{18})$

13. $\sqrt{2}(\sqrt{18} - \sqrt{3})$

14. $\sqrt{5}(\sqrt{15} + \sqrt{5})$

15. $(\sqrt{2} + 1)(\sqrt{3} + 1)$

16. $(\sqrt{3} + 3)(\sqrt{5} + 2)$

17. $(\sqrt{2} - \sqrt{3})(\sqrt{2} + \sqrt{3})$

18. $(\sqrt{7} + \sqrt{14})(\sqrt{7} - \sqrt{14})$

19. $(\sqrt{8} - \sqrt{2})(\sqrt{8} + \sqrt{2})$

20. $(\sqrt{20} - \sqrt{5})(\sqrt{20} + \sqrt{5})$

21. $(\sqrt{11} - \sqrt{7})(\sqrt{2} + \sqrt{5})$

22. $(\sqrt{13} - \sqrt{7})(\sqrt{3} + \sqrt{11})$

23. $(2\sqrt{3} + \sqrt{5})(3\sqrt{3} - 2\sqrt{5})$

24. $(\sqrt{7} - \sqrt{11})(2\sqrt{7} + 3\sqrt{11})$

25. $(\sqrt{5} + 2)^2$

26. $(\sqrt{11} - 1)^2$

27. $(\sqrt{21} - \sqrt{5})^2$

28. $(\sqrt{6} - \sqrt{2})^2$

29. $(2 + \sqrt[3]{6})(2 - \sqrt[3]{6})$

30. $(\sqrt[3]{3} + 6)(\sqrt[3]{3} - 6)$

31. $(2 + \sqrt[3]{2})(4 - 2\sqrt[3]{2} + \sqrt[3]{4})$

32. $(\sqrt[3]{3} - 1)(\sqrt[3]{9} + \sqrt[3]{3} + 1)$

33. $(3\sqrt{x} - \sqrt{5})(2\sqrt{x} + 1)$

34. $(4\sqrt{p} + \sqrt{7})(\sqrt{p} - 9)$

35. $(3\sqrt{r} - \sqrt{s})(3\sqrt{r} + \sqrt{s})$

36. $(\sqrt{k} + 4\sqrt{m})(\sqrt{k} - 4\sqrt{m})$

37. $(\sqrt[3]{2y} - 5)(4\sqrt[3]{2y} + 1)$

38. $(\sqrt[3]{9z} - 2)(5\sqrt[3]{9z} + 7)$

39. $(\sqrt{3x} + 2)(\sqrt{3x} - 2)$

40. $(\sqrt{6y} - 4)(\sqrt{6y} + 4)$

41. Concept Check A student incorrectly simplified the radical expression

$$6 - 4\sqrt{3} \quad \text{as} \quad 2\sqrt{3}.$$

WHAT WENT WRONG?

42. Concept Check A student rationalized the following denominator as shown.

$$\frac{5}{\sqrt[3]{2}} = \frac{5 \cdot \sqrt[3]{2}}{\sqrt[3]{2} \cdot \sqrt[3]{2}} = \frac{5\sqrt[3]{2}}{2} \quad \text{Incorrect}$$

WHAT WENT WRONG? Give the correct answer.

Rationalize each denominator. Assume that all variables represent positive real numbers. **See Examples 3 and 4.**

43. $\dfrac{7}{\sqrt{7}}$ **44.** $\dfrac{11}{\sqrt{11}}$ **45.** $\dfrac{15}{\sqrt{3}}$ **46.** $\dfrac{12}{\sqrt{6}}$

47. $\dfrac{\sqrt{3}}{\sqrt{2}}$ **48.** $\dfrac{\sqrt{7}}{\sqrt{6}}$ **49.** $\dfrac{9\sqrt{3}}{\sqrt{5}}$ **50.** $\dfrac{3\sqrt{2}}{\sqrt{11}}$

51. $\dfrac{-7}{\sqrt{48}}$ **52.** $\dfrac{-5}{\sqrt{24}}$ **53.** $\sqrt{\dfrac{7}{2}}$ **54.** $\sqrt{\dfrac{10}{3}}$

55. $-\sqrt{\dfrac{7}{50}}$ **56.** $-\sqrt{\dfrac{13}{75}}$ **57.** $\sqrt{\dfrac{24}{x}}$ **58.** $\sqrt{\dfrac{52}{y}}$

59. $\dfrac{-8\sqrt{3}}{\sqrt{k}}$ **60.** $\dfrac{-4\sqrt{13}}{\sqrt{m}}$ **61.** $-\sqrt{\dfrac{150m^5}{n^3}}$ **62.** $-\sqrt{\dfrac{98r^3}{s^5}}$

63. $\sqrt{\dfrac{288x^7}{y^9}}$ **64.** $\sqrt{\dfrac{242t^9}{u^{11}}}$ **65.** $\dfrac{5\sqrt{2m}}{\sqrt{y^3}}$

66. $\dfrac{2\sqrt{5r}}{\sqrt{m^3}}$ **67.** $-\sqrt{\dfrac{48k^2}{z}}$ **68.** $-\sqrt{\dfrac{75m^3}{p}}$

Simplify. Assume that all variables represent positive real numbers. **See Example 5.**

69. $\sqrt[3]{\dfrac{2}{3}}$ **70.** $\sqrt[3]{\dfrac{4}{5}}$ **71.** $\sqrt[3]{\dfrac{4}{9}}$ **72.** $\sqrt[3]{\dfrac{5}{16}}$ **73.** $\sqrt[3]{\dfrac{9}{32}}$

74. $\sqrt[3]{\dfrac{10}{9}}$ **75.** $-\sqrt[3]{\dfrac{2p}{r^2}}$ **76.** $-\sqrt[3]{\dfrac{6x}{y^2}}$ **77.** $\sqrt[3]{\dfrac{x^6}{y}}$ **78.** $\sqrt[3]{\dfrac{m^9}{q}}$

79. $\sqrt[4]{\dfrac{16}{x}}$ **80.** $\sqrt[4]{\dfrac{81}{y}}$ **81.** $\sqrt[4]{\dfrac{2y}{z}}$ **82.** $\sqrt[4]{\dfrac{7t}{s^2}}$

Rationalize each denominator. Assume that all variables represent positive real numbers and that no denominators are 0. **See Example 6.**

83. $\dfrac{3}{4 + \sqrt{5}}$ **84.** $\dfrac{4}{5 + \sqrt{6}}$ **85.** $\dfrac{6}{\sqrt{5} + \sqrt{3}}$

86. $\dfrac{12}{\sqrt{6} + \sqrt{3}}$ **87.** $\dfrac{\sqrt{8}}{3 - \sqrt{2}}$ **88.** $\dfrac{\sqrt{27}}{3 - \sqrt{3}}$

89. $\dfrac{2}{3\sqrt{5} + 2\sqrt{3}}$ **90.** $\dfrac{-1}{3\sqrt{2} - 2\sqrt{7}}$ **91.** $\dfrac{\sqrt{2} - \sqrt{3}}{\sqrt{6} - \sqrt{5}}$

92. $\dfrac{\sqrt{5} + \sqrt{6}}{\sqrt{3} - \sqrt{2}}$ **93.** $\dfrac{m - 4}{\sqrt{m} + 2}$ **94.** $\dfrac{r - 9}{\sqrt{r} - 3}$

95. $\dfrac{4}{\sqrt{x} - 2\sqrt{y}}$ **96.** $\dfrac{5}{3\sqrt{r} + \sqrt{s}}$ **97.** $\dfrac{\sqrt{x} - \sqrt{y}}{\sqrt{x} + \sqrt{y}}$

98. $\dfrac{\sqrt{a} + \sqrt{b}}{\sqrt{a} - \sqrt{b}}$ **99.** $\dfrac{5\sqrt{k}}{2\sqrt{k} + \sqrt{q}}$ **100.** $\dfrac{3\sqrt{x}}{\sqrt{x} - 2\sqrt{y}}$

Write each quotient in lowest terms. Assume that all variables represent positive real numbers. See Example 7.

101. $\dfrac{30 + 20\sqrt{6}}{10}$ **102.** $\dfrac{24 + 12\sqrt{5}}{12}$ **103.** $\dfrac{3 - 3\sqrt{5}}{3}$ **104.** $\dfrac{-5 + 5\sqrt{2}}{5}$

105. $\dfrac{16 - 4\sqrt{8}}{12}$ **106.** $\dfrac{12 - 9\sqrt{72}}{18}$ **107.** $\dfrac{6p + \sqrt{24p^3}}{3p}$ **108.** $\dfrac{11y - \sqrt{242y^5}}{22y}$

Extending Skills *Rationalize each denominator. Assume that all radicals represent real numbers and that no denominators are 0.*

109. $\dfrac{3}{\sqrt{x + y}}$ **110.** $\dfrac{5}{\sqrt{m - n}}$ **111.** $\dfrac{p}{\sqrt{p + 2}}$ **112.** $\dfrac{q}{\sqrt{5 + q}}$

Solve each problem.

113. The following expression occurs in a standard problem in trigonometry.

$$\frac{1}{\sqrt{2}} \cdot \frac{\sqrt{3}}{2} - \frac{1}{\sqrt{2}} \cdot \frac{1}{2}$$

Show that it simplifies to $\dfrac{\sqrt{6} - \sqrt{2}}{4}$. Then verify, using a calculator approximation.

114. The following expression occurs in a standard problem in trigonometry.

$$\frac{\sqrt{3} + 1}{1 - \sqrt{3}}$$

Show that it simplifies to $-2 - \sqrt{3}$. Then verify, using a calculator approximation.

RELATING CONCEPTS For Individual or Group Work (Exercises 115–118)

In calculus, it is sometimes desirable to rationalize the numerator. To rationalize a numerator, we multiply the numerator and the denominator by the conjugate of the numerator. For example,

$$\frac{6 - \sqrt{2}}{4} = \frac{\left(6 - \sqrt{2}\right)\left(6 + \sqrt{2}\right)}{4\left(6 + \sqrt{2}\right)} = \frac{36 - 2}{4\left(6 + \sqrt{2}\right)} = \frac{34}{4\left(6 + \sqrt{2}\right)} = \frac{17}{2\left(6 + \sqrt{2}\right)}.$$

Rationalize each numerator. Assume that all variables represent positive real numbers.

115. $\dfrac{6 - \sqrt{3}}{8}$ **116.** $\dfrac{2\sqrt{5} - 3}{2}$ **117.** $\dfrac{2\sqrt{x} - \sqrt{y}}{3x}$ **118.** $\dfrac{\sqrt{p} - 3\sqrt{q}}{4q}$

SUMMARY EXERCISES   *Performing Operations with Radicals and Rational Exponents*

> ### Conditions for a Simplified Radical
>
> 1. The radicand has no factor raised to a power greater than or equal to the index.
> 2. The radicand has no fractions.
> 3. No denominator contains a radical.
> 4. Exponents in the radicand and the index of the radical have greatest common factor 1.

Concept Check   *Give the reason why each radical is not simplified.*

1. $\sqrt{\dfrac{2}{5}}$ 2. $\sqrt[15]{x^5}$ 3. $\dfrac{5}{\sqrt[3]{10}}$ 4. $\sqrt[3]{x^5 y^6}$

Perform the indicated operations, and express each answer in simplest form. Assume that all variables represent positive real numbers.

5. $6\sqrt{10} - 12\sqrt{10}$ 6. $\sqrt{7}\left(\sqrt{7} - \sqrt{2}\right)$ 7. $\left(1 - \sqrt{3}\right)\left(2 + \sqrt{6}\right)$

8. $\sqrt{50} - \sqrt{98} + \sqrt{72}$ 9. $\left(3\sqrt{5} + 2\sqrt{7}\right)^2$ 10. $\dfrac{-3}{\sqrt{6}}$

11. $\dfrac{8}{\sqrt{7} + \sqrt{5}}$ 12. $\dfrac{1 - \sqrt{2}}{1 + \sqrt{2}}$ 13. $\left(\sqrt{5} + 7\right)\left(\sqrt{5} - 7\right)$

14. $\dfrac{1}{\sqrt{x} - \sqrt{5}}, \quad x \neq 5$ 15. $\sqrt[3]{8a^3 b^5 c^9}$ 16. $\dfrac{15}{\sqrt[3]{9}}$

17. $\dfrac{3}{\sqrt{5} + 2}$ 18. $\sqrt{\dfrac{3}{5x}}$ 19. $\dfrac{16\sqrt{3}}{5\sqrt{12}}$

20. $\dfrac{2\sqrt{25}}{8\sqrt{50}}$ 21. $\dfrac{-10}{\sqrt[3]{10}}$ 22. $\dfrac{\sqrt{6} + \sqrt{5}}{\sqrt{6} - \sqrt{5}}$

23. $\sqrt{12x} - \sqrt{75x}$ 24. $\left(5 - 3\sqrt{3}\right)^2$ 25. $\sqrt[3]{\dfrac{13}{81}}$

26. $\dfrac{\sqrt{3} + \sqrt{7}}{\sqrt{6} - \sqrt{5}}$ 27. $\dfrac{6}{\sqrt[4]{3}}$ 28. $\sqrt[3]{\dfrac{x^2 y}{x^{-3} y^4}}$

29. $\sqrt{12} - \sqrt{108} - \sqrt[3]{27}$ 30. $\dfrac{4^{1/2} + 3^{1/2}}{4^{1/2} - 3^{1/2}}$

31. $\sqrt[3]{16x^2} - \sqrt[3]{54x^2} + \sqrt[3]{128x^2}$ 32. $\left(1 - \sqrt[3]{3}\right)\left(1 + \sqrt[3]{3} + \sqrt[3]{9}\right)$

Simplify each expression. Write answers with positive exponents. Assume that all variables represent positive real numbers.

33. $3^{1/2} \cdot 3^{1/3}$ 34. $\left(\dfrac{x^2 y}{x^{-3} y^4}\right)^{1/3}$ 35. $\dfrac{x^{-2/3} y^{4/5}}{x^{-5/3} y^{-2/5}}$

36. $\left(\dfrac{x^{3/4} y^{2/3}}{x^{1/3} y^{5/8}}\right)^{24}$ 37. $\left(125x^3\right)^{-2/3}$ 38. $\left(3x^{-2/3} y^{1/2}\right)\left(-2x^{5/8} y^{-1/3}\right)$

10.6 Solving Equations with Radicals

OBJECTIVES

1 Solve radical equations using the power rule.
2 Solve radical equations that require additional steps.
3 Solve radical equations with indexes greater than 2.
4 Use the power rule to solve a formula for a specified variable.

VOCABULARY

☐ radical equation
☐ proposed solution
☐ extraneous solution

OBJECTIVE 1 Solve radical equations using the power rule.

An equation that includes one or more radical expressions with a variable in a radicand is a **radical equation.**

$$\sqrt{x-4}=8, \quad \sqrt{5x+12}=3\sqrt{2x-1}, \quad \sqrt[3]{6+x}=27 \qquad \text{Radical equations}$$

Solving radical equations involves a process that we have not yet seen, and it requires careful application. Notice that the equation $x = 1$ has only one solution. Its solution set is $\{1\}$. If we square both sides of this equation, we obtain $x^2 = 1$. This new equation has *two* solutions: -1 and 1. The solution of the original equation is also a solution of the "squared" equation. However, that equation has another solution, -1, that is *not* a solution of the original equation.

When solving equations with radicals, we use this idea of raising both sides to a power, which is an application of the **power rule.**

Power Rule for Solving a Radical Equation

If both sides of an equation are raised to the same power, all solutions of the original equation are also solutions of the new equation.

The power rule does not say that all solutions of the new equation are solutions of the original equation. They may or may not be. A value of the variable that appears to be a solution is a **proposed solution.** Such values that do not satisfy the original equation are **extraneous solutions.** They must be rejected.

⚠ CAUTION *When the power rule is used to solve an equation, every solution of the new equation must be checked in the original equation.*

NOW TRY EXERCISE 1

Solve $\sqrt{9x+7}=5$.

EXAMPLE 1 Using the Power Rule

Solve $\sqrt{3x+4}=8$.

$\left(\sqrt{a}\right)^2 = \sqrt{a}\cdot\sqrt{a}=a$ for $a \geq 0$.

$$\left(\sqrt{3x+4}\right)^2 = 8^2 \qquad \text{Use the power rule and square each side.}$$

$$3x+4 = 64 \qquad \text{Apply the exponents.}$$

$$3x = 60 \qquad \text{Subtract 4.}$$

$$x = 20 \qquad \text{Divide by 3.}$$

CHECK

$$\sqrt{3x+4}=8 \qquad \text{Original equation}$$

$$\sqrt{3\cdot 20+4}\overset{?}{=}8 \qquad \text{Let } x = 20.$$

$$\sqrt{64}\overset{?}{=}8 \qquad \text{Simplify.}$$

$$8=8 \checkmark \qquad \text{True}$$

NOW TRY ANSWER
1. $\{2\}$

Because 20 satisfies the *original* equation, the solution set is $\{20\}$. **NOW TRY** 🔄

Solving an Equation with Radicals

Step 1 **Isolate the radical.** Make sure that one radical term is alone on one side of the equation.

Step 2 **Apply the power rule.** Raise each side of the equation to a power that is the same as the index of the radical.

Step 3 **Solve** the resulting equation. If it still contains a radical, repeat Steps 1 and 2.

Step 4 **Check** all proposed solutions in the *original* equation. Discard any values that are not solutions of the original equation.

⚠ **CAUTION** Remember to check all proposed solutions (Step 4) or an incorrect solution set may result.

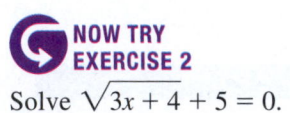

NOW TRY
EXERCISE 2
Solve $\sqrt{3x + 4} + 5 = 0$.

EXAMPLE 2 Using the Power Rule

Solve $\sqrt{5x - 1} + 3 = 0$.

Step 1	$\sqrt{5x - 1} = -3$	To isolate the radical on one side, subtract 3 from each side.
Step 2	$\left(\sqrt{5x - 1}\right)^2 = (-3)^2$	Square each side.
Step 3	$5x - 1 = 9$	Apply the exponents.
	$5x = 10$	Add 1.
	$x = 2$	Divide by 5.

Step 4 CHECK

Be sure to check the proposed solution.

$$\sqrt{5x - 1} + 3 = 0 \qquad \text{Original equation}$$
$$\sqrt{5 \cdot 2 - 1} + 3 \overset{?}{=} 0 \qquad \text{Let } x = 2.$$
$$\sqrt{9} + 3 \overset{?}{=} 0 \qquad \text{Evaluate the radicand.}$$
$$3 + 3 \overset{?}{=} 0 \qquad \text{Take the square root.}$$
$$6 = 0 \qquad \text{False}$$

This false result shows that the *proposed* solution 2 is *not* a solution of the original equation. It is extraneous. The solution set is $\varnothing$.   **NOW TRY** ↺

NOTE We could have determined after Step 1 that the equation in **Example 2** has no solution because the expression on the left cannot equal a negative number.

$$\sqrt{5x - 1} = -3 \qquad \begin{array}{l} \text{A square root radical} \\ \text{cannot be negative.} \end{array}$$

OBJECTIVE 2 Solve radical equations that require additional steps.

Recall the following rule for squaring a binomial.

$$(x + y)^2 = x^2 + 2xy + y^2$$

NOW TRY ANSWER
2. $\varnothing$

 **NOW TRY
EXERCISE 3**

Solve $\sqrt{16-x} = x + 4$.

EXAMPLE 3 Using the Power Rule (Squaring a Binomial)

Solve $\sqrt{4-x} = x + 2$.

Step 1 The radical is isolated on the left side of the equation.

Step 2 Square each side. On the right, $(x+2)^2 = x^2 + 2(x)(2) + 2^2$.

$$\left(\sqrt{4-x}\right)^2 = (x+2)^2 \quad \boxed{\text{Remember the middle term.}}$$

$$4 - x = x^2 + \underset{\underset{\text{Twice the product of 2 and } x}{\uparrow}}{4x} + 4$$

Step 3 The new equation is quadratic, so write it in standard form.

$$x^2 + 5x = 0 \qquad \text{Subtract 4, add } x, \text{ and interchange sides.}$$

$$x(x+5) = 0 \qquad \text{Factor.}$$

$$\boxed{\text{Set } \textit{each} \text{ factor equal to 0.}} \qquad x = 0 \quad \text{or} \quad x + 5 = 0 \qquad \text{Zero-factor property}$$

$$x = -5 \qquad \text{Solve.}$$

Step 4 Check each proposed solution in the original equation.

CHECK $\sqrt{4-x} = x + 2$ $\sqrt{4-x} = x + 2$

$\sqrt{4-0} \overset{?}{=} 0 + 2$ Let $x = 0$. $\sqrt{4-(-5)} \overset{?}{=} -5 + 2$ Let $x = -5$.

$\sqrt{4} \overset{?}{=} 2$ $\sqrt{9} \overset{?}{=} -3$

$2 = 2$ ✓   True $3 = -3$ False

The solution set is $\{0\}$. The proposed solution -5 is extraneous.   **NOW TRY**

 **NOW TRY
EXERCISE 4**

Solve
$\sqrt{x^2 - 3x + 18} = x + 3$.

EXAMPLE 4 Using the Power Rule (Squaring a Binomial)

Solve $\sqrt{x^2 - 4x + 9} = x - 1$.

Square each side. On the right, $(x-1)^2 = x^2 - 2(x)(1) + 1^2$.

$$\left(\sqrt{x^2 - 4x + 9}\right)^2 = (x-1)^2 \quad \boxed{\text{Remember the middle term.}}$$

$$x^2 - 4x + 9 = x^2 - \underset{\underset{\text{Twice the product of } x \text{ and } -1}{\uparrow}}{2x} + 1$$

$$-2x = -8 \qquad \text{Subtract } x^2 \text{ and 9. Add } 2x.$$

$$x = 4 \qquad \text{Divide by } -2.$$

CHECK $\sqrt{x^2 - 4x + 9} = x - 1$ Original equation

$\sqrt{4^2 - 4 \cdot 4 + 9} \overset{?}{=} 4 - 1$ Let $x = 4$.

$\sqrt{9} \overset{?}{=} 4 - 1$ Evaluate the radicand.

$3 = 3$ ✓   True

NOW TRY ANSWERS
3. $\{0\}$
4. $\{1\}$

The solution set is $\{4\}$.   **NOW TRY**

Solve

$$\sqrt{3x + 1} - \sqrt{x + 4} = 1.$$

EXAMPLE 5 Using the Power Rule (Squaring Twice)

Solve $\sqrt{5x + 6} + \sqrt{3x + 4} = 2$.

Isolate one radical on one side by subtracting $\sqrt{3x + 4}$ from each side.

$\sqrt{5x + 6} = 2 - \sqrt{3x + 4}$	Subtract $\sqrt{3x + 4}$.
$\left(\sqrt{5x + 6}\right)^2 = \left(2 - \sqrt{3x + 4}\right)^2$	Square each side.
$5x + 6 = 4 - 4\sqrt{3x + 4} + (3x + 4)$	Be careful here.

Remember the middle term.

Twice the product of 2 and $-\sqrt{3x + 4}$

The equation still contains a radical, so isolate the radical term on the right and square both sides again.

$5x + 6 = 8 - 4\sqrt{3x + 4} + 3x$	Combine like terms.
$2x - 2 = -4\sqrt{3x + 4}$	Subtract 8 and $3x$.
$x - 1 = -2\sqrt{3x + 4}$	Divide each term by 2 to obtain smaller numbers.
$(x - 1)^2 = \left(-2\sqrt{3x + 4}\right)^2$	Square each side again.
$x^2 - 2x + 1 = (-2)^2\left(\sqrt{3x + 4}\right)^2$	On the right, $(ab)^2 = a^2b^2$.
$x^2 - 2x + 1 = 4(3x + 4)$	Apply the exponents.
$x^2 - 2x + 1 = 12x + 16$	Distributive property
$x^2 - 14x - 15 = 0$	Standard form
$(x - 15)(x + 1) = 0$	Factor.
$x - 15 = 0 \quad \text{or} \quad x + 1 = 0$	Zero-factor property
$x = 15 \quad \text{or} \qquad x = -1$	Solve each equation.

Divide *each* term by 2.

CHECK First check the proposed solution 15. Then check -1.

$\sqrt{5x + 6} + \sqrt{3x + 4} = 2$	Original equation
$\sqrt{5(15) + 6} + \sqrt{3(15) + 4} \overset{?}{=} 2$	Let $x = 15$.
$\sqrt{81} + \sqrt{49} \overset{?}{=} 2$	Evaluate the radicands.
$9 + 7 \overset{?}{=} 2$	Take square roots.
$16 = 2$	False
$\sqrt{5x + 6} + \sqrt{3x + 4} = 2$	Original equation
$\sqrt{5(-1) + 6} + \sqrt{3(-1) + 4} \overset{?}{=} 2$	Let $x = -1$.
$\sqrt{1} + \sqrt{1} \overset{?}{=} 2$	Evaluate the radicands.
$1 + 1 \overset{?}{=} 2$	Take square roots.
$2 = 2 \checkmark$	True

The proposed solution 15 is extraneous and must be rejected, but -1 is valid. Thus, the solution set is $\{-1\}$.

NOW TRY

OBJECTIVE 3 Solve radical equations with indexes greater than 2.

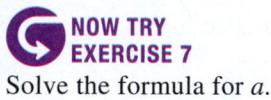
NOW TRY EXERCISE 6

Solve $\sqrt[3]{4x - 5} = \sqrt[3]{3x + 2}$.

EXAMPLE 6 Using the Power Rule for a Power Greater Than 2

Solve $\sqrt[3]{z + 5} = \sqrt[3]{2z - 6}$.

$$\left(\sqrt[3]{z + 5}\right)^3 = \left(\sqrt[3]{2z - 6}\right)^3 \qquad \text{Cube each side.}$$

$$z + 5 = 2z - 6 \qquad (\sqrt[3]{a})^3 = a$$

$$11 = z \qquad \text{Subtract } z. \text{ Add 6.}$$

CHECK

$$\sqrt[3]{z + 5} = \sqrt[3]{2z - 6} \qquad \text{Original equation}$$

$$\sqrt[3]{11 + 5} \overset{?}{=} \sqrt[3]{2 \cdot 11 - 6} \qquad \text{Let } z = 11.$$

$$\sqrt[3]{16} = \sqrt[3]{16} \checkmark \qquad \text{True}$$

The solution set is $\{11\}$.   NOW TRY

OBJECTIVE 4 Use the power rule to solve a formula for a specified variable.

NOW TRY EXERCISE 7

Solve the formula for a.

$$x = \sqrt{\frac{y + 2}{a}}$$

EXAMPLE 7 Solving a Formula from Electronics for a Variable

An important property of a radio-frequency transmission line is its **characteristic impedance**, represented by Z and measured in ohms. If L and C are the inductance and capacitance, respectively, per unit of length of the line, then these quantities are related by the following formula. Solve this formula for C.

$$Z = \sqrt{\frac{L}{C}}$$

> Our goal is to isolate C on one side of the equality symbol.

$$Z^2 = \left(\sqrt{\frac{L}{C}}\right)^2 \qquad \text{Square each side.}$$

$$Z^2 = \frac{L}{C} \qquad (\sqrt{a})^2 = a$$

$$CZ^2 = L \qquad \text{Multiply by } C.$$

$$C = \frac{L}{Z^2} \qquad \text{Divide by } Z^2.$$

NOW TRY

NOW TRY ANSWERS
6. $\{7\}$
7. $a = \dfrac{y + 2}{x^2}$

10.6 Exercises

FOR EXTRA HELP   ▶ **MyLab Math**

▶ *Video solutions for select problems available in MyLab Math*

Concept Check   *Check each equation to see if the given value for x is a solution.*

1. $\sqrt{3x + 18} - x = 0$
  (a) 6   (b) -3

2. $\sqrt{3x - 3} - x + 1 = 0$
  (a) 1   (b) 4

3. $\sqrt{x + 2} - \sqrt{9x - 2} = -2\sqrt{x - 1}$
  (a) 2   (b) 7

4. $\sqrt{8x - 3} = 2x$
  (a) $\dfrac{3}{2}$ (b) $\dfrac{1}{2}$

5. **Concept Check**   A student solved the following equation and claimed that 9 is a solution of the equation.

$$\sqrt{x} = -3$$

He received no credit for his answer. **WHAT WENT WRONG?** Give the correct solution set.

6. **Concept Check**   A student solved the following equation and obtained the proposed solutions $x = -3$ and $x = 6$.

$$\sqrt{3x + 18} = x$$

She gave $\{-3, 6\}$ as the solution set. **WHAT WENT WRONG?** Give the correct solution set.

Solve each equation. See Examples 1–4.

7. $\sqrt{x - 2} = 3$

8. $\sqrt{x + 1} = 7$

9. $\sqrt{6k - 1} = 1$

10. $\sqrt{7x - 3} = 6$

11. $\sqrt{4r + 3} + 1 = 0$

12. $\sqrt{5k - 3} + 2 = 0$

13. $\sqrt{3x + 1} - 4 = 0$

14. $\sqrt{5x + 1} - 11 = 0$

15. $4 - \sqrt{x - 2} = 0$

16. $9 - \sqrt{4x + 1} = 0$

17. $\sqrt{9x - 4} = \sqrt{8x + 1}$

18. $\sqrt{4x - 2} = \sqrt{3x + 5}$

19. $2\sqrt{x} = \sqrt{3x + 4}$

20. $2\sqrt{x} = \sqrt{5x - 16}$

21. $3\sqrt{x - 1} = 2\sqrt{2x + 2}$

22. $5\sqrt{4x + 1} = 3\sqrt{10x + 25}$

23. $x = \sqrt{x^2 + 4x - 20}$

24. $x = \sqrt{x^2 - 3x + 18}$

25. $x = \sqrt{x^2 + 3x + 9}$

26. $x = \sqrt{x^2 - 4x - 8}$

27. $\sqrt{9 - x} = x + 3$

28. $\sqrt{36 - x} = x + 6$

29. $\sqrt{5 - x} = x + 1$

30. $\sqrt{3 - x} = x + 3$

31. $\sqrt{6x + 7} = x + 2$

32. $\sqrt{4x + 13} = x + 4$

33. $\sqrt{k^2 + 2k + 9} = k + 3$

34. $\sqrt{x^2 - 3x + 3} = x - 1$

35. $\sqrt{z^2 + 12z - 4} + 4 - z = 0$

36. $\sqrt{x^2 - 15x + 15} + 5 - x = 0$

37. $\sqrt{r^2 + 9r + 15} - r - 4 = 0$

38. $\sqrt{m^2 + 3m + 12} - m - 2 = 0$

39. **Concept Check**   When solving the equation $\sqrt{3x + 4} = 8 - x$, a student wrote the following as her first step.

$$3x + 4 = 64 + x^2$$

WHAT WENT WRONG? Solve the given equation correctly.

40. **Concept Check**   When solving the equation $\sqrt{5x + 6} - \sqrt{x + 3} = 3$, a student wrote the following as his first step.

$$(5x + 6) + (x + 3) = 9$$

WHAT WENT WRONG? Solve the given equation correctly.

Solve each equation. See Example 5.

41. $\sqrt{k + 2} - \sqrt{k - 3} = 1$

42. $\sqrt{r + 6} - \sqrt{r - 2} = 2$

43. $\sqrt{2r + 11} - \sqrt{5r + 1} = -1$

44. $\sqrt{3x - 2} - \sqrt{x + 3} = 1$

45. $\sqrt{3p + 4} - \sqrt{2p - 4} = 2$

46. $\sqrt{4x + 5} - \sqrt{2x + 2} = 1$

47. $\sqrt{3 - 3p} - 3 = \sqrt{3p + 2}$

48. $\sqrt{4x + 7} - 4 = \sqrt{4x - 1}$

49. $\sqrt{2\sqrt{x + 11}} = \sqrt{4x + 2}$

50. $\sqrt{1 + \sqrt{24 - 10x}} = \sqrt{3x + 5}$

Solve each equation. See Example 6.

51. $\sqrt[3]{p-1} = 2$

52. $\sqrt[3]{x+8} = 3$

53. $\sqrt[3]{2x+5} = \sqrt[3]{6x+1}$

54. $\sqrt[3]{p+5} = \sqrt[3]{2p-4}$

55. $\sqrt[3]{2m-1} = \sqrt[3]{m+13}$

56. $\sqrt[3]{2k-11} = \sqrt[3]{5k+1}$

57. $\sqrt[3]{x^2+5x+1} = \sqrt[3]{x^2+4x}$

58. $\sqrt[3]{r^2+2r+8} = \sqrt[3]{r^2+3r+12}$

59. $\sqrt[4]{x+12} = \sqrt[4]{3x-4}$

60. $\sqrt[4]{z+11} = \sqrt[4]{2z+6}$

61. $\sqrt[3]{x-8} + 2 = 0$

62. $\sqrt[3]{r+1} + 1 = 0$

63. $\sqrt[4]{2k-5} + 4 = 0$

64. $\sqrt[4]{8z-3} + 2 = 0$

65. $\sqrt[3]{r^2+2r+8} = \sqrt[3]{r^2+3r+12}$

66. $\sqrt[3]{x^2+7x+2} = \sqrt[3]{x^2+6x+1}$

Extending Skills   *For each equation, write the expressions with rational exponents as radical expressions, and then solve, using the procedures explained in this section.*

67. $(2x-9)^{1/2} = 2 + (x-8)^{1/2}$

68. $(3w+7)^{1/2} = 1 + (w+2)^{1/2}$

69. $(2w-1)^{2/3} - w^{1/3} = 0$

70. $(x^2-2x)^{1/3} - x^{1/3} = 0$

Solve each formula for the specified variable. See Example 7. (Data from Cooke, N., and J. Orleans, Mathematics Essential to Electricity and Radio, *McGraw-Hill.)*

71. $Z = \sqrt{\dfrac{L}{C}}$ for L

72. $r = \sqrt{\dfrac{\mathcal{A}}{\pi}}$ for $\mathcal{A}$

73. $V = \sqrt{\dfrac{2K}{m}}$ for K

74. $V = \sqrt{\dfrac{2K}{m}}$ for m

75. $r = \sqrt{\dfrac{Mm}{F}}$ for M

76. $r = \sqrt{\dfrac{Mm}{F}}$ for F

To find the rotational rate N of a space station, the formula

$$N = \frac{1}{2\pi}\sqrt{\frac{a}{r}}$$

can be used. Here, a is the acceleration and r represents the radius of the space station in meters. To find the value of r that will make N simulate the effect of gravity on Earth, the equation must be solved for r, using the required value of N. (Data from Kastner, B., Space Mathematics, *NASA.)*

77. Solve the equation for the indicated variable.

   (a) for r **(b)** for a

78. If $a = 9.8$ m per sec^2, find the value of r (to the nearest tenth) using each value of N.

   (a) $N = 0.063$ rotation per sec   **(b)** $N = 0.04$ rotation per sec

10.7   Complex Numbers

OBJECTIVES

1 Simplify numbers of the form $\sqrt{-b}$, where $b > 0$.
2 Identify subsets of the complex numbers.
3 Add and subtract complex numbers.
4 Multiply complex numbers.
5 Divide complex numbers.
6 Simplify powers of i.

VOCABULARY
□ complex number
□ real part
□ imaginary part
□ nonreal complex number
□ pure imaginary number
□ complex conjugates

Recall that the set of real numbers includes many other number sets—the rational numbers, integers, and natural numbers, for example. In this section, we introduce a new set of numbers that includes the set of real numbers, as well as numbers that are even roots of negative numbers, like $\sqrt{-2}$.

OBJECTIVE 1 Simplify numbers of the form $\sqrt{-b}$, where $b > 0$.

The equation $x^2 + 1 = 0$ has no real number solution because any solution must be a number whose square is -1. In the set of real numbers, all squares are *nonnegative* numbers because the product of two positive numbers or two negative numbers is positive and $0^2 = 0$. To provide a solution of the equation

$$x^2 + 1 = 0,$$

we introduce a new number i.

Imaginary Unit i

The **imaginary unit i** is defined as follows.

$$i = \sqrt{-1}, \quad \text{and thus} \quad i^2 = -1$$

That is, i is the principal square root of -1.

We can use this definition to define any square root of a negative real number.

Meaning of $\sqrt{-b}$

For any positive real number b, $\sqrt{-b} = i\sqrt{b}$.

NOW TRY EXERCISE 1

Write each number as a product of a real number and i.

(a) $\sqrt{-49}$ **(b)** $-\sqrt{-121}$
(c) $\sqrt{-3}$ **(d)** $\sqrt{-32}$

EXAMPLE 1   Simplifying Square Roots of Negative Numbers

Write each number as a product of a real number and i.

(a) $\sqrt{-100} = i\sqrt{100} = 10i$

(b) $-\sqrt{-36} = -i\sqrt{36} = -6i$

(c) $\sqrt{-2} = i\sqrt{2}$

(d) $\sqrt{-54} = i\sqrt{54} = i\sqrt{9 \cdot 6} = 3i\sqrt{6}$

NOW TRY

! CAUTION   It is easy to mistake $\sqrt{2}i$ for $\sqrt{2i}$ with the i under the radical. For this reason, we usually write $\sqrt{2}i$ as $i\sqrt{2}$, as in the definition of $\sqrt{-b}$.

When finding a product such as $\sqrt{-4} \cdot \sqrt{-9}$, we cannot use the product rule for radicals because it applies only to *nonnegative* radicands.

For this reason, we change $\sqrt{-b}$ to the form $i\sqrt{b}$ before performing any multiplications or divisions.

NOW TRY ANSWERS
1. (a) $7i$ **(b)** $-11i$
   (c) $i\sqrt{3}$ **(d)** $4i\sqrt{2}$

**NOW TRY
EXERCISE 2**

Multiply.

(a) $\sqrt{-4} \cdot \sqrt{-16}$

(b) $\sqrt{-5} \cdot \sqrt{-11}$

(c) $\sqrt{-3} \cdot \sqrt{-12}$

(d) $\sqrt{13} \cdot \sqrt{-2}$

EXAMPLE 2 Multiplying Square Roots of Negative Numbers

Multiply.

(a) $\sqrt{-4} \cdot \sqrt{-9}$

First write all square roots in terms of i.

$$= i\sqrt{4} \cdot i\sqrt{9} \qquad \sqrt{-b} = i\sqrt{b}$$
$$= i \cdot 2 \cdot i \cdot 3 \qquad \text{Take square roots.}$$
$$= 6i^2 \qquad \text{Multiply.}$$
$$= 6(-1) \qquad \text{Substitute } -1 \text{ for } i^2.$$
$$= -6$$

(b) $\sqrt{-3} \cdot \sqrt{-7}$

First write all square roots in terms of i.

$$= i\sqrt{3} \cdot i\sqrt{7} \qquad \sqrt{-b} = i\sqrt{b}$$
$$= i^2\sqrt{3 \cdot 7} \qquad \text{Product rule}$$
$$= (-1)\sqrt{21} \qquad \text{Substitute } -1 \text{ for } i^2.$$
$$= -\sqrt{21} \qquad (-1)a = -a$$

(c) $\sqrt{-2} \cdot \sqrt{-8}$

$$= i\sqrt{2} \cdot i\sqrt{8} \qquad \sqrt{-b} = i\sqrt{b}$$
$$= i^2\sqrt{2 \cdot 8} \qquad \text{Product rule}$$
$$= (-1)\sqrt{16} \qquad i^2 = -1$$
$$= -4 \qquad \text{Take the square root.}$$

(d) $\sqrt{-5} \cdot \sqrt{6}$

$$= i\sqrt{5} \cdot \sqrt{6}$$
$$= i\sqrt{30}$$

NOW TRY

⚠ **CAUTION** Using the product rule for radicals *before* using the definition of $\sqrt{-b}$ gives an *incorrect* answer. **Example 2(a)** shows that

$$\sqrt{-4} \cdot \sqrt{-9} = -6, \qquad \text{Correct (Example 2(a))}$$

but

$$\sqrt{-4(-9)} = \sqrt{36} = 6. \qquad \text{Incorrect}$$

Thus,

$$\sqrt{-4} \cdot \sqrt{-9} \neq \sqrt{-4(-9)}.$$

**NOW TRY
EXERCISE 3**

Divide.

(a) $\dfrac{\sqrt{-72}}{\sqrt{-8}}$ **(b)** $\dfrac{\sqrt{-48}}{\sqrt{3}}$

EXAMPLE 3 Dividing Square Roots of Negative Numbers

Divide.

(a) $\dfrac{\sqrt{-75}}{\sqrt{-3}}$

First write all square roots in terms of i.

$$= \frac{i\sqrt{75}}{i\sqrt{3}}$$
$$= \sqrt{\frac{75}{3}} \qquad \frac{i}{i} = 1; \text{ Quotient rule}$$
$$= \sqrt{25} \qquad \text{Divide.}$$
$$= 5$$

(b) $\dfrac{\sqrt{-32}}{\sqrt{8}}$

$$= \frac{i\sqrt{32}}{\sqrt{8}} \qquad \sqrt{-32} = i\sqrt{32}$$
$$= i\sqrt{\frac{32}{8}} \qquad \text{Quotient rule}$$
$$= i\sqrt{4} \qquad \text{Divide.}$$
$$= 2i$$

NOW TRY

NOW TRY ANSWERS

2. **(a)** -8 **(b)** $-\sqrt{55}$
  (c) -6 **(d)** $i\sqrt{26}$

3. **(a)** 3 **(b)** $4i$

OBJECTIVE 2 Identify subsets of the complex numbers.

A new set of numbers, the *complex numbers,* is defined as follows.

Complex Number

If a and b are real numbers, then any number of the form

$$a + bi$$

Real part   Imaginary part

is a **complex number.** In the complex number $a + bi$, the number a is the **real part** and b is the **imaginary part.***

The following important concepts apply to a complex number $a + bi$.

1. If $b = 0$, then $a + bi = a$, which is a real number.

   Thus, the set of real numbers is a subset of the set of complex numbers. See **FIGURE 14.**

2. If $b \neq 0$, then $a + bi$ is a **nonreal complex number.**

   Examples: $7 + 2i$, $-1 - i$

3. If $a = 0$ and $b \neq 0$, then the nonreal complex number is a **pure imaginary number.**

   Examples: $3i$, $-16i$

A complex number written in the form $a + bi$ is in **standard form.** In this section, most answers will be given in standard form, but if $a = 0$ or $b = 0$, we consider answers such as a or bi to be in standard form.

The relationships among the various sets of numbers are shown in **FIGURE 14.**

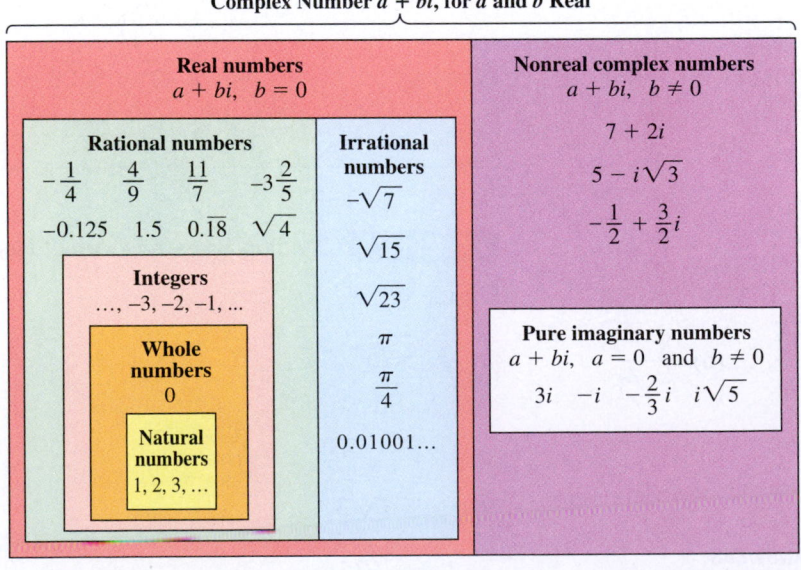

FIGURE 14

---

*Some texts define bi as the imaginary part of the complex number $a + bi$.

OBJECTIVE 3 Add and subtract complex numbers.

The commutative, associative, and distributive properties for real numbers are also valid for complex numbers.

To add complex numbers, we add their real parts and add their imaginary parts.

NOW TRY EXERCISE 4
Add.
(a) $(-3 + 2i) + (4 + 7i)$
(b) $(5 - i) + (-3 + 3i)$ $+ (6 - 4i)$

EXAMPLE 4 Adding Complex Numbers

Add.

(a) $(2 + 3i) + (6 + 4i)$

$= (2 + 6) + (3 + 4)i$ Commutative, associative, and distributive properties

$= 8 + 7i$ Add real parts. Add imaginary parts.

(b) $(4 + 2i) + (3 - i) + (-6 + 3i)$

$= [4 + 3 + (-6)] + [2 + (-1) + 3]i$ Commutative, associative, and distributive properties

$= 1 + 4i$ Add real parts. Add imaginary parts.   **NOW TRY**

To subtract complex numbers, we subtract their real parts and subtract their imaginary parts.

NOW TRY EXERCISE 5
Subtract.
(a) $(7 + 10i) - (3 + 5i)$
(b) $(5 - 2i) - (9 - 7i)$
(c) $(-1 + 12i) - (-1 - i)$

EXAMPLE 5 Subtracting Complex Numbers

Subtract.

(a) $(6 + 5i) - (3 + 2i)$

$= (6 - 3) + (5 - 2)i$ Commutative, associative, and distributive properties

$= 3 + 3i$ Subtract real parts. Subtract imaginary parts.

(b) $(7 - 3i) - (8 - 6i)$

$= (7 - 8) + [-3 - (-6)]i$

$= -1 + 3i$

(c) $(-9 + 4i) - (-9 + 8i)$

$= (-9 + 9) + (4 - 8)i$ Be careful.

$= 0 - 4i$

$= -4i$ **NOW TRY**

OBJECTIVE 4 Multiply complex numbers.

We multiply complex numbers in the same way that we multiply polynomials.

EXAMPLE 6 Multiplying Complex Numbers

Multiply.

(a) $4i(2 + 3i)$

$= 4i(2) + 4i(3i)$ Distributive property

$= 8i + 12i^2$ Multiply.

$= 8i + 12(-1)$ Substitute -1 for i^2.

$= -12 + 8i$ Standard form

NOW TRY ANSWERS
4. **(a)** $1 + 9i$ **(b)** $8 - 2i$
5. **(a)** $4 + 5i$ **(b)** $-4 + 5i$
(c) $13i$

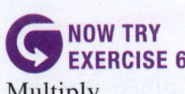

**NOW TRY
EXERCISE 6**
Multiply.

(a) $8i(3 - 5i)$

(b) $(7 - 2i)(4 + 3i)$

(b) $(3 + 5i)(4 - 2i)$

$= \underbrace{3(4)}_{\text{First}} + \underbrace{3(-2i)}_{\text{Outer}} + \underbrace{5i(4)}_{\text{Inner}} + \underbrace{5i(-2i)}_{\text{Last}}$ Use the FOIL method.

$= 12 - 6i + 20i - 10i^2$ Multiply.

$= 12 + 14i - 10(-1)$ Add imaginary parts; $i^2 = -1$

$= 12 + 14i + 10$ Multiply.

$= 22 + 14i$ Add real parts.

(c) $(2 + 3i)(1 - 5i)$

$= 2(1) + 2(-5i) + 3i(1) + 3i(-5i)$ FOIL method.

$= 2 - 10i + 3i - 15i^2$ Multiply.

$= 2 - 7i - 15(-1)$ Add imaginary parts; $i^2 = -1$

> Use parentheses around -1 to avoid errors.

$= 2 - 7i + 15$ Multiply.

$= 17 - 7i$ Add real parts.   **NOW TRY**

The two complex numbers $a + bi$ and $a - bi$ are **complex conjugates,** or simply *conjugates,* of each other. ***The product of a complex number and its conjugate is always a real number,*** as shown here.

$$(a + bi)(a - bi)$$

$= a^2 - abi + abi - b^2i^2$ FOIL method

$= a^2 - b^2(-1)$ Combine like terms; $i^2 = -1$

$= a^2 + b^2$ > The product eliminates i.

Example: $(3 + 7i)(3 - 7i) = 3^2 + 7^2 = 9 + 49 = 58$

OBJECTIVE 5 Divide complex numbers.

EXAMPLE 7 Dividing Complex Numbers

Divide.

(a) $\dfrac{8 + 9i}{5 + 2i}$

> $\dfrac{5 - 2i}{5 - 2i} = 1$

$= \dfrac{(8 + 9i)(5 - 2i)}{(5 + 2i)(5 - 2i)}$ Multiply numerator and denominator by $5 - 2i$, the conjugate of the denominator.

$= \dfrac{40 - 16i + 45i - 18i^2}{5^2 + 2^2}$ In the denominator, $(a + bi)(a - bi) = a^2 + b^2$.

$= \dfrac{40 + 29i - 18(-1)}{25 + 4}$ In the numerator, add imaginary parts; $i^2 = -1$

$= \dfrac{58 + 29i}{29}$ Multiply. Add real parts.
    Add in the denominator.

> Factor first. Then divide out the common factor.

$= \dfrac{29(2 + i)}{29}$ Factor the numerator.

$= 2 + i$ Lowest terms

NOW TRY ANSWERS
6. (a) $40 + 24i$ **(b)** $34 + 13i$

 **NOW TRY EXERCISE 7**

Find each quotient.

(a) $\dfrac{4 + 2i}{1 + 3i}$ (b) $\dfrac{5 - 4i}{i}$

(b) $\dfrac{1 + i}{i}$

$= \dfrac{(1 + i)(-i)}{i(-i)}$ Multiply numerator and denominator by $-i$, the conjugate of i.

$= \dfrac{-i - i^2}{-i^2}$ Use the distributive property in the numerator.
Multiply in the denominator.

$= \dfrac{-i - (-1)}{-(-1)}$ Substitute -1 for i^2.

> Use parentheses to avoid errors.

$= \dfrac{-i + 1}{1}$

$= 1 - i$ $\dfrac{a}{1} = a$ **NOW TRY**

OBJECTIVE 6 Simplify powers of i.

Powers of i can be simplified using the facts

$$i^2 = -1 \quad \text{and} \quad i^4 = (i^2)^2 = (-1)^2 = 1.$$

Consider the following powers of i.

$i^1 = i$ $i^5 = i^4 \cdot i = 1 \cdot i = i$

$i^2 = -1$ $i^6 = i^4 \cdot i^2 = 1(-1) = -1$

$i^3 = i^2 \cdot i = (-1) \cdot i = -i$ $i^7 = i^4 \cdot i^3 = 1 \cdot (-i) = -i$

$i^4 = i^2 \cdot i^2 = (-1)(-1) = 1$ $i^8 = i^4 \cdot i^4 = 1 \cdot 1 = 1$,   and so on.

Powers of i cycle through the same four outcomes

$$i, \quad -1, \quad -i, \quad \text{and} \quad 1$$

*because i^4 has the same multiplicative property as **1**. Also, any power of i with an exponent that is a multiple of **4** has value **1**. As with real numbers, $i^0 = 1$.*

 **NOW TRY EXERCISE 8**

Find each power of i.

(a) i^{16} (b) i^{21}

(c) i^{-6} (d) i^{-13}

EXAMPLE 8 Simplifying Powers of i

Find each power of i.

(a) $i^{12} = (i^4)^3 = 1^3 = 1$ $i^4 = 1$

(b) $i^{39} = i^{36} \cdot i^3 = (i^4)^9 \cdot i^3 = 1^9 \cdot (-i) = 1 \cdot (-i) = -i$

(c) $i^{-2} = \dfrac{1}{i^2} = \dfrac{1}{-1} = -1$

(d) $i^{-15} = \dfrac{1}{i^{15}} = \dfrac{1}{(i^4)^3 \cdot i^3} = \dfrac{1}{1^3 \cdot (-i)} = \dfrac{1}{-i}$

We divide by multiplying the numerator and denominator by i, the conjugate of $-i$.

$$\dfrac{1}{-i} = \dfrac{1(i)}{-i(i)} = \dfrac{i}{-i^2} = \dfrac{i}{-(-1)} = \dfrac{i}{1} = i$$ **NOW TRY**

NOTE In **Example 8(d)**, we could also multiply by a power of i that is a multiple of 4 (because $i^4 = 1$).

$$i^{-15} = i^{-15} \cdot i^{16} = i^1 = i \qquad i^{16} = 1$$

NOW TRY ANSWERS

7. (a) $1 - i$ (b) $-4 - 5i$

8. (a) 1 (b) i (c) -1 (d) $-i$

10.7 Exercises

FOR EXTRA HELP   **MyLab Math**

▶ *Video solutions for select problems available in MyLab Math*

Concept Check   *List all of the following sets to which each number belongs. A number may belong to more than one set.*

real numbers   pure imaginary numbers   nonreal complex numbers   complex numbers

1. $3 + 5i$

2. $-7i$

3. $\sqrt{2}$

4. $\dfrac{13}{3}$

5. $\sqrt{-49}$

6. $-\sqrt{-8}$

Concept Check   *Decide whether each expression is equal to* $1, -1, i,$ *or* $-i$.

7. $\sqrt{-1}$

8. $-\sqrt{-1}$

9. i^2

10. $-i^2$

11. $\dfrac{1}{i}$

12. $(-i)^2$

Write each number as a product of a real number and i. Simplify all radical expressions. **See Example 1.**

13. $\sqrt{-169}$

14. $\sqrt{-225}$

15. $-\sqrt{-144}$

16. $-\sqrt{-196}$

17. $\sqrt{-5}$

18. $\sqrt{-21}$

19. $\sqrt{-48}$

20. $\sqrt{-96}$

21. Concept Check   A student incorrectly multiplied as follows.
$$\sqrt{-9} \cdot \sqrt{-25} = \sqrt{225} = 15$$
WHAT WENT WRONG?   Find the product correctly.

22. Concept Check   A student incorrectly simplified $-\sqrt{-16}$ as follows.
$$-\sqrt{-16} = -(-4) = 4$$
WHAT WENT WRONG?   Simplify the expression correctly.

Multiply or divide as indicated. **See Examples 2 and 3.**

23. $\sqrt{-15} \cdot \sqrt{-15}$

24. $\sqrt{-19} \cdot \sqrt{-19}$

25. $\sqrt{-7} \cdot \sqrt{-15}$

26. $\sqrt{-3} \cdot \sqrt{-19}$

27. $\sqrt{-4} \cdot \sqrt{-25}$

28. $\sqrt{-9} \cdot \sqrt{-81}$

29. $\sqrt{-3} \cdot \sqrt{11}$

30. $\sqrt{-5} \cdot \sqrt{13}$

31. $\sqrt{5} \cdot \sqrt{-30}$

32. $\sqrt{-10} \cdot \sqrt{2}$

33. $\dfrac{\sqrt{-300}}{\sqrt{-100}}$

34. $\dfrac{\sqrt{-40}}{\sqrt{-10}}$

35. $\dfrac{\sqrt{-75}}{\sqrt{3}}$

36. $\dfrac{\sqrt{-160}}{\sqrt{10}}$

37. $\dfrac{-\sqrt{-64}}{\sqrt{-16}}$

38. $\dfrac{-\sqrt{-100}}{\sqrt{-25}}$

Add or subtract as indicated. Give answers in standard form. **See Examples 4 and 5.**

39. $(3 + 2i) + (-4 + 5i)$

40. $(7 + 15i) + (-11 + 14i)$

41. $(5 - i) + (-5 + i)$

42. $(-2 + 6i) + (2 - 6i)$

43. $(4 + i) - (-3 - 2i)$

44. $(9 + i) - (3 + 2i)$

45. $(-3 - 4i) - (-1 - 4i)$

46. $(-2 - 3i) - (-5 - 3i)$

47. $(-4 + 11i) + (-2 - 4i) + (7 + 6i)$

48. $(-1 + i) + (2 + 5i) + (3 + 2i)$

49. $[(7 + 3i) - (4 - 2i)] + (3 + i)$

50. $[(7 + 2i) + (-4 - i)] - (2 + 5i)$

Concept Check   *Fill in the blank with the correct response.*

51. Because $(4 + 2i) - (3 + i) = 1 + i$, using the definition of subtraction we can check this to find that

$$(1 + i) + (3 + i) = \underline{\hspace{1cm}}.$$

52. Because $\frac{-5}{2 - i} = -2 - i$, using the definition of division we can check this to find that

$$(-2 - i)(2 - i) = \underline{\hspace{1cm}}.$$

*Multiply. Give answers in standard form. **See Example 6.***

53. $(3i)(27i)$ **54.** $(5i)(125i)$ **55.** $(-8i)(-2i)$

56. $(-32i)(-2i)$ **57.** $5i(-6 + 2i)$ **58.** $3i(4 + 9i)$

59. $(4 + 3i)(1 - 2i)$ **60.** $(7 - 2i)(3 + i)$ **61.** $(4 + 5i)^2$

62. $(3 + 2i)^2$ **63.** $2i(-4 - i)^2$ **64.** $3i(-3 - i)^2$

65. $(12 + 3i)(12 - 3i)$ **66.** $(6 + 7i)(6 - 7i)$ **67.** $(4 + 9i)(4 - 9i)$

68. $(7 + 2i)(7 - 2i)$ **69.** $(1 + i)^2(1 - i)^2$ **70.** $(2 - i)^2(2 + i)^2$

Concept Check   *Answer each of the following.*

71. Let a and b represent real numbers.

    (a) What is the conjugate of $a + bi$?

    (b) If we multiply $a + bi$ by its conjugate, we obtain $\underline{\hspace{1cm}} + \underline{\hspace{1cm}}$, which is always a real number.

72. By what complex number should we multiply the numerator and denominator of $\frac{2 + i\sqrt{2}}{2 - i\sqrt{2}}$ to write the quotient in standard form?

    A. $\sqrt{2}$ **B.** $i\sqrt{2}$ **C.** $2 + i\sqrt{2}$ **D.** $2 - i\sqrt{2}$

*Divide. Give answers in standard form. **See Example 7.***

73. $\dfrac{2}{1 - i}$ **74.** $\dfrac{2}{1 + i}$ **75.** $\dfrac{8i}{2 + 2i}$

76. $\dfrac{-8i}{1 + i}$ **77.** $\dfrac{-7 + 4i}{3 + 2i}$ **78.** $\dfrac{-38 - 8i}{7 + 3i}$

79. $\dfrac{2 - 3i}{2 + 3i}$ **80.** $\dfrac{-1 + 5i}{3 + 2i}$ **81.** $\dfrac{3 + i}{i}$

82. $\dfrac{5 - i}{i}$ **83.** $\dfrac{3 - i}{-i}$ **84.** $\dfrac{5 + i}{-i}$

*Find each power of i. **See Example 8.***

85. i^{18} **86.** i^{26} **87.** i^{89} **88.** i^{48}

89. i^{38} **90.** i^{102} **91.** i^{43} **92.** i^{83}

93. i^{-5} **94.** i^{-17} **95.** i^{-20} **96.** i^{-27}

Ohm's law for the current I in a circuit with voltage E, resistance R, capacitive reactance X_c, and inductive reactance X_L is

$$I = \frac{E}{R + (X_L - X_c)i}.$$

Use this law to work each exercise.

97. Find I if $E = 2 + 3i$, $R = 5$, $X_L = 4$, and $X_c = 3$.

98. Find E if $I = 1 - i$, $R = 2$, $X_L = 3$, and $X_c = 1$.

RELATING CONCEPTS For Individual or Group Work (Exercises 99–102)

Some equations have nonreal complex solutions. **Work Exercises 99–102 in order,** to see how these nonreal complex solutions are related.

99. Consider the equation $x^2 - 2x + 26 = 0$. Show that each number is a solution of the equation.

(a) $1 + 5i$ (b) $1 - 5i$

100. From **Exercise 99,** the nonreal complex solutions of the equation

$$x^2 - 2x + 26 = 0$$

are $1 + 5i$ and $1 - 5i$. What do we call two complex numbers $a + bi$ and $a - bi$?

101. Show that $3 + 2i$ is a solution of the equation $x^2 - 6x + 13 = 0$.

102. Using the results of **Exercises 99–101,** make a conjecture about another nonreal complex solution of the equation $x^2 - 6x + 13 = 0$, and verify it.

Chapter 10   Summary

Key Terms

10.1

square root
positive (principal)
  square root
negative square root
radicand
radical
radical expression
perfect square
irrational number
cube root

fourth root
index (order)
perfect cube
perfect fourth power
square root function
cube root function

10.3

hypotenuse
legs (of a right triangle)
circle

center
radius

10.5

conjugates

10.6

radical equation
proposed solution
extraneous solution

10.7

complex number
real part
imaginary part
nonreal complex
  number
pure imaginary
  number
complex conjugates

New Symbols

$\sqrt{}$ radical symbol

$\sqrt[n]{a}$ radical; principal nth root of a

$\pm$ *"positive or negative"* or *"plus or minus"*

$\approx$ is approximately equal to

$a^{1/n}$ a to the power $\frac{1}{n}$

$a^{m/n}$ a to the power $\frac{m}{n}$

i imaginary unit

Test Your Word Power

See how well you have learned the vocabulary in this chapter.

1. A **radicand** is
   A. the index of a radical
   B. the number or expression under the radical symbol
   C. the positive root of a number
   D. the radical symbol.

2. The **Pythagorean theorem** states that, in a right triangle,
   A. the sum of the measures of the angles is 180°
   B. the sum of the lengths of the two shorter sides equals the length of the longest side
   C. the longest side is opposite the right angle
   D. the square of the length of the longest side equals the sum of the squares of the lengths of the two shorter sides.

3. A **hypotenuse** is
   A. either of the two shorter sides of a triangle
   B. the shortest side of a triangle
   C. the side opposite the right angle in a triangle
   D. the longest side in any triangle.

4. **Rationalizing the denominator** is the process of
   A. eliminating fractions from a radical expression
   B. changing the denominator of a fraction from a radical expression to a rational number
   C. clearing a radical expression of radicals
   D. multiplying radical expressions.

5. An **extraneous solution** is a value
   A. that does not satisfy the original equation
   B. that makes an equation true
   C. that makes an expression equal 0
   D. that checks in the original equation.

6. A **complex number** is
   A. a real number that includes a complex fraction
   B. a zero multiple of i
   C. a number of the form $a + bi$, where a and b are real numbers
   D. the square root of -1.

ANSWERS

1. B; *Example:* In $\sqrt{3xy}$, $3xy$ is the radicand.   2. D; *Example:* In a right triangle where $a = 6$, $b = 8$, and $c = 10$, $6^2 + 8^2 = 10^2$.

3. C; *Example:* In a right triangle where the sides measure 9, 12, and 15 units, the hypotenuse is the side opposite the right angle, with measure 15 units.

4. B; *Example:* To rationalize the denominator of $\dfrac{5}{\sqrt{3}+1}$, multiply both numerator and denominator by $\sqrt{3} - 1$ to obtain $\dfrac{5(\sqrt{3}-1)}{2}$.

5. A; *Example:* The proposed solution 2 is extraneous in $\sqrt{5x-1} + 3 = 0$.   6. C; *Examples:* -5 (or $-5 + 0i$), $7i$ (or $0 + 7i$), $\sqrt{2} - 4i$

Quick Review

CONCEPTS	EXAMPLES

10.1 Radical Expressions and Graphs

$\sqrt[n]{a} = b$ means $b^n = a$.

$\sqrt[n]{a}$ is the **principal nth root** of a.

$\sqrt[n]{a^n} = |a|$ if n is even. $\sqrt[n]{a^n} = a$ if n is odd.

Functions Defined by Radical Expressions
The square root function is
$$f(x) = \sqrt{x}.$$

The cube root function is
$$f(x) = \sqrt[3]{x}.$$

Find each root.

$$\sqrt{64} = 8 \quad \text{Principal square root}$$

$$-\sqrt{64} = -8 \qquad \sqrt{-64} \text{ is not a real number.}$$

$$\sqrt[4]{(-2)^4} = |-2| = 2 \qquad \sqrt[3]{-27} = -3$$

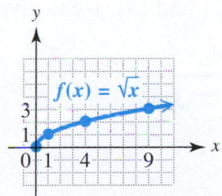

Square root function
Domain and range: $[0, \infty)$

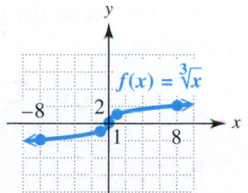

Cube root function
Domain and range: $(-\infty, \infty)$

CONCEPTS	EXAMPLES

10.2 Rational Exponents

$a^{1/n} = \sqrt[n]{a}$ whenever $\sqrt[n]{a}$ exists.

If m and n are positive integers with m/n in lowest terms, then

$$a^{m/n} = (a^{1/n})^m,$$ provided that $a^{1/n}$ is a real number.

All of the usual definitions and rules for exponents are valid for rational exponents.

Apply the rules for rational exponents.

$$81^{1/2} = \sqrt{81} = 9 \qquad -64^{1/3} = -\sqrt[3]{64} = -4$$

$$8^{5/3} = (8^{1/3})^5 = 2^5 = 32 \qquad (y^{2/5})^{10} = y^4$$

Write with positive exponents.

$$5^{-1/2} \cdot 5^{1/4} = 5^{-1/2+1/4} \qquad \frac{x^{-1/3}}{x^{-1/2}} = x^{-1/3-(-1/2)}$$

$$= 5^{-1/4} \qquad\qquad = x^{-1/3+1/2}$$

$$= \frac{1}{5^{1/4}} \qquad\qquad = x^{1/6}, \quad x > 0$$

10.3 Simplifying Radicals, the Distance Formula, and Circles

Product and Quotient Rules for Radicals

If n is a natural number and $\sqrt[n]{a}$ and $\sqrt[n]{b}$ are real numbers, then the following hold true.

$$\sqrt[n]{a} \cdot \sqrt[n]{b} = \sqrt[n]{ab} \quad \text{and} \quad \sqrt[n]{\frac{a}{b}} = \frac{\sqrt[n]{a}}{\sqrt[n]{b}} \quad \text{(where } b \neq 0\text{)}$$

Conditions for a Simplified Radical

1. The radicand has no factor raised to a power greater than or equal to the index.

2. The radicand has no fractions.

3. No denominator contains a radical.

4. Exponents in the radicand and the index of the radical have greatest common factor 1.

Simplify.

$$\sqrt{3} \cdot \sqrt{7} = \sqrt{21} \qquad \sqrt[5]{x^3 y} \cdot \sqrt[5]{xy^2} = \sqrt[5]{x^4 y^3}$$

$$\frac{\sqrt{x^5}}{\sqrt{x^4}} = \sqrt{\frac{x^5}{x^4}} = \sqrt{x}, \quad x > 0$$

$$\sqrt{18} = \sqrt{9 \cdot 2} = \sqrt{9} \cdot \sqrt{2} = 3\sqrt{2}$$

$$\sqrt[3]{54x^5 y^3} = \sqrt[3]{27x^3 y^3 \cdot 2x^2} = \sqrt[3]{27x^3 y^3} \cdot \sqrt[3]{2x^2} = 3xy\sqrt[3]{2x^2}$$

$$\sqrt{\frac{7}{4}} = \frac{\sqrt{7}}{\sqrt{4}} = \frac{\sqrt{7}}{2}$$

$$\sqrt[9]{x^3} = x^{3/9} = x^{1/3}, \quad \text{or} \quad \sqrt[3]{x}$$

Pythagorean Theorem

If a and b are the lengths of the shorter sides of a right triangle and c is the length of the longest side, then the following holds.

$$a^2 + b^2 = c^2$$

Find the length of the unknown side in the right triangle.

$$10^2 + b^2 = \left(2\sqrt{61}\right)^2$$

$$100 + b^2 = 4(61)$$

$$100 + b^2 = 244$$

$$b^2 = 144$$

$$b = 12$$

Distance Formula

The distance d between the points (x_1, y_1) and (x_2, y_2) is given by the following.

$$d = \sqrt{(x_2 - x_1)^2 + (y_2 - y_1)^2}$$

Find the distance between $(3, -2)$ and $(-1, 1)$.

$$d = \sqrt{(-1-3)^2 + [1-(-2)]^2} \qquad \text{Substitute.}$$

$$d = \sqrt{(-4)^2 + 3^2} \qquad \text{Subtract.}$$

$$d = \sqrt{16 + 9} \qquad \text{Square.}$$

$$d = \sqrt{25} \qquad \text{Add.}$$

$$d = 5 \qquad \text{Take the square root.}$$

CONCEPTS	EXAMPLES
Equation of a Circle A circle with center (h, k) and radius $r > 0$ has an equation of the form $$(x - h)^2 + (y - k)^2 = r^2.$$ A circle with center at the origin $(0, 0)$ and radius $r > 0$ has the following equation. $$x^2 + y^2 = r^2$$	Graph $(x + 2)^2 + (y - 3)^2 = 25$. This equation, which can be written $$[x - (-2)]^2 + (y - 3)^2 = 5^2,$$ is a circle with center $(-2, 3)$ and radius 5. 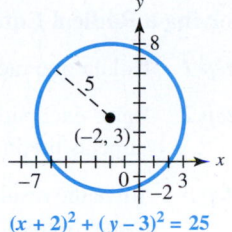 $(x + 2)^2 + (y - 3)^2 = 25$

10.4   Adding and Subtracting Radical Expressions

Add or subtract radical expressions using the distributive property. $$ac + bc = (a + b)c$$ ***Only radical expressions with the same index and the same radicand may be combined.***	Simplify. $$2\sqrt{28} - 3\sqrt{63} + 8\sqrt{112}$$ $$= 2\sqrt{4 \cdot 7} - 3\sqrt{9 \cdot 7} + 8\sqrt{16 \cdot 7}$$ $$= 2 \cdot 2\sqrt{7} - 3 \cdot 3\sqrt{7} + 8 \cdot 4\sqrt{7}$$ $$= 4\sqrt{7} - 9\sqrt{7} + 32\sqrt{7}$$ $$= (4 - 9 + 32)\sqrt{7}$$ $$= 27\sqrt{7}$$ $\left.\begin{array}{l}\sqrt{15} + \sqrt{30} \\ \sqrt{3} + \sqrt[3]{9}\end{array}\right\}$ Cannot be simplified further

10.5   Multiplying and Dividing Radical Expressions

Multiply binomial radical expressions using the FOIL method. Special product rules may apply. $$(x + y)(x - y) = x^2 - y^2$$ $$(x + y)^2 = x^2 + 2xy + y^2$$ $$(x - y)^2 = x^2 - 2xy + y^2$$	Perform the operations and simplify. $$(\sqrt{2} + \sqrt{7})(\sqrt{3} - \sqrt{6})$$ $$\quad \overset{F}{} \quad \overset{O}{} \quad \overset{I}{} \quad \overset{L}{}$$ $$= \sqrt{6} - 2\sqrt{3} + \sqrt{21} - \sqrt{42} \quad \sqrt{12} = 2\sqrt{3}$$ $(\sqrt{5} - \sqrt{10})(\sqrt{5} + \sqrt{10})$ $\quad\bigg|\quad$ $(\sqrt{3} - \sqrt{2})^2$ $= 5 - 10$ $\qquad\qquad\quad\bigg|\quad$ $= 3 - 2\sqrt{3} \cdot \sqrt{2} + 2$ $= -5$ $\qquad\qquad\qquad\bigg|\quad$ $= 5 - 2\sqrt{6}$
Rationalizing a Denominator Rationalize the denominator by multiplying both the numerator and the denominator by the same expression, one that will yield a rational number in the final denominator.	$$\frac{\sqrt{7}}{\sqrt{5}} = \frac{\sqrt{7} \cdot \sqrt{5}}{\sqrt{5} \cdot \sqrt{5}} = \frac{\sqrt{35}}{5}$$ $$\frac{4}{\sqrt{5} - \sqrt{2}} = \frac{4(\sqrt{5} + \sqrt{2})}{(\sqrt{5} - \sqrt{2})(\sqrt{5} + \sqrt{2})}$$ $$= \frac{4(\sqrt{5} + \sqrt{2})}{5 - 2} = \frac{4(\sqrt{5} + \sqrt{2})}{3}$$
To write a radical quotient in lowest terms, factor the numerator and denominator. Then divide out any common factor(s).	$$\frac{5 + 15\sqrt{6}}{10} = \frac{5(1 + 3\sqrt{6})}{5 \cdot 2} = \frac{1 + 3\sqrt{6}}{2}$$ Factor first, then divide.

CONCEPTS	EXAMPLES

10.6 Solving Equations with Radicals

Solving a Radical Equation

Step 1 Isolate one radical on one side of the equation.

Step 2 Raise each side of the equation to a power that is the same as the index of the radical.

Step 3 Solve the resulting equation. If it still contains a radical, repeat Steps 1 and 2.

Step 4 Check all proposed solutions in the *original* equation. Discard any values that are not solutions of the original equation.

Solve. $\sqrt{2x + 3} - x = 0$

$\sqrt{2x + 3} = x$ Add x.

$\left(\sqrt{2x + 3}\right)^2 = x^2$ Square each side.

$2x + 3 = x^2$ Apply the exponents.

$x^2 - 2x - 3 = 0$ Standard form

$(x - 3)(x + 1) = 0$ Factor.

$x - 3 = 0$ or $x + 1 = 0$ Zero-factor property

$x = 3$ or $x = -1$ Solve each equation.

A check shows that 3 is a solution, but -1 is extraneous (because it leads to $2 = 0$, a false statement). The solution set is $\{3\}$.

10.7 Complex Numbers

$i = \sqrt{-1}$,   and thus $i^2 = -1$.

For any positive real number b, $\sqrt{-b} = i\sqrt{b}$.

To multiply radicals with negative radicands, first change each factor to the form $i\sqrt{b}$ and then multiply. The same procedure applies to quotients.

Simplify.

$$\sqrt{-25} = i\sqrt{25} = 5i$$

$$\sqrt{-3} \cdot \sqrt{-27}$$

$$= i\sqrt{3} \cdot i\sqrt{27}$$ $\sqrt{-b} = i\sqrt{b}$

$$= i^2\sqrt{81}$$ Product rule

$$= -1 \cdot 9$$ $i^2 = -1$; Take the square root.

$$= -9$$ Multiply.

$$\frac{\sqrt{-18}}{\sqrt{-2}} = \frac{i\sqrt{18}}{i\sqrt{2}} = \sqrt{\frac{18}{2}} = \sqrt{9} = 3$$

Adding and Subtracting Complex Numbers

Add (or subtract) the real parts and add (or subtract) the imaginary parts.

Perform the operations.

$(5 + 3i) + (8 - 7i)$	$(5 + 3i) - (8 - 7i)$
$= 13 - 4i$	$= -3 + 10i$

Multiplying Complex Numbers

Multiply complex numbers using the FOIL method.

Multiply. $(2 + i)(5 - 3i)$

$$= 10 - 6i + 5i - 3i^2$$ Use the FOIL method.

$$= 10 - i - 3(-1)$$ Add imaginary parts; $i^2 = -1$

$$= 13 - i$$ Multiply. Add real parts.

Dividing Complex Numbers

Divide complex numbers by multiplying the numerator and the denominator by the conjugate of the denominator.

Divide. $\dfrac{20}{3 + i}$

$$= \frac{20(3 - i)}{(3 + i)(3 - i)}$$ Multiply both numerator and denominator by the conjugate of the denominator.

$$= \frac{20(3 - i)}{9 - i^2}$$ $(x + y)(x - y) = x^2 - y^2$

$$= \frac{20(3 - i)}{10}$$ $i^2 = -1$ and $9 - (-1) = 10$.

$$= 2(3 - i)$$ Divide out the common factor, 10.

$$= 6 - 2i$$ Distributive property

Chapter 10   Review Exercises

10.1 *Find each root.*

1. $\sqrt{100}$ **2.** $-\sqrt{289}$ **3.** $\sqrt[3]{216}$

4. $\sqrt[3]{-125}$ **5.** $-\sqrt[3]{27}$ **6.** $\sqrt[5]{-32}$

7. $\sqrt{x^2}$ **8.** $\sqrt[3]{x^3}$ **9.** $\sqrt[4]{x^{20}}$

10. Under what conditions is $\sqrt[n]{a}$ not a real number?

Graph each function, and give its domain and range.

11. $f(x) = \sqrt{x - 1}$ **12.** $f(x) = \sqrt[3]{x} + 4$

Use a calculator to approximate each radical to three decimal places.

13. $-\sqrt{47}$ **14.** $\sqrt[3]{-129}$ **15.** $\sqrt[4]{605}$

16. $\sqrt[4]{500}$ **17.** $-\sqrt[3]{500}$ **18.** $-\sqrt{28}$

Solve each problem.

19. The time t in seconds for one complete swing of a pendulum is given by

$$t = 2\pi\sqrt{\frac{L}{g}},$$

where L is the length of the pendulum in feet, and g, the force due to gravity, is about 32 ft per sec². Find the time of a complete swing of a 3-ft pendulum to the nearest tenth of a second.

20. Use Heron's formula

$$\mathcal{A} = \sqrt{s(s - a)(s - b)(s - c)},$$

where $s = \frac{1}{2}(a + b + c)$, to find the area of a triangle with sides of lengths 11 in., 13 in., and 20 in.

10.2 *Evaluate each exponential.*

21. $49^{1/2}$ **22.** $-121^{1/2}$ **23.** $16^{5/4}$ **24.** $-8^{2/3}$

25. $-\left(\dfrac{36}{25}\right)^{3/2}$ **26.** $\left(-\dfrac{1}{8}\right)^{-5/3}$ **27.** $\left(\dfrac{81}{10,000}\right)^{-3/4}$ **28.** $(-16)^{3/4}$

Write each exponential as a radical. Assume that all variables represent positive real numbers. Use the definition that takes the root first.

29. $10x^{1/2}$ **30.** $(3a + b)^{-5/3}$

Write each radical as an exponential and simplify. Assume that all variables represent positive real numbers.

31. $\sqrt{7^9}$ **32.** $\sqrt{11^6}$ **33.** $\sqrt[3]{r^{12}}$ **34.** $\sqrt[4]{z^4}$

Simplify each expression. Assume that all variables represent positive real numbers.

35. $5^{1/4} \cdot 5^{7/4}$

36. $\dfrac{96^{2/3}}{96^{-1/3}}$

37. $\dfrac{(a^{1/3})^4}{a^{2/3}}$

38. $\dfrac{y^{-1/3} \cdot y^{5/6}}{y}$

39. $\left(\dfrac{z^{-1}x^{-3/5}}{2^{-2}z^{-1/2}x} \right)^{-1}$

40. $r^{-1/2}(r + r^{3/2})$

Write each radical as an exponential and simplify. Leave answers in exponential form. Assume that all variables represent positive real numbers.

41. $\sqrt[5]{y} \cdot \sqrt[3]{y}$

42. $\dfrac{\sqrt{x^3}}{\sqrt{x^4}}$

43. $\dfrac{\sqrt{p^5}}{p^2}$

44. $\sqrt[4]{k^3} \cdot \sqrt{k^3}$

45. $\sqrt[3]{m^5} \cdot \sqrt[3]{m^8}$

46. $\dfrac{\sqrt[3]{t^5}}{\sqrt[6]{t}}$

47. $\sqrt{\sqrt{\sqrt{x}}}$

48. $\sqrt[3]{\sqrt[5]{x}}$

49. $\sqrt{\sqrt[6]{\sqrt[3]{x}}}$

50. The product rule does not apply to $3^{1/4} \cdot 2^{1/5}$. Why?

10.3   *Simplify. Assume that all variables represent positive real numbers.*

51. $\sqrt{6} \cdot \sqrt{11}$

52. $\sqrt{5} \cdot \sqrt{r}$

53. $\sqrt[3]{6} \cdot \sqrt[3]{5}$

54. $\sqrt[4]{7} \cdot \sqrt[4]{3}$

55. $\sqrt{20}$

56. $\sqrt{75}$

57. $-\sqrt{125}$

58. $\sqrt[3]{-108}$

59. $\sqrt{100y^7}$

60. $\sqrt[3]{64p^4q^6}$

61. $\sqrt[3]{108a^8b^5}$

62. $\sqrt[3]{632r^8t^4}$

63. $\sqrt{\dfrac{y^3}{144}}$

64. $\sqrt[3]{\dfrac{m^{15}}{27}}$

65. $\sqrt[3]{\dfrac{r^2}{8}}$

66. $\sqrt[4]{\dfrac{a^9}{81}}$

67. $\sqrt[6]{15^3}$

68. $\sqrt[4]{p^6}$

69. $\sqrt[3]{2} \cdot \sqrt[4]{5}$

70. $\sqrt{x} \cdot \sqrt[5]{x}$

71. Find the length c of the hypotenuse of the right triangle shown.

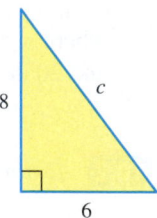

72. Find the distance between the points $(-4, 7)$ and $(10, 6)$.

Find the equation of a circle satisfying the given conditions.

73. Center $(0, 0)$, $r = 11$

74. Center $(-2, 4)$, $r = 3$

75. Center $(-1, -3)$, $r = 5$

76. Center $(4, 2)$, $r = 6$

Graph each circle. Identify the center and the radius.

77. $x^2 + y^2 = 25$

78. $(x + 3)^2 + (y - 3)^2 = 9$

79. $(x - 2)^2 + (y + 5)^2 = 9$

80. Why does the equation $x^2 + y^2 = -1$ have no points on its graph?

10.4 *Perform the indicated operations. Assume that all variables represent positive real numbers.*

81. $2\sqrt{8} - 3\sqrt{50}$

82. $8\sqrt{80} - 3\sqrt{45}$

83. $-\sqrt{27y} + 2\sqrt{75y}$

84. $2\sqrt{54m^3} + 5\sqrt{96m^3}$

85. $3\sqrt[3]{54} + 5\sqrt[3]{16}$

86. $-6\sqrt[4]{32} + \sqrt[4]{512}$

10.5 *Multiply, and then simplify.*

87. $\left(\sqrt{3} + 1\right)\left(\sqrt{3} - 2\right)$

88. $\left(\sqrt{7} + \sqrt{5}\right)\left(\sqrt{7} - \sqrt{5}\right)$

89. $\left(3\sqrt{2} + 1\right)\left(2\sqrt{2} - 3\right)$

90. $\left(\sqrt{13} - \sqrt{2}\right)^2$

91. $\left(\sqrt[3]{2} + 3\right)\left(\sqrt[3]{4} - 3\sqrt[3]{2} + 9\right)$

92. $\left(\sqrt[3]{4y} - 1\right)\left(\sqrt[3]{4y} + 3\right)$

Rationalize each denominator. Assume that all variables represent positive real numbers.

93. $\dfrac{\sqrt{6}}{\sqrt{5}}$

94. $\dfrac{-6\sqrt{3}}{\sqrt{2}}$

95. $\dfrac{3\sqrt{7p}}{\sqrt{y}}$

96. $\sqrt{\dfrac{11}{8}}$

97. $-\sqrt[3]{\dfrac{9}{25}}$

98. $\sqrt[3]{\dfrac{108m^3}{n^5}}$

99. $\dfrac{1}{\sqrt{2} + \sqrt{7}}$

100. $\dfrac{-5}{\sqrt{6} - 3}$

Write each quotient in lowest terms.

101. $\dfrac{2 - 2\sqrt{5}}{8}$

102. $\dfrac{-18 + \sqrt{27}}{6}$

10.6 *Solve each equation.*

103. $\sqrt{8x + 9} = 5$

104. $\sqrt{2x - 3} - 3 = 0$

105. $\sqrt{7x + 1} = x + 1$

106. $3\sqrt{x} = \sqrt{10x - 9}$

107. $\sqrt{x^2 + 3x + 7} = x + 2$

108. $\sqrt{x + 2} - \sqrt{x - 3} = 1$

109. $\sqrt[3]{5x - 1} = \sqrt[3]{3x - 2}$

110. $\sqrt[3]{1 - 2x} - \sqrt[3]{-x - 13} = 0$

111. $\sqrt[4]{x - 1} + 2 = 0$

112. $\sqrt[4]{x + 7} = \sqrt[4]{2x}$

Carpenters stabilize wall frames with a diagonal brace, as shown in the figure. The length of the brace is given by

$$L = \sqrt{H^2 + W^2}.$$

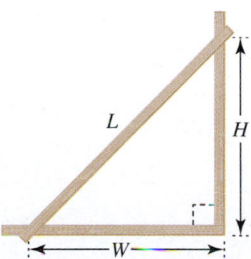

113. Solve this formula for H.

114. If the bottom of the brace is attached 9 ft from the corner and the brace is 12 ft long, how far up the corner post should it be nailed? Give the answer to the nearest tenth of a foot.

10.7 *Write each number as a product of a real number and i. Simplify.*

115. $\sqrt{-16}$

116. $\sqrt{-200}$

Perform the indicated operations. Give answers in standard form.

117. $(-2 + 5i) + (-8 - 7i)$

118. $(5 + 4i) - (-9 - 3i)$

119. $\sqrt{-5} \cdot \sqrt{-7}$

120. $\sqrt{-25} \cdot \sqrt{-81}$

121. $\dfrac{\sqrt{-72}}{\sqrt{-8}}$

122. $(2 + 3i)(1 - i)$

123. $(6 - 2i)^2$

124. $\dfrac{3 - i}{2 + i}$

Find each power of i.

125. i^{11}

126. i^{36}

127. i^{-10}

128. i^{-8}

Chapter 10   Mixed Review Exercises

Simplify. Assume that all variables represent positive real numbers.

1. $-\sqrt{169a^2b^4}$

2. $1000^{-2/3}$

3. $\dfrac{z^{-1/5} \cdot z^{3/10}}{z^{7/10}}$

4. $\sqrt[3]{54z^9t^8}$

5. $\sqrt{-49}$

6. $\dfrac{-1}{\sqrt{12}}$

7. $\sqrt[3]{\dfrac{12}{25}}$

8. i^{-1000}

9. $-5\sqrt{18} + 12\sqrt{72}$

10. $(4 - 9i) - (-1 + 2i)$

11. $\dfrac{\sqrt{50}}{\sqrt{-2}}$

12. $\dfrac{3 + \sqrt{54}}{6}$

13. $(3 + 2i)^2$

14. $8\sqrt[3]{x^3y^2} - 2x\sqrt[3]{y^2}$

15. $5i(3 - 7i)$

16. $\sqrt[3]{2} \cdot \sqrt[4]{5}$

17. $\dfrac{2\sqrt{z}}{\sqrt{z} - 2}$, $z \neq 4$

18. Find the perimeter of a triangular electronic highway road sign having the dimensions shown in the figure.

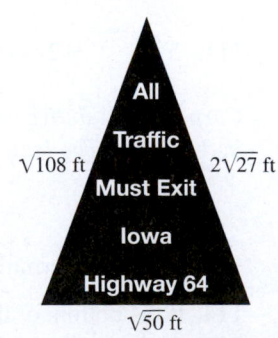

All

Traffic

$\sqrt{108}$ ft $2\sqrt{27}$ ft

Must Exit

Iowa

Highway 64

$\sqrt{50}$ ft

Solve each equation.

19. $\sqrt{x + 4} = x - 2$

20. $\sqrt[3]{2x - 9} = \sqrt[3]{5x + 3}$

21. $\sqrt{6 + 2x} - 1 = \sqrt{7 - 2x}$

22. $\sqrt{11 + 2x} + 1 = \sqrt{5x + 1}$

| Chapter 10 | Test | FOR EXTRA HELP | *Step-by-step test solutions are found on the Chapter Test Prep Videos available in* MyLab Math. |

▶ *View the complete solutions to all Chapter Test exercises in MyLab Math.*

Evaluate.

1. $-\sqrt{841}$

2. $\sqrt[3]{-512}$

3. $125^{1/3}$

Use a calculator to approximate each radical to three decimal places.

4. $\sqrt{478}$

5. $\sqrt[3]{-832}$

6. Graph the function $f(x) = \sqrt{x+6}$, and give its domain and range.

Simplify each expression. Leave answers in exponential form. Assume that all variables represent positive real numbers.

7. $\left(\dfrac{16}{25}\right)^{-3/2}$

8. $(-64)^{-4/3}$

9. $\dfrac{3^{2/5}x^{-1/4}y^{2/5}}{3^{-8/5}x^{7/4}y^{1/10}}$

10. $\left(\dfrac{x^{-4}y^{-6}}{x^{-2}y^3}\right)^{-2/3}$

11. $7^{3/4} \cdot 7^{-1/4}$

12. $\sqrt[3]{a^4} \cdot \sqrt[3]{a^7}$

13. Find the exact length of side b in the figure.

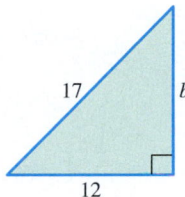

14. Find the distance between the points $(-4, 2)$ and $(2, 10)$.

15. Write an equation of a circle with center $(-4, 6)$ and radius 5.

16. Graph the circle $(x - 2)^2 + (y + 3)^2 = 16$. Identify the center and the radius.

Simplify. Assume that all variables represent positive real numbers.

17. $\sqrt{54x^5y^6}$

18. $\sqrt[4]{32a^7b^{13}}$

19. $\sqrt{2} \cdot \sqrt[3]{5}$

20. $3\sqrt{20} - 5\sqrt{80} + 4\sqrt{500}$

21. $\sqrt[3]{16t^3s^5} - \sqrt[3]{54t^6s^2}$

22. $\left(7\sqrt{5} + 4\right)\left(2\sqrt{5} - 1\right)$

23. $\left(\sqrt{3} - 2\sqrt{5}\right)^2$

24. $\dfrac{-5}{\sqrt{40}}$

25. $\dfrac{2}{\sqrt[3]{5}}$

26. $\dfrac{-4}{\sqrt{7} + \sqrt{5}}$

27. Write $\dfrac{6 + \sqrt{24}}{2}$ in lowest terms.

28. The following formula from physics relates the velocity V of sound to the temperature T.

$$V = \dfrac{V_0}{\sqrt{1 - kT}}$$

(a) Approximate V to the nearest tenth if $V_0 = 50$, $k = 0.01$, and $T = 30$.

(b) Solve the formula for T.

Solve each equation.

29. $\sqrt[3]{5x} = \sqrt[3]{2x - 3}$

30. $\sqrt{x + 6} = 9 - 2x$

31. $\sqrt{x + 4} - \sqrt{1 - x} = -1$

Perform the indicated operations. Give the answers in standard form.

32. $(-2 + 5i) - (3 + 6i) - 7i$ **33.** $(1 + 5i)(3 + i)$ **34.** $\dfrac{7 + i}{1 - i}$

35. Simplify i^{37}.

36. Answer *true* or *false* to each of the following.

(a) $i^2 = -1$ **(b)** $i = \sqrt{-1}$ **(c)** $i = -1$ **(d)** $\sqrt{-3} = i\sqrt{3}$

Chapters R–10   Cumulative Review Exercises

1. Complete the table of fraction, decimal, and percent equivalents.

	Fraction in Lowest Terms (or Mixed Number)	Decimal	Percent
(a)	$\frac{1}{25}$		
(b)			15%
(c)		0.8	
(d)	$1\frac{1}{4}$		

2. Find the value of each expression in the real number system.

(a) 4^2 **(b)** $(-4)^2$ **(c)** -4^2 **(d)** $-(-4)^2$ **(e)** $\sqrt{4}$ **(f)** $-\sqrt{4}$ **(g)** $\sqrt{-4}$

Perform the indicated operations.

3. $-4|2 - 7| - (-3)(-9)$

4. $\left(-\dfrac{5}{8} + \dfrac{1}{6}\right) - \left(-\dfrac{2}{3}\right)$

Solve each equation or inequality.

5. $3(x + 2) - 4(2x + 3) = -3x + 2$

6. $\dfrac{1}{3}x + \dfrac{1}{4}(x + 8) = x + 7$

7. $0.04x + 0.06(100 - x) = 5.88$

8. $-5 - 3(x - 2) < 11 - 2(x + 2)$

Solve each problem.

9. A piggy bank has 100 coins, all of which are nickels and quarters. The total value of the money is $17.80. How many of each denomination are there in the bank?

10. How many liters of pure alcohol must be mixed with 40 L of 18% alcohol to obtain a 22% alcohol solution?

11. Graph the equation $4x - 3y = 12$.

12. Find the slope of the line passing through the points $(-4, 6)$ and $(2, -3)$. Then find the equation of the line and write it in the form $y = mx + b$.

13. If $f(x) = 3x - 7$, find $f(-10)$.

Solve each system.

14. $3x - y = 23$
      $2x + 3y = 8$

15. $x + y + z = 1$
       $x - y - z = -3$
       $x + y - z = -1$

16. In 2017, sending five 2-oz letters and three 3-oz letters by first-class mail would have cost $6.23. Sending three 2-oz letters and five 3-oz letters would have cost $6.65. What was the rate for one 2-oz letter and one 3-oz letter? (Data from U.S. Postal Service.)

Perform the indicated operations.

17. $(3k^3 - 5k^2 + 8k - 2) - (4k^3 + 11k + 7) + (2k^2 - 5k)$

18. $(8x - 7)(x + 3)$

19. $\dfrac{6y^4 - 3y^3 + 5y^2 + 6y - 9}{2y + 1}$

Factor each polynomial completely.

20. $2p^2 - 5pq + 3q^2$ **21.** $3k^4 + k^2 - 4$ **22.** $x^3 + 512$

Perform each operation, and express the answer in lowest terms.

23. $\dfrac{y^2 + y - 12}{y^3 + 9y^2 + 20y} \div \dfrac{y^2 - 9}{y^3 + 3y^2}$

24. $\dfrac{1}{x + y} + \dfrac{3}{x - y}$

Simplify.

25. $\dfrac{\dfrac{-6}{x - 2}}{\dfrac{8}{3x - 6}}$

26. $\dfrac{\dfrac{1}{a} - \dfrac{1}{b}}{\dfrac{a}{b} - \dfrac{b}{a}}$

Solve.

27. $2x^2 + 11x + 15 = 0$ **28.** $5x(x - 1) = 2(1 - x)$

29. $\dfrac{x + 1}{x - 3} = \dfrac{4}{x - 3} + 6$ **30.** $\sqrt{3x - 8} = x - 2$

31. $2x + 4 < 10$ and $3x - 1 > 5$ **32.** $2x + 4 > 10$ or $3x - 1 < 5$

33. $|6x + 7| = 13$ **34.** $|2p - 5| \geq 9$

Simplify. Assume that all variables represent positive real numbers.

35. $27^{-2/3}$ **36.** $\sqrt[3]{16x^2y} \cdot \sqrt[3]{3x^3y}$

37. $\sqrt{50} + \sqrt{8}$ **38.** $\dfrac{1}{\sqrt{10} - \sqrt{8}}$

39. Find the distance between the points $(-4, 4)$ and $(-2, 9)$.

40. Express $\dfrac{6 - 2i}{1 - i}$ in standard form.

STUDY SKILLS REMINDER

Have you begun to prepare for your final exam? **Review Study Skill 10, *Preparing for Your Math Final Exam.***

11

QUADRATIC EQUATIONS, INEQUALITIES, AND FUNCTIONS

Quadratic functions, one of the topics of this chapter, have graphs that are *parabolas*. Cross sections of telescopes, satellite dishes, and automobile headlights form parabolas, as do the cables that support suspension bridges.

11.1 Solving Quadratic Equations by the Square Root Property

VOCABULARY
☐ quadratic equation
☐ second-degree equation

OBJECTIVE 1 Solve quadratic equations using the zero-factor property. (Review)

Recall that a *quadratic equation* is defined as follows.

> **Quadratic Equation**
>
> A **quadratic equation** (in x here) is an equation that can be written in the form
> $$ax^2 + bx + c = 0,$$
> where a, b, and c are real numbers and $a \neq 0$. The given form is called **standard form**.
>
> *Examples:* $4x^2 + 4x - 5 = 0$, $3x^2 = 4x - 8$ Quadratic equations (The first equation is in standard form.)

A quadratic equation is a **second-degree equation**—that is, an equation with a squared variable term and no terms of greater degree.

Recall that the **zero-factor property** can be used to solve some quadratic equations.

> **Zero-Factor Property**
>
> **If a and b are real numbers and if $ab = 0$, then $a = 0$ or $b = 0$.**
>
> That is, if the product of two numbers is 0, then at least one of the numbers must be 0. One number must be 0, but both *may* be 0.

NOW TRY EXERCISE 1

Solve each equation using the zero-factor property.

(a) $x^2 - x - 20 = 0$
(b) $x^2 = 36$

EXAMPLE 1 Solving Quadratic Equations Using the Zero-Factor Property

Solve each equation using the zero-factor property.

(a) $\qquad\qquad x^2 + 4x + 3 = 0$

$\qquad\qquad (x + 3)(x + 1) = 0$ Factor.

$\quad x + 3 = 0 \qquad \text{or} \qquad x + 1 = 0$ Zero-factor property

$\qquad x = -3 \qquad \text{or} \qquad\quad x = -1$ Solve each equation.

The solution set is $\{-3, -1\}$.

(b) $\qquad\qquad\qquad x^2 = 9$

$\qquad\qquad\qquad x^2 - 9 = 0$ Subtract 9.

$\qquad\quad (x + 3)(x - 3) = 0$ Factor.

$\quad x + 3 = 0 \qquad \text{or} \qquad x - 3 = 0$ Zero-factor property

$\qquad x = -3 \qquad \text{or} \qquad\quad x = 3$ Solve each equation.

The solution set is $\{-3, 3\}$.

NOW TRY

Not all quadratic expressions can easily be factored, so we develop other methods for solving quadratic equations.

OBJECTIVE 2 Solve equations of the form $x^2 = k$, where $k > 0$.

In **Example 1(b),** we might also have solved $x^2 = 9$ by noticing that x must be a number whose square is 9. Thus,

$$x = \sqrt{9} = 3 \quad \text{or} \quad x = -\sqrt{9} = -3.$$

This is generalized as the **square root property.**

Square Root Property

If k is a positive number and if $x^2 = k$, then

$$x = \sqrt{k} \quad \text{or} \quad x = -\sqrt{k}.$$

The solution set is $\left\{-\sqrt{k}, \sqrt{k}\right\}$, which can be written $\left\{\pm\sqrt{k}\right\}$.

(The symbol $\pm$ is read "*positive or negative*" or "*plus or minus.*")

EXAMPLE 2 Solving Quadratic Equations of the Form $x^2 = k$

Solve each equation. Write radicals in simplified form.

(a) $x^2 = 16$

By the square root property, if $x^2 = 16$, then

$$x = \sqrt{16} = 4 \quad \text{or} \quad x = -\sqrt{16} = -4.$$

Check each value by substituting it for x in the original equation. The solution set is

$$\{-4, 4\}, \quad \text{or} \quad \{\pm 4\}.$$

> This $\pm$ notation indicates *two* solutions, one *positive* and one *negative.*

(b) $x^2 = 5$

By the square root property, if $x^2 = 5$, then

$$x = \sqrt{5} \quad \text{or} \quad x = -\sqrt{5}.$$

> Don't forget the negative solution.

The solutions set is $\left\{-\sqrt{5}, \sqrt{5}\right\}$, or $\left\{\pm\sqrt{5}\right\}$.

(c)

$$5m^2 - 40 = 0$$

$$5m^2 = 40 \qquad \text{Add 40.}$$

$$m^2 = 8 \qquad \text{Divide by 5.}$$

> Don't stop here. Simplify the radicals

$$m = \sqrt{8} \quad \text{or} \quad m = -\sqrt{8} \qquad \text{Square root property}$$

$$m = 2\sqrt{2} \quad \text{or} \quad m = -2\sqrt{2} \qquad \sqrt{8} = \sqrt{4} \cdot \sqrt{2} = 2\sqrt{2}$$

CHECK Substitute each value in the original equation.

$5m^2 - 40 = 0$	$5m^2 - 40 = 0$
$5(2\sqrt{2})^2 - 40 \stackrel{?}{=} 0$ Let $m = 2\sqrt{2}$.	$5(-2\sqrt{2})^2 - 40 \stackrel{?}{=} 0$ Let $m = -2\sqrt{2}$.
$5(8) - 40 \stackrel{?}{=} 0$ Square $2\sqrt{2}$.	$5(8) - 40 \stackrel{?}{=} 0$ Square $-2\sqrt{2}$.
$40 - 40 \stackrel{?}{=} 0$ Multiply.	$40 - 40 \stackrel{?}{=} 0$ Multiply.
$0 = 0$ ✓ True	$0 = 0$ ✓ True

> $(2\sqrt{2})^2$
> $= 2^2 \cdot (\sqrt{2})^2$
> $= 4 \cdot 2$
> $= 8$

The solution set is $\left\{-2\sqrt{2}, 2\sqrt{2}\right\}$, or $\left\{\pm 2\sqrt{2}\right\}$.

**NOW TRY
EXERCISE 2**

Solve each equation. Write radicals in simplified form.

(a) $t^2 = 25$

(b) $x^2 = 13$

(c) $2x^2 - 5 = 35$

Galileo Galilei (1564–1642)

**NOW TRY
EXERCISE 3**

Tim is dropping roofing nails from the top of a roof 25 ft high into a large bucket on the ground. Use the formula in **Example 3** to determine how long it will take a nail dropped from 25 ft to hit the bottom of the bucket.

(d)
$$3x^2 + 5 = 11$$

$$3x^2 = 6 \qquad \text{Subtract 5.}$$

$$x^2 = 2 \qquad \text{Divide by 3.}$$

$$x = \sqrt{2} \quad \text{or} \quad x = -\sqrt{2} \qquad \text{Square root property}$$

The solution set is $\{-\sqrt{2}, \sqrt{2}\}$, or $\{\pm\sqrt{2}\}$.   **NOW TRY**

EXAMPLE 3   Using the Square Root Property in an Application

Galileo Galilei developed a formula for freely falling objects described by

$$d = 16t^2,$$

where d is the distance in feet that an object falls (disregarding air resistance) in t seconds, regardless of weight.

   If the Leaning Tower of Pisa is about 180 ft tall, use Galileo's formula to determine how long it would take an object dropped from the top of the tower to fall to the ground. (Data from www.brittanica.com)

$$d = 16t^2 \qquad \text{Galileo's formula}$$

$$180 = 16t^2 \qquad \text{Let } d = 180.$$

$$11.25 = t^2 \qquad \text{Divide by 16.}$$

$$t = \sqrt{11.25} \quad \text{or} \quad t = -\sqrt{11.25} \qquad \text{Square root property}$$

Time cannot be negative, so we discard $-\sqrt{11.25}$. Using a calculator,

$$\sqrt{11.25} \approx 3.4 \quad \text{so} \quad t \approx 3.4. \qquad \text{Round to the nearest tenth.}$$

The object would fall to the ground in 3.4 sec.   **NOW TRY**

OBJECTIVE 3   Solve equations of the form $(ax + b)^2 = k$, where $k > 0$.

In each equation so far, the exponent 2 appeared with a single variable as its base. We can extend the square root property to solve equations in which the base is a binomial.

EXAMPLE 4   Solving Quadratic Equations of the Form $(x + b)^2 = k$

Solve each equation, and check the solutions.

(a)   [Use $(x - 3)$ as the base.] $(x - 3)^2 = 16$

$$x - 3 = \sqrt{16} \quad \text{or} \quad x - 3 = -\sqrt{16} \qquad \text{Square root property}$$

$$x - 3 = 4 \qquad \text{or} \quad x - 3 = -4 \qquad \sqrt{16} = 4$$

$$x = 7 \qquad \text{or} \qquad x = -1 \qquad \text{Add 3.}$$

CHECK   Substitute each value in the original equation.

$(x - 3)^2 = 16$		$(x - 3)^2 = 16$	
$(7 - 3)^2 \stackrel{?}{=} 16$	Let $x = 7$.	$(-1 - 3)^2 \stackrel{?}{=} 16$	Let $x = -1$.
$4^2 \stackrel{?}{=} 16$	Subtract.	$(-4)^2 \stackrel{?}{=} 16$	Subtract.
$16 = 16$ ✓	True	$16 = 16$ ✓	True

The solution set is $\{-1, 7\}$.

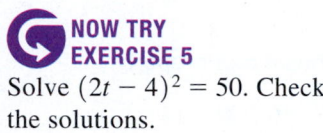
NOW TRY
EXERCISE 4
Solve $(x - 2)^2 = 32$, and check the solutions.

(b) $(x + 1)^2 = 6$

$x + 1 = \sqrt{6}$ or $x + 1 = -\sqrt{6}$ Square root property

$x = -1 + \sqrt{6}$ or $x = -1 - \sqrt{6}$ Add -1.

CHECK   Substitute each value in the original equation.

$(x + 1)^2 = 6$ $(x + 1)^2 = 6$

$(-1 + \sqrt{6} + 1)^2 \overset{?}{=} 6$ Let $x = -1 + \sqrt{6}$. $(-1 - \sqrt{6} + 1)^2 \overset{?}{=} 6$ Let $x = -1 - \sqrt{6}$.

$(\sqrt{6})^2 \overset{?}{=} 6$ Simplify. $(-\sqrt{6})^2 \overset{?}{=} 6$ Simplify.

$6 = 6$ ✓ True $6 = 6$ ✓ True

The solution set is $\{-1 - \sqrt{6}, -1 + \sqrt{6}\}$, or $\{-1 \pm \sqrt{6}\}$.   **NOW TRY**

NOW TRY
EXERCISE 5
Solve $(2t - 4)^2 = 50$. Check the solutions.

EXAMPLE 5   Solving a Quadratic Equation of the Form $(ax + b)^2 = k$

Solve $(3r - 2)^2 = 27$. Check the solutions.

$$(3r - 2)^2 = 27$$

$3r - 2 = \sqrt{27}$ or $3r - 2 = -\sqrt{27}$ Square root property

$3r - 2 = 3\sqrt{3}$ or $3r - 2 = -3\sqrt{3}$ $\sqrt{27} = \sqrt{9} \cdot \sqrt{3} = 3\sqrt{3}$

$3r = 2 + 3\sqrt{3}$ or $3r = 2 - 3\sqrt{3}$ Add 2.

$r = \dfrac{2 + 3\sqrt{3}}{3}$ or $r = \dfrac{2 - 3\sqrt{3}}{3}$ Divide by 3.

CHECK $(3r - 2)^2 = 27$

$$\left(3 \cdot \frac{2 + 3\sqrt{3}}{3} - 2\right)^2 \overset{?}{=} 27$$ Let $r = \dfrac{2 + 3\sqrt{3}}{3}$.

$$\left(2 + 3\sqrt{3} - 2\right)^2 \overset{?}{=} 27$$ Multiply.

$(ab)^2 = a^2b^2$ ⟶ $\left(3\sqrt{3}\right)^2 \overset{?}{=} 27$ Subtract.

$27 = 27$ ✓   True

The check of the other value is similar. The solution set is $\left\{\dfrac{2 \pm 3\sqrt{3}}{3}\right\}$.   **NOW TRY**

⚠ **CAUTION**   The solutions in **Example 5**,

$$\frac{2 + 3\sqrt{3}}{3} \quad \text{and} \quad \frac{2 - 3\sqrt{3}}{3},$$

are fractions that cannot be simplified because 3 is **not** a common factor in the numerator.

OBJECTIVE 4 Solve quadratic equations with solutions that are not real numbers.

If $k < 0$ in the equation $x^2 = k$, then there will be two nonreal complex solutions. Recall that the imaginary unit i is defined as

$$i = \sqrt{-1}, \quad \text{and thus} \quad i^2 = -1.$$

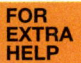 **NOW TRY
EXERCISE 6**

Solve each equation.
(a) $t^2 = -24$
(b) $(x + 4)^2 = -36$

EXAMPLE 6 Solving Quadratic Equations (Nonreal Complex Solutions)

Solve each equation.

(a) $x^2 = -15$

$$x = \sqrt{-15} \quad \text{or} \quad x = -\sqrt{-15} \qquad \text{Square root property}$$

$$x = i\sqrt{15} \quad \text{or} \quad x = -i\sqrt{15} \qquad \sqrt{-a} = i\sqrt{a}$$

The solution set is $\{i\sqrt{15}, -i\sqrt{15}\}$, or $\{\pm i\sqrt{15}\}$.

(b) $(x + 2)^2 = -16$

$$x + 2 = \sqrt{-16} \quad \text{or} \quad x + 2 = -\sqrt{-16} \qquad \text{Square root property}$$

$$x + 2 = 4i \quad \text{or} \quad x + 2 = -4i \qquad \sqrt{-16} = 4i$$

$$x = -2 + 4i \quad \text{or} \quad x = -2 - 4i \qquad \text{Add } -2.$$

NOW TRY ANSWERS
6. (a) $\{\pm 2i\sqrt{6}\}$
   (b) $\{-4 \pm 6i\}$

The solution set is $\{-2 + 4i, -2 - 4i\}$, or $\{-2 \pm 4i\}$.   **NOW TRY**

11.1 Exercises

FOR EXTRA HELP   ▶ **MyLab Math**

▶ *Video solutions for select problems available in MyLab Math*

Concept Check *Work each problem.*

1. An equation in the form $ax^2 + bx + c = 0$, where a, b, and c are real numbers and $a \neq 0$, is a(n) _____ equation, also called a(n) _____ degree equation. The greatest degree of the variable is _____.

2. Which of the following are quadratic equations?

   A. $x + 2y = 0$ **B.** $x^2 - 8x + 16 = 0$ **C.** $2x^2 - 5x = 3$ **D.** $x^3 + x^2 + 4 = 0$

3. Which quadratic equation is in standard form?

   A. $x^2 = 25$ **B.** $3x^2 - x = 4$ **C.** $(x - 5)^2 = 16$ **D.** $x^2 - x - 2 = 0$

4. The equation $x^2 = -9$ has solutions that (*are / are not*) real numbers. These solutions involve the imaginary unit i, which is defined as $i = $ _____. Thus, $i^2 = $ _____. For any positive real number a, we have $\sqrt{-a} = $ _____.

5. Concept Check Match each equation in Column I with the correct description of its solution in Column II.

I	II
(a) $x^2 = 12$ **(b)** $x^2 = -16$	**A.** Two nonreal complex solutions **B.** Two integer solutions
(c) $x^2 = \dfrac{25}{36}$ **(d)** $x^2 = 100$	**C.** Two irrational solutions **D.** Two rational solutions that are not integers

6. Concept Check   A student incorrectly solved the following equation as shown.

$$x^2 - x - 2 = 4$$

$$(x - 2)(x + 1) = 4 \qquad \text{Factor.}$$

$$x - 2 = 4 \quad \text{or} \quad x + 1 = 4 \qquad \text{Zero-factor property}$$

$$x = 6 \quad \text{or} \qquad x = 3 \qquad \text{Solve each equation.}$$

WHAT WENT WRONG?   Solve correctly and give the solution set.

7. Concept Check   A student solving $x^2 = 81$ wrote the solution set incorrectly as $\{9\}$. Her teacher did not give her full credit. The student argued that because $9^2 = 81$, her answer had to be correct. **WHAT WENT WRONG?** Give the correct solution set.

8. Concept Check   When solving a quadratic equation, a student obtained the solutions

$$x = \frac{3 + 2\sqrt{5}}{2} \quad \text{or} \quad x = \frac{3 - 2\sqrt{5}}{2},$$

and he wrote the solution set incorrectly as $\{3 \pm \sqrt{5}\}$. **WHAT WENT WRONG?**   Give the correct solution set.

Solve using the zero-factor property. ***See Example 1.***

9. $x^2 - x - 56 = 0$ **10.** $x^2 - 2x - 99 = 0$ **11.** $x^2 - 8x + 15 = 0$

12. $x^2 - 6x + 5 = 0$ **13.** $x^2 = 121$ **14.** $x^2 = 144$

15. $x^2 - 169 = 0$ **16.** $x^2 - 400 = 0$ **17.** $3x^2 - 13x = 30$

18. $5x^2 - 14x = 3$ **19.** $6x^2 + 19x + 10 = 0$ **20.** $8x^2 + 18x + 9 = 0$

Solve using the square root property. Simplify all radicals. ***See Example 2.***

21. $x^2 = 81$ **22.** $z^2 = 169$ **23.** $x^2 = 144$ **24.** $m^2 = 36$

25. $k^2 = 14$ **26.** $m^2 = 22$ **27.** $t^2 = 48$ **28.** $x^2 = 54$

29. $x^2 = \dfrac{25}{4}$ **30.** $m^2 = \dfrac{36}{121}$ **31.** $x^2 = 0.25$ **32.** $w^2 = 0.49$

33. $x^2 - 64 = 0$ **34.** $x^2 - 100 = 0$ **35.** $r^2 - 3 = 0$ **36.** $x^2 - 13 = 0$

37. $x^2 - 12 = 0$ **38.** $x^2 - 8 = 0$ **39.** $4x^2 - 72 = 0$ **40.** $2x^2 - 80 = 0$

41. $2t^2 + 7 = 61$ **42.** $3x^2 + 8 = 80$ **43.** $3x^2 - 8 = 64$ **44.** $2x^2 - 5 = 35$

45. $7x^2 = 4$ **46.** $2x^2 = 9$ **47.** $5x^2 + 4 = 8$ **48.** $7p^2 - 5 = 11$

Solve using the square root property. Simplify all radicals. ***See Examples 4 and 5.***

49. $(x - 3)^2 = 25$ **50.** $(x - 7)^2 = 16$ **51.** $(x - 4)^2 = 3$ **52.** $(x + 3)^2 = 11$

53. $(x - 8)^2 = 27$ **54.** $(p - 5)^2 = 40$ **55.** $(3x + 2)^2 = 49$ **56.** $(5t + 3)^2 = 36$

57. $(4x - 3)^2 = 9$ **58.** $(7z - 5)^2 = 25$ **59.** $(5 - 2x)^2 = 30$ **60.** $(3 - 2x)^2 = 70$

61. $(3k + 1)^2 = 18$ **62.** $(5z + 6)^2 = 75$ **63.** $\left(\dfrac{1}{2}x + 5\right)^2 = 12$

64. $\left(\dfrac{1}{3}m + 4\right)^2 = 27$ **65.** $\left(x - \dfrac{1}{8}\right)^2 = \dfrac{1}{64}$ **66.** $\left(x - \dfrac{1}{9}\right)^2 = \dfrac{1}{81}$

67. $\left(x - \dfrac{1}{3}\right)^2 = \dfrac{4}{9}$

68. $\left(x - \dfrac{1}{5}\right)^2 = \dfrac{16}{25}$

69. $\left(x + \dfrac{1}{4}\right)^2 = \dfrac{3}{16}$

70. $\left(x + \dfrac{1}{7}\right)^2 = \dfrac{11}{49}$

71. $(4x - 1)^2 - 48 = 0$

72. $(2x - 5)^2 - 180 = 0$

Solve each equation. (All solutions are nonreal complex numbers.) **See Example 6.**

73. $x^2 = -100$

74. $x^2 = -64$

75. $x^2 = -26$

76. $x^2 = -21$

77. $x^2 = -12$

78. $x^2 = -18$

79. $(x + 3)^2 = -4$

80. $(x - 5)^2 = -36$

81. $(r - 5)^2 = -3$

82. $(t + 6)^2 = -5$

83. $(6k - 1)^2 = -8$

84. $(4m - 7)^2 = -27$

Use Galileo's formula to solve each problem. Round answers to the nearest tenth. **See Example 3.**

85. The sculpture of American presidents at Mount Rushmore National Memorial is 500 ft above the valley floor. How long would it take a rock dropped from the top of the sculpture to fall to the ground? (Data from www.travelsd.com)

86. The Gateway Arch in St. Louis, Missouri, is 630 ft tall. How long would it take an object dropped from the top of the arch to fall to the ground? (Data from www.gatewayarch.com)

11.2   Solving Quadratic Equations by Completing the Square

OBJECTIVES

1 Solve quadratic equations by completing the square when the coefficient of the second-degree term is 1.

2 Solve quadratic equations by completing the square when the coefficient of the second-degree term is not 1.

3 Simplify the terms of an equation before solving.

OBJECTIVES 1 Solve quadratic equations by completing the square when the coefficient of the second-degree term is 1.

The methods we have studied so far are not enough to solve an equation such as

$$x^2 + 8x + 10 = 0.$$

We can use the square root property to solve *any* quadratic equation by writing it in the form

Square of a binomial → $(x + k)^2 = n.$ ← Constant

That is, we must write the left side of the equation as a perfect square trinomial that can be factored as $(x + k)^2$, the square of a binomial, and the right side must be a constant. This process is called **completing the square.**

Recall that the perfect square trinomial

$$x^2 + 10x + 25 \quad \text{can be factored as} \quad (x + 5)^2.$$

In the trinomial, the coefficient of x (the first-degree term) is 10 and the constant term is 25. If we take half of 10 and square it, we obtain the constant term, 25.

$$\begin{array}{cc} \text{Coefficient of } x & \text{Constant} \\ \downarrow & \downarrow \\ \left[\frac{1}{2}(10)\right]^2 = 5^2 = 25 \end{array}$$

Similarly, in $\quad x^2 + 12x + 36, \quad \left[\dfrac{1}{2}(12)\right]^2 = 6^2 = 36,$

and in $\quad m^2 - 6m + 9, \quad \left[\dfrac{1}{2}(-6)\right]^2 = (-3)^2 = 9.$

This relationship is true in general and is the idea behind completing the square.

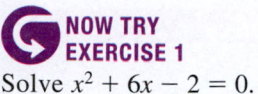

NOW TRY
EXERCISE 1
Solve $x^2 + 6x - 2 = 0$.

EXAMPLE 1 Solving a Quadratic Equation by Completing the Square ($a = 1$)

Solve $x^2 + 8x + 10 = 0$.

The trinomial on the left is nonfactorable, so this quadratic equation cannot be solved using the zero-factor property. It is not in the correct form to solve using the square root property. We can solve it by completing the square.

$$x^2 + 8x + 10 = 0 \quad \text{Original equation}$$

Only terms with variables remain on the left side.

$$x^2 + 8x = -10 \quad \text{Subtract 10.}$$

$$x^2 + 8x + \underline{\ ?\ } = -10 \quad \text{We must add a constant.}$$

Needs to be a perfect square trinomial

Take half the coefficient of the first-degree term, $8x$, and square the result.

$$\left[\tfrac{1}{2}(8)\right]^2 = 4^2 = 16 \longleftarrow \text{Desired constant}$$

We add this constant, 16, to *each* side of the equation and continue solving as shown.

$$x^2 + 8x + 16 = -10 + 16 \quad \text{Add 16 to each side.}$$

This is a key step.

$$(x + 4)^2 = 6 \quad \begin{array}{l}\text{Factor on the left.}\\ \text{Add on the right.}\end{array}$$

$$x + 4 = \sqrt{6} \qquad \text{or} \qquad x + 4 = -\sqrt{6} \quad \text{Square root property}$$

$$x = -4 + \sqrt{6} \qquad \text{or} \qquad x = -4 - \sqrt{6} \quad \text{Add } -4.$$

CHECK
$$x^2 + 8x + 10 = 0 \quad \text{Original equation}$$

Remember the middle term when squaring $-4 + \sqrt{6}$.

$$\left(-4 + \sqrt{6}\right)^2 + 8\left(-4 + \sqrt{6}\right) + 10 \overset{?}{=} 0 \quad \text{Let } x = -4 + \sqrt{6}.$$

$$16 - 8\sqrt{6} + 6 - 32 + 8\sqrt{6} + 10 \overset{?}{=} 0 \quad \text{Multiply.}$$

$$0 = 0 \ \checkmark \quad \text{True}$$

The check of $-4 - \sqrt{6}$ is similar. The solution set is

$$\left\{-4 + \sqrt{6},\ -4 - \sqrt{6}\right\}, \quad \text{or} \quad \left\{-4 \pm \sqrt{6}\right\}. \qquad \text{NOW TRY}$$

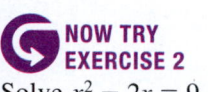

NOW TRY
EXERCISE 2
Solve $x^2 - 2x = 9$.

EXAMPLE 2 Solving a Quadratic Equation by Completing the Square ($a = 1$)

Solve $x^2 - 6x = 4$.

To complete the square on $x^2 - 6x$, take half the coefficient of x and square it.

$$\tfrac{1}{2}(-6) = -3 \quad \text{and} \quad (-3)^2 = 9$$

Add the result, 9, to each side of the equation.

$$x^2 - 6x = 4 \quad \text{Given equation}$$

$$x^2 - 6x + 9 = 4 + 9 \quad \text{Add 9.}$$

$$(x - 3)^2 = 13 \quad \text{Factor on the left. Add on the right.}$$

$$x - 3 = \sqrt{13} \qquad \text{or} \quad x - 3 = -\sqrt{13} \quad \text{Square root property}$$

$$x = 3 + \sqrt{13} \quad \text{or} \qquad x = 3 - \sqrt{13} \quad \text{Add 3.}$$

NOW TRY ANSWERS
1. $\left\{-3 \pm \sqrt{11}\right\}$
2. $\left\{1 \pm \sqrt{10}\right\}$

A check confirms that the solution set is $\left\{3 \pm \sqrt{13}\right\}$. 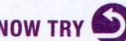 NOW TRY

> **Completing the Square to Solve** $ax^2 + bx + c = 0$ (Where $a \neq 0$)
>
> ***Step 1*** **Be sure the second-degree term has coefficient 1.**
> - If the coefficient of the second-degree term is 1, go to Step 2.
> - If it is not 1, but some other nonzero number a, divide each side of the equation by a.
>
> ***Step 2*** **Write the equation in correct form.** Make sure that all variable terms are on one side of the equality symbol and the constant term is on the other side.
>
> ***Step 3*** **Complete the square.**
> - Take half the coefficient of the first-degree term, and square it.
> - Add the square to each side of the equation.
> - Factor the variable side, which should be a perfect square trinomial, as the square of a binomial. Combine terms on the other side.
>
> ***Step 4*** **Solve** the equation using the square root property.

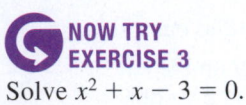

NOW TRY EXERCISE 3

Solve $x^2 + x - 3 = 0$.

EXAMPLE 3 Solving a Quadratic Equation by Completing the Square ($a = 1$)

Solve $x^2 + 5x - 1 = 0$.

The coefficient of the second-degree term is 1, so we begin with Step 2.

Step 2 $x^2 + 5x = 1$ Add 1 to each side.

Step 3 $\left[\frac{1}{2}(5)\right]^2 = \left(\frac{5}{2}\right)^2 = \frac{25}{4}$ Take half the coefficient of the first-degree term and square the result.

$$x^2 + 5x + \frac{25}{4} = 1 + \frac{25}{4}$$ Add the square to each side of the equation.

$$\left(x + \frac{5}{2}\right)^2 = \frac{29}{4}$$ Factor on the left. Add on the right.

Step 4 $x + \frac{5}{2} = \sqrt{\frac{29}{4}}$ or $x + \frac{5}{2} = -\sqrt{\frac{29}{4}}$ Square root property

$x + \frac{5}{2} = \frac{\sqrt{29}}{2}$ or $x + \frac{5}{2} = -\frac{\sqrt{29}}{2}$ $\sqrt{\frac{a}{b}} = \frac{\sqrt{a}}{\sqrt{b}}$

$x = -\frac{5}{2} + \frac{\sqrt{29}}{2}$ or $x = -\frac{5}{2} - \frac{\sqrt{29}}{2}$ Add $-\frac{5}{2}$.

$x = \frac{-5 + \sqrt{29}}{2}$ or $x = \frac{-5 - \sqrt{29}}{2}$ $\frac{a}{c} \pm \frac{b}{c} = \frac{a \pm b}{c}$

The solution set is $\left\{\frac{-5 \pm \sqrt{29}}{2}\right\}$.

NOW TRY

OBJECTIVE 2 Solve quadratic equations by completing the square when the coefficient of the second-degree term is not 1.

If a quadratic equation has the form

$$ax^2 + bx + c = 0, \quad \text{where} \quad a \neq 1,$$

we obtain 1 as the coefficient of x^2 by dividing each side of the equation by a.

NOW TRY ANSWER

3. $\left\{\frac{-1 \pm \sqrt{13}}{2}\right\}$

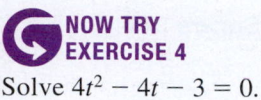

NOW TRY EXERCISE 4

Solve $4t^2 - 4t - 3 = 0$.

EXAMPLE 4 Solving a Quadratic Equation by Completing the Square ($a \neq 1$)

Solve $4x^2 + 16x - 9 = 0$.

Step 1 *Before we complete the square, the coefficient of x^2 must be 1,* not 4. We obtain 1 as the coefficient of x^2 here by dividing each side by 4.

$$4x^2 + 16x - 9 = 0 \qquad \text{Given equation}$$

The coefficient of x^2 must be 1.

$$x^2 + 4x - \frac{9}{4} = 0 \qquad \text{Divide by 4.}$$

Step 2 Write the equation so that all variable terms are on one side of the equation and all constant terms are on the other side.

$$x^2 + 4x = \frac{9}{4} \qquad \text{Add } \tfrac{9}{4}.$$

Step 3 Complete the square by taking half the coefficient of x, and squaring it.

$$\left[\tfrac{1}{2}(4)\right]^2 = 2^2 = 4$$

Add the result, 4, to each side of the equation.

$$x^2 + 4x + 4 = \frac{9}{4} + 4 \qquad \text{Add 4.}$$

$$(x + 2)^2 = \frac{25}{4} \qquad \text{Factor; } \tfrac{9}{4} + 4 = \tfrac{9}{4} + \tfrac{16}{4} = \tfrac{25}{4}$$

Step 4 Solve the equation using the square root property.

$$x + 2 = \sqrt{\frac{25}{4}} \quad \text{or} \quad x + 2 = -\sqrt{\frac{25}{4}} \qquad \text{Square root property}$$

$$x + 2 = \frac{5}{2} \quad \text{or} \quad x + 2 = -\frac{5}{2} \qquad \text{Take square roots.}$$

$$x = \frac{5}{2} - 2 \quad \text{or} \quad x = -\frac{5}{2} - 2 \qquad \text{Subtract 2.}$$

$$x = \frac{5}{2} - \frac{4}{2} \quad \text{or} \quad x = -\frac{5}{2} - \frac{4}{2} \qquad -2 = -\tfrac{4}{2}$$

$$x = \frac{1}{2} \quad \text{or} \quad x = -\frac{9}{2} \qquad \text{Subtract fractions.}$$

CHECK

$$4x^2 + 16x - 9 = 0$$

$$4\left(\frac{1}{2}\right)^2 + 16\left(\frac{1}{2}\right) - 9 \overset{?}{=} 0 \qquad \text{Let } x = \tfrac{1}{2}.$$

$$4\left(\frac{1}{4}\right) + 8 - 9 \overset{?}{=} 0$$

$$1 + 8 - 9 \overset{?}{=} 0$$

$$0 = 0 \ \checkmark \ \text{True}$$

$$4x^2 + 16x - 9 = 0$$

$$4\left(-\frac{9}{2}\right)^2 + 16\left(-\frac{9}{2}\right) - 9 \overset{?}{=} 0 \qquad \text{Let } x = -\tfrac{9}{2}.$$

$$4\left(\frac{81}{4}\right) - 72 - 9 \overset{?}{=} 0$$

$$81 - 72 - 9 \overset{?}{=} 0$$

$$0 = 0 \ \checkmark \ \text{True}$$

NOW TRY ANSWER

4. $\left\{-\frac{1}{2}, \frac{3}{2}\right\}$

The two values, $\frac{1}{2}$ and $-\frac{9}{2}$, check, so the solution set is $\left\{-\frac{9}{2}, \frac{1}{2}\right\}$.

NOW TRY

NOW TRY
EXERCISE 5
Solve $3x^2 + 12x - 5 = 0$.

EXAMPLE 5 Solving a Quadratic Equation by Completing the Square ($a \neq 1$)

Solve $2x^2 - 4x - 5 = 0$.

Divide each side by 2 to obtain 1 as the coefficient of the second-degree term.

$$x^2 - 2x - \frac{5}{2} = 0 \qquad \text{Divide by 2. (Step 1)}$$

$$x^2 - 2x = \frac{5}{2} \qquad \text{Add } \tfrac{5}{2}. \text{ (Step 2)}$$

$$\left[\tfrac{1}{2}(-2)\right]^2 = (-1)^2 = 1 \qquad \text{Complete the square. (Step 3)}$$

$$x^2 - 2x + 1 = \frac{5}{2} + 1 \qquad \text{Add 1 to each side.}$$

$$(x - 1)^2 = \frac{7}{2} \qquad \text{Factor on the left. Add on the right.}$$

$$x - 1 = \sqrt{\frac{7}{2}} \quad \text{or} \quad x - 1 = -\sqrt{\frac{7}{2}} \qquad \text{Square root property (Step 4)}$$

$$x = 1 + \sqrt{\frac{7}{2}} \quad \text{or} \qquad x = 1 - \sqrt{\frac{7}{2}} \qquad \text{Add 1.}$$

$$x = 1 + \frac{\sqrt{14}}{2} \quad \text{or} \qquad x = 1 - \frac{\sqrt{14}}{2} \qquad \sqrt{\frac{7}{2}} = \frac{\sqrt{7}}{\sqrt{2}} = \frac{\sqrt{7}}{\sqrt{2}} \cdot \frac{\sqrt{2}}{\sqrt{2}} = \frac{\sqrt{14}}{2}$$

Add the two terms in each solution as follows.

$$1 + \frac{\sqrt{14}}{2} = \frac{2}{2} + \frac{\sqrt{14}}{2} = \frac{2 + \sqrt{14}}{2}$$

$$1 - \frac{\sqrt{14}}{2} = \frac{2}{2} - \frac{\sqrt{14}}{2} = \frac{2 - \sqrt{14}}{2} \qquad 1 = \frac{2}{2}$$

The solution set is $\left\{\dfrac{2 \pm \sqrt{14}}{2}\right\}$.

NOW TRY

EXAMPLE 6 Solving a Quadratic Equation (Nonreal Complex Solutions)

Solve $4p^2 + 8p + 5 = 0$.

$$4p^2 + 8p + 5 = 0$$

The coefficient of the second-degree term must be 1.

$$p^2 + 2p + \frac{5}{4} = 0 \qquad \text{Divide by 4. (Step 1)}$$

$$p^2 + 2p = -\frac{5}{4} \qquad \text{Subtract } \tfrac{5}{4}. \text{ (Step 2)}$$

To complete the square, take half the coefficient of p, and square it.

$$p^2 + 2p + 1 = -\frac{5}{4} + 1 \qquad \begin{array}{l}\text{Complete the square. Add}\\ \left[\tfrac{1}{2}(2)\right]^2 = 1^2 = 1 \text{ to each side. (Step 3)}\end{array}$$

$$(p + 1)^2 = -\frac{1}{4} \qquad \text{Factor; } -\tfrac{5}{4} + 1 = -\tfrac{5}{4} + \tfrac{4}{4} = -\tfrac{1}{4}$$

NOW TRY ANSWER
5. $\left\{\dfrac{-6 \pm \sqrt{51}}{3}\right\}$

🔄 **NOW TRY**
EXERCISE 6
Solve $3t^2 - 12t + 15 = 0$.

$$p + 1 = \sqrt{-\frac{1}{4}} \quad \text{or} \quad p + 1 = -\sqrt{-\frac{1}{4}} \qquad \text{Square root property (Step 4)}$$

$$p + 1 = \frac{1}{2}i \quad \text{or} \quad p + 1 = -\frac{1}{2}i \qquad \sqrt{-\frac{1}{4}} = \frac{1}{2}i$$

$$p = -1 + \frac{1}{2}i \quad \text{or} \quad p = -1 - \frac{1}{2}i \qquad \text{Add } -1.$$

The solution set is $\left\{-1 \pm \frac{1}{2}i\right\}$.

NOW TRY 🔄

OBJECTIVE 3 Simplify the terms of an equation before solving.

🔄 **NOW TRY**
EXERCISE 7
Solve $(x - 5)(x + 1) = 2$.

EXAMPLE 7 Simplifying before Completing the Square

Solve $(x + 3)(x - 1) = 2$.

$$(x + 3)(x - 1) = 2$$

$$x^2 + 2x - 3 = 2 \qquad \text{Multiply using the FOIL method.}$$

$$x^2 + 2x = 5 \qquad \text{Add 3.}$$

$$x^2 + 2x + 1 = 5 + 1 \qquad \text{Add } \left[\frac{1}{2}(2)\right]^2 = 1^2 = 1.$$

$$(x + 1)^2 = 6 \qquad \text{Factor on the left. Add on the right.}$$

$$x + 1 = \sqrt{6} \quad \text{or} \quad x + 1 = -\sqrt{6} \qquad \text{Square root property}$$

$$x = -1 + \sqrt{6} \quad \text{or} \quad x = -1 - \sqrt{6} \qquad \text{Add } -1.$$

The solution set is $\left\{-1 \pm \sqrt{6}\right\}$.

NOW TRY 🔄

🔄 **NOW TRY**
EXERCISE 8
Solve $x(x + 5) = 3$.

EXAMPLE 8 Simplifying before Completing the Square

Solve $x(x + 7) = 2$.

$$x(x + 7) = 2$$

$$x^2 + 7x = 2 \qquad \text{Multiply.}$$

$$x^2 + 7x + \frac{49}{4} = 2 + \frac{49}{4} \qquad \text{Add } \left[\frac{1}{2}(7)\right]^2 = \left(\frac{7}{2}\right)^2 = \frac{49}{4}.$$

$$\left(x + \frac{7}{2}\right)^2 = \frac{57}{4} \qquad \text{Factor; } 2 + \frac{49}{4} = \frac{8}{4} + \frac{49}{4} = \frac{57}{4}$$

$$x + \frac{7}{2} = \sqrt{\frac{57}{4}} \quad \text{or} \quad x + \frac{7}{2} = -\sqrt{\frac{57}{4}} \qquad \text{Square root property}$$

$$x + \frac{7}{2} = \frac{\sqrt{57}}{2} \quad \text{or} \quad x + \frac{7}{2} = -\frac{\sqrt{57}}{2} \qquad \text{Quotient rule for radicals, } \sqrt{\frac{a}{b}} = \frac{\sqrt{a}}{\sqrt{b}}$$

$$x = -\frac{7}{2} + \frac{\sqrt{57}}{2} \quad \text{or} \quad x = -\frac{7}{2} - \frac{\sqrt{57}}{2} \qquad \text{Add } -\frac{7}{2}.$$

$$x = \frac{-7 + \sqrt{57}}{2} \quad \text{or} \quad x = \frac{-7 - \sqrt{57}}{2} \qquad \text{Add and subtract the fractions.}$$

The solution set is $\left\{\dfrac{-7 \pm \sqrt{57}}{2}\right\}$.

NOW TRY 🔄

NOW TRY ANSWERS

6. $\{2 \pm i\}$

7. $\{2 \pm \sqrt{11}\}$

8. $\left\{\dfrac{-5 \pm \sqrt{37}}{2}\right\}$

11.2 Exercises

  MyLab Math

▶ *Video solutions for select problems available in MyLab Math*

1. Concept Check   Which step is an appropriate way to begin solving the quadratic equation $2x^2 - 4x = 9$ by completing the square?

A. Add 4 to each side.   **B.** Factor the left side as $2x(x - 2)$.

C. Factor the left side as $x(2x - 4)$.   **D.** Divide each side by 2.

2. Concept Check   Which step is an appropriate way to begin solving the quadratic equation $x^2 + 12x = 13$ by completing the square?

A. Add 36 to each side.   **B.** Subtract 13 from each side.

C. Divide each side by x.   **D.** Add 6 to each side.

Complete each trinomial so that it is a perfect square. Then factor the trinomial. See Example 1.

3. $x^2 + 10x +$ _____   **4.** $x^2 + 16x +$ _____   **5.** $z^2 - 20z +$ _____

6. $x^2 - 32x +$ _____   **7.** $x^2 + 2x +$ _____   **8.** $m^2 - 2m +$ _____

9. $p^2 - 5p +$ _____   **10.** $x^2 + 3x +$ _____   **11.** $x^2 + \frac{1}{2}x +$ _____

12. $x^2 + \frac{1}{3}x +$ _____   **13.** $x^2 - 0.4x +$ _____   **14.** $x^2 - 0.8x +$ _____

Concept Check   *Solve each equation by completing the square.*

15. $x^2 + 4x = 1$

Take half the coefficient of x and square it.

$$\frac{1}{2} \cdot \text{\_\_\_\_} = 2, \text{ and } \text{\_\_\_\_}^2 = 4.$$

Now add _____ to each side of the equation.

$$x^2 + 4x + \text{\_\_\_\_} = 1 + 4$$

Factor and add.

$$(\text{_____})^2 = \text{\_\_\_\_}$$

Complete the solution.

16. $3x^2 + 5x - 2 = 0$

Divide each side by _____ to obtain

_____. Add $\frac{2}{3}$ to each side to

obtain _____.

Take half the coefficient of x and square it. Add _____ to each side of the equation.

$$x^2 + \frac{5}{3}x + \text{\_\_\_\_} = \frac{2}{3} + \frac{25}{36}$$

Factor and add.

$$(\text{_____})^2 = \text{\_\_\_\_}$$

Complete the solution.

Solve each equation by completing the square. See Examples 1–3.

17. $x^2 - 4x = -3$ **18.** $p^2 - 2p = 8$ **19.** $x^2 + 2x - 5 = 0$

20. $r^2 + 4r + 1 = 0$ **21.** $x^2 + 4x - 2 = 0$ **22.** $t^2 + 2t - 1 = 0$

23. $x^2 + 10x + 18 = 0$ **24.** $x^2 + 8x + 11 = 0$ **25.** $x^2 - 8x = -4$

26. $m^2 - 4m = 14$ **27.** $x^2 + 7x - 1 = 0$ **28.** $x^2 + 13x - 3 = 0$

Solve each equation by completing the square. See Examples 4, 5, 7 and 8.

29. $4x^2 + 4x = 3$ **30.** $9x^2 + 3x = 2$ **31.** $2k^2 + 5k - 2 = 0$

32. $3r^2 + 2r - 2 = 0$ **33.** $5x^2 - 10x + 2 = 0$ **34.** $2x^2 - 16x + 25 = 0$

35. $3x^2 - 9x + 5 = 0$ **36.** $6x^2 - 8x - 3 = 0$ **37.** $3k^2 + 7k = 4$

38. $2k^2 + 5k = 1$ **39.** $3w^2 - w = 24$ **40.** $4z^2 - z = 39$

41. $-x^2 + 2x = -5$ **42.** $-x^2 + 4x = 1$ **43.** $5x^2 + 6x - 11 = 0$

44. $7x^2 - 9x - 10 = 0$ **45.** $(x + 3)(x - 1) = 5$ **46.** $(x - 8)(x + 2) = 24$

47. $(r - 3)(r - 5) = 2$ **48.** $(x - 1)(x - 7) = 1$ **49.** $x(x - 3) = 1$

50. $x(x - 5) = 2$ **51.** $x(x + 3) = -1$ **52.** $x(x + 7) = -2$

53. $-3x^2 + 11x + 42 = 0$ **54.** $-9x^2 - 20x + 21 = 0$

55. $0.1x^2 - 0.2x - 0.1 = 0$ **56.** $0.1p^2 - 0.4p + 0.1 = 0$

   (*Hint:* First clear the decimals.)   (*Hint:* First clear the decimals.)

Extending Skills   *Solve each equation by completing the square. Give* (**a**) *exact solutions and* (**b**) *solutions rounded to the nearest thousandth.*

57. $3r^2 - 2 = 6r + 3$ **58.** $4p + 3 = 2p^2 + 2p$

59. $(x + 1)(x + 3) = 2$ **60.** $(x - 3)(x + 1) = 1$

Solve each equation. (All solutions are nonreal complex numbers.) **See Example 6.**

61. $m^2 + 4m + 13 = 0$ **62.** $t^2 + 6t + 10 = 0$

63. $m^2 + 6m + 12 = 0$ **64.** $x^2 + 10x + 27 = 0$

65. $3r^2 + 4r + 4 = 0$ **66.** $4x^2 + 5x + 5 = 0$

67. $-k^2 - 5k - 10 = 0$ **68.** $-x^2 - 3x - 8 = 0$

Extending Skills   *Solve for x. Assume that a and b represent positive real numbers.*

69. $x^2 - b = 0$ **70.** $x^2 = 4b$ **71.** $4x^2 = b^2 + 16$

72. $9x^2 - 25a = 0$ **73.** $(5x - 2b)^2 = 3a$ **74.** $x^2 - a^2 - 36 = 0$

RELATING CONCEPTS   For Individual or Group Work   (Exercises 75–80)

The Greeks had a method of completing the square geometrically in which they literally changed a figure into a square. For
example, to complete the square for $x^2 + 6x$, we begin with a
square of side x, as in the figure on the top. We add three rectangles of width 1 to the right side and the bottom to get a region
with area $x^2 + 6x$. To fill in the corner (complete the square),
we must add nine 1-by-1 squares as shown.

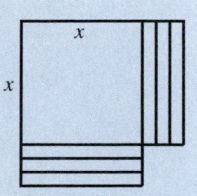

Work Exercises 75–80 in order.

75. What is the area of the original square?

76. What is the area of each strip?

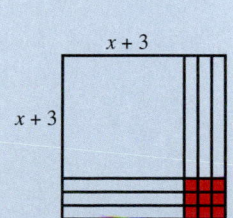

77. What is the total area of the six strips?

78. What is the area of each small square in the corner of the
   second figure?

79. What is the total area of the small squares?

80. What is the area of the new "complete" square?

11.3   Solving Quadratic Equations by the Quadratic Formula

OBJECTIVES

1 Derive the quadratic formula.

2 Solve quadratic equations using the quadratic formula.

3 Use the discriminant to determine number and type of solutions.

VOCABULARY

☐ quadratic formula
☐ discriminant

In this section, we complete the square to solve the general quadratic equation

$$ax^2 + bx + c = 0,$$

where a, b, and c are complex numbers and $a \neq 0$. The solution of this general equation gives a formula for finding the solution of *any* specific quadratic equation.

OBJECTIVE 1   Derive the quadratic formula.

We solve $ax^2 + bx + c = 0$ by completing the square (where $a > 0$) as follows.

$$ax^2 + bx + c = 0$$

$$x^2 + \frac{b}{a}x + \frac{c}{a} = 0 \qquad \text{Divide by } a. \text{ (Step 1)}$$

$$x^2 + \frac{b}{a}x = -\frac{c}{a} \qquad \text{Subtract } \tfrac{c}{a}. \text{ (Step 2)}$$

$$\left[\frac{1}{2}\left(\frac{b}{a}\right)\right]^2 = \left(\frac{b}{2a}\right)^2 = \frac{b^2}{4a^2} \qquad \text{Complete the square. (Step 3)}$$

$$x^2 + \frac{b}{a}x + \frac{b^2}{4a^2} = -\frac{c}{a} + \frac{b^2}{4a^2} \qquad \text{Add } \tfrac{b^2}{4a^2} \text{ to each side.}$$

$$\left(x + \frac{b}{2a}\right)^2 = \frac{b^2}{4a^2} + \frac{-c}{a} \qquad \begin{array}{l}\text{Factor on the left.}\\ \text{Rearrange the terms on the right.}\end{array}$$

$$\left(x + \frac{b}{2a}\right)^2 = \frac{b^2}{4a^2} + \frac{-4ac}{4a^2} \qquad \text{Write with a common denominator.}$$

$$\left(x + \frac{b}{2a}\right)^2 = \frac{b^2 - 4ac}{4a^2} \qquad \text{Add fractions.}$$

$$x + \frac{b}{2a} = \sqrt{\frac{b^2 - 4ac}{4a^2}} \quad \text{or} \quad x + \frac{b}{2a} = -\sqrt{\frac{b^2 - 4ac}{4a^2}} \qquad \begin{array}{l}\text{Square root property}\\ \text{(Step 4)}\end{array}$$

We can simplify $\sqrt{\dfrac{b^2 - 4ac}{4a^2}}$ as $\dfrac{\sqrt{b^2 - 4ac}}{\sqrt{4a^2}}$, or $\dfrac{\sqrt{b^2 - 4ac}}{2a}$.

The right side of each equation can be expressed as using the *t* symbol.

$$x + \frac{b}{2a} = \frac{\sqrt{b^2 - 4ac}}{2a} \quad \text{or} \quad x + \frac{b}{2a} = \frac{-\sqrt{b^2 - 4ac}}{2a}$$

$$x = \frac{-b}{2a} + \frac{\sqrt{b^2 - 4ac}}{2a} \quad \text{or} \quad x = \frac{-b}{2a} - \frac{\sqrt{b^2 - 4ac}}{2a}$$

If $a < 0$, the same two solutions are obtained.

$$x = \frac{-b + \sqrt{b^2 - 4ac}}{2a} \quad \text{or} \quad x = \frac{-b - \sqrt{b^2 - 4ac}}{2a}$$

This result is the **quadratic formula,** which is abbreviated using the $\pm$ symbol.

Quadratic Formula

The solutions of the equation $ax^2 + bx + c = 0$ (where $a \neq 0$) are given by

$$x = \frac{-b \pm \sqrt{b^2 - 4ac}}{2a}.$$

! CAUTION *In the quadratic formula, the square root is added to or subtracted from the value of $-b$ before dividing by 2a.*

OBJECTIVE 2 Solve quadratic equations using the quadratic formula.

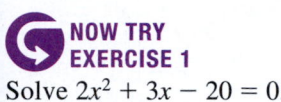
NOW TRY EXERCISE 1
Solve $2x^2 + 3x - 20 = 0$.

EXAMPLE 1 Using the Quadratic Formula (Two Rational Solutions)

Solve $6x^2 - 5x - 4 = 0$.

This equation is in standard form. Here a, the coefficient of the second-degree term, is 6, and b, the coefficient of the first-degree term, is -5. The constant c is -4.

$$x = \frac{-b \pm \sqrt{b^2 - 4ac}}{2a} \qquad \text{Quadratic formula}$$

$$x = \frac{-(-5) \pm \sqrt{(-5)^2 - 4(6)(-4)}}{2(6)} \qquad \text{Substitute } a = 6, b = -5, c = -4.$$

> Use parentheses and substitute carefully to avoid errors.

$$x = \frac{5 \pm \sqrt{25 + 96}}{12}$$

$$x = \frac{5 \pm \sqrt{121}}{12} \qquad \text{Add under the radical.}$$

$$x = \frac{5 \pm 11}{12} \qquad \text{Take the square root.}$$

There are two values represented, one from the $+$ sign and one from the $-$ sign.

$$x = \frac{5 + 11}{12} = \frac{16}{12} = \frac{4}{3} \quad \text{or} \quad x = \frac{5 - 11}{12} = \frac{-6}{12} = -\frac{1}{2}$$

Check each value in the original equation. The solution set is $\left\{ -\frac{1}{2}, \frac{4}{3} \right\}$.   **NOW TRY** ↻

NOTE When solving a quadratic equation, it is a good idea to try to factor the quadratic expression first. If it can be factored, then apply the zero-factor property.

$$6x^2 - 5x - 4 = 0 \qquad \text{See Example 1.}$$

$$(3x - 4)(2x + 1) = 0 \qquad \text{Factor.}$$

$$3x - 4 = 0 \quad \text{or} \quad 2x + 1 = 0 \qquad \text{Zero-factor property}$$

$$3x = 4 \quad \text{or} \qquad 2x = -1 \qquad \text{Solve each equation.}$$

$$x = \frac{4}{3} \quad \text{or} \qquad x = -\frac{1}{2} \qquad \text{Same solutions as in Example 1}$$

If it cannot be factored or if factoring is difficult, then use the quadratic formula.

NOW TRY ANSWER

1. $\left\{ -4, \frac{5}{2} \right\}$

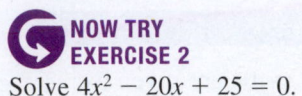
NOW TRY
EXERCISE 2
Solve $4x^2 - 20x + 25 = 0$.

EXAMPLE 2   Using the Quadratic Formula (One Rational Solution)

Solve $9x^2 + 12x + 4 = 0$.

The trinomial on the left side of the equality symbol can be factored, so this equation could be solved using the zero-factor property. However, if we did not recognize this and solved using the quadratic formula, our work might look similar to that below. The equation $9x^2 + 12x + 4 = 0$ is in standard form, so $a = 9$, $b = 12$, and $c = 4$.

$$x = \frac{-b \pm \sqrt{b^2 - 4ac}}{2a} \qquad \text{Quadratic formula}$$

$$x = \frac{-12 \pm \sqrt{12^2 - 4(9)(4)}}{2(9)} \qquad a = 9, b = 12, c = 4$$

$$x = \frac{-12 \pm \sqrt{144 - 144}}{18} \qquad \text{Simplify.}$$

$$x = \frac{-12 \pm 0}{18} \qquad \sqrt{0} = 0$$

$$x = -\frac{2}{3} \qquad \text{Write } \frac{-12}{18} \text{ in lowest terms.}$$

In this case, $b^2 - 4ac = 0$ because $9x^2 + 12x + 4$ is a perfect square trinomial. There is one *distinct* solution, $-\frac{2}{3}$. The solution set is $\left\{-\frac{2}{3}\right\}$. (Confirm this solution by factoring $9x^2 + 12x + 4$ and using the zero-factor property.)   **NOW TRY**

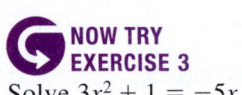
NOW TRY
EXERCISE 3
Solve $3x^2 + 1 = -5x$.

EXAMPLE 3   Using the Quadratic Formula (Two Irrational Solutions)

Solve $4x^2 = 8x - 1$.

Write the equation in standard form as $4x^2 - 8x + 1 = 0$.   This is a key step.

$$x = \frac{-b \pm \sqrt{b^2 - 4ac}}{2a} \qquad \text{Quadratic formula}$$

$$x = \frac{-(-8) \pm \sqrt{(-8)^2 - 4(4)(1)}}{2(4)} \qquad a = 4, b = -8, c = 1$$

$$x = \frac{8 \pm \sqrt{64 - 16}}{8} \qquad \text{Simplify in the numerator and denominator.}$$

$$x = \frac{8 \pm \sqrt{48}}{8} \qquad \text{Subtract under the radical.}$$

$$x = \frac{8 \pm 4\sqrt{3}}{8} \qquad \sqrt{48} = \sqrt{16} \cdot \sqrt{3} = 4\sqrt{3}$$

$$x = \frac{4(2 \pm \sqrt{3})}{4(2)} \qquad \text{Factor.}$$

$$x = \frac{2 \pm \sqrt{3}}{2} \qquad \text{Divide out the common factor 4 to write in lowest terms.}$$

The solution set is $\left\{\frac{2 \pm \sqrt{3}}{2}\right\}$.   **NOW TRY**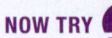

NOW TRY ANSWERS
2. $\left\{\frac{5}{2}\right\}$

3. $\left\{\frac{-5 \pm \sqrt{13}}{6}\right\}$

! CAUTION

1. *Before solving, every quadratic equation must be expressed in standard form $ax^2 + bx + c = 0$,* whether we use the zero-factor property or the quadratic formula.

2. *When writing solutions in lowest terms, be sure to factor first. Then divide out the common factor.* See the last two steps in **Example 3.**

NOW TRY EXERCISE 4
Solve $(x + 5)(x - 1) = -18$.

EXAMPLE 4 Using the Quadratic Formula (Two Nonreal Complex Solutions)

Solve $(9x + 3)(x - 1) = -8$.

This is a quadratic equation—when the first terms $9x$ and x are multiplied, we obtain a second-degree term, $9x^2$. We must write the equation in standard form.

$$(9x + 3)(x - 1) = -8$$
$$9x^2 - 6x - 3 = -8 \quad \text{Multiply.}$$
$$\text{Standard form} \rightarrow 9x^2 - 6x + 5 = 0 \quad \text{Add 8.}$$

From the equation $9x^2 - 6x + 5 = 0$, we identify $a = 9$, $b = -6$, and $c = 5$.

$$x = \frac{-b \pm \sqrt{b^2 - 4ac}}{2a} \quad \text{Quadratic formula}$$

$$x = \frac{-(-6) \pm \sqrt{(-6)^2 - 4(9)(5)}}{2(9)} \quad \text{Substitute.}$$

$$x = \frac{6 \pm \sqrt{-144}}{18} \quad \text{Simplify.}$$

$$x = \frac{6 \pm 12i}{18} \quad \sqrt{-144} = 12i$$

$$x = \frac{6(1 \pm 2i)}{6(3)} \quad \text{Factor first. Then divide out the common factor.} \quad \text{Factor.}$$

$$x = \frac{1 \pm 2i}{3} \quad \text{Divide out the common factor 6 to write in lowest terms.}$$

$$x = \frac{1}{3} \pm \frac{2}{3}i \quad \text{Standard form } a + bi \text{ for a complex number}$$

The solution set is $\left\{\frac{1}{3} \pm \frac{2}{3}i\right\}$.

NOW TRY

OBJECTIVE 3 Use the discriminant to determine number and type of solutions.

The solutions of the quadratic equation $ax^2 + bx + c = 0$ are given by

$$x = \frac{-b \pm \sqrt{b^2 - 4ac}}{2a}. \quad \leftarrow \text{Discriminant}$$

The expression under the radical symbol, $b^2 - 4ac$, is called the **discriminant** because it distinguishes among the number of solutions—one or two—and the type of solutions—rational, irrational, or nonreal complex—of a quadratic equation.

NOW TRY ANSWER
4. $\{-2 \pm 3i\}$

Using the Discriminant

If a, b, and c are integers in a quadratic equation $ax^2 + bx + c = 0$, then the discriminant $b^2 - 4ac$ can be used to determine the number and type of solutions of the equation as follows.

Discriminant $b^2 - 4ac$	Number and Type of Solutions
Positive, and the square of an integer	Two rational solutions
Positive, but not the square of an integer	Two irrational solutions
Zero	One rational solution
Negative	Two nonreal complex solutions

We can also use the discriminant to help decide how to solve a quadratic equation.

> *If a, b, and c are integers and the discriminant is a perfect square (including 0), then the equation can be solved using the zero-factor property. Otherwise, the quadratic formula should be used.*

EXAMPLE 5   Using the Discriminant

Find the discriminant. Use it to predict the number and type of solutions for each equation. Tell whether the equation can be solved using the zero-factor property, or if the quadratic formula should be used instead.

(a) $6x^2 - x - 15 = 0$

First identify the values of a, b, and c. Because $-x = -1x$, the value of b is -1. We find the discriminant by evaluating $b^2 - 4ac$.

$$b^2 - 4ac$$

Use parentheses and substitute carefully.

$$= (-1)^2 - 4(6)(-15) \qquad a = 6, b = -1, c = -15 \text{ (all integers)}$$

$$= 1 + 360 \qquad \text{Apply the exponent. Multiply.}$$

$$= 361 \qquad \text{Add.}$$

$$= 19^2, \quad \text{which is a perfect square.}$$

The discriminant 361 is a perfect square, so referring to the table we see that there will be two rational solutions. We can solve using the zero-factor property.

(b) $3x^2 - 4x = 5$

Write in standard form as $3x^2 - 4x - 5 = 0$.

$$b^2 - 4ac \qquad \text{Discriminant}$$

$$= (-4)^2 - 4(3)(-5) \qquad a = 3, b = -4, c = -5 \text{ (all integers)}$$

$$= 16 + 60 \qquad \text{Apply the exponent. Multiply.}$$

$$= 76 \qquad \text{Add.}$$

Because 76 is positive but *not* the square of an integer, the equation will have two irrational solutions. We solve using the quadratic formula.

NOW TRY
EXERCISE 5
Find the discriminant. Use it to predict the number and type of solutions for each equation. Tell whether the equation can be solved using the zero-factor property, or if the quadratic formula should be used instead.

(a) $8x^2 - 6x - 5 = 0$

(b) $9x^2 = 24x - 16$

(c) $3x^2 + 2x = -1$

NOW TRY ANSWERS

5. (a) 196; two rational solutions; zero-factor property
(b) 0; one rational solution; zero-factor property
(c) −8; two nonreal complex solutions; quadratic formula

(c) $4x^2 + x + 1 = 0$

 $x = 1x$, so $b = 1$.

$$b^2 - 4ac \qquad \text{Discriminant}$$
$$= 1^2 - 4(4)(1) \qquad a = 4, b = 1, c = 1 \text{ (all integers)}$$
$$= 1 - 16 \qquad \text{Apply the exponent. Multiply.}$$
$$= -15 \qquad \text{Subtract.}$$

Because the discriminant is negative, there will be two nonreal complex solutions. We solve using the quadratic formula.

(d) $4x^2 + 9 = 12x$ Write in standard form as $4x^2 - 12x + 9 = 0$.

$$b^2 - 4ac \qquad \text{Discriminant}$$
$$= (-12)^2 - 4(4)(9) \qquad a = 4, b = -12, c = 9 \text{ (all integers)}$$
$$= 144 - 144 \qquad \text{Apply the exponent. Multiply.}$$
$$= 0 \qquad \text{Subtract.}$$

The discriminant is 0, so there is only one rational solution. We solve using the zero-factor property.   **NOW TRY**

11.3 Exercises

FOR EXTRA HELP   ▶ **MyLab Math**

▶ *Video solutions for select problems available in MyLab Math*

Concept Check   *Answer each question.*

1. The documentation for an early version of Microsoft *Word* for Windows used the following for the quadratic formula. Was this correct? If not, correct it.

$$x = -b \pm \frac{\sqrt{b^2 - 4ac}}{2a} \qquad \text{Correct or incorrect?}$$

2. One patron wrote the quadratic formula, as shown here, on a wall at the Cadillac Bar in Houston, Texas. Was this correct? If not, correct it.

$$x = \frac{-b\sqrt{b^2 - 4ac}}{2a} \qquad \text{Correct or incorrect?}$$

3. Concept Check   A student solved $5x^2 - 5x + 1 = 0$ incorrectly as follows.

$$x = \frac{-(-5) \pm \sqrt{(-5)^2 - 4(5)(1)}}{2(5)} \qquad a = 5, b = -5, c = 1$$

$$x = \frac{5 \pm \sqrt{5}}{10} \qquad \text{Simplify.}$$

$$x = \frac{1}{2} \pm \sqrt{5} \qquad \text{Write in lowest terms.}$$

WHAT WENT WRONG?   Give the correct solution set.

4. Concept Check   A student incorrectly claimed that the equation $2x^2 - 5 = 0$ cannot be solved using the quadratic formula because there is no first-degree x-term. **WHAT WENT WRONG?**   Give the values of a, b, and c for this equation.

Use the quadratic formula to solve each equation. (All solutions for these equations are real numbers.) See Examples 1–3.

5. $x^2 - 8x + 15 = 0$ **6.** $x^2 + 3x - 28 = 0$ **7.** $6x^2 + 11x - 10 = 0$

8. $8x^2 + 10x - 3 = 0$ **9.** $4x^2 + 12x + 9 = 0$ **10.** $16x^2 + 40x + 25 = 0$

11. $36x^2 - 12x + 1 = 0$ **12.** $9x^2 - 6x + 1 = 0$ **13.** $2x^2 + 4x + 1 = 0$

14. $2x^2 + 3x - 1 = 0$ **15.** $2x^2 - 2x = 1$ **16.** $9x^2 + 6x = 1$

17. $x^2 + 18 = 10x$ **18.** $x^2 - 4 = 2x$ **19.** $4x^2 + 4x - 1 = 0$

20. $4r^2 - 4r - 19 = 0$ **21.** $2 - 2x = 3x^2$ **22.** $26r - 2 = 3r^2$

23. $\dfrac{x^2}{4} - \dfrac{x}{2} = 1$ **24.** $p^2 + \dfrac{p}{3} = \dfrac{1}{6}$ **25.** $-2t(t + 2) = -3$

26. $-3x(x + 2) = -4$ **27.** $(r - 3)(r + 5) = 2$ **28.** $(x + 1)(x - 7) = 1$

29. $(x + 2)(x - 3) = 1$ **30.** $(x - 5)(x + 2) = 6$ **31.** $p = \dfrac{5(5 - p)}{3(p + 1)}$

32. $x = \dfrac{2(x + 3)}{x + 5}$ **33.** $(2x + 1)^2 = x + 4$ **34.** $(2x - 1)^2 = x + 2$

Use the quadratic formula to solve each equation. (All solutions for these equations are nonreal complex numbers.) See Example 4.

35. $x^2 - 3x + 6 = 0$ **36.** $x^2 - 5x + 20 = 0$ **37.** $r^2 - 6r + 14 = 0$

38. $t^2 + 4t + 11 = 0$ **39.** $4x^2 - 4x = -7$ **40.** $9x^2 - 6x = -7$

41. $x(3x + 4) = -2$ **42.** $z(2z + 3) = -2$

43. $(2x - 1)(8x - 4) = -1$ **44.** $(x - 1)(9x - 3) = -2$

45. $(6x + 1)(x - 2) = (x + 2)(x - 5)$ **46.** $(4x + 3)(x - 1) = (x + 3)(x - 6)$

Find the discriminant. Use it to determine whether the solutions for each equation are

  A. *two rational numbers*   **B.** *one rational number*

  C. *two irrational numbers*   **D.** *two nonreal complex numbers.*

Tell whether the equation can be solved using the zero-factor property, or if the quadratic formula should be used instead. Do not actually solve. See Example 5.

47. $25x^2 + 70x + 49 = 0$ **48.** $4x^2 - 28x + 49 = 0$ **49.** $x^2 + 4x + 2 = 0$

50. $9x^2 - 12x - 1 = 0$ **51.** $3x^2 = 5x + 2$ **52.** $4x^2 = 4x + 3$

53. $3m^2 - 10m + 15 = 0$ **54.** $18x^2 + 60x + 82 = 0$

55. Find the discriminant for each quadratic equation. Use it to tell whether the equation can be solved using the zero-factor property, or the quadratic formula should be used instead. Then solve each equation.

  (a) $3x^2 + 13x = -12$ **(b)** $2x^2 + 19 = 14x$

56. Refer to the answers in **Exercises 47–54,** and solve the equation given in each exercise.

  (a) $25x^2 + 70x + 49 = 0$ **(Exercise 47)**   **(b)** $4x^2 - 28x + 49 = 0$ **(Exercise 48)**

  (c) $3x^2 = 5x + 2$ **(Exercise 51)**   **(d)** $4x^2 = 4x + 3$ **(Exercise 52)**

Extending Skills *Find the value of a, b, or c so that each equation will have exactly one rational solution. (Hint: The discriminant must equal 0 for an equation to have one rational solution.)*

57. $p^2 + bp + 25 = 0$ **58.** $r^2 - br + 49 = 0$ **59.** $am^2 + 8m + 1 = 0$

60. $at^2 + 24t + 16 = 0$ **61.** $9x^2 - 30x + c = 0$ **62.** $4m^2 + 12m + c = 0$

63. One solution of $4x^2 + bx - 3 = 0$ is $-\frac{5}{2}$. Find b and the other solution.

64. One solution of $3x^2 - 7x + c = 0$ is $\frac{1}{3}$. Find c and the other solution.

11.4   Equations That Lead to Quadratic Methods

OBJECTIVES

1 Solve rational equations that lead to quadratic equations.

2 Solve applied problems involving quadratic equations.

3 Solve radical equations that lead to quadratic equations.

4 Solve equations that are quadratic in form.

VOCABULARY
☐ quadratic in form

**NOW TRY
EXERCISE 1**

Solve $\dfrac{2}{x} + \dfrac{3}{x+2} = 1$.

NOW TRY ANSWER
1. $\{-1, 4\}$

OBJECTIVE 1 Solve rational equations that lead to quadratic equations.

A variety of nonquadratic equations can be written in the form of a quadratic equation and solved using the methods of this chapter.

EXAMPLE 1 Solving a Rational Equation That Leads to a Quadratic Equation

Solve $\dfrac{1}{x} + \dfrac{1}{x-1} = \dfrac{7}{12}$.

Clear fractions by multiplying each side by the least common denominator, $12x(x-1)$. (The domain is $\{x \mid x \text{ is a real number}, x \neq 0, 1\}$.)

$$12x(x-1)\left(\frac{1}{x} + \frac{1}{x-1}\right) = 12x(x-1)\left(\frac{7}{12}\right) \qquad \text{Multiply by the LCD.}$$

$$12x(x-1)\frac{1}{x} + 12x(x-1)\frac{1}{x-1} = 12x(x-1)\frac{7}{12} \qquad \text{Distributive property}$$

$$12(x-1) + 12x = x(x-1) \cdot 7 \qquad \text{Multiply.}$$

$$12x - 12 + 12x = 7x^2 - 7x \qquad \text{Distributive property}$$

$$24x - 12 = 7x^2 - 7x \qquad \text{Combine like terms.}$$

This trinomial is factorable. $\rightarrow$ $7x^2 - 31x + 12 = 0$ Standard form

$$(7x - 3)(x - 4) = 0 \qquad \text{Factor.}$$

$$7x - 3 = 0 \quad \text{or} \quad x - 4 = 0 \qquad \text{Zero-factor property}$$

$$x = \frac{3}{7} \quad \text{or} \qquad x = 4 \qquad \text{Solve each equation.}$$

These values are in the domain. Check them in the original equation. The solution set is $\left\{\frac{3}{7}, 4\right\}$.

NOW TRY

OBJECTIVE 2 Solve applied problems involving quadratic equations.

Some distance-rate-time (or motion) problems lead to quadratic equations. We use a form of the distance formula $d = rt$ to solve them.

NOW TRY
EXERCISE 2

A small fishing boat averages 18 mph in still water. It takes the boat $\frac{9}{10}$ hr to travel 8 mi upstream and return. Find the rate of the current.

Riverboat traveling upstream—the current slows it down.

FIGURE 1

EXAMPLE 2   Solving a Motion Problem

A riverboat for tourists averages 12 mph in still water. It takes the boat 1 hr, 4 min to travel 6 mi upstream and return. Find the rate of the current.

Step 1   **Read** the problem carefully.

Step 2   **Assign a variable.** Let x = the rate of the current.

The current slows down the boat as it travels upstream, so the rate of the boat traveling upstream is its rate in still water *less* the rate of the current, or $(12 - x)$ mph. See **FIGURE 1**.

Similarly, the current speeds up the boat as it travels downstream, so its rate downstream is $(12 + x)$ mph. Thus,

$$12 - x = \text{the rate upstream in miles per hour,}$$

and $$12 + x = \text{the rate downstream in miles per hour.}$$

	d	r	t
Upstream	6	$12 - x$	$\dfrac{6}{12 - x}$
Downstream	6	$12 + x$	$\dfrac{6}{12 + x}$

Complete a table. Use the distance formula, $d = rt$, solved for time t, $t = \frac{d}{r}$, to write expressions for t.

Step 3   **Write an equation.** We use the total time, 1 hr, 4 min, written as a fraction.

$$1 + \frac{4}{60} = 1 + \frac{1}{15} = \frac{16}{15} \text{ hr} \qquad \text{Total time}$$

The time upstream plus the time downstream equals $\frac{16}{15}$ hr.

$$
\underset{\dfrac{6}{12 - x}}{\text{Time upstream}} \quad + \quad \underset{\dfrac{6}{12 + x}}{\text{Time downstream}} \quad = \quad \underset{\dfrac{16}{15}}{\text{Total time}}
$$

Step 4   **Solve** the equation. The LCD is $15(12 - x)(12 + x)$.

$$15(12 - x)(12 + x)\left(\frac{6}{12 - x} + \frac{6}{12 + x}\right)$$

$$= 15(12 - x)(12 + x)\left(\frac{16}{15}\right)$$

Multiply by the LCD.

$$15(12 + x) \cdot 6 + 15(12 - x) \cdot 6 = (12 - x)(12 + x)16$$

Distributive property; Multiply.

$$90(12 + x) + 90(12 - x) = 16(144 - x^2) \qquad \text{Multiply.}$$

$$1080 + 90x + 1080 - 90x = 2304 - 16x^2 \qquad \text{Distributive property}$$

$$2160 = 2304 - 16x^2 \qquad \text{Combine like terms.}$$

$$16x^2 = 144 \qquad \text{Add } 16x^2\text{. Subtract 2160.}$$

$$x^2 = 9 \qquad \text{Divide by 16.}$$

$$x = 3 \quad \text{or} \quad x = -3 \qquad \text{Square root property}$$

NOW TRY ANSWER
2. 2 mph

Step 5   **State the answer.** The current rate cannot be -3, so the answer is 3 mph.

Step 6   **Check** that this value satisfies the original problem.

NOW TRY

PROBLEM-SOLVING HINT   Recall from earlier work that a person's work rate is $\frac{1}{t}$ part of the job per hour, where t is the time in hours required to do the complete job. Thus, the part of the job the person will do in x hours is $\frac{1}{t}x$.

EXAMPLE 3   Solving a Work Problem

It takes two carpet layers 4 hr to carpet a room. If each worked alone, one of them could do the job in 1 hr less time than the other. How long would it take each carpet layer to complete the job alone?

Step 1   **Read** the problem again. There will be two answers.

Step 2   **Assign a variable.**

Let x = the number of hours for the slower carpet layer to complete the job.

Then $x - 1$ = the number of hours for the faster carpet layer to complete the job.

The slower worker's rate is $\frac{1}{x}$, and the faster worker's rate is $\frac{1}{x-1}$. Together they can do the job in 4 hr. Complete a table as shown.

	Rate	Time Working Together	Fractional Part of the Job Done	
Slower Worker	$\frac{1}{x}$	4	$\frac{1}{x}(4)$	← Sum is 1 whole job.
Faster Worker	$\frac{1}{x-1}$	4	$\frac{1}{x-1}(4)$	←

Step 3   **Write an equation.**

Part done by slower worker   +   Part done by faster worker   =   1 whole job

$$\frac{4}{x} \quad + \quad \frac{4}{x-1} \quad = \quad 1$$

Step 4   **Solve** the equation. The LCD is $x(x-1)$.

$$x(x-1)\left(\frac{4}{x} + \frac{4}{x-1}\right) = x(x-1)(1) \qquad \text{Multiply by the LCD.}$$

$$(x-1) \cdot 4 + 4x = x(x-1) \qquad \text{Distributive property}$$

$$4x - 4 + 4x = x^2 - x \qquad \text{Distributive property}$$

$$x^2 - 9x + 4 = 0 \qquad \text{Standard form}$$

The trinomial on the left cannot be factored, so the equation cannot be solved using the zero-factor property. We use the quadratic formula.

$$x = \frac{-b \pm \sqrt{b^2 - 4ac}}{2a} \qquad \text{Quadratic formula}$$

$$x = \frac{-(-9) \pm \sqrt{(-9)^2 - 4(1)(4)}}{2(1)} \qquad a = 1, b = -9, c = 4$$

$$x = \frac{9 \pm \sqrt{65}}{2} \qquad \text{Simplify.}$$

$$x = \frac{9 + \sqrt{65}}{2} = 8.5 \quad \text{or} \quad x = \frac{9 - \sqrt{65}}{2} = 0.5 \qquad \text{Use a calculator. Round to the nearest tenth.}$$

NOW TRY
EXERCISE 3

Two electricians are running wire to finish a basement. One electrician could finish the job in 2 hr less time than the other. Together, they complete the job in 6 hr. How long (to the nearest tenth) would it take the slower electrician to complete the job alone?

NOW TRY
EXERCISE 4

Solve each equation.

(a) $x = \sqrt{9x - 20}$

(b) $x + \sqrt{x} = 20$

Step 5 **State the answer.** Only the solution 8.5 makes sense in the original problem.

If $x = 0.5$, then $x - 1 = 0.5 - 1 = -0.5,$ Time cannot be negative.

which cannot represent the time for the faster worker. The slower worker could do the job in 8.5 hr and the faster in $8.5 - 1 = 7.5$ hr.

Step 6 **Check** that these results satisfy the original problem.   NOW TRY

OBJECTIVE 3 Solve radical equations that lead to quadratic equations.

EXAMPLE 4 Solving Radical Equations That Lead to Quadratic Equations

Solve each equation.

(a) $x = \sqrt{6x - 8}$

This equation is not quadratic. However, squaring each side of the equation gives a quadratic equation that can be solved using the zero-factor property.

$$x^2 = \left(\sqrt{6x - 8}\right)^2 \qquad \text{Square each side.}$$

$$x^2 = 6x - 8 \qquad \left(\sqrt{a}\right)^2 = a$$

This trinomial is factorable. → $x^2 - 6x + 8 = 0$ Standard form

$$(x - 4)(x - 2) = 0 \qquad \text{Factor.}$$

$$x - 4 = 0 \quad \text{or} \quad x - 2 = 0 \qquad \text{Zero-factor property}$$

$$x = 4 \quad \text{or} \qquad x = 2 \qquad \text{Proposed solutions}$$

Squaring each side of a radical equation can introduce extraneous solutions. **All proposed solutions must be checked in the original (not the squared) equation.**

CHECK $x = \sqrt{6x - 8}$ $x = \sqrt{6x - 8}$

$4 \overset{?}{=} \sqrt{6(4) - 8}$ Let $x = 4$. $2 \overset{?}{=} \sqrt{6(2) - 8}$ Let $x = 2$.

$4 \overset{?}{=} \sqrt{16}$ $2 \overset{?}{=} \sqrt{4}$

$4 = 4$ ✓   True $2 = 2$ ✓   True

Both proposed solutions check, so the solution set is $\{2, 4\}$.

(b) $x + \sqrt{x} = 6$

$$\sqrt{x} = 6 - x \qquad \text{Isolate the radical on one side.}$$

$$\left(\sqrt{x}\right)^2 = (6 - x)^2 \qquad \text{Square each side.}$$

$$x = 36 - 12x + x^2 \qquad (x - y)^2 = x^2 - 2xy + y^2$$

$$x^2 - 13x + 36 = 0 \qquad \text{Write in standard form.}$$

$$(x - 4)(x - 9) = 0 \qquad \text{Factor.}$$

$$x - 4 = 0 \quad \text{or} \quad x - 9 = 0 \qquad \text{Zero-factor property}$$

$$x = 4 \quad \text{or} \qquad x = 9 \qquad \text{Proposed solutions}$$

CHECK $x + \sqrt{x} = 6$ $x + \sqrt{x} = 6$

$4 + \sqrt{4} \overset{?}{=} 6$ Let $x = 4$. $9 + \sqrt{9} \overset{?}{=} 6$ Let $x = 9$.

$6 = 6$ ✓   True $12 = 6$ False

Only the proposed solution 4 checks, so the solution set is $\{4\}$.   NOW TRY

OBJECTIVE 4 Solve equations that are quadratic in form.

A nonquadratic equation that can be written in the form

$$au^2 + bu + c = 0,$$

for $a \neq 0$ and an algebraic expression u, is **quadratic in form.**

Many equations that are quadratic in form can be solved more easily by defining and substituting a "temporary" variable u for an expression involving the variable in the original equation.

NOW TRY EXERCISE 5

Define a variable u in terms of x, and write each equation in the quadratic form

$$au^2 + bu + c = 0.$$

(a) $x^4 - 10x^2 + 9 = 0$

(b) $6(x+2)^2 - 11(x+2) + 4 = 0$

EXAMPLE 5 Defining Substitution Variables

Define a variable u in terms of x, and write each equation in the quadratic form $au^2 + bu + c = 0$.

(a) $x^4 - 13x^2 + 36 = 0$

Look at the two terms involving the variable x, ignoring their coefficients. Try to find one variable expression that is the square of the other. Here $x^4 = (x^2)^2$, so we can define $u = x^2$, and rewrite the original equation as a quadratic equation in u.

$$u^2 - 13u + 36 = 0 \qquad \text{Here, } u = x^2.$$

(b) $2(4x-3)^2 + 7(4x-3) + 5 = 0$

Because this equation involves both $(4x-3)^2$ and $(4x-3)$, we let $u = 4x - 3$.

$$2u^2 + 7u + 5 = 0 \qquad \text{Here, } u = 4x - 3.$$

(c) $2x^{2/3} - 11x^{1/3} + 12 = 0$

We apply a power rule for exponents, $(a^m)^n = a^{mn}$. Because $(x^{1/3})^2 = x^{2/3}$, we define $u = x^{1/3}$ and write the original equation as follows.

$$2u^2 - 11u + 12 = 0 \qquad \text{Here, } u = x^{1/3}. \qquad \textbf{NOW TRY}$$

EXAMPLE 6 Solving Equations That Are Quadratic in Form

Solve each equation.

(a)

$$x^4 - 13x^2 + 36 = 0 \qquad \text{See Example 5(a).}$$
$$(x^2)^2 - 13x^2 + 36 = 0 \qquad x^4 = (x^2)^2$$
$$u^2 - 13u + 36 = 0 \qquad \text{Let } u = x^2. \quad \text{(Quadratic in form)}$$
$$(u-4)(u-9) = 0 \qquad \text{Factor.}$$
$$u - 4 = 0 \quad \text{or} \quad u - 9 = 0 \qquad \text{Zero-factor property}$$
$$u = 4 \quad \text{or} \quad u = 9 \qquad \text{Solve. (Don't stop here.)}$$
$$x^2 = 4 \quad \text{or} \quad x^2 = 9 \qquad \text{Substitute } x^2 \text{ for } u.$$
$$x = \pm 2 \quad \text{or} \quad x = \pm 3 \qquad \text{Square root property}$$

Each value can be verified by substituting it into the original equation for x. The equation $x^4 - 13x^2 + 36 = 0$ is a fourth-degree equation and has four solutions, $-3, -2, 2, 3$.* The solution set is abbreviated $\{\pm 2, \pm 3\}$.

NOW TRY ANSWERS

5. (a) $u = x^2$; $u^2 - 10u + 9 = 0$
   (b) $u = x + 2$; $6u^2 - 11u + 4 = 0$

*In general, an equation in which an nth-degree polynomial equals 0 has n complex solutions, although some of them may be repeated.

NOW TRY
EXERCISE 6

Solve each equation.

(a) $x^4 - 17x^2 + 16 = 0$

(b) $x^4 + 4 = 8x^2$

(b)

$$4x^4 + 1 = 5x^2$$

$$4x^4 - 5x^2 + 1 = 0 \qquad \text{Standard form}$$

$$4(x^2)^2 - 5x^2 + 1 = 0 \qquad x^4 = (x^2)^2$$

$$4u^2 - 5u + 1 = 0 \qquad \text{Let } u = x^2.$$

$$(4u - 1)(u - 1) = 0 \qquad \text{Factor.}$$

$$4u - 1 = 0 \quad \text{or} \quad u - 1 = 0 \qquad \text{Zero-factor property}$$

$$u = \frac{1}{4} \quad \text{or} \quad u = 1 \qquad \text{Solve.}$$

> This is a key step.

$$x^2 = \frac{1}{4} \quad \text{or} \quad x^2 = 1 \qquad \text{Substitute } x^2 \text{ for } u.$$

$$x = \pm\frac{1}{2} \quad \text{or} \quad x = \pm 1 \qquad \text{Square root property}$$

Check that the solution set is $\left\{ \pm\frac{1}{2}, \pm 1 \right\}$.

(c)

$$x^4 = 6x^2 - 3$$

$$x^4 - 6x^2 + 3 = 0 \qquad \text{Standard form}$$

$$(x^2)^2 - 6x^2 + 3 = 0 \qquad x^4 = (x^2)^2$$

$$u^2 - 6u + 3 = 0 \qquad \text{Let } u = x^2.$$

The trinomial on the left is nonfactorable, so we cannot solve the equation using the zero-factor property. To solve, we use the quadratic formula.

$$u = \frac{-(-6) \pm \sqrt{(-6)^2 - 4(1)(3)}}{2(1)} \qquad a = 1, b = -6, c = 3$$

$$u = \frac{6 \pm \sqrt{24}}{2} \qquad \text{Simplify.}$$

$$u = \frac{6 \pm 2\sqrt{6}}{2} \qquad \sqrt{24} = \sqrt{4} \cdot \sqrt{6} = 2\sqrt{6}$$

$$u = \frac{2(3 \pm \sqrt{6})}{2} \qquad \text{Factor.}$$

$$u = 3 \pm \sqrt{6} \qquad \text{Divide out the common factor 2.}$$

$$x^2 = 3 + \sqrt{6} \quad \text{or} \quad x^2 = 3 - \sqrt{6} \qquad \text{Substitute } x^2 \text{ for } u.$$

> Find *both* square roots in each case.

$$x = \pm\sqrt{3 + \sqrt{6}} \quad \text{or} \quad x = \pm\sqrt{3 - \sqrt{6}} \qquad \text{Square root property}$$

The solution set contains four numbers and is written as follows.

$$\left\{ \pm\sqrt{3 + \sqrt{6}}, \pm\sqrt{3 - \sqrt{6}} \right\}$$

NOW TRY ANSWERS

6. (a) $\{ \pm 1, \pm 4 \}$

(b) $\left\{ \pm\sqrt{4 + 2\sqrt{3}}, \pm\sqrt{4 - 2\sqrt{3}} \right\}$

NOW TRY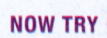

NOTE Expressions in equations like those in **Examples 6(a) and (b)** can be factored directly.

$$x^4 - 13x^2 + 36 = 0 \qquad \text{Example 6(a) equation}$$

$$(x^2 - 9)(x^2 - 4) = 0 \qquad \text{Factor.}$$

$$(x + 3)(x - 3)(x + 2)(x - 2) = 0 \qquad \text{Factor again.}$$

Using the zero-factor property gives the same solutions that we obtained in **Example 6(a)**. Equations that include nonfactorable quadratic expressions (as in **Example 6(c)**) must be solved using substitution and the quadratic formula.

Solving an Equation That Is Quadratic in Form by Substitution

Step 1 **Define a temporary variable** u, based on the relationship between the variable expressions in the given equation. Substitute u in the original equation and rewrite the equation in the form

$$au^2 + bu + c = 0.$$

Step 2 **Solve the quadratic equation obtained in Step 1** either by factoring the trinomial and applying the zero-factor property or by using the quadratic formula.

Step 3 **Replace u with the expression it defined in Step 1.**

Step 4 **Solve the resulting equations for the original variable.**

Step 5 **Check** all solutions by substituting them in the original equation. Write the solution set.

EXAMPLE 7 Solving Equations That Are Quadratic in Form

Solve each equation.

(a) $2(4x - 3)^2 + 7(4x - 3) + 5 = 0$

Step 1 Because of the repeated quantity $4x - 3$, substitute u for $4x - 3$.

$$2(4x - 3)^2 + 7(4x - 3) + 5 = 0 \qquad \text{See Example 5(b).}$$

$$2u^2 + 7u + 5 = 0 \qquad \text{Let } u = 4x - 3.$$

Step 2 $$(2u + 5)(u + 1) = 0 \qquad \text{Factor.}$$

$$2u + 5 = 0 \quad \text{or} \quad u + 1 = 0 \qquad \text{Zero-factor property}$$

$$\boxed{\text{Don't stop here.}} \quad u = -\frac{5}{2} \quad \text{or} \quad u = -1 \qquad \text{Solve for } u.$$

Step 3 $$4x - 3 = -\frac{5}{2} \quad \text{or} \quad 4x - 3 = -1 \qquad \text{Substitute } 4x - 3 \text{ for } u.$$

Step 4 $$4x = \frac{1}{2} \quad \text{or} \quad 4x = 2 \qquad \text{Solve for } x.$$

$$x = \frac{1}{8} \quad \text{or} \quad x = \frac{1}{2}$$

Step 5 Check that the solution set of the original equation is $\left\{\frac{1}{8}, \frac{1}{2}\right\}$.

NOW TRY
EXERCISE 7
Solve each equation.
(a) $6(x-4)^2 + 11(x-4)$
    $- 10 = 0$
(b) $2x^{2/3} - 7x^{1/3} + 3 = 0$

(b) $2x^{2/3} - 11x^{1/3} + 12 = 0$

Step 1   Because $x^{2/3} = (x^{1/3})^2$, we substitute u for $x^{1/3}$.

$$2x^{2/3} - 11x^{1/3} + 12 = 0 \qquad \text{See Example 5(c).}$$
$$2(x^{1/3})^2 - 11x^{1/3} + 12 = 0 \qquad x^{2/3} = (x^{1/3})^2$$
$$2u^2 - 11u + 12 = 0 \qquad \text{Let } u = x^{1/3}.$$

Step 2
$$(2u - 3)(u - 4) = 0 \qquad \text{Factor.}$$
$$2u - 3 = 0 \qquad \text{or} \qquad u - 4 = 0 \qquad \text{Zero-factor property}$$
$$u = \frac{3}{2} \qquad \text{or} \qquad u = 4 \qquad \text{Solve for } u.$$

Step 3
$$x^{1/3} = \frac{3}{2} \qquad \text{or} \qquad x^{1/3} = 4 \qquad \text{Substitute } x^{1/3} \text{ for } u.$$

Step 4
$$(x^{1/3})^3 = \left(\frac{3}{2}\right)^3 \qquad \text{or} \qquad (x^{1/3})^3 = 4^3 \qquad \text{Cube each side.}$$

$$x = \frac{27}{8} \qquad \text{or} \qquad x = 64 \qquad \text{Apply the exponents.}$$

Step 5   Because the original equation involves variables with rational exponents, check that neither of these solutions is extraneous. The solution set is $\left\{\frac{27}{8}, 64\right\}$.

NOW TRY

NOW TRY ANSWERS
7. (a) $\left\{\frac{3}{2}, \frac{14}{3}\right\}$ **(b)** $\left\{\frac{1}{8}, 27\right\}$

⚠ **CAUTION**   A common error when solving problems like those in **Examples 6 and 7** is to stop too soon. *Once we have solved for u, we must remember to substitute and solve for the values of the original variable.*

11.4 Exercises

FOR
EXTRA
HELP

 MyLab Math

▶ *Video solutions for select problems available in MyLab Math*

Concept Check   *Based on the discussion and examples of this section, give the first step to solve each equation. Do not actually solve.*

1. $\frac{14}{x} = x - 5$

2. $\sqrt{1 + x} + x = 5$

3. $(x^2 + x)^2 - 8(x^2 + x) + 12 = 0$

4. $3x = \sqrt{16 - 10x}$

5. Concept Check   Study this incorrect "solution."

$$x = \sqrt{3x + 4} \qquad \text{Square}$$
$$x^2 = 3x + 4 \qquad \text{each side.}$$
$$x^2 - 3x - 4 = 0$$
$$(x - 4)(x + 1) = 0$$
$$x - 4 = 0 \quad \text{or} \quad x + 1 = 0$$
$$x = 4 \quad \text{or} \qquad x = -1$$

Solution set: $\{4, -1\}$

WHAT WENT WRONG?   Give the correct solution set.

6. Concept Check   Study this incorrect "solution."

$$2(x - 1)^2 - 3(x - 1) + 1 = 0 \qquad \text{Let}$$
$$2u^2 - 3u + 1 = 0 \qquad u = x - 1.$$
$$(2u - 1)(u - 1) = 0$$
$$2u - 1 = 0 \quad \text{or} \quad u - 1 = 0$$
$$u = \frac{1}{2} \quad \text{or} \qquad u = 1$$

Solution set: $\left\{\frac{1}{2}, 1\right\}$

WHAT WENT WRONG?   Give the correct solution set.

Solve each equation. Check the solutions. See Example 1.

7. $\dfrac{14}{x} = x - 5$

8. $\dfrac{-12}{x} = x + 8$

9. $1 - \dfrac{3}{x} - \dfrac{28}{x^2} = 0$

10. $4 - \dfrac{7}{r} - \dfrac{2}{r^2} = 0$

11. $3 - \dfrac{1}{t} = \dfrac{2}{t^2}$

12. $1 + \dfrac{2}{x} = \dfrac{3}{x^2}$

13. $\dfrac{1}{x} + \dfrac{2}{x+2} = \dfrac{17}{35}$

14. $\dfrac{2}{m} + \dfrac{3}{m+9} = \dfrac{11}{4}$

15. $\dfrac{2}{x+1} + \dfrac{3}{x+2} = \dfrac{7}{2}$

16. $\dfrac{4}{3-p} + \dfrac{2}{5-p} = \dfrac{26}{15}$

17. $\dfrac{3}{2x} - \dfrac{1}{2(x+2)} = 1$

18. $\dfrac{4}{3x} - \dfrac{1}{2(x+1)} = 1$

19. $3 = \dfrac{1}{t+2} + \dfrac{2}{(t+2)^2}$

20. $1 + \dfrac{2}{3z+2} = \dfrac{15}{(3z+2)^2}$

21. $\dfrac{6}{p} = 2 + \dfrac{p}{p+1}$

22. $\dfrac{x}{2-x} + \dfrac{2}{x} = 5$

23. $1 - \dfrac{1}{2x+1} - \dfrac{1}{(2x+1)^2} = 0$

24. $1 - \dfrac{1}{3x-2} - \dfrac{1}{(3x-2)^2} = 0$

Concept Check *Answer each question.*

25. A boat travels 20 mph in still water, and the rate of the current is t mph.

    (a) What is the rate of the boat when it travels upstream?

    (b) What is the rate of the boat when it travels downstream?

26. It takes m hours to grade a set of papers.

    (a) What is the grader's rate (in job per hour)?

    (b) How much of the job will the grader do in 2 hr?

Solve each problem. Round answers to the nearest tenth as needed. See Examples 2 and 3.

27. In 4 hr, Kerrie can travel 15 mi upriver and come back. The rate of the current is 5 mph. Find the rate of her boat in still water.

Let $x =$ _____.

The rate traveling upriver (*against* the current) is _____ mph.

The rate traveling back downriver (*with* the current) is _____ mph.

Complete the table.

	d	r	t
Up			
Down			

Write an equation, and complete the solution.

28. Carlos can complete a certain lab test in 2 hr less time than Jaime can. If they can finish the job together in 2 hr, how long would it take each of them working alone?

Let $x =$ Jaime's time alone (in hours).

Then _____ = Carlos' time alone (in hours).

Complete the table.

	Rate	Time Working Together	Fractional Part of the Job Done
Carlos			
Jaime			

Write an equation, and complete the solution.

29. On a windy day William found that he could travel 16 mi downstream and then 4 mi back upstream at top speed in a total of 48 min. What was the top speed of William's boat if the rate of the current was 15 mph? (Let x represent the rate of the boat in still water.)

	d	*r*	*t*
Upstream	4	$x - 15$	
Downstream	16		

30. Vera flew for 6 hr at a constant rate. She traveled 810 mi with the wind, then turned around and traveled 720 mi against the wind. The wind speed was a constant 15 mph. Find the rate of the plane.

	d	*r*	*t*
With Wind	810		
Against Wind	720		

31. The distance from Jackson to Lodi is about 40 mi, as is the distance from Lodi to Manteca. Adrian drove from Jackson to Lodi, stopped in Lodi for a high-energy drink, and then drove on to Manteca at 10 mph faster. Driving time for the entire trip was 88 min. Find her rate from Jackson to Lodi. (Data from *State Farm Road Atlas.*)

32. Medicine Hat and Cranbrook are 300 km apart. Steve rides his Harley 20 km per hr faster than Mohammad rides his Yamaha. Find Steve's average rate if he travels from Cranbrook to Medicine Hat in $1\frac{1}{4}$ hr less time than Mohammad. (Data from *State Farm Road Atlas.*)

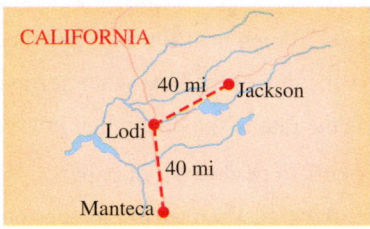

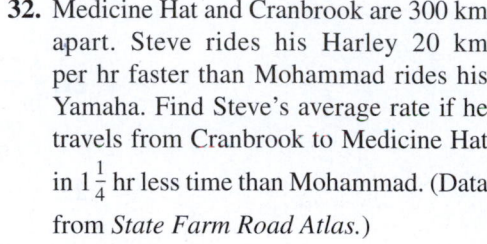

33. Working together, two people can cut a large lawn in 2 hr. One person can do the job alone in 1 hr less time than the other. How long would it take the faster worker to do the job? (Let x represent the time of the faster worker.)

	Rate	Time Working Together	Fractional Part of the Job Done
Faster Worker	$\frac{1}{x}$	2	
Slower Worker		2	

34. Working together, two people can clean an office building in 5 hr. One person takes 2 hr longer than the other to clean the building alone. How long would it take the slower worker to clean the building alone? (Let x represent the time of the slower worker.)

	Rate	Time Working Together	Fractional Part of the Job Done
Faster Worker			
Slower Worker	$\frac{1}{x}$		

35. Rusty and Nancy are planting flowers. Working alone, Rusty would take 2 hr longer than Nancy to plant the flowers. Working together, they do the job in 12 hr. How long would it have taken each person working alone?

36. Joel can work through a stack of invoices in 1 hr less time than Noel can. Working together they take $1\frac{1}{2}$ hr. How long would it take each person working alone?

37. A washing machine can be filled in 6 min if both the hot water and the cold water taps are fully opened. Filling the washer with hot water alone takes 9 min longer than filling it with cold water alone. How long does it take to fill the washer with cold water?

38. Two pipes together can fill a tank in 2 hr. One of the pipes, used alone, takes 3 hr longer than the other to fill the tank. How long would each pipe take to fill the tank alone?

Solve each equation. Check the solutions. ***See Example 4.***

39. $x = \sqrt{7x - 10}$ **40.** $z = \sqrt{5z - 4}$ **41.** $2x = \sqrt{11x + 3}$

42. $4x = \sqrt{6x + 1}$ **43.** $3x = \sqrt{16 - 10x}$ **44.** $4t = \sqrt{8t + 3}$

45. $t + \sqrt{t} = 12$ **46.** $p - 2\sqrt{p} = 8$ **47.** $x = \sqrt{\dfrac{6 - 13x}{5}}$

48. $r = \sqrt{\dfrac{20 - 19r}{6}}$ **49.** $-x = \sqrt{\dfrac{8 - 2x}{3}}$ **50.** $-x = \sqrt{\dfrac{3x + 7}{4}}$

Solve each equation. Check the solutions. ***See Examples 5–7.***

51. $x^4 - 29x^2 + 100 = 0$ **52.** $x^4 - 37x^2 + 36 = 0$

53. $4q^4 - 13q^2 + 9 = 0$ **54.** $9x^4 - 25x^2 + 16 = 0$

55. $x^4 + 48 = 16x^2$ **56.** $z^4 + 72 = 17z^2$

57. $(x + 3)^2 + 5(x + 3) + 6 = 0$ **58.** $(x - 4)^2 + (x - 4) - 20 = 0$

59. $3(m + 4)^2 - 8 = 2(m + 4)$ **60.** $(t + 5)^2 + 6 = 7(t + 5)$

61. $x^{2/3} + x^{1/3} - 2 = 0$ **62.** $x^{2/3} - 2x^{1/3} - 3 = 0$

63. $r^{2/3} + r^{1/3} - 12 = 0$ **64.** $3x^{2/3} - x^{1/3} - 24 = 0$

65. $4x^{4/3} - 13x^{2/3} + 9 = 0$ **66.** $9t^{4/3} - 25t^{2/3} + 16 = 0$

67. $2 + \dfrac{5}{3x - 1} = \dfrac{-2}{(3x - 1)^2}$ **68.** $3 - \dfrac{7}{2p + 2} = \dfrac{6}{(2p + 2)^2}$

69. $2 - 6(z - 1)^{-2} = (z - 1)^{-1}$ **70.** $3 - 2(x - 1)^{-1} = (x - 1)^{-2}$

The following exercises are not grouped by type. Solve each equation. (Exercises 83 and 84 require knowledge of complex numbers.) ***See Examples 1 and 4–7.***

71. $12x^4 - 11x^2 + 2 = 0$ **72.** $\left(x - \dfrac{1}{2}\right)^2 + 5\left(x - \dfrac{1}{2}\right) - 4 = 0$

73. $\sqrt{2x + 3} = 2 + \sqrt{x - 2}$ **74.** $\sqrt{m + 1} = -1 + \sqrt{2m}$

75. $2(1 + \sqrt{r})^2 = 13(1 + \sqrt{r}) - 6$ **76.** $(x^2 + x)^2 + 12 = 8(x^2 + x)$

77. $2m^6 + 11m^3 + 5 = 0$ **78.** $8x^6 + 513x^3 + 64 = 0$

79. $6 = 7(2w - 3)^{-1} + 3(2w - 3)^{-2}$ **80.** $x^6 - 10x^3 = -9$

81. $2x^4 - 9x^2 = -2$ **82.** $8x^4 + 1 = 11x^2$

83. $2x^4 + x^2 - 3 = 0$ **84.** $4x^4 + 5x^2 + 1 = 0$

SUMMARY EXERCISES   Applying Methods for Solving Quadratic Equations

We have introduced four methods for solving quadratic equations written in standard form $ax^2 + bx + c = 0$.

Method	Advantages	Disadvantages
Zero-factor property	This is usually the fastest method.	Not all polynomials are factorable. Some factorable polynomials are difficult to factor.
Square root property	This is the simplest method for solving equations of the form $(ax + b)^2 = c$.	Few equations are given in this form.
Completing the square	This method can always be used.	It requires more steps than other methods.
Quadratic formula	This method can always be used.	Sign errors may occur when evaluating $\sqrt{b^2 - 4ac}$.

Concept Check   *Decide whether the* zero-factor property, *the* square root property, *or the* quadratic formula *is most appropriate for solving each quadratic equation. Do not actually solve.*

1. $(2x + 3)^2 = 4$ **2.** $4x^2 - 3x = 1$ **3.** $x^2 + 5x - 8 = 0$

4. $2x^2 + 3x = 1$ **5.** $3x^2 = 2 - 5x$ **6.** $x^2 = 5$

Solve each quadratic equation by the method of your choice.

7. $p^2 = 7$ **8.** $6x^2 - x - 15 = 0$ **9.** $n^2 + 6n + 4 = 0$

10. $(x - 3)^2 = 25$ **11.** $\dfrac{5}{x} + \dfrac{12}{x^2} = 2$ **12.** $3x^2 = 3 - 8x$

13. $2r^2 - 4r + 1 = 0$ ***14.** $x^2 = -12$ **15.** $x\sqrt{2} = \sqrt{5x - 2}$

16. $x^4 - 10x^2 + 9 = 0$ **17.** $(2x + 3)^2 = 8$ **18.** $\dfrac{2}{x} + \dfrac{1}{x - 2} = \dfrac{5}{3}$

19. $t^4 + 14 = 9t^2$ **20.** $8x^2 - 4x = 2$ ***21.** $z^2 + z + 1 = 0$

22. $5x^6 + 2x^3 - 7 = 0$ **23.** $4t^2 - 12t + 9 = 0$ **24.** $x\sqrt{3} = \sqrt{2 - x}$

25. $r^2 - 72 = 0$ **26.** $-3x^2 + 4x = -4$ **27.** $x^2 - 5x - 36 = 0$

28. $w^2 = 169$ ***29.** $3p^2 = 6p - 4$ **30.** $z = \sqrt{\dfrac{5z + 3}{2}}$

*31. $\dfrac{4}{r^2} + 3 = \dfrac{1}{r}$ **32.** $2(3x - 1)^2 + 5(3x - 1) = -2$

---

*This exercise requires knowledge of complex numbers.

11.5 Formulas and Further Applications

OBJECTIVES

1. Solve formulas involving squares and square roots for specified variables.
2. Solve applied problems using the Pythagorean theorem.
3. Solve applied problems using area formulas.
4. Solve applied problems using quadratic functions as models.

OBJECTIVE 1 Solve formulas involving squares and square roots for specified variables.

EXAMPLE 1 Solving for Specified Variables

Solve each formula for the specified variable. Keep $\pm$ in the answer in part (a).

(a) $\quad w = \dfrac{kFr}{v^2} \quad$ for v The goal is to isolate v on one side.

$$v^2 w = kFr \qquad \text{Multiply by } v^2.$$

$$v^2 = \frac{kFr}{w} \qquad \text{Divide by } w.$$

$$v = \pm\sqrt{\frac{kFr}{w}} \qquad \text{Square root property}$$

Include both positive and negative roots.

$$v = \frac{\pm\sqrt{kFr}}{\sqrt{w}} \cdot \frac{\sqrt{w}}{\sqrt{w}} \qquad \text{Rationalize the denominator.}$$

$$v = \frac{\pm\sqrt{kFrw}}{w} \qquad \begin{array}{l}\sqrt{a}\cdot\sqrt{b}=\sqrt{ab};\\ \sqrt{a}\cdot\sqrt{a}=a\end{array}$$

(b) $\quad d = \sqrt{\dfrac{4\mathscr{A}}{\pi}} \quad$ for $\mathscr{A}$ The goal is to isolate $\mathscr{A}$ on one side.

$$d^2 = \frac{4\mathscr{A}}{\pi} \qquad \text{Square each side.}$$

$$\pi d^2 = 4\mathscr{A} \qquad \text{Multiply by } \pi.$$

$$\frac{\pi d^2}{4} = \mathscr{A}, \quad \text{or} \quad \mathscr{A} = \frac{\pi d^2}{4} \qquad \begin{array}{l}\text{Divide by 4.}\\ \text{Interchange sides.}\end{array} \qquad \text{NOW TRY}$$

NOW TRY EXERCISE 1

Solve each formula for the specified variable. Keep $\pm$ in the answer in part (a).

(a) $n = \dfrac{ab}{E^2}$ for E

(b) $S = \sqrt{\dfrac{pq}{n}}$ for p

EXAMPLE 2 Solving for a Specified Variable

Solve $s = 2t^2 + kt$ for t.

Because the given equation has terms with t^2 and t, write it in standard form $ax^2 + bx + c = 0$, with t as the variable instead of x.

$$s = 2t^2 + kt$$

$$0 = 2t^2 + kt - s \qquad \text{Subtract } s.$$

$$2t^2 + kt - s = 0 \qquad \text{Standard form}$$

We can solve this equation using the quadratic formula.

NOW TRY ANSWERS

1. **(a)** $E = \dfrac{\pm\sqrt{abn}}{n}$ **(b)** $p = \dfrac{nS^2}{q}$

**NOW TRY
EXERCISE 2**

Solve for r.

$$r^2 + 9r = -c$$

In the equation $2t^2 + kt - s = 0$, we have $a = 2$, $b = k$, and $c = -s$.

$$t = \frac{-b \pm \sqrt{b^2 - 4ac}}{2a}$$ Quadratic formula

$$t = \frac{-k \pm \sqrt{k^2 - 4(2)(-s)}}{2(2)}$$ In $2t^2 + kt - s = 0$, $a = 2$, $b = k$, and $c = -s$.

$$t = \frac{-k \pm \sqrt{k^2 + 8s}}{4}$$ Simplify.

The solutions are $t = \dfrac{-k + \sqrt{k^2 + 8s}}{4}$ and $t = \dfrac{-k - \sqrt{k^2 + 8s}}{4}$.   **NOW TRY**

OBJECTIVE 2 Solve applied problems using the Pythagorean theorem.

The Pythagorean theorem is represented by the equation

$$a^2 + b^2 = c^2.$$ See **FIGURE 2**.

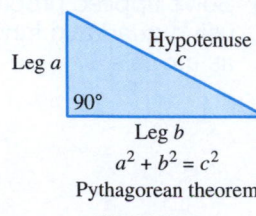

$a^2 + b^2 = c^2$
Pythagorean theorem

FIGURE 2

**NOW TRY
EXERCISE 3**

Matt is building a new barn, with length 10 ft more than width. While determining the footprint of the barn, he measured the diagonal as 50 ft. What will be the dimensions of the barn?

| **EXAMPLE 3**   Using the Pythagorean Theorem |

Two cars left an intersection at the same time, one heading due north, the other due west. Some time later, they were exactly 100 mi apart. The car headed north had gone 20 mi farther than the car headed west. How far had each car traveled?

Step 1   **Read** the problem carefully.

Step 2   **Assign a variable.**

Let $x =$ the distance traveled by the car headed west.

Then $x + 20 =$ the distance traveled by the car headed north.

See **FIGURE 3**. The cars are 100 mi apart, so the hypotenuse of the right triangle equals 100.

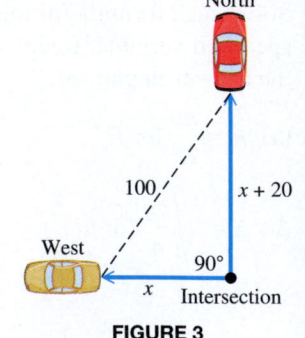

FIGURE 3

Step 3   **Write an equation.** Use the Pythagorean theorem.

$$a^2 + b^2 = c^2$$ Pythagorean theorem

$$x^2 + (x + 20)^2 = 100^2$$ See **FIGURE 3**.

$(x + y)^2 = x^2 + 2xy + y^2$

Step 4   **Solve.** $x^2 + x^2 + 40x + 400 = 10,000$ Square the binomial.

$$2x^2 + 40x - 9600 = 0$$ Standard form

$$x^2 + 20x - 4800 = 0$$ Divide by 2.

$$(x + 80)(x - 60) = 0$$ Factor.

$$x + 80 = 0 \quad \text{or} \quad x - 60 = 0$$ Zero-factor property

$$x = -80 \quad \text{or} \quad x = 60$$ Solve for x.

NOW TRY ANSWERS

2. $r = \dfrac{-9 \pm \sqrt{81 - 4c}}{2}$

3. 30 ft by 40 ft

Step 5   **State the answer.** Distance cannot be negative, so discard the negative solution. The required distances are 60 mi and $60 + 20 = 80$ mi.

Step 6   **Check.** Here $60^2 + 80^2 = 100^2$, as required.   **NOW TRY**

OBJECTIVE 3 Solve applied problems using area formulas.

NOW TRY EXERCISE 4

A football practice field is 30 yd wide and 40 yd long. A strip of grass sod of uniform width is to be placed around the perimeter of the practice field. There is enough money budgeted for 296 sq yd of sod. How wide will the strip be?

EXAMPLE 4 Solving an Area Problem

A rectangular reflecting pool in a park is 20 ft wide and 30 ft long. The gardener wants to plant a strip of grass of uniform width around the edge of the pool. She has enough seed to cover 336 ft^2. How wide will the strip be?

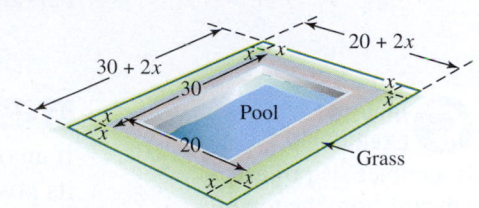

FIGURE 4

Step 1 **Read** the problem carefully.

Step 2 **Assign a variable.** The pool is shown in **FIGURE 4**.

Let x = the unknown width of the grass strip.

Then $20 + 2x$ = the width of the large rectangle (the width of the pool plus two grass strips),

and $30 + 2x$ = the length of the large rectangle.

Step 3 **Write an equation.** Refer to **FIGURE 4**.

$(30 + 2x)(20 + 2x)$ Area of large rectangle (length · width)

$30 \cdot 20,$ or 600 Area of pool (in square feet)

The area of the large rectangle minus the area of the pool should equal 336 ft^2, the area of the grass strip.

Area of large rectangle	−	area of pool	=	area of grass.
↓		↓		↓

$(30 + 2x)(20 + 2x) - 600 = 336$

Step 4 **Solve.**

$600 + 60x + 40x + 4x^2 - 600 = 336$ Multiply.

$4x^2 + 100x - 336 = 0$ Standard form

$x^2 + 25x - 84 = 0$ Divide by 4.

$(x + 28)(x - 3) = 0$ Factor.

$x + 28 = 0$ or $x - 3 = 0$ Zero-factor property

$x = -28$ or $x = 3$ Solve for x.

Step 5 **State the answer.** The width cannot be -28 ft, so the grass strip should be 3 ft wide.

Step 6 **Check.** If $x = 3$, we can find the area of the large rectangle (which includes the grass strip).

$(30 + 2 \cdot 3)(20 + 2 \cdot 3) = 36 \cdot 26 = 936$ ft^2 Area of pool and strip

The area of the pool is $30 \cdot 20 = 600$ ft^2, so the area of the grass strip is

$936 - 600 = 336$ ft^2,   as required.   **NOW TRY**

NOW TRY ANSWER
4. 2 yd

OBJECTIVE 4 Solve applied problems using quadratic functions as models.

Some applied problems can be modeled by *quadratic functions,* which for real numbers a, b, and c, can be written in the form

$$f(x) = ax^2 + bx + c \quad (\text{where } a \neq 0).$$

**NOW TRY
EXERCISE 5**

If an object is projected upward from the top of a 120-ft building at 60 ft per second, its position (in feet above the ground) is given by

$$s(t) = -16t^2 + 60t + 120,$$

where t is time in seconds after it was projected. When does it hit the ground (to the nearest tenth)?

EXAMPLE 5   Solving an Applied Problem Using a Quadratic Function

If an object is projected upward from the top of a 144-ft building at 112 ft per second, its position (in feet above the ground) is given by

$$s(t) = -16t^2 + 112t + 144,$$

where t is time in seconds after it was projected. When does it hit the ground?

When the object hits the ground, its distance above the ground is 0. We must find the value of t that makes $s(t) = 0$.

$s(t) = -16t^2 + 112t + 144$	Given model
$0 = -16t^2 + 112t + 144$	Let $s(t) = 0$.
$0 = t^2 - 7t - 9$	Divide by -16.
$t = \dfrac{-b \pm \sqrt{b^2 - 4ac}}{2a}$	Quadratic formula
$t = \dfrac{-(-7) \pm \sqrt{(-7)^2 - 4(1)(-9)}}{2(1)}$	Let $a = 1$, $b = -7$, and $c = -9$.
$t = \dfrac{7 \pm \sqrt{85}}{2}$	Simplify.
$t \approx \dfrac{7 \pm 9.2}{2}$	Approximate the square root to the nearest tenth.

Time cannot be negative, so discard -1.1.

$$t \approx 8.1 \quad \text{or} \quad t \approx -1.1 \qquad \text{Find the two solutions.}$$

The object will hit the ground about 8.1 sec after it is projected.   **NOW TRY**

EXAMPLE 6   Using a Quadratic Function to Model the CPI

The Consumer Price Index (CPI) is used to measure trends in prices for a "basket" of goods purchased by typical American families. This index uses a base period of 1982–1984, which means that the index number for that period is 100. The quadratic function

$$f(x) = -0.00833x^2 + 4.745x + 83.10$$

approximates the CPI for the years 1980–2016, where x is the number of years that have elapsed since 1980. (Data from Bureau of Labor Statistics.)

(a) Use the model to approximate the CPI for 2016.

For 2016, $x = 2016 - 1980 = 36$, so we find $f(36)$.

$f(x) = -0.00833x^2 + 4.745x + 83.10$	Given model
$f(36) = -0.00833(36)^2 + 4.745(36) + 83.10$	Let $x = 36$.
$f(36) = 243$	Nearest whole number

According to the model, the CPI for 2016 was 243.

NOW TRY ANSWER
5. 5.2 sec after it is projected

**NOW TRY
EXERCISE 6**

Refer to **Example 6.**

(a) Use the model to approximate the CPI for 2010, to the nearest whole number.

(b) In what year did the CPI reach 175? (Round down for the year.)

(b) In what year did the CPI reach 200?

Find the value of x that makes $f(x) = 200$.

$$f(x) = -0.00833x^2 + 4.745x + 83.10 \qquad \text{Given model}$$

$$200 = -0.00833x^2 + 4.745x + 83.10 \qquad \text{Let } f(x) = 200.$$

$$0 = -0.00833x^2 + 4.745x - 116.9 \qquad \text{Subtract 200.}$$

$$x = \frac{-4.745 \pm \sqrt{4.745^2 - 4(-0.00833)(-116.9)}}{2(-0.00833)}$$

Use $a = -0.00833$, $b = 4.745$, and $c = -116.9$ in the quadratic formula.

$$x = 25.8 \quad \text{or} \quad x = 543.8$$

Use a calculator. Round to the nearest tenth.

Rounding the first solution 25.8 down, the CPI first reached 200 in

$$1980 + 25 = 2005.$$

(Reject the solution $x = 543.8$, as this corresponds to a totally unreasonable year.)

NOW TRY ANSWERS

6. (a) 218   **(b)** 2000

NOW TRY

11.5 Exercises

FOR EXTRA HELP   ▶ **MyLab Math**

▶ *Video solutions for select problems available in MyLab Math*

Concept Check   *Answer each question.*

1. In solving a formula that has the specified variable in the denominator, what is the first step?

2. What is the first step in solving a formula like $gw^2 = 2r$ for w?

3. What is the first step in solving a formula like $gw^2 = kw + 24$ for w?

4. Why is it particularly important to check all proposed solutions to an applied problem against the information in the original problem?

For each triangle, solve for m in terms of the other variables (where $m > 0$).

5.

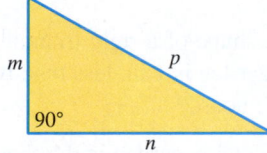

6.

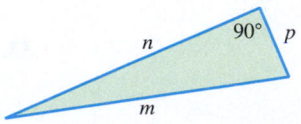

Solve each formula for the specified variable. (Leave $\pm$ in the answers as needed.) **See Examples 1 and 2.**

7. $d = kt^2$ for t **8.** $S = 6e^2$ for e **9.** $S = 4\pi r^2$ for r **10.** $s = kwd^2$ for d

11. $I = \dfrac{ks}{d^2}$ for d **12.** $R = \dfrac{k}{d^2}$ for d **13.** $F = \dfrac{kA}{v^2}$ for v

14. $L = \dfrac{kd^4}{h^2}$ for h **15.** $V = \dfrac{1}{3}\pi r^2 h$ for r **16.** $V = \pi r^2 h$ for r

17. $At^2 + Bt = -C$ for t **18.** $S = 2\pi rh + \pi r^2$ for r **19.** $D = \sqrt{kh}$ for h

20. $F = \dfrac{k}{\sqrt{d}}$ for d **21.** $p = \sqrt{\dfrac{k\ell}{g}}$ for ℓ **22.** $p = \sqrt{\dfrac{k\ell}{g}}$ for g

Extending Skills   *Solve each equation for the specified variable. (Leave ± in the answers.)*

23. $p = \dfrac{E^2 R}{(r + R)^2}$ for R (where $E > 0$)

24. $S(6S - t) = t^2$ for S

25. $10p^2c^2 + 7pcr = 12r^2$ for r

26. $S = vt + \dfrac{1}{2}gt^2$ for t

27. $LI^2 + RI + \dfrac{1}{c} = 0$ for I

28. $P = EI - RI^2$ for I

Solve each problem. When appropriate, round answers to the nearest tenth. **See Example 3.**

29. Find the lengths of the sides of the triangle.

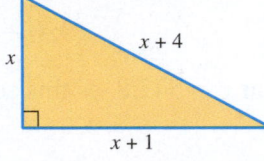

30. Find the lengths of the sides of the triangle.

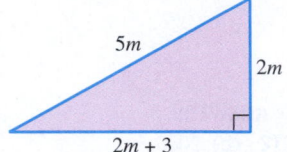

31. Two ships leave port at the same time, one heading due south and the other heading due east. Several hours later, they are 170 mi apart. If the ship traveling south traveled 70 mi farther than the other ship, how many miles did they each travel?

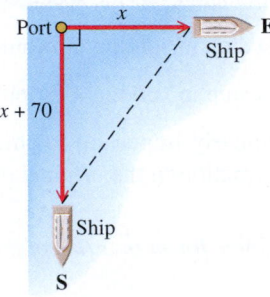

32. Deborah is flying a kite that is 30 ft farther above her hand than its horizontal distance from her. The string from her hand to the kite is 150 ft long. How high is the kite?

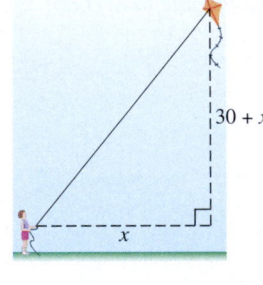

33. A game board is in the shape of a right triangle. The hypotenuse is 2 in. longer than the longer leg, and the longer leg is 1 in. less than twice as long as the shorter leg. How long is each side of the game board?

34. Manuel is planting a vegetable garden in the shape of a right triangle. The longer leg is 3 ft longer than the shorter leg, and the hypotenuse is 3 ft longer than the longer leg. Find the lengths of the three sides of the garden.

35. The diagonal of a rectangular rug measures 26 ft, and the length is 4 ft more than twice the width. Find the length and width of the rug.

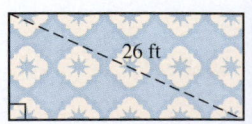

36. A 13-ft ladder is leaning against a house. The distance from the bottom of the ladder to the house is 7 ft less than the distance from the top of the ladder to the ground. How far is the bottom of the ladder from the house?

Solve each problem. See Example 4.

37. A club swimming pool is 30 ft wide and 40 ft long. The club members want an exposed aggregate border in a strip of uniform width around the pool. They have enough material for 296 ft². How wide can the strip be?

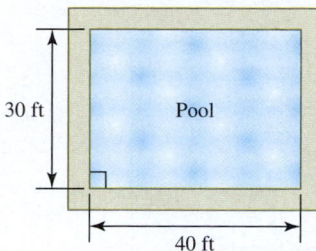

30 ft   Pool   40 ft

38. Lyudmila wants to buy a rug for a room that is 20 ft long and 15 ft wide. She wants to leave an even strip of flooring uncovered around the edges of the room. How wide a strip will she have if she buys a rug with an area of 234 ft²?

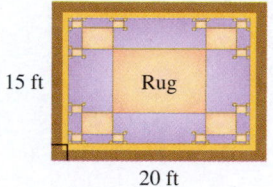

15 ft   Rug   20 ft

39. A rectangle has a length 2 m less than twice its width. When 5 m are added to the width, the resulting figure is a square with an area of 144 m². Find the dimensions of the original rectangle.

40. Mariana's backyard measures 20 m by 30 m. She wants to put a flower garden in the middle of the yard, leaving a strip of grass of uniform width around the flower garden. Mariana must have 184 m² of grass. Under these conditions, what will the length and width of the garden be?

41. A rectangular piece of sheet metal has a length that is 4 in. less than twice the width. A square piece 2 in. on a side is cut from each corner. The sides are then turned up to form an uncovered box of volume 256 in.³. Find the length and width of the original piece of metal.

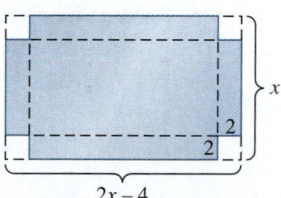

x 2   2 $2x - 4$

42. A rectangular piece of cardboard is 2 in. longer than it is wide. A square piece 3 in. on a side is cut from each corner. The sides are then turned up to form an uncovered box of volume 765 in.³. Find the dimensions of the original piece of cardboard.

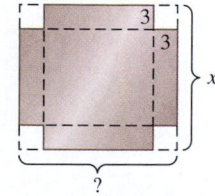

3   3 x ?

Solve each problem. When appropriate, round answers to the nearest tenth. See Example 5.

43. An object is projected directly upward from the ground. After t seconds its distance in feet above the ground is

$$s(t) = 144t - 16t^2.$$

After how many seconds will the object be 128 ft above the ground? (*Hint:* Look for a common factor before solving the equation.)

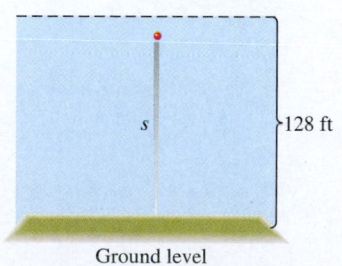

s 128 ft   Ground level

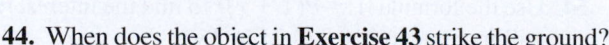

44. When does the object in **Exercise 43** strike the ground?

45. A ball is projected upward from the ground. Its distance in feet from the ground in t seconds is given by

$$s(t) = -16t^2 + 128t.$$

At what times will the ball be 213 ft from the ground?

46. A toy rocket is launched from ground level. Its distance in feet from the ground in t seconds is given by

$$s(t) = -16t^2 + 208t.$$

At what times will the rocket be 550 ft from the ground?

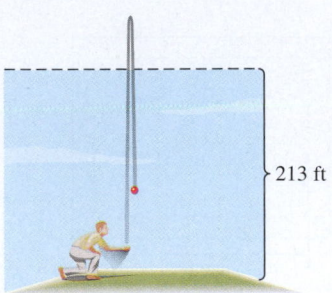

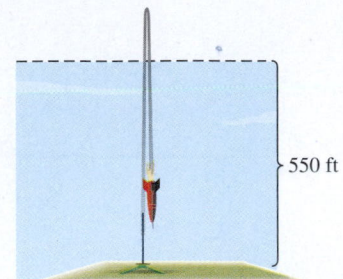

47. The following function gives the distance in feet a car going approximately 68 mph will skid in t seconds.

$$D(t) = 13t^2 - 100t$$

Find the time it would take for the car to skid 180 ft.

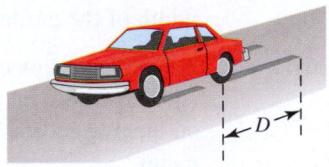

48. Refer to the function in **Exercise 47.** Find the time it would take for the car to skid 500 ft.

A ball is projected upward from ground level, and its distance in feet from the ground in t seconds is given by

$$s(t) = -16t^2 + 160t.$$

49. After how many seconds does the ball reach a height of 400 ft? Describe in words its position at this height.

50. After how many seconds does the ball reach a height of 425 ft? Interpret the mathematical result here.

Extending Skills   *Solve each problem using a quadratic equation.*

51. A certain bakery has found that the daily demand for blueberry muffins is $\frac{6000}{p}$, where p is the price of a muffin in cents. The daily supply is $3p - 410$. Find the price at which supply and demand are equal.

52. In one area the demand for Blu-ray discs is $\frac{1900}{P}$ per day, where P is the price in dollars per disc. The supply is $5P - 1$ per day. At what price, to the nearest cent, does supply equal demand?

53. The formula $A = P(1 + r)^2$ gives the amount A in dollars that P dollars will grow to in 2 yr at interest rate r (where r is given as a decimal), using compound interest. What interest rate will cause $2000 to grow to $2142.45 in 2 yr?

54. Use the formula $A = P(1 + r)^2$ to find the interest rate r at which a principal P of $10,000 will increase to $10,920.25 in 2 yr.

William Froude was a 19th century naval architect who used the following expression, known as the **Froude number,** in shipbuilding.

$$\frac{v^2}{g\ell}$$

This expression was also used by R. McNeill Alexander in his research on dinosaurs. (Data from "How Dinosaurs Ran," Scientific American.)

 Use this expression to find the value of v (in meters per second), given $g = 9.8$ m per sec². (Round to the nearest tenth.)

55. Rhinoceros: $\ell = 1.2$;

  Froude number = 2.57

56. Triceratops: $\ell = 2.8$;

  Froude number = 0.16

Recall that corresponding sides of similar triangles are proportional. Use this fact to find the lengths of the indicated sides of each pair of similar triangles. Check all possible solutions in both triangles. Sides of a triangle cannot be negative (and are not drawn to scale here).

57. Side AC

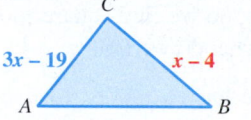

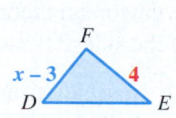

58. Side RQ

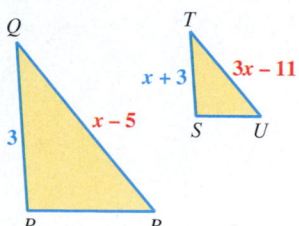

Total spending (in billions of dollars) in the United States from all sources on physician and clinical services for the years 2000–2016 are shown in the bar graph. One model for the data is the quadratic function

$$f(x) = -0.2901x^2 + 25.90x + 291.6.$$

Here, $x = 0$ represents 2000, $x = 2$ represents 2002, and so on. Use the graph and the model to work each of the following exercises. **See Example 6.**

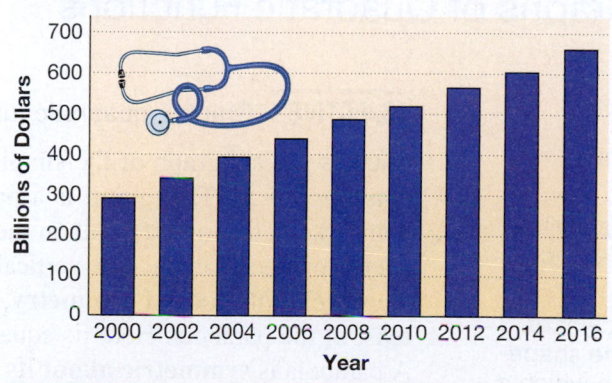

Spending on Physician and Clinical Services

Data from Centers for Medicare and Medicaid Services.

59. Approximate spending on physician and clinical services in 2014 to the nearest $10 billion using **(a)** the graph and **(b)** the model. How do the two approximations compare?

60. Repeat **Exercise 59** for the year 2008.

61. According to the model, in what year did spending on physician and clinical services first exceed $500 billion? (Round down for the year.)

62. Repeat **Exercise 61** for $400 billion.

RELATING CONCEPTS For Individual or Group Work (Exercises 63–66)

In the 1939 classic movie The Wizard of Oz, *Ray Bolger's character, the Scarecrow, wants a brain. When the Wizard grants him his "Th.D." (Doctor of Thinkology), the Scarecrow replies with the following statement.*

> *Scarecrow:* The sum of the square roots of any two sides of an isosceles triangle is equal to the square root of the remaining side.

His statement sounds like the formula for the Pythagorean theorem. To see why it is incorrect, **work Exercises 63–66 in order.**

63. To what kind of triangle does the Scarecrow refer in his statement? To what kind of triangle does the Pythagorean theorem actually refer?

64. In the Scarecrow's statement, he refers to square roots. In applying the formula for the Pythagorean theorem, do we find square roots of the sides? If not, what do we find?

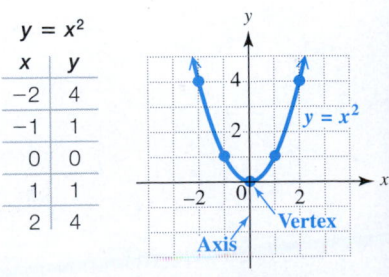

65. An isosceles triangle has two sides of equal length. Draw an isosceles triangle with two sides of length 9 units and remaining side of length 4 units. Show that this triangle does not satisfy the Scarecrow's statement. (This is a *counterexample* and is sufficient to show that his statement is false in general.)

66. Use wording similar to that of the Scarecrow, but state the Pythagorean theorem correctly.

11.6 Graphs of Quadratic Functions

OBJECTIVES

1. Graph a quadratic function.
2. Graph parabolas with horizontal and vertical shifts.
3. Use the coefficient of x^2 to predict the shape and direction in which a parabola opens.
4. Find a quadratic function to model data.

OBJECTIVE 1 Graph a quadratic function.

FIGURE 5 gives a graph of the simplest *quadratic function* $y = x^2$. This graph is a **parabola.** The point $(0, 0)$, the lowest point on the curve, is the **vertex** of the parabola. The vertical line through the vertex is the **axis of symmetry,** or simply the **axis,** of the parabola. Here its equation is $x = 0$. A parabola is **symmetric about its axis**—that is, if the graph were folded along the axis, the two portions of the curve would coincide.

$y = x^2$	
x	**y**
−2	4
−1	1
0	0
1	1
2	4

FIGURE 5

As **FIGURE 5** suggests, x can be any real number, so the domain of the function $y = x^2$, written in interval notation, is $(-\infty, \infty)$. Values of y are always nonnegative, so the range is $[0, \infty)$.

Quadratic Function

A function that can be written in the form

$$f(x) = ax^2 + bx + c$$

for real numbers a, b, and c, where $a \neq 0$, is a **quadratic function.**

Examples: $f(x) = x^2$, $f(x) = x^2 - 4x + 4$, $f(x) = -\frac{1}{2}x^2 + 3$

VOCABULARY
☐ parabola
☐ vertex
☐ axis of symmetry (axis)
☐ quadratic function

The graph of any quadratic function is a parabola with a vertical axis.

NOTE We use the variable y and function notation $f(x)$ interchangeably. Although we use the letter f most often to name quadratic functions, other letters can be used. We use the capital letter F to distinguish between different parabolas graphed on the same coordinate axes.

Parabolas have a special reflecting property that makes them useful in the design of telescopes, radar equipment, solar furnaces, and automobile headlights. (See **FIGURE 6**.)

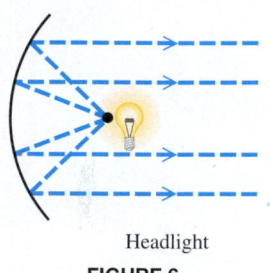

Headlight
FIGURE 6

OBJECTIVE 2 Graph parabolas with horizontal and vertical shifts.

Parabolas need not have their vertices at the origin, as does the graph of $f(x) = x^2$.

NOW TRY
EXERCISE 1
Graph $f(x) = x^2 - 3$. Give the vertex, axis, domain, and range.

EXAMPLE 1 Graphing a Parabola (Vertical Shift)

Graph $F(x) = x^2 - 2$.

If $x = 0$, then $F(x) = -2$, which gives the vertex $(0, -2)$. The graph of $F(x) = x^2 - 2$ has the same shape as that of $f(x) = x^2$ but is *shifted*, or *translated*, *down* 2 units. Every function value is 2 less than the corresponding function value of $f(x) = x^2$. Plotting points on both sides of the vertex gives the graph in **FIGURE 7**.

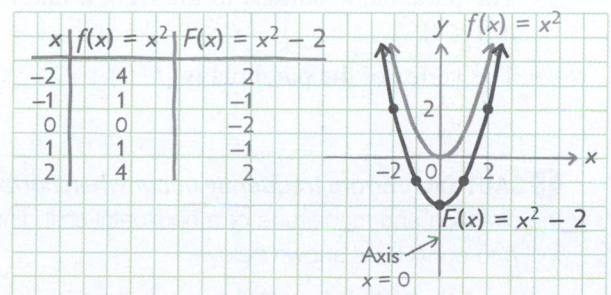

x	$f(x) = x^2$	$F(x) = x^2 - 2$
-2	4	2
-1	1	-1
0	0	-2
1	1	-1
2	4	2

$F(x) = x^2 - 2$
Vertex: $(0, -2)$
Axis of symmetry: $x = 0$
Domain: $(-\infty, \infty)$
Range: $[-2, \infty)$
The graph of $f(x) = x^2$ is shown for comparison.

FIGURE 7

NOW TRY ANSWER
1.
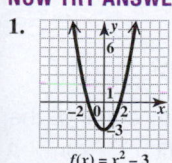
$f(x) = x^2 - 3$
vertex: $(0, -3)$; axis: $x = 0$;
domain: $(-\infty, \infty)$; range: $[-3, \infty)$

This parabola is symmetric about its axis $x = 0$, so the plotted points are "mirror images" of each other. Because x can be any real number, the domain is $(-\infty, \infty)$. The value of y (or $F(x)$) is always greater than or equal to -2, so the range is $[-2, \infty)$.

NOW TRY

Parabola with a Vertical Shift

The graph of $F(x) = x^2 + k$ is a parabola.

- The graph has the same shape as the graph of $f(x) = x^2$.
- The parabola is shifted up k units if $k > 0$, and down $|k|$ units if $k < 0$.
- The vertex of the parabola is $(0, k)$.

NOW TRY
EXERCISE 2

Graph $f(x) = (x + 1)^2$.
Give the vertex, axis, domain, and range.

EXAMPLE 2   Graphing a Parabola (Horizontal Shift)

Graph $F(x) = (x - 2)^2$.

If $x = 2$, then $F(x) = 0$, giving the vertex $(2, 0)$. The graph of

$$F(x) = (x - 2)^2$$

has the same shape as that of $f(x) = x^2$ but is shifted *to the right* 2 units. We plot several points on one side of the vertex and then use symmetry about the axis $x = 2$ to find corresponding points on the other side of the vertex.

If $x = 0$, for example, then

$$F(0) = (0 - 2)^2 = 4,$$

and the point $(0, 4)$ lies on the graph. The corresponding point two units on the "other" side of the axis $x = 2$ is the point $(4, 4)$, which also lies on the graph. See **FIGURE 8**.

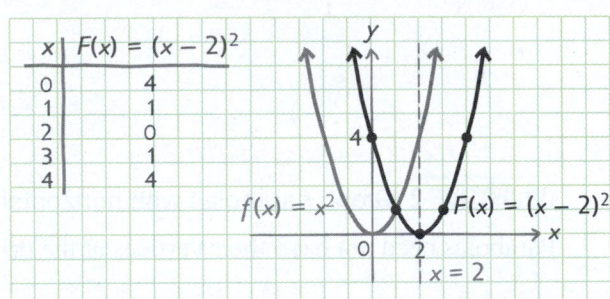

x	$F(x) = (x - 2)^2$
0	4
1	1
2	0
3	1
4	4

$F(x) = (x - 2)^2$
Vertex: $(2, 0)$
Axis of symmetry: $x = 2$
Domain: $(-\infty, \infty)$
Range: $[0, \infty)$

FIGURE 8

NOW TRY

Parabola with a Horizontal Shift

The graph of $F(x) = (x - h)^2$ is a parabola.

- The graph has the same shape as the graph of $f(x) = x^2$.
- The parabola is shifted to the right h units if $h > 0$, and to the left $|h|$ units if $h < 0$.
- The vertex of the parabola is $(h, 0)$.

⚠ **CAUTION**   *Errors frequently occur when horizontal shifts are involved.* To determine the direction and magnitude of a horizontal shift, find the value that causes the expression $x - h$ to equal 0, as shown below.

$F(x) = (x - 5)^2$	$F(x) = (x + 5)^2$
Because **+5** causes $x - 5$ to equal 0, the graph of $F(x)$ illustrates a shift	Because **−5** causes $x + 5$ to equal 0, the graph of $F(x)$ illustrates a shift
to the right 5 units.	**to the left 5 units.**

**NOW TRY
EXERCISE 3**

Graph $f(x) = (x + 1)^2 - 2$.
Give the vertex, axis, domain, and range.

EXAMPLE 3 Graphing a Parabola (Horizontal and Vertical Shifts)

Graph $F(x) = (x + 3)^2 - 2$.

This graph has the same shape as that of $f(x) = x^2$, but is shifted *to the left* 3 units (because $x + 3 = 0$ if $x = -3$) and *down* 2 units (because of the negative sign in -2). This gives the vertex $(-3, -2)$. Find and plot several ordered pairs, using symmetry as needed, to obtain the graph in **FIGURE 9**.

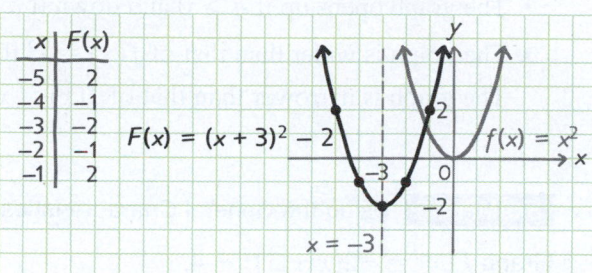

$F(x) = (x + 3)^2 - 2$
Vertex: $(-3, -2)$
Axis of symmetry: $x = -3$
Domain: $(-\infty, \infty)$
Range: $[-2, \infty)$

FIGURE 9

NOW TRY

Parabola with Horizontal and Vertical Shifts

The graph of $F(x) = (x - h)^2 + k$ is a parabola.

- The graph has the same shape as the graph of $f(x) = x^2$.
- The vertex of the parabola is (h, k).
- The axis of symmetry is the vertical line $x = h$.

OBJECTIVE 3 Use the coefficient of x^2 to predict the shape and direction in which a parabola opens.

Not all parabolas open up, and not all parabolas have the same shape as the graph of $f(x) = x^2$.

EXAMPLE 4 Graphing a Parabola That Opens Down

Graph $f(x) = -\frac{1}{2}x^2$.

This parabola is shown in **FIGURE 10**. The coefficient of x^2, $-\frac{1}{2}$, affects the shape of the graph—the $\frac{1}{2}$ makes the parabola wider $\left(\text{because the values of } \frac{1}{2}x^2 \text{ increase more slowly than those of } x^2\right)$, and the negative sign makes the parabola open down.

The graph is not shifted in any direction, so the vertex is $(0, 0)$. Unlike the parabolas graphed in **Examples 1–3**, the vertex $(0, 0)$ has the *greatest* function value of any point on the graph.

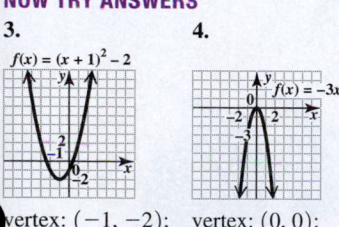
**NOW TRY
EXERCISE 4**

Graph $f(x) = -3x^2$.
Give the vertex, axis, domain, and range.

NOW TRY ANSWERS

3. $f(x) = (x + 1)^2 - 2$

vertex: $(-1, -2)$;
axis: $x = -1$;
domain: $(-\infty, \infty)$;
range: $[-2, \infty)$

4. $f(x) = -3x^2$

vertex: $(0, 0)$;
axis: $x = 0$;
domain: $(-\infty, \infty)$;
range: $(-\infty, 0]$

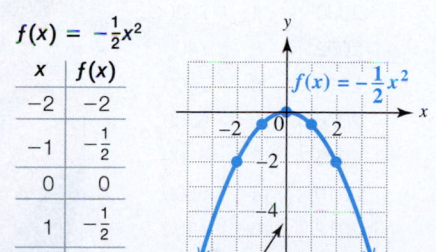

$f(x) = -\frac{1}{2}x^2$

x	$f(x)$
-2	-2
-1	$-\frac{1}{2}$
0	0
1	$-\frac{1}{2}$
2	-2

$f(x) = -\frac{1}{2}x^2$
Vertex: $(0, 0)$
Axis of symmetry: $x = 0$
Domain: $(-\infty, \infty)$
Range: $(-\infty, 0]$

FIGURE 10

NOW TRY

General Characteristics of the Graph of a Vertical Parabola

The graph of the quadratic function $F(x) = a(x - h)^2 + k$ (where $a \neq 0$) is a parabola.

- The vertex of the parabola is (h, k).
- The axis of symmetry is the vertical line $x = h$.
- The graph opens up if $a > 0$ and down if $a < 0$.
- The graph is wider than that of $f(x) = x^2$ if $0 < |a| < 1$.
  The graph is narrower than that of $f(x) = x^2$ if $|a| > 1$.

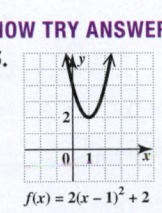

**NOW TRY
EXERCISE 5**

Graph $f(x) = 2(x - 1)^2 + 2$. Give the vertex, axis, domain, and range.

EXAMPLE 5   Using the General Characteristics to Graph a Parabola

Graph $F(x) = -2(x + 3)^2 + 4$.

The parabola opens down (because $a = -2$ and $-2 < 0$) and is narrower than the graph of $f(x) = x^2$ (because $|-2| = 2$ and $2 > 1$). This causes values of $F(x)$ to decrease more quickly than those of $f(x) = -x^2$.

The graph is shifted *to the left* 3 units (because $x + 3 = 0$ when $x = -3$) and *up* 4 units (because of the $+4$), which gives the vertex $(-3, 4)$. To complete the graph, we plotted the ordered pairs $(-4, 2)$ and, by symmetry, $(-2, 2)$. Symmetry can be used to find additional ordered pairs that satisfy the equation. See **FIGURE 11.**

$F(x) = -2(x + 3)^2 + 4$

x	$F(x)$
-5	-4
-4	2
-3	4
-2	2
-1	-4

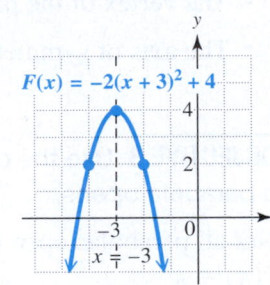

$F(x) = -2(x + 3)^2 + 4$
Vertex: $(-3, 4)$
Axis of symmetry: $x = -3$
Domain: $(-\infty, \infty)$
Range: $(-\infty, 4]$

FIGURE 11

NOW TRY

OBJECTIVE 4 Find a quadratic function to model data.

EXAMPLE 6   Modeling the Number of Multiple Births

The number of higher-order multiple births (triplets or more) in the United States is shown in the table. Here, x represents the number of years since 1996, and y represents the number of higher-order multiple births (to the nearest hundred).

Year	x	y
1996	0	5900
2000	4	7300
2002	6	7400
2004	8	7300
2006	10	6500
2008	12	6300
2010	14	5500
2012	16	4900
2014	18	4500

Data from National Center for Health Statistics.

NOW TRY ANSWER

5.

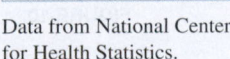

$f(x) = 2(x - 1)^2 + 2$

vertex: $(1, 2)$; axis: $x = 1$;
domain: $(-\infty, \infty)$; range: $[2, \infty)$

**NOW TRY
EXERCISE 6**

Using the points $(0, 5900)$, $(4, 7300)$, and $(12, 6300)$, find another quadratic model for the data on higher-order multiple births in **Example 6.** (Round values of a and b to the nearest tenth.)

Find a quadratic function that models the data in the table on the previous page.

A scatter diagram of the ordered pairs (x, y) is shown in **FIGURE 12**. The general shape suggested by the scatter diagram indicates that a parabola should approximate these points, as shown by the dashed curve in **FIGURE 13**. The equation for such a parabola would have a negative coefficient for x^2 because the graph opens down.

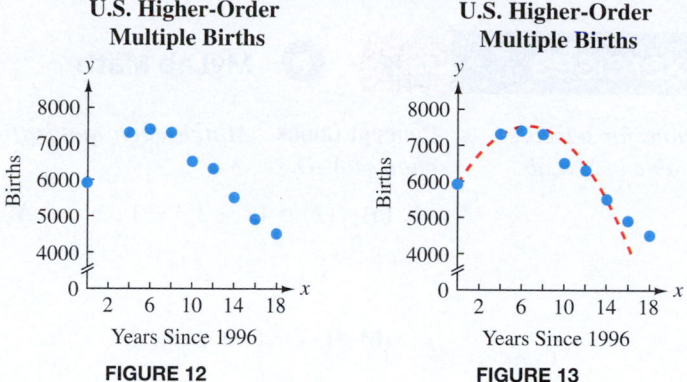

U.S. Higher-Order Multiple Births

Years Since 1996

FIGURE 12

U.S. Higher-Order Multiple Births

Years Since 1996

FIGURE 13

To find a quadratic function of the form

$$y = ax^2 + bx + c$$

that models, or *fits*, these data, we choose three representative ordered pairs from the table and use them to write a system of three equations.

$$(0, 5900), \quad (6, 7400), \quad \text{and} \quad (12, 6300) \qquad \text{Three ordered pairs } (x, y)$$

We substitute the x- and y-values from each ordered pair into the quadratic form $y = ax^2 + bx + c$ to obtain three equations.

$$
\begin{array}{lll}
a(0)^2 + b(0) + c = 5900 & \xrightarrow{\text{Simplify.}} & c = 5900 \qquad (1) \\
a(6)^2 + b(6) + c = 7400 & \longrightarrow & 36a + 6b + c = 7400 \qquad (2) \\
a(12)^2 + b(12) + c = 6300 & \longrightarrow & 144a + 12b + c = 6300 \qquad (3)
\end{array}
$$

To find the values of a, b, and c, we solve this system of three equations in three variables. From equation (1), $c = 5900$, so we substitute 5900 for c in equations (2) and (3) to obtain two equations in two variables.

$$
\begin{array}{lll}
36a + 6b + 5900 = 7400 & \longrightarrow & 36a + 6b = 1500 \qquad (4) \\
144a + 12b + 5900 = 6300 & \xrightarrow{\text{Subtract 5900.}} & 144a + 12b = 400 \qquad (5)
\end{array}
$$

We eliminate b from this system of equations in two variables by multiplying equation (4) by -2 and adding the results to equation (5).

$$
\begin{array}{ll}
-72a - 12b = -3000 & \text{Multiply equation (4) by } -2. \\
\underline{144a + 12b = 400} & (5) \\
72a = -2600 & \text{Add.} \\
a \approx -36.1 & \text{Use a calculator. Round to one decimal place.}
\end{array}
$$

We substitute -36.1 for a in equation (4) to find that $b \approx 466.6$. (Substituting in equation (5) will give $b \approx 466.5$ due to rounding procedures.) Using the values we found for a, b, and c, the model is

NOW TRY ANSWER
6. $y = -39.6x^2 + 508.4x + 5900$
(Answers may vary slightly due to rounding.)

$$y = \overset{a}{\underset{\downarrow}{-36.1}}x^2 + \overset{b}{\underset{\downarrow}{466.6}}x + \overset{c}{\underset{\downarrow}{5900}}.$$

NOW TRY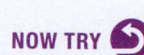

NOTE If we had chosen three different ordered pairs of data in **Example 6,** a slightly different, though similar, model would have resulted. (See **Now Try Exercise 6.**)

The *quadratic regression* feature on a graphing calculator can also be used to generate the quadratic model that best fits given data. See your owner's manual for details.

11.6 Exercises

 MyLab Math

 Video solutions for select problems available in MyLab Math

Concept Check *Match each quadratic function in parts (a)–(d) with its graph from choices A–D.*

1. (a) $f(x) = (x + 2)^2 - 1$

(b) $f(x) = (x + 2)^2 + 1$

(c) $f(x) = (x - 2)^2 - 1$

(d) $f(x) = (x - 2)^2 + 1$

A. 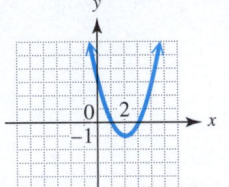   **B.**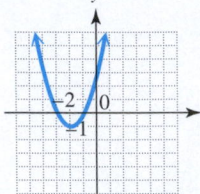

C. 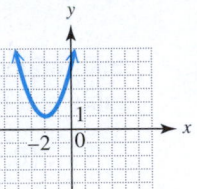   **D.**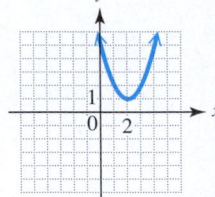

2. (a) $f(x) = -x^2 + 2$

(b) $f(x) = -x^2 - 2$

(c) $f(x) = -(x + 2)^2$

(d) $f(x) = -(x - 2)^2$

A. 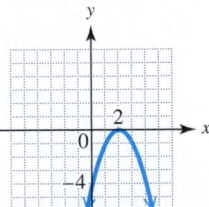   **B.**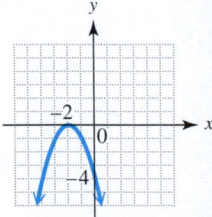

C. 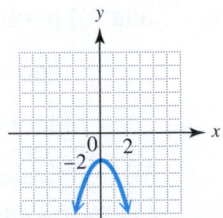   **D.**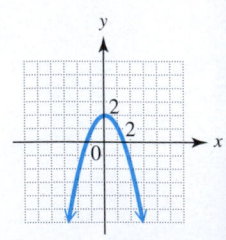

3. CONCEPT CHECK   Match each quadratic function in Column I with the description of the parabola that is its graph in Column II.

	I		**II**
(a)	$f(x) = (x - 4)^2 - 2$	**A.**	Vertex $(2, -4)$, opens down
(b)	$f(x) = (x - 2)^2 - 4$	**B.**	Vertex $(2, -4)$, opens up
(c)	$f(x) = (x + 4)^2 + 2$	**C.**	Vertex $(4, -2)$, opens down
(d)	$f(x) = -(x - 4)^2 - 2$	**D.**	Vertex $(4, -2)$, opens up
(e)	$f(x) = -(x - 2)^2 - 4$	**E.**	Vertex $(4, 2)$, opens down
(f)	$f(x) = -(x - 4)^2 + 2$	**F.**	Vertex $(-4, 2)$, opens up

4. **Concept Check**   For the quadratic function $f(x) = a(x - h)^2 + k$, in what quadrant is the vertex if the values of h and k are as follows?

(a) $h > 0, k > 0$ (b) $h > 0, k < 0$

(c) $h < 0, k > 0$ (d) $h < 0, k < 0$

Consider the value of a, and make the correct choice.

(e) If $a > 0$, then the graph opens (*up / down*).

(f) If $a < 0$, then the graph opens (*up / down*).

(g) If $|a| > 1$, then the graph is (*narrower / wider*) than the graph of $f(x) = x^2$.

(h) If $0 < |a| < 1$, then the graph is (*narrower / wider*) than the graph of $f(x) = x^2$.

*Identify the vertex of each parabola. **See Examples 1–4.***

5. $f(x) = -3x^2$

6. $f(x) = -4x^2$

7. $f(x) = \frac{1}{3}x^2$

8. $f(x) = \frac{1}{2}x^2$

9. $f(x) = x^2 + 4$

10. $f(x) = x^2 - 4$

11. $f(x) = (x - 1)^2$

12. $f(x) = (x + 3)^2$

13. $f(x) = (x + 3)^2 - 4$

14. $f(x) = (x + 5)^2 - 8$

15. $f(x) = -(x - 5)^2 + 6$

16. $f(x) = -(x - 2)^2 + 1$

For each quadratic function, tell whether the graph opens up *or* down *and whether the graph is* wider, narrower, *or the* same *shape as the graph of $f(x) = x^2$. **See Examples 4 and 5.***

17. $f(x) = -\frac{2}{5}x^2$

18. $f(x) = -2x^2$

19. $f(x) = 3x^2 + 1$

20. $f(x) = \frac{2}{3}x^2 - 4$

21. $f(x) = -4(x + 2)^2 + 5$

22. $f(x) = -\frac{1}{3}(x + 6)^2 + 3$

*Graph each parabola. Give the vertex, axis of symmetry, domain, and range. **See Examples 1–5.***

23. $f(x) = 3x^2$

24. $f(x) = \frac{1}{2}x^2$

25. $f(x) = -2x^2$

26. $f(x) = -\frac{1}{3}x^2$

27. $f(x) = x^2 - 1$

28. $f(x) = x^2 + 3$

29. $f(x) = -x^2 + 2$

30. $f(x) = -x^2 - 2$

31. $f(x) = (x - 4)^2$

32. $f(x) = (x + 1)^2$

33. $f(x) = (x + 2)^2 - 1$

34. $f(x) = (x - 1)^2 + 2$

35. $f(x) = (x - 1)^2 - 3$

36. $f(x) = (x + 1)^2 + 1$

37. $f(x) = 2(x - 2)^2 - 4$

38. $f(x) = 3(x - 2)^2 + 1$

39. $f(x) = \frac{1}{2}(x - 2)^2 - 3$

40. $f(x) = \frac{4}{3}(x - 3)^2 - 2$

41. $f(x) = -2(x + 3)^2 + 4$

42. $f(x) = -2(x - 2)^2 - 3$

43. $f(x) = -\frac{1}{2}(x + 1)^2 + 2$

44. $f(x) = -\frac{2}{3}(x + 2)^2 + 1$

Determine whether a linear function *or a* quadratic function *would be a more appropriate model for each set of graphed data. If linear, tell whether the slope should be* positive *or* negative. *If quadratic, tell whether the coefficient of* x^2 *should be* positive *or* negative. **See Example 6.**

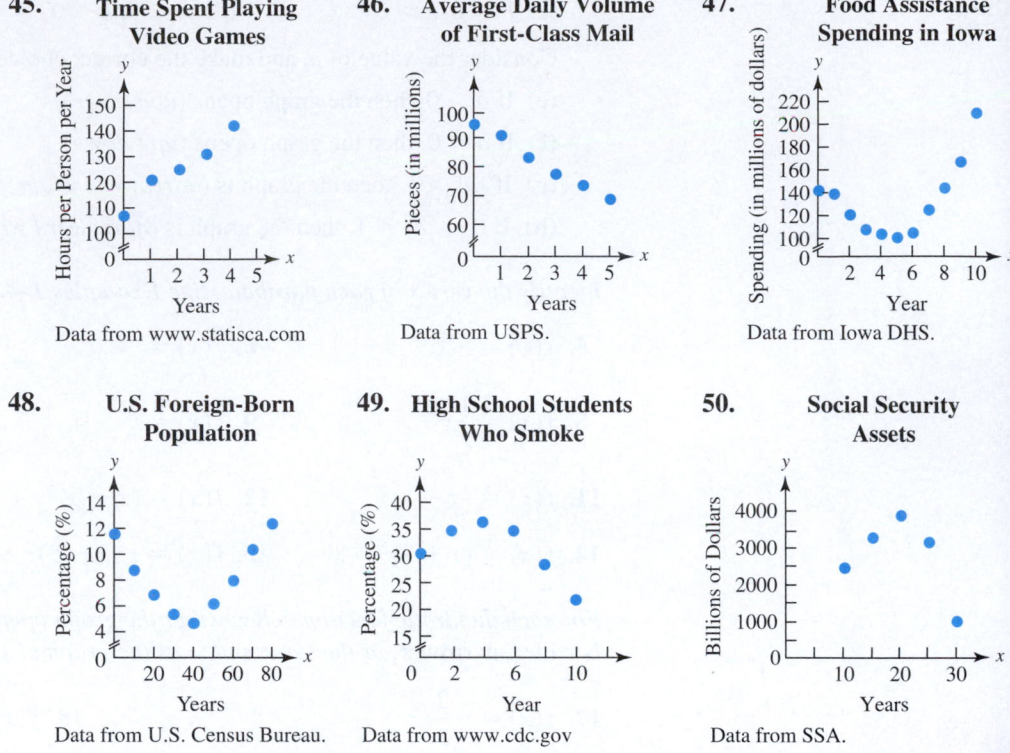

45.   **Time Spent Playing Video Games**
Hours per Person per Year / Years
Data from www.statisca.com

46.   **Average Daily Volume of First-Class Mail**
Pieces (in millions) / Years
Data from USPS.

47.   **Food Assistance Spending in Iowa**
Spending (in millions of dollars) / Year
Data from Iowa DHS.

48.   **U.S. Foreign-Born Population**
Percentage (%) / Years
Data from U.S. Census Bureau.

49.   **High School Students Who Smoke**
Percentage (%) / Year
Data from www.cdc.gov

50.   **Social Security Assets**
Billions of Dollars / Years
Data from SSA.

Solve each problem. See Example 6.

51. The number of U.S. households (in millions) with cable television service is shown in the table, where x represents the number of years since 2007 and y represents households.

Year	x	y
2007	0	95
2009	2	103
2011	4	105
2013	6	103
2015	8	100
2017	10	96

Data from Nielsen Media Research.

(a) Use the ordered pairs (x, y) to make a scatter diagram of the data.

(b) Would a linear or a quadratic function better model the data?

(c) Should the coefficient a of x^2 in a quadratic model $y = ax^2 + bx + c$ be positive or negative?

(d) Use the ordered pairs $(0, 95)$, $(4, 105)$, and $(10, 96)$ to find a quadratic function that models the data.

(e) Use the model from part (d) to approximate the number of U.S. households (in millions) with cable television service in 2009 and 2015. How well does the model approximate the actual data from the table?

52. The number (in thousands) of new, privately owned housing units completed in the United States is shown in the table. Here x represents the number of years since 2006, and y represents total housing units completed.

Year	x	y
2006	0	1980
2008	2	1120
2010	4	650
2012	6	650
2014	8	880
2016	10	1060

Data from U.S. Census Bureau.

(a) Use the ordered pairs (x, y) to make a scatter diagram of the data.

(b) Would a linear or a quadratic function better model the data?

(c) Should the coefficient a of x^2 in a quadratic model $y = ax^2 + bx + c$ be positive or negative?

(d) Use the ordered pairs $(0, 1980)$, $(4, 650)$, and $(8, 880)$ to find a quadratic function that models the data.

(e) Use the model from part (d) to approximate the number of such housing units (in thousands) completed in 2008 and 2012. How well does the model approximate the actual data from the table?

11.7 More about Parabolas and Their Applications

OBJECTIVES

1. Find the vertex of a vertical parabola.
2. Graph a quadratic function.
3. Use the discriminant to find the number of x-intercepts of a parabola with a vertical axis.
4. Use quadratic functions to solve problems involving maximum or minimum value.
5. Graph parabolas with horizontal axes.

NOW TRY EXERCISE 1

Find the vertex of the graph of

$$f(x) = x^2 + 2x - 8.$$

NOW TRY ANSWER

1. $(-1, -9)$

OBJECTIVE 1 Find the vertex of a vertical parabola.

When the equation of a parabola is given in the form $f(x) = ax^2 + bx + c$, there are two ways to locate the vertex.

1. Complete the square. (See **Examples 1 and 2.**)

2. Use a formula derived by completing the square. (See **Example 3.**)

EXAMPLE 1 Completing the Square to Find the Vertex ($a = 1$)

Find the vertex of the graph of $f(x) = x^2 - 4x + 5$.

We can express $x^2 - 4x + 5$ in the form $(x - h)^2 + k$ by completing the square on $x^2 - 4x$, as in the first section of this chapter. The process is slightly different here because we want to keep $f(x)$ alone on one side of the equation. Instead of adding the appropriate number to each side, we *add and subtract* it on the right.

$$f(x) = x^2 - 4x + 5$$

$$f(x) = (x^2 - 4x \qquad) + 5 \qquad \text{Group the variable terms.}$$

This is equivalent to adding 0. $\left[\frac{1}{2}(-4)\right]^2 = (-2)^2 = 4$ \qquad Square half the coefficient of the first-degree term.

$$f(x) = (x^2 - 4x + 4 - 4) + 5 \qquad \text{Add and subtract 4.}$$

$$f(x) = (x^2 - 4x + 4) - 4 + 5 \qquad \text{Bring } -4 \text{ outside the parentheses.}$$

$$f(x) = (x - 2)^2 + 1 \qquad \text{Factor. Combine like terms.}$$

The vertex of this parabola is $(2, 1)$.

NOW TRY

NOW TRY
EXERCISE 2

Find the vertex of the graph of

$$f(x) = -4x^2 + 16x - 10.$$

EXAMPLE 2   Completing the Square to Find the Vertex ($a \neq 1$)

Find the vertex of the graph of $f(x) = -3x^2 + 6x - 1$.

Because the x^2-term has a coefficient other than 1, we factor that coefficient out of the first two terms before completing the square.

$$f(x) = -3x^2 + 6x - 1$$

$$f(x) = (-3x^2 + 6x) - 1 \qquad \text{Group the variable terms.}$$

$$f(x) = -3(x^2 - 2x) - 1 \qquad \text{Factor out } -3.$$

$$f(x) = -3(x^2 - 2x \qquad) - 1 \qquad \text{Prepare to complete the square.}$$

$$\left[\tfrac{1}{2}(-2)\right]^2 = (-1)^2 = 1 \qquad \begin{array}{l}\text{Square half the coefficient} \\ \text{of the first-degree term.}\end{array}$$

$$f(x) = -3(x^2 - 2x + 1 - 1) - 1 \qquad \text{Add and subtract 1.}$$

Now bring -1 *outside the parentheses. Be sure to multiply it by* -3.

This is a key step. $\qquad f(x) = -3(x^2 - 2x + 1) + (-3)(-1) - 1 \qquad \text{Distributive property}$

$$f(x) = -3(x^2 - 2x + 1) + 3 - 1 \qquad \text{Multiply.}$$

$$f(x) = -3(x - 1)^2 + 2 \qquad \text{Factor. Combine like terms.}$$

The vertex is $(1, 2)$.   **NOW TRY**

We can complete the square to derive a formula for the vertex of the graph of the quadratic function $f(x) = ax^2 + bx + c$ (where $a \neq 0$).

$$f(x) = ax^2 + bx + c \qquad \text{Standard form}$$

$$f(x) = (ax^2 + bx) + c \qquad \text{Group the terms with } x.$$

$$f(x) = a\left(x^2 + \frac{b}{a}x \qquad \right) + c \qquad \text{Factor } a \text{ from the first two terms.}$$

$$\left[\tfrac{1}{2}\left(\tfrac{b}{a}\right)\right]^2 = \left(\tfrac{b}{2a}\right)^2 = \tfrac{b^2}{4a^2} \qquad \begin{array}{l}\text{Square half the coefficient} \\ \text{of the first-degree term.}\end{array}$$

$$f(x) = a\left(x^2 + \frac{b}{a}x + \frac{b^2}{4a^2} - \frac{b^2}{4a^2}\right) + c \qquad \text{Add and subtract } \tfrac{b^2}{4a^2}.$$

$$f(x) = a\left(x^2 + \frac{b}{a}x + \frac{b^2}{4a^2}\right) + a\left(-\frac{b^2}{4a^2}\right) + c \qquad \text{Distributive property}$$

$$f(x) = a\left(x^2 + \frac{b}{a}x + \frac{b^2}{4a^2}\right) - \frac{b^2}{4a} + c \qquad -\tfrac{ab^2}{4a^2} = -\tfrac{b^2}{4a}$$

$$f(x) = a\left(x + \frac{b}{2a}\right)^2 + \frac{4ac - b^2}{4a} \qquad \begin{array}{l}\text{Factor. Rewrite terms with} \\ \text{a common denominator.}\end{array}$$

$$f(x) = a\left[x - \left(\frac{-b}{2a}\right)\right]^2 + \frac{4ac - b^2}{4a} \qquad \begin{array}{l} f(x) = a(x - h)^2 + k; \\ \text{The vertex } (h, k) \text{ can be} \\ \text{expressed in terms of } a, b, \text{ and } c.\end{array}$$

$$\underbrace{\phantom{x - \left(\frac{-b}{2a}\right)}}_{h} \qquad \underbrace{\phantom{\frac{4ac - b^2}{4a}}}_{k}$$

The expression for k can be found by replacing x with $\frac{-b}{2a}$. Using function notation, if $y = f(x)$, then the y-value of the vertex is $f\left(\frac{-b}{2a}\right)$.

Vertex Formula

The graph of the quadratic function $f(x) = ax^2 + bx + c$ (where $a \neq 0$) has vertex

$$\left(\frac{-b}{2a}, f\left(\frac{-b}{2a} \right) \right).$$

The axis of symmetry of the parabola is the line having equation

$$x = \frac{-b}{2a}.$$

NOW TRY EXERCISE 3

Use the vertex formula to find the vertex of the graph of

$$f(x) = 3x^2 - 2x + 8.$$

EXAMPLE 3 Using the Formula to Find the Vertex

Use the vertex formula to find the vertex of the graph of $f(x) = x^2 - x - 6$.

The x-coordinate of the vertex of the parabola is given by $\frac{-b}{2a}$.

$$\frac{-b}{2a} = \frac{-(-1)}{2(1)} = \frac{1}{2} \leftarrow \begin{array}{l} a = 1, b = -1, \text{ and } c = -6 \\ x\text{-coordinate of vertex} \end{array}$$

The y-coordinate of the vertex of $f(x) = x^2 - x - 6$ is $f\left(\frac{-b}{2a} \right) = f\left(\frac{1}{2} \right)$.

$$f\left(\frac{1}{2} \right) = \left(\frac{1}{2} \right)^2 - \frac{1}{2} - 6$$

$$= \frac{1}{4} - \frac{1}{2} - 6$$

$$= -\frac{25}{4} \leftarrow y\text{-coordinate of vertex} \qquad \begin{array}{l} \frac{1}{4} - \frac{1}{2} - 6 = \frac{1}{4} - \frac{2}{4} - \frac{24}{4} \end{array}$$

The vertex is $\left(\frac{1}{2}, -\frac{25}{4} \right)$.

NOW TRY

OBJECTIVE 2 Graph a quadratic function.

Graphing a Quadratic Function $y = f(x)$

Step 1 **Determine whether the graph opens up or down.**
- If $a > 0$, then the parabola opens up.
- If $a < 0$, then it opens down.

Step 2 **Find the vertex.** Use the vertex formula or complete the square.

Step 3 **Find any intercepts.**
- To find the x-intercepts (if any), solve $f(x) = 0$.
- To find the y-intercept, evaluate $f(0)$.

Step 4 **Complete the graph.** Plot the points found so far. Find and plot additional points as needed, using symmetry about the axis.

NOW TRY ANSWER

3. $\left(\frac{1}{3}, \frac{23}{3} \right)$

NOW TRY
EXERCISE 4

Graph the quadratic function

$$f(x) = x^2 + 2x - 3.$$

Give the vertex, axis, domain, and range.

EXAMPLE 4   Graphing a Quadratic Function

Graph the quadratic function $f(x) = x^2 - x - 6$.

Step 1   From the equation, $a = 1$, so the graph of the function opens up.

Step 2   The vertex, $\left(\frac{1}{2}, -\frac{25}{4}\right)$, was found in **Example 3** using the vertex formula.

Step 3   Find any intercepts. The vertex, $\left(\frac{1}{2}, -\frac{25}{4}\right)$, is in quadrant IV and the graph opens up, so there will be two x-intercepts. Let $f(x) = 0$ and solve to find them.

$$f(x) = x^2 - x - 6$$
$$0 = x^2 - x - 6 \qquad \text{Let } f(x) = 0.$$
$$0 = (x - 3)(x + 2) \qquad \text{Factor.}$$
$$x - 3 = 0 \quad \text{or} \quad x + 2 = 0 \qquad \text{Zero-factor property}$$
$$x = 3 \quad \text{or} \qquad x = -2 \qquad \text{Solve each equation.}$$

The x-intercepts are $(3, 0)$ and $(-2, 0)$.

Find the y-intercept by evaluating $f(0)$.

$$f(x) = x^2 - x - 6$$
$$f(0) = 0^2 - 0 - 6 \qquad \text{Let } x = 0.$$
$$f(0) = -6 \qquad \text{Apply the exponent. Subtract.}$$

The y-intercept is $(0, -6)$.

Step 4   Plot the points found so far and additional points as needed using symmetry about the axis, $x = \frac{1}{2}$. The graph is shown in **FIGURE 14**.

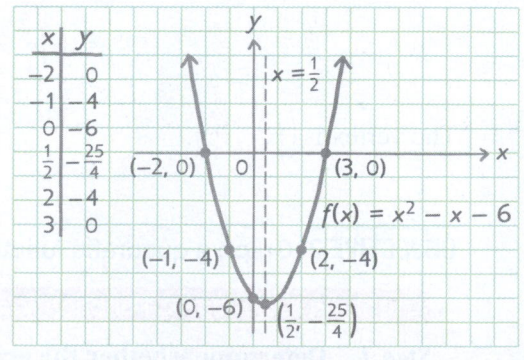

x	y
-2	0
-1	-4
0	-6
$\frac{1}{2}$	$-\frac{25}{4}$
2	-4
3	0

$f(x) = x^2 - x - 6$
Vertex: $\left(\frac{1}{2}, -\frac{25}{4}\right)$
Axis of symmetry: $x = \frac{1}{2}$
Domain: $(-\infty, \infty)$
Range: $\left[-\frac{25}{4}, \infty\right)$

FIGURE 14

 NOW TRY

NOW TRY ANSWER
4.

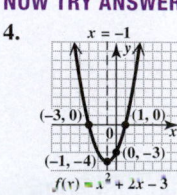

vertex: $(-1, -4)$; axis: $x = -1$;
domain: $(-\infty, \infty)$; range: $[-4, \infty)$

OBJECTIVE 3   Use the discriminant to find the number of x-intercepts of a parabola with a vertical axis.

Recall that the expression under the radical in $x = \dfrac{-b \pm \sqrt{b^2 - 4ac}}{2a}$, that is,

$$b^2 - 4ac, \qquad \text{Discriminant}$$

is the *discriminant* of the quadratic equation $ax^2 + bx + c = 0$ and that we can use it to determine the number of real solutions of a quadratic *equation*.

In a similar way, we can use the discriminant of a quadratic *function* to determine the number of x-intercepts of its graph. The three possibilities are shown in **FIGURE 15**.

1. If the discriminant is positive, the parabola will have two x-intercepts.

2. If the discriminant is 0, there will be only one x-intercept, and it will be the vertex of the parabola.

3. If the discriminant is negative, the graph will have no x-intercepts.

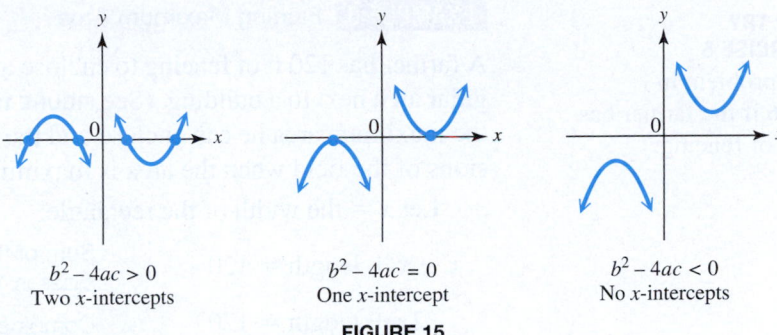

| $b^2 - 4ac > 0$ | $b^2 - 4ac = 0$ | $b^2 - 4ac < 0$ |
| Two x-intercepts | One x-intercept | No x-intercepts |

FIGURE 15

 NOW TRY EXERCISE 5

Find the discriminant and use it to determine the number of x-intercepts of the graph of each quadratic function.

(a) $f(x) = -2x^2 + 3x - 2$

(b) $f(x) = 3x^2 + 2x - 1$

(c) $f(x) = 4x^2 - 12x + 9$

EXAMPLE 5   Using the Discriminant to Determine Number of x-Intercepts

Find the discriminant and use it to determine the number of x-intercepts of the graph of each quadratic function.

(a) $f(x) = 2x^2 + 3x - 5$

$$
\begin{aligned}
b^2 - 4ac \qquad &\text{Discriminant} \\
= 3^2 - 4(2)(-5) \qquad &a = 2, b = 3, c = -5 \\
= 9 - (-40) \qquad &\text{Apply the exponent. Multiply.} \\
= 49 \qquad &\text{Subtract.}
\end{aligned}
$$

Because the discriminant is positive, the parabola has two x-intercepts.

(b) $f(x) = -3x^2 - 1$ (which can be written as $f(x) = -3x^2 + 0x - 1$)

$$
\begin{aligned}
b^2 - 4ac \\
= 0^2 - 4(-3)(-1) \qquad &a = -3, b = 0, c = -1 \\
= 0 - 12 \qquad &\text{Apply the exponent. Multiply.} \\
= -12 \qquad &\text{Subtract.}
\end{aligned}
$$

The discriminant is negative, so the graph has no x-intercepts.

(c) $f(x) = 9x^2 + 6x + 1$

$$
\begin{aligned}
b^2 - 4ac \\
= 6^2 - 4(9)(1) \qquad &a = 9, b = 6, c = 1 \\
= 36 - 36 \qquad &\text{Apply the exponent. Multiply.} \\
= 0 \qquad &\text{Subtract.}
\end{aligned}
$$

Because the discriminant is 0, the parabola has only one x-intercept (its vertex).

NOW TRY ANSWERS
5. (a) -7; none   (b) 16; two
   (c) 0; one

NOW TRY

OBJECTIVE 4 Use quadratic functions to solve problems involving maximum or minimum value.

The vertex of the graph of a quadratic function is either the highest or the lowest point on the parabola. It provides the following information.

1. The *y*-value of the vertex gives the maximum or minimum value of *y*.

2. The *x*-value tells where the maximum or minimum occurs.

> **NOW TRY**
> **EXERCISE 6**
> Solve the problem in
> **Example 6** if the farmer has
> only 80 ft of fencing.

EXAMPLE 6   Finding Maximum Area

A farmer has 120 ft of fencing to enclose a rectangular area next to a building. (See **FIGURE 16**.) Find the maximum area he can enclose and the dimensions of the field when the area is maximized.

FIGURE 16

Let *x* = the width of the rectangle.

$$x + x + \text{length} = 120 \qquad \text{Sum of the sides is 120 ft.}$$

$$2x + \text{length} = 120 \qquad \text{Combine like terms.}$$

$$\text{length} = 120 - 2x \qquad \text{Subtract } 2x.$$

The area $\mathcal{A}(x)$ is given by the product of the length and width.

$$\mathcal{A}(x) = (120 - 2x)x \qquad \text{Area = length } \cdot \text{ width}$$

$$\mathcal{A}(x) = 120x - 2x^2 \qquad \text{Distributive property}$$

$$\mathcal{A}(x) = -2x^2 + 120x \qquad \text{Standard form}$$

The graph of $\mathcal{A}(x) = -2x^2 + 120x$ is a parabola that opens down. To determine the maximum area, use the vertex formula to find the vertex of the parabola.

$$x = \frac{-b}{2a} = \frac{-120}{2(-2)} = \frac{-120}{-4} = 30 \qquad \text{Vertex formula; } a = -2, b = 120, c = 0$$

$$\mathcal{A}(30) = -2(30)^2 + 120(30) = -2(900) + 3600 = 1800$$

The vertex is $(30, 1800)$. The maximum area will be 1800 ft² when *x*, the width, is 30 ft and the length is $120 - 2(30) = 60$ ft.

NOW TRY ↺

EXAMPLE 7   Finding Maximum Height

If air resistance is neglected, a projectile on Earth shot straight upward with an initial velocity of 40 m per sec will be at a height *s* in meters given by

$$s(t) = -4.9t^2 + 40t,$$

where *t* is the number of seconds elapsed after projection. After how many seconds will it reach its maximum height, and what is this maximum height?

For this function, $a = -4.9$, $b = 40$, and $c = 0$. Use the vertex formula.

$$t = \frac{-b}{2a} = \frac{-40}{2(-4.9)} \approx 4.1 \qquad \text{Use a calculator. Round to the nearest tenth.}$$

NOW TRY ANSWER
6. The field should be 20 ft by 40 ft with maximum area 800 ft².

This indicates that the maximum height is attained at 4.1 sec.

**NOW TRY
EXERCISE 7**

A stomp rocket is launched from the ground with an initial velocity of 48 ft per sec so that its distance in feet above the ground after t seconds is

$$s(t) = -16t^2 + 48t.$$

Find the maximum height attained by the rocket and the number of seconds it takes to reach that height.

To find the maximum height at 4.1 sec, calculate $s(4.1)$.

$$s(t) = -4.9t^2 + 40t$$
$$s(4.1) = -4.9(4.1)^2 + 40(4.1) \qquad \text{Let } t = 4.1.$$
$$s(4.1) \approx 81.6 \qquad \text{Use a calculator. Round to the nearest tenth.}$$

The projectile will attain a maximum height of 81.6 m at 4.1 sec.   NOW TRY

⚠ **CAUTION** *Be careful when interpreting the meanings of the coordinates of the* ***vertex.*** *The first coordinate, x, gives the value for which the* function value, *y* or $f(x)$, *is a maximum or a minimum.*

OBJECTIVE 5 Graph parabolas with horizontal axes.

If x and y are interchanged in the equation

$$y = ax^2 + bx + c, \quad \text{the equation becomes} \quad x = ay^2 + by + c.$$

Because of the interchange of the roles of x and y, these parabolas are horizontal (with horizontal lines as axes of symmetry).

> **General Characteristics of the Graph of a Horizontal Parabola**
>
> The graph of an equation of the form
>
> $$x = ay^2 + by + c \quad \text{or} \quad x = a(y - k)^2 + h$$
>
> is a horizontal parabola.
>
> - The vertex of the parabola is (h, k).
> - The axis of symmetry is the horizontal line $y = k$.
> - The graph opens to the right if $a > 0$ and to the left if $a < 0$.

**NOW TRY
EXERCISE 8**

Graph $x = (y + 2)^2 - 1$. Give the vertex, axis, domain, and range.

EXAMPLE 8 Graphing a Horizontal Parabola ($a = 1$)

Graph $x = (y - 2)^2 - 3$. Give the vertex, axis, domain, and range.

This graph has its vertex at $(-3, 2)$ because the roles of x and y are interchanged. It opens to the right (the positive x-direction) because $a = 1$ and $1 > 0$, and has the same shape as $y = x^2$ (but situated horizontally).

To find additional points to plot, it is easiest to substitute a value for y and find the corresponding value for x. For example, let $y = 3$. Then

$$x = (3 - 2)^2 - 3 = -2, \quad \text{giving the point} \quad (-2, 3). \quad \text{Write the } x\text{-value first.}$$

Using symmetry, we can locate the point $(-2, 1)$. See **FIGURE 17**.

NOW TRY ANSWERS

7. 36 ft; 1.5 sec
8.

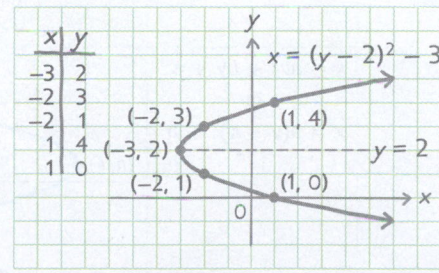

vertex: $(-1, -2)$; axis: $y = -2$;
domain: $[-1, \infty)$; range: $(-\infty, \infty)$

FIGURE 17

$x = (y - 2)^2 - 3$
Vertex: $(-3, 2)$
Axis of symmetry: $y = 2$
Domain: $[-3, \infty)$
Range: $(-\infty, \infty)$

NOW TRY

**NOW TRY
EXERCISE 9**

Graph $x = -3y^2 - 6y - 5$.
Give the vertex, axis, domain,
and range.

EXAMPLE 9   Graphing a Horizontal Parabola ($a \neq 1$)

Graph $x = -2y^2 + 4y - 3$. Give the vertex, axis, domain, and range.

$$x = -2y^2 + 4y - 3$$

$$x = (-2y^2 + 4y) - 3 \qquad \text{Group the variable terms.}$$

$$x = -2(y^2 - 2y) - 3 \qquad \text{Factor out } -2.$$

$$x = -2(y^2 - 2y \qquad) - 3 \qquad \text{Prepare to complete the square.}$$

$$\left[\tfrac{1}{2}(-2)\right]^2 = (-1)^2 = 1 \qquad \text{Square half the coefficient of the first-degree term.}$$

$$x = -2(y^2 - 2y + 1 - 1) - 3 \qquad \begin{array}{l}\text{Complete the square within the} \\ \text{parentheses. Add and subtract 1.}\end{array}$$

$$x = -2(y^2 - 2y + 1) + (-2)(-1) - 3 \qquad \text{Distributive property}$$

Be careful here.

$$x = -2(y - 1)^2 - 1 \qquad \text{Factor. Simplify.}$$

The vertex is $(-1, 1)$. Because of the negative coefficient -2 in $x = -2(y - 1)^2 - 1$, the graph opens to the left (the negative *x*-direction). The graph is narrower than the graph of $y = x^2$ because $|-2| = 2$, and $2 > 1$. See **FIGURE 18**.

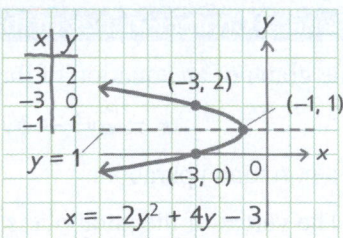
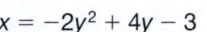

$x = -2y^2 + 4y - 3$
Vertex: $(-1, 1)$
Axis of symmetry: $y = 1$
Domain: $(-\infty, -1\,]$
Range: $(-\infty, \infty)$

FIGURE 18

NOW TRY

⚠ **CAUTION**   *Only quadratic equations that are solved for y (whose graphs are vertical parabolas) are functions.* The horizontal parabolas in **Examples 8 and 9** are *not* graphs of functions because they do not satisfy the conditions of the vertical line test.

▼ **Summary of Graphs of Parabolas**

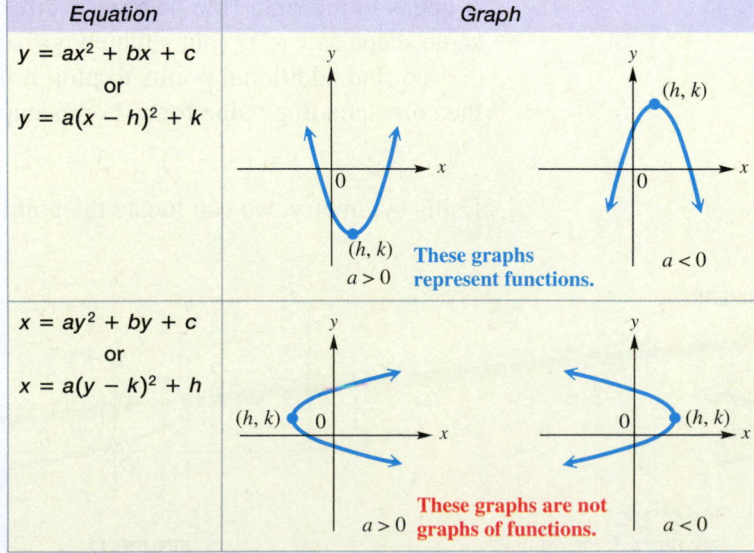

Equation	Graph
$y = ax^2 + bx + c$ or $y = a(x - h)^2 + k$	These graphs represent functions.
$x = ay^2 + by + c$ or $x = a(y - k)^2 + h$	These graphs are not graphs of functions.

NOW TRY ANSWER

9.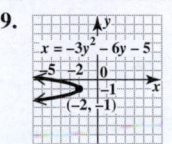

vertex: $(-2, -1)$; axis: $y = -1$;
domain: $(-\infty, -2\,]$; range: $(-\infty, \infty)$

11.7 Exercises

 MyLab Math

 Video solutions for select problems available in MyLab Math

Concept Check *Answer each question.*

1. How can we determine just by looking at the equation of a parabola whether it has a vertical or a horizontal axis?

2. Why can't the graph of a quadratic function be a parabola with a horizontal axis?

3. How can we determine the number of x-intercepts of the graph of a quadratic function without graphing the function?

4. If the vertex of the graph of a quadratic function is $(1, -3)$, and the graph opens down, how many x-intercepts does the graph have?

5. Which equations have a graph that is a vertical parabola?   A horizontal parabola?

    A. $y = -x^2 + 20x + 80$ **B.** $x = 2y^2 + 6y + 5$

    C. $x + 1 = (y + 2)^2$ **D.** $f(x) = (x - 4)^2$

6. Which of the equations in **Exercise 5** represent functions?

Find the vertex of each parabola. See Examples 1–3.

7. $f(x) = x^2 + 8x + 10$ **8.** $f(x) = x^2 + 10x + 23$

9. $f(x) = -2x^2 + 4x - 5$ **10.** $f(x) = -3x^2 + 12x - 8$

11. $f(x) = x^2 + x - 7$ **12.** $f(x) = x^2 - x + 5$

Find the vertex of each parabola. For each equation, decide whether the graph opens up, down, to the left, *or* to the right, *and whether it is* wider, narrower, *or the* same shape *as the graph of* $y = x^2$. *If it is a parabola with a vertical axis of symmetry, find the discriminant and use it to determine the number of x-intercepts.* **See Examples 1–3, 5, 8, and 9.**

13. $f(x) = 2x^2 + 4x + 5$ **14.** $f(x) = 3x^2 - 6x + 4$ **15.** $f(x) = -x^2 + 5x + 3$

16. $f(x) = -x^2 + 7x + 2$ **17.** $x = \dfrac{1}{3}y^2 + 6y + 24$ **18.** $x = \dfrac{1}{2}y^2 + 10y - 5$

Concept Check *Match each equation with its graph in choices A–F.*

19. $y = 2x^2 + 4x - 3$ **20.** $y = -x^2 + 3x + 5$ **21.** $y = -\dfrac{1}{2}x^2 - x + 1$

22. $x = y^2 + 6y + 3$ **23.** $x = -y^2 - 2y + 4$ **24.** $x = 3y^2 + 6y + 5$

A.
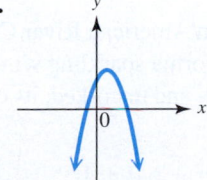

B.
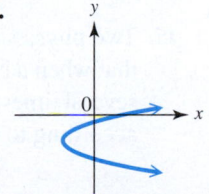

C.
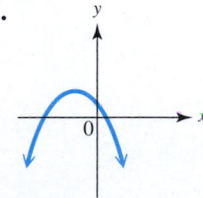

D.
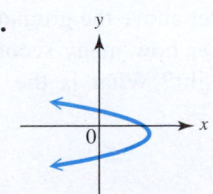

E.
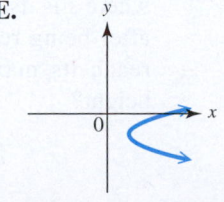

F.
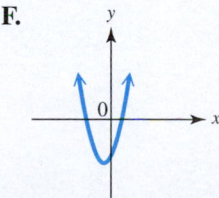

*Graph each parabola. Give the vertex, axis of symmetry, domain, and range. **See Examples 4, 8, and 9.***

25. $f(x) = x^2 + 8x + 10$

26. $f(x) = x^2 + 10x + 23$

27. $f(x) = x^2 + 4x + 3$

28. $f(x) = x^2 + 2x - 2$

29. $f(x) = -2x^2 + 4x - 5$

30. $f(x) = -3x^2 + 12x - 8$

31. $f(x) = -3x^2 - 6x + 2$

32. $f(x) = -2x^2 + 12x - 13$

33. $x = (y + 2)^2 + 1$

34. $x = (y + 3)^2 - 2$

35. $x = -(y - 3)^2 - 1$

36. $x = -(y - 2)^2 + 4$

37. $x = -\frac{1}{5}y^2 + 2y - 4$

38. $x = -\frac{1}{2}y^2 - 4y - 6$

39. $x = 3y^2 + 12y + 5$

40. $x = 4y^2 + 16y + 11$

*Solve each problem. **See Examples 6 and 7.***

41. Find the pair of numbers whose sum is 40 and whose product is a maximum. (*Hint:* Let x and $40 - x$ represent the two numbers.)

42. Find the pair of numbers whose sum is 60 and whose product is a maximum.

43. Polk Community College wants to construct a rectangular parking lot on land bordered on one side by a highway. It has 280 ft of fencing that is to be used to fence off the other three sides. What should be the dimensions of the lot if the enclosed area is to be a maximum? What is the maximum area?

44. Bonnie has 100 ft of fencing material to enclose a rectangular exercise run for her dog. One side of the run will border her house, so she will only need to fence three sides. What dimensions will give the enclosure the maximum area? What is the maximum area?

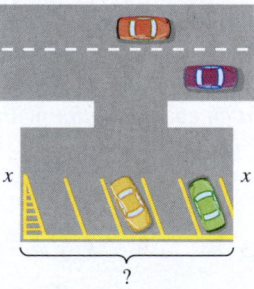

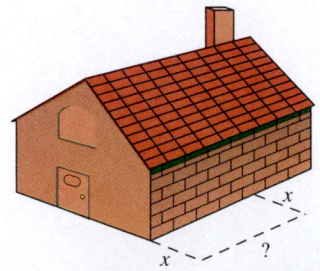

45. Two physics students from American River College find that when a bottle of California sparkling wine is shaken several times, held upright, and uncorked, its cork travels according to the function

$$s(t) = -16t^2 + 64t + 1,$$

where s is its height in feet above the ground t seconds after being released. After how many seconds will it reach its maximum height? What is the maximum height?

46. Professor Barbu has found that the number of students attending his intermediate algebra class is approximated by

$$S(x) = -x^2 + 20x + 80,$$

where x is the number of hours that the Campus Center is open daily. Find the number of hours that the center should be open so that the number of students attending class is a maximum. What is this maximum number of students?

47. Klaus has a taco stand. He has found that his daily costs are approximated by

$$C(x) = x^2 - 40x + 610,$$

where $C(x)$ is the cost, in dollars, to sell x units of tacos. Find the number of units of tacos he should sell to minimize his costs. What is the minimum cost?

48. Mohammad has a frozen yogurt cart. His daily costs are approximated by

$$C(x) = x^2 - 70x + 1500,$$

where $C(x)$ is the cost, in dollars, to sell x units of frozen yogurt. Find the number of units of frozen yogurt he must sell to minimize his costs. What is the minimum cost?

49. If an object on Earth is projected upward with an initial velocity of 32 ft per sec, then its height after t seconds is given by

$$s(t) = -16t^2 + 32t.$$

Find the maximum height attained by the object and the number of seconds it takes to hit the ground.

50. A projectile on Earth is fired straight upward so that its distance (in feet) above the ground t seconds after firing is given by

$$s(t) = -16t^2 + 400t.$$

Find the maximum height it reaches and the number of seconds it takes to reach that height.

51. The graph shows how the average retail price (in dollars per gallon) of regular unleaded gasoline in the United States changed between 2009 and 2016. The graph suggests that a quadratic function would be a good fit to the data. The data are approximated by the function

$$f(x) = -0.1171x^2 + 0.7768x + 2.296.$$

In the model, $x = 0$ represents 2009, $x = 1$ represents 2010, and so on.

(a) How could we have predicted that this quadratic model would have a negative coefficient for x^2, based only on the graph shown?

(b) Algebraically determine the vertex of the graph, with coordinates rounded to the nearest hundredth. Interpret the answer as it applies to this application.

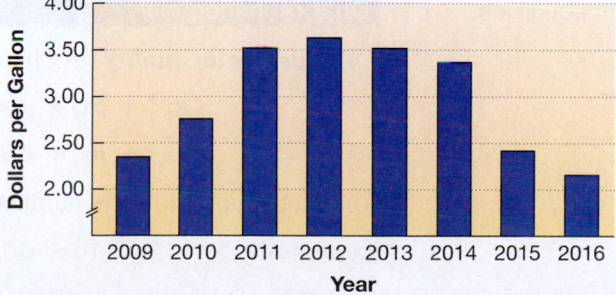

Data from International Energy Agency.

52. The graph shows how Social Security trust fund assets are expected to change and suggests that a quadratic function would be a good fit to the data. The data are approximated by the function

$$f(x) = -20.57x^2 + 758.9x - 3140.$$

In the model, $x = 10$ represents 2010, $x = 15$ represents 2015, and so on, and $f(x)$ is in billions of dollars.

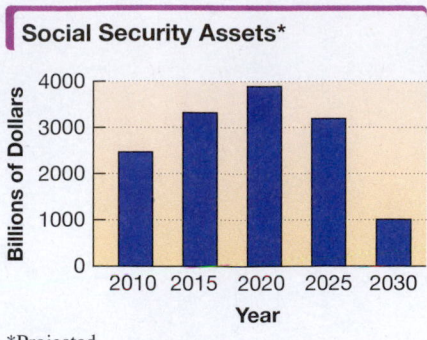

Social Security Assets*

*Projected

Data from Social Security Administration.

(a) How could we have predicted that this quadratic model would have a negative coefficient for x^2, based only on the graph shown?

(b) Algebraically determine the vertex of the graph. Express the x-coordinate to the nearest hundredth and the y-coordinate to the nearest whole number. Interpret the answer as it applies to this application.

Extending Skills *In each problem, find the following.*

(a) *A function $R(x)$ that describes the total revenue received*

(b) *The graph of the function from part (a)*

(c) *The number of unsold seats that will produce the maximum revenue*

(d) *The maximum revenue*

53. A charter flight charges a fare of $200 per person, plus $4 per person for each unsold seat on the plane. The plane holds 100 passengers. Let x represent the number of unsold seats. (*Hint:* To find $R(x)$, multiply the number of people flying, $100 - x$, by the price per ticket, $200 + 4x$.)

54. A charter bus charges a fare of $48 per person, plus $2 per person for each unsold seat on the bus. The bus has 42 seats. Let x represent the number of unsold seats. (*Hint:* To find $R(x)$, multiply the number riding, $42 - x$, by the price per ticket, $48 + 2x$.)

11.8 Polynomial and Rational Inequalities

OBJECTIVES

1 Solve quadratic inequalities.

2 Solve polynomial inequalities of degree 3 or greater.

3 Solve rational inequalities.

OBJECTIVE 1 Solve quadratic inequalities.

We can combine methods of solving linear inequalities and methods of solving quadratic equations to solve *quadratic inequalities.*

Quadratic Inequality

A **quadratic inequality** (in x here) is an inequality that can be written in the form

$$ax^2 + bx + c < 0, \qquad ax^2 + bx + c > 0,$$
$$ax^2 + bx + c \leq 0, \quad \text{or} \quad ax^2 + bx + c \geq 0,$$

where a, b, and c are real numbers and $a \neq 0$.

Examples: $x^2 - 3x - 10 < 0$, $6x^2 + x \geq 1$ Quadratic inequalities

VOCABULARY
☐ quadratic inequality
☐ rational inequality

**NOW TRY
EXERCISE 1**

Use the graph to solve each quadratic inequality.

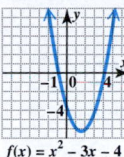

$f(x) = x^2 - 3x - 4$

(a) $x^2 - 3x - 4 > 0$
(b) $x^2 - 3x - 4 < 0$

One way to solve a quadratic inequality involves graphing the related quadratic function. This method is justified because the graph is *continuous*—that is, it has no breaks.

EXAMPLE 1   Solving Quadratic Inequalities by Graphing

Solve each inequality.

(a) $x^2 - x - 12 > 0$

We graph the related quadratic function $f(x) = x^2 - x - 12$. We are particularly interested in the *x*-intercepts, which are found by letting $f(x) = 0$ and solving the following quadratic equation.

$$x^2 - x - 12 = 0 \qquad \text{Let } f(x) = 0.$$
$$(x - 4)(x + 3) = 0 \qquad \text{Factor.}$$
$$x - 4 = 0 \quad \text{or} \quad x + 3 = 0 \qquad \text{Zero-factor property}$$
$$x = 4 \quad \text{or} \quad x = -3 \leftarrow \text{The } x\text{-intercepts are } (4, 0) \text{ and } (-3, 0).$$

The graph opens up because the coefficient of x^2 is positive. See **FIGURE 19(a)**. Notice that *x*-values less than -3 or greater than 4 result in *y*-values *greater than* 0. Thus, the solution set of $x^2 - x - 12 > 0$, written in interval notation, is

$$(-\infty, -3) \cup (4, \infty).$$

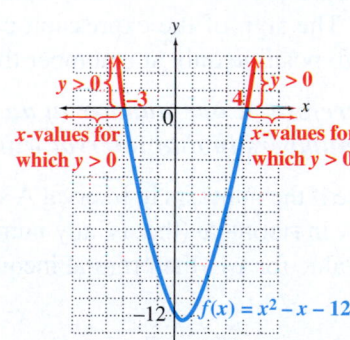

The graph is *above* the *x*-axis for
$(-\infty, -3) \cup (4, \infty)$.

(a)

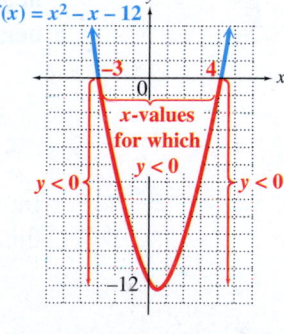

The graph is *below* the *x*-axis for
$(-3, 4)$.

(b)

FIGURE 19

(b) $x^2 - x - 12 < 0$

We want values of *y* that are *less than* 0. See **FIGURE 19(b)**. Notice from the graph that *x*-values between -3 and 4 result in *y*-values less than 0. Thus, the solution set of $x^2 - x - 12 < 0$, written in interval notation, is $(-3, 4)$.   **NOW TRY**

> **NOTE** If the inequalities in **Example 1** had used $\geq$ and $\leq$, the solution sets would have included the *x*-values of the intercepts, which make the quadratic expression equal to 0. They would have been written in interval notation as
>
> $$(-\infty, -3] \cup [4, \infty) \quad \text{and} \quad [-3, 4].$$
>
> Square brackets would indicate that the endpoints -3 and 4 are *included* in the solution sets.

NOW TRY ANSWERS
1. (a) $(-\infty, -1) \cup (4, \infty)$
   (b) $(-1, 4)$

Another method for solving a quadratic inequality uses the basic ideas of **Example 1** without actually graphing the related quadratic function.

EXAMPLE 2   Solving a Quadratic Inequality Using Test Values

Solve and graph the solution set of $x^2 - x - 12 > 0$.

Solve the quadratic equation $x^2 - x - 12 = 0$. (See **Example 1(a)**.)

$$x^2 - x - 12 = 0 \qquad \text{Let } f(x) = 0.$$
$$(x - 4)(x + 3) = 0 \qquad \text{Factor.}$$
$$x - 4 = 0 \quad \text{or} \quad x + 3 = 0 \qquad \text{Zero-factor property}$$
$$x = 4 \quad \text{or} \qquad x = -3 \qquad \text{Solve each equation.}$$

The numbers 4 and -3 divide a number line into Intervals A, B, and C, as shown in **FIGURE 20**. *Be careful to put the lesser number on the left.*

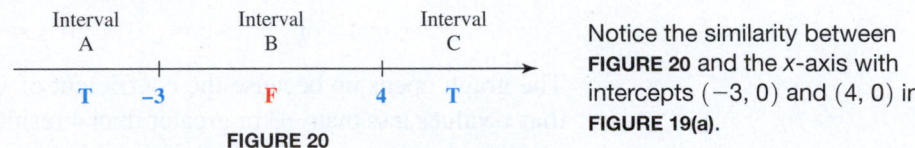

FIGURE 20

Notice the similarity between **FIGURE 20** and the x-axis with intercepts $(-3, 0)$ and $(4, 0)$ in **FIGURE 19(a)**.

The numbers 4 and -3 are the only values that make the quadratic expression $x^2 - x - 12$ equal to 0. All other numbers make the expression either positive or negative. The sign of the expression can change from positive to negative or from negative to positive only at a number that makes it 0.

Therefore, if one number in an interval satisfies the inequality, then all numbers in that interval will satisfy the inequality.

To see if the numbers in Interval A satisfy the inequality, choose any number from Interval A in **FIGURE 20** (that is, any number less than -3). We choose -5. Substitute this test value for x in the original inequality $x^2 - x - 12 > 0$.

$$x^2 - x - 12 > 0 \qquad \text{Original inequality}$$
$$(-5)^2 - (-5) - 12 \overset{?}{>} 0 \qquad \text{Let } x = -5.$$
$$25 + 5 - 12 \overset{?}{>} 0 \qquad \text{Simplify.}$$
$$18 > 0 \qquad \text{True}$$

> Use parentheses to avoid sign errors.

Because -5 satisfies the inequality, *all* numbers from Interval A are solutions, indicated by T in **FIGURE 20**.

Now try 0 from Interval B.

$$x^2 - x - 12 > 0 \qquad \text{Original inequality}$$
$$0^2 - 0 - 12 \overset{?}{>} 0 \qquad \text{Let } x = 0.$$
$$-12 > 0 \qquad \text{False}$$

The numbers in Interval B are *not* solutions, indicated by F in **FIGURE 20**.

Now try 5 from Interval C.

$$x^2 - x - 12 > 0 \qquad \text{Original inequality}$$
$$5^2 - 5 - 12 \overset{?}{>} 0 \qquad \text{Let } x = 5.$$
$$8 > 0 \qquad \text{True}$$

All numbers from Interval C are solutions, indicated by T in **FIGURE 20**.

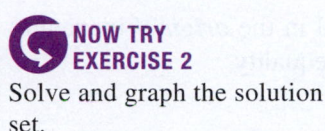

NOW TRY EXERCISE 2

Solve and graph the solution set.

$$x^2 + 2x - 8 > 0$$

Based on these results (shown by the colored letters in **FIGURE 20**), the solution set includes all numbers in Intervals A and C, as shown in **FIGURE 21**.

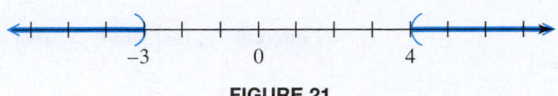

FIGURE 21

The solution set is written in interval notation as

$$(-\infty, -3) \cup (4, \infty).$$

-3 and 4 are **not** included because the symbol $>$ does not include equality.

This agrees with the solution set found in **Example 1(a)**.   **NOW TRY** ↺

Solving a Quadratic Inequality

Step 1 **Write the inequality as an equation and solve it.**

Step 2 **Use the solutions from Step 1 to determine intervals.** Graph the values found in Step 1 on a number line. These values divide the number line into intervals.

Step 3 **Find the intervals that satisfy the inequality.** Substitute a test value from each interval into the original inequality to determine the intervals that satisfy the inequality. All numbers in those intervals are in the solution set. A graph of the solution set will usually look like one of these.

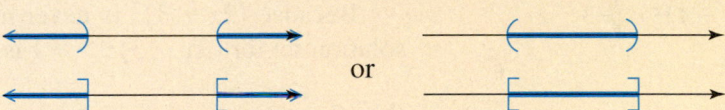

or

Step 4 **Consider the endpoints separately.** The values from Step 1 are included in the solution set if the inequality symbol is $\leq$ or $\geq$. They are not included if it is $<$ or $>$.

EXAMPLE 3   Solving a Quadratic Inequality

Step 1 Solve and graph the solution set of $2x^2 + 5x \leq 12$.

$$2x^2 + 5x = 12 \qquad \text{Related quadratic equation}$$

$$2x^2 + 5x - 12 = 0 \qquad \text{Standard form}$$

$$(2x - 3)(x + 4) = 0 \qquad \text{Factor.}$$

$$2x - 3 = 0 \quad \text{or} \quad x + 4 = 0 \qquad \text{Zero-factor property}$$

$$x = \frac{3}{2} \quad \text{or} \qquad x = -4 \qquad \text{Solve each equation.}$$

Step 2 The numbers $\frac{3}{2}$ and -4 divide a number line into three intervals. See **FIGURE 22**.

NOW TRY ANSWER

2.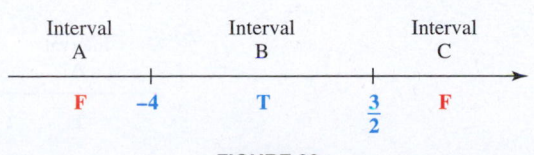

$(-\infty, -4) \cup (2, \infty)$

Interval A	Interval B	Interval C

F -4 T $\frac{3}{2}$ F

FIGURE 22

NOW TRY
EXERCISE 3
Solve and graph the solution set.

$$3x^2 - 11x \le 4$$

Steps 3 and 4   Substitute a test value from each interval in the *original* inequality $2x^2 + 5x \le 12$ to determine which intervals satisfy the inequality.

Interval	Test Value	Test of Inequality	True or False?
A	-5	$25 \le 12$	F
B	0	$0 \le 12$	T
C	2	$18 \le 12$	F

We use a table to organize this information. (Verify it.)

The numbers in Interval B are solutions. See **FIGURE 23**.

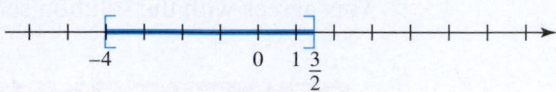

FIGURE 23

The solution set is the interval $\left[-4, \frac{3}{2}\right]$. -4 and $\frac{3}{2}$ are included because the symbol $\le$ includes equality.

NOW TRY

NOW TRY
EXERCISE 4
Solve each inequality.

(a) $(4x - 1)^2 > -3$

(b) $(4x - 1)^2 < -3$

EXAMPLE 4   Solving Special Cases

Solve each inequality.

(a) $(2x - 3)^2 > -1$

Because $(2x - 3)^2$ is never negative, it is always greater than -1. Thus, the solution set for $(2x - 3)^2 > -1$ is the set of all real numbers, $(-\infty, \infty)$.

(b) $(2x - 3)^2 < -1$

Using similar reasoning as in part (a), there is no solution for this inequality. The solution set is $\varnothing$.
NOW TRY

OBJECTIVE 2   Solve polynomial inequalities of degree 3 or greater.

EXAMPLE 5   Solving a Third-Degree Polynomial Inequality

Solve and graph the solution set of $(x - 1)(x + 2)(x - 4) \le 0$.

This *cubic* (third-degree) inequality can be solved by extending the zero-factor property to more than two factors. (Step 1)

$$(x - 1)(x + 2)(x - 4) = 0$$ Set the factored polynomial *equal* to 0.

$x - 1 = 0$ or $x + 2 = 0$ or $x - 4 = 0$ Zero-factor property

$x = 1$ or $x = -2$ or $x = 4$ Solve each equation.

Locate the numbers -2, 1, and 4 on a number line, as in **FIGURE 24**, to determine the Intervals A, B, C, and D. (Step 2)

NOW TRY ANSWERS

3.

$$\left[-\frac{1}{3}, 4\right]$$

4. (a) $(-\infty, \infty)$ **(b)** $\varnothing$

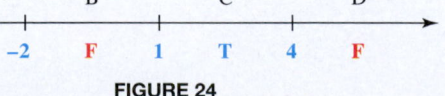

FIGURE 24

**NOW TRY
EXERCISE 5**

Solve and graph the solution set.

$(x + 4)(x - 3)(2x + 1) \leq 0$

Substitute a test value from each interval in the *original* inequality to determine which intervals satisfy $(x - 1)(x + 2)(x - 4) \leq 0$. (Step 3)

Interval	Test Value	Test of Inequality	True or False?
A	−3	−28 ≤ 0	T
B	0	8 ≤ 0	F
C	2	−8 ≤ 0	T
D	5	28 ≤ 0	F

The numbers in Intervals A and C are in the solution set. The three endpoints are included because the inequality symbol ≤ includes equality. (Step 4) See **FIGURE 25.**

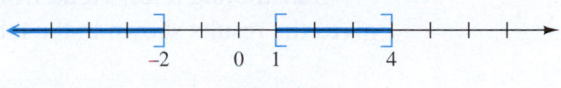

FIGURE 25

The solution set is the interval $(-\infty, -2] \cup [1, 4]$.

NOW TRY

OBJECTIVE 3 Solve rational inequalities.

Rational inequalities involve rational expressions and are solved similarly.

Solving a Rational Inequality

Step 1 **Write the inequality so that 0 is on one side** and there is a single fraction on the other side.

Step 2 **Determine the values that make the numerator or denominator equal to 0.**

Step 3 **Divide a number line into intervals.** Use the values from Step 2.

Step 4 **Find the intervals that satisfy the inequality.** Test a value from each interval by substituting it into the *original* inequality.

Step 5 **Consider the endpoints separately.** Exclude any values that make the denominator 0.

EXAMPLE 6 Solving a Rational Inequality

Solve and graph the solution set of $\dfrac{-1}{x - 3} > 1$.

Write the inequality so that 0 is on one side. (Step 1)

$$\frac{-1}{x - 3} - 1 > 0 \qquad \text{Subtract 1.}$$

$$\frac{-1}{x - 3} - \frac{x - 3}{x - 3} > 0 \qquad \text{Use } x - 3 \text{ as the common denominator.}$$

$$\frac{-1 - (x - 3)}{x - 3} > 0 \qquad \text{Write the left side as a single fraction.}$$

$$\frac{-1 - x + 3}{x - 3} > 0 \qquad \text{Distributive property}$$

Be careful with signs.

$$\frac{-x + 2}{x - 3} > 0 \qquad \text{Combine like terms in the numerator.}$$

NOW TRY ANSWER

5.

$(-\infty, -4] \cup \left[-\frac{1}{2}, 3\right]$

NOW TRY
EXERCISE 6

Solve and graph the solution set.

$$\frac{3}{x+1} > 4$$

The sign of $\frac{-x+2}{x-3}$ will change from positive to negative or negative to positive only at those values that make the numerator or denominator 0. The number 2 makes the numerator 0, and 3 makes the denominator 0. (Step 2) These two numbers, 2 and 3, divide a number line into three intervals. See **FIGURE 26**. (Step 3)

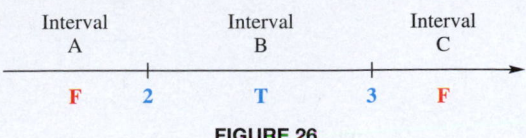

FIGURE 26

Substituting a test value from each interval in the *original* inequality, $\frac{-1}{x-3} > 1$, gives the results shown in the table. (Step 4)

Interval	Test Value	Test of Inequality	True or False?
A	0	$\frac{1}{3} > 1$	F
B	2.5	$2 > 1$	T
C	4	$-1 > 1$	F

The numbers in Interval B are solutions. This interval does not include 3 because it makes the denominator in the original inequality 0. The number 2 is not included either because the inequality symbol $>$ does not include equality. (Step 5) See **FIGURE 27**.

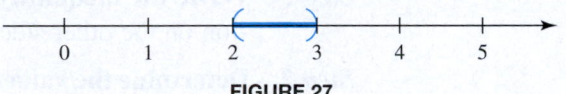

FIGURE 27

The solution set is the interval $(2, 3)$.

NOW TRY

> ⚠ **CAUTION** When solving a rational inequality, any number that makes the denominator 0 must be excluded from the solution set.

EXAMPLE 7   Solving a Rational Inequality

Solve and graph the solution set of $\frac{x-2}{x+2} \le 2$.

Write the inequality so that 0 is on one side. (Step 1)

$$\frac{x-2}{x+2} - 2 \le 0 \qquad \text{Subtract 2.}$$

$$\frac{x-2}{x+2} - \frac{2(x+2)}{x+2} \le 0 \qquad \text{Use } x+2 \text{ as the common denominator.}$$

$$\frac{x-2-2(x+2)}{x+2} \le 0 \qquad \text{Write as a single fraction.}$$

$$\frac{x-2-2x-4}{x+2} \le 0 \qquad \text{Distributive property}$$

Be careful with signs

$$\frac{-x-6}{x+2} \le 0 \qquad \text{Combine like terms in the numerator.}$$

NOW TRY ANSWER

6.

$-1 \quad -\frac{1}{4} \quad 0$

$\left(-1, -\frac{1}{4}\right)$

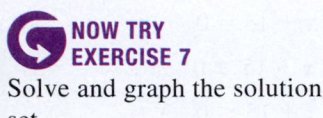

NOW TRY
EXERCISE 7

Solve and graph the solution set.

$$\frac{x-3}{x+3} \le 2$$

In the inequality $\frac{-x-6}{x+2} \le 0$, the number -6 makes the numerator 0, and -2 makes the denominator 0. (Step 2) These two numbers determine three intervals on a number line. See **FIGURE 28.** (Step 3)

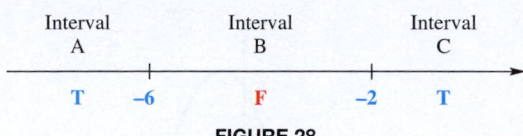

FIGURE 28

Substitute a test value from each interval in the *original* inequality $\frac{x-2}{x+2} \le 2$. (Step 4)

Interval	Test Value	Test of Inequality	True or False?
A	-8	$\frac{5}{3} \le 2$	T
B	-4	$3 \le 2$	F
C	0	$-1 \le 2$	T

The numbers in Intervals A and C are solutions. The number -6 satisfies the original inequality, but -2 does not because it makes the denominator 0. (Step 5) See **FIGURE 29.**

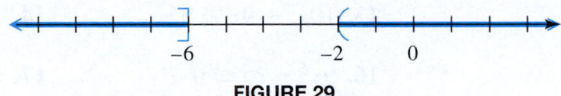

FIGURE 29

NOW TRY ANSWER

7.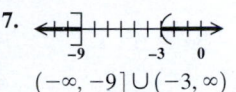

$(-\infty, -9] \cup (-3, \infty)$

The solution set is the interval $(-\infty, -6] \cup (-2, \infty)$.

NOW TRY

11.8 Exercises

FOR EXTRA HELP   ▶ **MyLab Math**

▶ *Video solutions for select problems available in MyLab Math*

1. Concept Check   The solution set of the inequality $x^2 + x - 12 < 0$ is the interval $(-4, 3)$. Without actually performing any work, give the solution set of the inequality

$$x^2 + x - 12 \ge 0.$$

2. Explain how to determine whether to include or to exclude endpoints when solving a quadratic or higher-degree inequality.

In each exercise, the graph of a quadratic function f is given. Use the graph to find the solution set of each equation or inequality. **See Example 1.**

3. (a) $x^2 - 4x + 3 = 0$

  (b) $x^2 - 4x + 3 > 0$

  (c) $x^2 - 4x + 3 < 0$

4. (a) $3x^2 + 10x - 8 = 0$

  (b) $3x^2 + 10x - 8 \ge 0$

  (c) $3x^2 + 10x - 8 < 0$

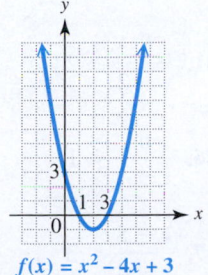

$f(x) = x^2 - 4x + 3$

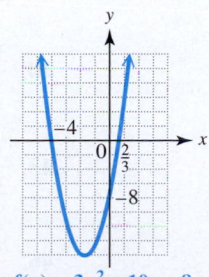

$f(x) = 3x^2 + 10x - 8$

5. (a) $-x^2 + 3x + 10 = 0$

(b) $-x^2 + 3x + 10 \geq 0$

(c) $-x^2 + 3x + 10 \leq 0$

6. (a) $-2x^2 - x + 15 = 0$

(b) $-2x^2 - x + 15 \geq 0$

(c) $-2x^2 - x + 15 \leq 0$

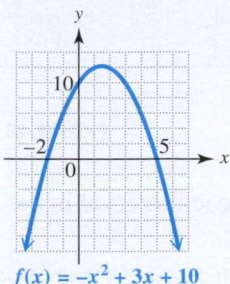

$f(x) = -x^2 + 3x + 10$

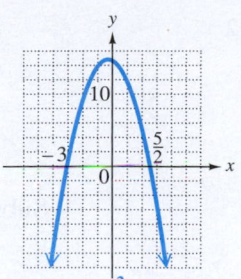

$f(x) = -2x^2 - x + 15$

Solve each inequality, and graph the solution set. **See Examples 2 and 3.** *(Hint: In Exercises 23 and 24, use the quadratic formula.)*

7. $(x + 1)(x - 5) > 0$

8. $(x + 6)(x - 2) > 0$

9. $(x + 4)(x - 6) < 0$

10. $(x + 4)(x - 8) < 0$

11. $x^2 - 4x + 3 \geq 0$

12. $x^2 - 3x - 10 \geq 0$

13. $10x^2 + 9x \geq 9$

14. $3x^2 + 10x \geq 8$

15. $4x^2 - 9 \leq 0$

16. $9x^2 - 25 \leq 0$

17. $6x^2 + x \geq 1$

18. $4x^2 + 7x \geq -3$

19. $z^2 - 4z \geq 0$

20. $x^2 + 2x < 0$

21. $3x^2 - 5x \leq 0$

22. $2z^2 + 3z > 0$

23. $x^2 - 6x + 6 \geq 0$

24. $3x^2 - 6x + 2 \leq 0$

Solve each inequality. **See Example 4.**

25. $(4 - 3x)^2 \geq -2$

26. $(7 - 6x)^2 \geq -1$

27. $(3x + 5)^2 \leq -4$

28. $(8x + 5)^2 \leq -5$

29. $(2x + 5)^2 < 0$

30. $(3x - 7)^2 < 0$

31. $(5x - 1)^2 \geq 0$

32. $(4x + 1)^2 \geq 0$

Solve each inequality, and graph the solution set. **See Example 5.**

33. $(x - 1)(x - 2)(x - 4) < 0$

34. $(2x + 1)(3x - 2)(4x + 7) < 0$

35. $(x - 4)(2x + 3)(3x - 1) \geq 0$

36. $(x + 2)(4x - 3)(2x + 7) \geq 0$

Solve each inequality, and graph the solution set. **See Examples 6 and 7.**

37. $\dfrac{x - 1}{x - 4} > 0$

38. $\dfrac{x + 1}{x - 5} > 0$

39. $\dfrac{2x + 3}{x - 5} \leq 0$

40. $\dfrac{3x + 7}{x - 3} \leq 0$

41. $\dfrac{8}{x - 2} \geq 2$

42. $\dfrac{20}{x - 1} \geq 1$

43. $\dfrac{3}{2x - 1} < 2$

44. $\dfrac{6}{x - 1} < 1$

45. $\dfrac{x - 3}{x + 2} \geq 2$

46. $\dfrac{m + 4}{m + 5} \geq 2$

47. $\dfrac{x - 8}{x - 4} < 3$

48. $\dfrac{2t - 3}{t + 1} > 4$

49. $\dfrac{4k}{2k-1} < k$

50. $\dfrac{r}{r+2} < 2r$

51. $\dfrac{2x-3}{x^2+1} \geq 0$

52. $\dfrac{9x-8}{4x^2+25} < 0$

53. $\dfrac{(3x-5)^2}{x+2} > 0$

54. $\dfrac{(5x-3)^2}{2x+1} \leq 0$

RELATING CONCEPTS For Individual or Group Work (Exercises 55–58)

A model rocket is projected vertically upward from the ground. Its distance s in feet above the ground after t seconds is given by the quadratic function

$$s(t) = -16t^2 + 256t.$$

Work Exercises 55–58 in order, to see how quadratic equations and inequalities are related.

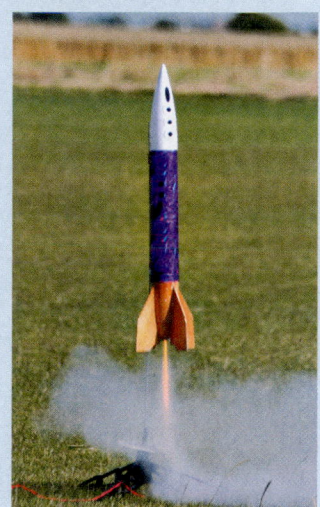

55. At what times will the rocket be 624 ft above the ground? (*Hint:* Let $s(t) = 624$ and solve the quadratic *equation.*)

56. At what times will the rocket be more than 624 ft above the ground? (*Hint:* Let $s(t) > 624$ and solve the quadratic *inequality.*)

57. At what times will the rocket be at ground level? (*Hint:* Let $s(t) = 0$ and solve the quadratic *equation.*)

58. At what times will the rocket be less than 624 ft above the ground? (*Hint:* Let $s(t) < 624$, solve the quadratic *inequality,* and observe the solutions in **Exercises 56 and 57** to determine the least and greatest possible values of t.)

Chapter 11   Summary

Key Terms

11.1
quadratic equation
second-degree equation

11.3
quadratic formula
discriminant

11.4
quadratic in form

11.6
parabola
vertex
axis of symmetry (axis)
quadratic function

11.8
quadratic inequality
rational inequality

Test Your Word Power

See how well you have learned the vocabulary in this chapter.

---

1. The **quadratic formula** is
 A. a formula to find the number of solutions of a quadratic equation
 B. a formula to find the type of solutions of a quadratic equation
 C. the standard form of a quadratic equation
 D. a general formula for solving any quadratic equation.

2. A **quadratic function** is a function that can be written in the form
 A. $f(x) = mx + b$, for real numbers m and b
 B. $f(x) = \dfrac{P(x)}{Q(x)}$, where $Q(x) \neq 0$
 C. $f(x) = ax^2 + bx + c$, for real numbers a, b, and c $(a \neq 0)$
 D. $f(x) = \sqrt{x}$, for $x \geq 0$.

3. A **parabola** is the graph of
 A. any equation in two variables
 B. a linear equation
 C. an equation of degree 3
 D. a quadratic equation in two variables.

4. The **vertex** of a parabola is
   A. the point where the graph intersects the y-axis
   B. the point where the graph intersects the x-axis
   C. the lowest point on a parabola that opens up or the highest point on a parabola that opens down
   D. the origin.

5. The **axis of symmetry** of a parabola is
   A. either the x-axis or the y-axis
   B. the vertical line (of a vertical parabola) or the horizontal line (of a horizontal parabola) through the vertex
   C. the lowest or highest point on the graph of a parabola
   D. a line through the origin.

6. A parabola is **symmetric about its axis** because its graph
   A. is near the axis
   B. is identical on each side of the axis
   C. looks different on each side of the axis
   D. intersects the axis.

ANSWERS

1. D; *Example:* The solutions of $ax^2 + bx + c = 0$ ($a \neq 0$) are given by $x = \dfrac{-b \pm \sqrt{b^2 - 4ac}}{2a}$. 2. C; *Examples:* $f(x) = x^2 - 2$,

$f(x) = (x + 4)^2 + 1$, $f(x) = x^2 - 4x + 5$ 3. D; *Example:* The quadratic equation $y = x^2$ has a parabola as its graph. 4. C; *Example:* The graph of $y = (x + 3)^2$ has vertex $(-3, 0)$, which is the lowest point on the graph. 5. B; *Example:* The axis of $y = (x + 3)^2$ is the vertical line $x = -3$.
6. B; *Example:* Because the graph of $y = (x + 3)^2$ is symmetric about its axis $x = -3$, the points $(-2, 1)$ and $(-4, 1)$ are on the graph.

Quick Review

CONCEPTS	EXAMPLES

11.1 Solving Quadratic Equations by the Square Root Property

Square Root Property
If x and k are complex numbers and $x^2 = k$, then

$$x = \sqrt{k} \quad \text{or} \quad x = -\sqrt{k}.$$

Solve $(x - 1)^2 = 8$.

$$x - 1 = \sqrt{8} \qquad \text{or} \qquad x - 1 = -\sqrt{8}$$

$$x = 1 + 2\sqrt{2} \quad \text{or} \qquad x = 1 - 2\sqrt{2}$$

Solution set: $\{1 + 2\sqrt{2}, 1 - 2\sqrt{2}\}$, or $\{1 \pm 2\sqrt{2}\}$

11.2 Solving Quadratic Equations by Completing the Square

Completing the Square
To solve $ax^2 + bx + c = 0$ (where $a \neq 0$), follow these steps.

Step 1 If $a \neq 1$, divide each side by a.

Step 2 Write the equation with the variable terms on one side and the constant on the other.

Step 3 Complete the square.
- Take half the coefficient of x and square it.
- Add the square to each side.
- Factor the perfect square trinomial, and write it as the square of a binomial. Combine terms on the other side.

Step 4 Use the square root property to solve.

Solve $2x^2 - 4x - 18 = 0$.

$$x^2 - 2x - 9 = 0 \qquad \text{Divide by 2.}$$

$$x^2 - 2x = 9 \qquad \text{Add 9.}$$

$$\left[\tfrac{1}{2}(-2)\right]^2 = (-1)^2 = 1$$

$$x^2 - 2x + 1 = 9 + 1 \qquad \text{Add 1.}$$

$$(x - 1)^2 = 10 \qquad \text{Factor. Add.}$$

$$x - 1 = \sqrt{10} \qquad \text{or} \quad x - 1 = -\sqrt{10} \qquad \text{Square root property}$$

$$x = 1 + \sqrt{10} \quad \text{or} \qquad x = 1 - \sqrt{10}$$

Solution set: $\{1 + \sqrt{10}, 1 - \sqrt{10}\}$, or $\{1 \pm \sqrt{10}\}$

11.3 Solving Quadratic Equations by the Quadratic Formula

Quadratic Formula
The solutions of $ax^2 + bx + c = 0$ (where $a \neq 0$) are given by

$$x = \frac{-b \pm \sqrt{b^2 - 4ac}}{2a}.$$

Solve $3x^2 + 5x + 2 = 0$.

$$x = \frac{-5 \pm \sqrt{5^2 - 4(3)(2)}}{2(3)} \qquad a = 3, b = 5, c = 2$$

$$x = \frac{-5 \pm 1}{6} \qquad \text{Simplify.}$$

$$x = -\frac{2}{3} \quad \text{or} \quad x = -1 \qquad \text{Two solutions, one from + and one from −}$$

Solution set: $\left\{-1, -\frac{2}{3}\right\}$

CONCEPTS	EXAMPLES

Using the Discriminant

The discriminant $b^2 - 4ac$ of $ax^2 + bx + c = 0$ (where a, b, and c are integers) can be used to determine the number and type of solutions.

Discriminant $b^2 - 4ac$	Number and Type of Solutions
Positive, and the square of an integer	Two rational solutions
Positive, but not the square of an integer	Two irrational solutions
Zero	One rational solution
Negative	Two nonreal complex solutions

For $x^2 + 3x - 10 = 0$, the discriminant is as follows.

$$b^2 - 4ac$$
$$= 3^2 - 4(1)(-10) \qquad a = 1, b = 3, c = -10$$
$$= 49 \qquad \text{There are two rational solutions.}$$
$$= 7^2 \qquad \text{Simplify.}$$

Because the discriminant is a perfect square, the quadratic equation can be solved using the zero-factor property.

11.4 Equations That Lead to Quadratic Methods

A nonquadratic equation that can be written in the form

$$au^2 + bu + c = 0,$$

for $a \neq 0$ and an algebraic expression u, is quadratic in form.

Solving an Equation Quadratic in Form by Substitution

Step 1 Define a temporary variable u.

Step 2 Solve the quadratic equation obtained in Step 1.

Step 3 Replace u with the expression it defined.

Step 4 Solve the resulting equations for the original variable.

Step 5 Check all solutions in the original equation.

Solve $3(x + 5)^2 + 7(x + 5) + 2 = 0$.

$$3u^2 + 7u + 2 = 0 \qquad \text{Let } u = x + 5.$$
$$(3u + 1)(u + 2) = 0 \qquad \text{Factor.}$$
$$3u + 1 = 0 \quad \text{or} \quad u + 2 = 0 \qquad \text{Zero-factor property}$$
$$u = -\frac{1}{3} \quad \text{or} \quad u = -2 \qquad \text{Solve for } u.$$
$$x + 5 = -\frac{1}{3} \quad \text{or} \quad x + 5 = -2 \qquad \text{Replace } u \text{ with } x + 5.$$
$$x = -\frac{16}{3} \quad \text{or} \quad x = -7 \qquad \text{Subtract 5.}$$

Solution set: $\left\{-7, -\frac{16}{3}\right\}$

11.5 Formulas and Further Applications

Solving a Formula for a Squared Variable

Case 1 **If the variable appears only to the second power:** Isolate the second-degree variable on one side of the equation, and then use the square root property.

Case 2 **If the variable appears to the first and second powers:** Write the equation in standard form, and then use the quadratic formula.

Solve $A = \dfrac{2mp}{r^2}$ for r. (*Case 1*)

$$r^2 A = 2mp \qquad \text{Multiply by } r^2.$$
$$r^2 = \frac{2mp}{A} \qquad \text{Divide by } A.$$
$$r = \pm\sqrt{\frac{2mp}{A}} \qquad \text{Square root property}$$
$$r = \frac{\pm\sqrt{2mpA}}{A} \qquad \text{Rationalize denominator.}$$

Solve $x^2 + rx = t$ for x. (*Case 2*)

$$x^2 + rx - t = 0 \qquad \text{Standard form}$$
$$x = \frac{-r \pm \sqrt{r^2 - 4(1)(-t)}}{2(1)} \qquad a = 1, b = r, c = -t$$
$$x = \frac{-r \pm \sqrt{r^2 + 4t}}{2} \qquad \text{Simplify.}$$

CONCEPTS	EXAMPLES

11.6 Graphs of Quadratic Functions

The graph of the quadratic function $F(x) = a(x - h)^2 + k$ (where $a \neq 0$) is a parabola.

- The vertex of the parabola is (h, k).
- The axis of symmetry is the vertical line $x = h$.
- The graph opens up if $a > 0$ and down if $a < 0$.
- The graph is wider than the graph of $f(x) = x^2$ if $0 < |a| < 1$ and narrower if $|a| > 1$.

Graph $f(x) = -(x + 3)^2 + 1$.

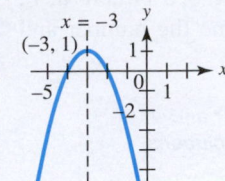

The graph opens down because $a < 0$.

Vertex: $(-3, 1)$

Axis of symmetry: $x = -3$

Domain: $(-\infty, \infty)$

Range: $(-\infty, 1]$

11.7 More about Parabolas and Their Applications

The vertex of the graph of $f(x) = ax^2 + bx + c$ (where $a \neq 0$) may be found by completing the square or by using the vertex formula $\left(\frac{-b}{2a}, f\left(\frac{-b}{2a} \right) \right)$.

Graphing a Quadratic Function

Step 1 Determine whether the graph opens up or down.

Step 2 Find the vertex.

Step 3 Find any intercepts.

Step 4 Find and plot additional points as needed.

Horizontal Parabolas

The graph of an equation of the form

$$x = ay^2 + by + c \quad \text{or} \quad x = a(y - k)^2 + h$$

is a horizontal parabola.

- The vertex of the parabola is (h, k)
- The axis of symmetry is the horizontal line $y = k$.
- The graph opens to the right if $a > 0$ and to the left if $a < 0$.

Horizontal parabolas do not represent functions.

Graph $f(x) = x^2 + 4x + 3$.

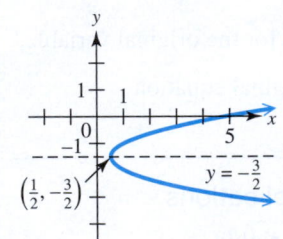

The graph opens up because $a > 0$.

Vertex: $(-2, -1)$

The solutions of $x^2 + 4x + 3 = 0$ are -1 and -3, so the x-intercepts are $(-1, 0)$ and $(-3, 0)$.

$f(0) = 3$, so the y-intercept is $(0, 3)$.

Axis of symmetry: $x = -2$

Domain: $(-\infty, \infty)$

Range: $[-1, \infty)$

Graph $x = 2y^2 + 6y + 5$.

The graph opens to the right because $a > 0$.

Vertex: $\left(\frac{1}{2}, -\frac{3}{2} \right)$

Axis of symmetry: $y = -\frac{3}{2}$

Domain: $\left[\frac{1}{2}, \infty \right)$

Range: $(-\infty, \infty)$

11.8 Polynomial and Rational Inequalities

Solving a Quadratic (or Higher-Degree Polynomial) Inequality

Step 1 Write the inequality as an equation and solve it.

Step 2 Use the values found in Step 1 to divide a number line into intervals.

Step 3 Substitute a test value from each interval into the *original* inequality to determine the intervals that belong to the solution set.

Step 4 Consider the endpoints separately.

Solve $2x^2 + 5x + 2 < 0$.

$$2x^2 + 5x + 2 = 0 \qquad \text{Related equation}$$

$$(2x + 1)(x + 2) = 0 \qquad \text{Factor.}$$

$$2x + 1 = 0 \quad \text{or} \quad x + 2 = 0 \qquad \text{Zero-factor property}$$

$$x = -\frac{1}{2} \quad \text{or} \qquad x = -2 \qquad \text{Solve each equation.}$$

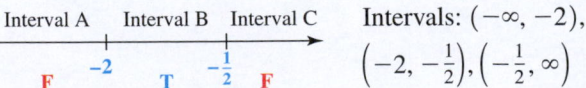

Intervals: $(-\infty, -2)$, $\left(-2, -\frac{1}{2} \right), \left(-\frac{1}{2}, \infty \right)$

Test values: -3 (Interval A), -1 (Interval B), 0 (Interval C); $x = -3$ makes the original inequality false, $x = -1$ makes it true, and $x = 0$ makes it false.

Solution set: $\left(-2, -\frac{1}{2} \right)$ (Endpoints are not included because $<$ does not include equality.)

CONCEPTS	EXAMPLES

Solving a Rational Inequality

Step 1 Write the inequality so that 0 is on one side and there is a single fraction on the other side.

Step 2 Determine the values that make the numerator or denominator 0.

Step 3 Use the values from Step 2 to divide a number line into intervals.

Step 4 Substitute a test value from each interval into the *original* inequality to determine the intervals that belong to the solution set.

Step 5 Consider the endpoints separately. Exclude any values that make the denominator 0.

Solve $\dfrac{x}{x+2} \geq 4$.

$$\dfrac{x}{x+2} - 4 \geq 0 \quad \text{Subtract 4.}$$

$$\dfrac{x}{x+2} - \dfrac{4(x+2)}{x+2} \geq 0 \quad \text{Write with a common denominator.}$$

$$\dfrac{x - 4x - 8}{x+2} \geq 0 \quad \text{Write as a single fraction; distributive property}$$

$$\dfrac{-3x - 8}{x+2} \geq 0 \quad \text{Subtract fractions.}$$

$-\dfrac{8}{3}$ makes the numerator 0, and -2 makes the denominator 0.

Intervals: $\left(-\infty, -\dfrac{8}{3}\right)$, $\left(-\dfrac{8}{3}, -2\right)$, $(-2, \infty)$

Test values: -4 from Interval A makes the original inequality false, $-\dfrac{7}{3}$ from Interval B makes it true, and 0 from Interval C makes it false.

Solution set: $\left[-\dfrac{8}{3}, -2\right)$ (The endpoint -2 is not included because it makes the denominator 0.)

Chapter 11   Review Exercises

11.1 *Solve each equation using the zero-factor property.*

1. $x^2 + 3x - 28 = 0$ 2. $2z^2 + 7z = 15$ 3. $r^2 - 169 = 0$

4. The High Roller observation wheel in Las Vegas has a height of about 168 m. Use the metric version of Galileo's formula,

$$d = 4.9t^2 \quad \text{(where } d \text{ is in meters)},$$

to find how long it would take a wallet dropped from the top of the High Roller to reach the ground. (Data from www.caesars.com)

Solve each equation using the square root property.

5. $t^2 = 121$ 6. $p^2 = 3$

7. $(2x + 5)^2 = 100$ *8. $(3x - 2)^2 = -25$

---

*This exercise requires knowledge of complex numbers.

11.2 *Solve each equation by completing the square.*

9. $x^2 + 4x = 15$

10. $2x^2 - 3x = -1$

11. $2z^2 + 8z - 3 = 0$

***12.** $4x^2 - 3x + 6 = 0$

11.3 *Solve each equation using the quadratic formula.*

13. $2x^2 + x - 21 = 0$

14. $x^2 + 5x = 7$

15. $(t + 3)(t - 4) = -2$

***16.** $2x^2 + 3x + 4 = 0$

***17.** $3p^2 = 2(2p - 1)$

18. $x(2x - 7) = 3x^2 + 3$

Find the discriminant and use it to predict whether the solutions to each equation are

A. *two rational numbers*

B. *one rational number*

C. *two irrational numbers*

D. *two nonreal complex numbers.*

Tell whether each equation can be solved using the zero-factor property, or the quadratic formula should be used instead. Do not actually solve.

19. (a) $x^2 + 5x + 2 = 0$

(b) $4t^2 = 3 - 4t$

20. (a) $4x^2 = 6x - 8$

(b) $9z^2 + 30z + 25 = 0$

11.4 *Solve each equation. Check the solutions.*

21. $\dfrac{15}{x} = 2x - 1$

22. $\dfrac{1}{n} + \dfrac{2}{n + 1} = 2$

23. $-2r = \sqrt{\dfrac{48 - 20r}{2}}$

24. $8(3x + 5)^2 + 2(3x + 5) - 1 = 0$

25. $2x^{2/3} - x^{1/3} - 28 = 0$

26. $p^4 - 10p^2 + 9 = 0$

Solve each problem. Round answers to the nearest tenth, as necessary.

27. Bahaa paddled a canoe 20 mi upstream, then paddled back. If the rate of the current was 3 mph and the total trip took 7 hr, what was Bahaa's rate?

28. Carol Ann drove 8 mi to pick up a friend, and then drove 11 mi to a mall at a rate 15 mph faster. If Carol Ann's total travel time was 24 min, what was her rate on the trip to pick up her friend?

29. An old machine processes a batch of checks in 1 hr more time than a new one. How long would it take the old machine to process a batch of checks that the two machines together process in 2 hr?

30. Zoran can process a stack of invoices 1 hr faster than Claude can. Working together, they take 1.5 hr. How long would the job take each person working alone?

	Rate	Time Working Together	Fractional Part of the Job Done
Zoran			
Claude			

11.5 *Solve each formula for the specified variable. (Leave $\pm$ in the answers as needed.)*

31. $k = \dfrac{rF}{wv^2}$ for v

32. $p = \sqrt{\dfrac{yz}{6}}$ for y

33. $mt^2 = 3mt + 6$ for t

---

*This exercise requires knowledge of complex numbers.

Solve each problem.

34. A large machine requires a part in the shape of a right triangle with a hypotenuse 9 ft less than twice the length of the longer leg. The shorter leg must be $\frac{3}{4}$ the length of the longer leg. Find the lengths of the three sides of the part.

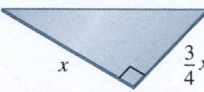

35. A square has an area of 256 cm². If the same amount is removed from one dimension and added to the other, the resulting rectangle has an area 16 cm² less. Find the dimensions of the rectangle.

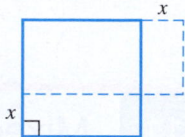

36. Allen wants to buy a mat for a photograph that measures 14 in. by 20 in. He wants to have an even border around the picture when it is mounted on the mat. If the area of the mat he chooses is 352 in.², how wide will the border be?

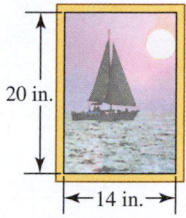

37. If a square piece of cardboard has 3-in. squares cut from its corners and then has the flaps folded up to form an open-top box, the volume of the box is given by the formula

$$V = 3(x - 6)^2,$$

where x is the length of each side of the original piece of cardboard in inches. What original length would yield a box with volume 432 in.³?

38. A searchlight moves horizontally back and forth along a wall. The distance of the light from a starting point at t minutes is given by the quadratic function

$$f(t) = 100t^2 - 300t.$$

How long will it take before the light returns to the starting point?

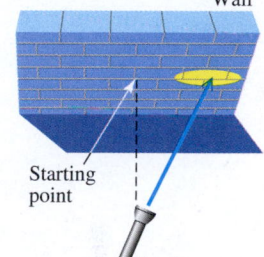

11.6, 11.7 *Identify the vertex of each parabola.*

39. $f(x) = -(x - 1)^2$

40. $f(x) = (x - 3)^2 + 7$

41. $x = (y - 3)^2 - 4$

42. $y = -3x^2 + 4x - 2$

Graph each parabola. Give the vertex, axis of symmetry, domain, and range.

43. $y = 2(x - 2)^2 - 3$

44. $f(x) = -2x^2 + 8x - 5$

45. $x = 2(y + 3)^2 - 4$

46. $x = -\frac{1}{2}y^2 + 6y - 14$

Solve each problem.

47. The height (in feet) of a projectile t seconds after being fired from Earth into the air is given by

$$f(t) = -16t^2 + 160t.$$

Find the number of seconds required for the projectile to reach maximum height. What is the maximum height?

48. Find the length and width of a rectangle having a perimeter of 200 m if the area is to be a maximum. What is the maximum area?

11.8 *Solve each inequality, and graph the solution set.*

49. $(x - 4)(2x + 3) > 0$

50. $x^2 + x \leq 12$

51. $(x + 2)(x - 3)(x + 5) \leq 0$

52. $(4x + 3)^2 \leq -4$

53. $\dfrac{6}{2z - 1} < 2$

54. $\dfrac{3t + 4}{t - 2} \leq 1$

Chapter 11   Mixed Review Exercises

Solve each equation or inequality.

1. $V = r^2 + R^2h$ for R

*2. $3t^2 - 6t = -4$

3. $(3x + 11)^2 = 7$

4. $S = \dfrac{Id^2}{k}$ for d

5. $(8x - 7)^2 \geq -1$

6. $2x - \sqrt{x} = 6$

7. $x^4 - 8x^2 = -1$

8. $\dfrac{-2}{x + 5} \leq -5$

9. $6 + \dfrac{15}{s^2} = -\dfrac{19}{s}$

10. $(x^2 - 2x)^2 = 11(x^2 - 2x) - 24$

11. $(r - 1)(2r + 3)(r + 6) < 0$

Work each problem.

12. Match each equation in parts (a) – (f) with the figure that most closely resembles its graph in choices A–F.

(a) $g(x) = x^2 - 5$

(b) $h(x) = -x^2 + 4$

(c) $F(x) = (x - 1)^2$

(d) $G(x) = (x + 1)^2$

(e) $H(x) = (x - 1)^2 + 1$

(f) $K(x) = (x + 1)^2 + 1$

A.
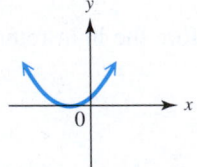

B.
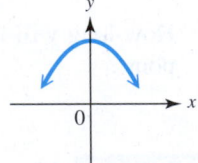

C.
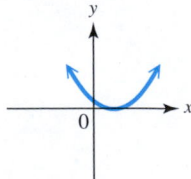

D.
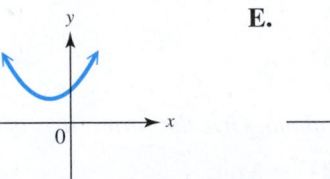

E.

F.
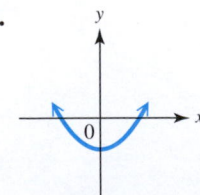

13. Graph $f(x) = 4x^2 + 4x - 2$. Give the vertex, axis of symmetry, domain, and range.

14. In 4 hr, Rajeed can travel 15 mi upriver and come back. The rate of the current is 5 mph. Find the rate of the boat in still water.

15. Two pieces of a large wooden puzzle fit together to form a rectangle with length 1 cm less than twice the width. The diagonal, where the two pieces meet, is 2.5 cm in length. Find the length and width of the rectangle.

---

*This exercise requires knowledge of complex numbers.

16. The percent of the U.S. population that was foreign-born during the years 1930–2010 can be modeled by the quadratic function

$$f(x) = 0.0043x^2 - 0.3245x + 11.53,$$

where $x = 0$ represents 1930, $x = 10$ represents 1940, and so on. (Data from U.S. Census Bureau.)

(a) The coefficient of x^2 in the model is positive, so the graph of this quadratic function is a parabola that opens up. Will the y-value of the vertex of this graph be a maximum or a minimum?

(b) According to the model, in what year during this period was the percent of foreign-born population a minimum? (Round down for the year.) Use the actual x-value of the vertex, to the nearest tenth, to find this percent, also to the nearest tenth.

| Chapter 11 | Test | **FOR EXTRA HELP** | *Step-by-step test solutions are found on the Chapter Test Prep Videos available in* **MyLab Math.** |

▶ *View the complete solutions to all Chapter Test exercises in MyLab Math.*

1. Solve $5x^2 + 13x = 6$ using the zero-factor property.

Solve each equation using the square root property or completing the square.

2. $t^2 = 54$ **3.** $(7x + 3)^2 = 25$ **4.** $x^2 + 2x = 4$

Solve each equation using the quadratic formula.

5. $2x^2 - 3x - 1 = 0$ ***6.** $3t^2 - 4t = -5$ **7.** $3x = \sqrt{\dfrac{9x + 2}{2}}$

*8. If k is a negative number, then which one of the following equations will have two nonreal complex solutions?

  A. $x^2 = 4k$ **B.** $x^2 = -4k$ **C.** $(x + 2)^2 = -k$ **D.** $x^2 + k = 0$

9. What is the discriminant for $2x^2 - 8x - 3 = 0$? How many and what type of solutions does this equation have? (Do not actually solve.)

Solve each equation by any method.

10. $3 - \dfrac{16}{x} - \dfrac{12}{x^2} = 0$ **11.** $4x^2 + 7x - 3 = 0$

12. $9x^4 + 4 = 37x^2$ **13.** $12 = (2n + 1)^2 + (2n + 1)$

14. Solve $S = 4\pi r^2$ for r. (Leave $\pm$ in your answer.)

Solve each problem.

15. Terry and Callie do word processing. For a certain prospectus, Callie can prepare it 2 hr faster than Terry can. If they work together, they can do the entire prospectus in 5 hr. How long will it take each of them working alone to prepare the prospectus? Round answers to the nearest tenth of an hour.

16. Qihong paddled a canoe 10 mi upstream and then paddled back to the starting point. If the rate of the current was 3 mph and the entire trip took $3\frac{1}{2}$ hr, what was Qihong's rate?

---

*This exercise requires knowledge of complex numbers.

17. Endre has a pool 24 ft long and 10 ft wide. He wants to construct a concrete walk around the pool. If he plans for the walk to be of uniform width and cover 152 ft², what will the width of the walk be?

18. At a point 30 m from the base of a tower, the distance to the top of the tower is 2 m more than twice the height of the tower. Find the height of the tower.

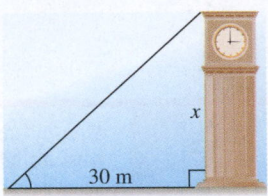

19. Which one of the following figures most closely resembles the graph of

$$f(x) = a(x - h)^2 + k \quad \text{if } a < 0, h > 0, \text{ and } k < 0?$$

A.

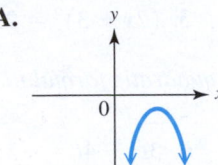

B.

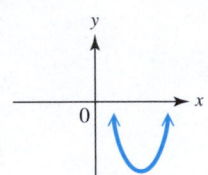

C.

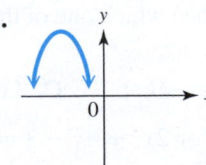

D.

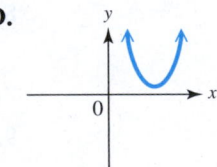

Graph each parabola. Identify the vertex, axis of symmetry, domain, and range.

20. $f(x) = \dfrac{1}{2}x^2 - 2$ **21.** $f(x) = -x^2 + 4x - 1$ **22.** $x = -(y - 2)^2 + 2$

23. Houston Community College is planning to construct a rectangular parking lot on land bordered on one side by a highway. The plan is to use 640 ft of fencing to fence off the other three sides. What should the dimensions of the lot be if the enclosed area is to be a maximum?

Solve each inequality, and graph the solution set.

24. $2x^2 + 7x > 15$ **25.** $\dfrac{5}{t - 4} \leq 1$

Chapters R–11 Cumulative Review Exercises

Perform the indicated operations.

1. $-8.84 - (-3.46)$

2. $\dfrac{7}{5} - \left(\dfrac{9}{10} - \dfrac{3}{2}\right)$

3. $|-12| + |13|$

4. Find 6% of 12.

5. Simplify $3 - 6(4^2 - 8)$.

6. Let $S = \left\{-\dfrac{7}{3}, -2, -\sqrt{3}, 0, 0.7, \sqrt{12}, \sqrt{-8}, 7, \dfrac{32}{3}\right\}$. List the elements of S that are elements of each set.

(a) Integers **(b)** Rational numbers **(c)** Real numbers **(d)** Complex numbers

Solve each equation or inequality.

7. $7 - (4 + 3t) + 2t = -6(t - 2) - 5$

8. $|6x - 9| = |-4x + 2|$

9. $2x = \sqrt{\dfrac{5x + 2}{3}}$

10. $\dfrac{3}{x - 3} - \dfrac{2}{x - 2} = \dfrac{3}{x^2 - 5x + 6}$

11. $(r - 5)(2r + 3) = 1$

12. $x^4 - 5x^2 + 4 = 0$

13. $-2x + 4 \le -x + 3$

14. $|3x - 7| \le 1$

15. $x^2 - 4x + 3 < 0$

16. $\dfrac{3}{p + 2} > 1$

Graph each relation. Decide whether or not y can be expressed as a function f of x, and if so, give its domain and range, and write using function notation.

17. $4x - 5y = 15$

18. $4x - 5y < 15$

19. $y = -2(x - 1)^2 + 3$

20. Find the slope and intercepts of the line with equation $-2x + 7y = 16$.

Write an equation for the specified line. Express each equation in slope-intercept form.

21. Passing through $(2, -3)$ and parallel to the line with equation $5x + 2y = 6$

22. Passing through $(-4, 1)$ and perpendicular to the line with equation $5x + 2y = 6$

Solve each system of equations.

23. $2x - 4y = 10$
$9x + 3y = 3$

24. $x + 2y = 5$
$4y = -2x + 8$

25. $x + y + 2z = 3$
$-x + y + z = -5$
$2x + 3y - z = -8$

26. The two top-grossing films of 2017 were *Star Wars: The Last Jedi* and *Beauty and the Beast*. The two films together grossed $1049 million. *Beauty and the Beast* grossed $41 million less than *Star Wars: The Last Jedi*. How much did each film gross? (Data from Box Office Mojo.)

Write with positive exponents only. Assume that variables represent positive real numbers.

27. $\left(\dfrac{x^{-3}y^2}{x^5y^{-2}}\right)^{-1}$

28. $\dfrac{(4x^{-2})^2(2y^3)}{8x^{-3}y^5}$

Perform the indicated operations.

29. $\left(\dfrac{2}{3}t + 9\right)^2$

30. Divide $4x^3 + 2x^2 - x + 26$ by $x + 2$.

Factor completely.

31. $24m^2 + 2m - 15$

32. $4t^2 - 100$

33. $8x^3 + 27y^3$

34. $9x^2 - 30xy + 25y^2$

Simplify. Express each answer in lowest terms. Assume denominators are nonzero.

35. $\dfrac{5x + 2}{-6} \div \dfrac{15x + 6}{5}$

36. $\dfrac{3}{2 - x} - \dfrac{5}{x} + \dfrac{6}{x^2 - 2x}$

37. $\dfrac{\dfrac{r}{s} - \dfrac{s}{r}}{\dfrac{r}{s} + 1}$

38. Two cars left an intersection at the same time, one heading due south and the other due east. Later they were exactly 95 mi apart. The car heading east had gone 38 mi less than twice as far as the car heading south. How far had each car traveled?

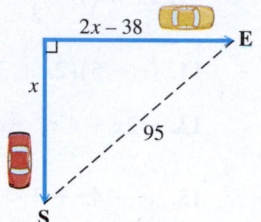

Simplify each radical expression.

39. $\sqrt[3]{\dfrac{27}{16}}$

40. $\dfrac{2}{\sqrt{7} - \sqrt{5}}$

STUDY SKILLS REMINDER

Have you begun to prepare for your final exam? **Review Study Skill 10,**
Preparing for Your Math Final Exam.

12

INVERSE, EXPONENTIAL, AND LOGARITHMIC FUNCTIONS

Compound interest earned on money, intensities of sounds, and population growth and decay are some examples of applications of *exponential* and *logarithmic functions*.

12.1 Inverse Functions

In this chapter we study two important types of functions, *exponential* and *logarithmic*. These functions are related: They are *inverses* of one another.

OBJECTIVE 1 Decide whether a function is one-to-one and, if it is, find its inverse.

Suppose we define the function

$$G = \{(-2, 2), (-1, 1), (0, 0), (1, 3), (2, 5)\}.$$

We can form another set of ordered pairs from G by interchanging the x- and y-values of each pair in G. We can call this set F, so

$$F = \{(2, -2), (1, -1), (0, 0), (3, 1), (5, 2)\}.$$

To show that these two sets are related as just described, F is called the *inverse* of G. For a function f to have an inverse, f must be a *one-to-one function*.

One-to-One Function

In a **one-to-one function,** each x-value corresponds to only one y-value, and each y-value corresponds to only one x-value.

VOCABULARY
☐ one-to-one function
☐ inverse of a function

The function in **FIGURE 1(a)** is one-to-one. The function shown in **FIGURE 1(b)** is not one-to-one because the y-value 7 corresponds to *two* x-values, 2 and 3. That is, the ordered pairs $(2, 7)$ and $(3, 7)$ both belong to the function.

STUDY SKILLS REMINDER

Make study cards to help you learn and remember the material in this chapter. **Review Study Skill 5, *Using Study Cards.***

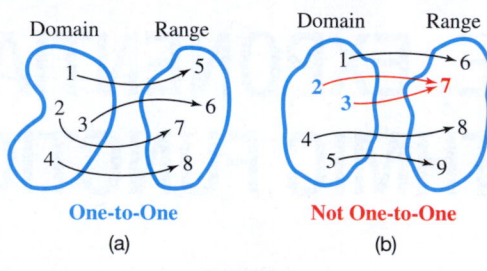

FIGURE 1

The *inverse* of any one-to-one function f is found by interchanging the components of the ordered pairs of f. The inverse of f is written $\boldsymbol{f^{-1}}$. Read f^{-1} as *"the inverse of f"* or *"f-inverse."*

⚠ **CAUTION** The symbol $f^{-1}(x)$ does **not** represent $\frac{1}{f(x)}$.

Inverse of a Function

The **inverse** of a one-to-one function f, written f^{-1}, is the set of all ordered pairs of the form (y, x), where (x, y) belongs to f. *The inverse is formed by interchanging x and y, so the domain of f becomes the range of f^{-1} and the range of f becomes the domain of f^{-1}.*

For inverses f and f^{-1}, it follows that for all x in their domains,

$$(f \circ f^{-1})(x) = x \quad \text{and} \quad (f^{-1} \circ f)(x) = x.$$

NOW TRY EXERCISE 1

Determine whether each function is one-to-one. If it is, find the inverse.

(a) $F = \{(0, 0), (1, 1),$
$(4, 2), (9, 3)\}$

(b) $G = \{(-1, -2), (0, 0),$
$(1, -2), (2, -8)\}$

(c) A Norwegian physiologist has developed a rule for predicting running times based on the time to run 5 km (5K). An example for one runner is shown here.

Distance	Time
1.5K	4:22
3K	9:18
5K	16:00
10K	33:40

Data from Stephen Seiler, Agder College, Kristiansand, Norway.

EXAMPLE 1 Finding Inverses of One-to-One Functions

Determine whether each function is one-to-one. If it is, find the inverse.

(a) $F = \{(3, 1), (0, 2), (2, 3), (4, 0)\}$

Every x-value in F corresponds to only one y-value, and every y-value corresponds to only one x-value, so F is a one-to-one function. The inverse function is found by interchanging the x- and y-values in each ordered pair.

$$F^{-1} = \{(1, 3), (2, 0), (3, 2), (0, 4)\}$$

The domain and range of F become the range and domain, respectively, of F^{-1}.

(b) $G = \{(-2, 1), (-1, 0), (0, 1), (1, 2), (2, 2)\}$

Each x-value in G corresponds to just one y-value. However, the y-value 1 corresponds to two x-values, -2 and 0. Also, the y-value 2 corresponds to both 1 and 2. Because some y-values correspond to more than one x-value, G is not one-to-one and does not have an inverse.

(c) The table shows the number of days in which the air in the Los Angeles–Long Beach–Anaheim metropolitan area exceeded air-quality standards in recent years.

Year	Number of Days Exceeding Standards
2010	4
2011	14
2012	37
2013	25
2014	28
2015	38
2016	25

Data from U.S. Environmental Protection Agency.

Let f be the function defined in the table, with the years forming the domain and the number of days exceeding air-quality standards forming the range. Then f is not one-to-one because in two different years (2013 and 2016) the number of days exceeding air-quality standards was the same, 25.

NOW TRY

OBJECTIVE 2 Use the horizontal line test to determine whether a function is one-to-one.

By graphing a function and observing the graph, we can use the *horizontal line test* to determine whether the function is one-to-one.

NOW TRY ANSWERS

1. (a) one-to-one;
$F^{-1} = \{(0, 0), (1, 1),$
$(2, 4), (3, 9)\}$
(b) not one-to-one
(c) one-to-one

Time	Distance
4:22	1.5K
9:18	3K
16:00	5K
33:40	10K

Horizontal Line Test

A function is one-to-one if every horizontal line intersects the graph of the function at most once.

The horizontal line test follows from the definition of a one-to-one function. Any two points that lie on the same horizontal line have the same y-coordinate. No two ordered pairs that belong to a one-to-one function may have the same y-coordinate. Therefore, no horizontal line will intersect the graph of a one-to-one function more than once.

**NOW TRY
EXERCISE 2**
Use the horizontal line test to
determine whether each graph
is the graph of a one-to-one
function.

(a)
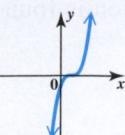

(b)
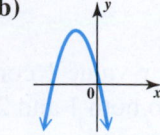

EXAMPLE 2   Using the Horizontal Line Test

Use the horizontal line test to determine whether each graph is the graph of a one-to-one function.

(a)
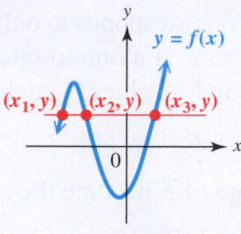

FIGURE 2

A horizontal line intersects the graph in **FIGURE 2** in more than one point. This function is not one-to-one.

(b)
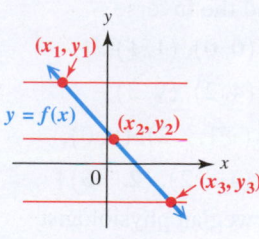

FIGURE 3

Every horizontal line will intersect the graph in **FIGURE 3** in exactly one point. This function is one-to-one. **NOW TRY**

OBJECTIVE 3   Find the equation of the inverse of a function.

The inverse of a one-to-one function is found by interchanging the x- and y-values of each of its ordered pairs. The equation of the inverse of a function $y = f(x)$ is found in the same way.

Finding the Equation of the Inverse of $y = f(x)$

For a one-to-one function f defined by an equation $y = f(x)$, find the defining equation of the inverse function f^{-1} as follows.

Step 1   Interchange x and y.

Step 2   Solve for y.

Step 3   Replace y with $f^{-1}(x)$.

EXAMPLE 3   Finding Equations of Inverses

Determine whether each equation defines a one-to-one function. If so, find the equation that defines the inverse.

(a) $f(x) = 2x + 5$

The graph of $y = 2x + 5$ is a nonvertical line, so by the horizontal line test, f is a one-to-one function. Find the inverse as follows.

$$f(x) = 2x + 5$$

$$y = 2x + 5 \qquad \text{Let } y = f(x).$$

Step 1 $\qquad x = 2y + 5 \qquad$ Interchange x and y.

Step 2 $\qquad 2y = x - 5 \qquad \begin{array}{l}\text{Subtract 5.}\\ \text{Interchange sides.}\end{array}\Big\}\;\text{Solve for } y.$

$$y = \frac{x - 5}{2} \qquad \text{Divide by 2.}$$

Step 3 $\qquad f^{-1}(x) = \dfrac{x - 5}{2} \qquad$ Replace y with $f^{-1}(x)$.

These are equivalent forms of the same equation.

$$f^{-1}(x) = \frac{x}{2} - \frac{5}{2}, \quad \text{or} \quad f^{-1}(x) = \frac{1}{2}x - \frac{5}{2} \qquad \frac{a - b}{c} = \frac{a}{c} - \frac{b}{c}$$

NOW TRY
EXERCISE 3
Determine whether each equation defines a one-to-one function. If so, find the equation that defines the inverse.

(a) $f(x) = 5x - 7$
(b) $f(x) = (x + 1)^2$
(c) $f(x) = x^3 - 4$

Thus, f^{-1} is a linear function. In the given function

$$f(x) = 2x + 5,$$

we start with a value of x, *multiply* by 2, and *add* 5. In the equation for the inverse

$$f^{-1}(x) = \frac{x - 5}{2}, \qquad \text{One form of } f^{-1}(x)$$

we *subtract* 5, and then *divide* by 2. This shows how an inverse is used to "undo" what a function does to the variable x.

(b) $y = x^2 + 2$

This equation has a vertical parabola as its graph, so some horizontal lines will intersect the graph at two points. For example, both $x = 1$ and $x = -1$ correspond to $y = 3$. See **FIGURE 4**.

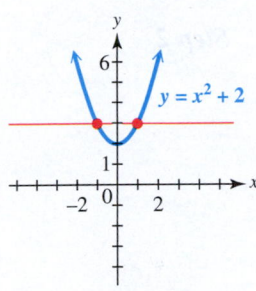

FIGURE 4

Because of the x^2-term in $y = x^2 + 2$, there are many pairs of x-values that correspond to the same y-value. The function is not one-to-one and does not have an inverse. If we try to find the equation of an inverse, we obtain the following.

$$y = x^2 + 2$$
$$x = y^2 + 2 \qquad \text{Interchange } x \text{ and } y.$$
$$y^2 = x - 2 \qquad \text{Solve for } y.$$

Remember both roots.
$$y = \pm \sqrt{x - 2} \qquad \text{Square root property}$$

The last step shows that there are two y-values for each choice of x in $(2, \infty)$—that is, for $x > 2$—so the function is not one-to-one. It does not have an inverse.

(c) $f(x) = (x - 2)^3$

A cubing function like this is one-to-one. See the graph in **FIGURE 5**.

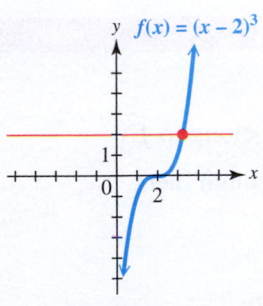

FIGURE 5

NOW TRY ANSWERS
3. **(a)** one-to-one function;
    $f^{-1}(x) = \frac{x + 7}{5}$, or
    $f^{-1}(x) = \frac{1}{5}x + \frac{7}{5}$
(b) not a one-to-one function
(c) one-to-one function;
    $f^{-1}(x) = \sqrt[3]{x + 4}$

$$f(x) = (x - 2)^3$$
$$y = (x - 2)^3 \qquad \text{Replace } f(x) \text{ with } y.$$

Step 1 $x = (y - 2)^3 \qquad \text{Interchange } x \text{ and } y.$

Step 2 $\sqrt[3]{x} = \sqrt[3]{(y - 2)^3} \qquad \text{Take the cube root on each side.}$

$\sqrt[3]{x} = y - 2 \qquad \sqrt[3]{a^3} = a \qquad$ Solve for y.

$y = \sqrt[3]{x} + 2 \qquad \text{Add 2. Interchange sides.}$

Step 3 $f^{-1}(x) = \sqrt[3]{x} + 2 \qquad \text{Replace } y \text{ with } f^{-1}(x).$

NOW TRY

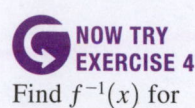

NOW TRY
EXERCISE 4
Find $f^{-1}(x)$ for

$$f(x) = \frac{x + 3}{x - 4}, \quad x \neq 4.$$

EXAMPLE 4 Using Factoring to Find an Inverse

It is shown in standard college algebra texts that the following function is one-to-one.

$$f(x) = \frac{x + 1}{x - 2}, \quad x \neq 2$$

Find $f^{-1}(x)$.

$$f(x) = \frac{x + 1}{x - 2}, \quad x \neq 2 \qquad \text{Given equation}$$

$$y = \frac{x + 1}{x - 2}, \quad x \neq 2 \qquad \text{Replace } f(x) \text{ with } y.$$

Step 1
$$x = \frac{y + 1}{y - 2}, \quad y \neq 2 \qquad \text{Interchange } x \text{ and } y.$$

Step 2
$$x(y - 2) = y + 1 \qquad \text{Multiply by } y - 2.$$
$$xy - 2x = y + 1 \qquad \text{Distributive property}$$
$$xy - y = 2x + 1 \qquad \text{Subtract } y. \text{ Add } 2x.$$
$$y(x - 1) = 2x + 1 \qquad \text{Factor out } y.$$
$$y = \frac{2x + 1}{x - 1} \qquad \text{Divide by } x - 1.$$

$$\left. \right\} \text{Solve for } y.$$

Step 3
$$f^{-1}(x) = \frac{2x + 1}{x - 1}, \quad x \neq 1 \qquad \begin{array}{l} \text{Replace } y \text{ with } f^{-1}(x). \\ \text{Note the restriction.} \end{array} \qquad \textbf{NOW TRY} \text{ } \textcircled{\small ↻}$$

OBJECTIVE 4 Graph f^{-1}, given the graph of f.

One way to graph the inverse of a function f whose equation is given follows.

Graphing the Inverse

Step 1 Find several ordered pairs that belong to f.

Step 2 Interchange x and y to obtain ordered pairs that belong to f^{-1}.

Step 3 Plot those points, and sketch the graph of f^{-1} through them.

We can also select points on the graph of f and *use symmetry* to find corresponding points on the graph of f^{-1}.

For example, suppose the point (a, b) shown in **FIGURE 6** belongs to a one-to-one function f. Then the point (b, a) belongs to f^{-1}. The line segment connecting the points (a, b) and (b, a) is perpendicular to, and cut in half by, the line $y = x$. The points (a, b) and (b, a) are "mirror images" of each other with respect to $y = x$.

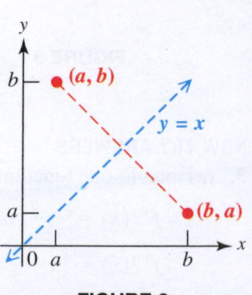

FIGURE 6

We can find the graph of f^{-1} from the graph of f by locating the mirror image of each point in f with respect to the line $y = x$.

NOW TRY
EXERCISE 5

Graph the inverse of the function labeled f in the figure.

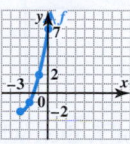

EXAMPLE 5   Graphing Inverses of Functions

Graph the inverse of each function labeled f in the figures.

(a) FIGURE 7(a) shows the graph of a one-to-one function f. The points

$$\left(-1, \tfrac{1}{2}\right), (0, 1), (1, 2), \text{ and } (2, 4) \qquad \text{Points on } f$$

lie on its graph. Interchange x and y to obtain ordered pairs that belong to f^{-1}.

$$\left(\tfrac{1}{2}, -1\right), (1, 0), (2, 1), \text{ and } (4, 2) \qquad \text{Points on } f^{-1}$$

Plot these points, and sketch the graph of f^{-1} through them. See FIGURE 7(b).

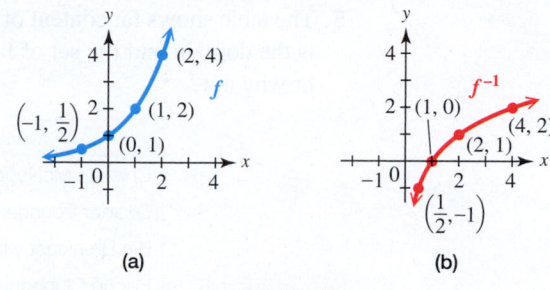

(a)   (b)

FIGURE 7

(b) Each function f (shown in blue) in FIGURES 8 and 9 is a one-to-one function.

Each inverse f^{-1} is shown in red. In both cases, the graph of f^{-1} is a reflection of the graph of f across the line $y = x$.

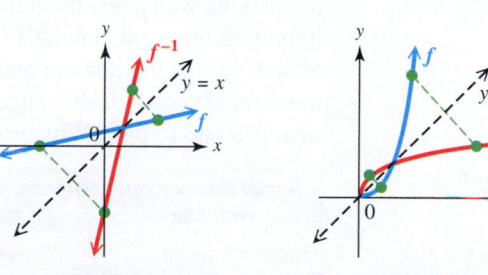

FIGURE 8   FIGURE 9   NOW TRY

NOW TRY ANSWER

5.

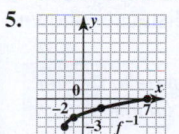

12.1 Exercises

FOR EXTRA HELP   **MyLab Math**

Video solutions for select problems available in MyLab Math

Concept Check   *Choose the correct response.*

1. If a function is made up of ordered pairs in such a way that the same y-value appears in a correspondence with two different x-values, then

A. the function is one-to-one

B. the function is not one-to-one

C. its graph does not pass the vertical line test

D. it has an inverse function associated with it.

2. Which equation defines a one-to-one function? Explain why the others do not, using specific examples.

A. $f(x) = x$

B. $f(x) = x^2$

C. $f(x) = |x|$

D. $f(x) = -x^2 + 2x - 1$

3. Only one of the graphs illustrates a one-to-one function. Which one is it? (**See Example 2.**)

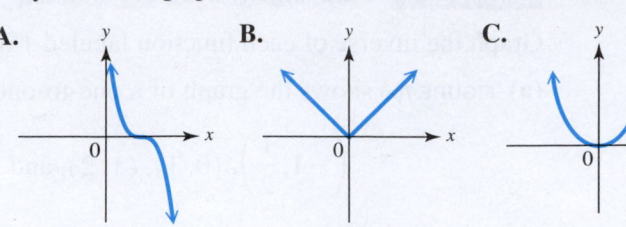

4. If a function f is one-to-one and the point (p, q) lies on the graph of f, then which point *must* lie on the graph of f^{-1}?

A. $(-p, q)$ **B.** $(-q, -p)$ **C.** $(p, -q)$ **D.** (q, p)

Answer each question.

5. The table shows fat content of various menu items at McDonald's. If the set of menu items is the domain and the set of fat contents is the range of a function, is it one-to-one? Why or why not?

Menu Item	Fat Content (in grams)
Chicken McNuggets (20 pieces)	60
Quarter Pounder with Cheese	43
Big Breakfast with Egg Whites	41
Bacon Clubhouse Burger	41
McFlurry with Reese's	37
Sausage Biscuit with Egg	33

Data from www.thebalancesmb.com

6. The table shows the most-visited social networking web sites, by number of visitors in millions, in June 2017. If the set of web sites is the domain and the set of numbers of visitors is the range of a function, is it one-to-one? Why or why not?

Social Networking Web Site	Number of Visitors (in millions)
Facebook	202
Instagram	121
Twitter	110
LinkedIn	103
Snapchat	95
Pinterest	86

Data from comScore, Inc.

7. The road mileage between Denver, Colorado, and several selected U.S. cities is shown in the table. If we consider this a function that pairs each city with a distance, is it one-to-one? How could we change the answer to this question by adding 1 mile to one of the distances shown?

City	Distance to Denver (in miles)
Atlanta	1398
Dallas	781
Indianapolis	1058
Kansas City, MO	600
Los Angeles	1059

8. The table lists caffeine amounts in several popular 12-oz sodas. If the set of sodas is the domain and the set of caffeine amounts is the range of a function, is it one-to-one? Why or why not?

Soda	Caffeine (in mg)
Mountain Dew	54
Diet Coke	46
Sunkist Orange Soda	41
Diet Pepsi-Cola	34
Coca-Cola Classic	34

Data from www.caffeineinformer.com

*Determine whether each function is one-to-one. If it is, find the inverse. **See Examples 1 and 3.***

9. $\{(3, 6), (2, 10), (5, 12)\}$

10. $\left\{(-1, 3), (0, 5), (5, 0), \left(7, -\dfrac{1}{2}\right)\right\}$

11. $\{(-1, 3), (2, 7), (4, 3), (5, 8)\}$

12. $\{(-8, 6), (-4, 3), (0, 6), (5, 10)\}$

13. $\{(0, 4.5), (2, 8.6), (4, 12.7)\}$

14. $\{(1, 5.8), (2, 8.8), (3, 8.5)\}$

15. $f(x) = x + 3$

16. $f(x) = x + 8$

17. $f(x) = -\dfrac{1}{2}x - 2$

18. $f(x) = -\dfrac{1}{4}x - 8$

19. $f(x) = 2x + 4$

20. $f(x) = 3x + 1$

21. $g(x) = -4x + 3$

22. $g(x) = -6x - 8$

23. $f(x) = 5$

24. $f(x) = -7$

25. $f(x) = \sqrt{x - 3}, \quad x \geq 3$

26. $f(x) = \sqrt{x + 2}, \quad x \geq -2$

27. $f(x) = \sqrt{x + 6}, \quad x \geq -6$

28. $f(x) = \sqrt{x - 4}, \quad x \geq 4$

29. $f(x) = 3x^2 + 2$

30. $f(x) = 4x^2 - 1$

31. $g(x) = (x + 1)^3$

32. $g(x) = (x - 4)^3$

33. $f(x) = x^3 - 4$

34. $f(x) = x^3 + 5$

*Each function is one-to-one. Find its inverse. **See Example 4.***

35. $f(x) = \dfrac{x + 4}{x + 2}, \quad x \neq -2$

36. $f(x) = \dfrac{x + 3}{x + 5}, \quad x \neq -5$

37. $f(x) = \dfrac{4x - 2}{x + 5}, \quad x \neq -5$

38. $f(x) = \dfrac{5x - 10}{x + 4}, \quad x \neq -4$

39. $f(x) = \dfrac{-2x + 1}{2x - 5}, \quad x \neq \dfrac{5}{2}$

40. $f(x) = \dfrac{-3x + 2}{3x - 4}, \quad x \neq \dfrac{4}{3}$

Concept Check *Let $f(x) = 2^x$. This function is one-to-one. Find each value.*

41. (a) $f(3)$

42. (a) $f(4)$

43. (a) $f(0)$

44. (a) $f(-2)$

(b) $f^{-1}(8)$

(b) $f^{-1}(16)$

(b) $f^{-1}(1)$

(b) $f^{-1}\left(\dfrac{1}{4}\right)$

Graphs of selected functions are given in the following exercises.

(a) *Use the horizontal line test to determine whether each function graphed is one-to-one.* ***See Example 2.***

(b) *If the function is one-to-one, graph its inverse.* ***See Example 5.***

45.

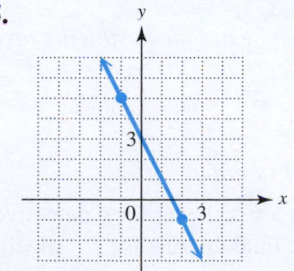

46.

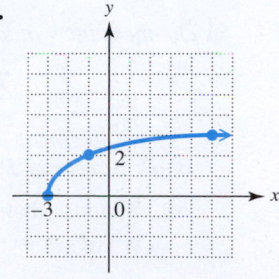

47.

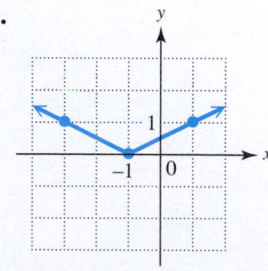

48.

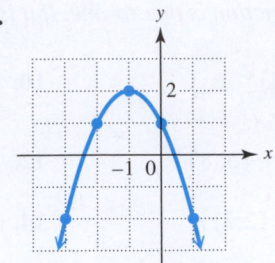

49.

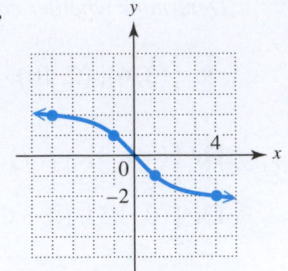

50.

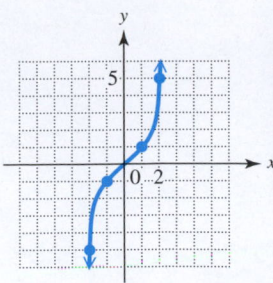

Each of the following functions is one-to-one. Graph the function as a solid line (or curve), and then graph its inverse on the same set of axes as a dashed line (or curve). Complete any tables to help graph the functions. See Example 5.

51. $f(x) = 2x - 1$ **52.** $f(x) = 2x + 3$ **53.** $f(x) = -4x$ **54.** $f(x) = -2x$

55. $f(x) = \sqrt{x}$,   **56.** $f(x) = -\sqrt{x}$,   **57.** $f(x) = x^3 - 2$ **58.** $f(x) = x^3 + 3$
  $x \geq 0$ $x \geq 0$

x	$f(x)$
0	
1	
4	

x	$f(x)$
0	
1	
4	

x	$f(x)$
-1	
0	
1	
2	

x	$f(x)$
-2	
-1	
0	
1	

RELATING CONCEPTS   **For Individual or Group Work** (Exercises 59–62)

Inverse functions can be used to send and receive coded information. A simple example might use the function

  $f(x) = 2x + 5.$ *(Note that it is one-to-one.)*

Suppose that each letter of the alphabet is assigned a numerical value according to its position, as follows.

A 1	**G** 7	**L** 12	**Q** 17	**V** 22
B 2	**H** 8	**M** 13	**R** 18	**W** 23
C 3	**I** 9	**N** 14	**S** 19	**X** 24
D 4	**J** 10	**O** 15	**T** 20	**Y** 25
E 5	**K** 11	**P** 16	**U** 21	**Z** 26
F 6				

This is an Enigma machine used by the Germans in World War II to send coded messages.

Using the function, the word ALGEBRA would be encoded as

  7   29   19   15   9   41   7,

because

  $f(A) = f(1) = 2(1) + 5 = 7,$ $f(L) = f(12) = 2(12) + 5 = 29,$ *and so on.*

The message would then be decoded using the inverse of f, which is $f^{-1}(x) = \dfrac{x - 5}{2}$.

  $f^{-1}(7) = \dfrac{7 - 5}{2} = 1 = A,$ $f^{-1}(29) = \dfrac{29 - 5}{2} = 12 = L,$ *and so on.*

Work Exercises 59–62 in order.

59. Suppose that you are an agent for a detective agency. Today's encoding function is $f(x) = 4x - 5$. Find the rule for f^{-1} algebraically.

60. You receive the following coded message today. (Read across from left to right.)

47 95 7 −1 43 7 79 43 −1 75 55 67 31 71 75 27

15 23 67 15 −1 75 15 71 75 75 27 31 51 23 71

31 51 7 15 71 43 31 7 15 11 3 67 15 −1 11

Use the letter/number assignment described on the previous page to decode the message.

61. Why is a one-to-one function essential in this encoding/decoding process?

62. Use $f(x) = x^3 + 4$ to encode your name, using the above letter/number assignment.

12.2 ## Exponential Functions

OBJECTIVES

1 Evaluate exponential expressions using a calculator.

2 Define and graph exponential functions.

3 Solve exponential equations of the form $a^x = a^k$ for x.

4 Use exponential functions in applications involving growth or decay.

VOCABULARY

☐ exponential function with base a
☐ asymptote
☐ exponential equation

NOW TRY EXERCISE 1

Use a calculator to approximate each exponential expression to three decimal places.

(a) $2^{3.1}$ **(b)** $2^{-1.7}$ **(c)** $2^{1/4}$

NOW TRY ANSWERS
1. (a) 8.574 **(b)** 0.308 **(c)** 1.189

OBJECTIVE 1 Evaluate exponential expressions using a calculator.

Consider the exponential expression 2^x for rational values of x.

$$2^3 = 8, \quad 2^{-1} = \frac{1}{2}, \quad 2^{1/2} = \sqrt{2}, \quad 2^{3/4} = \sqrt[4]{2^3} = \sqrt[4]{8} \qquad \text{Examples of } 2^x \text{ for rational } x$$

In more advanced courses it is shown that 2^x exists for all real number values of x, both rational and irrational. We can use a calculator to find approximations of exponential expressions that are not easily determined.

EXAMPLE 1 **Evaluating Exponential Expressions**

Use a calculator to approximate each exponential expression to three decimal places.

(a) $2^{1.6}$ **(b)** $2^{-1.3}$ **(c)** $2^{1/3}$

FIGURE 10 shows how a TI-84 Plus calculator approximates these values. The display shows more decimal places than we usually need, so we round to three decimal places as directed.

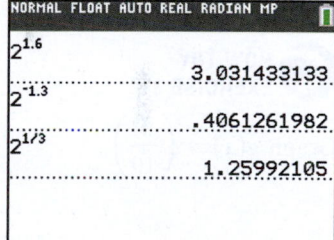

FIGURE 10

$$2^{1.6} \approx 3.031, \quad 2^{-1.3} \approx 0.406, \quad 2^{1/3} \approx 1.260$$

NOW TRY

OBJECTIVE 2 Define and graph exponential functions.

The definition of an exponential function assumes that a^x exists for all real numbers x.

Exponential Function

For $a > 0$, $a \neq 1$, and all real numbers x,

$$f(x) = a^x$$

defines the **exponential function with base a.**

**NOW TRY
EXERCISE 2**
Graph $f(x) = 4^x$.

When graphing an exponential function of the form $f(x) = a^x$, pay particular attention to whether $a > 1$ or $0 < a < 1$.

EXAMPLE 2   Graphing an Exponential Function ($a > 1$)

Graph $f(x) = 2^x$. Then compare it to the graph of $F(x) = 5^x$.

Choose some values of x, and find the corresponding values of $f(x) = 2^x$. Plotting these points and drawing a smooth curve through them gives the darker graph shown in **FIGURE 11**. This graph is typical of the graph of an exponential function of the form $f(x) = a^x$, where $a > 1$.

The larger the value of a, the faster the graph rises.

To see this, compare the graph of $F(x) = 5^x$ with the graph of $f(x) = 2^x$ in **FIGURE 11**. When graphing such functions, be sure to plot a sufficient number of points to see how rapidly the graph rises.

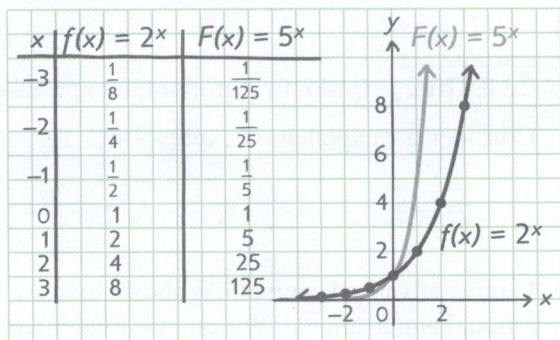

Exponential function with base $a > 1$

Domain: $(-\infty, \infty)$

Range: $(0, \infty)$

y-intercept: $(0, 1)$

The function is one-to-one, and its graph rises from left to right.

FIGURE 11

The vertical line test assures us that the graphs in **FIGURE 11** represent functions. These graphs show an important characteristic of exponential functions where $a > 1$.

As x gets larger, y increases at a faster and faster rate.   **NOW TRY**

**NOW TRY
EXERCISE 3**
Graph $g(x) = \left(\dfrac{1}{10}\right)^x$.

EXAMPLE 3   Graphing an Exponential Function ($0 < a < 1$)

Graph $g(x) = \left(\dfrac{1}{2}\right)^x$.

Find and plot some points on the graph. The graph in **FIGURE 12** is similar to that of $f(x) = 2^x$ (**FIGURE 11**) with the same domain and range, except that here *as x gets larger, y decreases.* This graph is typical of the graph of an exponential function of the form $f(x) = a^x$, where $0 < a < 1$.

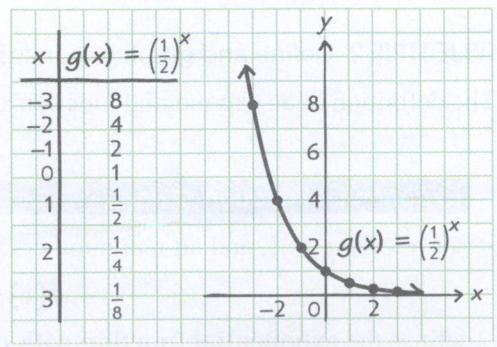

Exponential function with base $0 < a < 1$

Domain: $(-\infty, \infty)$

Range: $(0, \infty)$

y-intercept: $(0, 1)$

The function is one-to-one, and its graph falls from left to right.

FIGURE 12

NOW TRY ANSWERS

2. 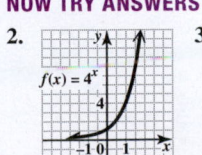   **3.**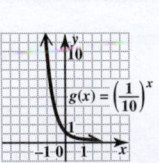

NOW TRY

⚠ **CAUTION** The graph of an exponential function of the form

$$f(x) = a^x$$

approaches the x-axis, but does *not* touch it. Recall that such a line is called an **asymptote**.

Characteristics of the Graph of $f(x) = a^x$

1. The graph contains the point $(0, 1)$, which is its y-intercept.

2. The function is one-to-one.

   • When $a > 1$, the graph *rises* from left to right. (See **FIGURE 11**.)

   • When $0 < a < 1$, the graph *falls* from left to right. (See **FIGURE 12**.)

   In both cases, the graph goes from the second quadrant to the first.

3. The graph approaches the x-axis but never touches it—that is, the x-axis is an asymptote.

4. The domain is $(-\infty, \infty)$, and the range is $(0, \infty)$.

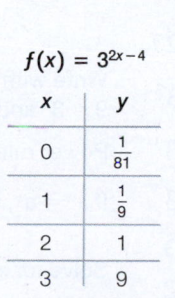

NOW TRY
EXERCISE 4
Graph $f(x) = 4^{2x-1}$.

EXAMPLE 4 Graphing a More Complicated Exponential Function

Graph $f(x) = 3^{2x-4}$.

Find some ordered pairs. We let $x = 0$ and $x = 2$ and find values of $f(x)$, or y.

$$f(x) = 3^{2x-4}$$
$$y = 3^{2(0)-4} \quad \text{Let } x = 0.$$
$$y = 3^{-4}$$
$$y = \frac{1}{81} \quad a^{-n} = \frac{1}{a^n}$$

$$f(x) = 3^{2x} - 4$$
$$y = 3^{2(2)-4} \quad \text{Let } x = 2.$$
$$y = 3^0$$
$$y = 1 \quad a^0 = 1$$

These ordered pairs, $\left(0, \frac{1}{81}\right)$ and $(2, 1)$, along with the other ordered pairs shown in the table, lead to the graph in **FIGURE 13**.

$f(x) = 3^{2x-4}$

x	y
0	$\frac{1}{81}$
1	$\frac{1}{9}$
2	1
3	9

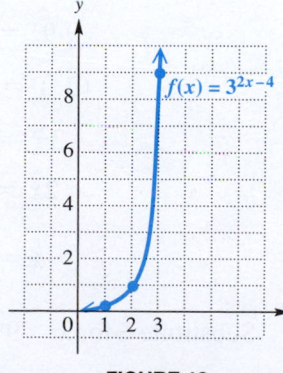

The graph of $f(x) = 3^{2x-4}$ is similar to the graph of $f(x) = 3^x$ except that it is shifted to the right and rises more rapidly.

NOW TRY ANSWER
4.

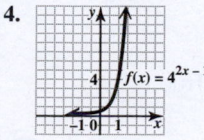

FIGURE 13

NOW TRY ↻

> **NOTE** *The two restrictions on the value of a in the definition of an exponential function f(x) = a^x are important.*
>
> 1. The restriction **a > 0** is necessary so that the function can be defined for *all* real numbers x. Letting a be negative, such as $a = -2$, and letting $x = \frac{1}{2}$ gives the expression $(-2)^{1/2}$, which is not a real number.
>
> 2. The restriction **a ≠ 1** is necessary because 1 raised to *any* power is equal to 1, resulting in the linear function $f(x) = 1$.

OBJECTIVE 3 Solve exponential equations of the form $a^x = a^k$ for *x*.

Up to this point, we have solved only equations that had the variable as a base, like $x^2 = 8$. In these equations, all exponents have been constants. An **exponential equation** is an equation that has a variable in an exponent, such as

$$9^x = 27.$$

We can use the following property to solve many exponential equations.

Property for Solving an Exponential Equation

For $a > 0$ and $a \neq 1$, if $a^x = a^y$ then $x = y$.

This property would not necessarily be true if $a = 1$.

Solving an Exponential Equation

Step 1 **Both sides must have the same base.** If the two sides of the equation do not have the same base, express each as a power of the same base if possible.

Step 2 **Simplify exponents** if necessary, using the rules of exponents.

Step 3 **Set exponents equal** using the property given in this section.

Step 4 **Solve** the equation obtained in Step 3.

NOW TRY
EXERCISE 5
Solve $8^x = 16$.

EXAMPLE 5   Solving an Exponential Equation

Solve $9^x = 27$.

$$9^x = 27$$

Step 1 $(3^2)^x = 3^3$ Write with the same base; $9 = 3^2$ and $27 = 3^3$.

Step 2 $3^{2x} = 3^3$ Power rule for exponents

Step 3 $2x = 3$ If $a^x = a^y$, then $x = y$.

Step 4 $x = \dfrac{3}{2}$ Solve for *x*.

NOW TRY ANSWER
5. $\left\{\frac{4}{3}\right\}$

CHECK Substitute $\frac{3}{2}$ for *x*: $9^x = 9^{3/2} = (9^{1/2})^3 = 3^3 = 27.$ ✓   True

The solution set is $\left\{\frac{3}{2}\right\}$.

NOW TRY ↻

**NOW TRY
EXERCISE 6**

Solve each equation.

(a) $3^{2x-1} = 27^{x+4}$

(b) $5^x = \dfrac{1}{625}$

(c) $\left(\dfrac{2}{7}\right)^x = \dfrac{343}{8}$

EXAMPLE 6 Solving Exponential Equations

Solve each equation.

(a) $4^{3x-1} = 16^{x+2}$

> Be careful multiplying the exponents.

$4^{3x-1} = (4^2)^{x+2}$ Write with the same base; $16 = 4^2$ (Step 1)

$4^{3x-1} = 4^{2x+4}$ Power rule for exponents (Step 2)

$3x - 1 = 2x + 4$ Set the exponents equal. (Step 3)

$x = 5$ Subtract $2x$. Add 1. (Step 4)

CHECK $4^{3x-1} = 16^{x+2}$

$4^{3(5)-1} \overset{?}{=} 16^{5+2}$ Substitute. Let $x = 5$.

$4^{14} \overset{?}{=} 16^7$ Perform the operations in the exponents.

$4^{14} \overset{?}{=} (4^2)^7$ $16 = 4^2$

$4^{14} = 4^{14}$ ✓ True

The solution set is $\{5\}$.

(b) $6^x = \dfrac{1}{216}$

$6^x = \dfrac{1}{6^3}$ $216 = 6^3$

$6^x = 6^{-3}$ Write with the same base; $\dfrac{1}{6^3} = 6^{-3}$.

$x = -3$ Set exponents equal.

CHECK   Substitute -3 for x.

$6^x = 6^{-3} = \dfrac{1}{6^3} = \dfrac{1}{216}$ ✓ True

The solution set is $\{-3\}$.

(c) $\left(\dfrac{2}{3}\right)^x = \dfrac{9}{4}$

$\left(\dfrac{2}{3}\right)^x = \left(\dfrac{4}{9}\right)^{-1}$ $\dfrac{9}{4} = \left(\dfrac{4}{9}\right)^{-1}$

$\left(\dfrac{2}{3}\right)^x = \left[\left(\dfrac{2}{3}\right)^2\right]^{-1}$ Write with the same base.

$\left(\dfrac{2}{3}\right)^x = \left(\dfrac{2}{3}\right)^{-2}$ Power rule for exponents

$x = -2$ Set exponents equal.

Check that the solution set is $\{-2\}$.

NOW TRY ANSWERS
6. (a) $\{-13\}$ **(b)** $\{-4\}$
  (c) $\{-3\}$

NOW TRY ↻

OBJECTIVE 4 Use exponential functions in applications involving growth or decay.

**NOW TRY
EXERCISE 7**
Use the function in **Example 7** to approximate average annual carbon dioxide concentration in 2010, to the nearest unit.

EXAMPLE 7 Applying an Exponential Growth Function

The graph in **FIGURE 14** shows average annual concentration of carbon dioxide (in parts per million) in the air. This concentration is increasing exponentially.

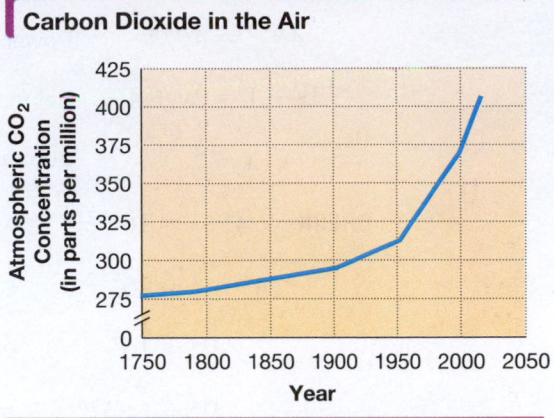

Carbon Dioxide in the Air

Data from National Oceanic and Atmospheric Administration.

FIGURE 14

The data in **FIGURE 14** are approximated by the exponential function

$$f(x) = 275.8 + 1.5172(1.0161)^x,$$

where x is number of years since 1750. Use this function to approximate average annual concentration of carbon dioxide in parts per million, to the nearest unit, for each year.

(a) 1900

Because x represents number of years since 1750, $x = 1900 - 1750 = 150$.

$$f(x) = 275.8 + 1.5172(1.0161)^x \qquad \text{Given function}$$

$$f(150) = 275.8 + 1.5172(1.0161)^{150} \qquad \text{Let } x = 150.$$

$$f(150) \approx 292 \qquad \text{Evaluate with a calculator.}$$

The concentration in 1900 was approximately 292 parts per million.

(b) 2000

$$f(x) = 275.8 + 1.5172(1.0161)^x \qquad \text{Given function}$$

$$f(250) = 275.8 + 1.5172(1.0161)^{250} \qquad x = 2000 - 1750 = 250$$

$$f(250) \approx 358 \qquad \text{Evaluate with a calculator.}$$

The concentration in 2000 was approximately 358 parts per million.   **NOW TRY**

EXAMPLE 8 Applying an Exponential Decay Function

Atmospheric pressure (in millibars) at a given altitude x, in meters, can be approximated by the exponential function

$$f(x) = 1038(1.000134)^{-x}, \quad \text{for values of } x \text{ between 0 and 10,000.}$$

Because the base is greater than 1 and the coefficient of x in the exponent is negative, function values decrease as x increases. This means that as altitude increases, atmospheric pressure decreases. (Data from Miller, A. and J. Thompson, *Elements of Meteorology,* Fourth Edition, Charles E. Merrill Publishing Company.)

NOW TRY ANSWER
7. 372 parts per million

NOW TRY EXERCISE 8

Use the function in **Example 8** to approximate the pressure at 6000 m, to the nearest unit.

(a) According to this function, what is the pressure at ground level?

$$f(x) = 1038(1.000134)^{-x} \qquad \text{Given function}$$

At ground level, $x = 0$.

$$f(0) = 1038(1.000134)^{-0} \qquad \text{Let } x = 0.$$

$$f(0) = 1038(1) \qquad a^0 = 1$$

$$f(0) = 1038 \qquad \text{Identity property}$$

The pressure is 1038 millibars.

(b) What is the pressure at 5000 m, to the nearest unit?

$$f(x) = 1038(1.000134)^{-x} \qquad \text{Given function}$$

$$f(5000) = 1038(1.000134)^{-5000} \qquad \text{Let } x = 5000.$$

$$f(5000) \approx 531 \qquad \text{Evaluate with a calculator.}$$

The pressure is approximately 531 millibars. **NOW TRY**

> **NOTE** The function in **Example 8** is equivalent to
>
> $$f(x) = 1038\left(\frac{1}{1.000134}\right)^x.$$
>
> In this form, the base a satisfies the condition $0 < a < 1$.

NOW TRY ANSWER

8. 465 millibars

12.2 Exercises

FOR EXTRA HELP

 MyLab Math

▶ *Video solutions for select problems available in MyLab Math*

Concept Check *Choose the correct response.*

1. For an exponential function $f(x) = a^x$, if $a > 1$, then the graph (*rises / falls*) from left to right.

2. For an exponential function $f(x) = a^x$, if $0 < a < 1$, then the graph (*rises / falls*) from left to right.

3. Which point lies on the graph of $f(x) = 3^x$?

    A. $(1, 0)$ **B.** $(3, 1)$ **C.** $(0, 1)$ **D.** $\left(\sqrt{3}, \frac{1}{3}\right)$

4. The asymptote of the graph of $f(x) = a^x$

    A. is the x-axis   **B.** is the y-axis

    C. has equation $x = 1$ **D.** has equation $y = 1$.

5. Which statement is true?

    A. The point $\left(\frac{1}{2}, \sqrt{5}\right)$ lies on the graph of $f(x) = 5^x$.

    B. $f(x) = 5^x$ is not a one-to-one function.

    C. The y-intercept of the graph of $f(x) = 5^x$ is $(0, 5)$.

    D. The graph of $y = 5^x$ rises at a faster rate than the graph of $y = 10^x$.

6. Which statement is false?

A. The domain of the function $f(x) = \left(\frac{1}{4}\right)^x$ is $(-\infty, \infty)$.

B. The graph of the function $f(x) = \left(\frac{1}{4}\right)^x$ has one x-intercept.

C. The range of the function $f(x) = \left(\frac{1}{4}\right)^x$ is $(0, \infty)$.

D. The point $(-2, 16)$ lies on the graph of $f(x) = \left(\frac{1}{4}\right)^x$.

Use a calculator to approximate each exponential expression to three decimal places. **See Example 1.**

7. $2^{1.9}$

8. $2^{2.7}$

9. $2^{-1.54}$

10. $2^{-1.88}$

11. $10^{0.3}$

12. $10^{0.5}$

13. $4^{1/3}$

14. $6^{1/5}$

15. $\left(\frac{1}{3}\right)^{1.5}$

16. $\left(\frac{1}{3}\right)^{2.4}$

17. $\left(\frac{1}{4}\right)^{-3.1}$

18. $\left(\frac{1}{4}\right)^{-1.4}$

Graph each exponential function. **See Examples 2–4.**

19. $f(x) = 3^x$

20. $f(x) = 5^x$

21. $g(x) = \left(\frac{1}{3}\right)^x$

22. $g(x) = \left(\frac{1}{5}\right)^x$

23. $f(x) = 4^{-x}$

24. $f(x) = 6^{-x}$

25. $f(x) = 2^{2x-2}$

26. $f(x) = 2^{2x+1}$

27. Concept Check A student incorrectly solved the following equation as shown.

$$2^x = 32 \quad \text{Given equation}$$

$$\frac{2^x}{2} = \frac{32}{2} \quad \text{Divide by 2.}$$

$$x = 16$$

WHAT WENT WRONG? Give the correct solution set.

28. Concept Check A student incorrectly solved the following equation as shown.

$$3^x = 81 \quad \text{Given equation}$$

$$3^x - 3 = 81 - 3 \quad \text{Subtract 3.}$$

$$x = 78$$

WHAT WENT WRONG? Give the correct solution set.

Solve each equation. **See Examples 5 and 6.**

29. $6^x = 36$

30. $8^x = 64$

31. $100^x = 1000$

32. $8^x = 4$

33. $16^x = 64$

34. $8^x = 32$

35. $4^{x-5} = 64^{2x}$

36. $125^{3x} = 5^{2x-7}$

37. $16^{2x+1} = 64^{x+3}$

38. $9^{2x-8} = 27^{x-4}$

39. $5^x = \dfrac{1}{125}$

40. $3^x = \dfrac{1}{81}$

41. $9^x = \dfrac{1}{27}$

42. $8^x = \dfrac{1}{32}$

43. $5^x = 0.2$

44. $10^x = 0.1$

45. $\left(\frac{3}{2}\right)^x = \dfrac{8}{27}$

46. $\left(\frac{4}{3}\right)^x = \dfrac{27}{64}$

47. $\left(\frac{5}{4}\right)^x = \dfrac{16}{25}$

48. $\left(\frac{3}{2}\right)^x = \dfrac{16}{81}$

The amount of radioactive material in an ore sample is given by the exponential function

$$A(t) = 100(3.2)^{-0.5t},$$

where $A(t)$ is the amount present, in grams, of the sample t months after the initial measurement.

49. How much radioactive material was present at the initial measurement? (*Hint: $t = 0$.*)

50. How much, to the nearest hundredth, was present 2 months later?

51. How much, to the nearest hundredth, was present 10 months later?

52. Graph the function on the axes as shown.

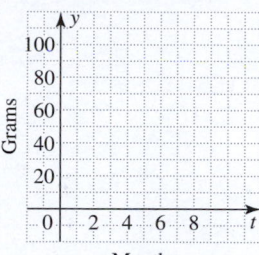

Months

A major scientific periodical published an article in 1990 dealing with the problem of global warming. The article was accompanied by a graph that illustrated two possible scenarios.

(a) *The warming might be modeled by an exponential function of the form*

$$f(x) = (1.046 \times 10^{-38})(1.0444^x).$$

(b) *The warming might be modeled by a linear function of the form*

$$g(x) = 0.009x - 17.67.$$

In both cases, x represents the year, and the function value represents the increase in degrees Celsius due to the warming. Use these functions to approximate the increase in temperature for each year, to the nearest tenth of a degree.

53. 2000 **54.** 2010 **55.** 2020 **56.** 2040

Solve each problem. See Examples 7 and 8.

57. The estimated number of monthly active Snapchat users (in millions) from 2013 to 2016 can be modeled by the exponential function

$$f(x) = 39.154(2.0585)^x,$$

where $x = 0$ represents 2013, $x = 1$ represents 2014, and so on. Use this model to approximate the number of monthly active Snapchat users in each year, to the nearest thousandth. (Data from Activate.)

(a) 2014 **(b)** 2015 **(c)** 2016

58. The number of paid music subscriptions (in millions) in the United States from 2010 to 2016 can be modeled by the exponential function

$$f(x) = 1.365(1.565)^x,$$

where $x = 0$ represents 2010, $x = 1$ represents 2011, and so on. Use this model to approximate the number of paid music subscriptions in each year, to the nearest thousandth. (Data from RIAA.)

(a) 2010 **(b)** 2013 **(c)** 2016

59. A small business estimates that the value $V(t)$ of a copy machine is decreasing according to the exponential function

$$V(t) = 5000(2)^{-0.15t},$$

where t is the number of years that have elapsed since the machine was purchased, and $V(t)$ is in dollars.

(a) What was the original value of the machine?

(b) What is the value of the machine 5 yr after purchase, to the nearest dollar?

(c) What is the value of the machine 10 yr after purchase, to the nearest dollar?

(d) Graph the function.

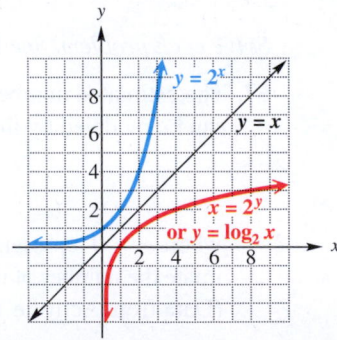

60. Refer to the exponential function in **Exercise 59.**

(a) When will the value of the machine be $2500? (*Hint:* Let $V(t) = 2500$, divide both sides by 5000, and use the method of **Example 5.**)

(b) When will the value of the machine be $1250?

12.3 Logarithmic Functions

OBJECTIVES

1 Define a logarithm.

2 Convert between exponential and logarithmic forms, and evaluate logarithms.

3 Solve logarithmic equations of the form $\log_a b = k$ for a, b, or k.

4 Use the definition of logarithm to simplify logarithmic expressions.

5 Define and graph logarithmic functions.

6 Use logarithmic functions in applications involving growth or decay.

OBJECTIVE 1 Define a logarithm.

The graph of $y = 2^x$ is the blue curve in **FIGURE 15.** Because $y = 2^x$ defines a one-to-one function, it has an inverse. Interchanging x and y gives

$$x = 2^y, \quad \text{the inverse of} \quad y = 2^x. \qquad \text{Roles of } x \text{ and } y \text{ are interchanged.}$$

The graph of the inverse is found by reflecting the graph of $y = 2^x$ across the line $y = x$. The graph of the inverse $x = 2^y$ is the red curve in **FIGURE 15.**

FIGURE 15

We can also write the equation of the red curve using a new notation that involves the concept of *logarithm.*

Logarithm

For all positive numbers a, where $a \neq 1$, and all positive real numbers x,

$$y = \log_a x \quad \text{is equivalent to} \quad x = a^y.$$

The abbreviation **log** is used for the word **logarithm.** Read $\log_a x$ as *"the logarithm of x with base a"* or *"the base a logarithm of x."* To remember the location of the base and the exponent in each form, refer to the following diagrams.

Logarithmic form: $\quad y = \log_a x$ $\qquad$ Exponential form: $\quad x = a^y$

Exponent → (on exponent) Base → (on base)

Meaning of $\log_a x$

A logarithm is an exponent. *The expression $\log_a x$ represents the exponent to which the base a must be raised to obtain x.*

OBJECTIVE 2 Convert between exponential and logarithmic forms, and evaluate logarithms.

We can use the definition of logarithm to carry out these conversions.

NOW TRY
EXERCISE 1

(a) Write $6^3 = 216$ in logarithmic form.

(b) Write $\log_{64} 4 = \frac{1}{3}$ in exponential form.

EXAMPLE 1 Converting between Exponential and Logarithmic Forms

The table shows several pairs of equivalent forms.

Exponential Form	Logarithmic Form
$3^2 = 9$	$\log_3 9 = 2$
$\left(\frac{1}{5}\right)^{-2} = 25$	$\log_{1/5} 25 = -2$
$10^5 = 100{,}000$	$\log_{10} 100{,}000 = 5$
$4^{-3} = \frac{1}{64}$	$\log_4 \frac{1}{64} = -3$

$y = \log_a x$
is equivalent to
$x = a^y$.

NOW TRY

NOW TRY
EXERCISE 2

Use a calculator to approximate each logarithm to four decimal places.

(a) $\log_2 7$ $\quad$ (b) $\log_5 8$
(c) $\log_{1/3} 12$ $\quad$ (d) $\log_{10} 18$

EXAMPLE 2 Evaluating Logarithms

Use a calculator to approximate each logarithm to four decimal places.

(a) $\log_2 5$ $\qquad$ (b) $\log_3 12$ $\qquad$ (c) $\log_{1/2} 12$ $\qquad$ (d) $\log_{10} 20$

FIGURE 16 shows how a TI-84 Plus calculator approximates the logarithms in parts (a)–(c).

```
NORMAL FIX4 AUTO REAL RADIAN MP
log₂(5)
                          2.3219
log₃(12)
                          2.2619
log₁/₂(12)
                         -3.5850
```

```
NORMAL FIX4 AUTO REAL RADIAN MP
log₁₀(20)
                          1.3010
log(20)
                          1.3010
```

FIGURE 16 $\qquad\qquad$ **FIGURE 17**

NOW TRY ANSWERS
1. (a) $\log_6 216 = 3$
   (b) $64^{1/3} = 4$
2. (a) 2.8074 $\quad$ (b) 1.2920
   (c) -2.2619 $\quad$ (d) 1.2553

FIGURE 17 shows the approximation for the expression $\log_{10} 20$ in part (d). Notice that the second display, which indicates log 20 (with no base shown), gives the same result. We shall see in a later section that when no base is indicated, the base is understood to be 10. A base 10 logarithm is a *common logarithm*.

NOW TRY

NOTE In a later section, we introduce another method of calculating logarithms like those in **Example 2(a)–(c)** using the *change-of-base rule*.

OBJECTIVE 3 Solve logarithmic equations of the form $\log_a b = k$ for a, b, or k.

A **logarithmic equation** is an equation with a logarithm in at least one term.

EXAMPLE 3 Solving Logarithmic Equations

Solve each equation.

(a) $\log_4 x = -2$ $\log_a x = y$ is equivalent to $x = a^y$.

$x = 4^{-2}$

$x = \dfrac{1}{4^2}$ Definition of negative exponent

$x = \dfrac{1}{16}$ Apply the exponent.

CHECK $\log_4 \dfrac{1}{16} = -2$ because $4^{-2} = \dfrac{1}{16}$. ✓

The solution set is $\left\{\dfrac{1}{16}\right\}$.

(b) $\log_{1/2}(3x + 1) = 2$

$3x + 1 = \left(\dfrac{1}{2}\right)^2$ This is a key step.   Write in exponential form.

$3x + 1 = \dfrac{1}{4}$ Apply the exponent.

$12x + 4 = 1$ Multiply each term by 4.

$12x = -3$ Subtract 4.

$x = -\dfrac{1}{4}$ Divide by 12. Write in lowest terms.

CHECK $\log_{1/2}(3x + 1) = 2$

$\log_{1/2}\left[3\left(-\dfrac{1}{4}\right) + 1\right] \overset{?}{=} 2$ Let $x = -\dfrac{1}{4}$.

$\log_{1/2} \dfrac{1}{4} \overset{?}{=} 2$ Simplify within parentheses.

$\left(\dfrac{1}{2}\right)^2 \overset{?}{=} \dfrac{1}{4}$ Write in exponential form.

$\dfrac{1}{4} = \dfrac{1}{4}$ ✓   True

The solution set is $\left\{-\dfrac{1}{4}\right\}$.

NOW TRY EXERCISE 3

Solve each equation.

(a) $\log_2 x = -5$

(b) $\log_{3/2}(2x-1) = 3$

(c) $\log_x 10 = 2$

(d) $\log_{125} \sqrt[3]{5} = x$

(c)
$$\log_x 3 = 2$$
$$x^2 = 3 \quad \text{Write in exponential form.}$$

Be careful here. $-\sqrt{3}$ is extraneous.

$$x = \pm\sqrt{3} \quad \text{Take square roots.}$$

Only the *principal* square root $\sqrt{3}$ satisfies the equation because the base must be a positive number.

CHECK
$$\log_x 3 = 2$$
$$\log_{\sqrt{3}} 3 \overset{?}{=} 2 \quad \text{Let } x = \sqrt{3}.$$
$$(\sqrt{3})^2 \overset{?}{=} 3 \quad \text{Write in exponential form.}$$
$$3 = 3 \; \checkmark \quad \text{True}$$

The solution set is $\{\sqrt{3}\}$.

(d)
$$\log_{49} \sqrt[3]{7} = x$$
$$49^x = \sqrt[3]{7} \quad \text{Write in exponential form.}$$
$$(7^2)^x = 7^{1/3} \quad \text{Write with the same base.}$$
$$7^{2x} = 7^{1/3} \quad \text{Power rule for exponents}$$
$$2x = \frac{1}{3} \quad \text{Set the exponents equal.}$$
$$x = \frac{1}{6} \quad \text{Divide by 2 }\left(\text{which is the same as multiplying by } \tfrac{1}{2}\right).$$

Check to verify that the solution set is $\left\{\tfrac{1}{6}\right\}$.

NOW TRY

OBJECTIVE 4 Use the definition of logarithm to simplify logarithmic expressions.

The definition of logarithm enables us to state several special properties.

Special Properties of Logarithms

For any positive real number b, where $b \neq 1$, the following hold true.

$$\log_b b = 1 \qquad\qquad \log_b 1 = 0$$
$$\log_b b^r = r \quad (r \text{ is real.}) \qquad b^{\log_b r} = r \quad (r > 0)$$

To prove the last statement, let $x = \log_b r$.
$$x = \log_b r$$
$$b^x = r \quad \text{Write in exponential form.}$$
$$b^{\log_b r} = r \quad \text{Replace } x \text{ with } \log_b r.$$

This is the statement to be proved.

NOW TRY ANSWERS

3. (a) $\left\{\tfrac{1}{32}\right\}$ (b) $\left\{\tfrac{35}{16}\right\}$
   (c) $\{\sqrt{10}\}$ (d) $\left\{\tfrac{1}{9}\right\}$

NOW TRY
EXERCISE 4
Use the special properties to evaluate each expression.

(a) $\log_{10} 10$ (b) $\log_8 1$
(c) $\log_{0.1} 1$ (d) $\log_3 3^9$
(e) $5^{\log_5 3}$ (f) $\log_3 81$

EXAMPLE 4 Using Special Properties of Logarithms

Use the special properties to evaluate.

(a) $\log_7 7 = 1$ $\log_b b = 1$ (b) $\log_{\sqrt{2}} \sqrt{2} = 1$ $\log_b b = 1$

(c) $\log_9 1 = 0$ $\log_b 1 = 0$ (d) $\log_{0.2} 1 = 0$ $\log_b 1 = 0$

(e) $\log_2 2^6 = 6$ $\log_b b^r = r$ (f) $\log_3 3^{-2.5} = -2.5$ $\log_b b^r = r$

(g) $4^{\log_4 9} = 9$ $b^{\log_b r} = r$ (h) $10^{\log_{10} 13} = 13$ $b^{\log_b r} = r$

(i) $\log_2 32$
$\quad = \log_2 2^5$ $32 = 2^5$
$\quad = 5$ $\log_b b^r = r$

(j) $\log_3 \dfrac{1}{3}$
$\quad = \log_3 3^{-1}$ Definition of negative exponent
$\quad = -1$ $\log_b b^r = r$

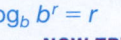 NOW TRY

OBJECTIVE 5 Define and graph logarithmic functions.

> **Logarithmic Function**
>
> If a and x are positive real numbers, where $a \neq 1$, then
>
> $$g(x) = \log_a x$$
>
> defines the **logarithmic function with base a.**

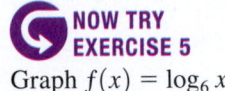

NOW TRY
EXERCISE 5
Graph $f(x) = \log_6 x$.

EXAMPLE 5 Graphing a Logarithmic Function ($a > 1$)

Graph $f(x) = \log_2 x$.

By writing $y = f(x) = \log_2 x$ in exponential form as $x = 2^y$, we can identify ordered pairs that satisfy the equation. It is easier to choose values for y and find the corresponding values of x. Plotting the points in the table of ordered pairs and connecting them with a smooth curve gives the graph in **FIGURE 18**. This graph is typical of logarithmic functions with base $a > 1$.

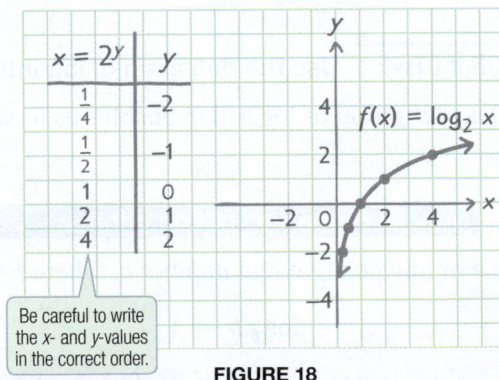

Be careful to write the x- and y-values in the correct order.

FIGURE 18

Logarithmic function with base $a > 1$

Domain: $(0, \infty)$

Range: $(-\infty, \infty)$

x-intercept: $(1, 0)$

The function is one-to-one, and its graph rises from left to right.

NOW TRY

EXAMPLE 6 Graphing a Logarithmic Function ($0 < a < 1$)

Graph $g(x) = \log_{1/2} x$.

We write $y = g(x) = \log_{1/2} x$ in exponential form as

$$x = \left(\frac{1}{2}\right)^y,$$

and then choose values for y and find the corresponding values of x. Plotting these points and connecting them with a smooth curve gives the graph in **FIGURE 19**. This graph is typical of logarithmic functions with base $0 < a < 1$.

NOW TRY ANSWERS
4. (a) 1   (b) 0   (c) 0
   (d) 9   (e) 3   (f) 4
5.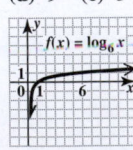

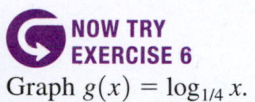
**NOW TRY
EXERCISE 6**
Graph $g(x) = \log_{1/4} x$.

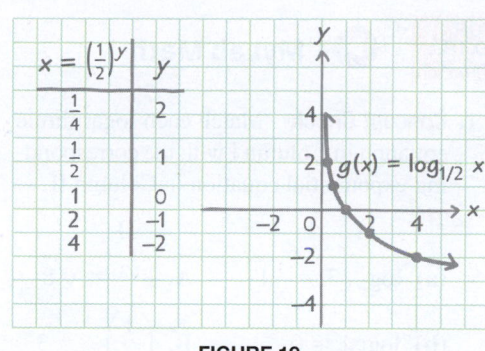

FIGURE 19

**Logarithmic function
with base $0 < a < 1$**

Domain: $(0, \infty)$

Range: $(-\infty, \infty)$

x-intercept: $(1, 0)$

The function is one-to-one, and its graph falls from left to right.

NOW TRY

Characteristics of the Graph of $g(x) = \log_a x$

1. The graph contains the point $(1, 0)$, which is its x-intercept.

2. The function is one-to-one.

   • When $a > 1$, the graph *rises* from left to right, from the fourth quadrant to the first. (See **FIGURE 18**.)

   • When $0 < a < 1$, the graph *falls* from left to right, from the first quadrant to the fourth. (See **FIGURE 19**.)

3. The graph approaches the y-axis, but never touches it—that is, the y-axis is an asymptote.

4. The domain is $(0, \infty)$, and the range is $(-\infty, \infty)$.

OBJECTIVE 6 Use logarithmic functions in applications involving growth or decay.

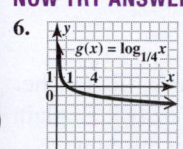
**NOW TRY
EXERCISE 7**

Suppose the gross national product (GNP) of a small country (in millions of dollars) is approximated by the logarithmic function

$G(t) = 15.0 + 2.00 \log_{10} t$,

where t is time in years since 2013. Approximate to the nearest tenth the GNP for each value of t.

(a) $t = 1$ **(b)** $t = 10$

EXAMPLE 7 Applying a Logarithmic Function

Barometric pressure in inches of mercury at a distance of x miles from the eye of a typical hurricane can be approximated by the logarithmic function

$$f(x) = 27 + 1.105 \log_{10} (x + 1).$$

(Data from Miller, A. and R. Anthes, *Meteorology*, Fifth Edition, Charles E. Merrill Publishing Company.)

Approximate the pressure 9 mi from the eye of the hurricane.

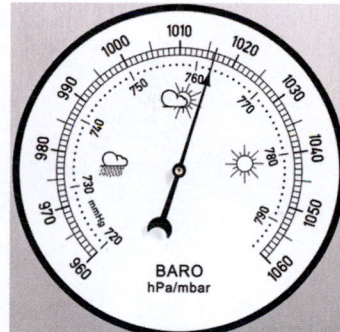

NOW TRY ANSWERS

6.

7. (a) $15.0 million
   (b) $17.0 million

$f(x) = 27 + 1.105 \log_{10} (x + 1)$

$f(9) = 27 + 1.105 \log_{10} (9 + 1)$ Let $x = 9$.

$f(9) = 27 + 1.105 \log_{10} 10$ Add inside parentheses.

$f(9) = 27 + 1.105(1)$ $\log_{10} 10 = 1$

$f(9) = 28.105$ Add.

The pressure 9 mi from the eye of the hurricane is 28.105 in.   **NOW TRY**

12.3 Exercises

 FOR EXTRA HELP

 MyLab Math

▶ *Video solutions for select problems available in MyLab Math*

1. Concept Check   Match each logarithmic equation in Column I with the corresponding exponential equation in Column II.

I	II
(a) $\log_{1/3} 3 = -1$	**A.** $8^{1/3} = \sqrt[3]{8}$
(b) $\log_5 1 = 0$	**B.** $\left(\dfrac{1}{3}\right)^{-1} = 3$
(c) $\log_2 \sqrt{2} = \dfrac{1}{2}$	**C.** $4^1 = 4$
(d) $\log_{10} 1000 = 3$	**D.** $2^{1/2} = \sqrt{2}$
(e) $\log_8 \sqrt[3]{8} = \dfrac{1}{3}$	**E.** $5^0 = 1$
(f) $\log_4 4 = 1$	**F.** $10^3 = 1000$

2. Concept Check   Match each logarithm in Column I with its corresponding value in Column II.

I	II
(a) $\log_4 16$	**A.** -2
(b) $\log_3 81$	**B.** -1
(c) $\log_3 \left(\dfrac{1}{3}\right)$	**C.** 2
(d) $\log_{10} 0.01$	**D.** 0
(e) $\log_5 \sqrt{5}$	**E.** $\dfrac{1}{2}$
(f) $\log_{13} 1$	**F.** 4

3. Concept Check   The domain of $f(x) = a^x$ is $(-\infty, \infty)$, while the range is $(0, \infty)$. Therefore because $g(x) = \log_a x$ is the inverse of f, the domain of g is _____, while the range of g is _____.

4. Concept Check   The graphs of both $f(x) = 3^x$ and $g(x) = \log_3 x$ rise from left to right. Which one rises at a faster rate as x gets large?

Write in logarithmic form. See Example 1.

5. $4^5 = 1024$ **6.** $3^6 = 729$ **7.** $\left(\dfrac{1}{2}\right)^{-3} = 8$ **8.** $\left(\dfrac{1}{6}\right)^{-3} = 216$

9. $10^{-3} = 0.001$ **10.** $36^{1/2} = 6$ **11.** $\sqrt[4]{625} = 5$ **12.** $\sqrt[3]{343} = 7$

13. $8^{-2/3} = \dfrac{1}{4}$ **14.** $16^{-3/4} = \dfrac{1}{8}$ **15.** $5^0 = 1$ **16.** $7^0 = 1$

Write in exponential form. See Example 1.

17. $\log_4 64 = 3$ **18.** $\log_2 512 = 9$ **19.** $\log_{12} 12 = 1$

20. $\log_{100} 100 = 1$ **21.** $\log_6 1 = 0$ **22.** $\log_\pi 1 = 0$

23. $\log_9 3 = \dfrac{1}{2}$ **24.** $\log_{64} 2 = \dfrac{1}{6}$ **25.** $\log_{1/4} \dfrac{1}{2} = \dfrac{1}{2}$

26. $\log_{1/8} \dfrac{1}{2} = \dfrac{1}{3}$ **27.** $\log_5 5^{-1} = -1$ **28.** $\log_{10} 10^{-2} = -2$

29. Concept Check   Match each logarithm in Column I with its value in Column II.

I	II
(a) $\log_8 8$	**A.** -1
(b) $\log_{16} 1$	**B.** 0
(c) $\log_{0.3} 1$	**C.** 1
(d) $\log_{\sqrt{7}} \sqrt{7}$	**D.** 0.1

30. Concept Check   When a student asked his teacher to explain how to evaluate

$$\log_9 3$$

without showing any work, his teacher told him to "*Think radically.*" Explain what the teacher meant by this hint.

Use a calculator to approximate each logarithm to four decimal places. **See Example 2.**

31. $\log_2 9$ **32.** $\log_2 15$ **33.** $\log_5 18$ **34.** $\log_5 26$

35. $\log_{1/4} 12$ **36.** $\log_{1/5} 27$ **37.** $\log_2 \dfrac{1}{3}$ **38.** $\log_2 \dfrac{1}{7}$

39. $\log_{10} 84$ **40.** $\log_{10} 126$ **41.** $\log 50$ **42.** $\log 90$

Solve each equation. **See Example 3.**

43. $x = \log_{27} 3$ **44.** $x = \log_{125} 5$ **45.** $\log_5 x = -3$

46. $\log_{10} x = -2$ **47.** $\log_x 9 = \dfrac{1}{2}$ **48.** $\log_x 5 = \dfrac{1}{2}$

49. $\log_x 125 = -3$ **50.** $\log_x 64 = -6$ **51.** $\log_{12} x = 0$

52. $\log_4 x = 0$ **53.** $\log_x x = 1$ **54.** $\log_x 1 = 0$

55. $\log_x \dfrac{1}{25} = -2$ **56.** $\log_x \dfrac{1}{10} = -1$ **57.** $\log_8 32 = x$

58. $\log_{81} 27 = x$ **59.** $\log_\pi \pi^4 = x$ **60.** $\log_{\sqrt{2}} \left(\sqrt{2}\right)^9 = x$

61. $\log_6 \sqrt{216} = x$ **62.** $\log_4 \sqrt{64} = x$ **63.** $\log_4 (2x + 4) = 3$

64. $\log_3 (2x + 7) = 4$ **65.** $\log_{1/3} (x - 4) = 2$ **66.** $\log_{1/2} (2x - 1) = 3$

Use the special properties of logarithms to evaluate each expression. **See Example 4.**

67. $\log_3 3$ **68.** $\log_8 8$ **69.** $\log_5 1$ **70.** $\log_{12} 1$

71. $\log_4 4^9$ **72.** $\log_5 5^6$ **73.** $\log_2 2^{-1}$ **74.** $\log_4 4^{-6}$

75. $6^{\log_6 9}$ **76.** $12^{\log_{12} 3}$ **77.** $8^{\log_8 5}$ **78.** $5^{\log_5 11}$

79. $\log_2 64$ **80.** $\log_2 128$ **81.** $\log_3 81$ **82.** $\log_3 27$

83. $\log_4 \dfrac{1}{4}$ **84.** $\log_6 \dfrac{1}{6}$ **85.** $\log_6 \sqrt[3]{6}$ **86.** $\log_9 \sqrt[3]{9}$

Graph each logarithmic function. **See Examples 5 and 6.**

87. $g(x) = \log_3 x$ **88.** $g(x) = \log_5 x$ **89.** $f(x) = \log_4 x$ **90.** $f(x) = \log_6 x$

91. $f(x) = \log_{1/3} x$ **92.** $f(x) = \log_{1/5} x$ **93.** $g(x) = \log_{1/4} x$ **94.** $g(x) = \log_{1/6} x$

Use the graph at the right to predict the value of $f(t)$ for the given value of t.

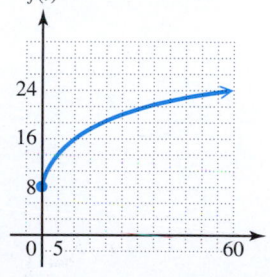

95. $t = 0$ **96.** $t = 10$ **97.** $t = 60$

98. Show that the points determined in **Exercises 95–97** lie on the graph of

$$f(t) = 8 \log_5 (2t + 5).$$

Concept Check   *Answer each question.*

99. Why is 1 not allowed as a base for a logarithmic function?

100. Why is a negative number not allowed as a base for a logarithmic function?

101. Why is $\log_a 1 = 0$ true for any value of a that is allowed as the base of a logarithm? Use a rule of exponents introduced earlier in the explanation.

102. Why is $\log_a a = 1$ true for any value of a that is allowed as the base of a logarithm?

*Solve each problem. **See Example 7.***

103. Sales (in thousands of units) of a new product are approximated by the logarithmic function

$$S(t) = 100 + 30 \log_3 (2t + 1),$$

where t is the number of years after the product is introduced.

(a) What were the sales, to the nearest unit, after 1 yr?

(b) What were the sales, to the nearest unit, after 13 yr?

(c) Graph $y = S(t)$.

104. A study showed that the number of mice in an old abandoned house was approximated by the logarithmic function

$$M(t) = 6 \log_4 (2t + 4),$$

where t is measured in months and $t = 0$ corresponds to January 2018. Find the number of mice in the house for each month.

(a) January 2018   **(b)** July 2018   **(c)** July 2020   **(d)** Graph $y = M(t)$.

105. An online sales company finds that its sales (in millions of dollars) are approximated by the logarithmic function

$$S(x) = \log_2 (3x + 1),$$

where x is the number of advertisements placed on a popular website. How many advertisements must be placed to earn sales of $4 million?

106. The population of deer (in thousands) in a certain area is approximated by the logarithmic function

$$f(x) = \log_5 (100x - 75),$$

where x is the number of years since 2017. During what year is the population expected to be 4 thousand deer?

RELATING CONCEPTS   For Individual or Group Work   (Exercises 107–110)

*To see how exponential and logarithmic functions are related, **work Exercises 107–110 in order.***

107. Complete the table of values, and sketch the graph of $y = 10^x$. Give the domain and range of the function.

x	y
−2	
−1	
0	
1	
2	

108. Complete the table of values, and sketch the graph of $y = \log_{10} x$. Give the domain and range of the function.

x	y
$\frac{1}{100}$	
$\frac{1}{10}$	
1	
10	
100	

109. Describe the symmetry between the graphs in **Exercises 107 and 108.**

110. What can we conclude about the functions

$$y = f(x) = 10^x \quad \text{and} \quad y = g(x) = \log_{10} x?$$

12.4 Properties of Logarithms

OBJECTIVES

1 Use the product rule for logarithms.
2 Use the quotient rule for logarithms.
3 Use the power rule for logarithms.
4 Use properties to write alternative forms of logarithmic expressions.

Logarithms were used as an aid to numerical calculation for several hundred years. Today the widespread use of calculators has made the use of logarithms for calculation obsolete. However, logarithms are still very important in applications and in further work in mathematics.

OBJECTIVE 1 Use the product rule for logarithms.

One way in which logarithms simplify problems is by changing a problem of multiplication into one of addition. For example, we know that

$$\log_2 4 = 2, \quad \log_2 8 = 3, \quad \text{and} \quad \log_2 32 = 5.$$

Therefore, we can make the following statements.

$$\log_2 32 = \log_2 4 + \log_2 8 \qquad 5 = 2 + 3$$

$$\log_2 (4 \cdot 8) = \log_2 4 + \log_2 8 \qquad 32 = 4 \cdot 8$$

This is an example of the product rule for logarithms.

Product Rule for Logarithms

If x, y, and b are positive real numbers, where $b \neq 1$, then the following holds true.

$$\log_b xy = \log_b x + \log_b y$$

That is, the logarithm of a product is the sum of the logarithms of the factors.

Examples: $\log_3 (4 \cdot 7) = \log_3 4 + \log_3 7$, $\log_{10} 8 + \log_{10} 9 = \log_{10} (8 \cdot 9)$

To prove this rule, let $m = \log_b x$ and $n = \log_b y$, and recall that

$\log_b x = m$ is equivalent to $b^m = x$ and $\log_b y = n$ is equivalent to $b^n = y$.

Now consider the product xy.

$xy = b^m \cdot b^n$	Substitute.
$xy = b^{m+n}$	Product rule for exponents
$\log_b xy = m + n$	Write in logarithmic form.
$\log_b xy = \log_b x + \log_b y$	Substitute for m and n.

The last statement is the result we wished to prove.

NOTE The word statement of the product rule can be restated by replacing the word "logarithm" with the word "exponent." The rule then becomes the familiar rule for multiplying exponential expressions:

The *exponent* of a product is *equal* to the sum of the *exponents* of the factors.

**NOW TRY
EXERCISE 1**

Use the product rule to rewrite each logarithm.

(a) $\log_{10} (7 \cdot 9)$

(b) $\log_5 11 + \log_5 8$

(c) $\log_5 (5x), \quad x > 0$

(d) $\log_2 t^3, \quad t > 0$

EXAMPLE 1　Using the Product Rule

Use the product rule to rewrite each logarithm. Assume $x > 0$.

(a) $\log_5 (6 \cdot 9)$

$= \log_5 6 + \log_5 9$ 　Product rule

(b) $\log_7 8 + \log_7 12$

$= \log_7 (8 \cdot 12)$ 　Product rule

$= \log_7 96$ 　Multiply.

(c) $\log_3 (3x)$

$= \log_3 3 + \log_3 x$ 　Product rule

$= 1 + \log_3 x$ 　$\log_b b = 1$

(d) $\log_4 x^3$

$= \log_4 (x \cdot x \cdot x)$ 　$x^3 = x \cdot x \cdot x$

$= \log_4 x + \log_4 x + \log_4 x$ 　Product rule

$= 3 \log_4 x$ 　Combine like terms.

NOW TRY

OBJECTIVE 2　Use the quotient rule for logarithms.

The rule for division is similar to the rule for multiplication.

Quotient Rule for Logarithms

If x, y, and b are positive real numbers, where $b \neq 1$, then the following holds true.

$$\log_b \frac{x}{y} = \log_b x - \log_b y$$

That is, the logarithm of a quotient is the difference of the logarithm of the numerator and the logarithm of the denominator.

Examples:　$\log_5 \frac{2}{3} = \log_5 2 - \log_5 3, \quad \log_7 3 - \log_7 5 = \log_7 \frac{3}{5}$

The proof of this rule is similar to the proof of the product rule.

**NOW TRY
EXERCISE 2**

Use the quotient rule to rewrite each logarithm.

(a) $\log_{10} \frac{7}{9}$

(b) $\log_4 x - \log_4 12, \quad x > 0$

(c) $\log_5 \frac{25}{27}$

(d) $\log_2 15 - \log_2 3$

EXAMPLE 2　Using the Quotient Rule

Use the quotient rule to rewrite each logarithm. Assume $x > 0$.

(a) $\log_4 \frac{7}{9}$

$= \log_4 7 - \log_4 9$ 　Quotient rule

(b) $\log_5 6 - \log_5 x$

$= \log_5 \frac{6}{x}$ 　Quotient rule

(c) $\log_3 \frac{27}{5}$

$= \log_3 27 - \log_3 5$ 　Quotient rule

$= 3 - \log_3 5$ 　$\log_3 27 = 3$

(d) $\log_6 28 - \log_6 7$

$= \log_6 \frac{28}{7}$ 　Quotient rule

$= \log_6 4$ 　$\frac{28}{7} = 4$ 　**NOW TRY**

NOW TRY ANSWERS

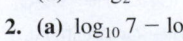

1. (a) $\log_{10} 7 + \log_{10} 9$
　(b) $\log_5 88$
　(c) $1 + \log_5 x$
　(d) $3 \log_2 t$

2. (a) $\log_{10} 7 - \log_{10} 9$
　(b) $\log_4 \frac{x}{12}$
　(c) $2 - \log_5 27$
　(d) $\log_2 5$

⚠ **CAUTION** *There is no property of logarithms to rewrite the logarithm of a sum.*

$$\log_b (x + y) \neq \log_b x + \log_b y$$

Also, $\log_b x \cdot \log_b y \neq \log_b xy$, and $\dfrac{\log_b x}{\log_b y} \neq \log_b \dfrac{x}{y}$.

OBJECTIVE 3 Use the power rule for logarithms.

Consider the exponential expression 2^3.

2^3 means $2 \cdot 2 \cdot 2$. The base 2 is used as a factor 3 times.

Similarly, the product rule can be extended to rewrite the logarithm of a power as the product of the exponent and the logarithm of the base.

$\log_5 2^3$	$\log_2 7^4$
$= \log_5 (2 \cdot 2 \cdot 2)$	$= \log_2 (7 \cdot 7 \cdot 7 \cdot 7)$
$= \log_5 2 + \log_5 2 + \log_5 2$	$= \log_2 7 + \log_2 7 + \log_2 7 + \log_2 7$
$= 3 \log_5 2$	$= 4 \log_2 7$

Furthermore, we saw in **Example 1(d)** that $\log_4 x^3 = 3 \log_4 x$. These examples suggest the following rule.

Power Rule for Logarithms

If x and b are positive real numbers, where $b \neq 1$, and if r is any real number, then the following holds true.

$$\log_b x^r = r \log_b x$$

That is, the logarithm of a number to a power equals the exponent times the logarithm of the number.

Examples: $\log_b m^5 = 5 \log_b m$, $\log_3 5^4 = 4 \log_3 5$

To prove the power rule, let $\log_b x = m$.

$b^m = x$	Write in exponential form.
$(b^m)^r = x^r$	Raise each side to the power r.
$b^{mr} = x^r$	Power rule for exponents
$\log_b x^r = mr$	Write in logarithmic form.
$\log_b x^r = rm$	Commutative property
$\log_b x^r = r \log_b x$	$m = \log_b x$ from above

This is the statement to be proved.

As a special case of the power rule, let $r = \dfrac{1}{p}$, so

$$\log_b \sqrt[p]{x} = \log_b x^{1/p} = \frac{1}{p} \log_b x.$$

Examples: $\log_b \sqrt[5]{x} = \log_b x^{1/5} = \dfrac{1}{5} \log_b x$, $\log_b \sqrt[3]{x^4} = \log_b x^{4/3} = \dfrac{4}{3} \log_b x$ $(x > 0)$

Another special case is

$$\log_b \frac{1}{x} = \log_b x^{-1} = -\log_b x. \qquad -a = -1 \cdot a$$

Example: $\log_9 \dfrac{1}{5} = \log_9 5^{-1} = -\log_9 5$

NOW TRY EXERCISE 3

Use the power rule to rewrite each logarithm. Assume $a > 0$, $x > 0$, and $a \neq 1$.

(a) $\log_7 5^3$ **(b)** $\log_a \sqrt{10}$

(c) $\log_3 \sqrt[4]{x^3}$ **(d)** $\log_4 \dfrac{1}{x^5}$

EXAMPLE 3   Using the Power Rule

Use the power rule to rewrite each logarithm. Assume $b > 0$, $x > 0$, and $b \neq 1$.

(a) $\log_5 4^2$

$= 2 \log_5 4$ Power rule

(b) $\log_b x^5$

$= 5 \log_b x$ Power rule

(c) $\log_b \sqrt{7}$

$= \log_b 7^{1/2}$ $\sqrt{x} = x^{1/2}$

$= \dfrac{1}{2} \log_b 7$ Power rule

(d) $\log_2 \sqrt[5]{x^2}$

$= \log_2 x^{2/5}$ $\sqrt[5]{x^2} = x^{2/5}$

$= \dfrac{2}{5} \log_2 x$ Power rule

(e) $\log_3 \dfrac{1}{x^4}$

$= \log_3 x^{-4}$ Definition of negative exponent

$= -4 \log_3 x$ Power rule

NOW TRY

We summarize the properties of logarithms from the previous section and this one.

Properties of Logarithms

If x, y, and b are positive real numbers, where $b \neq 1$, and r is any real number, then the following hold true.

Special Properties $\log_b b = 1$ $\log_b 1 = 0$

$\log_b b^r = r$ $b^{\log_b r} = r$ $(r > 0)$

Product Rule $\log_b xy = \log_b x + \log_b y$

Quotient Rule $\log_b \dfrac{x}{y} = \log_b x - \log_b y$

Power Rule $\log_b x^r = r \log_b x$

OBJECTIVE 4 Use properties to write alternative forms of logarithmic expressions.

EXAMPLE 4   Writing Logarithms in Alternative Forms

Use the properties of logarithms to rewrite each expression if possible. Assume that all variables represent positive real numbers.

(a) $\log_4 4x^3$

$= \log_4 4 + \log_4 x^3$ Product rule

$= 1 + 3 \log_4 x$ $\log_b b = 1$; Power rule

NOW TRY ANSWERS

3. **(a)** $3 \log_7 5$ **(b)** $\dfrac{1}{2} \log_a 10$

   (c) $\dfrac{3}{4} \log_3 x$ **(d)** $-5 \log_4 x$

**NOW TRY
EXERCISE 4**

Use the properties of logarithms to rewrite each expression if possible. Assume that all variables represent positive real numbers.

(a) $\log_3 9z^4$

(b) $\log_6 \sqrt{\dfrac{n}{3m}}$

(c) $\log_2 x + 3 \log_2 y - \log_2 z$

(d) $\log_5 (x + 10)$
$\quad + \log_5 (x - 10)$
$\quad - \dfrac{3}{5} \log_5 x, \quad x > 10$

(e) $\log_7 (49 + 2x)$

(b) $\log_7 \sqrt{\dfrac{m}{n}}$

$= \log_7 \left(\dfrac{m}{n}\right)^{1/2}$ — Write the radical expression with a rational exponent.

$= \dfrac{1}{2} \log_7 \dfrac{m}{n}$ — Power rule

$= \dfrac{1}{2} (\log_7 m - \log_7 n)$ — Quotient rule

(c) $\log_5 \dfrac{a^2}{bc}$

$= \log_5 a^2 - \log_5 bc$ — Quotient rule

$= 2 \log_5 a - \log_5 bc$ — Power rule

$= 2 \log_5 a - (\log_5 b + \log_5 c)$ — Product rule

$= 2 \log_5 a - \log_5 b - \log_5 c$ — Parentheses are necessary here.

(d) $4 \log_b m - \log_b n, \quad b \ne 1$

$= \log_b m^4 - \log_b n$ — Power rule

$= \log_b \dfrac{m^4}{n}$ — Quotient rule

(e) $\log_b (x + 1) + \log_b (2x + 1) - \dfrac{2}{3} \log_b x, \quad b \ne 1$

$= \log_b (x + 1) + \log_b (2x + 1) - \log_b x^{2/3}$ — Power rule

$= \log_b \dfrac{(x + 1)(2x + 1)}{x^{2/3}}$ — Product and quotient rules

$= \log_b \dfrac{2x^2 + 3x + 1}{x^{2/3}}$ — Multiply in the numerator.

(f) $\log_8 (2p + 3r)$ cannot be rewritten using the properties of logarithms. ***There is no property of logarithms to rewrite the logarithm of a sum.*** NOW TRY

In the next example, we use numerical values for $\log_2 5$ and $\log_2 3$. While we use the equality symbol to give these values, they are actually approximations because most logarithms of this type are irrational numbers. *We use $=$ with the understanding that the values are correct to four decimal places.*

EXAMPLE 5 Using the Properties of Logarithms with Numerical Values

Given that $\log_2 5 = 2.3219$ and $\log_2 3 = 1.5850$, use these values and the properties of logarithms to evaluate each expression.

(a) $\log_2 15$

$= \log_2 (3 \cdot 5)$ — Factor 15.

$= \log_2 3 + \log_2 5$ — Product rule

$= 1.5850 + 2.3219$ — Substitute the given values.

$= 3.9069$ — Add.

NOW TRY ANSWERS

4. (a) $2 + 4 \log_3 z$

(b) $\dfrac{1}{2}(\log_6 n - \log_6 3 - \log_6 m)$

(c) $\log_2 \dfrac{xy^3}{z}$ **(d)** $\log_5 \dfrac{x^2 - 100}{x^{3/5}}$

(e) cannot be rewritten

NOW TRY
EXERCISE 5

Given that $\log_2 7 = 2.8074$ and $\log_2 10 = 3.3219$, use these values and the properties of logarithms to evaluate each expression.

(a) $\log_2 70$ **(b)** $\log_2 0.7$

(c) $\log_2 49$

(b) $\log_2 0.6$

$= \log_2 \dfrac{3}{5}$ $0.6 = \frac{6}{10} = \frac{3}{5}$

$= \log_2 3 - \log_2 5$ Quotient rule

$= 1.5850 - 2.3219$ Substitute the given values.

$= -0.7369$ Subtract.

(c) $\log_2 27$

$= \log_2 3^3$ Write 27 as a power of 3.

$= 3 \log_2 3$ Power rule

$= 3(1.5850)$ Substitute the given value.

$= 4.7550$ Multiply.   **NOW TRY**

NOW TRY
EXERCISE 6

Determine whether each statement is *true* or *false*.

(a) $\log_2 16 + \log_2 16 = \log_2 32$

(b) $(\log_2 4)(\log_3 9) = \log_6 6^4$

EXAMPLE 6   Determining Whether Statements about Logarithms Are True

Determine whether each statement is *true* or *false*.

(a) $\log_2 8 - \log_2 4 = \log_2 4$

Evaluate each side.

$\log_2 8 - \log_2 4$	Left side	$\log_2 4$	Right side
$= \log_2 2^3 - \log_2 2^2$	Write 8 and 4 as powers of 2.	$= \log_2 2^2$	Write 4 as a power of 2.
$= 3 - 2$	$\log_b b^r = r$	$= 2$	$\log_b b^r = r$
$= 1$	Subtract.		

The statement is false because $1 \ne 2$.

(b) $\log_3 (\log_2 8) = \dfrac{\log_7 49}{\log_8 64}$

$\log_3 (\log_2 8)$	Left side	$\dfrac{\log_7 49}{\log_8 64}$	Right side
$= \log_3 (\log_2 2^3)$	Write 8 as a power of 2.	$= \dfrac{\log_7 7^2}{\log_8 8^2}$	Write 49 and 64 using exponents.
$= \log_3 3$	$\log_b b^r = r$	$= \dfrac{2}{2}$	$\log_b b^r = r$
$= 1$	$\log_b b = 1$	$= 1$	Simplify.

NOW TRY ANSWERS

5. (a) 6.1293   **(b)** −0.5145
   (c) 5.6148

6. (a) false   **(b)** true

The statement is true because $1 = 1$.   **NOW TRY**

12.4 Exercises

▶ **MyLab Math**

▶ *Video solutions for select problems available in MyLab Math*

Concept Check   *Determine whether each statement of a logarithmic property is* true *or* false. *If it is false, correct it by changing the right side of the equation.*

1. $\log_b x + \log_b y = \log_b (x + y)$

2. $\log_b \dfrac{x}{y} = \log_b x - \log_b y$

3. $\log_b xy = \log_b x + \log_b y$

4. $b^{\log_b r} = 1$

5. $\log_b b^r = r$

6. $\log_b x^r = \log_b rx$

7. Concept Check A student erroneously wrote $\log_a (x + y) = \log_a x + \log_a y$. When his teacher explained that this was indeed wrong, the student claimed that he had used the distributive property. **WHAT WENT WRONG?**

8. Concept Check Consider the following "proof" that $\log_2 16$ does not exist.

$$\log_2 16$$
$$= \log_2 (-4)(-4)$$
$$= \log_2 (-4) + \log_2 (-4)$$

The logarithm of a negative number is not defined, so the final step cannot be evaluated. Thus $\log_2 16$ does not exist. **WHAT WENT WRONG?**

Concept Check *Use the indicated rule of logarithms to complete each equation.*

9. $\log_{10} (7 \cdot 8)$ = _____ (product rule)

10. $\log_{10} \dfrac{7}{8}$ = _____ (quotient rule)

11. $3^{\log_3 4}$ = _____ (special property)

12. $\log_{10} 3^6$ = _____ (power rule)

13. $\log_3 3^9$ = _____ (special property)

14. $\log_3 9^2$ = _____ (special property)

Use the properties of logarithms to express each logarithm as a sum or difference of logarithms, or as a single logarithm if possible. Assume that all variables represent positive real numbers. **See Examples 1–4.**

15. $\log_7 (4 \cdot 5)$

16. $\log_8 (9 \cdot 11)$

17. $\log_5 \dfrac{8}{3}$

18. $\log_3 \dfrac{7}{5}$

19. $\log_4 6^2$

20. $\log_5 7^4$

21. $\log_3 \dfrac{\sqrt[3]{4}}{x^2 y}$

22. $\log_7 \dfrac{\sqrt[3]{13}}{pq^2}$

23. $\log_3 \sqrt{\dfrac{xy}{5}}$

24. $\log_6 \sqrt{\dfrac{pq}{7}}$

25. $\log_2 \dfrac{\sqrt[3]{x} \cdot \sqrt[5]{y}}{r^2}$

26. $\log_4 \dfrac{\sqrt[4]{z} \cdot \sqrt[5]{w}}{s^2}$

Use the properties of logarithms to write each expression as a single logarithm. Assume that all variables are defined in such a way that the variable expressions are positive, and bases are positive numbers not equal to 1. **See Examples 1–4.**

27. $\log_b x + \log_b y$

28. $\log_b w + \log_b z$

29. $\log_a m - \log_a n$

30. $\log_b x - \log_b y$

31. $(\log_a r - \log_a s) + 3 \log_a t$

32. $(\log_a p - \log_a q) + 2 \log_a r$

33. $3 \log_a 5 - 4 \log_a 3$

34. $3 \log_a 5 - \dfrac{1}{2} \log_a 9$

35. $\log_{10} (x + 3) + \log_{10} (x - 3)$

36. $\log_{10} (y + 4) + \log_{10} (y - 4)$

37. $\log_{10} (x + 3) + \log_{10} (x + 5)$

38. $\log_{10} (x + 4) + \log_{10} (x + 6)$

39. $3 \log_p x + \dfrac{1}{2} \log_p y - \dfrac{3}{2} \log_p z - 3 \log_p a$

40. $\dfrac{1}{3} \log_b x + \dfrac{2}{3} \log_b y - \dfrac{3}{4} \log_b s - \dfrac{2}{3} \log_b t$

To four decimal places, the values of $\log_{10} 2$ and $\log_{10} 9$ are

$$\log_{10} 2 = 0.3010 \quad \text{and} \quad \log_{10} 9 = 0.9542.$$

*Use these values and the properties of logarithms to evaluate each expression. DO NOT USE A CALCULATOR. **See Example 5.***

41. $\log_{10} 18$

42. $\log_{10} 4$

43. $\log_{10} \dfrac{2}{9}$

44. $\log_{10} \dfrac{9}{2}$

45. $\log_{10} 36$

46. $\log_{10} 162$

47. $\log_{10} \sqrt[4]{9}$

48. $\log_{10} \sqrt[5]{2}$

49. $\log_{10} 3$

50. $\log_{10} \dfrac{1}{9}$

51. $\log_{10} 9^5$

52. $\log_{10} 2^{19}$

Determine whether each statement is true *or* false. ***See Example 6.***

53. $\log_2 (8 + 32) = \log_2 8 + \log_2 32$

54. $\log_2 (64 - 16) = \log_2 64 - \log_2 16$

55. $\log_3 7 + \log_3 7^{-1} = 0$

56. $\log_3 49 + \log_3 49^{-1} = 0$

57. $\log_6 60 - \log_6 10 = 1$

58. $\log_3 8 + \log_3 \dfrac{1}{8} = 0$

59. $\dfrac{\log_{10} 7}{\log_{10} 14} = \dfrac{1}{2}$

60. $\dfrac{\log_{10} 10}{\log_{10} 100} = \dfrac{1}{10}$

12.5   Common and Natural Logarithms

OBJECTIVES

1 Evaluate common logarithms using a calculator.

2 Use common logarithms in applications.

3 Evaluate natural logarithms using a calculator.

4 Use natural logarithms in applications.

5 Use the change-of-base rule.

Logarithms are important in many applications in biology, engineering, economics, and social science. In this section we find numerical approximations for logarithms. Traditionally, base 10 logarithms were used most often because our number system is base 10. Logarithms to base 10 are **common logarithms,** and

$$\log_{10} x \text{ is abbreviated as } \log x,$$

where the base is understood to be 10.

OBJECTIVE 1   Evaluate common logarithms using a calculator.

We evaluate common logarithms using a calculator with a $\boxed{\text{LOG}}$ key.

EXAMPLE 1   Evaluating Common Logarithms

Evaluate each logarithm to four decimal places using a calculator as needed.

VOCABULARY
☐ common logarithm
☐ natural logarithm

(a) $\log 327.1 \approx 2.5147$

(b) $\log 437{,}000 \approx 5.6405$

(c) $\log 0.0615 \approx -1.2111$

(d) $\log 10^{6.1988} = 6.1988$

Special property $\log_b b^r = r$

**NOW TRY
EXERCISE 1**

Evaluate each logarithm to
four decimal places using a
calculator as needed.

(a) log 115

(b) log 539,000

(c) log 0.023

(d) log $10^{12.2139}$

In part (c), log 0.0615 ≈ −1.2111, which is a
negative result. *The common logarithm of a number between 0 and 1 is always negative* because the
logarithm is the exponent on 10 that produces the
number. In this case, we have

$$10^{-1.2111} \approx 0.0615.$$

If the exponent (the logarithm) were positive, the
result would be greater than 1 because $10^0 = 1$. The
graph in **FIGURE 20** illustrates these concepts.

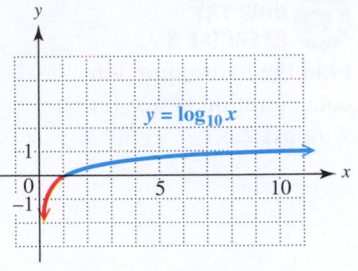
FIGURE 20

NOW TRY

OBJECTIVE 2 Use common logarithms in applications.

In chemistry, pH is a measure of the acidity or alkalinity of a solution. Water, for
example, has pH 7. In general, acids have pH numbers less than 7, and alkaline solutions have pH values greater than 7. **FIGURE 21** illustrates the pH scale.

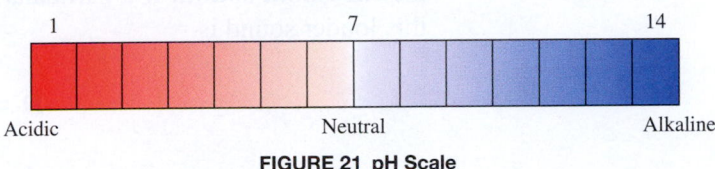
FIGURE 21 pH Scale

The **pH** of a solution is defined as

$$pH = -\log\,[H_3O^+],$$

where $[H_3O^+]$ is the hydronium ion concentration in moles per liter. It is customary
to round pH values to the nearest tenth.

**NOW TRY
EXERCISE 2**

Water taken from a wetland
has a hydronium ion
concentration of

3.4×10^{-5} mole per liter.

Find the pH value for the
water and classify the wetland
as a rich fen, a poor fen, or
a bog.

NOW TRY ANSWERS
1. **(a)** 2.0607 **(b)** 5.7316
   (c) −1.6383 **(d)** 12.2139
2. 4.5; poor fen

EXAMPLE 2 Using pH in an Application

Wetlands are classified as *bogs, fens, marshes,* and
swamps, on the basis of pH values. A pH value
between 6.0 and 7.5, such as that of Summerby
Swamp in Michigan's Hiawatha National Forest,
indicates that the wetland is a "rich fen." When the
pH is between 3.0 and 6.0, the wetland is a "poor
fen," and if the pH falls to 3.0 or less, it is a "bog."
(Data from Mohlenbrock, R., "Summerby Swamp,
Michigan," *Natural History.*)

Suppose that the hydronium ion concentration of a sample of water from a wetland
is 6.3×10^{-3} mole per liter. How would this wetland be classified?

pH = −log $[H_3O^+]$ Definition of pH

pH = −log (6.3×10^{-3}) Let $[H_3O^+] = 6.3 \times 10^{-3}$.

pH = −(log 6.3 + log 10^{-3})   Product rule

pH ≈ −[0.7993 − 3]   Use a calculator to approximate log 6.3; $\log_b b^r = r$

pH = −0.7993 + 3   Distributive property

pH ≈ 2.2   Add.

The pH is less than 3.0, so the wetland is a bog.

NOW TRY

Find the hydronium ion concentration of a solution with pH 2.6.

EXAMPLE 3 Finding Hydronium Ion Concentration

Find the hydronium ion concentration of drinking water with pH 6.5.

$$pH = -\log\left[H_3O^+\right] \quad \text{Definition of pH}$$

$$6.5 = -\log\left[H_3O^+\right] \quad \text{Let pH} = 6.5.$$

$$\log\left[H_3O^+\right] = -6.5 \quad \text{Multiply by } -1. \text{ Interchange sides.}$$

$$\left[H_3O^+\right] = 10^{-6.5} \quad \text{Write in exponential form, base 10.}$$

$$\left[H_3O^+\right] \approx 3.2 \times 10^{-7} \quad \text{Evaluate with a calculator.}$$

The hydronium ion concentration of drinking water is approximately 3.2×10^{-7} mole per liter.

NOW TRY

The loudness of sound is measured in a unit called a **decibel,** abbreviated **dB.** To measure with this unit, we first assign an intensity of I_0 to a very faint sound, called the **threshold sound.** If a particular sound has intensity I, then the decibel level of this louder sound is

$$D = 10\log\left(\frac{I}{I_0}\right).$$

The loudness of some common sounds are shown in the table. Any sound over 85 dB exceeds what hearing experts consider safe. Permanent hearing damage can be suffered at levels above 150 dB.

▼ Loudness of Common Sounds

Decibel Level	Example
60	Normal conversation
90	Rush hour traffic, lawn mower
100	Garbage truck, chain saw, pneumatic drill
120	Rock concert, thunderclap
140	Gunshot blast, jet engine
180	Rocket launching pad

Data from Deafness Research Foundation.

Find the decibel level to the nearest whole number of the sound from a jet engine with intensity I of

$$(6.312 \times 10^{13})I_0.$$

EXAMPLE 4 Measuring the Loudness of Sound

If music delivered through Bluetooth headphones has intensity I of $(3.162 \times 10^9)I_0$, find the average decibel level.

$$D = 10\log\left(\frac{I}{I_0}\right)$$

$$D = 10\log\left(\frac{(3.162 \times 10^9)I_0}{I_0}\right) \quad \text{Substitute the given value for } I.$$

$$D = 10\log\left(3.162 \times 10^9\right)$$

$$D \approx 95 \text{ dB} \quad \text{Evaluate with a calculator.} \quad \text{NOW TRY}$$

Leonhard Euler (1707–1783)

The number e is named
after Euler.

OBJECTIVE 3 Evaluate natural logarithms using a calculator.

Logarithms used in applications are often **natural logarithms,** which have as base the number **e.** The letter e was chosen to honor Leonhard Euler, who published extensive results on the number in 1748. It is an irrational number, so its decimal expansion never terminates and never repeats.

One way to see how e appears in an exponential situation involves calculating the values of $\left(1 + \frac{1}{x}\right)^x$ as x gets larger without bound. See the table.

x	$\left(1 + \frac{1}{x}\right)^x$
1	2
10	2.59374246
100	2.704813829
1000	2.716923932
10,000	2.718145927
100,000	2.718268237

These approximations are found using a calculator.

It appears that as x gets larger without bound, $\left(1 + \frac{1}{x}\right)^x$ approaches some number. This number is e.

Approximation for e

$$e \approx 2.718281828459$$

A calculator with an $\boxed{e^x}$ key can approximate powers of e.

$$e^2 \approx 7.389056099, \quad e^3 \approx 20.08553692, \quad \text{and} \quad e^{0.6} \approx 1.8221188$$

Logarithms with base e are called natural logarithms because they occur in natural situations that involve growth or decay.

The base e logarithm of x is written
ln x (read "*el en x*").

A graph of **y = ln x** is given in **FIGURE 22.**

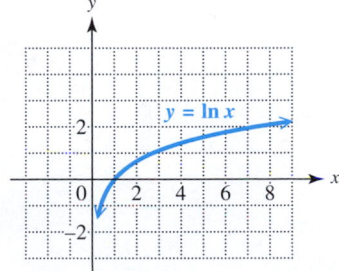

FIGURE 22

A calculator key labeled $\boxed{LN}$ is used to evaluate natural logarithms.

**NOW TRY
EXERCISE 5**

Evaluate each logarithm to four decimal places using a calculator as needed.

(a) ln 0.26   **(b)** ln 12

(c) ln 150   **(d)** ln $e^{5.8321}$

EXAMPLE 5 Evaluating Natural Logarithms

Evaluate each logarithm to four decimal places using a calculator as needed.

(a) ln 0.5841 ≈ −0.5377   **(b)** ln 0.9215 ≈ −0.0818

As with common logarithms, a number between 0 and 1 has a negative natural logarithm.

(c) ln 192.7 ≈ 5.2611   **(d)** ln $e^{4.6832}$ = 4.6832

Special property $\log_b b^r = r$

NOW TRY

NOW TRY ANSWERS

5. (a) −1.3471 **(b)** 2.4849
  (c) 5.0106 **(d)** 5.8321

OBJECTIVE 4   Use natural logarithms in applications.

NOW TRY
EXERCISE 6
Use the natural logarithmic
function in **Example 6** to
approximate the altitude
when atmospheric pressure is
600 millibars. Round to the
nearest hundred.

EXAMPLE 6   Applying a Natural Logarithmic Function

Altitude in meters that corresponds to an atmospheric pressure of x millibars can be approximated by the natural logarithmic function

$$f(x) = 51{,}600 - 7457 \ln x.$$

(Data from Miller, A. and J. Thompson, *Elements of Meteorology*, Fourth Edition, Charles E. Merrill Publishing Company.)

Use this function to find the altitude when atmospheric pressure is 400 millibars. Round to the nearest hundred.

$$f(x) = 51{,}600 - 7457 \ln x$$

$$f(400) = 51{,}600 - 7457 \ln 400 \qquad \text{Let } x = 400.$$

$$f(400) \approx 6900 \qquad\qquad\qquad \text{Evaluate with a calculator.}$$

Atmospheric pressure is 400 millibars at 6900 m.   NOW TRY

> **NOTE** In **Example 6,** the final answer was obtained using a calculator *without* rounding the intermediate values. In general, it is best to wait until the final step to round the answer. Otherwise, a buildup of round-off error may cause the final answer to have an incorrect final decimal place digit or digits.

OBJECTIVE 5   Use the change-of-base rule.

The change-of-base rule enables us to convert logarithms from one base to another.

Change-of-Base Rule

If $a > 0$, $a \neq 1$, $b > 0$, $b \neq 1$, and $x > 0$, then the following holds true.

$$\log_a x = \frac{\log_b x}{\log_b a}$$

To derive the change-of-base rule, let $\log_a x = m$.

$$\log_a x = m$$

$$a^m = x \qquad\qquad \text{Write in exponential form.}$$

$$\log_b a^m = \log_b x \qquad \text{Take the logarithm on each side.}$$

$$m \log_b a = \log_b x \qquad \text{Power rule}$$

$$(\log_a x)(\log_b a) = \log_b x \qquad \text{Substitute for } m.$$

$$\log_a x = \frac{\log_b x}{\log_b a} \qquad \text{Divide by } \log_b a.$$

NOW TRY ANSWER
6. 3900 m

This last statement is the change-of-base rule.

> **NOTE** Any positive number other than 1 can be used for base b in the change-of-base rule. Usually the only practical bases are e and 10, because calculators give logarithms for these two bases.

 NOW TRY
EXERCISE 7
Use the change-of-base rule to approximate each logarithm to four decimal places.
(a) $\log_2 7$ **(b)** $\log_{1/3} 12$

NOW TRY ANSWERS
7. (a) 2.8074 **(b)** -2.2619

EXAMPLE 7 Using the Change-of-Base Rule

Use the change-of-base rule to approximate $\log_5 12$ to four decimal places.

$$\log_5 12 = \frac{\log 12}{\log 5} \approx 1.5440$$

Either common or natural logarithms can be used.

$$\log_5 12 = \frac{\ln 12}{\ln 5} \approx 1.5440$$

NOW TRY

12.5 Exercises

FOR EXTRA HELP   **MyLab Math**

▶ *Video solutions for select problems available in MyLab Math*

Concept Check *Choose the correct response.*

1. What is the base in the expression $\log x$?
  A. x **B.** 1   **C.** 10   **D.** e

2. What is the base in the expression $\ln x$?
  A. x **B.** 1   **C.** 10   **D.** e

3. Given that $10^0 = 1$ and $10^1 = 10$, between what two consecutive integers is the value of $\log 6.3$?
  A. -1 and 0   **B.** 0 and 1   **C.** 6 and 7   **D.** 10 and 11

4. Given that $e^1 \approx 2.718$ and $e^2 \approx 7.389$, between what two consecutive integers is the value of $\ln 6.3$?
  A. 0 and 1   **B.** 1 and 2   **C.** 2 and 3   **D.** 6 and 7

Concept Check *Without using a calculator, give the value of each expression.*

5. $\log 10^{9.6421}$ **6.** $\ln e^{\sqrt{2}}$ **7.** $10^{\log \sqrt{3}}$

8. $e^{\ln 75.2}$ **9.** $\ln e^{-11.4007}$ **10.** $10 \ln e^4$

Use a calculator for most of the remaining exercises in this set.

Evaluate each logarithm to four decimal places. **See Examples 1 and 5.**

11. $\log 43$ **12.** $\log 98$ **13.** $\log 328.4$

14. $\log 457.2$ **15.** $\log 0.0326$ **16.** $\log 0.1741$

17. $\log (4.76 \times 10^9)$ **18.** $\log (2.13 \times 10^4)$ **19.** $\ln 7.84$

20. $\ln 8.32$ **21.** $\ln 0.0556$ **22.** $\ln 0.0217$

23. $\ln 388.1$ **24.** $\ln 942.6$ **25.** $\ln (8.59 \times e^2)$

26. $\ln (7.46 \times e^3)$ **27.** $\ln 10$ **28.** $\log e$

29. Concept Check Use a calculator to find approximations of each logarithm.
  (a) $\log 356.8$ **(b)** $\log 35.68$ **(c)** $\log 3.568$

  (d) Observe the answers and make a conjecture concerning the decimal values of the common logarithms of numbers greater than 1 that have the same digits.

30. Concept Check   Let k represent the number of letters in your last name.

(a) Use a calculator to find $\log k$.

(b) Raise 10 to the power indicated by the number found in part (a). What is the result?

(c) Explain why we obtained the answer found in part (b). Would it matter what number we used for k to observe the same result?

Suppose that water from a wetland area is sampled and found to have the given hydronium ion concentration. Is the wetland a rich fen, *a* poor fen, *or a* bog? **See *Example 2*.**

31. 3.1×10^{-5} **32.** 2.5×10^{-5} **33.** 2.5×10^{-2}

34. 3.6×10^{-2} **35.** 2.7×10^{-7} **36.** 2.5×10^{-7}

Find the pH *(to the nearest tenth) of the substance with the given hydronium ion concentration.* **See Example 2.**

37. Ammonia, 2.5×10^{-12} **38.** Egg white, 1.6×10^{-8}

39. Sodium bicarbonate, 4.0×10^{-9} **40.** Tuna, 1.3×10^{-6}

41. Grapes, 5.0×10^{-5} **42.** Grapefruit, 6.3×10^{-4}

Find the hydronium ion concentration of the substance with the given pH. **See Example 3.**

43. Human blood plasma, 7.4   **44.** Milk, 6.4

45. Human gastric contents, 2.0   **46.** Spinach, 5.4

47. Bananas, 4.6   **48.** Milk of magnesia, 10.5

Solve each problem. **See Examples 4 and 6.**

49. Managements of sports stadiums and arenas often encourage fans to make as much noise as possible. Find the average decibel level

$$D = 10 \log \left(\frac{I}{I_0} \right)$$

for each venue with the given intensity I.

(a) NFL fans, Kansas City Chiefs at Arrowhead Stadium: $I = (1.58 \times 10^{14})I_0$

(b) NBA fans, Sacramento Kings at Sleep Train Arena: $I = (3.9 \times 10^{12})I_0$

(c) MLB fans, Baltimore Orioles at Camden Yards: $I = (1.1 \times 10^{12})I_0$

(Data from www.baltimoresportsreport.com, www.guinessworldrecords.com)

50. Find the decibel level of each sound.

(a) noisy restaurant: $I = 10^8 I_0$

(b) farm tractor: $I = (5.340 \times 10^9)I_0$

(c) snowmobile: $I = 31,622,776,600 I_0$

(Data from The Canadian Society of Otolaryngology.)

51. The number of years, $N(x)$, since two independently evolving languages split off from a common ancestral language is approximated by

$$N(x) = -5000 \ln x,$$

where x is the percent of words (in decimal form) from the ancestral language common to both languages now. Find the number of years (to the nearest hundred years) since the split for each percent of common words.

(a) 85% (or 0.85)   **(b)** 35% (or 0.35)   **(c)** 10% (or 0.10)

52. The time t in years for an amount of money invested at an interest rate r (in decimal form) to double is given by

$$t(r) = \frac{\ln 2}{\ln (1 + r)}.$$

This is the **doubling time.** Find the doubling time to the nearest tenth for an investment at each interest rate.

(a) 2% (or 0.02)   **(b)** 5% (or 0.05)   **(c)** 8% (or 0.08)

53. The number of monthly active Twitter users (in millions) worldwide during the third quarter of each year from 2010 to 2017 is approximated by

$$f(x) = 25.6829 + 149.1368 \ln x,$$

where $x = 1$ represents 2010, $x = 2$ represents 2011, and so on. (Data from Twitter.)

(a) What does this model give for the number of monthly active Twitter users in 2012?

(b) According to this model, when did the number of monthly active Twitter users reach 300 million? (*Hint:* Substitute for $f(x)$, and then write the equation in exponential form to solve it.)

54. In the central Sierra Nevada of California, the percent of moisture that falls as snow rather than rain is approximated by

$$f(x) = 86.3 \ln x - 680,$$

where x is the altitude in feet.

(a) What percent of the moisture at 5000 ft falls as snow?

(b) What percent at 7500 ft falls as snow?

55. The approximate tax $T(x)$, in dollars per ton, that would result in an $x\%$ (in decimal form) reduction in carbon dioxide emissions is approximated by the **cost-benefit equation**

$$T(x) = -0.642 - 189 \ln (1 - x).$$

(a) What tax will reduce emissions 25%?

(b) Explain why the equation is not valid for $x = 0$ or $x = 1$.

56. The age in years of a female blue whale of length x in feet is approximated by

$$f(x) = -2.57 \ln \left(\frac{87 - x}{63} \right).$$

(a) How old is a female blue whale that measures 80 ft?

(b) The equation that defines this function has domain $24 < x < 87$. Explain why.

Use the change-of-base rule (with either common or natural logarithms) to approximate each logarithm to four decimal places. **See Example 7.**

57. $\log_3 12$ **58.** $\log_4 18$ **59.** $\log_5 3$ **60.** $\log_7 4$

61. $\log_{21} 0.7496$ **62.** $\log_{19} 0.8325$ **63.** $\log_{1/2} 5$ **64.** $\log_{1/3} 7$

65. $\log_3 \sqrt{2}$ **66.** $\log_6 \sqrt[3]{5}$ **67.** $\log_\pi e$ **68.** $\log_\pi 10$

12.6   Exponential and Logarithmic Equations; Further Applications

OBJECTIVES

1 Solve equations involving variables in the exponents.

2 Solve equations involving logarithms.

3 Solve applications of compound interest.

4 Solve applications involving base e exponential growth and decay.

General methods for solving these equations depend on the following properties.

> **Properties for Solving Exponential and Logarithmic Equations**
>
> For all real numbers $b > 0$, $b \neq 1$, and any real numbers x and y, the following hold true.
>
> **1.** If $x = y$, then $b^x = b^y$.
>
> **2.** If $b^x = b^y$, then $x = y$.
>
> **3.** If $x = y$, and $x > 0$, $y > 0$, then $\log_b x = \log_b y$.
>
> **4.** If $x > 0$, $y > 0$, and $\log_b x = \log_b y$, then $x = y$.

OBJECTIVE 1   Solve equations involving variables in the exponents.

VOCABULARY

☐ compound interest
☐ continuous compounding

**NOW TRY
EXERCISE 1**

Solve $5^x = 20$.
Approximate the solution to three decimal places.

EXAMPLE 1   Solving an Exponential Equation

Solve $3^x = 12$. Approximate the solution to three decimal places.

$$3^x = 12$$

$$\log 3^x = \log 12 \qquad \text{Property 3 (common logarithms):}$$
$$\text{If } x = y, \text{ then } \log_b x = \log_b y.$$

$$x \log 3 = \log 12 \qquad \text{Power rule}$$

$$\text{Exact solution} \longrightarrow x = \frac{\log 12}{\log 3} \qquad \text{Divide by log 3.}$$

$$\text{Decimal approximation} \longrightarrow x \approx 2.262 \qquad \text{Evaluate with a calculator.}$$

CHECK $3^x = 3^{2.262} \approx 12$ ✓   True

The solution set is $\{2.262\}$.   **NOW TRY**

⚠ CAUTION   Be careful: $\frac{\log 12}{\log 3}$ is **not** equal to log 4.

$$\frac{\log 12}{\log 3} \approx 2.262, \quad \text{but} \quad \log 4 \approx 0.6021.$$

> **NOTE**   In **Example 1**, we could have used Property 3 with natural logarithms.
>
>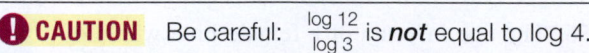
> $$\frac{\ln 12}{\ln 3} = \frac{\log 12}{\log 3} \approx 2.262 \qquad \text{(Verify this.)}$$

NOW TRY ANSWER
1. $\{1.861\}$

 **NOW TRY EXERCISE 2**

Solve $e^{0.12x} = 10$. Approximate the solution to three decimal places.

EXAMPLE 2 Solving an Exponential Equation (Base e)

Solve $e^{0.003x} = 40$. Approximate the solution to three decimal places.

$$e^{0.003x} = 40$$

$$\ln e^{0.003x} = \ln 40 \qquad \text{Property 3 (natural logarithms):}$$
$$\text{If } x = y, \text{ then } \ln x = \ln y.$$

$$0.003x \ln e = \ln 40 \qquad \text{Power rule}$$

$$0.003x = \ln 40 \qquad \ln e = \ln e^1 = 1$$

$$x = \frac{\ln 40}{0.003} \qquad \text{Divide by 0.003.}$$

$$x \approx 1229.626 \qquad \text{Evaluate with a calculator.}$$

CHECK $e^{0.003x} = e^{0.003(1229.626)} \approx 40$ ✓ True

The solution set is $\{1229.626\}$.  **NOW TRY**

 **NOW TRY EXERCISE 3**

Solve $3^{-x+5} = 8$. Approximate the solution to three decimal places.

EXAMPLE 3 Solving an Exponential Equation

Solve $7^{-x+4} = 17$. Approximate the solution to three decimal places.

$$7^{-x+4} = 17$$

$$\log 7^{-x+4} = \log 17 \qquad \text{Property 3 (common logarithms):}$$
$$\text{If } x = y, \text{ then } \log_b x = \log_b y.$$

$$(-x + 4) \log 7 = \log 17 \qquad \text{Power rule}$$

$$-x \log 7 + 4 \log 7 = \log 17 \qquad \text{Distributive property}$$

$$-x \log 7 = \log 17 - 4 \log 7 \qquad \text{Subtract 4 log 7.}$$

$$x = \frac{\log 17 - 4 \log 7}{-\log 7} \qquad \text{Divide by } -\log 7.$$

$$x \approx 2.544 \qquad \text{Evaluate with a calculator.}$$

CHECK $7^{-x+4} = 7^{-2.544+4} \approx 17$ ✓ True

The solution set is $\{2.544\}$. **NOW TRY**

> **General Method for Solving an Exponential Equation**
>
> Take logarithms having the same base on both sides and then use the power rule of logarithms or the special property $\log_b b^x = x$. **(See Examples 1–3.)**
>
> As a special case, if both sides can be written as exponentials with the same base, do so, and set the exponents equal.

OBJECTIVE 2 Solve equations involving logarithms.

EXAMPLE 4 Solving a Logarithmic Equation

Solve $\log_3 (4x + 1) = 4$.

$$\log_3 (4x + 1) = 4$$

$$4x + 1 = 3^4 \qquad \text{Write in exponential form.}$$

$$4x + 1 = 81 \qquad 3^4 = 81$$

$$4x = 80 \qquad \text{Subtract 1.}$$

$$x = 20 \qquad \text{Divide by 4.}$$

NOW TRY ANSWERS

2. $\{19.188\}$
3. $\{3.107\}$

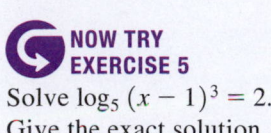

**NOW TRY
EXERCISE 4**
Solve $\log_6 (2x + 4) = 2$.

CHECK
$$\log_3 (4x + 1) = 4$$
$$\log_3 [4(20) + 1] \overset{?}{=} 4 \qquad \text{Let } x = 20.$$
$$\log_3 81 \overset{?}{=} 4 \qquad \text{Work inside the brackets.}$$
$$4 = 4 \; \checkmark \qquad \text{True; } \log_3 81 = 4$$

A true statement results, so the solution set is $\{20\}$.   **NOW TRY**

**NOW TRY
EXERCISE 5**
Solve $\log_5 (x - 1)^3 = 2$.
Give the exact solution.

EXAMPLE 5   Solving a Logarithmic Equation

Solve $\log_2 (x + 5)^3 = 4$. Give the exact solution.
$$\log_2 (x + 5)^3 = 4$$
$$(x + 5)^3 = 2^4 \qquad \text{Write in exponential form.}$$
$$(x + 5)^3 = 16 \qquad 2^4 = 16$$
$$x + 5 = \sqrt[3]{16} \qquad \text{Take the cube root on each side.}$$
$$x = -5 + \sqrt[3]{16} \qquad \text{Add } -5.$$
$$x = -5 + 2\sqrt[3]{2} \qquad \sqrt[3]{16} = \sqrt[3]{8 \cdot 2} = \sqrt[3]{8} \cdot \sqrt[3]{2} = 2\sqrt[3]{2}$$

CHECK
$$\log_2 (x + 5)^3 = 4 \qquad \text{Original equation}$$
$$\log_2 \left(-5 + 2\sqrt[3]{2} + 5\right)^3 \overset{?}{=} 4 \qquad \text{Let } x = -5 + 2\sqrt[3]{2}.$$
$$\log_2 \left(2\sqrt[3]{2}\right)^3 \overset{?}{=} 4 \qquad \text{Work inside the parentheses.}$$
$$\log_2 16 \overset{?}{=} 4 \qquad \left(2\sqrt[3]{2}\right)^3 = 2^3 \left(\sqrt[3]{2}\right)^3 = 8 \cdot 2 = 16$$
$$4 = 4 \; \checkmark \qquad \text{True; } \log_2 16 = 4$$

A true statement results, so the solution set is $\left\{-5 + 2\sqrt[3]{2}\right\}$.   **NOW TRY**

> **⚠ CAUTION**   Recall that the domain of $f(x) = \log_b x$ is $(0, \infty)$. *For this reason, always check that each proposed solution of an equation with logarithms yields only logarithms of positive numbers in the original equation.*

EXAMPLE 6   Solving a Logarithmic Equation

Solve $\log_2 (x + 1) - \log_2 x = \log_2 7$.
$$\log_2 (x + 1) - \log_2 x = \log_2 7$$

> Transform the left side to an expression with only *one* logarithm.

$$\log_2 \frac{x + 1}{x} = \log_2 7 \qquad \text{Quotient rule}$$
$$\frac{x + 1}{x} = 7 \qquad \text{Property 4: If } \log_b x = \log_b y, \text{ then } x = y.$$
$$x + 1 = 7x \qquad \text{Multiply by } x.$$
$$1 = 6x \qquad \text{Subtract } x.$$

> This proposed solution must be checked.

$$\frac{1}{6} = x \qquad \text{Divide by 6.}$$

NOW TRY ANSWERS
4. $\{16\}$
5. $\{1 + \sqrt[3]{25}\}$

We cannot take the logarithm of a *nonpositive* number, so both $x + 1$ and x must be positive here. If $x = \frac{1}{6}$, then this condition is satisfied.

**NOW TRY
EXERCISE 6**

Solve.

$\log_4 (2x + 13) - \log_4 (x + 1)$
$\quad = \log_4 10$

CHECK

$$\log_2 (x + 1) - \log_2 x = \log_2 7 \qquad \text{Original equation}$$

$$\log_2 \left(\frac{1}{6} + 1 \right) - \log_2 \frac{1}{6} \stackrel{?}{=} \log_2 7 \qquad \text{Let } x = \tfrac{1}{6}.$$

$$\log_2 \frac{7}{6} - \log_2 \frac{1}{6} \stackrel{?}{=} \log_2 7 \qquad \text{Add.}$$

$$\log_2 \frac{\frac{7}{6}}{\frac{1}{6}} \stackrel{?}{=} \log_2 7 \qquad \text{Quotient rule}$$

$$\frac{\frac{7}{6}}{\frac{1}{6}} = \frac{7}{6} \div \frac{1}{6} = \frac{7}{6} \cdot \frac{6}{1} = 7$$

$$\log_2 7 = \log_2 7 \quad \checkmark \qquad \text{True}$$

A true statement results, so the solution set is $\left\{ \frac{1}{6} \right\}$.  **NOW TRY**

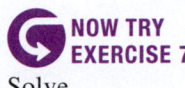

**NOW TRY
EXERCISE 7**

Solve.

$\log_4 (x + 2) + \log_4 2x = 2$

EXAMPLE 7 Solving a Logarithmic Equation

Solve $\log x + \log (x - 21) = 2$.

$$\log x + \log (x - 21) = 2$$

$$\log x(x - 21) = 2 \qquad \text{Product rule}$$

$$\text{(The base is 10.)} \quad x(x - 21) = 10^2 \qquad \text{Write in exponential form.}$$

$$x^2 - 21x = 100 \qquad \text{Distributive property; } 10^2 = 100$$

$$x^2 - 21x - 100 = 0 \qquad \text{Standard form}$$

$$(x - 25)(x + 4) = 0 \qquad \text{Factor.}$$

$$x - 25 = 0 \quad \text{or} \quad x + 4 = 0 \qquad \text{Zero-factor property}$$

$$x = 25 \quad \text{or} \qquad x = -4 \qquad \text{Solve each equation.}$$

The value -4 must be rejected as a solution because it leads to the logarithm of a negative number in the original equation.

$$\log (-4) + \log (-4 - 21) = 2 \qquad \text{The left side is undefined.}$$

Check that the only solution is 25, so the solution set is $\{25\}$. **NOW TRY**

⚠ **CAUTION** *Do not reject a potential solution just because it is nonpositive. Reject any value that leads to the logarithm of a nonpositive number.*

Solving a Logarithmic Equation

Step 1 **Transform the equation so that a single logarithm appears on one side.** Use the product rule or quotient rule of logarithms to do this.

Step 2 **Do one of the following.**

(a) **Use Property 4.**

If $\log_b x = \log_b y$, then $x = y$. (See Example 6.)

(b) **Write the equation in exponential form.**

If $\log_b x = k$, then $x = b^k$. (See Examples 4, 5, and 7.)

NOW TRY ANSWERS

6. $\left\{ \frac{3}{8} \right\}$

7. $\{2\}$

OBJECTIVE 3 Solve applications of compound interest.

We have solved simple interest problems using the formula $I = prt$. In most cases, interest paid or charged is **compound interest** (interest paid on both principal and interest). The formula for compound interest is an application of exponential functions. *In this text, monetary amounts are given to the nearest cent.*

Compound Interest Formula (for a Finite Number of Periods)

If a principal of P dollars is deposited at an annual rate of interest r compounded (paid) n times per year, then the account will contain

$$A = P\left(1 + \frac{r}{n}\right)^{nt}$$

dollars after t years. (In this formula, r is expressed as a decimal.)

NOW TRY EXERCISE 8

How much money will there be in an account at the end of 10 yr if $10,000 is deposited at 2.5% compounded monthly?

EXAMPLE 8 Solving a Compound Interest Problem for A

How much money will there be in an account at the end of 5 yr if $1000 is deposited at 3% compounded quarterly? (Assume no withdrawals are made.)

Because interest is compounded quarterly, $n = 4$.

$$A = P\left(1 + \frac{r}{n}\right)^{nt} \qquad \text{Compound interest formula}$$

$$A = 1000\left(1 + \frac{0.03}{4}\right)^{4 \cdot 5} \qquad \text{Substitute } P = 1000, r = 0.03 \text{ (because } 3\% = 0.03\text{), } n = 4, \text{ and } t = 5.$$

$$A = 1000(1.0075)^{20} \qquad \text{Simplify.}$$

$$A = 1161.18 \qquad \text{Evaluate with a calculator.}$$

The account will contain $1161.18.   **NOW TRY**

NOW TRY EXERCISE 9

Find the number of years, to the nearest hundredth, it will take for money deposited in an account paying 2% interest compounded quarterly to double.

EXAMPLE 9 Solving a Compound Interest Problem for t

Suppose inflation is averaging 3% per year. To the nearest hundredth of a year, how long will it take for prices to double? (This is the **doubling time** of the money.)

We want to find the number of years t for P dollars to grow to $2P$ dollars.

$$A = P\left(1 + \frac{r}{n}\right)^{nt} \qquad \text{Compound interest formula}$$

$$2P = P\left(1 + \frac{0.03}{1}\right)^{1t} \qquad \text{Let } A = 2P, r = 0.03, \text{ and } n = 1.$$

$$2 = (1.03)^t \qquad \text{Divide by } P. \text{ Simplify.}$$

$$\log 2 = \log (1.03)^t \qquad \text{Property 3}$$

$$\log 2 = t \log (1.03) \qquad \text{Power rule}$$

$$t = \frac{\log 2}{\log 1.03} \qquad \text{Interchange sides. Divide by } \log 1.03.$$

$$t \approx 23.45 \qquad \text{Evaluate with a calculator.}$$

NOW TRY ANSWERS
8. $12,836.92
9. 34.74 yr

Prices will double in 23.45 yr. To check, verify that $1.03^{23.45} \approx 2$.   **NOW TRY**

Interest can be compounded over various time periods per year, including

annually,   semiannually,   quarterly,   daily,   and so on.

If the value of n increases without bound, we have an example of **continuous compounding.**

Continuous Compound Interest Formula

If a principal of P dollars is deposited at an annual rate of interest r compounded continuously for t years, the final amount A on deposit is given by

$$A = Pe^{rt}.$$

NOW TRY EXERCISE 10

Suppose that $4000 is invested at 3% interest for 2 yr.

(a) How much will the investment grow to if it is compounded continuously?

(b) How long, to the nearest tenth of a year, would it take for the original investment to double?

EXAMPLE 10 Solving a Continuous Compound Interest Problem

In **Example 8** we found that $1000 invested for 5 yr at 3% interest compounded quarterly would grow to $1161.18.

(a) How much would this investment grow to if it is compounded continuously?

$A = Pe^{rt}$	Continuous compounding formula
$A = 1000e^{0.03(5)}$	Let $P = 1000$, $r = 0.03$, and $t = 5$.
$A = 1000e^{0.15}$	Multiply in the exponent.
$A = 1161.83$	Evaluate with a calculator.

The investment will grow to $1161.83 (which is $0.65 more than the amount in **Example 8** when interest was compounded quarterly).

(b) How long, to the nearest tenth of a year, would it take for the initial investment to triple?

We must find the value of t that will cause A to be $3(\$1000) = \$3000$.

$A = Pe^{rt}$	Continuous compounding formula
$3000 = 1000e^{0.03t}$	Let $A = 3P = 3000$, $P = 1000$, and $r = 0.03$.
$3 = e^{0.03t}$	Divide by 1000.
$\ln 3 = \ln e^{0.03t}$	Take natural logarithms.
$\ln 3 = 0.03t$	$\ln e^k = k$
$t = \dfrac{\ln 3}{0.03}$	Divide by 0.03. Interchange sides.
$t \approx 36.6$	Evaluate with a calculator.

It would take 36.6 yr for the original investment to triple.   **NOW TRY**

OBJECTIVE 4 Solve applications involving base e exponential growth and decay.

When situations involve growth or decay of a population, the amount or number of some quantity present at time t can be approximated by

$$f(t) = y_0 e^{kt}.$$

NOW TRY ANSWERS
10. (a) $4247.35
  (b) 23.1 yr

In this equation, y_0 is the amount or number present at time $t = 0$ and k is a constant.

NOW TRY
EXERCISE 11

Radium-226 decays according to the function

$$f(t) = y_0 e^{-0.00043t},$$

where t is time in years.

(a) If an initial sample contains $y_0 = 4.5$ g of radium-226, how many grams, to the nearest tenth, will be present after 150 yr?

(b) What is the half-life of radium-226? Round to the nearest year.

The continuous compounding of money is an example of exponential growth. In **Example 11,** we investigate exponential decay.

EXAMPLE 11   Applying an Exponential Decay Function

After a plant or animal dies, the amount of radioactive carbon-14 that is present disintegrates according to the natural logarithmic function

$$f(t) = y_0 e^{-0.000121t},$$

where t is time in years, $f(t)$ is the amount of the sample at time t, and y_0 is the initial amount present at $t = 0$.

(a) If an initial sample contains $y_0 = 10$ g of carbon-14, how many grams, to the nearest hundredth, will be present after 3000 yr?

$$f(3000) = 10e^{-0.000121(3000)} \approx 6.96 \text{ g} \qquad \text{Let } y_0 = 10 \text{ and } t = 3000 \text{ in the formula.}$$

(b) How long would it take to the nearest year for the initial sample to decay to half of its original amount? (This is the **half-life** of the sample.)

$$f(t) = y_0 e^{-0.000121t} \qquad \text{Exponential decay formula}$$

$$5 = 10e^{-0.000121t} \qquad \text{Let } f(t) = \tfrac{1}{2}(10) = 5 \text{ and } y_0 = 10.$$

$$\frac{1}{2} = e^{-0.000121t} \qquad \text{Divide by 10.}$$

$$\ln \frac{1}{2} = -0.000121t \qquad \text{Take natural logarithms; } \ln e^k = k$$

$$t = \frac{\ln \frac{1}{2}}{-0.000121} \qquad \text{Divide by } -0.000121. \text{ Interchange sides.}$$

$$t \approx 5728 \qquad \text{Evaluate with a calculator.}$$

NOW TRY ANSWERS
11. (a) 4.2 g   (b) 1612 yr

The half-life is 5728 yr.   **NOW TRY**

12.6 Exercises

FOR EXTRA HELP   ▶ **MyLab Math**

▶ *Video solutions for select problems available in MyLab Math*

Many of the problems in these exercises require a calculator.

Concept Check   *Determine whether common logarithms or natural logarithms would be a better choice to use for solving each equation. Do not actually solve.*

1. $10^{0.0025x} = 75$ 2. $10^{3x+1} = 13$

3. $e^{x-2} = 24$ 4. $e^{-0.28x} = 30$

Solve each equation. Approximate solutions to three decimal places. ***See Examples 1 and 3.***

5. $7^x = 5$ 6. $4^x = 3$ 7. $9^{-x+2} = 13$

8. $6^{-x+1} = 22$ 9. $3^{2x} = 14$ 10. $5^{3x} = 11$

11. $2^{x+3} = 5^x$ 12. $6^{x+3} = 4^x$ 13. $2^{x+3} = 3^{x-4}$

14. $4^{x-2} = 5^{3x+2}$ 15. $4^{2x+3} = 6^{x-1}$ 16. $3^{2x+1} = 5^{x-1}$

Solve each equation. Use natural logarithms. Approximate solutions to three decimal places when appropriate. See Example 2.

17. $e^{0.012x} = 23$ **18.** $e^{0.006x} = 30$ **19.** $e^{-0.205x} = 9$

20. $e^{-0.103x} = 7$ **21.** $\ln e^{3x} = 9$ **22.** $\ln e^{2x} = 4$

23. $\ln e^{0.45x} = \sqrt{7}$ **24.** $\ln e^{0.04x} = \sqrt{3}$ **25.** $\ln e^{-x} = \pi$

26. $\ln e^{2x} = \pi$ **27.** $e^{\ln 2x} = e^{\ln(x+1)}$ **28.** $e^{\ln(6-x)} = e^{\ln(4+2x)}$

Solve each equation. Give exact solutions. See Examples 4 and 5.

29. $\log_4 (2x + 8) = 2$ **30.** $\log_5 (5x + 10) = 3$

31. $\log_3 (6x + 5) = 2$ **32.** $\log_5 (12x - 8) = 3$

33. $\log_2 (2x - 1) = 5$ **34.** $\log_6 (4x + 2) = 2$

35. $\log_7 (x + 1)^3 = 2$ **36.** $\log_4 (x - 3)^3 = 4$

37. $\log_2 (x^2 + 7) = 4$ **38.** $\log_6 (x^2 + 11) = 2$

39. Concept Check Suppose that in solving a logarithmic equation having the term $\log (x - 3)$, we obtain the proposed solution 2. We know that our algebraic work is correct, so we give $\{2\}$ as the solution set. **WHAT WENT WRONG?**

40. Concept Check Suppose that in solving a logarithmic equation having the term $\log (3 - x)$, we obtain the proposed solution -4. We know that our algebraic work is correct, so we reject -4 and give $\varnothing$ as the solution set. **WHAT WENT WRONG?**

Solve each equation. Give exact solutions. See Examples 6 and 7.

41. $\log (6x + 1) = \log 3$ **42.** $\log (7 - 2x) = \log 4$

43. $\log_5 (3t + 2) - \log_5 t = \log_5 4$ **44.** $\log_2 (t + 5) - \log_2 (t - 1) = \log_2 3$

45. $\log 4x - \log (x - 3) = \log 2$ **46.** $\log (-x) + \log 3 = \log (2x - 15)$

47. $\log_2 x + \log_2 (x - 7) = 3$ **48.** $\log (2x - 1) + \log 10x = \log 10$

49. $\log 5x - \log (2x - 1) = \log 4$ **50.** $\log_3 x + \log_3 (2x + 5) = 1$

51. $\log_2 x + \log_2 (x - 6) = 4$ **52.** $\log_2 x + \log_2 (x + 4) = 5$

Solve each problem. See Examples 8–10.

53. Suppose that $2000 is deposited at 4% compounded quarterly.

    (a) How much money will be in the account at the end of 6 yr? (Assume no withdrawals are made.)

    (b) To one decimal place, how long will it take for the account to grow to $3000?

54. Suppose that $3000 is deposited at 3.5% compounded quarterly.

    (a) How much money will be in the account at the end of 7 yr? (Assume no withdrawals are made.)

    (b) To one decimal place, how long will it take for the account to grow to $5000?

55. What will be the amount A in an account with initial principal $4000 if interest is compounded continuously at an annual rate of 3.5% for 6 yr?

56. What will be the amount A in an account with initial principal $10,000 if interest is compounded continuously at an annual rate of 2.5% for 5 yr?

57. How long, to the nearest hundredth of a year, would it take an initial principal P to double if it were invested at 2.5% compounded continuously?

58. How long, to the nearest hundredth of a year, would it take $4000 to double at 3.25% compounded continuously?

59. Find the amount of money in an account after 12 yr if $5000 is deposited at 7% annual interest compounded as follows.

(a) Annually   (b) Semiannually   (c) Quarterly

(d) Daily (Use $n = 365$.)   (e) Continuously

60. Find the amount of money in an account after 8 yr if $4500 is deposited at 6% annual interest compounded as follows.

(a) Annually   (b) Semiannually   (c) Quarterly

(d) Daily (Use $n = 365$.)   (e) Continuously

61. How much money must be deposited today to amount to $1850 in 40 yr at 6.5% compounded continuously?

62. How much money must be deposited today to amount to $1000 in 10 yr at 5% compounded continuously?

Solve each problem. ***See Example 11.***

63. Revenues of software publishers in the United States for the years 2004–2016 can be modeled by the function

$$S(x) = 91.412e^{0.05195x},$$

where $x = 4$ represents 2004, $x = 5$ represents 2005, and so on, and $S(x)$ is in billions of dollars. Approximate, to the nearest unit, revenue for 2016. (Data from U.S. Census Bureau.)

64. Based on selected figures obtained during the years 1970–2015, the total number of bachelor's degrees earned in the United States can be modeled by the function

$$D(x) = 792{,}377e^{0.01798x},$$

where $x = 0$ corresponds to 1970, $x = 5$ corresponds to 1975, and so on. Approximate, to the nearest unit, the number of bachelor's degrees earned in 2015. (Data from U.S. National Center for Education Statistics.)

65. Suppose that the amount, in grams, of plutonium-241 present in a given sample is determined by the function

$$A(t) = 2.00e^{-0.053t},$$

where t is measured in years. Approximate the amount present, to the nearest hundredth, in the sample after the given number of years.

(a) 4   (b) 10   (c) 20   (d) What was the initial amount present?

66. Suppose that the amount, in grams, of radium-226 present in a given sample is determined by the function

$$A(t) = 3.25e^{-0.00043t},$$

where t is measured in years. Approximate the amount present, to the nearest hundredth, in the sample after the given number of years.

(a) 20   (b) 100   (c) 500   (d) What was the initial amount present?

67. A sample of 400 g of lead-210 decays to polonium-210 according to the function

$$A(t) = 400e^{-0.032t},$$

where t is time in years. Approximate answers to the nearest hundredth.

(a) How much lead will be left in the sample after 25 yr?

(b) How long will it take the initial sample to decay to half of its original amount?

68. The concentration of a drug in a person's system decreases according to the function

$$C(t) = 2e^{-0.125t},$$

where $C(t)$ is in appropriate units, and t is in hours. Approximate answers to the nearest hundredth.

(a) How much of the drug will be in the system after 1 hr?

(b) How long will it take for the concentration to be half of its original amount?

RELATING CONCEPTS For Individual or Group Work (Exercises 69–72)

Previously, we solved an equation such as $5^x = 125$ as follows.

$5^x = 125$	Original equation
$5^x = 5^3$	$125 = 5^3$
$x = 3$	Set exponents equal.

Solution set: $\{3\}$

The method described in this section can also be used to solve this equation. **Work Exercises 69–72 in order,** *to see how this is done.*

69. Take common logarithms on both sides, and write this equation.

70. Apply the power rule for logarithms on the left.

71. Write the equation so that x is alone on the left.

72. Use a calculator to find the decimal form of the solution. What is the solution set?

Chapter 12 Summary

Key Terms

12.1	**12.3**	**12.5**	**12.6**
one-to-one function	logarithm	common logarithm	compound interest
inverse of a function	logarithmic equation	natural logarithm	continuous compounding
	logarithmic function		
	with base a		

12.2

exponential function
    with base a
asymptote
exponential equation

New Symbols

$f^{-1}(x)$	inverse of $f(x)$	$\log x$	common (base 10) logarithm of x	$\ln x$	natural (base e) logarithm of x	e	a constant, approximately 2.718281828459
$\log_a x$	logarithm of x with base a						

Test Your Word Power

See how well you have learned the vocabulary in this chapter.

1. In a **one-to-one function**
 A. each *x*-value corresponds to only one *y*-value
 B. each *x*-value corresponds to one or more *y*-values
 C. each *x*-value is the same as each *y*-value
 D. each *x*-value corresponds to only one *y*-value and each *y*-value corresponds to only one *x*-value.

2. If *f* is a one-to-one function, then the **inverse** of *f* is
 A. the set of all solutions of *f*
 B. the set of all ordered pairs formed by interchanging the coordinates of the ordered pairs of *f*
 C. the set of all ordered pairs that are the opposite (negative) of the coordinates of the ordered pairs of *f*
 D. an equation involving an exponential expression.

3. An **exponential function** is a function defined by an expression of the form
 A. $f(x) = ax^2 + bx + c$ for real numbers a, b, c ($a \neq 0$)
 B. $f(x) = \log_a x$ for positive numbers a and x ($a \neq 1$)
 C. $f(x) = a^x$ for all real numbers x ($a > 0, a \neq 1$)
 D. $f(x) = \sqrt{x}$ for $x \geq 0$.

4. An **asymptote** is
 A. a line that a graph intersects just once
 B. a line that the graph of a function more and more closely approaches as the *x*-values increase or decrease without bound
 C. the *x*-axis or *y*-axis
 D. a line about which a graph is symmetric.

5. A **logarithmic function** is a function defined by an expression of the form
 A. $f(x) = ax^2 + bx + c$ for real numbers a, b, c ($a \neq 0$)
 B. $f(x) = \log_a x$ for positive numbers a and x ($a \neq 1$)
 C. $f(x) = a^x$ for all real numbers x ($a > 0, a \neq 1$)
 D. $f(x) = \sqrt{x}$ for $x \geq 0$.

6. A **logarithm** is
 A. an exponent
 B. a base
 C. an equation
 D. a polynomial.

ANSWERS

1. D; *Example:* The function $f = \{(0, 2), (1, -1), (3, 5), (-2, 3)\}$ is one-to-one. **2.** B; *Example:* The inverse of the one-to-one function f defined in Answer 1 is $f^{-1} = \{(2, 0), (-1, 1), (5, 3), (3, -2)\}$. **3.** C; *Examples:* $f(x) = 4^x$, $g(x) = \left(\frac{1}{2}\right)^x$ **4.** B; *Example:* The graph of $f(x) = 2^x$ has the *x*-axis ($y = 0$) as an asymptote. **5.** B; *Examples:* $f(x) = \log_3 x$, $g(x) = \log_{1/3} x$ **6.** A; *Example:* $\log_a x$ is the exponent to which a must be raised to obtain x; $\log_3 9 = 2$ because $3^2 = 9$.

Quick Review

CONCEPTS	EXAMPLES

12.1 Inverse Functions

Horizontal Line Test
A function is one-to-one if every horizontal line intersects the graph of the function at most once.

Find f^{-1} if $f(x) = 2x - 3$.

The graph of f is a nonhorizontal (slanted) straight line, so f is one-to-one by the horizontal line test and has an inverse.

Inverse Functions
For a one-to-one function f defined by an equation $y = f(x)$, find the defining equation of the inverse function f^{-1} as follows.

Step 1 Interchange *x* and *y*.

Step 2 Solve for *y*.

Step 3 Replace *y* with $f^{-1}(x)$.

$f(x) = 2x - 3$

$y = 2x - 3$ Let $y = f(x)$.

$x = 2y - 3$ Interchange *x* and *y*.

$x + 3 = 2y$ Add 3.

$y = \dfrac{x + 3}{2}$ Divide by 2. Interchange sides. } Solve for *y*.

$f^{-1}(x) = \dfrac{1}{2}x + \dfrac{3}{2}$ Replace *y* with $f^{-1}(x)$; $\frac{a+b}{c} = \frac{a}{c} + \frac{b}{c}$

CONCEPTS	EXAMPLES

In general, the graph of f^{-1} is the mirror image of the graph of f with respect to the line $y = x$. If the point (a, b) lies on the graph of f, then the point (b, a) lies on the graph of f^{-1}.

The graphs of a function f and its inverse f^{-1} are shown here.

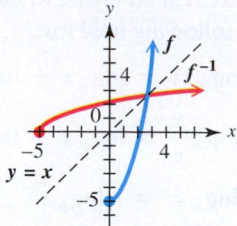

12.2 Exponential Functions

For $a > 0$, $a \neq 1$, and all real numbers x,

$$f(x) = a^x$$

defines the **exponential function with base a**.

Graph of $f(x) = a^x$

1. The graph contains the point $(0, 1)$, which is its y-intercept.

2. When $a > 1$, the graph rises from left to right.
   When $0 < a < 1$, the graph falls from left to right.

3. The x-axis is an asymptote.

4. The domain is $(-\infty, \infty)$, and the range is $(0, \infty)$.

$f(x) = 2^x$ defines the exponential function with base 2.

$F(x) = \left(\frac{1}{2}\right)^x$ defines the exponential function with base $\frac{1}{2}$.

x	$f(x) = 2^x$	$F(x) = \left(\frac{1}{2}\right)^x$
-2	$\frac{1}{4}$	4
-1	$\frac{1}{2}$	2
0	1	1
1	2	$\frac{1}{2}$
2	4	$\frac{1}{4}$

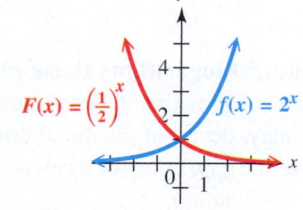

12.3 Logarithmic Functions

For all positive real numbers a, where $a \neq 1$, and all positive real numbers x,

$$y = \log_a x \quad \text{is equivalent to} \quad x = a^y.$$

Special Properties of Logarithms

For $b > 0$, where $b \neq 1$, the following hold true.

$$\log_b b = 1 \qquad \log_b 1 = 0$$

$$\log_b b^r = r \quad (r \text{ is real.}) \qquad b^{\log_b r} = r \quad (r > 0)$$

If a and x are positive real numbers, where $a \neq 1$, then

$$g(x) = \log_a x$$

defines the **logarithmic function with base a**.

Graph of $g(x) = \log_a x$

1. The graph contains the point $(1, 0)$, which is its x-intercept.

2. When $a > 1$, the graph rises from left to right.
   When $0 < a < 1$, the graph falls from left to right.

3. The y-axis is an asymptote.

4. The domain is $(0, \infty)$, and the range is $(-\infty, \infty)$.

$y = \log_2 x \quad \text{is equivalent to} \quad x = 2^y.$

Evaluate.

$$\log_3 3 = 1 \qquad \log_5 1 = 0$$

$$\log_5 5^6 = 6 \qquad 4^{\log_4 6} = 6$$

$g(x) = \log_2 x$ defines the logarithmic function with base 2.

$G(x) = \log_{1/2} x$ defines the logarithmic function with base $\frac{1}{2}$.

x	$g(x) = \log_2 x$	$G(x) = \log_{1/2} x$
$\frac{1}{4}$	-2	2
$\frac{1}{2}$	-1	1
1	0	0
2	1	-1
4	2	-2

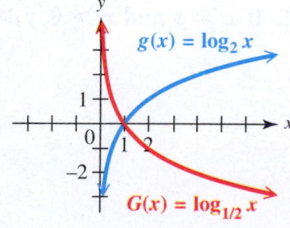

CONCEPTS	EXAMPLES

12.4 Properties of Logarithms

If x, y, and b are positive real numbers, where $b \neq 1$, and r is any real number, then the following hold true.

Product Rule $\log_b xy = \log_b x + \log_b y$

Quotient Rule $\log_b \dfrac{x}{y} = \log_b x - \log_b y$

Power Rule $\log_b x^r = r \log_b x$

Rewrite each logarithm.

$\log_2 3m = \log_2 3 + \log_2 m$ Product rule

$\log_5 \dfrac{9}{4} = \log_5 9 - \log_5 4$ Quotient rule

$\log_{10} 2^3 = 3 \log_{10} 2$ Power rule

12.5 Common and Natural Logarithms

Common logarithms (base 10) are used in applications such as pH and loudness of sound. Use the $\boxed{\text{LOG}}$ key of a calculator to evaluate common logarithms.

Use the formula $\mathbf{pH = -\log\,[H_3O^+]}$ to find the pH of grapes with hydronium ion concentration 5.0×10^{-5}.

$pH = -\log\,(5.0 \times 10^{-5})$ Substitute.

$pH = -(\log 5.0 + \log 10^{-5})$ Property of logarithms

$pH \approx 4.3$ Evaluate with a calculator.

Natural logarithms (base e) are most often used in applications of growth and decay, such as continuous compounding of money, decay of chemical compounds, and biological growth. Use the $\boxed{\text{LN}}$ key of a calculator to evaluate natural logarithms.

Use the following formula to find doubling time, to the nearest hundredth of a year, for an amount of money invested at 4%.

$t(r) = \dfrac{\ln 2}{\ln (1 + r)}$ Formula for doubling time (in years)

$t(0.04) = \dfrac{\ln 2}{\ln (1 + 0.04)}$ Let $r = 0.04$.

$t(0.04) \approx 17.67$ Evaluate with a calculator.

The doubling time is 17.67 yr.

Change-of-Base Rule

If $a > 0$, $a \neq 1$, $b > 0$, $b \neq 1$, and $x > 0$, then the following holds true.

$$\log_a x = \frac{\log_b x}{\log_b a}$$

Use the change-of-base rule to approximate $\log_3 17$.

$$\log_3 17 = \frac{\ln 17}{\ln 3} = \frac{\log 17}{\log 3} \approx 2.5789$$

12.6 Exponential and Logarithmic Equations; Further Applications

To solve exponential equations, use these properties (where $b > 0$, $b \neq 1$).

1. **If $b^x = b^y$, then $x = y$.**

2. **If $x = y$ and $x > 0$, $y > 0$, then $\log_b x = \log_b y$.**

Solve each equation.

$2^{3x} = 2^5$

$3x = 5$ Set the exponents equal.

$x = \dfrac{5}{3}$ Divide by 3.

Solution set: $\left\{\dfrac{5}{3}\right\}$

$5^x = 8$

$\log 5^x = \log 8$ Take common logarithms.

$x \log 5 = \log 8$ Power rule

$x = \dfrac{\log 8}{\log 5}$ Divide by log 5.

$x \approx 1.292$ Evaluate with a calculator.

Solution set: $\{1.292\}$

CONCEPTS	EXAMPLES
To solve logarithmic equations, use these properties (where $b > 0$, $b \neq 1$, $x > 0$, and $y > 0$). First use the product rule, quotient rule, power rule, or special properties as necessary to write the equation in the proper form.	Solve each equation.

<table>
<tr><td></td><td>$\log_3 2x = \log_3 (x + 1)$</td><td></td></tr>
<tr><td></td><td>$2x = x + 1$</td><td>Property 1</td></tr>
<tr><td></td><td>$x = 1$</td><td>Subtract x.</td></tr>
</table>

1. If $\log_b x = \log_b y$, then $x = y$.

2. If $\log_b x = k$, then $x = b^k$.

Always check that each proposed solution of a logarithmic equation yields only logarithms of positive numbers in the original equation.

A true statement results when 1 is substituted for x.
Solution set: $\{1\}$

$$\log_2 (3x - 1) = 4$$

$3x - 1 = 2^4$	Write in exponential form.
$3x - 1 = 16$	Apply the exponent.
$3x = 17$	Add 1.
$x = \dfrac{17}{3}$	Divide by 3.

A true statement results when $\frac{17}{3}$ is substituted for x.
Solution set: $\left\{\frac{17}{3}\right\}$

Chapter 12   Review Exercises

12.1 *Determine whether each graph is the graph of a one-to-one function.*

1.
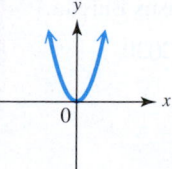

2.
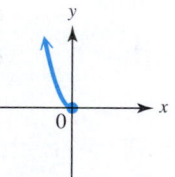

Determine whether each function is one-to-one. If it is, find its inverse.

3. $\{(-2, 4), (-1, 1), (0, 0), (1, 1), (2, 4)\}$

4. $\{(-2, -8), (-1, -1), (0, 0), (1, 1), (2, 8)\}$

5. $f(x) = -3x + 7$ **6.** $f(x) = \sqrt[3]{6x - 4}$ **7.** $f(x) = -x^2 + 3$

8. The table lists basic minimum wages in several states. If the set of states is the domain and the set of wage amounts is the range of a function, is it one-to-one? Why or why not?

State	Minimum Wage (in dollars)
Washington	11.50
Massachusetts	11.00
California	10.50
Arizona	10.50
Ohio	8.30
Iowa	7.25

Data from U.S. Department of Labor.

Each function graphed is one-to-one. Graph its inverse.

9.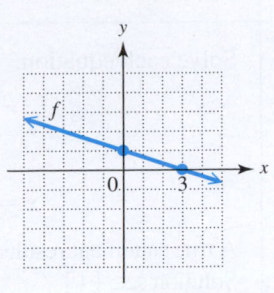

10.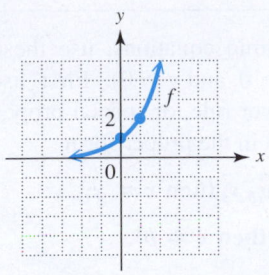

12.2 *Use a calculator to approximate each exponential expression to three decimal places.*

11. $5^{3.2}$

12. $\left(\dfrac{1}{5}\right)^{2.1}$

13. $8.3^{-1.2}$

Graph each exponential function.

14. $f(x) = 3^x$

15. $f(x) = \left(\dfrac{1}{3}\right)^x$

16. $f(x) = 2^{2x+3}$

Solve each equation.

17. $5^{2x+1} = 25$

18. $4^{3x} = 8^{x+4}$

19. $\left(\dfrac{1}{27}\right)^{x-1} = 9^{2x}$

20. The U.S. Hispanic population (in millions) can be approximated by

$$f(x) = 46.9 \cdot 2^{0.035811x},$$

where x represents the number of years since 2008. Use this function to approximate, to the nearest tenth, the Hispanic population in each year. (Data from U.S. Census Bureau.)

(a) 2015   **(b)** 2030

12.3 *Work each problem.*

21. Convert each equation to the indicated form.

    (a) Write in exponential form: $\log_5 625 = 4$.

    (b) Write in logarithmic form: $5^{-2} = 0.04$.

22. Fill in each blank with the correct response: The value of $\log_3 81$ is _____ . This means that if we raise _____ to the _____ power, the result is _____ .

Use a calculator to approximate each logarithm to four decimal places.

23. $\log_2 6$

24. $\log_7 3$

25. $\log_{10} 55$

Graph each logarithmic function.

26. $g(x) = \log_3 x$

27. $g(x) = \log_{1/3} x$

28. Use the special properties of logarithms to evaluate each expression.

    (a) $4^{\log_4 12}$ **(b)** $\log_9 9^{13}$ **(c)** $\log_5 625$

Solve each equation.

29. $\log_8 64 = x$ **30.** $\log_2 \sqrt{8} = x$ **31.** $\log_x \left(\dfrac{1}{49} \right) = -2$

32. $\log_4 x = \dfrac{3}{2}$ **33.** $\log_k 4 = 1$ **34.** $\log_6 x = -2$

A company has found that total sales, in thousands of dollars, are given by the function

$$S(x) = 100 \log_2 (x + 2),$$

where x is the number of weeks after a major advertising campaign was introduced.

35. What were total sales 6 weeks after the campaign was introduced?

36. Graph the function.

12.4 *Use the properties of logarithms to express each logarithm as a sum or difference of logarithms. Assume that all variables represent positive real numbers.*

37. $\log_4 3x^2$ **38.** $\log_5 \dfrac{a^3 b^2}{c^4}$

39. $\log_4 \dfrac{\sqrt{x} \cdot w^2}{z}$ **40.** $\log_2 \dfrac{p^2 r}{\sqrt{z}}$

Use the properties of logarithms to rewrite each expression as a single logarithm. Assume that all variables are defined in such a way that the variable expressions are positive, and bases are positive numbers not equal to 1.

41. $2 \log_a 7 - 4 \log_a 2$ **42.** $3 \log_a 5 + \dfrac{1}{3} \log_a 8$

43. $\log_b 3 + \log_b x - 2 \log_b y$ **44.** $\log_3 (x + 7) - \log_3 (4x + 6)$

12.5 *Evaluate each logarithm to four decimal places.*

45. $\log 28.9$ **46.** $\log 0.257$ **47.** $\ln 28.9$ **48.** $\ln 0.257$

Use the change-of-base rule (with either common or natural logarithms) to approximate each logarithm to four decimal places.

49. $\log_{16} 13$ **50.** $\log_4 12$ **51.** $\log_{1/4} 17$ **52.** $\log_\pi \sqrt{2}$

Find the pH (to the nearest tenth) of the substance with the given hydronium ion concentration.

53. Milk, 4.0×10^{-7} **54.** Crackers, 3.8×10^{-9}

Solve each problem.

55. If orange juice has pH 4.6, what is its hydronium ion concentration?

56. The magnitude of a star is given by the equation

$$M = 6 - 2.5 \log \frac{I}{I_0},$$

where I_0 is the measure of the faintest star and I is the actual intensity of the star being measured. The dimmest stars are of magnitude 6, and the brightest are of magnitude 1. Determine the ratio of intensities between stars of magnitude 1 and 3.

57. The concentration of a drug injected into the bloodstream decreases with time. The intervals of time T when the drug should be administered are given by

$$T = \frac{1}{k} \ln \frac{C_2}{C_1},$$

where k is a constant determined by the drug in use, C_2 is the concentration at which the drug is harmful, and C_1 is the concentration below which the drug is ineffective. (Data from Horelick, B. and S. Koont, "Applications of Calculus to Medicine: Prescribing Safe and Effective Dosage," *UMAP Module 202.*) Thus, if $T = 4$, the drug should be administered every 4 hr.

For a certain drug, $k = \frac{1}{3}$, $C_2 = 5$, and $C_1 = 2$. How often should the drug be administered? (*Hint:* Round down.)

12.6 *Solve each equation. Approximate solutions to three decimal places.*

58. $3^x = 9.42$ **59.** $2^{x-1} = 15$ **60.** $e^{0.06x} = 3$

Solve each equation. Give exact solutions.

61. $\log_3 (9x + 8) = 2$ **62.** $\log_5 (x + 6)^3 = 2$

63. $\log_3 (x + 2) - \log_3 x = \log_3 2$ **64.** $\log (2x + 3) = 1 + \log x$

65. $\log_4 x + \log_4 (8 - x) = 2$ **66.** $\log_2 x + \log_2 (x + 15) = \log_2 16$

Solve each problem.

67. Suppose \$20,000 is deposited at 4% annual interest compounded quarterly. How much will be in the account at the end of 5 yr? (Assume no withdrawals are made.)

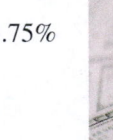

68. How much will \$10,000 compounded continuously at 3.75% annual interest amount to in 3 yr?

69. Which plan is better? How much more would it pay?

Plan A: Invest \$1000 at 4% compounded quarterly for 3 yr

Plan B: Invest \$1000 at 3.9% compounded monthly for 3 yr

70. Find the half-life of a radioactive substance that decays according to the function

$$Q(t) = A_0 e^{-0.05t}, \quad \text{where } t \text{ is in days, to the nearest tenth.}$$

Chapter 12   Mixed Review Exercises

Evaluate.

1. $\log_2 128$ **2.** $\log_{12} 1$ **3.** $\log_{2/3} \dfrac{27}{8}$ **4.** $5^{\log_5 36}$

5. $e^{\ln 4}$ **6.** $10^{\log e}$ **7.** $\log_3 3^{-5}$ **8.** $\ln e^{5.4}$

Solve.

9. $\log_3 (x + 9) = 4$ **10.** $\log_2 32 = x$ **11.** $\log_x \dfrac{1}{81} = 2$

12. $27^x = 81$ **13.** $2^{2x-3} = 8$ **14.** $5^{x+2} = 25^{2x+1}$

15. $\log_3 (x + 1) - \log_3 x = 2$ **16.** $\log (3x - 1) = \log 10$

17. $\log_4 (x + 2) - \log_4 x = 3$ **18.** $\ln (x^2 + 3x + 4) = \ln 2$

19. A machine purchased for business use **depreciates,** or loses value, over a period of years. The value of the machine at the end of its useful life is its **scrap value.** By one method of depreciation, the scrap value, S, is given by

$$S = C(1 - r)^n,$$

where C is original cost, n is useful life in years, and r is the constant percent of depreciation.

(a) Find the scrap value of a machine costing \$30,000, having a useful life of 12 yr and a constant annual rate of depreciation of 15%. Round to the nearest dollar.

(b) A machine has a "half-life" of 6 yr. Find the constant annual rate of depreciation, to the nearest percent.

20. One measure of the diversity of the species in an ecological community is the **index of diversity,** given by the logarithmic expression

$$-(p_1 \ln p_1 + p_2 \ln p_2 + \cdots + p_n \ln p_n),$$

where $p_1, p_2, \ldots, p_n$ are the proportions of a sample belonging to each of n species in the sample. (Data from Ludwig, J. and J. Reynolds, *Statistical Ecology: A Primer on Methods and Computing,* New York, John Wiley and Sons.)

Approximate the index of diversity to the nearest thousandth if a sample of 100 from a community produces the following numbers.

(a) 90 of one species, 10 of another   **(b)** 60 of one species, 40 of another

21. To solve the equation $5^x = 7$, we must find the exponent to which 5 must be raised in order to obtain 7. This is $\log_5 7$.

(a) Use the change-of-base rule and a calculator to find $\log_5 7$.

(b) Raise 5 to the number found in part (a). What is the result?

(c) Using as many decimal places as the calculator gives, write the solution set of $5^x = 7$.

22. Let m be the number of letters in your first name, and let n be the number of letters in your last name.

(a) Explain what $\log_m n$ means.   **(b)** Use a calculator to find $\log_m n$.

(c) Raise m to the power indicated by the number found in part (b). What is the result?

Chapter 12   Test

FOR EXTRA HELP — *Step-by-step test solutions are found on the Chapter Test Prep Videos available in* **MyLab Math**.

▶ *View the complete solutions to all Chapter Test exercises in MyLab Math.*

1. Decide whether each function is one-to-one.

   (a) $f(x) = x^2 + 9$

   (b)
   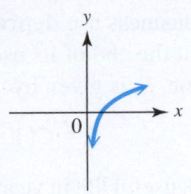

2. Find $f^{-1}(x)$ for the one-to-one function

   $$f(x) = \sqrt[3]{x + 7}.$$

3. Graph the inverse given the graph of $y = f(x)$.

   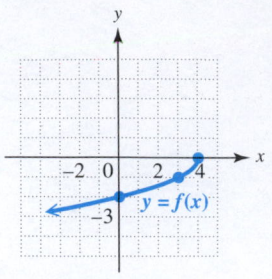

Graph each function.

4. $f(x) = 6^x$

5. $g(x) = \log_6 x$

6. Explain how the graph of the function in **Exercise 5** can be obtained from the graph of the function in **Exercise 4.**

Use the special properties of logarithms to evaluate each expression.

7. $7^{\log_7 9}$

8. $\log_3 3^6$

9. $\log_5 1$

Solve each equation. Give exact solutions.

10. $5^x = \dfrac{1}{625}$

11. $2^{3x-7} = 8^{2x+2}$

12. The atmospheric pressure (in millibars) at a given altitude x (in meters) is approximated by

   $$f(x) = 1013e^{-0.0001341x}.$$

   Use this function to approximate the atmospheric pressure, to the nearest unit, at each altitude.

   (a) 2000 m   **(b)** 10,000 m

13. Write in logarithmic form: $4^{-2} = 0.0625$.

14. Write in exponential form: $\log_7 49 = 2$.

Solve each equation.

15. $\log_{1/2} x = -5$

16. $x = \log_9 3$

17. $\log_x 16 = 4$

18. Fill in each blank with the correct response: The value of $\log_2 32$ is _____. This means that if we raise _____ to the _____ power, the result is _____.

Use the properties of logarithms to write each expression as a sum or difference of logarithms. Assume that variables represent positive real numbers.

19. $\log_3 x^2 y$

20. $\log_5 \left(\dfrac{\sqrt{x}}{yz} \right)$

Use the properties of logarithms to rewrite each expression as a single logarithm. Assume that all variables represent positive real numbers, and that bases are positive numbers not equal to 1.

21. $3 \log_b s - \log_b t$

22. $\dfrac{1}{4} \log_b r + 2 \log_b s - \dfrac{2}{3} \log_b t$

Evaluate each logarithm to four decimal places.

23. log 23.1   **24.** ln 0.82   **25.** $\log_6 45$

Work each problem.

26. Use the change-of-base rule to express $\log_3 19$ as described.

     (a) in terms of common logarithms   **(b)** in terms of natural logarithms

     (c) approximated to four decimal places

27. Solve $3^x = 78$, giving the solution to three decimal places.

28. Solve $\log_8 (x + 5) + \log_8 (x - 2) = 1$.

29. Suppose that $10,000 is invested at 3.5% annual interest, compounded quarterly.

     (a) How much will be in the account in 5 yr if no money is withdrawn?

     (b) How long, to the nearest tenth of a year, will it take for the initial principal to double?

30. Suppose that $15,000 is invested at 3% annual interest, compounded continuously.

     (a) How much will be in the account in 5 yr if no money is withdrawn?

     (b) How long, to the nearest tenth of a year, will it take for the initial principal to double?

Chapters R–12   Cumulative Review Exercises

1. Write each fraction as a decimal and a percent.

     (a) $\dfrac{1}{20}$ **(b)** $\dfrac{5}{4}$

2. Perform the indicated operations. Give answers in lowest terms.

     (a) $-\dfrac{7}{10} - \dfrac{1}{2}$ **(b)** $-\dfrac{8}{15} \div \dfrac{2}{3}$

3. Multiply or divide as indicated.

     (a) 37.5×100 **(b)** $37.5 \div 1000$

4. Let $S = \left\{ -\dfrac{9}{4}, -2, -\sqrt{2}, 0, 0.6, \sqrt{11}, \sqrt{-8}, 6, \dfrac{30}{3} \right\}$. List the elements of S that are members of each set.

     (a) Integers   **(b)** Rational numbers   **(c)** Irrational numbers

Simplify each expression, and solve each equation or inequality.

5. $|-8| + 6 - |-2| - (-6 + 2)$ **6.** $2(-5) + (-8)(4) - (-3)$

7. $7 - (3 + 4x) + 2x = -5(x - 1) - 3$ **8.** $2x + 2 \le 5x - 1$

Perform the indicated operations.

9. $(2p + 3)(3p - 1)$ **10.** $(4k - 3)^2$

11. $(3m^3 + 2m^2 - 5m) - (8m^3 + 2m - 4)$ **12.** Divide $15x^3 - x^2 + 22x + 8$ by $3x + 1$.

Factor.

13. $8x + x^3$

14. $24y^2 - 7y - 6$

15. $5z^3 - 19z^2 - 4z$

16. $16a^2 - 25b^4$

17. $8c^3 + d^3$

18. $16r^2 + 56rq + 49q^2$

Perform the indicated operations.

19. $\dfrac{(5p^3)^4(-3p^7)}{2p^2(4p^4)}$

20. $\dfrac{x^2 - 9}{x^2 + 7x + 12} \div \dfrac{x - 3}{x + 5}$

21. $\dfrac{2}{k + 3} - \dfrac{5}{k - 2}$

22. The graph projects that the number of international travelers to the United States will increase from 51.2 million in 2000 to 90.3 million in 2020.

(a) Is this the graph of a function?

(b) What is the slope of the line in the graph? Interpret the slope in the context of international travelers to the United States.

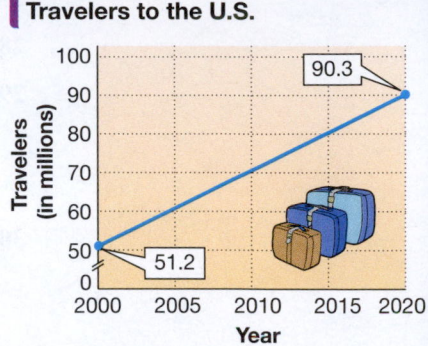

Travelers to the U.S.

Data from U.S. Department of Commerce.

23. Find an equation of the line passing through $(5, -1)$ and parallel to the line with equation $3x - 4y = 12$. Write the equation in slope-intercept form.

Solve each system.

24. $5x - 3y = 14$
$2x + 5y = 18$

25. $3x - 2y = 3$
$x = \dfrac{2}{3}y + 1$

26. $x + 2y + 3z = 11$
$3x - y + z = 8$
$2x + 2y - 3z = -12$

27. Candy worth $1.00 per lb is to be mixed with 10 lb of candy worth $1.96 per lb to obtain a mixture that will be sold for $1.60 per lb. How many pounds of the $1.00 candy should be used?

Number of Pounds	Price per Pound (in dollars)	Value (in dollars)
x	1.00	$1x$
	1.60	

Simplify.

28. $\sqrt{288}$

29. $2\sqrt{32} - 5\sqrt{98}$

30. $(5 + 4i)(5 - 4i)$

Solve each equation or inequality.

31. $|2x - 5| = 9$

32. $|4x + 2| > 10$

33. $\sqrt{2x + 1} - \sqrt{x} = 1$

34. $3x^2 - x - 1 = 0$

35. $x^2 + 2x - 8 > 0$

36. $x^4 - 5x^2 + 4 = 0$

37. $5^{x+3} = \left(\dfrac{1}{25}\right)^{3x+2}$

38. $\log_5 x + \log_5 (x + 4) = 1$

Graph.

39. $5x + 2y = 10$

40. $-4x + y \le 5$

41. $f(x) = \dfrac{1}{3}(x - 1)^2 + 2$

42. $f(x) = 2^x$

43. $f(x) = \log_3 x$

Work each problem.

44. Use properties of logarithms to write the following as a sum or difference of logarithms. Assume that variables represent positive real numbers.

$$\log \frac{x^3 \sqrt{y}}{z}$$

45. Let the number of bacteria present in a certain culture be given by

$$B(t) = 25{,}000e^{0.2t},$$

where t is time measured in hours, and $t = 0$ corresponds to noon. Approximate, to the nearest hundred, the number of bacteria present at each time.

(a) noon   **(b)** 1 P.M.   **(c)** 2 P.M.

(d) When will the population double?

STUDY SKILLS REMINDER

Have you begun to prepare for your final exam? **Review Study Skill 10,** *Preparing for Your Math Final Exam.*

13

NONLINEAR FUNCTIONS, CONIC SECTIONS, AND NONLINEAR SYSTEMS

An *ellipse,* one of a group of curves known as *conic sections,* has a special reflecting property responsible for "whispering galleries" like that in the Old House Chamber of the U.S. Capitol. We investigate ellipses in this chapter.

13.1   Additional Graphs of Functions

OBJECTIVE 1   Recognize graphs of the absolute value, reciprocal, and square root functions, and graph their translations.

The elementary function $f(x) = |x|$ is the **absolute value function.** This function pairs each real number with its absolute value. Its graph is shown in **FIGURE 1**.

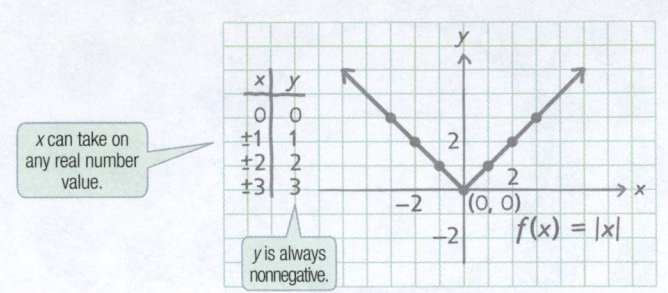

FIGURE 1

Recall that the **reciprocal function** $f(x) = \dfrac{1}{x}$ pairs every real number except 0 with its reciprocal. Its graph is shown in **FIGURE 2**. Because x can never equal 0, as x gets closer and closer to 0, $\dfrac{1}{x}$ approaches either ∞ or $-\infty$. Also, $\dfrac{1}{x}$ can never equal 0, and as x approaches ∞ or $-\infty$, $\dfrac{1}{x}$ approaches 0. The axes are the **asymptotes** for the function.

STUDY SKILLS REMINDER

Make study cards to help you learn and remember the material in this chapter. **Review Study Skill 5, *Using Study Cards.***

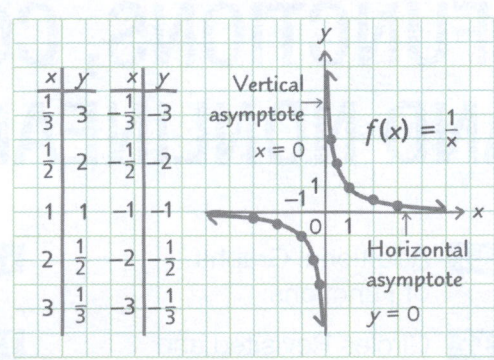

FIGURE 2

The **square root function** $f(x) = \sqrt{x}$, also introduced earlier, pairs every non-negative real number with its principal square root. See **FIGURE 3**.

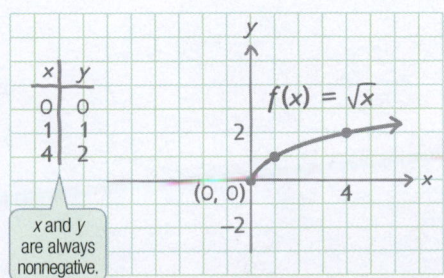

FIGURE 3

The graphs of these elementary functions can be shifted, or translated.

NOW TRY EXERCISE 1

Graph $f(x) = \frac{1}{x+3}$. Give the domain and range.

EXAMPLE 1 Applying a Horizontal Shift

Graph $f(x) = |x - 2|$. Give the domain and range.

If $x = 2$, then $f(x) = 0$, which gives the lowest point on the graph $(2, 0)$. The graph of $f(x) = |x - 2|$ has the same shape as that of $f(x) = |x|$ but is *shifted*, or *translated*, to the right 2 units, as shown in **FIGURE 4**.

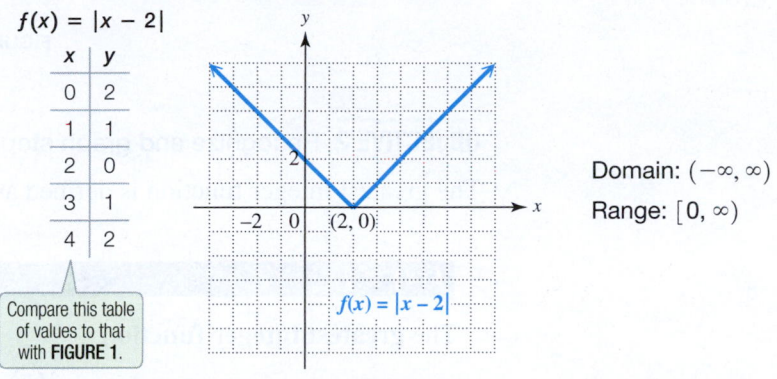

$f(x) = |x - 2|$

x	y
0	2
1	1
2	0
3	1
4	2

Compare this table of values to that with **FIGURE 1**.

Domain: $(-\infty, \infty)$

Range: $[0, \infty)$

FIGURE 4

NOW TRY

As seen in **Example 1,** the graph of

$\quad y = f(x + h)\quad$ is a *horizontal* translation of the graph of $\quad y = f(x)$.

In **Example 2,** we use the fact that the graph of

$\quad y = f(x) + k\quad$ is a *vertical* translation of the graph of $\quad y = f(x)$.

NOW TRY EXERCISE 2

Graph $f(x) = \sqrt{x} + 2$. Give the domain and range.

EXAMPLE 2 Applying a Vertical Shift

Graph $f(x) = \frac{1}{x} + 3$. Give the domain and range.

The graph is found by shifting the graph of $y = \frac{1}{x}$ up 3 units. See **FIGURE 5**.

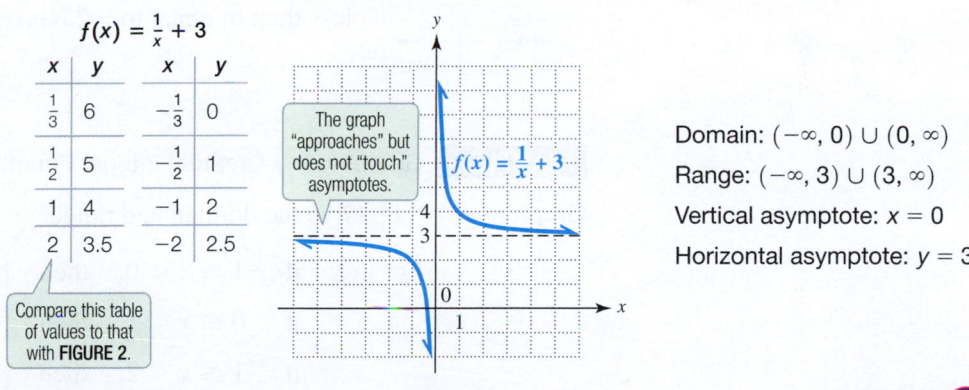

$f(x) = \frac{1}{x} + 3$

x	y	x	y
$\frac{1}{3}$	6	$-\frac{1}{3}$	0
$\frac{1}{2}$	5	$-\frac{1}{2}$	1
1	4	-1	2
2	3.5	-2	2.5

The graph "approaches" but does not "touch" asymptotes.

Compare this table of values to that with **FIGURE 2**.

Domain: $(-\infty, 0) \cup (0, \infty)$

Range: $(-\infty, 3) \cup (3, \infty)$

Vertical asymptote: $x = 0$

Horizontal asymptote: $y = 3$

FIGURE 5

NOW TRY

NOW TRY ANSWERS

1.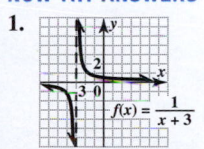

$f(x) = \frac{1}{x+3}$

domain: $(-\infty, -3) \cup (-3, \infty)$;
range: $(-\infty, 0) \cup (0, \infty)$

2.

$f(x) = \sqrt{x} + 2$

domain: $[0, \infty)$; range: $[2, \infty)$

EXAMPLE 3 Applying Both Horizontal and Vertical Shifts

Graph $f(x) = \sqrt{x + 1} - 4$. Give the domain and range.

This graph has the same shape as that of $f(x) = \sqrt{x}$ but is shifted to the left 1 unit (because $x + 1 = 0$ if $x = -1$) and down 4 units (because of the negative sign in -4). See **FIGURE 6** on the next page.

**NOW TRY
EXERCISE 3**

Graph $f(x) = |x + 1| - 3$.
Give the domain and range.

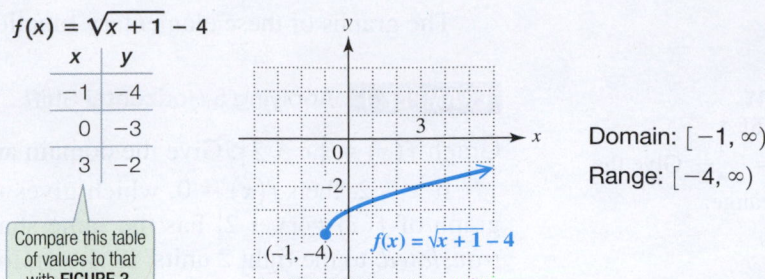

FIGURE 6

NOW TRY

Domain: $[-1, \infty)$
Range: $[-4, \infty)$

OBJECTIVE 2 Recognize and graph step functions.

The greatest integer function is defined as follows.

$$f(x) = [\![x]\!]$$

The **greatest integer function**

$$f(x) = [\![x]\!]$$

pairs every real number x with the greatest integer less than or equal to x.

**NOW TRY
EXERCISE 4**

Evaluate.

(a) $[\![5]\!]$ **(b)** $[\![-6]\!]$
(c) $[\![3.5]\!]$ **(d)** $[\![-4.1]\!]$

EXAMPLE 4 Finding the Greatest Integer

Evaluate.

(a) $[\![8]\!] = 8$ **(b)** $[\![-1]\!] = -1$ **(c)** $[\![0]\!] = 0$ If x is an integer, then $[\![x]\!] = x$.

(d) $[\![7.45]\!] = 7$ The greatest integer *less than or equal to* 7.45 is 7. This is like "rounding down."

(e) $[\![-2.6]\!] = -3$ Think of a number line with -2.6 graphed on it. Because -3 is to the *left of* (and therefore *less than*) -2.6, the greatest integer less than or equal to -2.6 is -3, **not** -2.

NOW TRY

EXAMPLE 5 Graphing the Greatest Integer Function

Graph $f(x) = [\![x]\!]$. Give the domain and range.

For $[\![x]\!]$, if $-1 \le x < 0$, then $[\![x]\!] = -1$;

if $0 \le x < 1$, then $[\![x]\!] = 0$;

if $1 \le x < 2$, then $[\![x]\!] = 1$;

if $2 \le x < 3$, then $[\![x]\!] = 2$;

if $3 \le x < 4$, then $[\![x]\!] = 3$,   and so on.

The graph, shown in **FIGURE 7** on the next page, consists of a series of horizontal line segments. In each segment, the left endpoint is included and the right endpoint is excluded. These segments continue infinitely following this pattern to the left and right. The appearance of the graph is the reason why this function is called a **step function.**

NOW TRY ANSWERS

3.

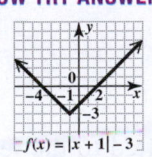

domain: $(-\infty, \infty)$; range: $[-3, \infty)$

4. (a) 5 **(b)** -6 **(c)** 3 **(d)** -5

**NOW TRY
EXERCISE 5**

Graph $f(x) = [\![x - 1]\!]$. Give the domain and range.

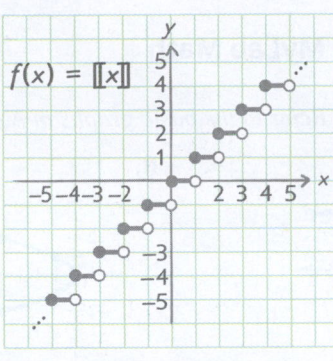

FIGURE 7

Greatest integer function

$$f(x) = [\![x]\!]$$

Domain: $(-\infty, \infty)$

Range: $\{\ldots, -3, -2, -1, 0, 1, 2, 3, \ldots\}$
(the set of integers)

The ellipsis points indicate that the graph continues indefinitely in the same pattern.

The graph of a step function also may be shifted. For example, the graph of

$$h(x) = [\![x - 2]\!] \quad \text{is the graph of } f(x) = [\![x]\!] \text{ shifted to the right 2 units.}$$

Similarly, the graph of

$$g(x) = [\![x]\!] + 2 \quad \text{is the graph of } f(x) \text{ shifted up 2 units.} \quad \textbf{NOW TRY} \text{ } \circlearrowleft$$

**NOW TRY
EXERCISE 6**

The cost of parking a car at an airport hourly parking lot is \$4 for the first hour and \$2 for each additional hour or fraction thereof. Let $y = f(x)$ represent the cost of parking a car for x hours. Graph $f(x)$ for x in the interval $(0, 5]$.

EXAMPLE 6   Applying a Greatest Integer Function

An overnight delivery service charges \$25 for a package weighing up to 2 lb. For each additional pound or fraction of a pound, there is an additional charge of \$3. Let $y = D(x)$ represent the cost to send a package weighing x pounds. Graph $D(x)$ for x in the interval $(0, 6]$.

For x in the interval $(0, 2]$, $y = 25$.

For x in the interval $(2, 3]$, $y = 25 + 3 = 28$.

For x in the interval $(3, 4]$, $y = 28 + 3 = 31$.

For x in the interval $(4, 5]$, $y = 31 + 3 = 34$.

For x in the interval $(5, 6]$, $y = 34 + 3 = 37$.

The graph, which is that of a step function, is shown in **FIGURE 8.**

NOW TRY ANSWERS

5.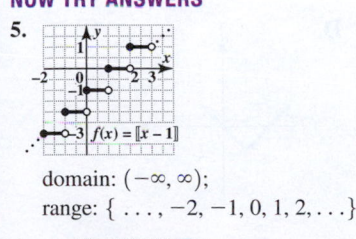

domain: $(-\infty, \infty)$;
range: $\{\ldots, -2, -1, 0, 1, 2, \ldots\}$

6.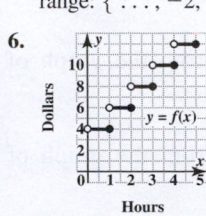

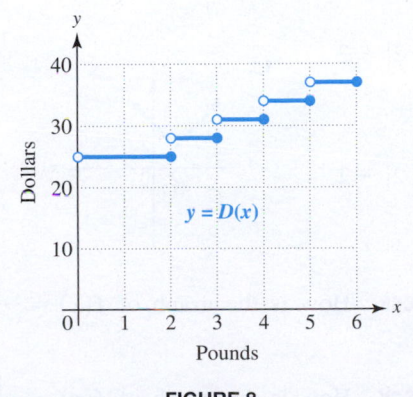

FIGURE 8

NOW TRY $\circlearrowleft$

13.1 Exercises

FOR EXTRA HELP   ▶ **MyLab Math**

▶ *Video solutions for select problems available in MyLab Math*

Concept Check *Refer to the basic graphs in A–F to answer each of the following.*

A.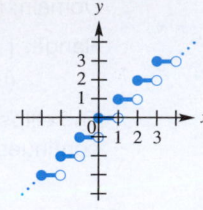

B.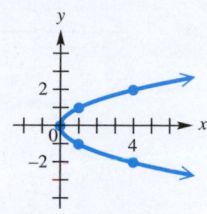

C.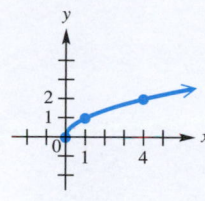

D.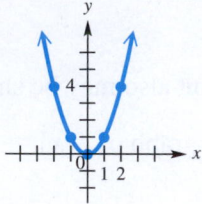

E.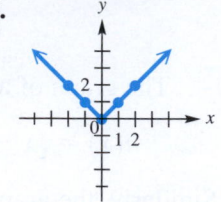

F.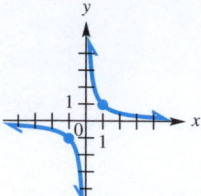

1. Which is the graph of $f(x) = |x|$? The lowest point on its graph has coordinates (_____, _____).

2. Which is the graph of $f(x) = x^2$? Give the domain and range.

3. Which is the graph of $f(x) = [\![x]\!]$? Give the domain and range.

4. Which is the graph of $f(x) = \sqrt{x}$? Give the domain and range.

5. Which is not the graph of a function? Why?

6. Which is the graph of $f(x) = \frac{1}{x}$? The lines with equations $x = 0$ and $y = 0$ are called its _____.

Concept Check *Without actually plotting points, match each function defined by the absolute value expression with its graph.*

7. $f(x) = |x - 2| + 2$ **A.**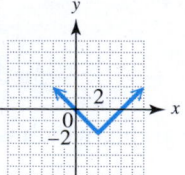

B.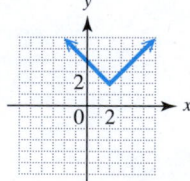

8. $f(x) = |x + 2| + 2$

9. $f(x) = |x - 2| - 2$ **C.**

D.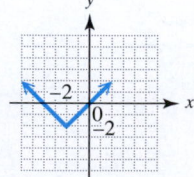

10. $f(x) = |x + 2| - 2$

11. Concept Check How is the graph of $f(x) = \frac{1}{x - 3} + 2$ obtained from the graph of $g(x) = \frac{1}{x}$?

12. Concept Check How is the graph of $f(x) = \frac{1}{x + 5} - 3$ obtained from the graph of $g(x) = \frac{1}{x}$?

Graph each function. Give the domain and range. See Examples 1–3.

13. $f(x) = |x + 1|$ **14.** $f(x) = |x - 1|$ **15.** $f(x) = \dfrac{1}{x} + 1$

16. $f(x) = \dfrac{1}{x} - 1$ **17.** $f(x) = \sqrt{x - 2}$ **18.** $f(x) = \sqrt{x + 5}$

19. $f(x) = \dfrac{1}{x - 2}$ **20.** $f(x) = \dfrac{1}{x + 2}$ **21.** $f(x) = \sqrt{x + 3} - 3$

22. $f(x) = \sqrt{x - 2} + 2$ **23.** $f(x) = |x - 3| + 1$ **24.** $f(x) = |x + 1| - 4$

25. Concept Check A student incorrectly evaluated $\llbracket -5.1 \rrbracket$ as -5. **WHAT WENT WRONG?** Evaluate correctly.

26. Concept Check A student incorrectly evaluated $\llbracket 3.75 \rrbracket$ as 4. **WHAT WENT WRONG?** Evaluate correctly.

Evaluate each expression. See Example 4.

27. $\llbracket 3 \rrbracket$ **28.** $\llbracket 18 \rrbracket$ **29.** $\llbracket 4.5 \rrbracket$ **30.** $\llbracket 8.7 \rrbracket$

31. $\left\llbracket \dfrac{1}{2} \right\rrbracket$ **32.** $\left\llbracket \dfrac{3}{4} \right\rrbracket$ **33.** $\left\llbracket \dfrac{8}{3} \right\rrbracket$ **34.** $\left\llbracket \dfrac{5}{2} \right\rrbracket$

35. $\llbracket -14 \rrbracket$ **36.** $\llbracket -5 \rrbracket$ **37.** $\llbracket -10.1 \rrbracket$ **38.** $\llbracket -6.9 \rrbracket$

Graph each step function. See Examples 5 and 6.

39. $f(x) = \llbracket x - 3 \rrbracket$ **40.** $f(x) = \llbracket x + 2 \rrbracket$

41. $f(x) = \llbracket x \rrbracket - 1$ **42.** $f(x) = \llbracket x \rrbracket + 1$

Solve each problem. See Example 6.

43. Suppose that postage rates are $0.55 for the first ounce, plus $0.24 for each additional ounce, and that each letter carries one $0.55 stamp and as many $0.24 stamps as necessary. Graph the function

$$y = p(x) = \text{the number of stamps}$$

on a letter weighing x ounces. Use the interval $(0, 5]$.

44. The cost of parking a car at an airport hourly parking lot is $3 for the first half-hour and $2 for each additional half-hour or fraction thereof. Graph the function

$$y = f(x) = \text{the cost of parking a car}$$

for x hours. Use the interval $(0, 2]$.

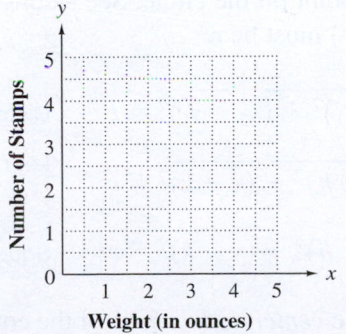

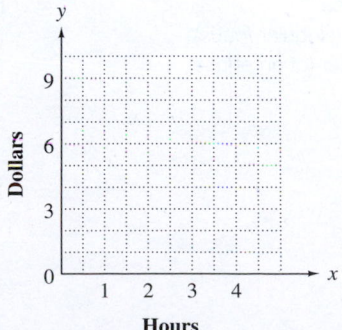

45. A certain long-distance carrier provides service between Podunk and Nowhereville. If x represents the number of minutes for the call, where $x > 0$, then the function f defined by

$$f(x) = 0.40[\![x]\!] + 0.75$$

gives the total cost of the call in dollars. Find the cost of a 5.5-minute call.

46. Total rental cost in dollars for a power washer, where x represents the number of hours with $x > 0$, can be represented by the function

$$f(x) = 12[\![x]\!] + 25.$$

Find the cost of a $7\frac{1}{2}$-hr rental.

13.2   Circles Revisited and Ellipses

OBJECTIVES

1 Graph circles.

2 Write an equation of a circle given its center and radius.

3 Determine the center and radius of a circle given its equation.

4 Recognize the equation of an ellipse.

5 Graph ellipses.

When an infinite cone is intersected by a plane, the resulting figure is called a **conic section.** The parabola is one example of a conic section. Circles, ellipses, and hyperbolas may also result. See **FIGURE 9.**

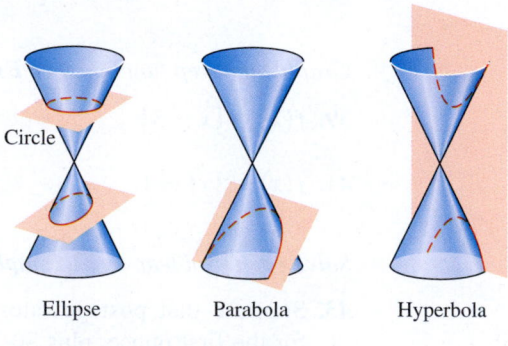

FIGURE 9

VOCABULARY

☐ conic section
☐ circle
☐ center (of a circle)
☐ radius
☐ ellipse
☐ foci (singular *focus*)
☐ center (of an ellipse)

OBJECTIVE 1 Graph circles.

Recall from our earlier work that a **circle** is a set of all points in a plane that lie a fixed distance from a fixed point. This fixed point is the **center** of the circle. The fixed distance is its **radius.**

Suppose that the point (h, k) is the center of a circle with radius r. We let (x, y) represent any point on the circle. See **FIGURE 10.** By definition, the distance between (h, k) and (x, y) must be r.

$$\sqrt{(x_2 - x_1)^2 + (y_2 - y_1)^2} = r \qquad \text{Distance formula}$$

$$\sqrt{(x - h)^2 + (y - k)^2} = r \qquad \begin{array}{l}\text{Let } (h, k) = (x_1, y_1) \text{ and} \\ (x, y) = (x_2, y_2).\end{array}$$

$$(x - h)^2 + (y - k)^2 = r^2 \qquad \text{Square each side.}$$

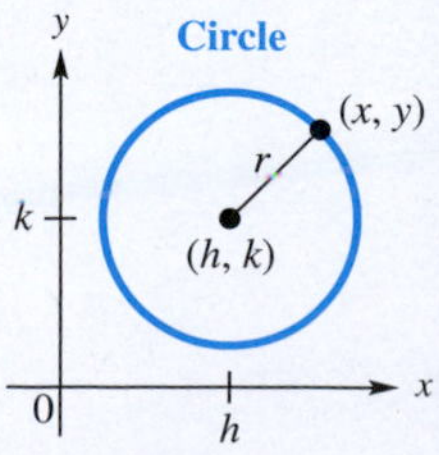

This result is the *center-radius form* of the equation of a circle.

FIGURE 10

Equation of a Circle (Center-Radius Form)

A circle with center (h, k) and radius r has an equation that can be written in the form

$$(x - h)^2 + (y - k)^2 = r^2 \quad \text{(where } r > 0\text{).}$$

If a circle has its center at the origin $(0, 0)$, then its equation becomes

$$x^2 + y^2 = r^2. \quad \text{Here, } h = 0 \text{ and } k = 0.$$

**NOW TRY
EXERCISE 1**

Find the center and radius of
each circle. Then graph the
circle.

(a) $(x - 2)^2 + (y - 3)^2 = 9$

(b) $x^2 + y^2 = 49$

EXAMPLE 1 Graphing Circles

Find the center and radius of each circle. Then graph the circle.

(a) $\quad (x - 3)^2 + (y + 1)^2 = 16$

> The radius r is the positive square root of 16.

$$(x - 3)^2 + [y - (-1)]^2 = 4^2 \quad (x - h)^2 + (y - k)^2 = r^2$$

$$\uparrow\uparrow\uparrow$$

$$hkr$$

The center (h, k) is $(3, -1)$, and the radius r is 4.

To graph the circle, plot the center $(3, -1)$, and then move right, left, up, and
down 4 units from the center to plot four points on the circle. Draw a smooth curve
through the four points. See **FIGURE 11**. (The center is *not* part of the graph.)

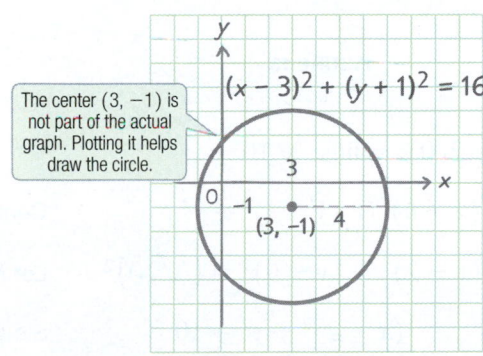

The center $(3, -1)$ is
not part of the actual
graph. Plotting it helps
draw the circle.

FIGURE 11

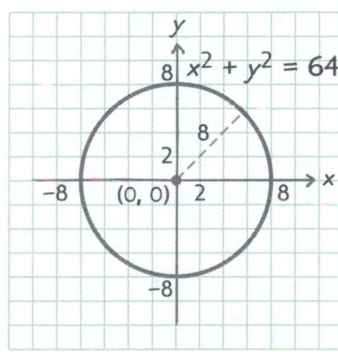

FIGURE 12

(b) $\quad x^2 + y^2 = 64$

$$(x - 0)^2 + (y - 0)^2 = 8^2 \quad \text{Here, } h = 0 \text{ and } k = 0.$$

The center is $(0, 0)$ and the radius is 8. The graph, which uses a scale of 2 on the axes,
is shown in **FIGURE 12**.

NOW TRY

NOW TRY ANSWERS

1. (a) center: $(2, 3)$; radius: 3

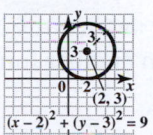

$(x - 2)^2 + (y - 3)^2 = 9$

(b) center: $(0, 0)$; radius: 7

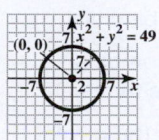

NOTE The x- and y-coordinates of the center of a circle are the values of x and y that
make each squared binomial in its center-radius form equal 0. Consider the equation

$$(x - 3)^2 + (y + 1)^2 = 16. \quad \text{See Example 1(a).}$$

If we let $x = 3$, then $(x - 3)^2$ equals 0. If we let $y = -1$, then $(y + 1)^2$ equals 0. Thus, the
center of the circle is $(3, -1)$.

OBJECTIVE 2 Write an equation of a circle given its center and radius.

NOW TRY
EXERCISE 2
Write the center-radius form
of each circle described. Then
graph the circle.

(a) Center: $(1, -2)$; radius: 3

(b) Center: $(0, 0)$; radius: $\sqrt{7}$

EXAMPLE 2 Writing Equations of Circles and Graphing Them

Write the center-radius form of each circle described. Then graph the circle.

(a) Center: $(-3, 4)$; radius: 2

Here, $(h, k) = (-3, 4)$ and $r = 2$.

$$(x - h)^2 + (y - k)^2 = r^2 \qquad \text{Center-radius form}$$

$$[x - (-3)]^2 + (y - 4)^2 = 2^2 \qquad \text{Substitute for } h, k, \text{ and } r.$$

> Watch signs here.

$$(x + 3)^2 + (y - 4)^2 = 4 \qquad \text{Simplify.}$$

See **FIGURE 13.**

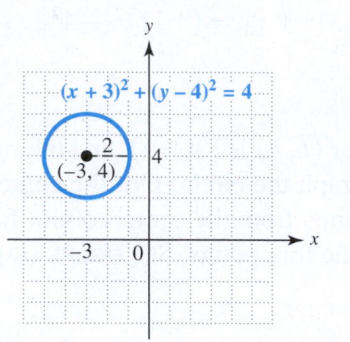

FIGURE 13   **FIGURE 14**

(b) Center: $(2, 0)$; radius: $\sqrt{10}$

$$(x - h)^2 + (y - k)^2 = r^2 \qquad \text{Center-radius form}$$

$$(x - 2)^2 + (y - 0)^2 = \left(\sqrt{10}\right)^2 \qquad \text{Let } h = 2, k = 0, \text{ and } r = \sqrt{10}.$$

$$(x - 2)^2 + y^2 = 10 \qquad \text{Simplify; } \left(\sqrt{a}\right)^2 = a.$$

The graph is shown in **FIGURE 14.** For the radius,

$$r = \sqrt{10} \approx 3.16. \qquad \text{Evaluate with a calculator.} \qquad \textbf{NOW TRY}$$

NOW TRY ANSWERS

2. **(a)** $(x - 1)^2 + (y + 2)^2 = 9$

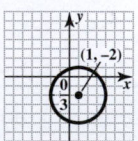

(b) $x^2 + y^2 = 7$

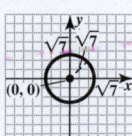

OBJECTIVE 3 Determine the center and radius of a circle given its equation.

If we begin with the center-radius form of the equation of a circle and rewrite it so that the binomials are expanded and the right side equals 0, we obtain another form of the equation. Consider the following.

$$(x + 3)^2 + (y - 4)^2 = 4 \qquad \text{See Example 2(a).}$$

> Remember the middle term when squaring each binomial.

$$x^2 + 6x + 9 + y^2 - 8y + 16 = 4 \qquad \text{Square each binomial.}$$

$$x^2 + y^2 + 6x - 8y + 21 = 0 \qquad \begin{array}{l}\text{Subtract 4. Combine}\\\text{and rearrange terms.}\end{array}$$

This result is a different, yet equivalent, form of the equation of the circle.

> **Equation of a Circle (General Form)**
>
> For some real numbers c, d, and e, an equation of the form
>
> $$x^2 + y^2 + cx + dy + e = 0$$
>
> may represent a circle.

In the above general form, both x^2- and y^2-terms have equal coefficients, here **1**. If the coefficients are equal but *not* 1, the above form can be obtained by dividing through by that coefficient.

NOW TRY
EXERCISE 3

Write the center-radius form of the circle with equation

$x^2 + y^2 - 8x + 10y - 8 = 0.$

Give the center and radius.

EXAMPLE 3 Completing the Square to Write the Center-Radius Form

Write the center-radius form of the circle with equation

$$x^2 + y^2 + 2x + 6y - 15 = 0.$$

Give the center and radius, and graph the circle.

To write the center-radius form, we complete the squares on x and y.

$$x^2 + y^2 + 2x + 6y - 15 = 0$$

$$x^2 + y^2 + 2x + 6y = 15$$ — Transform so that the constant is on the right.

$$(x^2 + 2x \qquad) + (y^2 + 6y \qquad) = 15$$ — Write in anticipation of completing the square.

$$\left[\tfrac{1}{2}(2)\right]^2 = 1 \qquad \left[\tfrac{1}{2}(6)\right]^2 = 9$$ — Square half the coefficient of each middle term.

$$(x^2 + 2x + 1) + (y^2 + 6y + 9) = 15 + 1 + 9$$ — Complete the squares on *both x* and *y.*

> Add 1 and 9 on *both* sides of the equation.

$$(x + 1)^2 + (y + 3)^2 = 25$$ — Factor. Add. Center-radius form

$$[x - (-1)]^2 + [y - (-3)]^2 = 5^2$$ — Rewrite to identify the center and radius.

The circle has center $(-1, -3)$ and radius 5. See **FIGURE 15**.   **NOW TRY**

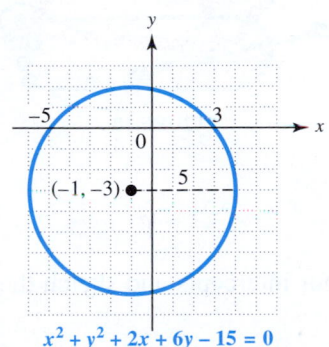

$$x^2 + y^2 + 2x + 6y - 15 = 0$$

FIGURE 15

> **NOTE** There are three possibilities for the graph of an equation of the form
>
> $$(x - h)^2 + (y - k)^2 = m, \quad \text{where } m \text{ is a constant.}$$
>
> **1.** If $m > 0$, then $r^2 = m$, and the graph of the equation is a circle with radius $\sqrt{m}$.
>
> **2.** If $m = 0$, then the graph of the equation is the single point (h, k).
>
> **3.** If $m < 0$, then no points satisfy the equation and the graph is nonexistent.

OBJECTIVE 4 Recognize the equation of an ellipse.

An **ellipse** is the set of all points in a plane the *sum* of whose distances from two fixed points is constant. These fixed points are the **foci** (singular: *focus*). The ellipse in **FIGURE 16** has foci $(c, 0)$ and $(-c, 0)$, with *x*-intercepts $(a, 0)$ and $(-a, 0)$ and *y*-intercepts $(0, b)$ and $(0, -b)$. It is shown in more advanced courses that $c^2 = a^2 - b^2$ for an ellipse of this type. The origin is the **center** of the ellipse.

NOW TRY ANSWER

3. $(x - 4)^2 + (y + 5)^2 = 49$; center: $(4, -5)$; radius: 7

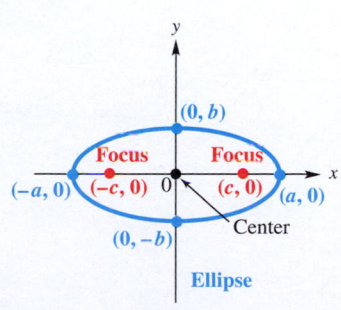

FIGURE 16

An ellipse has the following equation.

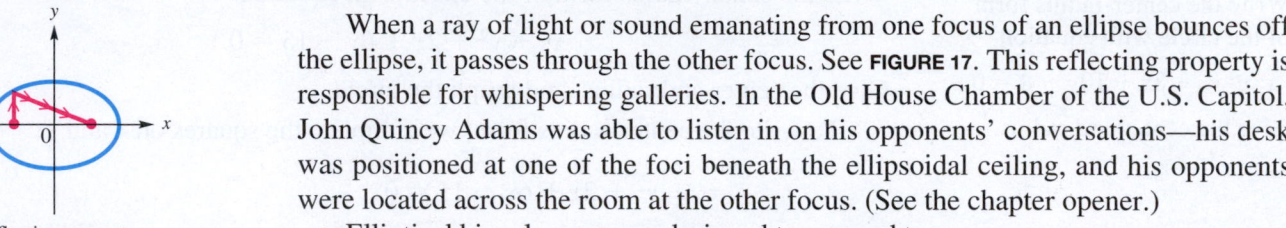

Equation of an Ellipse

An ellipse with x-intercepts $(a, 0)$ and $(-a, 0)$ and y-intercepts $(0, b)$ and $(0, -b)$ has an equation that can be written in the form

$$\frac{x^2}{a^2} + \frac{y^2}{b^2} = 1.$$

NOTE A circle is a special case of an ellipse, where $a^2 = b^2$.

When a ray of light or sound emanating from one focus of an ellipse bounces off the ellipse, it passes through the other focus. See **FIGURE 17**. This reflecting property is responsible for whispering galleries. In the Old House Chamber of the U.S. Capitol, John Quincy Adams was able to listen in on his opponents' conversations—his desk was positioned at one of the foci beneath the ellipsoidal ceiling, and his opponents were located across the room at the other focus. (See the chapter opener.)

Elliptical bicycle gears are designed to respond to the legs' natural strengths and weaknesses. At the top and bottom of the powerstroke, where the legs have the least leverage, the gear offers little resistance, but as the gear rotates, the resistance increases. This allows the legs to apply more power where it is most naturally available. See **FIGURE 18**.

Reflecting property
of an ellipse

FIGURE 17

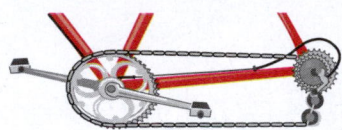

FIGURE 18

OBJECTIVE 5 Graph ellipses.

To graph an ellipse centered at the origin, we plot the four intercepts and sketch the ellipse through those points.

EXAMPLE 4 Graphing Ellipses

Graph each ellipse.

(a) $\dfrac{x^2}{49} + \dfrac{y^2}{36} = 1$

Here, $a^2 = 49$, so $a = 7$, and the x-intercepts are $(7, 0)$ and $(-7, 0)$. Similarly, $b^2 = 36$, so $b = 6$, and the y-intercepts are $(0, 6)$ and $(0, -6)$. Plotting the intercepts and sketching the ellipse through them gives the graph in **FIGURE 19**.

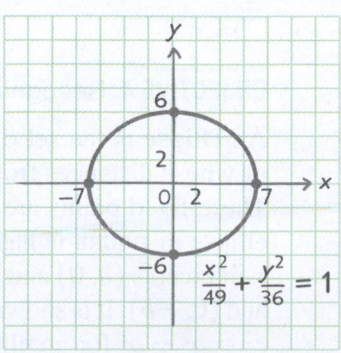

FIGURE 19

 **NOW TRY EXERCISE 4**

Graph $\dfrac{x^2}{16} + \dfrac{y^2}{25} = 1$.

(b) $\dfrac{x^2}{36} + \dfrac{y^2}{121} = 1$

The x-intercepts are $(6, 0)$ and $(-6, 0)$, and the y-intercepts are $(0, 11)$ and $(0, -11)$. Join these intercepts with the smooth curve of an ellipse. See **FIGURE 20**. (Note the scale of 2 on the axes.)

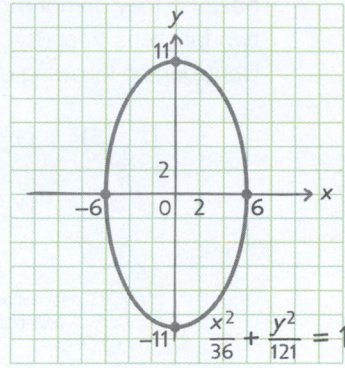

FIGURE 20

NOW TRY

🛑 **CAUTION** Hand-drawn graphs of ellipses are smooth curves that show symmetry with respect to the center. Again, the center is not part of the actual graph. It just provides help in drawing a more accurate graph.

 **NOW TRY EXERCISE 5**

Graph
$\dfrac{(x-3)^2}{36} + \dfrac{(y-4)^2}{4} = 1$.

EXAMPLE 5 Graphing an Ellipse Shifted Horizontally and Vertically

Graph $\dfrac{(x-2)^2}{25} + \dfrac{(y+3)^2}{49} = 1$.

Just as $(x - 2)^2$ and $(y + 3)^2$ would indicate that the center of a circle would be $(2, -3)$, so it is with this ellipse. **FIGURE 21** shows that the graph goes through the four points

$$(2, 4), \quad (7, -3), \quad (2, -10), \quad \text{and} \quad (-3, -3).$$

NOW TRY ANSWERS

4.
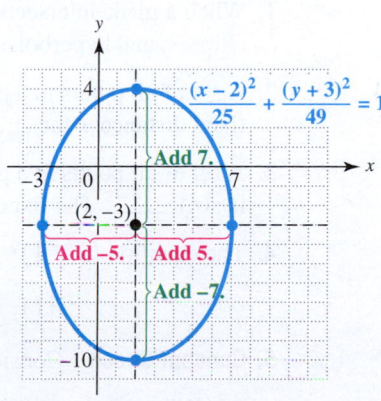

5.

$$\dfrac{(x-3)^2}{36} + \dfrac{(y-4)^2}{4} = 1$$

FIGURE 21

The x-values of these points are found by adding $\pm a = \pm 5$ to 2 (the x-value of the center). The y-values are found by adding $\pm b = \pm 7$ to -3 (the y-value of the center).

NOW TRY

Graphs of circles and ellipses are not graphs of functions—they fail the conditions of the vertical line test. The only conic section whose graph represents a function is the vertical parabola with equation $f(x) = ax^2 + bx + c$.

> **NOTE** A graphing calculator in function mode cannot directly graph a circle or an ellipse because they do not represent functions. We must first solve the equation for y to obtain two functions y_1 and y_2. The union of these two graphs is the graph of the entire figure. Consider the following equation of a circle.
>
> $$(x + 3)^2 + (y + 2)^2 = 25 \qquad \text{Solve the equation for } y.$$
>
> $$(y + 2)^2 = 25 - (x + 3)^2 \qquad \text{Subtract } (x + 3)^2.$$
>
> $$y + 2 = \pm\sqrt{25 - (x + 3)^2} \qquad \text{Take square roots.}$$
>
> Remember *both roots.* $\quad y = -2 \pm \sqrt{25 - (x + 3)^2} \qquad \text{Add } -2.$
>
> The two functions to be graphed are
>
> $$y_1 = -2 + \sqrt{25 - (x + 3)^2} \quad \text{and} \quad y_2 = -2 - \sqrt{25 - (x + 3)^2}.$$
>
> To get an undistorted screen, a **square viewing window** must be used. See **FIGURE 22**. The two semicircles seem to be disconnected because the calculator cannot show a true picture of the behavior at these points.

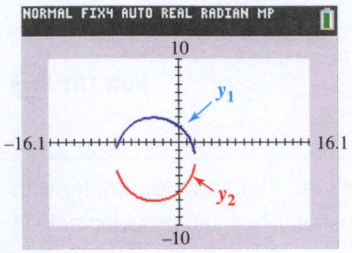

Square Viewing Window

FIGURE 22

13.2 Exercises

FOR EXTRA HELP

▶ **MyLab Math**

▶ *Video solutions for select problems available in MyLab Math*

Concept Check *Complete each statement. Choices may be used once, more than once, or not at all.*

circle	parabolas	radius	intercepts
foci	center	ellipse	conic sections

1. When a plane intersects an infinite cone at different angles, _____ such as _____, circles, ellipses, and hyperbolas result.

2. A set of all points in a plane that lie a fixed distance from a fixed point is a(n) _____. The fixed distance is the _____ and the fixed point is the _____.

3. A set of all points in a plane the sum of whose distances from two fixed points is constant is a(n) _____. The fixed points are the _____.

4. The equation $(x + 1)^2 + (y - 4)^2 = 4$ represents a(n) _____ with center

$$[(1, -4)/(1, 4)/(-1, 4)] \quad \text{and} \quad \underline{\hspace{1cm}} \text{ equal to 2.}$$

5. **Concept Check** Consider the circle whose equation is $x^2 + y^2 = 25$.

   (a) What are the coordinates of its center? **(b)** What is its radius?

   (c) Sketch its graph.

6. **Concept Check** Complete the following: The graph of a circle (*is/is not*) the graph of a function because it fails the conditions of the _____. The graph of an ellipse (*is/is not*) the graph of a function.

Concept Check   *Match each equation with the correct graph.*

7. $(x - 3)^2 + (y - 1)^2 = 25$ **A.** 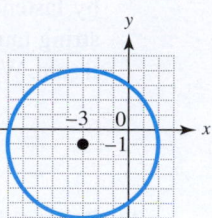   **B.**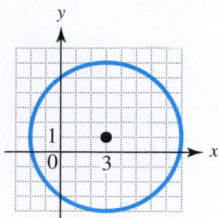

8. $(x + 3)^2 + (y - 1)^2 = 25$

9. $(x - 3)^2 + (y + 1)^2 = 25$ **C.** 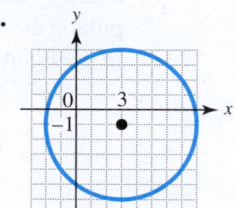   **D.**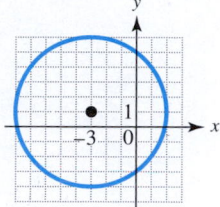

10. $(x + 3)^2 + (y + 1)^2 = 25$

Find the center and radius of each circle. Then graph the circle. **See Example 1.**

11. $x^2 + y^2 = 4$ **12.** $x^2 + y^2 = 1$

13. $x^2 + y^2 = 81$ **14.** $x^2 + y^2 = 36$

15. $(x - 5)^2 + (y + 4)^2 = 49$ **16.** $(x + 2)^2 + (y - 5)^2 = 16$

17. $(x + 1)^2 + (y + 3)^2 = 25$ **18.** $(x - 3)^2 + (y - 2)^2 = 4$

Write the center-radius form of each circle described. Then graph the circle. **See Examples 1 and 2.**

19. Center: $(0, 0)$; radius: 4   **20.** Center: $(0, 0)$; radius: 3

21. Center: $(-3, 2)$; radius: 3   **22.** Center: $(1, -3)$; radius: 4

23. Center: $(4, 3)$; radius: 5   **24.** Center: $(-3, -2)$; radius: 6

25. Center: $(-2, 0)$; radius: $\sqrt{5}$ **26.** Center: $(3, 0)$; radius: $\sqrt{13}$

27. Center: $(0, -3)$; radius: 7   **28.** Center: $(0, 4)$; radius: 4

Write the center-radius form of the circle with the given equation. Give the center and radius. (Hint: Divide each side by a common factor as needed.) **See Example 3.**

29. $x^2 + y^2 + 4x + 6y + 9 = 0$ **30.** $x^2 + y^2 - 8x - 12y + 3 = 0$

31. $x^2 + y^2 + 10x - 14y - 7 = 0$ **32.** $x^2 + y^2 - 2x + 4y - 4 = 0$

33. $3x^2 + 3y^2 - 12x - 24y + 12 = 0$ **34.** $2x^2 + 2y^2 + 20x + 16y + 10 = 0$

Write the center-radius form of the circle with the given equation. Give the center and radius, and graph the circle. **See Example 3.**

35. $x^2 + y^2 - 4x - 6y + 9 = 0$ **36.** $x^2 + y^2 + 8x + 2y - 8 = 0$

37. $x^2 + y^2 + 6x - 6y + 9 = 0$ **38.** $x^2 + y^2 - 4x + 10y + 20 = 0$

Concept Check   *Answer each question, and give a short explanation.* **See the Note following Example 3.**

39. How many points are there on the graph of $(x - 4)^2 + (y - 1)^2 = 0$?

40. How many points are there on the graph of $(x - 4)^2 + (y - 1)^2 = -1$?

41. Concept Check   A circle can be drawn on a piece of posterboard by fastening one end of a string with a thumbtack, pulling the string taut with a pencil, and tracing a curve, as shown in the figure. Why does this method work?

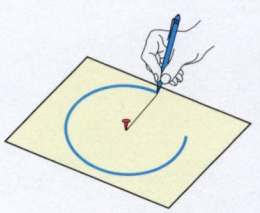

42. Concept Check   An ellipse can be drawn on a piece of poster-board by fastening two ends of a length of string with thumbtacks, pulling the string taut with a pencil, and tracing a curve, as shown in the figure. Why does this method work?

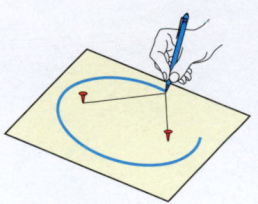

Graph each ellipse. See Examples 4 and 5.

43. $\dfrac{x^2}{9} + \dfrac{y^2}{25} = 1$

44. $\dfrac{x^2}{9} + \dfrac{y^2}{16} = 1$

45. $\dfrac{x^2}{36} + \dfrac{y^2}{16} = 1$

46. $\dfrac{x^2}{9} + \dfrac{y^2}{4} = 1$

47. $\dfrac{x^2}{16} + \dfrac{y^2}{4} = 1$

48. $\dfrac{x^2}{49} + \dfrac{y^2}{81} = 1$

49. $\dfrac{y^2}{25} = 1 - \dfrac{x^2}{49}$

50. $\dfrac{y^2}{9} = 1 - \dfrac{x^2}{16}$

51. $\dfrac{(x+1)^2}{64} + \dfrac{(y-2)^2}{49} = 1$

52. $\dfrac{(x-4)^2}{9} + \dfrac{(y+2)^2}{4} = 1$

53. $\dfrac{(x-2)^2}{16} + \dfrac{(y-1)^2}{9} = 1$

54. $\dfrac{(x+3)^2}{25} + \dfrac{(y+2)^2}{36} = 1$

Extending Skills   *A* lithotripter *is a machine used to crush kidney stones using shock waves. The patient is in an elliptical tub with the kidney stone at one focus of the ellipse. Each beam is projected from the other focus to the tub so that it reflects to hit the kidney stone. See the figure.*

55. Suppose a lithotripter is based on an ellipse with equation

$$\frac{x^2}{36} + \frac{y^2}{9} = 1.$$

How far from the center of the ellipse must the kidney stone and the source of the beam be placed? (*Hint:* Use the fact that $c^2 = a^2 - b^2$ because $a > b$ here.)

56. Rework **Exercise 55** if the equation of the ellipse is

$$9x^2 + 4y^2 = 36.$$

(*Hint:* Write the equation in fractional form by dividing each term by 36, and use $c^2 = b^2 - a^2$ because $b > a$ here.)

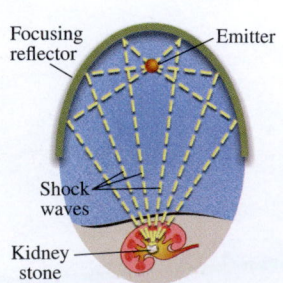

The top of an ellipse is illustrated in this depiction of how a lithotripter crushes a kidney stone.

Extending Skills *Solve each problem.*

57. An arch has the shape of half an ellipse. The equation of the ellipse, where x and y are in meters, is

$$100x^2 + 324y^2 = 32{,}400.$$

(a) How high is the center of the arch?

(b) How wide is the arch across the bottom?

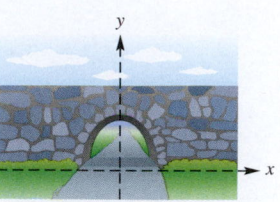

Not to scale

58. A one-way street passes under an overpass, which is in the form of the top half of an ellipse, as shown in the figure. Suppose that a truck 12 ft wide passes directly under the overpass. What is the maximum possible height of this truck?

15 ft

20 ft

Not to scale

59. Work each of the following.

(a) The Roman Colosseum is an ellipse with $a = 310$ ft and $b = \dfrac{513}{2}$ ft. Find the distance, to the nearest tenth, between the foci of this ellipse.

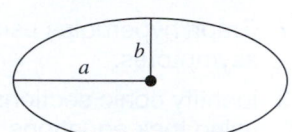

(b) The approximate perimeter of an ellipse is given by

$$P \approx 2\pi \sqrt{\frac{a^2 + b^2}{2}},$$

where a and b are the lengths given in part (a). Use this formula to find the approximate perimeter, to the nearest tenth, of the Roman Colosseum.

60. A satellite is in an elliptical orbit around Earth with least distance (**perigee**) of 160 km and greatest distance (**apogee**) of 16,000 km. Earth is located at a focus of the elliptical orbit. See the figure. (Data from *Space Mathematics*, Kastner, B., NASA.)

Find the equation of the ellipse. (*Hint:* Use the fact that $c^2 = a^2 - b^2$ here.)

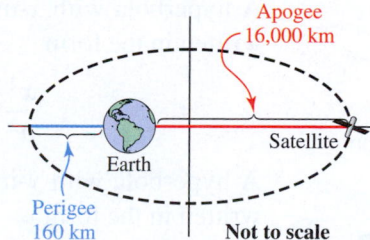

Apogee
16,000 km

Satellite

Earth

Perigee
160 km

Not to scale

Extending Skills *Solve each problem. See* **FIGURE 16** *and use the fact that $c^2 = a^2 - b^2$, where $a^2 > b^2$. Round answers to the nearest tenth. (Data from* Astronomy!, *Kaler, J. B., Addison-Wesley.)*

61. The orbit of Mars is an ellipse with the sun at one focus. For x and y in millions of miles, the equation of the orbit is

$$\frac{x^2}{141.7^2} + \frac{y^2}{141.1^2} = 1.$$

(a) Find the greatest distance (apogee) from Mars to the sun.

(b) Find the least distance (perigee) from Mars to the sun.

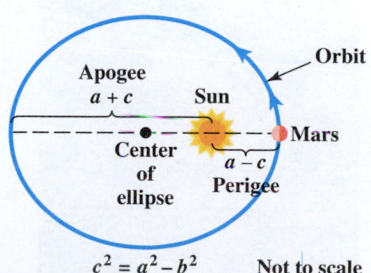

Apogee
$a + c$ Sun

Orbit

Center
of
ellipse

Mars

$a - c$

Perigee

$c^2 = a^2 - b^2$ Not to scale

62. The orbit of Venus around the sun (one of the foci) is an ellipse with equation

$$\frac{x^2}{5013} + \frac{y^2}{4970} = 1,$$

where x and y are measured in millions of miles.

(a) Find the greatest distance (apogee) between Venus and the sun.

(b) Find the least distance (perigee) between Venus and the sun.

13.3   Hyperbolas and Functions Defined by Radicals

OBJECTIVES

1 Recognize the equation of a hyperbola.

2 Graph hyperbolas using asymptotes.

3 Identify conic sections using their equations.

4 Graph generalized square root functions.

VOCABULARY

☐ hyperbola
☐ transverse axis
☐ fundamental rectangle
☐ asymptotes (of a hyperbola)
☐ generalized square root function

OBJECTIVE 1 Recognize the equation of a hyperbola.

A **hyperbola** is the set of all points in a plane such that the absolute value of the *difference* of the distances from two fixed points (the *foci*) is constant. The graph of a hyperbola has two parts, or *branches,* and two intercepts (or *vertices*) that lie on its axis, called the **transverse axis.**

The hyperbola in **FIGURE 23** has a horizontal transverse axis, with foci $(c, 0)$ and $(-c, 0)$ and x-intercepts $(a, 0)$ and $(-a, 0)$. (A hyperbola with vertical transverse axis would have its intercepts on the y-axis.)

A hyperbola centered at the origin has one of the following equations. (It is shown in more advanced courses that for a hyperbola, $c^2 = a^2 + b^2$.)

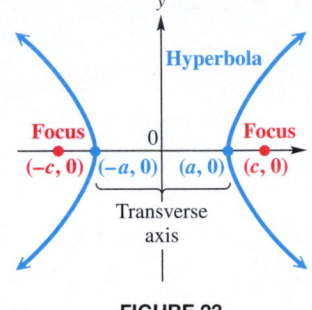

FIGURE 23

Equations of Hyperbolas

A hyperbola with x-intercepts $(a, 0)$ and $(-a, 0)$ has an equation that can be written in the form

$$\frac{x^2}{a^2} - \frac{y^2}{b^2} = 1. \qquad \text{Transverse axis on } x\text{-axis}$$

A hyperbola with y-intercepts $(0, b)$ and $(0, -b)$ has an equation that can be written in the form

$$\frac{y^2}{b^2} - \frac{x^2}{a^2} = 1. \qquad \text{Transverse axis on } y\text{-axis}$$

If we were to throw two stones into a pond, the ensuing concentric ripples would be shaped like a hyperbola. A cross-section of the cooling towers for a nuclear power plant is hyperbolic, as shown in the photo.

OBJECTIVE 2 Graph hyperbolas using asymptotes.

The two branches of the graph of a hyperbola approach a pair of intersecting straight lines, which are its *asymptotes*. (See **FIGURES 24** and **25** on the next page.) The asymptotes are useful for sketching the graph of the hyperbola.

Asymptotes of Hyperbolas

The extended diagonals of a rectangle, called the **fundamental rectangle,** with vertices (corners) at the points (a, b), $(-a, b)$, $(-a, -b)$, and $(a, -b)$ are the **asymptotes** of the hyperbolas

$$\frac{x^2}{a^2} - \frac{y^2}{b^2} = 1 \quad \text{and} \quad \frac{y^2}{b^2} - \frac{x^2}{a^2} = 1.$$

Using previous methods, we could show that the equations of these asymptotes are

$$y = \frac{b}{a}x \quad \text{and} \quad y = -\frac{b}{a}x.$$ Equations of the asymptotes of a hyperbola

Graphing a Hyperbola

Step 1   **Find and locate the intercepts.**

- At $(a, 0)$ and $(-a, 0)$ if the x^2-term has a positive coefficient
- At $(0, b)$ and $(0, -b)$ if the y^2-term has a positive coefficient

Step 2   **Find the fundamental rectangle.** Its vertices will be located at the points (a, b), $(-a, b)$, $(-a, -b)$, and $(a, -b)$.

Step 3   **Sketch the asymptotes.** The extended diagonals of the fundamental rectangle are the asymptotes of the hyperbola. They have equations $y = \pm\frac{b}{a}x$.

Step 4   **Draw the graph.** Sketch each branch of the hyperbola through an intercept, approaching (but not touching) the asymptotes.

NOW TRY EXERCISE 1

Graph $\dfrac{x^2}{25} - \dfrac{y^2}{9} = 1$.

EXAMPLE 1   Graphing a Horizontal Hyperbola

Graph $\dfrac{x^2}{16} - \dfrac{y^2}{25} = 1$.

Step 1   Here $a = 4$ and $b = 5$. The x-intercepts are $(4, 0)$ and $(-4, 0)$.

Step 2   The vertices of the fundamental rectangle are the four points

(a, b) $(-a, b)$ $(-a, -b)$ $(a, -b)$

$\downarrow\downarrow$ $\downarrow\downarrow$ $\downarrow\downarrow$ $\downarrow\downarrow$

$(4, 5)$, $(-4, 5)$, $(-4, -5)$,   and $(4, -5)$.

Steps 3 and 4   The equations of the asymptotes are $y = \pm\frac{b}{a}x$, or $y = \pm\frac{5}{4}x$. The hyperbola approaches these lines as x and y get larger in absolute value. See **FIGURE 24**.

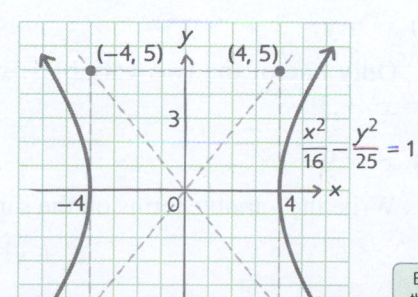

When graphing a hyperbola, the fundamental rectangle and the asymptotes are not part of the actual graph. They provide help in drawing a more accurate graph.

Be sure that the branches do not touch the asymptotes.

FIGURE 24

NOW TRY ANSWER

1.

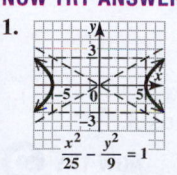

$\dfrac{x^2}{25} - \dfrac{y^2}{9} = 1$

NOW TRY

NOW TRY
EXERCISE 2

Graph $\dfrac{y^2}{9} - \dfrac{x^2}{16} = 1$.

EXAMPLE 2 Graphing a Vertical Hyperbola

Graph $\dfrac{y^2}{49} - \dfrac{x^2}{16} = 1$.

This hyperbola has y-intercepts $(0, 7)$ and $(0, -7)$. The asymptotes are the extended diagonals of the fundamental rectangle with vertices at

$$(4, 7), \quad (-4, 7), \quad (-4, -7), \quad \text{and} \quad (4, -7).$$

Their equations are $y = \pm \dfrac{7}{4} x$. See **FIGURE 25**.

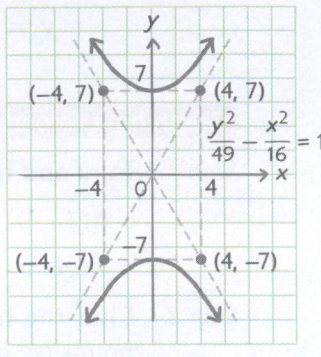

FIGURE 25

NOW TRY

OBJECTIVE 3 Identify conic sections using their equations.

Rewriting a second-degree equation in one of the forms given for ellipses, hyperbolas, circles, or parabolas makes it possible to identify the graph of the equation.

NOW TRY
EXERCISE 3

Identify the graph of each equation.

(a) $y^2 - 10 = -x^2$

(b) $y - 2x^2 = 8$

(c) $3x^2 + y^2 = 4$

EXAMPLE 3 Identifying the Graphs of Equations

Identify the graph of each equation.

(a) $9x^2 = 108 + 12y^2$

Both variables are squared, so the graph is either an ellipse or a hyperbola. (This situation also occurs for a circle, which is a special case of an ellipse.) To see which conic section it is, rewrite the equation so that the x^2- and y^2-terms are on one side of the equation and 1 is on the other side.

$$9x^2 - 12y^2 = 108 \qquad \text{Subtract } 12y^2.$$

$$\frac{x^2}{12} - \frac{y^2}{9} = 1 \qquad \text{Divide by 108.}$$

The subtraction symbol indicates that the graph of this equation is a hyperbola.

(b) $x^2 = y - 3$

Only one of the two variables, x, is squared, so this is the vertical parabola $y = x^2 + 3$.

(c) $x^2 = 9 - y^2$

Write the variable terms on the same side of the equation.

$$x^2 + y^2 = 9 \qquad \text{Add } y^2.$$

The graph of this equation is a circle with center at the origin and radius 3.

NOW TRY

NOW TRY ANSWERS

2.
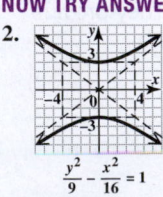

$\dfrac{y^2}{9} - \dfrac{x^2}{16} = 1$

3. **(a)** circle **(b)** parabola
   (c) ellipse

▼ SUMMARY OF CONIC SECTIONS

Equation	Graph	Description	Identification
$y = ax^2 + bx + c$ or $y = a(x - h)^2 + k$	 **Parabola**	It opens up if $a > 0$, down if $a < 0$. The vertex is (h, k).	It has an x^2-term. y is not squared.
$x = ay^2 + by + c$ or $x = a(y - k)^2 + h$	 **Parabola**	It opens to the right if $a > 0$, to the left if $a < 0$. The vertex is (h, k).	It has a y^2-term. x is not squared.
$(x - h)^2 +$ $(y - k)^2 = r^2$	 **Circle**	The center is (h, k), and the radius is r.	x^2- and y^2-terms have the same positive coefficient.
$\dfrac{x^2}{a^2} + \dfrac{y^2}{b^2} = 1$	 **Ellipse**	The x-intercepts are $(a, 0)$ and $(-a, 0)$. The y-intercepts are $(0, b)$ and $(0, -b)$.	x^2- and y^2-terms have different positive coefficients.
$\dfrac{x^2}{a^2} - \dfrac{y^2}{b^2} = 1$	 **Hyperbola**	The x-intercepts are $(a, 0)$ and $(-a, 0)$. The asymptotes are found from (a, b), $(a, -b)$, $(-a, -b)$, and $(-a, b)$.	x^2 has a positive coefficient. y^2 has a negative coefficient.
$\dfrac{y^2}{b^2} - \dfrac{x^2}{a^2} = 1$	 **Hyperbola**	The y-intercepts are $(0, b)$ and $(0, -b)$. The asymptotes are found from (a, b), $(a, -b)$, $(-a, -b)$, and $(-a, b)$.	y^2 has a positive coefficient. x^2 has a negative coefficient.

OBJECTIVE 4 Graph generalized square root functions.

Because they do not satisfy the conditions of the vertical line test, the graphs of horizontal parabolas and all circles, ellipses, and hyperbolas with horizontal or vertical axes do not satisfy the conditions of a function. However, by considering only a part of the graph of each of these, we have the graph of a function, as seen in **FIGURE 26**.

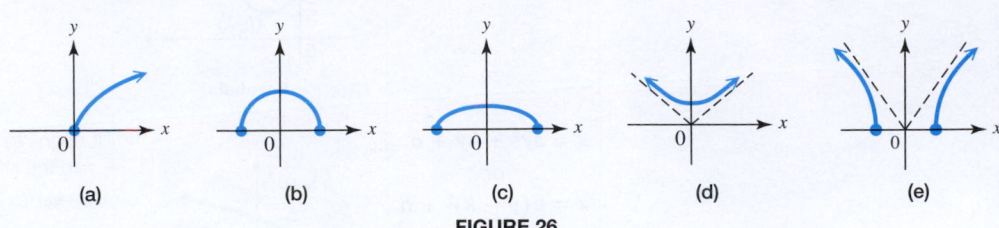

(a)   (b)   (c)   (d)   (e)

FIGURE 26

In parts (a)–(d) of **FIGURE 26**, the top portion of a conic section is shown (parabola, circle, ellipse, and hyperbola, respectively). In part (e), the top two portions of a hyperbola are shown. In each case, the graph is that of a function because the graph satisfies the conditions of the vertical line test.

Earlier in this chapter, we worked with the square root function $f(x) = \sqrt{x}$. To find equations for the types of graphs shown in **FIGURE 26**, we extend its definition.

> **Generalized Square Root Function**
>
> For an algebraic expression in x defined by u, where $u \geq 0$, a function of the form
>
> $$f(x) = \sqrt{u}$$
>
> is a **generalized square root function.**

**NOW TRY
EXERCISE 4**

Graph $f(x) = \sqrt{64 - x^2}$. Give the domain and range.

EXAMPLE 4 Graphing a Semicircle

Graph $f(x) = \sqrt{25 - x^2}$. Give the domain and range.

$$f(x) = \sqrt{25 - x^2} \qquad \text{Given function}$$

$$y = \sqrt{25 - x^2} \qquad \text{Replace } f(x) \text{ with } y.$$

$$y^2 = \left(\sqrt{25 - x^2}\right)^2 \qquad \text{Square each side.}$$

$$y^2 = 25 - x^2 \qquad \left(\sqrt{a}\right)^2 = a$$

$$x^2 + y^2 = 25 \qquad \text{Add } x^2.$$

This is the equation of a circle with center at $(0, 0)$ and radius 5.

Because $f(x)$, *or* y, *represents a principal square root in the original equation,* $f(x)$ *must be nonnegative. This restricts the graph to the upper half of the circle.*

See **FIGURE 27**. Use the graph and the vertical line test to verify that it is indeed a function. The domain is $[-5, 5]$, and the range is $[0, 5]$.

NOW TRY ANSWER

4.

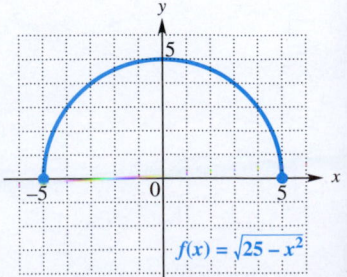

$f(x) = \sqrt{64 - x^2}$
domain: $[-8, 8]$; range: $[0, 8]$

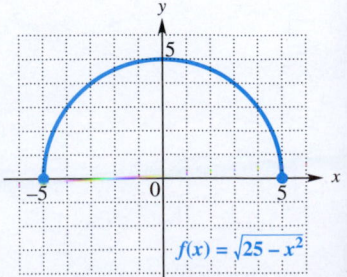

$f(x) = \sqrt{25 - x^2}$

FIGURE 27

NOW TRY

**NOW TRY
EXERCISE 5**

Graph $\dfrac{y}{4} = -\sqrt{1 - \dfrac{x^2}{9}}$.

Give the domain and range.

EXAMPLE 5	Graphing a Portion of an Ellipse

Graph $\dfrac{y}{6} = -\sqrt{1 - \dfrac{x^2}{16}}$. Give the domain and range.

$$\dfrac{y}{6} = -\sqrt{1 - \dfrac{x^2}{16}} \qquad \textcolor{blue}{\text{Given equation}}$$

$$\left(\dfrac{y}{6}\right)^2 = \left(-\sqrt{1 - \dfrac{x^2}{16}}\right)^2 \qquad \textcolor{blue}{\text{Square each side.}}$$

$$\dfrac{y^2}{36} = 1 - \dfrac{x^2}{16} \qquad \textcolor{blue}{\text{Apply the exponents.}}$$

$$\dfrac{x^2}{16} + \dfrac{y^2}{36} = 1 \qquad \textcolor{blue}{\text{Add } \dfrac{x^2}{16}.}$$

This is the equation of an ellipse with center at $(0, 0)$. The x-intercepts are $(4, 0)$ and $(-4, 0)$. The y-intercepts are $(0, 6)$ and $(0, -6)$.

Because $\dfrac{y}{6}$ equals a negative square root in the original equation, y must be nonpositive, restricting the graph to the lower half of the ellipse.

See **FIGURE 28.** The graph is that of a function. The domain is $[-4, 4]$, and the range is $[-6, 0]$.

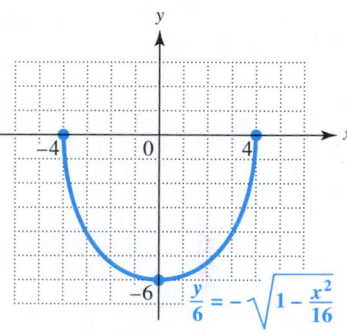

FIGURE 28

NOW TRY

NOW TRY ANSWER

5.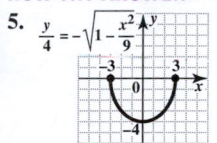

domain: $[-3, 3]$; range: $[-4, 0]$

13.3 Exercises

| FOR
EXTRA
HELP |

▶ **MyLab Math**

▶ *Video solutions for select problems available in MyLab Math*

Concept Check *Based on the discussions of ellipses in the previous section and of hyperbolas in this section, match each equation with its graph.*

1. $\dfrac{x^2}{25} + \dfrac{y^2}{9} = 1$ **A.**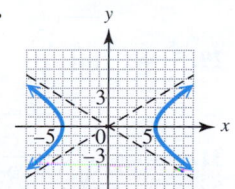

2. $\dfrac{x^2}{9} + \dfrac{y^2}{25} = 1$ **B.**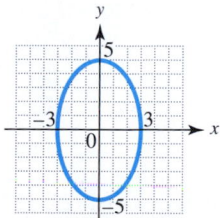

3. $\dfrac{x^2}{9} - \dfrac{y^2}{25} = 1$ **C.**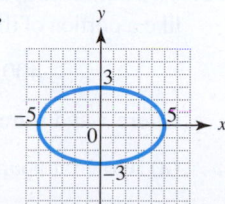

4. $\dfrac{x^2}{25} - \dfrac{y^2}{9} = 1$ **D.**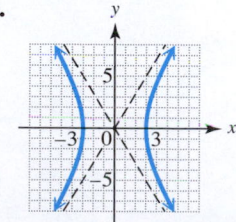

5. Concept Check   A student incorrectly described the graph of the equation

$$\frac{x^2}{9} - \frac{y^2}{16} = 1$$

as a vertical hyperbola with y-intercepts $(0, -4)$ and $(0, 4)$.   **WHAT WENT WRONG?**   Give the correct description of the graph.

6. Concept Check   A student incorrectly described the graph of the function

$$f(x) = \sqrt{49 - x^2}$$

as a circle centered at the origin with radius 7.   **WHAT WENT WRONG?**   Give the correct description of the graph.

Graph each hyperbola. See Examples 1 and 2.

7. $\dfrac{x^2}{16} - \dfrac{y^2}{9} = 1$

8. $\dfrac{x^2}{25} - \dfrac{y^2}{9} = 1$

9. $\dfrac{y^2}{4} - \dfrac{x^2}{25} = 1$

10. $\dfrac{y^2}{9} - \dfrac{x^2}{4} = 1$

11. $\dfrac{x^2}{25} - \dfrac{y^2}{36} = 1$

12. $\dfrac{x^2}{49} - \dfrac{y^2}{16} = 1$

13. $\dfrac{y^2}{9} - \dfrac{x^2}{9} = 1$

14. $\dfrac{y^2}{16} - \dfrac{x^2}{16} = 1$

Identify the graph of each equation as a parabola, circle, ellipse, *or* hyperbola, *and then sketch the graph. See Example 3.*

15. $x^2 - y^2 = 16$

16. $x^2 + y^2 = 16$

17. $4x^2 + y^2 = 16$

18. $9x^2 = 144 + 16y^2$

19. $y^2 = 36 - x^2$

20. $9x^2 + 25y^2 = 225$

21. $x^2 + 9y^2 = 9$

22. $x^2 - 2y = 0$

23. $x^2 = 4y - 8$

24. $y^2 = 4 + x^2$

Graph each generalized square root function. Give the domain and range. See Examples 4 and 5.

25. $f(x) = \sqrt{16 - x^2}$

26. $f(x) = \sqrt{9 - x^2}$

27. $f(x) = -\sqrt{36 - x^2}$

28. $f(x) = -\sqrt{25 - x^2}$

29. $\dfrac{y}{3} = \sqrt{1 + \dfrac{x^2}{9}}$

30. $\dfrac{y}{2} = \sqrt{1 + \dfrac{x^2}{4}}$

31. $y = -2\sqrt{1 - \dfrac{x^2}{9}}$

32. $y = -3\sqrt{1 - \dfrac{x^2}{25}}$

Extending Skills   *Solve each problem.*

33. Two buildings in a sports complex are shaped and positioned like a portion of the branches of the hyperbola with equation

$$400x^2 - 625y^2 = 250,000,$$

where x and y are in meters.

(a) How far apart are the buildings at their closest point?

(b) Find the distance d in the figure to the nearest tenth of a meter.

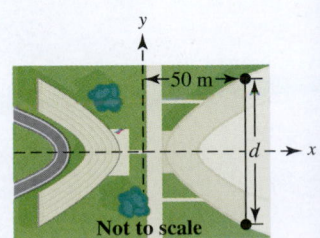

Not to scale

34. Using LORAN, a location-finding system, a radio transmitter at M sends out a series of pulses. When each pulse is received at transmitter S, it then sends out a pulse. A ship at P receives pulses from both M and S. A receiver on the ship measures the difference in the arrival times of the pulses. A special map gives hyperbolas that correspond to the differences in arrival times (which give the distances d_1 and d_2 in the figure). The ship can then be located as lying on a branch of a particular hyperbola.

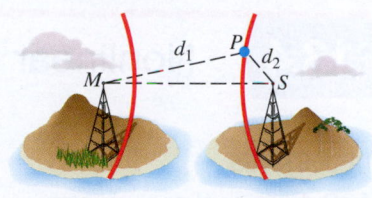

Suppose $d_1 = 80$ mi and $d_2 = 30$ mi, and the distance between transmitters M and S is 100 mi. Use the definition to find an equation of the hyperbola on which the ship is located.

Extending Skills *In rugby, after a try (similar to a touchdown in American football) the scoring team attempts a kick for extra points. The ball must be kicked from directly behind the point where the try was scored. The kicker can choose the distance but cannot move the ball sideways. It can be shown that the kicker's best choice is on the hyperbola with equation*

$$\frac{x^2}{g^2} - \frac{y^2}{g^2} = 1,$$

where $2g$ is the distance between the goal posts. Since the hyperbola approaches its asymptotes, it is easier for the kicker to estimate points on the asymptotes instead of on the hyperbola. (Data from Isaksen, Daniel C., "How to Kick a Field Goal." The College Mathematics Journal.)

35. What are the asymptotes of this hyperbola?

36. Why is it relatively easy to estimate the asymptotes?

RELATING CONCEPTS   For Individual or Group Work   (Exercises 37–40)

We have seen that the center of an ellipse may be shifted away from the origin. The same process applies to hyperbolas. For example, the hyperbola

$$\frac{(x + 5)^2}{4} - \frac{(y - 2)^2}{9} = 1,$$

has the same graph as

$$\frac{x^2}{4} - \frac{y^2}{9} = 1,$$

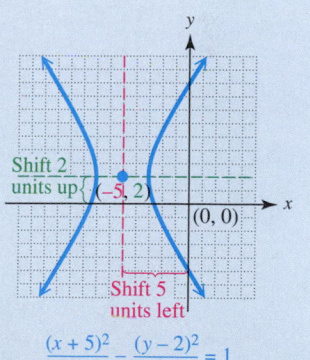

but it is centered at $(-5, 2)$, as shown at the right.
    Graph each hyperbola with center shifted away from the origin.

37. $\dfrac{(x - 2)^2}{4} - \dfrac{(y + 1)^2}{9} = 1$

38. $\dfrac{(x + 3)^2}{16} - \dfrac{(y - 2)^2}{25} = 1$

39. $\dfrac{y^2}{36} - \dfrac{(x - 2)^2}{49} = 1$

40. $\dfrac{(y - 5)^2}{9} - \dfrac{x^2}{25} = 1$

13.4   Nonlinear Systems of Equations

An equation in which some terms have more than one variable or a variable of degree 2 or greater is called a **nonlinear equation.** A **nonlinear system of equations** includes at least one nonlinear equation.

When solving a nonlinear system, it helps to visualize the types of graphs of the equations of the system to determine the possible number of points of intersection. For example, if a system includes two equations where the graph of one is a circle and the graph of the other is a line, then there may be zero, one, or two points of intersection, as illustrated in **FIGURE 29**.

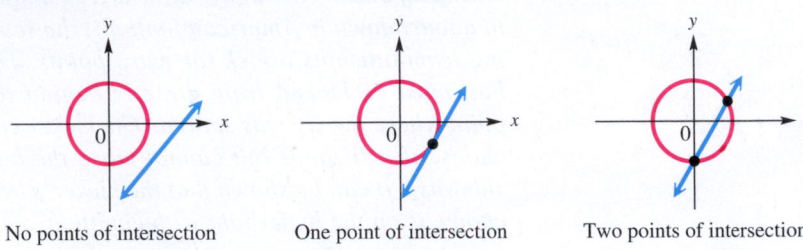

No points of intersection   One point of intersection   Two points of intersection

FIGURE 29

VOCABULARY

☐ nonlinear equation
☐ nonlinear system of equations

OBJECTIVE 1 Solve a nonlinear system using substitution.

We can usually solve a nonlinear system using the substitution method if one equation is linear.

EXAMPLE 1   Solving a Nonlinear System Using Substitution

Solve the system.

$$x^2 + y^2 = 9 \quad \text{(1)}$$
$$2x - y = 3 \quad \text{(2)}$$

The graph of equation (1) is a circle and the graph of equation (2) is a line, so the graphs could intersect in zero, one, or two points, as in **FIGURE 29**. We begin by solving the linear equation (2) for one of the two variables.

$$2x - y = 3 \qquad \qquad \text{(2)}$$
$$y = 2x - 3 \qquad \text{Solve for } y. \quad \text{(3)}$$

Then we substitute $2x - 3$ for y in the nonlinear equation (1).

$$x^2 + y^2 = 9 \qquad \text{(1)}$$
$$x^2 + (2x - 3)^2 = 9 \qquad \text{Let } y = 2x - 3.$$
$$x^2 + 4x^2 - 12x + 9 = 9 \qquad \text{Square } 2x - 3.$$
$$5x^2 - 12x = 0 \qquad \text{Combine like terms. Subtract 9.}$$
$$x(5x - 12) = 0 \qquad \text{Factor. The GCF is } x.$$
$$x = 0 \quad \text{or} \quad 5x - 12 = 0 \qquad \text{Zero-factor property}$$

Set *both* factors equal to 0.

$$x = \frac{12}{5} \qquad \text{Solve for } x.$$

**NOW TRY
EXERCISE 1**

Solve the system.

$$4x^2 + y^2 = 36$$
$$x - y = 3$$

Let $x = 0$ in equation (3), $y = 2x - 3$, to obtain $y = -3$. Then let $x = \frac{12}{5}$ in equation (3).

$$y = 2\left(\frac{12}{5}\right) - 3 \qquad \text{Let } x = \frac{12}{5}.$$

$$y = \frac{24}{5} - \frac{15}{5} \qquad \text{Multiply; } 3 = \frac{15}{5}$$

$$y = \frac{9}{5} \qquad \frac{a}{c} - \frac{b}{c} = \frac{a - b}{c}$$

The solution set of the system is $\left\{(0, -3), \left(\frac{12}{5}, \frac{9}{5}\right)\right\}$. The graph in **FIGURE 30** confirms the two points of intersection.

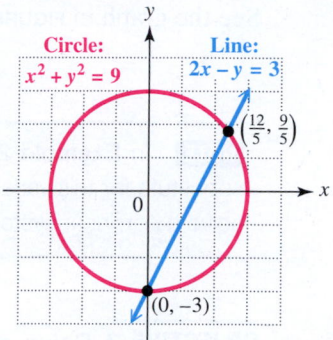

FIGURE 30

NOW TRY

EXAMPLE 2 Solving a Nonlinear System Using Substitution

Solve the system.

$$6x - y = 5 \qquad (1)$$
$$xy = 4 \qquad (2)$$

The graph of (1) is a line. It can be shown by plotting points that the graph of (2) is a hyperbola. Visualizing a line and a hyperbola indicates that there may be zero, one, or two points of intersection.

Since neither equation has a squared term, we can solve either equation for one of the variables and then substitute the result into the other equation. Solving $xy = 4$ for x gives $x = \frac{4}{y}$. We substitute $\frac{4}{y}$ for x in equation (1).

$$6x - y = 5 \qquad (1)$$

$$6\left(\frac{4}{y}\right) - y = 5 \qquad \text{Let } x = \frac{4}{y}.$$

$$\frac{24}{y} - y = 5 \qquad \text{Multiply.}$$

$$24 - y^2 = 5y \qquad \text{Multiply by } y, \ y \neq 0.$$

$$y^2 + 5y - 24 = 0 \qquad \text{Standard form}$$

$$(y - 3)(y + 8) = 0 \qquad \text{Factor.}$$

$$y - 3 = 0 \quad \text{or} \quad y + 8 = 0 \qquad \text{Zero-factor property}$$

$$y = 3 \quad \text{or} \qquad y = -8 \qquad \text{Solve each equation.}$$

NOW TRY ANSWER
1. $\left\{(3, 0), \left(-\frac{9}{5}, -\frac{24}{5}\right)\right\}$

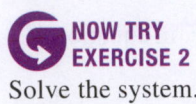

NOW TRY
EXERCISE 2
Solve the system.

$$xy = 2$$

$$x - 3y = 1$$

We substitute these values of y into $x = \dfrac{4}{y}$ to obtain the corresponding values of x.

If $y = 3$,   then $x = \dfrac{4}{3}$.

If $y = -8$,   then $x = -\dfrac{1}{2}$.

The solution set of the system is

$$\left\{ \left(\frac{4}{3}, 3 \right), \left(-\frac{1}{2}, -8 \right) \right\}.$$

Write the x-coordinates first in the ordered-pair solutions.

See the graph in **FIGURE 31**.

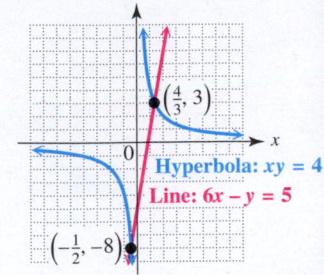

FIGURE 31

NOW TRY

NOTE   In **Example 2**, we could solve the *linear* equation for one of its variables and substitute for this variable in the *nonlinear* equation. There is often more than one way to solve a nonlinear system of equations.

OBJECTIVE 2 Solve a nonlinear system with two second-degree equations using elimination.

If a system consists of two second-degree equations, then there may be zero, one, two, three, or four solutions. **FIGURE 32** shows a case where a system consisting of a circle and a parabola has four solutions, all made up of ordered pairs of real numbers.

The elimination method is often used when both equations of a nonlinear system are second degree.

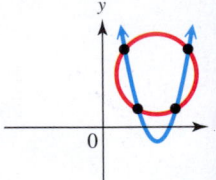

This system has four solutions because there are four points of intersection.

FIGURE 32

EXAMPLE 3   Solving a Nonlinear System Using Elimination

Solve the system.

$$x^2 + y^2 = 9 \qquad (1)$$

$$2x^2 - y^2 = -6 \qquad (2)$$

The graph of (1) is a circle, while the graph of (2) is a hyperbola. By analyzing the possibilities, we conclude that there may be zero, one, two, three, or four points of intersection. Adding the two equations will eliminate y.

$$
\begin{array}{lll}
x^2 + y^2 = 9 & & (1) \\
\underline{2x^2 - y^2 = -6} & & (2) \\
3x^2 = 3 & & \text{Add.} \\
x^2 = 1 & & \text{Divide by 3.} \\
x = 1 \quad \text{or} \quad x = -1 & & \text{Square root property}
\end{array}
$$

Each value of x gives corresponding values for y when substituted into one of the original equations. Using equation (1) is easier because the coefficients of the x^2- and y^2-terms are 1.

NOW TRY ANSWER
2. $\left\{ (-2, -1), \left(3, \frac{2}{3} \right) \right\}$

NOW TRY
EXERCISE 3

Solve the system.

$$x^2 + \quad y^2 = 16$$
$$4x^2 + 13y^2 = 100$$

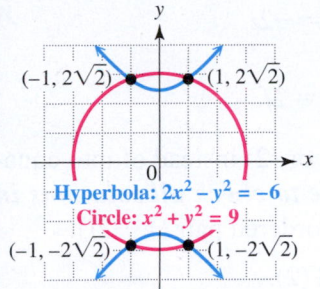

FIGURE 33

$$x^2 + y^2 = 9 \quad (1)$$
$$1^2 + y^2 = 9 \quad \text{Let } x = 1.$$
$$y^2 = 8$$
$$y = \sqrt{8} \quad \text{or} \quad y = -\sqrt{8}$$
$$y = 2\sqrt{2} \quad \text{or} \quad y = -2\sqrt{2}$$

$$x^2 + y^2 = 9 \quad (1)$$
$$(-1)^2 + y^2 = 9 \quad \text{Let } x = -1.$$
$$y^2 = 8$$
$$y = \sqrt{8} \quad \text{or} \quad y = -\sqrt{8}$$
$$y = 2\sqrt{2} \quad \text{or} \quad y = -2\sqrt{2}$$

The solution set is

$$\left\{ \left(1, 2\sqrt{2}\right), \left(1, -2\sqrt{2}\right), \left(-1, 2\sqrt{2}\right), \left(-1, -2\sqrt{2}\right) \right\}.$$

FIGURE 33 shows the four points of intersection.

NOW TRY

OBJECTIVE 3 Solve a nonlinear system that requires a combination of methods.

EXAMPLE 4   Solving a Nonlinear System Using a Combination of Methods

Solve the system.

$$x^2 + 2xy - y^2 = 7 \quad (1)$$
$$x^2 - y^2 = 3 \quad (2)$$

While we have not graphed equations like (1), its graph is a hyperbola. The graph of (2) is also a hyperbola. Two hyperbolas may have zero, one, two, three, or four points of intersection. We use the elimination method here in combination with the substitution method.

$$x^2 + 2xy - y^2 = \quad 7 \quad (1)$$
$$\underline{-x^2 \qquad\quad + y^2 = -3} \quad \text{Multiply (2) by } -1.$$

The x^2- and y^2-terms were eliminated.

$$2xy \qquad\quad = \quad 4 \quad \text{Add.}$$

Next, we solve $2xy = 4$ for one of the variables. We choose y.

$$2xy = 4$$

$$y = \frac{2}{x} \quad \text{Divide by } 2x. \quad (3)$$

Now, we substitute $y = \frac{2}{x}$ into one of the original equations.

$$x^2 - y^2 = 3 \quad \text{The substitution is easier in (2).}$$

$$x^2 - \left(\frac{2}{x}\right)^2 = 3 \quad \text{Let } y = \frac{2}{x}.$$

$$x^2 - \frac{4}{x^2} = 3 \quad \text{Square } \frac{2}{x}.$$

$$x^4 - 4 = 3x^2 \quad \text{Multiply by } x^2, x \neq 0.$$

$$x^4 - 3x^2 - 4 = 0 \quad \text{Subtract } 3x^2.$$

$$(x^2 - 4)(x^2 + 1) = 0 \quad \text{Factor.}$$

$$x^2 - 4 = 0 \quad \text{or} \quad x^2 + 1 = 0 \quad \text{Zero-factor property}$$

$$x^2 = 4 \quad \text{or} \quad x^2 = -1 \quad \text{Solve each equation.}$$

$$x = 2 \quad \text{or} \quad x = -2 \quad \text{or} \quad x = i \quad \text{or} \quad x = -i$$

NOW TRY ANSWER

3. $\left\{ \left(2\sqrt{3}, 2\right), \left(2\sqrt{3}, -2\right), \right.$
   $\left. \left(-2\sqrt{3}, 2\right), \left(-2\sqrt{3}, -2\right) \right\}$

NOW TRY EXERCISE 4

Solve the system.

$$x^2 + 3xy - y^2 = 23$$
$$x^2 - y^2 = 5$$

Substituting these four values of x into $y = \frac{2}{x}$ (equation (3)) gives the corresponding values for y.

If $x = 2,$ then $y = \frac{2}{2} = 1.$

If $x = -2,$ then $y = \frac{2}{-2} = -1.$

If $x = i,$ then $y = \frac{2}{i} = \frac{2}{i} \cdot \frac{-i}{-i} = -2i.$

> Multiply by the complex conjugate of the denominator. $i(-i) = 1$

If $x = -i,$ then $y = \frac{2}{-i} = \frac{2}{-i} \cdot \frac{i}{i} = 2i.$

If we substitute the x-values we found into equation (1) or (2) instead of into equation (3), we get extraneous solutions. ***It is always wise to check all solutions in both of the given equations.*** We show a check for the ordered pair $(i, -2i)$.

CHECK Let $x = i$ and $y = -2i$ in both equations (1) and (2).

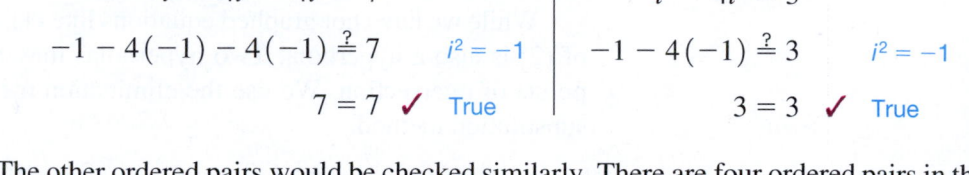

$$x^2 + 2xy - y^2 = 7 \quad \text{(1)}$$
$$i^2 + 2(i)(-2i) - (-2i)^2 \overset{?}{=} 7$$
$$i^2 - 4i^2 - 4i^2 \overset{?}{=} 7$$
$$-1 - 4(-1) - 4(-1) \overset{?}{=} 7 \qquad i^2 = -1$$
$$7 = 7 \; \checkmark \; \text{True}$$

$$x^2 - y^2 = 3 \quad \text{(2)}$$
$$i^2 - (-2i)^2 \overset{?}{=} 3$$
$$i^2 - 4i^2 \overset{?}{=} 3$$
$$-1 - 4(-1) \overset{?}{=} 3 \qquad i^2 = -1$$
$$3 = 3 \; \checkmark \; \text{True}$$

The other ordered pairs would be checked similarly. There are four ordered pairs in the solution set, two with real values and two with pure imaginary values. The solution set is

$$\{(2, 1), (-2, -1), (i, -2i), (-i, 2i)\}.$$

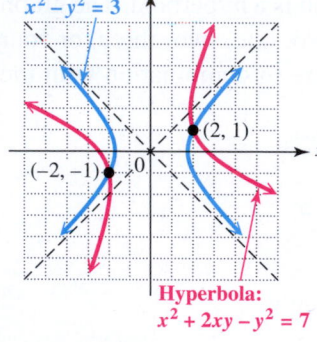

Hyperbola: $x^2 - y^2 = 3$

$(2, 1)$
$(-2, -1)$

Hyperbola: $x^2 + 2xy - y^2 = 7$

FIGURE 34

The graph of the system, shown in **FIGURE 34**, shows only the two real intersection points because the graph is in the real number plane. In general, if solutions contain nonreal complex numbers as components, they do not appear on the graph.

NOW TRY

NOTE It is not essential to visualize the number of points of intersection of the graphs in order to solve a nonlinear system. Sometimes we are unfamiliar with the graphs or, as in **Example 4,** there are nonreal complex solutions that do not appear as points of intersection in the real plane. Visualizing the graphs is only an aid to solving these systems.

NOW TRY ANSWER
4. $\{(3, 2), (-3, -2),$
   $(2i, -3i), (-2i, 3i)\}$

13.4 Exercises

FOR EXTRA HELP

▶ **MyLab Math**

▶ *Video solutions for select problems available in MyLab Math*

Concept Check   *Each sketch represents the graphs of a pair of equations in a system. How many ordered pairs of real numbers are in each solution set?*

1.
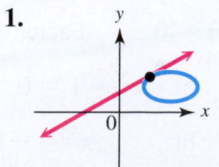

2.
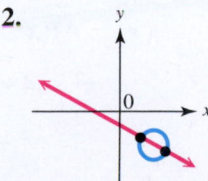

3.
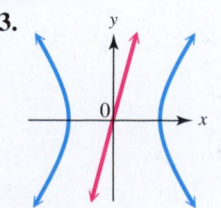

4.
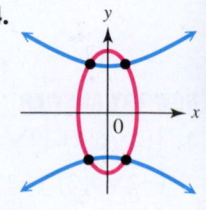

Concept Check *Suppose that a nonlinear system is composed of equations whose graphs are those described, and the number of points of intersection of the two graphs is as given. Make a sketch satisfying these conditions. (There may be more than one way to do this.)*

5. A line and a circle; no points

6. A line and a circle; one point

7. A line and a hyperbola; one point

8. A line and an ellipse; no points

9. A circle and an ellipse; four points

10. A parabola and an ellipse; one point

11. A parabola and an ellipse; four points

12. A parabola and a hyperbola; two points

Solve each system using the substitution method. **See Examples 1 and 2.**

13. $y = 4x^2 - x$
$y = x$

14. $y = x^2 + 6x$
$3y = 12x$

15. $y = x^2 + 6x + 9$
$x + y = 3$

16. $y = x^2 + 8x + 16$
$x - y = -4$

17. $x^2 + y^2 = 2$
$2x + y = 1$

18. $2x^2 + 4y^2 = 4$
$x = 4y$

19. $xy = 4$
$3x + 2y = -10$

20. $xy = -5$
$2x + y = 3$

21. $xy = -3$
$x + y = -2$

22. $xy = 12$
$x + y = 8$

23. $y = 3x^2 + 6x$
$y = x^2 - x - 6$

24. $y = 2x^2 + 1$
$y = 5x^2 + 2x - 7$

25. $2x^2 - y^2 = 6$
$y = x^2 - 3$

26. $x^2 + y^2 = 4$
$y = x^2 - 2$

27. $x^2 - xy + y^2 = 0$
$x - 2y = 1$

28. $x^2 - 3x + y^2 = 4$
$2x - y = 3$

Solve each system using the elimination method or a combination of the elimination and substitution methods. **See Examples 3 and 4.**

29. $3x^2 + 2y^2 = 12$
$x^2 + 2y^2 = 4$

30. $x^2 + 6y^2 = 9$
$4x^2 + 3y^2 = 36$

31. $5x^2 - 2y^2 = -13$
$3x^2 + 4y^2 = 39$

32. $x^2 + y^2 = 41$
$x^2 - y^2 = 9$

33. $2x^2 + 3y^2 = 6$
$x^2 + 3y^2 = 3$

34. $3x^2 + y^2 = 15$
$x^2 + y^2 = 5$

35. $2x^2 + y^2 = 28$
$4x^2 - 5y^2 = 28$

36. $x^2 + 3y^2 = 40$
$4x^2 - y^2 = 4$

37. $xy = 6$
$3x^2 - y^2 = 12$

38. $xy = 5$
$2y^2 - x^2 = 5$

39. $2x^2 = 8 - 2y^2$
$3x^2 = 24 - 4y^2$

40. $5x^2 = 20 - 5y^2$
$2y^2 = 2 - x^2$

41. $x^2 + xy + y^2 = 15$
$x^2 + y^2 = 10$

42. $2x^2 + 3xy + 2y^2 = 21$
$x^2 + y^2 = 6$

43. $x^2 + xy - y^2 = 11$
$x^2 - y^2 = 8$

44. $x^2 + xy - y^2 = 29$
$x^2 - y^2 = 24$

45. $3x^2 + 2xy - 3y^2 = 5$
$-x^2 - 3xy + y^2 = 3$

46. $-2x^2 + 7xy - 3y^2 = 4$
$2x^2 - 3xy + 3y^2 = 4$

Extending Skills   *Solve each problem using a nonlinear system.*

47. The area of a rectangular rug is 84 ft^2 and its perimeter is 38 ft. Find the length and width of the rug.

48. Find the length and width of a rectangular room whose perimeter is 50 m and whose area is 100 m^2.

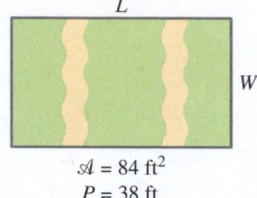

$\mathcal{A} = 84$ ft^2
$P = 38$ ft

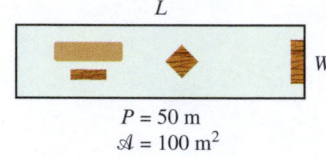

$P = 50$ m
$\mathcal{A} = 100$ m^2

49. A company has found that the price p (in dollars) of its scientific calculator is related to the supply x (in thousands) by the equation

$$px = 16.$$

The price is related to the demand x (in thousands) for the calculator by the equation

$$p = 10x + 12.$$

The **equilibrium price** is the value of p where demand equals supply. Find the equilibrium price and the supply/demand at that price. (*Hint:* Demand, price, and supply must all be positive.)

50. A company has determined that the cost y to make x (thousand) computer tablets is

$$y = 4x^2 + 36x + 20,$$

and that the revenue y from the sale of x (thousand) tablets is

$$36x^2 - 3y = 0.$$

Find the **break-even point,** where cost equals revenue.

13.5   Second-Degree Inequalities and Systems of Inequalities

OBJECTIVES

1   Graph second-degree inequalities.

2   Graph the solution set of a system of inequalities.

OBJECTIVE 1   Graph second-degree inequalities.

A **second-degree inequality** is an inequality with at least one variable of degree 2 and no variable of degree greater than 2.

Examples: $x^2 + y^2 > 9$, $y \leq 2x^2 - 4$, $x \geq y^2$

NOW TRY EXERCISE 1

Graph $x^2 + y^2 \geq 9$.

EXAMPLE 1   Graphing a Second-Degree Inequality

Graph $x^2 + y^2 \leq 36$.

The boundary of the inequality $x^2 + y^2 \leq 36$ is the graph of the equation $x^2 + y^2 = 36$, a circle with radius 6 and center at the origin, as shown in **FIGURE 35**.

The inequality $x^2 + y^2 \leq 36$ includes the points of the boundary (because the symbol $\leq$ includes equality) and either the points "outside" the circle or the points "inside" the circle. To decide which region to shade, we substitute any test point not on the circle.

$$x^2 + y^2 < 36 \quad \boxed{\text{We are testing the region.}}$$

$$0^2 + 0^2 \overset{?}{<} 36 \qquad \text{Use } (0, 0) \text{ as a test point.}$$

$$0 < 36 \qquad \text{True}$$

Because a true statement results, the original inequality includes the points *inside* the circle, the shaded region in **FIGURE 35**, and the boundary.

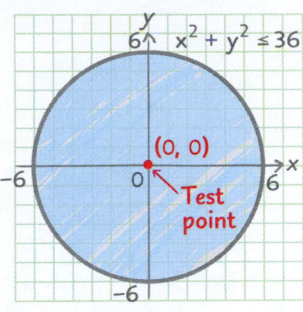

FIGURE 35

NOW TRY

NOTE   Because the substitution is easy, the origin is the test point of choice unless the graph actually passes through $(0, 0)$.

NOW TRY EXERCISE 2

Graph $y \geq -(x + 2)^2 + 1$.

EXAMPLE 2   Graphing a Second-Degree Inequality

Graph $y < -2(x - 4)^2 - 3$.

The boundary, $y = -2(x - 4)^2 - 3$, is a parabola that opens down with vertex $(4, -3)$. We must decide whether to shade the region "inside" or "outside" the parabola.

$$y < -2(x - 4)^2 - 3 \qquad \text{Original inequality}$$

$$0 \overset{?}{<} -2(0 - 4)^2 - 3 \qquad \text{Use } (0, 0) \text{ as a test point.}$$

$$0 \overset{?}{<} -32 - 3 \qquad \text{Simplify.}$$

$$0 < -35 \qquad \text{False}$$

Because the final inequality is a false statement, the points in the region containing $(0, 0)$ do not satisfy the inequality. As a result, the region inside (or below) the parabola is shaded in **FIGURE 36**. The parabola is drawn as a dashed curve since the points of the parabola itself do not satisfy the inequality.

CHECK   As additional confirmation, select a test point in the shaded region, such as $(4, -7)$, and substitute it into the original inequality.

$$y < -2(x - 4)^2 - 3$$

$$-7 \overset{?}{<} -2(4 - 4)^2 - 3 \qquad \text{Test } (4, -7).$$

$$-7 < -3 \ \checkmark \qquad \text{True}$$

A true statement results, so the correct region is shaded.

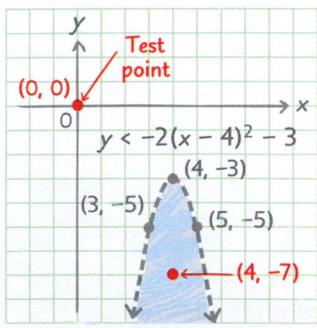

FIGURE 36

NOW TRY

NOW TRY ANSWERS
1.   2.

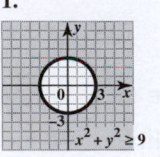   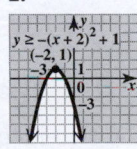

**NOW TRY
EXERCISE 3**

Graph $25x^2 - 16y^2 > 400$.

EXAMPLE 3 Graphing a Second-Degree Inequality

Graph $16y^2 \le 144 + 9x^2$.

$$16y^2 \le 144 + 9x^2$$

$$16y^2 - 9x^2 \le 144 \qquad \text{Subtract } 9x^2.$$

$$\frac{y^2}{9} - \frac{x^2}{16} \le 1 \qquad \text{Divide by 144.}$$

This form shows that the boundary is the hyperbola given by

$$\frac{y^2}{9} - \frac{x^2}{16} = 1.$$

Because the graph is a vertical hyperbola, the desired region will be either the region "between" the branches or the regions "above" the top branch and "below" the bottom branch. We choose $(0, 0)$ as a test point.

$$16y^2 < 144 + 9x^2$$

$$16(0)^2 \overset{?}{<} 144 + 9(0)^2 \qquad \text{Test } (0, 0).$$

$$0 < 144 \qquad \text{True}$$

Because a true statements results, we shade the region between the branches containing $(0, 0)$. See **FIGURE 37**.

NOW TRY

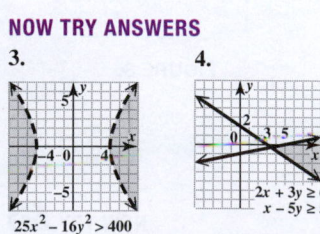

FIGURE 37

OBJECTIVE 2 Graph the solution set of a system of inequalities.

If two or more inequalities are considered at the same time, we have a **system of inequalities.** To find the solution set of the system, we find the intersection of the graphs (solution sets) of the inequalities in the system.

EXAMPLE 4 Graphing a System of Two Inequalities

Graph the solution set of the system.

$$x - y \le 4$$

$$x + 2y < 2$$

Both inequalities in the system are linear. We begin by graphing the solution set of $x - y \le 4$. The inequality includes the points of the boundary line $x - y = 4$ because the symbol $\le$ includes equality. The test point $(0, 0)$ leads to a true statement in $x - y < 4$, so we shade the region above the line. See **FIGURE 38**.

The graph of the solution set of $x + 2y < 2$ does not include the boundary line $x + 2y = 2$. Testing the point $(0, 0)$ leads to a true statement, indicating that we should shade the region below the dashed boundary line. See **FIGURE 39**.

**NOW TRY
EXERCISE 4**

Graph the solution set of the system.

$$2x + 3y \ge 6$$

$$x - 5y \ge 5$$

NOW TRY ANSWERS

3.   4.

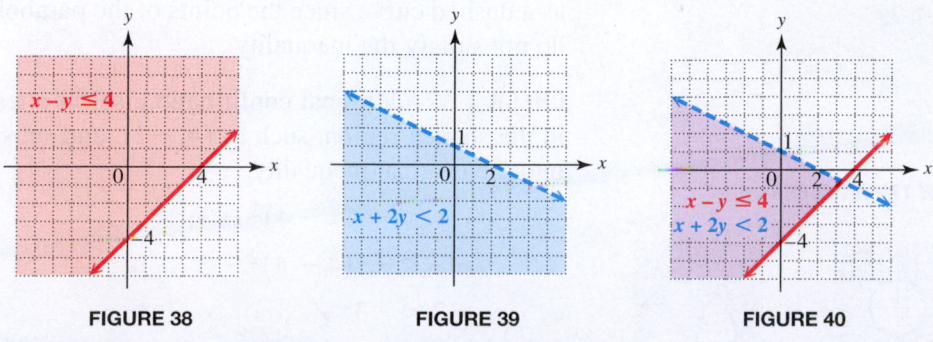

FIGURE 38   **FIGURE 39**   **FIGURE 40**

The solution set is the intersection of the graphs of the two inequalities. See the overlapping region in **FIGURE 40**, which includes one boundary line.

NOW TRY

NOW TRY EXERCISE 5

Graph the solution set of the system.

$$x^2 + y^2 \le 25$$
$$x + y \le 3$$

EXAMPLE 5 Graphing a System of Two Inequalities

Graph the solution set of the system.

$$2x + 3y > 6$$
$$x^2 + y^2 < 16$$

We begin by graphing the solution set of the linear inequality $2x + 3y > 6$. The boundary line is the graph of $2x + 3y = 6$ and is a dashed line because the symbol $>$ does not include equality. The test point $(0, 0)$ leads to a false statement in $2x + 3y > 6$, so we shade the region above the line, as shown in **FIGURE 41**.

The graph of $x^2 + y^2 < 16$ is the region inside of a dashed circle centered at the origin with radius 4. This is shown in **FIGURE 42**.

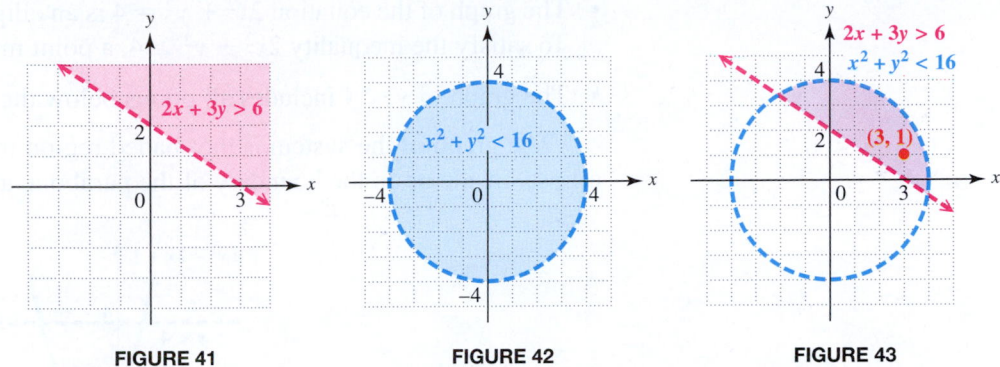

FIGURE 41   **FIGURE 42**   **FIGURE 43**

The graph of the solution set of the system is the intersection of the graphs of the two inequalities. The overlapping region in **FIGURE 43** is the solution set.

CHECK As additional confirmation, select a test point in the shaded region, such as $(3, 1)$, and substitute it into *both* inequalities.

$$2x + 3y > 6 \qquad\qquad x^2 + y^2 < 16$$
$$2(3) + 3(1) \overset{?}{>} 6 \quad \text{Test } (3,1). \qquad 3^2 + 1^2 \overset{?}{<} 16 \quad \text{Test } (3,1).$$
$$9 > 6 \ \checkmark \ \text{True} \qquad\qquad 10 < 16 \ \checkmark \ \text{True}$$

True statements result, so the correct region is shaded.   **NOW TRY**

NOW TRY EXERCISE 6

Graph the solution set of the system.

$$3x + 2y > 6$$
$$y \ge \frac{1}{2}x - 2$$
$$x \ge 0$$

EXAMPLE 6 Graphing a System of Three Inequalities

Graph the solution set of the system.

$$x + y < 1$$
$$y \le 2x + 3$$
$$y \ge -2$$

We graph each linear inequality on the same axes.

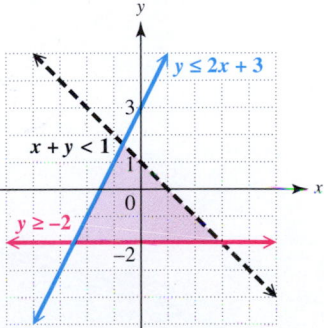

FIGURE 44

- The graph of $x + y < 1$ is the region that lies below the dashed line $x + y = 1$.
- The graph of $y \le 2x + 3$ is the region that lies below the solid line $y = 2x + 3$.
- The graph of $y \ge -2$ is the region above the solid horizontal line $y = -2$.

The test point $(0, 0)$ satisfies all three inequalities. The graph of the system, the intersection of these three graphs, is the triangular region enclosed by the three boundary lines in **FIGURE 44**, including two of its boundary lines.   **NOW TRY**

NOW TRY ANSWERS

5.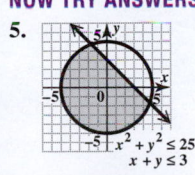

6.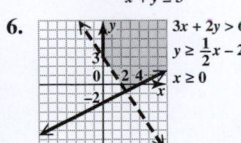

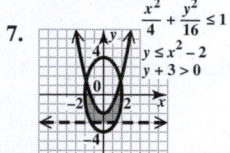

NOW TRY EXERCISE 7

Graph the solution set of the system.

$$\frac{x^2}{4} + \frac{y^2}{16} \le 1$$

$$y \le x^2 - 2$$

$$y + 3 > 0$$

EXAMPLE 7   Graphing a System of Three Inequalities

Graph the solution set of the system.

$$y \ge x^2 - 2x + 1$$

$$2x^2 + y^2 > 4$$

$$y < 4$$

We graph each inequality on the same axes.

- The graph of $y = x^2 - 2x + 1$ is a parabola with vertex at $(1, 0)$. Those points inside (or above) the parabola satisfy the condition $y > x^2 - 2x + 1$. Thus, the solution set of $y \ge x^2 - 2x + 1$ includes points on or inside the parabola.

- The graph of the equation $2x^2 + y^2 = 4$ is an ellipse. We draw it as a dashed curve. To satisfy the inequality $2x^2 + y^2 > 4$, a point must lie outside the ellipse.

- The graph of $y < 4$ includes all points below the dashed line $y = 4$.

The graph of the system is the shaded region in **FIGURE 45**, which lies outside the ellipse, inside or on the boundary of the parabola, and below the line $y = 4$.

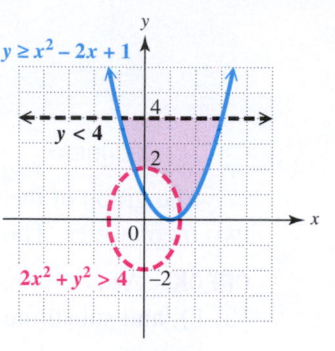

FIGURE 45

NOW TRY ANSWER

7.
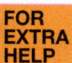

NOW TRY

13.5 Exercises

FOR EXTRA HELP   ▶ **MyLab Math**

▶ *Video solutions for select problems available in MyLab Math*

1. Concept Check   Match each inequality in parts (a)–(d) with its graph in choices A–D.

(a) $y \ge x^2 + 4$ **(b)** $y \le x^2 + 4$ **(c)** $y < x^2 + 4$ **(d)** $y > x^2 + 4$

A.

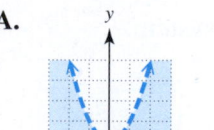

B.

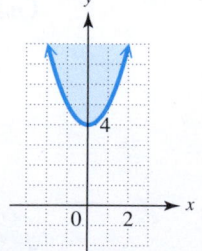

C.

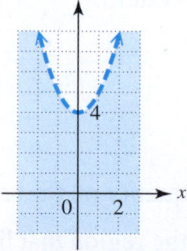

D.

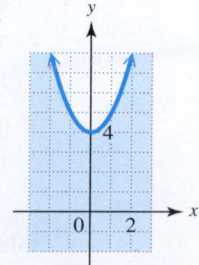

2. Concept Check   Match each system of inequalities in parts (a)–(d) with its graph in choices A–D.

(a) $x > -2$ **(b)** $x < -2$ **(c)** $x \geq 2$ **(d)** $x \leq 2$
   $y \leq 2$ $y \geq 2$ $y < -2$ $y > -2$

A.
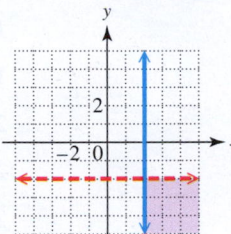

B.
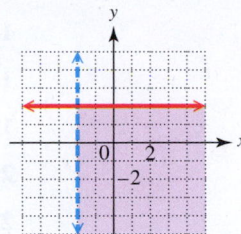

C.
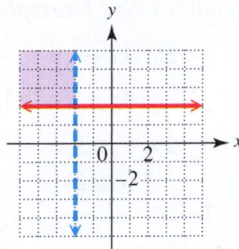

D.
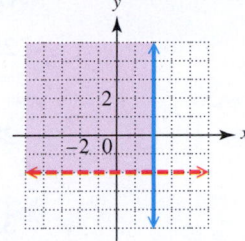

Concept Check   *Write a system of inequalities for the indicated region.*

3. Quadrant I, including the x- and y-axes

4. Quadrant IV, not including the x- and y-axes

5. Quadrant III, including the x-axis but not including the y-axis

6. Quadrant II, not including the x-axis but including the y-axis

7. Concept Check   Which one of the following is a description of the graph of the solution set of the following system?

$$x^2 + y^2 < 25$$
$$y > -2$$

A. All points outside the circle $x^2 + y^2 = 25$ and above the line $y = -2$

B. All points outside the circle $x^2 + y^2 = 25$ and below the line $y = -2$

C. All points inside the circle $x^2 + y^2 = 25$ and above the line $y = -2$

D. All points inside the circle $x^2 + y^2 = 25$ and below the line $y = -2$

8. Concept Check   Fill in each blank with the appropriate response. The graph of the solution set of the system

$$y > x^2 + 1$$
$$\frac{x^2}{9} + \frac{y^2}{4} > 1$$
$$y < 5$$

consists of all points _____ the parabola $y = x^2 + 1$, _____ the
                           (above / below)   (inside / outside)

ellipse $\dfrac{x^2}{9} + \dfrac{y^2}{4} = 1$, and _____ the line $y = 5$.
                                  (above / below)

Graph each inequality. ***See Examples 1–3.***

9. $y \geq x^2 - 2$

10. $y > x^2 - 1$

11. $2y^2 \geq 8 - x^2$

12. $y^2 \leq 4 - 2x^2$

13. $x^2 > 4 - y^2$

14. $x^2 \leq 16 - y^2$

15. $9x^2 > 16y^2 + 144$

16. $x^2 \leq 16 + 4y^2$

17. $9x^2 < 16y^2 - 144$

18. $y^2 > 4 + x^2$

19. $x^2 - 4 \geq -4y^2$

20. $4y^2 \leq 36 - 9x^2$

21. $x \leq -y^2 + 6y - 7$

22. $x \geq y^2 - 8y + 14$

23. $25x^2 \leq 9y^2 + 225$

24. $y^2 - 16x^2 \leq 16$

Graph each system of inequalities. ***See Examples 4–7.***

25. $2x + 5y < 10$
$\quad x - 2y < 4$

26. $3x - y > -6$
$\quad 4x + 3y > 12$

27. $5x - 3y \leq 15$
$\quad 4x + y \geq 4$

28. $4x - 3y \leq 0$
$\quad x + y \leq 5$

29. $x \leq 5$
$\quad y \leq 4$

30. $x \geq -2$
$\quad y \leq 4$

31. $x^2 + y^2 > 9$
$\quad y > x^2 - 1$

32. $x^2 - y^2 \geq 9$
$\quad \dfrac{x^2}{16} + \dfrac{y^2}{9} \leq 1$

33. $y > x^2 - 4$
$\quad y < -x^2 + 3$

34. $y \leq -x^2 + 5$
$\quad y \leq x^2 - 3$

35. $x + y > 1$
$\quad y \geq 2x - 2$
$\quad y \leq 4$

36. $3x - 4y \geq 12$
$\quad x + 3y > 6$
$\quad y \leq 2$

37. $x^2 + y^2 \geq 4$
$\quad x + y \leq 5$
$\quad x \geq 0$
$\quad y \geq 0$

38. $y^2 - x^2 \geq 4$
$\quad -5 \leq y \leq 5$

39. $y \leq -x^2$
$\quad y \geq x - 3$
$\quad y \leq -1$
$\quad x < 1$

40. $y < x^2$
$\quad y > -2$
$\quad x + y < 3$
$\quad 3x - 2y > -6$

Extending Skills   *The nonlinear inequality*

$$x^2 + y^2 \leq 4, \quad x \geq 0$$

is graphed in the figure. Only the right half of the interior of the circle and its boundary is shaded because of the restriction that x must be nonnegative. Graph each nonlinear inequality with the given restrictions.

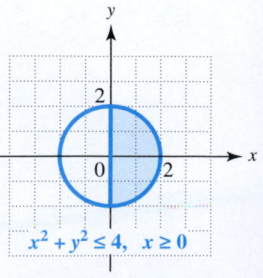

41. $x^2 + y^2 > 36, \quad x \geq 0$

42. $4x^2 + 25y^2 < 100, \quad y < 0$

43. $x < y^2 - 3, \quad x < 0$

44. $x^2 - y^2 < 4, \quad x < 0$

45. $4x^2 - y^2 > 16, \quad x < 0$

46. $x^2 + y^2 > 4, \quad y < 0$

47. $x^2 + 4y^2 \geq 1, \quad x \geq 0, y \geq 0$

48. $2x^2 - 32y^2 \leq 8, \quad x \leq 0, y \geq 0$

Chapter 13 Summary

Key Terms

13.1

absolute value function
reciprocal function
asymptote
square root function
greatest integer function
step function

13.2

conic section
circle
center (of a circle)
radius
ellipse
foci (singular *focus*)
center (of an ellipse)

13.3

hyperbola
transverse axis
fundamental rectangle
asymptotes (of a
  hyperbola)
generalized square root
  function

13.4

nonlinear equation
nonlinear system of
  equations

13.5

second-degree inequality
system of inequalities

New Symbols

$[\![x]\!]$ greatest integer less than or equal to x

Test Your Word Power

See how well you have learned the vocabulary in this chapter.

1. Conic sections are
  A. graphs of first-degree equations
  B. the result of two or more intersecting planes
  C. graphs of first-degree inequalities
  D. figures that result from the intersection of an infinite cone with a plane.

2. A **circle** is the set of all points in a plane
  A. such that the absolute value of the difference of the distances from two fixed points is constant
  B. that lie a fixed distance from a fixed point
  C. the sum of whose distances from two fixed points is constant
  D. that make up the graph of any second-degree equation.

3. An **ellipse** is the set of all points in a plane
  A. such that the absolute value of the difference of the distances from two fixed points is constant
  B. that lie a fixed distance from a fixed point
  C. the sum of whose distances from two fixed points is constant
  D. that make up the graph of any second-degree equation.

4. A **hyperbola** is the set of all points in a plane
  A. such that the absolute value of the difference of the distances from two fixed points is constant
  B. that lie a fixed distance from a fixed point
  C. the sum of whose distances from two fixed points is constant
  D. that make up the graph of any second-degree equation.

5. A **nonlinear equation** is an equation
  A. in which some terms have more than one variable or a variable of degree 2 or greater
  B. in which the terms have only one variable
  C. of degree 1
  D. of a linear function.

6. A **nonlinear system of equations** is a system
  A. with at least one linear equation
  B. with two or more inequalities
  C. with at least one nonlinear equation
  D. with at least two linear equations.

ANSWERS

1. D; *Example:* Parabolas, circles, ellipses, and hyperbolas are conic sections.   **2.** B; *Example:* The graph of $x^2 + y^2 = 9$ is a circle centered at the origin with radius 3.   **3.** C; *Example:* The graph of $\frac{x^2}{49} + \frac{y^2}{36} = 1$ is an ellipse centered at the origin with x-intercepts $(7, 0)$ and $(-7, 0)$ and y-intercepts $(0, 6)$ and $(0, -6)$.   **4.** A; *Example:* The graph of $\frac{x^2}{16} - \frac{y^2}{25} = 1$ is a horizontal parabola centered at the origin with vertices $(4, 0)$ and $(-4, 0)$.   **5.** A; *Examples:* $y = x^2 + 8x + 16, xy = 5, 2x^2 - y^2 = 6$ **6.** C; *Example:* $x^2 + y^2 = 2$
  $2x + y = 1$

Quick Review

CONCEPTS	**EXAMPLES**

13.1 Additional Graphs of Functions

Other Elementary Functions

- Absolute value function $f(x) = |x|$
- Reciprocal function $f(x) = \dfrac{1}{x}$
- Square root function $f(x) = \sqrt{x}$
- Greatest integer function $f(x) = [\![x]\!]$ (a step function)

Their graphs can be translated, as shown in the first three examples at the right.

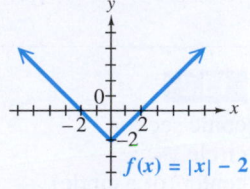
$f(x) = |x| - 2$

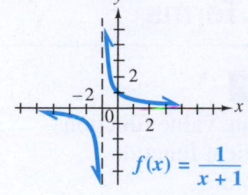
$f(x) = \dfrac{1}{x+1}$

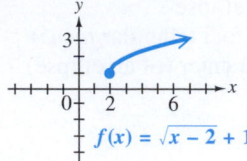
$f(x) = \sqrt{x-2} + 1$

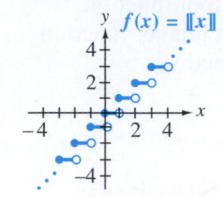
$f(x) = [\![x]\!]$

13.2 Circles Revisited and Ellipses

Circle

A circle with center (h, k), and radius $r > 0$ has an equation that can be written in the form

$$(x - h)^2 + (y - k)^2 = r^2. \quad \text{Center-radius form}$$

If its center is $(0, 0)$, then this equation becomes

$$x^2 + y^2 = r^2.$$

For some real numbers c, d, and e, an equation of the form

$$x^2 + y^2 + cx + dy + e = 0 \quad \text{General form}$$

may represent a circle. The x^2- and y^2-terms have equal coefficients, here 1.

Graph $(x + 2)^2 + (y - 3)^2 = 25$.

This equation, which can be written

$$[x - (-2)]^2 + (y - 3)^2 = 5^2,$$

represents a circle with center $(-2, 3)$ and radius 5.

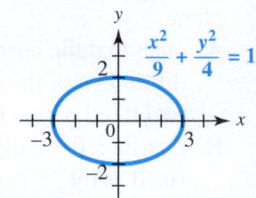
$(x + 2)^2 + (y - 3)^2 = 25$

The general form of the above equation is found as follows.

$$(x + 2)^2 + (y - 3)^2 = 25 \quad \text{Center-radius form}$$
$$x^2 + 4x + 4 + y^2 - 6y + 9 = 25 \quad \text{Square each binomial.}$$
$$x^2 + y^2 + 4x - 6y - 12 = 0 \quad \text{General form}$$

Ellipse

An ellipse with x-intercepts $(a, 0)$ and $(-a, 0)$ and y-intercepts $(0, b)$ and $(0, -b)$ has an equation that can be written in the form

$$\frac{x^2}{a^2} + \frac{y^2}{b^2} = 1.$$

Graph $\dfrac{x^2}{9} + \dfrac{y^2}{4} = 1$.

x-intercepts: $(3, 0)$ and $(-3, 0)$

y-intercepts: $(0, 2)$ and $(0, -2)$

$\dfrac{x^2}{9} + \dfrac{y^2}{4} = 1$

13.3 Hyperbolas and Functions Defined by Radicals

Hyperbola

A hyperbola centered at the origin has an equation that can be written in one of the following forms.

$$\frac{x^2}{a^2} - \frac{y^2}{b^2} = 1 \quad \text{or} \quad \frac{y^2}{b^2} - \frac{x^2}{a^2} = 1$$

x-intercepts $(a, 0)$ y-intercepts $(0, b)$
and $(-a, 0)$ and $(0, -b)$

The extended diagonals of the fundamental rectangle with vertices at the points (a, b), $(-a, b)$, $(-a, -b)$, and $(a, -b)$ are the asymptotes of these hyperbolas.

Graph $\dfrac{x^2}{4} - \dfrac{y^2}{4} = 1$.

x-intercepts: $(2, 0)$ and $(-2, 0)$

Vertices of the fundamental rectangle:
$(2, 2)$, $(-2, 2)$, $(-2, -2)$, and $(2, -2)$

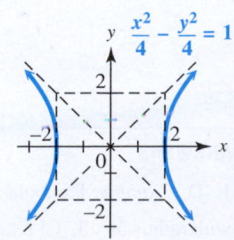
$\dfrac{x^2}{4} - \dfrac{y^2}{4} = 1$

CONCEPTS	EXAMPLES

Generalized Square Root Function

For an algebraic expression in x defined by u, where $u \geq 0$, a function of the form

$$f(x) = \sqrt{u}$$

is a generalized square root function.

Graph $y = -\sqrt{4 - x^2}$.

Square each side and rearrange terms.

$$x^2 + y^2 = 4$$

This equation has a circle as its graph. However, graph only the lower half of the circle because the original equation indicates that y cannot be positive.

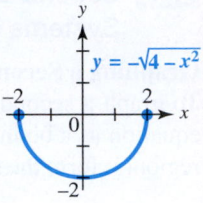

13.4 Nonlinear Systems of Equations

Solving a Nonlinear System

A nonlinear system can be solved by the substitution method, the elimination method, or a combination of the two.

Geometric Interpretation

If, for example, a nonlinear system includes two second-degree equations, then there may be zero, one, two, three, or four solutions of the system.

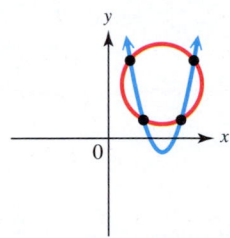

This system has four solutions because there are four points of intersection.

Solve the system.

$$x^2 + 2xy - y^2 = 14 \qquad (1)$$
$$x^2 - y^2 = -16 \qquad (2)$$

Multiply equation (2) by -1 and use the elimination method.

$$
\begin{aligned}
x^2 + 2xy - y^2 &= 14 \\
-x^2 \qquad\quad + y^2 &= 16 \\
\hline
2xy \qquad\quad &= 30 \\
xy &= 15
\end{aligned}
$$

Solve $xy = 15$ for y to obtain $y = \dfrac{15}{x}$, and substitute into equation (2).

$$x^2 - y^2 = -16 \qquad (2)$$

$$x^2 - \left(\frac{15}{x}\right)^2 = -16 \qquad \text{Let } y = \tfrac{15}{x}.$$

$$x^2 - \frac{225}{x^2} = -16 \qquad \text{Apply the exponent.}$$

$$x^4 + 16x^2 - 225 = 0 \qquad \begin{array}{l}\text{Multiply by } x^2.\\ \text{Add } 16x^2.\end{array}$$

$$(x^2 - 9)(x^2 + 25) = 0 \qquad \text{Factor.}$$

$$x^2 - 9 = 0 \quad \text{or} \quad x^2 + 25 = 0 \qquad \text{Zero-factor property}$$

$$x = \pm 3 \quad \text{or} \qquad x = \pm 5i \qquad \text{Solve each equation}$$

Substitute these values of x into $y = \dfrac{15}{x}$ to obtain the corresponding values of y.

If $x = 3$, then $y = \dfrac{15}{3} = 5$.

If $x = -3$, then $y = \dfrac{15}{-3} = -5$.

If $x = 5i$, then $y = \dfrac{15}{5i} = \dfrac{3}{i} \cdot \dfrac{-i}{-i} = -3i$.

If $x = -5i$, then $y = \dfrac{15}{-5i} = \dfrac{3}{-i} \cdot \dfrac{i}{i} = 3i$.

Solution set: $\{(3, 5), (-3, -5), (5i, -3i), (-5i, 3i)\}$

CONCEPTS	EXAMPLES
13.5 Second-Degree Inequalities and Systems of Inequalities	Graph $y \geq x^2 - 2x + 3$.
Graphing a Second-Degree Inequality To graph a second-degree inequality, graph the corresponding equation as a boundary, and use test points to determine which region(s) form the solution set. Shade the appropriate region(s).	The boundary is a parabola that opens up with vertex $(1, 2)$. Use $(0, 0)$ as a test point. Substituting into the inequality $y > x^2 - 2x + 3$ gives a false statement, $$0 > 3. \quad \text{False}$$ Shade the region inside (or above) the parabola. 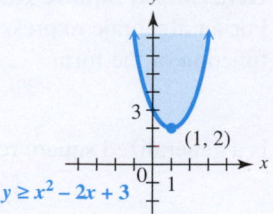
Graphing a System of Inequalities The solution set of a system of inequalities is the intersection of the graphs (solution sets) of the inequalities in the system.	Graph the solution set of the system. $$3x - 5y > -15$$ $$x^2 + y^2 \leq 25$$ Graph each inequality separately, on the same axes. The graph of the system is the region that lies below the dashed line *and* inside the solid circle. 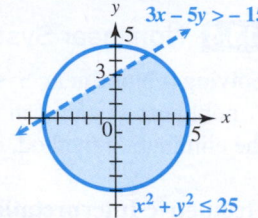

Chapter 13 **Review Exercises**

13.1 *Graph each function. Give the domain and range.*

1. $f(x) = |x + 4|$ **2.** $f(x) = \sqrt{x} + 3$

3. $f(x) = \dfrac{1}{x - 4}$ **4.** $f(x) = [\![-x]\!]$

Evaluate each expression.

5. $[\![12]\!]$ **6.** $\left[\!\!\left[2\dfrac{3}{4} \right]\!\!\right]$ **7.** $[\![-21]\!]$ **8.** $[\![-4.75]\!]$

13.2 *Write the center-radius form of each circle described.*

9. Center: $(0, 0)$; radius: 7   **10.** Center $(-2, 4)$; radius: 3

11. Center $(4, 2)$; radius: 6   **12.** Center: $(-5, 0)$; radius: $\sqrt{2}$

Write the center-radius form of the circle with the given equation. Give the center and radius.

13. $x^2 + y^2 + 6x - 4y - 3 = 0$ **14.** $x^2 + y^2 - 8x - 2y + 13 = 0$

15. $2x^2 + 2y^2 + 4x + 20y = -34$ **16.** $4x^2 + 4y^2 - 24x + 16y = 48$

17. Identify the graph of each equation as a *circle* or an *ellipse*.

     (a) $x^2 + y^2 = 121$ **(b)** $x^2 + 4y^2 = 4$

     (c) $\dfrac{x^2}{16} + \dfrac{y^2}{25} = 1$ **(d)** $3x^2 + 3y^2 = 300$

18. Match each equation of an ellipse in Column I with the appropriate intercepts in Column II.

I	II
(a) $36x^2 + 9y^2 = 324$	**A.** $(-3, 0), (3, 0), (0, -6), (0, 6)$
(b) $9x^2 + 36y^2 = 324$	**B.** $(-4, 0), (4, 0), (0, -5), (0, 5)$
(c) $\dfrac{x^2}{25} + \dfrac{y^2}{16} = 1$	**C.** $(-6, 0), (6, 0), (0, -3), (0, 3)$
(d) $\dfrac{x^2}{16} + \dfrac{y^2}{25} = 1$	**D.** $(-5, 0), (5, 0), (0, -4), (0, 4)$

Graph each equation.

19. $x^2 + y^2 = 16$

20. $(x + 3)^2 + (y - 2)^2 = 9$

21. $\dfrac{x^2}{16} + \dfrac{y^2}{9} = 1$

22. $\dfrac{x^2}{49} + \dfrac{y^2}{25} = 1$

13.3 *Graph each equation.*

23. $\dfrac{x^2}{16} - \dfrac{y^2}{25} = 1$

24. $\dfrac{y^2}{25} - \dfrac{x^2}{4} = 1$

25. $f(x) = -\sqrt{16 - x^2}$

26. Match each equation of a hyperbola in Column I with the appropriate intercepts in Column II.

I	II
(a) $\dfrac{x^2}{25} - \dfrac{y^2}{9} = 1$	**A.** $(-3, 0), (3, 0)$
(b) $\dfrac{x^2}{9} - \dfrac{y^2}{25} = 1$	**B.** $(0, -3), (0, 3)$
(c) $\dfrac{y^2}{25} - \dfrac{x^2}{9} = 1$	**C.** $(-5, 0), (5, 0)$
(d) $\dfrac{y^2}{9} - \dfrac{x^2}{25} = 1$	**D.** $(0, -5), (0, 5)$

Identify the graph of each equation as a parabola, circle, ellipse, *or* hyperbola.

27. $x^2 + y^2 = 64$

28. $y = 2x^2 - 3$

29. $y^2 = 2x^2 - 8$

30. $y^2 = 8 - 2x^2$

31. $x = y^2 + 4$

32. $x^2 - y^2 = 64$

13.4 *Answer each question.*

33. How many solutions are possible for a system of two equations whose graphs are a circle and a line?

34. How many solutions are possible for a system of two equations whose graphs are a parabola and a hyperbola?

Solve each system.

35. $2y = 3x - x^2$
$x + 2y = -12$

36. $y + 1 = x^2 + 2x$
$y + 2x = 4$

37. $x^2 + 3y^2 = 28$
$y - x = -2$

38. $xy = 8$
$x - 2y = 6$

39. $x^2 + y^2 = 6$
$x^2 - 2y^2 = -6$

40. $3x^2 - 2y^2 = 12$
$x^2 + 4y^2 = 18$

13.5 *Graph each inequality.*

41. $9x^2 \geq 16y^2 + 144$ **42.** $4x^2 + y^2 \geq 16$ **43.** $y < -(x + 2)^2 + 1$

44. Which one of the following is a description of the graph of $x^2 + y^2 > 4$?

    A. The region inside a circle with radius 4

    B. The region outside a circle with radius 4

    C. The region inside a circle with radius 2

    D. The region outside a circle with radius 2

Graph each system of inequalities.

45. $2x + 5y \leq 10$ **46.** $9x^2 \leq 4y^2 + 36$ **47.** $x + y \geq -2$
    $3x - y \leq 6$ $x^2 + y^2 \leq 16$ $y < x + 4$
                                                               $x < 1$

48. Match each system of inequalities in Column I with the quadrant that represents its graph in Column II.

<table>
<tr><td align="center" colspan="2">**I**</td><td align="center">**II**</td></tr>
<tr><td>**(a)** $x < 0$
$y > 0$</td><td>**(b)** $x > 0$
$y < 0$</td><td>**A.** Quadrant I
B. Quadrant II</td></tr>
<tr><td>**(c)** $x < 0$
$y < 0$</td><td>**(d)** $x > 0$
$y > 0$</td><td>**C.** Quadrant III
D. Quadrant IV</td></tr>
</table>

Chapter 13   Mixed Review Exercises

1. Find the center and radius of the circle with equation

$$x^2 + y^2 + 4x - 10y - 7 = 0.$$

2. *True* or *false:*   The graph of $y = -\sqrt{1 - x^2}$ is the graph of a function.

Graph.

3. $x^2 + 9y^2 = 9$ **4.** $x^2 - 9y^2 = 9$

5. $f(x) = \sqrt{4 - x}$ **6.** $x^2 + y^2 = 25$

7. $\dfrac{y^2}{4} - 1 \leq \dfrac{x^2}{9}$ **8.** $4y > 3x - 12$
                                                                 $x^2 < 16 - y^2$

Solve each system.

9. $y = x^2 - 84$ **10.** $x^2 - xy - y^2 = -5$
    $4y = 20x$ $x^2 - y^2 = -3$

Chapter 13   Test

▶ *View the complete solutions to all Chapter Test exercises in MyLab Math.*

Fill in each blank with the correct response.

1. For the reciprocal function defined by $f(x) = \frac{1}{x}$, _____ is the only real number not in the domain.

2. The range of the square root function $f(x) = \sqrt{x}$ is _____.

3. The range of $f(x) = [\![x]\!]$, the greatest integer function, is _____.

4. Match each function in parts (a)–(d) with its graph from choices A–D.

   (a) $f(x) = \sqrt{x - 2}$ **A.**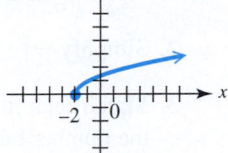

   (b) $f(x) = \sqrt{x + 2}$ **B.**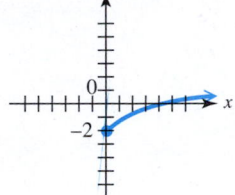

   (c) $f(x) = \sqrt{x} + 2$ **C.**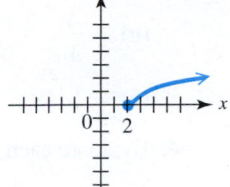

   (d) $f(x) = \sqrt{x} - 2$ **D.**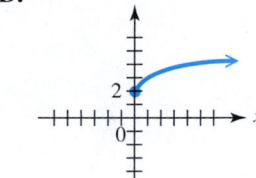

5. Graph $f(x) = |x - 3| + 4$. Give the domain and range.

6. Find the center and radius of the circle with equation $(x - 2)^2 + (y + 3)^2 = 16$. Sketch the graph.

7. Which one of the following equations is represented by the graph of a circle?

   A. $x^2 + y^2 = 0$ **B.** $x^2 + y^2 = -1$

   C. $x^2 + y^2 = x^2 - y^2$ **D.** $x^2 + y^2 = 1$

   For the equation that is represented by a circle, what are the coordinates of the center? What is the radius?

8. Find the center and radius of the circle with equation $x^2 + y^2 + 8x - 2y = 8$.

Graph.

9. $f(x) = \sqrt{9 - x^2}$ **10.** $4x^2 + 9y^2 = 36$

11. $16y^2 - 4x^2 = 64$ **12.** $\dfrac{y}{2} = -\sqrt{1 - \dfrac{x^2}{9}}$

Identify the graph of each equation as a parabola, hyperbola, ellipse, *or* circle.

13. $6x^2 + 4y^2 = 12$ **14.** $16x^2 = 144 + 9y^2$

15. $y^2 = 20 - x^2$ **16.** $4y^2 + 4x = 9$

Solve each system.

17. $2x - y = 9$
$xy = 5$

18. $x - 4 = 3y$
$x^2 + y^2 = 8$

19. $x^2 + y^2 = 25$
$x^2 - 2y^2 = 16$

Graph.

20. $y < x^2 - 2$

21. $x - y \le 2$
$x + 3y > 6$

22. $x^2 + 25y^2 \le 25$
$x^2 + y^2 \le 9$

Chapters R–13   Cumulative Review Exercises

1. Write $\frac{1}{100}$ as a decimal and as a percent.

2. Simplify $-|-1| - 5 + |4 - 10|$.

3. Match each number in Column I with the set (or sets) of numbers in Column II to which the number belongs.

	I		**II**
(a) 1	**(b)** 0	**A.** Natural numbers	**B.** Whole numbers
(c) $-\dfrac{2}{3}$	**(d)** π	**C.** Integers	**D.** Rational numbers
(e) $-\sqrt{49}$	**(f)** 2.75	**E.** Irrational numbers	**F.** Real numbers

4. Evaluate each expression in the complex number system.

(a) 6^2 **(b)** -6^2 **(c)** $(-6)^2$ **(d)** $\sqrt{36}$ **(e)** $-\sqrt{36}$ **(f)** $\sqrt{-36}$

Perform the indicated operations.

5. $(5y - 3)^2$

6. $\dfrac{8x^4 - 4x^3 + 2x^2 + 13x + 8}{2x + 1}$

7. $\dfrac{y^2 - 4}{y^2 - y - 6} \div \dfrac{y^2 - 2y}{y - 1}$

8. $\dfrac{5}{c + 5} - \dfrac{2}{c + 3}$

9. $\dfrac{p}{p^2 + p} + \dfrac{1}{p^2 + p}$

10. $\dfrac{1}{\dfrac{1}{x} - \dfrac{1}{y}}$

11. $4\sqrt[3]{16} - 2\sqrt[3]{54}$

12. $\dfrac{5 + 3i}{2 - i}$

Factor.

13. $12x^2 - 7x - 10$

14. $z^4 - 1$

15. $a^3 - 27b^3$

16. Henry and Lawrence want to clean their office. Henry can do the job alone in 3 hr, and Lawrence can do it alone in 2 hr. How long will it take them if they work together?

	Rate	Time Working Together	Fractional Part of the Job Done
Henry	$\frac{1}{3}$		
Lawrence			

Simplify. Assume all variables represent positive real numbers.

17. $\dfrac{(2a)^{-2}a^4}{a^{-3}}$

18. $\dfrac{3\sqrt{5x}}{\sqrt{2x}}$

Solve.

19. $4 - (2x + 3) + x = 5x - 3$

20. $-4x + 7 \geq 6x + 1$

21. $|5x| - 6 = 14$

22. $|2p - 5| > 15$

23. Find the slope of the line passing through the points $(2, 5)$ and $(-4, 1)$.

24. Find the equation of the line passing through the point $(-3, -2)$ and perpendicular to the graph of $2x - 3y = 7$.

Solve each system.

25. $3x - y = 12$
$2x + 3y = -3$

26. $x + y - 2z = 9$
$2x + y + z = 7$
$3x - y - z = 13$

27. $xy = -5$
$2x + y = 3$

28. Al and Bev traveled from their apartment to a picnic 20 mi away. Al traveled on his bike while Bev, who left later, took her car. Al's average rate was half of Bev's average rate. The trip took Al $\frac{1}{2}$ hr longer than Bev. What was Bev's average rate?

	d	r	t
Al	20		
Bev			

Solve.

29. $2\sqrt{x} = \sqrt{5x + 3}$

30. $10q^2 + 13q = 3$

31. $3x^2 - 3x - 2 = 0$

32. $2(x^2 - 3)^2 - 5(x^2 - 3) = 12$

33. $\log(x + 2) + \log(x - 1) = 1$

34. $F = \dfrac{kwv^2}{r}$ for v

35. If $f(x) = x^2 + 2x - 4$ and $g(x) = 3x + 2$, find the following.

(a) $(g \circ f)(1)$ (b) $(f \circ g)(x)$

36. If $f(x) = x^3 + 4$, find $f^{-1}(x)$.

37. Evaluate. (a) $3^{\log_3 4}$ (b) $e^{\ln 7}$

38. Use properties of logarithms to write $2\log(3x + 7) - \log 4$ as a single logarithm.

39. Give the domain and range of the function defined by $f(x) = |x - 3|$.

Graph.

40. $f(x) = -3x + 5$

41. $f(x) = -2(x - 1)^2 + 3$

42. $\dfrac{x^2}{25} + \dfrac{y^2}{16} \leq 1$

43. $f(x) = \sqrt{x - 2}$

44. $\dfrac{x^2}{4} - \dfrac{y^2}{16} = 1$

45. $f(x) = 3^x$

STUDY SKILLS REMINDER

Have you begun to prepare for your final exam? **Review Study Skill 10,** *Preparing for Your Math Final Exam.*

FURTHER TOPICS IN ALGEBRA

The number of ancestors in the family tree of a male honeybee forms the *Fibonacci sequence*

1, 1, 2, 3, 5, 8, . . . ,

where each term after the second is found by adding the two previous terms. This chapter covers *sequences, series,* and other topics in algebra.

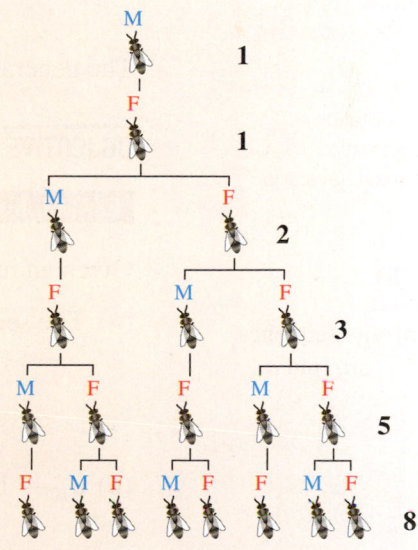

14.1 Sequences and Series

OBJECTIVES

1. Define infinite and finite sequences.
2. Find the terms of a sequence, given the general term.
3. Find the general term of a sequence.
4. Use sequences to solve applied problems.
5. Use summation notation to evaluate a series.
6. Write a series using summation notation.
7. Find the arithmetic mean (average) of a group of numbers.

VOCABULARY

☐ infinite sequence
☐ finite sequence
☐ terms of a sequence
☐ general term
☐ series
☐ summation notation
☐ index of summation
☐ arithmetic mean (average)

NOW TRY EXERCISE 1

Given an infinite sequence with $a_n = 5 - 3n$, find a_3.

OBJECTIVE 1 Define infinite and finite sequences.

In the Palace of the Alhambra, residence of the Moorish rulers of Granada, Spain, the Sultana's quarters feature an interesting architectural pattern.

There are 2 matched marble slabs inlaid in the floor, 4 walls, an octagon (8-sided) ceiling, 16 windows, 32 arches, and so on.

If this pattern is continued indefinitely, the set of numbers forms an *infinite sequence* whose *terms* are powers of 2.

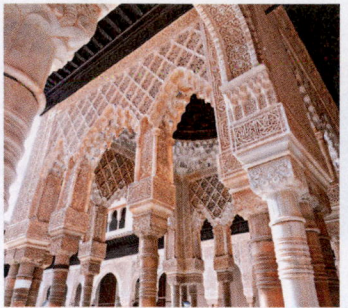

Sequence

An **infinite sequence** is a function with the set of positive integers (natural numbers)* as domain.

A **finite sequence** is a function with domain of the form $\{1, 2, 3, \ldots, n\}$, where n is a positive integer.

For any positive integer n, the function value of a sequence is written as a_n (read "*a sub-n*"). The function values

$$a_1, a_2, a_3, \ldots,$$

written in order, are the **terms** of the sequence, with a_1 the first term, a_2 the second term, and so on. The expression a_n, which defines the sequence, is its **general term.**

In the Palace of the Alhambra example, the first five terms of the sequence are

$$a_1 = 2, \quad a_2 = 4, \quad a_3 = 8, \quad a_4 = 16, \quad \text{and} \quad a_5 = 32.$$

The general term for this sequence is $a_n = 2^n$.

OBJECTIVE 2 Find the terms of a sequence, given the general term.

EXAMPLE 1 Writing the Terms of Sequences from the General Term

Given an infinite sequence with $a_n = n + \frac{1}{n}$, find the following.

(a) The second term of the sequence

$$a_2 = 2 + \frac{1}{2} = \frac{5}{2} \qquad \text{Replace } n \text{ in } a_n = n + \frac{1}{n} \text{ with 2.}$$

(b) $a_{10} = 10 + \frac{1}{10} = \frac{101}{10}$ $\quad 10 + \frac{1}{10} = \frac{10}{1} \cdot \frac{10}{10} + \frac{1}{10} = \frac{100}{10} + \frac{1}{10} = \frac{101}{10}$

(c) $a_{12} = 12 + \frac{1}{12} = \frac{145}{12}$ $\quad 12 + \frac{1}{12} = \frac{12}{1} \cdot \frac{12}{12} + \frac{1}{12} = \frac{144}{12} + \frac{1}{12} = \frac{145}{12}$

NOW TRY

NOW TRY ANSWER
1. $a_3 = -4$

*In this chapter we will use the terms *positive integers* and *natural numbers* interchangeably.

Graphing calculators can generate and graph sequences. See **FIGURE 1**. The calculator must be in dot mode so that the discrete points on the graph are not connected. *Remember that the domain of a sequence consists only of positive integers.*

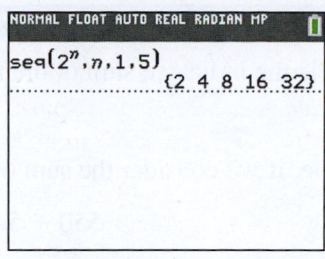

The first five terms of
the sequence $a_n = 2^n$

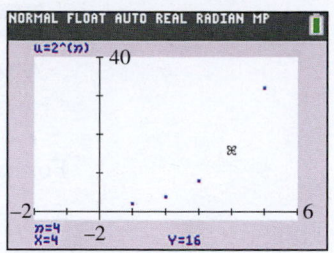

The first five terms of $a_n = 2^n$ are
graphed here. The display indicates
that the fourth term is 16; that is,
$a_4 = 2^4 = 16$.

FIGURE 1

OBJECTIVE 3 Find the general term of a sequence.

NOW TRY
EXERCISE 2
Determine an expression for
the general term a_n of the
sequence.

$$-3, 9, -27, 81, \ldots$$

EXAMPLE 2 Finding the General Term of a Sequence

Determine an expression for the general term a_n of the sequence.

$$5, 10, 15, 20, 25, \ldots$$

Notice that the terms are all multiples of 5. The first term is 5(1), the second is 5(2), and so on. The general term that will produce the given first five terms is

$$a_n = 5n.$$

NOW TRY

OBJECTIVE 4 Use sequences to solve applied problems.

Applied problems may involve *finite sequences.*

NOW TRY
EXERCISE 3
Chase borrows $8000 and
agrees to pay $400 monthly,
plus interest of 2% on the
unpaid balance from the
beginning of the first month.
Find the payments for the first
four months and the remaining
debt at the end of that period.

EXAMPLE 3 Using a Sequence in an Application

Saad borrows $5000 and agrees to pay $500 monthly, plus interest of 1% on the unpaid balance from the beginning of the first month. Find the payments for the first four months and the remaining debt at the end of that period.

The payments and remaining balances are calculated as follows.

First month	Payment: $500 + 0.01($5000) = $550
	Balance: $5000 − $500 = $4500
Second month	Payment: $500 + 0.01($4500) = $545
	Balance: $5000 − 2 · $500 = $4000
Third month	Payment: $500 + 0.01($4000) = $540
	Balance: $5000 − 3 · $500 = $3500
Fourth month	Payment: $500 + 0.01($3500) = $535
	Balance: $5000 − 4 · $500 = $3000

NOW TRY ANSWERS
2. $a_n = (-3)^n$
3. payments: $560, $552, $544,
   $536; balance: $6400

The payments for the first four months are

$$\$550, \$545, \$540, \$535$$

and the remaining debt at the end of the period is $3000.

NOW TRY

OBJECTIVE 5 Use summation notation to evaluate a series.

By adding the terms of a sequence, we obtain a *series*.

> **Series**
>
> A **series** is the indicated sum of the terms of a sequence.

For example, if we consider the sum of the payments listed in **Example 3,** namely,

$$550 + 545 + 540 + 535,$$

we have a series that represents the total payments for the first four months. Because a sequence can be finite or infinite, there are both finite and infinite series.

We use a compact notation, called **summation notation,** to write a series from the general term of the corresponding sequence. In mathematics, the Greek letter Σ **(sigma)** is used to denote summation. For example, the sum of the first six terms of the sequence with general term

$$a_n = 3n + 2$$

is written as

$$\sum_{i=1}^{6} (3i + 2).$$

The letter i is the **index of summation.** We read this as "*the sum from $i = 1$ to 6 of $3i + 2$.*" To find this sum, we replace the letter i in $3i + 2$ with 1, 2, 3, 4, 5, and 6, and add the resulting terms.

⚠ **CAUTION**   This use of i as the index of summation has no connection with the complex number i.

EXAMPLE 4 Evaluating Series Written Using Summation Notation

Write each series as a sum of terms and then find the sum.

(a) $\displaystyle\sum_{i=1}^{6} (3i + 2)$ Multiply and then add.

$$= (3 \cdot 1 + 2) + (3 \cdot 2 + 2) + (3 \cdot 3 + 2) \qquad \text{Replace } i \text{ with } 1, 2, 3, 4, 5, 6.$$
$$ + (3 \cdot 4 + 2) + (3 \cdot 5 + 2) + (3 \cdot 6 + 2)$$
$$= 5 + 8 + 11 + 14 + 17 + 20 \qquad \text{Simplify inside the parentheses.}$$
$$= 75 \qquad \text{Add.}$$

(b) $\displaystyle\sum_{i=1}^{5} (i - 4)$

$$= (1 - 4) + (2 - 4) + (3 - 4) + (4 - 4) + (5 - 4) \qquad i = 1, 2, 3, 4, 5$$
$$= -3 - 2 - 1 + 0 + 1 \qquad \text{Subtract.}$$
$$= -5 \qquad \text{Simplify.}$$

NOW TRY EXERCISE 4

Write the series as a sum of terms and then find the sum.

$$\sum_{i=1}^{5} (i^2 - 4)$$

(c) $\displaystyle\sum_{i=3}^{7} 3i^2$

$= 3(3)^2 + 3(4)^2 + 3(5)^2 + 3(6)^2 + 3(7)^2$ *i* = 3, 4, 5, 6, 7

$= 27 + 48 + 75 + 108 + 147$ Square, and then multiply.

$= 405$ Add.   **NOW TRY**

OBJECTIVE 6 Write a series using summation notation.

In **Example 4,** we started with summation notation and wrote each series using $+$ signs. Given a series, we can write it using summation notation by observing a pattern in the terms and writing the general term accordingly.

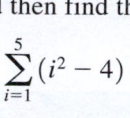

NOW TRY EXERCISE 5

Write each series using summation notation.

(a) $3 + 5 + 7 + 9 + 11$

(b) $-1 - 4 - 9 - 16 - 25$

EXAMPLE 5 Writing Series Using Summation Notation

Write each series using summation notation.

(a) $2 + 5 + 8 + 11$

First, find a general term a_n that will give these four terms for $a_1, a_2, a_3,$ and $a_4,$ respectively. Each term is one less than a multiple of 3, so try $3i - 1$.

$$3(1) - 1 = 2 \quad i = 1$$
$$3(2) - 1 = 5 \quad i = 2$$
$$3(3) - 1 = 8 \quad i = 3$$
$$3(4) - 1 = 11 \quad i = 4$$

(There may be other expressions that also work.) Because *i* ranges from 1 to 4,

$$2 + 5 + 8 + 11 = \sum_{i=1}^{4} (3i - 1).$$

(b) $8 + 27 + 64 + 125 + 216$

These numbers are the cubes of 2, 3, 4, 5, and 6, so the general term is i^3.

$$8 + 27 + 64 + 125 + 216 = \sum_{i=2}^{6} i^3$$

NOW TRY

OBJECTIVE 7 Find the arithmetic mean (average) of a group of numbers.

Arithmetic Mean or Average

The **arithmetic mean,** or **average,** of a set of values is symbolized $\bar{x}$ and is found by dividing their sum by the number of values.

$$\bar{x} = \frac{\displaystyle\sum_{i=1}^{n} x_i}{n}$$

The values of x_i represent the individual numbers in the group, and n represents the number of values.

NOW TRY ANSWERS

4. $-3 + 0 + 5 + 12 + 21 = 35$

5. (a) $\displaystyle\sum_{i=1}^{5} (2i + 1)$ **(b)** $\displaystyle\sum_{i=1}^{5} -i^2$

NOW TRY
EXERCISE 6

The table shows the top five American Quarter Horse States in 2016 based on the total number of registered Quarter Horses. To the nearest whole number, what is the average number of Quarter Horses registered per state in these top five states?

State	Number of Registered Quarter Horses
Texas	418,249
Oklahoma	164,265
California	114,623
Missouri	95,877
Montana	87,418

Data from American Quarter Horse Association.

NOW TRY ANSWER
6. 176,086 quarter horses

EXAMPLE 6   Finding the Arithmetic Mean (Average)

The table shows the number of FDIC-insured financial institutions for each year during the period from 2011 through 2017. What was the average number of institutions per year for this 7-yr period?

Year	Number of Institutions
2011	7523
2012	7255
2013	6950
2014	6669
2015	6358
2016	6068
2017	5797

Data from U.S. Federal Deposit Insurance Corporation.

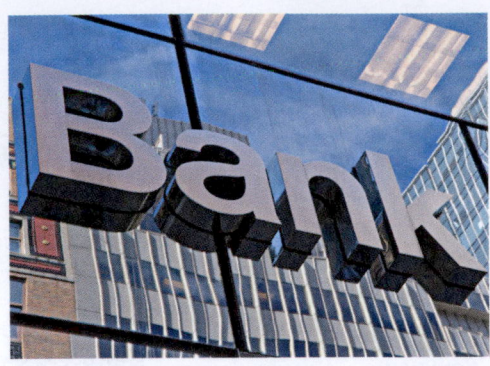

$$\bar{x} = \frac{\sum\limits_{i=1}^{7} x_i}{7}$$

Let $x_1 = 7523$, $x_2 = 7255$, and so on. There are 7 values in the group, so $n = 7$.

$$\bar{x} = \frac{7523 + 7255 + 6950 + 6669 + 6358 + 6068 + 5797}{7}$$

$$\bar{x} = 6660$$

The average number of institutions per year for this 7-yr period was 6660.

NOW TRY

14.1 Exercises

FOR EXTRA HELP

▶ **MyLab Math**

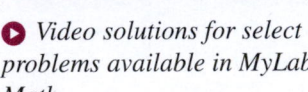
▶ *Video solutions for select problems available in MyLab Math*

STUDY SKILLS REMINDER
Make study cards to help you learn and remember the material in this chapter.
Review Study Skill 5, *Using Study Cards*.

Concept Check   *Fill in each blank with the correct response.*

1. The domain of an infinite sequence is _____.

2. In the sequence 3, 6, 9, 12, the term $a_3 =$ _____.

3. If $a_n = 2n$, then $a_{40} =$ _____.

4. If $a_n = (-1)^n$, then $a_{115} =$ _____.

5. The value of the sum $\sum\limits_{i=1}^{3} (i + 2)$ is _____.

6. The arithmetic mean of -4, -2, 0, 2, and 4 is _____.

Write the first five terms of each sequence. ***See Example 1.***

7. $a_n = n + 1$

8. $a_n = n + 4$

9. $a_n = \dfrac{n + 3}{n}$

10. $a_n = \dfrac{n + 2}{n}$

11. $a_n = 3^n$

12. $a_n = 2^n$

13. $a_n = -\dfrac{1}{n^2}$

14. $a_n = -\dfrac{2}{n^2}$

15. $a_n = 5(-1)^{n-1}$

16. $a_n = 6(-1)^{n+1}$

17. $a_n = n - \dfrac{1}{n}$

18. $a_n = n + \dfrac{4}{n}$

Find the indicated term for each sequence. See Example 1.

19. $a_n = -9n + 2$; a_8

20. $a_n = -3n + 7$; a_{12}

21. $a_n = \dfrac{3n + 7}{2n - 5}$; a_{14}

22. $a_n = \dfrac{5n - 9}{3n + 8}$; a_{16}

23. $a_n = (n + 1)(2n + 3)$; a_8

24. $a_n = (5n - 2)(3n + 1)$; a_{10}

Determine an expression for the general term a_n of each sequence. See Example 2.

25. $4, 8, 12, 16, \ldots$

26. $7, 14, 21, 28, \ldots$

27. $-8, -16, -24, -32, \ldots$

28. $-10, -20, -30, -40, \ldots$

29. $\dfrac{1}{3}, \dfrac{1}{9}, \dfrac{1}{27}, \dfrac{1}{81}, \ldots$

30. $\dfrac{2}{5}, \dfrac{2}{25}, \dfrac{2}{125}, \dfrac{2}{625}, \ldots$

31. $\dfrac{2}{5}, \dfrac{3}{6}, \dfrac{4}{7}, \dfrac{5}{8}, \ldots$

32. $\dfrac{1}{2}, \dfrac{2}{3}, \dfrac{3}{4}, \dfrac{4}{5}, \ldots$

Solve each applied problem by writing the first few terms of a sequence. See Example 3.

33. Horacio borrows $1000 and agrees to pay $100 plus interest of 1% on the unpaid balance each month. Find the payments for the first six months and the remaining debt at the end of that period.

34. Leslie is offered a new job with a salary of $20,000 + 2500n$ dollars per year at the end of the nth year. Write a sequence showing her salary at the end of each of the first 5 yr. If she continues in this way, what will her salary be at the end of the tenth year?

35. Suppose that an automobile loses $\frac{1}{5}$ of its value each year; that is, at the end of any given year, the value is $\frac{4}{5}$ of the value at the beginning of that year. If a car costs $20,000 new, what is its value at the end of 5 yr, to the nearest dollar?

36. A certain car loses $\frac{1}{2}$ of its value each year. If this car cost $40,000 new, what is its value at the end of 6 yr?

Write each series as a sum of terms and then find the sum. See Example 4.

37. $\displaystyle\sum_{i=1}^{5} (i + 3)$

38. $\displaystyle\sum_{i=1}^{6} (i + 9)$

39. $\displaystyle\sum_{i=1}^{3} (i^2 + 2)$

40. $\displaystyle\sum_{i=1}^{4} (i^3 + 3)$

41. $\displaystyle\sum_{i=1}^{6} (-1)^i \cdot 2$

42. $\displaystyle\sum_{i=1}^{5} (-1)^i \cdot i$

43. $\displaystyle\sum_{i=3}^{7} (i - 3)(i + 2)$

44. $\displaystyle\sum_{i=2}^{6} (i + 3)(i - 4)$

Write each series using summation notation. See Example 5.

45. $3 + 4 + 5 + 6 + 7$

46. $7 + 8 + 9 + 10 + 11$

47. $-2 + 4 - 8 + 16 - 32$

48. $-1 + 2 - 3 + 4 - 5 + 6$

49. $1 + 4 + 9 + 16$

50. $1 + 16 + 81 + 256$

51. Concept Check When asked to write the series $-1 - 4 - 9 - 16 - 25$ using summation notation, a student incorrectly wrote the following.

$$\sum_{t=1}^{5} (-i)^2$$

WHAT WENT WRONG? Give the correct summation notation.

52. Concept Check When asked to write the first five terms of the sequence defined by $a_n = 5n - 1$, a student incorrectly wrote the following.

$$4 + 9 + 14 + 19 + 24$$

WHAT WENT WRONG? Give the correct answer.

Find the arithmetic mean for each set of values. See Example 6.

53. 8, 11, 14, 9, 7, 6, 8

54. 10, 12, 8, 19, 23, 12

55. 5, 9, 8, 2, 4, 7, 3, 2, 0

56. 2, 1, 4, 8, 3, 7, 10, 8, 0

Solve each problem. See Example 6.

57. The number of mutual funds operating in the United States each year during the period 2012 through 2016 is given in the table. To the nearest whole number, what was the average number of mutual funds operating per year during the given period?

Year	Number of Mutual Funds
2012	8744
2013	8972
2014	9258
2015	9517
2016	9511

Data from Investment Company Institute.

58. The total assets of mutual funds operating in the United States, in billions of dollars, for each year during the period 2012 through 2016 are shown in the table. What were the average assets per year during this period?

Year	Assets (in billions of dollars)
2012	13,054
2013	15,049
2014	15,873
2015	15,650
2016	16,344

Data from Investment Company Institute.

14.2   Arithmetic Sequences

OBJECTIVES

1 Find the common difference of an arithmetic sequence.

2 Find the general term of an arithmetic sequence.

3 Use an arithmetic sequence in an application.

4 Find any specified term or the number of terms of an arithmetic sequence.

5 Find the sum of a specified number of terms of an arithmetic sequence.

OBJECTIVE 1 Find the common difference of an arithmetic sequence.

In this section, we introduce *arithmetic sequences*.

> **Arithmetic Sequence**
>
> An **arithmetic sequence,** or **arithmetic progression,** is a sequence in which each term after the first is found by adding a constant number to the preceding term.

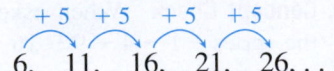

For example,   6, 11, 16, 21, 26, . . .   Arithmetic sequence

is an arithmetic sequence in which the difference between any two adjacent terms is always 5. The number 5 is the **common difference** of the arithmetic sequence. The common difference d is found by subtracting a_n from a_{n+1} in any such pair of terms.

$$d = a_{n+1} - a_n \qquad \text{Common difference}$$

 **NOW TRY EXERCISE 1**

Determine the common difference d for the arithmetic sequence.

$-4, -13, -22, -31, -40, \ldots$

EXAMPLE 1 Finding the Common Difference

Determine the common difference d for the arithmetic sequence.

$$-11, \ -4, \ 3, \ 10, \ 17, \ 24, \ldots$$

The sequence is arithmetic, so d is the difference between any two adjacent terms. We arbitrarily choose the terms 10 and 17.

$$d = a_{n+1} - a_n$$
$$d = 17 - 10 \qquad a_5 - a_4$$
$$d = 7 \qquad \text{Common difference}$$

Verify that *any* two adjacent terms would give the same result.   **NOW TRY**

 **NOW TRY EXERCISE 2**

Write the first five terms of the arithmetic sequence with first term 10 and common difference -8.

EXAMPLE 2 Writing the Terms of an Arithmetic Sequence

Write the first five terms of the arithmetic sequence with first term 3 and common difference -2.

$$
\begin{aligned}
a_1 &= \ 3 & &\text{First term} \\
a_2 &= \ 3 + (-2) = 1 & &\text{Add } d = -2. \\
a_3 &= \ 1 + (-2) = -1 & &\text{Add } -2. \\
a_4 &= -1 + (-2) = -3 & &\text{Add } -2. \\
a_5 &= -3 + (-2) = -5 & &\text{Add } -2.
\end{aligned}
$$

The first five terms are $3, \ 1, \ -1, \ -3, \ -5$.   **NOW TRY**

OBJECTIVE 2 Find the general term of an arithmetic sequence.

Generalizing from **Example 2,** if we know the first term a_1 and the common difference d of an arithmetic sequence, then the sequence is completely defined as

$$a_1, \quad a_2 = a_1 + d, \quad a_3 = a_1 + 2d, \quad a_4 = a_1 + 3d, \ldots.$$

Writing the terms of the sequence in this way suggests the following formula for a_n.

General Term of an Arithmetic Sequence

The general term a_n of an arithmetic sequence with first term a_1 and common difference d is given by the following.

$$a_n = a_1 + (n - 1)d$$

Because

$$a_n = a_1 + (n - 1)d \qquad \text{Formula for } a_n$$

$$a_n = dn + (a_1 - d) \qquad \text{Properties of real numbers}$$

is a linear function in n, any linear expression of the form $dn + c$, where d and c are real numbers, defines an arithmetic sequence.

NOW TRY
EXERCISE 3

Determine an expression for the general term of the arithmetic sequence. Then find a_{20}.

$$-5, \ 0, \ 5, \ 10, \ 15, \ldots$$

EXAMPLE 3   Finding the General Term of an Arithmetic Sequence

Determine an expression for the general term of the arithmetic sequence. Then find a_{2}.

$$-9, \ -6, \ -3, \ 0, \ 3, \ 6, \ldots$$

The first term is $a_1 = -9$. Begin by finding d.

$$d = -6 - (-9) \qquad \text{Let } d = a_2 - a_1.$$

$$d = 3 \qquad \text{Subtract.}$$

Now find a_n.

$$a_n = a_1 + (n-1)d \qquad \text{Formula for } a_n$$

$$a_n = -9 + (n-1)(3) \qquad \text{Let } a_1 = -9, \text{ and } d = 3.$$

$$a_n = -9 + 3n - 3 \qquad \text{Distributive property}$$

$$a_n = 3n - 12 \qquad \text{Combine like terms.}$$

The general term is $a_n = 3n - 12$. Now find a_{20}.

$$a_{20} = 3(20) - 12 \qquad \text{Let } n = 20.$$

$$a_{20} = 60 - 12 \qquad \text{Multiply.}$$

$$a_{20} = 48 \qquad \text{Subtract.} \qquad \textbf{NOW TRY}$$

OBJECTIVE 3   Use an arithmetic sequence in an application.

NOW TRY
EXERCISE 4

Ginny is saving money for her son's college education. She makes an initial contribution of $1000 and deposits an additional $120 each month for the next 96 months. Disregarding interest, how much money will be in the account after 96 months?

EXAMPLE 4   Applying an Arithmetic Sequence

Leonid's uncle decides to start a fund for Leonid's education. He makes an initial contribution of $3000 and deposits an additional $500 each month. How much will be in the fund after 24 months? (Disregard any interest.)

After n months, the fund will contain

$$a_n = 3000 + 500n \text{ dollars.} \qquad \text{Use an arithmetic sequence.}$$

To find the amount in the fund after 24 months, find a_{24}.

$$a_{24} = 3000 + 500(24) \qquad \text{Let } n = 24.$$

$$a_{24} = 3000 + 12{,}000 \qquad \text{Multiply.}$$

$$a_{24} = 15{,}000 \qquad \text{Add.}$$

The fund will contain $15,000 (disregarding interest) after 24 months.   **NOW TRY**

OBJECTIVE 4   Find any specified term or the number of terms of an arithmetic sequence.

The formula for the general term of an arithmetic sequence has four variables:

$$a_n, \quad a_1, \quad n, \quad \text{and} \quad d.$$

If we know the values of any three of these variables, we can use the formula to find the value of the fourth variable.

NOW TRY ANSWERS

3. $a_n = 5n - 10$; $a_{20} = 90$
4. $12,520

**NOW TRY
EXERCISE 5**

Find the indicated term for
each arithmetic sequence.

(a) $a_1 = 21, d = -3$; a_{22}

(b) $a_7 = 25, a_{12} = 40$; a_{19}

EXAMPLE 5 Finding Specified Terms in Arithmetic Sequences

Find the indicated term for each arithmetic sequence.

(a) $a_1 = -6, d = 12$; a_{15}

$$a_n = a_1 + (n-1)d \qquad \text{Formula for } a_n$$

$$a_{15} = a_1 + (15-1)d \qquad \text{Let } n = 15.$$

$$a_{15} = -6 + 14(12) \qquad \text{Let } a_1 = -6, d = 12.$$

$$a_{15} = 162 \qquad \text{Multiply, and then add.}$$

(b) $a_5 = 2, a_{11} = -10$; a_{17}

Any term can be found if a_1 and d are known. Use the formula for a_n.

$a_n = a_1 + (n-1)d$			$a_n = a_1 + (n-1)d$	
$a_5 = a_1 + (5-1)d$	Let $n = 5$.		$a_{11} = a_1 + (11-1)d$	Let $n = 11$.
$a_5 = a_1 + 4d$	Simplify.		$a_{11} = a_1 + 10d$	Simplify.
$2 = a_1 + 4d$	$a_5 = 2$		$-10 = a_1 + 10d$	$a_{11} = -10$

This gives a system of two equations in two variables, a_1 and d.

$$a_1 + 4d = 2 \qquad (1)$$

$$a_1 + 10d = -10 \qquad (2)$$

Multiply equation (2) by -1 and add to equation (1) to eliminate a_1.

$$a_1 + 4d = 2 \qquad (1)$$

$$\underline{-a_1 - 10d = 10} \qquad \text{Multiply (2) by } -1.$$

$$-6d = 12 \qquad \text{Add.}$$

$$d = -2 \qquad \text{Divide by } -6.$$

Now find a_1 by substituting -2 for d into either equation.

$$a_1 + 10d = -10 \qquad (2)$$

$$a_1 + 10(-2) = -10 \qquad \text{Let } d = -2 \text{ in (2).}$$

$$a_1 - 20 = -10 \qquad \text{Multiply.}$$

$$a_1 = 10 \qquad \text{Add 20.}$$

Use the formula for a_n to find a_{17}.

$$a_n = a_1 + (n-1)d \qquad \text{Formula for } a_n$$

$$a_{17} = a_1 + (17-1)d \qquad \text{Let } n = 17.$$

$$a_{17} = a_1 + 16d \qquad \text{Subtract.}$$

$$a_{17} = 10 + 16(-2) \qquad \text{Let } a_1 = 10, d = -2.$$

$$a_{17} = -22 \qquad \text{Multiply, and then add.}$$

NOW TRY

NOW TRY ANSWERS

5. (a) -42 (b) 61

NOW TRY
EXERCISE 6
Find the number of terms in
the arithmetic sequence.

$$1, \frac{4}{3}, \frac{5}{3}, 2, \ldots, 11$$

EXAMPLE 6   Finding the Number of Terms in an Arithmetic Sequence

Find the number of terms in the arithmetic sequence.

$$-8, -2, 4, 10, \ldots, 52$$

Let n represent the number of terms in the sequence. Here $a_n = 52$, $a_1 = -8$, and $d = -2 - (-8) = 6$, so we use the formula for a_n to find n.

$a_n = a_1 + (n-1)d$	Formula for a_n
$52 = -8 + (n-1)(6)$	Let $a_n = 52$, $a_1 = -8$, $d = 6$.
$52 = -8 + 6n - 6$	Distributive property
$52 = 6n - 14$	Simplify.
$66 = 6n$	Add 14.
$n = 11$	Divide by 6.

The sequence has 11 terms.   **NOW TRY**

OBJECTIVE 5   Find the sum of a specified number of terms of an arithmetic sequence.

To find a formula for the sum S_n of the first n terms of a given arithmetic sequence, we can write out the terms in two ways. We start with the first term, and then with the last term. Then we add the terms in columns.

$$S_n = a_1 + (a_1 + d) + (a_1 + 2d) + \cdots + [a_1 + (n-1)d]$$
$$\underline{S_n = a_n + (a_n - d) + (a_n - 2d) + \cdots + [a_n - (n-1)d]}$$
$$S_n + S_n = (a_1 + a_n) + (a_1 + a_n) + (a_1 + a_n) + \cdots + (a_1 + a_n)$$
$$2S_n = n(a_1 + a_n) \quad \text{There are } n \text{ terms of } a_1 + a_n \text{ on the right.}$$

Formula for S_n

$$S_n = \frac{n}{2}(a_1 + a_n) \qquad \text{Divide by 2.}$$

NOW TRY
EXERCISE 7
Evaluate the sum of the
first seven terms of the
arithmetic sequence in which
$a_n = 5n - 7$.

EXAMPLE 7   Finding the Sum of the First n Terms of an Arithmetic Sequence

Evaluate the sum of the first five terms of the arithmetic sequence in which $a_n = 2n - 5$.

Begin by evaluating a_1 and a_5.

$a_1 = 2(1) - 5$	$a_5 = 2(5) - 5$
$a_1 = -3$	$a_5 = 5$

Now evaluate the sum using $a_1 = -3$, $a_5 = 5$, and $n = 5$.

$S_n = \dfrac{n}{2}(a_1 + a_n)$	Formula for S_n
$S_5 = \dfrac{5}{2}(-3 + 5)$	Substitute.
$S_5 = 5$	Add, and then multiply.

NOW TRY

NOW TRY ANSWERS
6. 31
7. 91

It is possible to express the sum S_n of an arithmetic sequence in terms of a_1 and d, the quantities that define the sequence. We have established that

$$S_n = \frac{n}{2}(a_1 + a_n) \quad \text{and} \quad a_n = a_1 + (n-1)d.$$

We substitute the expression for a_n into the expression for S_n.

$$S_n = \frac{n}{2}(a_1 + [a_1 + (n-1)d]) \qquad \textcolor{blue}{\text{Substitute for } a_n.}$$

$$S_n = \frac{n}{2}[2a_1 + (n-1)d] \qquad \textcolor{blue}{\text{Combine like terms.}}$$

Sum of the First n Terms of an Arithmetic Sequence

The sum S_n of the first n terms of the arithmetic sequence with first term a_1, nth term a_n, and common difference d is given by either of the following.

$$S_n = \frac{n}{2}(a_1 + a_n) \qquad \text{or} \qquad S_n = \frac{n}{2}[2a_1 + (n-1)d]$$

The first formula is used when the first and last terms are known; otherwise, the second formula is used.

**NOW TRY
EXERCISE 8**

Evaluate the sum of the first nine terms of the arithmetic sequence having first term -8 and common difference -5.

EXAMPLE 8 Finding the Sum of the First n Terms of an Arithmetic Sequence

Evaluate the sum of the first eight terms of the arithmetic sequence having first term 3 and common difference -2.

The known values $a_1 = 3$, $d = -2$, and $n = 8$ appear in the second formula for S_n, so we use it.

$$S_n = \frac{n}{2}[2a_1 + (n-1)d] \qquad \textcolor{blue}{\text{Second formula for } S_n}$$

$$S_8 = \frac{8}{2}[2(3) + (8-1)(-2)] \qquad \textcolor{blue}{\text{Let } a_1 = 3, d = -2, n = 8.}$$

$$S_8 = 4[6 - 14] \qquad \textcolor{blue}{\text{Work inside the brackets.}}$$

$$S_8 = -32 \qquad \textcolor{blue}{\text{Subtract, and then multiply.}} \qquad \textbf{NOW TRY}$$

As mentioned earlier, a linear expression of the form

$$dn + c, \quad \text{where } d \text{ and } c \text{ are real numbers,}$$

defines an arithmetic sequence. For example, the sequences defined by $a_n = 2n + 5$ and $a_n = n - 3$ are arithmetic sequences. For this reason,

$$\sum_{i=1}^{n} (di + c)$$

represents the sum of the first n terms of an arithmetic sequence having first term $a_1 = d(1) + c = d + c$ and general term $a_n = d(n) + c = dn + c$. We can find this sum with the first formula for S_n, as shown in the next example.

NOW TRY ANSWER
8. -252

**NOW TRY
EXERCISE 9**

Evaluate $\displaystyle\sum_{i=1}^{11} (5i - 7)$.

EXAMPLE 9 Using S_n to Evaluate a Summation

Evaluate $\displaystyle\sum_{i=1}^{12} (2i - 1)$.

This is the sum of the first 12 terms of the arithmetic sequence having $a_n = 2n - 1$. We find this sum, S_{12}, using the first formula for S_n.

$$S_n = \frac{n}{2}(a_1 + a_n) \qquad \text{First formula for } S_n$$

$$S_{12} = \frac{12}{2}\big[\,(2(1) - 1) + (2(12) - 1)\,\big] \qquad \text{Let } n = 12,\ a_1 = 2(1) - 1,\ \text{and } a_{12} = 2(12) - 1.$$

$$S_{12} = 6(1 + 23) \qquad \text{Evaluate } a_1 \text{ and } a_{12}.$$

$$S_{12} = 6(24) \qquad \text{Add.}$$

$$S_{12} = 144 \qquad \text{Multiply.} \qquad \textbf{NOW TRY} \,\$$

NOW TRY ANSWER
9. 253

14.2 Exercises

FOR
EXTRA
HELP

 MyLab Math

▶ *Video solutions for select problems available in MyLab Math*

STUDY SKILLS REMINDER
Have you begun to prepare for your final exam? **Review Study Skill 10, *Preparing for Your Math Final Exam.***

Concept Check *Fill in each blank with the correct response.*

1. In an arithmetic sequence, if any term is subtracted from the term that follows it, the result is the common _____ of the sequence.

2. For the arithmetic sequence having $a_n = 2n + 4$, the term $a_3 = $ _____.

3. The sum of the first five terms of the arithmetic sequence 1, 6, 11, . . . is _____.

4. The number of terms in the arithmetic sequence 2, 4, 6, . . . , 100 is _____.

Concept Check *Fill in the blanks to complete the terms of each arithmetic sequence.*

5. $1, \dfrac{3}{2}, 2,$ _____ , _____ , _____

6. _____ , _____ , 1, 5, 9

7. 11, _____ , 21, _____ , 31

8. 7, _____ , _____ , −17, _____

If the given sequence is arithmetic, find the common difference d. If the sequence is not arithmetic, say so. See Example 1.

9. 1, 2, 3, 4, 5, . . .

10. 2, 5, 8, 11, . . .

11. 2, −4, 6, −8, 10, −12, . . .

12. 1, 2, 4, 7, 11, 16, . . .

13. 10, 5, 0, −5, −10, . . .

14. −6, −10, −14, −18, . . .

Write the first five terms of each arithmetic sequence. See Example 2.

15. $a_1 = 5, d = 4$

16. $a_1 = 6, d = 7$

17. $a_1 = -2, d = -4$

18. $a_1 = -3, d = -5$

19. $a_1 = \dfrac{1}{3}, d = \dfrac{2}{3}$

20. $a_1 = 0.25, d = 0.55$

Determine an expression for the general term of each arithmetic sequence. Then find a_{25}. See Example 3.

21. $a_1 = 2, d = 5$

22. $a_1 = 5, d = 3$

23. $3, \dfrac{15}{4}, \dfrac{9}{2}, \dfrac{21}{4}, \ldots$

24. $1, \dfrac{5}{3}, \dfrac{7}{3}, 3, \ldots$

25. $-3, 0, 3, \ldots$

26. $-10, -5, 0, \ldots$

Find the indicated term for each arithmetic sequence. See Example 5.

27. $a_1 = 4, d = 3; \quad a_{25}$

28. $a_1 = 1, d = -3; \quad a_{12}$

29. $2, 4, 6, \ldots; \quad a_{24}$

30. $1, 5, 9, \ldots; \quad a_{50}$

31. $a_{12} = -45, a_{10} = -37; \quad a_1$

32. $a_{10} = -2, a_{15} = 13; \quad a_3$

Find the number of terms in each arithmetic sequence. See Example 6.

33. $3, 5, 7, \ldots, 33$

34. $4, 8, 12, \ldots, 204$

35. $-7, -13, -19, \ldots, -157$

36. $4, 1, -2, \ldots, -32$

37. $\dfrac{3}{4}, 3, \dfrac{21}{4}, \ldots, 12$

38. $2, \dfrac{3}{2}, 1, \dfrac{1}{2}, \ldots, -5$

39. Concept Check A student incorrectly claimed that the common difference for the arithmetic sequence

$$-15, -10, -5, 0, 5, \ldots$$

is -5. **WHAT WENT WRONG?** Find the correct common difference.

40. Concept Check A student incorrectly claimed that there are 100 terms in the arithmetic sequence

$$2, 4, 6, 8, \ldots, 100.$$

WHAT WENT WRONG? How many terms are there?

Evaluate S_6 for each arithmetic sequence. See Examples 7 and 8.

41. $a_n = 3n - 8$

42. $a_n = 5n - 12$

43. $a_n = 4 + 3n$

44. $a_n = 9 + 5n$

45. $a_n = -\dfrac{1}{2}n + 7$

46. $a_n = -\dfrac{2}{3}n + 6$

47. $a_1 = 6, d = 3$

48. $a_1 = 5, d = 4$

49. $a_1 = 7, d = -3$

50. $a_1 = -5, d = -4$

51. $a = 0.9, d = -0.5$

52. $a_1 = 2.4, d = 0.7$

Use a formula for S_n to evaluate each series. See Example 9.

53. $\displaystyle\sum_{i=1}^{10} (8i - 5)$

54. $\displaystyle\sum_{i=1}^{17} (3i - 1)$

55. $\displaystyle\sum_{i=1}^{20} \left(\dfrac{3}{2}i + 4\right)$

56. $\displaystyle\sum_{i=1}^{11} \left(\dfrac{1}{2}i - 1\right)$

57. $\displaystyle\sum_{i=1}^{250} i$

58. $\displaystyle\sum_{i=1}^{2000} i$

Solve each problem. See Examples 4, 7, 8, and 9.

59. Nancy's aunt has promised to deposit \$1 in her account on the first day of her birthday month, \$2 on the second day, \$3 on the third day, and so on for 30 days. How much will this amount to over the entire month?

60. Billy's aunt deposits $2 in his savings account on the first day of February, $4 on the second day, $6 on the third day, and so on for the entire month. How much will this amount to? (Assume that it is a leap year.)

61. Suppose that Cherian is offered a job at $1600 per month with a guaranteed increase of $50 every six months for 5 yr. What will Cherian's salary be at the end of that time?

62. Malik is offered a job with a starting salary of $2000 per month with a guaranteed increase of $100 every four months for 3 yr. What will Malik's salary be at the end of that time?

63. A seating section in a theater-in-the-round has 20 seats in the first row, 22 in the second row, 24 in the third row, and so on for 25 rows. How many seats are there in the last row? How many seats are there in the section?

64. Constantin has started on a fitness program. He plans to jog 10 min per day for the first week and then to add 10 min per day each week until he is jogging an hour each day. In which week will this occur? What is the total number of minutes he will run during the first four weeks?

65. A child builds with blocks, placing 35 blocks in the first row, 31 in the second row, 27 in the third row, and so on. Continuing this pattern, can he end with a row containing exactly 1 block? If not, how many blocks will the last row contain? How many rows can he build this way?

66. A stack of firewood has 28 pieces on the bottom, 24 on top of those, then 20, and so on. If there are 108 pieces of wood, how many rows are there?

14.3   Geometric Sequences

OBJECTIVES

1 Find the common ratio of a geometric sequence.

2 Find the general term of a geometric sequence.

3 Find any specified term of a geometric sequence.

4 Find the sum of a specified number of terms of a geometric sequence.

5 Apply the formula for the future value of an ordinary annuity.

6 Find the sum of an infinite number of terms of a geometric sequence.

OBJECTIVE 1 Find the common ratio of a geometric sequence.

In an arithmetic sequence, each term after the first is found by *adding* a fixed number to the previous term. A *geometric sequence* is defined using multiplication.

> ### Geometric Sequence
>
> A **geometric sequence,** or **geometric progression,** is a sequence in which each term after the first is found by multiplying the preceding term by a nonzero constant.

We find the constant multiplier r, called the **common ratio,** by dividing any term a_{n+1} by the preceding term, a_n.

$$r = \frac{a_{n+1}}{a_n} \qquad \text{Common ratio}$$

For example,

$$\overset{\times 3 \;\; \times 3 \;\; \times 3 \;\; \times 3}{2, \;\; 6, \;\; 18, \;\; 54, \;\; 162, \ldots} \qquad \text{Geometric sequence}$$

is a geometric sequence in which the first term is 2 and the common ratio is 3.

$$r = \frac{6}{2} = \frac{18}{6} = \frac{54}{18} = \frac{162}{54} = 3 \;\leftarrow\; \frac{a_{n+1}}{a_n} = 3 \text{ for all } n.$$

VOCABULARY

☐ geometric sequence
  (geometric progression)
☐ common ratio
☐ annuity
☐ ordinary annuity
☐ payment period
☐ term of an annuity
☐ future value of an annuity

 **NOW TRY
EXERCISE 1**
Determine the common ratio r
for the geometric sequence.

$$\frac{1}{4}, -1, 4, -16, 64, \ldots$$

EXAMPLE 1 Finding the Common Ratio

Determine the common ratio r for the geometric sequence.

$$15, \frac{15}{2}, \frac{15}{4}, \frac{15}{8}, \ldots$$

To find r, choose any two successive terms and divide the second one by the first. We choose the second and third terms of the sequence.

$r = \dfrac{a_3}{a_2}$	Common ratio $r = \frac{a_{n+1}}{a_n}$
$r = \dfrac{\frac{15}{4}}{\frac{15}{2}}$	Substitute.
$r = \dfrac{15}{4} \div \dfrac{15}{2}$	Write as a division problem.
$r = \dfrac{15}{4} \cdot \dfrac{2}{15}$	Definition of division
$r = \dfrac{1}{2}$	Multiply. Write in lowest terms.

Any other two successive terms could have been used to find r. Additional terms of the sequence can be found by multiplying each successive term by $\frac{1}{2}$. **NOW TRY**

OBJECTIVE 2 Find the general term of a geometric sequence.

The general term a_n of a geometric sequence $a_1, a_2, a_3, \ldots$ is expressed in terms of a_1 and r by writing the first few terms.

$$a_1, \quad a_2 = a_1 r, \quad a_3 = a_1 r^2, \quad a_4 = a_1 r^3, \ldots, \quad a_n = a_1 r^{n-1}.$$

General Term of a Geometric Sequence

The general term a_n of the geometric sequence with first term a_1 and common ratio r is given by the following.

$$a_n = a_1 r^{n-1}$$

 **NOW TRY
EXERCISE 2**
Determine an expression
for the general term of the
geometric sequence.

$$\frac{1}{4}, -1, 4, -16, 64, \ldots$$

EXAMPLE 2 Finding the General Term of a Geometric Sequence

Determine an expression for the general term of the geometric sequence.

$$15, \frac{15}{2}, \frac{15}{4}, \frac{15}{8}, \ldots \quad \text{See Example 1.}$$

The first term is $a_1 = 15$ and the common ratio is $r = \frac{1}{2}$.

$a_n = a_1 r^{n-1}$	Formula for the general term a_n
$a_n = 15\left(\dfrac{1}{2}\right)^{n-1}$	Let $a_1 = 15$ and $r = \frac{1}{2}$.

NOW TRY ANSWERS
1. $r = -4$
2. $a_n = \frac{1}{4}(-4)^{n-1}$

It is not possible to simplify further because the exponent must be applied before the multiplication can be done. **NOW TRY**

NOW TRY
EXERCISE 3
Find the indicated term for each geometric sequence.

(a) $a_1 = 3$, $r = -2$; a_8

(b) $10, 2, \frac{2}{5}, \frac{2}{25}, \ldots$; a_7

OBJECTIVE 3 Find any specified term of a geometric sequence.

EXAMPLE 3   Finding Specified Terms in Geometric Sequences

Find the indicated term for each geometric sequence.

(a) $a_1 = 4$, $r = -3$; a_6

$$a_n = a_1 r^{n-1} \qquad \text{Formula for the general term } a_n$$

$$a_6 = a_1 \cdot r^{6-1} \qquad \text{Let } n = 6.$$

$$a_6 = 4(-3)^5 \qquad \text{Let } a_1 = 4, r = -3.$$

$$a_6 = -972 \qquad \boxed{\text{Evaluate } (-3)^5 \text{ and then multiply.}}$$

(b) $\frac{3}{4}, \frac{3}{8}, \frac{3}{16}, \ldots$; a_7

$$a_7 = \frac{3}{4}\left(\frac{1}{2}\right)^{7-1} \qquad \text{Let } a_1 = \frac{3}{4}, r = \frac{1}{2}, n = 7.$$

$$a_7 = \frac{3}{4}\left(\frac{1}{64}\right) \qquad \text{Apply the exponent.}$$

$$a_7 = \frac{3}{256} \qquad \text{Multiply.} \qquad \text{NOW TRY}$$

NOW TRY
EXERCISE 4
Write the first five terms of the geometric sequence having first term 25 and common ratio $-\frac{1}{5}$.

EXAMPLE 4   Writing the Terms of a Geometric Sequence

Write the first five terms of the geometric sequence having first term 5 and common ratio $\frac{1}{2}$.

$$a_1 = 5, \quad a_2 = 5\left(\frac{1}{2}\right) = \frac{5}{2}, \quad a_3 = 5\left(\frac{1}{2}\right)^2 = \frac{5}{4},$$

$$a_4 = 5\left(\frac{1}{2}\right)^3 = \frac{5}{8}, \quad a_5 = 5\left(\frac{1}{2}\right)^4 = \frac{5}{16}$$

Use $a_n = a_1 r^{n-1}$, with $a_1 = 5, r = \frac{1}{2}$, and $n = 1, 2, 3, 4, 5$.

NOW TRY

OBJECTIVE 4 Find the sum of a specified number of terms of a geometric sequence.

It is convenient to have a formula for the sum S_n of the first n terms of a geometric sequence. We can develop such a formula by first writing out S_n.

$$S_n = a_1 + a_1 r + a_1 r^2 + a_1 r^3 + \cdots + a_1 r^{n-1}$$

Next, we multiply both sides by $-r$.

$$-rS_n = -a_1 r - a_1 r^2 - a_1 r^3 - a_1 r^4 - \cdots - a_1 r^n$$

Now add.

$$\begin{aligned} S_n &= a_1 + a_1 r + a_1 r^2 + a_1 r^3 + \cdots + a_1 r^{n-1} \\ -rS_n &= \quad -a_1 r - a_1 r^2 - a_1 r^3 - \cdots - a_1 r^{n-1} - a_1 r^n \\ \hline S_n - rS_n &= a_1 \qquad\qquad\qquad\qquad\qquad\qquad\qquad - a_1 r^n \end{aligned}$$

$$S_n(1 - r) = a_1 - a_1 r^n \qquad \text{Factor on the left.}$$

$$S_n = \frac{a_1(1 - r^n)}{1 - r} \qquad \begin{array}{l}\text{Factor on the right.} \\ \text{Divide each side by } 1 - r.\end{array}$$

NOW TRY ANSWERS
3. (a) -384 (b) $\frac{2}{3125}$
4. $25, -5, 1, -\frac{1}{5}, \frac{1}{25}$

Sum of the First n Terms of a Geometric Sequence

The sum S_n of the first n terms of the geometric sequence having first term a_1 and common ratio r is given by the following.

$$S_n = \frac{a_1(1 - r^n)}{1 - r} \quad (\text{where } r \neq 1)$$

If $r = 1$, then $S_n = a_1 + a_1 + a_1 + \cdots + a_1 = na_1$.

Multiplying the formula for S_n by $\dfrac{-1}{-1}$ gives an alternative form.

$$S_n = \frac{a_1(1 - r^n)}{1 - r} \cdot \frac{-1}{-1} = \frac{a_1(r^n - 1)}{r - 1} \qquad \text{Alternative form}$$

NOW TRY
EXERCISE 5

Evaluate the sum of the first six terms of the geometric sequence with first term 4 and common ratio 2.

EXAMPLE 5   Finding the Sum of the First n Terms of a Geometric Sequence

Evaluate the sum of the first six terms of the geometric sequence having first term -2 and common ratio 3.

$$S_n = \frac{a_1(1 - r^n)}{1 - r} \qquad \text{Formula for } S_n$$

$$S_6 = \frac{-2(1 - 3^6)}{1 - 3} \qquad \text{Let } n = 6, a_1 = -2, r = 3.$$

$$S_6 = \frac{-2(1 - 729)}{-2} \qquad \text{Evaluate } 3^6. \text{ Subtract in the denominator.}$$

$$S_6 = -728 \qquad \text{Simplify.}$$

A series of the form

$$\sum_{i=1}^{n} a \cdot b^i$$

represents the sum of the first n terms of a geometric sequence having first term $a_1 = a \cdot b^1 = ab$ and common ratio b. The next example illustrates this form.

NOW TRY
EXERCISE 6

Evaluate $\displaystyle\sum_{i=1}^{5} 8\left(\frac{1}{2}\right)^i$.

EXAMPLE 6   Using the Formula for S_n to Evaluate a Summation

Evaluate $\displaystyle\sum_{i=1}^{4} 3 \cdot 2^i$.

Because the series is in the form $\displaystyle\sum_{i=1}^{n} a \cdot b^i$, it represents the sum of the first n terms of the geometric sequence with $a_1 = a \cdot b^1 = 3 \cdot 2^1 = 6$ and $r = b = 2$.

$$S_n = \frac{a_1(1 - r^n)}{1 - r} \qquad \text{Formula for } S_n$$

$$S_4 = \frac{6(1 - 2^4)}{1 - 2} \qquad \text{Let } n = 4, a_1 = 6, r = 2.$$

$$S_4 = \frac{6(1 - 16)}{-1} \qquad \text{Evaluate } 2^4. \text{ Subtract in the denominator.}$$

$$S_4 = 90 \qquad \text{Simplify.}$$

NOW TRY ANSWERS
5. 252
6. $\frac{31}{4}$, or 7.75

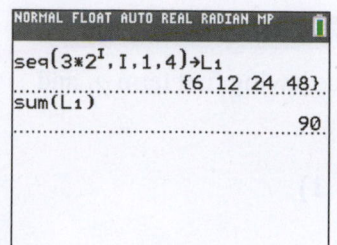

FIGURE 2

FIGURE 2 shows how a graphing calculator can store the terms in a list and then find the sum of these terms. The figure supports the result of **Example 6.**

OBJECTIVE 5 Apply the formula for the future value of an ordinary annuity.

A sequence of equal payments made over equal periods is an **annuity.** If the payments are made at the end of the period, and if the frequency of payments is the same as the frequency of compounding, the annuity is an **ordinary annuity.** The time between payments is the **payment period,** and the time from the beginning of the first payment period to the end of the last is the **term of the annuity.** The **future value of the annuity,** the final sum on deposit, is defined as the sum of the compound amounts of all the payments, compounded to the end of the term.

We state the following formula without proof.

Future Value of an Ordinary Annuity

The future value of an ordinary annuity is given by

$$S = R\left[\frac{(1 + i)^n - 1}{i}\right],$$

where

S is the future value,

R is the payment at the end of each period,

i is the interest rate per period,

and

n is the number of periods.

NOW TRY
EXERCISE 7

Work each problem. Give answers to the nearest cent.

(a) Billy deposits $600 at the end of each year into an account paying 2% per yr, compounded annually. Find the total amount on deposit after 18 yr.

(b) How much will be in Billy's account after 18 yr if he deposits $100 at the end of each month at 1.5% interest compounded monthly?

EXAMPLE 7   Applying the Formula for Future Value of an Annuity

Work each problem. Give answers to the nearest cent.

(a) Igor is an athlete who believes that his playing career will last 7 yr. He deposits $22,000 at the end of each year for 7 yr in an account paying 2% compounded annually. How much will he have on deposit after 7 yr?

Igor's payments form an ordinary annuity with $R = 22{,}000$, $n = 7$, and $i = 0.02$.

$$S = 22{,}000\left[\frac{(1 + 0.02)^7 - 1}{0.02}\right] \qquad \text{Substitute into the formula for } S.$$

$$S = 163{,}554.23 \qquad \text{Evaluate with a calculator.}$$

The future value of this annuity is $163,554.23.

(b) Amy has decided to deposit $200 at the end of each month in an account that pays interest of 2.8% compounded monthly for retirement in 20 yr. How much will be in the account at that time?

Because the interest is compounded monthly, $i = \frac{0.028}{12}$. Also, $R = 200$ and $n = 12(20)$. Substitute into the formula for S to find the future value.

$$S = 200\left[\frac{\left(1 + \frac{0.028}{12}\right)^{12(20)} - 1}{\frac{0.028}{12}}\right] = 64{,}245.50, \quad \text{or} \quad \$64{,}245.50$$

NOW TRY

NOW TRY ANSWERS
7. (a) $12,847.39
   (b) $24,779.49

OBJECTIVE 6 Find the sum of an infinite number of terms of a geometric sequence.

Consider an infinite geometric sequence such as

$$\frac{1}{3}, \frac{1}{6}, \frac{1}{12}, \frac{1}{24}, \frac{1}{48}, \ldots, \quad \text{with first term } \frac{1}{3} \text{ and common ratio } \frac{1}{2}.$$

The sum of the first two terms is

$$S_2 = \frac{1}{3} + \frac{1}{6} = \frac{1}{2}.$$

In a similar manner, we can find additional "partial sums."

$$S_3 = S_2 + \frac{1}{12} = \frac{1}{2} + \frac{1}{12} = \frac{7}{12} \approx 0.583$$

$$S_4 = S_3 + \frac{1}{24} = \frac{7}{12} + \frac{1}{24} = \frac{15}{24} = 0.625$$

$$S_5 = \frac{31}{48} \approx 0.64583$$

$$S_6 = \frac{21}{32} = 0.65625$$

$$S_7 = \frac{127}{192} \approx 0.6614583, \quad \text{and so on.}$$

Each term of the geometric sequence is less than the preceding one, so each additional term is contributing less and less to the partial sum. In decimal form (to the nearest thousandth), the first 7 terms and the 10th term are given in the table.

Term	a_1	a_2	a_3	a_4	a_5	a_6	a_7	a_{10}
Value	0.333	0.167	0.083	0.042	0.021	0.010	0.005	0.001

As the table suggests, the value of a term gets closer and closer to 0 as the number of the term increases. To express this idea, we say that as n increases without bound (written $n \to \infty$), the limit of the term a_n is 0, which is written

$$\lim_{n \to \infty} a_n = 0.$$

A number that can be defined as the sum of an infinite number of terms of a geometric sequence is found by starting with the expression for the sum of a finite number of terms.

$$S_n = \frac{a_1(1 - r^n)}{1 - r}$$

If $|r| < 1$, then as n increases without bound, the value of r^n gets closer and closer to 0. As r^n approaches 0, $1 - r^n$ approaches $1 - 0 = 1$. S_n approaches the quotient $\frac{a_1}{1 - r}$.

$$\lim_{n \to \infty} S_n = \lim_{n \to \infty} \frac{a_1(1 - r^n)}{1 - r} = \frac{a_1(1 - 0)}{1 - r} = \frac{a_1}{1 - r}$$

This limit is defined to be the *sum* of the terms of the infinite geometric sequence.

$$a_1 + a_1 r + a_1 r^2 + a_1 r^3 + \cdots = \frac{a_1}{1 - r}, \quad \text{if } |r| < 1$$

Sum of the Terms of an Infinite Geometric Sequence

The sum S of the terms of an infinite geometric sequence having first term a_1 and common ratio r, where $|r| < 1$, is given by the following.

$$S = \frac{a_1}{1 - r}$$

If $|r| \geq 1$, then the sum does not exist.

Now consider $|r| > 1$. For example, suppose the sequence is

$$6, \ 12, \ 24, \ldots, \ 3(2)^n, \ldots. \qquad \text{Here, } a_1 = 6 \text{ and } r = 2.$$

In this kind of sequence, as n increases, the value of r^n also increases and so does the sum S_n. Each new term adds a greater and greater amount to the sum, so there is no limit to the value of S_n. The sum S does not exist. Recall that if $r = 1$, then

$$S_n = a_1 + a_1 + a_1 + \cdots + a_1 = na_1.$$

 **NOW TRY**
EXERCISE 8
Evaluate the sum of the terms of the infinite geometric sequence having $a_1 = -4$ and $r = \frac{2}{3}$.

EXAMPLE 8   Finding the Sum of the Terms of an Infinite Geometric Sequence

Evaluate the sum of the terms of the infinite geometric sequence having $a_1 = 3$ and $r = -\frac{1}{3}$.

$$S = \frac{a_1}{1 - r} \qquad \textcolor{blue}{\text{Infinite sum formula}}$$

$$S = \frac{3}{1 - \left(-\frac{1}{3}\right)} \qquad \textcolor{blue}{\text{Let } a_1 = 3, r = -\frac{1}{3}.}$$

$$S = \frac{3}{\frac{4}{3}} \qquad \textcolor{blue}{\text{Subtract in the denominator;} \\ 1 - \left(-\frac{1}{3}\right) = 1 + \frac{1}{3} = \frac{3}{3} + \frac{1}{3} = \frac{4}{3}.}$$

$$S = 3 \div \frac{4}{3} \qquad \textcolor{blue}{\text{Write as a division problem.}}$$

$$S = 3 \cdot \frac{3}{4} \qquad \textcolor{blue}{\text{Definition of division}}$$

$$S = \frac{9}{4} \qquad \textcolor{blue}{\text{Multiply.}} \qquad \qquad \textbf{NOW TRY} \ \textcircled{\leftarrow}$$

In summation notation, the sum of an infinite geometric sequence is written as

$$\sum_{i=1}^{\infty} a_i.$$

For instance, the sum in **Example 8** would be written

$$\sum_{i=1}^{\infty} 3\left(-\frac{1}{3}\right)^{i-1}.$$

**NOW TRY
EXERCISE 9**

Evaluate $\displaystyle\sum_{i=1}^{\infty} \left(\frac{5}{8}\right)\left(\frac{3}{4}\right)^i$.

EXAMPLE 9 Evaluating an Infinite Geometric Series

Evaluate $\displaystyle\sum_{i=1}^{\infty} \left(\frac{2}{3}\right)\left(\frac{1}{2}\right)^i$.

This is the infinite geometric series

$$\frac{1}{3} + \frac{1}{6} + \frac{1}{12} + \cdots,$$

with $a_1 = \frac{1}{3}$ and $r = \frac{1}{2}$. Because $|r| < 1$, we find the sum as follows.

$$S = \frac{a_1}{1-r} \qquad \text{Infinite sum formula}$$

$$S = \frac{\frac{1}{3}}{1 - \frac{1}{2}} \qquad \text{Let } a_1 = \frac{1}{3}, r = \frac{1}{2}.$$

$$S = \frac{\frac{1}{3}}{\frac{1}{2}} \qquad \begin{array}{l}\text{Subtract in the denominator;}\\ 1 - \frac{1}{2} = \frac{2}{2} - \frac{1}{2} = \frac{1}{2}.\end{array}$$

$$S = \frac{2}{3} \qquad \text{Divide; } \frac{\frac{1}{3}}{\frac{1}{2}} = \frac{1}{3} \cdot \frac{2}{1} = \frac{2}{3}.$$

NOW TRY

NOW TRY ANSWER
9. $\frac{15}{8}$

14.3 Exercises

  MyLab Math

▶ *Video solutions for select problems available in MyLab Math*

Concept Check *Fill in each blank with the correct response.*

1. In a geometric sequence, if any term after the first is divided by the term that precedes it, the result is the common _____ of the sequence.

2. For the geometric sequence having $a_n = (-2)^n$, the term $a_5 = $ _____.

3. The sum of the first five terms of the geometric sequence 1, 2, 4, . . . is _____.

4. The number of terms in the geometric sequence 1, 2, 4, . . . , 2048 is _____.

Concept Check *Fill in the blanks to complete the terms of each geometric sequence.*

5. $\frac{1}{3}, -\frac{1}{9}, \frac{1}{27},$ _____, _____, _____

6. _____, _____, 1, 5, 25

7. 7, _____, 28, _____, 112

8. $-2,$ _____, _____, $-54,$ _____

If the given sequence is geometric, find the common ratio r. If the sequence is not geometric, say so. See Example 1.

9. 4, 8, 16, 32, . . .

10. 5, 15, 45, 135, . . .

11. $\frac{1}{3}, \frac{2}{3}, \frac{3}{3}, \frac{4}{3}, \cdots$

12. $\frac{5}{7}, \frac{8}{7}, \frac{11}{7}, 2, \cdots$

13. $1, -3, 9, -27, 81, \cdots$

14. $2, -8, 32, -128, \cdots$

15. $1, -\frac{1}{2}, \frac{1}{4}, -\frac{1}{8}, \cdots$

16. $\frac{2}{3}, -\frac{2}{15}, \frac{2}{75}, -\frac{2}{375}, \cdots$

Determine an expression for the general term of each geometric sequence. ***See Example 2.***

17. $-5, -10, -20, \ldots$ **18.** $-2, -6, -18, \ldots$ **19.** $-2, \dfrac{2}{3}, -\dfrac{2}{9}, \ldots$

20. $-3, \dfrac{3}{2}, -\dfrac{3}{4}, \ldots$ **21.** $10, -2, \dfrac{2}{5}, \ldots$ **22.** $8, -2, \dfrac{1}{2}, \ldots$

Find the indicated term for each geometric sequence. ***See Example 3.***

23. $a_1 = 2, r = 5; \quad a_{10}$ **24.** $a_1 = 1, r = 3; \quad a_{15}$

25. $\dfrac{1}{2}, \dfrac{1}{6}, \dfrac{1}{18}, \ldots; \quad a_{12}$ **26.** $\dfrac{2}{3}, \dfrac{1}{3}, \dfrac{1}{6}, \ldots; \quad a_{18}$

27. $a_2 = 18, a_5 = -486; \quad a_7$ **28.** $a_5 = 48, a_8 = -384; \quad a_{10}$

Write the first five terms of each geometric sequence. ***See Example 4.***

29. $a_1 = 2, r = 3$ **30.** $a_1 = 4, r = 2$ **31.** $a_1 = 5, r = -\dfrac{1}{5}$

32. $a_1 = 6, r = -\dfrac{1}{3}$ **33.** $a_1 = -4, r = 0.5$ **34.** $a_1 = -40, r = 0.25$

Evaluate the sum of the terms of each geometric sequence. In Exercises 39–44, give answers to the nearest thousandth. ***See Examples 5 and 6.***

35. $a_1 = -3, r = 4; \quad$ Find S_{10}.   **36.** $a_1 = -5, r = 7; \quad$ Find S_9.

37. $\dfrac{1}{3}, \dfrac{1}{9}, \dfrac{1}{27}, \dfrac{1}{81}, \dfrac{1}{243}$ **38.** $\dfrac{1}{4}, \dfrac{1}{16}, \dfrac{1}{64}, \dfrac{1}{256}, \dfrac{1}{1024}$

39. $-\dfrac{4}{3}, -\dfrac{4}{9}, -\dfrac{4}{27}, -\dfrac{4}{81}, -\dfrac{4}{243}, -\dfrac{4}{729}$ **40.** $\dfrac{5}{16}, -\dfrac{5}{32}, \dfrac{5}{64}, -\dfrac{5}{128}, \dfrac{5}{256}$

41. $\displaystyle\sum_{i=1}^{7} 4\left(\dfrac{2}{5}\right)^i$ **42.** $\displaystyle\sum_{i=1}^{8} 5\left(\dfrac{2}{3}\right)^i$

43. $\displaystyle\sum_{i=1}^{10} (-2)\left(\dfrac{3}{5}\right)^i$ **44.** $\displaystyle\sum_{i=1}^{6} (-2)\left(-\dfrac{1}{2}\right)^i$

Solve each problem. Round answers to the nearest cent. ***See Example 7.***

45. A father opened a savings account for his daughter on her first birthday, depositing $1000. Each year on her birthday he deposits another $1000, making the last deposit on her 21st birthday. If the account pays 1.5% interest compounded annually, how much is in the account at the end of the day on the daughter's 21st birthday?

46. A teacher puts $1000 in a retirement account at the end of each quarter $\left(\frac{1}{4} \text{ of a year}\right)$ for 15 yr. If the account pays 2.2% annual interest compounded quarterly, how much will be in the account at that time?

47. At the end of each quarter, a 50-year-old woman puts $1200 in a retirement account that pays 2% interest compounded quarterly. When she reaches age 60, she withdraws the entire amount and places it in a fund that pays 1% annual interest compounded monthly. From then on, she deposits $300 in the fund at the end of each month. How much is in the account when she reaches age 65?

48. At the end of each quarter, a 45-year-old man puts $1500 in a retirement account that pays 1.5% interest compounded quarterly. When he reaches age 55, he withdraws the entire amount and places it in a fund that pays 0.75% annual interest compounded monthly. From then on, he deposits $400 in the fund at the end of each month. How much is in the account when he reaches age 60?

*Find the sum, if it exists, of the terms of each infinite geometric sequence. **See Examples 8 and 9.***

49. $a_1 = 6, r = \dfrac{1}{3}$

50. $a_1 = 10, r = \dfrac{1}{5}$

51. $a_1 = 1000, r = -\dfrac{1}{10}$

52. $a_1 = 8800, r = -\dfrac{3}{5}$

53. $\displaystyle\sum_{i=1}^{\infty} \dfrac{9}{8}\left(-\dfrac{2}{3}\right)^i$

54. $\displaystyle\sum_{i=1}^{\infty} \dfrac{3}{5}\left(\dfrac{5}{6}\right)^i$

55. $\displaystyle\sum_{i=1}^{\infty} \dfrac{12}{5}\left(\dfrac{5}{4}\right)^i$

56. $\displaystyle\sum_{i=1}^{\infty} \left(-\dfrac{16}{3}\right)\left(-\dfrac{9}{8}\right)^i$

Extending Skills *Solve each application.*

57. When dropped from a certain height, a ball rebounds to $\dfrac{3}{5}$ of the original height. How high will the ball rebound after the fourth bounce if it was dropped from a height of 10 ft? Round to the nearest tenth.

58. A fully wound yo-yo has a string 40 in. long. It is allowed to drop, and on its first rebound it returns to a height 15 in. lower than its original height. Assuming that this "rebound ratio" remains constant until the yo-yo comes to rest, how far does it travel on its third trip up the string? Round to the nearest tenth.

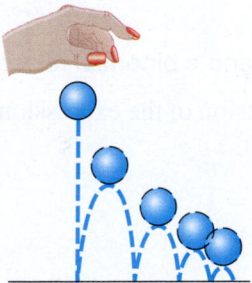

59. A ball is dropped from a height of 20 m, and on each bounce it returns to $\dfrac{3}{4}$ of its previous height. How far will the ball travel before it comes to rest? (*Hint:* Consider the sum of two sequences.)

60. A fully wound yo-yo is dropped the length of its 30-in. string. Each time it drops, it returns to $\dfrac{1}{2}$ of its original height. How far does it travel before it comes to rest? (*Hint:* Consider the sum of two sequences.)

61. A particular substance decays in such a way that it loses half its weight each day. In how many days will 256 g of the substance be reduced to 32 g? How much of the substance is left after 10 days?

62. A tracer dye is injected into a system with an ingestion and an excretion. After 1 hr, $\dfrac{2}{3}$ of the dye is left. At the end of the second hour, $\dfrac{2}{3}$ of the remaining dye is left, and so on. If one unit of the dye is injected, how much is left after 6 hr?

63. In a certain community, the consumption of electricity has increased about 6% per yr.

 (a) If the community uses 1.1 billion units of electricity now, how much will it use 5 yr from now? Round to the nearest tenth.

 (b) Find the number of years (to the nearest year) it will take for the consumption to double.

64. A growing community increases its consumption of electricity 2% per yr.

 (a) If the community uses 1.1 billion units of electricity now, how much will it use 5 yr from now? Round to the nearest tenth.

 (b) Find the number of years (to the nearest year) it will take for the consumption to double.

65. A machine depreciates by $\frac{1}{4}$ of its value each year. If it cost $50,000 new, what is its value after 8 yr?

66. A vehicle depreciates by 20% of its value each year. If it cost $35,000 new, what is its value after 6 yr?

67. The repeating decimal $0.99999\ldots$ can be written as the sum of the terms of a geometric sequence with $a_1 = 0.9$ and $r = 0.1$.

$$0.99999\ldots = 0.9 + 0.9(0.1) + 0.9(0.1)^2 + 0.9(0.1)^3 + 0.9(0.1)^4 + 0.9(0.1)^5 + \cdots$$

Because $|0.1| < 1$, this sum can be found from the formula $S = \dfrac{a_1}{1-r}$. Use this formula to find a more common way of writing the decimal $0.99999\ldots$.

68. If the result of **Exercise 67** seems hard to believe, look at it this way: Use long division to find the repeating decimal representations of $\frac{1}{3}$ and $\frac{2}{3}$. Then line up the decimals vertically and add them. The result must equal

$$\frac{1}{3} + \frac{2}{3} = 1.$$

14.4 The Binomial Theorem

OBJECTIVES

1 Expand a binomial raised to a power.

2 Find any specified term of the expansion of a binomial.

VOCABULARY

☐ Pascal's triangle

OBJECTIVE 1 Expand a binomial raised to a power.

Observe the expansion of the expression $(x + y)^n$ for the first six nonnegative integer values of n.

$(x + y)^0 = 1$ *Expansions of $(x + y)^n$*

$(x + y)^1 = x + y$

$(x + y)^2 = x^2 + 2xy + y^2$

$(x + y)^3 = x^3 + 3x^2y + 3xy^2 + y^3$

$(x + y)^4 = x^4 + 4x^3y + 6x^2y^2 + 4xy^3 + y^4$

$(x + y)^5 = x^5 + 5x^4y + 10x^3y^2 + 10x^2y^3 + 5xy^4 + y^5$

By identifying patterns, we can write a general expansion for $(x + y)^n$.

First, if n is a positive integer, each expansion after $(x + y)^0$ begins with x raised to the same power to which the binomial is raised. That is, the expansion of $(x + y)^1$ has a first term of x^1, the expansion of $(x + y)^2$ has a first term of x^2, and so on. Also, the last term in each expansion is y to this same power, so the expansion of

$$(x + y)^n$$

should begin with the term x^n and end with the term y^n.

The exponents on x decrease by 1 in each term after the first, while the exponents on y, beginning with y in the second term, increase by 1 in each succeeding term. Thus, the *variables* in the expansion of $(x + y)^n$ have the following pattern.

$$x^n, \quad x^{n-1}y, \quad x^{n-2}y^2, \quad x^{n-3}y^3, \quad \ldots, \quad xy^{n-1}, \quad y^n$$

This pattern suggests that the sum of the exponents on x and y in each term is n. For example, in the third term shown, the variable part is $x^{n-2}y^2$ and the sum of the exponents is $n - 2 + 2 = n$.

Now examine the pattern for the *coefficients* of the terms of the preceding expansions. Writing the coefficients alone in a triangular pattern gives **Pascal's triangle**, named in honor of the 17th-century mathematician Blaise Pascal.

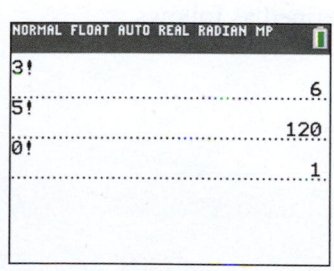

Blaise Pascal (1623–1662)

Pascal's Triangle

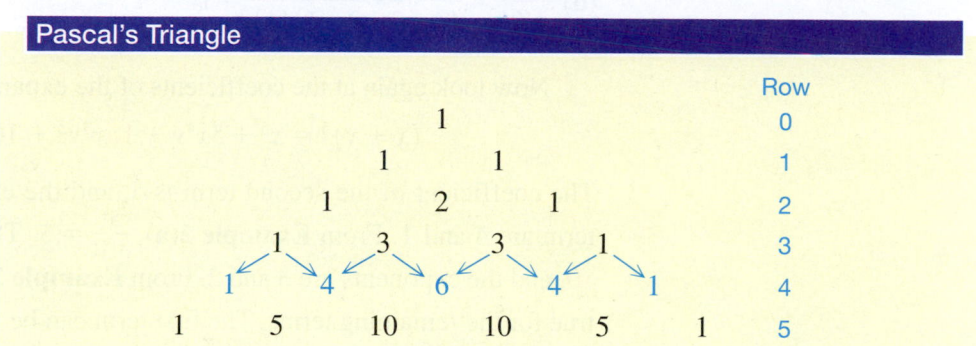

In this triangle, the first and last terms of each row are 1. Each number in the interior of the triangle is the sum of the two numbers just above it (one to the right and one to the left). For example, in row 4 of the triangle, 4 is the sum of 1 and 3, 6 is the sum of 3 and 3, and so on.

To obtain the coefficients for $(x + y)^6$, we need to attach row 6 to the table by starting and ending with 1, and adding pairs of numbers from row 5.

$$1 \quad 6 \quad 15 \quad 20 \quad 15 \quad 6 \quad 1 \qquad \text{Row 6}$$

We then use these coefficients to expand $(x + y)^6$.

$$(x + y)^6 = x^6 + 6x^5y + 15x^4y^2 + 20x^3y^3 + 15x^2y^4 + 6xy^5 + y^6$$

Although it is possible to use Pascal's triangle to find the coefficients in $(x + y)^n$ for any positive integer value of n, it is impractical for large values of n. A more efficient way to determine these coefficients uses the symbol ***n!*** (read ***"n factorial"***).

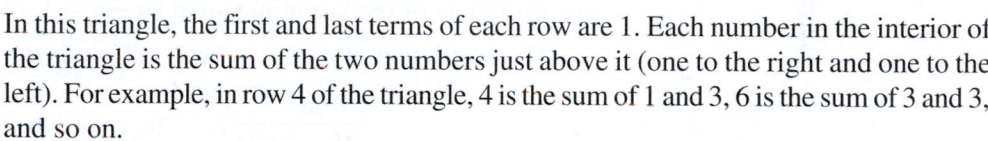

n Factorial (*n*!)

For any positive integer n,

$$n! = n(n - 1)(n - 2)(n - 3) \cdots (3)(2)(1).$$

By definition, **0! = 1.**

FIGURE 3

EXAMPLE 1 Evaluating Factorials

Evaluate each factorial.

(a) $3! = 3 \cdot 2 \cdot 1 = 6$ **(b)** $5! = 5 \cdot 4 \cdot 3 \cdot 2 \cdot 1 = 120$ **(c)** $0! = 1$

FIGURE 3 shows how a graphing calculator computes factorials.

NOW TRY

NOW TRY
EXERCISE 1
Evaluate 7!.

NOW TRY ANSWER
1. 5040

**NOW TRY
EXERCISE 2**
Find the value of each expression.

(a) $\dfrac{8!}{6!2!}$ (b) $\dfrac{8!}{5!3!}$

(c) $\dfrac{6!}{6!0!}$ (d) $\dfrac{6!}{5!1!}$

EXAMPLE 2   Evaluating Expressions Involving Factorials

Find the value of each expression.

(a) $\dfrac{5!}{4!1!} = \dfrac{5 \cdot 4 \cdot 3 \cdot 2 \cdot 1}{(4 \cdot 3 \cdot 2 \cdot 1)(1)} = 5$

(b) $\dfrac{5!}{3!2!} = \dfrac{5 \cdot 4 \cdot 3 \cdot 2 \cdot 1}{(3 \cdot 2 \cdot 1)(2 \cdot 1)} = \dfrac{5 \cdot 4}{2 \cdot 1} = 10$

(c) $\dfrac{6!}{3!3!} = \dfrac{6 \cdot 5 \cdot 4 \cdot 3 \cdot 2 \cdot 1}{(3 \cdot 2 \cdot 1)(3 \cdot 2 \cdot 1)} = \dfrac{6 \cdot 5 \cdot 4}{3 \cdot 2 \cdot 1} = 20$

(d) $\dfrac{4!}{4!0!} = \dfrac{4 \cdot 3 \cdot 2 \cdot 1}{(4 \cdot 3 \cdot 2 \cdot 1)(1)} = 1$ **NOW TRY** ↻

Now look again at the coefficients of the expansion

$$(x + y)^5 = x^5 + 5x^4y + 10x^3y^2 + 10x^2y^3 + 5xy^4 + y^5.$$

The coefficient of the second term is 5, and the exponents on the variables in that term are 4 and 1. From **Example 2(a)**, $\frac{5!}{4!1!} = 5$. The coefficient of the third term is 10, and the exponents are 3 and 2. From **Example 2(b)**, $\frac{5!}{3!2!} = 10$. Similar results are true for the remaining terms. The first term can be written as $1x^5y^0$, and the last term can be written as $1x^0y^5$. Then the coefficient of the first term should be $\frac{5!}{5!0!} = 1$, and the coefficient of the last term would be $\frac{5!}{0!5!} = 1$. (This is why 0! is *defined* to be 1.)

The coefficient of a term in $(x + y)^n$ in which the variable part is $x^r y^{n-r}$ is

$$\dfrac{n!}{r!(n - r)!}.$$ This is the binomial coefficient.

The **binomial coefficient** $\frac{n!}{r!(n - r)!}$ is often represented by the symbol $_nC_r$. This notation comes from the fact that if we choose *combinations* of n things taken r at a time, the result is given by that expression. We read $_nC_r$ as "**combinations of n things taken r at a time.**" Another common representation is $\binom{n}{r}$.

Formula for the Binomial Coefficient $_nC_r$

For nonnegative integers n and r, where $r \leq n$, $_nC_r$ is defined as follows.

$$_nC_r = \dfrac{n!}{r!(n - r)!}$$

EXAMPLE 3   Evaluating Binomial Coefficients

Evaluate each binomial coefficient.

(a) $_5C_4 = \dfrac{5!}{4!(5 - 4)!}$ Let $n = 5$, $r = 4$.

$= \dfrac{5!}{4!1!}$ Subtract.

$= \dfrac{5 \cdot 4 \cdot 3 \cdot 2 \cdot 1}{4 \cdot 3 \cdot 2 \cdot 1 \cdot 1}$ Definition of n factorial

$= 5$ Lowest terms

NOW TRY ANSWERS
2. (a) 28   (b) 56   (c) 1   (d) 6

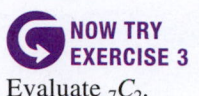

**NOW TRY
EXERCISE 3**
Evaluate $_7C_2$.

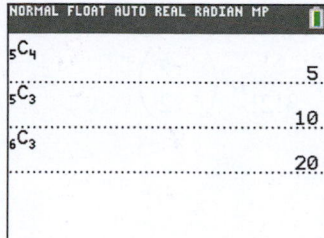

FIGURE 4

(b) $_5C_3 = \dfrac{5!}{3!(5-3)!} = \dfrac{5!}{3!2!} = \dfrac{5 \cdot 4 \cdot 3 \cdot 2 \cdot 1}{3 \cdot 2 \cdot 1 \cdot 2 \cdot 1} = 10$

(c) $_6C_3 = \dfrac{6!}{3!(6-3)!} = \dfrac{6!}{3!3!} = \dfrac{6 \cdot 5 \cdot 4 \cdot 3 \cdot 2 \cdot 1}{3 \cdot 2 \cdot 1 \cdot 3 \cdot 2 \cdot 1} = 20$

Binomial coefficients
will always be whole
numbers.

FIGURE 4 displays the binomial coefficients computed here.   **NOW TRY**

Our observations about the expansion of $(x + y)^n$ are summarized as follows.

- There are $n + 1$ terms in the expansion.

- The first term is x^n, and the last term is y^n.

- In each succeeding term, the exponent on x decreases by 1 and the exponent on y increases by 1.

- The sum of the exponents on x and y in any term is n.

- The coefficient of the term with $x^r y^{n-r}$ or $x^{n-r} y^r$ is $_nC_r$.

We now state the **binomial theorem,** or the **general binomial expansion.**

> **Binomial Theorem**
>
> For any positive integer n, $(x + y)^n$ is expanded as follows.
>
> $$(x + y)^n = x^n + \frac{n!}{1!(n-1)!}x^{n-1}y + \frac{n!}{2!(n-2)!}x^{n-2}y^2$$
>
> $$+ \frac{n!}{3!(n-3)!}x^{n-3}y^3 + \cdots + \frac{n!}{(n-1)!1!}xy^{n-1} + y^n$$
>
> The binomial theorem can be written using summation notation as follows.
>
> $$(x + y)^n = \sum_{r=0}^{n} \frac{n!}{r!(n-r)!}x^{n-r}y^r$$

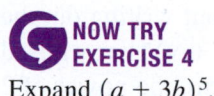

**NOW TRY
EXERCISE 4**
Expand $(a + 3b)^5$.

EXAMPLE 4 Using the Binomial Theorem

Expand $(2m + 3)^4$.

$(2m + 3)^4$

$= (2m)^4 + \dfrac{4!}{1!3!}(2m)^3(3) + \dfrac{4!}{2!2!}(2m)^2(3)^2 + \dfrac{4!}{3!1!}(2m)(3)^3 + 3^4$

$= 16m^4 + 4(8m^3)(3) + 6(4m^2)(9) + 4(2m)(27) + 81$

Remember:
$(ab)^m = a^m b^m.$

$= 16m^4 + 96m^3 + 216m^2 + 216m + 81$ **NOW TRY**

> **NOTE** The binomial coefficients for the first and last terms in **Example 4** are both 1. These values can be taken from Pascal's triangle or computed as follows.
>
> $\dfrac{4!}{0!4!} = 1$ Binomial coefficient of the first term
>
> $\dfrac{4!}{4!0!} = 1$ Binomial coefficient of the last term

NOW TRY ANSWERS
3. 21
4. $a^5 + 15a^4b + 90a^3b^2 +$
$270a^2b^3 + 405ab^4 + 243b^5$

**NOW TRY
EXERCISE 5**

Expand $\left(\dfrac{x}{3} - 2y\right)^4$.

EXAMPLE 5　Using the Binomial Theorem

Expand $\left(a - \dfrac{b}{2}\right)^5$.

$$\left(a - \dfrac{b}{2}\right)^5$$

$$= a^5 + \frac{5!}{1!4!}a^4\left(-\frac{b}{2}\right) + \frac{5!}{2!3!}a^3\left(-\frac{b}{2}\right)^2 + \frac{5!}{3!2!}a^2\left(-\frac{b}{2}\right)^3$$

$$+ \frac{5!}{4!1!}a\left(-\frac{b}{2}\right)^4 + \left(-\frac{b}{2}\right)^5$$

$$= a^5 + 5a^4\left(-\frac{b}{2}\right) + 10a^3\left(\frac{b^2}{4}\right) + 10a^2\left(-\frac{b^3}{8}\right)$$

$$+ 5a\left(\frac{b^4}{16}\right) + \left(-\frac{b^5}{32}\right)$$

> Notice that signs alternate positive and negative.

$$= a^5 - \frac{5}{2}a^4b + \frac{5}{2}a^3b^2 - \frac{5}{4}a^2b^3 + \frac{5}{16}ab^4 - \frac{1}{32}b^5$$

NOW TRY

⚠ **CAUTION**　When the binomial is a *difference* of two terms, as in **Example 5,** the signs of the terms in the expansion will alternate.

- Those terms with odd exponents on the second variable expression $\left(-\frac{b}{2}\text{ in } \textbf{Example 5}\right)$ will be negative.

- Those terms with even exponents on the second variable expression will be positive.

OBJECTIVE 2　Find any specified term of the expansion of a binomial.

Any single term of a binomial expansion can be determined without writing out the whole expansion. For example, if $n \geq 10$, then the 10th term of $(x + y)^n$ has y raised to the ninth power (because y has the power of 1 in the second term, the power of 2 in the third term, and so on).

The exponents on x and y in any term must have a sum of n, so the exponent on x in the 10th term is $n - 9$. The quantities 9 and $n - 9$ determine the factorials in the denominator of the coefficient. Thus, the 10th term of $(x + y)^n$ is

$$\frac{n!}{9!(n-9)!}x^{n-9}y^9.$$

kth Term of the Binomial Expansion

If $n \geq k - 1$, then the kth term of the expansion of $(x + y)^n$ is given by the following.

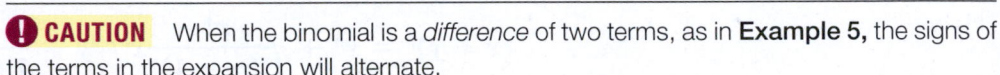

$$\frac{n!}{(k-1)![n-(k-1)]!}x^{n-(k-1)}y^{k-1}$$

NOW TRY ANSWER

5. $\dfrac{x^4}{81} - \dfrac{8x^3y}{27} + \dfrac{8x^2y^2}{3} - \dfrac{32xy^3}{3} +$

　　$16y^4$

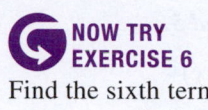

NOW TRY EXERCISE 6

Find the sixth term of the expansion of $(2m - n^2)^8$.

EXAMPLE 6 Finding a Single Term of a Binomial Expansion

Find the fourth term of the expansion of $(a + 2b)^{10}$.

In the fourth term, $2b$ has an exponent of $4 - 1 = 3$, and a has an exponent of $10 - 3 = 7$. The fourth term is determined as follows.

$$\frac{10!}{3!7!}(a^7)(2b)^3 \quad \boxed{\text{Parentheses MUST be used with } 2b.} \quad \text{Let } n = 10, x = a, y = 2b, \text{ and } k = 4.$$

$$= 120(a^7)(2^3b^3) \quad \text{Simplify the factorials; } (ab)^m = a^m b^m$$

$$= 120a^7(8b^3) \quad \text{Simplify.}$$

$$= 960a^7b^3 \quad \text{Multiply.}$$

NOW TRY

NOW TRY ANSWER
6. $-448m^3n^{10}$

14.4 Exercises

FOR EXTRA HELP ▶ **MyLab Math**

▶ *Video solutions for select problems available in MyLab Math*

Concept Check *Fill in each blank with the correct response.*

1. In each row of Pascal's triangle, the first and last terms are _____, and each number in the interior of the triangle is the _____ of the two numbers just above it (one to the right and one to the left).

2. The complete row of Pascal's triangle that begins with the terms 1, 4, is _____.

3. The first term in the expansion of $(x + y)^3$ is _____.

4. The last term in the expansion of $(x - y)^4$ is _____.

5. The value of $3!$ is _____.

6. The value of $0!$ is _____.

7. For any nonnegative integer n, the binomial coefficient ${}_nC_0$ is equal to _____.

8. For any nonnegative integer n, the binomial coefficient ${}_nC_n$ is equal to _____.

Evaluate each expression. See Examples 1–3.

9. $6!$	10. $4!$	11. $8!$	12. $9!$
13. $\dfrac{6!}{4!2!}$	14. $\dfrac{7!}{3!4!}$	15. $\dfrac{4!}{0!4!}$	16. $\dfrac{5!}{5!0!}$
17. $4! \cdot 5$	18. $6! \cdot 7$	19. ${}_6C_2$	20. ${}_7C_4$
21. ${}_{13}C_{11}$	22. ${}_{13}C_2$	23. ${}_{10}C_7$	24. ${}_{10}C_3$

Use the binomial theorem to expand each binomial. See Examples 4 and 5.

25. $(m + n)^4$	26. $(x + r)^5$	27. $(a - b)^5$
28. $(p - q)^4$	29. $(2x + 3)^3$	30. $(4x + 2)^3$
31. $\left(\dfrac{x}{2} - y\right)^4$	32. $\left(\dfrac{x}{3} - 2y\right)^5$	33. $(x^2 + 1)^4$
34. $(y^3 + 2)^4$	35. $(3x^2 - y^2)^3$	36. $(2p^2 - q^2)^3$

Write the first four terms of each binomial expansion. See Examples 4 and 5.

37. $(r + 2s)^{12}$ **38.** $(m + 3n)^{20}$ **39.** $(3x - y)^{14}$

40. $(2p - 3q)^{11}$ **41.** $(t^2 + u^2)^{10}$ **42.** $(x^2 + y^2)^{15}$

Find the indicated term of each binomial expansion. See Example 6.

43. $(2m + n)^{10}$;   fourth term   **44.** $(a + 3b)^{12}$;   fifth term

45. $\left(x + \dfrac{y}{2}\right)^8$;   seventh term   **46.** $\left(a + \dfrac{b}{3}\right)^{15}$;   eighth term

47. $(k - 1)^9$;   third term   **48.** $(r - 4)^{11}$;   fourth term

49. The middle term of $(x^2 + 2y)^6$ **50.** The middle term of $(m^3 + 2y)^8$

51. The term with x^9y^4 in $(3x^3 - 4y^2)^5$ **52.** The term with x^8y^2 in $(2x^2 + 3y)^6$

Chapter 14   Summary

Key Terms

14.1
infinite sequence
finite sequence
terms of a sequence
general term
series
summation notation

index of summation
arithmetic mean (average)

14.2
arithmetic sequence
    (arithmetic progression)
common difference

14.3
geometric sequence
    (geometric progression)
common ratio
annuity
ordinary annuity
payment period

term of an annuity
future value of an annuity

14.4
Pascal's triangle

New Symbols

a_n	nth term of a sequence	S_n	sum of first n terms of a sequence	$\displaystyle\sum_{i=1}^{\infty} a_i$	sum of an infinite number of terms	$_nC_r$ binomial coefficient (combinations of n things taken r at a time)
$\displaystyle\sum_{i=1}^{n} a_i$	summation notation with index i	$\displaystyle\lim_{n \to \infty} a_n$	limit of a_n as n increases without bound	$n!$	n factorial	

Test Your Word Power

See how well you have learned the vocabulary in this chapter.

1. An **infinite sequence** is
   A. the values of a function
   B. a function whose domain is the set of positive integers
   C. the sum of the terms of a function
   D. the average of a group of numbers.

2. A **series** is
   A. the sum of the terms of a sequence
   B. the product of the terms of a sequence
   C. the average of the terms of a sequence
   D. the function values of a sequence.

3. An **arithmetic sequence** is a sequence in which
   A. each term after the first is a constant multiple of the preceding term
   B. the numbers are written in a triangular array
   C. the terms are added
   D. each term after the first differs from the preceding term by a common amount.

4. A **geometric sequence** is a sequence in which
   A. each term after the first is a constant multiple of the preceding term
   B. the numbers are written in a triangular array
   C. the terms are multiplied
   D. each term after the first differs from the preceding term by a common amount.

5. The **common difference** is
   A. the average of the terms in a sequence
   B. the constant multiplier in a geometric sequence
   C. the difference between any two adjacent terms in an arithmetic sequence
   D. the sum of the terms of an arithmetic sequence.

6. The **common ratio** is
   A. the average of the terms in a sequence
   B. the constant multiplier in a geometric sequence
   C. the difference between any two adjacent terms in an arithmetic sequence
   D. the product of the terms of a geometric sequence.

ANSWERS

1. B; *Example:* The ordered list of numbers 3, 6, 9, 12, 15, . . . is an infinite sequence. **2.** A; *Example:* $3 + 6 + 9 + 12 + 15$, written in summation notation as $\sum_{i=1}^{5} 3i$, is a series. **3.** D; *Example:* The sequence $-3, 2, 7, 12, 17, . . .$ is arithmetic. **4.** A; *Example:* The sequence 1, 4, 16, 64, 256 , . . . is geometric. **5.** C; *Example:* The common difference of the arithmetic sequence $-3, 2, 7, 12, 17, \ . . .$ is 5, since $2 - (-3) = 5, 7 - 2 = 5$, and so on. **6.** B; *Example:* The common ratio of the geometric sequence 1, 4, 16, 64, 256, . . . is 4, since $\frac{4}{1} = \frac{16}{4} = \frac{64}{16} = \frac{256}{64} = 4$.

Quick Review

CONCEPTS	EXAMPLES

14.1 Sequences and Series

An infinite sequence is a function with the set of positive integers as domain.

A finite sequence is a function with domain of the form $\{1, 2, 3, . . . , n\}$, where n is a positive integer.

A series is the indicated sum of the terms of a sequence.

The infinite sequence $1, \frac{1}{2}, \frac{1}{3}, \frac{1}{4}, . . . , \frac{1}{n}$ has general term $a_n = \frac{1}{n}$.

The corresponding series is the *sum*

$$1 + \frac{1}{2} + \frac{1}{3} + \frac{1}{4} + \cdots + \frac{1}{n}.$$

14.2 Arithmetic Sequences

Assume that a_1 is the first term, a_n is the nth term, and d is the common difference in an arithmetic sequence.

Common Difference

$$d = a_{n+1} - a_n$$

nth Term

$$a_n = a_1 + (n - 1)d$$

Sum of the First n Terms

$$S_n = \frac{n}{2}(a_1 + a_n)$$

or

$$S_n = \frac{n}{2}[2a_1 + (n - 1)d]$$

Consider the following arithmetic sequence.

$$2, 5, 8, 11, . . .$$

$a_1 = 2$ a_1 is the first term.

$d = 3$ $d = a_2 - a_1$

(Any two successive terms could have been used.)

The tenth term is found as follows.

$a_{10} = 2 + (10 - 1)(3)$ Let $a_1 = 2$, $d = 3$, and $n = 10$.

$a_{10} = 2 + 9 \cdot 3$ Subtract inside the parentheses.

$a_{10} = 29$ Multiply, and then add.

The sum of the first ten terms can be found in either of two ways.

$S_{10} = \frac{10}{2}(2 + a_{10})$	$S_{10} = \frac{10}{2}[2(2) + (10 - 1)(3)]$
$S_{10} = 5(2 + 29)$	$S_{10} = 5(4 + 9 \cdot 3)$
$S_{10} = 5(31)$	$S_{10} = 5(4 + 27)$
$S_{10} = 155$	$S_{10} = 5(31)$
	$S_{10} = 155$

CONCEPTS	EXAMPLES

14.3 Geometric Sequences

Assume that a_1 is the first term, a_n is the nth term, and r is the common ratio in a geometric sequence.

Common Ratio

$$r = \frac{a_{n+1}}{a_n}$$

nth Term

$$a_n = a_1 r^{n-1}$$

Sum of the First n Terms

$$S_n = \frac{a_1(1 - r^n)}{1 - r} \qquad \text{(where } r \neq 1\text{)}$$

Future Value of an Ordinary Annuity

$$S = R\left[\frac{(1 + i)^n - 1}{i}\right],$$

where S is the future value, R is the payment at the end of each period, i is the interest rate per period, and n is the number of periods.

Sum of the Terms of an Infinite Geometric Sequence

$$S = \frac{a_1}{1 - r} \qquad \text{(where } |r| < 1\text{)}$$

Consider the following geometric sequence.

$$1, 2, 4, 8, \ldots$$

$a_1 = 1$ a_1 is the first term.

$r = 2$ $r = \frac{a_4}{a_3}$

(Any two successive terms could have been used.)

The sixth term is found as follows.

$a_6 = 1(2)^{6-1}$ Let $a_1 = 1$, $r = 2$, and $n = 6$.

$a_6 = 1(2)^5$ Subtract in the exponent.

$a_6 = 32$ Apply the exponent and multiply.

The sum of the first six terms is found as follows.

$S_6 = \dfrac{1(1 - 2^6)}{1 - 2}$ Substitute for a_1, r, and n.

$S_6 = \dfrac{1 - 64}{-1}$ Evaluate 2^6. Subtract in the denominator.

$S_6 = 63$ Simplify.

If \$5800 is deposited into an ordinary annuity at the end of each quarter for 4 yr and interest is earned at 2.4% compounded quarterly, how much will be in the account at that time?

$$R = \$5800, \quad i = \frac{0.024}{4} = 0.006, \quad n = 4(4) = 16$$

$$S = 5800\left[\frac{(1 + 0.006)^{16} - 1}{0.006}\right] = \$97,095.24 \qquad \text{Nearest cent}$$

Evaluate the sum S of the terms of an infinite geometric sequence with $a_1 = 1$ and $r = \frac{1}{2}$.

$$S = \frac{1}{1 - \frac{1}{2}} = \frac{1}{\frac{1}{2}} = 1 \div \frac{1}{2} = 1 \cdot \frac{2}{1} = 2$$

14.4 The Binomial Theorem

n Factorial ($n!$)
For any positive integer n,

$$n! = n(n - 1)(n - 2)(n - 3) \cdots (3)(2)(1).$$

By definition, $0! = 1$.

Binomial Coefficient

$$_nC_r = \frac{n!}{r!(n - r)!} \qquad \text{(where } r \leq n\text{)}$$

Evaluate $4!$.

$$4! = 4 \cdot 3 \cdot 2 \cdot 1 = 24$$

Evaluate $_5C_3$.

$$_5C_3 = \frac{5!}{3!(5 - 3)!} = \frac{5!}{3!2!} = \frac{5 \cdot 4 \cdot 3 \cdot 2 \cdot 1}{3 \cdot 2 \cdot 1 \cdot 2 \cdot 1} = 10$$

CONCEPTS	EXAMPLES

Binomial Theorem

For any positive integer n, $(x + y)^n$ is expanded as follows.

$(x + y)^n$

$= x^n + \dfrac{n!}{1!(n-1)!}x^{n-1}y + \dfrac{n!}{2!(n-2)!}x^{n-2}y^2$

$+ \dfrac{n!}{3!(n-3)!}x^{n-3}y^3 + \cdots + \dfrac{n!}{(n-1)!1!}xy^{n-1}$

$+ y^n$

Expand $(2m + 3)^4$.

$(2m + 3)^4$

$= (2m)^4 + \dfrac{4!}{1!3!}(2m)^3(3) + \dfrac{4!}{2!2!}(2m)^2(3)^2$

$+ \dfrac{4!}{3!1!}(2m)(3)^3 + 3^4$

$= 2^4m^4 + 4(2)^3m^3(3) + 6(2)^2m^2(9) + 4(2m)(27) + 81$

$= 16m^4 + 12(8)m^3 + 54(4)m^2 + 216m + 81$

$= 16m^4 + 96m^3 + 216m^2 + 216m + 81$

kth Term of the Binomial Expansion of $(x + y)^n$

If $n \geq k - 1$, then the kth term of the expansion of $(x + y)^n$ is given by the following.

$$\dfrac{n!}{(k-1)![n-(k-1)]!}x^{n-(k-1)}y^{k-1}$$

Find the eighth term of the expansion of $(a - 2b)^{10}$.

$\dfrac{10!}{7!3!}(a^3)(-2b)^7$ Let $n = 10$, $x = a$, $y = -2b$, and $k = 8$.

$= 120(a^3)(-2)^7b^7$ Simplify the factorials; $(ab)^m = a^mb^m$

$= 120a^3(-128b^7)$ Simplify.

$= -15{,}360a^3b^7$ Multiply.

Chapter 14 Review Exercises

14.1 *Write the first four terms of each sequence.*

1. $a_n = 2n - 3$ **2.** $a_n = \dfrac{n-1}{n}$ **3.** $a_n = n^2$

4. $a_n = \left(\dfrac{1}{2}\right)^n$ **5.** $a_n = (n+1)(n-1)$ **6.** $a_n = n(-1)^{n-1}$

Write each series as a sum of terms and then find the sum.

7. $\sum_{i=1}^{5} i^2$ **8.** $\sum_{i=1}^{6} (i+1)$ **9.** $\sum_{i=3}^{6} (5i-4)$

10. $\sum_{i=1}^{4} (i+2)^2$ **11.** $\sum_{i=1}^{6} 2^i$ **12.** $\sum_{i=4}^{7} \dfrac{i}{i+1}$

13. The table shows the worldwide number of electric vehicles, in thousands, in use from 2012 to 2017. To the nearest tenth, what was the average number of electric vehicles for this period?

Year	Electric Vehicles (in thousands)
2012	110
2013	220
2014	409
2015	727
2016	1186
2017	1928

Data from www.statista.com

14.2, 14.3   *Determine whether each sequence is arithmetic, geometric, or neither. If the sequence is arithmetic, find the common difference d. If it is geometric, find the common ratio r.*

14. $2, 5, 8, 11, \ldots$

15. $-6, -2, 2, 6, 10, \ldots$

16. $\dfrac{2}{3}, -\dfrac{1}{3}, \dfrac{1}{6}, -\dfrac{1}{12}, \ldots$

17. $-1, 1, -1, 1, -1, \ldots$

18. $64, 32, 8, \dfrac{1}{2}, \ldots$

19. $64, 32, 8, 1, \ldots$

20. The *Fibonacci sequence* begins $1, 1, 2, 3, 5, 8, \ldots$. What is the eleventh term of this sequence?

14.2   *Find the indicated term for each arithmetic sequence.*

21. $a_1 = -2, d = 5; \quad a_{16}$

22. $a_6 = 12, a_8 = 18; \quad a_{25}$

Determine an expression for the general term of each arithmetic sequence.

23. $a_1 = -4, d = -5$

24. $6, 3, 0, -3, \ldots$

Find the number of terms in each arithmetic sequence.

25. $7, 10, 13, \ldots, 49$

26. $5, 1, -3, \ldots, -79$

Evaluate S_8 for each arithmetic sequence.

27. $a_1 = -2, d = 6$

28. $a_n = -2 + 5n$

14.3   *Determine an expression for the general term of each geometric sequence.*

29. $-1, -4, -16, \ldots$

30. $\dfrac{2}{3}, \dfrac{2}{15}, \dfrac{2}{75}, \ldots$

Find the indicated term for each geometric sequence.

31. $2, -6, 18, \ldots; \quad a_{11}$

32. $a_3 = 20, a_5 = 80; \quad a_{10}$

Evaluate each sum if it exists.

33. $\displaystyle\sum_{i=1}^{5} \left(\dfrac{1}{4}\right)^i$

34. $\displaystyle\sum_{i=1}^{8} \dfrac{3}{4}(-1)^i$

35. $\displaystyle\sum_{i=1}^{\infty} 4\left(\dfrac{1}{5}\right)^i$

36. $\displaystyle\sum_{i=1}^{\infty} 2(3)^i$

14.4   *Use the binomial theorem to expand each binomial.*

37. $(2p - q)^5$

38. $(x^2 + 3y)^4$

39. $(3t^3 - s^2)^4$

40. Write the fourth term of the expansion of $(3a + 2b)^{19}$.

Chapter 14   Mixed Review Exercises

Find the indicated term and evaluate S_{10} for each sequence.

1. a_{10}: geometric; $-3, 6, -12, \ldots$

2. a_{40}: arithmetic; $1, 7, 13, \ldots$

3. a_{15}: arithmetic; $a_1 = -4, \quad d = 3$

4. a_9: geometric; $a_1 = 1, \quad r = -3$

Determine an expression for the general term of each arithmetic or geometric sequence.

5. 2, 8, 32, ...   **6.** 2, 7, 12, ...   **7.** 12, 9, 6, ...   **8.** 27, 9, 3, ...

Solve each problem.

9. When Faith's sled goes down the hill near her home, she covers 3 ft in the first second. Then, for each second after that, she goes 4 ft more than in the preceding second. If the distance she covers going down is 210 ft, how long does it take her to reach the bottom?

10. An ordinary annuity is set up so that $672 is deposited at the end of each quarter for 7 yr. The money earns 4.5% annual interest compounded quarterly. What is the future value of the annuity?

11. The school population in Middleton has been dropping 3% per yr. The current population is 50,000. If this trend continues, what will the population be in 6 yr?

12. A pump removes $\frac{1}{2}$ of the liquid in a container with each stroke. What fraction of the liquid is left in the container after seven strokes?

13. Consider the repeating decimal number 0.55555 . . . .

   (a) Write it as the sum of the terms of an infinite geometric sequence.

   (b) What is r for this sequence?

   (c) Find this infinite sum if it exists, and write it as a fraction in lowest terms.

14. Can the sum of the terms of the infinite geometric sequence defined by $a_n = 5(2)^n$ be found? Explain.

| Chapter 14 | Test | **FOR EXTRA HELP** | *Step-by-step test solutions are found on the Chapter Test Prep Videos available in* **MyLab Math.** |

▶ *View the complete solutions to all Chapter Test exercises in MyLab Math.*

Write the first five terms of each sequence described.

1. $a_n = (-1)^n + 1$ **2.** arithmetic, with $a_1 = 4$ and $d = 2$

3. geometric, with $a_4 = 6$ and $r = \frac{1}{2}$

Determine a_4 for each sequence described.

4. arithmetic, with $a_1 = 6$ and $d = -2$ **5.** geometric, with $a_5 = 16$ and $a_7 = 9$

Evaluate S_5 for each sequence described.

6. arithmetic, with $a_2 = 12$ and $a_3 = 15$ **7.** geometric, with $a_5 = 4$ and $a_7 = 1$

8. The first eight time intervals, in minutes, between eruptions of the Old Faithful geyser in Yellowstone National Park on July 19, 2018, are given here. Calculate the average number of minutes between eruptions to the nearest tenth. (Data from www.geysertimes.org)

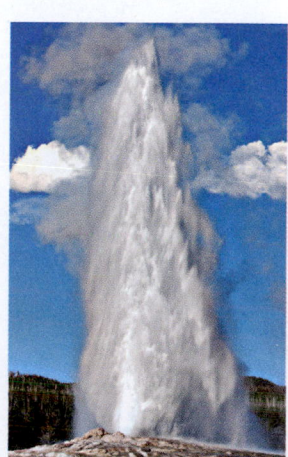

89, 94, 99, 85, 101, 89, 97, 86

9. If $4000 is deposited in an ordinary annuity at the end of each quarter for 7 yr and earns 6% interest compounded quarterly, how much will be in the account at the end of this term?

10. Under what conditions does an infinite geometric series have a sum?

Evaluate each sum if it exists.

11. $\displaystyle\sum_{i=1}^{5} (2i + 8)$ **12.** $\displaystyle\sum_{i=1}^{6} (3i - 5)$ **13.** $\displaystyle\sum_{i=1}^{500} i$

14. $\displaystyle\sum_{i=1}^{3} \frac{1}{2}(4^i)$ **15.** $\displaystyle\sum_{i=1}^{\infty} \left(\frac{1}{4}\right)^i$ **16.** $\displaystyle\sum_{i=1}^{\infty} 6\left(\frac{3}{2}\right)^i$

Evaluate each expression.

17. $8!$ **18.** $0!$ **19.** $\dfrac{6!}{4!2!}$ **20.** ${}_{12}C_{10}$

21. Expand $(3k - 5)^4$.

22. Write the fifth term of the expansion of $\left(2x - \dfrac{y}{3}\right)^{12}$.

Solve each problem.

23. Christian bought a new bicycle for $300. He agreed to pay $20 per month for 15 months, plus interest of 1% each month, on the unpaid balance. Find the total cost of the bicycle.

24. During the summer months, the population of a certain insect colony triples each week. If there are 20 insects in the colony at the end of the first week in July, how many are present by the end of September? (Assume exactly four weeks in a month.)

Chapters R–14 Cumulative Review Exercises

Simplify each expression.

1. $|-7| + 6 - |-10| - (-8 + 3)$ **2.** $-\dfrac{7}{30} + \dfrac{11}{45} - \dfrac{3}{10}$

Let $S = \left\{-\dfrac{8}{3}, 10, 0, \sqrt{13}, -\sqrt{3}, \dfrac{45}{15}, \sqrt{-7}, 0.82, -3\right\}$. *List the elements of S that are members of each set.*

3. Rational numbers   **4.** Irrational numbers

Solve each equation or inequality.

5. $9 - (5 + 3x) + 5x = -4(x - 3) - 7$ **6.** $\dfrac{x + 3}{12} - \dfrac{x - 3}{6} = 0$

7. $7x + 18 \leq 9x - 2$ **8.** $2x > 8$ or $-3x > 9$

9. $|4x - 3| = 21$ **10.** $|2x - 5| \geq 11$

11. $2x^2 + x = 10$ **12.** $6x^2 + 5x = 8$

13. $\dfrac{4}{x-3} - \dfrac{6}{x+3} = \dfrac{24}{x^2-9}$

14. $3^{2x-1} = 81$

15. $\log_8 x + \log_8 (x+2) = 1$

Simplify. Assume that all variables represent nonzero real numbers.

16. $\left(\dfrac{2}{3}\right)^{-2}$

17. $\dfrac{(3p^2)^3(-2p^6)}{4p^3(5p^7)}$

Perform the indicated operations.

18. $(4p + 2)(5p - 3)$

19. $(2m^3 - 3m^2 + 8m) - (7m^3 + 5m - 8)$

20. $(6t^4 + 5t^3 - 18t^2 + 14t - 1) \div (3t - 2)$ **21.** $(8 + 3i)(8 - 3i)$

Factor.

22. $6z^3 + 5z^2 - 4z$

23. $49a^4 - 9b^2$

24. $c^3 + 27d^3$

Simplify.

25. $\dfrac{x^2 - 16}{x^2 + 2x - 8} \div \dfrac{x - 4}{x + 7}$

26. $\dfrac{5}{p^2 + 3p} - \dfrac{2}{p^2 - 4p}$

27. $5\sqrt{72} - 4\sqrt{50}$

Solve each problem.

28. Find the slope of the line passing through $(4, -5)$ and $(-12, -17)$.

29. Find the standard form of the equation of the line passing through $(-2, 10)$ and parallel to the line with equation $3x + y = 7$.

30. Write the equation of a circle with center $(-5, 12)$ and radius 9.

Solve each system of equations.

31. $y = 5x + 3$
$2x + 3y = -8$

32. $x + 2y + z = 8$
$2x - y + 3z = 15$
$-x + 3y - 3z = -11$

33. $xy = -5$
$2x + y = 3$

34. Nuts worth \$3 per lb are to be mixed with 8 lb of nuts worth \$4.25 per lb to obtain a mixture that will be sold for \$4 per lb. How many pounds of the \$3 nuts should be used?

Graph.

35. $x - 3y = 6$

36. $4x - y < 4$

37. $f(x) = 2(x - 2)^2 - 3$

38. $g(x) = \left(\dfrac{1}{3}\right)^x$

39. $y = \log_{1/3} x$

40. $f(x) = \dfrac{1}{x - 3}$

41. $\dfrac{x^2}{9} + \dfrac{y^2}{25} = 1$

42. $x^2 - y^2 = 9$

Solve each problem.

43. Find $f^{-1}(x)$ if $f(x) = 9x + 5$.

44. Write the first five terms of the sequence with general term $a_n = 5n - 12$.

45. Find the sum of the first six terms of the arithmetic sequence with $a_1 = 8$ and $d = 2$.

46. Find the sum of the geometric series $15 - 6 + \dfrac{12}{5} - \dfrac{24}{25} + \cdots$.

47. Find the sum $\displaystyle\sum_{i=1}^{4} 3i$.

48. Evaluate. **(a)** $9!$ **(b)** $0!$ **(c)** $_{10}C_4$

49. Use the binomial theorem to expand $(2a - 1)^5$.

50. Find the fourth term in the expansion of $\left(3x^4 - \dfrac{1}{2}y^2\right)^5$.

Review of Exponents, Polynomials, and Factoring

(Transition from Beginning to Intermediate Algebra)

OBJECTIVES

1 Review the basic rules for exponents.

2 Review operations with polynomials.

3 Review factoring techniques.

OBJECTIVE 1 Review the basic rules for exponents.

Definitions and Rules for Exponents

For all integers m and n and all real numbers a and b for which the following are defined, these definitions and rules hold true.

		Examples
Product rule	$a^m \cdot a^n = a^{m+n}$	$7^4 \cdot 7^5 = 7^{4+5} = 7^9$
Zero exponent	$a^0 = 1$	$(-3)^0 = 1$
Negative exponent	$a^{-n} = \dfrac{1}{a^n}$	$5^{-3} = \dfrac{1}{5^3}$
Quotient rule	$\dfrac{a^m}{a^n} = a^{m-n}$	$\dfrac{2^2}{2^5} = 2^{2-5} = 2^{-3} = \dfrac{1}{2^3}$
Power rules (a)	$(a^m)^n = a^{mn}$	$(4^2)^3 = 4^{2\cdot3} = 4^6$
(b)	$(ab)^m = a^m b^m$	$(3k)^4 = 3^4 k^4$
(c)	$\left(\dfrac{a}{b}\right)^m = \dfrac{a^m}{b^m}$	$\left(\dfrac{2}{3}\right)^2 = \dfrac{2^2}{3^2}$
Negative-to-positive rules	$\dfrac{a^{-m}}{b^{-n}} = \dfrac{b^n}{a^m}$	$\dfrac{2^{-4}}{5^{-3}} = \dfrac{5^3}{2^4}$
	$\left(\dfrac{a}{b}\right)^{-m} = \left(\dfrac{b}{a}\right)^m$	$\left(\dfrac{4}{7}\right)^{-2} = \left(\dfrac{7}{4}\right)^2$

EXAMPLE 1 Applying Definitions and Rules for Exponents

Simplify. Write answers using only positive exponents. Assume that all variables represent nonzero real numbers.

(a) $(x^2 y^{-3})(x^{-5} y^7)$

$= (x^{2+(-5)})(y^{-3+7})$ Product rule

$= x^{-3} y^4$ Add exponents

$= \dfrac{1}{x^3} y^4$ Definition of negative exponent

$= \dfrac{y^4}{x^3}$ $\dfrac{1}{x^3} y^4 = \dfrac{1}{x^3} \cdot \dfrac{y^4}{1} = \dfrac{y^4}{x^3}$

(b) $\underset{\substack{\uparrow \\ \text{The base} \\ \text{is 5.}}}{-5^0}\ +\ \underset{\substack{\uparrow \\ \text{The base} \\ \text{is } -5.}}{(-5)^0}$ Notice the use of parentheses.

$= -1 + 1$ $a^0 = 1$

$= 0$ Add.

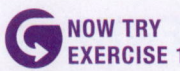
NOW TRY
EXERCISE 1

Simplify. Write answers using only positive exponents. Assume that all variables represent nonzero real numbers.

(a) $(m^{-8}n^4)(m^4n^{-3})$

(b) $-8^0 + 8^0$

(c) $\dfrac{(p^{-3}q)^4}{(p^2q^5)^2}$

(d) $\left(\dfrac{2x^{-2}y}{x^2y^{-4}}\right)^{-4}$

(c) $\dfrac{(t^5s^{-4})^2}{(t^{-3}s^5)^3}$

$= \dfrac{t^{10}s^{-8}}{t^{-9}s^{15}}$ Power rules (a) and (b)

$= \dfrac{t^{10}t^9}{s^{15}s^8}$ Negative-to-positive rule

$= \dfrac{t^{19}}{s^{23}}$ Product rule

(d) $\left(\dfrac{-3x^{-4}y}{x^5y^{-4}}\right)^{-2}$

$= \left(\dfrac{x^5y^{-4}}{-3x^{-4}y}\right)^2$ Negative-to-positive rule

$= \dfrac{x^{10}y^{-8}}{9x^{-8}y^2}$ Power rules (a), (b), and (c)

$= \dfrac{x^{18}}{9y^{10}}$ Quotient rule

(e) $(2x^2y^3z)^2(x^4y^2)^3$

$= (4x^4y^6z^2)(x^{12}y^6)$ Power rules (a) and (b)

$= 4x^{16}y^{12}z^2$ Product rule   **NOW TRY** ↻

OBJECTIVE 2 Review operations with polynomials.

Adding and Subtracting Polynomials

To add polynomials, add like terms.

To subtract polynomials, change the sign of each term in the subtrahend (second polynomial) and add the result to the minuend (first polynomial)—that is, add the *opposite* of each term of the second polynomial to the first polynomial.

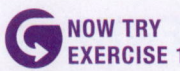
NOW TRY
EXERCISE 2

Add or subtract as indicated.

(a) $(3x^3 + x^2 - 5x - 6) +$
$(-6x^3 + 2x^2 + 4x - 1)$

(b) Subtract.

$4x^2 + 7x - 5$
$- (-5x^2 - 2x + 3)$

EXAMPLE 2 Adding and Subtracting Polynomials

Add or subtract as indicated.

(a) $(-4x^3 + 3x^2 - 8x + 2) + (5x^3 - 8x^2 + 12x - 3)$

$= (-4x^3 + 5x^3) + (3x^2 - 8x^2) + (-8x + 12x) + (2 - 3)$
Commutative and associative properties

$= (-4 + 5)x^3 + (3 - 8)x^2 + (-8 + 12)x + (2 - 3)$
Distributive property

$= x^3 - 5x^2 + 4x - 1$ Simplify.

(b) $-4(x^2 + 3x - 6) - (2x^2 - 3x + 7)$

$= -4(x^2 + 3x - 6) - 1(2x^2 - 3x + 7)$ $-a = -1a$

$= -4(x^2) - 4(3x) - 4(-6) - 1(2x^2) - 1(-3x) - 1(7)$ Distributive property

$= -4x^2 - 12x + 24 - 2x^2 + 3x - 7$ Multiply.

$= -6x^2 - 9x + 17$ Combine like terms.

(c) Subtract.

$\begin{array}{r} 2t^2 - 3t - 4 \\ - (-8t^2 + 4t - 1) \end{array}$ ⟶ $\begin{array}{r} 2t^2 - 3t - 4 \\ + (8t^2 - 4t + 1) \\ \hline 10t^2 - 7t - 3 \end{array}$ Change each sign.

Add.   **NOW TRY** ↻

NOW TRY ANSWERS

1. **(a)** $\dfrac{n}{m^4}$ **(b)** 0
   (c) $\dfrac{1}{p^{16}q^6}$ **(d)** $\dfrac{x^{16}}{16y^{20}}$
2. **(a)** $-3x^3 + 3x^2 - x - 7$
   (b) $9x^2 + 9x - 8$

Multiplying Polynomials

To multiply two polynomials, multiply each term of the second polynomial by each term of the first polynomial and add the products.

To multiply two binomials, use the FOIL method.

Also recall the special product rules. For x and y, the following hold true.

$$(x + y)^2 = x^2 + 2xy + y^2$$
$$(x - y)^2 = x^2 - 2xy + y^2$$
Square of a binomial

$$(x + y)(x - y) = x^2 - y^2$$
Product of a sum and difference of two terms

NOW TRY
EXERCISE 3

Find each product.

(a) $(6x - 5)(2x - 3)$

(b) $(4m - 3n)(4m + 3n)$

(c) $(7z + 1)^2$

(d) $(r + 3)(r^2 - 3r + 9)$

EXAMPLE 3   Multiplying Polynomials

Find each product.

(a) $(4y - 1)(3y + 2)$

| First | Outer | Inner | Last |
| terms | terms | terms | terms |

$$= 4y(3y) + 4y(2) - 1(3y) - 1(2) \quad \text{FOIL method}$$

$$= 12y^2 + 8y - 3y - 2 \quad \text{Multiply.}$$

$$= 12y^2 + 5y - 2 \quad \text{Combine like terms.}$$

(b) $(3x + 5y)(3x - 5y)$ $(ab)^2 = a^2b^2,\ \textbf{not}\ ab^2.$

$$= (3x)^2 - (5y)^2 \quad (x + y)(x - y) = x^2 - y^2$$

$$= 9x^2 - 25y^2 \quad (3x^2) = 3^2x^2 = 9x^2;\ (5y)^2 = 5^2y^2 = 25y^2$$

(c) $(2t + 3)^2$

$$= (2t)^2 + 2(2t)(3) + 3^2 \quad (x + y)^2 = x^2 + 2xy + y^2$$

$$= 4t^2 + 12t + 9 \quad \text{Remember the middle term.}$$

(d) $(5x - 1)^2$

$$= (5x)^2 - 2(5x)(1) + 1^2 \quad (x - y)^2 = x^2 - 2xy + y^2$$

$$= 25x^2 - 10x + 1 \quad (5x)^2 = 5^2x^2 = 25x^2$$

(e) $(3x + 2)(9x^2 - 6x + 4)$

$$
\begin{array}{r}
9x^2 - 6x + 4 \\
3x + 2 \\
\hline
18x^2 - 12x + 8 \\
27x^3 - 18x^2 + 12x \\
\hline
27x^3 \qquad\qquad + 8
\end{array}
$$

Multiply vertically.

$\longleftarrow$ $2(9x^2 - 6x + 4)$

$\longleftarrow$ $3x(9x^2 - 6x + 4)$

Add.

Be sure to write like terms in columns.

NOW TRY ANSWERS

3. (a) $12x^2 - 28x + 15$

(b) $16m^2 - 9n^2$

(c) $49z^2 + 14z + 1$

(d) $r^3 + 27$

The product is a sum of cubes, $27x^3 + 8$.

NOW TRY

OBJECTIVE 3 Review factoring techniques.

Recall that factoring involves writing a polynomial as a product.

Guidelines for Factoring a Polynomial

Question 1   **Is there a common factor other than 1?** If so, factor it out.

Question 2   **How many terms are in the polynomial?**

Two terms: The polynomial is a binomial. Is it a difference of squares or a sum or difference of cubes? If so, factor using the appropriate rule.

$$x^2 - y^2 = (x + y)(x - y)$$ Difference of squares
$$x^3 - y^3 = (x - y)(x^2 + xy + y^2)$$ Difference of cubes
$$x^3 + y^3 = (x + y)(x^2 - xy + y^2)$$ Sum of cubes

Three terms: The polynomial is a trinomial. Is it a perfect square trinomial? If so, factor as follows.

$$x^2 + 2xy + y^2 = (x + y)^2$$
$$x^2 - 2xy + y^2 = (x - y)^2$$ Perfect square trinomials

If the trinomial is not a perfect square trinomial, use one of the following methods.

- To factor $x^2 + bx + c$, find two integers whose product is c and whose sum is b, the coefficient of the middle term.

- To factor $ax^2 + bx + c$, find two integers having product ac and sum b. Use these integers to rewrite the middle term, and factor by grouping.

  Alternatively, use the FOIL method and try various combinations of the factors until the correct middle term is found.

Four terms: Try to factor by grouping.

Question 3   **Can any factors be factored further?** If so, factor them.

EXAMPLE 4 Factoring Polynomials

Factor each polynomial completely.

(a) $6x^2y^3 - 12x^3y^2$

$= 6x^2y^2(y) - 6x^2y^2(2x)$ $6x^2y^2$ is the greatest common factor.

$= 6x^2y^2(y - 2x)$ Distributive property

(b) $3x^2 - x - 2$

To find the factors, find two terms that multiply to give $3x^2$ (here $3x$ and x) and two terms that multiply to give -2 (here $+2$ and -1). Make sure that the sum of the outer and inner products in the factored form is the middle term of the trinomial, $-x$.

$3x^2 - x - 2$ factors as $(3x + 2)(x - 1)$.

NOW TRY
EXERCISE 4

Factor each polynomial completely.

(a) $10s^3t^6 + 5s^9t^2$

(b) $5x^2 - 20x - 60$

(c) $10t^2 + 13t - 3$

(d) $49x^2 + 42x + 9$

(e) $27x^3 - 1000$

(f) $mn - 2n + 5m - 10$

CHECK $(3x + 2)(x - 1)$

$= 3x(x) + 3x(-1) + 2(x) + 2(-1)$ Multiply using the FOIL method.

$= 3x^2 - x - 2$ ✓   Original polynomial

(c) $3x^2 - 27x + 42$

$= 3(x^2 - 9x + 14)$ Factor out the common factor.

$= 3(x - 7)(x - 2)$ Factor the trinomial.

(d) $100t^2 - 81$

$= (10t)^2 - 9^2$ Difference of squares

$= (10t + 9)(10t - 9)$ $x^2 - y^2 = (x + y)(x - y)$

(e) $4x^2 + 20xy + 25y^2$

The terms $4x^2$ and $25y^2$, which can be written as $(2x)^2$ and $(5y)^2$, are both perfect squares, so this trinomial might factor as a perfect square trinomial.

Try to factor $4x^2 + 20xy + 25y^2$ as $(2x + 5y)^2$.

CHECK   Take twice the product of the two terms in the squared binomial.

$2 \cdot 2x \cdot 5y = 20xy$ ⟵ Middle term of $4x^2 + 20xy + 25y^2$

Twice ⟶ First term   Last term

Because $20xy$ is the middle term of the trinomial, the trinomial is a perfect square.

$4x^2 + 20xy + 25y^2$ factors as $(2x + 5y)^2$.

(f) $1000x^3 - 27$

$= (10x)^3 - 3^3$ Difference of cubes

$= (10x - 3)[(10x)^2 + 10x(3) + 3^2]$ $x^3 - y^3 = (x - y)(x^2 + xy + y^2)$

$= (10x - 3)(100x^2 + 30x + 9)$ $(10x)^2 = 10^2x^2 = 100x^2$

(g) $6xy - 3x + 4y - 2$

Because there are four terms, try factoring by grouping.

$6xy - 3x + 4y - 2$

$= (6xy - 3x) + (4y - 2)$ Group the terms.

$= 3x(2y - 1) + 2(2y - 1)$ Factor each group.

$= (2y - 1)(3x + 2)$ Factor out $2y - 1$.

In the final step, factor out the greatest common factor, the binomial $2y - 1$.

NOW TRY

NOW TRY ANSWERS

4. **(a)** $5s^3t^2(2t^4 + s^6)$

   (b) $5(x - 6)(x + 2)$

   (c) $(5t - 1)(2t + 3)$

   (d) $(7x + 3)^2$

   (e) $(3x - 10)(9x^2 + 30x + 100)$

   (f) $(n + 5)(m - 2)$

⚠ **CAUTION**   When a polynomial has been factored, it is wise to do the following.

1. Check that the product of all the factors does indeed yield the original polynomial.

2. Check that the original polynomial has been factored **completely**.

A Exercises

 FOR EXTRA HELP   ▶ **MyMathLab**

▶ *Video solutions for select problems available in MyLab Math*

Concept Check   *Determine whether each statement is* true *or* false. *If false, correct the right-hand side of the statement.*

1. $(ab)^2 = ab^2$

2. $(5x)^3 = 5^3x^3$

3. $\left(\dfrac{4}{a}\right)^3 = \dfrac{4^3}{a}$ $(a \neq 0)$

4. $x^3 \cdot x^4 = x^7$

5. $xy^0 = 0$ $(y \neq 0)$

6. $(5^2)^3 = 5^6$

7. $-(-10)^0 = 1$

8. $3^{-1} = -\dfrac{1}{3}$

9. Concept Check   A friend incorrectly simplified

$$4^5 \cdot 4^2 \quad \text{as} \quad 16^7.$$

WHAT WENT WRONG? Give the correct answer.

10. Concept Check   A student incorrectly simplified

$$\frac{6^5}{3^2} \quad \text{as} \quad 2^3.$$

WHAT WENT WRONG? Give the correct answer.

Simplify each expression. Write the answers using only positive exponents. Assume that all variables represent positive real numbers. ***See Example 1.***

11. $(a^4b^{-3})(a^{-6}b^2)$

12. $(t^{-3}s^{-5})(t^8s^{-2})$

13. $(5x^{-2}y)^2(2xy^4)^2$

14. $(7x^{-3}y^4)^3(2x^{-1}y^{-4})^2$

15. $-6^0 + (-6)^0$

16. $(-12)^0 - 12^0$

17. $\dfrac{(2w^{-1}x^2y^{-1})^3}{(4w^5x^{-2}y)^2}$

18. $\dfrac{(5p^{-3}q^2r^{-4})^2}{(10p^4q^{-1}r^5)^{-1}}$

19. $\left(\dfrac{-4a^{-2}b^4}{a^3b^{-1}}\right)^{-3}$

20. $\left(\dfrac{r^{-3}s^{-8}}{-6r^2s^{-4}}\right)^{-2}$

21. $(7x^{-4}y^2z^{-2})^{-2}(7x^4y^{-1}z^3)^2$

22. $(3m^{-5}n^2p^{-4})^3(3m^4n^{-3}p^5)^{-2}$

23. Concept Check   A student added two polynomials vertically as follows.

$$\begin{array}{r} 4x^2 - 7x + 4 \\ + \underline{(2x^2 + 3x - 5)} \\ 6x^4 - 4x^2 - 1 \quad \text{Incorrect} \end{array}$$

WHAT WENT WRONG? Give the correct sum.

24. Concept Check   A student subtracted one polynomial from another vertically as follows.

$$\begin{array}{r} 15x^2 + 12 \\ - \underline{(7x^2 - 3)} \\ 8x^2 + 9 \quad \text{Incorrect} \end{array}$$

WHAT WENT WRONG? Give the correct difference.

Add or subtract as indicated. ***See Example 2.***

25. $(2a^4 + 3a^3 - 6a^2 + 5a - 12) + (-8a^4 + 8a^3 - 14a^2 + 21a - 3)$

26. $(-6r^4 - 3r^3 + 12r^2 - 9r + 9) + (8r^4 - 13r^3 - 14r^2 - 10r - 3)$

27. $(6x^3 - 12x^2 + 3x - 4) - (-2x^3 + 6x^2 - 3x + 12)$

28. $(10y^3 - 4y^2 + 8y + 7) - (7y^3 + 5y^2 - 2y - 13)$

29. $\quad 5x^2y + 2xy^2 + y^3$
$+ \underline{(-4x^2y - 3xy^2 + 5y^3)}$

30. $\quad 6ab^3 - 2a^2b^2 + 3b^5$
$+ \underline{(8ab^3 + 12a^2b^2 - 8b^5)}$

31. $\quad 6x^3 - 2x^2 + 3x - 1$
$- \underline{(-4x^3 + 2x^2 - 6x + 3)}$

32. $\quad -9y^3 - 2y^2 + 3y - 8$
$- \underline{(-8y^3 + 4y^2 + 3y + 1)}$

33. $3(5x^2 - 12x + 4) - 2(9x^2 + 13x - 10)$

34. $-4(2t^3 - 3t^2 + 4t - 1) - 3(-8t^3 + 3t^2 - 2t + 9)$

35. Concept Check A student multiplied incorrectly as follows.

$$(x + 4)^2 = x^2 + 16$$

WHAT WENT WRONG? Give the correct product.

36. Concept Check A student multiplied incorrectly as follows.

$$(x - 9)(x + 9) = x^2 + 81$$

WHAT WENT WRONG? Give the correct product.

Find each product. See Example 3.

37. $(3x + 1)(2x - 7)$
38. $(5z + 3)(2z - 3)$
39. $(4x - 1)(x - 2)$

40. $(7t - 3)(t - 4)$
41. $(4t + 3)(4t - 3)$
42. $(6x + 1)(6x - 1)$

43. $(2y^2 + 4)(2y^2 - 4)$
44. $(3b^3 + 2t)(3b^3 - 2t)$
45. $(4x - 3)^2$

46. $(9t + 2)^2$
47. $(6r + 5y)^2$
48. $(8m - 3n)^2$

49. $(c + 2d)(c^2 - 2cd + 4d^2)$
50. $(f + 3g)(f^2 - 3fg + 9g^2)$

51. $(4x - 1)(16x^2 + 4x + 1)$
52. $(5r - 2)(25r^2 + 10r + 4)$

53. $(7t + 5s)(2t^2 + 5st - s^2)$
54. $(8p + 3q)(2p^2 - 4pq + q^2)$

Concept Check *Match each polynomial in Column I with the method or methods for factoring it in Column II. The choices in Column II may be used once, more than once, or not at all.*

I	II
55. (a) $49x^2 - 81y^2$	**A.** Factor out the GCF.
(b) $125z^6 + 1$	**B.** Factor a difference of squares.
(c) $88r^2 - 55s^2$	**C.** Factor a difference of cubes.
(d) $64a^3 - 8b^9$	**D.** Factor a sum of cubes.
(e) $50x^2 - 128y^4$	**E.** The polynomial is prime.

I	II
56. (a) $ab - 5a + 3b - 15$	**A.** Factor out the GCF.
(b) $z^2 - 3z + 6$	**B.** Factor a perfect square trinomial.
(c) $25x^2 + 100$	**C.** Factor by grouping.
(d) $r^2 - 24r + 144$	**D.** Factor into two distinct binomials.
(e) $2y^2 + 36y + 162$	**E.** The polynomial is prime.

Factor each polynomial completely. ***See Example 4.***

57. $40ab - 16a$

58. $25xy - 15y$

59. $8x^3y^4 + 12x^2y^3 + 36xy^4$

60. $10m^5n + 4m^2n^3 + 18m^3n^2$

61. $x^2 - 2x - 15$

62. $x^2 + x - 12$

63. $2x^2 - 9x - 18$

64. $3x^2 + 2x - 8$

65. $4x^2 - 28x + 40$

66. $2x^2 - 18x + 36$

67. $36t^2 - 25$

68. $49r^2 - 9$

69. $16t^2 + 24t + 9$

70. $25t^2 + 90t + 81$

71. $4m^2p - 12mnp + 9n^2p$

72. $16p^2r - 40pqr + 25q^2r$

73. $x^3 + 1$

74. $x^3 + 27$

75. $8t^3 + 125$

76. $27s^3 + 64$

77. $t^6 - 125$

78. $w^6 - 27$

79. $t^4 - 1$

80. $r^4 - 81$

81. $5xt + 15xr + 2yt + 6yr$

82. $3am + 18mb + 2an + 12nb$

83. $6ar + 12br - 5as - 10bs$

84. $7mt + 35ms - 2nt - 10ns$

85. $4x^2 + 12xy + 9y^2 - 1$

86. $81t^2 + 36ty + 4y^2 - 9$

Synthetic Division

OBJECTIVES

1 Use synthetic division to divide by a polynomial of the form $x - k$.

2 Use the remainder theorem to evaluate a polynomial.

3 Use the remainder theorem to decide whether a given number is a solution of an equation.

OBJECTIVE 1 Use synthetic division to divide by a polynomial of the form $x - k$.

If a polynomial in x is divided by a binomial of the form $x - k$, a shortcut method called **synthetic division** can be used. Consider the following.

Polynomial Division	Synthetic Division
$$\begin{array}{r} 3x^2 + 9x + 25 \\ x-3\overline{)3x^3 + 0x^2 - 2x + 5} \\ \underline{3x^3 - 9x^2} \\ 9x^2 - 2x \\ \underline{9x^2 - 27x} \\ 25x + 5 \\ \underline{25x - 75} \\ 80 \end{array}$$	$$\begin{array}{r} 3 \quad 9 \quad 25 \\ 1-3\overline{)3 \quad 0 \quad -2 \quad 5} \\ \underline{3 \ -9} \\ 9 \quad -2 \\ \underline{9 \ -27} \\ 25 \quad 5 \\ \underline{25 \ -75} \\ 80 \end{array}$$

On the right above, exactly the same division is shown written without the variables. This is why it is *essential* to use 0 as a placeholder in synthetic division. All the numbers in color on the right are repetitions of the numbers directly above them, so we omit them to condense our work, as shown on the left below.

$$\begin{array}{r} 3 \quad 9 \quad 25 \\ 1-3\overline{)3 \quad 0 \quad -2 \quad 5} \\ -9 \\ 9 \quad -2 \\ -27 \\ 25 \quad 5 \\ -75 \\ 80 \end{array} \qquad \begin{array}{r} 3 \quad 9 \quad 25 \\ 1-3\overline{)3 \quad 0 \quad -2 \quad 5} \\ -9 \\ 9 \\ -27 \\ 25 \\ -75 \\ 80 \end{array}$$

The numbers in color on the left are again repetitions of the numbers directly above them. They too are omitted, as shown on the right above. If we bring the 3 in the dividend down to the beginning of the bottom row, the top row can be omitted because it duplicates the bottom row.

$$\begin{array}{r} 1-3\overline{)3 \quad 0 \quad -2 \quad 5} \\ \underline{-9 \ -27 \ -75} \\ 3 \quad 9 \quad 25 \quad 80 \end{array}$$

We omit the 1 at the upper left—it represents $1x$, which will always be the first term in the divisor. To simplify the arithmetic, we replace subtraction in the second row by addition. To compensate, we change the -3 at the upper left to its additive inverse, 3.

1013

Additive inverse of −3 → 3)$\overline{3 \quad 0 \quad -2 \quad 5}$

$$\begin{array}{r} 9 \quad 27 \quad 75 \end{array} \leftarrow \text{Signs changed}$$

$$3 \quad 9 \quad 25 \quad \textcolor{red}{80} \leftarrow \text{Remainder}$$

↓   ↓   ↓   ↓

The quotient is read from the bottom row. $3x^2 + 9x + 25 + \dfrac{80}{x-3}$ Remember to add $\dfrac{\text{remainder}}{\text{divisor}}$.

The first three numbers in the bottom row are the coefficients of the quotient polynomial with degree one less than the degree of the dividend. The last number gives the remainder.

Synthetic Division

Synthetic division is a shortcut procedure for polynomial division that eliminates writing the variable factors. It is used only when dividing a polynomial by a binomial of the form $x - k$.

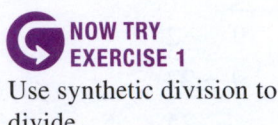

NOW TRY EXERCISE 1

Use synthetic division to divide.

$$\dfrac{4x^3 + 18x^2 + 19x + 7}{x+3}$$

EXAMPLE 1   Using Synthetic Division

Use synthetic division to divide.

$$\dfrac{5x^2 + 16x + 15}{x+2}$$ Remember that a fraction bar means division.

We change $x + 2$ into the form $x - k$ by writing it as

$$x + 2 = x - (-2), \quad \text{where } k = -2.$$

Now we write the coefficients of $5x^2 + 16x + 15$, placing -2 to the left.

$x + 2$ leads to -2. → $-2)\overline{5 \quad 16 \quad 15}$ ← Coefficients

$$-2)\overline{5 \quad 16 \quad 15}$$
$$\quad\quad -10$$
$$\quad 5$$

Bring down the 5, and multiply: $-2 \cdot 5 = -10$.

$$-2)\overline{5 \quad 16 \quad 15}$$
$$\quad\quad -10 \quad -12$$
$$\quad 5 \quad\;\; 6$$

Add 16 and -10, obtaining 6, and multiply 6 and -2 to obtain -12.

$$-2)\overline{5 \quad 16 \quad 15}$$
$$\quad\quad -10 \quad -12$$
$$\quad 5 \quad\;\; 6 \quad\;\;\; \textcolor{red}{3}$$ ← Remainder

Read the result from the bottom row.

Add 15 and -12, obtaining 3.

$$\dfrac{5x^2 + 16x + 15}{x+2} = 5x + 6 + \dfrac{3}{x+2}$$ Add $\dfrac{\text{remainder}}{\text{divisor}}$.

CHECK   We can check this result as with polynomial division.

$$(x+2)(5x+6) + 3 \qquad \text{Divisor} \times \text{Quotient} + \text{Remainder}$$
$$= 5x^2 + 6x + 10x + 12 + 3 \qquad \text{FOIL method}$$
$$= 5x^2 + 16x + 15 \;\checkmark \qquad \text{Dividend}$$

NOW TRY ANSWER

1. $4x^2 + 6x + 1 + \dfrac{4}{x+3}$

NOW TRY

NOW TRY
EXERCISE 2
Use synthetic division to divide.
$$\frac{-3x^4 + 13x^3 - 6x^2 + 31}{x - 4}$$

EXAMPLE 2 Using Synthetic Division with a Missing Term

Use synthetic division to divide.
$$\frac{-4x^5 + x^4 + 6x^3 + 2x^2 + 50}{x - 2}$$

This quotient could also be written as
$$(-4x^5 + x^4 + 6x^3 + 2x^2 + 50) \div (x - 2).$$

In long division form, the procedure is set up as follows.
$$x - 2 \overline{) -4x^5 + x^4 + 6x^3 + 2x^2 + 50}$$

Now use synthetic division.

$$\begin{array}{r|rrrrrr} 2) & -4 & 1 & 6 & 2 & 0 & 50 \\ & & -8 & -14 & -16 & -28 & -56 \\ \hline & -4 & -7 & -8 & -14 & -28 & -6 \end{array}$$

Use the steps given previously, first inserting a 0 for the missing x-term.

Read the result from the bottom row.

$$\frac{-4x^5 + x^4 + 6x^3 + 2x^2 + 50}{x - 2} = -4x^4 - 7x^3 - 8x^2 - 14x - 28 + \frac{-6}{x - 2}$$

NOW TRY

OBJECTIVE 2 Use the remainder theorem to evaluate a polynomial.

We can use synthetic division to evaluate polynomials. In the synthetic division of **Example 2,** where the polynomial was divided by $x - 2$, the remainder was -6.
Replacing x in the polynomial with 2 gives the following.

$$-4x^5 + x^4 + 6x^3 + 2x^2 + 50 \qquad \text{Dividend in \textbf{Example 2}}$$
$$= -4 \cdot 2^5 + 2^4 + 6 \cdot 2^3 + 2 \cdot 2^2 + 50 \qquad \text{Replace } x \text{ with 2.}$$
$$= -4 \cdot 32 + 16 + 6 \cdot 8 + 2 \cdot 4 + 50 \qquad \text{Evaluate the powers.}$$
$$= -128 + 16 + 48 + 8 + 50 \qquad \text{Multiply.}$$
$$= -6 \qquad \text{Add.}$$

This number, -6, is the same number as the remainder. Dividing by $x - 2$ produced a remainder equal to the result when x is replaced with 2. This always happens, as the **remainder theorem** states. This result is proved in more advanced courses.

> **Remainder Theorem**
>
> If the polynomial $f(x)$ is divided by $x - k$, then the remainder is equal to $f(k)$.

NOW TRY
EXERCISE 3
Let $f(x) = 3x^3 - 2x^2 + 5x + 30$. Use the remainder theorem to evaluate $f(-2)$.

EXAMPLE 3 Using the Remainder Theorem

Let $f(x) = 2x^3 - 5x^2 - 3x + 11$. Use the remainder theorem to evaluate $f(-2)$.
Divide $f(x)$ by $x - (-2)$, using synthetic division.

Value of $k \rightarrow$
$$\begin{array}{r|rrrr} -2) & 2 & -5 & -3 & 11 \\ & & -4 & 18 & -30 \\ \hline & 2 & -9 & 15 & -19 \end{array} \leftarrow \text{Remainder}$$

Thus, $f(-2) = -19.$

NOW TRY

NOW TRY ANSWERS
2. $-3x^3 + x^2 - 2x - 8 + \frac{-1}{x-4}$
3. -12

OBJECTIVE 3 Use the remainder theorem to decide whether a given number is a solution of an equation.

NOW TRY
EXERCISE 4
Use the remainder theorem to decide whether -4 is a solution of the equation.

$$5x^3 + 19x^2 - 2x + 8 = 0$$

EXAMPLE 4   Using the Remainder Theorem

Use the remainder theorem to decide whether -5 is a solution of the equation.

$$2x^4 + 12x^3 + 6x^2 - 5x + 75 = 0$$

If synthetic division gives a remainder of 0, then -5 is a solution. Otherwise, it is not.

Proposed solution $\longrightarrow$ $-5 \overline{)\begin{array}{ccccc} 2 & 12 & 6 & -5 & 75 \\ & -10 & -10 & 20 & -75 \\ \hline 2 & 2 & -4 & 15 & 0 \end{array}}$ $\leftarrow$ Remainder

Because the remainder is 0, the polynomial has value 0 when $k = -5$. So -5 is a solution of the given equation.

NOW TRY

The synthetic division in **Example 4** shows that $x - (-5)$ divides the polynomial with 0 remainder. Thus $x - (-5) = x + 5$ is a *factor* of the polynomial.

$$2x^4 + 12x^3 + 6x^2 - 5x + 75 \quad \text{factors as} \quad (x + 5)(2x^3 + 2x^2 - 4x + 15).$$

NOW TRY ANSWER
4. yes

The second factor is the quotient polynomial in the last row of the synthetic division.

B Exercises   FOR EXTRA HELP   ▶ **MyLab Math**

▶ *Video solutions for select problems available in MyLab Math*

Concept Check   *Choose the letter of the correct setup to perform synthetic division on the indicated quotient.*

1. $\dfrac{x^2 + 3x - 6}{x - 2}$

   A. $-2\overline{)\begin{array}{ccc} 1 & 3 & -6 \end{array}}$

   B. $-2\overline{)\begin{array}{ccc} -1 & -3 & 6 \end{array}}$

   C. $2\overline{)\begin{array}{ccc} 1 & 3 & -6 \end{array}}$

   D. $2\overline{)\begin{array}{ccc} -1 & -3 & 6 \end{array}}$

2. $\dfrac{x^3 - 3x^2 + 2}{x - 1}$

   A. $1\overline{)\begin{array}{ccc} 1 & -3 & 2 \end{array}}$

   B. $-1\overline{)\begin{array}{ccc} 1 & -3 & 2 \end{array}}$

   C. $1\overline{)\begin{array}{cccc} 1 & -3 & 0 & 2 \end{array}}$

   D. $1\overline{)\begin{array}{cccc} -1 & 3 & 0 & -2 \end{array}}$

Concept Check   *Fill in each blank with the appropriate response.*

3. In arithmetic, the result of the division

   $$\begin{array}{r} 3 \\ 5\overline{)19} \\ \underline{15} \\ 4 \end{array}$$

   can be written

   $$19 = 5 \cdot \underline{\hspace{1cm}} + \underline{\hspace{1cm}}.$$

4. In algebra, the result of the division

   $$\begin{array}{r} x + 3 \\ x - 1\overline{)x^2 + 2x + 3} \\ \underline{x^2 - x} \\ 3x + 3 \\ \underline{3x - 3} \\ 6 \end{array}$$

   can be written $x^2 + 2x + 3 =$

   $$(x - 1)(\underline{\hspace{1cm}}) + \underline{\hspace{1cm}}.$$

5. To perform the division $x - 3\overline{)x^3 + 6x^2 + 2x}$ using synthetic division, we begin by writing the following.

   $$\underline{\hspace{1cm}}\overline{)\begin{array}{cccc} 1 & \underline{\hspace{0.5cm}} & 2 & \underline{\hspace{0.5cm}} \end{array}}$$

6. Consider the following function.

$$f(x) = 2x^4 + 6x^3 - 5x^2 + 3x + 8$$

$$f(x) = (x - 2)(2x^3 + 10x^2 + 15x + 33) + 74$$

By inspection, we can state that $f(2) = $ _____.

7. Concept Check   A student attempted to divide

$$4x^3 + 2x^2 + 6 \quad \text{by} \quad x + 2$$

synthetically by setting up the division as follows.

$$-2 \overline{)4 \quad 2 \quad 6}$$

This is incorrect. **WHAT WENT WRONG?** Give the correct setup and the answer.

8. Concept Check   A student attempted to divide

$$4x^3 + 2x^2 + 6x \quad \text{by} \quad x + 2$$

synthetically by setting up the division as follows.

$$-2 \overline{)4 \quad 2 \quad 6}$$

This is incorrect. **WHAT WENT WRONG?** Give the correct setup and the answer.

Use synthetic division to divide. See Examples 1 and 2.

9. $\dfrac{x^2 - 6x + 5}{x - 1}$ **10.** $\dfrac{x^2 + 4x - 21}{x - 3}$ **11.** $\dfrac{4x^2 + 19x - 5}{x + 5}$

12. $\dfrac{3x^2 + 5x - 12}{x + 3}$ **13.** $\dfrac{2x^2 + 8x + 13}{x + 2}$ **14.** $\dfrac{4x^2 - 5x - 20}{x - 4}$

15. $\dfrac{x^2 - 3x + 5}{x + 1}$ **16.** $\dfrac{x^2 + 4x - 6}{x - 5}$ **17.** $\dfrac{4x^3 - 3x^2 + 2x - 3}{x - 1}$

18. $\dfrac{5x^3 - 6x^2 + 3x + 14}{x + 1}$ **19.** $\dfrac{x^3 + 2x^2 - 4x + 3}{x - 4}$ **20.** $\dfrac{x^3 - 3x^2 + 5x - 1}{x - 5}$

21. $\dfrac{2x^5 - 2x^3 + 3x^2 - 24x - 2}{x - 2}$ **22.** $\dfrac{3x^5 + x^4 - 84x^2 - 12x + 3}{x - 3}$

23. $\dfrac{-3x^5 - 3x^4 + 5x^3 - 6x^2 + 3}{x + 1}$ **24.** $\dfrac{-3x^5 + 2x^4 - 5x^3 - 6x^2 - 1}{x + 2}$

25. $\dfrac{x^5 + x^4 + x^3 + x^2 + x + 3}{x + 1}$ **26.** $\dfrac{x^5 - x^4 + x^3 - x^2 + x - 2}{x - 1}$

Use the remainder theorem to evaluate $f(k)$. See Example 3.

27. $f(x) = 2x^3 - 4x^2 + 5x - 3; \quad k = 2$ **28.** $f(x) = x^3 + 3x^2 - x + 5; \quad k = -1$

29. $f(x) = -x^3 - 5x^2 - 4x - 2; \quad k = -4$ **30.** $f(x) = -x^3 + 5x^2 - 3x + 4; \quad k = 3$

31. $f(x) = 2x^3 - 4x^2 + 5x - 33; \quad k = 3$ **32.** $f(x) = x^3 - 3x^2 + 4x - 4; \quad k = 2$

Use the remainder theorem to decide whether the given number is a solution of the equation. See Example 4.

33. $x^3 - 2x^2 - 3x + 10 = 0; \quad x = -2$ **34.** $x^3 - 3x^2 - x + 10 = 0; \quad x = -2$

35. $3x^3 + 2x^2 - 2x + 11 = 0; \quad x = -2$ **36.** $3x^3 + 10x^2 + 3x - 9 = 0; \quad x = -2$

37. $2x^3 - x^2 - 13x + 24 = 0;$ $x = -3$ **38.** $5x^3 + 22x^2 + x - 28 = 0;$ $x = -4$

39. $x^4 + 2x^3 - 3x^2 + 8x = 8;$ $x = -2$ **40.** $x^4 - x^3 - 6x^2 + 5x = -10;$ $x = -2$

RELATING CONCEPTS   For Individual or Group Work   (Exercises 41–46)

We can show a connection between dividing one polynomial by another and factoring the first polynomial. Let $f(x) = 2x^2 + 5x - 12$. **Work Exercises 41–46 in order.**

41. Factor $f(x)$.

42. Solve $f(x) = 0$.

43. Evaluate $f(-4)$.

44. Evaluate $f\left(\frac{3}{2}\right)$.

45. Complete the sentence: If $f(a) = 0$, then $x - $ _____ is a factor of $f(x)$.

46. Use the conclusion reached in **Exercise 45** to decide whether $x - 3$ is a factor of $g(x) = 3x^3 - 4x^2 - 17x + 6$. If so, factor $g(x)$ completely.

ANSWERS TO SELECTED EXERCISES

In this section we provide the answers that we think most students will obtain when they work the exercises using the methods explained in the text. If your answer does not look exactly like the one given here, it is not necessarily wrong. In many cases, there are equivalent forms of the answer that are correct. For example, if the answer section shows $\frac{3}{4}$ and your answer is 0.75, you have obtained the right answer, but written it in a different (yet equivalent) form. Unless the directions specify otherwise, 0.75 is just as valid an answer as $\frac{3}{4}$.

In general, if your answer does not agree with the one given in the text, see whether it can be transformed into the other form. If it can, then it is the correct answer. If you still have doubts, talk with your instructor.

R · PREALGEBRA REVIEW

Section R.1

1. true **3.** false; This is an improper fraction. Its value is 1.
5. false; The fraction $\frac{13}{39}$ is written in lowest terms as $\frac{1}{3}$. **7.** false;
Product refers to multiplication, so the product of 10 and 2 is 20.
9. C **11.** A **13.** prime **15.** composite; $2 \cdot 3 \cdot 5$
17. composite; $2 \cdot 2 \cdot 2 \cdot 2 \cdot 2 \cdot 2$ **19.** neither **21.** composite; $3 \cdot 19$
23. prime **25.** composite; $2 \cdot 2 \cdot 31$ **27.** composite; $2 \cdot 2 \cdot 5 \cdot 5 \cdot 5$
29. composite; $2 \cdot 7 \cdot 13 \cdot 19$ **31.** $\frac{1}{2}$ **33.** $\frac{5}{6}$ **35.** $\frac{3}{5}$ **37.** $\frac{1}{5}$ **39.** $\frac{6}{5}$
41. $1\frac{5}{7}$ **43.** $6\frac{5}{12}$ **45.** $7\frac{6}{11}$ **47.** $\frac{13}{5}$ **49.** $\frac{83}{8}$ **51.** $\frac{51}{5}$ **53.** $\frac{24}{35}$ **55.** $\frac{1}{20}$
57. $\frac{6}{25}$ **59.** $\frac{6}{5}$, or $1\frac{1}{5}$ **61.** 9 **63.** $\frac{65}{12}$, or $5\frac{5}{12}$ **65.** $\frac{38}{5}$, or $7\frac{3}{5}$ **67.** $\frac{21}{2}$, or
$10\frac{1}{2}$ **69.** $\frac{14}{27}$ **71.** $\frac{10}{3}$, or $3\frac{1}{3}$ **73.** 12 **75.** $\frac{1}{16}$ **77.** 10 **79.** 18
81. $\frac{35}{24}$, or $1\frac{11}{24}$ **83.** $\frac{84}{47}$, or $1\frac{37}{47}$ **85.** $\frac{11}{15}$ **87.** $\frac{2}{3}$ **89.** $\frac{8}{9}$ **91.** $\frac{29}{24}$, or $1\frac{5}{24}$
93. $\frac{107}{144}$ **95.** $\frac{43}{8}$, or $5\frac{3}{8}$ **97.** $\frac{101}{20}$, or $5\frac{1}{20}$ **99.** $\frac{5}{9}$ **101.** $\frac{2}{3}$ **103.** $\frac{1}{4}$
105. $\frac{17}{36}$ **107.** $\frac{67}{20}$, or $3\frac{7}{20}$ **109.** $\frac{11}{12}$ **111.** $\frac{32}{9}$, or $3\frac{5}{9}$ **113.** (a) $\frac{1}{2}$ (b) $\frac{1}{4}$
(c) $\frac{1}{3}$ (d) $\frac{1}{6}$ **115.** 6 cups **117.** $1\frac{1}{8}$ in. **119.** $\frac{9}{16}$ in. **121.** $618\frac{3}{4}$ ft
123. $5\frac{5}{24}$ in. **125.** 8 cakes (There will be some sugar left over.)
127. $16\frac{5}{8}$ yd **129.** $3\frac{3}{8}$ in. **131.** 4 million, or 4,000,000; 4.4 million, or
4,400,000 **133.** $\frac{2}{25}$ **135.** C

Section R.2

1. (a) 6 (b) 9 (c) 1 (d) 7 (e) 4 **3.** (a) 46.25 (b) 46.2
(c) 46 (d) 50 **5.** $\frac{4}{10}$ **7.** $\frac{64}{100}$ **9.** $\frac{138}{1000}$ **11.** $\frac{43}{1000}$ **13.** $\frac{3805}{1000}$
15. 143.094 **17.** 25.61 **19.** 15.33 **21.** 21.77 **23.** 81.716
25. 15.211 **27.** 116.48 **29.** 739.53 **31.** 0.006 **33.** 7.15 **35.** 2.8
37. 2.05 **39.** 1232.6 **41.** 5711.6 **43.** 94 **45.** 0.162 **47.** 1.2403
49. 0.02329 **51.** 1% **53.** $\frac{1}{20}$ **55.** $12\frac{1}{2}\%$, or 12.5% **57.** 0.25; 25%
59. $\frac{1}{2}$; 0.5 **61.** $\frac{3}{4}$; 75% **63.** 4.2 **65.** 2.25 **67.** 0.375 **69.** $0.\overline{5}$; 0.556

71. $0.1\overline{6}$; 0.167 **73.** 0.54 **75.** 0.07 **77.** 1.17 **79.** 0.024 **81.** 0.0625
83. 0.008 **85.** 79% **87.** 2% **89.** 0.4% **91.** 128% **93.** 40%
95. 600% **97.** $\frac{51}{100}$ **99.** $\frac{3}{20}$ **101.** $\frac{1}{50}$ **103.** $\frac{7}{5}$, or $1\frac{2}{5}$ **105.** $\frac{3}{40}$
107. 80% **109.** 14% **111.** $18.\overline{18}\%$ **113.** 225% **115.** $216.\overline{6}\%$
117. 160 **119.** 4.8 **121.** 109.2 **123.** $17.80; $106.80
125. $119.25; $675.75 **127.** 19.76 million, or 19,760,000 **129.** 5%

1 · THE REAL NUMBER SYSTEM

Section 1.1

1. false; $3^2 = 3 \cdot 3 = 9$ **3.** false; A number raised to the first power is that number, so $3^1 = 3$. **5.** false; $4 + 3(8 - 2)$ means $4 + 3(6)$, which simplifies to $4 + 18$, or 22. The common error leading to 42 is adding 4 to 3 and then multiplying by 6. One must follow the rules for order of operations. **7.** ①, ② **9.** ①, ③, ② **11.** ②, ④, ③, ① **13.** 49
15. 144 **17.** 64 **19.** 1000 **21.** 81 **23.** 1024 **25.** $\frac{1}{36}$ **27.** $\frac{16}{81}$
29. 0.36 **31.** 0.064 **33.** The multiplication should be performed before the addition. The correct value of the expression is 14. **35.** 32 **37.** 58
39. 22.2 **41.** 12 **43.** $\frac{49}{30}$, or $1\frac{19}{30}$ **45.** 13 **47.** 26 **49.** 4 **51.** 42
53. 5 **55.** 41 **57.** 95 **59.** 90 **61.** 14 **63.** 9 **65.** $3 \cdot (6 + 4) \cdot 2$
67. $10 - (7 - 3)$ **69.** $16 \le 16$; true **71.** $61 \le 60$; false **73.** $0 \ge 0$;
true **75.** $45 \ge 46$; false **77.** $66 > 72$; false **79.** $2 \ge 3$; false
81. $3 \ge 3$; true **83.** Five is less than seventeen; true **85.** Five is not equal to eight; true **87.** Seven is greater than or equal to fourteen; false
89. Fifteen is less than or equal to fifteen; true **91.** One-third is equal to three-tenths; false **93.** Two and five-tenths is greater than two and fifty-hundredths; false **95.** $15 = 5 + 10$ **97.** $9 > 5 - 4$
99. $16 \ne 19$ **101.** $\frac{1}{2} \le \frac{2}{4}$ **103.** $20 > 5$ **105.** $\frac{3}{4} < \frac{4}{5}$ **107.** $1.3 \le 2.5$
109. (a) $14.7 - 40 \cdot 0.13$ (b) 9.5 (c) 8.075; walking (5 mph)
(d) $14.7 - 55 \cdot 0.11$; 8.65; 7.3525, swimming **111.** Alaska, Texas, California, Idaho **113.** Alaska, Texas, California, Virginia, Idaho

Section 1.2

1. B **3.** A **5.** B **7.** The exponent refers only to the 4. The correct value is 80. **9.** (a) 11 (b) 13 **11.** (a) 16 (b) 24 **13.** (a) 16
(b) 26 **15.** (a) 64 (b) 144 **17.** (a) $\frac{5}{3}$ (b) $\frac{7}{3}$ **19.** (a) $\frac{7}{8}$ (b) $\frac{13}{12}$
21. (a) 52 (b) 114 **23.** (a) 25.836 (b) 38.754 **25.** (a) 24
(b) 28 **27.** (a) 12 (b) 33 **29.** (a) 6 (b) $\frac{9}{5}$ **31.** (a) $\frac{4}{3}$ (b) $\frac{13}{6}$
33. (a) $\frac{2}{3}$ (b) $\frac{22}{15}$ **35.** (a) 13 (b) 28 **37.** (a) 1 (b) $\frac{28}{17}$
39. (a) 3.684 (b) 8.841 **41.** $12x$ **43.** $x + 9$ **45.** $x - 2$
47. $7 - x$ **49.** $x - 8$ **51.** $\frac{18}{x}$ **53.** $6(x - 4)$ **55.** yes **57.** no
59. yes **61.** yes **63.** yes **65.** no **67.** $x + 8 = 18$; 10
69. $2x + 1 = 5$; 2 **71.** $16 - \frac{3}{4}x = 13$; 4 **73.** $3x = 2x + 8$; 8
75. expression **77.** equation **79.** equation **81.** 75 yr **83.** 78 yr

Section 1.3

1. 0 **3.** positive **5.** quotient; denominator **7.** 5, −5 **9. (a)** A
(b) A **(c)** B **(d)** B **11.** 4 **13.** 0 **15.** One example is $\sqrt{13}$.
There are others. **17.** true **19.** true **21.** false

In Exercises 23–27, answers will vary.

23. $\frac{1}{2}, \frac{5}{8}, 1\frac{3}{4}$ **25.** $-3\frac{1}{2}, -\frac{2}{3}, \frac{3}{7}$ **27.** $\sqrt{5}, \pi, -\sqrt{3}$

29. 2,216,602 **31.** −10,971 **33.** −39.73

35. (number line from −6 to 2) **37.** (number line from −6 to 4)

39. (number line from −4 to 4, marks −3.8, $-1\frac{5}{8}$, $\frac{1}{4}$, $2\frac{1}{2}$) **41. (a)** 3, 7 **(b)** 0, 3, 7 **(c)** −9, 0, 3, 7

(d) $-9, -1\frac{1}{4}, -\frac{3}{5}, 0, 0.\overline{1}, 3, 5.9, 7$ **(e)** $-\sqrt{7}, \sqrt{5}$ **(f)** All are real

numbers. **43. (a)** 11 **(b)** 0, 11 **(c)** 0, 11, −6

(d) $\frac{7}{9}, -2.\overline{3}, 0, -8\frac{3}{4}, 11, -6$ **(e)** $\sqrt{3}, \pi$ **(f)** All are real numbers.

45. (a) 2 **(b)** 2 **47. (a)** −8 **(b)** 8 **49. (a)** $\frac{3}{4}$ **(b)** $\frac{3}{4}$

51. (a) −5.6 **(b)** 5.6 **53.** 6 **55.** −12 **57.** $-\frac{2}{3}$ **59.** 3 **61.** −3

63. −11 **65.** $-\frac{2}{3}$ **67.** 4 **69.** $|-3.5|$ **71.** $-|-6|$ **73.** $|5-3|$

75. true **77.** true **79.** true **81.** false **83.** true **85.** false

87. Natural gas service, March to April **89.** Shelter, April to May

Section 1.4

1. negative   (number line showing −3, −2 from −5 to 0) **3.** negative (number line showing 2, −4 from −4 to 0)

5. −8; −6; 2 **7.** positive **9.** negative **11.** −8 **13.** −12

15. 2 **17.** −3 **19.** −9 **21.** 0 **23.** $-\frac{3}{5}$ **25.** $\frac{1}{2}$ **27.** $-\frac{19}{24}$

29. $-\frac{3}{4}$ **31.** 8.9 **33.** −6.01 **35.** 12 **37.** 5 **39.** 2 **41.** −9

43. 0 **45.** −7.7 **47.** −8 **49.** 0 **51.** −20 **53.** −3 **55.** −4

57. −8 **59.** −14 **61.** 10 **63.** −4 **65.** 4 **67.** $\frac{3}{4}$ **69.** $-\frac{11}{8}$, or

$-1\frac{3}{8}$ **71.** $\frac{15}{8}$, or $1\frac{7}{8}$ **73.** 11.6 **75.** −9.9 **77.** 10 **79.** −5 **81.** 11

83. −10 **85.** 22 **87.** −2 **89.** $\frac{37}{12}$, or $3\frac{1}{12}$ **91.** $-\frac{1}{4}$, or −0.25

93. −6 **95.** −12 **97.** −5.891 **99.** −5 + 12 + 6; 13

101. $[-19 + (-4)] + 14; -9$ **103.** $[-4 + (-10)] + 12; -2$

105. $[\frac{5}{7} + (-\frac{9}{7})] + \frac{2}{7}; -\frac{2}{7}$ **107.** 4 − (−8); 12 **109.** −2 − 8; −10

111. $[9 + (-4)] - 7; -2$ **113.** $[8 - (-5)] - 12; 1$ **115.** −9

117. +3 **119.** −4 **121.** −184 m **123.** 120°F **125.** −69°F

127. 17 **129. (a)** 3.6% **(b)** Americans spent more money than they
earned, which means they had to dip into savings or increase borrowing.

131. $19,900 **133.** $1045.55 **135.** $323.83 **137.** 16.38%

139. −12.18% **141.** 50,395 ft **143.** 1345 ft **145.** 136 ft

Section 1.5

1. positive **3.** negative **5.** positive **7.** 0 **9.** undefined; 0; Examples
include $\frac{1}{0}$, which is undefined, and $\frac{0}{1}$, which equals 0. **11.** −28 **13.** 30

15. 120 **17.** −33 **19.** 0 **21.** −2.38 **23.** $\frac{5}{6}$ **25.** $-\frac{1}{6}$ **27.** 6

29. −32, −16, −8, −4, −2, −1, 1, 2, 4, 8, 16, 32

31. −40, −20, −10, −8, −5, −4, −2, −1, 1, 2, 4, 5, 8, 10, 20, 40

33. −31, −1, 1, 31 **35.** 3 **37.** −7 **39.** 8 **41.** −6 **43.** −4

45. $\frac{32}{3}$, or $10\frac{2}{3}$ **47.** $-\frac{15}{16}$ **49.** 0 **51.** undefined **53.** −11

55. −2 **57.** −2 **59.** 35 **61.** 13 **63.** −22 **65.** 6 **67.** −18

69. 67 **71.** −8 **73.** 18 **75.** 3 **77.** 7 **79.** 4 **81.** −2 **83.** −1

85. 4 **87.** −3 **89.** 29 **91.** 47 **93.** 72 **95.** $-\frac{78}{25}$ **97.** 0

99. −5 **101.** 14 **103.** undefined **105.** 9 + (−9)(2); −9

107. −4 − 2(−1)(6); 8 **109.** 1.5(−3.2) − 9; −13.8

111. 12[9 − (−8)]; 204 **113.** $\frac{-12}{-5 + (-1)}; 2$ **115.** $\frac{15 + (-3)}{4(-3)}; -1$

117. $\frac{2}{5}[8 - (-1)]; 6$ **119.** 0.20(−5 · 6); −6 **121.** $(\frac{1}{2} + \frac{5}{8})(\frac{3}{5} - \frac{1}{3}); \frac{3}{10}$

123. $\frac{-\frac{1}{2}(\frac{3}{4})}{-\frac{2}{3}}; \frac{9}{16}$ **125.** $\frac{x}{3} = -3; -9$ **127.** x − 6 = 4; 10

129. x + 5 = −5; −10 **131. (a)** yes **(b)** no **133. (a)** no
(b) yes **135. (a)** yes **(b)** no **137. (a)** no **(b)** yes **139.** 42
140. 5 **141.** $8\frac{2}{5}$ **142.** $8\frac{2}{5}$ **143.** 4 **144.** 6 **145.** 2 **146.** $-12\frac{1}{2}$

SUMMARY EXERCISES Performing Operations with Real Numbers

1. −16 **2.** 4 **3.** 0 **4.** −24 **5.** −17 **6.** 76 **7.** −18 **8.** 90

9. 38 **10.** 1.02 **11.** 3.33 **12.** 25 **13.** 0 **14.** 8 **15.** −1

16. $\frac{6}{5}$, or $1\frac{1}{5}$ **17.** $\frac{17}{16}$, or $1\frac{1}{16}$ **18.** $-\frac{2}{3}$ **19.** 4 **20.** 5 **21.** −5

22. $\frac{5}{4}$, or $1\frac{1}{4}$ **23.** 9 **24.** $\frac{7}{10}$ **25.** $-\frac{7}{2}$, or $-3\frac{1}{2}$ **26.** 14 **27.** 13

28. undefined **29.** −4 **30.** $\frac{52}{37}$, or $1\frac{15}{37}$ **31.** 0 **32.** 4 **33.** −7

34. −3 **35.** −56 **36.** $\frac{1}{2}$ **37.** $-\frac{5}{13}$ **38.** 5 **39.** −1 **40.** 0

Section 1.6

1. (a) B **(b)** F **(c)** C **(d)** I **(e)** B **(f)** D, F **(g)** B **(h)** A
(i) G **(j)** H **3.** yes **5.** no **7.** no **9.** (large deposit) slip; large
(deposit slip) **11.** Subtraction is not associative. **13.** row 1: $-5, \frac{1}{5}$;
row 2: $10, -\frac{1}{10}$; row 3: $\frac{1}{2}, -2$; row 4: $-\frac{3}{8}, \frac{8}{3}$; row 5: $-x, \frac{1}{x}$; row 6: $y, -\frac{1}{y}$;
opposite; the same **15.** −15; commutative property **17.** 3;
commutative property **19.** 6; associative property **21.** 7; associative
property **23.** commutative property **25.** associative property
27. associative property **29.** inverse property **31.** inverse property
33. identity property **35.** commutative property **37.** distributive
property **39.** identity property **41.** distributive property **43.** 150
45. 2010 **47.** 400 **49.** 1400 **51.** 470 **53.** −9300 **55.** 11 **57.** 0
59. −0.38 **61.** 1 **63.** The student made a sign error in the second line.
The expression −3(4 − 6) means −3(4) − 3(−6), which simplifies to
−12 + 18, or 6. **65.** We must multiply $\frac{3}{4}$ by 1 in the form of a fraction,
$\frac{3}{3}$; $\frac{3}{4} \cdot \frac{3}{3} = \frac{9}{12}$ **67.** 85 **69.** 4t + 12 **71.** 7z − 56 **73.** −8r − 24
75. $-2x - \frac{3}{4}$ **77.** $-3x + \frac{4}{3}$ **79.** 12x + 10 **81.** −6x + 15
83. −4.8x − 0.72 **85.** −16y − 20z **87.** 24r + 32s − 40y
89. −24x − 9y − 12z **91.** −4t − 3m **93.** 5c + 4d **95.** 3q − 5r + 8s

Section 1.7

. B   **3.** C   **5.** The student made a sign error when applying the distributive property: $7x - 2(3 - 2x)$ means $7x - 2(3) - 2(-2x)$, which simplifies to $7x - 6 + 4x$, or $11x - 6$.   **7.** $4r + 11$ **9.** $21x - 28y$ **11.** $5 + 2x - 6y$ **13.** $32 + 12x$ **15.** $-7 + 3p$ **17.** $2 - 3x$ **19.** -12 **21.** 3   **23.** 1   **25.** -1 **27.** $\frac{1}{2}$ **29.** $\frac{2}{5}$ **31.** -0.5 **33.** 10   **35.** like   **37.** unlike   **39.** like   **41.** unlike   **43.** $13y$ **45.** $-9x$ **47.** $13b$ **49.** $7k + 15$ **51.** $-4y$ **53.** $2x + 6$ **55.** $14 - 7m$ **57.** $6.5x - 1.8$ **59.** $9.5x$ **61.** $-\frac{28}{3} - \frac{1}{3}t$ **63.** $9y^2$ **65.** $5p^2 - 14p^3$ **67.** $8x + 15$ **69.** $22 - 4y$ **71.** $-19p + 16$ **73.** $-t + 3$ **75.** $5x + 15$ **77.** $15 - 9x$ **79.** $-16y + 63$ **81.** $4r + 15$ **83.** $12k - 5$ **85.** $-\frac{3}{2}y + 16$ **87.** $-2x + 4$ **89.** $-\frac{14}{3}x - \frac{22}{3}$ **91.** $-23.7y - 12.6$ **93.** $-2k - 3$ **95.** $4x - 7$ **97.** $(4x + 8) + (3x - 2)$; $7x + 6$ **99.** $(5x + 1) - (x - 7)$; $4x + 8$ **101.** $(x + 3) + 5x$; $6x + 3$ **103.** $(13 + 6x) - (-7x)$; $13 + 13x$ **105.** $2(3x + 4) - (-4 + 6x)$; 12   **107.** $1000 + 5x$ (dollars)   **108.** $750 + 3y$ (dollars)   **109.** $1000 + 5x + 750 + 3y$ (dollars)   **110.** $1750 + 5x + 3y$ (dollars)

Chapter 1 Review Exercises

1. 625   **2.** $\frac{27}{125}$ **3.** $\frac{1}{64}$ **4.** 0.001   **5.** 27   **6.** 17   **7.** 4   **8.** 399   **9.** 4   **10.** 5   **11.** true   **12.** true   **13.** false   **14.** $13 < 17$ **15.** $5 + 2 \neq 10$ **16.** $\frac{2}{3} \geq \frac{4}{6}$ **17.** 30   **18.** 60   **19.** 14   **20.** 13   **21.** $x + 6$ **22.** $8 - x$ **23.** $6x - 9$ **24.** $12 + \frac{3}{5}x$ **25.** no   **26.** yes   **27.** $2x - 6 = 10$; 8   **28.** $4x = 8$; 2   **29.**

30.   **31.** rational numbers, real numbers   **32.** rational numbers, real numbers   **33.** natural numbers, whole numbers, integers, rational numbers, real numbers   **34.** irrational numbers, real numbers   **35.** -10 **36.** -9 **37.** $-\frac{3}{4}$ **38.** $-|23|$ **39.** true   **40.** true   **41.** true   **42.** true   **43.** (a) 9   (b) 9   **44.** (a) 0   (b) 0   **45.** (a) -6 (b) 6   **46.** (a) $\frac{5}{7}$ (b) $\frac{5}{7}$ **47.** 12   **48.** -3 **49.** -19 **50.** -7 **51.** -6 **52.** -4 **53.** -17 **54.** $-\frac{29}{36}$ **55.** -21.8 **56.** 14   **57.** -10 **58.** -19 **59.** -11 **60.** -1 **61.** 7   **62.** $-\frac{43}{35}$, or $-1\frac{8}{35}$ **63.** 10.31   **64.** -12 **65.** 10   **66.** -14.22 **67.** $(-31 + 12) + 19$; 0   **68.** $[-4 + (-8)] + 13$; 1   **69.** $-4 - (-6)$; 2   **70.** $[7 - (-5)] - 5$; 7   **71.** \$26.25   **72.** $-10°$F   **73.** $-$\$29   **74.** $-10°$ **75.** 38   **76.** 23,441.76   **77.** 36   **78.** -105 **79.** $\frac{1}{2}$ **80.** 10.08   **81.** -20 **82.** -10 **83.** -24 **84.** -35 **85.** 4   **86.** -20 **87.** $-\frac{3}{4}$ **88.** 11.3   **89.** undefined   **90.** 2   **91.** 1   **92.** 0   **93.** -18 **94.** 125   **95.** -423 **96.** undefined   **97.** $-4(5) - 9$; -29 **98.** $\frac{5}{6}[12 + (-6)]$; 5   **99.** $\frac{12}{8 + (-4)}$; 3   **100.** $\frac{-20(12)}{15 - (-15)}$; -8 **101.** $8x = -24$; -3 **102.** $\frac{x}{3} = -2$; -6 **103.** identity property   **104.** identity property   **105.** inverse property   **106.** inverse property   **107.** associative property   **108.** associative property   **109.** distributive property   **110.** commutative property   **111.** $7y + 14$ **112.** $-48 + 12t$ **113.** $6s + 15y$ **114.** $4r - 5s$ **115.** $11m$ **116.** $16p^2$ **117.** $16p^2 + 2p$ **118.** $-4k + 12$ **119.** $-2m + 29$ **120.** $4x - 3$ **121.** $-2(3x) - 7x$; $-13x$ **122.** $(5 + 4x) + 8x$; $5 + 12x$

Chapter 1 Mixed Review Exercises

1. 3; 3; $-\frac{1}{3}$ **2.** 12; -12; $\frac{1}{12}$ **3.** $-\frac{2}{3}; \frac{2}{3}; \frac{2}{3}$ **4.** 0.2; 0.2; 5   **5.** rational numbers, real numbers   **6.** 37   **7.** $\frac{8}{3}$, or $2\frac{2}{3}$ **8.** $-\frac{1}{24}$ **9.** 2   **10.** $-\frac{28}{15}$, or $-1\frac{13}{15}$ **11.** $-\frac{3}{2}$, or $-1\frac{1}{2}$ **12.** $\frac{25}{36}$ **13.** 16   **14.** 77.6   **15.** 11   **16.** $16t - 36$ **17.** $8x^2 - 21y^2$ **18.** 24   **19.** $-47°$F   **20.** 14,776 ft

Chapter 1 Test

[1.1] **1.** true   **2.** false   [1.3] **3.**

4. rational numbers, real numbers   **5.** $-|-8|$, or -8 **6.** -1.277 [1.5] **7.** $\frac{-6}{2 + (-8)}$; 1   [1.4, 1.5] **8.** 4   **9.** $-\frac{17}{6}$, or $-2\frac{5}{6}$ **10.** 2   **11.** 6   **12.** -20 **13.** 8   **14.** $\frac{30}{7}$, or $4\frac{2}{7}$ **15.** 11   [1.5] **16.** -70 **17.** 3   [1.3–1.5] **18.** 7000 m   **19.** 15   **20.** $-$\$0.61 trillion   [1.6] **21.** C   **22.** D   **23.** A   **24.** E   **25.** B   **26.** C   [1.7] **27.** $21x$ **28.** $-2.8t + 6.7$ **29.** $3x + \frac{1}{2}$ **30.** $15x - 3$

Chapters R and 1 Cumulative Review Exercises

[R.1] **1.** $\frac{5}{28}$ **2.** 6   **3.** $\frac{13}{48}$ [R.2] **4.** 26.78   **5.** 0.015   **6.** 0.0926   [R.1, R.2] **7.** $4\frac{1}{4}$ in.   **8.** \$56.85; \$322.15   [1.1] **9.** $105 \leq 65$; false   **10.** $8 > 2$; true   [1.2] **11.** no   **12.** yes   [1.3] **13.**

[1.4, 1.5] **14.** 0   **15.** -2.61 **16.** $-\frac{11}{8}$, or $-1\frac{3}{8}$ **17.** -2 **18.** 12   **19.** 1   **20.** undefined   [1.6] **21.** inverse property   **22.** associative property   [1.7] **23.** $3t - 5$ **24.** $13x - 27$ [1.4] **25.** 2698 ft

---

2   **LINEAR EQUATIONS AND INEQUALITIES IN ONE VARIABLE**

Section 2.1

1. equation; expression   **3.** equivalent equations   **5.** (a) expression; $x + 15$ (b) expression; $z + 7$ (c) equation; $\{-1\}$ (d) equation; $\{-17\}$ **7.** A, B   **9.** $\{12\}$ **11.** $\{31\}$ **13.** $\{-3\}$ **15.** $\{4\}$ **17.** $\{-9\}$ **19.** $\{6.3\}$ **21.** $\{-16.9\}$ **23.** $\left\{-\frac{3}{4}\right\}$ **25.** $\left\{\frac{1}{12}\right\}$ **27.** $\{-10\}$ **29.** $\{-13\}$ **31.** $\{10\}$ **33.** $\{10.1\}$ **35.** $\left\{\frac{4}{15}\right\}$ **37.** $\{-3\}$ **39.** $\{2\}$ **41.** $\{7\}$ **43.** $\{-4\}$ **45.** $\{-6\}$ **47.** $\{-5\}$ **49.** $\{-2\}$ **51.** $\{3\}$

53. $\{-3\}$ **55.** $\{0\}$ **57.** $\{2\}$ **59.** $\{-16\}$ **61.** $\{0\}$ **63.** $\{9\}$
65. $\{-7\}$ **67.** $\{2\}$ **69.** $\{13\}$ **71.** $\{-4\}$ **73.** $\{0\}$ **75.** $\{0\}$
77. $\left\{\frac{7}{15}\right\}$ **79.** $\{13\}$ **81.** $\{-2\}$ **83.** $\{7\}$ **85.** $\{-4\}$
87. $\{13\}$ **89.** $\{29\}$ **91.** $\{18\}$ **93.** $\{12\}$
95. Answers will vary. One example is $x - 6 = -8$.

Section 2.2

1. (a) multiplication property of equality **(b)** addition property
of equality **(c)** multiplication property of equality **(d)** addition
property of equality **3.** B **5.** $\frac{5}{4}$ **7.** 10 **9.** $-\frac{2}{9}$
11. -1 **13.** 6 **15.** -4 **17.** 0.12 **19.** -1 **21.** $\{6\}$
23. $\left\{\frac{15}{2}\right\}$ **25.** $\{-5\}$ **27.** $\{-4\}$ **29.** $\left\{-\frac{18}{5}\right\}$, or $\{-3.6\}$
31. $\{12\}$ **33.** $\{0\}$ **35.** $\{-12\}$ **37.** $\left\{\frac{3}{4}\right\}$ **39.** $\{40\}$
41. $\{-30\}$ **43.** $\{-2.4\}$ **45.** $\{3.5\}$ **47.** $\{-12.2\}$
49. $\{-48\}$ **51.** $\{72\}$ **53.** $\{-35\}$ **55.** $\{14\}$ **57.** $\{18\}$
59. $\left\{-\frac{27}{35}\right\}$ **61.** $\{3\}$ **63.** $\{-5\}$ **65.** $\{20\}$ **67.** $\{7\}$
69. $\{0\}$ **71.** $\left\{-\frac{3}{5}\right\}$ **73.** $\{18\}$ **75.** $\{-6\}$ **77.** $\{-3\}$
79. $\{-4\}$ **81.** Answers will vary. One example is $\frac{3}{2}x = -6$.

Section 2.3

1. Use the addition property of equality to subtract 8 from each side.
3. Clear parentheses by using the distributive property.
5. (a) identity; B **(b)** conditional; A **(c)** contradiction; C
7. $\{4\}$ **9.** $\{-5\}$ **11.** $\left\{\frac{5}{2}\right\}$ **13.** $\{4\}$ **15.** $\left\{-\frac{1}{2}\right\}$
17. $\{-3\}$ **19.** $\{5\}$ **21.** $\{0\}$ **23.** $\{-1\}$ **25.** $\left\{\frac{4}{3}\right\}$ **27.** $\left\{\frac{3}{2}\right\}$
29. $\left\{-\frac{5}{3}\right\}$ **31.** $\{0\}$ **33.** $\{$all real numbers$\}$ **35.** $\varnothing$ **37.** $\{0\}$
39. $\{5\}$ **41.** $\{8\}$ **43.** $\{0\}$ **45.** $\{$all real numbers$\}$ **47.** $\varnothing$
49. $\{$all real numbers$\}$ **51.** $\varnothing$ **53.** $11 - q$ **55.** $\frac{9}{x}$ **57.** $65 - h$
59. $x + 15; x - 5$ **61.** $25r$ **63.** $\frac{t}{5}$

Section 2.4

1. 8 **3.** 6 **5.** 12 **7.** 30 **9.** 35 **11.** 1 **13.** 72 **15.** 7
17. 10 **19.** 250 **21.** 9 **23.** 480 **25.** A **27.** $\{14\}$ **29.** $\{6\}$
31. $\{-2\}$ **33.** $\{-32\}$ **35.** $\{5\}$ **37.** $\{12\}$ **39.** $\varnothing$
41. $\{$all real numbers$\}$ **43.** $\{9\}$ **45.** $\{11\}$ **47.** $\{4\}$ **49.** $\{0\}$
51. $\{18\}$ **53.** $\{8\}$ **55.** $\{1\}$ **57.** $\{-2\}$ **59.** $\{-10\}$ **61.** $\{3\}$
63. $\left\{\frac{5}{4}\right\}$ **65.** $\{20\}$ **67.** $\varnothing$ **69.** $\{2\}$ **71.** $\{$all real numbers$\}$
73. $\{-15\}$ **75.** $\{120\}$ **77.** $\{6\}$ **79.** $\{15,000\}$

SUMMARY EXERCISES Applying Methods for Solving Linear Equations

1. equation; $\{-5\}$ **2.** expression; $7p - 14$ **3.** expression; $-3m - 2$
4. equation; $\{7\}$ **5.** equation; $\{0\}$ **6.** equation; $\left\{-\frac{96}{5}\right\}$
7. expression; $-10z - 6$ **8.** equation; $\{3\}$ **9.** expression; $-\frac{1}{6}x + 5$
10. expression; $2k - 11$ **11.** $\left\{\frac{7}{3}\right\}$ **12.** $\{4\}$ **13.** $\{-5.1\}$

14. $\{12\}$ **15.** $\{-25\}$ **16.** $\{-6\}$ **17.** $\{-6\}$ **18.** $\{-16\}$
19. $\{$all real numbers$\}$ **20.** $\{23.7\}$ **21.** $\{6\}$ **22.** $\{0\}$
23. $\{7\}$ **24.** $\{1\}$ **25.** $\{5\}$ **26.** $\varnothing$ **27.** $\varnothing$ **28.** $\{-10.8\}$
29. $\{25\}$ **30.** $\{$all real numbers$\}$ **31.** $\{3\}$ **32.** $\{1\}$
33. $\{-2\}$ **34.** $\left\{\frac{14}{17}\right\}$

Section 2.5

1. C; There cannot be a fractional number of cars. **3.** A; Distance
cannot be negative. **5.** 1; 16 (or 14); -7 (or -9); $x + 1$
7. complementary; supplementary **9.** $x + 9 = -26; -35$
11. $8(x + 6) = 104; 7$ **13.** $5x + 2 = 4x + 5; 3$
15. $3(x - 2) = x + 6; 6$ **17.** $\frac{3}{4}x + 6 = x - 4; 40$
19. $3x + (x + 7) = -11 - 2x; -3$ **21.** *Step 1:* Republicans;
Step 2: $x + 14$; Democrats; *Step 3:* $(x + 14) + x = 50$; Democrats: 32;
Republicans: 18 **23.** New York: 28 screens; Ohio: 24 screens
25. Democrats: 46; Republicans: 52 **27.** Beyoncé: \$169 million;
Guns N' Roses: \$131 million **29.** wins: 67; losses: 15 **31.** orange:
97 mg; pineapple: 25 mg **33.** 112 DVDs **35.** onions: 81.3 kg;
grilled steak: 536.3 kg **37.** 1950 Denver nickel: \$16.00; 1945
Philadelphia nickel: \$8.00 **39.** whole wheat: 25.6 oz; rye: 6.4 oz
41. American: 18 tickets; United: 11 tickets; Southwest: 26 tickets
43. shortest piece: 15 in.; middle piece: 20 in.; longest piece: 24 in.
45. gold: 46; silver: 37; bronze: 38 **47.** 36 million mi **49.** *A* and
B: 40°; *C*: 100° **51.** 68, 69 **53.** 101, 102 **55.** 10, 12 **57.** 17, 19
59. 10, 11 **61.** 18 **63.** 15, 17, 19 **65.** 18° **67.** 20° **69.** 39° **71.** 56

Section 2.6

1. The perimeter of a plane geometric figure is the measure of the outer
boundary of the figure. **3.** area **5.** perimeter **7.** area **9.** area
11. $P = 26$ **13.** $\mathcal{A} = 64$ **15.** $b = 4$ **17.** $t = 5.6$ **19.** $B = 14$
21. $r = 2.6$ **23.** $r = 10$ **25.** $\mathcal{A} = 50.24$ **27.** $r = 6$ **29.** $V = 150$
31. $V = 52$ **33.** $V = 7234.56$ **35.** $I = \$600$ **37.** $p = \$550$
39. $t = 1.5$ yr **41.** length: 18 in.; width: 9 in. **43.** length: 14 m;
width: 4 m **45.** shortest: 5 in.; medium: 7 in.; longest: 8 in.
47. two equal sides: 7 m; third side: 10 m **49.** perimeter: 5.4 m;
area: 1.8 m^2 **51.** 10 ft **53.** 154,000 ft^2 **55.** 194.48 ft^2; 49.42 ft
57. 23,800.10 ft^2 **59.** length: 36 in.; volume: 11,664 in.^{3} **61.** 48°, 132°
63. 70°, 110° **65.** 55°, 35° **67.** 30°, 60° **69.** 51°, 51° **71.** 105°, 105°

We give one possible answer for Exercises 73–109. There are other
correct forms.

73. $t = \dfrac{d}{r}$ **75.** $b = \dfrac{\mathcal{A}}{h}$ **77.** $d = \dfrac{C}{\pi}$ **79.** $H = \dfrac{V}{LW}$ **81.** $r = \dfrac{I}{pt}$
83. $h = \dfrac{2\mathcal{A}}{b}$ **85.** $h = \dfrac{3V}{\pi r^2}$ **87.** $b = P - a - c$ **89.** $W = \dfrac{P - 2L}{2}$
91. $m = \dfrac{y - b}{x}$ **93.** $y = \dfrac{C - Ax}{B}$ **95.** $r = \dfrac{M - C}{C}$ **97.** $a = \dfrac{P - 2b}{2}$
99. $b = 2S - a - c$ **101.** $F = \dfrac{9C + 160}{5}$ **103.** $y = -6x + 4$
105. $y = 5x - 2$ **107.** $y = \dfrac{3}{5}x - 3$ **109.** $y = \dfrac{1}{3}x - 4$
111. $W_1 + M = W_2 + N_2 + 1$ **112.** $M = W_2 + N_2 + 1 - W_1$

113. (a) 17 **(b)** 6 **(c)** 5 **114.** $M = -3$; A negative magic number indicates that Oakland has been eliminated from winning the division.

Section 2.7

1. compare; A, D **3. (a)** C **(b)** D **(c)** B **(d)** A **5.** 1 to 100; $\frac{1}{100}$; 0.01 **7.** 75 to 100; $\frac{75}{100}$; 0.75 **9.** 100 to 100; $\frac{100}{100}$; 1 **11.** $\frac{4}{3}$

13. $\frac{18}{55}$ **15.** $\frac{15}{2}$ **17.** $\frac{4}{15}$ **19.** $\frac{5}{6}$ **21.** 10 lb; $0.749 **23.** 64 oz; $0.047

25. 32 oz; $0.531 **27.** 32 oz; $0.056 **29.** 263 oz; $0.076 **31.** true

33. false **35.** true **37.** $\{35\}$ **39.** $\{7\}$ **41.** $\left\{\frac{45}{2}\right\}$ **43.** $\{1\}$

45. $\{2\}$ **47.** $\{-1\}$ **49.** $\{5\}$ **51.** $\left\{-\frac{31}{5}\right\}$ **53.** $\{-2\}$ **55.** $30.00

57. $8.75 **59.** $67.50 **61.** $56.40 **63.** 50,000 fish **65. (a)** 30 min

(b) yes **67.** 4 ft **69.** 2.7 in. **71.** 2.0 in. **73.** $2\frac{5}{8}$ cups

75. $352.17 **77.** $x = 4$ **79.** $x = 8$ **81.** $x = 22.5$; $y = 25.5$

83. (a)

Chair: x, Shadow 18; Pole 12, Shadow 4

(b) 54 ft **85.** $239 **87.** $285

89. (a) 2625 mg

(b) $\frac{125 \text{ mg}}{5 \text{ mL}} = \frac{2625 \text{ mg}}{x \text{ mL}}$ **(c)** 105 mL

91. 140.4 **93.** 700 **95.** 425

97. 8% **99.** 120% **101.** 80%

103. 28% **105.** 32% **107.** $3000 **109.** $15.00 **111.** 30

112. (a) $5x = 12$ **(b)** $\left\{\frac{12}{5}\right\}$ **113.** $\left\{\frac{12}{5}\right\}$ **114.** Both methods give the same solution set.

Section 2.8

1. A **3.** C **5.** D **7.** C **9.** Because the rate is given in miles per *hour*, the time must be in hours. So the distance is $45 \cdot \frac{1}{2} = \frac{45}{2}$, or $22\frac{1}{2}$ mi. **11.** 45 L **13.** $750 **15.** $17.50

17. 160 L **19.** $13\frac{1}{3}$ L **21.** 4 L **23.** 20 mL **25.** 4 L

27. $2100 at 5%; $900 at 4% **29.** $2500 at 6%; $13,500 at 5%

31. $1700 at 8%; $800 at 2% **33.** 10 nickels **35.** 49-cent stamps: 25; 21-cent stamps: 20 **37.** Arabian Mocha: 7 lb; Colombian Decaf: 3.5 lb

39. 530 mi **41.** 3.001 hr **43.** 9.18 m per sec **45.** 8.38 m per sec

47. $7\frac{1}{2}$ hr **49.** 5 hr **51.** $1\frac{3}{4}$ hr **53.** eastbound: 300 mph; westbound: 450 mph **55.** slower car: 40 mph; faster car: 60 mph

57. 75 hr **59.** $2280 **61.** $97.54 **63.** Bob: 7 yr old; Kevin: 21 yr old **65.** width: 3 ft; length: 9 ft

Section 2.9

1. $>$, $<$ (or $<$, $>$); $\geq$, $\leq$ (or $\leq$, $\geq$) **3.** $(0, \infty)$ **5.** $x > -4$ **7.** $x \leq 4$

9.

$(-\infty, 4]$

11.

$(-\infty, -3)$

13.

$(4, \infty)$

15.

$\left[-\frac{1}{2}, \infty\right)$

17.

$(-\infty, 0]$

19.

$(-2, \infty)$

21.

$[1, \infty)$

23.

$[5, \infty)$

25.

$(-\infty, -11)$

27. The inequality symbol must be reversed when multiplying or dividing by a negative number. **29.** When each side of an inequality is divided by a negative number, the direction of the inequality symbol must be reversed. The second and third lines of the solution should have $>$ symbols (instead of $<$) so that the last line is $x > -3$. The solution set is $(-3, \infty)$.

31.

$(-\infty, 6)$

33.

$[-10, \infty)$

35.

$(-\infty, -3)$

37.

$(-\infty, 0]$

39.

$(20, \infty)$

41.

$[-3, \infty)$

43.

$(-\infty, -3]$

45.

$(-1, \infty)$

47.

$[-5, \infty)$

49.

$(-\infty, 2]$

51.

$\left(-\infty, \frac{3}{2}\right)$

53.

$(-\infty, 1)$

55.

$(-\infty, 0]$

57.

$\left(-\frac{1}{2}, \infty\right)$

59.

$[4, \infty)$

61.

$[2, \infty)$

63.

$(-\infty, 32)$

65.

$(-\infty, 6]$

67.

$\left[\frac{5}{12}, \infty\right)$

69.

$(-21, \infty)$

71. $x \geq 18$ **73.** $x > 5$ **75.** $x \leq 20$ **77.** 83 or more

79. 80 or more **81.** more than 3.8 in. **83.** all numbers greater than 16

85. It is never less than $-13°$F. **87.** 32 or greater **89.** 12 min

91. $5x - 100$ **93.** $(5x - 100) - (125 + 4x)$, which simplifies to $x - 225$; $x > 225$ **95.** $-1 < x < 2$ **97.** $-1 < x \leq 2$

99.

$[8, 10]$

101.

$(0, 10]$

103.

$(-3, 4)$

105.

$(-2, 1]$

107.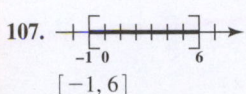
$[-1, 6]$

109.
$(1, 3)$

111.
$[3, 7)$

113.
$[-26, 6]$

115.
$[-3, 6]$

117.
$\left[-\frac{24}{5}, 0\right]$

119.
$(-6, 2)$

121.
$\left(-\frac{11}{6}, -\frac{2}{3}\right)$

123.
$\{4\}$

124.
$(4, \infty)$

125.
$(-\infty, 4)$

126. The graph would be the set of all real numbers.

Chapter 2 Review Exercises

1. $\{6\}$ **2.** $\{-12\}$ **3.** $\{7\}$ **4.** $\left\{\frac{2}{3}\right\}$ **5.** $\{11\}$ **6.** $\{17\}$
7. $\{5\}$ **8.** $\{-4\}$ **9.** $\{5\}$ **10.** $\{-12\}$ **11.** $\{15\}$ **12.** $\{4\}$
13. $\{$all real numbers$\}$ **14.** $\{-19\}$ **15.** $\{0\}$ **16.** $\{20\}$
17. $\varnothing$ **18.** $\{-1\}$ **19.** $-\frac{7}{2}$ **20.** Democrats: 67; Republicans: 51
21. Hawaii: 6425 mi^2; Rhode Island: 1212 mi^2 **22.** Seven Falls: 300 ft; Twin Falls: 120 ft **23.** oil: 4 oz; gasoline: 128 oz **24.** 11, 13
25. shortest piece: 13 in.; middle-sized piece: 20 in.; longest piece: 39 in.
26. 80° **27.** $h = 11$ **28.** $\mathcal{A} = 28$ **29.** $r = 4.75$ **30.** $V = 904.32$
31. 2 cm **32.** 42.2°; 92.8° **33.** 135°; 45° **34.** 100°; 100°

35. $h = \dfrac{\mathcal{A}}{b}$ **36.** $h = \dfrac{2\mathcal{A}}{b + B}$

For Exercises 37 and 38, there are other correct forms.

37. $y = -x + 11$ **38.** $y = \frac{3}{2}x - 6$ **39.** $\frac{3}{2}$ **40.** $\frac{5}{14}$ **41.** $\frac{3}{4}$
42. $\left\{\frac{7}{2}\right\}$ **43.** $\left\{-\frac{8}{3}\right\}$ **44.** $\left\{\frac{25}{19}\right\}$ **45.** $3.06 **46.** 375 km
47. 8 gold medals **48.** 18 oz; $0.249 **49.** 175% **50.** 2500
51. 3.75 L **52.** $5000 at 5%; $5000 at 3% **53.** 8.2 mph
54. $2\frac{1}{2}$ hr

55.
$[-4, \infty)$

56.
$(-\infty, 7)$

57.
$[-5, 6)$

58. B

59.
$[-3, \infty)$

60.
$(-\infty, 2)$

61.
$[3, \infty)$

62.
$[46, \infty)$

63.
$(-\infty, -5)$

64.
$(-\infty, -4)$

65.
$\left[-2, \frac{3}{2}\right]$

66.
$\left(\frac{4}{3}, 5\right]$

67. 88 or more **68.** all numbers less than or equal to $-\frac{1}{3}$

Chapter 2 Mixed Review Exercises

1. $\{7\}$ **2.** $r = \dfrac{I}{pt}$ **3.** $(-\infty, 2)$ **4.** $\{-9\}$ **5.** $\{70\}$ **6.** $\left\{\frac{13}{4}\right\}$
7. $\varnothing$ **8.** $\{$all real numbers$\}$ **9.** 4000 calories **10.** DiGiorno: $1014.6 million; Red Baron: $572.3 million **11.** 160 oz; $0.062
12. 24°, 66° **13.** 13 hr **14.** faster train: 80 mph; slower train: 50 mph
15. 44 m **16.** 50 m or less

Chapter 2 Test

[2.1–2.4] **1.** $\{-6\}$ **2.** $\{21\}$ **3.** $\varnothing$ **4.** $\{30\}$ **5.** $\{$all real numbers$\}$
6. $\left\{\frac{13}{4}\right\}$ [2.5] **7.** wins: 104; losses: 58 **8.** Hawaii: 4021 mi^2; Maui: 728 mi^2; Kauai: 551 mi^2 **9.** 50° **10.** 24, 26

[2.6] **11. (a)** $W = \dfrac{P - 2L}{2}$ **(b)** 18 **12.** $y = \frac{5}{4}x - 2$ (There are other correct forms.) **13.** 100°, 80° **14.** 75°, 75° [2.7] **15.** $\{6\}$

16. $\{-29\}$ **17.** 40% **18.** $264 **19.** 16 oz; $0.249 **20.** 2300 mi
[2.8] **21.** $8000 at 3%; $14,000 at 4.5% **22.** 4 hr
[2.9] **23. (a)** $x < 0$ **(b)** $-2 < x \le 3$

24.
$(-\infty, 11)$

25.
$[-3, \infty)$

26.
$(-\infty, 4]$

27.
$(-2, 6]$

28. 83 or more

Chapters R–2 Cumulative Review Exercises

[R.1] **1.** $\frac{3}{8}$ **2.** $\frac{3}{4}$ **3.** $\frac{31}{20}$ **4.** $\frac{551}{40}$, or $13\frac{31}{40}$ **5.** 6 **6.** $\frac{6}{5}$ [R.2] **7.** 21.53
8. 27.31 **9.** 30.51 **10.** 56.3 [1.4, 1.5] **11.** 1 **12.** 28 **13.** 1
14. 0 [1.2–1.5] **15.** $-\frac{19}{3}$ [1.6] **16.** distributive property
17. inverse property **18.** identity property [1.7] **19.** $m - 2$ **20.** $-p$
[2.1–2.4] **21.** $\{-1\}$ **22.** $\{-1\}$ **23.** $\{-12\}$ [2.7] **24.** $\{26\}$
[2.6] **25.** $y = -\frac{3}{4}x + 6$ (There are other correct forms.)
26. $c = P - a - b - B$

[2.9] **27.**
$(-\infty, 1]$

28.
$(-1, 2]$

[2.5] **29.** 4 cm; 9 cm; 27 cm [2.6] **30.** 12.42 cm [2.8] **31.** slower car: 40 mph; faster car: 60 mph [R.2] **32. (a)** 441,000 white cars **(b)** 399,000 silver cars **(c)** 273,000 red cars

3   **LINEAR EQUATIONS IN TWO VARIABLES**

Section 3.1

1. does; do not   **3.** II   **5.** 3   **7.** negative; negative   **9.** positive;
negative   **11.** If $xy < 0$, then either $x < 0$ and $y > 0$ or $x > 0$ and $y < 0$.
If $x < 0$ and $y > 0$, then the point lies in quadrant II. If $x > 0$ and $y < 0$,
then the point lies in quadrant IV.   **13.** between 2009 and 2010
15. 2011: 9%; 2012: 8%; decline: 1%   **17.** yes   **19.** yes   **21.** no
23. yes   **25.** yes   **27.** no   **29.** 17   **31.** -5 **33.** -4 **35.** -7
37. $8; 6; 3; (0, 8); (6, 0); (3, 4)$ **39.** $-9; 4; 9; (-9, 0); (0, 4); (9, 8)$
41. $12; 12; 12; (12, 3); (12, 8); (12, 0)$ **43.** $-10; -10; -10;$
$(4, -10); (0, -10); (-4, -10)$ **45.** $2; 2; 2; (9, 2); (2, 2); (0, 2)$
47. $-4; -4; -4; (-4, 4); (-4, 0); (-4, -4)$ **49.** The ordered pair
$(3, 4)$ represents the point 3 units to the right of the origin and 4 units up
from the x-axis. The ordered pair $(4, 3)$ represents the point 4 units to the
right of the origin and 3 units up from the x-axis.   **51.** $(2, 4)$; I
53. $(-5, 4)$; II   **55.** $(3, 0)$; no quadrant   **57.** $(4, -4)$; IV

59.–69. 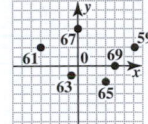   **71.** $-3; 6; -2; 4$ 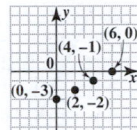   **73.** $-3; 4; -6; -\dfrac{4}{3}$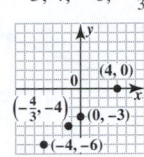

75. $-4; -4; -4; -4$ 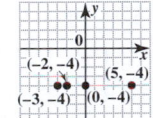   **77.** The points in each graph appear to lie on
a straight line.
79. **(a)** $(5, 95)$ **(b)** $(6, 110)$

81. **(a)** $(2010, 608), (2011, 845), (2012, 1056), (2013, 1228), (2014, 1393),$
$(2015, 1591), (2016, 1860)$ **(b)** $(2016, 1860)$ means that the worldwide number of monthly active Facebook users was 1860 million at the
end of the year 2016.

(c) 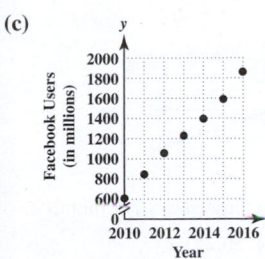   **(d)** The points lie approximately in a
linear pattern. The worldwide number
of monthly active Facebook users was
steadily increasing.

83. **(a)** $130; 117; 104; 91$ **(b)** $(20, 130), (40, 117), (60, 104), (80, 91)$
(c) 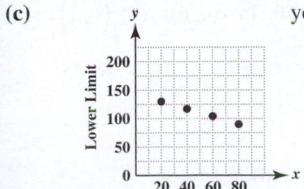   yes   **85.** between 130 and 170 beats per
minute; between 117 and 153 beats
per minute

Section 3.2

1. By; C; 0   **3.** **(a)** A   **(b)** C   **(c)** D   **(d)** B   **5.** x-intercept:
$(4, 0)$; y-intercept: $(0, -4)$ **7.** x-intercept: $(-2, 0)$; y-intercept: $(0, -3)$
9. **(a)** D   **(b)** C   **(c)** B   **(d)** A

11. $4; 4; 2$ 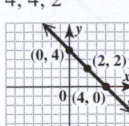   **13.** $1; 3; -1$ 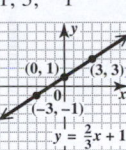   **15.** $-6; -2; -5$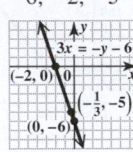

17. $(8, 0); (0, -8)$ **19.** $(4, 0); (0, -10)$ **21.** $(0, 0); (0, 0)$ **23.** $(2, 0);$
$(0, 4)$ **25.** $(6, 0); (0, -2)$ **27.** $(0, 0); (0, 0)$ **29.** $(4, 0)$; no y-intercept
31. no x-intercept; $(0, 2.5)$

33. 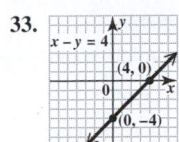   **35.**   **37.**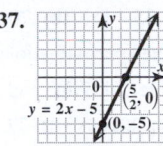

39. 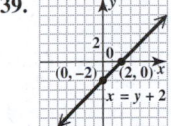   **41.** 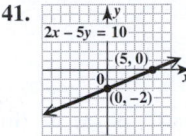   **43.**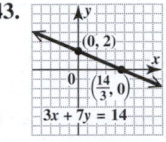

45. 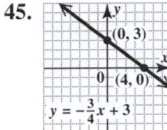   **47.** 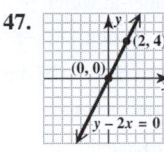   **49.**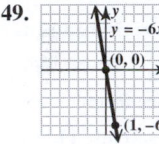

51. 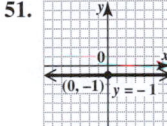   **53.** 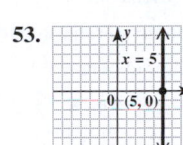   **55.**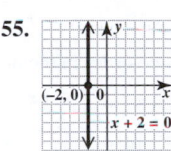

57. 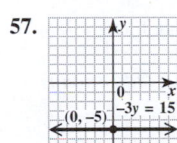   **59.** 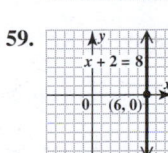   **61.**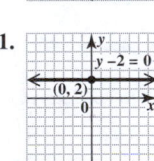

In Exercises 63–69, descriptions may vary.
63. The graph is a line with x-intercept $(-3, 0)$ and y-intercept $(0, 9)$.
65. The graph is a vertical line with x-intercept $(11, 0)$.   **67.** The
graph is a horizontal line with y-intercept $(0, -2)$.   **69.** The graph has
x- and y-intercepts $(0, 0)$. It passes through the points $(2, 1)$ and $(4, 2)$.

71. 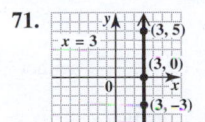 $x = 3$ **73.** 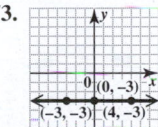 $y = -3$

75. **(a)** 121 lb, 143 lb, 176 lb   **77.** **(a)** \$62.50; \$100   **(b)** 200
(b) (62, 121), (66, 143),   **(c)** (50, 62.50), (100, 100),
(72, 176)   (200, 175)
(c)   **(d)**

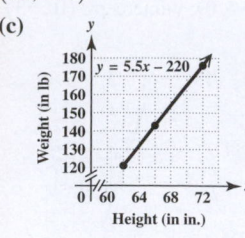   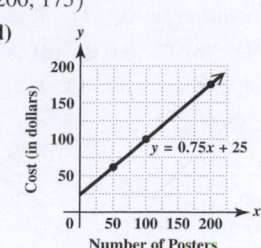

(d) 68 in.; 68 in.

79. **(a)** \$30,000   **(b)** \$15,000   **(c)** \$5000   **(d)** After 5 yr, the SUV
has a value of \$5000.   **81.** **(a)** 2000: 29.7 lb; 2008: 32.5 lb;
2016: 35.4 lb   **(b)** 2000: 29.8 lb; 2008: 32.4 lb; 2016: 36.6 lb
(c) The approximations using the equation are very close to the graph
estimates for 2000 and 2008. The approximation using the equation is
low for 2016.   **(d)** 37.5 lb; The approximation is 0.7 lb more than the
USDA projection.

Section 3.3

1. steepness; vertical; horizontal   **3.** **(a)** 6   **(b)** 4   **(c)** $\frac{6}{4}$, or $\frac{3}{2}$; slope of
the line   **(d)** Yes, it doesn't matter at which point we start. The slope
would be expressed as the ratio (quotient) of -6 and -4, which simpli-
fies to $\frac{3}{2}$.   **5.** **(a)** falls from left to right   **(b)** horizontal   **(c)** vertical
(d) rises from left to right   **7.** **(a)** negative   **(b)** zero   **9.** **(a)** positive
(b) negative   **11.** **(a)** zero   **(b)** negative   **13.** Because he found the
difference $3 - 5 = -2$ in the numerator, he should have subtracted in
the same order in the denominator to obtain $-1 - 2 = -3$. The correct
slope is $\frac{-2}{-3} = \frac{2}{3}$.   **15.** $\frac{8}{27}$ **17.** $-\frac{2}{3}$ **19.** 4   **21.** $-\frac{1}{2}$ **23.** 0   **25.** $\frac{5}{4}$
27. $\frac{3}{2}$ **29.** 0   **31.** -3 **33.** undefined   **35.** $\frac{1}{4}$ **37.** $-\frac{1}{2}$ **39.** 5
41. $\frac{1}{4}$ **43.** $\frac{3}{2}$ **45.** $\frac{3}{2}$ **47.** -1 **49.** $\frac{1}{2}$ **51.** 0   **53.** undefined

*In part (a) of Exercises 55 and 57, we used the intercepts. Other points
can be used.*

55. **(a)** (5, 0) and (0, 10); -2 **(b)** $y = -2x + 10$; -2 **57.** **(a)** (3, 0)
and (0, -5); $\frac{5}{3}$ **(b)** $y = \frac{5}{3}x - 5$; $\frac{5}{3}$
59. **(a)** 1   **(b)** $(-4, 0)$; (0, 4)   **61.** **(a)** $-\frac{1}{3}$ **(b)** $(-6, 0)$; (0, -2)
(c)   **(c)**

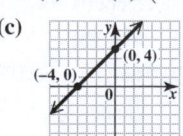   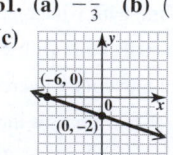

63. **(a)** -3 **(b)** $\frac{1}{3}$ **65.** A   **67.** $-\frac{2}{5}$; $-\frac{2}{5}$; parallel   **69.** $\frac{8}{9}$; $-\frac{4}{3}$; neither
71. $\frac{3}{5}$; $-\frac{5}{3}$; perpendicular   **73.** $\frac{4}{3}$; $\frac{4}{3}$; parallel   **75.** 5; $\frac{1}{5}$; neither
77. $\frac{2}{5}$; $-\frac{5}{2}$; perpendicular   **79.** **(a)** (2010, 19), (2016, 175)   **(b)** 26
(c) Shipments of tablets increased by 156 million units in 6 yr, or
26 million units per year.   **81.** 0.57   **82.** positive; increased
83. 0.57%   **84.** -0.18 **85.** negative; decreased   **86.** 0.18%

Section 3.4

1. m; $(0, b)$ **3.** **(a)** C   **(b)** B   **(c)** A   **(d)** D   **5.** y-axis
7. slope: $\frac{5}{2}$; y-intercept: (0, -4)   **9.** slope: -1; y-intercept: (0, 9)
11. slope: $\frac{1}{5}$; y-intercept: $\left(0, -\frac{3}{10}\right)$

13. 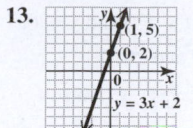   **15.** 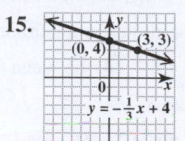   **17.**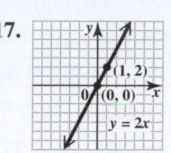

19. 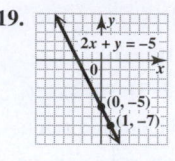   **21.** 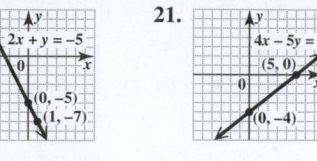   **23.**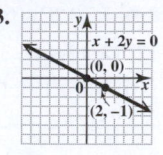

25. 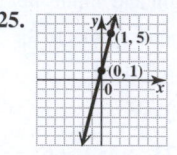   **27.** 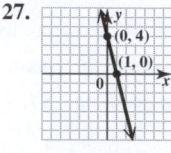   **29.**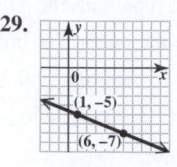

31. 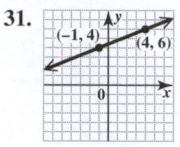   **33.** 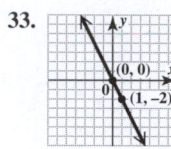   **35.**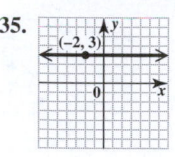

37. 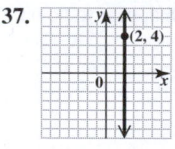   **39.** 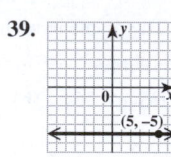   **41.**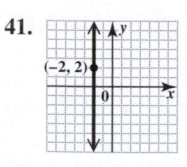

43. $y = 3x - 3$ **45.** $y = -x + 3$ **47.** $y = -\frac{1}{2}x + 2$
49. $y = 4x - 3$ **51.** $y = -x - 7$ **53.** $y = 2x - 7$ **55.** $y = -4x - 1$
57. $y = x - 6$ **59.** $y = \frac{3}{4}x + 4$ **61.** $y = 3$ **63.** $x = 2$ **65.** $x = 0$
67. $y = -6$
69. **(a)** 2   **(b)** (0, -1)   **71.** **(a)** $-\frac{1}{3}$ **(b)** (0, -2)
(c) $y = 2x - 1$ **(c)** $y = -\frac{1}{3}x - 2$
(d)   **(d)**

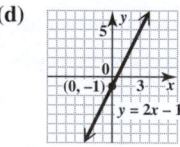   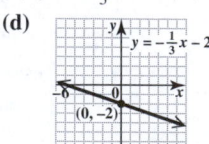

73. **(a)** 0.05; commission rate   **(b)** (0, 2000); base salary per month
(c) \$2500   **(d)** \$30,000   **75.** **(a)** \$400   **(b)** \$0.25
(c) $y = 0.25x + 400$ **(d)** \$425   **(e)** 1500   **77.** $y = -\frac{A}{B}x + \frac{C}{B}$
78. **(a)** $-\frac{2}{3}$ **(b)** 2   **(c)** $\frac{3}{7}$ **79.** $\left(0, \frac{C}{B}\right)$ **80.** **(a)** (0, 6)   **(b)** $\left(0, \frac{1}{2}\right)$
(c) (0, -3)

Section 3.5

1. m; (x_1, y_1) **3.** **(a)** D   **(b)** C   **(c)** B   **(d)** E   **(e)** A
5. A, B, D   **7.** $y = 5x + 2$ **9.** $y = x - 9$ **11.** $y = -3x - 4$
13. $y = -x + 1$ **15.** $y = \frac{2}{3}x + \frac{19}{3}$ **17.** $y = -\frac{4}{5}x + \frac{9}{5}$

19. In this text, an equation written in standard form has A, the coefficient

f x, greater than or equal to 0. In standard form, this equation would be

written $2x - y = -4$. **21.** (a) $y = x + 6$ (b) $x - y = -6$

23. (a) $y = \frac{1}{2}x + 2$ (b) $x - 2y = -4$ **25.** (a) $y = -\frac{3}{5}x - \frac{11}{5}$

(b) $3x + 5y = -11$ **27.** (a) $y = -\frac{9}{8}x - \frac{11}{8}$ (b) $9x + 8y = -11$

29. (a) $y = -\frac{1}{3}x + \frac{22}{9}$ (b) $3x + 9y = 22$ **31.** (a) $y = \frac{1}{3}x + \frac{4}{3}$

(b) $x - 3y = -4$ **33.** (a) $y = 3x - 9$ (b) $3x - y = 9$

35. (a) $y = -\frac{2}{3}x + \frac{4}{3}$ (b) $2x + 3y = 4$ **37.** (a) $y = \frac{5}{7}x + \frac{1}{7}$

(b) $5x - 7y = -1$ **39.** $y = 5x - 2$ **41.** $y = -2x - 3$

43. $y = 4x - 5$ **45.** $y = -3x + 14$ **47.** (a) $(1, 3310)$, $(2, 3340)$,

$(3, 3370)$, $(4, 3460)$, $(5, 3520)$

(b) 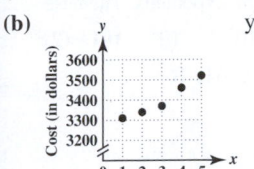 yes **(c)** $y = 60x + 3220$

(d) $\$3580$ $(x = 6)$

49. $y = 2.25x - 0.45$

51. $y = 0.2375x + 59.7$

53. $(0, 32)$; $(100, 212)$ **54.** $\frac{9}{5}$ **55.** $F - 32 = \frac{9}{5}(C - 0)$

56. $F = \frac{9}{5}C + 32$ **57.** $C = \frac{5}{9}(F - 32)$ **58.** $86°$

59. $10°$ **60.** $-40°$

SUMMARY EXERCISES Applying Graphing and Equation-Writing Techniques for Lines

1. (a) B (b) D (c) A (d) C **2.** A, B

3. 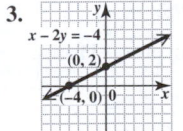 **4.** 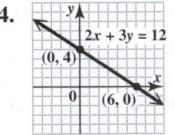 **5.**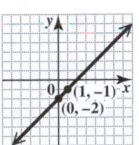

6. 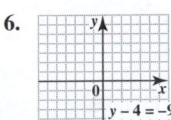 **7.** 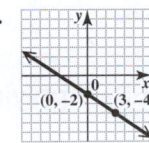 **8.**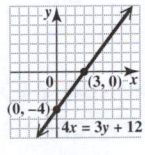

9. 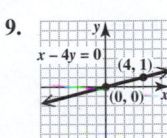 **10.** 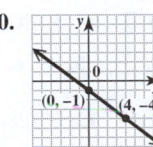 **11.**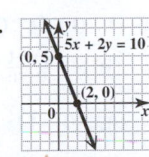

12. 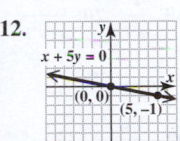 **13.** 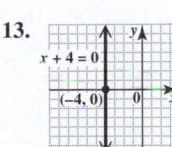 **14.**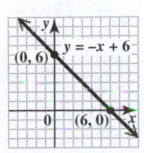

15. $y = -2x + 6$ **16.** $y = \frac{4}{3}x + 8$ **17.** $y = \frac{1}{2}x - 2$ **18.** $y = -\frac{2}{3}x + 5$

19. $y = -3x - 6$ **20.** $y = \frac{3}{2}x + 12$ **21.** $y = -4x - 3$ **22.** $x = 0$

23. $y = \frac{2}{3}x$ **24.** $y = -x - 4$ **25.** $y = x - 5$ **26.** $y = 0$

27. $y = \frac{5}{3}x + 5$ **28.** $y = -5x - 8$

Chapter 3 Review Exercises

1. from 2012 to 2013 and from 2015 to 2016 **2.** from 2014 to 2015

3. 2013: about 33 million trees; 2014: about 26 million trees

4. about 7 million trees **5.** -1; 2; 1 **6.** 2; $\frac{3}{2}$; $\frac{14}{3}$ **7.** 0; $\frac{8}{3}$; -9

8. 7; 7; 7 **9.** yes **10.** no **11.** yes **12.** no **13.** I **14.** II

15. none **16.** none **17.** $\left(-\frac{5}{2}, 0\right)$; $(0, 5)$

Graph for Exercises 13–16

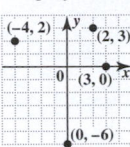 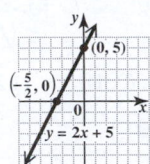

18. $\left(\frac{8}{3}, 0\right)$; $(0, 4)$ **19.** $(-4, 0)$; $(0, -2)$ **20.** $(-6, 0)$; no y-intercept

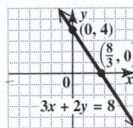 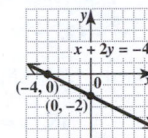 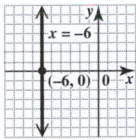

21. $-\frac{1}{2}$ **22.** undefined **23.** 3 **24.** 0 **25.** undefined

26. $\frac{3}{2}$ **27.** $-\frac{1}{3}$ **28.** $\frac{3}{2}$ **29.** (a) 2 (b) $\frac{1}{3}$ **30.** parallel

31. perpendicular **32.** neither **33.** $y = -x + \frac{2}{3}$ **34.** $y = -\frac{1}{2}x + 4$

35. $y = x - 7$ **36.** $y = \frac{2}{3}x + \frac{14}{3}$ **37.** $y = -\frac{3}{4}x - \frac{1}{4}$ **38.** $y = -\frac{1}{4}x + \frac{3}{2}$

39. $y = 1$ **40.** $x = -4$ **41.** $y = \frac{3}{2}x - 2$ **42.** $y = -\frac{1}{3}x + 1$

Chapter 3 Mixed Review Exercises

1. A **2.** C, D **3.** A, B, D **4.** D **5.** C **6.** B

7. $\left(-\frac{5}{2}, 0\right)$; $(0, -5)$; -2 **8.** $(0, 0)$; $(0, 0)$; $-\frac{1}{3}$

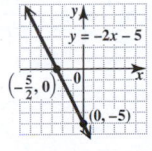 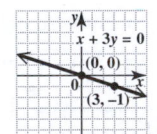

9. no x-intercept; $(0, 5)$; 0 **10.** $(-5, 0)$; no y-intercept; undefined slope

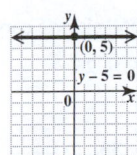 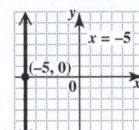

11. (a) $y = -\frac{1}{4}x - \frac{5}{4}$ (b) $x + 4y = -5$ **12.** (a) $y = -3x + 30$

(b) $3x + y = 30$ **13.** (a) $y = -\frac{4}{7}x - \frac{23}{7}$ (b) $4x + 7y = -23$

14. (a) $y = -5$ (b) $y = -5$ **15.** Because the graph rises from left to

right, the slope is positive. **16.** $(0, 81)$, $(6, 467)$ **17.** $y = 64.3x + 81$

18. 338 million members $(x = 4)$; It is a little high, as we might expect.

The actual data point lies slightly below the graph of the line.

Chapter 3 Test

[3.1] **1.** $-6; -10; -5$ **2.** no [3.2] **3.** To find the x-intercept, let $y = 0$, and to find the y-intercept, let $x = 0$. **4.** true

5. $(2, 0); (0, 6)$ **6.** $(0, 0); (0, 0)$ **7.** $(-3, 0)$; no y-intercept

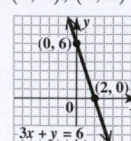

8. no x-intercept; $(0, 1)$ [3.2, 3.3] **9.** 1; $(0, -4)$

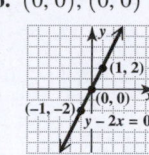

10.

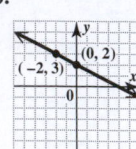

[3.3] **11.** $-\frac{8}{3}$ **12.** -2 **13.** $\frac{5}{2}$ **14.** 0

15. undefined [3.4, 3.5] **16.** $y = 2x + 6$

17. $y = \frac{5}{2}x - 4$ **18.** $y = -9x + 12$

19. $y = -\frac{3}{2}x + \frac{9}{2}$ [3.1, 3.5] **20.** The slope is negative because sales are decreasing.

21. (a) -5.3 (b) $y = -5.3x + 209$ (c) 124.2 thousand; The equation gives an approximation that is a little low. **22.** In 2015, worldwide snowmobile sales were 129.5 thousand.

Chapters R–3 Cumulative Review Exercises

[R.2] **1.** (a) 135.26 (b) 135.3 (c) 135 (d) 140 [R.1] **2.** $\frac{19}{18}$, or $1\frac{1}{18}$ **3.** $\frac{301}{40}$, or $7\frac{21}{40}$ **4.** 6 [R.2] **5.** 10.39 **6.** 0.016 **7.** 5.3026 [1.4] **8.** 7 [1.5] **9.** 13 **10.** $\frac{73}{18}$, or $4\frac{1}{18}$ [1.1–1.5] **11.** true **12.** -43 [1.6] **13.** distributive property [1.7] **14.** $-p + 2$ [2.6] **15.** $h = \frac{3V}{\pi r^2}$

[2.3] **16.** $\{-1\}$ **17.** $\{2\}$ [2.7] **18.** $\{-13\}$

[2.9] **19.** $(-2.6, \infty)$

20. $(0, \infty)$ **21.** $(-\infty, -4]$

[2.6] **22.** (a) $132°, 48°$ (b) $94°, 94°$ [R.2] **23.** 644,550 electric cars **24.** 120,853 electric cars [3.2] **25.** $(-4, 0); (0, 3)$ [3.3] **26.** $\frac{3}{4}$

[3.2] **27.**

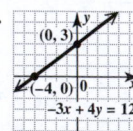

[3.3] **28.** perpendicular
[3.4, 3.5] **29.** $y = 3x - 11$ **30.** $y = 4$

---

4 EXPONENTS AND POLYNOMIALS

Section 4.1

1. false; $3^3 = 3 \cdot 3 \cdot 3 = 27$ **3.** false; $(a^2)^3 = a^{2 \cdot 3} = a^6$ **5.** false; $-2^2 = -(2 \cdot 2) = -4$ **7.** false; $(3x)^2 = 3^2 \cdot x^2 = 9x^2$ **9.** w^6 **11.** $\left(\frac{1}{2}\right)^6$ **13.** $(-4)^4$ **15.** $(-7y)^4$ **17.** In $(-3)^4$, -3 is the base. In -3^4, 3 is the base. $(-3)^4 = 81; -3^4 = -81$ **19.** base: 3; exponent: 5; 243 **21.** base: -3; exponent: 5; -243

23. base: -6; exponent: 2; 36 **25.** base: 6; exponent: 2; -36 **27.** base: $-6x$; exponent: 4 **29.** base: x; exponent: 4 **31.** 2; 5; 8^7 **33.** 5^8 **35.** 4^{12} **37.** t^{24} **39.** $-56r^7$ **41.** $42p^{10}$ **43.** $-30x^9$ **45.** $-30x^{11}$ **47.** The product rule does not apply. **49.** The product rule does not apply. **51.** 4^6 **53.** t^{20} **55.** 7^3r^3 **57.** $5^5x^5y^5$ **59.** 5^4 **61.** -5^5 **63.** $8q^3r^3$ **65.** $\frac{9^8}{5^8}$ **67.** $\frac{1}{8}$ **69.** $\frac{a^3}{b^3}$ **71.** $\frac{x^3}{64}$ **73.** $-\frac{32x^5}{y^5}$ **75.** $-8x^6y^3$ **77.** $9a^6b^4$ **79.** (a) negative (b) positive (c) positive (d) negative **81.** $\frac{5^5}{2^5}$ **83.** $\frac{9^5}{8^3}$ **85.** $2^{12}x^{12}$ **87.** $6^5x^{10}y^{15}$ **89.** x^{21} **91.** $4w^4x^{26}y^7$ **93.** $-r^{18}s^{17}$ **95.** $\frac{125a^6b^{15}}{c^{18}}$ **97.** (a) false (b) true (c) false (d) true **99.** Using power rule (a), multiply the exponents. The base stays the same. Simplify as follows: $(10^2)^3 = 10^{2 \cdot 3} = 10^6$. **101.** $12x^5$ **103.** $6p^7$ **105.** $125x^6$ **107.** \$304.16 **109.** \$1640.16

Section 4.2

1. negative **3.** negative **5.** positive **7.** 0 **9.** 11; 8; 3; 125 **11.** 1 **13.** 1 **15.** -1 **17.** -1 **19.** 0 **21.** 1 **23.** 0 **25.** 0 **27.** (a) B (b) C (c) D (d) B (e) E (f) B **29.** 2 **31.** $\frac{1}{64}$ **33.** 16 **35.** $\frac{49}{36}$ **37.** $\frac{1}{81}$ **39.** 3 **41.** $\frac{8}{15}$ **43.** $-\frac{7}{18}$ **45.** $\frac{7}{2}$ **47.** 125 **49.** $\frac{1}{9}$ **51.** $\frac{125}{9}$ **53.** 25 **55.** $\frac{1}{6^5}$ **57.** x^{15} **59.** 216 **61.** $2r^4$ **63.** $\frac{25}{64}$ **65.** $\frac{p^5}{q^8}$ **67.** r^9 **69.** $\frac{x^5}{6}$ **71.** $\frac{yz^2}{4x^3}$ **73.** $a + b$ **75.** $(x + 2y)^2$ **77.** 343 **79.** $\frac{1}{x^2}$ **81.** $\frac{64x}{9}$ **83.** $\frac{x^2z^4}{y^2}$ **85.** $6x$ **87.** $\frac{1}{m^{10}n^5}$ **89.** $\frac{1}{xyz}$ **91.** x^3y^9 **93.** $\frac{1}{2r}$ **95.** $-\frac{1}{4y}$ **97.** The student attempted to use the quotient rule with unequal bases. The correct way to simplify is $\frac{16^3}{2^2} = \frac{(2^4)^3}{2^2} = \frac{2^{12}}{2^2} = 2^{10} = 1024$. **99.** $\frac{a^{11}}{2b^5}$ **101.** $\frac{108}{y^5z^3}$ **103.** $\frac{9z^2}{400x^3}$

SUMMARY EXERCISES Applying the Rules for Exponents

1. (a) D (b) D (c) E (d) B (e) J (f) F (g) I (h) B (i) E (j) F **2.** $\frac{11}{30}$ **3.** 0 **4.** $\frac{61}{900}$ **5.** 1 **6.** -1 **7.** 0 **8.** $\frac{7}{24}$ **9.** 0 **10.** -1 **11.** $100,000x^7y^{14}$ **12.** $-128a^{10}b^{15}c^4$ **13.** $\frac{729w^3x^9}{y^{12}}$ **14.** $\frac{x^4y^6}{16}$ **15.** c^{22} **16.** $\frac{1}{k^4t^{12}}$ **17.** $y^{12}z^3$ **18.** $\frac{x^6}{y^5}$ **19.** $\frac{1}{z^2}$ **20.** $\frac{9}{r^2s^2t^{10}}$ **21.** $\frac{300x^3}{y^3}$ **22.** $\frac{3}{5x^6}$ **23.** x^8 **24.** $\frac{y^{11}}{x^{11}}$ **25.** $\frac{a^6}{b^4}$ **26.** $6ab$ **27.** 1 **28.** $\frac{343a^6b^9}{8}$ **29.** $\frac{27y^{18}}{4x^8}$ **30.** $\frac{1}{a^8b^{12}c^{16}}$ **31.** $\frac{x^{15}}{216z^9}$ **32.** $\frac{q}{8p^6r^3}$ **33.** x^6y^6 **34.** $\frac{343}{x^{15}}$ **35.** $\frac{9}{x^6}$ **36.** $5p^{10}q^9$ **37.** $\frac{r^{14}t}{2s^2}$ **38.** 1 **39.** $8p^{10}q$ **40.** $\frac{1}{mn^3p^3}$

Section 4.3

1. (a) C (b) A (c) B (d) D **3.** in scientific notation **5.** not in scientific notation; 5.6×10^6 **7.** not in scientific notation; 8×10^1

9. not in scientific notation; 4×10^{-3}

11. (a) $6; 4; 6.3; 4$ **(b)** $5; 2; 5.71; -2$ **13.** 5.876×10^9

15. 8.235×10^4 **17.** 7×10^{-6} **19.** 2.03×10^{-3} **21.** -1.3×10^7

23. -6×10^{-3} **25.** $750,000$ **27.** $5,677,000,000,000$

29. $1,000,000,000,000$ **31.** 6.21 **33.** 0.00078 **35.** 0.000000005134

37. -0.004 **39.** $-810,000$ **41.** 0.000002 **43.** 4.2×10^{42}

45. (a) 6×10^{11} **(b)** $600,000,000,000$ **47. (a)** 1.5×10^7

(b) $15,000,000$ **49. (a)** -6×10^4 **(b)** $-60,000$

51. (a) 2.4×10^2 **(b)** 240 **53. (a)** 6.3×10^{-2} **(b)** 0.063

55. (a) 3×10^{-4} **(b)** 0.0003 **57. (a)** -4×10 **(b)** -40

59. (a) 1.3×10^{-5} **(b)** 0.000013 **61. (a)** 5×10^2 **(b)** 500

63. (a) -3×10^6 **(b)** $-3,000,000$ **65. (a)** 2×10^{-7} **(b)** 0.0000002

67. $4.7\text{E}{-}7$ **69.** $2\text{E}7$ **71.** $1\text{E}1$ **73.** 1.04×10^8 **75.** 9.2×10^{-3}

77. 6×10^9 **79.** 1.5×10^{17} mi **81.** $\$3064$ **83.** $\$60,736$

85. 3.59×10^2, or 359 sec **87.** $\$109.02$ **89.** 5.580876×10^{18};

3.3485256×10^{20} **91.** The Valdivia earthquake was 10 times as intense as the Southern Sumatra earthquake. **92.** The Maule earthquake was 10 times as intense as the Gorkha earthquake. **93.** The Valdivia earthquake was 100 times as intense as the Farkhar earthquake. **94.** The Maule earthquake was 100,000 times as intense as the Mooreland earthquake. **95.** The Gorkha earthquake was 10,000 times as intense as the Mooreland earthquake.

Section 4.4

1. $4; 6$ **3.** 9 **5.** $19; 19$ **7.** 0 **9.** 6; one **11.** 1; one **13.** -19, -1; two **15.** $1, 8, 5$; three **17.** $2m^5$ **19.** $-r^5$ **21.** It cannot be simplified. **23.** $-5x^5$ **25.** $5p^9 + 4p^7$ **27.** 0 **29.** $-2y^2$ **31.** $-\frac{22}{15}tu^7$

33. already simplified; 4; binomial **35.** $11m^4 - 7m^3 - 3m^2$; 4; trinomial

37. x^4; 4; monomial **39.** $7; 0$; monomial **41.** $-13ab$; 2; monomial

43. The first term should be evaluated as $-(-2)^2 = -4$, not 4, The correct answer is -6. **45. (a)** 19 **(b)** -2 **47. (a)** 14 **(b)** -19

49. (a) 36 **(b)** -12 **51.** $5x^2 - 2x$ **53.** $5m^2 + 3m + 2$

55. $\frac{7}{6}x^2 - \frac{2}{15}x + \frac{5}{6}$ **57.** $6m^3 + m^2 + 4m - 14$ **59.** $3y^3 - 11y^2$

61. $4x^4 - 4x^2 + 4x$ **63.** $15m^3 - 13m^2 + 8m + 11$ **65.** $5m^2 - 14m + 6$

67. $4x^3 + 2x^2 + 5x$ **69.** $-11x^2 - 3x - 3$ **71.** $a^4 - a^2 + 1$

73. $5m^2 + 8m - 10$ **75.** $-6x^2 - 12x + 12$ **77.** -10 **79.** $4b - 5c$

81. $6x - xy - 7$ **83.** $c^4d - 5c^2d^2 + d^2$ **85.** $8x^2 + 8x + 6$

87. $2x^2 + 8x$ **89.** $8t^2 + 8t + 13$ **91. (a)** $23y + 5t$

(b) $25°, 67°, 88°$ **93.** $-7x - 1$

95. $0, -3, -4, -3, 0$

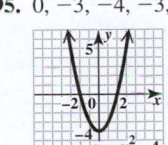

97. $7, 1, -1, 1, 7$

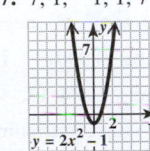

99. $0, 3, 4, 3, 0$

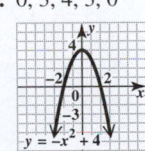

101. $4, 1, 0, 1, 4$

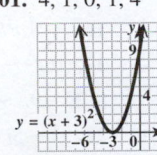

103. $5; 175$ **104.** $9; 63$ **105.** $6; 27$ **106.** $2.5; 130$

Section 4.5

1. (a) B **(b)** D **(c)** A **(d)** C **3.** distributive **5.** one

7. $15y^{11}$ **9.** $30a^9$ **11.** $15pq^2$ **13.** $-18m^3n^2$ **15.** $9y^{10}$

17. $-8x^{10}$ **19.** $6m^2 + 4m$ **21.** $-6p^4 + 12p^3$

23. $-16z^2 - 24z^3 - 24z^4$ **25.** $6y^3 + 4y^4 + 10y^7$

27. $28r^5 - 32r^4 + 36r^3$ **29.** $6a^4 - 12a^3b + 15a^2b^2$

31. $3m^2; 2mn; -n^3; 21m^5n^2 + 14m^4n^3 - 7m^3n^5$

33. $12x^3 + 26x^2 + 10x + 1$ **35.** $72y^3 - 70y^2 + 21y - 2$

37. $20m^4 - m^3 - 8m^2 - 17m - 15$

39. $6x^6 - 3x^5 - 4x^4 + 4x^3 - 5x^2 + 8x - 3$

41. $5x^4 - 13x^3 + 20x^2 + 7x + 5$ **43.** $3x^5 + 18x^4 - 2x^3 - 8x^2 + 24x$

45. first row: $x^2, 4x$; second row: $3x, 12$; Product: $x^2 + 7x + 12$

47. first row: $2x^3, 6x^2, 4x$; second row: $x^2, 3x, 2$; Product: $2x^3 + 7x^2 + 7x + 2$ **49. (a)** $2p; 3p; 6p^2$ **(b)** $2p; 7; 14p$ **(c)** $-5; 3p;$ $-15p$ **(d)** $-5; 7; -35$ **(e)** $6p^2 - p - 35$ **51.** $m^2 + 12m + 35$

53. $n^2 + 3n - 4$ **55.** $12x^2 + 10x - 12$ **57.** $81 - t^2$

59. $9x^2 - 12x + 4$ **61.** $10a^2 + 37a + 7$ **63.** $12 + 8m - 15m^2$

65. $20 - 7x - 3x^2$ **67.** $3t^2 + 5st - 12s^2$ **69.** $8xy - 4x + 6y - 3$

71. $15x^2 + xy - 6y^2$ **73.** $6y^5 - 21y^4 - 45y^3$ **75.** $-200r^7 + 32r^3$

77. $-3r^4 + 3r$ **79. (a)** $3y^2 + 10y + 7$ **(b)** $8y + 16$

81. $x^2 + 14x + 49$ **83.** $a^2 - 16$ **85.** $4p^2 - 20p + 25$

87. $25k^2 + 30kq + 9q^2$ **89.** $m^3 - 15m^2 + 75m - 125$

91. $8a^3 + 12a^2 + 6a + 1$ **93.** $-9a^3 + 33a^2 + 12a$

95. $56m^2 - 14m - 21$ **97.** $81r^4 - 216r^3s + 216r^2s^2 - 96rs^3 + 16s^4$

99. $6p^8 + 15p^7 + 12p^6 + 36p^5 + 15p^4$

101. $-24x^8 - 28x^7 + 32x^6 + 20x^5$ **103.** $6p^4 - \frac{5}{2}p^2q - \frac{25}{12}q^2$

105. $14x + 49$ **107.** $\pi x^2 - 9$ **109.** $30x + 60$

110. $30x + 60 = 600$; $\{18\}$ **111.** 10 ft by 60 ft **112.** 140 ft

113. $\$450$ **114.** $\$2870$

Section 4.6

1. The student neglected to include the term $2ab$. The correct answer is $a^2 + 2ab + b^2$. **3. (a)** $4x; 16x^2$ **(b)** $4x; 3; 24x$ **(c)** $3; 9$

(d) $16x^2 + 24x + 9$ **5.** $m^2 + 4m + 4$ **7.** $r^2 - 6r + 9$

9. $x^2 + 4xy + 4y^2$ **11.** $25p^2 + 20pq + 4q^2$ **13.** $16x^2 - 24x + 9$

15. $16a^2 + 40ab + 25b^2$ **17.** $36m^2 - \frac{48}{5}mn + \frac{16}{25}n^2$

19. $\frac{1}{4}x^2 + \frac{1}{3}x + \frac{1}{9}$ **21.** $2x^2 + 24x + 72$ **23.** $9t^3 - 6t^2 + t$

25. $48t^3 + 24t^2 + 3t$ **27.** $-16r^2 + 16r - 4$ **29.** $k^2 - 25$

31. $16 - 9t^2$ **33.** $25x^2 - 4$ **35.** $25y^2 - 9x^2$ **37.** $100x^2 - 9y^2$

39. $4x^4 - 25$ **41.** $\frac{9}{16} - x^2$ **43.** $81y^2 - \frac{4}{9}$ **45.** $25q^3 - q$

47. $-5a^2 + 5b^6$ **49.** $2k^2 - \frac{1}{2}$ **51.** $-x^2 + 1$ **53.** $x^3 + 3x^2 + 3x + 1$

55. $t^3 - 9t^2 + 27t - 27$ **57.** $r^3 + 15r^2 + 75r + 125$

59. $8a^3 + 12a^2 + 6a + 1$ **61.** $256x^4 - 256x^3 + 96x^2 - 16x + 1$

63. $81r^4 - 216r^3t + 216r^2t^2 - 96rt^3 + 16t^4$

65. $2x^4 + 6x^3 + 6x^2 + 2x$ **67.** $-4t^4 - 36t^3 - 108t^2 - 108t$

69. $x^4 - 2x^2y^2 + y^4$ **71.** 9999 **73.** $39,999$ **75.** $399\frac{3}{4}$

77. $\frac{1}{2}m^2 - 2n^2$ **79.** $9a^2 - 4$ **81.** $\pi x^2 + 4\pi x + 4\pi$

83. $x^3 + 6x^2 + 12x + 8$ **85.** $(a + b)^2$ **86.** a^2 **87.** $2ab$

88. b^2 **89.** $a^2 + 2ab + b^2$ **90.** They both represent the area of the entire large square. **91.** 1225 **92.** $30^2 + 2(30)(5) + 5^2$
93. 1225 **94.** They are equal.

Section 4.7

1. $10x^2 + 8$; 2; $5x^2 + 4$ **3.** The student wrote the second term of the quotient as $-12x$ rather than $-2x$. $\dfrac{6x^2 - 12x}{6} = \dfrac{6x^2}{6} - \dfrac{12x}{6} = x^2 - 2x$
5. $6p^4$; $18p^7$; $2p^2 + 6p^5$ **7.** $2m^2 - m$ **9.** $30x^3 - 10x + 5$
11. $4m^3 - 2m^2 + 1$ **13.** $4t^4 - 2t^2 + 2t$ **15.** $a^4 - a + \dfrac{2}{a}$
17. $-3p^2 - 2 + \dfrac{1}{p}$ **19.** $7r^2 - 6 + \dfrac{1}{r}$ **21.** $4x^3 - 3x^2 + 2x$
23. $-9x^2 + 5x + 1$ **25.** $2x + 8 + \dfrac{12}{x}$ **27.** $\dfrac{4x^2}{3} + x + \dfrac{2}{3x}$
29. $9r^3 - 12r^2 + 2r + \dfrac{26}{3} - \dfrac{2}{3r}$ **31.** $-m^2 + 3m - \dfrac{4}{m}$
33. $\dfrac{12}{x} - \dfrac{6}{x^2} + \dfrac{14}{x^3} - \dfrac{10}{x^4}$ **35.** $-4b^2 + 3ab - \dfrac{5}{a}$
37. $6x^4y^2 - 4xy + 2xy^2 - x^4y$ **39.** $x + 2$ **41.** $2y - 5$
43. $p - 4 + \dfrac{44}{p + 6}$ **45.** $6m - 1$ **47.** $2a - 14 + \dfrac{74}{2a + 3}$
49. $4x^2 - 7x + 3$ **51.** $4k^3 - k + 2$ **53.** $5y^3 + 2y - 3 + \dfrac{-5}{y + 1}$
55. $3k^2 + 2k - 2 + \dfrac{6}{k - 2}$ **57.** $2p^2 - 5p + 4 + \dfrac{6}{3p^2 + 1}$
59. $x^2 + 3x + 3$ **61.** $3y^2 - 2y + 2$ **63.** $2x^2 - 6x + 19 + \dfrac{-55}{x + 3}$
65. $r^2 + 2 + \dfrac{13}{r^2 - 4}$ **67.** $3x^2 + 3x - 1 + \dfrac{1}{x - 1}$ **69.** $y^2 - 3y + 9$
71. $a^2 + 5$ **73.** $x^2 - 4x + 2 + \dfrac{9x - 4}{x^2 + 3}$ **75.** $x^3 + 3x^2 - x + 5$
77. $\dfrac{3}{2}a - 10 + \dfrac{77}{2a + 6}$ **79.** $x^2 + \dfrac{8}{3}x - \dfrac{1}{3} + \dfrac{4}{3x - 3}$
81. $(x^2 + x - 3)$ units **83.** $(48m^2 + 96m + 24)$ units
85. $(5x^2 - 11x + 14)$ hours

Chapter 4 Review Exercises

1. 4^{11} **2.** $(-5)^{11}$ **3.** $-72x^7$ **4.** $10x^{14}$ **5.** 19^5x^5 **6.** -4^7y^7
7. $5p^4t^4$ **8.** $\dfrac{7^6}{5^6}$ **9.** $27x^6y^9$ **10.** t^{42} **11.** $36x^{16}y^4z^{16}$ **12.** $\dfrac{8m^9n^3}{p^6}$
13. -1 **14.** -1 **15.** 2 **16.** -2 **17.** $-\dfrac{1}{49}$ **18.** $\dfrac{64}{25}$ **19.** 64
20. $\dfrac{1}{81}$ **21.** $\dfrac{3}{4}$ **22.** $\dfrac{1}{36}$ **23.** x^2 **24.** y^7 **25.** $\dfrac{r^8}{81}$ **26.** $\dfrac{243}{p^3}$ **27.** $\dfrac{1}{a^3b^5}$
28. $72r^5$ **29.** 4.8×10^7 **30.** 2.8988×10^{10} **31.** 8.24×10^{-8}
32. -4.82×10^6 **33.** 24,000 **34.** 78,300,000 **35.** 0.000000897
36. -0.00076 **37.** (a) 8×10^2 (b) 800 **38.** (a) 5×10^0
(b) 5 **39.** (a) 4×10^6 (b) 4,000,000 **40.** (a) 2.5×10^{-2}
(b) 0.025 **41.** 100,720,000,000,000,000 **42.** 38,140,000
43. $3.50799552 \times 10^{16}$ **44.** 5.853×10^{11}
45. $20m^2$; 2; monomial **46.** $p^3 - p^2 - 4p$; 3; trinomial
47. $-8y^5 - 7y^4$; 5; binomial **48.** $-r^3 - 2r + 7$
49. $13x^3y^2 - 5xy^5 + 21x^2$ **50.** $a^3 + 4a^2$ **51.** $y^2 - 10y + 9$
52. $-13k^4 - 15k^2 + 18k$

53. 1, 4, 5, 4, 1 **54.** 10, 1, -2, 1, 10

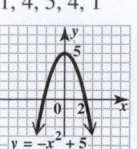 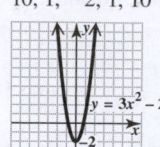

55. $a^3 - 2a^2 - 7a + 2$ **56.** $6r^3 + 8r^2 - 17r + 6$
57. $5p^5 - 2p^4 - 3p^3 + 25p^2 + 15p$ **58.** $m^2 - 7m - 18$
59. $6k^2 - 9k - 6$ **60.** $2a^2 + 5ab - 3b^2$ **61.** $12k^2 - 32kq - 35q^2$
62. $s^3 - 3s^2 + 3s - 1$ **63.** $a^2 + 8a + 16$ **64.** $4r^2 + 20rt + 25t^2$
65. $36m^2 - 25$ **66.** $25a^2 - 36b^2$ **67.** $r^3 + 6r^2 + 12r + 8$
68. $25t^3 - 30t^2 + 9t$ **69.** (a) Answers will vary. For example, let $x = 1$ and $y = 2$. $(1 + 2)^2 \neq 1^2 + 2^2$, because $9 \neq 5$. (b) Answers will vary. For example, let $x = 1$ and $y = 2$. $(1 + 2)^3 \neq 1^3 + 2^3$, because $27 \neq 9$. **70.** Find the third power of a binomial, such as $(a + b)^3$, as follows. $(a + b)^3 = (a + b)(a + b)^2 = (a + b)(a^2 + 2ab + b^2)$ $= a^3 + 2a^2b + ab^2 + a^2b + 2ab^2 + b^3 = a^3 + 3a^2b + 3ab^2 + b^3$
71. $x^6 + 6x^4 + 12x^2 + 8$ **72.** $\dfrac{4}{3}\pi x^3 + 4\pi x^2 + 4\pi x + \dfrac{4}{3}\pi$
73. $-\dfrac{5y^2}{3}$ **74.** $-2m^2n + mn + \dfrac{6n^3}{5}$ **75.** $y^3 - 2y + 3$
76. $-6r^5s - 3r^4 + \dfrac{2}{r^2s^5}$ **77.** $2r + 7$ **78.** $2a^2 + 3a - 1 + \dfrac{6}{5a - 3}$
79. $x^2 + 3x - 4$ **80.** $m^2 + 4m - 2$ **81.** $4x - 5$ **82.** $5y - 10$
83. $y^2 + 2y + 4$ **84.** $100x^4 - 10x^2 + 1$ **85.** $2y^2 - 5y + 4 + \dfrac{-5}{3y^2 + 1}$
86. $x^3 - 2x^2 + 4 + \dfrac{-3}{4x^2 - 3}$

Chapter 4 Mixed Review Exercises

1. 2 **2.** $\dfrac{216r^6p^3}{125}$ **3.** $144a^2 - 1$ **4.** $\dfrac{1}{16}$ **5.** $\dfrac{1}{256}$ **6.** $p - 3 + \dfrac{5}{2p}$
7. $\dfrac{2}{3m^3}$ **8.** $6k^3 - 21k - 6$ **9.** r^{13} **10.** $4r^2 + 20rs + 25s^2$
11. $y^2 + 5y + 1$ **12.** $10r^2 + 21r - 10$ **13.** $-y^2 - 4y + 4$
14. $\dfrac{5}{2} - \dfrac{4}{5xy} + \dfrac{3x}{2y^2}$ **15.** $10p^2 - 3p - 5$ **16.** $3x^2 + 9x + 25 + \dfrac{80}{x - 3}$
17. $49 - 28k + 4k^2$ **18.** $\dfrac{1}{x^4y^{12}}$ **19.** (a) $6x - 2$ (b) $2x^2 + x - 6$
20. (a) $20x^4 + 8x^2$ (b) $25x^8 + 20x^6 + 4x^4$

Chapter 4 Test

[4.1, 4.2] **1.** $\dfrac{1}{625}$ **2.** 2 **3.** $\dfrac{7}{12}$ **4.** -32 **5.** $\dfrac{216}{m^6}$ **6.** $9x^3y^5$
7. 8^5 **8.** x^2y^6 **9.** (a) positive (b) positive (c) negative
(d) positive (e) zero (f) negative [4.3] **10.** (a) 4.5×10^{10}
(b) 0.0000036 (c) 0.00019 **11.** (a) 1×10^3; 5.89×10^{12}
(b) 5.89×10^{15} mi [4.4] **12.** $-7x^2 + 8x$; 2; binomial
13. $4n^4 + 13n^3 - 10n^2$; 4; trinomial
14. 4, -2, -4, -2, 4 **15.** $-2y^2 - 9y + 17$

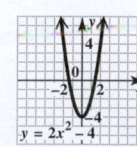

16. $-21a^3b^2 + 7ab^5 - 5a^2b^2$
17. $16r^2 - 19$ **18.** $-12t^2 + 5t + 8$
[4.5] **19.** $-27x^5 + 18x^4 - 6x^3 + 3x^2$
20. $t^2 - 5t - 24$ **21.** $8x^2 + 2xy - 3y^2$
[4.6] **22.** $25x^2 - 20xy + 4y^2$ **23.** $100v^2 - 9w^2$
[4.5] **24.** $2r^3 + r^2 - 16r + 15$ [4.6] **25.** $12x + 36$

26. $9x^2 + 54x + 81$ [4.7] **27.** $4y^2 - 3y + 2 + \dfrac{5}{y}$

28. $-3xy^2 + 2x^3y^2 + 4y^2$ **29.** $x - 2$ **30.** $3x^2 + 6x + 11 + \dfrac{26}{x-2}$

Chapters R–4 Cumulative Review Exercises

[R.1, R.2] **1.** $\dfrac{19}{24}$ **2.** $-\dfrac{1}{20}$ **3.** 0.000042 **4.** $31\frac{1}{4}$ yd^3 **5.** $\$1836$

[1.4, 1.5] **6.** -8 **7.** $\dfrac{1}{2}$ **8.** -4 [1.6] **9.** associative property

10. distributive property   [1.7] **11.** $-10x^2 + 21x - 29$

[2.1–2.3] **12.** $\left\{\dfrac{13}{4}\right\}$ **13.** $\varnothing$ [2.6] **14.** $r = \dfrac{d}{t}$ [2.7] **15.** $\{-5\}$

[2.1–2.4] **16.** $\{0\}$ **17.** $\{20\}$ **18.** $\{-12\}$ **19.** $\{$all real numbers$\}$

[2.5] **20.** exertion: 9443 calories; regulating body temperature: 1757 calories   [2.9] **21.** 11 ft, 22 ft   **22.** $\left(-\infty, -\dfrac{14}{5}\right)$ **23.** $[-4, 2)$

[3.2] **24.**

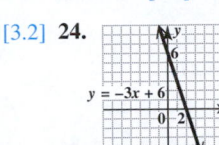

[3.3–3.5] **25.** (a) 1   (b) $y = x + 6$
[4.1, 4.2] **26.** 25   **27.** -16 **28.** $\dfrac{5}{4}$
29. 1   **30.** $\dfrac{2b}{a^{10}}$ [4.3] **31.** 10,800,000 km

[4.4] **32.** 4, 1, 0, 1, 4   **33.** $11x^3 - 20x^2 - x + 10$

[4.5] **34.** $63x^2 + 57x + 12$
[4.7] **35.** $y^2 - 2y + 6$

---

5 FACTORING AND APPLICATIONS

Section 5.1

1. product; multiplying   **3.** 4   **5.** 4   **7.** 6   **9.** 1   **11.** 8   **13.** $10x^3$

15. xy^2 **17.** 6   **19.** $6m^3n^2$ **21.** factored   **23.** not factored

25. The correct factored form is $18x^3y^2 + 9xy = 9xy(2x^2y + 1)$. If a polynomial has two terms, then the product of the factors must have two terms. $9xy(2x^2y) = 18x^3y^2$ is just one term.   **27.** $3m^2$ **29.** $2z^4$

31. $2mn^4$ **33.** $y + 2$ **35.** $a - 2$ **37.** $2 + 3xy$ **39.** $x(x - 4)$

41. $3t(2t + 5)$ **43.** $9m(3m^2 - 1)$ **45.** $m^2(m - 1)$ **47.** $8z^2(2z^2 + 3)$

49. $-6x^2(2x + 1)$ **51.** $5y^6(13y^4 + 7)$ **53.** no common factor (except 1)   **55.** $8mn^3(1 + 3m)$ **57.** $13y^2(y^6 + 2y^2 - 3)$

59. $-2x(2x^2 - 5x + 3)$ **61.** $9p^3q(4p^3 + 5p^2q^3 + 9q)$

63. $a^3(a^2 + 2b^2 - 3a^2b^2 + 4ab^3)$ **65.** $(x + 2)(c - d)$

67. $(m + 2n)(m + n)$ **69.** $(p - 4)(q^2 - 3)$ **71.** not in factored form; $(7t + 4)(8 + x)$ **73.** in factored form   **75.** not in factored form   **77.** The student should factor out -2, instead of 2, in the second step to obtain $x^2(x + 4) - 2(x + 4)$, which can be factored as $(x + 4)(x^2 - 2)$.   **79.** $(5 + n)(m + 4)$ **81.** $(2y - 7)(3x + 4)$

83. $(p + 4)(p + q)$ **85.** $(a - 2)(a + b)$ **87.** $(z + 2)(7z - a)$

89. $(3r + 2y)(6r - x)$ **91.** $(w + 1)(w^2 + 9)$ **93.** $(a + 2)(3a^2 - 2)$

95. $(x + 5)(10y + 1)$ **97.** $(3 - a)(4 - b)$ **99.** $(4m - p^2)(4m^2 - p)$

101. $(y + 3)(y + x)$ **103.** $(5 - 2p)(m + 3)$ **105.** $(z - 2)(2z - 3w)$

107. $(3r + 2y)(6r - t)$ **109.** $(1 + 2b)(a^5 - 3)$

111. commutative property   **112.** $2x(y - 4) - 3(y - 4)$

113. No, because the result is not a product. It is the difference of $2x(y - 4)$ and $3(y - 4)$.   **114.** $(y - 4)(2x - 3)$, or $(2x - 3)(y - 4)$; yes

Section 5.2

1. a and b must have different signs, one positive and one negative.

3. C   **5.** $a^2 + 13a + 36$ **7.** The greatest common factor must be included in the factorization. The correct factorization is $x(x + 7)(x - 4)$.

9. 1 and 48, -1 and -48, 2 and 24, -2 and -24, 3 and 16, -3 and -16, 4 and 12, -4 and -12, 6 and 8, -6 and -8; The pair with a sum of -19 is -3 and -16.   **11.** 1 and -24, -1 and 24, 2 and -12, -2 and 12, 3 and -8, -3 and 8, 4 and -6, -4 and 6; The pair with a sum of -5 is 3 and -8.   **13.** 20; 12; table entries: 2, 2, 12, 4, 4, 9; 10 and 2; $(y + 10)(y + 2)$ **15.** $p + 6$ **17.** $x + 11$ **19.** $x - 8$ **21.** $y - 5$

23. $x + 11$ **25.** $y - 9$ **27.** $(y + 8)(y + 1)$ **29.** $(b + 3)(b + 5)$

31. $(m + 5)(m - 4)$ **33.** $(y - 5)(y - 3)$ **35.** prime

37. $(z - 7)(z - 8)$ **39.** $(r - 6)(r + 5)$ **41.** $(a + 4)(a - 12)$

43. prime   **45.** $(x + 16)(x - 2)$ **47.** $(r + 2a)(r + a)$

49. $(x + y)(x + 3y)$ **51.** $(t + 2z)(t - 3z)$ **53.** $(v - 5w)(v - 6w)$

55. $(m + 6n)(m - 2n)$ **57.** $(a - 6b)(a - 3b)$ **59.** $4(x + 5)(x - 2)$

61. $2t(t + 1)(t + 3)$ **63.** $6z^2(z - 3)(z - 1)$

65. $-2x^4(x - 3)(x + 7)$ **67.** $5m^2(m^3 - 5m^2 + 8)$

69. $x(x - 4y)(x - 3y)$ **71.** $z^8(z - 7y)(z + 3y)$

73. $-a^3(a + 4b)(a - b)$ **75.** $mn(m - 6n)(m - 4n)$

77. $yz(y + 3z)(y - 2z)$ **79.** $(a + b)(x + 4)(x - 3)$

81. $(2p + q)(r - 9)(r - 3)$

Section 5.3

1. $(2t + 1)(5t + 2)$ **3.** $(3z - 2)(5z - 3)$ **5.** $(2s - t)(4s + 3t)$

7. (a) 2, 12, 24, 11   (b) 3, 8 (Order is irrelevant.)   (c) $3m$, $8m$

(d) $2m^2 + 3m + 8m + 12$ (e) $(2m + 3)(m + 4)$

(f) $(2m + 3)(m + 4) = 2m^2 + 8m + 3m + 12$, and combining like terms gives the original trinomial $2m^2 + 11m + 12$.   **9.** B   **11.** The student stopped too soon. He needs to factor out the common factor $4x - 1$ to obtain $(4x - 1)(4x - 5)$ as the correct answer.   **13.** B

15. A   **17.** $(4x + 4)$ cannot be a factor because its terms have a common factor of 4, but those of the trinomial do not. The correct factored form is $(4x - 3)(3x + 4)$.   **19.** $2a + 5b$ **21.** $x^2 + 3x - 4$; $x + 4$, $x - 1$, or $x - 1$, $x + 4$ **23.** $2z^2 - 5z - 3$; $2z + 1$, $z - 3$, or $z - 3$, $2z + 1$ **25.** $(3a + 7)(a + 1)$ **27.** $(2y + 3)(y + 2)$

29. $(3m - 1)(5m + 2)$ **31.** $(3s - 1)(4s + 5)$

33. $(5m - 4)(2m - 3)$ **35.** $(4w - 1)(2w - 3)$

37. $(4y + 1)(5y - 11)$ **39.** prime   **41.** $2(5x + 3)(2x + 1)$

43. $3(4x - 1)(2x - 3)$ **45.** prime   **47.** $(3x + 4)(x + 4)$

49. prime   **51.** $(12x - 5)(2x - 3)$ **53.** $4x(8x + 3)(x + 1)$

55. $-5x(2x + 7)(x - 4)$ **57.** $3n^2(5n - 3)(n - 2)$

59. $-q(5m + 2)(8m - 3)$ **61.** $y^2(5x - 4)(3x + 1)$

63. $(5a + 3b)(a - 2b)$ **65.** $(4s + 5t)(3s - t)$

67. $(3p + 4q)(4p - 3q)$ **69.** $(24y + 7x)(y - 2x)$

71. $2(24a + b)(a - 2b)$ **73.** $(4x + 3y)(6x + 5y)$

75. $x^2y^5(10x - 1)(x + 4)$ **77.** $4ab^2(9a + 1)(a - 3)$

79. $m^4n(3m + 2n)(2m + n)$ **81.** $-1(x + 7)(x - 3)$

83. $-1(3x + 4)(x - 1)$ **85.** $-1(a + 2b)(2a + b)$

87. $(8x^2 - 3)(3x^2 + 8)$ **89.** $(18x^2 - 5y)(2x^2 - 3y)$

91. $(m + 1)^3(5q - 2)(5q + 1)$ **93.** $(r + 3)^3(3x + 2y)^2$

95. $-4, 4$ **97.** $-11, -7, 7, 11$ **99.** $5 \cdot 7$ **100.** $-5(-7)$

101. The product of $3x - 4$ and $2x - 1$ is $6x^2 - 11x + 4$.

102. The product of $4 - 3x$ and $1 - 2x$ is $6x^2 - 11x + 4$.

103. The factors in Exercise 101 are the opposites of the factors in Exercise 102.   **104.** $(3 - 7t)(5 - 2t)$

Section 5.4

1. $1; 4; 9; 16; 25; 36; 49; 64; 81; 100; 121; 144; 169; 196; 225; 256; 289;$
$324; 361; 400$ **3.** A, D   **5.** The binomial $4x^2 + 16$ can be factored as $4(x^2 + 4)$. *After* any common factor is removed, a sum of squares (like $x^2 + 4$ here) *cannot* be factored.   **7.** $(y + 5)(y - 5)$

9. $(x + 12)(x - 12)$ **11.** prime   **13.** prime   **15.** $4(m^2 + 4)$

17. $(3r + 2)(3r - 2)$ **19.** $4(3x + 2)(3x - 2)$

21. $(14p + 15)(14p - 15)$ **23.** $(4r + 5a)(4r - 5a)$

25. $(9x + 7y)(9x - 7y)$ **27.** $6(3x + y)(3x - y)$ **29.** prime

31. $(2 + x)(2 - x)$ **33.** $(6 + 5t)(6 - 5t)$ **35.** $x(x^2 + 4)$

37. $x^2(x + 1)(x - 1)$ **39.** $(p^2 + 7)(p^2 - 7)$

41. $(x^2 + 1)(x + 1)(x - 1)$ **43.** $(p^2 + 16)(p + 4)(p - 4)$

45. B, C   **47.** 10   **49.** 9   **51.** $(w + 1)^2$ **53.** $(x - 4)^2$

55. prime   **57.** $2(x + 6)^2$ **59.** $(2x + 3)^2$ **61.** $(4x - 5)^2$

63. $4r(3r + 4)^2$ **65.** $(7x - 2y)^2$ **67.** $(8x + 3y)^2$

69. $2(5h - 2y)^2$ **71.** $z^2(25z^2 + 5z + 1)$ **73.** $1; 8; 27; 64; 125;$
$216; 343; 512; 729; 1000$ **75.** C, D   **77.** (a) neither of these

(b) perfect cube   (c) perfect square   (d) perfect square

(e) both of these   (f) perfect cube   **79.** $(a - 1)(a^2 + a + 1)$

81. $(m + 2)(m^2 - 2m + 4)$ **83.** $(y - 6)(y^2 + 6y + 36)$

85. $(k + 10)(k^2 - 10k + 100)$ **87.** $(3x - 4)(9x^2 + 12x + 16)$

89. $(5x + 2)(25x^2 - 10x + 4)$ **91.** $6(p + 1)(p^2 - p + 1)$

93. $5(x + 2)(x^2 - 2x + 4)$ **95.** $(y - 2x)(y^2 + 2yx + 4x^2)$

97. $2(x - 2y)(x^2 + 2xy + 4y^2)$ **99.** $(2p + 9q)(4p^2 - 18pq + 81q^2)$

101. $(3a + 4b)(9a^2 - 12ab + 16b^2)$

103. $(5t + 2s)(25t^2 - 10ts + 4s^2)$

105. $(2x - 5y^2)(4x^2 + 10xy^2 + 25y^4)$

107. $(3m^2 + 2n)(9m^4 - 6m^2n + 4n^2)$

109. $(5k - 2m^3)(25k^2 + 10km^3 + 4m^6)$

111. $(x + y)(x^2 - xy + y^2)(x^6 - x^3y^3 + y^6)$ **113.** $\left(p + \frac{1}{3}\right)\left(p - \frac{1}{3}\right)$

115. $\left(6m + \frac{4}{5}\right)\left(6m - \frac{4}{5}\right)$ **117.** $(x + 0.8)(x - 0.8)$

119. $\left(t + \frac{1}{2}\right)^2$ **121.** $(x - 0.9)^2$ **123.** $\left(x + \frac{1}{2}\right)\left(x^2 - \frac{1}{2}x + \frac{1}{4}\right)$

125. $4mn$ **127.** $(m - p + 2)(m + p)$

129. $(2x + 1 + y)(2x + 1 - y)$ **131.** $(3a - 8b + 2c)(3a - 8b - 2c)$

SUMMARY EXERCISES   Recognizing and Applying Factoring Strategies

1. G   **2.** H   **3.** A   **4.** B   **5.** E   **6.** I   **7.** C   **8.** F   **9.** I   **10.** E

11. $(a - 6)(a + 2)$ **12.** $(a + 8)(a + 9)$ **13.** $6(y - 2)(y + 1)$

14. $7y^4(y + 6)(y - 4)$ **15.** $6(a + 2b + 3c)$ **16.** $(m - 4n)(m + n)$

17. $(p - 11)(p - 6)$ **18.** $(z + 7)(z - 6)$ **19.** $(5z - 6)(2z + 1)$

20. $2(m - 8)(m + 3)$ **21.** $17xy(x^2y + 3)$ **22.** $5(3y + 1)$

23. $8a^3(a - 3)(a + 2)$ **24.** $(4k + 1)(2k - 3)$ **25.** $(z - 5a)(z + 2a)$

26. $50(z^2 - 2)$ **27.** $(x - 5)(x - 4)$ **28.** prime

29. $(3n - 2)(2n - 5)$ **30.** $(3y - 1)(3y + 5)$ **31.** $4(4x + 5)$

32. $(m + 5)(m - 3)$ **33.** $(3y - 4)(2y + 1)$ **34.** $(m + 9)(m - 9)$

35. $(6z + 1)(z + 5)$ **36.** $(12x - 1)(x + 4)$ **37.** $(2k - 3)^2$

38. $(8p - 1)(p + 3)$ **39.** $6(3m + 2z)(3m - 2z)$

40. $(4m - 3)(2m + 1)$ **41.** $(3k - 2)(k + 2)$

42. $(2a - 3)(4a^2 + 6a + 9)$ **43.** $7k(2k + 5)(k - 2)$

44. $(5 + r)(1 - s)$ **45.** $(y^2 + 4)(y + 2)(y - 2)$ **46.** prime

47. $8m(1 - 2m)$ **48.** $(k + 4)(k - 4)$ **49.** $(z - 2)(z^2 + 2z + 4)$

50. $(y - 8)(y + 7)$ **51.** prime   **52.** $9p^8(3p + 7)(p - 4)$

53. $8m^3(4m^6 + 2m^2 + 3)$ **54.** $(2m + 5)(4m^2 - 10m + 25)$

55. $(4r + 3m)^2$ **56.** $(z - 6)^2$ **57.** $(5h + 7g)(3h - 2g)$

58. $5z(z - 7)(z - 2)$ **59.** $(k - 5)(k - 6)$

60. $4(4p - 5m)(4p + 5m)$ **61.** $3k(k - 5)(k + 1)$

62. $(y - 6k)(y + 2k)$ **63.** $(10p + 3)(100p^2 - 30p + 9)$

64. $(4r - 7)(16r^2 + 28r + 49)$ **65.** $(2 + m)(3 + p)$

66. $(2m - 3n)(m + 5n)$ **67.** $(4z - 1)^2$

68. $(a^2 + 25)(a + 5)(a - 5)$ **69.** $3(6m - 1)^2$

70. $(10a + 9y)(10a - 9y)$ **71.** prime   **72.** $(2y + 5)(2y - 5)$

73. $8z(4z - 1)(z + 2)$ **74.** $5(2m - 3)(m + 4)$ **75.** $(8m - 5n)^2$

76. $(2 - q)(2 - 3p)$ **77.** $2(3a - 1)(a + 2)$

78. $6y^4(3y + 4)(2y - 5)$ **79.** prime   **80.** $4(2k - 3)^2$

81. $(4 + m)(5 + 3n)$ **82.** $12y^2(6yz^2 + 1 - 2y^2z^2)$

83. $(4k - 3h)(2k + h)$ **84.** $(2a + 5)(a - 6)$

85. $2(x + 4)(x^2 - 4x + 16)$ **86.** $15a^3b^2(3b^3 - 4a + 5a^3b^2)$

87. $(5y - 6z)(2y + z)$ **88.** $(m - 2)^2$ **89.** $(8a - b)(a + 3b)$

90. $5m^2(5m - 3n)(5m - 13n)$

Section 5.5

1. $ax^2 + bx + c$ **3.** 0; zero-factor   **5.** The term with greatest degree is greater than two. (It is *cubic*.)   **7.** (a) linear   (b) quadratic

(c) quadratic   (d) linear   **9.** Set each *variable* factor equal to 0 to obtain $2x = 0$ or $3x - 4 = 0$. The solution set is $\left\{0, \frac{4}{3}\right\}$.   **11.** $\{-5, 2\}$

13. $\left\{3, \frac{7}{2}\right\}$ **15.** $\left\{-\frac{1}{2}, \frac{1}{6}\right\}$ **17.** $\left\{-\frac{5}{6}, 0\right\}$ **19.** $\left\{0, \frac{4}{3}\right\}$ **21.** $\{6\}$

23. $\{-2, -1\}$ **25.** $\{1, 2\}$ **27.** $\{-8, 3\}$ **29.** $\{-1, 3\}$

31. $\{-2, -1\}$ **33.** $\{-4\}$ **35.** $\{10\}$ **37.** $\left\{-2, \frac{1}{3}\right\}$ **39.** $\left\{-\frac{4}{3}, \frac{1}{2}\right\}$

41. $\left\{-\frac{2}{3}\right\}$ **43.** $\left\{\frac{1}{5}\right\}$ **45.** $\{-3, 3\}$ **47.** $\left\{-\frac{7}{4}, \frac{7}{4}\right\}$ **49.** $\{-11, 11\}$

51. $\{-6, 0\}$ **53.** $\{0, 7\}$ **55.** $\left\{0, \frac{1}{2}\right\}$ **57.** $\{2, 5\}$ **59.** $\left\{-4, \frac{1}{2}\right\}$

61. $\{-17, 4\}$ **63.** $\{-4, 12\}$ **65.** $\{-9, -2\}$ **67.** $\left\{-\frac{7}{3}, 0, \frac{7}{3}\right\}$
69. $\{-2, 0, 4\}$ **71.** $\{-5, 0, 4\}$ **73.** $\left\{0, \frac{1}{2}, 4\right\}$ **75.** $\{-3, 0, 5\}$
77. $\left\{-\frac{5}{2}, \frac{1}{3}, 5\right\}$ **79.** $\left\{-\frac{7}{2}, -3, 1\right\}$ **81.** $\{-1, 3\}$ **83.** $\{-1, 3\}$
85. $\{3\}$ **87.** $\left\{-\frac{2}{3}, 4\right\}$ **89.** $\left\{-\frac{4}{3}, -1, \frac{1}{2}\right\}$ **91. (a)** 64; 144; 4; 6
(b) No time has elapsed, so the object hasn't fallen (been released) yet.

Section 5.6

1. Read; variable; equation; Solve; answer; Check; original
3. $\mathcal{A} = bh$; *Step 3:* $45 = (2x + 1)(x + 1)$; *Step 4:* $x = 4$ or $x = -\frac{11}{2}$;
Step 5: base: 9 units; height: 5 units; *Step 6:* $9 \cdot 5 = 45$ **5.** $\mathcal{A} = LW$;
Step 3: $80 = (x + 8)(x - 8)$; *Step 4:* $x = 12$ or $x = -12$; *Step 5:* length:
20 units; width: 4 units; *Step 6:* $20 \cdot 4 = 80$ **7.** length: 14 cm; width:
12 cm   **9.** base: 12 in.; height: 5 in.   **11.** height: 13 in.; width: 10 in.
13. length: 15 in.; width: 12 in.   **15.** mirror: 7 ft; painting: 9 ft
17. 20, 21   **19.** $-3, -2$ or 4, 5   **21.** 0, 1, 2 or 7, 8, 9   **23.** $-3, -1$ or
7, 9   **25.** 7, 9, 11   **27.** $-2, 0, 2$ or 6, 8, 10   **29.** 12 cm
31. length: 20 in.; width: 15 in.; diagonal: 25 in.   **33.** 12 mi   **35.** 8 ft
37. 112 ft   **39.** 256 ft   **41. (a)** 1 sec   **(b)** $\frac{1}{2}$ sec and $1\frac{1}{2}$ sec
(c) 3 sec   **(d)** The negative solution, -1, does not make sense because
t represents time, which cannot be negative in this situation.   **43.** 3 sec
45. (a) 10   **(b)** 297 million subscribers; The result obtained from the
model is slightly more than 296 million, the actual number for 2010.
(c) 16   **(d)** 390 million subscribers; The result is less than 396 million,
the actual number.   **(e)** 20   **(f)** 443 million subscribers   **47.** c^2
48. b^2 **49.** a^2 **50.** $a^2 + b^2 = c^2$; This is the equation of the
Pythagorean theorem.

Chapter 5 Review Exercises

1. $7(t + 2)$ **2.** $30z(2z^2 + 1)$ **3.** $-3x(x^2 - 2x - 1)$
4. $50m^2n^2(2n - mn^2 + 3)$ **5.** $(2y + 3)(x - 4)$
6. $(3y + 2x)(2y + 3)$ **7.** $(x + 3)(x + 2)$ **8.** $(y - 5)(y - 8)$
9. $(q + 9)(q - 3)$ **10.** $(r - 8)(r + 7)$ **11.** prime
12. $3x^2(x + 2)(x + 8)$ **13.** $-8p^3(p + 2)(p - 5)$
14. $(m + 3n)(m - 6n)$ **15.** $(y - 3z)(y - 5z)$
16. $(p + 12q)(p - 10q)$ **17.** $p^5(p - 2q)(p + q)$
18. $-3r^3(r + 3s)(r - 5s)$ **19.** r and $6r$, $2r$ and $3r$ **20.** Factor out z.
21. $(2k - 1)(k - 2)$ **22.** $(3r - 1)(r + 4)$ **23.** $(3r + 2)(2r - 3)$
24. $(5z + 1)(2z - 1)$ **25.** prime   **26.** $4x^3(3x - 1)(2x - 1)$
27. $-5y(3y + 2)(2y - 1)$ **28.** $(2a - 5b)(7a + 4b)$
29. $(3m - 5n)(m + 8n)$ **30.** $rs(5r + 6s)(2r + s)$ **31.** B
32. D   **33.** $(n + 7)(n - 7)$ **34.** $(5b + 11)(5b - 11)$
35. $(7y + 5w)(7y - 5w)$ **36.** $36(2p + q)(2p - q)$ **37.** prime
38. $(r - 6)^2$ **39.** $(3t - 7)^2$ **40.** $(m + 10)(m^2 - 10m + 100)$
41. $(5k + 4x)(25k^2 - 20kx + 16x^2)$ **42.** $(7x - 4)(49x^2 + 28x + 16)$
43. $(10 - 3x^2)(100 + 30x^2 + 9x^4)$
44. $(x - y)(x + y)(x^2 + xy + y^2)(x^2 - xy + y^2)$
45. $\left\{-\frac{3}{4}, 1\right\}$ **46.** $\left\{0, \frac{5}{2}\right\}$ **47.** $\{-3, -1\}$ **48.** $\{1, 4\}$
49. $\{3, 5\}$ **50.** $\left\{-\frac{4}{3}, 5\right\}$ **51.** $\left\{-\frac{8}{9}, \frac{8}{9}\right\}$ **52.** $\{0, 8\}$

53. $\{-1, 6\}$ **54.** $\{7\}$ **55.** $\{6\}$ **56.** $\{-3, 3\}$
57. $\left\{-2, -1, -\frac{2}{5}\right\}$ **58.** $\left\{-\frac{3}{8}, 0, \frac{3}{8}\right\}$ **59.** length: 10 ft; width: 4 ft
60. 5 ft   **61.** 26 mi   **62.** length: 6 m; width: 4 m   **63.** $-5, -4, -3$ or
5, 6, 7   **64. (a)** 256 ft   **(b)** 1024 ft   **65. (a)** 2013: 382 thousand cars;
2016: 1971 thousand cars   **(b)** 2013: The result is slightly less than
388 thousand, the actual number; 2016: The result is less than
2014 thousand, the actual number.

Chapter 5 Mixed Review Exercises

1. D   **2.** C   **3.** $(3k + 5)(k + 2)$ **4.** $(z - x)(z - 10x)$
5. $(y^2 + 25)(y + 5)(y - 5)$ **6.** $3m(2m + 3)(m - 5)$
7. prime   **8.** $2a^3(a + 2)(a - 6)$ **9.** $(3m + 4)(5m - 4p)$
10. $8abc(3b^2c - 7ac^2 + 9ab)$ **11.** $6xyz(2xz^2 + 2y - 5x^2yz^3)$
12. $(2r + 3q)(6r - 5)$ **13.** $(7t + 4)^2$
14. $(10a + 3)(100a^2 - 30a + 9)$ **15.** $\{0, 7\}$ **16.** $\{-5, 2\}$
17. $\left\{-\frac{2}{5}\right\}$ **18.** 15 m, 36 m, 39 m   **19.** 6 m   **20.** width: 10 m;
length: 17 m

Chapter 5 Test

[5.1–5.4] **1.** D   **2.** $6x(2x - 5)$ **3.** $m^2n(2mn + 3m - 5n)$
4. $(2x + y)(a - b)$ **5.** $(x + 3)(x - 8)$ **6.** $(2x + 3)(x - 1)$
7. $(5z - 1)(2z - 3)$ **8.** $3(x + 1)(x - 5)$ **9.** prime
10. prime   **11.** $(2 - a)(6 + b)$ **12.** $(3y + 8)(3y - 8)$
13. $(9a + 11b)(9a - 11b)$ **14.** $(x + 8)^2$ **15.** $(2x - 7y)^2$
16. $3t^2(2t + 9)(t - 4)$ **17.** $(r - 5)(r^2 + 5r + 25)$
18. $8(k + 2)(k^2 - 2k + 4)$ **19.** $(x^2 + 9)(x + 3)(x - 3)$
20. $(3x + 2y)(3x - 2y)(9x^2 + 4y^2)$ [5.5] **21.** $\{-3, 9\}$
22. $\left\{\frac{1}{2}, 6\right\}$ **23.** $\left\{-\frac{2}{5}, \frac{2}{5}\right\}$ **24.** $\{0, 9\}$ **25.** $\{10\}$ **26.** $\left\{-8, -\frac{5}{2}, \frac{1}{3}\right\}$
[5.6] **27.** $-2, -1$ **28.** 6 ft by 9 ft   **29.** 17 ft   **30.** \$13,987 billion

Chapters R–5 Cumulative Review Exercises

[R.1] **1.** $\frac{1}{2}$ **2.** $\frac{11}{18}$ **3.** $\frac{5}{9}$ **4.** $\frac{1}{2}$ [2.1–2.4] **5.** $\{0\}$ **6.** $\{0.05\}$
[1.7, 2.4] **7. (a)** equation; $\{6\}$ **(b)** expression; $\frac{1}{6}m - 1$
[2.6] **8.** $P = \dfrac{A}{1 + rt}$ **9.** $110°$ and $70°$ [2.5] **10.** gold: 9; silver: 8;
bronze: 6   [R.2, 2.7] **11.** 410; 340; 23%; 16%
[3.1] **12. (a)** negative; positive   **(b)** negative; negative
[3.2, 3.3] **13. (a)** $(-2, 0)$, $(0, -4)$ [3.3–3.5] **14. (a)** 135; A slope of
(b) -2 **(c)**   135 means that the number of digital
photos estimated to have been taken
worldwide increased by 135 billion
photos per year.   **(b)** (2014, 800)

[4.1, 4.2] **15.** $\frac{16}{9}$ **16.** 256   **17.** $\dfrac{1}{p^2}$ **18.** $\dfrac{1}{m^6}$

[4.4] **19.** $-4k^2 - 4k + 8$ [4.5] **20.** $45x^2 + 3x - 18$

[4.6] **21.** $9p^2 + 12p + 4$ [4.7] **22.** $4x^3 + 6x^2 - 3x + 10$

[5.2–5.4] **23.** $(2a - 1)(a + 4)$ **24.** $(4t + 3v)(2t + v)$ **25.** $(2p - 3)^2$

26. $(5r + 9t)(5r - 9t)$ [5.5] **27.** $\left\{-\frac{2}{3}, \frac{1}{2}\right\}$ [5.6] **28.** 5 m, 12 m, 13 m

6　RATIONAL EXPRESSIONS AND APPLICATIONS

Section 6.1

1. $3; 3; -5$　**3.** B, D　**5.** B　**7.** $-3; -3; -3; -6; \frac{3}{5}$

9. (a) $\frac{7}{10}$　(b) $\frac{8}{15}$　**11.** (a) 0　(b) -1　**13.** (a) $-\frac{64}{15}$　(b) undefined

15. (a) undefined　(b) $\frac{8}{25}$　**17.** (a) 0　(b) 0　**19.** (a) 0

(b) undefined　**21.** $x \neq 0$　**23.** $y \neq 0$　**25.** $x \neq 6$　**27.** $x \neq -\frac{5}{3}$

29. $m \neq -3, m \neq 2$　**31.** It is never undefined.　**33.** It is never

undefined.　**35.** numerator: $x^2, 4x$; denominator: $x, 4$　**37.** $\frac{3}{7}$　**39.** $\frac{3}{11}$

41. $3r^2$　**43.** $\frac{2}{5}$　**45.** $\frac{x-1}{x+1}$　**47.** $\frac{7}{5}$　**49.** $\frac{6}{7}$　**51.** $m-n$　**53.** $\frac{2}{t-3}$

55. $\frac{3(2m+1)}{4}$　**57.** $\frac{3m}{5}$　**59.** $\frac{3r-2s}{3}$　**61.** $\frac{1}{x+6}$　**63.** $\frac{x+3}{x-3}$

65. $\frac{13x}{7}$　**67.** $k-3$　**69.** $\frac{x-3}{x+1}$　**71.** $\frac{x+1}{x-1}$　**73.** $\frac{x+2}{x-4}$　**75.** $-\frac{3}{7t}$

77. $\frac{z-3}{z+5}$　**79.** $\frac{r+s}{r-s}$　**81.** $\frac{a+b}{a-b}$　**83.** $\frac{m+n}{2}$　**85.** $\frac{x^2+1}{x}$

87. $1-p+p^2$　**89.** x^2+3x+9　**91.** $-\frac{b^2+ba+a^2}{a+b}$　**93.** $\frac{k^2-2k+4}{k-2}$

95. $\frac{z+3}{z}$　**97.** $\frac{1-2r}{2}$　**99.** -1　**101.** $-(m+1)$　**103.** -1

105. It is already in lowest terms.　**107.** -2　**109.** $-x$

Answers may vary in Exercises 111, 113, and 115.

111. $\frac{-(x+4)}{x-3}, \frac{-x-4}{x-3}, \frac{x+4}{-(x-3)}, \frac{x+4}{-x+3}$

113. $\frac{-(2x-3)}{x+3}, \frac{-2x+3}{x+3}, \frac{2x-3}{-(x+3)}, \frac{2x-3}{-x-3}$

115. $\frac{-(3x-1)}{5x-6}, \frac{-3x+1}{5x-6}, \frac{3x-1}{-(5x-6)}, \frac{3x-1}{-5x+6}$

117. (a) 0　(b) 1.6　(c) 4.1　(d) The number of vehicles waiting

also increases.　**119.** x^2+3　**121.** Both yield $2x+3$.　**122.** Both

yield $2x+1$.　**123.** Both yield x^2+1.　**124.** Both yield x^2+2.

Section 6.2

1. (a) B　(b) D　(c) C　(d) A　**3.** $\frac{5}{12}$　**5.** $\frac{3a}{2}$　**7.** $\frac{40y^2}{3}$　**9.** $\frac{2}{c+d}$

11. $4(x-y)$　**13.** $\frac{t^2}{2}$　**15.** $\frac{x+3}{2x}$　**17.** $\frac{16q}{3p^3}$　**19.** $\frac{7}{r^2+rp}$

21. $\frac{z^2-9}{z^2+7z+12}$　**23.** $x-2; 3; x-2; 5; \frac{3}{4}$　**25.** $\frac{16}{13}$　**27.** 5

29. $-\frac{3}{2t^4}$　**31.** $\frac{1}{4}$　**33.** $-\frac{35}{8}$　**35.** $\frac{2(x+2)}{x(x-1)}$　**37.** $\frac{x(x-3)}{6}$

39. $\frac{5(x-4)}{x^2(x+4)}$　**41.** $\frac{-4t(t+1)}{t-1}$　**43.** $\frac{10}{9}$　**45.** $-\frac{3}{4}$　**47.** $-\frac{9}{2}$

49. $\frac{p+4}{p+2}$　**51.** -1　**53.** $-\frac{m+2}{m+1}$　**55.** $\frac{(2x-1)(x+2)}{x-1}$

57. $\frac{(k-1)^2}{(k+1)(2k-1)}$　**59.** $\frac{4k-1}{3k-2}$　**61.** $\frac{m+4p}{m+p}$　**63.** $\frac{m+6}{m+3}$　**65.** $\frac{y+3}{y+4}$

67. $\frac{m}{m+5}$　**69.** $\frac{r+6s}{r+s}$　**71.** $\frac{(q-3)^2(q+2)^2}{q+1}$　**73.** $\frac{3-a-b}{2a-b}$

75. $-\frac{(x+y)^2(x^2-xy+y^2)}{3y(y-x)(x-y)}$, or $\frac{(x+y)^2(x^2-xy+y^2)}{3y(x-y)^2}$

77. $\frac{x+10}{10}$　**79.** $\frac{5xy^2}{4q}$

Section 6.3

1. B　**3.** The factor x should appear in the LCD the *greatest* number of

times it appears in any single denominator, not the *total* number of times.

The correct LCD is $50x^4$.　**5.** $5; 5$; one; $5; 2; 5; 50$　**7.** 60　**9.** 1800

11. x^5　**13.** $30p$　**15.** $84r^5$　**17.** $50m^4$　**19.** $15a^5b^3$　**21.** r^9t^3

23. $(x+1)(x-1)$　**25.** $12p(p-2)$　**27.** $28m^2(3m-5)$

29. $30(b-2)$　**31.** $18(r-2)$　**33.** $c-d$ or $d-c$

35. $m-3$ or $3-m$　**37.** $p-q$ or $q-p$　**39.** $2(x+1)(x-1)$

41. $3(x-4)^2$　**43.** $12p(p+5)^2$　**45.** $8(y+2)(y+1)$

47. $k(k+5)(k-2)$　**49.** $a(a+6)(a-3)$

51. $(p+3)(p+5)(p-6)$　**53.** $(k+3)(k-5)(k+7)(k+8)$

55. $\frac{20}{55}$　**57.** $\frac{-45}{9k}$　**59.** $\frac{60m^2k^3}{32k^4}$　**61.** $\frac{57z}{6z-18}$　**63.** $\frac{-4a}{18a-36}$

65. $\frac{14(z-2)}{z(z-3)(z-2)}$　**67.** $\frac{(t-r)(4r-t)}{t^3-r^3}$　**69.** $\frac{2y(z-y)(y-z)}{y^4-z^3y}$, or

$\frac{-2y(y-z)^2}{y^4-z^3y}$　**71.** $\frac{36r(r+1)}{(r-3)(r+2)(r+1)}$　**73.** $\frac{ab(a+2b)}{2a^3b+a^2b^2-ab^3}$

75. 7　**76.** 1　**77.** identity property of multiplication　**78.** 7　**79.** 1

80. identity property of multiplication

Section 6.4

1. E　**3.** C　**5.** B　**7.** G　**9.** *Each* term in the numerator of the second

expression must be subtracted. Using parentheses will help avoid this

error. The correct answer is $\frac{x-1}{x+5}$.　**11.** $\frac{2}{3}$　**13.** $\frac{11}{m}$　**15.** $\frac{4}{y+4}$

17. 1　**19.** 4　**21.** $\frac{m-1}{m+1}$　**23.** b　**25.** x　**27.** $y-6$　**29.** $\frac{1}{x-3}$

31. -1　**33.** $\frac{17}{30}$　**35.** $\frac{3z+5}{15}$　**37.** $\frac{10-7r}{14}$　**39.** $\frac{-3x-2}{4x}$　**41.** $\frac{61}{28t}$

43. $\frac{x+1}{2}$　**45.** $\frac{5x+9}{6x}$　**47.** $\frac{7-6p}{3p^2}$　**49.** $\frac{-k-8}{k(k+4)}$　**51.** $\frac{x+4}{x+2}$

53. $\frac{6m^2+23m-2}{(m+2)(m+1)(m+5)}$　**55.** $\frac{4y^2-y+5}{(y+1)^2(y-1)}$　**57.** $\frac{3}{t}$

59. $m-2; 2-m$　**61.** $\frac{-2}{x-5}$, or $\frac{2}{5-x}$　**63.** -4

65. $\frac{-5}{x-y^2}$, or $\frac{5}{y^2-x}$　**67.** $\frac{x+y}{5x-3y}$, or $\frac{-x-y}{3y-5x}$

69. $\frac{-6}{4p-5}$, or $\frac{6}{5-4p}$　**71.** 3　**73.** $\frac{-m-n}{2(m-n)}$

75. $\frac{-x^2+6x+11}{(x+3)(x-3)(x+1)}$　**77.** $\frac{-5q^2-13q+7}{(3q-2)(q+4)(2q-3)}$　**79.** $y-7$

81. $\frac{7x+31}{x+4}$　**83.** $\frac{-5x+13}{4x}$　**85.** $\frac{2x^2+6x}{(x-7)(x-3)}$, or $\frac{2x(x+3)}{(x-7)(x-3)}$

87. $\dfrac{2a + 21}{3(a - 2)}$ **89.** $\dfrac{x - 8}{2(x - 3)}$, or $\dfrac{8 - x}{2(3 - x)}$ **91.** $\dfrac{1}{x - 2}$

93. $\dfrac{9r + 2}{r(r + 2)(r - 1)}$ **95.** $\dfrac{2(x^2 + 3xy + 4y^2)}{(x + y)(x + y)(x + 3y)}$,

or $\dfrac{2(x^2 + 3xy + 4y^2)}{(x + y)^2(x + 3y)}$ **97.** $\dfrac{15r^2 + 10ry - y^2}{(3r + 2y)(6r - y)(6r + y)}$

99. **(a)** $\dfrac{9k^2 + 6k + 26}{5(3k + 1)}$ **(b)** $\frac{1}{4}$ **101.** $\dfrac{8000 + 10x}{49(101 - x)}$

Section 6.5

1. division   **3.** **(a)** $6; \frac{1}{6}$ **(b)** $12; \frac{3}{4}$ **(c)** $\frac{1}{6} \div \frac{3}{4}$ **(d)** $\frac{2}{9}$ **5.** Choice D
is correct, because every sign has been changed in the fraction. This

means it was multiplied by $\frac{-1}{-1} = 1$.   **7.** -6 **9.** $\frac{31}{50}$ **11.** $\dfrac{1}{xy}$

13. $\dfrac{1}{6pq}$ **15.** $\dfrac{2a^2b}{3}$ **17.** $\dfrac{m(m + 2)}{3(m - 4)}$ **19.** $\dfrac{2}{x}$ **21.** $\dfrac{8}{x}$ **23.** $\dfrac{a^2 - 5}{a^2 + 1}$

25. $\frac{31}{50}$ **27.** $\dfrac{y^2 + x^2}{xy(y - x)}$ **29.** $\dfrac{40 - 12p}{85p}$, or $\dfrac{4(10 - 3p)}{85p}$ **31.** $\dfrac{5y - 2x}{3 + 4xy}$

33. $\dfrac{a - 2}{2a}$ **35.** $\dfrac{z - 5}{4}$ **37.** $\dfrac{-m}{m + 2}$ **39.** $\dfrac{x + 8}{-x + 7}$

41. $\dfrac{x^2 + y^2}{x^2 - y^2}$, or $\dfrac{x^2 + y^2}{(x + y)(x - y)}$ **43.** $\dfrac{ab}{b + a}$ **45.** $\dfrac{3m(m - 3)}{(m - 1)(m - 8)}$

47. $\dfrac{2x - 7}{3x + 1}$ **49.** $\dfrac{y + 4}{y - 8}$ **51.** $\dfrac{30}{(a + b)(a - b)}$ **53.** $\dfrac{x + y}{x^2 + xy + y^2}$

55. The negative exponents are on terms, not factors. Terms with
negative exponents cannot be simply moved across a fraction bar.

57. $\dfrac{x^2y^2}{y^2 + x^2}$ **59.** $\dfrac{y^2 + x^2}{xy^2 + x^2y}$, or $\dfrac{y^2 + x^2}{xy(y + x)}$ **61.** $\dfrac{p^2 + k}{p^2 - 3k}$ **63.** $\dfrac{1}{2xy}$

65. $\dfrac{x - 3}{x - 5}$ **67.** $\frac{5}{3}$ **69.** $\frac{13}{2}$ **71.** $\dfrac{19r}{15}$ **73.** $\dfrac{\frac{3}{8} + \frac{5}{6}}{2}$ **74.** $\frac{29}{48}$ **75.** $\frac{29}{48}$

76. Answers will vary.   **77.** 6   **78.** $\frac{1}{3}$ **79.** -3 **80.** $-\frac{7}{5}$

Section 6.6

1. 12   **3.** xyz **5.** expression; $\frac{43}{40}x$ **7.** equation; $\left\{\frac{40}{43}\right\}$ **9.** expression;

$-\frac{1}{10}y$ **11.** equation; $\{-10\}$ **13.** $\{-6\}$ **15.** $\{-15\}$ **17.** $\{7\}$

19. $\{-6\}$ **21.** $\{-15\}$ **23.** $\{-5\}$ **25.** $\{-14\}$ **27.** $\{9\}$

29. $\{12\}$ **31.** $\{5\}$ **33.** $\{1\}$ **35.** $\{2\}$ **37.** $x \neq -2, x \neq 0$

39. $x \neq -3, x \neq 4, x \neq -\frac{1}{2}$ **41.** $x \neq -9, x \neq 1, x \neq -2, x \neq 2$

43. $\left\{\frac{1}{4}\right\}$ **45.** $\left\{-\frac{3}{4}\right\}$ **47.** $\{-5\}$ **49.** $\varnothing$ **51.** $\{3\}$ **53.** $\{-2, 12\}$

55. $\left\{-\frac{1}{5}, 3\right\}$ **57.** $\left\{-\frac{3}{5}, 3\right\}$ **59.** $\{3\}$ **61.** $\{-4\}$ **63.** $\varnothing$

65. $\{-1\}$ **67.** $\{-3\}$ **69.** $\left\{x \mid x \neq \pm\frac{4}{3}\right\}$ **71.** $\{-6\}$ **73.** $\{0\}$

75. $\left\{\frac{7}{2}\right\}$ **77.** $\left\{-6, \frac{1}{2}\right\}$ **79.** $\{6\}$ **81.** This is an expression, *not* an
equation. The student multiplied by the LCD, 14, instead of writing each

coefficient with the LCD. The answer is $\frac{31}{14}t$.   **83.** $F = \dfrac{ma}{k}$ **85.** $a = \dfrac{kF}{m}$

87. $y = mx + b$ **89.** $R = \dfrac{E - Ir}{I}$, or $R = \dfrac{E}{I} - r$ **91.** $\mathcal{A} = \dfrac{h(B + b)}{2}$

93. $a = \dfrac{2S - ndL}{nd}$, or $a = \dfrac{2S}{nd} - L$ **95.** $y = \dfrac{xz}{x + z}$

97. $t = \dfrac{rs}{rs - 2s - 3r}$, or $t = \dfrac{-rs}{-rs + 2s + 3r}$ **99.** $z = \dfrac{3y}{5 - 9xy}$, or

$z = \dfrac{-3y}{9xy - 5}$ **101.** $t = \dfrac{2x - 1}{x + 1}$, or $t = \dfrac{-2x + 1}{-x - 1}$ **103.** **(a)** $x \neq -3$

(b) $x \neq -1$ **(c)** $x \neq -3, x \neq -1$ **104.** $\dfrac{15}{2x}$ **105.** If $x = 0$,

the divisor R is equal to 0, and division by 0 is undefined.

106. $(x + 3)(x + 1)$ **107.** $\dfrac{7}{x + 1}$ **108.** $\dfrac{11x + 21}{4x}$ **109.** $\varnothing$

110. We know that -3 is not allowed because P and R are undefined for
$x = -3$.

SUMMARY EXERCISES   Simplifying Rational Expressions vs. Solving Rational Equations

1. expression; $\dfrac{10}{p}$ **2.** expression; $\dfrac{y^3}{x^3}$ **3.** expression; $\dfrac{1}{2x^2(x + 2)}$

4. equation; $\{9\}$ **5.** equation; $\{39\}$ **6.** expression;

$\dfrac{5k + 8}{k(k - 4)(k + 4)}$ **7.** expression; $\dfrac{y + 2}{y - 1}$ **8.** expression; $\dfrac{t - 5}{3(2t + 1)}$

9. expression; $\dfrac{13}{3(p + 2)}$ **10.** equation; $\left\{-1, \frac{12}{5}\right\}$ **11.** equation;

$\left\{\frac{1}{7}, 2\right\}$ **12.** expression; $\dfrac{16}{3k}$ **13.** expression; $\dfrac{7}{12z}$ **14.** equation; $\{13\}$

15. expression; $\dfrac{3m + 5}{(m + 3)(m + 2)(m + 1)}$ **16.** expression; $\dfrac{k + 3}{5(k - 1)}$

17. equation; $\varnothing$ **18.** equation; $\varnothing$ **19.** expression; $\dfrac{t + 2}{2(2t + 1)}$

20. equation; $\{-7\}$

Section 6.7

1. into a headwind: $(m - 5)$ mph; with a tailwind: $(m + 5)$ mph

3. $\frac{1}{10}$ job per hr   **5.** **(a)** the amount   **(b)** $5 + x$ **(c)** $\dfrac{5 + x}{6} = \dfrac{13}{3}$

7. x represents the original numerator; $\dfrac{x + 3}{(x + 6) + 3} = \dfrac{5}{7}; \dfrac{12}{18}$

9. x represents the original numerator; $\dfrac{x + 2}{3x - 2} = 1; \dfrac{2}{6}$

11. x represents the number; $\frac{1}{6}x = x + 5; -6$ **13.** x represents

the quantity; $x + \frac{3}{4}x + \frac{1}{2}x + \frac{1}{3}x = 93; 36$ **15.** 18.809 min

17. 326.307 m per min   **19.** 3.371 hr   **21.** $\dfrac{8}{4 - x} = \dfrac{24}{4 + x}$ **23.** 8 mph

25. 32 mph   **27.** 165 mph   **29.** 3 mph   **31.** 18.5 mph

33. $\frac{1}{8}t + \frac{1}{6}t = 1$, or $\frac{1}{8} + \frac{1}{6} = \frac{1}{t}$ **35.** 12 hr   **37.** $2\frac{2}{5}$ hr   **39.** 10 hr

41. 3 hr   **43.** 10 mL   **45.** $\frac{15}{8}$ hr, or $1\frac{7}{8}$ hr

46. **(a)** The player multiplied: $5 \cdot 3 = 15$.

(b) The player added: $5 + 3 = 8$.   **(c)** The player added the two times
and divided by 2—that is, he averaged the times: $\dfrac{5 + 3}{2} = 4$.

47. Solve the equation $\frac{1}{a}x + \frac{1}{b}x = 1$ to obtain $x = \frac{a \cdot b}{a + b}$.

48. 10 hr; 10 hr; The same answer results.

Chapter 6 Review Exercises

1. (a) $-\frac{4}{7}$　**(b)** -16　**2. (a)** $\frac{11}{8}$　**(b)** $\frac{13}{22}$　**3. (a)** undefined　**(b)** 1

4. (a) undefined　**(b)** $\frac{1}{2}$　**5.** $x \neq 3$　**6.** $y \neq 0$　**7.** $k \neq -5, k \neq -\frac{2}{3}$

8. $m \neq -1, m \neq 3$　**9.** $\frac{b}{3a}$　**10.** -1　**11.** $\frac{-(2x+3)}{2}$　**12.** $\frac{2p + 5q}{5p + q}$

13. $\frac{x+3}{y+4}$　**14.** $x^2 + x + 1$

Answers may vary in Exercises 15 and 16.

15. $\frac{-(4x-9)}{2x+3}, \frac{-4x+9}{2x+3}, \frac{4x-9}{-(2x+3)}, \frac{4x-9}{-2x-3}$

16. $\frac{-(8-3x)}{3-6x}, \frac{-8+3x}{3-6x}, \frac{8-3x}{-(3-6x)}, \frac{8-3x}{-3+6x}$

17. $\frac{72}{p}$　**18.** 2　**19.** $\frac{2}{3m^6}$　**20.** $\frac{3}{2}$　**21.** $\frac{5}{8}$　**22.** $\frac{r+4}{3}$

23. $\frac{3a-1}{a+5}$　**24.** $\frac{y-2}{y-3}$　**25.** $\frac{p+5}{p+1}$　**26.** $\frac{3z+1}{z+3}$　**27.** $\frac{2(p+2q)}{p-2q}$

28. 2　**29.** 96　**30.** $108y^4$　**31.** $m(m+2)(m+5)$

32. $(x+3)(x+1)(x+4)$　**33.** $\frac{35}{56}$　**34.** $\frac{40}{4k}$　**35.** $\frac{15a}{10a^4}$　**36.** $\frac{-54}{18-6x}$

37. $\frac{15y}{50-10y}$　**38.** $\frac{4b(b+2)}{(b+3)(b-1)(b+2)}$　**39.** $\frac{15}{x}$　**40.** $-\frac{2}{p}$

41. $\frac{4k-45}{k(k-5)}$　**42.** $\frac{28+11y}{y(7+y)}$　**43.** $\frac{-2-3m}{6}$　**44.** $\frac{3(16-x)}{4x^2}$

45. $\frac{7a+6b}{(a-2b)(a+2b)}$　**46.** $\frac{-k^2-6k+3}{3(k+3)(k-3)}$　**47.** $\frac{5z-16}{z(z+6)(z-2)}$

48. $\frac{-13p+33}{p(p-2)(p-3)}$　**49.** $\frac{10}{13}$　**50.** $\frac{a}{b}$　**51.** $\frac{4(y-3)}{y+3}$

52. $\frac{xw+1}{xw-1}$　**53.** $\frac{(q-p)^2}{pq}$　**54.** $(x+4)(x+3)$, or $x^2 + 7x + 12$

55. $\frac{1-r-t}{1+r+t}$　**56.** $\frac{y+x}{xy}$　**57.** $\left\{\frac{35}{6}\right\}$　**58.** $\{-16\}$　**59.** $\{-4\}$

60. $\varnothing$　**61.** $\{0\}$　**62.** $\{3\}$　**63.** $t = \frac{Ry}{m}$　**64.** $s = br - t$

65. $d = \frac{b-ac}{a}$, or $d = \frac{b}{a} - c$　**66.** $t = \frac{rs}{s-r}$　**67.** $\frac{20}{15}$　**68.** $\frac{3}{18}$

69. 2.020 hr　**70.** 10 mph　**71.** $3\frac{1}{13}$ hr　**72.** 2 hr

Chapter 6 Mixed Review Exercises

1. $\frac{m+7}{(m-1)(m+1)}$　**2.** $8p^2$　**3.** $\frac{1}{6}$　**4.** 3　**5.** $\frac{z+7}{(z+1)(z-1)^2}$

6. $\frac{-t-1}{(t+2)(t-2)}$, or $\frac{t+1}{(2+t)(2-t)}$　**7.** $\frac{x-7}{2x+3}$　**8.** $\{4\}$　**9.** $\{-2, 3\}$

10. $\{2\}$　**11.** $\{2\}$　**12.** $v = at + w$　**13.** 3　**14.** 2 hr

15. 150 km per hr

Chapter 6 Test

[6.1]　**1. (a)** $\frac{11}{6}$　**(b)** undefined　**2.** $x \neq -2, x \neq 4$

3. (Answers may vary.) $\frac{-(6x-5)}{2x+3}, \frac{-6x+5}{2x+3}, \frac{6x-5}{-(2x+3)}, \frac{6x-5}{-2x-3}$

4. $-3x^2y^3$　**5.** $\frac{3a+2}{a-1}$　[6.2]　**6.** $\frac{25}{27}$　**7.** $\frac{3k-2}{3k+2}$　**8.** $\frac{a-1}{a+4}$

9. $\frac{x-5}{3-x}$　[6.3]　**10.** $150p^5$　**11.** $(2r+3)(r+2)(r-5)$

12. $\frac{240p^2}{64p^3}$　**13.** $\frac{21}{42m-84}$　[6.4]　**14.** 2　**15.** $\frac{-14}{5(y+2)}$

16. $\frac{-x^2+x+1}{3-x}$, or $\frac{x^2-x-1}{x-3}$　**17.** $\frac{-m^2+7m+2}{(2m+1)(m-5)(m-1)}$

[6.5]　**18.** $\frac{2k}{3p}$　**19.** $(x-5)(x-3)$, or $x^2 - 8x + 15$　**20.** $\frac{-2-x}{4+x}$

[6.6]　**21.** $\left\{-\frac{1}{2}, 1\right\}$　**22.** $\left\{-\frac{1}{2}, 5\right\}$　**23.** $\varnothing$　**24.** $\left\{-\frac{1}{2}\right\}$

25. $D = \frac{dF-k}{F}$, or $D = d - \frac{k}{F}$　[6.7]　**26.** $2\frac{2}{9}$ hr　**27.** 3 mph　**28.** -4

Chapters R–6 Cumulative Review Exercises

[R.1]　**1.** $\frac{71}{28}$　**2.** $\frac{5}{24}$　[R.2]　**3.** -24.166　**4.** 5.76　[R.1, 1.7]　**5.** 2

[2.3]　**6.** $\{17\}$　[2.6]　**7.** $b = \frac{2\mathcal{A}}{h}$　[2.7]　**8.** $\left\{-\frac{2}{7}\right\}$

[2.9]　**9.**　　**10.**　

　　$[-8, \infty)$　　　　　　$(4, \infty)$

[3.2, 3.3]　**11. (a)** $(-3, 0)$　**(b)** $(0, -4)$　**12.** $-\frac{4}{3}$

[3.1, 3.2]　**13.**　　[4.4]　**14.**　

[4.1, 4.2]　**15.** $\frac{1}{16x^7}$　**16.** $\frac{1}{m^6}$　[4.4]　**17.** $k^2 + 2k + 1$

[4.6]　**18.** $4a^2 - 4ab + b^2$　[4.5]　**19.** $3y^3 + 8y^2 + 12y - 5$

[4.7]　**20.** $6p^2 + 7p + 1 + \frac{3}{p-1}$　[5.3]　**21.** $(4t+3v)(2t+v)$

[5.4]　**22.** $(4x^2+1)(2x+1)(2x-1)$　[5.5]　**23.** $\{-3, 5\}$

24. $\left\{5, -\frac{1}{2}, \frac{2}{3}\right\}$　[5.6]　**25.** 6 m　**26.** -2 or -1　[6.1]　**27.** A　**28.** D

[6.4]　**29.** $\frac{3r+28}{7r}$　**30.** $\frac{7}{15(q-4)}$　[6.2]　**31.** $\frac{7(2z+1)}{24}$

[6.5]　**32.** $\frac{195}{29}$　[6.6]　**33.** $\left\{\frac{21}{2}\right\}$　**34.** $\{-2, 1\}$　[6.7]　**35.** $1\frac{1}{5}$ hr

7 LINEAR EQUATIONS, GRAPHS, AND SYSTEMS

Section 7.1

1. (a) I (b) III (c) II (d) IV (e) none (f) none

3. The student interchanged the x- and y-coordinates. To plot this point correctly, move from 0 to the left 4 units on the x-axis and then up 2 units parallel to the y-axis.

5. $-4; -3; -2; -1; 0$ 7. $\frac{5}{2}; 5; \frac{3}{2}; 1$

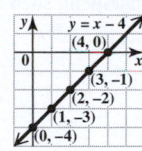 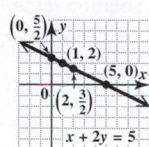

9. (a) C (b) D (c) B (d) A

11. $(6,0); (0,4)$ 13. $(6,0); (0,-2)$ 15. $(0,0); (0,0)$

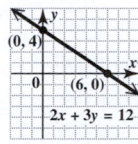 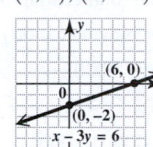 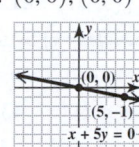

17. $(0,0); (0,0)$ 19. none; $(0,5)$ 21. $(2,0)$; none

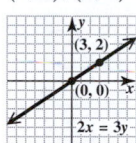 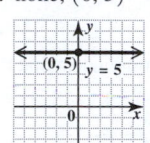 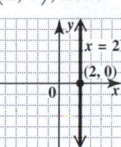

23. (a) $(-2,0); (0,3)$ (b) B   25. (a) none; $(0,-1)$ (b) A
(c) 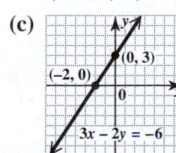   (c)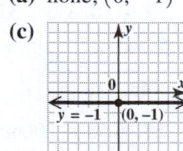

27. $(-5,-1)$ 29. $\left(\frac{9}{2}, -\frac{3}{2}\right)$ 31. $\left(0, \frac{11}{2}\right)$ 33. $(2.1, 0.9)$ 35. A, B, D, F

37. (a) 2 (b) 0 (c) undefined (d) $-\frac{1}{3}$ (e) 1 (f) -4
(g) -1 (h) $\frac{7}{4}$ 39. The x-values in the denominator should be subtracted in the same order as the y-values in the numerator. The denominator should be $2-(-4)$. The correct slope is $\frac{2}{6} = \frac{1}{3}$.

41. (a) 8 (b) rises   43. (a) $\frac{5}{6}$ (b) rises   45. (a) 0
(b) horizontal   47. (a) $-\frac{1}{2}$ (b) falls   49. (a) undefined
(b) vertical   51. (a) -1 (b) falls   53. -2 55. $\frac{4}{3}$
57. $-\frac{5}{2}$ 59. undefined

In part (a) of Exercises 61 and 63, we used the intercepts. Other points can be used.

61. (a) $(4,0)$ and $(0,-8)$; 2 (b) $y=2x-8$; 2 (c) $A=2, B=-1$; 2
63. (a) $(-3,0)$ and $(0,-3)$; -1 (b) $y=-x-3$; -1
(c) $A=1, B=1$; -1

65. $-\frac{1}{2}$ 67. 4   69. undefined

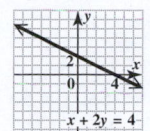 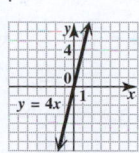 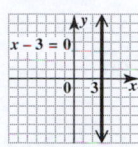

71. 0   73. 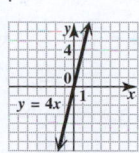   75.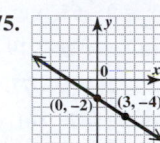

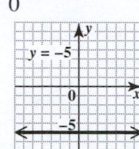

77. 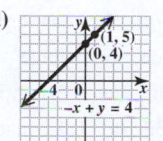   79.   81.

83.   85. parallel   87. perpendicular   89. neither
91. parallel   93. neither   95. perpendicular
97. perpendicular

99. $-\$4000$ per year; The value of the machine is decreasing $\$4000$ per year during these years.   101. 0% per year (or no change); The percent of pay raise is not changing—it is 3% per year during these years.   103. (a) In 2016, there were 396 million wireless subscriber connections in the U.S. (b) 16 (c) The number of subscribers increased by an average of 16 million per year from 2011 to 2016.   105. (a) -9 theaters per yr (b) The negative slope means that the number of drive-in theaters decreased by an average of 9 per year from 2010 to 2017.   107. $\$0.04$ per year; The price of a gallon of gasoline increased by an average of $\$0.04$ per year from 2000 to 2016.   109. -16 million digital cameras per year; The number of digital cameras sold decreased by an average of 16 million per year from 2010 to 2016.

Section 7.2

1. A   3. A   5. $3x+y=10$ 7. A   9. C   11. H   13. B
15. $y=5x+15$ 17. $y=-\frac{2}{3}x+\frac{4}{5}$ 19. $y=x-1$
21. $y=\frac{2}{5}x+5$ 23. $y=\frac{2}{3}x+1$ 25. $y=-x-2$ 27. $y=2x-4$
29. $y=-\frac{3}{5}x+3$

31. (a) $y=x+4$ (b) 1   33. (a) $y=-\frac{6}{5}x+6$ (b) $-\frac{6}{5}$
(c) $(0,4)$ (c) $(0,6)$
(d)   (d)

35. (a) $y=\frac{4}{5}x-4$ (b) $\frac{4}{5}$ 37. (a) $y=-\frac{1}{2}x-2$ (b) $-\frac{1}{2}$
(c) $(0,-4)$ (c) $(0,-2)$
(d)   (d)

39. (a) $y = -2x + 18$ **(b)** $2x + y = 18$ **41. (a)** $y = -\frac{3}{4}x + \frac{5}{2}$
(b) $3x + 4y = 10$ **43. (a)** $y = \frac{1}{2}x + \frac{13}{2}$ **(b)** $x - 2y = -13$
45. (a) $y = 4x - 12$ **(b)** $4x - y = 12$ **47. (a)** $y = 1.4x + 4$
(b) $7x - 5y = -20$ **49.** $2x - y = 2$ **51.** $x + 2y = 8$ **53.** $y = 5$
55. $x = 7$ **57.** $y = -3$ **59.** $2x - 13y = -6$ **61.** $y = 5$
63. $x = 9$ **65.** $y = -\frac{3}{2}$ **67.** $y = 8$ **69.** $x = 0.5$ **71.** $y = 0$
73. (a) $y = 3x - 19$ **(b)** $3x - y = 19$ **75. (a)** $y = \frac{1}{2}x - 1$
(b) $x - 2y = 2$ **77. (a)** $y = -\frac{1}{2}x + 9$ **(b)** $x + 2y = 18$
79. (a) $y = 7$ **(b)** $y = 7$ **81.** $y = 45x$; $(0, 0)$, $(5, 225)$, $(10, 450)$
83. $y = 3.75x$; $(0, 0)$, $(5, 18.75)$, $(10, 37.50)$ **85.** $y = 150x$; $(0, 0)$,
$(5, 750)$, $(10, 1500)$ **87. (a)** $y = 140x + 18.50$ **(b)** $(5, 718.50)$;
The cost for 5 tickets and the delivery fee is $718.50. **(c)** $298.50
89. (a) $y = 41x + 99$ **(b)** $(5, 304)$; The cost for a 5-month
membership is $304. **(c)** $591 **91. (a)** $y = 90x + 25$ **(b)** $(5, 475)$;
The cost of the plan for 5 months is $475. **(c)** $2185
93. (a) $y = -79.7x + 716$; Sales of e-readers in the United States
decreased by $79.7 million per year from 2013 to 2016.
(b) $636.3 million **(c)** $556.6 million; The result is a little high,
as we might expect. The graphed line lies above the actual data point.
95. (a) The slope would be positive because spending on home health
care is increasing over these years. **(b)** $y = 3.5x + 36$ **(c)** $88.5 billion;
It is very close to the actual value.

Section 7.3

1. system of linear equations; same **3.** inconsistent; no; independent
5. dependent; consistent; infinitely many **7.** The ordered pair $(1, -2)$
is not a solution of the system because it is not a solution of the second
equation, $2x + y = 4$. **9.** B; The ordered-pair solution must be in
quadrant II. **11. (a)** B **(b)** C **(c)** D **(d)** A **13. (a)** $\varnothing$
(b) $\{(x, y) \mid 2x - y = 4\}$ **15.** no **17.** yes **19.** yes **21.** no
23. yes **25.** no
27. $\{(4, 2)\}$ **29.** $\{(0, 4)\}$ **31.** $\{(4, -1)\}$

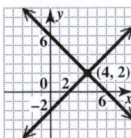 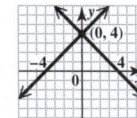 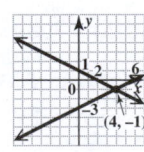

In Exercises 33–49, we do not show the graphs.

33. $\{(1, 3)\}$ **35.** $\{(x, y) \mid 3x + y = 5\}$; dependent equations
37. $\{(0, -3)\}$ **39.** $\{(-4, 0)\}$ **41.** $\varnothing$; inconsistent system
43. $\{(x, y) \mid 3x - y = -6\}$; dependent equations **45.** $\{(4, -3)\}$
47. $\varnothing$; inconsistent system **49.** $\{(0, 0)\}$ **51. (a)** The equations have
graphs that intersect in one point. **(b)** one solution **(c)** neither
53. (a) The equations have graphs that are the same line. **(b)** infinite
number of solutions **(c)** dependent equations **55. (a)** The equations
have graphs that intersect in one point. **(b)** one solution **(c)** neither
57. (a) The equations have graphs that are parallel lines. **(b)** no solution
(c) inconsistent system **59. (a)** The equations have graphs that
are parallel lines. **(b)** no solution **(c)** inconsistent system

61. (a) The equations have graphs that are the same line.
(b) infinite number of solutions **(c)** dependent equations
63. (a) 2012–2014 **(b)** 2015; about $1200 million **(c)** (2015, 1200)
65. 40; 30 **66.** (40, 30) **67.** Supply exceeds demand.
68. Demand exceeds supply.

Section 7.4

1. The student must find the value of y and write the solution as an
ordered pair. The solution set is $\{(3, 0)\}$. **3.** A false statement, such as
$0 = 3$, occurs. **5.** $\{(3, 9)\}$ **7.** $\{(7, 3)\}$ **9.** $\{(-4, 8)\}$
11. $\{(3, -2)\}$ **13.** $\{(-3, 0)\}$ **15.** $\{(0, 5)\}$
17. $\{(x, y) \mid 3x - y = 5\}$ **19.** $\varnothing$ **21.** $\{(x, y) \mid 2x - y = -12\}$
23. $\varnothing$ **25.** $\{(0, 0)\}$ **27.** $\{(\frac{1}{4}, -\frac{1}{2})\}$ **29.** $\{(2, -3)\}$
31. $\{(2, -4)\}$ **33.** $\{(-2, 1)\}$ **35.** $\{(10, 4)\}$ **37.** $\{(4, -9)\}$
39. $\{(5, 0)\}$ **41.** To find the total cost, multiply the number
of bicycles (x) by the cost per bicycle ($400), and add the fixed
cost ($5000). Thus, $y_1 = 400x + 5000$ gives this total cost (in dollars).
42. $y_2 = 600x$ **43.** $y_1 = 400x + 5000$, $y_2 = 600x$; solution set:
$\{(25, 15,000)\}$ **44.** 25; 15,000; 15,000

Section 7.5

1. true **3.** A false statement indicates that the solution set is $\varnothing$.
5. $\{(4, 6)\}$ **7.** $\{(-1, -3)\}$ **9.** $\{(5, 3)\}$ **11.** $\{(-2, 3)\}$
13. $\{(\frac{1}{2}, 4)\}$ **15.** $\{(-3, 4)\}$ **17.** $\{(3, -6)\}$ **19.** $\{(-3, 2)\}$
21. $\{(0, 4)\}$ **23.** $\{(-4, 0)\}$ **25.** $\{(0, 0)\}$ **27.** $\{(7, 4)\}$
29. $\{(-6, 5)\}$ **31.** $\{(0, 3)\}$ **33.** $\{(3, 0)\}$ **35.** $\varnothing$
37. $\{(x, y) \mid x - 3y = -4\}$ **39.** $\{(x, y) \mid x + 3y = 6\}$ **41.** $\varnothing$
43. $\{(-\frac{5}{7}, -\frac{2}{7})\}$ **45.** $\{(\frac{1}{8}, -\frac{5}{6})\}$ **47.** $\{(11, 15)\}$ **49.** $\{(13, -\frac{7}{5})\}$
51. $\{(6, -4)\}$ **53.** $6.55 = 2006a + b$ **54.** $8.65 = 2016a + b$
55. $6.55 = 2006a + b$, $8.65 = 2016a + b$; solution set:
$\{(0.21, -414.71)\}$ **56. (a)** $y = 0.21x - 414.71$ **(b)** 8.44 ($8.44);
This is just $0.01 greater than the actual figure.

SUMMARY EXERCISES Applying Techniques for Solving Systems of Linear Equations

1. (a) Use substitution because the second equation is solved for y.
(b) Use elimination because the coefficients of the y-terms are opposites.
(c) Use elimination because the equations are in $Ax + By = C$ form with
no coefficients of 1 or -1. Solving by substitution would involve fractions.
2. System B is easier to solve by substitution because the second equation
is already solved for y. **3.** $\{(3, 12)\}$ **4.** $\{(-3, 2)\}$ **5.** $\{(\frac{1}{3}, \frac{1}{2})\}$
6. $\varnothing$ **7.** $\{(3, -2)\}$ **8.** $\{(-1, -11)\}$ **9.** $\{(x, y) \mid 2x - 3y = 5\}$
10. $\{(9, 4)\}$ **11.** $\{(\frac{45}{31}, \frac{4}{31})\}$ **12.** $\{(4, -5)\}$ **13.** $\varnothing$ **14.** $\{(\frac{22}{13}, -\frac{23}{13})\}$
15. $\{(0, 0)\}$ **16.** $\{(3, 0)\}$ **17.** $\{(x, y) \mid 3x + y = 7\}$ **18.** $\{(2, -3)\}$
19. $\{(24, -12)\}$ **20.** $\{(3, 2)\}$ **21.** $\{(-4, 2)\}$ **22.** $\{(5, 3)\}$

Section 7.6

1. Answers will vary. Some possible answers are **(a)** two perpendicular walls and the ceiling in a normal room, **(b)** the floors of three different levels of an office building, and **(c)** three pages of a book (because they intersect in the spine). **3.** The statement means that when -1 is substituted for x, 2 is substituted for y, and 3 is substituted for z in the three equations, the resulting three statements are true. **5.** 4 **7.** $\{(3, 2, 1)\}$
9. $\{(1, 4, -3)\}$ **11.** $\{(0, 2, -5)\}$ **13.** $\{(1, 0, 3)\}$ **15.** $\{(1, \frac{3}{10}, \frac{2}{5})\}$
17. $\{(-\frac{7}{3}, \frac{22}{3}, 7)\}$ **19.** $\{(-12, 18, 0)\}$ **21.** $\{(0.8, -1.5, 2.3)\}$
23. $\{(4, 5, 3)\}$ **25.** $\{(2, 2, 2)\}$ **27.** $\{(\frac{8}{3}, \frac{2}{3}, 3)\}$ **29.** $\{(-1, 0, 0)\}$
31. $\{(-4, 6, 2)\}$ **33.** $\{(-3, 5, -6)\}$ **35.** $\varnothing$; inconsistent system
37. $\{(x, y, z) \mid x - y + 4z = 8\}$; dependent equations **39.** $\{(3, 0, 2)\}$
41. $\{(x, y, z) \mid 2x + y - z = 6\}$; dependent equations **43.** $\{(0, 0, 0)\}$
45. $\varnothing$; inconsistent system **47.** $\{(2, 1, 5, 3)\}$ **49.** $\{(-2, 0, 1, 4)\}$
51. $2a + b + c = -5$ **52.** $a - c = 1$ **53.** $3a + 3b + c = -18$
54. $a = 1, b = -7, c = 0$ **55.** $x^2 + y^2 + x - 7y = 0$

Section 7.7

1. **(a)** 6 oz **(b)** 15 oz **(c)** 24 oz **(d)** 30 oz **3.** $1.99x$
5. **(a)** $(10 - x)$ mph **(b)** $(10 + x)$ mph **7.** **(a)** 220 ft **(b)** $\dfrac{d}{44}$ sec
9. length: 78 ft; width: 36 ft **11.** wins: 102; losses: 60 **13.** AT&T: $163.8 billion; Verizon: $125.1 billion **15.** $x = 40$ and $y = 50$, so the angles measure 40° and 50°. **17.** Red Sox: $360.66; Indians: $179.44 **19.** ribeye: $30.30; salmon: $21.60 **21.** Busch Gardens: $90; Universal Studios: $110 **23.** general admission: 76; with student ID: 108 **25.** 25% alcohol: 6 gal; 35% alcohol: 14 gal **27.** nuts: 14 kg; cereal: 16 kg **29.** 2%: $1000; 4%: $2000 **31.** pure acid: 6 L; 10% acid: 48 L **33.** $1.75-per-lb candy: 7 lb; $1.25-per-lb candy: 3 lb **35.** train: 60 km per hr; plane: 160 km per hr **37.** freight train: 50 km per hr; express train: 80 km per hr **39.** boat: 21 mph; current: 3 mph **41.** plane: 300 mph; wind: 20 mph **43.** $x + y + z = 180$; 70°, 30°, 80° **45.** 20°, 70°, 90° **47.** 12 cm, 25 cm, 33 cm **49.** gold: 7; silver: 6; bronze: 6 **51.** upper level: 1170; center court: 985; floor: 130 **53.** bookstore A: 140; bookstore B: 280; bookstore C: 380 **55.** first chemical: 50 kg; second chemical: 400 kg; third chemical: 300 kg **57.** wins: 46; losses: 23; overtime losses: 13 **59.** box of fish: 8 oz; box of bugs: 2 oz; box of worms: 5 oz

Chapter 7 Review Exercises

1. $5; \frac{10}{3}; 2; \frac{14}{3}$

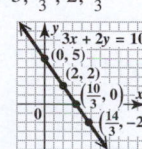

2. $-6; 5; -5; 6$

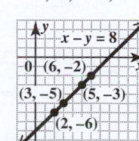

3. $(3, 0); (0, -4)$

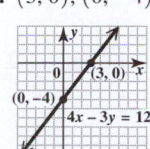

4. $(\frac{28}{5}, 0); (0, 4)$

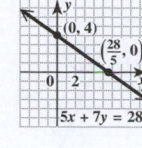

5. $(10, 0); (0, 4)$

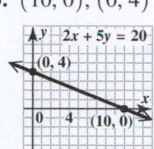

6. $(8, 0); (0, -2)$

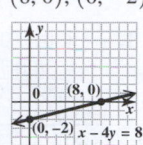

7. $(0, 2)$ **8.** $(-\frac{9}{2}, \frac{3}{2})$ **9.** $-\frac{7}{5}$ **10.** $-\frac{1}{2}$ **11.** undefined **12.** 2
13. $\frac{3}{4}$ **14.** 0 **15.** $-\frac{1}{3}$ **16.** $\frac{2}{3}$ **17.** $-\frac{1}{3}$ **18.** -1 **19.** -3
20. **(a)** positive **(b)** negative **(c)** undefined **(d)** 0
21. perpendicular **22.** parallel **23.** 12 ft **24.** $1057 per year
25. **(a)** $y = \frac{3}{5}x - 8$ **(b)** $3x - 5y = 40$ **26.** **(a)** $y = -\frac{1}{3}x + 5$
(b) $x + 3y = 15$ **27.** **(a)** $y = -9x + 13$ **(b)** $9x + y = 13$
28. **(a)** $y = \frac{7}{5}x + \frac{16}{5}$ **(b)** $7x - 5y = -16$ **29.** **(a)** $y = 4x - 26$
(b) $4x - y = 26$ **30.** **(a)** $y = -\frac{5}{2}x + 1$ **(b)** $5x + 2y = 2$
31. $y = 12$ **32.** $x = 2$ **33.** $x = 0.3$ **34.** $y = 4$
35. **(a)** $y = 229x + 2829$ **(b)** $11,760 **36.** **(a)** $y = 1007.5x + 29,822.5$;
The cost at private 4-year universities increased by an average of $1007.50 per year from 2013 to 2017. **(b)** $44,935 **37.** yes
38. no **39.** $\{(3, 1)\}$ **40.** $\{(0, -2)\}$ **41.** $\varnothing$
42. $\{(x, y) \mid x - 2y = 2\}$ **43.** It would be easiest to solve for x in the second equation because its coefficient is -1. No fractions would be involved. **44.** The false statement $0 = 5$ is an indication that the system has no solution. The solution set is $\varnothing$. **45.** $\{(2, 1)\}$ **46.** $\{(3, 5)\}$
47. $\{(x, y) \mid x + 3y = 6\}$ **48.** $\{(6, 4)\}$ **49.** C **50.** **(a)** 2 **(b)** 9
51. $\{(7, 1)\}$ **52.** $\{(-4, 3)\}$ **53.** $\{(\frac{10}{7}, -\frac{9}{7})\}$ **54.** $\varnothing$
55. $\{(-4, 1)\}$ **56.** $\{(x, y) \mid 2x - 3y = 0\}$ **57.** $\{(9, 2)\}$
58. $\{(8, 9)\}$ **59.** $\{(2, 1)\}$ **60.** $\{(-3, 2)\}$ **61.** $\{(1, -5, 3)\}$
62. $\varnothing$; inconsistent system **63.** $\{(1, 2, 3)\}$ **64.** $\{(5, -1, 0)\}$
65. $\{(x, y, z) \mid 3x - 4y + z = 8\}$; dependent equations
66. $\{(0, 0, 0)\}$ **67.** length: 200 ft; width: 85 ft **68.** plane: 300 mph; wind: 20 mph **69.** $6-per-lb nuts: 30 lb; $3-per-lb candy: 70 lb
70. 8% solution: 5 L; 20% solution: 3 L **71.** gold: 12; silver: 8; bronze: 21
72. 85°, 35°, 60° **73.** 10%: $40,000; 6%: $100,000; 5%: $140,000
74. Mantle: 54; Maris: 61; Berra: 22

Chapter 7 Mixed Review Exercises

1. parallel **2.** perpendicular **3.** -0.575 lb per year; Per capita consumption of potatoes decreased by an average of 0.575 lb per year from 2008 to 2016. **4.** $y = -0.575x + 37.8$ **5.** $y = 3x$ **6.** $x + 2y = 6$
7. Answers will vary.

(a)

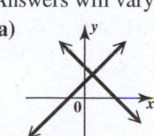

(b)

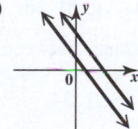

(c)

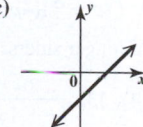

8. B; The second equation is already solved for y. **9.** $\{(0, 4)\}$
10. $\{(5, 3)\}$ **11.** $\{(3, -1)\}$ **12.** $\{(x, y) \mid x + 2y = 48\}$
13. $\{(1, 2, 3)\}$ **14.** $\{(1, 0, -1)\}$ **15.** 20 L **16.** United States: 121; China: 70; Great Britain: 67

Chapter 7 Test

[7.1] 1. $(-3, 0)$; $(0, 4)$　　**2.** $(2, 0)$; none　　**3.** $(0, 0)$; $(0, 0)$

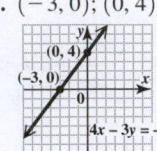
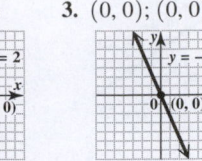

4. $\frac{1}{2}$　**5.** $\frac{3}{2}$; $\left(\frac{13}{3}, 0\right)$; $\left(0, -\frac{13}{2}\right)$　**6.** B　**7.** perpendicular

8. -438 farms per yr; The number of farms decreased an average of 438 farms per year from 2000 to 2016.　**[7.2] 9. (a)** $y = -\frac{2}{5}x + 3$

(b) $2x + 5y = 15$　**10. (a)** $y = -\frac{1}{2}x + 2$　**(b)** $x + 2y = 4$

11. (a) $y = -5x + 19$　**(b)** $5x + y = 19$　**12.** $y = 14$　**13.** $x = 5$

14. (a) $y = -\frac{3}{5}x - \frac{11}{5}$　**(b)** $y = -\frac{1}{2}x - \frac{3}{2}$

15. (a) $y = 142.75x + 45$　**(b)** \$901.50

[7.3] 16. $\{(6, 1)\}$　　**[7.4, 7.5] 17.** $\{(6, -4)\}$　**18.** $\left\{\left(-\frac{9}{4}, \frac{5}{4}\right)\right\}$

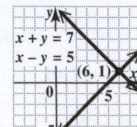

19. $\{(x, y) \mid 12x - 5y = 8\}$; dependent equations

20. $\{(3, 3)\}$　**21.** $\{(0, -2)\}$

22. $\varnothing$; inconsistent system

[7.6] 23. $\left\{\left(-\frac{2}{3}, \frac{4}{5}, 0\right)\right\}$　**24.** $\{(3, -2, 1)\}$　**25.** $\varnothing$

[7.7] 26. *Captain America:* \$408.1 million; *Deadpool:* \$363.1 million

27. slower car: 45 mph; faster car: 75 mph　**28.** 20% solution: 4 L; 50% solution: 8 L　**29.** 25°, 55°, 100°　**30.** Orange Pekoe: 60 oz; Irish Breakfast: 30 oz; Earl Grey: 10 oz

Chapters R–7 Cumulative Review Exercises

[R.1, R.2] 1. (a) $\frac{11}{12}$　**(b)** $\frac{2}{3}$　**(c)** 4.75　**2. (a)** 8　**(b)** 150　**(c)** 0.125

3. (a) 0.01; 1%　**(b)** $\frac{1}{2}$; 50%　**(c)** $\frac{3}{4}$; 0.75　**(d)** 1; 100%

[1.3, 1.4] 4. always true　**5.** never true　**6.** sometimes true; For example, $3 + (-3) = 0$, but $3 + (-1) = 2$ and $2 \neq 0$.　**[1.1, 4.1] 7.** 9

8. -9　**9.** 9　**[1.5, 1.6] 10.** -39　**[2.3, 2.4] 11.** $\left\{\frac{7}{6}\right\}$

12. $\{11\}$　**[2.6] 13.** $x = \dfrac{d - by}{a}$　**[2.9] 14.** $(1, \infty)$

[2.5] 15. pennies: 35; nickels: 29; dimes: 30　**16.** 46°, 46°, 88°

[3.4, 7.2] 17. $y = 6$　**18.** $x = 4$　**[3.3, 7.1] 19.** $-\frac{4}{3}$　**20.** $\frac{3}{4}$

[3.5, 7.2] 21. $4x + 3y = 10$

[3.4, 7.1] 22.

[4.6] 23. $49x^2 + 42xy + 9y^2$

[4.7] 24. $2x^2 + x + 3$

[4.4] 25. $x^3 + 12x^2 - 3x - 7$

[4.3] 26. (a) 4.638×10^{-4}　**(b)** 566,000

[5.1–5.4] 27. $(2w + 7)(8w - 3)$　**28.** $(10x^2 + 9)(10x^2 - 9)$

29. $(2p + 3)(4p^2 - 6p + 9)$　**[5.5] 30.** $\left\{\frac{1}{3}\right\}$　**[5.6] 31.** 4 ft

32. longer siders: 18 in.; distance between: 16 in.

[6.4] 33. $\dfrac{6x + 22}{(x + 1)(x + 3)}$　**[6.2] 34.** $\dfrac{(x + 3)^2}{3x}$

[6.5] 35. 6　**[6.6] 36.** $\{5\}$　**[7.4, 7.5] 37.** $\{(3, -3)\}$

38. $\{(x, y) \mid x - 3y = 7\}$　**[7.6] 39.** $\{(5, 3, 2)\}$

[7.3] 40. (a) $x = 8$, or 800 items; \$3000　**(b)** about \$400

8　INEQUALITIES AND ABSOLUTE VALUE

Section 8.1

1. D　**3.** B　**5.** F　**7. (a)** $x < 100$　**(b)** $100 \le x \le 129$

(c) $130 \le x \le 159$　**(d)** $160 \le x \le 189$　**(e)** $x \ge 190$

9. The student divided by 4, a *positive* number. Reverse the symbol only when multiplying or dividing by a *negative* number. The solution set is $[-16, \infty)$.

11. $[16, \infty)$

13. $(7, \infty)$

15. $(-\infty, -4)$

17. $(-4, \infty)$

19. $(-\infty, -40]$

21. $(-\infty, 4]$

23. $(-\infty, -10]$

25. $(7, \infty)$

27. $\left(-\infty, -\frac{15}{2}\right)$

29. $(-\infty, -7)$

31. $\left[\frac{1}{2}, \infty\right)$

33. $[2, \infty)$

35. $(3, \infty)$

37. $(-\infty, 4)$

39. $\left(-\infty, \frac{23}{6}\right]$

41. $\left(-\infty, \frac{76}{11}\right)$

43. $(-\infty, \infty)$

45. $\varnothing$

47. $(1, 11)$

49. $[-14, 10]$

51. $[-5, 6]$

53. $\left[-\frac{14}{3}, 2\right]$

55. $\left(-\frac{1}{3}, \frac{1}{9}\right]$

57. $\left[-1, \frac{5}{2}\right]$

59. $(-4, 8)$

61. $\left[-\frac{1}{2}, \frac{35}{2}\right]$

63.

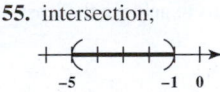

$\left[\frac{6}{5}, \frac{16}{5}\right]$

Section 8.2

1. true **3.** false; The union is $(-\infty, 7) \cup (7, \infty)$. **5.** $\{4\}$, or D
7. $\{1, 3, 5\}$, or B **9.** $\varnothing$ **11.** $\{1, 2, 3, 4, 5, 6\}$, or A **13.** $\{1, 3, 5, 6\}$

15.

17.

19.

$(-3, 2)$

21.

$(-\infty, 2]$

23. $\varnothing$
25.

$[5, 9]$

27.

$(-3, -1)$

29.

$(-\infty, 4]$

31.

33.

35.

$(-\infty, 8]$

37.

$[-2, \infty)$

39.

$(-\infty, \infty)$

41.

$(-\infty, -5) \cup (5, \infty)$

43.

$(-\infty, -1) \cup (2, \infty)$

45.

$(-\infty, \infty)$

47. $[-4, -1]$ **49.** $[-9, -6]$
51. $(-\infty, 3)$ **53.** $[3, 9]$

55. intersection;

$(-5, -1)$

57. union;

$(-\infty, 4)$

59. union;

$(-\infty, 0] \cup [2, \infty)$

61. intersection;

$[4, 12]$

63. {Tuition and fees} **65.** {Tuition and fees, Board rates, Dormitory charges} **67.** 160 ft; 150 ft; 220 ft; 120 ft **69.** Maria, Joe **71.** none of them **73.** Maria, Joe **75.** (a) A: $185 + x \ge 585$; B: $185 + x \ge 520$; C: $185 + x \ge 455$ (b) A: $x \ge 400$; 89%; B: $x \ge 335$; 75%; C: $x \ge 270$; 60% **76.** $520 \le 185 + x \le 584$; $335 \le x \le 399$; $75\% \le$ average $\le 89\%$ **77.** A: $105 + x \ge 585$; $x \ge 480$; impossible; B: $105 + x \ge 520$; $x \ge 415$; 93%; C: $105 + x \ge 455$; $x \ge 350$; 78% **78.** $455 \le 105 + x \le 519$; $350 \le x \le 414$; $78\% \le$ average $\le 92\%$

Section 8.3

1. E; C; D; B; A **3.** (a) one (b) two (c) none **5.** $\{-12, 12\}$
7. $\{-5, 5\}$ **9.** $\{-6, 12\}$ **11.** $\{-5, 6\}$ **13.** $\left\{-3, \frac{11}{2}\right\}$ **15.** $\left\{-\frac{19}{2}, \frac{9}{2}\right\}$
17. $\left\{\frac{7}{3}, 3\right\}$ **19.** $\{12, 36\}$ **21.** $\{-12, 12\}$ **23.** $\{-10, -2\}$
25. $\left\{-\frac{32}{3}, 8\right\}$ **27.** $\{-75, 175\}$

29.

$(-\infty, -3) \cup (3, \infty)$

31.

$(-\infty, -4] \cup [4, \infty)$

33.

$(-\infty, -25] \cup [15, \infty)$

35.

$\left(-\infty, -\frac{12}{5}\right) \cup \left(\frac{8}{5}, \infty\right)$

37.

$(-\infty, -2) \cup (8, \infty)$

39.

$\left(-\infty, -\frac{9}{5}\right] \cup [3, \infty)$

41. (a)

(b)

43.

$[-3, 3]$

45.

$(-4, 4)$

47.

$(-25, 15)$

49.

$\left[-\frac{12}{5}, \frac{8}{5}\right]$

51.

$[-2, 8]$

53.

$\left(-\frac{9}{5}, 3\right)$

55.

$(-\infty, -5) \cup (13, \infty)$

57.

$\left(-\frac{13}{3}, 3\right)$

59.

$\{-6, -1\}$

61.

$\left[-\frac{10}{3}, 4\right]$

63.

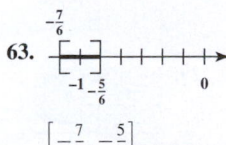

$\left[-\frac{7}{6}, -\frac{5}{6}\right]$

65.

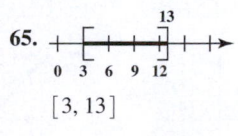

$[3, 13]$

67. $\left(-\infty, \frac{1}{2}\right] \cup \left[\frac{7}{6}, \infty\right)$ **69.** $\left\{-\frac{5}{3}, \frac{11}{3}\right\}$ **71.** $(-\infty, -20) \cup (40, \infty)$
73. $\{-5, 1\}$ **75.** $\{3, 9\}$ **77.** $\{0, 20\}$ **79.** $\{-5, 5\}$ **81.** $\{-5, -3\}$
83. $(-\infty, -3) \cup (2, \infty)$ **85.** $[-10, 0]$ **87.** $(-\infty, 20] \cup [30, \infty)$
89. $\left\{-\frac{5}{3}, \frac{1}{3}\right\}$ **91.** $\{-1, 3\}$ **93.** $\left\{-3, \frac{5}{3}\right\}$ **95.** $\left\{-\frac{1}{3}, -\frac{1}{15}\right\}$ **97.** $\left\{-\frac{5}{4}\right\}$
99. $(-\infty, \infty)$ **101.** $\varnothing$ **103.** $\left\{-\frac{1}{4}\right\}$ **105.** $\varnothing$ **107.** $(-\infty, \infty)$
109. $\left\{-\frac{3}{7}\right\}$ **111.** $\left\{\frac{2}{5}\right\}$ **113.** $(-\infty, \infty)$ **115.** $\varnothing$ **117.** between 60.8 and 67.2 oz, inclusive **119.** between 31.36 and 32.64 oz, inclusive
121. $(-0.05, 0.05)$ **123.** $(2.74975, 2.75025)$ **125.** between 6.8 and 9.8 lb **127.** $|x - 1000| \le 100$; $900 \le x \le 1100$
129. 814.8 ft **130.** Bank of America Center

131. Williams Tower, Bank of America Center, Texaco Heritage Plaza, 609 Main at Texas, Enterprise Plaza, Centerpoint Energy Plaza, 1600 Smith St., Fulbright Tower **132. (a)** $|x - 814.8| \geq 95$ **(b)** $x \geq 909.8$ or $x \leq 719.8$ **(c)** JPMorgan Chase Tower, Wells Fargo Plaza **(d)** It makes sense because it includes all buildings *not* listed in the answer to **Exercise 131.**

SUMMARY EXERCISES Solving Linear and Absolute Value Equations and Inequalities

1. $\{12\}$ **2.** $\{-5, 7\}$ **3.** $\{7\}$ **4.** $\left\{-\frac{2}{5}\right\}$ **5.** $\varnothing$ **6.** $(-\infty, -1]$

7. $\left[-\frac{2}{3}, \infty\right)$ **8.** $\{-1\}$ **9.** $\{-3\}$ **10.** $\left\{1, \frac{11}{3}\right\}$ **11.** $(-\infty, 5]$

12. $(-\infty, \infty)$ **13.** $\{2\}$ **14.** $(-\infty, -8] \cup [8, \infty)$ **15.** $\varnothing$ **16.** $(-\infty, \infty)$

17. $(-5.5, 5.5)$ **18.** $\left\{\frac{13}{3}\right\}$ **19.** $\left\{-\frac{96}{5}\right\}$ **20.** $(-\infty, 32]$ **21.** $(-\infty, -24)$

22. $\left[\frac{3}{4}, \frac{15}{8}\right]$ **23.** $\left\{\frac{7}{2}\right\}$ **24.** $\{60\}$ **25.** $\{$all real numbers$\}$ **26.** $(-\infty, 5)$

27. $\left[-\frac{9}{2}, \frac{15}{2}\right]$ **28.** $\{24\}$ **29.** $\left\{-\frac{1}{5}\right\}$ **30.** $\left(-\infty, -\frac{5}{2}\right]$ **31.** $\left[-\frac{1}{3}, 3\right]$

32. $[1, 7]$ **33.** $\left\{-\frac{1}{6}, 2\right\}$ **34.** $\{-3\}$ **35.** $(-\infty, -1] \cup \left[\frac{5}{3}, \infty\right)$

36. $\left\{\frac{3}{8}\right\}$ **37.** $\left\{-\frac{5}{2}\right\}$ **38.** $(-6, 8)$ **39.** $(-\infty, -4) \cup (7, \infty)$

40. $(1, 9)$ **41.** $(-\infty, \infty)$ **42.** $\left\{\frac{1}{3}, 9\right\}$ **43.** $\{$all real numbers$\}$

44. $\left\{-\frac{10}{9}\right\}$ **45.** $\{-2\}$ **46.** $\varnothing$ **47.** $(-\infty, -1) \cup (2, \infty)$ **48.** $[-3, -2]$

Section 8.4

1. (a) yes **(b)** yes **(c)** no **(d)** yes **3. (a)** no **(b)** no **(c)** no **(d)** yes **5.** solid; below **7.** dashed; above **9.** $\leq$ **11.** $>$

13. 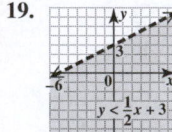 **15.** 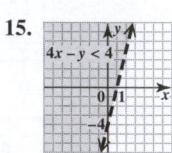 **17.**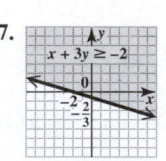

19. 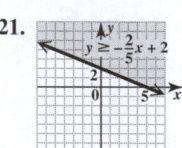

19. 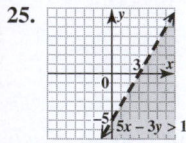 **21.** 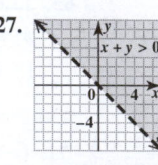 **23.**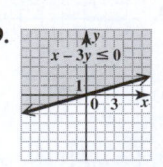

25. 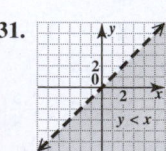 **27.** 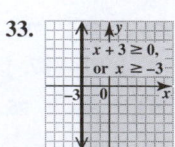 **29.**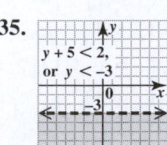

31. 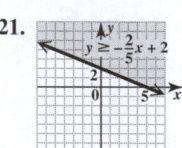 **33.** 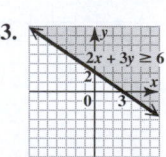 **35.**

37. $2; (0, -4); 2x - 4$; solid; above; $\geq$; $\geq 2x - 4$

39. C **41.** B **43.** This is quadrant I, including the x- and y-axes.

45. This is quadrant IV, not including the x- and y-axes.

47. (a) no **(b)** no **(c)** yes **49. (a)** no **(b)** yes **(c)** no

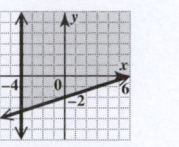 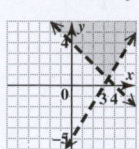

51. 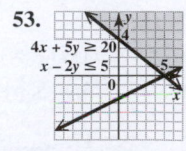 **53.** **55.**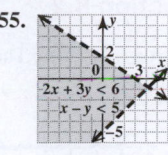

57. 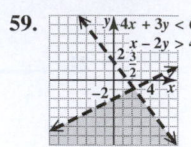 **59.** **61.**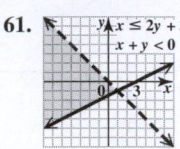

63. 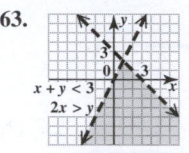 **65.** 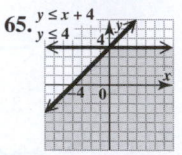 **67.**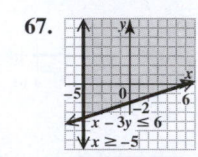

69. 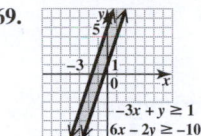 **71.** 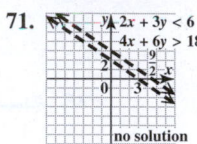 no solution **73.**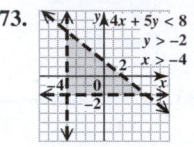

75. 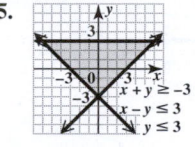 **77.** 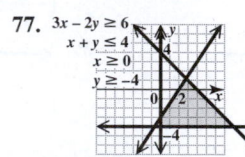 **79.** $x \leq 200,$ $x \geq 100,$ $y \geq 3000$

80. 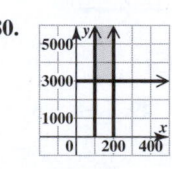 **81.** $C = 50x + 100y$ **82.** Some examples are $(100, 5000)$, $(150, 3000)$, and $(150, 5000)$. The corner points are $(100, 3000)$ and $(200, 3000)$. **83.** The least value occurs when $x = 100$ and $y = 3000$. **84.** The company should use 100 workers and manufacture 3000 units to achieve the least possible cost.

Chapter 8 Review Exercises

1. $(-9, \infty)$ **2.** $(-\infty, -3]$ **3.** $\left(\frac{3}{2}, \infty\right)$ **4.** $[-3, \infty)$ **5.** $[3, 5)$

6. $\left(\frac{59}{31}, \infty\right)$ **7.** $\{a, c\}$ **8.** $\{a\}$ **9.** $\{a, c, e, f, g\}$ **10.** $\{a, b, c, d, e, f, g\}$

11. 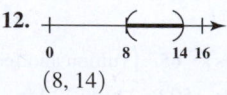 $(6, 9)$ **12.** 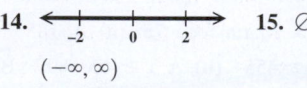 $(8, 14)$

13. $(-\infty, -3] \cup (5, \infty)$ **14.** $(-\infty, \infty)$ **15.** $\varnothing$

16. $(-\infty, -2] \cup [7, \infty)$ **17.** $(-3, 4]$ **18.** $(-\infty, 2]$ **19.** $(4, \infty)$

20. $(1, \infty)$ **21.** $\{-7, 7\}$ **22.** $\{-11, 7\}$

23. $\left\{-\frac{1}{3}, 5\right\}$ **24.** $\varnothing$ **25.** $\{0, 7\}$ **26.** $\left\{-\frac{3}{2}, \frac{1}{2}\right\}$ **27.** $\left\{-\frac{3}{4}, \frac{1}{2}\right\}$

28. $\left\{-\frac{1}{2}\right\}$ **29.** $(-14, 14)$ **30.** $[-1, 13]$ **31.** $[-3, -2]$

32. $(-\infty, \infty)$ **33.** $\varnothing$ **34.** $(-\infty, \infty)$ **35.** between 46.56 and 49.44 oz, inclusive   **36.** between 31.68 and 32.32 oz, inclusive

37. 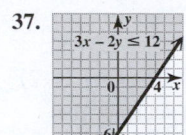   **38.** 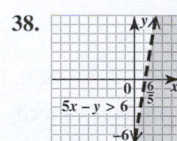   **39.**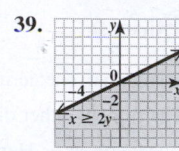

40. 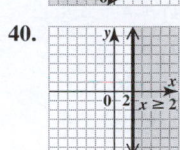   **41.** 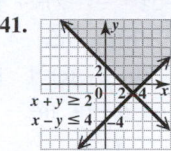   **42.**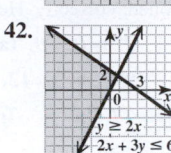

43. 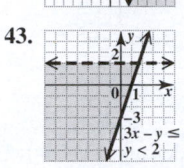   **44.** B

Chapter 8 Mixed Review Exercises

1. $(-2, \infty)$ **2.** $[-2, 3)$ **3.** $(-\infty, \infty)$ **4.** $(-\infty, 2]$ **5.** $\left\{-\frac{7}{3}, 1\right\}$

6. $[-16, 10]$ **7.** $\left(-\infty, \frac{14}{17}\right)$ **8.** $\left(-3, \frac{7}{2}\right)$ **9.** $\left(-\infty, -\frac{13}{5}\right) \cup (3, \infty)$

10. $(-\infty, \infty)$ **11.** $\left\{-4, -\frac{2}{3}\right\}$ **12.** $\left\{1, \frac{11}{3}\right\}$ **13.** $(6, 8)$

14. $(-\infty, -2] \cup [7, \infty)$ **15.** D

16.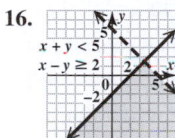

Chapter 8 Test

[8.1] **1.**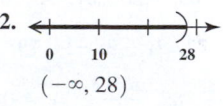

$[1, \infty)$

2. (number line)

$(-\infty, 28)$

3. 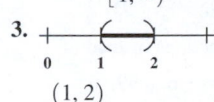   **4.** C   [8.2] **5.** (a) $\{1, 5\}$

$(1, 2)$

(b) $\{1, 2, 5, 7, 9, 12\}$ **6.** $[2, 9)$

7. $(-\infty, 3) \cup [6, \infty)$ [8.3] **8.** $\left[-\frac{5}{2}, 1\right]$

9. $\left(-\infty, -\frac{7}{6}\right) \cup \left(\frac{17}{6}, \infty\right)$ **10.** $\{1, 5\}$ **11.** $\left(\frac{1}{3}, \frac{7}{3}\right)$ **12.** $\varnothing$ **13.** $\left\{-\frac{5}{3}, 3\right\}$

14. $\left\{-\frac{5}{7}, \frac{11}{3}\right\}$ **15.** (a) $\varnothing$ (b) $(-\infty, \infty)$ (c) $\varnothing$

[8.4] **16.** 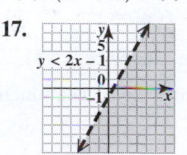   **17.** 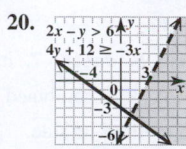   **18.** C

19. 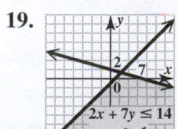   **20.** (graph)

Chapters R–8 Cumulative Review Exercises

[R.1, R.2] **1.** (a) D, I   (b) C, H   (c) G   (d) A, F   (e) B   (f) E
[1.3] **2.** (a) A, B, C, D, F   (b) B, C, D, F   (c) D, F   (d) C, D, F
(e) E, F   (f) D, F   [1.1] **3.** 32   [1.3, 1.4] **4.** 0   [1.7] **5.** -7
[2.3] **6.** $\{-65\}$ **7.** $\{$all real numbers$\}$ [2.6] **8.** $t = \dfrac{A - p}{pr}$
[2.9, 8.1] **9.** $(-\infty, 6)$ [R.2, 2.7] **10.** 3500; 1950; 36%; 31%
[3.3, 7.1] **11.** $-\frac{4}{3}$ **12.** 0   [3.4, 3.5, 7.2] **13.** (a) $y = -4x + 15$
(b) $4x + y = 15$ **14.** (a) $y = 4x$ (b) $4x - y = 0$
[3.2, 7.1] **15.** (graph)   [8.4] **16.**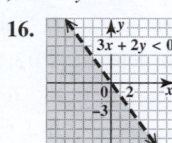

[7.1, 7.2] **17.** (a) -0.44 gal per yr; Per capita consumption of whole milk decreased by an average of 0.44 gal per year from 1970 to 2015.   (b) $y = -0.44x + 25.3$ (c) 12.1 gal

[4.1, 4.2] **18.** $\dfrac{8m^9 n^3}{p^6}$ **19.** $\dfrac{y^7}{x^{13} z^2}$ [4.4] **20.** $2x^2 - 5x + 10$

[4.5] **21.** $15x^2 + 7xy - 2y^2$ [4.7] **22.** $4xy^4 - 2y + \dfrac{1}{x^2 y}$

23. $m^2 - 2m + 3$ [5.2, 5.3] **24.** $(m + 8)(m + 4)$

[5.4] **25.** $(5t^2 + 6)(5t^2 - 6)$ **26.** $(9z + 4)^2$

[6.2] **27.** $\dfrac{x + 1}{x}$ **28.** $(t + 5)(t + 3)$, or $t^2 + 8t + 15$

[6.4] **29.** $\dfrac{-2x - 14}{(x + 3)(x - 1)}$ [6.5] **30.** -21 [5.5] **31.** $\{-2, -1\}$

[6.6] **32.** $\{19\}$ [7.3–7.5] **33.** $\{(3, 2)\}$ **34.** $\varnothing$

[7.6] **35.** $\{(1, 0, -1)\}$ [7.7] **36.** length: 42 ft; width: 30 ft

[8.2] **37.** $(-4, 4)$ **38.** $(-\infty, 0] \cup (2, \infty)$

[8.3] **39.** $\left\{-\frac{1}{3}, 1\right\}$ **40.** $\left(-\infty, -\frac{8}{3}\right] \cup [2, \infty)$

9   RELATIONS AND FUNCTIONS

Section 9.1

1. relation; ordered pairs   **3.** domain; range   **5.** independent variable; dependent variable   **7.** $\{(2, -2), (2, 0), (2, 1)\}$ **9.** $\{(1960, 0.76),$ $(1980, 2.69), (2000, 5.39), (2016, 8.65)\}$ **11.** $\{(A, 4), (B, 3), (C, 2),$ $(D, 1), (F, 0)\}$

In Exercises 13–17, answers will vary.

13. (graph)   **15.** $\{(-1, -3), (0, -1), (1, 1), (3, 3)\}$
17. $y = 2$ **19.** function; domain: $\{5, 3, 4, 7\}$;
range: $\{1, 2, 9, 6\}$ **21.** not a function; domain: $\{2, 0\}$; range: $\{4, 2, 5\}$

23. function; domain: $\{-3, 4, -2\}$; range: $\{1, 7\}$
25. not a function; domain: $\{1, 0, 2\}$; range: $\{1, -1, 0, 4, -4\}$
27. not a function; domain: $\{1\}$; range: $\{5, 2, -1, -4\}$ **29.** function; domain: $\{4, 2, 0, -2\}$; range: $\{-3\}$ **31.** function; domain: $\{2, 5, 11, 17, 3\}$; range: $\{1, 7, 20\}$ **33.** function; domain: $\{-2, 0, 3\}$; range: $\{2, 3\}$ **35.** function; domain: $(-\infty, \infty)$; range: $(-\infty, \infty)$

37. not a function; domain: $\{-2\}$; range: $(-\infty, \infty)$ **39.** function; domain: $(-\infty, \infty)$; range: $\{-2\}$ **41.** not a function; domain: $(-\infty, 0]$; range: $(-\infty, \infty)$ **43.** function; domain: $(-\infty, \infty)$; range: $(-\infty, 4]$ **45.** not a function; domain: $[-4, 4]$; range: $[-3, 3]$ **47.** not a function; domain: $(-\infty, \infty)$; range: $[2, \infty)$ **49.** function; $(-\infty, \infty)$ **51.** function; $(-\infty, \infty)$ **53.** function; $(-\infty, \infty)$ **55.** not a function; $[0, \infty)$ **57.** not a function; $(-\infty, \infty)$ **59.** function; $(-\infty, \infty)$ **61.** function; $(-\infty, \infty)$ **63.** function; $(-\infty, 0) \cup (0, \infty)$ **65.** function; $(-\infty, 4) \cup (4, \infty)$ **67.** function; $\left(-\infty, -\frac{1}{2}\right) \cup \left(-\frac{1}{2}, \infty\right)$ **69.** not a function; $[1, \infty)$ **71.** function; $(-\infty, 0) \cup (0, \infty)$ **73.** (a) yes (b) domain: $\{2013, 2014, 2015, 2016, 2017\}$; range: $\{52.8, 52.6, 53.2, 53.7\}$ (c) 52.6; 2017 (d) Answers will vary. Two possible answers are (2014, 52.6) and (2017, 53.7).

Section 9.2

1. $f(x)$; function; domain; x; f of x (or "f at x") **3.** line; -2; linear; $-2x + 4$; -2; 3; -2 **5.** 4 **7.** 13 **9.** -11 **11.** 4 **13.** -296 **15.** 3 **17.** 2.75 **19.** $-3p + 4$ **21.** $3x + 4$ **23.** $-3x - 2$ **25.** $-6t + 1$ **27.** $-\pi^2 + 4\pi + 1$ **29.** $-3x - 3h + 4$ **31.** $-\frac{p^2}{9} + \frac{4p}{3} + 1$ **33.** (a) -1 (b) -1 **35.** (a) 2 (b) 3 **37.** (a) 15 (b) 10 **39.** (a) 4 (b) 1 **41.** (a) 3 (b) -3 **43.** (a) -3 (b) 2 **45.** (a) 2 (b) 0 (c) -1 **47.** (a) $f(x) = -\frac{1}{3}x + 4$ (b) 3 **49.** (a) $f(x) = 3 - 2x^2$ (b) -15 **51.** (a) $f(x) = \frac{4}{3}x - \frac{8}{3}$ (b) $\frac{4}{3}$

53. domain: $(-\infty, \infty)$; range: $(-\infty, \infty)$
55. domain: $(-\infty, \infty)$; range: $(-\infty, \infty)$
57. domain: $(-\infty, \infty)$; range: $(-\infty, \infty)$

59. domain: $(-\infty, \infty)$; range: $(-\infty, \infty)$
61. domain: $(-\infty, \infty)$; range: $\{-4\}$
63. domain: $(-\infty, \infty)$; range: $\{0\}$

65. x-axis **67.** (a) 11.25 (dollars) (b) 3 is the value of the independent variable, which represents a package weight of 3 lb. $f(3)$ is the value of the dependent variable, which represents the cost to mail a 3-lb package. (c) $18.75; f(5) = 18.75$ **69.** (a) $f(x) = 12x + 100$ (b) 1600; The cost to print 125 t-shirts is $1600. (c) 75; $f(75) = 1000$; The cost to print 75 t-shirts is $1000. **71.** (a) 1.1 (b) 4 (c) -1.2 (d) $(0, 3.5)$ (e) $f(x) = -1.2x + 3.5$ **73.** (a) $[0, 100]$; $[0, 3000]$ (b) 25 hr; 25 hr (c) 2000 gal (d) $f(0) = 0$; The pool is empty at time 0. (e) $f(25) = 3000$; After 25 hr, there are 3000 gal of water in the pool. **75.** (a) 194.53 cm (b) 177.29 cm (c) 177.41 cm (d) 163.65 cm **77.** Because it falls from left to right, the slope is negative.

78. $-\frac{3}{2}$ **79.** $-\frac{3}{2}; \frac{2}{3}$ **80.** $\left(\frac{7}{3}, 0\right)$ **81.** $\left(0, \frac{7}{2}\right)$ **82.** $f(x) = -\frac{3}{2}x + \frac{7}{2}$ **83.** $-\frac{17}{2}$ **84.** $\frac{23}{3}$ **85.** $2; f(1) = -\frac{3}{2}(1) + \frac{7}{2}$, which simplifies on the right to $f(1) = -\frac{3}{2} + \frac{7}{2}$. This gives $f(1) = 2$.

Section 9.3

1. polynomial; one; terms; powers **3.** C **5.** 0; 1; 2; $(0, 0)$, $(1, 1)$, $(2, 2)$ **7.** The student either did not substitute correctly or did not apply the exponent correctly. Here $f(-2) = -(-2)^2 + 4 = -(4) + 4 = 0$, so $f(-2) = 0$. **9.** (a) -10 (b) 8 (c) -4 **11.** (a) 8 (b) -10 (c) 0 **13.** (a) 8 (b) 2 (c) 4 **15.** (a) 7 (b) 1 (c) 1 **17.** (a) 8 (b) 74 (c) 6 **19.** (a) -11 (b) 4 (c) -8

21.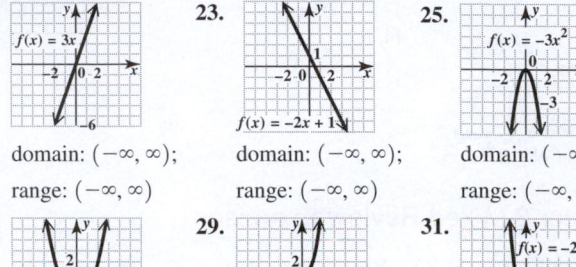
domain: $(-\infty, \infty)$; range: $(-\infty, \infty)$
23. domain: $(-\infty, \infty)$; range: $(-\infty, \infty)$
25. domain: $(-\infty, \infty)$; range: $(-\infty, 0]$

27.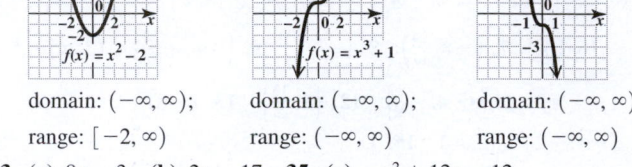
domain: $(-\infty, \infty)$; range: $[-2, \infty)$
29. domain: $(-\infty, \infty)$; range: $(-\infty, \infty)$
31. domain: $(-\infty, \infty)$; range: $(-\infty, \infty)$

33. (a) $8x - 3$ (b) $2x - 17$ **35.** (a) $-x^2 + 12x - 12$ (b) $9x^2 + 4x + 6$ **37.** $f(x)$ and $g(x)$ can be any two polynomials that have a sum of $3x^3 - x + 3$, such as $f(x) = 3x^3 + 1$ and $g(x) = -x + 2$. **39.** $x^2 + 2x - 9$ **41.** 6 **43.** $x^2 - x - 6$ **45.** 6 **47.** -33 **49.** 0 **51.** $-\frac{9}{4}$ **53.** $-\frac{9}{2}$ **55.** (a) $P(x) = 8.49x - 50$ (b) $799 **57.** $10x^2 - 2x$ **59.** $2x^2 - x - 3$ **61.** $8x^3 - 27$ **63.** $2x^3 - 18x$ **65.** -20 **67.** 32 **69.** 36 **71.** 0 **73.** $\frac{35}{4}$ **75.** $5x - 1; 0$ **77.** $2x - 3; -1$ **79.** $4x^2 + 6x + 9; \frac{3}{2}$ **81.** $\frac{x^2 - 9}{2x}$, $x \neq 0$ **83.** $-\frac{5}{4}$ **85.** $\frac{x - 3}{2x}$, $x \neq 0$ **87.** 0 **89.** $-\frac{35}{4}$ **91.** $\frac{7}{2}$ **93.** B **95.** A **97.** The student multiplied the functions instead of composing them. The correct answer is $(f \circ g)(x) = -6$. **99.** 6 **101.** 83 **103.** 53 **105.** 13 **107.** $2x^2 + 11$ **109.** $2x - 2$ **111.** $\frac{97}{4}$ **113.** 8 **115.** 1 **117.** 9 **119.** 1 **121.** $(f \circ g)(x) = 63{,}360x$; It computes the number of inches in x miles. **123.** (a) $s = \frac{x}{4}$ (b) $y = \frac{x^2}{16}$ (c) 2.25 **125.** (a) $g(x) = \frac{1}{2}x$ (b) $f(x) = x + 1$ (c) $(f \circ g)(x) = \frac{1}{2}x + 1$ (d) $(f \circ g)(60) = 31$; The sale price is $31. **127.** $(\mathcal{A} \circ r)(t) = 4\pi t^2$; This is the area of the circular layer as a function of time.

Section 9.4

1. increases; decreases **3.** direct **5.** direct **7.** inverse **9.** inverse **11.** inverse **13.** direct **15.** joint **17.** combined **19.** The perimeter of a square varies directly as the length of its side. **21.** The surface area of a sphere varies directly as the square of its radius. **23.** The area of a triangle varies jointly as the lengths of its base and height.

25. $4; 2; 4\pi; \frac{4}{3}\pi; \frac{1}{2}; \frac{1}{3}\pi$ **27.** $A = kb$ **29.** $h = \dfrac{k}{t}$ **31.** $M = kd^2$

33. $I = kgh$ **35.** 36 **37.** $\frac{16}{9}$ **39.** 0.625 **41.** $\frac{16}{5}$ **43.** $222\frac{2}{9}$

45. $3.92 **47.** 8 lb **49.** 450 cm^3 **51.** 256 ft **53.** $106\frac{2}{3}$ mph

55. 100 cycles per sec **57.** $21\frac{1}{3}$ foot-candles **59.** $420 **61.** 11.8 lb

63. 448.1 lb **65.** 68,600 calls **67.** Answers will vary. **69.** $(0, 0)$,

$(1, 3.75)$ **70.** 3.75 **71.** $y = 3.75x + 0$, or $y = 3.75x$

72. $a = 3.75, b = 0$ **73.** It is the price per gallon and the slope of

the line. **74.** It can be written in the form $y = kx$ (where $k = a$).

The value of a is the constant of variation.

Chapter 9 Review Exercises

1. not a function; domain: $\{-4, 1\}$; range: $\{2, -2, 5, -5\}$

2. function; domain: $\{9, 11, 4, 17, 25\}$; range: $\{32, 47, 69, 14\}$

3. function; domain: $[-4, 4]$; range: $[0, 2]$ **4.** not a function;

domain: $(-\infty, 0]$; range: $(-\infty, \infty)$ **5.** function; domain: $(-\infty, \infty)$;

linear function **6.** not a function; domain: $(-\infty, \infty)$ **7.** not a function;

domain: $[0, \infty)$ **8.** function; domain: $(-\infty, 6) \cup (6, \infty)$

9. -6 **10.** -8.52 **11.** -8 **12.** $-2k^2 + 3k - 6$

13. $f(x) = 2x^2$; 18 **14.** C **15.** It is a horizontal line. **16. (a)** yes

(b) domain: $\{1960, 1970, 1980, 1990, 2000, 2010, 2015\}$; range:

$\{69.7, 70.8, 73.7, 75.4, 76.8, 78.7, 78.8\}$ **(c)** Answers will vary.

Two possible answers are $(1960, 69.7)$ and $(2010, 78.7)$. **(d)** 73.7; In 1980,

life expectancy at birth was 73.7 yr. **(e)** 2000 **17.** -11 **18.** 4 **19.** 7

20. 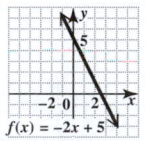 **21.** 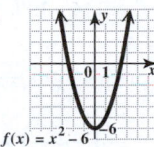 **22.**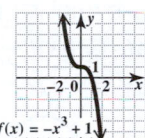

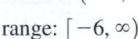

domain: $(-\infty, \infty)$;   domain: $(-\infty, \infty)$;   domain: $(-\infty, \infty)$;

range: $(-\infty, \infty)$ range: $[-6, \infty)$ range: $(-\infty, \infty)$

23. (a) $5x^2 - x + 5$ **(b)** -9 **24. (a)** $-5x^2 + 5x + 1$ **(b)** 11

25. (a) $36x^3 - 9x^2$ **(b)** -45 **26. (a)** $4x - 1, \ x \neq 0$ **(b)** 7

27. (a) $75x^2 + 220x + 160$ **(b)** 1495 **(c)** 20 **28. (a)** $15x^2 + 10x + 2$

(b) 167 **(c)** 42 **29.** C **30.** 430 mm **31.** 5.59 vibrations per sec

32. 22.5 ft^3

Chapter 9 Mixed Review Exercises

1. domain: $\{14, 91, 75, 23\}$; range: $\{9, 70, 56, 5\}$; not a function; 75 in

the domain is paired with two different values, 70 and 56, in the range.

2. (a) -1 **(b)** -2 **(c)** 2 **(d)** $(-\infty, \infty)$; $(-\infty, \infty)$ **3.** -2 **4.** 24

5. -112 **6.** 1 **7.** 94 **8.** 13 **9.** 12 ft^2 **10.** 32.97 in.

Chapter 9 Test

[9.1] **1.** D **2.** C **3.** domain: $[0, \infty)$; range: $(-\infty, \infty)$

4. domain: $\{0, -2, 4\}$; range: $\{1, 3, 8\}$ [9.2] **5.** 0; $-a^2 + 2a - 1$

6. 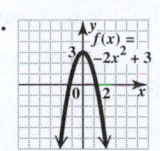 [9.3] **7.** 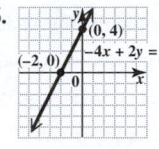 **8.**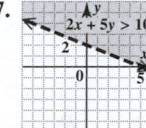

domain: $(-\infty, \infty)$;   domain: $(-\infty, \infty)$;   domain: $(-\infty, \infty)$;

range: $(-\infty, \infty)$ range: $(-\infty, 3]$ range: $(-\infty, \infty)$

9. (a) -18 **(b)** $-2x^2 + 12x - 9$ **(c)** $-2x^2 + 2x - 3$ **(d)** -7

10. (a) $x^3 + 4x^2 + 5x + 2$ **(b)** 0 **11. (a)** $x + 2, \ x \neq -1$ **(b)** 0

12. (a) 23 **(b)** $3x^2 + 11$ **(c)** $9x^2 + 30x + 27$ [9.4] **13.** 200 amps

14. 0.8 lb

Chapters R–9 Cumulative Review Exercises

[R.1, R.2, 1.4, 1.5] **1.** $-\frac{11}{12}$ **2.** $\frac{5}{8}$ **3.** $-\frac{6}{5}$ **4.** -5.67 **5.** 0.0525

6. 4360 [1.1, 4.1] **7. (a)** 25 **(b)** -25 **(c)** 25 [1.5] **8.** -199

[2.3] **9.** $\left\{-\frac{15}{4}\right\}$ [2.9, 8.1] **10.** $\left(-\infty, \frac{240}{13}\right]$ [2.6] **11.** 6 m

[2.8] **12.** $4000 at 4%; $8000 at 3% [3.3, 7.1] **13.** $-\frac{3}{2}$ **14.** $-\frac{3}{4}$

[3.5, 7.2] **15.** $y = -\frac{3}{2}x + \frac{1}{2}$

[3.2, 7.1] **16.** [8.4] **17.**

x-intercept: $(-2, 0)$;

y-intercept: $(0, 4)$

[4.1, 4.2] **18.** $\dfrac{m}{n}$ [4.4] **19.** $4y^2 - 7y - 6$ [4.5] **20.** $12f^2 + 5f - 3$

[4.6] **21.** $\frac{1}{16}x^2 + \frac{5}{2}x + 25$ [4.7] **22.** $x^2 + 4x - 7$

[5.3] **23.** $(2x + 5)(x - 9)$ [5.4] **24.** $(2p + 5)(4p^2 - 10p + 25)$

[6.1] **25.** $\dfrac{y + 4}{y - 4}$ [6.2] **26.** $\dfrac{a(a - b)}{2(a + b)}$ **27.** $\dfrac{2(x + 3)}{(x + 2)(x^2 + 3x + 9)}$

[6.4] **28.** 3 [5.5] **29.** $\left\{-\frac{7}{3}, 1\right\}$ [6.6] **30.** $\{-4\}$ [8.3] **31.** $\left\{\frac{2}{3}, 2\right\}$

32. $(-\infty, -2] \cup \left[\frac{2}{3}, \infty\right)$ [7.4, 7.5] **33.** $\{(-1, 3)\}$ [7.6] **34.** $\{(-2, 3, 1)\}$

[9.1] **35.** function; domain: $(-\infty, \infty)$; range: $(-\infty, \infty)$

[9.2] **36. (a)** $\frac{5}{3}x - \frac{8}{3}$ **(b)** -1 [9.3] **37. (a)** $2x^3 - 2x^2 + 6x - 4$

(b) $2x^3 - 4x^2 + 2x + 2$ **(c)** -14 **(d)** $x^4 + 2x^2 - 3$ [9.4] **38.** $9.92

| 10 | **ROOTS, RADICALS, AND ROOT FUNCTIONS** |

Section 10.1

1. 1; 4; 9; 16; 25; 36; 49; 64; 81; 100; 121; 144; 169; 196; 225; 256; 400;

625; 900; 2500 **3.** true **5.** false; Zero has only one square root.

7. true **9.** a must be positive. **11.** a must be negative.

13. $-3, 3$ **15.** $-8, 8$ **17.** $-13, 13$ **19.** $-\frac{5}{14}, \frac{5}{14}$ **21.** $-30, 30$

23. There is no real number that can be squared to obtain -4. Therefore,

$\sqrt{-4}$ is not a real number. **25.** 1 **27.** 7 **29.** 10 **31.** -4 **33.** -16

35. $\frac{2}{5}$ **37.** $-\frac{12}{11}$ **39.** 0.8 **41.** -0.2 **43.** It is not a real number.

45. It is not a real number. **47.** 19 **49.** 19 **51.** $\frac{2}{3}$ **53.** $3x^2 + 4$

55. rational; 5   **57.** irrational; 5.385   **59.** rational; -8

61. irrational; -17.321 **63.** It is not a real number.   **65.** irrational; 34.641   **67.** 9 and 10   **69.** 7 and 8   **71.** -7 and -6 **73.** 4 and 5

75. 1; 8; 27; 64; 125; 216; 343; 512; 729; 1000   **77.** The student divided 27 by the index, 3, instead of taking the cube root. $\sqrt[3]{27} = 3$ because $3^3 = 27$.   **79.** 1   **81.** 5   **83.** 9   **85.** -3 **87.** -6 **89.** 2   **91.** $\frac{2}{3}$ **93.** $-\frac{6}{5}$ **95.** 0.5   **97.** 3   **99.** 5   **101.** 6   **103.** It is not a real number.   **105.** -3 **107.** 2   **109.** -4

In Exercises 111–117, we give the domain and then the range.

111.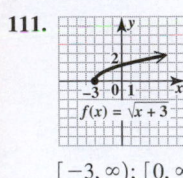

$[-3, \infty)$; $[0, \infty)$

113.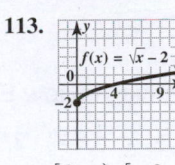

$[0, \infty)$; $[-2, \infty)$

115.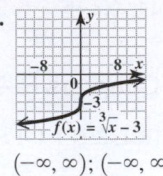

$(-\infty, \infty)$; $(-\infty, \infty)$

117.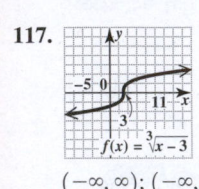

$(-\infty, \infty)$; $(-\infty, \infty)$

119. 12   **121.** 10   **123.** 2   **125.** -9 **127.** -5 **129.** $|x|$ **131.** $|z|$ **133.** x **135.** x^5 **137.** $|x|^5$ (or $|x^5|$)

139. When rounding to three decimal places, we look at the digit in the fourth decimal place. If it is 5 or greater, we round up. The correct answer is $\sqrt{19} \approx 4.359$.

141. 97.381   **143.** 16.863   **145.** -9.055 **147.** 7.507

149. 3.162   **151.** 1.885   **153. (a)** 1,183,000 cycles per sec **(b)** 118,000 cycles per sec   **155.** 10 mi   **157.** 1.732 amps

159. 437,000 mi^2 **161.** 330 m^2

Section 10.2

1. C   **3.** A   **5.** H   **7.** B   **9.** D   **11.** In $27^{1/3}$, the base and exponent should not be multiplied. The denominator of the rational exponent is the index of the radical, $\sqrt[3]{27}$. The correct answer is 3.   **13.** 13   **15.** 9

17. 2   **19.** $\frac{8}{9}$ **21.** -3 **23.** It is not a real number.   **25.** 1000

27. 27   **29.** -1024 **31.** 16   **33.** $\frac{1}{8}$ **35.** $\frac{1}{512}$ **37.** $\frac{9}{25}$ **39.** $\frac{27}{8}$

41. $\sqrt{10}$ **43.** $\left(\sqrt[4]{8}\right)^3$ **45.** $5\left(\sqrt[3]{x}\right)^2$ **47.** $9\left(\sqrt[8]{q}\right)^5 - \left(\sqrt[3]{2x}\right)^2$

49. $\dfrac{1}{\left(\sqrt[5]{x}\right)^3}$ **51.** $\left(\sqrt[3]{2y + x}\right)^2$ **53.** $15^{1/2}$ **55.** 64   **57.** 64

59. x **61.** x^{10} **63.** 9   **65.** 4   **67.** y **69.** $x^{5/12}$ **71.** $k^{2/3}$

73. $x^3 y^8$ **75.** $\dfrac{1}{x^{10/3}}$ **77.** $\dfrac{1}{m^{1/4} n^{3/4}}$ **79.** $\dfrac{y^{17/3}}{x^8}$ **81.** $\dfrac{c^{11/3}}{b^{11/4}}$ **83.** $\dfrac{q^{5/3}}{9p^{7/2}}$

85. $p + 2p^2$ **87.** $k^{7/4} - k^{3/4}$ **89.** $6 + 18a$ **91.** $-5x^2 + 5x$

93. $x^{17/20}$ **95.** $t^{8/15}$ **97.** $\dfrac{1}{x^{3/2}}$ **99.** $y^{5/6} z^{1/3}$ **101.** $m^{1/12}$ **103.** $x^{1/24}$

105. $\sqrt{a^2 + b^2} = \sqrt{3^2 + 4^2} = 5$; $a + b = 3 + 4 = 7$; $5 \neq 7$

107. 4.5 hr   **109.** 19.0°; The table gives 19°.   **111.** 4.2°; The table gives 4°.

Section 10.3

1. D   **3.** B   **5.** D   **7.** The student "dropped" the index, 3. The correct product is $\sqrt[3]{65}$.   **9.** $\sqrt{9}$, or 3   **11.** $\sqrt{36}$, or 6   **13.** $\sqrt{30}$ **15.** $\sqrt{14x}$

17. $\sqrt{42pqr}$ **19.** $\sqrt[3]{10}$ **21.** $\sqrt[3]{14xy}$ **23.** $\sqrt[4]{33}$ **25.** $\sqrt[4]{6xy^2}$

27. This expression cannot be simplified by the product rule.   **29.** $\dfrac{8}{11}$

31. $\dfrac{\sqrt{3}}{5}$ **33.** $\dfrac{\sqrt{x}}{5}$ **35.** $\dfrac{p^3}{9}$ **37.** $-\dfrac{3}{4}$ **39.** $\dfrac{\sqrt[3]{r^2}}{2}$ **41.** $-\dfrac{3}{x}$ **43.** $\dfrac{1}{x^3}$

45. $\sqrt{12}$ can be simplified further. The *greatest* perfect square factor that divides into 48 is 16, not 4. Thus, $\sqrt{48} = \sqrt{16 \cdot 3} = \sqrt{16} \cdot \sqrt{3} = 4\sqrt{3}$.

47. $2\sqrt{3}$ **49.** $12\sqrt{2}$ **51.** $-4\sqrt{2}$ **53.** $-2\sqrt{7}$ **55.** This radical cannot be simplified further.   **57.** $4\sqrt[3]{2}$ **59.** $2\sqrt[3]{5}$ **61.** $-2\sqrt[3]{2}$

63. $-4\sqrt[4]{2}$ **65.** $2\sqrt[5]{2}$ **67.** $-3\sqrt[5]{2}$ **69.** $2\sqrt[6]{2}$ **71.** $6k\sqrt{2}$

73. $12xy^4\sqrt{xy}$ **75.** $11x^3$ **77.** $-3t^4$ **79.** $-10m^4z^2$ **81.** $5a^2b^3c^4$

83. $\frac{1}{2}r^2t^5$ **85.** $5x\sqrt{2x}$ **87.** $-10r^5\sqrt{5r}$ **89.** $x^3y^4\sqrt{13x}$ **91.** $2z^2w^3$

93. $-2zt^2\sqrt[3]{2z^2t}$ **95.** $3x^3y^4$ **97.** $-3r^3s^2\sqrt[4]{2r^3s}$ **99.** $\dfrac{y^5\sqrt{y}}{6}$

101. $\dfrac{x^5\sqrt[3]{x}}{3}$ **103.** $4\sqrt{3}$ **105.** $\sqrt{5}$ **107.** $x^2\sqrt{x}$ **109.** $x\sqrt[5]{x^3}$

111. $\sqrt[6]{432}$ **113.** $\sqrt[12]{6912}$ **115.** $\sqrt[6]{x^5}$ **117.** 5   **119.** $8\sqrt{2}$

121. $2\sqrt{14}$ **123.** 13   **125.** $9\sqrt{2}$ **127.** $\sqrt{17}$ **129.** 5

131. $6\sqrt{2}$ **133.** $\sqrt{5y^2 - 2xy + x^2}$ **135. (a)** B   **(b)** C   **(c)** D   **(d)** A   **137.** $x^2 + y^2 = 144$ **139.** $(x + 4)^2 + (y - 3)^2 = 4$

141. $(x + 8)^2 + (y + 5)^2 = 5$

143.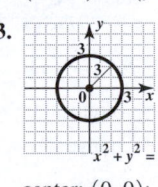

$x^2 + y^2 = 9$
center: $(0, 0)$; radius: 3

145.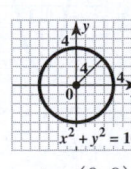

$x^2 + y^2 = 16$
center: $(0, 0)$; radius: 4

147.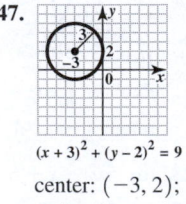

$(x + 3)^2 + (y - 2)^2 = 9$
center: $(-3, 2)$; radius: 3

149.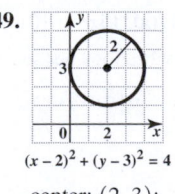

$(x - 2)^2 + (y - 3)^2 = 4$
center: $(2, 3)$; radius: 2

151. 42.0 in.   **153.** 581   **155.** $d = 1.224\sqrt{h}$ **157.** 12.5 mi

158. (a) 3 units, 4 units   **(b)** If we let $a = 3$, $b = 4$, and $c = 5$, then the Pythagorean theorem is satisfied because $3^2 + 4^2 = 5^2$ is a true statement.   **159. (a)** $(a + b)^2$, or $a^2 + 2ab + b^2$

(b) $c^2 + 2ab$ **(c)** Subtract $2ab$ from each side to obtain $a^2 + b^2 = c^2$.

160. (a) $\frac{1}{2}(a + b)(a + b)$ **(b)** PWX: $\frac{1}{2}ab$; PZY: $\frac{1}{2}ab$; PXY: $\frac{1}{2}c^2$

(c) $\frac{1}{2}(a + b)(a + b) = \frac{1}{2}ab + \frac{1}{2}ab + \frac{1}{2}c^2$; When simplified, we obtain the equivalent equation $a^2 + b^2 = c^2$.

Section 10.4

1. B **3.** The terms 3 and $3xy$ are not like terms and cannot be combined. The expression $(3 + 3xy)\sqrt[3]{xy^2}$ cannot be simplified further. **5.** -4 **7.** $8\sqrt{10}$ **9.** $-\sqrt{5}$ **11.** $7\sqrt{3}$ **13.** The expression cannot be simplified further. **15.** $14\sqrt[3]{2}$

17. $24\sqrt{2}$ **19.** $20\sqrt{5}$ **21.** $4\sqrt{2x}$ **23.** $-11m\sqrt{2}$ **25.** $5\sqrt[4]{2}$ **27.** $7\sqrt[3]{2}$ **29.** $2\sqrt[3]{x}$ **31.** $-7\sqrt[3]{x^2y}$ **33.** $-x\sqrt[3]{xy^2}$ **35.** $19\sqrt[4]{2}$

37. $x\sqrt[4]{xy}$ **39.** $9\sqrt[4]{2a^3}$ **41.** $(4 + 3xy)\sqrt[3]{xy^2}$ **43.** $4t\sqrt[3]{3st} - 3s\sqrt{3st}$

45. $4x\sqrt[3]{x} + 6x\sqrt[4]{x}$ **47.** $2\sqrt{2} - 2$ **49.** $\dfrac{5\sqrt{5}}{6}$ **51.** $\dfrac{7\sqrt{2}}{6}$

53. $\dfrac{5\sqrt{2}}{3}$ **55.** $5\sqrt{2} + 4$ **57.** $\dfrac{5 + 3x}{x^4}$ **59.** $\dfrac{30\sqrt{2} - 21}{14}$

61. $\dfrac{m\sqrt[3]{m^2}}{2}$ **63.** $\dfrac{3x\sqrt[3]{2} - 4\sqrt[3]{5}}{x^3}$ **65. (a)** $\sqrt{7}$ **(b)** 2.645751311

(c) 2.645751311 **(d)** equal **67.** A; 42 m **69.** $\left(12\sqrt{5} + 5\sqrt{3}\right)$ in.

71. $\left(24\sqrt{2} + 12\sqrt{3}\right)$ in. **73.** $2\sqrt{106} + 4\sqrt{2}$

Section 10.5

1. E **3.** A **5.** D **7.** $3\sqrt{6} + 2\sqrt{3}$ **9.** $6 - 4\sqrt{3}$

11. $20\sqrt{2}$ **13.** $6 - \sqrt{6}$ **15.** $\sqrt{6} + \sqrt{2} + \sqrt{3} + 1$

17. -1 **19.** 6 **21.** $\sqrt{22} + \sqrt{55} - \sqrt{14} - \sqrt{35}$ **23.** $8 - \sqrt{15}$

25. $9 + 4\sqrt{5}$ **27.** $26 - 2\sqrt{105}$ **29.** $4 - \sqrt[3]{36}$

31. 10 **33.** $6x + 3\sqrt{x} - 2\sqrt{5x} - \sqrt{5}$ **35.** $9r - s$

37. $4\sqrt[3]{4y^2} - 19\sqrt[3]{2y} - 5$ **39.** $3x - 4$ **41.** Because 6 and $4\sqrt{3}$ are not like terms, they cannot be combined. The expression $6 - 4\sqrt{3}$ cannot be simplified further. **43.** $\sqrt{7}$ **45.** $5\sqrt{3}$ **47.** $\dfrac{\sqrt{6}}{2}$

49. $\dfrac{9\sqrt{15}}{5}$ **51.** $-\dfrac{7\sqrt{3}}{12}$ **53.** $\dfrac{\sqrt{14}}{2}$ **55.** $-\dfrac{\sqrt{14}}{10}$ **57.** $\dfrac{2\sqrt{6x}}{x}$

59. $\dfrac{-8\sqrt{3k}}{k}$ **61.** $\dfrac{-5m^2\sqrt{6mn}}{n^2}$ **63.** $\dfrac{12x^3\sqrt{2xy}}{y^5}$ **65.** $\dfrac{5\sqrt{2my}}{y^2}$

67. $-\dfrac{4k\sqrt{3z}}{z}$ **69.** $\dfrac{\sqrt[3]{18}}{3}$ **71.** $\dfrac{\sqrt[3]{12}}{3}$ **73.** $\dfrac{\sqrt[3]{18}}{4}$ **75.** $-\dfrac{\sqrt[3]{2pr}}{r}$

77. $\dfrac{x^2\sqrt[3]{y^2}}{y}$ **79.** $\dfrac{2\sqrt[4]{x^3}}{x}$ **81.** $\dfrac{\sqrt[4]{2yz^3}}{z}$ **83.** $\dfrac{3(4 - \sqrt{5})}{11}$

85. $3\left(\sqrt{5} - \sqrt{3}\right)$ **87.** $\dfrac{6\sqrt{2} + 4}{7}$ **89.** $\dfrac{2(3\sqrt{5} - 2\sqrt{3})}{33}$

91. $2\sqrt{3} + \sqrt{10} - 3\sqrt{2} - \sqrt{15}$ **93.** $\sqrt{m} - 2$

95. $\dfrac{4(\sqrt{x} + 2\sqrt{y})}{x - 4y}$ **97.** $\dfrac{x - 2\sqrt{xy} + y}{x - y}$ **99.** $\dfrac{5\sqrt{k}(2\sqrt{k} - \sqrt{q})}{4k - q}$

101. $3 + 2\sqrt{6}$ **103.** $1 - \sqrt{5}$ **105.** $\dfrac{4 - 2\sqrt{2}}{3}$ **107.** $\dfrac{6 + 2\sqrt{6p}}{3}$

109. $\dfrac{3\sqrt{x + y}}{x + y}$ **111.** $\dfrac{p\sqrt{p + 2}}{p + 2}$ **113.** Each expression is approximately equal to 0.2588190451. **115.** $\dfrac{33}{8(6 + \sqrt{3})}$ **116.** $\dfrac{11}{2(2\sqrt{5} + 3)}$

117. $\dfrac{4x - y}{3x(2\sqrt{x} + \sqrt{y})}$ **118.** $\dfrac{p - 9q}{4q(\sqrt{p} + 3\sqrt{q})}$

SUMMARY EXERCISES   Performing Operations with Radicals and Rational Exponents

1. The radicand is a fraction, $\frac{2}{5}$. **2.** The exponent in the radicand and the index of the radical have greatest common factor 5. **3.** The denominator contains a radical, $\sqrt[3]{10}$. **4.** The radicand has two factors, x and y, that are raised to powers greater than the index, 3.

5. $-6\sqrt{10}$ **6.** $7 - \sqrt{14}$ **7.** $2 + \sqrt{6} - 2\sqrt{3} - 3\sqrt{2}$

8. $4\sqrt{2}$ **9.** $73 + 12\sqrt{35}$ **10.** $\dfrac{-\sqrt{6}}{2}$ **11.** $4\left(\sqrt{7} - \sqrt{5}\right)$

12. $-3 + 2\sqrt{2}$ **13.** -44 **14.** $\dfrac{\sqrt{x} + \sqrt{5}}{x - 5}$ **15.** $2abc^3\sqrt[3]{b^2}$

16. $5\sqrt[3]{3}$ **17.** $3\left(\sqrt{5} - 2\right)$ **18.** $\dfrac{\sqrt{15x}}{5x}$ **19.** $\dfrac{8}{5}$ **20.** $\dfrac{\sqrt{2}}{8}$

21. $-\sqrt[3]{100}$ **22.** $11 + 2\sqrt{30}$ **23.** $-3\sqrt{3x}$ **24.** $52 - 30\sqrt{3}$

25. $\dfrac{\sqrt[3]{117}}{9}$ **26.** $3\sqrt{2} + \sqrt{15} + \sqrt{42} + \sqrt{35}$ **27.** $2\sqrt[4]{27}$

28. $\dfrac{x\sqrt[3]{x^2}}{y}$ **29.** $-4\sqrt{3} - 3$ **30.** $7 + 4 \cdot 3^{1/2}$, or $7 + 4\sqrt{3}$

31. $3\sqrt[3]{2x^2}$ **32.** -2 **33.** $3^{5/6}$ **34.** $\dfrac{x^{5/3}}{y}$ **35.** $xy^{6/5}$ **36.** $x^{10}y$

37. $\dfrac{1}{25x^2}$ **38.** $\dfrac{-6y^{1/6}}{x^{1/24}}$

Section 10.6

1. (a) yes **(b)** no **3. (a)** yes **(b)** no **5.** There is no solution. The radical expression, which is nonnegative, cannot equal a negative number. The solution set is $\varnothing$. **7.** $\{11\}$ **9.** $\left\{\frac{1}{3}\right\}$ **11.** $\varnothing$ **13.** $\{5\}$

15. $\{18\}$ **17.** $\{5\}$ **19.** $\{4\}$ **21.** $\{17\}$ **23.** $\{5\}$ **25.** $\varnothing$ **27.** $\{0\}$

29. $\{1\}$ **31.** $\{-1, 3\}$ **33.** $\{0\}$ **35.** $\varnothing$ **37.** $\{1\}$ **39.** We cannot just square each term. The right side should be $(8 - x)^2 = 64 - 16x + x^2$. The correct first step is $3x + 4 = 64 - 16x + x^2$. The solution set is $\{4\}$.

41. $\{7\}$ **43.** $\{7\}$ **45.** $\{4, 20\}$ **47.** $\varnothing$ **49.** $\left\{\frac{5}{4}\right\}$ **51.** $\{9\}$ **53.** $\{1\}$

55. $\{14\}$ **57.** $\{-1\}$ **59.** $\{8\}$ **61.** $\{0\}$ **63.** $\varnothing$ **65.** $\{-4\}$

67. $\{9, 17\}$ **69.** $\left\{\frac{1}{4}, 1\right\}$ **71.** $L = CZ^2$ **73.** $K = \dfrac{V^2m}{2}$

75. $M = \dfrac{r^2F}{m}$ **77. (a)** $r = \dfrac{a}{4\pi^2N^2}$ **(b)** $a = 4\pi^2N^2r$

Section 10.7

1. nonreal complex, complex   **3.** real, complex   **5.** pure imaginary, nonreal complex, complex   **7.** i **9.** -1 **11.** $-i$ **13.** $13i$ **15.** $-12i$ **17.** $i\sqrt{5}$ **19.** $4i\sqrt{3}$ **21.** It is incorrect to use the product rule for radicals before using the definition of $\sqrt{-b}$. The correct product is -15.   **23.** -15 **25.** $-\sqrt{105}$ **27.** -10 **29.** $i\sqrt{33}$ **31.** $5i\sqrt{6}$ **33.** $\sqrt{3}$ **35.** $5i$ **37.** -2 **39.** $-1+7i$ **41.** 0 **43.** $7+3i$ **45.** -2 **47.** $1+13i$ **49.** $6+6i$ **51.** $4+2i$ **53.** -81 **55.** -16 **57.** $-10-30i$ **59.** $10-5i$ **61.** $-9+40i$ **63.** $-16+30i$ **65.** 153 **67.** 97 **69.** 4 **71.** (a) $a-bi$ (b) $a^2; b^2$ **73.** $1+i$ **75.** $2+2i$ **77.** $-1+2i$ **79.** $-\frac{5}{13}-\frac{12}{13}i$ **81.** $1-3i$ **83.** $1+3i$ **85.** -1 **87.** i **89.** -1 **91.** $-i$ **93.** $-i$ **95.** 1 **97.** $\frac{1}{2}+\frac{1}{2}i$ **99.** (a) Substitute $1+5i$ for x in the equation. A true statement results—that is, $(1+5i)^2 - 2(1+5i) + 26$ will simplify to 0 when the operations are applied. Thus, $1+5i$ is a solution.   (b) Substituting $1-5i$ for x in the equation results in a true statement, indicating that $1-5i$ is a solution.   **100.** They are complex conjugates.   **101.** Substituting $3+2i$ for x in the equation results in a true statement, indicating that $3+2i$ is a solution.   **102.** $3-2i$; Substituting $3-2i$ for x in the equation results in a true statement, indicating that $3-2i$ is a solution.

Chapter 10 Review Exercises

1. 10 **2.** -17 **3.** 6 **4.** -5 **5.** -3 **6.** -2 **7.** $|x|$ **8.** x **9.** $|x|^5$ (or $|x^5|$)   **10.** $\sqrt[n]{a}$ is not a real number if n is even and a is negative.

11.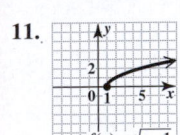
domain: $[1, \infty)$;
range: $[0, \infty)$

12.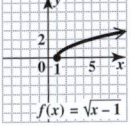
domain: $(-\infty, \infty)$;
range: $(-\infty, \infty)$

13. -6.856
14. -5.053
15. 4.960 **16.** 4.729
17. -7.937
18. -5.292
19. 1.9 sec
20. 66 in.^{2} **21.** 7

22. -11 **23.** 32 **24.** -4 **25.** $-\frac{216}{125}$ **26.** -32 **27.** $\frac{1000}{27}$ **28.** It is not a real number.   **29.** $10\sqrt{x}$ **30.** $\frac{1}{(\sqrt[3]{3a+b})^5}$

31. $7^{9/2}$ **32.** 1331 **33.** r^4 **34.** z **35.** 25 **36.** 96 **37.** $a^{2/3}$ **38.** $\frac{1}{y^{1/2}}$ **39.** $\frac{z^{1/2}x^{8/5}}{4}$ **40.** $r^{1/2}+r$ **41.** $y^{8/15}$ **42.** $\frac{1}{x^{1/2}}$ **43.** $p^{1/2}$ **44.** $k^{9/4}$ **45.** $m^{13/3}$ **46.** $t^{3/2}$ **47.** $x^{1/8}$ **48.** $x^{1/15}$ **49.** $x^{1/36}$ **50.** The product rule for exponents applies only if the bases are the same.

51. $\sqrt{66}$ **52.** $\sqrt{5r}$ **53.** $\sqrt[3]{30}$ **54.** $\sqrt[4]{21}$ **55.** $2\sqrt{5}$ **56.** $5\sqrt{3}$ **57.** $-5\sqrt{5}$ **58.** $-3\sqrt[3]{4}$ **59.** $10y^3\sqrt{y}$ **60.** $4pq^2\sqrt[3]{p}$ **61.** $3a^2b\sqrt[3]{4a^2b^2}$ **62.** $2r^2t\sqrt[3]{79r^2t}$ **63.** $\frac{y\sqrt{y}}{12}$ **64.** $\frac{m^5}{3}$ **65.** $\frac{\sqrt[3]{r^2}}{2}$ **66.** $\frac{a^2\sqrt[4]{a}}{3}$

67. $\sqrt{15}$ **68.** $p\sqrt{p}$ **69.** $\sqrt[12]{2000}$ **70.** $\sqrt[10]{x^7}$ **71.** 10 **72.** $\sqrt{197}$ **73.** $x^2+y^2=121$ **74.** $(x+2)^2+(y-4)^2=9$ **75.** $(x+1)^2+(y+3)^2=25$ **76.** $(x-4)^2+(y-2)^2=36$

77.
center: $(0, 0)$;
radius: 5

78.
$(x+3)^2+(y-3)^2=9$
center: $(-3, 3)$;
radius: 3

79.
$(x-2)^2+(y+5)^2=9$
center: $(2, -5)$;
radius: 3

80. It is impossible for the sum of the squares of two real numbers to be negative.   **81.** $-11\sqrt{2}$ **82.** $23\sqrt{5}$ **83.** $7\sqrt{3y}$ **84.** $26m\sqrt{6m}$ **85.** $19\sqrt[3]{2}$ **86.** $-8\sqrt[4]{2}$ **87.** $1-\sqrt{3}$ **88.** 2 **89.** $9-7\sqrt{2}$ **90.** $15-2\sqrt{26}$ **91.** 29 **92.** $2\sqrt[3]{2y^2}+2\sqrt[3]{4y}-3$ **93.** $\frac{\sqrt{30}}{5}$ **94.** $-3\sqrt{6}$ **95.** $\frac{3\sqrt{7py}}{y}$ **96.** $\frac{\sqrt{22}}{4}$ **97.** $-\frac{\sqrt[3]{45}}{5}$ **98.** $\frac{3m\sqrt[3]{4n}}{n^2}$ **99.** $\frac{\sqrt{2}-\sqrt{7}}{-5}$ **100.** $\frac{5(\sqrt{6}+3)}{3}$ **101.** $\frac{1-\sqrt{5}}{4}$ **102.** $\frac{-6+\sqrt{3}}{2}$ **103.** $\{2\}$ **104.** $\{6\}$ **105.** $\{0, 5\}$ **106.** $\{9\}$ **107.** $\{3\}$ **108.** $\{7\}$ **109.** $\left\{-\frac{1}{2}\right\}$ **110.** $\{14\}$ **111.** $\varnothing$ **112.** $\{7\}$ **113.** $H=\sqrt{L^2-W^2}$ **114.** 7.9 ft   **115.** $4i$ **116.** $10i\sqrt{2}$ **117.** $-10-2i$ **118.** $14+7i$ **119.** $-\sqrt{35}$ **120.** -45 **121.** 3 **122.** $5+i$ **123.** $32-24i$ **124.** $1-i$ **125.** $-i$ **126.** 1 **127.** -1 **128.** 1

Chapter 10 Mixed Review Exercises

1. $-13ab^2$ **2.** $\frac{1}{100}$ **3.** $\frac{1}{z^{3/5}}$ **4.** $3z^3t^2\sqrt[3]{2t^2}$ **5.** $7i$ **6.** $-\frac{\sqrt{3}}{6}$ **7.** $\frac{\sqrt[3]{60}}{5}$ **8.** 1 **9.** $57\sqrt{2}$ **10.** $5-11i$ **11.** $-5i$ **12.** $\frac{1+\sqrt{6}}{2}$ **13.** $5+12i$ **14.** $6x\sqrt[3]{y^2}$ **15.** $35+15i$ **16.** $\sqrt[12]{2000}$ **17.** $\frac{2\sqrt{z}(\sqrt{z}+2)}{z-4}$ **18.** $\left(12\sqrt{3}+5\sqrt{2}\right)$ ft   **19.** $\{5\}$ **20.** $\{-4\}$ **21.** $\left\{\frac{3}{2}\right\}$ **22.** $\{7\}$

Chapter 10 Test

[10.1] **1.** -29 **2.** -8 [10.2] **3.** 5 [10.1] **4.** 21.863 **5.** -9.405

6.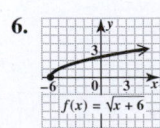
$f(x)=\sqrt{x+6}$
domain: $[-6, \infty)$;
range: $[0, \infty)$

[10.2] **7.** $\frac{125}{64}$ **8.** $\frac{1}{256}$ **9.** $\frac{9y^{3/10}}{x^2}$ **10.** $x^{4/3}y^6$ **11.** $7^{1/2}$ **12.** $a^{11/3}$ [10.3] **13.** $\sqrt{145}$ **14.** 10 **15.** $(x+4)^2+(y-6)^2=25$

16.

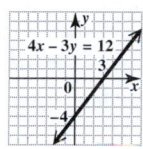

$(x-2)^2 + (y+3)^2 = 16$
center: $(2, -3)$;
radius: 4

17. $3x^2y^3\sqrt{6x}$ **18.** $2ab^3\sqrt[4]{2a^3b}$ **19.** $\sqrt[6]{200}$

[10.4] **20.** $26\sqrt{5}$ **21.** $(2ts - 3t^2)\sqrt[3]{2s^2}$

[10.5] **22.** $66 + \sqrt{5}$ **23.** $23 - 4\sqrt{15}$

24. $-\dfrac{\sqrt{10}}{4}$ **25.** $\dfrac{2\sqrt[3]{25}}{5}$

26. $-2(\sqrt{7} - \sqrt{5})$ **27.** $3 + \sqrt{6}$ [10.6] **28. (a)** 59.8

(b) $T = \dfrac{V_0{}^2 - V^2}{-V^2 k}$, or $T = \dfrac{V^2 - V_0{}^2}{V^2 k}$ **29.** $\{-1\}$ **30.** $\{3\}$

31. $\{-3\}$ [10.7] **32.** $-5 - 8i$ **33.** $-2 + 16i$ **34.** $3 + 4i$

35. i **36. (a)** true **(b)** true **(c)** false **(d)** true

Chapters R–10 Cumulative Review Exercises

[R.2] **1. (a)** $0.04; 4\%$ **(b)** $\dfrac{3}{20}; 0.15$ **(c)** $\dfrac{4}{5}; 80\%$ **(d)** $1.25; 125\%$

[1.1, 4.1, 10.1] **2. (a)** 16 **(b)** 16 **(c)** -16 **(d)** -16 **(e)** 2

(f) -2 **(g)** It is not a real number. [1.4, 1.5] **3.** -47 [R.1, 1.4] **4.** $\dfrac{5}{24}$

[2.3, 2.4] **5.** $\{-4\}$ **6.** $\{-12\}$ **7.** $\{6\}$ [2.9, 8.1] **8.** $(-6, \infty)$

[2.8] **9.** 36 nickels; 64 quarters **10.** $2\dfrac{2}{39}$ L

[3.2, 7.1] **11.**

[3.3–3.5, 7.1, 7.2] **12.** $-\dfrac{3}{2}; y = -\dfrac{3}{2}x$

[9.2] **13.** -37 [7.4, 7.5] **14.** $\{(7, -2)\}$

[7.6] **15.** $\{(-1, 1, 1)\}$

[7.7] **16.** 2-oz letter: $0.70; 3-oz letter: $0.91

[4.4] **17.** $-k^3 - 3k^2 - 8k - 9$ [4.5] **18.** $8x^2 + 17x - 21$

[4.7] **19.** $3y^3 - 3y^2 + 4y + 1 + \dfrac{-10}{2y + 1}$

[5.2–5.4] **20.** $(2p - 3q)(p - q)$ **21.** $(3k^2 + 4)(k - 1)(k + 1)$

22. $(x + 8)(x^2 - 8x + 64)$ [6.2] **23.** $\dfrac{y}{y + 5}$ [6.4] **24.** $\dfrac{4x + 2y}{(x + y)(x - y)}$

[6.5] **25.** $-\dfrac{9}{4}$ **26.** $\dfrac{-1}{a + b}$ [5.5] **27.** $\left\{-3, -\dfrac{5}{2}\right\}$ **28.** $\left\{-\dfrac{2}{5}, 1\right\}$

[6.6] **29.** $\varnothing$ [10.6] **30.** $\{3, 4\}$ [8.2] **31.** $(2, 3)$ **32.** $(-\infty, 2) \cup (3, \infty)$

[8.3] **33.** $\left\{-\dfrac{10}{3}, 1\right\}$ **34.** $(-\infty, -2] \cup [7, \infty)$ [10.2] **35.** $\dfrac{1}{9}$

[10.3] **36.** $2x\sqrt[3]{6x^2y^2}$ [10.4] **37.** $7\sqrt{2}$ [10.5] **38.** $\dfrac{\sqrt{10} + 2\sqrt{2}}{2}$

[10.3] **39.** $\sqrt{29}$ [10.7] **40.** $4 + 2i$

11 **QUADRATIC EQUATIONS, INEQUALITIES, AND FUNCTIONS**

Section 11.1

1. quadratic; second; two **3.** D **5. (a)** C **(b)** A **(c)** D **(d)** B

7. The equation is also true for -9. The solution set is $\{-9, 9\}$, or $\{\pm 9\}$.

9. $\{-7, 8\}$ **11.** $\{3, 5\}$ **13.** $\{\pm 11\}$ **15.** $\{\pm 13\}$ **17.** $\left\{-\dfrac{5}{3}, 6\right\}$

19. $\left\{-\dfrac{5}{2}, -\dfrac{2}{3}\right\}$ **21.** $\{\pm 9\}$ **23.** $\{\pm 12\}$ **25.** $\{\pm\sqrt{14}\}$

27. $\{\pm 4\sqrt{3}\}$ **29.** $\left\{\pm\dfrac{5}{2}\right\}$ **31.** $\{\pm 0.5\}$ **33.** $\{\pm 8\}$

35. $\{\pm\sqrt{3}\}$ **37.** $\{\pm 2\sqrt{3}\}$ **39.** $\{\pm 3\sqrt{2}\}$ **41.** $\{\pm 3\sqrt{3}\}$

43. $\{\pm 2\sqrt{6}\}$ **45.** $\left\{\pm\dfrac{2\sqrt{7}}{7}\right\}$ **47.** $\left\{\pm\dfrac{2\sqrt{5}}{5}\right\}$

49. $\{-2, 8\}$ **51.** $\{4 \pm \sqrt{3}\}$ **53.** $\{8 \pm 3\sqrt{3}\}$

55. $\left\{-3, \dfrac{5}{3}\right\}$ **57.** $\left\{0, \dfrac{3}{2}\right\}$ **59.** $\left\{\dfrac{5 \pm \sqrt{30}}{2}\right\}$

61. $\left\{\dfrac{-1 \pm 3\sqrt{2}}{3}\right\}$ **63.** $\{-10 \pm 4\sqrt{3}\}$ **65.** $\left\{0, \dfrac{1}{4}\right\}$

67. $\left\{-\dfrac{1}{3}, 1\right\}$ **69.** $\left\{\dfrac{-1 \pm \sqrt{3}}{4}\right\}$ **71.** $\left\{\dfrac{1 \pm 4\sqrt{3}}{4}\right\}$ **73.** $\{\pm 10i\}$

75. $\{\pm i\sqrt{26}\}$ **77.** $\{\pm 2i\sqrt{3}\}$ **79.** $\{-3 \pm 2i\}$ **81.** $\{5 \pm i\sqrt{3}\}$

83. $\left\{\dfrac{1}{6} \pm \dfrac{\sqrt{2}}{3}i\right\}$ **85.** 5.6 sec

Section 11.2

1. D **3.** $25; (x + 5)^2$ **5.** $100; (z - 10)^2$ **7.** $1; (x + 1)^2$

9. $\dfrac{25}{4}; \left(p - \dfrac{5}{2}\right)^2$ **11.** $\dfrac{1}{16}; \left(x + \dfrac{1}{4}\right)^2$ **13.** $0.04; (x - 0.2)^2$

15. $4; 2; 4; 4; x + 2; 5; \{-2 \pm \sqrt{5}\}$ **17.** $\{1, 3\}$

19. $\{-1 \pm \sqrt{6}\}$ **21.** $\{-2 \pm \sqrt{6}\}$ **23.** $\{-5 \pm \sqrt{7}\}$

25. $\{4 \pm 2\sqrt{3}\}$ **27.** $\left\{\dfrac{-7 \pm \sqrt{53}}{2}\right\}$ **29.** $\left\{-\dfrac{3}{2}, \dfrac{1}{2}\right\}$

31. $\left\{\dfrac{-5 \pm \sqrt{41}}{4}\right\}$ **33.** $\left\{\dfrac{5 \pm \sqrt{15}}{5}\right\}$ **35.** $\left\{\dfrac{9 \pm \sqrt{21}}{6}\right\}$

37. $\left\{\dfrac{-7 \pm \sqrt{97}}{6}\right\}$ **39.** $\left\{-\dfrac{8}{3}, 3\right\}$ **41.** $\{1 \pm \sqrt{6}\}$

43. $\left\{-\dfrac{11}{5}, 1\right\}$ **45.** $\{-4, 2\}$ **47.** $\{4 \pm \sqrt{3}\}$ **49.** $\left\{\dfrac{3 \pm \sqrt{13}}{2}\right\}$

51. $\left\{\dfrac{-3 \pm \sqrt{5}}{2}\right\}$ **53.** $\left\{-\dfrac{7}{3}, 6\right\}$ **55.** $\{1 \pm \sqrt{2}\}$

57. (a) $\left\{\dfrac{3 \pm 2\sqrt{6}}{3}\right\}$ **(b)** $\{-0.633, 2.633\}$

59. (a) $\{-2 \pm \sqrt{3}\}$ **(b)** $\{-3.732, -0.268\}$ **61.** $\{-2 \pm 3i\}$

63. $\{-3 \pm i\sqrt{3}\}$ **65.** $\left\{-\dfrac{2}{3} \pm \dfrac{2\sqrt{2}}{3}i\right\}$ **67.** $\left\{-\dfrac{5}{2} \pm \dfrac{\sqrt{15}}{2}i\right\}$

69. $\{\pm\sqrt{b}\}$ **71.** $\left\{\pm\dfrac{\sqrt{b^2 + 16}}{2}\right\}$ **73.** $\left\{\dfrac{2b \pm \sqrt{3a}}{5}\right\}$

75. x^2 **76.** x **77.** $6x$ **78.** 1 **79.** 9 **80.** $(x + 3)^2$, or $x^2 + 6x + 9$

Section 11.3

1. No. The fraction bar should extend under the term $-b$. The correct

formula is $x = \dfrac{-b \pm \sqrt{b^2 - 4ac}}{2a}$. **3.** The last step is wrong. Because

5 is not a common factor in the numerator, the fraction cannot be

simplified. The solution set is $\left\{\dfrac{5 \pm \sqrt{5}}{10}\right\}$. **5.** $\{3, 5\}$ **7.** $\left\{-\dfrac{5}{2}, \dfrac{2}{3}\right\}$

9. $\left\{-\frac{3}{2}\right\}$ **11.** $\left\{\frac{1}{6}\right\}$ **13.** $\left\{\frac{-2 \pm \sqrt{2}}{2}\right\}$ **15.** $\left\{\frac{1 \pm \sqrt{3}}{2}\right\}$

17. $\{5 \pm \sqrt{7}\}$ **19.** $\left\{\frac{-1 \pm \sqrt{2}}{2}\right\}$ **21.** $\left\{\frac{-1 \pm \sqrt{7}}{3}\right\}$

23. $\{1 \pm \sqrt{5}\}$ **25.** $\left\{\frac{-2 \pm \sqrt{10}}{2}\right\}$ **27.** $\{-1 \pm 3\sqrt{2}\}$

29. $\left\{\frac{1 \pm \sqrt{29}}{2}\right\}$ **31.** $\left\{\frac{-4 \pm \sqrt{91}}{3}\right\}$ **33.** $\left\{\frac{-3 \pm \sqrt{57}}{8}\right\}$

35. $\left\{\frac{3}{2} \pm \frac{\sqrt{15}}{2}i\right\}$ **37.** $\{3 \pm i\sqrt{5}\}$ **39.** $\left\{\frac{1}{2} \pm \frac{\sqrt{6}}{2}i\right\}$

41. $\left\{-\frac{2}{3} \pm \frac{\sqrt{2}}{3}i\right\}$ **43.** $\left\{\frac{1}{2} \pm \frac{1}{4}i\right\}$ **45.** $\left\{\frac{4}{5} \pm \frac{2\sqrt{6}}{5}i\right\}$

47. 0; B; zero-factor property   **49.** 8; C; quadratic formula

51. 49; A; zero-factor property   **53.** -80; D; quadratic formula

55. (a) 25; zero-factor property; $\left\{-3, -\frac{4}{3}\right\}$ **(b)** 44; quadratic formula;

$\left\{\frac{7 \pm \sqrt{11}}{2}\right\}$ **57.** -10 or 10   **59.** 16   **61.** 25   **63.** $b = \frac{44}{5}; \frac{3}{10}$

Section 11.4

1. Multiply by the LCD, x.   **3.** Substitute a variable for $x^2 + x$.

5. The proposed solution -1 does not check. The solution set is $\{4\}$.

7. $\{-2, 7\}$ **9.** $\{-4, 7\}$ **11.** $\left\{-\frac{2}{3}, 1\right\}$ **13.** $\left\{-\frac{14}{17}, 5\right\}$

15. $\left\{-\frac{11}{7}, 0\right\}$ **17.** $\left\{\frac{-1 \pm \sqrt{13}}{2}\right\}$ **19.** $\left\{-\frac{8}{3}, -1\right\}$

21. $\left\{\frac{2 \pm \sqrt{22}}{3}\right\}$ **23.** $\left\{\frac{-1 \pm \sqrt{5}}{4}\right\}$ **25. (a)** $(20 - t)$ mph

(b) $(20 + t)$ mph   **27.** the rate of her boat in still water; $x - 5$; $x + 5$;

row 1 of table: 15, $x - 5$, $\frac{15}{x - 5}$; row 2 of table: 15, $x + 5$, $\frac{15}{x + 5}$;

$\frac{15}{x - 5} + \frac{15}{x + 5} = 4$; 10 mph   **29.** 25 mph   **31.** 50 mph

33. 3.6 hr   **35.** Rusty: 25.0 hr; Nancy: 23.0 hr   **37.** 9 min

39. $\{2, 5\}$ **41.** $\{3\}$ **43.** $\left\{\frac{8}{9}\right\}$ **45.** $\{9\}$ **47.** $\left\{\frac{2}{5}\right\}$ **49.** $\{-2\}$

51. $\{\pm 2, \pm 5\}$ **53.** $\left\{\pm 1, \pm \frac{3}{2}\right\}$ **55.** $\{\pm 2, \pm 2\sqrt{3}\}$

57. $\{-6, -5\}$ **59.** $\left\{-\frac{16}{3}, -2\right\}$ **61.** $\{-8, 1\}$ **63.** $\{-64, 27\}$

65. $\left\{\pm 1, \pm \frac{27}{8}\right\}$ **67.** $\left\{-\frac{1}{3}, \frac{1}{6}\right\}$ **69.** $\left\{-\frac{1}{2}, 3\right\}$ **71.** $\left\{\pm \frac{\sqrt{6}}{3}, \pm \frac{1}{2}\right\}$

73. $\{3, 11\}$ **75.** $\{25\}$ **77.** $\left\{-\sqrt[3]{5}, -\frac{\sqrt[3]{4}}{2}\right\}$ **79.** $\left\{\frac{4}{3}, \frac{9}{4}\right\}$

81. $\left\{\pm \frac{\sqrt{9 + \sqrt{65}}}{2}, \pm \frac{\sqrt{9 - \sqrt{65}}}{2}\right\}$ **83.** $\left\{\pm 1, \pm \frac{\sqrt{6}}{2}i\right\}$

SUMMARY EXERCISES   Applying Methods for Solving Quadratic Equations

1. square root property   **2.** zero-factor property   **3.** quadratic formula

4. quadratic formula   **5.** zero-factor property   **6.** square root property

7. $\{\pm \sqrt{7}\}$ **8.** $\left\{-\frac{3}{2}, \frac{5}{3}\right\}$ **9.** $\{-3 \pm \sqrt{5}\}$ **10.** $\{-2, 8\}$

11. $\left\{-\frac{3}{2}, 4\right\}$ **12.** $\left\{-3, \frac{1}{3}\right\}$ **13.** $\left\{\frac{2 \pm \sqrt{2}}{2}\right\}$ **14.** $\{\pm 2i\sqrt{3}\}$

15. $\left\{\frac{1}{2}, 2\right\}$ **16.** $\{\pm 1, \pm 3\}$ **17.** $\left\{\frac{-3 \pm 2\sqrt{2}}{2}\right\}$ **18.** $\left\{\frac{4}{5}, 3\right\}$

19. $\{\pm \sqrt{2}, \pm \sqrt{7}\}$ **20.** $\left\{\frac{1 \pm \sqrt{5}}{4}\right\}$ **21.** $\left\{-\frac{1}{2} \pm \frac{\sqrt{3}}{2}i\right\}$

22. $\left\{-\frac{\sqrt[3]{175}}{5}, 1\right\}$ **23.** $\left\{\frac{3}{2}\right\}$ **24.** $\left\{\frac{2}{3}\right\}$ **25.** $\{\pm 6\sqrt{2}\}$

26. $\left\{-\frac{2}{3}, 2\right\}$ **27.** $\{-4, 9\}$ **28.** $\{\pm 13\}$ **29.** $\left\{1 \pm \frac{\sqrt{3}}{3}i\right\}$

30. $\{3\}$ **31.** $\left\{\frac{1}{6} \pm \frac{\sqrt{47}}{6}i\right\}$ **32.** $\left\{-\frac{1}{3}, \frac{1}{6}\right\}$

Section 11.5

1. Find a common denominator, and then multiply both sides by the common denominator.   **3.** Write it in standard form (with 0 on one side, in decreasing powers of w).   **5.** $m = \sqrt{p^2 - n^2}$ **7.** $t = \frac{\pm \sqrt{dk}}{k}$

9. $r = \frac{\pm \sqrt{S\pi}}{2\pi}$ **11.** $d = \frac{\pm \sqrt{skI}}{I}$ **13.** $v = \frac{\pm \sqrt{kAF}}{F}$

15. $r = \frac{\pm \sqrt{3\pi Vh}}{\pi h}$ **17.** $t = \frac{-B \pm \sqrt{B^2 - 4AC}}{2A}$ **19.** $h = \frac{D^2}{k}$

21. $\ell = \frac{p^2 g}{k}$ **23.** $R = \frac{E^2 - 2pr \pm E\sqrt{E^2 - 4pr}}{2p}$

25. $r = \frac{5pc}{4}$ or $r = -\frac{2pc}{3}$ **27.** $I = \frac{-cR \pm \sqrt{c^2 R^2 - 4cL}}{2cL}$

29. 7.9, 8.9, 11.9   **31.** eastbound ship: 80 mi; southbound ship: 150 mi

33. 8 in., 15 in., 17 in.   **35.** length: 24 ft; width: 10 ft   **37.** 2 ft

39. 7 m by 12 m   **41.** 20 in. by 12 in.   **43.** 1 sec and 8 sec

45. 2.4 sec and 5.6 sec   **47.** 9.2 sec   **49.** It reaches its *maximum* height at 5 sec because this is the only time it reaches 400 ft.

51. $1.50   **53.** 0.035, or 3.5%   **55.** 5.5 m per sec   **57.** 5 or 14

59. (a) $600 billion   **(b)** $597 billion; They are about the same.

61. 2008   **63.** isosceles triangle; right triangle   **64.** no; The formula for the Pythagorean theorem involves the *squares* of the sides, not the square roots.

65.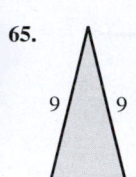
$\sqrt{9} + \sqrt{9} = 3 + 3 = 6$;
$\sqrt{4} = 2$;
$6 \neq 2$

66. The sum of the squares of the two shorter sides (legs) of a right triangle is equal to the square of the longest side (hypotenuse).

Section 11.6

1. (a) B **(b)** C **(c)** A **(d)** D **3. (a)** D **(b)** B **(c)** F **(d)** C
(e) A **(f)** E **5.** $(0, 0)$ **7.** $(0, 0)$ **9.** $(0, 4)$ **11.** $(1, 0)$
13. $(-3, -4)$ **15.** $(5, 6)$ **17.** down; wider **19.** up; narrower
21. down; narrower

23. 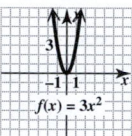   **25.** 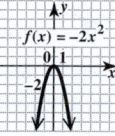   **27.**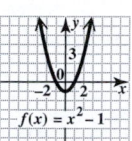

vertex: $(0, 0)$;   vertex: $(0, 0)$;   vertex: $(0, -1)$;
axis: $x = 0$;   axis: $x = 0$;   axis: $x = 0$;
domain: $(-\infty, \infty)$;   domain: $(-\infty, \infty)$;   domain: $(-\infty, \infty)$;
range: $[0, \infty)$ range: $(-\infty, 0]$ range: $[-1, \infty)$

29. 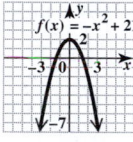   **31.** 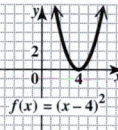   **33.**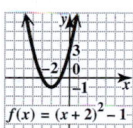

vertex: $(0, 2)$;   vertex: $(4, 0)$;   vertex: $(-2, -1)$;
axis: $x = 0$;   axis: $x = 4$;   axis: $x = -2$;
domain: $(-\infty, \infty)$;   domain: $(-\infty, \infty)$;   domain: $(-\infty, \infty)$;
range: $(-\infty, 2]$ range: $[0, \infty)$ range: $[-1, \infty)$

35. 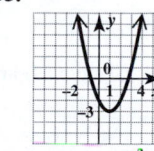   **37.** 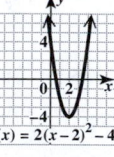   **39.**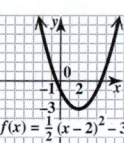

$f(x) = (x-1)^2 - 3$
vertex: $(1, -3)$;   vertex: $(2, -4)$;   vertex: $(2, -3)$;
axis: $x = 1$;   axis: $x = 2$;   axis: $x = 2$;
domain: $(-\infty, \infty)$;   domain: $(-\infty, \infty)$;   domain: $(-\infty, \infty)$;
range: $[-3, \infty)$ range: $[-4, \infty)$ range: $[-3, \infty)$

41.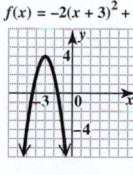
$f(x) = -2(x+3)^2 + 4$
vertex: $(-3, 4)$;
axis: $x = -3$;
domain: $(-\infty, \infty)$;
range: $(-\infty, 4]$

43. $f(x) = -\frac{1}{2}(x+1)^2 + 2$
vertex: $(-1, 2)$;
axis: $x = -1$;
domain: $(-\infty, \infty)$;
range: $(-\infty, 2]$

45. linear function; positive
47. quadratic function; positive
49. quadratic function; negative

51. (a)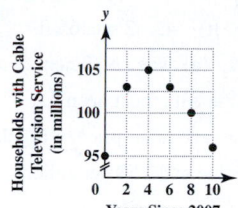

(b) quadratic function
(c) negative
(d) $y = -0.4x^2 + 4.1x + 95$
(e) 2009: 101.6 million households; 2015: 102.2 million households; The model approximates the data fairly well.

Section 11.7

1. If x is squared, it has a vertical axis. If y is squared, it has a horizontal axis. **3.** Find the discriminant of the function. If it is positive, there are two x-intercepts. If it is 0, there is one x-intercept (at the vertex), and if it is negative, there is no x-intercept. **5.** A, D are vertical parabolas. B, C are horizontal parabolas. **7.** $(-4, -6)$ **9.** $(1, -3)$

11. $\left(-\frac{1}{2}, -\frac{29}{4}\right)$ **13.** $(-1, 3)$; up; narrower; no x-intercepts

15. $\left(\frac{5}{2}, \frac{37}{4}\right)$; down; same shape; two x-intercepts **17.** $(-3, -9)$; to the right; wider **19.** F **21.** C **23.** D

25. $f(x) = x^2 + 8x + 10$ **27.** 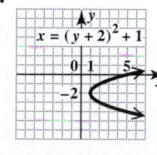   **29.** $f(x) = -2x^2 + 4x - 5$

vertex: $(-4, -6)$;   vertex: $(-2, -1)$;   vertex: $(1, -3)$;
axis: $x = -4$;   axis: $x = -2$;   axis: $x = 1$;
domain: $(-\infty, \infty)$;   domain: $(-\infty, \infty)$;   domain: $(-\infty, \infty)$;
range: $[-6, \infty)$ range: $[-1, \infty)$ range: $(-\infty, -3]$

31. $f(x) = -3x^2 - 6x + 2$ **33.** 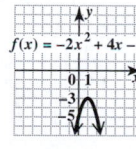   **35.** $x = -(y-3)^2 - 1$

vertex: $(-1, 5)$; $x = (y+2)^2 + 1$
axis: $x = -1$;   vertex: $(1, -2)$;   vertex: $(-1, 3)$;
domain: $(-\infty, \infty)$;   axis: $y = -2$;   axis: $y = 3$;
range: $(-\infty, 5]$ domain: $[1, \infty)$;   domain: $(-\infty, -1]$;
                        range: $(-\infty, \infty)$ range: $(-\infty, \infty)$

37. $x = -\frac{1}{5}y^2 + 2y - 4$

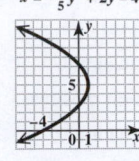

vertex: $(1, 5)$;
axis: $y = 5$;
domain: $(-\infty, 1]$;
range: $(-\infty, \infty)$

39.

$x = 3y^2 + 12y + 5$

vertex: $(-7, -2)$;
axis: $y = -2$;
domain: $[-7, \infty)$;
range: $(-\infty, \infty)$

41. 20 and 20 **43.** 140 ft by 70 ft; 9800 ft^2 **45.** 2 sec; 65 ft
47. 20 units; $210 **49.** 16 ft; 2 sec **51. (a)** The coefficient of x^2 is
negative because a parabola that models the data must open
down. **(b)** $(3.32, 3.58)$; In 2012, the average retail price of regular
unleaded gasoline reached a maximum value of $3.58 per gallon.
53. (a) $R(x) = (100 - x)(200 + 4x)$
$$= 20,000 + 200x - 4x^2$$

(b)

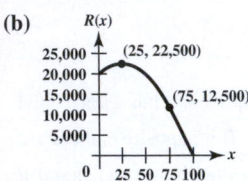

$R(x)$

25,000 (25, 22,500)
20,000
15,000
10,000 (75, 12,500)
5,000

0   25 50 75 100

(c) 25 **(d)** $22,500

Section 11.8

1. $(-\infty, -4] \cup [3, \infty)$ **3. (a)** $\{1, 3\}$ **(b)** $(-\infty, 1) \cup (3, \infty)$
(c) $(1, 3)$ **5. (a)** $\{-2, 5\}$ **(b)** $[-2, 5]$ **(c)** $(-\infty, -2] \cup [5, \infty)$

7.
$(-\infty, -1) \cup (5, \infty)$

9.
$(-4, 6)$

11.
$(-\infty, 1] \cup [3, \infty)$

13.
$\left(-\infty, -\frac{3}{2}\right] \cup \left[\frac{3}{5}, \infty\right)$

15.
$\left[-\frac{3}{2}, \frac{3}{2}\right]$

17.
$\left(-\infty, -\frac{1}{2}\right] \cup \left[\frac{1}{3}, \infty\right)$

19.
$(-\infty, 0] \cup [4, \infty)$

21.
$\left[0, \frac{5}{3}\right]$

23.
$(-\infty, 3 - \sqrt{3}] \cup [3 + \sqrt{3}, \infty)$

25. $(-\infty, \infty)$ **27.** $\varnothing$
29. $\varnothing$ **31.** $(-\infty, \infty)$

33.
$(-\infty, 1) \cup (2, 4)$

35.
$\left[-\frac{3}{2}, \frac{1}{3}\right] \cup [4, \infty)$

37.
$(-\infty, 1) \cup (4, \infty)$

39.
$\left[-\frac{3}{2}, 5\right)$

41.
$(2, 6]$

43.
$\left(-\infty, \frac{1}{2}\right) \cup \left(\frac{5}{4}, \infty\right)$

45.
$[-7, -2)$

47.
$(-\infty, 2) \cup (4, \infty)$

49.
$\left(0, \frac{1}{2}\right) \cup \left(\frac{5}{2}, \infty\right)$

51.
$\left[\frac{3}{2}, \infty\right)$

53.
$\left(-2, \frac{5}{3}\right) \cup \left(\frac{5}{3}, \infty\right)$

55. 3 sec and 13 sec **56.** between 3 sec
and 13 sec **57.** at 0 sec (the time when it
is initially projected) and at 16 sec (the time
when it hits the ground) **58.** between 0
and 3 sec and between 13 and 16 sec

Chapter 11 Review Exercises

1. $\{-7, 4\}$ **2.** $\left\{-5, \frac{3}{2}\right\}$ **3.** $\{\pm 13\}$ **4.** 5.9 sec **5.** $\{\pm 11\}$
6. $\{\pm \sqrt{3}\}$ **7.** $\left\{-\frac{15}{2}, \frac{5}{2}\right\}$ **8.** $\left\{\frac{2}{3} \pm \frac{5}{3}i\right\}$ **9.** $\{-2 \pm \sqrt{19}\}$ **10.** $\left\{\frac{1}{2}, 1\right\}$
11. $\left\{\frac{-4 \pm \sqrt{22}}{2}\right\}$ **12.** $\left\{\frac{3}{8} \pm \frac{\sqrt{87}}{8}i\right\}$ **13.** $\left\{-\frac{7}{2}, 3\right\}$
14. $\left\{\frac{-5 \pm \sqrt{53}}{2}\right\}$ **15.** $\left\{\frac{1 \pm \sqrt{41}}{2}\right\}$ **16.** $\left\{-\frac{3}{4} \pm \frac{\sqrt{23}}{4}i\right\}$
17. $\left\{\frac{2}{3} \pm \frac{\sqrt{2}}{3}i\right\}$ **18.** $\left\{\frac{-7 \pm \sqrt{37}}{2}\right\}$ **19. (a)** 17; C; quadratic formula
(b) 64; A; zero-factor property **20. (a)** -92; D; quadratic formula
(b) 0; B; zero-factor property **21.** $\left\{-\frac{5}{2}, 3\right\}$ **22.** $\left\{-\frac{1}{2}, 1\right\}$
23. $\{-4\}$ **24.** $\left\{-\frac{11}{6}, -\frac{19}{12}\right\}$ **25.** $\left\{-\frac{343}{8}, 64\right\}$ **26.** $\{\pm 1, \pm 3\}$
27. 7 mph **28.** 40 mph **29.** 4.6 hr **30.** Zoran: 2.6 hr; Claude: 3.6 hr
31. $v = \frac{\pm \sqrt{rFkw}}{kw}$ **32.** $y = \frac{6p^2}{z}$ **33.** $t = \frac{3m \pm \sqrt{9m^2 + 24m}}{2m}$
34. 9 ft, 12 ft, 15 ft **35.** 12 cm by 20 cm **36.** 1 in. **37.** 18 in.
38. 3 min **39.** $(1, 0)$ **40.** $(3, 7)$ **41.** $(-4, 3)$ **42.** $\left(\frac{2}{3}, -\frac{2}{3}\right)$

43.

$y = 2(x - 2)^2 - 3$

vertex: $(2, -3)$;
axis: $x = 2$;
domain: $(-\infty, \infty)$;
range: $[-3, \infty)$

44.

$f(x) = -2x^2 + 8x - 5$

vertex: $(2, 3)$;
axis: $x = 2$;
domain: $(-\infty, \infty)$;
range: $(-\infty, 3]$

45.

$x = 2(y + 3)^2 - 4$

vertex: $(-4, -3)$;
axis: $y = -3$;
domain: $[-4, \infty)$;
range: $(-\infty, \infty)$

46.

vertex: $(4, 6)$;

axis: $y = 6$;

domain: $(-\infty, 4]$;

range: $(-\infty, \infty)$

$x = -\frac{1}{2}y^2 + 6y - 14$

47. 5 sec; 400 ft

48. length: 50 m; width: 50 m; maximum area: 2500 m²

49.

$\left(-\infty, -\frac{3}{2}\right) \cup (4, \infty)$

50.

$[-4, 3]$

51.

$(-\infty, -5] \cup [-2, 3]$

52. $\varnothing$

53.

$\left(-\infty, \frac{1}{2}\right) \cup (2, \infty)$

54.

$[-3, 2)$

Chapter 11 Mixed Review Exercises

1. $R = \dfrac{\pm\sqrt{Vh - r^2h}}{h}$ **2.** $\left\{1 \pm \dfrac{\sqrt{3}}{3}i\right\}$

3. $\left\{\dfrac{-11 \pm \sqrt{7}}{3}\right\}$ **4.** $d = \dfrac{\pm\sqrt{SkI}}{I}$ **5.** $(-\infty, \infty)$

6. $\{4\}$ **7.** $\left\{\pm\sqrt{4 + \sqrt{15}}, \pm\sqrt{4 - \sqrt{15}}\right\}$

8. $\left(-5, -\dfrac{23}{5}\right]$ **9.** $\left\{-\dfrac{5}{3}, -\dfrac{3}{2}\right\}$ **10.** $\{-2, -1, 3, 4\}$

11. $(-\infty, -6) \cup \left(-\dfrac{3}{2}, 1\right)$ **12. (a)** F **(b)** B **(c)** C **(d)** A

(e) E **(f)** D

13.

vertex: $\left(-\dfrac{1}{2}, -3\right)$;

axis: $x = -\dfrac{1}{2}$;

domain: $(-\infty, \infty)$;

range: $[-3, \infty)$

$f(x) = 4x^2 + 4x - 2$

14. 10 mph **15.** length: 2 cm; width: 1.5 cm

16. (a) minimum **(b)** 1967; 5.4%

Chapter 11 Test

[11.1, 11.2] **1.** $\left\{-3, \dfrac{2}{5}\right\}$ **2.** $\{\pm 3\sqrt{6}\}$ **3.** $\left\{-\dfrac{8}{7}, \dfrac{2}{7}\right\}$ **4.** $\{-1 \pm \sqrt{5}\}$

[11.3] **5.** $\left\{\dfrac{3 \pm \sqrt{17}}{4}\right\}$ **6.** $\left\{\dfrac{2}{3} \pm \dfrac{\sqrt{11}}{3}i\right\}$ [11.4] **7.** $\left\{\dfrac{2}{3}\right\}$

[11.1] **8.** A   [11.3] **9.** discriminant: 88; There are two irrational solutions.

[11.1–11.4] **10.** $\left\{-\dfrac{2}{3}, 6\right\}$ **11.** $\left\{\dfrac{-7 \pm \sqrt{97}}{8}\right\}$ **12.** $\left\{\pm\dfrac{1}{3}, \pm 2\right\}$

13. $\left\{-\dfrac{5}{2}, 1\right\}$ [11.5] **14.** $r = \dfrac{\pm\sqrt{\pi S}}{2\pi}$ [11.4] **15.** Terry: 11.1 hr; Callie: 9.1 hr **16.** 7 mph   [11.5] **17.** 2 ft **18.** 16 m   [11.6, 11.7] **19.** A

20.

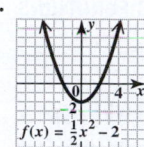

vertex: $(0, -2)$;

axis: $x = 0$;

domain: $(-\infty, \infty)$;

range: $[-2, \infty)$

$f(x) = \frac{1}{2}x^2 - 2$

21. $f(x) = -x^2 + 4x - 1$

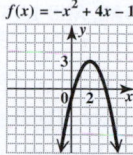

vertex: $(2, 3)$;

axis: $x = 2$;

domain: $(-\infty, \infty)$;

range: $(-\infty, 3]$

22.

vertex: $(2, 2)$;

axis: $y = 2$;

domain: $(-\infty, 2]$;

range: $(-\infty, \infty)$

$x = -(y - 2)^2 + 2$

23. 160 ft by 320 ft

[11.8] **24.**

$(-\infty, -5) \cup \left(\dfrac{3}{2}, \infty\right)$

25.

$(-\infty, 4) \cup [9, \infty)$

Chapters R–11 Cumulative Review Exercises

[R.1, R.2, 1.3, 1.4] **1.** -5.38 **2.** 2 **3.** 25 **4.** 0.72   [1.1, 1.7] **5.** -45

[1.3, 10.7] **6. (a)** $-2, 0, 7$ **(b)** $-\dfrac{7}{3}, -2, 0, 0.7, 7, \dfrac{32}{3}$

(c) All are real except $\sqrt{-8}$. **(d)** All are complex numbers.

[2.3] **7.** $\left\{\dfrac{4}{5}\right\}$ [8.3] **8.** $\left\{\dfrac{11}{10}, \dfrac{7}{2}\right\}$ [10.6] **9.** $\left\{\dfrac{2}{3}\right\}$ [6.6] **10.** $\varnothing$

[11.2, 11.3] **11.** $\left\{\dfrac{7 \pm \sqrt{177}}{4}\right\}$ [11.4] **12.** $\{\pm 1, \pm 2\}$

[2.9, 8.1] **13.** $[1, \infty)$ [8.3] **14.** $\left[2, \dfrac{8}{3}\right]$ [11.8] **15.** $(1, 3)$ **16.** $(-2, 1)$

[3.2, 7.1, 9.1, 9.2]

17.

function;

domain: $(-\infty, \infty)$;

range: $(-\infty, \infty)$;

$4x - 5y = 15$

$f(x) = \dfrac{4}{5}x - 3$

[8.4, 9.1]

18.

$4x - 5y < 15$

not a function

[11.6] **19.**

function;

domain: $(-\infty, \infty)$;

range: $(-\infty, 3]$;

$y = -2(x - 1)^2 + 3$

$f(x) = -2(x - 1)^2 + 3$

[3.2, 3.3, 7.1] **20.** $m = \dfrac{2}{7}$; x-intercept: $(-8, 0)$; y-intercept: $\left(0, \dfrac{16}{7}\right)$

[7.2] **21.** $y = -\dfrac{5}{2}x + 2$ **22.** $y = \dfrac{2}{5}x + \dfrac{13}{5}$ [7.3-7.5] **23.** $\{(1, -2)\}$

24. $\varnothing$ [7.6] **25.** $\{(3, -4, 2)\}$ [7.7] **26.** *Star Wars: The Last Jedi:* $545 million; *Beauty and the Beast:* $504 million

[4.1, 4.2] **27.** $\dfrac{x^8}{y^4}$ **28.** $\dfrac{4}{xy^2}$ [4.6] **29.** $\dfrac{4}{9}t^2 + 12t + 81$

[4.7] **30.** $4x^2 - 6x + 11 + \dfrac{4}{x + 2}$ [5.1–5.4] **31.** $(4m - 3)(6m + 5)$

32. $4(t + 5)(t - 5)$ **33.** $(2x + 3y)(4x^2 - 6xy + 9y^2)$ **34.** $(3x - 5y)^2$

[6.2] **35.** $-\dfrac{5}{18}$ [6.4] **36.** $-\dfrac{8}{x}$ [6.5] **37.** $\dfrac{r - s}{r}$

[11.5] **38.** southbound car: 57 mi; eastbound car: 76 mi

[10.3] **39.** $\dfrac{3\sqrt[3]{4}}{4}$ [10.5] **40.** $\sqrt{7} + \sqrt{5}$

12   INVERSE, EXPONENTIAL, AND LOGARITHMIC FUNCTIONS

Section 12.1

1. B   **3.** A   **5.** It is not one-to-one. The Big Breakfast with Egg Whites and the Bacon Clubhouse Burger are paired with the same fat content, 41 g.   **7.** Yes, it is one-to-one. Adding 1 to 1058 would make two distances be the same, so the function would not be one-to-one.

9. $\{(6, 3), (10, 2), (12, 5)\}$ **11.** not one-to-one

13. $\{(4.5, 0), (8.6, 2), (12.7, 4)\}$ **15.** $f^{-1}(x) = x - 3$

17. $f^{-1}(x) = -2x - 4$ **19.** $f^{-1}(x) = \dfrac{x - 4}{2}$, or $f^{-1}(x) = \dfrac{1}{2}x - 2$

21. $g^{-1}(x) = \dfrac{-x + 3}{4}$, or $g^{-1}(x) = -\dfrac{1}{4}x + \dfrac{3}{4}$ **23.** not one-to-one

25. $f^{-1}(x) = x^2 + 3, \ x \geq 0$ **27.** $f^{-1}(x) = x^2 - 6, \ x \geq 0$

29. not one-to-one   **31.** $g^{-1}(x) = \sqrt[3]{x} - 1$ **33.** $f^{-1}(x) = \sqrt[3]{x + 4}$

35. $f^{-1}(x) = \dfrac{-2x + 4}{x - 1}, \ x \neq 1$ **37.** $f^{-1}(x) = \dfrac{-5x - 2}{x - 4}, \ x \neq 4$

39. $f^{-1}(x) = \dfrac{5x + 1}{2x + 2}, \ x \neq -1$ **41.** (a) 8 (b) 3   **43.** (a) 1 (b) 0

45. (a) one-to-one   **47.** (a) not one-to-one   **49.** (a) one-to-one

(b) 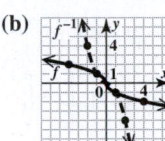   (b)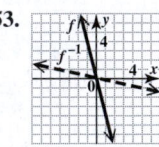

51.   **53.**   **55.**

x	f(x)
0	0
1	1
4	2

57.

x	f(x)
−1	−3
0	−2
1	−1
2	6

59. $f^{-1}(x) = \dfrac{x + 5}{4}$, or $f^{-1}(x) = \dfrac{1}{4}x + \dfrac{5}{4}$

60. MY CALCULATOR IS THE GREATEST THING SINCE SLICED BREAD.   **61.** If the function were not one-to-one, there would be ambiguity in some of the characters, as they could represent more than one letter.   **62.** Answers will vary. For example, Jane Doe is 1004 5 2748 129 68 3379 129.

Section 12.2

1. rises   **3.** C   **5.** A   **7.** 3.732   **9.** 0.344   **11.** 1.995   **13.** 1.587
15. 0.192   **17.** 73.517

19. 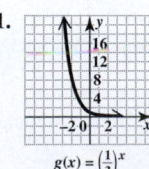   **21.**   **23.**

25.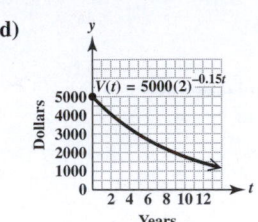
$f(x) = 2^{2x - 2}$

27. The division in the second step does not lead to x on the left side. By expressing each side using the same base, $2^x = 2^5$, we obtain the correct solution set, $\{5\}$.   **29.** $\{2\}$ **31.** $\left\{\dfrac{3}{2}\right\}$ **33.** $\left\{\dfrac{3}{2}\right\}$

35. $\{-1\}$ **37.** $\{7\}$ **39.** $\{-3\}$ **41.** $\left\{-\dfrac{3}{2}\right\}$ **43.** $\{-1\}$

45. $\{-3\}$ **47.** $\{-2\}$ **49.** 100 g   **51.** 0.30 g   **53.** (a) 0.6°C
(b) 0.3°C   **55.** (a) 1.4°C   (b) 0.5°C   **57.** (a) 80.599 million users
(b) 165.912 million users   (c) 341.530 million users
59. (a) $5000   (b) $2973   (c) $1768

(d)

$V(t) = 5000(2)^{-0.15t}$

Section 12.3

1. (a) B   (b) E   (c) D   (d) F   (e) A   (f) C   **3.** $(0, \infty); (-\infty, \infty)$

5. $\log_4 1024 = 5$ **7.** $\log_{1/2} 8 = -3$ **9.** $\log_{10} 0.001 = -3$

11. $\log_{625} 5 = \dfrac{1}{4}$ **13.** $\log_8 \dfrac{1}{4} = -\dfrac{2}{3}$ **15.** $\log_5 1 = 0$ **17.** $4^3 = 64$

19. $12^1 = 12$ **21.** $6^0 = 1$ **23.** $9^{1/2} = 3$ **25.** $\left(\dfrac{1}{4}\right)^{1/2} = \dfrac{1}{2}$

27. $5^{-1} = 5^{-1}$ **29.** (a) C   (b) B   (c) B   (d) C   **31.** 3.1699

33. 1.7959   **35.** −1.7925   **37.** −1.5850   **39.** 1.9243   **41.** 1.6990

43. $\left\{\dfrac{1}{3}\right\}$ **45.** $\left\{\dfrac{1}{125}\right\}$ **47.** $\{81\}$ **49.** $\left\{\dfrac{1}{5}\right\}$ **51.** $\{1\}$

53. $\{x \,|\, x > 0, x \neq 1\}$ **55.** $\{5\}$ **57.** $\left\{\dfrac{5}{3}\right\}$ **59.** $\{4\}$ **61.** $\left\{\dfrac{3}{2}\right\}$

63. $\{30\}$ **65.** $\left\{\dfrac{37}{9}\right\}$ **67.** 1   **69.** 0   **71.** 9   **73.** −1   **75.** 9

77. 5   **79.** 6   **81.** 4   **83.** −1   **85.** $\dfrac{1}{3}$

87. 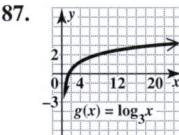   **89.** 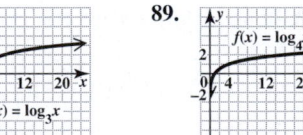   **91.**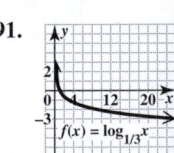

93. 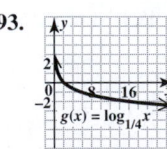   **95.** 8   **97.** 24   **99.** Because every real number power of 1 equals 1, if $y = \log_1 x$, then $x = 1^y$ and so $x = 1$ for every y. This contradicts the definition of a function.

101. $x = \log_a 1$ is equivalent to $a^x = 1$. The only value of x that makes $a^x = 1$ is 0. (Recall that $a \neq 1$.)

103. (a) 130 thousand units   (b) 190 thousand units

(c)   **105.** 5 advertisements

$S(t) = 100 + 30 \log_3 (2t + 1)$

107. $\frac{1}{100}$; $\frac{1}{10}$; 1; 10; 100 **108.** -2; -1; 0; 1; 2

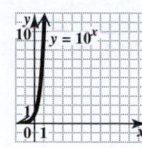

domain: $(-\infty, \infty)$; range: $(0, \infty)$

domain: $(0, \infty)$; range: $(-\infty, \infty)$

109. The graphs have symmetry across the line $y = x$.

110. They are inverses.

Section 12.4

1. false; $\log_b x + \log_b y = \log_b xy$ **3.** true **5.** true **7.** In the notation $\log_a (x + y)$, the parentheses do not indicate multiplication. They indicate that $x + y$ is the result of raising a to some power.

9. $\log_{10} 7 + \log_{10} 8$ **11.** 4 **13.** 9 **15.** $\log_7 4 + \log_7 5$

17. $\log_5 8 - \log_5 3$ **19.** $2 \log_4 6$ **21.** $\frac{1}{3} \log_3 4 - 2 \log_3 x - \log_3 y$

23. $\frac{1}{2} \log_3 x + \frac{1}{2} \log_3 y - \frac{1}{2} \log_3 5$ **25.** $\frac{1}{3} \log_2 x + \frac{1}{5} \log_2 y - 2 \log_2 r$

27. $\log_b xy$ **29.** $\log_a \frac{m}{n}$ **31.** $\log_a \frac{rt^3}{s}$ **33.** $\log_a \frac{125}{81}$

35. $\log_{10} (x^2 - 9)$ **37.** $\log_{10} (x^2 + 8x + 15)$ **39.** $\log_p \frac{x^3 y^{1/2}}{z^{3/2} a^3}$

41. 1.2552 **43.** -0.6532 **45.** 1.5562 **47.** 0.2386 **49.** 0.4771 **51.** 4.7710 **53.** false **55.** true **57.** true **59.** false

Section 12.5

1. C **3.** B **5.** 9.6421 **7.** $\sqrt{3}$ **9.** -11.4007 **11.** 1.6335 **13.** 2.5164 **15.** -1.4868 **17.** 9.6776 **19.** 2.0592 **21.** -2.8896 **23.** 5.9613 **25.** 4.1506 **27.** 2.3026 **29. (a)** 2.552424846 **(b)** 1.552424846 **(c)** 0.552424846 **(d)** The whole number parts will vary, but the decimal parts are the same. **31.** poor fen **33.** bog **35.** rich fen **37.** 11.6 **39.** 8.4 **41.** 4.3 **43.** 4.0×10^{-8} **45.** 1.0×10^{-2} **47.** 2.5×10^{-5} **49. (a)** 142 dB **(b)** 126 dB **(c)** 120 dB **51. (a)** 800 yr **(b)** 5200 yr **(c)** 11,500 yr **53. (a)** 189.53 million monthly active Twitter users **(b)** 2015 **55. (a)** \$54 per ton **(b)** If $x = 0$, then $\ln (1 - x) = \ln 1 = 0$, so $T(x)$ would be negative. If $x = 1$, then $\ln (1 - x) = \ln 0$, but the domain of $\ln x$ is $(0, \infty)$. **57.** 2.2619 **59.** 0.6826 **61.** -0.0947 **63.** -2.3219 **65.** 0.3155 **67.** 0.8736

Section 12.6

1. common logarithms **3.** natural logarithms **5.** $\{0.827\}$ **7.** $\{0.833\}$ **9.** $\{1.201\}$ **11.** $\{2.269\}$ **13.** $\{15.967\}$ **15.** $\{-6.067\}$ **17.** $\{261.291\}$ **19.** $\{-10.718\}$ **21.** $\{3\}$ **23.** $\{5.879\}$ **25.** $\{-\pi\}$, or $\{-3.142\}$ **27.** $\{1\}$ **29.** $\{4\}$ **31.** $\{\frac{2}{3}\}$ **33.** $\{\frac{33}{2}\}$ **35.** $\{-1 + \sqrt[3]{49}\}$ **37.** $\{\pm 3\}$ **39.** 2 cannot be a solution because $\log (2 - 3) = \log (-1)$, and -1 is not in the domain of $\log x$. **41.** $\{\frac{1}{3}\}$ **43.** $\{2\}$ **45.** $\varnothing$

47. $\{8\}$ **49.** $\{\frac{4}{3}\}$ **51.** $\{8\}$ **53. (a)** \$2539.47 **(b)** 10.2 yr **55.** \$4934.71 **57.** 27.73 yr **59. (a)** \$11,260.96 **(b)** \$11,416.64 **(c)** \$11,497.99 **(d)** \$11,580.90 **(e)** \$11,581.83 **61.** \$137.41 **63.** \$210 billion **65. (a)** 1.62 g **(b)** 1.18 g **(c)** 0.69 g **(d)** 2.00 g **67. (a)** 179.73 g **(b)** 21.66 yr **69.** $\log 5^x = \log 125$ **70.** $x \log 5 = \log 125$ **71.** $x = \frac{\log 125}{\log 5}$ **72.** $\frac{\log 125}{\log 5} = 3$; $\{3\}$

Chapter 12 Review Exercises

1. not one-to-one **2.** one-to-one **3.** not one-to-one

4. $\{(-8, -2), (-1, -1), (0, 0), (1, 1), (8, 2)\}$ **5.** $f^{-1}(x) = \frac{x - 7}{-3}$, or $f^{-1}(x) = -\frac{1}{3}x + \frac{7}{3}$ **6.** $f^{-1}(x) = \frac{x^3 + 4}{6}$ **7.** not one-to-one

8. This function is not one-to-one because two states in the list have minimum wage \$10.50.

9. 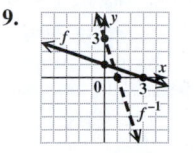 **10.** 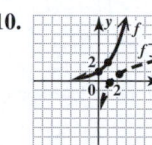 **11.** 172.466 **12.** 0.034 **13.** 0.079

14. 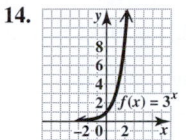 **15.** 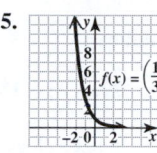 **16.**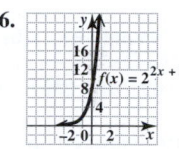

17. $\{\frac{1}{2}\}$ **18.** $\{4\}$ **19.** $\{\frac{3}{7}\}$ **20. (a)** 55.8 million people **(b)** 81.0 million people **21. (a)** $5^4 = 625$ **(b)** $\log_5 0.04 = -2$ **22.** 4; 3; fourth; 81 **23.** 2.5850 **24.** 0.5646 **25.** 1.7404

26. 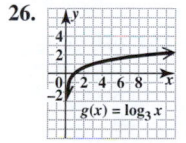 **27.** 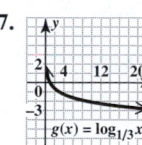 **28. (a)** 12 **(b)** 13 **(c)** 4 **29.** $\{2\}$ **30.** $\{\frac{3}{2}\}$ **31.** $\{7\}$ **32.** $\{8\}$ **33.** $\{4\}$ **34.** $\{\frac{1}{36}\}$

35. \$300,000

36. 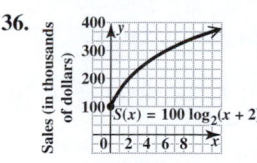 **37.** $\log_4 3 + 2 \log_4 x$ **38.** $3 \log_5 a + 2 \log_5 b - 4 \log_5 c$ **39.** $\frac{1}{2} \log_4 x + 2 \log_4 w - \log_4 z$

40. $2 \log_2 p + \log_2 r - \frac{1}{2} \log_2 z$ **41.** $\log_a \frac{49}{16}$ **42.** $\log_a 250$

43. $\log_b \frac{3x}{y^2}$ **44.** $\log_3 \frac{x + 7}{4x + 6}$ **45.** 1.4609 **46.** -0.5901 **47.** 3.3638

48. -1.3587 **49.** 0.9251 **50.** 1.7925 **51.** -2.0437 **52.** 0.3028 **53.** 6.4 **54.** 8.4 **55.** 2.5×10^{-5} **56.** Magnitude 1 is about 6.3 times as intense as magnitude 3. **57.** every 2 hr **58.** $\{2.042\}$ **59.** $\{4.907\}$ **60.** $\{18.310\}$ **61.** $\{\frac{1}{9}\}$ **62.** $\{-6 + \sqrt[3]{25}\}$ **63.** $\{2\}$ **64.** $\{\frac{3}{8}\}$ **65.** $\{4\}$ **66.** $\{1\}$ **67.** \$24,403.80 **68.** \$11,190.72 **69.** Plan A is better because it would pay \$2.92 more. **70.** 13.9 days

Chapter 12 Mixed Review Exercises

1. 7 **2.** 0 **3.** −3 **4.** 36 **5.** 4 **6.** e **7.** −5 **8.** 5.4

9. $\{72\}$ **10.** $\{5\}$ **11.** $\left\{\frac{1}{9}\right\}$ **12.** $\left\{\frac{4}{3}\right\}$ **13.** $\{3\}$ **14.** $\{0\}$

15. $\left\{\frac{1}{8}\right\}$ **16.** $\left\{\frac{11}{3}\right\}$ **17.** $\left\{\frac{2}{63}\right\}$ **18.** $\{-2, -1\}$ **19. (a)** \$4267 **(b)** 11%

20. (a) 0.325 **(b)** 0.673 **21. (a)** 1.209061955
(b) 7 **(c)** $\{1.209061955\}$ **22.** Answers will vary. Suppose
the name is Jeffery Cole, with $m = 7$ and $n = 4$. **(a)** $\log_7 4$ is
the exponent to which 7 must be raised to obtain 4.
(b) 0.7124143742 **(c)** 4

Chapter 12 Test

[12.1] **1. (a)** not one-to-one **(b)** one-to-one **2.** $f^{-1}(x) = x^3 - 7$

3. 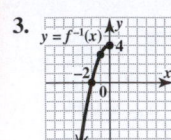 [12.2] **4.** 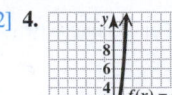 [12.3] **5.**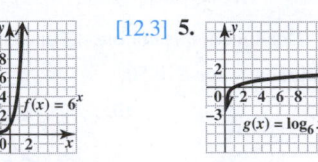

6. Interchange the x- and y-values of the ordered pairs, because the functions
are inverses. **7.** 9 **8.** 6 **9.** 0 [12.2] **10.** $\{-4\}$ **11.** $\left\{-\frac{13}{3}\right\}$

12. (a) 775 millibars **(b)** 265 millibars [12.3] **13.** $\log_4 0.0625 = -2$

14. $7^2 = 49$ **15.** $\{32\}$ **16.** $\left\{\frac{1}{2}\right\}$ **17.** $\{2\}$ **18.** 5; 2; fifth; 32

[12.4] **19.** $2\log_3 x + \log_3 y$ **20.** $\frac{1}{2}\log_5 x - \log_5 y - \log_5 z$

21. $\log_b \dfrac{s^3}{t}$ **22.** $\log_b \dfrac{r^{1/4}s^2}{t^{2/3}}$ [12.5] **23.** 1.3636 **24.** −0.1985

25. 2.1245 **26. (a)** $\dfrac{\log 19}{\log 3}$ **(b)** $\dfrac{\ln 19}{\ln 3}$ **(c)** 2.6801

[12.6] **27.** $\{3.966\}$ **28.** $\{3\}$ **29. (a)** \$11,903.40 **(b)** 19.9 yr

30. (a) \$17,427.51 **(b)** 23.1 yr

Chapters R–12 Cumulative Review Exercises

[R.2] **1. (a)** 0.05; 5% **(b)** 1.25; 125% [R.1, 1.4] **2. (a)** $-\frac{6}{5}$

(b) $-\frac{4}{5}$ [R.2] **3. (a)** 3750 **(b)** 0.0375 [1.3] **4. (a)** $-2, 0, 6, \frac{30}{3}$ (or 10)

(b) $-\frac{9}{4}, -2, 0, 0.6, 6, \frac{30}{3}$ (or 10) **(c)** $-\sqrt{2}, \sqrt{11}$

[1.3–1.5] **5.** 16 **6.** −39 [2.3] **7.** $\left\{-\frac{2}{3}\right\}$ [2.9, 8.1] **8.** $[1, \infty)$

[4.5] **9.** $6p^2 + 7p - 3$ [4.6] **10.** $16k^2 - 24k + 9$

[4.4] **11.** $-5m^3 + 2m^2 - 7m + 4$ [4.7] **12.** $5x^2 - 2x + 8$

[5.1] **13.** $x(8 + x^2)$ [5.2, 5.3] **14.** $(3y - 2)(8y + 3)$

15. $z(5z + 1)(z - 4)$ [5.4] **16.** $(4a + 5b^2)(4a - 5b^2)$

17. $(2c + d)(4c^2 - 2cd + d^2)$ **18.** $(4r + 7q)^2$

[4.1, 4.2] **19.** $-\dfrac{1875p^{13}}{8}$ [6.2] **20.** $\dfrac{x + 5}{x + 4}$ [6.4] **21.** $\dfrac{-3k - 19}{(k + 3)(k - 2)}$

[3.3, 7.1, 9.1] **22. (a)** yes **(b)** 1.955; The number of travelers shows an
increase of an average of 1.955 million per year during these years.

[7.2] **23.** $y = \frac{3}{4}x - \frac{19}{4}$ [7.3–7.5] **24.** $\{(4, 2)\}$

25. $\{(x, y) \mid 3x - 2y = 3\}$ [7.6] **26.** $\{(1, -1, 4)\}$

[7.7] **27.** 6 lb [10.3] **28.** $12\sqrt{2}$ [10.4] **29.** $-27\sqrt{2}$

[10.7] **30.** 41 [8.3] **31.** $\{-2, 7\}$ **32.** $(-\infty, -3) \cup (2, \infty)$

[10.6] **33.** $\{0, 4\}$ [11.2, 11.3] **34.** $\left\{\dfrac{1 \pm \sqrt{13}}{6}\right\}$

[11.8] **35.** $(-\infty, -4) \cup (2, \infty)$ [11.4] **36.** $\{\pm 1, \pm 2\}$

[12.2] **37.** $\{-1\}$ [12.6] **38.** $\{1\}$

[3.2, 7.1] **39.** 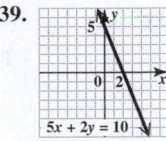 [8.4] **40.**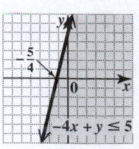

[11.6] **41.** $f(x) = \frac{1}{3}(x - 1)^2 + 2$ 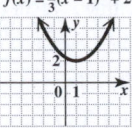 [12.2] **42.**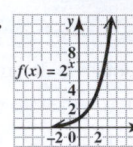

[12.3] **43.** 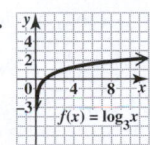 [12.4] **44.** $3\log x + \frac{1}{2}\log y - \log z$

[12.6] **45. (a)** 25,000 bacteria
(b) 30,500 bacteria **(c)** 37,300 bacteria
(d) in 3.5 hr, or at 3:30 P.M.

---

13 NONLINEAR FUNCTIONS, CONIC SECTIONS, AND NONLINEAR SYSTEMS

Section 13.1

1. E; 0; 0 **3.** A; $(-\infty, \infty)$; $\{\ldots, -2, -1, 0, 1, 2, \ldots\}$ **5.** B; It does
not satisfy the conditions of the vertical line test. **7.** B **9.** A

11. Shift the graph of $g(x) = \dfrac{1}{x}$ to the right 3 units and up 2 units.

13.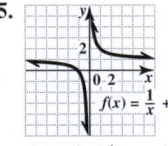
domain: $(-\infty, \infty)$;
range: $[0, \infty)$

15.
domain: $(-\infty, 0) \cup (0, \infty)$;
range: $(-\infty, 1) \cup (1, \infty)$

17.
domain: $[2, \infty)$;
range: $[0, \infty)$

19.
domain: $(-\infty, 2) \cup (2, \infty)$;
range: $(-\infty, 0) \cup (0, \infty)$

21.
domain: $[-3, \infty)$;
range: $[-3, \infty)$

23.
domain: $(-\infty, \infty)$;
range: $[1, \infty)$

25. $[\![x]\!]$ means the greatest integer *less than* or equal to x. Because −6 is
less than −5.1, the greatest integer less than −5.1 is −6. **27.** 3

29. 4 **31.** 0 **33.** 2 **35.** -14 **37.** -11

39.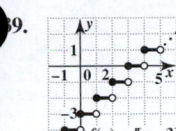

41.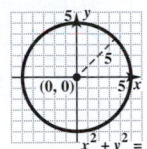
$f(x) = [\![x]\!] - 1$

43.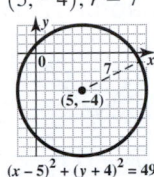

45. $2.75

Section 13.2

1. conic sections; parabolas **3.** ellipse; foci

5. (a) $(0, 0)$ **(b)** 5 **(c)**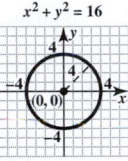
$x^2 + y^2 = 25$

7. B **9.** C

11. $(0, 0)$; $r = 2$ **13.** $(0, 0)$; $r = 9$ **15.** $(5, -4)$; $r = 7$

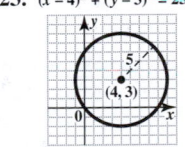
$x^2 + y^2 = 4$

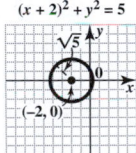
$x^2 + y^2 = 81$

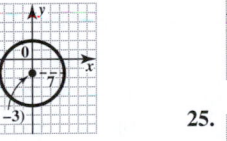
$(x - 5)^2 + (y + 4)^2 = 49$

17. $(-1, -3)$; $r = 5$ **19.** $x^2 + y^2 = 16$ **21.** $(x + 3)^2 + (y - 2)^2 = 9$

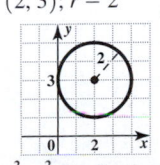
$(x + 1)^2 + (y + 3)^2 = 25$

23. $(x - 4)^2 + (y - 3)^2 = 25$ **25.** $(x + 2)^2 + y^2 = 5$ **27.** $x^2 + (y + 3)^2 = 49$

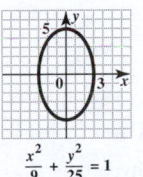

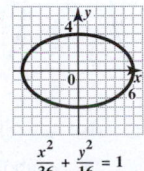

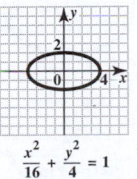

29. $(x + 2)^2 + (y + 3)^2 = 4$; $(-2, -3)$; $r = 2$

31. $(x + 5)^2 + (y - 7)^2 = 81$; $(-5, 7)$; $r = 9$

33. $(x - 2)^2 + (y - 4)^2 = 16$; $(2, 4)$; $r = 4$

35. $(x - 2)^2 + (y - 3)^2 = 4$; $(2, 3)$; $r = 2$

37. $(x + 3)^2 + (y - 3)^2 = 9$; $(-3, 3)$; $r = 3$

$x^2 + y^2 - 4x - 6y + 9 = 0$

$x^2 + y^2 + 6x - 6y + 9 = 0$

39. one point; The only ordered pair that satisfies the equation is $(4, 1)$.

41. The thumbtack acts as the center, and the length of the string acts as the radius.

43.
$\frac{x^2}{9} + \frac{y^2}{25} = 1$

45.
$\frac{x^2}{36} + \frac{y^2}{16} = 1$

47.
$\frac{x^2}{16} + \frac{y^2}{4} = 1$

49.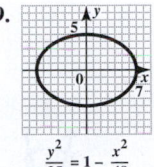
$\frac{y^2}{25} = 1 - \frac{x^2}{49}$

51.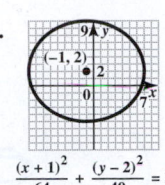
$\frac{(x + 1)^2}{64} + \frac{(y - 2)^2}{49} = 1$

53.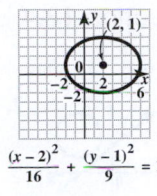
$\frac{(x - 2)^2}{16} + \frac{(y - 1)^2}{9} = 1$

55. $3\sqrt{3}$ units **57. (a)** 10 m **(b)** 36 m **59. (a)** 348.2 ft
(b) 1787.6 ft **61. (a)** 154.7 million mi **(b)** 128.7 million mi

Section 13.3

1. C **3.** D **5.** Because the coefficient of the x^2-term is positive, this is a horizontal hyperbola with x-intercepts $(-3, 0)$ and $(3, 0)$.

7.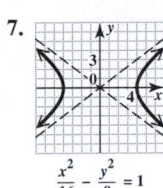
$\frac{x^2}{16} - \frac{y^2}{9} = 1$

9.
$\frac{y^2}{4} - \frac{x^2}{25} = 1$

11.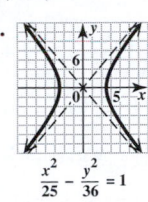
$\frac{x^2}{25} - \frac{y^2}{36} = 1$

13.
$\frac{y^2}{9} - \frac{x^2}{9} = 1$

15. hyperbola
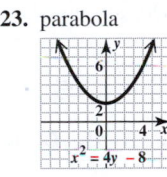
$x^2 - y^2 = 16$

17. ellipse
$4x^2 + y^2 = 16$

19. circle
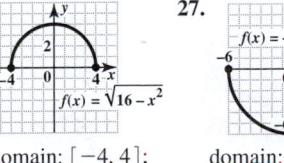
$y^2 = 36 - x^2$

21. ellipse
$x^2 + 9y^2 = 9$

23. parabola
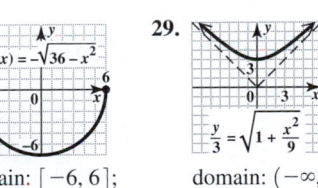
$x^2 = 4y - 8$

25.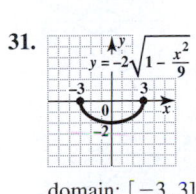
$f(x) = \sqrt{16 - x^2}$
domain: $[-4, 4]$;
range: $[0, 4]$

27.
$f(x) = -\sqrt{36 - x^2}$
domain: $[-6, 6]$;
range: $[-6, 0]$

29.
$\frac{y}{3} = \sqrt{1 + \frac{x^2}{9}}$
domain: $(-\infty, \infty)$;
range: $[3, \infty)$

31.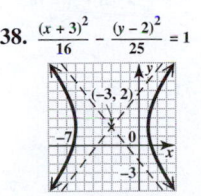
$y = -2\sqrt{1 - \frac{x^2}{9}}$
domain: $[-3, 3]$;
range: $[-2, 0]$

33. (a) 50 m **(b)** 69.3 m **35.** $y = \pm x$

37. $\frac{(x - 2)^2}{4} - \frac{(y + 1)^2}{9} = 1$

38. $\frac{(x + 3)^2}{16} - \frac{(y - 2)^2}{25} = 1$

39. $\frac{y^2}{36} - \frac{(x - 2)^2}{49} = 1$

40. $\frac{(y - 5)^2}{9} - \frac{x^2}{25} = 1$

Section 13.4

1. one   **3.** none

5. 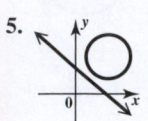   **7.** 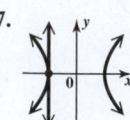   **9.** 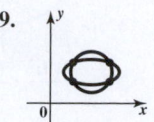   **11.**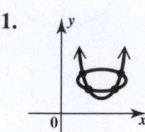

13. $\left\{(0,0),\left(\frac{1}{2},\frac{1}{2}\right)\right\}$ **15.** $\{(-6,9),(-1,4)\}$

17. $\left\{\left(-\frac{1}{5},\frac{7}{5}\right),(1,-1)\right\}$ **19.** $\left\{(-2,-2),\left(-\frac{4}{3},-3\right)\right\}$

21. $\{(-3,1),(1,-3)\}$ **23.** $\left\{\left(-\frac{3}{2},-\frac{9}{4}\right),(-2,0)\right\}$

25. $\{(-\sqrt{3},0),(\sqrt{3},0),(-\sqrt{5},2),(\sqrt{5},2)\}$

27. $\left\{\left(\frac{\sqrt{3}}{3}i,-\frac{1}{2}+\frac{\sqrt{3}}{6}i\right),\left(-\frac{\sqrt{3}}{3}i,-\frac{1}{2}-\frac{\sqrt{3}}{6}i\right)\right\}$

29. $\{(-2,0),(2,0)\}$ **31.** $\{(1,3),(1,-3),(-1,3),(-1,-3)\}$

33. $\{(-\sqrt{3},0),(\sqrt{3},0)\}$

35. $\{(-2\sqrt{3},-2),(-2\sqrt{3},2),(2\sqrt{3},-2),(2\sqrt{3},2)\}$

37. $\left\{\left(i\sqrt{2},-3i\sqrt{2}\right),\left(-i\sqrt{2},3i\sqrt{2}\right),\left(-\sqrt{6},-\sqrt{6}\right),\left(\sqrt{6},\sqrt{6}\right)\right\}$

39. $\{(-2i\sqrt{2},-2\sqrt{3}),(-2i\sqrt{2},2\sqrt{3}),(2i\sqrt{2},-2\sqrt{3}),(2i\sqrt{2},2\sqrt{3})\}$

41. $\left\{\left(-\sqrt{5},-\sqrt{5}\right),\left(\sqrt{5},\sqrt{5}\right)\right\}$

43. $\{(-3,-1),(3,1),(-i,3i),(i,-3i)\}$

45. $\{(i,2i),(-i,-2i),(2,-1),(-2,1)\}$ **47.** length: 12 ft; width: 7 ft

49. $20; $\frac{4}{5}$ thousand or 800 calculators

Section 13.5

1. (a) B   **(b)** D   **(c)** A   **(d)** C   **3.** $x \geq 0$ **5.** $x < 0$ **7.** C
$\qquad\qquad\qquad\qquad\qquad\qquad\qquad\ \ y \geq 0 \qquad\ y \leq 0$

9. 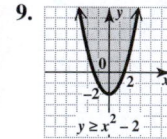   **11.** 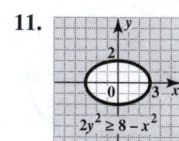   **13.**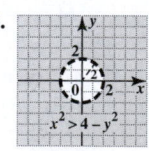

15. 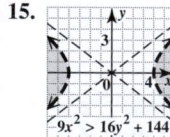   **17.** 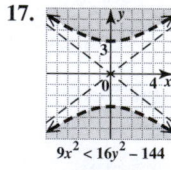   **19.**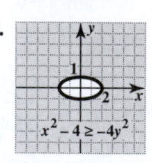

21. 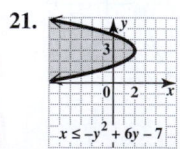   **23.** 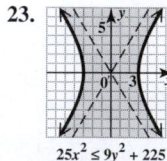   **25.**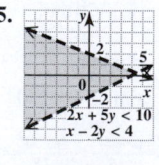

27. 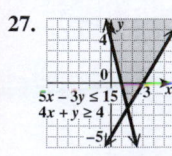   **29.** 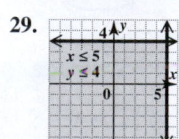   **31.**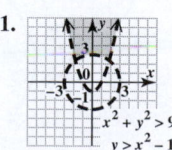

33. 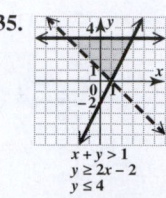   **35.** 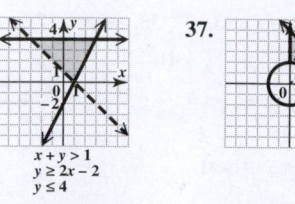   **37.**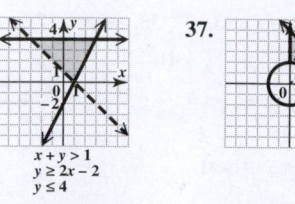

39. 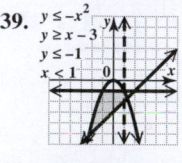   **41.** 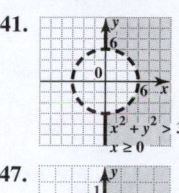   **43.**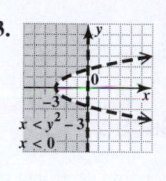

45. 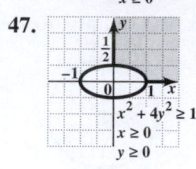   **47.**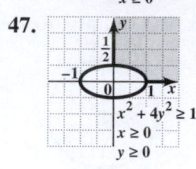

Chapter 13 Review Exercises

1. 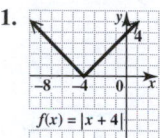   **2.**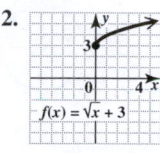

domain: $(-\infty,\infty)$;   domain: $[0,\infty)$;

range: $[0,\infty)$ range: $[3,\infty)$

3. 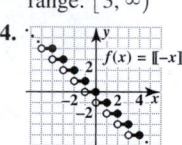   **4.**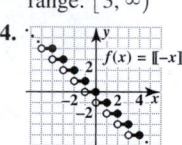

domain: $(-\infty,4)\cup(4,\infty)$;   domain: $(-\infty,\infty)$;

range: $(-\infty,0)\cup(0,\infty)$ range: $\{\ldots,-2,-1,0,1,2,\ldots\}$

5. 12   **6.** 2   **7.** -21 **8.** -5 **9.** $x^2+y^2=49$

10. $(x+2)^2+(y-4)^2=9$ **11.** $(x-4)^2+(y-2)^2=36$

12. $(x+5)^2+y^2=2$ **13.** $(x+3)^2+(y-2)^2=16;\ (-3,2);\ r=4$

14. $(x-4)^2+(y-1)^2=4;\ (4,1);\ r=2$

15. $(x+1)^2+(y+5)^2=9;\ (-1,-5);\ r=3$

16. $(x-3)^2+(y+2)^2=25;\ (3,-2);\ r=5$

17. (a) circle   **(b)** ellipse   **(c)** ellipse   **(d)** circle

18. (a) A   **(b)** C   **(c)** D   **(d)** B

19. 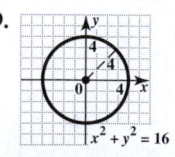   **20.** 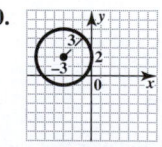   **21.**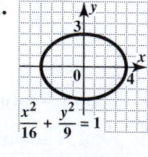

22. 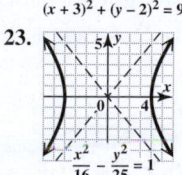   **23.** 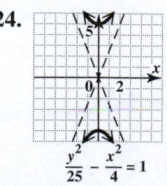   **24.**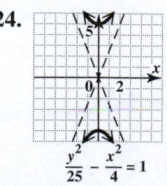

25. 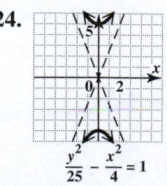   **26. (a)** C   **(b)** A   **(c)** D   **(d)** B

27. circle   **28.** parabola   **29.** hyperbola

30. ellipse   **31.** parabola   **32.** hyperbola

33. 0, 1, or 2   **34.** 0, 1, 2, 3, or 4

35. $\{(6, -9), (-2, -5)\}$ **36.** $\{(1, 2), (-5, 14)\}$

37. $\{(4, 2), (-1, -3)\}$ **38.** $\{(-2, -4), (8, 1)\}$

39. $\{(-\sqrt{2}, 2), (-\sqrt{2}, -2), (\sqrt{2}, -2), (\sqrt{2}, 2)\}$

40. $\{(-\sqrt{6}, -\sqrt{3}), (-\sqrt{6}, \sqrt{3}), (\sqrt{6}, -\sqrt{3}), (\sqrt{6}, \sqrt{3})\}$

41. 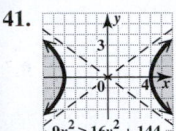   **42.** 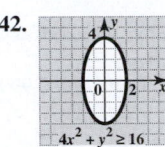   **43.** 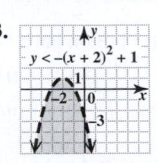   **44.** D

45. 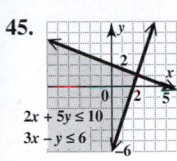   **46.** 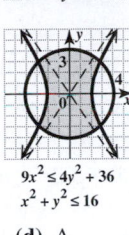   **47.**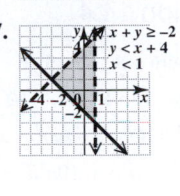

48. (a) B   **(b)** D   **(c)** C   **(d)** A

Chapter 13 Mixed Review Exercises

1. $(-2, 5); r = 6$ **2.** true

3. 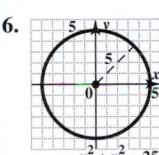   **4.** 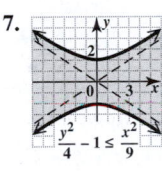   **5.**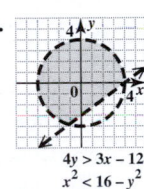

6. 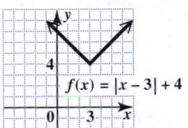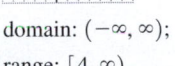   **7.** 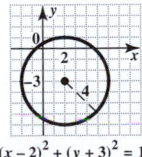   **8.**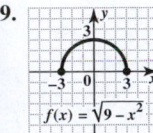

9. $\{(-7, -35), (12, 60)\}$ **10.** $\{(-1, -2), (1, 2), (-2i, i), (2i, -i)\}$

Chapter 13 Test

[13.1] **1.** 0   **2.** $[0, \infty)$ **3.** $\{\ldots, -2, -1, 0, 1, 2, \ldots\}$

4. (a) C   **(b)** A   **(c)** D   **(d)** B

5. 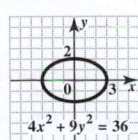   [13.2] **6.**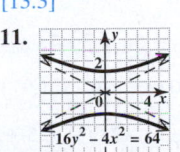

domain: $(-\infty, \infty)$;
range: $[4, \infty)$

$(x-2)^2 + (y+3)^2 = 16$
$(2, -3); r = 4$

7. D; $(0, 0); r = 1$ **8.** $(-4, 1); r = 5$

[13.3]   [13.2]   [13.3]

9. 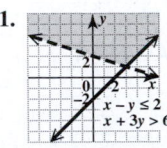   **10.** 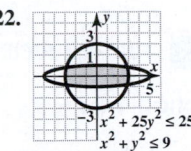   **11.**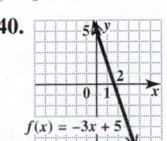

---

12. 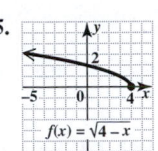   **13.** ellipse   **14.** hyperbola   **15.** circle

16. parabola   [13.4] **17.** $\left\{\left(-\frac{1}{2}, -10\right), (5, 1)\right\}$

18. $\left\{(-2, -2), \left(\frac{14}{5}, -\frac{2}{5}\right)\right\}$

19. $\left\{(-\sqrt{22}, -\sqrt{3}), (-\sqrt{22}, \sqrt{3}), (\sqrt{22}, -\sqrt{3}), (\sqrt{22}, \sqrt{3})\right\}$

[13.5]

20. 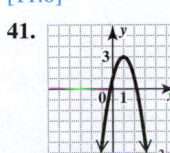   **21.** 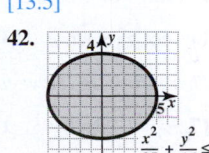   **22.**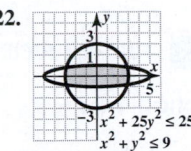

Chapters R–13 Cumulative Review Exercises

[R.2] **1.** 0.01; 1%   [1.3, 1.4] **2.** 0   [1.3] **3. (a)** A, B, C, D, F

(b) B, C, D, F   **(c)** D, F   **(d)** E, F   **(e)** C, D, F   **(f)** D, F

[1.1, 4.1, 10.1, 10.7] **4. (a)** 36   **(b)** -36 **(c)** 36   **(d)** 6   **(e)** -6

(f) $6i$ [4.6] **5.** $25y^2 - 30y + 9$ [4.7] **6.** $4x^3 - 4x^2 + 3x + 5 + \dfrac{3}{2x + 1}$

[6.2] **7.** $\dfrac{y - 1}{y(y - 3)}$ [6.4] **8.** $\dfrac{3c + 5}{(c + 5)(c + 3)}$ **9.** $\dfrac{1}{p}$

[6.5] **10.** $\dfrac{xy}{y - x}$ [10.4] **11.** $2\sqrt[3]{2}$ [10.7] **12.** $\dfrac{7}{5} + \dfrac{11}{5}i$

[5.2, 5.3] **13.** $(3x + 2)(4x - 5)$ [5.4] **14.** $(z^2 + 1)(z + 1)(z - 1)$

15. $(a - 3b)(a^2 + 3ab + 9b^2)$ [6.7] **16.** $1\frac{1}{5}$ hr   [4.1, 4.2] **17.** $\dfrac{a^5}{4}$

[10.5] **18.** $\dfrac{3\sqrt{10}}{2}$ [2.3] **19.** $\left\{\frac{2}{3}\right\}$ [2.9, 8.1] **20.** $\left(-\infty, \frac{3}{5}\right]$

[8.3] **21.** $\{-4, 4\}$ **22.** $(-\infty, -5) \cup (10, \infty)$ [3.3, 7.1] **23.** $\frac{2}{3}$

[7.2] **24.** $3x + 2y = -13$ [7.3–7.5] **25.** $\{(3, -3)\}$

[7.6] **26.** $\{(4, 1, -2)\}$ [13.4] **27.** $\left\{(-1, 5), \left(\frac{5}{2}, -2\right)\right\}$

[7.7] **28.** 40 mph   [10.6] **29.** $\varnothing$ [5.5] **30.** $\left\{\frac{1}{5}, -\frac{3}{2}\right\}$

[11.2, 11.3] **31.** $\left\{\dfrac{3 \pm \sqrt{33}}{6}\right\}$ [11.4] **32.** $\left\{\pm\dfrac{\sqrt{6}}{2}, \pm\sqrt{7}\right\}$

[12.6] **33.** $\{3\}$ [11.5] **34.** $v = \dfrac{\pm\sqrt{rFkw}}{kw}$ [9.3] **35. (a)** -1

(b) $9x^2 + 18x + 4$ [12.1] **36.** $f^{-1}(x) = \sqrt[3]{x - 4}$

[12.4, 12.5] **37. (a)** 4   **(b)** 7   [12.4] **38.** $\log\dfrac{(3x + 7)^2}{4}$

[13.1] **39.** domain: $(-\infty, \infty)$; range: $[0, \infty)$

[9.2]   [11.6]   [13.5]

40. 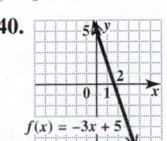   **41.** 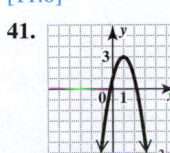   **42.**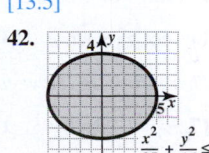

[13.1]   [13.3]   [12.2]

43.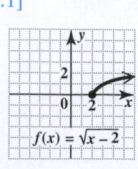

44.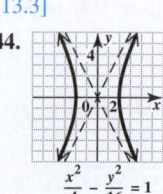

45.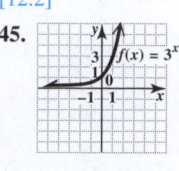

31. $5, -1, \frac{1}{5}, -\frac{1}{25}, \frac{1}{125}$ **33.** $-4, -2, -1, -0.5, -0.25$ **35.** $-1,048,575$

37. $\frac{121}{243}$ **39.** -1.997 **41.** 2.662 **43.** -2.982 **45.** \$24,470.52

47. \$74,156.06   **49.** 9   **51.** $\frac{10,000}{11}$ **53.** $-\frac{9}{20}$

55. The sum does not exist.   **57.** 1.3 ft   **59.** 140 m   **61.** 3 days; $\frac{1}{4}$ g

63. (a) 1.5 billion units   (b) 12 yr   **65.** \$5005.65

67. $\frac{a_1}{1-r} = \frac{0.9}{1-0.1} = \frac{0.9}{0.9} = 1$; Therefore, $0.99999\ldots = 1$.

14   FURTHER TOPICS IN ALGEBRA

Section 14.1

1. the set of all positive integers (natural numbers)   **3.** 80

5. 12   **7.** 2, 3, 4, 5, 6   **9.** $4, \frac{5}{2}, 2, \frac{7}{4}, \frac{8}{5}$ **11.** 3, 9, 27, 81, 243

13. $-1, -\frac{1}{4}, -\frac{1}{9}, -\frac{1}{16}, -\frac{1}{25}$ **15.** $5, -5, 5, -5, 5$ **17.** $0, \frac{3}{2}, \frac{8}{3}, \frac{15}{4}, \frac{24}{5}$

19. -70 **21.** $\frac{49}{23}$ **23.** 171   **25.** $4n$ **27.** $-8n$ **29.** $\frac{1}{3^n}$ **31.** $\frac{n+1}{n+4}$

33. \$110, \$109, \$108, \$107, \$106, \$105; \$400   **35.** \$6554

37. $4 + 5 + 6 + 7 + 8 = 30$ **39.** $3 + 6 + 11 = 20$

41. $-2 + 2 - 2 + 2 - 2 + 2 = 0$ **43.** $0 + 6 + 14 + 24 + 36 = 80$

Answers may vary for Exercises 45–49.

45. $\sum\limits_{i=1}^{5} (i+2)$ **47.** $\sum\limits_{i=1}^{5} 2^i(-1)^i$ **49.** $\sum\limits_{i=1}^{4} i^2$

51. In summation notation, each replacement for i should be squared first before the negative symbol is applied. The correct answer is $\sum\limits_{i=1}^{5} -i^2$.

53. 9   **55.** $\frac{40}{9}$ **57.** 9200 mutual funds

Section 14.2

1. difference   **3.** 55   **5.** $\frac{5}{2}; 3; \frac{7}{2}$ **7.** 16; 26   **9.** $d = 1$

11. not arithmetic   **13.** $d = -5$ **15.** 5, 9, 13, 17, 21

17. $-2, -6, -10, -14, -18$ **19.** $\frac{1}{3}, 1, \frac{5}{3}, \frac{7}{3}, 3$ **21.** $a_n = 5n - 3$; 122

23. $a_n = \frac{3}{4}n + \frac{9}{4}$; 21   **25.** $a_n = 3n - 6$; 69   **27.** 76   **29.** 48

31. -1 **33.** 16   **35.** 26   **37.** 6   **39.** The student subtracted terms in the wrong order. Subtracting the first term from the second term, we find that $d = -10 - (-15) = -10 + 15 = 5$.   **41.** 15   **43.** 87   **45.** $\frac{63}{2}$

47. 81   **49.** -3 **51.** -2.1 **53.** 390   **55.** 395   **57.** 31,375

59. \$465   **61.** \$2100 per month   **63.** 68 seats; 1100 seats

65. no; 3 blocks; 9 rows

Section 14.3

1. ratio   **3.** 31   **5.** $-\frac{1}{81}; \frac{1}{243}; -\frac{1}{729}$ **7.** 14; 56   **9.** $r = 2$

11. not geometric   **13.** $r = -3$ **15.** $r = -\frac{1}{2}$

There are alternative forms of the answers in Exercises 17–21.

17. $a_n = -5(2)^{n-1}$ **19.** $a_n = -2\left(-\frac{1}{3}\right)^{n-1}$ **21.** $a_n = 10\left(-\frac{1}{5}\right)^{n-1}$

23. 3,906,250   **25.** $\frac{1}{354,294}$ **27.** -4374 **29.** 2, 6, 18, 54, 162

Section 14.4

1. 1; sum   **3.** x^3 **5.** 6   **7.** 1   **9.** 720   **11.** 40,320

13. 15   **15.** 1   **17.** 120   **19.** 15   **21.** 78   **23.** 120

25. $m^4 + 4m^3n + 6m^2n^2 + 4mn^3 + n^4$

27. $a^5 - 5a^4b + 10a^3b^2 - 10a^2b^3 + 5ab^4 - b^5$

29. $8x^3 + 36x^2 + 54x + 27$ **31.** $\frac{x^4}{16} - \frac{x^3y}{2} + \frac{3x^2y^2}{2} - 2xy^3 + y^4$

33. $x^8 + 4x^6 + 6x^4 + 4x^2 + 1$ **35.** $27x^6 - 27x^4y^2 + 9x^2y^4 - y^6$

37. $r^{12} + 24r^{11}s + 264r^{10}s^2 + 1760r^9s^3$

39. $3^{14}x^{14} - 14(3^{13})x^{13}y + 91(3^{12})x^{12}y^2 - 364(3^{11})x^{11}y^3$

41. $t^{20} + 10t^{18}u^2 + 45t^{16}u^4 + 120t^{14}u^6$ **43.** $120(2^7)m^7n^3$

45. $\frac{7x^2y^6}{16}$ **47.** $36k^7$ **49.** $160x^6y^3$ **51.** $4320x^9y^4$

Chapter 14   Review Exercises

1. $-1, 1, 3, 5$ **2.** $0, \frac{1}{2}, \frac{2}{3}, \frac{3}{4}$ **3.** 1, 4, 9, 16   **4.** $\frac{1}{2}, \frac{1}{4}, \frac{1}{8}, \frac{1}{16}$

5. 0, 3, 8, 15   **6.** $1, -2, 3, -4$ **7.** $1 + 4 + 9 + 16 + 25 = 55$

8. $2 + 3 + 4 + 5 + 6 + 7 = 27$ **9.** $11 + 16 + 21 + 26 = 74$

10. $9 + 16 + 25 + 36 = 86$ **11.** $2 + 4 + 8 + 16 + 32 + 64 = 126$

12. $\frac{4}{5} + \frac{5}{6} + \frac{6}{7} + \frac{7}{8} = \frac{2827}{840}$ **13.** 763.3 thousand electric vehicles

14. arithmetic; $d = 3$ **15.** arithmetic; $d = 4$ **16.** geometric; $r = -\frac{1}{2}$

17. geometric; $r = -1$ **18.** neither   **19.** neither   **20.** 89   **21.** 73

22. 69   **23.** $a_n = -5n + 1$ **24.** $a_n = -3n + 9$ **25.** 15   **26.** 22

27. 152   **28.** 164   **29.** $a_n = -1(4)^{n-1}$ **30.** $a_n = \frac{2}{3}\left(\frac{1}{5}\right)^{n-1}$

31. 118,098   **32.** 2560 or -2560 **33.** $\frac{341}{1024}$ **34.** 0   **35.** 1

36. The sum does not exist.   **37.** $32p^5 - 80p^4q + 80p^3q^2 - 40p^2q^3 + 10pq^4 - q^5$ **38.** $x^8 + 12x^6y + 54x^4y^2 + 108x^2y^3 + 81y^4$

39. $81t^{12} - 108t^9s^2 + 54t^6s^4 - 12t^3s^6 + s^8$ **40.** $7752(3)^{16}a^{16}b^3$

Chapter 14   Mixed Review Exercises

1. $a_{10} = 1536; S_{10} = 1023$ **2.** $a_{40} = 235; S_{10} = 280$

3. $a_{15} = 38; S_{10} = 95$ **4.** $a_9 = 6561; S_{10} = -14,762$ **5.** $a_n = 2(4)^{n-1}$

6. $a_n = 5n - 3$ **7.** $a_n = -3n + 15$ **8.** $a_n = 27\left(\frac{1}{3}\right)^{n-1}$

9. 10 sec   **10.** \$21,973.00   **11.** approximately 42,000 people

12. $\frac{1}{128}$ **13.** (a) $0.5 + 0.05 + 0.005 + 0.0005 + \cdots$ (b) 0.1   (c) $\frac{5}{9}$

14. No, the sum cannot be found, because $r = 2$. This value of r does not satisfy $|r| < 1$.

Chapter 14 Test

[14.1] **1.** 0, 2, 0, 2, 0   [14.2] **2.** 4, 6, 8, 10, 12   [14.3] **3.** 48, 24, 12, 6, 3

[14.2] **4.** 0   [14.3] **5.** $\frac{64}{3}$ or $-\frac{64}{3}$ [14.2] **6.** 75   [14.3] **7.** 124 or 44

[14.1] **8.** 92.5 min   [14.3] **9.** $137,925.91   **10.** It has a sum if $|r| < 1$.

[14.2] **11.** 70   **12.** 33   **13.** 125,250   [14.3] **14.** 42   **15.** $\frac{1}{3}$

16. The sum does not exist.   [14.4] **17.** 40,320   **18.** 1   **19.** 15

20. 66   **21.** $81k^4 - 540k^3 + 1350k^2 - 1500k + 625$ **22.** $\dfrac{14,080x^8y^4}{9}$

[14.1] **23.** $324   [14.3] **24.** 3,542,940 insects

Chapters R–14 Cumulative Review Exercises

[R.1, 1.3, 1.4] **1.** 8   **2.** $-\frac{13}{45}$ [1.3] **3.** $-\frac{8}{3}$, 10, 0, $\frac{45}{15}$ (or 3), 0.82, -3

4. $\sqrt{13}, -\sqrt{3}$ [2.3] **5.** $\left\{\frac{1}{6}\right\}$ [2.4] **6.** $\{9\}$ [2.9, 8.1] **7.** $[10, \infty)$

[8.2] **8.** $(-\infty, -3) \cup (4, \infty)$ [8.3] **9.** $\left\{-\frac{9}{2}, 6\right\}$

10. $(-\infty, -3] \cup [8, \infty)$ [5.5] **11.** $\left\{-\frac{5}{2}, 2\right\}$

[11.2, 11.3] **12.** $\left\{\dfrac{-5 \pm \sqrt{217}}{12}\right\}$ [6.6] **13.** $\varnothing$ [12.2] **14.** $\left\{\frac{5}{2}\right\}$

[12.6] **15.** $\{2\}$ [4.1, 4.2] **16.** $\frac{9}{4}$ **17.** $-\dfrac{27p^2}{10}$

[4.5] **18.** $20p^2 - 2p - 6$ [4.4] **19.** $-5m^3 - 3m^2 + 3m + 8$

[4.7] **20.** $2t^3 + 3t^2 - 4t + 2 + \dfrac{3}{3t - 2}$ [10.7] **21.** 73

[5.2, 5.3] **22.** $z(3z + 4)(2z - 1)$ [5.4] **23.** $(7a^2 + 3b)(7a^2 - 3b)$

24. $(c + 3d)(c^2 - 3cd + 9d^2)$ [6.2] **25.** $\dfrac{x + 7}{x - 2}$

[6.4] **26.** $\dfrac{3p - 26}{p(p + 3)(p - 4)}$ [10.4] **27.** $10\sqrt{2}$

[3.3, 7.1] **28.** $\frac{3}{4}$ [7.2] **29.** $3x + y = 4$

[10.3, 13.2] **30.** $(x + 5)^2 + (y - 12)^2 = 81$

[7.3–7.5] **31.** $\{(-1, -2)\}$ [7.6] **32.** $\{(2, 1, 4)\}$

[13.4] **33.** $\left\{(-1, 5), \left(\frac{5}{2}, -2\right)\right\}$ [7.7] **34.** 2 lb

[3.2, 7.1] **35.** 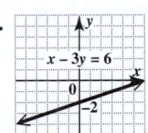   [8.4] **36.**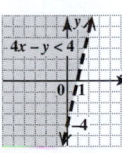

[11.6] **37.** 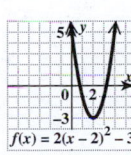   [12.2] **38.**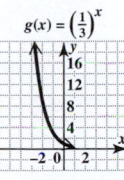

[12.3] **39.** 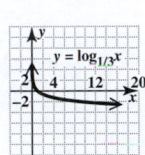   [13.1] **40.**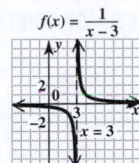

[13.2] **41.** 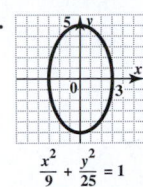   [13.3] **42.**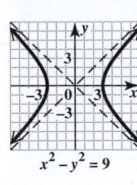

[12.1] **43.** $f^{-1}(x) = \dfrac{x - 5}{9}$, or $f^{-1}(x) = \dfrac{1}{9}x - \dfrac{5}{9}$

[14.1] **44.** $-7, -2, 3, 8, 13$ [14.2] **45.** 78   [14.3] **46.** $\frac{75}{7}$

[14.1] **47.** 30   [14.4] **48.** (a) 362,880   (b) 1   (c) 210

49. $32a^5 - 80a^4 + 80a^3 - 40a^2 + 10a - 1$ **50.** $-\dfrac{45x^8y^6}{4}$

APPENDIX A   REVIEW OF EXPONENTS, POLYNOMIALS, AND FACTORING

1. false; $(ab)^2 = a^2b^2$ **3.** false; $\left(\dfrac{4}{a}\right)^3 = \dfrac{4^3}{a^3}$ **5.** false; $xy^0 = x \cdot 1 = x$

7. false; $-(10)^0 = -(1) = -1$ **9.** The bases should not be multiplied: $4^5 \cdot 4^2 = 4^7$.   **11.** $\dfrac{1}{a^2b}$ **13.** $\dfrac{100y^{10}}{x^2}$ **15.** 0

17. $\dfrac{x^{10}}{2w^{13}y^{15}}$ **19.** $-\dfrac{a^{15}}{64b^{15}}$ **21.** $\dfrac{x^{16}z^{10}}{y^6}$ **23.** Only the coefficients of the like terms should be added, *not* the exponents. The correct sum is $6x^2 - 4x - 1$.   **25.** $-6a^4 + 11a^3 - 20a^2 + 26a - 15$

27. $8x^3 - 18x^2 + 6x - 16$ **29.** $x^2y - xy^2 + 6y^3$

31. $10x^3 - 4x^2 + 9x - 4$ **33.** $-3x^2 - 62x + 32$

35. The student squared each term instead of multiplying the binomial times itself. The correct product is $x^2 + 8x + 16$.

37. $6x^2 - 19x - 7$ **39.** $4x^2 - 9x + 2$ **41.** $16t^2 - 9$

43. $4y^4 - 16$ **45.** $16x^2 - 24x + 9$ **47.** $36r^2 + 60ry + 25y^2$

49. $c^3 + 8d^3$ **51.** $64x^3 - 1$ **53.** $14t^3 + 45st^2 + 18s^2t - 5s^3$

55. (a) B   (b) D   (c) A   (d) A, C   (e) A, B   **57.** $8a(5b - 2)$

59. $4xy^3(2x^2y + 3x + 9y)$ **61.** $(x + 3)(x - 5)$

63. $(2x + 3)(x - 6)$ **65.** $4(x - 5)(x - 2)$ **67.** $(6t + 5)(6t - 5)$

69. $(4t + 3)^2$ **71.** $p(2m - 3n)^2$ **73.** $(x + 1)(x^2 - x + 1)$

75. $(2t + 5)(4t^2 - 10t + 25)$ **77.** $(t^2 - 5)(t^4 + 5t^2 + 25)$

79. $(t^2 + 1)(t + 1)(t - 1)$ **81.** $(5x + 2y)(t + 3r)$

83. $(6r - 5s)(a + 2b)$ **85.** $(2x + 3y - 1)(2x + 3y + 1)$

APPENDIX B   SYNTHETIC DIVISION

1. C   **3.** 3; 4   **5.** 3; 6; 0   **7.** The student neglected to include 0 as the coefficient of the linear term. The correct setup is $-2\overline{)4 \quad 2 \quad 0 \quad 6}$ and the correct answer is $4x^2 - 6x + 12 + \dfrac{-18}{x+2}$.   **9.** $x - 5$

11. $4x - 1$ **13.** $2x + 4 + \dfrac{5}{x+2}$ **15.** $x - 4 + \dfrac{9}{x+1}$

17. $4x^2 + x + 3$ **19.** $x^2 + 6x + 20 + \dfrac{83}{x-4}$

21. $2x^4 + 4x^3 + 6x^2 + 15x + 6 + \dfrac{10}{x-2}$

23. $-3x^4 + 5x^2 - 11x + 11 + \dfrac{-8}{x+1}$ **25.** $x^4 + x^2 + 1 + \dfrac{2}{x+1}$

27. 7   **29.** -2 **31.** 0   **33.** yes   **35.** no   **37.** yes   **39.** no

41. $(2x - 3)(x + 4)$ **42.** $\left\{-4, \dfrac{3}{2}\right\}$ **43.** 0   **44.** 0   **45.** a

46. Yes, $x - 3$ is a factor; $g(x) = (x - 3)(3x - 1)(x + 2)$

STUDY SKILLS
p. S-1 Borchee/E+/Getty Images; **p. S-2** Originoo stock/123RF; **p. S-4** G-stockstudio/Shutterstock; **p. S-6** Rawpixel/Shutterstock; **p. S-8** Studio DMM Photography, Designs & Art/Shutterstock; **p. S-10** Doyeol (David) Ahn/Alamy Stock Photo

CHAPTER R
p. 11 Arek Malang/Shutterstock; **p. 15** Ryan McVay/Stockbyte/Getty Images

CHAPTER 1
p. 27 Pearson Education, Inc.; **p. 35** Cathy Yeulet/123RF; **p. 42** David Pereiras/Shutterstock; **p. 50** Kurt Kleemann/Shutterstock; **p. 62** Terry McGinnis; **p. 63** Africa Studio/Shutterstock; **p. 65** GaryLHampton/iStock/Getty Images; **p. 75** Belchonock/123RF; **p. 80** Famveldman/123RF; **p. 101** Sascha Burkard/Shutterstock; **p. 102** Callie Daniels

CHAPTER 2
p. 103 Callie Daniels; **p. 134** PCN Photography/Alamy Stock Photo; **p. 142** Terry McGinnis; **p. 143** (top) Belchonock/123RF, (bottom) Brandon Alms/Shutterstock; **p. 156** ZUMA Press Inc./Alamy Stock Photo; **p. 158** Yulia Davidovich/123RF; **p. 161** Education & Exploration 3/Alamy Stock Photo; **p. 163** Shutterstock; **p. 166** (top) Jim West/Alamy Stock Photo, (bottom left) Chokkicx/iStock/360/Getty Images, (bottom right) Spass/Fotolia; **p. 167** Tovovan/Shutterstock; **p. 168** Sumire8/Fotolia; **p. 169** Chris Cooper-Smith/Alamy Stock Photo; **p. 170** Stuart Jenner/Shutterstock; **p. 174** (top) Reiulf Gronnevik/Shutterstock, (bottom) Robynrg/Shutterstock; **p. 178** Wavebreakmedia/Shutterstock; **p. 179** Linda Richards/Alamy Stock Photo; **p. 181** Josh randall/Shutterstock; **p. 189** Kali9/Getty Images; **p. 194** Sasi Ponchaisang/123RF; **p. 195** Rob/Fotolia

CHAPTER 3
p. 207 Sergunt/Fotolia; **p. 208** DreamPictures/Tetra Images, LLC/Alamy Stock Photos; **p. 212** Georgios Kollidas/Alamy Stock Photo; **p. 214** Maksym Topchii/123RF; **p. 218** Air Images/Shutterstock; **p. 226** Wavebreakmedia/Shutterstock; **p. 253** Koksharov Dmitry/Shutterstock; **p. 256** Matt Benoit/Shutterstock; **p. 259** Rawpixel.com/Shutterstock; **p. 261** Rebecca Fisher/Fotolia

CHAPTER 4
p. 271 Nikonomad/Fotolia; **p. 280** Imagesbavaria/123RF; **p. 294** Studio 72/Shutterstock; **p. 297** (top) Photoroad/123RF, (bottom) Albert Barr/Shutterstock; **p. 298** Deepspace/Shutterstock; **p. 309** Terry McGinnis; **p. 339** NASA

CHAPTER 5
p. 343 Alexandre Solodov/Fotolia; **p. 382** Nickolae/Fotolia; **p. 390** Petr Jilek/Shutterstock; **p. 393** Offscreen/Shutterstock; **p. 395** Marmaduke St. John/Alamy Stock Photo; **p. 400** Goran Bogicevic/123RF; **p. 401** Elenabsl/123RF; **p. 407** Stephen Barnes/Transport/Alamy Stock Photo

CHAPTER 6
p. 411 HodagMedia/Shutterstock; **p. 444** Denis Tabler/Fotolia; **p. 470** Tommaso Altamura/123RF; **p. 472** Johnny Habell/Shutterstock; **p. 474** (top left) Sakda saetiew/123RF, (top right) Sirtravelalot/Shutterstock, (bottom) Alexandr Shevchenko/Shutterstock; **p. 475** MelindaFawver/Fotolia; **p. 477** Llandrea/Fotolia

CHAPTER 7
p. 489 Mmphotos/Fotolia; **p. 514** Kazzland Inc./Shutterstock; **p. 516** Podfoto/Shutterstock; **p. 519** (top) Siraphol/123RF, (bottom) Karen roach/123RF; **p. 520** Wavebreakmedia/Shutterstock; **p. 538** Vadim Ponomarenko/Fotolia; **p. 557** Pavol Kmeto/Shutterstock; **p. 558** Blickwinkel/Alamy Stock Photo; **p. 565** Marie C Fields/Shutterstock; **p. 567** AndreAnita/Shutterstock; **p. 568** (top) Chris Whitehead/Cultura Creative (RF)/Alamy Stock Photo, (bottom) Thomas M Perkins/Shutterstock; **p. 570** Callie Daniels; **p. 571** Dpa picture alliance/Alamy Stock Photo; **p. 572** Callie Daniels

CHAPTER 8
p. 587 Leonid Shcheglov/Shutterstock; **p. 601** Jeayesy/123RF; **p. 603** Michaeljung/Shutterstock; **p. 604** Chad McDermott/Shutterstock; **p. 611** Leonid Shcheglov/Shutterstock; **p. 615** Katherine Martin/123RF; **p. 625** Derek Meijer/Alamy Stock Photo; **p. 634** National Atomic Museum Foundation

CHAPTER 9
p. 635 Michelangeloop/Shutterstock; **p. 653** Rido/Shutterstock; **p. 655** Neelsky/Shutterstock; **p. 662** Gorbelabda/Shutterstock; **p. 666** Anton Starikov/Shutterstock; **p. 668** Terry McGinnis; **p. 674** Dana Bartekoske/123RF; **p. 677** Dmitry Deshevykh/Alamy Stock Photo; **p. 678** Cathy Yeulet/123RF

CHAPTER 10
p. 687 Giulio Meinardi/Fotolia; **p. 699** Stoupa/Shutterstock; **p. 708** Maria Moroz/Fotolia; **p. 721** (dog) Anneka/Shutterstock, (screen) Cobalt/Fotolia; **p. 722** Kevin Kipper/Alamy Stock Photo; **p. 745** Johnson Space Center/NASA; **p. 765** Scol22/Fotolia

CHAPTER 11
p. 767 Sarot Chamnankit/123RF; **p. 770** Portrait of Galileo (ca. 1853), Charles Knight. Steel engraving. Library of Congress Prints and Photographs Division [LC-DIG-pga-38085]; **p. 791** Phovoir/Shutterstock; **p. 804** Micro10x/Shutterstock; **p. 809** Look Die Bildagentur der Fotografen GmbH/Alamy Stock Photo; **p. 810** Steven J. Everts/123RF; **p. 814** Zurijeta/Shutterstock; **p. 818** Andriy Popov/123RF; **p. 819** Alexwhite/Shutterstock; **p. 828** Kaband/Shutterstock; **p. 829** Bonniej/E+/Getty Images; **p. 839** Hartrockets/iStock/Getty Images

CHAPTER 12
p. 851 Chien321/Shutterstock; **p. 853** Frontpage/Shutterstock; **p. 860** Mark Higgins/Shutterstock; **p. 869** Bernhard Staehli/Shutterstock; **p. 870** Maksym Bondarchuk/123RF; **p. 875** Per Tillmann/Fotolia; **p. 878** Prisma by Dukas Presseagentur GmbH/Alamy Stock Photo; **p. 887** Romrodphoto/Shutterstock; **p. 889** Georgios Kollidas/Fotolia; **p. 892** (top) Jim West/Alamy Stock Photo, (bottom) Silvano Rebai/Fotolia; **p. 893** WaterFrame/Alamy Stock Photo; **p. 901** Sergey Nivens/Shutterstock; **p. 902** Armadillo Stock/Shutterstock; **p. 908** Shutterstock; **p. 910** Syda Productions/Shutterstock; **p. 911** IndustryAndTravel/Alamy Stock Photo

CHAPTER 13
p. 917 Terry McGinnis; **p. 921** Samuel Acosta/Shutterstock; **p. 934** Petr Student/Shutterstock; **p. 941** Faiz Azizan/Shutterstock

CHAPTER 14
p. 965 Callie Daniels; **p. 966** Madeinitaly4k/Shutterstock; **p. 970** Frank11/Shutterstock; **p. 979** Blend Images/Shutterstock; **p. 980** Creativa Images/Shutterstock; **p. 988** Paul Hakimata Photography/Shutterstock; **p. 991** Atlaspix/Alamy Stock Photo; **p. 1001** Jtbaskinphoto/Shutterstock; **p. 1003** Baloncici/123RF

INDEX

Triangles and Angles

Right Triangle
Triangle has one 90° (right) angle.

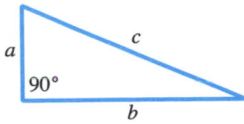

Pythagorean Theorem (for right triangles)

$a^2 + b^2 = c^2$

Isosceles Triangle
Two sides are equal.

$AB = BC$

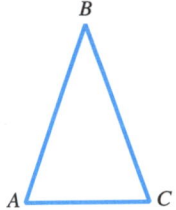

Equilateral Triangle
All sides are equal.

$AB = BC = CA$

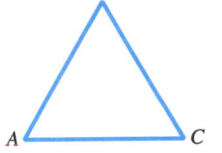

Sum of the Angles of Any Triangle

$A + B + C = 180°$

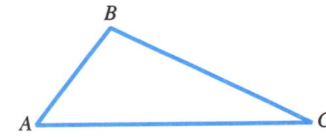

Similar Triangles
Corresponding angles are equal. Corresponding sides are proportional.

$A = D, B = E, C = F$

$$\frac{AB}{DE} = \frac{AC}{DF} = \frac{BC}{EF}$$

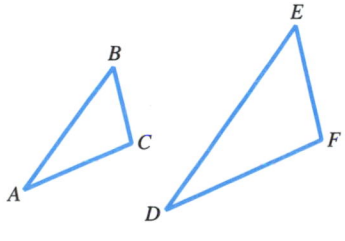

Right Angle
Measure is 90°.

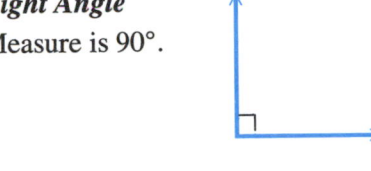

Straight Angle
Measure is 180°.

Complementary Angles
The sum of the measures of two complementary angles is 90°.

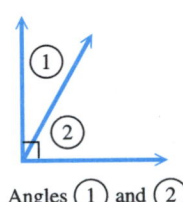

Angles ① and ② are complementary.

Supplementary Angles
The sum of the measures of two supplementary angles is 180°.

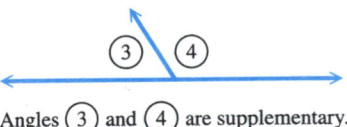

Angles ③ and ④ are supplementary.

Vertical Angles
Vertical angles have equal measures.

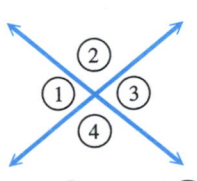

Angle ① = Angle ③
Angle ② = Angle ④

Geometry Formulas

Figure	Formulas	Illustration
Square	Perimeter: $\quad P = 4s$ Area: $\quad \mathcal{A} = s^2$	
Rectangle	Perimeter: $\quad P = 2L + 2W$ Area: $\quad \mathcal{A} = LW$	
Triangle	Perimeter: $\quad P = a + b + c$ Area: $\quad \mathcal{A} = \dfrac{1}{2}bh$	
Parallelogram	Perimeter: $\quad P = 2a + 2b$ Area: $\quad \mathcal{A} = bh$	
Trapezoid	Perimeter: $\quad P = a + b + c + B$ Area: $\quad \mathcal{A} = \dfrac{1}{2}h(b + B)$	
Circle	Diameter: $\quad d = 2r$ Circumference: $\quad C = 2\pi r$ $\qquad\qquad\qquad C = \pi d$ Area: $\quad \mathcal{A} = \pi r^2$	

Geometry Formulas

Figure	Formulas	Illustration
Cube	Volume: $V = e^3$ Surface area: $S = 6e^2$	
Rectangular Solid	Volume: $V = LWH$ Surface area: $S = 2HW + 2LW + 2LH$	
Right Circular Cylinder	Volume: $V = \pi r^2 h$ Surface area: $S = 2\pi rh + 2\pi r^2$ (Includes both circular bases)	
Cone	Volume: $V = \dfrac{1}{3}\pi r^2 h$ Surface area: $S = \pi r\sqrt{r^2 + h^2} + \pi r^2$ (Includes circular base)	
Right Pyramid	Volume: $V = \dfrac{1}{3}Bh$ B = area of the base	
Sphere	Volume: $V = \dfrac{4}{3}\pi r^3$ Surface area: $S = 4\pi r^2$	

Other Formulas

Distance: $d = rt$ (r = rate or speed, t = time)

Percent: $p = br$ (p = percentage, b = base, r = rate)

Temperature: $F = \dfrac{9}{5}C + 32$ $C = \dfrac{5}{9}(F - 32)$

Simple Interest: $I = prt$ (p = principal or amount invested, r = rate or percent, t = time in years)